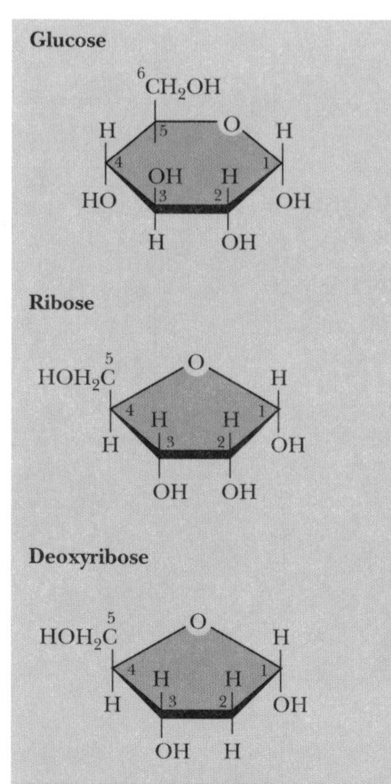
Glucose
Ribose
Deoxyribose

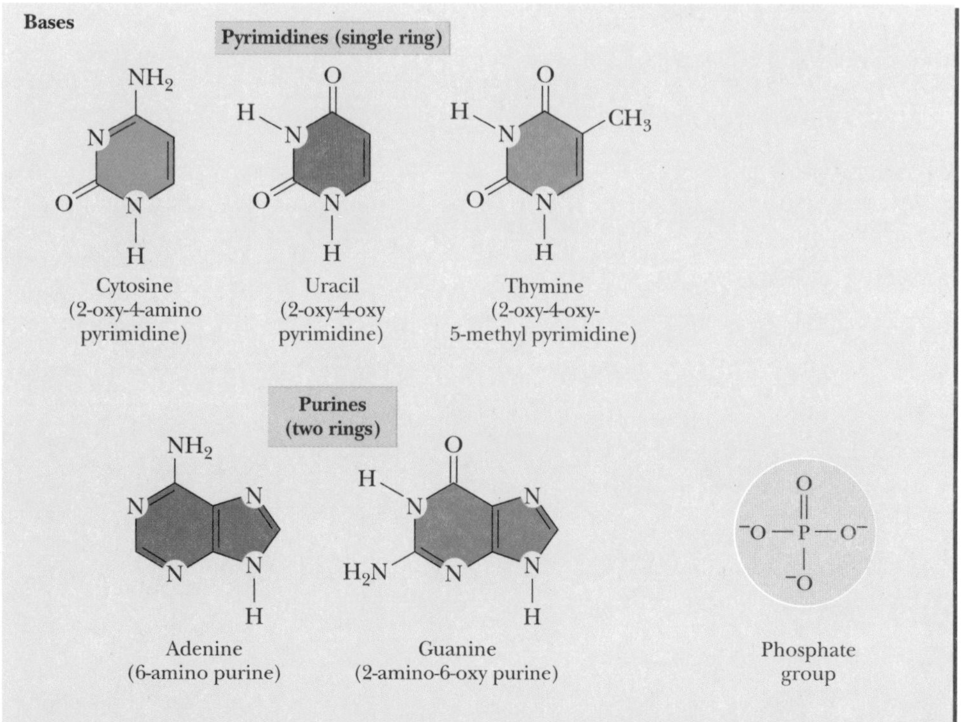
Bases
Pyrimidines (single ring)
Cytosine
(2-oxy-4-amino
pyrimidine)
Uracil
(2-oxy-4-oxy
pyrimidine)
Thymine
(2-oxy-4-oxy-
5-methyl pyrimidine)
Purines
(two rings)
Adenine
(6-amino purine)
Guanine
(2-amino-6-oxy purine)
Phosphate
group

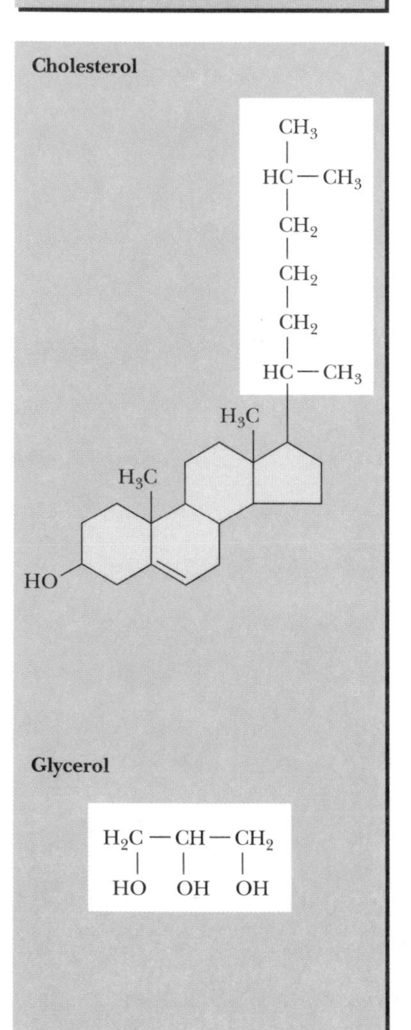
Cholesterol
Glycerol

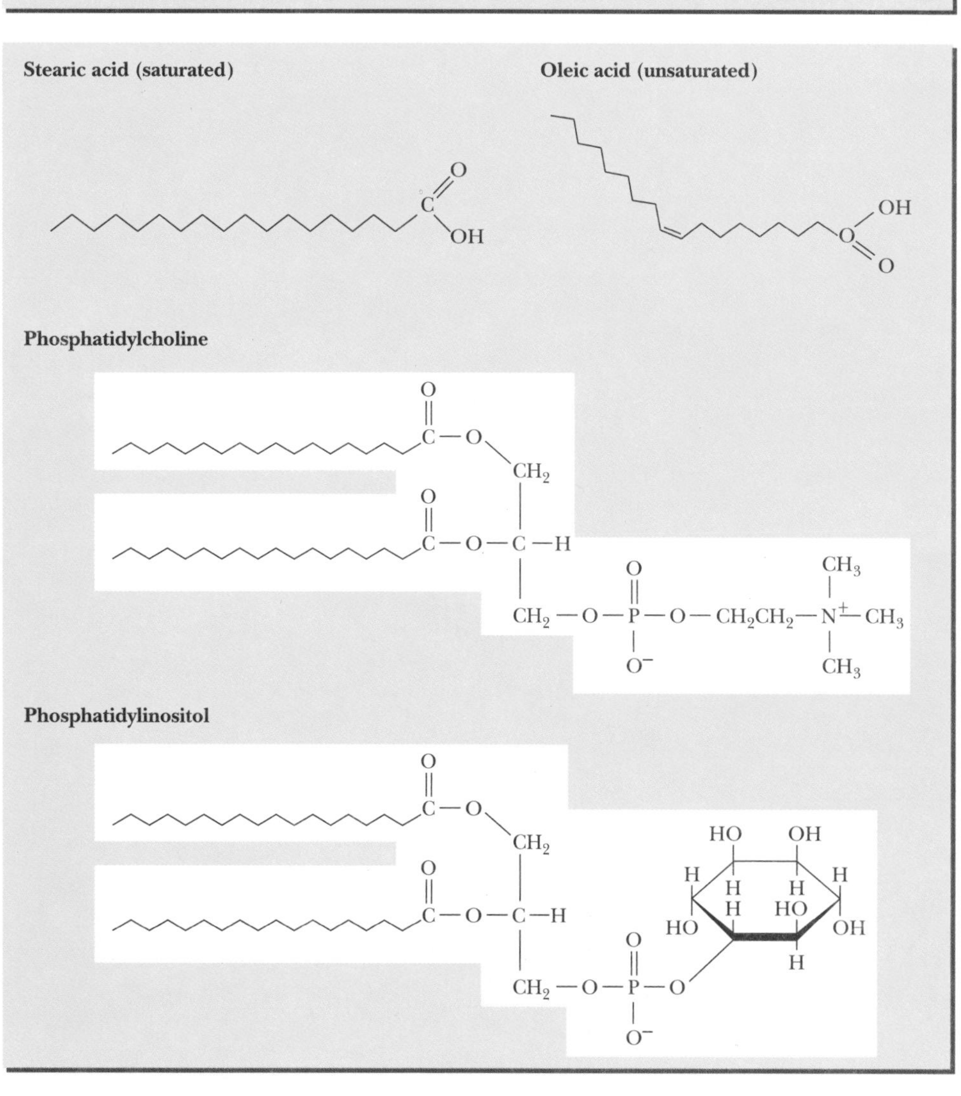
Stearic acid (saturated)
Oleic acid (unsaturated)
Phosphatidylcholine
Phosphatidylinositol

Asking About Cells

ALLAN J. TOBIN

University of California, Los Angeles

RICHARD E. MOREL

Smith College

SAUNDERS COLLEGE PUBLISHING

Harcourt Brace College Publishers

Fort Worth Philadelphia San Diego New York Orlando Austin
San Antonio Toronto Montreal London Sydney Tokyo

Asking About Cells

Text Typeface: New Baskerville
Compositor: York Graphic Services, Inc.
Executive Editor: Julie Levin Alexander, Edith Beard Brady
Developmental Editors: Richard Morel, Lee Marcott
Managing Editor: Carol Field
Project Editor: Elizabeth Ahrens
Copy Editor: Donna Walker
Manager, Art & Design: Carol Bleistine
Art Director: Caroline McGowan
Art & Design Coordinator: Kathleen Flanagan
Text Designer: Donna Wickes
Cover Designer: Lawrence R. Didona
Text Artwork: J/B Woolsey Associates
Layout Artist: Anne Muldrow
Photo Research: Linda Sykes, PhotoSearch, Inc.; Amy Ellis Dunleavy
Senior Production Manager: Charlene Catlett Squibb
Manager of Production: Joanne Cassetti
Senior Production Manager: Sue Westmoreland
Senior Vice President/Director of EDP: Tim Frelick

Cover Credit: Fibrin threads and red blood cells together form a blood clot to bind wounds. © Boehringer Ingelheim International GmbH, photo Lennart Nilsson, THE INCREDIBLE MACHINE, National Geographic Society
Title Page Credit: Nancy Kedersha. Microfilaments of natural killer cells.
Part Opener Credits: I, James Dennis/CNRI/Phototake NYC; II, Nancy Kedersha; III, Biophoto Associates/Science Source/Photo Researchers; IV, A. B. Dowsett/Science Photo Library/Photo Researchers; V, Synaptek Scientific Products/Science Photo Library/Photo Researchers
Lab Tour Credits: Donna Coveney, MIT (Horvitz); Ro Kampman (Goodenough); Susan Abmayr (Abmayr); Michael Smith (Smith); James Townsel (Townsel)

Printed in the United States of America

ASKING ABOUT CELLS
0-03-098018-6

Library of Congress Catalog Card Number: 95-67686

89012345 032 10 98765432

DEDICATION

to our partners,
Janet Hadda
Beth Morel

and our children,
Adam Tobin
David Tobin
Eric Morel
Jon Morel
Lucy Morel

PREFACE

Cell biology has become the lead actor in the drama that is modern biology—versatile, evocative, and charismatic. Exciting new techniques have given cell biology the power to advance our understanding of processes ranging from photosynthesis and pheromone action to muscle contraction and muscular dystrophy. More powerful microscopes have revealed amazing details in the organization and action of cells: Our cover photograph, for example, shows how a network of clotting proteins prevents the loss of red blood cells from a wound. Likewise, molecular biology tells us which molecules are responsible for initiating the formation of blood clots.

To many biologists, the prominence of cell biology is just as it should be. Since the 1925 appearance of the third edition of E. B. Wilson's landmark book, *The Cell in Development and Inheritance,* many biologists have seen cell biology as the central actor in the drama of biological and medical discovery. Yet in the undergraduate curriculum, cell biology has, until recently, usually played only a supporting role. Now, however, undergraduate curricula may include cell biology as a first course.

Cell biology continues its attention to subcellular structure, but it is increasingly involved in issues of broad social concern, such as cancer, infection, immunity, brain function, and agriculture. Again and again, cell biology has revealed the macroscopic importance of initially obscure cellular processes in mammals, flowering plants, bacteria, yeasts, worms, and flies. For students and scientists alike, the excitement of learning more about cell biology comes from understanding the process by which researchers discover the connections between cells and the rest of biology. Our book tries to convey the fervor and the energy of inquiry, as reflected in its title, *Asking About Cells.*

One of the most engaging enterprises of contemporary cell biology is the continuing attempt to understand complex processes in terms of known molecular structures. But the details can be overwhelming: Even as the teaching of cell biology has moved forward in the curriculum to sophomore and sometimes freshman year, cell biology textbooks have become longer and heavier.

We have tried to avoid writing an encyclopedia of cell biology. Instead, we have concentrated on presenting a picture of the questing and questioning nature of cell biology. We have, of course, also offered many molecular and cellular details—certainly enough to prepare students for higher-level courses or for the Medical College Admissions Test. But our main focus is on such larger issues as "What do engaged cell biologists argue about? How do the latest techniques allow clever researchers to ask innovative questions? What are the most important unanswered questions?" Our goals are both to introduce the facts and concepts of cell biology and to show that they derive from a process of inquiry that undergraduates can easily understand and even emulate. The process of discovery is not mysterious: Future discoveries will, in fact, depend on some of the very people who will be reading this book.

We mean, above all, for the book to be friendly and engaging to our undergraduate readers. We want them to understand that mere facts are not the center of science: New techniques may show that once-accepted "facts" are wrong, and new concepts may completely change the context that makes a "fact" more or less important. We want students to know that questions are often more important than answers. And we want them to savor the zest of cell biological research.

■ ORGANIZATION

The book consists of 24 chapters, grouped into five parts. Each chapter begins with a list of "Key Concepts" that tie together the material within the chapter, and each ends with a set of questions designed to help the student review and integrate concepts from the chapter. In our illustration program, we have presented both actual photographic examples of cells and cellular structures and detailed drawings of molecular assemblies. We have also used schematic drawings to summarize temporal sequences in cellular processes and in experimental programs.

Each chapter has at least one boxed essay. The essays introduce new experimental advances or show how the information discussed in the chapter has led to new insights of social relevance, usually involving medicine or agriculture. In the essays that accompany Chapter 1, "What Are 'Peer-Reviewed Publications,' and Why Are They So Important to Scientists?" and "How to Read a Scientific Paper," we introduce students to the significance of scientific publications. We hope that instructors will appropriately reinforce this point by discussing recent research papers. We provide an annotated list of suggested papers in the Instructor's Manual, and we will update these suggestions.

In both essays and text, we have repeatedly returned to the topic of cancer cells—a subject of particular concern to society in general and to cell biologists in particular. Although some textbooks treat cancer in a separate chapter, usually at the end of the book, we have chosen to refer to cellular processes relevant to cancer as points of contrast in many different contexts. Examples include our discussions of the cell cycle (Chapter 9); tumor viruses and oncogenes (Essay, Chapter 9); telomeres, senescence, and cancer (Essay, Chapter 13); retroviruses (Chapter 16); programmed cell death (Essay, Chapter 20); and cancer as a failure of the immune system (Chapter 22).

Each of the five parts of the book begins with a "Lab Tour," which introduces an active research scientist. Each interview focuses on the scientist's research and memorable moments of discovery, with a view to conveying some of the human intensity that accompanies the often faceless facts that appear in textbooks.

Part I, "Introduction: The Chemistry of Cells" (Chapters 1–4), begins with an introductory chapter, "The Challenges of Cell Biology." In this chapter, we introduce the principles of cell biology and present a view of scientific inquiry that derives from our own experience as researchers. Again, we stress that research is a human quest, not the disembodied operation of a mysterious monastic order. Chapters 2–4 introduce the small building blocks of cells, macromolecules, and the thermodynamic and kinetic constraints on chemical transformations.

Part II, "The Functional Organization of Cells" (Chapters 5–8), introduces the techniques of cell biology and shows what they have revealed about subcellular organization. We emphasize relatively new discoveries about cytoskeletal elements and membrane-bounded compartments within eukaryotic cells. Our treatment of cellular respiration and photosynthesis stresses chemiosmosis and the importance of the subcellular organization in ATP production.

Part III, "The Continuity of Cellular Information" (Chapters 9–13), introduces cell reproduction (Chapter 9), transmission genetics (Chapters 10 and 11), and molecular genetics. In our discussion of Mendel's Laws, we emphasize their roots in chromosomal behavior, rather than their basis in probability theory. Our discussion of genes and DNA replication focuses on concrete examples, emphasizing the specific experiments by which researchers discovered general principles.

Part IV, "Information Flow and Its Manipulation" (Chapters 14–17), presents our current understanding of how information in DNA gives rise to RNAs, proteins, and cellular structures. We show how general knowledge of information flow and specific information about regulatory signals has led both to an understanding of unconventional genetic systems, such as viruses and plasmids, and to biologists' newfound ability to manipulate genes, cells, and organisms.

Part V, "Cell Specialization, Integration, and Evolution" (Chapters 18–24), shows how cellular and molecular knowledge has given biologists an increasingly profound understanding of complex multicellular processes, including movement (Chapter 18), hormonal signaling (Chapter 19), animal and plant development (Chapters 20 and 21), the immune response (Chapter 22), neural integration (Chapter 23), and prebiotic evolution (Chapter 24).

■ ACKNOWLEDGMENTS

We are grateful to our many friends and colleagues who have contributed to our endeavor:

to our families, who provided support and inspiration;

to Dan Lynch and Nancy Roseman (Williams College), who wrote the "Questions for Review and Understanding" that appear at the end of each chapter;

to George Fleck (Smith College) for his valuable chemical counsel; Elaine Tobin (UCLA) for her help with plant biology; David Tobin (UCSF) for his help in reorganizing the section on information flow; Beth Morel for proofreading galleys and pages; and Jennie Dusheck for her continuing counsel and ideas;

to Pouneh Beizai and Diep Nguyen for their help in preparing essays and in finding engaging pedagogic examples;

to John Woolsey, Betty Woolsey, Laura Colangelo, and Regina Santoro, who provided both drawings and structural insights;

to Linda Sykes and Amy Ellis Dunleavy for finding the photographs that illustrate our contemporary views of cells and cellular structures, and to Donald Lovett, Trenton State University, for calculating the scale bars that appear on many of the photographs;

to our teachers—John Edsall, Howard E. Evans, Mary Monahan Frain, Ed Herbert, Vernon Ingram, Ephraim Katchalsky Katzir, George Laties, Matthew Meselson, K. D. Roeder, Walter Rosenblith, Chester Roys, Hans-Lukas Teuber, Milton Wexler, and George Zink, who inspired our love of scientific inquiry;

to our reviewers, who have encouraged us and brought new insights to both our science and our pedagogy: Susan Abmayr (Pennsylvania State University), Richard Adler (Woods Hole Marine Biological Laboratory), Howard Arnott (University of Texas at Arlington), Ellen Baker (University of Nevada, Reno), Tobias Baskin (University of Missouri, Columbia), James Brammer (North Dakota State University), Carol Brenner (Woods Hole Marine Biological Laboratory), David Cochrane (Tufts University), Stanley Cordilis (Smith College), Nancy Dengler (University of Toronto), Ernest DuBrul (University of Toledo), Richard Elinson (University of Toronto), George Fleck (Smith College), Edward Florance (Lewis and Clark College), Dana García (Southwest Texas State University), Ursula Goodenough (Washington University), John Greenwood (University of Guelph), J. K. Haynes (Morehouse College), Stephen Heideman (Michigan State University), Gae Kovalick (Miami University), Hallie Krider (University of Connecticut), Daniel Lynch (Williams College), Debra Meuler (Cardinal Stritch College), Melissa Michael (University of Illinois at Champaign-Urbana), Jon Minden (Carnegie Mellon University), Beth Mullin (University of Tennessee), Donald Mykles (Colorado State University), Thomas McKnight (Texas A&M University), James Pushnick (California State University, Chico), Maurice Ringuette (University of Toronto), Donna Ritch (University of Wisconsin, Green Bay), Elaine Rubenstein (Skidmore College), David Scicchitano (New York University), Julian Shepherd (Binghamton University), and Pat Williamson (Amherst College);

and, especially, to Julie Alexander, Liz Widdicombe, Jane Sanders Wood, Lee Marcott, and Beth Ahrens, all of Saunders College Publishing, who actually made the book happen.

Allan J. Tobin
Richard E. Morel
December 1995

ABOUT THE AUTHORS

Allan J. Tobin is presently Professor of Neuroscience and Professor of Neurology at UCLA. He is director of the UCLA Brain Research Institute and the Scientific Director of the Hereditary Disease Foundation. As Scientific Director, he has encouraged the application of cell biology and molecular genetics to disorders of the brain, and he helped organize the consortium that identified the gene responsible for Huntington's disease. His undergraduate work at MIT was in literature and biology, followed by doctoral training at Harvard University in Biophysics, with an emphasis on physical biochemistry. He did postdoctoral work on cell surface molecules at Weizmann Institute of Science and on erythroid development at MIT. He has an active research laboratory, which employs cellular and molecular methods to study the production and action of GABA, the major inhibitory signal in the brain. These studies, he hopes, may eventually lead to new therapeutic approaches to epilepsy, Huntington's disease, and juvenile diabetes. Dr. Tobin is the recipient of a Jacob Javits Neuroscience Investigator Award from the National Institute of Neurological Disorders and Stroke.

Dr. Tobin has taught introductory courses in cell and molecular biology since 1971, first at Harvard University, then at UCLA. An innovative and interactive teacher, he received a Faculty Teaching and Service Award from the Biology Department at UCLA. Like the textbook he has co-written, his teaching style is marked by a continuing interest in examining and chronicling the process of scientific inquiry.

Richard E. Morel is a professional science writer and editor. He holds an appointment as Research Associate in the Chemistry Department at Smith College. During his career as an editor, he has developed art and text for some of the most successful science texts in college publishing. He was educated in biology at Tufts University where his research interests centered on nerve cells and insect behavior. His current research is related to the biological significance of self-ordering structures found in far from equilibrium thermodynamic systems. His questions concern how these phenomena relate to biodiversity and the possible existence and characteristics of extraterrestrial life forms.

CONTENTS OVERVIEW

TABLE OF CONTENTS

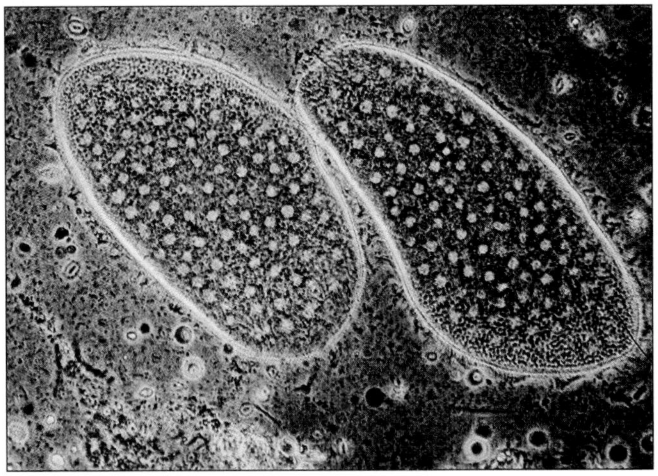

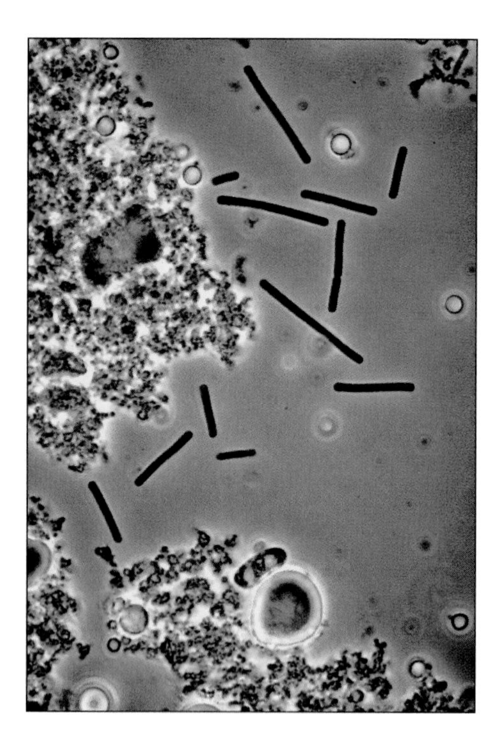

CHAPTER 3
Macromolecules 60

CHAPTER 4
Directions and Rates of Biochemical Processes 87

PART II

The Functional Organization of Cells 117

CHAPTER 5
How Do Biologists Study Cells? And What Have They Learned About Subcellular Organization? 120

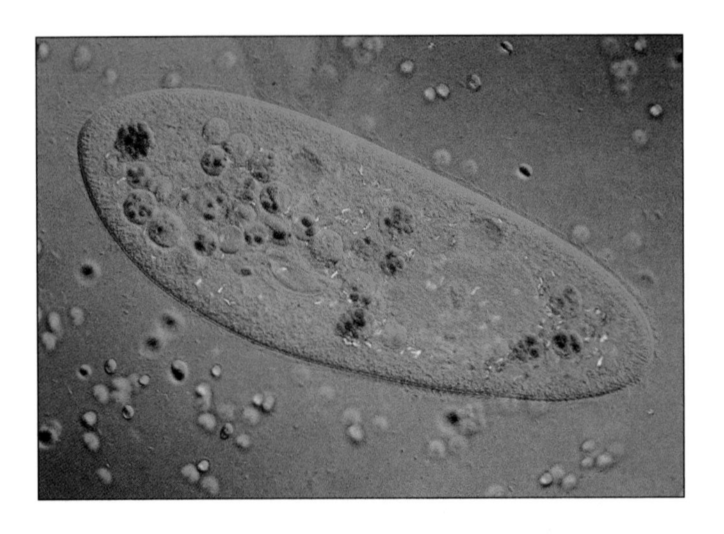

CHAPTER 6
Cell Membranes, Skeletons, and Surfaces 151

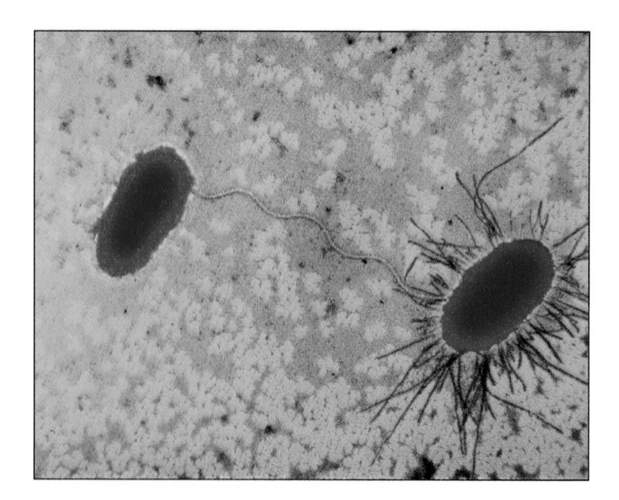

CHAPTER 8

How Do Cells Get Energy from the Sun? Photosynthesis 225

PART III

The Continuity of Cellular Information 249

CHAPTER 9

Cell Reproduction 252

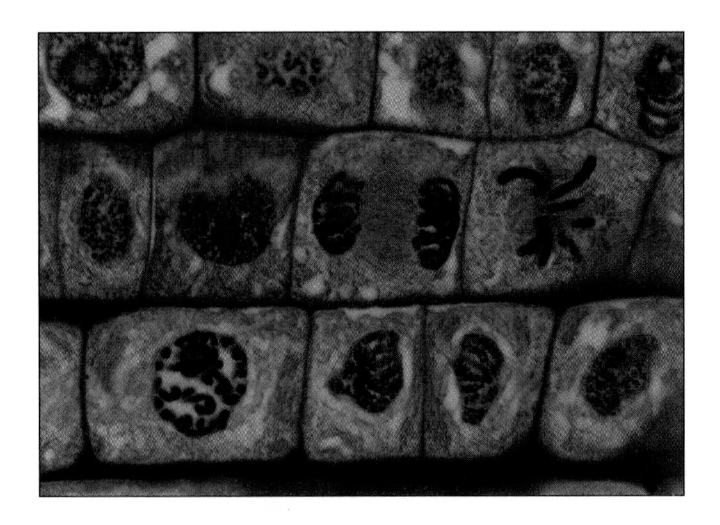

CHAPTER 10

Chromosomes and the Life Cycle: The Role of Meiosis in Sexual Reproduction 276

CHAPTER 11

Inheritance: Genes and Chromosomes 294

CHAPTER 12

What Is a Gene? 316

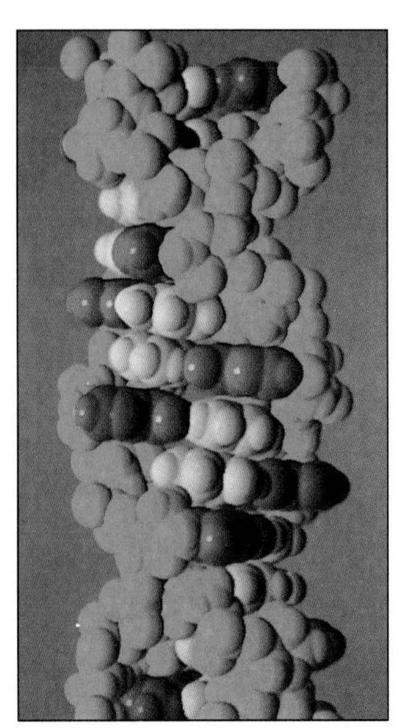

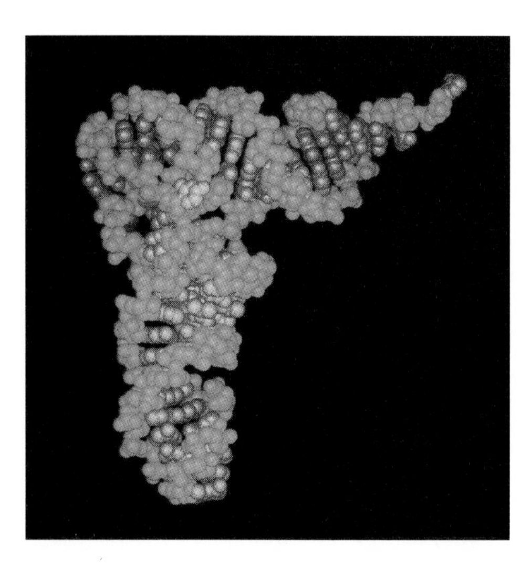

BLACK HILLS
MARATHON
2:39.32

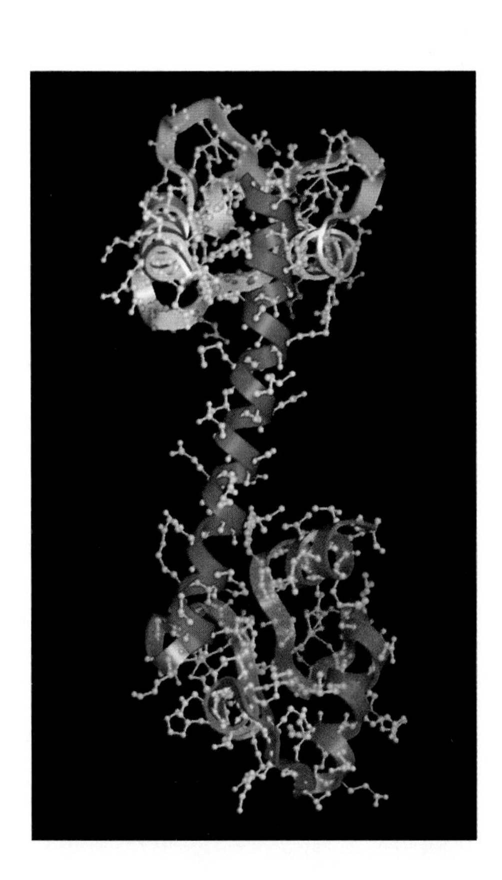

CHAPTER 20
Cellular Contributions to Animal Development 539

CHAPTER 21
Growth and Development of Flowering Plants 576

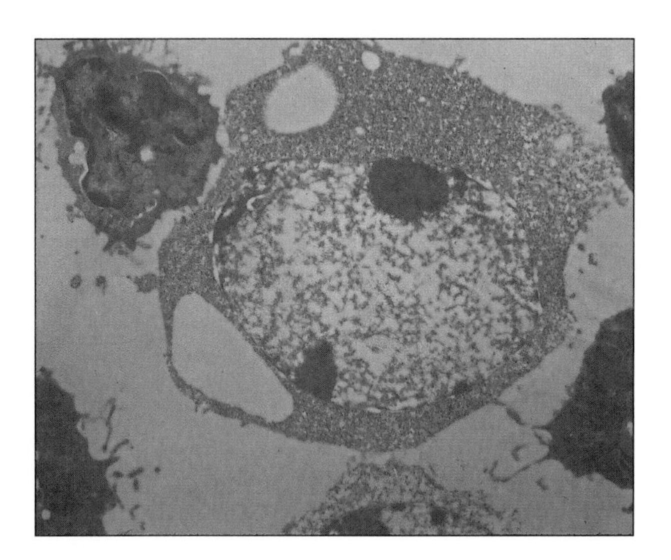

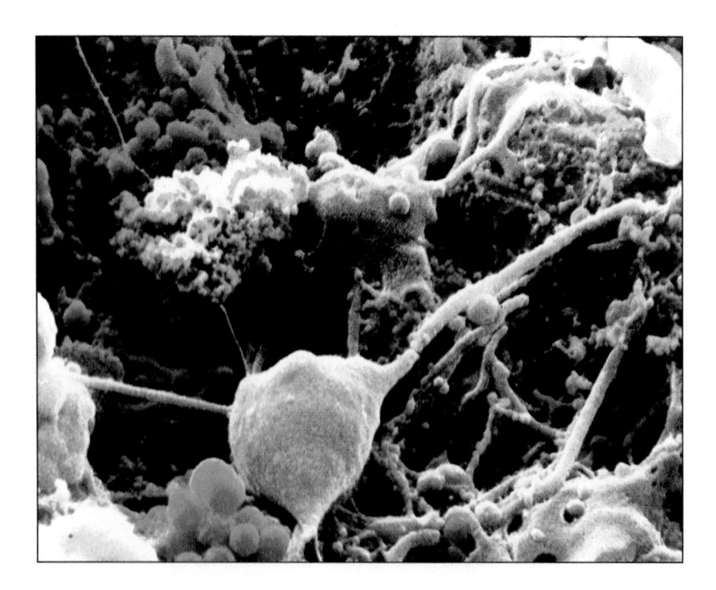

SYNOPSIS OF ICON AND COLOR USE IN ILLUSTRATIONS

The following symbols and colors are used in this text to help in illustrating structures, reactions, and biochemical principles.

Elements

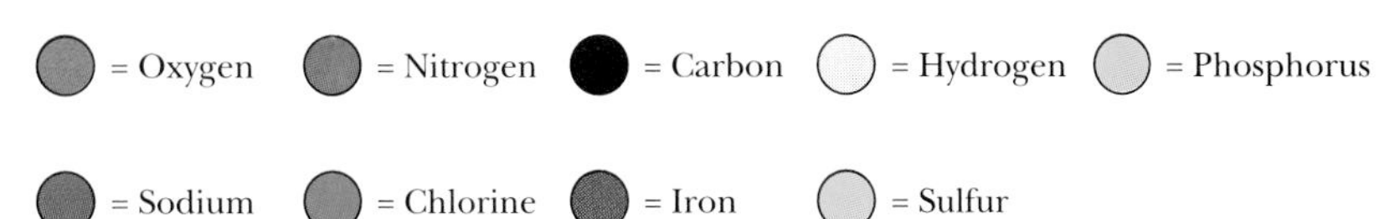

Electronegativity icons

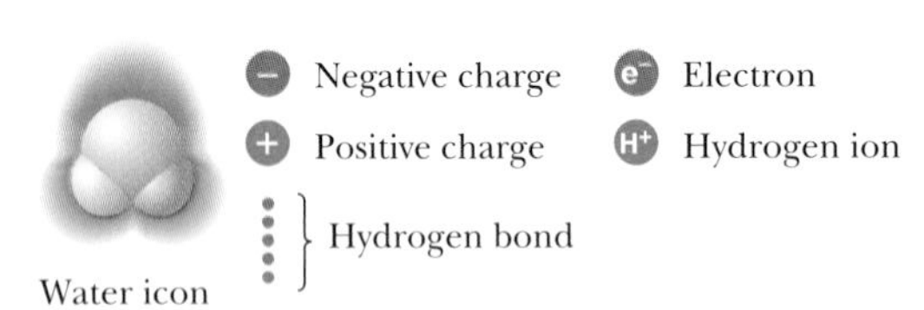

Small molecules and groups that are common reactants or products in many biochemical reactions are symbolized by the following icons:

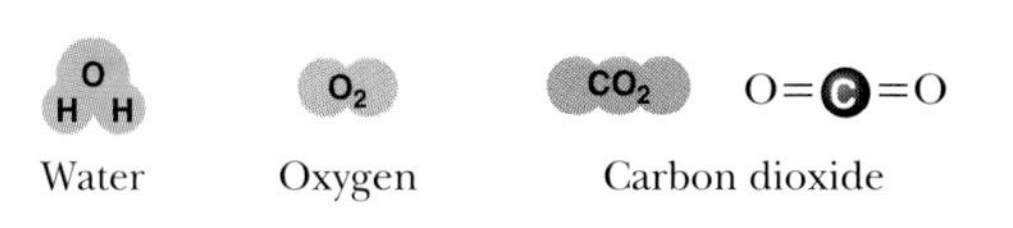

All sugars

Icons representing "energy" and metabolism molecules:

Cell organelles:

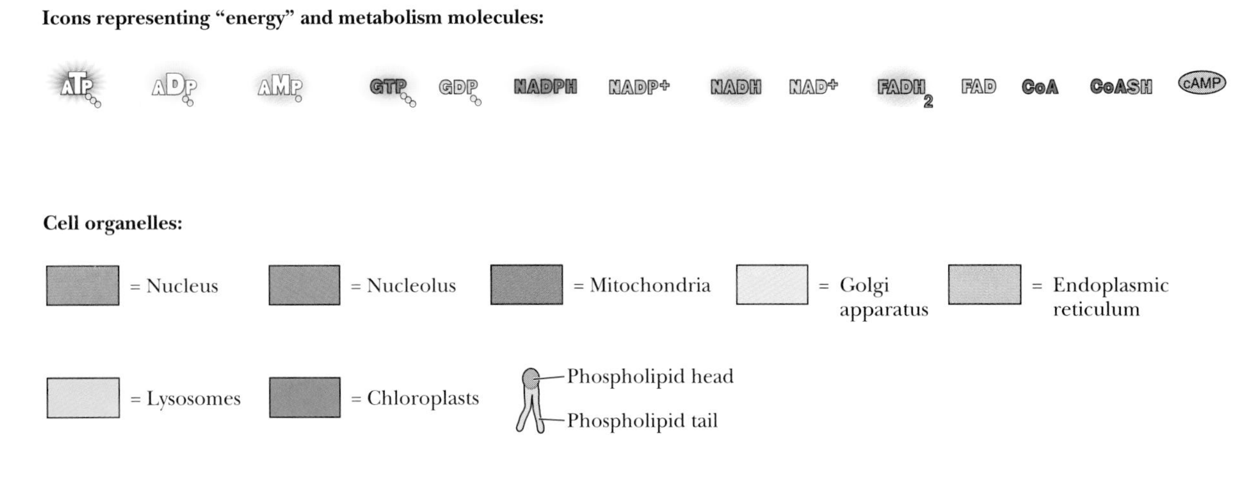

Nucleotide bases:

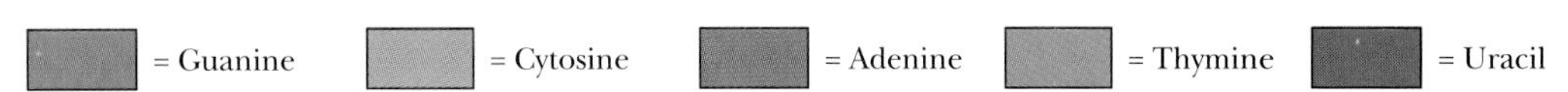

Animal cell **Plant cell**

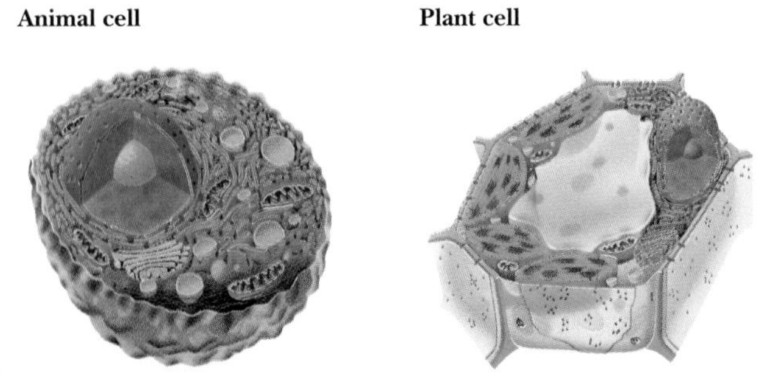

Asking About Cells

PART I

Introduction: The Chemistry of Cells

Chicken skeletal muscle.

LAB TOUR

NAME: H. ROBERT HORVITZ
TITLE: PROFESSOR OF BIOLOGY
ADDRESS: MASSACHUSETTS INSTITUTE OF TECHNOLOGY, CAMBRIDGE, MA 02139

■ ***How did you come to study biology?***

I suppose we could start with my butterfly collection in my first year of high school, but I would say that that drove me more away from biology than towards it. As an undergraduate, one of my ex-roommates told me that biology was more than collecting butterflies and dissecting formaldehyde-treated dead animals. He said that it had become quite analytical. So with basically no knowledge about and no experience in biology, I decided to go to graduate school in the area of biology.

■ ***What is the major question that you and your laboratory are asking?***

There are actually three questions. The first is a question of development, or how one starts with a single cell, a fertilized egg, and generates from that fertilized egg a vast number of cells of many different types that somehow get together and form the organisms that we know, animals and plants. The second is a question of behavior: How do nervous systems control what animals can and do do? That question relates to how a nervous system is wired and how it processes information. Both of these questions relate to the third: How can these processes go wrong in the context of a human being and cause disease? What are the biological bases of human disease?

In terms of what we actually do, the disease that we've put the most effort into is the human degenerative disease known as Lou Gehrig's disease, for the baseball player, and more precisely as Amyotrophic Lateral Sclerosis or ALS. This is a disease in which a particular class of nerve cells—motor neurons that drive muscles—die, and by doing so cause the patient to lose control over muscular activity. Ultimately what happens is that patients die because their muscles can't drive breathing.

■ ***Why have the big questions remained important to you?***

The big questions remain important because they are not solved. That response might surprise a lot of people, because there has been such a revolution in biology during the last 30 years that the level of understanding is profound. Nonetheless, if you had to design something to make a human being from an egg, you still wouldn't know how to do it. In the area of human disease it's a little different, because some diseases are mechanistically understood. But there are still a lot of diseases out there. The ones that I'm most interested in are neurodegenerative diseases, like ALS, and cancer, which is a disease of development.

■ ***What is your strategy as you pursue answers to the questions you have posed?***

The strategy that we've taken is to start with the simplest things and move step by step in a forward direction. My own choice is not to study people, who develop in a very complex way, and who behave in even more complex and incomprehensible ways. They're very interesting, but people are difficult objects for study. I decided to work on the simplest animal that undergoes a profound development, generating many cell and tissue types including a nervous system and a musculature and an intestine and so forth. This animal is a microscopic worm, known as *Caenorhabditis elegans.*

■ ***Why do you use this organism in your research?***

It's very easy to use in the context of genetic experiments. Geneticists are concerned with rare events, mutations that occur in one in a million organisms. Growing a million mice is difficult and expensive and requires a lot of room. Growing a million worms is easy. It requires just a few petri plates. Also, the worms are transparent. This allows us to view a living animal and see every single cell. It turns out that there are only 959 cells in its body, and that number is precise and constant from animal to animal. By contrast, a human being has 10^{12} to 10^{14} cells in the body. Furthermore, you can observe the cells in *C. elegans* from the one-cell stage, so you can watch the one cell as it divides to two, and the two as they divide to four, so we know the complete developmental history of every cell in the animal. In addition, by making very thin slices and looking at them in the electron microscope, every cell can be studied structurally, and this has yielded a wiring diagram of the nervous system. We can look at every nerve cell and see how it connects to other nerve cells and to muscles. *C. elegans* is the pilot project for the human genome project. It's the first animal for which there is an essentially complete physical map of the genome, and its DNA is now being sequenced in its entirety.

■ ***Describe some typical experimental steps in your research.***

Well, let's say we want to know what controls the behavior of egg-laying. We would study egg-laying and determine, as we have, that the environment controls egg-laying, that there are conditions in the environment that cause eggs to be laid at a random rate, and conditions that cause eggs to be laid at a slower rate. Then we would analyze that behavior, for example, with the laser microbeam. We take a laser and shoot it through a microscope and in this way we can do surgery on single cells in living animals. By shooting out cells that we suspect might be involved in the behavior, we can see which of the cells really are involved in the behavior. We get a phenotype: "This is what happens when that cell isn't there."

One thing that is worth commenting on is that people often think that you do experiments to prove what you know. If every experiment you did proved what you knew, nothing would be exciting. The excitement comes when something is unanticipated. And, to me, the biggest excitement comes when something is anticipated but turns out to be wrong. I think there is a great circle in experimental biology. Because one has a question—"How is egg-laying controlled?" We do an experiment. We get a result. And from that, we can define new experiments.

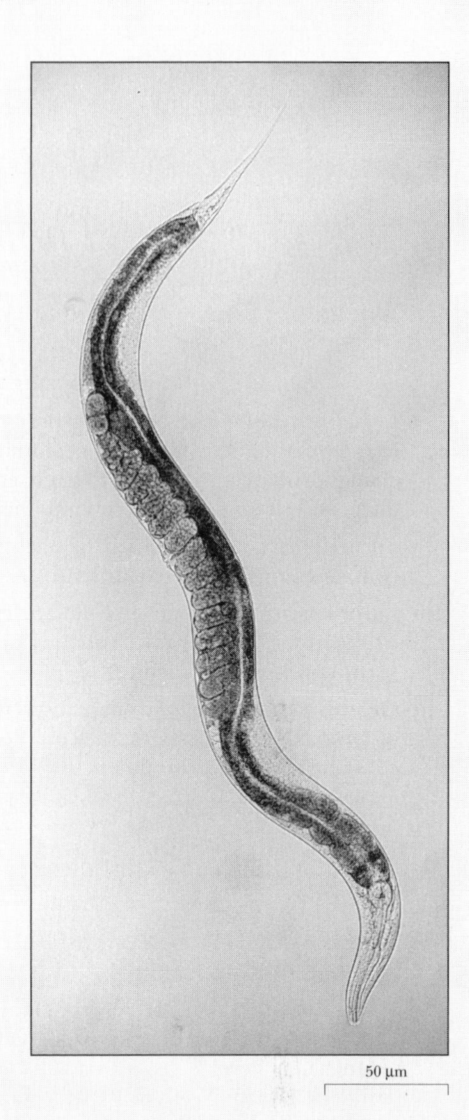

Caenorhabditis elegans. (G. J. Beitel and H. R. Horvitz)

CHAPTER 1

The Challenges of Cell Biology

KEY CONCEPTS

- Living organisms consist of well-ordered parts, obtain energy from their surroundings, perform chemical reactions, change with time, respond to their environments, reproduce, and share parts of a common history.
- All organisms are composed of one or more cells, which contain common subcellular structures.
- Eukaryotic cells contain subcellular structures surrounded by membranes, but prokaryotic cells have no such membrane-surrounded structures.
- Organisms and cells have adaptations that resist the effects of environmental changes by acting to restore a relatively constant internal environment.
- A model is a simplified view of a particular process, generally based on observations.
- Identifying causes and effects requires experiments, not just observations.
- Models lead to hypotheses that can be tested by experiment.
- Hypotheses are most valuable if they lead to predictions that can be disproved by experiment.
- Statistics allows researchers to evaluate experiments that give varying results.
- The generation of hypotheses often depends on bringing together information and ideas from separate fields.

OUTLINE

ESSAYS

Looking at our surroundings through the telescope and the microscope has fundamentally changed the ways that we think about ourselves. The telescope showed that the earth is just a part of the universe, not its center. The microscope showed that humans are, in many ways, just another species, recognizably similar to the millions of other species on our planet.

Revolutions in both astronomy and biology began in 1609, when Galileo began to devise new ways of producing high-quality lenses. Galileo's observations with his new telescope supported the view that the earth revolves around the sun, rather than the reverse, as people had previously thought. His eloquent challenge to official doctrine, however, put him in trouble with the authorities of his time. In 1633, the 69-year-old Galileo was forced to deny the conclusions of his life's work and to spend the last 8 years of his life under house arrest.

The microscope has stirred up less trouble than the telescope, but its contributions have been no less revolutionary. Anton van Leeuwenhoek, who sold drapes and pins in 17th-century Holland, became fascinated with the power of lenses to reveal new forms of life. Everywhere Leeuwenhoek looked—in the water of ponds and rain barrels, in the human mouth and intestine—he found new forms of life (like those in Figure 1-1) swimming about. He was the first person to see and to draw bacteria.

With his tiny lenses, some no larger than a pinhead, Leeuwenhoek showed that life is not limited to the organisms of everyday experience. Because Leeuwenhoek did not share the secrets of microscope making, no one else (other than those who used his microscopes) was to see such small organisms for another 200 years, although other microscopes were also built elsewhere. Leeuwenhoek's lenses, however, revealed a new world, one that people wanted to see. Both the Queen of England and the Tsar of Russia peered through his lenses.

Other microscopes eventually revealed that every organism is composed of enclosed compartments called cells. The word "cell" originally referred to the small apartments of a monastery, nunnery, or prison. Thomas Hooke, a 17th-century English physicist and astronomer, coined the biological use of "cell" when he saw the dead walls of cells in a piece of cork through a microscope in the collection of the Royal Society of London. All organisms—including bacteria and lilacs, tortoises and hares—share this fundamental principle of organization, as illustrated in Figure 1-2.

CELLS ARE THE FUNDAMENTAL LIVING UNIT OF ALL ORGANISMS

Microscopes have forced biologists to ponder our definition of life ever more closely. Many of the things that Leeuwenhoek called "animalcules" were single-celled organisms, which we now call protozoa or protists. These free-living cells display the characteristics of life listed in Figure 1-3. They are (a) highly organized; (b) obtain energy from their surroundings; (c) perform chemical reactions; (d) change with time; (e) respond to their environments; and (f) reproduce. They also display the last listed characteristic of life: (g) they appear to be related to more complex forms of life, such as ourselves.

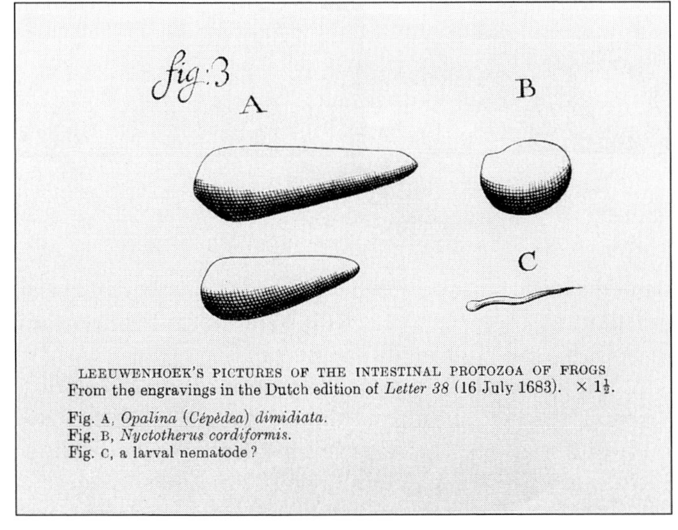

(a)

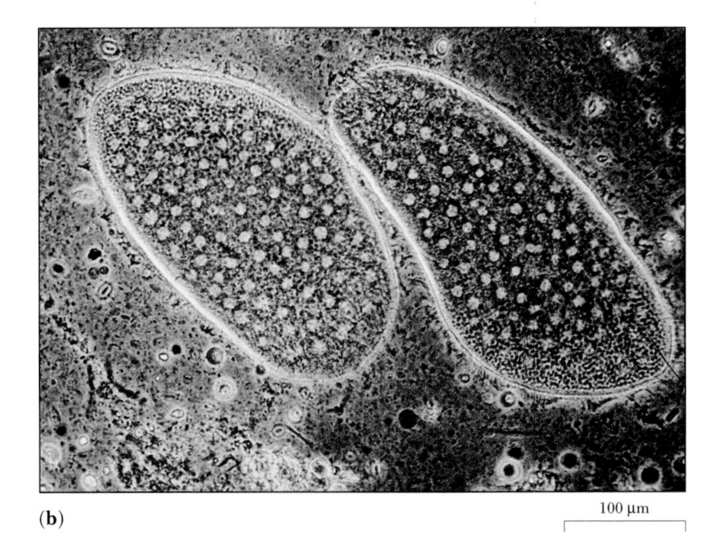

Figure 1-1 "Animalcules," seen for the first time through Leeuwenhoek's lenses, are highly organized single-celled organisms. **(a)** Leeuwenhoek's drawings; **(b)** *Opalina,* as seen with a modern microscope. *(a, College of Physicians of PA; b, Biophoto Associates/Photo Researchers)*

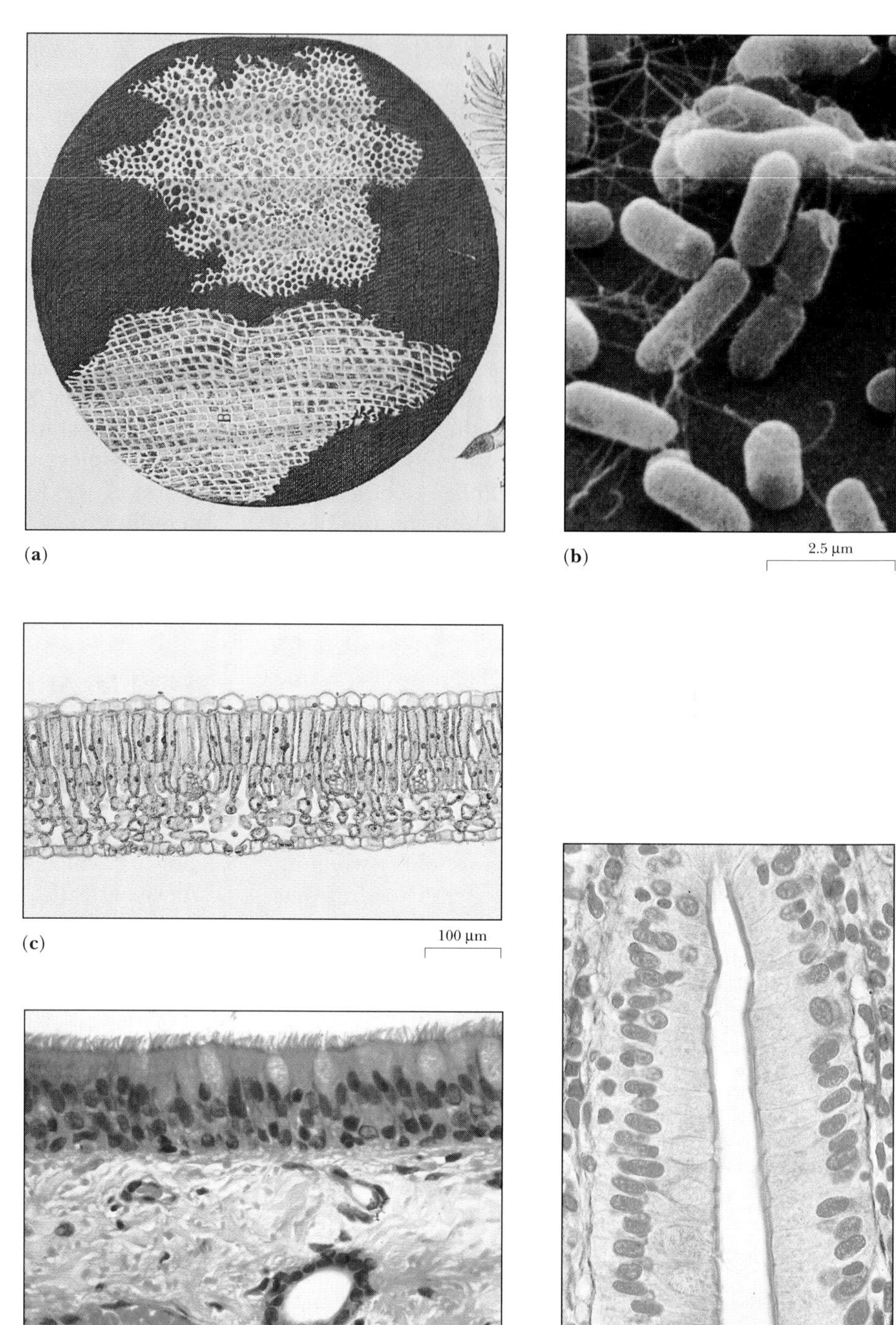

Figure 1-2 Cells are the basic organizational unit of all organisms. **(a)** Hooke's drawing (published in 1665) of cellular organization in a piece of cork; the drawing reminded Hooke of cells in a monastery. **(b)** *E. coli,* common bacteria in the human gut; each organism is a single cell. **(c)** Cross section of a lilac leaf. **(d)** Cells lining the lung of a mammal. **(e)** Cells lining the intestine of a mammal; notice the resemblance and differences between the cells lining the lung and those lining the intestine.
(a, Leonard Lessin/Peter Arnold; b, David M. Phillips/Visuals Unlimited; c, Runk/Schoenberger/Grant Heilman; d, Biophoto Associates/Photo Researchers; e, Manfred Kage/Peter Arnold)

Further improvements in the microscope in the early 19th century revealed that cells have an internal structure. Armed with better microscopes, biologists examined more and more tissues from both plants and animals. Wherever they looked, they found cells that resembled the free-living cells seen by Leeuwenhoek. Every cell appeared to be independently alive. In 1839 botanist Matthias Schleiden and zoologist Theodor Schwann proposed that *all* plants and *all* animals consist of cells.

But where do cells come from? How do the organisms in a pond or rain barrel get there? Where do the new masses of cells in cancers come from? How does an inflamed tissue become filled with white blood cells? Understanding the origin of cells is tremendously important to both biology and medicine.

The physician, statesman, and anthropologist Rudolf Virchow studied these questions in great detail in the mid–19th century. He observed individual cells dividing, each giving rise to two daughter cells. Virchow showed that cells always come from other cells: Cells in ponds descend from other cells, cancers arise by the uncontrolled division of an organism's own cells, and white blood cells accumulate in an inflamed tissue from the circulating blood. Virchow's work demonstrated that cells never arise from

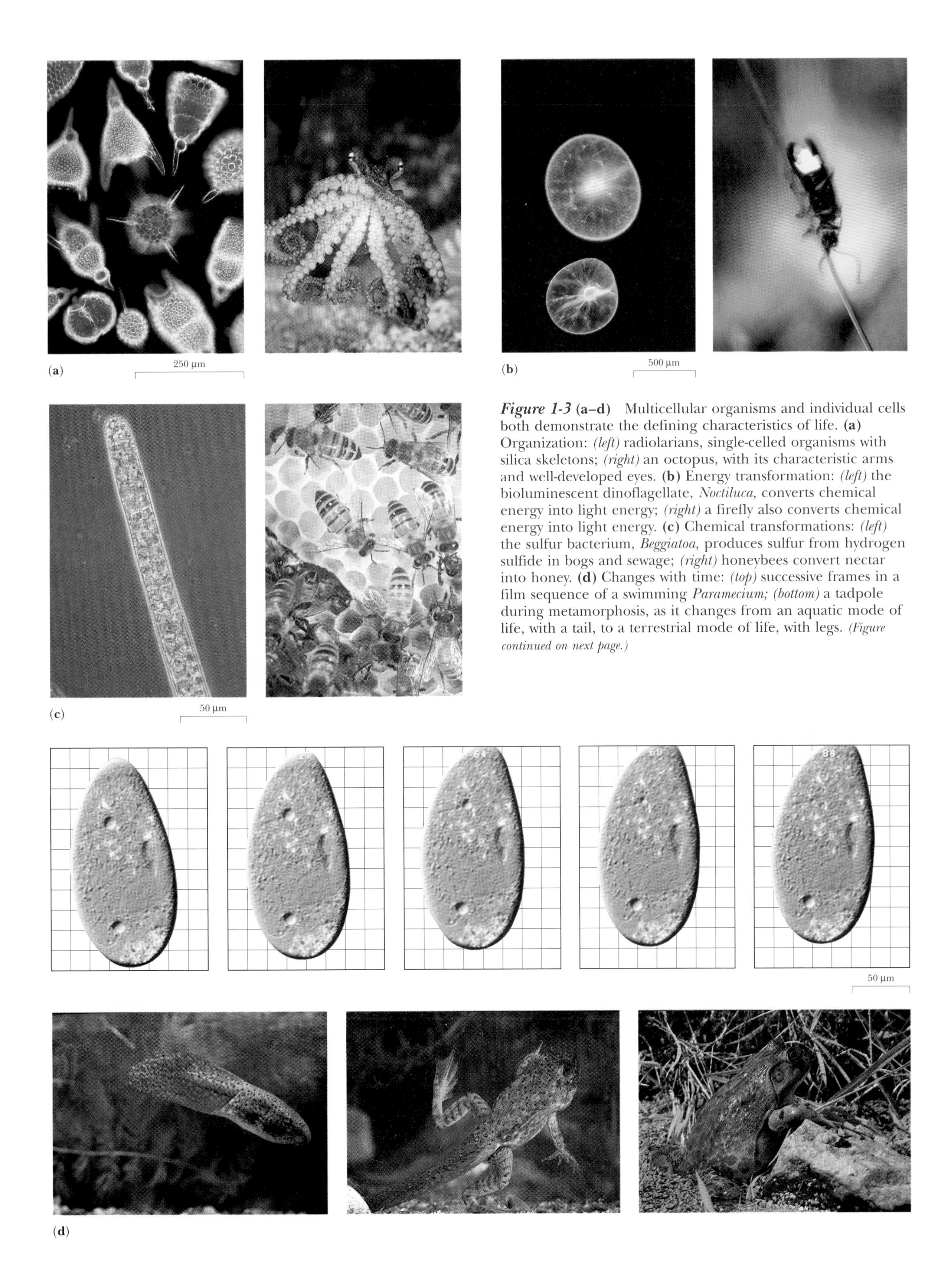

Figure 1-3 **(a–d)** Multicellular organisms and individual cells both demonstrate the defining characteristics of life. **(a)** Organization: *(left)* radiolarians, single-celled organisms with silica skeletons; *(right)* an octopus, with its characteristic arms and well-developed eyes. **(b)** Energy transformation: *(left)* the bioluminescent dinoflagellate, *Noctiluca,* converts chemical energy into light energy; *(right)* a firefly also converts chemical energy into light energy. **(c)** Chemical transformations: *(left)* the sulfur bacterium, *Beggiatoa,* produces sulfur from hydrogen sulfide in bogs and sewage; *(right)* honeybees convert nectar into honey. **(d)** Changes with time: *(top)* successive frames in a film sequence of a swimming *Paramecium; (bottom)* a tadpole during metamorphosis, as it changes from an aquatic mode of life, with a tail, to a terrestrial mode of life, with legs. *(Figure continued on next page.)*

Figure 1-3 **(e–g)** **(e)** Responsiveness to environment: *(left)* amebas of the cellular slime mold, *Dictyostelium discoideum,* aggregating into a multicellular form that produces spores; *(right)* a plant growing toward light. **(f)** Reproduction: *(left)* a yeast cell dividing; *(right)* a brood of goslings, following a goose. **(g)** Continuity: *(left)* limestone deposits more than 2 billion years old are nearly identical in form to stromatolites formed by contemporary cyanobacteria; *(right)* a fossil of an ancient prawn, from the Jurassic period (about 250 million years ago), next to a contemporary relative. *(a, left, Manfred Kage/Peter Arnold, right, E. R. Degginger/Photo Researchers; b, left, Manfred Kage/Peter Arnold, right, Gregory K. Scott/Photo Researchers; c, left, Paul W. Johnson/Biological Photo Service, right, W. J. Weber/Visuals Unlimited; d, top, Karl Aufderheide, Texas A&M University, bottom, (1) Patrice/Visuals Unlimited, (2, 3) Joe McDonald/Visuals Unlimited; e, left, Cabisco/Visuals Unlimited, right, Runk/Schoenberger/Grant Heilman; f, left, Charles Hoffman, Boston College, right, Lester Christman/Visuals Unlimited; g, left, Sinclair Stammers/Photo Researchers, right, (1) A. D. Copley/Visuals Unlimited, (2) Tom McHugh/Photo Researchers)*

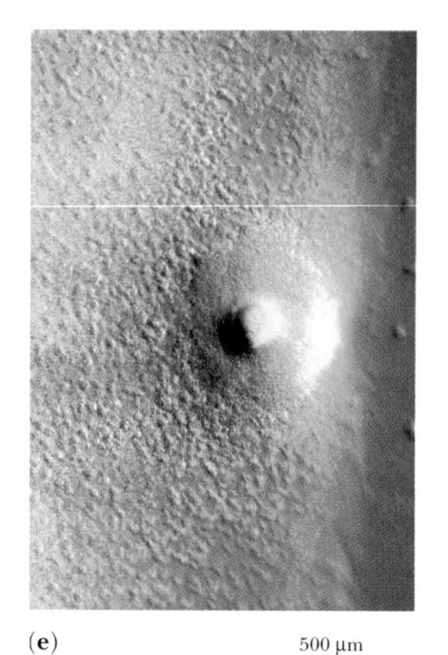

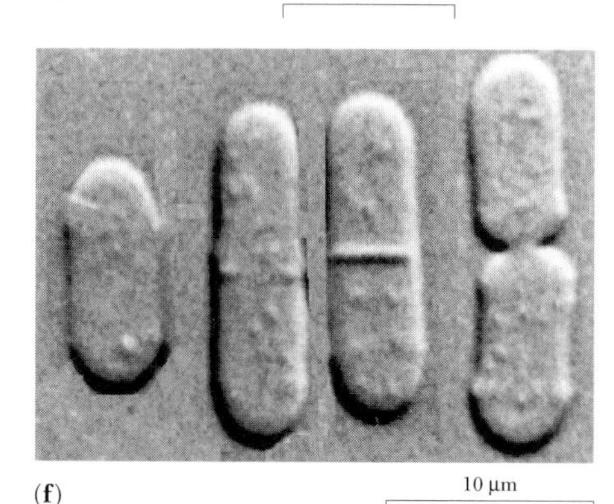

noncellular material. Virchow also revolutionized medicine by showing that diseases result from changes in specific kinds of cells.

The works of Schleiden, Schwann, and Virchow are summarized in what is now called the **Cell Theory:**

1. All organisms are composed of one or more cells.
2. Cells themselves are alive and are the basic living unit of function and organization of all organisms.
3. All cells come from other cells.

Organisms from All Species Are Made of Cells

Most biologists now divide all organisms into six large categories, called kingdoms, as illustrated in Figure 1-4. The

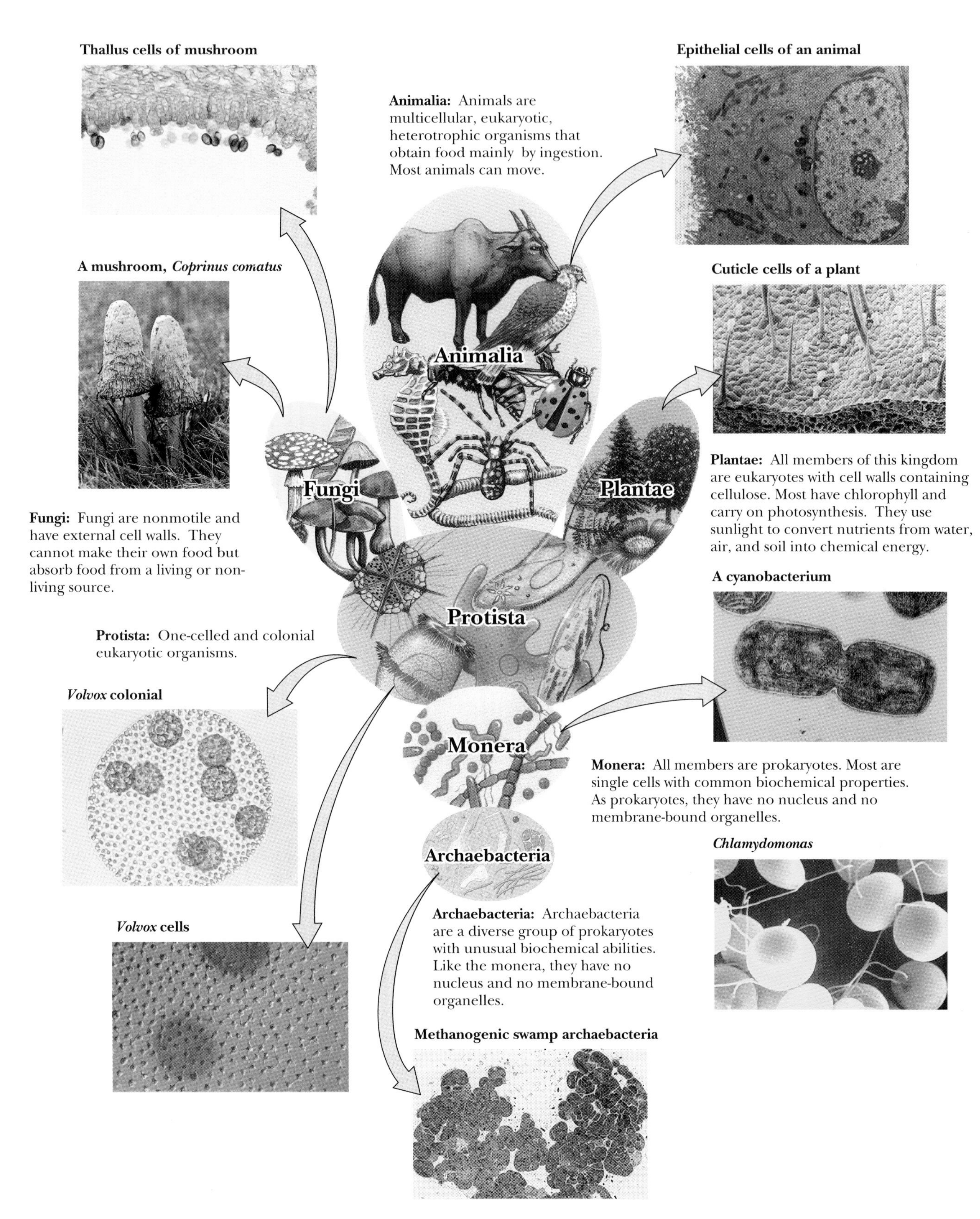

Figure 1-4 Cells of the six kingdoms all contain a nucleus or nucleoid, cytosol, and a surrounding membrane. The membrane can be seen in transmission electron micrographs. The photos shown depict the following: A **moneran,** *Agmenellum quadruplicatum,* a cyanobacterium; cyanobacteria have no nucleus or other subcellular organelles that are surrounded by membranes. An **archaebacterium,** *Methanosarcina barkeri,* a methane-producing organism isolated from sewage; note the absence of a nucleus. A **protist,** *Volvox,* a colonial green alga, at the light microscope level and individual cells of a *Volvox* seen with a Nomarski interference microscope; note that each cell in a *Volvox* colony resembles the single-celled alga *Chlamydomonas,* a protist. A **fungus,** the shaggy mane mushroom, *Coprinus comatus,* and a micrograph showing individual cells of this mushroom. Individual cells in a **plant,** a geranium leaf; scanning electron micrograph. Individual cells in an **animal,** a single cell from the lining of the gut; transmission electron micrograph. *(clockwise from top right: David Phillips/Photo Researchers; C. Y. Shih, R. Kessel/Visuals Unlimited; Henry Aldrich/Visuals Unlimited; R. Kessel, G. Shih/Visuals Unlimited; Henry Aldrich/Visuals Unlimited; Walker England/Visuals Unlimited; James W. Richardson/Visuals Unlimited; Dick Poe/Visuals Unlimited; R. Knauft/Photo Researchers)*

(a)

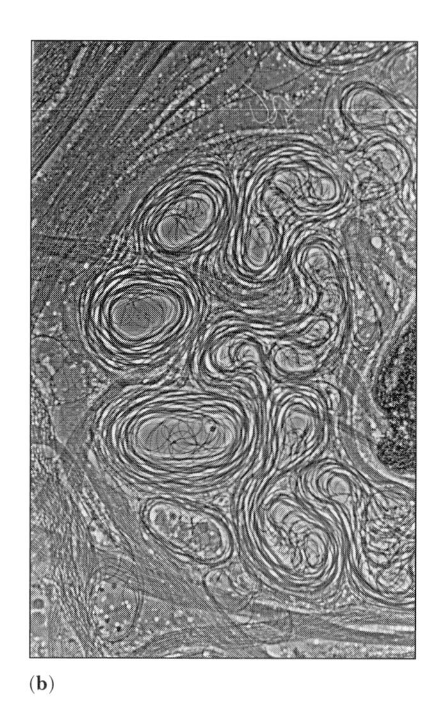

(b)

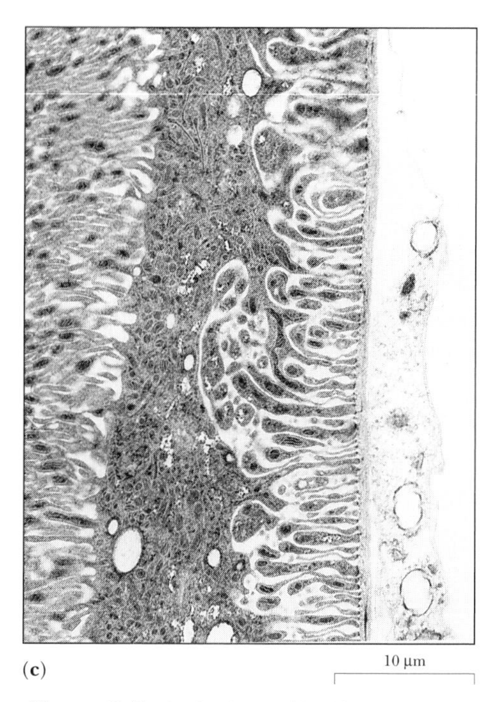

(c)

(d)

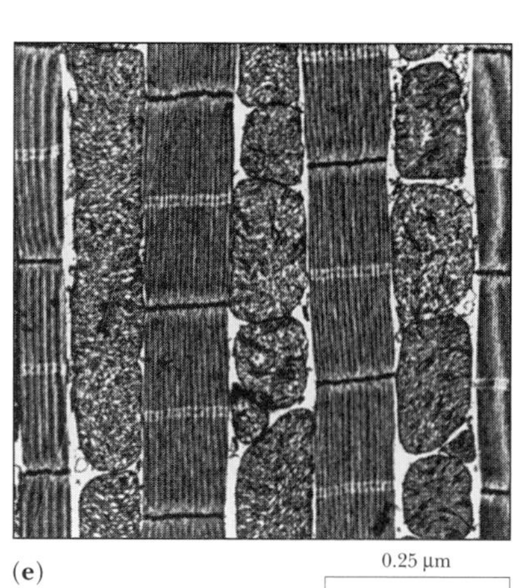

(e)

Figure 1-5 A single multicellular organism contains many distinctive cell types, here illustrated within the fruit fly *Drosophila,* a favorite experimental organism. **(a)** *Drosophila;* scanning electron micrograph; **(b)** sperm cells; **(c)** Malpighian tubule cells, which function like a kidney; **(d)** omatidia, the organizational units of the *Drosophila* eye, are each composed of 8 photoreceptor cells, 7 of which are visible from this top view; **(e)** flight muscle. *(a, Dr. Jeremy Burgess/Science Photo Library/Photo Researchers; b, Margaret Fuller, Stanford University; c, Dr. Helen Skaer, Dept. of Human Anatomy, Oxford University; d, Tania Wolff, University of California, Berkeley; e, Mary Reedy, Duke University Medical Center)*

most familiar of these kingdoms include the organisms that are always multicellular—**Animalia** (animals) and **Plantae** (plants). **Fungi** include both multicellular organisms (like mushrooms) and single-celled organisms (like baker's yeast). A fourth kingdom, the **Protista,** consists mostly of single-celled organisms but also contains some related multicellular species as well. The protists include algae, water molds, slime molds, and protozoa (the "animalcules" of Leeuwenhoek). The fifth kingdom, the **Monera,** consists mostly of single-celled organisms with distinctive properties, including bacteria and cyanobacteria (formerly called blue-green algae). Recently, a growing number of biologists have added a sixth kingdom to the list, the **Archaebacteria,** a diverse group of ancient, single-celled organisms with distinctive biochemical features, which we discuss more in Chapter 24.

Just as organisms are wonderfully diverse, so are cells (as illustrated in Figure 1-5). The appearance of a cell often reflects its specialization to perform a particular function. We can see some examples of such specialization in the microscopic picture of human skin in Figure 1-6: nerve cells that carry touch information, epithelial cells that serve as a barrier to the outside world, and cells that produce hairs. Plant cells, such as those illustrated in Figure 1-7, are similarly specialized: Some cells are specialized to transform energy from sunlight into energy-rich sugar molecules, while other cells are specialized for the distribution of these molecules to the rest of the plant. Mammals have at least 200 recognizably different cell types.

Viruses Are Not Cells and Are Not Alive

In the last four decades, powerful microscopes, called *electron microscopes* (discussed in Chapter 5), have revealed the structure of disease-causing agents called **viruses.** (See Figure 1-8 and Chapter 18.) Viruses, however, are not cells,

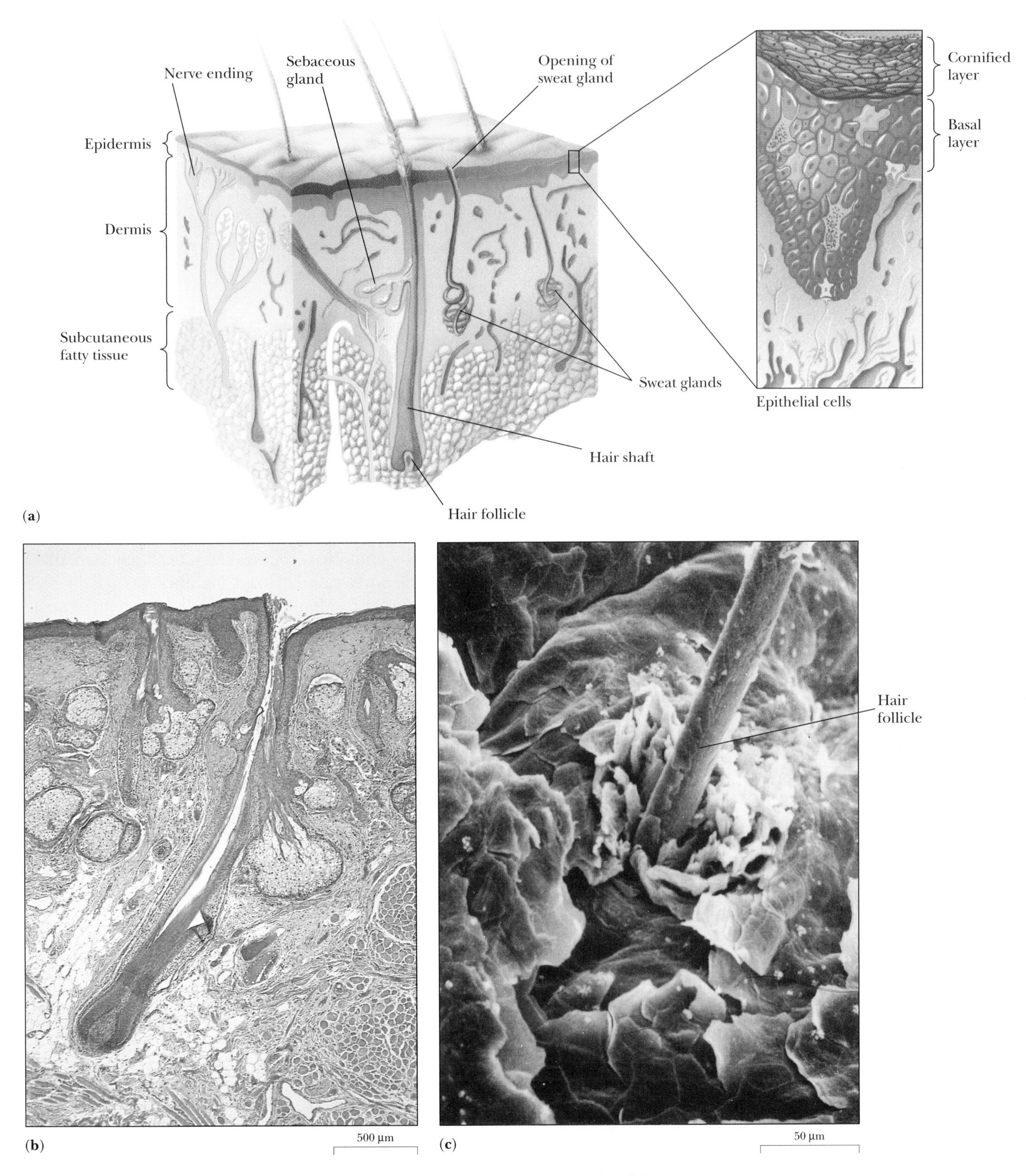

Figure 1-6 Specialized cells have exaggerated versions of common properties. **(a)** Cross section of human skin, showing several types of specialized cells, including the endings of a nerve cell, which carries information in the form of electrical impulses; epithelial cells, which form a tightly connected surface that serves as a barrier to infection; and hair follicles, which assemble structural proteins to form a hair; **(b)** cross section of skin seen with a light microscope; **(c)** the surface of the skin, seen with a scanning electron microscope, showing cells on the skin's surface and a hair follicle. *(b, G. W. Willis, MD/Biological Photo Service; c, Fred Hossler/Visuals Unlimited)*

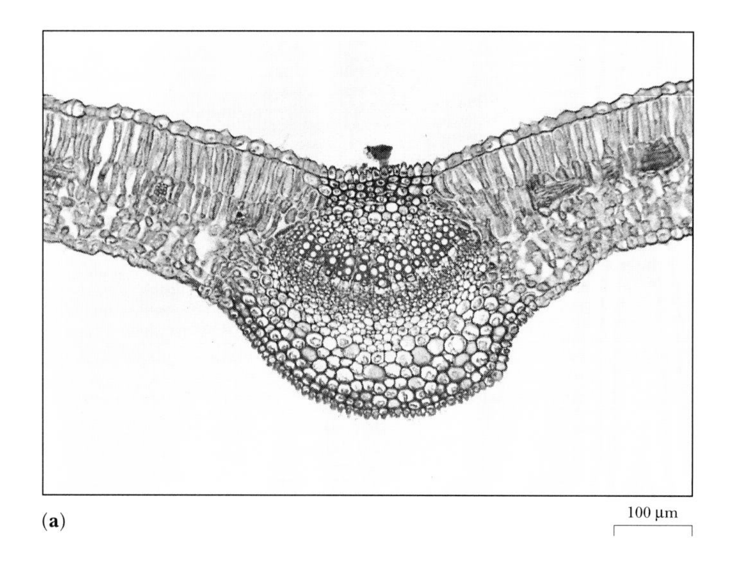

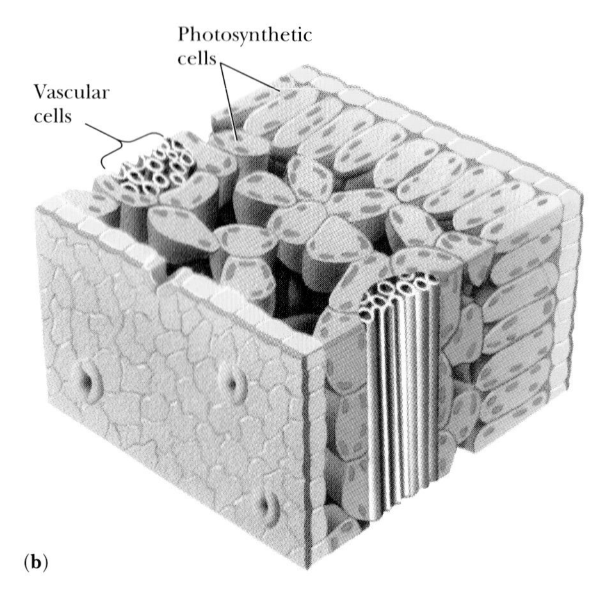

Figure 1-7 Specialized cells in a lilac leaf, showing photosynthetic and nonphotosynthetic cells **(a)** as seen with a light microscope; **(b)** drawing. *(a, Bruce Iverson)*

and they do not fulfill all the definitions of life. Viruses must depend on cells to obtain energy and perform chemical reactions.

Cells Contain Common Subcellular Structures

Beneath the superficial diversity of cell types is a basic unity. With an electron microscope, biologists have found that all cells contain recognizable subcellular structures, including a fairly standardized membrane that encloses the cell's contents. Membrane-bounded subcellular structures are called **organelles** [little organs].

Some subcellular structures are present in all types of cells, but others are more limited in their distribution. All cells show one of two basic plans, called eukaryotic and prokaryotic. **Eukaryotic** cells [Greek, *eu* = true +

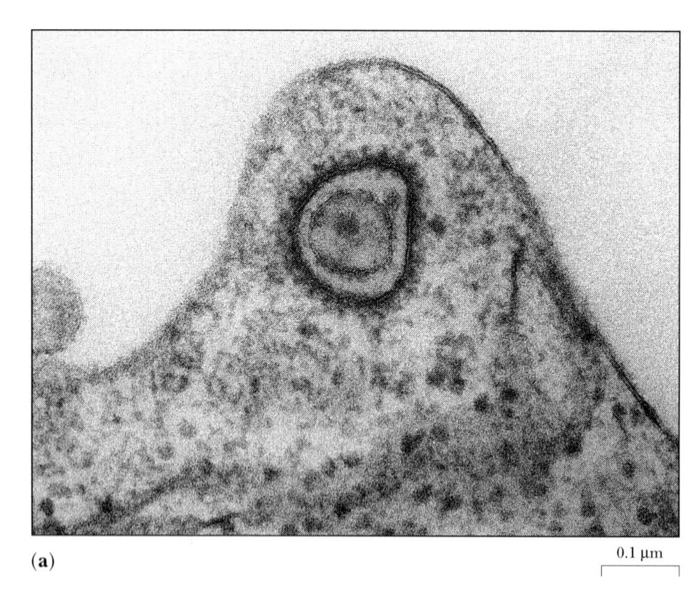

Figure 1-8 Viruses are not cellular and are not alive. **(a)** The human immunodeficiency virus (HIV), which causes AIDS, in the process of budding out of a white blood cell; **(b)** isolated adenovirus. *(a, Hans Gelderblom/Visuals Unlimited; b, Phototake NYC)*

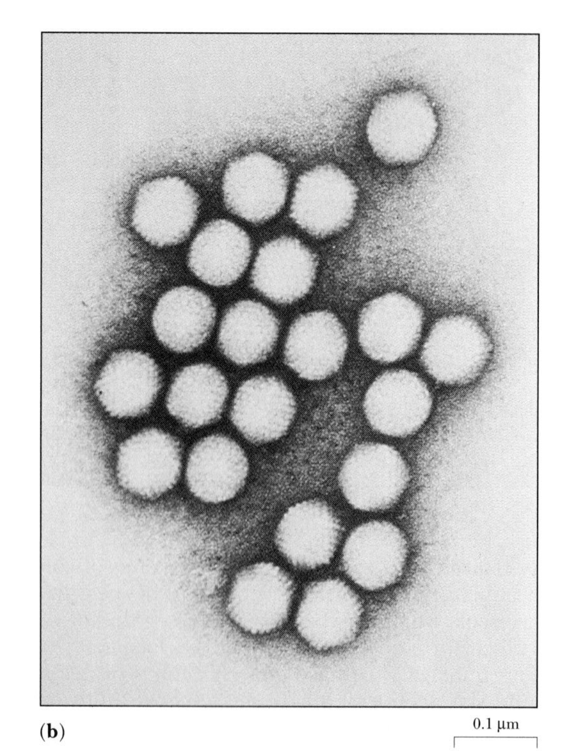

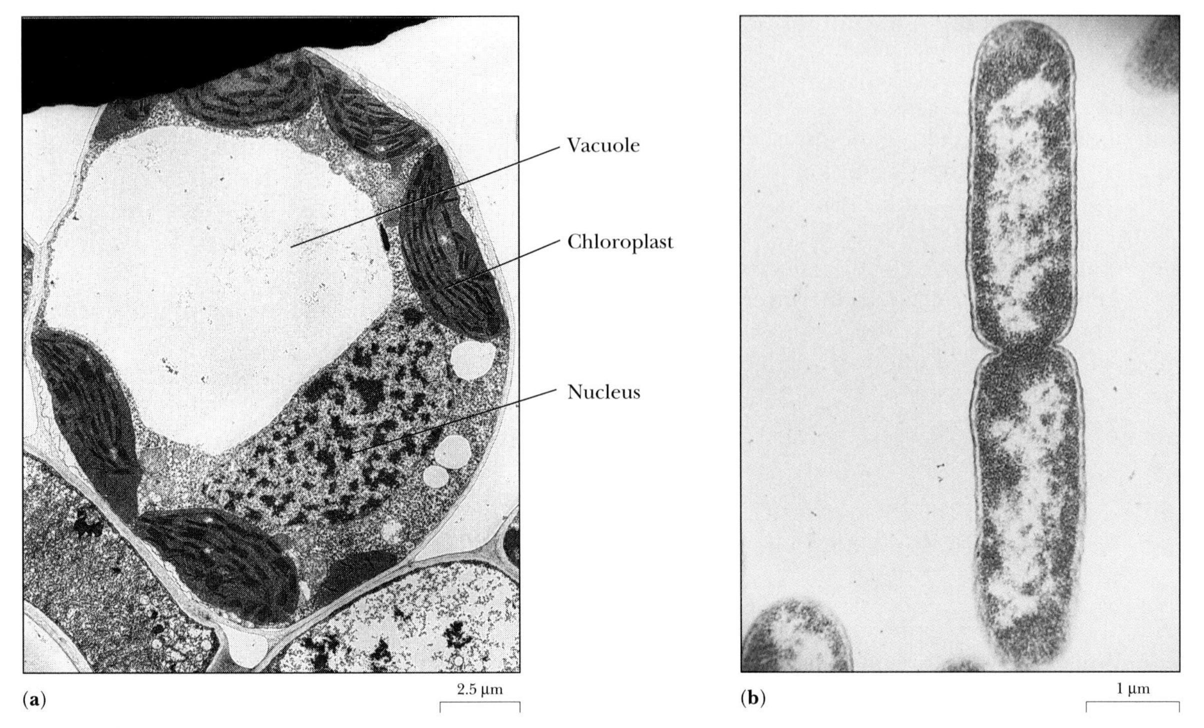

Figure 1-9 **(a)** Cell within a plant leaf, showing common organelles at a higher magnification than in Figure 1-7; transmission electron micrograph; **(b)** *E. coli,* showing lack of internal membranes surrounding the nucleoid region; transmission electron micrograph. *(a, Dr. G. F. Leedale/Photo Researchers; b, Kwang Shing Kim/Peter Arnold)*

karyon = nucleus] contain a central organelle called a nucleus, which is surrounded by a double membrane. Eukaryotic cells also contain other specific types of organelles, as shown in Figure 1-9a.

In contrast, **prokaryotic** cells [Greek, *pro* = before] (illustrated in Figure 1-9b) contain neither a nucleus nor membrane-bounded organelles. The Monera are all prokaryotes. Although we (as eukaryotic organisms) tend to view eukaryotic cells as more advanced, prokaryotes are, in some ways, more successful: You have 100 times more (10^{15}) prokaryotic bacteria living in your own digestive tract than you have eukaryotic human cells in your whole body (10^{13})! As we shall see, eukaryotic cells almost certainly evolved from prokaryotic ancestors. (See Chapter 24.)

Despite the many types of cells, there are only 20 or so kinds of organelles, which we discuss in Chapter 5. The diversity of cell types appears to result from combinations and arrangements of the same kinds of smaller units. The specialization of cells almost always depends on the exaggeration of common properties. For example, the cells lining the intestines have elaborately folded surface membranes that increase enormously their ability to absorb food molecules, as illustrated in Figure 1-10.

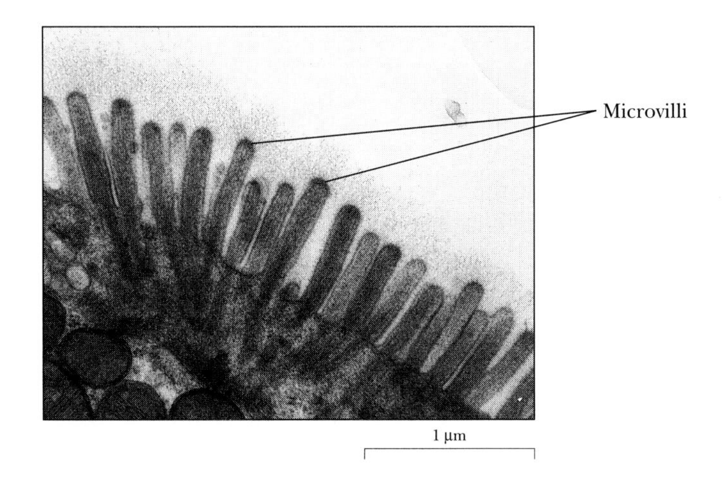

Figure 1-10 Cell specialization. The lining of the human intestine, showing elaborately folded membranes (microvilli). The folding increases the surface area through which the lining can absorb food molecules. Also notice the organelles, surrounded by membranes, in the bottom left; these are mitochondria, which provide the cell with most of its energy. *(David M. Phillips/Visuals Unlimited)*

All Cells and Organisms Resemble One Another at the Molecular Level

Careful observations have established a major theme concerning the chemical nature of cells and organisms. Cells are made almost entirely from only four basic kinds of molecules—lipids, sugars, amino acids, and nucleotides. (See Chapter 2.) At the molecular level, then, cells are more alike than they appear under the electron microscope. The diversity of cells is far greater than the diversity of their molecular components. This diversity arises from the huge number of ways in which the basic molecular components can be arranged.

Modern Biology Seeks to Explain the Properties of Organisms and Cells in Terms of Their Underlying Structures

The effort to understand the whole by understanding the parts is called **reductionism.** From whole organisms to molecules, for every level of organization, biologists seek to explain structure and function in terms of the next finer level of structure. For example, we can understand the action of arm muscles by analyzing levels of organization from whole muscles to the cellular and subcellular structures of the muscles themselves. The reductionist approach has been immensely successful in understanding a wide variety of biological phenomena. The most strident reductionists argue that *all* the properties of an object are totally explainable in terms of the properties of its components.

Critics of reductionism argue, however, that life is not as simple as reductionists would like. Complex structures interact in ways that cannot be predicted from knowledge of the component parts. The whole is greater than the sum of its parts, just as a symphony is much more than a collection of individual notes and chords, and a film is more than a progression of still photographs. Figure 1-11 illustrates this same principle for a painting. At higher magnifications, we lose the sense of the whole picture, and we see only a rather abstract pattern of dots.

To many people, the most interesting and beautiful aspects of biology involve **emergent properties,** characteristics that arise only at the more complex levels of organization. Although it is certainly true that organisms and

(a)

(b)

(c)

Figure 1-11 The "big picture" may sometimes disappear in common details, as in the painting, "The Seine River at Herblay," by Maxamilien Luce, shown here at successively higher magnifications in **(a), (b),** and **(c)**. The original painting measures 80 × 50 cm (about 30 × 20 inches). *(Musée d'Orsay, Réunion de Musées Nationaux)*

cells must all obey the laws of physics and chemistry, physics and chemistry *by themselves* provide little insight into such phenomena as thought, perception, sexuality, and fear. As we see throughout this book, however, physics and chemistry do help explain how cells and organisms operate. Most biologists therefore accept both the power and the limitations of reductionism.

CELLULAR ORGANIZATION ALLOWS THE MAINTENANCE OF A CONSTANT INTERNAL ENVIRONMENT, AS WELL AS CELL SPECIALIZATION AND REPLACEMENT

Although cells and organisms are constantly in flux, their survival depends on their ability to maintain a relatively constant state. The more stable the internal environment of a cell or a multicellular organism, the more likely it is to survive and reproduce. Failure to respond appropriately to environmental changes is likely to mean an early death and fewer descendants. Any mechanism that helps maintain the internal environment in the face of outside changes therefore provides a selective advantage in evolutionary time. The 19th-century French physiologist, Claude Bernard, summarized the immense value of these mechanisms: "The constancy of the internal environment is the necessary condition for free life." The process of achieving a relatively stable internal environment is called **homeostasis** [Greek, *homeo* = like, similar + *stasis* = standing]. Cells, like organisms, counteract changes in their environments, such as fluctuations in the concentrations of individual molecules, that might otherwise interfere with the processes of life.

Despite its derivation [from *stasis* = standing], the term "homeostasis" does not refer to a static condition but to a dynamic process. Homeostasis always involves two stages: (1) detecting deviations from the stable state, and (2) responding to such deviations.

Once a cell or an organism detects deviations, it often uses a strategy called **negative feedback** that neutralizes the effects of external changes. The household thermostat provides a familiar nonliving example of negative feedback. When the temperature of a room falls below the set temperature, the thermostat activates a heater so that the room warms up. When the room again reaches the set temperature, the heater is turned off. The heater stays off until the temperature again falls below the thermostat setting. Although the temperature of the room varies slightly as the heater cycles on and off, the effect of this mechanism is to keep it nearly constant.

The cells of all organisms also regulate their internal environments. Cells make, destroy, and pump chemicals in or out in order to maintain their characteristic compositions and to sustain their normal activities.

The Need to Regulate the Internal Environment Limits Cell Size

Almost all eukaryotic cells are about the same size, usually from 10 to 100 micrometers (μm). Prokaryotic cells are almost always smaller, ranging from 0.4 to 5 μm. No cells are smaller than 0.4 μm, and few are larger than 100 μm. (Some animal eggs are much larger, but most of their volume is taken up with food stores, and some cells of plants, animals, and protists are much longer.) These limitations in part reflect the common evolutionary origin of all cells on earth. But cells are not just an historical accident: Subdivision provides an organism with several important advantages.

For the biochemical machinery of cells to function, the cell must maintain a relatively constant internal environment. Otherwise, the cell's molecular machinery would not adopt the particular shapes needed for biological activity. (See Chapters 3 and 4.) In particular, cells must regulate pH (acidity) and salt concentrations. In addition, cells must be able to take in and use energy-rich molecules and get rid of waste products.

All the materials that enter or leave a cell must somehow get through the cell membrane. Membranes contain specialized "gates" and "pumps" that together regulate the cell's contents by selectively admitting, accumulating, or excluding specific molecules and ions. (See Chapters 6 and 23.) The membrane must keep pace with the demands of such regulation, but it can do so only as long as the cell's volume is not too large.

What happens as a cell gets bigger? Think, for a minute, about the relationship between the surface area and the volume of any object. If the linear dimension (the side) of a cube (like that in Figure 1-12) doubles, its surface area increases by a factor of 4 and its volume by a factor of 8. The larger the cube, the smaller is the **surface-to-volume ratio,** the amount of surface area for each bit of volume. For each 1000 μm^3 of volume, the smaller cube of Figure 1-12 has 60 μm^2 of surface and the larger cube has only 30 μm^2 of surface.

Because larger objects have smaller surface-to-volume ratios than smaller objects of the same shape, larger cells have proportionally less membrane with which to regulate their internal space. Larger cells have a harder time obtaining nutrients, getting rid of wastes, and regulating the internal concentrations of ions and molecules.

Knowing that size affects the regulation of a cell's internal environment, we can now ask why eukaryotic cells are generally so much larger than prokaryotic cells. The answer is that eukaryotic cells have special adaptations—including especially large amounts of internal membranes—that increase their surface areas. Large plant

(**a**) Surface area vs. volume in a cube: As the cube's size increases, there is less and less surface area for each unit of volume.

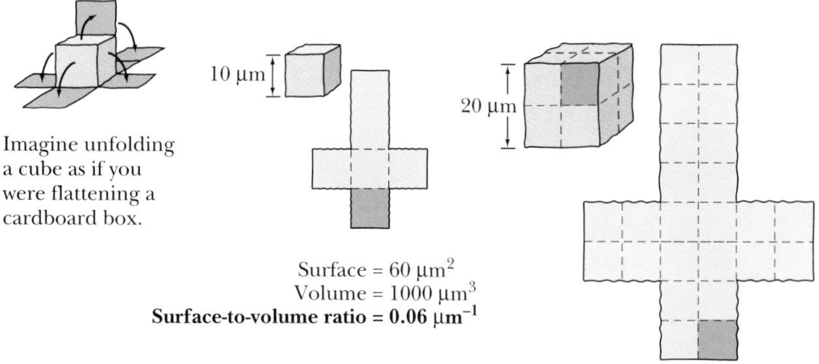

Surface = 240 μm^2
Volume = 8000 μm^3
Surface-to-volume ratio = 0.03 μm^{-1}

(**b**) Instead of growing larger by increasing cell size, organisms grow larger by producing more cells, keeping the high surface-to-volume ratios.

1 cell

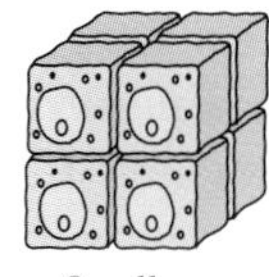

8 cells

For all 8 cells, surface-to-volume ratio is the same.

Conclusion: Smaller cells have more surface area per volume than do larger cells.

Figure 1-12 The relationship between surface and volume in a cube. (**a**) A cube 10 μm on a side has a surface area of 60 square micrometers (60 μm^2) and a volume of 1000 cubic micrometers (1000 μm^3); a cube 20 μm on a side has a surface area of 240 μm^2 and a volume of 8000 μm^3. (**b**) Cellular organization allows each cell to have a high surface-to-volume ratio, even in a large organism.

cells, for example, are usually long and thin, increasing their surface-to-volume ratio, as illustrated in Figure 1-13a. Many eukaryotic cells have convoluted surface membranes, and almost all have elaborate internal membrane systems. Indeed, some eukaryotic cells, such as the cone cell from the retina of a mammalian eye shown in Figure 1-13b, have 50 times more membrane inside than on the surface.

A Multicellular Organism Can Lose and Replace Individual Cells

Organisms, unlike nonliving things, constantly repair and renew themselves. While the material of a nonliving thing stays the same, an organism constantly replaces and remolds itself, exchanging old atoms and molecules for new

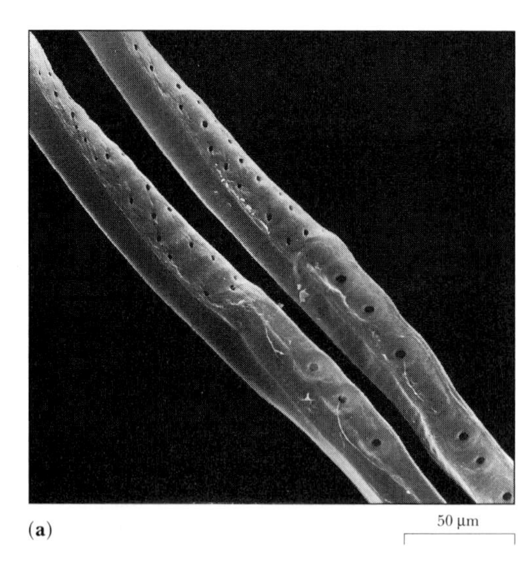

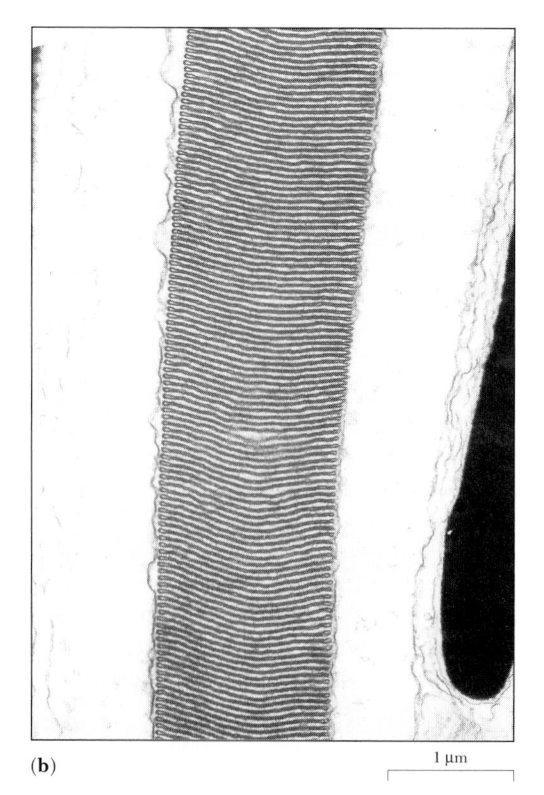

Figure 1-13 Large cells with increased external or internal surface. (**a**) Long thin cells, such as the cells that form the transport systems in a woody plant, have a higher surface-to-volume ratio than spherical cells; (**b**) a cone cell from the mammalian retina showing enormous amounts of internal membrane. *(a, courtesy of W. A. Côté, Jr., N. C. Brown Center for Ultrastructure Studies, SUNY; b, John Dowling, Harvard University)*

ones. All the while, its form and internal properties stay almost constant.

Cells can often live and die independently of the whole organism. In fact, the death of some cells is part of the normal program of development. Cells of the skin, blood, and intestines are continuously replaced during your life. The advantage of such replacement is that the life of a multicellular organism can extend far beyond the life of an individual cell.

Individual Cells May Specialize for Different Tasks

Individual cells may perform distinctive tasks. Your red blood cells carry oxygen, while muscle cells produce mechanical movement. Cells in the roots of a tree absorb nutrients from the soil, while cells in the leaves harvest the energy of sunlight.

In almost all multicellular organisms, individual cells express only part of the genetic information stored in their nuclei. Just as the division of labor makes human societies more productive, the specialization of cells can allow an organism to function efficiently as a "cellular city."

Specialization is important even within individual cells. Membranes divide eukaryotic cells into many compartments, and membrane-bounded organelles can independently regulate their own flows of materials and energy. Each kind of organelle is specialized to accomplish particular tasks, such as the production of ATP or the degradation of wastes.

In summary, the multicellular organization of large organisms brings three general advantages: the ability to regulate the internal environment, the capacity to replace individual cells, and the flexibility to develop specialized cells that perform different tasks. These three characteristics help explain why each of us consists of 10^{13} cells rather than a single large blob.

HOW DO CELL BIOLOGISTS SEEK ANSWERS TO QUESTIONS?

When we ask how an organism (or an automobile or a computer) works, we generally want to know how particular structures are used. The challenge for cell biologists is to explain each biological process in terms of participating cells and subcellular structures. To see how to approach such a problem, we consider the cellular and molecular underpinnings of two familiar exercises—the chin-up and the push-up. What cellular structures change as the arms bend and unbend? How do cells convert the energy provided by food into movement?

The first step in understanding is usually looking (although hearing, smelling, and touching may also contribute to our biological observations). As we have already seen, one of the greatest leaps in biological understanding—realizing that all organisms are cells or groups of cells—depended on the improved microscopes that became available in the 19th century. The first goal of many biologists is to observe, as carefully as possible, the structures involved in a process and how they change. On the basis of such observations we try to picture how changes in structures can bring about a particular biological process. Scientists call such a picture a **model**—a simplified view of how the components of a structure operate.

A model is useful for two reasons. First, it can make predictions that can be tested by experiment. If the predictions are correct, we can have some confidence that we have understood more than we did previously. If the predictions are incorrect, we must modify or reject our model. Second, a model can also suggest how other similar processes work. A successful model may provide a common explanation for many processes. The model that we will develop for muscle movement applies not just to the muscles of the arm but to hundreds of other human muscles, as well as to muscles of other species.

A Model of Chin-Ups and Push-Ups Pictures the Movements of Muscles and Bones

A chin-up, shown in Figure 1-14a, requires the forceful bending *(flexion)* of the arm at the elbow, and a push-up, shown in Figure 1-14b, requires the forceful unbending *(extension)* of the forearm. During a chin-up one can feel the bulging of the *biceps,* the major muscle on the front side of the upper arm, as shown in Figure 1-14a. In a push-up (as shown in Figure 1-14b), another major muscle becomes apparent—the *triceps,* which lies on the back side of the upper arm. As the triceps bulges during a push-up, the biceps becomes less prominent. Similarly, the triceps becomes less prominent as the biceps bulges in a chin-up.

(a)

(b)

Figure 1-14 How do the muscles power movement? A chin-up depends on the action of the biceps muscle, while a push-up depends on the action of the triceps. *(Michael Newman/Photo Edit)*

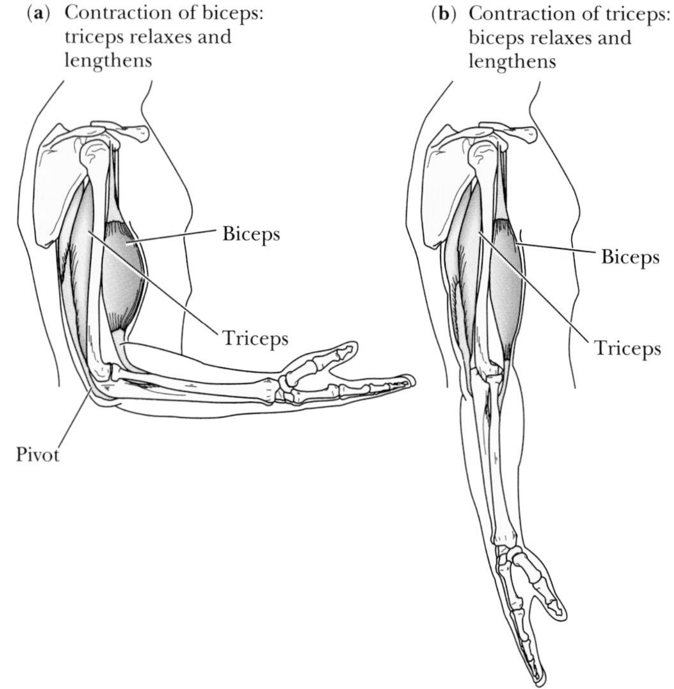

Figure 1-15 The skeleton and muscles of the human arm. **(a)** The contraction of biceps powers the flexion of the arm, while **(b)** the contraction of the triceps powers extension.

We can guess that the bulging of these muscles provides the power needed for movement—the biceps for flexion and the triceps for extension. To understand the mechanism of movement, however, we need to know in more detail how these muscles are attached to the bones of the arm. We can get this anatomical information only from careful dissection and examination of muscles and bones. We then may form a simplified model.

As we can see in Figure 1-15, the skeleton of the upper arm consists of a single long bone that is attached to the shoulder and to the two bones of the forearm. On the front of the arm the biceps muscle runs from one part of the shoulder to one of the bones of the forearm. On the back side, the triceps runs from another bone of the shoulder to the second bone of the forearm.

Now we can make a simple model of the flexion and extension of the arm. In our model, sketched in Figure 1-15, we picture the forearm as a lever that pivots on the elbow. To flex, the biceps contracts and the triceps relaxes (Figure 1-15a); to extend, the triceps contracts and the biceps relaxes (Figure 1-15b).

Our model predicts that the lengths of the biceps and triceps will change during extension and flexion: Each muscle will shorten as it contracts and lengthen as it relaxes. Careful observation shows that this is exactly what happens. The shortening of each muscle leads directly to the bulging that we can feel or see. The model explains why the biceps bulges during chin-ups and the triceps during push-ups. It also suggests a more general picture of how animals move their limbs. For every movement, some muscles contract and others relax.

The biceps and the triceps are an example of an **antagonistic pair,** two muscles that accomplish opposing movements. One member of the pair relaxes while the other contracts. In addition to allowing us to understand the action of the arm, this model immediately helps us make sense of the organization of the muscles of the limbs in the rest of the body. We can try to generalize by asking whether all muscles involved in body movements are arranged in antagonistic pairs. The answer appears to be yes.

Identifying Causes and Effects Requires Experiments, Not Just Observations

Does a change in the length of a muscle move bones, or does the reorientation of bones change the length of the muscle? Or do both bones and muscle change in response to a third set of changes, for example in the nerves?

To establish which events are causes and which are effects, we cannot rely on observations; we must do experiments. We must interfere with the normal arrangement of muscles, bones, and nerves and see if our model can predict what will happen. If shortening the biceps causes flexion, then the muscle alone, without the rest of the body, should be able to move the bones. Our model predicts that if we somehow stimulate the biceps to contract, the arm will flex. If our model correctly predicts the results of our experiment, we can say that we understand the mechanism of this movement.

This represents a practical problem. How can we possibly devise such an experiment? We cannot really separate biceps, triceps, and the bones of the arm from the rest of the body. Biologists try to find material for their experiments that will be particularly suitable for manipulation. Whenever they can, experimenters prefer to use materials (or "preparations") that others have previously studied. This allows them to take practical advantage of specific knowledge and to compare their results with those of other researchers who have used the same experimental preparation.

One of history's most influential biologists formulated and tested a model of limb movement. The scientist was Galen, a Roman physician and philosopher of the second century. Galen made careful observations of gladiators' wounds and dissected a variety of animals, from frogs to monkeys. On the basis of his observations, Galen shrewdly (but incorrectly) proposed that muscle contraction was caused by the nerves' pumping "animal spirits" into the muscles. These animal spirits, Galen said, caused the muscle to bulge and consequently to shorten. On the face of it, this was a reasonable hypothesis. It was particularly appealing because it suggested an easy way to understand the mechanism of movement.

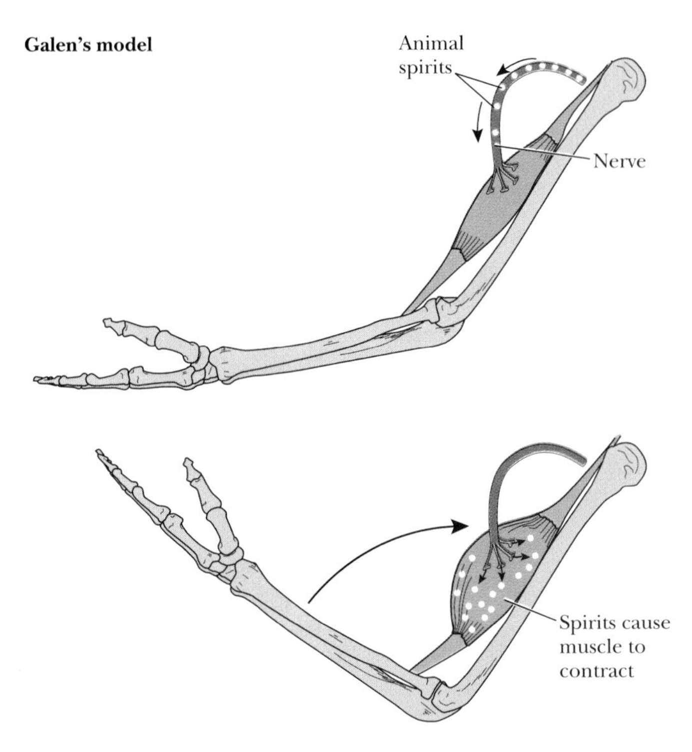

Figure 1-16 Galen's model of arm movement, in which nerves, rather than muscles, provide the power for movement.

Galen's model, shown in Figure 1-16, portrayed nerves rather than muscles as the structures responsible for movement. Is this true? Do muscles merely respond passively to the action of nerves, or do they themselves act? Distinguishing these two alternatives required examination of the activity of a muscle that was freed of nerves. If an isolated muscle were still able to contract upon proper stimulation, we would know that Galen's idea was wrong. Following this reasoning, Galen himself removed nerves from muscles and showed that they would not contract. He interpreted the failure of isolated muscle to contract as support for his idea about "animal spirits." In fact, the failure to contract was due to his failure to use the right stimulus.

Only after the discovery of electricity in the late 18th century did anyone find a proper stimulus for muscle contraction. In 1798, the Italian physiologist Luigi Galvani demonstrated that electricity could stimulate the contraction of a frog's leg muscles. Galvani's experiments showed that muscle action does not depend exclusively on nerves. In fact, as discussed in Chapter 23, nerves do contribute to muscle contraction by providing electrical stimulation.

Frog muscles are much smaller than human muscles, but their appearance is similar. At the microscopic level the two are almost indistinguishable. What is true for the leg muscle of the frog is generally true for muscles involved in bodily movements in other animals, including the muscles we need for chin-ups and push-ups. On the basis of experiments with frog muscles, we can conclude that, after electrical stimulation, muscles themselves cause movement.

Knowing that muscles cause movement, we can now say with some confidence that we understand the flexion and extension of the human arm. Our model is not much more complicated than the working of a seesaw or any other simple lever. The model can be easily summarized: (1) Movement results from the contraction of muscles, and (2) muscles are arranged in pairs, so that the action of one is opposed by the action of the other.

Answering one set of questions always leads to many others. For example, we may also want to understand our model at a more detailed level: How does the muscle itself work? What are the cellular and subcellular bases of muscle contraction?

A More Refined Model of Muscle Contraction Requires Looking at a Finer Level

A microscope allows us to see the common organization of all vertebrate muscles involved in movement. Such muscles consist of bundles of thin fibers. Each of these fibers is a single cell, about 10–40 μm in diameter and up to several centimeters long. Each cell contains long cylindrical structures, called *myofibrils,* each about 1–2 μm in diameter. Regular stripes, called **striations,** cross each fiber, as illustrated in Figure 1-17. When a muscle contracts, the distance between the striations decreases. This observation suggests that the contraction of a muscle depends on the movement of smaller structures.

The electron microscope reveals that each myofibril is made of still smaller fibers, called *filaments,* as shown in Figure 1-18. The basic event of muscle movement is the relative movement of two sets of microscopic filaments. (See Chapter 18.) As was the case for muscle action on the larger scale, observation with a microscope suggested a model. Again, testing the model required experiments.

The Hungarian biochemist Albert Szent-Gyorgyi provided the preparation needed for such experiments. Szent-Gyorgyi treated a rabbit muscle in such a way that he removed most of its cellular structures but left microscopic striations intact. He then added a solution of **adenosine triphosphate (ATP),** a molecule that provides energy for many biochemical processes in all organisms, as discussed throughout this book. The result of adding ATP was that the striations moved closer together, just as they do when an intact muscle contracts. Szent-Gyorgyi's preparation has allowed biologists to study the biochemical basis of muscle contraction.

Szent-Gyorgyi's experiments established that the basis of muscle contraction lies in the action of ATP upon the striated structures that can be observed in the microscope. We can now begin to see how the operation of specific biological structures may underlie and explain complex biological processes. Many of the questions that we ask in this book concern the relationships of structures

(a) **Myofibril relaxed**

Z-line 2.5 μm Z-line

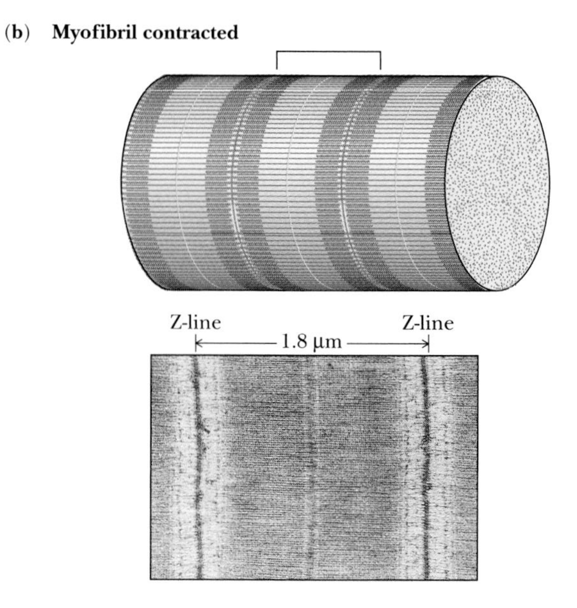

Figure 1-17 Each fiber of striated (voluntary) muscle is a single long cell with many nuclei. Each cell contains many myofibrils, each about 1–2 μm in diameter. The distance between striations changes during muscle contraction, suggesting that the movement depends on the movement of molecules. **(a)** A relaxed myofibril; **(b)** a contracted myofibril. *(a, b, James Dennis/Phototake/NYC)*

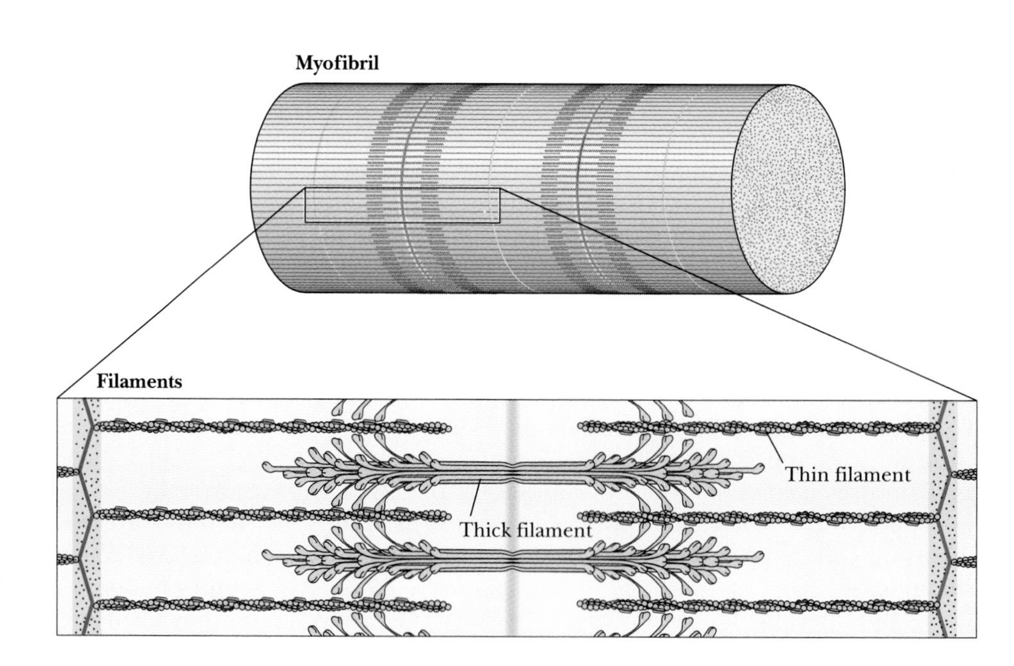

Figure 1-18 Each myofibril consists of two sets of microscopic filaments.

and processes. The questions are often the same: What are the involved structures? What are they made of? How do they capture energy? What controls their operation? The answers to our questions may lie at either a more or a less detailed level of biological structure. The goal, however, remains the same—to understand processes in terms of specific structures.

HOW IS SCIENTIFIC INQUIRY DIFFERENT FROM CURIOSITY?

Our discussion of muscle action illustrates how scientists go about trying to understand the world. The beginning of scientific inquiry is curiosity: Something in the world commands our attention, and we begin to ask questions.

Sometimes our attention is captured by pure wonder or by arresting beauty. In other cases, we are stimulated by a practical need, perhaps to cure a disease or to increase the yield of a crop.

Natural curiosity such as this is the beginning of science. Albert Einstein, for example, attributed his scientific achievements to his "slow development." Einstein said that he never ceased to ask the questions that most people ask only as children, but he continued to ask such questions when he had the power to answer some of them. Many scientists pride themselves on their child-like capacity for asking lots of questions, but scientific inquiry is not merely asking questions. Science is curiosity controlled and channeled. Scientists try to limit themselves to questions to which they can seek answers by making models and doing experiments.

Because science has been so successful in increasing knowledge, many philosophers and historians have tried to understand exactly how scientists work. Many have referred to a systematic **scientific method,** a manner of formulating, testing, and eliminating hypotheses. Most scientists, however, do not think of themselves as following the rules of a single "method." Rather, scientists pursue knowledge in a variety of individually creative ways. We can nonetheless give a general description of how most scientists go about learning about the world.

Models Lead to Hypotheses

A scientist's first step is to focus his or her curiosity on a single question or small set of questions. The next task is to think how to answer them. This often means recognizing resemblances to more familiar structures or processes, as in the similarity of an arm and a lever. Such similarities lead to a model of how a process works or how a structure is assembled.

A model simplifies and offers new insight. We saw the arm-lever resemblance by ignoring all the parts of the arm except the four bones and two muscles. A model must also be consistent with other scientific knowledge and ways of thinking.

One of the biggest joys in pursuing science comes when we can explain a complex and confusing phenomenon with a simple model. But model building is also inherently frustrating. First, a model can never be as complete as the real world. Second, knowledge is always incomplete: A good model of muscle action, for example, had to await the discovery of electricity. A provisional model, even if inadequate, allows a scientist to formulate an **hypothesis,** an informed guess about the way a process works or a structure is organized.

Hypotheses Lead to Predictions About the Results of Experiments

An hypothesis not only takes into account what is already known but also leads to specific predictions about what will happen in a new, experimental situation. Galen's model of muscle action, for example, led to his hypothesis that muscle contraction requires the pumping of animal spirits from nerves. This hypothesis in turn led to a concrete and testable prediction: If nerves are removed from a muscle, the muscle will not contract. When an experiment turns out as predicted, scientists become more confident about the hypothesis and the underlying model. If an experiment turns out differently, however, the hypothesis is likely to be wrong.

Notice that experiments can never really *prove* hypotheses. The experimenter could obtain the predicted result even if the model were incorrect. Galen's prediction was correct—muscle without nerve did not contract. But his model (as we now know) was wrong, because a muscle can in fact contract without a nerve, if it is stimulated electrically.

An experiment can *disprove* an hypothesis if the predicted results are not obtained. Even when a prediction is not correct, however, a scientist may still hold on to an hypothesis and an underlying model. Such a scientist may not just be stubborn: An experiment may not give the predicted result for reasons that have nothing to do with the hypothesis being tested. For example, the tissue may be dead, or a process may occur only at a particular temperature. Or the prediction might be wrong because of faulty logic or because the simplified model neglected something important.

When Is an Hypothesis Valuable?

We can quickly generate many hypotheses to explain any observation. Scientists need ways of deciding which hypotheses to ponder. In general, an hypothesis is valuable only when someone can devise an experiment that could *disprove* it if it were incorrect. If there is no conceivable way of disproving an hypothesis, then it presents no opportunity to know whether the underlying model is correct. The hypothesis that the movement of limbs depends on the pumping of fluid from the nerves to the muscle is testable. But the hypothesis that limb movement depends only on the spiritual state of the experimental animal is not testable.

When an hypothesis makes correct predictions about experimental results, scientists naturally gain confidence both in the hypothesis and in the model that suggested it. The more a model can explain, the more important it becomes to other scientists.

The model of arm flexion and extension is particularly valuable because it allows us to understand many other movements. The model suggests many related hypotheses, for example, that antagonistic pairs of muscles are responsible for the movements of legs, fingers, or toes. For each kind of movement, there is another set of testable hypotheses. When a number of such tests give consistent results, however, scientists are often ready to accept a generalized hypothesis and a model. Establishing such generalizations from a pattern of specific examples is called **inductive reasoning.**

Once scientists have worked sufficiently hard to test general models for a group of related structures or processes, they may decide that they can accept a fairly broad set of hypotheses. They may refer to this set of related hypotheses as a **theory,** a system of statements and ideas that explains a group of facts or phenomena, such as "atomic theory" or "evolutionary theory."

Scientists and nonscientists use the word "theory" differently. To a nonscientist, a theory is an untested idea or speculation; to a scientist, it is a set of interconnected, rigorously tested hypotheses. Sometimes this difference in word usage confuses nonscientists, who are left thinking that scientific theories are much more tenuous than they actually are.

How Do Scientists Know What Experiments to Do?

Discussions of scientific method often present experimental tests as more or less repetitive exercises. The design of experiments for hypothesis testing, however, is perhaps the greatest challenge to a scientist's ingenuity and the greatest source of excitement. This challenge makes research exciting. Every scientist has his or her own stories to tell about clever experiments, but perhaps the most famous story concerns Otto Loewi's elegant test of the hypothesis that nerves control muscle contractions by producing a chemical messenger.

Loewi, an early 20th-century German physiologist, reasoned that if a nerve acts through a chemical signal, a muscle ought to respond to the right chemical (if enough of it were present) even when the nerve itself is not present. Loewi had been working with the vagus nerve, which controls the heart rate. When Loewi electrically stimulated a frog's vagus nerve, the heart rate slowed. Loewi wanted to distinguish between two alternative hypotheses: (1) that the slowing results from a direct connection of the nerve to the heart, and (2) that the slowing depends on a chemical produced and secreted by the vagus nerve in response to its electrical activity.

Loewi isolated two frog hearts, one with the vagus nerve still attached, the other without the attached nerve. He stimulated the vagus nerve of the first heart, collected the fluid surrounding it, and transferred the fluid to the heart with no nerve connections. The second heart slowed down, supporting the hypothesis that the slowing was a response to some substance produced by the vagus nerve. Loewi called this substance "vagus stuff," and we now know it as *acetylcholine.*

As described, Loewi's experiment seems straightforward, almost routine. But the design of the experiment was not. Much later, Loewi himself told of the origin of his experimental design:

> *The night before Easter Sunday of 1920 I awoke, turned on the light, and jotted down a few notes on a tiny slip of thin paper. Then I fell asleep again. It occurred to me at six o'clock in the morning that during the night I had written down something most important, but I was unable to decipher the scrawl. The next night, at three o'clock the idea returned. It was the design of an experiment to determine whether or not the hypothesis of chemical transmission that I had uttered seventeen years ago was correct. I got up immediately, went to the laboratory, and performed a simple experiment on a frog heart according to the nocturnal design.*

Loewi's work was the beginning of a new level of understanding of communication, not only between nerves and muscles but also within the nervous system itself. The hypothesis that nerves secrete chemical signals turned out to be general. For example, the speeding up of the heart as well as the other physiological responses to excitement or fear is mediated by a chemical called adrenaline or epinephrine.

Loewi's 1920 experiment was justly acclaimed, and Loewi later received the Nobel Prize. The story of his discovery is a dramatic illustration of the difficulty of good hypothesis testing and of the importance of creativity in experimental design.

How Do Scientists Evaluate Experiments That Give Varying Results?

Up to this point we have talked as if an experiment yields the same results each time it is performed—that electrical stimulation of a muscle always results in contraction, that a solution of ATP always decreases the spacing of muscle striations, that "vagus stuff" (acetylcholine) always slows down a frog heart. In fact, however, experimental results are not always the same. There is always variation in nature, even in simple experiments. Scientists must therefore evaluate their results carefully and quantitatively.

In order for an experiment to say something about an hypothesis, observed changes must result from the proposed cause. They must not depend on chance variations alone or on factors that are unconnected with the hypothesis. To evaluate an experiment, a researcher must compare measurements after an experimental manipulation with similar measurements in a *control,* in which conditions are the same except that no experimental manipulation has been done. For example, to evaluate the hypothesis that ATP causes muscle movement, Szent-Gyorgyi needed to compare the spacing of muscle striations after adding a solution of ATP with the spacing after adding water without dissolved ATP. The choice of controls is extremely important because an hypothesis generally aims to distinguish causes from irrelevant factors. An ideal experimental system allows a researcher to change one condition at a time in order to connect measured effects to specific causes.

To evaluate an hypothesis, a scientist must be able to tell whether an experimental manipulation produces a result different from the control situation. But what happens when the results are not exactly the same each time the experiment is performed? How, for example, would a muscle researcher deal with variations in the distances between striations from experiment to experiment?

Statistics is the science of collecting and analyzing such numerical data. Statistics applies probability theory, a branch of theoretical mathematics, to practical problems, not only in experimental science but also in economics, census taking, opinion polling, and quality control in manufacturing. Among the questions that statistics addresses are two that are especially important for analyzing experiments: (1) How closely does a set of measurements represent the actual value of a quantity? (2) Is a quantity measured after an experimental manipulation significantly different from the control situation?

If, for example, we measure the distances between the striations in a muscle, we obtain slightly different values for each measurement. After a number of such measurements, we can express both the average value of the spacing and the variability of our measurements. Statistics allows us to estimate the likelihood that the spacing before and after the addition of ATP actually changes or is simply different among striations. To have a significant result, we must see differences between the experimental and control conditions which are greater than the variability in the measurements themselves. Statistics provides the tools for estimating the significance of experimental measurements.

In teaching biology to beginning students, instructors and textbooks naturally stress experimental results whose statistical significance is well established. This book therefore makes little use of formal statistics. For experimental scientists, however, statistical methods are crucial for proper hypothesis testing. Because of the great variability and diversity of organisms, statistics is particularly important to biologists.

Hypothesis testing presents two practical problems: (1) how to design or choose a good experiment, and (2) how to tell whether the results are significant. Learning how to solve these two problems is the major task in becoming a professional scientist. The craft of hypothesis testing is discussed in all science courses, but it is properly learned only by actually doing experiments. Students get a taste of this process in the laboratory exercises of introductory courses. Like all crafts, however, hypothesis testing is best learned by a more sustained apprenticeship; this experience is the most important part of a scientist's training.

WHERE DO HYPOTHESES COME FROM?

If we describe hypothesis *testing* as the craft of science, then the art of science is hypothesis *creation*. One could think about science as a collection of formal procedures for gaining knowledge by testing hypotheses. Then the source of the hypotheses would make no difference: Scientific understanding in this view would depend only on how well the hypotheses are tested, not on where they come from. A moment's thought, however, reveals that this view is naive. Such tests require time, talent, and luck. Not all possible hypotheses can ever be tested, and testing takes place only when the deviser of an hypothesis can do a critical experiment to test the hypothesis or can convince someone else to do so.

In many cases an hypothesis follows from other tested hypotheses or from the application of a successful model or theory to a new situation. Much of the steady pace of scientific progress depends on this kind of activity, which leads to a continual refinement and expansion of accepted theory. Much of this work represents the most fascinating kind of puzzle solving: The scientist, already having a good idea of what he or she is looking for, proposes and tests hypotheses to reveal, piece by piece, a picture that no one else has yet seen. In this way scientific knowledge continuously evolves. Isaac Newton best summarized this process in talking about his own great work: "If I have seen farther than others it is because I have stood on the shoulders of giants."

There is another way of creating hypotheses. Wonderful moments of inspiration punctuate the history of science. The most arresting image of such a moment is that of Archimedes running naked through the streets shouting, "Eureka! Eureka!" ("I have found it!"). Archimedes, who lived in the third century BC in Syracuse in Sicily, had been pondering in his bath a problem put to him by the king: Was the king's crown really made of gold? Archimedes noticed that the full tub overflowed as he stepped in, and he hypothesized that equal weights of different materials might cause different amounts of water to overflow from a full container. He was then able to show that every material has a characteristic density, or weight for a given volume. He could show that the king's crown was indeed made of gold. He determined the crown's density by weighing it and by determining its volume from the amount of water it displaced. Archimedes later refined and generalized his hypotheses on density and buoyancy.

What is the role of such moments of inspiration in science? The essential part of inspiration is often the bringing together of realms of thought whose relationship is not obvious. Often such syntheses occur, as in the cases of Loewi and Archimedes, in a moment of relaxation, distraction, or even sleep. We have only a few stories identifying precise moments at which such connections occur. Many historians of science argue, however, that the important leaps of scientific understanding have resulted from borrowing ideas, methods, and images from other fields of endeavor, both scientific and nonscientific.

The discovery of oxygen provides a striking example of the importance of interdisciplinary connections. The

ESSAY

WHAT ARE "PEER-REVIEWED PUBLICATIONS" AND WHY ARE THEY SO IMPORTANT TO SCIENTISTS?

Stories about scientific advances appear with increasing frequency in newspapers and on television. The media usually address the social or practical implications of a new discovery—how someone's research may lead to a cure for cancer or diabetes, how it may explain the differences between men and women, how it throws new light on the death of the dinosaurs or the madness of monarchs, or how it may lead to a more humane mousetrap.

Professional scientists are often as interested in these public presentations as nonscientific readers, and most read or watch such stories. But scientists are suspicious not only of a reporter's rendering of research findings but even of verbal presentations by other scientists, made in laboratories, over coffee, and in public scientific meetings. Scientists regard only one type of communication as acceptable currency for the advancement of scientific knowledge: a "peer-reviewed" paper in a scientific journal. "Peer review" means that, before publication, the paper is evaluated by other scientists, usually people who work in the same area of research and who are able to evaluate the reported techniques, logic, and the relationship to other work in the field.

Most scientists resist talking about their work to journalists until a peer-reviewed paper appears in print. Most print and electronic journalists time their reports of scientific discoveries to appear only when a scientific paper is actually published.

What does a peer reviewer do? First, a reviewer must decide whether the paper is sound: Is it logical? Does it answer the question posed? A particularly important question about any paper is whether it gives enough detail so that another researcher could reproduce the experiments in another laboratory. This question is important because the essence of science is reproducibility. Answers must be answers for everyone, not a private truth. While scientific dishonesty does occur and occasionally makes headlines, it cannot persist for very long; science is intrinsically self-correcting.

Finally, a reviewer, or perhaps a journal's editor, must decide whether the paper is interesting. Does it address a question that concerns other scientists? If the paper is sound but uninteresting, it is unlikely to appear in a widely read journal.

A *scientific journal*—a periodical that contains articles on a particular subject—differs from a *magazine* [from the French *magasin* = storehouse], which assembles stories, photographs, and articles, often on a variety of subjects. While authors always want their works to be read and publishers want their publica-

development of mechanical pumps in the 17th century made it possible to remove air from containers and to show that air was necessary for life. At that time air was still regarded as a single, indivisible substance, so it came as a surprise when later in the century John Mayow found that when a burning candle or a live mouse was put into a closed container, the candle went out and the mouse died after only a fraction of the air was depleted. Mayow also found that if a mouse and candle were placed in the same container, each ran out of air and "expired" more rapidly than when alone. This suggested that they were both using the same component of the air.

When a candle burns, it disappears. Until the late 18th century, scientists pictured the combustion of a candle or a piece of wood as the loss of something. They called this lost substance "phlogiston."

Scientists also knew that when a metal was heated in the presence of air, it lost its shininess and other characteristic properties. Perhaps, they reasoned, phlogiston is lost in all three cases—when metals are heated, when a candle burns, and when a mouse breathes. This theory was particularly appealing because it explained why metals are similar to each other, even though their ores are so different—perhaps the metals, but not the ores, all contained phlogiston.

The difficulty with the Phlogiston Theory emerged when scientists began to weigh metals carefully before and after combustion. They found that metals weigh more after they have presumably lost their phlogiston! Phlogiston must therefore have a negative weight. To the French chemist Antoine Lavoisier a negative weight for phlogiston was totally unacceptable. Instead he hypothesized that

tions to be bought, most scientific journals have a small and self-selected readership. A few journals, particularly *Nature* and *Science,* play to a much wider audience, which includes interested lay people as well as scientists. These publications also carry nontechnical news stories about science, scientists, technology, and public policy. The scientific articles in these journals are still subject to peer review, without which they would not be considered authentic.

Throughout this book, we show how scientists' curiosity is channeled into experiments that address interesting questions. In some cases, the end point of such an experiment is a single scientific paper that answers a single question. More often, however, each question leads to new questions, and each research group publishes related papers that move toward an answer. Papers from different groups may be put together to establish an answer, or different groups may report alternative experimental approaches that lead to conflicting conclusions. The published papers may then become a kind of public scientific debate.

So how does a scientist decide whether she or he is ready to publish? The decision to publish usually reflects a researcher's judgment that other workers in the field will want to know the results and that the result will influence their own next questions. But there is no standard answer to the question of when to publish, and researchers vary greatly in what they consider publishable. By submitting a paper for publication, however, a researcher is saying that he or she is willing to reach a conclusion based on the data summarized in that paper. In submitting and publishing a paper, the authors are also saying that they take both credit and responsibility for the conclusions.

Some scientists prefer to publish many short papers, keeping their colleagues and competitors aware of what they are doing, and making sure that their employers know that they are working hard. In fact, many academic scientists are explicitly told, "Publish or perish!" meaning that they will not keep their jobs if they do not publish regularly. Many researchers, however, prefer to publish relatively few papers, each of which may represent many years of work leading to a single coherent story, which (they hope) will change the way that others answer or approach a particular question.

Your instructors may ask you to read a number of scientific papers in order for you to see how the progress of science actually occurs. The accompanying essay, "How to Read a Scientific Paper," offers some help in approaching such an assignment.

combustion was sustained by an invisible gas, which he named oxygen. The material of a burning candle only seems to disappear: Invisible products must be made. Indeed, such products can be easily detected by the proper techniques.

The point here is that Lavoisier saw the world differently from his contemporaries and predecessors. As a landlord in 18th-century France, he spent a good deal of his time managing a huge estate, in addition to performing his chemical and agricultural experiments. Perhaps he took the weight gains and losses more seriously than others because of his experience in accounting.

Unfortunately, Lavoisier's noble position cost him his head during the Reign of Terror that followed the French Revolution. In connecting the chemistry of combustion with that of animal respiration, however, he did much to establish the principle that the rules of chemistry apply to both living and nonliving matter. This principle is basic to modern biology and continues to be a rich source of new hypotheses about the living world.

■ HOW CAN STUDENTS LEARN TO ASK PRODUCTIVE QUESTIONS ABOUT CELLS?

We have seen how scientists refine questions, formulate models, and test hypotheses. In the rest of this book, we examine how researchers have come to their present understanding of cells and molecules, and we also discuss the areas where more research is needed. Throughout these discussions, we first outline the questions that motivate sci-

(Text continues on page 28.)

ESSAY

HOW TO READ A SCIENTIFIC PAPER

Almost every scientific paper adheres strictly to a traditional format: Title, Authors, Abstract (Summary), Introduction, Materials and Methods, Results, Discussion, Bibliography. Some journals abbreviate and deemphasize Methods and Bibliography, and many journals impose a word limit on papers.

Scientists read papers with differing degrees of attention and skepticism. Sometimes we read just to find out what is happening in a field that is tangential to our own, and our reading is rather casual, not unlike the way we might read a magazine. Often, however, we want to learn the results and interpretation of other researchers working on a particular problem of intense interest.

At our most critical, scientists may bring a kind of intellectual *machismo* to a published paper. In this mode, we are ready to assume that authors, reviewers, and editors have all had major lapses in thinking, and only our own critical sensibility can prevent our friends and students from sinking into a scientific swamp. We focus on the data presented in the Results section, usually presented as graphs, tables, or photographs. We ask whether the data themselves are believable, and whether the authors' interpretation of their own data is correct. We may even say, "I *never* read the Introduction or Discussion of a paper. I only want the facts, so I only read the Results." Or even, "I only look at the graphs and figures."

Looking at the data carefully is important, as is learning to evaluate the data independently of the authors own interpretation. In a Ph.D. Qualifying Examination, for example, a graduate student may be given a paper without a Discussion section and asked to write the Discussion himself or herself. But most of the time, the context of the paper, set forward in the Introduction and elaborated in the Discussion, is what makes a paper most interesting and stimulating. While many students learn that they should first read the Abstract of a paper to learn what it says before deciding whether to read the entire paper, the Introduction is actually a better starting point, since the Introduction frames the scientific questions that the paper addresses.

What papers should you take the time to read carefully? When you are searching for information about a topic of interest, you will find it useful to alternate between cursory readings of papers from which you just want to know conclusions or methods and critical readings of crucial or controversial papers.

For a paper that you, or your instructor, decide you should read carefully, we can provide some general guidelines. First, understand that you may have to read a paper several times before it begins to come into sharp focus. At first encounter, many papers seem obscure and unnecessarily detailed. Your first task is to discover the paper's main point: What question is it addressing, and what answers does it offer? Set yourself the task of actually writing down the main question asked in the paper and your own one- or two-sentence summary of the paper's conclusions.

Then state the general experimental strategy, the methods by which the authors addressed their question in a way that leads to the answer. Resist getting involved in too much detail while examining the methods—just state the general approach.

Next, observe how the experimental approach works for one or two experiments—summarized by the figures in a paper. Ultimately, you will want to be able to discuss the data presented in each of the figures and tables: What are the authors measuring? How do the data contribute to the conclusion? Do you believe the data? Do you trust the conclusion? Usually, you will not be able to make confident judgments about the believability of the data until you have actually had some related laboratory experience.

Finally, place the paper in the context of other things that you have been learning. Why do you think this paper is important enough to read it critically, rather than just glancing at its conclusions? How does the paper relate to information that you learned elsewhere or to the "big picture" developed in this text or in your class's lectures?

Remember, just because something is published—either in a textbook or in a peer-reviewed journal article—does not mean it represents the final truth about a subject. Scientific understanding changes, sometimes dramatically, as new techniques allow the collection of new information and as new insights accumulate both within a given area and from other fields. On the other hand, a new technique—even one that is used by many critical scientists in many laboratories—can sometimes mislead a whole field, be-

What Is the Function of Each of the Parts of a Paper?

Part	Function	Maximally critical reading	More relaxed reading
Title	Tells what the paper is about	Allows decision of whether to read more	Allows decision of whether to read more
Authors	Tells who did the work and who assumes responsibility for it	Not particularly important	Important: some authors are consistently interesting, no matter what their subject
Abstract	Summarizes the results of the paper, and sometimes the interpretation	Important: most salient facts in one place	Not as important as the frame of the paper in the introduction
Introduction	Sets the framework of the paper: why it is important or interesting	Not important: the reader should be able to place the paper in context	The most important part of the paper
Materials and Methods	Gives details of materials used and of experimental methods	Worth detailed attention; tells exactly how experiments were done; the place to look for weaknesses in approach	Important only when methods are not standard or paper is otherwise unbelievable; usually obscure to someone not working in the field
Results	Reports what the researchers actually found; data may be in graphs, tables, or photographs	Ultimately the most important section of the paper; "just the facts"	Important: a careful reader will evaluate whether the results actually support the stated hypothesis and if they also support alternative views or raise additional questions
Discussion	Discusses two sets of issues: (1) the adequacy of the experiments themselves and (2) the relationship of the results to other work in the field	Not so important: the reader should evaluate the data and place them in context	The best window into the author's context, revealing the level of confidence in the conclusions
Bibliography	Lists other papers relevant to experiments or conclusions	Where to find details of methods and context	Source of additional information

cause of an unnoticed artifact. A prominent scientist once embarrassedly described the status of a particular field that had been misled by a flawed technique, saying, "Unfortunately, the scientific literature is far ahead of our understanding."

Still, peer-reviewed papers are the currency of scientific progress. Students and researchers alike must go to such papers for information and understanding, but they must continue to be skeptical, even after a paper is published. Science in general is self-correcting, especially in fields like cell biology, where new understanding can lead to valuable practical applications, or disastrous consequences, in medicine or agriculture. In science, a misunderstanding, artifact, or deception generally cannot persist for long, especially as more people employ a skeptical but respectful approach to reading the scientific literature.

entists to undertake their research, and then we see how researchers set about answering these questions. In most cases, we start with observations and questions and then proceed to discuss models, hypotheses, and experiments.

Although this book is entitled *Asking About Cells,* it (like any textbook) also *tells* about cells. The current appreciation of the workings and origins of cells depends on understanding not only the basic principles of biology but also the rules of physics and chemistry. In Part I of this book, we see how organisms are composed of well-ordered parts and how they perform chemical reactions. In Part II we see how cells obtain energy from their surroundings and describe the molecular and cellular structures responsible for cell reproduction. In Parts III and IV we trace the flow of information from genes to proteins and show how molecular biologists have learned to manipulate the properties of cells and even of organisms. And in Part V, we explore the workings of several types of specialized cells, discuss the development of multicellular animals and plants from single cells, and examine current views of the evolutionary origins of cells.

The scope of cell biology is enormous, and the task at first seems overwhelming. However, even the most complicated process can be understood in relatively simple terms. This understanding requires work, but the inevitable rewards include a finer appreciation of our places in our exquisite living world.

Throughout the book we see again and again that it is people who ask questions and who have insights. These insights often depend on the peculiarities of a particular observation, the insight of an hypothesis, or the cleverness of an experiment. Cell biology is an exciting human endeavor, one in which teachers and textbook writers enthusiastically want to engage their students.

SUMMARY

The development of the microscope led to the discovery that all organisms consist of cells. Cells are the smallest units of life: They are highly organized, obtain energy from their surroundings, perform chemical reactions, change with time, respond to their environments, and reproduce. Cells always arise from other cells.

Biologists divide all organisms into six kingdoms, and all organisms from all kingdoms consist of one or more cells. Although cells are highly diverse, they contain common subcellular structures, such as an outer membrane. At a cellular level, the major distinction among organisms is between eukaryotes, whose cells contain a nucleus and other internal membrane-bounded structures, and prokaryotes, whose cells do not contain such structures. At a molecular level, however, all organisms resemble one another and contain the same types of molecules.

The evolution of multicellular organisms has given rise to cell specialization and permits the replacement of individual cells rather than of the whole organism.

To answer questions about cells, researchers try to develop a picture, a model, of how a particular process works. The contraction of muscle provides a particularly nice example of how macroscopic processes, such as a chin-up, depend on cellular structures, such as the protein filaments within muscle cells. For scientists, the process of gaining knowledge requires experimentation that can test hypotheses. Researchers try to design experiments that disprove a particular hypothesis. The formulation of interesting and provocative hypotheses, as well as of clever ways of testing those hypotheses, depends on the imagination and creativity of individual scientists.

KEY TERMS

adenosine triphosphate (ATP)
Animalia
antagonistic pair
Archaebacteria
Cell Theory
emergent properties
eukaryotic cell
Fungi
homeostasis
hypothesis
inductive reasoning
model
Monera
negative feedback
organelle
Plantae
prokaryotic cell
Protista
reductionism
scientific method
striation
surface-to volume ratio
theory
virus

QUESTIONS FOR REVIEW AND UNDERSTANDING

1. Cells are made up of what four basic kinds of molecules?
2. Explain why homeostasis is necessary. What aspects of the cell does this process maintain?
3. What technique would you use to determine whether a cell should be classified as a eukaryote or a prokaryote? What are the criteria you would use in making the decision? On what basis could you differentiate between an animal and a plant cell?
4. Contrast viruses and bacteria in relation to the seven characteristics of free-living cells.
5. What four things were revealed about the structure of organisms by observing cells under the microscope?
6. Why is having a large surface-to-volume ratio a problem for cells? What are some ways that large cells solve the problem of a large surface-to-volume ratio?
7. Define the terms "model," "hypothesis," and "theory." Explain how they relate to one another.
8. What are the three hypotheses that make up the Cell Theory?
9. What is meant by the phrase "cell specialization"? Give two examples.
10. What are three advantages of being a multicellular organism?
11. If all cells of a multicellular organism contain the same genetic information, what accounts for cell specialization?
12. What is the reductionist approach to understanding biological phenomena? Are there limitations to this approach?
13. Why is performing a statistical analysis so important in the evaluation of data generated by any kind of research, be it in the biological sciences or economics?
14. Construct a model that would assist you in forming an hypothesis to explain how plants obtain energy for growth. What are some experiments you could do that would prove or disprove your hypothesis? Specifically, design an experiment to test that plants require oxygen.
15. How might you explain the fact that two separate laboratories that are working from the same model but unique hypotheses arrive at conflicting data?
16. If peer review is such a stringent process, how can it be that the "truth" being told in a published paper sometimes changes, is disproved, or is sometimes not reproducible? Can you think of any examples of "scientific truth" that were later shown to be false?

SUGGESTED READINGS

BODENHEIMER, F. S., *The History of Biology,* Dawson, London, 1958.

BRADBURY, S., *The Evolution of the Microscope,* Pergammon Press, New York, 1967.

KOESTLER, A., *On Creativity,* Macmillan, New York, 1964. A creative and controversial author discusses the role of inspiration in artistic and scientific creation.

KUHN, T. S., *The Structure of Scientific Revolutions,* 2nd edition, University of Chicago Press, Chicago, 1970. A highly influential book, which discusses the process by which communities of scientists change their views.

MARGULIS, L. and K. V. Schwartz, *Five Kingdoms: An Illustrated Guide to the Phyla of Life on Earth,* 2nd edition, W. H. Freeman, New York, 1988. An accessible introduction to the kingdoms.

POPPER, K., *Conjectures and Refutations: The Growth of Scientific Knowledge,* 2nd edition, Basic Books, New York, 1965. The most influential 20th century philosopher of science presents his view of the "scientific method."

CHAPTER 2

Small Molecules of the Cell

KEY CONCEPTS

- Almost all molecules within cells are built from six types of atoms: hydrogen, carbon, nitrogen, oxygen, phosphorus, and sulfur.
- The distribution of electrons within atoms determines the way they combine to form molecules.
- The distribution of electrons within molecules determines the way they interact with other molecules.
- Molecules with uneven charge distributions are polar and interact with other polar molecules. Within cells, the most important molecular interactions involve water molecules.
- Almost all cellular structures are built from combinations of the same 35 small molecules. These include lipids, sugars, amino acids, and nucleotides.

OUTLINE

ESSAYS

Biologists have always wondered about the relationship between the materials that make up living and nonliving things. In the second century, for example, Galen argued that the function of the heart was to give color and vitality to the blood in the same way that a furnace can transform a dull ore into a bright, useful metal. Galen was trying to explain a biological process in terms of *chemistry*, the study of the properties and transformations of matter. Perhaps the greatest achievement of 20th-century biology has been to understand the chemistry that underlies cellular processes—from the inheritance of a gene to the contraction of a muscle.

Just as engineers can assemble dazzling and diverse computers from a few kinds of chips, organisms can produce extraordinary structures from a small set of standard parts. In this chapter, we describe 35 small molecules (or "building blocks") from which most cellular components are built. These molecules obey the same rules as the molecules of nonliving matter, with their special properties—especially their interactions with water—depending on the distribution of electrons. From a chemist's point of view, two aspects of living matter are particularly intriguing—the impoverished molecular repertoire and the diversity of structures built from the same few small molecules.

THE DISTRIBUTION OF ELECTRONS WITHIN ATOMS EXPLAINS HOW COVALENT BONDS HOLD MOLECULES TOGETHER

Early in this century, physicists discovered that an atom consists of a small, positively charged nucleus surrounded by diffuse volumes of space inhabited by negatively charged electrons. Each electron moves about within a limited portion of the atom's space, called an **orbital** [to distinguish it from a planet's restricted "orbit"]. Like clouds, orbitals have characteristic sizes and shapes, even though their boundaries are fuzzy.

Electron orbitals are grouped into **shells** whose electrons have nearly equal energy. The atoms most important in cells use only three shells. The first of these can contain only two electrons, while (in biologically important molecules) the second and third can contain up to eight.

Atoms Form Molecules by Sharing Electrons

Molecules are specific combinations of individual atoms. The atoms are held together by **covalent bonds,** shared arrangements of electrons. The formation of covalent bonds is best understood with the help of a generalization called the **octet rule,** which states that an atom is particularly stable and chemically unreactive when its outermost shell is full, meaning (usually) that it contains eight electrons. An atom that already contains a full outer shell (like neon or argon) is chemically unreactive. Most atoms, however, do not have full outer shells but can acquire them in one of two ways: by sharing electrons to form a covalent bond, or by gaining or losing electrons to form an **ion,** an atom (or a molecule) with a net electrical charge, the result of a different number of electrons and protons.

The molecules that we introduce in this chapter are built from six types of atoms: hydrogen (H), carbon (C), nitrogen (N), oxygen (O), phosphorus (P), and sulfur (S). (A useful way of remembering these six biologically important atoms is to rearrange their symbols into the pronounceable "SPONCH.") Organisms are so closely associated with the compounds of carbon that all carbon-containing compounds are called *organic* molecules (even when they have nothing to do with organisms). **Organic chemistry** is the study of the structures and reactions of carbon compounds. **Biochemistry** (or biological chemistry) is the study of the structures and reactions that actually occur in living organisms.

The outer shells of the SPONCH atoms determine their ability to form molecules. Carbon, for example, has four electrons in its outer shell. In Figure 2-1a, each of these electrons is shown as a small dot contained in a volume of space. To fill its outer shell with eight electrons, a

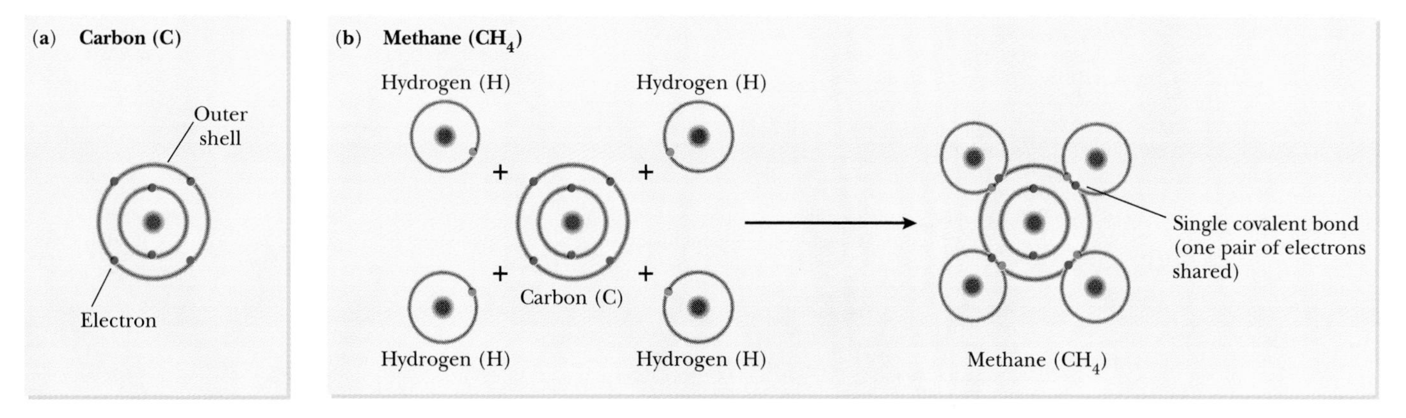

Figure 2-1 **(a)** A carbon atom; **(b)** the four electrons in carbon's outer shell form covalent bonds with hydrogen atoms in a methane molecule.

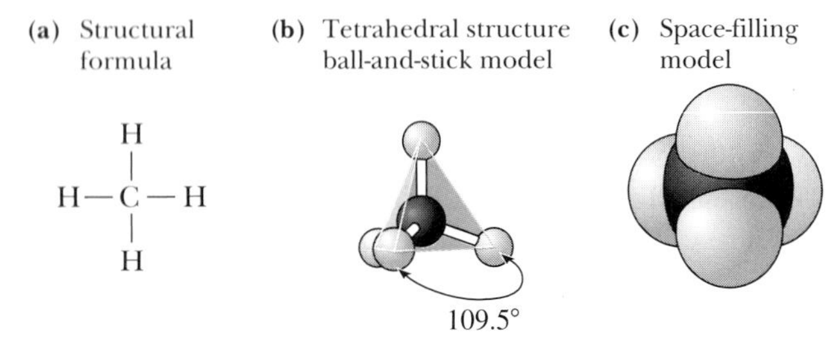

Figure 2-2 The three-dimensional structure of methane, showing single covalent bonds between carbon and hydrogen atoms. **(a)** Structural formula, each line represents a pair of shared electrons; **(b)** "ball-and-stick" representation, showing the angle between two CH bonds; **(c)** space-filling model, the four bonds point toward the corners of a tetrahedron, and the angle between any two bonds is 109.5°. In the standard two-dimensional drawing of an organic molecule, the convention is that bonds drawn vertically actually point away from the viewer, behind the plane of the page, while bonds drawn horizontally point out at the viewer. Although some people are able to grasp molecular shapes from flat drawings, three-dimensional models help most people understand molecular structure more clearly.

carbon atom must share an additional four electrons with other atoms, forming four covalent bonds, as in the case of methane (CH_4), the simplest organic compound, shown in Figure 2-1b.

Similarly, nitrogen's outer shell has five electrons, and oxygen and sulfur have six. Nitrogen may acquire an octet by sharing three additional electrons to form three covalent bonds, while oxygen and sulfur can form two covalent bonds. Hydrogen atoms are a special case: Hydrogen contains only one electron, and a complete outer shell contains only two electrons. Hydrogen obeys a "duet" rather than the octet rule. Phosphorus is discussed on page 46.

The covalent bonds within a molecule have fixed directions and lengths. Every molecule has a definite size and shape, which chemists and physicists have determined by techniques such as x-ray crystallography, discussed in Chapter 3 (p. 73). The lengths and directions of chemical bonds are the same from molecule to molecule. The distance and bond angle between a carbon atom and a hydrogen atom in the most complex organic molecule is the same as in methane, whose three-dimensional structure is shown in Figure 2-2. Similarly, the angles between bonds are the same in a wide range of compounds.

Two atoms may share more than one pair of electrons. In ethylene, a molecule important in the ripening of fruit, the two carbon atoms are held together by a **double covalent bond,** in which two atoms share two pairs of electrons, as shown in Figure 2-3a. In sharing two electron pairs with each other, each carbon atom can combine with only two (rather than three) additional hydrogen atoms. The geometry of the double bond is also distinct from that of a single bond, as illustrated in Figure 2-3, with all six atoms (two C and four H) of ethylene lying in the same plane. Some molecules, like those of acetylene (shown in Figure 2-3b), contain **triple covalent bonds,** in which two atoms share three pairs of electrons, but these compounds are not common in cells.

Because bond lengths and angles are so constant, we can predict much about the structure of a molecule by making a three-dimensional model. Chemists and biologists can also represent the structures of molecules on computer screens.

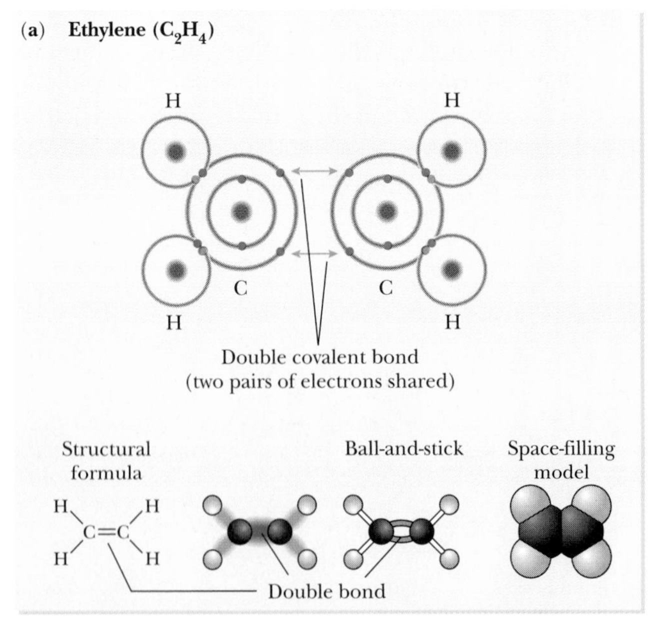

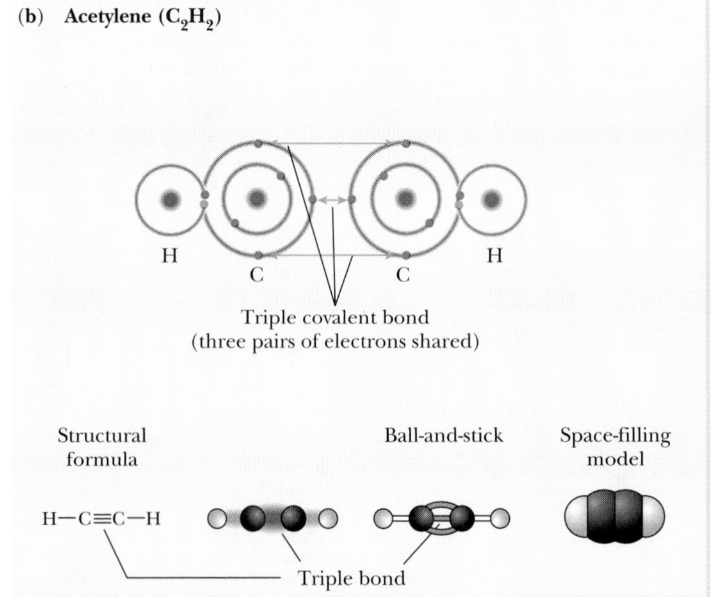

Figure 2-3 Carbon atoms make double and triple covalent bonds by sharing two and three pairs of electrons. **(a)** Ethylene; **(b)** acetylene.

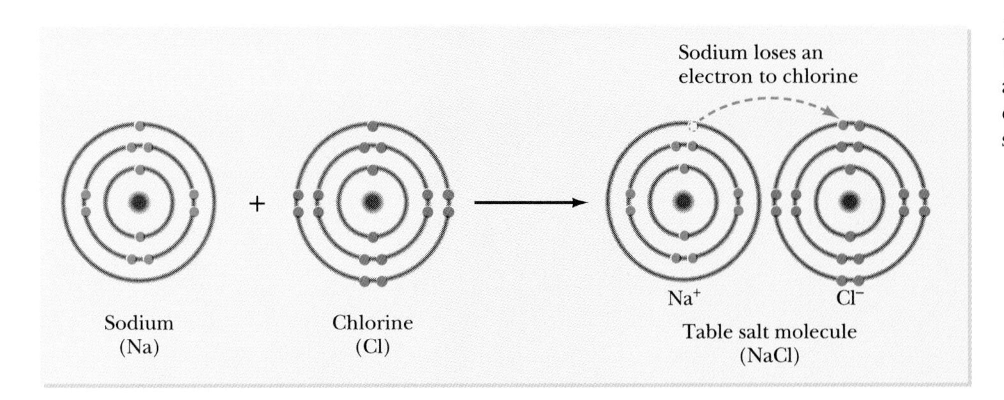

Figure 2-4 Sodium (Na^+) and chloride (Cl^-) ions form as a sodium atom loses an electron to form a stable octet and a chlorine atom gains an electron to form a stable octet.

Noncovalent Interactions Depend on Electron Distributions

The properties of electron orbitals determine a molecule's geometry and chemical reactivity. The distribution of electrons also determines a molecule's ability to participate in **noncovalent interactions,** associations with other molecules and ions.

The most familiar noncovalent interactions are **electrostatic interactions,** the attraction or repulsion of charges. In salts such as sodium chloride, the attractive interactions between sodium and chlorine atoms are often called "ionic bonds." Like the carbon and hydrogen atoms of methane, the atoms in sodium chloride have stable electron configurations, with octets in their outer shells. Instead of sharing electrons with each other as in covalent compounds, however, the sodium and chlorine atoms lose and gain electrons to form positively and negatively charged ions (abbreviated Na^+ and Cl^-), as shown in Figure 2-4. The positively charged sodium ions and the negatively charged chlorine ions are held together by simple electrical attraction.

When sodium metal and chlorine gas come into contact, a violent chemical reaction occurs. During this reaction each sodium atom (with one electron in its outer shell) loses that electron, making a previously inner shell into an outer octet. In losing an electron, sodium becomes a **cation,** a positively charged ion [so called because it moves toward a negatively charged electrode, or cathode]. At the same time, chlorine, with seven electrons in its outer shell, captures an electron and gains an outer octet to become an **anion,** a negatively charged ion [so called because it moves toward a positively charged electrode, or anode].

Electrical forces hold together ions of opposite charge, forming a complex of positive and negative ions, rather than a true molecule. In a salt crystal, such as that shown in Figure 2-5, each cation is surrounded by anions and each anion by cations. In solution, however, each ion is surrounded by water molecules.

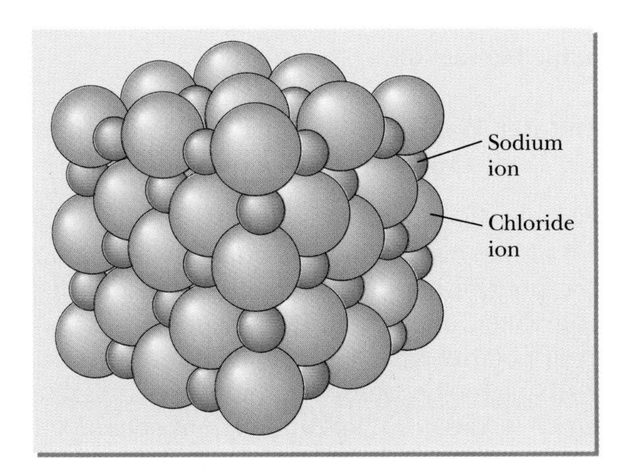

Figure 2-5 In a crystal of sodium chloride electrostatic interactions hold sodium and chloride ions together.

Individual Types of Atoms Have Different Tendencies to Gain or Lose Electrons

Chlorine has a high tendency to gain an electron to form the chloride ion, while sodium has a high tendency to lose an electron to form the sodium ion. Carbon, on the other hand, tends to share its outer electrons to form covalent bonds.

The American chemist Linus Pauling realized that the atoms of each element have a characteristic **electronegativity,** the tendency of atoms to gain electrons. Fluorine, oxygen, and chlorine have the greatest tendencies to gain electrons, and they are said to be the most electronegative atoms. Sodium, potassium, and lithium are among the least electronegative elements—they have the greatest tendency to lose electrons.

Knowing the electronegativity of two atoms allows one to predict whether a bond between them is covalent or ionic: The larger the difference in the electronegativities of two atoms, the more likely they are to form an ionic rather than a covalent bond. Sodium and chloride have a large difference in electronegativities and form ionic bonds. Carbon and nitrogen, on the other hand, have similar electronegativities and are likely to form covalent bonds.

Even in a covalent bond, atoms may not share electrons equally. When atoms differ in electronegativity, they do not share electrons equally. Instead, the diffuse clouds

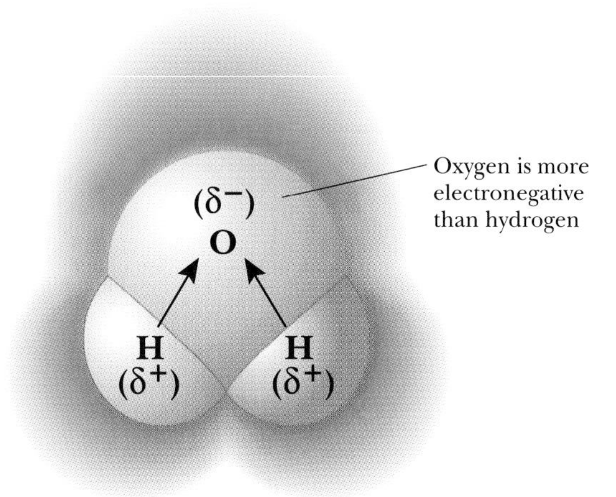

Figure 2-6 The unequal charge distribution in a water molecule. The covalent bonds between O and H atoms are polar, with electrons more concentrated around the oxygen nucleus. "δ" is the symbol for a partial charge.

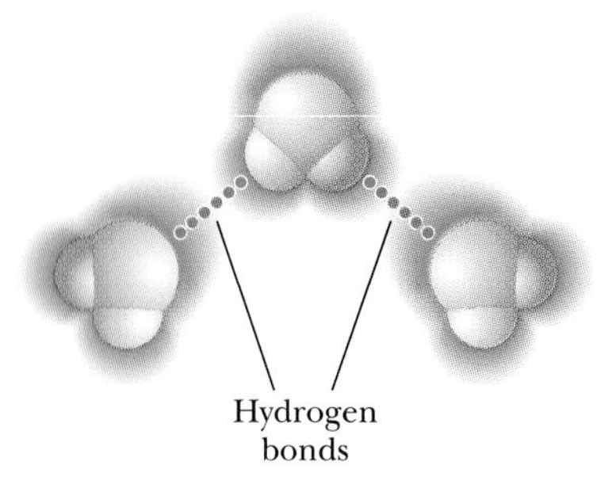

Figure 2-7 Hydrogen bonds between adjacent water molecules. The hydrogen atoms of one water molecule are attracted to the oxygen atom of another water molecule.

of shared electrons tilt toward the more electronegative atoms. In the familiar water molecule, for example, the shared electrons are slightly more concentrated around the oxygen nucleus than around the two hydrogen nuclei, as shown in Figure 2-6. This results in a slight negative charge on the oxygen atom and a slight positive charge on the two hydrogen atoms. Covalent bonds between atoms that differ in electronegativity have a slightly ionic character. On the other hand, the bonds between two carbon atoms or even between a carbon and a hydrogen atom are almost purely covalent; that is, the electrons are evenly shared.

Molecules with uneven distributions of electrical charge are said to be **polar** because they have positive and negative ends (or poles). A covalent bond in which the two atoms do not share electrons equally is said to be polar. Molecules with approximately uniform charge distributions are said to be **nonpolar.** When a polar molecule like water comes close to an ion or to another polar molecule, its negative pole points toward the nearby positive charge or positive pole, and its positive pole toward a nearby negative charge or negative pole.

THE POLARITY (OR NONPOLARITY) OF MOLECULES UNDERLIES THEIR NONCOVALENT INTERACTIONS

The forces that hold atoms together in covalent bonds are strong, and breaking them requires lots of energy. Electrostatic interactions in salt crystals ("ionic bonds") are also strong. Most of the chemistry that concerns biologists, however, happens in or around water, and we need to understand the kinds of interactions that molecules can have with each other in an **aqueous** (watery) environment. Each of these interactions is weak compared with a covalent bond; that is, it takes less energy to break. But the sum of many weak interactions is responsible for the complex structures within and between cells.

Hydrogen Bonds Can Hold Polar Molecules Together

When a hydrogen atom attaches to a highly electronegative atom like oxygen or nitrogen, the resulting covalent bond is polar. In this case the hydrogen atom acquires a slight positive charge. Such a hydrogen atom can then participate in a **hydrogen bond,** a weak attraction to a negatively charged atom in another molecule, as shown in Figure 2-7. The most common hydrogen bonds are those between water molecules (discussed below), but other hydrogen bonds also play a critical role in the structure of proteins and DNA.

Hydrophobic Interactions—the Avoidance of Water—Lead to the Association of Nonpolar Molecules

Nonpolar molecules, that is, molecules with equal charge distributions, are said to be **hydrophobic** [Greek, *hydro* = water + *phobos* = fear], meaning that they avoid associations with water. The resulting coming together of nonpolar molecules, illustrated in Figure 2-8, is called **hydrophobic interaction.**

The van der Waals Attraction Pulls Two Molecules Toward Each Other If They Are Already Very Close

All molecules have constantly fluctuating charge distributions. At a given instant, these fluctuations may temporarily produce positive and negative poles even in a nonpolar molecule. Such a temporarily polar molecule may, at its instant of polarity, induce a change in an adjacent molecule, so that it too becomes temporarily polar, as illustrated in Figure 2-9. Van der Waals interactions operate only over very short distances, and they generally reinforce the hydrophobic interactions among nonpolar molecules in water.

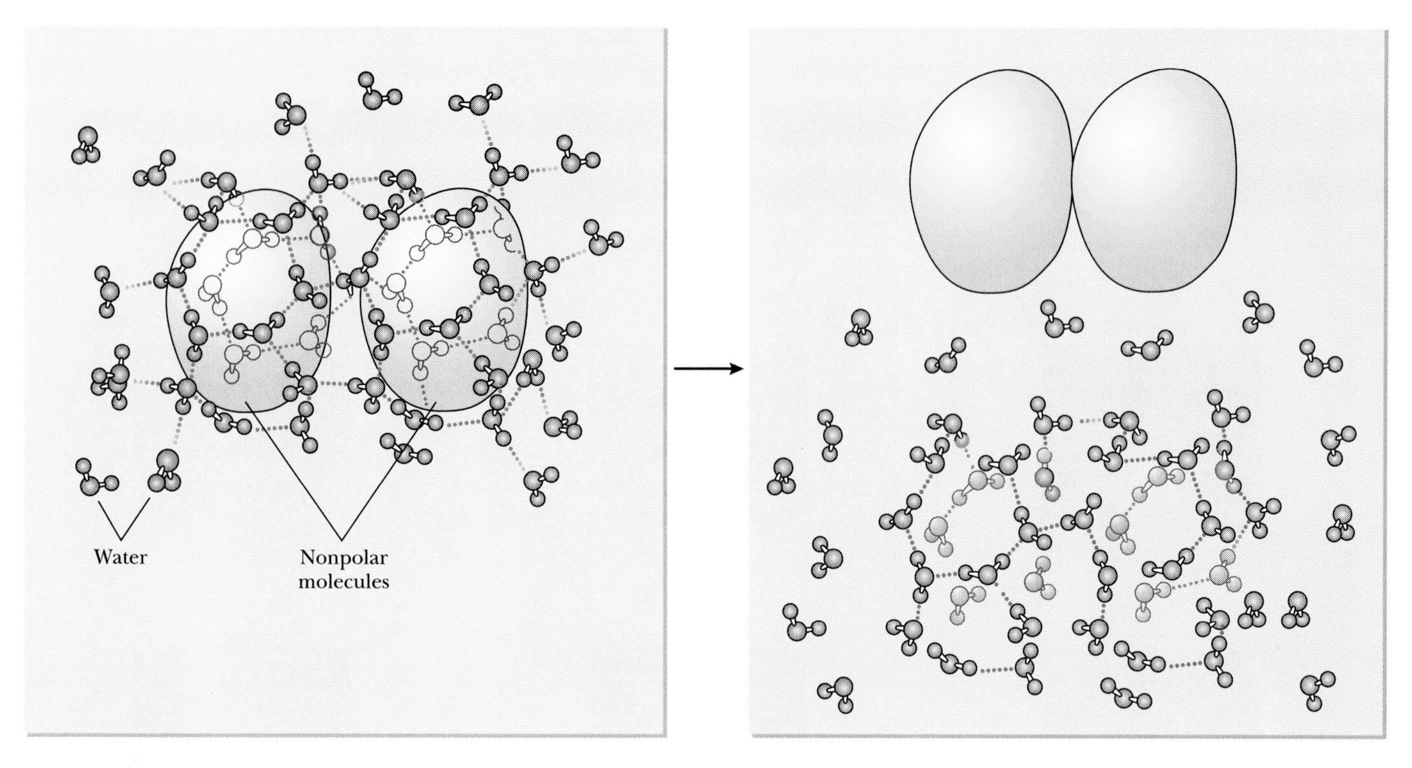

Figure 2-8 Hydrophobic interactions. Nonpolar molecules cannot form hydrogen bonds with water. They associate with one another, avoiding interactions with water.

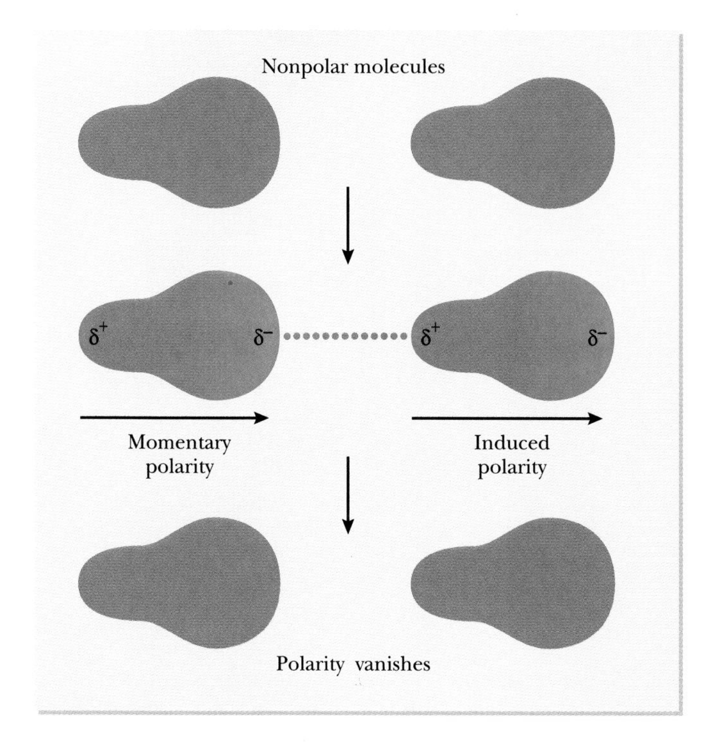

Figure 2-9 Van der Waals interactions attract nonpolar molecules to one another over short distances. A momentary polarity in one molecule causes a redistribution of electrons and induced polarity in an adjacent molecule.

The properties of polar and nonpolar molecules play a key role in cells. The tendency of nonpolar molecules to separate from water underlies the formation of the membrane boundaries that separate the insides and outsides of cells. In eukaryotic cells, such membranes are fundamental to the establishment of subcellular compartments within cells.

THE POLARITY OF THE WATER MOLECULE EXPLAINS ITS UNIQUE ROLE IN CELLS

Water is the most important polar molecule, both within cells and in their external environments. Water molecules interact strongly with each other and with other charged and polar molecules. These interactions and the lack of interactions with nonpolar molecules are the most important factors in establishing biological structures.

Because water is so common, people often do not realize just how unusual it is. Water's special properties make it uniquely fit for its important role in life. Among these special properties are (1) greater density as a liquid than as a solid, allowing ice to float and pond life below to survive the winter; (2) cohesion and high surface tension, allowing

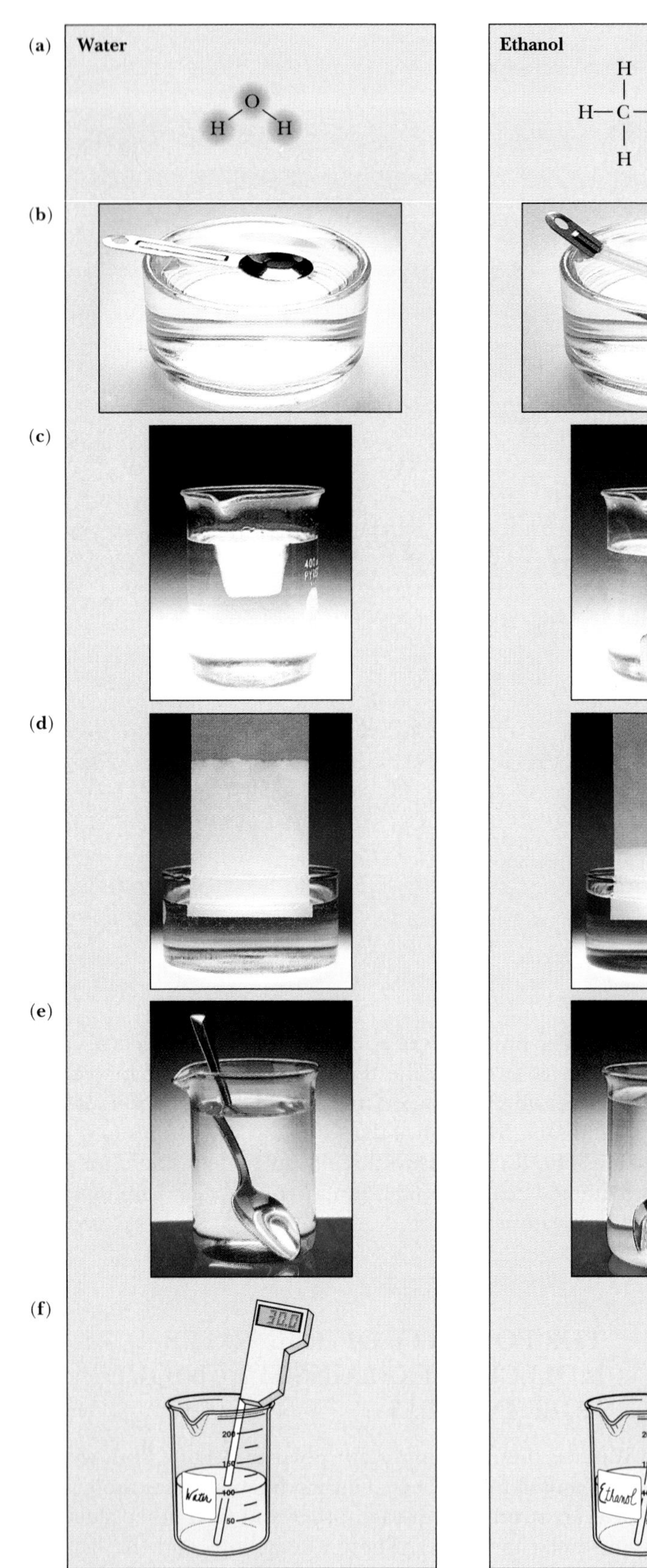

Figure 2-10 Contrasting physical properties of water, ethanol, and oleic acid. **(a)** Molecular structure: water is more polar than either ethanol or oleic acid; **(b)** cohesion: the attraction of water molecules to one another resists the sinking of the spoon; the spoon sinks in ethanol or in oleic acid; **(c)** density: the density of the liquid and solid phases; ice floats, but ethanol ice and solid oleic acid do not; **(d)** adhesion: water forms stronger interactions with the polar groups in chromatography paper, so water rises farther than ethanol or oleic acid; **(e)** ability to act as a solvent: table sugar (sucrose) dissolves more readily in water than in ethanol or oleic acid; **(f)** heat capacity: the same amount of applied heat raises the temperature of water less than it raises the temperature of ethanol or oleic acid. *(b–e, Charles D. Winters)*

the maintenance of tall columns of water in the transport systems of trees; (3) adhesion, allowing a close interaction between water and many biological and nonbiological surfaces; (4) a melting (freezing) point that allows our water-based life to exist in every ocean and on every continent on this planet; (5) high heat capacity, permitting organisms to be insulated from rapid changes in temperature; and (6) excellence as a solvent, providing a medium in which ions and polar molecules readily dissolve.

We can see some of the ways in which water is remarkable by comparing its characteristics with those of ethanol (the alcohol in wines, beers, and liquors) and of oleic acid (the major substance in olive oil), as illustrated in Figure 2-10. The properties of water depend upon the polarity of the water molecule, with the oxygen atom having a slight negative charge and attracting the positively charged hydrogen atoms of other water molecules. The hydrogen atoms participate in hydrogen bonds between water molecules. Ethanol and oleic acid are less able to form hydrogen bonds with other molecules. Each water molecule can make hydrogen bonds with up to four other water molecules. In ice all these bonds form, as shown in Figure 2-11. In liquid water, however, the bonds are constantly changing, and each water molecule is more mobile. Hydrogen bonds rapidly form, break, and reform in different orientations.

Water Is a Powerful Solvent

A **solvent** is a substance in which other substances dissolve, and a **solute** is a substance that dissolves within the solvent. Water is an especially good solvent for the ions and polar molecules of cells, although not for the nonpolar molecules that make up cell membranes. Table sugar (sucrose), for example, dissolves readily in water, less well in weakly polar ethanol, and hardly at all in nonpolar olive oil.

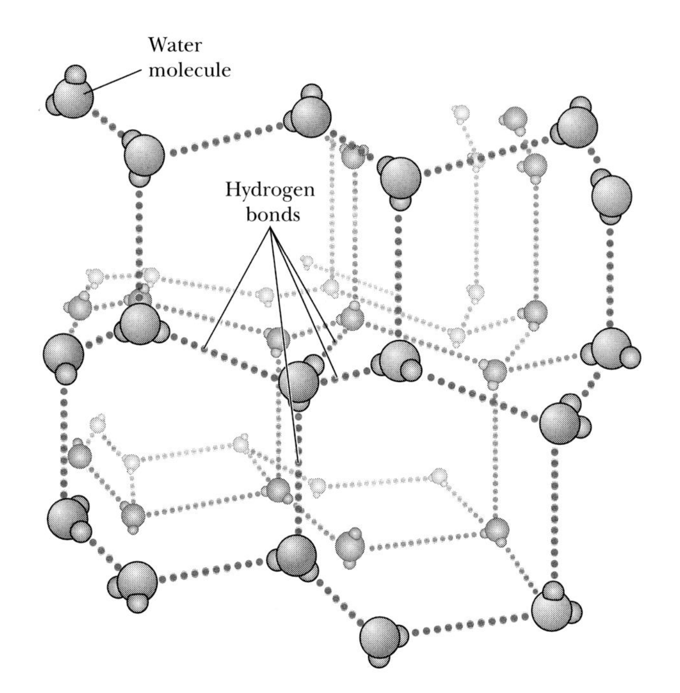

Figure 2-11 In ice, each water molecule forms hydrogen bonds to four other water molecules.

Salts, which are composed of ions, are also extremely soluble in water. Even when ions cannot make hydrogen bonds with water molecules, the polar water molecules orient so that their negatively charged ends are close to the dissolved cations and their positively charged ends are closer to the dissolved anions. Water molecules form a kind of shell around each ion, as shown in Figure 2-12. This shell shields positive and negative ions from each other, allowing them to move farther away from each other than they can in their solid form (their crystals), where they have a regular arrangement.

In cells, ions and molecules move from place to place dissolved in water. Many of these movements occur in tiny channels in the membranes that separate individual cells and subcellular compartments.

Hydrogen bonds also contribute to the excellence of water as a solvent. Again, hydrogen bonds can form whenever a hydrogen atom forms a covalent bond with a more electronegative atom, such as oxygen or nitrogen. Compounds with such bonds are usually polar. Sucrose and other polar solutes form hydrogen bonds and dissolve readily in water.

The polar properties of water explain why some molecules are hydrophilic and others are hydrophobic. **Hydrophilic** molecules interact strongly with water molecules, while nonpolar, hydrophobic molecules do not. Oil, a mixture of nonpolar molecules, does not readily mix with water.

Many biologically important molecules, including oleic acid and its close relative palmitic acid (shown in Figure 2-13a), contain both hydrophilic head and hydrophobic tail regions and are said to be **amphipathic** [Greek, *amphi* = both + *pathos* = feeling]. When an amphipathic molecule is in water, its hydrophobic regions avoid the water, and its hydrophilic regions establish hydrogen bonds with water, forming a **micelle,** a cluster of amphipathic molecules, as shown in Figure 2-13b. Mayonnaise, for example, consists of tiny oil droplets suspended in a watery solution, roughly the same arrangement as in the fat-containing cell (or adipocyte).

Detergents are amphipathic molecules that interact with both water and hydrophobic molecules. These interactions not only allow detergents to remove grease from dirty dishes, but also allow cell biologists to produce solutions of cell extracts for chemical studies.

Water Participates in Many Biochemical Reactions

Water is present at high concentrations wherever there is life, and many metabolic processes produce or use water.

Figure 2-12 When a salt dissolves in water, water molecules surround ions, with O atoms attracted to positive ions, such as sodium (Na^+), and H atoms attracted to negative ions, such as chloride (Cl^-).

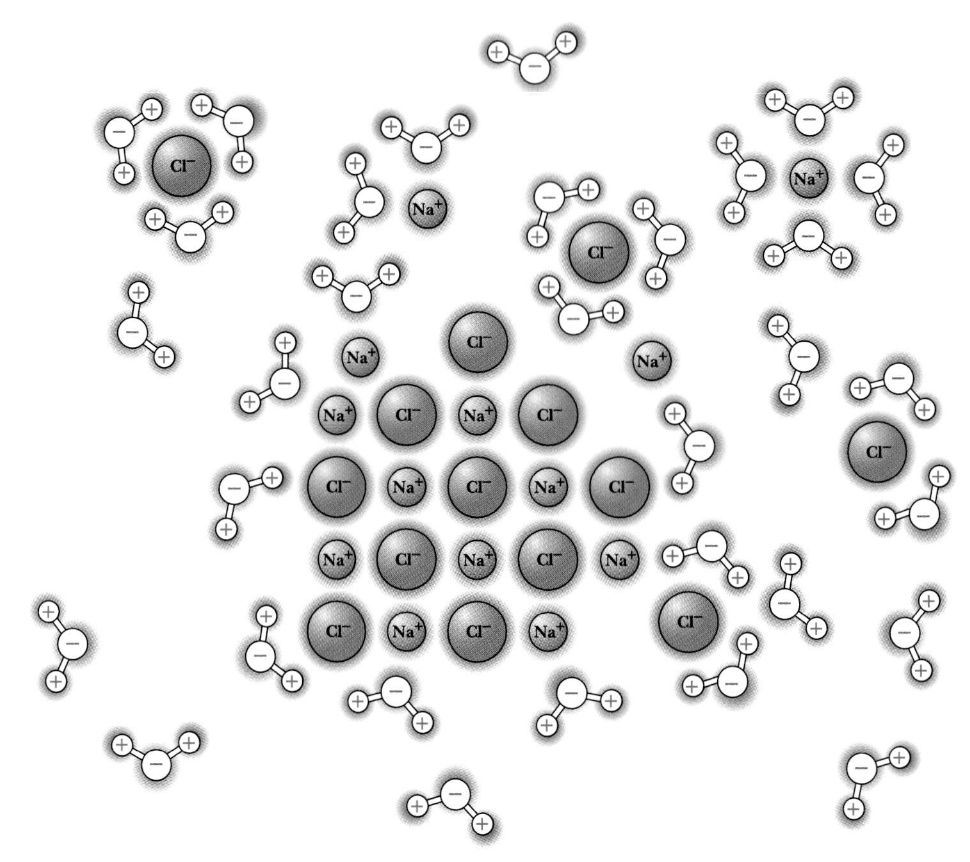

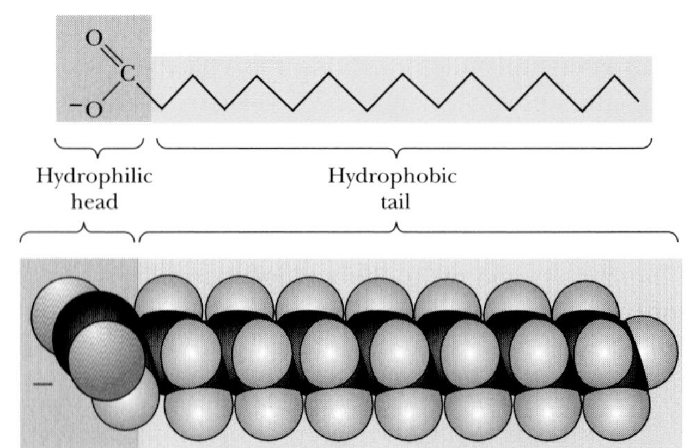

Figure 2-13 Amphipathic molecules contain a hydrophobic tail and a hydrophilic head. **(a)** Palmitic acid; **(b)** a micelle, formed as water molecules cluster around the hydrophilic head and the tails form a central hydrophobic region.

Breaking down large molecules in food, for example, requires the addition of water molecules, as discussed in Chapter 3. The resulting small molecules act as fuel for cellular processes or as building blocks for cellular structures.

Besides serving as the medium in which most biochemical reactions occur, water itself participates in many reactions. Just as water's physical properties depend on its polarity, so do its chemical properties. The slight negative charge of the oxygen atom in water produces a tendency for it to approach positively charged atoms in other molecules. The covalent bond between oxygen and one of the hydrogen atoms may then break as oxygen forms a new covalent bond. The abandoned hydrogen atom then leaves as a hydrogen ion, which quickly combines with a water molecule, as described below.

Water Molecules Continuously Dissociate into Hydrogen Ions and Hydroxide Ions

The electrons in the chemical bonds of a water molecule are closer to the oxygen atom than to hydrogen. When a hydrogen atom participates in a hydrogen bond with another water molecule, it also becomes closely associated with a second oxygen atom. The hydrogen nucleus can effectively jump back and forth between the oxygen atoms of two adjacent molecules, as sketched in Figure 2-14. At one instant a hydrogen ion can be covalently bonded to the oxygen atom of one molecule and hydrogen-bonded to the oxygen of another molecule. At the next instant covalent and hydrogen bonds may be reversed. Because of this ability, Albert Szent-Gyorgyi said that water is "the only molecule that can turn around without turning around."

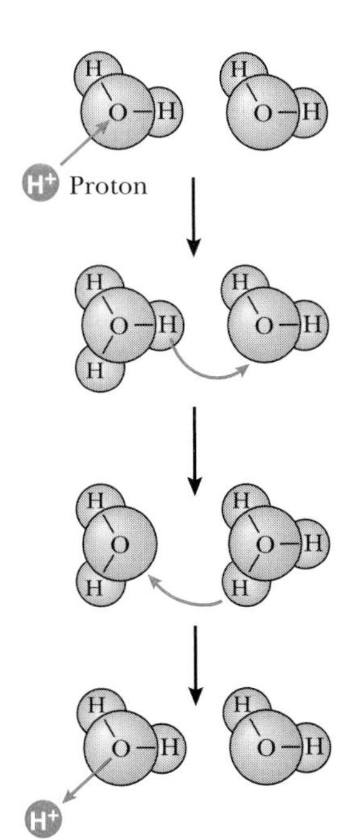

Figure 2-14 A hydrogen nucleus can jump back and forth between two hydrogen-bonded water molecules.

On average, each water molecule has two hydrogen atoms and one oxygen atom, and it remains uncharged. A small fraction of water molecules, however, have only one hydrogen atom, and a similar fraction have an extra hydrogen. A deprived water molecule is called a hydroxide ion (OH^-) and has a charge of -1. A recipient water molecule becomes a **hydronium ion** (H_3O^+) and has a charge of $+1$, as shown in Figure 2-15. We can represent this exchange of hydrogen ions by a chemical equation:

$$H_2O + H_2O \longrightarrow OH^- + H_3O^+$$

For the sake of brevity, however, we can leave one of the water molecules out of the equation. We therefore represent the shuttling of hydrogen ions as the splitting (or dissociation) of water into a hydroxide ion (OH^-) and a naked **hydrogen ion** (H^+):

$$H_2O \longrightarrow H^+ + OH^-$$

The opposite charges of hydrogen and hydroxide ions ensure that they do not get far. They quickly recombine (reassociate) with other hydroxide and hydrogen ions to form uncharged water molecules again:

$$H^+ + OH^- \longrightarrow H_2O$$

Water, then, constantly dissociates into hydrogen and hydroxide ions, and the ions constantly reassociate to form water. We can represent both processes in a single chemical equation, using a double arrow to show that the reaction proceeds continuously in both directions:

$$H_2O \rightleftharpoons H^+ + OH^-$$

At a given instant only a small fraction of water molecules are dissociated into hydrogen and hydroxide ions. The dissociation and association balance one another, so the concentrations of hydrogen and hydroxide ions stay constant. This balance between forward and reverse reactions is an example of chemical equilibrium, discussed further in Chapter 4.

At equilibrium, the dissociation of water into hydrogen ions and hydroxide ions exactly balances the association of hydrogen and hydroxide ions to form water. In pure water the concentration of hydrogen ions exactly equals that of hydroxide ions.

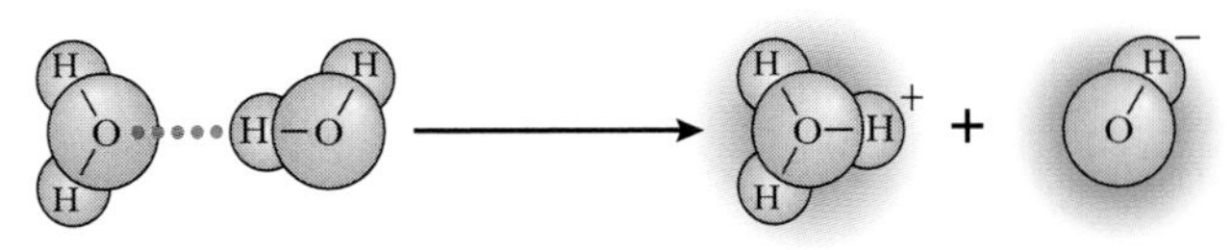

Figure 2-15 A hydrogen nucleus can jump from one water molecule to another, forming a hydronium ion (H_3O^+) and a hydroxide ion (OH^-).

pH Is a Measure of the Concentration of Hydrogen Ions in a Solution

Chemists often express the concentration of a dissolved substance as **molarity,** or moles of the dissolved substance per liter of solution. A **mole** is the amount of a substance in grams equal to its molecular mass. For example, the molecular mass of ethanol (the alcohol of wine and beer) is 46, so a mole of ethanol is 46 grams. One mole of any substance contains 6×10^{23} (Avogadro's number) molecules. A 1 molar (1 M) solution of ethanol in water contains 46 grams of ethanol in 1 liter of water, and contains 6×10^{23} molecules of ethanol.

The concentrations of hydrogen and hydroxide ions in pure water are each 10^{-7} M, meaning that 1 liter contains one ten-millionth of a mole of hydrogen (and hydroxide) ions, or about 6×10^{16} of each. But when water contains other substances, hydrogen and hydroxide ions may differ in their concentrations. In vinegar, for example, there are about 1 million (10^6) times more hydrogen ions than hydroxide ions, and in stomach acid, about 1 trillion (10^{12}) times more.

As the concentration of hydrogen ions increases, the concentration of free hydroxide ions decreases. The equilibrium between water and its ions is such that the molar concentration of hydrogen ions (designated $[H^+]$) times the molar concentration of hydroxide ions (designated $[OH^-]$) is always 10^{-14}:

$$[H^+] \times [OH^-] = 10^{-14}$$

Once we specify the concentration of either ion, we can calculate the concentration of the other. In vinegar, the concentration of hydrogen ions is about 10^{-4} M, so the concentration of hydroxide ions must be $10^{-14}/10^{-4} = 10^{-10}$ M.

Because the concentrations of hydrogen and hydroxide ions can vary widely in different solutions, chemists have defined a more manageable scale, called **pH,** based on the logarithm (to the base 10) of the molar hydrogen ion concentration:

$$\text{pH} = \log_{10} 1/[H^+] = -\log [H^+]$$

where $[H^+]$ is again the molar concentration of the hydrogen ion. By using logarithms, the pH scale specifies concentrations of hydrogen ions (and therefore of hydroxide ions) without using exponential notation. Figure 2-16 shows the pH of some familiar compounds. The pH of pure water is $-\log(10^{-7}) = 7$, and that of vinegar is $-\log(10^{-4}) = 4$. A 10-fold change in concentration means a change of only one pH unit.

Why Is pH Important to Cells?

Many chemical reactions in organisms involve exchanges of hydrogen ions. For example, when you exercise, your muscles produce carbon dioxide (CO_2), which your blood carries to your lungs where it is exhaled. Most of the carbon dioxide in the blood, however, combines with water to form carbonic acid (H_2CO_3). Most of the carbonic acid molecules release a hydrogen ion to produce bicarbonate (HCO_3^-) ions, which dissolve readily in the blood:

$$CO_2 + H_2O \longrightarrow H_2CO_3 \longrightarrow H^+ + HCO_3^-$$

When the bicarbonate ions reach the lungs, they reacquire a hydrogen ion and reform carbon dioxide and water:

$$H^+ + HCO_3^- \longrightarrow H_2CO_3 \longrightarrow CO_2 + H_2O$$

A molecule (or part of a molecule) that can give up a hydrogen ion is called an **acid.** A molecule (or part of a molecule) that can accept a hydrogen ion is called a **base.** Because a hydrogen ion is the same as a proton, an acid is often defined as a proton donor and a base as a proton acceptor. An acid dissolved in water contributes hydrogen ions to water molecules and raises the concentration of hydrogen ions, decreasing the pH. Conversely, bases dissolved in water remove hydrogen ions from water molecules and increase pH.

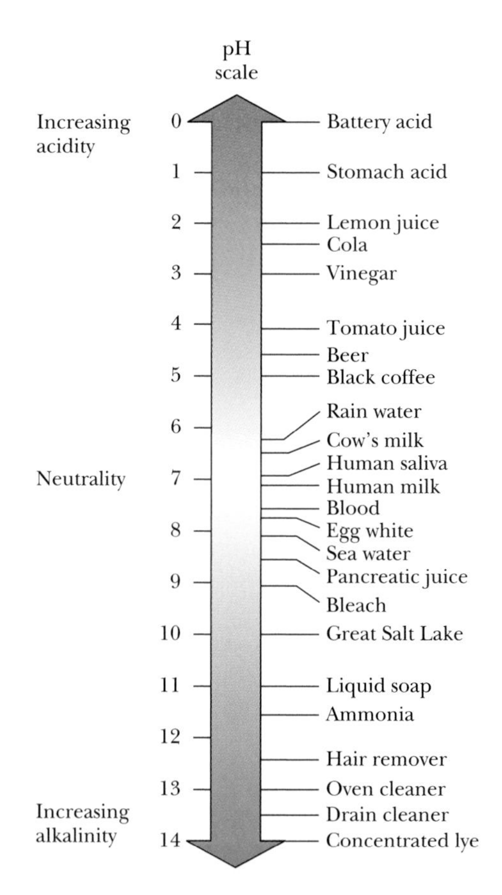

Figure 2-16 The pH scale, noting the pH of some common solutions.

ESSAY

LIFE BELOW PH 7

Organisms that live in water usually require a pH near 7, the approximate pH of cellular interiors and the pH at which natural selection has favored an appropriate mix of acidic and basic groups in functional proteins. For thousands of years people have killed unwanted organisms in their food by pickling—that is, by soaking meats or vegetables in vinegar, a weak acid. Some microorganisms (such as those pictured in Figure 1a), however, thrive in acidic environments. As lactobacilli bacteria grow in milk, they reduce the pH of their environment. This acidity not only prevents the growth of most other organisms but also induces the coagulation of the milk into buttermilk, sour cream, cheese, or yogurt (depending on the species of lactobacilli). The characteristic flavors of these milk products also develop from the action of the bacteria.

Other microorganisms can tolerate still lower pH. Bacteria growing at pH below 3 have been isolated from the acid hot springs of Yellowstone National Park, as well as from the refuse piles of coal mines and from rivers and ponds polluted with industrial wastes. These bacteria, as well as others that live at extremes of pH, appear to have special adaptations that maintain their internal pH near 7 in spite of their acidic or basic external environment. (See Figure 1b.)

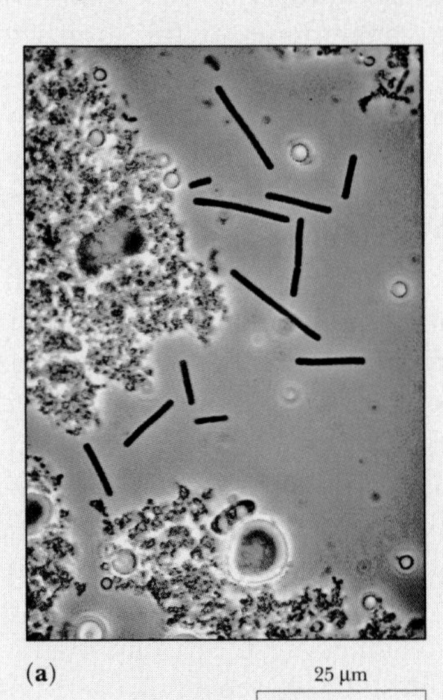

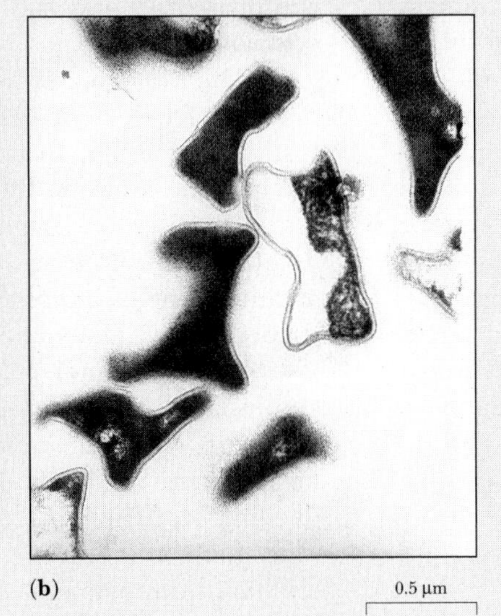

Figure 1 *Bacteria that thrive in acid. **(a)** Lactobacilli in yogurt; **(b)** thermoacidophiles that live in an acid hot spring in Yellowstone Park.* *(a, John Walsh/Photo Researchers; b, Dennis W. Grogan, University of Cincinnati)*

Yet another remarkable property of water may not be apparent: It can act either as an acid (by donating a hydrogen ion, leaving behind a hydroxide ion) or as a base (by accepting a proton to form a hydronium ion). In addition, the hydroxide ion, always present in equilibrium with water, is an extremely strong base, able to accept a hydrogen ion to form water.

Changing the pH of a solution also affects the properties of other molecules. The characteristic sour taste of acids results from the protons they contribute to the molecules on the tongue that report taste to the brain. Similarly, basic solutions are slippery to the touch because they change the characteristics of molecules on the skin.

When a molecule donates or accepts a hydrogen ion, its net charge changes, as in the case of carbonic acid and bicarbonate. This change alters the molecule's interactions with other molecules and ions. In a large molecule (such as the proteins discussed in Chapter 3), small groups of atoms within the larger molecule can independently act as acids and bases. Changes in the pH of the solution change the charge on these groups and influence the way individual parts of the same molecule interact with one another.

Because pH dramatically affects both the structure and the chemical reactivity of most biologically important molecules, cells have mechanisms that maintain nearly constant pH. The interiors of cells are almost always about pH 7.

Organisms use two kinds of devices to maintain constant pH—chemical and physiological. The chemical strategy is to employ **buffers,** molecules that easily interconvert between acidic and basic forms by donating or accepting hydrogen ions.

In the blood the main buffer is the carbonic acid/bicarbonate system we just described. The equilibrium between carbonic acid and bicarbonate helps maintain a constant pH. When the digestion of food produces extra hydrogen ions in the blood, bicarbonate ions can absorb them by becoming carbonic acid, and the pH of the blood changes little. Similarly, hydrogen ions provided by carbonic acid can "soak up" hydroxide ions added to the blood. In both cases the presence of the buffer minimizes changes in the total concentration of hydrogen ions.

Buffers like carbonic acid/bicarbonate reduce but do not eliminate pH changes. To maintain the constant internal environment necessary for life, cells rely on physiological adaptations, either in the cells themselves or, in multicellular organisms, of the organism as a whole. Within all cells, energy-driven hydrogen ion pumps maintain cytoplasmic pH. Eukaryotic cells, which have subcellular organelles, maintain separate compartments with differing pH. (In Chapters 7 and 8, we discuss the importance of these pH differences.)

In animals, the maintenance of blood pH, for example, is crucial for life: If the pH of the blood were to decrease until it was very slightly acidic (to pH 6.95), coma and death would soon follow. At slightly more basic pH (pH 7.7), muscle spasms and convulsions would begin. These disasters are prevented by a number of physiological adaptations, which include the regulation of the heart rate, changes in the rate and depth of breathing, and the secretion of hydrogen ions by the kidneys.

CELLS BUILD ALMOST ALL THEIR STRUCTURES FROM FOUR CLASSES OF BUILDING BLOCKS—LIPIDS, SUGARS, AMINO ACIDS, AND NUCLEOTIDES

If we analyze the sizes of all the molecules in a cell, we discover a surprising generalization. Cells have many small molecules, with molecular weights (M_rs) less than 300, and many large molecules, with M_rs greater than 10,000. But they have relatively few molecules with intermediate sizes.

We refer to the large molecules as **macromolecules** [Greek, *macro* = large], discussed in Chapter 3. Macromolecules consist of small molecules joined together in long chains. Macromolecules, then, are all **polymers** [Greek, *poly* = many + *meros* = part], molecules that consist of smaller molecules linked together. Each of the component parts is called a **monomer** [Greek, *mono* = single].

Cells contain many types of small molecules, but almost all of them are closely related to the 35 discussed in this chapter. These 35 molecules, listed in a table inside the front cover of this book, serve as universal building blocks for macromolecules in all organisms. In addition, they also play many biological roles—for example, in storing energy and carrying signals between cells. We may think of small molecules as letters of an alphabet and macromolecules as sentences composed from these letters. Cells are a strange alphabet soup, with free-floating letters and sentences, but with relatively few words standing alone.

What Determines the Biological Properties of an Organic Molecule?

The biological functions of molecules are determined by their structures. Because water is the universal medium of life, the attraction or avoidance of water by molecules or parts of molecules is arguably the most important determinant of cellular organization. The interactions of molecules with water and with each other in turn depend on the properties and the arrangements of the atoms that compose them.

The biological roles played by a small molecule or a macromolecule depend on shape and charge distribution. Just as the parts of a car or a bicycle must fit together for the machine to work, the molecules in a cell must also interact. We therefore begin by looking at the properties that determine the interactions among small molecules. The arrangement of atoms within a molecule determines its shape, as well as its chemical and biological properties.

Whether a molecule is hydrophobic, hydrophilic, or amphipathic in turn depends on the polarity of individual bonds within the molecule. Polar bonds between atoms mean unequal charge distributions and provide an opportunity for the molecule to form hydrogen bonds with water molecules. Polar molecules are more soluble in water than nonpolar molecules. The polarity of each bond in turn depends on the electronegativity of the participating atoms. The biological roles of a molecule therefore depend not only on its shape and the number and kind of atoms, but also on the specific arrangements of atoms.

To illustrate the importance of bond polarity, let us consider two simple molecules shown in Figure 2-17—ethane, a component of natural gas, and ethanol. Each of these molecules contains two carbon atoms and six hydrogen atoms; ethanol also has a single atom of oxygen.

Ethane contains only nonpolar bonds, one between the two carbons (C—C) and six between carbon and hydrogen (C—H). Ethanol, on the other hand, contains two polar bonds, one between carbon and oxygen (C—O) and one between oxygen and hydrogen (O—H). The unequal charge distribution in the O—H bond allows ethanol molecules to form hydrogen bonds with other ethanol molecules and with water, so ethanol is highly hydrophilic. In

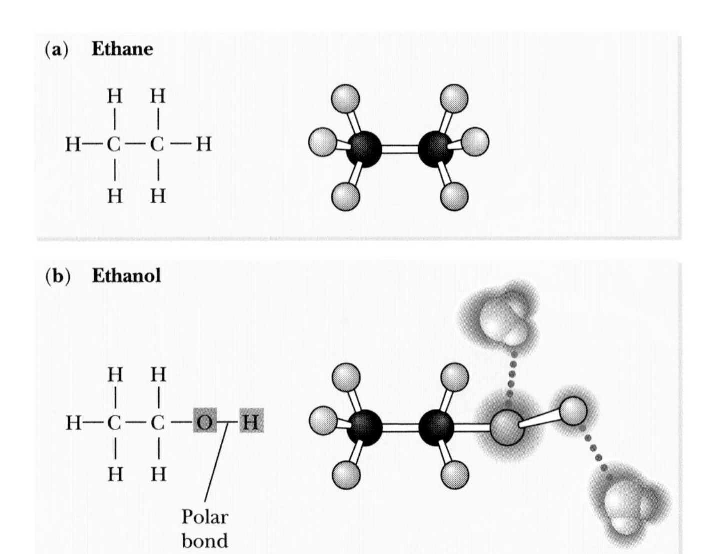

Figure 2-17 The structures of **(a)** ethane and **(b)** ethanol. Ethanol contains a polar group (OH), allowing it to form hydrogen bonds with water.

contrast, ethane molecules do not interact much even with each other. Ethane boils at a much lower temperature (−89°C) than ethanol (which boils at 79°C). At room temperature ethane is a gas and ethanol is a liquid.

Biologically Important Molecules Contain a Small Number of Functional Groups

Many of the characteristics of ethanol, including the ability to form hydrogen bonds, depend on the presence of its O—H grouping. Other molecules that contain an O—H group attached to a carbon atom have similar properties.

In analyzing the structures of many organic molecules, chemists have repeatedly found standard small groupings of atoms attached to carbon atoms. These arrangements, called **functional groups,** determine many characteristics of organic molecules. Familiarity with just the four functional groups shown in Figure 2-18 is enough to begin to understand the chemical properties of our 35 building blocks. For this discussion, recall that carbon can acquire a full outer shell by sharing four electrons, nitrogen by sharing three, oxygen by sharing two, and hydrogen by sharing one.

As we've said, the O—H functional group, called a **hydroxyl** group, allows molecules that contain it to form hydrogen bonds. In addition to the hydroxyl group, shown in Figure 2-18a, two other functional groups involve polar bonds between carbon and oxygen atoms. The **carbonyl** group (shown in Figure 2-18b) consists of a carbon atom attached to an oxygen atom by a double bond (C═O), that is, by sharing four rather than two electrons. The **carboxyl** group (shown in Figure 2-18c) contains two oxygen

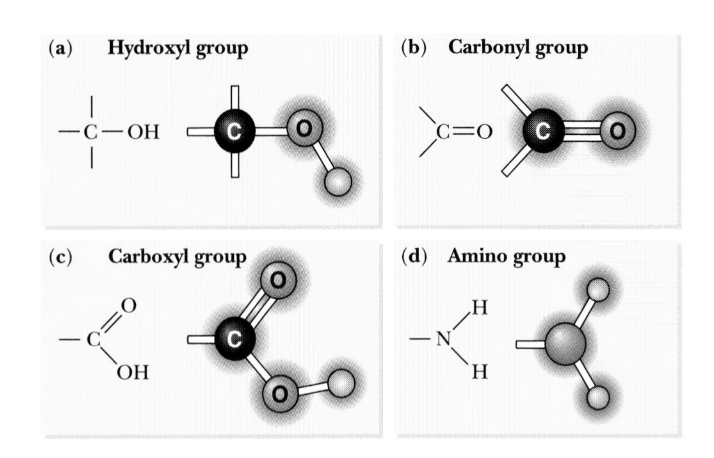

Figure 2-18 Functional groups important in establishing the properties of biological molecules. **(a)** Hydroxyl group; **(b)** carbonyl group; **(c)** carboxyl group; **(d)** amino group.

atoms rather than one. The carbon atom forms a double bond with one oxygen atom and a single bond with the other. The carboxyl carbon shares three of its four outer electrons with oxygen atoms. One oxygen atom shares two of its outer electrons with the carbon atom, and the second oxygen atom shares one with a carbon atom and the other with a hydrogen atom.

Figure 2-18 shows the structures of hydroxyl, carbonyl, and carboxyl groups. Notice that a carbon atom attached to a hydroxyl group has three electrons to share with other atoms. In ethanol, for example, the carbon that is connected to the hydroxyl group is also connected to another carbon atom and to two hydrogen atoms. In contrast, the carbonyl carbon has only two electrons left to form other bonds, and the carboxyl carbon has only one. In fact, we may speak of a carbon atom as "losing" electrons in going from a hydroxyl group to a carbonyl group to a carboxyl group.

Ethanol, acetaldehyde, and acetic acid are three related organic molecules that differ in their physical and chemical properties. Each of these molecules contains two carbon atoms, but each has a different functional group. The characteristics of each molecule depend on its functional groups. Ethanol burns in air, as does acetaldehyde but not acetic acid. Acetaldehyde, which has a fruity smell, is chemically reactive and is widely used as a starting material in the chemical industry. Acetic acid is familiar as a principal component of vinegar.

The other important functional group in biological molecules is the **amino** group, which consists of a nitrogen atom and two hydrogen atoms (NH_2), as shown in Figure 2-18d. The nitrogen atom acquires a stable octet by sharing three electrons, two electrons with hydrogen atoms and a third with a carbon atom.

The nitrogen atom of the amino group also contains an unshared pair of electrons. At cellular pH (about pH 7), the amino group associates with a hydrogen ion

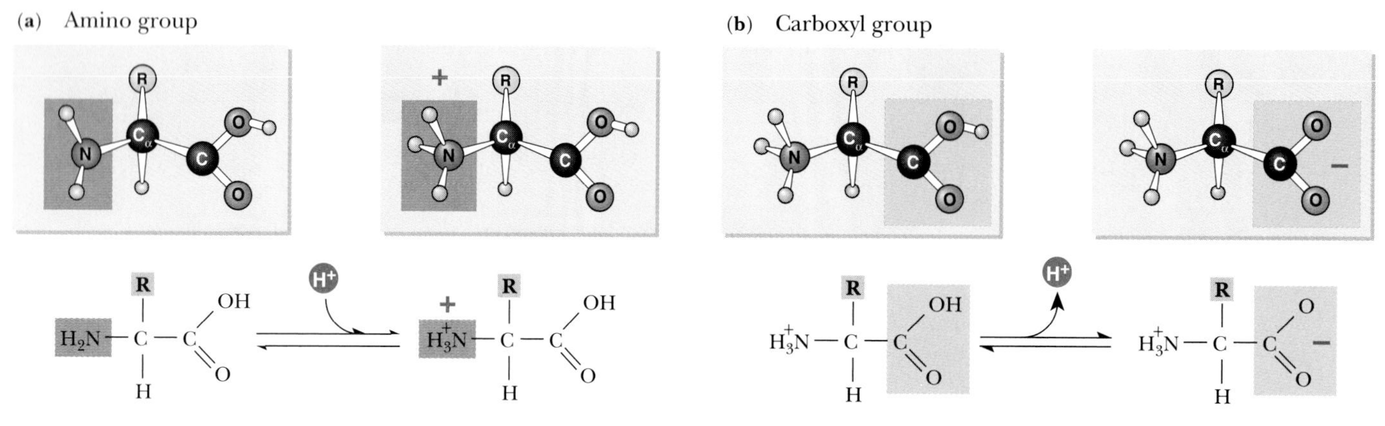

Figure 2-19 **(a)** An amino group is a base: It can acquire a proton and a positive charge (NH_3^+). **(b)** A carboxyl group is an acid: It can give up a proton and acquire a negative charge.

(H^+), as shown in Figure 2-19a. The NH_2 group becomes positively charged and is now written NH_3^+.

When an amino nitrogen acquires a hydrogen ion, the amino group is acting as a base. In a roughly parallel manner (shown in Figure 2-19b), the oxygen-attached hydrogen of a carboxyl group can detach to give a free hydrogen ion. The carboxyl group acquires an extra electron (from hydrogen) and a net negative charge. This is another example of how functional groups give character to molecules: An amino group acts as a base, while a carboxyl group acts as an acid.

Knowledge of the functional groups described above allows us to understand the importance of the associations of atoms within molecules. But molecules can be composed of the same atoms arranged in different ways. **Isomers** are molecules that contain the same atoms arranged differently. **Structural isomers** have the same atoms, but they are grouped in different ways to produce different functional groups, giving widely differing properties. Compare, for example, the properties of the two molecules shown in Figure 2-20, ethanol and its structural isomer, dimethyl ether. Ethanol contains the polar hydroxyl group, but dimethyl ether does not. Ethanol is a common product of cells: It is highly soluble in water and has a well-known intoxicating effect on humans and other animals. Dimethyl ether is not found in organisms, is much less soluble in water, and has little biological effect.

Carbon Atoms May Form Chains of Any Length

Carbon atoms may form single, double, or triple bonds with other carbon atoms. A common organic structure is a **hydrocarbon chain,** a chain of connected carbon atoms, with hydrogen atoms sharing other available outer-shell electrons. For example, stearic acid, a component of cell membranes, consists of a hydrocarbon chain with 17 carbon atoms attached to a carboxyl group, as illustrated in Figure 2-21.

Hydrocarbon chains can also form rings. Because carbon can form covalent bonds only at fixed angles, most carbon-containing rings have either five or six atoms. The members of a ring sometimes include other types of atoms, especially oxygen and nitrogen, as in glucose and dimethylpyrazine (chocolate) rings, shown in Figure 2-22a and c.

The atoms of a ring are connected by single covalent bonds, with each atom in the ring connected to its two neighbors by bonds that make an angle of about 109°. The ring atoms do not lie in the same plane but zig and zag to form a structure that looks rather like a chair. But in some ring-type compounds, all the atoms lie together in

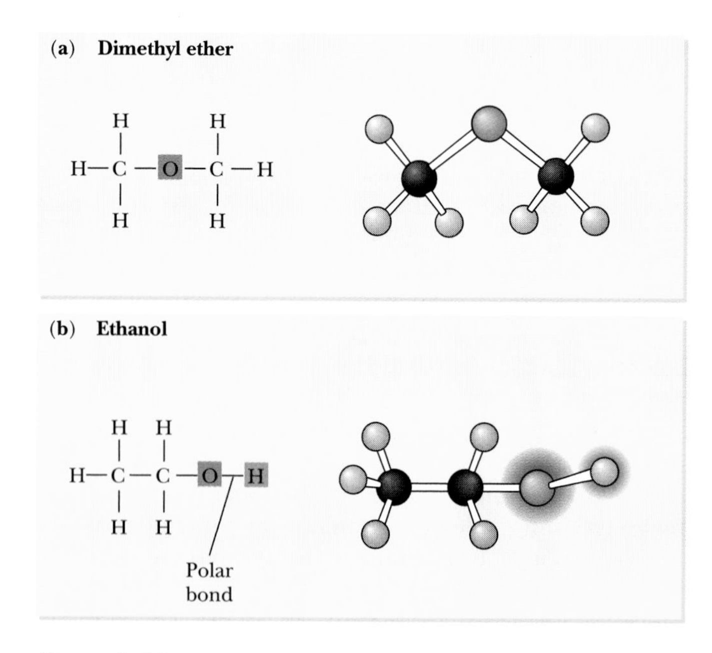

Figure 2-20 **(a)** A molecule of dimethyl ether, a nonpolar molecule, consists of the same atoms as ethanol, a polar molecule, shown in **(b).** Differing arrangements of atoms into functional groups gives isomers differing properties.

ESSAY

MOLECULES WITH SIMILAR SHAPES CAN MIMIC ONE ANOTHER: AMPHETAMINE AND NOREPINEPHRINE

The general arrangements of atoms in a molecule determine many of its properties—whether it is polar or nonpolar, whether it undergoes oxidation or reduction. But the specific arrangement of atoms determines the molecule's size and shape. In many cases, the shape and charge distribution of a molecule—much more than its chemical properties—determines its biological effects. In many cases, for example, synthetic molecules (that is, molecules made in the laboratory and never found in nature) can have shapes and charge distributions so similar to those of natural compounds that they can mimic their biological effects.

A particularly dramatic example of this mimicry is the case of amphetamine, whose three-dimensional structure (shown in Figure 1) closely resembles norepinephrine, a powerful chemical signal in the hormone system and the nervous system. Norepinephrine helps coordinate the "fight or flight" reaction, by which animals (including humans) respond to threatening situations.

During World War II, flyers and soldiers on all sides used amphetamine to keep alert during night missions. After 1945, amphetamine became a widely abused drug, with many instances of addiction in Japan and the United States. Amphetamine is dangerous because the brain cannot regulate the effects of amphetamine as it regulates its responses to norepinephrine. Amphetamine therefore has severe disorganizing effects and, even in the short term, leads to an inevitable depressive crash.

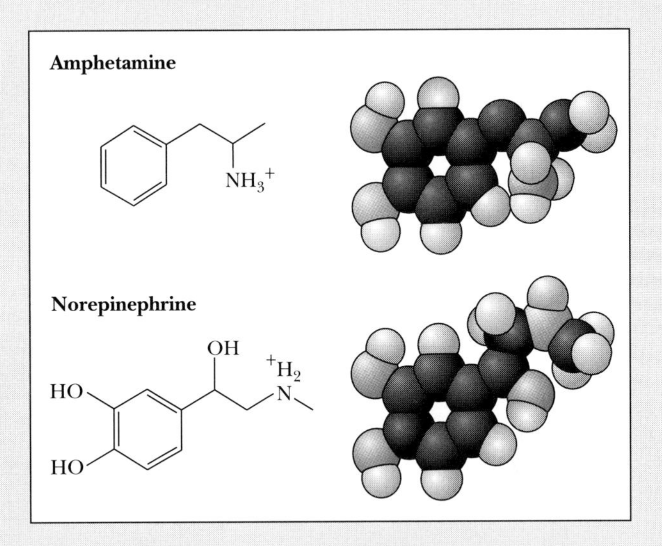

Figure 1 The similar shapes of norepinephrine, a natural signalling molecule, and amphetamine, a synthetic and widely abused drug.

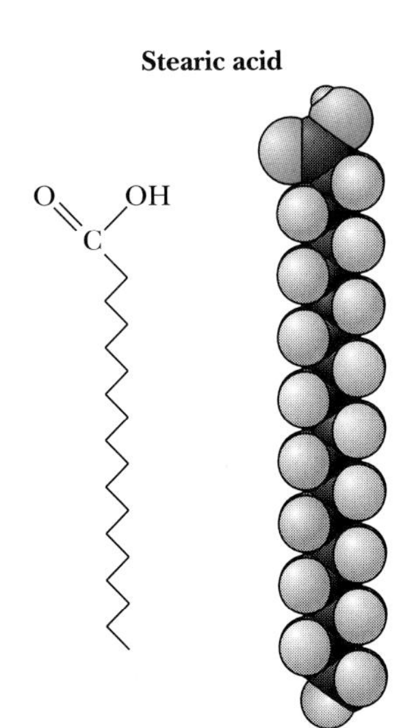

Figure 2-21 A molecule of stearic acid, containing 18 C atoms—17 in a hydrocarbon chain and one in the carboxyl group—all held together with single bonds.

the same plane, as in benzene, shown in Figure 2-22b. Molecules with benzene-like planar rings are called **aromatic** compounds. The rings in aromatic compounds usually consist of five or six atoms linked together by sharing their valence electrons not just between two atoms but among all the members of the ring. This electron arrangement makes aromatic compounds especially stable and less likely to react than chemists had originally expected. Some biologically important compounds containing aromatic rings (such as phenylalanine, shown in Figure 2-22d) have only carbon atoms, while others have one or two nitrogen atoms.

We are now in a position to understand the properties of many organic molecules, including the 35 on our list. Think of each molecule as a skeleton of carbon atoms

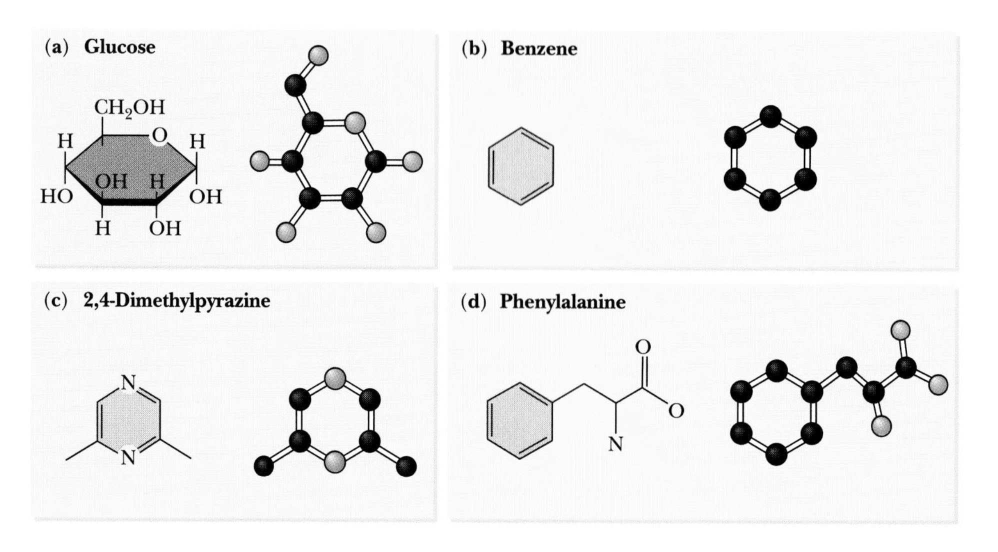

Figure 2-22 Carbon, nitrogen, and oxygen atoms in six-membered rings. **(a)** Glucose; in the actual structure, the C atoms zig and zag, distorting the ring so that it resembles a chair; **(b)** benzene; **(c)** dimethylpyrazine (chocolate); **(d)** phenylalanine. Hydrogen atoms are not shown, but every carbon atom shares four electrons. In the benzene ring, for example, each C atom is attached to one H atom.

to which hydrogen atoms or functional groups may be attached. Molecules (or parts of molecules) consisting only of carbon and hydrogen atoms (hydrocarbon chains) are nonpolar and hydrophobic. Hydrophilic molecules (or parts of molecules) always contain polar functional groups. Table sugar (sucrose), for example, dissolves easily in water because it contains many hydroxyl groups.

Sulfur and Phosphorus Also Form Important Functional Groups in Biological Molecules

Functional groups containing sulfur and phosphorus are also important in biological molecules. Sulfur, like oxygen, can share two electrons. In the **sulfhydryl group** (shown in Figure 2-23a), sulfur forms one bond with a hydrogen atom and another with a carbon atom. Like a hydroxyl group, the sulfhydryl group is polar. The sulfhydryl group gives the amino acid cysteine (shown in Figure 2-23b) the special property of being able to combine with another sulfhydryl group to form a **disulfide bond** (shown in Figure 2-23c), in which two sulfur atoms form a covalent bond with one another.

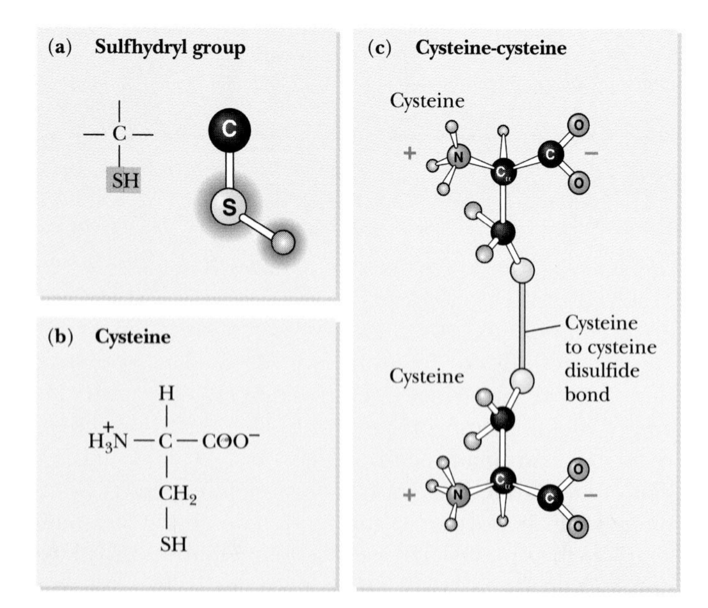

Figure 2-23 The sulfhydryl group in cysteine can form a covalent disulfide bond with the sulfhydryl group of another cysteine. **(a)** Sulfhydryl group; **(b)** cysteine; **(c)** a disulfide bond between two cysteine molecules.

Phosphorus, like nitrogen, has five electrons in its outer shell. In organisms, phosphorus is almost always bonded to four atoms of oxygen. In order that all five atoms have stable electron octets, they need to acquire a total of three additional electrons. In phosphoric acid (H_3PO_4), shown in Figure 2-24, these electrons come from three hydrogen atoms that form bonds with the oxygen atoms. The hydrogens can, however, all dissociate as hydrogen ions, leaving behind their electrons in a **phosphate** group, with an excess of three negative charges (PO_4^{-3}).

In a **phosphoric acid ester,** such as adenosine monophosphate, a hydroxyl group attached to a carbon atom replaces one of the oxygen atoms, attaching a phosphate group to a carbon atom, as illustrated in Figure 2-24c. At the pH of cells, the phosphoric acid esters give up hydrogen ions to form negatively charged phosphate-containing organic molecules, as illustrated in Figure 2-24c.

BIOLOGICAL BUILDING BLOCKS FALL INTO FOUR CLASSES, EACH WITH CHARACTERISTIC FUNCTIONAL GROUPS

All the large molecules (macromolecules) of a cell are built from combinations and polymers of the small molecule building blocks. These small molecules fall into four classes:

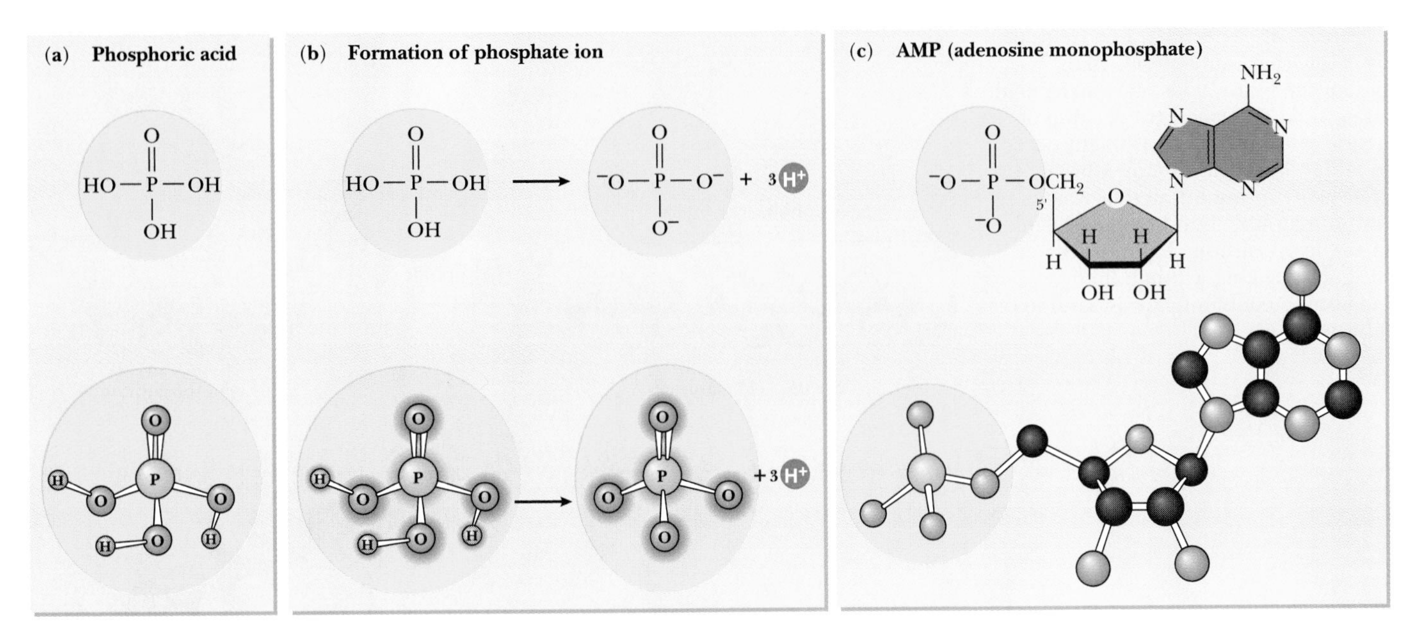

Figure 2-24 Phosphoric acid loses hydrogen ions to form a phosphate ion. **(a)** Phosphoric acid; **(b)** phosphate ion; **(c)** adenosine monophosphate (AMP), a phosphate ester.

1. **Lipids** are compounds that are less soluble in water than in nonpolar solvents like olive oil.
2. **Sugars** are molecules that have the equivalent of one molecule of water (that is two hydrogen atoms and one oxygen atom) for every atom of carbon.
3. **Amino acids** are molecules that contain both amino and carboxyl groups.
4. **Nucleotides** are molecules that each consist of a nitrogen-containing aromatic ring compound, a sugar, and a molecule of phosphoric acid.

As we see in more detail in Chapter 3, sugars form polymers called **polysaccharides.** Similarly, nucleotides form polymers called **nucleic acids,** and amino acids form polymers called **proteins.** Lipids do not form polymers, but they do assemble into large structures. Lipids are the major components of the membranes that enclose cells and cell compartments.

We may think of the building blocks of both lipids and macromolecules as letters in different alphabets. Just as sentences in English, Greek, Japanese, and Hebrew are written in different alphabets, the different classes of cellular molecules consist of "sentences" that use different biochemical alphabets. In the rest of this chapter we discuss the characteristics of the letters of each biochemical alphabet.

Lipids Include a Variety of Nonpolar Compounds

Many familiar compounds from the kitchen—oils, fats, and floor wax—are lipids. The defining characteristic of lipids is their solubility in nonpolar solvents such as gasoline, cleaning fluid, and olive oil. The low solubility of lipids in water is due to their low content of polar functional groups. Lipid molecules contain few oxygen or nitrogen atoms but instead consist mostly of carbon and hydrogen atoms. The limited solubility of lipids in water explains why removing a butter or gravy stain often requires dry cleaning—that is, treatment with nonpolar solvents.

Lipids contain more chemical energy per gram than other biological molecules, and lipids often serve as energy stores. Because of their low solubility in water, lipids are especially important components of the membranes that separate cells from their external environments and, in eukaryotic cells, cell compartments from each other.

Not all lipids, however, are totally hydrophobic. Some lipid molecules are amphipathic: They contain a polar functional group as well as nonpolar chains of carbon and hydrogen atoms.

One class of amphipathic lipids is the **fatty acids,** three examples of which we have already mentioned, stearic acid, oleic acid, and palmitic acid. A fatty acid consists of a long hydrocarbon chain ending in a carboxyl group, as shown in Figure 2-25a. The carboxyl group can form hydrogen bonds with water molecules, so that one end of the fatty acid molecule is hydrophilic and the other (hydrocarbon) end is hydrophobic. The carboxyl group also makes the molecule an acid because it can give up a hydrogen ion to water. In cells, carboxyl groups are usually ionized, so that they are even more hydrophilic than in their uncharged form.

Fatty acids can differ in both the number of carbon

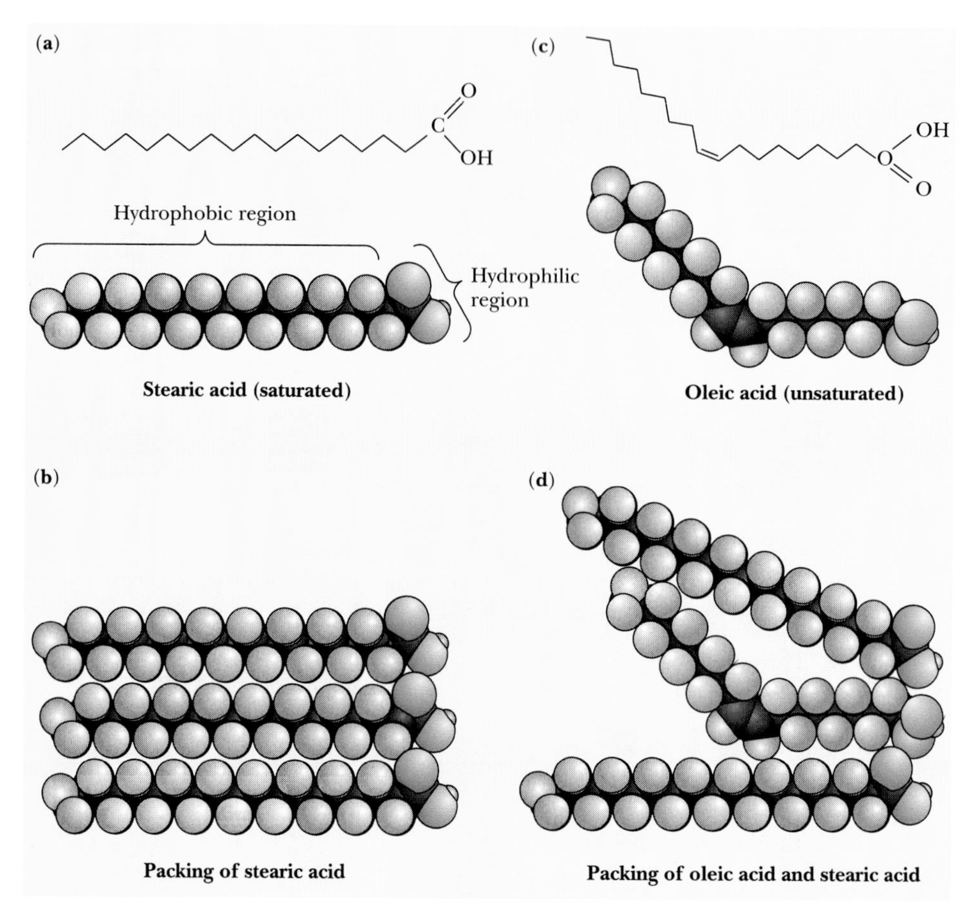

Figure 2-25 Saturated fatty acids can pack together more closely than unsaturated fatty acids. **(a)** Stearic acid, a saturated fatty acid; **(b)** packing of stearic acid molecules, as might occur in a lipid layer within a membrane; **(c)** oleic acid, an unsaturated fatty acid, with one double bond causing a kink in the hydrocarbon chain; **(d)** the irregular packing within a lipid membrane containing an unsaturated fatty acid.

atoms they contain and the relative numbers of single and double bonds. The hydrocarbon chain in a fatty acid consists of carbon atoms linked either to other carbon atoms or to hydrogen atoms. When all the bonds between carbon atoms are single bonds, the hydrocarbon chain contains the maximum number of hydrogen atoms and is said to be **saturated,** as is stearic acid. When a fatty acid contains at least one double bond between two carbon atoms, it is said to be **unsaturated,** as is oleic acid (Figure 2-25c), because it can accept two more hydrogen atoms per double bond. A fatty acid with many double bonds is said to be **polyunsaturated.**

Single bonds between carbon atoms allow parts of the hydrocarbon chain to rotate freely. This flexibility allows chains of neighboring molecules to attract each other in van der Waals interactions. The presence of double bonds, however, reduces the flexibility of a hydrocarbon chain and restricts interactions between molecules, as illustrated in Figure 2-25b and d. The result is that saturated hydrocarbon chains interact more than unsaturated chains and solidify more easily. That is, saturated fatty acids have a higher melting point than unsaturated fatty acids.

Most of the Fatty Acids in Organisms Are Chemically Combined with Glycerol

Glycerol, shown in Figure 2-26a, contains three hydroxyl groups. Each hydroxyl group can attach by means of a dehydration reaction to the carboxyl carbon of a fatty acid to form an ester linkage. In a **triacylglycerol** (or **triglyceride**), shown in Figure 2-26b, all three hydroxyl groups are attached to fatty acids, so that glycerol acquires three long hydrocarbon tails. When a triacylglycerol forms, the hydrophilic properties of the carboxyl group of the fatty acid and of the hydroxyl groups of the glycerol are lost, so triacylglycerols are even less soluble in water than fatty acids or glycerol.

A triacylglycerol that contains only saturated hydrocarbon chains is solid at room temperature and is called a **fat.** A triacylglycerol that is polyunsaturated is likely to be a liquid at room temperature and is called an oil. Animals contain more fats and plants more oils. Most of the fat deposits of animals and the oils of plants consist of triacylglycerols.

In a **phospholipid,** shown in Figure 2-27, only two hydroxyl groups of glycerol attach to fatty acids, so there

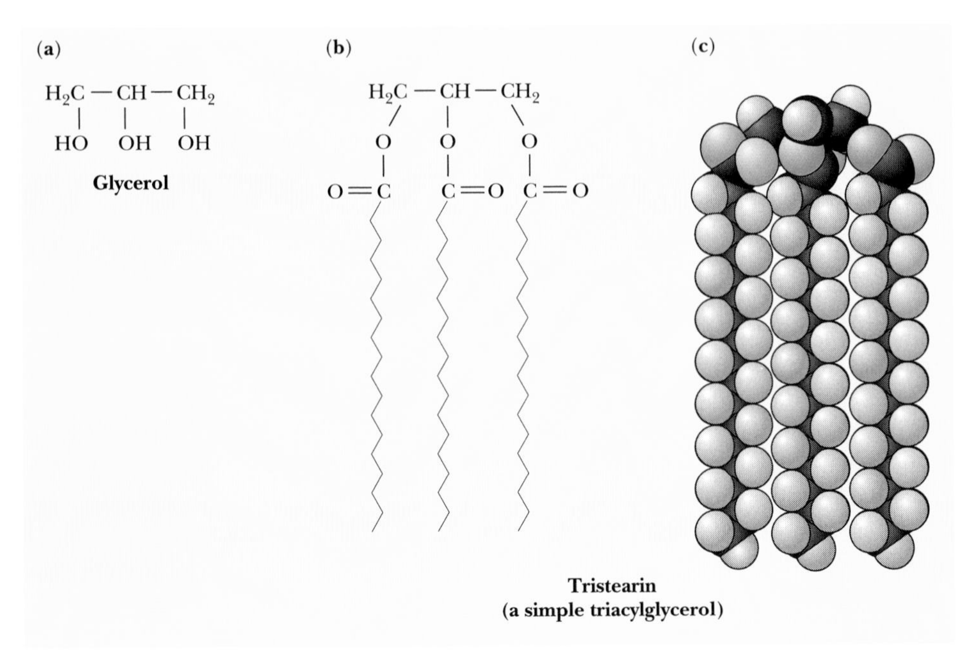

Figure 2-26 A triacylglycerol (triglyceride) is formed from glycerol and three fatty acids. **(a)** Glycerol; **(b)** tristearin, which consists of three molecules of stearic acid each linked to an oxygen atom from a hydroxyl group of glycerol; **(c)** a space-filling model of tristearin.

(a) **Phosphatidylcholine**

Hydrophobic tail

Hydrophilic head

O
C—O
CH2
O
C—O—C—H
O CH3
CH2—O—P—O—CH2CH2—N+—CH3
O− CH3

(b) **Phosphatidylinositol**

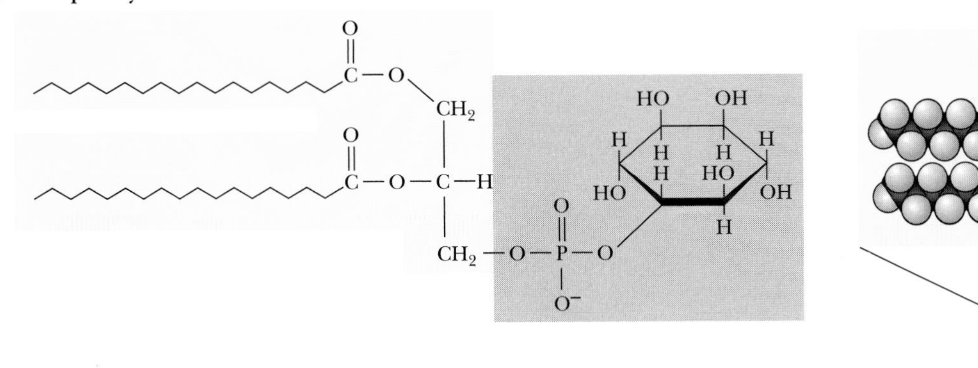

Phospholipid icon

Figure 2-27 A phospholipid consists of a hydrophilic head, with a charged phosphate group, and a hydrophobic tail, which contains two fatty acids. **(a)** Phosphatidylcholine; **(b)** phosphatidylinositol.

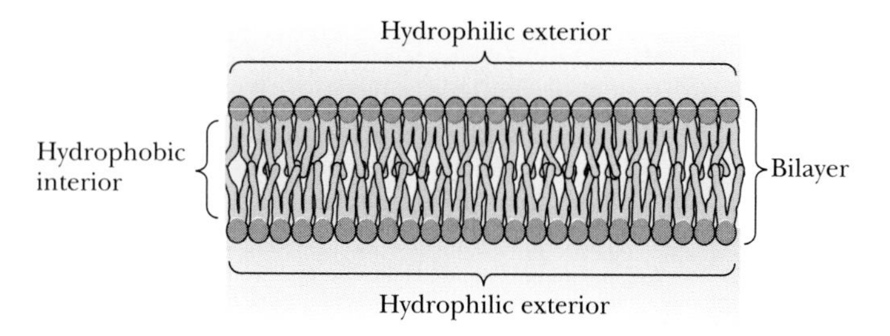

Figure 2-28 Membranes form from two phospholipid layers lying with their hydrophilic heads pointing outward and their hydrophobic tails forming the membrane's interior.

are only two tails. The third hydroxyl group of glycerol instead attaches to a phosphate group, which in turn attaches to another polar molecule. Phospholipids are thus highly amphipathic. Each has a hydrophobic "tail," consisting of the hydrocarbon chains of two fatty acids, and a hydrophilic "head," which contains a charged phosphate group.

In membranes, the hydrophobic tails associate with each other to form an oily interior, and the hydrophilic heads associate with surrounding water molecules, as shown in Figure 2-28. The membrane boundaries between cells and cell compartments consist mostly of phospholipids. The membranes have a hydrophobic interior and a hydrophilic exterior.

Steroids, Another Class of Lipids, Consist of Four Interconnected Rings

Individual types of **steroids** differ in the functional groups that attach to a common hydrocarbon ring structure, as shown in Figure 2-29a. Many of the hormones responsible for sexual development and reproductive functions in animals are steroids. Small differences in the attached functional groups can make enormous differences in a steroid's biological properties. For example, testosterone (whose structure is shown in Figure 2-29c)—the hormone almost entirely responsible for inducing the genes characteristic of male sexual differentiation—differs by only a few atoms from the female hormone progesterone (shown in Figure 2-29d). The starting point for the synthesis of all other steroids is **cholesterol,** a hydrocarbon with the same pattern of four interconnected rings whose structure is shown in Figure 2-29b.

Sugars Contain Many Hydroxyl Groups

Both sugars and polysaccharides are **carbohydrates,** compounds that contain the equivalent of one water molecule (one oxygen atom and two hydrogen atoms) for every carbon atom. Most of the carbon atoms in carbohydrates are attached to both a hydrogen atom and a hydroxyl group. Each of these hydroxyl groups can form a hydrogen bond with a water molecule, accounting for the high solubility of most carbohydrates in water.

Simple sugars, or **monosaccharides** [Greek, *mono* = one + *saccharine* = sugar), may contain from three to nine carbon atoms. Figure 2-30 shows three monosaccharides—glucose, ribose, and deoxyribose—that are important components of biological molecules.

The most common monosaccharide, glucose (whose structure is shown in Figure 2-30a), contains six carbon atoms and is therefore called a **hexose** [Greek, *hex* = six]. In contrast, ribose and deoxyribose (Figure 2-30b,c), which are components of nucleotides, contain five carbons and are called **pentoses** [Greek, *pente* = five].

Sugars and polysaccharides are among the most important energy storage molecules in cells. Glucose serves as the main carrier of energy in the blood of animals. Vir-

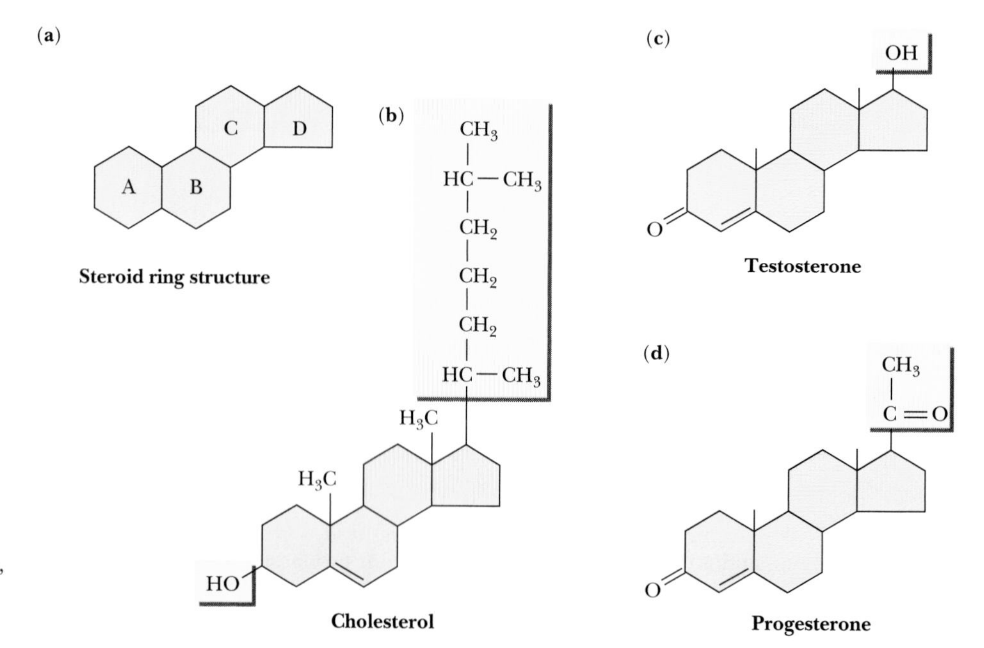

Figure 2-29 Steroid molecules are derivatives of cholesterol. **(a)** The basic four-ring structure of steroids; **(b)** the structure of cholesterol; **(c)** testosterone, a male sex hormone; **(d)** progesterone, a female sex hormone.

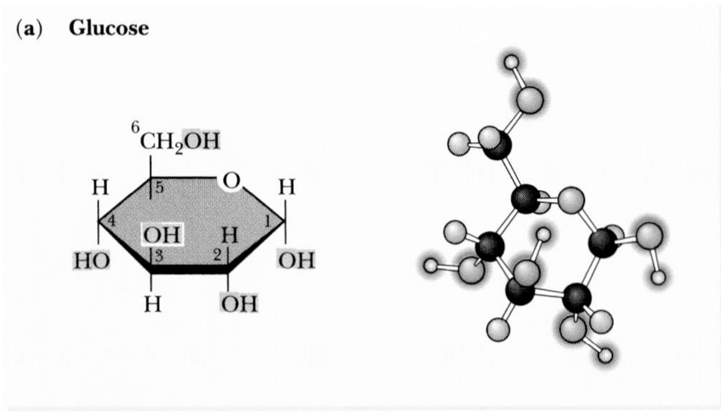

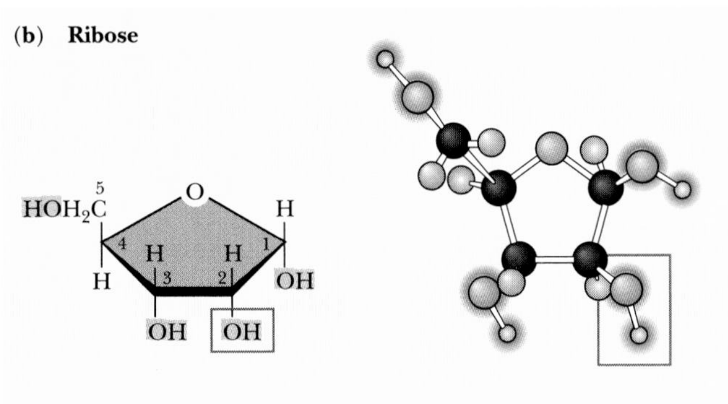

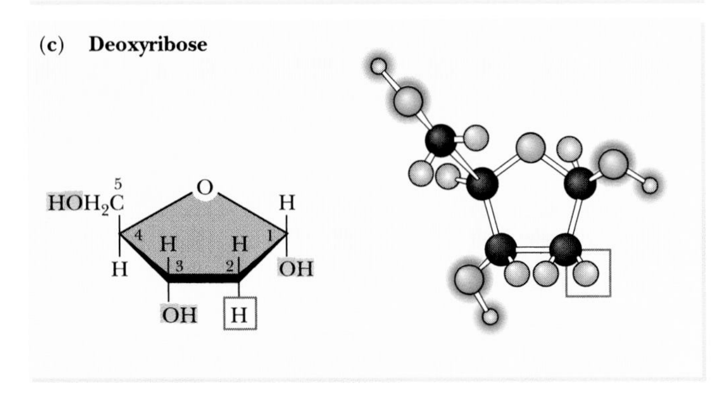

Figure 2-30 Simple sugars (monosaccharides). **(a)** Glucose, a six-carbon sugar; the starting point for much of energy metabolism; **(b)** ribose, a five-carbon sugar that is a component of ATP and of RNA; **(c)** deoxyribose, a five-carbon sugar derived from ribose, a component of DNA.

tually all the interrelated biochemical reactions that produce energy are in some way connected to the synthesis and breakdown of glucose. All cells and organisms have special adaptations that keep the concentration of glucose nearly constant.

Amino Acids Contain Both Carboxyl and Amino Groups

All 20 amino acids on our list are variations of the same basic structure. As shown for alanine in Figure 2-31, each amino acid contains a carboxyl group, which makes it an acid, and an amino group, which also makes it a base. Both

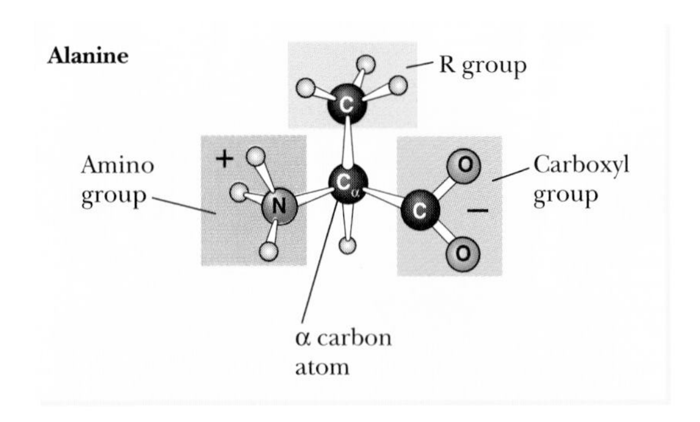

Figure 2-31 An amino acid—here alanine—contains a carboxyl group (here on the right), an amino group (here on the left), a hydrogen atom, and a side chain (R group) (here pointing up), all attached to a single carbon atom, called the α carbon.

these functional groups are attached to the same carbon atom, called the α (alpha) carbon.

Each of the amino acids has a characteristic group of atoms, called a side chain or R group, attached (like the amino group, the carboxyl group, and a hydrogen atom) to the α-carbon atom. Figure 2-32 shows the 20 amino acids found in proteins. Side chains may contain hydrocarbon chains of different lengths, as well as different functional groups. Some side chains are hydrophobic, some hydrophilic. Some normally have a positive charge and some a negative charge. The properties of the side chains establish the individual character of each amino acid.

Several amino acids and their derivatives themselves serve specialized biological functions. Some are neurotransmitters, chemicals secreted by nerve cells to signal nerve and other cells. Some nerve cells, for instance, cause other nerve cells to transmit impulses by secreting glutamic acid. Other nerve cells secrete a compound derived from the amino acid tyrosine, called norepinephrine (or noradrenaline). (See Essay, page 45.)

A Nucleotide Has Three Parts, Each with Different Functional Groups

A nucleotide consists of a **nitrogenous base,** a sugar (ribose or deoxyribose), and one or more phosphate groups. Each nitrogenous base (so called because it contains nitrogen and can accept hydrogen ions) contains one or two aromatic rings of carbon and nitrogen atoms, with attached hydrogen atoms. Also attached to these rings are polar functional groups that can form hydrogen bonds with water or (as discussed in Chapter 13) with other nitrogenous bases. Five nitrogenous bases, whose structures are shown in Figure 2-33a, are on our list of building blocks. A **nucleoside** consists of a nitrogenous base attached to a sugar by a covalent bond between a carbon atom of the sugar and a nitrogen atom of the base, as shown in Figure 2-33b. A **nucleotide** (illustrated in Figure 2-33c) is a nucleoside

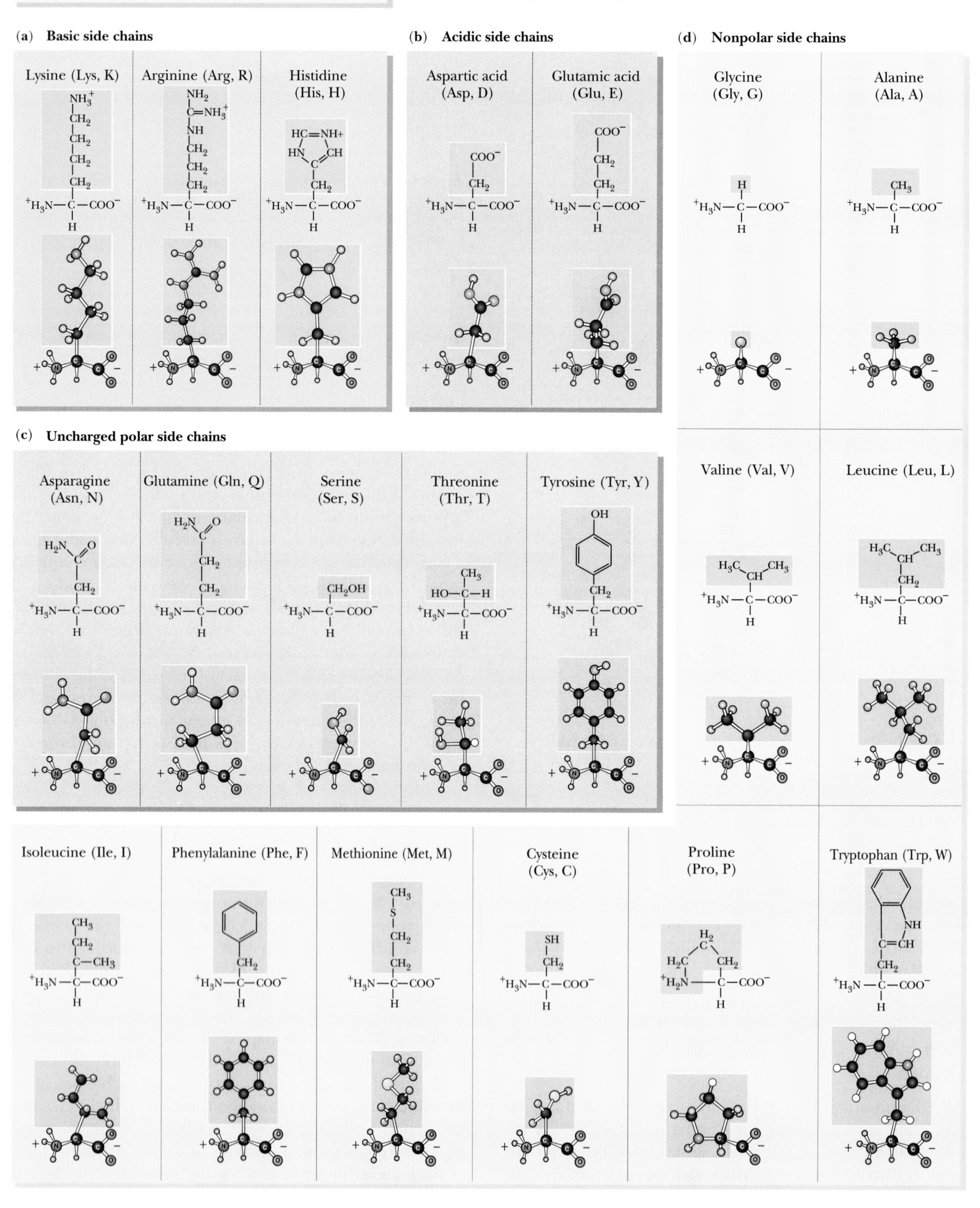

Figure 2-32 The 20 amino acids used in proteins, along with their standard 3-letter and 1-letter abbreviations. (Actually, one of the 20, proline, is an "imino" acid because its amino group is part of a ring structure.)

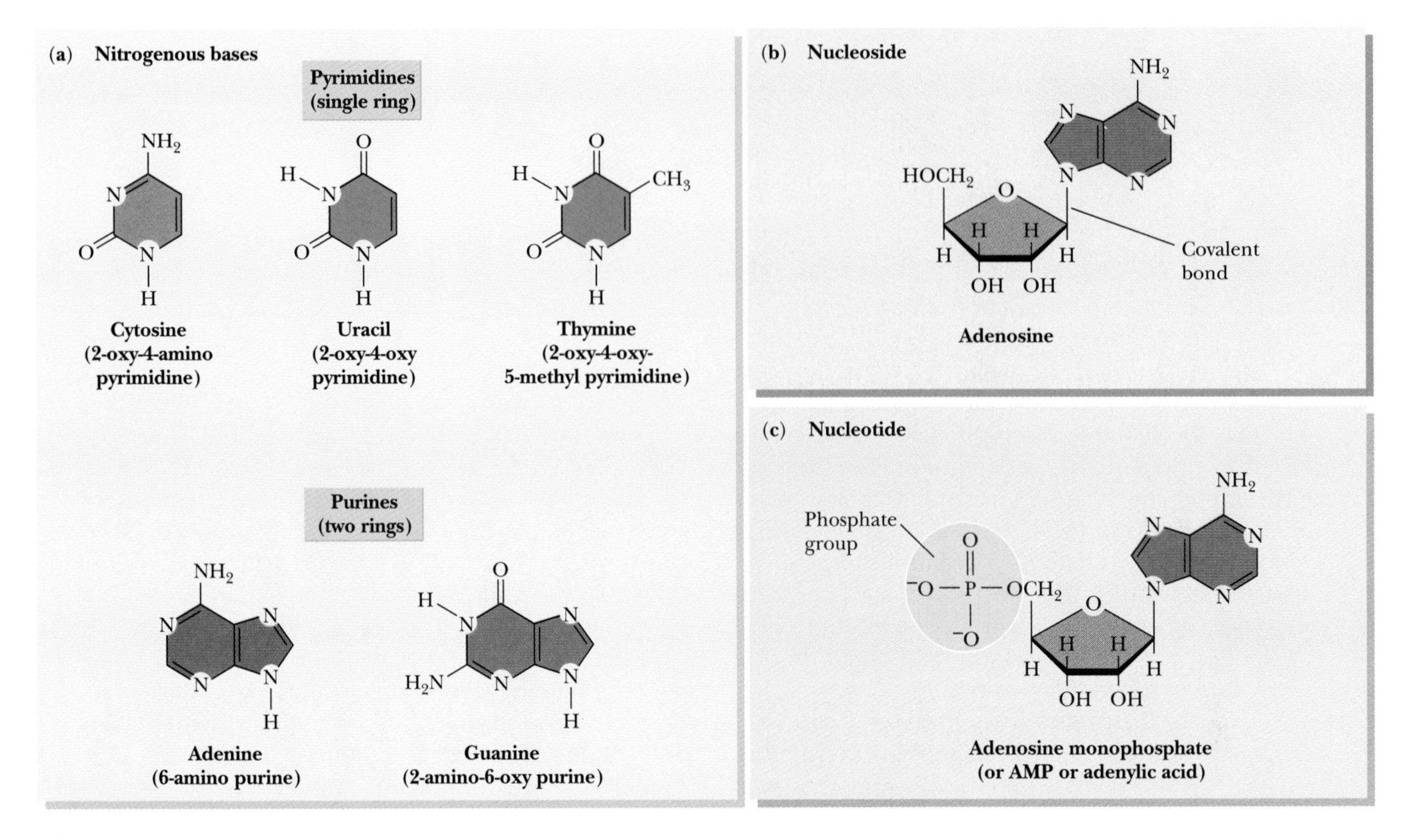

Figure 2-33 Nitrogenous bases, nucleosides, and nucleotides. **(a)** Nitrogenous bases may be pyrimidines, with single rings of 2 N and 4 C atoms, or purines with double rings of 4 N and 5 C atoms. **(b)** Adenosine, a nucleoside; nucleosides are nitrogenous bases linked to a sugar, here a ribose molecule. **(c)** AMP, a nucleotide; nucleotides are nitrogenous bases linked to a sugar linked to a phosphate group.

linked to a phosphoric acid group. At the pH of a cell, phosphoric acid is ionized to form the negatively charged phosphate ion, so each nucleotide has a negative charge.

As shown in Figure 2-33a, there are two classes of nitrogenous bases: **pyrimidines,** which contain a single ring of atoms, and **purines,** which contain two interlocking rings. Our list contains two purines—adenine and guanine—and three pyrimidines—cytosine, uracil, and thymine. Thymine and uracil are almost identical, differing only in the presence of a methyl (CH_3) group in thymine.

Like sugars and amino acids, nucleotides themselves serve important biological roles, especially in the packaging and transport of energy and in the regulation of a cell's chemical transformations. Free nucleotides may contain one, two, or three phosphate groups linked to one another; adenosine triphosphate (ATP) (whose structure is shown in Figure 2-34) contains three phosphate groups. The major source of energy for all biological processes in all organisms is the energy released by breaking the phosphate-phosphate bonds of ATP.

Why Are Biological Structures Made from So Few Building Blocks?

Given the enormous diversity of organic compounds, the relatively small number of compounds used by organisms

(Text continues on page 56.)

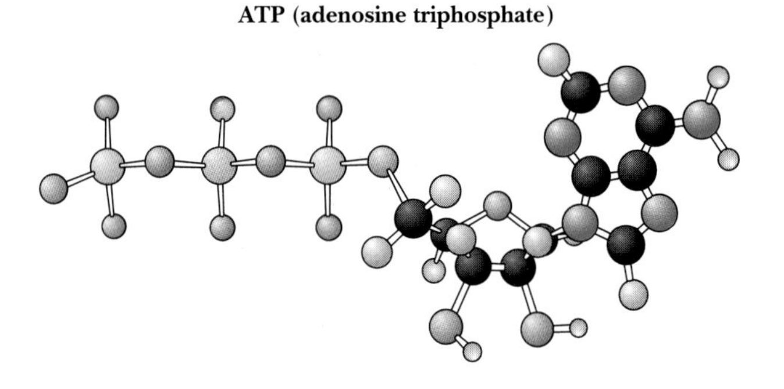

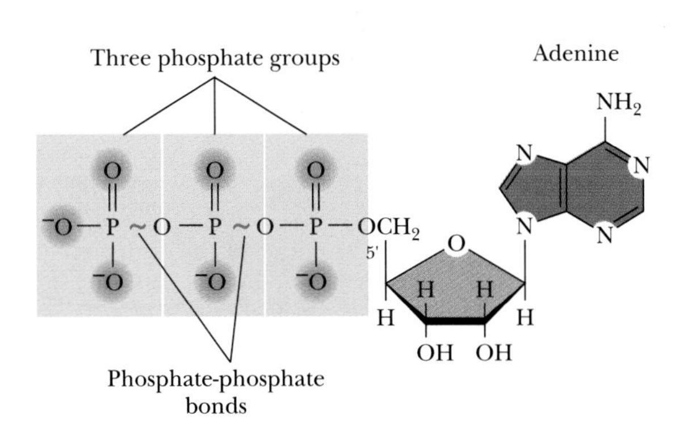

Figure 2-34 ATP (adenosine triphosphate), the energy currency of all cells, is a nucleotide consisting of the nitrogenous base adenine, the sugar ribose, and three phosphate groups.

ESSAY

OPTICAL ISOMERS AND LIFE

Organisms use only a minuscule fraction of all molecules that they could conceivably make from combinations of the same atoms. Molecules with the same atoms and functional groups are called **stereoisomers,** and they differ only in the spatial arrangements of their atoms. Two stereoisomers cannot superimpose, no matter how much they twist and turn. Thus, even when two molecules contain the same functional groups, they may still be isomers, and organisms can usually distinguish between them.

Sometimes organisms use stereoisomers to good advantage, as in the case of the **geometrical isomers.** In order to understand geometrical isomers, we need to compare the properties of single and double bonds between two carbon atoms. When two atoms are connected by a single bond, the groups attached to them can rotate freely, as two loosely connected wheels can rotate about the same axle. If two atoms are connected by a double bond, however, no rotation is possible. The attached groups remain on one or the other side of the bond. When two attached groups extend in the same direction from the two sides of a double bond, the groups are said to be in the *cis* [Latin, on this side] configuration; when the groups extend in opposite directions they are said to be in the *trans* [Latin, across] configuration.

Changing the arrangement of atoms around a double bond from *cis* to *trans* requires first breaking and then remaking the bond. Such a change, from one geometrical isomer to another, is the primary chemical event in vision, as shown in Figure 1. Light converts a *cis* isomer of retinal (a derivative of vitamin A) to *trans*. The *trans* form no longer fits into the pocket occupied by the *cis* retinal within the visual pig-

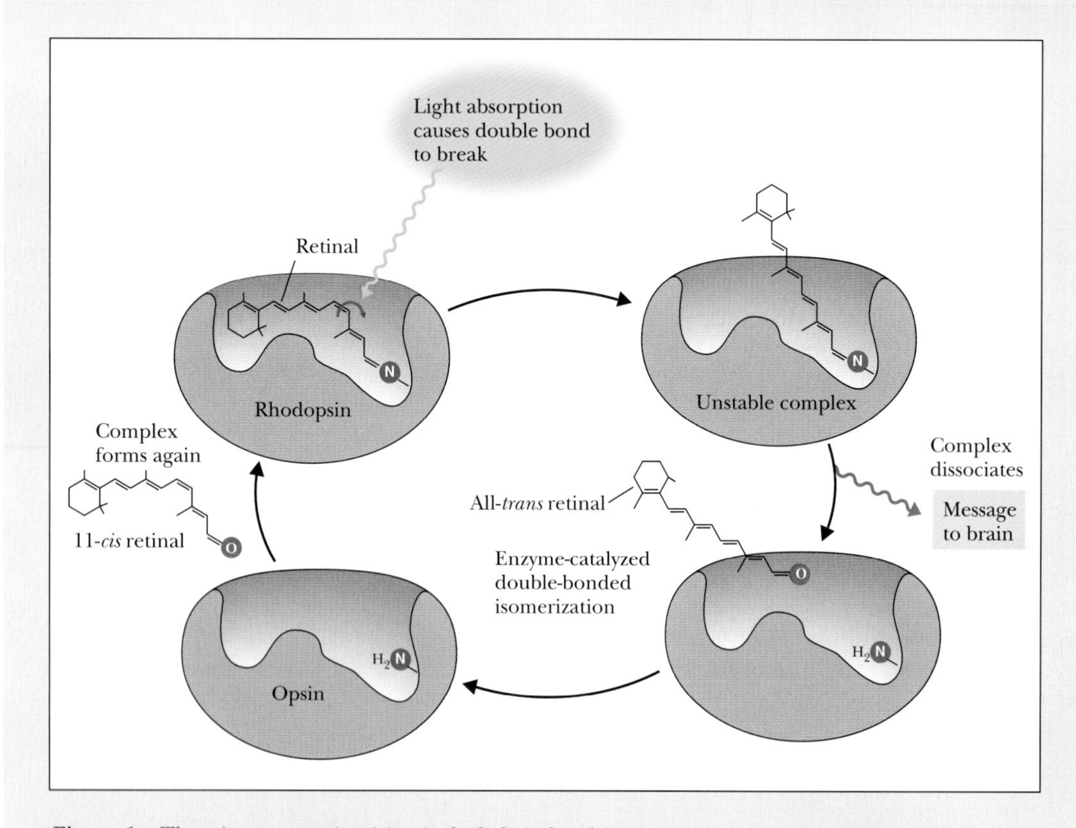

Figure 1 *The primary event in vision is the light-induced isomerization of retinal, in an unsaturated hydrocarbon chain. The resulting shape change causes retinal to be released from its binding protein, rhodopsin. The retinal-free rhodopsin, opsin, participates in the sending of a message to the brain.*

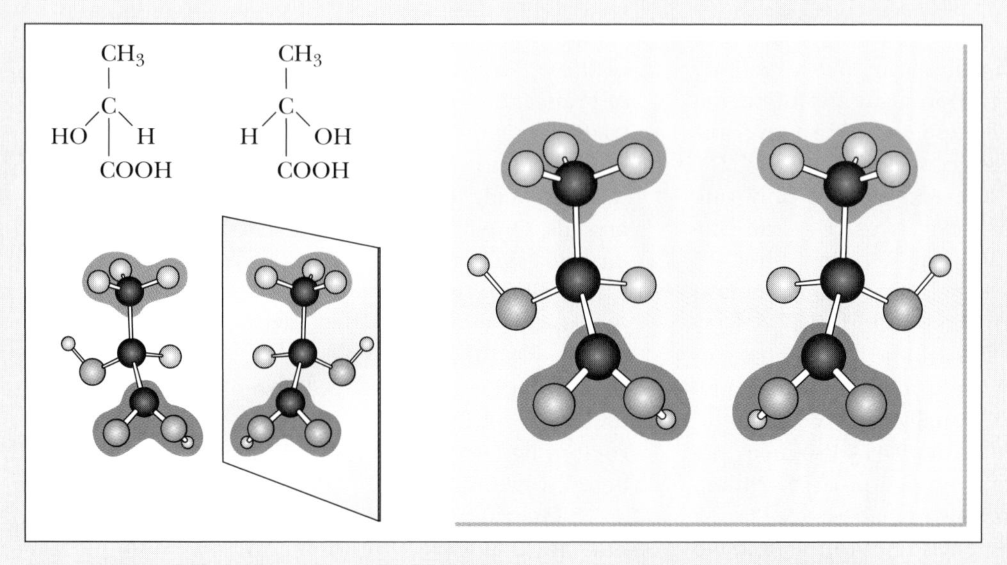

Figure 2 *Optical isomers are mirror images of each other and cannot be superimposed on one another by any rotation. Optical isomers can occur whenever a compound has an asymmetric C atom, a C atom attached to four different chemical groupings. Here the four groups are (1) a carboxyl group (COOH), (2) a hydrogen atom (H), (3) a hydroxyl group (OH), and (4) a methyl group (CH_3).*

ment rhodopsin. The resulting change in the rhodopsin molecule stimulates a message to the brain, resulting in the perception of light.

Another kind of stereoisomer, which does not depend on the presence of double bonds, is more common in organisms than geometrical isomers. These stereoisomers, called **optical isomers,** schematically illustrated in Figure 2, occur when four different groupings of atoms are attached to a single carbon atom. When this happens, the four groups can lie in ways that are never identical, no matter how much twisting occurs around the bonds. Molecules that are mirror images of one another, like left and right gloves, are examples of optical isomers.

To understand optical isomers requires thinking in three dimensions. If the groups of the molecule were really arranged in a plane, as they are on a page, it would be possible to rotate them about single bonds to obtain the original arrangement. But now think of the central carbon atom as a tennis ball surrounded by four balloons of different colors. If any two balloons are now exchanged, the resulting arrangement is the mirror image of the starting one. The new arrangement cannot be converted to the original by rotation about a single bond. It is the three-dimensional arrangement of the single bonds around a carbon atom that makes optical isomers possible.

A carbon atom surrounded by four different groups is said to be **asymmetric.** A molecule may have more than one asymmetric carbon. Let us take the six-carbon sugar glucose as an example. Figure 3 shows glucose in its linear (nonring) form in order to em-

(Essay continues on next page.)

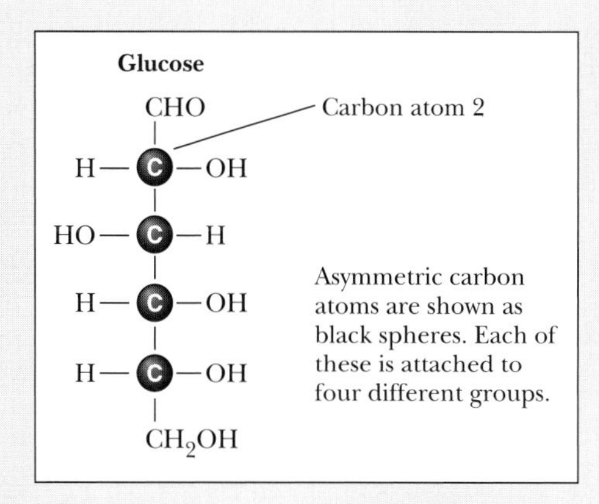

Figure 3 *Glucose, a six-carbon sugar, has four asymmetric carbon atoms. Carbon atom 2, for example, is attached to the following four groups: (1) CHO, (2) H, (3) OH, and (4) the remaining four C atoms with their attached H's and OH's.*

ESSAY *continued*

phasize the asymmetry of the carbon atoms. Glucose is one of our 35 small molecules, one that we will encounter many times. A six-carbon sugar has four asymmetric carbons. At each of these, two alternative spatial arrangements are possible, so there are $2 \times 2 \times 2 \times 2 = 2^4 = 16$ possible alternatives. But only one particular stereoisomer, glucose, is used as the biochemical crossroads of the cell. Of the other 15 stereoisomers, only one—galactose—is common enough in organisms to appear at all in this book.

The spatial arrangement around asymmetric carbon atoms determines how light interacts with a molecule. Asymmetric carbons actually can "rotate" light. Imagine the following experiment: (1) Produce "polarized" light by passing it through a polarizing filter, such as are used in polarizing sun glasses. (2) Place a second polarizing filter at right angles to the first. No light passes through the second filter. (3) Place a solution of glucose between the two filters. To extinguish the polarized light again, you must turn the second filter. The glucose solution rotated the plane of polarization of the light.

The extraordinary thing about "optical rotation" is that it occurs only within biological molecules. The connection between optical rotation and life was the first major discovery of the French scientist Louis Pasteur. In 1848, at the age of 26, Pasteur announced his discovery to the Paris Academy of Science.

Pasteur had been studying two distinct substances that had the same chemical properties. One of these, tartaric acid, is formed in the fermentation of grapes; the other, racemic acid, is produced industrially. Only the first can rotate polarized light.

When Pasteur looked carefully at crystals of racemic acid, he found that there were two types of crystals. One of these types looked just like crystals of tartaric acid, and the other was its mirror image. With a magnifying glass and a pair of tweezers, Pasteur painstakingly separated the two types of crystal. Pasteur showed that the crystals that looked like tartaric acid crystals could rotate polarized light and could be used as food for microorganisms. The other crystals rotated light in the opposite direction and could not be used by organisms.

The rotation of polarized light does not seem to have any biological significance. Why, then, should optical activity be associated with life? Optical rotation reveals something about molecular structure—that molecules can exist as mirror images of one another and that these mirror images have different properties. Optical isomers also present different shapes to the molecular machinery of cells. During evolution, organisms have acquired the ability to process only a limited number of small molecules. The ability to recognize these molecules depends crucially on the precise arrangement of their atoms. To cells, two optical isomers are not equivalent, and, in general, organisms use only one.

comes as a surprise. Most of the material of all organisms consists of combinations of the 35 building blocks. The surprise is a pleasant one, both to the biologist who wants to explore cells and design experiments and to the student who needs to learn about the structures. But it should also make us wonder just why there should be so few building blocks. Why, for example, should every single organism on this planet use the same 20 amino acids to make proteins, and why should almost every organism use the sugar glucose as its principal energy source?

The extremely limited number of molecules used to form all the wonderfully diverse forms of life supports the idea that all life on earth has descended from a common origin. (See Essay: Optical Isomers and Life.) The same molecules are used again and again because organisms have transmitted the genetic information for using the same molecules to accomplish the tasks of life. Thirty-five building blocks seem to have been enough.

SUMMARY

Most cell components are built from only about 35 small molecules, which serve as building blocks for the assembly of larger molecules. These 35 molecules consist of only six types of atoms: hydrogen, carbon, nitrogen, oxygen, phosphorus, and sulfur. The atoms in a molecule are held together by covalent bonds, formed by the sharing of electrons between neighboring atoms.

The ability of an atom to form a covalent bond depends on its electrons, which are arranged in shells that are each centered on the atom's nucleus. Covalent bonds form as two atoms share two, four, or six electrons, in a way that produces an outer shell of two electrons for hydrogen and eight electrons for other atoms.

Atoms and molecules can interact noncovalently; for example, unlike charges attract and like charges repel. Many biologically important molecules are uncharged but have uneven distributions of electrons, resulting in polar molecules. Polar molecules interact more strongly with each other and with water than do nonpolar molecules. Water molecules and other polar molecules often participate in hydrogen bonds, in which a slightly positive hydrogen atom forms a bridge between itself and a negatively charged atom. Nonpolar molecules tend to associate with other nonpolar molecules in hydrophobic interactions.

The polarity of water molecules and their ability to form hydrogen bonds with one another explain water's unique properties, including its density, cohesion, adhesion, melting (freezing) point, high heat capacity, and its excellence as a solvent. Water molecules also continuously dissociate into hydrogen and hydroxide ions. The concentration of hydrogen ions, usually given as the pH of a solution, affects the structure and chemical reactivity of most biologically important molecules.

The properties of the building blocks depend on specific functional groups, particular arrangements of atoms that appear in many molecules. Different arrangements of the same atoms, called isomers, result in different functional groups and different molecular properties.

The building blocks, from which most cellular structures are made, fall into four classes: lipids, sugars, amino acids, and nucleotides. Each class of building block contains characteristic functional groups. In addition to small molecular building blocks, cells also contain macromolecules, which are long chains of small molecule building blocks.

KEY TERMS

acid
amino
amino acid
amphipathic
anion
aqueous
aromatic
asymmetric
base
biochemistry
buffer
carbohydrate
carbonyl
carboxyl
cation
cholesterol
covalent bond
detergent
disulfide bond
double covalent bond
electronegativity
electrostatic interaction
fat
fatty acid
functional group
geometrical isomer
glycerol
hexose
hydrocarbon chain
hydrogen bond
hydrogen ion
hydronium ion
hydrophilic
hydrophobic
hydrophobic interaction
hydroxyl
ion
isomer
lipid
macromolecule
micelle
molarity
mole
molecule
monomer
monosaccharide
nitrogenous base
noncovalent interaction
nonpolar
nucleic acid
nucleoside
nucleotide
octet rule
optical isomer
orbital
organic chemistry
pentose
pH
phosphate
phospholipid
phosphoric acid ester
polar
polymer
polysaccharide
polyunsaturated
protein
purine
pyrimidine
saturated
shell
solute
solvent
stereoisomer
steroid
structural isomer
sugar
sulfhydryl group
triacylglycerol (triglyceride)
triple covalent bond
unsaturated
van der Waals attraction

QUESTIONS FOR REVIEW AND UNDERSTANDING

1. Complete the following: Combinations of individual atoms are ____________. These atoms share electrons in their ____________ bonds. The attraction or repulsion of charges is known as ____________ interactions and gives rise to ____________ bonds. A positively charged ion is a ____________ while a negatively

charged ion is an ________. Together, these two ions form a ________, which dissociates in water.

The tendency of atoms to gain electrons is referred to as ________. Molecules composed of atoms differing in electronegativity are said to be ________ because they have an uneven charge distribution. In contrast to a ________ bond, which requires a hydrogen atom having a slight positive charge, ________ interactions occur between nonpolar molecules. ________ molecules interact with water while ________ molecules do not. ________ molecules contain both polar and nonpolar regions.

Large molecules consisting of smaller molecules linked together are called ________. The properties of these molecules are determined in large part by their constituent functional groups, including ________, ________, ________, and ________ groups. Two other functional groups common to biological systems are ________ groups, which contain sulfur, and ________ groups, which contain phosphorus.

The four classes of small molecules common in cells include ________, ________, ________, and ________. While three of these are used to construct macromolecules (polymers), ________ do not, but instead can aggregate to form structures such as membranes. ________ is a lipid that is the precursor to ________ hormones such as testosterone. Nitrogenous bases having a single ring of atoms are ________ and those having two rings are ________. A molecule having a nitrogenous base and a sugar is a ________. If the molecule also has a phosphate group it is a ________. Many energy-requiring processes in cells use the nucleotide ________ as an immediate energy source.

2. List the six elements common to biological molecules. Give an example of a biologically important molecule that contains one or more of the six elements.
3. Since all naturally occurring carbon-containing compounds found on earth are the direct or indirect products of photosynthesis (see Chapter 8), what does this say about the field of organic chemistry?
4. Reconcile the electron structure of carbon and its capacity to form covalent bonds, including double bonds.
5. What are the consequences of electronegativity differences between atoms comprising a molecule?
6. Explain why nonpolar molecules exhibit hydrophobic interactions, or more precisely, are excluded from polar environments such as water.
7. What are six properties of water that allow it to support life?
8. Consider the following problems and calculations concerning pH:
   (a) In pure water, why must the concentrations of hydrogen ions and hydroxide ions be equal? Why is the pH of pure water 7?
   (b) What is a buffer? Explain in your own words how a buffer can maintain constant pH.
   (c) The tolerable range of blood pH is 6.95 to 7.7. What is the tolerable range of hydrogen ion concentration?
9. Why can we say that the polarity or hydrophobicity of molecules is critical in determining cell organization?
10. Why can an amino acid act as either a base or an acid?
11. Construct a table of the four classes of building blocks, listing for each one:
    (a) the constituent functional groups
    (b) whether it is hydrophilic, hydrophobic, or amphipathic
    (c) the macromolecules or cellular structures containing it
12. Distinguish between saturated, unsaturated, and polyunsaturated fatty acids. Given that cooking oils contain unsaturated fatty acids attached to glycerol, but butter and animal fat contain primarily saturated fatty acids, explain the physical effects of double bonds in these lipids.
13. Describe in your own words how the amphipathic nature of a phospholipid makes it ideally suited to form a membrane bilayer.
14. Fatty acids, phospholipids, and cholesterol differ greatly in chemical structure but are all classified as lipids. Why?
15. What is the defining chemical feature of carbohydrates?
16. With your knowledge of optical isomers, explain how you would determine whether a sample of a pure amino acid (for example, alanine) was synthesized chemically or biologically. Why could such a determination not be made on a sample of the amino acid glycine?
17. As we shall see in other chapters, the oxidation of a molecule (that is, a decrease in the ratio of hydrogen atoms to carbon atoms or an increase in the ratio of oxygen atoms to carbon atoms) is accompanied by a release of energy. How is this consistent with the fact that ethanol and acetaldehyde readily burn (oxidize) in air, releasing heat, but acetic acid does not?

SUGGESTED READINGS

GARRETT, R. H. and C. M. Grisham, *Biochemistry,* Saunders College Publishing, Philadelphia, 1995. An up-to-date biochemistry textbook.

HENDERSON, L. J., *The Fitness of the Environment,* Macmillan, New York, 1913 (republished 1970, P. Smith, Gloucester, MA). A classic discussion of the properties of water and the significance for life.

KOTZ, J. and K. Purcell, *Chemistry and Chemical Reactivity,* 3rd edition, Saunders College Publishing, Philadelphia, 1996. A well-written introduction to modern chemistry.

LOEWY, A. G., P. Sievkevitz, J. R. Menninger, and J. A. N. Gallant, *Cell Structure and Function: An Integrated Approach,* 3rd edition, Saunders College Publishing, Philadelphia, 1991. Cell biology from a physical and chemical perspective. Chapter 3 covers much of the material in this chapter.

STRYER, L., *Biochemistry,* 3rd edition, W. H. Freeman, New York, 1988. Now considered a classic introduction to biochemistry, beautifully illustrated. Chapter 1 covers much of the material in this chapter.

CHAPTER 3

Macromolecules

KEY CONCEPTS

- Macromolecules are formed from their component building blocks by condensation reactions that remove the equivalent of one molecule of water.
- Each kind of protein molecule has a distinct sequence and characteristic properties.
- The three-dimensional structure of a protein depends on interactions of amino acid residues.

OUTLINE

Macromolecules are large molecules that contain thousands or even millions of atoms. About 20% of the weight of a living cell consists of macromolecules: polysaccharides, proteins, and nucleic acids. Each class of macromolecule plays a number of different roles. Polysaccharides store energy, form protective coats around cells, and furnish structural support. Nucleic acids store, transmit, and help to interpret genetic information. Proteins serve both as passive structural supports and as molecular machines that perform most of the activities of cells.

This chapter describes the covalent structure of all three classes of macromolecules and then discusses the relationship of the covalent structure to the three-dimensional structure of proteins. We concentrate on proteins in this chapter because many of the structures and processes described in the next few chapters depend on the properties of proteins. Many biologists consider proteins to be the most important class of macromolecules and note that the word "protein" derives from the Greek *proteios,* meaning primary. Other biologists, however, stress the importance of the information-carrying molecules DNA and RNA, whose structures and cellular roles are discussed in Chapters 13 and 14.

MACROMOLECULES ARE COVALENTLY LINKED CHAINS OF BUILDING BLOCKS

Cells do not produce macromolecules by assembling individual atoms, but by linking together the ready-made building blocks described in Chapter 2. Recall that sugars are the building blocks of polysaccharides, amino acids the building blocks of proteins, and nucleotides the building blocks of nucleic acids.

To form each type of macromolecule, components are linked together by similar but distinct chemical reactions, described later in this chapter. In the case of proteins and nucleic acids, the components are linked end to end to form long, unbranched chains, like a child's string of "pop beads." In the case of polysaccharides, however, the chains of building blocks are often branched. In unbranched chains, the building blocks themselves may have branches of atoms, but the backbone chain remains unbranched.

Macromolecular chains fold to form complex three-dimensional shapes. Most of the biological roles performed by macromolecules depend on the shapes of the folded chains.

For All Three Types of Macromolecules, Building Blocks Join by Eliminating a Water Molecule

In each type of macromolecule the building blocks are held together by a distinctive type of covalent bond. These bonds must be strong enough that the macromolecules remain stable for some time.

Cells use a large variety of biochemical reactions and specialized machinery to build macromolecules, but the same basic principle applies in every case. Small molecules link into chains, and the formation of each link eliminates two hydrogen atoms and one oxygen atom—the equivalent of a molecule of water. The changed building block, now part of a larger molecule, is called a **residue.** The linking of building blocks, as illustrated in Figure 3-1a, is achieved by a **condensation** (or **dehydration condensation**) reaction. The whole process of putting together a macromolecule is sometimes called polymerization.

On the other hand, cells must be able to disassemble the macromolecules relatively easily. Animals, for example, must be able to use building blocks from the macromolecules they eat, and all cells must be able to recycle the building blocks of their own macromolecules as they grow and change. We ourselves continually break down and remake the proteins of our skin, muscles, and even bones.

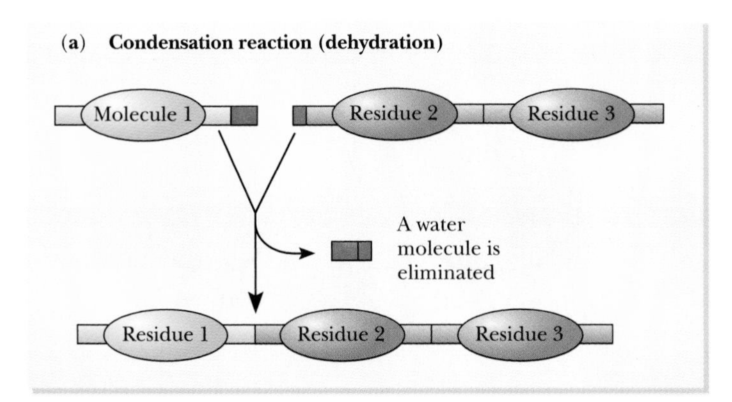

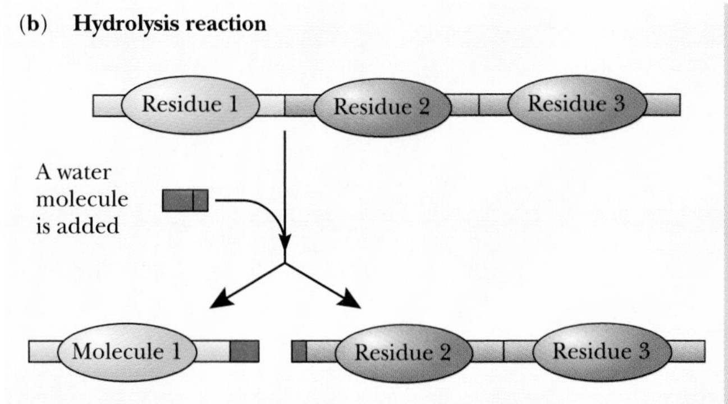

Figure 3-1 Biological building blocks form macromolecules by dehydration condensation reactions. **(a)** Two small molecules link together, eliminating a water molecule; **(b)** addition of a water molecule (hydrolysis) splits apart two condensed small molecules.

Taking macromolecules apart into their components reverses the process. One molecule of water is added to each residue as it becomes detached from the larger structure. This breaking-down process, diagrammed in Figure 3-1b, is called **hydrolysis** [Greek, *hydro* = water + *lysis* = breaking].

Both assembly and breakdown occur in cells. But what determines whether a macromolecule is broken down into its component building blocks or building blocks are assembled into macromolecules? This problem is an example of the difficulties that cells must overcome to be able to exist at all. A collection of unconnected building blocks is more stable than an assembled macromolecule because it requires energy to make a macromolecule. This means that, given enough time, a macromolecule is hydrolyzed into its component parts. (See Chapter 4.) Inside cells, however, spontaneous hydrolysis occurs only slowly, so macromolecules and the larger structures that they can form are relatively stable.

How Do Sugar Monomers Form Polysaccharides?

Two sugar molecules can join together by forming a **glycosidic bond,** in which an oxygen atom forms a bridge between carbon atoms on two sugar molecules. A glycosidic bond forms as the hydroxyl group on one sugar molecule reacts with a carbon of a second sugar molecule, displacing a hydroxyl group on the second molecule. This condensation reaction eliminates one hydroxyl group and one hydrogen atom, the equivalent of a water molecule.

Disaccharides contain two sugar molecules and just one glycosidic bond. Two glucose molecules, for example, can link to form maltose (malt sugar), as shown in Figure 3-2a; the linking of glucose and fructose forms sucrose (table sugar), as shown in Figure 3-2b. Glucose is the form in which energy is distributed in animals, while disaccharides (especially sucrose) are the usual energy-transporting molecules in plants.

Polysaccharides may consist of hundreds or thousands of monosaccharide residues linked together by glycosidic bonds. Some polysaccharides—starch in plants and glycogen in animals, both shown in Figure 3-3a, b (page 64)—are commonly used to store energy. The polysaccharides of potatoes and pasta are a rich source of calories in the diet of many university students. Other polysaccharides have important structural and protective roles for organisms and cells (especially those of prokaryotes, fungi, and plants). The most common of these is cellulose, the major structural material of plant cell walls and of the fibers of wood, cotton, and paper.

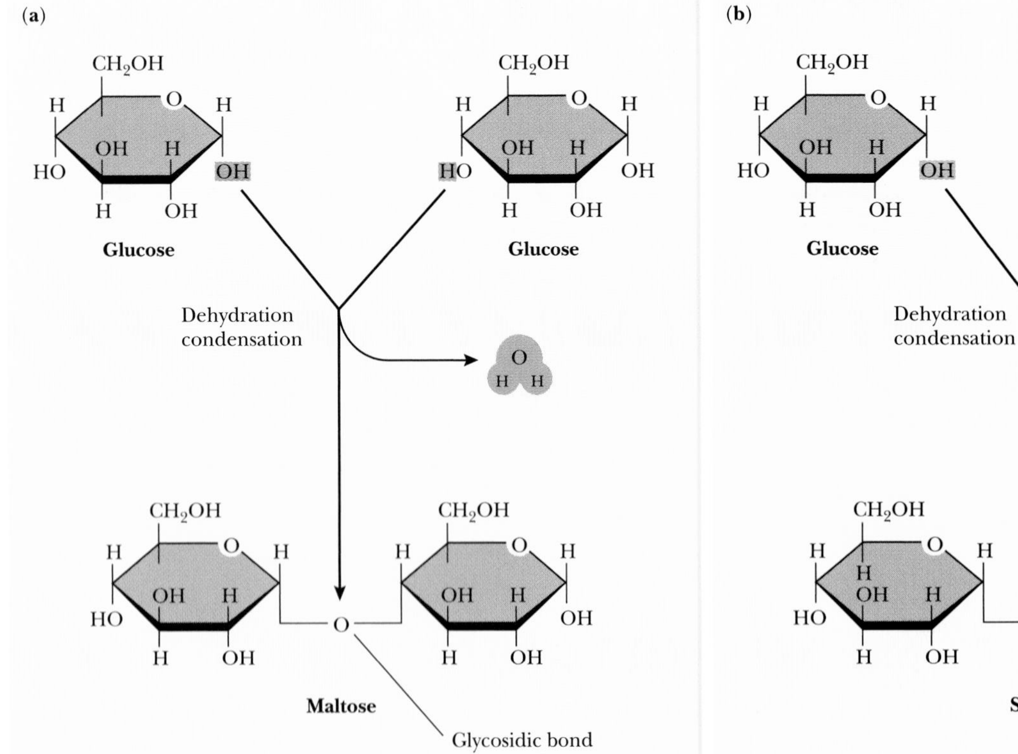

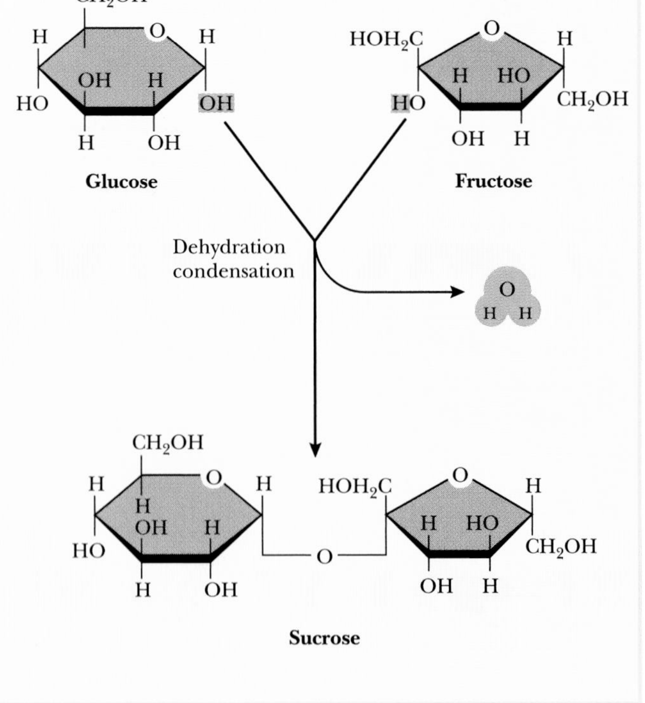

Figure 3-2 Formation of two disaccharides: **(a)** maltose, formed from two glucose molecules; **(b)** sucrose, formed from glucose and frutose. (Cells perform this reaction in a more elaborate way.)

Polysaccharides May Contain Branched as Well as Unbranched Chains

The degree of branching of a polysaccharide varies: The glycogen of animals, for example, is usually more branched than the starches of plants. Although different sugars and sugar derivatives are components of many polysaccharides, the most common polysaccharides—starch, cellulose, and glycogen—consist of glucose residues alone.

Both the three-dimensional structure and the biological properties of polysaccharides depend both on the kind of monosaccharide residues of which they are composed and on the way in which these residues are joined. For example, both cellulose and some starches consist of long, unbranched chains of glucose residues. As we can see in Figure 3-3a and c, however, the arrangement of the glycosidic links between residues in cellulose differs from that in starch.

The Hydrolysis of Polysaccharides by Cells Depends on Specific Molecular Catalysts

Enzymes are large molecules, almost always proteins, that accelerate the rates of specific chemical reactions. (See Chapter 4.) Many cells have enzymes, called **glycosidases,** which allow the rapid digestion of starch or glycogen. Few animal cells, however, have enzymes that can break down the distinctive glycosidic linkage in cellulose, shown in Figure 3-3c. As a result, the bonds between the residues in cellulose cannot be digested by the enzymes found in most animals, and cellulose therefore has no nutritional value to most animals.

The food industry has used cellulose and its derivatives to produce food additives that have no calories—such as milk-shake thickeners and no-calorie pasta. Another result of the geometry of the glycosidic bonds of cellulose is that wood, books, and clothes do not easily rot: Few organisms produce enzymes that can break them down. Cows derive energy from the cellulose of grass only because one of their stomachs, full of cellulose-digesting bacteria, serves as a digestion vessel. Similarly, cellulose-digesting bacteria that live in the digestive tracts of termites allow them to digest the cellulose in wood.

How Do Amino Acids Form Proteins?

Like polysaccharides, proteins are composed of chains of building-block residues joined together by dehydration condensation reactions. The building blocks in proteins are amino acids, and they are linked by peptide bonds (shown in Figure 3-4), in which the carboxyl carbon of one amino acid molecule joins to the nitrogen atom in the amino group of the next. When a peptide bond forms, the carboxyl group loses a hydrogen atom and an oxygen atom, and the amino group loses a hydrogen atom—again the equivalent of a molecule of water.

Chains of amino acid residues are called **polypeptides** and are always linear, never branched. A protein may consist of a single polypeptide or of several polypeptides. In proteins with more than one chain, the individual chains are held together tightly and specifically, so that the protein behaves like a single molecule. The chains may or may not be identical to one another.

Proteins play many roles. They can function as structural supports, identification tags, protectors against invasion by foreign organisms or molecules, carriers of ions and small molecules, and signalling and sensing devices. All of the chemical activities of a cell are regulated by enzymes, almost all of which are also proteins.

The three-dimensional shape of a protein determines its biological activity. As we discuss in some detail later in this chapter, a protein's shape depends upon its particular sequence of amino acid residues.

Antibodies Bind to Specific Molecular Shapes

Protein molecules called **antibodies** are able to bind tightly to specific molecules or parts of molecules. As discussed in Chapter 22, each type of antibody has a distinct amino acid sequence and can bind to a particular molecular shape. This ability allows antibodies to participate in the defense against infection by binding and helping to inactivate proteins of viruses or bacteria.

Antibodies have been an enormous practical boon. Experimental animals can be made to produce antibodies against particular molecules, for example, against the protein human chorionic gonadotrophin (HCG), a hormone that is made only after the onset of pregnancy. Antibodies against HCG can be isolated and used as the basis of a pregnancy test. Antibodies to other biological molecules provide important means of identifying those molecules in cells and cell extracts.

Enzymes Recognize the Shapes of Molecules Whose Reactions They Accelerate

As discussed in Chapter 4, enzymes speed up biochemical reactions by binding to specific molecules, in much the same way that antibodies bind to specific molecules. In the case of enzymes, however, the result is the speeding up of a chemical transformation. In the case of the enzyme urease, for example, the enzyme binds to urea, a prevalent component of urine. The enzyme helps the urea break down into carbon dioxide (CO_2) and ammonia (NH_3).

Specific Enzymes Cut Peptide Bonds Either Randomly or at Particular Places

The hydrolysis of proteins usually takes place only slowly within cells. In the digestive process, however, proteins must be broken down into their component amino acids, which may then be reused, converted to other amino acids, or utilized as sources of energy. This digestion depends on

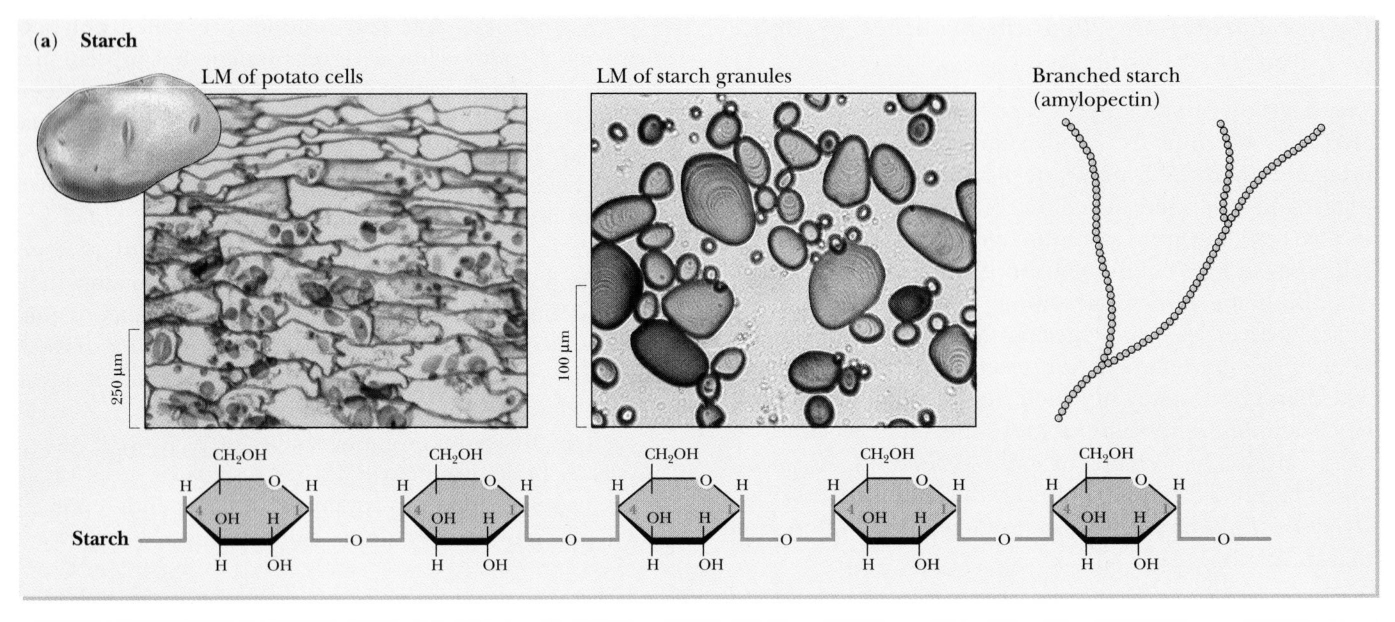
(a) Starch
LM of potato cells
LM of starch granules
Branched starch (amylopectin)
250 μm
100 μm
CH₂OH
H
O
OH
H
Starch
OH

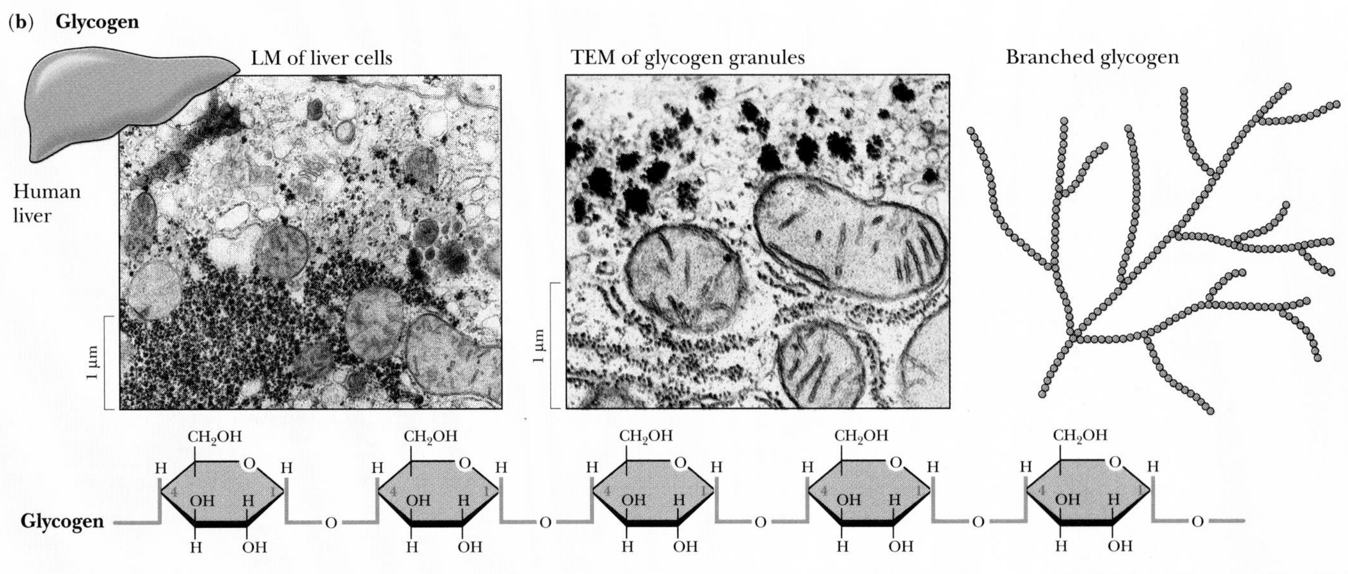
(b) Glycogen
LM of liver cells
TEM of glycogen granules
Branched glycogen
Human liver
1 μm
1 μm
CH₂OH
H
O
OH
Glycogen

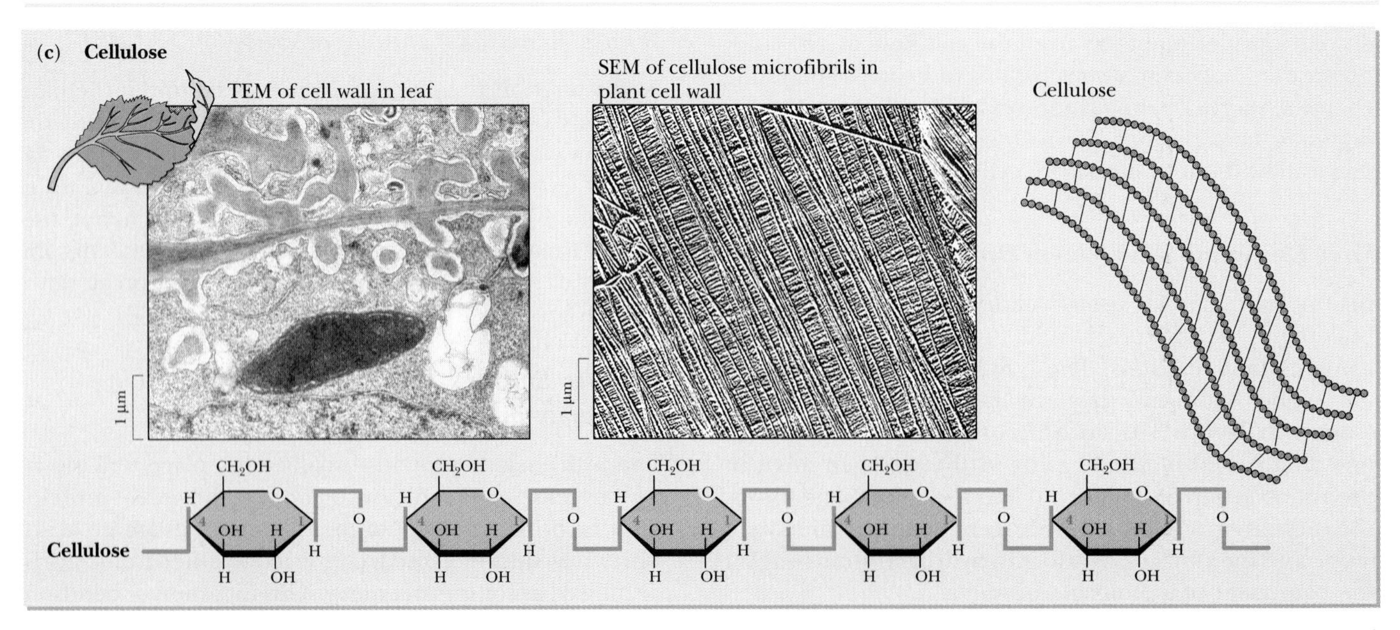
(c) Cellulose
TEM of cell wall in leaf
SEM of cellulose microfibrils in plant cell wall
Cellulose
1 μm
1 μm
CH₂OH
H
O
OH
Cellulose

◀ ***Figure 3-3*** Starch, glycogen, and cellulose are all polysaccharides formed from glucose, but with different arrangements of glycosidic bonds. Animals have enzymes that hydrolyze starch and glycogen, but few animals can hydrolyze cellulose. **(a)** Starch (amylopectin): most glucose residues are linked between carbon atoms 1 and 4, with few bonds between carbon atoms 1 and 6 (not shown), creating branches in the chain; *(left)* cells in a potato, seen with a light microscope (LM); starch grains are stained purple; *(center)* isolated starch grains in a potato; *(right)* branched chain of amylopectin, with about one (1—6) branch per 30 (1—4) linkages. **(b)** Glycogen: most glucose residues are linked between carbon atoms 1 and 4, with some between carbon atoms 1 and 6 (not shown), creating branches; *(left)* liver cells, seen with a light microscope, showing glycogen granules; *(center)* glycogen granules in liver, seen with a transmission electron microscope (TEM); *(right)* branched chain of glycogen, with about one (1—6) branch per 10 (1—4) linkages. **(c)** Cellulose: glucose residues are linked between carbon atoms 1 and 4, but with a different orientation, called a β-linkage, from starch and glycogen; *(left)* the cell wall of a zinnia leaf; *(center)* cellulose fibrils in a plant cell; *(right)* unbranched chains of cellulose seen with a scanning electron microscope (SEM). *(a, left, Cabisco/Visuals Unlimited, center, Runk/Schoenberger/Grant Heilman; b, left, Biophoto Associates/Photo Researchers; center, Don Fawcett/Visuals Unlimited; c, left, Dr. Jeremy Burgess/Photo Researchers, center, Biophoto Associates/Photo Researchers)*

the presence of **proteases** (also called peptidases), digestive enzymes that accelerate the hydrolysis of proteins. Some proteases hydrolyze peptide bonds irrespective of the amino acid side groups on the neighboring amino acid residues. Other proteases require the presence of a specific amino acid. Trypsin, a common protease produced in the mammalian pancreas and secreted into the intestine, for example, cuts only next to a positively charged amino acid side group (those of lysine or arginine), as illustrated in Figure 3-5. Chymotrypsin, another mammalian protease, preferentially cuts next to an aromatic amino acid residue.

How Do Nucleotides Form Nucleic Acids?

The nucleic acids, DNA and RNA, are composed of long chains of nucleotide residues. Chains of nucleotides, called **polynucleotides,** are always unbranched (with a very few exceptions). Again, joining of the building blocks de-

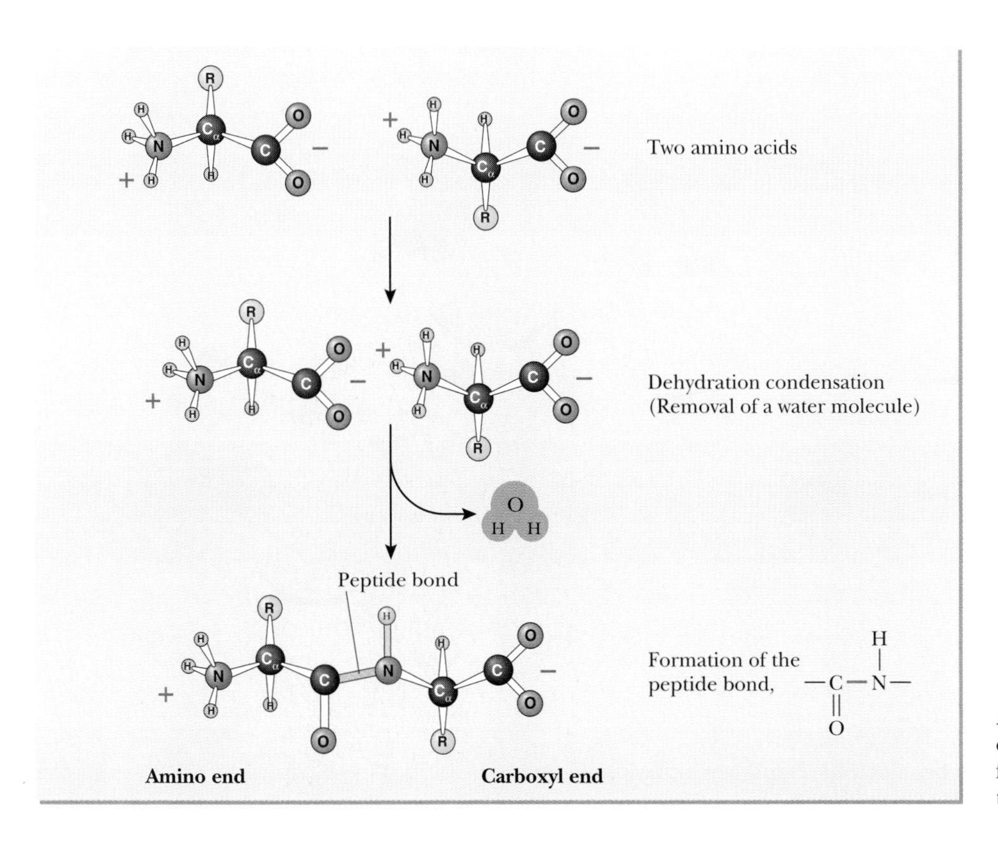

Figure 3-4 Dehydration condensation of two amino acids to form a peptide bond. (Cells perform this reaction in a more elaborate way.)

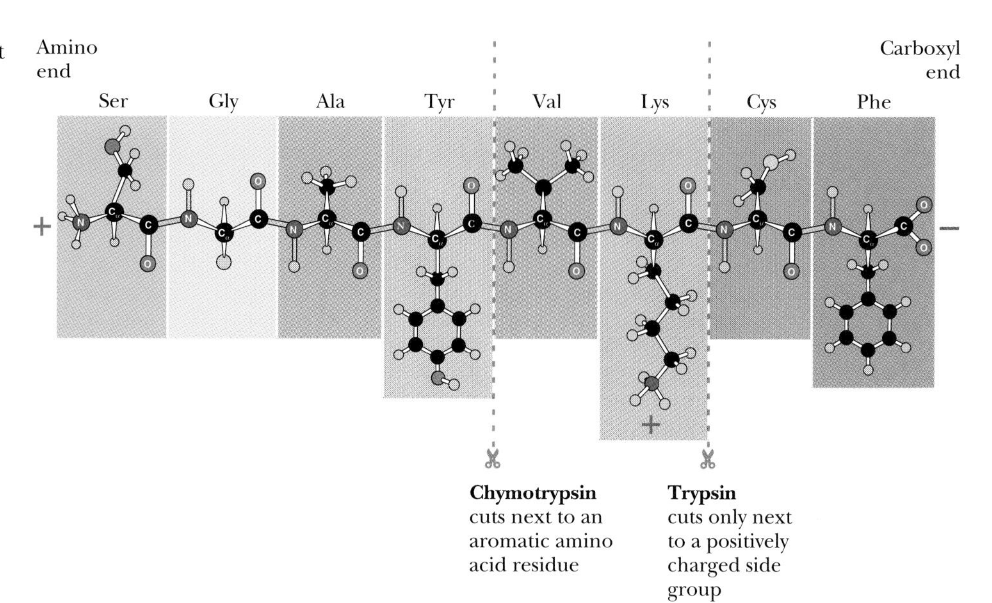

Figure 3-5 Individual proteases digest polypeptides by cutting at different places. Chymotrypsin hydrolyzes a peptide following an aromatic side chain, while trypsin cuts after a positively charged side chain.

pends on dehydration condensation reactions. The links (shown in Figure 3-6) are called **phosphodiester** bonds and are formed by the phosphate group of one nucleotide attaching to a carbon atom in the sugar component of another nucleotide. Forming the bond eliminates one hydroxyl group from the sugar and one hydrogen atom from the phosphate group, the equivalent of one molecule of water.

DNA and RNA are both assembled from their nucleotide building blocks into linear chains, which fold into specific three-dimensional shapes. Chapter 13 discusses both the three-dimensional structure of DNA and how important it is in determining its biological function. Here we say only that DNA usually consists of two polynucleotide chains held together by hydrogen bonds, and a molecule of RNA usually consists of a single polynucleotide chain.

Like polysaccharides and polypeptides, nucleic acids are subject to hydrolysis to their component building blocks. Most of the degradation of nucleic acids that occurs in cells depends on specific enzymes, called **nucle-**

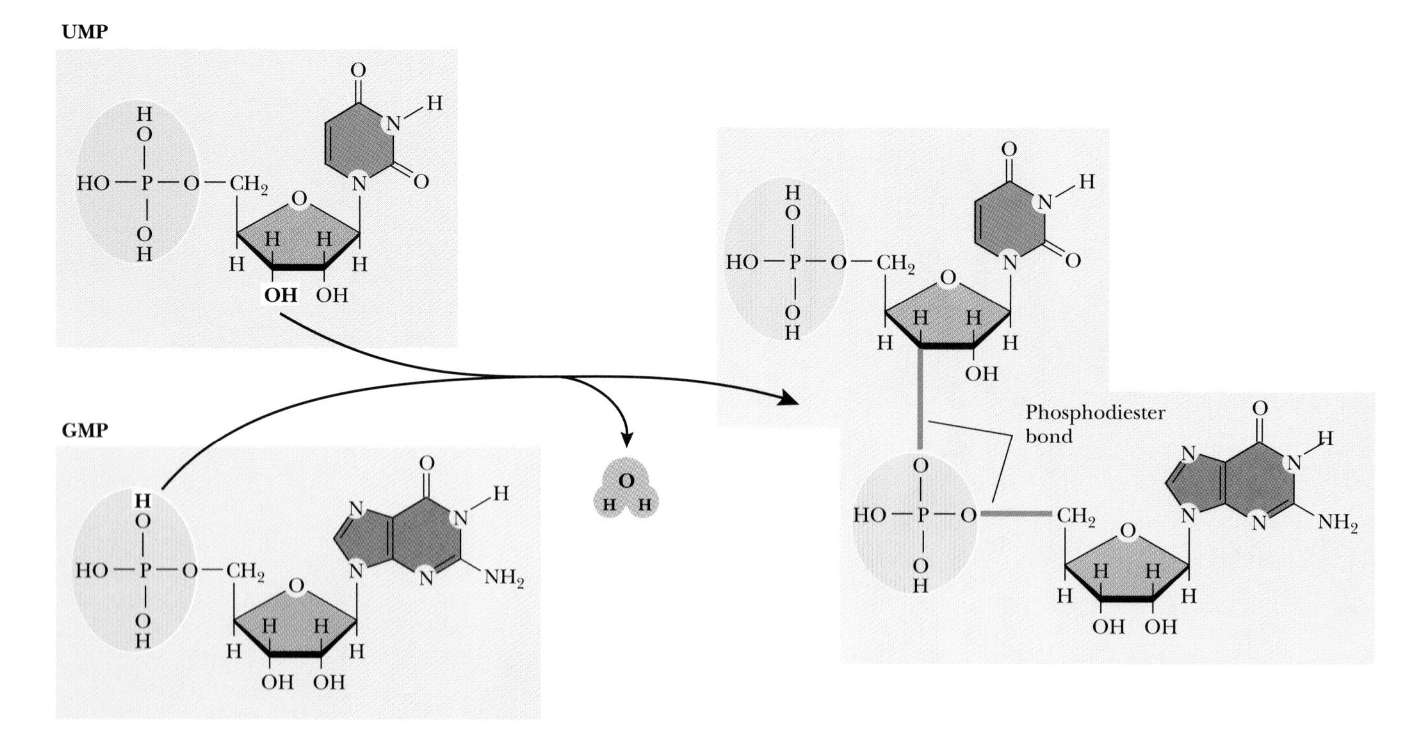

Figure 3-6 Dehydration condensation of two nucleotides to form a phosphodiester bond. (Cells perform this reaction in a more elaborate way.)

ases. Some nucleases act at random, while others cut nucleic acids only after nucleotide residues containing a certain nitrogenous base. For example, ribonuclease A, a major enzyme produced in the mammalian pancreas and secreted into the intestine, cuts only after cytosine or uridine residues in RNA.

HOW DO BIOLOGISTS STUDY THE STRUCTURE OF MACROMOLECULES?

The chains that make up polysaccharides, proteins, and nucleic acids are not infinitely flexible but fold in particular ways to form biologically active structures. The folding is determined in large part by the sequence of residues in their chains.

The number and ordering of residues in polysaccharides make much less difference than in proteins or nucleic acids. The number of glucose residues in a molecule of starch, for example, does not affect its ability to store energy. In contrast, even a single changed amino acid residue in a protein can destroy its functional abilities. When such a change is genetically programmed, all the proteins of that type are altered in a way that can make the difference between life and death. Much of a cell's machinery is devoted to avoiding accidental alterations of sequence changes in proteins and nucleic acids.

The sequences of amino acids in proteins and of nucleotides in nucleic acids establish both their structure and their function. Some individuals within a species may have only a single difference in sequence, which leads to a nonfunctional protein or an aberrant gene. In the case of functions needed for life, the change may be fatal, as in the case of a variant sequence that is responsible for one form of Tay-Sachs disease, a fatal childhood disease. Or a sequence variation may merely lead to one of the countless differences we see among species or among individuals of a single species.

Widely Used Techniques Allow Biologists to Determine the Sequences of Building Blocks in Proteins and Nucleic Acids

By studying the differences in the three-dimensional structures of naturally occurring variants of proteins or genes, biologists have come to understand more about the biological importance of the sequences of macromolecules. Using new techniques, biologists can actually create desired variant sequences of proteins and nucleic acids (as we see in Chapter 17), greatly accelerating the study of the relationships among sequence, structure, and function.

Several methods are now available to determine the sequences of proteins and nucleic acids. Each method employs the same basic steps: (1) purification of a single kind of polypeptide or nucleic acid chain from all the other components of the starting tissue; (2) cutting the purified chain into shorter lengths, typically about 100 amino acid residues for proteins or several hundred nucleotide residues of DNA; (3) determining the sequence of residues of each fragment, starting from one end and working toward the other; and (4) fitting together the sequences of the individual fragments to give the sequence of the whole chain.

Automated techniques have allowed biologists to determine the sequences of thousands of proteins and lengths of DNA (Figure 3-7). The U.S. Department of Energy and the National Institutes of Health, in cooperation with similar European and Japanese organizations, have embarked on a massive program, called the Human Genome Project, which plans to determine the sequence

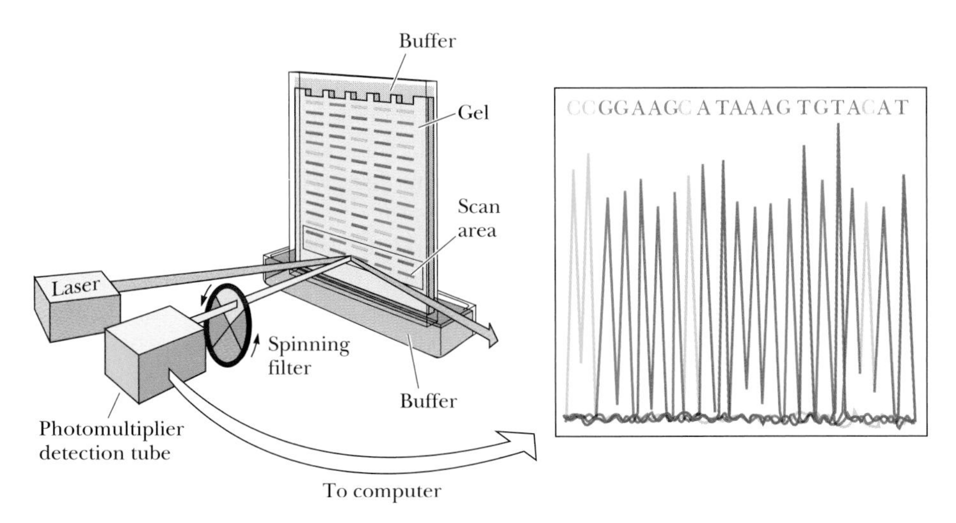

Figure 3-7 Automated sequencing of DNA. A single machine can determine the sequence of thousands of nucleotides in DNA in a single day.

of all the DNA in a human cell. This project should give new information about the tens of thousands of genes that make up the human genetic plan.

Biologists Can Also Make Proteins and Nucleic Acids in the Laboratory

Biologists can now produce any sequence of amino acid residues or nucleotide residues in the laboratory. For technical reasons, chemical methods have most often been restricted to producing polypeptides and polynucleotides that are relatively small, usually containing fewer than 50 residues. In addition to chemical methods, however, biologists can also manipulate genes so that bacteria (or other cells) can become living "factories" for custom-made proteins or nucleic acids. We discuss these techniques in Chapter 17.

By making macromolecules of known sequences, scientists have learned much about the relationships of chemical structure and biological function. First, if a molecule of known structure, made in the laboratory, has the same function as a molecule made by cells, then biochemists can be sure that the molecule they have painstakingly purified from cells is really the one that performs a particular biochemical role. For example, the protein called growth hormone has the same structure and biological effects whether it is isolated from the organs that normally produce it or is produced in a laboratory.

In addition, laboratory techniques can produce macromolecules that vary in particular known ways. By comparing the structures and biological activities of variant molecules, researchers can study the contributions of individual residues to the unaltered molecule. In growth hormone, for example, changing one amino acid residue may have little effect, while changing another may completely alter its structure and abolish its biological activity.

Such studies have led to a greater understanding of the relationships of residue sequences, three-dimensional structure, and function. For example, in a protein the changes that make the smallest differences (the most "conservative" changes) are those that do not alter the general character of amino acid side groups. Substitutions of one hydrophobic residue for another and of one negatively charged side chain for another are examples of "conservative" changes. On the other hand, replacing a hydrophobic side chain with a hydrophilic one, or a negatively charged residue with one that is positively charged, may destroy the folding of the polypeptide and make it biologically useless.

Human growth hormone is only one example of a useful protein that industrial biologists have made available in greater amounts than can practically be had from natural sources. With ever-increasing frequency, the newspapers announce yet another useful product that derives from the application of recent biological techniques. These products include hormones, antibiotics, insecticides, and many other proteins with medical, agricultural, and commercial uses. (See Chapter 17.)

Each Kind of Protein Has a Distinct Sequence and Characteristic Properties

Since the middle of the 19th century, biochemists have known that proteins have unusual properties. They know, for example, that they are large, because, unlike dissolved salt or sugar, dissolved proteins do not diffuse through the tiny pores in a parchment container. Furthermore, while heating tends to make most substances more soluble, heated proteins tend to come out of solution, as anyone who has boiled an egg may attest. (The white of an uncooked egg is a watery solution of proteins.) For a long time, then, proteins have been regarded as unconventional substances that do not appear to obey the same chemical rules as ordinary compounds.

Until the 1920s, most biochemists thought that proteins were large, relatively nonspecific aggregates of smaller molecules. Two lines of evidence, however, convinced them that proteins are well-defined molecules that can perform previously unsuspected roles. The first line of evidence came from the experiments of the Swedish physical chemist Theodor Svedberg, who showed that specific proteins have characteristic molecular weights. To do this, Svedberg devised an instrument called an ultracentrifuge, with which he observed that each type of protein moves as if it has a large but definite molecular weight, rather than just being random polymers of amino acids.

The second line of evidence derived from the work of the American biochemist James Sumner. Sumner obtained pure crystals of the enzyme urease from extracts of jack beans. The ability to form crystals, Sumner knew, meant that urease is a well-defined molecule. Sumner showed that the urease crystals, when redissolved, could still catalyze the breakdown of urea into carbon dioxide and ammonia. He also showed that the urease is composed of polypeptide chains: When hydrolyzed in strong acid, urease yields amino acids. Sumner's work, like Svedberg's, occurred at a time when most biochemists thought that proteins were not definable molecules, but random polymers with no biological functions other than to give cells a gel-like interior (as the dissolved protein collagen gives substance to gelled desserts). Sumner's work showed that (1) urease is a true molecule (because it forms crystals) and (2) urease is a protein (because the pure enzyme is composed of amino acids).

Before Sumner, biochemists knew that enzymes are cellular components that could catalyze specific chemical reactions, but they did not know the chemical identities of enzymes. Indeed, many prominent biochemists doubted that enzymes were proteins even after Sumner's work. They changed their minds only after several other enzymes were also shown to be proteins (that is, to consist of amino acids held together by peptide bonds).

The ultimate proof that proteins have definite structures, however, came only with the demonstration, in the early 1950s, that insulin has a definite sequence of amino acid residues. This demonstration depended on the pioneering work of Frederick Sanger in Cambridge, England. Sanger was largely responsible for developing the first techniques for sequencing proteins. More remarkably, at the time that Sanger began his ambitious project to determine insulin's sequence, many biochemists doubted that his question would have any answer at all, because they thought that proteins did not have fixed sequences. Sanger persisted, however, and received a Nobel Prize in recognition of this work; Sanger later received a second Nobel Prize for developing the most widely used method for sequencing DNA.

The unique sequence of a protein dictates its three-dimensional structure, which in turn determines its function. In the case of insulin or growth hormone, the function of the protein is to act as a signaling molecule; in the case of urease, the function is to act as an enzyme.

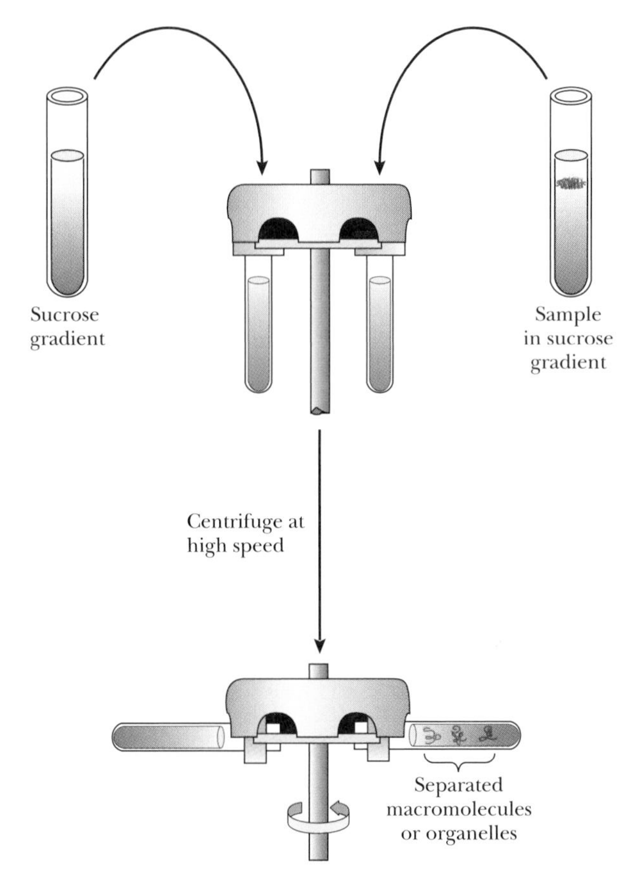

Figure 3-8 Centrifugation can separate macromolecules and organelles according to size and shape. The centrifuge can spin a tube containing a solution of macromolecules up to 80,000 rpm, generating forces of more than 100,000 times gravity. Large molecules travel more rapidly than small molecules toward the bottom of the tube.

How Can One Obtain a Single Type of Protein to Study?

Today, we understand that all proteins consist of definite sequences of amino acid residues. The three-dimensional structure and the function of each protein derive directly from the specific amino acid sequence. Determining the three-dimensional structure, however, is difficult. The important first step is the isolation of the protein away from other compounds.

Differences among proteins in amino acid sequences establish differences in size, charge, and other properties, such as the ability to bind to small molecules. The most useful contemporary methods for separating proteins—centrifugation, chromatography, and electrophoresis—are based on these properties.

During each step in any separation procedure, the researcher monitors the functional activity of the desired protein, for example, by an **enzyme assay,** which measures the ability of the protein to catalyze a particular reaction, or by a **binding assay,** which measures the ability of the protein to bind to a specific small molecule. At the same time, the researcher measures the total amount of protein present, often using dyes that bind specifically to proteins. As the purification of a protein proceeds, the **specific activity,** the amount of enzyme or binding activity per gram of total protein, increases until it reaches a maximum when the protein is completely homogeneous.

Centrifugation separates proteins (as well as other macromolecules and subcellular structures) according to size (and shape) by subjecting them to high gravitational fields in a spinning tube within an ultracentrifuge, as shown in Figure 3-8. In a widely used method of separation, the experimenter places a sample containing a protein mixture at the top of a tube containing a sucrose gradient, a mixture with a higher sucrose concentration at the bottom of the tube and a lower concentration at the top. This gradient minimizes the mixing of the separated proteins during and after centrifugation. Larger molecules tend to move more rapidly than smaller molecules, although the rate of movement depends also on shape. Compact molecules move more rapidly than extended molecules of the same molecular mass.

Electrophoresis separates charged molecules (such as proteins) according to their ability to move in an electrical field. In general, the more highly charged molecules move most rapidly (positively charged molecules toward the negative electrode and negatively charged molecules toward the positive electrode). Size and shape are also important in determining the rate of movement. More compact molecules move more rapidly than extended ones, and larger molecules tend to move more slowly than smaller ones. The most commonly used method of electrophoresis is illustrated in Figure 3-9, which shows an apparatus that allows electrophoresis to occur in a solution contained within a jelly-like substance called polyacry-

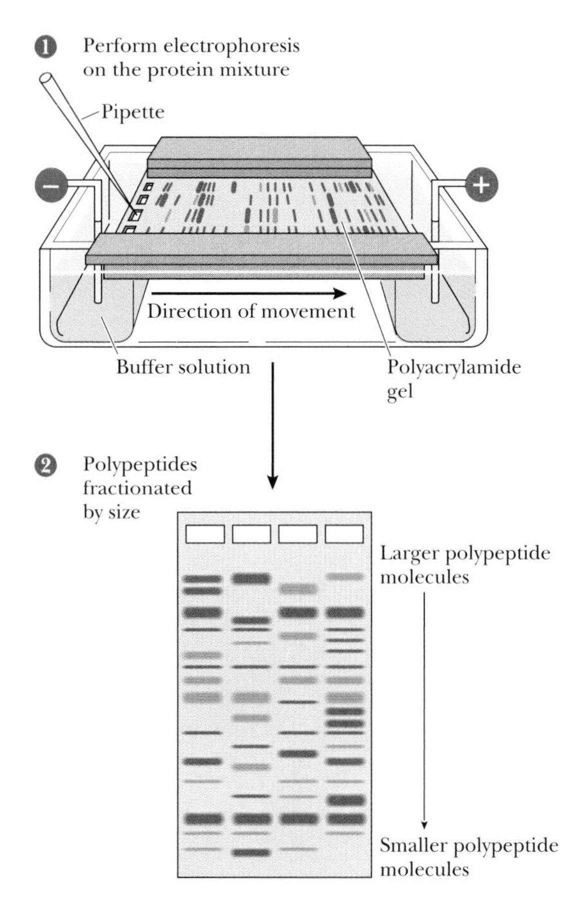

Figure 3-9 Electrophoresis is used to separate macromolecules according to charge, shape, and, most often, size. Most frequently, negatively charged DNA or RNA, or polypeptides complexed with a negatively charged detergent, move toward the positive electrode. Larger molecules move more slowly than smaller molecules, because the supporting gel retards them.

lamide. Like a sucrose gradient, the polyacrylamide gel minimizes the mixing of the separated molecules.

Perhaps the most frequently used method for separating individual polypeptides is a type of electrophoresis in which the solution contains a detergent called sodium dodecyl sulfate (SDS, a common ingredient in shampoos). Because the polar head of this detergent has a negative charge, all polypeptides become negatively charged as the detergent's nonpolar tail binds to the nonpolar side groups of a protein. The negatively charged detergent-protein complexes all move toward the positive electrode. The larger polypeptides, however, move more slowly, and the net result is a separation according to size.

Proteins separated by electrophoresis are detected by their ability to bind to specific protein-binding dyes or to silver ions, which bind to the sulfhydryl groups of proteins. Specific proteins can be detected by antibodies, which can be tagged with a radioactive label, a fluorescent dye, or other marking devices. In the separation, the desired protein is detected after electrophoresis by blotting the contents of the electrophoresis gel onto a paper-like support and incubating this "blot" with a specific antibody—a technique called **immunoblotting**, or **western blotting** [because the method derives from a similar blotting method, called "Southern blotting" after its inventor, Edward Southern].

A variation of electrophoresis, called **isoelectric focusing,** allows molecules to migrate in an electrical field in such a way that they concentrate ("focus") at a point where they cease to have a net electrical charge. A technique called **two-dimensional electrophoresis,** the sequential application of isoelectric focusing and SDS electrophoresis, can separate a mixture of proteins into several thousand components, as illustrated in Figure 3-10. The two-dimensional separations are more effective than electrophoresis alone because they distinguish among polypeptides with the same (or similar) molecular sizes.

The final method discussed here is **chromatography,** illustrated in Figure 3-11, which separates molecules according to their relative affinities for a stationary support, called the *stationary phase,* and a moving solution, called the *mobile phase.* Most often, the stationary phase is some insoluble material packed into a glass or metal column, and the method is called **column chromatography.** Molecules that have little affinity for the stationary material stay in the moving solution and quickly emerge from the column, where they can be collected into test tubes, for example, with a device called a fraction collector. Materials that have greater affinity for the stationary phase emerge more slowly from the column and may even require changing the mobile phase solution.

A widely used contemporary chromatographic technique is **high performance liquid chromatography (HPLC),** in which high pressure drives the mobile phase over a finely divided stationary phase in a metal column. HPLC has been applied to proteins, small peptides, and a wide variety of small molecules. A commonly used application is the separation of amino acids, which can now be done in less than 1 hour.

A frequently used type of chromatography, called **ion exchange chromatography,** uses a stationary phase with a net charge. If the stationary phase has a positive charge, then more negatively charged molecules stick more tightly and move more slowly down the column. Gradually increasing the salt concentration of the mobile phase weakens the ionic bonds to the stationary phase, allowing tightly bound material to emerge.

Another widely used method, called **affinity chromatography,** makes use of the specific binding of a protein to a small molecule, which is bound to an insoluble material and packed into a column. Biochemists have used affinity chromatography, for example, to isolate proteins that have high binding affinities for ATP by coupling a derivative of ATP to a stationary phase. ATP-binding proteins attach to the stationary phase and are released by the addition of free ATP. In **immunoaffinity chromatography,** an antibody is attached to the stationary phase, and the specif-

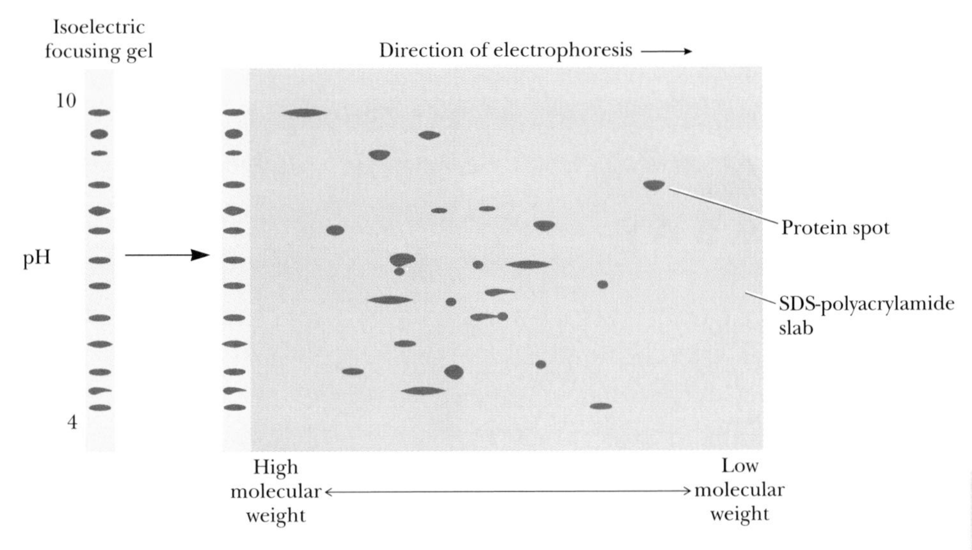

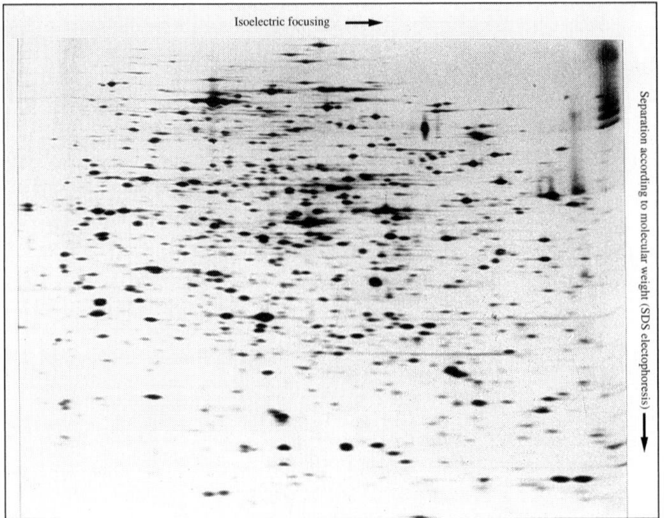

Figure 3-10 Two-dimensional electrophoresis can separate molecules according to charge, size, and shape. This technique, here applied to an extract of *E. coli* cells, is mostly used to separate same-sized molecules according to charge. (*Patrick O'Farrell, UCSF*)

ically bound protein is eluted by changing the solution and weakening the attachment to the antibody.

Purification of a protein is necessary before biochemists can determine its sequence and begin to study its functional and structural properties. Despite the huge repertoire of powerful separation techniques, however, the purification of each protein still requires resourcefulness and ingenuity. By now, biochemists have purified and studied several thousand proteins, each of which has characteristic structural and functional properties. In the next section, we look at how researchers determine the three-dimensional structures of proteins.

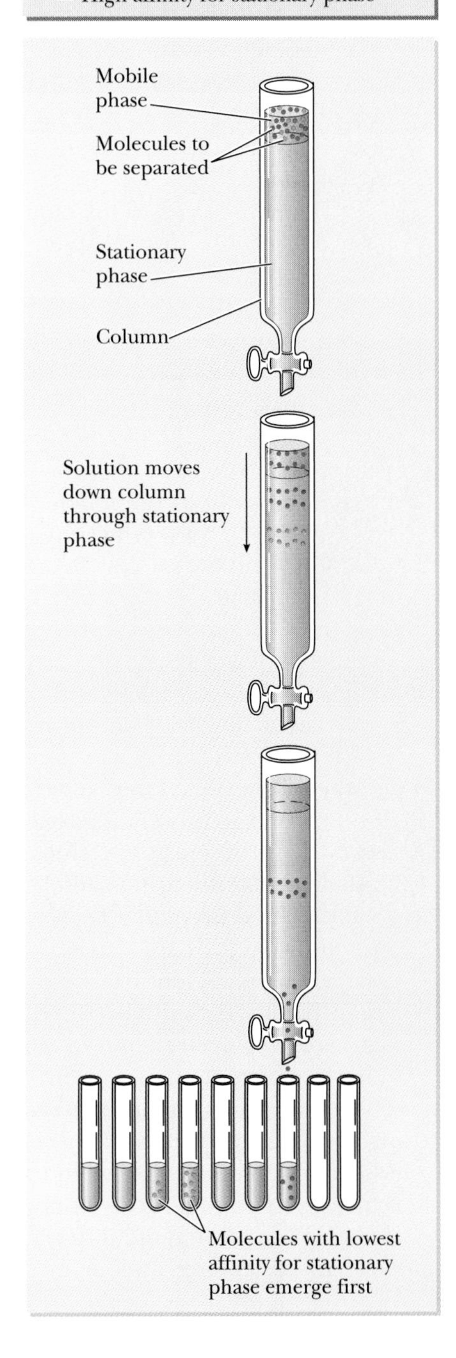

Figure 3-11 Chromatography can separate molecules according to their relative affinities for a stationary phase. The stationary phase may interact either specifically or nonspecifically with molecules in the mobile phase. In the case of affinity chromatography, the stationary phase may contain molecules that bind to specific macromolecules in the solution.

ESSAY

HEME, A SMALL ORGANIC MOLECULE, HELPS PROTEINS INTERACT WITH OXYGEN

While all proteins consist of amino acids linked together in polypeptide chains, many proteins also contain other organic molecules, called *prosthetic groups,* which may be bound by covalent or noncovalent bonds. Such groups allow a protein to perform some chemical function that it could not have done otherwise. A widely used prosthetic group is *heme,* whose structure is shown in Figure 1. All the atoms in the heme molecule lie in a single plane, with an iron atom held in the middle and sharing electrons with four nitrogen atoms.

Heme is crucial to the handling of oxygen in almost every cell in the biosphere. The extraordinary electron affinity of oxygen has allowed life to flourish. Almost all eukaryotic cells, and many prokaryotic cells as well, derive most of their energy from the highly regulated movement of electrons from carbon and hydrogen to oxygen atoms. (See Chapter 7.) But oxygen is also a poison: The uncontrolled movement of the electrons to oxygen can also destroy cellular machinery, just as the direct movement of electrons from iron to oxygen can rust the mechanism of a bicycle.

Heme

Figure 1 *Structure of a heme molecule. The iron (Fe^{2+}) in the center of the ring complex can bind directly to an oxygen molecule (O_2). Note that all the atoms lie in the same plane.*

Cells and organisms make wide use of iron (and a few other types of metal atoms) to handle the flow of electrons to oxygen atoms. As electrons flow from organic molecules to oxygen, iron (Fe) atoms participate in a bucket brigade. An electron may combine with a ferric ion (Fe^{3+}) to form a ferrous ion (Fe^{2+}). The ferrous ion (Fe^{2+}) can then pass the electron to another electron carrier and revert to a ferric ion (Fe^{3+}). Such a controlled passage could not happen, however, if iron ions were present simply in solution; in that case all the iron would "rust," losing electrons to oxygen and becoming Fe^{3+}. Instead, almost all the iron atoms in cells are part of heme groups.

Heme is seldom free within cells, but is almost always bound to a protein. Dozens of cellular proteins participate in electron movements and in the distribution of oxygen atoms. As in the case of free iron ions in solution, the iron ions in heme may be either Fe^{2+} or Fe^{3+}. The tendency of the heme iron to gain or lose electrons depends on the particular protein to which the heme group is bound, so different heme proteins can serve quite different functions.

In vertebrates and in many invertebrates, the main oxygen-carrying protein is hemoglobin, the major protein in blood. It is the iron-containing heme that gives hemoglobin and blood its red color. (The degradation of heme is responsible for most of the color in body wastes.) The protein part of the hemoglobin molecule, called *globin* holds the heme in such a way that it prevents the Fe^{2+} from losing electrons to oxygen. The result is that hemoglobin can serve as an oxygen "raft," binding oxygen and releasing it, without any transfer of electrons.

Other prosthetic groups (and other metal ions) contribute to the function of many other proteins. Particularly interesting are prosthetic groups that participate in enzyme reactions by reversibly binding to specific groups in organic molecules. Such binding allows the enzyme to transfer a group between larger molecules in a manner that resembles hemoglobin's transfer of oxygen from lungs to muscles. Many of these prosthetic groups are derived from *vitamins,* organic molecules that many animals (including humans) have lost the ability to make and that must therefore come from the diet.

HOW DOES A PROTEIN MOLECULE MAINTAIN A PRECISE THREE-DIMENSIONAL STRUCTURE?

The long chains of amino acid residues in a protein molecule fold to form a specific three-dimensional shape. The sequence of the side groups in large part determines the way the chain folds. Nonpolar side groups, as shown in Figure 3-12a for example, help form the hydrophobic regions inside a protein molecule. Charged or polar side groups as shown in Figure 3-12b, on the other hand, are hydrophilic and form electrostatic interactions and hydrogen bonds with water, with other groups of the same protein, or with other molecules. Before we examine these interactions in more detail, we first look at how biochemists have learned about the structures of proteins.

Most proteins, including enzymes, are **globular proteins,** relatively compact molecules roughly spherical in shape. (See Figure 3-13a.) A typical globular protein may have a diameter of 4–10 nm, while the small molecules discussed in Chapter 2 typically have a diameter of about 0.5–1 nm. Many structural proteins, like those of hair, tendons, and skin, have a far more elongated shape and are called **fibrous proteins.** Figure 3-13b shows an electron micrograph of collagen, the most common fibrous protein in animals.

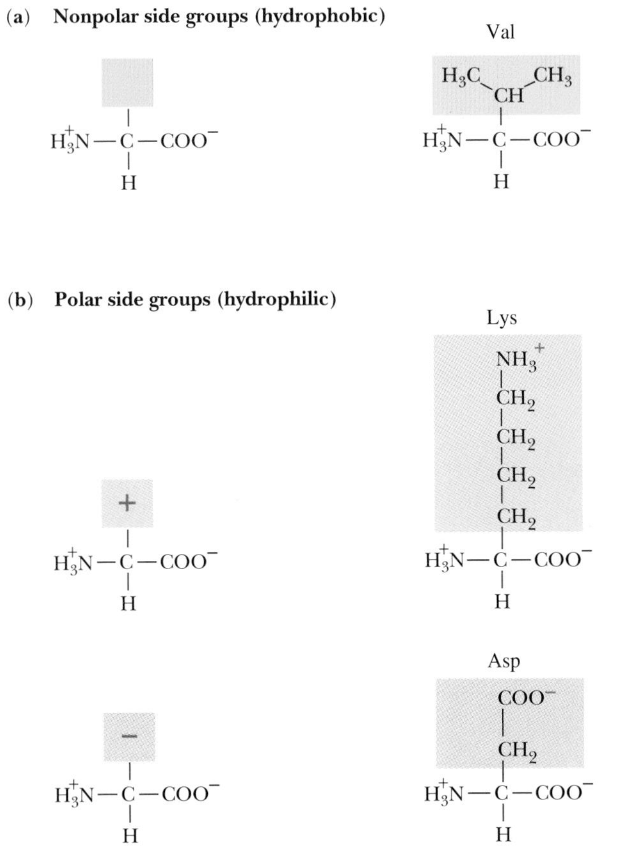

The function of a protein molecule depends on its three-dimensional structure. Our first question, then, is how to visualize the actual three-dimensional structures of these tiny marvels.

(a) **Nonpolar side groups (hydrophobic)**

$H_3^+N-C(H)-COO^-$ Val: $(H_3C)(CH_3)CH-C(H)(H_3^+N)-COO^-$

(b) **Polar side groups (hydrophilic)**

(+) $H_3^+N-C(H)-COO^-$ Lys: $NH_3^+-CH_2-CH_2-CH_2-CH_2-C(H)(H_3^+N)-COO^-$

(−) $H_3^+N-C(H)-COO^-$ Asp: $COO^--CH_2-C(H)(H_3^+N)-COO^-$

Figure 3-12 **(a)** Nonpolar side groups help form hydrophobic regions inside a protein molecule. **(b)** Polar, charged, side groups are hydrophilic and form electrostatic interactions and hydrogen bonds with water and other charged groups on the same protein or other molecules.

X-ray Crystallography Has Revealed the Three-Dimensional Structure of Many Proteins

X-ray crystallography is the analysis of the diffraction or scattering of x-rays by **crystals,** solids that are enclosed by geometrically regular faces, such as those shown in Figure 3-14. By the end of the 19th century, scientists suspected that the external symmetry of crystals resulted from the regular arrangement of atoms or molecules within them. Proof of this view came in 1912 from the work of the German physicist Max von Laue. Laue showed that a crystal can **diffract,** or scatter, x-rays in such a way that they form a pattern of spots on an appropriately placed photographic film. Laue and other physicists understood the meaning of such spots immediately: X-rays, like visible light, are electromagnetic waves. The spots result from the interference of waves scattered by the crystal's atoms, the crest of one wave canceling the trough of another in a regular pattern that reflects the precise arrangement of atoms, as diagrammed in Figure 3-15.

Physicists had observed and studied similar interference patterns with visible light for more than 100 years. They knew that such interference occurs only when the spacing between the scattering objects is about the same as the wavelength of scattered light. Laue's experiment showed that the distance between the regularly arranged atoms of a crystal is about the same as the wavelength of the x-rays.

Almost immediately after Laue's discovery, two British physicists, William Henry Bragg and his son Lawrence Bragg, showed that analysis of x-ray diffraction patterns can reveal the precise atomic arrangements within a crystal. The Braggs' work served as the basis for determining all of the macromolecular structures discussed in the rest of this chapter.

Each atom in a crystal contributes to a pattern of spots, such as those shown in Figure 3-16, generated by an impinging beam of x-rays. The analysis of such complicated x-ray diffraction patterns is currently the only way to determine the detailed three-dimensional structure of a protein.

Crystallographers can calculate the structure of the individual molecules of the crystal from the intensities of these spots. For a complicated structure like a protein, researchers must gather and analyze huge amounts of data, tasks made much more feasible with the development of

Figure 3-13 Globular and fibrous proteins. **(a)** Globular proteins, such as myoglobin, are roughly spherical in shape. **(b)** Fibrous proteins are more extended and often serve structural roles; collagen, for example, strengthens skin, tendons, and blood vessels. (*a, Joel Berendzen/Los Alamos National Lab, NM; b, Dr. David Birke/Tufts University*)

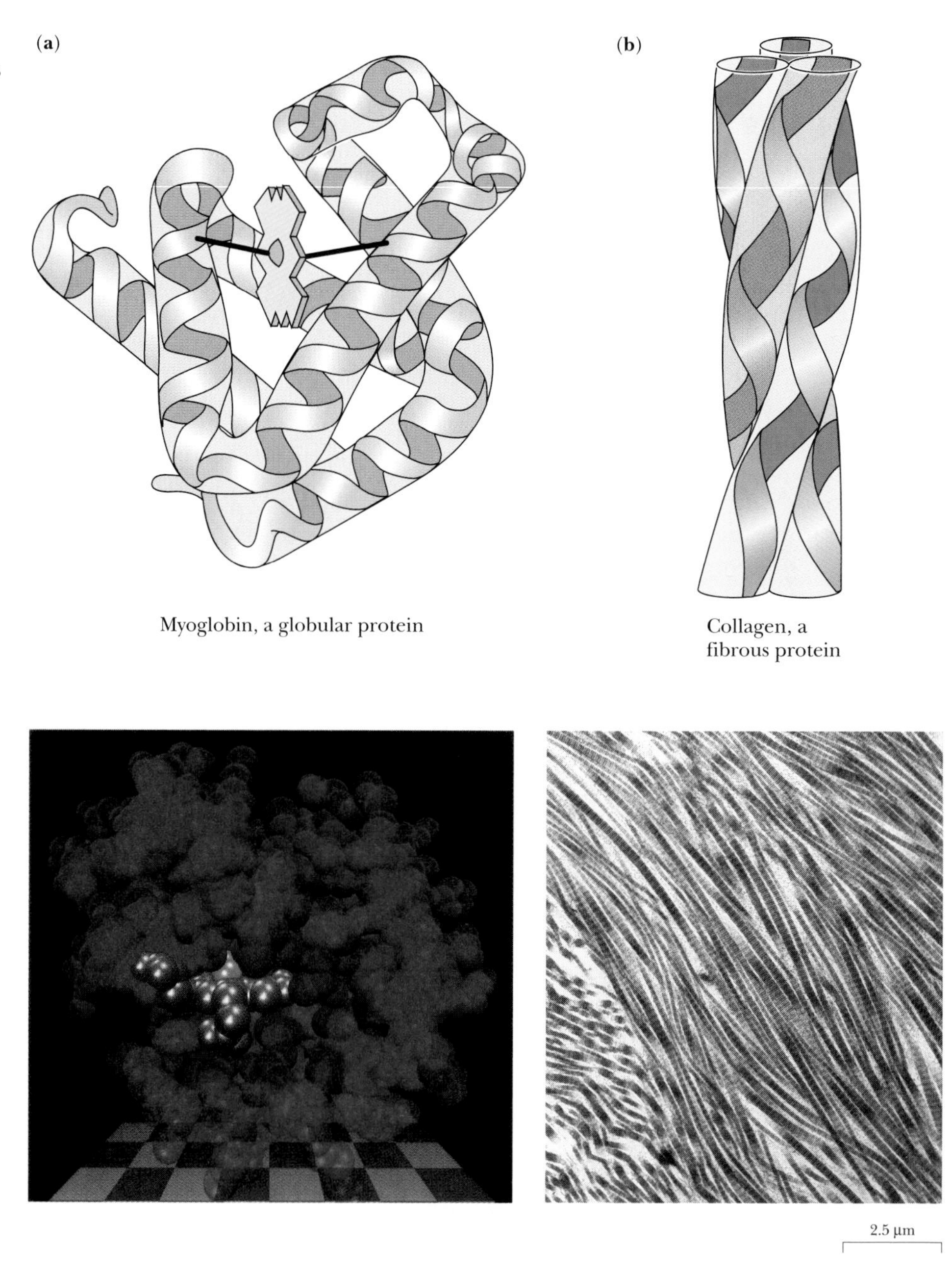

(a) 500 μm

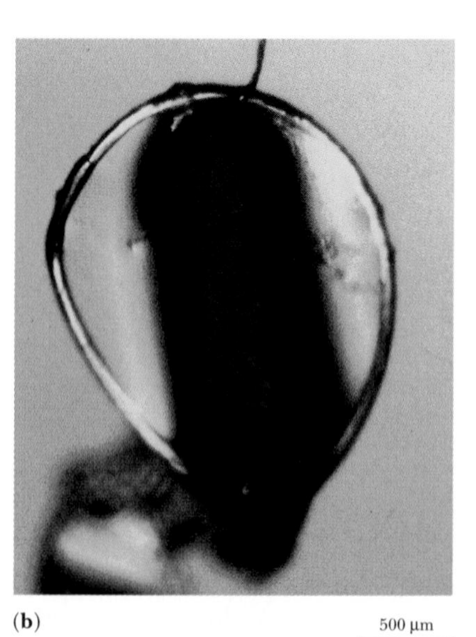

Figure 3-14 Crystals formed by the regular packing of ions or molecules: **(a)** crystals of table salt (sodium chloride); **(b)** a crystal of myoglobin, an oxygen-binding protein. (*a, Charles M. Falco/Photo Researchers; b, Joel Berendzen/Los Alamos National Lab, NM*)

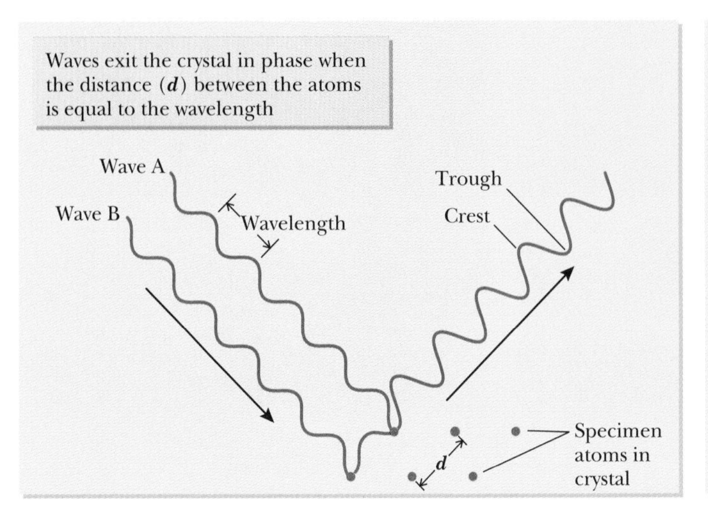

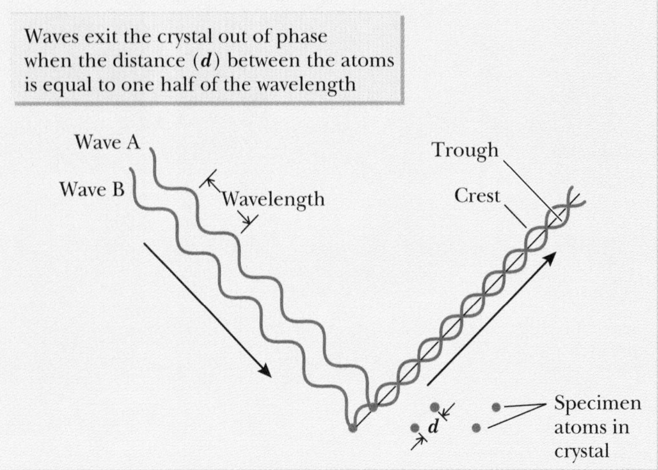

Figure 3-15 X-rays, like light, can behave like a wave. In some situations, the trough of one wave, emanating from one atom, cancels the crest of another wave, emanating from another atom. In other situations, the waves reinforce one another. These phenomena can give rise to a pattern of spots, such as those shown in Figure 3-16. The application of x-ray diffraction to study the structure of biologically important molecules was started in Cambridge, England, by William and Lawrence Bragg, father and son. The two shared the Nobel Prize in 1915. The younger Bragg had to leave the front during the First World War to accept the prize.

automated equipment and high-speed computers. The complete structure of a protein requires more than 25,000 measurements and equations with more than 1 billion terms! The rapid advance of computer technology has greatly accelerated the analysis of crystallographic data, and crystallographers have determined the complete structures of more than 100 proteins.

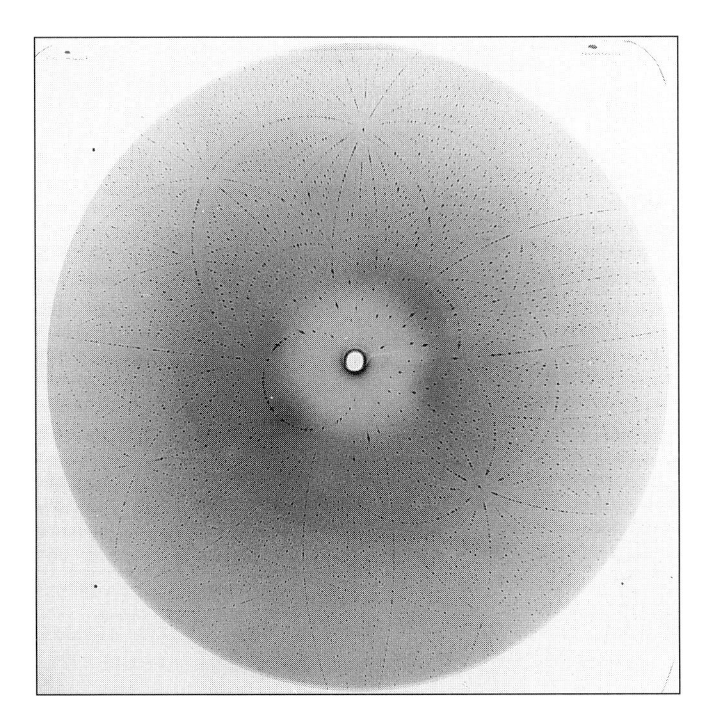

Figure 3-16 X-ray diffraction pattern from a myoglobin crystal, such as that shown in Figure 3-14b. (*Joel Berendzen/Los Alamos National Lab, NM*)

The most complete description of the structure of a protein is a three-dimensional space-filling model. Space-filling models, such as that shown in Figure 3-13a, give particularly useful information about the surface of a protein molecule but reveal little of how the inside interactions establish the outside shape. In addition, such models are extremely expensive and difficult to circulate to interested scientists and students.

Protein structures are therefore most often shown as skeletons of all or some of the atoms, as shown in Figure 3-17a. The model may be represented in perspective, in a series of cross-sections, or in stereo photographs that present slightly different images to each eye. Different kinds of representations help us to understand different aspects of structure and function: how the backbone folds, how the side chains interact, how features on the surface interact with other molecules. New computer graphics techniques as illustrated in Figure 3-17b and c, originally developed for designers and architects, have allowed scientists and students to view models of proteins from virtually any perspective.

The Three-Dimensional Structure of a Protein Depends on Interactions of Amino Acid Residues

Biochemists have defined four levels of protein structure, as illustrated in Figure 3-18. The **primary structure** of a protein is the linear sequence of amino acid residues in each polypeptide chain. The fundamental defining char-

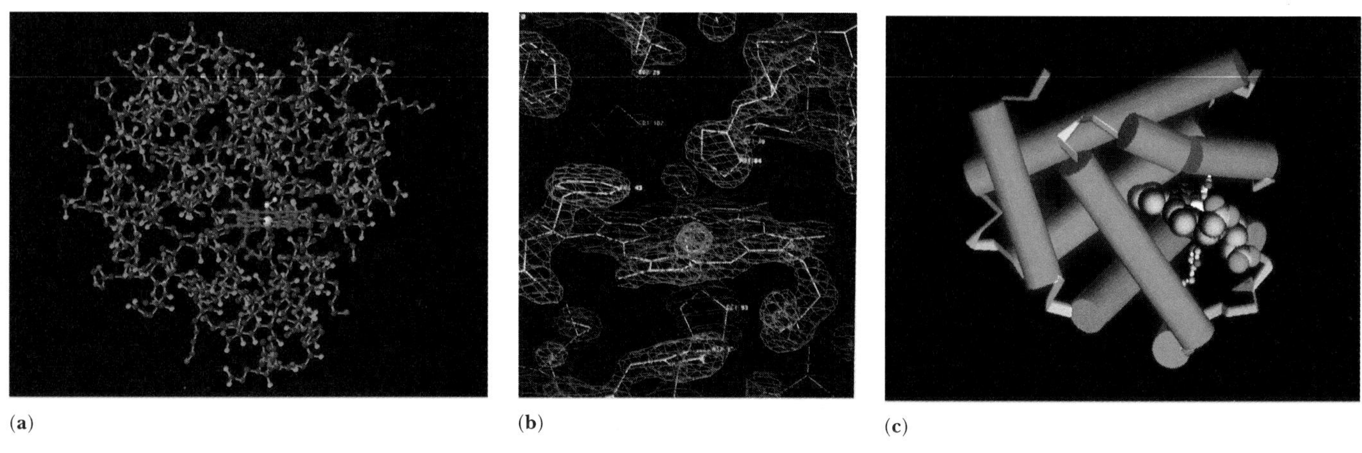

(a) (b) (c)

Figure 3-17 Models of a myoglobin molecule. **(a)** Computer-generated model, showing 3-dimensional arrangement of all atoms; computer can rotate the model to give different perspectives. **(b)** A detailed representation of atoms in the vicinity of the oxygen-binding region of myoglobin; the model shows the heme group, discussed in the accompanying Essay (p. 72). **(c)** A ribbon model of myoglobin, showing the arrangement of the oxygen-binding heme molecule and of helical structures (shown in blue). *(a, Ken Eward/Science Source, Photo Researchers; b, Joel Berendzen/Los Alamos National Lab, NM; c, Dr. Arthur Lesk/Photo Researchers)*

acteristic of a polypeptide chain is its sequence of amino acids. We might expect some regularities of structure to result from the regular repetition of peptide bonds, and we discuss the two main examples (the α helix and the β sheet) later in this chapter. These regular local structures, which are almost always confined to relatively short lengths of a polypeptide chain, are examples of a protein's **secondary structure,** and result from regular hydrogen bonding within adjacent stretches of the polypeptide backbone.

A protein's **tertiary structure** is the complete arrangement of all its atoms. Biochemists also speak of the **conformation** of a polypeptide chain—the way the whole polypeptide chain folds, the equivalent of the tertiary structure. (In contrast, the secondary structure refers to mostly local interactions.) When a protein contains more than one polypeptide chain, the relationship among the separate chains is called its **quaternary structure.**

The arrangement of atoms in a protein is a compromise between the tendency of the backbone to form regular secondary structures and the tendency of individual side chains to interact in more complicated ways. In globular proteins, about 35% (on average) of the amino acid residues in proteins are in regular arrangements. The rest are found in irregular, highly individualized arrangements. Together, regular and irregular structures form compact units, each of which is called a **domain,** which usually consists of 50–350 amino acid residues, as shown in Figure 3-19. A single protein may have one or several domains.

Scale Models Allowed the Prediction of the Regular Structures of Protein Molecules

A repeating structure tends to adopt the form of a helix, the arrangement exemplified by a spiral staircase. In a regular helix, every repeating unit has the same interactions with its neighbors, just as in the spiral staircase each step is identical and has the same relationships to the steps above and below.

Both the regular and irregular structures in proteins depend on the same four kinds of noncovalent interactions: (1) charge interactions ("ionic bonds"); (2) hydrogen bonds; (3) hydrophobic interactions; and (4) van der Waals interactions. Although ionic bonds, hydrophobic interactions, and van der Waals interactions are important in determining a protein's three-dimensional structure, the greatest insights into the regular arrangements within proteins came from considering only the hydrogen bonds that could be formed by the atoms of the polypeptide backbone—the repetitive N—C—C—N—C—C—N—C—C—···.

Even before x-ray analysis provided any information about the actual structure of proteins, the chemist Linus Pauling and his co-workers in California correctly predicted the kinds of helix that a polypeptide might adopt. They arrived at their suggestions not by calculations but by the simple (and, to some, outrageous) procedure of building a scale model of a polypeptide chain. They based their scale model of a protein on measurements of the arrangements of atoms in amino acids and small polypeptides that had been synthesized in the laboratory.

Literally within hours of reading Pauling's 1951 paper, Max Perutz, a biochemist and x-ray crystallographer in Cambridge, England, was able to confirm one of its predictions from the x-ray analyses he had already performed. Pauling's approach correctly predicted common arrangements of polypeptide chains. He also showed that scientists could deduce some of the properties of a macromolecule from the properties of its building blocks. This astonishing finding has had tremendous influence on the later development of molecular biology and biochemistry,

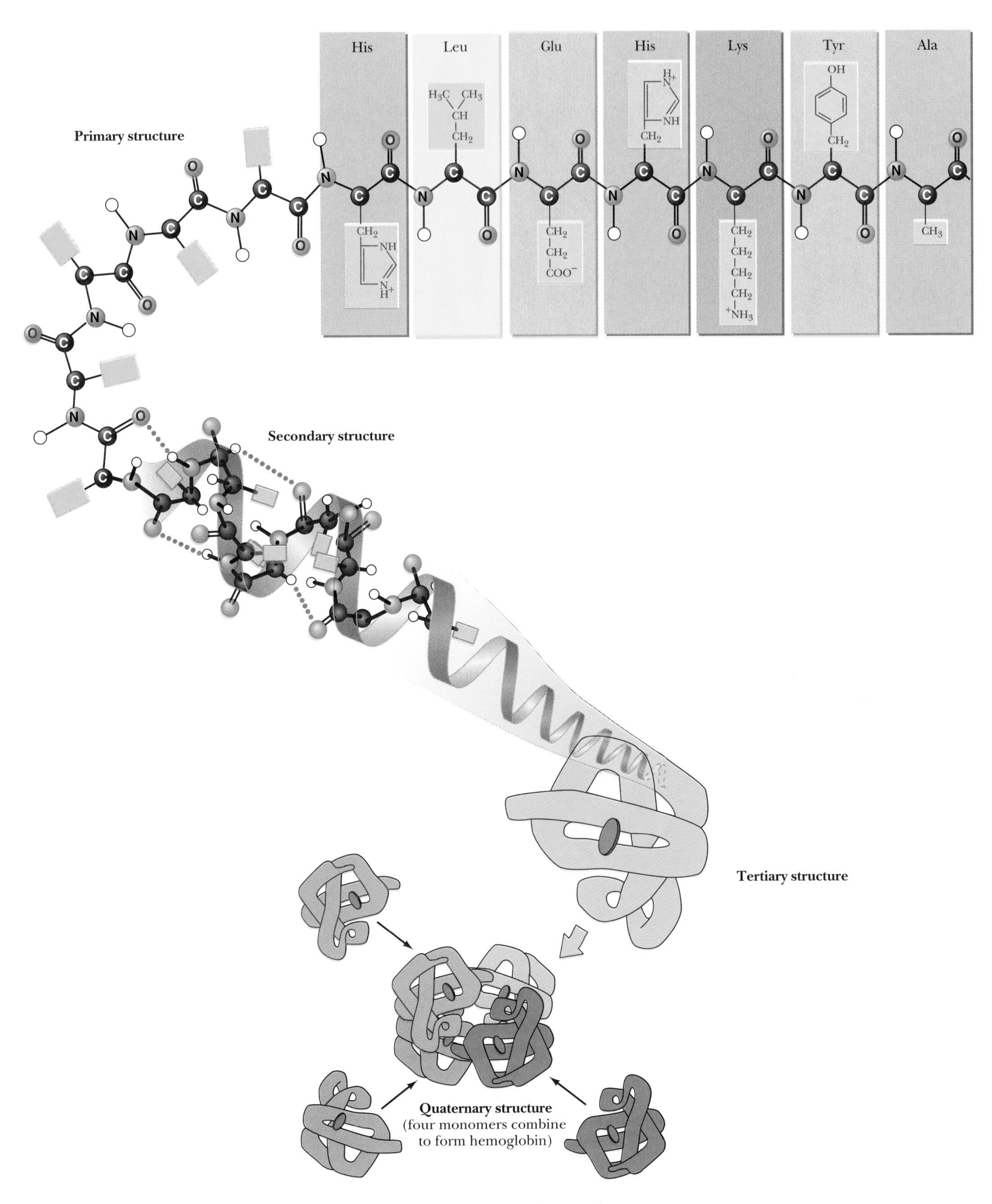

Figure 3-18 Primary, secondary, tertiary, and quaternary structure of a protein.

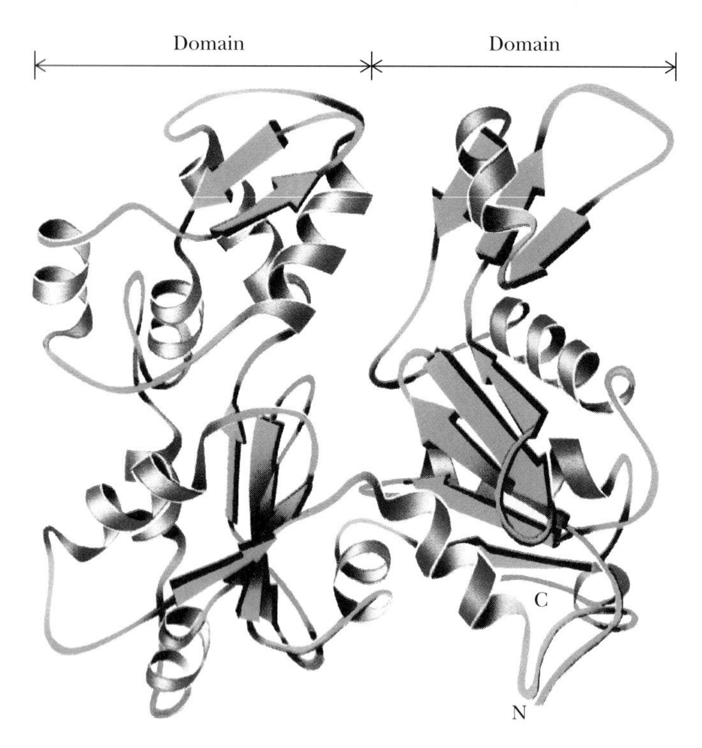

Figure 3-19 Domains within an actin molecule. Each domain is thought to fold independently of other domains. "N" marks the amino-terminal end of the polypeptide and "C" its carboxyl-terminal end.

greatly simplifying the way scientists think about macromolecules. Most spectacularly, Pauling's approach led directly to understanding the structure and function of DNA, discussed in Chapter 13.

Pauling's scale models showed the basis of the secondary structures of proteins. Some arrangements of a polypeptide backbone allow a regular pattern of hydrogen bonds between the slightly positive amino group of one amino acid residue and the slightly negative carbonyl group of another residue. In the ***α* helix** (shown in Figure 3-20), the first secondary structure proposed, every carbonyl group is hydrogen-bonded to the amino group four amino acids farther down the chain. Each residue is turned 100 degrees from the last and advanced 0.15 nanometers (nm) along the axis. Each 360° turn contains 3.6 amino acid residues, and there are 0.54 nm between turns. The side chains point away from the backbone.

α helices are extremely common in proteins. Myoglobin, the oxygen-binding protein of red muscle, is the champion, with about 70% of its amino acid residues in *α* helix. Most enzymes and other globular proteins contain substantial amounts of *α* helix. A typical *α* helix segment is a short rod, usually less than 27 residues (4 nm) long. Some fibrous proteins, such as the keratins of hair and wool, contain much longer *α*-helical regions.

A second common secondary structure, shown in Figure 3-21a, is called the ***β* structure.** Here the amino and carbonyl groups are hydrogen bonded to other polypeptide chains or to distant regions of the same chain folded back on itself. *β* structures are the main feature of the

Figure 3-20 An *α* helix, first proposed by Linus Pauling on the basis of regular hydrogen bonds between atoms in the polypeptide backbone. The carboxyl group of residue 1 forms a hydrogen bond to the amino group of residue 5; the carboxyl group of residue 2 to the amino group of residue 6, etc. Pauling had already made major contributions to chemistry, summarized in his influential book, *The Nature of the Chemical Bond,* and he was largely responsible for the current concept of electronegativity, which we discuss in Chapter 2. He won two Nobel Prizes, one for his scientific work and one for his efforts on behalf of world peace, which led to the 1964 international treaty outlawing atmospheric tests of nuclear weapons.

(**a**) An α helix

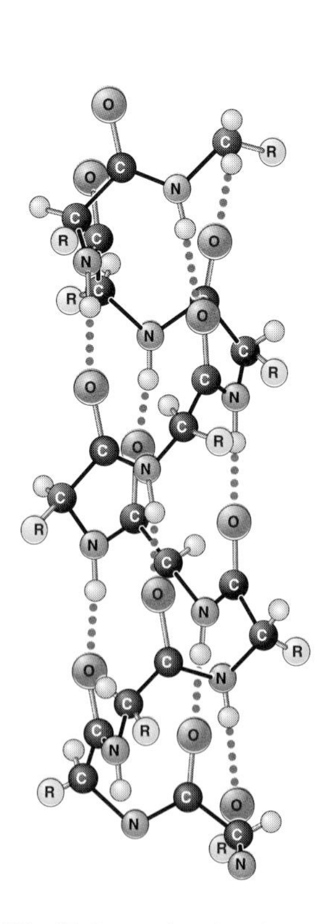

(**b**) Hydrogen bonds stabilize the helix structure

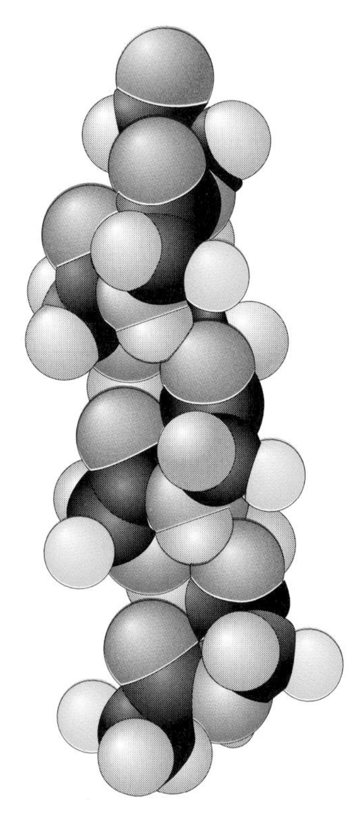

(**c**) Space-filling model

(a)

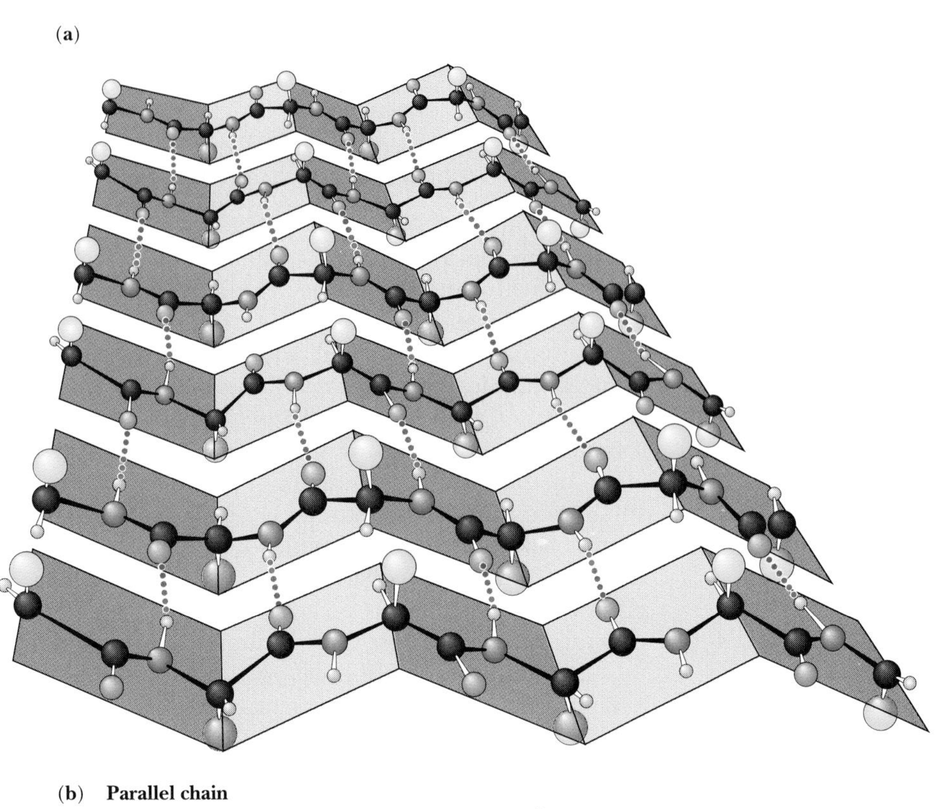

(b) Parallel chain

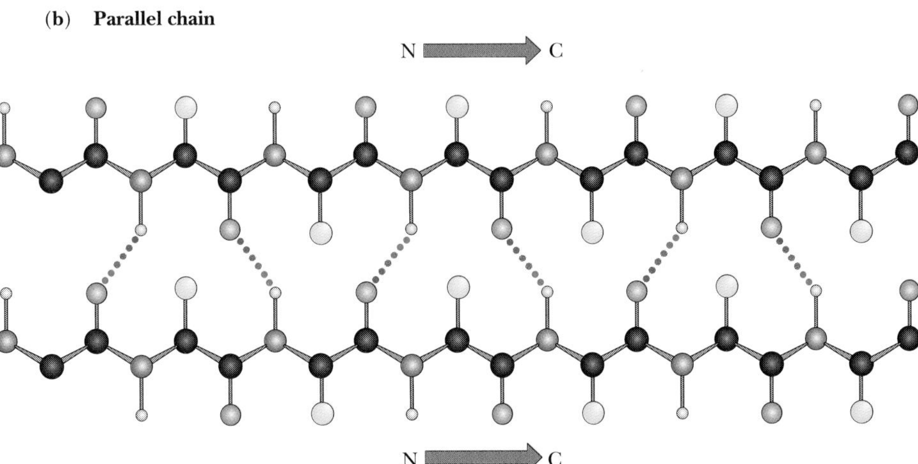

(c) Antiparallel chain

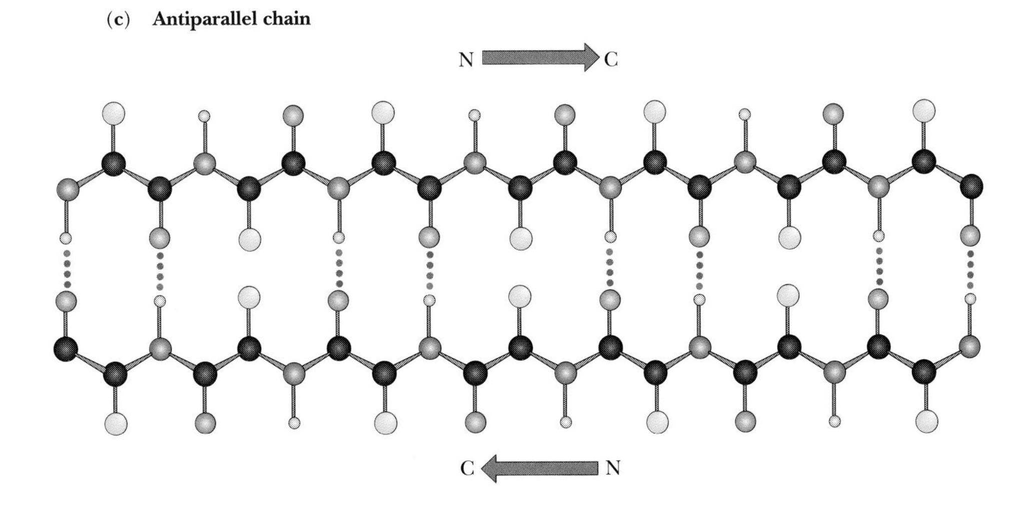

Figure 3-21 Two forms of β structure **(a),** also predicted by Pauling; **(b)** hydrogen bonds between amino groups and carboxyl groups of polypeptide backbones running in the same (parallel) direction; **(c)** hydrogen bonds between amino groups and carboxyl groups of polypeptide backbones running in opposite (antiparallel) directions. Both β structures can be thought of as helices with each residue turned 180° and advanced 0.35 nm.

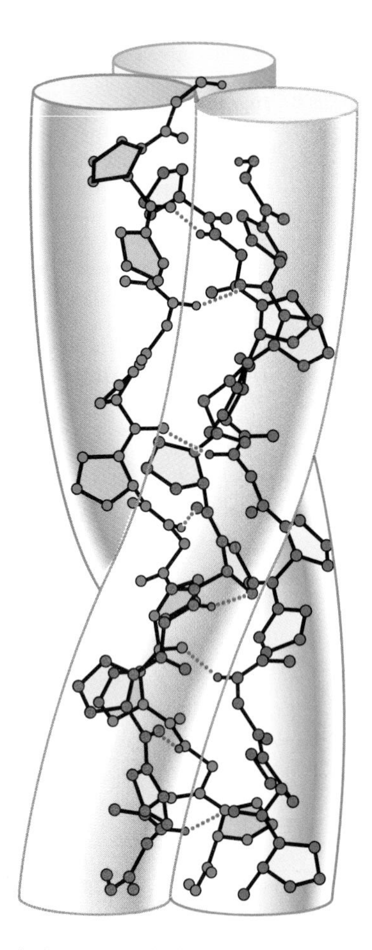

Figure 3-22 A three-stranded collagen helix.

proteins of silk. Many globular proteins also contain β structures, with two to five parallel or antiparallel sections of a single chain forming a **β sheet,** as illustrated in Figure 3-21b and c.

The frequent occurrence of α and β structures in proteins has led biochemists to devise another way of portraying protein structures, called the **ribbon model.** In the ribbon model an α helix is depicted as a coil, while β sheets are sets of arrows. The ribbon model stresses the secondary structures within a globular protein (see Figure 3-19).

A third type of regular structure, shown in Figure 3-22, is called the **collagen helix.** The collagen helix is found in only a few proteins, mainly in the structural protein collagen, which makes up about 25% of all animal protein. Different types of collagen contribute to skin, tendons, bone, ligaments, and, in smaller amounts, almost every other organ in the body. The collagen helix consists of three polypeptide chains wound around each other.

α helices and β structures are the major regular arrangements found in most proteins. These repeating forms (along with that of collagen) occur commonly in proteins that provide mechanical support within organisms. These simple patterns allow such fibrous proteins to function as standardized building materials, comparable to the standardized cuts of lumber or sizes of bricks with which a contractor assembles a house or a factory.

Both the Secondary and Tertiary Structures of a Polypeptide Depend on the Sequence of Amino Acid Residues

Each of the repeated forms is stabilized by different patterns of hydrogen bonding along the polypeptide backbone. But why should one type of helix be found in tendons, another in hair, and still another in silk? The answer must lie in the particular arrangements of amino acid side groups in each protein because the peptide backbone is always the same and the side groups vary among proteins. In the regular regions of structural proteins, the amino acid sequence is extremely regular. In the β structures of silk protein, for example, every other amino acid is glycine, and in the collagen helix, every third residue is proline. In these fibrous proteins, the regular secondary structures extend almost throughout the entire molecule.

In contrast to the fibrous proteins, globular proteins have no striking long-range regularities of sequence. Their intricate pattern of folding is highly individual. The uniqueness of each globular protein is not surprising because each protein must interact specifically with other molecules: Enzymes are associated with molecules whose transformations they catalyze, antibodies bind to their targets, and structural proteins interact with other structural proteins. The irregular but precise folding of polypeptide chains results from interactions of the side chains of amino acid residues.

Each side group has its particular characteristics. Some are charged, some polar, some nonpolar. Some can serve as donors in hydrogen bonds, others as acceptors. In addition, two residues of the amino acid cysteine can form a covalent disulfide bond with each other, attaching two distant points in a polypeptide chain in a kind of "spot weld," as shown in Figure 3-23.

As noted earlier in this chapter, the stability of each conformation depends on noncovalent interactions. The formation of a hydrogen bond stabilizes a conformation, as does the coming together of positively and negatively charged side groups. On the other hand, bringing together two like-charged side groups destabilizes a conformation, as does the placement of a charged or polar side group in a nonpolar, hydrophobic environment.

Singly, any of these interactions contributes only weakly to the stability of a given conformation. Together, however, they establish a three-dimensional structure that is extremely stable in a given chemical environment. Because the internal environment of cells is relatively constant, each protein maintains its characteristic conformation. Subtle changes in the cellular environment (for example, in the concentration of calcium ions) can bring about a new conformation with altered biological activity. The sensitivity of protein conformation to environmental factors is the key to the regulation of biological processes.

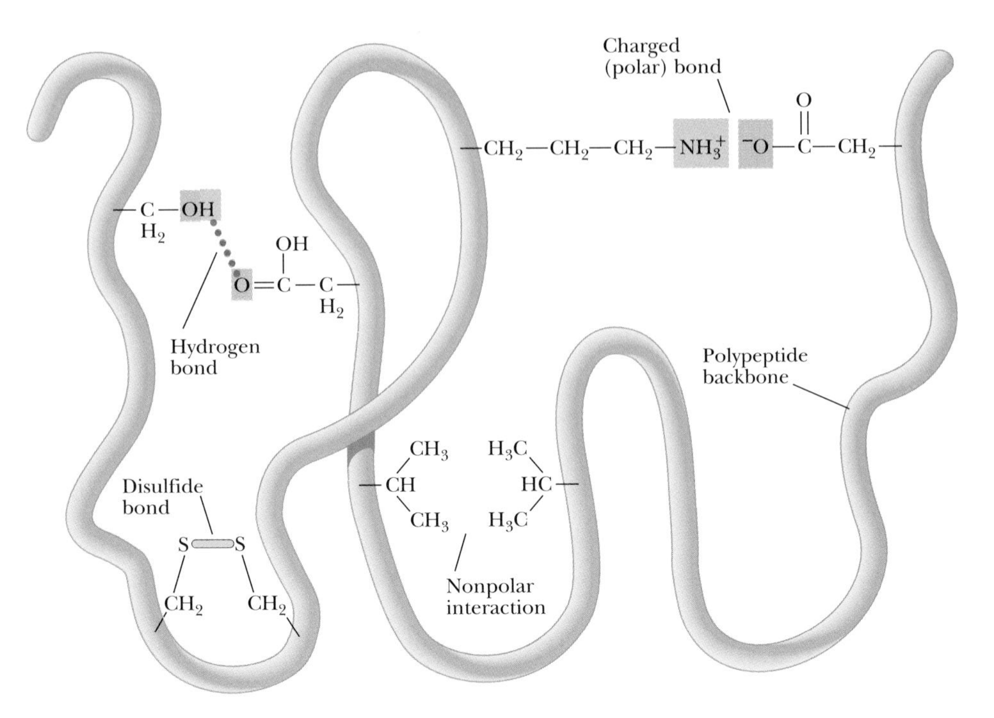

Figure 3-23 Side groups can be involved in hydrogen bonds, nonpolar interactions, or polar bonds. A disulfide bond provides a "spot weld" between two polypeptide chains or between segments of a single polypeptide chain.

Are the Interactions of Amino Acid Residues Enough to Determine Protein Structure?

Interactions of the functional groups—the polypeptide backbone and the amino acid side groups, both among themselves and with water—stabilize the conformation of a protein. But these interactions might not be enough to determine the precise three-dimensional structure of a protein. We might imagine, for example, that the conformation of a polypeptide depends not only on its amino acid sequence but also on the presence of some other "instructional" molecule that bends or molds the polypeptide into a particular shape. How can we determine whether the conformation is really established by the amino acid sequence?

The critical experiment, shown in Figure 3-24, was to unfold an enzyme and allow it to fold up again. In the 1950s, biochemist Christian Anfinsen and his co-workers at Harvard Medical School isolated the enzyme ribonuclease, which speeds up the hydrolysis of RNA. The **native** form of a protein is the three-dimensional structure that has biological activity. In the case of ribonuclease, the native form is the conformation that helps to hydrolyze RNA into nucleotides.

Anfinsen and his colleagues denatured ribonuclease; that is, they disrupted its native structure so that it lost its enzymatic activity. They did this by treating the enzyme with a solution that interfered with hydrogen bonds and hydrophobic interactions and also broke the disulfide spot welds between cysteine side chains.

(Text continues on page 84.)

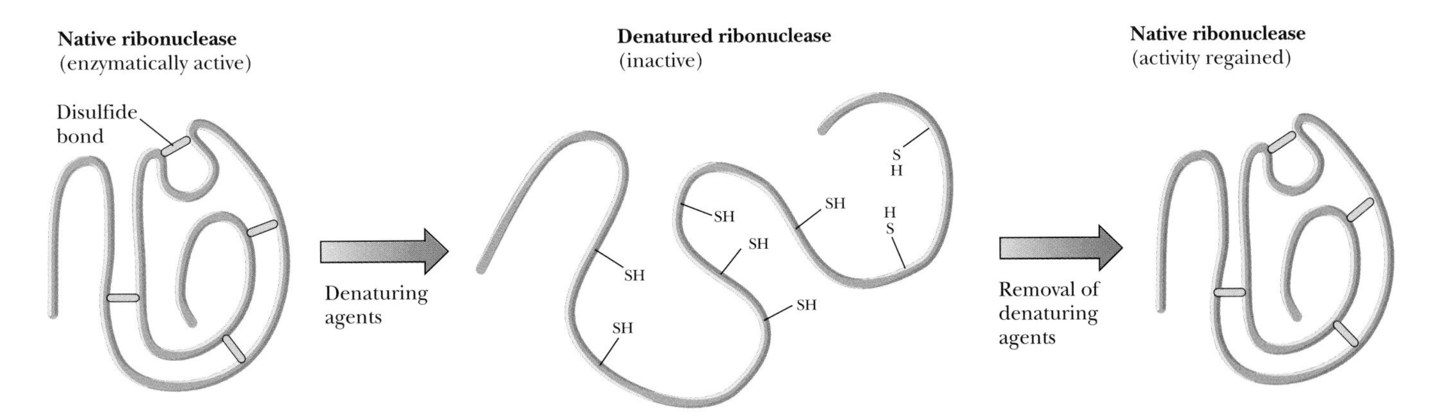

Figure 3-24 Christian Anfinsen broke the disulfide bonds and disrupted the noncovalent bonds that maintained the three-dimensional structure of ribonuclease. After the denaturing agents were removed, the same noncovalent and disulfide bonds reformed, and enzyme activity was restored. This experiment suggested that the amino acid sequence of a protein determines its three-dimensional structure, without the need for additional guidance.

ESSAY

HOW CAN HEMOGLOBIN DELIVER OXYGEN BOTH TO A MUSCLE AND TO A DEVELOPING FETUS?

In air-breathing vertebrates, hemoglobin—present within red blood cells—binds oxygen in the lungs. Red blood cells containing *oxyhemoglobin* then move with the blood throughout the body, releasing oxygen to muscles and other tissues and regenerating *deoxyhemoglobin.* But how does the hemoglobin "know" when to bind and when to release oxygen?

The answer lies in the differences in oxygen concentration between lungs and muscle. In order to describe these differences, scientists speak of the *partial pressure* of oxygen, a measure of oxygen concentration. To understand the idea of partial pressure, recall that air is a mixture of gases. About 21% of the molecules are oxygen (O_2), about 0.03% are CO_2, and almost all the rest are nitrogen (N_2). The total pressure exerted by the atmosphere—which we can measure with a barometer—is the sum of all these partial pressures. At sea level, the total pressure exerted by the atmosphere is enough to support a column of mercury 760 mm high (about 30 inches). We say that the pressure is 760 torr, after Torricelli, the inventor of the mercury barometer. The partial pressure of oxygen is 21% of 760 torr, or about 160 torr.

To understand hemoglobin action, researchers have studied the binding of oxygen to hemoglobin at different partial pressures of oxygen. Binding is fairly easy to study, because oxyhemoglobin has a different color from deoxyhemoglobin, as illustrated in Figure 1. The oxygen-loaded blood of the arteries is bright red, while the relatively depleted blood of the veins is slightly blue-red. With an appropriate measuring instrument, called a *spectrophotometer,* researchers can determine the precise fraction of oxy- and deoxyhemoglobin by measuring the fraction of light absorbed at a particular wavelength.

A typical experiment begins with a sample of blood in a tube. The researcher removes air from the tube with a vacuum pump, thereby converting oxyhemoglobin to deoxyhemoglobin. Now she admits a measured amount of air and determines the fraction of oxyhemoglobin. The result is a curve, called an *oxygen dissociation curve,* shown in Figure 2. For the blood from adult humans, the fraction of oxyhemoglobin increases from a partial pressure of about 10 torr to about 60 torr. Above 60 torr, the fraction of oxyhemoglobin does not increase much; the hemoglobin is effectively saturated with oxygen. The shape of the whole curve vaguely resembles the letter "*S*."

How does the oxygen dissociation curve relate to the delivery of oxygen in the body? To approach this question, biologists first determine the concentration of oxygen in the fluids of the lungs and of the tissues, expressing the concentrations as the partial pressures of oxygen in air that would have produced those concentrations of dissolved oxygen. In the fluids of the lungs, for example, the oxygen concentration is the partial pressure of oxygen in the lung itself—about 100 torr. This is less than the partial pressure of oxygen in the air because only part of the air in the lungs is refreshed with each breath. At this concentration, essentially all the hemoglobin is oxyhemoglobin.

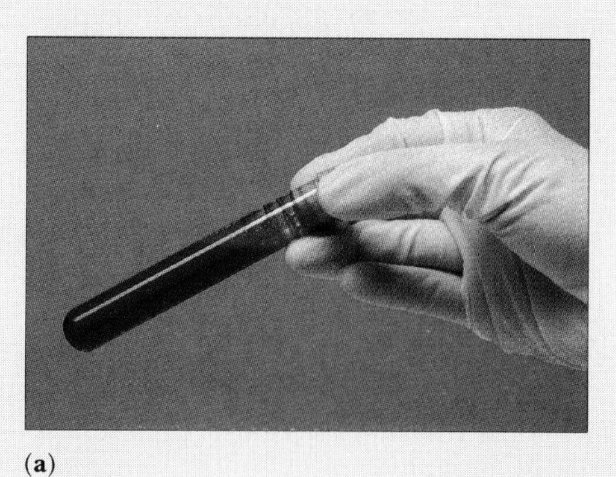

(a)

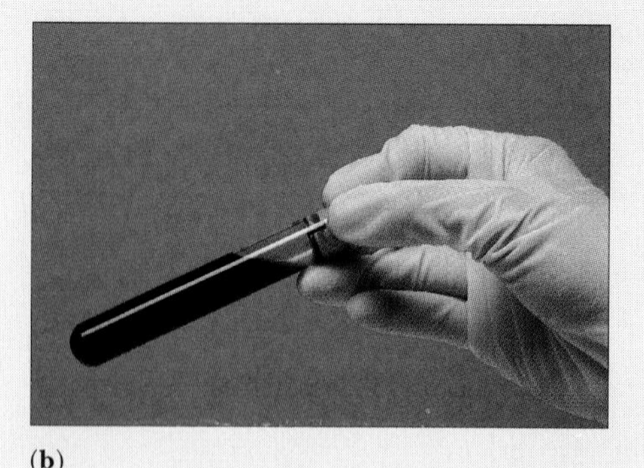

(b)

Figure 1 *Hemoglobin changes color when it combines with oxygen.* ***(a)*** *Blood with oxyhemoglobin is bright red;* ***(b)*** *blood with mostly deoxyhemoglobin is bluish.* (© Matt Meadows)

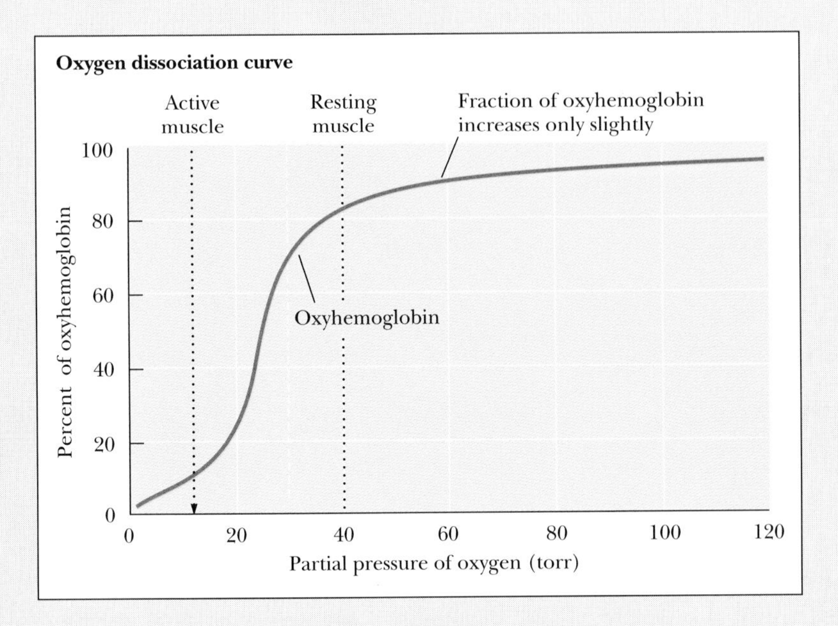

Figure 2 *The oxygen dissociation curve of adult hemoglobin. As the partial pressure of oxygen increases, from the level present in muscles to the level present in the lungs, hemoglobin binds more oxygen. As the circulation returns blood to the muscles, the hemoglobin releases oxygen, where it drives the energy-producing flow of electrons.*

Tissues that use oxygen have lower oxygen concentrations than the lungs. A resting muscle, for example, has an oxygen concentration of 40 torr. At this concentration, only about 75% of the hemoglobin is oxyhemoglobin. As the blood moves from lungs to tissues, then, it gives up about 25% of its oxygen.

During vigorous activity, the oxygen concentration in a muscle may be much lower. So when oxygen and blood come to equilibrium, fewer hemoglobin molecules have oxygen bound—that is, more of the oxyhemoglobin molecules have let go of their passengers. Blood releases more oxygen to active tissues than to resting ones.

Another factor, pH, contributes to the release of oxygen in active tissues. Cells that are especially active may exceed their ordinary supply of oxygen. They then depend (temporarily) on glycolysis to provide ATP. The result is the accumulation of lactic acid, which decreases the pH. (See Chapter 7.) The lowered pH decreases the affinity of hemoglobin for oxygen, shifting the oxygen binding curve to the right. The dependence of oxygen binding on pH (called the *Bohr effect*) increases the unloading of hemoglobin to active tissues.

An important aspect of oxygen binding involves the interaction of the hemoglobin molecule with a small organic molecule called 2,3-bisphosphoglycerate, or BPG, shown in Figure 3. Most cells have only small concentrations of BPG, but mammalian red cells have almost exactly as many molecules of BPG as of hemoglobin. Physiologists long regarded the presence of BPG as a complicating curiosity, and it was not until the 1960s that researchers realized that BPG binds specifically to hemoglobin and changes its oxygen binding properties.

BPG decreases the oxygen binding of hemoglobin. If we study the binding of oxygen to purified adult hemoglobin, we find that its oxygen affinity is much higher than that of blood. In the presence of BPG, however, the hemoglobin behaves as it does in blood itself. Were it not for BPG, then, hemoglobin would be saturated with oxygen even in the tissues, so it would not be useful for oxygen delivery.

When oxygen is less available—for example, at high altitudes or in lung disease—BPG increases and the oxygen binding curves shift to the right. Even with such a shift, hemoglobin is nearly saturated with oxygen in the lungs. But the shift means less oxyhemoglobin in the tissues and more efficient oxygen delivery.

A fetus has a special problem in obtaining oxygen, because it depends not on its own, but on its mother's lungs. Maternal blood carries oxygen from

(Essay continues on next page.)

ESSAY *continued*

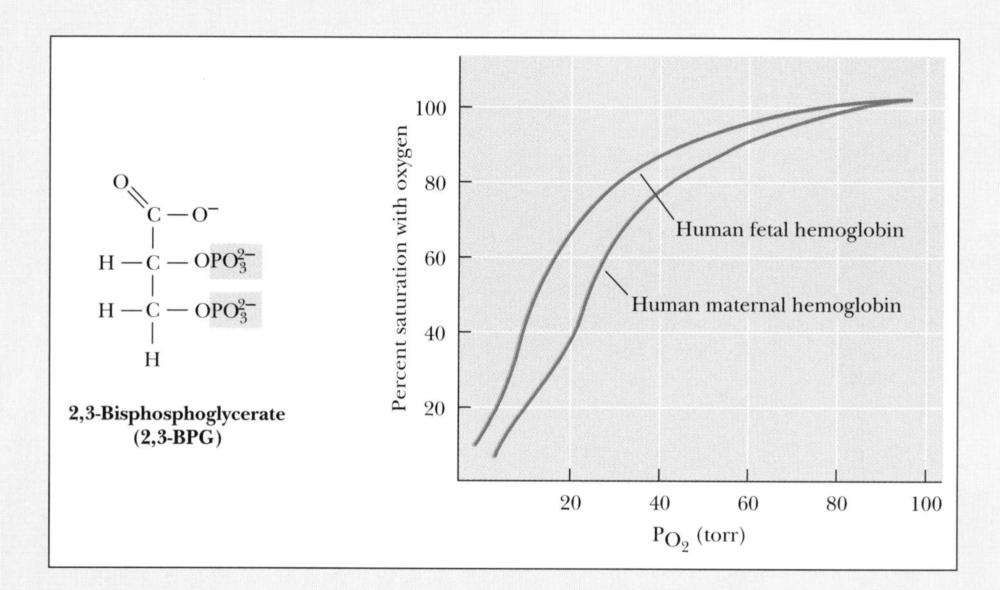

Figure 3 *The structure of 2,3-bisphosphoglycerate, which decreases the oxygen affinity of adult hemoglobin but not of fetal hemoglobin.*

lungs to tissues, including the placenta, a structure in the uterus where maternal and fetal circulation come into close contact. In the placenta, oxygen moves from the maternal to the fetal blood. For this to happen, the fetal blood must have a higher affinity for oxygen than the maternal blood.

Indeed the fetal blood does have a greater affinity for oxygen. The fetus has a hemoglobin whose structure is distinct from that of adults. The major molecular difference between fetal and adult hemoglobins (in humans) is the failure of fetal hemoglobin to bind BPG. The oxygen binding curve of fetal blood thus resembles that of adult hemoglobin, without BPG. The higher oxygen affinity means that the fetal hemoglobin can acquire oxygen from maternal blood and deliver it efficiently to the tissues of the developing fetus.

The efficient passing of oxygen from adult, maternal hemoglobin to fetal hemoglobin illustrates how subtly a polypeptide can influence the properties of an attached prosthetic group. Not only do the adult and fetal hemoglobin polypeptides prevent the oxidation of iron from Fe^{2+} to Fe^{3+}, but they specifically alter the oxygen dissociation curves, allowing the growing fetus to enjoy an environment that is rich in oxygen.

The question was whether the enzyme molecule would regain its enzymatic activity if the denaturing agents were removed. The answer was yes, biological activity returned. The enzyme "renatured": It regained its activity and its native form. This renaturation did not require other sources of energy or other molecules to guide the protein's folding. The amino acid sequence of the protein appeared to be sufficient to determine its three-dimensional structure and biological activity.

Anfinsen's experiment suggested that proteins spontaneously fold into their biologically active forms. The free energy required for this ordering must come from the interactions of the amino acids themselves. (See Chapter 4.) Many protein molecules self-organize; they form their active structures without the need for additional energy or guidance. Other proteins fold properly only in the presence of special helper proteins, called "chaperones."

The instructions to build a complex three-dimensional protein are contained within its one-dimensional structure—the not-yet-folded polypeptide chain. The amino acid sequence of a polypeptide in turn depends on another linear structure—a chain of nucleotides in DNA. We can therefore see how a one-dimensional gene can specify a three-dimensional structure.

SUMMARY

Small building blocks assemble into different kinds of macromolecules by the same kind of general reaction, called a dehydration condensation reaction, because a molecule of water is eliminated each time two building blocks come together. The breaking down of a macromolecule into its constituent building blocks is called hydrolysis because it involves the addition of a molecule of water as each small molecule is removed.

The properties of macromolecules depend on the sequence in which the component building blocks assemble. Biochemists can determine the sequences of amino acids in a protein or of nucleotides in DNA. They can also make molecules with specific sequences in the laboratory. A major goal of biochemistry is to understand, for any macromolecule, the relationship of sequence, three-dimensional structure, and biological function.

Every kind of protein has a distinct sequence and characteristic properties. These properties permit their isolation, usually by a combination of several methods. The most often used purification methods are centrifugation (separation according to size and shape in a high gravitational field); electrophoresis (separation according to charge, size, and shape in an electrical field); and chromatography (separation according to the relative affinities for a stationary support and a mobile solution).

The folding of a polypeptide into a precise three-dimensional structure generally requires energy. This energy comes from the free energy of noncovalent interactions among the atoms of both the polypeptide backbone and the amino acid side groups. X-ray crystallography can reveal the three-dimensional structures of proteins and other molecules. Polypeptides can form both regular and irregular structures. Such structures are stabilized by four kinds of noncovalent interactions—charge interactions, hydrogen bonds, hydrophobic interactions, and van der Waals interactions. The regular structures within protein molecules include α helices, β structures, and collagen helices. The interactions among the atoms of the polypeptide may completely determine a protein's structure in a given environment.

KEY TERMS

affinity chromatography
α helix
antibody
β sheet
β structure
binding assay
centrifugation
chromatography
collagen helix
column chromatography
condensation
conformation
crystal
dehydration condensation
diffract
domain
electrophoresis
enzyme
enzyme assay
fibrous protein
globular protein
glycosidase
glycosidic bond
high performance liquid chromatography (HPLC)
hydrolysis
immunoaffinity chromatography
immunoblotting
ion exchange chromatography
isoelectric focusing
native protein
nuclease
phosphodiester
polynucleotide
polypeptide
primary structure
protease
quaternary structure
residue
ribbon model
secondary structure
specific activity
tertiary structure
two-dimensional separation
western blotting

QUESTIONS FOR REVIEW AND UNDERSTANDING

1. What are the three major classes of macromolecules and what are their functions in the cell?
2. Assembly of macromolecules is accomplished via a ________ reaction, which eliminates a molecule of ________. This reaction links together individual subunits, or ________, which together form the long ________. The ________ store energy and can also form part of the structural support of the cell. The cell is able to harvest this energy via a ________ reaction, which cleaves the ________ bond. Polysaccharides, which are made of units of ________, are the only polymers that can be ________; all other macromolecules are ________ in structure. Proteins, which can serve as structural supports as well as performing many of the functions of the cell, are polymers of ________, which are held together via ________ bonds. Proteins can be hydrolyzed by enzymes known as

______. Genetic information is stored in polymers of ______ residues, which are held together by ______ bonds. This bond can be cleaved by ______.

3. Define the term "specific activity."
4. Describe and explain four methods of separating or purifying proteins.
5. Explain why cellulose can serve as a low-calorie food additive for humans. What organisms find such food nutritionally useful and why?
6. Define the terms "hydrophilic" and "hydrophobic" and explain how they relate to the structures formed by proteins.
7. Both antibodies and enzymes are proteins, but in what way are they different?
8. Explain why, in terms of their functions, the order and identity of residues are less critical in polysaccharides than in proteins and nucleic acids.
9. Of the three macromolecules, proteins may be the most versatile. Discuss some of the jobs proteins do.
10. Describe the four levels of structure that occur in protein folding. Name the chemical bonds involved and describe how each contributes to the formation of these structures.
11. Discuss how the structures of proteins that function as structural components differ from those that act as enzymes. How are these different structures created?
12. Why is it critical for a protein or a region of DNA to be pure for sequencing?
13. How would you go about experimentally testing an hypothesis that suggests that a disease such as sickle cell anemia is inherited and is caused by a single amino acid change? Describe the techniques and materials you would use.
14. Discuss how Linus Pauling's model building and Max Perutz' x-ray crystallography data came together in a classic example of how models and hypotheses inform one another.
15. Adopt the philosophy of a reductionist and defend the way we study proteins today. How was this view successful in earlier studies that demonstrated that proteins carried specific information and were not simply random polymers?

SUGGESTED READINGS

BRANDON, C. and J. Tooze, *Introduction to Protein Structure,* Garland, New York, 1991. A fine introduction to protein structure.

DICKERSON, R. E. and I. Geis, *Proteins: Structure, Function, and Evolution,* Benjamin-Cummings, Menlo Park, 1982. An introduction to protein structure with exceptional illustrations.

DICKERSON, R. E. and I. Geis, *Hemoglobin: Structure, Function, Evolution, and Pathology,* Benjamin-Cummings, Menlo Park, 1983. A close look at the hemoglobin molecule.

DOOLITTLE, R. F., "Proteins," *Scientific American,* October 1985. A well illustrated introduction to protein structure. Part of an entire issue of *Scientific American* devoted to the molecules of life.

GARRETT, R. H. and C. M. Grisham, *Biochemisty,* Saunders College Publishing, Philadelphia, 1995. Chapter 4 covers much of the material in this chapter.

STRYER, L., *Biochemistry,* 3rd edition, W. H. Freeman, New York, 1988. Chapters 2 and 3 cover much of the material in this chapter.

VOET, D. and J. G. Voet, *Biochemistry,* John Wiley & Sons, New York, 1990. Another fine biochemistry textbook.

C H A P T E R 4

Directions and Rates of Biochemical Processes

KEY CONCEPTS

- The Second Law of Thermodynamics predicts the direction of any process. Less energy becomes available for work as a reaction or process occurs.
- Chemical reactions depend on the random movements of molecules. The rate of a chemical reaction depends on the activation energy needed to rearrange chemical bonds.
- Enzymes speed up biochemical reactions by lowering their activation energies. Enzymes orient substrates, strain substrates so that they approximate the reaction's transition state, and, in many cases, form transient covalent bonds with the substrate.
- Cells and organisms regulate their own metabolism by changing the shapes and activities of enzymes.

OUTLINE

ESSAYS

Understanding a biological or chemical process (or any process, for that matter) requires more than noticing changes in structures. We also want to know the answers to two general questions: (1) *Which way* (what will happen)? and (2) *how fast* (when)? In this chapter, we see how scientists approach these questions for biochemical reactions. Answering these two questions requires different kinds of thinking involving two fields of study, called thermodynamics and kinetics.

Every event, every reaction, every process is accompanied by transformation of **energy,** the capacity to do work, to move an object against an opposing force. A car moves forward as it converts the chemical energy of gasoline into the energy of motion; a cell moves (or changes its molecular composition) as it harnesses the chemical energy stored in ATP. The direction of any energy transformation depends on general rules derived from the study of **thermodynamics** [Greek, *thermo* = heat + *dynamis* = power, force], the study of the relationships among different forms of energy. From the rules of thermodynamics, a scientist can predict *which way* a process or a chemical reaction will run (its *direction*) without knowing how it takes place. We know, for example, that water always flows downhill, even if we do not know whether it gets there through a pipe, a stream, or a waterfall.

On the other hand, to know *how fast* a reaction occurs we must know something about the process itself. Water moves downhill more quickly in a waterfall than in a gentle stream and faster through a big pipe than through a small one. The study of the *rates* of reactions is called **kinetics** [Greek, *kinetikos* = moving].

(a) Compressing a spring

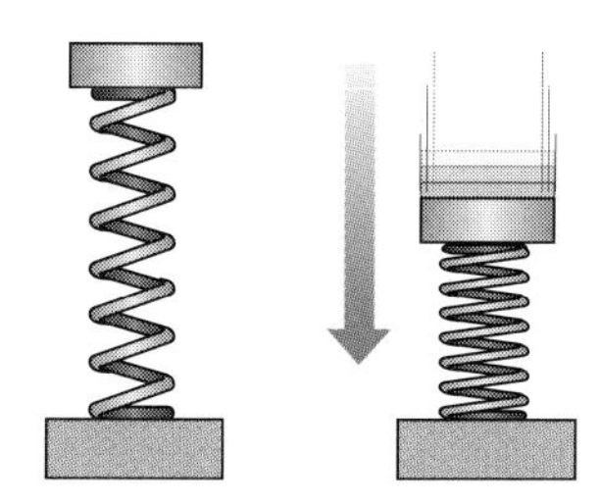

(b) Compressing phosphate groups

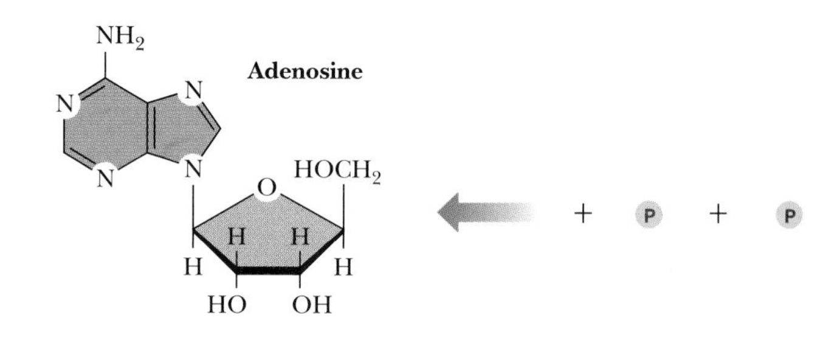

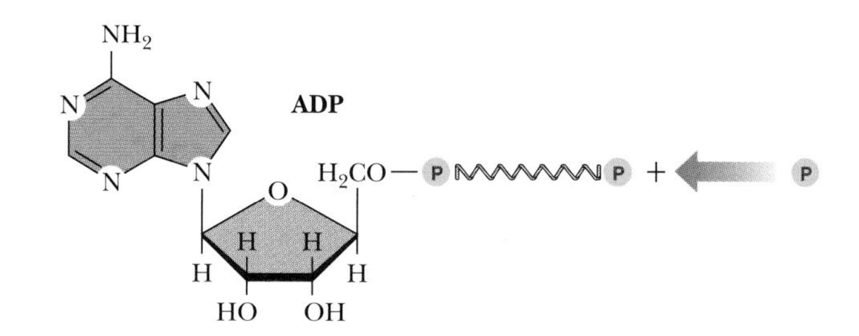

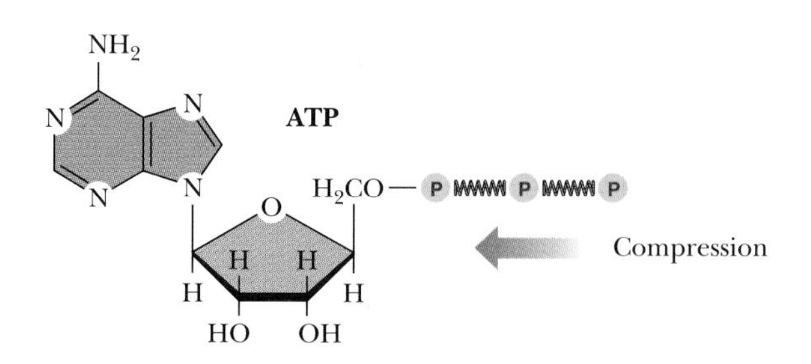

Figure 4-1 Potential energy stored in **(a)** the compression of a spring; **(b)** the squeezing together of negatively charged phosphate groups in ATP.

WHICH WAY DOES A PROCESS PROCEED?

Everyone knows the direction of time's arrow: A waterfall goes down, not up; a balloon bursts but does not spontaneously reassemble; a rubber ball bounces down and not up a flight of stairs. Events flow toward lower energy states or toward more disorder. Biological processes also have direction: A muscle contracts; a fertilized egg begins to divide; dead cells decay.

The key to predicting direction lies in understanding the conversions of energy. In any process, energy changes form. The study of thermodynamics has provided rules that govern all energy changes, both biological and nonbiological. These rules establish time's arrow, the direction of events.

Energy Changes Form But Is Never Lost

Physicists define **work** as the movement of an object against a force. A physics student can calculate, for example, the amount of work required to lift a brick or a ball or a pail of water against the force of gravity, or to compress a spring, as illustrated in Figure 4-1a. All these tasks are forms of **mechanical work.** Cell biologists are also concerned with other forms of work: **chemical work,** the formation of chemical bonds; **concentration work,** bringing a substance to a different concentration in one region from that in the surroundings; and **electrical work,** changing the separation of charges.

Rearranging atoms in a chemical reaction is also a form of work, as illustrated by the formation of ATP (adenosine triphosphate), the energy-rich molecule introduced in Chapter 1. ATP is a nucleotide that consists of an adenosine molecule linked to three phosphate groups, as illustrated in Figure 4-1b. Making a molecule of ATP requires the squeezing together of three negatively charged groups, working against an electrical force.

Both a compressed spring and the electrical repulsion within an ATP molecule can perform further work of their own. Releasing the spring can propel a ball, and splitting ATP can move a muscle, converting the work of compressing the spring or of forming ATP into **kinetic energy,** the energy of moving objects. **Heat** is a form of kinetic energy—that of moving molecules. **Potential energy** is a general term for energy that can ultimately be converted into kinetic energy. Potential energy can exist in many forms, including gravitational, electrical, and chemical. A compressed spring and the compressed charges of an ATP molecule both contain potential energy.

Scientists and engineers first began to understand the relationship among different forms of energy in the late 18th century. The stimulus for the development of thermodynamics was the widening use of the steam engine, which converted the chemical energy of coal into mechanical energy. The same rules that apply to steam engines apply to the energy conversions in organisms, such as the contraction of a muscle or the swimming of a protozoan.

One of the early and major conclusions of thermodynamics was that *the total amount of energy stays constant in any process.* This statement is called the **First Law of Thermodynamics.** Energy is neither lost nor gained—it only changes form. A cell lives by using the available energy in food or sunlight to carry out the energy-requiring processes of organization and reproduction.

In Any Process, Less Energy Becomes Available for Work

The **Second Law of Thermodynamics** states that, *although the total energy in the universe does not change, less and less energy remains available to do work.* The Second Law is the basis for the direction of time's arrow. That is, we perceive time as proceeding as the world moves from order to disorder, from more usable energy to less. As time passes, for example, a car converts more and more of the energy stored in gasoline into motion and heat, and less and less fuel remains in the tank.

Let us see how these two laws apply to the movement of the *Paramecium* shown in Figure 4-2. Covering the surface of the *Paramecium* are rows of filamentous protein assemblies called **cilia** [singular, **cilium,** Latin, eyelid, from the hairlike appearance of a cilium]. The cilia perform work: They move the *Paramecium* through the surrounding medium by acting like little oars.

The energy for this work comes from the chemical energy in ATP. The ATP molecule gives up part of its stored energy by splitting off one of its three phosphate groups.

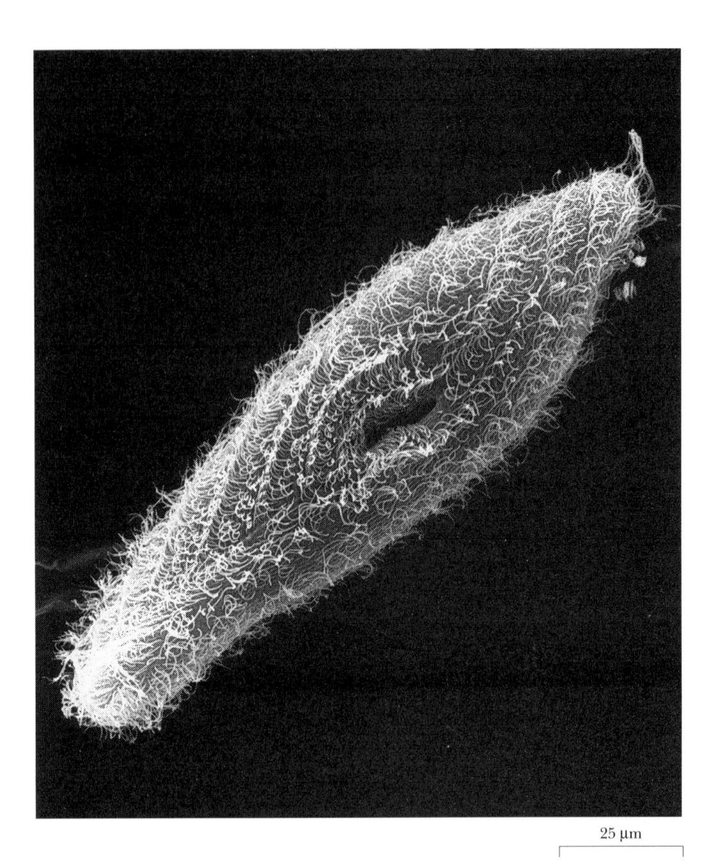

Figure 4-2 Cilia, which line the surface of *Paramecium,* act as little oars that propel this single-celled organism through its liquid surroundings. *(Karl Aufderheide/Visuals Unlimited)*

The remaining molecule has only two phosphate groups and is called **adenosine diphosphate,** or **ADP** [Greek, *di* = two]. ADP is still a strained molecule with stored energy, but it has less energy than ATP. The energy released performs work: It changes the arrangement of the proteins in the cilia and causes movement. Once the ATP is used up, the *Paramecium* can no longer move unless it takes in more energy (from food) and makes more ATP. (Cell movement is discussed further in Chapters 6 and 18.)

Because the First Law says that the total amount of energy does not change, the chemical energy released by ATP exactly equals the work done by the moving *Paramecium* plus any heat released in the process. According to the Second Law, the energy released from the splitting of ATP is greater than the work done in movement.

Cell movement converts only a fraction of the energy into work, the remaining energy going into heat (the kinetic energy of molecules). This apparent waste is not undesirable, however, because it establishes the direction of the whole process. The reverse process—the production of ATP from heat—does not happen because random molecular movements only rarely act in a concerted manner.

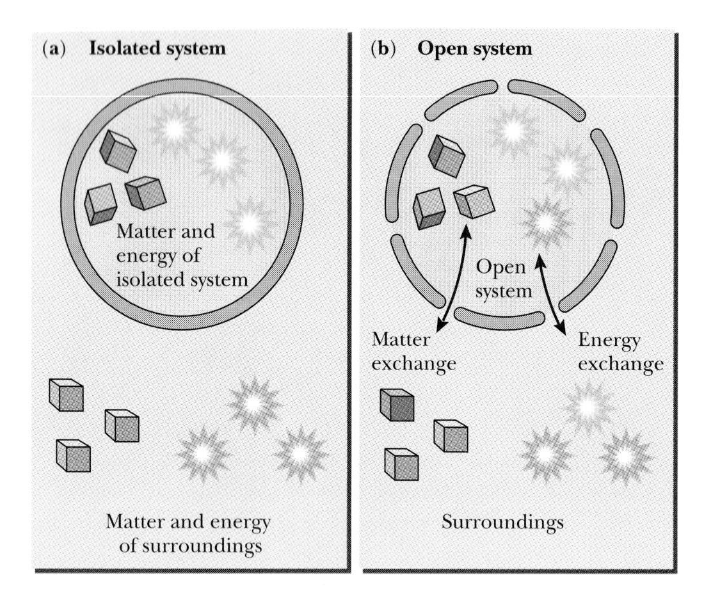

Figure 4-3 Open and isolated systems: **(a)** In an isolated system matter and energy *do not* exchange with matter and energy outside; **(b)** in an open system, matter and energy *do* exchange.

The First Law applies both to the universe as a whole and to an **isolated system,** any region that does not exchange matter or energy with its surroundings. As illustrated in Figure 4-3, however, organisms and cells are actually **open systems** because they can exchange both materials and energy with their surroundings. Strictly speaking, the laws of thermodynamics apply only to isolated systems, but in fact they also help explain the workings of open systems, including cells and organisms.

The First Law implies that all forms of energy—heat and work, chemical, electrical, and mechanical—are interchangeable and can be measured in the same units. The measure of energy that we use in this book is the **calorie,** the amount of energy needed to raise the temperature of 1 gram of water 1°C. The energy contained in food is often reported in Calories (with a capital C). One Calorie is equal to 1000 calories, or 1 kilocalorie (kcal). The daily diet of an average adult in the United States usually contains about 3000 kcal of energy.

The Second Law tells us that, in both biological and nonbiological processes, less and less energy remains available for work. Heat, despite its ability to power steam engines, is actually the least useful form of energy. For many processes, then, the answer to *"which way?"* is that a process goes in the direction that releases rather than absorbs heat. A red-hot steel ingot sitting at room temperature cools down; it does not heat up.

For a chemical reaction, however, the predictions of the Second Law are more complicated than for a mechanical system such as a steel ingot. The formation of any chemical bond between two atoms releases energy, while the breaking of a chemical bond requires energy. But the amount of energy released by forming a chemical bond varies, so the overall change in energy in a reaction depends on which bonds are broken and which bonds are formed. In addition, chemical reactions, like other processes, generate (or absorb) heat. In splitting ATP, for example, the formation of new chemical bonds releases more energy than the breaking of bonds, so the net result (as described later in this chapter) is the release of energy, partly as heat.

To predict the direction of a chemical reaction, we need an overall measure of the availability of energy. **Free energy** is a measure of available energy under the conditions of a biochemical reaction. Free energy is abbreviated ***G,*** after the American thermodynamicist J. Willard *G*ibbs, who single-handedly worked out the theory of chemical thermodynamics in the 1870s. Every chemical reaction has a measurable change in free energy, abbreviated ΔG, where Δ (delta) is the mathematical symbol for the change in a quantity; so $\Delta G = G_{products} - G_{reactants}$. The Second Law indicates whether any process occurs *spontaneously,* that is, without any external input of energy. In any spontaneous reaction, less energy remains available for work after the reaction, meaning that free energy decreases: $G_{products}$ must be less than $G_{reactants}$, and ΔG must be negative.

In any process (spontaneous or not), the change in free energy depends only on the initial and final states. The change (ΔG) does not depend on the way the process occurs or on how long it takes. A ball on a staircase may fall down the whole staircase in one big bounce or in many smaller bounces. The ball may sit unmoving on the top step for years until it is jostled by an earthquake, a passing truck, or a curious child. Similarly, thermodynamics cannot predict how long it will take for a muscle cell to break down ATP into ADP and phosphate. All that thermodynamics tells us is that, if the ball ever moves, it will eventually bounce down (not up) and that ATP will eventually form ADP and phosphate.

Processes tend to go "downhill," from higher to lower free energy. A process is said to be **exergonic** (Figure 4-4a) when free energy decreases and **endergonic** when free energy increases, as shown in Figure 4-4b. An exergonic reaction has a *negative* ΔG and occurs spontaneously (given enough time). An endergonic reaction has a *positive* ΔG and does *not* occur spontaneously.

Exergonic chemical reactions usually produce more low-energy chemical bonds, while the formation of high-energy chemical bonds requires an endergonic reaction. To restate the Second Law in another way, an exergonic reaction occurs spontaneously, while an endergonic reaction requires energy from some other source, that is, from some exergonic reaction.

Rearrangements of Chemical Bonds May Require or Release Energy

When two isolated atoms form a chemical bond, energy is released. The tighter the bond, the more energy is released in its formation. But chemical reactions rarely form

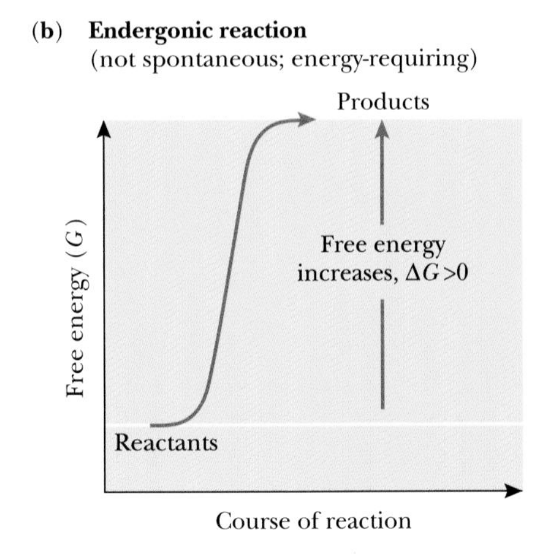

Figure 4-4 Exergonic and endergonic reactions: **(a)** Free energy decreases in an exergonic reaction, and the reaction will occur spontaneously; **(b)** free energy increases in an endergonic reaction, and the reaction will not occur spontaneously.

bonds from unbonded atoms. Instead, a chemical reaction leads to the breaking of some bonds (in the reactants), and the formation of other bonds (in the products). The conversion of ATP to ADP (Figure 4-5) illustrates this point. The actual reaction is

$$\text{ATP} + \text{water} \longrightarrow \text{ADP} + \text{phosphate}$$

Here, the oxygen atom of a water molecule displaces an oxygen atom in the third phosphate group of ATP. The phosphate group leaves the ATP molecule, leaving behind an ADP molecule and acquiring another H atom on one of its oxygen atoms. Two bonds break: a P—O bond in ATP and an O—H in water. And a new bond forms: a P—O bond in phosphate. The new P—O bond is more stable than the old one. So the free energy of the products (ADP and phosphate) is less than the free energy of the reactants (ATP and water), meaning that ADP has less capacity to perform work than ATP. The ΔG for this reaction is negative.

The direction of the reaction favors the breaking of less stable bonds and the forming of more stable bonds. Biochemists speak of **high-energy bonds** as relatively unstable chemical bonds that give up energy as new, more

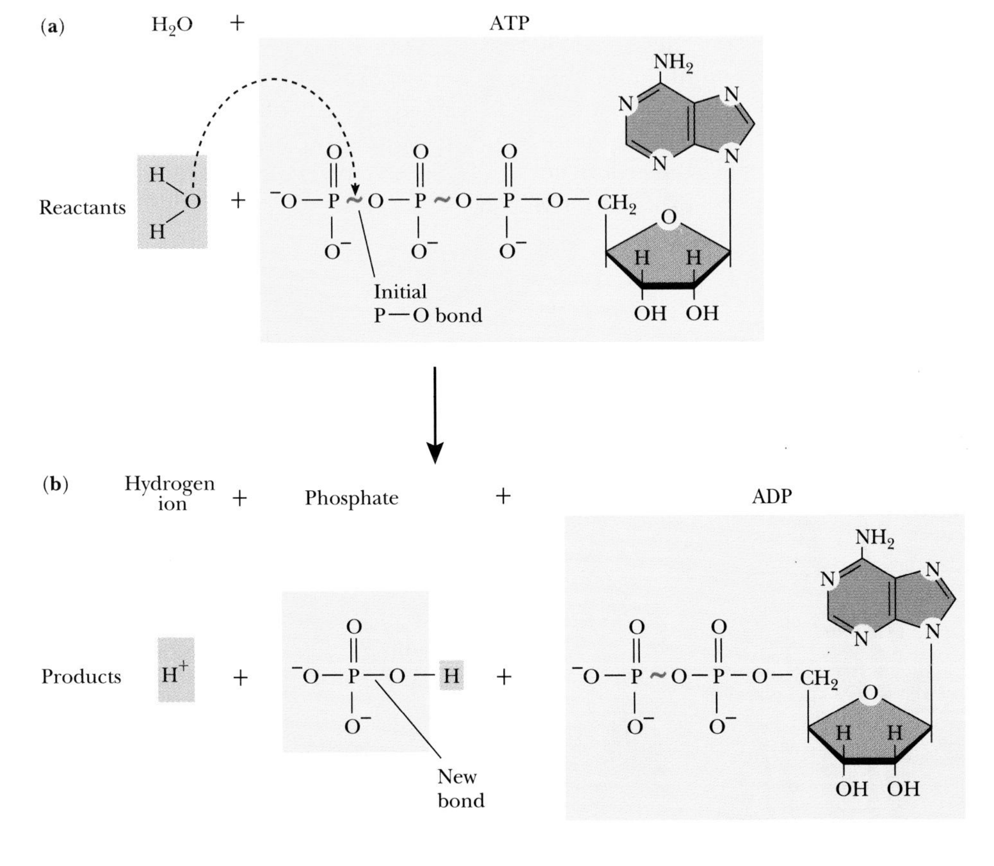

Figure 4-5 Splitting of a high-energy bond in ATP. **(a)** Attack of a water molecule breaks the bond between two phosphate groups; **(b)** the hydrolysis of the phosphate bond allows the negatively charged phosphate group to move away from the negatively charged ADP molecule, releasing energy.

stable bonds form. In contrast, **low-energy bonds** are more stable. The bonds that hold the second and third phosphates of ATP are high-energy bonds. Biochemists sometimes represent such high-energy bonds as squiggles, as shown in Figure 4-5.

Chemists (and biochemists) can measure the free energy changes associated with the making and breaking of chemical bonds. These changes determine the direction of a chemical reaction. As we discuss later in this chapter, free energy change is a function of both the particular molecules involved and their concentrations. To compare different chemical reactions, then, we must specify the standard free energy change, the free energy change at some standard concentrations. For biochemists, the standard free energy change is designated $\Delta G^{0\prime}$ [pronounced "delta G nought prime"], where the superscript ^{0} means that the free energy change is calculated for standard concentrations (1 mole/liter) of reactants and products, under certain defined conditions (25°C and atmospheric pressure). The ′ [prime] means that the free energy change is determined for pH 7, that is, at a hydrogen ion concentration of 10^{-7} M, instead of at 1 M, as is the case for the other reactants and products. For example, $\Delta G^{0\prime}$ for the conversion of ATP into ADP and phosphate is −7.3 kcal/mole. The minus sign shows that the reaction is exergonic, that the free energy decreases.

Exergonic Processes Can Drive Endergonic Processes

We can see how an exergonic process can provide the energy for an endergonic process by using the example of two balls on a staircase. Imagine (as illustrated in Figure 4-6) that we arrange ropes and pulleys so that the falling of one ball lifts another. If the balls are of equal weight and the steps of equal height (as in Figure 4-6a), the process could easily go in either direction: The falling of ball A could lift ball B, or the falling of ball B could lift ball A. On the other hand, if ball A were heavier than ball B (as in Figure 4-6b), then it would lose more energy in dropping a step than ball B would gain in going up a step. The excess energy would always drive the system in the same direction—the lifting of ball B.

The Second Law requires only that the combined process (the falling of A and the lifting of B) is exergonic. The linking of the endergonic reaction (the lifting of the lighter B) to an exergonic reaction (the falling of the heavier A) not only allows the endergonic reaction to occur but ensures that it always takes place in the same direction (A down and B up). Such linkage (or "coupling") allows an exergonic reaction to drive an endergonic reaction. Coupling is common in cellular reactions, as we see in Chapters 7 and 8. The benefit of coupling is that it fixes the direction of the combined reaction; the cost of coupling is that less free energy is available from the exergonic reaction to do other work.

Changes in free energy of coupled reactions also determine the direction of biological processes. In each case, the coupled biochemical reactions must—all together—be exergonic. For most of the biological processes discussed in this book, the driving exergonic reaction is the splitting of ATP.

How does a cell accomplish the coupling of endergonic and exergonic reactions? The answer lies in the ability of cells to speed up particular chemical reactions selectively. Just as the arrangement of a rope and pulley determines whether the falling of a ball will lift another, cellular mechanisms can couple two chemical reactions. The agents by which cells accomplish this coupling are large molecules called **enzymes,** which speed up specific chemical reactions, as discussed on page 99. Enzymes do not alter the free energy change in a reaction, but each kind of enzyme speeds up a particular reaction. For example, an enzyme can couple two reactions by allowing the product of an endergonic reaction to participate in an exergonic reaction, so the two reactions together are ex-

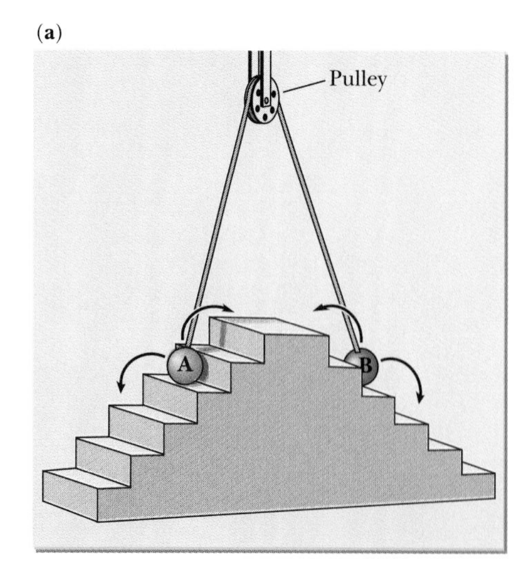

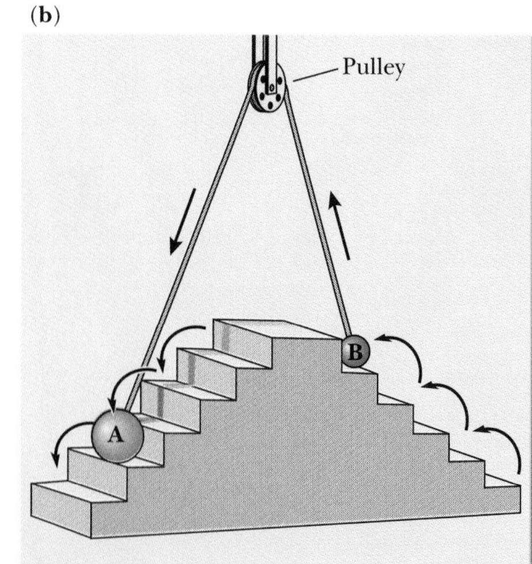

Figure 4-6 Coupling of endergonic and exergonic processes. **(a)** With balls of equal weight, the reaction can go in either direction: energy released in the exergonic falling of A can be coupled to the endergonic lifting of B, or energy released in the exergonic falling of B can be coupled to the endergonic lifting of A; **(b)** when A is heavier than B, A will always fall and B, if coupled, will always rise. The exergonic falling of A establishes the direction of the coupled reaction.

ergonic. In a swimming *Paramecium,* for example, an enzyme (called dynein) couples the hydrolysis of ATP to the movement of cilia.

The breakdown of food molecules within cells releases more free energy than the production of ATP, so thermodynamics predicts that, given the proper coupling, cells will make ATP. But how do food molecules get their energy in the first place? The answer is that the energy must be provided from elsewhere. The synthesis of sugars such as glucose and of other energy-rich food molecules must be coupled to some exergonic process. This energy comes from powerful nuclear reactions in the sun, resulting in sunlight, which ultimately powers life on our planet. We discuss in Chapter 8 just how this is done.

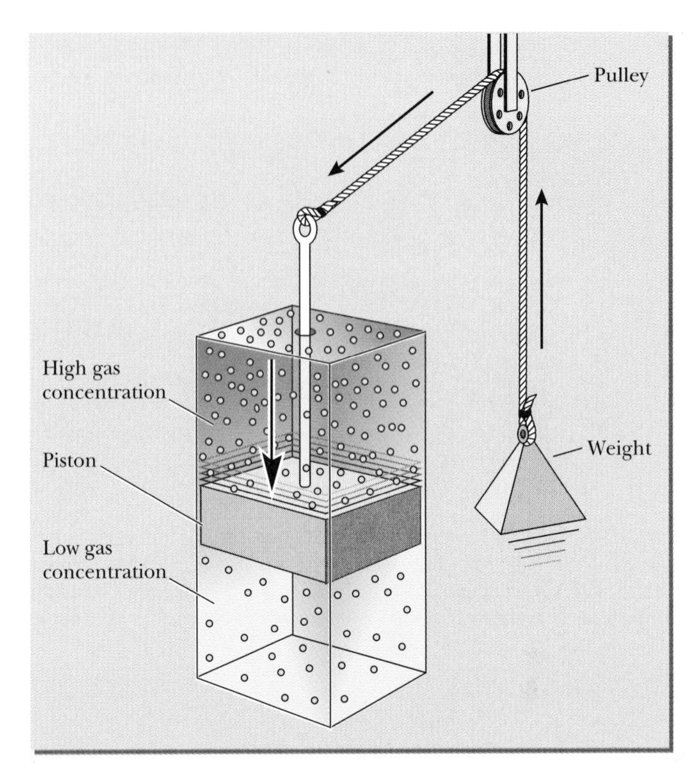

Figure 4-7 Differences in concentration as potential energy: The higher concentration of gas molecules in the upper chamber can perform work, here by lifting a weight.

How Does the Concentration of a Substance Affect Its Free Energy?

A popular way of stating the Second Law of Thermodynamics is that the order of the universe is always decreasing. Any process that converts an orderly arrangement of the world to a less orderly one can perform work. Differences in concentration of any substance (for example, between the inside and outside of a cell) represent a more ordered arrangement than the homogeneous distribution of that substance. Such concentration differences can drive useful work. For example, as shown in Figure 4-7, the differences in the concentration of gas molecules on the two sides of a box could be used to drive a piston, which in turn could lift a weight. Within cells, concentration differences (particularly of hydrogen ions) are an important source of energy, and such concentration differences are actually used to produce ATP.

The close relationship between order and energy leads to an important statement about cells (and about life in general): Cells (and organisms) need energy to maintain organization. A simple example of the increased order within a cell is the difference in concentrations of sodium ions (Na^+) inside and outside of red blood cells. The Na^+ concentration outside a red blood cell (in the liquid part of the blood) is 0.145 M (or 145 millimolar = 145 mM). Inside the cell, the concentration is 12 mM. The cell maintains this difference in concentration by pumping Na^+ ions across the cell membrane, using ATP as an energy source.

Cells must perform work to keep their internal environments constant. When no energy supply is available to perform this work, the differences between the inside and the outside of the cell disappear. Such a failure to maintain a constant internal environment means that the cell dies.

Because the movement of molecules from a more concentrated to a less concentrated state can perform work, free energy also depends on concentration. This means that the actual free energy change in any chemical reaction depends not only on bond energies but on the concentrations of both reactants and products.

Imagine a chemical reaction in which a single reactant A is converted to a single product B:

$$A \rightleftharpoons B$$

If we begin with a solution of A alone, B accumulates as the reaction proceeds. At first, the free energy of A is greater than that of B because of its much greater concentration, as illustrated in Figure 4-8a and b. As the reaction proceeds, the difference in energy between the reactants and the products decreases. The reaction becomes less and less "downhill," until, finally, the free energy of B equals the free energy of A (and ΔG is zero). At that point, the reaction has reached **equilibrium,** the point at which no further net conversion of reactants and products takes place.

A specific example of a chemical reaction involving a single reactant and product is the conversion, diagrammed in Figure 4-9, of 3-phosphoglycerate (3PG) to 2-phosphogylcerate (2PG), whose cellular significance is discussed in Chapter 7. We can guess (correctly) that the free energy of a 1 M solution of 2PG is greater than that of a 1 M solution of 3PG because the negatively charged phosphate group is closer to the negatively charged carboxyl group in 2PG than in 3PG. So $\Delta G^{0\prime}$, the standard free

(a)

Free energy (G) ⟶

Early in the reaction, ΔG is large

As reaction proceeds, ΔG decreases

At equilibrium, $\Delta G = G - G_{eq} = 0$

ΔG

ΔG

Progress of reaction ⟶ $\Delta G = 0$

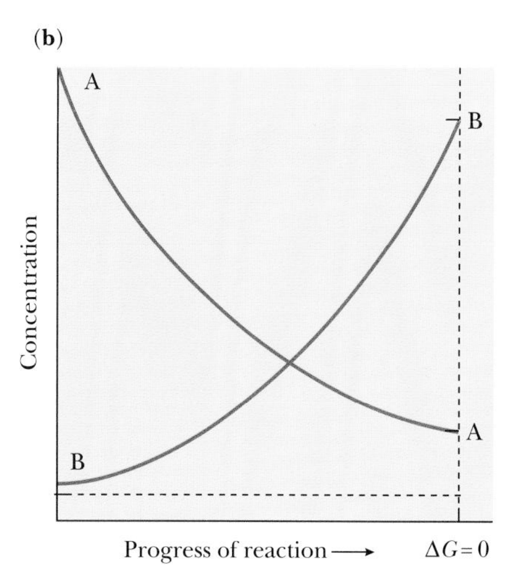

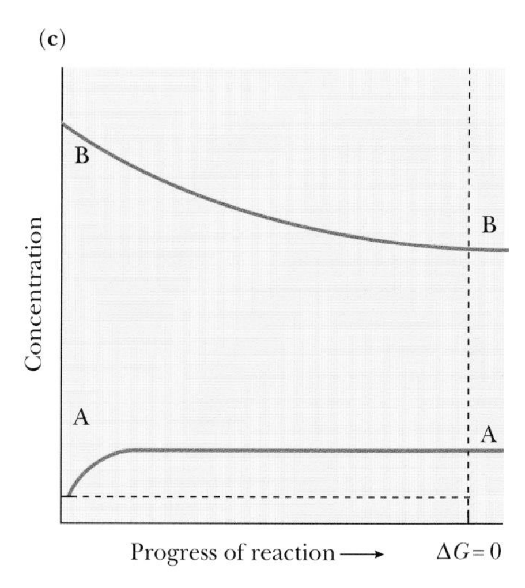

Figure 4-8 As a reaction proceeds, free energy decreases. At equilibrium, $G_A = G_B$ and $\Delta G = 0$. **(a)** Free energy decreases as the reaction progresses; **(b, c)** concentration changes in the reaction A ⟶ B. In (b), the initial concentration of A is much greater than B; in (c), the initial concentration of B is higher than at equilibrium.

energy change, is positive for the forward reaction (3PG ⟶ 2PG). If, however, we begin with a solution of 2PG, the conversion to 3PG has a negative free energy change.

Thermodynamicists, starting with Gibbs, have determined the quantitative relationship between a compound's free energy and its concentration. Biochemists usually use G' (and $\Delta G'$), meaning the free energy (and free energy changes) in aqueous solution at pH 7, atmospheric pressure, and 25°C. We know from the statements earlier in this chapter that free energy increases with concentration. The actual relationship is

$$G_A' = G_A^{0\prime} + RT \ln [A]$$

where G' is the free energy of substance A, $G_A^{0\prime}$ is its standard free energy (in a 1 M solution at 1 atmosphere pressure and 25°C), R is the gas constant (1.987 cal/mol/°K), T is the absolute temperature (°K, usually 298°K = 25°C), and ln [A] is the natural logarithm of its molar concentration. Similarly,

$$G_B' = G_B^{0\prime} + RT \ln [B]$$

For the reaction A ⇌ B, then,

$$\Delta G = G_B - G_A = G_B' - G_A' + RT \ln [B] - RT \ln [A]$$

Since $G_B' - G_A' = \Delta G^{0\prime}$, and $\ln [B] - \ln [A] = \ln \frac{[B]}{[A]}$

$$\Delta G = \Delta G^{0\prime} + RT \ln \frac{[B]}{[A]}$$

At equilibrium (as previously discussed and as illustrated in Figure 4-8), $\Delta G = 0$. So, for this reaction, the concentrations of A and B at equilibrium—$[A]_{eq}$ and $[B]_{eq}$—are given by

$$\Delta G = 0 = \Delta G^{0\prime} + RT \ln \frac{[B]_{eq}}{[A]_{eq}}$$

or

$$\Delta G^{0\prime} = -RT \ln \frac{[B]_{eq}}{[A]_{eq}}$$

3-Phosphoglycerate (3PG)		2-Phosphoglycerate (2PG)
COO^- / HCOH / $CH_2O\ PO_3^{2-}$	⇌	COO^- / $HCO\ PO_3^{2-}$ / CH_2OH

$\Delta G^{0\prime} = +1.1$ kcal

Figure 4-9 The conversion of 3-phosphoglycerate (3PG) to 2-phosphoglycerate (2PG) is endergonic, because the negatively charged phosphate moves closer to the negatively charged carboxyl group. $\Delta G^{0\prime} = +1.1$ kcal, corresponding to a final ratio of 3PG to 2PG of about 6 : 1.

For the conversion of 3PG to 2PG, $\Delta G^{0\prime}$ is 1.1 kcal/mol, so that at equilibrium, the relative concentrations are given by

$$\ln \frac{[\mathrm{B}]_{\mathrm{eq}}}{[\mathrm{A}]_{\mathrm{eq}}} = -\frac{\Delta G^{0\prime}}{RT} = -\frac{(1100 \text{ cal/mol})}{(1.987 \text{ cal/mol/°K})(298\text{°K})} = 1.858$$

So $[2\mathrm{PG}]/[3\mathrm{PG}] = [\mathrm{B}]_{\mathrm{eq}}/[\mathrm{A}]_{\mathrm{eq}} = e^{-1.858} = 0.156$. That is, at equilibrium, the concentration of 3PG is about six times that of 2PG.

This last concentration ratio (the concentration of product divided by the concentration of reactant) is the reaction's **equilibrium constant (K_{eq})**. When there is more than one product or reactant, the equilibrium constant is

$$K_{\mathrm{eq}} = \frac{[\text{Product 1}] \times [\text{Product 2}] \times \cdots}{[\text{Reactant 1}] \times [\text{Reactant 2}] \times \cdots}$$

When a reaction involves more than one product or reactant, the formula is slightly more complicated, but that situation is beyond the scope of our discussion. The reactions with which we are most concerned usually involve no more than two reactants and two products. In every case, the same relationship exists between the equilibrium constant and the free energy change:

$$\Delta G^{0\prime} = -RT \ln K_{\mathrm{eq}}$$

The more negative the standard free energy change of the reaction, the greater the difference in the concentrations of the reactants and products at equilibrium. When, for example, $\Delta G^{0\prime}$ is -2.7 kcal, the concentration of B (the product) at equilibrium is about 100 times that of A (the reactant). If the reaction were still more exergonic, it would proceed still more to the right, and at equilibrium the ratio of B to A would be still higher. If, on the other hand, the reaction is less exergonic, a smaller fraction of the reactants is converted to products when the reaction reaches equilibrium. A reaction with a positive $\Delta G^{0\prime}$ gives a greater concentration of reactants than products at equilibrium, as was the case for the conversion of 3PG into 2PG.

Thermodynamics predicts where every equilibrium lies. But it gives no information about how long it takes to get there.

Disorder (Entropy) Increases in Spontaneous Processes

The Second Law says that the amount of available energy in an isolated system can only decrease. This means that the uncoordinated motion of atoms and molecules cannot spontaneously change back into motion in a single direction. For example, the random motion of the molecules of a rubber ball increases as the ball bounces. We would not expect all the molecules in the ball to start moving in a single direction to propel the ball higher than the preceding bounce. Nor would we expect the random movements of molecules to coordinate in such a way that ADP and phosphate would spontaneously form ATP and water. A ball will bounce higher, and ATP will form, only if we link these endergonic processes to exergonic processes. Examples of such linkage would be the ball's collision with a swinging bat or the coupling of ATP synthesis to the capture of light energy from the sun.

According to the Second Law, uncoordinated motion is more probable than coordinated motion. It is thus more likely that the pieces of a burst balloon will fly apart than that they will spontaneously reassemble. It is more likely that the books in an office will become disorganized than that they will remain neatly shelved in alphabetical order. The Second Law is a formal statement that disorder is more probable than order. The amount of disorder in the universe or in any isolated system tends to increase.

Physicists have defined a formal measure of disorder, called **entropy (S)**. Entropy has a high value when objects are disordered or distributed at random and a low value when they are ordered. Because disorder and random motion are more probable, the tendency of entropy to increase can be used to produce work, as in the device in Figure 4-7 that harnesses the differences in the concentrations of gases.

In calculating the free energy of a system, we must take entropy into account. Changes in entropy ultimately determine the direction of many chemical reactions. There is a trade-off, however, between entropy and energy, summarized in the equation:

$$\Delta G = \Delta H - T\Delta S$$

Where ΔH is the change in the heat content (or **enthalpy**) in going from reactants to products (roughly a measure of the change in bond energies), ΔS is the change in entropy or degree of disorganization, and T is the absolute temperature (in °K). A chemical reaction may be spontaneous (that is, exergonic, with ΔG less than zero) when it leads to (1) the formation of more stable chemical bonds; (2) an increase in entropy; or (3) a combination of both. It is the sum of these two changes that matters, so a decrease in ΔH may overcome a decrease in entropy in the same reaction, or an increase in entropy may outweigh an increase in ΔH.

The randomness or disorder in any isolated system (or in the universe as a whole) increases in any process, but order may be purchased by the expenditure of energy. The books in an office can be restored to alphabetical or-

Figure 4-10 A transient ordered structure, called a Bénard cell, that maximizes entropy production. Highly ordered structures may arise locally and transiently, even as entropy (disorder) increases in the universe. Cells and organisms are still more complicated examples of temporary local order in the face of globally increasing entropy. *(E. L. Koschmieder, University of Texas at Austin)*

der only by expending energy to put them in order on their proper shelves.

The Second Law of Thermodynamics establishes the direction of time's arrow and predicts that eventually the universe will run down, as useful energy becomes more and more unavailable and entropy increases. A remarkable feature of the world around us, however, is the widespread (but transient) appearance of highly ordered structures, such as cells and organisms or the inorganic example shown in Figure 4-10. Each of these structures is an open system, and each maintains temporary local order by exporting entropy to the rest of the universe.

Systems tend to produce entropy at a predictable rate, which is why people long used the consumption of a burning candle to measure time. Lars Onsager, a prominent thermodynamicist in the first half of the 20th century, noted that systems actually increase entropy at the maximum possible rate. Locally organized structures are devices for maximizing entropy production. Extending this view, we may view cells and the evolution of life as a way of accelerating the production of entropy because life processes move matter from states of low entropy to states of higher entropy.

WHAT DETERMINES THE VELOCITY OF A CHEMICAL REACTION?

Thermodynamics does not make any predictions about the rate at which a process occurs. In order to understand *how fast* a biochemical reaction occurs, we need to have a model of molecular interactions.

To help understand chemical reactions, we may borrow a fanciful but reasonably accurate model of randomly bouncing molecules. Picture molecules as tiny frogs, like those in Figure 4-11, constantly jumping about at random. Imagine individual frogs (molecules) jumping around at random in a box labeled "compartment A" and an adjacent empty box labeled "compartment B." Some frogs naturally jump into the B compartment. If compartment B is lower than compartment A, the frogs have a harder time

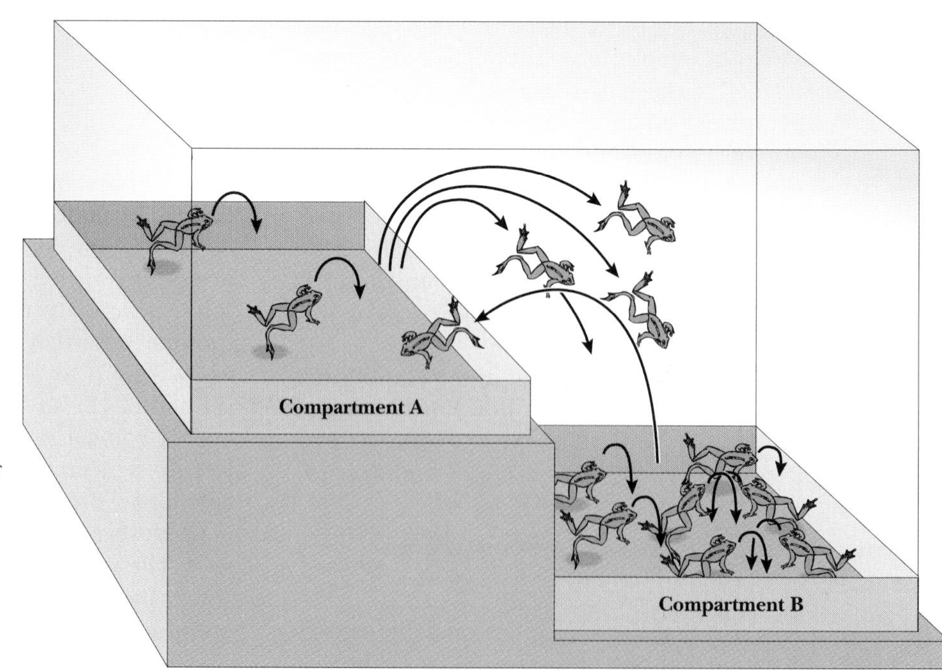

Figure 4-11 The random bouncing of molecules, represented as the random jumping of frogs. Initially, frogs are placed in compartment A; they then begin to jump over the barrier; at equilibrium more frogs end up in compartment B, which has a lower energy than compartment A.

going from B to A than from A to B. Once they arrive at the lower level, fewer are able to jump back to the A compartment—only those with sufficient jumping power, that is, those with higher kinetic energy.

The Random Bouncing of Molecules Helps Explain Chemical Equilibrium

The ideas that molecules are constantly in motion and that their behavior explains many of the properties of matter were developed in the 1870s by the Austrian physicist Ludwig Boltzmann. Boltzmann was the first to realize that temperature is a measure of the average kinetic energy of moving molecules. The higher the temperature, the faster molecules are moving on average, although at any given temperature, molecules have a range of kinetic energies. The range of kinetic energy in individual molecules means that, in our frog model, the frogs have a range of jumping ability: At any given temperature, some frogs jump higher than others. The greater the temperature, the more frogs are able to jump over a fixed barrier, like that pictured in Figure 4-11.

As time goes on, the frogs (molecules) come to equilibrium. That is, the number of frogs or molecules in each chamber remains the same because the number leaving each chamber equals the number entering that chamber. Each minute just as many frogs jump from A into B as jump from B into A. For this to happen, however, there must be more frogs in B than in A because a smaller fraction of the frogs in B will be able to make the uphill jump. The lower position of the B compartment represents the lower energy of the B compartment and means that more frogs will end up in the B compartment than in the A compartment. This model leads to the same conclusion as our thermodynamic view—that molecules move in the direction of lower free energy.

The jumping frog model also allows us to understand how the concentrations of reactants and products influence chemical equilibrium. The number of frogs jumping each minute is proportional to the total number of frogs in the compartment. We can call the fraction of frogs that make the jump from A to B in any minute k_{AB}. The number of frogs that jump from A to B in any given minute is $k_{AB} \times [A]$, where [A] is the concentration of frogs in A. The fraction of frogs making the reverse jump is k_{BA}, so the number of frogs that jump from B to A is $k_{BA} \times [B]$. At equilibrium, the same number of frogs jumps in each direction, so

$$k_{AB}\,[A] = k_{BA}\,[B]$$

or

$$\frac{[B]}{[A]} = \frac{k_{AB}}{k_{BA}}$$

Recall that the equilibrium constant (K_{eq}) of a chemical reaction is the concentration of the product [B] divided by the concentration of the reactant [A], so

$$K_{eq} = \frac{[B]}{[A]} = \frac{k_{AB}}{k_{BA}}$$

(As mentioned previously, when there is more than one product, the numerator is the product of their concentrations; when there is more than one reactant, the denominator is the product of their concentrations.)

The jumping frogs provide a way of understanding the relationship of the equilibrium constant to free energy change (although the thermodynamic conclusions do not depend on the specifics of the model). The greater the difference in height between A and B, the more frogs end up in the B compartment. If the standard free energy change is zero (that is, if the height of the two chambers is the same), the equilibrium constant is 1, and there will be the same number of frogs or molecules in each compartment.

How Does Molecular Motion Help Explain Reaction Rates?

Thermodynamics predicts the direction of a process, independent of the precise mechanism. Thermodynamics says only whether something will happen eventually. Because cells and organisms do not live forever, this kind of forecasting is not enough. We also want to know how long a process will take.

To determine how rapidly a process occurs, we must understand its mechanism. We need to know what actually happens, what "path" a process takes, not just the beginning and end states. For chemical reactions, we must go beyond discussing changes in bond energies and concentrations. We must try to understand the behavior of atoms and molecules. Although we have discussed the molecular interpretations of free energy and entropy, thermodynamic predictions do not depend on the behavior of atoms and molecules. Predicting the *rates* of processes, however, depends upon understanding molecular behavior.

Molecules and atoms are constantly in motion. When a substance absorbs heat, its temperature increases and the atoms and molecules jostle about more rapidly. This understanding of the movement of molecules and atoms is the basis of kinetic theory. Kinetic theory allows us to understand and predict both the rates of chemical reactions and the dependence of these rates on temperature and other environmental factors. This understanding permits us to appreciate the exquisite ways in which cells and organisms regulate their activities, speeding up different reactions under different conditions. After a meal, for example, our liver cells convert glucose and other sugars from our food into polysaccharides. Later, the same cells break down the stored polysaccharides into glucose, which circulates in the blood and provides energy to other cells in the body.

Chemical Reactions Require Molecular Movement

Every process, no matter how much it is favored thermodynamically, needs a "push" to get started. In chemical reactions, the energy comes from the random movements of molecules. Appreciation of the nature of these molecular "pushes" allowed scientists to understand the factors that determine *how fast* chemical processes occur.

The experiments that convinced scientists of the importance of molecular movement examined the rates of reactions at different temperatures. The higher the temperature (that is, the greater the energy of molecular movement), the more rapidly a reaction proceeds.

To understand the factors that influence the rates of chemical reactions, let us go back to the jumping frog model. We can represent the difficulty in starting a chemical reaction as the height of the barrier between the two compartments. In order to get from compartment A to compartment B, the frogs must first jump over the barrier, shown in Figure 4-12. The molecules or frogs must have a minimum jumping ability (a minimum energy) in order to do this. The minimum energy needed for a process to occur is called the **activation energy** and is usually abbreviated as $\Delta G^{\ddagger}$ (pronounced "delta *G* double dagger").

The basis of kinetic theory is the constant bouncing about of molecules. At any given temperature, the average energy of all the molecules in a system is fixed. The average kinetic energy of these molecular movements is directly proportional to the absolute temperature.

The number of molecules that are able to react depends on the temperature and the height of the activation energy barrier. The higher the temperature, the greater the fraction of molecules with enough energy to jump the barrier. The lower the activation energy, the greater the fraction of molecules that can jump the lower barrier, and the more rapidly equilibrium is achieved.

Laboratory and industrial chemists can speed up a specific chemical reaction by using a **catalyst,** a substance that lowers the activation energy of a reaction and is not itself consumed in that reaction. Cells and organisms also use catalysts, called **enzymes.** Enzymes are large molecules, almost always proteins. By using enzymes to lower the activation energy for specific reactions, cells can greatly speed up their reaction, even at the modest temperatures at which they live. Without enzymes, most of the chemical reactions necessary for life would occur too slowly to be of any use.

Although a catalyst speeds up a specific chemical reaction, each reaction produces the same free energy change (which is independent of the reaction rate). Because enzymes and other catalysts do not affect the free energy change of a reaction, they do not change the relative concentrations of reactants and products at equilibrium; they only decrease the time required to attain equilibrium.

The presence of an active enzyme allows a cell to use molecules such as ATP in specific ways, coupling the exergonic splitting of ATP to such endergonic reactions as cell movement, the synthesis of a macromolecule, or the pumping of ions through a membrane. Cells control their chemistry and their other activities by regulating the production and activity of individual enzymes.

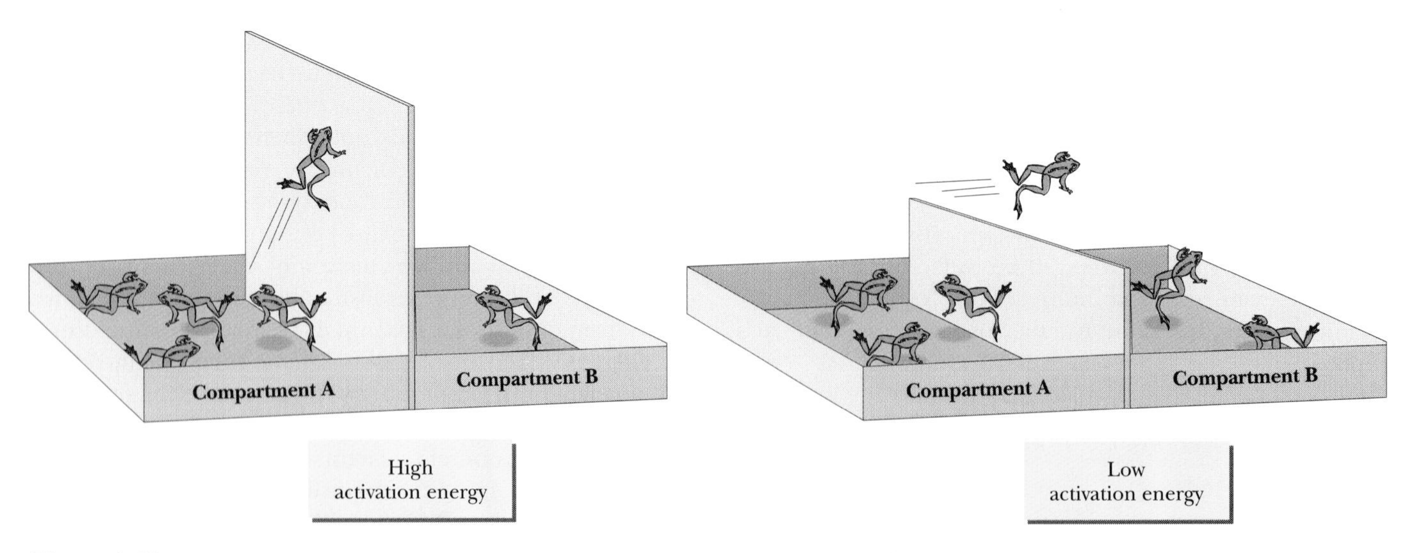

Figure 4-12 The height of the barrier between compartments A and B corresponds to the activation energy of a chemical reaction. The higher the barrier, the fewer frogs (or molecules) jump between compartments per second.

HOW DO ENZYMES ACT AS CATALYSTS?

Enzymes accelerate specific chemical reactions by specifically binding to the reacting molecules, which are called **substrates.** A substrate is usually (but not always) much smaller than an enzyme. The noncovalent (and sometimes covalent) interactions between enzyme and substrate are the basis for the remarkable properties of enzymes. These properties include their *catalytic power,* typically speeding up a reaction by a factor of 10^6 or more); their *specificity,* each enzyme (with a few exceptions) catalyzing only a single chemical reaction; and, in many cases, the *regulation* of cellular reactions in response to changing cellular conditions.

The substrate binds to an **active site,** a groove or cleft on the enzyme's surface, as represented in Figure 4-13a. The binding of a substrate to an enzyme depends on their having *complementary surfaces,* which fit together so that the "bumps" on one match the "hollows" on the other. Biochemists have often compared the matching of the surfaces of enzyme and substrate with the matching of a lock and a key. (We discuss a more modern view of enzyme-substrate interactions later in this chapter.)

The binding of substrate to enzyme results in a lowering of the activation energy for a particular chemical reaction. The substrate temporarily binds to the enzyme using a combination of many weak noncovalent bonds—hydrogen bonds, ionic bonds, hydrophobic interactions, and van der Waals attractions—and sometimes also by a covalent bond (as in a case discussed later in the chapter). In many cases, binding to the substrate changes the conformation of the enzyme itself, as illustrated in Figure 4-13b. When this occurs, the interaction of enzyme and substrate is called an **induced fit** (as if a lock changed to fit the key).

Binding of enzyme and substrate is reversible, and the reactant or product molecules quickly leave the active site and re-enter the solution. Some enzymes bind, process, and release substrate molecules more than 100,000 times a second. But, in order for catalysis to occur, the enzyme and substrate must closely interact, so that the rate of any biochemical reaction depends not only on the temperature and the concentration of reactants but also on the amount of enzyme present.

The Formation of an Enzyme-Substrate Complex Accounts for the Kinetics and Concentration Dependence of Enzymatic Reactions

What determines the **velocity** (the rate of appearance of product) of an enzymatic reaction? In an uncatalyzed chemical reaction, the velocity is proportional to the concentration of each reactant, as discussed on page 97 and shown diagrammatically in Figure 4-14a. For enzymatic reactions at particular enzyme concentrations, however, the reaction rate also increases in proportion to substrate concentration, but only to a certain point, where it reaches a maximum velocity as shown in Figure 4-14b. The maximum velocity is proportional to the amount of enzyme present. In 1913, the German chemist Leonor Michaelis proposed that the maximum velocity of an enzymatic reaction results from the necessary formation of an **enzyme-substrate complex,** the association of enzyme and substrate that we have already discussed. Michaelis's deduction preceded by half a century the physical demonstration of an enzyme-substrate complex using x-ray diffraction techniques.

Together with his assistant, Maud Menten, Michaelis formulated an extremely useful analysis of the reaction velocity (abbreviated V) of an enzymatic reaction, called

(Text continues on page 103.)

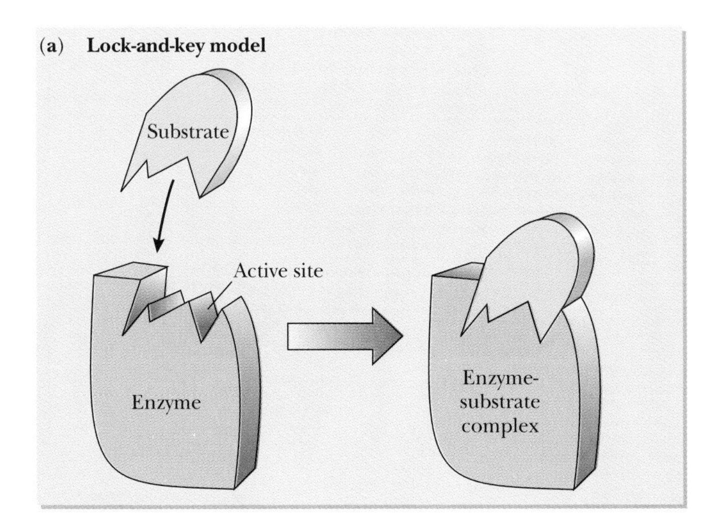

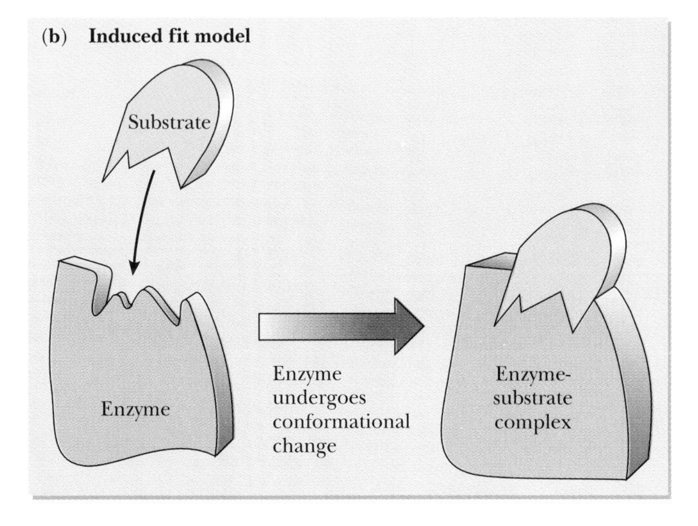

Figure 4-13 The binding of an enzyme to a substrate to form an enzyme-substrate complex. **(a)** The substrate fits into a complementary surface on the enzyme; **(b)** the binding of the substrate changes the structure of the enzyme.

ESSAY

ALMOST ALL ENZYMES ARE PROTEINS, BUT SOME ARE RNA

"Don't waste clean thinking on dirty enzymes!" This excellent advice continually rings in the ears of biochemists and biologists. But obtaining a "clean" enzyme—a pure preparation of a single type of catalytic molecule—is not easy. After all, cells contain thousands of types of macromolecules: Isolating just the one responsible for a particular type of catalysis is an immense challenge.

The first enzyme to be isolated in pure form was *urease,* which converts urea—a nitrogen-containing waste product—into ammonia and carbon dioxide, as shown in Figure 1a. Many plants use this reaction to obtain nitrogen from animal waste, and the enzyme is present in relatively high concentrations, for example, in beans. In the mid-1920s, James Sumner, a young professor at Cornell, purified urease from jack beans by allowing a concentrated and filtered extract to form crystals. The formation of crystals depends on the regular bonding patterns between molecules, so crystals usually contain only a single substance. These crystals, Sumner found, were almost entirely protein, and, when dissolved in water, they were powerful catalysts for the breakdown of urea. Sumner concluded that urease, an enzyme, is a protein.

Sumner's conclusions contradicted those of Richard Willstätter, an eminent German biochemist and Nobel laureate. (Willstätter's prestige and imposing presence contrasted markedly with those of the much younger Sumner, who had lost an arm in a shooting accident as a teenager.) Willstätter had also purified an enzyme, called *invertase,* which splits sucrose into glucose and fructose, as shown in Figure 1b. But he could not find any protein in his preparations. We now realize that Willstätter's assay for protein was simply not sensitive enough, but Willstätter concluded that enzymes "do not belong to any of the known groups of complicated cellular molecules." Instead, he argued, enzymes are small molecules somehow embedded in protein gels.

Willstätter's prestige was so great that Sumner's work was viewed with great skepticism. But two other biochemists, John Northrop and Moses Kunitz, at the Rockefeller Institute in New York, went on to crystallize and study a number of digestive enzymes. Their

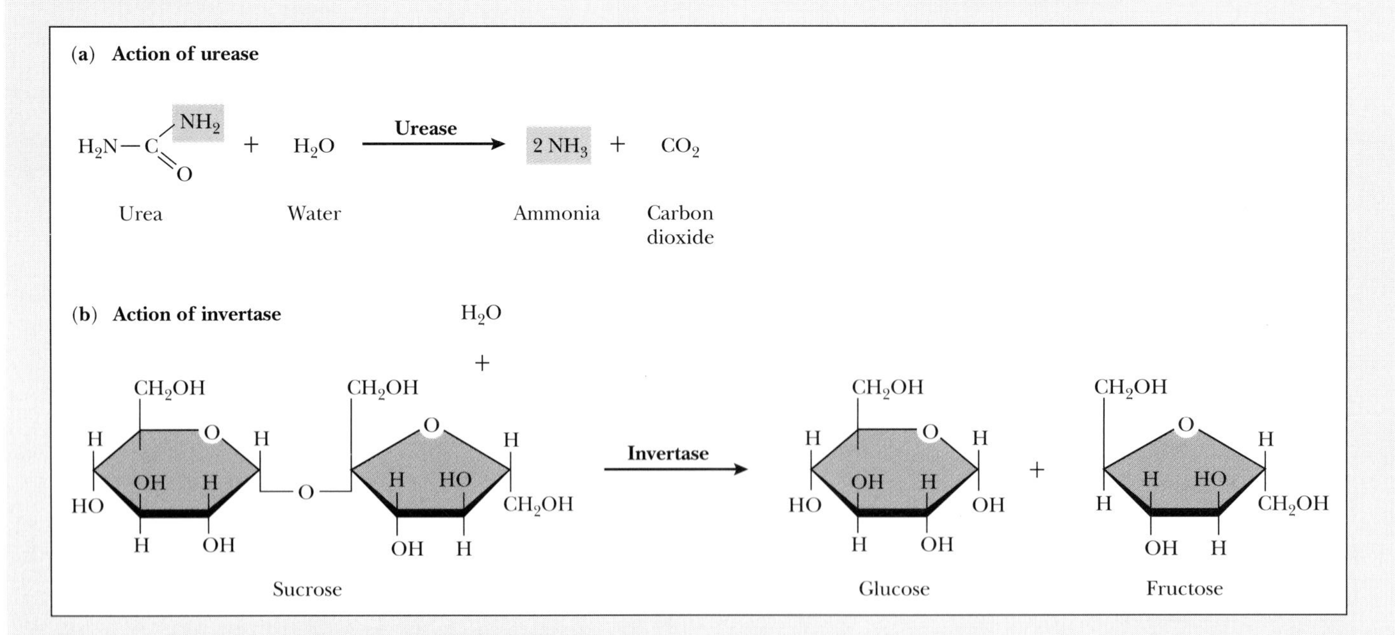

Figure 1 *Reactions catalyzed by urease and invertase.* ***(a)*** *Urease catalyzes the conversion of urea into ammonia and carbon dioxide;* ***(b)*** *invertase catalyzes the conversion of sucrose into glucose and fructose.*

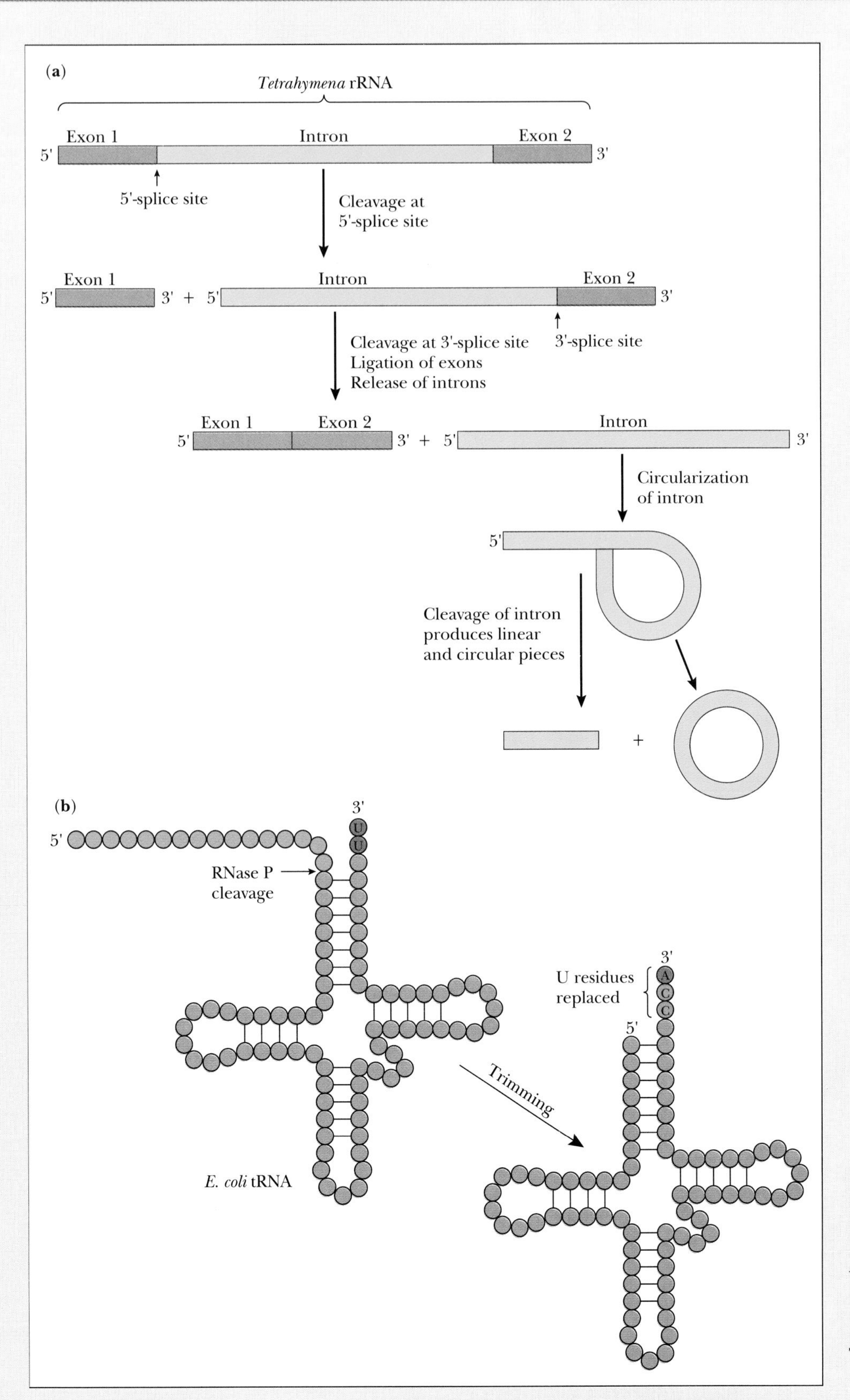

Figure 2 *Reactions catalyzed by naturally occurring ribozymes.* ***(a)*** *The self-catalyzed removal of a segment of an RNA molecule in* Tetrahymena; ***(b)*** *the trimming of an* E. coli *tRNA molecule.*

(continued)

ESSAY *continued*

work showed that these enzymes were proteins: Not only did the proteins crystallize, but enzymatic activity was always associated with single kinds of macromolecules during electrophoresis and centrifugation. (See Chapter 3, pp. 69–71.) In 1946, James Sumner and John Northrop shared the Nobel Prize for the discovery that enzymes are proteins.

This conclusion is not the end of the story, however. In the early 1980s, two young investigators studying extremely different organisms found evidence that RNA molecules could be cut by enzymes even in the absence of proteins, as illustrated in Figure 2. Tom Cech at the University of Colorado showed that a particular type of RNA molecule in the protist *Tetrahymena* accomplished the removal of a stretch of 419 nucleotide residues, apparently by itself—with no protein. About the same time, Sidney Altman at Yale showed that another type of RNA molecule, this time in *E. coli,* could be trimmed by an enzyme preparation that lacked any protein. In each case, enzyme catalysis did not depend on a protein but only on RNA molecules. These enzymatic RNA molecules are now called *ribozymes.* This discovery—though initially greeted with skepticism—was particularly exciting since increasing numbers of biologists had accepted the idea, discussed in Chapter 24, that the earliest genes were made of RNA. Cech's and Altman's discoveries established that RNA could not only store information but even participate in its processing.

Ribozymes are again a subject of much excitement, now in the realm of biotechnology, which applies biological knowledge to problems of medicine and agriculture. Cech, Altman, and many others are developing methods, as illustrated in Figure 3, for using ribozymes to destroy specific kinds of RNAs—such as those involved in the action of the human immunodeficiency virus, which causes AIDS.

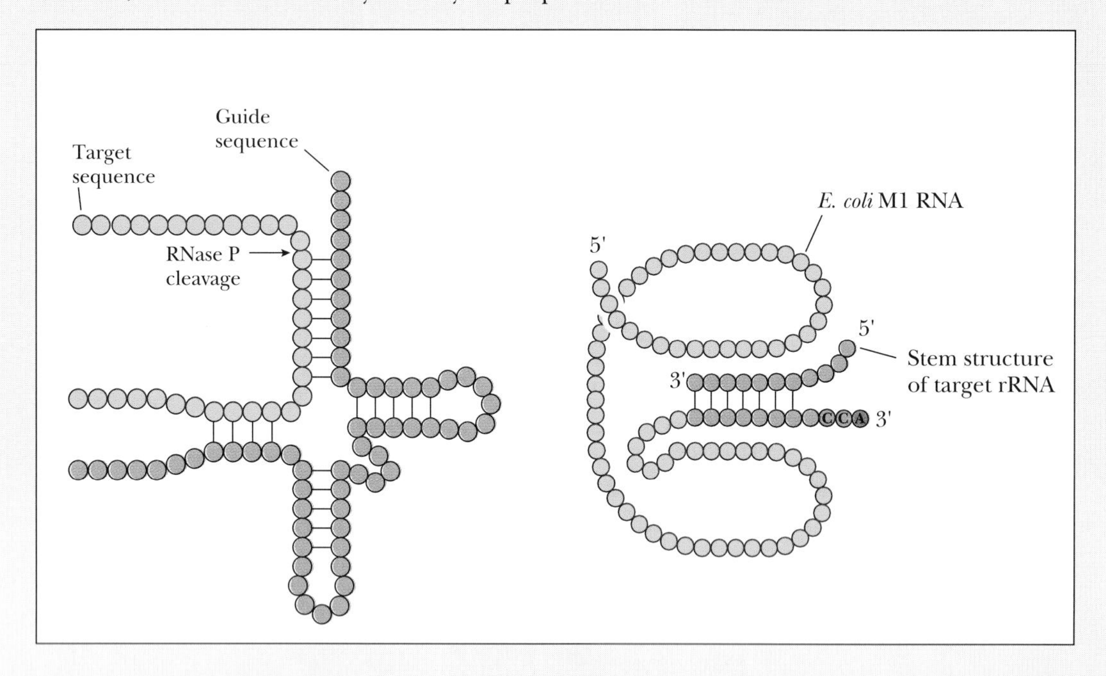

Figure 3 *RNase P recognizes structures in which two RNA segments have formed a "stem" held together by hydrogen bonds.* ***(a)*** *A natural substrate for RNase P, with a "guide" sequence on the right and the "target" sequence on the left;* ***(b)*** *an engineered "guide" sequence directs RNase P to cut at a specified place in a natural target, such as a viral RNA.*

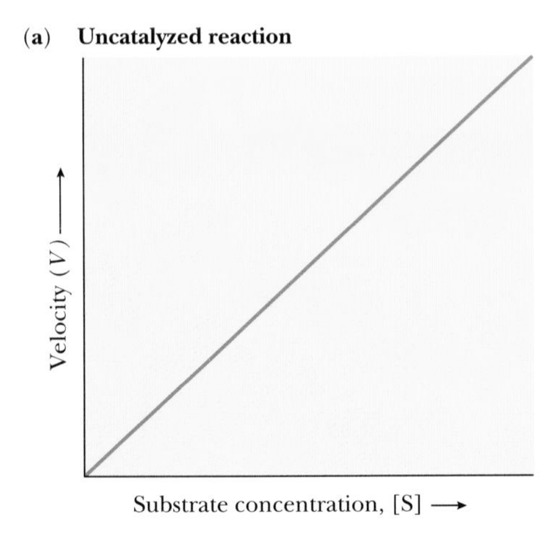

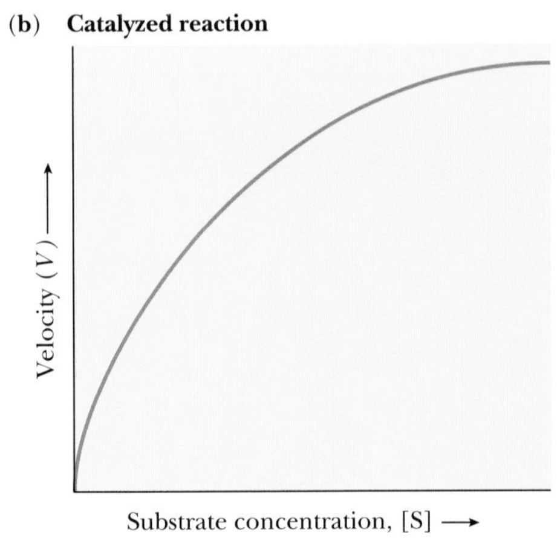

Figure 4-14 Dependence of reaction velocity *(V)* on substrate concentrations [S]. **(a)** In an uncatalyzed reaction, *V* increases linearly with substrate concentration; **(b)** in a catalyzed reaction, such as an enzymatic reaction, *V* reaches a maximum when the catalyst (enzyme) is saturated.

Michaelis-Menten kinetics, based on the necessity for the formation of an enzyme-substrate complex. The conclusion of this analysis was that

$$V = \frac{V_{max}\ [S]}{[S] + K_M}$$

where V is the reaction velocity (typically in mol of product produced per minute), [S] is the molar concentration of substrate, and K_M is a constant (called the **Michaelis constant**) that is equal to the substrate concentration at which the reaction velocity is half of its maximum value. The Michaelis-Menten equation fits experimental data for many enzyme reactions, such as that shown in Figure 4-14b.

The Michaelis-Menten equation is based on the simple idea that free enzyme (E) and substrate (S) form a complex (ES)

$$E + S \longrightarrow ES$$

which can then dissociate to produce either (1) free enzyme and product (P) or (2) free enzyme and substrate.

$$ES \longrightarrow E + P$$

or

$$ES \longrightarrow E + S$$

Putting these together and assigning rate constants, we have

$$E + S \underset{k_2}{\overset{k_1}{\rightleftharpoons}} ES \overset{k_3}{\rightleftharpoons} E + P$$

At early times in the reaction, the recombination of E and P to reform ES is negligible.

The velocity of the enzymatic reaction, the rate of P formation, is proportional to the concentration of ES (ES), with a rate constant of k_3.

$$V = k_3\ [ES]$$

Similarly, the rate of formation of the ES complex is proportional to the concentrations of E and S with a rate constant of k_1.

$$\text{rate of ES formation} = k_1\ [E]\ [S]$$

and the total rate of ES breakdown is

$$k_2[ES] + k_3[ES] = (k_2 + k_3)[ES]$$

At the *steady state,* when the relative concentrations of free enzyme and ES complex are not changing, the rate of ES formation is equal to the rate of ES breakdown, so

$$k_1\ [E]\ [S] = (k_2 + k_3)\ [ES]$$

Rearranging, we can solve for [ES],

$$[ES] = [E]\ [S]\left(\frac{k_1}{k_2 + k_3}\right) = \frac{[E]\ [S]}{K_M}$$

where K_M is $(k_2 + k_3)/k_1$.

For this equation to be useful, however, we need to know the relationship between the concentration of ES ([ES]), free enzyme ([E]), and the total enzyme concentration ($[E_T]$).

$$[E_T] = [E] + [ES]$$

or

$$[E] = [E_T] - [ES]$$

(a)

V_{max}

❸ At high [S] ([S] >> K_M), V is independent of [S]

❷ At intermediate [S] ([S] ≈ K_M), V is dependent on [S]

❶ At low [S] ([S] << K_M), V is directly proportional to [S]

Velocity (V)

Substrate concentration, [S]

[E] is constant

(b)

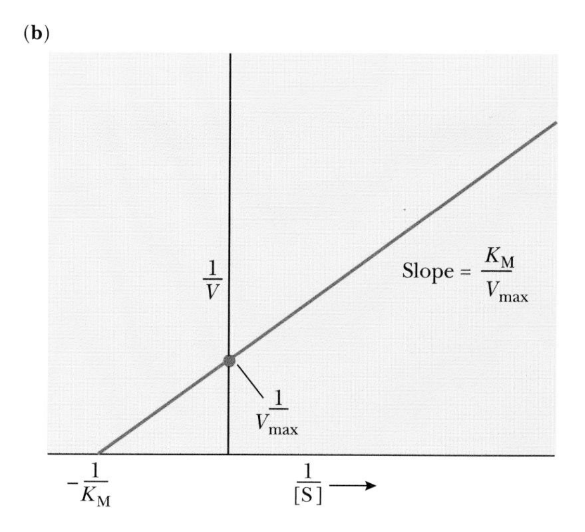

Figure 4-15 The Michaelis-Menten equation predicts the dependence of reaction velocity (V) on substrate concentration ([S]). **(a)** The graph of V as a function of [S] is a hyperbola. **(b)** The Michaelis-Menten equation predicts a straight line, if one plots $1/V$ versus $1/[S]$, a method devised by Lineweaver and Burk. The slope of the line is K_M/V_{max} and the y-intercept is $1/V_{max}$.

So we may now write,

$$[ES] = \frac{[E]\ [S]}{K_M} = \frac{([E_T] - [ES])\ [S]}{K_M}$$

and

$$[ES] = \frac{[E_T]\ [S]}{K_M\left(\dfrac{1}{1 + [S]/K_M}\right)} = [E_T]\ \frac{[S]}{[S] + K_M}$$

Since $V = k_3$ [ES], we may define the maximum velocity, V_{max}, as the velocity when all the enzyme is present as ES complex, that is when [ES] = [E_T]:

$$V_{max} = k_3\ [E_T]$$

This leads directly to the Michaelis-Menten equation

$$V = \frac{V_{max}\ [S]}{[S] + K_M} = V_{max}\ \frac{1}{1 + K_M/[S]}$$

The Michaelis-Menten equation makes the following predictions, noted in Figure 4-15a:

1. At low substrate concentrations (when $[S] \ll K_M$), the reaction velocity is directly proportional to substrate concentration:

$$V \approx \frac{V_{max}\ [S]}{K_M}$$

2. At intermediate substrate concentrations (when $[S] \approx K_M$), it makes a quantitative and testable prediction of the relationship between velocity and substrate concentration—that the reaction velocity depends on substrate concentration in the way shown by Figures 4-14 and 4-15.
3. At high substrate concentrations (when $[S] \gg K_M$), the reaction velocity is independent of substrate concentration:

$$V \approx V_{max}$$

To evaluate the Michaelis-Menten model for any enzyme reaction, biochemists study the dependence of reaction velocity upon substrate concentration. For hundreds of enzymes, the Michaelis-Menten predictions hold, providing experimental support for the formation of an enzyme-substrate complex as a rate-limiting step in enzymatic catalysis.

How Does an Enzyme Lower the Activation Energy of a Chemical Reaction?

The formation of noncovalent bonds between enzyme and substrate provides the free energy needed to form a more ordered state. This new arrangement of substrate lowers the activation energy for a reaction in at least three ways: (1) by orienting two or more substrate molecules to increase the chances of their interaction (Figure 4-16a); (2) by straining the covalent bonds of the substrate molecules (Figure 4-16b); and (3) by allowing the enzyme itself to

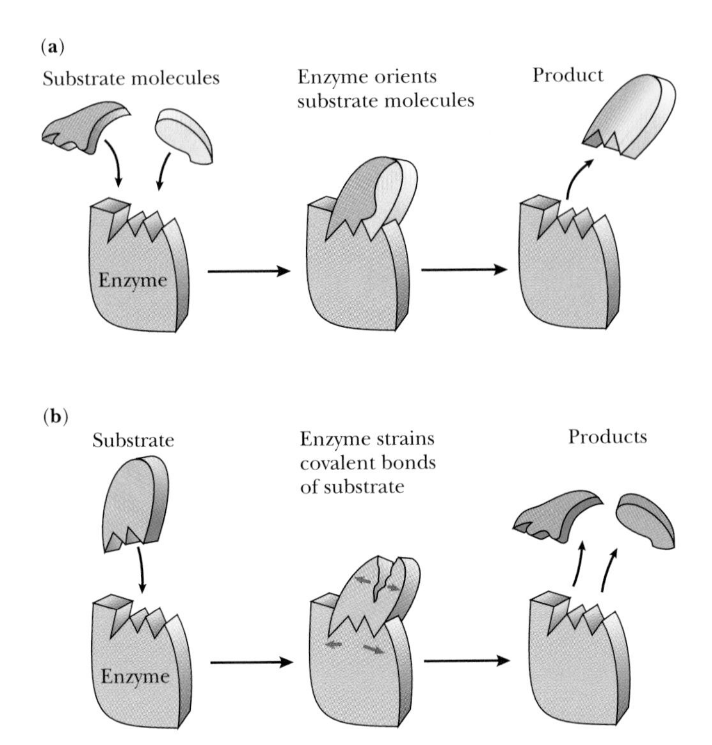

Figure 4-16 Two of the ways in which an enzyme can catalyze a reaction: **(a)** by orienting two substrate molecules, increasing the chance of their interaction; **(b)** by straining a substrate molecule, increasing the chance of reaction.

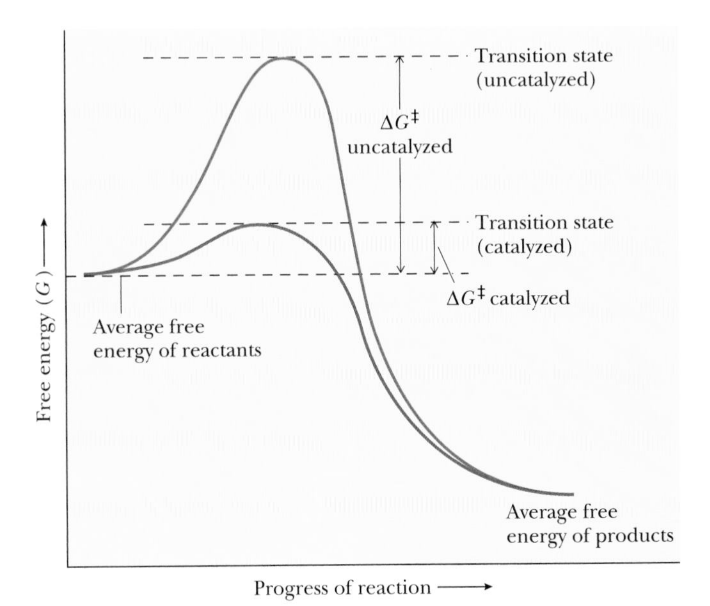

Figure 4-17 An enzyme lowers the activation energy ($\Delta G^{\ddagger}$) of a reaction by stabilizing the transition state.

participate in the chemical reaction (Figure 4-19, page 107). In all cases, however, the enzyme itself remains unchanged after the completion of the reaction.

Substrate molecules bouncing around in solution bump into each other at random, but few of these collisions lead to a reaction. Enzymes orient the substrates: They bring the reactants together in a way that favors reaction.

When a substrate molecule binds to an enzyme, the substrate's covalent bonds may be strained as it maximizes its noncovalent interactions with the enzyme molecule. Such strained bonds break more easily. The distorted form of a substrate represents a **transition state,** a molecular form intermediate between the starting reactant and the final product. The transition state exists for only a brief moment.

The transition state has a higher free energy than either the reactant or the product because the bonds are strained. At the same time, however, the presence of the bonds between the enzyme and the transition state stabilizes the transition state, lowering the activation energy ($\Delta G^{\ddagger}$) required for the reaction to proceed, as illustrated in Figure 4-17. The net effect of the interactions between enzyme and substrate is to lower the amount of additional energy needed to get over the activation energy barrier.

Enzymes may contribute directly to the chemical reaction, for example, by lending or temporarily accepting an atom or an ion. This change strains the substrate molecule, bringing it closer to its transition state. Again, the net effect is to lower the activation energy barrier to the reaction.

Since the late 1960s, x-ray diffraction studies have revealed much not only about the three-dimensional structures of many enzymes, but also about the way in which they interact with substrate molecules. These studies often show exactly how enzymes orient and strain substrate molecules and help explain why chemical or genetic modifications of particular amino acid residues in an enzyme interfere with its catalytic abilities.

The protein-digesting enzyme *chymotrypsin,* a protease, provides an illustration. Chymotrypsin hydrolyzes peptide bonds, but usually only after an aromatic amino acid (tyrosine, phenylalanine, or tryptophan) in the protein that is being digested. Figure 4-18a is a representation of the tertiary structure of chymotrypsin, showing the locations of two residues that interact with the substrate. These residues—His 57 and Ser 195—are relatively far apart on the polypeptide chain but are brought close together by the chain's twists. Other residues with nonpolar side chains form a hydrophobic pocket for the aromatic side chain of the substrate.

During the process of catalysis, the hydroxyl oxygen of Ser 195 attaches to the carboxyl carbon of the substrate. This new bond changes the character of the substrate's peptide bond: The carbon's bonds no longer form a *planar* arrangement but adopt a tetrahedral arrangement. As illustrated in Figure 4-19, the transition state then rearranges, leaving the carboxyl carbon still attached to Ser 195. The nitrogen atom of the peptide bond then acquires a hydrogen ion from His 57 to form an amino group, as shown in Figure 4-19. The peptide fragment with the free

(a) Chymotrypsin tertiary structure

36 64 109 82 87 75 C 245 49 60 115 105 29 25 149 5 240 154 N 1 His 57 94 Ser 195 234 213 20 190 146 98 219 227 127 184 130 178 161 223 174 170

(b) Chymotrypsin tertiary structure

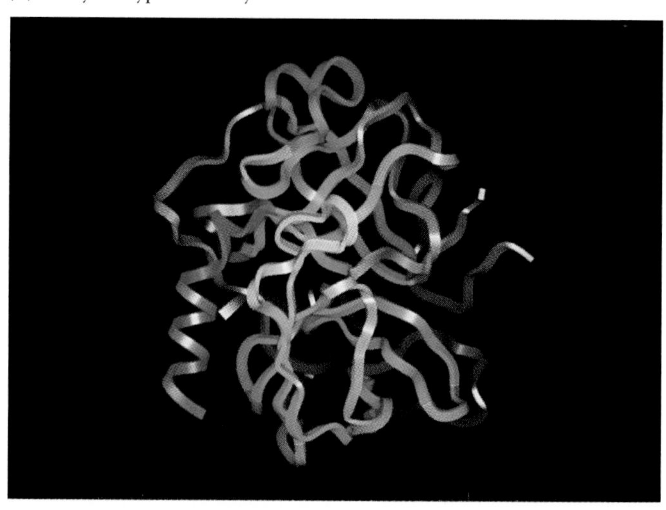

Figure 4-18 The tertiary structure of chymotrypsin, a protease. **(a)** A representation of the course of chymotrypsin's polypeptide chain, showing the amino terminal residue (#1, marked N), the carboxy terminal (#245, marked C), and two amino acid residues that participate in the enzymatic reaction (His 57 and Ser 195); residues 147 and 148 are missing because they were removed by another protease to activate chymotrypsin from its precursor molecule, chymotrypsinogen; **(b)** a three-dimensional representation of the chymotrypsin molecule. *(b, Melinda M. Whaley)*

amino group (the "downstream" peptide fragment) then diffuses away from the active site, leaving the "upstream" fragment still attached (by the terminal carboxyl group) to Ser 195. Finally, a water molecule, guided by its attachment to His 57, hydrolyzes the bond between Ser 195 and the remaining peptide fragment, as shown in Figure 4-19f, g, and h.

Chymotrypsin illustrates the three ways in which an enzyme can accelerate a biochemical reaction: (1) the *orientation* of the substrate and a water molecule; (2) the *straining* of the peptide bond toward the transition state; and (3) the *direct participation* of the enzyme in the reaction.

Environmental Conditions Affect the Rates of Enzymatic Reactions

In our previous discussion of reaction kinetics, we saw that chemical reactions occur more rapidly at higher temperatures because more molecules possess enough kinetic energy to get over the activation energy barrier. But increasing the temperature of an enzymatic reaction increases the rate only to a point. As the temperature increases, the bonds that maintain the enzyme's structure (hydrogen bonds, hydrophobic interactions, charge interactions, and van der Waals interactions) begin to break, as the polypeptide chain acquires greater kinetic (thermal) energy. The enzyme **denatures**—it loses its activity. Most organisms therefore cannot survive temperatures much higher than 40°C. (Some organisms live at higher temperatures, but they can do so only if their enzymes have evolved to be unusually stable.) Similarly, an enzyme's ability to catalyze a reaction depends on other environmental conditions, including pH, the concentration of specific ions, and the presence of molecules that affect its structure.

HOW DO CELLS AND ORGANISMS REGULATE THEIR OWN METABOLISM?

The complex network of biochemical conversions within a cell is collectively called **metabolism.** Cells perform metabolic reactions differently at different times—for example, before and after a meal. They can obtain energy when nutrients are available or break down their own stores when they are not. Often they can make a particular building block that is not available from their surroundings, or

(Text continues on page 110.)

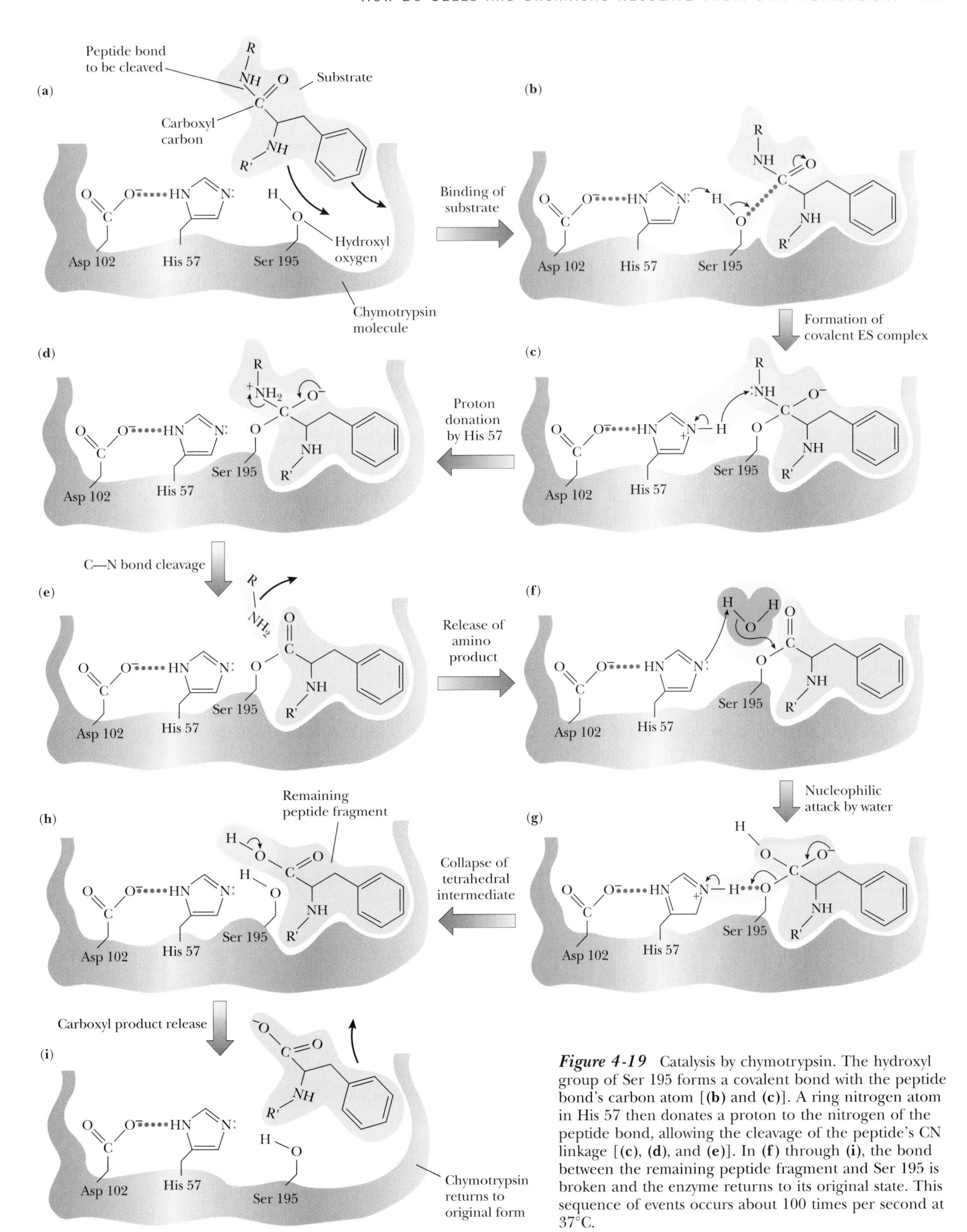

Figure 4-19 Catalysis by chymotrypsin. The hydroxyl group of Ser 195 forms a covalent bond with the peptide bond's carbon atom [(**b**) and (**c**)]. A ring nitrogen atom in His 57 then donates a proton to the nitrogen of the peptide bond, allowing the cleavage of the peptide's CN linkage [(**c**), (**d**), and (**e**)]. In (**f**) through (**i**), the bond between the remaining peptide fragment and Ser 195 is broken and the enzyme returns to its original state. This sequence of events occurs about 100 times per second at 37°C.

ESSAY

BLOOD CLOTS AND THE THERAPEUTIC USES OF ENZYMES

Animals continually suffer breaks in their skins and their blood vessels. Such wounds give microorganisms a chance to enter and disperse through the body. They can also result in a life-threatening loss of blood and reduction of blood pressure. Happily, the body has a number of adaptations—collectively called *hemostasis*—that prevent blood loss. Hemostasis involves at least three interconnected processes that begin immediately after injury:

1. *vasoconstriction,* the contraction of smooth muscles in damaged blood vessels
2. *platelet aggregation,* the attachment of small cells, called platelets, to the site of injury
3. *blood clot formation,* the assembly of protein networks that prevent further blood loss through the wound

Both the formation and the dissolution of blood clots depend on the sequential action of specialized enzymes. Detailed knowledge about these enzymes has revolutionized the understanding and treatment of heart attacks and strokes.

A blood clot can form in a tube of blood outside the body, as illustrated in Figure 1a. Still more remarkably, clotting occurs even in the absence of any blood cells. The ability to clot must therefore be a property of substances dissolved in the *plasma,* the liquid, noncellular part of the blood (Figure 1b). Biochemists have studied intensively the process of clotting and have identified a set of proteins responsible for clot formation.

The clot itself consists of a network of *fibrin,* a fibrous protein. But fibrin and fibrin fibers are not present in unclotted blood, otherwise blood would not be able to flow. Where, then, does fibrin come from?

Fibrin derives from *fibrinogen,* a precursor protein in the plasma. Fibrinogen becomes fibrin as a result of the action of *thrombin,* an enzyme closely related to chymotrypsin that digests away part of the fibrinogen polypeptide. But what prevents thrombin from acting inappropriately, causing fibrin formation and clotting when there is no damage?

The answer is that active thrombin is not usually present in plasma. Instead, plasma contains *prothrombin,* a precursor of thrombin. At the site of tissue damage, another enzyme, called *Factor X* ("Factor ten"), converts inactive prothrombin to active thrombin, initiating the clot. Factor X is itself normally inactive, but

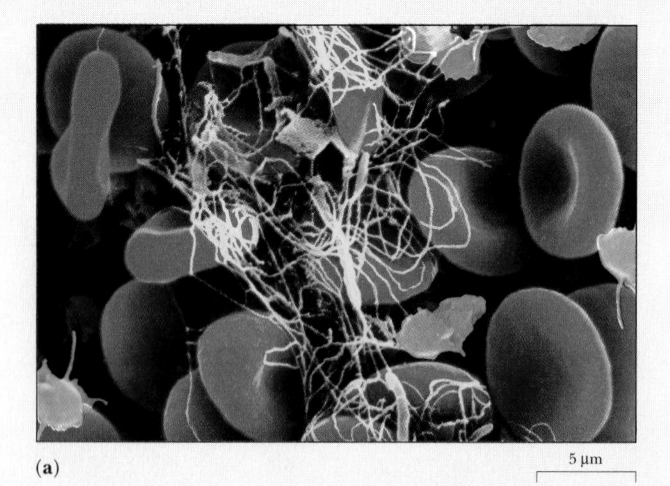

(a)

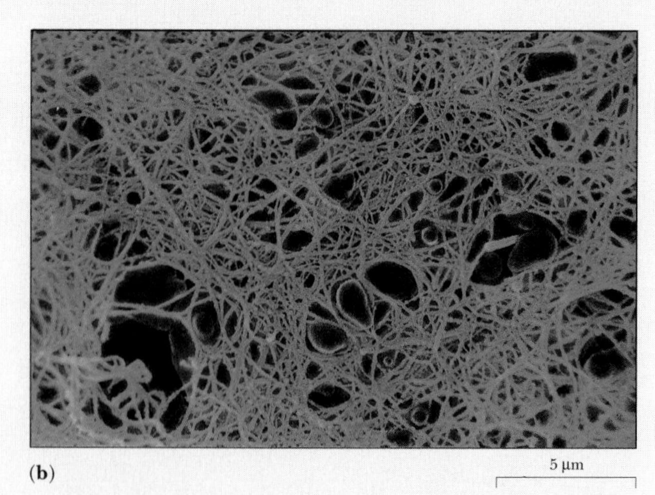

(b)

Figure 1 *Blood clots, composed of fibrin filaments, can form outside the body, even in the absence of blood cells:* ***(a)*** *a blood clot showing trapped red blood cells in a mesh of fibrin filaments;* ***(b)*** *a clot formed in plasma alone, with no blood cells present.* *(a, Dr. Dennis Kunkel/Phototake NYC; b, David Phillips/Photo Researchers)*

other enzymes can convert it to an active form at the site of damage. This conversion depends either on an enzyme made by surrounding damaged tissue, or by two additional plasma proteins, *Factor VIII* and *Factor IX*. Activation of Factor IX in turn depends on *Factor XI*, which depends on *Factor XII*, which (finally!) is directly dependent on the collagen exposed during vessel damage.

The various steps form a complex cascade of events that lead to clot formation, as diagrammed in Figure 2. At each step, a single molecule of one enzyme activates many molecules for the next step. The

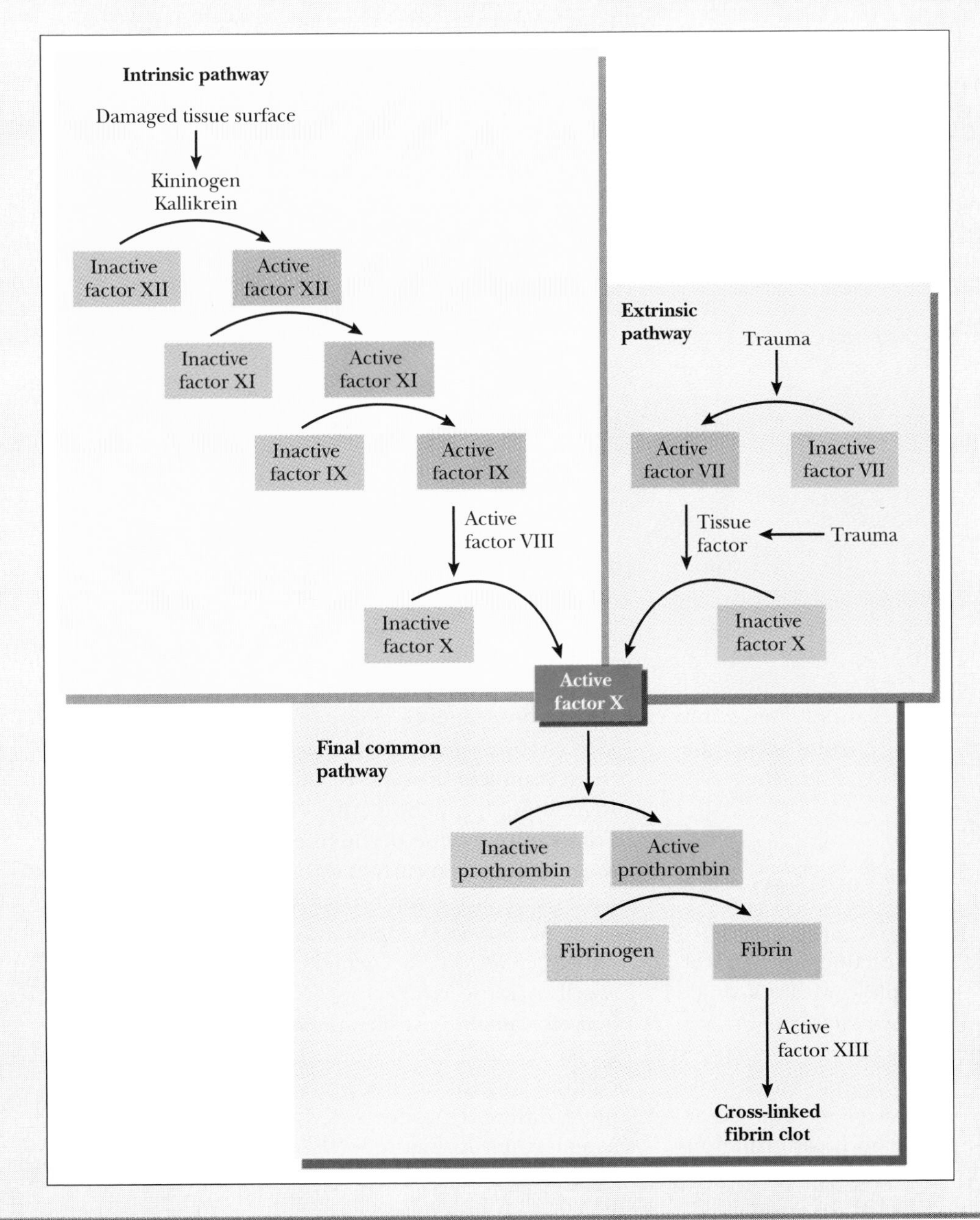

Figure 2 *The blood clotting cascade. The activated form of one factor catalyzes the activation of the next factor, finally resulting in the formation of fibrin from fibrinogen.*

(continued)

ESSAY *continued*

result is a tremendous multiplication of the initial stimulus. But this sensitivity poses another danger—that clots can form inappropriately and clog the blood vessels.

The clotting system is so powerful that it requires an equally powerful antagonistic system, an anticlotting system, to keep the blood from clotting inappropriately. Formation of clots within the circulation can cause serious damage to the brain (a stroke), the heart (a heart attack), or other tissues.

Like the clotting system, the anticlotting system consists of a multiplying cascade of enzymes, diagrammed in Figure 3. The last enzyme in the cascade is *plasmin,* which digests the bonds between fibrin molecules, dissolving the clot.

Plasmin derives from *plasminogen,* a precursor protein in the plasma. The activation of plasminogen depends either on the action of plasma enzymes or on *tissue plasminogen activators,* enzymes produced by tissues that activate the anticlotting system. The uterus, for example, contains high levels of a tissue plasminogen activator, which prevents the clotting of menstrual blood.

Recombinant DNA techniques now allow the production of relatively large amounts of tissue plasminogen activators (tPA). Administration of tPA appears to be valuable in the treatment of heart attacks. By stimulating the rapid action of plasmin, tPA helps restore normal blood delivery to the heart. In a clinical trial completed in 1990, for example, the rapid administration of tPA (or an equivalent protelytic enzyme) to heart attack victims reduced the mortality by a factor of 4, from about 35% to less than 9%.

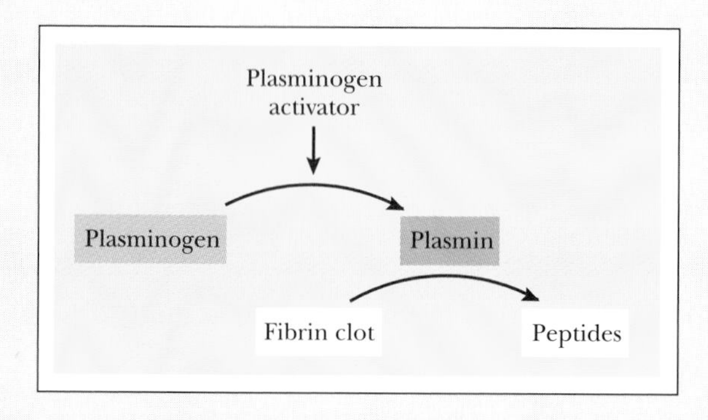

Figure 3 *The activation of plasmin by plasminogen activator.*

they can stop making a building block when it becomes available. If a particular amino acid is abundant, for example, it is wasteful to spend hard-won ATP to make more.

Cells adjust their metabolism by regulating the amounts and activities of enzymes. In doing so, they both adjust the overall flow of energy and channel energy into performing individual tasks, such as the production of specific building blocks.

Sequences of Enzymatic Reactions Form Cycles and Paths

Thousands of different chemical reactions—each catalyzed by a different enzyme—take place within a single cell. But why should there be so many reactions?

The network of biochemical reactions is so complex for at least four reasons: (1) Some of the chemical transformations of cells are so intricate that they can be accomplished only in several steps; (2) many transformations (especially those of **anabolism,** the synthesis of large molecules from smaller ones) are endergonic and require coupling to exergonic reactions, such as the breakdown of ATP to ADP; (3) many exergonic transformations (especially those of **catabolism,** the breakdown of food molecules to smaller molecules) produce more energy than the cell can store all at once; more reactions each with a smaller free energy change allow more efficient step-by-step capture of energy for later use; and (4) the intermediate products in some reaction sequences serve as starting materials for the synthesis of needed building blocks. For example, some of the intermediate steps in the breakdown of sugars to carbon dioxide and water are important precursors of the amino acids.

Sequences of enzymatic reactions may be arranged in either cycles or branched paths. Because they allow a cell to reuse relatively scarce materials, cycles are particularly valuable, as illustrated in Figure 4-20 for the tricarboxylic acid cycle, which is discussed in Chapter 7. In contrast, organisms use branched paths whenever two or more different products derive from the same precursor, as illustrated in Figure 4-21 for the synthesis, from aspartate, of asparagine, methionine, threonine, lysine, and isoleucine.

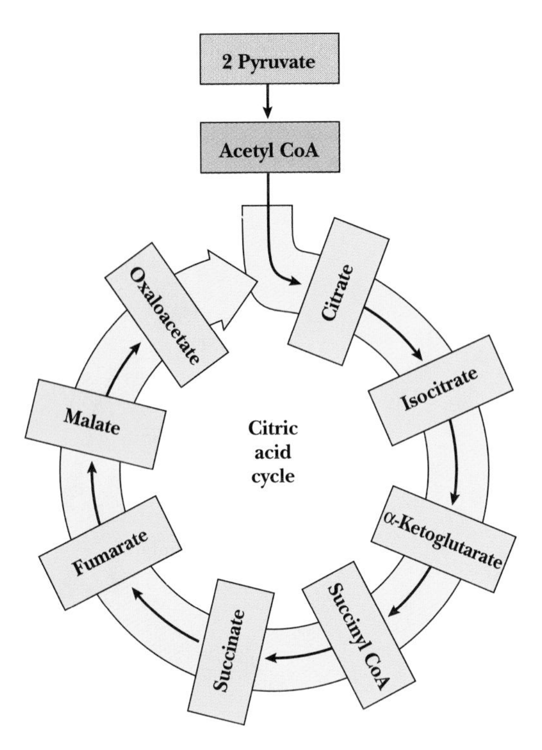

Figure 4-20 Arrangement of enzymatic reactions to form a cycle, here the citric acid cycle, discussed in Chapter 7.

The regulation of cycles is different from that of branched paths. Inactivating one enzyme in a cycle turns off the whole cycle. Inactivating one enzyme after a branch point, however, may still allow one set of products to be made, even when others are no longer produced.

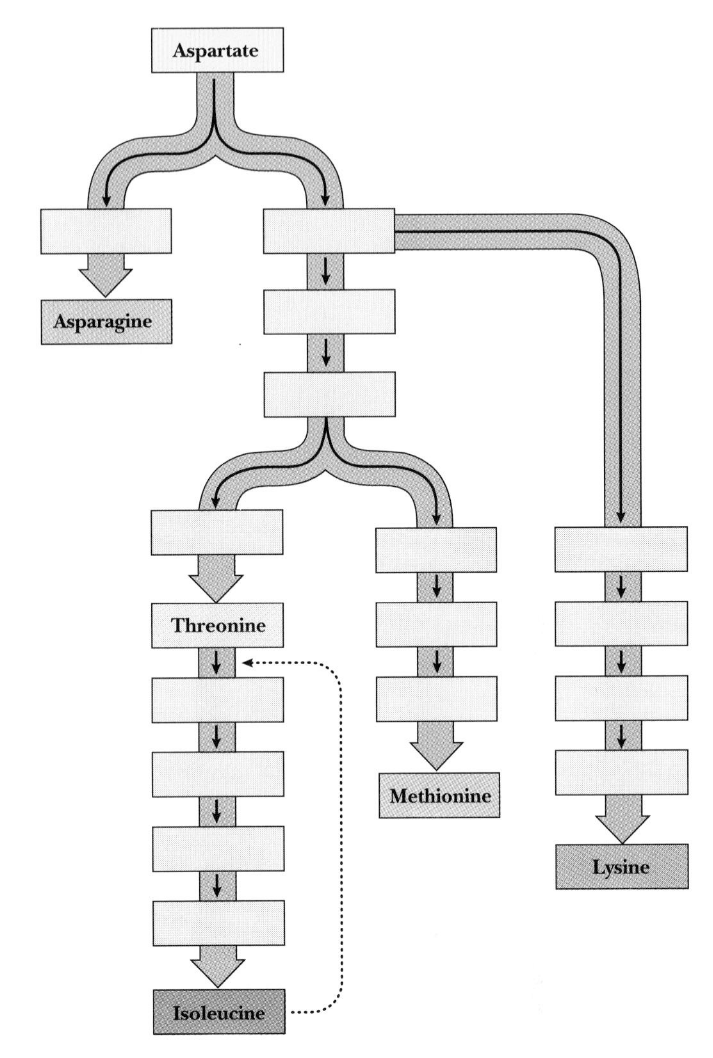

Figure 4-21 Arrangement of enzymatic reactions to form a branched pathway, to form many products from the same precursors.

How Does a Cell Turn an Enzyme Off or On?

How do cells themselves selectively regulate enzymes in response to their needs? In many cases, cells do not waste energy in making building blocks that they do not require, nor do they make enzymes to metabolize molecules that are not present.

An example of such regulation is the synthesis of the amino acid isoleucine by bacteria. When isoleucine is not available, cells make it from another molecule called threonine, in a sequence of five enzymatic reactions. When enough isoleucine has accumulated, the pathway from threonine shuts down. Isoleucine, the end product of the pathway, specifically inhibits the first step of this pathway. This **feedback inhibition** is one of many examples of homeostasis in metabolism. Molecules other than end products may also specifically change the activity of enzymes—often inhibiting enzyme action, sometimes increasing enzymatic activity.

How does an enzyme regulate its activity? All enzymes recognize the shape of their substrates. In the laboratory, enzymes can often be "fooled" into binding molecules that are similar to their substrates in size and shape. Such impersonators, called *inhibitors,* may block or change the active site, as shown in Figure 4-22a, and prevent the enzyme from acting. This kind of inhibition is called **steric inhibition** because it has to do with shape. Steric inhibition can often be overcome by increased substrate concentration, in which case the inhibitor is called a **competitive inhibitor.** As shown in Figure 4-22b, a competitive inhibitor does not affect the V_{max} of an enzyme, but does change its apparent K_M.

Noncompetitive inhibitors, in contrast, bind to enzymes at locations other than the active site. Lead and other heavy metals, for example, inhibit many enzymes by changing their overall shapes. Noncompetitive inhibitors cannot be overcome by increased substrate concentration, so studies of enzyme kinetics reveal alterations in both V_{max} and K_M, as shown in Figure 4-22b.

Some compounds form stable covalent bonds with enzymes and act as **irreversible inhibitors.** Chemical weapons, such as the nerve gas DFP (diisopropylfluorophosphate, shown in Figure 4-23), are examples of such irreversible inhibitors. DFP binds to the serine hydroxyl

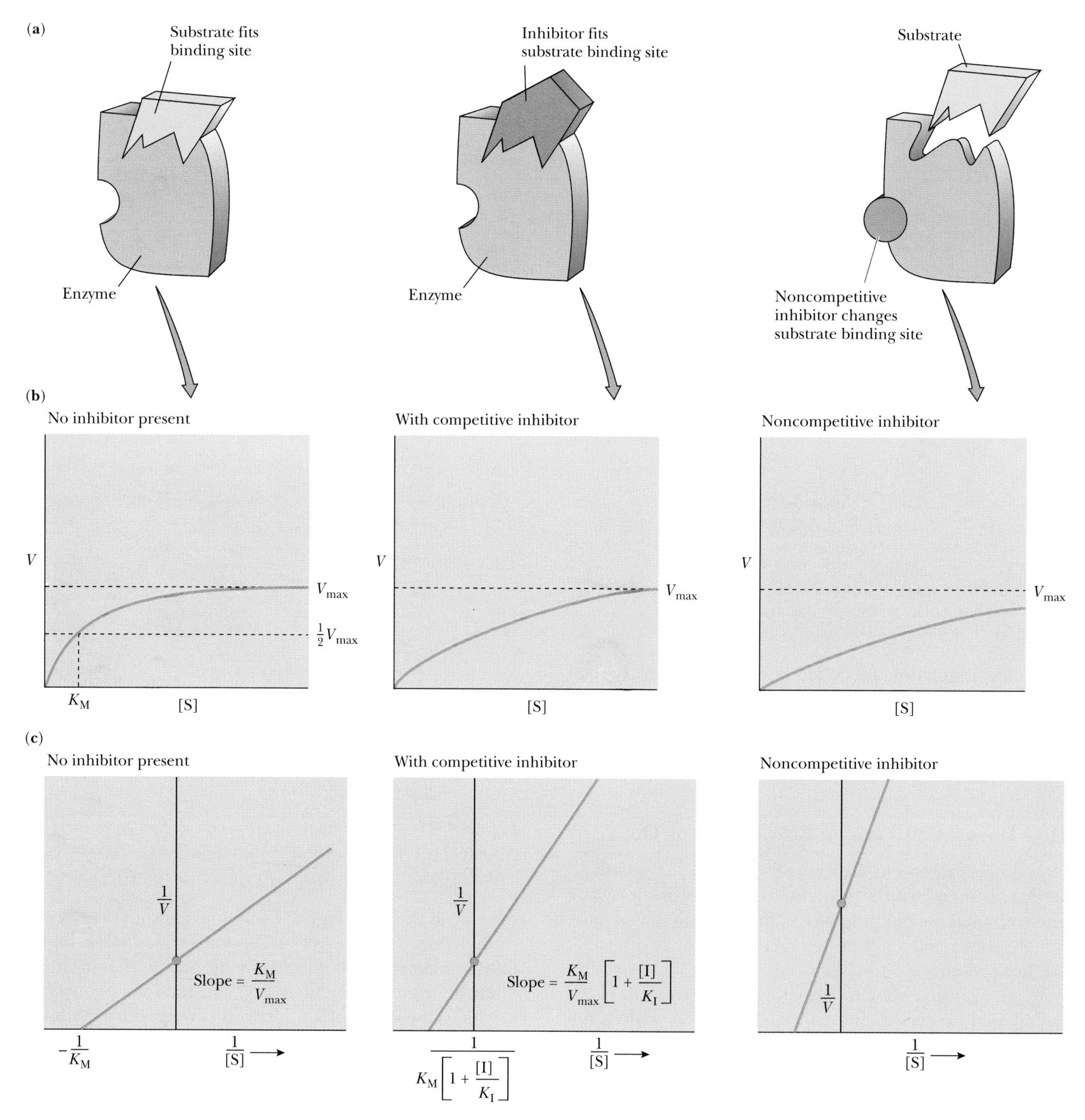

Figure 4-22 Specific molecules can inhibit an enzyme in differing ways. **(a)** A steric inhibitor may directly compete with the substrate in binding to the active site; a noncompetitive inhibitor may bind to another site on the enzyme molecule. **(b)** A competitive inhibitor does not alter the V_{max} of the enzyme, but a noncompetitive inhibitor reduces V_{max}. **(c)** The effects of competitive and noncompetitive inhibition are often easier to visualize in a Lineweaver-Burk plot.

groups in the active site of enzymes like chymotrypsin; its action as a nerve gas arises because one of the enzymes that it inhibits (acetylcholinesterase) is responsible for breaking down a signalling molecule in the brain (acetylcholine).

Any molecule or ion that changes the activity of an enzyme is called an **effector.** When it binds to a site different from the active site, it is called an **allosteric effector** (because it acts other than sterically), as illustrated in Figure 4-24.

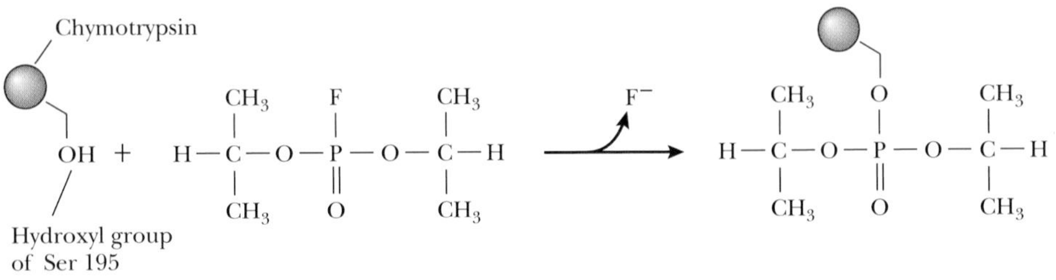

Chymotrypsin

Hydroxyl group of Ser 195

$$\text{OH} + \text{H}-\underset{\text{CH}_3}{\overset{\text{CH}_3}{\text{C}}}-\text{O}-\underset{\text{O}}{\overset{\text{F}}{\text{P}}}-\text{O}-\underset{\text{CH}_3}{\overset{\text{CH}_3}{\text{C}}}-\text{H} \xrightarrow{\ \text{F}^-\ } \text{H}-\underset{\text{CH}_3}{\overset{\text{CH}_3}{\text{C}}}-\text{O}-\underset{\text{O}}{\overset{\text{O}}{\text{P}}}-\text{O}-\underset{\text{CH}_3}{\overset{\text{CH}_3}{\text{C}}}-\text{H}$$

Diisopropylfluorophosphate

Diisopropylphosphoryl derivative of chymotrypsin

Figure 4-23 Diisopropylflurophosphate (DFP), a nerve gas, binds to the hydroxyl group of Ser 195 in chymotrypsin.

What is the mechanism of allosteric regulation? An enzyme can adopt a huge number of different possible shapes. For most enzymes, the most stable shape is the most biologically active—the expected result of natural selection for functional activity. The binding of an allosteric effector, however, may change the enzyme's shape to one that is more or less active. We can therefore picture an allosteric enzyme as having two alternative shapes. An allosteric effector switches an enzyme from one form to the other. It thus "pulls" or "pushes" the enzyme into a more or less active shape.

The molecules within a cell could undergo countless thermodynamically favored chemical reactions. The only reactions that happen rapidly enough to contribute to the cell's economy, however, are those catalyzed by enzymes. Such reactions include those that provide cells with the energy needed to maintain order in the face of the Second Law and that accomplish many of the other tasks of living. Not only do enzymes serve as catalysts for specific reactions, but they also regulate the rates of reactions in response to environmental changes. A major goal of contemporary biology is to understand the ways that enzymes work, in particular the relationship of an enzyme's function to its structure.

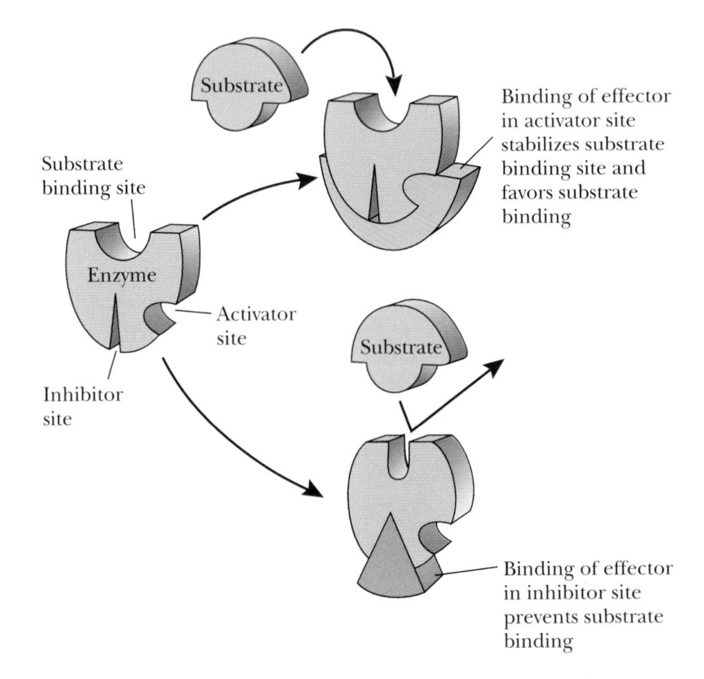

Figure 4-24 An allosteric effector binds to a site other than the substrate-binding site. Effector binding may either increase or decrease enzyme activity.

■ SUMMARY

All processes—biological and nonbiological—are accompanied by changes in the form of energy, which are subject to the laws of thermodynamics. The First Law of Thermodynamics states that the total amount of energy in an isolated system does not change. The Second Law of Thermodynamics states that the total amount of energy in a isolated system becomes less and less available for work: More and more of it is converted to heat.

The laws of thermodynamics allow scientists to predict whether any given process will occur spontaneously. In biochemical reactions, the making and breaking of chemical bonds between atoms lead to changes in available energy. A process occurs spontaneously if the energy available to do work—the free energy—decreases. A reaction is exergonic if free energy decreases and endergonic if free energy increases. Exergonic reactions occur spontaneously, but thermodynamics provides no information about how quickly they occur. By coupling endergonic processes to exergonic reactions, organisms can accomplish "uphill reactions." Cells achieve such coupling by using catalysts called enzymes, which are almost always protein molecules. In many cases, cells derive needed energy from splitting the high-energy bonds of ATP.

The free energy change in a chemical reaction depends on the concentrations of reactants and products, as well as on the number of chemical bonds that are made and broken. Reactions tend to go in the direction of increased disorder, or entropy, in which the concentrations of molecules become more uniform. The creation and maintenance of order by a cell require a decrease in order outside the cell.

To determine how long a process takes requires

knowledge of how it happens. The study of kinetics allows scientists to predict the rates of chemical reactions from the properties of molecules. The basis of kinetics is that molecules are constantly in motion, with the average energy of a group of molecules determined by the temperature. We can imagine a chemical equilibrium as molecules jumping back and forth between two compartments.

In organisms enzymes speed up specific chemical reactions. Enzymes and other catalysts do not affect the free energy changes of chemical reactions. They increase the speed at which equilibrium is approached, but they do not affect the character of the equilibrium.

Enzymes accelerate particular chemical reactions by lowering their activation energy. They accomplish this by binding to the reacting molecules via noncovalent interactions. The activity of an enzyme is highly sensitive to environmental conditions that can affect its shape. An enzyme may orient two or more substrate molecules or strain the bonds of a single substrate molecule. Some enzymes actually participate in the chemical reaction by temporarily giving or receiving atoms or ions.

Cells can regulate the activity of many enzymes. This allows them to control their metabolism in response to different environmental conditions. Steric inhibitors may block the enzyme's active site, while allosteric effectors can change the enzyme's shape and activity.

KEY TERMS

activation energy
active site
adenosine diphosphate (ADP)
allosteric effector
anabolism
calorie
catabolism
catalyst
chemical work
cilia
competitive inhibitor
concentration work
denaturation
effector
electrical work
endergonic
energy
enthalpy
entropy (S)
enzyme
enzyme-substrate complex
equilibrium constant (K_{eq})
exergonic
feedback inhibition
First Law of Thermodynamics
free energy (G)
heat
high-energy bond
induced fit
irreversible inhibitor
isolated system
kinetic energy
kinetics
low-energy bond
mechanical work
metabolism
Michaelis constant (K_M)
Michaelis-Menten kinetics
noncompetitive inhibitor
open system
potential energy
Second Law of Thermodynamics
steric inhibition
substrate
thermodynamics
transition state
velocity
work

QUESTIONS FOR REVIEW AND UNDERSTANDING

1. Define and contrast the following pairs of terms:
   (a) thermodynamics vs. kinetics
   (b) kinetic energy vs. potential energy
   (c) endergonic reaction vs. exergonic reaction
   (d) enthalpy vs. entropy
   (e) catalyst vs. enzyme
   (f) substrate vs. active site
   (g) competitive inhibition vs. noncompetitive inhibition
   (h) catabolism vs. anabolism
   (i) Michaelis constant (K_M) vs. maximum velocity (V_{max})
2. State, in your own words, the First and Second Laws of Thermodynamics.
3. Why are organisms (or cells) considered open systems?
4. When applying the laws of thermodynamics to a cell or an organism, why must its surroundings also be taken into consideration?
5. What is the unit of energy commonly used by biologists?
6. What is meant when we say a reaction is "spontaneous"?
7. The concept of coupled reactions is very important to the understanding of energy flow in cellular processes, as we will see in later chapters. Consider the following questions:
   (a) What is meant by "coupling"?

(b) What are two requirements for a coupled reaction?

(c) In general terms, why are coupled reactions so important to cells?

8. Why is it appropriate that enzymes are referred to as "biological catalysts"?
9. Support the statement, "A cell at equilibrium is dead."
10. Describe the relationship between the free energy change of a reaction and the equilibrium constant of a reaction, in mathematical terms and in your own words.
11. You have set up an experiment to determine the value of the standard free energy change, $\Delta G^{0\prime}$ for the reaction:

Dihydroxyacetone phosphate $\longrightarrow$ Glyceraldehyde phosphate

Initially, the reaction mixture contains each compound at a concentration of 1.0 M, and the reaction mixture is held at standard temperature, pressure, and pH for a very long time to allow the reaction mixture to reach equilibrium.

(a) What is the equilibrium constant (K_{eq}) if the concentrations of the the two compounds at equilibrium are:

	Dihydroxyacetone phosphate	Glyceraldehyde phosphate
(i)	1.0 M	1.0 M
(ii)	1.9 M	0.1 M
(iii)	0.0002 M	1.998 M

(b) Knowing the gas constant (R) is 1.987 cal/mol/°K and the temperature of the reaction (T) is 298°K, calculate the standard free energy change ($\Delta G^{0\prime}$) for the reaction using the equilibrium constants determine in part (a) for the three hypothetical results.

(*Note:* The results of this experiment given here and, therefore, the calculated values of $\Delta G^{0\prime}$, are hypothetical. The actual value of $\Delta G^{0\prime}$ for this reaction is +1.8 kcal/mol.)

12. What types of changes in the enthalpy and the entropy of a reaction are typical if the reaction is exergonic (spontaneous)? What if the reaction is endergonic?
13. Many reactions that are exergonic (spontaneous) do not occur in the absence of a catalyst or an enzyme. Explain why this is so.
14. Define in your own words the term "activation energy." What is the affect of catalysts and enzymes on the activation energy of a reaction?
15. Suppose you found that the experimental reaction given in Question 11 was very slow to reach equilibrium, and so you decided to add an appropriate enzyme to speed up the reaction and shorten the time of the experiment. Would the addition of the enzyme alter K_{eq} or $\Delta G^{0\prime}$? Explain your answer.
16. List three important properties of enzymes.
17. What is the transition state? How is an enzyme's function related to the formation of the transition state?
18. Given that enzyme structure and activity are affected by environmental conditions, suggest why a high fever (above 105°F or 40°C) is potentially lethal.
19. Allosteric regulation is an important means of controlling the activity of specific enzymes and the metabolic pathways in which they are found. Suggest how feedback inhibition (see Figure 4-22) may involve the endproduct of a pathway acting as an allosteric effector.
20. Many drugs and related pharmaceutical agents act as inhibitors of specific enzymes. Speculate on the information you would want to know in order to rationally design a new pharmaceutical agent.

SUGGESTED READINGS

BLUM, H. F., *Time's Arrow and Evolution,* 3rd edition, Princeton University Press, Princeton, N.J., 1968. A close but accessible look at the operation of thermodynamics in biology.

EDSALL, J. T. and H. Gutfreund, *Biothermodynamics: The Study of Biochemical Processes at Equilibrium,* Wiley, New York, 1983.

GARRETT, R. H. and C. M. Grisham, *Biochemistry,* Saunders College Publishing, Philadelphia, 1995. Chapters 11 and 12 discuss enzyme kinetics and regulation.

KLOTZ, I. M., *Energy Changes in Biochemical Reactions,* Academic Press, New York, 1967.

MONOD, J., *Chance and Necessity,* Alfred Knopf, New York, 1971.

MOROWITZ, H. J., *Entropy for Biologists: An Introduction to Thermodynamics,* Academic Press, New York, 1970.

STRYER, L., *Biochemistry,* 3rd edition, W. H. Freeman, New York, 1988. Chapters 8 and 9 concern the action of enzymes.

PART II

The Functional Organization of Cells

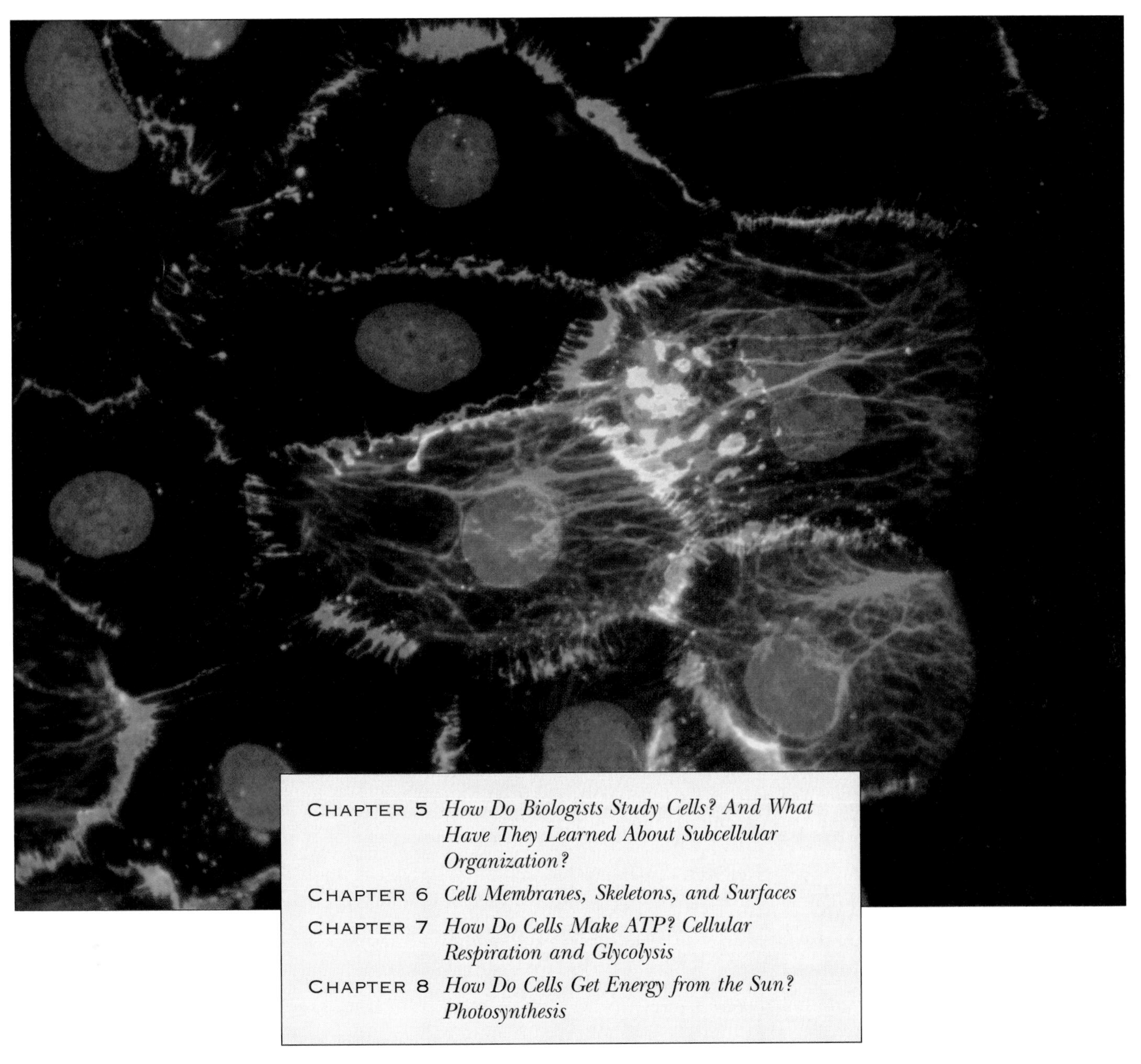

Cultured human skin carcinoma cells stained with an antibody to keratin.

LAB TOUR

NAME: URSULA GOODENOUGH
TITLE: PROFESSOR OF BIOLOGY
PRESIDENT, AMERICAN SOCIETY OF CELL BIOLOGY
ADDRESS: WASHINGTON UNIVERSITY, ST. LOUIS, MO 63130

■ ***What is the major question that you and your laboratory have been asking?***

We are attempting to understand why eukaryotic organisms have sex.

■ ***How did you come to this question?***

The topic has always intrigued me. When I was a graduate student, I first watched mating in my experimental organism *Chlamydomonas reinhardii,* a single-celled green alga. Most people in the laboratory used mating merely as means to genetic analysis, to find out how genes were linked to one another on chromosomes. But I found myself intrigued by the elaborate cell biology of *Chlamydomonas* mating. I wondered why such a simple organism had evolved so complicated a mechanism, especially since it could also reproduce asexually, and sex was only an optional part of its life cycle. I began to realize that optional sex had evolved in many single-celled organisms and wondered what it was good for and what maintains it.

■ ***Why has this question remained important to you?***

As the tools of cell biology and molecular biology became more sophisticated, I found that I could pursue the same set of questions at a more and more refined level. While I still don't have a final "answer" to the question, I have continued to pose more questions, to have more ideas about why things are the way they are, and to formulate more models and hypotheses to test.

■ ***What is your general strategy as you pursue answers to the questions you have posed?***

As in every field, every time I got an answer to a question, there were still more questions; every time I walked through a door, I found eight more closed doors in the next room. I kept finding that there were always more questions than answers. The trick has always been to decide which of the eight new doors is the most profitable to open and then to ignore the other seven.

For example, I isolated my first mutant, nonmating strain of *Chlamydomonas* in 1971. My laboratory is still using this strain in 1995 and still learning things from it. Along the way, we learned first that this mutant lacks a surface glycoprotein involved in recognizing gametes of the opposite sex. From genetic experiments, we also learned that the mutation is linked to a "mating type" region of DNA, which determines sex in this organism. Using recombinant DNA techniques, we have now cloned the mating type region and have identified the mutation—the precise DNA alteration in the mutant strain. By sequencing the gene containing this mutation, we have deduced the amino acid sequence of the sexual recognition protein. Now we are asking a new set of questions by looking at the DNA of different, but closely related species. We hope to learn which aspects of sexual recognition have remained the same during evolution and which have changed.

■ ***Why do you study this organism?***

First, it's a beautiful organism. Its

mating reaction is extremely elegant and illustrates many of the most important general processes of cells and molecules—regulated gene expression, cell-cell recognition, actin polymerization, and membrane fusion. In addition, *Chlamydomonas* has a sexual cycle that is readily elicited in the laboratory and subjected to genetic analysis.

Chlamydomonas is an excellent organism for genetics; matings are easy to accomplish for classical genetic analysis, and cells can also be transformed with recombinant DNAs to directly assess the effects of normal and mutant genes. Furthermore, ecologists have identified a number of closely related species, which allows us also to ask evolutionary questions.

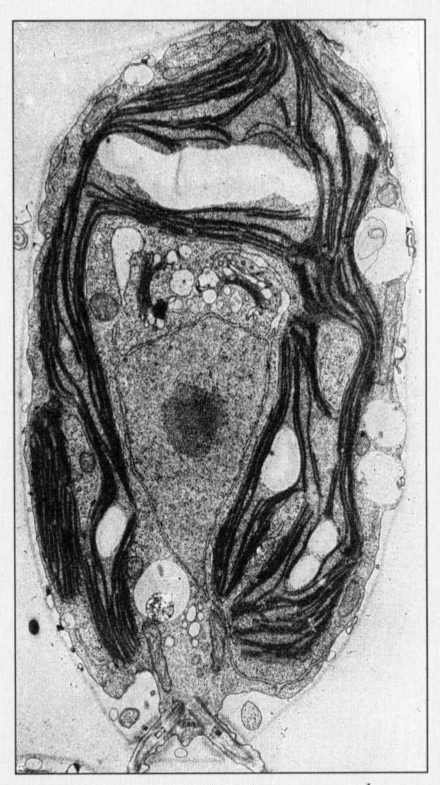

Chlamydomonas, *the experimental organism studied in Professor Goodenough's lab.* (Ursula Goodenough, Washington University)

■ ***What instruments do you use in your work?***

Because few other labs are studying the mating process, we have had to do almost every kind of cellular and molecular study, ranging from all available forms of microscopy (including fluorescence microscopy and electron microscopy) to protein purification, genetic analysis, and gene cloning.

■ ***Describe a particularly exciting day in the life of the laboratory.***

I can describe a relatively recent "eureka" experience.

Our genetic analyses had identified many mating mutations that mapped to a single region of DNA, called the "mating type locus." Our data suggested that the mating type locus contained a lot of genes, and our task was to find out what exactly was different between the two mating types, which we call "plus" and "minus" to avoid any temptation for anthropomorphism.

Using recombinant DNA, we had identified a marker gene, which was in the region of the mating type locus, though still at some distance. A postdoctoral fellow in the lab, Patrick Ferris, then undertook to "walk" to the mating locus itself. The trick was to find, in a recombinant DNA library, a long piece of DNA that had both the marker and a piece of DNA that lay in the direction of the mating locus. After isolating such a DNA, the next step was to use the end of that piece to find another piece of DNA that moved another "step"' toward the mating locus. Each step in this walk was about 20,000 nucleotides (20 kilobases).

After three years of hard work, Patrick had walked over a million bases (1.2 megabases). After each step, he compared the newly characterized region in plus and minus strains. Each time they looked the same. One morning, however, Patrick excitedly came into my office to say that he was "getting something really weird." The last step of his walk had yielded a region of DNA in which the plus and minus strains were very different from one another. This was what we had been looking for, but we were surprised by the extent of the differences; the DNA was highly rearranged and had many of the properties of sex chromosomes found in other species. Our continuing task now is to find out just how many genes are in this region and what they do.

■ ***In addition to all your scientific work, you are President of the American Society of Cell Biology (ASCB). What have your goals been as president of ASCB?***

ASCB is a dynamic and actively growing society. It has a life independent of the life of its president or of any officer. Aside from its traditional role in promoting communication among cell biologists, its most important activities include the recruitment of minorities to science and increasing scientific literacy by promoting better science education in elementary schools and high schools.

ASCB is also trying to educate members of Congress about the importance of biomedical funding. It is crucial that federal funding continue to support both basic and targeted research. If we look at the history of science, we find that both basic and targeted research have made vital contributions to what we know and to how we approach problems of human disease and of feeding our population. My own work is not targeted but basic, but I know that it contributes to a better understanding of fundamental cellular processes. Such understanding is crucial to developing new solutions to medical and societal problems.

CHAPTER 5

How Do Biologists Study Cells? And What Have They Learned About Subcellular Organization?

KEY CONCEPTS

- Every cell has a boundary, a set of genes, and a cell body.
- Biologists use light microscopes and electron microscopes to magnify images of cells and cell components. They use stains and optical techniques to achieve contrast among different parts of cells.
- Immunocytochemistry, microsurgery, and cell fractionation allow biologists to study the structures and functions of subcellular components.
- Internal membranes divide eukaryotic cells into distinct compartments, each with a characteristic structure and function.

OUTLINE

The first way to learn about cells is to look at them. Each major improvement in microscopy has led to both a burst of new knowledge and a host of new questions. The development of the microscope in the 17th century made it possible to see cells for the first time. The improvements of the early 19th century revealed that all organisms are composed of cells. Starting with the "Golden Age of Microscopy" in the late 19th century, biologists began to see the elaborate adaptations of cells to specific functions. In the mid-20th century, the development of new types of microscopes again revolutionized the study of cells by revealing the rich internal structure of cytoplasm, nuclei, and even membranes.

WHEN ARE CELLS AND THEIR COMPONENTS VISIBLE?

We can see an object only when it is big enough and when it stands out from its background. First, an object must be big enough to affect the signals from our eyes to our brains. In order to see an object the size of a cell, we must use lenses to magnify the image. The ratio of the size of the image to the size of the object itself is called the **magnification.** Figure 5-1 shows how increased magnification reveals details unsuspected with the naked eye.

But even when we can see an object (or a magnified photograph of it), we still may not be able to see it clearly enough to learn anything. When two objects (or two details within the same object) are too close together, we cannot even tell whether we are looking at one or two objects. The **resolution** of a microscope (or any other optical instrument, including your eye) is the minimum distance between two objects that allows them to form distinct images. Resolution is thus a measure of image detail. Your eyes have a resolution of about 100 μm (0.1 mm). As illustrated in Figure 5-2, a microscope such as the one you would use in a student laboratory may have a maximum resolution of about 1 μm, enough to distinguish individual prokaryotic cells and to resolve some internal structure in eukaryotic cells, while a microscope in a research laboratory might have a maximum resolution of 250–500 nanometers (nm).

Visibility depends not only on magnification and resolution but on **contrast,** an object's ability to absorb more or less light than its surroundings. A transparent object—one that absorbs light exactly as much as its surroundings—is invisible. Microscopists have devised ingenious ways of increasing contrast. For example, in the mid-19th century, biologists began to treat their microscope samples with the new organic dyes made by chemists for the central European textile industry. They found that different dyes specifically bound to different parts of the cell—one to the nucleus, another to the membrane, another to parts of the cytoplasm. Dyes that bind differently to cell components and thus increase contrast are called *stains.*

Most of the time, the process of staining cells also kills them, so biologists had a hard time studying processes in living cells. Beginning in the 1930s, however, ingenious optical techniques allowed microscopic examination of unstained, living cells, as we discuss later in this chapter.

The Light Microscope Reveals the Organization of Cells in Tissues but Little of Their Internal Structure

Figure 5-3 shows the workings of a light microscope. Because such a light microscope contains several lenses, it is

(a)

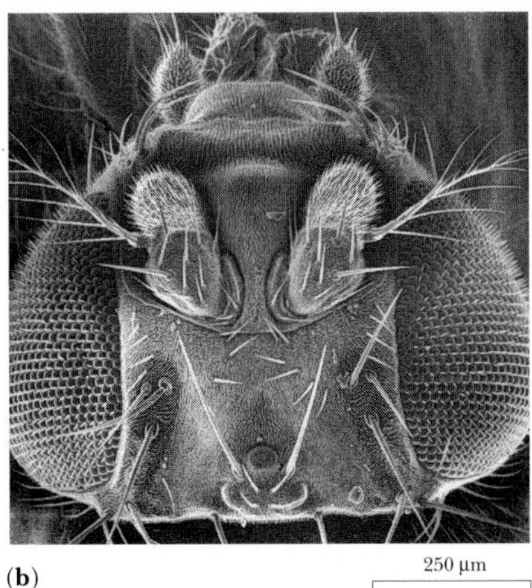
(b)

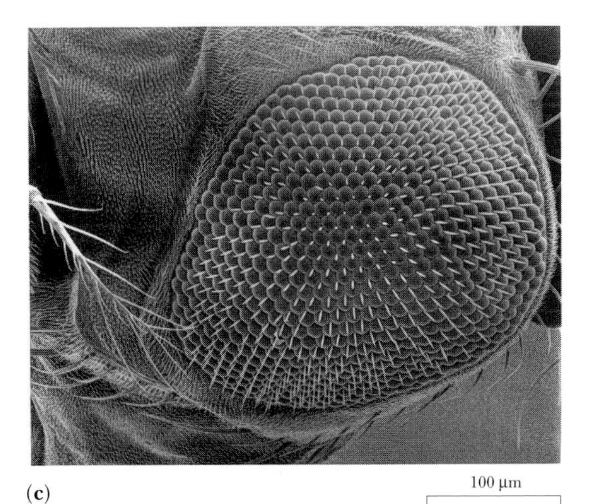
(c)

Figure 5-1 Increased magnification reveals details not visible to the naked eye. **(a)** *Drosophila melanogaster,* a widely used experimental organism; **(b)** higher magnification, showing details of the fly's head; **(c)** still higher magnification, showing that the *Drosophila* eye consists of many identical units, called ommatidia. *(a, Dr. Jeremy Burgess/Science Photo Library/Photo Researchers; b, c, Dr. Dennis Kunkel/Phototake)*

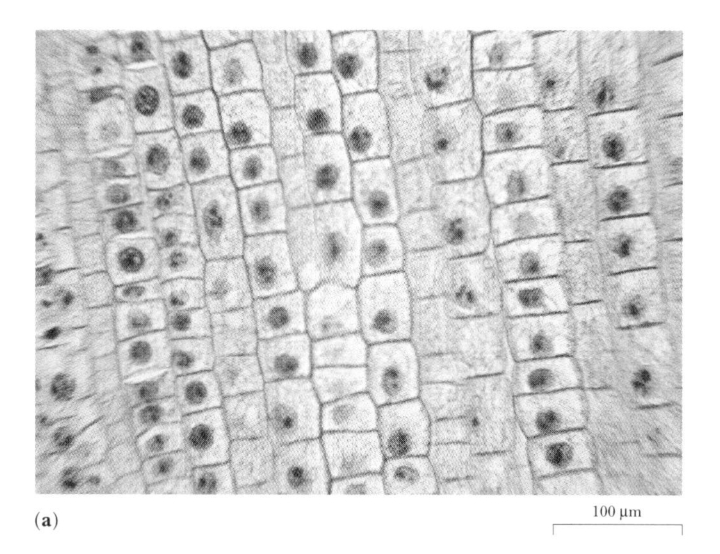

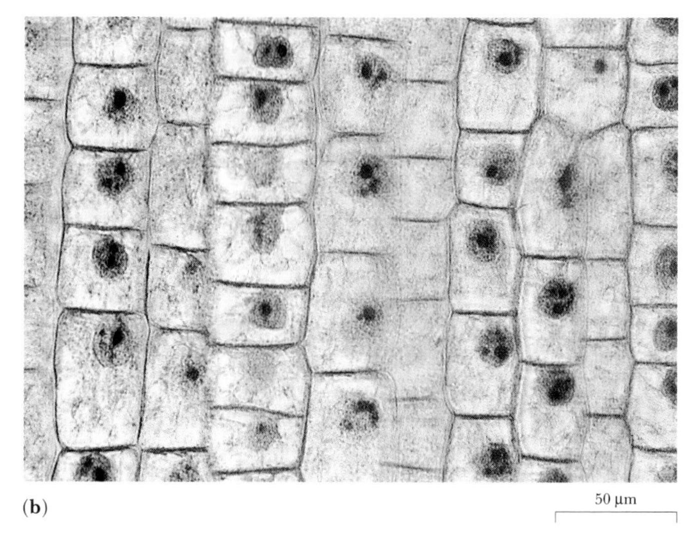

Figure 5-2 Photographs of an onion skin taken through microscopes with different resolutions: **(a)** through an inexpensive student microscope; **(b)** through a microscope with better lenses and finer resolution—note the subcellular detail. *(Charles D. Winters)*

often called a **compound microscope.** It consists of a *tube* with lenses at each end, a movable *stage* (or platform) that holds the observed specimen, a *light source* and a *condenser lens* that provide even illumination of the specimen, and a mechanism for moving the tube to focus the light that passes through the specimen. The lens near the eye is called the *ocular,* and the lens at the other end of the tube, near the object being examined, is called the *objective.* Both the ocular and the objective may consist of two or more individual lenses that act as a unit.

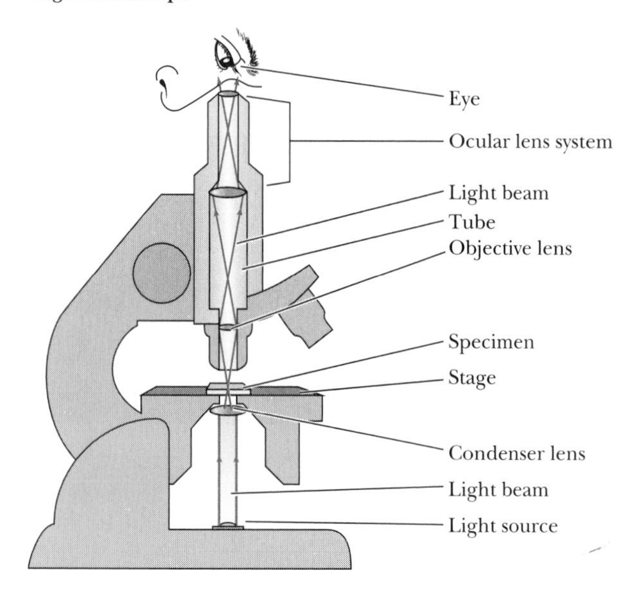

Figure 5-3 The path of light in a light microscope.

The magnification of the light microscope is the product of the magnification of the objective and that of the ocular. A single light microscope often has three objective lenses (for example, with magnifications of 10, 40, and 100×) mounted on a rotating turret, so that the viewer can easily change magnifications. Ocular lenses often contribute another 10-fold magnification, so the final image may be 100–1000 times the size of the object viewed.

The resolution of a light microscope depends both on the properties of light and on the design of the objective lens. One of the many problems that lens makers must confront is **chromatic aberration,** the failure of light of different colors (wavelengths) to focus at the same point. Modern objective lenses consist of layers of several different types of materials, which compensate for chromatic aberration. In the last 25 years, computers have greatly assisted in the design of well-corrected lenses, making them less expensive and more generally available.

Even the best lenses, however, have a resolution limited by the nature of light. Light is composed of waves and cannot resolve structures much smaller than half its wavelength. The wavelength of visible light ranges from about 0.4 μm (for blue light) to about 0.7 μm (for red light). For objects smaller than about 0.2 μm, then, light microscopes simply do not reveal the details of subcellular structures.

Electron Microscopes Reveal the Internal Structures of Cells

Cell biologists have overcome the resolution limits of light microscopes with microscopes that use electrons instead of visible light. These microscopes, called **electron microscopes,** depend on the wave-like properties of moving electrons.

Light and electrons each have both wave-like and particle-like properties. Wavelength and energy are intimately related, with higher-energy electrons having shorter wavelengths. In a 100,000-volt electron microscope, for example, the wavelength of an electron is 0.004 nm, or 0.000004 μm, less than the diameter of an individual atom. Because of these short wavelengths, the potential resolving power of an electron microscope is far greater than that of any light microscope.

The lenses of an electron microscope are electromagnets that bend the path of electrons as glass lenses bend the path of visible light. Because of various problems, both with the design of electron microscopes and with their application to studying biological structures, the actual resolution for looking at cells is about 2 nm, about 100 times better than the best light microscope. An electron microscope therefore reveals previously unobservable **ultrastructure,** or subcellular structure, which was once invisible.

Biologists commonly use two kinds of electron microscope—the **transmission electron microscope (TEM)** and the **scanning electron microscope (SEM).** In each case the electrons form an image on a phosphorescent screen, similar to that of a television set. Figure 5-4a and b compare the views of a *Paramecium* with a light microscope and

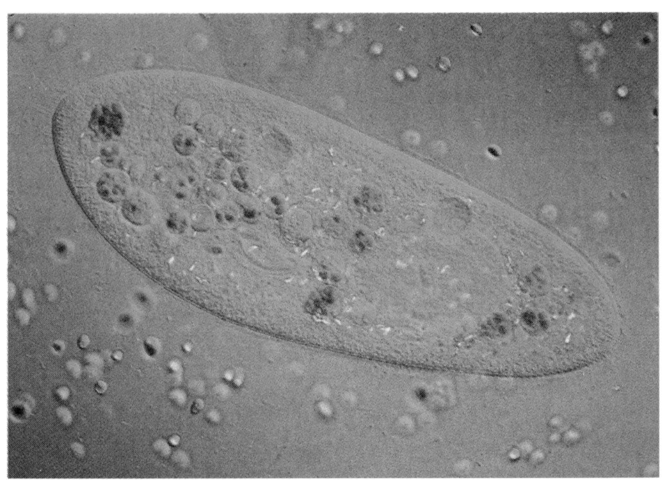

(a) 50 μm

(b) 0.5 μm

(c) 25 μm

Figure 5-4 Views of *Paramecium* with different types of microscopes. **(a)** With a light microscope, the fringe coating the cell's surface is rows of cilia, which propel this protist through the water; **(b)** with a transmission electron microscope, showing three rows of cilia in cross section; notice the highly ordered arrangement of hollow tubules that run along the axis of each cilium; **(c)** with a scanning electron microscope, which reveals surface details, including the protruding cilia. *(a, Eric Grave/Phototake; b, David Phillips/Visuals Unlimited; c, Dr. Dennis Kunkel/Phototake)*

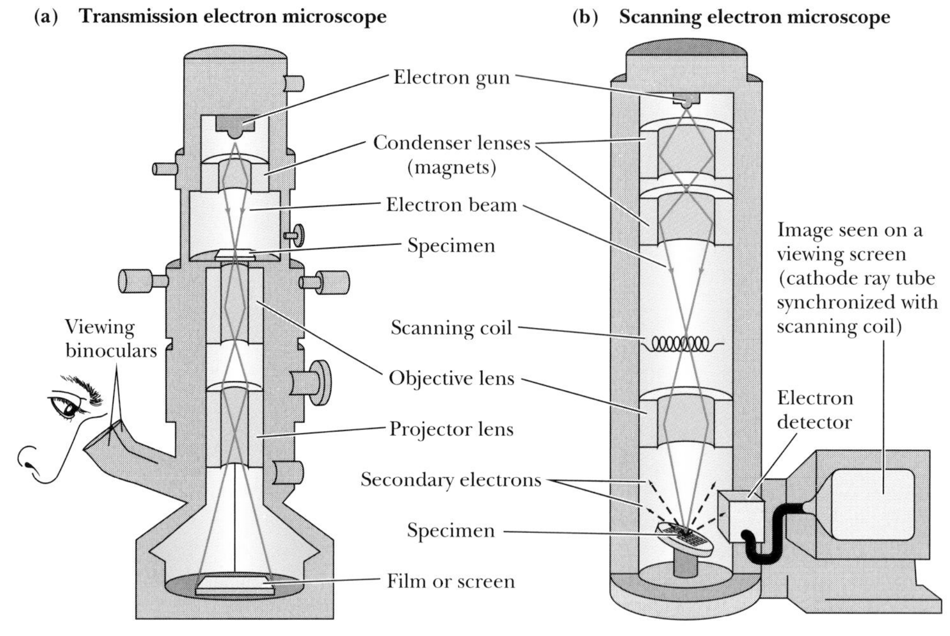

Figure 5-5 Transmission and scanning electron microscopes: **(a)** The transmission electron microscope works more or less in the same way as a light microscope, but uses electrons and magnetic lenses instead of light and optical lenses; its arrangement is upside down compared with the light microscope of Figure 5-3. **(b)** The scanning electron microscope bombards a specimen with electrons and then measures the scattering of the electrons from the specimen's surface.

a transmission electron microscope. Figure 5-4c shows the same organism as seen with a scanning electron microscope.

The design of the TEM, shown in Figure 5-5a, resembles that of a light microscope—a tube containing a set of three lenses. The TEM's tube differs, however, in three ways: (1) It also contains the specimen and the condenser; (2) it is kept at a high vacuum because air molecules would scatter the electrons; and (3) it is upside down. Electrons emerge from their source at the top of a tube in much the same way that light flows from a source in a light microscope. An electromagnetic condenser lens focuses the electron beam on the specimen. After passing through the specimen, the beam passes through two more electromagnetic lenses, which magnify the image. Finally, the electrons form an image on the phosphorescent screen or on a photographic plate.

In the SEM, shown in Figure 5-5b, the electron source forms a narrow beam of electrons which moves back and forth across the specimen. Wherever the beam hits the surface of the specimen, it causes the surface to emit more electrons, called secondary electrons. These secondary electrons then produce a detailed, almost three-dimensional, picture of the specimen's surface.

Seeing the Details of Biological Structures Requires Special Sample Preparation

Microscopists often cut biological samples into slices, called *sections.* Using sections may avoid two problems: (1) If a specimen is too thick, light or electrons simply do not get through; and (2) a thick sample can produce a confusion of superimposed images of different objects. Sections for light microscopy are typically 1–10 μm thick. For the TEM, sections may be as thin as 50 nm. New high-voltage TEMs use more than 1 million volts to accelerate electrons, instead of the 40,000 to 100,000 volts used in conventional electron microscopes. These microscopes allow cell biologists to examine sections up to 1 μm thick with high resolution. Because the electron beam of the SEM does not pass through the sample, specimen thickness does not pose the same problems in the SEM as it does for light microscopy and TEM.

Handling thin sections of fragile biological materials is tricky. (Think about how thinly you can slice a tomato before the slices fall apart.) Before cutting sections into parts or further handling the cells, microscopists subject the tissue to *fixation,* a process that keeps the sections from falling apart by binding cellular components together. The primary purpose of fixation, however, is to preserve the internal structures of cells. Fixation often involves treatments that make covalent bonds between cellular molecules.

Tissues to be cut into thin sections are usually embedded in liquid wax or plastic, which is then solidified. When they solidify, these substances allow the tissue to hold together in thin sections and to maintain its organization even in the high vacuum inside the electron microscope.

Different stains must be used for light microscopy and electron microscopy. Light microscopy requires colored stains that cause specific cellular structures to absorb light of a specific color more than the surrounding material. Electron microscopy requires stains that alter the paths of electrons through some structures more than others. Many stains for electron microscopes contain heavy metal atoms. A long-time favorite is osmium tetroxide, which also fixes the material as it stains. Sometimes otherwise invisible molecules or surfaces are coated with heavy metals (such as platinum), as shown in Figure 5-6.

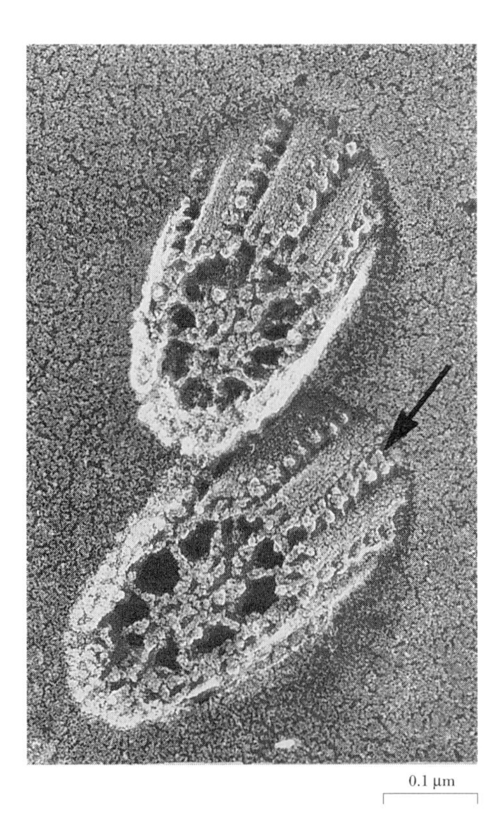

Figure 5-6 Heavy metal atoms serve as effective stains for transmission electron microscopy because of their ability to scatter electrons. Here, the surface of a green alga, *Chlamydomonas,* has been shadowed with a film of platinum, revealing the fine structure of the flagella. *(Ursula Goodenough, Washington University)*

Can Microscopists Observe Living Cells?

How can we be sure that a structure within a cell seen in the microscope is real, rather than a result of fixation, sectioning, and staining? This question haunts microscopists. They must be especially wary of **artifacts,** remnants of what the experimenter does, rather than of the cell itself. To convince themselves and others that a structure is real, microscopists use two major criteria: (1) The structure must be visible when they prepare a specimen with different kinds of fixation and staining, and (2) more convincingly, it must be visible in living cells.

Only since the 1930s have microscopic techniques allowed the detailed observation of living cells. In conventional light microscopy, contrast depends on the absorbance of light by different cellular structures. But most materials in unstained cells hardly absorb visible light. All molecules, however, do interact with light waves. As a result of such interactions, different substances change the speed of light to different extents. Light passes through the cell nucleus, for example, more slowly than it does through the watery cytoplasm. The light passing through the nucleus becomes "out of phase" with parallel waves that pass through the cytoplasm. Special optical arrangements, including **phase contrast** and **differential interference contrast** microscopy (also called Nomarski microscopy, after its inventor), reveal phase differences among different cell components without the use of stains. These techniques permit the observation of living cells, as illustrated in Figure 5-7.

Computers allow the electronic enhancement of images, vastly amplifying small differences in light intensity

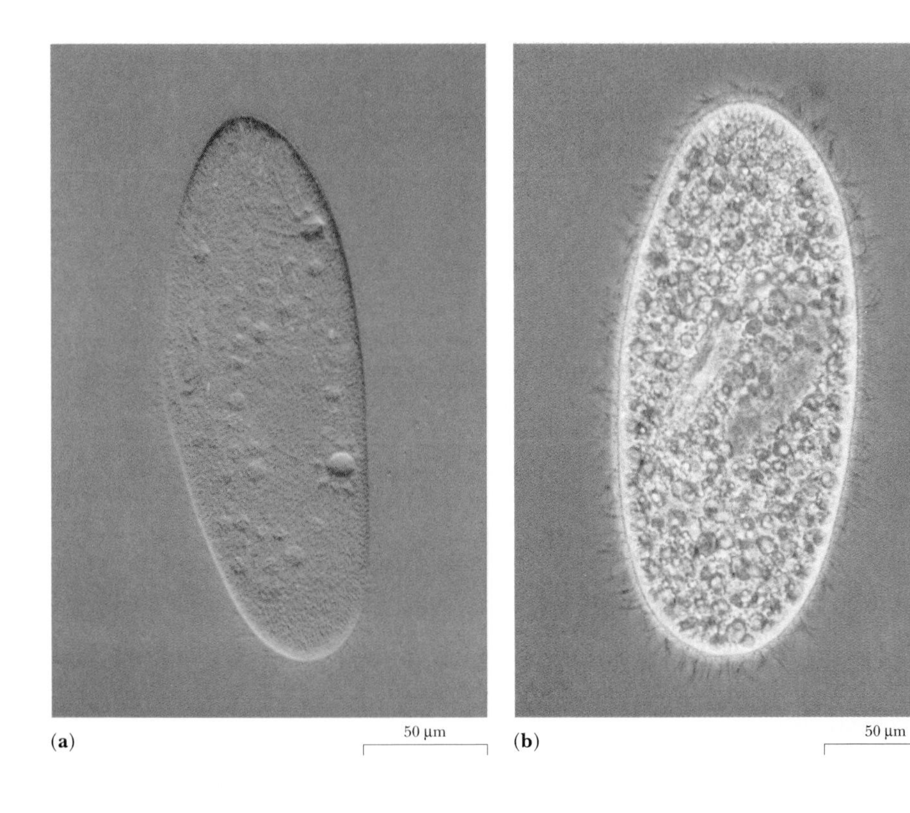

Figure 5-7 Special optical arrangements permit the detailed observation of living cells, such as *Paramecium.* **(a)** Differential interference contrast microscopy (also called "Nomarski optics"); notice the depressions within the cytoplasm, which are ingested alga. **(b)** Phase contrast microscopy; notice the fringe of waving cilia. *(Dr. David J. Patterson/Science Photo Library/Photo Researchers)*

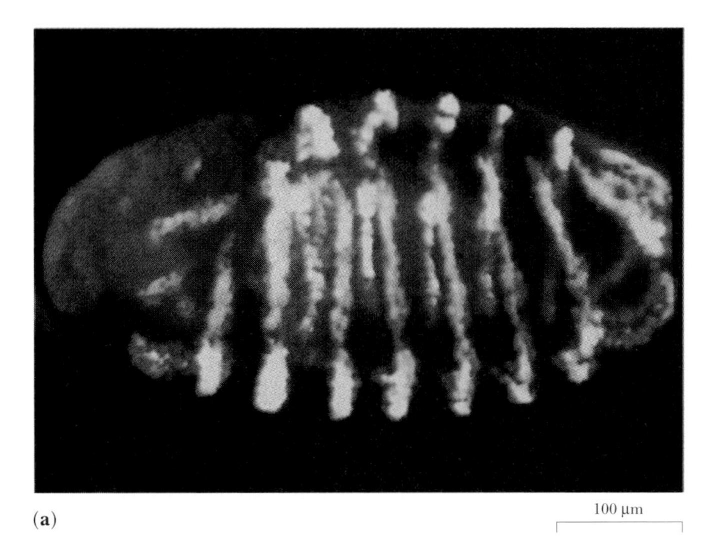

(a)

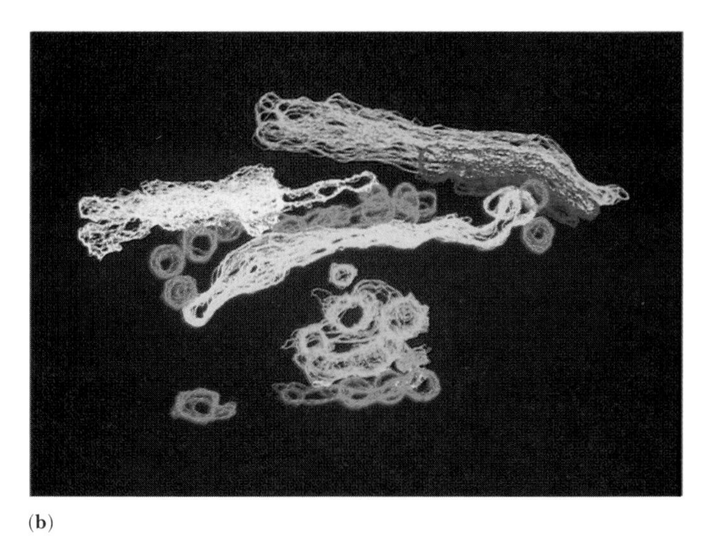

(b)

Figure 5-8 Computers have revolutionized microscopy by enhancing conventional images, controlling the illumination of a specimen, and allowing the assembly of two-dimensional images into a three-dimensional structure. **(a)** A *Drosophila* embryo seen with a confocal scanning microscope, an instrument that controls illumination by a fine laser beam and produces a thin "optical section" of the specimen. **(b)** A three-dimensional image of a Golgi apparatus constructed from dozens of optical images, assembled by computer-based image processing. *(a, Paddock/Custom Medical Stock Photo; b, Mark S. Ladinsky, University of Colorado, Boulder)*

and allowing the observation of unstained cells, as shown in Figure 5-8. In addition, computer processing can also produce three-dimensional pictures of cells by controlling the relative positions of the sample and the beam of light or of electrons.

Other methods for looking at living cells not only allow microscopists to verify the structures seen in stained tissue, but also allow biologists to study cell behavior: movement, shape changes, and internal rearrangements. Single-celled organisms—bacteria and protozoa—are easy to watch as they swim, tumble, mate, or defend themselves. But how can we look at plant and animal cells, which are usually part of thick tissues?

Biologists Can Observe Individual Living Cells from Multicellular Organisms

An experimenter can break plant or animal tissue into single cells in a number of ways. Some enzymes digest the proteins (and in the case of plants, the cell walls) that help hold cells together, and some organic chemicals tear away the metal ions needed for intercellular adhesion. In some cases, just forcing a tissue through the narrow opening of a syringe breaks it into single cells. After breaking up tissue by one or more of these methods, the cell biologist may filter out the remaining multicellular chunks of tissue and begin to watch individual cells. If the cells' environment is suitable, the cells not only stay alive but even grow and divide.

What kind of environment is "suitable"? To stay alive, cells require a proper balance of salts, a source of energy, and a supply of chemicals that they cannot make themselves. Many bacteria require little in the way of chemicals other than a few salts, sources of nitrogen (N), sulfur (S), and phosphorus (P) atoms, and glucose. The glucose provides both energy and a source of carbon (C), hydrogen (H), and oxygen (O) atoms. Some bacteria and plant cells derive energy from sunlight and can live on even simpler sources of SPONCH atoms, along with a few salts.

Animal cells, however, are much fussier because they are unable to make all their own building blocks. They therefore require more kinds of nutrients, as well as a variety of vitamins. Until the early 1970s, biologists had not figured out exactly what kind of a diet animal cells required. Culture media therefore included some crude biological extract—usually blood serum (the liquid part of the blood left behind after the removal of cells)—as first used in the early 20th century by the American developmental biologist Ross Harrison. Now, however, researchers know the precise dietary needs of most kinds of animal cells and grow them in culture media with completely defined recipes. Different kinds of cells often have slightly varying requirements. For example, some cells require the addition of an iron-binding protein called transferrin, while other cells do not.

Most normal animal cells also need a solid surface on which to grow and divide. In an animal this surface is provided by other cells or by proteins and polysaccharides. In the laboratory, the surface to which cells attach is specially treated plastic. Interestingly, cancer cells differ from normal cells in being able to grow without attaching to a surface. Indeed, this ability of cancer cells to grow in abnormal conditions is part of the reason that they can invade tissues in different parts of the body.

Increasingly powerful microscopic methods have revealed more and more about the internal organization of cells. But the microscope alone cannot show what these subcellular structures do or how they work.

HOW DO BIOLOGISTS STUDY THE FUNCTIONS OF SUBCELLULAR STRUCTURES?

Cell biologists make their first guesses about function simply by comparing the ultrastructure of different specialized cells. Cells known to export large amounts of protein, such as the pancreas cell in Figure 5-9a (which is specialized for the export of digestive enzymes) and the root cap cell of Figure 5-9b (which is specialized for the production of mucilage), are packed with internal membranes. This suggests that the membranes are involved in secretion of newly made proteins. Cells known to take in and digest food particles or foreign organisms, such as the cell shown in Figure 5-9c (a type of white blood cell), are packed with membrane-bounded organelles called lysosomes, which suggests that these organelles perform some digestive function.

Once a cell biologist formulates an hypothesis concerning the function of an observed subcellular structure,

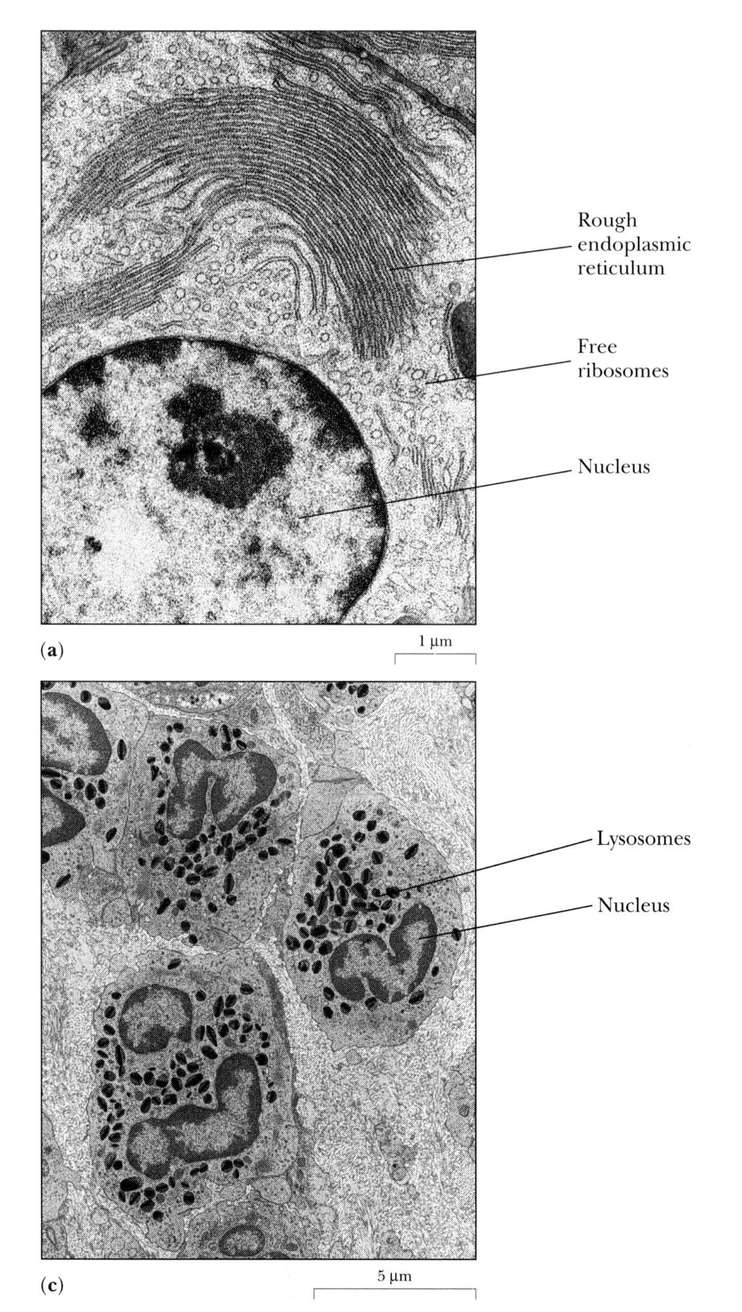

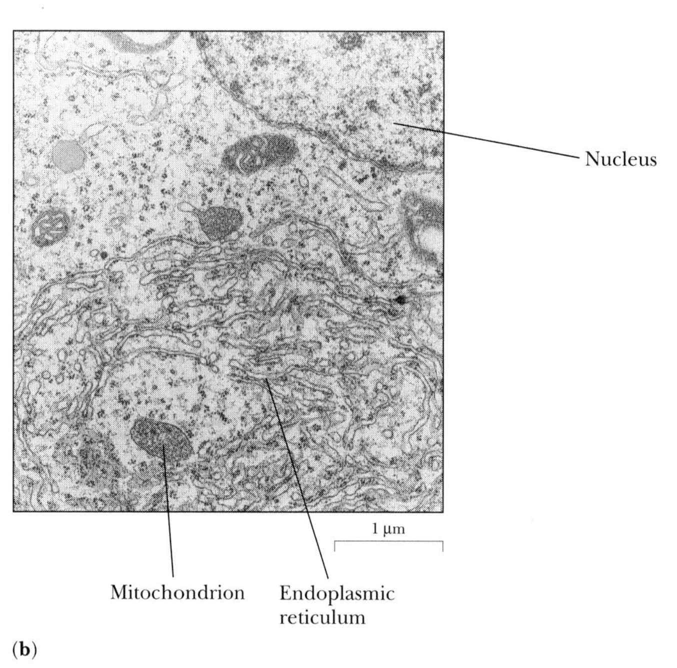

Figure 5-9 Cells specialized for export have extensive internal membranes, suggesting that these membranes participate in the export process; cells specialized for digestion contain membrane-bounded sacs full of digestive enzymes. **(a)** The nucleus and part of the cytoplasm of an acinar cell from a pancreas, specialized for the export of digestive enzymes into the stomach; notice the extensive network of membranes, studded with little particles; this is the rough endoplasmic reticulum. **(b)** The nucleus and cytoplasm of a corn root cap cell, specialized for export of mucilage; notice the network of internal membranes in the endoplasmic reticulum. **(c)** White blood cells, specialized for digestion of foreign particles (like invading bacteria); notice the cytoplasmic sacs, called lysosomes, which contain digestive enzymes. *(a, Don W. Fawcett/Visuals Unlimited; b, Biophoto Associates/Photo Researchers; c, David M. Phillips/The Population Council/Science Source/Photo Researchers)*

he or she needs to formulate experimental tests of the hypothesis. Three techniques have been extraordinarily useful in such tests: (1) **immunocytochemistry** [Greek, *immunis* = free (referring to the immune response) + *cyto* = cell], also called **immunohistochemistry** [Greek, *histos* = tissue], a method that reveals the cellular locations of particular molecules of known function; (2) **microsurgery,** the physical manipulation of subcellular structures; and (3) **subcellular fractionation,** the bulk isolation of specific subcellular structures.

Immunocytochemistry Uses Antibodies to Locate Individual Kinds of Molecules Within Cells

One way of determining the function of an organelle is to establish that it contains a specific kind of molecule with a known function. To do this, cell biologists take advantage of the natural defenses of animals to foreign molecules. A vertebrate, such as a rabbit or a mouse, responds to an injected **antigen** (a molecule recognized as foreign by the animal) by producing specific blood proteins, called **antibodies,** which bind specifically to a particular antigen and tag it for destruction. The production of antibodies is part of a vertebrate's **immune response,** which helps protect it against infection. (See Chapter 22.)

The polypeptide chains of an antibody fold to form a surface that recognizes specific shapes on the surface of the antigen. If a solution of specific antibodies is placed on a properly prepared tissue section, the antibodies bind to antigen molecules in the section. The distribution of the antibodies can reveal the distribution of a particular molecule. Immunocytochemistry uses antibodies to locate molecules within cells and tissues.

By labeling antibodies with a fluorescent dye, a cell biologist can determine the cellular distribution of the antigen they recognize (as in the photograph in Figure 5-10) after stimulating fluorescence with light focused on the specimen. For electron microscopic studies, detection of the antibodies depends on using metal-tagged proteins that stick specifically to antibodies.

The *confocal scanning microscope* allows fluorescence microscopy with much higher resolution than in conventional microscopes. It employs a special optical arrangement that focuses a laser beam at a single point within a thick specimen. The microscope detects emitted light from that point while excluding light from other points. By scanning the incident light across a plane, the microscope builds a sharp image of a thin "optical section" of a biological specimen. The result of this technological innovation has been to provide high-resolution photographs of the distribution of specific antigens.

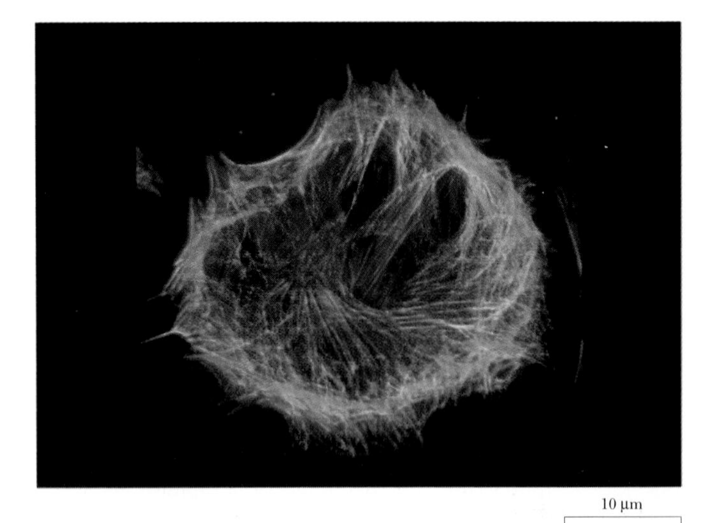

Figure 5-10 Immunocytochemistry can reveal the subcellular distribution of specific proteins. Here antibodies to two components of the cytoskeleton reveal the distribution of these two proteins. Actin filaments are shown in red and vimentin filaments in green. The high resolution of the two components depends on the use of a confocal microscope. *(Leica, Inc., Deerfield, IL)*

Cellular Surgery Allows Cell Biologists to Examine the Functioning of Subcellular Structures in Altered Environments

Microsurgical experiments with the nucleus of the marine alga *Acetabularia,* also called "the mermaid's wineglass," provide a classic example of the way biologists have learned about subcellular functions. *Acetabularia* is a huge single cell, 5 to 9 cm long, consisting of a "foot," a stalk, and an umbrella-like cap. The nucleus is always in the foot and may be isolated from the rest of the cell by cutting the stalk. After this operation, only the fragment containing the nucleus can regenerate a whole new cell. The stalk and the cap cannot. This experiment, first performed by the German biologist Joachim Hämmerling in the 1950s, supports the idea that the nucleus is the director of cellular activity.

In other experiments (Figure 5-11), more sophisticated microsurgery has allowed biologists to remove nuclei from one cell and transfer them to another. The result of such experiments is that the recipient cells acquire characteristics specified by the transferred nucleus. All these experiments are consistent with the nucleus serving an executive role within the cell.

Subcellular Fractionation Allows the Study of Cell Components

The most definitive way of showing what a cell component does is to study its biochemical or mechanical operation when it is isolated from the rest of the cell. This approach

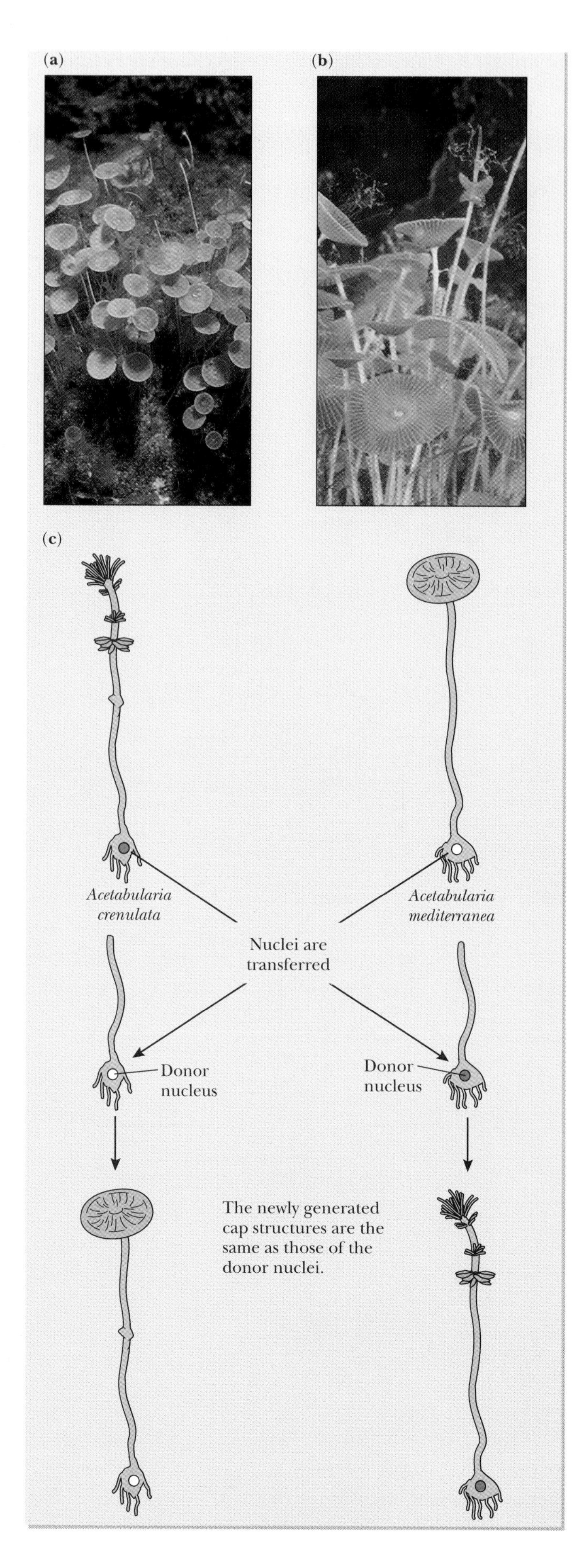

Figure 5-11 *Acetabularia,* the "mermaid's wineglass," is a huge single cell. The cap of **(a)** *Acetabularia crenulata* is smaller and more wrinkled than that of **(b)** *Acetabularia mediterranea.* **(c)** Transfer of the nucleus of *A. mediterranea* to the cytoplasm of *A. crenulata* results in the growth of a *mediterranea*-type cap. The same result is obtained when the nucleus of *A. crenulata* is transferred to the cytoplasm of *A. mediterranea*—a *crenulata* cap develops, showing that genetic information about cap shape resides in the nucleus. *(a, John Forsythe/Visuals Unlimited; b, L. L. Sims/Visuals Unlimited)*

has been extremely rewarding despite two major problems: (1) It is usually hard to isolate a pure component, and (2) it is often difficult to get an isolated component to function normally away from its normal environment.

A popular method of subcellular fractionation separates cell components on the basis of their sizes and densities. To accomplish this the experimenter gently grinds the tissue and breaks open the cells. The resulting extract is then put into a special test tube and then into a centrifuge, which generates a high force by spinning at high speeds. (See Chapter 3, p. 69.) The faster the centrifuge spins, the greater is the force. A small desk-top centrifuge creates forces up to about 1000 g (that is, 1000 times the force of gravity). A larger centrifuge generates forces up to 20,000 g. In a specially designed **ultracentrifuge,** tubes may spin at 80,000 rpm, subjecting their contents to forces up to 500,000 g.

Each time an extract is centrifuged, it is separated into two fractions—the **pellet,** which moves to the bottom of the tube and contains the heavier particles, and the **supernatant,** which remains in suspension and contains the lighter particles and dissolved small molecules. The supernatant can then be transferred to another tube and spun at higher speed. By **differential centrifugation,** successive centrifugations at increasing speeds, the experimenter obtains pellets that contain subcellular components of decreasing sizes, as illustrated in Figure 5-12—first nuclei and assorted cell debris, then membrane-bounded organelles, then sacs of membranes, then large molecular complexes. Most macromolecules fail to pellet and remain in the supernatant.

Centrifugation can also separate cell components according to buoyant density rather than according to size. Just as your body floats higher in the salt water of the ocean than in the fresh water of a lake, each cell component sinks or floats differently in solutions with different amounts of salts or sugars. A popular method for separating organelles of different densities is to layer a cell extract on top of a centrifuge tube containing a specially prepared solution of sucrose (table sugar). In the tube the concentration of sucrose is highest at the bottom and decreases in a graded way to the top in an arrangement called a **density gradient,** as shown earlier in Figure 3-8. Upon

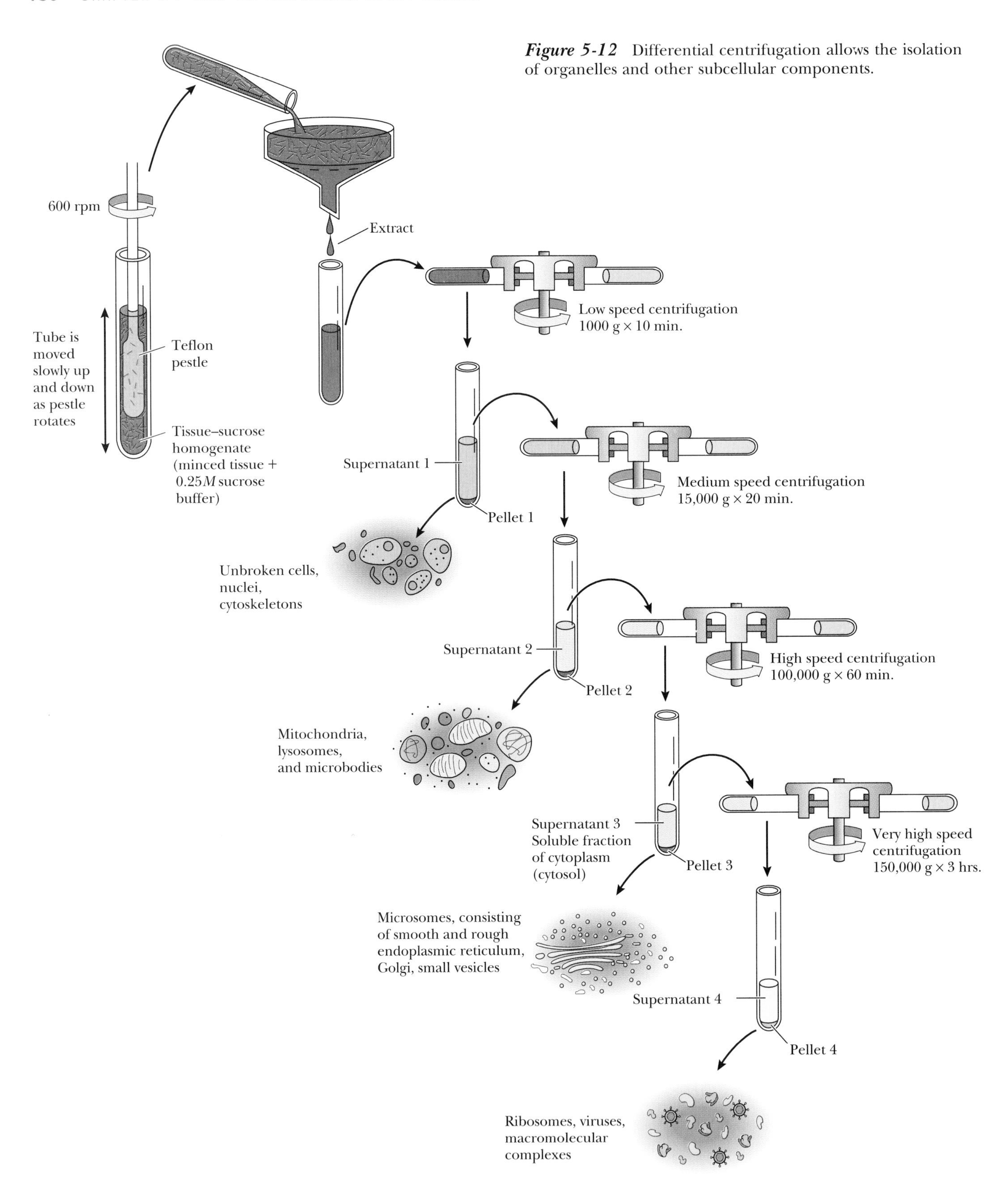

Figure 5-12 Differential centrifugation allows the isolation of organelles and other subcellular components.

centrifugation, each cell component moves through the density gradient. When a component reaches the point where its own density equals that of the sucrose solution, it stops. If it moves beyond this point, its density is less than that of its surroundings, and it floats up.

At each step of the cell fractionation, the experimenter can obtain cellular components and examine them with a light microscope or electron microscope. The trick is to minimize the disruption of cell components during the fractionation, so that the organelles and other cellular components remain recognizable and functional.

Subcellular fractionation and electron microscopy have allowed cell biologists to obtain purified cell components and to study both their chemical compositions and their functions. For example, differential centrifugation allows the isolation of lysosomes from white blood cells. Sometimes, however, obtaining pure and intact components has been difficult. Subcellular fractionation disturbs the intricate system of membranes inside eukaryotic cells, breaking them into fragments that immediately reseal to form little sacs, or vesicles. Even though such vesicles may not represent actual components of living cells, they can often carry out the same biochemical processes as the membranes from which they derived in the intact cell.

Even when purification has been possible, the real goal is to get the component to perform its suspected function ***in vitro*** [Latin = in glass, that is, in a test tube, in contrast to ***in vivo*** = in life]. Biochemists and cell biologists have been extremely clever in designing cell-free systems from specific cellular components, in which cellular processes, such as the synthesis of proteins or DNA, can go on. This approach has led to an understanding of the role of subcellular components in providing energy to the cell, harvesting the energy of sunlight, digesting food, secreting waste, and synthesizing proteins.

INTERNAL MEMBRANES DIVIDE EUKARYOTIC CELLS INTO DISTINCT COMPARTMENTS

All cells from all organisms have three common features: (1) a boundary that separates the inside of the cell from the rest of the world; (2) a set of genetic instructions; and (3) a cell body. The **plasma membrane** is a cell's boundary. A highly organized and responsive structure, the plasma membrane not only defines the limits of a cell but also helps to regulate the cell's internal environment. The membrane selectively admits and excretes specific molecules and ions.

The genetic instructions for each cell are contained in DNA, which is confined to a restricted part of the cell. In all eukaryotes, most of each cell's DNA is within a membrane-bounded structure called the **nucleus,** as illustrated in Figure 5-13a. In prokaryotes, the DNA is not surrounded by a membrane but still occupies a limited region of the cell, called a **nucleoid**. In eukaryotes some organelles also have their own DNA. (See Chapter 16.)

In eukaryotes, the part of the cell outside the nucleus but inside the membrane is called the **cytoplasm.** For a long time, biologists thought of the cytoplasm as a more or less homogeneous jelly-like substance. In the last 40 years, however, improved ways of looking at cells have revealed (as illustrated in Figures 5-4 and 5-13) that the cytoplasm contains a host of small structures, called **organelles,** each of which performs a specialized task. In eukaryotic cells many organelles are enclosed by membranes, which isolate the contents of the organelle from the rest of the cytoplasm.

The greatest surprise revealed by the TEM was the huge amounts of internal membrane in eukaryotic cells. These membranes divide cells into seven distinct compartments, as shown in Figure 5-14: the nucleus, the cytosol, the endoplasmic reticulum, the Golgi apparatus, lysosomes, peroxisomes (see Figure 5-22), and mitochondria. A plant cell, such as that in Figure 5-14b, contains an additional type of compartment, the plastid.

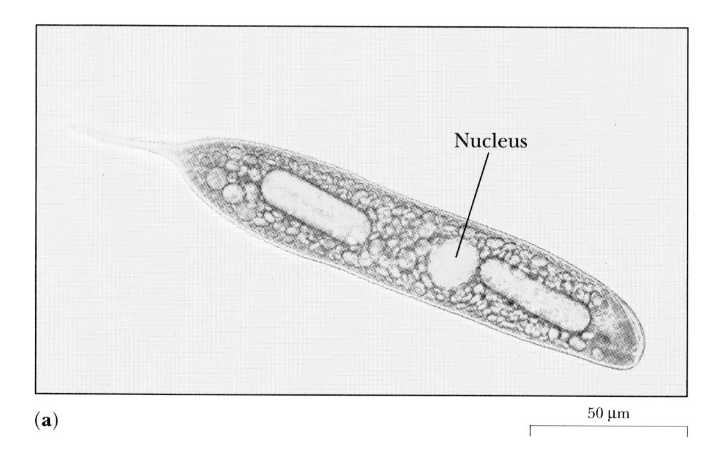

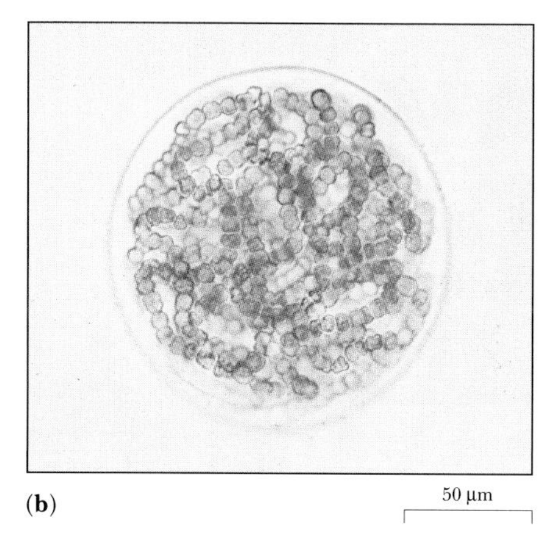

Figure 5-13 Both eukaryotic and prokaryotic cells are surrounded by a similar plasma membrane. The DNA of a eukaryotic cell is (almost all) in a membrane-bounded nucleus; the DNA of a prokaryotic cell lies in a region called the *nucleoid,* which is not surrounded by a membrane. **(a)** A single-celled eukaryote, *Euglena,* a green alga; **(b)** a colony of a prokaryote, *Nostoc,* a cyanobacterium (formerly called a blue-green alga). *(a, Alex Rakosky/Custom Medical Stock Photo; b, Winston Patnode/Photo Researchers)*

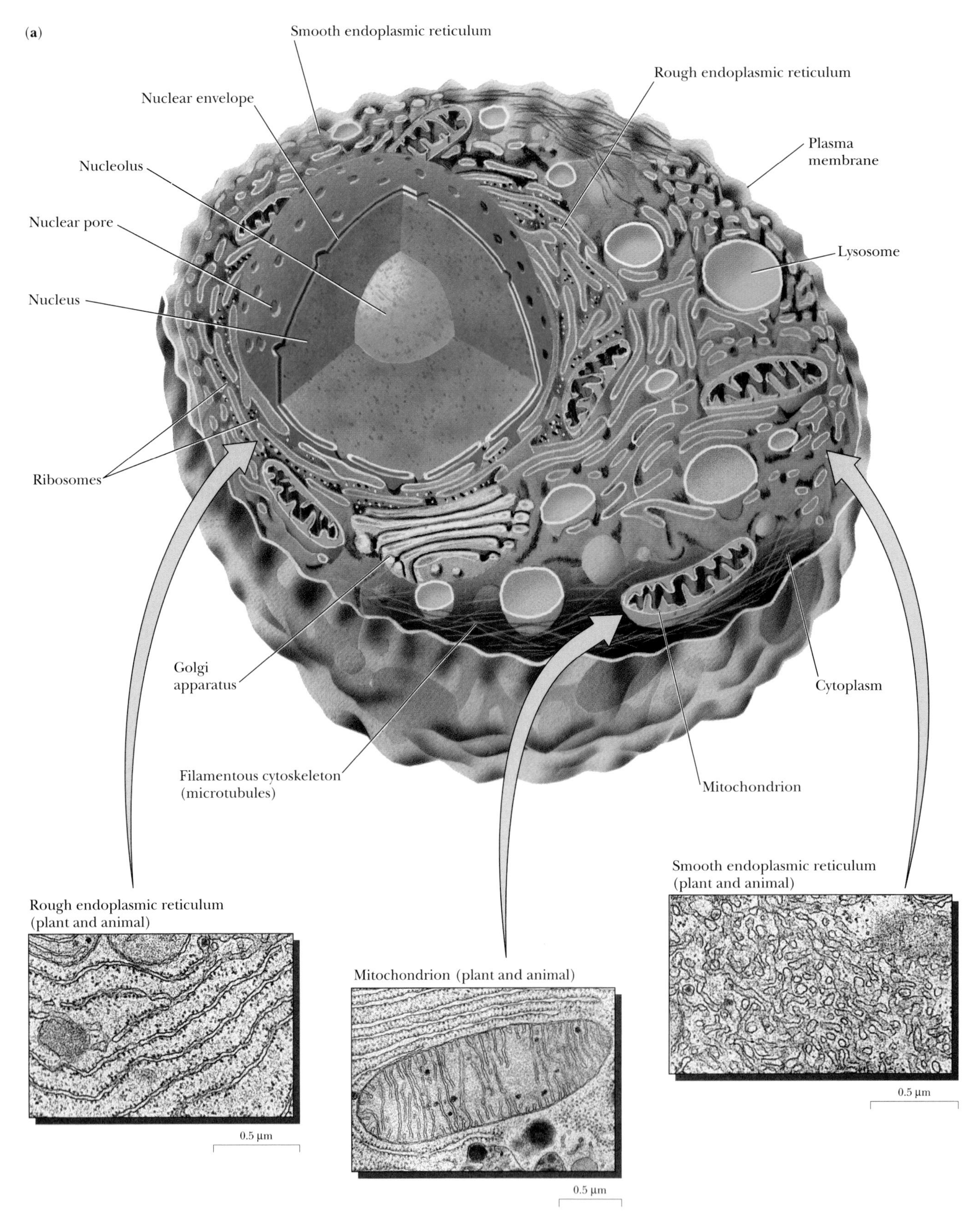

Figure 5-14 Composite (generalized) eukaryotic cells, with distinct membrane-bounded compartments: **(a)** an animal cell; **(b)** a plant cell. *(RER, SER: R. Bolender-D. Fawcett/Visuals Unlimited; mitochondrion: Keith R. Porter/Photo Researchers; chloroplast, nucleus: R. Howard Berg/Visuals Unlimited; Golgi apparatus: Dr. Dennis Kunkel/Phototake NYC)*

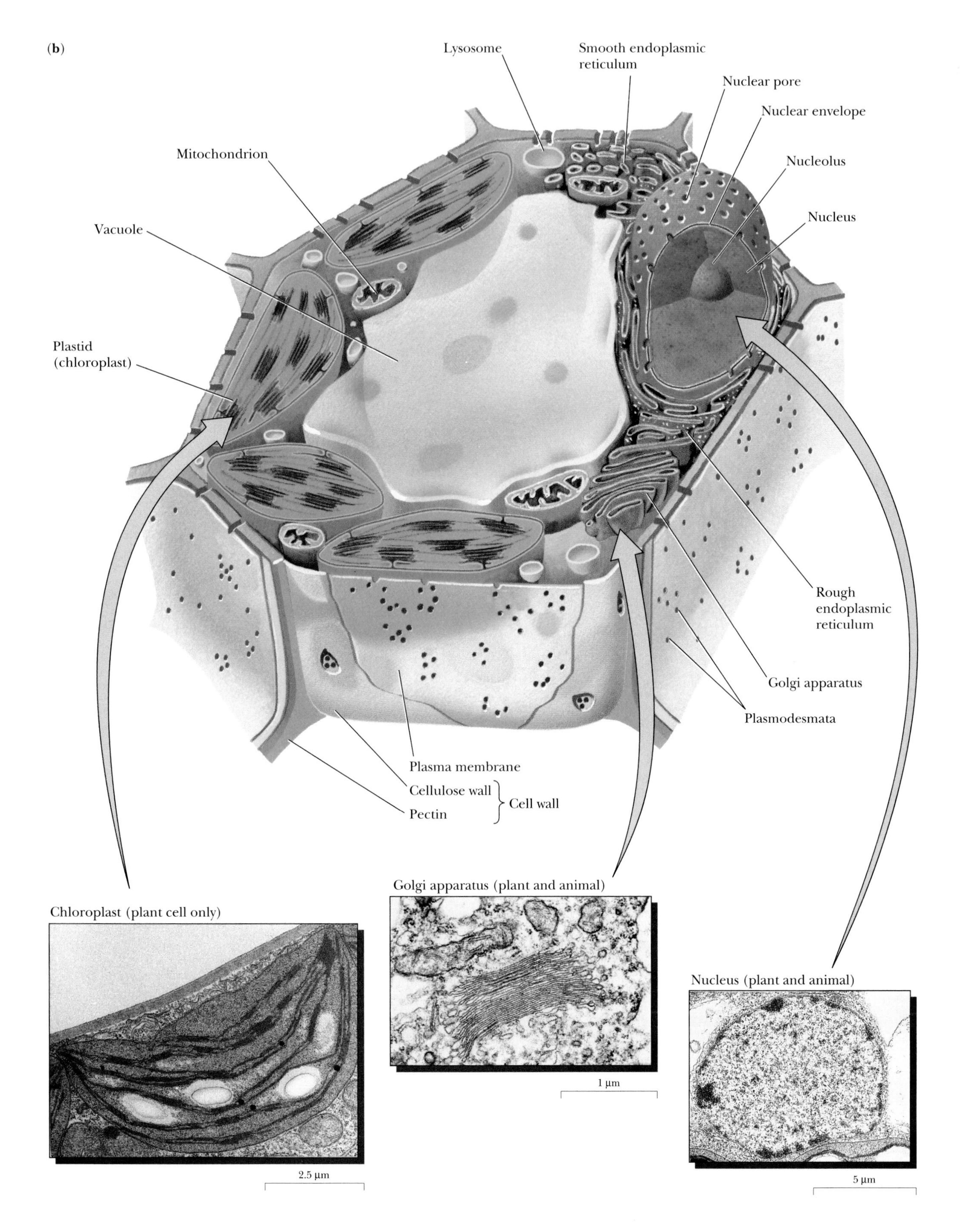
(b)
Lysosome
Smooth endoplasmic reticulum
Nuclear pore
Nuclear envelope
Nucleolus
Nucleus
Mitochondrion
Vacuole
Plastid (chloroplast)
Rough endoplasmic reticulum
Golgi apparatus
Plasmodesmata
Plasma membrane
Cellulose wall
Pectin
Cell wall
Chloroplast (plant cell only)
2.5 μm
Golgi apparatus (plant and animal)
1 μm
Nucleus (plant and animal)
5 μm

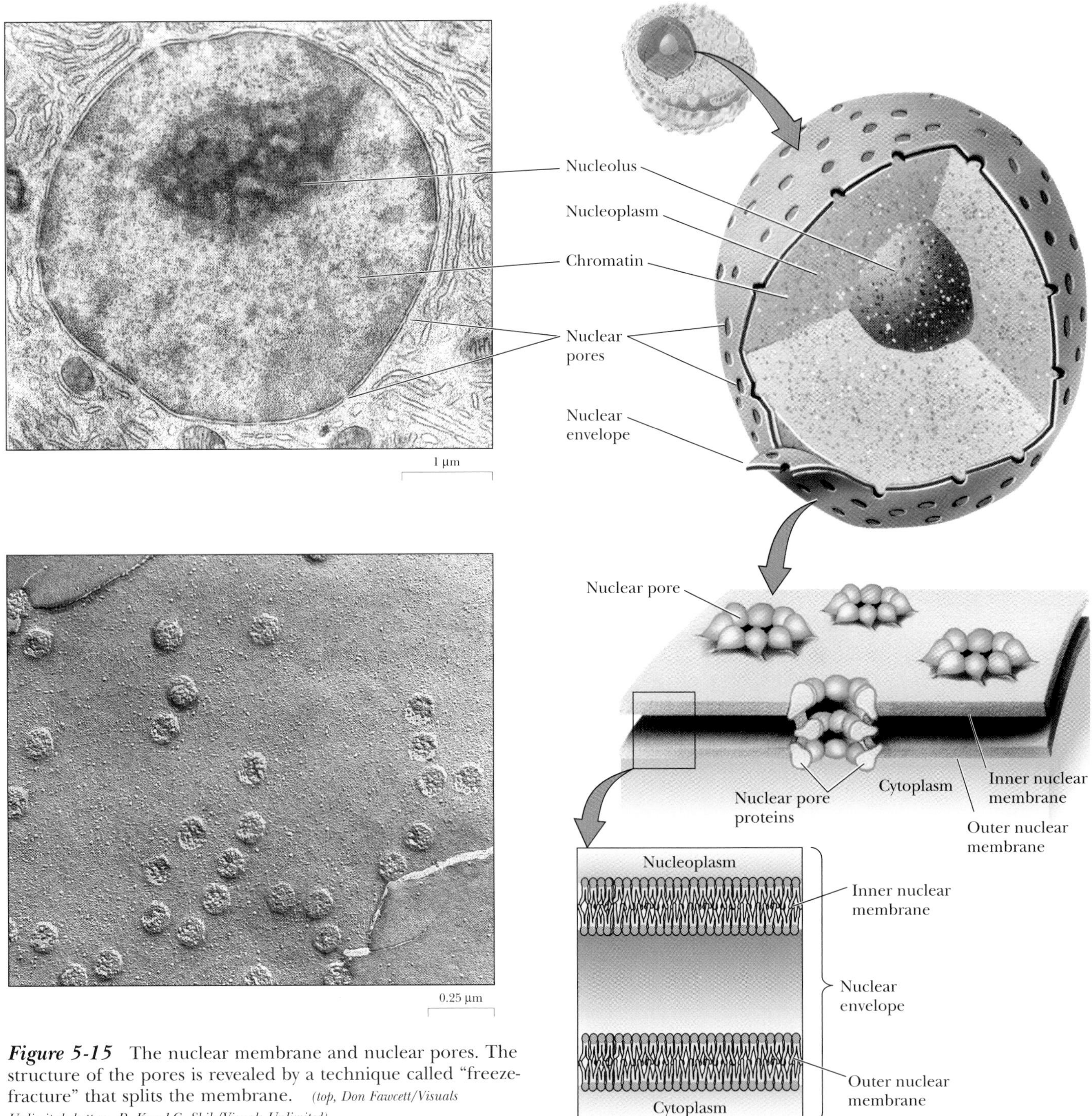

Figure 5-15 The nuclear membrane and nuclear pores. The structure of the pores is revealed by a technique called "freeze-fracture" that splits the membrane. *(top, Don Fawcett/Visuals Unlimited; bottom, R. Kessel-G. Shih/Visuals Unlimited)*

Eukaryotic cells also contain a variety of **vesicles** and **vacuoles,** membrane-bounded sacs without any obvious internal structure which differ only in size. The vesicles of animal cells are generally about 100 nm in diameter. In plant cells, vacuoles are especially large and may occupy up to 95% of the volume of the cell. Prokaryotic cells do not usually have any membrane-bounded internal compartments.

To varying extents, cell biologists can isolate the different compartments by subcellular fractionation and study them in cell-free systems. Such experiments have allowed cell biologists to learn much about the functions of each type of cell compartment.

The Nucleus Is the Most Prominent Compartment in a Eukaryotic Cell

The nucleus usually makes up 5–10% of the volume of the cell. In nondividing cells, commonly used stains produce a diffuse pattern in the nucleus. This pattern results from the binding of stains to **chromatin** [Greek, *chroma* = color], a complex of DNA and protein. In dividing cells, however, staining reveals that DNA lies in distinct pieces called **chromosomes** [Greek, *chroma* = color + *soma* = body].

With the electron microscope, we can see that the nucleus's boundary, shown in Figure 5-15, is a double membrane called the **nuclear envelope.** The two membranes are separated by about 20–40 nm. In many places, however, the two membranes come together to form interruptions called **nuclear pores,** which (as can be seen in Figure 5-15) form channels between the **nucleoplasm** (the contents of the nucleus) and the cytoplasm.

As discussed earlier in this chapter (p. 128), the nucleus serves as the "chief executive officer" of the cell, while the rest of the cell is its "administrative domain." The nucleus contains the genetic information (coded in DNA) that is passed from generation to generation of organisms and from one cell to both its "daughter" cells. In Part III of this book, we consider how DNA carries genetic information, how cells use this information, and how they pass it to their daughter cells.

Most nuclei have conspicuous bodies called **nucleoli** (singular, **nucleolus**) [Latin = a small nucleus]. Each nucleolus contains a **nucleolar organizer,** a region of one or more chromosomes that contains the genes for the RNA within the ribosomes, the structures on which protein synthesis occurs.

The Cytosol Is the Cytoplasm That Lies Outside Membrane-Bounded Organelles

The part of the cytoplasm not contained in membrane-bounded organelles is called the **cytosol.** Most of a cell's biochemical transformations occur within the cytosol. Running through the cytosol is a complicated network of protein fibers, called the **cytoskeleton** (which is also visible in Figures 5-4 and 5-13). The cytoskeleton, which is discussed in more detail in Chapter 6, gives the cell its shape, holds organelles in place, and participates in cell movement.

Cell biologists usually isolate the cytosol from the organelles suspended in it simply by breaking cells open and removing the organelles by centrifugation. The cytosol contains thousands of different kinds of enzymes responsible for producing building blocks, processing small molecules to derive energy, and synthesizing proteins. Although the cytosol is aqueous, about 20% of its weight is protein, making its consistency somewhat gelatinous.

Within the cytosol are organelles, vesicles and vacuoles, and smaller structures that are not enclosed by membranes, as can be seen in Figure 5-16a. These include granules of glycogen and droplets of stored fat. The TEM also shows many little round dots, about 15–30 nm in diameter, called **ribosomes** [Latin, *soma* = body + "ribo" because they contain *ribo*nucleic acid (RNA)].

Ribosomes (Figure 5-16b) are complexes of RNA molecules and protein molecules. They consist of two separate assemblies, called ribosomal subunits. They are indispensable for protein synthesis both in cells and in cell-free systems. They are the structures on which protein molecules are assembled, and they contain the enzymes needed to form peptide bonds. When ribosomes lie in the

(a)

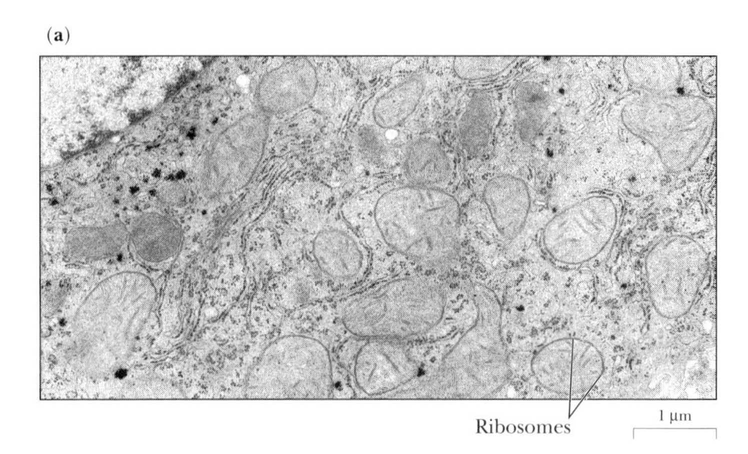

(b)

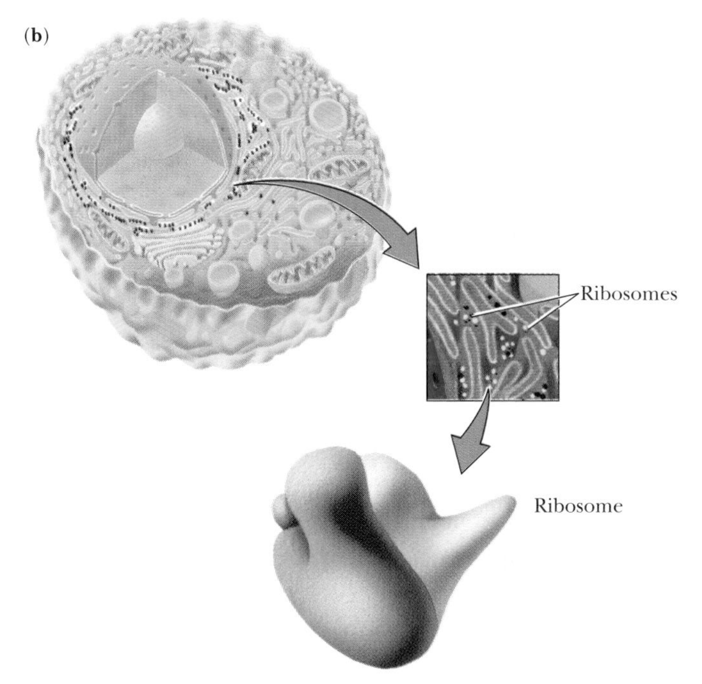

Figure 5-16 **(a)** A liver cell showing free ribosomes and membrane-bound ribosomes (rough ER). **(b)** Ribosomes are complex assemblies of RNA and protein molecules. *(a, Dr. Dennis Kunkel/Phototake)*

cytosol, they are called **free ribosomes.** Ribosomes may also be associated with a cell's internal membranes, in which case they are called **membrane-bound ribosomes.**

The Endoplasmic Reticulum Is a Folded Membrane That Encloses an Extensive Space Inside a Cell

In a single microscopic section, the **endoplasmic reticulum (ER)** [Greek, *endon* = within + *plasmein* = to mold + Latin, *reticulum* = network] appears as an extensive and convoluted network, as shown in Figure 5-17. By looking at the ER in successive ("serial") sections, microscopists gradually became convinced that the ER is actually a single sheet of membrane enclosing a single **lumen,** or enclosed space [Latin = light, an opening]. The ER mem-

(Text continues on page 138.)

ESSAY

NUCLEAR PORE COMPLEXES SERVE AS GATEKEEPERS AND ESCORTS

Almost all the genetic information of eukaryotic cells lies within the nucleus, in DNA. The nucleus also contains many RNA molecules, all made in the nucleus, and many proteins, all made in the cytoplasm. As we will discuss in Chapter 14, RNA molecules (messenger RNAs or mRNAs) carry information that specifies the amino acid sequences of proteins. These RNA molecules must move from the nucleus to the cytoplasm, where they attach to the protein-synthesizing ribosomes. Besides transporting information-carrying mRNA, the nucleus also exports assembled ribosomal subunits, made within the nucleolus, as well as other RNAs needed for protein synthesis.

In addition to the outgoing traffic, the nucleus must also import proteins—the proteins of the ribosomes that will later be exported, enzymes needed to copy DNA, other proteins that compress DNA into compact structures, enzymes that copy ("transcribe") DNA into RNA, proteins that regulate the process of transcription, and proteins responsible for the subnuclear structures seen in Figure 1. Subnuclear structures include the nucleoli, a fibrous network called the *nuclear matrix,* and *nuclear pores complexes.* Each nucleus has about 3000–4000 nuclear pore complexes, which allow the selective passage of molecules into and out of the nucleus.

Figure 1 *A freeze-fracture electron micrograph of the nucleus of a human cell line (HeLa), showing the nuclear membrane and several nuclear pores.* (Biophoto Associates/Photo Researchers)

Nuclear proteins move from the cytoplasm into the nucleus. Proteins imported into the nucleus have "address tags," short amino acid sequences (4–8 residues long) that somehow tell the cell to direct them to the nucleus. The present view is that nuclear pore complexes play an important role in directing traffic into and out of the nucleus. In addition to serving as a gatekeeper, moreover, the nuclear pore complex appears to assist directly in both the outward transport of RNAs and ribosomal subunits and the inward transport of proteins.

Transport between the nucleus and cytoplasm differs from that between the cytoplasm and other membrane-bounded organelles. In the case of most other organelles, proteins first pass through the membrane of the endoplasmic reticulum in an unfolded form, usually following a "leader" sequence composed of hydrophobic amino acid residues. The nucleus, however, communicates with the cytoplasm through aqueous pores, which allow proteins and ribosomal subunits to pass, apparently in a fully folded form. The electron microscope reveals that the pores include both the inner and outer nuclear membranes, with the outer membrane continuous with the endoplasmic reticulum, as seen in Figure 2.

In the last decade, biochemists and microscopists have been able to isolate intact pores from detergent extracts of nuclear membranes. Biochemical analysis reveals that the nuclear pore complexes consist of more than 100 different proteins, with a total molecular size of about 125 million. (For comparison, a eukaryotic ribosome has a molecular size of about 4 million.) Researchers have been able to isolate a few of these proteins and have identified others from studies of mutant yeasts that cannot make fully functional pores.

In the electron microscope, the pores have a striking octagonal symmetry and an apparent diameter of about 100 nm. Using computer reconstructions of high resolution scanning and transmission electron micrographs, researchers have produced stunning pictures of nuclear pores, such as that shown in Figure 2a. Figure 2b shows a recent model of a nuclear pore

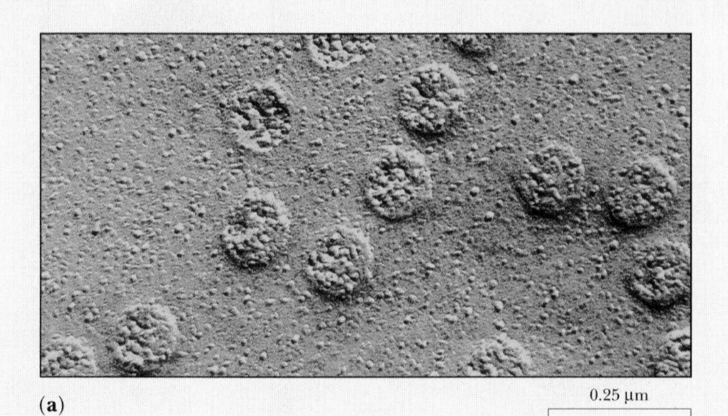

Figure 2 *The nuclear pore complex: **(a)** a high resolution micrograph of a nuclear pore complex, taken with a technique called "field-emission in-lens scanning electron microscopy"; **(b)** a model of the nuclear pore complex.* (*a, Richard Kessel/Visuals Unlimited*)

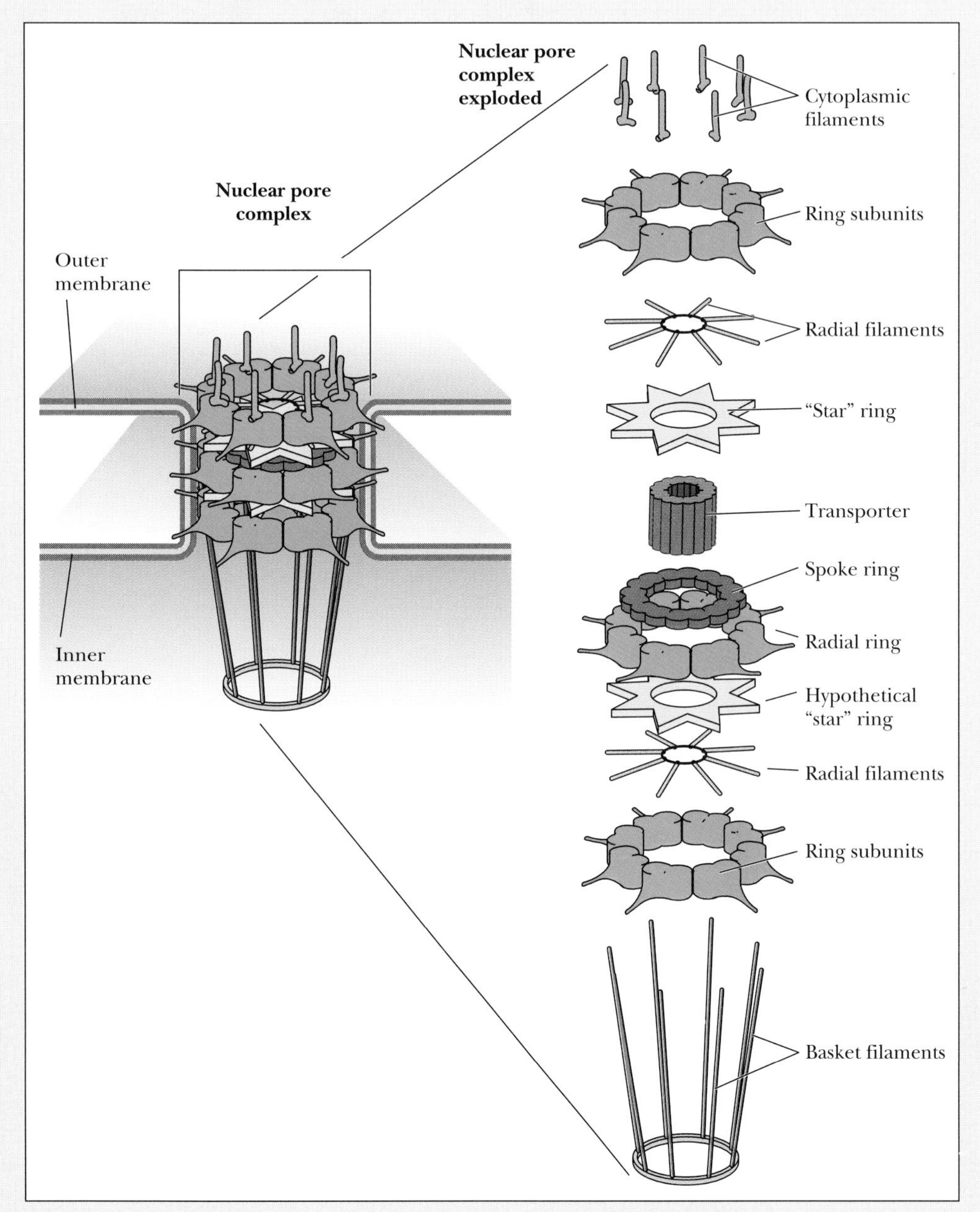

(*continued*)

ESSAY *continued*

complex, based on microscopic studies of pore complexes after mild treatments with detergents and proteases.

While the apparent diameter of the nuclear pore is about 50 nm, the actual functional diameter of the pore is much narrower. To determine the functional diameter of the pore, researchers attached particles of colloidal gold to proteins and RNAs and injected the labeled macromolecules into nuclei and cytoplasm. They found that small proteins (with molecular sizes less than 20,000) diffuse relatively rapidly between nucleus and cytoplasm, but that most proteins with molecular sizes larger than 60,000 cannot pass at all. From studies of the rates of entry and exit, researchers concluded that the functional diameter of the nuclear pore is about 9 nm.

Much larger molecules and molecular complexes can pass through the pore however. Nucleoplasmin is a large nuclear protein that contains specific nuclear targeting sequences. When attached to gold particles, nucleoplasmin can move easily through the pores, as illustrated in Figure 3. Similarly, gold-labeled RNA molecules, injected into the nucleus, can exit through the pores. Such studies have led to the hypothesis that some proteins of the nuclear pore complex form a diaphragm that can dilate to admit molecules as large as 20 nm in diameter. Understanding the structure and the dynamics of nuclear pore complexes is an active area of current research for cell biologists.

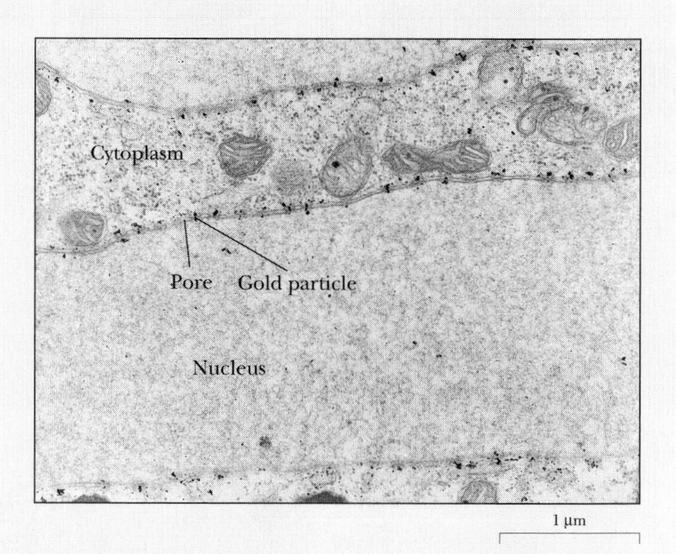

Figure 3 *The large nuclear protein, nucleoplasmin, can move through nuclear pores, though it is considerably larger than the cut-off size for freely diffusing particles. Here the black dots are colloidal gold particles that have been attached to proteins that are too big to pass through the nuclear pores.* *(Carl Feldherr, University of Florida)*

brane is also continuous with the outer membrane of the nucleus. In many eukaryotic cells the ER lumen makes up about 15% of the cell volume, and the membranes of the ER more than half the total membrane.

The ER consists of two parts that are distinct in both appearance and function. The **rough ER,** seen in Figure 5-17a, is studded with ribosomes on the cytosolic side of the membrane. The **smooth ER,** seen in Figure 5-18b, has no ribosomes. Rough and smooth ER are both present in most eukaryotic cells, but some specialized cells may have particularly large amounts of one or the other. Cells specialized for the export of proteins, such as the plasma cell in Figure 5-18a, have large amounts of rough ER. Cells specialized for lipid production, such as the cells that produce large amounts of steroid hormones (shown in Figure 5-18b), contain large amounts of smooth ER.

The rough ER can be isolated by subcellular fractionation and its function studied *in vitro.* As suspected, the rough ER is the site of synthesis of exported proteins. Almost all of the proteins made by the ribosomes on the rough ER end up outside the cell or on its external surface. As these exported proteins are made, they immediately pass to the ER lumen. There most of them are modified by the addition of carbohydrates to become **glycoproteins** [Greek, *glykis* = sweet]. In contrast, proteins that are destined for the cytosol do not contain attached carbohydrates and are made by free ribosomes.

Free ribosomes and the ribosomes of the rough ER are identical. As discussed in Chapter 14, the presence of special hydrophobic sequences in an exported protein or a membrane protein causes the newly made proteins and the ribosomes that make them to attach to the ER. The protein then moves directly into the ER lumen as soon as it is made.

The smooth ER is devoted mostly to the synthesis and metabolism of lipids. Enzymes on the smooth ER of liver cells also help animals get rid of toxic chemicals by converting them to a water-soluble form.

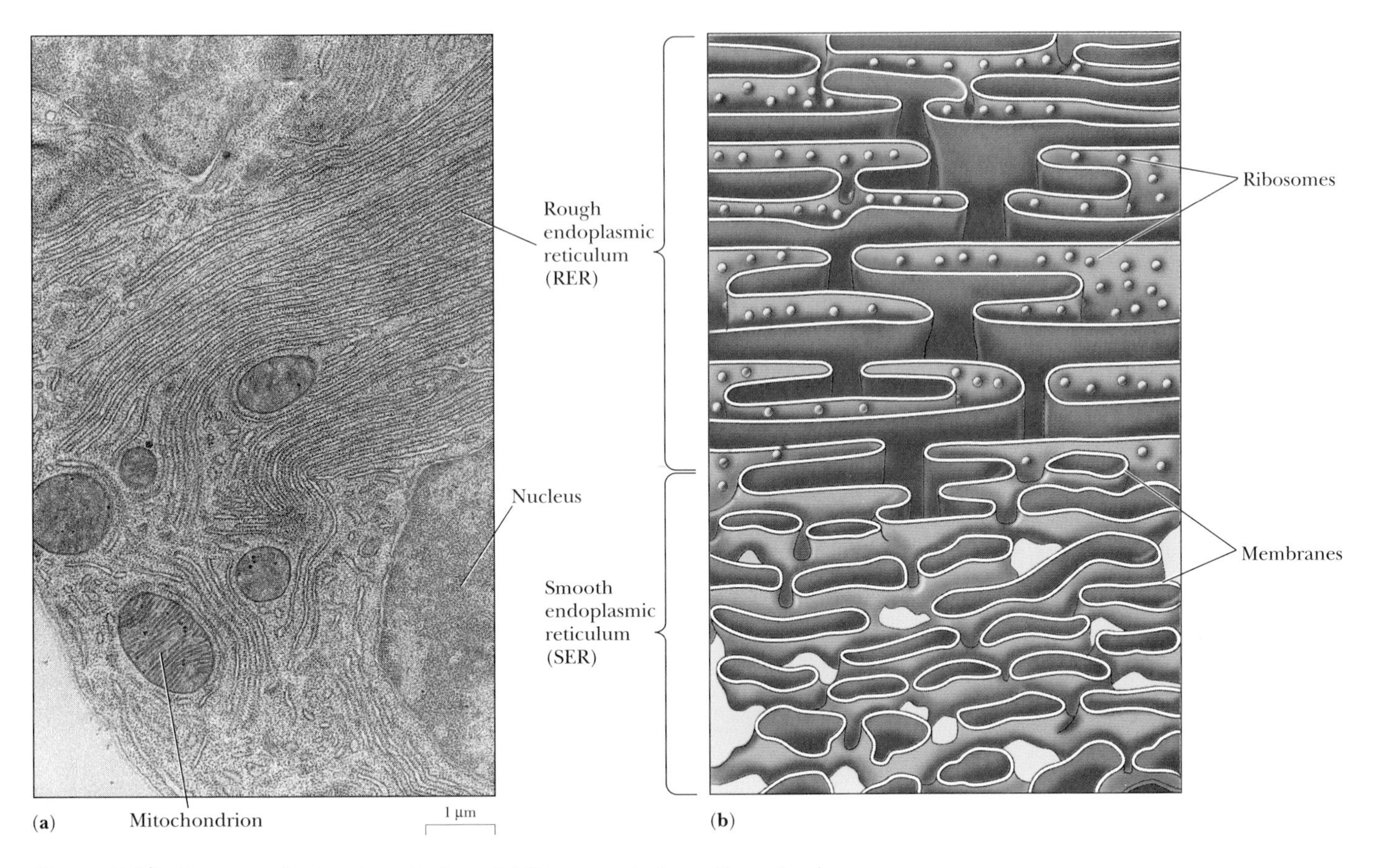

Figure 5-17 Rough endoplasmic reticulum. **(a)** Photograph shows the extensive rough ER in an acinar cell from a bat pancreas, highly specialized for the production of digestive enzymes; **(b)** drawing of rough and smooth ER. *(a, K. R. Porter/Photo Researchers)*

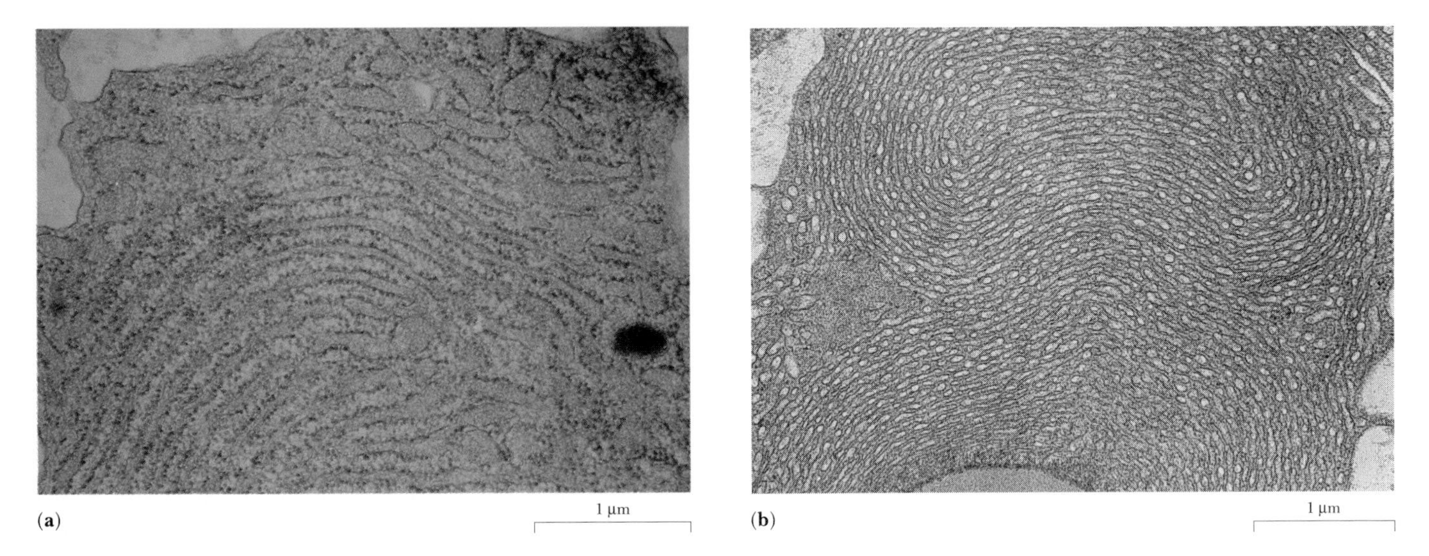

Figure 5-18 **(a)** Rough endoplasmic reticulum in a plasma cell; **(b)** smooth endoplasmic reticulum in a steroid-producing cell. *(a, Dr. R. Alexley/Peter Arnold; b, Don W. Fawcett/Photo Researchers)*

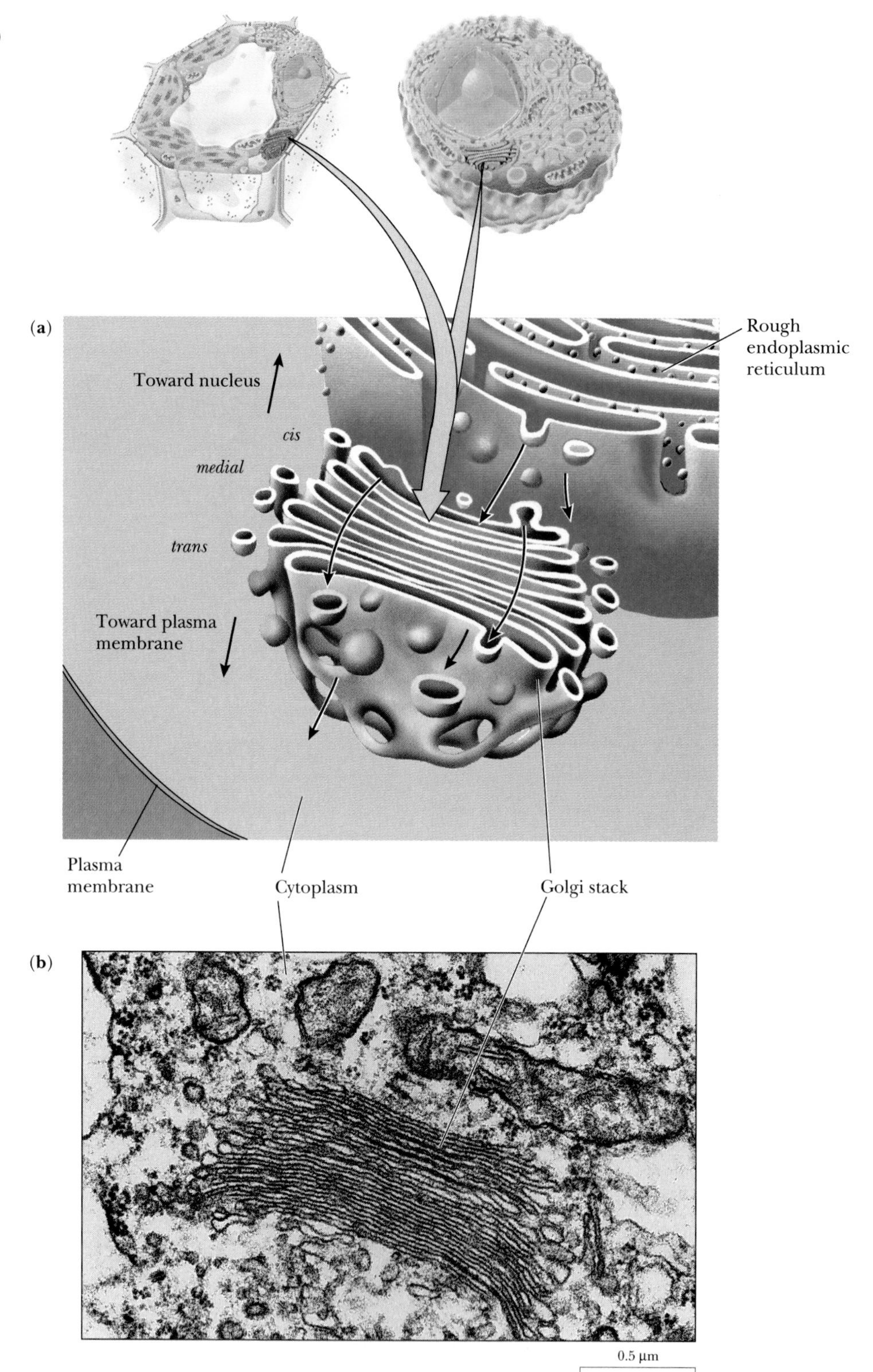

Figure 5-19 The Golgi apparatus: **(a)** directional organization; **(b)** the Golgi apparatus in a cell specialized for the secretion of mucus into the intestine. *(b, Dennis Kunkel/Phototake NYC)*

The Golgi Apparatus Directs the Flow of Newly Made Proteins

The **Golgi apparatus** (or Golgi complex), as illustrated in Figure 5-19, appears in the TEM as sets of flattened discs, resembling stacked dinner plates, usually near the nucleus. Each stack is about 1 μm in diameter and typically consists of about six discs (although, in single-celled eukaryotes, a single stack may have more than 30 discs).

Each stack has three distinct regions, which cell biologists call ***cis, medial,*** and ***trans,*** with the *cis* region usually nearest the nucleus and the *trans* region usually nearest the plasma membrane, as illustrated in Figure 5-19a. Between the *cis* side of the Golgi stack and the adjacent ER is a less regular structure called the *cis* Golgi network. Between the *trans* side of the Golgi stack and the plasma membrane is the *trans* Golgi network. Surrounding the stacks of flat discs are some smaller, spherical vesicles.

Every eukaryotic cell has a Golgi apparatus, but the total number of stacks varies among different cell types. Cells that secrete large amounts of glycoproteins (such as the cells shown in Figure 5-19b, which secrete mucus into the intestines) have particularly large numbers of Golgi stacks.

Cell biologists have learned about the role of the Golgi apparatus and the rough ER in cells that secrete large amounts of proteins by labeling newly made glycoproteins with radioactive amino acids or carbohydrates. Among the most influential experiments of this type were those performed by George Palade and Marilyn Farquhar, then at Rockefeller University. This work showed that proteins destined for export appear first in the rough ER and then in the Golgi apparatus, as shown in Figure 5-20a. The small vesicles near the *cis* face of the Golgi apparatus shuttle glycoproteins from the rough ER to the Golgi stack. During their passage through the Golgi apparatus, glycoproteins have their carbohydrates modified and are then packaged into membrane-bounded **secretory vesicles.** The secretory vesicles are always on the *trans* side of the Golgi apparatus, nearest to the plasma membrane. They then fuse with the plasma membrane, discharging their contents to the outside of the cell by the process of **exocytosis,** the transport of materials contained within a vesicle to the outside of a cell. We discuss the process in more detail in Chapter 6.

The Golgi apparatus participates intimately in the process of export. It is also involved in the formation of membrane-bounded structures that stay in the cell, such as lysosomes, which we discuss later in this chapter. The Golgi apparatus appears to function both as a packaging center and as traffic director.

How does the Golgi apparatus order the flow of proteins to different destinations? For some proteins the answer to this question appears to lie in modifications of the

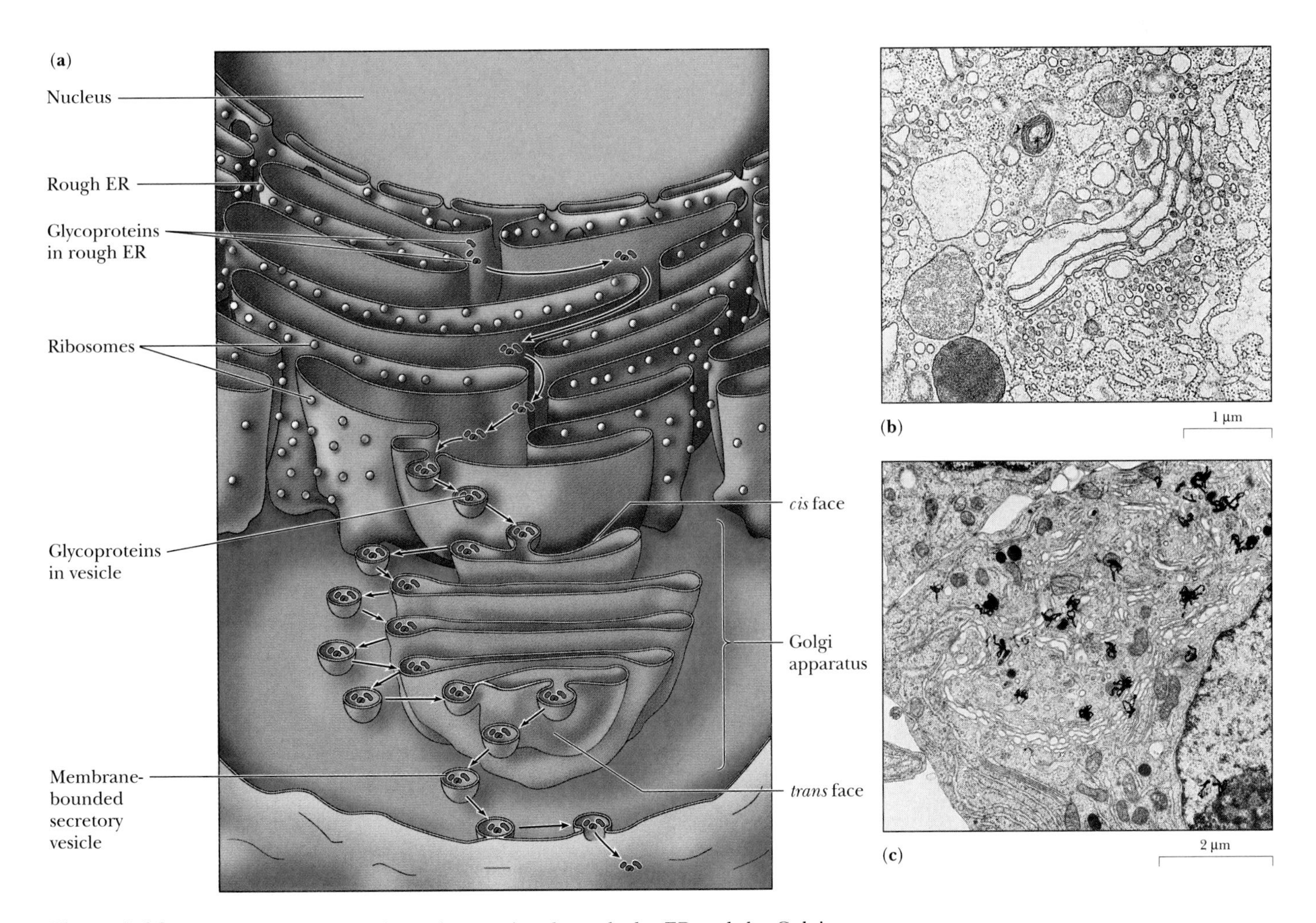

Figure 5-20 The secretion of newly made proteins through the ER and the Golgi apparatus. (**a**) Diagram of pathway. (**b**) The Golgi region from a pancreatic acinar cell; these highly specialized cells, which were intensely studied by George Palade and Marilyn Farquhar, have provided many insights into the mechanisms of protein export; Palade has said that the cells had "the effect of the song of a mermaid: Irresistible and half transparent. . . . Reasonable working hypotheses were already suggested by the structural organization itself." (**c**) Movement of radioactively labeled proteins (dark spots) within a pancreatic acinar cell. *(b, c, George Palade and Marilyn Farquhar)*

carbohydrates originally attached to glycoproteins in the ER lumen. These proteins have amino acid sequences that serve as "zip codes," signaling the machinery of the Golgi apparatus to add specific "tags" that direct them to appropriate targets.

Lysosomes Function as Digestion Vessels

Lysosomes are small membrane-bounded vesicles found in all cells. They vary widely in size, but all contain hydrolytic enzymes that break down proteins, nucleic acids, sugars, lipids, and other complex molecules. The large vacuole of a plant cell contains similar enzymes, and in many ways it resembles a giant lysosome.

Lysosomes are particularly numerous in cells that actively perform phagocytosis. Granulocytes, the white blood cells that engulf and digest foreign invaders and cellular debris, are packed with lysosomes. The lysosomes of granulocytes are visible even in the light microscope because they specifically take up certain stains so that they appear as colored granules. With an electron microscope, cell biologists can distinguish two types of lysosomes, which can be seen in Figure 5-21: *primary lysosomes,* which are nearly spherical and do not contain any debris, and *secondary lysosomes,* which are larger and more irregular and contain material that is in the process of being digested.

Granulocytes engulf extracellular material by surrounding it with plasma membrane. (See Chapter 6.) The resulting vesicle, called an **endosome**, fuses with a primary lysosome, as shown in Figure 5-21. Endosomes are the intermediate transporters of virtually all particles taken into a cell from outside.

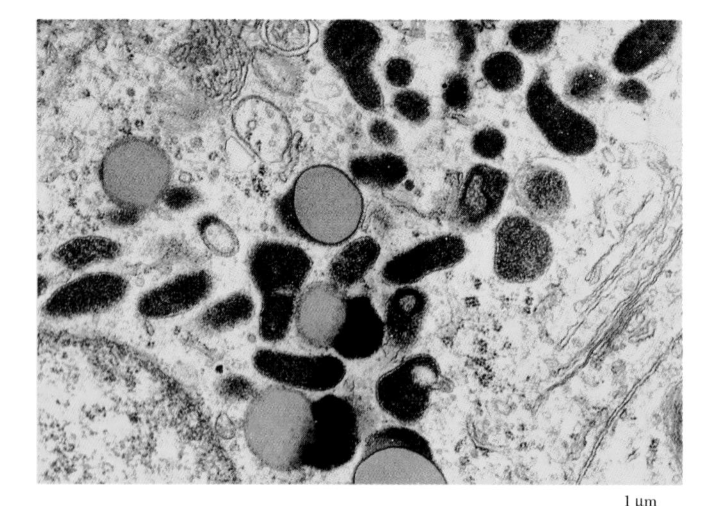

Figure 5-21 Primary and secondary lysosomes. Secondary lysosomes form as primary lysosomes fuse with a membrane-bounded vesicle formed during endocytosis, as discussed in Chapter 6. *(Don Fawcett/Visuals Unlimited)*

Fusion of an endosome and a primary lysosome produces a secondary lysosome, in which enzymes digest the engulfed material. The membranes of lysosomes, which are themselves resistant to digestion, keep their hydrolytic enzymes from digesting the cytosol's own proteins. If the lysosome membrane breaks down, however, the cell begins to consume itself and dies.

Lysosomes form from membranes derived from the Golgi apparatus. Their digestive enzymes are made (in inactive form) on the ribosomes of the rough ER. They are packaged into lysosomes after passing through the ER lumen, where each lysosomal enzyme acquires a special attached sugar. This sugar provides the lysosomal proteins with an address tag, so that the Golgi apparatus directs them to lysosomes rather than exporting the proteins to the outside of the cell. The digestive enzymes become active only after entering the acidic environment (about pH 5) within the lysosomal membrane.

Peroxisomes Produce Peroxide and Metabolize Small Organic Molecules

Another kind of membrane-bounded vesicle, **peroxisomes** (shown in Figure 5-22), contains enzymes that use oxygen to break down various molecules by pathways that produce hydrogen peroxide (H_2O_2), hence their name. Peroxisomes are widespread in all kinds of eukaryotes.

The peroxisomes of different kinds of cells contain different sets of enzymes. For example, specialized peroxisomes (called **glyoxisomes**) in the seeds of some plants provide energy for the growing plant embryo by breaking down stored fats. When yeast cells grow in a medium rich in alcohol, they produce large numbers of peroxisomes, which contain an enzyme that breaks down alcohol. In some cases, peroxisomes can occupy half the volume of the cell.

Peroxisomes are especially abundant in cells that are actively engaged in the synthesis, storage, or breakdown of lipids. Their diameters can range from 0.15 to more than 1 μm. All peroxisomes, however, produce hydrogen peroxide, which is highly reactive and can easily damage cells. They therefore all also contain a protective enzyme (called catalase) that breaks down hydrogen peroxide to form water and oxygen.

We would expect the enzymes of peroxisomes to be directed to these organelles in much the same way as those of lysosomes, but this does not seem to be the case. Free ribosomes, rather than rough ER, produce at least some peroxisome enzymes. Proteins destined for the peroxisomes carry a three-amino-acid address label—Ser-Lys-Leu.

Mitochondria Break Down Small Organic Molecules and Capture Their Energy in ATP

Mitochondria [singular, **mitochondrion;** Greek, *mitos* = thread + *chondrion* = a grain], shown in Figure 5-23, are

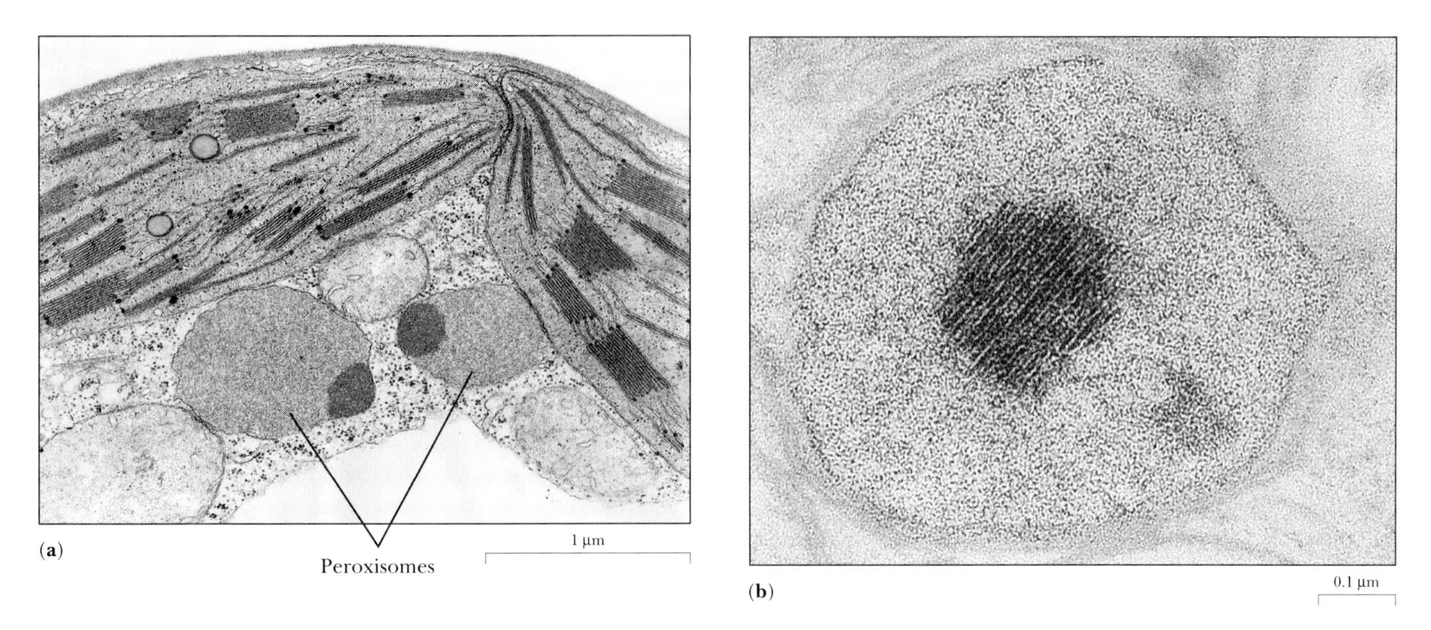

Figure 5-22 Peroxisomes. **(a)** TEM showing two chloroplasts with two peroxisomes next to them; **(b)** peroxisome in a liver cell. *(a, Biological Photo Service; b, H. D. Fahimi, Universitat Heidelberg)*

one of the most prominent organelles in most eukaryotic cells. Typically 0.5–1 μm in diameter and 5–10 μm long, they can barely be seen in the light microscope, although their presence has long been known because they actively take up certain dyes.

The TEM reveals the mitochondrion's characteristic ultrastructure. Two membranes, 6–10 nm apart, separate the mitochondrion's **matrix** (its innermost chamber) from the cytosol. The outer membrane of a mitochondrion is smooth, but the inner membrane forms elaborate folds,

(Text continues on page 146.)

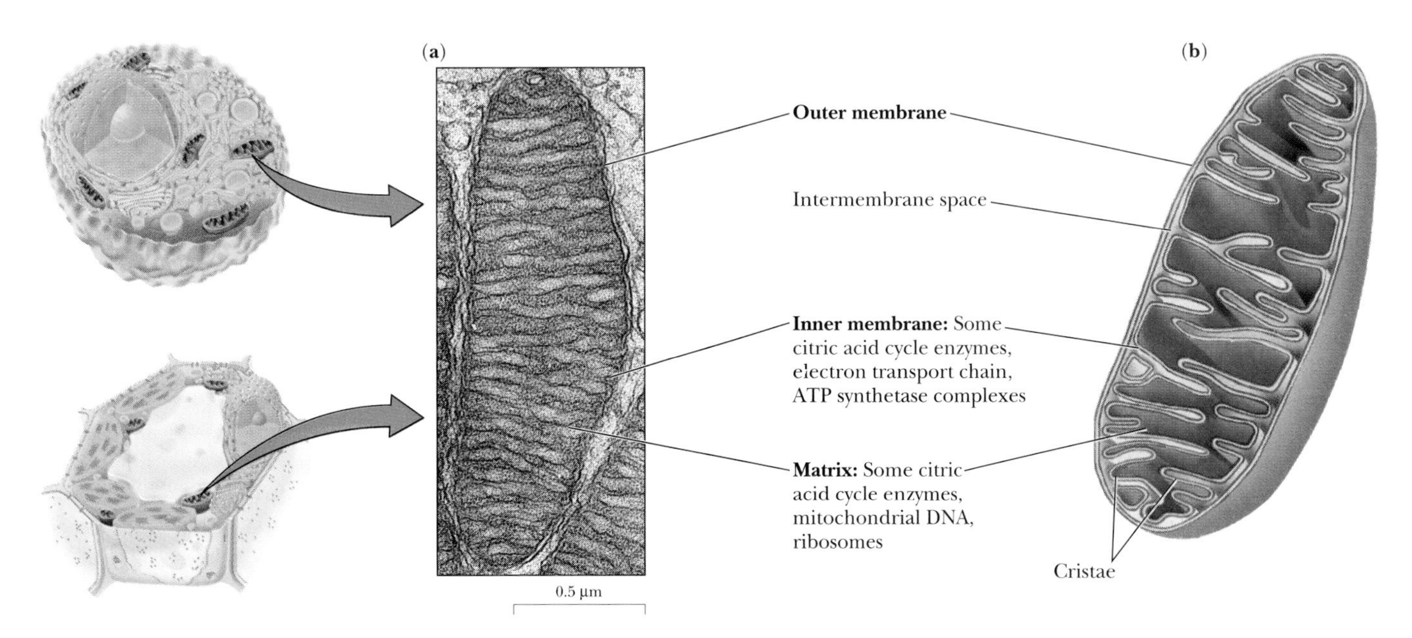

Figure 5-23 Mitochondria have an outer membrane and an inner membrane; the inner membrane folds to form *cristae,* which extend into the mitochondrial *matrix.* *(D. W. Fawcett/Photo Researchers)*

ESSAY

LYSOSOMAL DISEASES

Besides digesting materials taken into a cell by endocytosis, lysosomes perform a major role in the normal turnover of cellular constituents. In addition to proteases and nucleases, then, lysosomes contain an elaborate set of enzymes that break down complex molecules found throughout the body. The lipid components of brain membranes are particularly challenging substrates for lysosomal enzymes, and digestion of a single type of molecule may require the sequential action of many enzymes.

Genetic deficiencies affecting any of these individual enzymes can lead to devastating neurological diseases. The hallmark of these *lysosomal diseases,* also called *lysosomal storage diseases,* is the accumulation of abnormal lysosomes, such as those shown in Figure 1, filled with molecules that cannot be digested into recyclable components. One of the most common of the lysosomal diseases in the U.S. is *Tay-Sachs disease,* a genetic disorder particularly prevalent among the descendants of Eastern European Jews. Infants with Tay-Sachs disease appear normal for much of the first year, but then they develop mental retardation, weakness, and blindness. They die by age 3 or 4.

The brains of children who have died from Tay-Sachs disease have swollen lysosomes, which contain high levels of a lipid molecule called ganglioside GM_2. Gangliosides are lipids derived not from glycerol (Figure 2a) but from *sphingosine,* an amine with a long hydrocarbon chain, shown in Figure 2b. Sphingosine is coupled to a fatty acid (in an amide linkage) to form *ceramide* (Figure 2c), and ceramide is coupled to glucose to form *glucocerebroside* (Figure 2d). GM_2 and other gangliosides form by the attachment of specific sugars to glucocerebroside (Figure 2e). Gangliosides are particularly prominent components of the "gray matter" of the brain, where they make up about 6% of the lipids.

Normally, gangliosides are continuously degraded and replaced, using the pathway shown in Figure 3. Tay-Sachs patients lack an enzyme, called hexosaminidase, which removes the sugar N-acetylglucosamine from GM_2. Lacking this enzyme, children with Tay-Sachs disease accumulate GM_2 in nerve cells, causing the associated damage and death.

Other lysosomal diseases lead to the accumulation of other intermediate products of ganglioside degradation. Gaucher's disease, for example, leads to the accumulation of glucocerebroside, but this compound does not accumulate in the brain as much as does GM_2 in Tay-Sachs, presumably because of the action of other brain enyzmes. Instead, Gaucher's patients may survive for decades, though glucocerebroside does accumulate in and enlarge both the liver and spleen. In a few cases, physicians have successfully treated Gaucher's disease by administering a purified enzyme. The enzyme, engineered to be taken up by

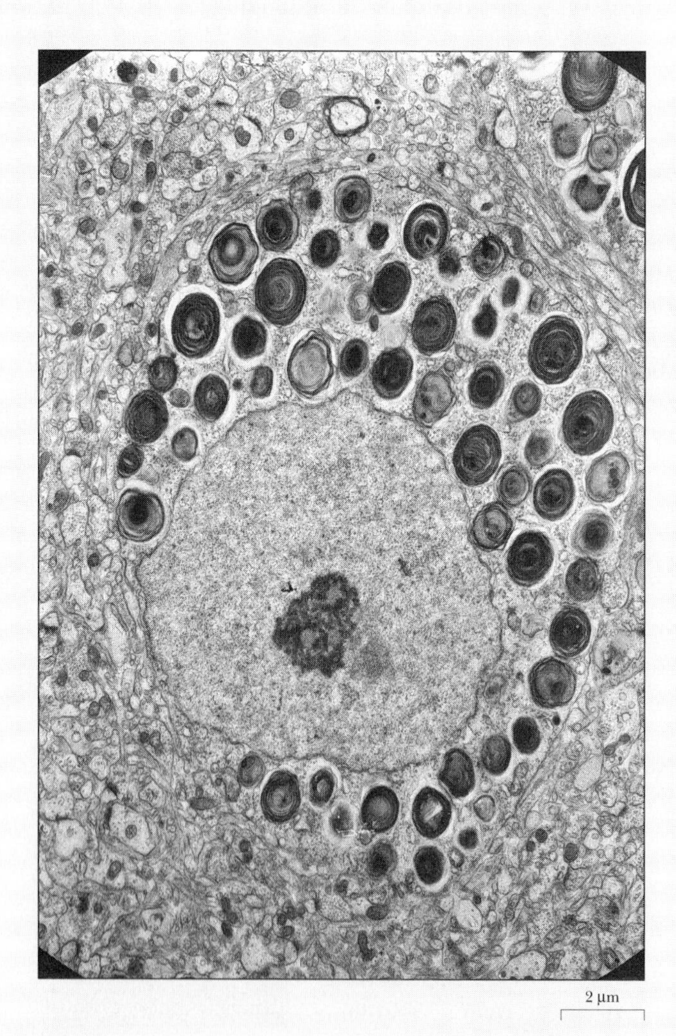

Figure 1 *Lysosomes swollen with undigested products of metabolism. These lysosomes come from the brain of a child who died of Tay-Sachs disease.* (Dr. Robert D. Terry)

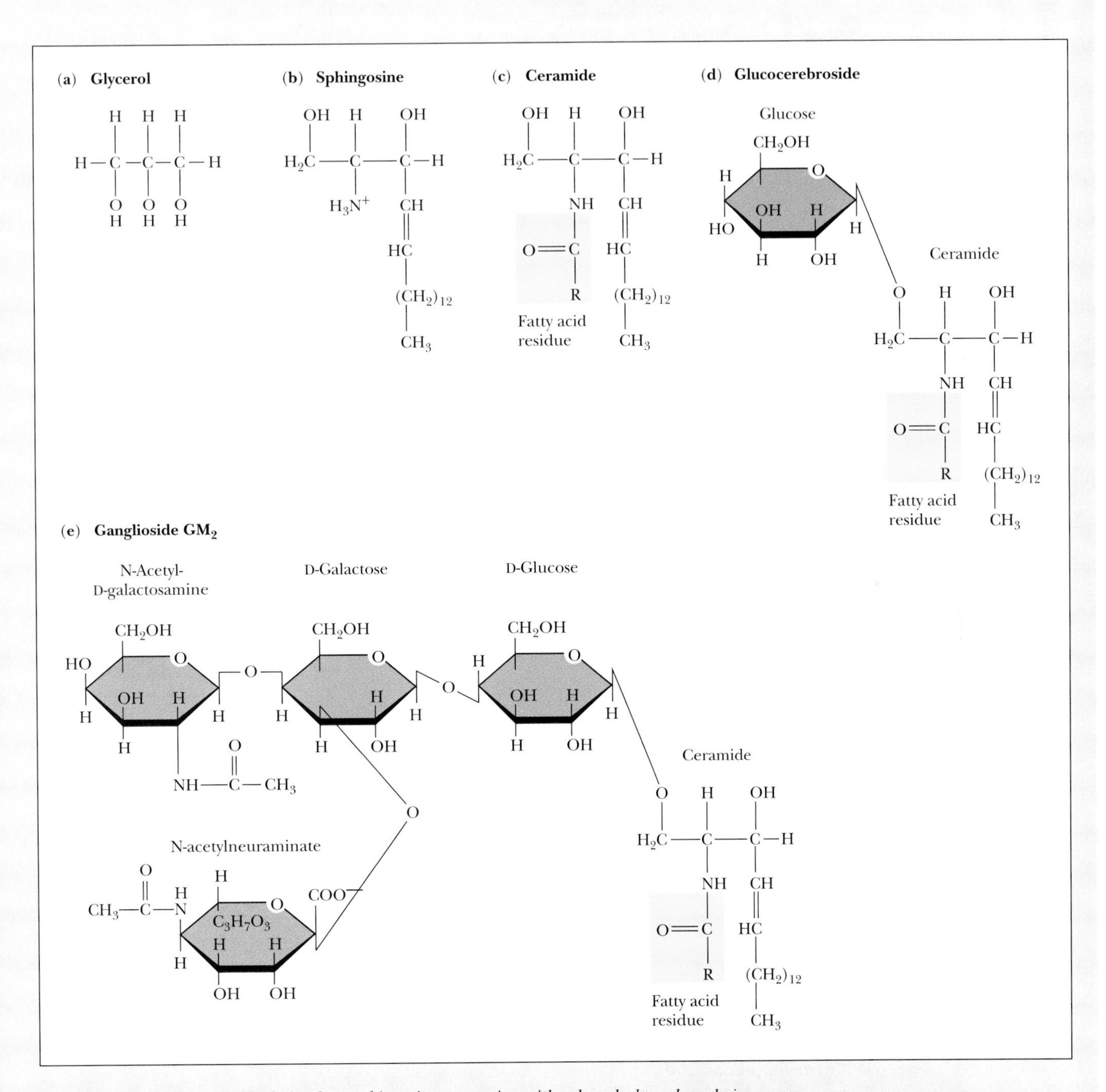

Figure 2 *Many brain lipids derive from sphingosine, an amine with a long hydrocarbon chain, rather than from glycerol:* ***(a)*** *glycerol;* ***(b)*** *sphingosine;* ***(c)*** *ceramide;* ***(d)*** *glucocerebroside;* ***(e)*** *ganglioside GM$_2$.*

(continued)

ESSAY *continued*

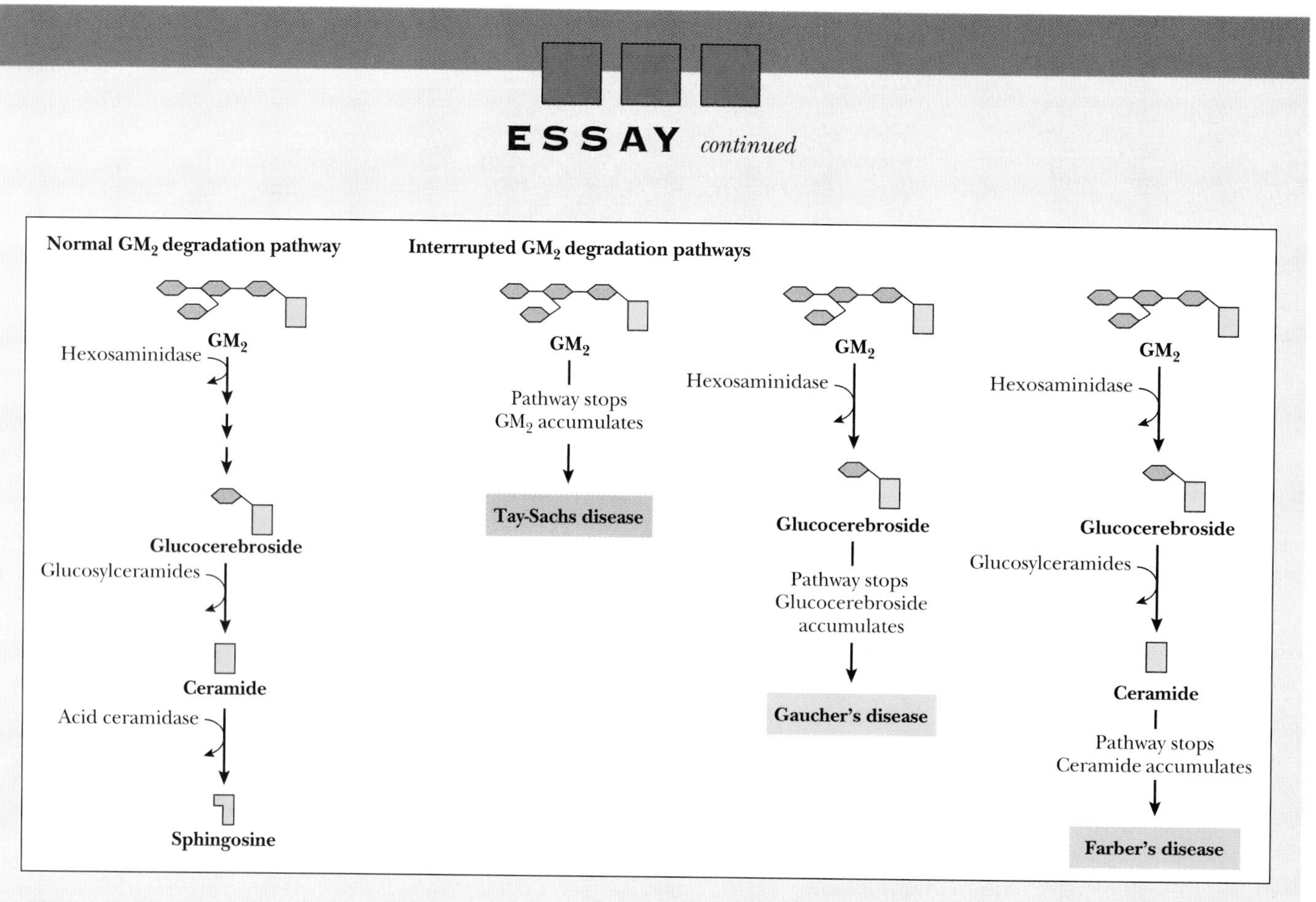

Figure 3 The pathway of ganglioside degradation in normal lysosomes. Sugars are sequentially removed from ganglioside GM_2, producing glucocerebroside, ceramide, and sphingosine. The enzymatic defect in Tay-Sachs disease leads to the accumulation of GM_2. Another enzymatic defect, in Gaucher's disease, leads to the accumulation of glucocerebroside; still another, in Farber's disease leads to the accumulation of ceramide.

lysosomes, has reduced the damage to liver and spleen in these patients. Genetically manipulated cells with the same enzyme defect have been successfully repaired by treatment with a gene that encodes an active enzyme, and many researchers think that Gaucher's disease and other lysosomal diseases are likely to be particularly susceptible to gene therapy, a topic that we will discuss in Chapter 17.

The enzyme deficiencies in Tay-Sachs disease and other lysosomal diseases can also be detected in carriers, who have half the level of enzyme activity as noncarriers. (See Chapter 12.) Tay-Sachs screening programs for carriers are common. Couples who have a risk of conceiving a Tay-Sachs embryo can also elect to assay the embryo for the enzyme deficiency.

called **cristae.** As shown in Figure 5-23, the two mitochondrial membranes divide each mitochondrion into four distinct locations: (1) the outer membrane; (2) the intermembrane space; (3) the inner membrane; and (4) the matrix, the space within the inner mitochondrial membrane.

Mitochondria come in a variety of sizes and shapes. A liver cell may contain several thousand mitochondria, making up about a quarter of the cell volume. Cells that use particularly large amounts of energy, such as those of heart muscle, may contain still more. Time-lapse photography of living cells with a phase contrast microscope reveals that mitochondria may be constantly in motion, changing shape, fusing with each other, and dividing.

Cell biologists can easily isolate mitochondria by cell fractionation and can study their role in cellular function. They are the powerhouses of the cell, producing most of the ATP in nearly all eukaryotic cells. The production of

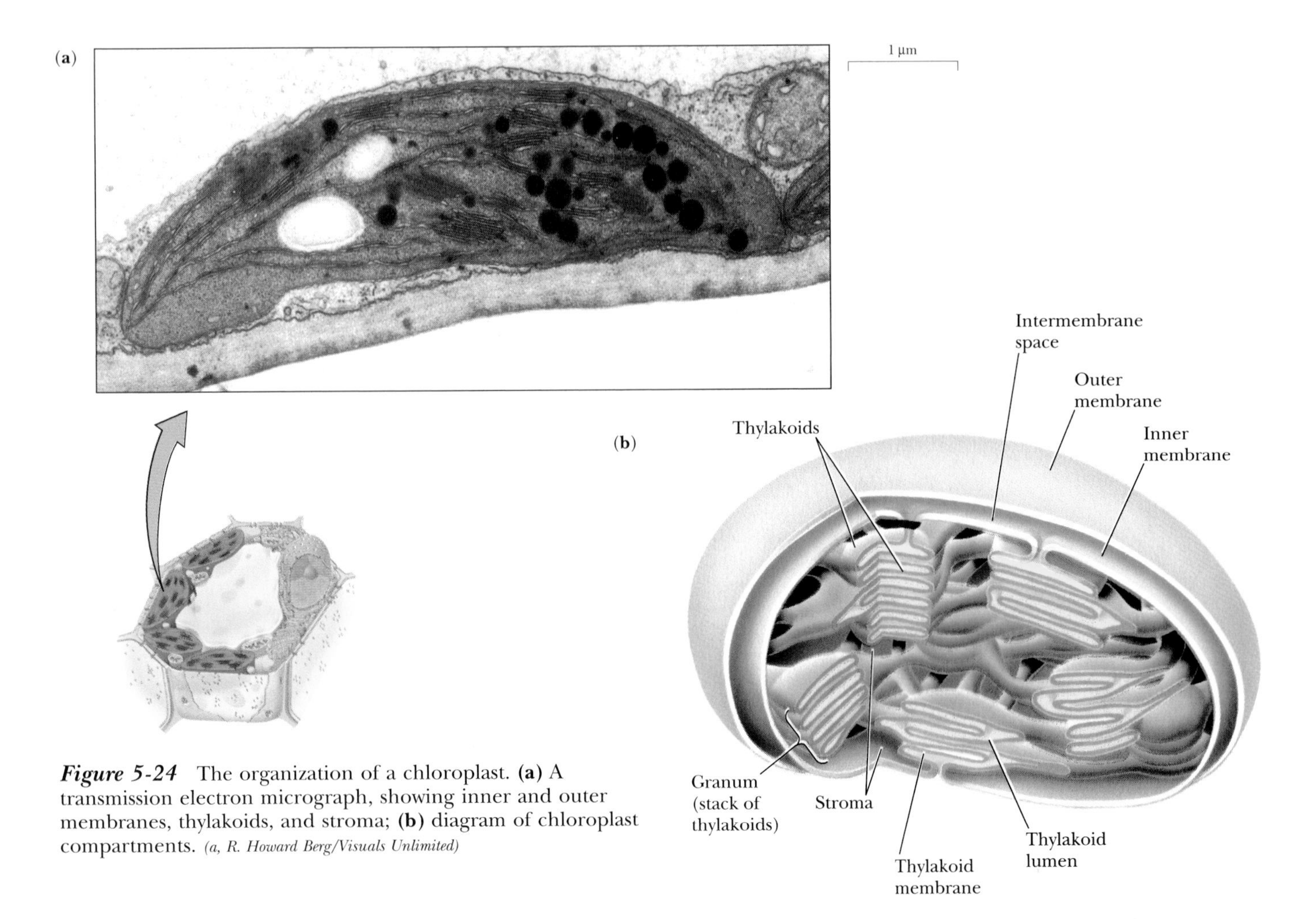

Figure 5-24 The organization of a chloroplast. **(a)** A transmission electron micrograph, showing inner and outer membranes, thylakoids, and stroma; **(b)** diagram of chloroplast compartments. *(a, R. Howard Berg/Visuals Unlimited)*

ATP by mitochondria depends on the availability of oxygen and on the organization of their internal and external membranes. (See Chapter 7.)

Plastids Are Organelles Found Only in Plants and Protists

Plastids, like mitochondria and nuclei, are surrounded by a double membrane. Plants may have at least four types of plastids, shown in Figures 5-24 and 5-25: chloroplasts, chromoplasts, amyloplasts, and proplastids.

The most important type of plastid is the **chloroplast,** as seen in Figure 5-24, a large green organelle, found only in plants and photosynthetic protists, that uses the energy of sunlight to build energy-rich sugar molecules in the process of **photosynthesis** [Greek, *photo* = light + *syntithenai* = to put together]. Chloroplasts vary from species to species in size, shape, and number per cell.

The TEM has revealed that a chloroplast, like a mitochondrion, has an intricate internal structure of folded membranes. These internal membranes, however, are not continuous with the inner membrane of the chloroplast envelope. Instead, as shown in Figure 5-24b, the innermost membranes lie in stacks of flattened discs, called **thylakoids** [Greek, *thylakos* = sac + *oides* = like]. Each of these stacks is called a **granum** [Latin = grain; plural, **grana**].

The double membrane of the chloroplasts together with the membrane of the thylakoids divides the chloroplast into six distinct locations, as shown in Figure 5-24: (1) the outer membrane; (2) the intermembrane space; (3) the inner membrane; (4) the **stroma,** the space inside the inner membrane but outside the thylakoids; (5) the thylakoid membrane; and (6) the interior of the thylakoids.

A leaf cell typically contains several dozen disc-like chloroplasts. A chloroplast is typically 2–4 μm thick and 5–10 μm in diameter and contains many grana. A granum consists of ten or more stacked thylakoids, together resembling a pile of coins.

Chloroplasts, shown in Figure 5-24, can be isolated by subcellular fractionation. They carry out the reactions of photosynthesis, discussed in Chapter 8. Chloroplasts in plants also serve as storage sites for starches. The sugars produced during photosynthesis may be assembled into starches within the chloroplast and temporarily stored there.

Figure 5-25 shows other plastids. Chromoplasts contain the pigments that give yellow, orange, and red color

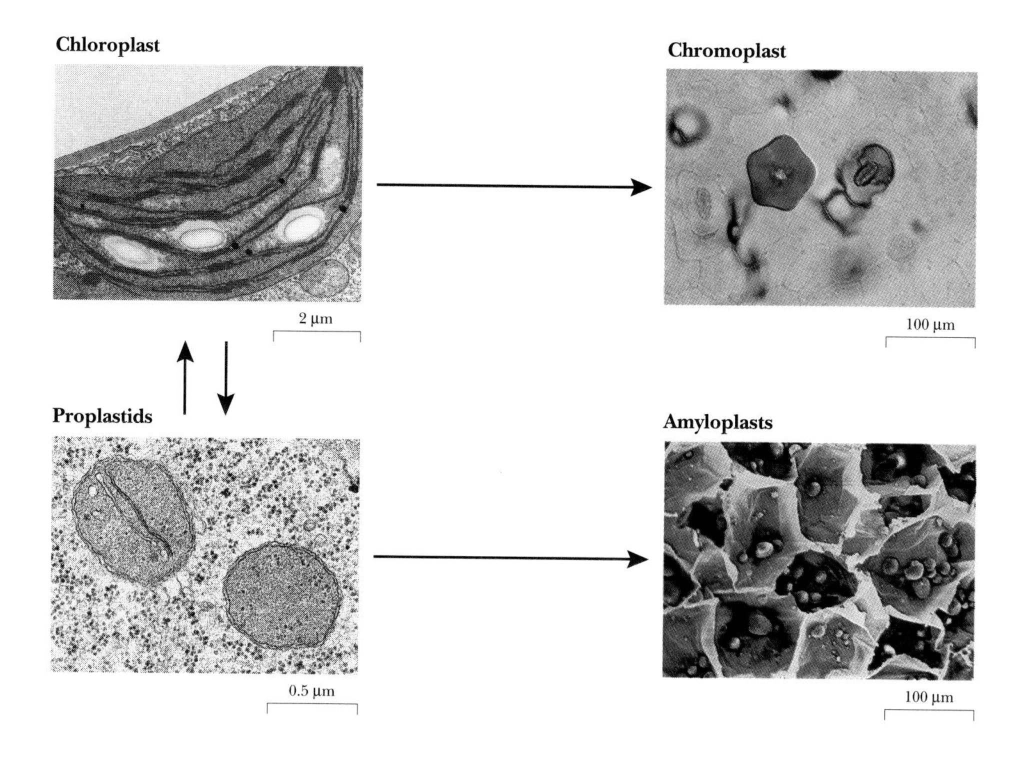

Figure 5-25 Plastids in plant cells. *(chloroplast: R. Howard Berg/Visuals Unlimited; proplastid: E. H. Newcomb, University of Wisconsin/BPS: amyloplast: Science Photo Library/Custom Medical Stock Photo; chromoplast: Richard Green/Photo Researchers)*

to many fruits and flowers, as well as to roots such as carrots and to the beautiful maple leaves of a New England autumn. Chromoplasts arise from chloroplasts by the orderly breakdown of chlorophyll and thylakoid membranes, followed by a reshaping of the inner membrane. **Amyloplasts**, which contain large granules of starch, do not derive directly from chloroplasts, but from the common precursor of all plastids, which is called a **proplastid**. Amyloplasts are storage sites for starch and are especially abundant in potatoes and turnips, as well as in many seeds.

Why Do Cells Have So Much Membrane?

The TEM has revealed that membranes are a prominent component in all cells. In prokaryotes, cell size is ultimately limited by the ability of the membrane to exchange materials between inside and outside. In eukaryotes, the presence of internal membranes allows cells to achieve greater size. These internal membranes appear to dominate cellular organization in eukaryotes.

In the last 20 years, cell biologists and biochemists have intensively studied membrane structure and function. This work has led to an appreciation of four major roles of cellular membranes: (1) Membranes are boundaries that separate the inside from the outside and internal compartments from one another; (2) membranes serve as a "workbench" for a variety of biochemical reactions, especially those involving the metabolism of lipids and the secretion of proteins; (3) membranes regulate the contents of the spaces they enclose (as discussed in more detail in Chapter 9); and (4) membranes participate intimately in energy conversions (as discussed in Chapters 7 and 8).

■ SUMMARY

Seeing the internal structures of a cell requires sufficient magnification, resolution, and contrast. The resolution of light microscopes is limited to about half the wavelength of visible light (about 0.2 μm). Smaller objects, however, can be seen with the electron microscope, which can resolve biological objects as small as 2 nm. The transmission electron microscope works much like a light microscope, showing the distribution of dyes (stains) that bind to specific cell components, while the scanning electron microscope reveals the details of the surfaces of objects. Phase and differential interference contrast microscopy allows cell biologists to observe living cells. Individual cells of plants and animals can be grown in defined chemical conditions and on special surfaces.

To study the functions of cell components, biologists have used techniques of microsurgery, immunocytochemistry, and subcellular fractionation. Subcellular fractionation depends on centrifugation, which can separate different classes of cell components in an artificial gravitational field. The function of cell components can often be duplicated and studied in a test tube.

Membranes divide eukaryotic cells into many compartments, which make characteristic contributions to the cellular economy. The nucleus contains the cell's genetic instructions and serves as the chief executive officer of the cell. The cytosol is the part of the cytoplasm outside the membrane-bounded organelles. It contains most of the enzymes that carry out the metabolism of the cell. Within the cytoplasm are up to six well defined kinds of membrane-bounded organelles (endoplasmic reticulum, Golgi apparatus, lysosomes, peroxisomes, mitochondria, and plastids), as well as vesicles and vacuoles, which are also bounded by membranes but whose structures and functions vary widely.

The endoplasmic reticulum (ER) is an extensive network of membranes that encloses a space called the ER lumen. Part of the ER—the rough ER—is studded with ribosomes and makes proteins destined mainly for export to the outside or the outer surface of the cell. Part of the ER—the smooth ER—contains no ribosomes and appears to function in the synthesis of lipids.

The Golgi apparatus is a set of discs, each of which is enclosed by a membrane. It serves as the traffic director in the cell, helping to package some proteins for export and others for inclusion in other membrane-bounded organelles.

Lysosomes and peroxisomes are small membrane-bounded organelles that are responsible for digestion. They dispose of both material taken in from outside the cell and the cell's own debris.

Mitochondria and chloroplasts are the largest organelles outside the nucleus. Each is bounded by a double membrane, and each contains complicated internal membranes. Mitochondria are the powerhouses of cells, making most of the cell's ATP by breaking down small organic molecules in the presence of oxygen. Chloroplasts are found only in photosynthetic cells. They are the sites of photosynthesis.

KEY TERMS

amyloplast
antibody
antigen
artifact
chloroplast
chromatic aberration
chromatin
chromosome
cis, medial, and *trans* Golgi
compound microscope
contrast
cristae
cytoplasm
cytoskeleton
cytosol
density gradient
differential centrifugation
differential interference contrast
electron microscope
endoplasmic reticulum (ER)
endosome
exocytosis
free ribosome
glycoprotein
glyoxisome
Golgi apparatus
granum
immune response
immunocytochemistry
immunohistochemistry
in vitro
in vivo
lumen
lysosome
magnification
matrix
membrane-bound ribosome
microsurgery
mitochondrion
nuclear envelope
nuclear pore
nucleoid
nucleolar organizer
nucleolus
nucleoplasm
nucleus
organelle
pellet
peroxisome
phase contrast
photosynthesis
plasma membrane
plastid
proplastid
resolution
ribosome
rough ER
scanning electron microscope (SEM)
secretory vesicle
smooth ER
stroma
subcellular fractionation
supernatant
thylakoid
transmission electron microscope (TEM)
ultracentrifuge
ultrastructure
vacuole
vesicle

QUESTIONS FOR REVIEW AND UNDERSTANDING

1. The three parameters that contribute to the ability to visualize an object are: magnification, resolution, and contrast. Define each and explain how they contribute to our ability to see an object.
2. What is the definition of an artifact?
3. When using a compound microscope, can the magnification of an object be deduced by the objective lens setting alone? Explain your answer.
4. Describe four different kinds of microscopes. Include the advantages or disadvantages of each in terms of

their ability to view biological specimens. Explain why different microscopes must be used to view objects of different sizes.

5. Does what one sees in a microscope always represent what is really present? Explain how the preparation of a sample can give rise to misleading observations in both compound and electron microscopes. How do scientists correct for artifacts?
6. Why is it that animal cells, derived from a multicellular organism, have more complex nutrient requirements than bacteria?
7. Briefly explain the concept of immunocytochemistry. What kind of information is obtained by this technique?
8. Contrast density gradient and differential centrifugation. Why are these techniques useful in isolating cellular components? How could you demonstrate that each procedure was successful in isolating a particular cellular component?
9. What is *in vivo* versus *in vitro?* Why is developing a cell-free system important to the study of cellular processes?
10. Draw a typical eukaryotic cell and label all of the compartments and organelles.
11. What is the function of each of the following cellular components?

lysosome	plasma membrane
nuclear envelope	cytoskeleton
smooth endoplasmic reticulum	ribosome
rough endoplasmic reticulum	chloroplast
mitochondrion	Golgi apparatus

13. Describe a process that governs protein localization and movement in the cell.
14. What allows eukaryotic cells to be larger than prokaryotic cells?
15. Why is it important for a scientist to obtain a pure sample of a cellular structure as part of the process of determining what its function might be?

■ SUGGESTED READINGS

ALBERTS, B., D. Bray, J. Lewis, M. Raff, K. Roberts, and J. D. Watson, *Molecular Biology of the Cell,* 3rd edition, Garland, New York, 1994. An authoritative and accessible compendium of current knowledge about cells and cellular mechanisms. See Chapters 10 and 12 for much of the material covered in this chapter.

DE DUVE, C., *A Guided Tour of the Living Cell,* Scientific American Books, New York, 1984. A masterful view of the insides of cells, by one of the pioneers of modern cell biology.

LODISH, H., D. Baltimore, A. Berk, S. L. Zipursky, P. Matsudaira, and J. Darnell, *Molecular Cell Biology,* 3rd edition, Scientific American Books, New York, 1995. See Chapter 5 for much of the material covered in this chapter.

CHAPTER 6

Cell Membranes, Skeletons, and Surfaces

KEY CONCEPTS

- Membranes are organized arrangements of lipids and proteins.
- Membranes regulate the chemical composition of the spaces they enclose.
- The outer surfaces of cells are responsible for their interactions with other cells and with other molecules in their environments.
- The cytoskeleton contributes to the internal organization of eukaryotic cells.
- Cells of all kingdoms have both common and distinctive internal features.

OUTLINE

ESSAYS

Cells pose extraordinary challenges to the experimental ingenuity of biologists. They are highly organized, as we have already seen, and we would like to know just how each of their parts helps them live. But once a cell is disrupted, for example to study just one of its parts, the cell is no longer alive. Still, many of a cell's components manage to function after isolation, and, as we saw in Chapter 5, cell biologists have learned much about their organization and workings. In this chapter, we discuss the current understanding of membranes, which define the surface of the cell and enclose many of its internal components, and of the cytoskeleton, which gives structure and shape to most eukaryotic cells.

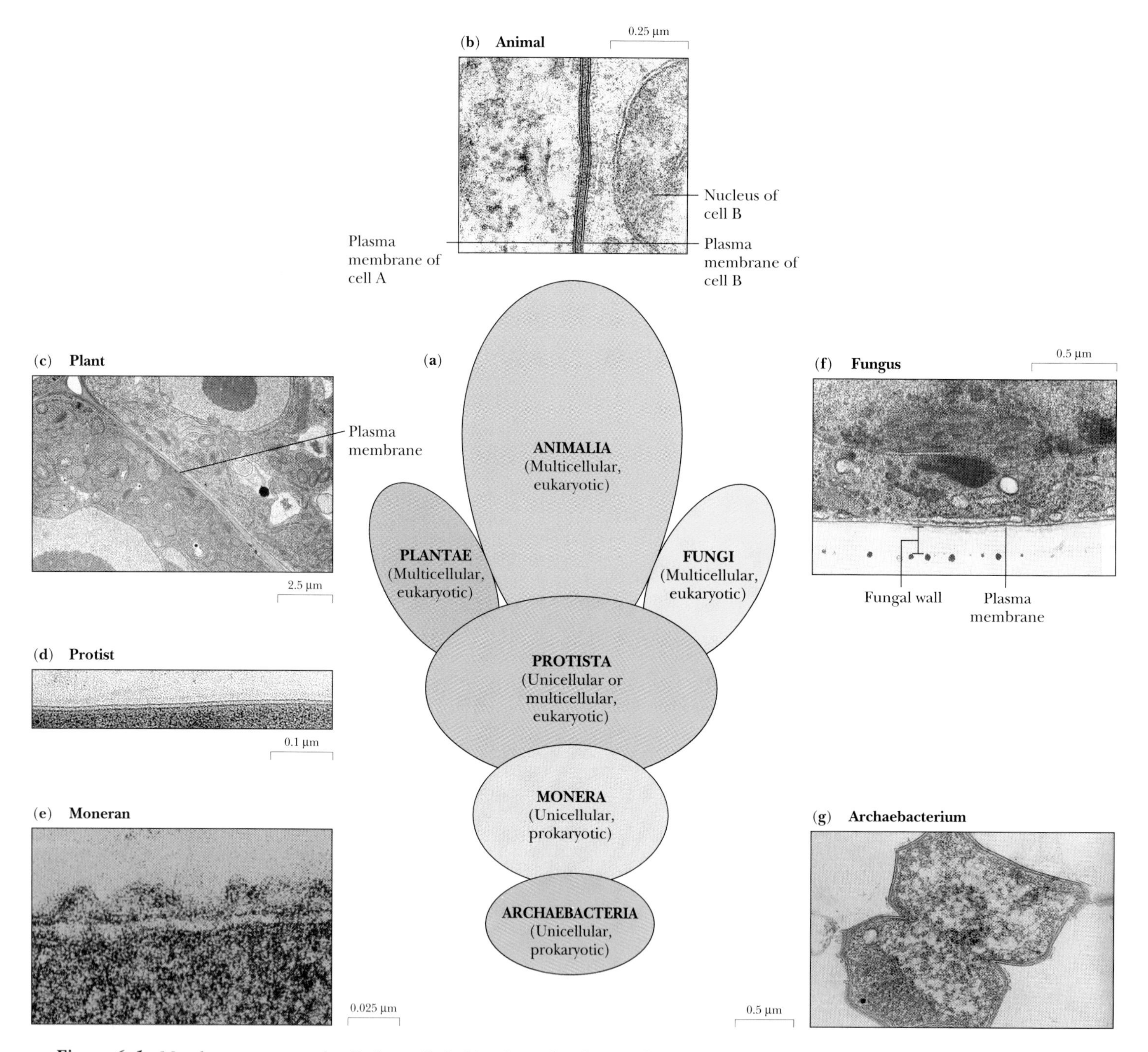

Figure 6-1 Membranes surround cells from all six kingdoms; in all cases the membranes' cross sections resemble railroad tracks 7–10 nm apart. **(a)** The six kingdoms; **(b)** the membranes of two adjacent cells in an animal; **(c)** two adjacent plant cells; **(d)** a protist; **(e)** a bacterial cell; **(f)** a fungal cell; **(g)** an archaebacterium. *(b, Don Fawcett/Photo Researchers; c, Myron C. Ledbetter/Biophoto Associates/Photo Researchers; d, J. Beveridge/Visuals Unlimited; e, Biophoto Associates/Photo Researchers; f, Dr. Charles Mims, University of Georgia; g, Dr. Karl Stetter, Dr. Heinz Schlesner, Kiel University, Germany)*

MEMBRANES ARE ORGANIZED ARRANGEMENTS OF LIPIDS AND PROTEINS

Every cell is surrounded by a plasma membrane. The membrane defines the limits of the cell and separates it from the outside world. Eukaryotic cells also have large amounts of internal membranes, as we have already seen in Chapter 5.

As illustrated in Figure 6-1, the plasma membranes of cells from all kingdoms look pretty much the same. Under a transmission electron microscope, the plasma membrane appears in cross-section as two dark lines, 7–10 nm apart, which resemble tiny railroad tracks, with a lighter zone between them. In three dimensions the membrane consists of two parallel layers. An important question, then, concerns the structure of these layers. What molecules make up the layers and what organizes them into the same characteristic structure in all organisms?

Two specialized types of cells have been extremely valuable in studying the molecular organization of cell membranes. The first of these is the mammalian red blood cell, a micrograph of which is shown in Figure 6-2a. These cells, which lose their nuclei before they enter the blood stream, are ideally suited for membrane studies because they lack almost all other subcellular structure. Their lack of nuclei and internal organelles enables them to deform more easily and to squeeze through the circulatory system. Red blood cells have no internal membranes, no organelles, and few types of macromolecules other than hemoglobin, the red oxygen-binding protein. Biochemists can obtain essentially pure membranes by breaking the red blood cells to release their contents and then using a centrifuge to recover their membranes, called "ghosts," shown in Figure 6-2c.

The second specialized cell that has been important in determining the structure of membranes is shown in Figure 6-3. Called a Schwann cell, it wraps extensions of

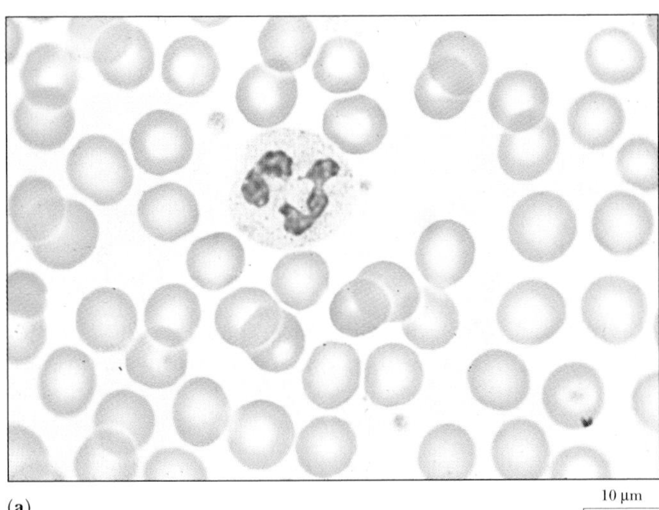

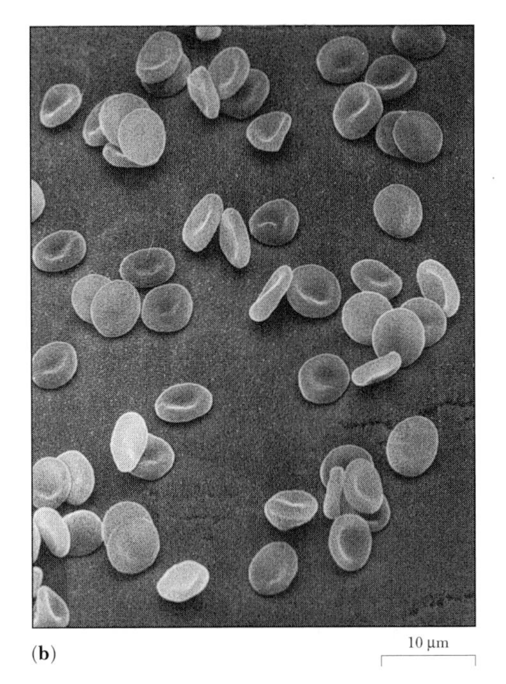

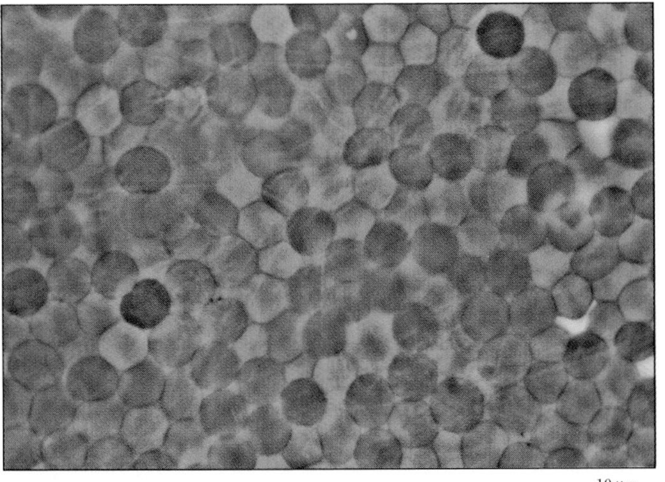

Figure 6-2 Red blood cells from mammals lack nuclei and subcellular organelles. Their membranes are easily isolated and studied. **(a)** A light micrograph of human blood, showing the absence of nuclei in the red blood cells; **(b)** a scanning electron micrograph of red blood cells; **(c)** red blood cell ghosts, membranes left behind after the cells' contents have emptied. *(a, Biophoto Associates/Photo Researchers; b, K. R. Porter/Photo Researchers; c, Robert Knauft/Photo Researchers)*

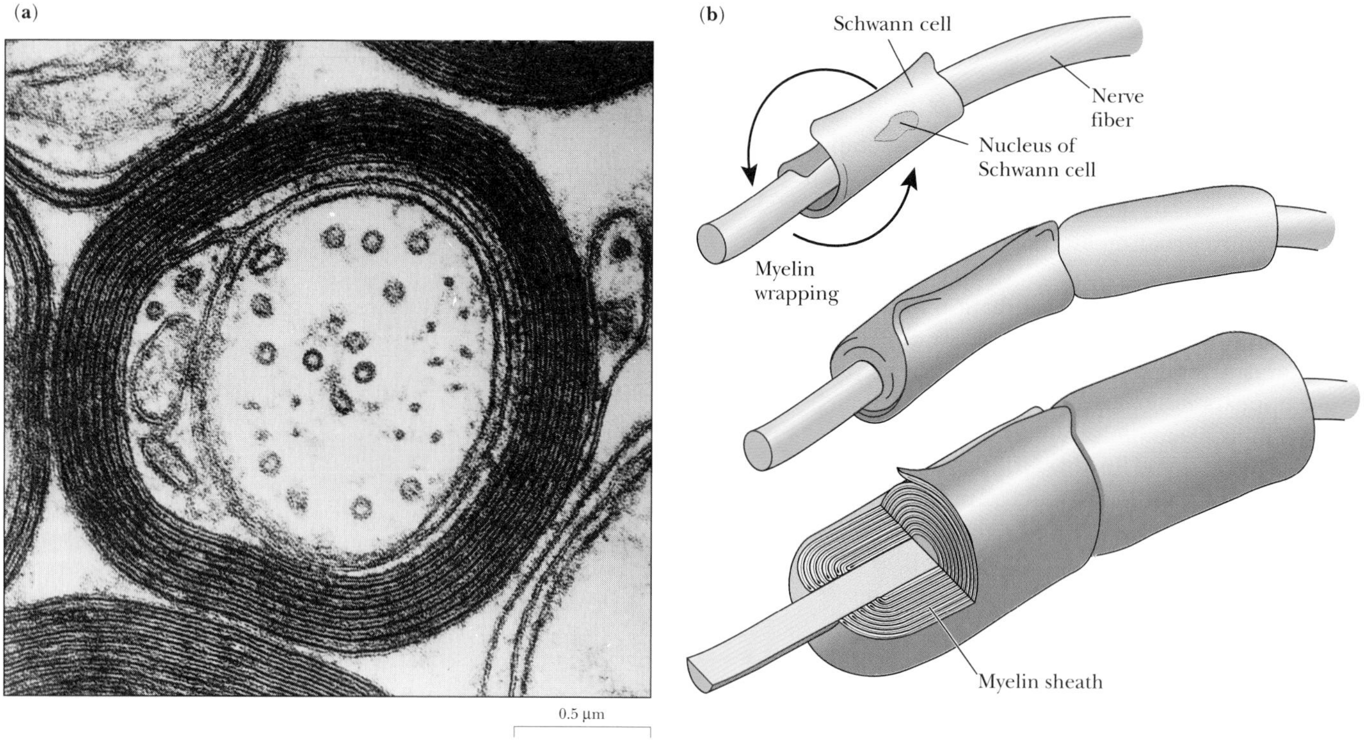

(a) (b)

Figure 6-3 An insulating myelin sheath surrounds many nerve fibers. The myelin consists of stacks of membranes formed as a Schwann cell wraps around the nerve fiber. **(a)** Cross section through a myelinated nerve fiber; each membrane layer has the same railroad-track appearance as a single cell membrane; **(b)** the wrapping process. *(a, C. S. Raine/Visuals Unlimited)*

its plasma membrane around a nerve fiber to form an insulating structure called a **myelin** sheath. As shown in the electron micrograph in Figure 6-3a, a myelin sheath consists of a stack of membranes, each of which looks like the plasma membranes of other cell types.

What Kinds of Molecules Do Membranes Contain?

The first insights into the molecular structure of membranes came from studies of red blood cell ghosts. Biochemical analysis showed that these membranes consist mostly of lipids and proteins, as well as some carbohydrates.

In the 1920s, biochemists determined that a red cell ghost contains just enough lipid to enclose the cell with a continuous layer two molecules thick. Further investigation revealed that the most prevalent lipids in the membrane are phospholipids. As discussed in Chapter 2, phospholipids are highly amphipathic; that is, they have a hydrophilic "head" and a hydrophobic "tail." When pure phospholipids are placed into water, they can spontaneously organize into sheets that are two layers thick. The oily tails of the lipids point toward the interior of the sheets, and their hydrophilic heads point toward the outside, as shown in Figure 6-4a. The question, then, is whether such phospholipid bilayers are the basis of biological membranes. And, if so, where do the membrane proteins lie?

By the mid-1930s, biologists were convinced that biological membranes were formed from lipid bilayers. Because they knew that some membrane proteins were found on the membrane outside the cell and other proteins were found associated with the membrane inside of the cell, they hypothesized that the membrane proteins were spread on the inner and outer surfaces of the lipid bilayer, in contact with the polar heads of the phospholipid, as illustrated in Figure 6-4b. But this turns out to be not quite correct.

Current understanding of the actual arrangement of lipids and proteins in biological membranes came in part from studies of myelin structure. Researchers were able to determine that, as suspected, each membrane consists of a lipid bilayer, but the interior of the bilayer appeared to contain some protein as well.

Membrane Proteins Have Distinctive Structures and Modes of Synthesis

Combining the biochemical studies of proteins and lipids, on the one hand, with structural studies using electron mi-

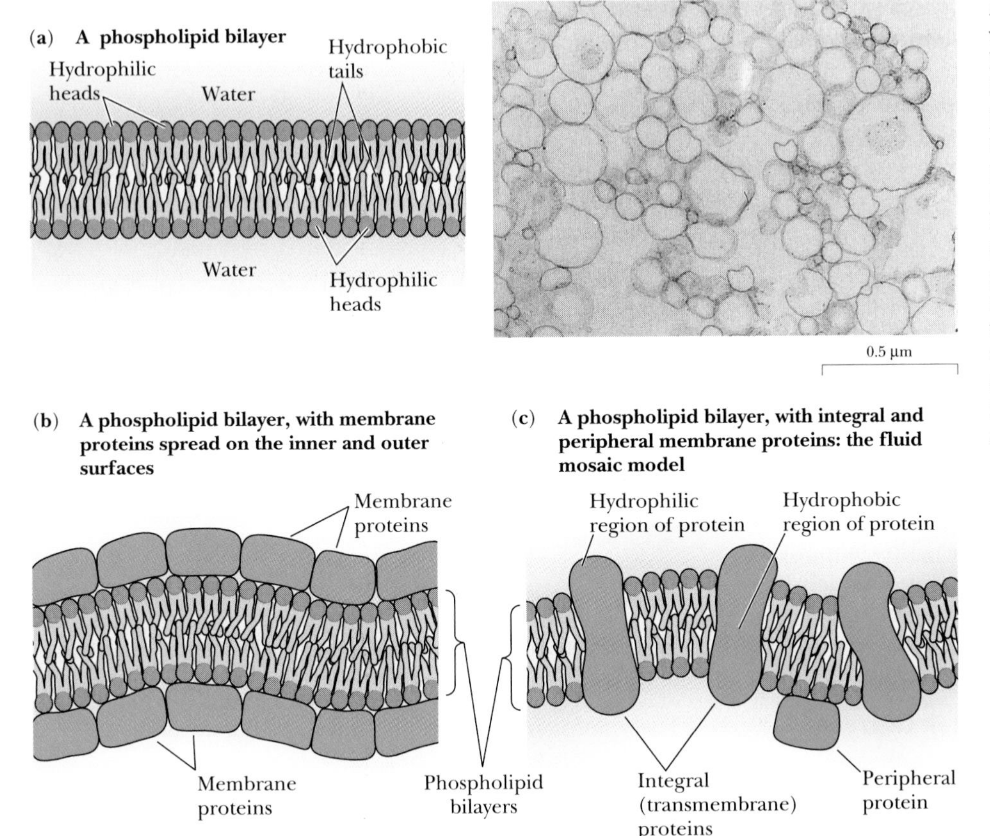

Figure 6-4 Phospholipids placed in water can form a bilayer with a hydrophobic interior and two hydrophilic surfaces, but where do the proteins reside in a biological membrane? **(a)** A diagram of a phospholipid bilayer and a micrograph of vesicles, each surrounded by a bilayer of pure phospholipid; **(b)** one model of a biological membrane, with proteins spread on the inside and outside surfaces; once the accepted view, but wrong; **(c)** currently accepted view of a biological membrane, with many membrane proteins extending through the phospholipid bilayer. *(a, David Deamer, University of California, Santa Cruz)*

croscopy and x-ray diffraction, on the other, cell biologists came to the model summarized in Figure 6-4c. According to this model, some membrane proteins (called **peripheral proteins**) are located only on the outer and inner surfaces of the membrane; others (called **integral** or **transmembrane proteins**) span the lipid bilayer.

Transmembrane proteins are amphipathic, with both polar (hydrophilic) and nonpolar (hydrophobic) regions. The polar regions interact with the aqueous solutions of the cytoplasm and the cell exterior and with the polar heads of the lipids. The nonpolar regions interact with the hydrophobic interior of the membrane.

Cells produce both transmembrane proteins and the peripheral proteins of the membrane's outer surface as if they were exported proteins. As discussed in Chapter 5 (p. 138), these proteins are made in the rough endoplasmic reticulum (ER) and pass into the ER lumen. There and in the Golgi apparatus these proteins acquire sugars and become glycoproteins. In transmembrane proteins, sugars are present only in those portions that lie on the outside surface of the membrane. Sorting machinery within the Golgi apparatus directs these proteins to the cell membrane. In contrast, the proteins that associate exclusively with the membrane's inner surface are produced by free ribosomes and arrive at the membrane through the cytoplasm.

Many membrane proteins also contain other modifications: Some have phosphate groups added to particular amino acid side chains (usually serine, threonine, or tyrosine). Other proteins become attached to small nonpolar molecules that anchor them to the lipids of the bilayer.

The Two Leaflets of the Lipid Bilayer Differ from One Another

Cell biologists refer to the two "faces" of the bilayer—the **cytosolic** face, which touches the cytosol, and the **exoplasmic** face, which contacts the outside of the cell. In the case of the internal membranes that surround a eukaryotic organelle, the exoplasmic face touches the *inside* of the organelle.

One way of examining the different composition of the cytosolic and exoplasmic faces of a membrane makes use of digestive enzymes called phospholipases. When added to the outside of a cell, phospholipases can hydrolyze phospholipids in the exoplasmic but not the cytosolic layer (because they do not enter the cytoplasm). If, however, phospholipases are added to a cell homogenate, they digest phospholipids on both faces of the membrane. By comparing the products of phospholipase action on intact cells and on homogenized cells, cell biologists have shown that the two layers have different chemical compositions. These experiments are actually a little more complicated, because membrane fragments reseal to form vesicles; some of these vesicles are right side out and some are inside out. But these experiments have revealed, for example, that in red blood cell ghosts, phosphatidylcholine is located primarily within the exoplasmic face, while phosphatidyl-ethanolamine is located primarily within the cytoplasmic face.

The two faces also differ in the way they interact with membrane proteins, so that every integral membrane protein has a unique orientation. Because (as discussed later in the chapter) many membrane proteins regulate the passage of ions and molecules into and out of cells, the direction in which they lie is biologically important.

Carbohydrates are also asymmetrically distributed: They are almost entirely attached to the proteins on the exoplasmic face. Some carbohydrates are also components of **glycolipids,** which consist of carbohydrates attached to certain lipids.

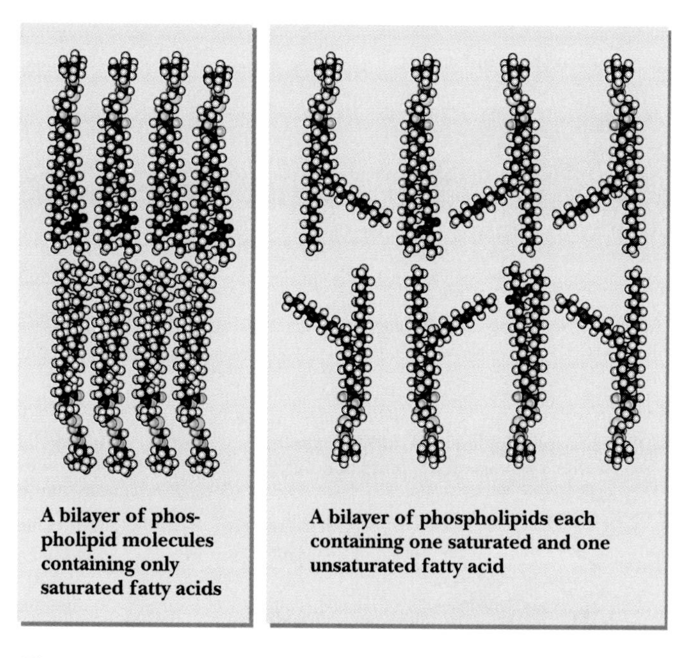

Figure 6-5 Phospholipids with more saturated fatty acids form more stable associations within the hydrophobic interior of a membrane than those with more unsaturated fatty acids. The more stable associations reduce the fluidity of the membrane.

The Fluid Mosaic Model Summarizes Current Understanding of Biological Membranes

The cell membrane has a consistency similar to that of light machine oil, such as you might use to lubricate a bicycle. Nonpolar molecules can diffuse rapidly both through the membrane and within each of the two layers of the membrane.

What about individual molecules of the membrane itself? Are they able to move about? To answer this question for the membrane's phospholipid molecules, biochemists tagged certain molecules with a functional group that could be detected by a technique called *electron spin resonance spectroscopy.* Researchers could then study the rotational movement of tagged molecules that had been incorporated into the membrane. These experiments demonstrated that phospholipid molecules within the cell membrane can move relatively freely within each of the membrane's layers.

Biochemists have used electron spin resonance spectroscopy to study the influence of different types of phospholipids on membrane fluidity, a measure of the ability of substances to move within the membrane. Membranes made from phospholipids that contain fatty acids with saturated chains are less fluid than those containing unsaturated fatty acids. We can understand this difference in fluidity in terms of the differing flexibility of individual molecules of saturated and unsaturated fatty acids, as mentioned in Chapter 2 (p. 48) and illustrated in Figure 6-5.

Membrane fluidity also depends on components other than the fatty acids of phospholipids. Cholesterol (as shown in Figure 6-6) stiffens and strengthens the membrane by bracing the nonpolar tails and the polar heads. Increased levels of saturated fats and cholesterol make membranes less fluid.

Membranes contain proteins as well as lipids, and we would also like to know how easily they move. To study the movement of proteins within the membrane, biochemists tagged membrane proteins with a fluorescent dye. They then used a laser to provide an intense spot of light for a brief time that bleached the dye in a small region of the membrane, as illustrated in Figure 6-7. They then used a microscope to detect the appearance of new fluorescent molecules in that area, as proteins tagged with unbleached dye entered from adjacent areas. These experiments showed that about half of the proteins in a membrane were able to diffuse freely, while about half were tightly bound to particular positions within the membrane.

Experiments with electron spin resonance and fluorescence bleaching showed that both lipid and protein molecules can move about relatively easily within the individual leaflets of the membrane bilayer. Other experiments show, however, that molecules do not easily "flip-flop" between the two layers.

The observed fluidity of biological membranes is consistent with a model of membrane structure proposed in 1972 by two biochemists at the University of California, San Diego, Jonathan Singer and Garth Nicholson. Their model, called the **Fluid Mosaic Model** (which is the basis of Figure 6-8), stresses that proteins and phospholipid molecules can move within each leaflet of the lipid bilayer unless they are restricted by special interactions.

We can summarize the Fluid Mosaic Model as follows:

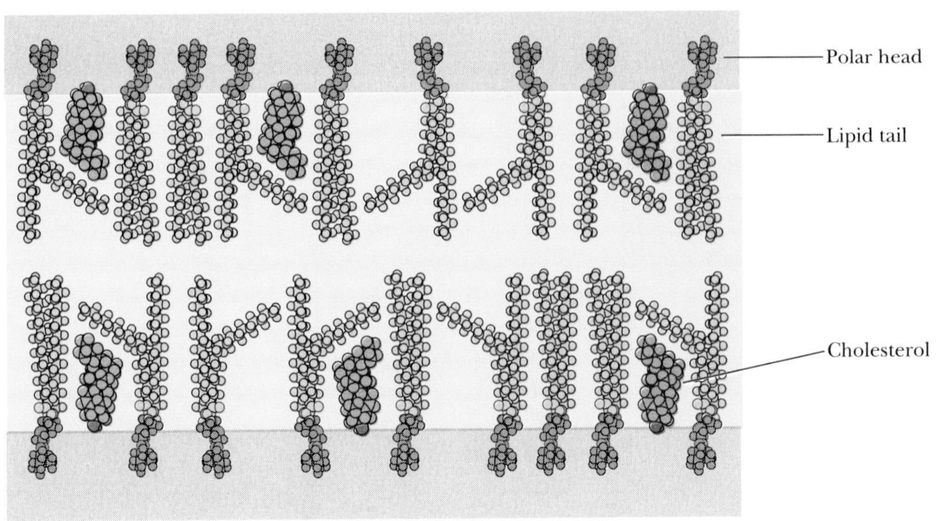

Figure 6-6 Cholesterol interacts with the hydrophobic tails of phospholipids, stiffening and strengthening membranes.

1. The basic structure of the membrane is a lipid bilayer, with two sheets of phospholipids arranged tail to tail.
2. Proteins are dispersed throughout the membrane, like the individual pieces of stone or glass in a mosaic pattern. Proteins contribute to membrane structure and function. Some proteins span the membrane, while others are confined to the inner or outer surface.
3. The membrane is fluid: Most protein and lipid molecules can move freely within each of the two leaflets of the membrane, although they cannot easily move between the two leaflets.
4. The lipid bilayer serves as a hydrophobic barrier, confining hydrophilic molecules to the inside or the outside of a cell (or cell compartment).
5. Some membrane proteins serve as channels, carriers, and pumps for transporting specific molecules across the membrane (discussed later in this chapter).

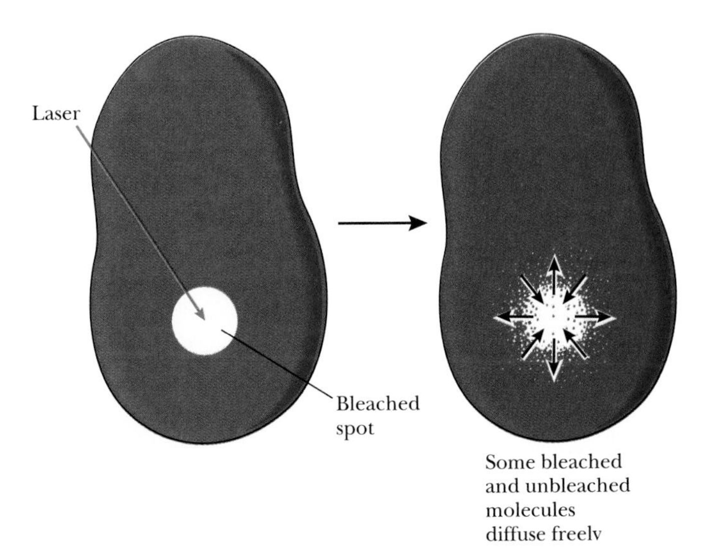

Figure 6-7 A fluorescence bleaching experiment: Membrane proteins are tagged with a fluorescent dye, and the dye is bleached in a small spot. Unbleached dye diffuses back into the bleached area, revealing the mobility of proteins within the membrane's surface.

One technique, called freeze-fracture (illustrated in Figure 6-9), provides a particularly pretty way of viewing the mosaic arrangement of membrane proteins. In this method, a sharp blow (with a fine blade) to a quickly frozen tissue section can fracture the membrane between the two lipid layers, exposing the protein molecules that span the two layers. In red blood cell membranes, most of the small scattered particles are molecules of a transmembrane protein that serves as a channel for chloride (Cl^-) and bicarbonate (HCO_3^-) ions.

HOW DO MEMBRANES REGULATE THE CHEMICAL COMPOSITION OF THE SPACES THEY ENCLOSE?

Biological membranes are **selectively permeable,** meaning that they allow the passage of some ions and molecules (especially of water) much more rapidly than others. Proteins, for example, cannot directly move through membranes at all. Biological membranes also contain molecules that actively participate in the regulation of the internal contents of cells and organelles. Biologists have identified many such membrane molecules and are now seeking to understand just how they work. Before we can discuss the role of membranes in regulating a cell's internal environment, however, we must first discuss the rules that govern the movement of water and dissolved substances.

Water Moves from Solutions with Higher Potential Energy to Solutions of Lower Potential Energy

As discussed in Chapter 4, the Second Law of Thermodynamics states that energy inevitably becomes less and less

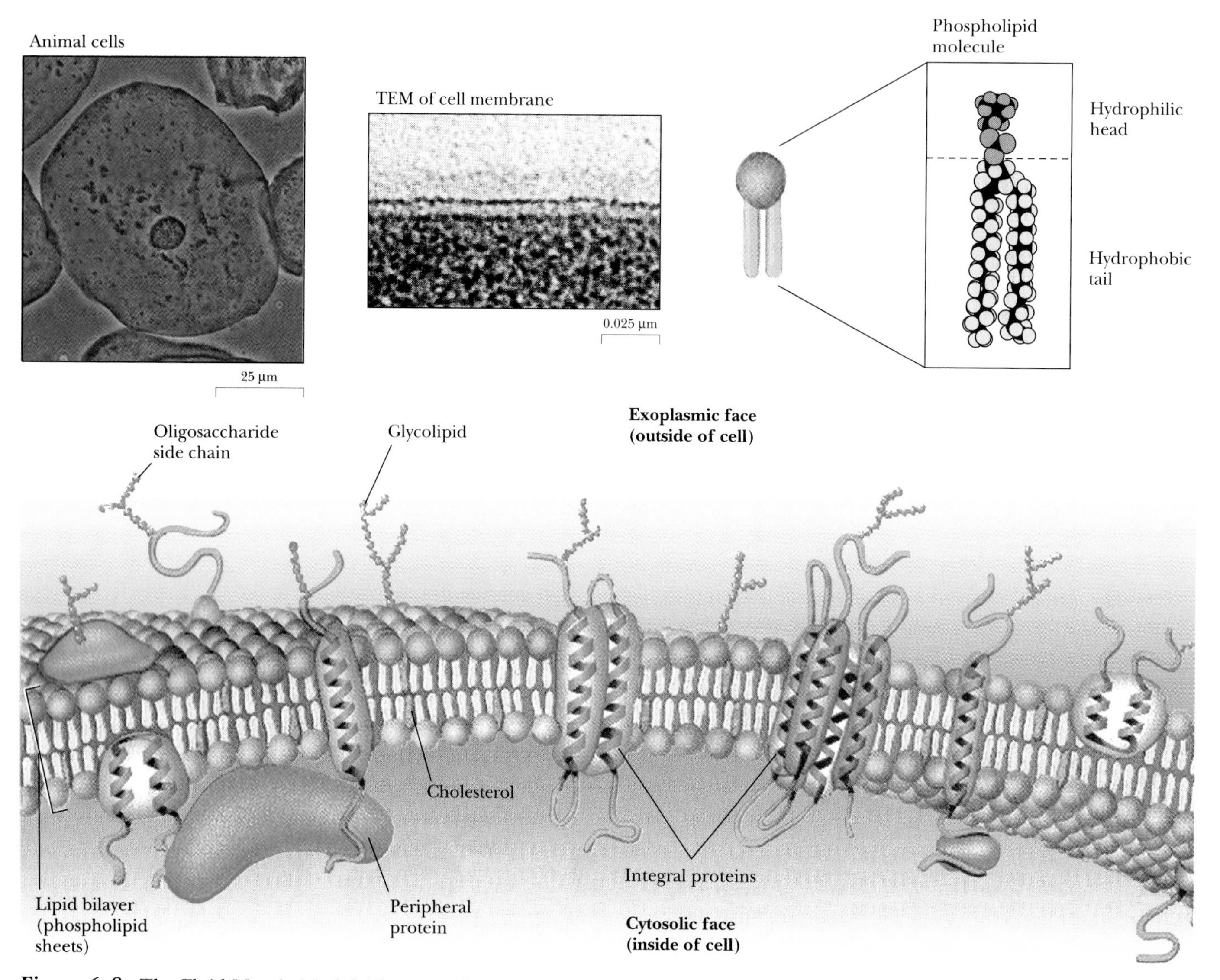

Figure 6-8 The Fluid Mosaic Model. Transmembrane ("integral") proteins usually pass through the lipid bilayer as an α-helix consisting of about 20 residues with nonpolar side chains. The proteins and lipids on the exoplasmic face are often linked to oligosaccharides. *(left, Jim Solliday/Biological Photo Service; center, Biophoto Associates)*

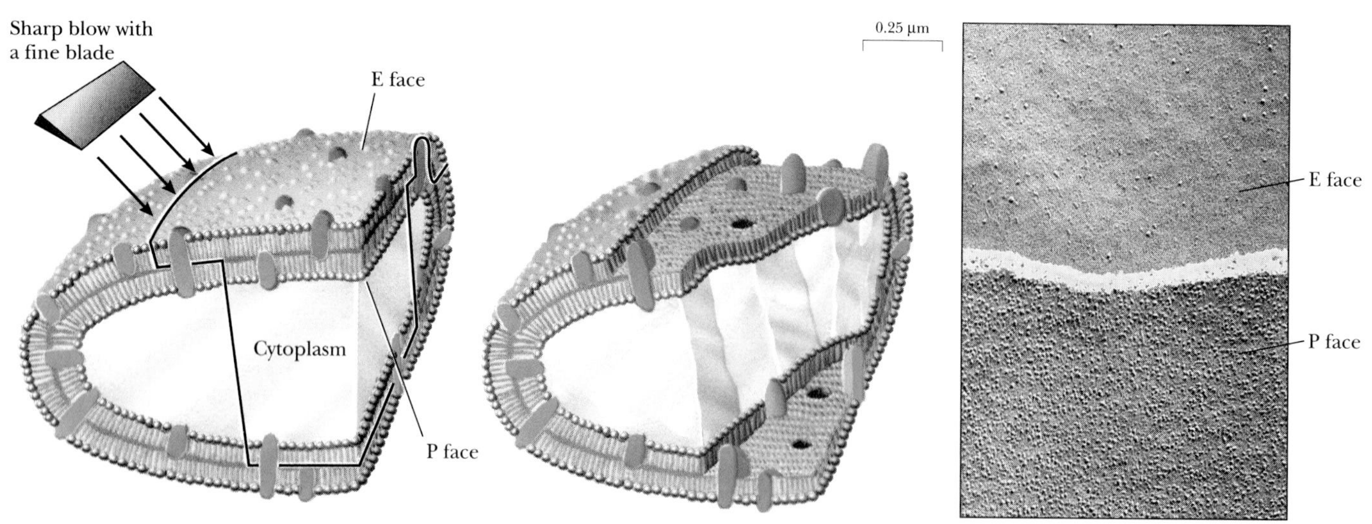

Figure 6-9 Freeze fracture and electron microscopy reveal transmembrane protein molecules. The diagram shows how a sharp knife splits the hydrophobic interior of a membrane bilayer, revealing protein molecules on two faces of the membrane: the P (protoplasmic) face, whose outside touches the cytosol ("protoplasm"), and the E (external) face, whose outside touches the cell's exterior. The density of proteins associated with the P face is much higher than those associated with the E face. *(Don Fawcett/Photo Researchers)*

available to do work. The Second Law predicts that a ball will not spontaneously roll uphill but will spontaneously roll downhill, in the direction of lower free energy. The Second Law predicts the direction of any process whose initial and final states differ in free energy.

The Second Law applies to all kinds of energy. Consider, for example, the differences in energy between pure water and a salt solution. At a given temperature and pressure, the free energy of a salt solution is always less than the free energy of pure water because some of the water's free energy is used in the process of dissolving the salt. Just as a stream inevitably flows downhill, water molecules inevitably move from pure water into a salt solution or into a cell, whose cytosol is essentially a solution of salts and molecules in water.

Because the movement of water from place to place is work (that is, movement against a resisting force), we may speak of the potential energy of water. Plant biologists have introduced the term **water potential,** the potential energy of water per liter, which we use for our discussion here. Water potential is a measure of the tendency of water to flow from a place with higher potential to a place with lower potential.

Water potential predicts the direction that water will move in response to differences in free energy of the water in separated areas of a cell or organism, just as height predicts the direction an object will move in response to gravity. Water potential (abbreviated by the Greek letter ψ) is highest in pure water and lower in any solution.

Differences in Water Potential Cause Water to Move Across Membranes

When a selectively permeable membrane separates two solutions with different water potentials, water flows in the direction that reduces the difference. Water flows in the direction that lowers the concentration difference between the two sides of the membrane as long as both solutions are under the same pressure. This makes sense in terms of our discussion in Chapter 4: As water flows from the high water potential of pure water to the lower water potential of a solution, it increases the system's entropy; the system maximizes its entropy by equalizing the concentration of the solute in both compartments.

The flow of water across a selectively permeable membrane as a result of concentration differences is called **osmosis** [Greek, *osmos* = push, thrust]. Water can also flow across a membrane as a result of differences in mechanical pressure, which contributes importantly to water flow in plants. Water potential, then, consists of two components, osmotic potential and pressure potential.

Researchers can measure the osmotic pressure of a solution with a device called an osmometer, illustrated in Figure 6-10a. In an osmometer, a membrane separates two compartments with different water potentials. The selectively permeable membrane allows water molecules but not solute molecules to pass. In a typical experiment, pure water would be put into one compartment (compartment 2 in the osmometer shown in Figure 6-10a) and a solution

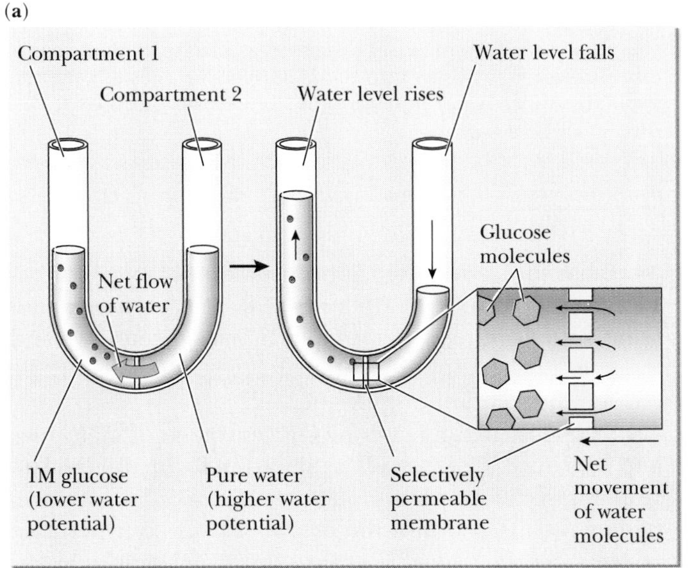

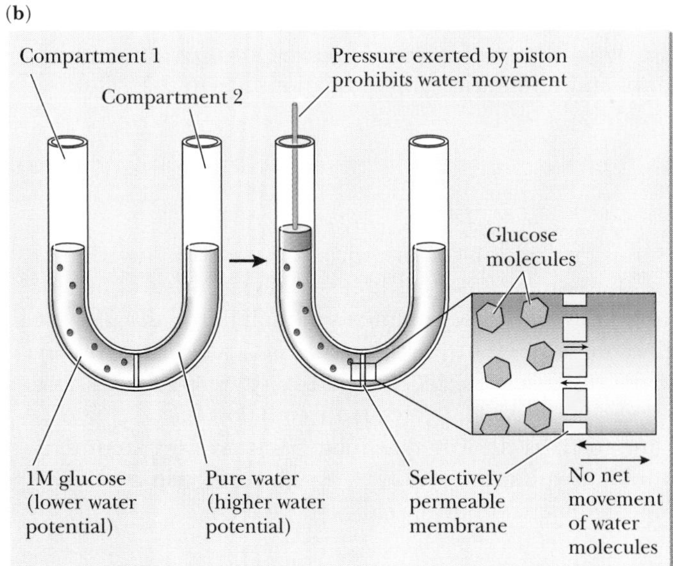

Figure 6-10 An osmometer: **(a)** the dissolved glucose molecules in compartment 1 lower the water potential, so water flows from compartment 2, raising the level on the left side; **(b)** the piston on top of compartment 1 prevents water movement; the pressure needed to prevent the water movement is equal to the osmotic pressure generated by the difference in water potentials between the two compartments.

containing, for example, 1 M glucose into the other compartment (compartment 1 in Figure 6-10a).

The pure water in compartment 2 has a greater water potential than the solution in compartment 1, and it begins to move through the membrane into compartment 1. As water flows into compartment 1, it begins to fill the vertical tube, creating a pressure head that resists the further flow of water. The height of the water in the tube is a measure of the work done by the water in compartment 2 and of its water potential.

Osmotic potential is exactly equivalent to pressure potential. In an osmometer, such as that illustrated in Figure 6-10b, for example, a piston that exerts mechanical pressure can prevent the flow of water into compartment 1. By measuring the mechanical pressure required to resist flow, osmometers can determine osmotic potential.

The potential energy of a solution (its capacity to do work) decreases with increasing solute concentration. By convention, the water potential of pure water is taken as zero. Because flow is away from a compartment of pure water, water potentials of solutions are negative numbers. The greater the concentration of solute molecules in a solution, the more negative its water potential; that is, water has a greater tendency to flow into a concentrated solution than into a dilute solution.

Measurements with osmometers show that osmotic potential is directly proportional to the concentration of solute molecules. As long as there are no opposing pressures, the osmotic potential is equal to the water potential. In 1887, Jacobus van't Hoff, a Dutch chemist (and winner of the first Nobel Prize in Chemistry), discovered this simple relationship between osmotic pressure (the pressure generated by differences in osmotic potential in an osmometer) and solute concentrations. He also noted the dependence of osmotic pressure on absolute temperature and summarized these relationships in the following equation:

$$\pi = R\,T\,\frac{n}{V}$$

Here, π is the osmotic pressure, n/V the concentration of the solution in moles per liter (n is the number of ions or molecules and V the volume of the solution), and T is the absolute temperature. R is a constant, called the gas constant, discussed in more detail in your chemistry and physics courses. The van't Hoff equation has exactly the same form as the law that relates the pressure, temperature, and volume of an ideal gas (a relationship you will repeatedly encounter in chemistry courses).

Osmosis Occurs in Cells (and Membrane-Bounded Compartments) as It Does in Osmometers

We may now begin to describe how water and other molecules flow into and out of cells. When a solution outside a cell has the same solute concentration as the cell's interior (as represented in Figure 6-11a), the solution is said to be **isotonic** [Greek, *isos* = equal + *tonos* = tension], and there is no net flow of water. But what happens if we change the solution outside the cell?

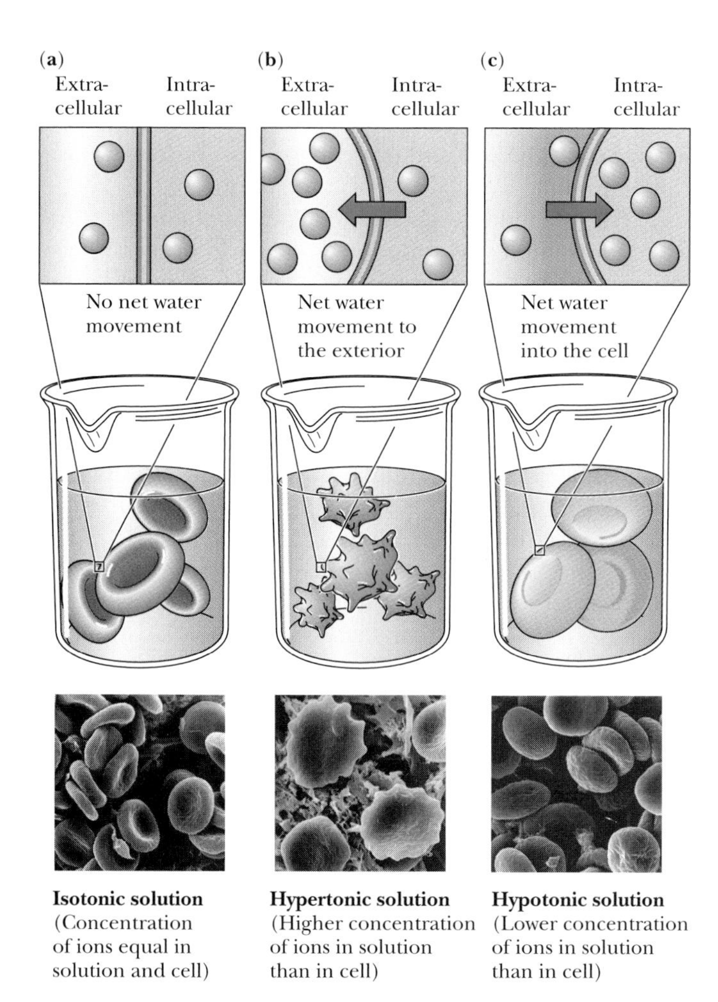

Figure 6-11 Red blood cells act as osmometers: **(a)** in an isotonic solution, the water potential outside the cell equals that inside the cell, and there is no net flow; **(b)** in a hypertonic solution, the water potential outside the cell is lower than that inside the cell and water flows out of the cell, causing the shriveled ("crenated") shapes; **(c)** in a hypotonic solution, the water potential inside the cell is lower than that outside, and water flows into the cell; here the cells are slightly swollen; a more hypotonic solution would lead to cell lysis. *(Dennis Kunkel/Phototake NYC)*

Suppose we increase the concentration of molecules and ions outside of the cell, as shown in Figure 6-11b. The resulting solution, whose total concentration of solutes is higher than that within the cell, is said to be **hypertonic** [Greek, *hyper* = above]. Because the hypertonic solution has a lower water potential than the cell's interior, and because the selectively permeable membrane allows water to pass freely, water flows out of the cell and the cell shrivels (as seen in the figure). When enough water flows out that

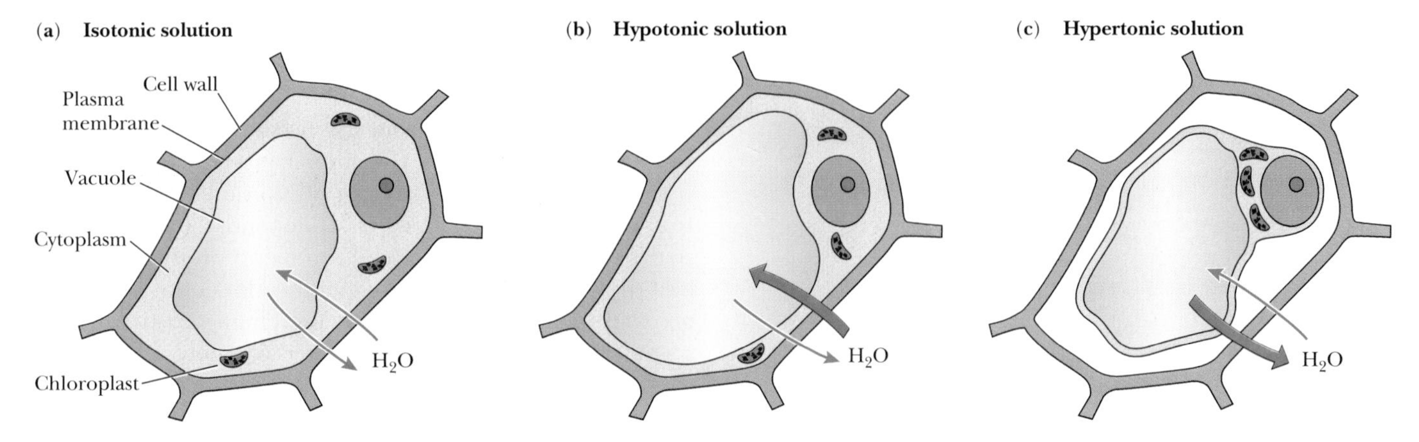

Figure 6-12 Plant cells act as osmometers: **(a)** in an isotonic solution, there is no net flow of water; **(b)** in a hypotonic solution, the vacuole fills, but the cell cannot expand because of the rigid cell wall; **(c)** in a hypertonic solution, water flows out of the central large vacuole, causing the cytoplasm to shrink away from the rigid cell wall ("plasmolysis").

the new concentration of molecules within the cell is again the same as outside the cell, no more water flows.

Conversely, we may decrease the solute concentration outside the cell, creating a **hypotonic** [Greek, *hypo* = under] solution, which has a higher water potential than the cell's interior. Water then flows into the cell, causing it to expand, as shown in Figure 6-11c. Water flows in until the water potential is equal on both sides of the membrane. But the difference in free energy on the two sides of the membrane may be too great for this equalization to occur, for example, if the cells are suspended in pure water. In this case, the water flowing into the cell can cause the cell to burst and release its contents. This method is the way that cell biologists usually prepare red blood cell ghosts.

When plant cells, such as the cell shown in Figure 6-12a, are exposed to a hypotonic solution, as in Figure 6-12b, water moves into the cell but does not accumulate in the cytoplasm. Instead, the entering water passes from the cytoplasm into a membrane-bounded vacuole. The expansion of the vacuole pushes the surrounding cytoplasm against the plant cell's rigid wall. Conversely, when an onion skin is placed in a hypertonic solution (Figure 6-12c), water flows out of the vacuole and the cell separates from the cell wall, a phenomenon called **plasmolysis** [Greek, *plasma* = form + *lysis* = loosening].

In a hypotonic solution, the pressure exerted on the cell by the cell wall eventually prevents the further flow of water into the cell. Water actually continues to flow equally in both directions through the cell membrane, but there is no net water flow.

The pressure of a plant cell's contents against its cell wall is called **turgor pressure** [Latin, *turgor* = swelling]. Turgor pressure provides the principal mechanical support of nonwoody plants.

What Determines the Direction and Rate of Movement Through a Selectively Permeable Membrane?

We can ask the same two questions about the passage of molecules and ions through a membrane as we asked about chemical reactions: "Which way?" and "How fast?" (See Chapter 4.) In addressing the first question, membrane researchers distinguish between **passive transport** (as diagrammatically shown in Figure 6-13a), which occurs spontaneously without the expenditure of energy, and **active transport** (as diagrammatically shown in Figure 6-13b), which requires energy.

Active transport moves molecules or ions from a region of lower free energy to a region of higher free energy. (See Chapter 4 for a discussion of free energy and concentration.) For uncharged molecules, passive transport means that molecules move from a region of higher concentration to a region of lower concentration; active transport of uncharged molecules means movement from a region of lower concentration to a region of higher concentration. For ions, active transport means movement against an electrochemical gradient. (See below and Chapter 23.) The answer to the question "Which way?" is different for passive transport and active transport.

For Passive Transport, Thermodynamics Predicts the Relative Concentrations of a Substance on the Two Sides of a Membrane

For uncharged molecules, passive transport results in the uniform distribution of a substance on the two sides of a membrane. For charged substances, however, the distribution also depends on the charge distribution across the membrane. As discussed in Chapter 23, the interior of a cell usually has an excess of negative charge, so passively

(a) **Passive transport**

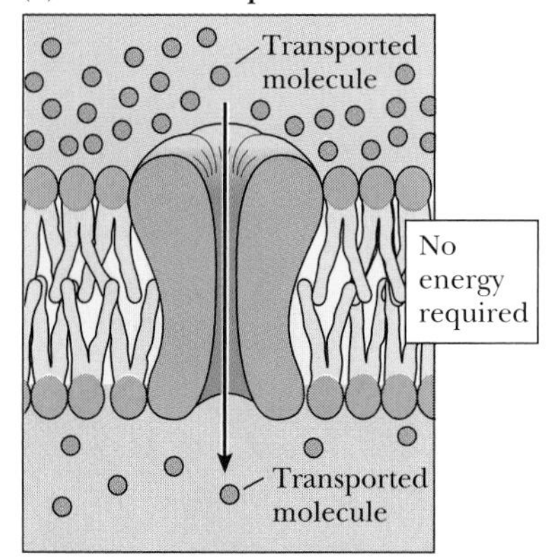

(b) **Active transport**

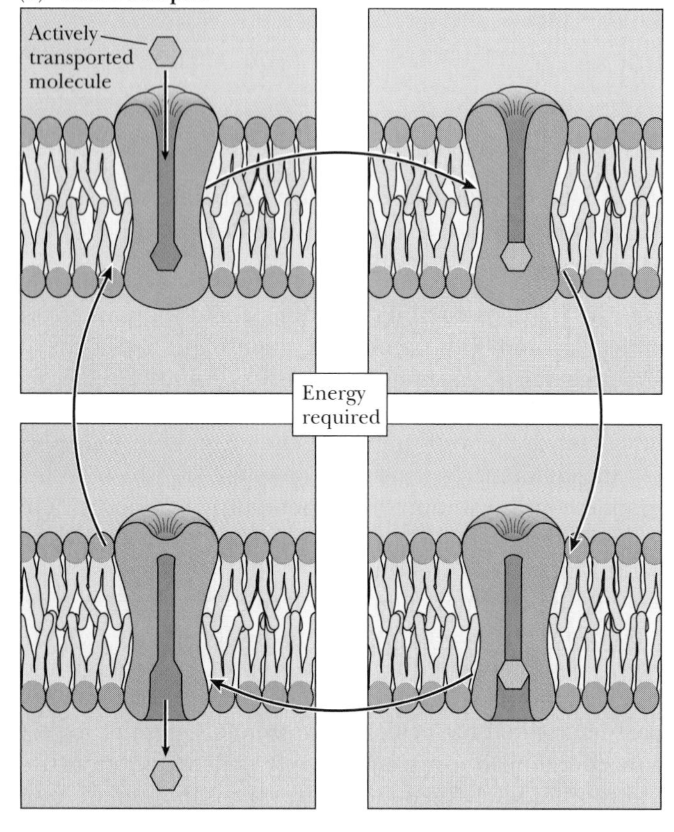

Figure 6-13 Passive and active transport. **(a)** Passive transport occurs spontaneously, but may be accelerated by membrane proteins that act as channels or carriers; **(b)** active transport requires a carrier protein that links transport to the expenditure of energy.

transported cations (each with a positive charge) have higher concentrations inside cells, and passively transported anions (each with a negative charge) have higher concentrations outside cells.

Voltage, a measure of electrical potential energy, depends on the charge difference between two points (in our case, across a membrane). The greater the voltage across a membrane, the more work may be accomplished by allowing a charged particle to move from higher potential energy to lower potential energy. For any ion, then, the concentration ratio across the cell membrane depends directly on the voltage across the membrane, as we see in Chapter 23. Deviations from the predicted concentration ratio indicate that active transport has occurred.

Even for passively transported substances, the answer to the question "How fast?" is "It depends." Just as the kinetics of a biochemical reaction depend on the presence and the properties of an enzyme, the kinetics of transport across a membrane depend on the characteristics of the membrane. Cell biologists have used a variety of methods to study the transport of individual substances across the membrane. In one widely used approach, the researcher places a small amount of radioactively labeled ion or compound on one side of a membrane and measures the rate of its appearance on the other side.

Such experiments have shown that most small organic molecules move across membranes in a way that makes their concentrations more uniform on the two sides of the membrane. That is, these molecules are passively transported. Passive transport, however, occurs at different rates for different molecules and ions. For example, nonpolar molecules generally move across the membrane more rapidly than polar molecules, and small polar molecules move across membranes much faster than large polar molecules.

Most small nonpolar molecules move across the membrane by simple **diffusion** [Latin, *diffundere* = to pour out], in which random movements lead to a uniform distribution of molecules both within a solution and on the two sides of a membrane. The rate of diffusion of a compound in a solution also depends on its size and shape. Diffusion of a compound across a membrane depends on other molecular characteristics, such as polarity, with more nonpolar molecules more able to diffuse through the membrane's lipid bilayer.

Some Passively Transported Molecules Move Across Membranes Much Faster Than Would Be Expected on the Basis of Diffusion Alone

Facilitated diffusion, an increased rate of passive transport, depends on the action of specific transporter molecules within the membrane. In most cells (including red blood cells), for example, glucose is subject to such facilitated diffusion, even though its final concentration is consistent with passive transport. (That is, the final concentrations are equal inside and outside the cell.)

Biochemists have identified the protein, called glucose permease, responsible for glucose transport in red blood cells. Glucose permease is a transmembrane protein consisting of about 400 amino acid residues. To show that it is responsible for the facilitated diffusion of glucose, researchers studied the rates of entry of radioactive glucose into **liposomes,** vesicles produced in the laboratory that are surrounded by phospholipid bilayers. Pure liposomes scarcely allow the entry of glucose, but when the purified

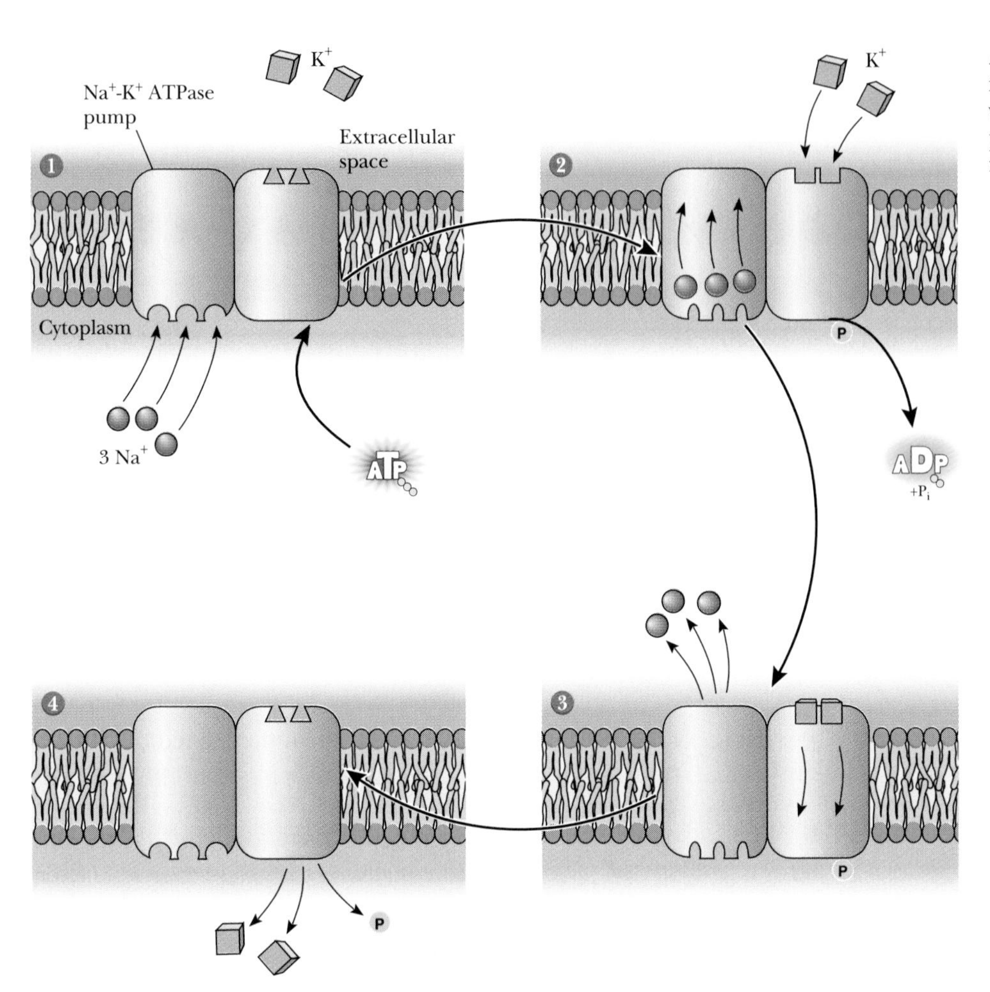

Figure 6-14 A hypothetical scheme for the action of Na^+-K^+ ATPase, which actively transports sodium and potassium ions across the cell membrane.

glucose permease is incorporated into the liposome membrane, the rate of glucose entry increases dramatically, showing that the protein performs its suspected function.

Transport proteins like glucose permease bind to the molecules they transport, in much the same way that an enzyme binds to its substrate, and somehow carry them across the membrane (as diagrammed in Figure 6-13a). As in the case of an enzyme, one can measure the dependence of the transport rate as a function of substrate concentration. Such experiments show that the kinetics of transport resemble those derived for enzymes by Michaelis and Menten. That is, the rate of passive transport increases with the concentration of the transported molecule, but only up to a certain point. (See Chapter 4, p. 103.)

The plateauing of transport rate suggests that a transporter protein becomes saturated at high concentrations of the transported molecule, in much the same way that an enzyme molecule becomes saturated at high substrate concentrations. In red blood cell membranes, the measured K_M for the transport of D-glucose (the molecule from which virtually all cells derive energy) — the concentration of glucose at which transport is half its maximum value — is 1.5 mM. The glucose concentration in the blood generally varies from about 4.4 mM (after a fast) to 6.6 mM (after a meal), so the glucose transporter in red blood cells nearly always functions close to its maximal rate.

Like an enzyme, transporters bind specifically to certain molecules, so glucose permease transports only sugars whose structure is similar to that of glucose. The K_M for the transport of galactose is 30 mM, while the K_M for L-glucose (from which we cannot derive energy) is greater than 3000 mM. These higher values of K_M mean that glucose permease binds galactose less avidly than it binds glucose; it binds L-glucose hardly at all.

How Can a Membrane Accomplish Active Transport?

To transport a substance actively across a membrane requires a special "pump" mechanism that somehow couples transport to an energy source (such as ATP). A number of ATP-driven pumps drive the active transport of ions across the plasma membrane.

The active transport of ions is said to be **electrogenic** when it leads to the accumulation of charge on one side of a membrane. An important active transporter, called the **Na^+-K^+ ATPase** (or the "sodium-potassium pump"), simultaneously transports sodium ions (Na^+) out of cells and potassium ions (K^+) into cells, using the energy of ATP. This transporter is electrogenic because for each ATP expended, it carries two K^+ ions in and three Na^+ ions out; that is, three positive charges move out of the cell for each two that move into the cell. The Na^+-K^+ ATPase binds to ATP and Na^+ ions on the inside of the plasma membrane and to K^+ ions on the outside of the membrane, as diagrammed in Figure 6-14. As in the case of glucose per-

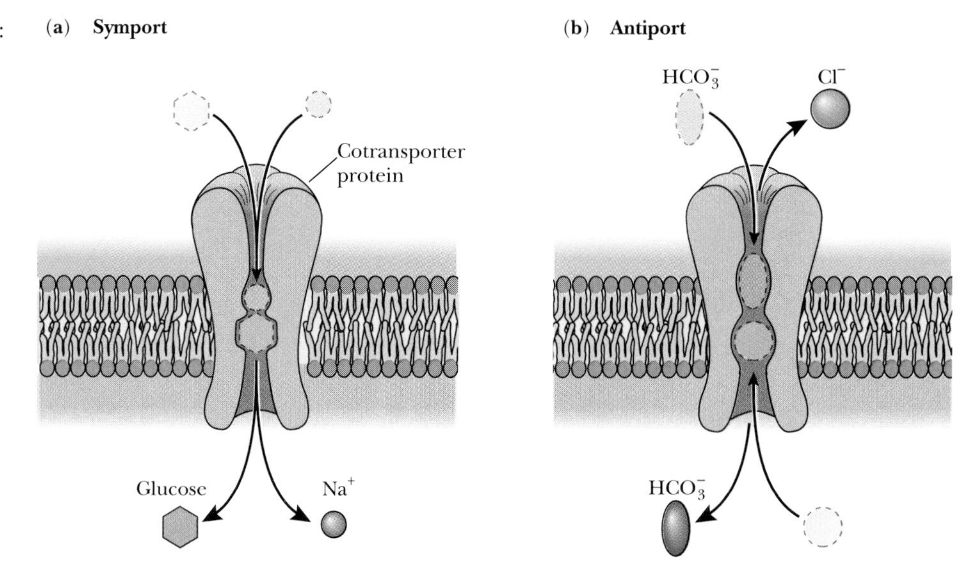

Figure 6-15 Two types of cotransport: **(a)** symport of glucose and Na^+ ions in the small intestine; **(b)** antiport of Cl^- and HCO_3^- in red blood cells.

mease, the Na^+-K^+ ATPase has been purified. When inserted into liposomes, the pure protein (which consists of two polypeptide chains) transports Na^+ and K^+ in opposite directions, using the energy obtained by hydrolyzing ATP.

Somewhat surprisingly, many other examples of active transport do not directly depend on ATP. Instead, the cell couples the "downhill" transport of one substance—usually hydrogen (H^+), sodium (Na^+), or (in plant cells) potassium (K^+) ions—to the "uphill" transport of another. In other words, the cell harnesses the free energy of an already existing **concentration gradient,** a difference in the concentration of a substance [so called because the concentration changes in a *graded* manner, in this case across a membrane].

The cells that line the small intestine, for example, actively take up glucose, using energy stored in the concentration difference of Na^+ ions (as illustrated in Figure 6-15a). Glucose transport by these cells depends on active transport, in contrast to the passive transport of glucose elsewhere, for example in red blood cells. The active transporter permits intestinal cells to accumulate glucose to a higher concentration than is present extracellularly in the gut. The coupled transport of two substances is called **cotransport** and depends on specific transmembrane protein molecules, called cotransporters, which bind to both transported molecules and ions. Because the glucose and Na^+ ions move in the same direction across the membrane, the process is called **symport** [Greek, *sym* = together with]. In other cases, cotransport involves an **antiport** mechanism, in which the transported materials move in opposite directions, as in the case of chloride (Cl^-) and bicarbonate (HCO_3^-) ions in red blood cells (illustrated in Figure 6-15b).

Membranes, then, are important regulators of a cell's contents. They allow water to pass freely, with the net movement dependent on the relative water potential inside and outside the cell. They selectively admit and expel molecules and ions. In some cases, they speed the passive transport of specific molecules with transport proteins like glucose permease. In other cases, they harness the energy of ATP or of existing concentration gradients to accumulate or exclude particular molecules or ions.

Although we have spoken almost entirely about plasma membranes, the internal membranes of eukaryotic cells are equally important in regulating the contents of the spaces they enclose. As we see in Chapters 7 and 8, the membranes of mitochondria and chloroplasts are crucial to the production of ATP.

MEMBRANE FUSIONS ALLOW THE UPTAKE OF MACROMOLECULES AND EVEN LARGER PARTICLES

Biological membranes can rapidly change shape. The ruptured membranes of red cells, for example, can reseal to form intact membranes ("ghosts") that are remarkably able to regulate their internal environments by pumping ions and molecules. Similarly, when cells are broken open for subcellular fractionation, the internal membranes of the endoplasmic reticulum are disrupted, and membrane fragments quickly reseal to form **microsomes,** vesicles derived from fragments of the ER membranes.

The fusion and reorganization of membranes with one another are not simple processes, however. The outside surfaces of most biological membranes carry a net negative charge and repel each other. Yet the ability of membranes to reseal is an important part of many cellular functions—including cell division, when the plasma membrane of the parent cell is distributed to the two daughter cells. Although biologists do not yet know the mechanisms involved, they have identified a number of proteins that help bring membranes together for fusion by binding to sites on each membrane.

(a)

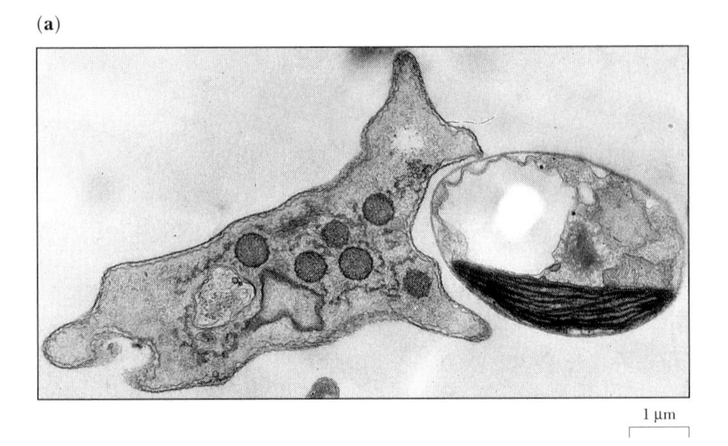

(b)

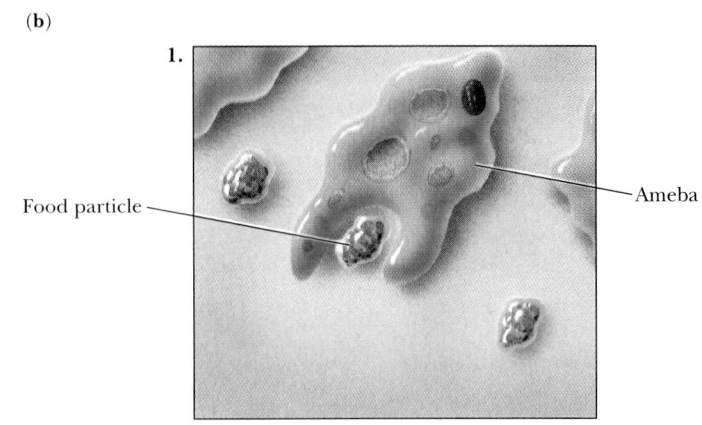

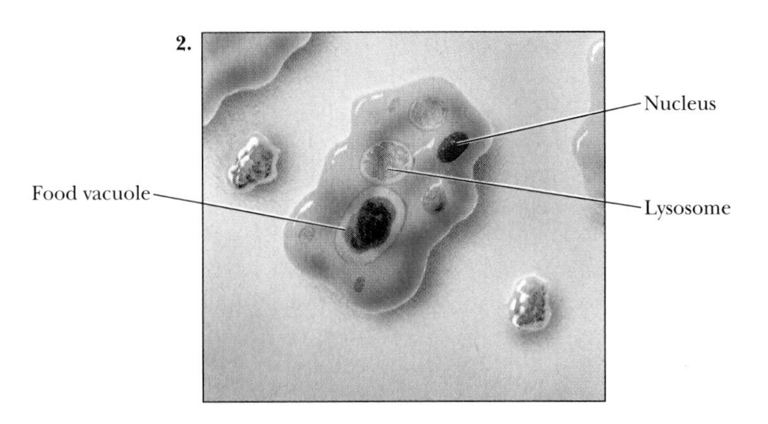

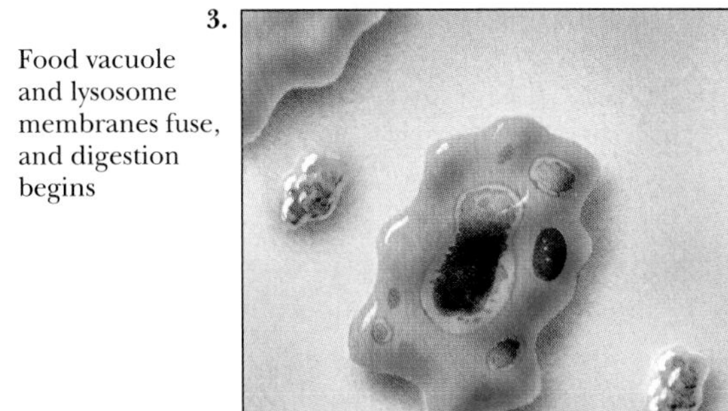

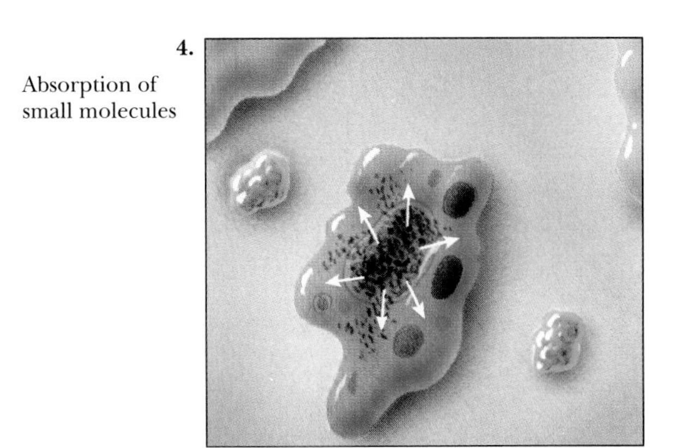

Figure 6-16 Phagocytosis: **(a)** an ameba ingesting an alga by phagocytosis; **(b)** a diagram of the process of ingestion. *(a, Biophoto Associates)*

Cells Can Engulf External Particles by Extending Plasma Membrane

The kind of membrane reorganization best known to biology students (because it is easily seen with a light microscope) is **phagocytosis** [Greek, *phagein* = to eat + *kytos* = hollow vessel], in which the cell's membrane surrounds a relatively large solid particle, such as a microorganism or cell debris. As shown in Figure 6-16, part of the cell membrane surrounds the particle. The membrane then fuses to form a large vesicle or vacuole, which then travels to the interior of the cell. The vesicle later fuses with the membranes that surround lysosomes, organelles that contain digestive enzymes. (See Chapter 5, p. 142.) Amebas and many other single-celled eukaryotes feed primarily by phagocytosis. In animals, phagocytosis occurs mostly in a few types of specialized cells, whose role is not to take in food but to engulf and destroy assorted cell debris as well as bacteria.

Phagocytosis depends on interactions with specific proteins on the cell's surface. In the case of the scavenger cells of our own bodies, for example, a specific cell surface protein (called the F_c receptor) recognizes a part of the antibodies (called the F_c region) after antibodies have attached to an invading virus or bacteria (as illustrated in Figure 6-17).

Cells Can Take in External Particles and Molecules by Folding and Fusing Plasma Membrane

Phagocytosis requires the expansion of the plasma membrane to engulf an external particle. In contrast, **endocytosis** [Greek, *endon* = within] is the process of taking in materials from outside a cell in vesicles (endosomes) that arise by the inward folding ("invagination") of the plasma membrane. The pouch-like vesicles formed in endocyto-

(Text continues on page 168.)

ESSAY

HOW DO PHAGOCYTES ENGULF AND DESTROY INVADING MICROORGANISMS?

Many single-celled eukaryotes, such as amebas, use phagocytosis as a source of food. Animals, however, use phagocytosis mostly to rid themselves of invading microorganisms. Specialized "professional phagocytes" include certain types of circulating white blood cells as well as macrophages that reside in tissues all over the body.

Because phagocytosis provides an important defense against infection and because the light microscope reveals so many details about it, cell biologists have been studying the mechanisms of phagocytosis for almost 100 years. In the past 10 to 20 years, molecular biologists have also contributed to our current understanding of how phagocytic cells recognize, engulf, and destroy their targets.

The first step in phagocytosis is the preparation of the target microorganism, by coating it with proteins. This preparation process is called *opsonization* [Latin, *opsonare* = to buy provisions, that is, to prepare for eating], and the coating proteins are called *opsonins.* The most frequently used opsonins are antibodies, which bind to components of the target cell's surface. The surfaces of professional phagocytic cells, like macrophages, contain opsonin receptors; when the operative opsonins are antibodies, the receptors are called F_c receptors, because they recognize a specific fragment (F_c) of the antibody molecule. The interaction between the opsonins coating the target and the opsonin receptors on the phagocytosing cell allows a "zippering" of the target and the phagocyte membranes, as illustrated in Figure 1.

Zippering occurs, however, only with the active participation of the phagocyte. Temporarily poisoning the phagocyte prevents phagocytosis, which continues only after the poison is removed. Removing the opsonins from part of the target cell also stops the process, demonstrating that opsonization does not just trigger phagocytosis but is required for actual internalization.

What are the cellular processes required for internalization? The most important one involves the polymerization of actin monomers into actin filaments, which drives the extension of cellular "arms" called *pseudopods,* as illustrated in Figure 2. These pseudopods actively aid in the engulfment process. A number of laboratories are actively trying to understand the process by which contact between opsonin and opsonin receptors initiates actin polymerization. Recent work has shown that opsonin receptors, like the F_c receptors, trigger the addition of a phosphate group to specific membrane proteins. These proteins then regulate the interactions of actin filaments and plasma membranes that lead to engulfment.

After a phagocytic cell surrounds its target microorganism, it encloses it in membrane, to form a *phagosome.* The phagosome then fuses with a primary lysosome to form a secondary lysosome, or *phagolysosome,* within which the microorganism is destroyed and digested by lysosomal enzymes.

An intense area of cell biological research asks the question, "What actually kills the captured microorganism?" At least two distinct mechanisms appear to contribute—the formation of toxic oxygen products, particularly the superoxide ion O_2^-, and the action of specific antimicrobial agents found in the lysosomes of professional phagocytes. Among the most potent of the antimicrobial agents are small polypeptides called *defensins,* which contain 29 to 35 amino acid residues with a net positive charge. Defensins initially contained within the lysosomes apparently move into the membrane of the engulfed microorganism, where they assemble to form pores, killing the invader by preventing it from regulating its own environment. The hydrolytic enzymes of the lysosome do the rest, and soon nothing is left but universal building blocks.

Ongoing research on defensins is addressing not only such basic questions as how lysosomes prevent their own destruction by defensins but also questions about the possible extracellular use of defensins and defensin-like peptides as antibiotics.

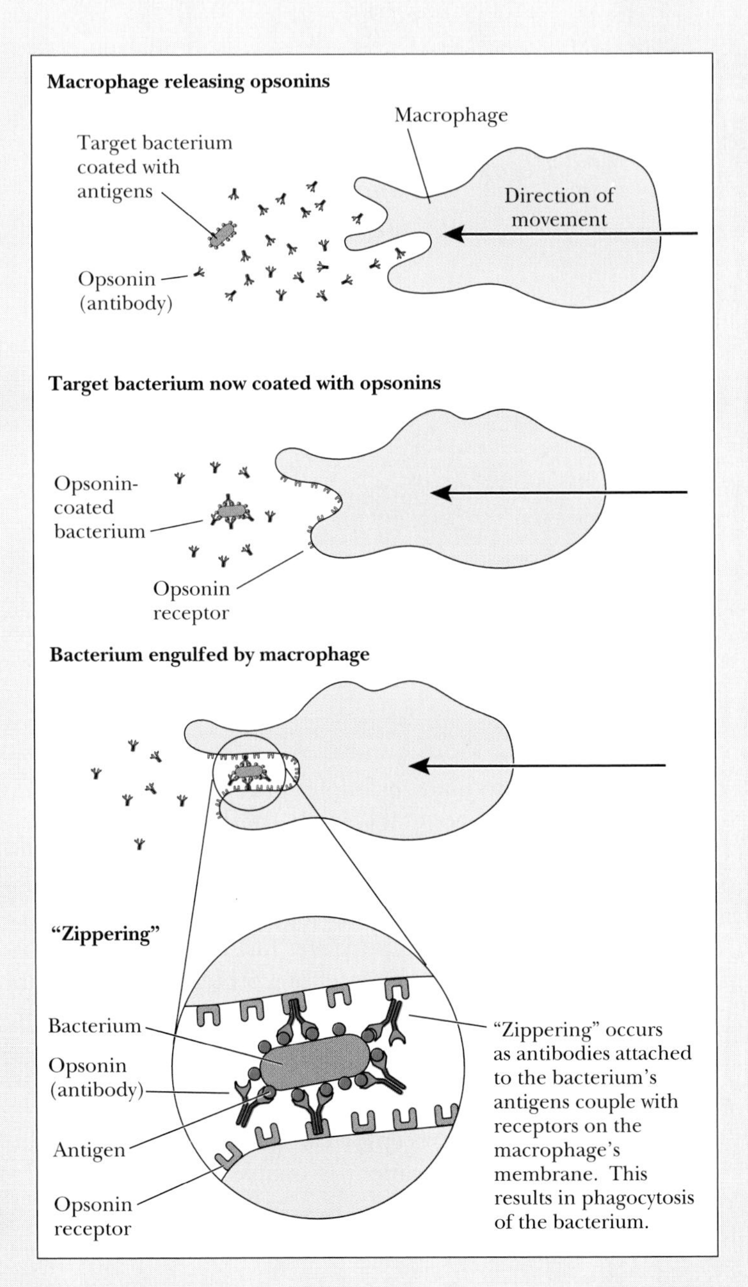

Figure 1 *"Zippering" of the membranes of a phagocyte and a target microorganism that is coated with an opsonin, such as an antibody.*

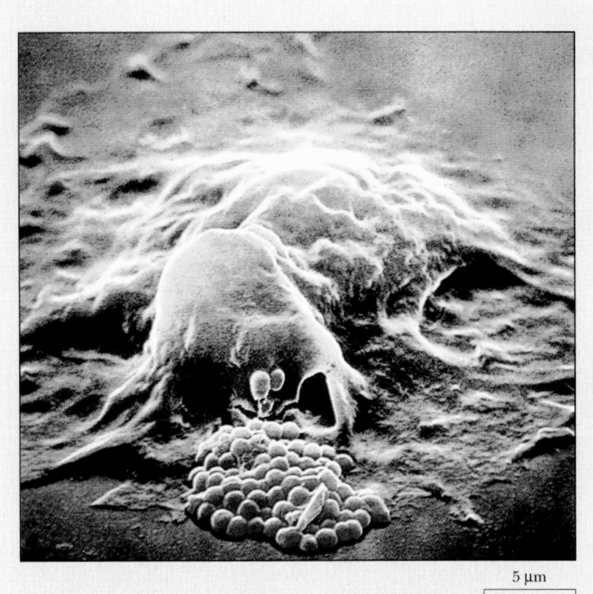

Figure 2 *Pseudopods extend around the foreign microorganism. Extension requires the polymerization of actin into actin filaments.* *(Lennart Nilsson)*

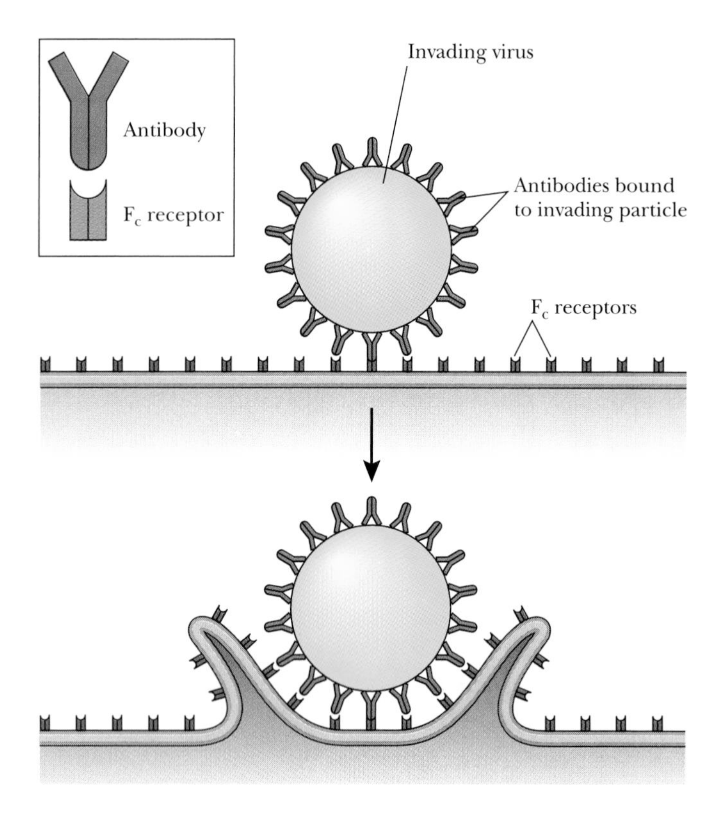

Figure 6-17 Specific proteins on the cell surface (F_c receptors) can aid phagocytosis of microorganisms that invade a mammal's body. The F_c receptor on the surface of a white blood cell recognizes a part of each antibody molecule that attaches to the foreign microbe.

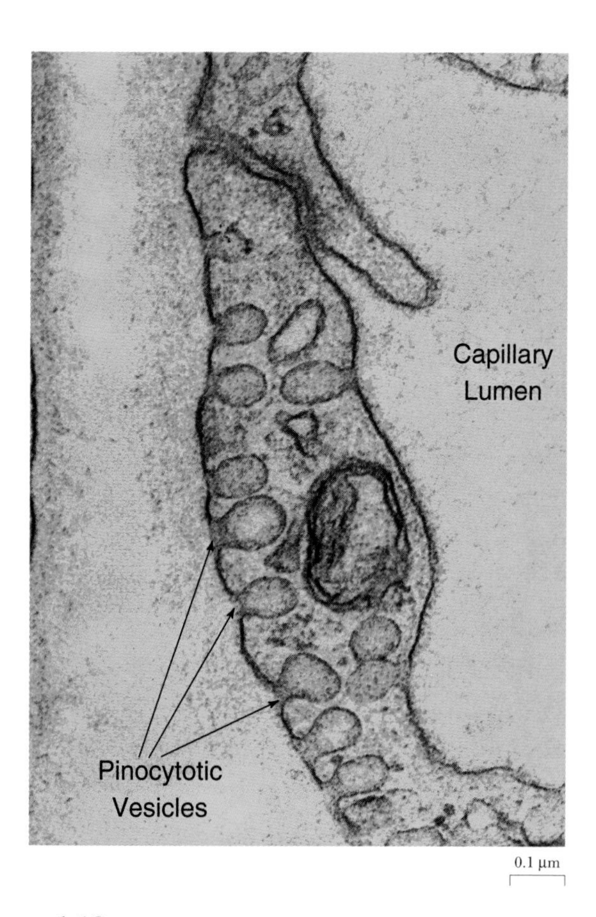

Figure 6-18 Pinocytosis, a transmission electron micrograph. *(Don Fawcett/Visuals Unlimited)*

sis are smaller (usually about 0.1 μm in diameter) than those formed in phagocytosis (which may be as large as 1–2 μm in diameter). Cell biologists now recognize two types of endocytosis: **pinocytosis** [Greek, *pinein* = to drink], the nonspecific uptake of bits of liquid and dissolved molecules (shown in Figure 6-18), and **receptor-mediated endocytosis** (illustrated for the special case of cholesterol uptake in Figure 6-19), the uptake of specific substances that are recognized by receptor proteins on the plasma membrane. Both types of endocytosis, like phagocytosis, depend on membrane fusion, after the "lips" of the membrane draw together.

Receptor-mediated endocytosis has been the subject of much study in recent years, partly because it is important to the regulation of cholesterol uptake by cells throughout the vertebrate body. Most of the cholesterol in the blood is associated with a carrier called **low density lipoprotein (LDL).** Alterations in the structure of the membrane receptor for LDL are responsible for *familial hypercholesterolemia,* a disease characterized by high blood cholesterol and increased susceptibility to heart attacks.

Complexes of LDL and cholesterol bind to an integral membrane protein (called the LDL receptor) and enter cells in a series of steps that cell biologists now know are common to many examples of receptor-mediated endocytosis. In the general case, we speak of a molecule that binds specifically to another molecule (such as a receptor) as a **ligand.** After the binding of a ligand, the receptors in the membrane cluster together and begin to form a depression on the cell surface. Just below the now clustered receptors appears a layer of a protein called **clathrin,** which forms first a **coated pit** and then a coated vesicle, as illustrated in Figure 6-19. For receptor-mediated endocytosis to occur, the membrane receptors must attach to the coated pit. In one type of hypercholesterolemia, for example, the LDL receptor cannot properly bind to the coated pit, and the patient is unable to regulate cholesterol levels.

Cells Can Also Export Molecules by Folding and Fusing Their Plasma Membranes

Exocytosis (which we have already mentioned in Chapter 5, p. 141) is the export of molecules from a cell by a process that (as shown in Figure 6-20) is approximately the reverse of pinocytosis. Molecules to be exported are surrounded by membranes that move to the cell surface. They then

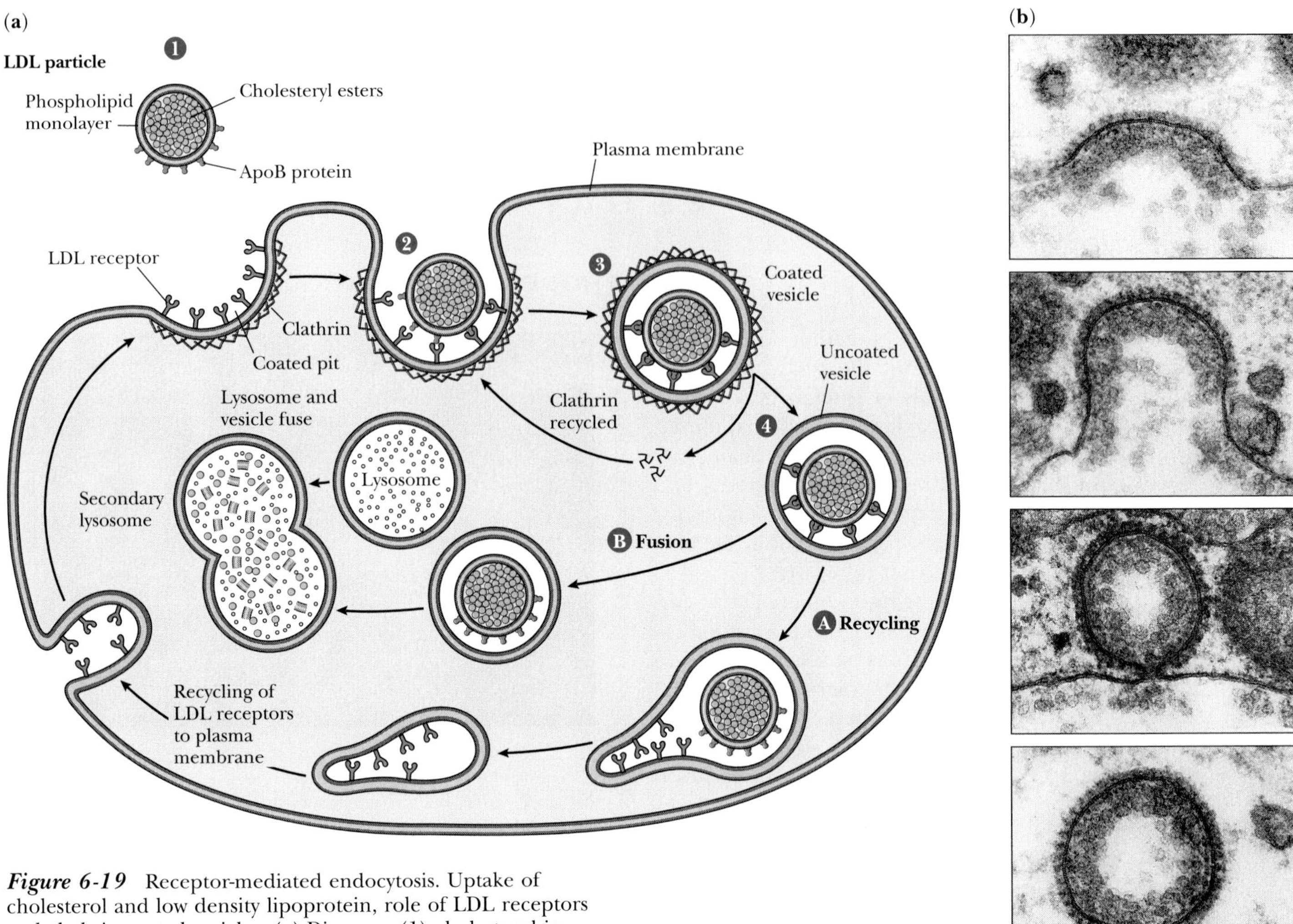

Figure 6-19 Receptor-mediated endocytosis. Uptake of cholesterol and low density lipoprotein, role of LDL receptors and clathrin-coated vesicles. **(a)** Diagram: (1) cholesterol in the blood is surrounded by a phospholipid protein complex forming an LDL particle; (2) a protein on the surface, called ApoB, binds to specific receptors on the outer surface of the plasma membrane, above the clathrin-coated region; (3) the clathrin-coated region of the membrane folds to form a clathrin-coated vesicle containing an LDL particle; (4) the clathrin detaches from the vesicle and recycles to the membrane; the uncoated vesicle then follows one of two paths: A, the recycling of LDL receptors to the plasma membrane, or B, fusion. **(b)** Electron micrographs. *(b, M. M. Perry and A. B. Gilbert, J. Cell Sci. 39:257–272, 1979)*

fuse with the plasma membrane, releasing their contents to the outside of the cell. Molecules exported by exocytosis include proteins that function outside of cells (such as digestive enzymes) and chemical signals (such as hormones and neurotransmitters). In some cases, such as the secretion of digestive enzymes by pancreas cells, secretion goes on continuously *(constitutive secretion)*. In other cases, such as the secretion of hormones and neurotransmitters, secretion depends on a specific cue *(regulated secretion)*.

Figure 6-20 Exocytosis and endocytosis.

ESSAY

HOW DO MICROBIAL PATHOGENS STOW AWAY?

Scavenger white blood cells, together with macrophages dispersed throughout the body, are highly effective in ridding animals of foreign microorganisms. The immune system, discussed in Chapter 22, provides additional defense mechanisms against infection. Yet microorganisms can cause serious and long-lasting disease, and some pathogens even live for long periods as intracellular parasites within the very cells that are specialized to get rid of them.

The bacterium that causes tuberculosis, *Mycobacterium tuberculosis,* shown in Figure 1, is a good example of a pathogen that subverts normal phagocytosis. Macrophages engulf *Mycobacterium,* but the bacteria-containing phagosome never fuses with lysosomes. Instead, a viable bacterium stays within the phagosome, its components undigested. In contrast, a phagosome containing a dead *Mycobacterium* fuses normally, and the bacterial macromolecules are degraded by lysosomal enzymes.

Mycobacterium resists lysosomal fusion by preventing acidification. While the interior of lysosomes is normally acidic, with a pH below 5, *Mycobacterium*-containing phagosomes have a pH above 6.3. Somehow the *Mycobacterium* prevents the action of the pump responsible for maintaining a low lysosomal pH, allowing it to reside for long periods as an intracellular parasite.

Another intracellular parasite, *Trypanosoma cruzi,* shown in Figure 2, subverts lysosomal function in a different manner. In humans, *Trypanosoma cruzi* causes Chagas' disease, a chronic condition affecting millions of people in Latin America. The symptoms of Chagas' disease include fever, lassitude, and enlargement of the spleen and lymph nodes; Charles Darwin is thought to have suffered from Chagas' disease, which he picked up during his South American voyage between 1831 and 1836. Some scholars have speculated

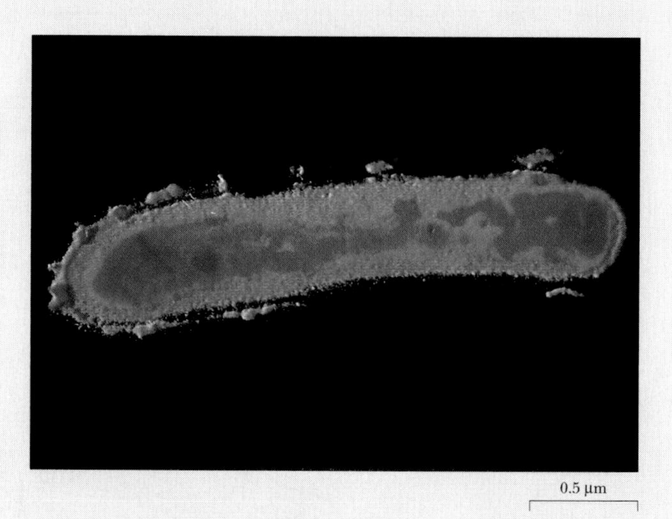

Figure 1 Mycobacterium tuberculosis, *the bacterium that causes tuberculosis, can live for long times within a phagosome without ever being subject to death and degradation by lysosomal enzymes. The secret of its survival is that it somehow prevents acidification.* (Institut Pasteur/CNRI/Phototake)

THE OUTER SURFACES OF CELLS ARE RESPONSIBLE FOR THEIR INTERACTIONS WITH OTHER CELLS AND WITH OTHER MOLECULES IN THEIR ENVIRONMENTS

Many of the properties of the surface of a cell depend on membrane proteins that extend into the space outside the cell. These proteins (as well as some of the lipids on the exoplasmic face) are rich in carbohydrates and are visible with stains that specifically attach to carbohydrates (as illustrated in Figure 6-21a). Such a densely staining zone is visible outside most eukaryotic cells and is called the **glycocalyx** [Greek, *glykos* = sweet + Latin, *calix* = cup], or cell coat. The cell coat contains molecules that allow specific types of cells to recognize and associate with each other.

Cell Walls and Extracellular Matrix Lie Outside the Cell Membrane

Additional cell surface structures lie outside the membrane itself. All plant cells and most prokaryotes, for example, have an external rigid structure, made of carbohydrates and proteins, called the **cell wall,** as shown in Figure 6-21b. Many animal cells are also enclosed in a complex network of extracellular molecules, both carbohydrates and proteins, which is called the **extracellular matrix,** shown in Figure

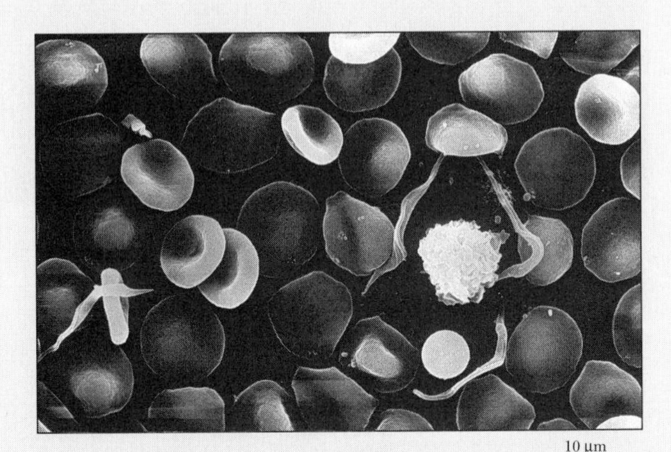

Figure 2 Trypanosoma cruzi, *the protist that causes Chagas' disease, attracts lysosomes to the point of its cellular entry. The membrane it gathers to surround itself allows it to enter its host by phagocytosis. Soon it pokes a hole in the phagosome membrane and becomes a free intracellular parasite.* *(Dr. Elizabeth Robbins, NYU Medical Center)*

that Darwin's slowness in writing *On the Origin of Species,* first published in 1859, resulted from the effects of *Trypanosoma cruzi.*

Trypanosomes are protists and live as intracellular parasites, reproducing within a wide variety of cell types. The host's cells include both macrophages and cells, like the fibroblasts and epithelial cells, that are not "professional phagocytes." In many cases, the invading form of the trypanosome may be nearly as big as its host cell, raising the problem of how to surround it with membrane. Ordinarily, the membrane surrounding a phagocytosed microorganism derives from the plasma membrane, but in the case of trypanosomes, the amount of plasma membrane is simply not enough. Instead, as the trypanosome begins to enter a host cell, it attracts lysosomes to the site of entry. Their membranes, rather than plasma membrane, become the boundary of the phagosome. But soon thereafter, the trypanosome manages to poke a hole in the surrounding membrane and enter the cytoplasm as a free parasite.

Both *Mycobacterium tuberculosis* and *Trypanosoma cruzi* subvert the normal defense mechanisms of the cell to their own advantage. In each case, however, the initial subversion is only a first step. For a pathogen to live and reproduce at the expense of the hosts requires additional interactions with normal cellular mechanisms. Research on these pathogens is therefore important not only because of the desire to treat specific diseases but also because it continues to reveal much about normal cellular mechanisms. Conversely, medical research is unlikely to find new treatments and cures for such diseases as tuberculosis and Chagas' disease without continuing basic research on cellular and molecular mechanisms.

6-21c. Specific components of the extracellular matrix bind to specific transmembrane receptor proteins, collectively called the **integrins.**

In animals, tissues have different ratios of extracellular matrix to cells, as illustrated in Figure 6-22. **Connective tissue** consists of a relatively sparse population of cells within a bed of extracellular matrix. Connective tissue, such as bone and cartilage, withstands mechanical stress largely by virtue of the extracellular matrix. In contrast, **epithelial tissue** [Greek, *epi* = upon + *thele* = nipple], which consists of cells tightly linked together to form a sheet, has little extracellular matrix. The cells, rather than the extracellular matrix, must withstand mechanical stress by virtue of their internal cytoskeletal proteins, together with proteins that join the cells to one another and to the extracellular matrix. In epithelial tissue, most of the extracellular matrix is present in the underlying **basal lamina,** a thin, tough molecular carpet to which the epithelial sheet attaches.

Multicellular Organisms Depend on Communication Among Cells

Multicellular organisms depend on cooperation among individual cells, usually by some form of chemical signalling in which the activities of one cell influence those of another in the same organism.

In plants, fungi, and multicellular protists, direct

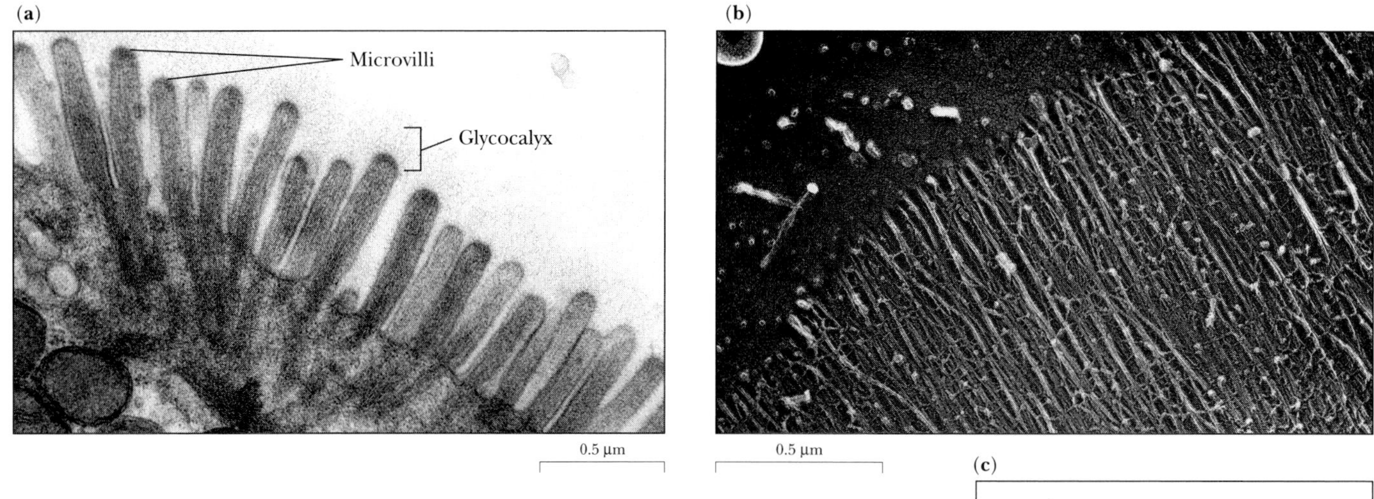

Figure 6-21 The outer surfaces of cells: **(a)** the glycocalyx or cell coat, the fuzzy area outside the microvilli on these cells from a mammalian intestinal tract; **(b)** a plant cell wall, showing network of cellulose fibers; **(c)** fibers of the extracellular matrix proteins bind to specific transmembrane receptor proteins, called integrins. *(a, David Phillips/Visuals Unlimited; b, Keith Roberts, John Inness Center, Norwich, England)*

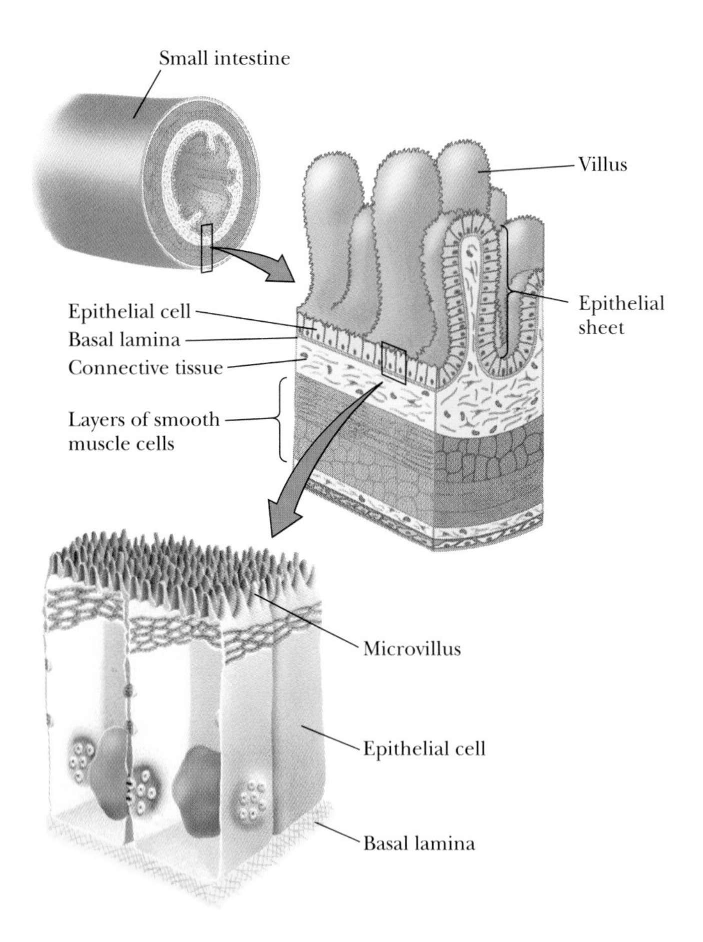

Figure 6-22 Connective tissue consists of a sparse population of cells embedded in extracellular matrix. In epithelial tissue, cells are tightly linked together, and most of the extracellular matrix is in the basal lamina.

communication between cells is limited by the presence of the cell wall, although the cell wall does not limit the diffusion of small molecules that serve as chemical signals. In some cases, however, divisions between cells are absent, so that many nuclei share a single cytoplasm. This strategy for cooperation, however, limits the possibilities of cell specialization, because all the nuclei are exposed to the same cytoplasmic environment, and all the cytoplasmic materials tend to mix together.

Individual Cells of Plants and Animals May Have Direct Communication with One Another

The connections between two adjacent cells depend on **communicating junctions,** membrane-associated structures that allow small molecules to pass freely between two adjacent cells.

Plant cells are both cemented together and separated from each other by rigid cell walls. Small molecules pass easily through cell walls, but larger molecules are stopped. The living cells of plants (as well as of fungi and multicellular protists) are nonetheless in intimate contact through **plasmodesmata** (singular, **plasmodesma**) [Greek, *plassein* = to mold + *desmos* = to bond], fine intercellular channels 50–100 nm in diameter. Plasmodesmata, shown in Figure 6-23, allow large molecules to pass directly from

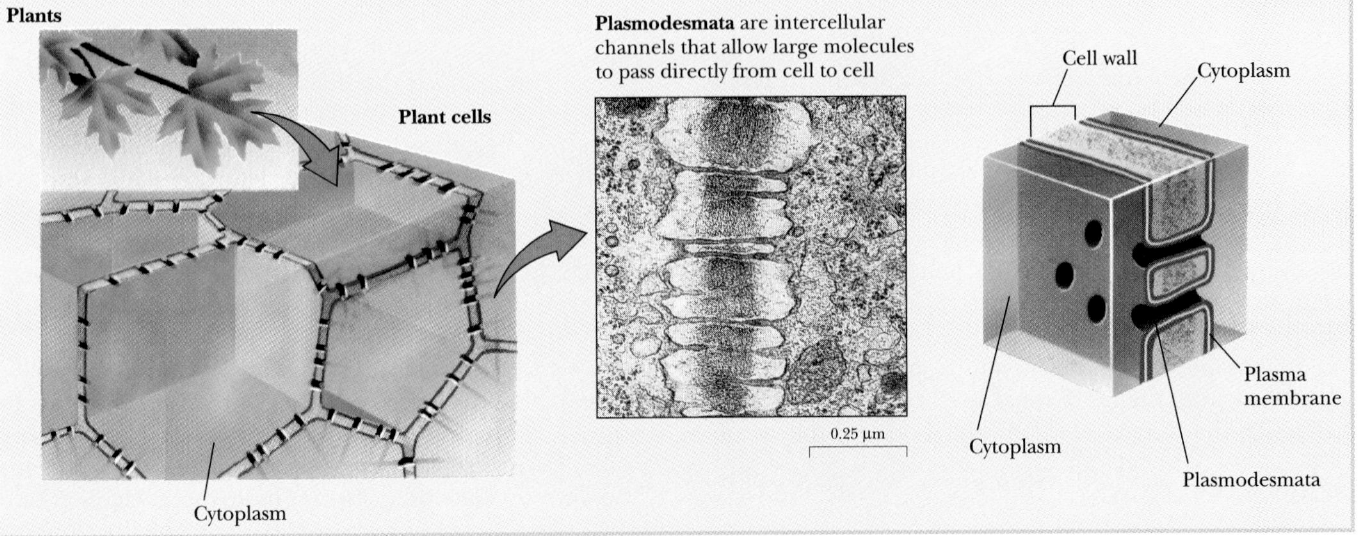

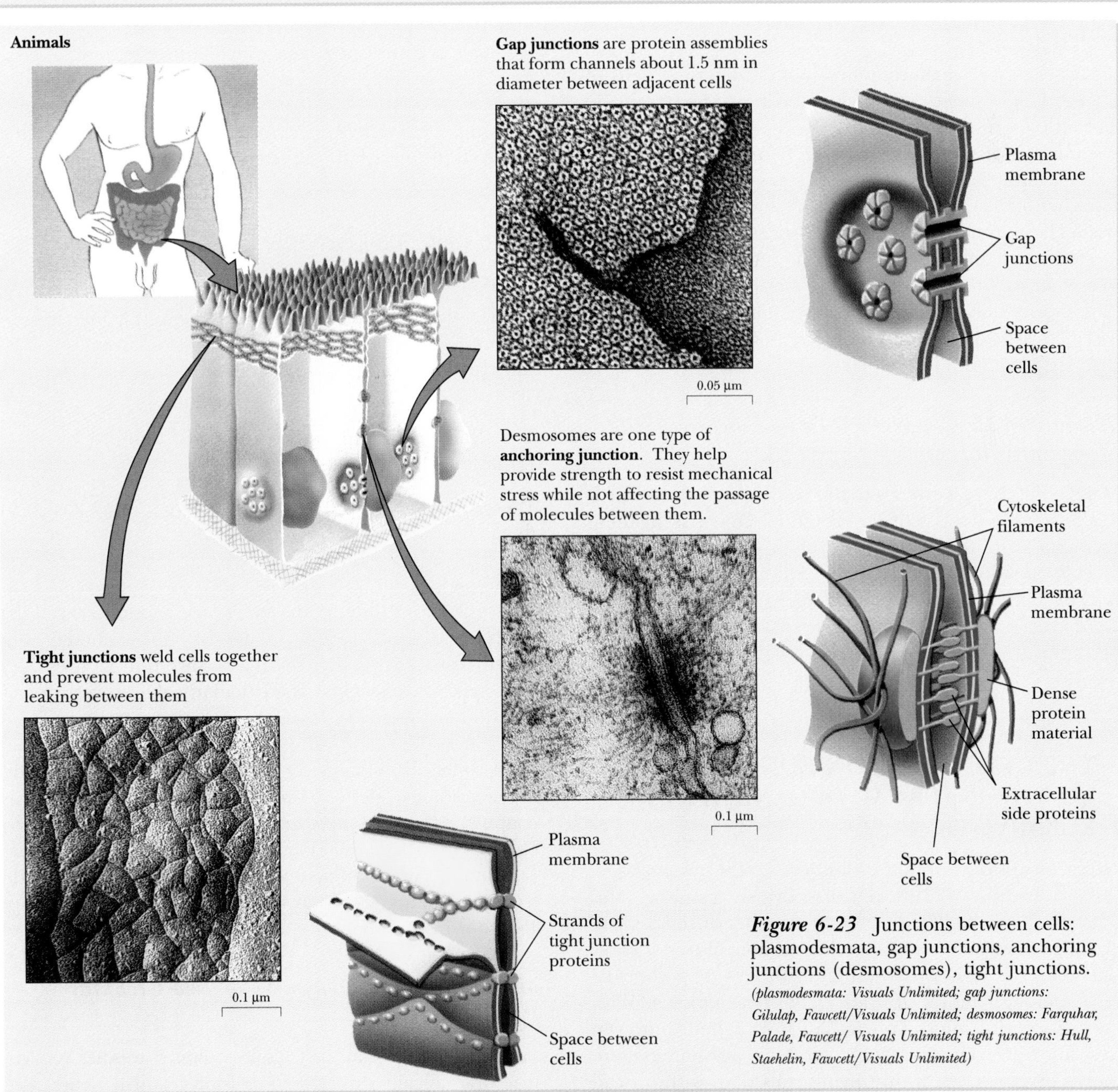

Figure 6-23 Junctions between cells: plasmodesmata, gap junctions, anchoring junctions (desmosomes), tight junctions. *(plasmodesmata: Visuals Unlimited; gap junctions: Gilulap, Fawcett/Visuals Unlimited; desmosomes: Farquhar, Palade, Fawcett/ Visuals Unlimited; tight junctions: Hull, Staehelin, Fawcett/Visuals Unlimited)*

cell to cell. Their number varies widely among different cell types. When more plasmodesmata are present, materials are passed more rapidly from one cell to the next.

Animal cells do not have plasmodesmata but have two other forms of communicating junctions—chemical synapses, specialized junctions in nerve and muscle cells, which are discussed in Chapter 23, and **gap junctions,** protein assemblies that form channels about 1.5 nm in diameter between adjacent cells. As illustrated in Figure 6-23, the membranes within the gap junctions are only 2–4 nm apart, instead of the more usual 10–20 nm. The gap junction's channels consist of a specialized protein called connexin.

Anchoring Junctions Attach Cells to Each Other or to Extracellular Matrix

Several types of **anchoring junctions** connect the cells of an epithelial sheet. Anchoring junctions help provide strength to resist mechanical stress while not affecting the passage of molecules between them. Electron microscopists have distinguished several types of anchoring junctions between cells and between cells and extracellular matrix. These junctions consist of specialized membrane proteins that, on the extracellular side, attach to other cells or to extracellular matrix. On the intracellular side of the membrane, they attach to specific components of the cytoskeleton, as shown for a desmosome in Figure 6-23 and discussed later in this chapter.

Specialized Cell Junctions Insulate the Two Sides of an Epithelial Sheet

Many epithelia separate two fluid-containing cavities, such as the blood vessels and the intestine. One task of these epithelia is to keep the two fluids from mixing. **Impermeable junctions** (also called occluding junctions) not only weld cells together, but also prevent any molecules from leaking between them. In vertebrates the main kind of impermeable junction is called a **tight junction** (Figure 6-23).

■ THE CYTOSKELETON CONTRIBUTES TO THE INTERNAL ORGANIZATION OF EUKARYOTIC CELLS

While examining pollen grains under the microscope, Robert Brown, a 19th-century British surgeon and botanist, discovered the jerky movements of small particles, which we now call **Brownian motion.** We now know that these movements result from the random thermal movements of molecules described in Chapter 4. With more powerful microscopes than Brown had, however, we can now see that such random movements are much less apparent within living cells than within dead cells.

The reason for the highly reduced Brownian movements in living cytoplasm is the active maintenance of a dense cellular network of protein fibers collectively known as the **cytoskeleton.** The cytoskeleton, which is responsible for the jelly-like consistency of the cytoplasm, became visible (as illustrated in Figure 6-24) only after the development of electron microscope techniques in the 1950s. It contributes to the large-scale order of the cytoplasm and its component organelles.

Transmission electron microscopy and immunocytochemistry (as shown in Figure 6-25) have revealed the distinctive distributions of the three types of filaments that make up the cytoskeleton: microtubules, actin filaments, and intermediate filaments. These three elements are comparable to the girders, cables, and bolts that support, maintain, and enclose large buildings.

The name "cytoskeleton" is slightly misleading, however, because the cytoskeleton does more than give mechanical support. (In plant cells, in fact, most of the mechanical strength of cells comes from the cell wall, not the cytoskeleton.) Unlike the skeleton of an animal, the cytoskeleton also provides the force for most of the movements of cells, including the transport of materials from one place to another within the cell, changes in overall shape, cellular swimming, and cellular crawling. Some of the specialized proteins of muscle filaments are closely related to proteins within the cytoskeleton. While muscle filaments are essentially permanent structures, however, the cytoskeleton is always changing.

The Thickest Elements of the Cytoskeleton Are Hollow Cylinders

Microtubules, shown in Figure 6-25a, are about 25 nm in diameter and of variable lengths, up to several micrometers. Unlike the other elements of the cytoskeleton, microtubules do not interconnect with one another, but extend in straight lines from fixed places within a cell. Like struts of an airplane or the bones of our own bodies, microtubules are hollow cylinders (as shown in Figures 6-25 and 6-26d), a form that is particularly able to resist compression and bending forces.

Microtubules are the major components of the machinery that distributes the chromosomes during cell division and, like actin filaments (described later in this chapter), they contribute to cell movements and to changes in cell shape. Microtubules are also the principal components of permanent structures (cilia and eukaryotic flagella) that allow cells to swim or to move fluids. (See p. 187 later in this chapter.)

Microtubules Are Polymers of Two Globular Protein Molecules

As discussed in detail in Chapter 9, microtubules reorganize dramatically during cell division to form clearly visible structures that microscopists described more than a

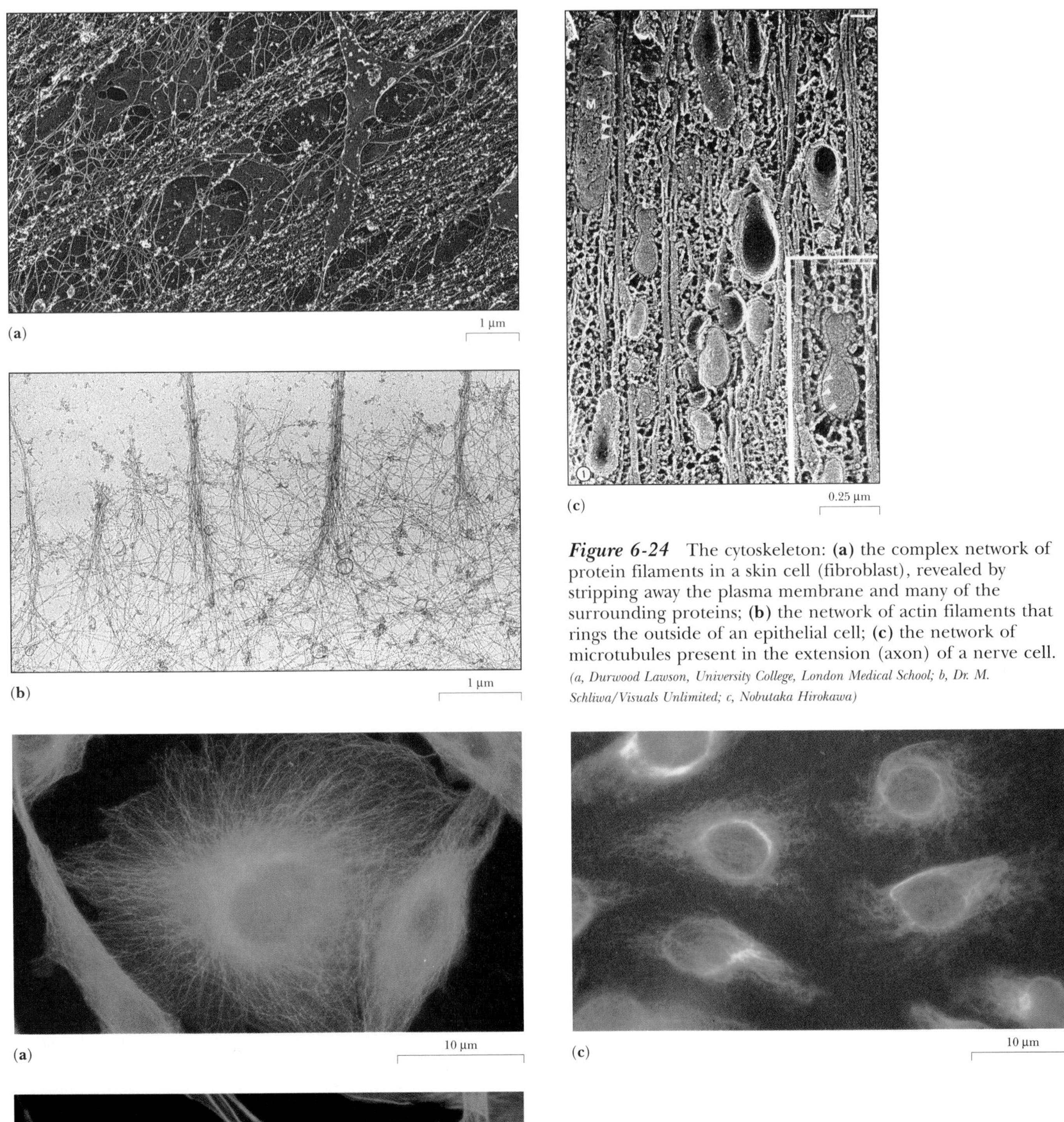

Figure 6-24 The cytoskeleton: **(a)** the complex network of protein filaments in a skin cell (fibroblast), revealed by stripping away the plasma membrane and many of the surrounding proteins; **(b)** the network of actin filaments that rings the outside of an epithelial cell; **(c)** the network of microtubules present in the extension (axon) of a nerve cell. *(a, Durwood Lawson, University College, London Medical School; b, Dr. M. Schliwa/Visuals Unlimited; c, Nobutaka Hirokawa)*

(b)
10 µm

Figure 6-25 The cytoskeleton's three types of protein filaments have distinctive distributions, here revealed in cultured BHK (baby hamster kidney) cells by immunocytochemistry with specific antibodies. **(a)** Microtubules; **(b)** actin filaments; **(c)** intermediate filaments. *(a, b, Dr. Gopal Murti/Science Photo Library/Photo Researchers; c, K.G. Murti/Visuals Unlimited)*

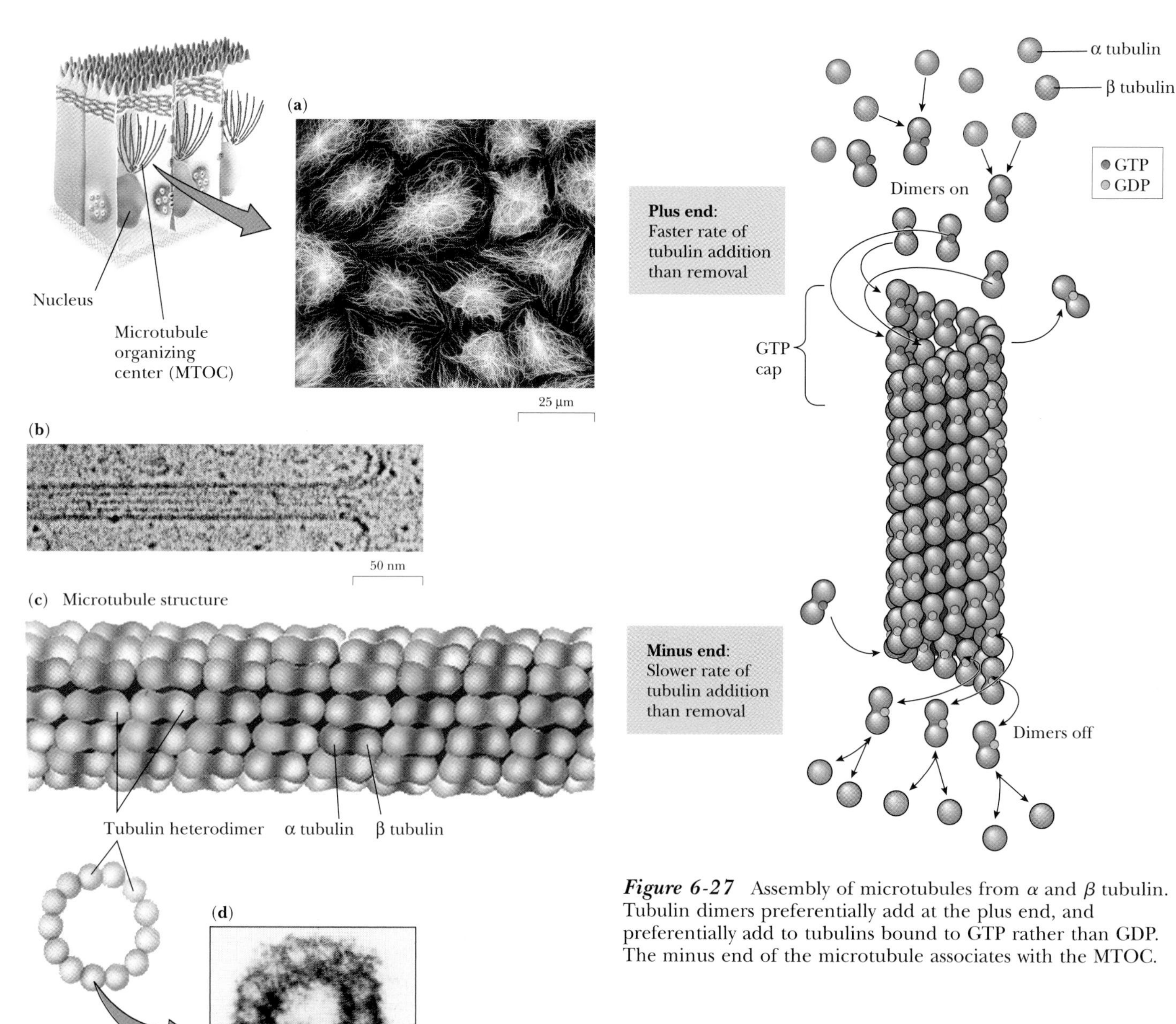

Figure 6-27 Assembly of microtubules from α and β tubulin. Tubulin dimers preferentially add at the plus end, and preferentially add to tubulins bound to GTP rather than GDP. The minus end of the microtubule associates with the MTOC.

Figure 6-26 The organization of microtubules: **(a)** microtubules emanating from the microtubule organizing center (MTOC); **(b)** a TEM of a single microtubule; **(c)** microtubules consist of helical assemblies of α and β tubulin; **(d)** a cross section of a microtubule, showing that it is a hollow cylinder. *(a, R. O. Hynes/Photo Researchers; b, Dr. Eva-Maria Mendelkow, Max-Planck-Gesellschaft; d, Paul R. Burton, University of Kansas/BPS)*

century ago. The microtubules of the cytoskeleton were not detected until much later, after the development of a new fixation technique that prevented their disintegration.

Colchicine, long used for the treatment of gout, disrupts microtubule assembly during cell division. We now know that this disruption results from the binding of colchicine to individual microtubule proteins, preventing their polymerization.

Cell biologists used the ability of colchicine to bind to the components of microtubules to identify the component proteins in cell extracts. These biochemical studies led to the isolation of two similar globular proteins, called α and β **tubulin.** Each tubulin has a molecular size of 55,000, about 450 amino acid residues. Vertebrates actually have six types each of α and β tubulin, while baker's yeast has only a single type of tubulin. All the tubulins, however, appear to be closely related, with their amino acid sequences differing only slightly.

Isolated α and β tubulins spontaneously associate into a dimer with a molecular size of 110,000. Under the proper conditions, the dimers can assemble *in vitro* to form long, lattice-like structures that, under the electron microscope, look just like microtubules seen within cells (Figure 6-26c and d). Successful polymerization in a test tube was difficult to achieve because researchers had to discover

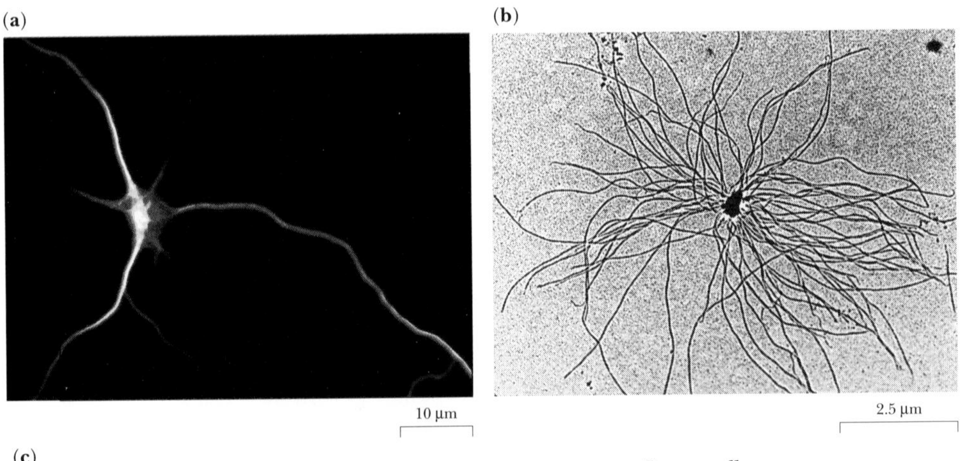

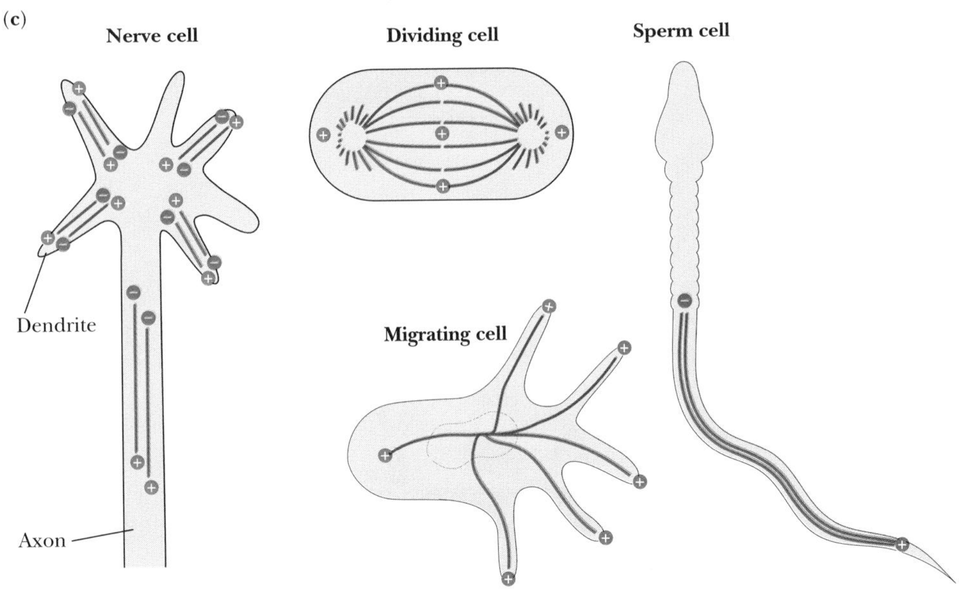

Figure 6-28 Microtubules emanating from the MTOC establish directional routes for cell traffic. **(a)** MTOC and microtubules in a neuron; **(b)** MTOC and microtubules in a migrating fibroblast; **(c)** directional routes and polarities in different cell types and processes. *(a, Peter Baas, Wenqian Yu, David Sharp, University of Wisconsin, Madison; b, Tim Mitchinson, UCSF)*

that assembly required a warmer temperature (37°C) than was usually used for biochemical experiments (4°C), as well as the removal of Ca^{2+} ions and the addition of guanosine triphosphate (GTP). One molecule of GTP binds to each tubulin dimer; after assembly into a microtubule, the GTP is eventually split to guanosine diphosphate (GDP).

Microtubules have an intrinsic polarity, whether they are observed *in vivo* or formed *in vitro,* with the two ends designated *plus* and *minus.* Tubulin dimers can add onto or leave from either end, but growth is more rapid at the plus end, as shown in Figure 6-27. Microtubules that contain GTP in the most recently added tubulin dimers are likely to grow faster than those with GDP. In a rapidly growing microtubule, assembly can occur faster than GTP hydrolysis, so the plus end may contain a "cap" with many GTP-tubulin dimers, as illustrated in Figure 6-27. Removal of the GTP cap with a microbeam of ultraviolet light leads to a rapid disassembly and shrinkage, showing the importance of the cap in determining the dynamics of microtubule growth.

In most cells, the minus ends of the microtubules emanate from **microtubule organizing centers (MTOC),** shown in Figure 6-26a and Figure 6-28a and b. In the cytoplasm of most eukaryotic cells, the major MTOC is near the nucleus in a structure called the **centrosome,** which (in many cells) also contains a distinctive organelle called the **centriole.** The MTOCs of the centrosome play an important role in cell division, as discussed in Chapter 9. Isolated MTOCs also serve as seeds for the polymerization of tubulin *in vitro.* The polarity of the microtubules also establishes the routes for cell components that use the microtubules as distribution tracks for movement away from (or toward) the cell center, as illustrated in Figure 6-28c.

Other Proteins Bind to Microtubules and Modify Their Organization and Function

Biochemical studies of microtubules have identified more than a dozen **microtubule-associated proteins** (or **MAPs**), which bind to microtubules and influence their organization. MAPs fall into two broad classes—fibrous MAPs and

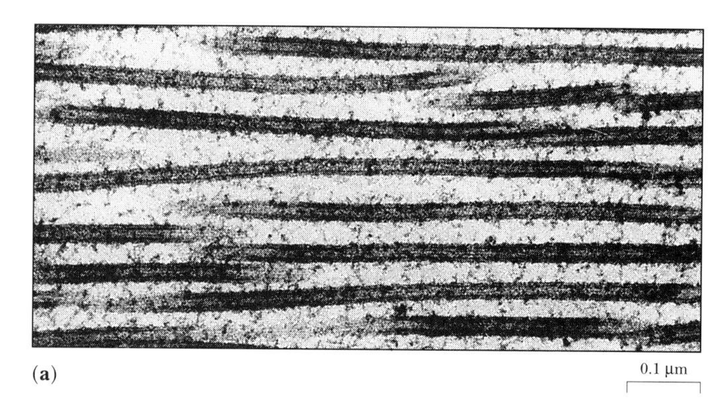

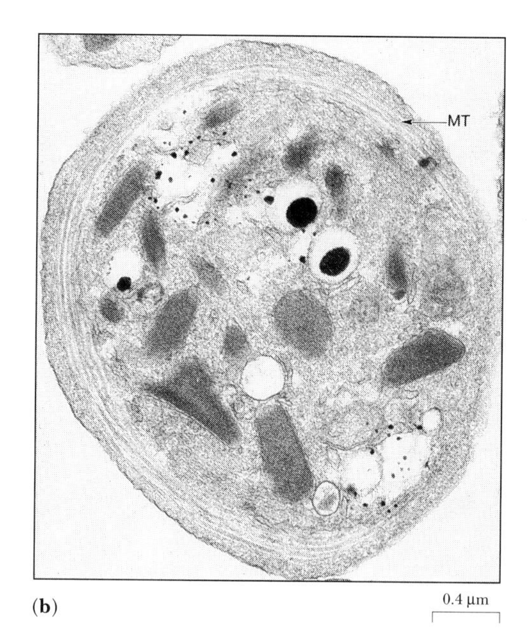

Figure 6-29 Microtubule associated proteins (MAPs) cross-link adjacent microbutules into bundles. **(a)** Microtubule bundles in a platelet; **(b)** presence of a MAP in a microtubule bundle, revealed by immunocytochemistry; antibody molecules, labeled with gold particles, bind to a specific MAP. *(a, Dr. Don Fawcett/Photo Researchers; b, Dr. Fern Tablin, UC, Davis)*

motors. *Fibrous MAPs* cross-link adjacent microtubules into bundles (as shown in Figure 6-29a). The bundling of microtubules is important in establishing the shapes of many cells, as in the case of the blood platelets shown in Figure 6-29b, where a coiled loop presses against the plasma membrane. In plant cells, the orientation of cellulose fibrils is determined by the orientation of microtubule assemblies. Disruption of the microtubular assemblies with colchicine leads to a disorganized pattern of cellulose fibrils.

Motors are microtubule-associated proteins that generate movement, using energy obtained by splitting ATP. One motor molecule, called **kinesin,** moves toward the plus end of a microtubule (as shown in Figure 6-30a), while another, called **dynein,** moves toward the minus end (as shown in Figure 6-30b). We discuss protein motors further in Chapter 18. One end of each motor molecule binds to the microtubule, and the other can bind to a vesicle or other cell component, so as the motor molecule moves, it can carry organellar freight down the microtubular track.

The Thin Filaments of the Cytoskeleton Consist of Actin, a Globular Protein Closely Related to the Actin in Muscle Fibers

Actin filaments, also called **microfilaments,** are shown in Figure 6-31a. Actin filaments, only about 7 nm in diameter, are the most flexible elements of the cytoskeleton.

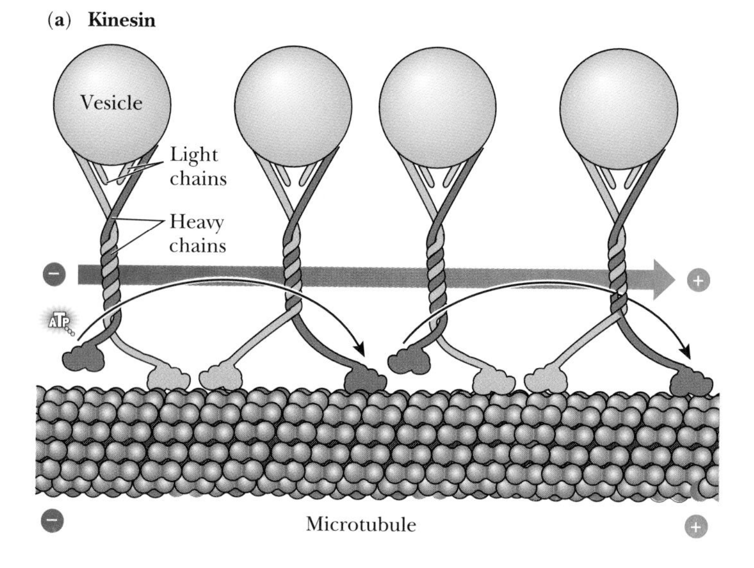

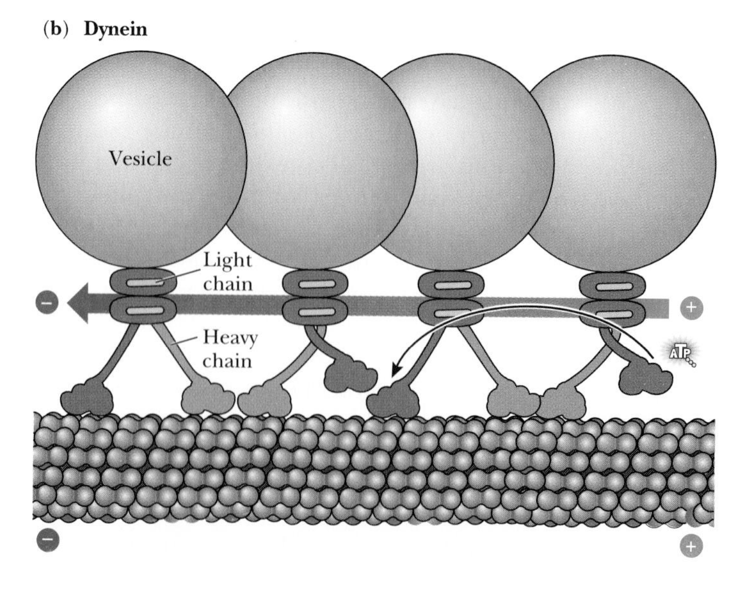

Figure 6-30 Motor molecules generate movement along microtubules. **(a)** Kinesin molecules move toward plus ends; **(b)** dynein molecules move toward minus ends.

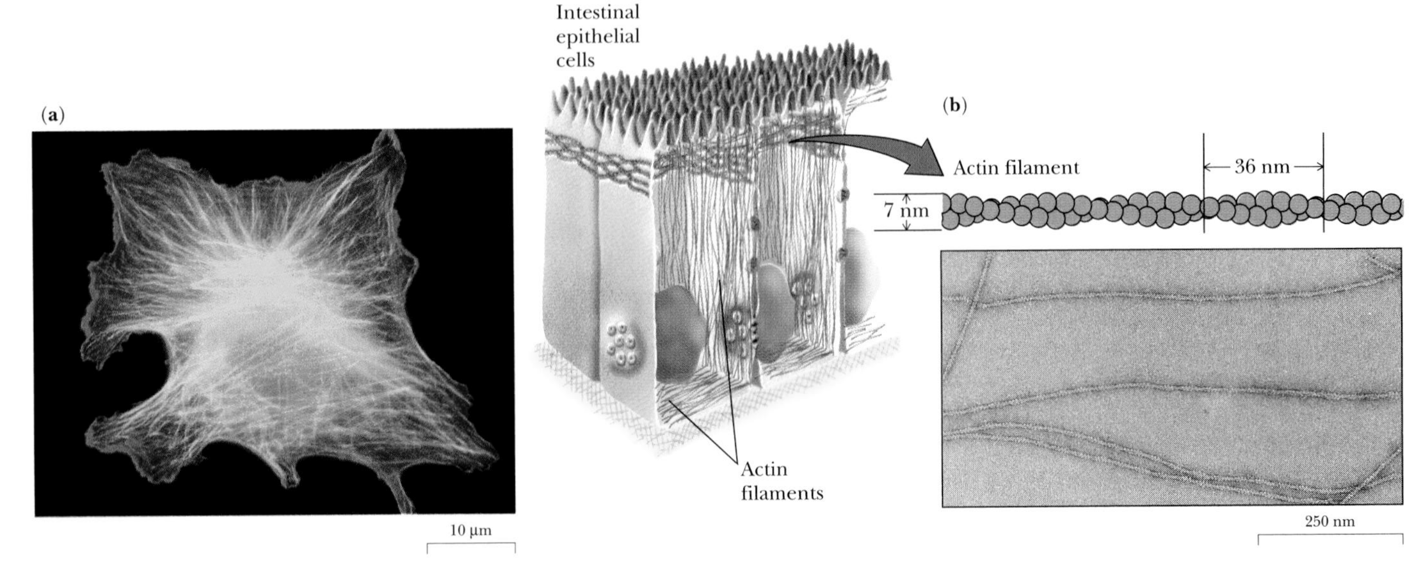

Figure 6-31 Assemblies of actin filaments: **(a)** the cortex of an epithelial cell, just below the plasma membrane with the immunofluorescence showing the distribution of actin filaments; **(b)** an actin filament (TEM). *(a, M. Schliwa/Visuals Unlimited; b, Ueli Aebi, Maurice E. Müller-Institut)*

They form a meshwork just below a cell's surface, as shown in Figure 6-31b. They can also assemble in parallel arrays that extend from the cell surface, as in the hair cells of the inner ear, which detect minuscule air movements of sound waves.

Actin filaments are so named because they consist mostly of **actin,** a protein that is highly similar to the protein first identified as making up the thin filaments of muscle (mentioned in Chapter 1). Antibodies to muscle actin recognize the thin filaments within essentially all eukaryotic cells.

Each actin molecule is a globular protein consisting of 375 amino acid residues, with a molecular size of about 42,000. Actin molecules from diverse species are extremely similar to one another, suggesting an ancient and successful evolutionary invention. In vertebrates, however, the actin within muscles has an amino acid sequence distinct from that of the actin in the cytoplasm of all cells (including the muscle cells themselves). A mammal, for example, has six different types of actin, two of which (called β and γ actin) are present in all cells, while four are present in different types of muscle cells. Yet all of these actins are recognizably similar, differing in only about 6% of their amino acid residues.

In muscle, actin filaments are permanent structures that interact with other proteins to produce the force leading to contraction. In nonmuscle cells, actin filaments are constantly being built and destroyed. They may form cross-linked cables that provide mechanical support within cells. They may also associate with other proteins to furnish "muscle-like" contractile forces. The proteins that associate with actin filaments in generating a pulling force have been compared with the winches that pull cables.

Actin Molecules Can Form Actin Filaments in a Test Tube

Even in the test tube, isolated actin molecules (called *G-* or *globular* actin), can self-assemble with other G-actin molecules to form a two-stranded helix (called *F-* or *filamentous* actin), similar to the actin filament seen in Figure 6-31b. Cell biologists have studied this polymerization of actin monomers into actin polymers (F-actin) for almost 50 years, using both biochemical and microscopic techniques. These studies have established that each actin filament has a visible direction and that one end (called the *plus* end) grows more rapidly. That is, the plus end has a greater chance of adding more actin monomers, while the *minus* end has a greater chance of losing monomers. The result is that, as illustrated in Figure 6-32b, actin polymerization tends to move actin filaments in a single direction.

ATP Affects the Movements of Actin Filaments

Each actin monomer can bind a molecule of ATP, which can then by hydrolyzed to ADP (forming ADP-actin), as illustrated in Figure 6-32. The rate of addition of monomers is greater for ATP-actin than for ADP-actin, so, in the presence of ATP, an actin filament grows longer, providing one way in which energy stored in ATP can bring about movement.

But the best-documented effects of ATP on actin-dependent movements depend on the interaction of actin filaments with another protein, called **myosin.** Myosin is a long two-headed protein consisting of six polypeptide chains, two heavy and four light, as illustrated in Figure 6-33a. Like actin, myosin is a major component of mus-

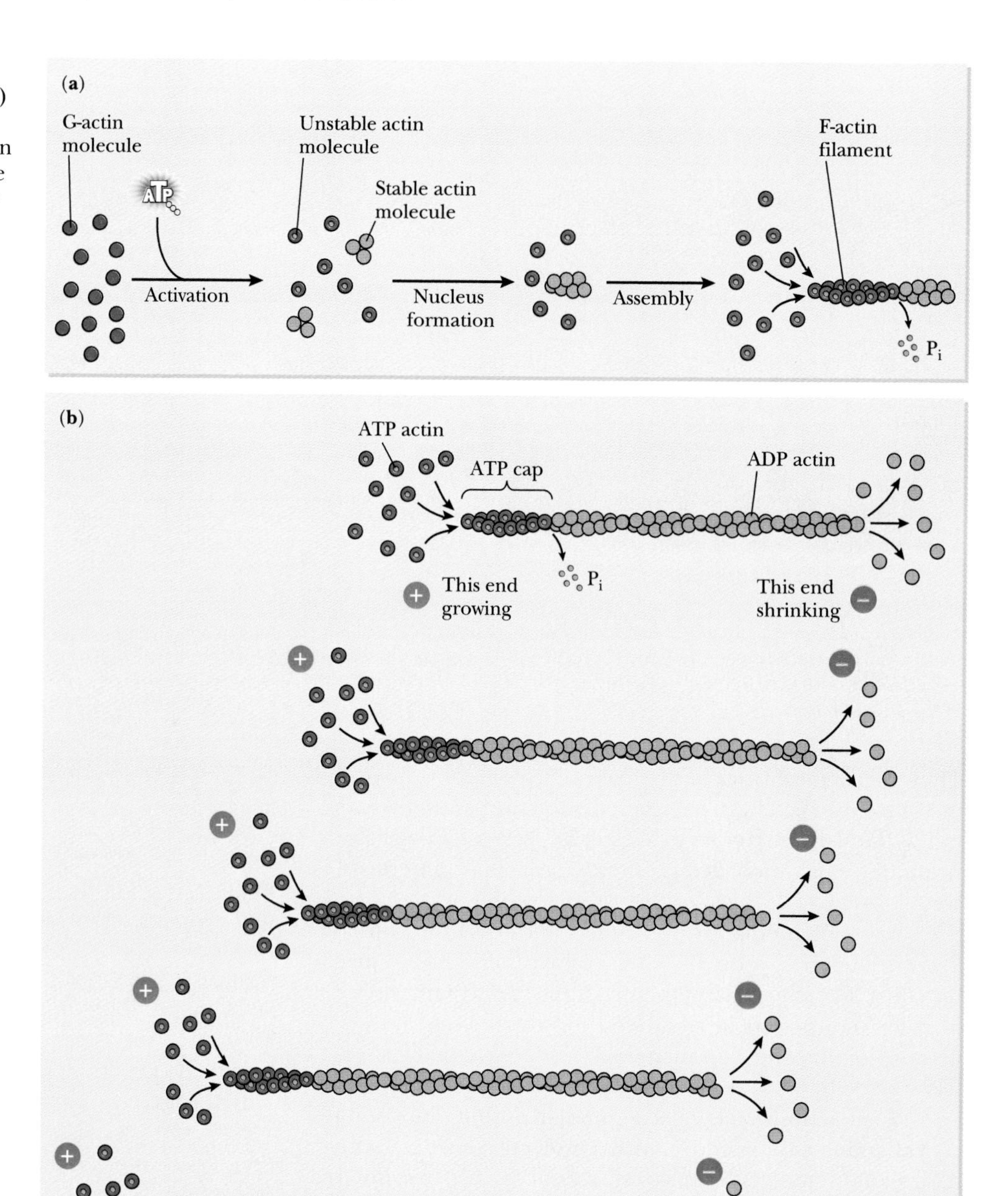

Figure 6-32 Actin filaments can assemble spontaneously *in vitro.* **(a)** Activation of actin monomers (G-actin) by ATP allows polymerization into fibrous actin (F-actin); **(b)** like microtubules, actin filaments grow at the plus end.

cle, where it was first discovered and characterized. (See Chapter 1 and Chapter 18.)

Myosin heads can bind to actin; see Figure 6-33c. Cell biologists actually use head-containing fragments of myosin molecules to demonstrate the directionality of actin filaments, as illustrated in Figure 6-33c. The arrowhead in the picture points toward the minus end of the actin filament. Myosin heads also serve as an **ATPase,** an enzyme that hydrolyzes ATP into ADP and phosphate. When associated with actin, the myosin ATPase converts the energy stored in ATP into movement of the myosin molecules relative to the actin filaments, as discussed in more detail in Chapter 18.

A Variety of Actin-Binding Proteins Affects the Forms of Actin Filaments

Proteins other than myosin also interact specifically with actin. At least 20 such proteins have been described, each of which has a distinctive effect on the organization or growth of actin filaments. A number of these effects are summarized in Figure 6-34. Some actin-binding proteins,

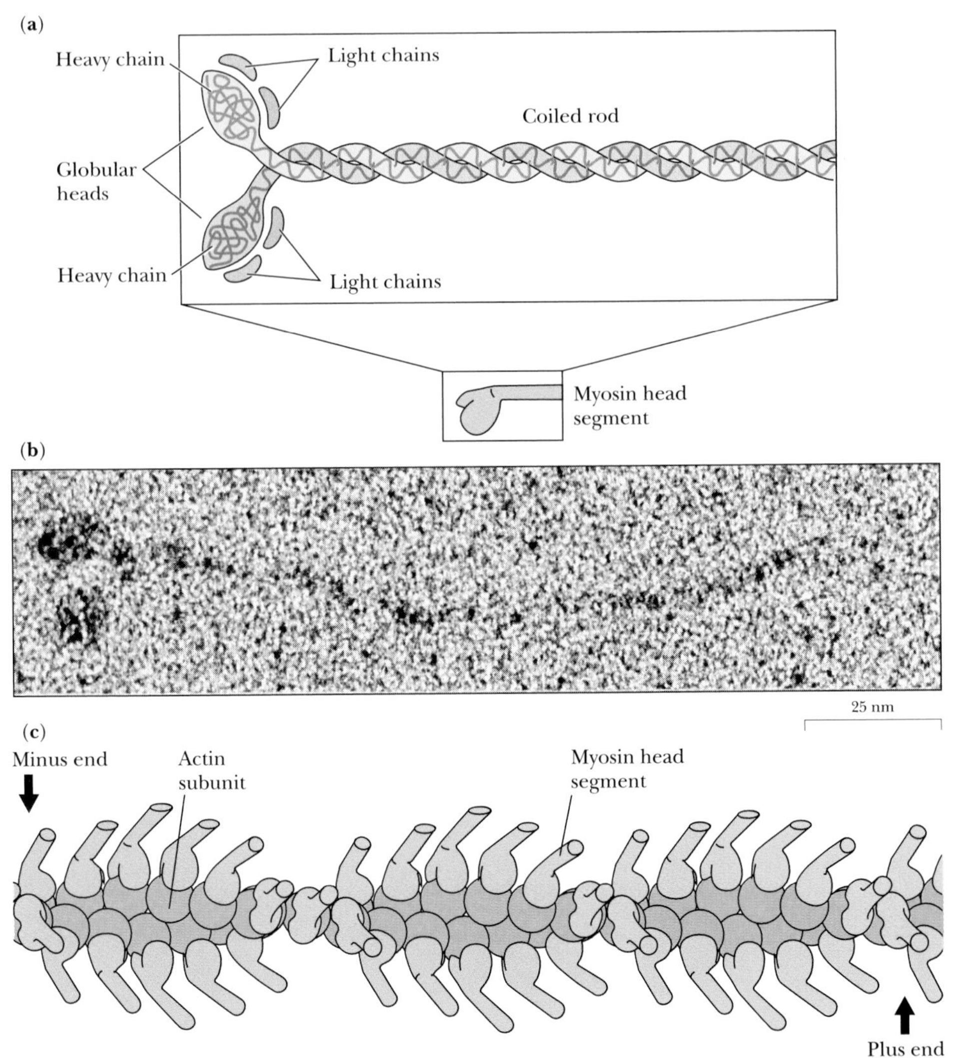

Figure 6-33 Many actin-dependent movements depend on the interactions of actin filaments with myosin. **(a)** A myosin molecule, consisting of six polypeptide chains (four light and two heavy), forming a long, two-headed protein; **(b)** TEM of a myosin molecule; **(c)** assembly of actin and myosin. *(b, courtesy of Henry Slayter, Harvard Medical School)*

for example, promote the shrinking of actin filaments by binding specifically to and sequestering unpolymerized G-actin. Others promote the loose association of actin filaments to form the gel that gives the cytoplasm the viscous quality that slows Brownian movements. Still others bind together actin filaments into the tight parallel arrangements.

Actin filaments are often anchored to the cell surface, allowing them to provide the force for movement, shape changes, and phagocytosis. As in the case of membrane transport studies, red blood cell ghosts have provided an opportunity to study the organization of actin filaments in the cell cortex.

As shown in Figure 6-35, the inside surface of a red cell membrane consists of a dense network of proteins, including actin. The major protein within this mesh is called *spectrin* [*specter* = phantom, ghost], because it was discovered in a red blood cell "ghost." Spectrin is a large protein that forms a thin molecule about 200 nm long. As shown in Figure 6-35, spectrin molecules are thought to lie on the inner surface of the membrane, anchored to integral membrane proteins either directly or by other linking proteins (such as *ankyrin,* also shown in Figure 6-35). Actin filaments bind to spectrin, which secures them to the membrane's inner face. Spectrin and ankyrin are also present in other cells, where they interact with actin to establish cell shape.

Actin filaments may cause cell movement and changes in cell shape by three different mechanisms: changes in polymerization of actin monomers, rearrangements of myosin-actin interactions that depend on the action of myosin ATPase, and alterations in the cross-linking of actin filaments by actin-binding proteins. We have more to say about these mechanisms in Chapters 18 and 20.

Intermediate Filaments Consist of Fibrous Proteins That Differ Among Cell Types

Intermediate filaments, shown in Figure 6-36, are usually 8–10 nm in diameter, thinner than microtubules but thicker than actin filaments. They are especially prominent in parts of cells that are subject to mechanical stress, suggesting that they strengthen cells and tissues. They do

Effect of protein	Protein	Comparative shapes, sizes, and molecular masses	Schematic of interaction with actin
Form filaments	Actin	370 × 43 kD/μm	Minus end; Plus end; Preferred subunit addition
Attach sides of filaments to plasma membrane	Spectrin	2× 265 kD plus 2 × 260 kD	
Slide filaments	Myosin-II	2 × 260 kD	ATP, ADP
Move vesicles on filaments	Myosin-I	150 kD	ATP, ADP
Bundle filaments	Fimbrin	68 kD	14 nm
	α-actinin	2 ×100 kD	40 nm
Strengthen filaments	Tropomyosin	2 × 35 kD	
Cross-link filaments into gel	Filamin	2 × 270 kD	
Fragment filaments	Gelsolin	90 kD	Ca^{2+}
Sequester G-actin monomers	Thymosin	5 kD	

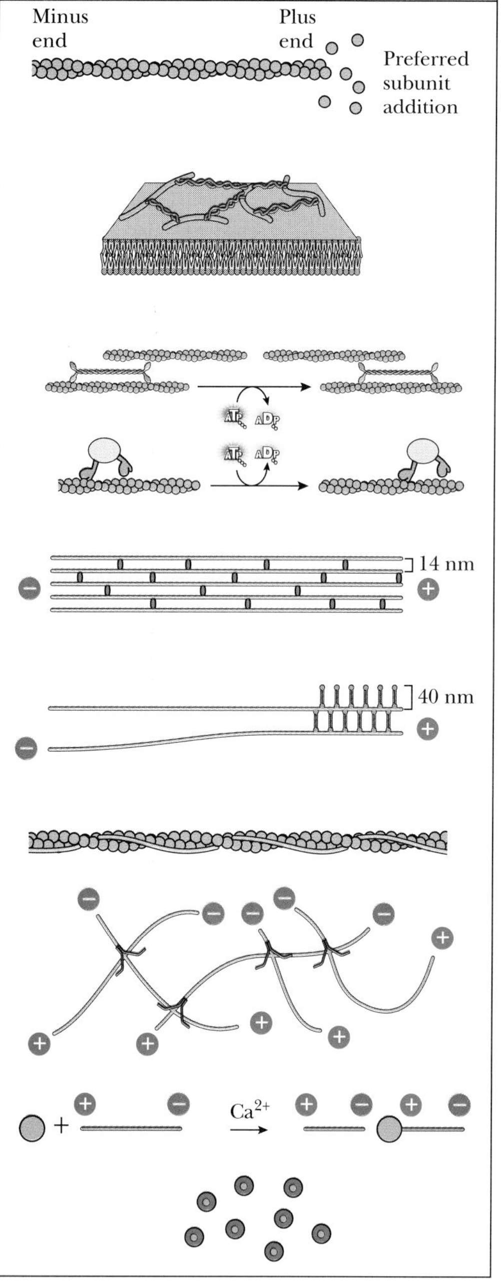

Figure 6-34 Some actin-binding proteins, showing how they interact with actin.

not appear to contribute to cell movement but form a basket-like structure that stabilizes cell shape and resists deformation. Intermediate filament proteins are all fibrous proteins. As sketched in Figure 6-36a, all contain a globular head, a globular tail, and a fibrous center—an α helical region, about 45–50 nm long, consisting of about 310 amino acid residues. Two such fibrous molecules wrap around each other, and these dimers then assemble into tetramers, which link end to end into protofilaments. The protofilaments associate and overlap to form rope-like filaments, as shown in Figure 6-36a. The overlapping fibrous elements of intermediate filaments are largely responsible for the resistance to tensile forces that pull on epithelial sheets.

A specialized intermediate filament protein—keratin—is the major structural component of wool, hair, fin-

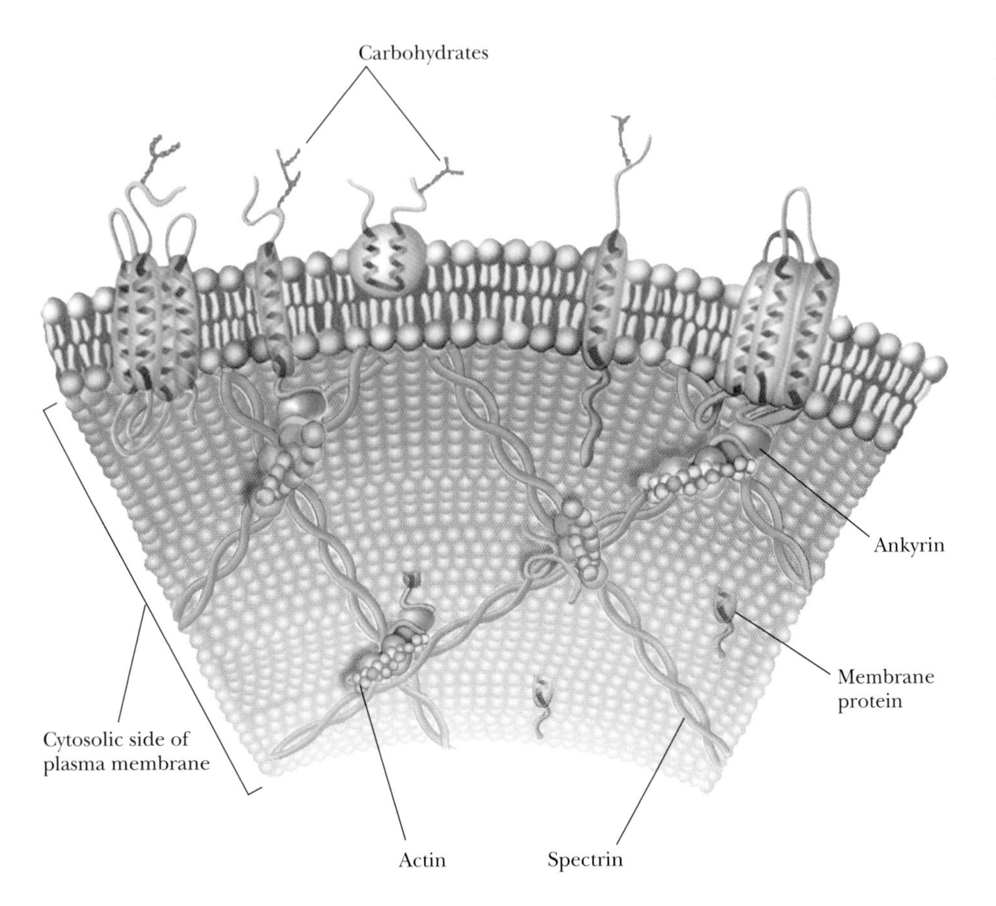

Figure 6-35 The molecular organization of the cortex of a red blood cell.

gernails, and the outer layers of the skin. Keratin filaments are also important in reinforcing the sheets formed by epithelial cells.

Another set of three intermediate filament proteins—called nuclear lamins—are present in all eukaryotic cells and form the **nuclear lamina,** a fibrous sheath just below the nuclear envelope, as shown in Figure 6-36b. The dissociation and association of the nuclear lamins are responsible for the breakdown and reassembly of the nucleus during cell division.

Yet another set of intermediate filaments—the neurofilaments—is present in nerve cells and stabilizes the structures of these highly elongated cells. When nerve cells change shape in response to injury or disease or to the normal signals of development, the neurofilaments are partially digested by a specialized protease called *calpain,* whose activity is controlled by the availability of Ca^{2+} ions.

In contrast to microtubules and actin filaments, which are "dynamically unstable" and polar, intermediate filaments are all stable and have no intrinsic polarity. While tubulin and actin require a nucleotide for polymerization, intermediate filament proteins do not. Finally, while the diverse forms of microtubules and actin filaments depend largely on associated proteins, intermediate filament proteins are themselves diverse and do not associate extensively with other proteins.

Anchoring Junctions Are Points of Membrane Attachment of Cytoskeletal Elements

Anchoring junctions provide continuity between the cytoskeletal proteins of two adjacent cells or a cell and the extracellular matrix. **Adhering junctions** connect to actin filaments on the inner surface of the plasma membrane, as shown in Figure 6-37. Transmembrane proteins, called *cadherins,* associate with each other on the extracellular side of the membrane. In epithelial sheets, cadherins form a continuous *adhesion belt* around each cell in the sheet.

Adhering junctions also anchor actin filaments within a cell to extracellular matrix at specialized membrane regions called **focal contacts,** also shown in Figure 6-37. The transmembrane proteins that make intracellular contact with actin filaments and extracellular contact with matrix are members of the integrin family.

Other anchoring junctions consist of transmembrane proteins that attach to a cell's intermediate filaments. **Desmosomes** were shown in Figure 6-23. They serve as rivets between cells. On the intracellular side, a desmosome consists of a button-like protein assembly that associates with intermediate filaments, usually keratin. The transmembrane proteins in desmosomes are members of the cadherin family, which, on the extracellular side, link the two participating cells to one another.

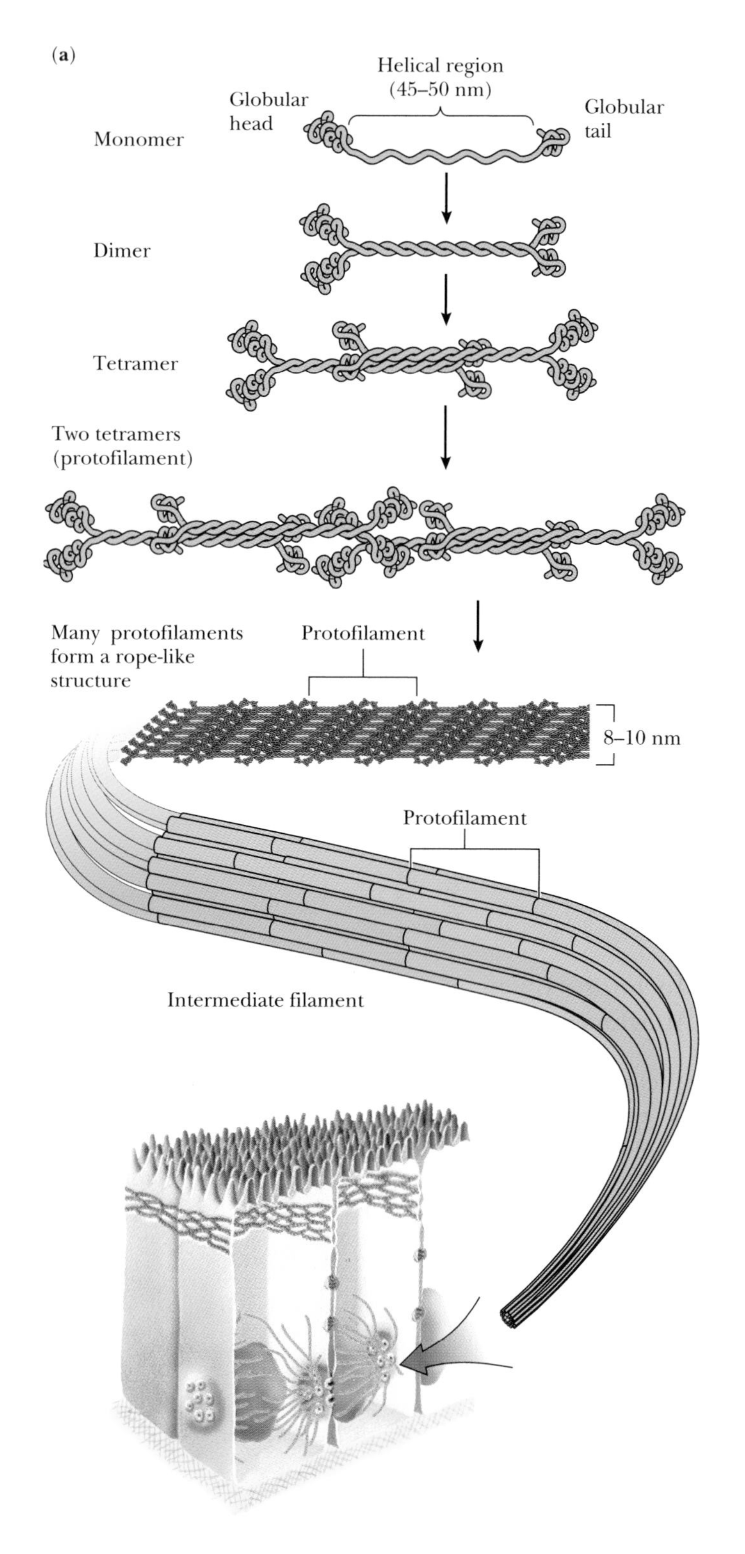

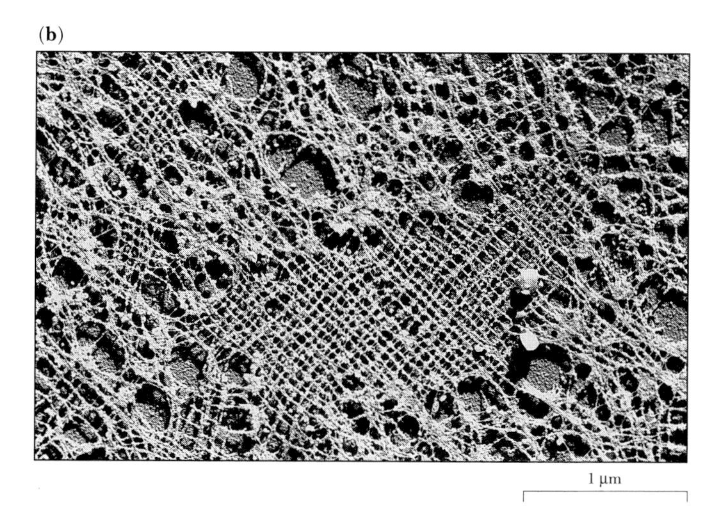

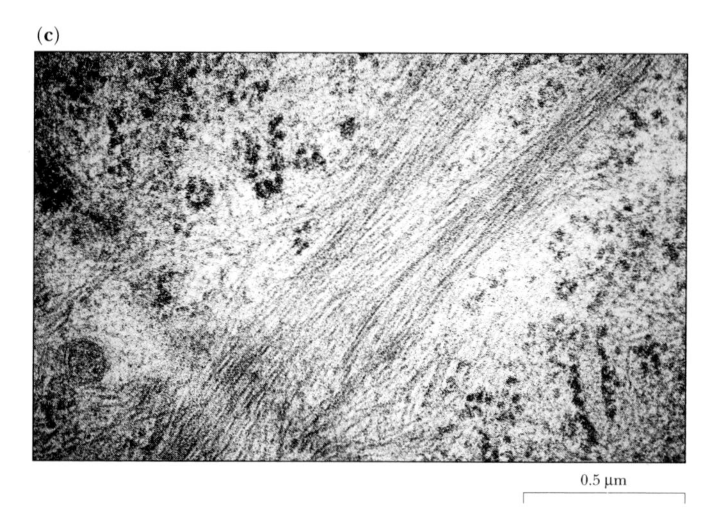

Figure 6-36 **(a)** The organization of intermediate filaments in an epithelial cell; subunits assemble to form a filament; **(b)** TEM of the nuclear lamina; **(c)** TEM of intermediate filaments. (*b, Ueli Aebi, Maurice E. Müller-Institut; c, Dr. R. Alexley/Peter Arnold*)

Hemidesmosomes, shown in Figure 6-38, attach the intermediate filaments of a cell's extracellular matrix. Despite the similarity of their names, desmosomes and hemidesmosomes are distinctive structures, with cadherins serving as the transmembrane proteins in desmosomes and integrins in hemidesmosomes.

CELLS OF ALL KINGDOMS HAVE BOTH COMMON AND DISTINCTIVE FEATURES

The cells within each kingdom and even within a single multicellular organism are enormously diverse. Our initial survey, however, examines the principal features of just one cell type of one species of each kingdom. The cells shown are not "typical" because no typical cells exist, but they illustrate some of the common and distinguishing features of cells from each kingdom.

An important distinction among organisms concerns the way they obtain the organic molecules they need to use as building blocks and as energy sources. Organisms that can make their own organic molecules from simple

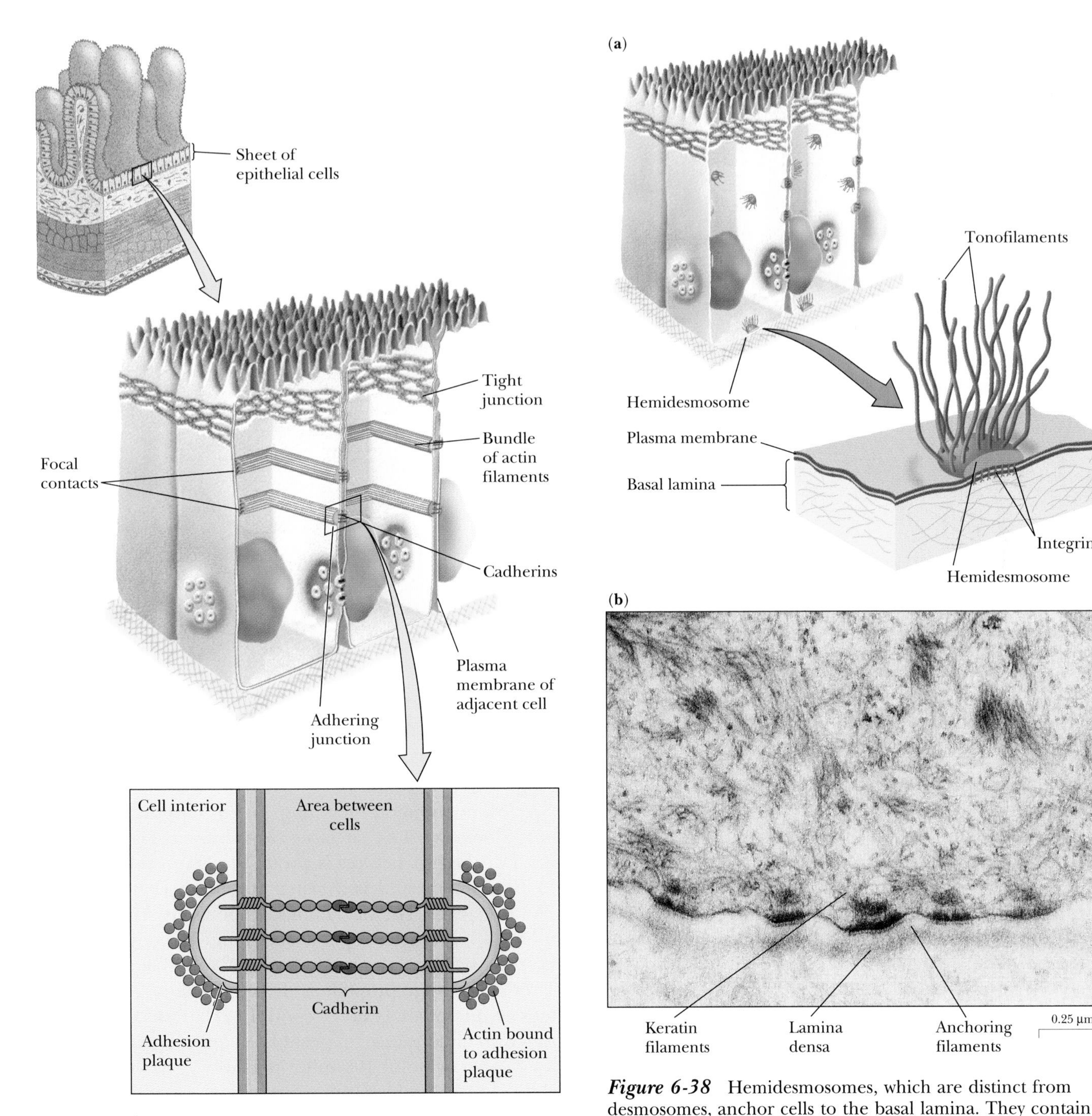

Figure 6-37 Adhering junctions—a type of anchoring junction—between cells in an epithelial sheet. Focal contacts consist of transmembrane proteins that, in the cytoplasm, attach to actin filaments.

Figure 6-38 Hemidesmosomes, which are distinct from desmosomes, anchor cells to the basal lamina. They contain transmembrane proteins that, in the cytoplasm, attach to intermediate filaments. *(b, Dr. Qian-Chun Yu and Dr. Elaine Fuchs)*

inorganic compounds (like carbon dioxide, water, and ammonia) are called **autotrophs** [Greek, *autos* = self + *trophos* = feeder]. Most autotrophs derive their energy from sunlight, although some prokaryotes can use more obscure sources such as inorganic chemicals. Organisms that derive their energy from organic molecules made by other organisms are called **heterotrophs** [Greek, *heteros* = other]. Some monera and some protists are autotrophs, as are almost all plants. All fungi and animals are heterotrophs.

Prokaryotes Are Single Cells Without Nuclei or Internal Membrane-Bounded Organelles

All prokaryotes share two common properties: They lack a nucleus and membrane-bounded organelles. Until re-

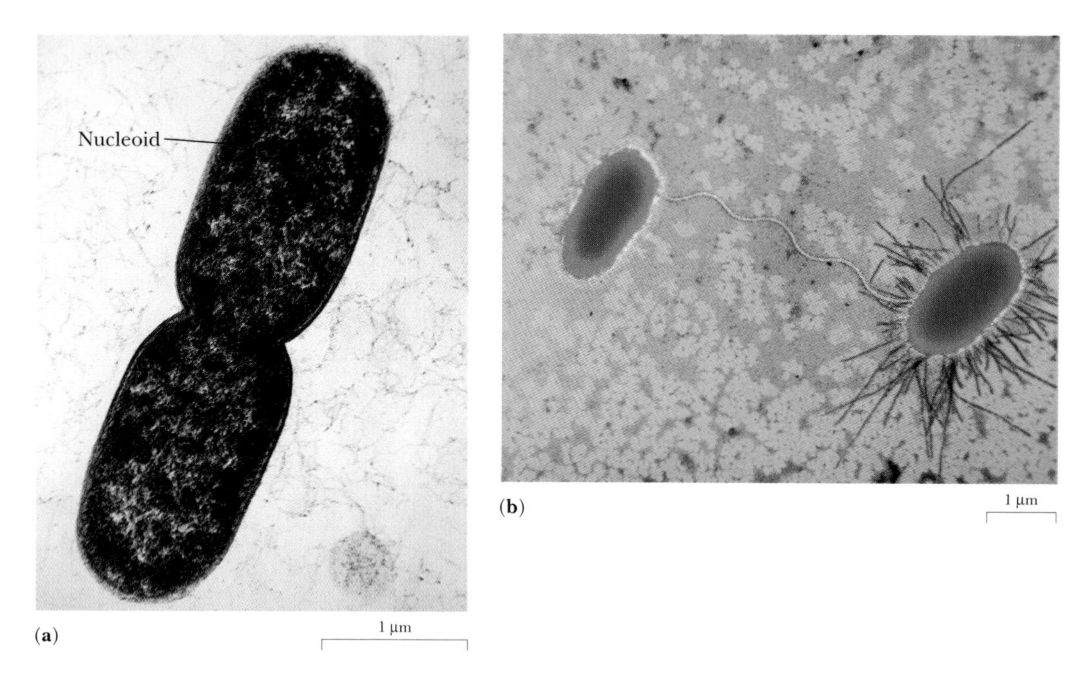

Figure 6-39 *Escherichia coli,* a prokaryote and a favorite experimental organism: **(a)** The DNA-containing nucleoid stains differently from the rest of the cell, but it is not a true nucleus, surrounded by a membrane; **(b)** the extensions from the surface of *E. coli* include flagelli, which propel the cell through its liquid surroundings and a pilus, through which conjugating bacteria exchange DNA. *(a, Dr. Tony Brain/Science Photo Library/ Photo Researchers; b, Dr. Dennis Kunkel/Phototake NYC)*

cently almost all biologists included all the prokaryotes in the kingdom Monera. Recently, however, many biologists have begun to consider the prokaryotes as at least two kingdoms—the **eubacteria** [Greek, *eu* = true], the commonly occurring prokaryotes that live in water or soil or within larger organisms, and the **archaebacteria** [Greek, *archein* = to begin], which live in particularly inhospitable environments, like acid hot springs, bogs, salt ponds, and the ocean depths. Archaebacteria and eubacteria differ from each other in their metabolic abilities, the composition of their membranes, and the structure of their ribosomes. (See Chapter 24.)

The most studied of all organisms (other than humans) is a eubacterium—*Escherichia coli* **(*E. coli*)**, shown in Figure 6-39. *E. coli* is a common resident in the human gut and (because of its rapid growth) a favorite experimental organism for thousands of research and industrial biologists. *E. coli* is the same size as most prokaryotes, about 2 μm long. With the transmission electron microscope we can see that *E. coli,* like all other cells, is surrounded by a membrane. Like all but the smallest prokaryotes, it is surrounded by a cell wall a few nanometers thick. The cell wall gives a cell shape; in *E. coli* the cell wall makes the organism more like a rod than a sphere.

The DNA of *E. coli,* like that of other prokaryotes, lies in the nucleoid, a region of the cell that stains differently from the rest of its contents. *E. coli's* nucleoids, isolated by subcellular fractionation, each contain a single molecule of DNA about 3 million nucleotide residues long.

Many monera, including *E. coli,* have a number of long fibers on their surfaces (as shown in Figure 6-39). Each of these structures is about 10–20 nm thick and up to 10 μm long and is a called **flagellum** [Latin = whip; plural *flagella*]. The flagellum's proportions are similar to those of a half-inch rope 40 feet long. In contrast to eukaryotic flagella, which bend, prokaryotic flagella rotate. They act as huge propellers, driving bacteria at speeds of tens of micrometers per second. In relation to the bacteria's own dimensions, this would be equivalent to a 15-foot-long car moving at 100 miles per hour. The bacterium's ability to move allows it to swim toward specific chemicals that it can use for food.

Like all other organisms, *E. coli* have ribosomes. Its ribosomes, as well as those of other monera, are significantly smaller than those of eukaryotes. Differences in ribosome structure have helped biologists tell which prokaryotic species are more closely related to one another, leading to a view of early evolution discussed in Chapter 24.

Protists Are (Mostly) Single-Celled Eukaryotes

Unlike the archaebacteria and eubacteria, the protists are eukaryotes, with nuclei and other membrane-bounded cytoplasmic compartments. Most protists are single cells, al-

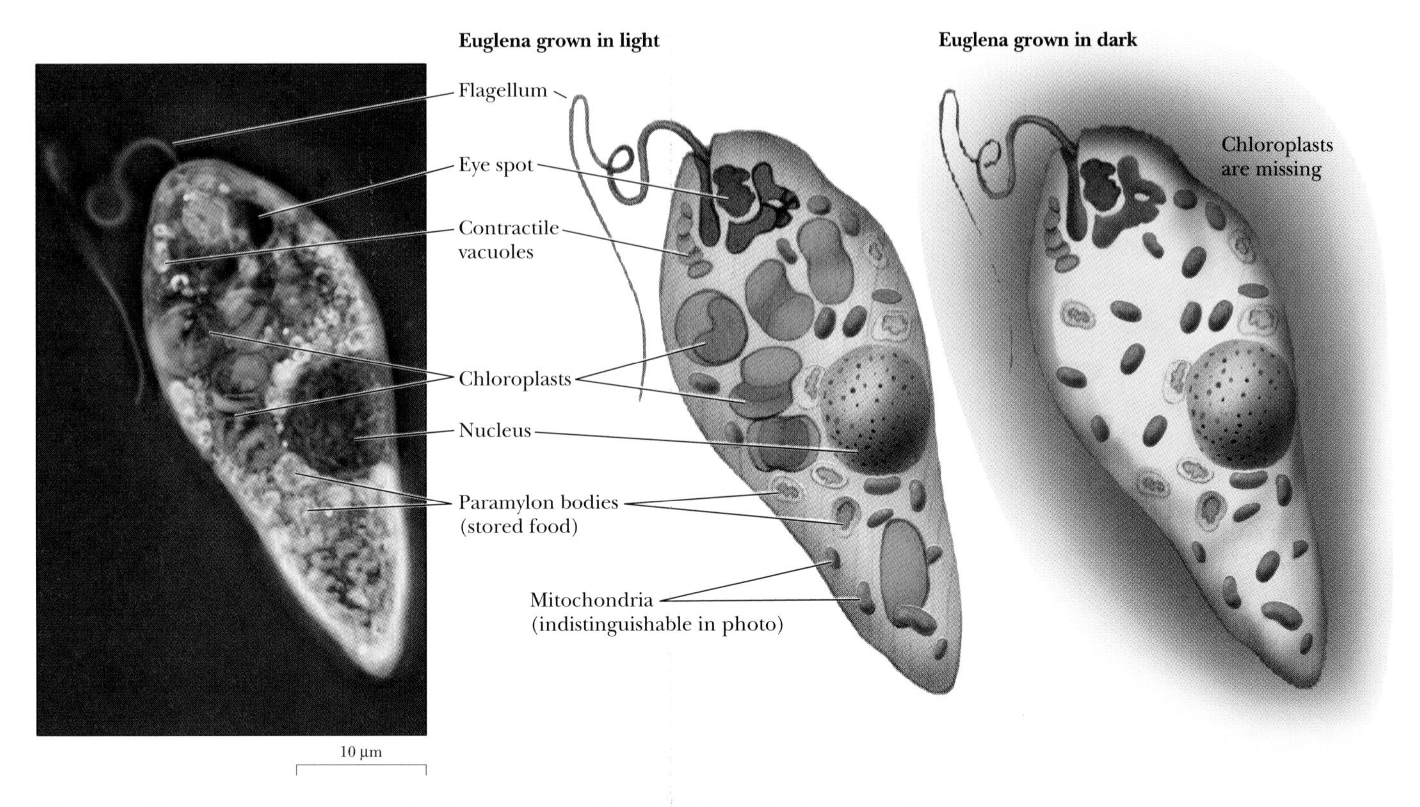

Figure 6-40 *Euglena gracilis,* a protist, is an autotroph in the light and a heterotroph in the dark. A light-grown *Euglena* has chloroplasts and can perform photosynthesis. A dark-grown *Euglena* has no chloroplasts. *(Biophoto Associates/Photo Researchers)*

though some closely related species are multicellular. Protists are not always simple organisms, however; they include some of the most complex cells on our planet.

The example discussed here is the relatively simple *Euglena gracilis,* which usually grows in fresh water ponds or lakes. *Euglena,* shown in Figure 6-40, illustrates many features of the protists and of eukaryotic cells in general. *Euglena* has a nucleus, internal membranes, a cytoskeleton, mitochondria, and chloroplasts.

Euglena illustrates the difficulty of classifying protists as plants or animals. Grown in the light, *Euglena* is an autotroph: It can derive its energy from sunlight and make its own organic molecules from carbon dioxide, ammonia, and a few salts. Under these conditions *Euglena* is green and contains chloroplasts. Grown in the dark, however, *Euglena* is white and has no chloroplasts. It lives as a heterotroph, depending on organic molecules made by other organisms. After a few days in the light, however, chloroplasts reappear and again perform photosynthesis. Chloroplasts and the ability to transform the energy of sunlight into chemical bonds go together. *Euglena* contains mitochondria in both the light and the dark.

Another feature of *Euglena* that is common in protists, plants, and animals is the eukaryotic flagellum. This structure is distinct from the flagellum of prokaryotes. When these structures are long and each cell has only a few, they are generally called flagella; when they are short and each cell has many (arranged in rows, as in the *Tetrahymena* shown in Figure 6-41a), they are instead called **cilia.**

In the electron microscope, both cilia and flagella have the same characteristic cross section. As illustrated in Figure 6-41a, each consists of a ring of nine sets of microtubules, surrounding two more in the center. The microtubule organizing center of a cilium or flagellum also has a nine-membered ring of microtubular structures and is called a **basal body** (Figure 6-41c).

Protists and some cells of animals (including sperm) and of nonflowering plants use flagella or cilia to propel themselves. Cilia on the surfaces of the digestive, respiratory, and reproductive tracts push fluids along those surfaces (as shown in Figure 6-41b).

Fungi Differ from Other Eukaryotes in the Way They Obtain Food

Fungi are always heterotrophic. They derive their food from other organisms, either living or dead, by secreting digestive enzymes onto their hosts and then absorbing the products of this digestion through specialized root-like structures. Fungi are also distinct from other eukaryotes in never having cells with cilia or flagella.

Saccharomyces cerevisiae, brewer's yeast or budding yeast (shown in Figure 6-42a), is widely used by bread makers

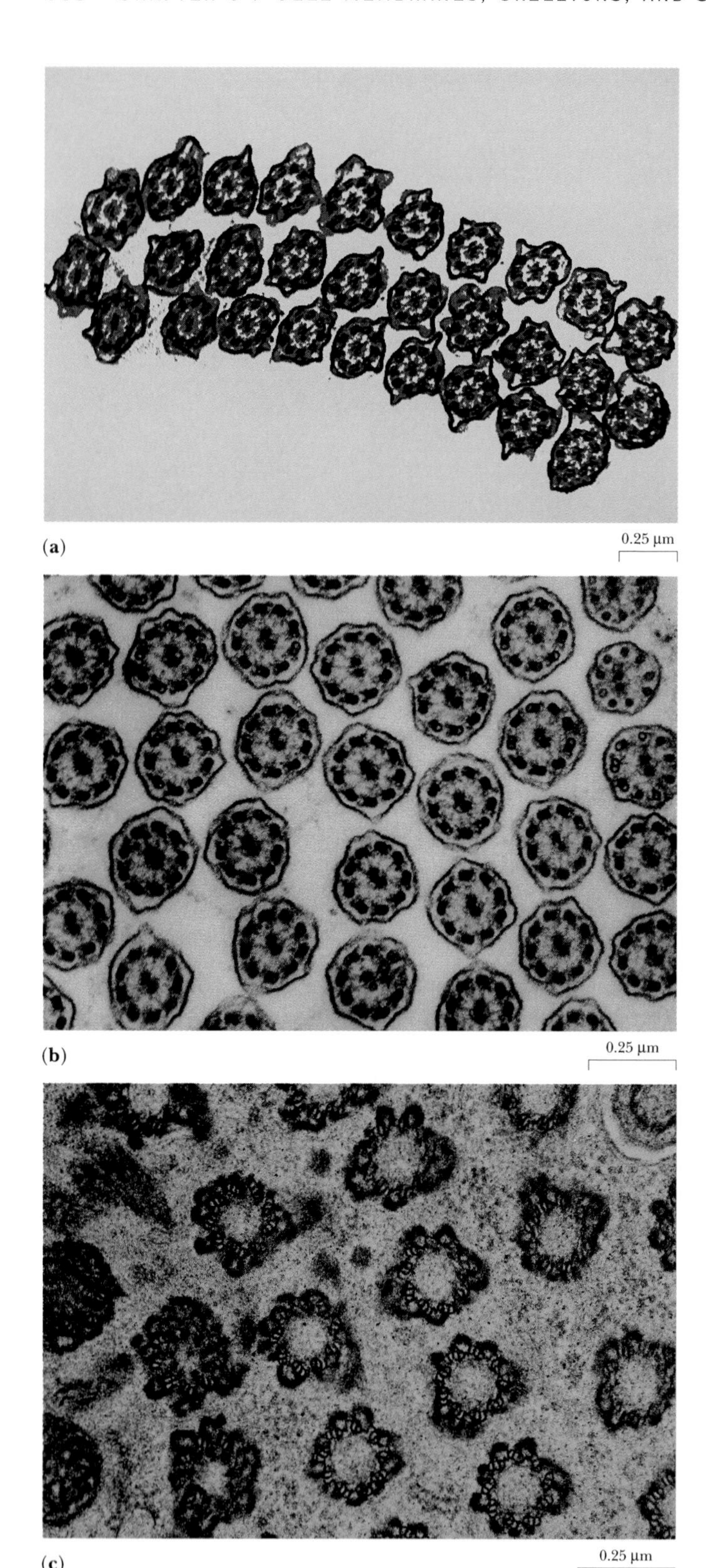

Figure 6-41 Eukaryotic flagella and cilia, both composed of microtubules, are distinct from the flagella of prokaryotes. **(a)** Cross section of cilia from *Tetrahymena;* note the "9 + 2" structure. **(b)** Cross section of cilia from a mammal's oviduct; note the same "9 + 2" structure as in *Tetrahymena.* **(c)** Cross section of a basal body from oviduct; notice the 9-fold symmetry. *(a, Scott Camazine, Sharon Bilotta-Best/Photo Researchers; b, c, David M. Phillips/Visuals Unlimited)*

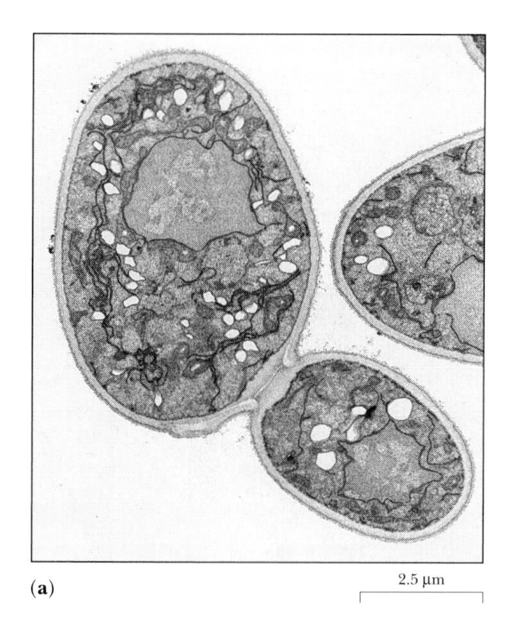

Figure 6-42 Fungi: **(a)** *Saccharomyces cerevisiae,* brewer's yeast, or budding yeast, TEM; **(b)** a morel mushroom. *(a, Biophoto Associates/Photo Researchers; b, Matt Meadows/Peter Arnold)*

and cell biologists. The morel mushrooms, shown in Figure 6-42b, are less widely used experimentally but are cherished for their subtle flavor. Like other fungi, yeast cells do not have plastids, but they do have nuclei, mitochondria, and the other types of membrane-bounded organelles. Also like other fungi, yeast cells are surrounded by a cell wall made of carbohydrates. Yeast is a single-celled organism, like many but not all fungi.

Plants Are Multicellular and Autotrophic

All plants are multicellular. Plant cells are extremely diverse, both among different species and even within a single plant. They nonetheless share many characteristics—including cell walls and vacuoles. In (almost) every plant, at least some cells have chloroplasts.

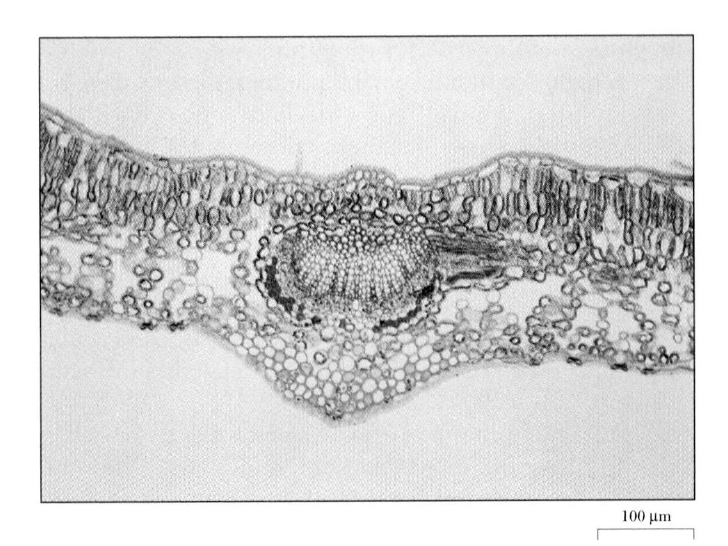

Figure 6-43 Plants: cross section through a leaf, showing red, photosynthesizing parenchymal cells. *(James Mauseth, University of Texas, Austin)*

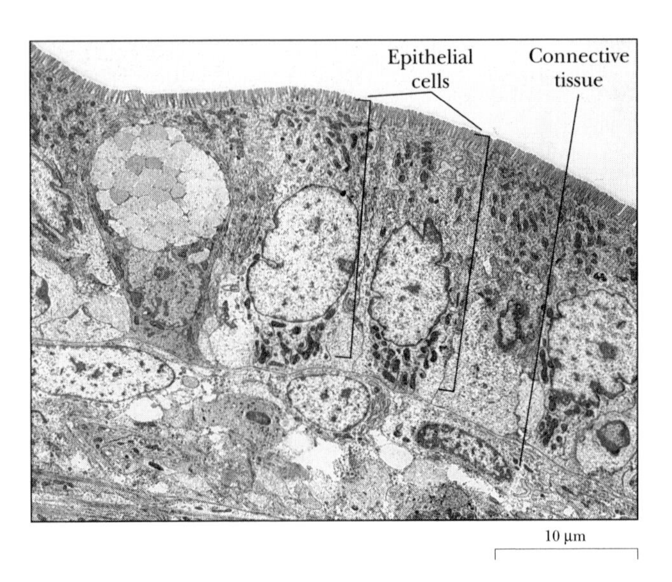

Figure 6-44 Animals: a cross section of the intestinal lining of a rat. *(Biophoto Associates/Photo Researchers)*

The dandelion is a familiar plant in city, suburb, and countryside. Dandelion cells demonstrate some of the common features and the diversity of plant cells. Some of its cells form transporting vessels and cease dividing, while other cells are not obviously specialized and keep on dividing actively. A cross section of a leaf (as in Figure 6-43) shows several types of cells. The most numerous cells in the leaf, called **parenchymal cells,** have thin cell walls, chloroplasts, and large vacuoles and perform photosynthesis. As in many other plant cells, the vacuole may take up a large fraction of the cytoplasm.

Animal Cells Lack Cell Walls, Chloroplasts, and Large Vacuoles

The cells of animals are even more diverse than those of plants. Here we discuss only four types of cells in one animal—the laboratory rat *(Rattus rattus).*

A cross section through the rat's intestine shows four major types of cells found in a mammal, two of which are seen in Figure 6-44: (1) epithelial cells, tightly linked together to form a sheet (as discussed above); (2) cells of connective tissue, which are not tightly connected to each other but are embedded in the extracellular matrix; (3) muscle cells, whicweh contain long protein fibers and produce movement; and (4) nerve cells, which are elongated and specialized for communication. Animal cells contain most of the subcellular structures found in the other eukaryotic kindoms, except for chloroplasts, cell walls, and large vacuoles.

SUMMARY

Cells from all six kingdoms are highly organized. Eukaryotic cells are larger than prokaryotic cells and contain large amounts of internal membranes that divide the cytosol into distinct compartments, each with separate functions.

All cells are surrounded by membranes. Membranes consist mainly of lipids and proteins. The lipids, mostly phospholipid, are arranged in two layers (a "bilayer"). Their hydrophobic tails make up the interior of the membrane, and their hydrophilic heads point toward the cytoplasm and the outside of the cell. Some proteins interact with the external or internal surface of the bilayer and with the hydrophilic cytoplasm or cell exterior. Other proteins pass all the way through the membrane. The membrane is fluid: Both protein and lipid molecules can move freely within it. This picture of membrane structure is called the Fluid Mosaic Model.

Membranes are selectively permeable: They allow some molecules and ions to pass in and out freely and confine others to the inside or the outside. Some molecules and ions move passively across the plasma membrane and attain the same concentration inside and outside the cell.

Others are actively transported: They come to different concentrations inside and outside of cells. Active transport requires the expenditure of energy.

Membrane fusions allow the uptake of macromolecules and even larger particles. Certain cells can engulf large particles by extending their plasma membranes in a process called phagocytosis. Most cells are also capable of endocytosis, taking in materials by the folding in of the membrane. Endocytosis may be nonspecific or may depend on specific receptors on the membrane.

The outer surfaces of cells are responsible for their interactions with other cells and with other molecules in their environments. Cell walls enclose most prokaryotes and many eukaryotic cells (although not animal cells). Animal cells are often surrounded with extracellular matrix, a complex of carbohydrates and proteins.

Multicellular organisms depend on coordination of activities among many cells. Plants and animals have a variety of specialized structures by which cells interact. Some of these structures—plasmodesmata in plants and gap junctions in animals—allow the free passage of some molecules between adjoining cells. Other specialized junctions between cells provide mechanical strength to sheets of cells. Tight junctions weld some cells together and prevent the passage of molecules through them.

Together with internal membranes, the cytoskeleton provides internal organization for eukaryotic cells. It consists of three sets of filaments: microtubules, actin filaments, and intermediate filaments. In most cells the cytoskeleton consists of structures that are continually assembled and disassembled. Some cells have their permanent structure composed of elements of the cytoskeleton. Cilia and eukaryotic flagella are composed of microtubules. The fibers responsible for muscle contraction contain actin filaments that are closely related to the actin filaments of nonmuscle cells.

Organisms that can make their own organic molecules from simple inorganic compounds are called autotrophs, while organisms that derive their energy from organic molecules made by other organisms are called heterotrophs. Prokaryotes—both archaebacteria and eubacteria—are single cells without nuclei or membrane-bounded organelles. Protists are (mostly) single-celled eukaryotes. Fungi differ from other eukaryotes in the way they obtain food. Plants are multicellular and autotrophic. Animal cells lack cell walls, chloroplasts, and large vacuoles. Cells of all kingdoms thus have both common and distinctive features.

KEY TERMS

actin
actin filament
active transport
adhering junction
anchoring junction
antiport
archaebacteria
ATPase
autotroph
basal body
basal lamina
Brownian motion
cell wall
centriole
centrosome
cilium
clathrin
coated pit
communicating junction
concentration gradient
connective tissue
cortex
cotransport
cytoplasmic face
cytoskeleton
desmosome
diffusion
dynein
E. coli
electrogenic
endocytosis
epithelial cell
epithelium
eubacteria
exocytosis
exoplasmic face
extracellular matrix
facilitated diffusion
flagellum
Fluid Mosaic Model
focal contact
gap junction
glycocalyx
glycolipid
hemidesmosome
heterotroph
hypertonic
hypotonic
impermeable junction
integral protein
integrin
intermediate filament
isotonic
kinesin
ligand
liposome
low density lipoprotein (LDL)
microfilament
microsomes
microtubule
microtubule-associated protein (MAP)
microtubule organizing center (MTOC)
motor
myelin
myosin
Na^+-K^+ ATPase
nuclear lamina
osmosis
parenchymal cell
passive transport
peripheral proteins
phagocytosis
pinocytosis
plasmodesma
plasmolysis
receptor-mediated endocytosis
selectively permeable
symport
tight junction
transmembrane protein
tubulin
turgor pressure
water potential

QUESTIONS FOR REVIEW AND UNDERSTANDING

1. Complete the following by filling in the blanks:

   Proteins that span the lipid bilayer of a membrane and so protrude from both sides are referred to as __________ proteins, while __________ proteins are those that associate (by electrostatic interactions or other means) with the membrane surface. Proteins and lipids having carbohydrates as part of their structure are called __________ and __________, respectively. The lipids and proteins that comprise the membrane generally exhibit mobility within the plane of the membrane. Membrane __________ is the term used to describe this mobility of membrane components.

   Biological membranes exhibit __________ __________, allowing certain molecules to pass through membranes more readily than others. Although water is a polar molecule, it can penetrate membranes with relative ease. The movement of water across a semipermeable membrane in response to differences in concentration of solute molecules is __________. Some molecules pass through membranes from high concentration to low with the assistance of specific proteins, a process called __________ __________. Other proteins, membrane "pumps," can actually use energy to transport molecules across membranes from low concentration to high by the process of __________ __________. Such pumps may use __________ as an energy source or rely on an existing __________ __________. The latter process is referred to as __________.

   In addition to the important role(s) played by membranes in regulating the inside of the cell, important cellular constituents are found outside of the cell. The __________ is involved in cell recognition. Many cells are surrounded by an __________ __________. Others attach to a __________ __________ that anchors them in place. Plant cells are surrounded by a __________ __________ that minimizes contact between them. However, neighboring plant cells have __________, channels of communication that allow even large molecules to pass. Some animal cells have functionally analogous channels called __________ __________. __________ __________ are formed between cells by fusion of the outer leaflets of the plasma membrane of adjacent cells. When these form in a sheet of cells, the sheet becomes permeable only to molecules that can pass through the cells themselves, rather than between or around cells. Adhering junctions, formed by proteins called __________, are responsible for joining adjacent cells. To anchor cells to the extracellular matrix, transmembrane proteins of the integrin family form __________ __________. These various points of membrane attachment are called __________ __________.

   Inside the cell, the __________ contributes to the ordering and properties of the cytoplasm. It is composed of three different major types of filaments, including __________, __________ filaments, and __________ filaments.

2. Define and contrast the following terms:
   (a) hypertonic vs. hypotonic vs. isotonic
   (b) passive transport vs. active transport
   (c) symport vs. antiport
   (d) endocytosis vs. exocytosis
   (e) pinocytosis vs. receptor-mediated endocytosis
   (f) adhering junction vs. tight junction
   (g) MAP vs. MTOC
   (h) kinesin vs. dynein
   (i) G-actin vs. F-actin
   (j) autotrophs vs. heterotrophs
   (k) prokaryotes vs. eukaryotes
3. Describe in your own words the Fluid Mosaic Model of membrane structure.
4. Why must transmembrane proteins have polar and nonpolar regions? Where would you expect to find polar regions and nonpolar regions of such a protein?
5. Describe the respective routes to the plasma membrane used by integral membrane proteins and by peripheral proteins located on the cytoplasmic surface of the plasma membrane.
6. Describe in your own words how phospholipases have been used in experiments to determine the distribution of lipids between the two faces (leaves) of a membrane.
7. Cholesterol is an important component of the plasma membrane of animal cells. What is cholesterol's effect on membrane properties? Based on this, can you speculate why it is logical that cholesterol is found especially in the plasma membrane?
8. What are the two components contributing to water potential?
9. Define "turgor" and propose how a decrease in turgor pressure in plant cells results in the wilting of a plant.
10. The rate of simple diffusion of a molecule across a membrane is related to the size, shape, and polarity of the molecule. Given that the interior of a membrane bilayer is hydrophobic, predict the relative rate of diffusion for the following six molecules and support your ranking of the molecules' permeabilities with a brief explanation: Na^+, CH_3OH, glucose, O_2, fatty acid (palmitic acid), and glutamic acid.

11. What characteristics of enzymes (described in Chapter 4) are also relevant to transport proteins such as glucose permease?
12. Describe in your own words the defining characteristics of electrogenic pumps.
13. Glucose is taken up by cells via cotransport (symport) with Na^+. The Na^+ concentration is higher outside of the cell because it is pumped out of the cell by the Na^+-K^+ ATPase. What does this say about the ultimate source of energy for transporting glucose and other selected molecules by cotransport?
14. Describe the role of membrane fusion in endocytosis and exocytosis.
15. What is the role of the receptor in receptor-mediated endocytosis? Can you speculate how a virus might use this mechanism to enter the cytoplasm of a cell?
16. Plasmodesmata are found between plant cells that are derived from a common parental cell by cell division. What does this suggest about the process of plasmodesmata formation?
17. What are three functions facilitated by microtubules? What conditions are required for tubulin polymerization? Describe the importance of the GTP cap in determining the polarity of microtubule growth.
18. From your knowledge of bioenergetics, explain why motors associated with microtubules require ATP.
19. What is the suggested role of intermediate filaments?
20. What are the two prokaryotic kingdoms?
21. It is thought that eukaryotes evolved from archaebacteria (see Chapter 24). If this is true, what might you find if you compared the structures of ribosomes from eukaryotes, archaebacteria, and eubacteria?
22. How is the flagellum of a bacterial cell different from the flagellum or cilium of a eukaryotic cell?
23. Compare the characteristics of cells from the five kingdoms discussed in this chapter, noting their similarities and differences.

SUGGESTED READINGS

ALBERTS, B., D. Bray, J. Lewis, M. Raff, K. Roberts, and J. D. Watson, *Molecular Biology of the Cell,* 3rd edition, Garland, New York, 1994. See Chapters 16 and 19 for much of the material covered in this chapter.

BRAY, D., *Cell Movements,* Garland, New York, 1992. A clear and concise book about cell movement and the molecules responsible for it.

DE DUVE, C., *A Guided Tour of the Living Cell,* Scientific American Books, New York, 1984.

LODISH, H., D. Baltimore, A. Berk, S. L. Zipursky, P. Matsudaira, and J. Darnell, *Molecular Cell Biology,* 3rd edition, Scientific American Books, New York, 1995. Chapter 22 concerns the cytoskeleton and cell movements.

CHAPTER 7

How Do Cells Make ATP? Cellular Respiration and Glycolysis

KEY CONCEPTS

- Cellular respiration converts energy into ATP by controlled oxidations.
- Oxidative phosphorylation couples electron transport to the production of high-energy phosphate bonds in ATP, in a process that depends on the intact organization of mitochondria.
- Glycolysis derives energy by rearranging the molecules of glucose to form two molecules of a three-carbon compound called pyruvate.
- Respiring cells convert two of the three carbon atoms of pyruvate into the activated form of acetic acid, called acetyl CoA.
- The citric acid cycle removes eight electrons from the two carbon atoms of acetyl CoA. These electrons reduce two types of electron carriers—NAD^+ and FAD—in eight enzymatic steps.
- Heterotrophs digest the macromolecules of food into smaller molecules that can enter cells and participate in energy metabolism.

OUTLINE

ESSAYS

If you spend the next 24 hours studying cell biology, you will use the energy of about 10^{25} molecules of ATP, about 40 kg. (You would use about the same amount of ATP in a few hours if you were to run a marathon.) Almost every time one of your cells performs an endergonic reaction, it derives the needed energy from the splitting of ATP, the universal currency of biological energy.

One of the most important goals of biologists and biochemists has therefore been to learn how cells convert the energy of sunlight or food into the high-energy bonds of ATP. Until the early 1960s, biochemists believed that they could understand ATP formation solely in terms of the rearrangements of chemical bonds in ordinary reactions catalyzed by enzymes in aqueous solutions. Decades of ingenious research produced detailed maps of such enzyme-catalyzed transformations, many of which made or used ATP. But researchers were still not able to explain how most ATP was made.

In the end, the clue to the major route of ATP formation did not come from the properties of enzyme reactions (although enzymes are certainly important), but from the properties of membranes and subcellular organelles. In producing ATP, cells undeniably obey the rules of chemistry, but they do so in a way that eluded most biochemists, who had not understood the significance of the subcellular organization discussed in Chapter 5.

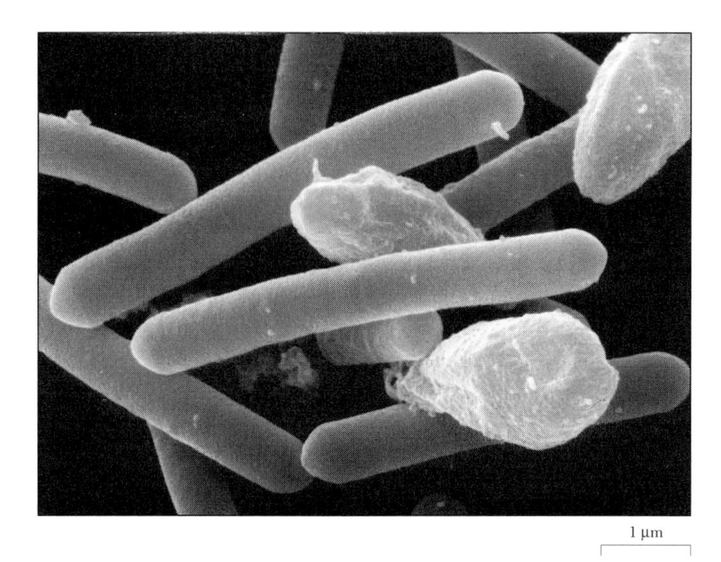

Figure 7-1 The food-poisoning bacteria, *Clostridium botulinum*, lives without oxygen. *(U. S. Department of Agriculture)*

CELLULAR RESPIRATION CONVERTS ENERGY INTO ATP BY CONTROLLED OXIDATIONS

Almost all eukaryotic cells (and many prokaryotic cells as well) are **aerobic;** that is, they require oxygen. Aerobic organisms produce most of their ATP by **respiration,** the oxygen-dependent extraction of energy from food molecules. In eukaryotes, most of the steps of cellular respiration occur in mitochondria. Cells that are most active in respiration tend to have the greatest number of mitochondria.

Even organisms that also produce ATP by photosynthesis (plants, some protists, and some prokaryotes) also can make ATP by respiration. In the dark and in the winter (for plants that lose their leaves), plants, like animals, depend on cellular respiration for most of their ATP.

Some prokaryotes live and produce ATP **anaerobically,** without oxygen, such as the food-poisoning bacteria, *Clostridium botulinum,* shown in Figure 7-1. Some eukaryotes can also live anaerobically for some time, such as the yeast shown in Figure 7-2. We will see how anaerobic life occurs, but only after we have first discussed respiration. Most of our discussion centers on the extraction of energy from glucose, the starting point for the same set of reactions in all organisms.

Respiration Transfers the Electrons of Glucose to Other Molecules

Most energy-producing reactions in cells are **reduction-oxidation (redox) reactions,** which transfer electrons from one molecule (the **reducing agent,** or electron donor) to another (the **oxidizing agent,** or electron acceptor), as shown in Figure 7-3a. Oxidation and reduction reactions always go together: The oxidation of one atom, molecule, or ion provides the electrons for the reduction of another atom, molecule, or ion.

Compounds Differ in Their Tendency to Accept or Donate Electrons

As we saw in Chapter 2, individual types of atoms have different affinities for electrons. Recall that chemists de-

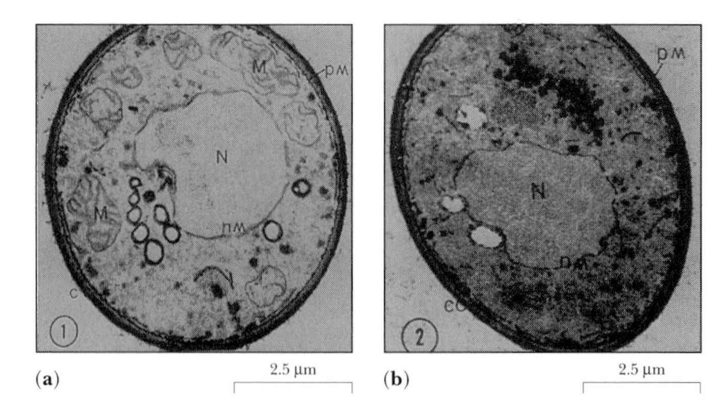

Figure 7-2 Brewer's yeast, *Saccharomyces cerevisiae,* can live either **(a)** aerobically or **(b)** anaerobically. *(Carolyn Damsky,* Journal of Cell Biology, *43: 173–179, 1969)*

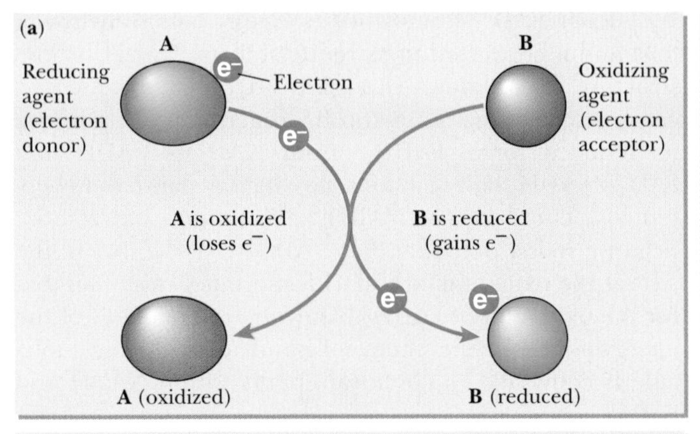

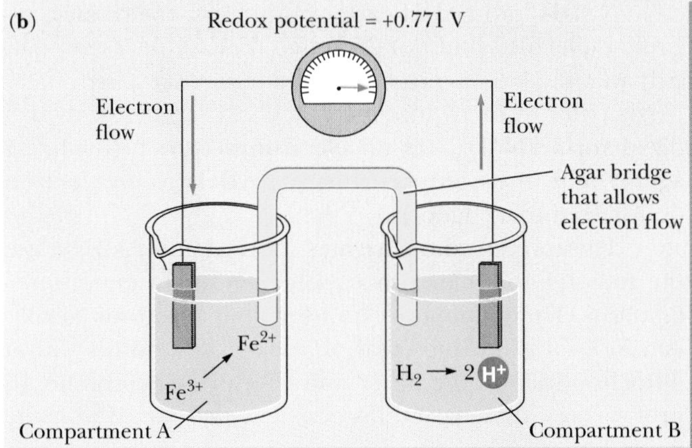

(c)

Oxidation and reduction potentials of various biological compounds

Oxidation	Redox potentials (in volts)	ΔG^0
Acetaldehyde $\longrightarrow$ acetate + $2H^+$ + $2e^-$	–0.58	112
Isocitrate $\longrightarrow$ α-ketoglutarate + CO_2 + $2H^+$ + $2e^-$	–0.38	73
β-Hydroxybutyrate $\longrightarrow$ acetoacetate + $2H^+$ + $2e^-$	–0.346	67
NADPH + H^+ $\longrightarrow$ $NADP^+$ + $2H^+$ + $2e^-$	–0.324	65
NADH + H^+ $\longrightarrow$ NAD^+ + $2H^+$ + $2e^-$	–0.320	62
Ethanol $\longrightarrow$ acetaldehyde + $2H^+$ + $2e^-$	–0.197	38
Lactate $\longrightarrow$ pyruvate + $2H^+$ + $2e^-$	–0.185	35
Malate $\longrightarrow$ oxaloacetate + $2H^+$ + $2e^-$	–0.166	32
Succinate $\longrightarrow$ fumarate + $2H^+$ + $2e^-$	–0.031	6
2 cytochrome $b_{K(red)}$ $\longrightarrow$ 2 cytochrome $b_{K(ox)}$ + $2e^-$	+0.030	–6
Coenzyme Q_{red} $\longrightarrow$ coenzyme Q_{ox} + $2H^+$ + $2e^-$	+0.10	–19
2 cytochrome c_{red} $\longrightarrow$ 2 cytochrome c_{ox} + $2e^-$	+0.254	–49
2 cytochrome $a_{3(red)}$ $\longrightarrow$ 2 cytochrome $a_{3(ox)}$ + $2e^-$	+0.385	–74
H_2O $\longrightarrow$ $\frac{1}{2}O_2$ + $2H^+$ + $2e^-$	+0.816	–157

Figure 7-3 A reduction-oxidation (redox) reaction: **(a)** A, the electron donor or reducing agent, donates an electron to B, the electron acceptor or oxidizing agent; A becomes oxidized and B becomes reduced; **(b)** for any redox reaction, chemists can measure a redox potential, which reflects differences in the tendency of two substances to donate electrons; **(c)** redox potentials and ΔG^0 for the oxidation of some biologically important compounds.

fine **electronegativity** as the tendency of atoms to gain electrons.

The apparatus shown in Figure 7-3b measures the **redox potential,** the differing tendencies of two redox reactions, one in each compartment, to move electrons from electron donor (reducing agent) to electron acceptor (oxidizing agent). If the reaction in compartment A has a greater tendency to donate electrons, electrons move from B to A. A voltmeter measures the redox potential, designated E_0', in volts, a measure of potential energy per charge.

Chemists have measured the redox potentials for the transfer of electrons between many specific compounds or atoms, such as the transfer of an electron from one type of iron ion, Fe^{+2} (ferrous ion), to another type of iron ion, Fe^{+3} (ferric ion):

$$Fe^{+3} + e^- \longrightarrow Fe^{+2}$$

This reaction is actually only a *half reaction* because the electrons must be supplied from somewhere else. All redox reactions are therefore paired, with one half reaction supplying electrons for the other half reaction.

Chemists calculate all redox potentials for the reduction of the oxidized form of a given molecule or ion as if the electrons are always supplied by a standard half reaction, shown in Figure 7-3b, involving the oxidation of a hydrogen atom (half a molecule of H_2, hydrogen gas). [Recall that a hydrogen atom consists just of a proton (H^+) and an electron (e^-).]

$$\tfrac{1}{2}H_2 \longrightarrow H^+ + e^-$$

This reaction is arbitrarily said to have a redox potential of zero. Pairing this half reaction with the reduction of ferric ion, chemists determine that the redox potential for the $Fe^{+3} \rightarrow Fe^{+2}$ half reaction is 0.77 volt.

A strong *oxidizing* agent, such as O_2, has a high tendency to gain electrons. The half reaction

$$\tfrac{1}{2}O_2 + 2\,H^+ + 2\,e^- \longrightarrow H_2O$$

has a high positive redox potential (0.82 volt). In contrast, a strong *reducing* agent, with a strong tendency to donate electrons, generates strongly negative redox potentials. For example, the compound NADH (discussed later in this chapter) is a strong reducing agent. The half reaction that converts the oxidized form of NADH (called NAD^+) to its reduced form has a negative redox potential, −0.32 volt.

In respiring cells, the ultimate electron acceptor is O_2. We can convert redox potentials into free energy differences (ΔG^0) for the oxidation of biologically important reducing agents (like NADH) to the oxidized form, as shown in Figure 7-3c. We use ΔG in most of our discussions of redox reactions, although many biochemists prefer to speak of redox potentials. The two measures are completely equivalent because as electrons move, energy becomes more or less available to perform work.

Redox Potentials Predict the Direction of Electron Flow

The direction of electron movement is from half reactions with negative redox potentials to those with more positive redox potentials. As noted above, the $O_2 \rightarrow H_2O$ half reaction has a strongly positive redox potential. Another way of saying this is that ΔG is negative for the movement of electrons to oxygen, and such electron movement occurs spontaneously. (See Chapter 4.)

Both redox potentials and ΔG predict that electrons ultimately move from glucose and other energy-rich molecules to oxygen. Their path, however, takes them via intermediate electron acceptors with intermediate redox potentials and free energies.

Respiration ultimately depends on the properties of the oxygen atom, in particular on its high affinity for electrons. Electrons from the chemical bonds of food molecules combine with oxygen and hydrogen ions to form water. The leftover carbon atoms from the starting molecules ultimately become carbon dioxide (CO_2). The fate of glucose is summarized by the following equation:

$$C_6H_{12}O_6 \text{ (glucose)} + 6\ O_2 \longrightarrow 6\ CO_2 + 6\ H_2O$$

This overall reaction is strongly exergonic with a standard free energy change ($\Delta G^{0\prime}$) for this reaction of −686 kcal per mole of glucose.

Respiration transfers 24 electrons from glucose to oxygen (four per carbon atom). As discussed later in this chapter, the electrons move from glucose and its derivatives in two sets of reactions, called glycolysis and the citric acid cycle. For 20 of these 24 electrons, the first electron acceptor is a molecule called **NAD^+** [*n*icotinamide *a*denine *d*inucleotide]. For the other four electrons of glucose, the oxidizing agent is **FAD** [*f*lavin *a*denine *d*inucleotide]. The structures of both NAD^+ and FAD and their reduced forms are shown in Figure 7-4.

Each NAD^+ accepts two electrons (and a hydrogen ion) and is converted to its reduced form, **NADH** (whose structure is also shown in Figure 7-4). Similarly, FAD accepts two electrons (and two hydrogen ions) to produce its reduced form, **$FADH_2$**. Both $NAD^+ \rightarrow NADH$ and $FAD \rightarrow FADH_2$ half reactions have high negative redox potentials. The $NAD^+ \rightarrow NADH$ half reaction has a more negative redox potential than $FAD \rightarrow FADH_2$; that is, the ΔG for the oxidation of NADH is more negative than that for the oxidation of $FADH_2$. Altogether about 90% of the energy stored in the chemical bonds of a glucose molecule is converted to chemical energy within NADH and $FADH_2$.

NADH and $FADH_2$ are examples of **coenzymes,** organic molecules (but not proteins) that are necessary participants in certain enzyme reactions. NAD^+, especially, participates in many enzyme reactions, serving in its oxidized form (NAD^+) as an electron acceptor (oxidizing agent) and in its reduced form (NADH) as an electron donor (reducing agent).

These reduced coenzymes (NADH and $FADH_2$) are the major initial electron transfer agents in energy metabolism. They ultimately transfer their electrons to oxygen in a series of steps that, in eukaryotes, occurs within mitochondria, on the inner mitochondrial membrane. In prokaryotes, these reactions occur in association with the plasma membrane.

The two electrons from each molecule of reduced coenzyme ultimately reduce a single oxygen atom. Together with the addition of two hydrogen ions from the surrounding water, the result is the formation of one molecule of water. The flow of electrons to oxygen is steeply downhill, that is, highly exergonic. Natural selection has favored the evolution of molecules and mechanisms that efficiently capture this energy. The next question, then, is how cells couple the downhill flow of electrons to the uphill production of the high-energy bonds of ATP.

How Does the Flow of Electrons Lead to the Formation of High-Energy Phosphate Bonds in ATP?

Oxidative phosphorylation is the process that couples the oxidation of NADH and $FADH_2$ to the production of high-energy phosphate bonds in ATP. The strategy for the efficient capture of energy is to break the big waterfall (the free energy released by directly oxidizing NADH with O_2) into a gentler cascade (a series of steps, each with a smaller ΔG).

The electrons from NADH and $FADH_2$ shuttle to their ultimate acceptor (O_2) through a series of other molecules, called **electron carriers.** The pathway of the electrons (from one carrier to another) is called the **electron transport chain** (or **respiratory chain**). Biochemists compare the respiratory chain to a "bucket brigade" that car-

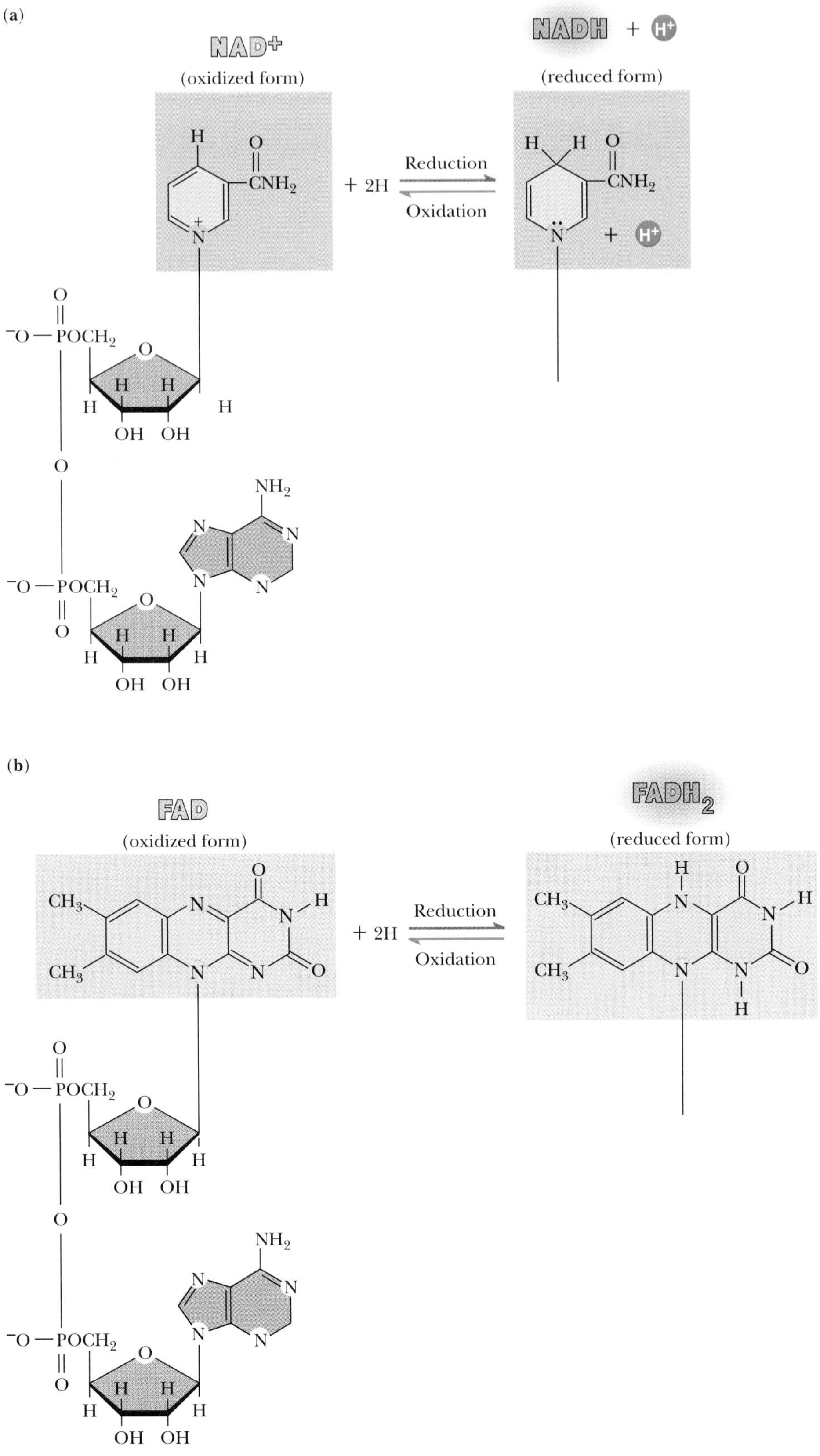

(a)

NAD^+ (oxidized form) + 2H ⇌ (Reduction / Oxidation) NADH + H^+ (reduced form)

(b)

FAD (oxidized form) + 2H ⇌ (Reduction / Oxidation) $FADH_2$ (reduced form)

Figure 7-4 NAD^+ and FAD are important electron acceptors in cells.

ries electrons to oxygen. Each electron carrier passes its electrons to the next carrier in the line. So a reduced carrier becomes oxidized as it gives up its electrons, and the next carrier becomes reduced as it receives electrons. Electrons move in one direction because successive carriers have ever-increasing electron affinities, so successive transfers are exergonic.

Which Molecules Serve as Electron Carriers?

In 1925, the British biochemist David Keilin discovered that cellular respiration depends on a set of specific proteins that change color as they accept or donate electrons. These proteins, called **cytochromes** [Greek, *kytos* = hollow vessel + *chroma* = color] have a reddish color that changes hue as they change from their oxidized to their reduced forms, as shown in Figure 7-5a. In all the cytochromes, the color derives from **heme,** an iron-containing organic molecule (shown in Figure 7-5b) that also gives hemoglobin its color. Heme, like NAD^+ and FAD, may donate and accept electrons. As it does so, the charge on its iron atom changes between +2 (Fe^{+2}) and +3 (Fe^{+3}).

(a)

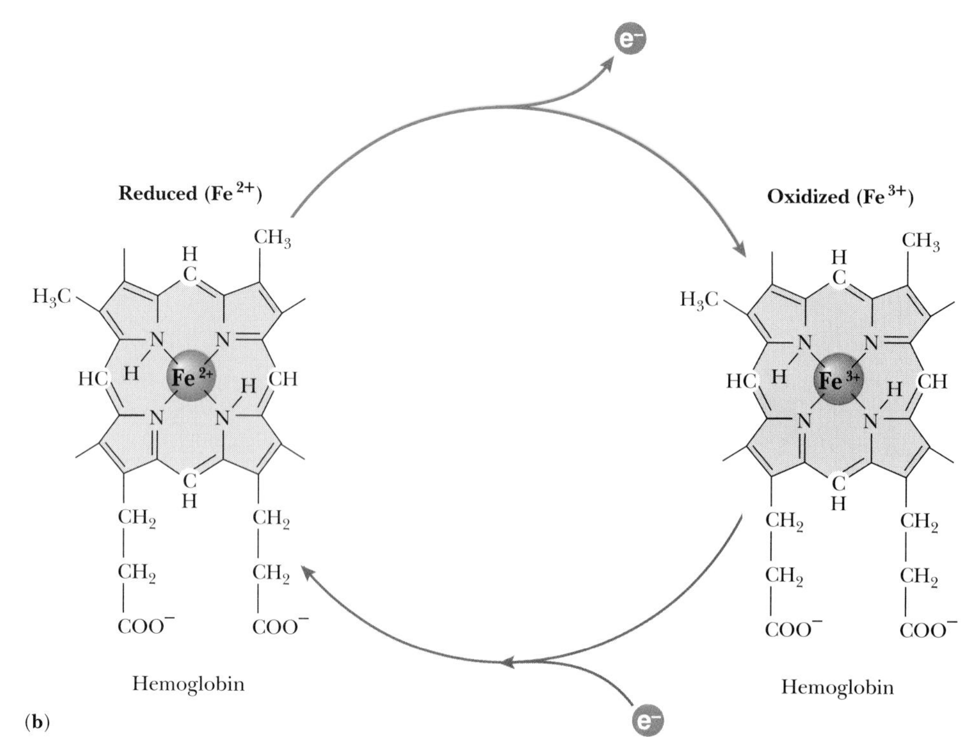

(b)

Figure 7-5 As a cytochrome c changes from its oxidized (Fe^{3+}) to its reduced (Fe^{2+}) state, its color changes: **(a)** reduced *(left)* and oxidized *(right)* cytochrome c; **(b)** chemical structures of the heme group in cytochrome c. *(a, Dick Fish, Smith College)*

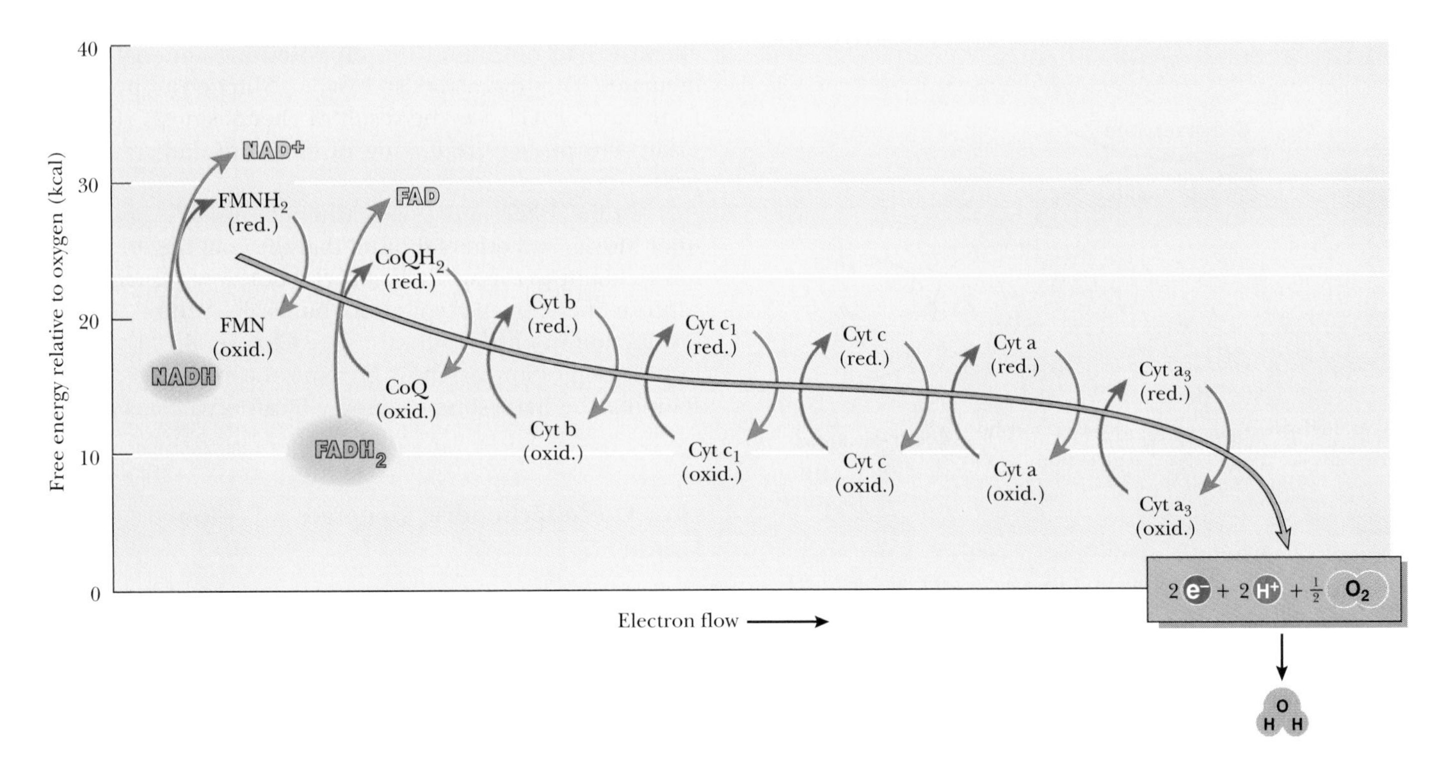

Figure 7-6 The electron transport chain, the path of electrons from NADH to oxygen. The high affinity of oxygen for electrons pulls electrons through the chain, so that NADH is oxidized to NAD^+ and $FADH_2$ to FAD.

Keilin and others showed that cytochromes in the test tube could accept electrons from NADH. Cytochromes can also accept electrons from each other, but only in a fixed order. By comparing the flow of electrons in laboratory experiments with isolated cytochromes and whole cells, biochemists worked out the electron transport chain, the path of electrons from NADH through the cytochromes and other intermediates to oxygen (Figure 7-6). They could also predict where electrons could enter the chain from $FADH_2$. In addition to five distinct cytochromes (called b, c_1, c, a, and a_3), two small molecules—called FMN (for *f*lavin *mono*nucleotide) and Q (for ubi*q*uinone, or coenzyme *Q*)—act as intermediate electron carriers.

Altogether about 40 proteins participate in the electron transport chain. Probably the most important of these, perhaps even more than the cytochromes, are a class of at least six proteins called **iron-sulfur proteins.** As in the case of the cytochromes, an electron moving through the chain temporarily reduces an iron atom in the active center of the protein. Each of these proteins contains an **iron-sulfur center,** which contains two or four iron atoms and an equal number of sulfur atoms; despite the presence of more than one iron atom, each protein can accept only one electron at a time. Less is known about the iron-sulfur centers than about the cytochromes because redox changes cannot be studied by changes in absorbance and require more sophisticated experiments with a technique called electron spin resonance spectroscopy.

How Do Cells Harvest the Energy of Electron Transport?

After the discovery of the electron transport chain, the big question still remained: How do cells harvest the energy released by the flow of electrons? Biochemists looked for clues in **substrate-level phosphorylations,** reactions (like that illustrated in Figure 7-7) in which a redox reaction is coupled to the production of a high-energy phosphate group that is eventually transferred to ADP to form ATP. Perhaps, they reasoned, the oxidation of NADH was similarly coupled to the formation of a compound that contained a high-energy phosphate bond (like 1,3-bisphosphoglycerate in Figure 7-7).

The proposed schemes, based on this example, were entirely reasonable and were dutifully studied by several generations of investigators and biochemistry students. But no one was able to find any of the proposed intermediates. The experts were convinced that intermediates with high-energy phosphates must exist but concluded that they were so unstable that they fell apart before they could be isolated in an experiment.

The situation was this: The electron transport chain could pass electrons in a test tube, but no ATP could be generated. Isolated mitochondria could perform both electron transport and ATP synthesis, but as soon as the mitochondria were disrupted (by adding tiny amounts of detergents), ATP synthesis stopped even though electron flow continued. In retrospect, these experiments clearly

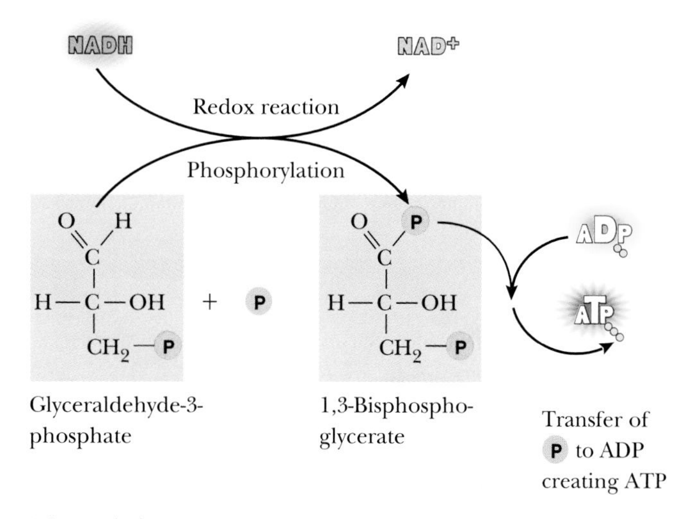

Figure 7-7 A "substrate-level" phosphorylation couples a redox reaction to the production of a high energy phosphate. Here the phosphorylation of glyceraldehyde-3-phosphate is coupled with its oxidation from an aldehyde to an acid, producing 1,3-bisphosphoglycerate. This reaction served as an unsuccessful model for ATP production in oxidative phosphorylation in mitochondria.

showed the importance of intact mitochondrial organization for the production of ATP.

The puzzle was resolved in the early 1960s by Peter Mitchell, an English biochemist. Mitchell had been studying the way bacteria pump hydrogen ions across their membranes. The inside of a bacterium has a higher pH (a lower H^+ concentration) than the medium in which it is growing. The difference in H^+ concentration across a membrane is called a **proton gradient.** As discussed in Chapter 2, a so-called hydrogen ion, H^+, is actually a hydronium ion, H_3O^+, so it is not really accurate to refer to it as a proton. Nonetheless, "proton" is a convenient and accepted shorthand.

Mitchell showed that bacteria could use the energy stored in this proton gradient to transport other substances into the cell (as discussed for other cotransporters in Chapter 6, p. 164). In 1961, he suggested that mitochondria might, like bacteria, be able to make use of the energy stored in a proton gradient. The idea that a proton gradient could be used to make ATP was, at the time, thoroughly unconventional. (Another example of Mitchell's unorthodoxy was the site at which he performed his research—a manor house in the country, where Mitchell lived and worked after he left his university position.)

Mitchell's hypothesis had two parts: (1) The flow of electrons through the electron transport chain establishes a proton gradient across the mitochondrial membrane, and (2) some special molecular machinery in the membrane captures the energy stored in the proton gradient by making ATP. Mitchell's hypothesis suggested why biochemists had not found the supposed high-energy intermediates: They did not exist. Instead, Mitchell argued, the formation of ATP was the result of **chemiosmosis** [Greek, *osmos* = to push], the linking of chemical and transport processes.

In the 1960s and 1970s, Mitchell, his colleague Jennifer Moyle, and others showed that the coupling of a proton gradient to the synthesis of ATP not only explains the capture of energy in respiration, but is also central to ATP production in photosynthesis. (See Chapter 8.) Mitchell's radical suggestion has now become the basis for understanding the harvesting of energy by all organisms.

How Do Mitochondria Generate a Proton Gradient?

Recall that a mitochondrion has two membranes, so that (as shown in Figure 7-8) a mitochondrion provides four distinct locations for the participants in cellular respiration: the outer membrane, the intermembrane space, the inner membrane, and the **mitochondrial matrix,** the space within the inner mitochondrial membrane. The matrix accounts for about two thirds of the volume of a typical mitochondrion. Cell fractionation techniques have allowed biochemists to separate isolated mitochondria into individual components and to study the biochemical function of each.

Most of a cell's NADH and $FADH_2$ are produced within the mitochondrial matrix (by the enzymes of the citric acid cycle, discussed later in this chapter). The electron transport chain lies within the inner membrane. The orientation of the carriers within the inner membrane is such that, as the electron carriers pass electrons to one another, protons (hydrogen ions) move across the inner mitochondrial membrane from the matrix into the intermembrane space. This movement of hydrogen ions out of the matrix establishes a pH difference, or proton gradient, across the inner membrane, between the matrix and the intermembrane space, as shown in Figure 7-8c. Careful measurements with intact mitochondria confirmed that a proton gradient really exists across the inner mitochondrial membrane, with the pH about one unit lower outside than inside. (No such proton gradient exists across the outer mitochondrial membrane: The pH of the intermembrane space is the same as the pH of the cytosol.) The movement of electrons through the electron transport chain (in the inner mitochondrial membrane) causes protons to move *out* of the mitochondrial matrix.

Biochemical experiments have demonstrated that even fragments of inner membranes can pump protons. In these studies, mitochondria were disrupted by ultrasound and their inner membranes allowed to reassemble spontaneously into vesicles, as diagrammed in Figure 7-9b. Proton pumping occurred when electrons flowed, for example, from NADH to O_2. When there was no electron

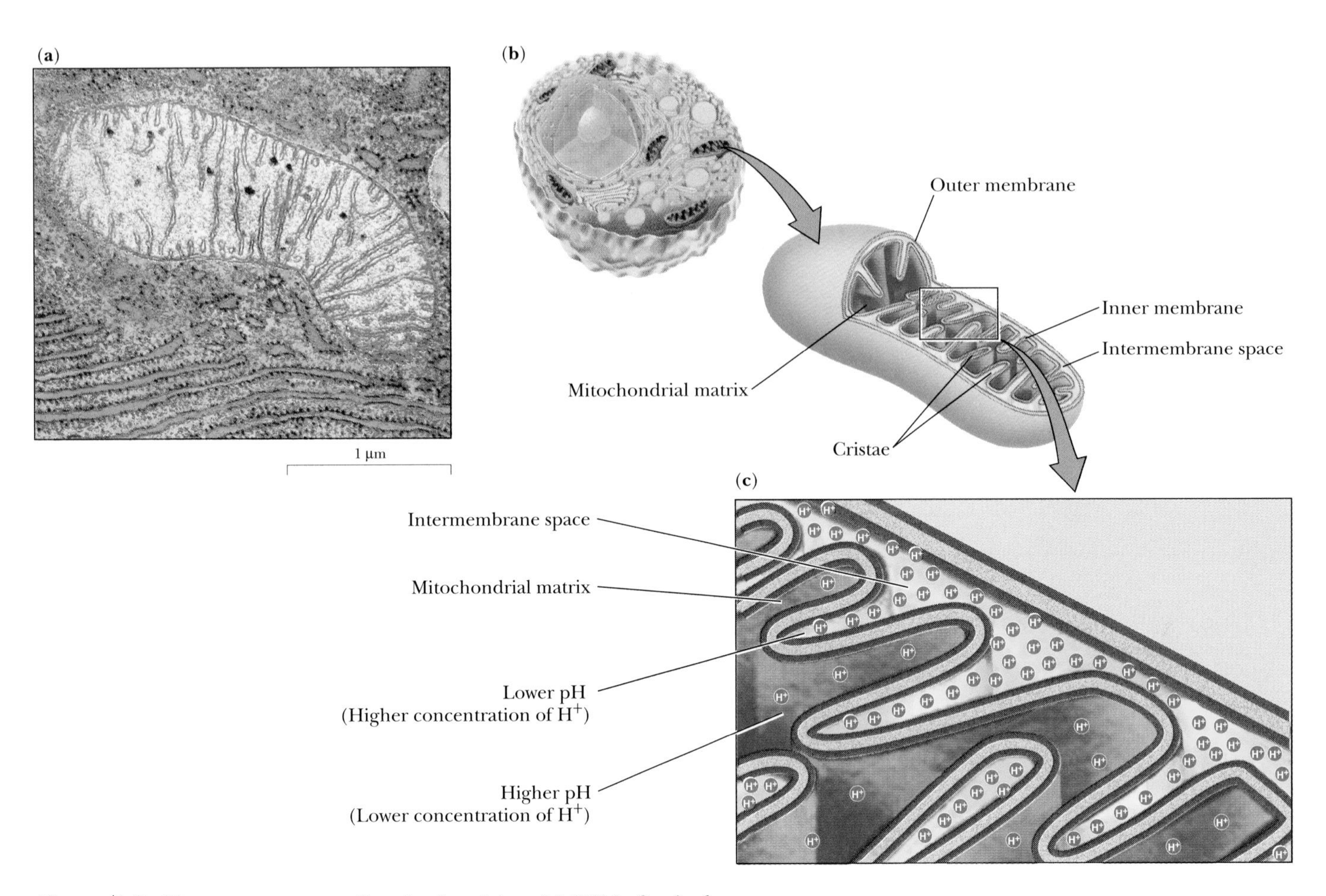

Figure 7-8 The compartments of a mitochondrion. (**a**) TEM of a single mitochondrion; (**b**) mitochondrial compartments; (**c**) differing pH within mitochondrial compartments. *(a, B. King/BPS)*

flow, for example, if no NADH was added, or in the presence of an inhibitor, there was no proton pumping. These experiments established the ability of the electron transport chain to pump protons.

By taking apart the inner membrane still further, biochemists found four distinct electron-transporting complexes in the inner mitochondrial membrane. Three of these complexes can move protons across the inner membrane as they pass electrons part of the way down the electron transport chain, as shown in Figure 7-10. The remaining complex, complex II, which is responsible for the initial oxidation of $FADH_2$, transfers electrons but does not move hydrogen ions.

Each of the respiratory complexes has a fixed orientation in the membrane, so that it pumps protons only in a single direction. The orientation of the complexes establishes the direction of the proton gradient. For example, one of the respiratory complexes, called cytochrome oxidase, passes electrons from cytochrome c to oxygen, again moving protons from the mitochondrial matrix into the intermembrane space. Purified cytochrome oxidase complex, even when added to vesicles produced from artificial membranes (not from mitochondria), transfers electrons from reduced cytochrome c to oxygen and simultaneously moves hydrogen ions across the vesicle's membrane.

The outer mitochondrial membrane allows the easy passage of small molecules and ions (including hydrogen ions), so the intermembrane space has the same pH as the bulk of the cytosol, about pH 7. As the electron transport chain pumps hydrogen ions out of the matrix, the pH of the matrix increases to about pH 8. The inner membrane is not generally permeable to H^+ ions, so the pH difference is stable as long as electron flow continues.

In addition to the pH gradient established by proton pumping, there is a difference of charge across the inner membrane because pumping the positive hydrogen ions out of the matrix leaves behind an excess of negative ions in the matrix. (That is, the proton pump is *electrogenic,* leading to a voltage across the membrane; see Chapter 6, p. 163.) The proton pump generates an **electrochemical gradient,** a double gradient composed of a *chemical* gradi-

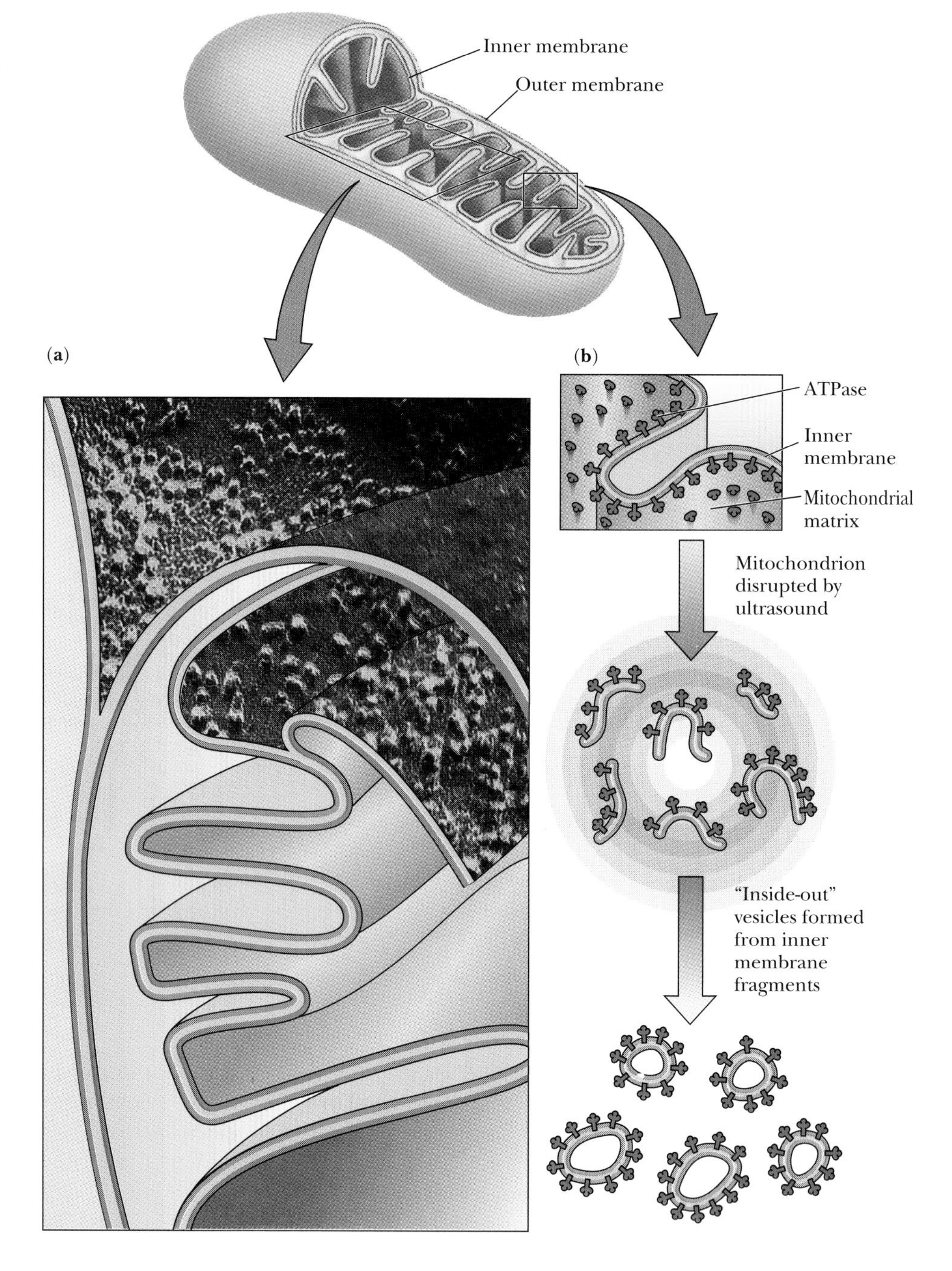

Figure 7-9 The distinctness of the inner and outer mitochondrial membranes. (**a**) Freeze fracture TEMs of inner and outer membranes; (**b**) the formation of "inside-out" vesicles of inner membrane fragments. *(a, Lester Packer, University of California, Berkeley)*

ent (the difference in hydrogen ion concentration, or pH) and an *electrical* gradient (the difference in charge). Both components of the electrochemical gradient favor the return of hydrogen ions to the matrix—hydrogen ions tend to flow in such a way as to minimize both the difference in concentration (that is, down the chemical gradient) and the difference in charge (that is, down the electrical gradient).

How Does the Flow of Hydrogen Ions Back into the Matrix Cause the Synthesis of ATP?

Once hydrogen ions are pumped out of the matrix, there is only one way back in—through special channels in the otherwise impermeable inner membrane. As the H^+ ions flow "downhill" through the channels into the matrix, they are moving down the electrochemical gradient and their free energy decreases. A protein complex called ATP synthase, **coupling factor,** or the **F_0-F_1 complex,** uses the flow of protons to drive the synthesis of ATP, as illustrated in Figure 7-11a. The role of the coupling factor is similar to that of a turbine that converts the flow of water or gas to electrical energy.

Electron microscopists can actually see coupling factors as components of the inner mitochondrial membrane, as shown in Figure 7-11b. Each looks like a miniature lol-

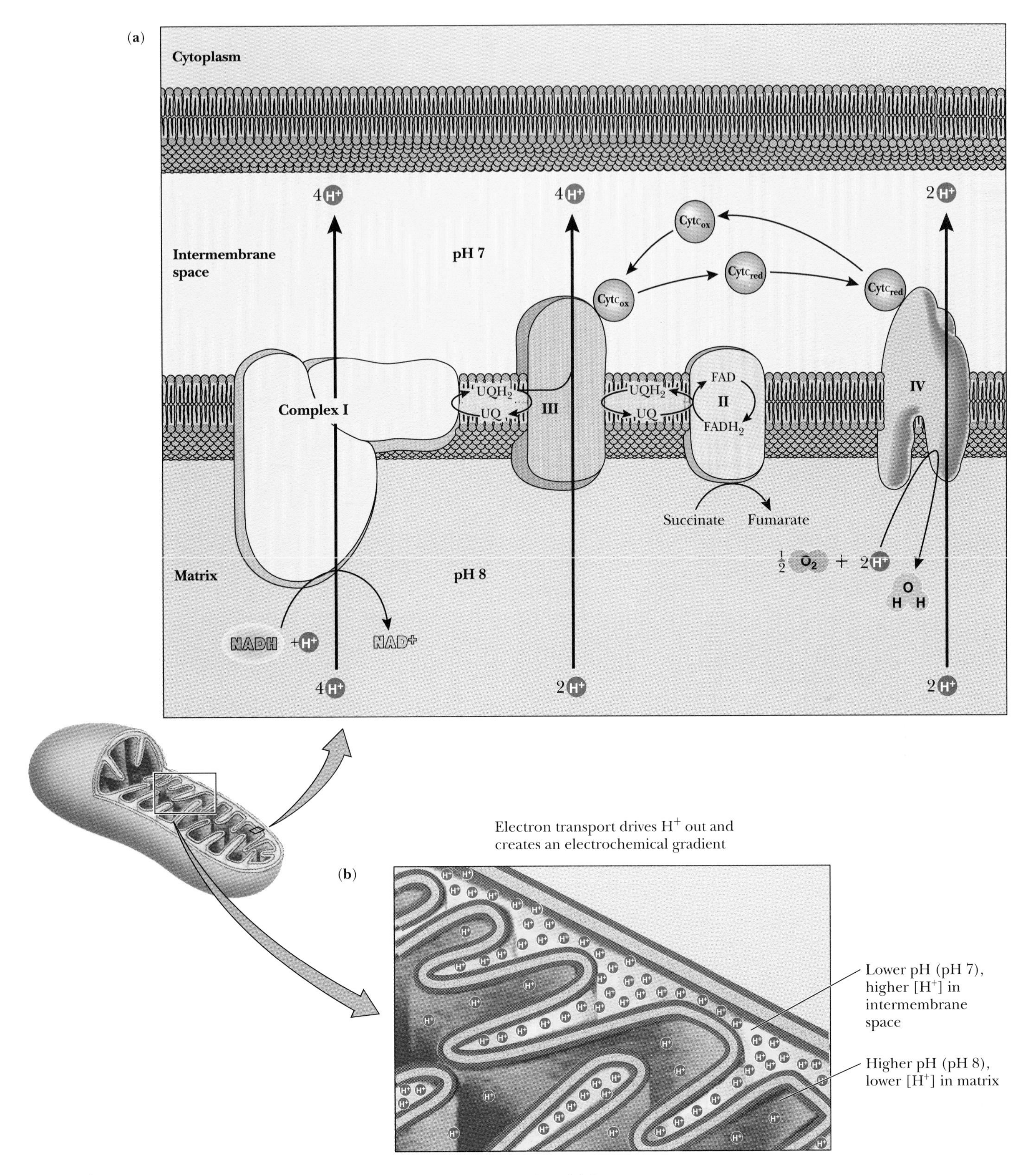

Figure 7-10 Electron transport complexes within the inner mitochondrial membrane move electrons down the respiratory chain and protons across the membrane from the matrix to the intermembrane space.

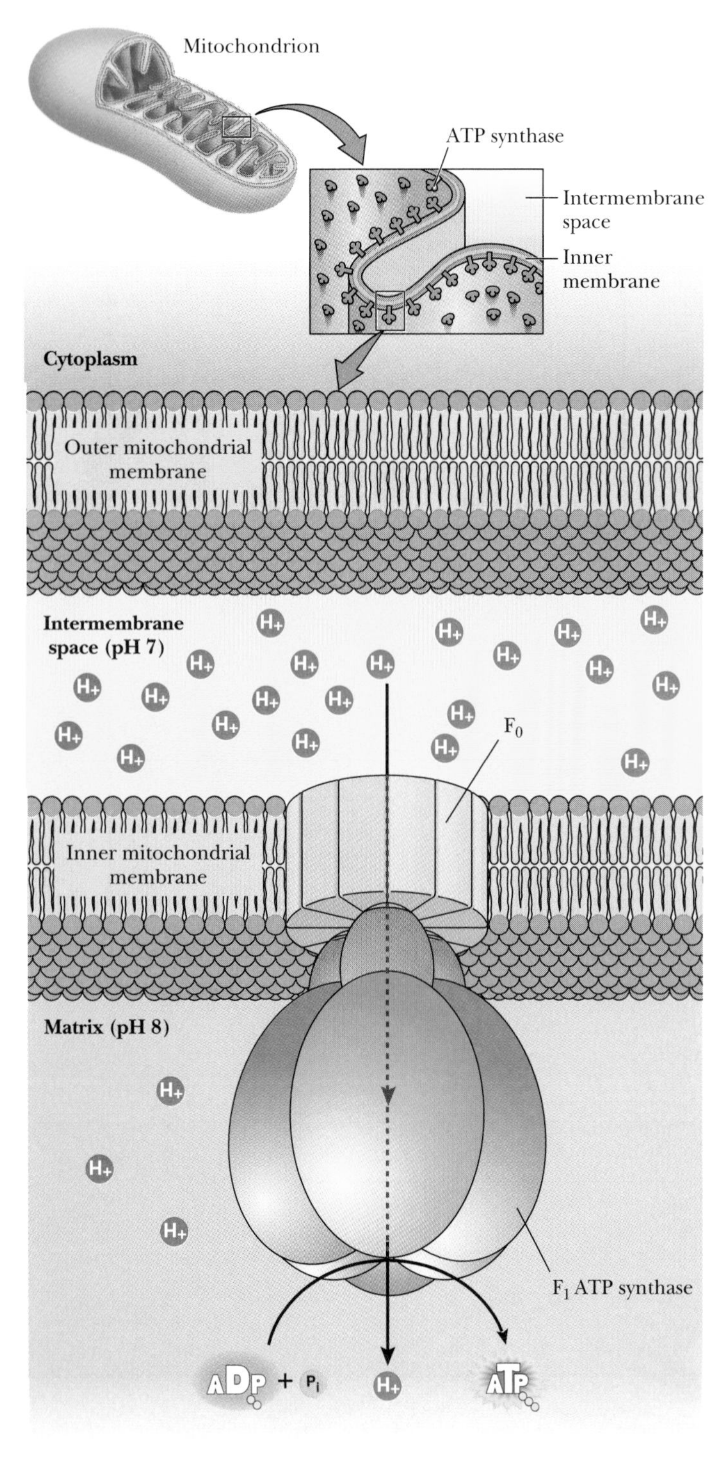

(a)

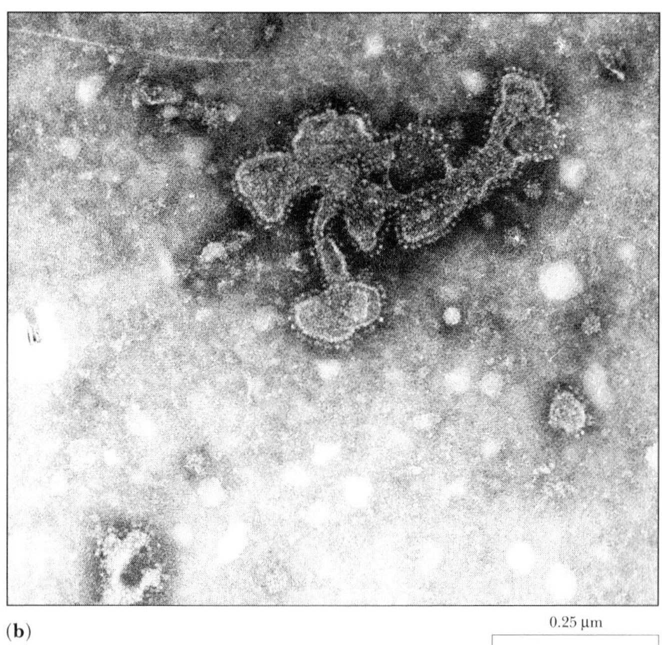

(b)

Figure 7-11 The F_0-F_1 particles harness energy stored in the proton gradient across the inner mitochondrial membrane. (**a**) F_0 serves as a channel that allows protons to move through the membrane, and F_1 is an ATP synthase that makes ATP as the protons pass; (**b**) TEM of the F_0-F_1 particles. *(b, R. Bhatnagar/Visuals Unlimited)*

lypop protruding from the surface of the inner mitochondrial membrane into the mitochondrial matrix. Biochemists have also been able to isolate these complexes and study their structures. As shown in Figure 7-11a, each contains two major components: The stick, which traverses the inner membrane, is called F_0, and the top is called F_1.

In experiments similar to those described above, fragments of inner mitochondrial membrane formed inside-out vesicles from which F_0-F_1 complexes protrude. Gentle shaking removes F_1 from the membrane, leaving F_0 behind. After this treatment, the inside-out vesicles still allow protons to flow down an electrochemical gradient, suggesting that F_0 functions as the **proton channel,** the route that hydrogen ions follow as they flow back into the matrix. But the removal of F_1 abolishes the ability of the vesicles to make ATP. Replacing the F_1 fragment allows ATP synthesis to resume. So F_1 appears to be responsible for making ATP and is called **ATP synthase.** We do not yet know, however, exactly how the F_0-F_1 complex harnesses the proton flow to create new high-energy phosphate bonds in ATP.

Are Proton Pumping and ATP Synthesis Really Separate Processes?

Even before the discovery that the F_0-F_1 complex is responsible for coupling proton flow to ATP synthesis, biochemists became convinced of the chemiosmotic hypothesis because of experiments with photosynthetic bacteria from the salt flats near San Francisco Bay. These bacteria provided a particularly elegant demonstration of the separateness of proton pumping and ATP synthesis. A purple protein called *bacteriorhodopsin* pumps protons across the bacterial membrane in response to light, as part of the process by which the bacteria capture light energy in pho-

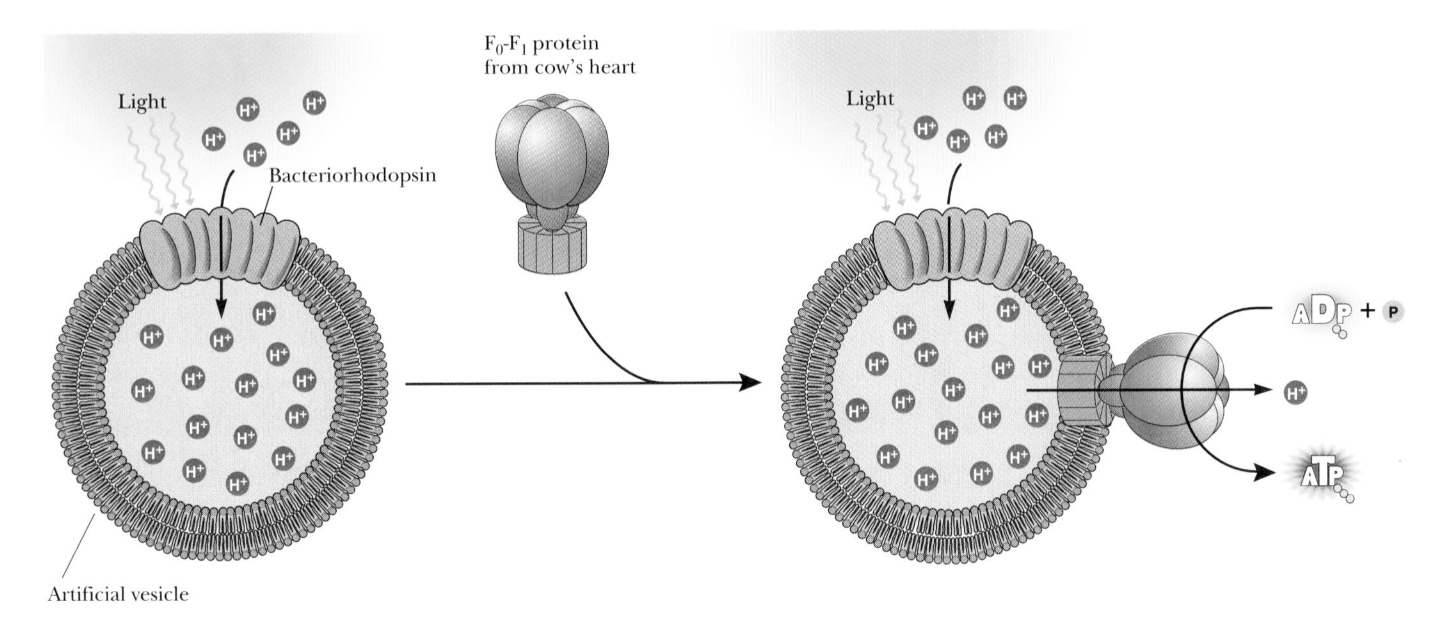

Figure 7-12 Diagram of the experiment that convinced everyone that a proton gradient can generate ATP. Energy for establishing the proton gradient came from the absorption of light by bacteriorhodopsin, and the F_0-F_1 particles came from a cow's heart.

tosynthesis. Even when this protein is put into artificial membrane vesicles, light causes the purple protein to pump protons into the vesicles. This establishes a pH gradient between the inside and the outside, as shown in Figure 7-12.

To the artificial bacteriorhodopsin vesicles researchers then added F_0-F_1 protein isolated from the mitochondria of a cow's heart, as shown in Figure 7-12. This complex allowed protons to flow out of the vesicles and also caused the production of ATP. This experiment showed that the F_0-F_1 complex can harness the energy of proton flow even in a highly artificial situation, with separate components from two extraordinarily distinctive sources—a rare tiny prokaryote and a familiar large eukaryote.

The Chemiosmotic Hypothesis, summarized in Figure 7-13, explains how cells transfer the energy from reduced coenzymes (NADH and $FADH_2$) to ATP: (1) The electron transport chain creates an electrochemical gradient across the inner mitochondrial membrane, and (2) the F_0-F_1 complex couples the flow of protons down the gradient to the synthesis of ATP. The main point is that the creation of the proton gradient and the generation of high-energy bonds in ATP are separate processes—a point established in the experiments with bacteriorhodopsin and cow F_0-F_1 complex. In Chapter 8, we see that proton gradients not only can produce ATP in mitochondria, but also drive ATP synthesis in chloroplasts during photosynthesis. In addition, the energy stored in proton gradients can also drive the transport of other ions as well as the movement of bacterial flagella.

In the rest of this chapter we see how the breakdown of glucose (and other food molecules) frees electrons for the production of NADH and $FADH_2$, the starting materials for oxidative phosphorylation. The responsible reactions fall into three stages (as illustrated in Figure 7-14): (1) the conversion of the six-carbon glucose into two three-carbon molecules; (2) the conversion of each three-carbon compound into a two-carbon compound; and (3) the processing of the two-carbon compounds to carbon dioxide. Each of these stages produces reduced cofactors (NADH and $FADH_2$). The first and third stages also include phosphate transfers that generate additional high-energy phosphate bonds.

GLYCOLYSIS IS THE FIRST STAGE IN THE EXTRACTION OF ENERGY FROM GLUCOSE

The first stage in the metabolism of glucose, called **glycolysis** [Greek, *glykys* = sweet (referring to sugar) + *lyein* = to loosen], is a set of ten chemical reactions. These reactions convert glucose, a sugar with six carbon atoms, into two molecules of **pyruvate,** with three carbon atoms each. They also convert some of the energy of glucose into two high-energy bonds of ATP and some into two energy-rich molecules of NADH. All ten reactions take place in the cytosol, that is, outside the mitochondria.

The most remarkable fact about glycolysis is that *all* organisms accomplish it in exactly the same way; that is,

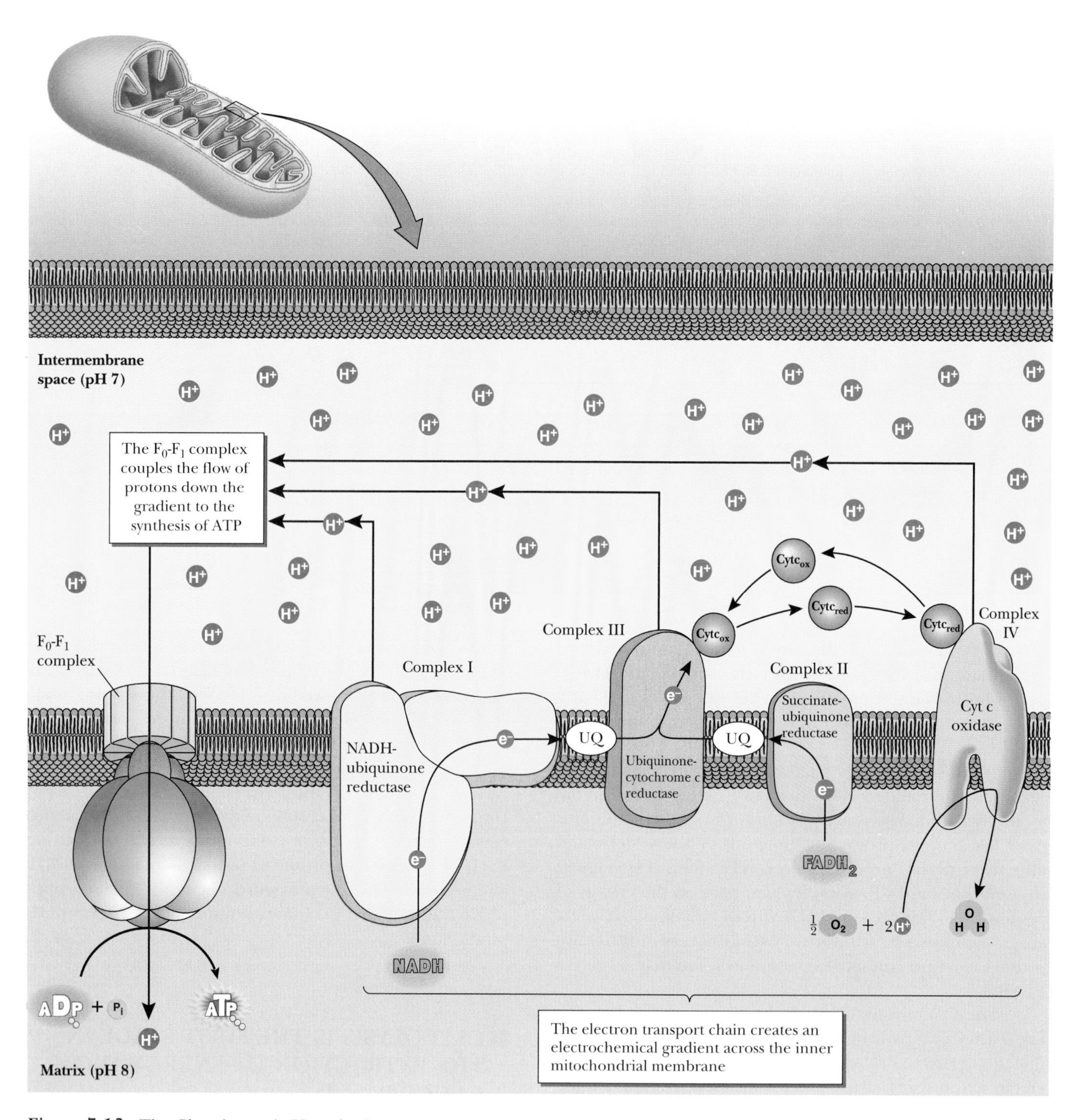

Figure 7-13 The Chemiosmotic Hypothesis.

all organisms perform the same ten reactions. This universality suggests that the common ancestors of all present-day organisms could perform glycolysis, and that the enzymes that catalyze the ten reactions must already have evolved at least 3 billion years ago.

Our major question about glycolysis is how cells capture energy in ATP and NADH. Like other biochemical pathways, glycolysis proceeds in small steps. Some of these steps transfer phosphate groups within molecules and between molecules, eventually leading to the net synthesis of ATP. One step also leads to the production of NADH. The success of glycolysis (the reason that it has persisted for 3 billion years) lies in its ability to couple downhill exergonic reactions to the uphill (endergonic) synthesis of ATP and NADH.

As glycolysis rearranges the atoms of glucose into

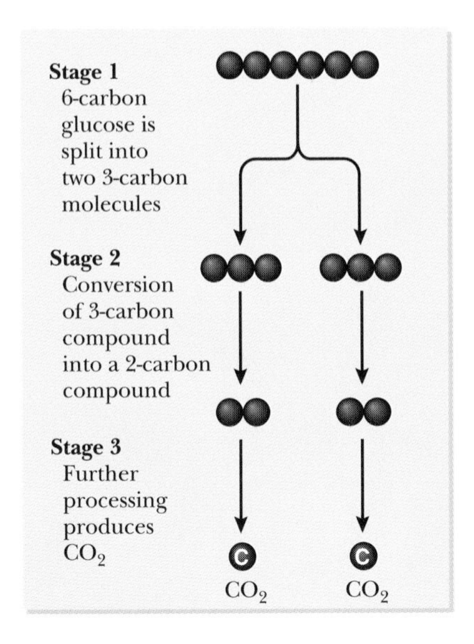

Figure 7-14 Three stages in the breakdown of glucose.

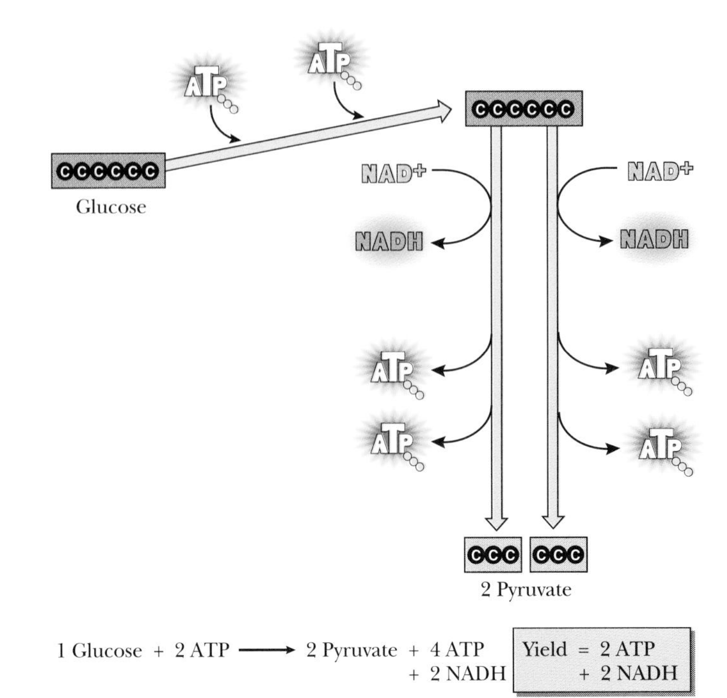

Figure 7-15 Glycolysis rearranges the atoms of glucose into two molecules of pyruvate.

two molecules of pyruvate, it also converts two molecules of NAD^+ to NADH and two molecules of ADP to ATP. The overall reaction, summarized schematically in Figure 7-15, is

$$\text{glucose} + 2\,\text{ADP} + 2\,\text{phosphate}\ (P_i) + 2\,\text{NAD}^+ \longrightarrow 2\,\text{pyruvate} + 2\,\text{ATP} + 2\,\text{NADH} + 2\,\text{H}^+ + 2\,\text{H}_2\text{O}$$

The overall free energy change for this reaction (under the conditions within a cell) is −18.3 kcal per mole of glucose. Even though the downhill reaction (glucose → pyruvate) is coupled to two uphill reactions (ADP → ATP and $NAD^+ \rightarrow NADH$), glycolysis is still exergonic and therefore occurs spontaneously.

Glycolysis consists of ten steps, which are shown in Figure 7-16. These steps are of just five types:

First Type: *phosphate transfers*—in some cases a phosphate group moves from one carbon atom in a compound to another, but in most cases a phosphate group moves from ATP to the carbon atom of a glucose-derived molecule; when phosphate is transferred from ATP, the enzyme involved is called a *kinase.*

Second Type: *redox reactions,* in which electrons are transferred from a reducing agent to an oxidizing agent; in many cases the net effect is to remove electrons as hydrogen atoms (that is, as electrons plus protons), in which case the enzyme is called a *dehydrogenase;* almost always, the removed electrons end up in NADH or $FADH_2$.

Third Type: *isomerizations* (each catalyzed by an *isomerase)* in which the atoms within a molecule are rearranged.

Fourth Type: *dehydrations,* in which a water molecule is removed.

Fifth Type: a *cleavage reaction,* in which a six-carbon compound is split into two three-carbon compounds.

All five types of chemical reactions in glycolysis are common to many enzyme pathways, so in learning about glycolysis you are also learning some of the most important reactions of metabolism. We can divide the ten chemical reactions of glycolysis into three parts: first, conversion of the six-carbon glucose into two three-carbon molecules; second, oxidation of the three-carbon molecules and reduction of NAD^+; and third, further oxidation of the new three-carbon molecules to pyruvate.

Part 1 (steps 1–5), the conversion of the six-carbon glucose molecule into two three-carbon molecules, requires five separate reactions, each catalyzed by a different enzyme. The process would be an uphill one, but it is coupled to the transfer of two phosphate groups from two ATP molecules to glucose at steps 1 and 3. The energy realized from splitting the high-energy phosphates pushes the reactions in the direction of the three-carbon product, glyceraldehyde-phosphate. At the end of part 1, glycolysis has split a six-carbon molecule into two three-carbon molecules and has consumed rather than harvested energy.

Part 2 of glycolysis (steps 6 and 7), the oxidation of the three-carbon molecules of glyceraldehyde-phosphate (the reaction illustrated in Figure 7-7), consists of two

(Text continues on page 210.)

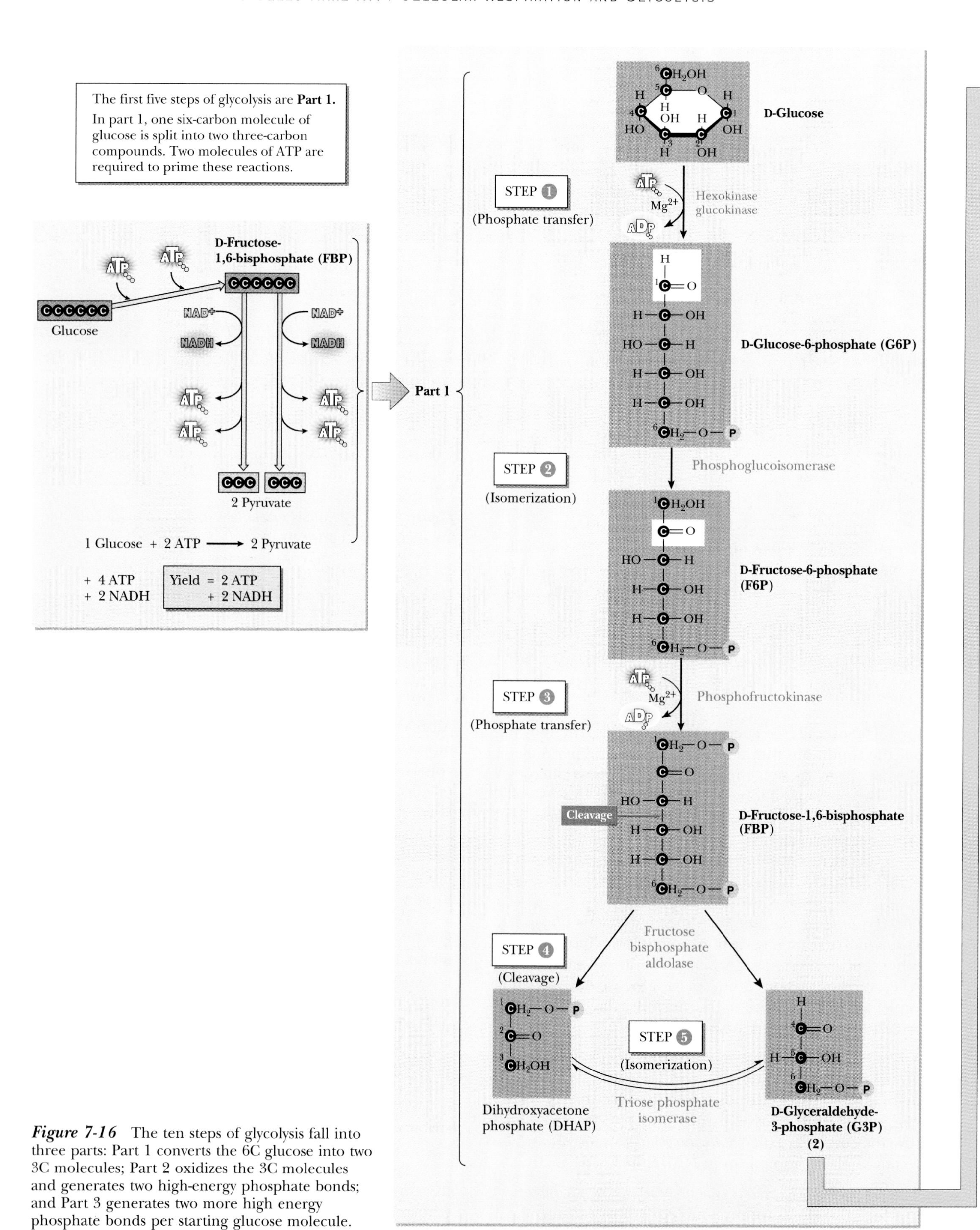

Figure 7-16 The ten steps of glycolysis fall into three parts: Part 1 converts the 6C glucose into two 3C molecules; Part 2 oxidizes the 3C molecules and generates two high-energy phosphate bonds; and Part 3 generates two more high energy phosphate bonds per starting glucose molecule.

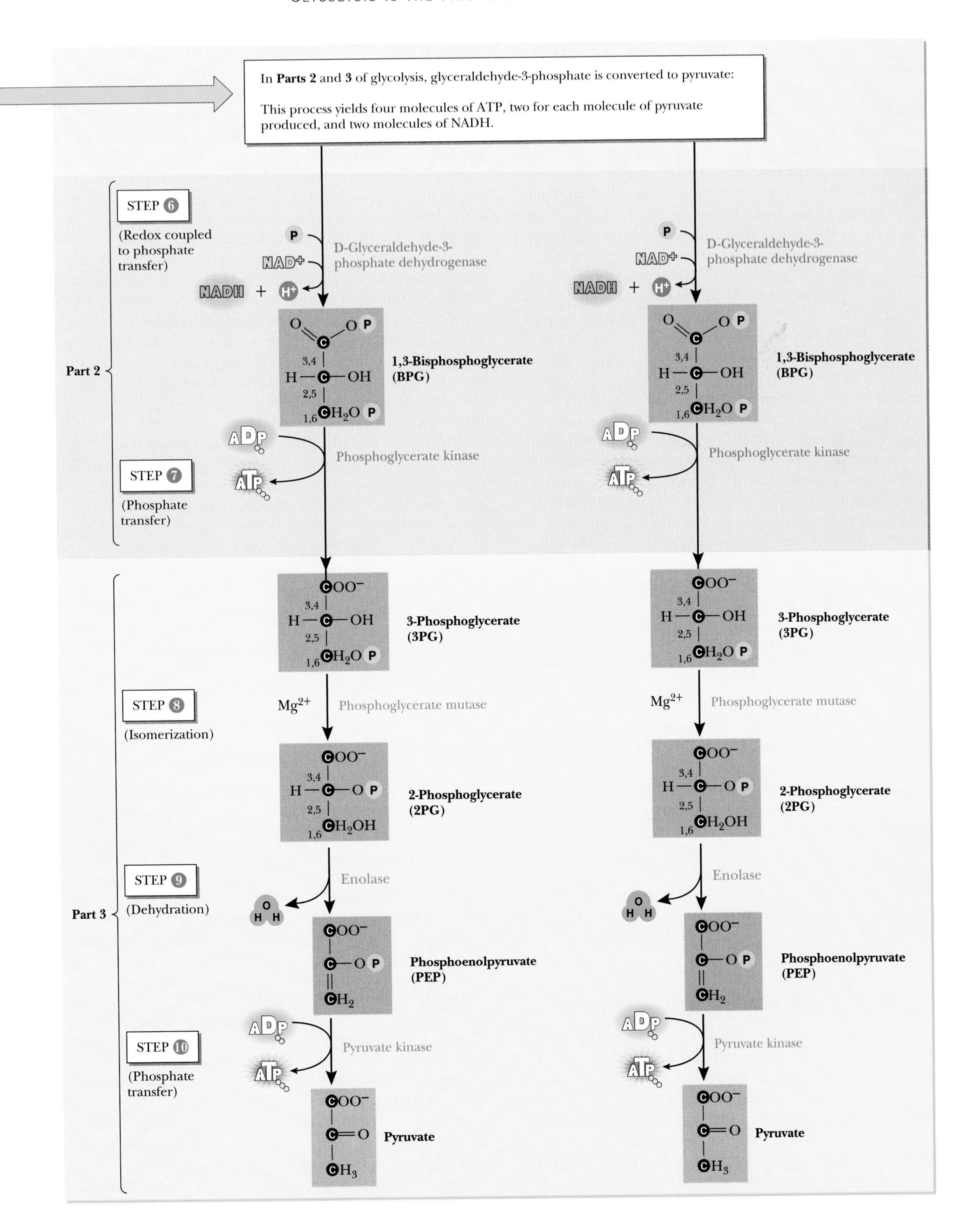

In **Parts 2** and **3** of glycolysis, glyceraldehyde-3-phosphate is converted to pyruvate:
This process yields four molecules of ATP, two for each molecule of pyruvate produced, and two molecules of NADH.
Part 2
STEP 6
(Redox coupled to phosphate transfer)
P
NAD+
NADH + H+
D-Glyceraldehyde-3-phosphate dehydrogenase
1,3-Bisphosphoglycerate (BPG)
STEP 7
(Phosphate transfer)
ADP
ATP
Phosphoglycerate kinase
Part 3
3-Phosphoglycerate (3PG)
STEP 8
(Isomerization)
Mg2+
Phosphoglycerate mutase
2-Phosphoglycerate (2PG)
STEP 9
(Dehydration)
Enolase
Phosphoenolpyruvate (PEP)
STEP 10
(Phosphate transfer)
Pyruvate kinase
Pyruvate

steps. For each molecule of glyceraldehyde-phosphate, these reactions generate one ATP as well as one molecule of NADH. So for each starting glucose molecule, part 2 yields two ATPs and two NADHs, recovering the ATP investment of part 1.

The first of the two steps in part 2 (step 6) oxidizes glyceraldehyde-phosphate, with NAD^+ serving as the oxidizing agent. In doing so it uses a phosphate group from the surrounding solution to create a new high-energy phosphate in the three-carbon molecule bisphosphoglycerate. The next step (step 7) transfers this high-energy phosphate group to ADP to form ATP and 3-phosphoglycerate.

The reactions of part 2 require both phosphate ions (P_i) and NAD^+. If either is missing, the reactions of part 2 simply cannot occur. Cells must therefore be able to recycle both of these. Phosphate is usually not a problem because it is present at fairly high concentrations in most cells and is continuously released as ATP is split. In aerobic organisms, the regeneration of NAD^+ generally requires transfer of NADH's electrons to the electron transport chain and therefore usually depends on the availability of oxygen. In the absence of oxygen (and in anaerobes), however, the cell must regenerate NAD^+ by other means, discussed later in this chapter.

Part 3 of glycolysis (steps 8–10) converts phosphoglycerate to pyruvate and transfers its phosphate to ADP, thus forming an additional ATP molecule for each phosphoglycerate (that is, two ATP molecules for each glucose). Since part 2's reactions already recovered the two ATP molecules invested in part 1, the net gain of ATP during glycolysis is two ATPs per molecule of glucose.

The overall effect of glycolysis, then, is to convert one molecule of glucose into two molecules of pyruvate. Besides making pyruvate, glycolysis produces two molecules of NADH and a net gain of two molecules of ATP. Without oxidative phosphorylation, glycolysis captures only about 15 kcal of energy per mole of glucose (2×7.3 kcal/mole ATP), only a small bite out of a much larger apple. The pyruvate molecules still contain lots of energy. This energy is extracted only in the next stages of energy metabolism, discussed later in this chapter. First, however, we look at how cells can use glycolysis to obtain energy even in the absence of oxygen.

How Can Glycolysis Continue Without Oxygen?

To sustain glycolysis, a cell must regenerate NAD^+ from NADH. Just how it does this depends on the species of organism and on whether oxygen is present.

In the presence of oxygen, most cells can oxidize NADH to NAD^+, using the reactions of oxidative phosphorylation to deliver electrons to oxygen. In the absence of oxygen, however, cells must regenerate NAD^+ by somehow using pyruvate as an oxidizing agent. Another way of saying this is that cells must use NADH to reduce pyruvate. Their manner of doing so depends on what enzymes they have, which in turn depends on their genetic legacy.

Many microorganisms live by **fermentation** [Latin, *fervere* = to boil], the anaerobic extraction of energy from organic compounds. Yeast cells, for example, perform fermentation in the brewing of beer or wine and in the rising of bread, converting the glucose into ethanol and carbon dioxide. To do this they use all the reactions of glycolysis that we have discussed and then use the accumulated NADH to reduce pyruvate and form ethanol and carbon dioxide. In the process, NADH is oxidized back to NAD^+, as shown in Figure 7-17a. Yeast (as shown in Figure 7-2a and b) is a **facultative anaerobe;** that is, it can live either anaerobically (by fermentation) or aerobically (using oxidative phosphorylation), while other microorganisms, such as *Clostridium botulinum* (shown in Figure 7-1), are **obligate anaerobes,** meaning that they grow only in the absence of oxygen.

Animals, including humans, often suffer a temporary lack of oxygen in some of their cells—for example, in muscle cells during vigorous exercise when the amount of oxygen needed to oxidize NADH exceeds the ability of the blood to deliver it. (See Essay, "Sprints, Dives, and Marathons.") In these cases, the cells can regenerate NAD^+ in the absence of oxygen. Instead of producing carbon dioxide and ethanol (like a yeast cell), however, muscle cells make lactate, a compound with three carbon atoms (shown in Figure 7-17b). It is the accumulation of lactate in muscles that causes leg cramps after a vigorous sprint.

How Do Biochemists Study the Reactions of Glycolysis?

Understanding a metabolic pathway such as glycolysis means learning about each reaction separately and then seeing how they all fit together. This is usually impossible in an intact cell because the products of one reaction go quickly on to the next step. To learn what is happening, biochemists must be able to stop each reaction and analyze what has happened. They usually do this in two ways: (1) by treating cells or cell extracts with an enzyme inhibitor to stop progress through the pathway; and (2) by separating individual enzymes from one another so that they can study one reaction at a time *in vitro* (outside the cell). Biochemists can then determine the structures of the products of each reaction and can even follow the fate of individual atoms by using compounds that are labeled with radioisotopes.

What determines the direction and the rates of the individual steps of the pathway? Biochemists have determined the standard free energy change ($\Delta G^{0\prime}$) of each reaction in glycolysis *in vitro*. At first glance, these measure-

ESSAY

SPRINTS, DIVES, AND MARATHONS

Respiration yields at least 15 times more ATP than glycolysis alone, and animals depend on oxygen for most of their energy. After the entrance of pyruvate into the citric acid cycle, oxygen serves as the ultimate acceptor of electrons from NADH. Similarly, the operation of the citric acid cycle requires that NADH and $FADH_2$ pass their electrons to oxygen through the respiratory chain. Extracting energy therefore depends on the delivery of oxygen to the sites where these processes occur. For land vertebrates, continuous energy production requires efficient lungs (to take in air) and heart (to pump blood from the lungs to the muscles).

Although training and experience can improve the performance of the systems that deliver oxygen, we and other animals often need to spend energy more quickly than our oxygen supply allows—for a pedestrian to avoid an inattentive driver, for an outfielder to chase a baseball, for a predator to pursue prey, or for prey to elude a predator. Several energy reservoirs allow such bursts of energy: (1) Muscles have stores of ATP and other compounds with high-energy phosphate bonds, enough to allow some work to continue without any new ATP production; (2) muscles can use blood glucose and stored glycogen to produce ATP by anaerobic glycolysis, producing lactate as the end product.

Figure 1 The red-eared turtle, *Chrysemys scripta elegans*, can stay under water for two weeks at a time, relying on glycolysis for energy production. *(William Weber/Visuals Unlimited)*

The length of time that an organism can depend on anaerobic glycolysis for ATP production is limited both by experience and by genetic capacity. Humans can manage only a few minutes, but some animals such as whales, seals, and other divers can live by glycolysis alone for much longer times. The fresh water red-eared turtle shown in Figure 1, for example, can stay under water for as long as 2 weeks!

After anaerobic exertion, animals must get rid of the accumulated lactate and restore their supplies of glucose and glycogen. Most of the time the animals can recover more energy from the accumulated lactate by reconverting it into pyruvate and using it for respiration. This depends on the availability of oxygen, which permits the lactate to enter the citric acid cycle. In vertebrates the new production of glucose goes on in the liver. Some of the lactate made by anaerobic glycolysis travels via the blood to the liver, where it can be recycled into glucose. The liver then resupplies the glucose reservoir of the blood.

Anaerobic glycolysis provides animals with a way of borrowing energy. Mammals repay this "oxygen debt" by heavy breathing and increased circulation after the loan. The loan, however, is only short term: The supply of oxygen limits sustained exercise. We can therefore sprint much faster than we can run long distances: An untrained school child can run 17 miles per hour in a 100-meter dash, but even champion athletes can run only about 11.5 miles per hour in a 26-mile marathon.

Nonmammalian vertebrates—fish, amphibia, and reptiles (like the red-eared turtle)—can get on without air for much longer periods, deriving most or all of their energy from glycolysis. Instead of using oxygen to recycle the accumulated lactate, they may excrete the lactate in their urine. This represents a huge waste of energy, but it can allow these animals to function and feed in environments that could otherwise not sustain them.

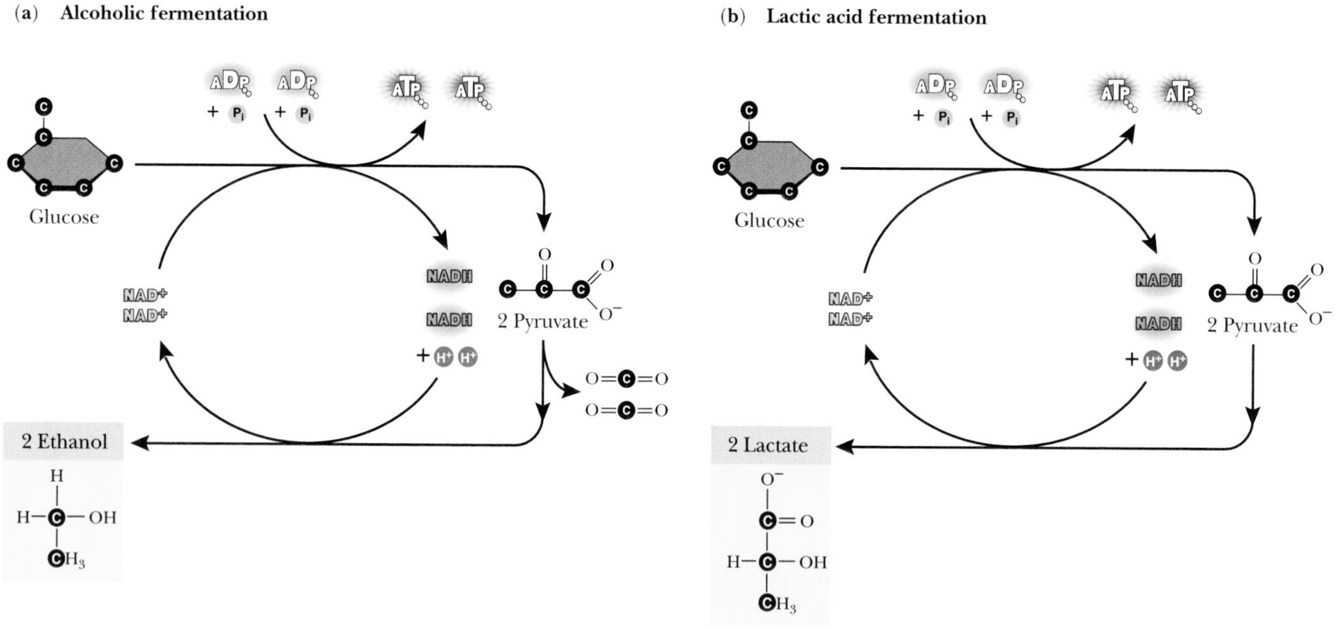

Figure 7-17 Fermentation, extraction of energy in the absence of oxygen, requires the oxidation of NADH by pyruvate or another organic compound. **(a)** In alcoholic fermentation, the ultimate product is ethanol; **(b)** in lactic acid fermentation, the ultimate product is lactic acid (lactate).

ments are a little troublesome: Of the ten steps of glycolysis, only four are exergonic (steps 1, 3, 7, and 10) under standard conditions. Figure 7-18 shows the free energy changes of each step, when the intermediates of glycolysis are at their typical cellular concentrations. We can see that most of the reactions are actually exergonic in a cell, but how do the endergonic reactions occur?

The answer lies in the coupling of subsequent reactions in the pathway. The product of one reaction immediately enters the next reaction, so that exergonic reactions "pull" the endergonic reactions that precede them.

What regulates the overall rate of glycolysis? In the cell the overall process of glycolysis is highly exergonic. This means that glycolysis proceeds spontaneously and irreversibly. Because all the needed enzymes are present in cells, the reactions should also proceed quickly. How, then, does the cell keep all its glucose from being immediately converted to pyruvate?

The conclusion is that glycolysis must be regulated. The questions are: Where? and How? In general, biochemical pathways are regulated at the first step that is unique to the pathway. This strategy ensures that molecules are not needlessly processed when they could be used for alternate purposes. The commitment to glycolysis takes place in step 3, and the enzyme phosphofructokinase is the site of the major regulation.

We can see the effect of ATP concentration on the rate of production of fructose bisphosphate (step 3). When ATP concentration is low, the reaction rate depends on substrate concentration, just as predicted by the Michaelis-Menten model (see Chapter 4, p. 103), as shown in Figure 7-19a. At high ATP concentrations, however, the reaction rate is less, as shown in Figure 7-19b, suggesting that ATP regulates the enzyme's activity.

We would expect glycolysis to proceed rapidly when the cell is low in ATP and to proceed slowly when there is already abundant ATP. This is indeed the case. Notice that ATP is both a substrate for phosphofructokinase and a regulator. The inhibition occurs at a separate binding site, so ATP is an *allosteric* inhibitor. (See Chapter 4, p. 112.) The result of this allosteric interaction is that, when ATP is not needed, glycolysis slows, and less ATP is made.

THE FORMATION OF ACETYL CoA IS THE SECOND STAGE IN THE EXTRACTION OF ENERGY FROM GLUCOSE

Pyruvate, the product of glycolysis, still contains a lot of energy. This energy is extracted through respiratory processes that occur in the mitochondria. After pyruvate arrives in the mitochondria, it undergoes an oxidation reaction in which NAD^+ serves as the electron acceptor. This reaction, shown in Figure 7-20a, converts pyruvate to a derivative of acetate, a two-carbon compound. In this compound, called **acetyl CoA,** acetate is linked by a high-energy bond to a molecule called coenzyme A, whose

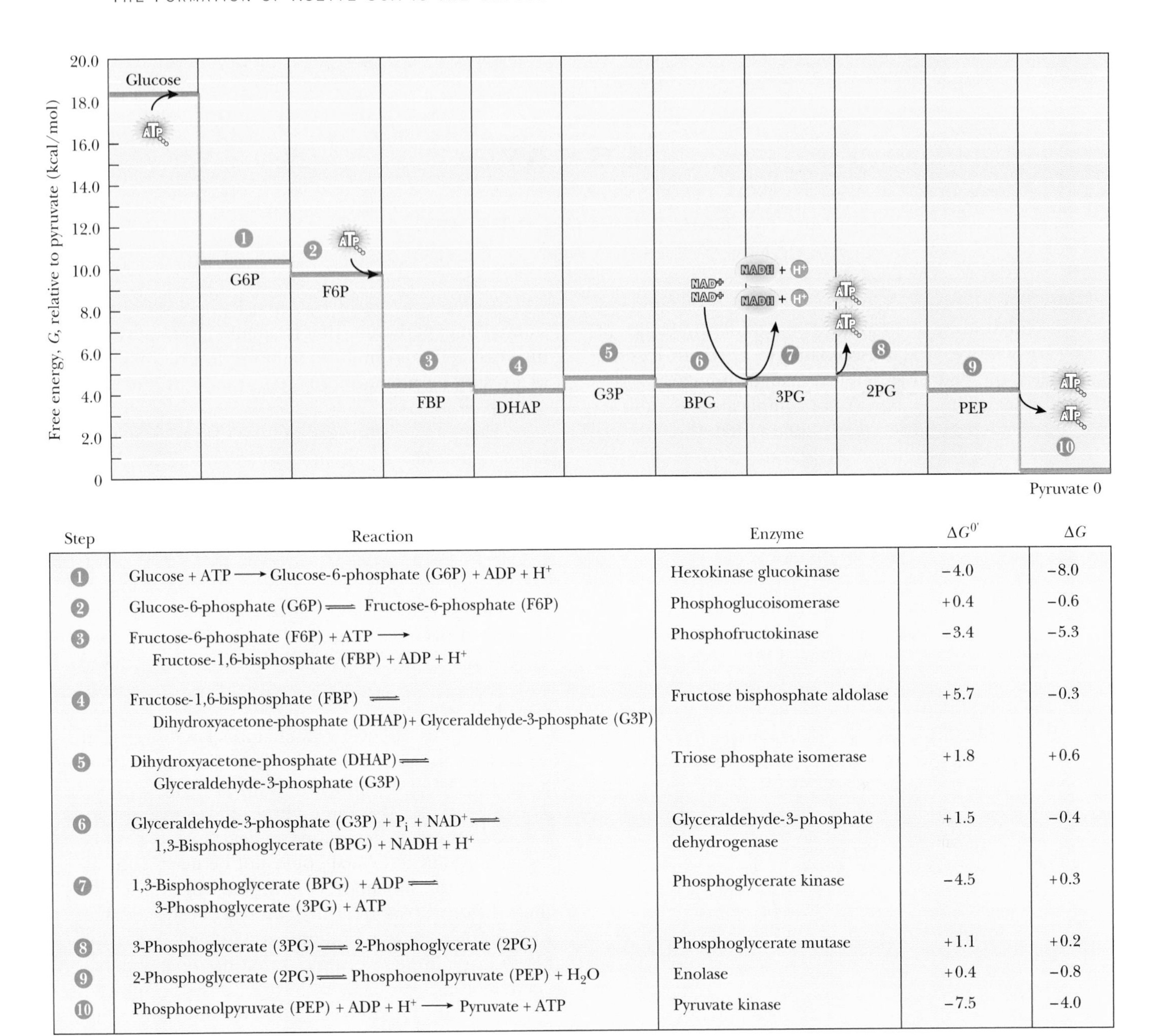

Step	Reaction	Enzyme	$\Delta G^{0'}$	ΔG
1	Glucose + ATP ⟶ Glucose-6-phosphate (G6P) + ADP + H^+	Hexokinase glucokinase	−4.0	−8.0
2	Glucose-6-phosphate (G6P) ⇌ Fructose-6-phosphate (F6P)	Phosphoglucoisomerase	+0.4	−0.6
3	Fructose-6-phosphate (F6P) + ATP ⟶ Fructose-1,6-bisphosphate (FBP) + ADP + H^+	Phosphofructokinase	−3.4	−5.3
4	Fructose-1,6-bisphosphate (FBP) ⇌ Dihydroxyacetone-phosphate (DHAP)+ Glyceraldehyde-3-phosphate (G3P)	Fructose bisphosphate aldolase	+5.7	−0.3
5	Dihydroxyacetone-phosphate (DHAP) ⇌ Glyceraldehyde-3-phosphate (G3P)	Triose phosphate isomerase	+1.8	+0.6
6	Glyceraldehyde-3-phosphate (G3P) + P_i + NAD^+ ⇌ 1,3-Bisphosphoglycerate (BPG) + NADH + H^+	Glyceraldehyde-3-phosphate dehydrogenase	+1.5	−0.4
7	1,3-Bisphosphoglycerate (BPG) + ADP ⇌ 3-Phosphoglycerate (3PG) + ATP	Phosphoglycerate kinase	−4.5	+0.3
8	3-Phosphoglycerate (3PG) ⇌ 2-Phosphoglycerate (2PG)	Phosphoglycerate mutase	+1.1	+0.2
9	2-Phosphoglycerate (2PG) ⇌ Phosphoenolpyruvate (PEP) + H_2O	Enolase	+0.4	−0.8
10	Phosphoenolpyruvate (PEP) + ADP + H^+ ⟶ Pyruvate + ATP	Pyruvate kinase	−7.5	−4.0

Figure 7-18 Free energy changes during glycolysis.

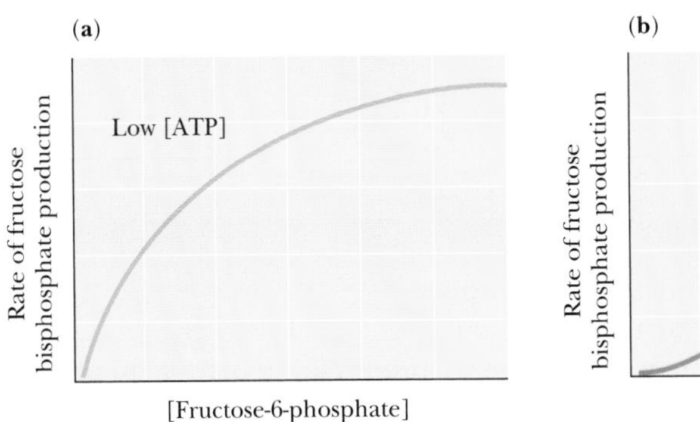

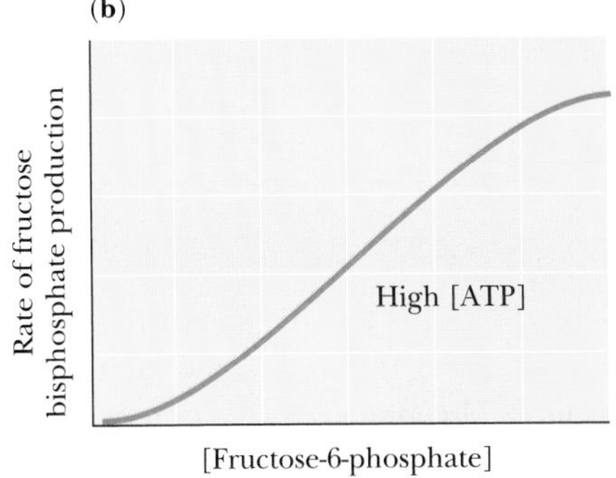

Figure 7-19 The enzymatic activity of phosphofructokinase depends on allosteric regulation by ATP. This reaction regulates the overall rate of glycolysis. **(a)** At low ATP concentrations, the enzyme is more active and glycolysis is rapid; **(b)** at high ATP concentrations (when the cell may not require more ATP), enzyme activity is low and glycolysis is slow.

ESSAY

EARLY DISCOVERIES AND CONTROVERSIES ABOUT GLYCOLYSIS

Because of ethanol's intoxicating effects on humans, careful studies of its production evidently began at least 8000 years ago. According to Egyptian legend, Osiris, the god of agriculture, taught humans to prepare beer. The recipe was not complicated: Barley was germinated in buried pots, water added, the mixture exposed to yeasts in the air, and the mixture allowed to ferment. Notwithstanding the legend, devising this procedure must have required patient experimentation to avoid the production of beer that tasted like vinegar.

Biochemists can describe fermentation as a set of chemical transformations. Although we now accept that organisms carry out chemical reactions, many chemists were at first scandalized by the idea that fermentation—a chemical transformation—was performed by organisms. In 1839 the most prominent early biochemists, Wöhler and Liebig, published an anonymous paper mocking the idea that organisms performed chemical reactions. Part of the paper reads

> *Beer broken up in water is resolved by [the microscope] into innumerable small spheres. . . . When placed in sugar water it can be seen that these are the eggs of animals; they swell, burst, and there develop small animals which multiply with incredible rapidity in a most unprecedented way. The form of these animals differs from that of the 600 species already described; it is the shape of a distilling flask without the condenser. The tube of the still-head is a kind of sucking snout covered internally with fine cracks; although teeth and eyes are not to be seen, one can distinguish a stomach, intestine, the anus (a rose-pink spot), and the organs of urine secretion. . . . The animals suck in sugar, which can clearly be seen in the stomach. It is immediately digested, and the digestion is followed by excretion. They excrete alcohol from the intestine and carbon dioxide from the urine organs. The urine bladder in the full condition is shaped like a champagne bottle. . . .*
> [from Thimann, K. V., *The Life of Bacteria*, 2nd edition, Macmillan, New York, 1963]

Twenty years later, however, Pasteur definitively showed in experiments that fermentation requires living yeast cells and that the amount of yeast increases as a result. Fermentation, Pasteur said, was "life without air."

How can cells bring about the chemical changes of fermentation? To discover this required the ability to carry out the cellular reactions *outside* the cell. This finally became possible in 1897 when two brothers named Büchner were attempting to preserve a yeast extract to sell as a medicine. To prepare their product for market, they tried adding sugar, a common preservative in jams and jellies, to prevent their extract from decomposing. To their surprise, their extract began to froth. This was the first clear demonstration of a complicated biological process occurring outside a cell.

After this discovery biochemists could begin to study fermentation in detail. They found that fermentation requires two kinds of materials in the yeast extract: enzymes (meaning "in yeast")—large molecules that could be inactivated by heat; and coenzymes—smaller molecules that could not be inactivated by heat. By 1940, biochemists had determined each of the individual steps in fermentation. They showed that the steps of yeast fermentation were identical to those of glycolysis in muscle, except that the final product is ethanol in fermentation and lactic acid in muscle. Since 1940, the enzymes catalyzing the individual steps in glycolysis have been isolated and their properties studied. These enzymes couple the rearrangement of atoms in glucose to the generation of ATP.

---

structure is shown in Figure 7-20b. The third carbon of pyruvate is released as CO_2.

The formation of acetyl CoA is complex: The reaction removes a molecule of carbon dioxide as well as two electrons and two hydrogen ions from pyruvate. The enzyme that catalyzes this reaction, called pyruvate dehydrogenase (shown in Figure 7-21), is one of the largest known, consisting of 72 polypeptide chains.

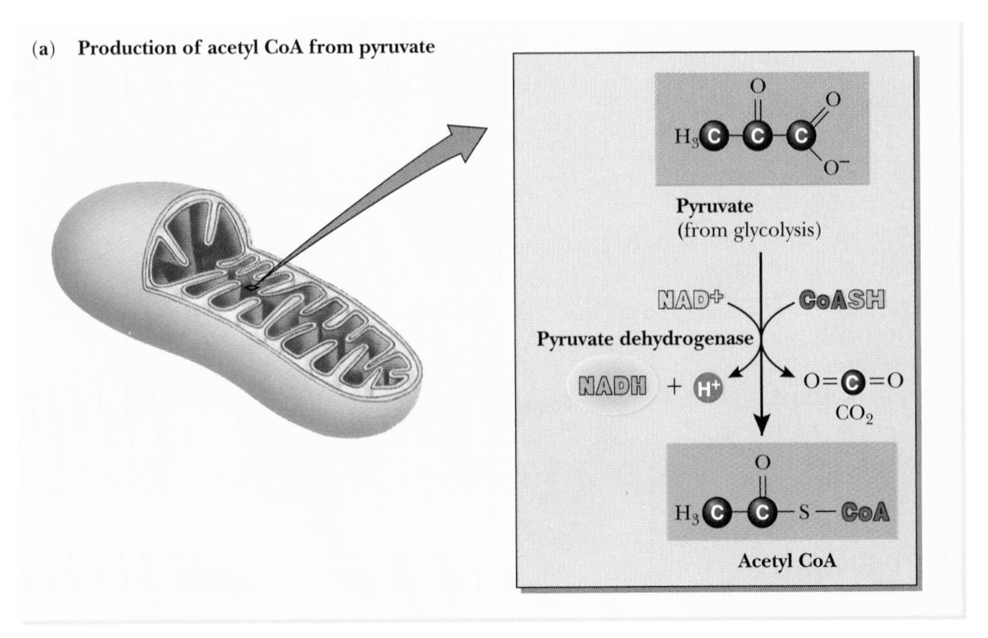

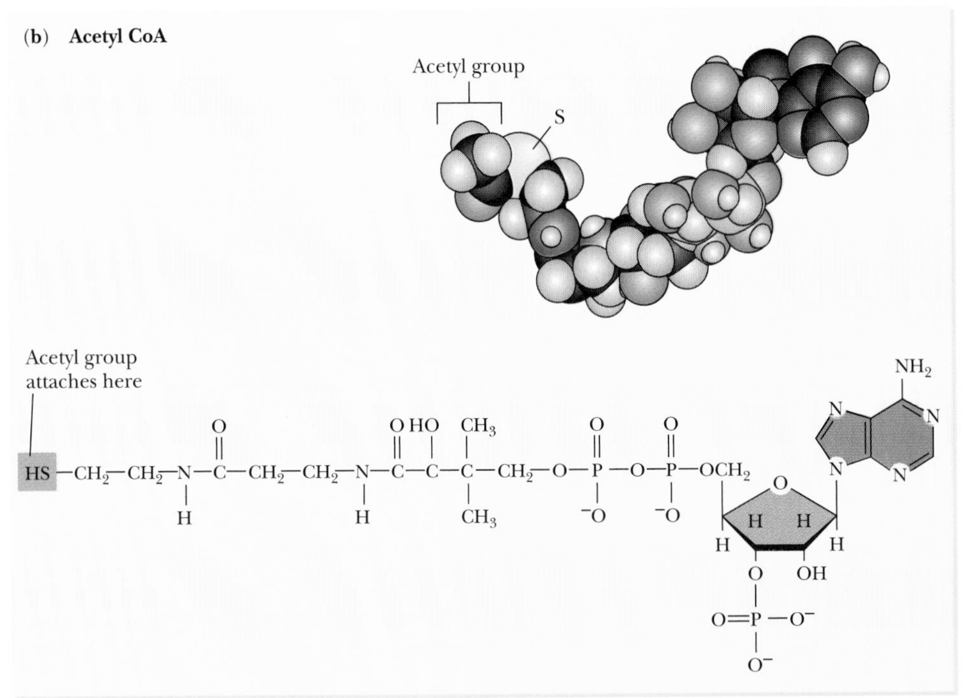

Figure 7-20 In the second stage of energy extraction, pyruvate is oxidized to form acetyl CoA. (**a**) The enzymatic formation of acetyl CoA; (**b**) the structure of acetyl CoA.

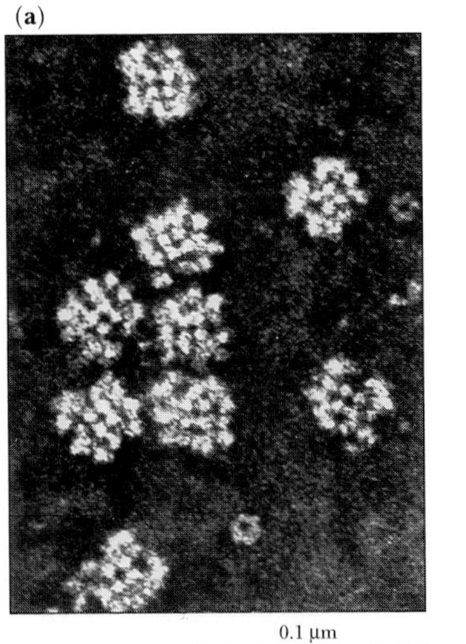

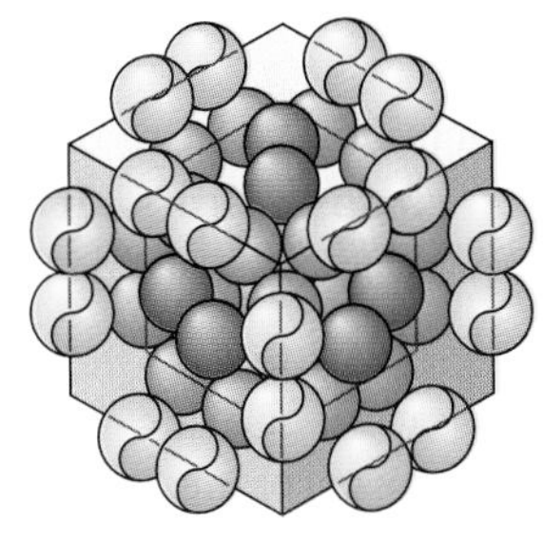

Figure 7-21 Pyruvate dehydrogenase, one of the largest known enzyme complexes, consists of 72 polypeptide chains. *(a, Lester J. Reed, University of Texas at Austin)*

Acetyl CoA also comes from sources other than pyruvate, including the breakdown of fatty acids and some amino acids. Acetyl CoA has only two principal fates, however: (1) It can enter the third stage of energy metabolism and generate more ATP, as discussed below, or (2) it can be used to synthesize new lipids, either fatty acids or cholesterol. When a cell's ATP levels are high, the rates of respiration and oxidative phosphorylation decrease, and the cell uses acetyl CoA (derived from carbohydrates, fats, and proteins) to make new fats. This redirection of acetyl CoA is the reason that we accumulate fat when we consume more energy-rich molecules than we use.

HOW DOES A CELL EXTRACT ENERGY FROM ACETYL CoA?

In most organisms acetyl CoA enters directly into a set of reactions called the **citric acid cycle,** also called **Krebs's cycle,** which ultimately converts the carbon atoms of acetyl CoA into carbon dioxide. We discuss the reasons for these two names later in this chapter. The citric acid cycle is a nearly universal set of reactions, which, in eukaryotes, take place within the mitochondrial matrix.

The citric acid cycle removes eight electrons from the two carbons of acetyl CoA. NAD^+ accepts six of these electrons, and FAD accepts the other two. The citric acid cycle also forms one high-energy phosphate bond, but, unlike the other reactions we have discussed, the phosphate acceptor is guanosine diphosphate (GDP), rather than ADP. The resulting GTP has the same amount of free energy as ATP.

The overall equation for the citric acid cycle, also summarized in Figure 7-22, is

$$\text{acetyl CoA} + 3\ NAD^+ + FAD + GDP + P_i + 2\ H_2O \longrightarrow 2\ CO_2 + 3\ NADH + FADH_2 + GTP + 2\ H^+ + CoA$$

Oxygen is absolutely essential for the citric acid cycle to proceed. Only when oxygen is available can the cell regenerate NAD^+ and FAD and allow the cycle to continue.

The citric acid cycle not only provides energy, but, as discussed later in this chapter, it also serves as the major route for the synthesis of many amino acids and other small molecules. Because it is a cycle rather than a path, changes in the rate of any single reaction speed up or slow down the whole cycle. (See Chapter 4, p. 110.) The flow of molecules through the cycle responds not only to the energy needs of a cell but also to the availability of amino acids.

Figure 7-23 summarizes the steps of the citric acid cycle. These eight reactions fall into six types, three of which are also used in glycolysis (p. 207): (1) *phosphate transfer;* (2) *dehydration;* (3) *redox reactions;* (4) *condensation,* the joining of two molecules (here the two-carbon acetyl CoA to the four-carbon oxaloacetate) to form a single molecule; (5) *hydration,* the addition of a water molecule; and (6) *decarboxylation,* the removal of a carbon dioxide molecule. The central events of the citric acid cycle are its four oxidation-reduction reactions, which employ two kinds of electron acceptors—NAD^+ and FAD.

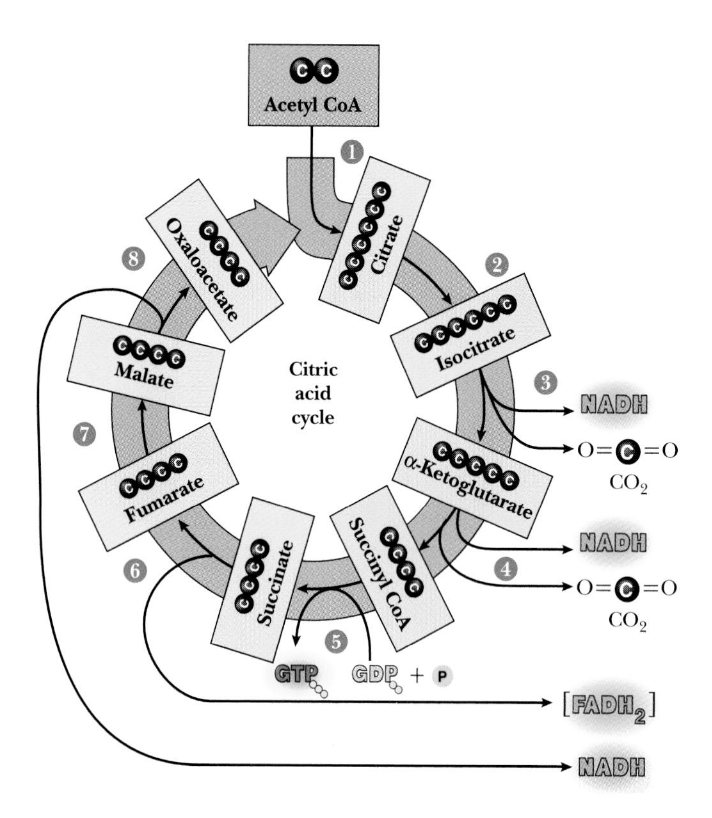

Figure 7-22 The Citric Acid Cycle. A single molecule of acetyl CoA generates 3 NADH, 1 $FADH_2$, 1 GTP, and 2 CO_2.

The reactions of the cycle fall into three parts: (1) steps 1 and 2, the incorporation of acetyl CoA into citrate (citric acid), a six-carbon compound that gives the cycle its name, followed by its rearrangement into isocitrate, an isomer; (2) steps 3–5, the conversion of isocitrate into a four-carbon compound called succinate, which is linked to CoA; and (3) steps 6–8, the production of a four-carbon compound, called **oxaloacetate,** which can combine with the acetyl group from acetyl CoA to regenerate citrate.

Each Turn of the Citric Acid Cycle Transfers Eight Electrons from Acetyl CoA to Produce NADH and $FADH_2$

Each revolution of the citric acid cycle begins with the condensation of oxaloacetate and acetyl CoA to form citrate. Within each turn, citrate loses two carbon atoms as carbon dioxide. Notice, however, that the carbon atoms of the carbon dioxide from each turn are not those that entered the cycle as acetyl CoA. Nor is the oxaloacetate re-

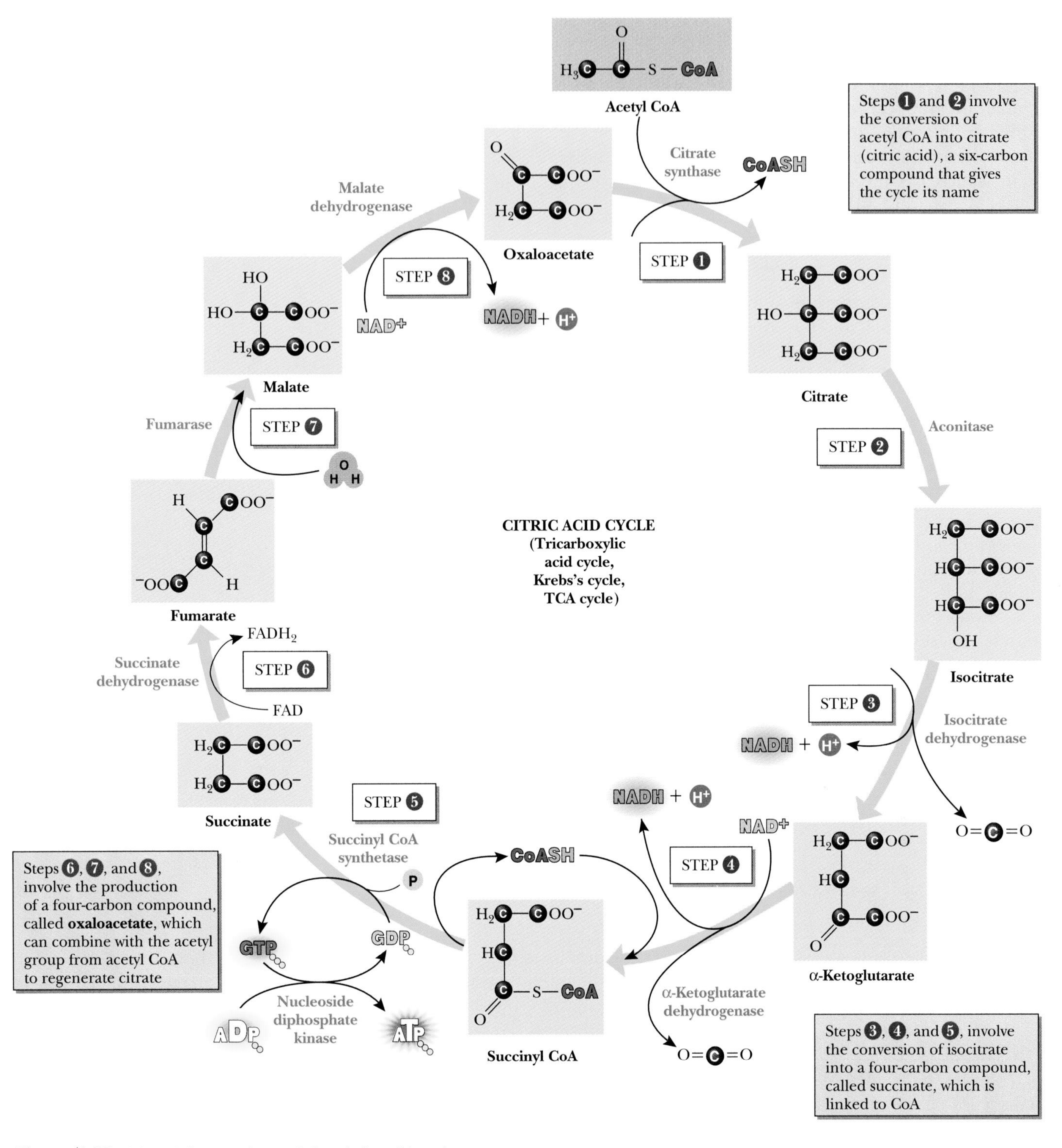

Figure 7-23 The eight reactions of the citric acid cycle.

generated by the cycle the same one that started the cycle. In each turn of the cycle citrate loses a total of eight electrons, six to NAD^+ and two to FAD.

How Do We Know That the Citric Acid Cycle Is a Cycle?

Biochemists tried for some time to understand how cells obtain energy from organic molecules in the presence of oxygen. The Hungarian biochemist Albert Szent-Gyorgyi (mentioned in Chapter 1) studied the uptake of oxygen in extracts of pigeon breast muscles. These extracts were particularly active in using oxygen. To his surprise, Szent-Gyorgyi found that certain four-carbon compounds stimulated the uptake of oxygen far more than he expected. He calculated the amount of oxygen required for the complete oxidation of each added compound, but he found that more oxygen was used than complete oxidation would require. The reason for this discrepancy, we now realize, is that the four-carbon compounds that Szent-Gyorgyi added are intermediates in the citric acid cycle—they increase the rate at which acetyl CoA enters the cycle. This in turn increases the amount of NADH and $FADH_2$ produced and therefore the amount of oxygen needed to regenerate NAD^+ and FAD.

By 1937, biochemists knew that pigeon muscle extracts could convert citric acid to these same four-carbon compounds. The missing link and the final interpretation were provided in 1938 by Hans Krebs, a refugee from Nazi Germany working in England. He showed that the same muscle extracts could make citric acid from a four-carbon compound called oxaloacetate and a product of pyruvate (which we now know to be acetyl CoA).

Krebs's crucial contribution was the idea that the conversion of acetate to carbon dioxide involves a cycle. Interestingly, 6 years earlier, while still in Germany, Krebs had discovered another metabolic cycle involved in the disposal of nitrogen-containing compounds. Because the two cycles intersect, the two have been called "Krebs's bicycle."

Krebs's citric acid cycle explained Szent-Gyorgyi's earlier observation. Because the four-carbon molecules could go around the cycle more than once, they stimulated more oxygen consumption than they would have if they were merely oxidized themselves. Krebs showed how the particular four-, five-, and six-carbon compounds fit together in a single cycle. He also showed that the same chemical transformations occurred both in cells and in test tubes.

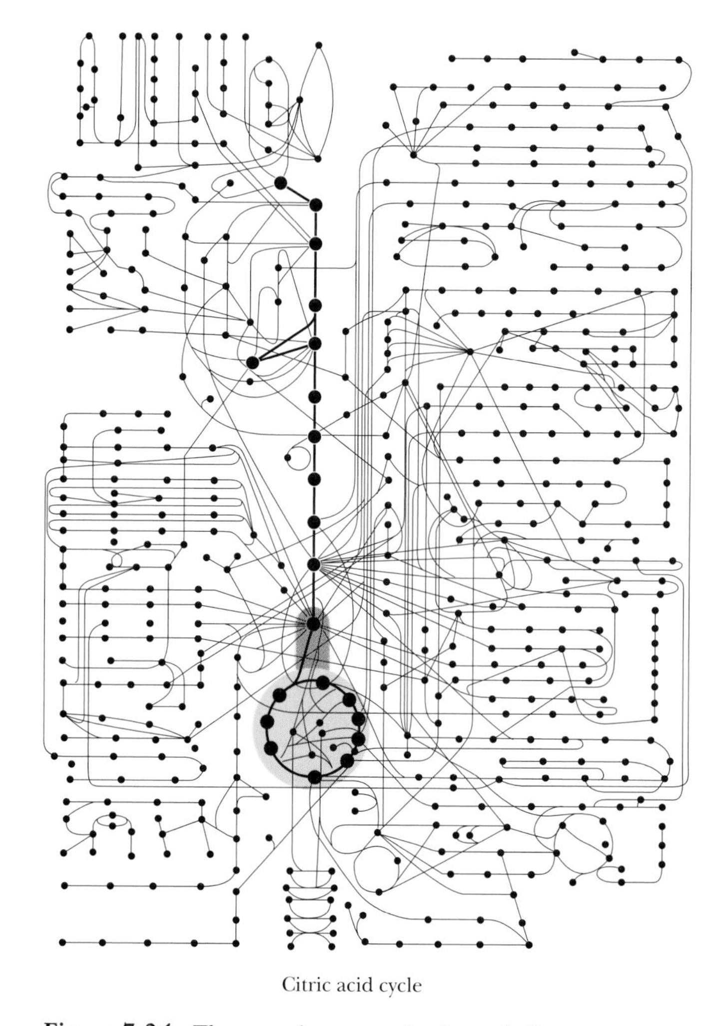

Figure 7-24 The complex network of metabolism.

The Citric Acid Cycle Intersects with Other Biochemical Pathways

The citric acid cycle lies at the center of the complex network of biochemical conversions (represented in Figure 7-24) that are collectively called metabolism. Metabolism is divided into two classes of pathways—**catabolism** [Greek, *cata* = down + *ballein* = to throw], the breakdown of complex molecules, such as those in food, and **anabolism** [Greek, *ana* = up + *ballein* = to throw], the synthesis of complex molecules. In general, catabolism produces energy and usually involves oxidation—the removal of electrons and the production of NADH. Anabolism usually consumes energy and involves reduction, thus using up NADH and other reducing agents.

Cells regulate the rates at which pathways or sets of pathways operate. High levels of ATP favor anabolism, while low levels favor catabolism. When ATP is high but the products of anabolism are abundant, the energy in food is stored for future use—as carbohydrates, proteins, or fats. In spite of the importance of glucose and other carbohydrates in energy metabolism, the preferred form of energy storage in animals is fat.

The end product of the catabolism of fats is acetyl CoA. Acetyl CoA from fat catabolism enters the citric acid cycle just as the acetyl CoA produced by pyruvate dehydrogenase.

Fats are more reduced compounds than sugars; that is, for each carbon atom, fats provide more electrons for oxidative phosphorylation. Fats therefore contain more free energy—about 9 kcal per gram of fat, instead of about 4 kcal per gram of carbohydrate or protein. In addition, fats are less polar and less hydrophilic than carbohydrates, which may bind up to twice their weight in water. As a result of these factors, each gram of stored fat contains about six times as much energy as each gram of stored glycogen. A 50-kg (110-lb) woman, for example, has fuel reserves of about 70,000 kcal in fats, about 18,000 kcal in proteins (mostly in muscle), about 400 kcal in glycogen, and about 30 kcal in glucose. Her fat stores, however, weigh only about 8 kg (18 lb): To store the same amount of energy as carbohydrates (including their bound water), she would need to weigh at least 90 kg (200 lb)!

HOW MUCH USABLE ENERGY CAN A CELL HARVEST FROM A MOLECULE OF GLUCOSE?

As electrons move through the electron transport chain, H^+ ions move from the mitochondrial matrix into the intermembrane space. And, as the protons flow back into the matrix, they drive the formation of high-energy bonds in ATP. The transfer of a pair of electrons from each NADH to oxygen yields enough free energy to produce three high-energy phosphate bonds, that is, to make three molecules of ATP. In contrast, the transfer of a pair of electrons from each molecule of $FADH_2$ to oxygen is enough to provide energy to make two molecules of ATP. According to these numbers, the eight molecules of NADH made in a mitochondrion from each glucose molecule could therefore yield 24 ATPs and the two molecules of $FADH_2$ an additional four ATPs. (See Table 7-1.)

Actual measurements in isolated mitochondria, however, suggest a slightly lower yield of ATP. The experiments show that the oxidation of each NADH leads to the movement of 10 H^+ ions across the inner membrane and the synthesis of only 2.5 molecules of ATP. Similar experiments show that each $FADH_2$ yields 1.5 ATPs. So, for each molecule of glucose, under conditions likely to represent those within living cells, the NADH produced in the mitochondria yields 20 (instead of 24) ATPs, and the $FADH_2$ yields 3 (instead of 4) ATPs.

Glycolysis produces two molecules of NADH. These are made in the cytoplasm and do not readily cross the inner mitochondrial membrane. Instead, the electrons of cytoplasmic NADH usually shuttle across the inner mitochondrial membrane to form $FADH_2$ in the matrix. Because each $FADH_2$ can produce only 1.5 ATPs, the shuttling of electrons from cytoplasmic NADH results in a lower ATP yield than the NADH produced in the mitochondria themselves. Heart and liver cells, however, have an alternate electron shuttle across the inner membrane, allowing the formation of NADH in the matrix. In this case, the yield from the cytoplasmic NADH produced by glycolysis is the same as that from mitochondrial NADH. The two molecules of NADH made during glycolysis thus produce either 3 (2×1.5 ATP per $FADH_2$) or 5 (2×2.5 ATP per NADH) molecules of ATP, depending on the tissue.

Two ATPs are also produced in glycolysis and two GTPs in the mitochondria by phosphate-transfer reactions. Because the high-energy phosphates of the GTPs readily produce ATP, we may say that substrate-level phosphorylation produces an additional four ATPs per molecule of glucose, two from glycolysis and two from the citric acid cycle.

Respiration thus produces 30 (or 32) molecules of ATP for each molecule of glucose, depending on the tissue. Earlier estimates of ATP yield were based on the expected yield of three ATPs per NADH and two per $FADH_2$, giving a total yield of 36 (or 38) ATPs per molecule of glucose. All but four of the ATP molecules produced from a glucose molecule are produced by oxidative phosphorylation. In the absence of oxygen, glycolysis alone can produce only two molecules of ATP.

The ATP molecules produced in the mitochondria travel across the mitochondrial membranes to the cytoplasm, and ADP molecules from the cytoplasm move into the mitochondria. Within the mitochondria, the rates of the citric acid cycle and oxidative phosphorylation depend on the relative amounts of ATP and ADP, in much the same way as in the overall rate of glycolysis. (See p. 212.) Both the first enzyme of the citric acid cycle and the F_0-F_1 ATP synthetase are more active when ATP within the

Table 7-1 ***Energy Capture in Respiration***

Source	ATP/GTP produced by substrate level phosphorylation	NADH/ $FADH_2$ produced	Maximum number of high-energy phosphates from oxidative phosphorylation	Maximum total number of high-energy phosphates	Cellular site
Glycolysis	2ATP	2 NADH	6	8	Cytoplasm
Pyruvate dehydrogenase		2 NADH	6	6	Mitochondrial matrix
Citric acid cycle	2 GTP	6 NADH 2 $FADH_2$	22	24	Mitochondrial matrix
Totals					
Cytoplasm	2 ATP	2 NADH	6	8	
Mitochondion	2 GTP	8 NADH 2 $FADH_2$	28	30	

ESSAY

THE METABOLISM OF ALCOHOL

For thousands of years and in many societies, ethanol has provided a popular means of changing mood and behavior. In addition to the immediate dangers of intoxication, ethanol can cause permanent cellular damage in heavy drinkers. Knowledge of the metabolic pathways described in this chapter allows us to understand how this happens.

In humans, only liver cells can metabolize ethanol. There NAD^+ oxidizes ethanol, first to acetaldehyde and then to acetic acid, as shown in Figure 1. The abundantly generated NADH then donates its electrons to the electron transport chain. Oxidative phosphorylation occurs without attendant citric acid cycle operation. Sustained use of large amounts of alcohol leads to strange-looking mitochondria, such as those shown in Figure 2, the result of active oxidative phosphorylation and the shut-off of the citric acid cycle. Carbohydrates that would ordinarily enter the citric acid cycle are instead converted to fat. The fat is secreted by the liver cells into the blood in little vesicles produced by the endoplasmic reticulum.

After a few years, liver cells begin to fill with fat and to cease functioning. The liver becomes less and less able to deal with the demands of the consumed alcohol. Eventually it may become heavily scarred, as illustrated in Figure 3—a condition called cirrhosis. Cirrhosis of the liver is in fact the seventh leading cause of death in the United States. This disease illustrates that animals have evolved to obtain energy from a large variety of chemicals. Biochemical homeostasis keeps the cells supplied with ATP and reducing power. Other sources of energy are not wasted but stored sensibly. But animals have also evolved within a relatively narrow set of conditions. They (we) cannot tolerate too large a variation of diet for an extended period.

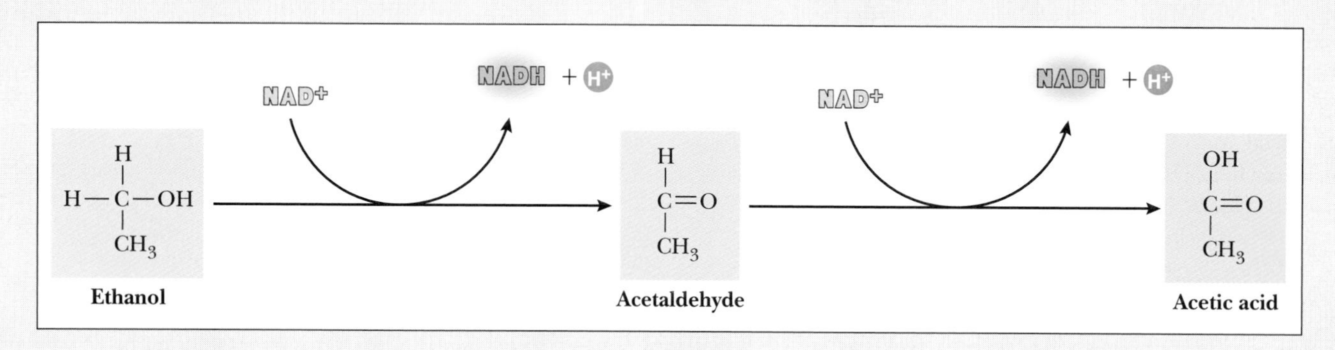

Figure 1 The oxidation of ethanol to acetaldehyde and acetic acid, which can occur in the human liver.

mitochondria is low and ADP is high. Because of such regulation, your body catabolizes sugars and fats 5 to 10 times more rapidly when you are exercising than when you are resting.

One last point about energy yield: The free energy for the hydrolysis of ATP is often stated to be −7.3 kcal per mole. This number is the free energy under standard conditions, when ATP, ADP, and phosphate are all present at equal concentrations. In cells, however, the concentration of ATP is 5 to 10 times that of ADP, which further favors the production of ADP. (As discussed in Chapter 4, increased concentration of a reactant favors the formation of more product; see p. 93). The free energy of hydrolysis is actually about −12 kcal per mole, and the efficiency of energy extraction from glucose under cellular conditions is about 50%. For comparison, the efficiency of an electric motor or a gasoline engine is about 10% to 20%.

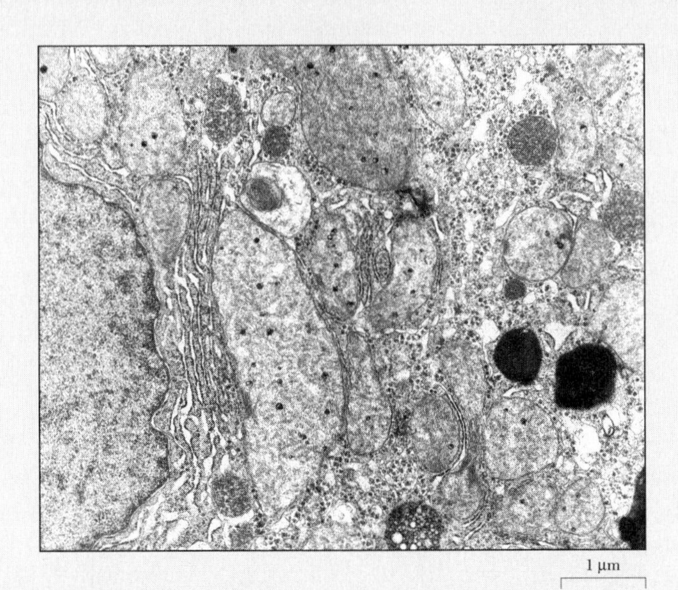

Figure 2 Abnormal mitochondria formed in the liver of an alcoholic. *(Dr. Antonio Chedid, Chicago Medical School)*

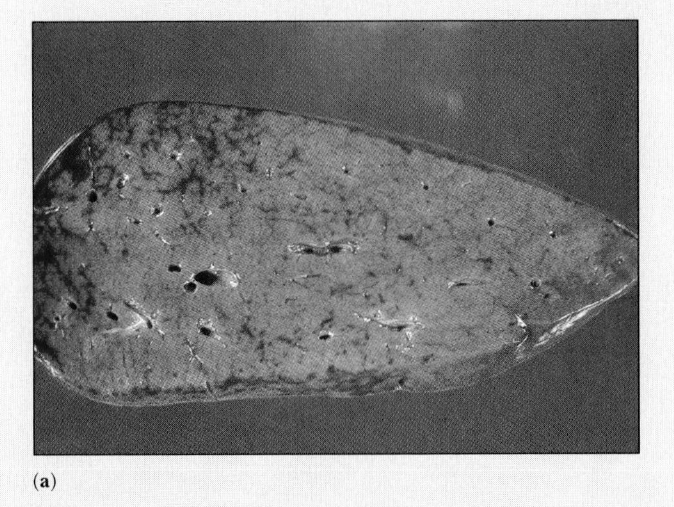

(a)

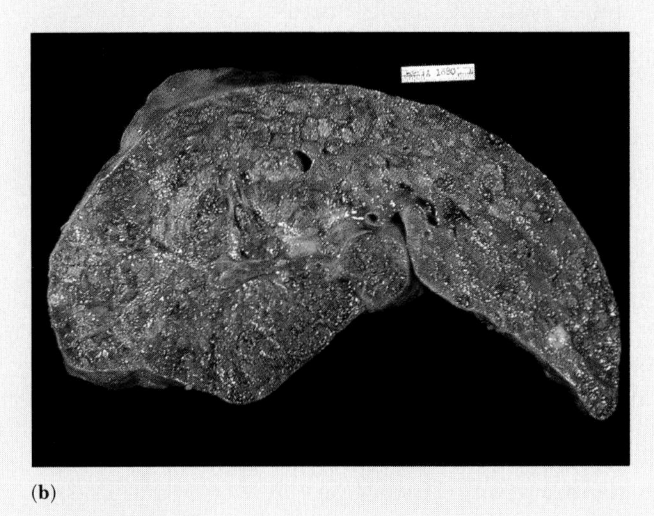

(b)

Figure 3 Chronic alcoholism leads to cirrhosis, the heavy scarring of the liver: **(a)** a normal human liver; **(b)** a cirrhotic liver from a chronic alcoholic. *(a, Biophoto Associates/Photo Researchers; b, Martin Rotker/Photo Researchers)*

HOW DO HETEROTROPHS EXTRACT ENERGY FROM MACROMOLECULES?

Before they can enter the catabolic pathways we have described, macromolecules must be hydrolyzed into monomers (such as glucose). **Digestion** is the process of hydrolyzing large molecules into smaller units—proteins to amino acids, polysaccharides to glucose and other simple sugars, and fats to fatty acids and glycerol. Digestion almost always takes place *outside* of cells (or at least outside the cytosol). In animals and fungi, most digestion takes place through the action of secreted enzymes, in the intestinal tract of animals or in space surrounding a fungus. In heterotrophic protists and in some animal cells, digestion at first seems to occur inside cells following phagocytosis or pinocytosis. (See Chapter 6, p. 165.) However, the lysosomal membrane separates digestive enzymes from the rest of the cell, and even in these cells we can still consider digestion inside the lysosome to be "extracellular."

Digestion differs from the rest of catabolism in two ways: (1) It does not provide energy, and (2) it does not occur in a fixed sequence. The accomplishment of digestion is the generation of energy-rich small molecules, which, unlike macromolecules, can move across the plasma membrane into the cytosol. Inside the cell, systems of enzymes transform these molecules and harness the energy from their chemical bonds into the chemical bonds of ATP.

Extracellular digestive enzymes hydrolyze macromolecules and fats to their component building blocks. Although they usually do not act in a fixed order, digestive enzymes nonetheless cooperate with each other. One enzyme, for example, may cut peptide bonds next to positively charged amino acid residues, another next to nonpolar residues. Once the products of digestion enter the cytosol, however, they enter the various metabolic pathways—some catabolic, to provide more ATP, and some anabolic, to make new molecules.

SUMMARY

Almost all eukaryotic cells (and many prokaryotic cells as well) are aerobic; that is, they live in oxygen-containing air or water. Aerobic organisms produce most of their ATP by respiration, the oxygen-dependent extraction of energy from food molecules. Most energy-producing reactions in cells are reduction-oxidation (redox) reactions, which transfer electrons from one molecule (the reducing agent) to another (the oxidizing agent).

Altogether, respiration transfers 24 electrons from glucose to oxygen in a complex set of reactions. For 20 of these electrons, the first acceptor is NAD^+, which can accept two electrons and a hydrogen ion to give its reduced form, called NADH. The other four electrons of glucose are transferred to another oxidizing agent, called FAD, whose reduced form is called $FADH_2$. These reduced coenzymes are the major energy storage molecules in energy metabolism. They ultimately transfer their electrons to oxygen in a series of steps that, in eukaryotes, occurs in mitochondria, within the inner mitochondrial membrane. In prokaryotes, these reactions occur in association with the plasma membrane.

Oxidative phosphorylation is the process that couples the oxidation of NADH and $FADH_2$ to the production of high-energy phosphate bonds in ATP. The electrons from NADH and $FADH_2$ shuttle to oxygen, their ultimate acceptor, through a series of electron carriers. The pathway of the electrons (from one carrier to another) is called the electron transport chain (or respiratory chain). Biochemists have worked out the precise path of electrons from NADH to oxygen.

Researchers looked for clues to understanding oxidative phosphorylation in phosphate-transfer reactions in which a redox reaction is coupled to the production of high-energy phosphate groups. But oxidative phosphorylation operates by a different mechanism, summarized in the Chemiosmotic Hypothesis: (1) The flow of electrons through the electron transport chain pumps protons out of the matrix, establishing an electrochemical gradient across the inner mitochondrial membrane; and (2) some special molecular machinery in the membrane captures the energy of this gradient in ATP.

The orientation of the electron carrier complexes establishes the direction of the proton gradient. Once H^+ ions are pumped out of the matrix, there is only one way back in—through special channels in the otherwise impermeable membrane. As the H^+ ions flow "downhill" through the channel, down the electrochemical gradient, their free energy decreases. The F_0-F_1 complex (coupling factor) uses the flow of protons to drive the synthesis of ATP.

Glycolysis is the first stage in the metabolism of glucose. Glycolysis is a set of ten chemical reactions that convert glucose, a sugar with six carbon atoms, into two smaller molecules, called pyruvate, with three carbon atoms each. These reactions also convert some of the energy in the chemical bonds of glucose into two high-energy bonds of ATP. All ten reactions take place in the cytosol, that is, outside the mitochondria.

Many microorganisms obtain energy from anaerobic glycolysis, or fermentation. Yeast cells, for example, perform fermentation in the brewing of beer or wine and in the rising of bread, converting glucose into ethanol and carbon dioxide, using NADH to reduce pyruvate to form ethanol and converting NADH back to NAD^+.

In eukaryotic cells, energy metabolism (other than glycolysis) takes place in the mitochondria. Pyruvate enters the mitochondria, where respiration proceeds. Pyruvate then undergoes an oxidation reaction, in which NAD^+ serves as the electron acceptor. In acetyl CoA, acetate is linked, via a high-energy bond, to a molecule called coenzyme A.

The citric acid cycle, which takes place within the mitochondrial matrix, ultimately converts the carbon atoms of acetyl CoA into carbon dioxide. The citric acid cycle is a nearly universal set of reactions.

The citric acid cycle not only provides energy, but also serves as the major route for the synthesis of many amino acids and other small molecules. The flow of molecules through the cycle responds not only to the energy needs of a cell, but also to the availability of amino acids.

The citric acid cycle lies at the center of the complex network of biochemical conversions that are collectively called metabolism. Metabolism is divided into two classes of pathways—catabolism, the breakdown of complex molecules such as those in food, and anabolism, the synthesis of complex molecules. In general, catabolism produces energy and anabolism requires energy. Catabolism usually involves oxidation—the removal of electrons and the production of NADH—while anabolism involves reduction and the using up of NADH and related reducing agents.

Respiration produces 30 (or 32) molecules of ATP for each molecule of glucose. All but four of these ATP molecules are produced by oxidative phosphorylation. In the absence of oxygen, glycolysis alone can produce only two molecules of ATP.

KEY TERMS

acetyl CoA
aerobic
anabolism
anaerobic
ATP synthase
catabolism
chemiosmosis
citric acid cycle
coenzyme
coupling factor
cytochrome
digestion
electrochemical gradient
electron carrier
electron transport chain
electronegativity
F_0-F_1 complex
facultative anaerobe
FAD
$FADH_2$
fermentation
glycolysis
heme
iron-sulfur center
iron-sulfur protein
Krebs's cycle
mitochondrial matrix
NAD^+
NADH
obligate anaerobe
oxaloacetate
oxidative phosphorylation
oxidizing agent
proton channel
proton gradient
pyruvate
redox potential
reducing agent
reduction-oxidation reaction
respiration
respiratory chain
substrate-level phosphorylation

QUESTIONS FOR REVIEW AND UNDERSTANDING

1. Define and contrast the following pairs of terms:
   (a) aerobic vs. anaerobic
   (b) oxidation vs. reduction
   (c) electronegativity vs. redox potential
   (d) oxidizing agent vs. reducing agent
   (e) NAD^+ vs. FAD
   (f) facultative anaerobe vs. obligate anaerobe
   (g) catabolism vs. anabolism
   (h) substrate-level phosphorylation vs. oxidative phosphorylation
   (i) chemical gradient vs. electrical gradient
   (j) digestion vs. catabolism
2. Describe the relationship between ΔG for a redox reaction and the difference in reduction potentials between the reducing agent and the oxidizing agent in mathematical terms and in your own words.
3. What is the role of the electron transport chain in the redox reaction between NADH and O_2?
4. Why do electrons move in only one direction in the electron transport chain?
5. What is actually responsible for the electron transfer ability of cytochromes?
6. How do iron-sulfur proteins differ from cytochromes? What is actually responsible for the electron transfer ability of FeS proteins?
7. What is meant by "hydrogen ion (or proton) gradient"? What two "component" gradients contribute to this electrochemical gradient?
8. Describe the respective roles of the F_0 and F_1 subunits of the ATP synthase.
9. Explain why NADH yields 2.5 ATP while $FADH_2$ yields 1.5 ATP when each is oxidized by the electron transport chain. Take into account the reduction potentials of the nucleotide coenzymes and the proton pumping activities of the membrane complexes.
10. If a proton gradient is all that is required for ATP synthesis (along with a functional ATP synthase complex!) propose an experiment that would be as simple as possible but demonstrate the role of the proton gradient.
11. Outline the major points of the chemiosmotic theory proposed by Mitchell to explain the coupling of electron transport and oxidative phosphorylation.
12. Recalling the concept of coupling and coupled reactions given in Chapter 4, explain why we can say that:
    (a) ATP synthesis is coupled to electron transport.

(b) The formation of a proton gradient is coupled to electron transport.

(c) The synthesis of ATP is coupled to dissipation of the proton gradient.

13. If the reaction

$$NADH + H^+ + \tfrac{1}{2}O_2 \longrightarrow NAD^+ + H_2O$$

has a value for $\Delta G = -53$ kcal/mol and the transfer of two moles of electrons through the respiratory chain contributes sufficiently to the proton gradient to allow for the synthesis of approximately 2.5 moles of ATP (ΔG of ATP formation is +12 kcal/mol), what is the efficiency (i.e., energy captured/energy released) of oxidative phosphorylation?

14. Dinitrophenol (DNP) is a compound known as an "uncoupler" because it uncouples ATP synthesis from electron transport. That is, in the presence of DNP, electron transport will operate but ATP synthesis is reduced or stopped. DNP does this by acting as a proton carrier, transferring protons across the mitochondrial membrane from high concentration to low, and so by-passing the ATP synthase. Earlier in this century, DNP was actually prescribed by doctors as a diet pill. Speculate on why DNP would result in weight loss, i.e., utilization of energy (fat) stores. This practice was later stopped when patients died from even minor overdoses. Speculate on the cause of death from an overdose of DNP. (*Hint:* Keep in mind that relieving the proton gradient releases energy.)

15. What are the principal products of glycolysis?

16. Name the three "parts" of glycolysis.

17. If the conversion of pyruvate to lactate or ethanol does not yield any energy, why do anaerobic cells convert pyruvate to these products?

18. What is the yield of ATP when a molecule of glucose is broken down during glycolysis under aerobic conditions when NADH is oxidized by the electron transport chain? What is the yield of ATP under anaerobic conditions?

19. Why do we say that phosphofructokinase is an allosteric enzyme (defined in Chapter 4), and that ATP is an allosteric regulator of the enzyme, in addition to being a substrate?

20. Considering the role of phosphofructokinase in glycolysis and the contribution of glycolysis to the oxidation of glucose, why is it "logical" that ATP is an allosteric regulator of phosphofructokinase?

21. What are the three sources of acetyl CoA in cells? What are the possible fates of acetyl CoA?

22. Why is the conversion of pyruvate to acetyl CoA referred to as an "oxidative decarboxylation"?

23. What are the principal products of the citric acid cycle? What are the respective fates of these products?

24. A variety of enzymes termed dehydrogenases are found in the citric acid cycle. What is the common feature of these enzymes?

25. Explain why oxygen (O_2) is essential for the citric acid cycle to proceed, even though it is not used in the citric acid cycle.

26. If the ΔG of glucose $\rightarrow CO_2 + H_2O$ is -686 kcal/mol and the ΔG of ADP + $P_i \rightarrow$ ATP is 12 kcal/mol, what is the efficiency of aerobic respiration?

27. Why is fat ideally suited for energy storage in animals?

28. All biochemical pathways, whether catabolic or anabolic (biosynthetic), are exergonic. Why must this be the case? How does coupling (see Chapter 4) allow cells to drive pathways?

29. Why is a cell biologist's definition of respiration, as given in this chapter, consistent with a physiologist's definition of respiration, which is the uptake of oxygen and the release of carbon dioxide and water?

SUGGESTED READINGS

ALBERTS, B., D. Bray, J. Lewis, M. Raff, K. Roberts, and J. D. Watson, *Molecular Biology of the Cell,* 3rd edition, Garland, New York, 1994. See Chapter 2 on energy and ATP production and Chapter 14 on mitochondria.

GARRETT, R. H. and C. M. Grisham, *Biochemistry,* Saunders College Publishing, Philadelphia, 1995. Chapters 17–20 cover the material in this chapter.

LODISH, H., D. Baltimore, A. Berk, S. L. Zipursky, P. Matsudaira, and J. Darnell, *Molecular Cell Biology,* 3rd edition, Scientific American Books, New York, 1995. Chapter 17 covers the material in this chapter.

STRYER, L., *Biochemistry,* 3rd edition, W. H. Freeman, New York, 1988. Chapters 13–17 cover the material in this chapter.

VOET, D. and J. G. Voet, *Biochemistry,* John Wiley & Sons, New York, 1990. Another fine biochemistry textbook.

CHAPTER 8

How Do Cells Get Energy from the Sun? Photosynthesis

KEY CONCEPTS

- Carbon dioxide and water are the raw materials of photosynthesis. Sunlight provides the energy.
- Specialized photosynthetic machinery absorbs light and transforms light energy into chemical energy.
- Photosynthesis requires two distinct but interacting sets of light-dependent reactions.
- The energy of sunlight, first captured in the chemical bonds of NADPH and ATP, is initially stored in the chemical bonds of carbohydrates.
- The efficiency of photosynthesis depends both on a plant's genetic capacity and on its environmental conditions.

OUTLINE

ESSAYS

Life on earth ultimately depends on light energy from the sun. Every endergonic reaction that you are performing at this moment is driven by chemical energy somehow stolen from sunlight. **Photosynthesis** [Greek, *photos* = light + *syntithenai* = to put together] is the process that transduces light energy into the energy of chemical bonds. In this chapter, we ask how cells—both of microorganisms and of plants—accomplish photosynthesis. How does photosynthesis capture light energy? How does it store the captured energy? What molecules and subcellular structures participate?

WHERE DO PLANTS GET THEIR RAW MATERIALS?

Aristotle, the most influential of the Greek philosophers, was one of the first to confront the question of where plants get their raw materials. He suggested that a plant's roots extract materials from the earth in much the same way that an animal's stomach extracts materials from its food. Aristotle's view was not illogical: Farmers have long known that even the richest soils lose the ability to support vegetation. Only centuries later did they learn that the soil provides plants with nitrogen, phosphorus, and other essential elements (or minerals), but not with the carbon, hydrogen, and oxygen that make up most of organic compounds.

Scientists became convinced that Aristotle's view was incorrect as a result of a spectacular experiment performed in the early 17th century by the Belgian physician and alchemist, Johann Baptista van Helmont. As shown in Figure 8-1, Van Helmont planted a 5-pound willow sapling in 200 pounds of soil, which he had carefully dried in a furnace and placed in a large pot. He watered the growing tree as needed and surrounded it with a metal plate to keep dust from falling into the pot. After five years, he again weighed the tree and the soil. The tree then weighed 169 pounds and the soil only 2 ounces less than the starting 200 pounds. Van Helmont concluded that "164 pounds of wood, bark, and roots arose out of water only." To van Helmont this transformation of water seemed similar to the transformation of elements that alchemists had long sought to accomplish. To us, however, with an understanding of modern chemistry, the transformation is yet another example of the rearrangement of atoms to form new compounds, in this case compounds that make the parts of the tree.

Van Helmont's experiment made him realize the importance of water in plants, but he apparently did not think about the role of the air. The first suggestion that the atmosphere contributed to the stuff of plants came later—and not from chemists or alchemists but from microscopists. With the new microscopes developed in the early 17th century an Englishman and an Italian almost simultaneously discovered **stomata** [singular, **stoma;** Greek = mouth] (also called **stomates**), minute pores in the leaves of plants (shown in Figure 8-2) that allow air to pass to the interior of the leaves. The activities of plants, they speculated, somehow must depend on interactions with air. By the mid-18th century, scientists had concluded, "vegetation is planted in the air as well as in the earth."

What Do Plants and Air Do for Each Other?

The next big clue to the interaction between plants and air came from experiments, already mentioned in Chapter 1 (p. 24), in which a mouse was placed into a closed container. After some time the mouse died. In the 1770s,

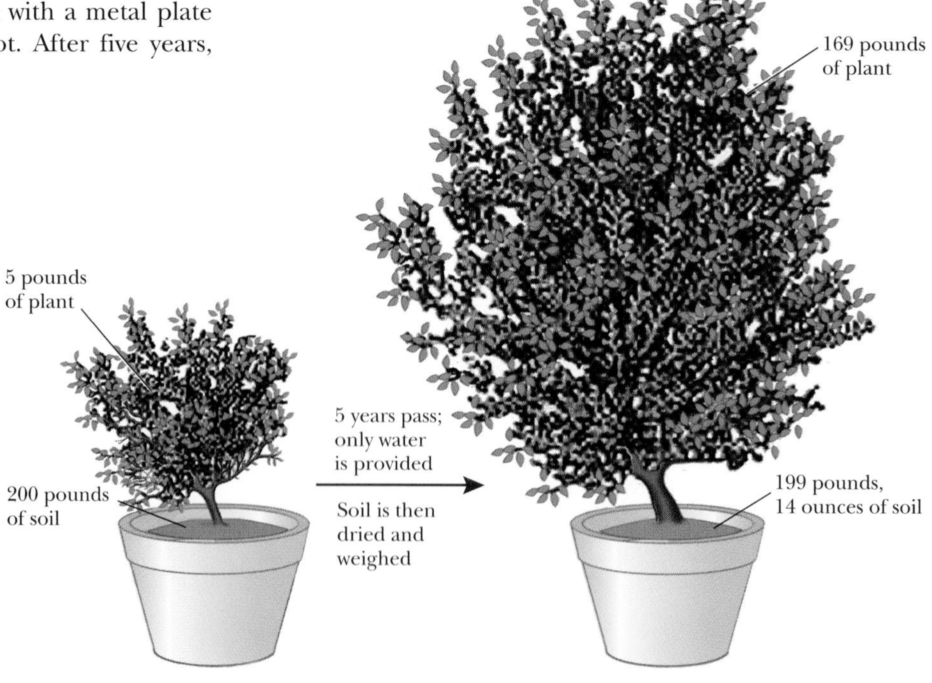

Figure 8-1 Van Helmont's experiment showed the importance of water and the unimportance of the soil in providing bulk materials for plant growth.

Figure 8-2 The discovery of stomata suggested that plants use gases from the air as well as water from the ground. **(a)** Stomata in a plant leaf, conventional light microscopy; **(b)** a scanning electron micrograph showing stomata on the lower surface of a parsley leaf. *(a, Mike Peres, RBP/Custom Medical Stock Photo; b, Dr. L. M. Beidler/Photo Researchers)*

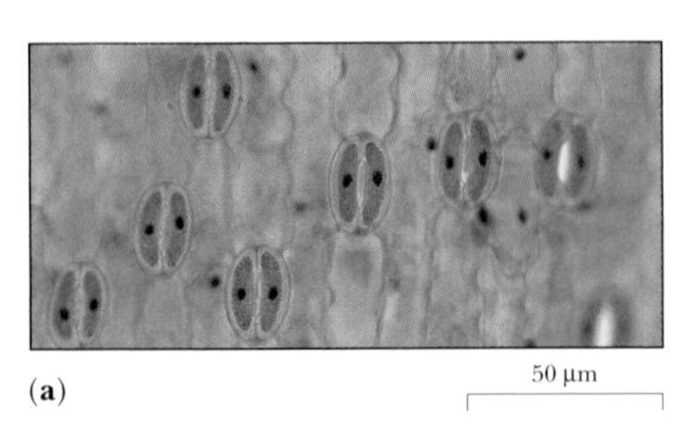

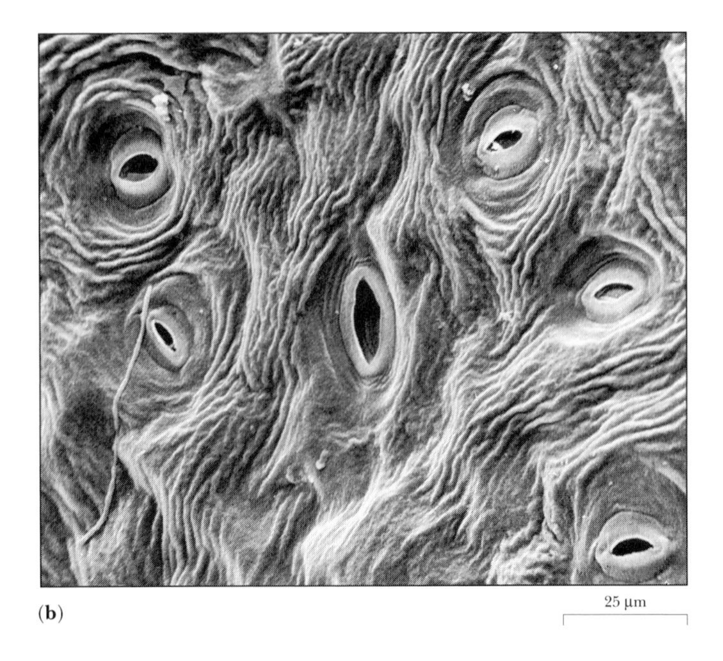

the English clergyman Joseph Priestley showed that a plant could restore the air in the jar and keep the mouse alive. Plants put back what animals (or burning candles) take out. This discovery was nothing less than the discovery of oxygen, whose existence (as discussed in Chapter 1) Lavoisier had independently found at about the same time. Credit for the discovery usually goes to both men.

Scientists soon learned that the material that plants restore to the air is oxygen and what they take out is carbon dioxide. During photosynthesis, as shown in Figure 8-3, plants take carbon dioxide from the air (and water from the ground) and release oxygen as a waste product. Plants also respire, however; as shown in Figure 8-3, they take in oxygen and release carbon dioxide. Respiration provides the energy for most of their metabolism.

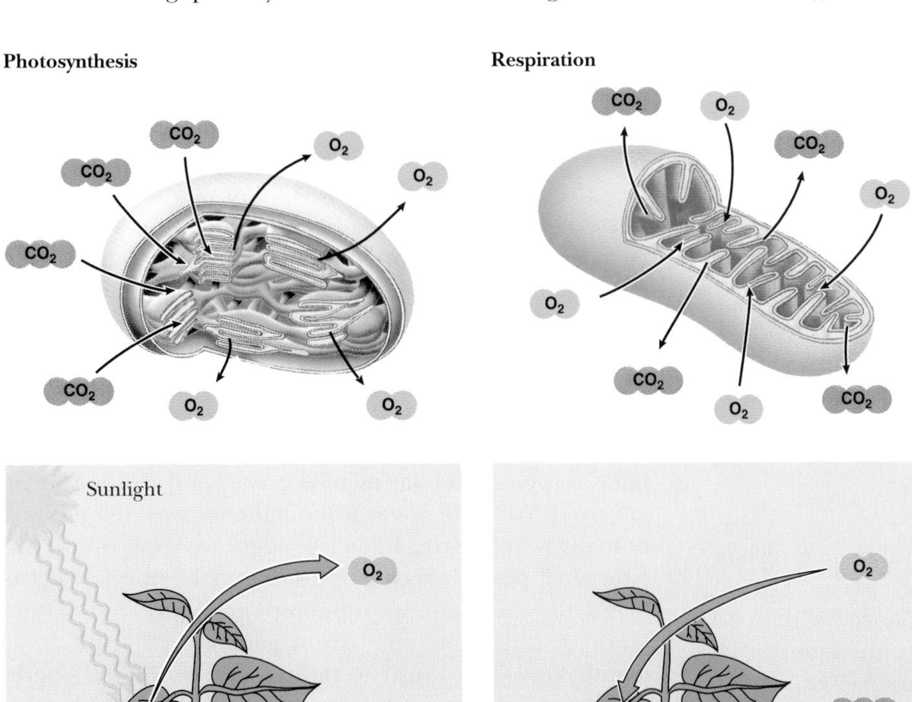

Figure 8-3 In the light, plants perform photosynthesis. They use carbon dioxide from the air and water from the ground to make carbohydrates. In both the dark and the light, they derive energy by respiration, starting with the carbohydrates they produced in the light.

Where Do the Atoms Go in Photosynthesis?

Now that we know the raw materials of photosynthesis, we can summarize the overall chemical transformation in the following equation:

$$\text{carbon dioxide} + \text{water} \xrightarrow{\text{light}} \text{glucose} + \text{oxygen}$$

or

$$6\,CO_2 + 6\,H_2O \xrightarrow{\text{light}} C_6H_{12}O_6 + 6\,O_2$$

Notice that this is the exact reversal of the overall equation for respiration, given in Chapter 7 (p. 196). Another way of writing the equation for photosynthesis is

$$CO_2 + H_2O \longrightarrow (CH_2O) + O_2$$

Here (CH_2O) is the general formula for a carbohydrate (as discussed in Chapter 2), with each carbon atom associated with the equivalent of one molecule of water.

This equation suggested an hypothesis for how photosynthesis might occur: Light splits carbon dioxide into oxygen, which is expelled into the atmosphere, and carbon, which combines with water to form carbohydrate. Although this idea was appealing, it also turned out to be wrong.

The first realization that this view of photosynthesis was wrong came from studies of photosynthetic bacteria. In the early 1930s, a graduate student at Stanford University named Cornelius van Niel was examining photosynthesis in bacteria that used hydrogen sulfide (H_2S) instead of water as a raw material. Instead of producing oxygen as a by-product, these bacteria produce sulfur. Van Niel reasoned that the basic strategy of photosynthesis was likely to be the same wherever it was found, and he suggested that water (H_2O) and hydrogen sulfide (H_2S) serve the same biochemical role in different organisms. Oxygen, he said, must come from water, just as sulfur comes from hydrogen sulfide.

$$CO_2 + H_2S \longrightarrow CH_2O + S$$

This insight suggests that light splits water, not carbon dioxide.

Conclusive proof for this idea came more than 10 years later, when scientists could observe the fate of oxygen from water that contained the heavy isotope ^{18}O. Plants provided with normal carbon dioxide and $H_2{}^{18}O$ produce heavy oxygen, *not* heavy carbohydrate:

$$CO_2 + H_2{}^{18}O \longrightarrow {}^{18}O_2 + (CH_2O)$$

This experiment, illustrated in Figure 8-4, conclusively demonstrated that water, not CO_2, is split by sunlight. But how do plants channel light energy into this chemical reaction?

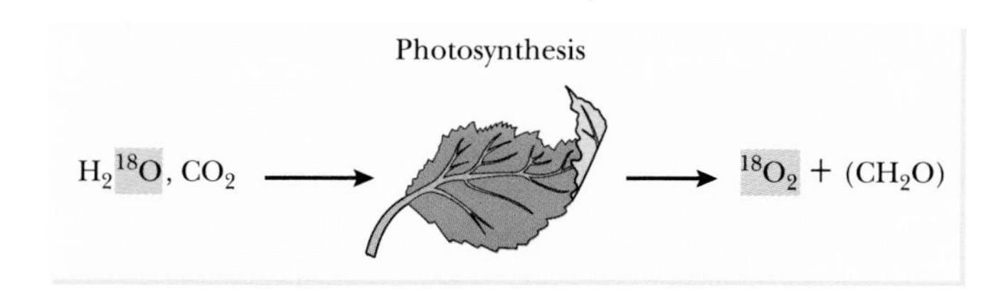

Figure 8-4 If photosynthesis split CO_2 into C and O_2, then the ^{18}O of $H_2{}^{18}O$ would have ended up in carbohydrate. Instead, the experiment with $H_2{}^{18}O$ showed that photosynthesis splits water, releasing the ^{18}O from $H_2{}^{18}O$ as free $^{18}O_2$.

HOW DO PLANTS COLLECT ENERGY FROM THE SUN?

Julius Mayer, one of the discoverers of the First Law of Thermodynamics, elegantly summarized the business of photosynthesis: "Nature has put itself the problem of how to catch in flight light streaming to the earth and to store the most elusive of all powers in rigid form."

What Is Light?

Light is a form of **electromagnetic radiation,** a form of energy transmitted through space as periodically changing electrical and magnetic forces. Light waves are oscillations of forces, rather than (like water waves) oscillations of matter. The most important wave-like property of light is its **wavelength**—the distance between the crests of two successive waves, as shown in Figure 8-5. The light of our everyday experience is a mixture of many wavelengths, as also shown in Figure 8-5.

Our eyes and brains allow us to see only those wavelengths ranging from about 400 to about 750 nm—what we call **visible light.** We see different wavelengths of visible light as different colors. For example, we see 400-nm light as violet, 500-nm as blue-green, and 600-nm as orange-red. What we see as white light consists of a mixture of many wavelengths. In fact the light we see is only a tiny part of all possible wavelengths that make up the electromagnetic spectrum, represented in Figure 8-6. Other kinds of radiation range from gamma rays (with wavelengths less than 1 nm) to radio waves (with wavelengths measured in meters or even kilometers).

Unlike all other forms of energy, light can move through empty space. It is always moving. Light stops moving only when some object absorbs it, and then it is no longer light but some other kind of energy, such as heat. The rug, curtains, and furniture of your room absorb the morning sunlight and transform light energy into heat. In photosynthesis, specialized cellular machinery absorbs light and transforms light energy into chemical energy.

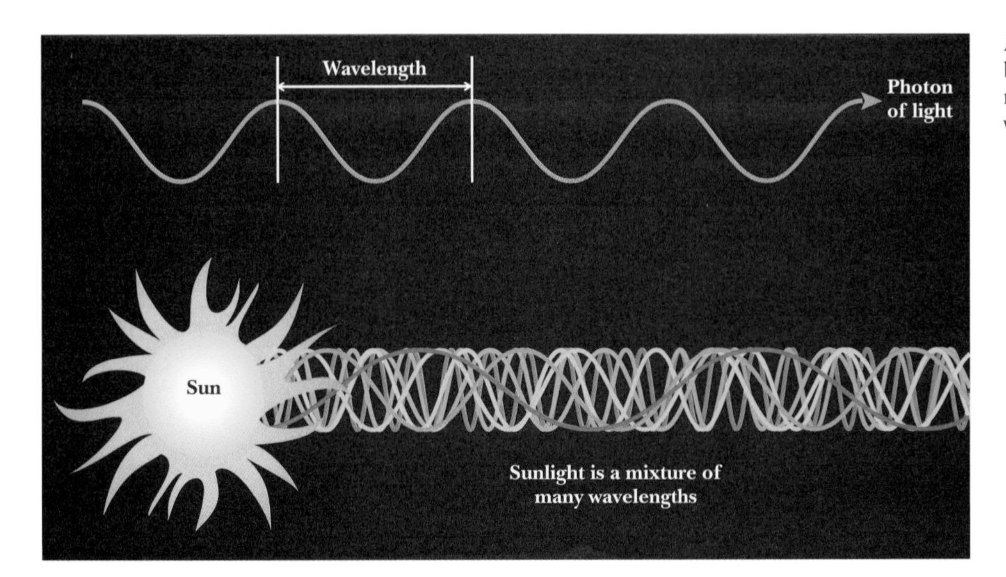

Figure 8-5 Light has properties of both particles and waves. Sunlight is a mixture of light with many wavelengths.

Ordinary rules cannot describe light, which behaves both like a particle and like a wave. Like a particle, light travels in straight lines. Like the waves in a rain puddle, light has troughs and crests that interfere with each other. Nothing in our ordinary experience, however, provides a good model for the behavior of light. We are stuck talking separately about "wave-like" and "particle-like" properties.

To understand the interaction of light and matter, we must talk of light's particle-like and wave-like properties at the same time. A light particle, or **photon,** is a package of energy. The amount of energy in each photon depends entirely on the wavelength of light. (This statement sounds strange, because we must speak of the wavelength of a particle!) Photons of shorter wavelengths (toward the blue end of the visible spectrum) have more energy than photons of wavelengths toward the longer (red) end of the visible spectrum. An atom or a molecule can absorb a photon of light if the energy of the photon matches the energy needed to boost one of its electrons from one energy level to a higher one, as shown in Figure 8-7. Brighter light contains more photons than dimmer light (for example, a 150-watt light bulb emits more photons than a 50-watt light bulb.) The brighter light can excite more atoms or molecules, but only if it has the proper wavelengths.

We see visible light because it excites electrons in specialized molecules in our eyes. In other situations, absorbing a photon can produce an electrical signal, for example, in a photographer's exposure meter or video camera. In photosynthesis, light excites electrons in particular molecules within photosynthesizing cells of plants, protists, or prokaryotes, thereby increasing the free energy of the absorbing molecules. The photosynthetic machinery captures some of this free energy and transduces it into high-energy chemical bonds.

How Much Energy Does a Photon Contain?

For each photon, a short wavelength means a high energy, and a long wavelength means lower energy. Blue light has about twice as much energy per photon as red light.

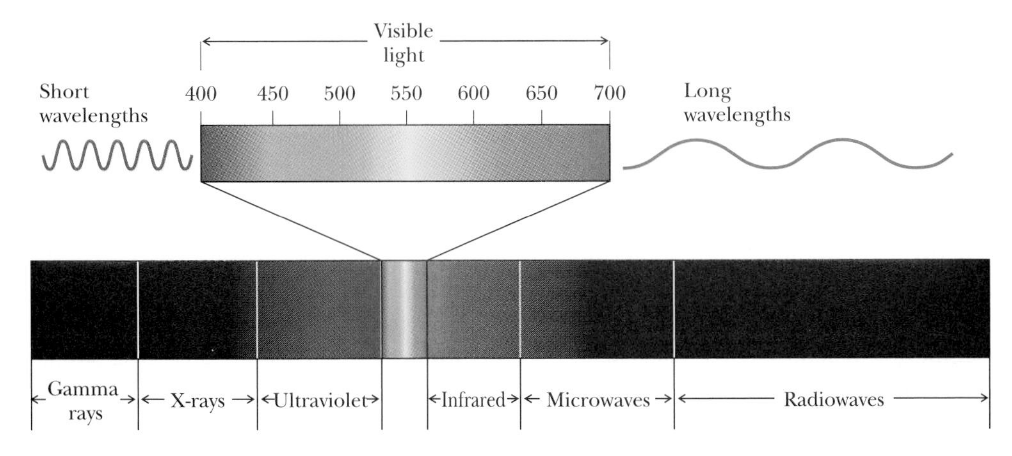

Figure 8-6 Humans can see light with wavelengths between 400 and 700 nm; different wavelengths are seen as different colors.

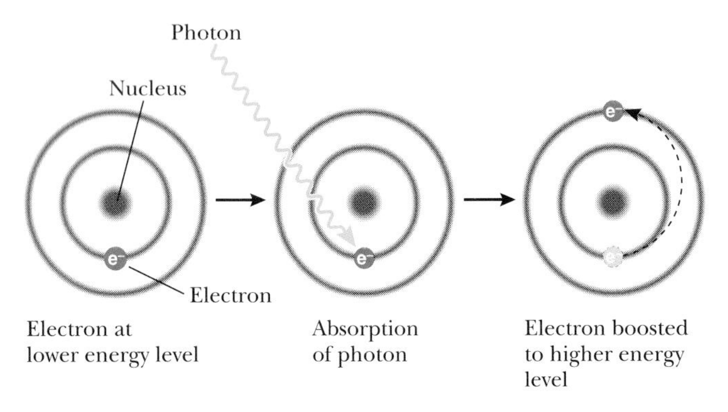

Figure 8-7 The absorption of a photon boosts the energy of an electron within an atom or molecule. But any particular atom or molecule can absorb only certain wavelengths—those corresponding to the fixed energy differences between electron orbitals.

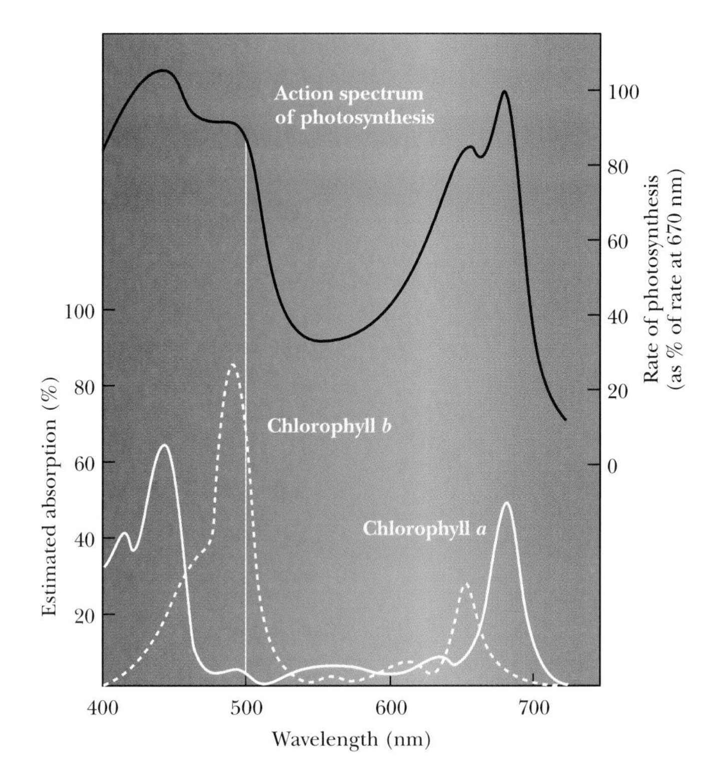

Figure 8-8 The action spectrum of photosynthesis roughly follows the absorption spectrum of chlorophyll, but not exactly.

Most of the photons that reach the earth's surface have wavelengths in the visible range. Most of the more energetic photons of ultraviolet ("beyond violet") light are absorbed by ozone in the upper atmosphere. And most of the less energetic photons of infrared ("below red") light are absorbed by water vapor and carbon dioxide in the atmosphere.

Not surprisingly, evolution has produced molecules that can make use of the energies provided by photons that best penetrate the atmosphere. Life has evolved in a way that takes fullest advantage of available energy. The energy contained in a visible photon (35–70 kcal per mole of photons) is enough to produce the chemical reactions of vision and photosynthesis.

The Relative Effectiveness of Light with Different Wavelengths Can Reveal Which Molecules Are Responsible for a Light-Dependent Process in Living Cells

Modern instruments can quickly determine the **absorption spectrum** of a solution, the relative amounts of light of different wavelengths that the solution absorbs. The wavelengths of absorbed light reveal the differences in the energy levels of electrons in the absorbing atoms or molecules. Most of a cell's molecules absorb only ultraviolet light, because the energy required to move their electrons to a higher energy level is relatively high. Some molecules, however, have smaller energy differences between their electron arrangements. These molecules are called **pigments,** molecules that absorb visible light and have color when seen by human eyes.

A solution of isolated **chlorophyll,** the green pigment of plant leaves, absorbs red and blue photons and allows photons of intermediate (green) wavelengths to pass or to be reflected. These photons excite pigment molecules in our eyes, and we perceive a green color. Similarly, the red color of hemoglobin results from the selective absorption of blue and green photons. The transmitted longer wavelength (red) light excites other pigment molecules in our eyes, and we see red.

For any light-dependent process, we can also determine an **action spectrum,** the relative effectiveness of different wavelengths in promoting a specific process. For example, we can measure the amount of oxygen produced by photosynthesis with lights of different colors. The action spectrum of photosynthesis roughly follows the absorption spectrum of chlorophyll. This correspondence suggests that the absorption of light by chlorophyll may be responsible for photosynthesis.

More careful measurements of the action spectrum of photosynthesis, however, show that the action spectrum does not exactly correspond to the absorption spectrum of chlorophyll, as shown in Figure 8-8. Light with a wavelength of 500 nm (blue-green) can cause the production of oxygen, showing that photosynthesis is occurring, even though light at a wavelength of 500 nm is not absorbed by chlorophyll. It was reasonable to look for a role for pigments that absorb 500 nm light.

These other pigments, with the expected absorbance spectrum, are indeed present in leaves. They are the **carotenoids,** yellow or orange pigments that absorb blue-green light. Carotenoids are the pigments responsible for the beautiful colors of autumn leaves. Comparing the action spectrum for photosynthesis with the absorption of purified carotenoids shows that carotenoids participate in photosynthesis. We now know that they absorb blue-green light and transfer the energy of their excited electrons to chlorophyll.

By studying how much oxygen is produced as the amount of light is increased, researchers have found that photosynthesis reaches a maximum when only a small fraction of the leaf's chlorophyll molecules are excited. In fact, most of the chlorophyll molecules do not participate directly in photosynthesis. Instead, most are associated with carotenoids to form **antenna complexes,** which trap light and transfer the energy to the much smaller number of chlorophyll molecules that actually participate in photosynthesis. These participating chlorophylls are part of another complex, called the **photochemical reaction center,** which actually converts the captured light energy to chemical energy. X-ray diffraction studies of the photochemical reaction center from the purple bacterium mentioned in Chapter 7 (pp. 204–205) have revealed the exact arrangement of chlorophyll and other pigment molecules.

Transfers of energy within the reaction center occur quickly, within 10^{-12} to 10^{-9} second. Researchers have devised powerful new techniques to study these rapid reactions. In one approach, for example, a laser provides a burst of light that lasts only 1 picosecond (10^{-12} second). A device called a *picosecond absorbance spectrometer* can detect the rapid changes in the absorption spectra of individual pigments and has revealed the path of each excited electron in the reaction center, as diagrammed in Figure 8-9.

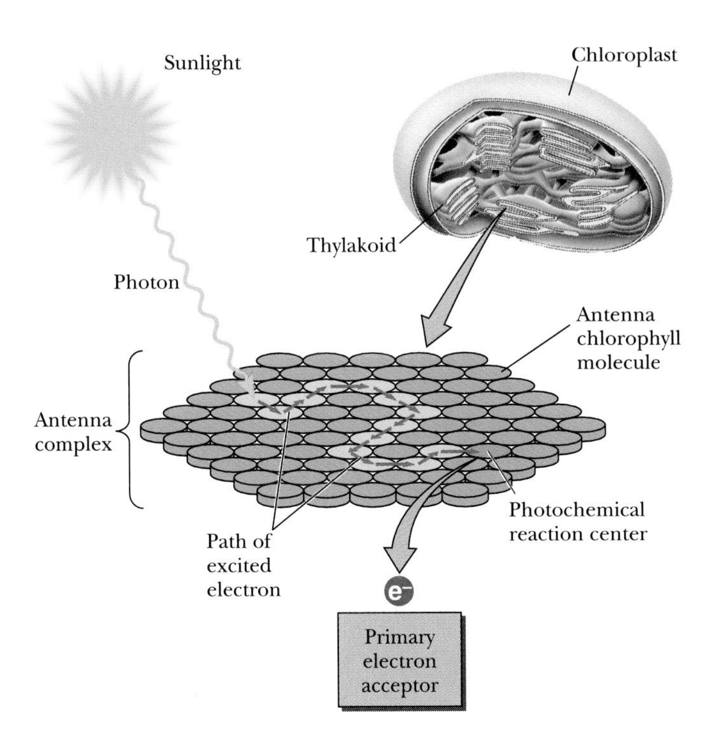

Figure 8-9 An antenna complex traps light. Its photochemical reaction center converts light energy to chemical energy.

Light Excites Electrons in Two Types of Reaction Centers

We can now describe a fairly simple model of photosynthesis: (1) Chlorophyll and carotenoids absorb light; (2) these pigments transfer the captured energy to the reaction center; (3) the energy splits oxygen from water and somehow forms chemical bonds.

If this simple model were correct, each absorbed photon would contribute equally to photosynthesis, which is the case in some prokaryotes. Photosynthesis in plants, however, appears to operate by a more complicated mechanism: Careful comparisons of the action spectrum and the absorption spectrum show that photons of some wavelengths are not effective in promoting photosynthesis, even when they are absorbed. As we can see in Figure 8-8, the efficiency drops when the wavelength of the absorbed light is longer than 680 nm. But red light (with a wavelength of 700 nm or longer) promotes rapid photosynthesis when shorter wavelength light (with a wavelength of 680 nm) also illuminates the plant. When both 680-nm and 700-nm light shine on a plant, photosynthesis is also far more efficient than with 680-nm light alone. Analysis of the differences between the action spectrum of photosynthesis and the absorption spectrum of carotenoids and chlorophyll suggests that photosynthesis in plants uses light in more than one way.

The synergistic effect of the 680-nm and 700-nm light suggests that effective photosynthesis requires two distinct but interacting sets of light-dependent reactions, which depend on light of different wavelengths. One set of reactions, called **photosystem I,** best absorbs and uses light with wavelengths of about 700 nm. The other set, called **photosystem II,** requires light with wavelengths of about 680 nm. In contrast, photosynthetic bacteria have only a single photosystem, which leads to the creation of a proton gradient across the cell membrane. The existence of the two photosystems explains why plants grown in artificial light grow better when lit with both a fluorescent bulb and an incandescent bulb, which differ slightly in their emitted wavelengths, than with either bulb alone. Later in this chapter we discuss how the two photosystems each contribute to harnessing light energy.

Absorbance of Light by the Two Photosystems Can Generate Proton Gradients and NADPH, a Strong Electron Donor

The two photosystems lie in distinct particles that can be isolated by subcellular fractionation. Both are within chloroplasts, where they are associated with the **thylakoid membrane,** which delineates stacked vesicles, as shown in Figure 8-10. Each stack is called a **granum** [plural, *grana*]. Each photosystem particle consists of chlorophyll molecules together with carotenoids, specific proteins, and

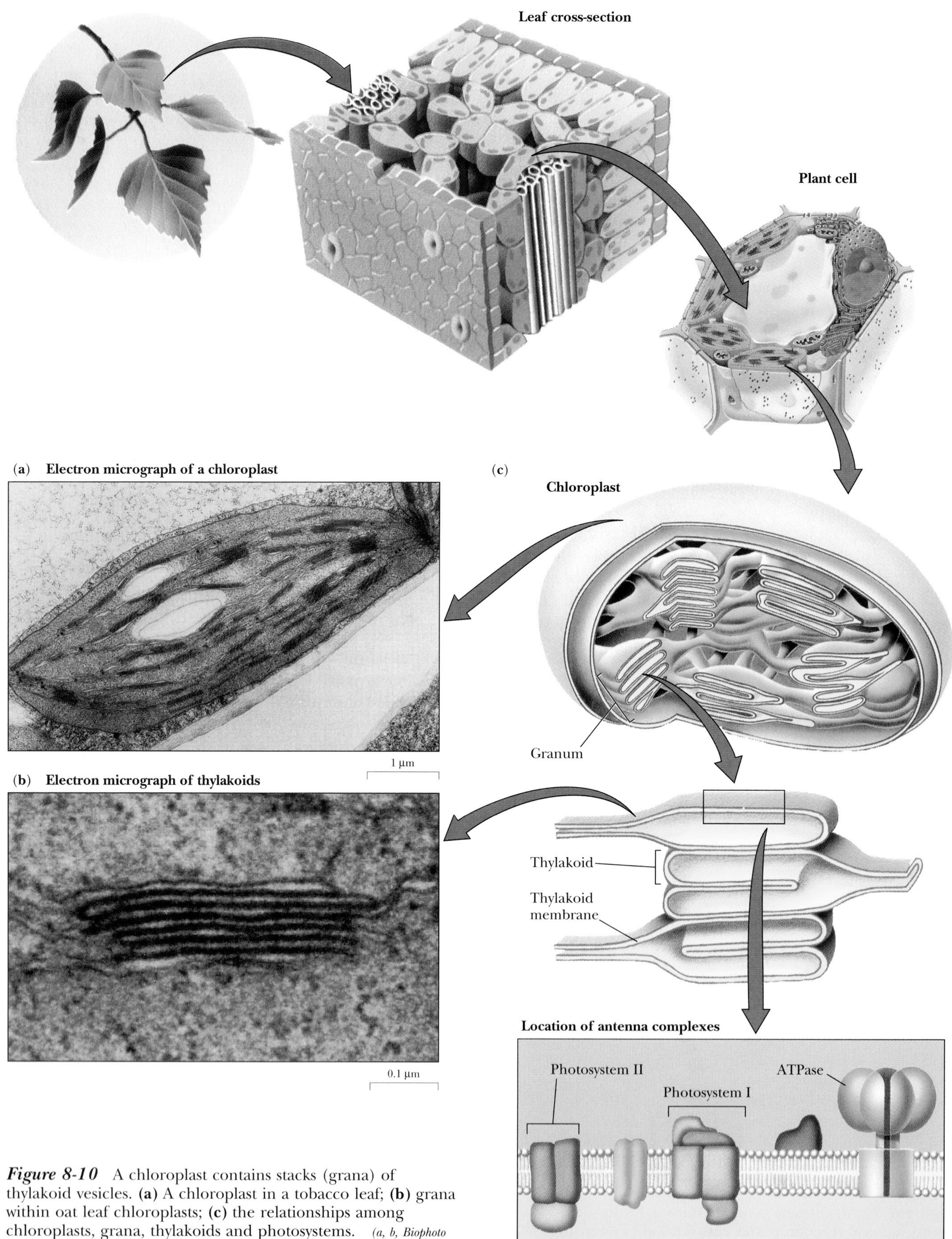

Figure 8-10 A chloroplast contains stacks (grana) of thylakoid vesicles. **(a)** A chloroplast in a tobacco leaf; **(b)** grana within oat leaf chloroplasts; **(c)** the relationships among chloroplasts, grana, thylakoids and photosystems. *(a, b, Biophoto Associates)*

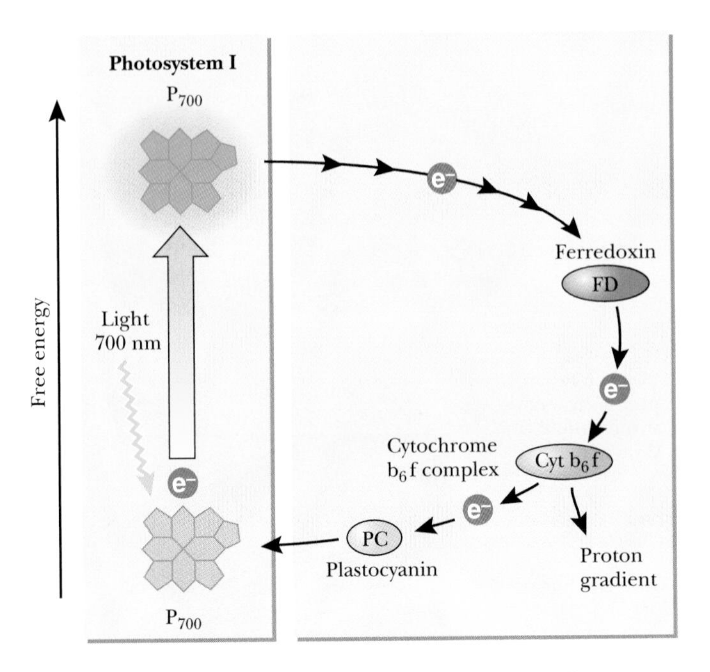

Figure 8-11 The fate of an excited P_{700} electron (in photosystem I) in cyclic photophosphorylation.

other molecules and ions. Each performs its own biochemical reactions. Both photosystems are required for plant photosynthesis.

The chlorophyll in the reaction center of photosystem I absorbs light most efficiently at 700 nm and is called P_{700} [where P stands for pigment]. When P_{700} receives energy from the antenna complex, its free energy increases and it becomes a strong reducing agent (electron donor) (Figure 8-11). The excited P_{700} readily gives up a high-energy electron to a series of iron-sulfur proteins. (See Chapter 7, p. 199.) With the transfer of its excited electron, the P_{700} loses most of the free energy it had acquired from the absorbed photon. Because it has now given up an electron, P_{700} is in its oxidized state.

The last iron-sulfur protein in the electron transport chain of photosystem I (called *ferredoxin*) can deliver its newly acquired, high-energy electron to one of two electron acceptors—a cytochrome complex that resembles the mitochondrion's cytochrome b-c complex (p. 199) or a coenzyme ($NADP^+$)closely related to the mitochondrion's NAD^+ (discussed later in this chapter).

The Excited Electrons of Photosystem I Can Move in a Circle Back to Chlorophyll and Produce ATP

One of the two electron transport routes associated with photosystem I leads the excited electron back to oxidized P_{700}. This cycle restores P_{700} to its original state and makes it ready to absorb another photon.

The electron acceptor here, as shown in Figure 8-11, is a complex called the *cytochrome b_6-f complex.* This complex then passes the electron to a copper-containing protein called *plastocyanin,* which then uses it to reduce oxidized P_{700}, completing the cycle. As in mitochondria, the passage of electrons through the cytochrome complex results in a proton gradient across a membrane.

As we noted above, and as schematically shown in Figure 8-12, the chloroplast has much in common with the mitochondrion—an outer membrane, an intermembrane space, an inner membrane, and an enclosed region, called the **stroma** [Latin = mattress], which corresponds to the mitochondrial matrix. The chloroplast has an additional compartment, however: the inside of the thylakoid, separated from the stroma by the thylakoid membrane. In chloroplasts, the electron transport chain is located in the thylakoid membrane (rather than in the inner membrane), and the chain pumps protons from the stroma into the thylakoid (rather than into the intermembrane space, as in mitochondria).

Like the inner mitochondrial membrane, the thylakoid membrane contains turbine-like ATP synthase molecules, which produce ATP as the protons flow back down their gradient out of the thylakoid lumen into the stroma. The chloroplast ATP synthase lies on the stroma side of the membrane, just as mitochondrial ATP synthase lies on the matrix side of the cristae. ATP is made in the stroma, just where it will be used for glucose synthesis.

The net effect of the electron route just described is the production of ATP from the energy of absorbed light, with no other net changes in the chloroplast or cell. This pathway is therefore called **cyclic photophosphorylation.**

The Excited Electrons of Photosystem I Can Also Produce the Strong Reducing Agent, NADPH

The alternate electron path in photosystem I, shown in Figure 8-13, does not produce ATP. Instead, the energy of the excited electron of P_{700} is used to make a strong reducing agent—NADPH (*n*icotinamide *a*denine *d*inucleotide *p*hosphate), a derivative of NADH (Figure 8-14). (The oxidized form of NADPH is called $NADP^+$.) NADPH then contributes directly to the production of glucose.

NADPH and $NADP^+$ are used in the building of high-energy molecules such as glucose (anabolism), whereas NADH and NAD^+ are used in the pathways that harvest energy, such as glycolysis and respiration (catabolism). By using NADPH and NADH in different pathways, cells can separately regulate anabolism and catabolism. This independence would be harder to achieve if the breakdown and the synthesis were just the reverse of one another.

The noncyclic route of photosystem I supplies most of the reducing power for sugar synthesis. But after it has given up its excited electron to form NADPH, P_{700} needs another electron before it can again serve as an electron

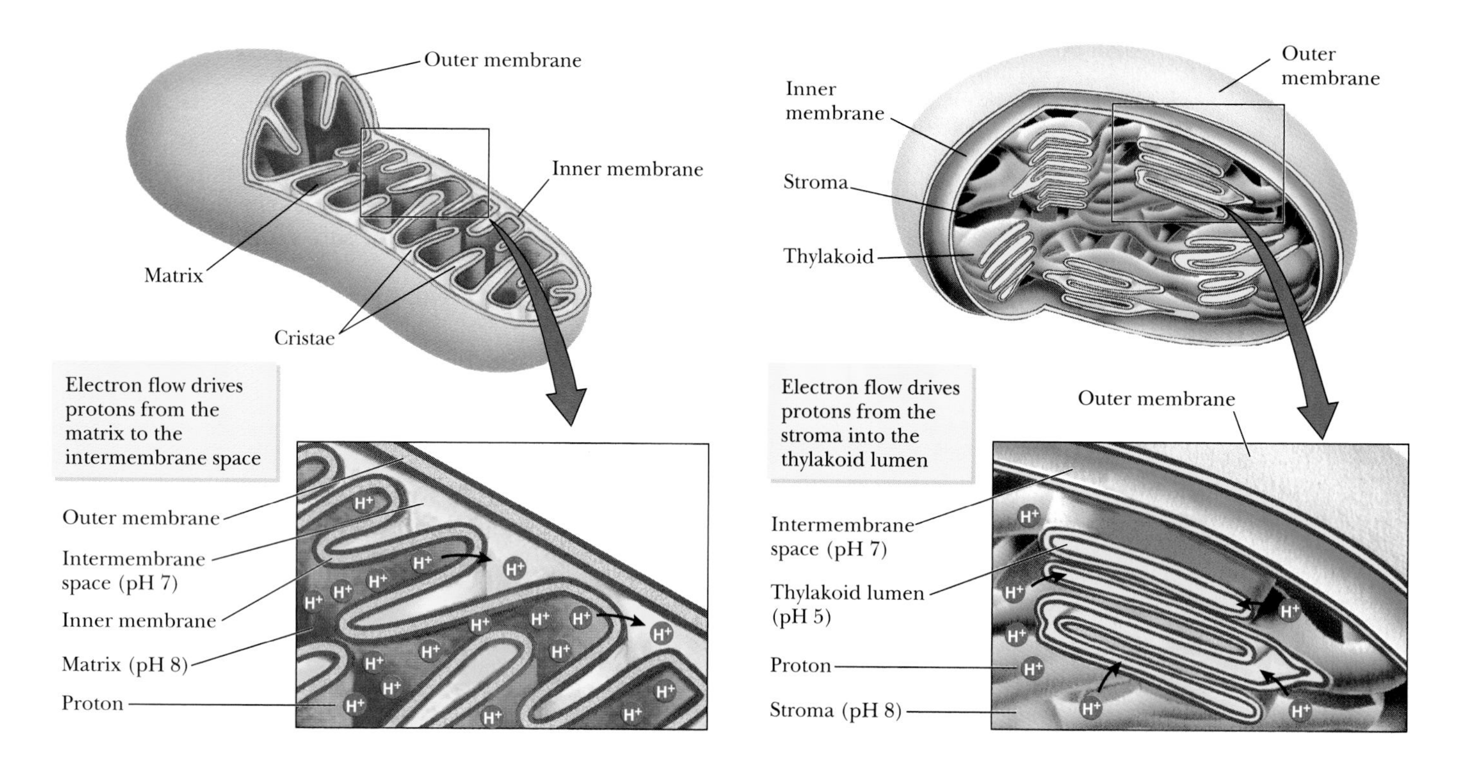

Figure 8-12 Comparison of proton movements and ATP generation in a mitochondrion and a chloroplast.

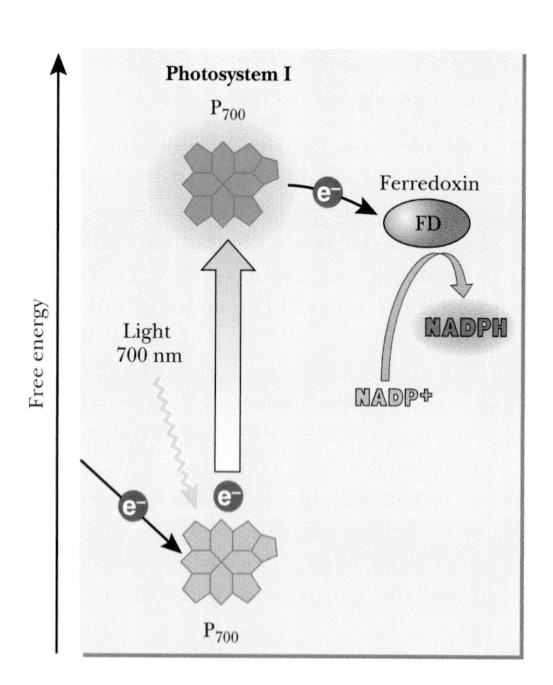

Figure 8-13 The fate of an excited P_{700} electron (in photosystem I) in noncyclic electron flow—the production of NADPH.

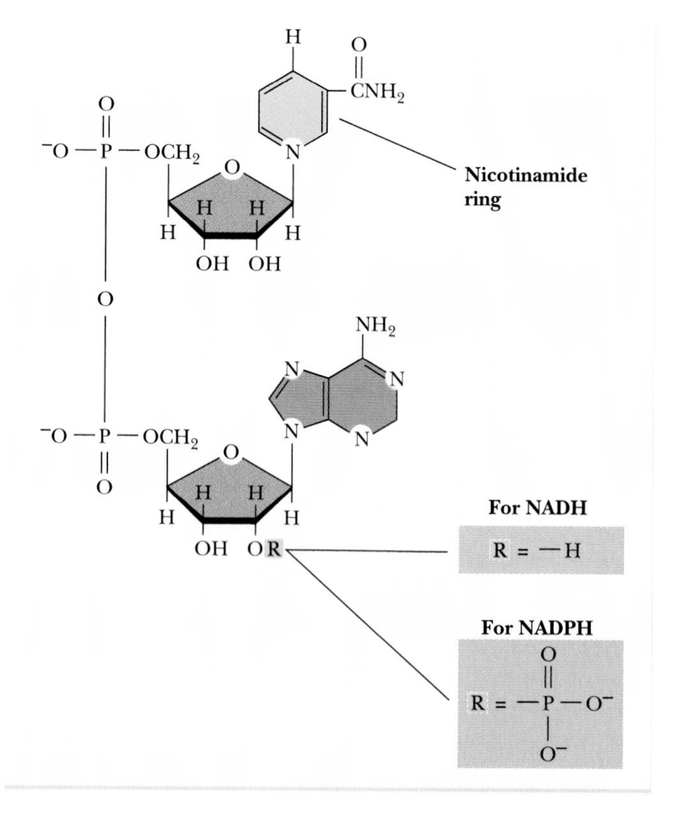

Figure 8-14 The structure of NADH and NADPH.

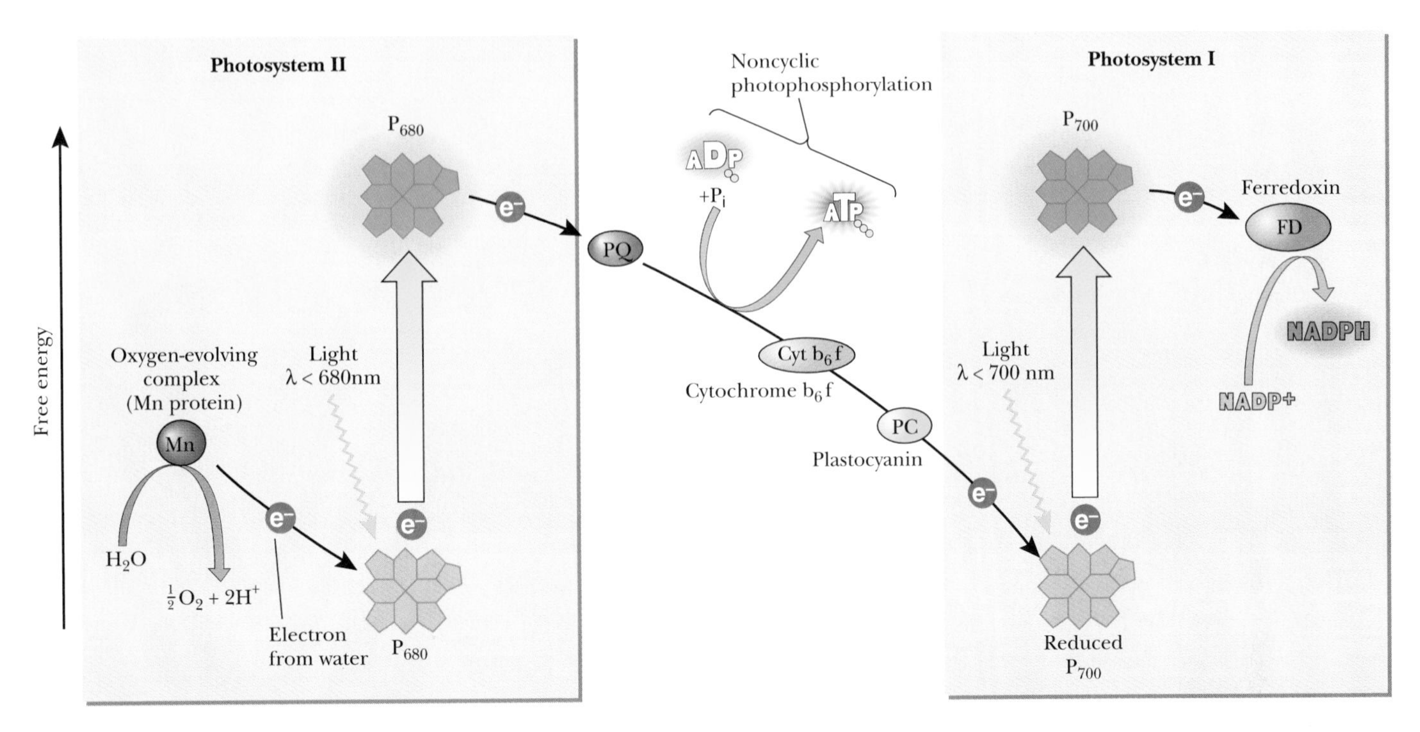

Figure 8-15 The fate of an excited P_{680} electron (in photosystem II). The excited P_{680} electron reduces oxidized P_{700} (in photosystem I). The oxygen evolving complex regenerates the reduced form of P_{680}.

donor. That is, some reducing agent must contribute electrons to oxidized P_{700}.

Photosystem II Provides Electrons to Reduce P_{700}

The chlorophyll molecules associated with the reaction center of photosystem II absorb light best at 680 nm and are called P_{680}. Like excited P_{700}, excited P_{680} is a powerful reducing agent that can transfer a high-energy electron to an electron acceptor, as shown in Figure 8-15. Electrons move from excited P_{680}, through the electron transport chain of photosystem II, to oxidized P_{700} (produced by photosystem I). By regenerating reduced P_{700}, the operation of photosystem II permits photosystem I to keep going even when there is no cyclic flow of electrons.

Photosystem II also produces ATP as protons flow down the gradient generated by its electron transport chain. This ATP production is called **noncyclic photophosphorylation,** to distinguish it from the ATP production in the cyclic path of photosystem I.

But we have the same question all over again: If the electron from excited P_{680} is transferred to oxidized P_{700}, how is the reduced form of P_{680} regenerated? The answer is that electrons come from water and result in the production of oxygen, as shown in Figure 8-15, using a protein called the **oxygen evolving complex.** This protein, which is unusual in containing manganese ions, coordinates the transfer of four electrons needed to oxidize a single molecule of water to oxygen. Oxidized P_{680} is a powerful electron acceptor—more powerful even than oxygen itself, so it is able to pull electrons away from the oxygen in water and split the water molecule. Photosystem II produces the oxygen released by photosynthesis.

The Light-Dependent Reactions of Photosynthesis Transform Light Energy into the Chemical Bonds of NADPH and ATP

Figure 8-16 shows how photosystems I and II together make up the light-dependent reactions of photosynthesis. The zig-zag path of the electron, which is apparent in Figure 8-15, is often described as the "Z-scheme of photosynthesis." Figure 8-16 summarizes the pathway of electron flow in the context of the physical make-up of the thylakoid membrane. The results of the absorption of light by the two photosystems are (1) the splitting of water into oxygen and hydrogen ions; (2) the production of ATP; and (3) the production of NADPH. Oxygen is essentially a waste product and is released into the atmosphere. ATP and NADPH remain mostly in the chloroplast stroma, where they contribute to the light-independent reactions of photosynthesis—the synthesis of glucose. The light-independent reactions used to be called the "dark reaction," but this term is no longer used because the reactions occur in the light as well as in the dark.

How efficient is this process, which Albert Szent-Gyorgyi characterized as "a little electric current driven by

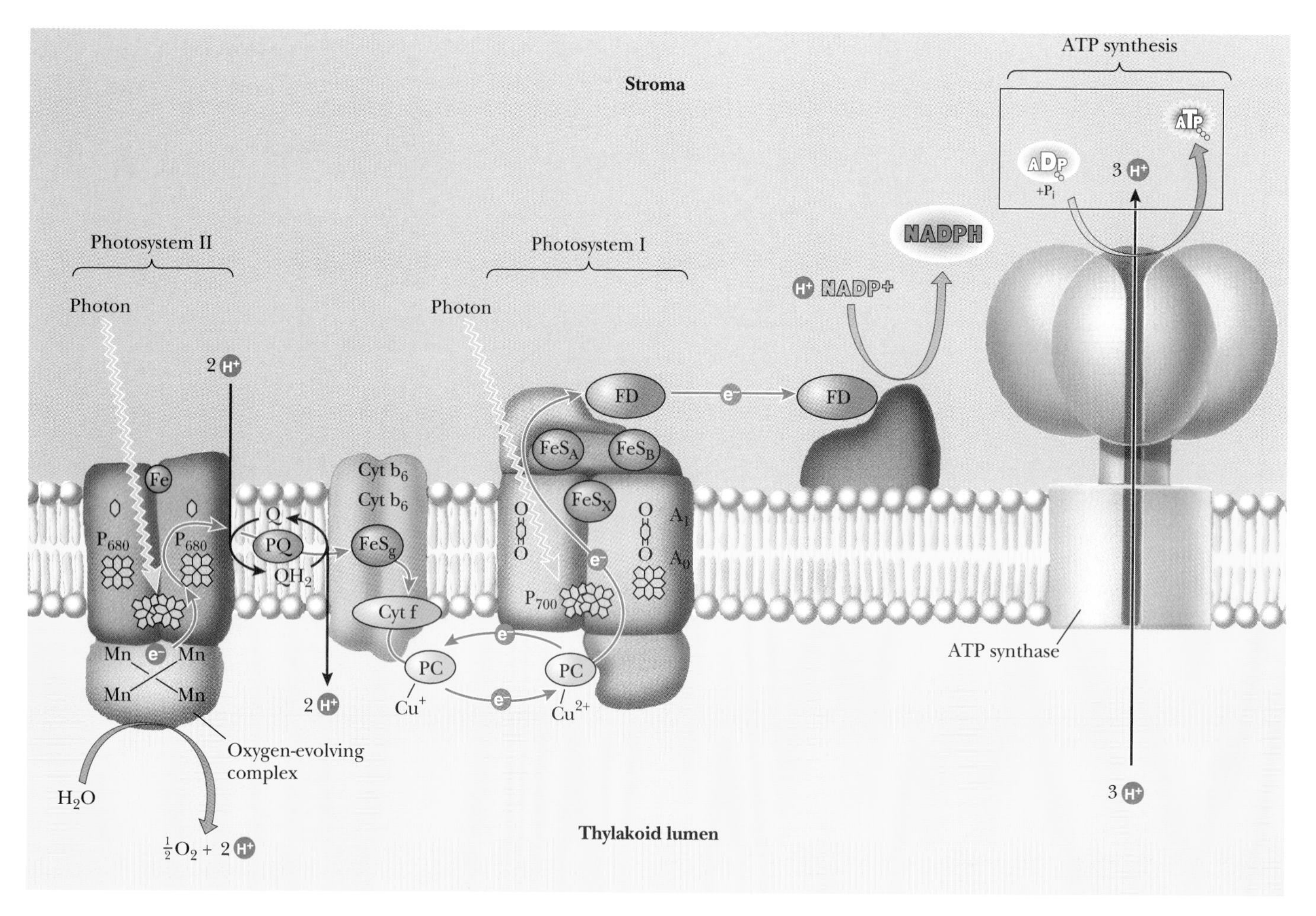

Figure 8-16 Summary of electron flow from photosystem II to photosystem I, in the context of the thylakoid membrane. The proton gradient generated by the electron flow from photosystem II to photosystem I drives ATP synthesis by the ATP synthase complex shown on the right. Cyclic photophosphorylation by photosystem I can also generate a proton gradient and ATP.

the sunshine"? The reduction of each molecule of $NADP^+$ to NADPH requires two electrons. To provide each excited electron for this electron path, photosystems I and II must each absorb a photon. To provide the two electrons, then, requires the absorption of four photons, two by each photosystem. The production of a molecule of NADPH therefore requires four photons of light. As this pair of electrons flows from photosystem II to photosystem I, it pumps enough protons to produce about one molecule of ATP. So, in noncyclic photophosphorylation, the energy of four photons is transformed into the chemical energy of one molecule of NADPH and one molecule of ATP.

The energy stored in the chemical bonds of NADPH and ATP is quickly stored in glucose. As we see later in this chapter, however, the synthesis of each molecule of glucose requires 18 molecules of ATP and 12 molecules of NADPH, that is, 1.5 molecules of ATP for each molecule of NADPH. But noncyclic photophosphorylation gives an ATP/NADPH ratio of 1. The additional ATP comes from cyclic photophosphorylation. Cells change the ratios of ATP to NADPH production by varying the relative rates of noncyclic and cyclic photophosphorylation.

Because glucose is uncharged and highly soluble, it can be more easily stored in the form of starch, which can often form an insoluble deposit in chloroplasts. Alternatively, the cell may convert the glucose into sucrose, a solution of which moves throughout the plant.

HOW DO PLANTS MAKE GLUCOSE?

Glucose (and the molecules made on the way to making glucose) provide plants with their most important sources of energy and building blocks that contain carbon, hydrogen, and oxygen. (As noted before, plants must obtain their nitrogen, phosphorus, sulfur, and other atoms from the soil.) We already know that the carbon, hydrogen, and oxygen atoms for glucose come from the carbon dioxide

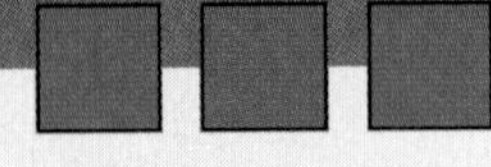

ESSAY

HOW DO HERBICIDES KILL PLANTS?

Herbicides are chemicals that kill plants but not animals. In most cases, the ability of a particular chemical to act as an herbicide was known long before its site of chemical action, but subsequent research has, in many cases, revealed exactly how herbicides act.

Some herbicides selectively kill different types of plants. For example, 2,4-D—a synthetic compound that mimics the plant hormone auxin and disrupts plant development—is more effective in killing C3 plants than C4 plants, allowing farmers to kill C3 weeds while preserving their corn. (See p. 244.) Most herbicides, however, do not distinguish among plants but between plants and animals. Not surprisingly, then, most herbicides act by interfering with processes that occur within chloroplasts.

Some herbicides act on metabolic pathways that are unique to plants. The widely used herbicide glyphosate (Roundup®), for example, interferes with the synthesis of aromatic amino acids (phenylalanine, tyrosine, and tryptophan). Glyphosate does not affect aromatic amino acids in animals because the synthesizing pathways are completely different from those in plants.

A number of herbicides, whose structures are shown in Figure 1, act by interfering directly with photosynthetic electron transport. Atrazine and diuron [also called DCMU or 3-(3,4-dichlorophenyl)-1,-1-dimethylurea] both interfere with electron transport in photosystem II. Both of these compounds prevent the transfer of electrons from excited P_{680} to the cytochrome b_6-f complex. The site of interference is a thylakoid protein, called QB, which is involved in transferring electrons from a quinone (Q) molecule within the P_{680} reaction center to a molecule called plastoquinone (PQ), which then carries electrons to the cytochrome b_6-f complex, as illustrated in Figure 2. Atrazine and diuron bind to QB and prevent electron transfer.

Corn farmers have used atrazine and related compounds heavily, since corn contains enzymes that makes them tolerant. In contrast, diuron is as toxic to corn as it is to weeds. But in the last decades, many species of weeds have developed tolerance to atrazine. Researchers have been able to identify the precise cause of this resistance. Atrazine-resistant weeds have an altered QB protein. In nonresistant weeds, QB binds both to quinone and to atrazine, but in the resistant weeds QB binds to quinone but not to atrazine. In one case, the mutant QB protein, which consists of about 300 amino acid residues, has a single amino acid change, which prevents atrazine binding and thereby makes the weeds unresponsive to the herbicide.

Another widely used photosynthetic inhibitor is paraquat, whose structure is also shown in Figure 1. Paraquat and related compounds accept electrons from photosystem I and transfer them directly to O_2 to produce free radicals such as the superoxide ion (O_2^-) and the hydroxy ion (OH^+). These radicals are highly reactive. They rapidly attack thylakoid membranes, destroying their ability to sustain photophos-

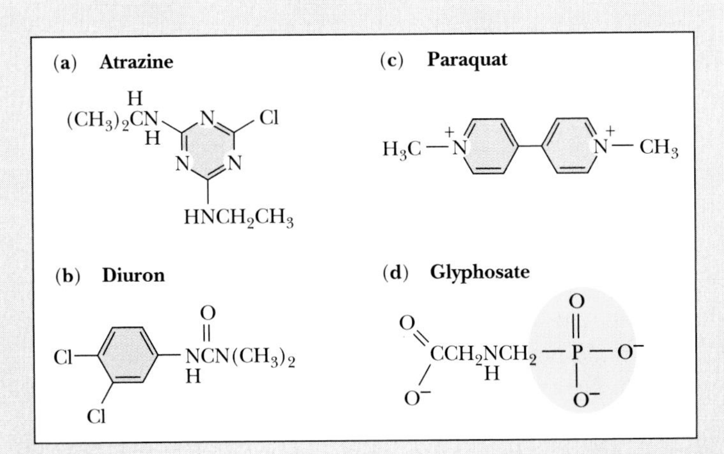

Figure 1 Structures of some herbicides: **(a)** atrazine, which blocks the transfer of electrons in photosystem II: **(b)** diuron, which blocks electron transfer at the same point as atrazine; **(c)** paraquat, which transfers electrons from photosystem I to O_2 molecules, creating superoxide and hydroxy radicals; **(d)** glyphosate (Roundup®), which blocks the synthesis of aromatic amino acids in plants but not in animals.

(continued)

ESSAY *(continued)*

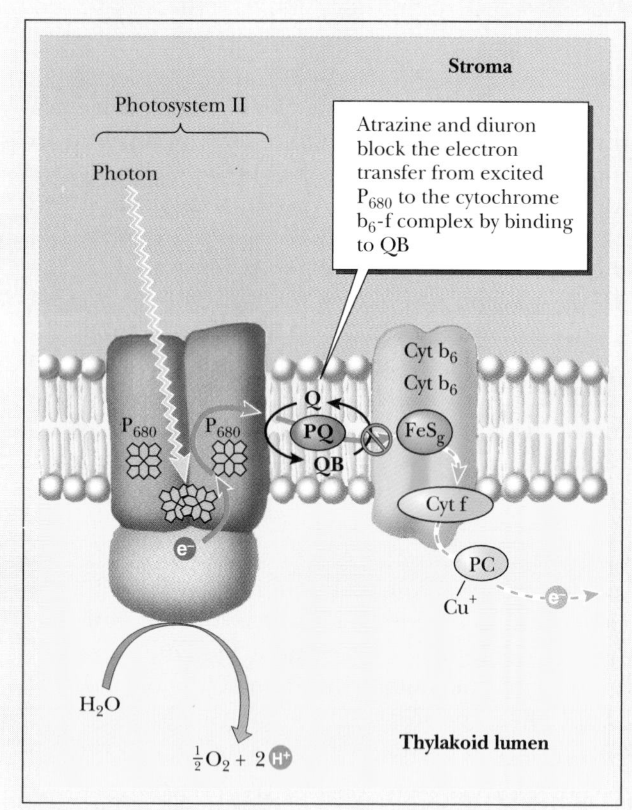

Figure 2 Details of electron transfer in photosystem II. Excitation of the P_{680} reaction center leads to the transfer of electrons from P_{680} to quinone (Q). These electrons are transferred to plastoquinone (PQ), which carries them to the cytochrome b_6-f complex. Atrazine and diuron block the transfer from Q to PQ.

phorylation and leading to plant death. Because paraquat serves as an electron shuttle—taking electrons from photosystem I and transferring them to O_2—it continues to do damage for as long as photosystem I is operating. It is particularly effective at low doses. Unfortunately, however, paraquat is also toxic to humans and must be used with caution.

Herbicides have, in many cases, been a great boon to agriculture. But experience with atrazine and paraquat has suggested two notes of caution: (1) weeds can evolve resistance to herbicides; and (2) even when chemicals act selectively on plants, they can also affect humans and other animals.

A number of herbicide-producing companies are now using genetic engineering techniques to produce herbicide-resistant crop plants, starting with genes that confer herbicide resistance, such as that responsible for the atrazine-resistant QB protein. (See Chapter 17.) These companies argue that such herbicide-resistant crops will allow the selective destruction of weeds by application of a particular herbicide, so farmers will not have to cultivate their crops at all. Most environmentalists, however, are highly critical of this strategy, since it will ultimately increase herbicide use. The wide use of herbicides allows them to enter the food chain and to be much more concentrated in human tissues than in the plants to which they were originally applied.

in the atmosphere and the water in the soil. From our discussion of the light-dependent reactions, we know that the carbon atoms of carbon dioxide are reduced by electrons that ultimately come from water. This reduction depends on the transfer of electrons from water to P_{680}, their excitation by light, and their transfer to NADPH. Because the overall reaction of photosynthesis is the reverse of the overall reaction of respiration, we know from Chapter 7 that the synthesis of glucose from carbon dioxide and water must be highly endergonic:

$$6\,CO_2 + 6\,H_2O \longrightarrow C_6H_{12}O_6 + 6\,O_2$$
$$\Delta G^{0\prime} = +686 \text{ kcal/mole}$$

Just as the individual reactions in the breakdown of glucose divide the huge downhill process into steps whose energy can be harnessed, so the individual reactions of glucose synthesis allow the energy of light to be harnessed a little bit at a time. The energy "hill" is climbed by a set of small steps.

What Happens to Carbon Dioxide in Photosynthesis?

The discovery in 1940 of the radioactive isotope ^{14}C provided a means to answer this question. Martin Kamen and Samuel Ruben, then working at the University of California's Berkeley campus, immediately realized that $^{14}CO_2$ could be used to investigate the chemical transformations of the light-independent photosynthetic reaction. In 1945, Melvin Calvin and Andrew Benson, also at Berkeley, began the crucial experiments, shown in Figure 8-17a, which represented one of the first important contributions that radioisotopes made to biological research.

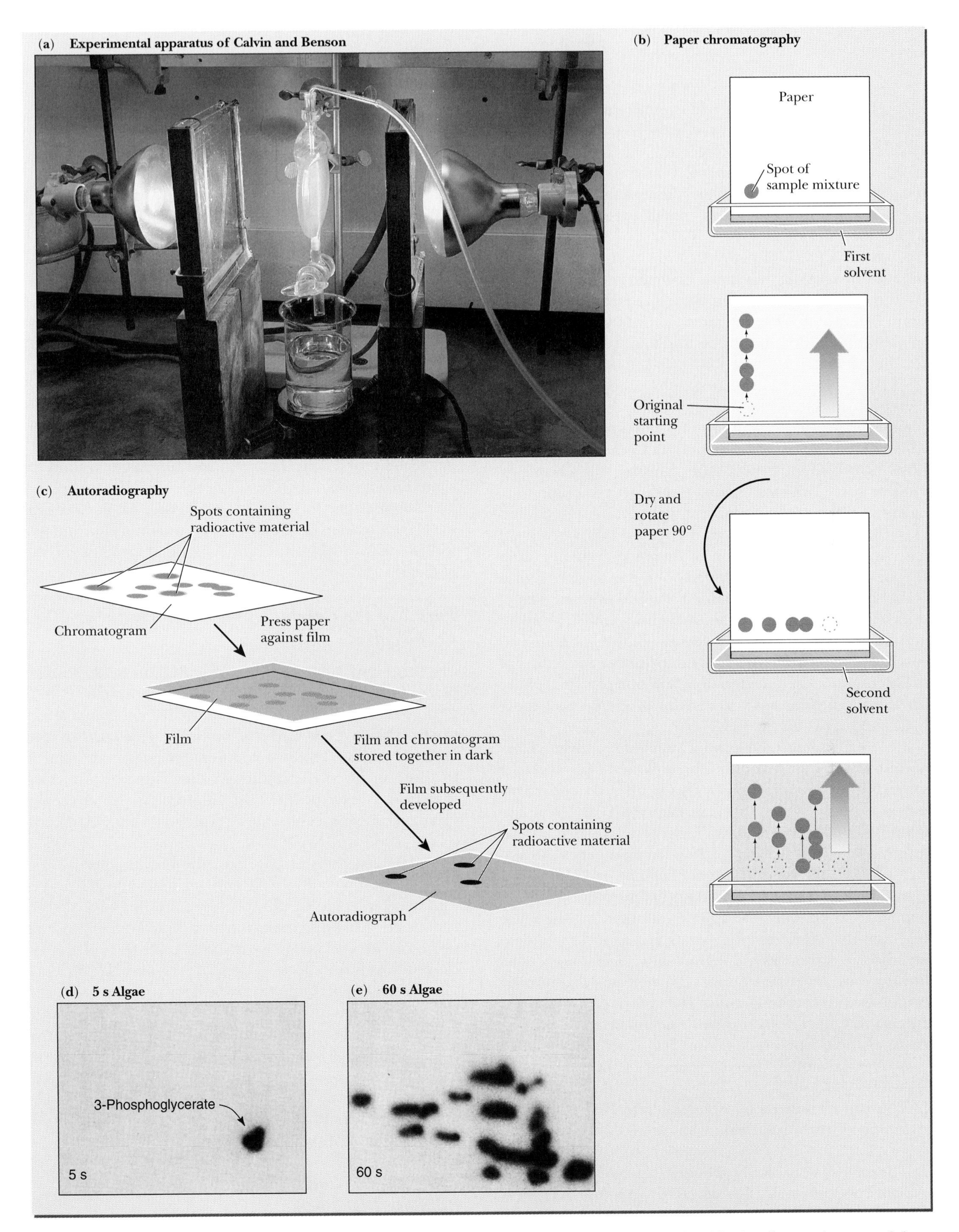

Figure 8-17 Analysis of the products of photosynthesis, as performed by Calvin and Benson. **(a)** $^{14}CO_2$ is added to algae growing in bright light; the reactions are stopped and the algae extracted after different times; **(b)** compounds in the extracts are separated from one another by two-dimensional paper chromatography; **(c)** the radioactive products are detected by autoradiography; **(d)** after 5 seconds, most of the radioactivity is in a single spot, corresponding to 3-phosphoglycerate; **(e)** at later times, the pattern becomes more complicated, reflecting the subsequent conversion of 3-phosphoglycerate to other compounds. *(a, Melvin Calvin, U. C. Berkeley; d, e, Dr. J. A. Bassam)*

Calvin and Benson injected a small amount of $^{14}CO_2$ into a flask of algae that were actively growing in light. Quickly, sometimes within 5 seconds, they killed the algae by adding alcohol and then determined the fate of the radioactive carbon atoms. They next analyzed the compounds made by the algae by paper chromatography, a technique (shown in Figure 8-17b) that separates compounds on a piece of paper that has been dipped into a mixture of solvents that move gradually along the paper. As the solvents move over the paper, individual compounds move at different rates, according to their solubilities in the different solvents and their relative affinities for the paper. (See Chapter 3, p. 70.)

After paper chromatography, the Berkeley researchers detected the radioactive compounds by a technique called **autoradiography,** illustrated in Figure 8-17c, which detects radioactive compounds by their ability to expose a photographic film. After pressing the chromatographic paper, with its separated radioactive compounds, against a photographic film, the film became black wherever ^{14}C was present, as seen in Figure 8-17c.

In his later description of these experiments (after the work had been recognized with the awarding of a Nobel Prize), Calvin commented that the primary data of these experiments was contained "in the number, position, and intensity—that is, radioactivity—of the blackened areas. The paper ordinarily does not print out the names of these compounds, unfortunately, and our principal chore for the succeeding ten years was to properly label those blackened areas on the film."

The first detectable product was a three-carbon compound. When Calvin and Benson collected algae within 5 seconds after the injection of $^{14}CO_2$, all the radioactivity was in a single spot, as seen in Figure 8-17d. They later identified that spot as 3-phosphoglycerate, which is also an intermediate in glycolysis (p. 209). At later times the pattern of spots became much more complicated (as seen in Figure 8-17e), and it took a while to figure out just what happened when. Calvin and Benson worked out the set of reactions that converted the carbon atoms of carbon dioxide into the carbon atoms of glucose. This set of reactions forms a cycle which regenerates the molecule that initially combines with carbon dioxide. The cycle is called CO_2 fixation; the Calvin-Benson cycle, after its principal discoverers; or the Carbon Reduction Cycle (CRC).

The Calvin-Benson cycle consists of three parts:

1. The production of phosphoglycerate from the combining of carbon dioxide with an acceptor molecule.

2. The conversion of phosphoglycerate into glyceraldehyde phosphate in two reactions that use ATP and NADPH produced in the light reactions.

3. The regeneration of the initial carbon dioxide acceptor.

In Part 1 of the Calvin-Benson Cycle, Carbon Dioxide Combines with a Five-Carbon Molecule and Forms Two Three-Carbon Molecules

The first reaction of the Calvin-Benson cycle, shown in Figure 8-18, is the production of phosphoglycerate (PG). The acceptor of carbon dioxide in this reaction is a five-carbon compound called ribulose bisphosphate (RuBP). Together they form an unstable six-carbon compound, which immediately breaks down into two molecules of phosphoglycerate.

The enzyme catalyzing this reaction, ribulose bisphosphate carboxylase (Rubisco) shown in Figure 8-19, has a challenging task—it is almost solely responsible for the recapture of carbon dioxide into the biosphere. The enzyme performs this job rather slowly, processing only about three molecules per second (compared with 1000 per second for a typical enzyme and 10,000 per second for some champions). As a result of this sluggishness, plants typically produce lots of this carboxylase, so that Rubisco may be more than 50% of the total protein in chloroplasts. Rubisco is probably the most abundant protein on the planet.

Part 2 of the Calvin-Benson Cycle Produces Carbohydrate

The next two steps of the Calvin-Benson cycle convert phosphoglycerate into glyceraldehyde phosphate (G3P), a three-carbon sugar. Notice that the overall composition of glyceraldehyde is $C_3H_6O_3$, or $(CH_2O)_3$, so it is indeed a carbohydrate. The two reactions of part 2 are similar to those in glycolysis (although they run in the reverse directions). (See p. 209.) In chloroplasts, however, the electron-carrying coenzyme is not NAD^+ but $NADP^+$, in its reduced form NADPH.

The reactions of part 2 of the Calvin-Benson cycle require one molecule of ATP and one molecule of NADPH for each molecule of phosphoglycerate. Because the reactions involving one molecule of carbon dioxide produce two molecules of phosphoglycerate, each turn of the cycle requires two molecules each of ATP and NADPH.

A portion of the glyceraldehyde phosphate produced by the continuous activity of the cycle enters the cytoplasm, where it serves as a precursor of glucose by using reactions that are essentially the reverse of those of the first part of glycolysis. It also serves as a precursor of fats and amino acids. Some of the glyceraldehyde phosphate remains in the chloroplast, where it regenerates ribulose bisphosphate, the task of part 3 of the Calvin-Benson cycle.

Part 3 of the Calvin-Benson Cycle Regenerates Ribulose Bisphosphate

The regeneration of the acceptor for carbon dioxide is the messy part of the Calvin-Benson cycle, involving seven dif-

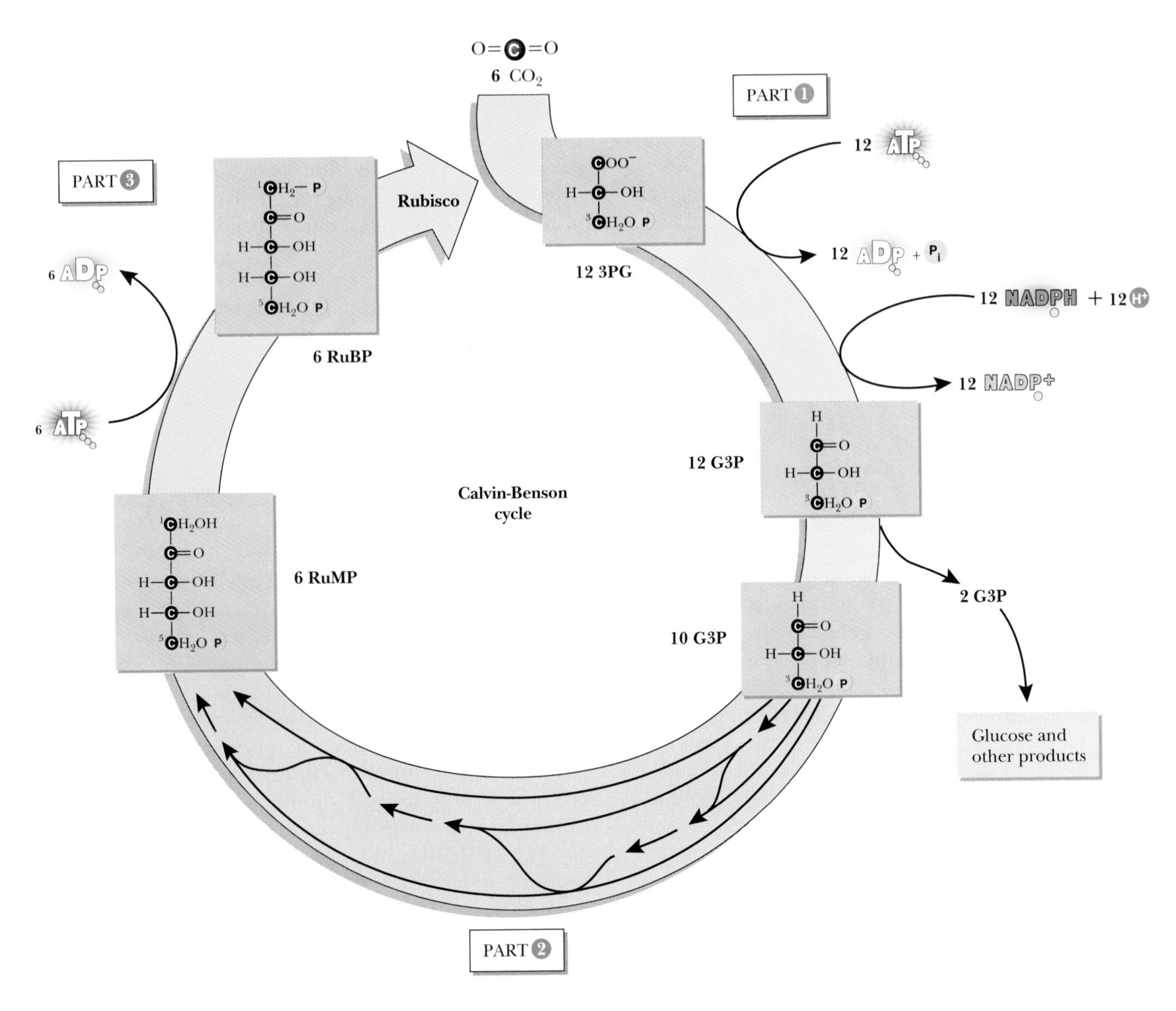

Figure 8-18 The Calvin-Benson cycle. In part 1, CO_2 combines with RuBP to produce 3PG, an acid; in part 2, G3P is converted to a carbohydrate; part 3 regenerates more RuBP for the next turn of the cycle.

ferent enzymatic steps (which are beyond the scope of this book). The end result is the production of the five-carbon ribulose bisphosphate from the three-carbon glyceraldehyde phosphate. These reactions move atoms about in a series of isomerization reactions. The last step of part 3 is the phosphate transfer, in which ATP contributes a second phosphate group to form ribulose bisphosphate.

How Much ATP and NADPH Are Required to Make a Molecule of Glucose?

Each molecule of glucose requires that six molecules of carbon dioxide enter the Calvin-Benson cycle. As the cycle turns six times, it produces 12 molecules of three-carbon glyceraldehyde phosphate. Of these, two form glu-

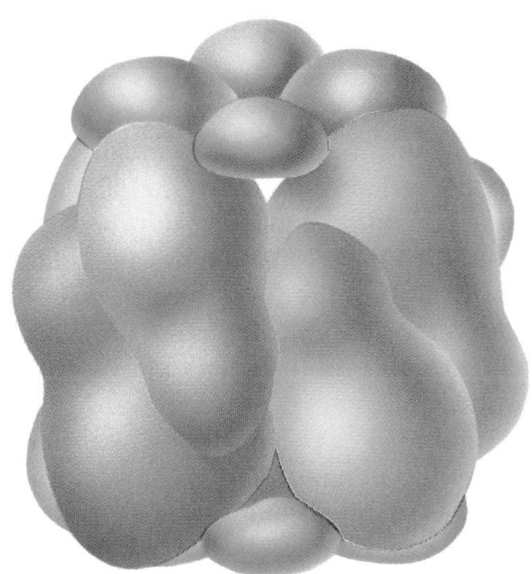

Figure 8-19 Ribulose bisphosphate carboxylase (Rubisco), the most abundant enzyme in the biosphere, converts RuBP and CO_2 into 3PG.

cose, and the remaining ten (which contain 30 carbon atoms) regenerate six molecules of five-carbon ribulose bisphosphate (again containing 30 carbon atoms).

Each turn of the Calvin-Benson cycle requires one molecule of ribulose bisphosphate and two molecules each of ATP and NADPH. Each turn produces two molecules of glyceraldehyde phosphate. Six turns, then, use 12 molecules each of ATP and NADPH.

In addition, because the production of each ribulose bisphosphate molecule requires a molecule of ATP, the production of six molecules requires six ATPs. The total cost of making a molecule of glucose from 6 molecules of carbon dioxide is 18 molecules of ATP and 12 molecules of NADPH, or 3 molecules of ATP and 2 of NADPH for each carbon atom fixed.

The energy driving the Calvin-Benson cycle comes from the high-energy bonds of NADPH and ATP, which formed during the light-dependent reactions. The 12 molecules of NADPH require the absorption of 48 photons of light by photosystems I and II. The same 48 photons also provide energy for 12 of the needed molecules of ATP. The additional six ATP molecules generated by cyclic photophosphorylation require another 24 photons (approximately).

When the light used has a wavelength of 600 nm, one mole of photons carries about 48 kcal of energy, 72 moles of photons contain 3500 kcal (72 moles $\times$ 48 kcal/mole). The free energy change for the production of a molecule of glucose from carbon dioxide and water is +686 kcal/mole. So the overall efficiency of photosynthesis is about 686/3500 = 20%.

WHAT DETERMINES THE PRODUCTIVITY OF PHOTOSYNTHESIS?

The factors that influence plant productivity are not only scientifically interesting, but are also crucially important to feeding our entire planet. We have already seen that the wavelength of light influences the efficiency of photosynthesis: Light that cannot be absorbed cannot contribute to photosynthesis. The intensity of light is also important. Plants do not produce as much carbohydrate in the shade as in the light. The length of the day, the length of the growing season, and angle at which sunlight enters the atmosphere all influence how much light energy is available to an individual plant or to an entire ecosystem.

Other environmental conditions influence the amount of light reaching a leaf as well. In London, for example, pollution and fog reduce the total sunlight reaching the ground by half. In a tropical rain forest, as many as 15 or 20 layers of vegetation may absorb more than 95% of the light before it reaches the ground.

Photosynthesis requires water and carbon dioxide as well as light. The roots of land plants take water from the soil and transport it to the leaves. Some of this water is used in photosynthesis, but most of the water evaporates into the surrounding air, passing through the stomata. (See p. 227.) This continuous movement of water through the plant provides both a means of transporting nutrients from the soil to other parts of the plant and a way of regulating the plant's temperature through evaporation.

The stomata serve not only as escape routes for water vapor, but also as exits for oxygen and entrances for carbon dioxide. These multiple functions of the stomata create a problem, however, when the weather is hot and dry: The stomata close to save water, but this reduces the amount of carbon dioxide available for photosynthesis.

Low carbon dioxide concentrations also present a problem when plants grow close together and deplete atmospheric carbon dioxide. Many greenhouse operators, for example, must supply additional carbon dioxide from gas cylinders to replenish the enclosed air. The problem of low carbon dioxide concentrations is especially crucial because Rubisco requires a relatively high concentration of carbon dioxide to work. The reason for this requirement is that, at low concentration of carbon dioxide, Rubisco preferentially converts ribulose bisphosphate into one molecule of phosphoglycerate and a two-carbon compound called phosphoglycolate. This reaction is the beginning of an oxygen-dependent process called **photorespiration,** diagrammed in Figure 8-20, which ultimately produces a molecule of carbon dioxide and a molecule of the amino acid serine. Photorespiration does not produce any ATP or NADH, and most plant biologists think that it provides no benefit to the plant. It causes plants to waste as much as half of the carbohydrates they produce.

Some plants have evolved a surprising solution to the problem of photorespiration. In these plants, the first reaction of carbon dioxide does not produce the three-carbon molecule phosphoglycerate, but a four-carbon compound (oxaloacetate), as shown in Figure 8-21a. One of the special enzymes uses NADPH to convert oxaloacetate to

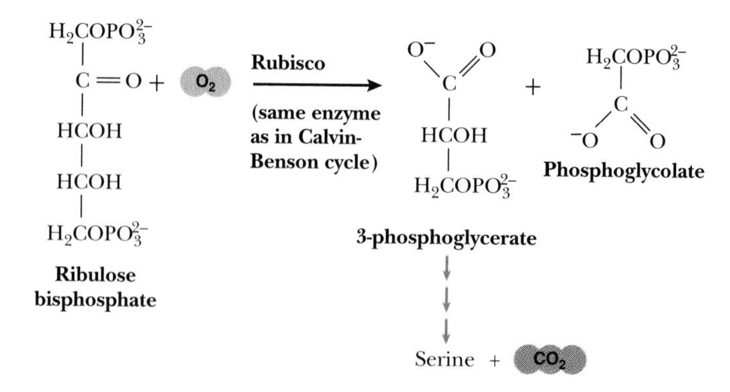

Figure 8-20 Photorespiration: At low CO_2 concentrations, Rubisco preferentially converts RuBP into phosphoglycerate and phosphoglycolate.

ESSAY

HOW IS THE ENERGY OF CAPTURED PHOTONS BEST USED?

About 1–2% of the light energy that falls on a plant is ultimately stored in the chemical bonds of simple carbohydrates and of more complex molecules like cellulose, starch, and proteins. Since feeding the planet will require increasing amounts of foods, plant scientists have spent considerable time trying to increase the yield of photosynthesis. But even with the most efficient of currently available crops, such as wheat, it seems unlikely that the earth will be able to support more than 5–6 billion people on a totally vegetarian diet.

Among the ways that plant scientists have been trying to increase the food supply, reducing photorespiration is a major goal. Plants that use the C4 pathway for carbon fixation lose less of the captured energy, but researchers are looking for ways of decreasing photorespiration in C3 plants as well. A particularly promising and surprising finding has been that spraying a dilute solution of methanol on cabbages doubles the size of the resulting heads. The biochemical mechanism of methanol action is at this point unknown, but is a subject of active investigation.

Other researchers are trying to select plants that put more of the captured energy into the edible portions of a plant. In a corn plant, for example, this means selectively diverting the products of photosynthesis to seeds rather than to cobs, stalks, and leaves. Still other studies address the effects of growing plants in artificially high levels of carbon dioxide, relying on closed chambers or on pipes that carry CO_2 into the field. Some people even argue that the "greenhouse effect" will significantly increase crop yields by providing increased CO_2 in the atmosphere as well as increased temperatures.

Despite the pressure to increase the edible yield of photons, many researchers are also studying ways of diverting plant metabolism to make industrially useful products. Almost all of the oils produced by plants, for example, consist of just six fatty acids, but plants can make hundreds of other fatty acids as well. The genetic engineering techniques described in Chapter 17 have made it possible to transfer the ability to make exotic fatty acids to crop plants that are easily farmed. For example, ricinoleic acid, the oil in the castor plant (*Ricinus communis*), long known for its cathartic effects, is extremely useful as a lubricant, and ricinoleic acid produced in a crop plant would find a large market. Similarly, other oils now present only in undomesticated plants are useful as detergents, coatings, lubricants, and cosmetics.

Figure 1 The leaf cells of a genetically engineered plant make granules of a biodegradable plastic, polyhydroxybutyrate (PHB). *(Yves Poinier, Carnegie Institution)*

A particularly surprising product, produced in genetically engineered plants, is a plastic, polyhydroxybutyrate (PHB). As shown in Figure 1, researchers have transferred two bacterial genes into a widely used experimental plant *Arabidopsis thaliana*, which we will again encounter in Chapter 22. These two enzymes convert products of photosynthesis into PHB, which accumulates in small granules. This experiment shows the possibility of directly converting solar energy into a useful industrial material.

(a)

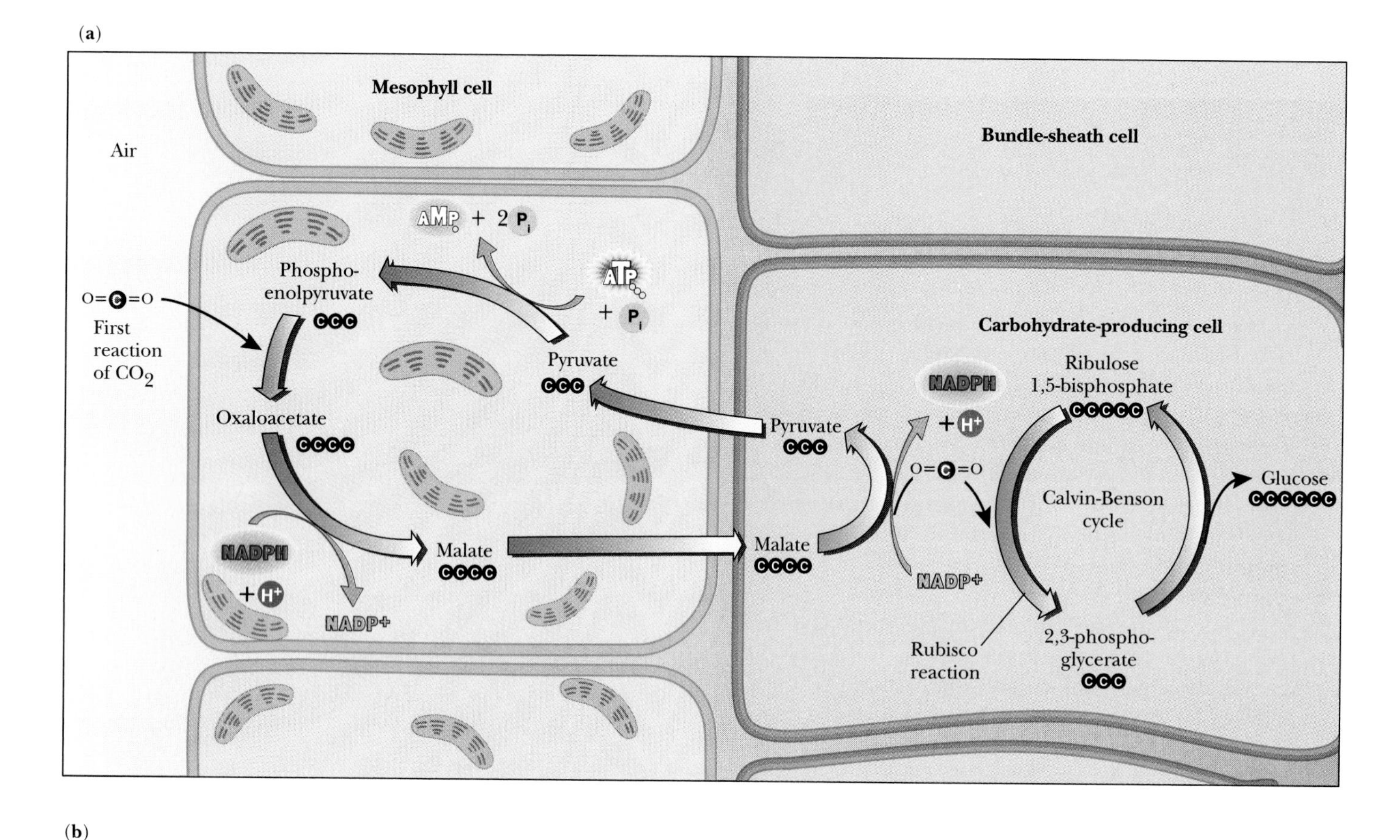

(b)

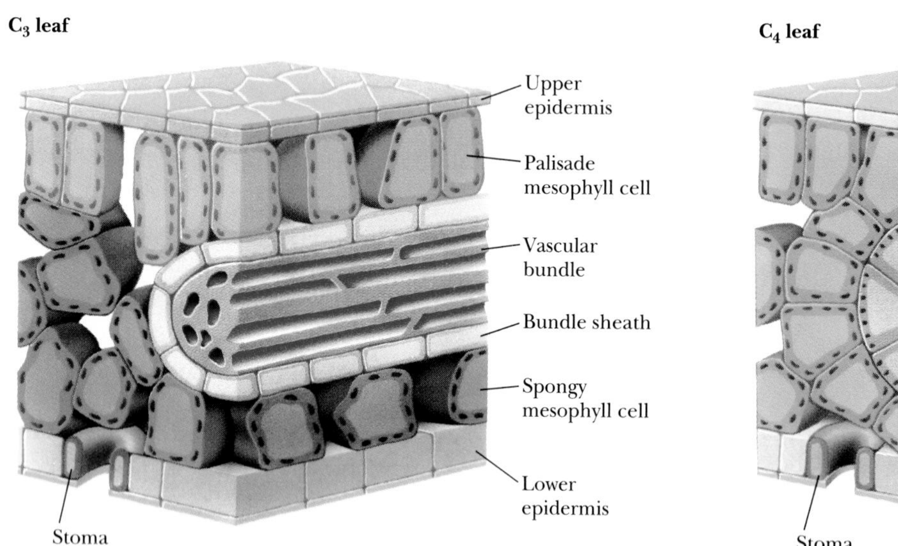

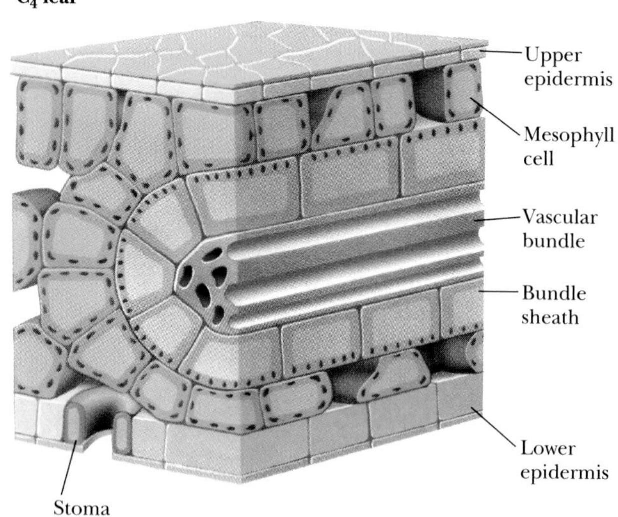

Figure 8-21 The special anatomy of C4 plants, such as corn and sugar cane, minimizes photorespiration. **(a)** CO_2 combines with a four-carbon compound in mesophyll cells; carbohydrate synthesis occurs in bundle sheath cells, where little photorespiration can occur; **(b)** the comparative cellular anatomy of a C3 plant, in which CO_2 fixation and carbohydrate synthesis occur in the same cells, and a C4 plant.

another four-carbon compound, malate, which moves into carbohydrate-producing cells. There it breaks down to release carbon dioxide, which immediately enters the Calvin-Benson cycle (via the Rubisco reaction).

Plants with this special pathway and structure are called C4 plants, while most plants, which take up carbon dioxide directly into phosphoglyceric acid, are called C3 plants. C4 plants include corn and sugar cane, which can grow in hot and dry environments. Compared with the leaves of C3 plants, the leaves of plants with the C4 pathway have a special cellular anatomy, shown in Figure 8-21b. The C4 enzymes and the C4-associated anatomy are adaptations that reduce water loss through the stomata while providing enough carbon dioxide to support a high rate of photosynthesis. The uphill transport of carbon dioxide requires energy. Each molecule of carbon dioxide moved by the C4 cycle costs two high-energy phosphate bonds from ATP.

Plants with the C4 pathway have another advantage besides being able to take up carbon dioxide even at low concentrations: They hold on better to the sugars they have made, while C3 plants waste much of their hard-won carbohydrates in photorespiration. Carbon dioxide inhibits photorespiration, and at high carbon dioxide concentrations Rubisco produces only phosphoglycerate. Because the carbon dioxide pump of C4 plants provides them with high concentrations of carbon dioxide, they have little photorespiration. As long as there is enough light to provide the extra ATP requirement of C4 plants, they are far more efficient than C3 plants.

In hot, dry climates, C4 plants predominate. In temperate zones with less light and more water, C3 plants have the advantage. Many admirers of Kentucky bluegrass (a C3 plant and the most popular landscaping plant in the United States) are frustrated by the selective advantage of C4 plants. As the summer wears on, often becoming hot and dry, their beautiful single crop lawns are taken over by less attractive, yellow-green crabgrass (a C4 plant).

The C4 path represents an important adaptation. Plants that thrive in hot dry climates, the succulents (which include the cactuses), have a related but independent scheme, called crassulacean acid metabolism, for reducing water loss during the acquisition of carbon dioxide. C4 plants are more important for our discussion here because they include important crop plants, such as sugar cane and corn. Such C4 crops are especially efficient in capturing solar energy, converting as much as 8% of the energy falling on a field to the chemical bonds in carbohydrates. As oil and coal reserves diminish, some countries have begun to produce alcohol as fuel for cars and trucks from such C4 plants as corn and sugar cane.

The C4 pathway and associated anatomy represent an important adaptation to hot and dry climates. Species with these genetic endowments have advantages in certain environments over species limited to the C3 path. But for any individual plant, the ability to produce carbohydrates is determined by its immediate environment—its supply of light, carbon dioxide, water, and other nutrients. Both genes and environment determine the phenotype and the success of an individual organism.

SUMMARY

Green plants and other photosynthetic organisms harvest the sun's energy and convert it into the chemical energy of carbohydrates. The raw materials for photosynthesis are water and carbon dioxide. The products are carbohydrates and oxygen.

Photosynthesis uses light energy to split water to generate oxygen and high-energy electrons. These electrons provide the reducing power to convert carbon dioxide to carbohydrates. Some prokaryotic organisms do not use water in photosynthesis but derive their reducing power from other compounds, such as hydrogen sulfide.

Light has both wave-like and particle-like properties. Each particle of light, called a photon, carries a fixed amount of energy, which depends upon its wavelength. To have an effect, light must first be absorbed. Plants contain two sets of pigments that absorb light, the chlorophylls and the carotenoids. The energy of the absorbed light increases the energy of electrons in the absorbing molecules. In plants, energy absorbed by a large number of chlorophyll and carotenoid molecules is transferred to one of two kinds of reaction centers, each of which absorbs light with characteristic wavelengths. For one center, called P_{700}, this wavelength is 700 nm; for the other, called P_{680}, it is 680 nm. Photosynthesis requires that both P_{700} and P_{680} absorb light to produce excited electrons.

Each reaction center participates in a separate "photosystem," located on the thylakoid membrane of the chloroplast. P_{700} is part of photosystem I, and P_{680} is part of photosystem II. The excited electrons of each photosystem are transferred to special electron acceptors, leaving the reaction center in an oxidized state. The excited electrons of each photosystem make different contributions to the harnessing of energy.

Excited electrons of photosystem I can flow back to reduce the oxidized P_{700}, via an electron transport chain that generates a proton gradient, which in turn can generate ATP. The end result of this path—called cyclic photophosphorylation—is the regeneration of P_{700} and the production of ATP.

Alternatively, the excited electrons of photosystem I can flow down a different electron transport chain to produce NADPH, which is used directly in the synthesis of carbohydrates. This noncyclic path requires that P_{700} be reduced by an excited electron of photosystem II.

The electron transport chain of photosystem II carries an excited electron from excited P_{680} to the oxidized form of P_{700}. As the electron moves down the chain, it generates a proton gradient and the consequent formation of ATP. The net result is noncyclic photophosphorylation.

The electrons needed to regenerate reduced P_{680}

come from water. The removal of four electrons from water results in the production of oxygen (O_2) and four H^+ ions.

The reduction of each molecule of $NADP^+$ to NADPH requires two electrons and the absorption of four photons (two by each photosystem). At the same time, about one molecule of ATP is formed.

The light-independent reactions of photosynthesis are those that produce carbohydrates by using the ATP and NADPH formed in the light-dependent reactions. The set of reactions that produces glucose from carbon dioxide, NADPH, and ATP is called the Calvin-Benson cycle. Each carbon dioxide molecule converted to carbohydrate requires the expenditure of two molecules of NADPH and three molecules of ATP. The production of a molecule of glucose from 6 carbon dioxide molecules requires 72 photons of light.

Several environmental factors limit the efficiency of photosynthesis. There must be sufficient supplies of light, water, and carbon dioxide. Some plants, called C4 plants, have evolved a special mechanism that increases the efficiency of carbon dioxide uptake. These plants first take up carbon dioxide to form a four-carbon molecule, which is then converted to another four-carbon molecule and shuttled to the cells that perform the Calvin-Benson cycle. The net result is the ATP-dependent pumping of carbon dioxide. These C4 plants are more efficient producers of carbohydrates when carbon dioxide is limiting and when there is enough light to provide the extra energy needed for the carbon dioxide pump. The efficiency of photosynthesis thus depends both on the genetic capacity of the plant and on its environmental conditions.

KEY TERMS

absorption spectrum
action spectrum
antenna complex
autoradiography
carotenoid
chlorophyll
cyclic photophosphorylation
electromagnetic radiation
granum
noncyclic photophosphorylation
oxygen evolving complex
photochemical reaction center
photon
photorespiration
photosynthesis
photosystem I
photosystem II
pigment
stomata
stroma
thylakoid membrane
visible light
wavelength

QUESTIONS FOR REVIEW AND UNDERSTANDING

1. Define and contrast the following terms:
   (a) absorbtion spectrum vs. action spectrum
   (b) antenna complex vs. photochemical reaction center
   (c) photosystem I vs. photosystem II
   (d) $NADP^+$ vs. NAD^+
   (e) cyclic photophosphorylation vs. noncyclic photophosphorylation
   (f) C4 plant vs. C3 plant
   (g) stroma vs. thylakoids
2. How did studies of sulfur-producing bacteria lead to a greater understanding of photosynthesis in plants?
3. How does a photon of blue light differ from a photon of red light?
4. What is the difference between a pigment molecule and other biological molecules?
5. What role do carotenoids play in photosynthesis?
6. Describe in your own words the synergistic effect of light of 680 nm wavelength and 700 nm wavelength on photosynthetic electron transport and photophosphorylation.
7. How does the cytochrome b_6f complex contribute to photophosphorylation?
8. What is the role of the oxygen evolving complex in photosynthetic electron transport?
9. Why is it significant that the oxidized form of P_{680} has a very positive reduction potential, i.e., can readily accept electrons? Why is it significant that the excited form of P_{700} has a very negative reduction potential, i.e., can readily donate electrons?
10. Why is a combination of cyclic and noncyclic photophosphorylation "optimal" with respect to the synthesis of glucose?
11. List the three parts of the Calvin-Benson cycle.
12. What is the role of the enzyme Rubisco? Why is it so abundant in plant tissues?
13. What are the two principal products of the light reactions of photosynthesis? How are they used in the dark reactions of photosynthesis?

14. What are the two fates of glyceraldehyde phosphate produced in part 2 of the Calvin-Benson cycle?
15. Why is the movement of water through a plant useful?
16. Explain what is meant by the statement, "Photorespiration salvages phosphoglycolate following the utilization of oxygen, rather than carbon dioxide, by Rubisco."
17. What is the advantage to plants of having the C4 pathway? What is the disadvantage? How does this influence the geographical distribution of C3 and C4 plants?
18. Compare the processes of mitochondrial oxidative phosphorylation (see Chapter 7) and the light-dependent reactions of photosynthesis. Consider the fates of nucleotide coenzymes, ATP synthesis, energy (reduction potential) changes, and the role of oxygen.
19. Explain how the chemiosmotic theory (see Chapter 7) applies to photosynthetic electron transport and photophosphorylation.
20. Why do you think the development of photosystem II by photosynthetic organisms is considered a key event in biological evolution and the earth's history?
21. If a few algal cells are placed in a flask containing just tap water and the flask is illuminated, the cells will divide and grow, increasing their mass several-fold over a few days. What biomolecules must they be creating? What fuel must they be using for ATP synthesis (see Chapter 7)? What does this tell us about the biosynthetic capabilities of plants (algae)? Can other organisms (bacteria, fungi, animals) grow in tap water or other media not having a pre-existing source of reduced carbon?

SUGGESTED READINGS

ALBERTS, B., D. Bray, J. Lewis, M. Raff, K. Roberts, and J. D. Watson, *Molecular Bilogy of the Cell,* 3rd edition, Garland, New York, 1994. See Chapter 14 on chloroplasts.

GARRETT, R. H. and C. M. Grisham, *Biochemistry,* Saunders College Publishing, Philadelphia, 1995. Chapter 22 covers much of the material in this chapter.

LODISH, H., D. Baltimore, A. Berk, S. L. Zipursky, P. Matsudaira, and J. Darnell, *Molecular Cell Biology,* 3rd edition, Scientific American Books, New York, 1995. Chapter 18 covers the material in this chapter.

MAUSETH, J. D., *Botany, An Introduction to Plant Biology,* 2nd edition, Saunders College Publishing, Philadelphia, 1995. Chapter 10 deals with photosynthesis.

RAVEN, P. H., R. F. Evert, and S. E. Eichhorn, *Biology of Plants,* 5th edition, Worth Publishers, New York, 1992. Chapter 7 deals with photosynthesis.

STRYER, L., *Biochemistry,* 3rd edition, W. H. Freeman, New York, 1988. Chapter 22 covers much of the material in this chapter.

PART III

The Continuity of Cellular Information

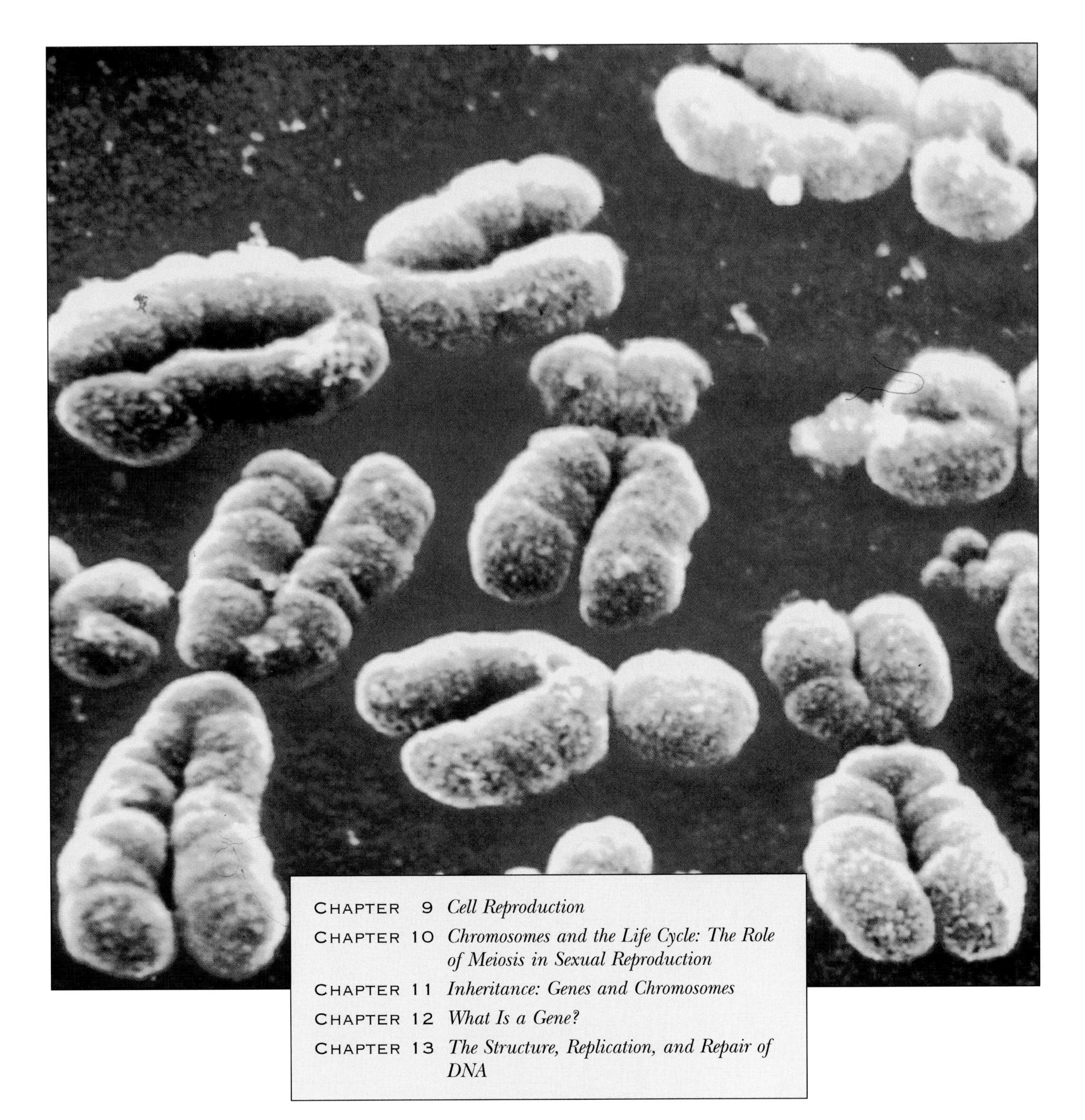

SEM of human chromosomes.

LAB TOUR

NAME: SUSAN M. ABMAYR
TITLE: ASSISTANT PROFESSOR OF MOLECULAR GENETICS
ADDRESS: THE PENNSYLVANIA STATE UNIVERSITY,
UNIVERSITY PARK, PA 16802-4500

■ ***What is the major question that you and your laboratory are asking?***

We want to ask, "What is the cascade of decisions that a cell makes in order to become a particular cell type?" More specifically, "How does a cell decide to be a muscle cell rather than any other kind of cell?"

■ ***How did you choose this question?***

Before I started graduate school, I was interested in gene transcription, particularly in how an individual gene gets turned on or off. While in school, I studied the ways by which viruses appropriate the transcriptional machinery so that the cell preferentially expresses the virus's genes rather than its own genes, following an infection. These studies were entirely *in vitro,* in a test tube. During that time, I was always being asked, "How do you know that genes work the same way in an organism as in a test tube?" I decided that, in order to study how genes actually worked in organisms, I would need to use genetics. I also decided I was more interested in development than in viruses.

At the time, people didn't know nearly as much about transcription as we now do, but already a transcription factor critical in muscle development had been identified in mice. This molecule, called MyoD, was exactly the sort of molecule I wanted to work with, but in a species that was accessible to genetic studies. I and a colleague therefore set out to find the corresponding gene in *Drosophila.*

■ ***Why has the overall question remained important to you?***

The studies became addictive. Every time I found a piece of the puzzle, I wanted to find more pieces. Every bit of understanding made me realize that there was still more to understand.

■ ***What is your strategy as you pursue answers to the questions you have posed?***

Right now, our approach combines genetics and molecular biology. We've isolated a couple of genes, and we're going back to make specific mutations in those genes and look at the effect of those mutations on muscle development. We've also found mutations, in unknown genes, which affect muscle development, and we're trying to identify and clone the genes that are mutated. So sometimes we have a gene and want to find out what it does, while other times we start by knowing what a gene does and wanting to find out what it is.

■ ***What other people work with you in your research program?***

Besides myself, there are three graduate students, a technician, and several undergraduates. The undergraduates make mutations and characterize them genetically and phenotypically, by looking through a microscope for muscle defects. We then try to "rescue" the mutants by genetic engineering—by producing an embryo that contains a nonmutant form of the mutated gene.

Professor Abmayr examining and propagating her lab's collection of Drosophila *mutations.*

■ ***What organisms do you use in your resesarch? Why those organisms?***

We work entirely with *Drosophila* because it's so approachable genetically. Ultimately, we would like to think that when we identify new genes in *Drosophila,* their mammalian counterparts will likewise be important for mammalian development.

■ ***What instruments do you use in your research?***

We use lots of microscopes. Our single most utilized tool is a high-powered light microscope equipped with differential interference contrast (Nomarski) optics, which we use to look at muscle structure. We also use the standard equipment of a molecular biology laboratory—centrifuges, water baths, and electrophoresis setups for the separation of DNA fragments and for DNA sequencing. Because we cannot store *Drosophila* in the freezer and need to maintain living stocks of about 500 types of mutant flies, we have lots of incubators. We keep most of these incubators at 25°C, so the flies mature in 10–12 days, but we keep some at 18°C to slow development to 20–24 days.

■ ***What are some typical experimental steps in your research?***

Right now we have a fly with a deletion, or deficiency, in one of its chromosomes. Several genes are missing, including one that we are interested in. We've set up a series of genetic crosses that allow us to determine whether new mutations lie within the deleted region. We can then go back to the original stock, recover the particular mutation, and use the microscope to study the effects of that mutation in embryos that have been bred to be homozygotes. At the same time, we clone and determine the sequence of the particular gene of interest, for example, the fly version of *MyoD,* which is called *Nautilus.* Finally, we try to rescue the mutation by engineering the cloned gene into a mobile genetic element that can carry the gene into the embryo of a mutant fly.

■ ***Please describe a day in which something happened in your laboratory that was especially exciting.***

Two weeks ago, we found that certain flies had no muscle defects, even though they're missing DNA in a particular region of the chromosome that we've been studying. If we removed an additional small region, then the muscles developed abnormally. This was especially exciting because the *Nautilus* gene lies exactly in that small region. We are now trying to rescue the mutants by adding back the cloned *Nautilus* gene.

■ ***How do new questions arise from week to week or from month to month?***

Usually, as soon as we get an exciting result, we think of ten other experiments to do. Sometimes, [though], we need to step back and reevaluate. For example, the mutation being studied by Mary Erickson, a graduate student in the lab, lay very near to the *Nautilus* gene, and produced abnormal muscle, so we thought the mutation lay in the *Nautilus* gene itself. We isolated 11 point mutations in Mary's gene, and tried to rescue them with [a nondefective] *Nautilus* gene, but we couldn't. So Mary's gene is not *Nautilus,* it just happens to lie close by. This conclusion was difficult to reach. For a long time we thought that the mutations were in the *Nautilus* gene, but we just weren't able to prove it. Ultimately, it was the strength of the genetic approach that told us that Mary's mutation was not *Nautilus.* Once we were able to establish this unexpected fact, it told us that we had uncovered a new gene that was critical for muscle development.

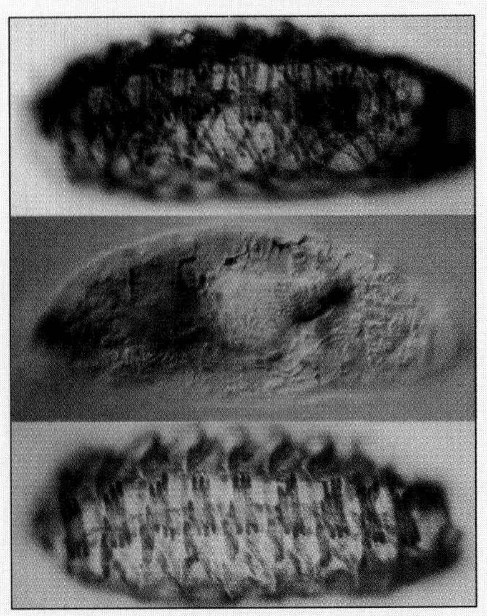

The top panel shows the musculature of the Drosophila *embryo, the organism that is studied in the lab. The mutant embryo in the middle panel does not develop muscles due to a defect in an important gene. In the bottom panel, a normal copy of this gene has been put back into the mutant embryo and muscles can now develop.*

CHAPTER 9

Cell Reproduction

KEY CONCEPTS

- Cell reproduction occurs in an orderly sequence of events called the cell cycle.
- Cell reproduction involves three distinct but interconnected processes: the duplication of genetic information in the nucleus, its equal distribution to each of two daughter cells, and the division of the cytoplasm.
- In eukaryotic cells, DNA forms compact complexes with proteins. During mitosis, these complexes become even more compact and are visible as distinct chromosomes.
- During mitosis microtubules and other subcellular components bring about chromosome movements that distribute one copy of each chromosome to the daughter cells.
- Eukaryotic cells, ranging from yeasts to humans, use similar molecular mechanisms to regulate passage through the cell cycle.
- Interactions with other cells and extracellular molecules regulate cell division in multicellular organisms.

OUTLINE

Cells are the fundamental *living* units of all organisms. An organism may consist of a single cell or many cells—an adult human, for example, contains about 10^{13} cells. All the cells of a multicellular organism, however, originate from a single cell. The continuation of life ultimately depends on cell reproduction.

In the 5 minutes or so that it will take you to read this page, your body will produce about a billion new cells. Most of these will replace battered and dying cells of your skin, intestines, and blood. Each newly made cell contains a copy of the genetic information you inherited from your parents as well as the molecular machinery needed to interpret that information. Each daughter cell contains not only a copy of the parent cell's genes, but also its membranes, organelles, macromolecules, and small molecules.

Cell reproduction is so common that it is easy to forget how extraordinary it is. Our wonder at this copying process is even more heightened by our everyday experience with nonbiological copying—the recording of music, the filming and taping of movies and television programs, the printing of books, and even the duplication of class assignments. In each case, complicated machinery carries out a complex but ordered process. But cell reproduction is even more remarkable than these other processes because the cell is both its own copying machine and the thing being copied. Still more—it is the builder of the next copier.

Cells reproduce themselves in a process called **cell division,** in which a parent cell gives rise to two **daughter cells** that carry the same genetic information as the parent cell. All organisms (and all cells) use DNA as their genetic material: The sequence of nucleotides in DNA contains coded information about cellular molecules and processes. (Later, we will see just how this information is decoded.) One of the main tasks of cell reproduction is to **replicate** [Latin, *replicare* = to fold back], or copy, the cell's DNA. Each DNA molecule consists of two polynucleotide strands. Each daughter DNA molecule contains one of the parent chains and a newly synthesized chain. In Chapter 13, we see exactly how this is accomplished.

After DNA replication, the cell's machinery must distribute the same coded information to each of the two daughter cells. When only one parent DNA molecule is involved (as in most bacteria), such equal distribution is relatively simple. When several (or many) DNA molecules are involved (as in all eukaryotes), the distribution process is more elaborate because each daughter cell must receive a copy of all the genetic information.

Cell division also splits the rest of the cell's contents between the two daughter cells. Cell reproduction, then, accomplishes three tasks—the replication of DNA, the equal distribution of the DNA to the two daughter cells, and the splitting of the other cell components into two (usually) equivalent daughter cells, each surrounded by its own plasma membrane.

In a single-celled organism, cell reproduction leads directly to a population increase. In a multicellular organism, however, cell reproduction does not increase the number of organisms but contributes to the growth and maintenance of a single organism. To the dismay of lawnkeepers, for example, a dandelion root can regenerate leaves and flowers even after the exposed part of the plant is removed. More happily, children quickly replace the skin cells that they scrape and tear in their inevitable bangs and falls.

HOW DOES A DIVIDING CELL ENSURE THAT ITS DAUGHTER CELLS INHERIT THE SAME GENETIC INFORMATION?

We can describe the process of cell reproduction in terms of three m's—materials, machinery, and memory: the small molecules that carry energy or serve as building blocks, the organelles and macromolecular structures needed to carry on cellular processes, and the information, contained in DNA, for building cellular machinery from the building blocks. Before a cell divides, it usually doubles its materials, machinery, and memory. When it divides, each daughter cell usually receives an equal share of each.

The most important task of cell reproduction is to propagate the information encoded in DNA. Without the memory encoded in DNA, the cell's materials and machinery are useless. The precise doubling of DNA uses some of the cell's most elaborate molecular mechanisms, which we discuss at length in Chapter 13. In this chapter, we see how dividing cells coordinate the duplication and distribution of DNA within the **cell cycle,** the orderly sequence of events that accomplish cell reproduction.

How Do Prokaryotic Cells Divide?

Cell division in prokaryotes is fairly straightforward. Prokaryotes divide by **binary fission,** in which a cell pinches in two, distributing its materials and molecular machinery more or less evenly to the two daughter cells. As is the case for all cells, special mechanisms duplicate and distribute the molecular memories encoded in DNA.

In most prokaryotes genetic information is stored in a single molecule of DNA, which, like the *E. coli* DNA shown in Figure 9-1a, usually forms a closed loop (always referred to as a circle, although it never really looks like a true circle). Before the cell divides, it replicates its DNA. Each of the two resulting daughter DNA molecules attaches to the membrane fold at which binary fission begins. As the cell divides, the two DNA molecules move apart so that each daughter cell receives a single (double-stranded) molecule of DNA, as shown in Figure 9-1b and c. The attachment of each DNA molecule to the plasma membrane accomplishes the distribution of one DNA molecule to each daughter cell.

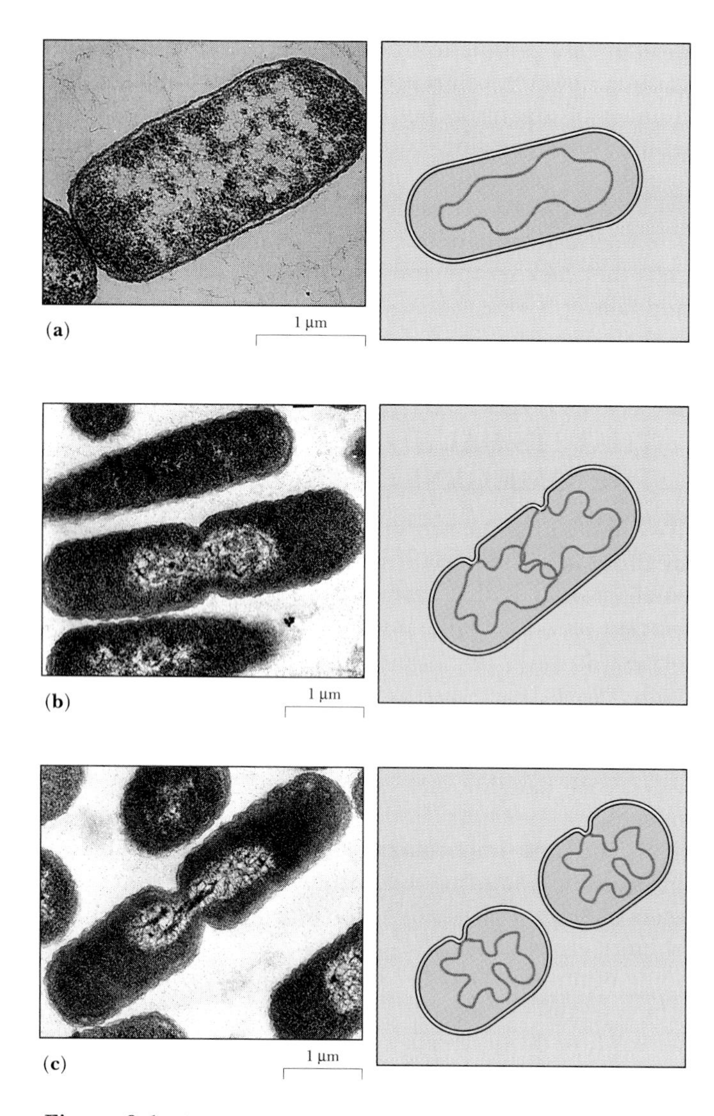

Figure 9-1 Cell reproduction in *E. coli.* **(a)** *E. coli* DNA is a closed loop, here just beginning to replicate. **(b, c)** As a cell divides, the two molecules of DNA, each attached to the cell membrane, move apart. *(a, Biology Media/Photo Researchers; b, c, BPS)*

Microscopic Studies Reveal That Eukaryotic Cell Division Is a Complex Process

The internal membranes and compartments of eukaryotic cells make the process of cell reproduction far more complicated than in prokaryotes. Just the existence of a nucleus poses a problem: DNA cannot easily be distributed to daughter cells as long as it is surrounded by the nuclear envelope. Finally, the amount of DNA in a eukaryotic cell may be 1000 times or more that in a prokaryotic cell—far more than in any single prokaryotic DNA molecule.

A light microscope reveals cells in various stages of the cell cycle in an onion root. In Figure 9-2, for example, we can easily see individual cells in the process of dividing their nucleus and cytoplasm by staining with a dye that binds to DNA.

We can observe the doubling and distribution of cellular memory. We can see that, in some cells, the material of the nucleus has become organized into discrete bodies called **chromosomes** [Greek, *chroma* = color + *soma* = body, because they are stained by certain dyes]. The cells of each species contain a characteristic number of chromosomes: Humans have 46, carrots have 18, and the fruit fly *Drosophila melanogaster* has 8. With very few exceptions, every cell in your body has copies of the same 46 chromosomes.

In the second half of the 19th century, microscopists had improved their ability to see structures within cells, partly because of improvements in lens design by Ernst Abbe, a professor of physics at the University of Jena in Germany and later the owner of the Zeiss Optical Works. Some of the most important advances in microscopy, however, came from improved contrast as researchers began to use both natural dyes (like the blue dye hematoxylin, extracted from the logwood tree of Central America) and the new synthetic dyes (such as the red dye eosin) that had been developed for the burgeoning textile industry. As shown in Figure 9-2, cells stained with hematoxylin and eosin (a combination that microscopists still commonly use) show the nucleus to be full of **chromatin** [Greek, *chroma* = color], a material that consists of DNA and proteins and appears grainy after staining. Chromatin is present in almost all eukaryotic cells, while chromosomes are visible only in cells that are actively dividing.

By 1880, microscopists had described thread-like chromosomes in a wide variety of animal and plant cells. But no one knew whether these structures were actually present in living cells or were artifacts caused by tissue preparation (as discussed in Chapter 6). In 1882, however, Walther Flemming, a German physician and microscopist, showed that the chromosomes were visible not only in tis-

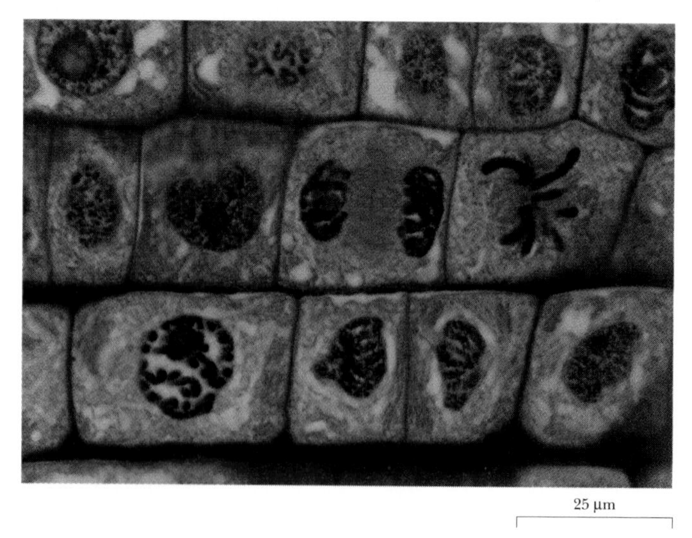

Figure 9-2 Dividing cells in an onion root, stained with a dye that binds to DNA. Notice the distinct chromosomes in some cells and the diffuse chromatin in others. *(M. I. Walker/Photo Researchers)*

sue that had been fixed and stained but also in living salamander larvae.

Because he could observe chromosomes in living cells, Flemming was also the first to describe the sequence of chromosomal movements. Because stained cells gave much better views of internal cell structure than unstained cells, Flemming regarded the stained cells as a set of still photographs. From these, he pieced together the first description of **mitosis** [Greek, *mitos* = thread], the process of the equal distribution of chromosomes during cell division.

One of the first events that Flemming noted was the gathering of the chromatin into visible threads (the chromosomes). He also observed a star-like object just outside the nucleus. Flemming called this structure an **aster** [Latin = star]; we now know it to contain a microtubule organizing center. The aster lies at the center of a set of thin threads, much finer than the chromosomes, which we now know to be microtubules.

Between mitoses, a cell duplicates its aster, and, during mitosis the two daughter asters migrate to opposite poles of the cell. While the asters are migrating, the chromosomes first line up in a row in the middle of the cell, and then each one splits into two. Each of the two half-chromosomes then migrates toward the asters. Finally, the chromosomes lose their thready appearance, and the nuclei regain their grainy texture, with no visible chromosomes.

Only later did biologists realize the importance of the chromosome dance that Flemming had observed. By the early years of the 20th century, biologists understood that the chromosomes contained the cell's genetic instructions. By 1952, they knew that the DNA in the chromosomes is the genetic material.

On the basis of such microscopic observations, biologists distinguish three stages in the process of cell division: (1) **mitosis,** the division of the chromosomes; (2) **cytokinesis** [Greek, *kytos* = hollow vessel + *kinesis* = movement], the division of the cytoplasm and formation of two separate plasma membranes; and (3) **interphase,** the rest of the life of a cell, when the chromosomes are not condensed and the cytoplasm is not dividing. Once biologists knew the importance of DNA, however, they realized that the microscope alone did not reveal all that happens during cell reproduction.

A Cell Has Different Amounts of DNA at Different Phases of the Cell Cycle

In some special cases, such as the rapid cell divisions right after fertilization in many species, mitosis and cytokinesis occur in as little as half an hour. For many plant and animal cells, however, the cell cycle usually takes about a day. Mitosis and cytokinesis, the most visible portions of the cell cycle, may occupy only 1 or 2 hours of this time, with interphase lasting 20 hours or more. Although microscopy

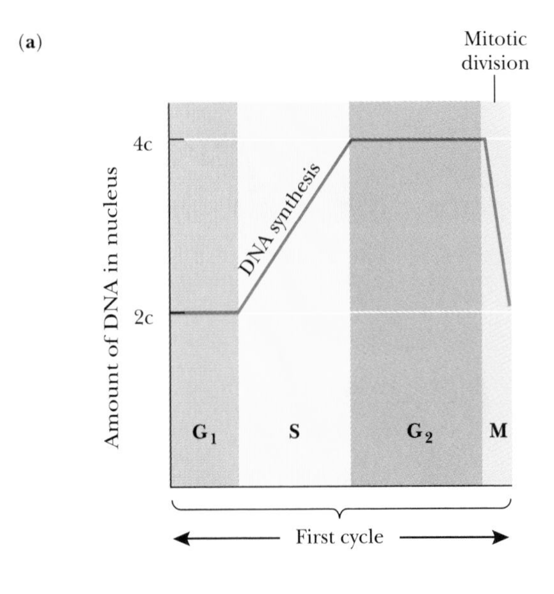

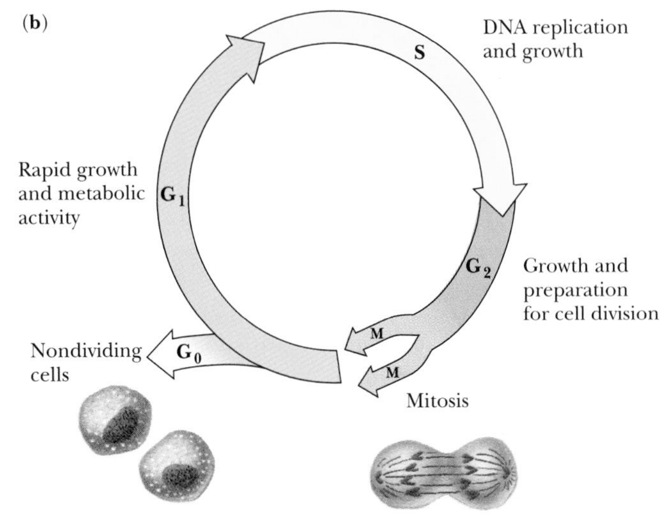

Figure 9-3 The eukaryotic cell cycle consists of the visually distinct processes of mitosis and cytokinesis and the apparently homogeneous interphase. But interphase actually consists of three distinct phases: G_1, S, and G_2. G_1 is the period of growth before DNA replication, S the period of DNA replication, and G_2 the period after DNA replication during which the cell prepares for mitosis and cytokinesis. **(a)** Changes in the cell's DNA content; 2c is the amount of DNA in a nondividing somatic cell; **(b)** the phases of the cell cycle.

alone does not reveal much about the interphase part of the cell cycle, cells go through a set of well-defined stages in a highly ordered and tightly regulated manner.

The amount of DNA in a cell provides a useful way of dividing the cell cycle into phases (as illustrated in Figure 9-3); **M,** *m*itosis, which here includes cytokinesis; $\mathbf{G_1}$, the *g*ap between the completion of M and the beginning of DNA replication (also called the first growth phase); **S,** the period of DNA replication, or DNA *s*ynthesis, when a cell doubles its DNA content; and $\mathbf{G_2}$, the *g*ap between the completion of DNA synthesis and the beginning of M (of the next cell cycle). This description divides the period

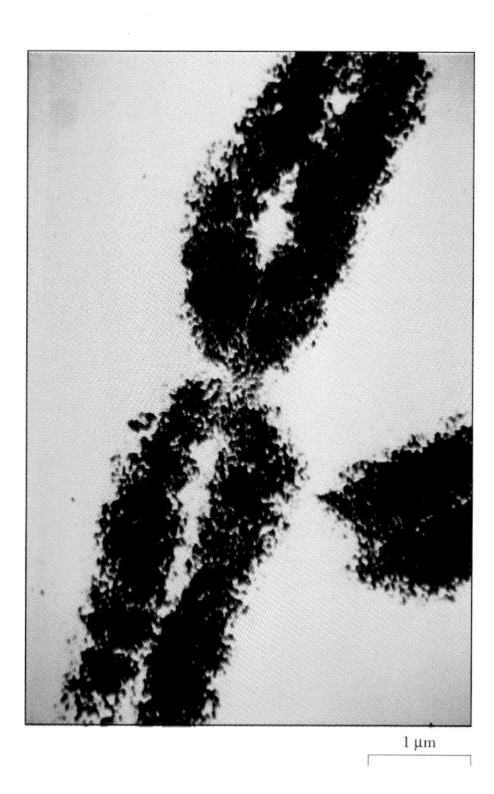

Figure 9-4 A human chromosome at the close of G_2, just before the beginning of mitosis; transmission electron micrograph. *(Science VU/Visuals Unlimited)*

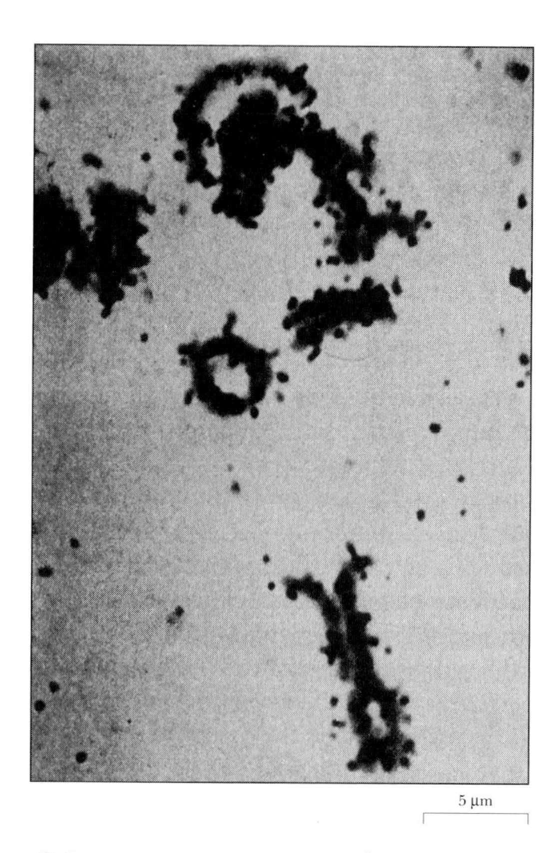

Figure 9-5 A radioactive nucleoside (^{3}H-thymidine) is incorporated into the DNA of a growing bean root during the S phase of the cell cycle. Cells that are in the S phase during the labeling period have radioactive chromosomes, revealed by autoradiography. *(J. H. Taylor,* Proceedings of the National Academy of Science, *43: 122, 1957, courtesy New York Academy of Medicine)*

called "interphase" into three separate phases—G_1, S, and G_2. During G_1, a cell accumulates the materials and machinery needed for DNA synthesis (during S). During G_2, it prepares for mitosis and cytokinesis: It assembles the molecular machinery needed to sort the chromosomes and partition the cells.

In each phase of the cell cycle, the chromosomes have distinctive characteristics. At the beginning of mitosis, when the chromosomes first become visible, each chromosome consists of two separate but connected bodies, called **chromatids,** which are shown in Figure 9-4. The two chromatids are joined at a single point called the **centromere.** Each chromatid (we now know) contains a long molecule of double-stranded DNA complexed with proteins. Each centromere contains DNA with special sequences that are repeated many times, together with a characteristic set of proteins. The two **sister chromatids** (the two chromatids that make up a single chromosome) are duplicate copies of the same genetic information.

At the end of mitosis each chromosome consists of a single chromatid. Each chromatid duplicates during the S phase, and the amount of DNA doubles (see Figure 9-3a). In the G_2 phase each chromosome already consists of two paired sister chromatids, although these become visible only at the beginning of mitosis. During M, each daughter cell receives one of the two sister chromatids from each chromosome. In G_1 each chromosome consists of a single chromatid, although the chromatids have re-dispersed into chromatin, and (as in G_2 and S) we cannot see them in a microscope.

Researchers can determine when a cell is in S by studying the cellular uptake of radioactive DNA precursors. In this experiment a radioactively labeled nucleoside, ^{3}H-thymidine, is added to growing root tips of a bean plant. Inside the cell, enzymes convert the radioactive compound into thymidine triphosphate, which can be incorporated into DNA during the S phase of the cell cycle. Figure 9-5 shows how radioactive chromosomes are revealed by autoradiography.

The amount of DNA in each cell is an index of its phase in the cell cycle. One way to measure the DNA content of individual cells is to expose cells to a fluorescent dye that binds to DNA, as illustrated in Figure 9-6. The amount of fluorescence from each cell is a measure of its DNA content and its phase in the cell cycle.

Most of the work of cell reproduction occurs during the visually unspectacular G_1 phase, when the cell doubles most of its components (other than DNA). If the cell runs out of nutrients, it stops the cycle in G_1 phase. Some specialized cells, like nerve cells, altogether withdraw from the cell cycle during G_1 and enter a state called $\mathbf{G_0}$. Nerve

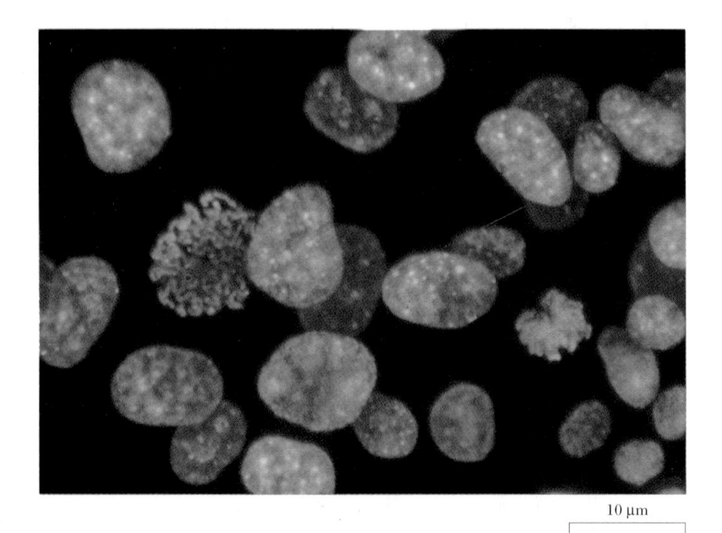

Figure 9-6 The DNA content in dividing cells changes with the phase of the cell cycle. Cells exposed to a DNA-binding fluorescent dye show different amounts of fluorescence of chromosomes and chromatin, depending on the phase of the cell cycle. *(David Phillips/Visuals Unlimited)*

cells can undergo extensive growth and specialization during G_0. In many cells, such as white blood cells, however, G_0 represents a quiescent state with only low levels of protein synthesis and other cellular activities.

HOW DOES MITOSIS DISTRIBUTE ONE COPY OF EACH CHROMOSOME TO EACH DAUGHTER CELL?

The major task of mitosis is to distribute one chromatid from each chromosome to each daughter cell. Remarkably, all eukaryotic cells accomplish mitosis with the same chromosomal ballet that Flemming first showed more than 100 years ago. Modern optical techniques allow biologists to watch the whole process in living cells (far more conveniently than could Walther Flemming) with a phase-contrast or an interference contrast microscope, as in Figure 9-7. In addition, immunocytochemistry with antibodies to specific molecules has allowed researchers to follow the movements of individual participants in the mitotic process.

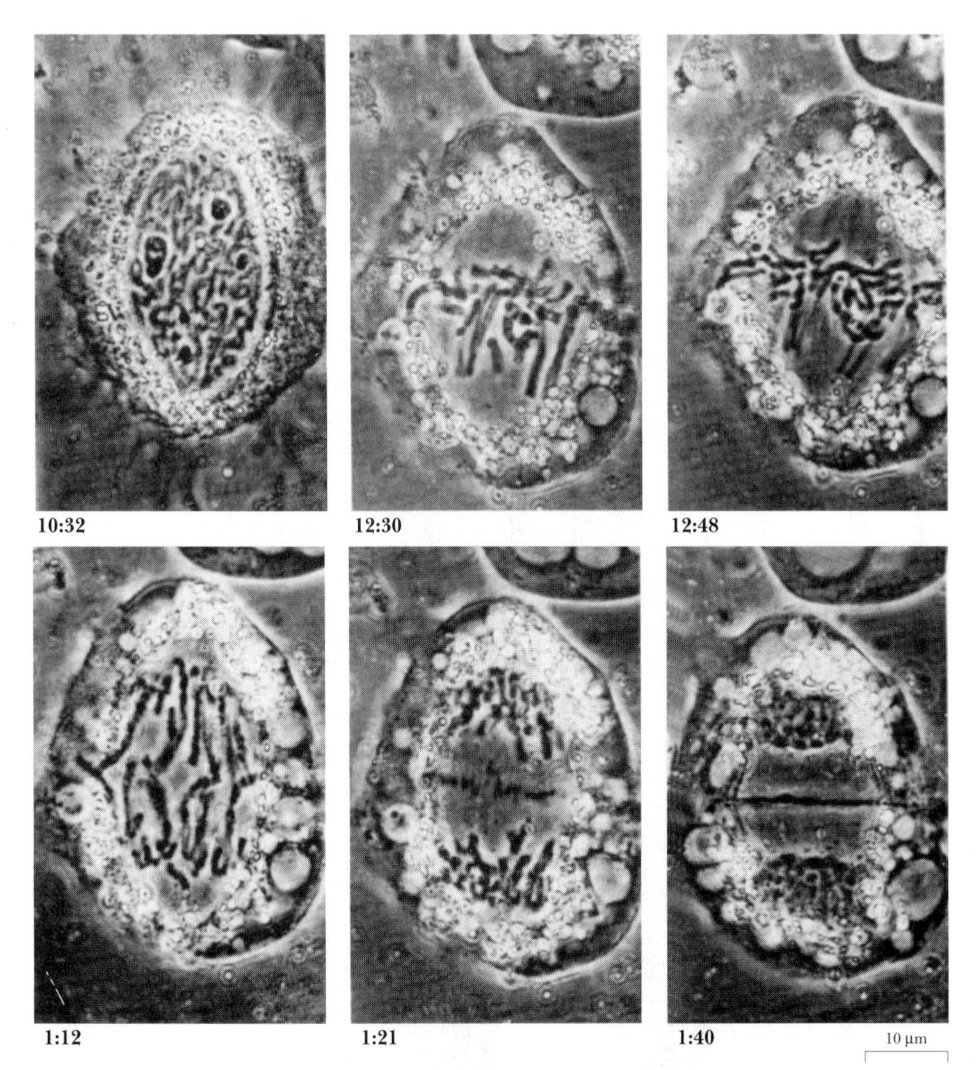

Figure 9-7 Modern microscopic techniques allow biologists to observe mitosis in living cells, here in a lily (*Haemanthus*), which has unusually large chromosomes. The chromosomes behave in exactly the same way as can be seen in stained cells, showing that the appearance and apparent movement of the chromosomes did not result from artifacts of staining. *(Dr. Andrew Bajer/University of Oregon)*

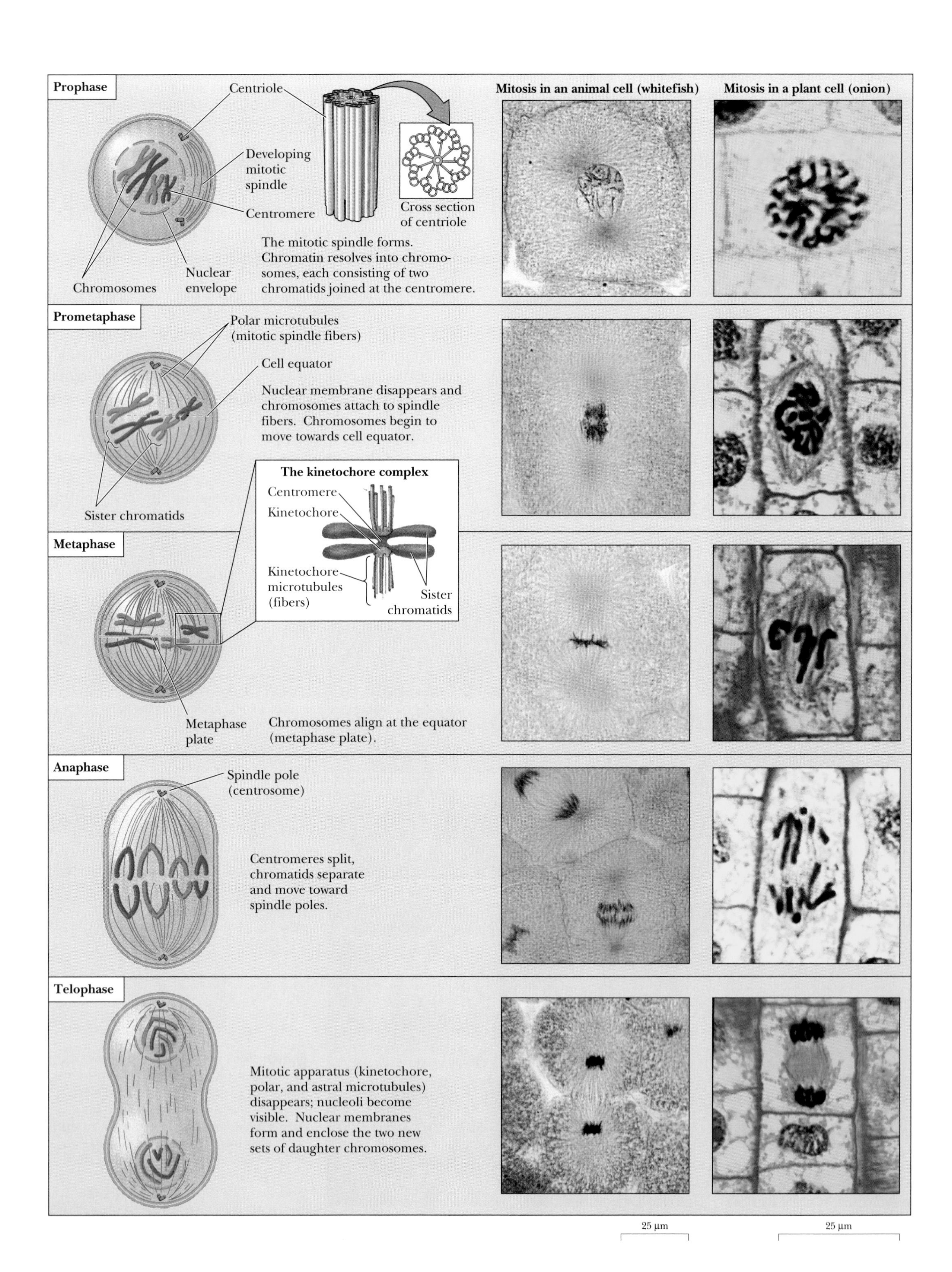
Prophase
Centriole
Developing
mitotic
spindle
Centromere
Cross section
of centriole
Chromosomes
Nuclear
envelope
The mitotic spindle forms.
Chromatin resolves into chromo-
somes, each consisting of two
chromatids joined at the centromere.
Mitosis in an animal cell (whitefish)
Mitosis in a plant cell (onion)
Prometaphase
Polar microtubules
(mitotic spindle fibers)
Cell equator
Nuclear membrane disappears and
chromosomes attach to spindle
fibers. Chromosomes begin to
move towards cell equator.
Sister chromatids
The kinetochore complex
Centromere
Kinetochore
Kinetochore
microtubules
(fibers)
Sister
chromatids
Metaphase
Metaphase
plate
Chromosomes align at the equator
(metaphase plate).
Anaphase
Spindle pole
(centrosome)
Centromeres split,
chromatids separate
and move toward
spindle poles.
Telophase
Mitotic apparatus (kinetochore,
polar, and astral microtubules)
disappears; nucleoli become
visible. Nuclear membranes
form and enclose the two new
sets of daughter chromosomes.
25 μm
25 μm

Mitosis Is a Continuous Process, But Biologists Distinguish Five Phases

For many years, biologists have divided mitosis into four phases. In recent years, however, many cell biologists have preferred to divide the process into five phases, as shown in Figure 9-8: prophase, prometaphase, metaphase, anaphase, and telophase. All of these phases are somewhat arbitrary because mitosis is a continuous process, but the events described always take place in exactly the same order and always end with G_1, the beginning of the next interphase.

During Prophase, Chromosomes Condense and the Mitotic Spindle Forms

During the first phase of mitosis, **prophase** [Greek, *pro* = before], the diffusely stained chromatin resolves into discrete chromosomes, each consisting of two chromatids, joined together at the centromere, as shown in Figure 9-8. As the chromosomes become visible within the nucleus, the nucleoli disappear. This process takes about 10–15 minutes in mammalian cells in culture.

At the same time a new, football-shaped structure, called a **mitotic spindle,** develops outside the nucleus. The mitotic spindle contains the machinery that moves the chromatids apart. The electron microscope reveals that the spindle contains prominent bands of microtubules, 25-nm hollow tubes of protein, identical in appearance to those of the cytoskeleton, as well as of cilia and flagella. (See Chapter 6.)

In animal cells (and in some other eukaryotes as well), each pole of the mitotic spindle contains a pair of **centrioles,** small cylindrical organelles about 0.2 μm in diameter and 0.4 μm long, that lie at right angles to one another. Electron microscopic examination of a centriole in cross-section shows nine clustered groups of microtubules lying around the circumference, with faint spokes radiating inward, rather like a cartwheel (as the inset shows in Figure 9-8).

Microtubules emanate from each pole of the spindle. Some, the **polar microtubules,** run toward the equator. Other microtubules, called **astral microtubules,** extend outward from each centriole to form an aster, as first noticed by Flemming. The cells of vascular plants, however, usually have neither centrioles nor asters. Mitosis in these plants is said to be **anastral,** in contrast to **astral mitosis** in animals and in nonvascular plants. By the end of prophase, the spindle has started to elongate, and the two poles (with or without asters) begin to move apart.

◀ ***Figure 9-8*** Stages of mitosis, shown diagrammatically, correspond to stages observed in the microscope. The components of the mitotic apparatus are labeled; note the appearance and locations of the mitotic spindle, sister chromatids, centriole, centrosome, kinetochores, centromere, kinetochore microtubules, polar microtubules, metaphase plate, and asters. *(Whitefish photos: John Cunningham/Visuals Unlimited; onion photos: top, Cabisco/Visuals Unlimited, middle (3), Robert Calentine/Visuals Unlimited, bottom, Jack M. Bostrack/Visuals Unlimited)*

During Prometaphase the Nuclear Membrane Disappears and the Chromosomes Attach to the Spindle Fibers

The next phase of mitosis is called **prometaphase** [Greek, *meta* = middle] (previously called early metaphase). Prometaphase begins with the disappearance of the nuclear membrane, which allows the chromosomes to attach to the mitotic spindle. In mammalian cells, this process usually takes 10–20 minutes.

The inset drawing in Figure 9-8 reveals how the chromosomes attach to the spindle microtubules. Each chromatid develops a **kinetochore,** a specialized disc-shaped structure that attaches the mitotic spindle to the centromere. Some of the polar microtubules attach to kinetochores (and thus to chromosomes) and become **kinetochore microtubules.** By the end of prometaphase, then, the animal cells have three sets of microtubules, all of which originate at a spindle pole: (1) kinetochore microtubules (which run from a pole to a kinetochore); (2) polar microtubules (which run from each pole toward the equator); and (3) astral microtubules (which extend from each pole without attaching to any other visible structure). Again, vascular plant cells contain kinetochore and polar microtubules, but no astral microtubules. After the spindle apparatus has formed during prometaphase, the attached chromosomes begin to move along the spindle toward the cell's equator.

Chromosomes Align During Metaphase

By the end of **metaphase,** the next (and longest) stage of mitosis, the chromosomes have moved halfway between the two poles of the spindle, where they accumulate in a disc, called the **metaphase plate.** All the chromosomes lie (usually for about 1 hour) in a single plane at the equator (at right angles to the spindle fibers). At this stage, we can best see the individual characteristics of the chromosomes, which (among other things) may differ in size and in the placement of the centromere.

Chromatids Separate During Anaphase

Anaphase [Greek, *ana* = up, again], the most dramatic stage of mitosis, is the time of chromatid separation. All at once, the chromatids of each chromosome simultaneously begin to move in opposite directions away from the metaphase plate. All the chromatids move at the same speed (about 1 μm per minute) toward the poles of the spindle, and the whole process lasts for only about 5–10 minutes. The centromere of each chromosome splits so that each chromatid contains its own centromere connected to kinetochore microtubules. The centromeres appear to lead the way, with the rest of the moving chromatid lagging behind. This pattern suggests that the kinetochore microtubules are pulling the chromatid toward the poles.

The appearance of the chromosomes at anaphase differs among species, depending on the placement of the centromere (and kinetochore) on the chromosome. If the centromere is at or near the middle of a chromosome (as is the case for many human chromosomes), the chromosome is said to be **metacentric** and looks like a V. If the centromere is much closer to one end than the other (as is the case for other human chromosomes), the chromosome is said to be **acrocentric** and looks like a bent capital L. If the centromere is at one end of each chromosome (as is the case for mouse chromosomes), the chromosomes are said to be **telocentric.**

Even as the chromosomes move toward the spindle poles, the poles themselves are moving farther apart, often doubling the distance between them. The mechanism that accomplishes the polar separation is distinct from that responsible for chromosome movement toward the poles, and it is sensitive to disruption by different cellular poisons. To distinguish between these two processes, cell biologists often refer to movement of the chromosomes as **anaphase A** and the movement of the poles as **anaphase B.** By the end of anaphase, the two processes, often acting simultaneously, have brought about the separation of the two sets of chromatids to positions around the two spindle poles, some 10–20 μm apart.

The Mitotic Apparatus Disappears During Telophase

During **telophase** [Greek, *telos* = end], the last phase of mitosis, the mitotic apparatus (including kinetochore, polar, and astral microtubules) disperses. The chromosomes lose their distinct identities, once more assuming the diffuse appearance of chromatin. Finally, nuclear membranes form and enclose the two new sets of daughter chromosomes, and the nucleoli again become visible. Each daughter nucleus has the same number of chromosomes as the parent nucleus, but now each chromosome consists of a single chromatid. Each chromosome replicates during the next S phase.

What Propels the Chromosomes During Mitosis?

We have described three sets of movements whose molecular mechanisms we would like to understand: (1) the movement of the chromosomes toward the equator during prometaphase to form the metaphase plate; (2) the movement of the separated chromatids toward the poles during anaphase A; and (3) the movement of the spindle poles away from each other during prophase and again during anaphase B. The chromosomes themselves do not play active roles in the movements of mitosis. Rather, the major players are the microtubules, and understanding the mechanics of mitosis requires an appreciation of how microtubules assemble and fall apart.

Other cell components also contribute to mitosis, including the kinetochores and microtubule-associated proteins (MAPs). In many species, including all animals and many protists, but not plants, the centrioles also play an important role. In order to determine how these molecules bring about the various migrations of chromosomes and spindle poles, cell biologists have not only examined dividing cells with the light and electron microscopes but have also tried to recreate the same processes in the test tube, using purified molecular components.

Microtubules and MAPs Bring About Chromosomal Movements

The microtubules of the mitotic spindle consist mostly of α- and β-tubulin as illustrated in Figure 6-27, arranged in 13 parallel chains. The spindle microtubules do not form the triplets and doublets like the more permanent microtubules in cilia and flagella.

Spindle microtubules rapidly appear and disappear and, like the microtubules of the cytoskeleton, are said to be dynamically unstable: Unless they are actually growing or stabilized by specific "cap" proteins, they begin to shrink by losing tubulin molecules. In fact, the average lifetime of a microtubule is only about 1 minute. By studying the assembly of tubulin into microtubules *in vitro* [Latin = in glass, that is, in a test tube], cell biologists have come to understand much about the way cells regulate the assembly and disassembly of microtubules during mitosis. (See Chapters 6 and 18.)

Microtubule organizing centers (MTOCs) promote microtubule formation—both in cells and *in vitro*—by blocking their *minus* ends and allowing the accumulation (or, under other conditions, the loss) of tubulins at their *plus* ends. MTOCs isolated by subcellular fractionation stimulate such directed growth *in vitro.*

Microtubule-associated proteins (MAPs) also participate in mitosis. (See Chapter 6.) Fibrous MAPs bind tightly to tubulins, stabilizing microtubules and allowing them to interact with other cellular components. The motor MAPs, **kinesin** and **dynein,** use microtubules like railroad tracks, with kinesin carrying materials toward the plus end and dynein toward the minus end.

Microtubule Organizing Centers Play a Crucial Role in the Assembly and Disassembly of the Mitotic Spindle

During interphase each cell contains a single MTOC, called the **cell center** or **centrosome,** a complex of proteins that (in animal cells) includes a pair of centrioles. (Plant cells have centrosomes, but no centrioles.) In an interphase cell, microtubules radiate from the centrosome toward the cell's surface, forming part of the cell's cytoskeleton.

As a cell enters prophase, the centrosome (with or without a centriole) duplicates and forms two centrosomes. New centrosomes come only from old centrosomes. When centrioles are present, each centriole somehow stimulates the formation of a second centriole, which lies at right angles to it. The microtubules of the cell's cytoskeleton (previously assembled from the centrosome) fall apart and release tubulin molecules into the cytoplasm. About 100 million of these tubulin molecules recycle, adding to the microtubules that emanate from the duplicated centrosomes. At the end of mitosis, the spindle apparatus disassembles, and some of the tubulin molecules then reassemble into cytoskeleton.

During prophase, the polar microtubules grow between the two centrosomes, with their two plus ends overlapping. This overlap evidently helps to stabilize these microtubules until the end of metaphase. During anaphase and telophase, the polar microtubules increase in length, forcing apart the two poles of the spindle appartaus. Part of the energy for this process comes from the splitting of ATP and may depend on a form of dynein associated with the polar microtubules.

During metaphase, kinetochores capture the plus ends of some microtubules whose minus ends lie in the centrosomes at the poles, as shown in Figure 9-9. During anaphase, the plus ends of the kinetochore microtubules begin to shorten. The kinetochore, however, maintains contact between the centromere and the attached microtubule, and this shortening is thought to provide at least part of the force that pulls the chromosomes to the poles.

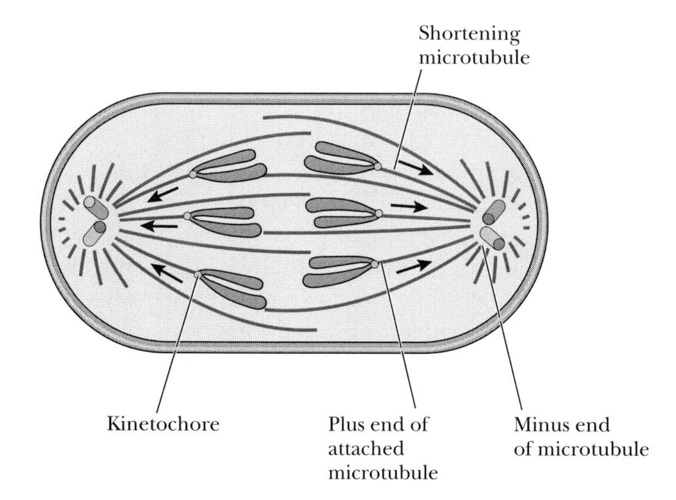

Figure 9-9 During metaphase, the plus ends of some spindle microtubules begin to attach to kinetochores, which are in turn attached to the centromeres.

Three Kinds of Experiments Demonstrate the Importance of Microtubules in Mitosis

Cell biologists have studied the role of microtubules (1) by using drugs that interfere with microtubule assembly, (2) by disrupting the mitotic spindle with a finely focused laser, and (3) by selectively labeling newly assembled (as opposed to pre-existing) microtubules. An example of the first approach is the use of colchicine, a product of the meadow saffron plant *Colchicum autumnale*, which has been used medicinally for 3000 years. Colchicine blocks mitosis by binding to tubulin molecules and thereby inhibiting their assembly into microtubules. After colchicine treatment, chromosomes condense normally but do not complete their movements. Instead, the condensed chromosomes form an array that resembles a metaphase plate. Treatment with colchicine (or with other similar drugs, like colcemide and nocadazole, which are now more widely used) provides a useful way of obtaining condensed chromosomes, which can then be counted and distinguished from one another.

Laser microsurgery has demonstrated the importance of the kinetochore microtubules. In one experiment, for example, the researcher cut the kinetochore fibers that attach one sister chromatid to one pole at metaphase. The result was that both sister chromatids moved toward the opposite pole, thus demonstrating that the kinetochore microtubules ordinarily pull the two chromatids in opposite directions.

To label newly made microtubules, researchers first isolate tubulin molecules (not assembled into microtubules) and chemically couple each tubulin to a small organic molecule called biotin. They then inject these labeled tubulin molecules into cells undergoing mitosis. One experiment that selectively labeled kinetochore fibers, for example, showed that tubulin molecules specifically add to the kinetochore ends during metaphase and then move toward the poles.

Application of the same technique to the chromosome movements of anaphase showed that tubulin molecules are lost from kinetochore microtubules as the chromosomes migrate toward the poles. The question remains, however, whether the gain or loss of tubulin molecules is the cause of chromosome movement or merely the result. ATP-splitting motor molecules are also present in the vicinity of the kinetochores, and they may be responsible for chromosome movements. (See Chapter 18.) Figure 9-10 shows two models of how the kinetochore and microtubules may interact to cause chromosome movement.

The construction and operation of the mitotic machinery seem to depend on interactions both among tubulin molecules and between tubulin molecules and other proteins of the spindle, centrosomes, and kinetochores. Cell biologists still do not fully understand the mechanism of mitosis, and the identities and roles of the components of the mitotic machinery are the subject of intense research.

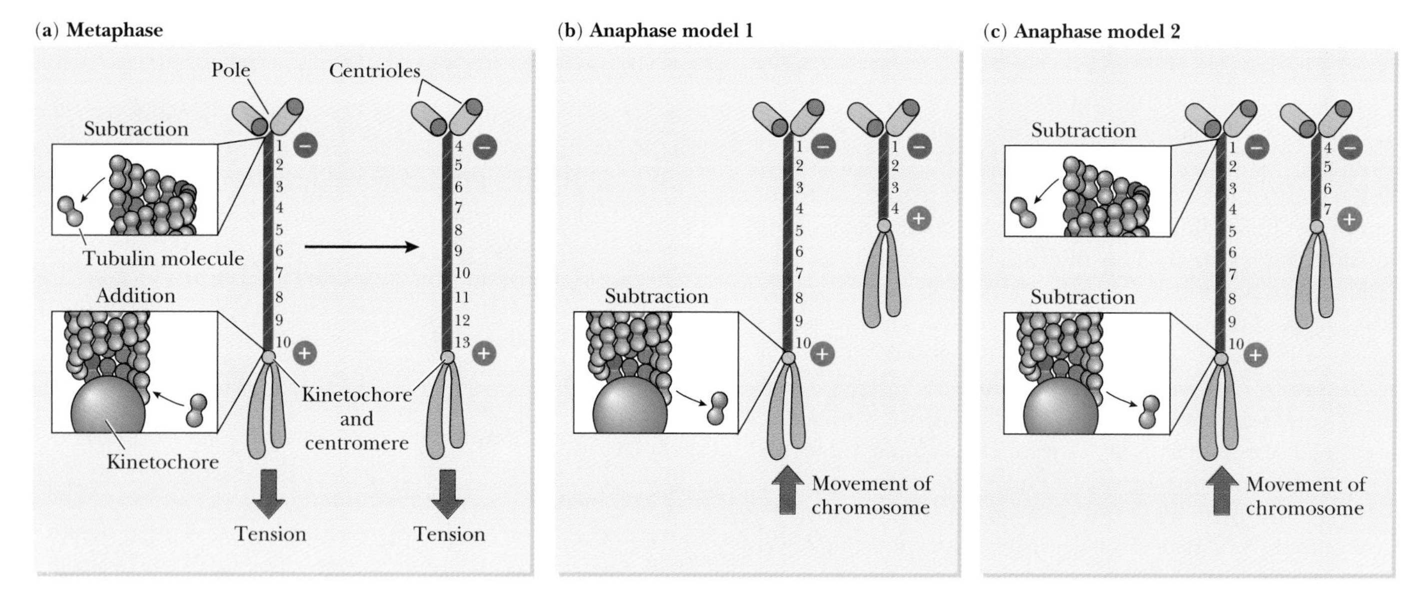

Figure 9-10 Alternative models for the role of tubulin polymerization within the kinetochore: Is microtubule disassembly a cause or a result of chromosome movement?

HOW DOES A CELL FIT ALL ITS DNA INTO A NUCLEUS?

Based on the information stored in its DNA, the nucleus issues molecular orders for the construction of cell machinery. When a cell divides, the cell must duplicate this information and pass it on to its two daughters.

Each molecule of DNA contains two polynucleotide strands, wrapped around each other in a double-helix. (See Chapters 3 and 6.) This rope-like structure is so fine that a DNA molecule long enough to circle the earth at the equator would weigh only slightly more than 1 mg (less than a grain of sand)! In contrast, if fine thread 100 μm (1/250 inch) in diameter were to circle the earth, the needed length would weigh well over 600 pounds. The DNA in each of your cells weighs about 6×10^{-12} grams and has a total length of about 2 meters. A cell must therefore solve a major problem: how to fit this long molecule into a nucleus only 5 μm or so in diameter without the long threads of DNA getting tangled as the cell copies its instructions.

During mitosis a single chromosome is about 1–10 μm long, while its DNA may be 1,000–10,000 times longer. Fitting DNA into so small a structure depends mostly on its tight interactions with specialized proteins. The most abundant of these are the **histones** [Greek, *histos* = web], a set of small, positively charged proteins. Most cells have five types of histones (called H1, H2A, H2B, H3, and H4), each of which may be present at about 60 million copies per cell.

Histones and DNA Interact to Form Particles Called Nucleosomes, Which Resemble Beads on a String

The positive charge of histones comes from their unusually high content of the basic amino acids, arginine and lysine. At cellular pH, the basic side chains of these amino acids attract hydrogen ions and give a net positive charge to each of the histone molecules. As shown in Figure 9-11, the positively charged histones bind tightly to DNA, whose phosphate groups give it a negative charge. DNA and four of the five histones associate with each other to form a structure, shown in Figure 9-12a, b, and c, whose appearance in the electron microscope resembles beads on a string. Each bead is called a **nucleosome** and is a DNA-histone complex, about 11 nm in diameter. Each nucleosome contains a 146 nucleotide–long stretch of DNA and eight histone molecules (two each of H2A, H2B, H3, and H4). Histone H1 also binds to DNA, but not in the nucleosomes themselves. The formation of nucleosomes is the first stage in the packing of DNA into a nucleus.

This method of packing DNA must have evolved long ago because histones from all eukaryotes resemble one another. For example, histone H4 of cows differs in only two amino acid residues (out of 102) from that of peas. This extraordinary similarity suggests that the structure of histone H4 has remained almost unchanged for a billion years or more, and that natural selection has tolerated few variations in histone structure during evolution.

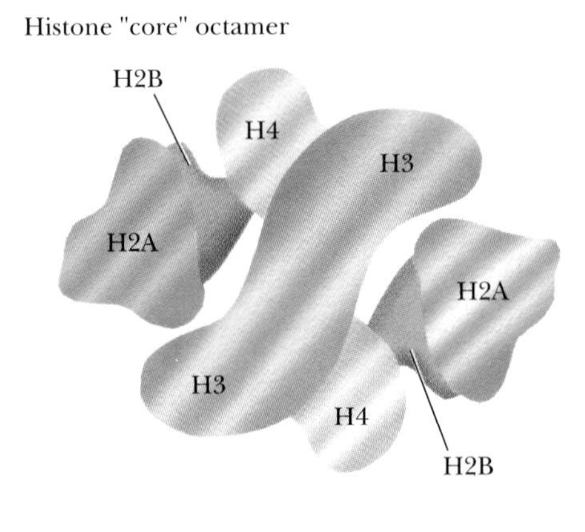

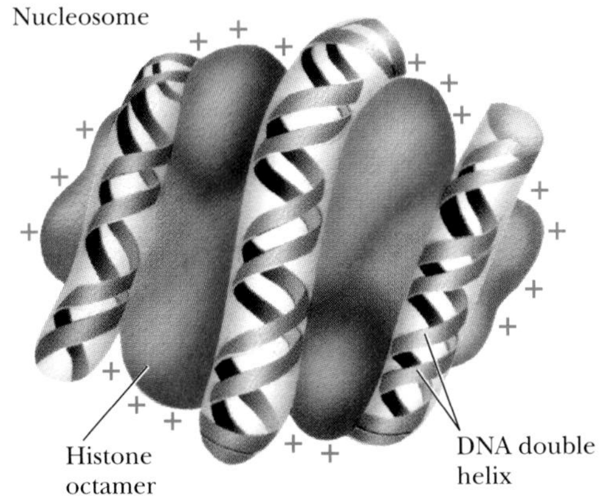

Figure 9-11 Histones, which are positively charged proteins, bind to DNA, which is negatively charged. Eight histones and their associated DNA form a nucleosome.

DNA, Histones, and Other Proteins Fold to Form Still More Compact Structures

Histone H1 appears to be responsible for holding together groups of nucleosomes to form a hollow tube, or solenoid, that is 30 nm in diameter, as shown in Figure 9-12d and e. As a result of this packing, the DNA of an average human chromosome could be condensed from an average length of 5 cm to about 1 mm—a terrific achievement, but still not enough to fit into a 5-μm nucleus.

We still do not know exactly how DNA is made more compact. In some specialized cells, microscopists can see loops extending from chromosomes, such as those in Figure 9-12f and g. Many biologists now think that such loops are common to the chromatin of all cells. Perhaps some other proteins stabilize the coiling of the 30-nm fiber.

During mitosis, the DNA is even more condensed in chromosomes than it is in nondividing cells. DNA and protein form still more loops. In each mitotic chromosome these loops cluster in characteristic ways.

The specific pattern of DNA clustering results in **chromosome banding,** the pattern, visible in a microscope, that results from the selective binding of certain dyes, as first discovered by the Swedish biochemist Torbjörn Caspersson in 1971. The banding patterns, such as those shown in Figure 9-13, allow microscopists to distinguish among individual chromosomes—a feat that allows the detection of many chromosomal abnormalities.

Only Limited Regions of DNA Can Replicate at Any One Time

Before a cell can enter mitosis, it must have two chromatids, that is, two copies of each chromosome. It must not only produce two copies of each chromosome's DNA, but the DNA must associate with proteins to form nucleosomes and more elaborately folded structures. Electron micrographs, such as that in Figure 9-14, show the doubling of a nucleosome fiber.

There is just not room in the nucleus, however, for all the highly condensed DNA to unpack and replicate at the same time. Instead, limited regions of DNA replicate as a unit and then reform into chromatin. The whole process of chromosome doubling, the S phase, takes about 8–10 hours in most eukaryotic cells. After DNA has replicated, both old and newly made histones associate with the newly made DNA molecules.

■ HOW DOES A CELL DIVIDE ITS CYTOPLASM?

So far we have discussed how a cell distributes the two copies of its genetic information. But we still need to describe cytokinesis—the process by which a dividing cell partitions its cytoplasm, including the two new nuclei that form during mitosis.

Cytokinesis and mitosis are separate processes, and they depend on different molecular machinery. In some organisms, for example, mitosis sometimes occurs without cytokinesis, leading to two or more nuclei within a single cell, as in the early fruit fly embryo and in specialized tissues of a plant seed. Similarly, treatment of dividing cells with cytochalasin B, a poison made by a fungus, prevents cytokinesis without stopping mitosis. Almost always, however, cytokinesis accompanies mitosis, usually beginning during anaphase and finishing shortly after the end of telophase.

What Molecular Structures Accomplish Cytokinesis in Animal Cells?

During cytokinesis, the plane of cell division is always perpendicular to the axis of the mitotic spindle, so that the separation between cells forms around the spindle's equator, as shown in Figure 9-15a and b. Cytokinesis, however, uses actin filaments to achieve its movements instead of microtubules (which are used in mitosis). (See Chapter 6, p. 179.) In early anaphase, dividing cells form a **contrac-**

(a)

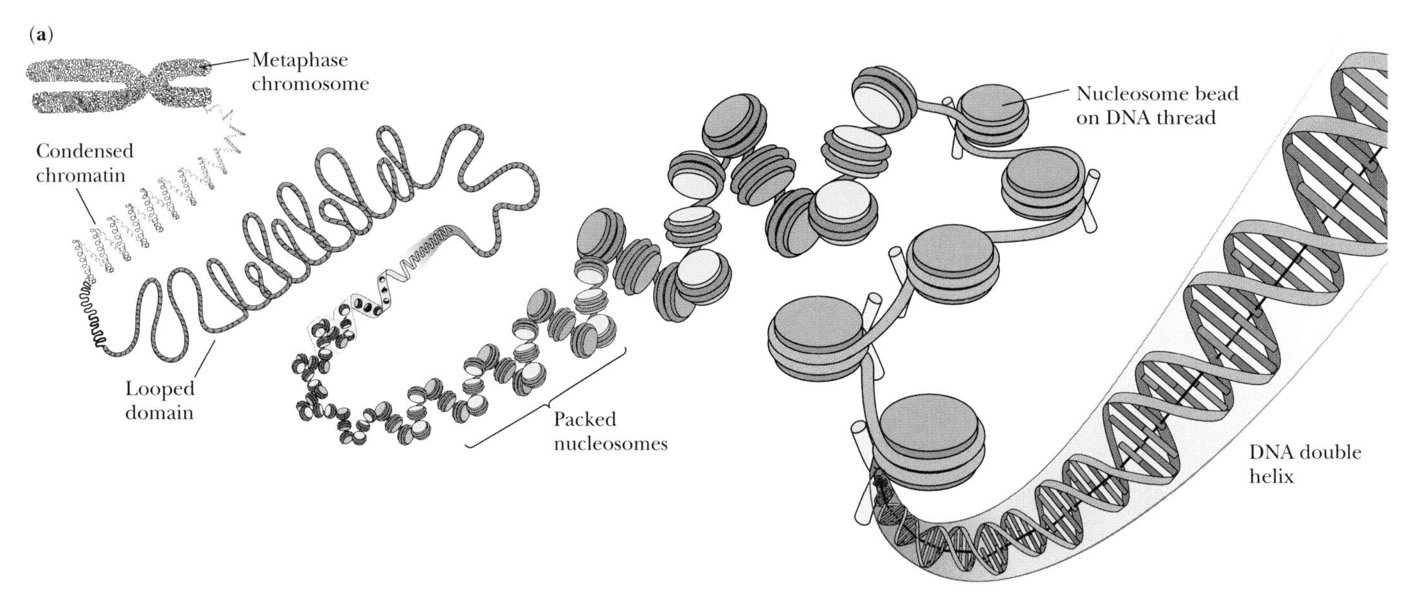

(b) "Beads on a string" chromatin form

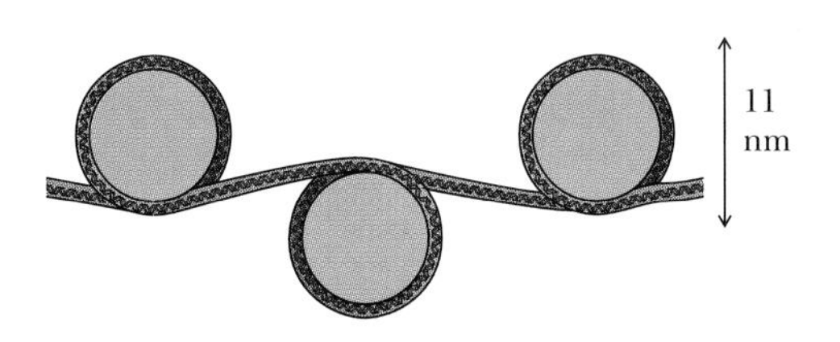

(c)

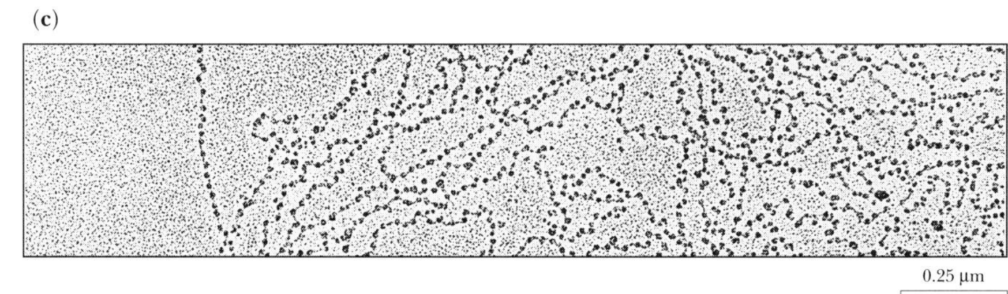

0.25 µm

(d) Solenoid (six nucleosomes per turn)

(e)

0.1 µm

(f) Loops (50 turns per loop)

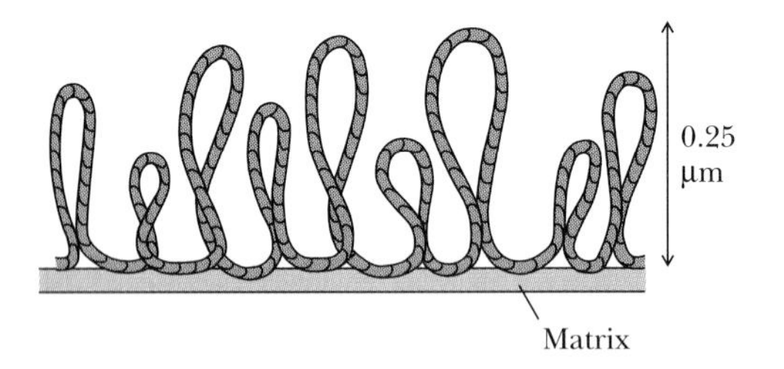

(g)

25 µm

Figure 9-12 DNA in a cell is tightly folded, both in an interphase cell and in a cell undergoing mitosis. **(a)** The folding in a metaphase chromosome; **(b)** nucleosomes associated with DNA resemble beads on a string; **(c)** TEM of nucleosomes; **(d)** DNA-nucleosome strings coil to form a 30 nm fiber; **(e)** TEM of 30 nm chromatin fiber; **(f)** a chromatin fiber folds and forms loops; **(g)** TEM of looped chromosomes. *(c, S. L. McKnight and O. L. Miller,* Cell: *449–454, 1976; e, Barbara Hamkalo, UC, Irvine; g, Joseph Gall, Carnegie Institution)*

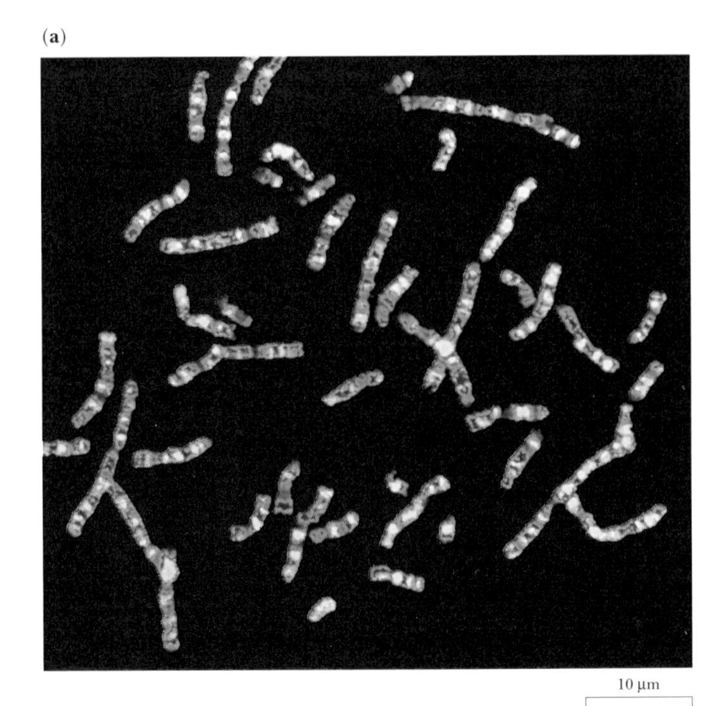

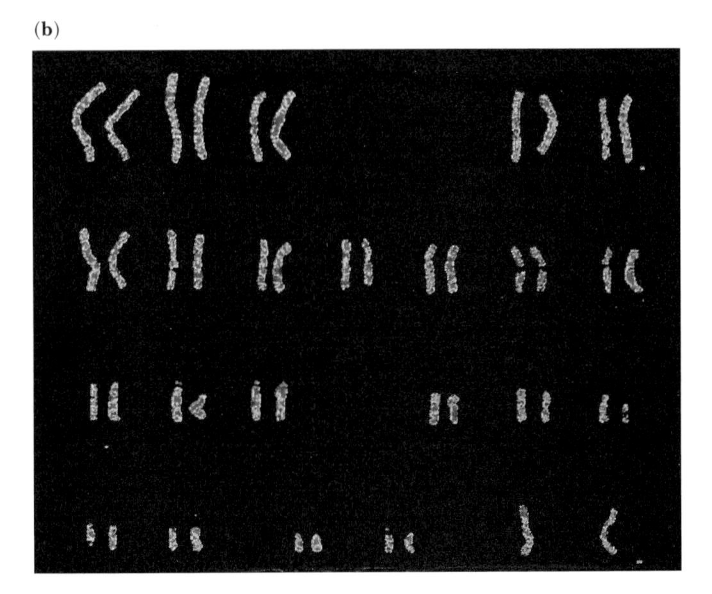

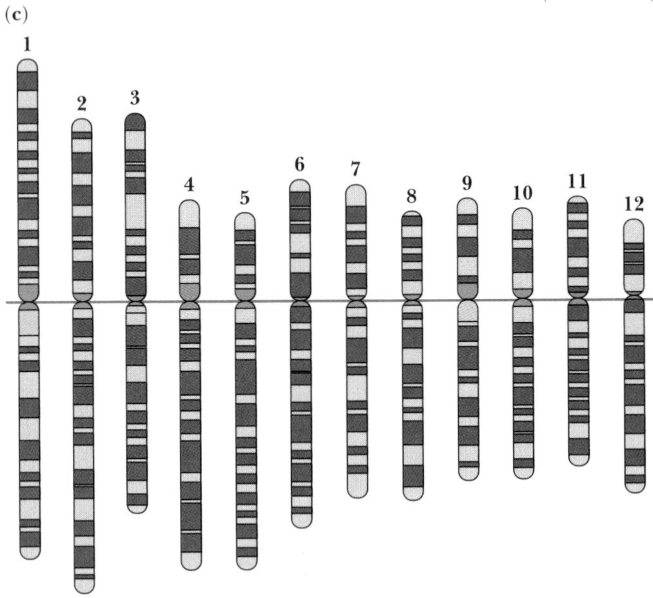

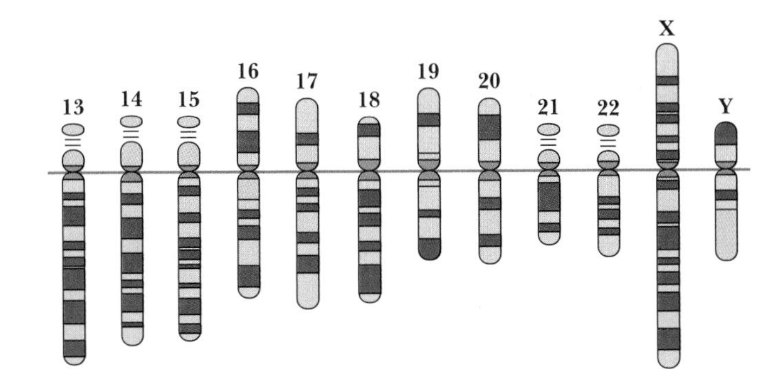

Figure 9-13 Certain dyes reveal a banding pattern in chromosomes that results from different clustering of DNA and chromatin proteins. Each chromosome has a distinctive pattern, allowing cell biologists to distinguish them and to notice that they come in matched pairs. **(a)** A human cell, showing chromosome banding; **(b,c)** a human karyotype, showing individual human chromosomes in order of descending size. *(a, b, Custom Medical Stock)*

tile ring, a bundle of actin filaments that surrounds the dividing cell. The contractile ring works like a purse string, eventually pinching the cell into two parts (Figure 9-15c).

In animal cells, the force for this movement comes from a mechanism similar to that of muscle contraction, in which interacting actin and myosin molecules use energy released from the hydrolysis of ATP. As in the case of muscle, calcium ions play an important role in regulating the contraction. (See Chapter 18.) Like the mitotic spindle, the contractile ring is a transient structure, in contrast with the permanent contractile structures of muscle cells.

As the contractile ring tightens, the plasma membrane pulls inward to form a deepening groove, called a **cleavage furrow** (shown in Figure 9-15c). As telophase proceeds, the two daughter cells become almost completely separated, but a thin connection between the daughter cells, the **midbody** (shown in Figure 9-16), persists until the end of mitosis. The midbody is packed with microtubules from the spindle apparatus. Finally, the microtubules disassemble, and the midbody breaks, leaving the daughter cells completely separated.

As the two daughter cells separate, they must increase their total amount of plasma membrane (because two cells have more surface area than a single cell of the same vol-

(a)

Contractile ring of actin filaments

(b)

Cleavage furrow

(c)

250 μm

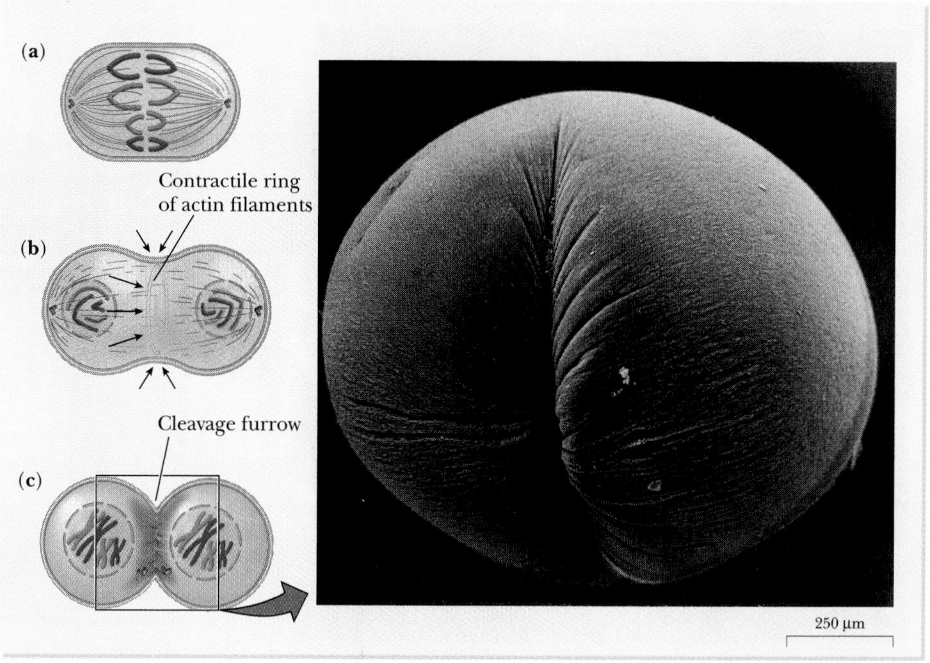

Figure 9-15 During cytokinesis, a contractile ring of actin filaments pinch the cell into two parts. **(a)** At anaphase, the chromosomes move apart, but stay within a single cell; **(b)** in early telophase, the daughter cells begin to pinch apart; **(c)** in late telophase, the daughter cells become almost separated. On the right is an SEM of a cleavage furrow, which is formed by the contractile ring of actin filaments located inside of the cell. *(David M. Phillips/Visuals Unlimited)*

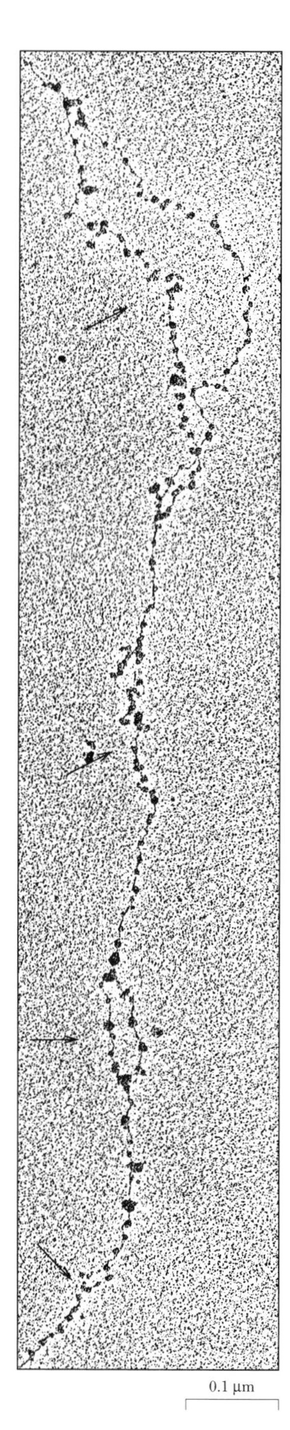

Figure 9-14 When DNA replicates, it remains associated with histones. Here, a chromatin fiber in an early *Drosophila* embryo has just replicated, and there are nucleosomes both in the still unreplicated single fiber and in the newly replicated double fiber. *(Victoria Foe)*

ume). The additional membrane material is immediately available because the parent cell has made extra membrane during interphase and accumulates it before entering mitosis.

How Does a Dividing Plant Cell Build a New Cell Wall Between Its Daughters?

Because of its rigid cellulose wall, a plant cell cannot simply pinch in two. Thus, cytokinesis in plant cells requires the building of new cell walls between the daughter cells.

The beginnings of the new partition are visible in telophase. The new wall begins as a small, flattened disc, called the early cell plate, in the center of the space between the two daughter nuclei, as shown in Figure 9-17. The disc grows to become a **cell plate,** which eventually seals the two daughters off from one another. The electron microscope reveals that the early cell plate is closely associated with a set of microtubules, called the **phragmoplast** [Greek, *phragmos* = fence + *plasma* = mold, form], which extends between the two cells at right angles to the cell plate.

Where do the materials come from to build the cell plate? Associated with the phragmoplast are tiny vesicles (which are visible in Figure 9-17b) that originated in the Golgi apparatus. These vesicles, which contain the cell wall precursors, travel down the fibers of the phragmoplast to the midpoint, where they fuse to form the early cell plate. After the early cell plate is formed, the microtubules disassemble. New microtubules then attach to the outside of the early cell plate. Again, more vesicles travel to the disc. After several such cycles, enough material accumulates so that the cell plate forms a complete cross wall between the two cells. The two cells remain connected through continuous regions of the smooth endoplasmic reticulum, called **plasmodesmata.**

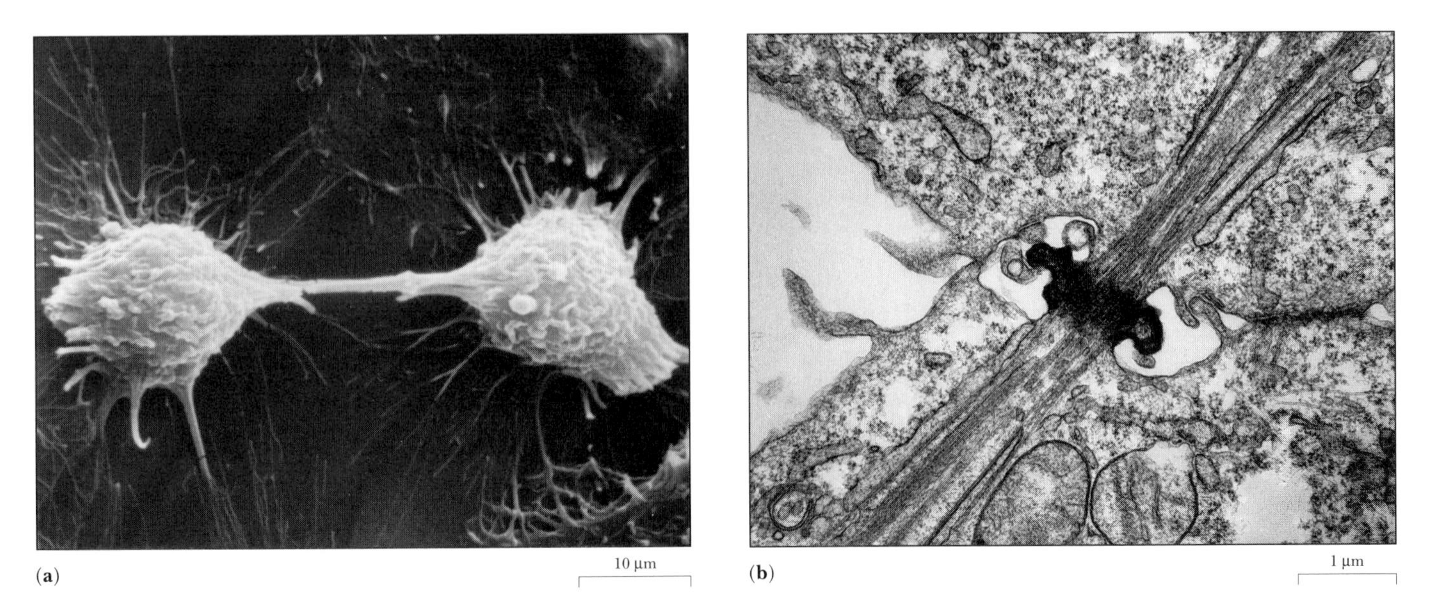

Figure 9-16 Near the end of telophase, the midbody still connects the two daughter cells. **(a)** Scanning EM of two daughter cells nearly separated; **(b)** TEM of midbody, showing overlapping microtubules extending from the two daughter cells.
(a, C. Y. Shih, R. Kessel/Visuals Unlimited; b, David M. Phillips/Visuals Unlimited)

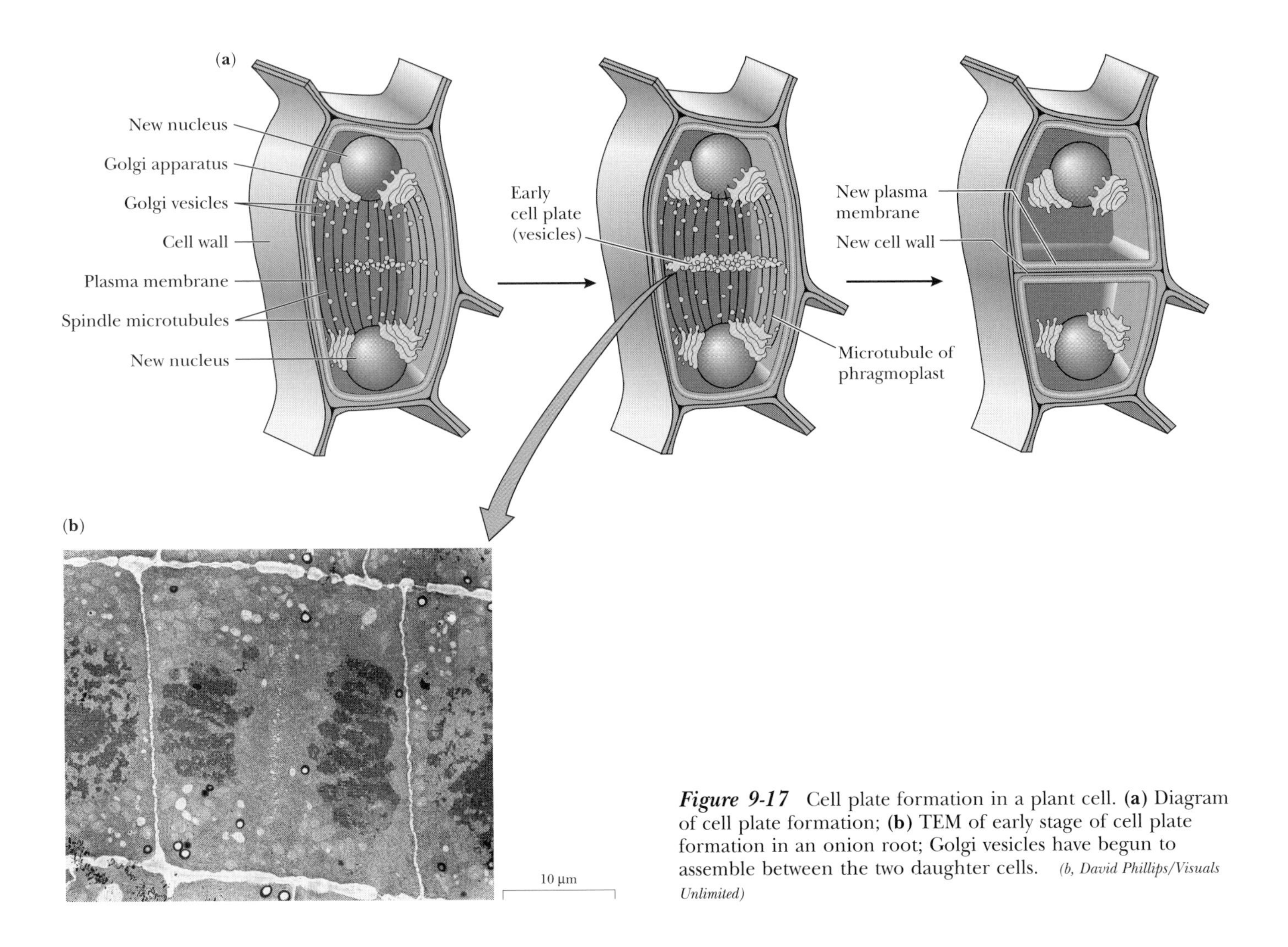

Figure 9-17 Cell plate formation in a plant cell. **(a)** Diagram of cell plate formation; **(b)** TEM of early stage of cell plate formation in an onion root; Golgi vesicles have begun to assemble between the two daughter cells. *(b, David Phillips/Visuals Unlimited)*

HOW DOES A CELL REGULATE PASSAGE THROUGH THE CELL CYCLE?

Cells must somehow sense when to divide. In the previous part of this chapter, we saw how cells accomplish the equal distribution of chromosomes by mitosis and the division of cytoplasmic components by cytokinesis. But how do cells know when to begin the cell cycle? And how do they coordinate its separate parts?

When Does a Cell Divide?

Most cells divide only after they have first doubled their mass. Otherwise, daughter cells would get smaller with every generation. Most cells therefore divide only after they somehow sense that they have reached a critical size and have enough nutrients. We do not know how eukaryotic cells accomplish this sensing, but once they do, they become irreversibly committed to cell division. Once a cell proceeds beyond a certain point in G_1, the cell proceeds through the rest of the cycle, including mitosis and cytokinesis. This "point of no return" is called **Start** or the **restriction (R) point.** Arrival at Start depends heavily on the environment of the cell—on the availability of nutrients and on signals from other cells. Once the cell has gone through Start, however, the rest of the cycle proceeds independently of the extracellular environment.

The early embryos of many animal species are exceptions to the general rule that cells double their mass before dividing. Many embryos derived from large eggs begin life after fertilization with a series of rapid **cleavage divisions,** which partition the embryo without increasing its mass. A frog egg, for example, is a single cell about 1 mm in diameter. After fertilization, the cleavage divisions rapidly divide the cytoplasm into cells that are more nearly the size of cells in the adult, about 10–20 μm in diameter. These smaller cells are then free to follow different developmental paths. (See Chapter 20.) The rapid cell divisions in the embryos of frogs, sea urchins, and many marine invertebrates have made them valuable experimental organisms for studying the coordination of the cell cycle.

Apparently Universal Mechanisms Coordinate Events of DNA Synthesis, Mitosis, and Cytokinesis

Cell reproduction requires the coordination of three cycles, each of which depends on separate subcellular components: (1) the duplication and packaging of DNA, which depend on specific nuclear enzymes and on histones and other chromosomal proteins (see Chapter 13); (2) the duplication of the centrosomes and the operation of the mitotic spindle apparatus, which depend on the dynamic properties of microtubules, kinetochores, and MTOCs; and (3) cytokinesis, which depends on actin and other muscle-like proteins of the contractile ring. Recent research has shown that eukaryotes from yeasts to humans use almost identical mechanisms to coordinate the events of these three processes. Our discussion here focuses on the mechanism by which cells become committed to enter the cell cycle.

A Regulatory Factor Called MPF Was First Identified in Vertebrate Cells

Many cell cycle researchers have studied mammalian cells growing in artificial conditions in a plastic dish. Their studies have often depended on the ability to produce **synchronous cell populations,** cells that are all at the same stage of the cell cycle. One way of producing such a population is to use an inhibitor that reversibly blocks DNA synthesis, stopping the cell cycle just before S. After the inhibitor is removed, all the cells begin S at the same time and continue in step through the rest of the cell cycle.

Such synchronous cell populations allowed researchers to show that cells undergoing mitosis contain a factor, called **MPF,** for **mitosis promoting factor** (originally called "**m**aturation **p**romoting **f**actor" because of its effect on the maturation of frog eggs) that triggers the events of mitosis. In one experiment, for example, researchers artificially induced the fusion of two cells, allowing their cytoplasms to mix while their nuclei remained separate. The resulting cell is a **heterokaryon** [Greek, *heteros,* different + *karyon* = kernel, nucleus], a cell that has two nuclei from different sources. A heterokaryon made from a cell in mitosis (M) and a cell in G_1 results, as shown in Figure 9-18 in the premature condensation of the G_1 chromatin into

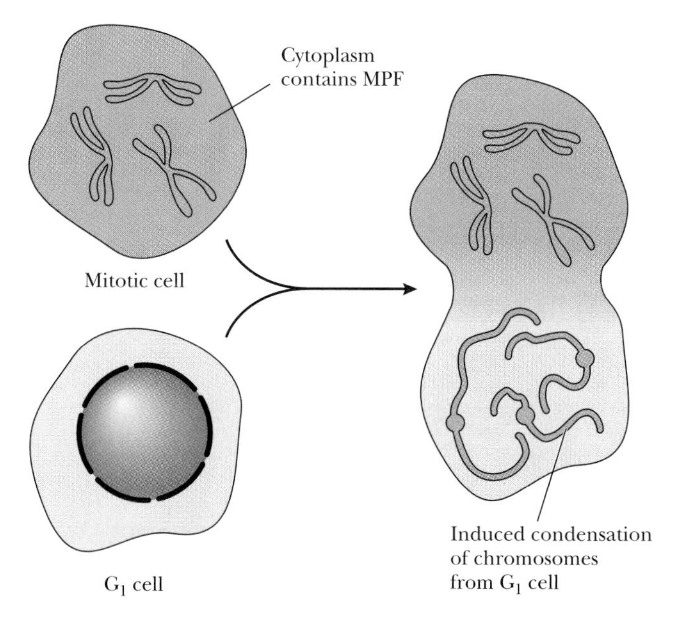

Figure 9-18 By artificially fusing a cell in mitosis with a cell in G_1, researchers demonstrated that the mitotic cell produces a factor that actively induces chromosome condensation.

chromosomes (with only one chromatid each)—the result of prematurely exposing the nucleus of the G_1 cell to the MPF in the M cell's cytoplasm. The nuclear membrane of the G_1 cell also disappeared prematurely.

Other researchers have purified MPF from frog eggs and shown that it is a **protein kinase,** an enzyme that transfers the phosphate group from an ATP to a protein. One of the proteins that receives a phosphate group as a result of MPF action is histone H1, which, as discussed previously, is directly involved in the packing of DNA. The role of MPF in stimulating mitosis, however, is not fully known and is still a subject of active research.

Studies of the Cell Cycle in Yeasts Have Identified Specific Regulators of the Cell Cycle

In addition to research on the cell cycle in vertebrate cells, many informative experiments have studied cell division in two distantly related species of yeast—**fission yeast,** *Schizosaccharomyces pombe,* and **budding yeast,** *Saccharomyces cerevisiae* (also called baker's yeast and brewer's yeast). Although both species are yeasts, they are thought to have had their last common ancestor more than 1 billion years ago, and their modes of cell division are quite different. *S. pombe* is a rod-shaped cell that divides by fission into two daughter cells of equal size. In contrast, *S. cerevisiae* divides by forming a small bud, so the two daughter cells differ in size. Both yeasts divide rapidly (doubling their number in about 2 hours). As researchers studied cell division in these yeasts, they established that the cell cycle always proceeds in a fixed order: For example, mitosis never begins before DNA synthesis is complete. Moreover, both yeasts have distinctive appearances at each stage of the cell cycle; in fission yeast, for example, the length of the cell is a good indicator of the stage of the cycle.

Researchers studying the cell cycle in yeast have experimentally produced **cell division cycle (cdc) mutants,** which cannot complete the cell division cycle because of alterations (mutations) in certain genes. Despite their distinctive patterns of cell division, the genes known to control cell division work the same way in these two yeasts. One of these genes specifies the structure of a protein that stimulates passage through Start. Closely related versions of this same protein are specified by a gene called ***CDC28*** in fission yeast and ***cdc2*** in budding yeast. A mutant yeast that cannot make the normal form of this protein cannot reproduce. (See Figure 9-19.)

Although frogs and starfish are only remotely related to the two yeasts, the *CDC28/cdc2* protein, like MPF, is a protein kinase—suggesting that starfish, frogs, and yeasts may regulate the cell cycle in similar ways. In fact, the amino acid sequences of the *CDC28* and *cdc2* proteins are recognizably similar to that of a polypeptide that is part of vertebrate MPF. In one experiment, researchers even showed that a human MPF polypeptide can stimulate yeast to pass through Start! This experiment shows that yeast and humans use virtually identical signals to coordinate at least one part of their cell cycles. It also suggests that the molecular mechanism for regulating the cell cycle is likely to have evolved well over a billion years ago, about the time that eukaryotic cells first appeared.

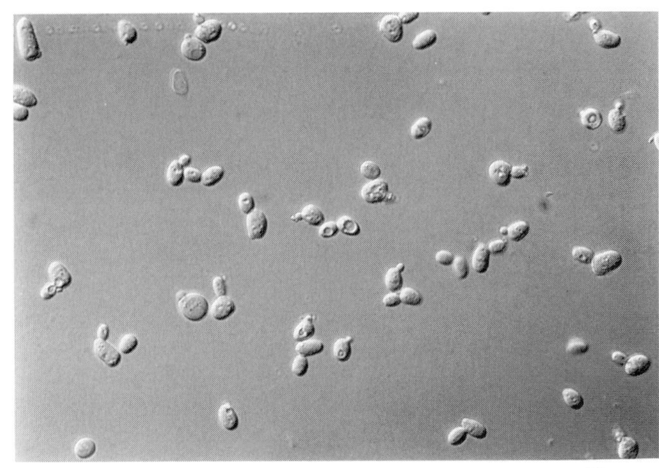

(a)

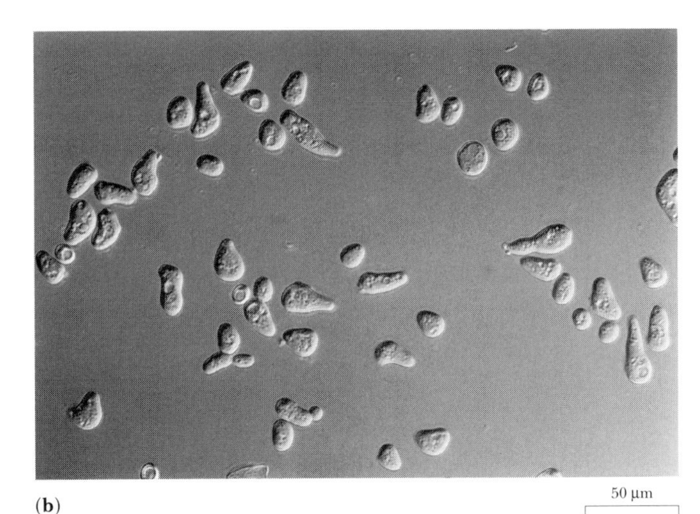

(b)

Figure 9-19 A cell-cycle mutant of the budding yeast, *Saccharomyces cerevisiae.* **(a)** Normal division, with functional *CDC28* kinase; **(b)** abnormal division with nonfunctional *CDC28* kinase. *(Steve Reed, Scripps Research Institute)*

How Do CDC28, cdc2, and MPF Change Their Activity During the Cell Cycle?

Other experiments (with yeast, mammalian cells in culture, and the embryos of frogs, clams, and sea urchins) have shown that the coordination of the cell cycle involves several regulatory factors and at least three different kinds of mechanisms. Cell biologists refer to these three regulatory mechanisms as *dominoes, checkpoints,* and *clocks.* Some cell cycle events, such as detachment of the two chro-

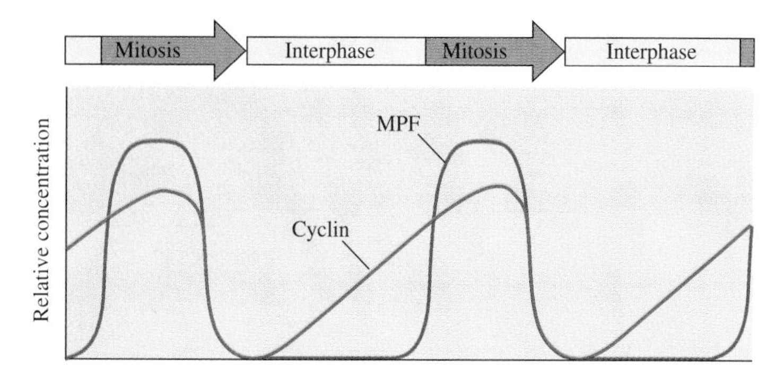

Figure 9-20 The concentration of cyclin and the activity of MPF change regularly through the cell cycle. These concentration changes are thought to regulate a cell's passage through the cell cycle.

matids, directly lead to the next event, chromosome movement toward the poles, just as one falling domino in a line knocks over the next. Other events in the cell cycle depend on the cell's checking that another process has been completed: For example, a cell checks that DNA synthesis has finished before beginning mitosis.

Other cell cycle events are subject to regulation by the clock-like cycling of regulatory molecules. The amount of active MPF, for example, increases and decreases during the cell cycle, as shown in Figure 9-20. Another protein, called **cyclin,** regularly increases and decreases in concentration during the rapid cell divisions that occur in early embryos. Although cyclin was first identified in the embryos of clams and starfish, more recent experiments have found it also in fission yeast, where (as in animal cells) it interacts with *cdc2*. Current evidence demonstrates that cyclin physically interacts with the animal's version of the *CDC28/cdc2* protein to make active MPF. The active MPF then stimulates mitosis.

A new and attractive view of cell cycle regulation is that changing levels of MPF and cyclin entirely determine a cell's passage through the cycle. A powerful approach to this question came from an experiment in which researchers add a sperm nucleus (whose membranes have been removed) to a concentrate of frog egg cytoplasm. Within 20 minutes, the sperm nucleus becomes surrounded with a new nuclear envelope and begins DNA synthesis. After 40 minutes, the nuclear envelope breaks down and mitosis begins. As in the case of ordinary frog development, cyclins accumulate during interphase and disappear at the end of mitosis.

To show that the increase in cyclin concentration could by itself stimulate mitosis, the researchers treated the egg concentrate so that it could not synthesize any of its own proteins. They did this by destroying the concentrate's messenger RNA, which (as we discuss in Chapter 14) specifies the sequences of amino acid residues in newly made proteins. After they destroyed the mRNA, they specifically stimulated cyclin production by adding cyclin messenger RNA. The nuclei then entered mitosis again, showing that cyclin by itself is enough to stimulate mitosis.

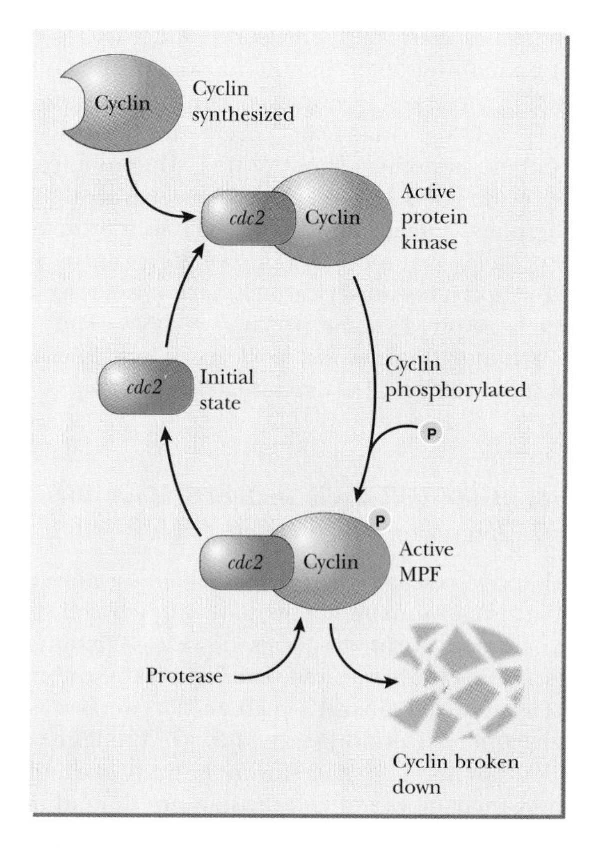

Figure 9-21 A current model of the interaction between cyclin and *cdc2* to form MPF during the cell cycle.

Later experiments showed that the destruction of cyclin is necessary to leave mitosis. Synthesis of cyclin allowed the nuclei to enter mitosis, but if the cyclin was not destroyed they never got beyond metaphase, and chromosomes were stuck at the metaphase plate. The periodic increase and decrease of cyclin was both necessary and sufficient for cell cycling in this experimental system. Figure 9-21 illustrates a current model of the interaction between cyclin and *cdc2* that seems to cause cell cycling.

INTERACTIONS WITH OTHER CELLS AND EXTRACELLULAR MOLECULES ALSO REGULATE CELL DIVISION IN MULTICELLULAR ORGANISMS

If cell division continued without stopping, single-celled organisms would soon run out of space and food. Consider a yeast cell with a cycle time of 2 hours. After 2 hours, it would have produced 2 cells; after 4 hours (2 generations), 2^2 (4) cells. After 1 day (12 generations), one cell would have grown into 2^{12} (about 4000) cells. After just 4 days of unperturbed growth, the single cell would have divided 48 times and produced 2^{48} (about 10^{15}) descen-

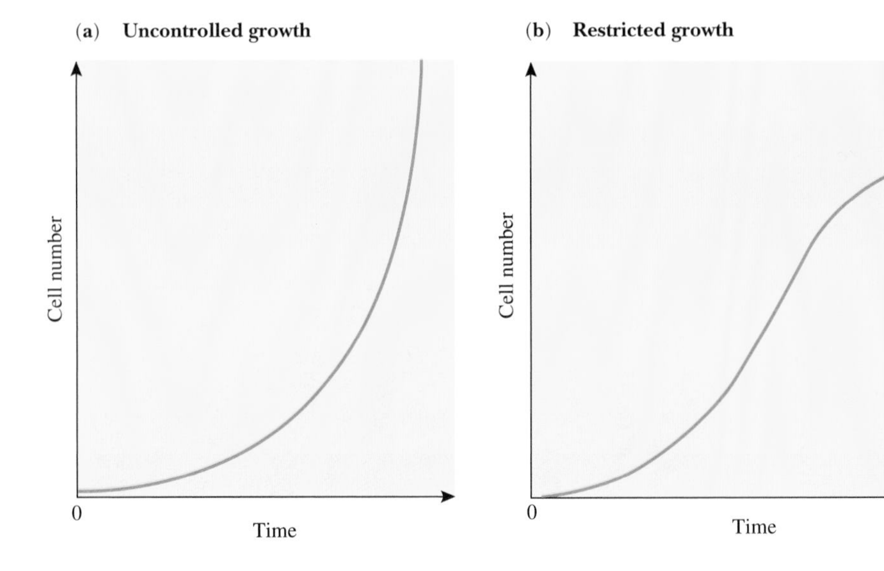

Figure 9-22 Unrestricted cell division would result in an exponential increase in cell number. External factors (like food supply) and internal regulation, however, restrict growth so that any cell population reaches a plateau. **(a)** Unrestricted exponential growth; **(b)** restricted growth, a more realistic portrayal of the increase in cell numbers.

dants, about 100 times more cells than are contained in your body. Figure 9-22a graphs the progress of such unbridled cell division.

In multicellular organisms, the failure to regulate cell division (which happens, for example, in cancer) leads to the loss of needed cooperation among cells, tissues, and organs. Cells divide rapidly during periods of growth and more slowly (or not at all) in mature organisms. At least two general mechanisms help prevent runaway cell division—cell senescence (aging) and growth control.

Cell Senescence Limits the Number of Times a Cell Can Divide

The more times a cell divides (at least under conditions of laboratory culture), the more likely it is to enter G_0 and withdraw from the cell cycle. An average cell taken from a newborn baby, for example, can divide about 50 times in a standard culture medium. But cells taken from an 80 year old stop cycling after about 30 divisions. Researchers do not know the mechanism of cell senescence but think that it may involve differences in the response to **growth factors,** extracellular proteins that stimulate cell division.

How Do Normal Cells Determine When to Stop Dividing?

A second set of regulatory mechanisms—those involved in growth control—prevents delinquent division by allowing the cell cycle to proceed under some conditions and to stop under others. As an example of growth control, think about what happens when you cut yourself. The cells on the cut edges of your skin begin to divide and fill the space left by the wound. The cells divide until the two edges again touch. Once the wound heals, cell division ceases.

In artificial culture, cells can regulate their division in much the same way. Cells growing in an appropriate growth medium attach to the bottom of the culture dish and divide until they form a confluent monolayer, that is, a single layer of cells that occupies the whole surface. Then they stop dividing.

If we now make a "wound" in the monolayer by scraping away a swath of cells, the cells on the margins start moving into the wound area and begin to divide. Division stops only after the space is filled and the monolayer again is confluent.

We can summarize the growth control of normal cells in culture with the following rules: (1) Cells stop dividing during G_1 (at Start) when they run out of free space on which to spread—that is, when neighboring cells all touch each other; and (2) cells that have proceeded beyond Start begin to divide when contact with their neighbors ceases. This kind of growth control is called **contact inhibition of cell division.**

How does contact inhibition work? Its mechanism is not known, but it must involve signaling from the cell surface (which is either in contact or not in contact with other cells) to the nucleus (which either proceeds or does not proceed with the cell cycle).

Besides sensitivity to contact with other cells, growth control also depends upon a cell's response to other external signals, including growth factors and steroid hormones. Some of these molecular signals act on the cell surface, while others diffuse through the membrane and act directly on molecules in the cytoplasm. Some of the signals on the cell surface directly affect the organization of the cytoskeleton.

Cancer Cells Fail to Show Normal Growth Control

Cancer cells fail to regulate their cell cycles normally. For example, cancer cells, unlike normal cells, do not exhibit

ESSAY

TUMOR VIRUSES, ONCOGENES, AND GROWTH CONTROL

Crucial insights on growth control have come from the study of tumor viruses, viruses that can cause normal cells to grow as cancer cells (Figure 1). Tumor viruses damage growth control in cells growing either in culture or in susceptible animals. The viruses accomplish their subversion by adding copies of their own genes to the genes of their hosts.

The first known tumor virus, Rous sarcoma virus, was discovered in 1910 by Peyton Rous. This virus contains enough genetic information for only four genes. Three of these genes provide the virus with the special molecules necessary for its own multiplication. Only one gene—called *src*—directly affects the growth of the virus's host cells. The protein whose blueprints are contained in the *src* gene is an enzyme that acts on many components of the cell, subverting the host cell's ability to regulate its growth. The *src* enzyme does this by functioning as a protein kinase, transferring phosphate groups from ATP to the tyrosine residues of proteins, especially those present at the points of contact between cells.

The *src* gene is an **oncogene** [Greek, *onkos* = bulk, tumor], a gene that can change normal cells into cancer-like cells. Molecular biologists have now identified more than 25 oncogenes from other tumor viruses. Many oncogenes appear to code for protein kinases similar to the *src* protein. The products of other oncogenes also disrupt the regulation of cell reproduction, some by producing growth factors that act on the cell surface, others by interacting directly

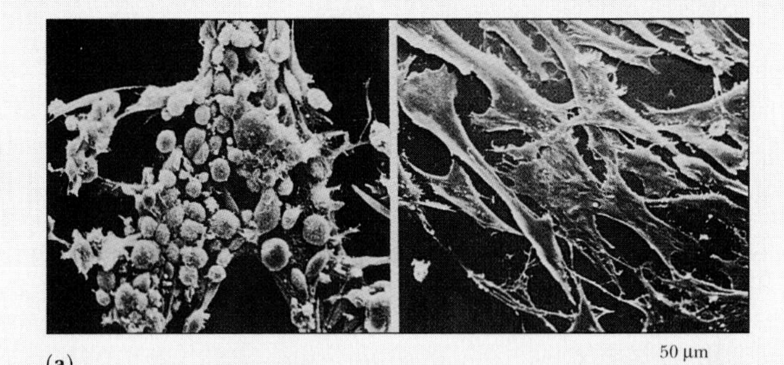

(a)

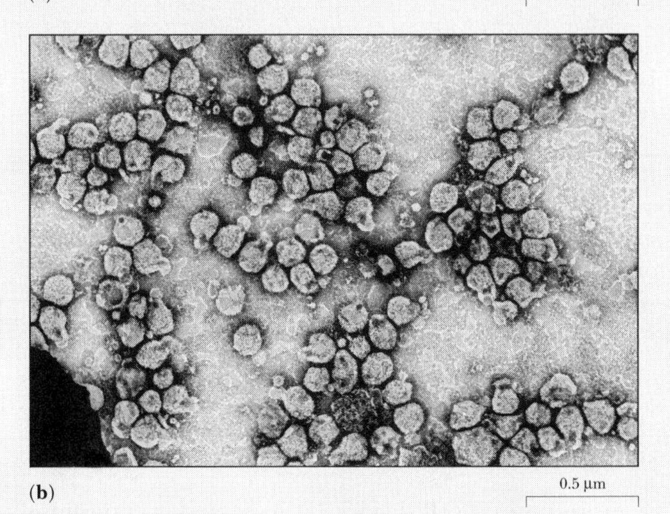

(b)

Figure 1 Cells infected with Rous sarcoma virus. **(a)** Infected cells (left) grow like cancer cells, piling up on top of one another (uninfected cells are shown on the right); **(b)** TEM of Rous sarcoma virus.
(a, G. Steven Martin/Visuals Unlimited; b, K. G. Murti/Visuals Unlimited)

contact inhibition in cell culture. Instead of forming a confluent monolayer, they pile up on top of one another and grow to much higher densities than normal cells.

Cancer cells are generally less fastidious than normal cells: They often can grow without sticking to a surface (as normal cells require) and without the normal levels of specific growth factors. They also can produce tumors when injected into susceptible animals. Because of the medical importance of cancer, scientists have paid enormous attention to questions related to growth regulation in both normal and cancer cells. We discuss these issues in more detail in Chapters 16 and 21.

■ SUMMARY

Cells reproduce themselves in a process called cell division, in which a parent cell gives rise to two daughter cells with the same genetic information in the form of DNA. Dividing cells duplicate and distribute DNA during an orderly sequence of events called the cell cycle.

Prokaryotes divide by a simple process called bi-

with genes in the nucleus, still others by unknown mechanisms.

How do oncogenes arise during evolution? The emerging answer is that oncogenes of cancer viruses arose from normal cellular genes that somehow were "captured" by invading viruses. Many viruses indeed have the ability to incorporate a copy of part of its host's DNA into its own genetic information. When the captured DNA contained information that stimulates cell division, then infection by a virus could trigger inappropriate cell division. Cancer viruses cause exactly such uncontrolled division. Scientists believe that the captured genes stimulate division because they are themselves involved in normal regulation of cell reproduction.

For example, epidermal growth factor (or EGF), a protein that stimulates cell division of certain cells, stimulates the activity of a protein kinase (called the EGF receptor) that transfers phosphate groups from ATP to the tyrosine residues of several proteins. Some of these proteins are identical to the proteins acted on by the kinases specified by oncogenes. In many cases, the oncogenes of a cancer virus differ slightly from the oncogenes within a normal cell.

The oncogenes within normal cells are called **proto-oncogenes.** Viral oncogenes are always closely related to proto-oncogenes and almost certainly evolved from them. The presence of proto-oncogenes in normal cells suggests that abnormal growth control in cancer is closely related to the normal regulation of cell division. Apparently, tumor viruses merely provide an overdose of normal cellular proteins that ordinarily function in controlling normal cell division.

Recently, cancer researchers have identified another class of genes—**anti-oncogenes** or **tumor suppressor genes**—which antagonize the action of oncogenes and are intimately involved in the normal regulation of cell growth. One such gene specifies the structure of a protein called RB. RB is normally present in the nucleus of cells as a phosphorylated protein whose level of phosphorylation varies during the cell cycle. RB operates at Start, reporting the cell's readiness to enter the S phase. Cancer researchers think that this protein ordinarily participates in the regulation of the cell cycle. When a tumor suppressor protein is missing or altered, cell growth can run amok: Alterations in RB, for example, cause retinoblastoma, a rare eye tumor that appears in young children. Both RB and the products of several proto-oncogenes appear to participate in normal cell regulation, but too much activity of a proto-oncogene product or too little activity of a tumor suppressor can lead to cancer. So any environmental agent that causes such changes in proto-oncogenes or in tumor suppressors can cause cancer.

nary fission, which distributes one copy of the parent cell's DNA to each daughter cell. Cell reproduction is more complicated in eukaryotic cells, whose DNA is enclosed in a membrane-bounded nucleus. Chromosomes are discrete bodies visible in dividing cells with a light microscope.

Cell biologists divide the cell cycle into five phases, called mitosis, cytokinesis, G_1, S, and G_2. The process of evenly distributing chromosomes to daughter cells is called mitosis. The process of dividing a cell's cytoplasm is called cytokinesis. During G_1 a cell accumulates the materials and machinery necessary to replicate its DNA; during S, it synthesizes DNA, resulting in two copies (replicas) of the parent cell's DNA; during G_2, it prepares for mitosis and cytokinesis. G_1, S, and G_2 together are called interphase.

Just before a cell begins mitosis, each chromosome consists of two chromatids. Mitosis is a continuous process, but biologists distinguish five phases: prophase, prometaphase, metaphase, anaphase, and telophase. During prophase, chromosomes condense and the mitotic apparatus forms. During prometaphase, the nuclear membrane disappears and the chromosomes attach to the spindle fibers. During metaphase, the chromosomes align in a disc called a metaphase plate. During anaphase, the two chromatids that make up each chromosome separate. During telophase, the mitotic apparatus disappears and new daughter nuclei form.

During interphase the cell duplicates its single centrosome (microtubule organizing center). During prophase, a mitotic apparatus assembles with microtubules emerging from each centrosome to form a mitotic spindle. Microtubules connect to each chromatid in a structure called a kinetochore. The mitotic apparatus is responsible for chromosome movements during mitosis.

The DNA of eukaryotic cells is highly folded. Protein molecules called histones bind to DNA and form particles called nucleosomes, which resemble beads on a string. DNA, histones, and other proteins fold to form still more compact structures. The orderly packing of DNA produces a characteristic pattern of chromosome banding in each metaphase chromosome.

During cytokinesis, the plane of cell division is always perpendicular to the axis of the mitotic spindle. The pinching movement of cytokinesis results from the action of a contractile ring, made of actin filaments, which surrounds the dividing cell. As the contractile ring contracts, it pulls the membrane inward to form a cleavage furrow, which deepens and separates the two daughter cells.

A plant cell cannot simply pinch in two because of its rigid cell wall. Cytokinesis in plant cells requires a special mechanism for building a new cell wall from materials made in the Golgi apparatus.

Cells actively regulate passage through the cell cycle. One regulatory protein, called mitosis promoting factor, or MPF, stimulates cells to enter mitosis. MPF is a protein kinase that transfers phosphate groups from ATP to proteins, including histone H1. MPF isolated from animal cells is similar to a yeast protein that stimulates cells to continue their cell cycles, suggesting that all eukaryotes may regulate cell cycles in a similar manner.

Regulated cell division in multicelled organisms depends on cell senescence and growth control. Cell senescence limits the number of times a cell can divide. Growth control regulates cell reproduction according to external conditions. Many cells exhibit contact inhibition of cell division, in which cells divide only when they are not in contact with their neighbors. Growth control may also involve growth factors, extracellular proteins that stimulate cell division. Cancer cells fail to show normal growth control.

■ KEY TERMS

acrocentric
anaphase
anaphase A
anaphase B
anastral mitosis
anti-oncogene
aster
astral microtubule
astral mitosis
binary fission
budding yeast
cdc2/CDC28
cell center
cell cycle
cell division
cell division cycle (cdc) mutants
cell plate
cell senescence
centriole
centromere
centrosome
chromatid
chromatin
chromosome
chromosome banding
cleavage division
cleavage furrow
contact inhibition of cell division
contractile ring
cyclin
cytokinesis
daughter cell
dynein
fission yeast
G_0
G_1 phase
G_2 phase
growth factors
heterokaryon
histone
interphase
kinesin
kinetochore
kinetochore microtubule
M phase
metacentric
metaphase
metaphase plate
midbody
mitosis
mitosis promoting factor (MPF)
mitotic spindle
nucleosome
oncogene
phragmoplast
plasmodesmata
polar microtubule
prometaphase
prophase
protein kinase
proto-oncogene
replicate
restriction (R) point
S phase
senescence
sister chromatid
Start
synchronous cell population
telocentric
telophase
tumor suppressor gene

QUESTIONS FOR REVIEW AND UNDERSTANDING

1. Define and contrast the following terms:
   (a) chromosome vs. chromatin
   (b) mitosis vs. cytokinesis
   (c) G_1 vs. G_2
   (d) G_0 vs. G_1
   (e) mitotic spindle vs. centriole
   (f) polar vs. astral vs. kinetochore microtubules
   (g) astral mitosis vs. anastral mitosis
   (h) kinetochore vs. centromere
   (i) metacentric vs. acrocentric vs. telocentric
   (j) histone vs. nucleosome
   (k) contractile ring vs. cleavage furrow
   (l) oncogene vs. tumor suppressor gene
   (m) proto-oncogene vs. oncogene
2. What are the three processes associated with cell reproduction?
3. Why is cell reproduction important even in tissues, organs, or individuals that are not growing?
4. Define what is meant by "cell cycle."
5. What is the role of the membrane in prokaryotic cell division?
6. What are some of the factors that make eukaryotic cell division more complex than prokaryotic cell division?
7. Why is the process of mitosis not observed in prokaryotes?
8. What is meant by "interphase"?
9. Explain why a "half chromosome" (a chromatid) as observed during mitosis is equivalent to a chromosome in a cell during interphase.
10. Describe how ^{3}H-thymidine is used to detect DNA replication during S phase. Why is thymidine better than other nucleosides (such as cytosine or guanosine) for following DNA synthesis?
11. What is the source of the microtubules (tubulin molecules) for constructing the mitotic spindle?
12. What are three experimental approaches used to study the mitotic spindle? How has each approach contributed to our understanding of the role of microtubules in chromosome movement?
13. What is the nature of the interaction between histones and DNA?
14. What is the evidence suggesting that histone structure has been highly conserved through evolution?
15. Describe and contrast cytokinesis in plant cells and animal cells.
16. Why is Start considered the "point of no return"?
17. What is the evidence that MPF plays a role in promoting mitosis?
18. What is the relationship between *CDC28/cdc2* and MPF? How is cyclin involved?
19. Why is it significant that human MPF can stimulate yeast to progress through the cell cycle?
20. Explain, in general terms, why contact inhibition must involve signaling from the cell surface to the nucleus.
21. In recent years, many significant discoveries pertinent to cancer have been made by researchers examining cell cycle control in yeast. Describe in your own words how these two areas are related.

SUGGESTED READINGS

MITCHISON T. J., "Mitosis—From Molecules to Machine," *American Zoologist* 29:523–535, 1989. A truly outstanding review of the mechanics of mitosis.

MOORE, J. A., "Science As a Way of Knowing—Genetics," *American Zoologist* 26:583–747, 1986. This article is part of a multivolume work, *Science As a Way of Knowing,* a project of the Education Committee of the American Society of Zoologists. Pages 619–626 of this article deal with mitosis. The articles by Murray and Mitchison also come from this series.

MURRAY, A.W., "The Cell Cycle," *American Zoologist* 29:511–522, 1989. A nice discussion of the relationship between cell cycle regulation in yeast and in vertebrates.

PINES, J., "Cell Proliferation and Control," *Current Opinion in Cell Biology* 4:144–148, 1992. A review of the current research concerning the initiation of the S phase of the cell cycle.

SCIENCE 246, 1989. Most of this issue is devoted to advances in understanding the cell cycle.

WATSON, J. D., M. Gilman, J. Witkowski, and M. Zoller, *Recombinant DNA,* 2nd edition, Scientific American Books, New York, 1992. Chapter 18, "Oncogenes and Anti-oncogenes," pp. 335–367: the roles of viral and cellular genes in cancer and in normal growth control. Chapter 19, "Molecular Analysis of the Cell Cycle," pp. 369–388: an excellent discussion of how recombinant DNA and immunological techniques have demonstrated the close connections between cell cycle control in yeasts and animal cells.

CHAPTER 10

Chromosomes and the Life Cycle: The Role of Meiosis in Sexual Reproduction

KEY CONCEPTS

- Sexually reproducing eukaryotes have pairs of homologous chromosomes.
- Meiosis distributes one chromosome from each pair to each daughter cell.
- At fertilization the chromosomes of the two gametes combine to form a zygote with the same number of chromosomes as the parents.
- The behavior of chromosomes underlies the rules of inheritance.

OUTLINE

ESSAY

Every individual inherits genetic plans that distinguish it from individuals of other species. The mechanisms of inheritance must maintain this constancy. In Chapter 9, we saw how individual cells within an organism distribute the genetic information in their chromosomes to their daughter cells. In this chapter, we see how sexually reproducing organisms pass on a complete set of chromosomes to their offspring. We also preview the relationship of chromosome management during the sexual life cycle and the laws of heredity that we discuss in Chapter 11.

Figure 10-1 Genetic variation in a natural population of snails. *(John Cunningham/Visuals Unlimited)*

SEXUAL REPRODUCTION PROMOTES GENETIC DIVERSITY

While all the individuals within a species share many common characteristics, they also differ from one another, as illustrated by the variant forms of the snails shown in Figure 10-1. In all plants and animals (and many other species as well), most of the genetic diversity among individuals comes from combining genes inherited separately from the two parents.

Sexual reproduction produces offspring that have inherited genetic information from two parents rather than one. Because the genes from each parent are likely to differ, sexual reproduction provides new combinations of genes in every individual in every generation. In contrast, **asexual reproduction** produces offspring with genes from a single parent. Asexual reproduction allows far fewer new combinations of genes than does sexual reproduction.

A genetically diverse population is far more likely to survive in the face of changing environmental conditions than one with limited diversity. A field of genetically diverse corn plants (*Zea mays*), the most important crop plant in the U.S., contains plants that differ not only in height and grain yield, but also in the ability to resist drought and insect pests. In contrast, genetically identical corn plants may be bred to produce a high yield under normal agricultural conditions. In the face of varying conditions, however, a genetically identical crop may be destroyed altogether, while the genetically diverse crop survives. Corn breeders are therefore constantly working to introduce new traits into agriculturally valuable varieties. The sources of diversity must often be distantly related wild species, such as *Zea diploperrenis,* because much of the natural diversity of *Zea mays* has been lost.

Genetic variation arises from **mutations,** random, spontaneous changes in the genetic information coded in the sequence of nucleotides of DNA. Sexual reproduction accomplishes the "field testing" of new mutations by allowing the formation of new combinations of new and old variations. Most of the time a new mutation reduces the ability of an individual to reproduce, but occasionally a mutation increases the chances of that individual's survival and reproduction. Sexual reproduction, then, both permits populations to be diverse and provides the raw material for natural selection and evolution.

For species that can reproduce either sexually or asexually, asexual reproduction usually occurs only when environmental conditions are stable. Plants may reproduce sexually or asexually. Similarly, while all animals can produce sexually, some, like the hydra, can also reproduce by budding or by regenerating a whole organism from a small part. If such asexual reproduction can occur even in complex organisms, why should organisms spend time and energy on sexual reproduction?

We have some hint of the importance of sexually generated diversity from studies of species that actually change their mode of reproduction in different environments. When food and space are abundant, for example, the water flea shown in Figure 10-2 can proliferate rapidly by asexual means, producing large numbers of almost exactly identical individuals. When resources are strained and uncertain, however, it begins to reproduce sexually. In these circumstances, the genetic diversity provided by sexual reproduction becomes important because it increases the chance that at least some individuals, with slightly different characteristics, will survive. Sexual reproduction not only replaces individuals, but also produces populations in which some individuals may be better suited to life in varying environments.

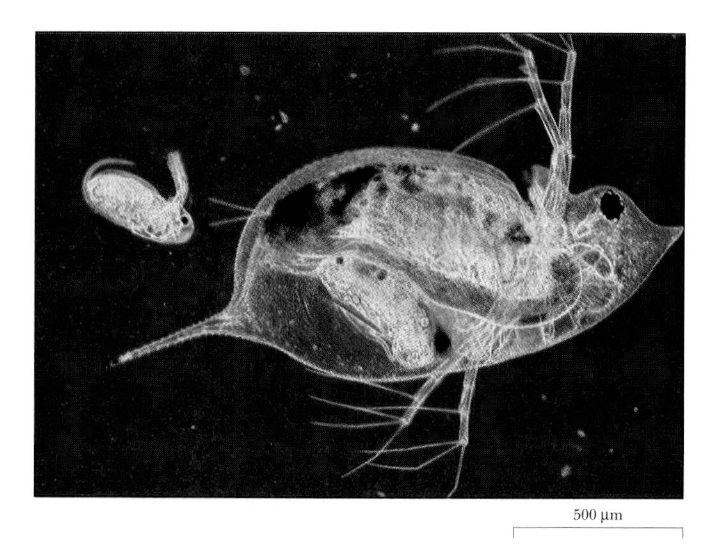

Figure 10-2 *Daphnia,* a water flea, can proliferate asexually as long as resources are abundant. When food is scarce, it reproduces sexually. *(M. I. Walker/Science Source/Photo Researchers)*

HOW DO SEXUALLY REPRODUCING ORGANISMS KEEP THE SAME NUMBER OF CHROMOSOMES FROM GENERATION TO GENERATION?

During asexual reproduction in eukaryotes, mitosis ensures that each offspring inherits a complete set of chromosomes. But sexual reproduction raises a new problem in chromosome management: How can each new organism receive copies of chromosomes from both parents and not double the total number of chromosomes in each generation?

The answer is that each parent contributes half of the genetic material. Chromosomes of sexually reproducing organisms exist in pairs, and each parent contributes one chromosome to each pair of chromosomes. We begin our discussion by showing the existence of such chromosome pairs in our own species.

Let us look carefully at the chromosomes within a dividing skin cell from a healthy woman, shown in Figure 10-3a. By cutting out photographs of the individual chromosomes from this figure and arranging them in order of decreasing size, we could produce a tidy picture of all 46 chromosomes, such as that shown in Figure 10-3b. Such a picture shows the cell's **karyotype** [Greek, *karyon* = kernel or nucleus + *typos* = stamp], its chromosomal makeup.

Notice in Figure 10-3 that for each chromosome with a characteristic size and banding pattern, there is another with the same properties. The two matching chromosomes are called **homologous chromosomes,** and the matching members of each pair are called **homologs.** We can therefore think of the 46 chromosomes of a human female as

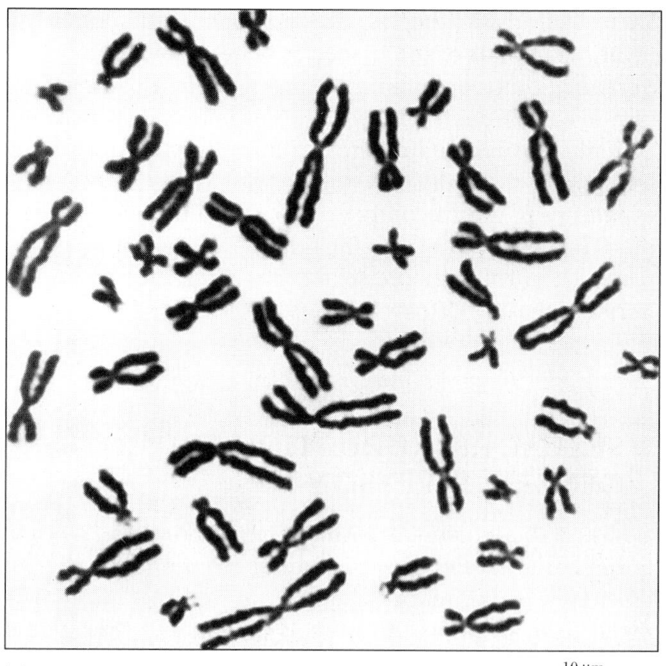

(**a**)

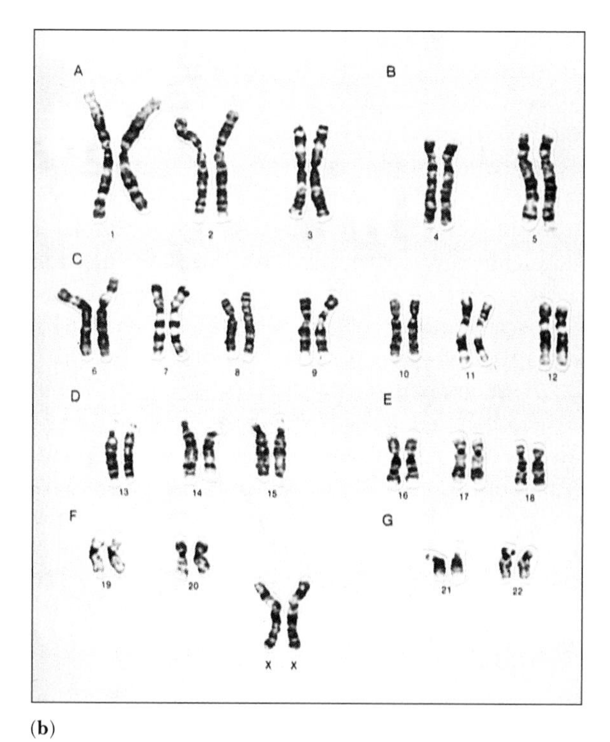

(**b**)

Figure 10-3 The mitotic chromosomes in a skin cell from a human female. This cell has been treated with colchicine, blocking mitosis in metaphase, and stained with a dye that reveals a characteristic banding pattern for each chromosome. **(a)** Chromosomes at the metaphase plate; **(b)** a karyotype, produced by cutting out photographs of individual chromosomes and arranging them according to size and banding pattern. *(a, Biophoto Associates/Photo Researchers; b, Leonard Lessin/Peter Arnold)*

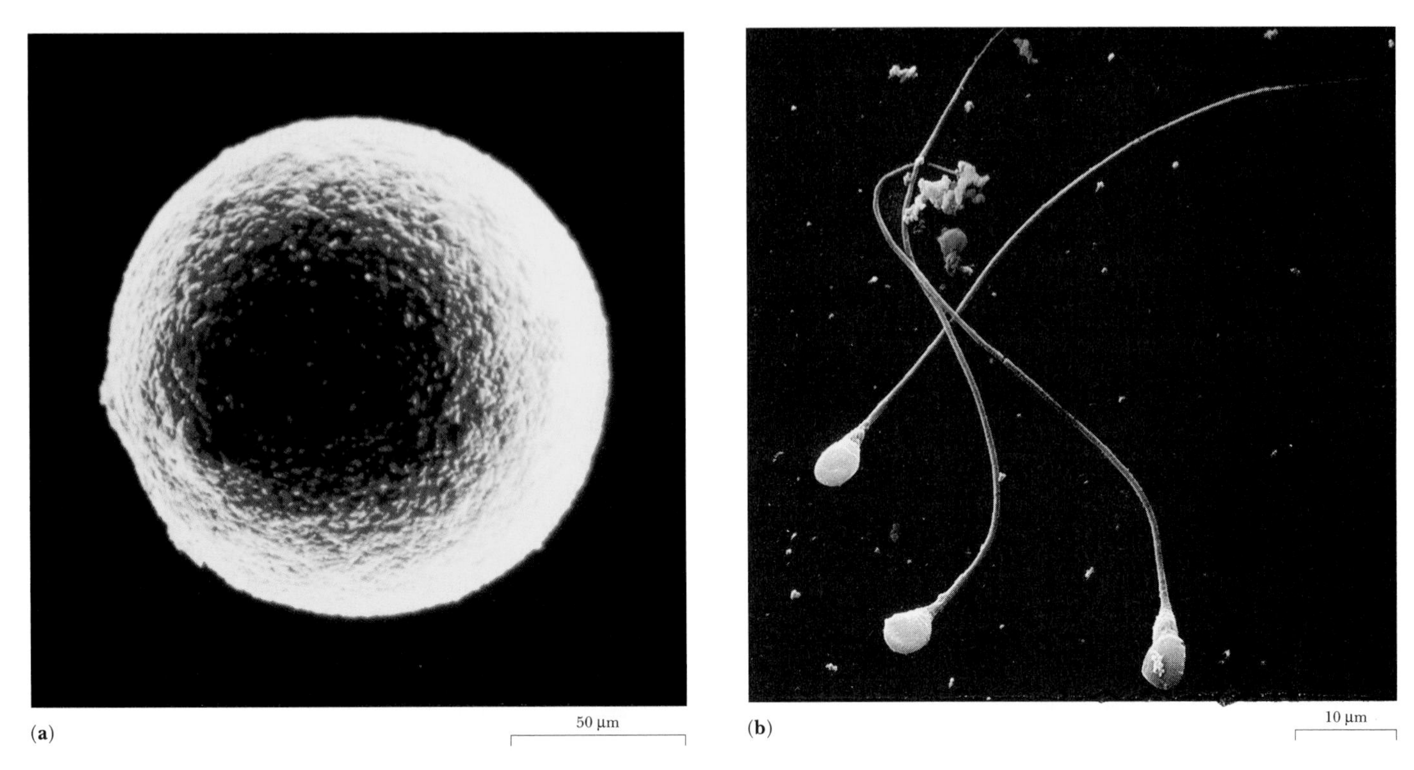

Figure 10-4 **(a)** Human egg; **(b)** a human sperm. *(a, John Giannicchi/Science Source/Photo Researchers; b, David Phillips/Photo Researchers)*

two sets of 23 distinct chromosomes. (Human males have a slightly different karyotype, shown in Figure 10-13, which we discuss shortly.) Cells that contain two sets of chromosomes are said to be **diploid** [Greek, *di* = double + *ploion* = vessel]. One complete set of chromosomes (that is, one homolog of each pair) comes from each parent.

Sexually reproducing parents pass chromosomes to their offspring via specialized reproductive cells called **gametes** [Greek, *gamos* = marriage]. In many species, including all animals and plants, the male and female gametes are distinct from one another. Each female gamete, called an **egg** or an **ovum** [Latin = egg; plural, **ova**], is larger and usually nonmotile, while each male gamete, called a **sperm** or **spermatozoan** [Greek, *sperma* = seed + *zoos* = living], is smaller than the egg and motile. In humans, for example, the egg (shown in Figure 10-4a) is nonmotile and about 140 μm in diameter, while the motile sperm (shown in Figure 10-4b) consists of a tail about 70 μm long and the nucleus-containing head, only 5 μm in diameter.

The gametes and the cells from which they arise are called **germ cells** or the **germ line.** Most of the cells in a multicelled organism, however, do not give rise to gametes or undergo meiosis and are called **somatic cells** [Greek, *soma* = body], or the **soma**. Genetic instructions pass to subsequent generations exclusively through the germ line.

Each gamete is **haploid** [Greek, *haploos* = single], meaning that it carries a single set of chromosomes. Unfortunately, the word "haploid" sounds as if it means that cells contain "half" a chromosome set rather than a complete single set; to avoid this possible confusion, some people refer to the single set of chromosomes as "monoploid." But how does each parent accomplish the needed splitting of chromosomes into two haploid sets and provide just one homolog from each pair?

The haploid gametes arise from diploid cells by **meiosis** [Greek, *meioun* = to make smaller], a process that distributes chromosomes so that each of four daughter cells receives one chromosome from each homologous pair. The similarity of the word "meiosis" to "mitosis" has been another unfortunate source of confusion to many generations of biology students. The two processes are indeed related, using some of the same cellular machinery, but they have very different results.

The First Cell of the New Generation Has Two Sets of Chromosomes

The beginning of a new generation occurs at **fertilization,** the union of the two haploid gametes to form a diploid cell, called a **zygote** [Greek = yoke], as illustrated in Figure 10-5. As discussed in Chapters 20 (for animals) and 21

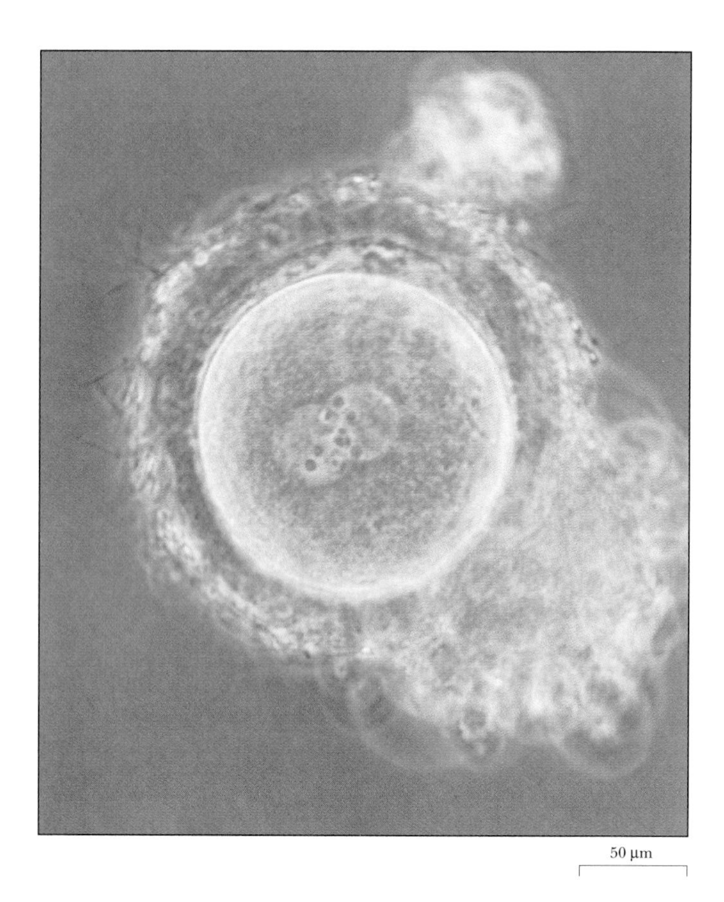

Figure 10-5 Joining of sperm and egg nuclei to form a human zygote. The two nuclei fuse to form the zygote nucleus. *(© 1991 Richard G. Rawlins/Custom Medical Stock)*

(for plants), fertilization depends heavily on adaptations of the gametes. A spermatozoan is specialized for movement and the entrance of its nucleus into the egg's cytoplasm. The egg, in turn, responds to the entry of the sperm nucleus into its cytoplasm by moving the sperm and egg nuclei together and fusing the two haploid sets of chromosomes. The egg is also often specialized to provide the materials and machinery for early development. Both gametes, however, contribute equally to the zygote's genetic endowment.

Gamete formation requires other adaptations as well. Special structures produce the gametes and bring them together for fertilization. In animals, for example, the gamete-producing organs, where meiosis occurs, are called **gonads** [Greek, *gonos* = seed]. In males, **testes** produce sperm, while in females the **ovaries** produce eggs. In mammals, fertilization takes place within the female's reproductive system and requires sexual intercourse. Mammals have evolved elaborate structural and behavioral adaptations that bring gametes together. Plants, fungi, and protists have their own diverse adaptations, but fertilization always accomplishes the same thing: The life cycle of sexually reproducing organisms combines two haploid sets of chromosomes into one diploid set of chromosomes (as illustrated in Figure 10-6).

If the gametes were diploid instead of haploid, each zygote would have twice as many chromosomes as the cells of the previous generation. The accomplishment of meiosis, then, is that it produces gametes with complete hap-

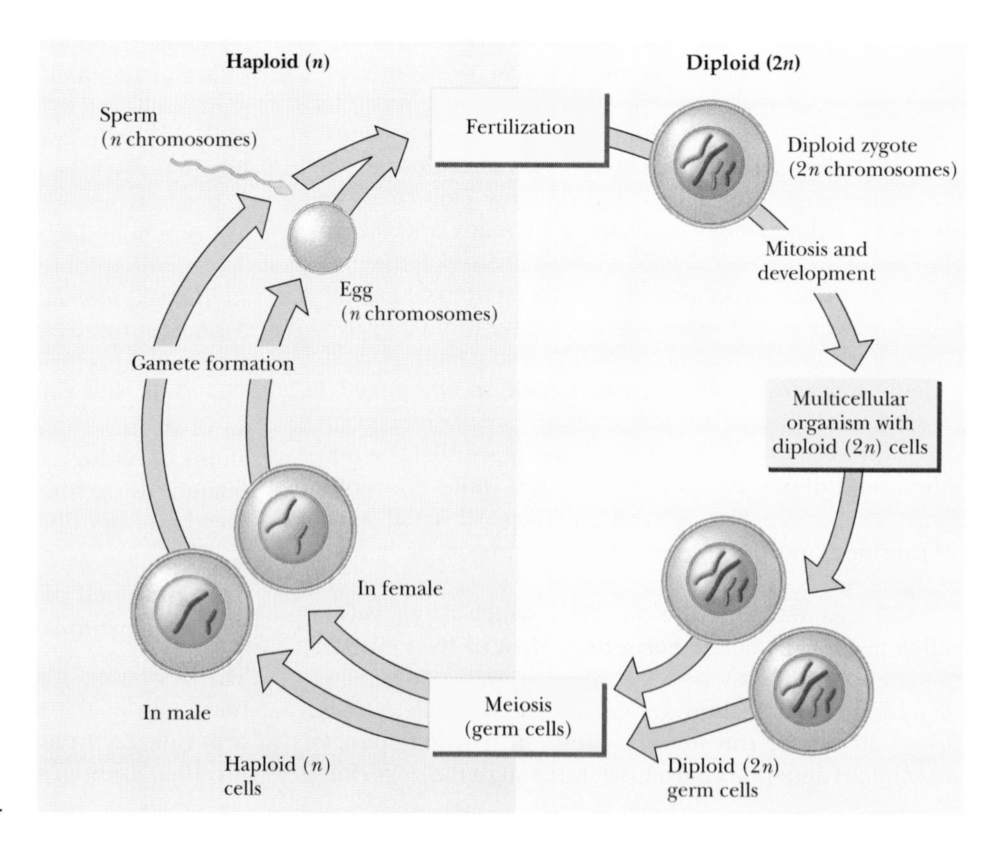

Figure 10-6 Fertilization combines two haploid sets of chromosomes to produce a diploid set of chromosomes.

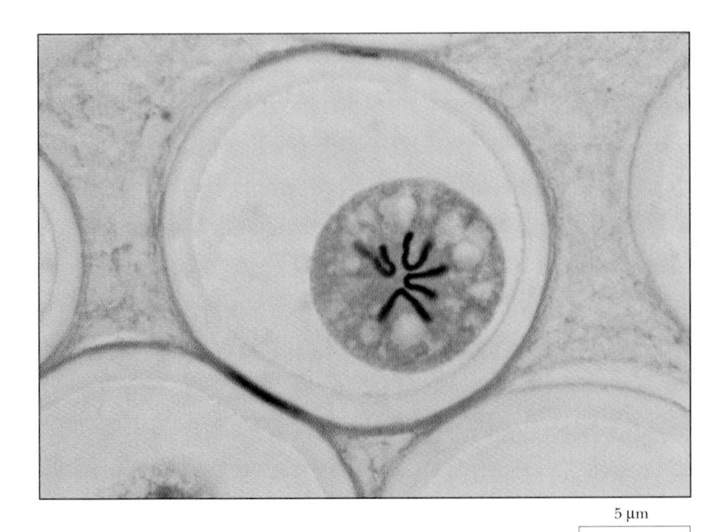

Figure 10-7 The roundworm *Ascaris* has only two pair of chromosomes. In the late 19th century, the German embryologist Theodor Boveri first described meiosis from his studies of gamete formation in *Ascaris.* *(Custom Medical Stock)*

loid sets of chromosomes by taking one chromosome from each homologous pair. Meiosis in all sexually reproducing eukaryotes occurs in essentially the same way, in an elaborate chromosomal ballet that (as we will see) both resembles and differs from mitosis.

How Does Meiosis Distribute Chromosomes to the Gametes?

One of the first descriptions of meiosis came from studies of the roundworm *Ascaris,* whose somatic cells contain only four chromosomes, two pairs of homologous chromosomes, as shown in Figure 10-7.

Figure 10-8 shows the overall process of meiosis, which consists of two sequential divisions, called **meiosis I** and **meiosis II,** which occur only in cells of the germ line during the production of gametes. Like mitosis, meiosis is a continuous process, which biologists divide into a series of steps for purposes of discussion. DNA replication occurs before meiosis I (as it does before mitosis in somatic cells), so that at the beginning of meiosis, each chromosome consists of two sister chromatids. Meiosis II immediately follows meiosis I, but without intervening DNA replication. At the end of meiosis II, then, each chromosome consists of just one chromatid.

Biologists divide each meiotic division into four phases, illustrated in Figure 10-8: prophase, metaphase, anaphase, and telophase. (We do not usually distinguish a separate prometaphase, as we do for mitosis.) In general outline, each stage of meiosis resembles the corresponding stage of mitosis: During each prophase, chromosomes condense, and the nuclear membrane breaks down; during each metaphase, chromosomes move to the equator of the spindle apparatus; during each anaphase, chromosomes move toward the poles; and during each telophase, the nuclear membrane reforms. But the end results of meiosis and mitosis are crucially different: Each mitotic division produces two daughter cells, each with the same diploid chromosome set as the parent cell, while the two divisions of meiosis produce four daughter cells, each with a haploid set of chromosomes.

Meiosis does not always produce four gametes. For example, in plants and other organisms that have alternation of generations, the products of meiosis are spores (which can divide mitotically) rather than gametes (which cannot). In contrast, during meiosis in female mammals, the cytoplasm divides unequally, so that only one of the four products of meiosis is actually an egg. By this inequality, the egg receives almost all of the cytoplasm, which provides the nutrients and cellular machinery for the zygote formed at fertilization. The other three products of meiosis are cytoplasm-poor "polar bodies."

The major difference between mitosis and meiosis is the absence of DNA replication before prophase of meiosis II. The DNA content of the haploid cell is half that of a corresponding diploid cell (in G_1), consistent with each chromosome containing a single chromatid after the completion of meiosis. Figure 10-8b shows the changing DNA content of cells undergoing mitosis and meiosis.

Homologous Chromosomes Form Pairs During Prophase I

The central event of meiosis is **synapsis** [Greek = union], the pairing of homologous chromosomes in prophase I, shown in Figure 10-8 and in more detail in Figure 10-9. Following synapsis, the chromatids of each chromosome do *not* separate, but stay together in meiosis I. In meiosis I, the homologous *chromosomes* separate during anaphase, while in mitosis (and in meiosis II) it is the *sister chromatids* that separate. Synapsis is the process that allows the sorting of the homologous chromosomes into two haploid sets.

Prophase I occupies more than 90% of the total time of meiosis, and cell biologists have traditionally divided the prophase of meiosis I into five subphases, as illustrated in Figure 10-9a. As in mitotic prophase, chromosomes condense, the nucleoli and the nuclear membrane disappear, and the spindle apparatus begins to form. The major difference from mitotic prophase is that, after the chromosomes begin to condense, the homologous pairs come together in synapsis. The homologous chromosomes align.

Each united chromosome pair consists of four chromatids, called a **tetrad** (as shown in Figure 10-9b). Because the two sister chromatids of each chromosome are so close to one another, the tetrads appear to consist of two rather than four components, so the chromosome pairs are also

(a)

Meiosis

Meiosis is a continuous process consisting of two sequential divisions, meiosis I and meiosis II. Chromosomes are distributed so that each of the four daughter cells receives one chromosome from each homologous pair.

Meiosis I

Homologous chromosomes separate. One chromosome from each pair goes to each daughter cell.

Meiosis II

Sister chromatids separate and are distributed among daughter cells, which results in four new cells, each having haploid nuclei.

Prophase I

Homologous chromosomes form pairs and are sorted into two haploid sets (synapsis).

Nuclear envelope

Prophase II

Daughter cells resulting from meiosis I

Nuclear membrane begins to break down in daughter cells. Chromosomes attach to newly assembled spindle fibers.

Metaphase I

Tetrad

Nuclear envelope disappears. Each tetrad migrates to the equator. The two centromeres are attached to spindle fibers from opposite poles.

Equator

Bivalent

Metaphase II

Equator

Chromosomes line up across the equator of the spindle apparatus, with each centromere attached to spindle fibers from both poles.

Centromeres

Anaphase I

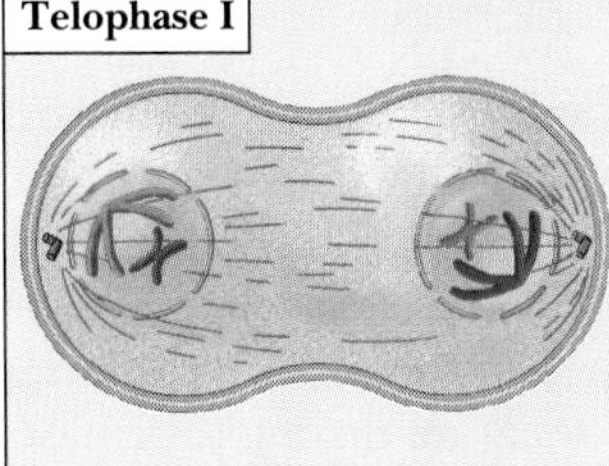

Spindle fibers (microtubules)

Sister chromatids (bivalents) move to the same pole.

Centriole

Anaphase II

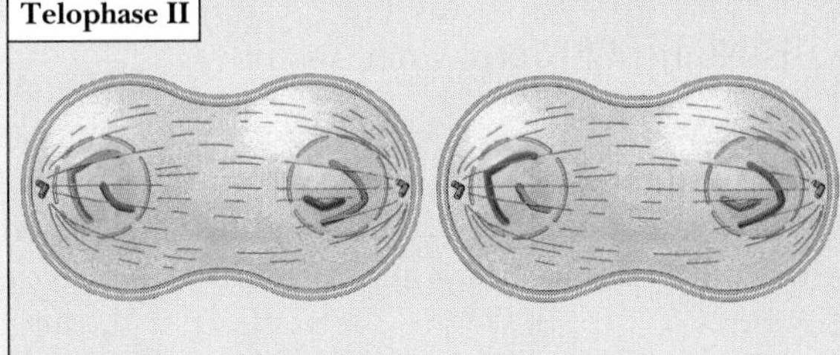

The chromosomes split and the chromatids move to each pole.

Telophase I

Chromosomes separate into two groups. Nuclear envelope reappears, and cytoplasm divides. The two resulting daughter cells, which contain one chromosome from each pair, move into meiosis II.

Telophase II

Nuclear membrane reforms. The four resulting daughter cells each have a complete haploid set of chromosomes, each derived from a single chromatid.

(b)

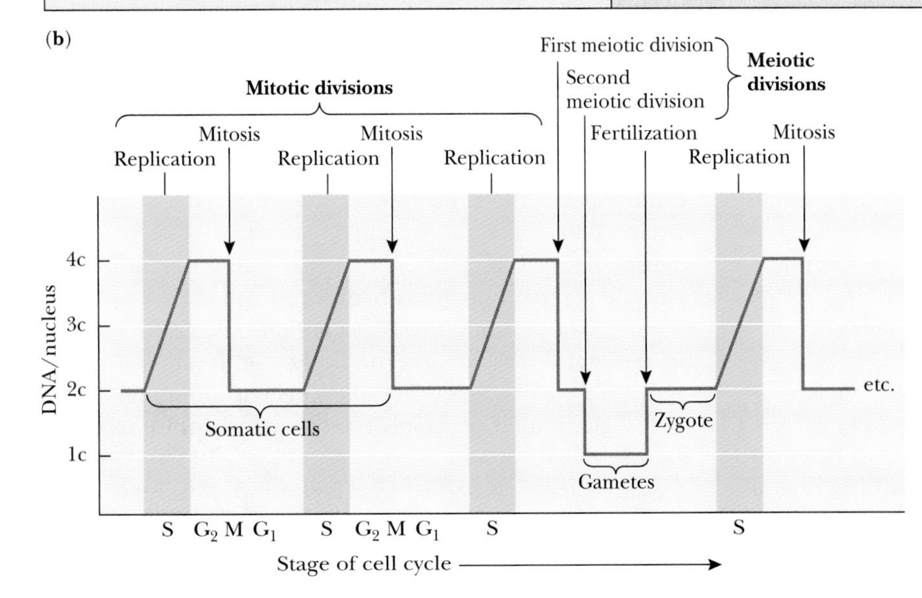

Figure 10-8 Meiosis, an overview: **(a)** stages of the two meiotic divisions; **(b)** the amount of DNA per cell changes as cells undergo mitosis, meiosis, and fertilization.

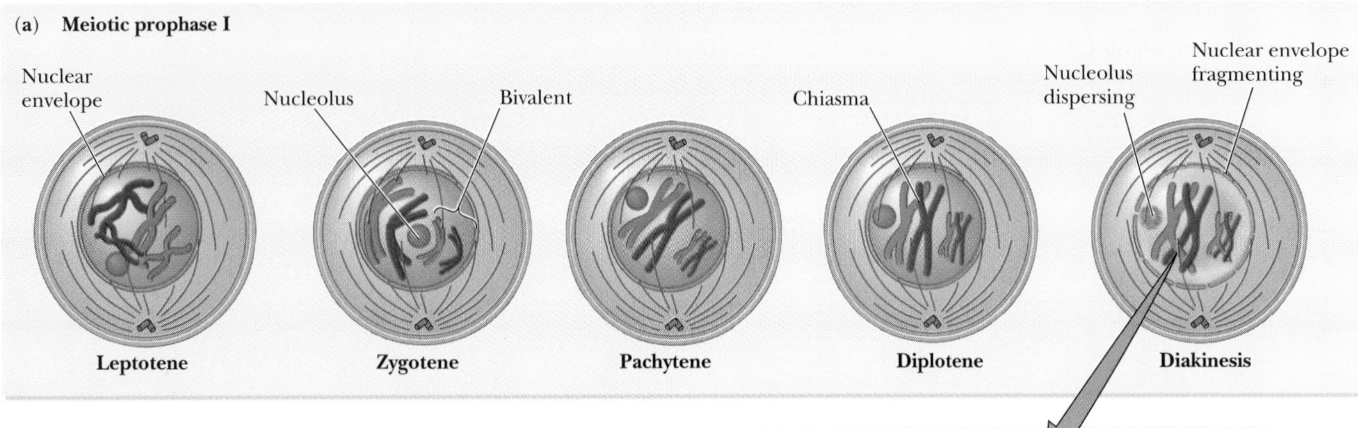

Figure 10-9 During prophase I of meiosis, homologous chromosomes pair and crossing over occurs. **(a)** Subphases of prophase I; **(b)** crossing over, the exchange of chromosome segments during prophase I.

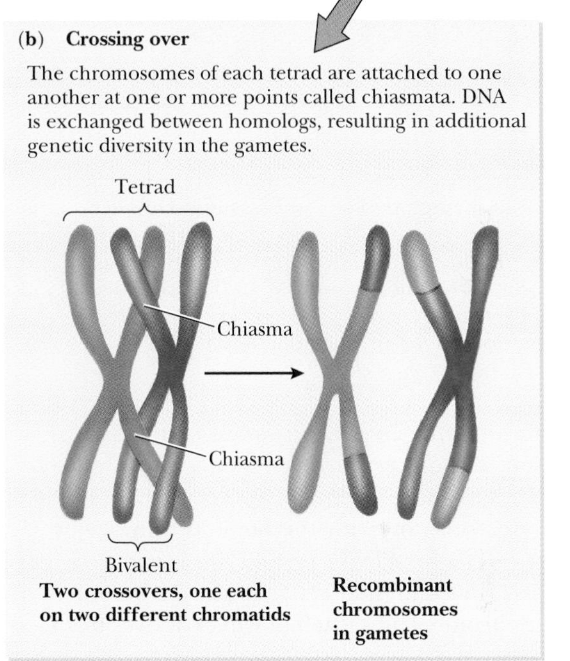

called **bivalents.** In Figure 10-9b, we can see that the chromosomes of each tetrad are attached to one another at one or more points, each of which is called a **chiasma** [Greek = cross; plural, **chiasmata**]. The chiasmata are the sites of exchange of DNA between homologs, as shown in Figure 10-9b, resulting in additional genetic diversity (recombinant chromosomes) in the gametes. (See Chapter 11.)

The extended duration of prophase I in some species is important for the proper building of an egg. In a frog's egg, for example, prophase I may last several years, during which time the egg makes and stores RNA for later use by the embryo. The embryo's early development after fertilization often depends on RNAs, proteins, and energy-rich lipids and carbohydrates, which have accumulated during meiosis. Only later can the developing embryo fend for itself.

In some species (including our own) meiosis may actually stop for extended times during prophase of meiosis I. In human females, meiosis begins before birth and then arrests during prophase I. In some cells, meiosis may resume some 40 years later.

During the Rest of Meiosis I, One Chromosome from Each Pair Goes to Each Daughter Cell

The next movement of the meiotic dance is metaphase I. Each tetrad migrates to the equator, as illustrated in Figure 10-8a. The two centromeres are attached to spindle fibers from opposite poles, as shown in Figure 10-10a.

Notice, as illustrated in Figure 10-8a, that in meiosis the sister chromatids go to the *same* pole during anaphase I. In contrast, the two chromatids go to *opposite* poles in mitosis, as shown in Figure 10-14 (p. 290). Telophase I and cytokinesis rapidly follow. The chromosomes stay partly condensed, and the daughter cells of meiosis I (now haploid) move right into meiosis II without any more DNA synthesis. Here again, meiosis differs from mitosis, where DNA synthesis must precede the next cell cycle.

Meiosis II Distributes Sister Chromatids to Daughter Cells

Meiosis II, as shown in Figure 10-8a, resembles mitosis except that each cell starts with a haploid (rather than a diploid) number of chromosomes. The specific accomplishment of meiosis II is the separation of sister chromatids: Each chromosome of the daughter cell derives from a single chromatid.

Prophase II is brief because the chromosomes are still mostly condensed from meiosis I. (See Figure 10-8b.) If a nuclear membrane has reappeared during telophase I, it breaks down during prophase II. The chromosomes attach to newly assembled spindle fibers. During metaphase II, the chromosomes line up across the equator of the spin-

(a) **Spindle attachment in metaphase of meiosis I**

Sister chromatids of homolog I
Sister chromatids of homolog II
Kinetochores of homolog I
Kinetochores of homolog II (one not visible)
Pole of spindle
Tetrad

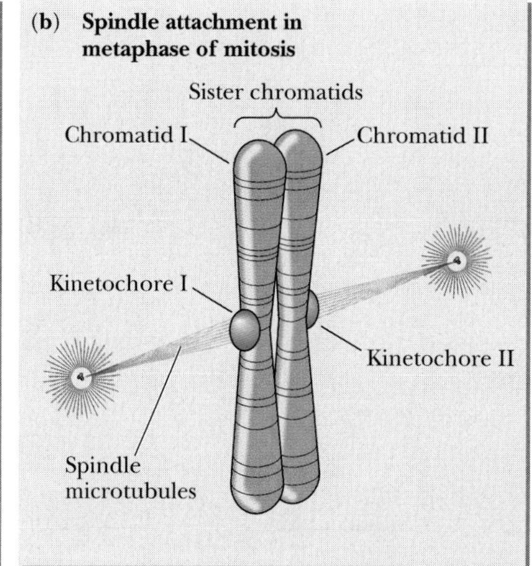

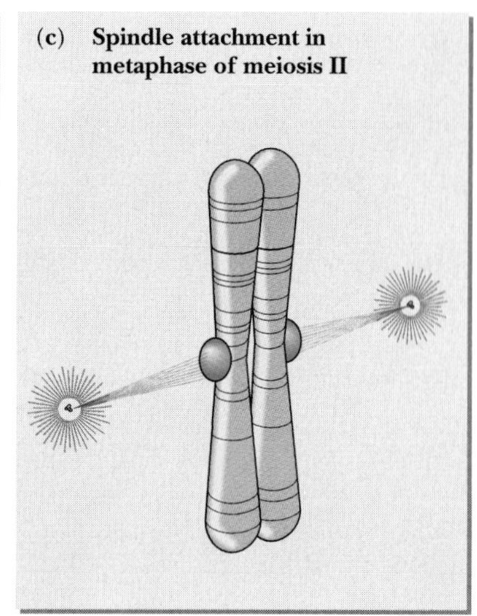

Figure 10-10 During anaphase I, sister chromatids go to the same pole, while in mitosis and in anaphase II, sister chromatids go to opposite poles. **(a)** Spindle attachment in metaphase I of meiosis; fused kinetochores are responsible for the two sister chromatids going to the same pole during anaphase I. **(b)** Spindle attachment in metaphase of mitosis; the kinetochores of the two sister chromatids are not fused, and the two sister chromatids move to opposite poles during anaphase. **(c)** Spindle attachment during metaphase II of meiosis; as in mitosis, the two sister chromatids move to opposite poles during anaphase II.

dle apparatus, with each centromere connected to spindle fibers from both poles. During anaphase II, the chromosomes split—one chromatid moves to each pole (just as in mitosis, but now with the haploid number of chromosomes). The nuclear membrane again forms in telophase II. The daughter cells each have a complete haploid set of chromosomes, each of which derives from a single chromatid.

The moving apart of two homologous chromosomes during anaphase I and of two sister chromatids in anaphase II is called **disjunction.** The failure of these separations to proceed normally is called **nondisjunction.** Nondisjunction results in a gamete having too many or too few chromosomes (that is, more or fewer than the haploid number). After fertilization, the resulting zygote also has too many or too few chromosomes (that is, more or fewer than the diploid number). A cell or individual with an abnormal number of chromosomes is said to be **aneuploid,** while one with the correct number of chromosomes is said to be **euploid.** One of the most common consequences of nondisjunction in humans is Down syndrome, which we discuss in the accompanying essay.

Meiosis Employs Much of the Molecular Machinery of Mitosis, as Well as Additional Structures

As in mitosis, the chromosome movements of meiosis depend on the interactions among the microtubules of the spindle apparatus, the kinetochores, and the centrosome. (See Chapter 9.) During meiosis I, however, the kinetochores of the two sister chromatids are fused (as shown in Figure 10-10a) and behave as a single kinetochore, so that, at anaphase I, the two sister chromatids of each chromosome migrate together to a single pole. During meiosis II, the kinetochores behave as they do in mitosis, and the two sister chromatids separate, as shown in Figure 10-10b and c.

Regulation of meiosis uses some of the same molecular machinery that regulates the mitotic cell cycle. MPF, for example, serves as a trigger for meiosis as well as for mitosis. (In fact, MPF—now called mitosis-promoting factor—was first called maturation-promoting factor, referring to its ability to promote meiosis in frogs.) But meiosis also depends on other molecular structures that are not used in mitosis.

One of the characteristic differences between mitosis and meiosis is the formation, during prophase I of meiosis, of the **synaptonemal complex** (shown in Figure 10-11), in which proteins hold homologous chromosomes together at synapsis. The proteins of the synaptonemal complex somehow keep the paired homologs in register with each other, so that corresponding segments of the paternal and maternal chromosomes are aligned. Within the synaptonemal complex are other large assemblies of proteins, called **recombination nodules,** which assist in the formation of chiasmata between corresponding segments of the two homologs. Understanding just how the synaptonemal complex and the recombination nodules work is an important goal of contemporary research on meiosis.

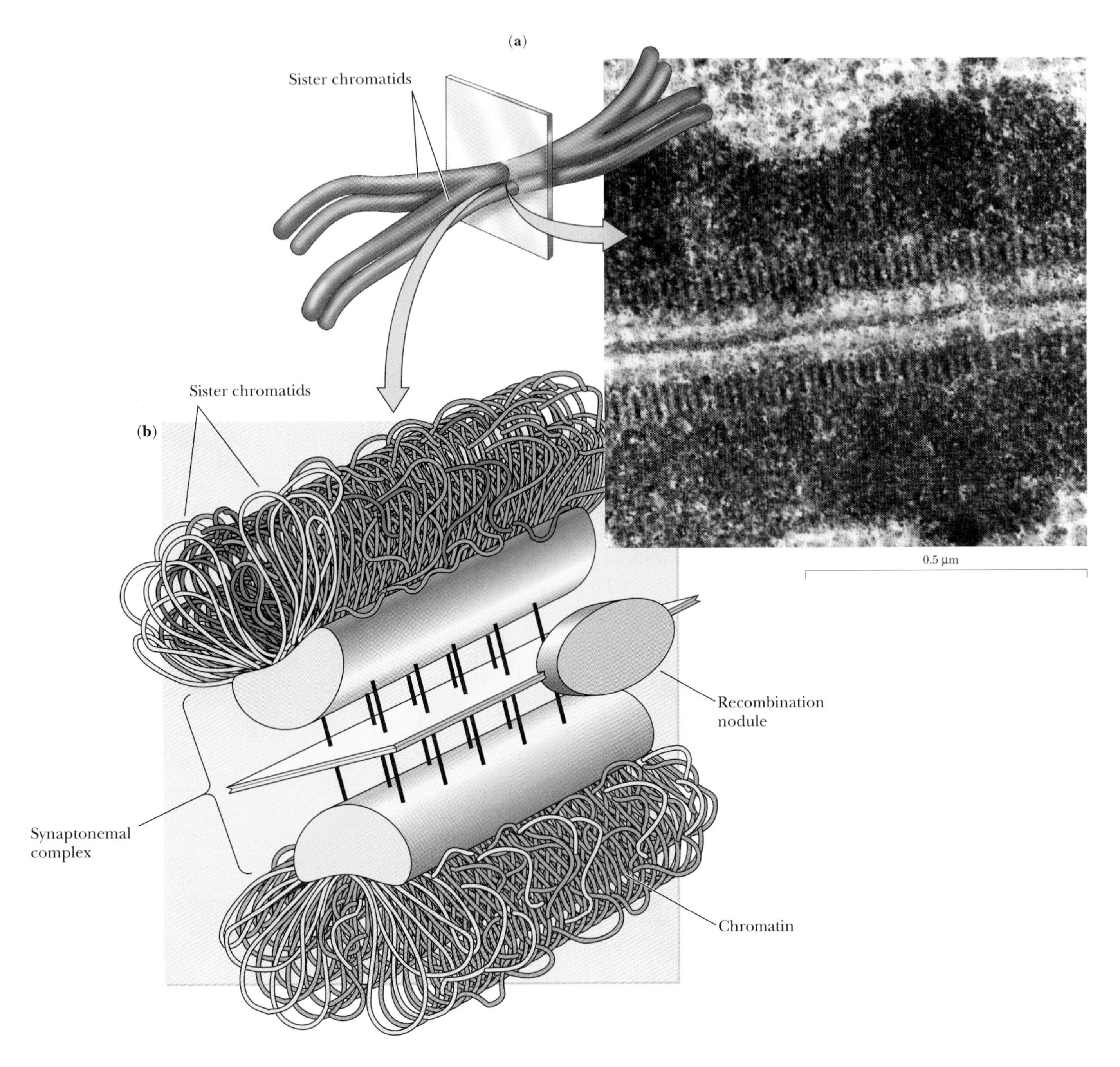

Figure 10-11 During prophase I of meiosis, proteins hold homologous chromosomes together in a structure called the synaptonemal complex. **(a)** An electron micrograph of a synaptonemal complex of two hamster chromosomes; **(b)** diagram of a synaptonemal complex, showing a recombination nodule. *(a, D. VonWettstein)*

Random Assortment of Homologous Chromosomes During Metaphase I Creates New Chromosome Combinations in Gametes

How do the tetrads align themselves during metaphase I? Do all the maternally derived chromosomes go to one pole and the paternally derived chromosomes to the other? If so, meiosis would more or less recreate the same chromosome sets generation after generation, although formation of chiasmata leads to some shuffling of maternal and paternal information.

In fact, however, the assortment of chromosomes is random, with each pair of homologs aligning independently during metaphase I. This means that meiosis is a major source of genetic diversity. As shown in Figure 10-12, two pairs of chromosomes can form four (2^2) kinds of gametes. Similarly, three pairs of chromosomes can form eight (2^3) kinds. A single human germ cell, with 23 pairs

(Text continues on page 288.)

ESSAY

WHAT HAPPENS WHEN MEIOSIS GOES WRONG?

More than 30% of all human zygotes do not survive to birth. They die as embryos, usually in the first 3 months of pregnancy. Such embryo deaths are called *spontaneous abortions,* or miscarriages. Most of these embryos die because they have abnormal numbers of chromosomes. A few zygotes with chromosome abnormalities, however, do develop into embryos that survive to birth. Most of these children have serious health problems and may die in early childhood.

An example of a chromosomal abnormality that results from abnormal meiosis is *Down syndrome* (also called Down's syndrome), after the mid-19th century English physician who first described its physical symptoms. Down syndrome is a disorder that leads to mental retardation and the abnormal development of the face, heart, and other parts of the body. Because the abnormalities are relatively mild (compared with those of children with other chromosomal abnormalities), fetuses and children with Down syndrome usually survive at least into their teens and sometimes much longer. Only rarely, however, do Down syndrome patients have children of their own.

After the development of modern karyotyping methods in the mid-1950s, researchers were able to determine that Down syndrome is almost always associated with a chromosomal abnormality, shown in Figure 1—a **trisomy** [Greek, *tri* = three + *soma* = body], the presence of three, rather than two, copies of a chromosome, in this case of chromosome 21. About 80% of the embryos with this chromosomal abnormality do not survive until birth, but about 20% do survive.

How does such a chromosomal abnormality arise? The failure, as illustrated schematically in Figure 2, is nondisjunction, most commonly occurring during meiosis I. If, for example, the two chromosomes 21 fail to separate during anaphase I, then two of the four possible gametes produced during meiosis each have two copies instead of one, and the other two have no copies of chromosome 21. When these gametes unite with normal gametes at fertilization, they produce abnormal zygotes—some with three copies of chromosome 21, and some with only one copy. Those with only one copy of chromosome 21 survive poorly, and only a few such cases have been reported.

The chance of a woman under 30 giving birth to a Down syndrome child is less than 1 in 1000; the

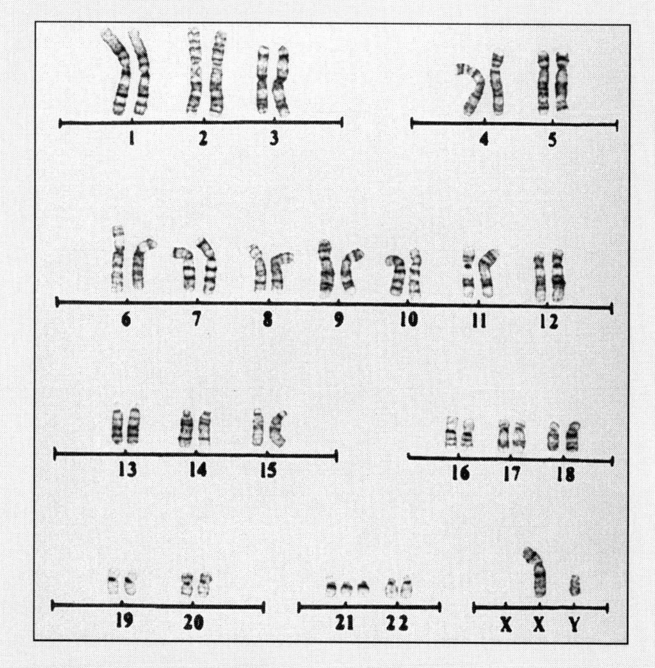

Figure 1 Down syndrome karyotype: Notice the presence of three copies rather than two copies of chromosome 21. *(courtesy of Dr. Leonard Sciorra)*

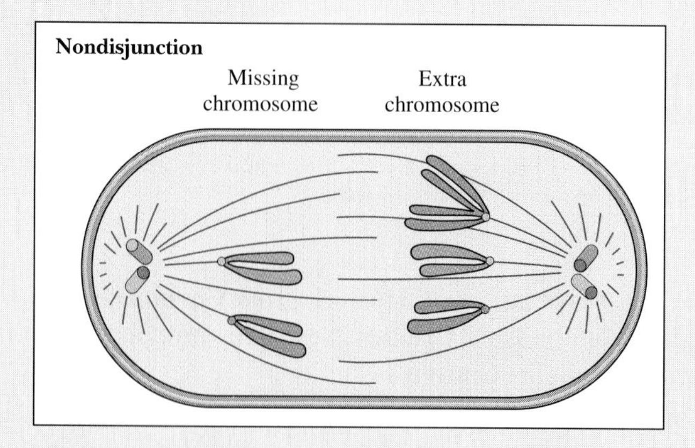

Figure 2 Nondisjunction during meiosis I or II can produce a gamete with a missing chromosome or an extra copy of a chromosome.

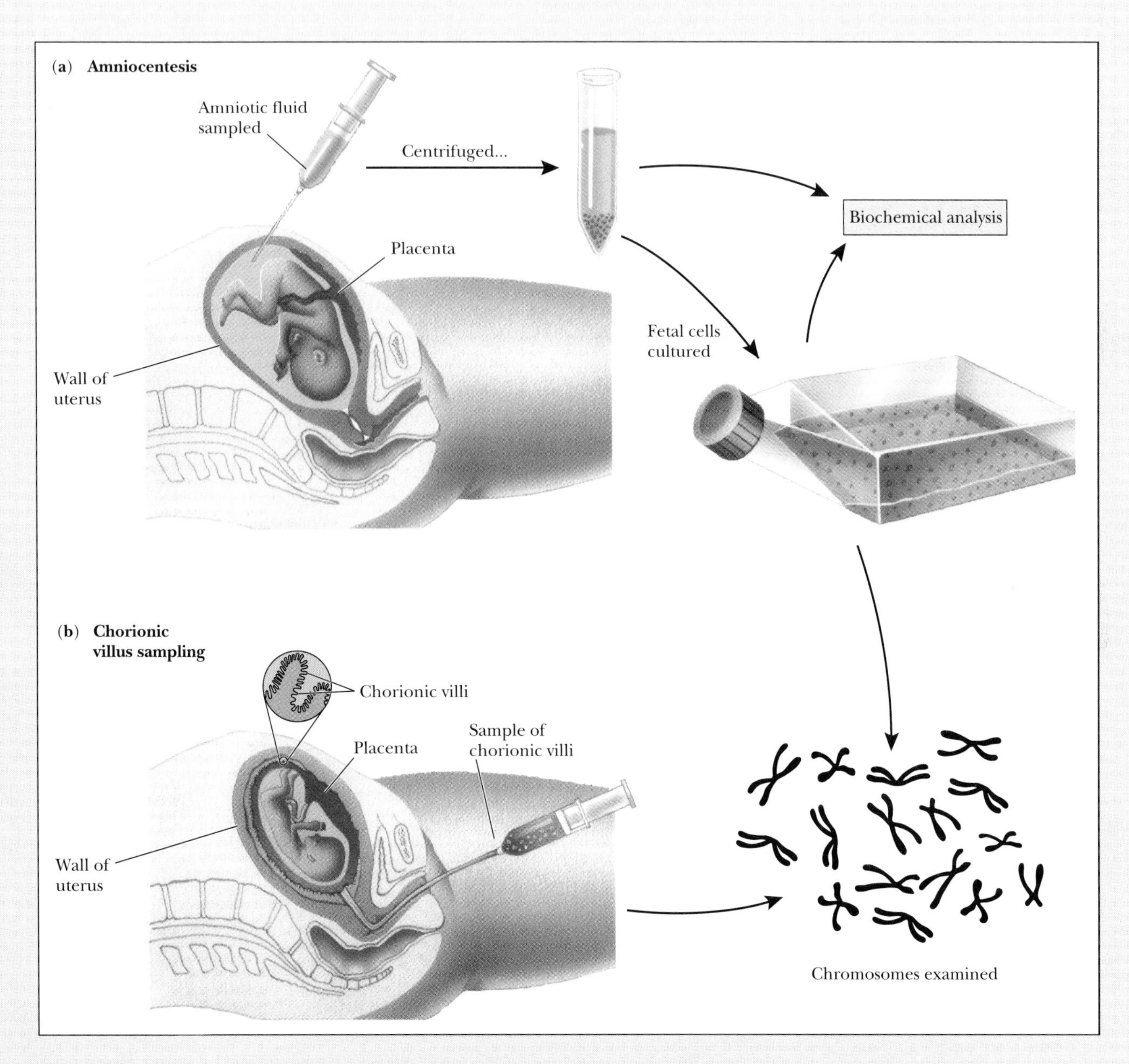

Figure 3 Amniocentesis and chorionic villus sampling allow the detection of chromosomal abnormalities. **(a)** Amniocentesis; **(b)** chorionic villus sampling.

(continued)

ESSAY *continued*

chance increases to over 1 in 100 for mothers over 40. Overall in the United States, about one baby in 700 has Down syndrome. Children with Down syndrome are most frequently born to older mothers. Cell biologists think that the chances of nondisjunction (at least for chromosome 21) go up with maternal age because of the particular pattern of meiosis. Human females begin meiosis well before birth, but (as mentioned on p. 283) it arrests in the midst of prophase I. Nondisjunction of chromosome 21 may be the result of interrupting the complicated process of meiosis for so long a time. About 20% of the time, however, the extra chromosome 21 in a Down syndrome child comes from the father.

Karyotype analysis can allow couples to learn early in pregnancy if they have produced a zygote with a chromosomal abnormality. The most common method of determining the karyotype involves sampling cells in the *amniotic fluid,* [Greek, *amnion* = membrane around a fetus], which surrounds the fetus, in a procedure called *amniocentesis* [Greek, *centes* = puncture], as illustrated in Figure 3a. Amniocentesis is usually done between the 14th and 16th weeks after conception. To determine the karyotype, the cytogeneticist grows the cells in culture, blocks mitosis with colchicine or another drug, and prepares the karyotype. A newer procedure, diagrammed in Figure 3b, called *chorionic villus sampling,* uses fetal cells present in the placenta, the tissue that carries nutrients and oxygen from the mother's blood to the fetus and carries wastes from the fetus back to the mother. Chorionic villus sampling may be done earlier in pregnancy than amniocentesis. Some couples prefer this procedure because information becomes available sooner.

Cells obtained by amniocentesis or by chorionic villus sampling can also provide other information about the genetic makeup of the zygote. Besides looking for chromosomal abnormalities by karyotype analysis, genetics clinics can now detect a variety of genetic diseases—including sickle cell disease, Tay-Sachs disease, and muscular dystrophy—in the cells of an early embryo by using the techniques described in subsequent chapters.

of chromosomes, can therefore form 2^{23} (about 8 million) kinds of gametes. A single couple, then, could produce more than 64 trillion (8 million eggs × 8 million sperm) genetically different offspring. All this diversity comes from the random alignment and subsequent separation of tetrads in metaphase I—without even including the new combinations of genes that result from chiasmata formation. No wonder sexually reproducing species (including our own) are so wonderfully diverse!

In Some Animals, One Sex Has One Pair of Nonhomologous Chromosomes

In protists, fungi, plants, and many animals, both males and females have complete sets of homologous chromosomes. In all mammals (including humans) and birds, and in a number of other animals (including *Drosophila*), however, males and females have distinctive chromosome sets. As shown in a human karyotype in Figure 10-13, male mammals have two nonmatching chromosomes.

Other animals also have a nonhomologous pair of chromosomes in one sex, but not always in the male. In birds, for example, all the male's chromosomes are in pairs, but the female has a nonhomologous pair.

Instead of having 23 homologous pairs of chromosomes, a cell from a human male has 22 pairs of homologous chromosomes plus two nonhomologous chromosomes. One of these two unmatched chromosomes has the same size and banding pattern as the 23rd pair of chromosomes in female cells. The other unmatched chromosome is much smaller than any of the other chromosomes seen in the female karyotype.

The chromosomes that differ between males and females (in humans, the 23rd pair of chromosomes) are called the **sex chromosomes;** the other chromosomes (the 22 pairs that, in humans, are the same in men and women) are called **autosomes.** In mammals, the two sex chromosomes are called X and Y. An individual with both an X and a Y chromosome is male, and an individual with two X chromosomes is a female. The determination of sexual

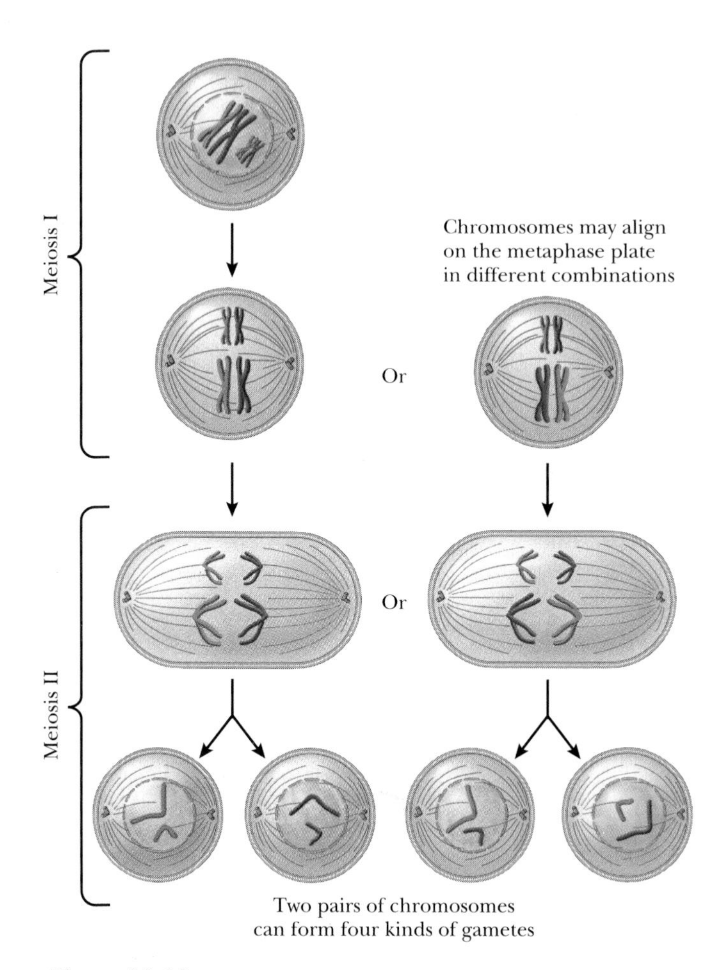

Figure 10-12 The random assortment of homologous chromosomes during meiosis means that two pairs of chromosomes can form four types of gametes.

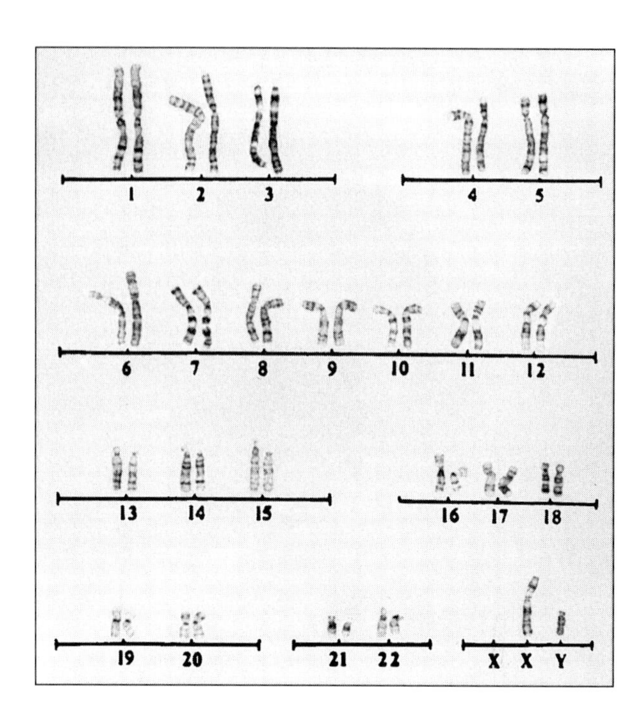

Figure 10-13 Male mammals have two nonmatching chromosomes. Cells from a human male, for example, have 22 pairs of homologous chromosomes, called autosomes, and two sex chromosomes, called X and Y. Chromosomes from a human female, shown in Figure 10-3, have 23 pairs of homologous chromosomes—22 pairs of autosomes and a pair of X chromosomes. *(Leonard Sciorra)*

phenotype (maleness versus femaleness) in mammals depends heavily on the presence of the Y chromosome, but, as discussed in Chapter 22, the story is more complicated.

In both females and males, the two sex chromosomes (XX in females and XY in males) pair and segregate during meiosis in the same way as homologous pairs of autosomes. In women, then, each egg contains 22 autosomes plus an X chromosome. In men, each sperm also contains 22 autosomes plus either an X chromosome or a Y chromosome. When a sperm carrying a Y chromosome fertilizes an egg, the resulting zygote is XY and therefore male; when a fertilizing sperm carries an X chromosome, the resulting zygote is XX and therefore female.

How can X and Y pair in synapsis if they differ so much in size and form? The reason is that they share a region that is homologous. In spite of their striking differences in size and form, they still behave as homologs during meiosis.

THE BEHAVIOR OF CHROMOSOMES UNDERLIES THE RULES OF INHERITANCE

In Chapter 11 we examine the rules that govern the inheritance of genetic variations. We can, however, preview these rules from our knowledge of chromosome movements during meiosis and from the fact that the chromosomes contain genetic information in the form of DNA.

The realization that chromosomes behave like genes—in specifying inherited traits—came in the first years of the 20th century, largely from the work of Walter Sutton, then a graduate student at Columbia University in the laboratory of E. B. Wilson, the most distinguished cell biologist in the United States. Sutton was studying the chromosomes of a grasshopper (*Brachystola*) during gamete formation and fertilization. Because the chromosomes have distinctive sizes, Sutton was able to demonstrate that the chromosomes of somatic cells occur in pairs, which separate during the course of meiosis.

Sutton was aware of the just rediscovered laws of heredity, worked out 35 years earlier by Gregor Mendel. (See Chapter 11.) Sutton realized that the chromosome behavior he observed could explain the laws of heredity

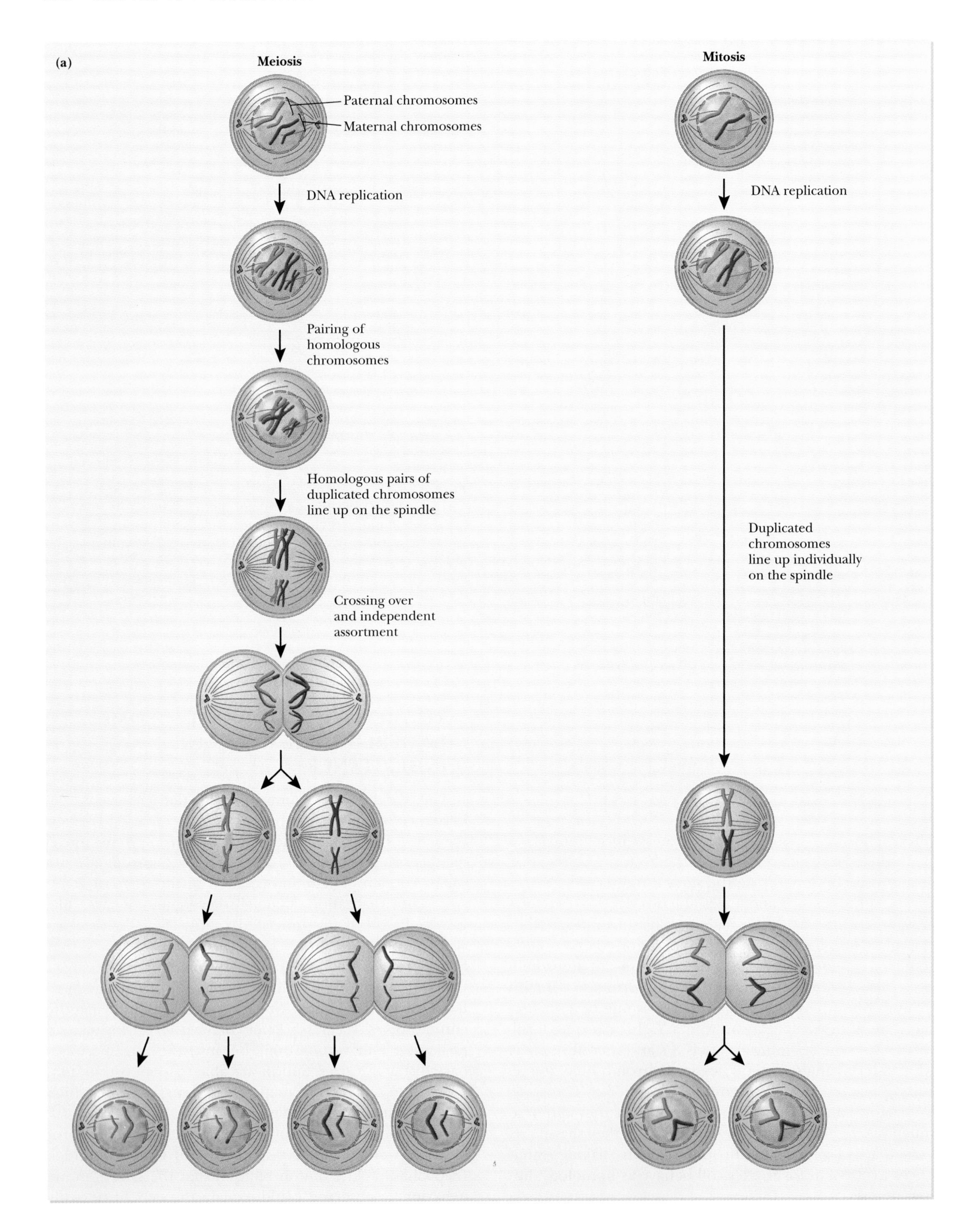
(a)
Meiosis
Paternal chromosomes
Maternal chromosomes
DNA replication
Pairing of
homologous
chromosomes
Homologous pairs of
duplicated chromosomes
line up on the spindle
Crossing over
and independent
assortment
Mitosis
DNA replication
Duplicated
chromosomes
line up individually
on the spindle

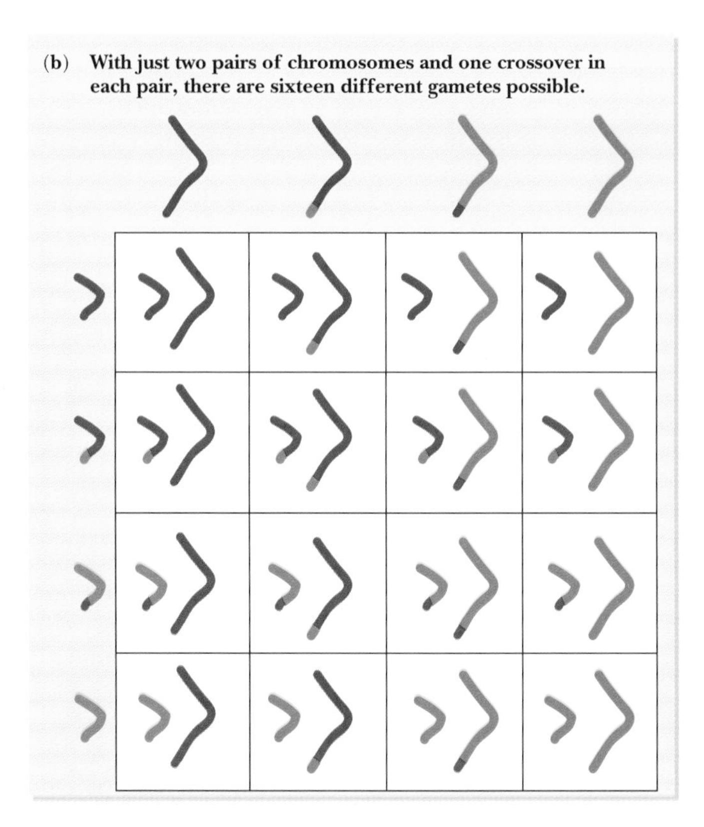

Figure 10-14 **(a)** A comparison of meiosis and mitosis. **(b)** Gametes that are possible with 2 chromosomes and 1 crossover.

if (as we now know) genes were physically a part of chromosomes.

Sutton argued that meiosis was responsible for new combinations of genes in the next generation. The reason for such new combinations, Sutton said, was that, in forming haploid sets of chromosomes during meiosis, a cell did not distinguish between the maternally and paternally derived homologs. A gamete (or haploid spore) could contain combinations of variant genes that were not present in either parent.

We now see how Sutton's insights help us understand the chromosomal basis of the laws of inheritance, discussed in Chapter 11. Sutton excitedly reported his ideas to his mentor, Wilson, saying that he had discovered "why the yellow dog is yellow." Wilson was not immediately receptive, but he soon became convinced. The rest of the biology world was slower to accept Sutton's formulation, however. Sutton himself left graduate school without completing a Ph.D. He returned to Columbia 2 years later to study medicine, a career which he successfully pursued until his death at age 39.

We can now summarize the rules of inheritance that we can take from Sutton's work and our subsequently gained knowledge of chromosome behavior.

1. Just as diploid cells have two copies of each chromosome, so they have two copies of each gene. Each copy may be slightly different from the other and, for example, specify alternative versions of the same trait.
2. When genes that specify two different characteristics lie on different chromosomes, meiosis can bring about **genetic recombination,** associations of variant genes that did not exist in the parents.
3. Genetic recombination also results from **crossing over,** the exchange of chromatid segments that occurs in chiasmata. After crossing over, a new chromosome contains some genes from each of the homologs (that is, originally from each of the two parents). These new combinations of genes then stay together through the rest of meiosis. The chromosomes of each gamete, then, can differ from the chromosomes in the gametes of the previous generation.

Chapter 11 discusses these points in much greater detail. Figure 10-14 presents a comparison between meiosis and mitosis and points out the two major sources of the reassortment of genetic material during meiosis: crossing over and independent assortment.

SUMMARY

A sexually reproducing organism receives genetic information from two parents rather than one. This genetic information is contained in two sets of chromosomes, one set from each parent. The somatic cells of an organism—those that do not contribute to the next generation—all have two copies of each chromosome and are said to be diploid. The gametes are the reproductive cells that carry information to the next generation. The gametes and the cells from which they derive are called germ cells. Each gamete has only one copy of each chromosome and is said to be haploid.

Male and female gametes come together at fertilization to form a zygote. Fertilization restores the diploid chromosome number.

During the life cycle of sexually reproducing organisms, the process of meiosis randomly selects one chromosome from each pair. Meiosis generates combinations of genes that are different from those present in the parents and thus account for much of the genetic diversity of individuals within a species.

Meiosis consists of two cell divisions (meiosis I and meiosis II), each of which consists of prophase, metaphase, anaphase, and telophase. During prophase I homologous chromosomes form precisely aligned pairs in a process called synapsis. During the rest of meiosis I one chromosome from each pair (with two chromatids) goes to each daughter cell. Meiosis II distributes the sister chromatids of each chromosome to daughter cells.

During synapsis, homologous chromosomes pair with one another and exchange segments in a process called crossing over. Sometimes homologous chromosomes fail to separate after synapsis. This failure—called nondisjunction—leads to the production of gametes with an abnormal number of chromosomes.

In mammals, males have two unmatched sex chromosomes, called X and Y. Human males have 22 homologous pairs of autosomes, while human females have 22 pairs of autosomes and a pair of X chromosomes. The X and Y chromosomes have homologous regions and synapse during meiosis.

In diploid cells that have two copies of each chromosome, there are two copies of each gene. When genes that specify two different characteristics lie on different chromosomes, meiosis can bring about genetic recombination. In meiosis, genetic recombination also results from crossing over.

KEY TERMS

aneuploid
asexual reproduction
autosome
bivalent
chiasma (chiasmata)
crossing over
diploid
disjunction
egg
euploid
fertilization
gamete
genetic recombination
germ cell
germ line
gonad
haploid
homolog
homologous chromosome
karyotype
meiosis
meiosis I
meiosis II
mutation
nondisjunction
ovary
ovum
recombination nodule
sex chromosome
sexual reproduction
somatic cell
sperm
synapsis
synaptonemal complex
testis
tetrad
trisomy
zygote

QUESTIONS FOR REVIEW AND UNDERSTANDING

1. Define and contrast the following terms:
   (a) haploid vs. diploid
   (b) germ cells vs. somatic cells
   (c) ovum vs. spermatozoan
   (d) sexual reproduction vs. asexual reproduction
   (e) autosomes vs. sex chromosomes
   (f) homologous chromosomes vs. sister chromatids
   (g) aneuploid vs. euploid
2. What could you learn about an individual by simply examining his or her karyotype?
3. Describe several specific adaptations of a sperm and an egg. How does the structure of each match its function?
4. Explain why meiosis is necessary in sexual reproduction.
5. Discuss the statement: "Mutations are always detrimental to an organism and ultimately to a species."
6. How are homologus chromosomes similar to each other? How are they different?
7. Contrast the following steps in mitosis and meiosis:
   mitotic prophase vs. prophase I
   mitotic metaphase vs. metaphase I
   mitotic metaphase vs. metaphase II
8. How are maternal and paternal chromosomes segregated during meiosis versus mitosis?
9. Are one's offspring always more fit in terms of survival? Explain, in general terms, why or why not.
10. Describe in your own words what occurs during synapsis.
11. Why is synapsis such a critical event in meiosis?
12. What is a chiasma? How do events at the chiasmata impact genetic diversity?
13. Describe the steps of meiosis I and explain what they accomplish. Do the same for meiosis II. Which part of meiosis actually accomplishes the reduction in chromosome number?
14. Discuss the genetic and evolutionary advantages of sexual reproduction.
15. Consider that there are organisms that do not regu-

larly undergo sexual reproduction. Argue the advantages and disadvantages of an asexual reproductive strategy.

16. Propose a model that would explain why a particular genetic trait is seen only in males and not females. Propose a model that would explain why a given genetic trait could occur more frequently in females.
17. With the development of large agribusinesses and the decline of the family farm comes the planting of fields with genetically identical seed. Discuss the pros and cons of this agricultural practice.
18. Summarize all the steps and mechanisms of meiosis that contribute to the generation of genetic diversity.
19. Describe and discuss the medical procedures a woman over the age of 40 who is considering having a child might undergo. Explain why a physician would recommend these tests.
20. It is clear that the ability to detect the presence of a variety of genetic diseases in a fetus is only going to improve. Consider the social and moral implications of this technology.

SUGGESTED READINGS

ALBERTS, B., D. Bray, J. Lewis, M. Raff, K. Roberts, and J. D. Watson, *Molecular Biology of the Cell,* 3rd edition, Garland Publishing, New York, 1994. Chapter 20, "Germ Cells and Fertilization," discusses much of the material in this chapter.

GRIFFITHS, A. J. F., J. H. Miller, D. T. Suzuki, R. C. Lewontin, and W. M. Gelbart, *An Introduction to Genetic Analysis,* 5th edition, W. H. Freeman and Company, New York, 1993. Chapter 3, "Chromosome Theory of Inheritance," discusses much of the material in this chapter.

MOORE, J. A., "Genetics," *American Zoologist* 26:583–747, 1986.

CHAPTER 11

Inheritance: Genes and Chromosomes

KEY CONCEPTS

- Genetics explains the inheritance of variations.
- The inheritance of genes resembles the inheritance of chromosomes during meiosis and fertilization.
- Gregor Mendel established the principles of genetics even before biologists had discovered chromosome movements.
- Genes lie on chromosomes.
- A chromosome contains many genes.

OUTLINE

ESSAYS

Genetics, the study of inheritance, must explain both genetic constancy (why offspring resemble their parents) and the origin of genetic variation (why offspring differ from their parents). In this chapter, we discuss **transmission genetics,** which deals with patterns of inheritance. In Chapters 12–17, we discuss **molecular genetics,** which studies how DNA carries genetic instructions and how cells carry out these instructions. In this chapter we show that the behavior of chromosomes during meiosis and fertilization does much to explain the observed inheritance of variation between generations.

■ VARIATIONS AMONG INDIVIDUALS DEPEND ON DIFFERENCES IN GENES AND IN ENVIRONMENT

Transmission genetics deals mostly with the inheritance of variations. Until the 1970s, when molecular biologists learned how to determine the sequence of bases in DNA, geneticists could study only variations in **phenotype,** the set of an organism's observable properties. A **phenotypic trait** is a single aspect of phenotype in which individuals may vary. Figure 11-1 illustrates variations in an easily observed phenotypic trait—the height of corn plants.

Individuals vary in phenotype because of differences in their genes and in their environments. For example, the height of the corn plants in Figure 11-1 depends not only on their genetic inheritance but also on the amount of received water, the intensity of the absorbed sunlight, and amount of applied fertilizer. It makes little sense to study the genetic control of the height of a plant if it lacks water. One of the most difficult tasks of genetics is to separate the environmental and genetic influences on phenotype. Many experimental studies of genetics concern inherited traits, such as hair color or seed shape, that do not depend heavily on environmental variation.

(a) (b) (c) (d)

Figure 11-1 Phenotype depends on genes and on environment. **(a)** A dwarf variety of corn, unable to make a hormone called gibberellin; **(b)** the same dwarf variety, after spraying with gibberellin; **(c)** the common, non-dwarf variety, treated with gibberellin; **(d)** a common variety of corn plant grown under standard conditions. *(Courtesy of B. O. Phinney, University of California, Los Angeles)*

Geneticists Study the Inheritance of Both Products and Plans

Phenotype ultimately reflects the presence of particular cellular products, often enzymes or other proteins. The plans for these products lie in the organism's genes. Geneticists use the word **genotype** to refer to the genes (that is, the plans) present in a particular organism or cell. When we speak about a specific trait like a plant's height, genotype usually has a more restricted meaning—the gene or genes that control that one trait. The plants shown in Figure 11-1a and b, for example, do not contain a gene needed to make a compound called gibberellin, which stimulates plant growth. (See Chapter 21.)

Because sexually reproducing organisms are diploid, they have two copies of most genes on two homologous chromosomes. The corn plants in Figure 11-1a and b have two defective genes, and we might designate their genotype (for this trait) as ga^-/ga^-.

In this chapter, we focus our attention on the rules that govern the inheritance of variant phenotypes. Although (as we see in Chapter 12) geneticists can now directly study the inheritance of genes themselves, some of the most interesting genetic questions concern not the plans but the products. An organism's phenotype (a reflection of the products present) is not only interesting to us as fellow organisms, but also determines the way that the organism gets on in the world.

The Same Laws of Inheritance Apply to All Sexually Reproducing Organisms

Agricultural breeders have long studied inherited variations, but in such practical traits as the yields of meat, milk, and grain. Indeed, the beginnings of civilization, some 12,000 years ago, depended on the success of such agricultural breeders in producing grains that could be sown and harvested according to plan. Early farmers also bred sheep and goats for the production of wool and milk, as well as cats, dogs, and horses. Ancient documents attest to the importance of heredity: For example, the 4000-year-old Egyptian model shown in Figure 11-2 illustrates genetic traits in cattle. Ancient Babylonian tablets show how to cross-pollinate date palms; and the Bible portrays Jacob as a master animal breeder.

Figure 11-2 The ancient Egyptians were successful cattle breeders. This miniature stable, which dates from about 2000 BC, shows longhorn cattle. Other cattle breeds had short horns or no horns. *(Metropolitan Museum of Art, Rogers Fund and Edward S. Harkness Gift, 1920)*

In trying to understand the rules of genetics, researchers have examined the transmission of such phenotypic variations as sizes, shapes, and colors. The best recommendations for an experimental organism are rapid reproduction (to minimize the time between generations) and the easy identification of many phenotypic traits that are not heavily dependent on environmental variation (to allow the rapid study of many genes). Favorite organisms for this research have included peas (which vary in the shapes of seeds and the color of flowers) and the fruit fly (*Drosophila*), which can vary in eye color and in the shapes and lengths of wings, legs, and antennae.

No matter how useful an experimental organism is in elucidating general rules, humans are ultimately interested in the characteristics of our own species. Geneticists and physicians have therefore also studied the inheritance of variations in humans as well as in experimental animals. They have studied both well-defined traits, such as inherited diseases, height, and eye color, and more vaguely defined characteristics whose genetic bases are highly questionable, such as "criminality" and "feeblemindedness."

As we have already stated repeatedly and discuss in detail in Chapter 12, genes consist of specific sequences of nucleotides in DNA. Genetic variation therefore ultimately depends on **mutations,** changes in the nucleotide sequences of DNA.

All sexually reproducing eukaryotes follow the genetic rules described in this chapter. There are some notable exceptions to these rules, including the inheritance of genes carried in the DNA of mitochondria and chloroplasts (discussed in Chapter 16) and the life cycle of the social insects (not discussed in this book). The reasons for this unity are that (1) all organisms use DNA as the genetic material; (2) the DNA of all eukaryotic organisms is organized into chromosomes; (3) almost all chromosomes exist in homologous pairs at some time during a sexual life cycle; and (4) homologous chromosomes behave in the same ways during meiosis and fertilization in all eukaryotes.

THEORIES OF PREFORMATION AND BLENDING FAILED TO EXPLAIN OBSERVED PATTERNS OF INHERITANCE

Historically, biologists had no trouble inventing theories to explain how organisms stay the same between generations: A favorite theory was that miniature versions of future organisms were already preformed in sperm or eggs, so that the properties of all future generations were already determined.

Before the 17th century, most biologists had accepted Aristotle's theory that the mother contributes merely the "substance" of a child, while the father determines the "form." Philosophers (male philosophers) even compared the presumed creative action of the sperm to the divine creation of the universe from formless matter! Acknowledgment of the equal role of females came only from breeders of ornamental plants, who found that both parental varieties contributed equally to the next generation, no matter which variety provided the pollen (sperm).

In the face of such evidence, most biologists reasoned that the phenotypes of offspring somehow result from the "blending" of the properties of the parents. Indeed, we are not surprised when the mating of a large and a small dog produces puppies that grow to intermediate sizes or when crossing a red-flowering plant with a white-flowering plant, such as the snapdragons shown in Figure 11-3, produces a plant with pink flowers. The color of the offspring appeared (at first) to be the result of blending, as if two cans of paint were poured together and vigorously stirred. The blending model of inheritance works well for some traits, especially those like height or weight, which we now know to depend on many genes.

The blending model, however, does not account for the reappearance of unblended characteristics in later generations—for example, pink flowers giving rise to red and white offspring. The blending model also raised a serious problem in understanding the mechanism of evolution. If blending occurred all the time, variation would eventually disappear into a grey average for each species,

Figure 11-3 Red, white, and pink snapdragons. Snapdragon flower color appears to result from a kind of "blending" inheritance: Red snapdragons crossed with white snapdragons produce plants with pink flowers. *(John D. Cunningham/Visuals Unlimited)*

and natural selection would not have any inherited variations on which to act. Blending inheritance and natural selection, then, were incompatible theories.

GENES BEHAVE LIKE CHROMOSOMES

In Chapter 10, we discussed the behavior of chromosomes during meiosis and fertilization. We also know (and discuss in more detail in Chapter 12) that genes are contained in chromosomes. We are therefore in a better position to understand genetics than were earlier biologists. Armed with our knowledge, we can understand why the blending model of inheritance is wrong and explain the basis of the laws of inheritance.

The key to understanding genetics in sexual organisms is to realize that each gene is present in two copies. Recall that we inferred this double dosage from the existence of homologous pairs of chromosomes. The observed patterns of the transmission of genes confirm this deduction.

A Diploid Cell Has Two Copies of Every Gene

Let us consider the inheritance of flower color in the snapdragon, as illustrated in Figure 11-4. Suppose we obtain two varieties of snapdragons, one that produces only white flowers and the other only red. We can show that each of these varieties is **true-breeding** for flower color; that is, all the progeny (offspring) of the red-flowered variety produce red flowers generation after generation, and all the progeny of the white-flowered variety produce white flowers.

The simplest model we can make of flower color in snapdragons is that color depends on a single gene that can exist in two variant versions, or **alleles.** One allele (which we call R) specifies red flowers, and another allele (which we call R') specifies white flowers. One might imagine, for example, that the R gene allows the plant to make a red pigment molecule, while the R' gene allows the making of a white pigment.

From our previous discussions, we know that genes are on chromosomes and that each diploid cell has two copies of each chromosome. So each diploid cell in a snapdragon plant would have two copies of the color gene, one copy on each member of a pair of chromosomes. True-breeding varieties contain two copies of just one of these alleles. An organism is said to be **homozygous** for an allele when it has two copies of the same allele. An organism is **heterozygous** when it contains two different alleles for a single gene. We would say, then, that the genotype of a homozygous red snapdragon is RR and the genotype of the homozygous white snapdragon is $R'R'$. The genotype of the heterozygous pink snapdragon is RR'.

Ordinarily, snapdragons self-fertilize. Each flower produces both pollen (sperm) and egg cells, and the struc-

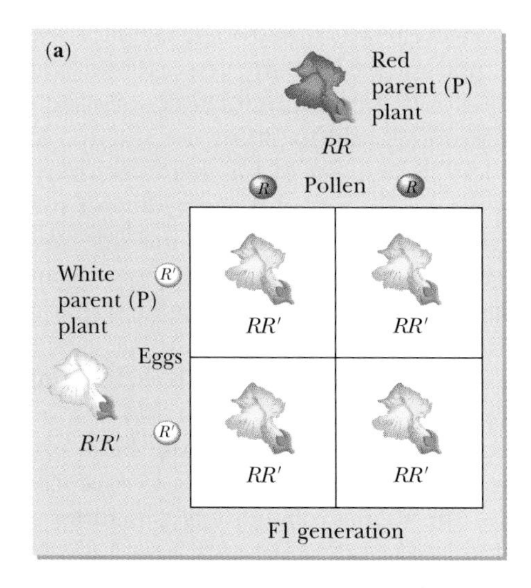

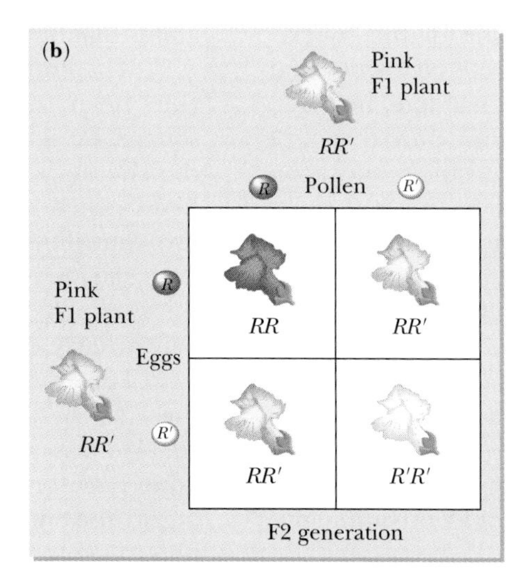

Figure 11-4 Inheritance of flower color in snapdragons. RR plants produce red flowers, $R'R'$ plants produce white flowers, and RR' heterozygotes produce pink flowers. This method of representation is called a Punnett square.

ture of the flower permits the pollen to fertilize eggs within the same flower. A breeder can, however, prevent self-fertilization by removing the pollen-carrying structures and artificially fertilizing the flower with pollen from another variety. This procedure of breeding two genetically distinct organisms is called **cross-breeding** (or **crossing**), and the progeny of such a cross are called **hybrids.** In snapdragons, the results of crossing a true-breeding (homozygous) red snapdragon with a true-breeding (homozygous) white snapdragon are heterozygous hybrids with pink flowers.

The original parents in a cross are called the **parental,** or **P** generation, and the progeny are the **first filial** [Latin = son] or **F1** generation. The **F2 generation** are the progeny of the F1 generation. In the snapdragon cross, then, the parents are *RR* and *R′R′* homozygotes, and the F1 generation are heterozygous hybrids. The genotypes of the hybrids are all *RR′*, and their phenotypes are pink flowers.

What Are the Genotypes and Phenotypes of the F2 Generation?

A typical question that a geneticist would ask is, "What color flowers will a snapdragon grower obtain by crossing these hybrids with one another?" We know that meiosis produces gametes that have just one chromosome from each pair of homologous chromosomes. So each snapdragon gamete receives just one version (allele) of the color gene, either *R* or *R′*. The genotype of the F1 hybrids is *RR′*, so half of all the gametes contain the *R* allele and half the *R′* allele. We can now make some guesses about the distribution of genotypes in the F2 generation.

A convenient way to make these calculations relies on the checkerboard representation, shown in Figure 11-4, invented by an early 20th-century British geneticist named Reginald Punnett. In a Punnett square we represent each kind of gamete produced by one parent along the top of the square and each kind of gamete produced by the other parent along the left side of the square. Within the small squares we can write the genotypes that would be produced by each combination. Note, however, that a particular genotype can result from more than one combination of gametes. For example, in the snapdragon cross, the *RR′* genotype is equivalent to the *R′R* genotype—that is, the flowers are pink no matter which allele comes from which parent.

The Punnett square (Figure 11-4b) shows that the three possible genotypes of the F2 generation of our snapdragon experiment should be in the ratio 1:2:1. Because the pink phenotype of the heterozygote is distinct from the phenotypes of either homozygote (red or white), we can directly count the ratios of the genotypes in the F2 generation. As predicted, such counts of F2 offspring give a ratio close to 1:2:1.

The Rules of Probability Predict the Ratios of Genotypes and Phenotypes

Probability is the number of times an event actually occurs divided by the number of opportunities it could have occurred. In the most familiar example, the probability of a tossed coin turning up "heads" is 1 (the number of sides that are heads) divided by 2 (the total number of sides). Similarly, the probability of an *RR′* heterozygote producing an *R*-bearing gamete is 1/2.

Numerical predictions of the ratios of geneotypes in genetic crosses depend on two simple rules: the **product rule,** which states that the chance of two independent events taking place together is the product of their probabilities, and the **sum rule,** which states that the probability that either of two separate events will occur is the sum of their individual probabilities.

The product rule allows us to calculate the probability of two independent coin tosses both coming up heads: That probability is the product of the probabilities of each coin coming up heads, that is, $1/2 \times 1/2 = 1/4$. We can also calculate the probability of two six-sided dice both coming up "1": $1/6 \times 1/6 = 1/36$. Similarly, we can calculate the probability of two *RR′* heterozygotes in the F1 generation producing an *RR* offspring in the F2 generation. For this to happen, both gametes must carry the *R* allele, so the probability is $1/2 \times 1/2 = 1/4$. The probability of an *R′R′* offspring is also 1/4.

The sum rule (together with the product rule) allows us to calculate the probability that a pair of dice will come up to a total of, say, 5. There are four ways to obtain a total of five: Die A could be "1" and B could be "4"; A could be "2" and B "3"; A could be "3" and B "2"; or A could be "4" and B "1". By the product rule, the probability of each combination is 1/36; by the sum rule the probability of getting a "5" is $1/36 + 1/36 + 1/36 + 1/36 = 4/36 = 1/9$. Similarly, the chance of F1 heterozygotes (with an *RR′* genotype) producing F2 offspring with *RR′* genotype is 1/2—the sum of the probability that the pollen will carry *R* and the egg cell *R′* (1/4) and the probability that the pollen will carry *R′* and the egg cell *R* (1/4).

The Punnett square represents the product and sum rules in an easily visualized manner. In forming a Punnett square, we use the product rule to fill in the squares of the checkerboard: The probability of the genotype specified in each box is the product of the probability of the two contributing gametes, given on the top and side. We then calculate the total probability of each genotype from the sum rule, just as we calculated the probability that a pair of dice would total "5."

For example, we can see in Figure 11-4b that 1/4 of the F2 generation progeny will be *R′R′* and 1/4 will be *RR*. We also see the two ways of obtaining the heterozygote. Each of these has a probability of 1/4, so (according to the sum rule) the total probability of obtaining a heterozygote is $1/4 + 1/4 = 1/2$.

GREGOR MENDEL ESTABLISHED THE PRINCIPLES OF GENETICS BEFORE BIOLOGISTS HAD DISCOVERED THE CHROMOSOME MOVEMENTS OF MEIOSIS

Today, the inheritance of flower color in snapdragons makes sense because of our knowledge of the relationship of chromosomes and genes. But would the explanation have seemed as simple if we did not already have a model in mind?

In fact, the principles of genetics were discovered in the mid-19th century by Gregor Mendel, an Austrian monk who studied the inheritance of variations in pea plants, well before a model was available. But biologists ignored Mendel's conclusions until after they began to appreciate the importance of chromosome movements, at the very beginning of this century.

Mendel's Insights Came from Careful Counting

Gregor Mendel was the first person to realize the significance of the relative numbers of progeny with differing phenotypes after a genetic cross. The organism that he studied most successfully was the common garden pea, which he cultivated in his monastery's garden in Brünn (Brno), which is now in the Czech Republic.

Mendel studied seven different traits, illustrated in Figure 11-5: round versus wrinkled seeds, yellow versus green seeds, purple versus white flowers, axial versus terminal flowers, tall versus dwarf stems, inflated versus constricted pods, and green versus yellow pods. Each of these traits showed the same pattern of inheritance. Mendel chose traits in which the different variants were clearly visible and distinct.

In each case, Mendel first showed that the parental stocks were true-breeding. For example, self-fertilization of plants with round seeds always produced progeny with round seeds. Mendel then prevented self-fertilization and performed cross-fertilization, as illustrated in Figure 11-6. His crosses involved two stocks that differed in a single phenotypic characteristic, such as purple versus white flowers.

The Characteristics That Mendel Studied in Peas Did Not Blend in the F1 Generation

The F1 hybrids that Mendel studied in peas did not have an intermediate phenotype, as did the hybrid (pink) snapdragons. Instead, his heterozygous F1 hybrid had a phenotype that was indistinguishable from one of the ho-

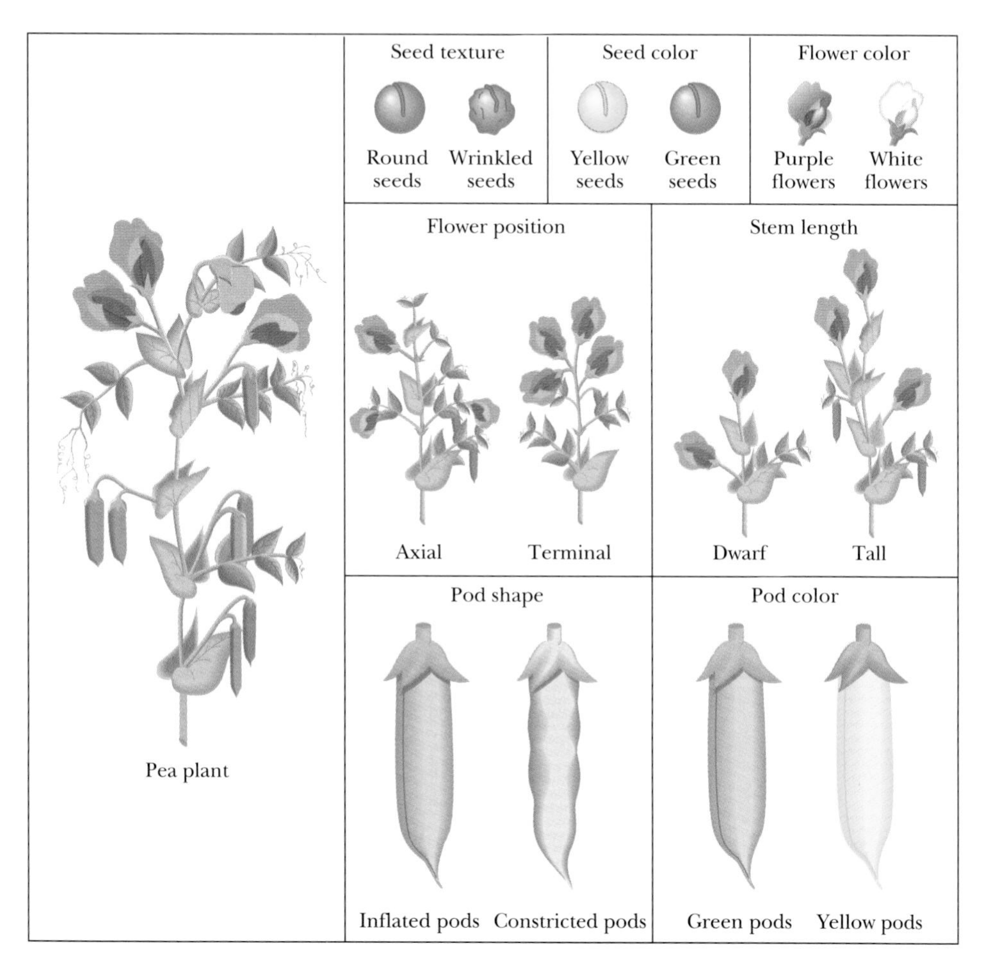

Figure 11-5 Gregor Mendel studied these seven traits in pea plants.

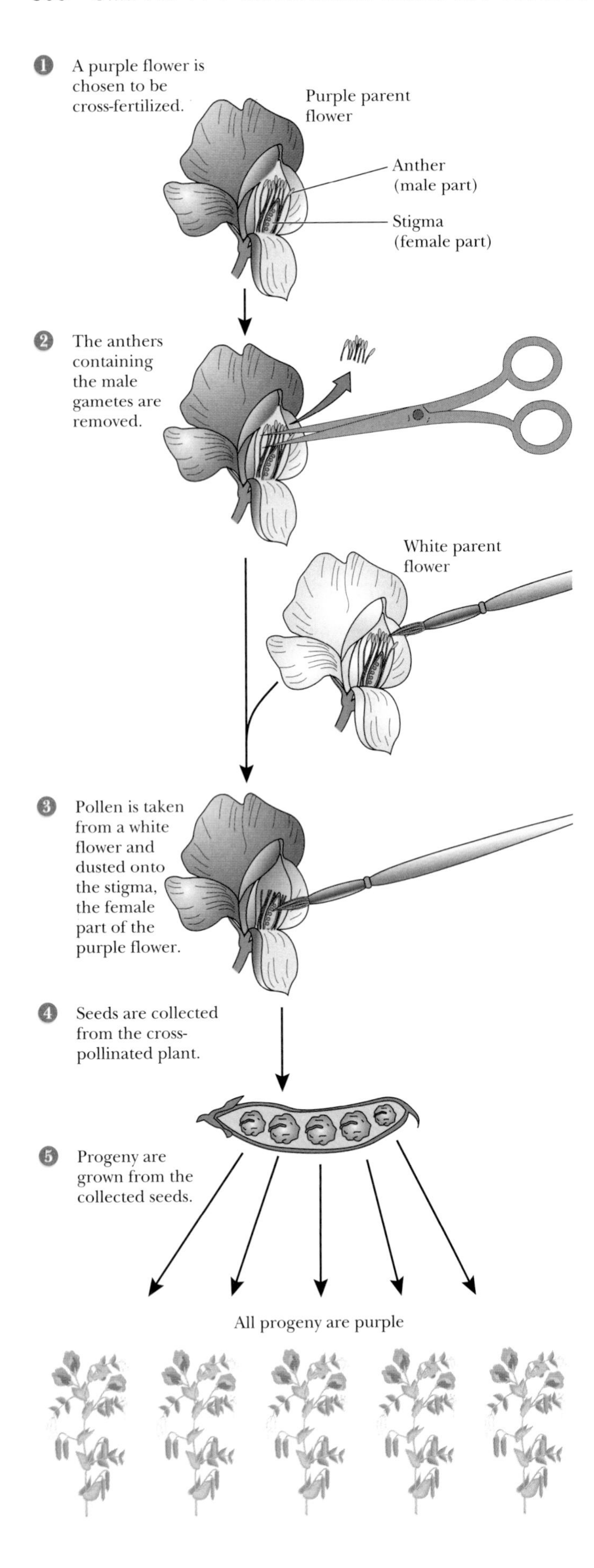

Figure 11-6 Mendel's study of inheritance in peas required that he prevent self-fertilization in order to perform only the desired crosses by cross pollination.

mozygous parents. For example, when Mendel crossed true-breeding peas with purple flowers with true-breeding white-flowered peas, all the hybrids had purple flowers (as illustrated in the Punnett square in Figure 11-7a). We say that an allele is **dominant** when it alone determines the phenotype of a heterozygote, as in the case of the purple allele in peas. An allele is **recessive** when it contributes nothing to the phenotype of a heterozygote, as in the case of the white allele of peas.

When two alleles each contribute to the phenotype, as in the case of the R and R' alleles of snapdragons (which together gave a color different from either homozygote), the alleles are said to be **codominant.** The occurrence of codominance originally provided evidence for blending inheritance, but (as we said previously) blending inheritance cannot explain the reappearance of red and white snapdragons after crossing the hybrids. By convention, we begin the names of dominant and codominant alleles with a capital letter, and those of recessive alleles with a small letter.

The F1 hybrids all had the same phenotype in all seven of Mendel's crosses: Plants that produced round peas crossed with wrinkled pea stocks gave F1 hybrids with round peas; tall pea stocks crossed with short pea stocks give tall F1 hybrids; and so on. The purple flower allele is therefore dominant to the white flower allele, the round allele is dominant to the wrinkled allele, and the tall allele dominant to the short allele.

The F2 Generation Had Characteristic Ratios of Phenotypes

Mendel allowed the F1 plants to self-fertilize, and he then *counted* the F2 progeny. In every case the phenotype specified by the dominant allele accounted for 3/4 of the progeny.

We can understand the 3:1 ratio in terms of the behavior of chromosomes in meiosis and fertilization. Just as in the case of the snapdragon cross described earlier, one quarter of the progeny were homozygotes for each allele, and half were heterozygotes, as illustrated in the Punnett square in Figure 11-7b.

In each of Mendel's crosses, one allele was dominant and one recessive, so the heterozygotes of the F2 generation (like the F1 hybrids) had the same phenotype as plants homozygous for the dominant allele. (That is, they had the same phenotype despite their different genotypes.) So, in the case of the purple versus white cross, 1/4 were homozygous for the purple allele, 1/4 were homozygous for the white allele, and 1/2 were heterozygous. Because

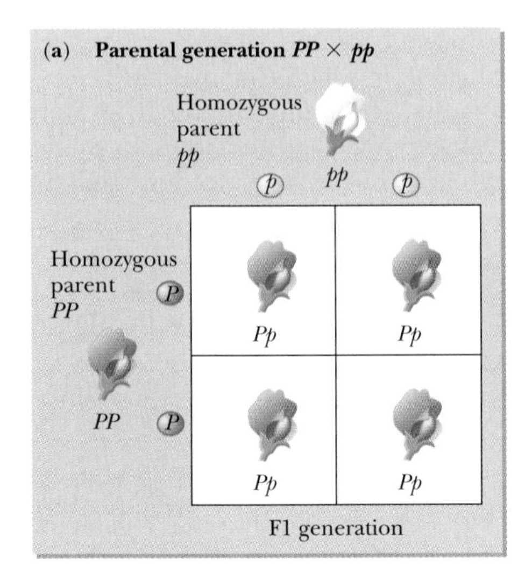

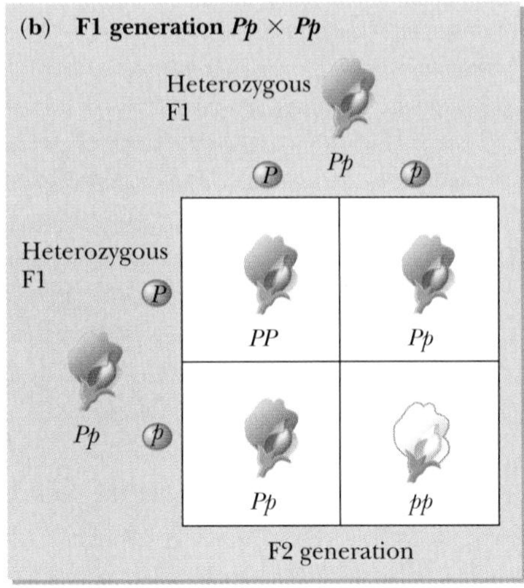

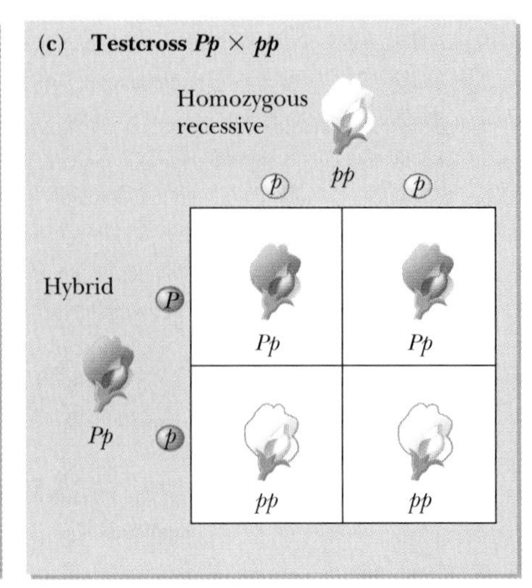

Figure 11-7 Punnett squares showing three crosses of purple- and white-flowered peas. **(a)** homozygous white (*pp*) versus homozygous purple (*PP*); **(b)** heterozygous purple (*Pp*) versus heterozygous purple (*Pp*); **(c)** a testcross, homozygous white (*pp*) versus heterozygous purple (*Pp*).

ESSAY

CHARLES DARWIN ALSO TRIED TO UNDERSTAND THE PRINCIPLES OF INHERITANCE, BUT FAILED

Charles Darwin, probably the greatest biologist of the 19th century, also studied the inheritance of variations. But Darwin failed where Mendel succeeded. Darwin's defeat nonetheless illustrates an important genetic concept—the often unequal contributions of alleles to phenotype.

Darwin crossed true-breeding snapdragons with flowers of two distinct shapes—normal, irregular flowers and peloric [Greek, *peloros* = monstrous], symmetrical flowers. All the F1 progeny of this cross had normal flowers; not one flower was peloric.

Darwin correctly inferred that the genetic plans for normal flowers somehow suppressed the plans for peloric flowers. Today we would say that the allele that produces normal flowers is dominant and the allele for the peloric flowers is recessive.

Darwin then allowed the F1 snapdragons (with normal flowers) to self-fertilize. Of the 125 F2 plants, 88 had normal flowers, while 37 had peloric flowers. Darwin did not know how to interpret these results. He concluded that the "latent character" of the peloric flowers had "gained strength by the intermission of a generation." Today, such an explanation seems vague, especially from the founder of modern biology.

Why did Darwin go astray? First, he apparently did not try to formulate a model that he could test for other genes and other aspects of phenotype. Second, he evidently did not think enough about the possible meaning of the numbers.

Based on our discussion of the genetics of snapdragon color, we can immediately understand the problem of flower shape. In the case of the color, the red and white alleles were codominant. In the case of flower shape, if one allele is dominant, then we can calculate the fraction of F2 progeny that would have each phenotype. The 1:2:1 ratio of genotypes in the F2 generation should give a 3:1 ratio of phenotypes, just about what Darwin observed!

the purple allele is dominant, the heterozygotes had the same phenotype as the purple homozygote. So 3/4 (1/4 + 1/2) were purple, and 1/4 were white.

Mendel had to make sense of the data without the knowledge that we now possess. He did this just by considering the possible meanings of the ratios of phenotypes in his breeding experiments.

Mendel's first conclusion is called the **Principle of Segregation,** or **Mendel's First Law:** *A sexually reproducing organism has two "determinants" (or genes, in modern terms) for each characteristic; these two copies* ***segregate*** *(or separate) during the production of gametes.* Mendel's analysis showed that recessive traits did not disappear, as the blending model suggested. Furthermore, Mendel could use the rules of probability to predict the ratios in the F2 generation of progeny with dominant and recessive characteristics.

Mendel Established the Principle of Segregation by Showing That Hybrids Produce Gametes That Carry Recessive Alleles

To test the Principle of Segregation, Mendel performed a new kind of cross, called a **testcross.** Instead of allowing the F1 heterozygotes to self-fertilize, he crossed them with the parental stock that contained only the recessive allele (for example, the true-breeding plants that produce white flowers, as shown in Figure 11-7c). In terms of phenotypes alone, this cross would look just like the original cross—a purple-flowered plant with a white-flowered plant. But Mendel knew that the purple-flowered F1 hybrids included two different "determinants" (genotypes) and should therefore produce two kinds of gametes, one with the purple allele, and the other with the white allele. The plants with white flowers, on the other hand, would produce only gametes with the white allele. He correctly predicted that half the progeny should have white flowers and half purple flowers.

Mendel's testcross showed that breeding does not "strengthen" or in any way change genes. Instead it supported Mendel's idea that genes are unchanging and indivisible, the genetic equivalent of an atom. We now know this view is not quite correct because genes do change, but they don't change in each generation, as Mendel's contemporaries (including Charles Darwin) thought. The reason that F1 hybrids often resemble just one of their parents is not that breeding changes the genes, but that one of the alleles is dominant.

Mendel Established the Rules That Govern the Inheritance of Two Variant Characteristics

In one of his next experiments, Mendel took two true-breeding stocks that differed in both the color and the form of the peas they produced. One parent produced yellow, round peas; the other green and wrinkled peas. Mendel already knew that yellow and round alleles were dominant to the green and wrinkled, and, as expected, he found that the F1 progeny of the cross produced all yellow and round peas. He then determined the inheritance of the characteristics in the F2 generation that resulted from crossing the yellow and round F1 hybrids.

Again, we can diagram this experiment by using a Punnett square. As diagrammed in Figure 11-8, the allele specifying round peas is *R*, that for wrinkled peas *r;* the allele specifying yellow peas is *Y*, that for green peas *y*. The genotypes of the parents are therefore *RRYY* and *rryy*.

In this cross, each parent can produce only a single kind of gamete: *RRYY* plants produce only *RY* gametes, and *rryy* plants produce only *ry* gametes. The genotype of all the F1 plants must therefore be *RrYy*, producing a round yellow phenotype.

What happens in the F2 generation? Mendel allowed the F1 to self-pollinate and then counted 556 pea plants in the F2: Of these 315 produced yellow and round peas (representing the action of the dominant alleles, *R* and *Y*), and only 32 were green and wrinkled (and must therefore be homozygous for both *r* and *y*). The rest represented new phenotypes, unlike either of the parental stocks: 101 were yellow and wrinkled, and 108 were green and round. The two crosses had produced new combinations of genes, illustrating again how sexual reproduction provides diversity in a population. But how did Mendel make sense of the numbers?

First, he noted that the ratios of the characters associated with each gene were in a ratio of about 3:1, just as in his earlier experiments with only a single variant trait. In the F2 generation there were three times as many yellow as green peas (416:140) and three times as many round as wrinkled (423:133). Mendel realized that the two traits were behaving independently: Yellowness or greenness did not affect roundness or wrinkledness (and vice versa).

Mendel could explain the data if he assumed that each F1 plant produces equal numbers of four kinds of gametes, representing all the possible combinations of two alleles of the two genes: *ry*, *Ry*, *rY*, and *RY*. The Punnett square of Figure 11-8b correctly predicts the distribution of genotypes and phenotypes of the F2 generation of this cross. It shows that nine of the sixteen combinations of gametes produce plants with yellow round peas, but only one combination produces plants with green wrinkled peas. The overall ratio of phenotypes predicted by this analysis is 9:3:3:1.

The correspondence of Mendel's experiments to this prediction established the **Principle of Independent Assortment,** sometimes called **Mendel's Second Law.** This principle (in modern language) states that *the alleles for one gene segregate independently of the alleles of another gene.*

Notice that independent assortment of genes corresponds exactly to the independent segregation of chro-

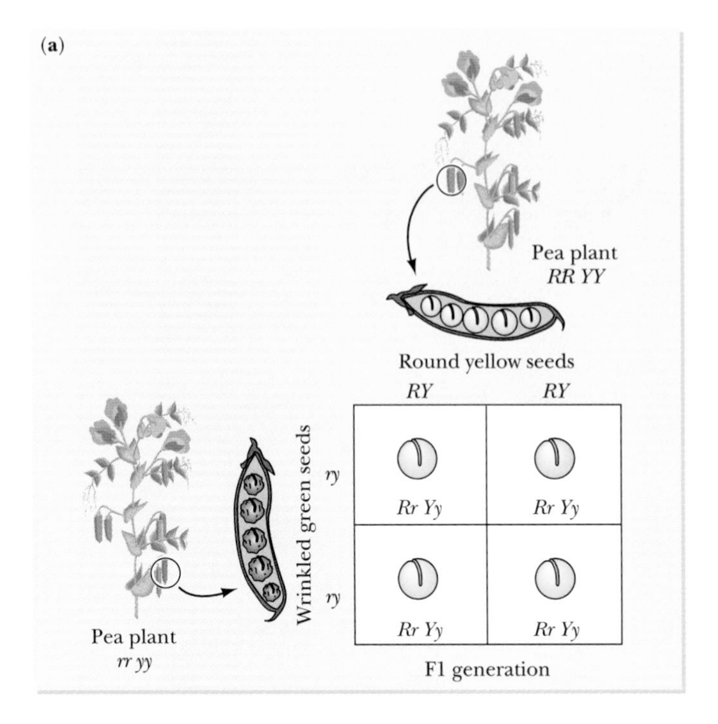

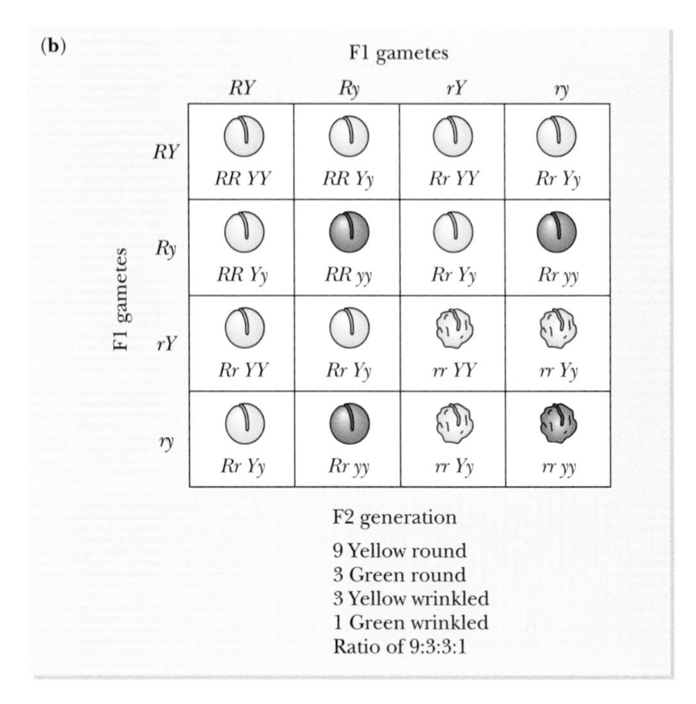

Figure 11-8 A "two-factor" cross. **(a)** Mendel crossed plants that produced round yellow peas with plants that produced wrinkled green peas. **(b)** He then crossed the resulting F1 hybrids to produce the F2 generation.

mosomes during meiosis, as illustrated in Figure 11-9. (See Chapter 10.) Two paternal (or two maternal) chromosomes have only a 50:50 chance of staying together during gamete formation. For example, you have two versions of chromosome 1, one from your mother, which we call 1M, and one from your father, which we call 1P. Similarly, you have two versions of chromosome 2, 2M and 2P. The probability that 1M will stay with 2M during meiosis as you produce gametes is 1/2; the probability that 1M will move with 2P during meiosis is also 1/2. The independent assortment of homologous chromosomes during meiosis underlies the independent assortment of genes and thereby results in genetic recombination (the association of alleles that did not exist in the parents).

Mendel verified his understanding of the genotypes of the gametes produced by the F1 hybrids by doing another testcross, shown in Figure 11-10 (p. 306), with plants that produce green wrinkled peas. Because both *r* (for wrinkled peas) and *y* (for green peas) are recessive, the genotype of the plants used in the testcross must be *rryy*. The only gametes produced by these plants must be *ry*.

If the genotype of the F1 plants is really *RrYy* and the two genes assort independently, then the four kinds of gametes (*RY*, *rY*, *Ry*, and *ry*) should be made in equal numbers, as shown in Figure 11-10. When these gametes form zygotes with *ry* gametes, we expect equal numbers of the four phenotypes, in contrast to the 9:3:3:1 ratio expected in the case of self-fertilization of the F1 plants. Again, Mendel obtained the expected ratios, confirming his understanding of the process.

EXPERIMENTS WITH FRUIT FLIES ESTABLISHED THAT GENES LIE ON CHROMOSOMES

The fruit fly *Drosophila melanogaster* has been the favorite organism for genetic research since 1909, when it was introduced by embryologist and geneticist Thomas Hunt Morgan. Morgan's laboratory at Columbia University in New York was the center of genetic research for more than 25 years.

Morgan's fruit flies have many advantages for genetic research: (1) They breed rapidly, growing from eggs to sexually mature adults in less than 2 weeks; (2) they are prolific, with each female laying hundreds of eggs a day; (3) the adults are small (about 3 mm long) and easy to maintain; (4) they are intricately and precisely built organisms, with dozens of identifiable traits under genetic control. An important part of Morgan's motivation for choosing *Drosophila* was that it was readily available and inexpensive to maintain: Morgan reportedly collected flies on a ripe

Figure 11-9 The behavior of genes in Mendel's two-factor cross corresponds to that of chromosomes in meiosis.

A dihybrid cross

Diploid parental genotypes

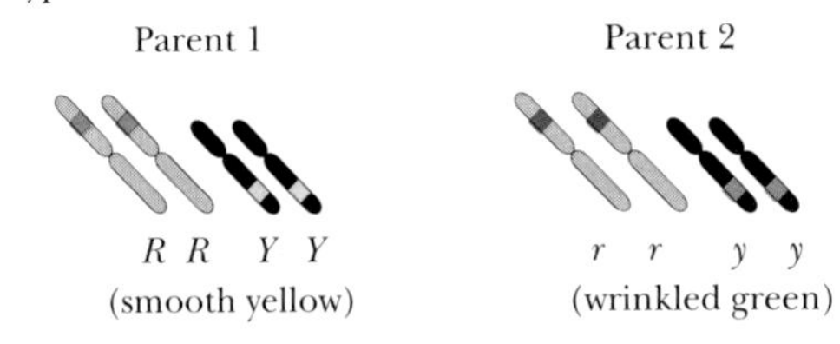

F1 generation

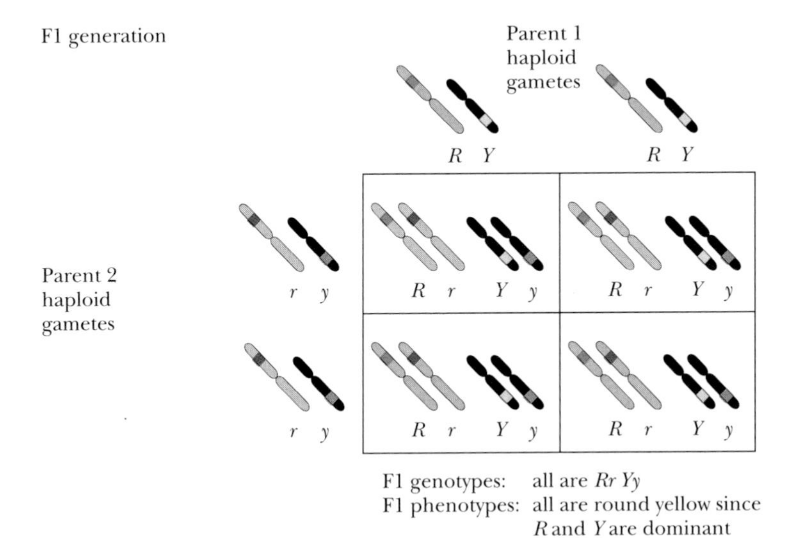

F1 genotypes: all are *Rr Yy*
F1 phenotypes: all are round yellow since *R* and *Y* are dominant

F2 generation

F1 parent 1 haploid gametes

F1 parent 2 haploid gametes

	R Y	*R y*	*r Y*	*r y*
R Y	*R R Y Y*	*R R Y y*	*R r Y Y*	*R r Y y*
R y	*R R Y y*	*R R y y*	*R r Y y*	*R r y y*
r Y	*R r Y Y*	R r Y y	r r Y Y	*r r Y y*
r y	*R r Y y*	*R r y y*	*r r Y y*	*r r y y*

F2 genotypes:

F2 phenotypes:

$\frac{1}{16}$ (*RR YY*) + $\frac{2}{16}$ (*Rr YY*) + $\frac{2}{16}$ (*RR Yy*) + $\frac{4}{16}$ (*Rr Yy*) = $\frac{9}{16}$ smooth yellow seeds

$\frac{1}{16}$ (*RR yy*) + $\frac{2}{16}$ (*Rr yy*) = $\frac{3}{16}$ smooth green seeds

$\frac{1}{16}$ (*rr YY*) + $\frac{2}{16}$ (*rr Yy*) = $\frac{3}{16}$ wrinkled yellow seeds

$\frac{1}{16}$ (*rr yy*) = $\frac{1}{16}$ wrinkled green seeds

ESSAY

WHY WAS MENDEL IGNORED?

Mendel presented his work to the Brünn Society for the Study of Natural Sciences in 1865 and published a formal report in the *Transactions* of the Society the next year. His experiments and analysis are among the most elegant and important in the whole history of science, but his work was completely ignored until 1900, 16 years after his death. Then it was independently rediscovered by three botanists, Hugo DeVries in the Netherlands, Karl Correns in Germany, and Erich Tschermak in Austria. Each of these researchers was independently studying the inheritance of variations in plants, DeVries in the American evening primrose and Correns and Tschermak in garden peas. As each began to prepare his results for publication, he independently discovered Mendel's neglected paper.

Why had so many scientists missed the point? Why didn't Mendel get the attention he deserved? Part of the answers to these questions must have to do with Mendel's credibility as a scientist. He was a monk in an obscure monastery, the son of a peasant. He had studied 2 years at the University of Vienna, and he had been a substitute teacher of Greek and mathematics in a high school near the monastery. He never was able, however, to pass the licensing examination to become a teacher, and he received his lowest marks in geology and biology!

Some of the condescending attitudes that Mendel encountered are revealed in his correspondence with the distinguished botanist, Carl von Nägeli, Professor at the University of Munich. Mendel had sent Nägeli a copy of his paper and asked Nägeli's advice on further work. Nägeli replied, "It seems to me that your experiments . . . are far from being completed and are indeed only just beginning."

At Nägeli's request, Mendel sent him 140 carefully labeled packages of peas, but Nägeli evidently never used them. Nägeli also suggested that Mendel undertake to repeat his experiments with hawkweed, one of Nägeli's own favorite plants for breeding experiments. Mendel, however, could not obtain the same results with hawkweed (because it reproduces most of the time without fertilization). For Mendel the encounter with the distinguished professor was disastrous, leading him to doubt the validity of his own revolutionary insights and experiments.

Was the rejection of Mendel's work by Nägeli and others mere snobbery? Probably not. It seems more likely that Nägeli and other biologists were unable to understand Mendel's calculations and abstractions. Biologists were not used to counting, measuring, and thinking mathematically. In addition, mid-19th century biologists knew nothing of chromosomes and their movements. They were, it seems, simply not prepared to follow Mendel's reasoning. To 20th-century biologists, now more mathematically inclined, Mendel's 1865 paper stands out as one of the greatest achievements in the entire history of biology.

One lesson here, even for students entering the 21st century, is that biologists must not be afraid of numbers. Understanding genetics requires mathematics, although only a little arithmetic and two rules of probability.

The other lesson to learn from Mendel's experience is that a young person's new ideas often face resistance from established experts, even when the new ideas are correct. Acceptance of new ideas often requires the passage of time. Max Planck, one of the founders of quantum physics, once quipped, "The only way for new ideas to gain acceptance is for adherents of the old ideas to die." Planck's view is harsh, but it points out the need both for open minds on the part of critics and for convincing arguments on the part of proponents.

pineapple outside his window and kept them in small milk bottles "borrowed" from his home milk delivery man.

Thousands of geneticists and perhaps hundreds of thousands of undergraduate biology students have studied inheritance in *Drosophila,* showing again and again that traits such as eye color, body color, wing shape, and bristle length all obey Mendel's laws. Many strange phenotypes, such as white eyes, or a leg replacing an antenna, as shown in Figure 11-11, can be shown to behave in "mendelian" fashion.

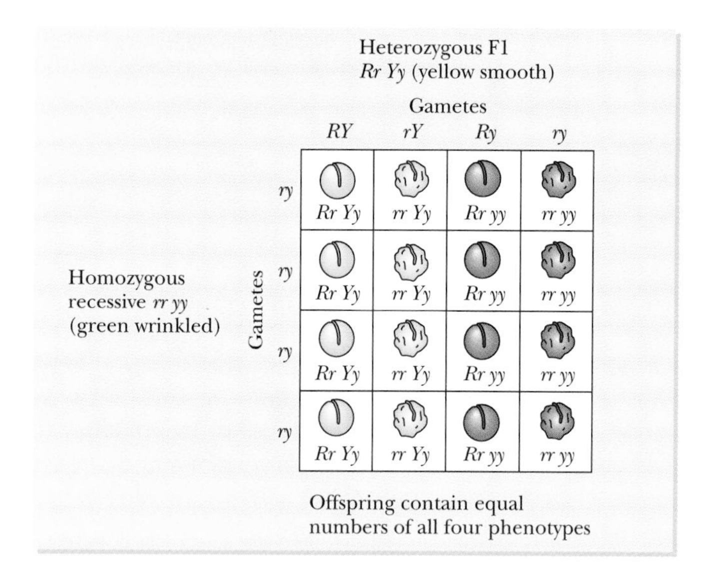

Figure 11-10 A testcross allowed Mendel to distinguish genotypes of the products of his two-factor cross.

Genes on the *Drosophila* X Chromosome Show Different Patterns of Inheritance in Males and Females

The detailed genetic analysis of *Drosophila* has allowed geneticists to understand much about the relationships between genes and chromosomes. *Drosophila* males have three pairs of homologous chromosomes plus two unpaired sex chromosomes, called (as in mammals) X and Y, while *Drosophila* females have two X chromosomes. Any gene that lies on the X chromosome, then, is present in two copies in females but in only one copy in males. Males are therefore said to be **hemizygous** for genes on the X chromosome; that is, they have only one (instead of two) copies of genes that lie on the X chromosome. As an example, one of Morgan's early discoveries concerned a gene for eye color, which, he showed, lies on the X chromosome.

Most fruit flies have red eyes. This phenotype is called "wild type." One day, however, Calvin Bridges, a Columbia undergraduate hired to wash the dirty milk bottles in Morgan's laboratory, noticed a male fly with white eyes. Morgan then crossed the white-eyed male with red-eyed females, in order to show that the allele for white eyes behaved like other genes, as shown in Figure 11-12a. As expected, he found that all the F1 progeny had red eyes, so the *white-eyed* allele (w) must be recessive and the corresponding red allele (W) dominant.

Morgan then crossed the F1 males and females and obtained an overall ratio of 3:1 of red-eyed flies to white-eyed flies, but all of the white-eyed flies were male, as shown in Figure 11-12b! Half of the F2 males had white eyes, but all of the females had red eyes. What was going on? Morgan hypothesized that the gene for white versus red eyes was on the X chromosome.

The diagram of Morgan's experiment with Punnett squares in Figure 11-12 designates the dominant red allele as X^W and the recessive white allele as X^w. We see that the F1 males and females have different genotypes. The males have a Y chromosome and the red eye allele on their single X chromosome. The females have one X chromosome with the white allele and one with the red allele. Males are hemizygous for the gene on the X chromosome.

The F1 males are hemizygous for the red allele, while the F1 females are heterozygous—having one copy of each allele. In the F2 generation, the males must receive their Y chromosomes from their fathers and their X chromosomes from their mothers. Therefore, half have X^W and have red eyes, and half have X^w and have white eyes. Because the females receive one of their X chromosomes

(a)

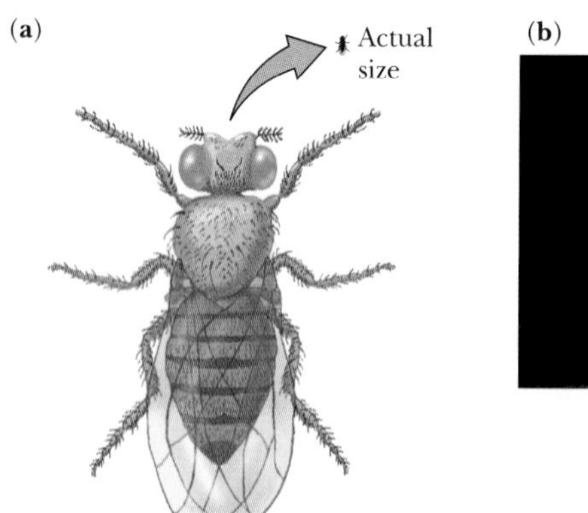

(b)

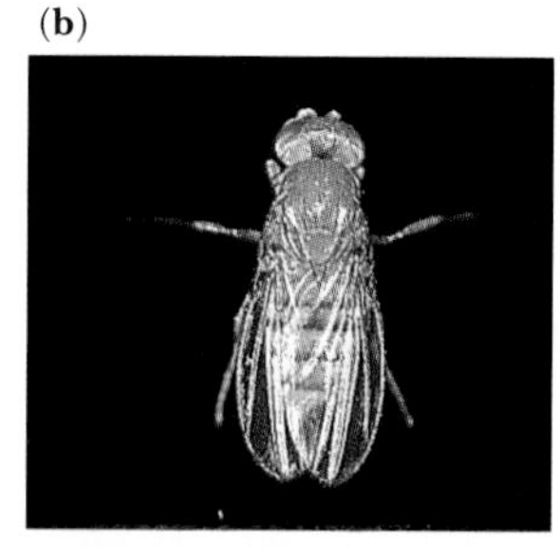

(c)

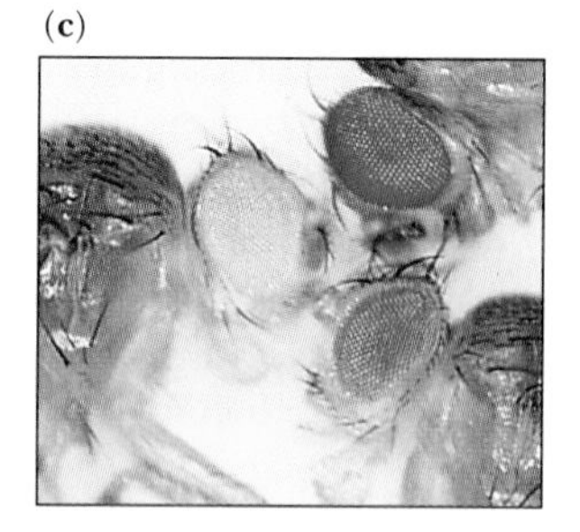

(d)

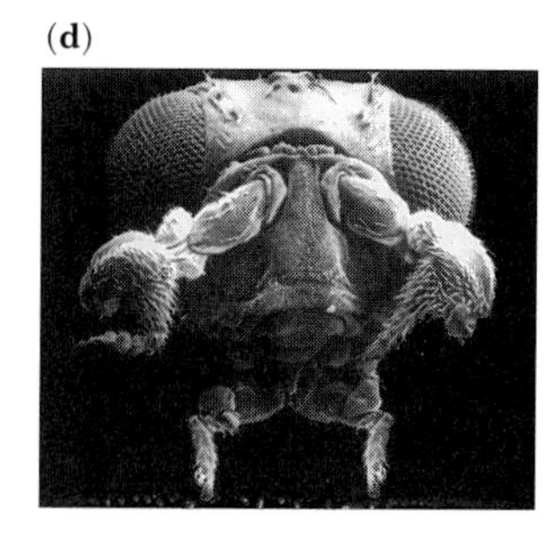

Figure 11-11 **(a, b)** *Drosophila melanogaster,* a favorite organism for genetic analysis; **(c)** an eye-color mutant, called white, first discovered by Calvin Bridges, an undergraduate who washed glassware in Morgan's laboratory; **(d)** a mutant called *antennapedia,* which has legs growing where its antennae should be. *(b, Biology Media/Photo Researchers; c, Robert W. Levis, Fred Hutchinson Cancer Research Center; d, Thomas Kaufman, Howard Hughes Medical Center, Bloomington, IN)*

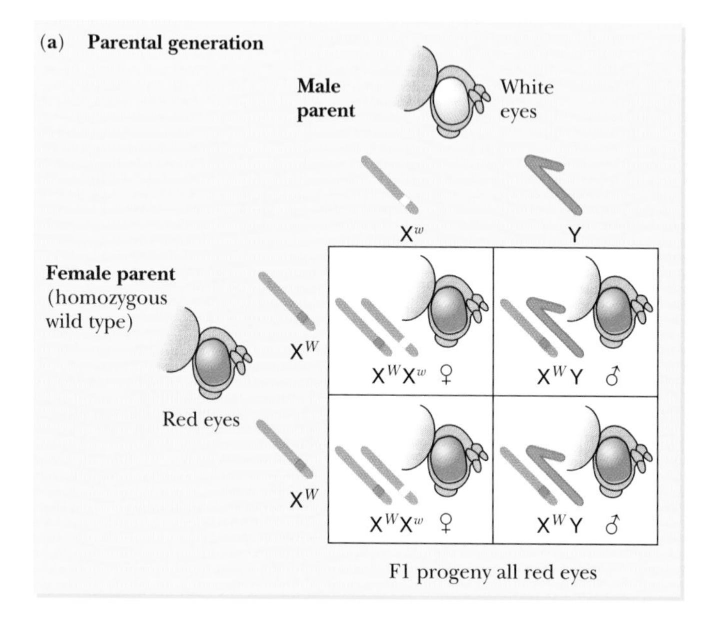

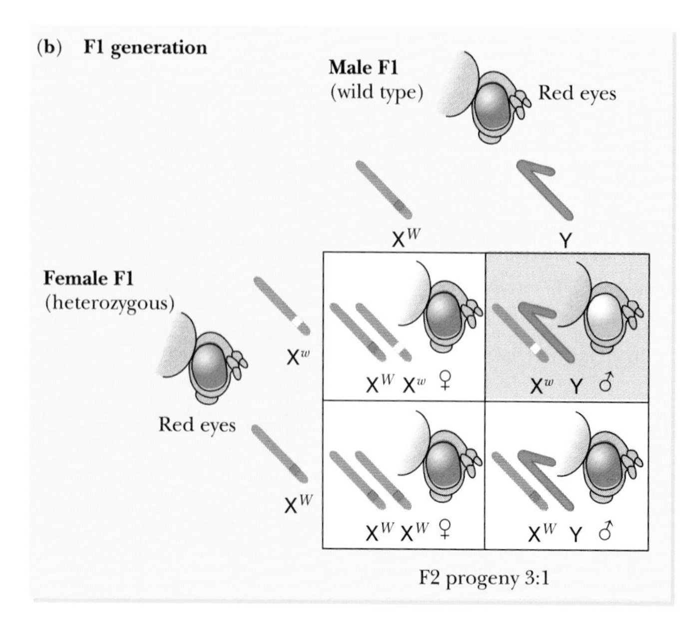

Figure 11-12 Sex linkage: The *white* gene is on the X chromosome. **(a)** Crossing white-eyed males with wild-type (red-eyed) females produces F1 progeny with red-eyed males and red-eyed females, since the red allele *(W)* is dominant to the white allele *(w)*. **(b)** An F1 cross; crossing a female heterozygous for *white* to a wild-type male produces females with red eyes, but half the males have white eyes.

from each parent, all the females receive the dominant X^W from their fathers and have red eyes. Of these, however, half should be heterozygous and half homozygous for X^W (Figure 11-12b).

Morgan checked this analysis by performing a testcross, mating the F1 females (which should have been heterozygotes) with the original white-eyed male. His results were exactly as he predicted (as illustrated in the Punnett square of Figure 11-13): Half of the males and half of the females had white eyes. This result established that the white gene lay on the X chromosome.

Still more convincing proof came in 1916, when Calvin Bridges—now Morgan's graduate student—analyzed the results of nondisjunction, abnormal meioses that resulted in the production of gametes with either two X chromosomes or with no X chromosomes at all. Bridges showed that he could interpret his data according to a model in which the white mutation was a recessive allele that produced a white-eyed phenotype only when no wild-type allele was present on another X chromosome in the same individual. Bridges' results, which appeared as the very first paper in the now prestigious journal *Genetics,* was entitled, "Nondisjunction as a Proof of the Chromosomal Theory of Heredity."

The gene studied in this experiment is one of almost 1000 genes in *Drosophila* that are sex-linked. **Sex-linked genes** have different patterns of inheritance in males and females because they are contained on chromosomes that also determine the gender of the offspring. Because the X chromosome contains many more genes than the Y chromosome, sex linkage nearly always refers to genes on the X chromosome. The work of Morgan and Bridges on white-eyed flies, as well as others performed by Morgan and his students, provided the most convincing early evidence that genes are on chromosomes, just as Sutton (another Columbia undergraduate) had proposed a decade earlier.

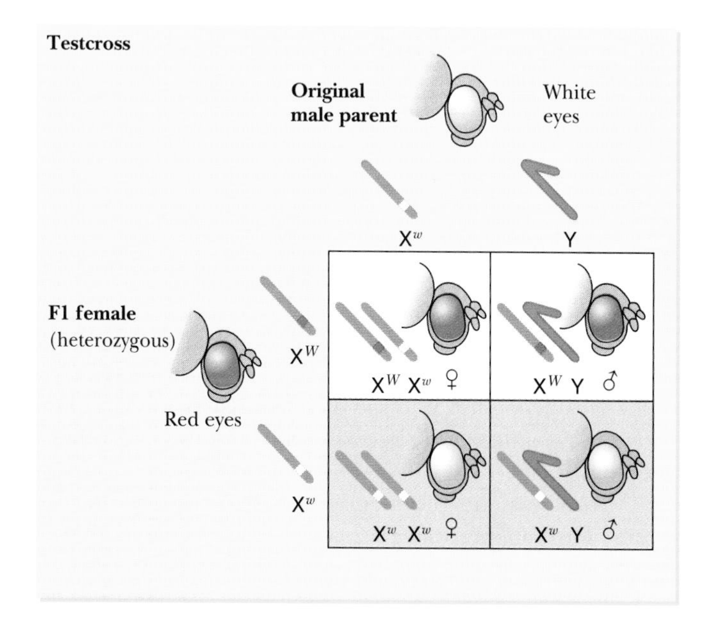

Figure 11-13 A testcross to a white-eyed male demonstrates that the F1 females used in the Figure 11-12 cross are heterozygous for the white gene on the X chromosome. Males are hemizygous.

A Chromosome Has Many Genes

As we have already said, the behavior of genes closely parallels the behavior of chromosomes. Knowing how chromosomes move during meiosis makes Mendel's laws easier to understand: The separation of homologous chromosomes accounts for the Principle of Segregation, and the independent assortment of chromosomes accounts for the Principle of Independent Assortment. Morgan's work opened the door for a new kind of question: Just *where* on the chromosome does a particular gene lie?

There are many more genes than there are chromosomes. At least 1000 genes lie on the *Drosophila* X chromosome, and the human X chromosome may contain 10,000 or more. The Principle of Independent Assortment depends on the independent assortment of chromosomes, but it cannot hold for genes that are physically linked to one another because they reside on the same chromosome. Mendel was therefore extremely fortunate (or consciously selective) in choosing to study seven traits that happened to lie on the seven different chromosomes in pea plants. If Mendel had chosen an eighth trait to study, he may well have found that it did not assort independently. Sutton realized this problem in 1903, but Morgan and his students were the first geneticists to address it specifically. In doing so, they further established the chromosomal basis of inheritance.

To investigate this problem, Morgan crossed two types of flies: **wild-type** individuals, those whose characteristics are most represented in wild populations, with red eyes and wings of a more or less standard length, and another type of fly that had purple eyes, which depends on a different gene from that for white eyes, and short wings, called vestigial wings. Morgan had previously shown that purple-eyed flies with vestigial wings bred true, and he inferred that the purple, vestigial phenotype results from the homozygous condition of two genes. Morgan named the purple allele *pr* and the vestigial allele *vg;* he called the corresponding wild-type alleles pr^+ and vg^+. So the genotype of the purple vestigial flies was *pr pr vg vg.* Each such fly could produce eggs or sperm carrying both the *pr* and the *vg* allele.

When crossed with true-breeding (homozygous) wild-type flies (as shown in Figure 11-14a), the purple vestigial flies produced phenotypically normal F1 flies. This shows that *pr* and *vg* are recessive alleles, and pr^+ and vg^+ are dominant.

To assess the independent assortment of the two genes, Morgan performed a testcross: He mated the F1 flies with purple vestigial flies. If the two genes were to assort independently, this cross would yield equal numbers of all four possible phenotypes, as shown in the Punnett square of Figure 11-14b—purple eyes, vestigial wings; purple eyes, normal wings; red eyes, vestigial wings; and red eyes, normal wings. Instead, Morgan obtained a radically different result: Almost 90% of the flies had the same phenotypes as the two original parents—purple eyes, vestigial wings and red eyes, normal wings. The remaining 10% of the flies were divided into purple eyes, normal wings and red eyes, vestigial wings.

Morgan deduced that the two genes whose alleles produced purple eyes and vestigial wings did not assort independently. Instead, they behaved as if they were on the same chromosome. Because the *vg* and *pr* alleles come from the same parent, they mostly stay together in the gametes. Similarly, the vg^+ and pr^+ alleles stay together 90% of the time. Instead of independent assortment, then, genes on the same chromosome behave almost like a single gene.

Morgan's experiments showed that genes assort independently only when they are on separate chromosomes (or, as discussed later in this chapter, when they are far apart on the same chromosome). Otherwise genes on the same chromosomes tend to travel together during the formation of gametes. The tendency of two or more genes to segregate together is called **genetic linkage.** Morgan and his students found that fly mutants fell into four distinct **linkage groups,** sets of genes that do not assort independently. Instead of independently assorting (as expected from Mendel's Second Law), parental combinations from within each linkage group tend to stay together, as did *vg* and *pr* in the cross just described.

Morgan realized that the four linkage groups corresponded to the four pairs of homologous chromosomes in *Drosophila,* shown in Figure 11-15. The group of sex-linked traits (controlled by genes on the X chromosome), which contains the *white-eyed* gene, also contains, for example, a gene called *bar,* which reduces the normally round eyes to a narrow bar. The *vestigial* and *purple* genes lie in a distinct linkage group (Group II). Another linkage group had many fewer genes on it than the others and corresponded to the tiny chromosome pair seen in the microscope. Just as the behavior of chromosomes is the basis of Mendel's laws, it also explains the existence of linkage groups, which at first seemed to be deviations from Mendel's laws.

Genetic Recombination Can Arise from Independent Assortment or from Crossing Over

The Principle of Independent Assortment applies to genes on different chromosomes. In contrast, genes on the same chromosome (if they are close enough to one another) segregate together during the formation of gametes; that is, they are genetically linked. Such linkage explains why most of the flies in the cross shown in Figure 11-14b resemble one or the other parental phenotype. But why do some flies in the testcross have purple eyes and normal wings? Genetic recombination (as defined in Chapter 10, p. 291) produces new combinations of alleles, combinations not present in the parents. Recombinant gametes

(a) F1 generation

Wild type parent
$pr^+ pr^+ \; vg^+ vg^+$

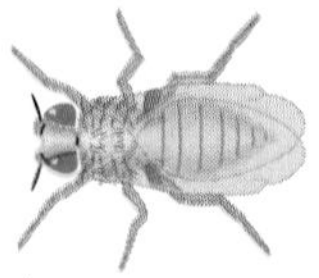

Homozygous purple/vestigial parent
$pr\,pr \; vg\,vg$

	$pr^+ \; vg^+$	$pr^+ \; vg^+$
$pr \; vg$	$pr^+ pr \; vg^+ vg$	$pr^+ pr \; vg^+ vg$
$pr \; vg$	$pr^+ pr \; vg^+ vg$	$pr^+ pr \; vg^+ vg$

All heterozygous genotype
All wild type phenotype

(b) Testcross

F1 heterozygous parent
$pr^+ pr \; vg^+ vg$

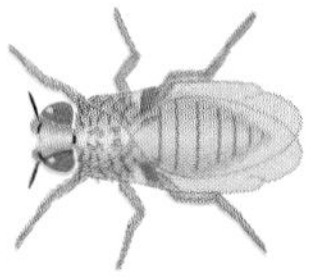

Homozygous purple/vestigial parent
$pr\,pr \; vg\,vg$

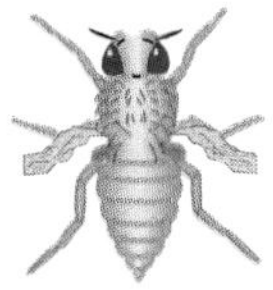

	$pr^+ \; vg^+$	$pr^+ \; vg$	$pr \; vg^+$	$pr \; vg$
$pr \; vg$	$pr^+ pr \; vg^+ vg$	$pr^+ pr \; vg\,vg$	$pr\,pr \; vg^+ vg$	$pr\,pr \; vg\,vg$
$pr \; vg$	$pr^+ pr \; vg^+ vg$	$pr^+ pr \; vg\,vg$	$pr\,pr \; vg^+ vg$	$pr\,pr \; vg\,vg$
$pr \; vg$	$pr^+ pr \; vg^+ vg$	$pr^+ pr \; vg\,vg$	$pr\,pr \; vg^+ vg$	$pr\,pr \; vg\,vg$
$pr \; vg$	$pr^+ pr \; vg^+ vg$	$pr^+ pr \; vg\,vg$	$pr\,pr \; vg^+ vg$	$pr\,pr \; vg\,vg$

	Wild type phenotype	Red eye vestigial wing	Purple eye wild type wing	Purple eye vestigial wing
Predicted from Mendel's Law of independent assortment	25%	25%	25%	25%
Observed	45%	5%	5%	45%

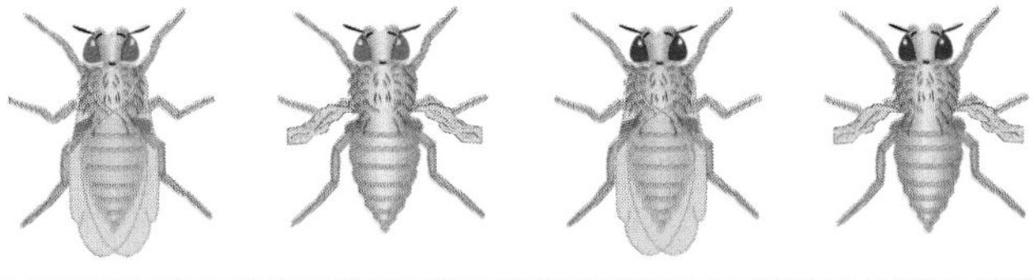

Figure 11-14 Establishment of genetic linkage between the vestigial and the purple genes. **(a)** Predicted genotypes of the F1 generation resulting from a cross between wild type and purple vestigial parents. **(b)** Testcross of the F1 hybrids. Mendel's Law of Independent Assortment predicts that 25% of the testcross progeny would have each of the phenotypes shown, but the fraction of recombinant phenotypes (red *vestigial* and *purple* normal-wing) was much less than expected because the *vestigial* and *purple* genes are close to each other on the same chromosome.

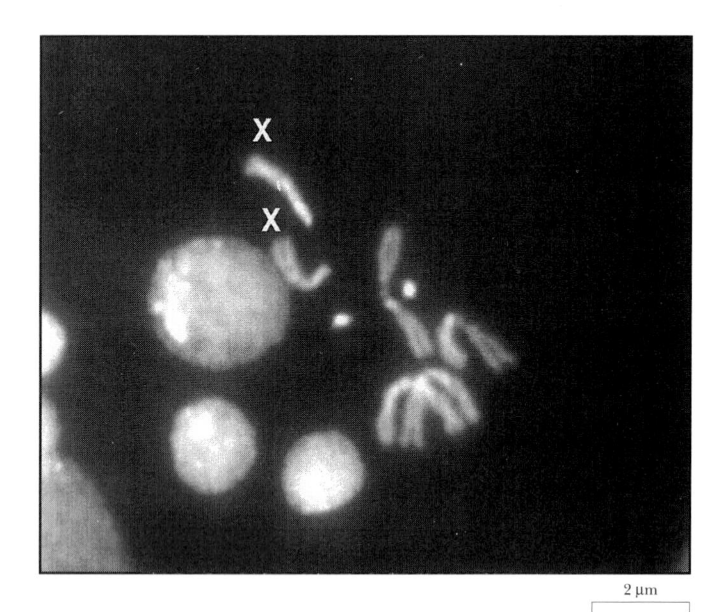

Figure 11-15 The four chromosomes of *Drosophila melanogaster.* *(Sharyn Endow, Duke Medical Center [Komma and Endow 1995, Haploidy and androgenesis in* Drosophila. *PNAS, in press])*

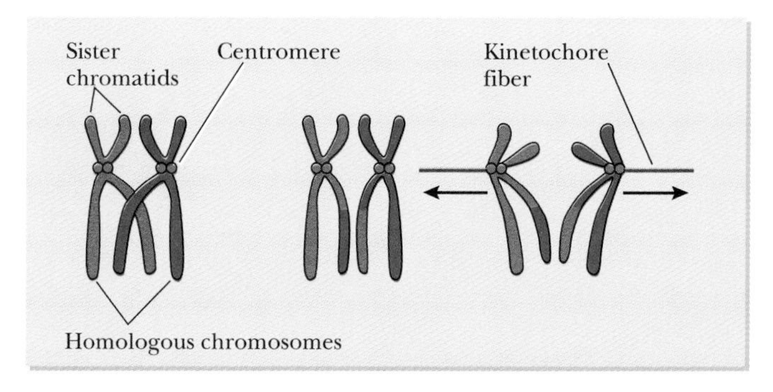

Figure 11-16 Crossing over, a type of genetic recombination, occurs during prophase I of meiosis. Breakage and rejoining of single chromatids on paired homologous chromosomes.

may arise from the independent assortment of individual chromosomes. And they can also arise, Morgan showed, from the exchange of material between the homologous chromosomes.

We can actually observe these exchanges during prophase of meiosis I, when homologous chromosomes are paired with one another (p. 282). **Crossing over,** one type of genetic recombination, involves the breakage and rejoining of single chromatids of homologous chromosomes (Figure 11-16), as first shown in 1931 by Harriet Creighton and Barbara McClintock at Cornell University, who compared the pattern of inheritance of alleles in corn with the behavior of chromosomes. Creighton and McClintock took advantage of homologous chromosomes that differed in their microscopic appearance. They used these chromosomes to show that genetic recombination involves actual physical exchanges between paired homologous chromosomes.

The Frequency of Recombination Between Two Genes Reflects the Physical Distance Between Them

The physical explanation of genetic recombination makes a strong prediction about the locations of genes and the probability of recombination: The farther apart two genes are on the chromosome, the greater the likelihood of crossing over and genetic recombination. The discoverer of this insight was Alfred Sturtevant, another undergraduate in Morgan's laboratory. Sturtevant realized that he could make a **genetic map** (or **linkage map**) that summarized the distances between genes. He expressed the genetic distance between two genes as the chance that recombination would occur between them. He defined a **map unit** as the distance between two genes that would produce 1% recombinant gametes (and 99% parental gametes). In tribute to Morgan, geneticists now often use the term "centiMorgan," or cM, as a measure of genetic distance, where 1 cM = 1 map unit. Geneticists refer to the position of a gene on a chromosome as its **locus** [plural, **loci;** Latin = place]. The distance between the *vg* locus and the *pr* locus is about 10 map units (10 cM), meaning that 10% of the offspring of the cross we discussed above are recombinant—with purple eyes and normal wings or with normal eyes and vestigial wings.

In the same way, geneticists have mapped thousands of genes on all four *Drosophila* chromosomes. Other species follow the same genetic rules as *Drosophila,* and geneticists have made similar maps for organisms ranging from bacteria (and even bacterial viruses) to humans. These maps again confirm the view that genes are on chromosomes and that they have a physical reality.

Visible Features of Chromosomes Demonstrate the Physical Basis of Genetic Maps

Drosophila has another special feature that has allowed it to contribute to understanding the relationship between genes and chromosomes. Certain *Drosophila* cells undergo DNA replication without accompanying cell division. Each of these cells acquires giant **polytene chromosomes** [Greek, *poly* = many + *tainia* = ribbon], which consist of about 1000 chromatids aligned in parallel. Along each polytene chromosome are dark bands interrupted by lighter regions, as shown in Figure 11-17a. The bands (about 5000) contain most of the cell's DNA. These bands provide the visible landmarks for a physical map of each chromosome.

The landmarks on giant chromosomes allowed geneticists to determine the physical location of genes. In certain mutations, called **deletions,** part of a chromosome

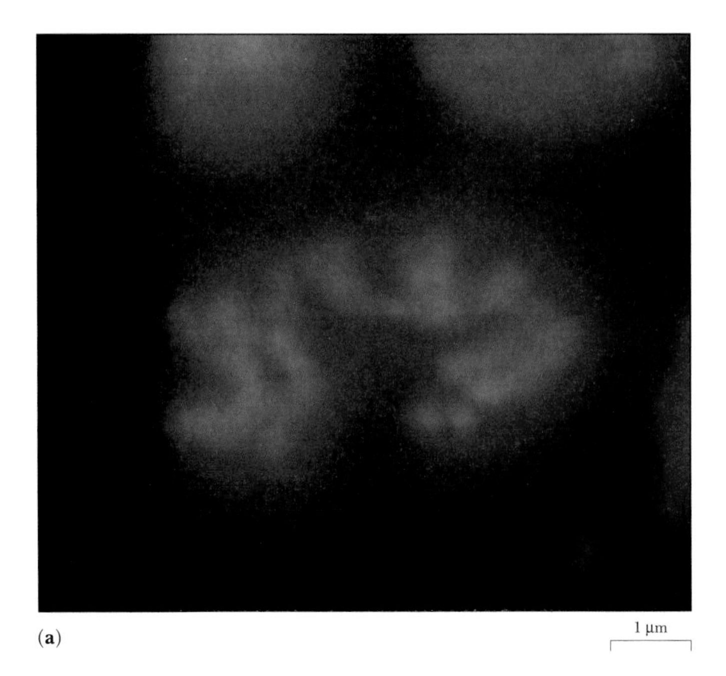

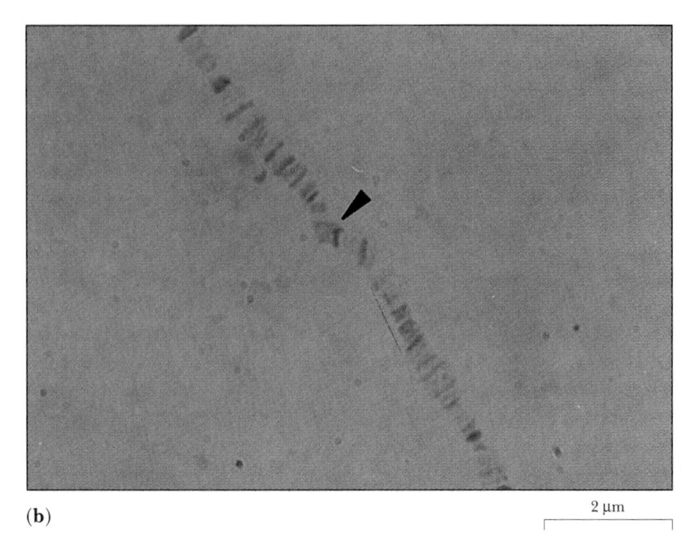

Figure 11-17 Chromosomes from *Drosophila* salivary glands: **(a)** wild-type chromosomes; **(b)** a deletion in the *vestigial* region of polytene chromosome II. *(a, Lisa Zimmerman, St. John's University; b, Johnson Chen, St. John's University)*

is missing, as shown in Figure 11-17b. Such mutations provided geneticists with a way of visualizing physical distances on a chromosome and correlating these differences with the genetic distances on a linkage map.

Careful examination of polytene chromosomes reveals the physical location of many deletion mutants. By combining such physical information with the linkage maps derived from crosses, geneticists can assign genes to particular bands on the giant chromosomes, again demonstrating that genes are not abstractions but real physical objects.

THE RELATIONSHIP BETWEEN GENOTYPE AND PHENOTYPE IS OFTEN COMPLEX

So far this chapter has revealed that understanding inheritance is simpler than anyone might have imagined. All the genetics we have discussed can be explained by rules that reflect the presence of genes on chromosomes and the behavior of chromosomes during meiosis and fertilization. But nature is often more complex, even when the underlying rules are simple. In this section, we examine some of the complexities of transmission genetics. We see that a single gene may affect many phenotypic traits, and that phenotype often depends on the interaction of more than one gene.

A Single Gene Can Affect Many Aspects of Phenotype

Phenylketonuria (PKU) is a human disease that results from the absence of the enzyme phenylalanine hydroxylase, which converts phenylalanine to tyrosine. With the path to tyrosine blocked, excess phenylalanine is converted to another compound, phenylpyruvic acid (a "phenylketone," hence the name of the disease), which accumulates in the blood and urine. PKU is a genetic disease caused by the absence of a functional gene for phenylalanine hydroxylase. The PKU gene is recessive; that is, one good copy of the gene (that is, one wild-type allele) can provide the information to produce sufficient enzyme to avoid phenylketone accumulation.

The accumulation of phenylalanine in PKU patients leads to diverse symptoms—mental retardation, reduced head size, and light hair color. Untreated children may never learn to walk, often have severe convulsions, and usually die before age 30. These all result from the failure to produce a single enzyme. This capacity of a single gene to affect many aspects of phenotype is called **pleiotropy** [Greek, *pleios* = more + *trope* = turning].

Many (perhaps most) genes in humans and in other organisms are pleiotropic. For example, an allele of the gene for the β-globin polypeptide leads to a disease called **sickle cell disease.** The sickle cell disease allele directs the production of an abnormal β-globin called β^S-globin, which combines with a normal α-globin to produce an abnormal hemoglobin and red blood cells with a sickle shape (Figure 11-18). Sickle cell disease also brings on many other changes: fatigue, decreased numbers of red blood cells, and damage to bones, heart, kidneys, lungs, and spleen.

PKU Illustrates the Importance of Interactions Between Genes and Environment

Studies of PKU led to a breakthrough in the diagnosis and treatment of human genetic diseases. Infants with PKU can be identified at birth by an inexpensive chemical test. In

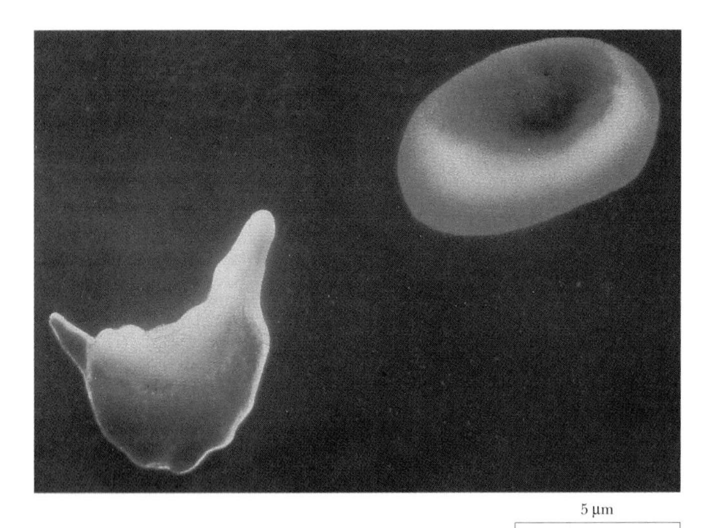

Figure 11-18 A sickled red blood cell next to a red blood cell with normal appearance. *(Roseman/Custom Medical Stock)*

most states this test is compulsory for all newborns because about 1 in 15,000 children have PKU. Children with PKU then receive a special diet that is low in phenylalanine. They receive enough phenylalanine and tyrosine to support protein synthesis and other needs, but too little to accumulate and cause damage. With this treatment PKU children usually develop normally.

PKU illustrates the importance of the interaction between genes and environment in establishing phenotype. A PKU child fed a normal diet develops a severely abnormal phenotype. Fed a special diet low in phenylalanine, however, the child develops normally. No matter what the genotype of an organism, its phenotype ultimately depends on both genes and environment.

Some Phenotypic Characters Depend on More than One Gene

Phenotype also often depends on the interactions of more than one gene. Height, for example, has a clear genetic component. Children are more likely to be tall if their parents are tall, short if their parents are short. Mutations that affect the function of the thyroid or pituitary gland (both of which participate in the regulation of growth) produce offspring with below-average height. But height also depends on a variety of unknown genes. Height is therefore said to be a **polygenic** trait; that is, more than one gene contributes to its inherited variation.

Environment and genes both contribute to height determination. In particular, nutrition is an especially important determinant of height: A malnourished child does not grow to 6 feet, no matter how tall his or her parents.

Most phenotypic traits are polygenic. In some cases, such as the genes that govern height, we know little about the identities and action of each of the genes. In other cases, we know a great deal about how one gene modifies the expression of another. For example, coat color in mice, the subject of decades of research in mouse genetics, depends on the action of at least five genes. One gene determines the distribution of pigment within individual hairs; a second determines the color of pigment; a third allows or prevents the expression of the pigment genes; a fourth determines the intensity of the pigmentation; and a fifth controls the presence or absence of spots on the mouse's coat. The gene that regulates the intensity of pigmentation is called a **modifier gene** because it regulates the expression of another, separate gene. An **epistatic gene** is a type of modifier gene that limits the expression of another gene, for example, the coat color gene that determines whether pigmentation occurs at all.

Epistasis refers only to the effect of a gene on another gene, not on a different allele on the homologous chromosome. Interactions between two alleles result in incomplete dominance, as in the case of snapdragon color, where the R and R' alleles together produce a phenotype (pink flowers) that is distinct from the phenotype produced by either RR or $R'R'$ homozygotes.

Both modifier genes and environmental factors can affect the expression of genes, complicating genetic analysis. Understanding such interactions is particularly difficult in humans because geneticists must rely only on the histories of existing families.

Geneticists working with any organism, however, are likely to encounter examples of variable **expressivity,** the range of phenotypes associated with a given genotype. For example, a relatively common human disease, neurofibromatosis, is caused by a dominant allele of a gene located on human chromosome 17. The expressivity of the disease-causing allele varies: In some people it causes blindness or horribly disfiguring growths, while in others the only symptom is a coffee-colored spot.

To complicate genetic analysis still further, two individuals with the same genotype at a given locus may differ in their phenotype. Geneticists therefore speak of a genotype's **penetrance,** the fraction of individuals with that genotype who show an associated phenotype. Some people known to carry the neurofibromatosis allele show no phenotypic abnormalities at all, and the gene is said to be incompletely penetrant.

Coat color in mammals is an aspect of phenotype for which the genetics and biochemistry are well understood. For many phenotypic traits, however, we do not even know how many genes are involved. Many such traits are said to be **quantitative characters,** meaning that they may be described numerically within a range of values rather than as clear-cut alternatives. Examples of quantitative characters are the sizes of mice, humans, or tomatoes.

The familiar contributions of polygenes to quantitative characters helped establish the old idea of blending inheritance. Now, however, we can explain the inheritance of many quantitative characters in terms of the independent assortment of many independent genes. That is, the inheritance of quantitative characters is often polygenic.

The inheritance of quantitative characters is particularly complicated because of the importance of environmental factors. For example, musical talent, high grades, healthy teeth, or wealth may each run in families. Do these facts suggest that these qualities are quantitative genetic characters? No, such a conclusion is not warranted unless environmental factors are all the same. The children of rich parents are likely to be rich themselves for nongenetic reasons, such as inheriting money! Similarly, environmental influences (such as a family's cultural interests, its diet, or a community's fluorinated water supply) are likely to contribute strongly to academic achievements, musical ability, and dental health.

A Single Gene May Have Multiple Alleles

So far we have spoken mostly as if each gene can have only two alleles—a dominant allele and a recessive allele. But every gene may exist in many alternate forms, each of which contains a distinct sequence of nucleotides that specifies a distinct polypeptide. Some of the specified polypeptides cannot function at all, while others function with reduced or somehow altered capacity.

A widely used example of multiple alleles is that responsible for ABO blood types. This locus has three alleles, which result in three different versions of a surface marker on red blood cells.

Perhaps the best-documented example of a multiple allele system is the gene for β-globin (which combines with α-globin to form a hemoglobin molecule). The β^S allele of the β-globin gene is responsible, when homozygous, for sickle cell disease. Researchers have also found more than 100 other alleles of this gene, each of which codes for a slightly different polypeptide.

In homozygotes, some of the β-globin alleles (like β^S) can have devastating effects. Many alleles, however, have little or no effect on phenotype. In fact, some heterozygotes are better able than the wild-type allele (called β^A) to survive and reproduce in some environments. For example, a β^S-globin/β^A-globin heterozygote is (for reasons not completely understood) much more likely to survive an attack of malaria than a β^A/β^A homozygote. Multiple alleles, present at different levels in different populations, can help a species to survive in a variety of environments.

SUMMARY

Genetics, the study of inheritance, must explain the inheritance of variations. Variations in the appearance of organisms arise from differences in the genes and environments of individual organisms. Inherited variations derive from differences in genes, which ultimately derive from mutations, changes in the sequences of nucleotides in DNA. Mutations may arise spontaneously or may be induced by radiation or chemicals.

Despite the obvious resemblance between parents and their offspring, the laws of genetics were long obscure because geneticists were limited to studying the inheritance of phenotypes rather than of genes themselves. Until the work of Mendel, which was done in the mid-19th century but not appreciated until the early 20th century, most biologists thought that the characteristics of offspring represented a blending of the characteristics of their parents. The key to understanding genetics was the realization that most genes are present in two copies, one from each parent, and that—most of the time—these genes do not change as they pass from generation to generation.

Mendel's experiments made use of true-breeding variants of pea plants, that is, plants whose offspring had the same characteristics as the parents. Mendel systematically examined the consequences of cross-breeding such variants. The basic principles of genetics allow geneticists to predict the types and numbers of different offspring that are likely to be produced by two parents.

The Principle of Segregation states that the two copies of a gene separate during the production of gametes. The Principle of Independent Assortment states that the distribution of the two copies of one gene during the formation of gametes is independent of the distribution of other genes. These two principles, sometimes called Mendel's Laws, explain the inheritance of many aspects of form and function in diploid organisms. Species whose genetics have been especially well studied include peas (which Mendel himself used) and *Drosophila* (the common fruit fly), intensely examined by Morgan and his students.

The laws of genetics are consistent with the behavior of chromosomes during meiosis and fertilization. Compelling evidence that genes are on chromosomes first came from the analysis of genes on the X chromosome of *Drosophila*. Genes that are close together on the same chromosome do not assort independently but demonstrate genetic linkage. Later studies established genetic maps of *Drosophila* and many other organisms. Such maps place genes on chromosomes and explain deviations from the Principle of Independent Assortment. The giant polytene chromosomes of *Drosophila* have permitted geneticists to show that the distances on the genetic maps of chromosomes correspond to the physical arrangement of genes on chromosomes.

KEY TERMS

allele
codominant
cross-breeding (crossing)
crossing over
deletion
dominant
Drosophila
epistatic gene
expressivity
F2 generation
first filial (F1) generation
genetic linkage
genetic map (or linkage map)
genetics
genotype
hemizygous
heterozygous
homozygous
hybrid
linkage group
locus (loci)
map unit
Mendel's First Law
Mendel's Second Law
modifier gene
molecular genetics
mutation
parental (P) generation
penetrance
phenotype
phenotypic trait
phenylketonuria (PKU)
pleiotropy
polygenic
polytene chromosome
Principle of Independent Assortment
Principle of Segregation
probability
product rule
quantitative character
recessive
segregate
sex-linked gene
sickle cell disease
sum rule
testcross
transmission genetics
true-breeding
wild-type

QUESTIONS FOR REVIEW AND UNDERSTANDING

1. Fill in the blanks to complete the following statements.
   An individual's ____________ is a consequence of the interaction of environmental factors and the individual's ____________. A ____________ gene is one that may influence more than one phenotypic characteristic. Conversely, a phenotypic trait is ____________ if it is under the control of more than one gene.
   A particular "version" of a gene is referred to as an ____________. An individual is ____________ if it possesses two identical versions of the gene, while an individual is ____________ if it contains two different versions of a gene. In a heterozygous individual, a ____________ allele will determine a phenotype while a ____________ allele will not. In other cases, ____________ alleles may both contribute to the phenotype. A particular genotype, even if it is controlled by a dominant allele, may be associated with a range of phenotypes as a consequence of ____________. On the other hand, the ____________ of a particular genotype refers to the frequency at which an associated phenotype is observed.
   Although Mendel stated in his Second Law that alleles for one gene segregate independently of the alleles for another gene, it was later demonstrated in Morgan's laboratory that certain genes tend to segregate together during gamete formation. This is referred to as ____________, and such a group of ____________ genes corresponds to a ____________.
2. Say whether the following statements are TRUE or FALSE. If false, provide a brief answer explaining why.
   (a) The frequency of an allele in a given population is related to its survival value.
   (b) A dominant allele is more abundant in a given population.
   (c) A dominant allele has greater survival value.
3. Considering the symptoms of phenylketonuria as a phenotype, how can this phenotype be influenced by the environment? What does this suggest regarding the contributions of diet, exercise, and "healthy living" to reducing the likelihood that an individual will contract a disease for which he/she may have a "genetic predisposition"?
4. How can a particular phenotype, such as height, be influenced by various genotypes, i.e., can you propose how various genes (or the products of those genes) may influence an individual's growth and his/her adult height?
5. The sickle cell allele is considered recessive because heterozygous individuals carrying one sickle cell allele do not exhibit obvious symptoms of sickle cell anemia; only homozygous individuals do. However, heterozygous individuals are less susceptible to malarial infection, and electrophoretic analysis of hemoglobin from heterozygous individuals demonstrates the presence of both wild-type and sickle cell hemoglobin.
   (a) What does this tell us about recessive alleles and their expression?
   (b) Argue in favor of the sickle cell and wild-type alleles being considered codominant.
   (c) What do the points stated above say about the determination of phenotype?
6. Cystic fibrosis (CF) is a common inherited disorder (1 in 2500 newborns of European extraction are affected) involving a particular gene coding for a transmem-

brane chloride channel protein. The CF allele is recessive, and heterozygous individuals (carriers) having one wild-type and one CF allele do not display any symptoms.

(a) Using a Punnett square, determine the probability that a child born to two carriers will be affected. What is the probability that a child born to the two carriers will not carry the CF allele?

(b) Using a Punnett square, determine the expected ratios of the genotypes (and associated phenotypes) of offspring produced by a mother who exhibits symptoms of sickle cell anemia and is a carrier of the CF allele and a father who is a carrier of both the sickle cell allele and the CF allele.

7. What are two mechanisms of genetic recombination? Describe how they differ.
8. Provide a mechanistic explanation of genetic linkage and how it can alter the expected ratios of genotypes (and phenotypes) in a cross.
9. Genetic or linkage mapping is an extremely valuable tool, allowing the relative position of a gene on a chromosome (its locus) to be determined from the percentage of recombinant gametes, and hence offspring, produced. Determine the arrangement of the genes A, B, C, D, and E on a chromosome, given the following values for frequency of recombination of genes.

Genes	% Recombination
A & C	80
B & D	40
C & E	45
A & D	10
B & E	15

SUGGESTED READINGS

GRIFFITHS, A. J. F., J. H. Miller, D. T. Suzuki, R. C. Lewontin, and W. M. Gelbart, *An Introduction to Genetic Analysis,* 5th edition, W. H. Freeman and Company, New York, 1993. Chapters 4 and 5 cover much of the material in this chapter.

LODISH, H., D. Baltimore, A. Berk, S. L. Zipursky, P. Matsudaira, and J. Darnell, *Molecular Cell Biology,* 3rd edition, Scientific American Books, New York, 1995. Chapter 8, "Genetic Analysis in Cell Biology," gives a contemporary view of Mendelian genetics.

CHAPTER 12

What Is a Gene?

KEY CONCEPTS

- A mutation may produce an organism that lacks a particular enzyme or that makes a polypeptide with an altered amino acid sequence.
- Experiments with bacteria and bacterial viruses showed that genes are made of DNA.
- Each gene (with some exceptions) specifies the structure of a single polypeptide.

OUTLINE

ESSAYS

Defining a gene remains a problem not only to students studying for biology examinations but also to scientists searching for biological molecules. To Mendel (in 1865), a gene was a "determinant," an abstraction that somehow determined the choice of alternative traits. To Sutton and Morgan (in the early years of the 20th century) genes were parts of physical structures, the chromosomes, which microscopists could actually see.

But biologists still did not know what genes actually are or how cells execute and replicate the information that they contain. That knowledge came only when geneticists tried to understand the function and the structure of genes in chemical terms. In this chapter, we see how researchers established what genes do and what they are made of. We see that genes specify the sequence of amino acids in protein molecules and that they are made of DNA. In Chapter 13, we see that the three-dimensional structure of DNA is especially favorable for replication and repair. In Chapter 14, we see how the plans contained in genes lead to the synthesis of specific protein products.

GENES AFFECT BIOCHEMICAL PROCESSES AS WELL AS MORPHOLOGICAL PROPERTIES

During the second half of the 19th century, biologists came to accept the idea that organisms perform chemical reactions. This was not an obvious idea because living organisms at first seem to violate basic chemical rules—like the apparent creation and preservation of order. Recall (from the mocking 1839 essay quoted in Chapter 7), for example, the hostility to the idea that fermentation is a chemical process.

By the beginning of the 20th century, however, many biologists and physicians began to focus on chemical pathways, asking questions about the sequence of chemical reactions responsible for building or degrading particular biological molecules. This work used the same techniques that some of the initially hostile organic chemists had developed years earlier. For decades, chemists had applied such analytical and synthetic techniques to more conventional chemistry, including the production of the new dyes that changed both the textile industry and cell biology. But, for the first half of the 20th century, most geneticists did not think of genes in biochemical terms, as we now must.

Inherited Human Diseases Provided the First Hints About the Products of Genes

The first hint of the biochemical importance of genes came in 1902 from the work of an English physician, Archibald Garrod. Among Garrod's patients at the Hospital for Sick Children in London was a baby whose diapers developed black urine stains. The baby's disease, alkaptonuria, was first described in 1649 and named for the dark alkapton bodies, which formed after the urine was exposed to air. The precursor of the alkapton bodies was identified by standard techniques of analytical organic chemistry to be homogentisic acid (HA), which oxidizes to form the alkapton bodies.

Garrod, working just 2 years after the rediscovery of Mendel's paper, thought that alkaptonuria might be an inherited disease because the baby's parents were first cousins. He realized that alkaptonuria behaves as if it were caused by a recessive allele, with each parent producing gametes with the same allele inherited from one of their common grandparents.

Like the amino acids phenylalanine and tyrosine, HA contains a benzene ring, and Garrod suggested that homogentisic acid is ordinarily an intermediate in their metabolism. The tissues of individuals with normal phenotypes have an enzyme called HA oxidase. This enzyme catalyzes the oxidation of homogentisic acid, shown in Figure 12-1, so that it is ultimately excreted as carbon diox-

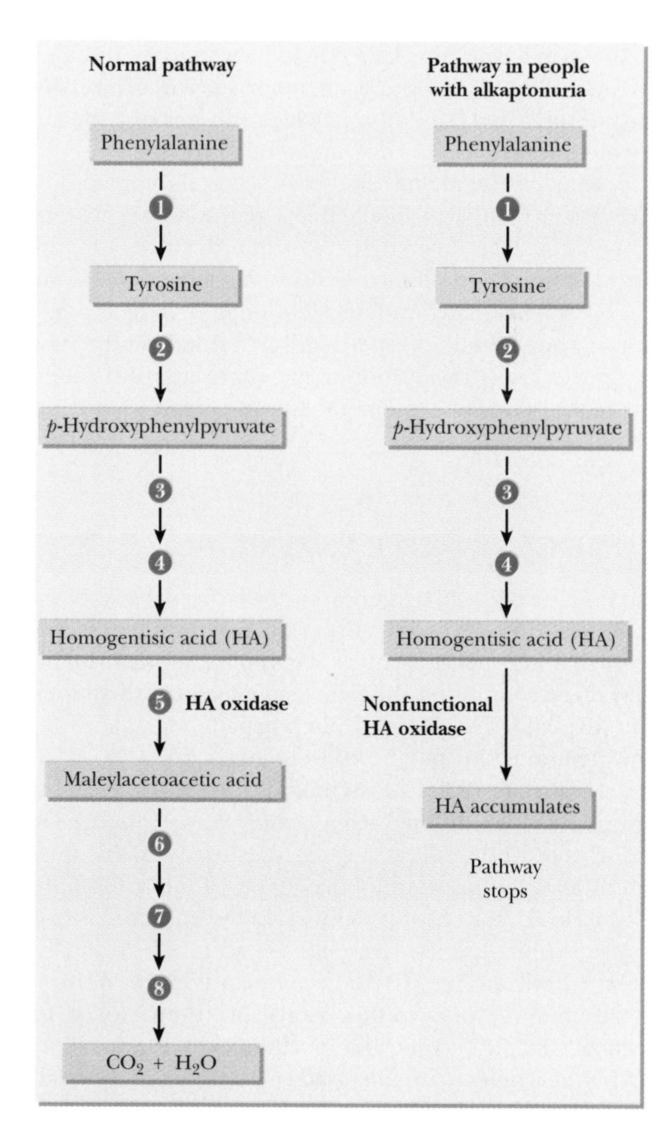

Figure 12-1 Phenylalanine and tyrosine, amino acid building blocks, are degraded by a common pathway, which is blocked in alkaptonuria.

ide and water. Alkaptonuria patients, however, cannot process homogentisic acid, and it accumulates in their urine.

Garrod realized that the absence of HA oxidase could produce such a condition. This enzyme deficiency, he concluded, was the result of the recessive allele whose inheritance he had charted. An alkaptonuria patient, he realized, had two copies of a defective allele of the gene for HA oxidase.

Inherited Diseases Allowed Dissection of Metabolic Pathways and Suggested the Nature of Genes

Other inherited conditions, such as albinism (lack of skin pigmentation), also result from failures of normal biochemical pathways. In many cases, the patient's inability to process a chemical intermediate can lead to the accumulation of compounds that are normally found only in small amounts. Often such compounds appear in the urine, leading to distinctive colors (as in alkaptonuria) or smells (as in a condition called maple syrup urine disease). Garrod called alkaptonuria an **inborn error of metabolism,** an inherited condition in which the lack of a specific enzyme interrupts a normal metabolic process.

Analysis of such inborn errors of metabolism allowed researchers to dissect biochemical pathways because intermediates accumulate that are ordinarily present in only trace amounts. An inborn error may stop a metabolic process at a normally undetectable step, as Garrod put it, "just as when a film . . . is brought to a standstill the moving figures are left with foot in air." Only later did geneticists realize that inborn errors also reveal just what genes do.

ONE GENE—ONE ENZYME

Garrod's insight—that genes control the activity of enzymes—at first seemed to apply only to a few rare diseases in which abnormal substances accumulate in the urine. In all of these conditions, the failure to make a particular enzyme explains the observed phenotypic abnormality. But how general is this picture of what genes do?

In the late 1930s, two Stanford University geneticists, George Beadle (who had been a student in Morgan's laboratory) and Edward Tatum, set out to answer this question by looking for other inborn errors of metabolism. Instead of looking in hospital clinics, however, they chose to study a simple fungus—the pink bread mold, *Neurospora crassa,* shown in Figure 12-2. Beadle and Tatum induced mutations in *Neurospora* with x-rays and then looked for mutants that altered specific biochemical pathways. Such biochemical defects are the mold counterparts of Garrod's inborn errors of metabolism. They were relatively easy to produce because *Neurospora* is a haploid organism and has only one copy of each gene, whereas most inborn errors

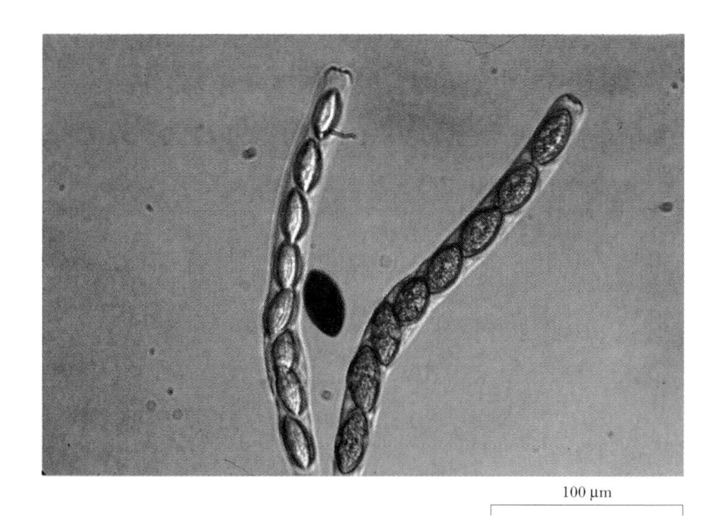

Figure 12-2 *Neurospora crassa,* a common bread mold. *(James Richardson/Visuals Unlimited)*

of metabolism in humans occur only when both copies of a gene are mutant, which usually happens only in inbred populations, in first-cousin marriages, and as the result of incest.

Metabolic Mutants of *Neurospora* Cannot Grow in a Simple Medium

Wild-type *Neurospora* can grow on an extremely simple medium (called a minimal medium), which contains sugar, ammonia, a few salts, and biotin, a vitamin. Beadle and Tatum identified a large number of mutant *Neurospora* molds that could not grow on such a minimal medium. These mutants were **auxotrophs,** meaning that they could not synthesize all the compounds needed to grow on a minimal medium. Many of these auxotrophs, however, could grow on an enriched medium, in which the minimal medium was supplemented with amino acids, nucleotides, and vitamins.

Beadle and Tatum reasoned that each auxotrophic mutation destroyed the mold's ability to synthesize some needed compound. Adding the compound to the medium should therefore allow the mold to grow. After isolating many such mutants, Beadle and Tatum were often able to find a specific compound that would, when added to the minimal medium, permit the *Neurospora* to grow. One mutant strain, for example, grew if the minimal medium were enriched just with vitamin B_6, while other strains grew only in the presence of a particular amino acid, for example, arginine.

Altogether Beadle and Tatum found seven mutant strains that would grow only if they added arginine. Perhaps, they reasoned, the mutations in the arginine-requiring *Neurospora* had destroyed the activity of individual enzymes in the pathway for the synthesis of arginine.

Without added arginine, the mold would die; with added arginine, the mold would live, reproduce, and form a visible colony.

Genetic Analysis Established That the Products of at Least Seven *Neurospora* Genes Contribute to Arginine Synthesis

Did all the seven arginine-requiring mutants (with mutant genes called *arg*-1, *arg*-2, *arg*-3, etc.) result from a defect in the same gene? To answer this question, Beadle and Tatum applied a simple genetic analysis. They took advantage of the ability of *Neurospora* to reproduce sexually as well as asexually. As illustrated in Figure 12-3, under the right conditions, two *Neurospora* cells can unite to form a zygote, the equivalent of fertilization in a plant or animal. The zygote immediately undergoes meiosis to form four cells, which each go through one mitotic division to produce eight *ascospores* (all haploid), so called because they are enclosed in a tiny case called an *ascus* [Greek = bladder]. All the meiotic products of a single zygote are present in a single ascus.

The first kind of experiment that Beadle and Tatum performed, shown in Figure 12-4a, analyzed a cross be-

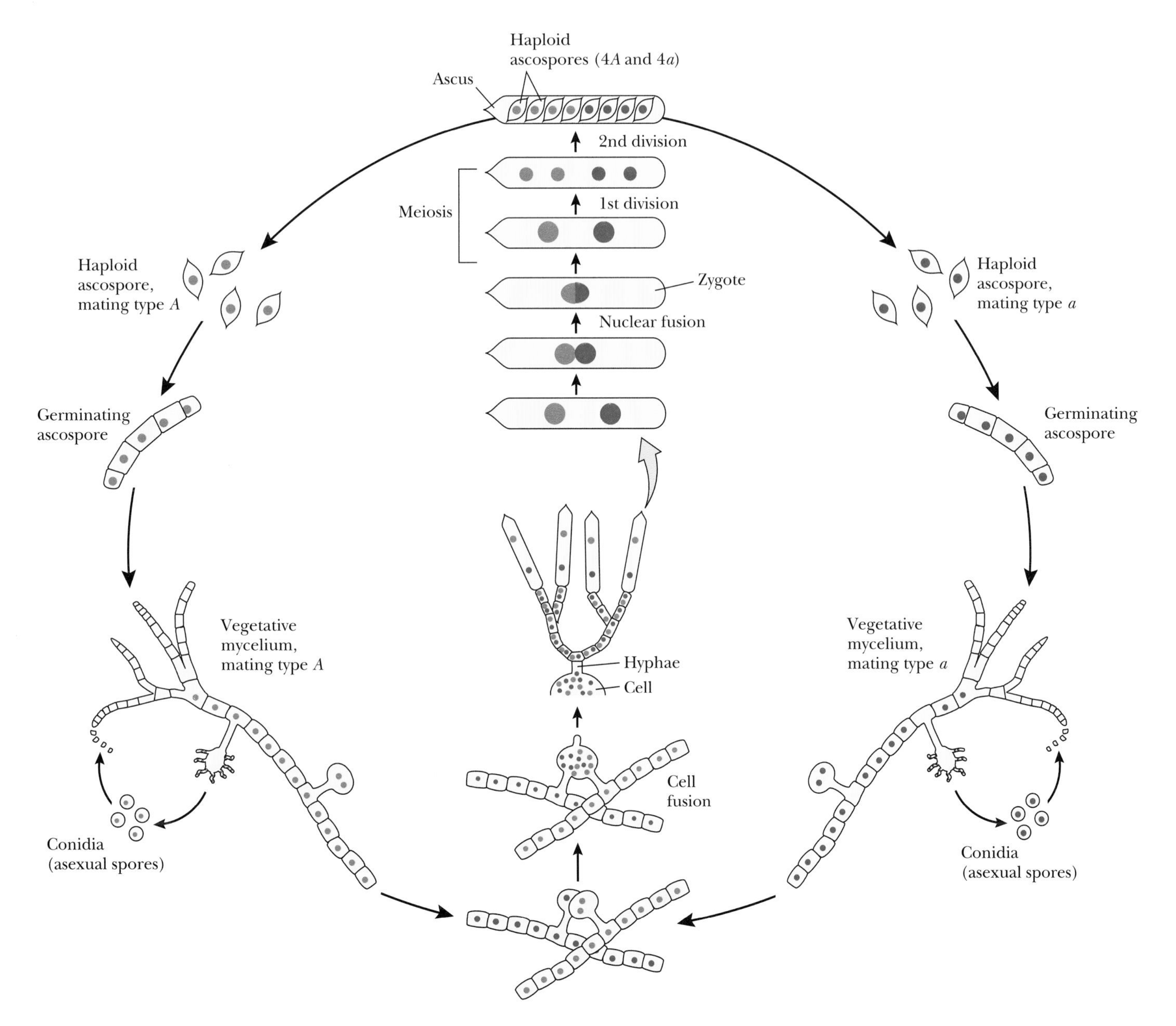

Figure 12-3 The ability of *Neurospora* to reproduce sexually allowed Beadle and Tatum to perform genetic analyses of mutations. Meiosis produces haploid ascospores, which produce multicellular, haploid mycelia. When two mycelia with different mating types (*A* and *a*) come into contact, as shown in the bottom center, the haploid cells can fuse and their nuclei can mix. Nuclei of the two mating types can then form a zygote that can immediately undergo meiosis to form more haploid ascospores.

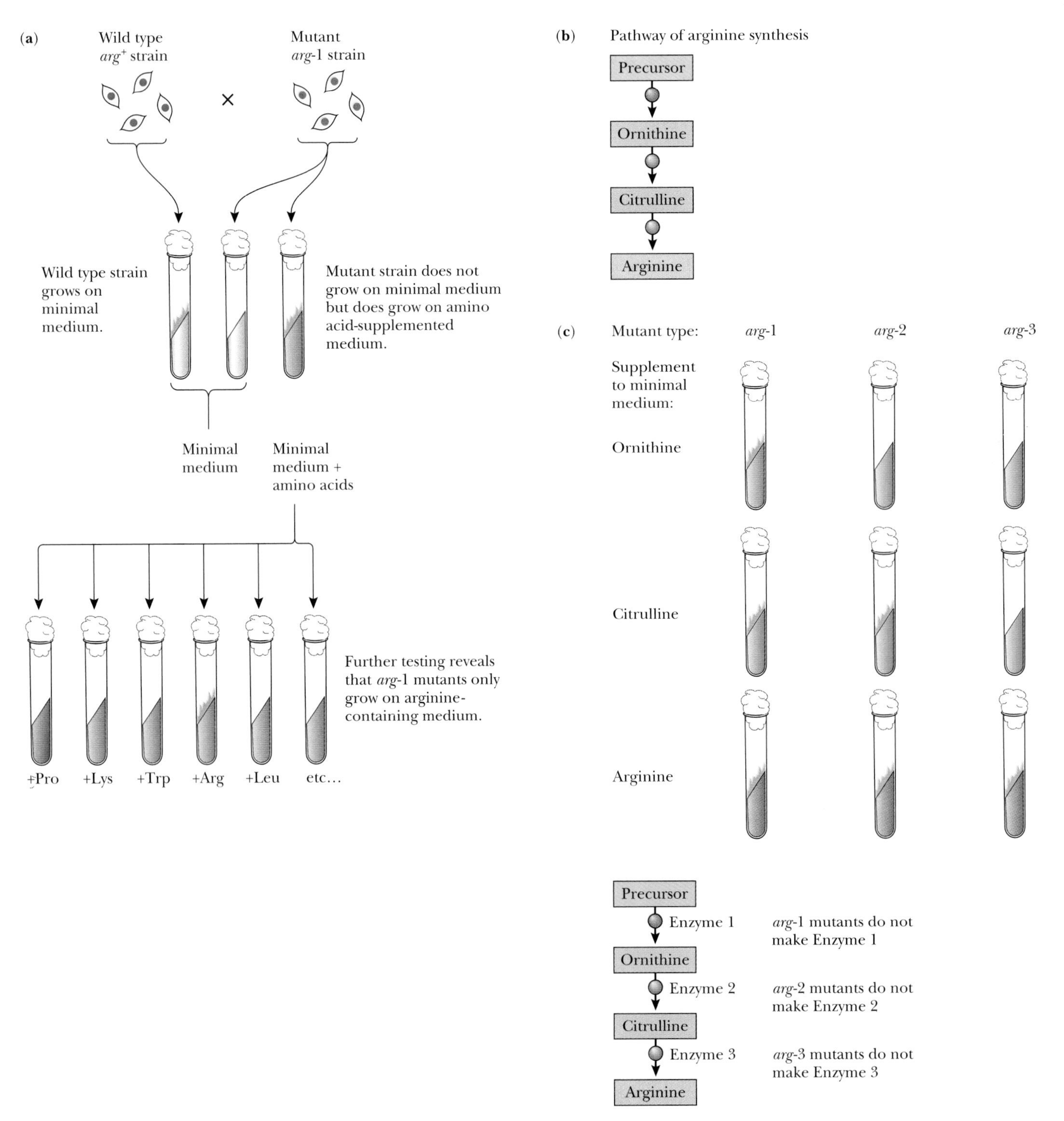

Figure 12-4 Beadle and Tatum showed that different *arg* mutants of *Neurospora* lacked distinct enzymes. They analyzed the ability of mycelia derived from single ascospores to grow on different types of media. **(a)** Wild-type strain grows on minimal medium, but strain called *arg*-1 grows only if minimal medium is supplemented with arginine; **(b)** arginine synthesis pathway; ornithine is converted to citrulline, which is converted to arginine; **(c)** one arginine-requiring mutant (*arg*-1) can grow on ornithine or citrulline instead of arginine; another (*arg*-2) can grow on citrulline but not on ornithine; while another (*arg*-3) can grow only on arginine itself. Beadle and Tatum concluded that each mutant lacks a different enzyme in the pathway of arginine synthesis.

tween a mutant strain (for example, one containing the *arg*-1 mutation) and a wild-type strain (which is designated arg^+. After fertilization, meiosis, and ascospore formation, they found that four of the eight ascospores in the ascus could grow on minimal medium, while four required added arginine. This is consistent with Mendel's Principle of Segregation and suggests that a single mutant gene is responsible for the arginine requirement in each strain.

They then analyzed crosses between two different arginine auxotrophs (for example, with the *arg*-1 and *arg*-2 mutations). If *arg*-1 and *arg*-2 are alleles of the same gene, then all the offspring of an *arg*-1 × *arg*-2 cross should require arginine. If, on the other hand, the mutations are in different genes, then some of the offspring should grow on minimal medium. They found that the seven mutations responsible for the arginine requirement are all separate genes. This result suggested to them that arginine synthesis in *Neurospora* involves at least seven different enzymes, each controlled by a separate gene.

Three *Neurospora* Genes Specify Three Enzymes That Catalyze Specific Reactions in the Arginine Pathway

Beadle and Tatum were able to relate their genetic findings to biochemical studies done in the early 1930s by Hans Krebs (the same biochemist who contributed so much to understanding the citric acid cycle, as discussed in Chapter 7). Krebs found out the last three steps in arginine synthesis (in vertebrates, at least) are: (1) the synthesis of a compound called ornithine; (2) the conversion of ornithine to citrulline; and (3) the conversion of citrulline to arginine.

With this scheme in mind, Beadle and Tatum examined the ability of their seven arginine auxotrophs to grow on ornithine and citrulline, as well as on arginine. Figure 12-4c shows the growth properties of three mutants (*arg*-1, *arg*-2, and *arg*-3). The *arg*-1 mutant grew when arginine was added, but it also grew in the presence of ornithine and citrulline, suggesting that the genetic defect was in the ability to produce ornithine and not in the ability to convert ornithine to arginine. In contrast, the *arg*-2 mutant grew in the presence of arginine or citrulline, but not in the presence of only ornithine, suggesting that it could produce arginine from citrulline, but not citrulline from ornithine alone. The *arg*-3 mutant required arginine; neither citrulline nor ornithine would do. The *arg*-3 mutant, then, could not convert citrulline or ornithine to arginine.

These results suggested that the last steps in the synthesis of arginine in *Neurospora* resembled the corresponding steps in vertebrates, as illustrated in Figure 12-4c. Each step in the synthesis of arginine is catalyzed by a different enzyme. The combination of biochemical and genetic reasoning allowed Beadle and Tatum to deduce that the *arg*-1 mutation must lie in the gene for enzyme 1 in Figure 12-4c. Similarly, *arg*-2 and *arg*-3 correspond to the genes for enzymes 2 and 3.

Beadle and Tatum's work helped unravel previously unknown biochemical pathways because the mutations blocked the pathways at different steps and allowed biochemical analysis, just as Garrod had appreciated two generations earlier. Still more importantly, however, Beadle and Tatum provided a definitive link between genetics and biochemistry. They summarized and generalized their work in the memorable phrase "one gene—one enzyme." That is, each gene somehow specifies the structure and production of a single enzyme.

■ ONE GENE—ONE POLYPEPTIDE

The generalization of Beadle and Tatum established the character of many genes. But do genes do other things besides specifying enzymes? Are enzymes the only link between genotype and phenotype?

Analysis of Sickle Cell Disease Showed That Genes Can Specify Proteins Other than Enzymes

The answers to these questions came with the unraveling of the molecular basis of **sickle cell disease** (also called sickle cell anemia), a blood disorder, introduced in Chapter 11, p. 311, that gets its name from the curled appearance of red blood cells in sickle cell patients. The distorted blood cells clump and interfere with the circulation, causing pain, heart problems, and brain damage. The body's defense mechanisms eliminate the troubling cells as rapidly as they can, bringing about a severe anemia. Patients with sickle cell disease almost always die young.

Sickle cell disease is a genetic disease. Like alkaptonuria, it results from the action of a recessive allele. Unlike alkaptonuria, however, it is both serious and common—with the mutant allele carried in almost 1 of every 30 African Americans. People who are homozygous for the sickle cell allele have sickle cell disease, a generally fatal condition, but people who are heterozygous for the sickle cell gene are much more likely to survive an infection with the malaria parasite, one of the world's major health problems.

In 1949 Linus Pauling (whose work on protein structure is discussed in Chapter 3) showed that sickle cell disease is a disease of **hemoglobin,** the red oxygen-binding protein of the blood. As illustrated in Figure 12-5, Pauling and his collaborators used electrophoresis to show that the hemoglobin of sickle cell patients (called HbS) differs from the usual adult hemoglobin, HbA.

The parents of sickle cell patients have both HbS and HbA, but the patients lack HbA entirely. A simple explanation of the disease emerged: (1) A recessive allele codes for an abnormal protein, HbS; (2) heterozygotes have genes for both normal and abnormal proteins; (3) in the presence of HbA, HbS does not cause the disease, so heterozygotes are healthy; (4) homozygotes for the recessive allele make only HbS; (5) in the absence of HbA, HbS causes sickle cell disease. Because differences in the gene had caused differences in the hemoglobin protein, not in an enzyme, Pauling's work suggested that genes control proteins. "One gene—one enzyme" became "one gene—one protein."

(**a**) Perform electrophoresis on hemoglobins isolated from red blood cells

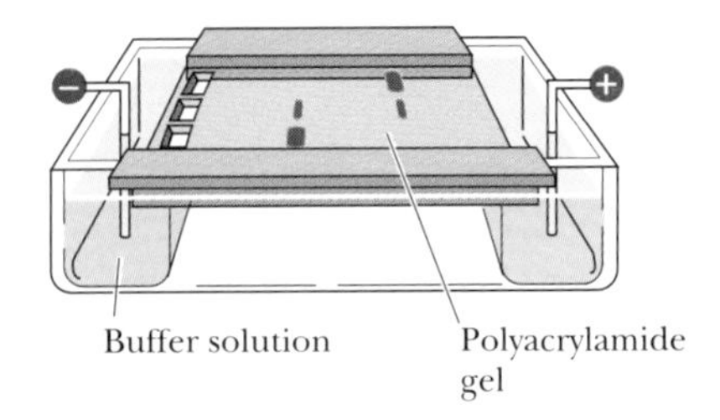

(**b**) Correlation of hemoglobin type with disease status

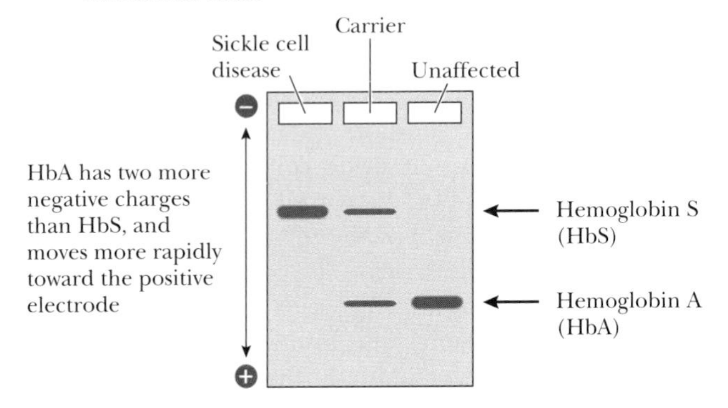

Figure 12-5 The hemoglobin in patients with sickle cell disease differs from that in unaffected people.
(a) Electrophoretic analysis of hemoglobins; electrophoresis detects differences in the sizes and charges of molecules by the way they move in an electric field; HbA and HbS have the same size but HbA is more negatively charged. **(b)** Correlation of hemoglobin type with disease status; sickle cell patient has only HbS, while a carrier (heterozygote) has both HbS and HbA.

What Exactly Is Different About Hemoglobin S?

Exactly how does an abnormal gene change a protein? To answer this question, Vernon Ingram, a young biochemist working at Cambridge University in 1956, used new methods of protein chemistry—electrophoresis and chromatography—to analyze the structural differences between HbS and HbA.

Hemoglobin consists of two types of polypeptide chains, called α-globin and β-globin. As diagrammed in Figure 12-6, each hemoglobin molecule has two α-globin polypeptides and two β-globin polypeptides. Ingram showed that only the β-globin polypeptides are different in HbS and HbA. They differ only at a single residue in the β-chain's entire amino acid sequence. Amino acid residue 6, ordinarily glutamic acid in HbA, is valine in HbS. This difference alters the charge of the hemoglobin molecule. This charge difference changes the movement of HbS (or, in Ingram's experiments, of HbS fragments) during electrophoresis.

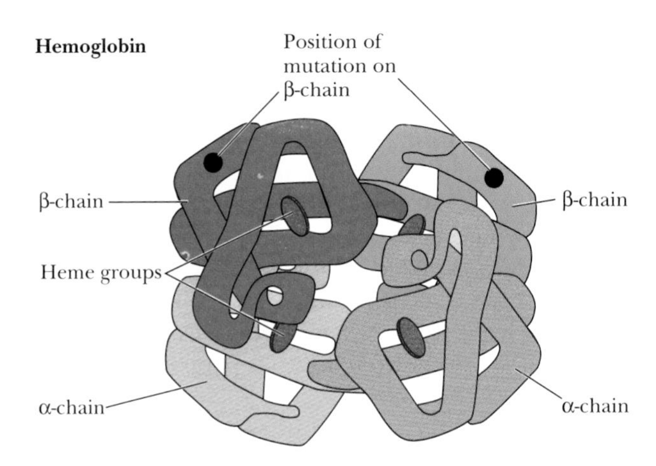

Figure 12-6 Hemoglobin consists of four polypeptide chains, 2 α- and 2 β-chains. The mutation responsible for sickle cell disease changes one amino acid in each β-chain.

This charge difference also alters the way that hemoglobin molecules pack into the crowded red cell. The result is that the HbS tends to crystallize within the cell, leading to the distortion of the cell that gives the disease its name. This work suggested that genes somehow control the amino acid sequence of proteins.

Ingram found that the sickle cell allele produces abnormalities only in the β-globin polypeptide chain, never in the α-globin chain. Further research revealed that the gene for another hereditary blood disease that produces abnormalities in the α-globin chain is completely unlinked to the sickle cell gene. These findings suggested that each gene is responsible for the production not necessarily of an entire protein (in this case, hemoglobin), but of a polypeptide chain. "One gene—one protein" became "one gene—one polypeptide."

This generalization is not complete, however. Some genes, for example those that specify the sequence of RNA molecules, do not contain information for amino acid sequences. And the sequences of some polypeptides, such as those involved in the immune response, are determined by more than one gene.

Even with these exceptions, the generalization means that a gene usually specifies the structure of a single polypeptide chain. But how does this statement help us understand the relationship of genotype and phenotype? The answer is that the polypeptide specified by a gene may be part of an enzyme like HA oxidase (the enzyme missing in alkaptonuria), a component of the cytoskeleton, or a protein that regulates the passage of ions through a cell membrane (as in the case of cystic fibrosis, the most common genetic disease among people of Northern European origin). Each of these polypeptides may play a part in determining an organism's phenotype. In Chapter 14, we see how a gene is "read" into a polypeptide. But before we can understand this process, we must look closely at what a gene is.

ESSAY

REPAIRING THE GENETIC DEFECT IN CYSTIC FIBROSIS CELLS

Cystic fibrosis (CF) is the most common genetic disease among Americans of Northern European origin. In this group, it affects approximately 1 in 2500 children, with about 1000 new cases each year in the U.S. Like alkaptonuria, cystic fibrosis is a recessive disorder, resulting from the inability to produce a specific polypeptide. About 5% of Caucasian Americans are carriers (heterozygotes).

In the case of cystic fibrosis, the missing polypeptide is a membrane protein that transports chloride ions across the plasma membrane of epithelial cells. The protein, which was named before its function was understood, is called CFTR (cystic fibrosis transmembrane conductance regulator). In 70% of CF patients, the defective protein lacks a single amino acid residue. This defect leads to a blockage in the movement of CFTR to the membrane, where it would normally function. Other CF patients have other defects, but the result is always a failure of chloride transport across the plasma membrane.

Failed chloride transport affects epithelial cells that line ducts in the pancreas, sweat glands, liver, and lungs. In the sweat glands, the result is characteristically salty sweat, a hallmark of CF that provides a reliable diagnosis. Northern European folklore refers to the sad prognosis for a child with salty sweat: "Woe to that child which when kissed on the forehead tastes salty: He is bewitched and soon must die." In the pancreas, the result of abnormal chloride transport is the blocking of the ducts that carry digestive enzymes to the gut. Pancreatic failure was once a major cause of death in CF patients, but it no longer is because patients can supply themselves with digestive enzymes in capsules taken when they eat.

Today, death of CF patients usually results from infections in the lungs, where the failure of chloride transport leads to the accumulation of a thick mucus in which bacterial infections can flourish. Current therapies include pounding the back or chest to release the thick mucus and clear the airways, antibiotics to control infections, and the inhalation of the enzyme DNase. The introduction of DNase into the airways digests the DNA released by dying epithelial cells. Because DNA molecules are so long and thin, they contribute greatly to the stickiness of the mucus, and their destruction by DNase is effective in reducing the thickness of the mucus and the chance of bacterial infection.

The available therapies for CF address the results, rather than the causes of the sticky mucus. But the new molecular understanding of the disease has suggested another therapeutic strategy—adding functional CFTR molecules to the epithelial cells that lack them. Providing CFTR protein directly is not considered feasible because protein delivery methods are not efficient. But many researchers are confident that delivery of the CFTR gene to epithelial cells will be possible. Already researchers have succeeded in restoring CFTR function in laboratory cultures of cells taken from CF patients, using recombinant DNAs that contain the CFTR gene.

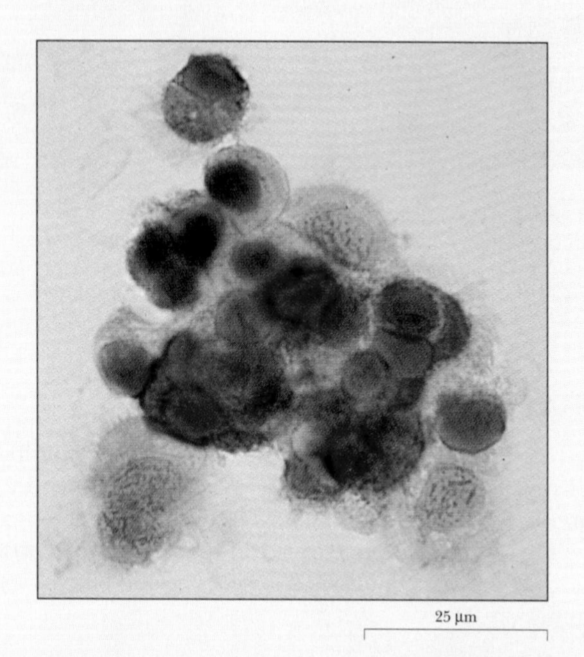

An approach to gene therapy. The red stain indicates that these lung cells of a rat are expressing the normal human allele of the gene responsible for cystic fibrosis. A modified virus that infects lung cells, but cannot reproduce in them, was used as the vector to introduce the human gene. *(courtesy of Melissa A. Rosenfeld and Ronald G. Crystal, National Heart, Lung and Blood Institute; National Institutes of Health. From* Cell *Vol. 68, no. 1, cover. Copyright by* Cell *Press)*

But the difficult part is to put the CFTR gene into cells in a living patient. Researchers are actively approaching this goal by placing the CFTR gene into a common virus, called adenovirus, which infects epithelial cells of the airways. Such engineered viruses have indeed been able to provide airway cells in experimental animals with functional CFTR molecules. At this time, however, results with human CF patients have been only moderately encouraging. Nonetheless, many physicians and biologists think that CF is likely to be the first major success for gene therapy.

Figure 12-7 Two strains of *Streptococcus pneumoniae,* "pneumococcus." The R (rough) strain is not virulent, while the S (smooth) strain is virulent. *(Avery, MacLeod, and McCarty,* Journal of Experimental Biology, *Vol. 79, 1944, New York Academy of Medicine)*

WHAT ARE GENES MADE OF?

Research on an amazing variety of organisms—including humans, peas, flies, and bread mold—has contributed to our understanding of what genes *do.* To learn what genes *are,* we turn to work on bacteria and the viruses that prey on them. The experiments we describe provided the proof that genes are DNA.

Extracts of Bacteria Can Change the Genetic Properties of Other Bacteria

Before general improvements in hygiene and the development of antibiotics, bacterial pneumonia, caused by a bacterium called *Streptococcus pneumoniae* (commonly referred to as pneumococcus), was an extremely serious disease. Research originally directed at its prevention led directly to the discovery of the nature of genes.

One approach to preventing bacterial pneumonia was to develop a vaccine against pneumococci. In 1928 Frederick Griffith, an English bacteriologist, was trying to provoke the immune defenses of mice against pneumococcus with injections of bacteria. Griffith worked with two strains of pneumococcus, one of which (called the S strain) produced pneumonia, while the other (called the R strain) did not.

When strains of pneumococcus with R and S phenotypes grow in artificial media, the virulent (disease-causing) S strain grows into glistening smooth colonies, while the nonvirulent R strain forms rough-looking colonies, as shown in Figure 12-7. We now know that the virulent S bacteria form smooth colonies because they produce a polysaccharide capsule. This capsule prevents their digestion by phagocytic white blood cells in the blood and makes them effective in causing pneumonia. The R bac-

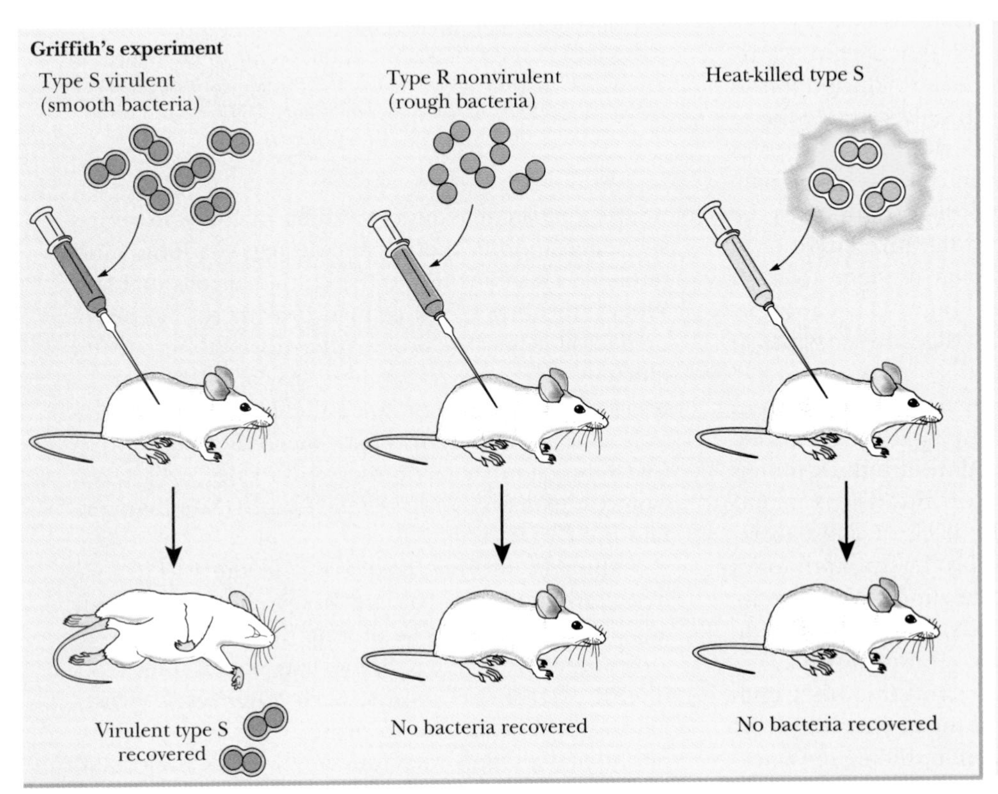

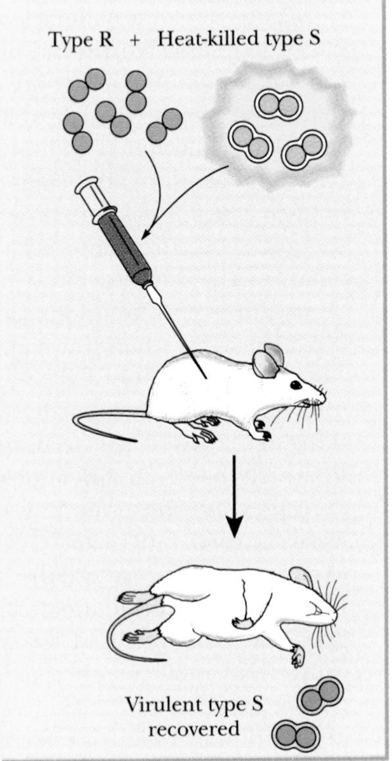

Figure 12-8 The experiment of Frederick Griffith, showing that an extract of a virulent strain of pneumococcus can "transform" a nonvirulent strain.

teria are mutants—they lack an enzyme needed to produce the polysaccharide capsule. They are much more susceptible to phagocytosis and therefore do not produce pneumonia.

Griffith tried to produce immunity to pneumonia by injecting mice with different combinations of live and heat-killed pneumococcus, as diagrammed in Figure 12-8. As expected, mice injected with live S strain bacteria got pneumonia and died, while mice injected with live R strain bacteria or with heat-killed S bacteria survived. Griffith was surprised, however, with the results of injecting a mouse with a mixture of live R strain bacteria and heat-killed S strain bacteria: The mice died. Moreover, the blood of the sick mice contained live S bacteria! Something in the heat-killed S bacteria had *transformed* the R bacteria into virulent S bacteria. That "something," called "the transforming factor," must be responsible for the change in the properties of the bacteria.

What Is the Transforming Factor?

After Griffith's work, other scientists found that they could convert R strains into S strains in a test tube as well as in a mouse. Extracts of S strains, dead or alive, could transform R strains. Once the R strains were transformed, moreover, their descendants inherited the properties of S strains. **Transformation,** as we now understand it, is the transfer of one or more genes from one organism to another. Although experiments using transformation now go on in thousands of laboratories, Griffith's work was the first example and provided the first opportunity to learn what genes are actually made of.

In the early 1930s, Oswald Avery at the Rockefeller Institute in New York set out to determine the chemical basis of the transformation of pneumococcus. With his coworkers Colin MacLeod and Maclyn McCarty, Avery carefully isolated different kinds of molecules from heat-killed S bacteria and tested to see which component is responsible for transformation, as diagrammed in Figure 12-9.

At the time, almost everyone expected that the transforming factor would turn out to be a protein because proteins were already known to perform so many cellular functions. Yet, surprisingly, transformation had occurred even when the bacteria had been heat-killed—a process that denatures proteins. Avery and his colleagues worked almost 10 years on this problem: They isolated various substances from heat-killed S strain bacteria and then tested each substance to see if it could transform R strain bacteria. They treated each isolated substance with a battery of enzymes that would hydrolyze proteins into amino acids, polysaccharides into sugars, and nucleic acids into nucleotides. To their (and everyone else's) surprise, they concluded (in 1944) that the transforming factor is DNA: Only the DNA fraction of the cells produces transformation, and the transforming activity is destroyed by an enzyme (DNAase) that hydrolyzes DNA into nucleotides.

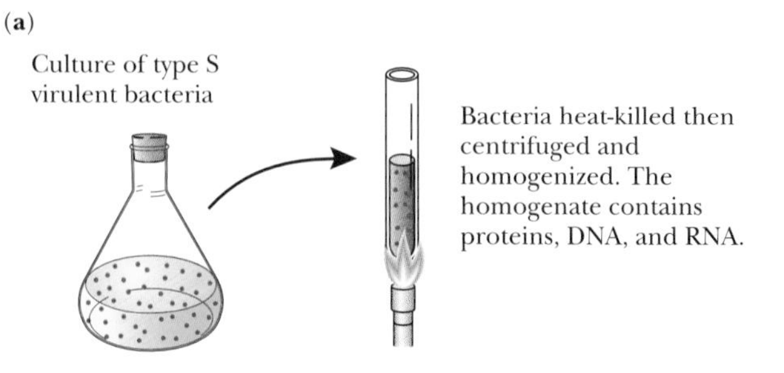

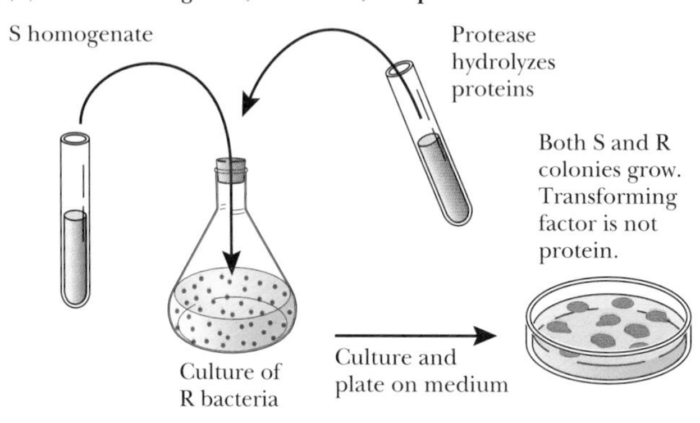

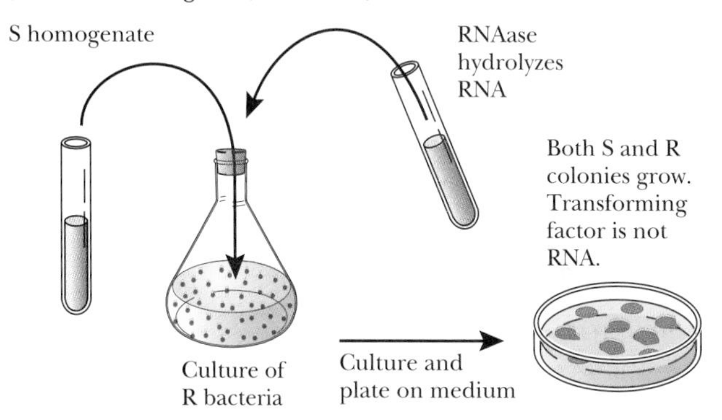

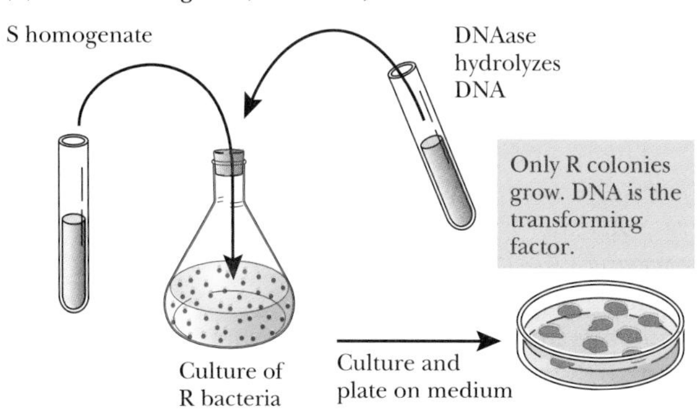

Figure 12-9 Identification of the "transforming factor" as DNA. **(a)** Extraction of proteins, DNA, and RNA from virulent (S) bacteria; **(b)** protease treatment does not prevent transformation, showing that the transforming factor is not a protein; **(c)** RNAase treatment does not prevent transformation, showing that the transforming factor is not RNA; **(d)** DNAase treatment prevents transformation, showing that the transforming factor is DNA.

ESSAY

EXPERIMENTAL METHODS FOR INTRODUCING DNA INTO CELLS

The discovery that DNA could transform a nonvirulent strain of bacteria into a strain that causes pneumonia not only established the identity of DNA as the genetic material, but also served as a model for experiments that are now routine in thousands of cell biology laboratories. In these experiments, academic and industrial researchers systematically introduce specific pieces of DNA into both prokaryotic and eukaryotic cells. The introduced DNA is seldom the entire DNA of a particular bacteria (as in the case of the pneumococcus experiments) but is rather recombinant DNA sequences that have been deliberately manipulated, as discussed in Chapter 17.

Such experiments have allowed researchers to learn the role of specific DNA sequences in encoding a protein sequence and in designating the amount and intracellular address of that protein. Similar experiments have also allowed researchers to alter specific properties of the DNA-treated cells, much as Griffiths and Avery did with pneumococcus but with a much wider repertoire of cell types and possible outcomes. In a growing number of cases, researchers have been able to restore a normal function to cells that suffer from a genetic defect. (See Essay: "Repairing the Genetic Defect in Cystic Fibrosis Cells.")

Because of the power of DNA-based cellular engineering, molecular biologists have expended considerable effort in figuring out the best ways of accomplishing transformation, that is, for introducing DNA into cells. Because many early experiments focused on the problem of how viruses inject their DNA into cells, researchers often call the transfer of any naked DNA "transfection," to distinguish it from "infection," the takeover of a cell by an intact virus. (See Chapter 17.)

Despite the early successes with the transformation of pneumococcus, the task of finding standard ways of transferring genes has not been simple. Cells do not naturally take up foreign DNA, and they have substantial barriers to prevent the subversion of their own genetic programs.

Notwithstanding the difficulties, cell biologists now routinely use four methods of transferring DNA into cells—microinjection, calcium phosphate precipitation, liposome fusion, and electroporation. In addition, many laboratories accomplish gene transfer by using recombinant DNA techniques to put a specific piece of DNA into a virus, so that the proteins responsible for viral infection help provide an efficient route for introducing the engineered DNA.

The most direct, but also the most limited method is microinjection with a specially prepared fine glass needle. Because cells are so small, and, for eukaryotic cells, DNA must be targeted to the nucleus one cell at a time, direct injection of DNA is limited in its application. The procedure is mostly used in ex-

The prejudice against DNA was so pervasive that biologists did not generally accept Avery's conclusion for almost a decade. Few people thought that DNA is capable of carrying information because it was considered too simple a molecule, composed as it is of only four nucleotides. Biochemists and biologists thought that DNA merely stabilized the structures within the nucleus. In the 1940s, however, another group of researchers began work with an even more unusual set of genes, those of a bacterial virus. This work finally convinced the scientific world of DNA's importance.

What Is a Virus?

In 1913, even before Griffith's work on pneumococcus, two bacteriologists—Frederick Twort and Felix d'Herelle—separately discovered that bacteria themselves can be killed by a tiny infectious agent called a **bacteriophage** [Greek, *bakterion* = little rod + *phagein* = to eat], or phage. D'Herelle, a young French microbiologist, was at the time trying to stop a locust epidemic in Mexico by spreading bacteria that would cause the locust to contract diarrhea. The "bacteria eaters" were nuisances that destroyed his bacterial cultures. D'Herelle and others later thought that bacteriophages might in fact provide a way of killing disease-causing bacteria—a prospect that was especially exciting in those days before the discovery of penicillin and other antibiotics. Unfortunately, however, no one ever found a way to use bacteriophages to destroy bacteria in infected organisms.

D'Herelle and Twort did not know what bacteriophages actually are, only that they were small enough to

periments requiring high efficiency but relatively few transformed cells, such as the production of transgenic embryos. (See Chapter 17.)

Perhaps the most widely used method of transfection, calcium phosphate precipitation, depends on the ability of cells to take up DNA that has coprecipitated with calcium phosphate. In this method, the researcher prepares pure DNA in a phosphate buffer and adds a solution of calcium chloride ($CaCl_2$). Calcium phosphate precipitates from this solution and carries DNA along with it in a fine precipitate. Incubation of the DNA-calcium phosphate mixture with targeted cells results in the uptake of DNA into a sizable fraction of cells.

Up to 50% of the cells may take up DNA-calcium phosphate coprecipitates by endocytosis. But only a small fraction of these cells actually incorporate the transfected DNA into their own nuclear DNA. The cells that do take up and express transfected DNA, however, will take up quite a bit. Researchers have learned to select those cells that express the foreign DNA by adding to it a specific gene for a trait that can be selected. A common trick, for example, is to "cotransfect" cells with DNA that specifies antibiotic resistance. A common selection method, for example, employs the eukaryotic antibiotic G418, closely related to the bacterial antibiotic neomycin. The neomycin resistance gene specifies an enzyme that inactivates both neomycin and G418. Eukaryotic cells grown in G418 will ordinarily die, but cells that have taken up the neomycin resistance gene will survive. Those cells, it turns out, express both the antibiotic resistance and the properties specified by the other, cotransfected DNA.

The two other methods—liposome fusion and electroporation—circumvent the need for transfected DNA to go through the cell's endocytosis machinery. Liposomes that contain DNA can fuse with the plasma membrane and release DNA directly into the cytosol. Electroporation allows DNA to enter the cytosol by using transient electrical currents to poke transient holes in the plasma membrane. In each case, some cells allow the exogenous DNA to move successfully through the cytosol into the nucleus, where it can be expressed, along with a selectable gene such as the gene that encodes resistance to G418.

The ability to alter a cell's characteristics by transferring specific bits of DNA underscores the conclusions reached in the much earlier studies of human metabolic disease, of *Neurosopora* mutants, of pneumococcus transformation, and of viral infection: that genes are made of DNA and that each gene specifies the sequence of a single polypeptide chain.

pass through the filters they used to remove bacteria. D'Herelle and Twort both found that these filterable agents could reproduce themselves within their bacterial hosts and therefore seemed to be alive.

Bacteriophages are not actually alive, however. They are not organisms but **viruses,** molecular assemblies capable of reproducing themselves only as intracellular parasites within the cells of host organisms. Viruses are not cells, they cannot reproduce outside of a host cell, and they do not themselves obtain energy from their environments. A single virus can infect a cell and subvert the cell's molecular machinery to make more viruses.

In the 1940s, Max Delbruck and Salvador Luria, both European refugees working in the United States, began to study the genetics of the bacteriophages that infect the common intestinal bacterium, *Escherichia coli (E. coli).* Because bacteriophages reproduce so rapidly, they were, Delbruck and Luria decided, the ideal material for genetic studies. A bacterium that has been infected with a single bacteriophage breaks open after about 20 minutes and releases some 200 new phage particles. Each of these new phages can then infect another bacterium. If the bacteria are growing on a solid culture medium, the phages create a clear spot (a plaque) on the otherwise uniform bacterial "lawn," as shown in Figure 12-10. As the phages multiply, their descendants may eventually consume almost the entire lawn.

Bacterial Viruses Provided More Evidence That Genes Are Made of DNA

Delbruck, Luria, and the other researchers whom they attracted to work on bacteriophages catalogued many in-

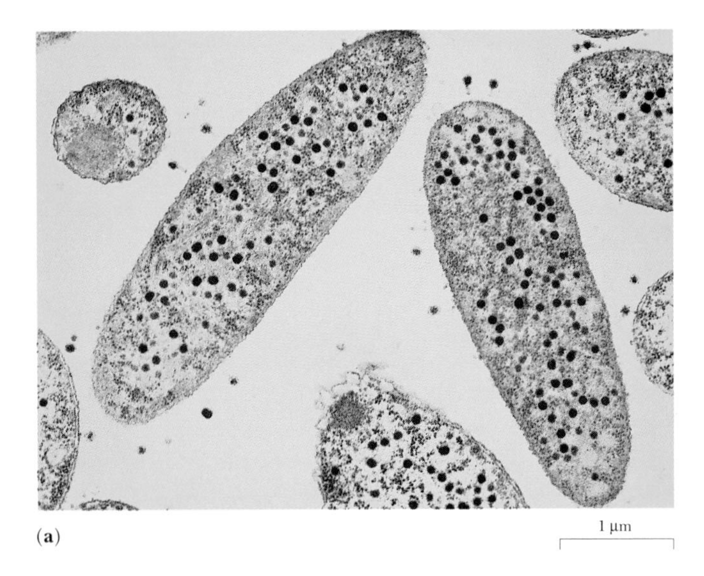

(a)

(b)

Figure 12-10 Bacteriophages infect and destroy bacteria, releasing more bacteriophages. The characteristics of the plaques formed by the destroyed bacteria depend on the genes of the bacteriophage. **(a)** TEM of an *E. coli* bacterium infected by T4 bacteriophages; **(b)** clear plaques formed by bacteriophages in a "lawn" of *E. coli.* *(a, M. Maeder/Biozentrum, University of Basel/Science Photo Library/Photo Researchers; b, R. Humbert/BPS)*

herited variants. They produced mutant phages with distinct phenotypes. For example, some phages produced clear plaques (as shown in Figure 12-10b), while others produced cloudy (or turbid) plaques, and some phages infected one strain of *E. coli* but not another. Such variant phenotypes of each mutant are inherited: Phage particles isolated from turbid plaques themselves produce turbid plaques. Bacteriophages, then, like organisms, have genes. Alleles of bacteriophage genes specify alternative phenotypes such as plaque morphology.

Bacteriophages consist of only two types of macromolecules—proteins and nucleic acid, with complexes of proteins surrounding a molecule of nucleic acid, usually DNA. Their genes must therefore be either protein or DNA.

The nature of phage genes was settled by an experiment performed in 1952 by Alfred Hershey and Martha Chase as part of their long-term studies of bacteriophage genetics. Although Avery had shown that DNA codes the phenotype of pneumococci, he could not show that this transforming factor is actually inherited. Hershey and Chase therefore conducted their experiment specifically to discover if DNA is also the hereditary material in bacteriophages.

Hershey and Chase were among the first geneticists to use radioisotopes in their experiments. The availability of radioisotopes had increased greatly after World War II, partly as a result of the work done to develop the atomic bomb. As diagrammed in Figure 12-11, Hershey and Chase grew bacteria in ^{32}P-labeled phosphate or in ^{35}S-labeled sulfate. Recall that proteins contain sulfur but no phosphorus and that nucleic acids have phosphorus but no sulfur. Bacteria grown in ^{35}S-labeled sulfate incorporate ^{35}S into the cysteine and methionine of their proteins, while bacteria grown in ^{32}P-labeled phosphate incorporate ^{32}P into the phosphate backbones of their DNA and RNA.

Hershey and Chase then infected radioactively labeled bacteria with T2 bacteriophages. The descendants of the infecting phages became radioactively labeled. Phages grown in ^{32}P-labeled bacteria acquired ^{32}P-labeled DNA; phages grown in ^{35}S-labeled bacteria acquired ^{35}S-labeled proteins.

Hershey and Chase then used the radioactive phage to infect new, nonradioactive bacteria. They waited just long enough for the infection to begin and then interrupted the process with a kitchen blender. The blender knocked off the empty phage particles outside the bacteria and separated these particles from the bacteria. They then examined the infected bacteria to see what part of the phage had entered—the ^{32}P-labeled DNA or the ^{35}S-labeled protein.

As illustrated in Figure 12-11, Hershey and Chase found that the labeled DNA entered the infected cell, while the labeled protein was removed from the cell's surface with the blender. They also showed that some new phages—the phage offspring produced within the bacteria and released when the cells lysed—contained ^{32}P, but never ^{35}S. They therefore concluded that the genes of the bacteriophage are made of DNA, not protein.

The phage group had convinced many geneticists that the genes of bacteriophages must be similar to the genes of organisms. The Hershey-Chase experiment convinced a large group of influential biologists in a way that the Avery paper, 8 years earlier, had not, that DNA is the

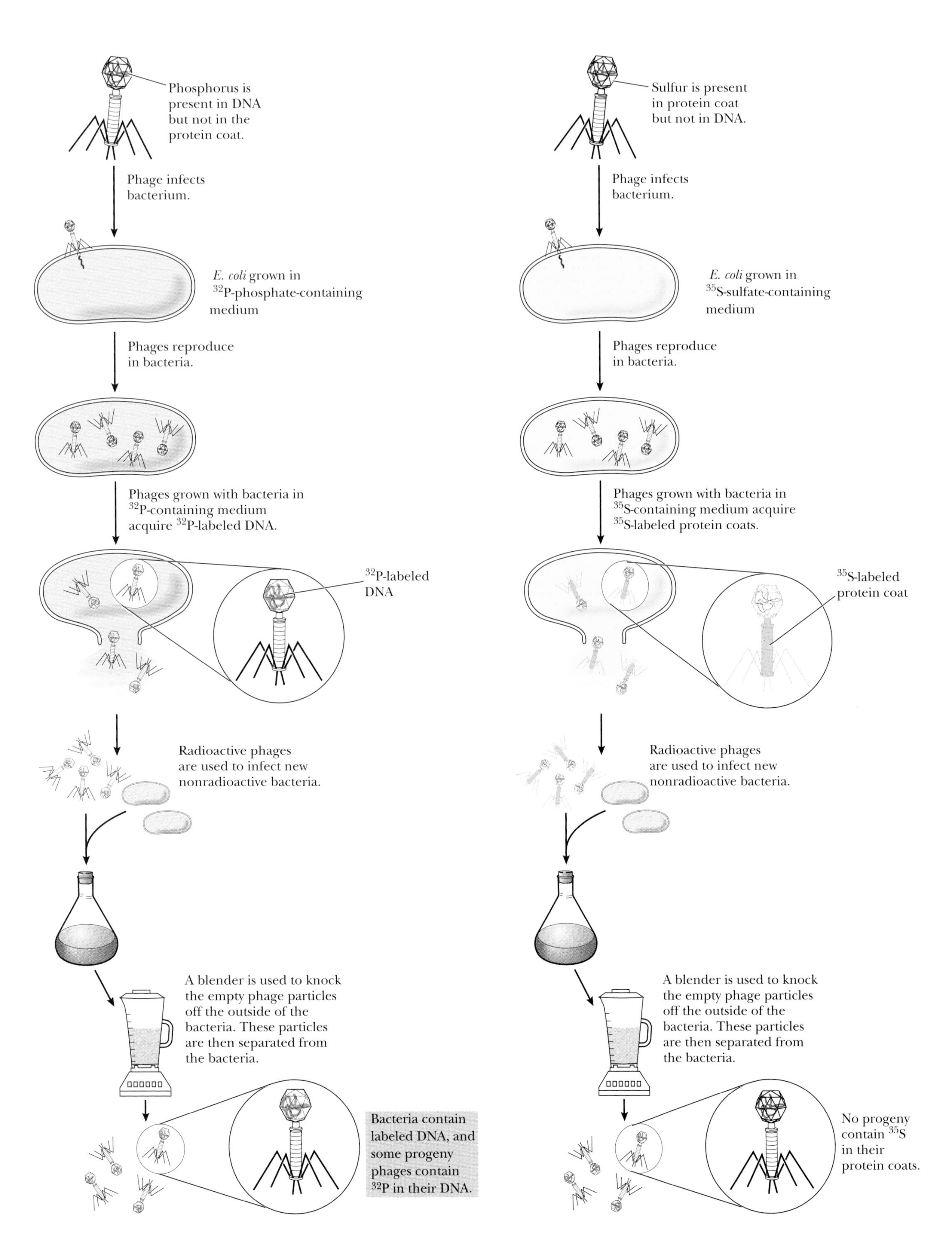

Figure 12-11 The Hershey-Chase experiment.

repository of inherited genetic information. After 1952, nearly everyone agreed that DNA is the genetic material.

Within a few years after scientists accepted the centrality of DNA, researchers came to understand much about its structure and replication. In the next two decades, they learned about how the information stored in DNA is expressed in the form of proteins and how cells tightly regulate both the replication and the expression of specific genes. Now, researchers have already determined the arrangement and structure of all the genes present in the **genome**—the collection of all the DNA—in several bacterial species, and they are within sight of learning the structure of the entire human genome.

■ SUMMARY

Many genes contain the information for the structure and expression of enzymes. When an organism cannot make a functional enzyme, it cannot carry out the biochemical conversions catalyzed by the enzyme. The result is often the failure to produce a needed compound and the buildup of some intermediate in the biochemical pathway leading to that compound. Patients with alkaptonuria, for example, lack the capacity to process tyrosine, and they accumulate a compound that turns the urine black.

By studying similar interruptions in biochemical pathways in the bread mold *Neurospora,* Beadle and Tatum reached the conclusion that each gene determines the structure of a single enzyme ("one gene—one enzyme"). Many genes do not determine the structure of enzymes but of other proteins, such as hemoglobin. A better statement of gene function, then, is "one gene—one polypeptide."

Different alleles of a gene produce different versions of the same protein. An alteration in the gene coding for the β-globin polypeptide leads to hemoglobins with different structures and properties. Sickle cell disease results from the homozygous occurrence of an allele that alters the structure of β-globin at a single amino acid residue.

The conclusion that genes are made of DNA came from work on bacteria and bacterial viruses. The ability of the pneumonia bacteria to cause disease can be transferred to a nonvirulent strain by DNA from the virulent strain. Bacteriophages also transmit genetic instructions to their offspring through DNA.

■ KEY TERMS

auxotroph
bacteriophage
genome
hemoglobin
inborn error of metabolism
sickle cell disease
transformation
virus

■ QUESTIONS FOR REVIEW AND UNDERSTANDING

1. What is an inborn error of metabolism?
2. Define and contrast the following terms:
   (a) gene vs. allele
   (b) auxotroph vs. autotroph
   (c) bacteriophage vs. virus
   (d) plaque vs. lawn
3. Why do you suppose that many found it difficult to accept DNA as the genetic material?
4. Considering how many metabolic pathways are required for survival, why is it that we see relatively few inborn errors of metabolism in the population at large?
5. Explain why the study of mutants is so useful in our knowledge of biological processes.
6. Discuss the experimental and historical context of each of the following phrases:
   (a) "one gene—one enzyme"
   (b) "one gene—one protein"
   (c) "one gene—one polypeptide"
7. Defend the following statement: "Diploid organisms have an advantage over haploid organisms."
8. Why are organisms that can undergo both sexual and asexual reproduction so useful in the study of the genetics of biochemical processes?
9. Why were bacteriophages chosen to be an experimental organism in determining the chemical nature of genetic material?
10. How did Beadle and Tatum determine that they had seven unique strains of *Neurospora,* which all required arginine?

11. You have isolated a mutant which you name *arg*-8. Upon crossing this new mutant with wild type you find that none of the eight ascospores can grow on minimal media. How do you interpret these results?
12. What properties would you look for in selecting an organism for genetic analysis?
13. Discuss how genotype is related to phenotype.
14. If Avery had grown S strain bacteria in ^{32}P and then mixed those labeled bacteria with R strain bacteria, what would you expect to happen to some of the radioactive label? What would the phenotype be of all bacteria that contained the radioactivity?
15. Using the tools of biochemistry that you have learned (see Chapter 5), how do you suppose Avery and his co-workers isolated the components of the S strain bacteria? What were those components?
16. What methods could you use to determine whether an individual is a carrier for sickle cell disease?
17. Summarize the experimental evidence that demonstrated that DNA is the genetic material.

■ SUGGESTED READING

GRIFFITHS, A. J. F., J. H. Miller, D. T. Suzuki, R. C. Lewontin, and W. M. Gelbart, *An Introduction to Genetic Analysis,* 5th edition, W. H. Freeman and Company, New York, 1993. Chapter 12 covers much of the material in this chapter.

C H A P T E R 13

The Structure, Replication, and Repair of DNA

KEY CONCEPTS

- DNAs from different organisms may have different proportions of nucleotides, but the amount of adenine always equals the amount of thymine and the amount of guanine always equals the amount of cytosine.
- DNA consists of two polynucleotide strands, which form a double helix.
- The two strands of DNA contain complementary versions of the same information. During DNA replication, each strand directs the synthesis of another complementary strand.
- DNA synthesis uses energy obtained by splitting high-energy phosphate bonds in nucleotide triphosphates.
- Enzyme systems in cells correct most but not all mistakes in DNA.
- Researchers can use the reactions of DNA synthesis to determine the sequence of nucleotides in DNA.

OUTLINE

ESSAYS

The experiments described in Chapter 12 established that DNA is the genetic material. The next questions were (1) what is the structure of DNA? and (2) how does it work? In this chapter, we discuss the three-dimensional structure of DNA and how this structure helps explain the replication of DNA. We also discuss the mechanisms that have evolved to minimize errors in DNA replication.

WHAT IS THE CHEMICAL STRUCTURE OF DNA?

The discovery of the structure of DNA is one of the most influential advances in the history of biology. Knowing the structure of DNA has allowed molecular biologists not only to explain further how genes work, but also to change and manipulate genes almost at will. In addition, the discovery of DNA structure is perhaps the best-documented story in the history of science because many of the participants have written extensively and passionately about both the science and the scientists.

What Do All DNA Molecules Have in Common?

Studies of DNA depended on the ability to isolate DNA from several different sources. The Swiss biochemist Friedrich Miescher, who discovered DNA in 1869, obtained it from the pus (white blood cells) in the bandages of soldiers wounded in the Austro-Hungarian War, as well as from fish sperm. Later researchers isolated DNA from other sources and found that whatever its source, DNA has the same chemical properties. The chemical structure of DNA was discovered in the 1920s by the biochemist P. A. Levene, a Russian immigrant to the United States. Levene showed that, like other macromolecules, DNA consists of building blocks linked together in long chains (as discussed in Chapter 3). In the case of DNA (and RNA), these building blocks are nucleotides, and they are linked together by phosphodiester bonds, as shown in Figure 13-1.

The phosphodiester bonds in DNA (and in RNA) connect the sugar components with phosphate groups. Each polynucleotide has a "backbone" consisting of sugar—phosphate—sugar—phosphate—etc. From this backbone hang the nitrogenous (nitrogen-containing) bases,

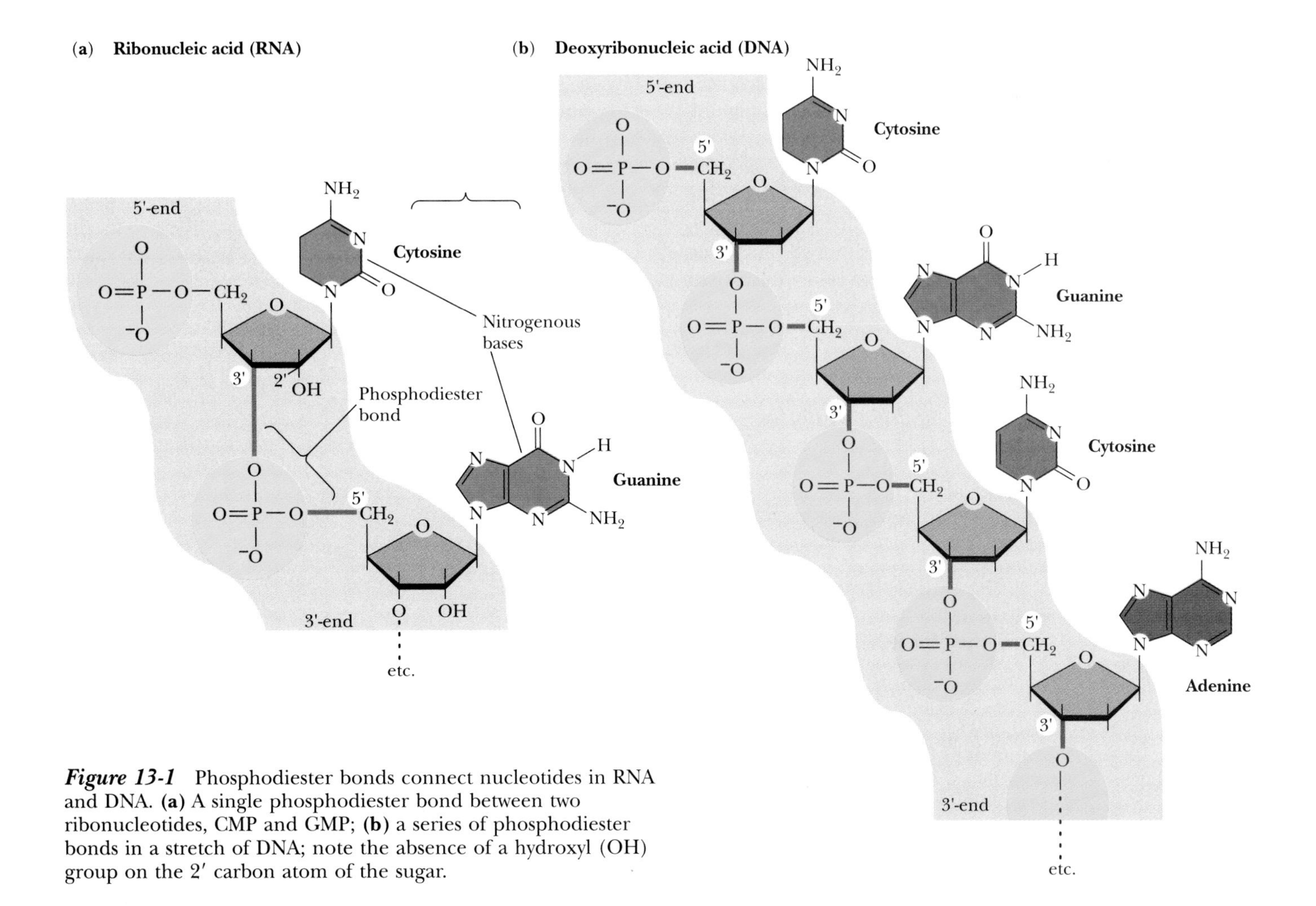

Figure 13-1 Phosphodiester bonds connect nucleotides in RNA and DNA. **(a)** A single phosphodiester bond between two ribonucleotides, CMP and GMP; **(b)** a series of phosphodiester bonds in a stretch of DNA; note the absence of a hydroxyl (OH) group on the 2′ carbon atom of the sugar.

as illustrated in Figure 13-1. Recall that the formation of links between building blocks in a macromolecule eliminates the equivalent of one water molecule from each link, so that each unit in a chain of nucleotides, sugars, or amino acids is called a residue.

Each polynucleotide chain has a direction. That is, we can distinguish between its two ends. In the backbone, carbon atom 3 (called the 3′ carbon to distinguish it from the carbon atoms in the nitrogenous base) of the deoxyribose molecule of one nucleotide residue is attached via a phosphate group to carbon 5 (the 5′carbon) of the sugar in the next residue. The bases are attached to carbon 1 (the 1′ carbon) of each sugar. We speak of the two ends of a polynucleotide as the **5′-end** (the nucleotide residue whose 5′carbon is not attached to another residue but to a phosphate or a hydroxyl group) and the **3′-end** (the nucleotide residue whose 3′ carbon is not attached to another residue but to a phosphate or a hydroxyl group).

DNA contains four types of nucleotide residues, which differ only in the nitrogenous bases attached to the deoxyribose. Two of these bases are *pyrimidines;* that is, each consists of a six-membered ring, with four carbon and two nitrogen atoms. The other two bases are *purines;* they contain a nine-membered double ring, with five carbon and four nitrogen atoms. In DNA the purine bases are adenine and guanine, and the pyrimidine bases are cytosine and thymine. (Recall from Chapter 4 that RNA contains three of these four bases, with uracil replacing thymine as the fourth base.)

In DNA and RNA, each nucleotide residue consists of a nitrogenous base + sugar + phosphate. Table 13-1 gives the full names of the nucleotides that contain each base, but we generally refer to each residue by a single letter abbreviation: A, G, C, U, or T. In RNA, U is substituted for T. This convention not only saves space; it also emphasizes that the nucleotides are indeed letters in the DNA or RNA alphabet.

Like other macromolecules, DNA can be broken down (hydrolyzed) into its component parts, accompanied by the addition of the equivalent of a water molecule to each nucleotide residue. DNA is more resistant to hydrolysis than most macromolecules. This stability serves it well as it carries information between generations. Acid solutions, however, break DNA into small molecules, as do certain enzymes. A **nuclease** is any enzyme that catalyzes the hydrolysis of a nucleic acid; an enzyme that breaks down DNA is called a DNase, and an enzyme that breaks down RNA is called an RNase.

Are All DNA Molecules the Same?

After hydrolyzing DNA into its component nucleotides, Levene identified its nitrogenous bases as A, G, C, and T. He was unable, however, to determine the relative amounts of each base accurately, so he incorrectly concluded that DNA contained almost equal amounts of each base. Levene hypothesized that the basic structure of DNA was a repeating "tetranucleotide" (a sequence of four nucleotides).

In the 1940s, another immigrant biochemist working in the U.S., Erwin Chargaff, developed a new method for analyzing the base composition of DNA. Chargaff found that the DNAs of different species have different base compositions; for example, in cow DNA 29% of the bases are A and 21% are G, while in crab DNA 47% of the bases are A and only 3% are G. These differences disproved Levene's hypothesis that all DNA contained equal numbers of all four bases.

Table 13-1 ***The Nucleotides of RNA and DNA***

Bases	RNA Nucleotide	DNA Deoxynucleotide
Purines		
Adenine (A)	Adenosine monophosphate (AMP)	Deoxyadenosine monophosphate (dAMP)
Guanine(G)	Guanosine monophosphate (GMP)	Deoxyguanosine monophosphate (dGMP)
Pyrimidines		
Cytosine (C)	Cytidine monophosphate (CMP)	Deoxycytidine monophosphate (dCMP)
Uracil (U)	Uridine monophosphate (UMP)	—
Thymine (T)	—	Thymidine monophosphate (dTMP)

Chargaff also made a curious observation, summarized in what came to be called "Chargaff's rules." Although the proportions of the bases can vary widely from species to species, the amount of A is always equal to the amount of T, the amount of G is always equal to the amount of C, and the ratio of A plus T to G plus C is constant within a species. Neither Chargaff nor anyone else at the time understood the significance of this finding, but within a few years it contributed to the working out of the relationship between the structure and function of DNA.

Why Did Many Biologists Think That DNA Was Not Important?

Avery's 1944 paper showing that DNA is responsible for the transformation of pneumonia-causing bacteria should have convinced other scientists that DNA is the genetic material. (See Chapter 12, p. 325.) Still, most biologists continued to regard DNA as merely a structural component of the chromosomes.

An important reason for their skepticism was that they accepted Levene's view of DNA as a tetranucleotide. If DNA were really composed of the same sequence repeated over and over, it would not be able to carry useful information. Before Chargaff's work, all DNA seemed to be identical. And if DNA were really the genetic material, people thought, the DNA of bacteria should somehow differ from the DNA of butterflies or bananas.

Chargaff's measurements showed that such species differences do exist—that the DNAs of butterflies and bacteria are not identical—and destroyed the tetranucleotide hypothesis. Biologists therefore were no longer as resistant to the idea that DNA might carry information. So by the time Hershey and Chase published their famous experiment in 1952, people were far more ready than they were in 1944 to accept DNA as the genetic material. (See Chapter 12, p. 328.)

HOW DID RESEARCHERS DETERMINE THE THREE-DIMENSIONAL STRUCTURE OF DNA?

In the last few decades scientists have determined the atomic structures of several hundred macromolecules. Much of their success has depended on the steady improvement in technique, including the use of high-speed computers, new methods of calculating results, and more powerful x-ray cameras. In the early 1950s, however, the major advances in understanding the structure of macromolecules came from realizing the importance of helices. Recall (from Chapter 3) that a helix is a spiral arrangement of component units along a chain, with each unit advanced and rotated from the previous unit—as are the stairs in a spiral staircase. (See p. 78, Figure 3-20.)

By correctly deducing the arrangement of atoms in the α helix of proteins, Linus Pauling changed the way scientists thought about macromolecules. In particular, Pauling introduced the idea that the atoms within the building blocks of macromolecules have more or less the same arrangements as they have when in isolated small molecules. This insight permitted Pauling and his followers to build scale models of macromolecules, with the atoms arranged according to the structures deduced for their component parts. This method led directly to the most important discovery of 20th-century biology—the structure of DNA.

DNA Is a Helix

Much of the DNA story took place in the early 1950s in Cambridge, England, and 50 miles away in London. The principals in the Cambridge part of the DNA story were James Watson, an American postdoctoral fellow in his early 20s, and Francis Crick, a physicist in his mid-30s who had recently turned to the study of biological structures. Both worked at the Cavendish Laboratories in Cambridge, in the same building as Max Perutz and John Kendrew, pioneers in the application of x-ray diffraction to macromolecules. Watson and Crick were convinced that DNA was important and that its structure would help explain how it works. The problem was how to determine its structure. In London, two independent laboratories at King's College were working on the structure of DNA—that of Maurice Wilkins, a physicist turned biochemist, and that of Rosalind Franklin, a physical chemist whose previous work had concerned x-ray diffraction studies of carbon fibers in coal.

The only possible method to solve the three-dimensional structure of DNA was x-ray diffraction, but there were two difficulties: (1) X-ray diffraction required crystals of DNA, and these were hard to make; and (2) even the best crystals could not give enough information to deduce a structure. Because DNA is a long, thin molecule, it was impossible to obtain true DNA crystals. DNA crystals became available only much later, in the 1960s, when researchers produced crystals from short pieces of DNA synthesized in the laboratory. In the initial studies of DNA structure, however, the crystallographers had to settle for studying fibers of DNA, in which hundreds of millions of DNA molecules lie parallel to one another, such as the one shown in Figure 13-2a. In the early 1950s, the best DNA fibers were those being studied by Rosalind Franklin.

Before working on DNA, Crick had worked on understanding the way that a helix would interact with x-rays. That work, already published, had shown that a helix should generate a cross-shaped pattern. The obvious cross in Franklin's x-ray films immediately suggested that DNA must be a helix. Crick's previous work allowed both him and Franklin to deduce the dimensions of the helix from Franklin's unpublished photograph, as illustrated in Figure 13-2b. It is 2 nm in diameter, and each full 360° turn

(a)

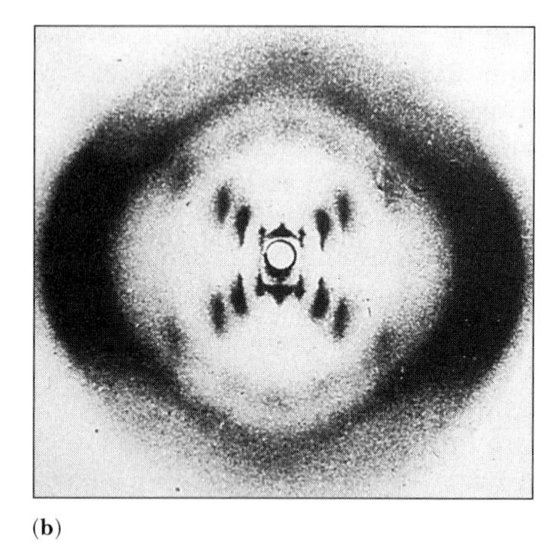

(b)

Figure 13-2 **(a)** DNA fibers consist of many parallel DNA molecules. **(b)** While not true crystals, they produce characteristic diffraction patterns, such as this one, originally obtained by Rosalind Franklin at King's College in London. *(a, C. Case/Visuals Unlimited; b, CSHL Archives/Peter Arnold)*

of the spiral is 3.4 nm long. Within this 3.4 nm, the data suggested, were ten units of 0.34 nm each. Even knowing this, Watson and Crick still had to figure out how the units of the helix—the nucleotide residues—were arranged.

Watson and Crick used Pauling's approach of model building, starting with the known structures of each nucleotide. They knew, for example, that the atoms of the purines and pyrimidines should all lie in the same plane. They had more trouble, however, in figuring out what to do with the sugars and phosphates.

The key to solving the structure came from a biological insight: The genetic material had to duplicate itself. Watson and Crick realized that a helix could fill this function if it consisted of two chains rather than one chain (or strand) of nucleotides. In such a double helix, the two strands could come apart, and each could direct the assembly of a second strand. This double structure is sometimes called a duplex.

The x-ray data told them that the helix had an unexpected property. Rotating the helix through 180 degrees (turning it upside down, as illustrated in Figure 13-3a) produces an equivalent helix. This could not possibly happen if the helix were single or triple, so Watson and Crick deduced that the helix was double, with the two chains running in antiparallel (opposite) directions. Finally, they realized that only a double helix would give the density of atoms consistent with the measured density of DNA.

The most troubling part of this otherwise dazzling story is the use made of Franklin's unpublished data without her knowledge. She did show her photographs to Watson when he visited London, but Watson and Crick obtained additional information about her work from Wilkins and from Franklin's official, but unpublished, annual report that Crick obtained from Perutz. When, 10 years later, Watson and Crick (Figure 13-3b) received a Nobel Prize for their work, they shared it with Wilkins. By that time, Franklin had died of cancer, at the age of 37.

The Double Helix Explains Chargaff's Rules

Watson and Crick still had to figure out how to build a model of a double helix starting with nucleotide building blocks. If DNA carried genetic information, it should be able to accommodate any sequence of nucleotides, just as each line of print on this page must be able to accommodate any sequence of letters.

In order to account for the x-ray data, the helix had to have a constant diameter. Watson and Crick initially thought that the bases must go on the outside and the consistently repeating units, the sugars and phosphates, on the inside. This arrangement would allow the helix to have a constant diameter, no matter what the sequence of bases. Such an arrangement, with the repeating components on the inside and the varying components on the outside, was the plan of the helix of proteins. In the α helix, the constant polypeptide backbone was on the inside and the variable amino acid side chains on the outside.

Watson and Crick soon realized, however, that they could also build a model of a double helix with the bases on the inside if they established a single rule: Each step in the helix must consist of two bases in the same plane—a purine (A or G) and a pyrimidine (T or C). By having a larger, two-ring purine always in the same "step" as a smaller one-ring pyrimidine, the helix could have a constant diameter. This insight allowed them to begin to explain the origin of Chargaff's rules: Every time G is found in one strand of the helix, C is found in the other; similarly A is always paired with T.

Watson and Crick showed that the arrangement of atoms in the bases is such that A is perfectly arranged to make two hydrogen bonds to T, while G can form three hydrogen bonds with C. Base pairing via hydrogen bonds

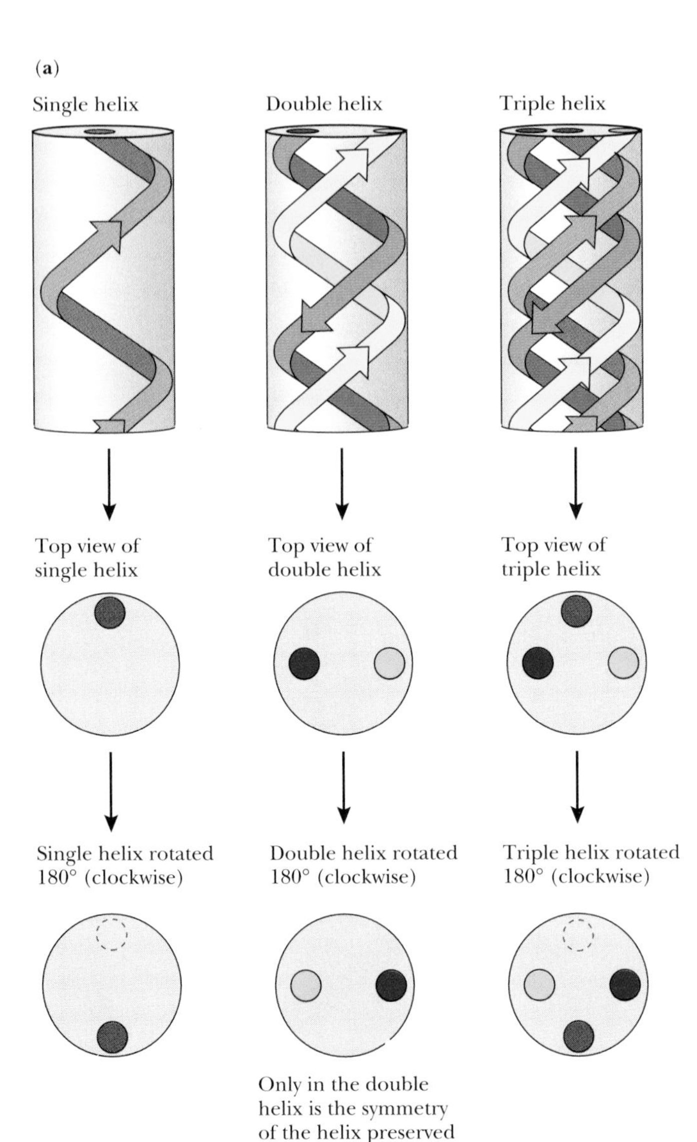

Figure 13-3 X-ray diffraction data indicated that the DNA helix could be rotated through 180° without changing its form, suggesting that it was a double helix with the two chains running in opposite directions. **(a)** The rotation of a double helix is consistent with the x-ray data, but the rotation of a single or triple helix is not. **(b)** Watson, Crick, and their model of DNA structure. *(b, CSHL/Peter Arnold)*

keeps the two strands of the helix together, as diagrammed in Figure 13-4.

Still more importantly, the pairing rule establishes the basis of DNA's ability to replicate itself. The purine A, for example, ordinarily forms coordinated hydrogen bonds only with the pyrimidine T, and not with G, C, or another A. Because a base on one strand can match with only one other base, the sequence of bases of each strand can direct the assembly of bases in the other strand. In their famous paper, Watson and Crick coyly commented on this property: "It has not escaped our notice that the specific pairing we have postulated immediately suggests a possible copying mechanism for the genetic material." Each strand has the capacity to direct the synthesis of the other, with exactly the same sequence, in the same way that a photographer can make a positive print from a negative or a negative from a positive print.

Before we look at the copying mechanism, let us look more closely at the structure of DNA. We may imagine DNA as a twisted ladder, as shown in Figure 13-5. Each rung of the ladder consists of a pair of bases (AT or GC). These rungs lie on top of one another. The bases, how-

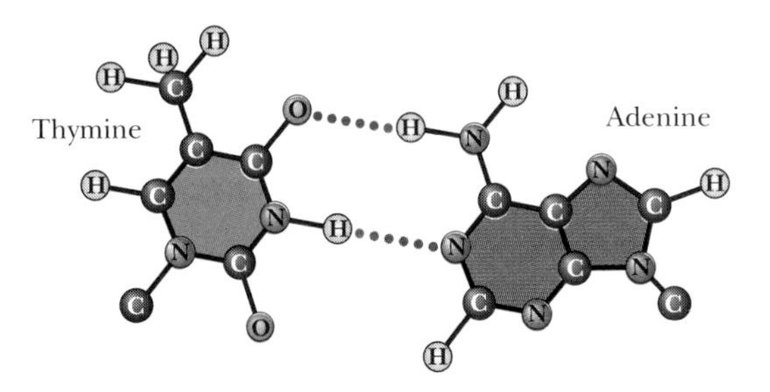

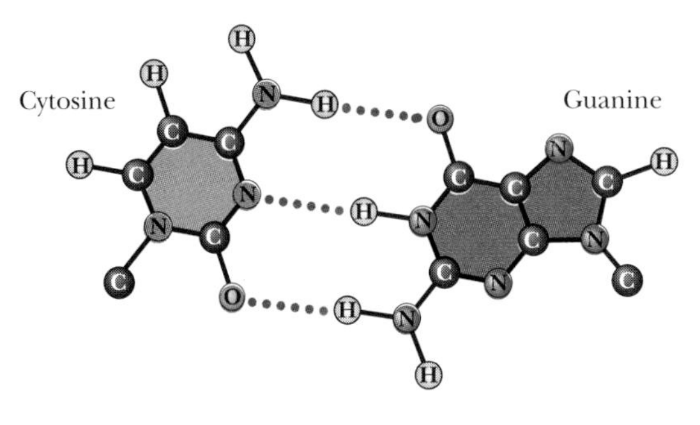

Figure 13-4 Hydrogen bonds between the bases of DNA. A and T can form two hydrogen bonds; G and C can form three.

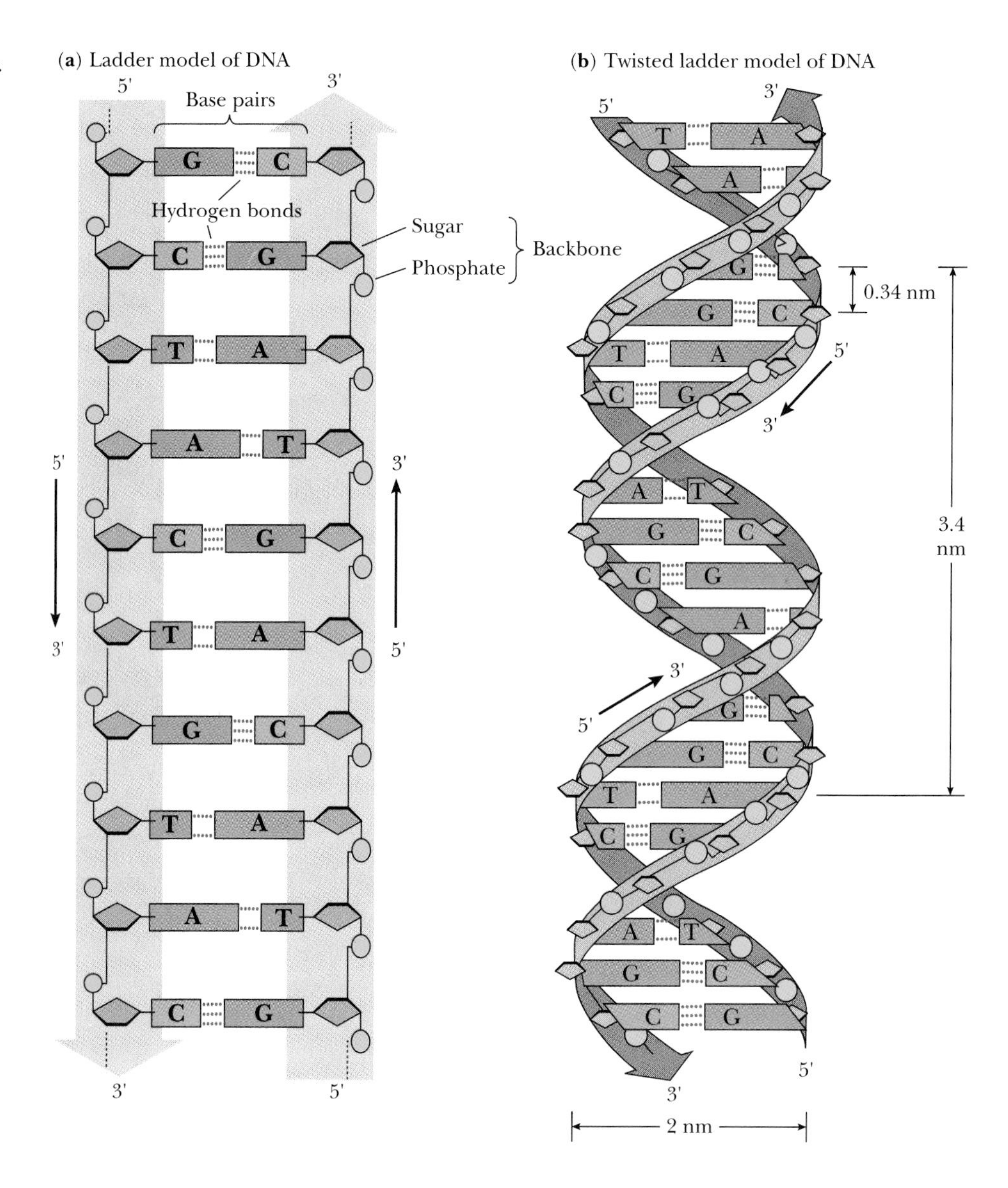

Figure 13-5 DNA is a twisted ladder. **(a)** An untwisted ladder with two antiparallel and complementary polynucleotide chains; **(b)** a twisted ladder.

ever, are more like flat dishes than the steps of a real ladder. Each twist of the ladder contains 10 rungs (or dishes), each of which is 0.34 nm apart. The uprights of the ladder are the sugars and phosphates of the two polynucleotide strands, running in opposite directions. If we draw an arrow from the 5′-end to the 3′-end of each strand, one arrow points up the helix, the other down.

A single molecule of DNA may consist of thousands or millions of nucleotide pairs, but the diameter of the ladder is always only 2 nm. As we said in Chapter 10, DNA is an "exquisitely thin molecule," which must be specially folded to fit into cells.

Subsequent work on the structure of DNA has convinced everyone that genes have a physical reality that can even be seen with a scanning tunneling microscope, as illustrated in Figure 13-6. Although DNA always obeys the basic rules of base pairing, its structure is not always the same as in the Watson-Crick model. DNA appears to be a dynamic molecule that most often has the form that Wat-

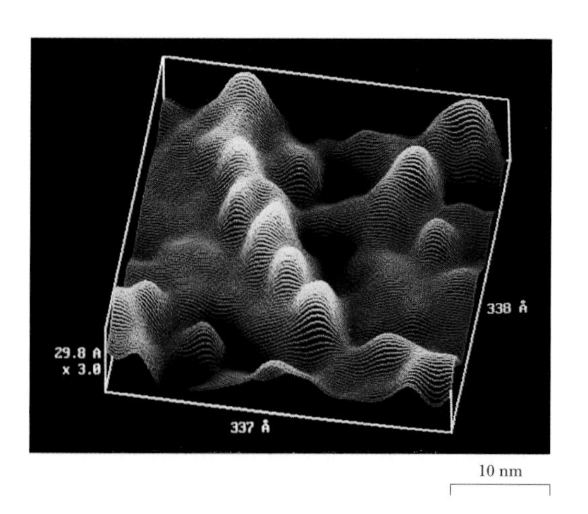

Figure 13-6 DNA visualized by a scanning tunneling microscope. The yellow ridges running down the left side of the photograph represent sequential turns of the double helix. *(Lawrence Livermore Laboratory/Photo Researchers)*

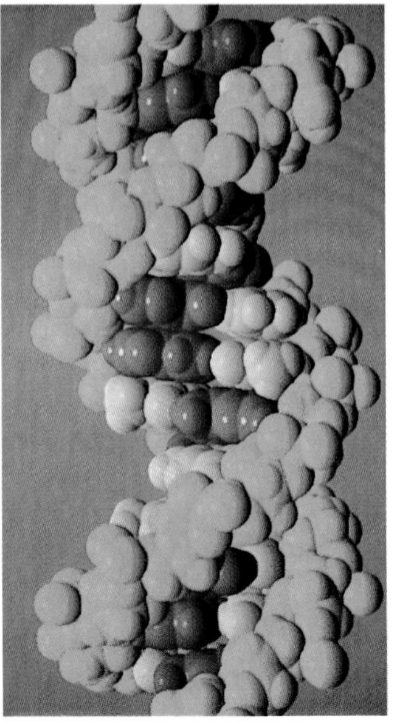

(a)

(b)

Figure 13-7 DNA may have different forms in addition to the most prevalent B-DNA form, of the Watson-Crick model. **(a)** A computer model of B-DNA, with the phosphodiester backbone in blue, the bases of one strand in yellow, and the bases of the other strand in pink; note major and minor grooves along the helix; proteins that bind to DNA usually interact within the major groove. **(b)** An alternative form of DNA, called Z-DNA; note that the helix is left-handed and that the grooves have all but disappeared. *(a, b, Richard Feldmann/Phototake NYC)*

son and Crick studied (called B-DNA), shown in Figure 13-7a, but is sometimes found in other forms, which include Z-DNA, illustrated by the computer representation of Figure 13-7b.

The Watson-Crick model of DNA has allowed scientists to answer both questions we posed at the beginning of this chapter—what is the structure of DNA and how does it work? The rules of base pairing in DNA mean that the sequence of one polynucleotide strand completely specifies the sequence of the second strand. The two strands are **complementary,** meaning that they fit together to make a unified whole—the double helix. This complementarity is the basis for understanding both how a chromosome replicates itself during cell division and how genes actually work.

The sequence of bases in DNA, like the sequence of letters on this page, can contain coded information. In Chapter 14, we see how cells make use of such information. Here we examine how dividing cells copy it. The basis for this DNA replication is that the sequence of residues in one chain fixes the sequence of residues in the other. The two strands of the DNA double helix contain equivalent copies of the same information.

During DNA Replication, Each Strand of DNA Remains Intact as It Directs the Synthesis of a Complementary Strand

The Watson-Crick model of DNA structure suggests a model of DNA replication. Picture a molecule of DNA coming apart into two complementary single strands. If the needed materials and the proper machinery were available, each strand could then direct the synthesis of a new complementary strand. The result would be the formation of two daughter DNA molecules, one from each parent strand, each a replica of the parent DNA.

To determine whether this is what really happens, two young scientists at the California Institute of Technology—Matthew Meselson and Frank Stahl—devised a particularly elegant experiment in the late 1950s. Meselson and Stahl realized that if the Watson-Crick model of DNA replication were correct, each daughter DNA molecule would contain one old strand and one new strand. According to this model, DNA replication would be **semiconservative;** that is, half of each parent molecule would be present in each daughter molecule, as illustrated in Figure 13-8a.

One can imagine other ways in which DNA could replicate, however. For example, a cell might contain a molecular copying machine that would produce complete copies of the parent DNA molecule, as illustrated in Figure 13-8b. This mechanism would be **conservative;** that is, the parent molecules would remain intact, and both strands of the descendant molecules would be newly assembled. In another possible model of DNA replication, the **dispersive** model illustrated in Figure 13-8c, each daughter molecule would contain interspersed pieces of old (parental) DNA and newly polymerized DNA.

Meselson and Stahl's experiment was "elegant" because it distinguished among these models in a direct and unambiguous manner. (Scientists often use the word "elegant" in appreciation of an experiment's achievement and restraint—getting a job done without excessive diversions.) Meselson and Stahl earned this accolade because their experiment distinguished among the possibilities by inventing a new method just suited to the problem—a method that could distinguish between the parental DNA and newly made DNA.

Like Hershey and Chase, Meselson and Stahl took advantage of cells to incorporate isotopically labeled compounds into their DNA. Instead of using a radioactive isotope, however, Meselson and Stahl used a heavy, nonradioactive isotope, ^{15}N. As we see later in this chapter, using the heavy isotope allowed them to separate old and new DNA. They grew bacteria for many generations in ^{15}N-ammonia ("heavy" ammonia), so that the nitrogenous bases in the bacterial DNA (as well as other nitrogen-containing molecules) became "heavy"—that is, denser than molecules containing ^{14}N. They then used an ultracentrifuge to separate the heavy molecules from the light ones and

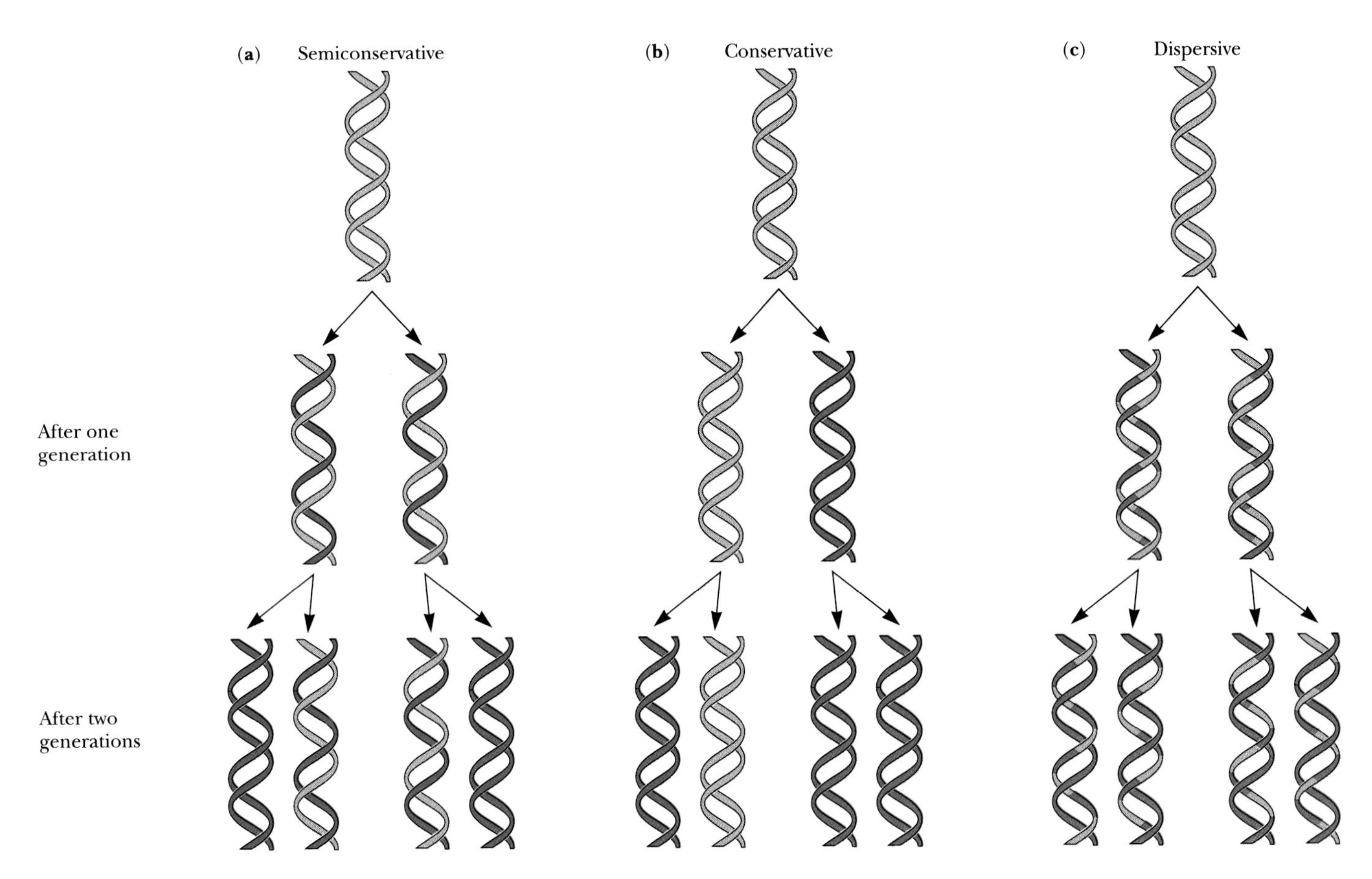

Figure 13-8 Alternative models of DNA replication. **(a)** In the semiconservative model, each strand of the double helix directs the synthesis of the second strand. **(b)** In the conservative model, each double helix directs the synthesis of a second double helix. **(c)** In the dispersive model, parts of the parental strands are incorporated into the newly made strands. The blue strands represent the newly made DNA.

examined how the DNA density changed as the bacteria divided.

Meselson and Stahl's technique to separate the heavy and light molecules requires some explanation. In an ultracentrifuge, large molecules whose density is higher than that of the solution in the tube move more quickly toward the bottom of the tube, to the part of the tube farthest from the axis of rotation. Molecules that are sufficiently small do not settle because of their rapid thermal motion. For molecules and ions of intermediate size, settling in the centrifugal field is only partly offset by thermal motion, and the centrifugal field generates a concentration gradient.

When placed in a strong centrifugal field (say 100,000 times gravity), cesium chloride forms a concentration gradient in which the solution's varying density spans the densities of ^{14}N-DNA and ^{15}N-DNA. Now think about what happens as the DNA moves through the tube. As it moves, the solution it displaces becomes denser and denser. At the point where the density of the cesium chloride solution equals the density of DNA, the DNA cannot move any farther and forms a band there. Another way of saying this is that when a DNA molecule has the same density as the surrounding solution, it does not settle any farther in the tube but concentrates at that density (Figure 13-9a). Because ^{15}N-DNA is denser than ^{14}N-DNA, the band for ^{15}N-DNA is lower in the tube than that of ^{14}N-DNA.

Meselson and Stahl adjusted the gradient (by varying the total concentration of the cesium chloride and the speed of the centrifuge) so that heavy and light DNAs, which differ in density only by about 1%, stop far enough away from each other that they can be separately visualized. The optical system of the analytical ultracentrifuge allowed Meselson and Stahl to measure the amount of DNA at each place in the gradient, as illustrated in Figure 13-9b.

Meselson and Stahl compared the density of DNA at different times after bacteria were transferred from ^{15}N-

Figure 13-9 The Meselson-Stahl experiment gave results consistent with the semiconservative model of DNA replication, but not with the conservative or dispersive models. ▶

(**a**)
Tubes of DNA and CsCl before centrifugation

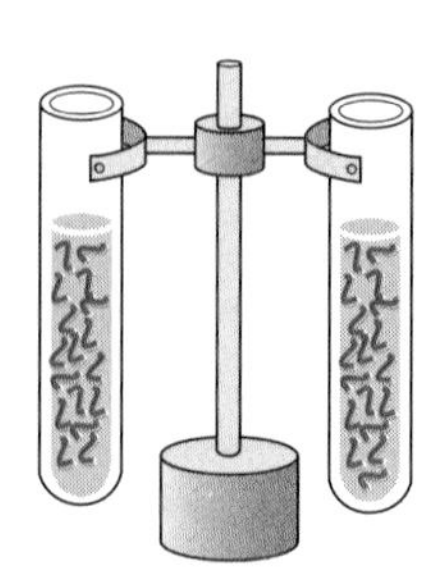

As the DNA moves down the tube, the solution it displaces becomes denser and denser.

When a DNA molecule has the same density as the surrounding solution, it will concentrate at that density.

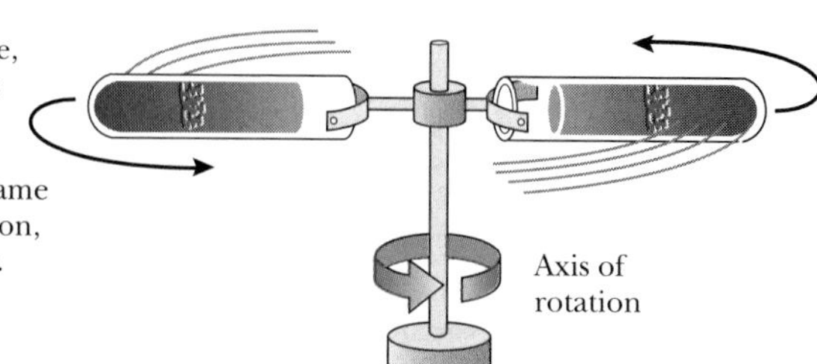

(**b**) Grow bacteria in ^{15}N (heavy) medium.

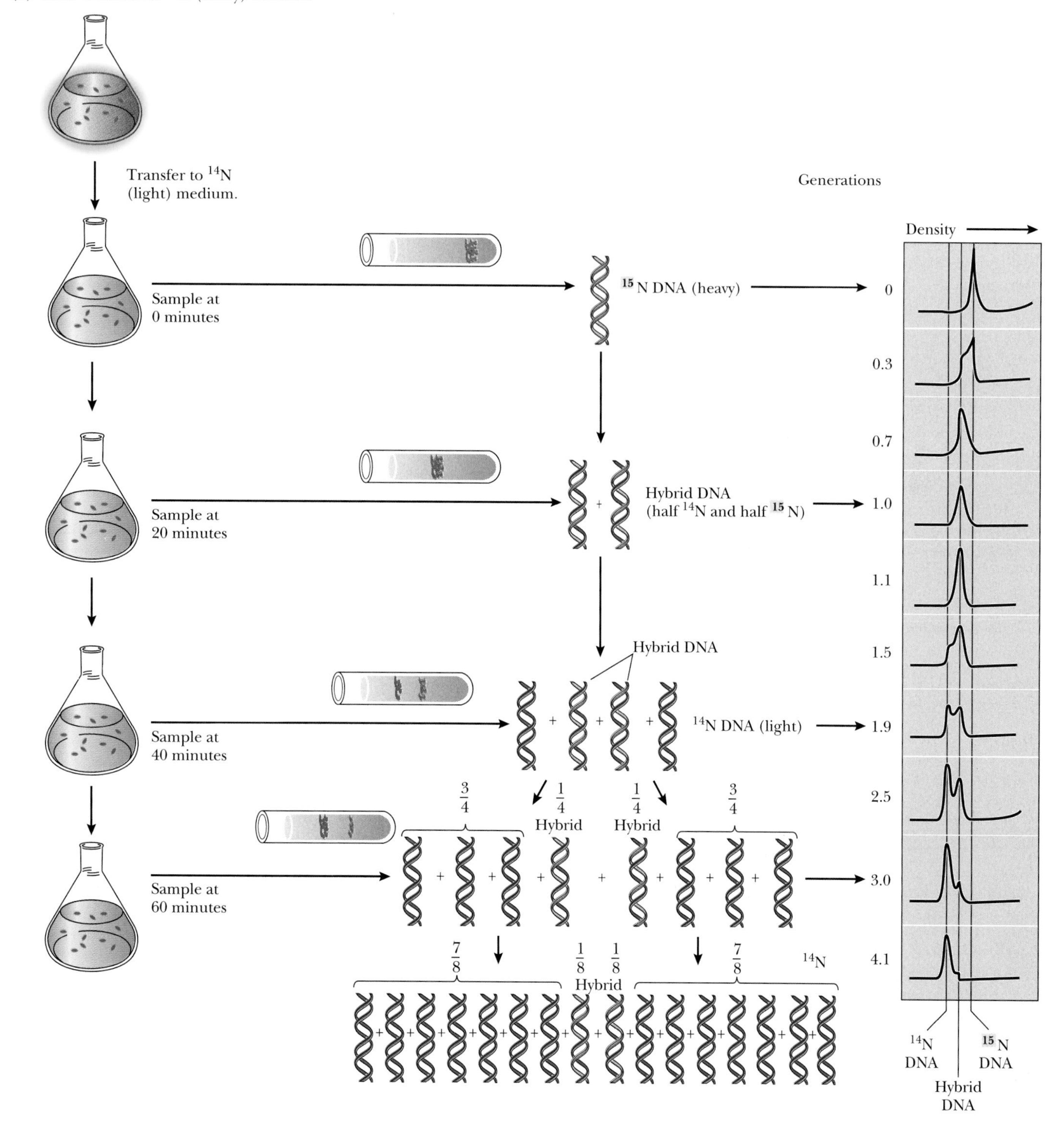

ammonia to ^{14}N-ammonia. After one round of DNA replication—that is, after one bacterial generation— the DNA had a density halfway between ^{15}N-DNA and ^{14}N-DNA—exactly what they expected from a hybrid molecule consisting of one old (^{15}N) and one new (^{14}N) strand. After two generations, half the molecules were all new (all ^{14}N) and half were of the intermediate density, again, exactly what the semiconservative model predicts.

Meselson and Stahl's experiment gave results consistent with the semiconservative model of DNA replication and ruled out the conservative model. The dispersive model could also have generated similar data, but only with the further assumption that half of each parent molecule is present in each daughter molecule. Ultimately this model was also disproved by examining the densities of the separated DNA strands. After one generation, the semiconservative model predicts that each strand will be either all ^{14}N or all ^{15}N, while the dispersive model predicts that each strand will contain a mixture of ^{14}N and ^{15}N. When this experiment was done, each strand was either all heavy or all light, again supporting the semiconservative model.

The complementarity of the two strands of DNA is not merely a curiosity. It is the basis for the replication of genetic information.

■ HOW DOES ONE DNA STRAND DIRECT THE SYNTHESIS OF ANOTHER?

Our discussion has so far focused on how DNA passes information to daughter DNA molecules. But what is the mechanism? How do cells actually produce new DNA? As in the case of other biochemical reactions, DNA replication depends on enzymes. The full process of DNA replication requires at least a dozen enzymes. Even now, more than 35 years after DNA synthesis was first accomplished outside of a cell in a laboratory experiment, biochemists still do not know all the details of DNA replication. The basic steps of the process, however, are well understood.

DNA Synthesis Uses Energy Obtained by Splitting High-Energy Phosphate Bonds in Nucleoside Triphosphates

The assembly of DNA involves putting together many small molecules (nucleotides) into a larger one, a process called **polymerization** [Greek, *polys* = many + *meros* = part]. The enzyme that strings together the nucleotides is called **DNA polymerase.** Recall (from Chapter 3) that the joining (condensation) of two nucleotide building blocks to form a phosphodiester bond removes the equivalent of a water molecule. Conversely, the breaking of DNA into nucleotides is an example of hydrolysis, the splitting of two residues with the addition of a water molecule.

The polymerization of nucleotides, however, is not just the reverse of hydrolysis. Hydrolysis is a spontaneous reaction with a negative free energy change, while polymerization has a positive free energy change and requires energy input to occur. Furthermore, in order to achieve DNA replication, polymerization of nucleotides in each strand must occur in fixed order, somehow guided by the order of bases in the complementary strand.

Figure 13-10 outlines the actual polymerization reaction as a G residue is added to a growing polynucleotide chain. Notice that in the illustration the nucleotide to be added enters the reaction as deoxyguanosine triphosphate

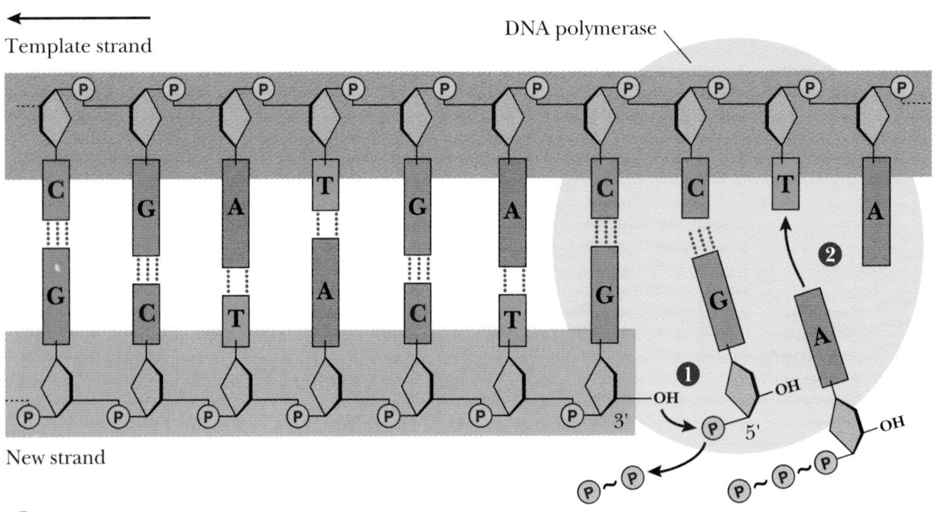

Figure 13-10 The formation of a phosphodiester bond. Here dGTP adds, opposite a C, to the 3′-end of a growing DNA strand.

(dGTP). The first phosphate group (on the 5′ carbon) of the dGTP reacts with the 3′-OH of the growing polynucleotide chain, splitting the bond between the first and the second phosphate groups of dGTP and releasing a **pyrophosphate** molecule, which consists of two phosphate groups linked together in a high-energy bond.

Recall from Chapter 3 that nucleoside triphosphates like dGTP contain two high-energy phosphate bonds, designated with a squiggle in Figure 13-10. So when one of these splits during polymerization, a second still remains. When the high-energy bond of pyrosphosphate is split, still more free energy is released. The splitting of an extra phosphate-phosphate bond is common to the biosynthesis of nucleic acids. It provides necessary input of free energy to the reactions and keeps the overall synthetic process exergonic. As discussed in Chapter 4, many cellular processes (including DNA synthesis) use enzymes that couple downhill (energetically favorable) reactions to uphill (energetically unfavorable) reactions, allowing the net reaction to occur spontaneously.

DNA Polymerase Catalyzes the Ordered Addition of Nucleotides to Each DNA Strand

The Watson-Crick model for DNA replication suggests how each strand of DNA serves as a **template,** or guide, for the assembly of a complementary sequence. Each base in the template strand serves as a guide for the insertion of the complementary base in the growing strand. Our understanding of just how the template works and how the new chains are assembled depends heavily on the work of Arthur Kornberg and his colleagues at Stanford University. By 1957, Kornberg had demonstrated that extracts of bacteria could catalyze DNA synthesis. This achievement allowed him and his colleagues to take the process apart: They first isolated DNA polymerase and studied the reactions it catalyzed, and they later studied the roles of a number of other enzymes that contribute to DNA replication.

Somewhat surprisingly, Kornberg found that DNA polymerase can add nucleotides only to a **primer,** an already existing polynucleotide. As shown in Figure 13-11,

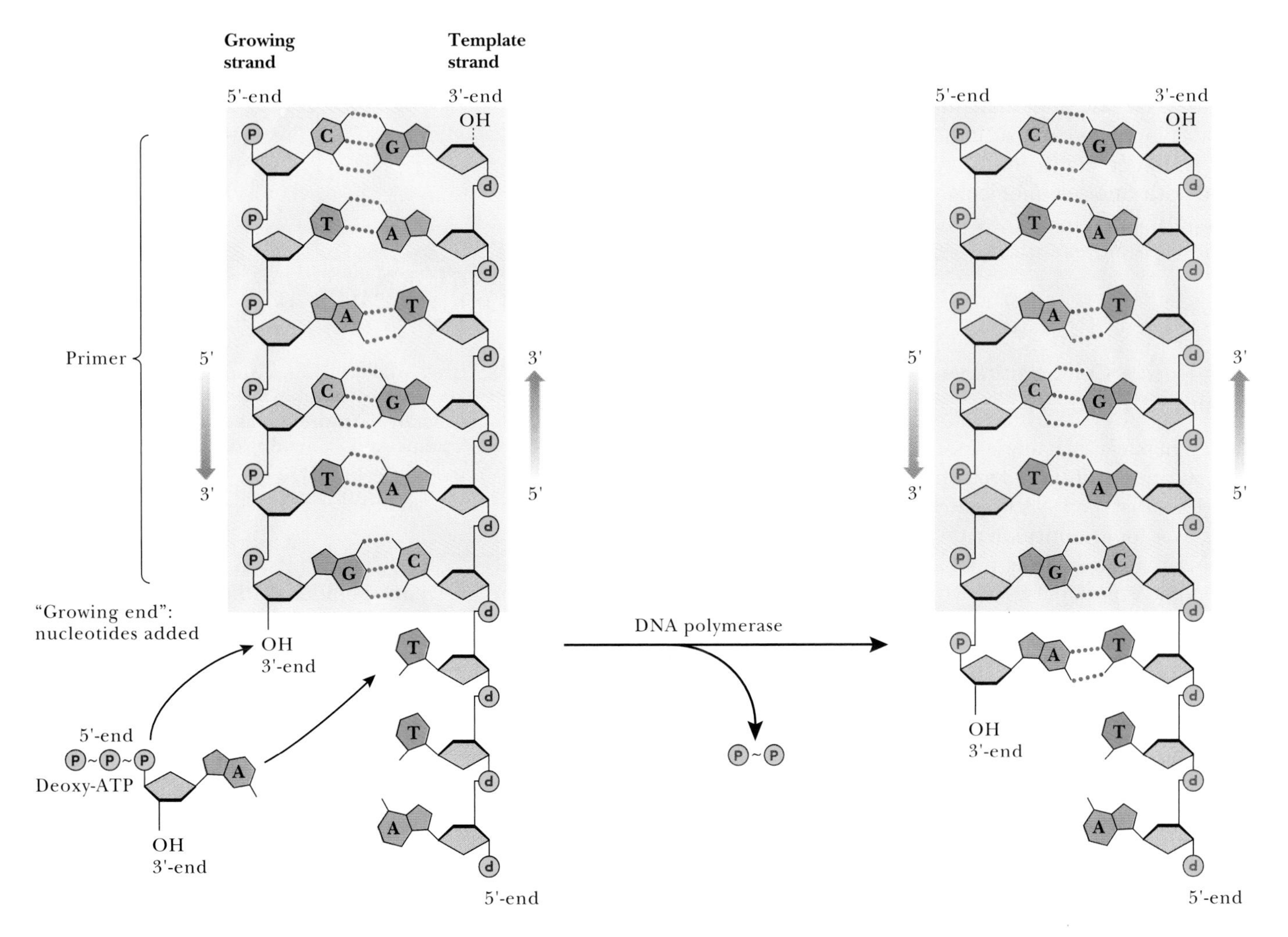

Figure 13-11 DNA polymerase adds a complementary nucleotide only to the 3′-end of a growing strand, called a primer.

DNA polymerase adds nucleotides to the 3′-end of the existing polynucleotide, so that the polynucleotide grows in a single direction—from its 5′-end toward its 3′-end. Elongation occurs as the nucleotide at the 3′-end of the growing chain attaches to the closest phosphate group of an unattached nucleoside triphosphate.

The particular nucleotide added at each step depends upon the base sequence of the template strand. Each added nucleotide makes hydrogen bonds to the next base on the template strand. So each strand extends in a single direction, starting with the sequence closest to the 5′-end of the primer and extending from its 3′-end.

Once a free 3′-end exists (Figure 13-11), we can see how the process will continue. But if DNA can add nucleotides only to the 3′-end of existing polynucleotides, how can the replication process ever begin? The resolution of this paradox is that DNA replication does not begin with a DNA primer, but with a temporary RNA primer.

The RNA primer arises from the action of **primase,** an RNA polymerase that copies short stretches (fewer than 10 nucleotides) of DNA into RNA without requiring a primer. Primase is said to be a "DNA-dependent RNA polymerase" because it uses DNA as a template for making RNA. DNA polymerase can use this RNA primer's 3′-end to begin synthesizing the new strand of DNA, adding nucleotides, step by step, matching the template and creating a complementary strand. Finally, another growing DNA strand enters (from the right in Figure 13-12), associated with another molecule of DNA polymerase. The DNA polymerase hydrolyzes the primer one nucleotide at a time, as it moves with the second growing DNA chain.

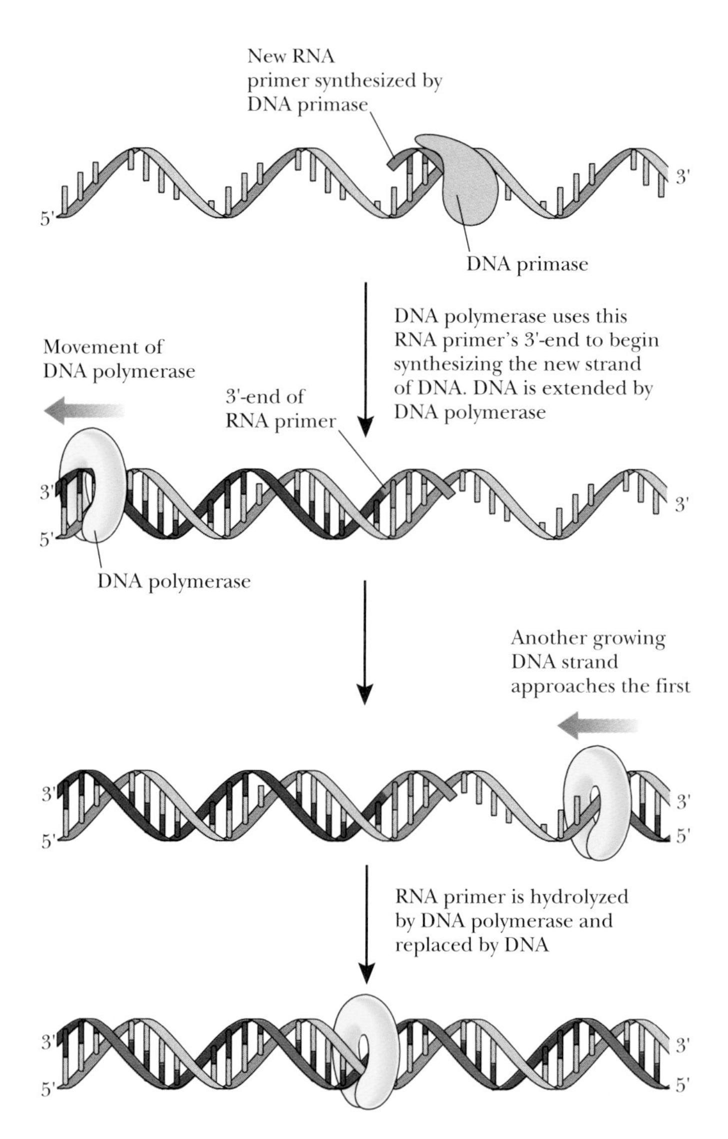

Figure 13-12 DNA primase (shown in blue) copies short stretches (< 10 nucleotides) of one DNA strand into an RNA primer. DNA polymerase (shown here as a yellow donut) then extends the RNA primer at its 3′-end. Notice that extension is always in the 5′ → 3′ direction, but in this figure (unlike other figures) the 3′-end of the growing strand is to the left.

How Does a Cell Simultaneously Copy Both Strands of DNA?

Researchers can actually see cellular DNA in the process of replication with the electron microscope. As shown in Figure 13-13a, the DNA in a dividing tissue culture cell appears as a long, thin molecule punctuated by "bubbles." As DNA replicates, these bubbles expand. The outside of the bubble is parental DNA, while the interior of the bubble is replicated DNA, consisting of one newly made strand.

Each bubble consists of two **replication forks,** Y-shaped regions of DNA where the two strands of the helix have come apart, as diagrammed in Figure 13-13b. At the replication fork, each strand begins to direct the assembly of a new complementary strand. As replication proceeds, the forks move away from each other. In bacteria the replication forks move away from each other at about 500 nucleotide pairs per second, and, in mammals, at about 50 nucleotide pairs per second. The slower movement in mammals may reflect the more elaborate organization of eukaryotic chromosomes, as discussed in Chapter 9, or the enzymatic abilities of the eukaryotic, as opposed to the prokaryotic, DNA polymerase.

Think about the problem of replication in terms of what we know about DNA polymerase. In any replication fork, one of the strands grows by adding complementary bases at the 3′-end, while the other, apparently, must add bases at its 5′-end. How does DNA replication proceed in both strands at the same location (the replication fork) if each strand can grow only from the 5′- to the 3′-end?

We can see some of the problems of DNA synthesis by following a single replication fork, as in Figure 13-13c. Imagine the fork as a Y lying on its side, with its jaws opening to the left. As replication proceeds, the DNA in the stem of the Y is opened and replicated; as the jaws of the

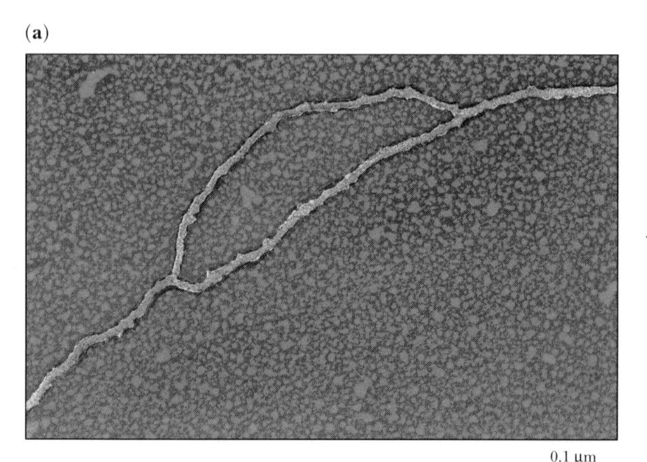

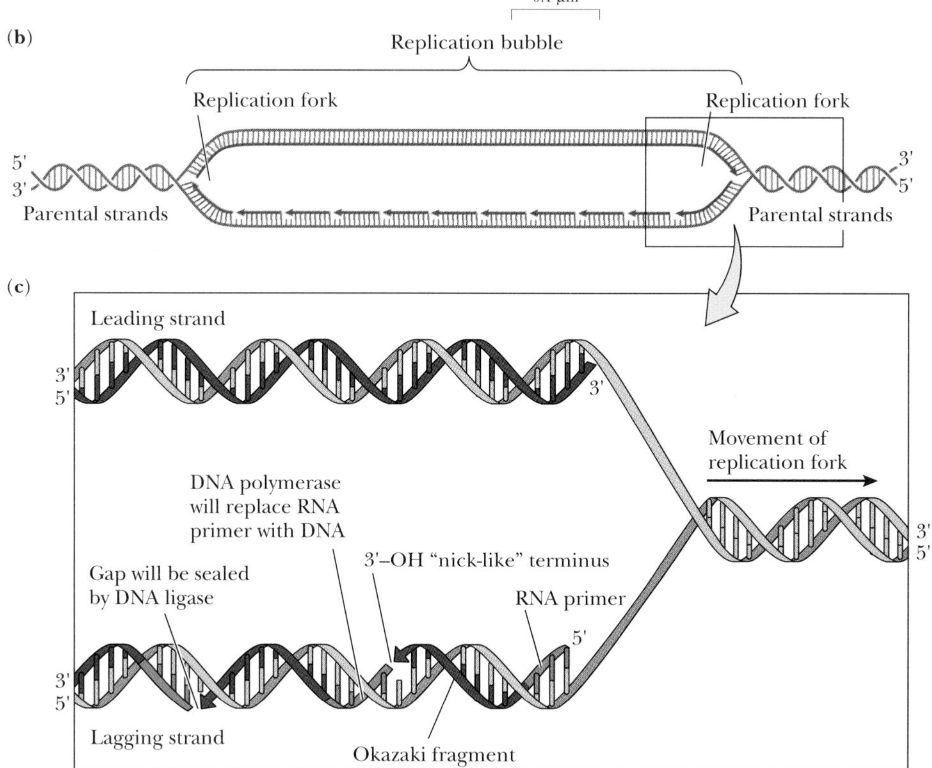

Figure 13-13 Replicating eukaryotic DNA appears as a long thin molecule punctuated by bubbles. **(a)** An electron micrograph of replicating DNA from a human cell grown in tissue culture; **(b)** each bubble consists of two replication forks, which move in opposite directions; **(c)** in the replication fork, one strand ("the leading strand") adds nucleotides continuously while the other ("the lagging strand") adds discontinuous "Okazaki fragments," each originally joined to an RNA primer. DNA ligase then stitches fragments together. *(a, Dr. Gopal Murti/Science Photo Library/Custom Medical Stock)*

Y open more, the whole Y-shaped fork moves toward the right.

One of the new DNA strands (the upper strand in Figure 13-13c) extends continuously, as DNA polymerase adds nucleotides to its 3′-end. This newly made strand is called the **leading strand,** because it extends continuously, as shown in Figure 13-13c. The synthesis of the other new strand, called the **lagging strand,** is not so simple. As the fork moves to the right, nucleotides must be added to its 5′-end, a task that DNA polymerase cannot perform. How is this strand made?

The answer is that the lagging strand is produced discontinuously, as shown in Figure 13-13c. After the moving fork has exposed about 1000 nucleotides of template for the lagging strand, primase produces a short piece of complementary RNA. This RNA serves as a primer for DNA polymerase, which attaches nucleotides to its 3′-end. This process produces fragments, called **Okazaki fragments,** after their discoverer Rejii Okazaki. Each Okazaki fragment contains a short stretch of RNA connected to about 1000 nucleotides of a DNA strand. DNA polymerase removes the RNA primer and replaces it with DNA sequence. Another enzyme, called **DNA ligase,** stitches the fragments together by catalyzing the formation of phosphodiester bonds between the two fragments.

How Does DNA Synthesis Begin?

A cell must regulate DNA synthesis precisely so that all its genes replicate once and only once during each cell cycle. *E. coli,* the organism whose DNA synthesis has been

ESSAY

TELOMERES, SENESCENCE, AND CANCER

The end of a chromosome poses a special problem for the machinery of DNA replication. DNA polymerase works in a single direction, extending new DNA strands only at their 3′-ends, so the copying of the lagging strand is necessarily discontinuous. Almost all of the lagging strand is ultimately replicated, using RNA primers and producing Okazaki fragments that are stitched together by DNA ligase. But the 5′-end of the lagging strand is never copied in conventional DNA replication, and the ends of a DNA molecule shorten with each round of cell division. While the amount of this shortening is only about 12 bases, the incompleteness of replication renders the ends of a chromosome especially susceptible to degradation.

The loss of DNA with cell division could be fatal if genes near chromosome ends were damaged. But cells have special mechanisms to protect the ends of chromosomes and to ensure their complete replication. Each chromosome ends not with a conventionally replicated DNA sequence but with a special structure called a *telomere.* In vertebrate cells, for example, each telomere consists of tens of thousands of copies of the same DNA sequence (CCCTAA on one strand, TTAGGG on the other) repeated again and again.

Each time a cell divides, it is the telomeres that become shorter. When one or more telomere reaches a minimum length, cell division stops, preventing the otherwise inescapable damage to genes that lie inside the telomeres. So each cell is effectively limited to a fixed number of cell divisions unless it has some special way of repairing the nibbled telomeres.

A number of researchers are trying to understand how the progressively shortening telomere triggers the shutdown of cell division. One current hypothesis is that the loss of a telomere, or of a gene just inside a telomere, leads to the induction of tumor-suppressor genes. One of the tumor-suppressor genes inhibits the cyclin-dependent activation of cell division. The result is the cessation of the cell cycle.

Cells that divide without apparent limit are able to repair the telomeres, using an enzyme called *telomerase,* which replaces the missing repeated sequences of each telomere. Such effectively immortal cells include cancer cells and the cells of the germ line (those that give rise to gametes—sperm and eggs—that produce the next generation). Other cells, including most of the cells that make up the body, lack telomerase. Many researchers think that they may be able to stop the unregulated cell division of cancer by devising ways of inhibiting telomerase.

---

most studied, has a single piece of DNA that forms a continuous loop and is conventionally called *circular* DNA. In *E. coli,* DNA replication always begins at the same place in the circle, at a site called the **origin of replication,** as shown in Figure 13-14. Two replication forks move in opposite directions away from the origin of replication, and DNA synthesis stops when the fork that is moving clockwise meets the counterclockwise fork.

Several *E. coli* proteins participate in the beginning of DNA replication. These proteins recognize and bind to the origin of replication. They bend the DNA and open the double helix. An enzyme called **helicase** uses energy from ATP to untwist the helix, while another enzyme called **DNA gyrase** prevents the accumulation of kinks at the stems of the replication forks, as shown in Figure 13-15. A **single-strand binding protein** binds to the unwound DNA and keeps open the jaws of the replication fork. The remaining steps in replication are the same as those we have already discussed: production of RNA primers, extension of leading and lagging strands, the hydrolysis of the RNA primers, and the stitching together of the Okazaki fragments with DNA ligase.

In eukaryotic cells, DNA replication is more involved. The DNA of each cell contains thousands of replication origins. Groups of 20–50 origins, called **replication units,** form replication forks at the same time. Within a replication unit, the forks move in opposite directions from each origin until they encounter a fork from an adjacent origin.

The replication of a eukaryotic chromosome (during the S phase of the cell cycle) typically takes about 8 hours, with individual replication units active at different

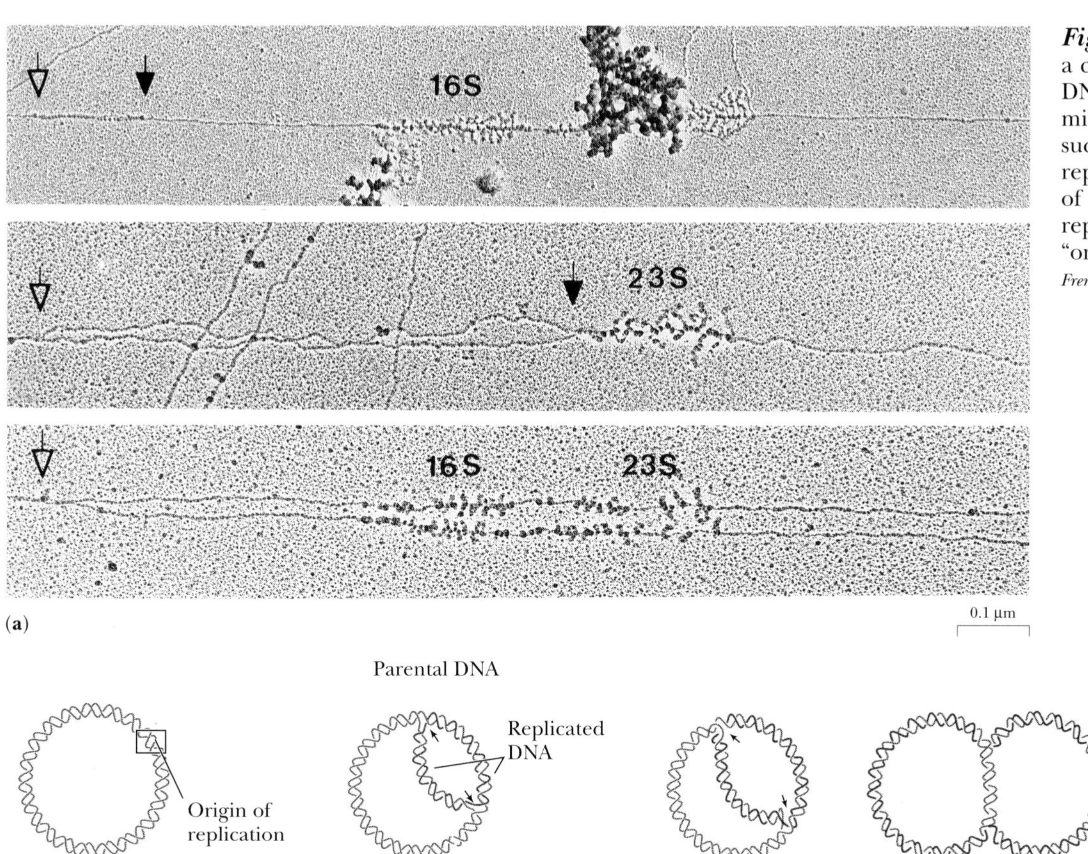

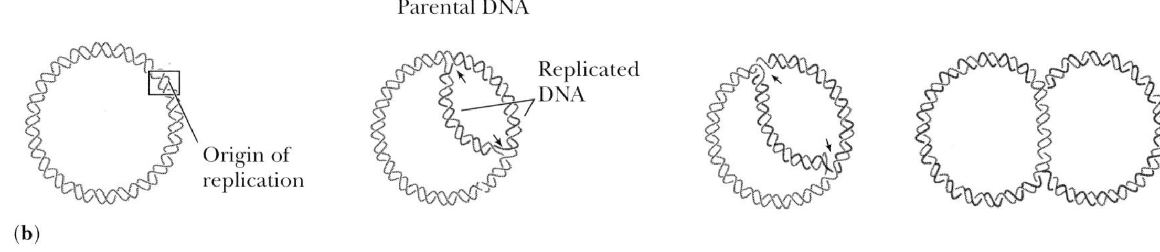

Figure 13-14 Replication of a circular molecule of *E. coli* DNA: **(a)** electron micrographs of DNA at successive stages of replication; **(b)** interpretation of micrographs, with replication starting at the "origin of replication." *(a, Sarah French, University of Virginia)*

(a)

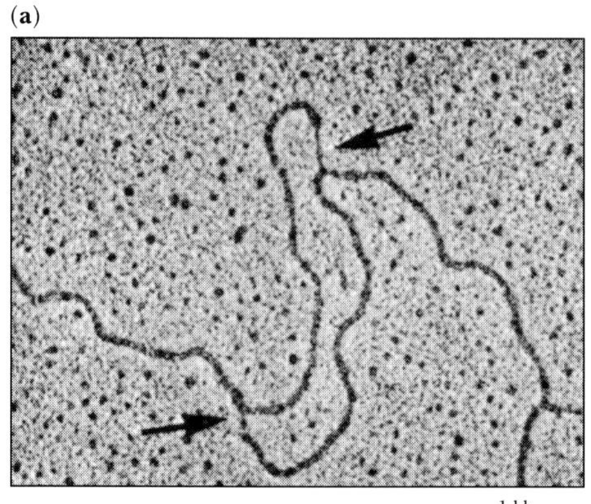

Figure 13-15 More detailed view of a replication fork. **(a)** TEM of a replication fork; **(b)** diagram of the proteins and the nucleic acids present at a replication fork. *(a, David Hogness and Henry Kriegstein, Stanford University School of Medicine,* Proceedings of the National Academy of Science, *71 #1, January 1974, pp. 135–9)*

(b)

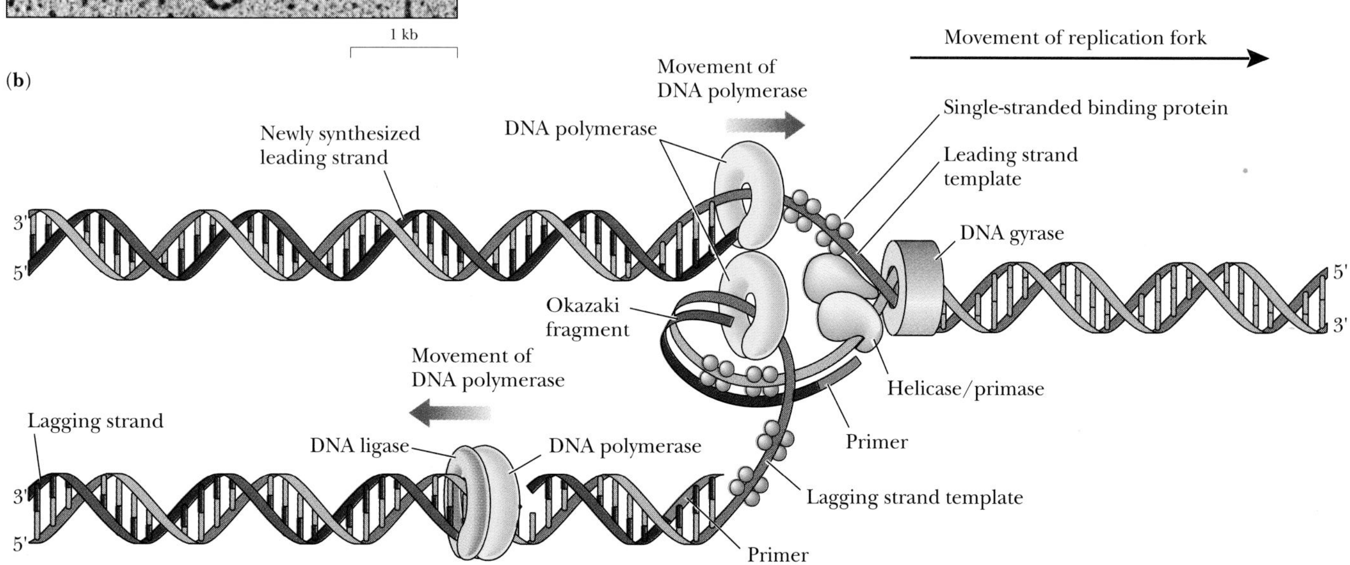

times during the whole process. After a unit has finished replicating, it does not replicate again until the next S phase. Researchers do not yet know exactly how cells prevent over-replication.

MOLECULAR BIOLOGISTS USE DNA POLYMERASE TO MAKE LARGE AMOUNTS OF SPECIFIC DNA SEQUENCES BY THE POLYMERASE CHAIN REACTION

Understanding how genes work ultimately requires knowing what nucleotide sequences they contain. The recombinant DNA methods discussed in Chapter 17 provide ways of purifying and amplifying individual genes or parts of genes. Since the 1970s, such methods have—among other things—allowed researchers to determine nucleotide sequences.

Recombinant DNA techniques depend on the ability to program bacteria (or other cells) to use their own DNA replication machinery to copy rearranged pieces of DNA, often spliced together from a variety of sources. Since the late 1980s, researchers have been able to make millions of copies of DNA sequences in a test tube without using whole living cells, by using a DNA polymerase purified from a bacteria. The method is called the **polymerase chain reaction,** or **PCR,** because it specifically and repetitively copies a segment of DNA between two defined nucleotide sequences. The reaction first makes two copies of the sequence, then copies those copies to make 4 copies, then 8, then 16, and so on. A single copy of a chosen sequence can be amplified into millions or billions of copies in less than a day.

The process depends on the ability to make specific polynucleotides that correspond to sequences at each end of the sequence to be copied. These synthetic polynucleotides are typically about 20 nucleotide residues long and are called oligonucleotides [Greek, *oligos* = few]. Each oligonucleotide serves as a primer for DNA polymerase, first forming a double-stranded DNA with the complementary sequences in the target DNA, in much the same way that RNA primers form hybrid duplexes in cellular DNA replication. Instead of relying on helicase to unwind the DNA, however, the experimenter carefully heats the DNA so that the two strands come apart. The primers bind to complementary sequences—one on each strand, as shown in Figure 13-16. DNA polymerase then copies each strand until the experimenter (or the machine) stops the reaction by again raising the temperature.

The increased temperature also separates the template strand and the newly made complementary copy. Lowering the temperature then allows DNA polymerase to go back to work, again starting with the oligonucleotide primers. The only region that is copied—again and again with further cycles—is the segment flanked by the two

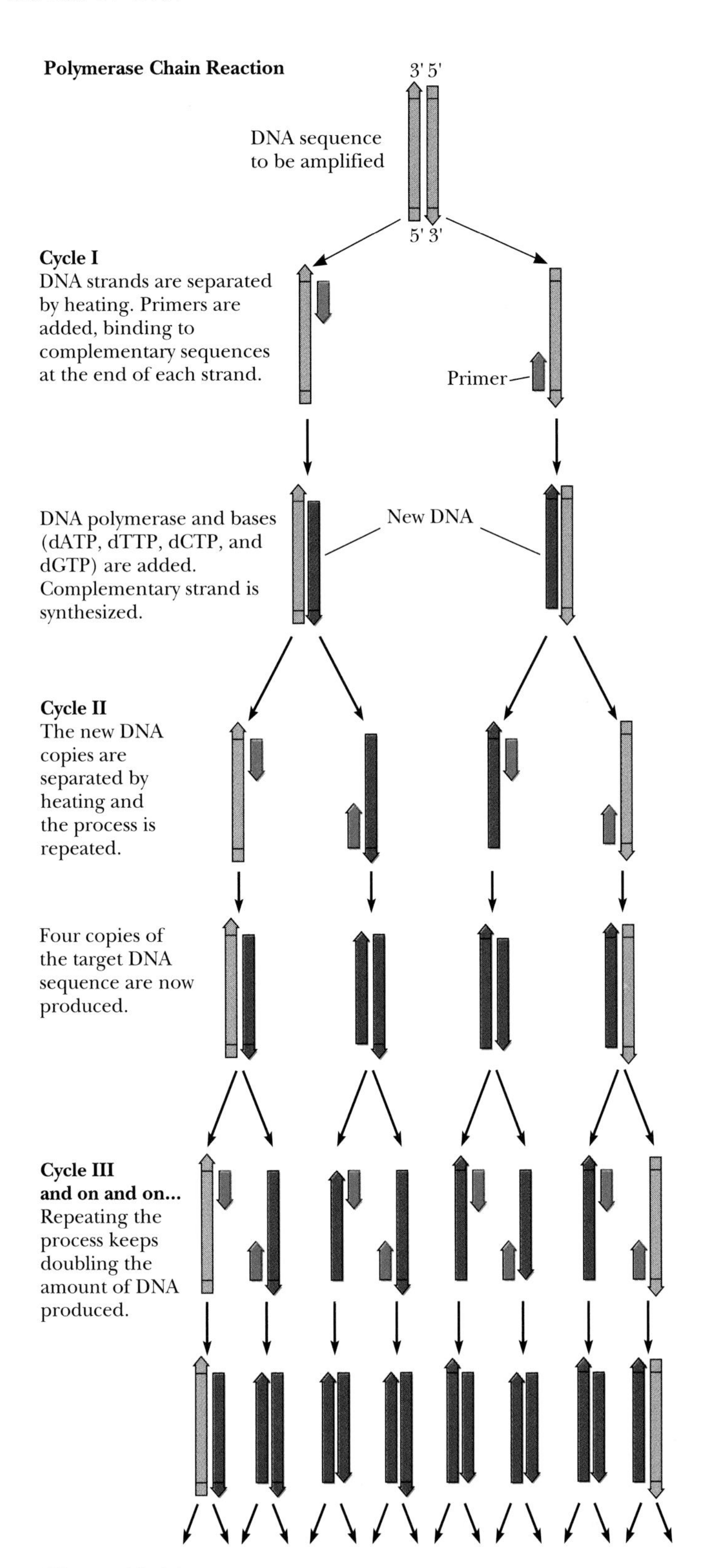

Figure 13-16 The polymerase chain reaction (PCR).

primers. The result is the specific amplification of this segment. Specialized machines, called thermal cyclers, can automatically vary the temperature for the different stages of the polymerase chain reaction. The amplified segment can then be used for a variety of purposes, including sequence determination, mutation detection, and studies of interactions with specific proteins.

■ HOW DO CELLS HANDLE MISTAKES IN THE NUCLEOTIDE SEQUENCE OF DNA?

Any molecule—including DNA—is subject to chemical change from random (thermal) collisions. For example (see Figure 13-17a), the bond between the purine bases in DNA and the backbone sugar is slightly unstable, sometimes leading to **depurination,** the loss of a purine base. In each of your cells, some 5000–10,000 A and G nucleotide residues are lost every day. Another spontaneous change, shown in Figure 13-17b, is **deamination,** the removal of an amino group from adenine, guanine, or cytosine. The most frequent of these changes (which occur about 100 times per day in every human cell) is the deamination of cytosine to produce uracil—a base that is not ordinarily present in DNA.

In addition to these changes, DNA is also subject to damage from radiation (such as ultraviolet light or x-rays) and to mistakes made by DNA polymerase. Ultraviolet light produces a characteristic change, the formation of **thymine dimers,** two adjacent T residues in the same chain linked together, as shown in Figure 13-17c.

Cells must be able to repair damage to DNA if they are to maintain the continuity of life. This is especially important for the germ line cells that pass information to the next generation, but even mutations in somatic cells

(Text continues on page 352.)

(a) Depurination

etc. Thymine Purine base Guanine Cytosine OH Guanine Thymine Cytosine

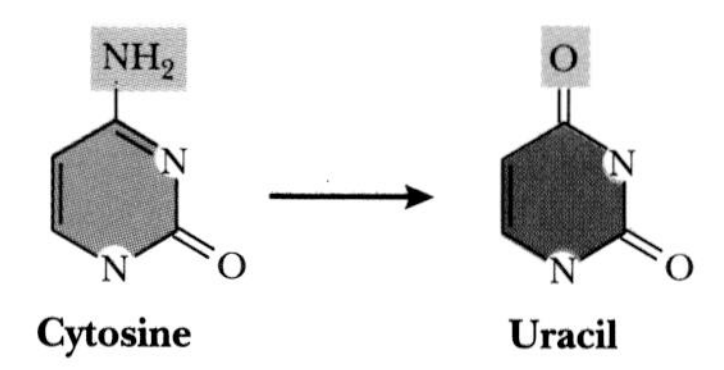

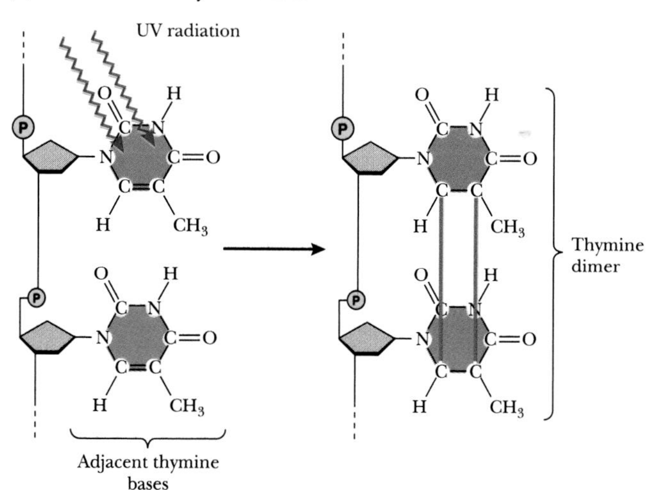

Figure 13-17 The most frequent changes in DNA: **(a)** depurination; **(b)** deamination; and **(c)** thymine dimer formation.

ESSAY

FORENSIC USES OF PCR

The haploid human genome contains some 3 billion (3×10^9) base pairs of DNA. Only about 5% of this DNA (about 1.5×10^8 base pairs) codes for proteins, with the rest of the DNA thought to lie in special sequences at centromeres and telomeres, between genes, and in sequences within genes that do not encode proteins. The protein-coding sequences of DNA vary among individuals, giving rise to genetic variations in molecules ranging from hormones to hemoglobins, as discussed in Chapters 11 and 12.

Notwithstanding these allelic variations, protein-coding DNA sequences are highly conserved: Natural selection eliminates mutations that result in nonviable, or even less viable phenotypes. The 95% of the DNA that does not encode protein is much less subject to natural selection, however, and this noncoding DNA contains far more sequence variation. Estimates of the level of variation in noncoding DNA suggest that a human is heterozygous at 1 base pair out of 500: That is, the chromosome 1 that you inherited from your mother (which contains about 2.5×10^8 base pairs) differs from the chromosome 1 that you inherited from your father at about 500,000 base pairs! This level of variation means that every individual is genetically distinct from every other, except an identical twin, and DNA sequencing could provide a reliable means of distinguishing every individual in a population.

Variations in DNA sequence have been crucial for sorting out patterns of inheritance in natural populations, including humans, in which genetic experimentation is impossible. Geneticists now usually identify sites of variant sequences using the polymerase chain reaction (PCR), and they have now mapped more than 15,000 variable sites in the human genome. These sites provide signposts ("markers") for the entire human genome, and any genetic locus can now be placed on a map whose resolution is about 200,000 base pairs.

A remarkable amount of variability within the human genome occurs as variations in the number of dinucleotide repeats (such as CACACA.../ TGTGTG...) Larger sequences are also repeated a varying number of times in the genomes of individuals, and researchers have termed such polymorphic sites *hypervariable* sites or VNTRs (variable number of tandem repeats). Dinucleotide VNTRs are particularly common and easy to detect. They occur about 100,000 times in the human genome, about every 50,000 to 100,000 base pairs. Each such block has, on average, abut 25 repeats, with a range between 4 and 40. By using PCR primers that flank a given dinucleotide repeat, an investigator can determine the number of repeats at a particular site, as illustrated in Figure 1. Studies of a small number of such sites can distinguish any individual from any other. Some sites are particularly variable, so only a relatively small number of sites may be necessary to distinguish two individuals and to track the inheritance of a given chromosome from parent to offspring.

The most publicized use of PCR-based identification has been to identify individuals suspected of a crime. The forensic power of this technique depends on the highly individualized patterns of repeats and the ability of PCR to detect such repeats from a minuscule amount of DNA, for example, the DNA remaining in a few hairs at the scene of a crime or, in the case of rape, from a vaginal smear.

The use of DNA-based evidence, however, has been much debated in U.S. courts. Establishing nonidentity will always be easier than establishing identity. Much of the discussion on establishing identity has focused on two points: (1) whether the whole population is as variable as prosecutors claim and (2) whether DNA gathered at a crime scene can remain secure and uncontaminated by other DNAs gathered in an investigation or present in a testing laboratory. Researchers agree that the level of DNA sequence variation is adequate to distinguish among individuals if investigators examine enough VNTRs, but expert witnesses have differed in their calculations of the number of tests needed to establish identity.

An issue of greater concern to the courts has been the handling of DNA evidence. Given the extreme sensitivity of PCR-based analysis, some juries have been convinced that careless handling could result in spurious identifications.

Even as PCR-based analyses are increasingly common for the identification of individuals and the diagnosis of disease-causing genes, some geneticists, ethicists, and policy makers have raised serious questions about the ethics of using DNA for identification. Who, they ask, has the right to know about any one else's genes? Should an insurance company or an employer be told about a person's genetic risks? Should they be allowed to discriminate in the writing of insurance or in promotion decisions on the basis of, for example, an increased likelihood of developing cancer? These questions will almost certainly be the subject of important policy debates in the next decade, as more people have their DNA collected for purposes of definitive identification, even if they are not suspected of a crime.

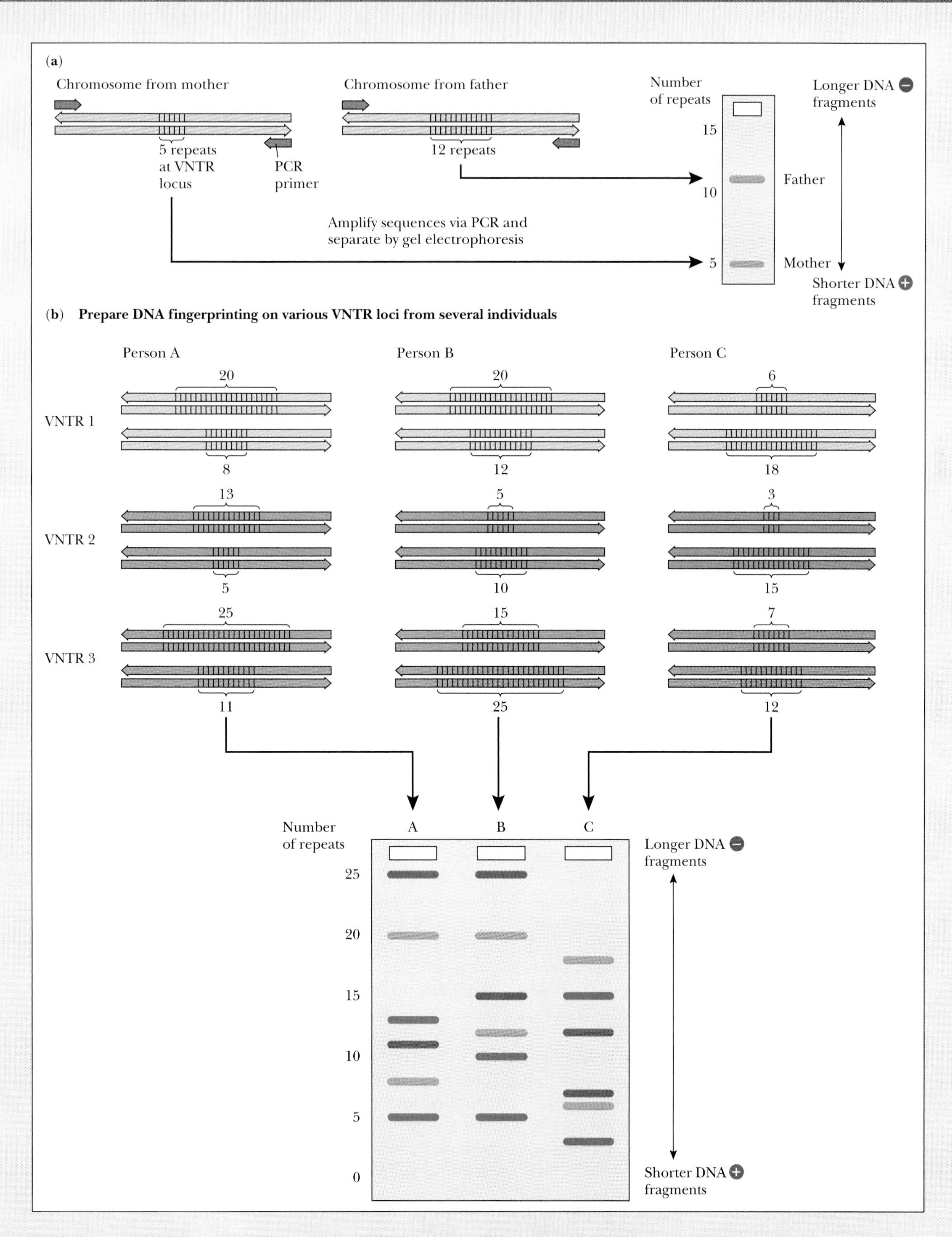

Figure 1 Detection of genetic variants by PCR. **(a)** Using PCR primers that flank a particular VNTR, investigators can distinguish alleles with different numbers of repeats; **(b)** simultaneous analysis of several VNTR loci.

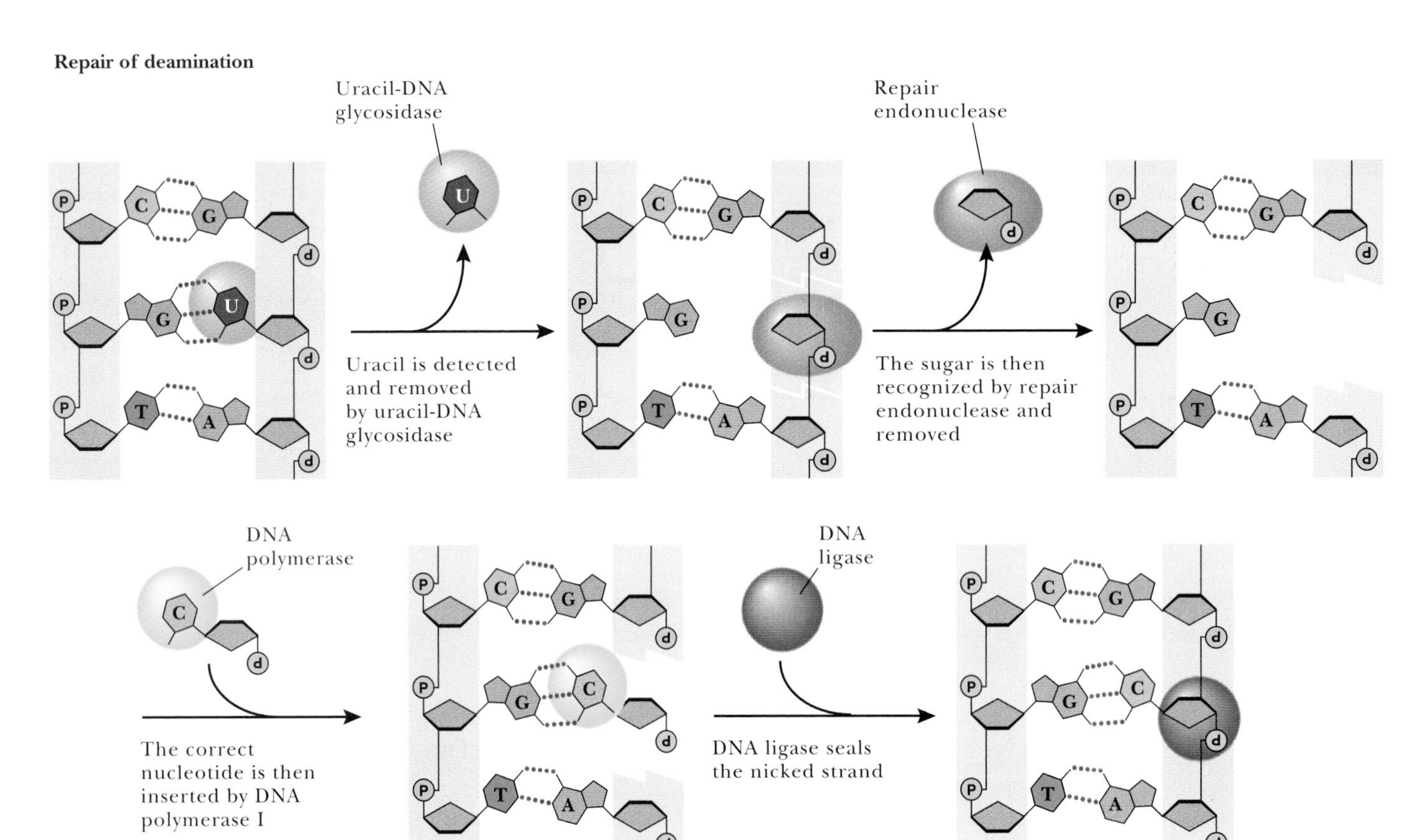

Figure 13-18 Repair of deamination. Uracil-DNA glycosidase recognizes the strange base (deoxy-U) and removes it.

are now thought to be dangerous, in many cases leading to the development of cancers. (See Chapter 9.)

Both prokaryotic and eukaryotic cells have sophisticated mechanisms to repair damage to DNA and to correct mistakes in replication. These mechanisms are crucial for the stability of genetic information. They depend upon the duplex structure of DNA and the presence of enzymes that recognize commonly occurring mistakes.

In each case, an enzyme detects something wrong in one strand of the DNA and removes the error, as shown in Figure 13-18. For example, when the deamination of cytosine produces uracil, an enzyme called uracil–DNA glycosidase recognizes the strange base and removes it, as shown in Figure 13-18. Similar enzymes recognize depurinated nucleotides or thymine dimers.

Once the correction enzymes discard the offending residues, DNA polymerase goes to work. It copies the information in the intact second strand and creates a sequence that is properly matched. DNA ligase then seals the gap, and the original sequence is restored. More than 50 enzymes "proofread" DNA for errors—unusual bases, gaps, and bulges. In both prokaryotes and eukaryotes, the DNA polymerase responsible for error correction is distinct from the enzyme responsible for replication during the normal cell cycle.

HOW DO RESEARCHERS DETERMINE THE SEQUENCE OF NUCLEOTIDES IN DNA?

Current molecular biological techniques allow scientists to prepare large amounts of individual pieces of DNA and to study both their chemical structure and their biological function. One of the first questions in such studies is "what is the sequence of nucleotides in a given piece of DNA?"

The most widely used method for determining DNA sequences depends on the use of the DNA polymerase. This method was developed in Cambridge, England, by Fred Sanger in the late 1970s; he was also the first to determine the sequence of a protein. Sanger's method is diagrammed in Figure 13-19. The method depends on the way that DNA synthesis occurs: (1) DNA synthesis is semiconservative; (2) a newly made DNA strand grows in a single direction, adding one nucleotide residue at a time to the 3′-end; (3) the nucleotide sequence of the template strand determines the sequence of nucleotides in the growing strand; (4) DNA polymerase requires a primer polynucleotide; and (5) the copying of a single strand into a new complementary strand is catalyzed by a single enzyme, DNA polymerase.

Another method for DNA sequencing, developed by

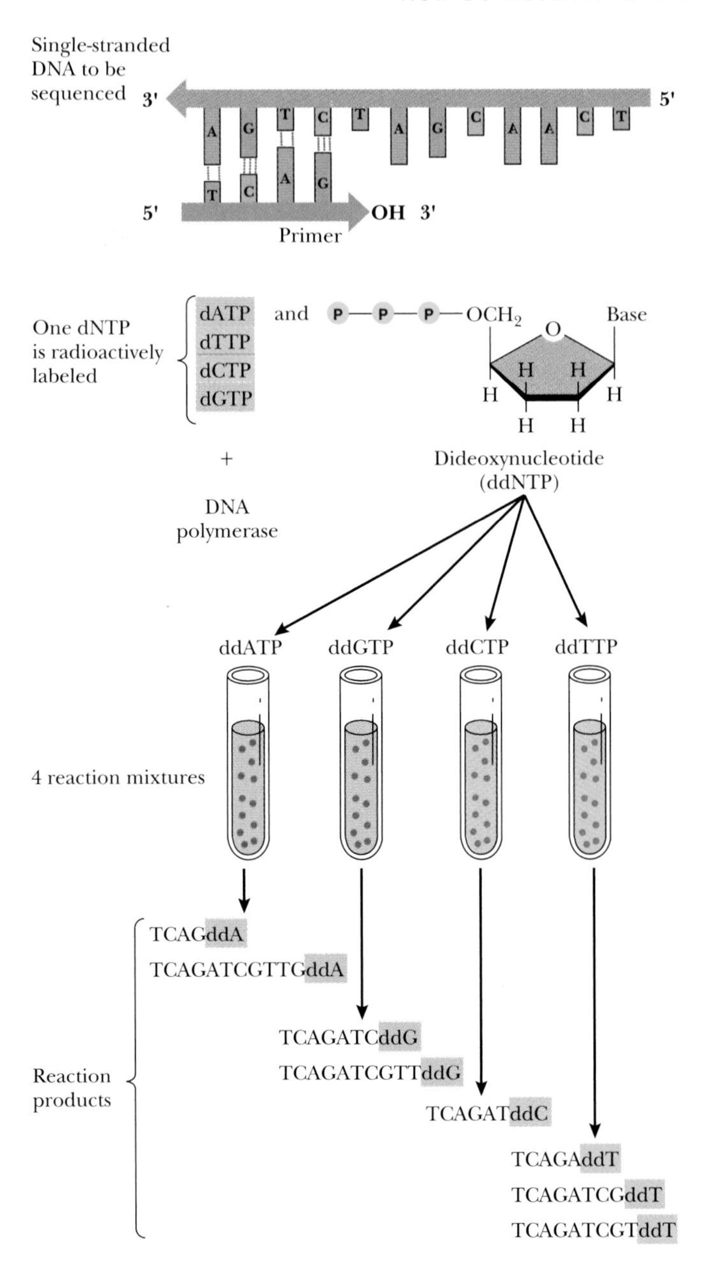

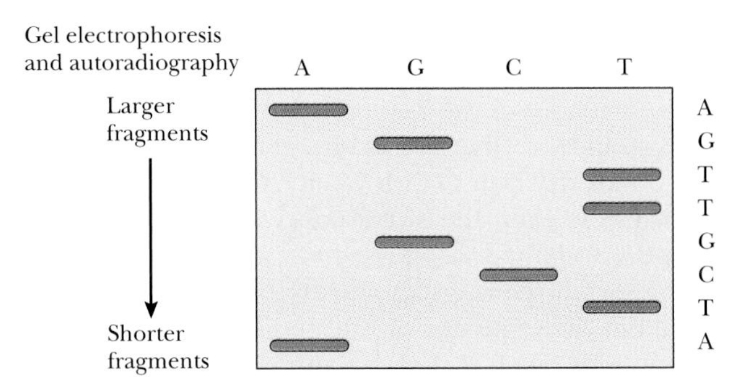

Reading sequence bottom to top: –A–T–C–G–T–T–G–A

Its complement is the original template strand (3' → 5'): –T–A–G–C–A–A–C–T

Result:

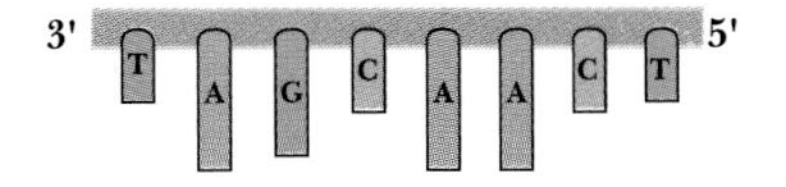

Figure 13-19 DNA sequencing by the dideoxy method of Fred Sanger. Notice the lack of a 3′-OH group on the dideoxynucleotide.

Alan Maxam and Walter Gilbert at Harvard University, does not depend on DNA-synthesizing enzymes, but on the ability of chemical agents to cleave DNA at specific residues. This method, however, is no longer widely used. The sequencing of DNA by either method depends on the ability to separate DNA fragments of different lengths by electrophoresis in long, thin gels of polyacrylamide. (See Chapter 3 for a discussion of electrophoresis.)

In the Sanger procedure, the first step is to heat DNA so that the two strands come apart. As shown in Figure 13-19, Sanger began the sequence determination with a specific complementary primer. In the presence of deoxynucleoside triphosphates, DNA polymerase would now have all the necessary materials and could extend the primer until it produced a complete copy of the template strand.

Sanger devised a way of stopping the polymerase reaction at different places. He did this by replacing a fraction of the nucleoside triphosphates with a nucleoside triphosphate that would not allow further polymerization.

Recall that a strand extends by adding a nucleotide to its 3′-hydroxyl group. Imagine what would happen if we replaced some of a deoxyribonucleoside triphosphate with a **dideoxynucleoside triphosphate,** a similar molecule that lacks a 3′-hydroxyl group. In Figure 13-19, one example shows the replacement of deoxyadenosine triphosphate with dideoxyadenosine triphosphate. Each time the DNA polymerase sees a T in the template strand, it places an A in the new strand. But if the selected A has no 3′-hydroxyl group, the chain stops growing at that position. Some DNA molecules stop after the first A in the new strand, some after the second, some after the third, and so on. The result is a collection of DNA fragments, all of which end in A. As shown in Figure 13-19, electrophoresis can then separate these fragments according to size.

Similarly, we can stop polymerization after the other three nucleotides, by using dideoxy-derivatives of C, G, and T. In each case, electrophoresis can separate the resulting collections of fragments by size, as shown in Figure 13-19. The four sets of bands in Figure 13-19 allow the experimenter to read the sequence of the newly made DNA. The shortest fragments are at the bottom of each lane of the gel and represent the sequence closest to the primer. These are closest to the 5′-end of the new strand and complementary to the 3′-end of the template strand. Figure 13-20 shows some real data and their interpretation.

Until the late 1980s, the dideoxy method for determining DNA sequences relied on radioactive labeling of DNA produced by DNA polymerase. More recently, however, several modifications have allowed the tagging of newly made DNA strands with fluorescent dyes. Commer-

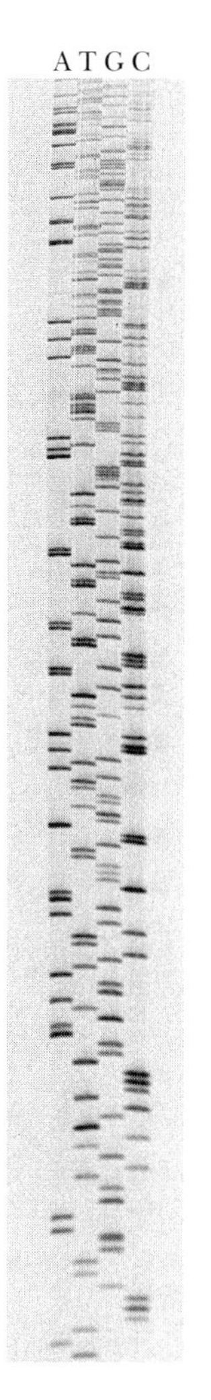

Figure 13-20 Some actual data from a DNA sequencing gel. Reading from the 5′-end (bottom of the gel), the sequence is TATCCCGTTGGAAGGTCGTCTGCTCCCTGGA AGTAG. . . . *(James D. Colandene, University of Virginia)*

cial machines are now available that can detect this fluorescence and automatically measure the sizes of the DNA strands that contain it, as shown in Figure 13-20. The power of these automatic machines is such that many molecular biologists believe that, by sometime early in the 21st century, we will have determined the entire sequence of all 3 billion nucleotides of human DNA.

The sequencing of DNA provides important information. It allows the detection of genetic changes associated with mutations and with diseases, such as sickle cell disease. The ability to determine DNA sequences rapidly has allowed molecular biologists to study how changes in macromolecular structure can affect function. In addition, new methods allow the selective alteration of nucleotide sequences in genes and the placing of genes into cells to study the effects of these changes. Such methods allow biochemists to study the relationship between the amino acid sequence of a polypeptide and its enzymatic function.

SUMMARY

Knowing the structure of DNA has allowed molecular biologists not only to explain further how genes work, but also to change and manipulate genes almost at will. DNA consists of nucleotide residues held together by phosphodiester bonds between the third (3′) carbon of the sugar of one nucleotide residue with the fifth (5′) carbon of the sugar of the next nucleotide residue. Each polynucleotide chain has a direction with a 5′-end (which has no neighboring residue attached to its 5′ carbon) and a 3′-end (which has no neighboring residue attached to its 3′ carbon).

DNA contains four kinds of nucleotides, which differ only in the nitrogen-containing bases attached to their sugar components. The bases that distinguish the nucleotides are either of two purines (adenine and guanine, abbreviated A and G) or two pyrimidines (cytosine and thymine, abbreviated C and T). Careful analysis of DNA, summarized in Chargaff's rules, from many species showed that the amount of C always equals the amount of G, and the amount of A equals the amount of T.

In cells, DNA consists of two polynucleotide strands twisted around each other in a double helix. X-ray diffraction studies of DNA and the building of molecular models allowed James Watson and Francis Crick to determine the structure of the double helix in the early 1950s. The two strands of DNA are held together by specific hydrogen bonds that pair G with C and A with T. These specific pairings explain the equal concentrations of G and C, and of A and T.

The specific pairings also mean that the information contained in the sequence of nucleotides is equivalent in the two strands. Each strand can direct the synthesis of a complementary strand in DNA replication. The experiment of Meselson and Stahl showed that DNA replication takes place in a semiconservative manner, with each strand remaining intact as it directs the synthesis of a new polynucleotide strand.

The enzyme that catalyzes DNA synthesis is called DNA polymerase. DNA polymerase assembles nucleotides into sequences specified by one strand of DNA, the template strand. The enzyme, however, can add nucleotides only to a pre-existing primer molecule. DNA polymerase adds nucleotides only to the 3′-end of the primer and continues to add nucleotides to the 3′-end of the growing polynucleotide.

In cells DNA replication occurs at replication forks, Y-shaped regions of DNA where the two strands of the helix have come apart. Nucleotides add directly to the 3′-end of one strand in the fork, called the leading strand. The other strand, however, is produced discontinuously, in fragments of about 1000 nucleotides. Each of these fragments starts with a short stretch of RNA, which serves as a primer. Another enzyme, called DNA ligase, joins the fragments together as DNA replication proceeds.

Special sequences in DNA serve as origins of replication, where DNA synthesis always begins. The sequences bind to proteins that open and untwist the double helix.

DNA molecules continuously suffer inevitable damage, particularly from the loss of purine bases. Cells have sophisticated mechanisms to repair such damage and to correct mistakes made during DNA replication.

KEY TERMS

3′-end
5′-end
complementary
conservative replication
deamination
depurination
dideoxynucleoside triphosphate
dispersive replication
DNA gyrase
DNA ligase
DNA polymerase
helicase
lagging strand
leading strand
nuclease
Okazaki fragment
origin of replication
polymerase chain reaction (PCR)
polymerization
primase
primer
pyrophosphate
replication fork
replication unit
semiconservative replication
single-strand binding protein
template
thymine dimer

QUESTIONS FOR REVIEW AND UNDERSTANDING

1. What are the four building blocks of DNA? Which are purines and which are pyrimidines?
2. What are the chemical components of a phosphodiester bond? What are the chemical components of a nucleotide?
3. What is the "backbone" of the polynucleotide chain composed of?
4. How can you differentiate between RNA and DNA?
5. Describe at least two differences between B-DNA and Z-DNA.
6. What are some of the ways that DNA can become damaged?
7. What would an RNA-dependent DNA polymerase do?
8. You have been determining the base composition of a new organism and your data indicate that the amount of adenine equals the amount of guanine present. Should you throw away your data and start over? Explain your answer.
9. How, chemically, can you distinguish one end of a polynucleotide chain from the other?
10. As the complementary strand is assembled, what kinds of chemical bonds are being formed? What molecule is eliminated as a result of phosphodiester bond formation?
11. Describe what stops DNA polymerase from extending the complementary strand in a sequencing reaction.
12. Why were Rosalind Franklin's data so critical to the determination of the structure of DNA?
13. What is the significance of "Chargaff's rules"? How did they influence Watson and Crick when they built their model of the DNA molecule?
14. Discuss, in general terms, how the chemistry and structure of DNA makes it so suitable for its biological function of carrying and replicating information.
15. How was semiconservative replication of DNA demonstrated? If conservative replication was correct, what would the data have been?
16. Describe the process which initiates DNA replication.
17. Draw a DNA bubble and mark how the strands would be labeled in a heavy isotope experiment like the one Meselson and Stahl performed.
18. Contrast what occurs along the leading versus the lagging strand during DNA replication.
19. What is the purpose of a polymerase chain reaction? How might one use the product of such a reaction?
20. What are the characteristics of the bands at the top of a DNA sequencing gel?
21. Discuss how sequencing and selective gene alteration might be useful in the study of the relationship between genotype and phenotype.

SUGGESTED READINGS

ALBERTS, B., D. Bray, J. Lewis, M. Raff, K. Roberts, and J. D. Watson, *Molecular Biology of the Cell,* 3rd edition, Garland, New York, 1994. Chapter 6 deals with the mechanisms of DNA replication and DNA repair.

LODISH, H., D. Baltimore, A. Berk, S. L. Zipursky, P. Matsudaira, and J. Darnell, *Molecular Cell Biology,* 3rd edition, Scientific American Books, New York, 1995. Chapter 10 concerns DNA replication and repair.

PART IV

Information Flow and Its Manipulation

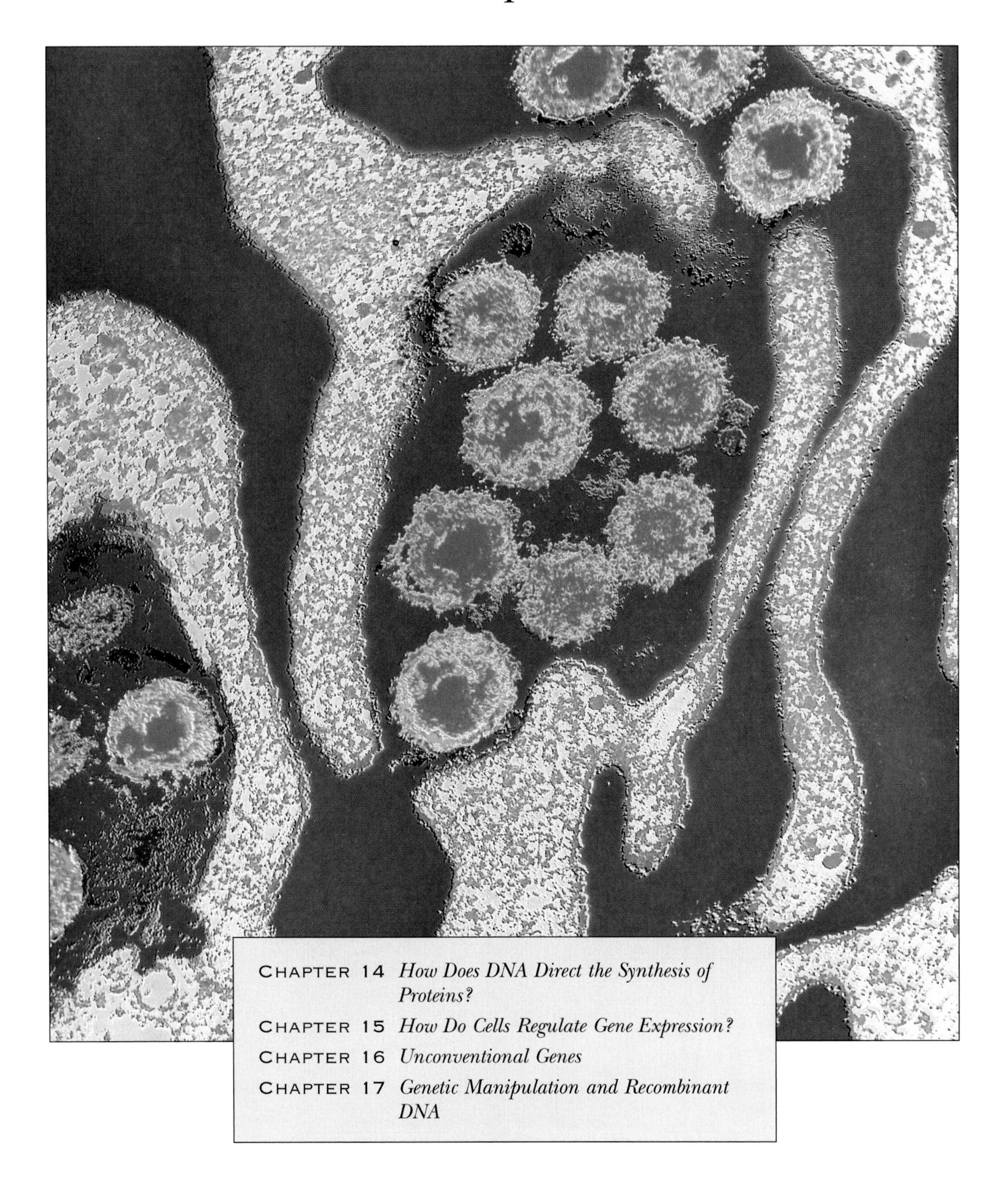

TEM of human herpes virus type 6 infecting a human cell. Virus particles can be seen in red.

LAB TOUR

NAME: MICHAEL SMITH
TITLE: PETER WALL DISTINGUISHED PROFESSOR OF BIOTECHNOLOGY AND UNIVERSITY PROFESSOR
ADDRESS: UNIVERSITY OF BRITISH COLUMBIA, VANCOUVER, B.C., CANADA V6T 1Z3

■ ***You are most famous (and received the Nobel Prize) for having invented site-directed mutagenesis, a widely used method for deliberately altering a particular sequence in DNA. How have you been using this technique? What is the major question that you and your laboratory are asking?***

My laboratory is now in a "winding-down" phase, since I am 64 years old. For many years, however, we have been applying the technology of site-directed mutagenesis to study the relationship between the structure of a protein and its biological function. We have been systematically replacing amino acid residues in certain proteins in order to determine which residues are important for function.

■ ***Why has the overall question remained important to you?***

With questions about the relationship of protein structure and function, one doesn't get answers right away. You start with some ideas, and the first thing to find out is that your initial ideas are wrong. Then you need to try other approaches. It takes some time to pin down what is actually going on. Many models only give you part of the information, and the details are usually much more subtle than you think at first.

■ ***How did you come to choose these questions?***

We began our work in the late 1960s with the idea of using oligonucleotides synthesized in the laboratory to isolate DNA fragments with sequences that were complementary to the oligonucleotide, a process known as affinity chromatography. Just about then, the whole technology of recombinant DNA came along. From our previous work, it became obvious that we could use synthetic oligonucleotides to identify recombinant DNAs that contain specific sequences.

In 1975–76, I was in Fred Sanger's laboratory in Cambridge, England, when he developed his first version of rapid DNA sequencing and applied it to the bacteriophage ϕX-174. The discovery that ϕX DNA contained overlapping genes made us think about developing precise genetic tools to dissect gene action, and we [Clyde Hutchinson and I] began to talk about ideas for site-directed mutagenesis.

I knew from our model studies that a short oligonucleotide, nine nucleotides long, could form a duplex even if there was a mismatch in the middle. A duplex with such a mismatch was potentially mutagenic, if we could think of a way to accomplish it. It was slow work because it then took so long—six months—to synthesize an oligonucleotide. Few laboratories made oligonucleotides in those days, and I was lucky to have a laboratory that was already doing it. I didn't set out to devise the method in 1968. I was lucky to be in Cambridge at the right time, with the ϕX-174 sequence DNA available and Clyde Hutchinson to talk to, and lucky to have a laboratory that was doing chemical synthesis.

Michael Smith in his laboratory holding an autoradiograph of a DNA sequencing gel.

■ ***What organisms do you use in your research? Why those organisms?***

We use *E. coli* and yeast. To do a thorough study with many mutants, we need to have an organism in which we can propagate mutated genes readily and produce proteins in adequate amounts.

■ ***What instruments do you use in your research, and how do they contribute to the experimental pursuit of your questions?***

The instrument that has revolutionized our work is the automated oligonucleotide synthesizer. Neither Kary Mullis nor I would have won the Nobel Prize in 1993 without the synthesizer because, to succeed, technology must have broad application. If producing an oligonucleotide had remained a major task that took several months, mutagenesis would not have been much used.

■ ***What other people work with you in your research program?***

Like most people who are successful in science, I have been lucky to have had great people in my laboratory. It is invidious to pick one out of the 75–100 with whom I have worked in the course of my career, but one critical person who was involved in the work that won the Nobel Prize was Dr. Shirley Gillam, who worked in the laboratory for over ten years in the 1970s, and now has her own laboratory in the Department of Pathology. While developing the method for site-directed mutagenesis, I had such confidence in Shirley's exceptional talent that I knew if an experiment didn't work, it was because it was a bad idea, not because it was not done well.

■ ***How does your work fit into the broad context of cell biology?***

Our studies of the relationship between the structure and function of proteins is part of a broad assault that is going on in many laboratories. The immediate and most important goal is to determine the sequence of the many genes within genomes of humans and other species. Once this is done, we'll be faced with all these gene sequences without knowing what they do. Obtaining this information will be the foundation of biology for the next half century. One approach to understanding what unknown genes do will be to use site-directed mutagenesis.

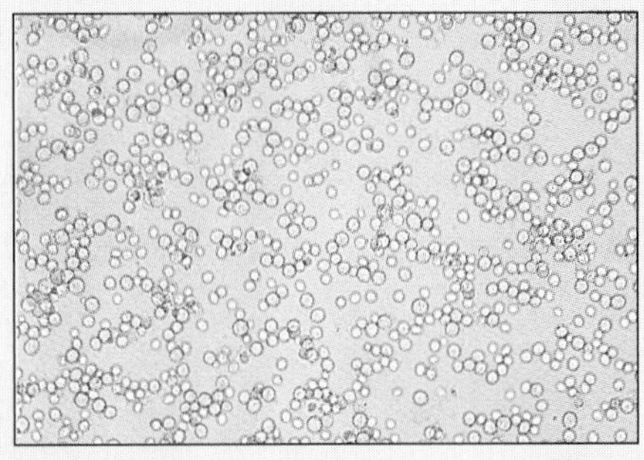

Dr. Smith has worked extensively with the yeast Saccharomyces cerevisiae. *(Courtesy James Mauseth, University of Texas, Austin)*

■ ***In addition to your scientific work, you have been involved in a number of educational and other projects. What have your goals been in this other work?***

I want nonscientists to understand why science is important, and I want people to understand that scientists who do biotechnology are not ogres. To this end, I tend to accept invitations to speak from organizations like the Rotary Club, and I also serve on the advisory board of a foundation that supports the research of young scientists working on the central nervous system and mental illness. As to the money from the Nobel Prize, I gave it to three organizations that do important work: One fosters research in schizophrenia, one helps women in science and technology, and a third educates elementary school teachers about science and the scientific method. There is much unjustified fear of science in society, and this fear is rooted in widespread ignorance of the nature of science and technology. I think it is important to get at the roots of this fear in the context of elementary education.

CHAPTER 14

How Does DNA Direct the Synthesis of Proteins?

KEY CONCEPTS

- DNA and RNA use a 4-letter alphabet, while proteins use a 20-letter alphabet.
- The ability of cell extracts to perform protein synthesis in a test tube allowed researchers to understand how information flows from messenger RNA to polypeptide.
- RNA polymerase produces RNA molecules by copying one strand of DNA, starting at a promoter sequence and ending at a terminator sequence.
- The RNA products of eukaryotic transcription often undergo extensive modification.
- Translation of messenger RNA requires correct initiation, elongation, and termination of polypeptide synthesis.
- In eukaryotic cells, special amino acid sequences help specify the destinations of newly made polypeptides.

OUTLINE

ESSAYS

In Chapter 13, we saw that the sequence of nucleotides in a gene somehow specifies the sequence of amino acids in a protein. The function of a protein—its ability to serve as an enzyme, a structural element, a transporter, or a motor—depends on its three-dimensional structure, which in turn depends on its amino acid sequence. Our questions in this chapter center on the cellular processes that manufacture proteins.

The complicated process of building a protein from the blueprints in DNA results from a series of chemical reactions that biologists now understand in great detail. These reactions are subject to the same thermodynamic laws as other chemical processes. The same kinds of forces drive them. No special force is needed to build the machinery of life.

Think for a moment about the production of a protein as you would about the production of a car. A full understanding of each process should summarize (1) the flow of information from design to working drawings to finished product and (2) the character and contributions of each stage in the assembly line. For a car, we want to know how workers and their machines execute detailed plans by putting together pieces of metal, plastic, and semiconductors. For a protein, we have the same questions about intracellular assembly. First, however, we examine the basic flow of information stored in DNA.

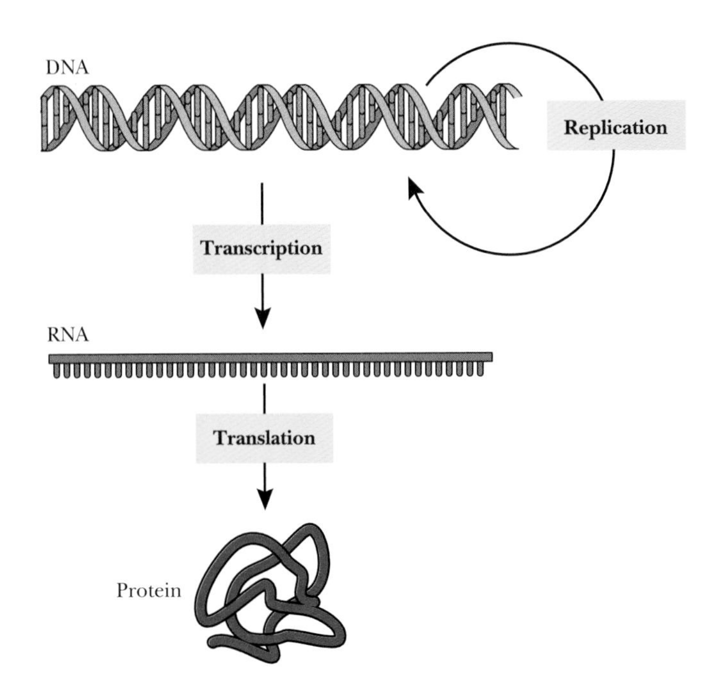

Figure 14-1 The "Central Dogma" of molecular biology: DNA specifies RNA, which specifies protein; DNA can also specify more DNA.

HOW DOES GENETIC INFORMATION FLOW FROM DNA TO RNA TO PROTEINS?

By the mid-1950s, biologists had concluded that genes specify the sequences of amino acids in proteins and that genes are made of DNA. The next question, then, was how a sequence of nucleotides in DNA specifies a sequence of amino acids in a polypeptide.

In the early 1960s, molecular biologists discovered that RNA is always an intermediate in the production of proteins. Francis Crick, one of the discoverers of the DNA double helix, summarized cellular information flow in a statement that he facetiously called the **Central Dogma** of molecular biology: "DNA specifies RNA, which specifies proteins." Figure 14-1 schematically illustrates the Central Dogma, noting that DNA also specifies more DNA, as explained in Chapter 13. Before we discuss the molecular machinery responsible for this flow of information, we will examine the general rules by which cells convert information from a nucleotide alphabet to an amino acid alphabet.

DNA and RNA Use a 4-Letter Alphabet, While Proteins Use a 20-Letter Alphabet

Molecular biologists speak of the nucleotide sequences in DNA and RNA and the amino acid sequences of polypeptides as two different languages. The languages of DNA and RNA are written with alphabets of 4 nucleotides; the polypeptide alphabet has 20 amino acid letters.

Recall that RNA, like DNA, is a polynucleotide and that it differs from DNA in three ways: (1) The backbone sugar is ribose instead of deoxyribose; (2) uracil (U) replaces thymine (T) as one of the pyrimidine bases; and (3) RNA is usually single stranded, whereas DNA is usually double stranded. (See Chapter 5.)

As illustrated in Figure 14-2, however, uracil can form a Watson-Crick base pair with adenine in exactly the same way as does thymine. The RNA alphabet (A, G, C, and U) is functionally the same as the DNA alphabet (A, G, C, and T) and totally different from the protein alphabet (the 20 amino acids). Molecular biologists call the production of RNA from DNA **transcription,** while the conversion of information from RNA into polypeptides is called **translation,** as illustrated in Figure 14-1.

Although the DNA and protein alphabets are different, we know that a close relationship, called the **Genetic Code,** exists between the letters of the two alphabets. Like the International Morse Code or the signing alphabet, the genetic code specifies the "meaning" in one alphabet of each set of symbols in the other alphabet. Because of the strict relationship of letters in the two alphabets, the problem of translating from the DNA language into the polypeptide language is much less difficult than, say, translating from English into Greek.

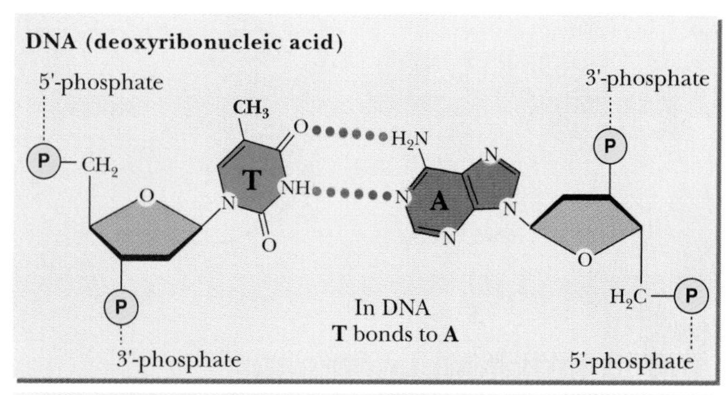

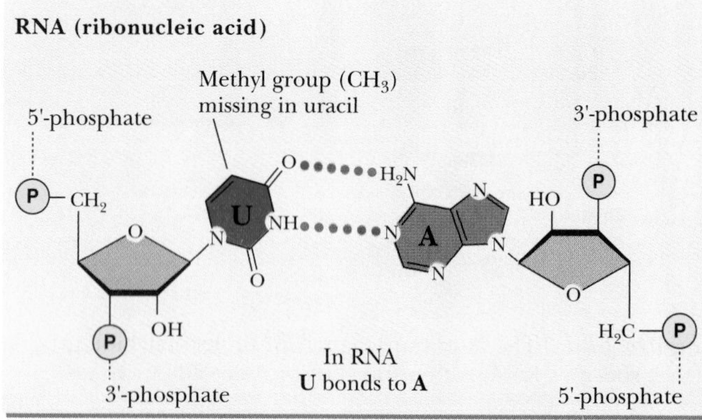

Figure 14-2 In DNA, A pairs with T; in RNA, A pairs with U.

First letter (5'-end)	Second letter: U	C	A	G	Third letter (3'-end)
U	UUU Phe	UCU Ser	UAU Tyr	UGU Cys	U
	UUC	UCC	UAC	UGC	C
	UUA Leu	UCA	UAA Stop	UGA Stop	A
	UUG	UCG	UAG Stop	UGG Trp	G
C	CUU Leu	CCU Pro	CAU His	CGU Arg	U
	CUC	CCC	CAC	CGC	C
	CUA	CCA	CAA Gln	CGA	A
	CUG	CCG	CAG	CGG	G
A	AUU Ile	ACU Thr	AAU Asn	AGU Ser	U
	AUC	ACC	AAC	AGC	C
	AUA	ACA	AAA Lys	AGA Arg	A
	AUG Met	ACG	AAG	AGG	G
G	GUU Val	GCU Ala	GAU Asp	GGU Gly	U
	GUC	GCC	GAC	GGC	C
	GUA	GCA	GAA Glu	GGA	A
	GUG	GCG	GAG	GGG	G

= Chain termination codon (stop)

= Initiation codon

Figure 14-3 The Genetic Code. Of the 64 possible triplet codons, 61 specify amino acids, and 3 signify "stop." The Met codon, AUG, can also specify "start," as discussed later.

How Many Nucleotides Specify a Single Amino Acid?

A change (mutation) in a single nucleotide in the DNA of a particular gene often leads to a single amino acid substitution in the corresponding protein, as first demonstrated for hemoglobin S in sickle cell disease. (See Chapter 12, p. 322.) The most straightforward explanation for this is that the nucleotide letters of DNA somehow code for the letters of the protein. Deciphering the genetic code began in the 1950s, initially as an intellectual exercise in cryptography. An impressive group of scientists, many of them trained as physicists, attempted to solve the puzzle through logic and through the then limited knowledge of the amino acid sequences of a few proteins.

This group's first question concerned the nature of the nucleotide "word" that specified a single amino acid. If each nucleotide corresponded to a single amino acid, then DNA could specify only four different amino acids. Specifying more than four amino acids from only four nucleotides requires that each nucleotide word must be a combination of nucleotides. Two nucleotides could specify a total of 16 (4×4) different amino acids. Because there are 20 amino acids in proteins, there must be at least three nucleotides per nucleotide word, which could specify a total of 64 ($4 \times 4 \times 4$) different amino acids. Each nucleotide word is called a **codon,** a group of nucleotides which specifies a single amino acid residue. The actual genetic code for nuclear genes is shown in Figure 14-3. By convention, codons are stated as the sequences found on mRNA.

Experimental confirmation that each codon consists of a triplet of nucleotide residues came in 1961 from studies with **acridine dyes,** chemical mutagens that cause mutations by inserting themselves into the backbone of the DNA double helix, as illustrated in Figure 14-4. Each inserted dye molecule leads to the insertion or deletion of a single nucleotide residue. Working in Cambridge, England, Francis Crick and Sydney Brenner showed that a single acridine-induced insertion would cause a mutation of the bacteriophage gene that they were studying. A second such insertion would still be nonfunctional, but a third single-base insertion in the same gene restored nearly normal function.

These results supported the hypothesis that each codon consists of three nucleotides. Each insertion, Crick and Brenner reasoned, caused a shift in **reading frame,** the grouping of nucleotides into codons that specify an amino acid sequence. As illustrated in Figure 14-5, a shift of one nucleotide or of two nucleotides in a triplet code leads to a complete garbling of the message. In contrast,

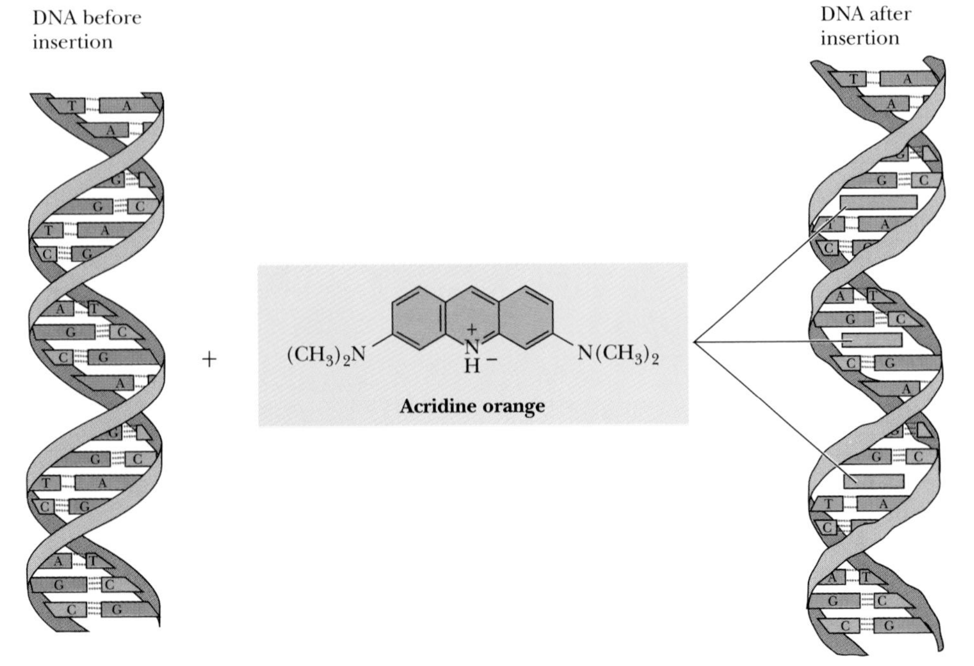

Figure 14-4 A molecule of acridine dye can insert into the DNA double helix, leading to the insertion of an extra nucleotide when DNA replicates.

insertions of three nucleotides can restore most of the message's original sense, even though it is slightly different from the original.

On the basis of these arguments and limited experimental evidence, molecular biologists concluded that a triplet of nucleotides in DNA specifies each amino acid residue in a protein. The next task was to "crack" the code—to assign each possible codon to a specific amino acid. To accomplish this, logic itself was not enough: The code was solved only after researchers could perform experiments with protein synthesis in a test tube. The solution of the Genetic Code depended on the ability to test the products specified by nucleotide sequences that were artificially synthesized. We describe these experiments later in this chapter, but we must first introduce the cell's relevant machinery.

Does DNA Itself Direct Protein Synthesis?

In order to reconstitute protein synthesis *in vitro*, experimenters had to know the responsible cell components. Because a sequence of codons in DNA specifies the sequence of amino acids in a protein, we might expect that DNA is involved directly in the linking of amino acids. But this is not the case.

The first hint that DNA is not directly involved in protein synthesis came from studies of immature red blood cells. In mammals red blood cells lose their nuclei, and therefore their DNA, as they enter the circulation from the bone marrow (where they first develop). Even after they lose their nuclei, however, they continue to make hemoglobin, as demonstrated by their ability to incorporate added amino acids into hemoglobin. In experiments, researchers isolated immature red blood cells from rabbits and incubated the cells in a mixture of amino acids, one of which (methionine) contained a radioactive isotope (^{35}S).

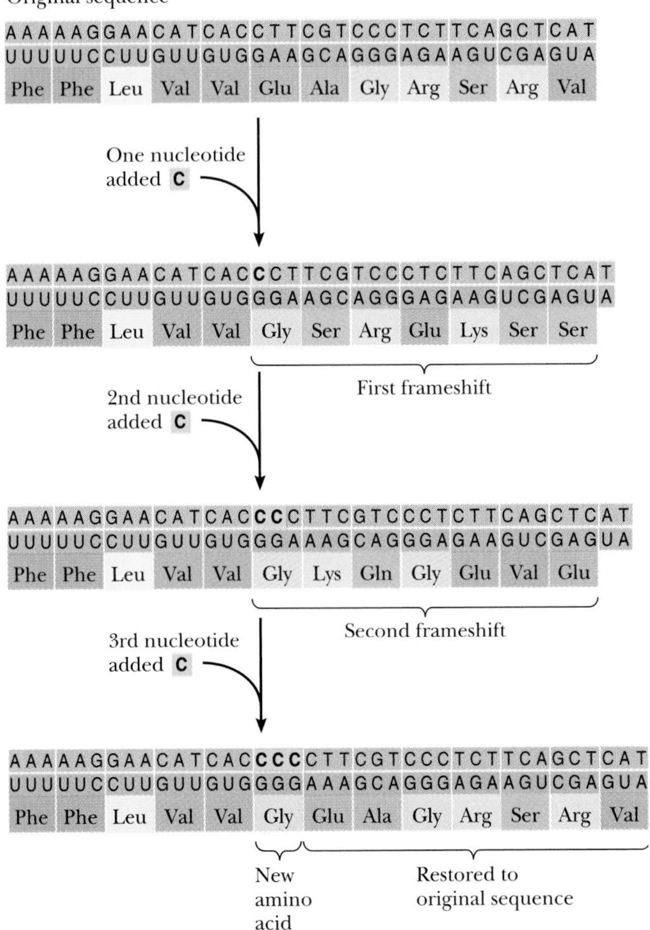

Figure 14-5 Insertion of a single nucleotide changes a Glu codon (GAA) to a Gly codon (GGA); it also shifts all the "downstream" codons. When two more nucleotides are inserted, the grouping of nucleotides ("reading frame") returns to the original, and the sequence downstream from the inserted nucleotides becomes normal again.

The cells incorporated some of the labeled methionine into newly made protein (mostly hemoglobin), showing that protein synthesis does not depend directly on the nucleus and that it does not depend directly on DNA.

But what components of the cytoplasm are responsible for protein synthesis? To answer this question, biochemists learned to make cell extracts that could convert radioactive amino acids into proteins in the test tube *(in vitro)*. Using the subcellular fractionation techniques described in Chapter 8, researchers could separate the components of these extracts and then combine them in different ways. As suspected from the studies of red blood cells, neither the nuclei of eukaryotic cells nor the nucleoids of prokaryotic cells are required for protein synthesis. The biggest subcellular components that are absolutely necessary are the *ribosomes*, assemblies of RNA and proteins that are the sites of protein synthesis. Ribosomes are not enough, however; protein synthesis also requires other cell components that are much smaller than ribosomes.

CELLS TRANSCRIBE INFORMATION FROM DNA INTO RNA AND TRANSLATE INFORMATION FROM RNA INTO PROTEINS

Subcellular fractionation established that DNA does not directly contribute to protein synthesis. How, then, can DNA specify polypeptide sequences?

To transmit its genetic orders, DNA relies on molecules that carry its information to the ribosomes, where protein synthesis occurs. The carriers of this information are **messenger RNA (mRNA)** molecules—RNA molecules whose sequence, copied from DNA, contains the information for building a polypeptide chain. As in DNA, a triplet codon in an mRNA specifies one amino acid.

Translation occurs on the ribosomes—the molecular machines that perform protein synthesis. As discussed in more detail later in this chapter, translation involves the use of "adapter" molecules, which match mRNA codons to individual amino acids. These adapters are members of another class of RNAs—**transfer RNAs (tRNAs),** each of which becomes linked to a particular amino acid. As the ribosome moves along the strand of mRNA, tRNA molecules—each linked to a specific amino acid—attach to the mRNA. The order of the tRNAs, and therefore the order of the amino acids, depends on the order of codons in the mRNA because each tRNA contains a codon-recognition sequence that forms Watson-Crick base pairs with an mRNA codon. A complete polypeptide ultimately forms according to the DNA blueprint and the mRNA working plans.

To summarize, protein synthesis requires at least the following components: (1) mRNAs, which carry genetic information from DNA; (2) ribosomes, which serve as the workbenches of polypeptide synthesis; (3) tRNAs, which serve as translators of each codon; and (4) additional enzymes and other proteins that are discussed later in this chapter. Before we talk more about the mechanism of protein synthesis, however, we look at how the ability to reconstitute protein synthesis from these cellular components allowed researchers to "crack" the Genetic Code.

Translating Artificial mRNAs in Cell Extracts Helped to Solve the Genetic Code

In 1961, Marshall Nirenberg and his collaborators at the National Institutes of Health in Bethesda, Maryland, were studying protein synthesis in extracts of bacteria. Nirenberg and J. Heinrich Matthei incubated their extracts with poly U, an artificially produced RNA that consisted entirely of U residues, rather than with mRNA isolated from a bacterial extract. They then examined the incorporation of different radioactive amino acids into protein. To their surprise (because the poly U was originally intended as a control that would not stimulate protein synthesis), the extracts that received poly U did produce a polypeptide. The product was strange, however, because it consisted of just one amino acid—phenylalanine—repetitively linked together by peptide bonds, as shown in Figure 14-6a. Subsequent similar experiments showed that another synthetic RNA, poly A, stimulated the production of a unique polypeptide, but in this case, the only amino acid incorporated was lysine. Similarly, poly C directed the synthesis of a polypeptide that contained only proline.

Nirenberg's experiments depended on the ability to use a bacterial enzyme to assemble nucleotides into artificial RNAs, so he could produce RNAs that had only a single type of codon. For example, the only triplet codon in poly U is UUU, the only codon in poly A is AAA, and the only codon in poly C is CCC. Nirenberg concluded that the codon UUU specifies phenylalanine, AAA lysine, and CCC proline.

Nirenberg could also produce RNAs with mixtures of codons. For example, an RNA made to contain a mixture of U and A residues could contain eight possible codons (UUU, AUU, UAU, UUA, UAA, AAU, AUA, and AAA), but these would be arranged at random within the polynucleotide chain. The resulting polypeptide would consist of random arrangements of phenylalanine, isoleucine, tyrosine, leucine, asparagine, and lysine but no other amino acids. But information of this sort was not enough to solve the code; the solution ultimately depended on the ability to make RNAs with completely known sequences.

The needed synthetic methods came in the early 1960s from the work of Gobind Khorana, an organic chemist at the University of Wisconsin. Khorana had developed new methods for linking nucleotides together. In one experiment, for example, he made small units of two or three nucleotides and put them together in long chains.

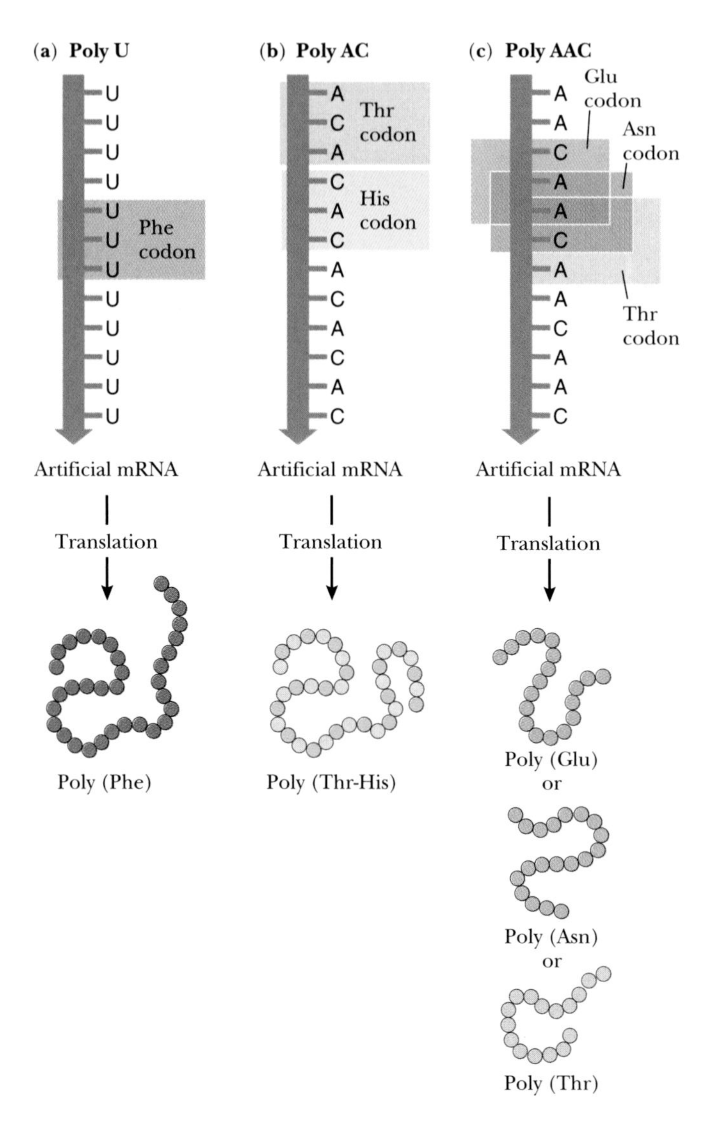

Figure 14-6 Artificial mRNAs, made in the laboratory, can specify polypeptides. **(a)** Poly U contains only one codon, UUU = Phe; **(b)** poly AC contains two codons, ACA = Thr and CAC = His; **(c)** poly AAC contains three codons, CAA = Glu, AAC = Asn, and ACA = Thr.

One such artificial RNA was poly (AC), which contained the sequence ACACACACAC. . . .

Khorana found that poly (AC) stimulated the production of a polypeptide of alternating threonine and histidine residues, as shown in Figure 14-6b. As also shown in Figure 14-6, the repeating triplet poly (AAC) contained the sequence AACAACAACAACAAC. . . . Poly (AAC) directed the synthesis of three distinct polypeptides—polymers of asparagine, threonine, and glutamine, as shown in Figure 14-7c. Khorana understood that each type of polymer resulted from the three different reading frames, which resulted in the different grouping of triplet codons—AAC AAC AAC AAC . . . or A ACA ACA ACA . . . or AA CAA CAA CAA CAA. . . . ACA specifies threonine, consistent with experiments using the ACACACACA polymer, while AAC specifies asparagine and CAA glutamine. From a series of such artificial mRNAs Khorana and his colleagues established a dictionary for many of the codons.

These experiments with artificial mRNAs demonstrated that mRNA is read sequentially, one codon after another. That is, each nucleotide is read as a part of only one codon, rather than having overlapping codons. The sequence AACAACAAC . . . is read AAC (Asn) AAC (Asn), AAC (Asn), *not* AAC (Asn) ACA (Thr) CAA (Gln) . . .

These experiments also confirmed that each codon consists of three nucleotides. Note in the example above that ACACACACA . . . pattern produced a polypeptide chain of two alternating amino acids. If each codon consisted of four nucleotides, the sequence would have been read ACAC ACAC, or CACA CACA, and the experiment would have yielded a polypeptide chain composed of only one repeating amino acid.

The use of artificial mRNAs of known sequence quickly led to an almost complete dictionary of the genetic code, but some codons were still missing. The final cracking of the Genetic Code depended on another method, developed by Nirenberg and Philip Leder in 1964. Using appropriate cell extracts, Nirenberg and Leder linked tRNAs to their corresponding amino acids, as shown in Figure 14-7. They then showed that each "charged" tRNA, a tRNA linked to its corresponding amino acid, would bind to ribosomes when the correct triplet codon was present, even if the codon was not part of mRNA or a large, synthetic RNA molecule. For example, if they added the synthetic trinucleotide UUU to a mixture of ribosomes and charged tRNAs, only tRNA molecules containing radioactively labeled lysine would attach to ribosomes. By 1966, this method had filled the remaining holes in the codon dictionary: The Genetic Code was solved.

The Genetic Code Is Degenerate, Contains Nonsense, and Is Nearly Universal

Experiments with *in vitro* protein-synthesizing systems led to the codon dictionary shown in Figure 14-3. A striking characteristic of the Code is that most amino acids can be specified by more than one codon. The Code is therefore said to be **degenerate** (a term borrowed from quantum physics), meaning that several codons may have the same meaning. This apparent redundancy within the Genetic Code was expected from the fact that there are 64 possible nucleotide triplets, but only 20 amino acids used in proteins.

Of the 64 triplets, 61 specify amino acids, while three do not. These three codons (UAA, UAG, and UGA) are called **nonsense codons** because they do not specify any amino acid. They are also called **stop codons** because they normally signal the end of a polypeptide chain.

The Genetic Code is **universal,** meaning that it is used by all species. For example, the precursors of red

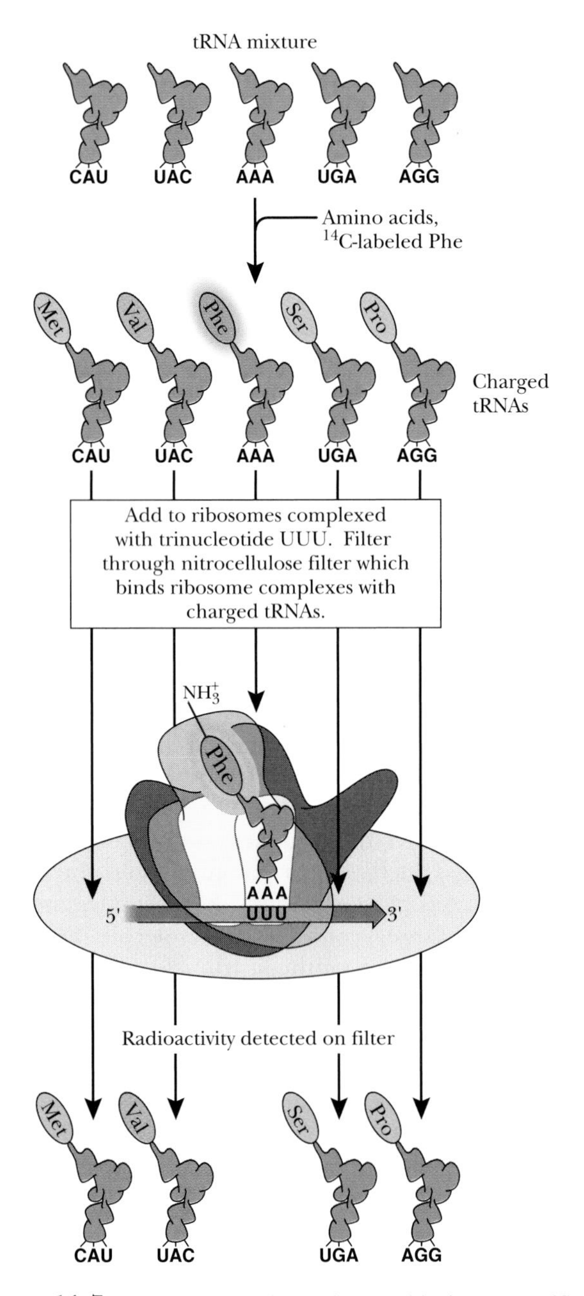

Figure 14-7 Each nucleotide triplet can bind to a specific aminoacyl-tRNA. UUU forms base pairs with the AAA "anticodon" of $tRNA^{Phe}$. In a mixture of aminoacyl-tRNAs, only Phe-$tRNA^{Phe}$ will bind to UUU.

blood cells, which are highly specialized for the production of hemoglobin, contain high levels of the mRNAs for α- and β-globin, the polypeptides that together form the hemoglobin molecule. When purified globin mRNA from rabbit red cell precursors is added to the protein-synthesizing machinery of wheat germ (ribosomes, tRNAs, and soluble enzymes and other proteins), the wheat machinery begins to make rabbit globin! This experiment again confirms that each mRNA contains the information for a particular sequence of amino acids. It also shows that the same Genetic Code is used in both wheat and rabbits.

Several exceptions to the universality of the Genetic Code, however, are present in the protein-synthesizing machinery within mitochondria, with two differences even between the mitochondria of yeast and those of mammals. These differences, not seen outside the mitochondria of any organisms, are consistent with two hypotheses that we discuss further in Chapter 24: (1) that mitochondria are descended from ancient prokaryotes that ceased to exist independently about 1.5 billion years ago and (2) that yeast and mammalian mitochondria had independent origins.

Experiments with *in vitro* protein synthesis solved the Genetic Code. They also established the Central Dogma: Genetic information flows from DNA to RNA to proteins. We now turn our attention to the mechanisms of this information flow, starting with transcription, the production of RNA.

HOW DOES RNA POLYMERASE TRANSCRIBE DNA INTO RNA?

RNA polymerase is the enzyme responsible for transcribing DNA into RNA. Using RNA polymerase isolated from cell extracts, researchers have discovered not only how RNA polymerase copies the nucleotide sequences of DNA, but also how signals within DNA designate the beginning and end of each RNA molecule.

Like DNA polymerase, RNA polymerase assembles each polynucleotide by adding nucleotides only to the 3′-end of a growing chain. In contrast to DNA polymerase, however, RNA polymerase does not require a primer and copies only one strand of DNA, called the **template** strand, as illustrated in Figure 14-8.

Transcription Involves RNA Polymerase Binding, Initiation, Elongation, and Termination

Figure 14-8 outlines the steps of transcription—initiation, elongation, and termination. **Initiation,** the start of synthesis, requires first that RNA polymerase bind to DNA and separate the two strands in such a way that it copies one (the template strand) but not the other, as shown in Figure 14-8a and b. A base pair then forms between a base in the template strand and a ribonucleotide triphosphate, also shown in Figure 14-8b. Notice that this first nucleotide retains its three phosphate groups (attached to the 5′ carbon atom of the ribose) as the RNA chain grows on the OH group attached to its 3′ carbon.

Elongation, the adding of additional nucleotide residues, occurs (as shown in Figure 14-8c) as an appropriate nucleotide triphosphate pairs with the next base on the template strand and RNA polymerase catalyzes the formation of a phosphodiester bond. The RNA strand grows from its 5′-end to its 3′-end, as it copies the template DNA strand from its 3′-end to its 5′-end. The addition of each

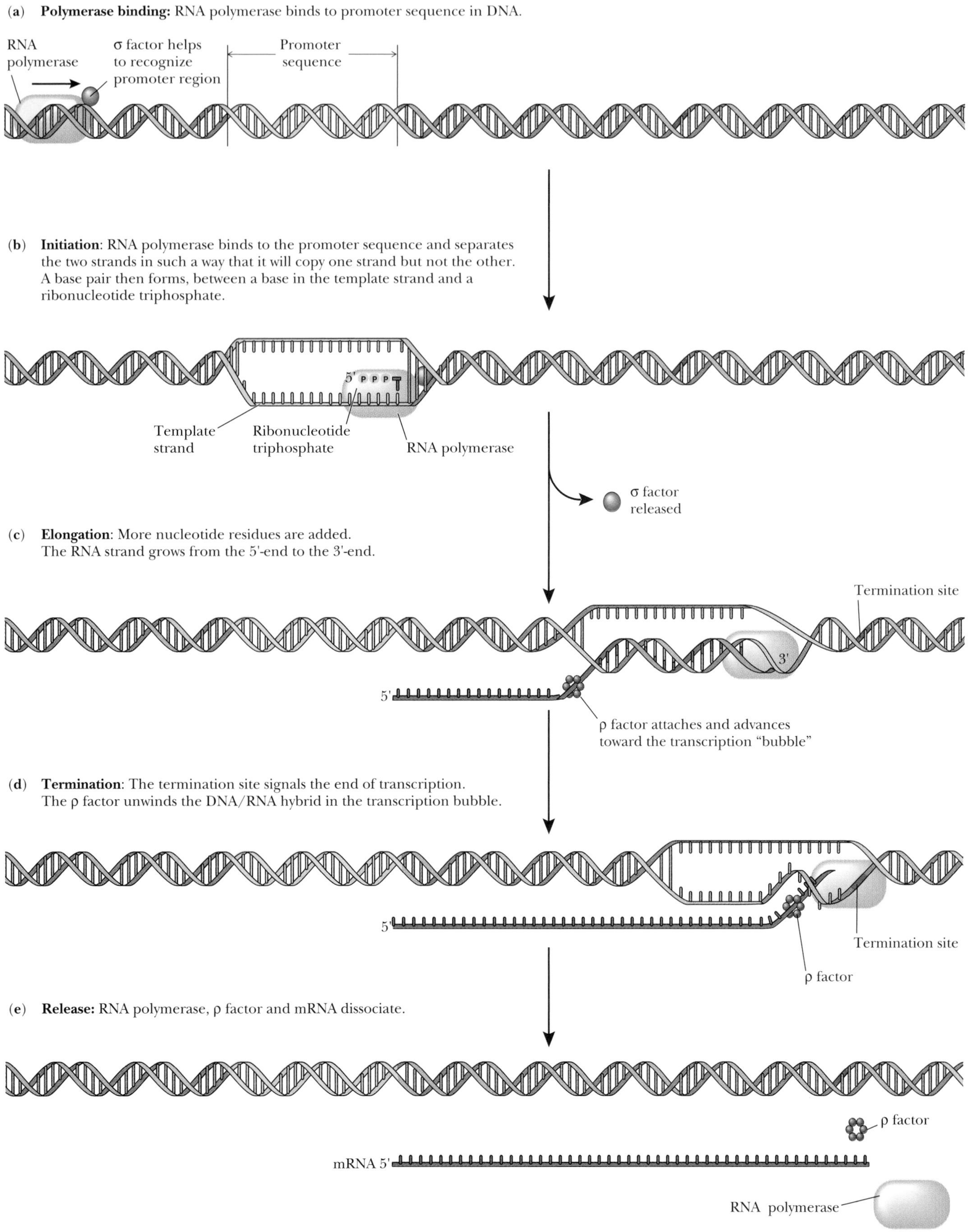

(a) **Polymerase binding:** RNA polymerase binds to promoter sequence in DNA.

RNA polymerase

σ factor helps to recognize promoter region

Promoter sequence

(b) **Initiation**: RNA polymerase binds to the promoter sequence and separates the two strands in such a way that it will copy one strand but not the other. A base pair then forms, between a base in the template strand and a ribonucleotide triphosphate.

5' P P P

Template strand

Ribonucleotide triphosphate

RNA polymerase

σ factor released

(c) **Elongation**: More nucleotide residues are added. The RNA strand grows from the 5'-end to the 3'-end.

Termination site

3'

5'

ρ factor attaches and advances toward the transcription "bubble"

(d) **Termination**: The termination site signals the end of transcription. The ρ factor unwinds the DNA/RNA hybrid in the transcription bubble.

5'

Termination site

ρ factor

(e) **Release:** RNA polymerase, ρ factor and mRNA dissociate.

ρ factor

mRNA 5'

RNA polymerase

Figure 14-8 Transcription in bacteria. **(a)** Binding of RNA polymerase; **(b)** initiation at the promoter sequence, assisted by the σ (sigma) factor; **(c)** elongation; **(d)** termination, here illustrated for a gene whose termination requires the ρ (rho) factor; for many genes, termination does not require ρ; **(e)** release of RNA polymerase, ρ, and mRNA.

nucleotide requires that the DNA duplex unwind a little more, a process helped along by the formation of a duplex between the newly made RNA and the DNA template strand, which extends for 10–12 bases behind the most recently added residue. As the RNA polymerase moves down the DNA, however, the growing RNA molecule detaches from the template and the DNA helix again forms. The new RNA molecule has exactly the same sequence of bases as the nontemplate strand of DNA, except that wherever DNA has a T, the RNA has a U.

Termination, the ending of chain growth, occurs when the RNA polymerase detaches from the DNA template and the growing RNA chain, as illustrated in Figure 14-8d. Specific sequences in DNA, called **termination sites,** signal the end of transcription. In many prokaryotic genes, the DNA sequence at the termination site specifies either a self-complementary sequence in the RNA causing the RNA to fold back on itself in a hairpin duplex, or a stretch of U's. Termination mechanisms in eukaryotes are still being elucidated.

The stretch of U's in the RNA forms an unstable RNA-DNA duplex, because each AU base pair has only two hydrogen bonds, while a GC pair has three. Mutations that reduce the size of the U stretch or reduce the ability to form a hairpin decrease the probability of chain termination resulting in an oversized RNA.

Termination also occurs at termination sites that do not have these characteristic features. These terminations require additional protein factors, the best studied of which is called **rho** (ρ). Other proteins, called **antiterminators,** can allow elongation to continue through DNA sequences that could otherwise serve as termination sites.

How Does RNA Polymerase Know Where to Begin and in Which Direction to Transcribe?

As mentioned previously, RNA polymerase does not require a primer. RNA polymerase gains access to the template strand by locally unwinding and separating the two strands of DNA at the **promoter,** the DNA region where RNA polymerase first binds as it begins transcription. The binding of the RNA polymerase to the promoter region determines which of the two DNA strands will be transcribed and therefore the direction in which the RNA polymerase will move along the chromosome.

The ability of RNA polymerase to bind and unwind DNA depends on untranscribed sequences in the region of the promoter. When RNA polymerase is mixed with DNA from the promoter region, it forms a tight complex. When this DNA-protein mixture is exposed to chemical agents that break DNA, the nucleotides within the promoter are protected from destruction, while nucleotides outside the promoter are not protected. When the chemically treated DNA is then subjected to electrophoretic separation such as that used for DNA sequencing, the protected nucleotides within the promoter leave a characteristic "footprint" in the banding pattern.

Comparisons of the DNA sequences in the promoter regions of many genes reveal shared **consensus sequences** (also called "sequence motifs" or "boxes" because of the way they are marked in diagrams)—characteristic nucleotide arrangements that are common to many genes (usually in many species). The promoter region of many bacterial genes, for example, contains two consensus sequences of about six nucleotide pairs each, with about 25 nucleotide pairs between them, as shown in Figure 14-9. One of these consensus sequences is called the **Pribnow box** (in bacteria), after its discoverer, or the **TATA box** (in eukaryotes), after its first four nucleotide residues. The actual **initiation site,** the first nucleotide actually copied into RNA, lies about 10 nucleotide pairs in the 3′ ("downstream") direction from the Prinbnow box in bacteria and about 25–30 nucleotide pairs downstream from the TATA box in eukaryotes..

Using recombinant DNA techniques (described in Chapter 17), researchers can change any DNA sequence at will. Such directed changes in one of the identified consensus sequences usually reduce the rate at which RNA polymerase begins transcription. Such mutations are said to "alter the strength of the promoter."

Molecular Biologists Define "Upstream" and "Downstream" with Respect to the Nontemplate Strand

The direction of RNA synthesis during transcription, like the direction of DNA replication, is always from 5′ to 3′, with additional nucleotides added at the 3′-end. RNA polymerase reads the template strand of DNA in the 3′ to 5′ direction while catalyzing the elongation of RNA in the 5′ to 3′ direction. By convention, however, molecular biologists describe the direction of transcription in terms of the nontemplate strand, that is, the strand of DNA whose

Figure 14-9 RNA polymerase binds to specific sequences in DNA. Shown here is the consensus sequence for RNA polymerase binding—nucleotides protected from chemical degradation by RNA polymerase.

Pribnow box; Initiation site

Consensus sequence: T C T T G A C A T ···[about 25 bp]··· T A T A A T ···[about 10 bp]··· A

sequence is the same as that of the mRNA. The promoter is said to be **"upstream"**—that is, in a 5′ direction—from the termination site. Similarly, the termination site is said to be **"downstream"** from the promoter site.

Different Types of RNA Polymerase Recognize Different Promoters

The RNA polymerase from *E. coli,* the first RNA polymerase to be characterized biochemically, consists of four polypeptide chains, two of which are identical and called α, while the other two are called β and β' (to indicate that they are distinct from α and similar but not identical to one another). This RNA polymerase is designated as $\alpha_2\beta\beta'$. The $\alpha_2\beta\beta'$ assembly by itself can catalyze the formation of phosphodiester bonds, but it cannot initiate transcription. Specific strong attachment to the promoter region requires an accessory protein, called **sigma (σ),** which associates with the other polypeptides and allows binding to the promoter and the initiation of transcription, as shown in Figure 14-8.

Bacteria produce a variety of σ factors that recognize different consensus sequences. For example, when a culture of *E. coli* encounters a **heat shock,** a sudden increase in temperature (say to 50°C), the bacteria quickly begin to make large amounts of about 17 proteins that apparently contribute to surviving the new and adverse temperature. This switch in the pattern of protein synthesis depends on the production of a new σ factor (called σ^{32}), which recognizes consensus sequences distinct from the σ factor that is normally present in *E. coli.*

Eukaryotic cells have at least five RNA polymerases. One type of these polymerases is limited to mitochondria and one to chloroplasts, while three are found in nuclei. The polymerases of mitochondria and chloroplasts resemble those of bacteria, but those of nuclei consist of at least ten different polypeptides. The amino acid sequences of the two largest of these polypeptides, however, resemble the sequences of the β polypeptide of *E. coli,* suggesting that they have a common evolutionary origin.

Nuclear RNA polymerases—which are classified as polymerases I, II, and III—transcribe different sets of genes. Polymerase I, which is located in the nucleolus, is responsible for the synthesis of the large RNAs of the ribosomes. Polymerase II produces mRNA and its precursors. Polymerase III makes a large set of small, stable RNAs (generally smaller than 300 nucleotides), which contribute to protein synthesis (as components of the ribosomes or as the tRNA adapters, discussed later in this chapter), or to the processing of mRNA precursors. Each kind of RNA polymerase starts transcription at distinct sets of promoters.

Unlike prokaryotic RNA polymerases, however, eukaryotic nuclear RNA polymerases do not themselves recognize the nucleotide sequences within promoters. Instead, they depend on the cooperation of other proteins, called transcription factors. For example, one transcription factor for RNA polymerase II (called TFIID) binds to the TATA box, which exists in most eukaryotic genes about 25 nucleotides away from the first nucleotide to be copied into RNA. The interaction of RNA polymerases with transcription factors is a major factor in determining which genes are expressed in which cells and under what conditions. Identifying these factors and understanding their interactions with specific DNA sequences and with RNA polymerase is a major activity of contemporary molecular biologists. We have more to say about this in Chapter 15.

HOW DOES EUKARYOTIC TRANSCRIPTION DIFFER FROM THAT IN PROKARYOTES?

In prokaryotes, RNA polymerase can directly produce a functional mRNA. A bacterial mRNA binds to ribosomes and begins to direct protein synthesis even as it is still being transcribed from DNA. In eukaryotes, however, such direct coupling of transcription and translation is impossible because RNA is made in the nucleus and proteins are made in the cytoplasm.

Eukaryotic transcription and translation are separated by much more than the nuclear membrane. In eukaryotes, no mature RNA (an RNA molecule that functions in protein synthesis) is identical to the corresponding **primary transcript,** the RNA first transcribed in the nucleus. Instead, the primary transcripts made by all three nuclear RNA polymerases undergo **post-transcriptional processing,** a series of chemical modifications that convert the primary transcript to a mature (functional) RNA. Post-transcriptional processing usually includes cutting (hydrolysis) of the primary transcript and may also involve the splicing of RNA fragments, addition or removal of chemical groups (methyl and phosphate groups), selective addition of nucleotides not encoded by DNA, and even substitution of specific nucleotide residues. Among other results of post-transcriptional processing, almost all functional RNAs in eukaryotic cells are shorter than the primary transcripts from which they derive.

Three Mature Ribosomal RNAs Derive from a Single Primary Transcript Produced by RNA Polymerase I

Ribosomes are macromolecular assemblies, with molecular masses of several million. (See Chapter 5) Ribosomes of all species consist of two subunits, called the large subunit and the small subunit. Each subunit consists of 1–3 rRNA molecules and dozens of protein molecules. Figure 14-11 compares the composition of large and small subunits from prokaryotes and eukaryotes The ribosomes of mitochondria and chloroplasts from eukaryotes resemble those

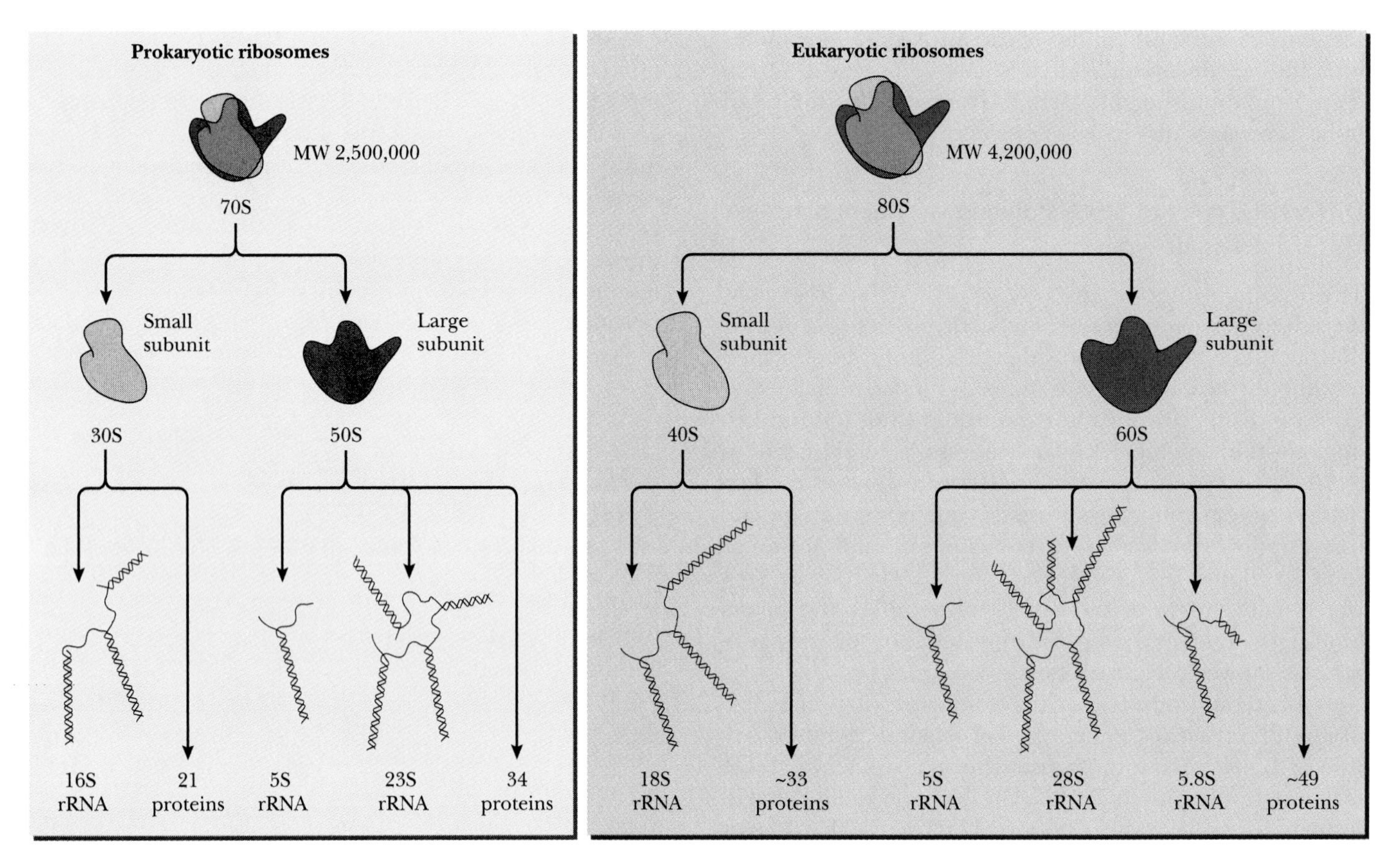

Figure 14-10 Ribosomes from prokaryotes differ from those of eukaryotes in the total size of the ribosomes, in the sizes of ribosomal RNAs, and in the numbers of proteins associated with each ribosomal subunit. The names of the ribosomes, ribosomal subunits, and ribosomal RNAs are given as S values (70S, 80S, etc.), referring to the speed (in units called "Svedbergs") with which these molecules move in a centrifugal field.

of prokaryotes, again supporting the idea that energy organelles evolved from prokaryotic endosymbionts (as discussed in Chapters 6 and 24).

The *E. coli* ribosome, for example, has a total molecular size of 2.5×10^6 (corresponding to a mass of 4×10^{-18} g) and consists of three molecules of RNA and 55 types of protein molecules. The cytoplasmic ribosomes of eukaryotic cells (which we call eukaryotic ribosomes), however, are larger and more complicated than prokaryotic ribosomes. They have a molecular size of 4.2×10^6 and consist of four molecules of RNA and some 82 types of protein.

Within any species, the rRNA molecules in cytoplasmic ribosomes are the same from ribosome to ribosome no matter what types of protein they are producing. In contrast, every mRNA specifies the amino acid sequence of a specific polypeptide, so a cell contains 10^3–10^5 different mRNAs, but only a few rRNAs. These rRNAs are named according to the rate at which they sediment in an ultracentrifuge. As noted in Figure 14-10, prokaryotic ribosomes have one rRNA, called 16S rRNA, in their small subunits, and two rRNAs, called 5S and 23S rRNAs, in their large subunits. The small subunits of eukaryotic ribosomes also have one rRNA, which is slightly larger than the prokaryotic equivalent and is called 18S, while the large subunits contains three (rather than two) rRNAs—called 5S, 28S, and 5.8S rRNA.

The four rRNAs of eukaryotic cytoplasmic ribosomes are transcribed within the nucleus from genes that are present in more than 100 copies in the genome of mammals. RNA polymerase III transcribes 5S RNA, while RNA polymerase I transcribes the other three rRNAs. Remarkably, however, the three rRNAs transcribed by RNA polymerase I (5.8S, 18S, and 28S) all derive from a single giant precursor molecule, called 45S pre-rRNA.

The first evidence for the existence of this precursor came from experiments that studied the sizes of newly made RNAs in cultured mammalian cells. After incubating cells for different times with ^{3}H-uridine, a radioactive precursor of RNA, researchers analyzed the sizes of the newly made, radioactive RNAs by ultracentrifugation. After long incubations, the most prominently labeled RNAs were the 18S and 28S rRNAs, which make up more than 90% of the total RNA within cells. After short incubations, however, most of the radioactivity was in 45S RNA, a molecule that is present in such low overall concentrations that researchers were not previously aware of its existence.

What happened to the 45S RNA during the longer

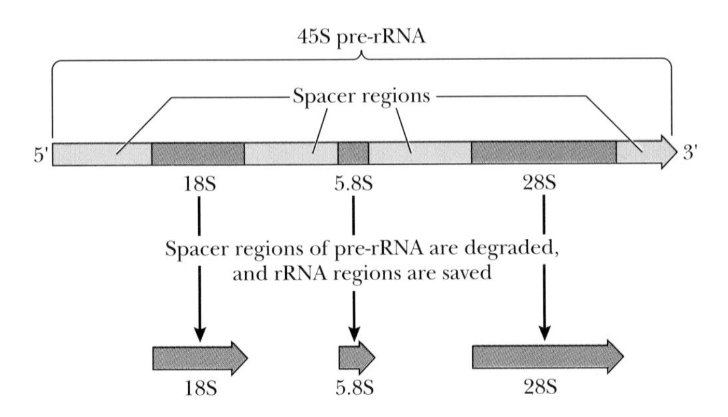

Figure 14-11 Three ribosomal RNAs—18S, 5.8S, and 28S—all derive from the same precursor. In the precursor pre-rRNA, the three sequences are separated from each other by "spacers," which are rapidly degraded after the rRNA sequences are cut out.

incubation periods? To study the fate of the 45S RNA, researchers then performed a "pulse-chase" experiment. They exposed the cells to radioactively labeled uridine (^{3}H-uridine) for only a few minutes, enough to label 45S RNA. They then stopped the further incorporation of the label by adding actinomycin D, an inhibitor of RNA synthesis. The labeled 45S RNA disappeared, and the label then appeared in several smaller RNAs (41S, 32S, and 20S)—and, eventually, in 5.8S, 18S, and 28S rRNAs. These experiments led to a picture, diagrammed in Figure 14-11, of post-transcriptional processing of rRNA precursors, involving cuts at specific points in the rRNA precursors. The biggest surprise of these experiments was that the processing of a single RNA molecule (the 45S pre-rRNA) yields three different RNA components of a functional ribosome.

Studies of rDNA (the gene from which the 45S rRNA is transcribed) confirmed the conclusion that a single precursor contains the sequences for three rRNAs. Later studies in a number of species also established the organizational principles shared by all eukaryotic rDNAs: (1) Many copies (usually 100–200) of the gene for the ribosomal RNA precursor are present in tandem arrays on several chromosomes, with **untranscribed spacer DNA** lying between each of the repeated copies; (2) in all species examined, the precursor gene contains rRNA sequences in the same order, 18S, then 5.8S, then 28S; (3) the length of the transcribed precursor is larger than the sum of the lengths of the three mature rRNAs, with 20% to 50% of the precursor consisting of **transcribed spacer RNA,** RNA segments between the mature rRNA sequences which are rapidly destroyed within the nucleus; (4) the lengths of the untranscribed spacer DNA and the transcribed spacer RNAs vary from species to species, but the organization in each species is always identical.

The processing of pre-rRNA takes place in the nucleolus. Formation of the nucleolus within the nucleus depends on interactions with the rRNA genes. The pre-rRNA genes act as a **nucleolar organizer,** loops of DNA (sometimes from different chromosomes) that come together after mitosis and serve as the site where the nucleolus reforms. Recall that nucleoli disappear during mitotic prophase. The nucleolus itself (as discussed in Chapter 5) is a distinctive part of the nucleus where rRNA is transcribed from DNA, processed, and assembled with proteins to form ribosomal subunits. The newly assembled ribosomal subunits then exit from the nucleus via nuclear pores.

The Formation of Transfer RNAs Requires Processing of a Primary Transcript Produced by RNA Polymerase III

RNA polymerase III, like RNA polymerase I, transcribes a relatively small set of genes whose final products are RNAs rather than proteins. The RNAs made by RNA polymerase III are all short (less than 300 bases) and relatively abundant. They include 5S rRNA (which is a component of the large ribosomal subunit), tRNAs (a set of about 40 molecules that serve as adapter molecules in protein synthesis, as discussed later in this chapter), and a set of small nuclear RNAs (snRNAs, which participate in the processing of mRNA precursors produced by RNA polymerase II). In some species at least, the 5S RNA that appears in ribosomes is identical to the primary transcript of the 5S genes. tRNAs, however, are highly modified after transcription.

Mature tRNAs are 75–80 bases long, but their corresponding primary transcripts are up to 40 bases longer. The primary transcripts are trimmed to size. In some cases, moreover, nuclear machinery removes a piece of RNA in the middle of the primary transcript and splices the two remaining pieces together. Other nuclear enzymes modify about 10% of the bases in each tRNA molecule, for example changing some uridine residues into a nucleotide called "pseudouridine." Finally, another enzyme removes the three nucleotide residues at the 3′-end of every tRNA and replaces them with the residues CCA.

These complex processes—trimming, splicing, base modification, and CCA addition—all depend on nuclear enzymes. The nucleus of a frog oocyte, a large cell that eventually matures into an egg, can correctly process an injected precursor of a yeast tRNA. This suggests that both the signals for processing in the tRNA precursor and the machinery for processing in the nucleus are likely to be universal among eukaryotes.

The Formation of Mature mRNAs Requires Processing of Primary Transcripts Produced by RNA Polymerase II

In eukaryotes, all genes that encode polypeptides (as opposed to genes that specify types of RNA that are not translated into proteins) are transcribed by RNA polymerase II.

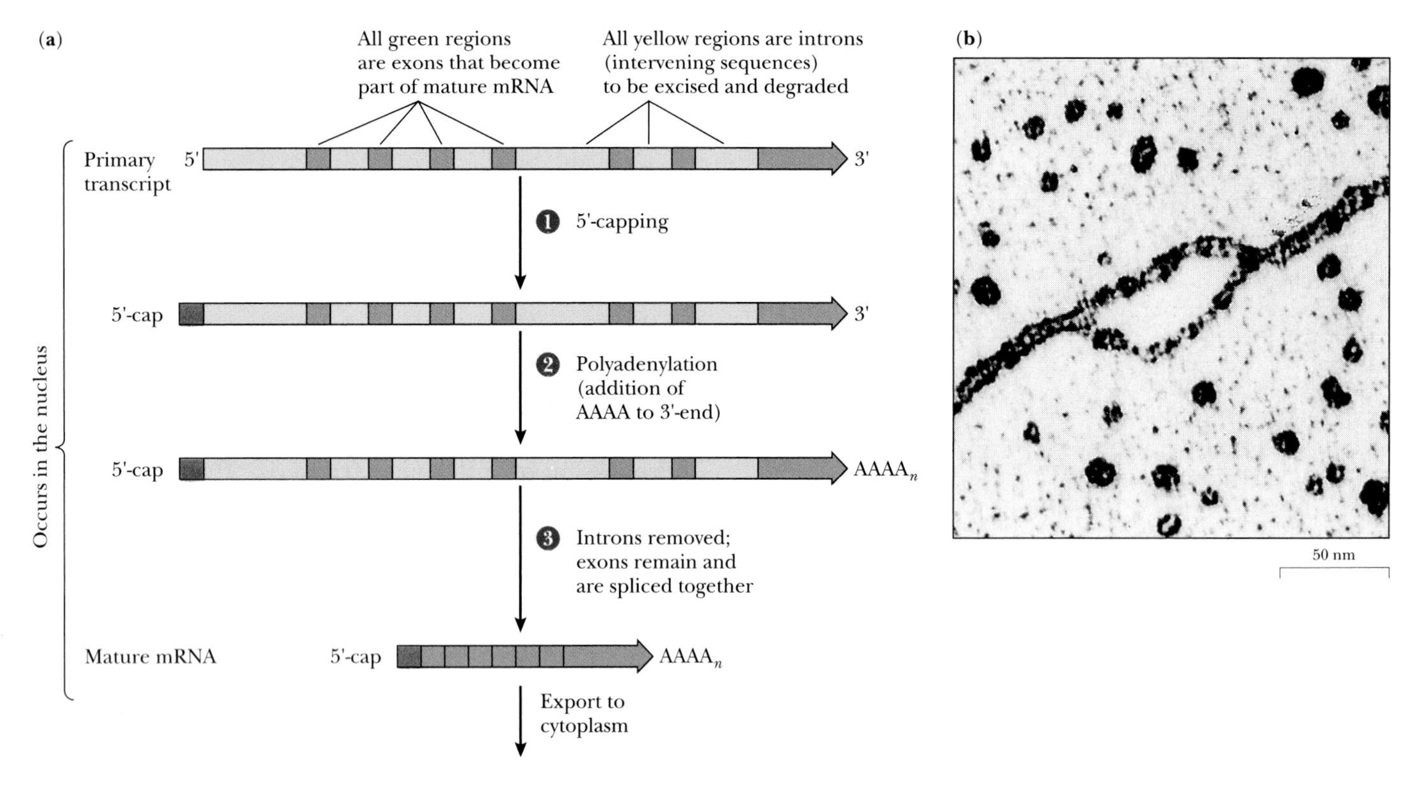

Figure 14-12 Production of a mature messenger RNA from a primary transcript. **(a)** Post-transcriptional events include capping, polyadenylation, and exon splicing. **(b)** Electron micrograph of an intron within the pre-mRNA for β-globin. *(b, Jeff Ross, McArdle Laboratory)*

The mRNAs that are ultimately translated into polypeptides, however, are never identical to the primary transcripts but result from post-transcriptional processing of those transcripts. The term **pre-mRNA** refers to both the primary transcript and its partly modified, not yet mature products. The collection of pre-mRNAs in the nucleus is called **heterogeneous nuclear RNA** (or **hnRNA**). While pre-mRNAs may be tens of thousands of bases long, most mature eukaryotic mRNAs are 1000–3000 bases long.

Before a mature mRNA exits from the nucleus, it undergoes a number of modifications, which are shown in Figure 14-12: (1) "capping" the 5′-end of the RNA with a modified GTP; (2) adding a string of A residues (polyadenylation) to the 3′-end of most mRNAs; and (3) removing large pieces of RNA from the transcript and splicing the remainder together. In addition, a recently discovered process, called RNA editing, can add, remove, or change the sequence of nucleotides in particular regions of RNA, thereby changing the protein-coding properties of that RNA.

Almost all eukaryotic genes contain **introns** (or **intervening sequences**), long stretches of DNA that are transcribed into RNA but are excised before the primary transcript matures into functional mRNA. The sequences present in mature mRNAs (most of which encode polypeptides) are called **exons.** The discovery of introns and exons, in 1977, was particularly surprising because it demonstrated that genes contain, and are actually interrupted by, sequences that do not code for proteins.

A single pre-mRNA molecule may contain many introns, ranging in size from about 80 to more than 10,000 nucleotides. These introns help account for part of the 98% of human DNA that does not directly specify polypeptide sequences.

Mature mRNA usually results from an elaborate process that removes introns and splices together the remaining exons, like so many pieces of film or magnetic tape. DNA and the primary RNA transcript also contain sequences that specify where the splices are to occur. The pattern of splicing of a single transcript can vary from tissue to tissue.

As discussed in Chapter 24, the first genes were probably RNA rather than DNA, and present genes probably resulted from the combination of small coding regions. Like contemporary eukaryotic genes, the earliest genes, we now think, were mosaics of exons and introns.

RNA Modification and Splicing Depend on Nuclear Machinery and Specific Signals Within Pre-mRNAs

Within the nucleus, one enzyme caps the 5′-end of a pre-mRNA, while another adds a poly A tail to its 3′-end, usually about 20 bases downstream from the sequence AAUAAA. The removal of introns and the splicing of ex-

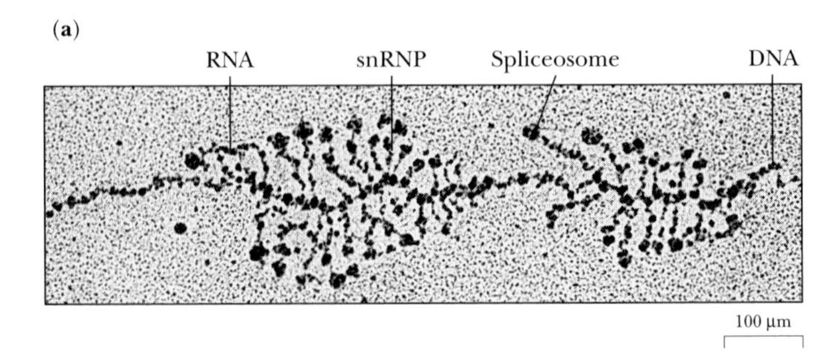

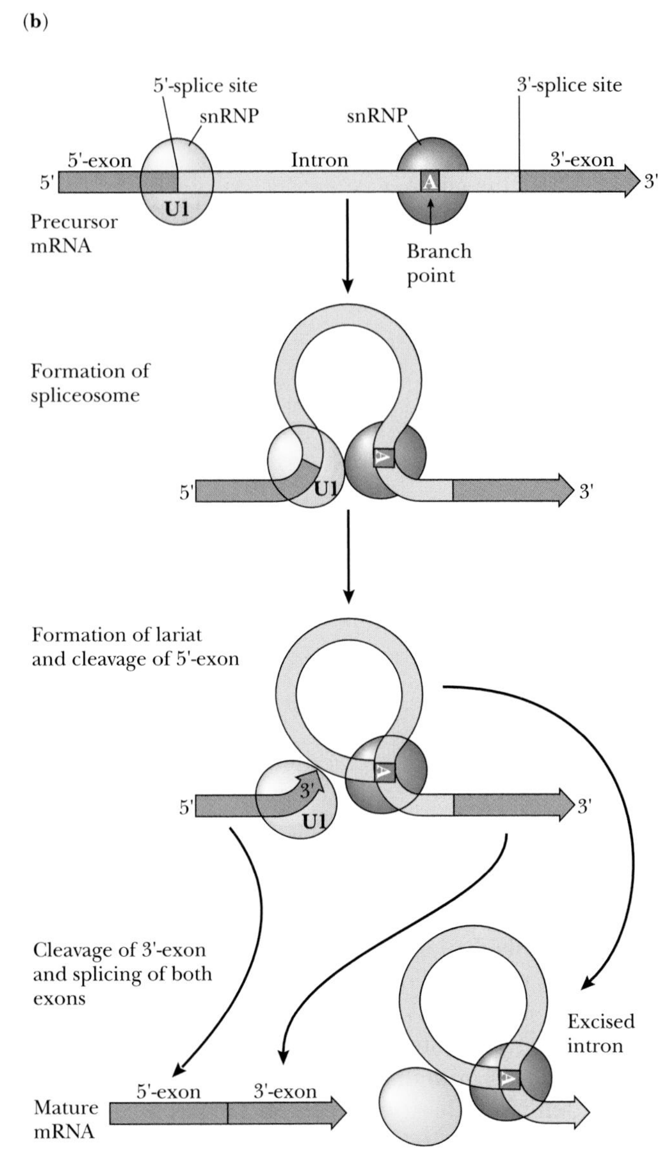

Figure 14-13 The mechanics of exon splicing in pre-mRNA. **(a)** An electron micrograph of a spliceosome, the protein-RNA complex in which exon splicing occurs; **(b)** steps in removing an intron. *(a, Yvonne Osheim)*

ons, however, depend on more elaborate complexes, called **spliceosomes,** illustrated in Figure 14-13a. Each spliceosome contains a pre-mRNA and several smaller complexes (called **snRNPs**—small nuclear ribonucleoprotein particles—and pronounced "snurps") of proteins and small nuclear RNAs (some of those transcribed by RNA polymerase III).

The splicing of a pre-mRNA takes place in several steps (illustrated in Figure 14-13b), which researchers can now study in cell extracts as well as in intact cells, aided by the ability to produce pre-RNAs in a test tube, using recombinant DNA techniques. Even before transcription is completed, one of the small nuclear RNAs (called U1) forms a duplex with complementary sequences at the 5′-end of an intron, facilitating the first cut by the U1 snRNP. At the same time (or possibly before) another snRNP attaches to a **branch point,** an A residue about 25 bases upstream from the 3′ splice site. After the 5′ cut, other snRNPs hold the 3′-end of the first exon in the spliceosome, while the 5′-end of the intron forms a **lariat** structure (because it resembles a rope-twirler's lariat) at the branch point. Finally, the snRNPs attached to the branch point facilitate the cut at the 3′-end of the intron and the splicing together of the two exons.

While splicing of pre-mRNAs (and of pre-tRNAs and pre-rRNAs) almost always requires the participation of specific enzymes and other proteins (such as those of the snRNPs), some RNA molecules are able to perform **self-splicing,** removal of intervening sequences without the participation of other molecules. The first known example of such self-splicing was in the production of ribosomal RNAs in the protozoan *Tetrahymena thermophila.* The process is illustrated in Figure 14-14. Pre-rRNA from *Tetrahymena,* as well as a number of other RNA precursors from other organisms, acts as an enzyme that cuts and splices itself. The demonstration that an RNA (as well as a protein) could act as an enzyme led to the reformulation of accepted ideas about early evolution, as discussed in Chapter 24.

PROTEIN SYNTHESIS IN A TEST TUBE ALLOWED RESEARCHERS TO DISSECT THE SYNTHETIC MACHINERY

Much as the head of a tape deck converts the information in a magnetic tape to a sequence of sounds, ribosomes convert the information in a molecule of mRNA to a sequence of amino acid residues in a polypeptide. As mentioned above, the ability to translate mRNAs into polypeptides in a cell extract, such as wheat germ extracts, allowed researchers to deduce that the genetic code is universal. Because researchers were also able to subdivide the extracts into definable biochemical components, they also allowed detailed studies of the mechanisms of protein synthesis. Our current understanding of this process comes from two kinds of experiments: (1) isolating individual components of the translation machinery and putting

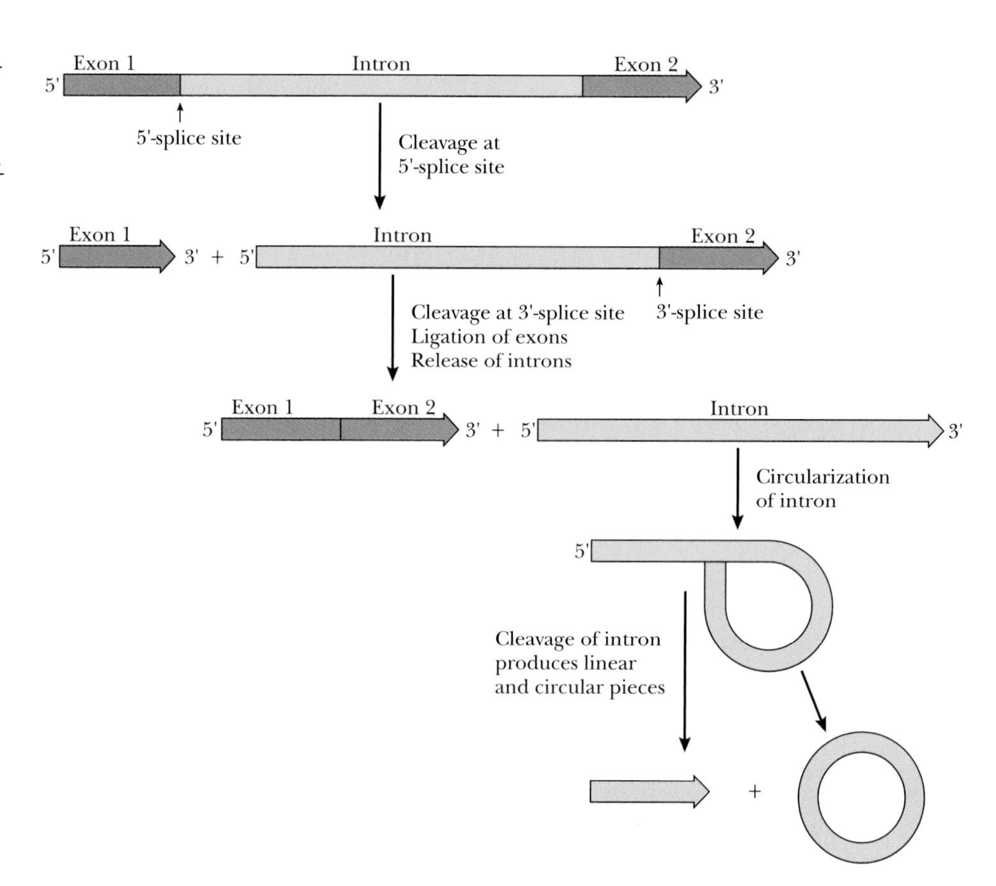

Figure 14-14 The mechanics of self-splicing in the production of ribosomal RNA in *Tetrahymena.* This process occurs without the help of proteins, with the RNA acting as a self-cutting enzyme. First it cuts itself at the 5′-side of the intron; next the 3′-side of the intron is cut and the two exons joined; the intron then circularizes and cleaves into a linear and a circular fragment.

them back together into a functional system; and (2) blocking individual steps in the translation with specific **antibiotics,** substances produced by microorganisms that kill (or interfere with) other microorganisms. (See Essay, "Antibiotics and Protein Synthesis.")

Studies of protein synthesis in both cells and cell extracts identified three quite distinct processes, each of which was subject to inhibition by different antibiotics: *initiation,* the attachment of mRNA to the ribosomes, the protein-synthesizing factories; *elongation,* the addition of amino acid residues to a growing polypeptide; and *termination,* the release of the completed polypeptide chain. Before we discuss the overall process in detail, we examine some of the characteristics of the participants.

Ribosomes Link Amino Acids to Make Peptide Bonds

While the corresponding ribosomal proteins from distantly related species are quite different from each other, ribosomal RNAs from species of all kingdoms are similar, with the same patterns of internal base pairing to form double-stranded stems and intervening loops, as illustrated in Figure 14-15. The similarity of these structures in rRNAs suggests that there may be a strong selection for particular structures to play important functional roles and that the rRNAs themselves (rather than their associated proteins) may be the catalysts of peptide bond formation.

Ribosomes catalyze a complicated set of reactions that leads to the formation of peptide bonds between amino acids, in a sequence dictated by mRNA. Their main tasks are to read the information in mRNA and to translate it into a sequence of amino acid residues.

In addition, however, ribosomes must also recognize where to start ("initiate") and where to stop ("terminate") the production of each polypeptide. To see the importance of proper initiation, for example, consider the translation of the mRNA sequence AUG CAU GCA AUG CAU GCA. If a ribosome started reading at the first nucleotide, it would assemble the peptide methionine-histidine-alanine-methionine-histidine-alanine. If, however, the ribosome started with the second nucleotide, the reading frame would shift and the ribosome would read the sequence UGC AUG CAA UGC AUG into cysteine-methionine-glutamine-cysteine-methionine. Similarly, initiation on the third nucleotide would result in reading GCA UGC AAU GCA UGC as alanine-cysteine-asparagine-alanine-cysteine.

mRNAs fix the correct reading frame with signals that say where to "start" and "stop." These signals are the punctuation of the letters of the mRNA message. An important part of the ribosome's job is to recognize and act on these signals. In bacteria, at least, the RNA of the smaller ribosomal subunit forms complementary base pairs with a sequence in mRNA that lies before the start site for translation. This binding helps start translation at the correct site.

ESSAY

ANTIBIOTICS AND PROTEIN SYNTHESIS

Antibiotics are usually complex organic molecules that compromise and ultimately kill cells. Many antibiotics are natural products, made by bacteria and fungi to dispose of their competitors. The medical uses of antibiotics began at least 2500 years ago, when the Chinese began to use the moldy curd of soybeans to treat boils and similar infections. In the 20th century the pursuit of antibiotics as a treatment for human disease started with Alexander Fleming. In 1928 Fleming discovered that the growth of bacteria in one of his experiments had been stopped by the presence of a green mold (Figure 1). Were it not for the significance of Fleming's insight, we might say that the presence of mold in a culture of bacteria was the result of his poor technique.

Fleming identified the mold as a species of *Penicillium*. Later other scientists isolated the active substance and named it penicillin. Penicillin and its many derivatives have remained the most important of the antibiotics. They act by interfering with cell wall synthesis in many bacteria. The battlefields of World War II were the testing grounds for penicillin, not only in halting infectious diseases but in fighting infections after surgery. Prior to World War II and the advent of penicillin, more soldiers died from infections than from the wounds themselves.

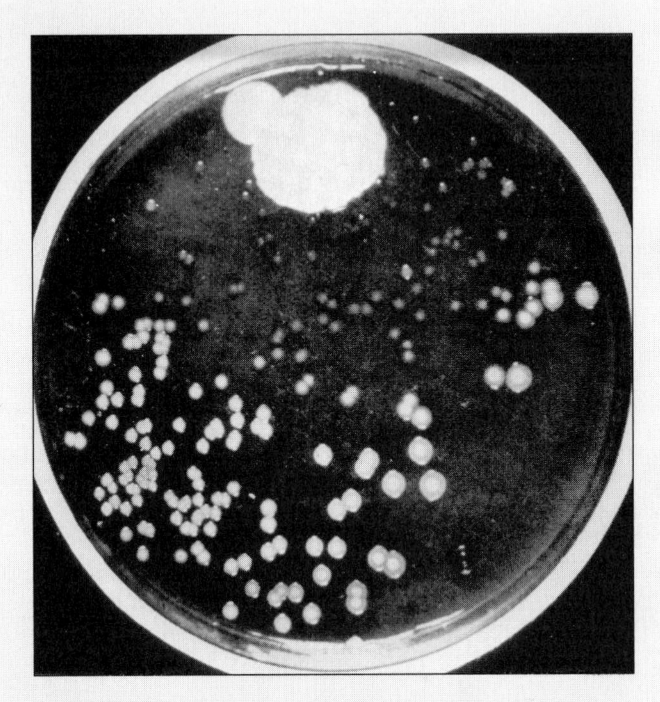

Figure 1 Alexander Fleming discovered that something—now called penicillin—made by the mold *Penicillium* stops the growth of bacteria. *(E. Chain and H. Florey,* Endeavor *3:9, 1944, courtesy New York Academy of Medicine)*

After World War II, penicillin and other antibiotics became increasingly common in civilian medicine, with large numbers of antibiotics isolated from many microorganisms. Dozens of these are now commercially produced and commonly used to treat bacterial and fungal infections both in humans and in livestock. In fact, antibiotic usage is so widespread that many researchers and physicians believe that they are overused, because many new strains of antibiotic-resistant microorganisms have spread rapidly.

Many antibiotics act by interfering with protein synthesis, and they have been extremely important tools for dissecting the steps of protein synthesis. For example, each of the steps of elongation is subject to inhibition by a different set of antibiotics: In prokaryotic cells tetracycline interferes with positioning, chloramphenicol with peptide bond formation, and erythromycin with translocation.

Most useful in fighting bacterial infections are antibiotics that interfere with prokaryotic protein synthesis without affecting eukaryotic protein synthesis. For example, while erythromycin blocks translocation of the growing peptide chain in bacteria, it does not inhibit this step during protein synthesis on eukaryotic ribosomes (although another antibiotic—cycloheximide—does). This specificity reflects differences in the proteins of prokaryotic and eukaryotic ribosomes.

Antibiotics have dramatically decreased the deaths caused by infectious disease. They not only can cure an initial infection, but they also can prevent long-term consequences, such as the rheumatic fever that can result from persistent throat infections. In addition, killing the bacteria that make one person sick prevents the spread of an infection. Finally, the ability to deal with infections has made surgery safer and allowed the development of operations ranging from the removal of cancers to the replacement of vital organs.

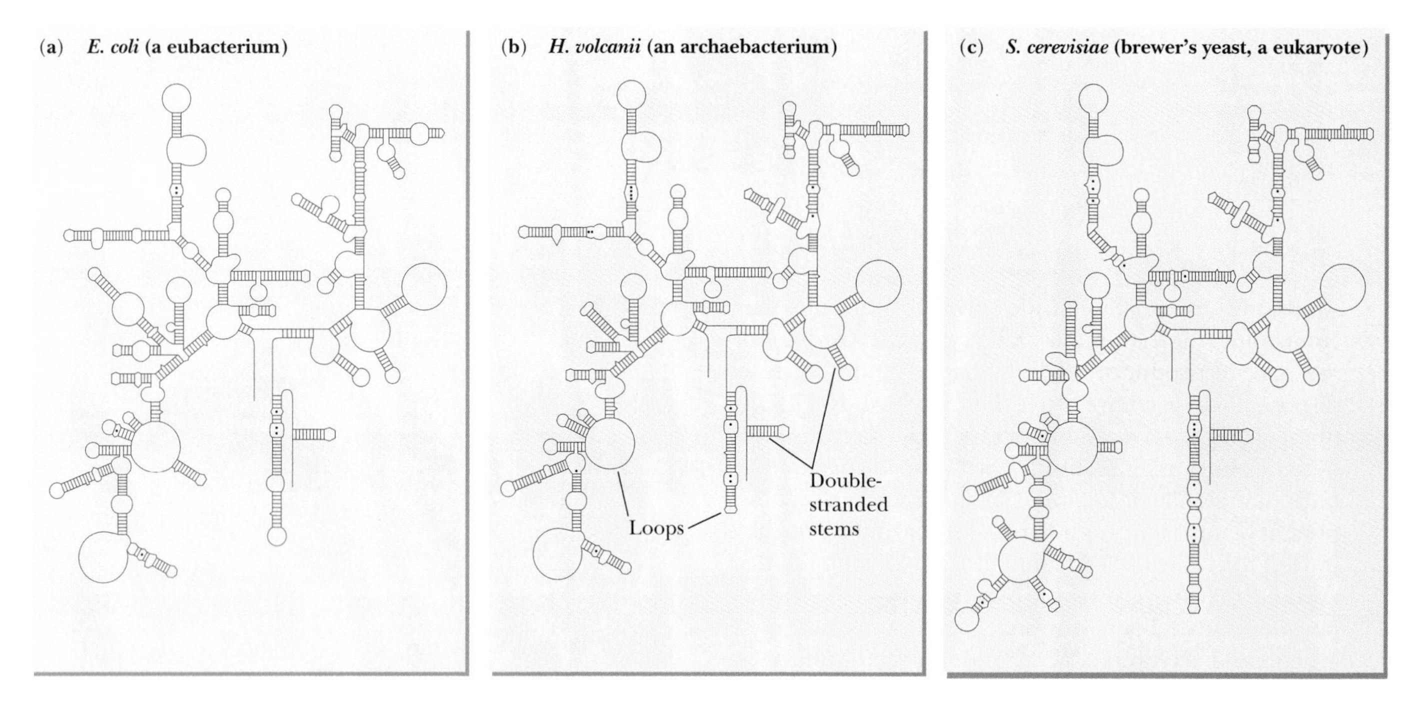

Figure 14-15 Ribosomal RNAs from all kingdoms have similar stem-and-loop structures, suggesting strong evolutionary conservation, consistent with a catalytic role for rRNA. **(a)** Small rRNA from a eubacterium; **(b)** small rRNA from an archaebacterium; **(c)** small rRNA from a fungus.

Both prokaryotic and eukaryotic ribosomes appear to work in the same manner. In both eukaryotes and prokaryotes, protein synthesis requires the participation of both ribosomal subunits. The large subunit contains the machinery needed to make peptide bonds, while the small subunit initiates the process in the correct reading frame. Each ribosome contains two grooves, one that binds to mRNA and a second that accommodates the growing polypeptide chain.

Transfer RNAs Serve as Adapters That Allow Amino Acids to Be Assembled in the Right Order

Once molecular biologists realized that mRNAs dictate the sequences of amino acids in polypeptides, they tried to imagine how this could occur. An early idea was that each codon in an mRNA molecule might bind a specific amino acid. No one, however, could find any evidence of such binding.

The actual explanation, proposed by Francis Crick, is that the amino acids are first linked to specific adapter molecules, which can interact more specifically with the different codons in mRNA. Subsequent work showed that the adapters are tRNAs.

A clever experiment has demonstrated that the ordering of amino acid residues depends on recognition of the tRNAs, rather than on recognition of the amino acids themselves. Recall that each type of tRNA links to a single amino acid, and we can refer to each tRNA according to the corresponding amino acid; a tRNA for cysteine, for example, is called $tRNA^{cys}$. Researchers prepared cysteine attached to its specific tRNA (cys-$tRNA^{cys}$) and then chemically converted the cysteine into alanine to form ala-$tRNA^{cys}$, that is, alanine attached to the tRNA for cysteine. They next added this incorrectly loaded tRNA to a protein-synthesizing extract. The extract produced incorrect polypeptides: Wherever cysteine was supposed to be inserted, there was an alanine instead. This result shows that the translation of information in mRNA depends on the interactions of mRNA and adapter tRNAs rather than on interactions between mRNA and amino acid molecules.

How Many tRNAs Are There?

Each tRNA consists of 75 to 85 nucleotide residues. Every known tRNA can fold upon itself to form four base-paired "stems" separated by four "loops," segments with unpaired nucleotides. In a drawing of tRNA, each molecule resembles a cloverleaf, as shown in Figure 14-16.

Researchers predicted the cloverleaf structure of tRNAs by examining their nucleotide sequences and noticing that every tRNA had a structure that permitted the formation of short duplexes held together by complementary bases, as illustrated in Figure 14-16a. X-ray diffraction studies of tRNA structure confirmed the existence of these predictions and also showed that the whole mol-

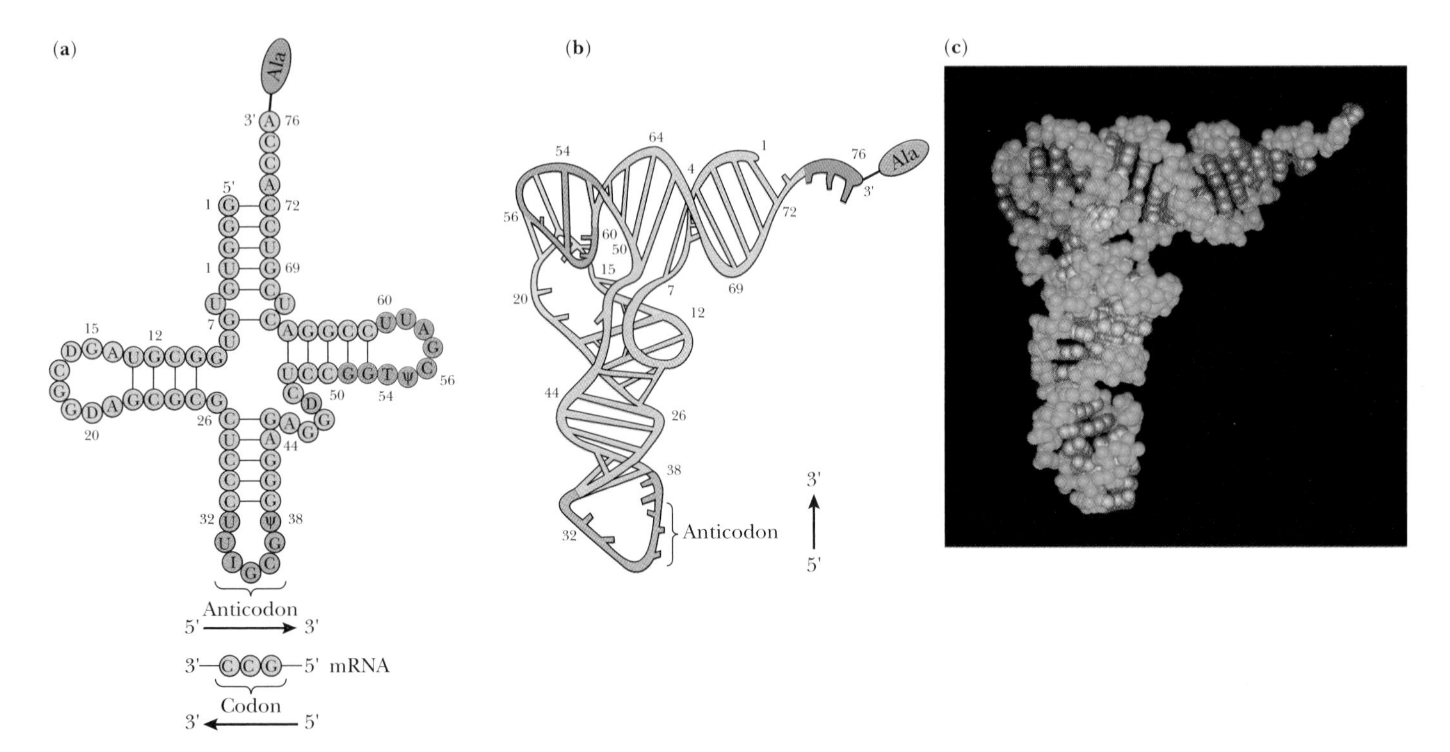

Figure 14-16 The structure of the tRNA for alanine ($tRNA^{Ala}$). **(a)** Stem and loop structure, portrayed as a clover leaf. In addition to the conventional bases in RNA (A, G, C, and U), tRNAs contain several kinds of modified bases, not found in other RNAs; these include D (dihydrouridine), I (inosine), T (thymine), and ψ (pseudouridine). In $tRNA^{Ala}$, I in the anticodon pairs with C in the codon. **(b)** Three-dimensional structure of $tRNA^{Ala}$, shown diagrammatically; shape resembles an inverted L. **(c)** A computer representation of the three-dimensional structure of $tRNA^{Ser}$; serine-binding site is highlighted in yellow; anticodon is highlighted in red. Notice that, in this figure, the 5′-to-3′ orientation of the tRNA is from left-to-right, so that the 5′-to-3′ orientation of the mRNA is from right-to-left. *(c, Melinda M. Whaley)*

ecule bends into a still more intricate structure that resembles the letter L (as shown in Figure 14-16b). One of the loops in each tRNA molecule contains an **anticodon,** a sequence of three nucleotides in tRNA that forms specific base pairs with the corresponding codon sequence in mRNA.

The amino acid corresponding to the codon is attached to one end (the 3′-end) of the tRNA, as shown in Figure 16a and b. At each step in the reading of mRNA, the ribosome moves the next charged tRNA into place, as the tRNA anticodon binds to the complementary codon sequence in the mRNA.

There is at least one tRNA for each amino acid. We might have expected one for each of the 64 ($4 \times 4 \times 4$) triplets (actually 61, because three codons are "nonsense"). Actually, every species has more than 20 tRNAs but fewer than 61: 30–40 types in bacteria and up to 50 in plants and animals.

Notice also in Figure 14-3 that equivalent codons (ones that specify the same amino acid) often differ only in the third nucleotide position. For example, UCU, UCC, UCA, and UCG all specify serine, and CCU, CCC, CCA, and CCG all code for proline. Organisms have fewer than 61 tRNAs because the matching of anticodon and codon may not perfectly obey the base-pairing rules of Watson and Crick. **Wobble,** nonstandard base pairing, between the first base in the anticodon and the third base of the codon, allows some tRNAs to recognize more than one codon for the same amino acid, as illustrated in Figure 14-17. In these cases, a single type of tRNA can recognize all four equivalent codons, because the matching of codon and anticodon depends on two rather than three bases. Wobble allows organisms to get by with fewer types of tRNA.

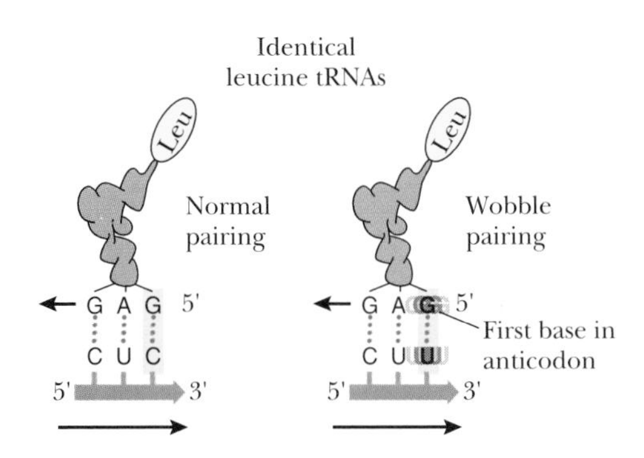

Figure 14-17 The same molecule of $Leu\text{-}tRNA^{Leu}$ can form base pairs with two codons in mRNA, CUC and CUU, because of "wobble" between the third base of the codon and the first base of the anticodon.

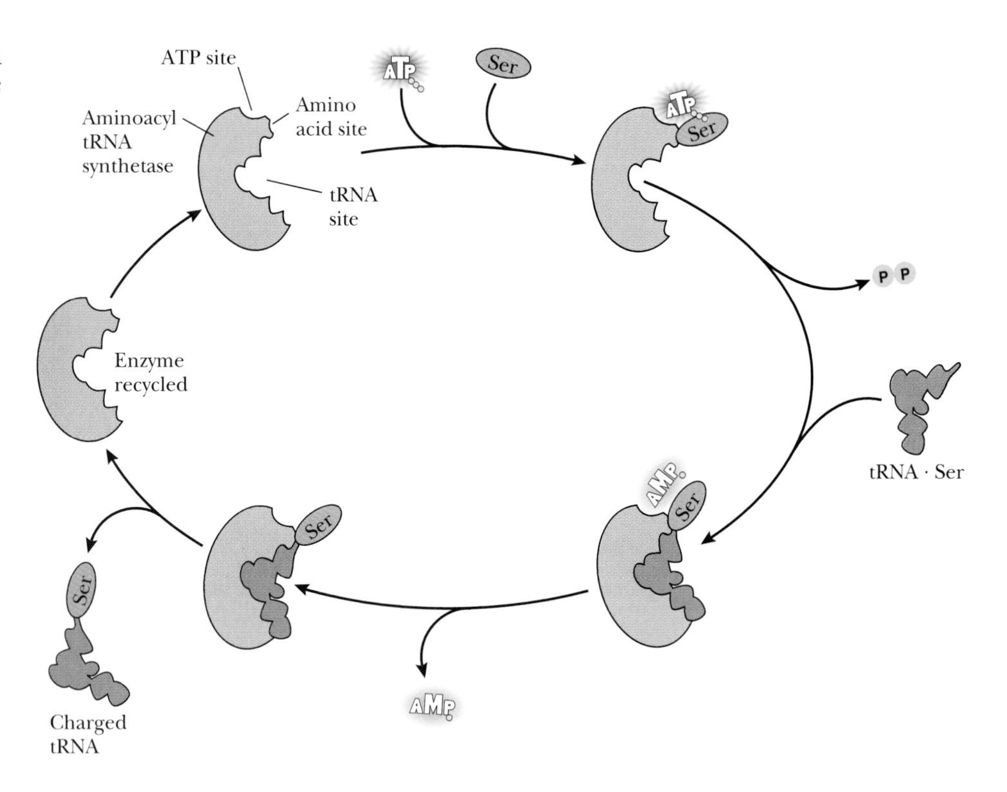

Figure 14-18 Each aminoacyl-tRNA synthetase links a specific tRNA (here $tRNA^{Ser}$) to its corresponding amino acid (here Ser) to form a "charged" tRNA (aminoacyl-tRNA, here Ser-$tRNA^{Ser}$). The energy for the linkage comes from the hydrolysis of ATP to AMP.

Aminoacyl-tRNA Synthetases Servie as Decoders

Transfer RNA molecules convert information from the language of nucleotides to that of amino acids. The genetic code is the dictionary for this translation. The actual work of decoding, however, depends upon a set of decoding enzymes, called **aminoacyl-tRNA synthetases,** which link tRNAs to their corresponding amino acids to form "charged" tRNAs—tRNAs joined to their amino acids by high-energy bonds, as illustrated in Figure 14-18.

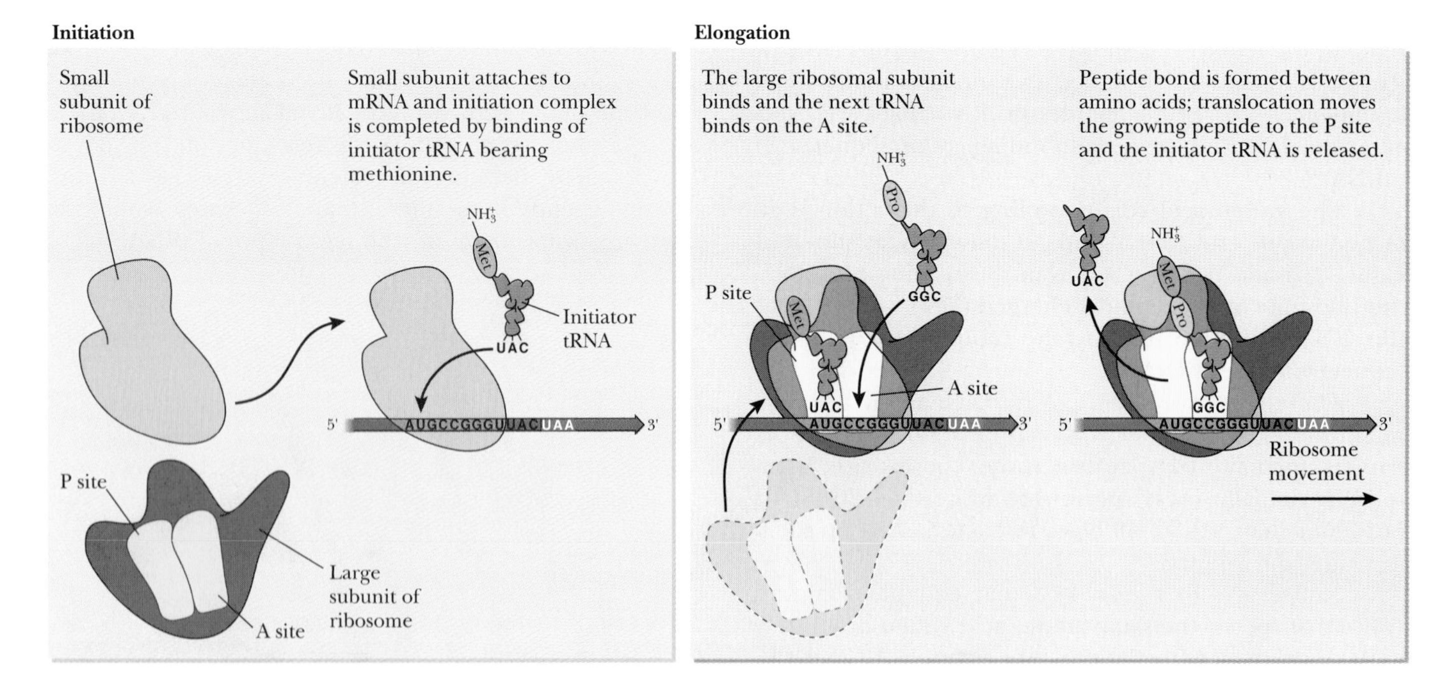

Figure 14-19 Overview of the steps of translation: initiation, elongation, and termination. Note that in mRNA the 5′-to-3′ direction is from left to right, while in the tRNAs, the 5′-to-3′ direction is from right to left. The P and A sites bind to aminoacyl-tRNAs.

These tRNA adapters then carry amino acids into the positions specified by the codons in mRNA. But how do the tRNAs and amino acids recognize each other?

The task is to match the unique shape and charge distribution of each amino acid with the shape and charge distribution of the appropriate tRNA. The tRNAs differ from one another not only in their anticodons, but also in other parts of their sequences. These sequence differences lead to differences in three-dimensional structures, which can be recognized by the decoding enzymes. Each aminoacyl tRNA synthetase has three binding sites: One holds an ATP, the second a distinct type of tRNA, and the third the corresponding amino acid. The enzyme then catalyzes the specific linking of the amino acid to the 3′-end of the appropriate tRNA, as shown in Figure 14-18. The resulting molecule is called an **aminoacyl-tRNA.**

Forming a bond between an amino acid and a tRNA molecule requires the expenditure of energy from ATP. In fact, each aminoacyl-tRNA requires the splitting of two high-energy phosphate bonds, as shown in Figure 14-18. Aminoacyl-tRNAs are high-energy compounds whose energy contributes to the formation of peptide bonds during protein synthesis.

This energy input is important because the hydrolysis of a polypeptide can occur spontaneously (without energy input), but polypeptide synthesis cannot. The required additional energy to form each peptide bond comes from the high-energy bond of aminoacyl-tRNA, whose energy is derived from ATP.

HOW DO RIBOSOMES TRANSLATE AN mRNA INTO A POLYPEPTIDE?

Figure 14-19 provides an overview of translation that resulted from research carried out in dozens of laboratories in the 1960s and 1970s. The process consists of three broad stages: initiation, elongation, and termination.

Initiation Aligns the mRNA in the Proper Reading Frame

Initiation, the first step of translation, begins with the attachment of the small subunit of the ribosome to mRNA and to an initiator tRNA, which binds to the first codon in an mRNA molecule. In both prokaryotic and eukaryotic cells, initiation involves recognition of mRNA sequences just upstream from the codon for the first amino acid in the polypeptide (the amino-terminal residue). Initiation also requires the participation of nonribosomal proteins called **initiation factors (IFs)**, which then facilitate the subsequent attachment of the large ribosomal subunit, as illustrated in Figure 14-20.

But where does a ribosome begin? Correctly starting translation is especially important because, as discussed above, a mistake of a single nucleotide could lead to a complete misreading of the information. The actual initiation site in mRNA is always the same—the codon AUG, which specifies the amino acid methionine. This means

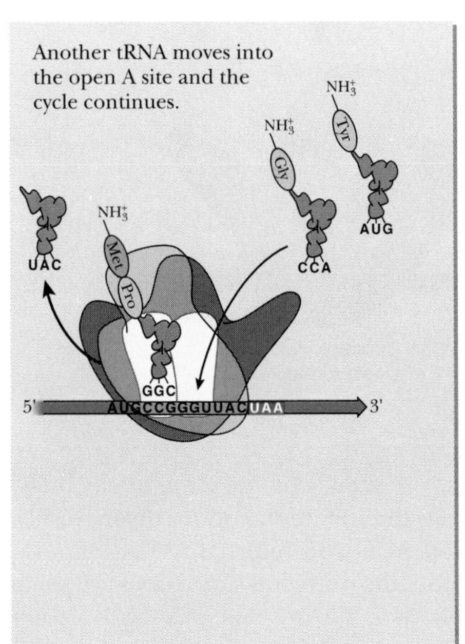

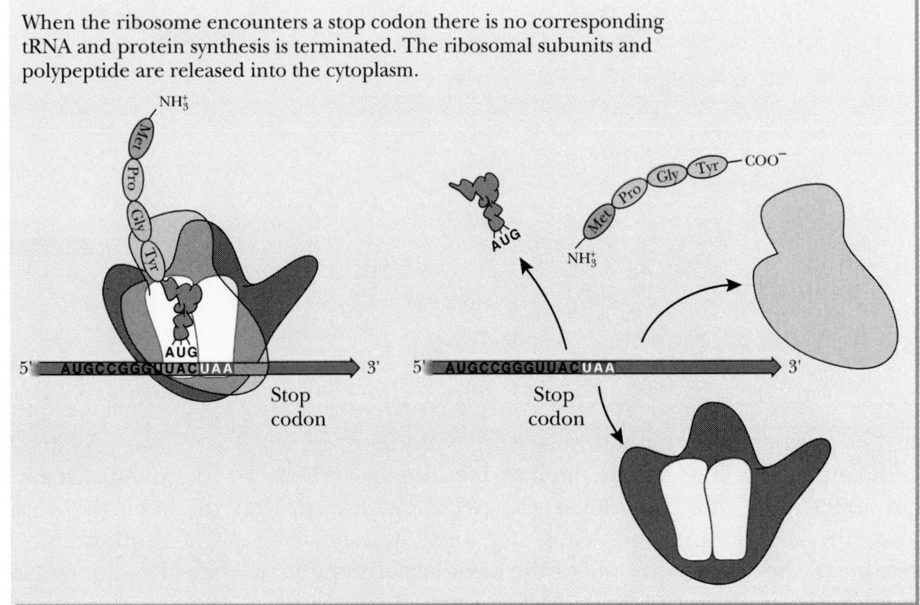

Figure 14-20 Steps in the initiation phase of protein synthesis.

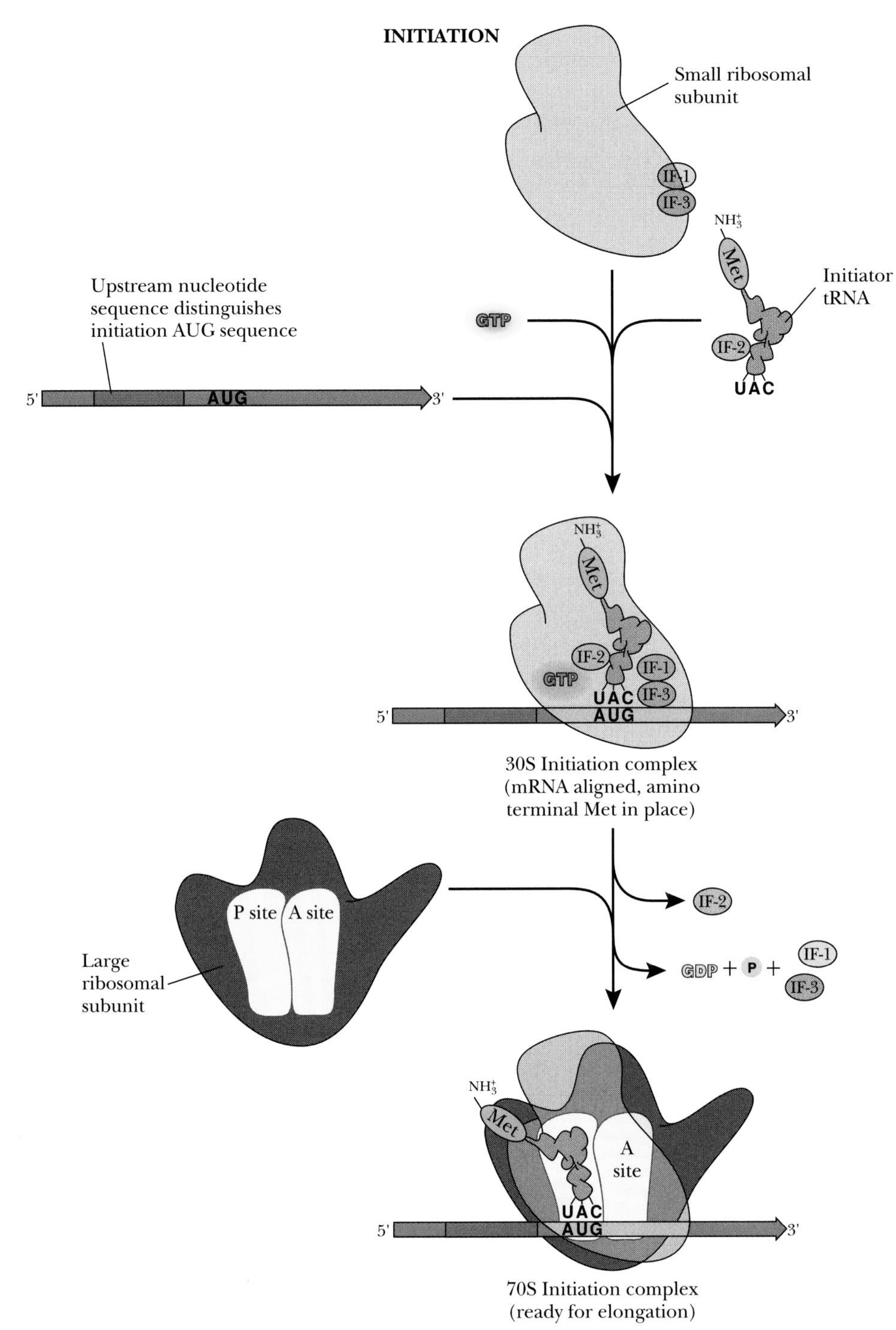

that *all* polypeptides originally start with methionine. Most cellular proteins, however, do not have methionine at their amino-terminal end. In most cases, a special enzyme clips away the methionine and leaves the next amino acid residue as the amino terminal of the cellular polypeptide.

Notice in Figure 14-3 that AUG is the only codon for methionine. How, then, can the ribosome distinguish between the AUG that means "start" and the AUGs that specify methionines in the middle of a polypeptide? The answer to this question differs in prokaryotes and eukaryotes.

In prokaryotes, a sequence of about six nucleotides, slightly upstream from the initiation codon, distinguishes between the initiation AUG and internal AUGs. Because this sequence lies before the initiation site, it does not specify any amino acids. Rather, it forms base pairs with a complementary sequence in the rRNA of the small ribosomal subunit.

In eukaryotes, on the other hand, no such specific base pairing occurs. Instead, each ribosome attaches to the 5′-end of the mRNA and moves along until it encounters the first AUG, where translation begins. These upstream nucleotides do not code for amino acids but somehow mark the site of initiation.

In both prokaryotes and eukaryotes, the small ribosomal subunit binds a special initiator tRNA before it attaches to the initiation site. The initiator tRNA contains an anticodon complementary to AUG and is already linked to the methionine (Met) that will occupy the first position of the new polypeptide, as shown in Figure 14-20. This binding of initiator tRNA depends on a special protein, called initiation factor 2 (IF2 in prokaryotes, eIF2 in eukaryotes).

The initiation process aligns the mRNA so that it is properly read and moves the amino-terminal methionine into place. After the initiator tRNA and its attached methionine are in place, elongation of the polypeptide chain begins.

Elongation: How Do Ribosomes Make Peptide Bonds?

Elongation, unlike initiation, depends upon the large ribosomal subunit, which contains two special pockets for tRNAs. The interaction of tRNAs with the ribosomes stabilizes the specific (but weak) binding of the anticodon of tRNA to the corresponding codon in mRNA. The two sites on the ribosome are called the P site [for *p*eptide] and the A site [for *a*mino acid], as diagrammed in Figure 14-20. The P and A sites bind two tRNAs, one at each site, aligned to adjacent codons in the mRNA. Elongation consists of three steps: (1) positioning of the next aminoacyl-tRNA at the A site, (2) peptide bond formation, and (3) translocation of the growing peptide to the P site.

During the initiation stage, the initiator tRNA and its attached methionine bind directly to the P site. The first step of elongation, as shown in Figure 14-21, is the **positioning** (binding) of a second tRNA (specified by the next codon in the mRNA) and its attached amino acid (here proline) to the vacant A site on the large ribosomal subunit.

When both sides are filled, **peptide bond formation,** the second step of elongation, occurs, as shown in Figure 14-21. After forming the peptide bond, the tRNA in the P site, no longer bound to an amino acid, falls away from the ribosome. It diffuses back into the cytoplasm, where it can be "recharged"—relinked to its specific amino acid

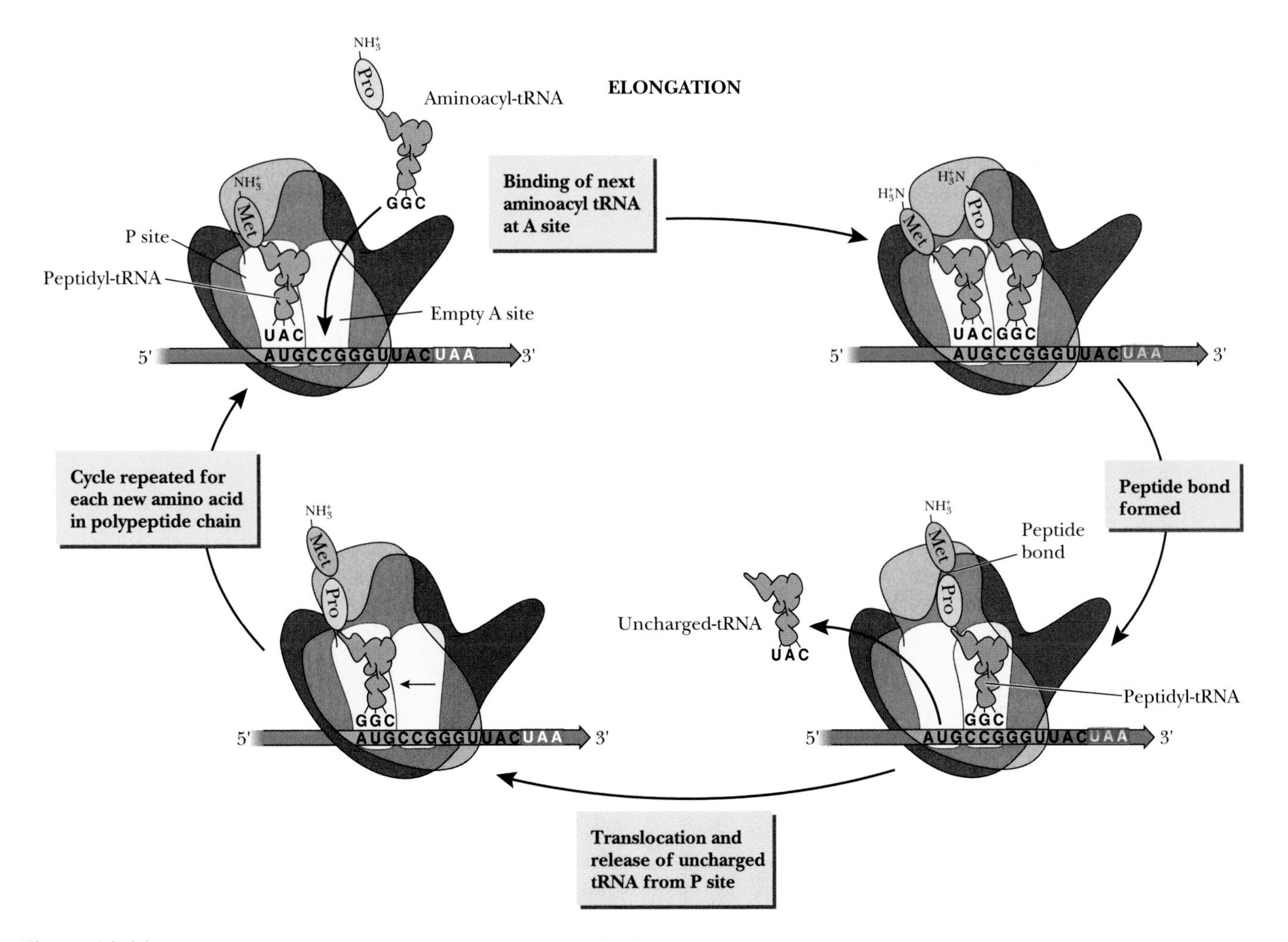

Figure 14-21 Steps in the elongation phase of protein synthesis.

by aminoacyl-tRNA synthetase. Because the amino acid it carries has become part of the polypeptide, the tRNA in the A site, formerly attached to a single amino acid, is now linked to the growing polypeptide chain.

In the third step of the elongation cycle, called **translocation,** this peptidyl-tRNA (with its attached peptide) moves to the P site, as shown in Figure 14-21. The interaction of the tRNA with the mRNA and with the P site holds the attached polypeptide to the ribosome.

For each added amino acid, these three steps repeat, with the appropriate aminoacyl-tRNA positioned according to the next codon in mRNA. In this way, the ribosome "reads" an mRNA from its initiation site near its 5′-end toward its 3′-end, and the resulting polypeptide is assembled from its amino-terminal residue to its carboxy-terminal residue. Each cycle takes about one fifth of a second, so the synthesis of an average-sized polypeptide (containing 100–300 amino acid residues) requires 20 to 60 seconds.

Each step of elongation requires the splitting of two more high-energy phosphate bonds, one for binding aminoacyl-tRNA to the A site and one for moving the ribosome to the next codon. Recall also that the formation of each aminoacyl-tRNA required two high-energy bonds. All together, then, each peptide bond requires four high-energy bonds. In most cells, the making of proteins consumes more energy than the making of any other kind of molecules.

As soon as a ribosome has moved far enough away from the start site, another ribosome has room to attach and begin translation, so several ribosomes, spaced as little as 80 nucleotides apart, are attached to a single mRNA at one time. This complex, shown in Figure 14-22, is called a **polyribosome,** or **polysome.**

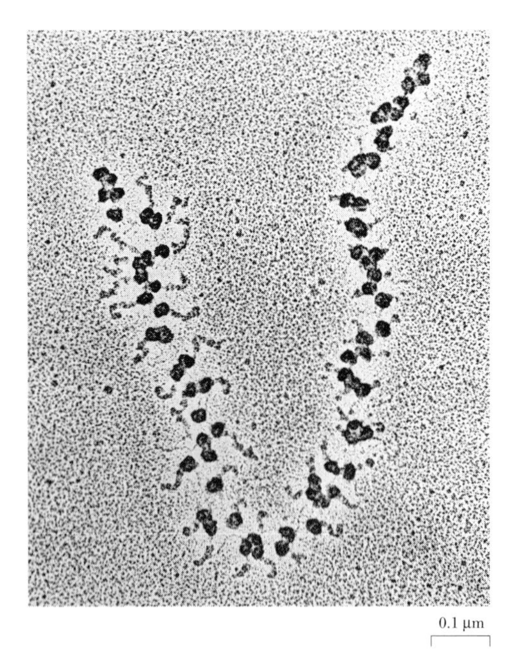

Figure 14-22 Electron micrograph of a large polysome. *(Oscar L. Miller, Jr., University of Virginia)*

Termination: How Does Polypeptide Synthesis Stop?

The appearance of a stop codon in mRNA signals termination, the completion of a polypeptide chain. When one of the three stop codons (UAA, UAG, or UGA) reaches the A site on the ribosome, there is no corresponding tRNA. Instead cytoplasmic proteins, called **release factors** bind to form a termination complex, as shown in Figure 14-23. The enzyme that produces peptide bonds during elongation now splits the link between the polypeptide and the tRNA in the P site. Its only bond to the ribosome now broken, the completed polypeptide is released the ribosome. The ribosomal subunits then dissociate from each other, available for another round of initiation and elongation.

Most of the time the mRNA continues to be translated by the other attached ribosomes of the polysome cluster (Figure 14-22). A given mRNA may be translated thousands of times before it is ultimately hydrolyzed by cytoplasmic nucleases. Different mRNAs have different lifetimes, ranging from minutes to days, determined in part by the structure of the mRNA itself. Different types of mRNA may direct the synthesis of widely varying amounts of their corresponding polypeptides.

■ HOW DO CELLS REGULATE THE ACTIVITY AND LOCATION OF A POLYPEPTIDE AFTER ITS TRANSLATION?

Even as mRNAs are translated, the resulting polypeptides begin to fold into their three-dimensional conformations. Later, they may also assemble into complexes with other polypeptide chains, either of the same or of different structure. As discussed in Chapters 3 and 4, the folding of a polypeptide into a three-dimensional structure and its association with other polypeptides depend both on the genetically determined amino acid sequence and on the environment, for example, on the surrounding pH, the concentrations of specific ions, and the presence of specific allosteric effectors (small molecules that alter a protein's conformation and activity, as discussed in Chapter 4).

Each polypeptide is also subject to covalent modifications, such as the formation of disulfide bonds between cysteine side chains and specific cleavage by proteases, as illustrated in Figure 14-24a for **insulin,** a protein hormone that regulates cellular glucose uptake. An even more dra-

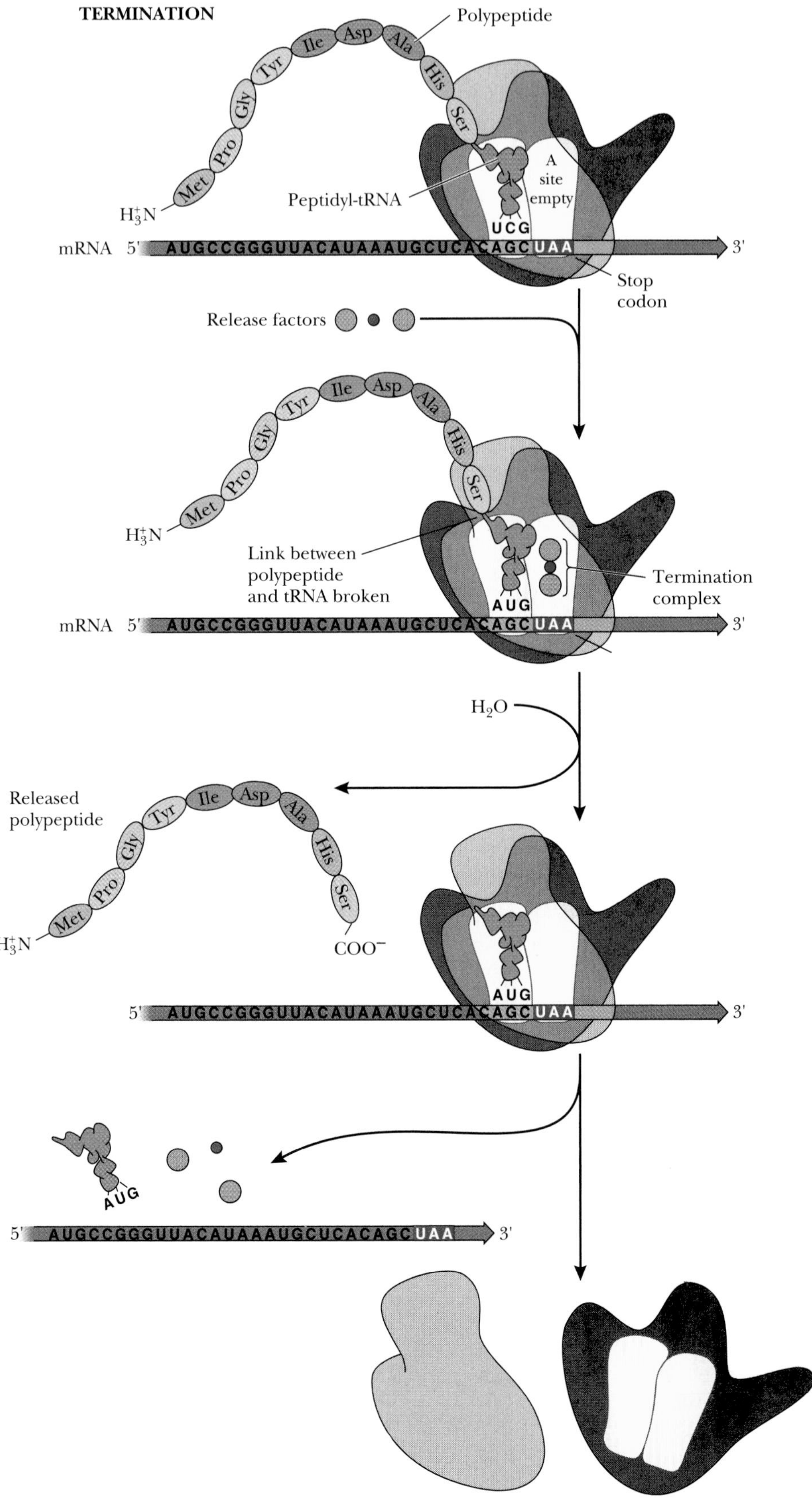

Figure 14-23 Steps in the termination phase of protein synthesis.

ESSAY

CHAPERONES HELP PREVENT PROMISCUOUS PROTEIN FOLDING

A major problem for the biotechnology industry is the difficulty in producing properly folded proteins. This problem is particularly severe for secreted proteins, such as monoclonal antibodies, hormones, and hydrolytic enzymes, which fold within the lumen of the endoplasmic reticulum. Biotechnology researchers did not fully anticipate the problem of proper protein folding because, since the 1950s, molecular biologists have accepted the generalization that the sequence of amino acid residues in a polypeptide determines its ultimate three-dimensional structure. (See Chapter 3, p. 80.)

While it is true that a pure protein in a dilute solution will ultimately adopt a single, thermodynamically determined conformation, newly made proteins left to themselves usually do not fold properly—either in the preparative vats of biotechnologists or in the ER lumen of a single cell. At the high protein concentrations (about 200 mg/ml) within the ER lumen, for example, polypeptides are likely to form intermolecular noncovalent and disulfide bonds with one another, rather than develop the intramolecular interactions that stabilize the native conformation. The task of restricting promiscuous intermolecular interactions falls to special proteins called *chaperones,* which bind to unfolded polypeptide segments and catalyze their folding into a native conformation.

Two classes of chaperone proteins have been especially well studied. These classes had been previously identified as *heat shock proteins,* proteins whose synthesis is dramatically increased by exposure, even brief exposure, to elevated temperature. The two classes of chaperones correspond to groups of heat shock proteins called Hsp60, with approximate molecular sizes of 60,000, and Hsp70, with approximate molecular sizes of 70,000. The Hsp70 family is particularly conserved in evolution, with 50% of the amino acids in *E. coli* Hsp70 identical to the corresponding human proteins.

One member of the Hsp70 family, called *binding protein,* or *BiP,* is particularly abundant in the ER lumen of eukaryotic cells. BiP binds to the amino terminal segment of a nascent polypeptide as it enters the ER lumen and blocks illicit interactions. As the bound polypeptide segment folds into its native conformation, BiP dissociates from it. BiP may then bind to subsequently entering segments of the same polypeptide, as it enters the ER lumen.

The Hsp70-class of chaperones, including BiP, specifically recognize and bind to exposed hydrophobic side groups in a polypeptide segment of about seven amino acid residues. In the final conformation of an active protein, hydrophobic side groups are usually buried in the protein's interior, so the chaperone binds to a region of the protein that is incorrectly folded.

The timing of BiP binding and release depends on BiP's ability to bind ATP and also to act as a slow ATPase. It is ADP-BiP that binds to the unfolded polypeptide segment. As the polypeptide segment folds properly, the ADP is displaced by ATP, which triggers BiP's release of the nascent polypeptide. BiP then catalyzes the hydrolysis of ATP to ADP, and the BiP-ADP binds to another unfolded segment. The timing between binding events depends on the rapidity of ATP hydrolysis: The faster the hydrolysis, the more rapidly the next segment is bound. The result of the binding to and unbinding from BiP is that the polypeptide chain has more time to fold without the distraction of other polypeptides in the protein-rich lumen.

After individual segments have properly folded, other proteins within the lumen contribute to proper folding of the whole polypeptide and the formation of the proper disulfide bonds. An Hsp60-type chaperone, for example, can form a large barrel-like structure, which allows a completed polypeptide to fold in isolation from other proteins. Other proteins that participate in the final folding of a polypeptide include *protein disulfide isomerase,* which catalyzes the exchange of disulfide bonds and accelerates the formation of the most thermodynamically stable disulfides, and *peptidyl prolyl isomerase,* which allows the twisting of a polypeptide chain around the otherwise fixed peptide bond that connects to a proline residue. The final folding of a polypeptide within the ER lumen, then, depends not just on the thermodynamics of noncovalent and disulfide interactions but also on the participation of several other proteins, some of which consume ATP, to accelerate a protein's access to its thermodynamically favored structure.

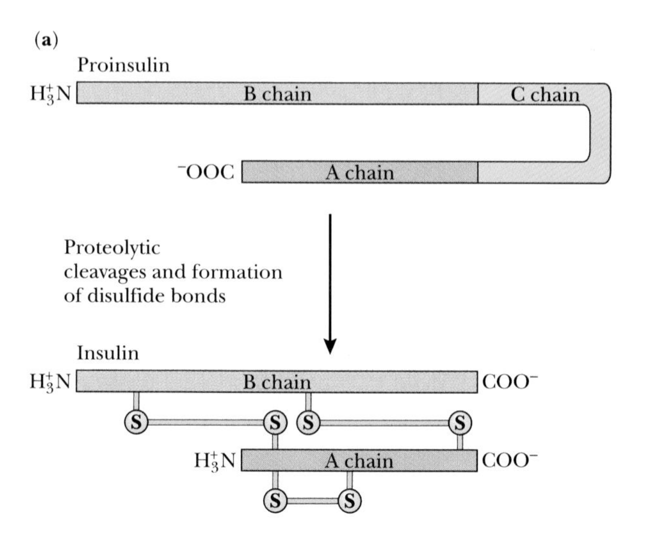

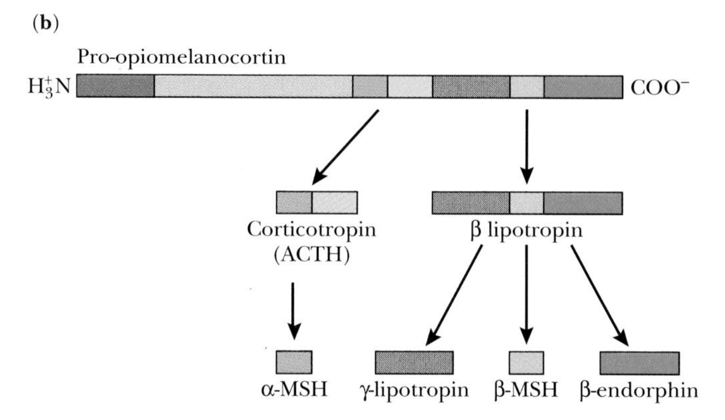

Figure 14-24 Post-translational processing of protein precursors. **(a)** Insulin derives from proinsulin after two proteolytic cleavages (which release the C peptide) and the formation of three disulfide bonds. **(b)** At least six biologically active peptides derive from a common precursor, called pro-opiomelanocortin.

matic illustration of the importance of specific polypeptide cleavage comes from studies of the 134 amino acid residue polypeptide **proopiomelanocortin (POMC),** illustrated in Figure 14-24b. POMC is a precursor of at least seven biologically active peptides, including adrenocorticotropic hormone (ACTH); β-endorphin; β- and γ-lipotropic hormones; and α-, β-, and γ-melanocyte-stimulating hormones (MSH).

Still other covalent modifications of polypeptides include the addition of sugars (glycosylation), phosphate groups (phosphorylation), methyl groups (methylation), and fatty acid groups (acylation), as well as linkage to peptides or to membrane lipids. These covalent changes often affect the conformation of individual polypeptides, their association with other polypeptides, and their biological function. As discussed in Chapter 19, the biological activity of many proteins depends upon their regulated phosphorylation and dephosphorylation. Here we discuss covalent changes responsible for the selective targeting of newly made proteins to different intracellular addresses.

How Do Cells Direct Newly Made Proteins to Different Intracellular and Extracellular Addresses?

Most of the proteins produced on the ribosomes stay in the cytosol, but many must find their way to various organelles, to the plasma membrane, or to the outside of cells. Cell biologists have learned much about the mechanisms that direct newly made polypeptides to secretory vesicles or to different intracellular addresses. Although many questions remain to be answered, these studies have demonstrated that newly made proteins contain or acquire address labels that target them to different parts of the cell.

Clear evidence that some polypeptides contain address labels within their sequences came from studies of secreted proteins, such as the digestive enzymes made in the pancreas and exported into the digestive tract. Gunther Blobel, a cell biologist at Rockefeller University in New York, prepared the mRNA for such a protein and added it to a test-tube translation system similar to the wheat germ extract discussed previously. The resulting polypeptide, however, was longer than the normally secreted protein.

The extra length of the polypeptide produced *in vitro* came from a **signal peptide,** a polypeptide segment of 16–30 amino acid residues at the amino-terminal end of the newly made protein. A typical signal peptide has one or more positively charged amino acid residues (histidine, lysine, or arginine) near the amino terminal, adjacent to another 6–12 hydrophobic residues.

According to Blobel's **signal hypothesis,** illustrated in Figure 14-25, this hydrophobic segment directs the growing polypeptide chain through the membrane of the endoplasmic reticulum into the ER lumen. Because the growing polypeptide is still attached to ribosomes, the ribosomes themselves become attached to the ER membrane, creating the rough ER. The ribosomes of the rough ER are identical in structure to those in the cytoplasm: They become associated with membrane only because of the hydrophobic segments of the growing polypeptide.

Subsequent experiments have established that the signal hypothesis is correct, not only for exported proteins, but also for other proteins made on the rough ER, including membrane proteins and lysosomal proteins. *In vitro* protein synthesis in the presence of ER fragments ("microsomes") results in the trapping of newly made secretory proteins in membrane vesicles, with their signal peptides removed. In the absence of microsomes, the signal peptide is not removed, even if microsomes are added later, suggesting that the targeting of secreted proteins is *cotranslational;* that is, it occurs while translation is still in progress.

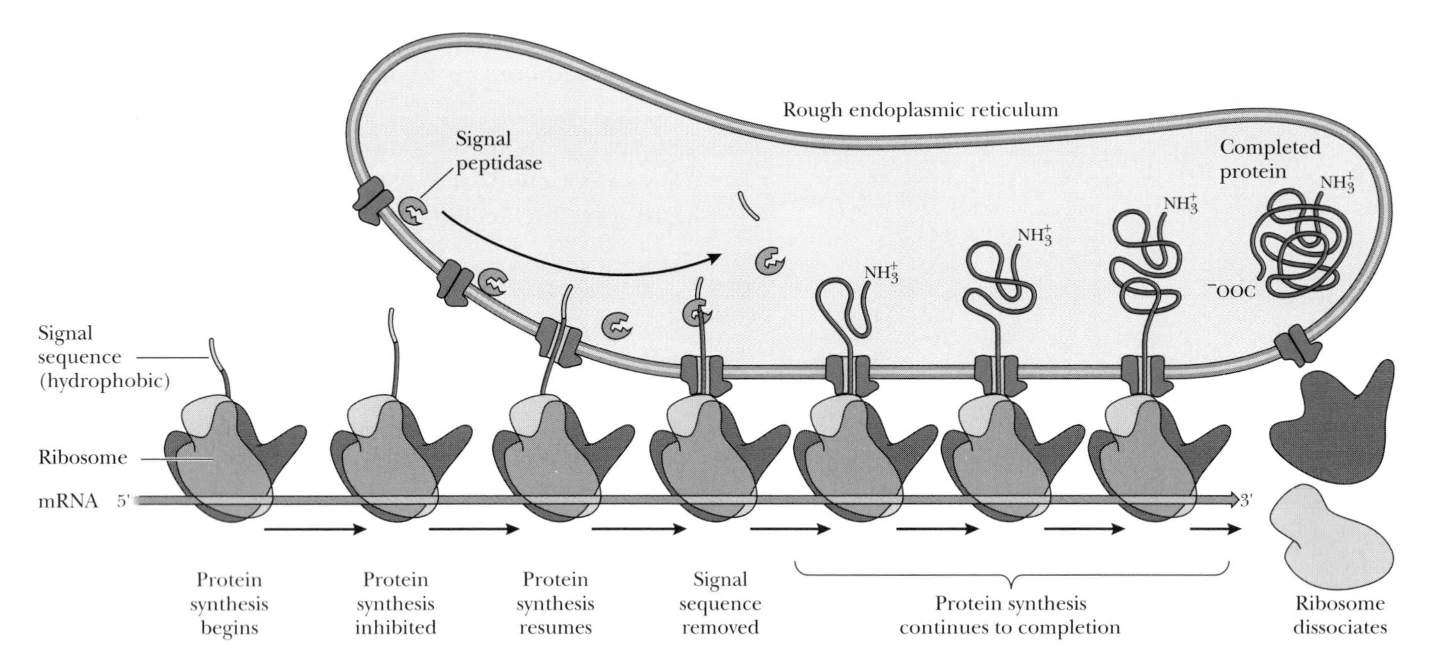

Figure 14-25 The Signal Hypothesis: A hydrophobic segment at the amino-terminal end directs the newly made polypeptide into the lumen of the rough endoplasmic reticulum; within the lumen, a signal peptidase removes the signal.

Another experiment demonstrated the importance of the signal peptide by making use of recombinant DNA techniques. Researchers engineered the production of hybrid polypeptide, with a signal peptide at the amino terminal and globin, a normally cytoplasmic protein, at the carboxy terminal. The result was the transport of globin to the ER lumen (a target to which it would not ordinarily have gone).

Once inside the lumen, the signal peptide is almost always removed by **signal peptidase,** an enzyme confined to the ER lumen. The growing polypeptide follows its new amino terminal into the lumen. From the ER lumen, the protein is further modified and packaged into vesicles. The vesicles then deliver the protein to the Golgi apparatus and, in the case of exported proteins, into secretory vesicles, as discussed in Chapter 6.

Like secreted proteins, proteins destined for the plasma membrane, the lysosomes, or the ER itself are also synthesized on membrane-bound ribosomes of the rough ER, and they also use a hydrophobic signal peptide to direct them to the ER lumen. So the amino acid sequence of a polypeptide contains not only the information needed to specify its three-dimensional structure, but in the case of polypeptides made on the rough ER, a part of the amino acid sequence serves as an address tag specifying its intracellular or extracellular destination.

In contrast to the polypeptides produced by the membrane-attached ribosomes of the rough ER, polypeptides destined for cytosol, nucleus, mitochondria, chloroplasts, or peroxisomes are made on free (cytosolic) ribosomes. In addition, as discussed in Chapter 16, a few mitochondrial and chloroplast polypeptides derive from mRNAs that are transcribed and translated within the energy organelles themselves.

While the targeting of proteins made on membrane-bound ribosomes is cotranslational, the targeting of proteins made on free ribosomes is **post-translational;** that is, it occurs after translation is completed. In each documented case of a polypeptide made on free ribosomes and targeted to one of these organelles, researchers have been able to identify a signal sequence, also called a **transit peptide,** to distinguish it from signal sequences in polypeptides derived to the ER lumen. Polypeptides directed to mitochondria usually contain a targeting sequence of 20–60 amino acid residues at the amino terminal, which is removed after arrival within the mitochondria. Distinctive targeting sequences direct mitochondrial proteins to the matrix or the intermembrane space, as illustrated in Figure 14-26. Nuclear proteins contain one of several localization sequences, which typically carry a net positive charge, in marked contrast to the signal sequences of secreted proteins, which are highly hydrophobic. Unlike transit peptides, nuclear localization signals are an integral part of the mature protein.

Receptor Proteins for Signal Sequences Contribute to Intracellular Targeting

What directs polypeptides and appropriate signal peptides to, for example, the ER lumen? A complex of proteins and RNA, called the **signal recognition particle,** or **SRP,** recognizes the signal sequence on the peptide and shuttles

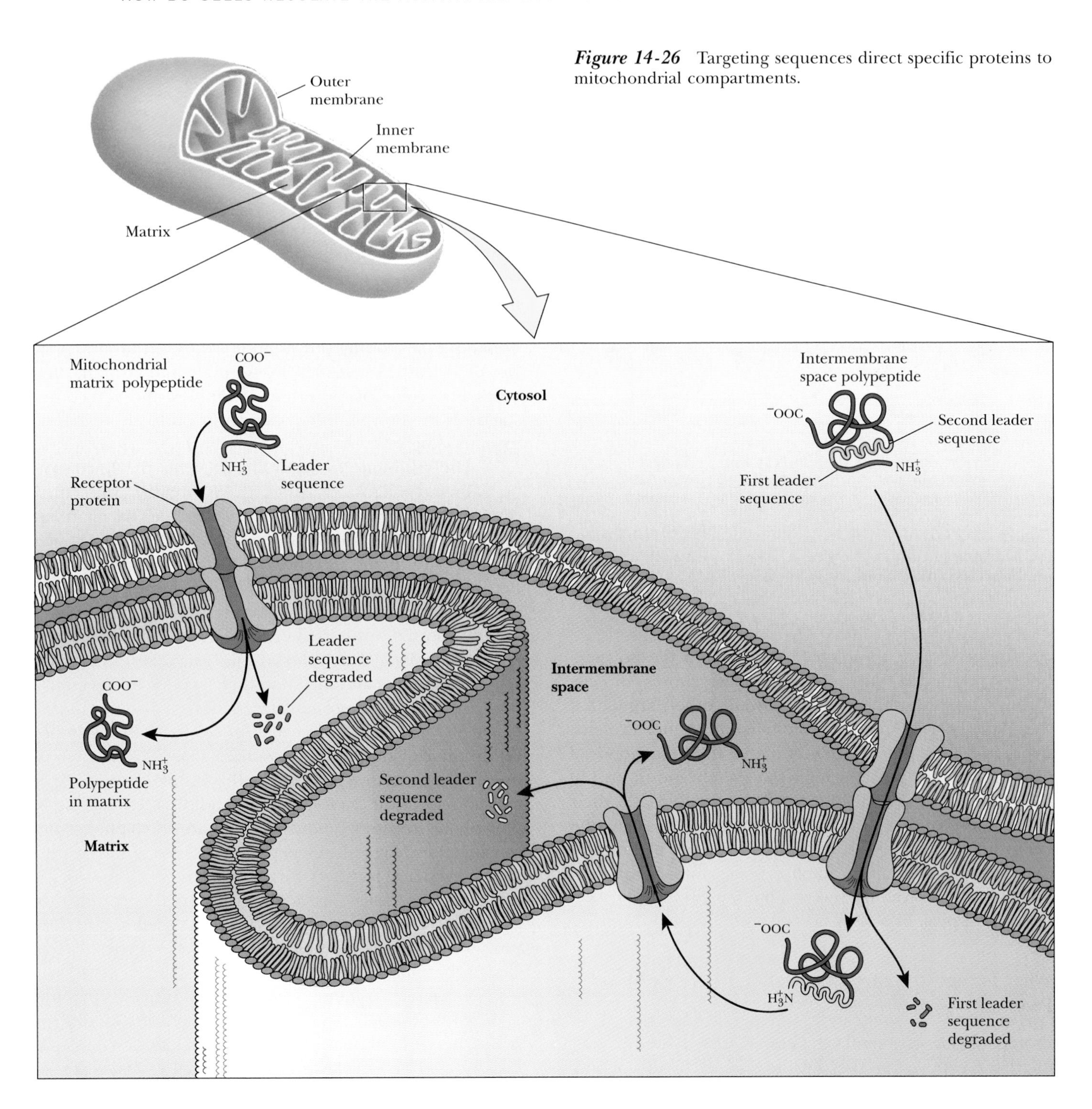

Figure 14-26 Targeting sequences direct specific proteins to mitochondrial compartments.

the marked polypeptide to the ER membrane, as shown in Figure 14-27. There a membrane protein, called the **docking protein,** or the **SRP receptor,** pulls the polypeptide into the lumen. The signal peptide apparently moves into the membrane with its amino-terminal end oriented toward the cytoplasmic face, with a hairpin turn exposed in the ER lumen. Signal peptidase cleaves the still growing chain, leaving the new amino terminal within the lumen, as shown in Figure 14-27.

Researchers have characterized both SRP and the SRP receptor: SRP consists of six polypeptides and a 300-base RNA, while the SRP receptor consists of two polypeptides. Together, these two complexes initiate the movement of the growing polypeptide into microsomes *in vitro.* Another protein on the ER membrane, the **signal sequence receptor,** binds the signal sequence after its release by SRP and the SRP receptor.

Despite recent advances in characterizing target sequences and their receptors, the mechanisms by which proteins move across organellar membranes are still unknown. Some researchers argue that the growing chain (when the polypeptide is produced on membrane-bound

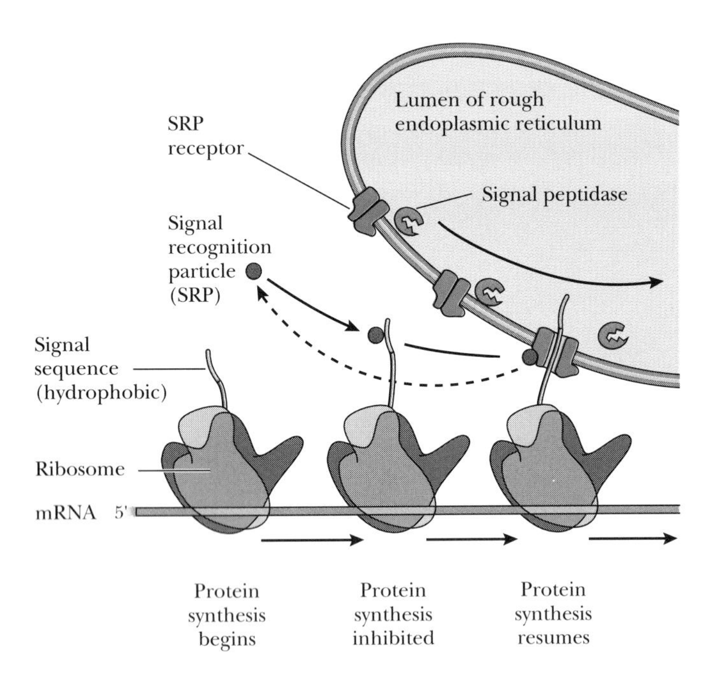

Figure 14-27 The signal recognition particle (SRP) binds to the hydrophobic signal, allowing it to bind to the SRP receptor on the rough endoplasmic reticulum.

ribosomes) or the completed chain (when the polypeptide is produced on free ribosomes) moves directly through the phospholipid bilayer, while others argue that the chain passes through a protein-lined channel. Proteins that pass across the ER membrane or the mitochondrial membrane must pass through the membrane in an unfolded form. In the ER lumen, they must fold into their proper conformations before further movement to the Golgi apparatus. In the mitochondria, special proteins called **chaperonins** aid in stabilizing the unfolded state of mitochondrial proteins and in allowing the proper folding once a protein arrives at its target.

What Is the Role of Glycosylation of Newly Made Protein?

Proteins made on membrane-bound ribosomes and destined for export, for the outside surface of the plasma membrane, and for lysosomes are usually **glycoproteins;** that is, they contain covalently attached sugars. Proteins made on free ribosomes are almost never glycoproteins. **Glycosylation,** the process of adding sugars, takes place in the ER lumen and in the Golgi apparatus. Most of the action takes place in the Golgi apparatus, with the *cis*, *medial*, and *trans* regions making distinctive contributions to the final pattern of glycosylation, as illustrated in Figure 14-28a.

Eukaryotic cells perform two types of glycosylation: (1) O-linked, where a sugar is attached to the oxygen of the hydroxyl group of serine or threonine, as illustrated in Figure 14-28b; and (2) N-linked, where a sugar is attached to the amide nitrogen of asparagine, as illustrated in Figure 14-28c. O-linked sugar complexes are generally small, while N-linked complexes may be much larger.

While O-linked sugar complexes vary in size and in the kinds of added sugars, all the N-linked complexes have a common oligosaccharide core: two molecules of N-acetylglucosamine, one of which is attached to the asparagine nitrogen, and three molecules of mannose. This oligosaccharide derives from a more complex precursor, a branched oligosaccharide with 14 sugar molecules attached by a high-energy bond to a long-chain lipid called **dolichol.** This oligosaccharide complex is transferred from dolichol to a growing polypeptide in the ER lumen. A series of at least 11 enzymes, in assembly line fashion, then trims away some of the original complex and adds other sugars as the polypeptide passes from ER to *cis*- to *medial*- to *trans*-Golgi.

The antibiotic tunicamycin prevents the formation of N-linked oligosaccharides. Cells treated with tunicamycin make proteins without N-linked oligosaccharides. For some proteins, this blockage prevents transport to the Golgi apparatus and secretion, while for others it does not. Because of the variable effects of tunicamycin, cell biologists have concluded that N-glycosylation is not necessary for transport, sorting to alternative subcellular addresses, or secretion, but that N-linked oligosaccharides can help stabilize proteins and facilitate their proper folding.

In the case of proteins targeted to lysosomes in animal cells, however, N-glycosylation is important. Enzymes within the *cis*-Golgi add a phosphate group to carbon-6 of one of the mannose residues in the N-linked complex. A component of the *trans*-Golgi membrane, the mannose-6-phosphate receptor, binds this mannose-6-phosphate, causing these tagged polypeptides to be segregated into sorting vesicles targeted for lysosomes. The interior of these sorting vesicles is acidic, causing the mannose-6-phosphate receptor to release the lysosomal protein for transport to the lysosomes, while the receptor itself recycles to the *trans*-Golgi.

Improper targeting to the lysosomes occurs in a human disease, called *I cell disease,* in which one of the enzymes needed for mannose phosphorylation is missing. Lacking mannose-6-phosphate, lysosomal enzymes are secreted from the cell rather than targeted to intracellular lysosomes. Fibroblasts from I cell disease patients have large vesicles with material that would ordinarily be subject to degradation by lysosomal enzymes.

Mannose-6-phosphate receptors are also present on the plasma membrane, where they can bind to extracellular molecules that contain mannose-6-phosphate. Such molecules can then be taken up by endocytosis, as discussed in Chapter 6, and directed to lysosomes. Recently researchers have shown that they can partially replace the defective lysosomal enzyme (hexosaminidase A) in fibroblasts cultured from patients with Tay-Sachs disease. Understanding lysosomal targeting may then contribute

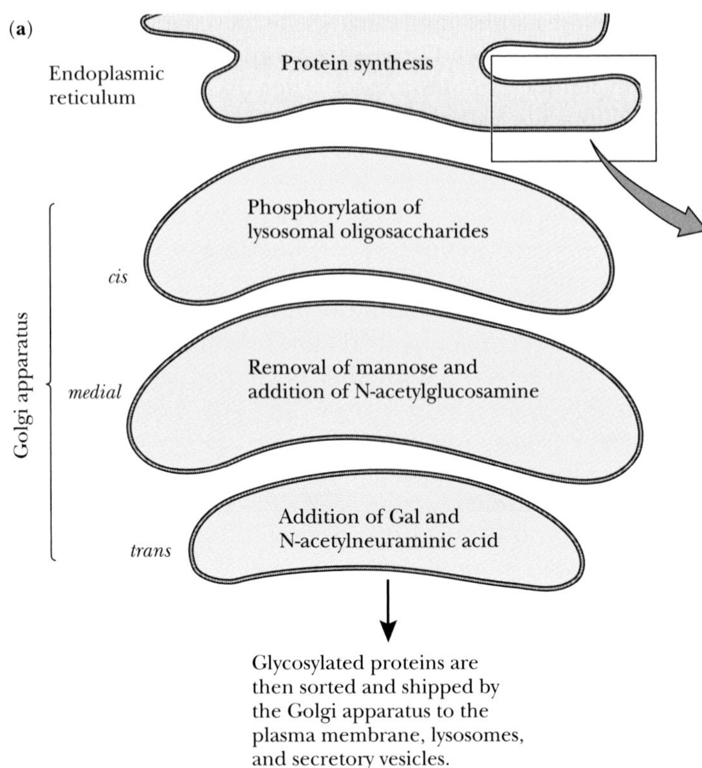

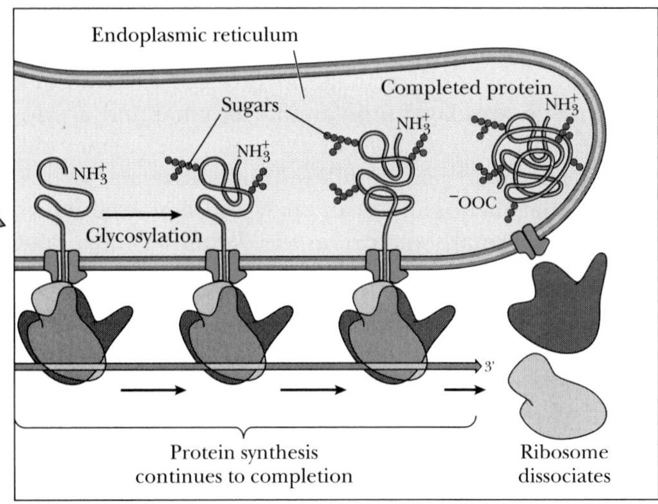

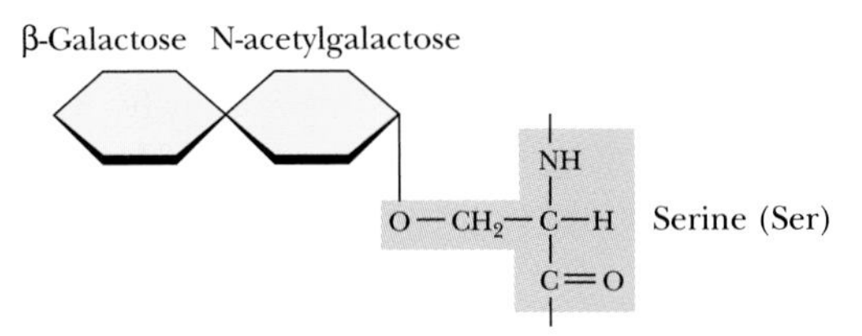

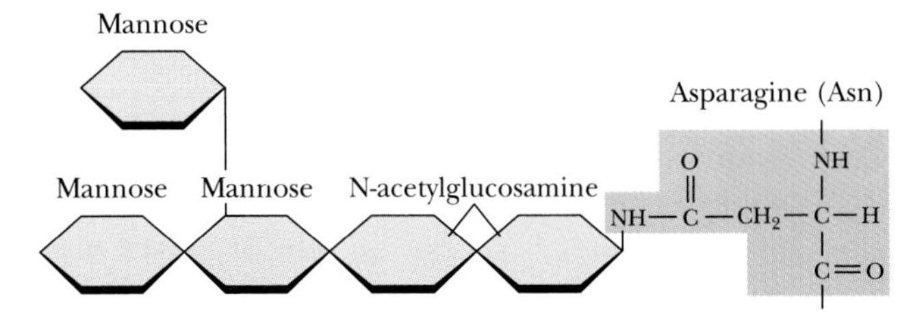

Figure 14-28 Glycosylation: **(a)** Different sugars are added in different compartments of the Golgi complex; **(b)** O-linked oligosaccharides; **(c)** N-linked oligosaccharides.

to devising new types of medical treatment, such as enzyme replacement in Tay-Sachs disease.

The Proper Orientation of Membrane Proteins Depends on Membrane-Spanning Sequences

Membrane proteins are produced on rough ER, that is, on ribosomes associated with the ER membrane. Unlike proteins targeted to the inside of secretory vesicles or lysosomes, however, membrane proteins do not fully pass into the ER lumen, but remain embedded in the ER membrane. They then move, as integral parts of the transport vesicle membranes, to the Golgi apparatus, where they are glycosylated on segments that will ultimately lie on the extracellular face of the plasma membrane.

Most integral membrane proteins have one or more **membrane-spanning segments,** sequences of 20–25 hydrophobic amino acid residues that can form an α helix extending through the membrane. Detailed examination of the structures and orientations of integral membrane proteins showed that in many proteins with a single such membrane-spanning segment (such as the LDL receptor, discussed in Chapter 6), the amino terminal lies on the exoplasmic face of the membrane (that is, originally pointing into the ER lumen), while the carboxy terminal lies on the cytoplasmic face. This orientation is exactly what we would expect for a polypeptide that originally contained a signal sequence (as does the LDL receptor).

But what keeps the protein from moving all the way through the membrane, as the secreted proteins do? The answer to this question is that once the membrane-spanning segment moves into the membrane, it prevents further movement. The result is that the amino terminal stays inside the cell and the carboxy terminal outside. The membrane-spanning region acts as a **topogenic sequence,** a polypeptide segment that determines the orientation within a membrane. In the case of the LDL receptor, the topogenic sequence stops the further transfer of the growing polypeptide and anchors the polypeptide within the membrane.

Experimental verification of the role of the topogenic sequence came from an experiment (again using recombinant DNA techniques) in which cultured cells produced a protein that lacked the membrane-spanning sequence. The result was that the polypeptide moved all the way into the lumen of the ER and was secreted from the cell, showing the importance of the identified *stop-transfer membrane-anchor* sequence.

How does a cell manage to produce a membrane protein with an exoplasmic amino terminal and a cytoplasmic carboxy terminal? In this case, the membrane-spanning segment plays a different role: It serves both as a membrane anchor and as an uncleaved signal sequence. There is no signal sequence at the amino terminal, and SRP binds instead to the internal membrane-spanning sequence. The internal signal sequence orients, as shown, with the amino terminal outside the cell, but no cleavage occurs. The polypeptide continues to extend into the ER lumen, where glycosylation occurs.

Still other proteins (such as the glucose transporter and the receptors for many hormones and neurotransmitters) contain more than one membrane-spanning domain. The same rules seem to apply to all membrane-spanning proteins: (1) If there is a cleavable amino-terminal signal sequence, then the amino terminal of the polypeptide moves into the ER lumen and the mature protein is oriented with its amino terminal on the exoplasmic face (facing the lumen) of the ER membrane; (2) if there is no cleavable signal sequence, an internal membrane-spanning region acts as an uncleaved signal-anchor sequence, while the next membrane-spanning region acts as a stop-transfer and anchor sequence. Each membrane-spanning region serves as a topogenic sequence, together fixing the orientation of the membrane protein.

The Elaborate Mechanisms of Protein Sorting Expend Energy to Achieve Intracellular Order

As we have noted many times since Chapter 4, the laws of thermodynamics imply that the achievement of order involves the expenditure of energy. The synthesis of a polypeptide imposes a strict order on formerly dispersed amino acids, using energy from ATP and GTP hydrolysis. In contrast, the folding of a protein into a specific three-dimensional conformation usually does not require more externally provided energy, but is a spontaneous process; that is, the free energy of the folded protein is less than the free energy of the unfolded protein.

Similarly, the sorting of proteins into different intracellular compartments requires the expenditure of energy, with substantial amounts of energy consumed just to move proteins into or across lipid membranes. In some cases, we know where and how the energy is expended. For example, the large unit of the SRP receptor binds GTP and uses the energy from GTP hydrolysis to force the dissociation of SRP and the growing polypeptide chain. In other cases, however, we do not yet know how energy is captured to drive nonspontaneous processes, even when we know, for example, that ATP is the source.

Contemporary studies of protein sorting address questions about energy transformations and molecular mechanisms. Among the powerful tools now available for such studies are (1) mutations in more than 25 genes that prevent protein secretion from yeast cells and (2) cell-free extracts that permit *in vitro* studies of the transfer of polypeptides among cellular compartments. A combination of genetic, molecular, and cellular approaches is rapidly increasing our understanding of how newly made polypeptides assemble with other macromolecules and lipids to create the elaborate subcellular organization of a eukaryotic cell.

SUMMARY

DNA and RNA carry information in the sequence of nucleotide residues. Protein synthesis translates this information into a sequence of amino acid residues. The Genetic Code is the dictionary that specifies the relationship of the nucleotide sequences in DNA and RNA to the amino acid sequences of proteins.

A codon is a group of three nucleotides that specify a single amino acid residue. There are 64 possible codons. Of these, 61 specify the 20 amino acids that make up proteins, while three codons signal the end of a polypeptide.

Cells transcribe information from DNA into RNA and translate information from RNA into protein. RNA polymerase copies information from one strand of DNA into RNA. The starting point for transcription is called a promoter. Prokaryotic cells usually have a single RNA polymerase, while eukaryotic cells have three kinds of RNA polymerase that recognize distinct promoters and produce distinctive products. Polymerases I and III produce RNA products, while polymerase II produces mRNA, the RNA that actually directs the synthesis of polypeptides.

In prokaryotes, RNA polymerase produces mature RNAs directly. In eukaryotes, however, cells modify the initially transcribed RNAs into functional products. For pre-mRNA, modifications include the cutting and splicing of the original transcription product to form a patchwork mRNA. The mRNA consists of sequences called exons, while the discarded RNA fragments are called introns.

Ribosomes are the molecular factories that assemble amino acids into polypeptides. Several ribosomes may associate with a single mRNA molecule to form a polyribosome (or polysome). Before ribosomes can assemble a sequence of amino acids according to the information encoded in an mRNA, amino acids must first be linked to adaptor molecules called transfer RNAs (tRNAs). Each tRNA contains an anticodon, a specific nucleotide sequence that forms base pairs with a codon in mRNA.

The pairing of specific tRNAs with their appropriate amino acids is the task of enzymes called aminoacyl tRNA synthetases. Each enzyme recognizes one amino acid and the corresponding tRNA and links them together with a high-energy bond.

The translation of information from mRNA into the sequence of amino acid residues in a polypeptide occurs in three distinct steps, called initiation, elongation, and termination. Initiation, which depends on the action of the smaller subunit of the ribosome, positions the mRNA in the correct reading frame, always starting with AUG as the codon for methionine. Elongation consists of three steps: (1) putting the next amino acid in place, (2) forming a peptide bond, and (3) moving the ribosome to read the next codon. The process of elongation involves the larger subunit of the ribosome, which contains two sites for binding tRNAs. Termination occurs when the ribosome encounters a stop codon.

Eukaryotic cells sort newly made polypeptides to different addresses such as the cytosol, organelles, membranes, or locations outside of the cell. The destination of a polypeptide often depends on the presence of signal peptides at its amino terminal. Polypeptides destined for the outside of a cell move into the lumen of the rough endoplasmic reticulum. There an enzyme usually clips off the signal peptide, and other enzymes add a sugar complex to specific amino acid sequences.

KEY TERMS

acridine dye
aminoacyl-tRNA
aminoacyl-tRNA synthetase
antibiotic
anticodon
antiterminator
branch point
Central Dogma
chaperonin
codon
consensus sequence
degenerate
docking protein (or SRP receptor)
dolichol
downstream
elongation
exon
Genetic Code
glycoprotein
glycosylation
heat shock
heterogeneous nuclear RNA (hnRNA)
initiation
initiation factor (IF)
initiation site
insulin
intron (or intervening sequence)
lariat
membrane-spanning segment
messenger RNA (mRNA)
nonsense codon
nucleolar organizer
peptide bond formation
polyribosome (or polysome)
positioning
post-transcriptional processing
post-translational
pre-mRNA
Pribnow box
primary transcript
pro-opiomelanocortin (POMC)
promoter
reading frame
release factor
rho (ρ)
RNA polymerase
self-splicing
sigma (σ)
signal hypothesis
signal peptidase
signal peptide
signal recognition particle (SRP)
signal sequence receptor
snRNP
spliceosome
SRP receptor (or docking protein)
stop codon
TATA box
template
termination
termination site
topogenic sequence
transcribed spacer RNA
transcription
transfer RNA (tRNA)
transit peptide
translation
translocation
universal genetic code
upstream
untranscribed spacer DNA
wobble

QUESTIONS FOR REVIEW AND UNDERSTANDING

1. List the four functions a protein may have. What ultimately determines the function of a protein?
2. List four ways proteins can be covalently modified.
3. Define and contrast the following terms:
   (a) transcription vs. translation
   (b) DNA vs. RNA
   (c) nucleotide vs. amino acid
   (d) codon vs. open reading frame
   (e) DNA polymerase vs. RNA polymerase
   (f) promoter vs. terminator
   (g) upstream vs. downstream
   (h) intron vs. exon
   (i) transit peptide vs. signal peptide
4. What is the Central Dogma? Who proposed it? What two other discoveries, or models, is this person known for?

5. How might you distinguish the 5′-end from the 3′-end of an RNA molecule?
6. How was it demonstrated that the Genetic Code is universal?
7. What factors determine the tertiary and quaternary structure of proteins?
8. Can a single nucleotide in mRNA participate in one or more codons? Cite evidence for your answer.
9. Do all codons specify an amino acid? Explain.
10. What does it mean that the Genetic Code is degenerate? Why was it expected that it would be degenerate?
11. Describe the two pieces of evidence that suggested that a triplet of nucleotides specifies one amino acid.
12. What is the significance of wobble?
13. List the different kinds of RNA molecules. Describe their functions.
14. Do amino acids play a direct role in ordering themselves in a polypeptide chain? What evidence supports your answer?
15. Eukaryotes can have as many as five unique RNA polymerases. Discuss the properties and functions of each. How do the different eukaryotic RNA polymerases recognize their respective DNA target sequences?
16. What are consensus sequences? What is their significance? Explain how they are identified experimentally.
17. How is it determined which strand of the DNA double helix serves as the template for transcription?
18. The study of transcription factors is one of the more highly competitive fields in molecular biology today. Why do you suppose this is so?
19. Describe how pre-mRNA is modified to become mature mRNA.
20. Explain how a pulse chase experiment might be done to show the existence of post-transcriptional processing of RNA.
21. If you obtained the DNA and mature mRNA sequence originating from a given eukaryotic gene, would they be complementary? Explain your answer.
22. Describe the following processes of transcription: initiation, elongation, and termination.
23. Describe the following processes of translation: initiation, elongation, and termination.
24. What are the energy-requiring steps of translation? How much energy is used in each?
25. Describe how prokaryotes and eukaryotes determine which AUG of an mRNA molecule is the initiation site for translation.
26. Where and when does peptide bond formation occur?
27. What does the signal peptide do?
28. How was the function of the signal peptide demonstrated directly?
29. What are chaperonins?
30. Contrast the fate of proteins translated on the rough ER with those translated on free ribosomes.
31. In the case of an integral membrane protein:
    (a) Why does the carboxy terminus lie oriented toward the cytoplasmic face?
    (b) What are the characteristics of membrane-spanning segments?
    (c) What is the function of a topogenic sequence?
32. Once scientists have completed sequencing the human genome, will all the sequence information they obtain correspond to polypeptides? Explain your answer.
33. Discuss the parallel between the use of mutants and the use of antibiotics in the study of biological processes.

SUGGESTED READINGS

ALBERTS, B., D. Bray, J. Lewis, M. Raff, K. Roberts, and J. D. Watson, *Molecular Biology of the Cell*, 3rd edition, Garland, New York, 1994. Part of Chapter 6 deals with the basic mechanisms of RNA and protein synthesis. Chapters 12 and 13 deal with intracellular compartments and protein sorting and with intracellular transport.

LODISH, H., D. Baltimore, A. Berk, S. L. Zipursky, P. Matsudaira, and J. Darnell, *Molecular Cell Biology*, 3rd edition, Scientific American Books, New York, 1995. Chapter 4 concerns the basic mechanisms of RNA and protein synthesis.

STRYER, L., *Biochemistry*, 4th edition, W. H. Freeman and Company, New York, 1995. Chapter 5 deals with basic mechanisms of RNA and protein synthesis.

CHAPTER 15

How Do Cells Regulate Gene Expression?

KEY CONCEPTS

- Changes in DNA can lead to changes in the structures or the amounts of specific proteins.
- Genetic changes in the amount of a specific protein can result from changes in the gene encoding that protein or in genes that specify regulatory proteins.
- Cell specialization in a multicellular organism depends on gene regulation, with individual genes expressed at different levels in different types of cells.
- Many regulatory genes specify proteins that affect transcription by interacting with specific sequences in DNA.
- A single regulatory gene can affect the expression of many other genes.
- Eukaryotic cells have a number of mechanisms that allow them to alter the sequence and amounts of individual mRNAs.

OUTLINE

ESSAYS

Chapter 14 discussed how cells convert the information in the DNA of a given gene into a polypeptide with a specific amino acid sequence. We also saw that the amino acid sequence of a protein not only determines its three-dimensional structure but also specifies its intracellular address. But not all cells make all proteins at all times, just as not all the water taps in your home are flowing at all times. The production of individual proteins, like the flow of water in your kitchen sink and washing machine, varies according to need.

Regulation saves the energy and materials that would otherwise be used for unneeded protein synthesis. Regulated gene expression also establishes the relative concentrations of interacting proteins. In multicellular organisms, regulation is indispensable to the division of labor among cells, because specialization depends on the ability of different types of cells to produce different proteins.

This chapter discusses the many ways in which prokaryotic and eukaryotic cells regulate the expression of individual genes. Information about gene regulation has come both from studies of mutations that affect gene regulation, especially in bacteria, and from biochemical and molecular manipulation of gene regulation *in vitro.* We begin by re-examining the nature of mutations, using the information about protein synthesis discussed in Chapter 14.

MUTATIONS MAY AFFECT EITHER THE STRUCTURE OF PROTEINS OR THEIR REGULATION OF EXPRESSION

All genetic variation ultimately depends on **mutations,** permanent changes in the sequence of DNA, and the first task of genetics was to explain the rules that govern the inheritance of such variations. (See Chapter 11.) Our focus here is on the origin and consequences of mutations: How do mutations arise? And how do they influence phenotype?

The word "mutation" was first used by Hugo DeVries, whose work on the American evening primrose led to the rediscovery of Mendel's laws (and Mendel's paper). DeVries's work was also the immediate stimulus for Thomas Hunt Morgan's search for *Drosophila* mutations. (See Chapter 11.) Morgan especially wanted to know the relationship between these developmental changes and the changes that occur during evolution. Upon learning of DeVries's work, Morgan thought that mutations might hold the key to such changes, and he therefore changed the direction of his research. Morgan's work with *Drosophila,* and thousands of studies since, have shown that mutations can produce many effects on phenotype, ranging from none at all to major changes in form.

A Mutation May Occur Anywhere in DNA, But Its Impact Depends on Where It Occurs

Changes in DNA sequence occur continuously. Changes can initially result from random (thermal) molecular collisions, from mistakes in the replication process (for example, from a failure to form a correct base pair), or from chemical changes in the structure of a nucleotide. The rate of mutation (the number of changes per nucleotide pair per generation) depends both on how fast the sequence changes first occur and on the efficiency with which cells repair these errors. (See Chapter 13.)

The rate of mutation is roughly the same for any piece of chromosomal DNA, but some sequences, called *mutational hot spots,* are more likely than others to mutate. In bacteria, for example, the average rate of mutations during ordinary DNA replication is about 1 mutation in 10 million nucleotide pairs (10^{-7}) per generation, but some sequences mutate 25 times more rapidly and others 25 times more slowly. Even the highest rates of mutation, however, are far less than the error levels to which we are accustomed in everyday life. A good typist, for example, may make only one mistake per 1000 to 10,000 characters, that is, 10,000 to 1000 times more frequently than an average bacterial cell replicating its DNA.

We may distinguish three different types of mutation, according to their effects on phenotype: **silent mutations,** changes in DNA sequence that have no effect on phenotype; **structural mutations,** which lead to alterations in the amino acid sequence of a polypeptide; and **regulatory mutations,** which affect the amount (rather than the structure) of a polypeptide. This distinction is valuable but sometimes blurry. For example, a mutation may appear to be silent only because researchers have not yet found an effect on phenotype. Furthermore (as we discuss later in this chapter), many regulatory mutations are actually structural mutations in genes for regulatory proteins.

Most silent mutations are changes that do not alter the structure of a polypeptide. Many such mutations lie in DNA sequences that do not code for a polypeptide (for example, in an intron sequence). Many silent mutations, however, do lie within a coding region, but (because of the redundancy of the Genetic Code) they do not lead to a change in the specified amino acid residue.

The mutations that are most easily understood are structural, such as those illustrated in Figure 15-1 for the *E. coli* enzyme, tryptophan synthetase. A structural mutation leads to an altered sequence of amino acids, often resulting in a lack of enzyme activity. A regulatory mutation, on the other hand, may result from many types of alterations—for example, in the DNA sequences that are recognized by the transcription machinery or in the structures of proteins that interact with sequences in DNA or RNA.

The altered proteins that result from structural mutations may have different effects on phenotype. In some

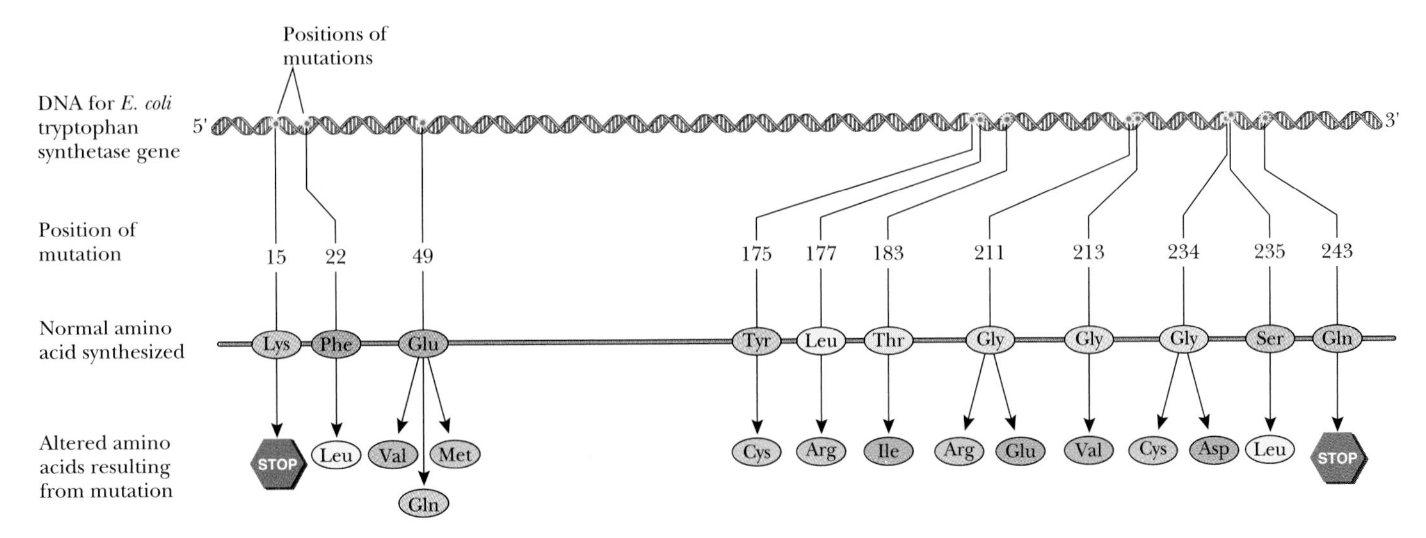

Figure 15-1 Structural mutations in the gene for *E. coli* tryptophan synthetase. Missense mutations lead to the substitution of one amino acid for another, while nonsense mutations lead to a prematurely terminated polypeptide. The numbers indicate the codon (and amino acid) number in which a mutation has occurred.

cases, the protein may be able to function more or less normally, and the phenotype is changed little. At the other extreme, the mutation may be a **null mutation,** in which the resulting protein is never synthesized in a functional form. In still other cases, the DNA change leads to impaired but nonetheless functional proteins, as in sickle cell disease: As long as the hemoglobin S does not precipitate, it functions well as an oxygen carrier.

Conditional Mutants Are Valuable Experimental Tools

Some DNA alterations are **conditional mutations,** in which the resulting protein may be able to function under some conditions but not others. **Temperature-sensitive mutations,** one kind of conditional mutation, encode proteins that function normally at one temperature (called the *permissive temperature,* usually relatively low), but not at a second temperature (the *restrictive temperature,* generally higher than the permissive temperature). In one *Drosophila* mutant, the fly functions normally at 20°C but becomes paralyzed at 30° as a protein fails to regulate ion transport across nerve cell membranes. Temperature sensitivity results when an amino acid substitution in a protein reduces its thermal stability, so that it denatures more readily than the wild-type protein.

Regulation of gene expression—ensuring that the right amount of each protein is made at the right time—is crucial to survival. Recall, for example, the genes whose products are responsible for coordinating the events of the cell cycle. (See Chapter 9.) Mutations in these genes can prevent the completion of the cell cycle, so such mutations never persist in naturally growing populations. Finding these genes required a search for temperature-sensitive mutations. The mutant cells could divide at the permissive temperature but not at the restrictive temperature. The use of these conditional mutations allowed cell biologists to study the regulation of the eukaryotic cell cycle in more detail than they otherwise could have.

Mutations Can Change a Single Base or Extend over Many Bases

We can also categorize mutations on the basis of the size of the changes in DNA. Geneticists often distinguish between **point mutations** (single-base mutations), which change one nucleotide pair (or, in some cases, a few nucleotide pairs), and **chromosomal mutations,** which affect relatively large regions of chromosomes. A point mutation in a structural gene (shown in Figure 15-2) may be a **base substitution,** the replacement of one base (nucleotide) by another; an **insertion,** the addition of one or more nucleotides; or a **deletion,** the removal of one or more nucleotides. A base substitution may result in a **missense mutation,** which changes the codon for one amino acid residue into a codon for another amino acid residue, or a **nonsense mutation,** which changes the codon for an amino acid residue into a nonsense or termination codon and can lead to a truncated polypeptide.

Insertions and deletions may produce a **frame-shift mutation,** illustrated in Figure 15-2, the altered grouping of nucleotides into codons. As discussed in Chapter 14, the addition or subtraction of a single base wreaks havoc in the ribosomes' reading of mRNA downstream from the mutation. The result is often a string of missense changes, followed by a nonsense (stop) codon.

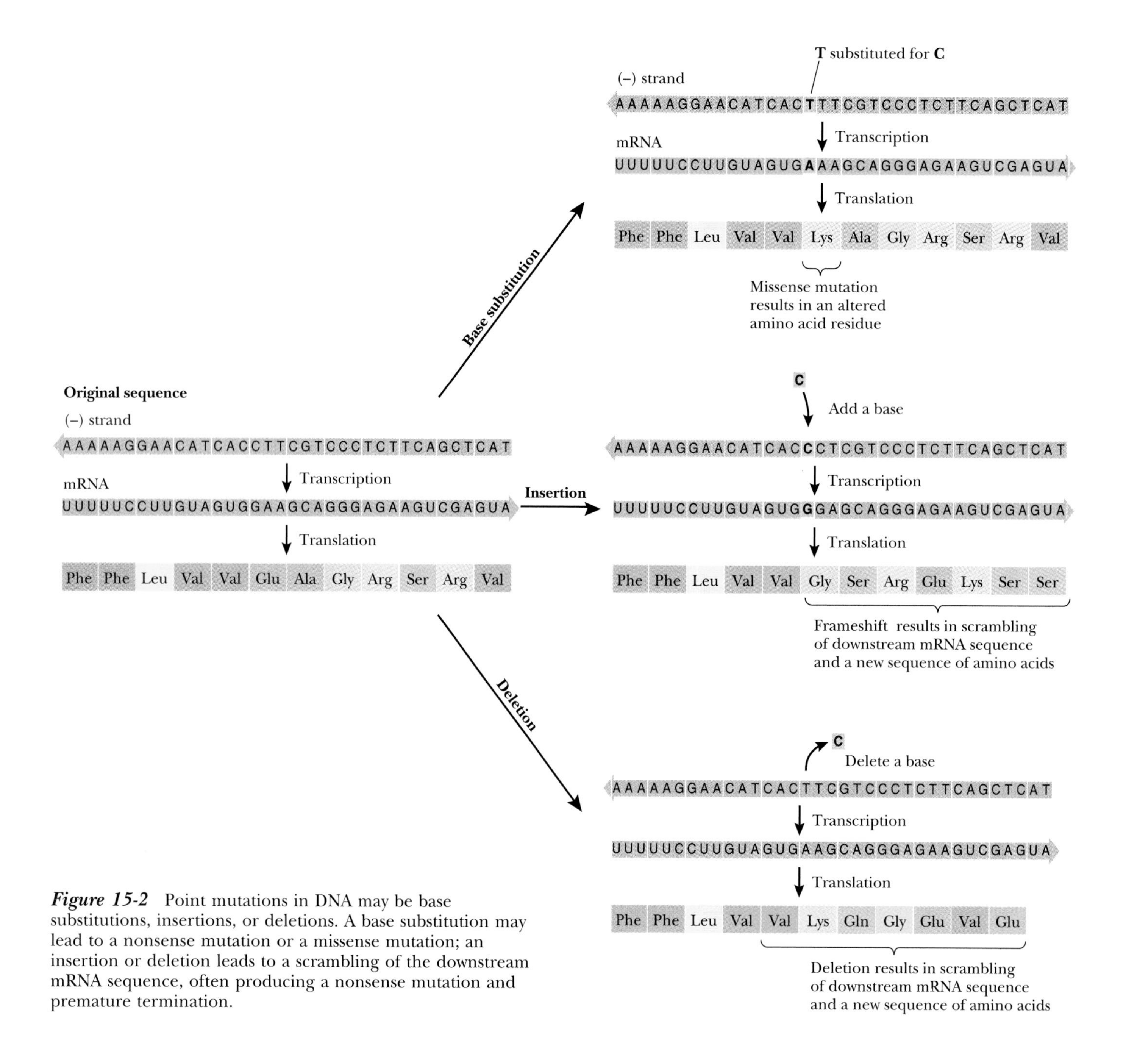

Figure 15-2 Point mutations in DNA may be base substitutions, insertions, or deletions. A base substitution may lead to a nonsense mutation or a missense mutation; an insertion or deletion leads to a scrambling of the downstream mRNA sequence, often producing a nonsense mutation and premature termination.

Chromosomal mutations are larger than point mutations and may affect large regions of chromosomes (or even whole chromosomes). Chromosomal mutations include insertions and deletions (which are larger than a few nucleotides), as well as **translocations,** in which part of one chromosome is moved to another chromosome; **inversions,** in which a segment of a chromosome is turned 180 degrees from its normal orientation; and **duplications,** in which part of a chromosome appears twice. Many insertions arise from unconventional genetic systems, such as viruses and transposons, as discussed in Chapter 16. **Aneuploidy** (abnormal chromosome number, as discussed on p. 284) is regarded as another type of chromosomal mutation. Aneuploidy includes **trisomy,** the presence of three copies of one chromosome in a diploid cell, and **monosomy,** the presence of only a single copy.

Mutagens are agents that increase the rate of mutation. Ultraviolet light, high-energy radiation from x-rays or radioactive decay, and many chemicals are mutagens. Each mutagen has characteristic effects. Ultraviolet light, for example, causes the linking together of adjacent T residues in the same DNA strand; x-rays cause breakage and incorrect rejoining of DNA, leading to rearrangements; acridine dyes cause insertion or deletion of single nucleotide residues; and other chemicals increase the base-pairing errors.

X-rays are particularly effective producers of chromosomal mutations, although these mutations also arise spontaneously or after exposure to certain chemicals. Polytene chromosomes of *Drosophila* allow the visualization of a variety of chromosomal mutations.

A Second Mutation Can Sometimes Restore a Normal Phenotype

Point mutations are much more susceptible than chromosomal mutations to **reversion,** a subsequent mutation that restores the previous version of the sequence. Geneticists usually detect reversions as second mutations that restore the previous (unmutated) phenotype. Reversion mutations, however, may not be exact restorations of the previous nucleotide sequence. For example, the reversion may have a nucleotide sequence that is not identical to the original but specifies the same amino acid.

In some cases an apparent reversion is actually a **suppressor mutation,** a mutation in one gene that prevents the phenotypic changes caused by a mutation in another gene. Among the best-studied suppressors is a strain of *E. coli* with a mutation in a gene for a tyrosine tRNA. The mutation, in the anticodon loop, allows the $tRNA^{Tyr}$ to recognize the UAG stop codon. The suppressor tRNA prevents the premature termination of polypeptides with UAG missense mutations. Because UAG also functions as a stop codon, however, the suppressor mutation also leads to some improperly long polypeptides.

PROKARYOTIC REGULATION: MUTATIONS HAVE REVEALED MANY MECHANISMS BY WHICH PROKARYOTES REGULATE GENE EXPRESSION

Regulatory mutations may affect both an organism's ability to respond to environmental stimuli and, in multicellular organisms, the pattern of cell specialization during development. In this section we examine regulatory mutations that affect gene regulation in prokaryotes.

Prokaryotes May Transcribe Several Genes into a Single Messenger RNA

The ability to respond to the environment can be a matter of life or death. It is not surprising, then, that organisms have evolved sensitive molecular mechanisms that ensure the production of enzymes when they need them. The first insights into such molecular mechanisms came from studies of a prokaryote, *E. coli.*

A rich growth medium directly provides both energy sources and building blocks. As long as the medium contains high enough concentrations of amino acids, bacteria devote most of their energy to rapid division and do not squander energy making amino acids that are readily available from their environment.

If no amino acids are available, however, the bacteria must make them from scratch, usually taking the needed carbon atoms from sugar in the medium and nitrogen atoms from ammonia or another nitrogen-containing compound. To accomplish the synthesis of the needed amino acids, *E. coli* produces enzymes that are not present when they grow in a rich medium. Making histidine from simple C- and N-containing compounds requires the participation of seven different enzymes.

Genetic analysis of *E. coli* mutants that cannot make histidine showed that the genes encoding the seven enzymes required for histidine synthesis lie right next to each other on the *E. coli* chromosome. Molecular studies of these genes led to the surprising conclusion that all seven genes are transcribed into a single huge molecule of mRNA. A set of genes transcribed into a single mRNA (in prokaryotes) is called an **operon.** An mRNA that encodes more than a single polypeptide is said to be **polycistronic.**

Making a single mRNA from several related genes ensures that all the enzymes needed to make a particular final product (like histidine) are turned on or off at the same time. This is one strategy for achieving *coordinate regulation,* the simultaneous expression of several genes. Coordinately regulated genes often specify protein products with related functions. When bacteria need to make histidine, RNA polymerase copies all seven genes for the seven enzymes in the pathway of histidine biosynthesis. When the medium contains enough histidine, the bacteria do not make this RNA, and none of the seven genes is expressed.

The Lactose Operon Is the Best-Understood Example of Transcriptional Regulation in Prokaryotes

Most of our first knowledge about transcriptional control in prokaryotes came from studies of mutations in the *E. coli* genes responsible for the ability to grow on lactose ("milk sugar"). At least in the laboratory, glucose is the usual source of energy and carbon atoms. Many bacteria, including *E. coli,* however, can also grow on lactose, the major sugar in milk. As shown in Figure 15-3, lactose is a disaccharide consisting of glucose and galactose held together by a glycosidic bond.

When sufficient glucose is present, bacteria do not use the enzymes of lactose utilization. When glucose is absent and lactose is present, however, bacteria produce three proteins needed to derive energy from lactose: the enzyme β-galactosidase (which splits lactose into galactose and glucose) and two other proteins, called permease and acetylase. The genes encoding these three proteins lie next to each other in *E. coli* DNA and form the **lactose** (or ***lac***) **operon.** As in the case of the histidine operon, transcrip-

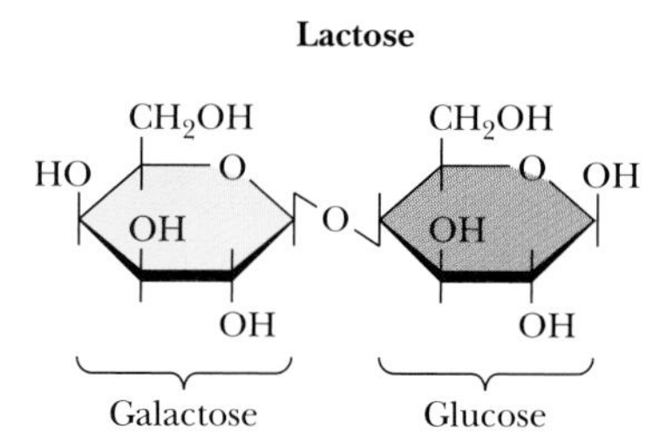

Figure 15-3 Lactose ("milk sugar") is a disaccharide consisting of one molecule each of galactose and glucose.

tion of the *lac* operon gives a polycistronic mRNA, which encodes all three proteins.

When glucose is not present, lactose stimulates (or induces) the expression of the *lac* operon. When no lactose is present, each bacterial cell has only about 3 molecules of β-galactosidase; after induction there may be as many as 3000 molecules of the enzyme, about 3% of the cell's protein. Biochemists can now study the regulation of transcription in the *lac* operon *in vitro.* Long before these experiments were possible, however, François Jacob and Jacques Monod, microbiologists at the Institut Pasteur in Paris, had deduced much about this regulation from purely genetic experiments.

In normal *E. coli,* β-galactosidase and the other *lac* operon proteins are **inducible;** that is, bacteria make high levels of the proteins only in the presence of inducer (in this case, lactose or a similar molecule). Some of Jacob and Monod's mutants, however, were **noninducible,** meaning that they *never* made β-galactosidase; other mutants were **constitutive,** meaning that they made high levels of β-galactosidase even when lactose was not present.

In classifying their mutants, Jacob and Monod distinguished between structural mutations (changes in the DNA that codes for the protein's amino acid sequence) and regulatory mutations (which control the level of the gene's expression). Using techniques of bacterial genetics, Jacob and Monod discovered two distinct regulatory sites, whose positions on the genetic map were just next to the structural gene for β-galactosidase.

Lac Operon Regulation Depends on the Interaction of an Inducer with a Negative Regulator

One of the two regulatory sites discovered by Jacob and Monod is the *i* gene (for *i*nducible). Later work showed that the *i* gene is the structural gene for the ***lac* repressor,** a protein that prevents the expression of the *lac* operon. The *i* gene is thus both a regulatory gene (for the *lac* operon) and a structural gene (for the *lac* repressor). The *lac* repressor is an example of a **negative regulator,** a molecule that reduces transcription when it binds to a specific site in DNA.

Under normal growth conditions where little or no lactose is present, wild-type *E. coli* makes little β-galactosidase. The *lac* repressor prevents the transcription of the *lac* operon because the *lac* repressor binds to the **operator** region, the DNA sequence in an operon to which a repressor protein binds. The *lac* operator is a sequence of 21 nucleotides just upstream from the structural gene for β-galactosidase, as shown in Figure 15-4a.

The *lac* operator is the second regulatory site discovered by Jacob and Monod. The operator overlaps with the promoter of the operon. When the *lac* repressor binds to the operator, it prevents RNA polymerase from transcribing the structural gene.

How does this system of repressor and operator allow the lactose to induce gene expression from the *lac* operon? The key to understanding the induction of the *lac* operon is the realization that the *lac* repressor can bind both to the operator and, when lactose is present, to lactose, as shown in Figure 15-4b. The *lac* repressor exists in two distinct conformations, so that when it binds to lactose it cannot bind to the operator. This is another example of an allosteric effector changing the activity of a protein. (See Chapter 6, p. 112.)

Lactose prevents the repressor from repressing. It therefore induces the transcription of the operon. The action of lactose in removing the repressor from the *lac* operator and allowing the movement of RNA polymerase has been fancifully compared with the role of a peanut used to coax an elephant from a train track so that a locomotive can pass.

With this picture in mind, we are in a good position to explain the phenotypes of the mutants isolated by Jacob and Monod. First consider the effects of changes in the *lac i* gene. Changes that interfere with the binding of the repressor to the operator give constitutive (uncontrolled) expression. Transcription can still be turned off, however, by adding wild-type repressor—either in a test tube experiment or by genetic manipulations that produce bacteria with two copies of the *lac i* gene.

Changes that interfere with the binding of lactose to the repressor, on the other hand, give a noninducible phenotype. In this case, adding more repressor does not help because some repressor that is unresponsive to lactose is always able to bind to the operator and prevent transcription.

What is the effect of a change in the operator region? Any change that prevents the binding of the repressor also leads to constitutive expression. Here, again, added repressor cannot block transcription because there is no place for it to bind.

The *lac* Operon Also Responds to a Positive Regulator

Bacteria also use **positive regulators,** which increase the rate of transcription. One such positive regulator, whose interaction with the *lac* operon is shown in Figure 15-5a

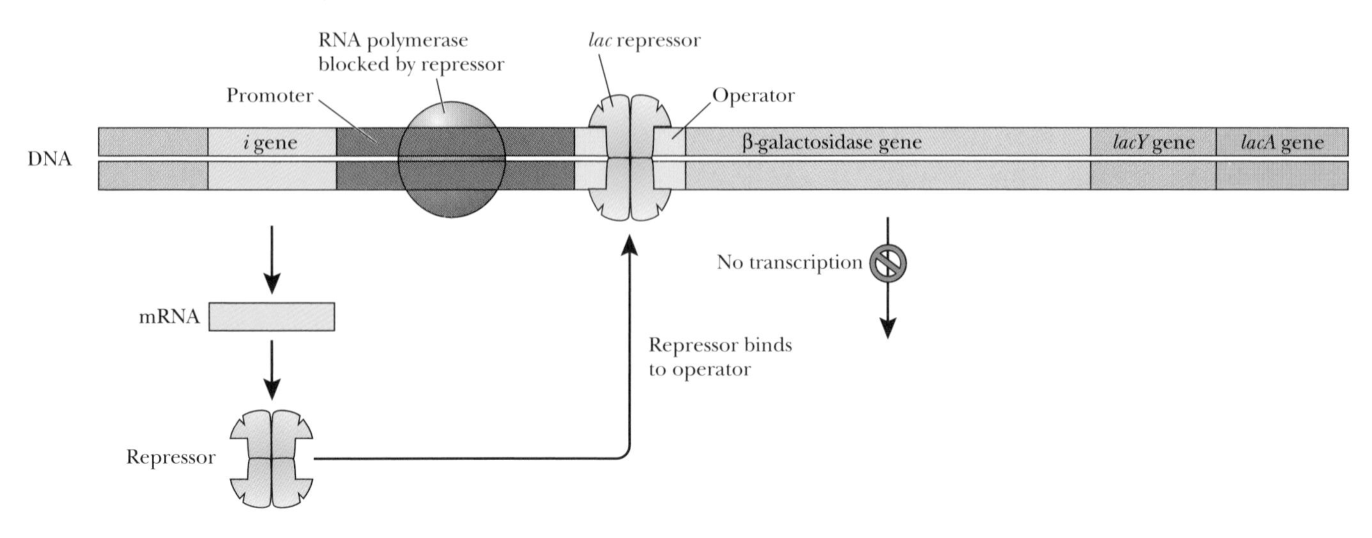

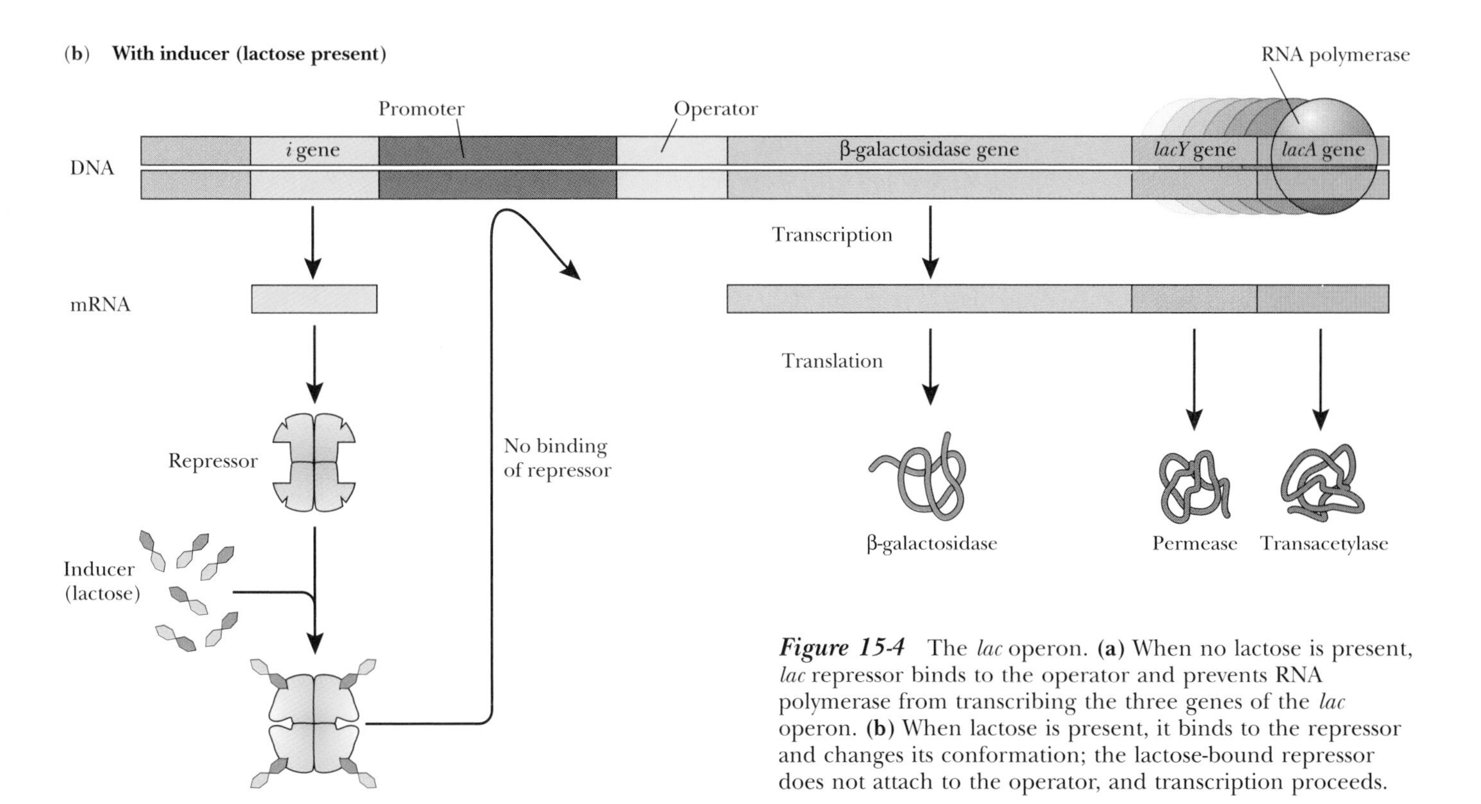

Figure 15-4 The *lac* operon. **(a)** When no lactose is present, *lac* repressor binds to the operator and prevents RNA polymerase from transcribing the three genes of the *lac* operon. **(b)** When lactose is present, it binds to the repressor and changes its conformation; the lactose-bound repressor does not attach to the operator, and transcription proceeds.

and b, is called **CAP,** or **catabolite activator protein.** CAP binds to the promoter region of the *lac* operon.

Like the *lac* repressor, CAP has two binding sites—one for a particular sequence in DNA and one for a small molecule. The small molecule that binds to CAP is **cyclic AMP,** a nucleotide that is unusual in containing an internal phosphodiester bond, as shown in Figure 15-5c. When cyclic AMP binds to CAP, CAP binds to DNA in the promoter region. This binding somehow increases the rate at which RNA polymerase binds, stimulating the synthesis of RNA. Cyclic AMP and cyclic GMP are also important regulators in eukaryotic cells, as discussed later in this chapter and in Chapter 19.

CAP gets its name because its action depends in part on the availability of glucose or glucose breakdown products ("catabolites"). In bacteria, cyclic AMP is present at high concentrations only when glucose is in short supply. When glucose is plentiful, RNA polymerase transcribes the *lac* operon only slowly, even in the presence of lactose.

Bacteria respond quickly to changes in their envi-

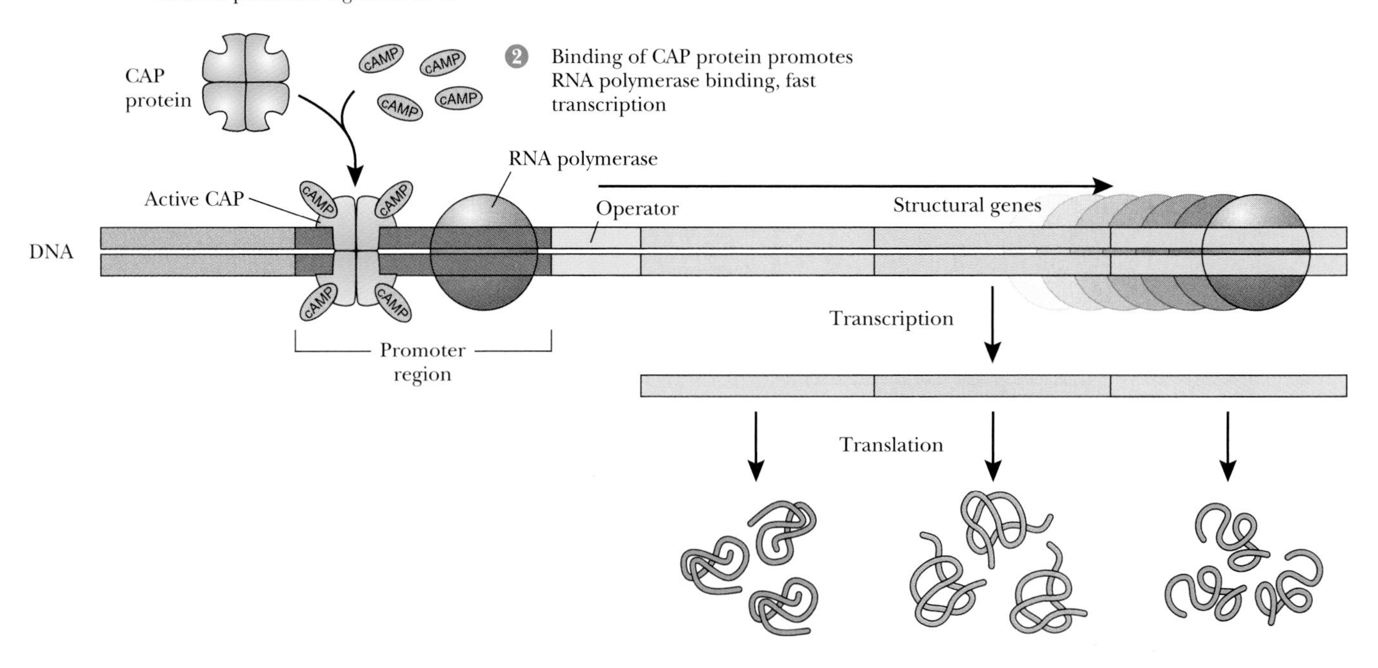

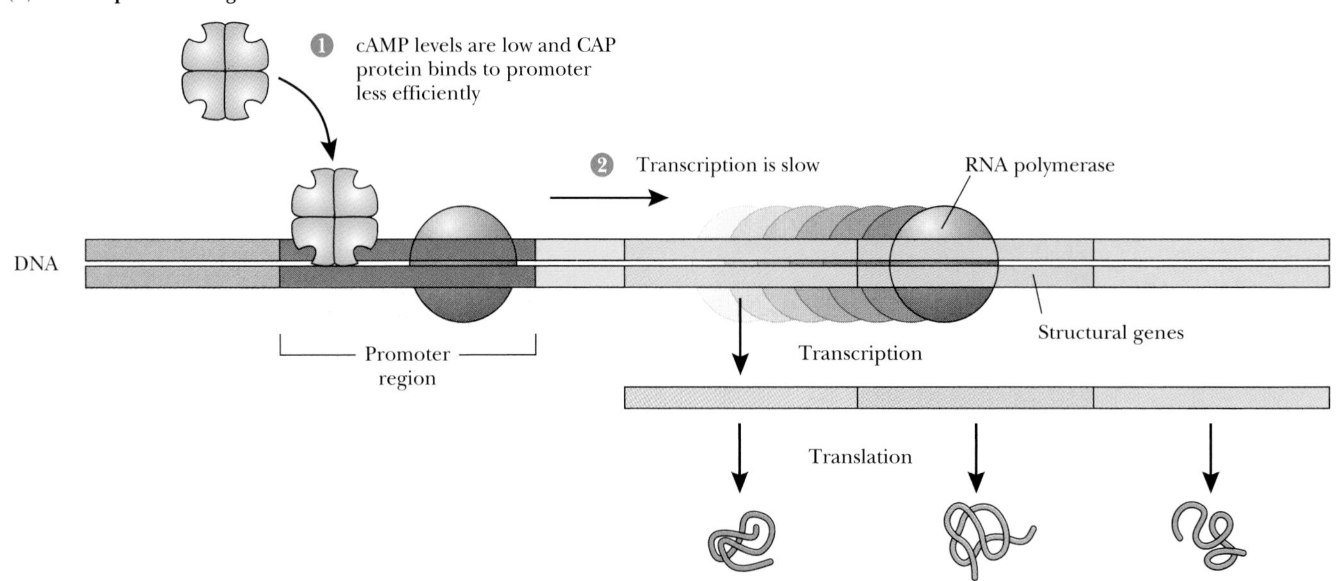

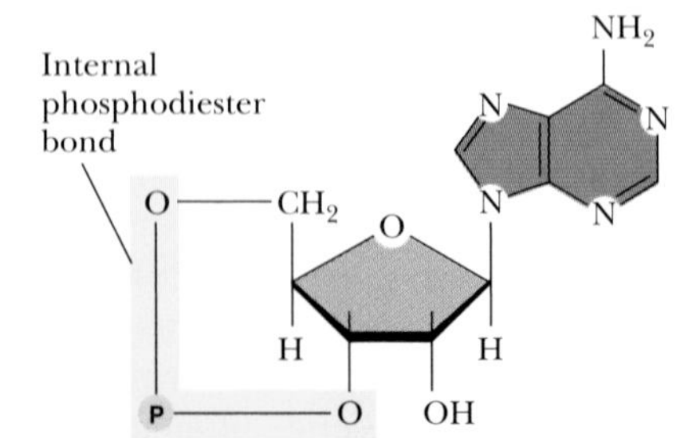

Figure 15-5 CAP (catabolite activator protein) is a positive regulator of *lac* operon transcription. **(a)** When no glucose is present, cyclic AMP levels are high; cyclic AMP activates CAP, which can then bind to the promoter region and stimulate transcription. **(b)** When glucose is present, cyclic AMP levels are low, and CAP does not bind to the promoter; transcription proceeds only slowly. **(c)** Structure of cyclic AMP.

ronment. For example, when lactose is present but glucose is not, *E. coli* quickly begins to make β-galactosidase. Similarly, when glucose becomes available or lactose is exhausted, they almost cease production of β-galactosidase within minutes. The rapidity of these responses depends not only on the regulation of transcription but also on the rapid destruction of the *lac* operon mRNA. Like most prokaryotic mRNAs, *lac* mRNA has a lifetime of only a few minutes.

CAP regulates the transcription not only of the *lac* operon but also of other genes whose protein products are enzymes required to obtain energy from molecules other than glucose. CAP provides an example of how a single regulator can affect the transcription of genes that are physically separate from one another.

Prokaryotes Can Regulate Gene Expression at Levels Other than Transcriptional Initiation

Two other examples of gene regulation in prokaryotes deserve mention here: **attenuation,** regulated premature termination, and **translational control,** alterations in the rate at which specific mRNAs are translated.

Attenuation was first discovered in the *E. coli trp* operon, a set of five structural genes whose protein products are responsible for the synthesis of tryptophan. As with the *lac* operon, transcription of the *trp* operon depends on environmental conditions. But the direction of the regulation is different: Whereas transcription of the *lac* operon *increases* when lactose is present (because the induced enzymes metabolize it), transcription of the *trp* operon *decreases* when tryptophan is present (because there is no need to synthesize it). Like the *lac* operon, the *trp* operon is subject to negative regulation by the *trp* repressor, but the *trp* repressor binds to the *trp* operator only when tryptophan is present. The result of this arrangement is the *repression* of transcription in the presence of tryptophan and *expression* in the absence of tryptophan. With this additional twist, transcriptional regulation of the *trp* operon appears to resemble that of the *lac* operon.

But researchers could not understand a new set of mutants, in which a deletion between the operator and the first gene in the operon (*trp E* in Figure 15-6a) led to higher than normal levels of tryptophan-producing enzymes, even in the presence of tryptophan. Further studies revealed that the region deleted in these mutants include an *attenuator site,* where transcription terminates before the gene is fully transcribed into RNA. The attenuator site also occurs in the middle of a *leader RNA,* a sequence transcribed from the region between the promoter and the *trp E* structural gene. The sequence of the leader RNA suggested that it could fold into a hairpin, with Watson-Crick base pairing between segments 3 and 4, shown in Figure 15-6b. Such a hairpin can stop RNA polymerase from continuing and cause the termination of transcription; under these conditions, the bacteria produce little *trp* mRNA. High concentrations of tryptophan activate the *trp* repressor and allow only low levels of transcriptional initiation. The expression of the *trp* operon is further limited by attenuation of this low level of transcription.

When the tryptophan concentration decreases, however, *trp* repressor is no longer activated, and termination at the attenuator site ceases, as shown in Figure 15-6a. The mechanism for this latter change depends on the ability of the leader RNA to serve as a functional mRNA, initiating protein synthesis at its AUG even as transcription is still going on. The explanation for the effect of low tryptophan concentrations involves the leader mRNA's 10th and 11th codons, both of which specify tryptophan. When the tryptophan concentration is low, so is that of the tryptophan aminoacyl-tRNA, and the ribosome on the leader mRNA stalls—unable to proceed without the next tRNA. The result of this stalling (as illustrated in Figure 15-6c) is that the ribosome covers region 1, prevents the formation of the termination hairpin, and allows the transcription of the rest of the *trp* operon.

The combination of regulation by controlled termination and controlled binding of the *trp* repressor allows changes (up to a factor of 700) in the rates of production of the tryptophan synthetic enzymes. This mechanism for controlled termination depends on the close coupling of transcription and translation in prokaryotes. Such a mechanism would be impossible in eukaryotes because transcription takes place in the nucleus and translation in the cytoplasm.

Our final example of bacterial regulation is the translational control of the synthesis of ribosomal proteins. The genes encoding ribosomal proteins in bacteria lie in at least 20 operons, but the synthesis of all the proteins is tightly coordinated. The mechanism for this regulation includes the binding of some ribosomal proteins to ribosomal protein mRNAs. In one case, illustrated in Figure 15-7, this binding can block the translation of a polycistronic mRNA that encodes (along with other polypeptides) that same ribosomal protein; in another case, a ribosomal protein prevents the translation of an mRNA that encodes other ribosomal proteins.

The regulation of ribosomal protein synthesis depends on the higher affinity of the regulating ribosomal protein for ribosomal RNA (rRNA) than for mRNA. Binding to the mRNA therefore happens only when the protein is present in excess of the rRNA. Then its excess concentration serves as a signal to turn down its synthesis, achieving feedback regulation at the translational level.

(a)

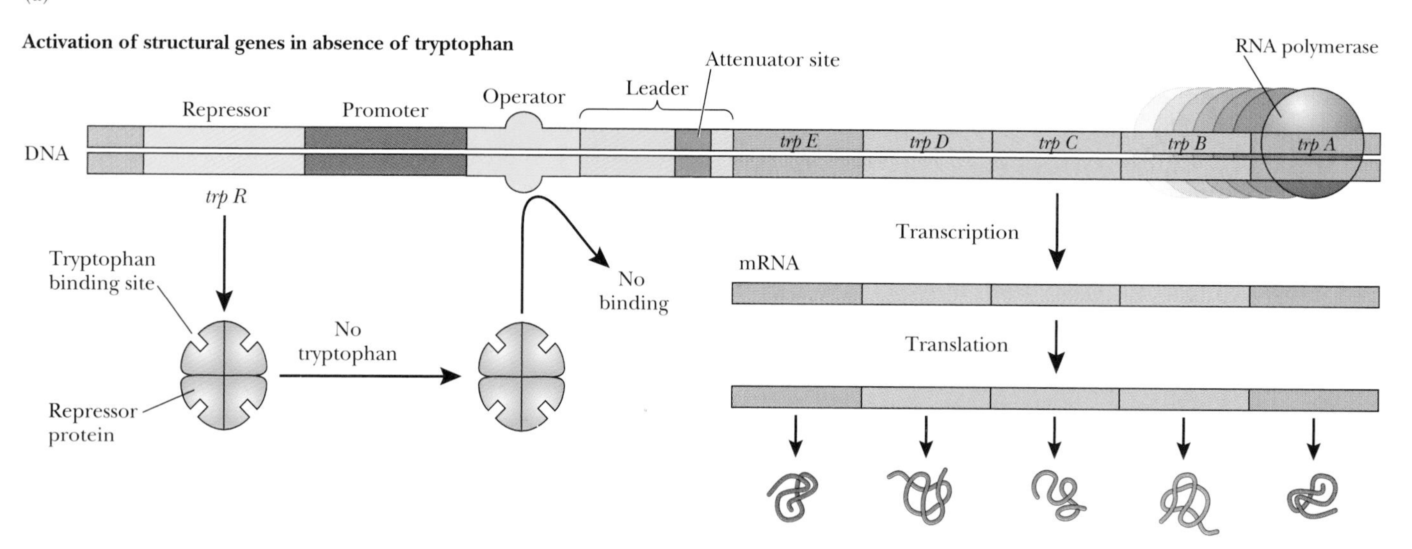
Activation of structural genes in absence of tryptophan
RNA polymerase
Attenuator site
Leader
Operator
Repressor
Promoter
DNA
trp E
trp D
trp C
trp B
trp A
trp R
Transcription
Tryptophan binding site
mRNA
No binding
No tryptophan
Translation
Repressor protein

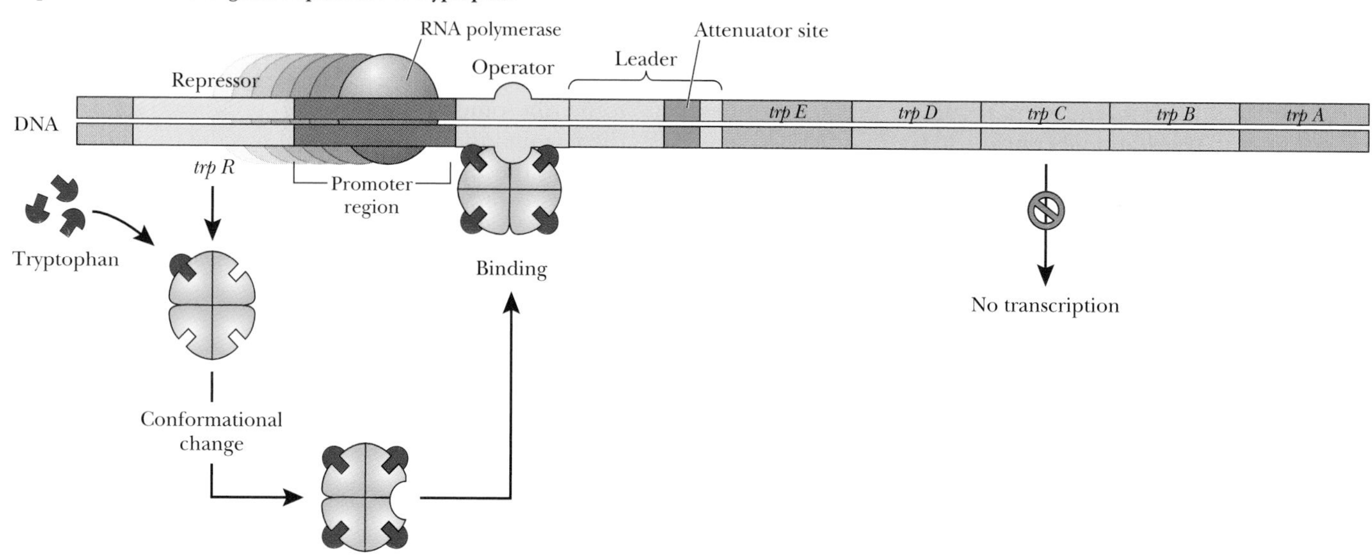
Repression of strucural genes in presence of tryptophan
RNA polymerase
Attenuator site
Leader
Operator
Repressor
DNA
trp E
trp D
trp C
trp B
trp A
trp R
Promoter region
Tryptophan
Binding
No transcription
Conformational change

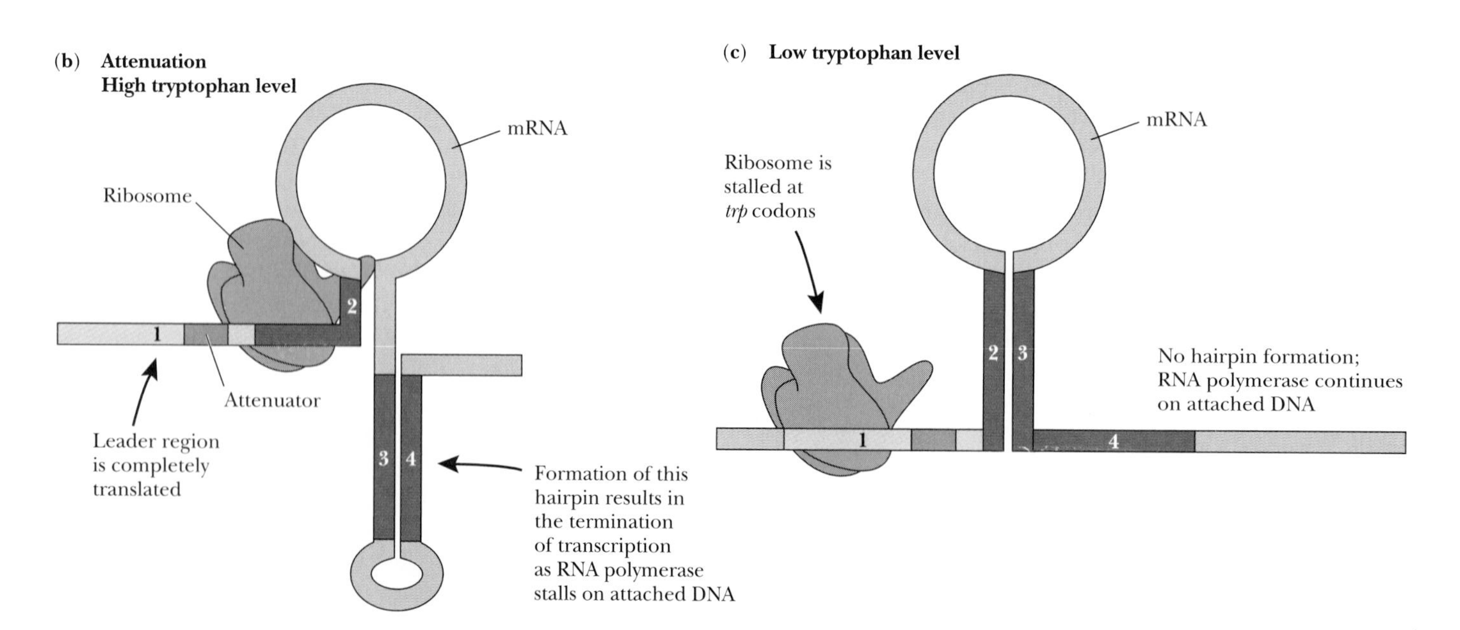
(b) Attenuation
High tryptophan level
mRNA
Ribosome
2
1
Attenuator
3
4
Leader region is completely translated
Formation of this hairpin results in the termination of transcription as RNA polymerase stalls on attached DNA
(c) Low tryptophan level
mRNA
Ribosome is stalled at trp codons
2
3
No hairpin formation; RNA polymerase continues on attached DNA
1
4

◀ ***Figure 15-6*** The tryptophan operon. **(a)** When no tryptophan is present, the repressor cannot bind to the *trp* operator; transcription proceeds, and the cell can make its own tryptophan. When tryptophan is present, the tryptophan-bound repressor attaches to the operator, and transcription stops. **(b)** Attenuation: At high levels of tryptophan, a hairpin formed in the newly made RNA, while it is still being transcribed from DNA, leads to the premature termination of transcription and the production of just a short leader RNA. **(c)** At low levels of tryptophan, the hairpin does not form in the newly made RNA, and the whole operon is transcribed.

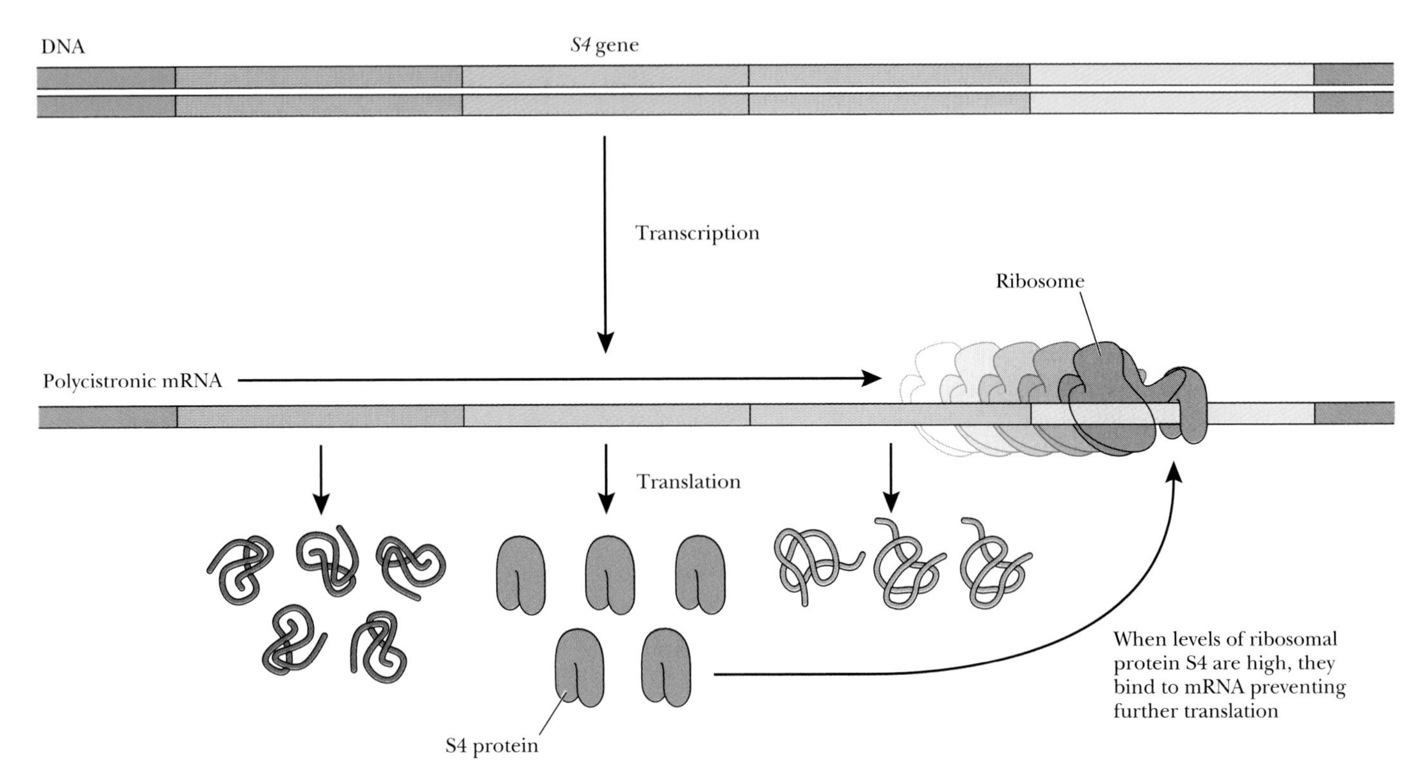

Figure 15-7 Translational control of ribosomal protein synthesis. The ribosomal protein S4 binds to the polycistronic mRNA encoding ribosomal proteins.

EUKARYOTIC REGULATION: REGULATED GENE EXPRESSION CAN OCCUR AT MANY LEVELS

Eukaryotes face the same kinds of regulatory challenges as prokaryotes. Both must respond to their environments, and both must coordinate the expression of related genes. But eukaryotes and prokaryotes differ in several important respects.

Until recently, molecular biologists knew far more about regulation in prokaryotes than in eukaryotes. New techniques (especially those based on recombinant DNA) have allowed the accumulation of detailed knowledge of DNA structure and expression in eukaryotic cells. The result has been an explosion of information about previously unknown mechanisms of eukaryotic gene regulation. The conclusion from these studies is that eukaryotes control gene expression at more levels than do prokaryotes. A eukaryotic cell can alter the accessibility of a gene to RNA polymerase, the rate of transcription, the pattern of RNA splicing, the stability of the RNA, and the rate of translation.

First, the nuclear membrane of eukaryotes prevents the physical coupling of transcription and translation. This physical separation may make it necessary for most eukaryotic mRNAs to have longer lifetimes than prokaryotic mRNAs. While most prokaryotic mRNAs have lifetimes of minutes, most eukaryotic mRNAs have lifetimes measured in hours or days, meaning that rapid changes in mRNA levels are not as common in eukaryotes as in prokaryotes. Some eukaryotic mRNAs, however, especially those encoding proteins that regulate cell growth and gene expression, are short-lived. The additional stability of eukaryotic mRNAs depends upon the modifications of pre-mRNAs within the nucleus—capping, splicing, and poly A addition. (See Chapter 14.) These additional steps also provide additional opportunities for regulation.

Second, eukaryotes generally have much more DNA than prokaryotes. Each human sperm or egg cell contains about 1000 times as much DNA as in a single *E. coli* bacterium. To fit into the nucleus, moreover, the DNA of eukaryotic cells is elaborately complexed with proteins to form chromatin. (See Chapter 13.) The compacted structure of DNA provides additional problems and opportunities for regulation because any given gene may be more or less accessible to the transcriptional machinery.

Third, eukaryotes have many intracellular compartments and must have elaborate mechanisms for the proper distribution of newly made proteins. The correct targeting of molecules to the nucleus is especially important for regulation of chromatin structure and gene expression.

Fourth, eukaryotic mRNAs are not polycistronic. Each mRNA directs the synthesis of a single polypeptide. In some special cases, a polypeptide may be cut into smaller peptides with differing functions. But the coordinate regulation of related genes does not generally depend on their sharing a common mRNA.

Finally, most of our favorite eukaryotes are multicellular, with dramatic differences in the gene expression among different cell types. Multicellular organisms must regulate the expression of groups of genes whose products are all expressed in a particular cell type or at a particular time during development. A developing red blood cell, for example, must simultaneously express the genes for both α and β globin, but not for the proteins of muscle filaments. Similarly, plants cells specialized for transporting the products of photosynthesis make different proteins from cells in the tip of a growing shoot.

Plants and animals both undergo development in which the zygote develops first into an **embryo** [Greek, *en* = in + *bryein* = to be full of], a set of early developmental stages in which the organism differs from its mature form. (See Chapters 20 and 21.) Regulatory mechanisms must ensure that the cells of the developing organism make the appropriate proteins at the appropriate times.

In our own lives, perhaps the most interesting changes in gene expression occur in our nerve cells in response to changes in our experience. As we learn, we also change the pattern of gene expression within the cells of our brains. Understanding eukaryotic gene regulation should help us to understand not only cell specialization but also the even more complex processes of development, learning, and possibly thought itself.

Multicellularity: Do All Cells Have the Same Genetic Information?

One way in which cells might express different sets of genes would be to distribute genes during development so that different types of specialized cells had different sets of genes. If this mechanism were to operate, muscle cells would not express hemoglobin because they do not have hemoglobin genes.

Many biologists once favored this idea of differential gene loss because some organisms (such as the parasitic worm, *Ascaris,* shown in Figure 15-8) actually lose chromosomes in the course of development. Studies of *Ascaris* were influential in establishing the importance of chromosomes in the first place, largely because of the work of the German cell biologist Theodor Boveri, who was also the first to show that sperm and ovum contribute equal numbers of chromosomes to the zygote. Ironically, however, chromosome behavior in *Ascaris* misled many developmental biologists into thinking that chromosome loss was a general mechanism for cell *differentiation*—the process by which types of cells become specialized and different from one another.

Examination of the chromosomes and measurement of the amount of DNA in each cell of a multicellular organism convinced biologists that all cells had the same amount of DNA and that the selective disposal of DNA was

(a)

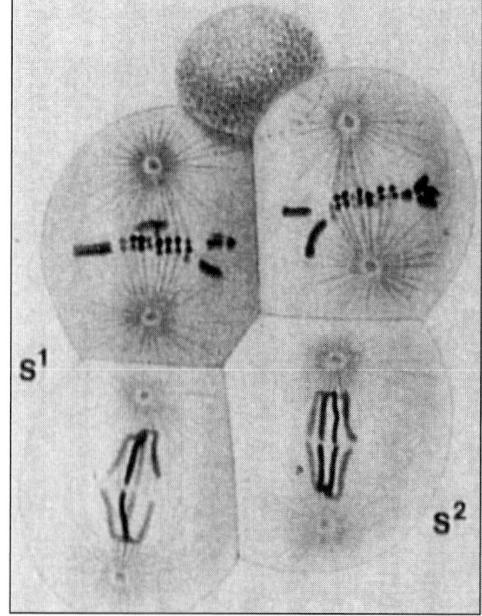

(b)

Figure 15-8 Chromosome loss in *Ascaris.* **(a)** A mature *Ascaris* worm, an intestinal parasite; **(b)** chromosome loss during the early development of an *Ascaris* embryo. *(a, Martin Rotker/Phototake NYC; b, Eric Davidson,* Gene Expression in Early Development, *3rd edition, Academic Press, 1986)*

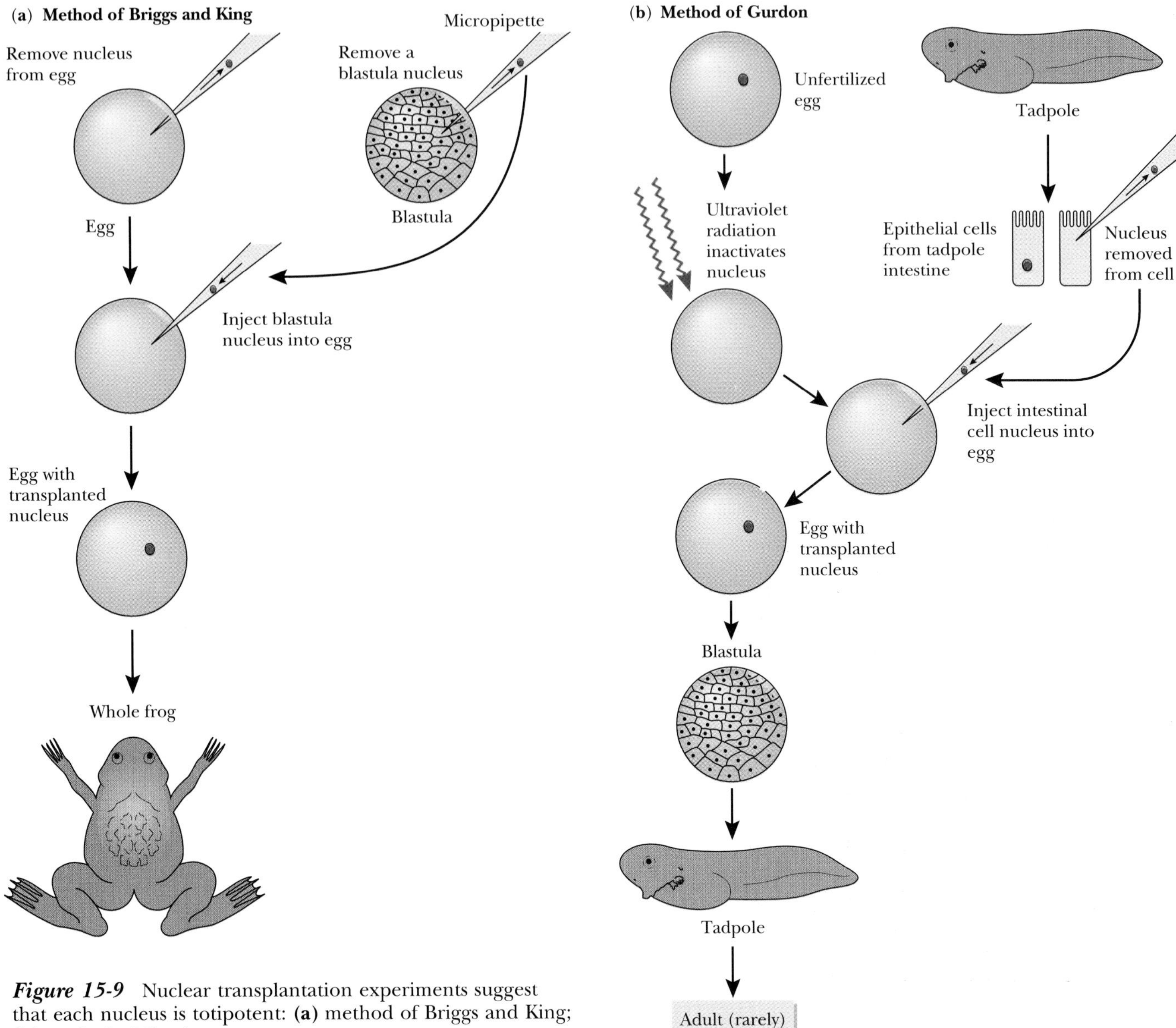

Figure 15-9 Nuclear transplantation experiments suggest that each nucleus is totipotent: **(a)** method of Briggs and King; **(b)** method of Gurdon.

probably *not* the mechanism by which cells accomplish selective gene expression. But the possibility remained that genes were permanently changed during the course of development, perhaps by some process of selective mutation. Further experiments, however, supported the idea that differentiated cells have the same genetic capacity. Ultimately, we now think, development depends on the activation and repression of genes.

Nuclear Transplantation Allowed the Direct Testing for Gene Loss During Animal Development

In the 1950s developmental biologists began to approach the question of gene loss experimentally. The American embryologists Robert Briggs and Thomas King devised a method for removing the nucleus from a frog egg before fertilization. Using a fine glass tube, they performed a **nuclear transplantation** experiment, in which they injected a nucleus from another cell into an egg whose own nucleus had been removed, as shown in Figure 15-9a. The egg then developed under the control of the transplanted nucleus.

Briggs and King soon discovered that nuclei from an early frog embryo could each program the development of a whole frog. That is, each nucleus was **totipotent**—it could direct the development of a whole organism. This finding meant that Briggs and King could produce a group of genetically identical frogs, each frog derived from a single nucleus within the same embryo. Each frog made in this way constitutes a **clone** [Greek, *klon* = twig], a population of genetically identical individuals or cells descended from a single ancestor. Because all the cells of the donor embryo derived from a single cell (the zygote), *all* the frogs derived from a single embryo are also a clone! This experiment stimulated a rash of science fiction based on its possible (but unlikely) application to humans. Interestingly, novelists have tended to choose historical figures for perpetuation by cloning, while journalists have preferred movie stars and athletes.

Briggs and King then examined the ability of nuclei at later stages to direct frog development. They found that this ability declined with development, and they were unable to produce a swimming tadpole from nuclei taken after a certain embryonic stage (except when they used nuclei from germ-line cells, that is, from the progenitors of gametes). Briggs and King concluded that somatic nuclei lose their capacity to direct complete development, perhaps indicating the permanent loss or inactivation of some genes.

While Briggs and King did their experiments on the American frog *(Rana catesbiana)*, the English biologist John Gurdon was working with the South African clawed frog *(Xenopus laevis)*, using a slightly different method of nuclear transplantation, shown in Figure 15-9b. Gurdon used nuclei from differentiated cells taken from the intestines of tadpoles. These, he found, could generate all types of cells present in a swimming tadpole. In a few cases, the transplanted intestinal nuclei could give rise to a fertile adult.

Gurdon's results strongly suggested that the nuclei of differentiated cells continue to be totipotent. Gurdon argued that the apparent loss of totipotency observed by Briggs and King and in many of his own experiments resulted from the extreme demands of the experimental procedure. Some critics, however, have argued that Gurdon's few totipotent nuclei might have derived from a minority population of undifferentiated cells.

The argument is at present unresolved: Biologists do not know whether all somatic nuclei remain totipotent. We do know, however, that even the nuclei of differentiated cells retain the capacity to express more genes than they normally do. Contemporary biologists almost all consider development to be largely a question of gene regulation rather than of selective gene loss.

Totipotency Is Easier to Demonstrate in Plant Cells than in Animal Cells

Almost anyone who has a houseplant knows how to produce whole new plants from cuttings, and many plants of commercial importance are propagated entirely from cuttings and graftings. (See Chapter 21). Even a single cell from a fully developed plant can give rise to a whole new plant, as most convincingly demonstrated by the carrot cloning experiment summarized in Figure 15-10.

The cloning of a carrot was achieved by F. C. Steward and his colleagues at Cornell University in 1958. From a slice of carrot, Steward grew disorganized colonies of cells in cultures nourished by coconut milk. Single cells derived from these colonies gave rise to groups of cells that resembled little roots. Steward transferred some of these to an agar plate and found that they could actually produce young "plantlets." These, in turn, could grow into whole new carrot plants. If it had been left alone, the progenitor cell of these new plants would have been part of the root transport system. By putting the cell into successively new environments, however, Steward succeeded in demonstrating that the cell was totipotent.

Nuclear transplantation experiments in animals and cloning of plant cells strongly suggest that differentiated cells of plants and animals contain all the genes necessary to build a whole organism. In a few cases, however, re-

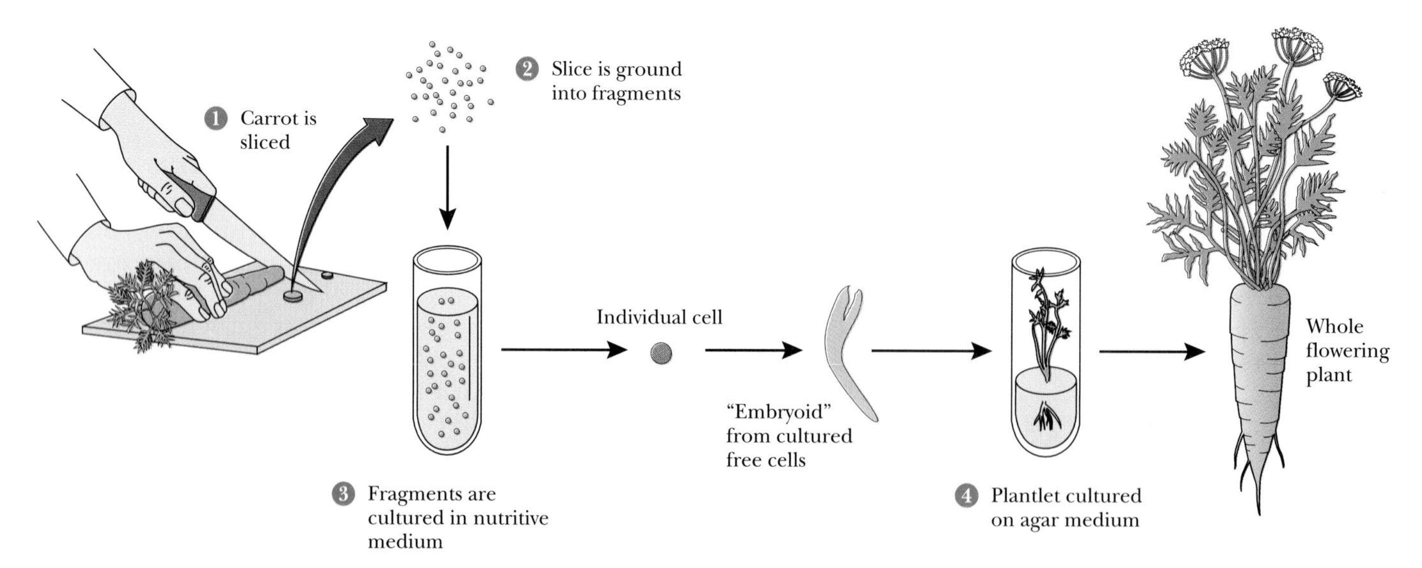

Figure 15-10 The development of a whole carrot plant from a single cell demonstrates cellular totipotency.

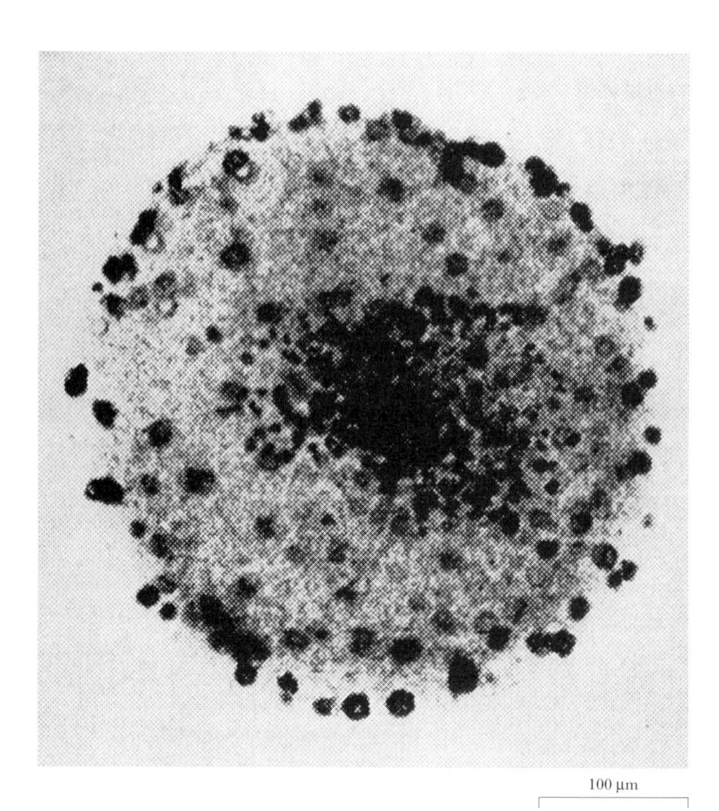

Figure 15-11 Ribosomal DNA amplification in *Xenopus* oocytes. This exceptional mechanism prepares oocytes for rapid protein synthesis after fertilization; it is not a general mechanism for cell specialization, though it was once suspected to be. *(D. D. Brown and I. B. Dawid,* Science, *Vol. 160, 1968, p. 272, courtesy New York Academy of Medicine)*

searchers have demonstrated that the genes of some cells within a multicellular organism systematically differ from those of other cells. The best studied examples of such gene changes in somatic cells are **gene rearrangements,** the cutting and splicing of segments of specific genes during development, which occur in the immune system. (See Chapter 22.) Another example is **gene amplification,** the specific multiplication of the genes for rRNA during frog embryonic development, illustrated in Figure 15-11.

Using molecular biological techniques, researchers have examined a large number of genes to see whether gene rearrangement or gene amplification represents a generally used mechanism for the regulation of gene activity. In all the genes outside those involved in the immune response and the development of germ cells, however, neither gene rearrangement nor gene amplification appears to play a part in gene regulation. These examples appear not to represent any general mechanism for cell specialization.

Development requires **differential gene expression,** the production of differing amounts of individual proteins (and RNAs) in different types of cells and at different times of development. The actual set of genes expressed in a given cell must depend on something besides its DNA, presumably (as in the case of prokaryotes) on cues from the cell's environment. The response to such cues involves a wide variety of regulatory mechanisms, which we now discuss.

Individual Genes Differ in Their Accessibility to RNA Polymerase

Most of a eukaryotic cell's DNA is complexed with proteins in such a way that it is not accessible for transcription. While a prokaryote transcribes almost all its DNA into RNA, a eukaryotic cell typically transcribes only about 20% of its DNA into RNA.

In eukaryotes, both transcribed and nontranscribed genes are organized into nucleosomes (as described in Chapter 13). The organization of the chromatin is such that genes are often accessible for transcription only in cells that are actively transcribing those genes. Harold Weintraub, a molecular biologist who worked at Princeton and at the Hutchinson Cancer Research Center in Seattle, demonstrated differences in gene accessibility in a particularly clever and simple way.

Weintraub added deoxyribonuclease (DNase), an enzyme that digests DNA, to the nuclei of two types of chicken cells—red cell precursors that were making globin mRNA and oviduct cells that were making the protein ovalbumin. DNase digests globin genes in red blood cell precursors, but not in oviduct cells. Conversely, the ovalbumin gene was accessible to DNase digestion in oviduct but not in red cell precursors. These experiments showed that genes not being transcribed by RNA polymerase are also not accessible to DNase digestion, while active genes are accessible to both RNA polymerase and DNase. Weintraub found no such differences when he did the same experiments with purified DNA from red cell precursors and oviduct, suggesting that differences in accessibility depend on the interactions of DNA and proteins within the nucleus.

Transcriptional Regulation Depends on the Interaction of Specific DNA Sequences with Nuclear Proteins

As in prokaryotes, most gene regulation in eukaryotes is at the transcriptional level, although post-transcriptional regulation is often important as well. Also, as in prokaryotes, the best-studied cases of transcriptional regulation depend on the interaction of specific DNA sequences with regulatory proteins. The major differences between transcriptional regulation in prokaryotes and eukaryotes derive from the presence of the nuclear membrane, so that only nuclear proteins can contribute to transcriptional regulation, and from the much larger amount of DNA, so that the expression of a given gene usually depends on more than a single DNA-protein interaction.

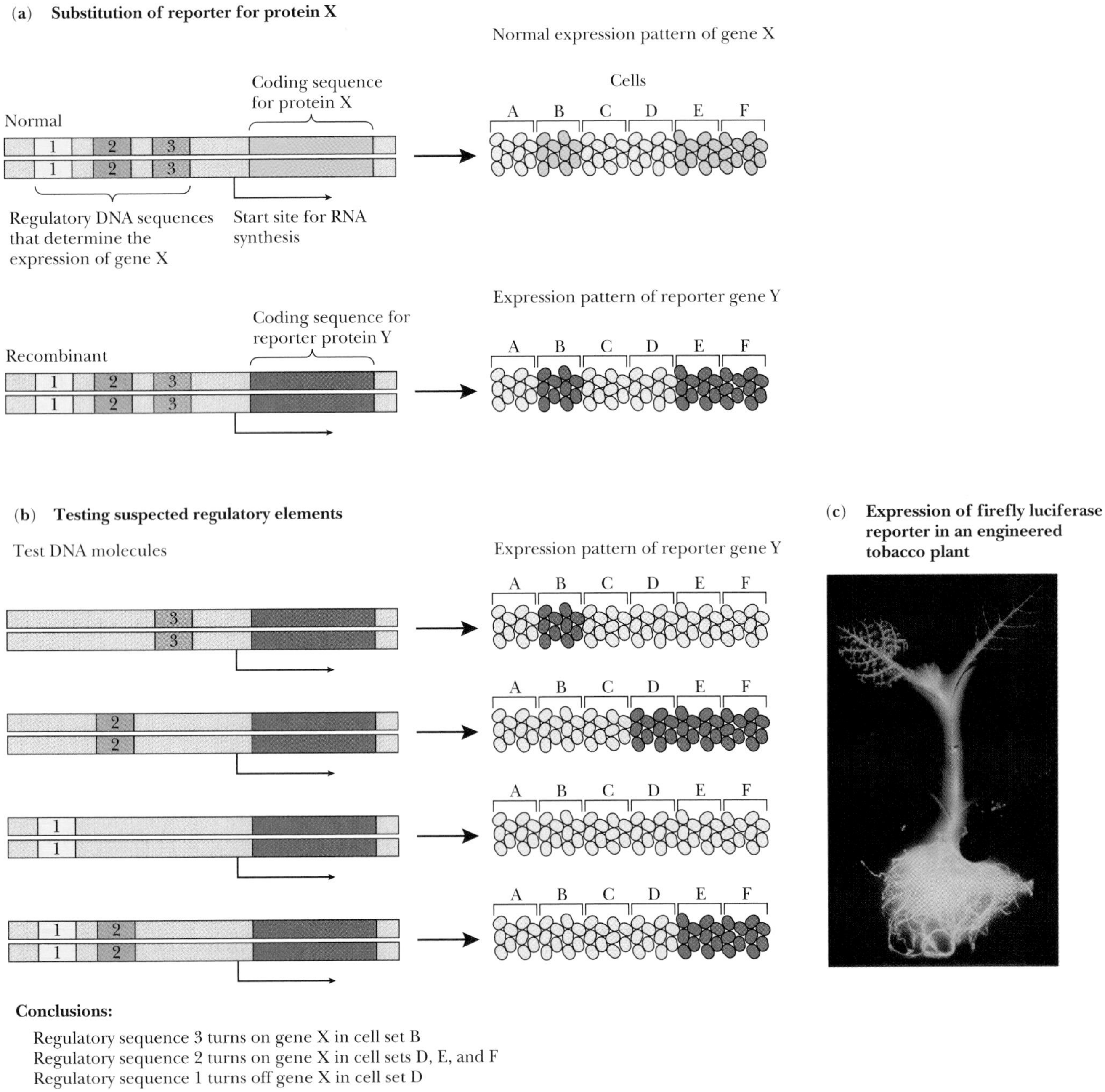

(a) **Substitution of reporter for protein X**

Normal expression pattern of gene X

Cells

A B C D E F

Coding sequence for protein X

Normal

1 2 3

Regulatory DNA sequences that determine the expression of gene X

Start site for RNA synthesis

Expression pattern of reporter gene Y

Coding sequence for reporter protein Y

Recombinant

1 2 3

A B C D E F

(b) **Testing suspected regulatory elements**

Test DNA molecules

Expression pattern of reporter gene Y

A B C D E F

(c) **Expression of firefly luciferase reporter in an engineered tobacco plant**

Conclusions:

Regulatory sequence 3 turns on gene X in cell set B
Regulatory sequence 2 turns on gene X in cell sets D, E, and F
Regulatory sequence 1 turns off gene X in cell set D

Figure 15-12 Identification of regulatory sequences in recombinant DNAs that contain suspected regulatory elements coupled to the coding sequence for a reporter protein. Commonly used reporters are β-galactosidase and firefly luciferase. **(a)** Substitution of the coding sequence for a reporter for the coding sequence of protein X often directs reporter expression in exactly the same pattern as that of protein X. **(b)** Coupling of individual DNA fragments to the reporter can allow the identification of regulatory elements needed for expression in a specific set of cells. **(c)** Expression of a firefly luciferase reporter in an engineered tobacco plant. *(c, Keith Wood/Visuals Unlimited)*

Elucidating the nature of these interactions has depended in large part on recombinant DNA techniques, which allow the creation of hybrid DNA molecules. (See Chapter 17.) A commonly used strategy for identifying the DNA sequences ("regulatory elements") that control gene expression is to build hybrid genes that contain a suspected regulatory region of one gene and the coding region of another gene. The product of the second gene serves as a **reporter,** a measure of the ability of a given test sequence to regulate expression in a particular type of cell, as shown in Figure 15-12a. The bacterial enzyme β-galactosidase is a frequently used reporter because its expression is easy to detect. Other useful reporters are chloramphenicol acetyltransferase, which catalyzes an easily assayed reaction, and firefly luciferase shown in Figure 15-12c, whose presence is revealed by luminescence in the presence of an appropriate energy source. The level of reporter activity in cells containing a hybrid gene indicates the effectiveness of a given test sequence in regulating gene expression.

Researchers usually test regulatory sequences for their ability to drive the expression of a reporter in cells that have taken up the recombinant DNA, as shown in Figure 15-12b. Techniques are also available for producing **transgenic** plants or animals, each of which contains a particular recombinant DNA in all its cells, as illustrated in Figure 15-12c. Either in cells or in transgenic organisms, researchers can examine the effects on reporter protein production of different DNA segments, usually starting with the DNA sequences flanking the 5′-end of the coding sequence. Once they identify suspected regulatory segments, researchers try to identify the nucleotide residues responsible for regulation, usually by testing the effects of changes produced by *site-directed mutagenesis,* which specifically alters a chosen nucleotide sequence. (See Chapter 17.)

Among the DNA sequences identified in this way are (a) **promoters,** which specify the starting point and direction of transcription; (b) **response elements,** called *cis-regulatory elements* [Latin, *cis* = on the same side, because the regulatory element is on the same piece of DNA that is being regulated] or **enhancers,** because they usually stimulate the transcription of neighboring DNA, without themselves serving as promoters; and (c) **termination signals.**

Promoters, response elements, and termination signals usually consist of 6–20 base pairs of DNA, which bind specific regulatory proteins, called **transcription factors** or *trans-acting factors* [Latin, *trans* = across, because they regulate the expression of other molecules]. Most eukaryotic promoters, for example, contain the sequence "TATA" (also found in the prokaryotic Pribnow box, mentioned elsewhere, p. 368), embedded in a longer consensus sequence, about 25 base pairs 5′ to the start site of transcription. This TATA box specifically binds a transcription factor, TFIID, which interacts with other transcription factors and with RNA polymerase II, as diagrammed in Figure 15-13a.

While many response elements of eukaryotic genes

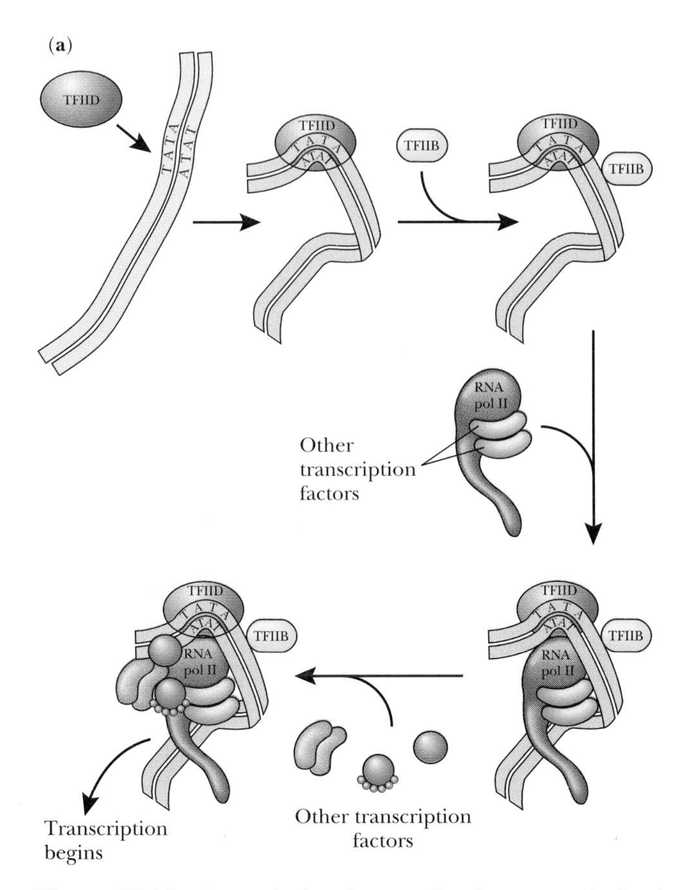

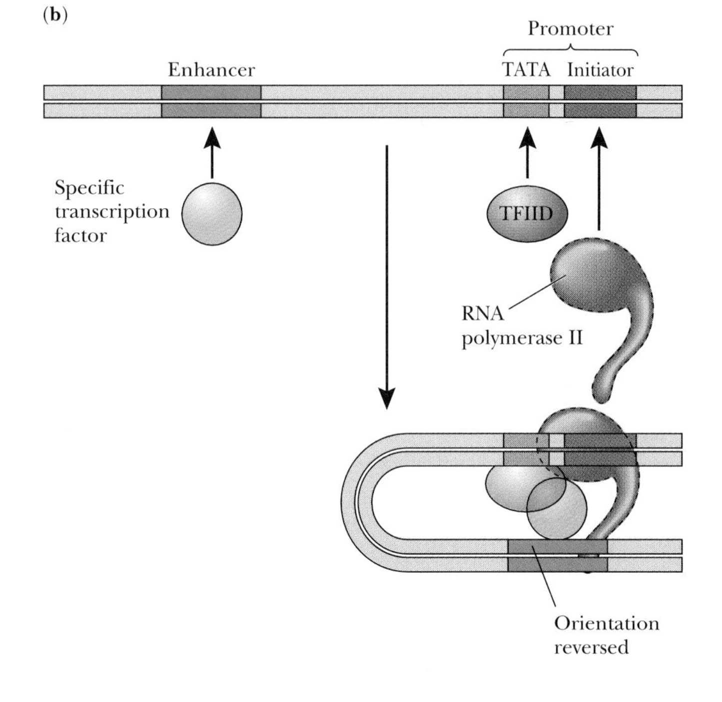

Figure 15-13 Transcription factors stimulate transcription by interacting with specific sequences in DNA and with other proteins. **(a)** The transcription factor TFIID binds to the TATA box of genes transcribed by RNA polymerase II; other transcription factors and RNA polymerase II can then bind to the DNA-protein complex. **(b)** Transcription factors often bind to sites relatively distant from the transcription start site; separated DNA-protein complexes may interact with one another, locally bending the DNA in a region of active transcription.

ESSAY

TRANSCRIPTIONAL REGULATION DEPENDS ON PROTEINS THAT BEND DNA

Both the response to environmental cues and the specialization of cells in multicellular organisms depend heavily on the regulated transcription of individual genes. In mammals, for example, proper development depends on the precise turning on and turning off of some 100,000 genes, a process that has been compared to the blinking of 100,000 light bulbs at just the right times and places and at just the right intensities.

While researchers have already identified scores of transcription factors, regulation cannot possibly depend on having a different transcription factor for each regulated gene. Instead, the transcription of a single gene usually depends on some specific combination of transcription factors. Since the number of possible combinations far exceeds the number of individual factors, the distinctive regulation of individual genes is much more imaginable given the limited number of factors that could be encoded by the genome.

Investigators have documented many examples of cooperation among transcription factors. For hundreds of genes, researchers have identified DNA sequences that are responsible for transcriptional regulation and the specific proteins that bind to these sequences. As such information has accumulated, a major question has remained: How do proteins that may bind to segments of DNA that lie thousands of base pairs away from each other all affect the transcription of a single gene by RNA polymerase?

The answer to this question appears to lie in the ability of certain DNA-binding proteins to bend the DNA double-helix into a structure that allows physical contacts between DNA-bound proteins that are separated even by thousands of base pairs, as illustrated in Figure 1. The TATA-binding protein (TBP), for example, binds to the TATA sequence in the promoter region of genes transcribed by RNA polymerase II. X-ray diffraction analyses of complexes between TBP and DNA show that TBP binding bends the DNA molecule by 70°. Further x-ray diffraction analysis has shown that the TBP binds to the outside of the bend, while the transcription factor TFIIB binds to the inside of the bend. Some 50 other proteins bind to the complex, in a manner that depends on the presence of the bent DNA.

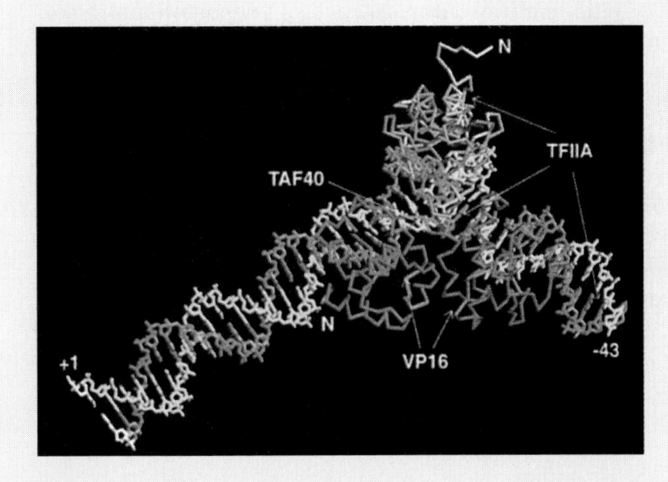

Figure 1 The binding of regulatory proteins to DNA causes the bound DNA to assume an L shape. Arrows show where some other transcription factors bind.
(Science, *Vol. 270, p. 1587 from Nikolov et al.,* Nature, *Vol. 377, pp. 119–128)*

TBP is one of a number of known "architectural proteins," which bend DNA and thereby bring together other proteins that would otherwise not be in physical contact. These close interactions among transcription factors are responsible for the combinatorial power of transcriptional regulation.

are, like those in prokaryotes, close to the start sites of transcription, some enhancers may lie 1000 bases or more away. This long-distance action suggests that activation of transcription may sometimes involve a physical rearrangement of DNA, as diagrammed in Figure 15-13b. Also in contrast to prokaryotic genes, enhancers may lie either upstream or downstream from the genes they regulate, and their ability to regulate transcription is maintained even if their orientation is reversed, as illustrated in Figure 15-13b.

The binding of a transcription factor to a regulatory element can either affect the immediate rate of transcription (for example, by increasing or decreasing the probability of initiation by the RNA polymerase complex) or produce a stable change in chromatin structure, which (as discussed previously) can alter the accessibility of a particular gene to the transcriptional machinery. In the case of such structural alterations, the continuous presence of the trans-acting factor might not be required; regulation may require only an initial local refolding of the DNA.

Often, however, the rate of transcription must change in response to relatively rapid changes in a cell's environment, as in the case of *metallothionein,* a protein that tightly binds several metal ions, including zinc, copper, and mercury. Metallothionein apparently helps protect organisms against metal poisoning and also provides them with reservoirs of zinc and copper ions. The presence of heavy metals stimulates metallothionein synthesis. Like many other proteins in mammals, metallothionein is present in some tissues (such as the liver) but not in others (such as the brain).

The increased metallothionein after exposure to heavy metals depends on an increased rate of transcription. Certain steroid hormones also increase the rate of transcription of the metallothionein gene. Both metals and steroids bind to distinct nuclear transcription factors, each of which changes the rate of transcription.

Several factors, then, determine the rate of transcription of the metallothionein gene (just as many factors can influence the expression of the *lac* operon). And, just as CAP coordinates the expression of many genes in *E. coli,* so steroid receptors can influence the expression of many genes in mammals.

Many Transcription Factors Have Common Structural Features

Researchers have used several methods to detect the specific binding of transcription factors to specific DNA sequences. One popular method is called a **footprinting** experiment: It detects the presence of proteins bound to DNA by showing that a bound protein prevents the chemical or enzymatic cleavage of a specific DNA sequence, as illustrated in Figure 15-14a. Another method of showing that a particular DNA sequence binds to a protein is a **gel retardation assay,** which detects protein binding to DNA by comparing the electrophoretic mobility of a specific DNA fragment in the presence or absence of protein; as shown in Figure 15-14b, the bound protein slows the movement of the DNA fragment in the electrical field. Finally, **DNA affinity chromatography** allows a researcher to purify a DNA-binding protein by virtue of its ability to bind to a specific DNA sequence, as shown in Figure 15-14c. These methods have led to the isolation and characterization of a number of transcription factors. Still other transcription factors have been identified as the products of genes responsible for normal development in *Drosophila,* as shown in Figure 15-15, or in the intensively studied roundworm *Caenorhabditis elegans,* shown in the Lab Tour on p. 3.

Almost all of the transcription factors so far identified have one of three distinctive structural designs (as illustrated in Figure 15-16)—called **helix-turn-helix** (containing three α-helices separated by short turns), **zinc finger** (which contains zinc atoms bound to the side chains of either four cysteine residues or two cysteine and two histidine residues), and **amphipathic helix** (which contains α-helices with nonpolar side chains extending from one side, allowing the formation of dimers). The steroid receptor that regulates metallothionein transcription, along with a large set of other steroid receptors, contains zinc fingers, while the eukaryotic transcription factor that binds cyclic AMP has an amphipathic helix, as discussed later in this chapter. Another classification scheme for transcription factors subdivides the known transcription factors into 12 distinctive families, 5 of which are listed in Table 15-1. Each of these 12 families is the subject of much ongoing research.

A transcription factor usually has several structural domains—(a) a *DNA-binding domain,* which recognizes a specific DNA regulatory sequence; (b) an *effector domain* (or *transactivating domain*), which can interact with the transcription machinery or with another transcription factor, in many cases forming a dimer with another molecule of the same type (a homodimer) or of another type (a heterodimer); and, in some cases, (c) a *ligand-binding domain,* which binds to a small molecule such as a steroid hormone; the binding of the small hormone molecule changes the conformation and the activity of the factor, for example, by permitting the DNA-binding domain to recognize a specific DNA sequence.

Many, perhaps most, transcription factors appear to act as dimers—either homodimers or heterodimers with another member of the same family. The response elements recognized by such dimeric factors are often symmetrical, so that the sequence (reading from 5′ to 3′) of each strand of DNA is the same within the element. These symmetrical sequences are often called *palindromes* (which are sentences with similar symmetry, as in "Madam, I'm Adam"). The symmetry of response elements and of the dimeric transcription factors may explain why transcriptional regulation does not depend on the orientation of enhancer elements.

Intracellular Signaling Molecules Can Activate Transcription

The best studied eukaryotic example of transcriptional regulation by an effector molecule involves the action of cyclic AMP, which affects the transcription of a number of genes. Examination of the response elements that bring about the cyclic AMP effects has revealed a similar 8-base long DNA sequence, called the **cyclic AMP response element (CRE),** in a number of genes. The transcription fac-

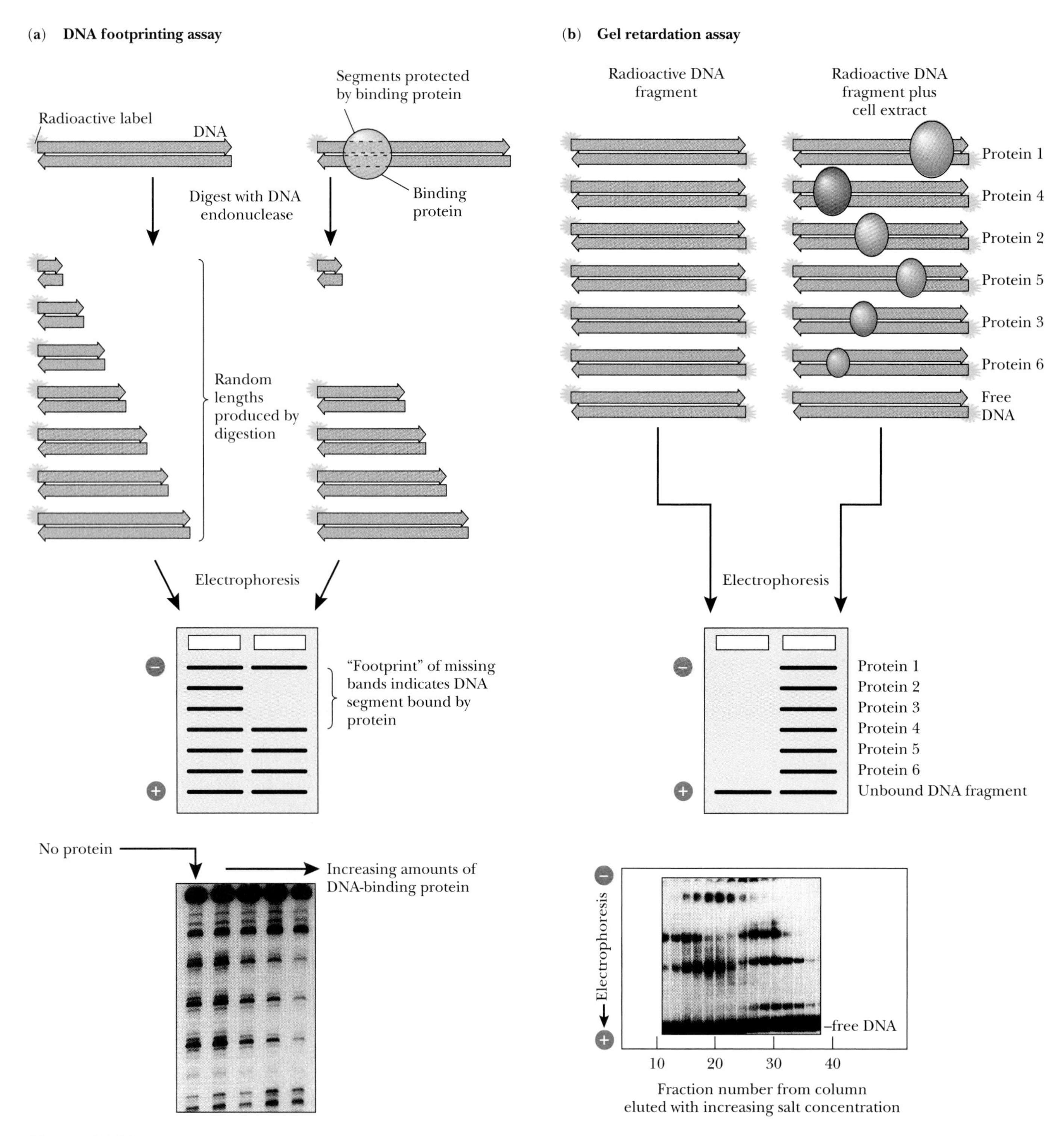

***Figure 15-14*(a, b)** Methods for detecting DNA-binding proteins. **(a)** DNA footprinting: A bound protein protects a labeled DNA fragment from digestion, so certain DNA fragment lengths are missing from the otherwise random mixture. **(b)** Gel retardation assay: Proteins bound to a DNA fragment slow its migration in an electric field, giving rise to additional bands after electrophoresis; below, individual DNA-binding proteins, isolated from cell extracts by chromatography, slow electrophoresis to differing extents. *(footprint photo: Michael Carey, Ph.D., UCLA Medical School; gel retardation data: C. Scheidereit, A. Hequy, and R. G. Roeder,* Cell, *51:783–793, 1987, courtesy New York Academy of Medicine)*

tor that binds to CRE is called CREB (*c*yclic-AMP *r*esponse *e*lement *b*inding protein) and is a member of the amphipathic helix family: Each CREB molecule contains a characteristic **leucine zipper,** formed by the association of two amphipathic α-helices, with leucine side groups forming hydrophobic interactions and promoting dimerization.

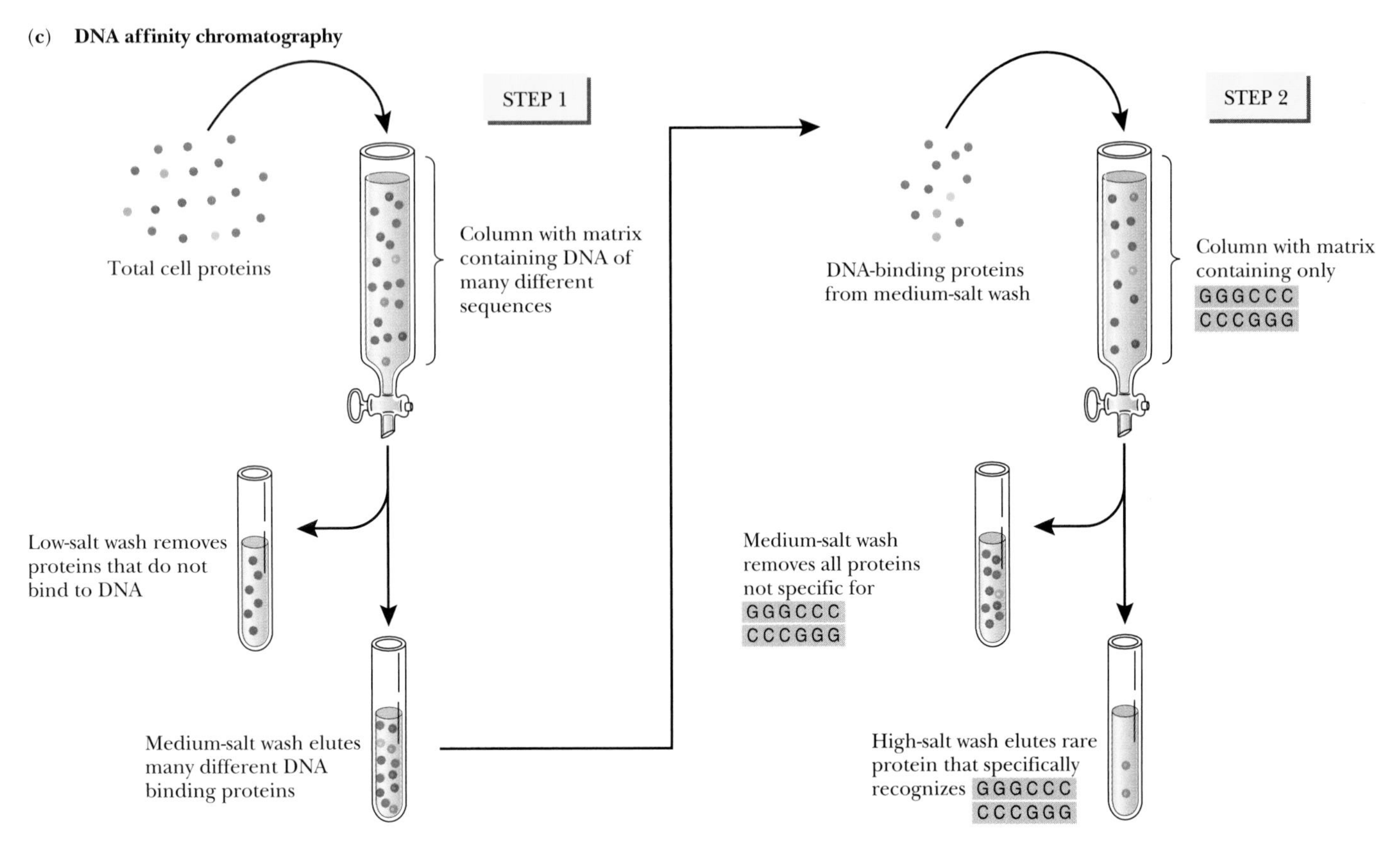

***Figure 15-14*(c)** DNA affinity chromatography: The first column isolates a mixture containing a large number of different DNA-binding proteins; the second column isolates proteins that specifically bind to the sequence GGGCCC.

In eukaryotic cells, cyclic AMP serves as a "second messenger" whose intracellular concentration changes in response to a variety of hormones and transmitters. (See Chapter 19.) Cyclic AMP does not itself bind to CREB, as it does to the bacterial CAP. Instead it affects the binding of CREB to DNA by stimulating its phosphorylation by **protein kinase A,** an enzyme that transfers a phosphate group from ATP to specific proteins and whose activity depends on the presence of cyclic AMP. Cyclic AMP also leads to the phosphorylation of many other cellular proteins, often changing their functional activity in response to extracellular signals—hence the term "second messenger." In its phosphorylated form, CREB binds to CRE and stimulates transcription. Other cellular proteins that are phosphorylated in response to cyclic AMP mediate other (nontranscriptional) effects in response to increased levels of cyclic AMP.

How Complex Is the Problem of Transcriptional Regulation?

In a multicellular organism, each gene may be expressed in a manner that depends on cell type and on developmental stage, as well as on internal and external environmental cues. It is not surprising, then, that transcriptional regulation is complex. In the last decade, researchers have discovered at least six types of complexity in transcriptional regulation: (1) the existence of a large number of transcription factors, perhaps 5000 in mammals; (2) the combinations of several types of response elements in in-

Figure 15-15 Some mutations in genes encoding transcription factors cause abnormalities in development, as seen in this *bithorax* mutant of *Drosophila,* which has two sets of wings rather than one. *(E. B. Lewis, California Institute of Technology, Pasadena)*

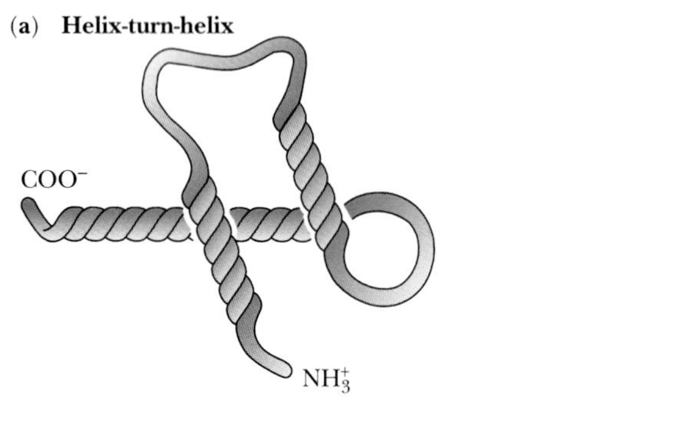

(b) Zinc finger

Zinc bound to two cysteine and two histidine residues:

Zinc bound to four cysteine residues:

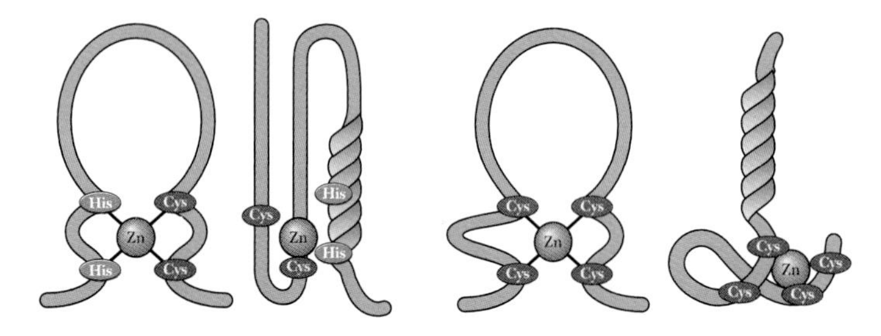

(c) Amphipathic helix

H_3^+N

NH_3^+

Leu

Nonpolar side chains (such as leucine) allow formation of a dimer

COO⁻ COO⁻

Figure 15-16 Structural designs common to many transcription factors: **(a)** helix-turn-helix; **(b)** zinc finger; **(c)** amphipathic helix, here a "leucine zipper."

Table 15-1 *Five Families of Transcription Factors*

Class and family of factor	Examples	Distinguishing characteristics
Helix-turn-helix		
Homeobox	*Antennapedia, bithorax*	Highly basic domain of about 60 amino acid residues; distantly related to helix-turn-helix regulators of bacterial transcription
POU	Pit-1 (responsible for cell-specific activation of prolactin and growth hormone); Oct-2 (binds to response element in an antibody gene); Unc-86 (mutations in which lead to uncoordinated movements of *C. elegans*)	Each contains a homeobox but with about 20/60 amino acid residues matching those of the homeobox factors; additional common segment of about 80 amino acid residues
Zinc finger		
Nuclear receptors	Receptors for steroid hormone, thyroid hormone, and retinoic acid	Zinc binding site formed by four cysteine residues; act as dimers
Amphipathic helix		
Leucine zipper	CREB	α-Helix with leucine in every seventh position; forms heterodimer through hydrophobic interactions
Helix-loop-helix	MyoD (a transcription factor that can stimulate the differentiation of a fibroblast line into muscle cells)	Each contains two conserved α-helical regions of 12–15 amino acid residues, connected by a nonconserved "loop"; helices are amphipathic with hydrophobic side chains extending from one side

dividual genes, providing targets for several transcription factors—for example, the response elements for both steroids and metal ions in the metallothionein gene; (3) interactions among members of the same factor family—for example, the formation of a dimer between two factors via a leucine zipper; (4) interactions among members of different families; (5) covalent modifications, such as the phosphorylation of CREB; and (6) the regulated partitioning of transcription factors—for example, a factor may become sequestered in the cytoplasm so that it cannot reach its target in the nucleus.

Researchers have estimated that 10% of genes in *E. coli* encode regulatory proteins, and 1% of genes in *C. elegans* encode members of the homeobox family, 1 of 12 identified families of transcription factors. Gene regulation in the mammalian nervous system, for example, is likely to be at least as complex as that in bacteria or worms, so we might expect mammals to have at least 10^3–10^4 transcription factors and a corresponding number of response elements. While these numbers seem awfully big, even 10^4 transcription factors could not individually specify the sets of 10^4–10^5 kinds of protein molecules in each of 10^{12} nerve cells in the human brain; much less could they specify their regulation in response to developmental or experiential cues. The diverse patterns of gene expression in multicellular organisms must therefore result from *combinations* of transcription factors, at least some of which must be responsive to environmental factors.

A Single Transcriptional Regulator May Affect Many Genes

The ability to coordinate the expression of the many genes in the same cell type is a fundamental requirement for an organism with specialized cells, tissues, and organs. Molecular biologists have therefore expected to find "master genes" that control the expression of such gene sets.

Perhaps the best example is MyoD1. This gene encodes a transcription factor that turns on the expression of many genes needed to make a muscle cell. Artificial introduction of this gene into skin cells can convert them into muscle cells. Many molecular biologists are now searching for similar master genes that control the development of other cell types.

Sex determination in mammals illustrates the widespread effects of regulation by a single transcription factor. In male mammals, a gene on the Y chromosome (called the **sex-determining region**) encodes a regulatory protein that influences gene expression during the embryonic development of the gonads. (See Essay: Chromosomes and Sexual Equality.) In the absence of this regulatory protein, the embryonic precursors of the gonads organize themselves into ovaries, which eventually produce ova (eggs) and female steroid hormones (estrogen and progesterone). In the presence of this regulatory protein, however, the gonad precursor cells organize into testes, which soon begin to produce the male steroid hormone testosterone. The mode of action of the regulatory factor derived from the sex-determining region is still unknown, but the actions of testosterone and its receptor are well understood.

Testosterone stimulates the formation of the male genitalia (penis, scrotum, and the internal ducts that carry sperm). Later, at puberty, testosterone stimulates the development of male secondary sexual characteristics—male distribution of fat and body hair, deepening of the voice, and patterns of growth and even behavior.

Three medical conditions dramatically illustrate the role of testosterone in sexual differentiation. Like men, women make testosterone, but at lower levels and in the adrenal glands rather than in the nonexistent testes. In a condition called *adrenal virilism,* women with excessive adrenal gland activity—for example, accompanying adrenal cancer—produce high levels of testosterone and develop male secondary sexual characteristics. This condition goes away when the adrenal tumor is surgically reduced or hormonally inhibited.

Fetal masculinization is the development of male characteristics in a fetus carried by a woman treated for cancer with a testosterone-like compound while pregnant. This condition provides a still more striking example of the influence of nongenetic factors on sexual phenotype. The presence of testosterone induces the formation of male genitalia, even in XX embryos. The danger of such masculinization, now realized, has led to avoiding such steroid treatments in pregnant women.

Finally, a condition called *testicular feminization* leads to another abnormality of sexual differentiation—a female phenotype in a chromosomal (XY) male. (See Essay.) These individuals produce testosterone, but it does not cause male sexual differentiation. These individuals therefore develop into women, with female external genitalia and secondary sexual characteristics. These women, however, do not have ovaries. Instead they have testes, still in the body cavity, rather than descended (because there is no scrotum). Essentially all other aspects of their phenotype are female, despite the presence of the Y chromosome, testes, and testosterone. We now know that the testicular feminization mutants fail to make the testosterone receptor—a transcription factor of the steroid receptor group—which binds both to testosterone and to target response elements in DNA, thereby influencing gene expression.

Abnormalities of sexual determination provide a wealth of illustrations of the interactions of genes and environment to yield phenotype. We see that the conventional rule (XX = female; XY = male) is only part of the story. Sexual phenotype has many aspects—including gonads, internal genitalia, external genitalia, and secondary sexual characteristics. Each of these aspects of phenotype depends on both genes and chemical environment. With respect to sexual behavior, environment (including social environment and psychological experience) clearly also contribute.

(Text continues on page 418.)

ESSAY

CHROMOSOMES AND SEXUAL EQUALITY

Male and female mammals have different sex chromosomes: Each diploid cell in a female has a pair of X chromosomes, while males have an unpaired X and Y. But how do the X and Y chromosomes influence sexual development?

The sex chromosomes of mammals pair and move apart (disjoin) during normal meiosis, so that each gamete ultimately receives one and only one sex chromosome. When the sex chromosomes fail to move apart properly (that is, when nondisjunction occurs), however, each gamete may receive two sex chromosomes or none at all. When such an abnormal gamete participates in fertilization, the result is an aneuploid zygote, with an abnormal number of sex chromosomes and an aberrant phenotype.

A zygote that receives no X chromosomes at all does not survive. This is not surprising, given that the X chromosome contains thousands of genes, while the Y chromosome contains relatively few. Many individuals with sex chromosome abnormalities, however, do grow to maturity. Although most are infertile, they still have recognizably male or female characteristics. In humans, for example, three such abnormalities are relatively common—XO (that is, an individual with a single X chromosome), called Turner's syndrome; XXY, called Klinefelter's syndrome; and XYY.

What is the relationship between the sex chromosomes and sexual phenotype? We could imagine, for example, either that "femaleness" results from the presence of two X chromosomes instead of one or that "maleness" results from the presence of a Y chromosome, independent of the number of X chromosomes. In mammals, the presence of the Y chromosome suffices to give a "male" phenotype. A person with XXY sex chromosomes is male but infertile. In contrast, an XXY fruitfly looks female.

What does the Y chromosome do in mammals? Recently, researchers have learned that the Y chromosome has a relatively small region (called the *sex-determining region*) that is responsible for the determination of sexual phenotype. Researchers have identified the gene responsible for sex determination by analyzing translocation mutants, in which the sex-determining region has moved from the Y to the X chromosome. An individual with this translocation is XX but has a male phenotype.

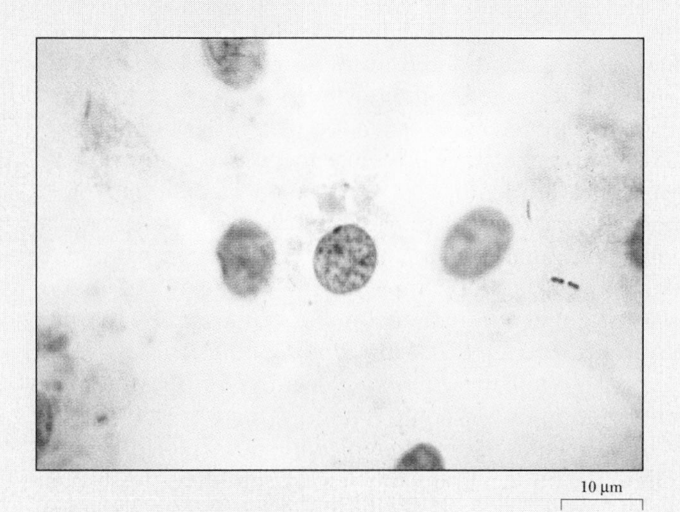

Figure 1 Barr bodies in human epithelial cells. In female mammals, each cell has one inactivated X chromosome, which appears as a condensed object on the edge of the nucleus. *(Michael Abbey/Photo Researchers)*

The chromosomal determination of sex ensures nearly equal numbers of males and females, but the chromosomal differences between males and females also pose special problems. Chapter 12 discussed one of these problems in talking about the genetics of sex-linked characteristics: Male mammals have no back-up copy of genes to furnish correct instructions when genes on the X chromosome are defective. Boys are therefore far more susceptible than girls to diseases ranging from color-blindness to hemophilia.

Another problem concerns the genetic differences between males and females. About 10% of human genes are present in two copies in women but in only one copy in men. Yet the phenotypic differences between men and women are relatively small (although we may be especially aware of them). Even most genes on the X chromosome are expressed to the same extent in men and women. How is this equalization accomplished?

In *Drosophila,* both X chromosomes are active in females, but each gene is only half as active as it would be in a male. This regulation, whose mechanism is still unknown, is called *dosage compensation.*

Mammals have evolved a different mechanism

(a)

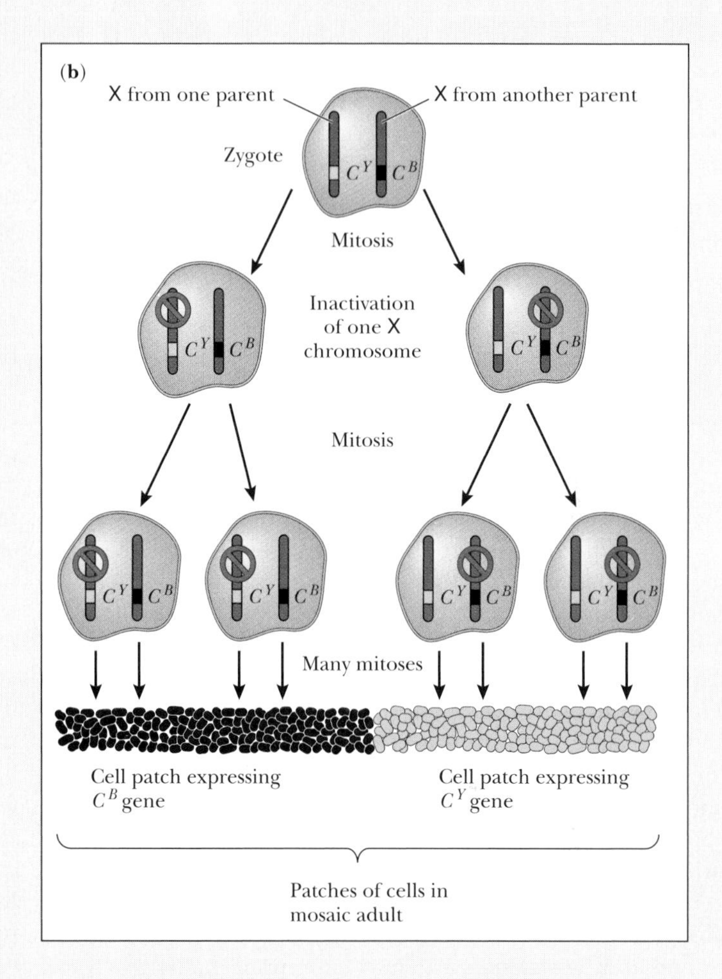

Figure 2 X chromosome activation. **(a)** A calico cat results from X chromosome inactivation; **(b)** one X chromosome has an allele that specifies black pigment, whereas the other X chromosome has an allele that specifies yellow pigment. *(a, Frederic Jacana/Photo Researchers)*

for dealing with X chromosomes. Although a female mammal has two X chromosomes in every cell, only one is functional. During early embryonic development, one of the two X chromosomes becomes condensed and inactive. Even during interphase, the condensed X chromosome, called a Barr body, may be seen at the outer edge of the nucleus (Figure 1).

Calico cats, such as the one shown in Figure 2a, beautifully illustrate the consequences of X chromosome inactivation. A gene on the X chromosome contributes to the establishment of coat color. As illustrated in Figure 2b, the C^Y allele of this gene produces a yellow (or orange) coat, the C^B allele a black coat. Calico cats are C^Y/C^B heterozygotes. In some cells the X chromosome bearing the yellow allele is active, while in others the black allele is active. Once an X chromosome is inactivated in a cell, it remains inactive in all the descendants of that cell. Each yellow (or orange) patch, then, consists of the descendants of a single cell in which the C^B-containing X chromosome was inactivated. In a calico cat, X chromosome inactivation happens early in development, when there are only a few dozen cells that will eventually make the cat's fur, so there are only a small number of patches.

X chromosome inactivation also happens in humans. In women heterozygous for an enzyme encoded by a gene on the X chromosome, different forms of the enzyme are present in different cells. All female mammals are genetic mosaics: Different X chromosomes are active in different cells. In kangaroos and other marsupials, females also inactivate one of their X chromosomes, but it is always the X chromosome derived from the father that is inactivated. The maternal and paternal X chromosomes of humans (and of other placental mammals) have an equal chance of inactivation.

Abnormalities of sexual phenotype illustrate that single regulatory molecules can affect many specific aspects of phenotype in multicellular eukaryotes. They also illustrate that regulation can occur at several sequential steps: The presence of the Y chromosome triggers the development of testes, which in turn produce testosterone. The action of testosterone in turn depends on the proper expression of another gene, that encoding the testosterone receptor protein, a single transcription factor.

How Can More than One Polypeptide Arise from a Single Gene? Alternative Splicing, Polyadenylation, and Editing of Primary Transcripts

Most eukaryotic genes are mosaics of exons and introns. In the usual picture of mRNA production, transcription begins at a single promoter and yields a single primary transcript containing all the introns and exons. Molecular machinery within the nucleus adds a 5′ cap and a 3′ tail and then cuts and splices the primary transcripts to produce mature mRNAs without introns.

In addition to the coding sequence—the nucleotide sequence that specifies the sequence of amino acids in a polypeptide—each mRNA contains additional nucleotides both "upstream" (toward the 5′-end) and "downstream" (toward the 3′-end) from the coding sequence. These sequences, called the 5′ untranslated region (or 5′ UTR) and the 3′ untranslated region (or 3′ UTR) of the mRNA, help determine the stability of each mRNA and the rate at which it binds to ribosomes.

Although many genes appear to produce only one primary transcript and a unique mature mRNA, other genes produce many alternative forms of primary transcripts and mature mRNAs. Primary transcripts may start or end at different sites and may have more than one site for the addition of poly A. Furthermore, different splicing events may string together different subsets of exons from a single primary transcript. In such a case, one mRNA's exon becomes another mRNA's intron, thus fogging definitions. Most investigators now use "exon" to refer to any segment of DNA or RNA that codes for a polypeptide sequence segment in *any* cell.

A single stretch of DNA may encode a variety of polypeptides. Sometimes, a single cell may simultaneously express these different polypeptides; in other cases, the pattern of transcription and splicing is subject to developmental regulation. After geneticists spent decades arriving at the definition summarized as "one gene—one polypeptide chain," subsequent studies of gene expression conclusively showed that one gene can encode several polypeptide chains, occasionally with distinct functions. Most molecular biologists now accept a modified definition of **gene** to be "a DNA sequence that is transcribed as a single unit and encodes one set of closely related products."

Multicellular organisms provide many examples in which the same segment of DNA encodes several polypeptide chains. Two contrasting examples are especially instructive: (a) the gene for NCAM (neural cell adhesion molecules), a family of related proteins responsible for many cell-cell interactions during development, and (b) the gene for two apparently unrelated peptides—calcitonin, a Ca^{2+}-regulating peptide hormone in the thyroid gland, and CGRP (calcitonin gene-related peptide), a signalling polypeptide in the brain. Alternative splicing of NCAM results in alternative proteins that either do or do not contain a transmembrane domain (Figure 15-17a), and the two products differ in their manner of association with the cell membrane.

In contrast, the alternative expression of genetic information in the calcitonin-CGRP gene is more complicated (as shown in Figure 15-17b). The pre-mRNA for calcitonin and CGRP contains six exons, five introns, and two sites for polyadenylation. In the thyroid, the transcript is polyadenylated at the end of the fourth exon, whereas in neurons, poly A lies at the end of the sixth exon. Splicing in the thyroid yields calcitonin mRNA, which consists of exons 1, 2, 3, and 4; in nerve cells, however, exon 4 is removed and exons 5 and 6 are attached to the mRNA. The two mRNAs each encode polypeptides that are precursors of the biologically active peptides. Exons 1 and 6 do not encode translated polypeptide sequences. Post-translational proteolytic processing in both thyroid and brain removes the amino acid sequences encoded by exons 2 and 3, so the final calcitonin peptide in thyroid derives entirely from exon 4 and CGRP in brain from exon 5.

Research on trypanosomes, parasitic protozoa that cause sleeping sickness (illustrated in Figure 15-18a), has led to the discovery of two additional, totally unexpected processes that alter the protein-encoding ability of RNAs —***trans* RNA splicing** and **RNA editing.** All the mRNAs in trypanosomes have a common sequence at their 5′ ends. This common sequence is transcribed from a single gene that is distinct from those that specify the coding region of each mRNA molecule. After the two parts of each mRNA are separately made, special enzymes link them together in a process called *trans* RNA splicing.

The mitochondria of trypanosomes, like those of other eukaryotes, contain their own DNA and produce a number of mRNAs. (See Chapter 16.) But in trypanosome mitochondria, mature mRNAs arise in an unconventional way: Transcription gives rise to two sets of RNAs, as shown in Figure 15-18b—mRNA precursors and two types of short "guide RNAs," 40–80 nucleotides long, each ending in a string of U residues at the 3′-end. The 5′-end of one guide RNA forms a duplex with part of an mRNA precursor. But the match is not perfect, with extra Us present at many places in the guide, but not in the pre-mRNA. Complex editing machinery removes U residues from the 3′-end of the guide RNA and inserts them into the pre-mRNA, changing the reading frame and the coding sequence of the mRNA.

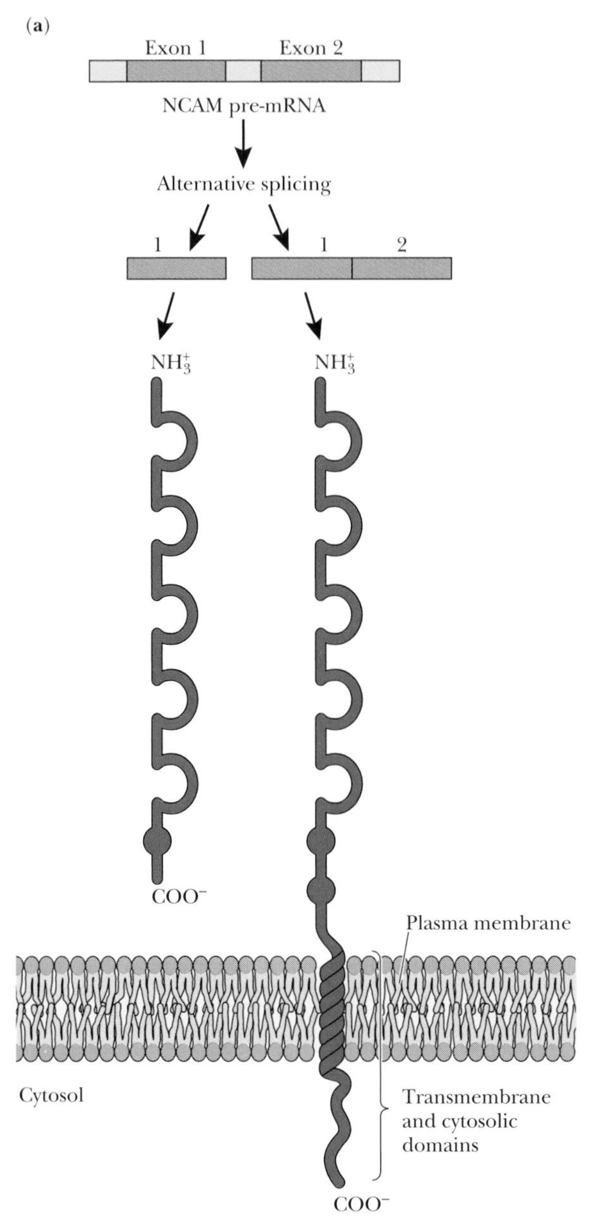

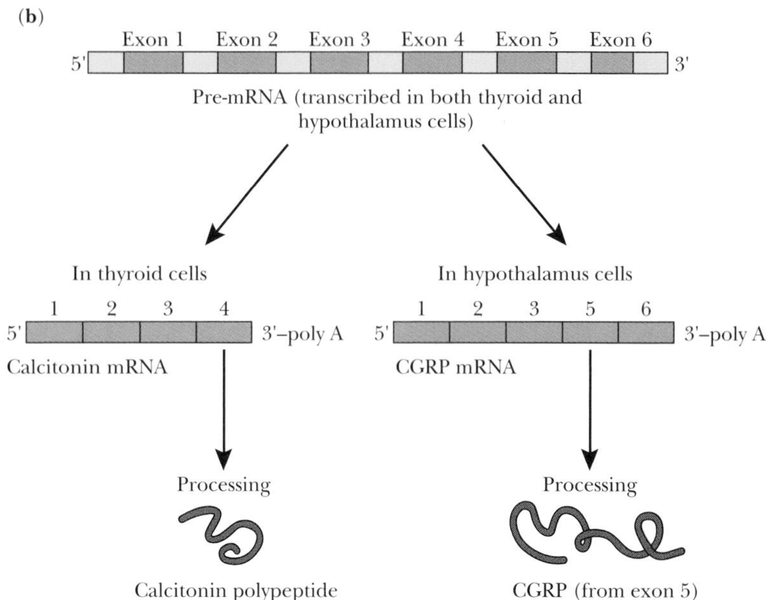

Figure 15-17 Alternative splicing of mRNA precursors. **(a)** NCAM pre-mRNA can be alternatively spliced to give a form with or without a transmembrane domain; **(b)** the pre-mRNA for calcitonin and CGRP can be alternatively spliced to produce either a calcitonin polypeptide or CGRP.

After RNA editing was discovered in trypanosome mitochondria, researchers found other forms of RNA editing in other eukaryotes. In plant mitochondria, for example, nearly every mRNA is edited to some extent; while there are no insertions or deletions, many Cs are changed to U. In mammals, RNA editing has so far been detected only rarely, although in one case RNA editing appears to be responsible for a change in the ion selectivity of a receptor protein in the brain.

Individual mRNAs Differ in Their Rates of Translation and Degradation

Once mature mRNAs arrive in the cytoplasm, their ability to direct protein synthesis is subject to further regulation through alterations in (1) the efficiency of translation and (2) the rapidity of their degradation. A particularly well-studied example of the translational control of gene expression is the mRNA contained in the sea urchin egg prior to fertilization. The unfertilized sea urchin egg contains many apparently mature mRNAs, all synthesized during the process of egg formation in the mother, but translation is extremely slow until after fertilization.

The Rate of Translation of Individual mRNAs is Subject to Regulation

An example of more specific translational regulation is provided by *ferritin,* a protein that serves as an intracellular iron depot. The translation of ferritin mRNA *increases* when free iron atoms are present within cells. A protein, called a **translation repressor protein,** can bind to sequences in the 5′ untranslated region of ferritin mRNA. Binding occurs, however, only when free iron is absent or in low concentrations, as shown in Figure 15-19a. When iron is present at high concentration, the conformation of the repressor changes, no binding occurs, and translation proceeds, as shown in Figure 15-19b.

The Degradation of Individual mRNAs May also Change in Response to Cellular Environment

Another iron-related protein provides a nice example. The *transferrin receptor* contributes to the uptake of iron from the outside to the inside of cells, and its production is re-

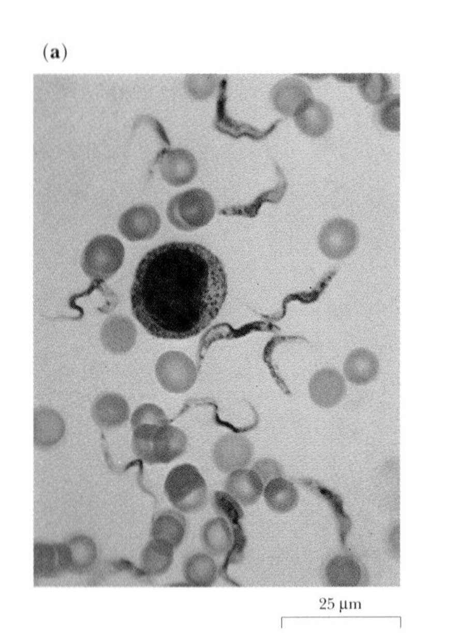

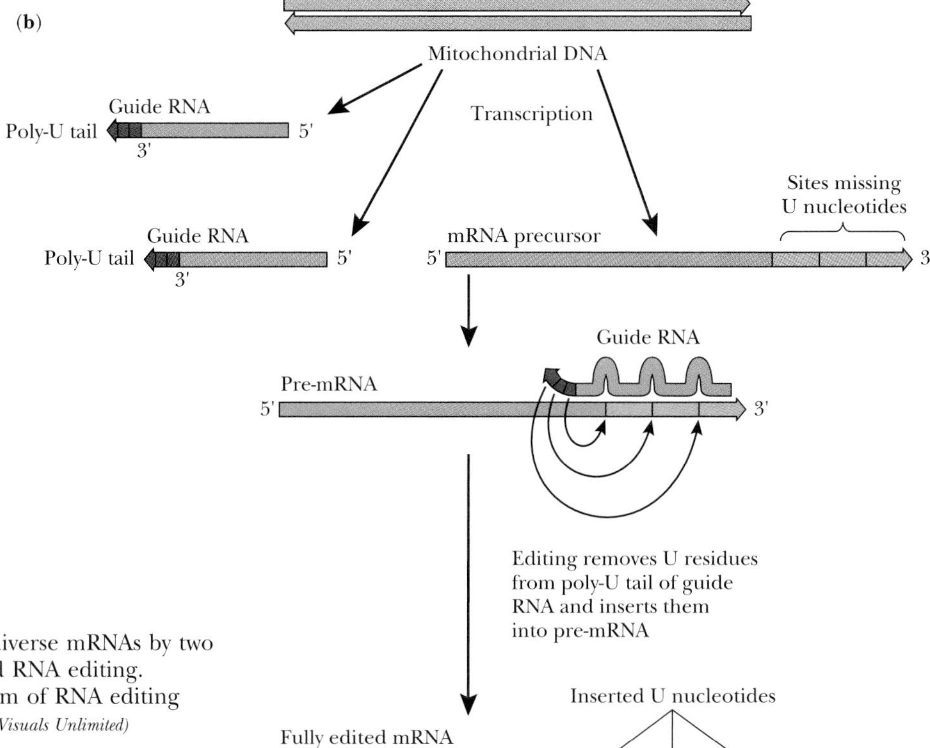

Figure 15-18 Trypanosomes generate diverse mRNAs by two unusual processes, trans-RNA splicing and RNA editing. **(a)** Trypanosomes in blood; **(b)** mechanism of RNA editing in trypanosome mitochondria. *(a, Cabisco/Visuals Unlimited)*

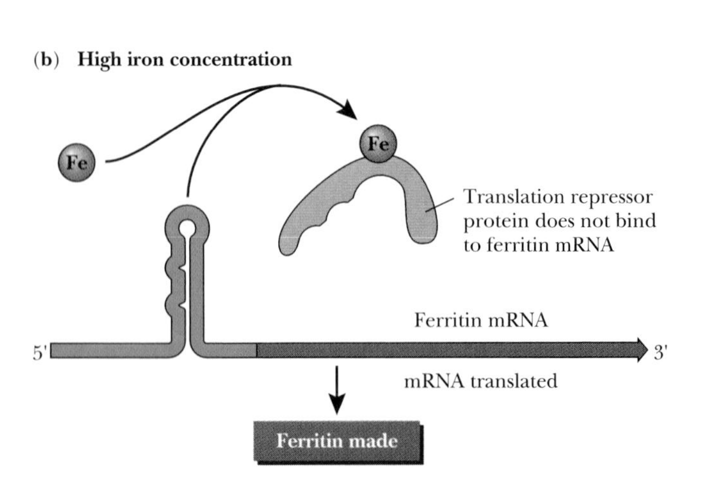

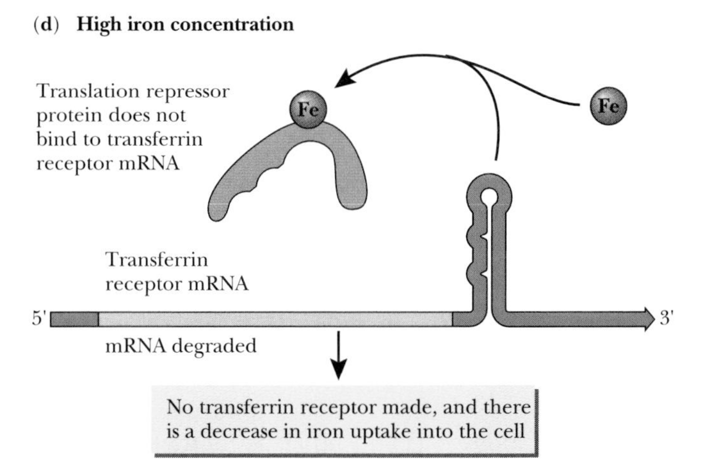

duced when intracellular iron is high. The same protein involved in the translational repression of ferritin synthesis can bind to a specific RNA sequence in the 3′ untranslated region of transferrin receptor mRNA, as illustrated in Figure 15-19c and d. This binding inhibits the action of a ribosome-associated ribonuclease, thereby stabilizing the mRNA. When the intracellular iron concentration is low, the protein binds to the mRNA; the result is a net increase in transferrin receptor synthesis. When intracellular iron concentration is high, the protein does not bind to the mRNA; the result is a decrease in production of transferrin receptor and a decrease in iron movement into the cells.

Finally, the cytoskeletal protein tubulin provides another example of the regulated stability of mRNA. Microtubules form by the ordered association of two related cytoplasmic proteins, α and β tubulin. (See Chapter 6.) When excess $\alpha\beta$ tubulin dimers accumulate in the cytoplasm, however, they activate a ribosome-associated nuclease that specifically digests the mRNA on ribosomes that are producing tubulin polypeptides. The result of this mechanism is the feedback regulation of tubulin production, adjusting synthesis downward when there are excess tubulin dimers and allowing full production when the concentration of tubulin dimers is low.

Eukaryotic cells can regulate gene expression at many levels (as shown schematically in Figure 15-20). A eukaryotic cell can alter the accessibility of a gene to RNA polymerase, the rate of transcription, the pattern of RNA splicing, and the rate of translation. While researchers know how hundreds of genes regulate their expression, many questions remain, and eukaryotic gene regulation is an especially active and challenging field.

◄
Figure 15-19 Regulation of translation and mRNA stability for ferritin, an iron-binding protein. **(a)** At low iron concentrations, an iron-binding translation repressor protein associates with a sequence in the 5′ untranslated region of ferritin mRNA, preventing translation. **(b)** At high iron concentrations, the translation repressor protein binds iron, undergoes a conformational change and dissociates from the ferritin mRNA; ferritin mRNA can now be translated. **(c)** The same translation repressor protein also binds to a similar sequence in the 3′ untranslated region of the mRNA for transferrin receptor; at low iron concentrations, the binding of the protein protects the transferrin receptor mRNA from degradation, increasing the level of ferritin receptor synthesis and the iron uptake into the cell. **(d)** At high iron concentrations, the translation repressor protein no longer binds to the receptor mRNA, and the mRNA is more rapidly degraded.

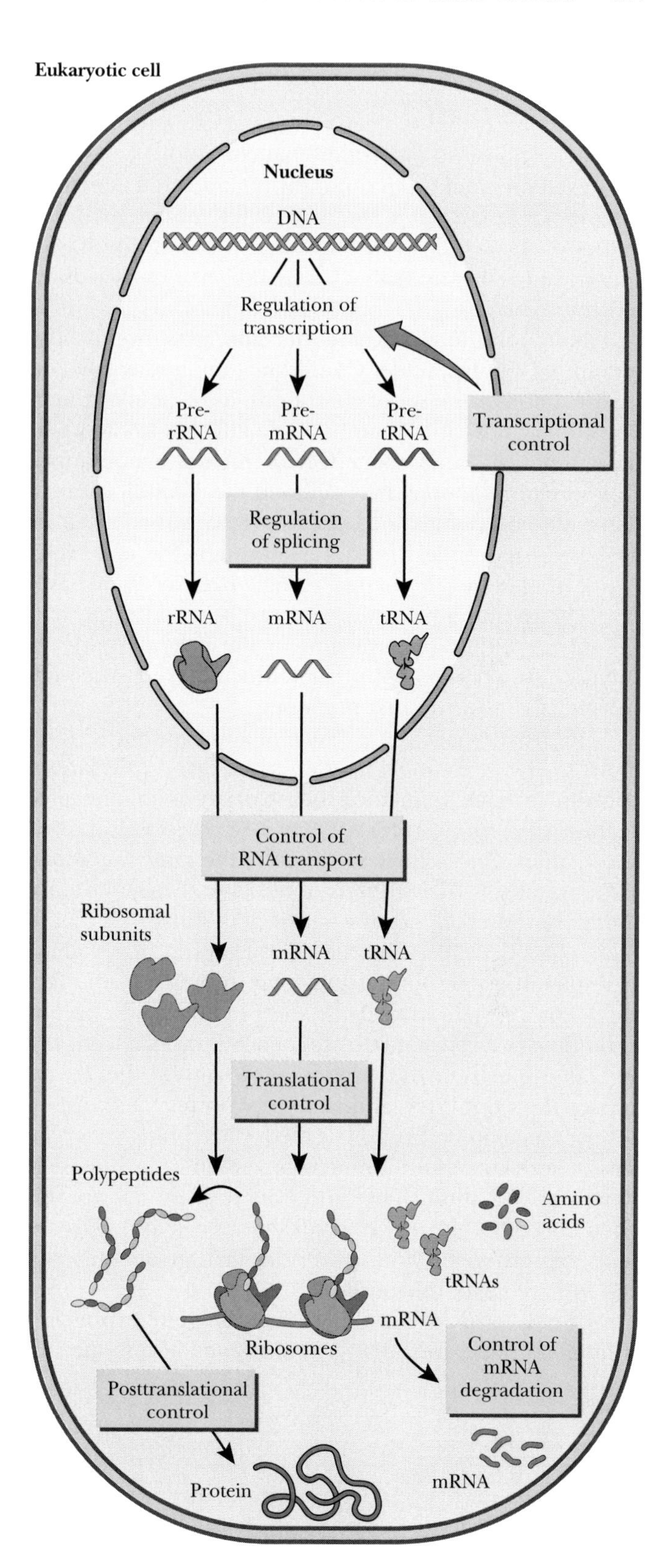

Figure 15-20 Summary of informational levels at which eukaryotic gene regulation can occur.

SUMMARY

Gene regulation allows a cell to change its pattern of proteins in response to environmental conditions or as part of a developmental program. Information about gene regulation has come from the analysis of mutations, from studies of the synthesis of individual RNAs and polypeptides within cells, and from *in vitro* studies of transcription and translation.

Mutations may affect either the structure or the amount of a polypeptide. Conditional mutations have little effect under one set of conditions; for example, a temperature-sensitive mutation has little effect at low temperature but a pronounced effect at higher temperature. Conditional mutations have allowed the identification of genes that affect functions needed for cell survival, such as those responsible for passage through the cell cycle.

A mutation can change a single base or many bases in DNA. An additional mutation can restore the normal phenotype, either by reversion to the original base sequence or by serving a suppressor, which prevents the phenotypic effects of the first mutation.

In prokaryotes, the absence of a nuclear membrane allows the close coupling of transcription and translation. Coordinate gene regulation often depends on the production of a single mRNA from a set of closely linked genes, all of which may contribute to the same metabolic pathway, such as the synthesis of histidine. Analysis of mutations in the *lac* operon, a set of genes involved in the metabolism of lactose, showed the role of specific proteins in regulating gene expression. One protein, the *lac* repressor, is a negative regulator and prevents transcription by binding to a specific DNA sequence upstream from the regulated genes. In the presence of lactose, the *lac* repressor does not bind, and transcription may proceed. A positive regulatory protein, called CAP, binds to cyclic AMP, whose concentration increases when little glucose is available. CAP stimulates transcription of the *lac* operon when no glucose is present. Prokaryotes can also regulate gene expression by premature termination of transcription or by control of translation.

A eukaryotic cell is able to regulate gene expression even more elaborately than a prokaryote because the nuclear membrane uncouples transcription and translation. Eukaryotic cells generally have much more DNA than prokaryotes, and many multicellular eukaryotes show distinctive patterns of gene expression in different cell types. Nuclear transplantation experiments suggest that all cells in a multicellular organism have the same genetic information, so cell specialization depends on regulated gene expression.

Eukaryotic gene regulation can occur at many levels. A eukaryotic cell can alter the accessibility of a gene to RNA polymerase, the rate of transcription, the pattern of RNA splicing, and the rate of translation. Transcriptional regulation depends on the interaction of specific regulatory proteins with specific DNA sequences, often just upstream from the DNA that specifies the polypeptide sequence. Recombinant DNA techniques allow the linking of suspected regulatory sequences in DNA to a reporter gene, which encodes an easily detected protein. By studying the pattern of reporter expression, researchers can identify the DNA sequences needed for transcription within a particular type of cell. By studying the interaction of these sequences with nuclear proteins, researchers can identify factors responsible for the regulation of transcription of a specific gene.

Transcription factors fall into distinct classes, each with common structural features. Many transcription factors have one of three structural designs, called helix-turn-helix, zinc finger, or amphipathic helix. Some transcription factors can bind to small molecules, such as steroids, allowing regulation of transcription by small molecules. Some transcriptional factors act on many genes, allowing the coordination of complex programs such as muscle development or sex determination.

Eukaryotic mRNAs arise from intron-containing precursors. In some cases, a single gene may give rise to distinctive mRNAs by alternative splicing patterns of exons. Other changes in the sequences or amounts of specific polypeptides may result from differences in polyadenylation, mRNA degradation, translational efficiency, or from post-transcriptional alterations in the base sequence of RNA, called RNA editing.

KEY TERMS

amphipathic helix
aneuploidy
attenuation
base substitution
CAP, or catabolite activator protein
chromosomal mutation
clone
conditional mutation
constitutive
CRE, or cyclic AMP response element
cyclic AMP
deletion
differential gene expression
DNA affinity chromatography
duplication
embryo
enhancer
footprinting
frame-shift mutation
gel retardation assay
gene (a modified definition)
gene amplification
gene rearrangement
helix-turn-helix

inducible
insertion
inversion
lac operon
lac repressor
leucine zipper
missense mutation
monosomy
mutagen
mutation
negative regulator
noninducible
nonsense mutation
nuclear transplantation
null mutation
operator
operon
point mutation
polycistronic
positive regulator
promoter
protein kinase A
regulatory mutation
reporter
response element
reversion
RNA editing
sex-determining region
silent mutation
structural mutation
suppressor mutation
temperature-sensitive mutation
termination signal
totipotent
transcription factor
transgenic
translational control
translational repressor protein
translocation
trans RNA splicing
trisomy
zinc finger

QUESTIONS FOR REVIEW AND UNDERSTANDING

1. Define and contrast the following terms:
   (a) silent mutation vs. structural mutation vs. regulatory mutation
   (b) conditional mutation vs. temperature-sensitive mutation
   (c) permissive temperature vs. restrictive temperature
   (d) point mutation vs. chromosomal mutation
   (e) missense mutation vs. nonsense mutation
   (f) noninducible mutant vs. constitutive mutant
   (g) promoter vs. enhancer
   (h) *cis*-regulatory elements vs. transcription factors (*trans*-acting factors)
   (i) footprinting vs. gel retardation assay
   (j) CRE vs. CREB
2. Given what you know about the Genetic Code, explain why a point mutation lying within a coding region may not result in a change in the amino acid sequence of the polypeptide.
3. Suggest why, generally speaking, frameshift mutations such as insertions and deletions are more deleterious than base substitutions.
4. Describe in your own words the four commonly found types of chromosomal mutations.
5. Give an example of a regulatory mutation that is actually a structural mutation. Contrast this to a true regulatory mutation.
6. Contrast the mechanism by which a reversion restores a normal phenotype to the mechanism by which a suppressor mutation restores a normal phenotype.
7. Why can sunlight be considered a mutagen?
8. Describe in your own words how an operon, yielding a polycistronic mRNA, allows for coordinate regulation in prokaryotes.
9. Why can we say that the *i* gene for the *lac* repressor is both a regulatory gene and a structural gene? Why is it significant that the repressor can bind to either the operator or lactose, but not both simultaneously?
10. Describe the role of CAP in coordinating the utilization of lactose and glucose.
11. How is the *trp* operon and its repressor similar to the *lac* operon? How is it different?
12. Describe the role of the hairpin structure in attenuation. Why is this useful for regulation in prokaryotes? Although it is described here in relation to tryptophan synthesizing enzymes, it is also employed as a means of regulating the production of enzymes needed to synthesize other amino acids such as histidine, leucine, or phenylalanine. What codons might you expect to find in abundance in the leader RNA regions for the mRNA specifying enzymes needed for, say, phenylalanine biosynthesis?
13. In general terms, state how coordinate regulation in eukaryotes differs from that in prokaryotes.
14. What is meant by differential gene expression and why is it required for development?
15. Regarding Weintraub's experiment using DNase, consider the following questions.
   (a) Why would we expect the globin gene to be accessible only in red blood cell precursors and the ovalbumin gene to be accessible only in oviduct cells?
   (b) Why is it significant that the DNase digested both genes when DNA samples purified from the two cell types were tested?
   (c) Picturing the process described in this experiment on a molecular level, why might the results of this experiment be different if the DNase were a much larger protein molecule than the RNA poly-

merase? What if DNase were much smaller than RNA polymerase?

16. Describe in your own words how reporter genes are used to characterize regulatory elements.
17. Why do the properties of enhancers suggest that a rearrangement of the DNA occurs during activation of transcription?
18. Why can we say that steroid hormone receptors are transcription factors?
19. What are three "structural designs" found in transcription factors?
20. Describe the function of each of the three structural domains typically found on a transcription factor.
21. Why would an inhibitor of kinase activity affect cAMP's ability to activate transcription of specific genes?
22. It is now understood that the combination of transcription factors present in a cell is important in regulating gene expression. Why is this logical from a mathematical standpoint? Why is it rational from a biological standpoint?
23. Why does alternative splicing lead us to reconsider the general rule "one gene, one polypeptide"? How did it lead us to revise our definition of a gene?
24. What is a Barr body? Why do we say that all female mammals are "genetic mosaics"?

■ SUGGESTED READINGS

ALBERTS, B., D. Bray, J. Lewis, M. Raff, K. Roberts and J. D. Watson, *Molecular Biology of the Cell,* 3rd edition, Garland, New York, 1994. Chapter 9 deals with the control of gene expression.

LODISH, H., D. Baltimore, A. Berk, S. L. Zipursky, P. Matsudaira, and J. Darnell, *Molecular Cell Biology,* 3rd edition, Scientific American Books, New York, 1995. Chapter 11 concerns the regulation of transcriptional initiation.

STRYER, L., *Biochemistry,* 4th edition, W. H. Freeman and Company, New York, 1995. Chapters 36 and 37 deal with gene regulation in prokaryotes and eukaryotes.

WATSON, J. D., M. Gilman, J. Witkowski, and M. Zoller, *Recombinant DNA,* 2nd edition, Scientific American Books, New York, 1992. Chapter 9 deals with the control of gene expression in eukaryotes.

CHAPTER 16

Unconventional Genes

KEY CONCEPTS

- A virus is an assembly of protein and nucleic acid that can reproduce only within a living cell.
- Viruses are examples of mobile genes. Some mobile genes can replicate separately from other types of DNA, while others must first integrate into another DNA molecule.
- The genomes of both prokaryotes and eukaryotes contain DNA that originally derived from mobile genes.
- Energy organelles—mitochondria and chloroplasts—contain independent genetic systems and probably evolved from ancient associations of prokaryotes.

OUTLINE

ESSAYS

Since the beginning of the 20th century, geneticists have developed methods for studying both the information within a gene and its chromosomal address. In some respects, a gene's address is more important than its information in determining inheritance patterns because geographically distant genes are more likely to recombine than are nearby genes.

Some genes, however, are **unconventional;** they do not have a stable cellular address. Some (those of viruses) may never have a cellular address and sometimes are not even made of DNA; some (those of chloroplasts and mitochondria) have addresses that are not in the nucleus; and some **(mobile genes)** have addresses that change without notice. This chapter deals with a heterogeneous collection of such unconventional genes. In all cases, even when a cellular address is only transient, the replication and expression of unconventional genes can occur only within a host cell.

Consideration of unconventional genes has changed cell biology: It has given fundamental insights into diseases ranging from the common cold to cancer to AIDS; it has led to new views of the origin of eukaryotic cells and of life itself; and (as discussed in Chapter 17) it has led directly to the extraordinary growth of the biotechnology industry.

■ WHAT IS A VIRUS?

The most familiar and socially relevant examples of unconventional genes are viruses, the agents of many human diseases. A **virus** is an assembly of nucleic acid (DNA or RNA) and proteins (and occasionally other components, such as lipids or carbohydrates) that can reproduce only within a living cell. A virus does not itself take energy from its environment or perform metabolic reactions. No known virus, for example, makes ATP or ribosomes. Most biologists do not consider viruses to be alive. (See Chapter 1, p. 10.)

Viruses use the biochemical machinery of their hosts to replicate their genetic material and build their proteins. Viruses can never reproduce or express their genes without the machinery of their hosts. Viruses, therefore, must reproduce within cells.

The word *virus* has a curious history. It comes from the Latin, and means "slimy liquid or poison." Long used as a synonym for "venom," in the last century "virus" came to mean any agent associated with death and disease.

By the late 19th century, Louis Pasteur and others had shown that bacteria and other small cells not only cause many infectious diseases but are also responsible for the souring of milk, wine, and beer and for the spoiling of food. Finding ways to prevent bacterial infection became a high priority for medical, scientific, and commercial reasons. Pasteur himself developed a way of sterilizing food with heat, a process now called pasteurization. Pasteur's colleague, Charles Chamberland, developed two other ways of eliminating microorganisms—heating solutions under pressure to high temperatures in a device called an autoclave, and forcing solutions through a filter that removed even the smallest cells.

These sterilization methods prevented food spoilage and the transmission of bacterial infections. But some infectious agents managed to get through the bacterial filters and were therefore called "filterable viruses." As microbiologists identified the specific bacteria responsible for many diseases, the filterable infectious agents were simply called viruses.

In the 1890s, within a few years after Chamberland's filter became available, two men—the Russian pathologist Dmitri Iwanowski and the Dutch microbiologist Martinus Beijerinck—demonstrated that a filterable virus was responsible for a disease of tobacco leaves. Because the diseased leaves have a patchy, "mosaic" appearance before they die, the disease is called tobacco mosaic disease, and the virus is now called tobacco mosaic virus.

Other microbiologists soon discovered that viruses are the causes of several diseases, including yellow fever (an important human disease carried by mosquitoes) and hoof-and-mouth disease (a disease of cattle). In 1911, Francis Peyton Rous, an American microbiologist, was the first person to identify a **tumor virus,** a virus that causes its host cells to lose their normal ability to regulate cell division. Rous's virus, now called Rous sarcoma virus, caused chickens to develop tumors within their connective tissues. Other scientists did not realize the full importance of Rous's discovery until much later, and Rous—happily still alive at the time—received a Nobel Prize in 1966.

Viruses infect not only plants and animals but even bacteria. In 1915, the English bacteriologist Frederick Twort discovered a **bacteriophage** [Greek, *phagein* = to eat], a virus that infects bacteria. Two years later, Felix d'Herelle also demonstrated that a filterable virus could kill his bacterial cultures.

All of these discoveries convinced biologists that some disease-causing agents are smaller than the pores in a bacterial filter (about 0.5 μm). No one, however, knew what viruses actually are, although most people thought that they are just cells too small to be stopped by a filter or seen in a microscope. This view dissolved in 1935, when Wendel Stanley, a biochemist working at the Rockefeller Institute in New York, produced crystals (precisely ordered arrays) of tobacco mosaic virus. This was confusing indeed. A crystal forms only when identical molecules arrange themselves in a regular lattice (as in the case of the familiar salt and sugar crystals on the breakfast table). How could an infectious agent—something that was assumed to be alive—form a crystal? We certainly couldn't imagine a crystal of bacteria or of mice!

The implication was that a virus is just a large molecule and that viral diseases are similar to chemical poisoning. But poisonous molecules do not replicate, and tobacco mosaic virus from a dissolved crystal could be passed indefinitely from plant to plant.

We now understand that viruses are large, regular molecular complexes that can sometimes form crystals. By analyzing such crystals with x-ray diffraction techniques, researchers have been able to determine the structure of some viruses in great detail. The complexes consist of proteins and nucleic acid (sometimes DNA, sometimes RNA), and the nucleic acid component contains the virus's genes. Some of a virus's genes direct the synthesis of proteins that make up the virus itself, while other genes contribute to its ability to divert the host cell's machinery to make more virus. Viruses provide our first examples of unconventional genetic systems.

The Electron Microscope Has Revealed the Structure of Many Viruses

With the development of the electron microscope in the 1940s, scientists could actually see viruses. The electron microscope revealed that viruses are large particles, smaller than cells but larger than single protein molecules. Figure 16-1, for example, shows the rod-like particles of tobacco mosaic virus, while Figure 16-2 shows the more nearly spherical particles of some other viruses. Viruses range in size from smallpox virus (about 250 nm in diameter) to poliovirus (about 27 nm in diameter), with some viruses as small as 20 nm in diameter.

Viruses have many shapes, but most have one of two forms: a long helix, such as tobacco mosaic virus, shown in Figure 16-2a; or a compact nearly spherical particle, such as poliovirus, shown in Figure 16-2b. This nearly spherical particle is actually an icosahedron, a symmetrical solid with 20 faces that meet at 12 vertices.

Other viruses have more complex forms. The bacteriophage T4, shown in Figures 16-2c and f, is among the most complex of viruses. With an icosahedral head and an intricate tail structure, it looks and functions like a tiny inoculation syringe, injecting the bacteriophage's DNA into its bacterial host.

The outside surface of a virus is either a **capsid,** a coat of protein, or a **membrane envelope.** Many animal viruses, such as HIV and pox virus, shown in Figures 16-2d and e, have envelopes that derive from the membranes of their hosts together with proteins specified by the virus's own genes.

The combined use of the electron microscope and the ultracentrifuge (see Chapter 5) has allowed virologists to analyze the chemical composition of isolated viruses. From these studies we now know that viruses consist almost entirely of proteins and nucleic acid. The nucleic acid—either DNA or RNA, depending on the virus—serves as the virus's genetic material and contains the genetic information for most, it not all, of the virus's proteins. Recent x-ray diffraction studies (described in Chapter 3) have revealed the precise three-dimensional structure of the proteins and the nucleic acids of several viruses, including the bacteriophage T4 shown in Figure 16-3.

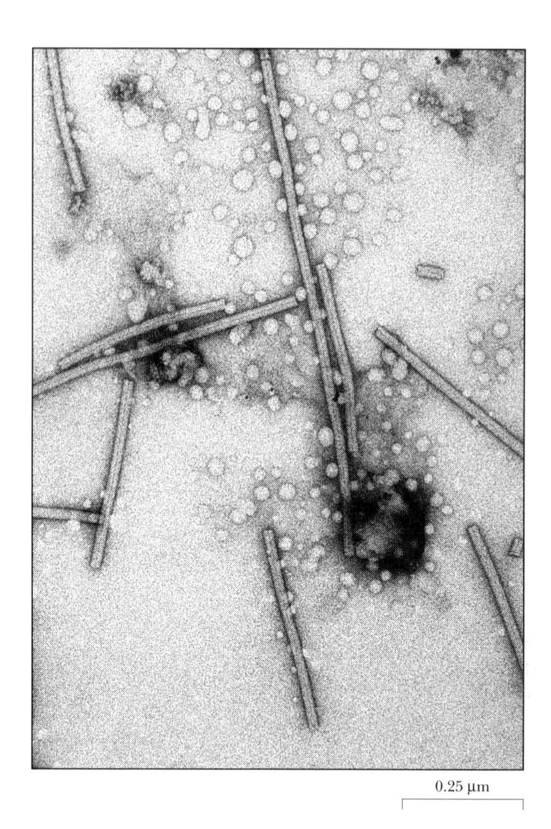

Figure 16-1 The electron microscope has revealed the structures of many viruses including tobacco mosaic virus. *(Jack D. Griffith, University of North Carolina)*

Although Viruses Are Not Alive, They Are Still Subject to Natural Selection

Viruses are adapted to specific hosts, whose cellular machinery they commandeer. The **host range** of a virus is the set of hosts that a particular virus can infect. For example, bacteriophage T4 infects *E. coli* but not other bacteria; in fact, T4's host range is limited to certain strains of *E. coli.*

Infection with a virus usually reduces the ability of the host to reproduce. Hosts that evolve ways to protect against viral infection are therefore more likely to survive and reproduce, so potential hosts may have evolved specific and general mechanisms that resist viral infection. Many strains of *E. coli,* for example, produce **restriction enzymes,** enzymes that cut (and thus render inoperable) foreign DNA that enters the cell, to protect against viral infection. We discuss these enzymes in greater detail in Chapter 17.

As hosts evolve resistance to a virus, natural selection operates on the virus itself. As its host range shrinks, variant viruses that are nonetheless able to propagate them-

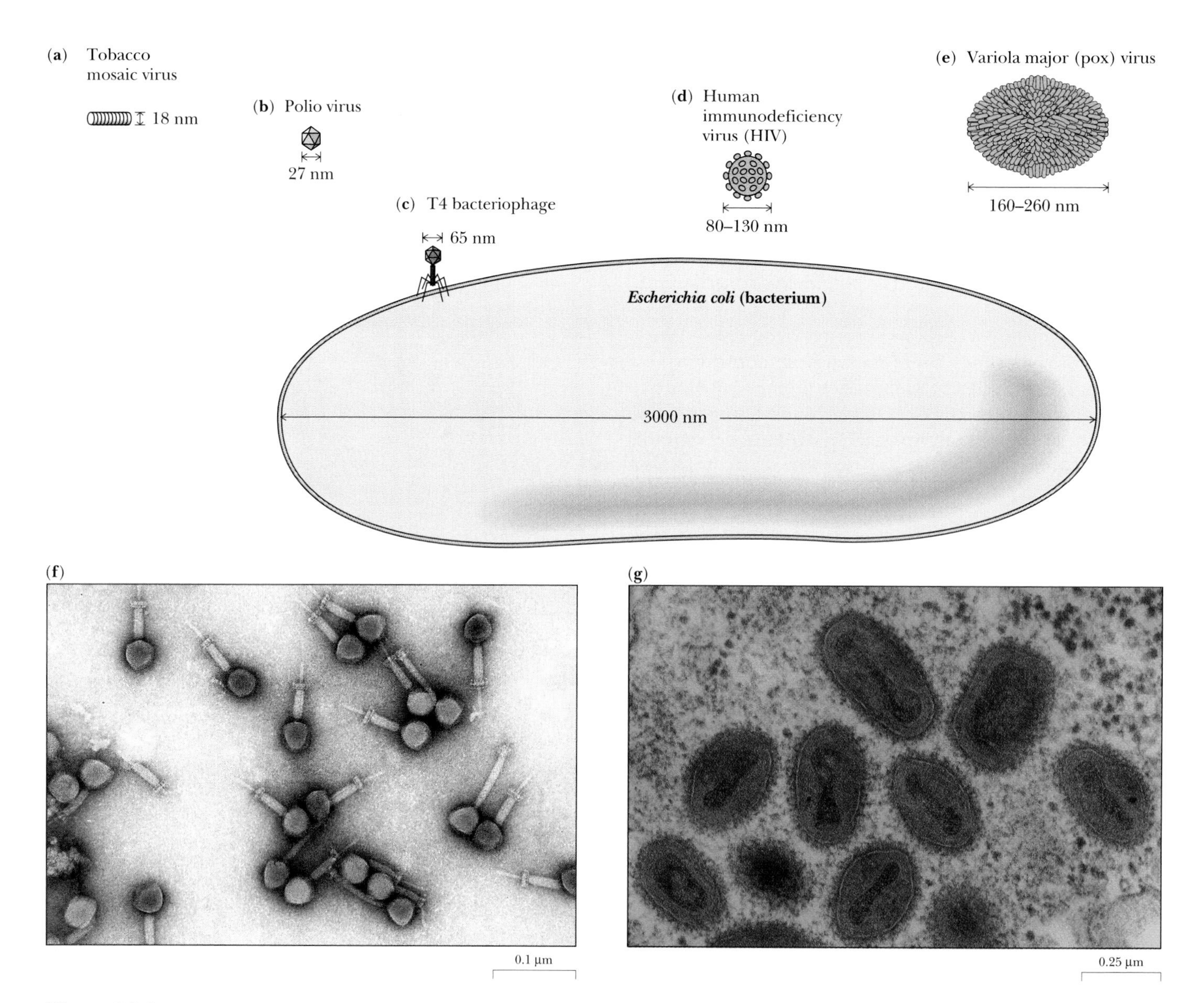

Figure 16-2 Viruses may have different sizes and shapes: **(a)** tobacco mosaic virus; **(b)** polio virus; **(c)** bacteriophage T4; **(d)** human immunodeficiency virus (HIV); **(e)** pox virus; **(f)** electron micrograph of bacteriophage T4; **(g)** electron micrograph of a pox virus; like HIV, pox virus is surrounded by a membrane envelope. *(f, Dr. Thomas Broker/Phototake NYC; g, Hans Gelderblom/Visuals Unlimited)*

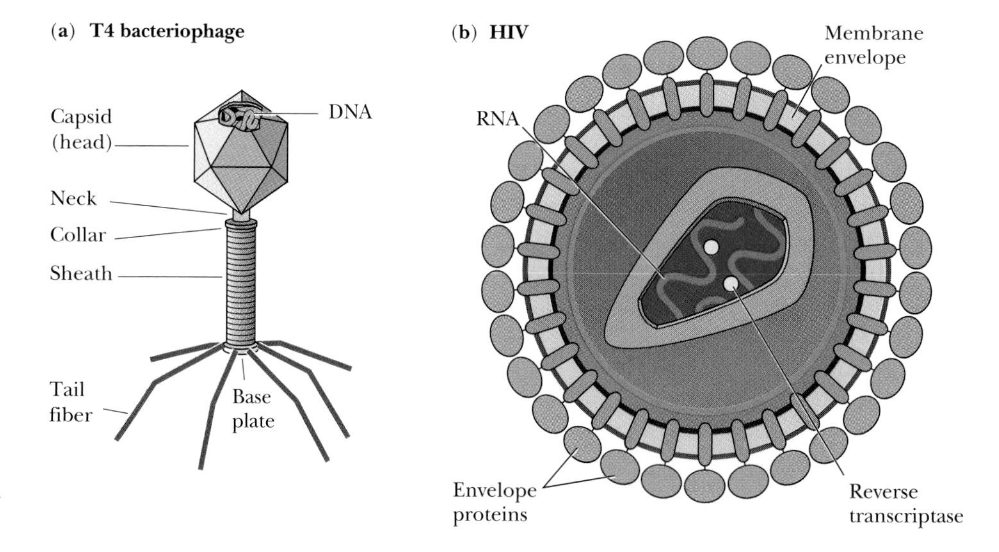

Figure 16-3 Structural studies of viruses have revealed the precise arrangements of nucleic acids and proteins. **(a)** T4 bacteriophage, a DNA virus; **(b)** HIV, an RNA virus.

selves are most represented in the next generations. For example, viruses that infect *E. coli* have DNA lacking the target sequence for the restriction enzyme that would otherwise destroy their DNA.

Natural selection has driven viral reproductive cycles to become more and more efficient. Cycles are generally rapid, with lots of progeny viruses produced in each infected cell. Viral genes have evolved regulatory sequences that are recognized by the host's transcription and translation machinery, even in preference to the regulatory sequences in the host's own genes. These signals drive the production of new viral components at the expense of the host cell.

Keeping their genetic material to a minimum has allowed viruses to speed up their cycle times. Some viral genomes have become incredibly streamlined. For example, in adenovirus, a small icosahedral virus that infects the respiratory tract, different mRNAs arise from the same precursor from different patterns of RNA splicing.

Perhaps the most extraordinarily efficient use of genetic material, however, occurs in the bacteriophage ϕX-174 ("phi-X"). This virus has **overlapping genes,** a single stretch of DNA that contains the information for distinct polypeptides, as illustrated in Figure 16-4. In all conventional genes, the translation machinery reads codons in only one reading frame, for example, grouping nucleotides 123 456 789. . . . The mRNA from ϕX-174, however, is read (by *E. coli* ribosomes) not only as 123 456 789. . . , but also as 234 567. . . .

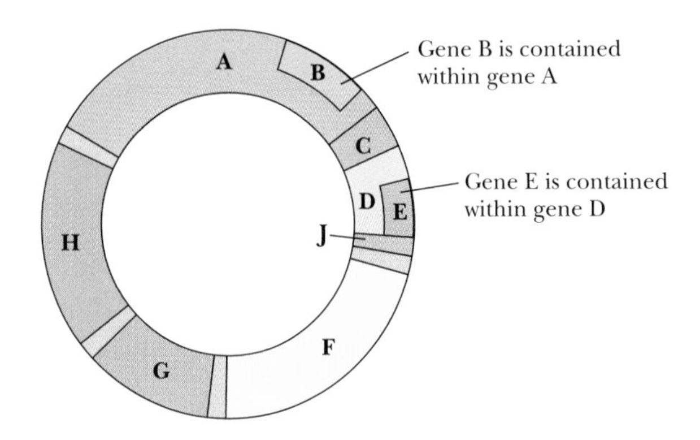

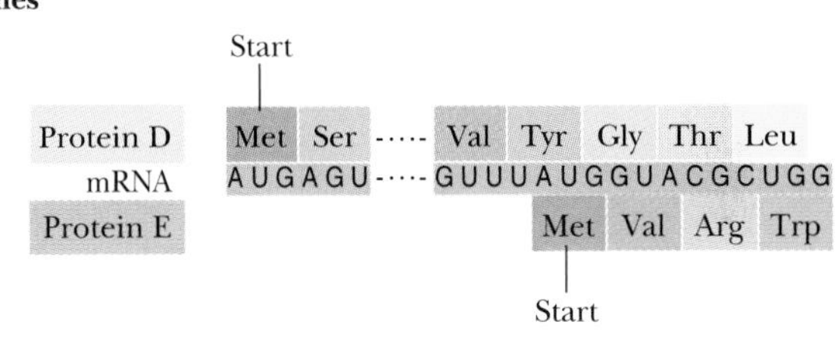

Figure 16-4 Bacteriophage ϕX-174 has overlapping reading frames that lead, in two cases, to the production of two proteins from a single stretch of DNA. **(a)** Gene B is contained within gene A, and gene E within gene D; **(b)** the overlapping reading frames of genes D and E.

How Can the Ability to Grow Viruses in Laboratory Cell Cultures Allow Researchers to Study Their Reproductive Mechanisms?

Viruses can infect cells in laboratory cultures. The ability to propagate viruses in the laboratory allows virologists to study the reproductive mechanisms of individual viruses.

Although viruses can make more viruses in many different ways, their reproductive cycles always include the stages shown in Figure 16-5a: (1) attachment to the host cell; (2) entry of viral nucleic acid into the cell; (3) synthesis of proteins specified by the virus's genes; (4) replication of the virus's DNA (or RNA); (5) assembly of new virus particles; and (6) release of the new viruses. Each stage in the reproductive cycle depends both on components of the host cell and on the molecules of the virus itself.

Attachment Depends on Interaction Between a Viral Protein and Molecules on the Host Cell's Membrane

Specific proteins on the surface of the virus (the viral coat) bind to a molecule on the surface of the host cell, often a membrane protein. The attachment site may be on all the host's cells or just on a special set of cells. Among human viruses, some, such as the influenza virus (which causes flu), specifically infect cells of the respiratory tract; other, such as measles and chickenpox viruses, infect skin cells; still others, like rabies and polio, attach specifically to nerve cells. HIV (human immunodeficiency virus), the virus responsible for causing AIDS, primarily infects cells of the immune system and the brain (Figure 16-6).

Viral Nucleic Acid Enters the Host Cell by Mechanisms That Depend on the Available Machinery of Both the Host and the Virus

Bacteriophage T4, for example, contains an enzyme that digests away part of a bacterial cell wall, allowing the phage to inject its DNA into its host. Other viruses are taken into cells by receptor-mediated endocytosis, as described in Chapter 6. In this case, a protein on the viral surface triggers the fusion of the viral envelope with the vesicle's membrane, thereby releasing the viral genes into the cytoplasm, as illustrated in Figure 16-5b. Still other viruses, including HIV, are enclosed in membranes captured as they exit an infected cell; these viruses can enter their next host as the virus's membrane envelope fuses with the host cell's plasma membrane.

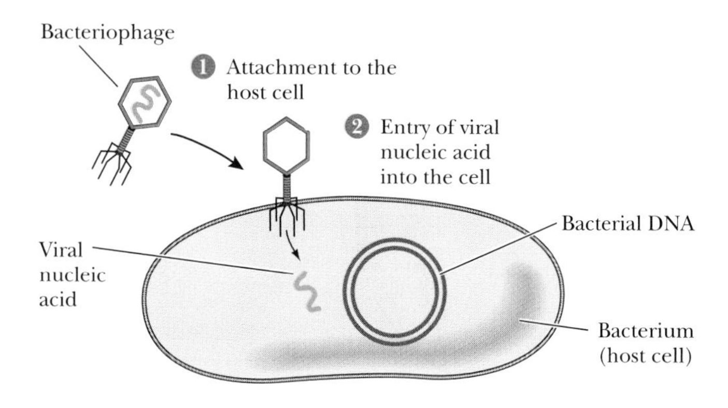

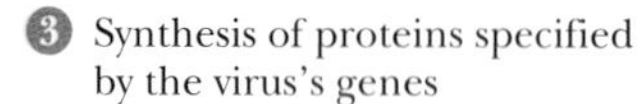

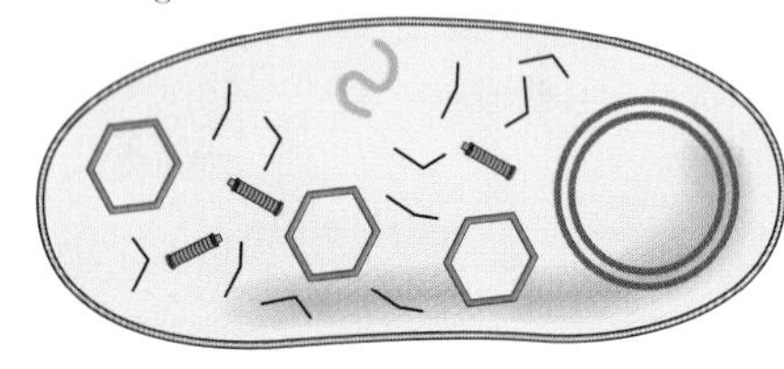

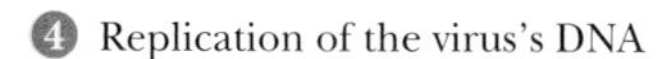

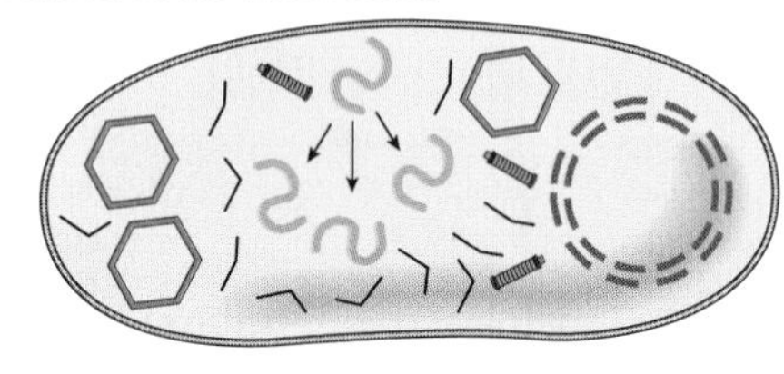

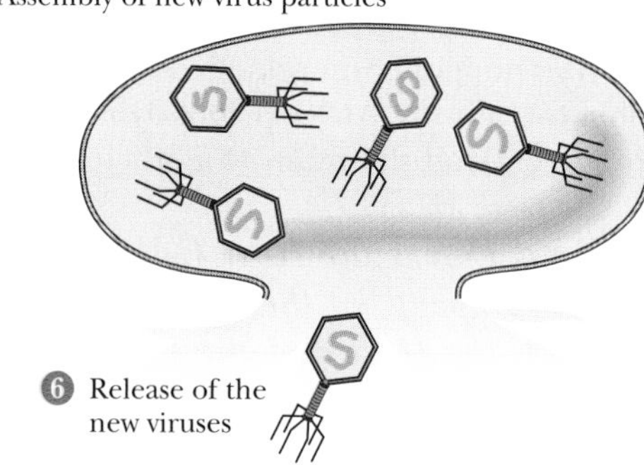

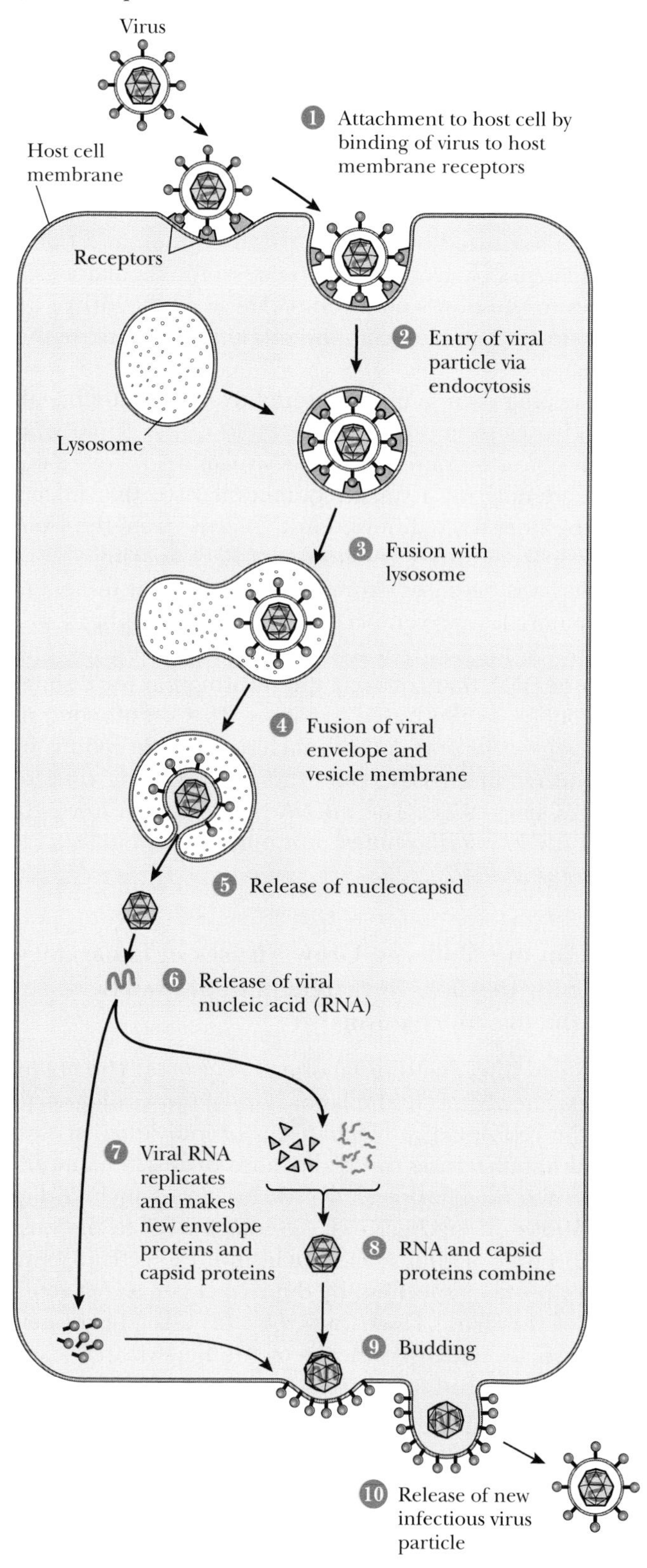

Figure 16-5 The reproductive cycle of **(a)** a nonenveloped virus and **(b)** an enveloped virus.

Once Inside the Host Cell, the Virus's Genes Direct the Synthesis of Viral Proteins

Some viruses take over their host's protein-synthesizing machinery by employing an enzyme that destroys the host's own DNA. Other viruses rely on promoters and other sequences that ensure rapid replication, transcription, and translation. For most viruses, infection results in the preferential production of the virus's (rather than the host cell's) proteins.

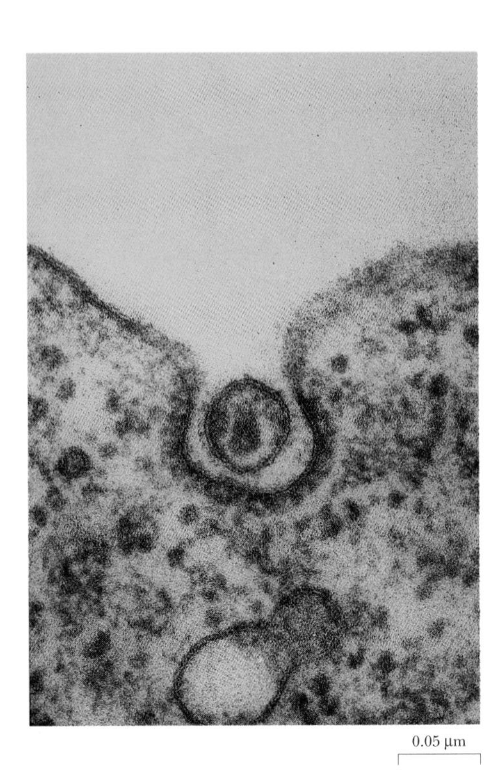

Figure 16-6 HIV surrounded by the membrane of a cell that is about to be infected. Note the clathrin pit. *(Hans Gelderblom/ Visuals Unlimited)*

Viral Proteins Spontaneously Assemble with the Viral Nucleic Acids to Form New Virus Particles, Which Are Released Outside the Host Cell

For a bacteriophage, the whole cycle may take only 20 minutes and may produce 100 or more new bacteriophages from each "parent" bacteriophage. After these new viruses are produced, a viral enzyme causes host cell **lysis** [Greek, *lysis* = loosening], the breaking of the host cell's membranes. The ruptured cell releases the newly made viruses, which are ready to infect more host cells. Viruses that destroy their host cells in this way are called **lytic viruses.** Not all viruses cause cell lysis, however. Some viral life cycles are compatible with survival of the host cell, as discussed in the next section.

The Genes of Some Viruses Can Integrate into the Chromosomal DNA of the Host Cell

Killing the host cell is not the only strategy for a virus's reproductive success. Elimination of all host cells, after all, would ultimately mean the end of the line for the virus. Some viruses have a reproductive cycle similar to that of a lytic virus but can be released without destroying their hosts. Many viruses, however, use an alternative strategy for survival: The genetic information of these viruses actually becomes part of the host cell's own DNA and is reproduced by host cells along with their own genetic material when the cells reproduce.

The Bacteriophage λ (Lambda) Has Alternative Reproductive Cycles in its E. coli *Hosts*

Lambda is said to be a **lysogenic virus,** meaning that the virus either can reproduce in a **lytic cycle,** in which it destroys its host cell, similarly to T4, or the virus can lie in a dormant state that can subsequently be thrown into a lytic cycle. In its dormant, or "peacetime," form, a lysogenic virus integrates its DNA into the host's own DNA to form a **provirus** (or in the case of a bacteriophage, a **prophage**) [Greek, *pro* = before].

A provirus does not lyse its host cell, and it cannot infect other cells. Treatment with environmental insults, such as ultraviolet light or certain chemicals, however, can trigger the escape of the viral DNA and entry into a lytic cycle. Therefore, the cell containing the provirus is part of the viruses **lysogenic cycle** [Greek, *lysis* = loosen + *genos* = offspring) (see Figure 16-7a) because it can generate the destructive lytic cycle shown in Figure 16-7b.

When phage λ is present as a prophage, the host cell does not make viral proteins. The reason for this is that λ DNA directs the synthesis of a powerful repressor, called the **lambda repressor,** or **cI protein,** which prevents the transcription of λ genes in much the same way as described in Chapter 15 for the *lac* repressor. Environmental insults that threaten to destroy the host, such as ultraviolet light, also induce an "escape" mechanism in the prophage. These triggers cause a repression of the λ repressor (by another negative transcriptional factor called **cro protein**), allowing the expression of the genes within λ DNA. The expression of these genes leads to a full lytic cycle, including the destruction of the host cell and the release of new λ bacteriophages from the lysed cell.

Tumor Viruses, Like Lysogenic Bacteriophages, Can Incorporate Their DNA into Their Host's Chromosomes

Tumor viruses keep in step with their hosts, rather than multiplying as rapidly as possible and killing their host cells in a lytic cycle. After infection, the genes of most tumor viruses become integrated into the DNA of their hosts. The virus uses the host's enzymes to express one or more of its genes. The resulting viral proteins then alter the host's ability to control its own reproductive cycle.

Viral genes that affect growth control in their host cells are called **oncogenes** (or viral oncogenes). (See Chapter 12.) The oncogenes of tumor viruses are closely related to chromosomal genes (called **cellular oncogenes** or **proto-oncogenes**), which normally control growth and differentiation. The expression of an oncogene leads to the increased proliferation of the host cell and the formation of a tumor. Because its host cell produces more progeny than cells subject to normal growth control, the expression of an oncogene provides a selective advantage to the virus: Each daughter cell contains the virus's as well as the host's genes.

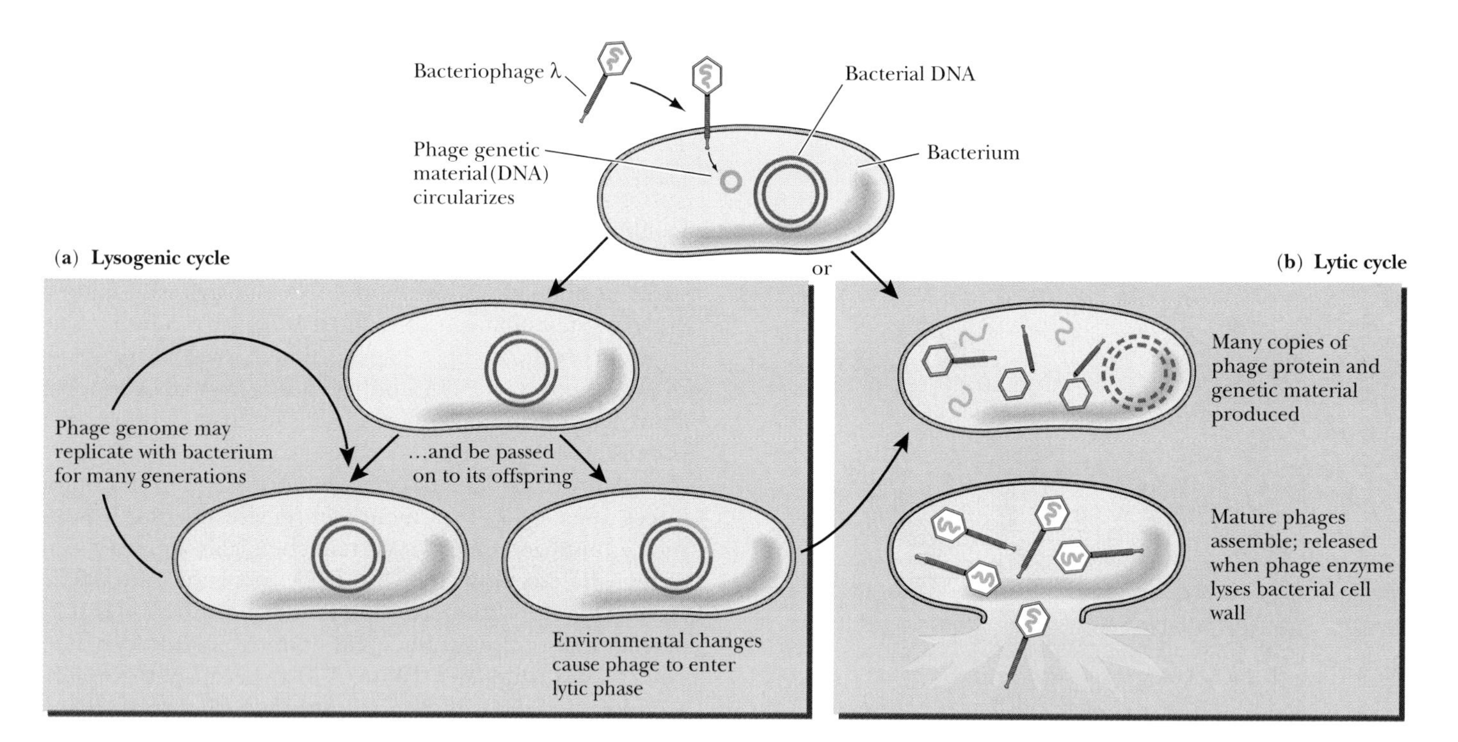

Figure 16-7 The bacteriophage λ can propagate itself in two distinct ways, by **(a)** the lysogenic cycle or **(b)** the lytic cycle.

A Virus Serves as a Vehicle for Propagating Viral Genes

This statement sounds a little ridiculous, like Samuel Butler's idea that "a hen is only an egg's way of producing another egg." Still, it helps us to understand the evolution both of viruses and of other mobile genes as well. We have seen two general strategies by which viruses propagate themselves: **autonomous replication,** in which the virus's DNA remains separate from the DNA of the host cell, and **integrated replication,** as in the lysogenic cycle or the reproductive cycle of tumor viruses, in which the virus's DNA becomes integrated into the host cell's DNA and is replicated along with that of the host. **Episomes** are viruses or other genetic systems that can propagate either autonomously or as an integrated part of the host's chromosome. The bacteriophage λ is an example of an episome.

Many Viruses Use Genetic Material Other than Double-Stranded DNA

So far we have talked mostly about viruses whose genes are double-stranded DNA, just like the genes of their hosts. But as scientists isolated and studied more viruses, they learned that many viruses do not contain double-stranded DNA but have some other nucleic acid instead. Some, like ϕX-174 for example, have DNA with only one strand. Although the discovery of single-stranded viruses was initially a big surprise, the replication of these viruses still follows the Watson-Crick model. The replication of ϕX-174 DNA, for example, starts with copying the single strand into a complementary strand to make a double-stranded "replicative form," as illustrated in Figure 16-8. The replicative form then replicates, using Watson-Crick pairing rules.

How Do RNA Viruses Replicate Their Genetic Material?

Other viruses, including those causing polio, influenza, AIDS, and many animal tumors, use RNA as their genetic material. How do genes function if they are made of RNA rather than DNA? They must use special enzymes, encoded in the virus's genome. The strategy for replication varies with the virus but always uses the Watson-Crick rules to make a complementary strand.

Virologists have classified RNA viruses according to their replicative strategies. Virologists call a coding (mRNA-like) strand a "plus" strand, so "plus-strand viruses" contain RNA that can itself function as mRNA. Immediately after entering a cell, such viruses start directing protein

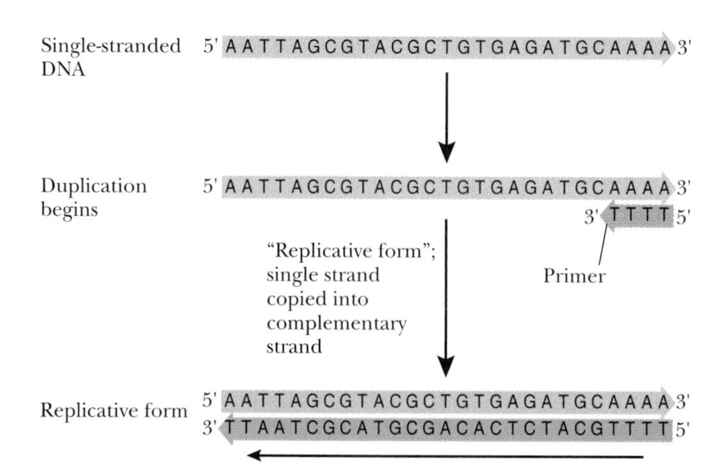

Figure 16-8 The bacteriophage ϕX-174, which contains single-stranded DNA, replicates its DNA by first making a double-stranded replicative form, which can then replicate like other double-stranded DNAs.

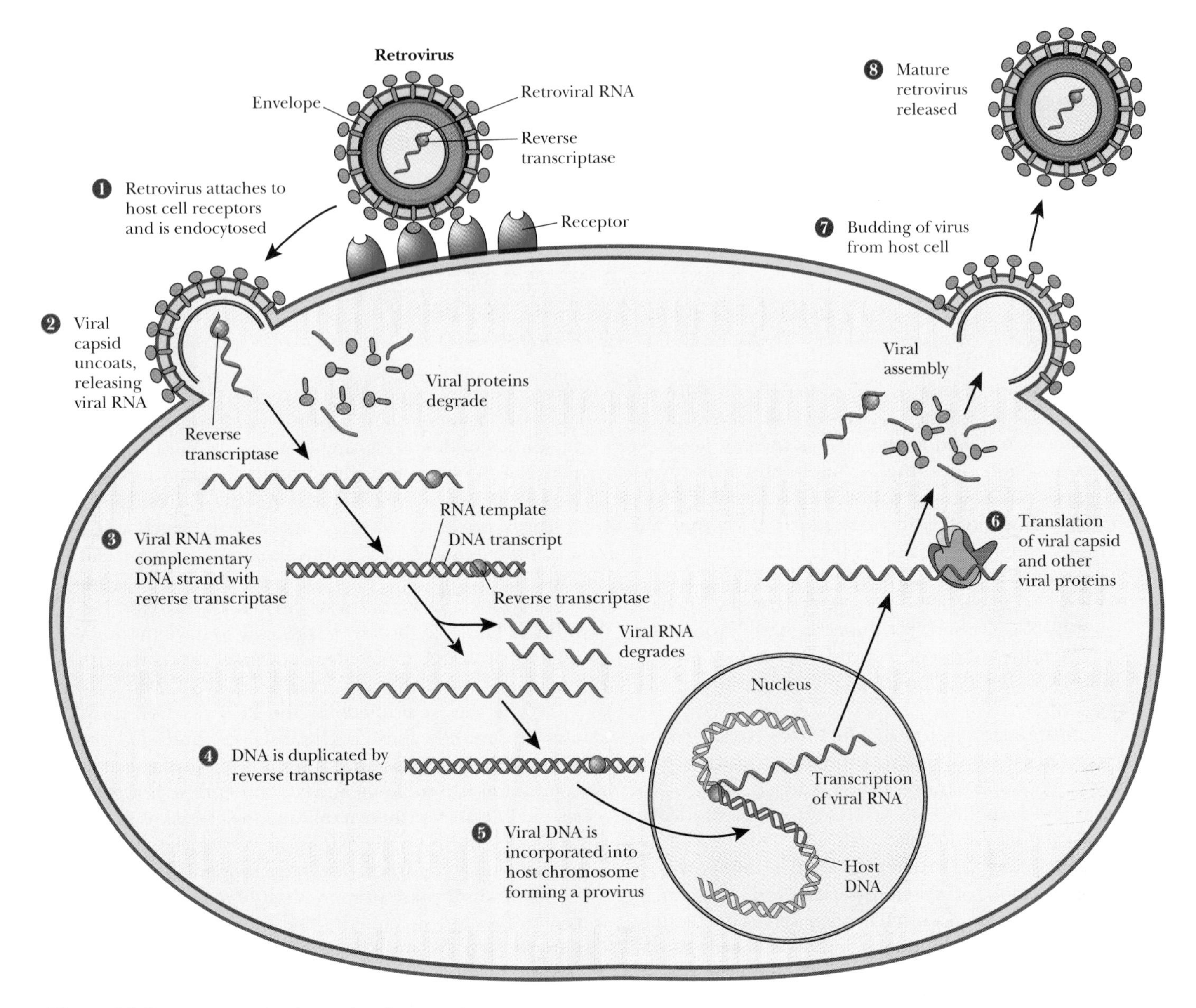

Figure 16-9 The reproductive cycle of a retrovirus.

synthesis. In other RNA viruses, called "minus-strand viruses," a viral enzyme first copies the RNA into a complementary (plus) strand that functions as mRNA.

RNA Tumor Viruses Pose Special Problems

By the mid-1960s virologists had discovered many of the ways that RNA viruses can reproduce. But they could not understand the reproductive cycle of RNA tumor viruses, such as Rous sarcoma virus.

Tumor viruses permanently change not only the host cell but also all of its descendants. Somehow the viral genes are integrated into the host cell's DNA. DNA tumor viruses could function by inserting their genes directly into the DNA of their hosts, permanently changing their genetic makeup. But how could an RNA tumor virus function? Certainly single-stranded RNA could not insert directly into chromosomal DNA.

The answer was first suggested in the 1960s by Howard Temin, a virologist at the University of Wisconsin. Perhaps, Temin argued, the RNA is copied into DNA, and the DNA is then inserted into the host cell's genome. Many scientists found this suggestion heretical because it seemed to violate the Central Dogma: "DNA specifies RNA, which specifies proteins." (See Chapter 14, p. 361.)

In 1970, however, Temin and David Baltimore, a researcher at MIT, independently discovered that RNA tumor viruses contain an enzyme, called **reverse transcriptase,** which copies RNA into DNA. The enzyme, formally called RNA-dependent DNA polymerase, received its more commonly used name because the direction of information flow in these viruses is backward from the normal transcription, where information goes from DNA to RNA. Viruses that use reverse transcriptase in this way are called **retroviruses.** The discovery of reverse transcriptase supported Temin's hypothesis.

Subsequent work has shown exactly how retroviruses can subvert their host cells to produce more retrovirus, as shown in Figure 16-9. Reverse transcriptase, the product of a viral gene that is packaged within the viral coat, en-

ESSAY

COMBINING REVERSE TRANSCRIPTASE AND PCR TO ANALYZE GENE EXPRESSION

The discovery that retroviruses contain an RNA-dependent DNA polymerase, or reverse transcriptase, has been a tremendous boon to studies of gene expression. Soon after the isolation of reverse transcriptase from a retrovirus, investigators used it to copy globin mRNA into a single-stranded DNA that was complementary to the mRNA template. The resulting DNA, called cDNA (complementary DNA) can be made with radioactively labeled nucleotides. Such labeled cDNAs are now routinely used as "probes" for specific mRNA sequences in cell extracts or even in thin tissue sections prepared for microscopy. (See Chapter 17.)

More recently, investigators have combined the use of reverse transcriptase with the polymerase chain reaction in a technique called RT-PCR, which produces and amplifies cDNAs derived from a particular tissue or even from individual cells. This type of experiment allows researchers to analyze differences in the expression of specific genes even from small amounts of tissue. In some cases, it has been possible to study the levels of individual mRNAs in cells known to be performing a particular role, for example, in nerve cells that play a defined role in a neural circuit.

The strategy for RT-PCR is straightforward: (1) RNA is extracted from a specific tissue, or from the contents of a single cell; (2) the extracted RNA is copied into cDNA by reverse transcriptase; and (3) specific cDNAs are amplified by PCR and detected after electrophoretic separation.

Like other DNA polymerases, reverse transcriptase does not begin a DNA strand from scratch but adds nucleotides to a primer. When reverse transcriptase is used to copy a set of mRNAs, the primers must match a sequence in each RNA to be copied. Most often, researchers now use random mixtures of oligonucleotides, reasoning that one or more components of the mixture will be complementary to some part of each mRNA. Sometimes, however, investigators will use a primer consisting just of oligo-T, which forms a complementary duplex with the poly A sequence at the 3′-end of mRNA. Other times, researchers will use an oligonucleotide that is specific for a single kind of mRNA. The goal, however, is usually to have the population of cDNA molecules accurately represent the population of mRNA molecules in the extract.

The sets of primers for the PCR reaction must also be carefully chosen. The usual method is to use primers that correspond to known mRNAs whose concentrations are to be measured. Sometimes, however, researchers may wish to amplify all the cDNAs as much as possible to the same extent. In these cases, investigators usually incorporate special sequences into the reverse transcriptase primer. A sequence complementary to the added sequence can then be used to amplify all the resulting cDNAs by PCR. In this case, the second PCR primer may correspond to a sequence tagged on to the 3′-end of the original cDNA.

RT-PCR has become a standard tool in the analysis of mRNA populations in small pieces of tissue and in single cells. Researchers have used RT-PCR to document changes in specific mRNAs during development and in response to injury or experimental manipulation. Application of RT-PCR to studies of the nervous system promises to reveal cellular differences associated with the roles of individual cells in specific neural circuits and changes that occur with learning or with sensory experience.

ters the cell along with the virus's RNA. It copies the RNA first into single-stranded and then into double-stranded DNA. The double-stranded DNA then integrates into the host cell's DNA, so that the virus's genes are replicated along with those of the host. As in the case of DNA-containing tumor viruses, many (but not all) retroviruses also contain an oncogene, whose expression can disrupt the host's growth regulation and lead to tumor formation.

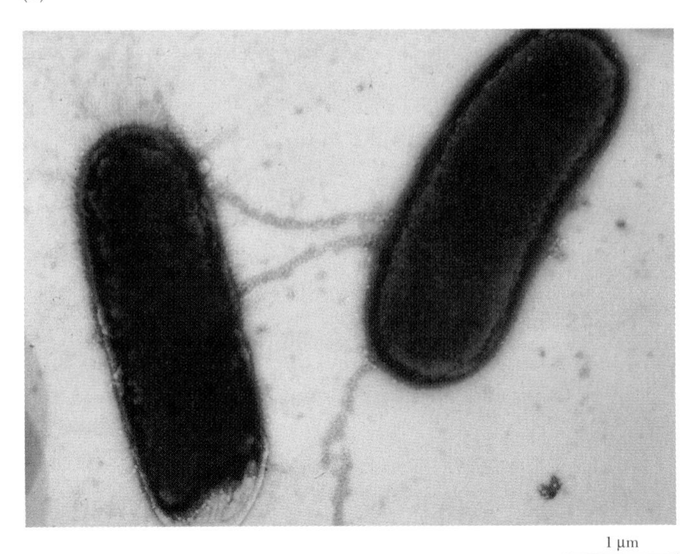

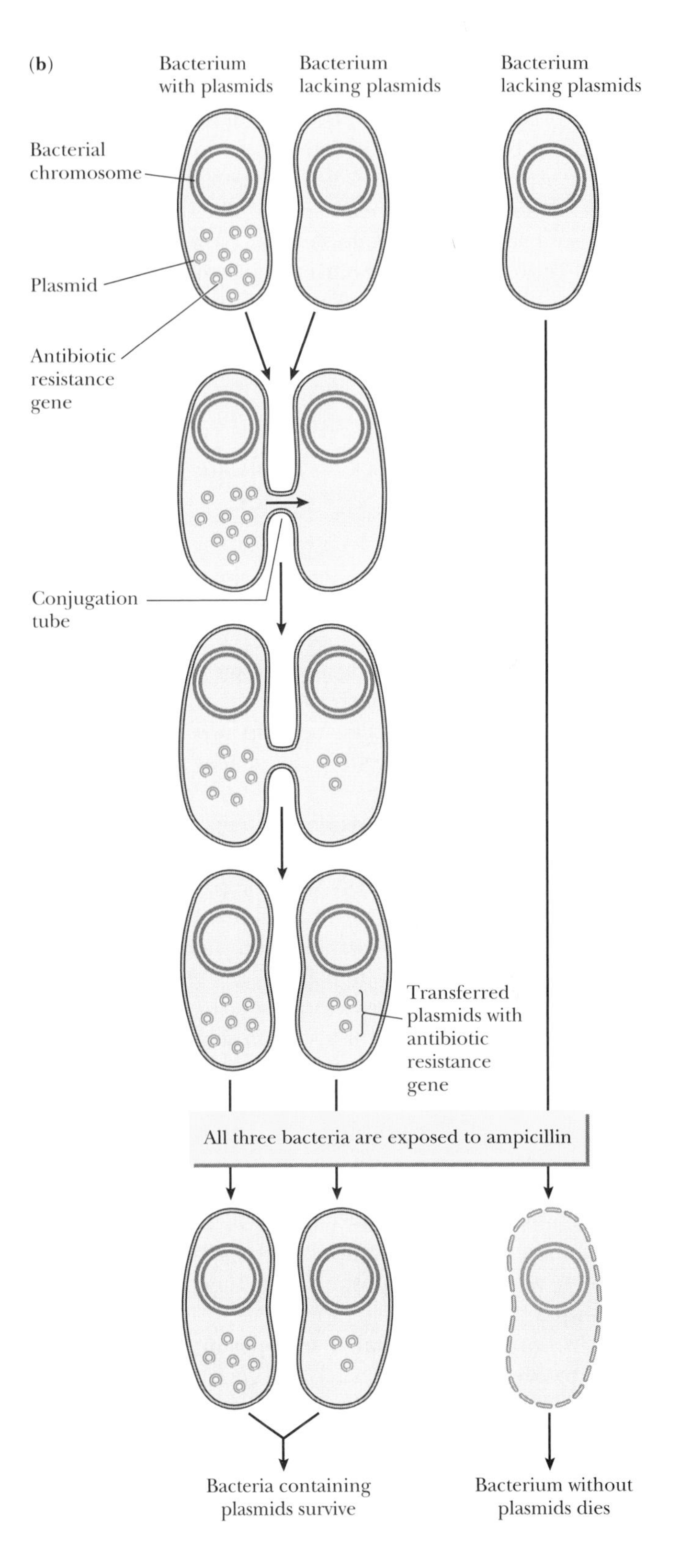

Figure 16-10 During conjugation, two bacteria exchange genetic material. **(a)** Two *E. coli* bacteria conjugating; **(b)** transfer of plasmids carrying a gene for ampicillin resistance allows bacteria to survive in the presence of ampicillin, which would otherwise kill them. Each bacterium may have multiple copies of the same plasmid, some of which may pass between cells along with bacterial DNA. *(a, Fred Marsik/Visuals Unlimited)*

MOBILE GENES CONTRIBUTE TO THE GENETIC DIVERSITY OF BACTERIA

Viral infection is one way that new genetic information can be introduced into a bacterial cell. We now discuss other mobile genes in bacteria, including **plasmids,** which can replicate autonomously; **transposable elements,** which can replicate only in integrated form (as part of a host's chromosome); and **F-factors,** a kind of episome (meaning that they can replicate either autonomously or in integrated form). These different types of mobile genes can quickly bring new genetic information to a bacterial cell.

Plasmids Are Circular DNAs That Replicate Autonomously

A plasmid, a small piece of circular DNA, can replicate independently because it contains an origin of replication (the DNA sequence where DNA replication begins, as described for *E. coli* DNA in Chapter 13, p. 346). Although plasmids usually replicate in concert with their hosts, they can replicate even when host DNA synthesis has stopped. Plasmids, unlike viruses, do not produce protein coats that would allow them to move easily between cells. They can move from one bacterial host to another, however, during the process of **conjugation,** when two temporarily attached bacteria exchange genetic material, as shown in Figure 16-10.

Plasmids often contain genes that bring their hosts some advantage over competitors. For example, some plasmids—called **resistance factors,** or **R-factors**—contain genes for enzymes that inactivate antibiotics. Other plasmids contain genes for the production of toxins that kill other bacteria, for resistance to heavy metal poisoning, or for the metabolism of unusual compounds (such as those found in oil spills). Besides these naturally occurring plasmids, molecular biologists have produced thousands of **recombinant** plasmids, which contain combinations of genes never found in nature. We discuss the production and uses of such recombinant DNAs in Chapter 17.

Transposable Elements Can Replicate Only When Integrated into a Host Cell's DNA

Transposable elements are mobile DNA sequences whose replication depends not only on the energy and the machinery of its host, but also on an origin of replication of the DNA into which they integrate. A transposable element, unlike an episome or a prophage, contains a gene for an enzyme called **transposase,** which catalyzes insertion into new sites. Bacteria contain two kinds of transposable elements—insertion sequences and complex transposons.

The simplest transposable elements, called **insertion sequences,** or **simple transposons,** as illustrated in Figure 16-11, are short lengths of DNA, no longer than a few thousand base pairs, that can move from place to place in the genome. Copies of simple transposons may also move, so that they multiply within their host's DNA. At least six different insertion sequences are present in the genome of *E. coli,* and several identical insertion sequences are present 5–10 times within each *E. coli* genome.

The limited information in a simple transposon specifies transposase, but nothing else. Unlike an R-factor, a simple transposon does not appear to convey any selective advantages to its host, although transposons in general serve to increase genetic variability by helping to rearrange the genome and by interrupting genes. As far as we can now tell, a simple transposon's only function is perpetuating itself.

In contrast, a **complex transposon** contains other genes besides transposase, such as a gene for antibiotic resistance. Some complex transposons contain insertion sequences, which assist in their movement from one site to another, as shown in Figure 16-12. Transposons that contain antibiotic resistance genes can move to plasmid DNA as well as to the host cell's chromosomal DNA, allowing a single plasmid (for example, one that carries a penicillin resistance gene) to accumulate genes for resistance to more than one antibiotic (for example, to tetracycline as well as to penicillin). The ability of complex transposons to move between plasmid and host DNA helps explain the awful mystery of how antibiotic resistance spreads so rapidly among bacteria.

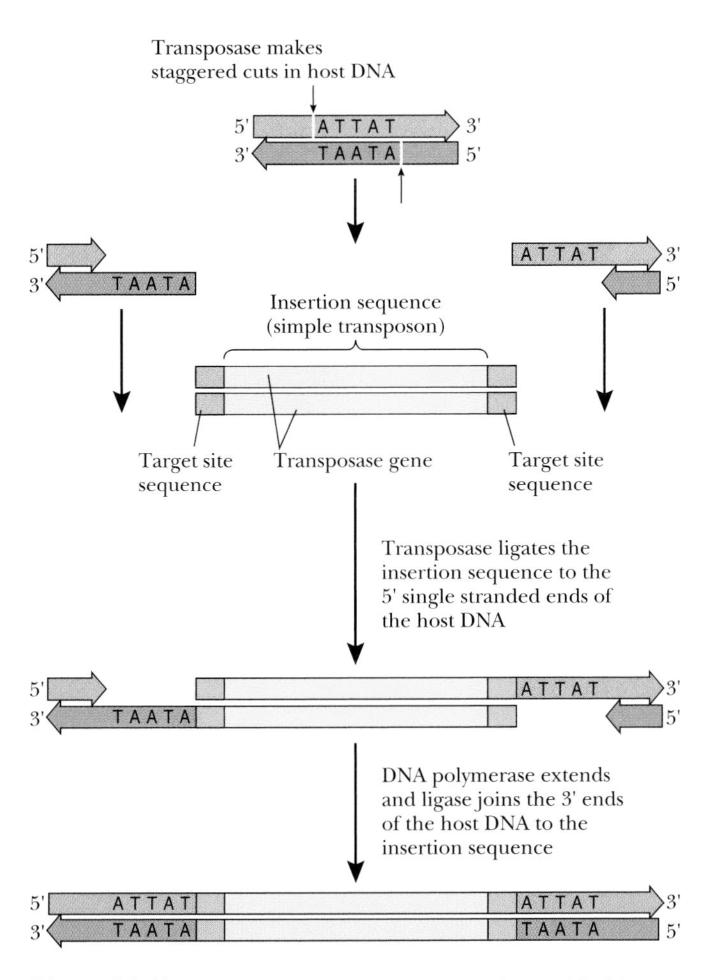

Figure 16-11 Transposase, whose sequence is specified by a gene within an insertion sequence (or simple transposon), catalyzes the insertion of the sequence into new sites.

F-Factors Can Replicate Either Autonomously or After Insertion into Bacterial DNA

The F-factor (or fertility factor) of *E. coli* is an episome that contains genes whose products are necessary for bacterial conjugation. Each F-factor contains at least 25 genes, many of which provide information for polypeptides of the **pilus,** the long appendage that serves as the means of

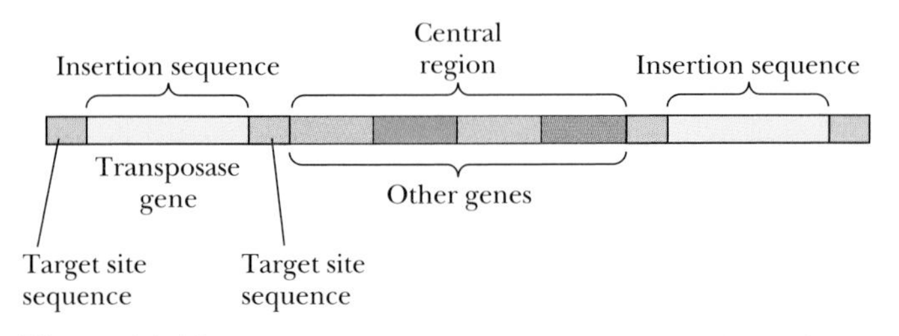

Figure 16-12 Complex transposons consist of two insertion sequences surrounding one or more other genes, such as genes that specify resistance to an antibiotic.

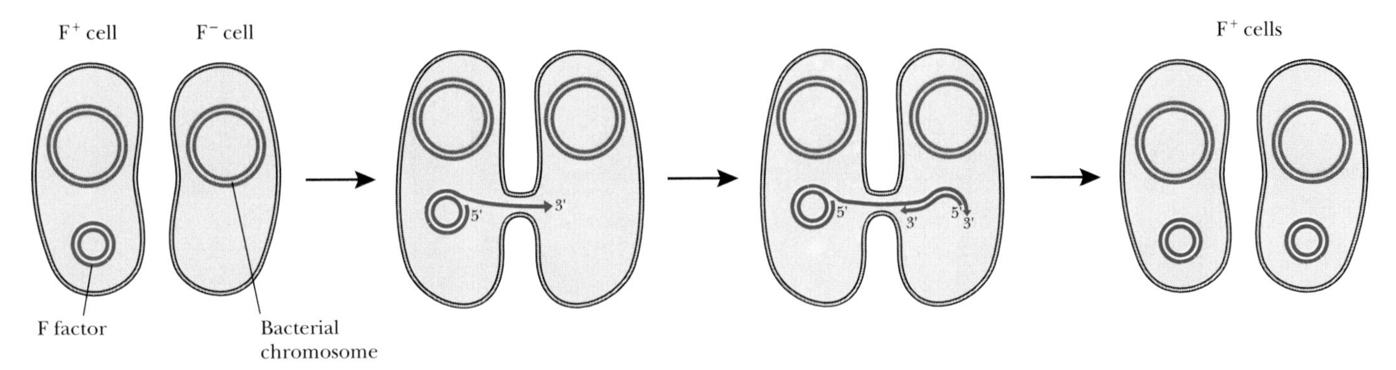

Figure 16-13 An F-factor can replicate by itself within a cell or can be transferred from one cell to another during conjugation. When a cell acquires an F-factor, it becomes F^+.

attachment of conjugating bacteria and a conduit for the transfer of DNA. "Fertility" in bacteria refers to the ability of a cell that contains the F-factor to transfer some of its genes to another cell during conjugation.

The F-factor can replicate independently because, like a plasmid, it has its own origin of replication. During conjugation, a cell may transfer one or more copies of the independent F-factor to a cell that did not previously contain an F-factor. A cell that contains an F-factor is called F^+, while a cell that lacks an F factor is F^-, as shown in Figure 16-13. During conjugation, an F^+ bacterium donates genetic material to its F^- partner and the recipient also becomes F^+.

Occasionally an F-factor integrates into *E. coli* DNA, where it replicates as part of the *E. coli* genome rather than as an independent unit. An integrated F-factor allows a copy of the whole *E. coli* DNA to move through the pilus during conjugation. The genes of the integrated F-factor (like those of an autonomous F-factor) direct the production of a pilus. When conjugation occurs, part of the F-factor moves into the pilus, pulling the rest of newly replicated *E. coli* DNA into the F^- cell, as shown in Figure 16-14. Bacterial geneticists have determined the order of genes in *E. coli* by measuring the time required for individual genes to enter the F^- host, as illustrated in Figure 16-14.

An *E. coli* cell that contains an integrated F-factor is called an **Hfr** bacterium ["*h*igh *f*requency of *r*ecombination"]. The name derives from the ability of the integrated F-factor to stimulate the transfer of genetic information and to produce new combinations of genes in the F^- cell. As a result of the transfer, the F^- cell gains an F-factor and so becomes F^+. It also gains new genes, which can recombine with the genes of the recipient F^- cell. This transfer represents a primitive type of sexuality, quite distinct from the sexuality of multicellular organisms: Mating in multicellular organisms does not convert females into males. The integrated F-factor conveys a selective advantage both to itself and to its host DNA. During conjugation it propagates both itself and the genes of its host, and, by creating new combinations of *E. coli* genes, it also increases the genetic variability.

The Properties of Mobile Genes Suggest the Evolutionary Origin of Viruses

Scientists believe viruses originated as fragments of cellular DNA that acquired the ability to replicate independently and to carry with them the genes specifying the apparatus for getting them in and out of cells. Knowing what we now know about mobile genes, we can imagine that mobile genes first appeared as simple insertion sequences that could move from place to place in the genome. Insertion sequences acquired a huge selective advantage when they associated with genes that benefited the host cell—like the ability to break down poisons or antibiotics or to make a toxin that killed competitors. These genes attained more independence when they acquired the ability to replicate independently as plasmids. Finally, some antonomously reproducing plasmids could have acquired genes for proteins such as viral coat proteins, which allowed them to move efficiently from cell to cell, rather than depending on rare conjugation events.

■ EUKARYOTIC CELLS ALSO HAVE MOBILE GENES

The study of mobile genes in eukaryotic DNA has led to still more surprises. We have already mentioned, for example, that tumor virus DNA can integrate into the genome of a host cell. Each kind of DNA tumor virus, however, can also exist independently of the host's DNA (although usually in a different type of cell). DNA tumor viruses, then, are examples of eukaryotic mobile genes that resemble bacterial episomes—able to replicate either autonomously or as integrated parts of a host chromosome.

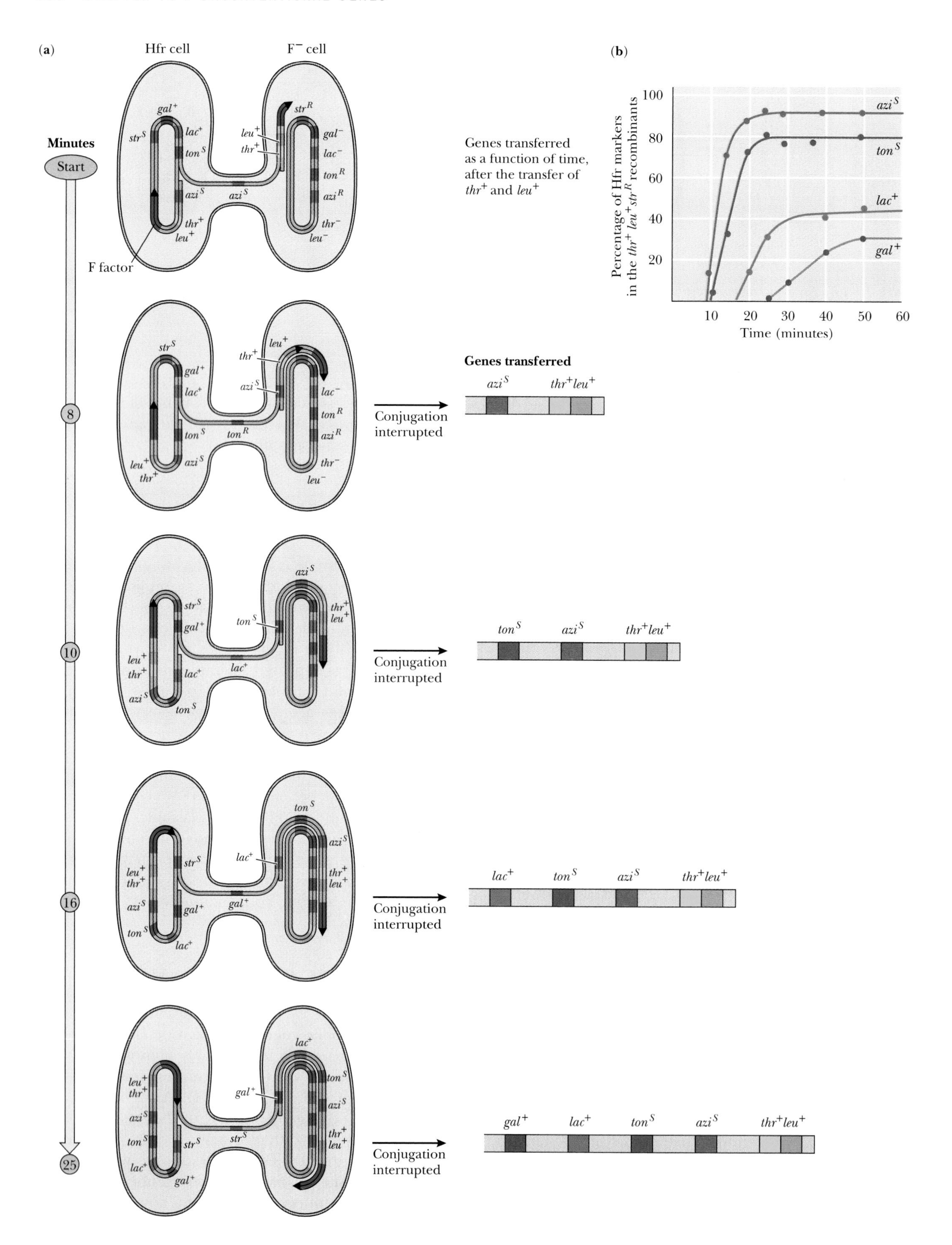

(a)
Hfr cell
F⁻ cell
Minutes
Start
8
10
16
25
F factor
Genes transferred as a function of time, after the transfer of thr⁺ and leu⁺
Genes transferred
Conjugation interrupted
azi^S thr⁺leu⁺
ton^S azi^S thr⁺leu⁺
lac⁺ ton^S azi^S thr⁺leu⁺
gal⁺ lac⁺ ton^S azi^S thr⁺leu⁺
(b)
Percentage of Hfr markers in the thr⁺ leu⁺ str^R recombinants
Time (minutes)
azi^S
ton^S
lac⁺
gal⁺

◀ ***Figure 16-14*** Gene transfer between conjugating *E. coli.* An F-factor has integrated into *E. coli* DNA in the bacterium on the left, converting the bacterium into an Hfr (high frequency of recombination) bacterium. Part of the F-factor DNA leads the way through the pilus, into the recipient F^- cell, pulling a copy of the Hfr DNA with it. By disrupting conjugation at different times, researchers can deduce the order of gene transfer.

Other eukaryotic mobile genes, however, appear to have no autonomous existence but can move from place to place within the genome. The dappled colors of "Indian corn," for example (Figure 16-15), result from the random jumping of a mobile gene from one chromosomal location to another, in much the same way as a bacterial insertion sequence. Insertion of one of these DNA fragments next to certain genes prevents that gene's expression and produces colorless kernels.

Barbara McClintock, a geneticist at the Cold Spring Harbor Laboratory on Long Island, New York, began to study the genetics of corn in the 1940s. McClintock was the first to realize that the unexpected patterns of inheritance resulted from jumping genes. Only 20 years later did biologists begin to realize the importance of McClintock's conclusions and how broadly they applied to mobile genes in other organisms, including bacteria. For this work, McClintock received a Nobel Prize in 1983.

Figure 16-15 Mobile genes are responsible for the dappled coloring of Indian corn. Such "jumping genes" in corn were the subject of important studies by Barbara McClintock. *(Michael Gadomski/Photo Researchers)*

Scientists now realize that both prokaryotes and eukaryotes contain mobile genes. Even mobile genes do not move frequently, however, so geneticists can map them on chromosomes. Given enough time, however, their map locations may change.

Do Mobile Genes Benefit Their Hosts?

Many mobile genes are copies of stationary genes, which stay in their original places, so mobile genes often contribute significantly to the sizes of genomes. Some genes, however, have become so proficient at multiplying themselves that they have taken over as much as 10% of the genome of some eukaryotes!

It costs energy to replicate DNA. We might therefore think that natural selection would lead to organisms with stripped-down genomes, as in viruses. Instead, however, some organisms have huge amounts of DNA. The genomes of some salamanders contain 30 times more DNA than the human genome, which, in turn, contains about 1000 times more DNA than that of *E. coli.*

Some biologists regard mobile genes as the ultimate illustration of a **selfish gene,** a sequence of DNA whose only role is to reproduce itself. A selfish gene may have no effect whatever on the phenotype of the cell or organism in which it lies. This argument suggests that the relationship between mobile genes and their hosts resembles **commensalism,** an association between individuals of two species in which one organism benefits without harming the other one.

Other biologists argue that the relationship between mobile genes and their hosts resembles **parasitism,** an association that benefits only one participant—in this case the mobile gene—and often harms the other. Only by waiting long enough, say the proponents of this view, are we able to see the deleterious effects of such parasitism.

Still other biologists argue that mobile genes have evolved so that they somehow serve to increase the evolutionary success of their hosts. Otherwise, this argument goes, the replication of the mobile genes would consume energy and decrease the reproductive success of their hosts. These biologists argue that the relationship between mobile genes and their hosts resembles **mutualism,** an association between organisms of two species from which both organisms benefit.

Arguments about the costs and benefits of mobile genes have also led to suggestions about the origins of viruses. We can imagine, for example, that animal viruses "captured" cellular genes involved in regulating cell growth and division. These genes evolved into oncogenes, which caused the cells containing tumor virus DNA to outgrow cells that do not contain viral oncogenes.

We do not know how different mobile genes will ultimately fare during future evolution. We can only try to

ESSAY

PRIONS

Kuru, a disease found only among the Fore highlanders of Papua New Guinea, is one of a number of fatal human and animal diseases characterized by spongy holes in the brain. Kuru, which causes a loss of motor coordination and cognitive deficits, appears to have spread by ritual cannibalism; now that the practice has stopped, the disease has disappeared. A more common human disease, with similar losses of coordination and cognition, is Creuzfeldt-Jakob disease. Creuzfeldt-Jakob disease appears in the United States in about 1 person out of a million, although some families have an apparently genetic form with much higher incidence. A still more common disease in sheep, scrapie, leads to a similar lack of coordination and a similar spongy appearance to the brain.

All of these diseases are experimentally transmissible. Injection of brain extracts from affected individuals can bring on each disease in experimental animals. In a few tragic cases, doctors unwittingly transmitted Creuzfeldt-Jakob disease in treating children for dwarfism, because the brains of affected patients were included in the preparation of brain tissue for extracting human growth hormone.

These diseases have all the hallmarks of virally transmitted diseases, except for one point: Treating the infectious brain extracts with ultraviolet light or with nucleases—treatments that would destroy DNA or RNA in a virus—does not alter the extracts' ability to cause disease in experimental animals. Biochemical purification of the infectious agent from brain extracts has led to the surprising conclusion that a small protein, of about 250 amino acid residues, is responsible for transmission. This protein, extensively studied by Stanley Prusiner and his colleagues at the University of California, San Francisco, is called the *prion protein,* or PrP, where *prion* refers to "proteinaceous infectious particles."

Given what we know about viral diseases, most investigators have suspected that kuru, Creuzfeldt-Jakob disease, and scrapie are viral diseases, but all attempts to find the responsible viruses have so far been fruitless. In contrast, Prusiner's conclusion that these diseases depend on infectious proteins has gained increasing support. In the mid-1980s, Prusiner and his collaborators obtained a partial amino acid sequence of PrP and used this information to produce recombinant DNAs that encode Prp, employing techniques that we discuss in Chapter 17. If PrP were a component of a disease-causing virus, then the recombinant DNA should have allowed the identification of that virus. Instead, however, the PrP gene was found as a normal gene in the chromosomes of mice, humans, and every other mammal whose genes have been examined.

Not only do mice and humans have PrP genes, they normally produce PrP in their brains. The PrP in diseased brains, however, was more resistant to protease digestion, suggesting that it had an altered structure. Reseachers found the basis of this altered structure in another prion-related human condition, called Gerstmann-Sträussler-Scheinker (GSS) disease. In GSS patients, the gene encoding PrP has a mutation that leads to an altered PrP sequence, with a leucine replacing a proline at position 102 in the protein sequence.

Prusiner's group has used genetic engineering techniques, described in Chapter 17, to produce transgenic mice that make the mutated PrP that is present in the brains of GSS patients. These mice also develop a disease similar to GSS, with the characteristic spongy appearance of the brain. And extracts from the brains of these mice contain prions—that is, the extracts trigger the disease in healthy mice that contain the mutant PrP gene. So far, however, these extracts have not caused disease in mice that contain only the wild-type gene.

In the last decade researchers have established the involvement of PrP with a set of neurodegenerative diseases in humans, sheep, cows, and mink. The ability of mutated PrP to cause similar diseases in experimentally manipulated mice has suggested that an alteration in PrP structure can cause neurodegeneration. But serious questions remain: Many researchers are still not convinced that mutant PrP can permanently alter the structure of normal PrP in a manner that causes disease.

understand how mobile genes may or may not contribute to the phenotype and survival of their hosts. We do know that most mobile genes in eukaryotes do not interfere with sequences that specify proteins. In this respect, at least, they do not damage their hosts (although in some cases a mobile gene can cause damage, for example, by interrupting a needed functional gene).

In addition to mobile genes whose movements can be studied in the laboratory, the chromosomes of many species contain much DNA that may have derived from genes that were once mobile but no longer are. Some formerly mobile genes appear to benefit their hosts. Centromeres, for example, contain clusters of highly repeated DNA sequences, which almost certainly resulted from the proliferation of mobile genes. These centromere sequences now appear, however, to contribute to the pairing of chromosomes during meiosis. Other DNA sequences are found scattered in many places through the genome, some as many as 100,000 times.

MITOCHONDRIA AND CHLOROPLASTS CONTAIN THEIR OWN DNA

Both chloroplasts and mitochondria contain DNA. Almost every eukaryotic cell therefore carries at least one genetic system independent of the nucleus. Cell fractionation techniques have allowed biologists to isolate mitochondria and chloroplasts and to examine their contents. They found that all mitochondria and chloroplasts contain DNA molecules distinct from those of the nucleus.

In a mammalian cell, each mitochondrion contains 5–10 identical DNA molecules, each about 16,000 nucleotides long. Because a single cell may have hundreds of mitochondria, mitochondrial DNA may comprise up to 1% of the total cell DNA. As shown in Figure 16-16a, each mitochondrial DNA molecule is **circular,** meaning that it forms a continuous loop, like the DNA of bacteria and many viruses.

Plant mitochondria also contain circular DNA molecules, but usually of several different kinds. The total length of mitochondrial DNA in plants is 30–100 times more than that in animals.

Chloroplasts contain DNA as well, as shown in Figure 16-16b. Each species has a single kind of chloroplast DNA. The chloroplast DNA is circular and usually about 150,000 nucleotides long. In higher plants, chloroplast DNA comprises about 15% of the total DNA in a cell.

What Is the Function of the DNA in Mitochondria and Chloroplasts?

The reproduction of mitochondria and chloroplasts depends on cellular machinery specified by genetic plans in chromosomal DNA in the cell nucleus. Just the replication and expression of organelle DNA, for example, requires about 90 proteins, in most cases far more than the organelle's DNA can itself encode. Most of these proteins are specified by the nuclear genome and made on cytoplasmic ribosomes.

The DNA of mitochondria and chloroplasts contains information for many of their own components. None of the products of these genes ever leaves the organelles. For example, both mitochondria and chloroplasts contain an

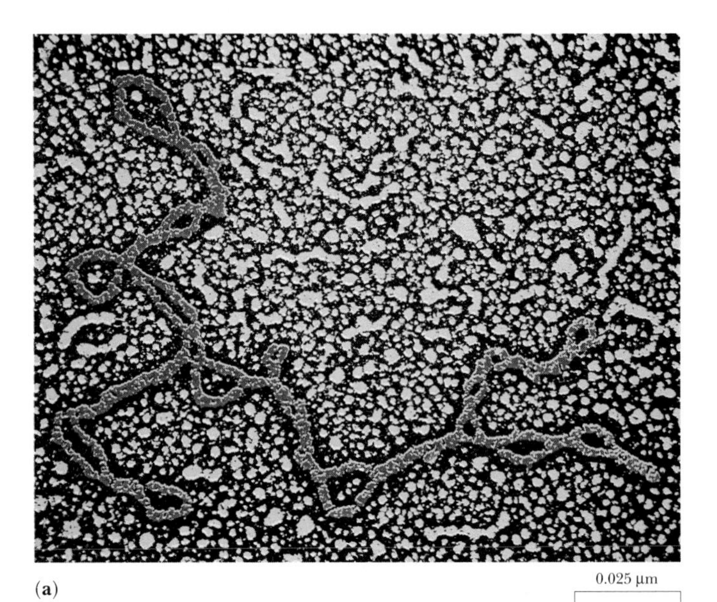

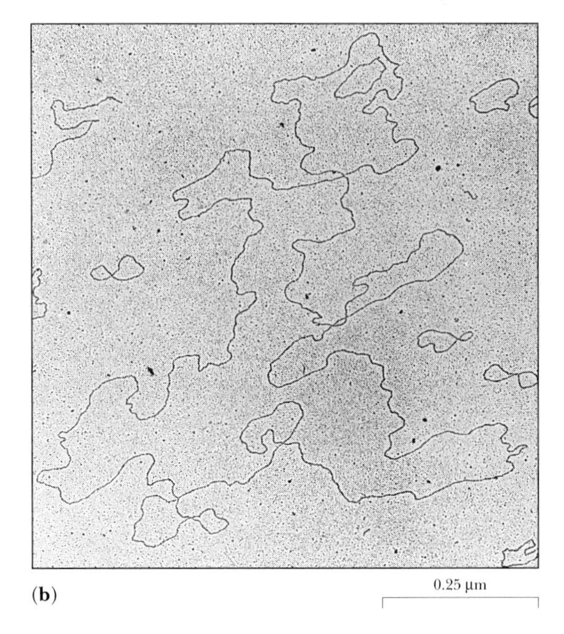

Figure 16-16 Circular DNA in energy organelles: **(a)** in a mitochondrion; **(b)** in a chloroplast. Chloroplast DNA is 10 times longer than that in animal mitochondria. The smaller circular DNAs are from a DNA virus and were added as size markers. *(a, CNRI/Phototake NYC; b, Dr. Richard Rolodner, Dana-Farber Cancer Institute, Boston, MA)*

enzyme complex (F_0-F_1) that harnesses the energy of proton gradients to make ATP. (See Chapters 7 and 8.) Each F_0-F_1 complex contains nine different kinds of polypeptides. The DNA of animal mitochondria contains the information for two of these polypeptides, while the mitochondrial DNA of plants and yeast specify three. Nuclear DNA specifies the sequences of the remaining F_0-F_1 polypeptides: Animal nuclear DNA specifies seven F_0-F_1 polypeptides, while plant nuclear DNA specifies six. In chloroplasts, however, the organelle DNA contains genes for six of the polypeptides. In both mitochondria and chloroplasts, then, the ability to make ATP depends on information in both organellar and nuclear DNA.

In addition to DNA, mitochondria and chloroplasts contain complete systems for transcription and translation, including ribosomes, tRNAs, and an RNA polymerase whose structure is remarkably similar to that of bacteria. As in the case of ATP synthetase, some components of the machinery of protein synthesis are made in the cytoplasm, and other components are made within the organelles.

To find out the identities of the genes in mitochondrial DNA, Fred Sanger first applied his new method for DNA sequencing (described in Chapter 14) to the 16,569 nucleotides of human mitochondrial DNA. These studies, whose results are summarized in Figure 16-17, revealed the presence of 13 protein-coding regions; two of these encode polypeptides of ATP synthetase, three encode polypeptides of cytochrome oxidase, and one encodes a polypeptide of cytochrome b. Mitochondrial DNA also contains genes for two ribosomal RNAs that are similar to, but smaller than, the ribosomal RNAs of bacteria. Mitochondrial DNA also contains genes for 22 kinds of tRNAs. These tRNAs are essential to mitochondrial protein synthesis because no cytoplasmic tRNAs enter the mitochondria.

Mitochondrial protein synthesis gets along with fewer tRNAs than cytoplasmic protein synthesis, which employs 31 different tRNAs. The smaller number of mitochondrial tRNAs is sufficient, however, because they generally allow more wobble than cytoplasmic tRNAs (that is, a single anticodon can form duplexes with codons that vary in the third position, as discussed in Chapter 14, p. 377). The less rigorous base pairing with mRNA codons also means that mitochondrial protein synthesis has a greater error rate than that in the cytoplasm.

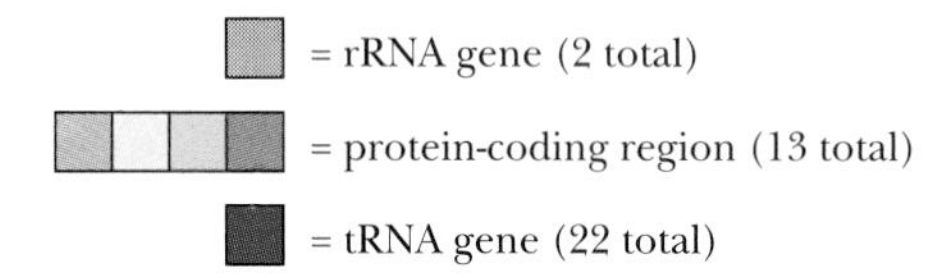

Total length of genome = 16,569 base pairs

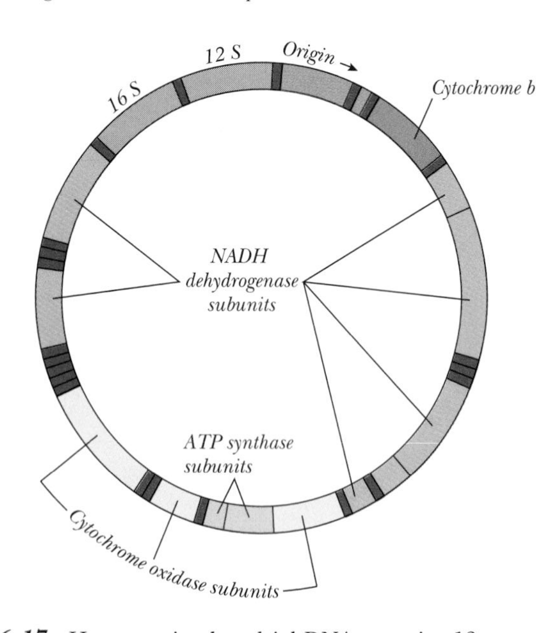

Figure 16-17 Human mitochondrial DNA contains 13 protein-coding regions, as well as genes for two ribosomal RNAs and 22 tRNAs.

Studies of the DNAs and RNAs of Energy Organelles Support Their Derivation from Ancient Prokaryotes

Like prokaryotic DNAs, the DNAs of energy organelles are circular. And the ribosomes of mitochondria and chloroplasts are more like bacterial ribosomes than like the ribosomes of eukaryotic cytoplasm. These and many other similarities between the energy organelles and prokaryotes support the **endosymbiotic theory** for the origin of chloroplasts and mitochondria, which holds that present-day energy organelles are descended from free-living prokaryotes that were engulfed but not digested by early eukaryotes. (See Chapter 24.)

The genetic code used in mitochondrial protein synthesis is slightly different from that used in either prokary-

	Codon	UGA	AUA	CUA	AGA/AGC
	Universal code	STOP	Ile	Leu	Arg
Mitochondrial codes	Mammalian mitochondria	Trp	Met	Leu	STOP
	Drosophila mitochondria	Trp	Met	Leu	Ser
	Yeast mitochondria	Trp	Met	Thr	Arg
	Plant mitochondria	STOP	Ile	Leu	Arg

Figure 16-18 The genetic code in mitochondria differs from the "universal code" used by prokaryotes and by cytoplasmic ribosomes of eukaryotes.

otes or eukaryotic cytoplasm, with four codon assignments that differ from that in the "universal" code. (See Figure 16-18.) For example, in cytoplasmic protein synthesis, the codon UGA is a stop codon, while in the mitochondria of animals and fungi UGA specifies tryptophan.

If mitochondria derived from free-living prokaryotes, their different genetic code suggests an ancient origin, before the establishment of the otherwise universal genetic code. Plant mitochondria seem to have had an independent origin because their DNAs are so much larger than the mitochondrial DNAs of animals and fungi and because, as shown in Figure 16-18, they have distinctive codon assignments. In plant mitochondria, for example, UGA is a stop codon as it is in the cytoplasm, unlike its specification of tryptophan in animal and fungal mitochondria. (See Chapter 24 for more discussion of the origin of mitochondria and chloroplasts.)

SUMMARY

Unconventional genetic systems are DNA (and sometimes RNA) sequences that replicate and function apart from chromosomal DNA. These include mobile genes such as viruses, plasmids, transposable elements, and F-factors, which can move from cell to cell or from place to place within a chromosome, and the DNA of mitochondria and chloroplasts. All of these unconventional genes depend on cellular machinery specified by chromosomal DNA.

A virus is an assembly of nucleic acid and proteins that can reproduce within a living cell. Viruses were first discovered as disease-causing agents that could pass through filters that would remove bacteria. Viruses have many properties of life, but they cannot generate their own energy-transfer molecules and cannot form the building blocks needed for the synthesis of their components. Most biologists therefore do not consider them to be alive.

The electron microscope has revealed the structure of many viruses to be one of two general forms—a long helix, or an icosahedron—a nearly spherical particle. X-ray diffraction studies have revealed the detailed structure of several viruses.

Although viruses are not alive, they still evolve by natural selection. Viruses are able to take over the machinery of their host's cells and produce more viruses. The growth of viruses in laboratory cell cultures has allowed researchers to study their reproductive cycles. These cycles consist of six steps: attachment to the host cell, entry of viral nucleic acid into the cell, synthesis of viral proteins, replication of viral nucleic acid, assembly of new virus particles, and release of new viruses.

Many viruses kill their hosts in the process of replicating themselves. These viruses are called lytic viruses. Sometimes, however, viral genes can integrate into the chromosomal DNA of their hosts. The integrated virus, called a provirus, does not immediately produce more virus. Induction of a provirus begins another lytic cycle, which produces more virus particles. Cells that contain a provirus are said to be lysogenic.

RNA serves as the genetic material in some viruses. In some RNA viruses, called plus-strand viruses, the viral RNA can function as RNA to direct polypeptide synthesis. In minus-strand viruses, an enzyme copies the viral RNA into a complementary strand that functions as mRNA. In retroviruses, such as the viruses that cause AIDS and several types of tumors, the enzyme reverse transcriptase copies viral RNA into DNA, which then integrates into the host cell's chromosomal DNA.

Mobile genes in bacteria include plasmids, which can replicate only autonomously; transposable elements, which can replicate only in integrated form; and F-factors, which—like some viruses—are episomes (that is, they can replicate either autonomously or in integrated form). Some transposable elements, called insertion sequences, appear to carry information only for their own perpetuation, while other transposable elements, called complex transposons, may carry genes for antibiotic resistance. F-factors carry genes responsible for the transfer of genetic information between bacteria.

Eukaryotic cells also have mobile genes, including viruses. For example, some mobile genes, which resemble bacterial insertion sequences, move within the chromosomal DNA of corn and cause the dappled colors of Indian corn.

Biologists disagree about the evolutionary significance of mobile genes. Some say that mobile genes do not affect their hosts at all; others say they reduce the evolutionary fitness of their hosts; and still others say they benefit their hosts.

Energy organelles—chloroplasts and mitochondria—have their own DNA. The circular form of these DNAs and the genes they contain suggest that they derived from ancient prokaryotes that lived within ancient eukaryotic cells.

KEY TERMS

autonomous replication
bacteriophage
capsid
cellular oncogene
cI protein
circular DNA
commensalism
complex transposon
conjugation

cro protein
endosymbiotic theory
episome
F-factor
Hfr
host range
insertion sequence
integrated replication
lambda repressor
lysis
lysogenic cycle
lysogenic virus
lytic virus
membrane envelope
mobile gene
mutualism
oncogene
overlapping gene
parasitism
pilus
plasmid
prophage
proto-oncogene
provirus
recombinant
resistance factor, or R-factor
restriction enzyme
retrovirus
reverse transcriptase
selfish gene
simple transposon
transposable element
transposase
tumor virus
unconventional gene
virus

QUESTIONS FOR REVIEW AND UNDERSTANDING

1. What is meant by the term "unconventional gene"?
2. What is a virus?
3. What are viruses composed of?
4. How would you go about separating bacteria from viruses in the laboratory?
5. Contrast the terms commensalism, parasitism, and mutualism.
6. List four ways viruses have evolved to become very efficient at reproduction.
7. What are the six stages of the viral reproductive cycle?
8. Give an example of how a virus evolved or changes in order to be able to survive and replicate in a given host. Give an example of how a bacteria evolves or changes in order to be able to survive and replicate in a given environment.
9. What is a lytic as compared to a lysogenic virus?
10. Why is it not to a virus's benefit to always destroy its host?
11. What are oncogenes? Where are they found?
12. Why don't all RNA viruses require the enzyme reverse transcriptase in order to replicate?
13. Explain what the three kinds of bacterial mobile genes are.
14. What is required for a genetic element to be able to replicate independently?
15. What is a simple transposon as compared to a complex transposon?
16. What genes are encoded by human mitochondrial DNA?
17. What is the endosymbiotic theory? What evidence supports this theory?
18. What are the potential consequences of conjugation?
19. Why is it a good thing, from a human point of view, that viruses do not readily change their host range?
20. How do mobile genetic elements potentially contribute to the phenotype or survival of the host?

SUGGESTED READINGS

LODISH, H., D. Baltimore, A. Berk, S. L. Zipursky, P. Matsudaira, and J. Darnell, *Molecular Cell Biology*, 3rd edition, Scientific American Books, New York, 1995. Chapter 19 discusses the DNAs of mitochondria and chloroplasts.

STRYER, L., *Biochemistry*, 4th edition, W. H. Freeman and Company, New York, 1995. Chapter 32 deals with gene rearrangements.

WATSON, J. D., M. Gilman, J. Witkowski, and M. Zoller, *Recombinant DNA*, 2nd edition, Scientific American Books, New York, 1992. Chapter 10 deals with movable genes.

CHAPTER 17

Genetic Manipulation and Recombinant DNA

KEY CONCEPTS

- Genetic manipulation depends on choosing organisms with desirable genetic properties from a population that contains natural or induced variation.
- Recombinant DNA techniques allow the production of relatively large amounts of pure genes and gene products.
- Recombinant DNAs can reprogram both prokaryotes and eukaryotes to have new properties.

OUTLINE

ESSAYS

Genetic manipulation began some 10,000 years ago, when farmers learned how to select varieties of grain that could provide a more stable source of food. Their success allowed the peoples of the Middle East, northwestern China, and southern Mexico to live in larger groups and to develop villages and cities. At the same time animal breeders succeeded in selecting animals for such desirable qualities as the production of wool and milk. By 5000 years ago, the peoples of the Middle East had mastered enough microbiology to ferment their grain to produce beer and to convert their milk into cheese and yogurt.

The genetic manipulation of plants and animals depended primarily on careful selection of which individuals to breed. Once breeders had chosen desirable varieties of plants or animals, the trick was to preserve their genetic traits. They did this by inbreeding: They allowed individuals with the desired qualities to breed only with each other. The Bible provides an early example—Jacob kept his spotted goats apart from the nonspotted goats of his father-in-law. The manipulation of microorganisms also depended on selection. Brewers, for example, learned to choose culture conditions that allowed the growth of microorganisms that would produce alcohol, but not acetic acid; this choice allowed them to produce beer and not vinegar.

Plants and microorganisms can be propagated from a single parent to give genetically identical offspring. Many indoor gardeners, for example, propagate geranium plants from cuttings of a single favorite, as shown in Figure 17-1. All the descendant plants form a clone (genetically identical offspring) of the original plant. (See Chapter 15, p. 405.) In contrast, a group of plants grown from most seeds is not a clone, because (as Mendel showed) the seeds are not genetically identical to either parent: Sexual reproduction means (usually) that each descendant inherits only half its genes from each of its parents.

Figure 17-1 A clone of geranium plants. *(Lefever/Grushow/Grant Heilman)*

GENETIC MANIPULATION DEPENDS HEAVILY ON CHOOSING CLONES WITH DESIRABLE PROPERTIES

Beer brewing, cheesemaking, and agriculture itself are early examples of **biotechnology,** the use of living organisms for practical purposes. Modern biotechnology depends heavily on microorganisms that are chosen for their desirable genetic qualities. A particularly important example is the production of penicillin.

Recall (from the Essay, "Antibiotics and Protein Synthesis," in Chapter 14) that Alexander Fleming discovered penicillin as the result of a chance contamination of one of his bacterial cultures. Fleming isolated the mold responsible for the death of his bacteria and grew the mold in pure cultures. This mold produced enough penicillin to prevent bacterial growth in his cultures, but not enough to use for medical purposes. Fleming himself seems never to have considered any use for penicillin except to improve the culturing of bacteria in the laboratory.

In 1938, 9 years after Fleming's discovery, two Oxford University scientists—Ernst Chain, a German biochemist, and his professor, Howard Florey, an Australian pathologist—realized that penicillin might be able to play a role in controlling diseases caused by bacteria. Within 2 years, Florey and Chain announced that penicillin, even in very small amounts, could protect mice from an otherwise fatal injection of bacteria. But in order to obtain even a microgram of penicillin, Florey's team had to process thousands of liters of mold culture. To make penicillin in the amounts needed to control a human disease, scientists had to develop ways of improving the yield. They did this both by finding better growth conditions and by choosing strains of mold that could produce more penicillin from each gram of mold.

Genetic Manipulation Makes Use of Both Natural and Induced Variation

Choosing strains (or clones) with desired properties from a large collection is called **screening.** To find better penicillin producers, microbiologists screened soil samples from all over the world. The most promising one came from the U.S. Department of Agriculture Laboratory in Peoria, Illinois.

Microbiologists in government laboratories, universities, and pharmaceutical companies then subjected the Peoria strain to mutagens—x-rays, ultraviolet light, and chemicals. After each round of mutagenesis, they chose the strains with the highest yield of penicillin. The result of this work was the isolation of a strain that produces nearly 100 times the amount of penicillin of the original Peoria strain. Together with improvements in the conditions of culture, such genetic manipulations allowed ever-increasing production of this important drug, in amounts that could help large numbers of people. Some historians have even said that the availability of penicillin to Allied troops during World War II may ultimately have decided the outcome of the war.

Knowing Specific Biochemical Pathways Can Allow More Efficient Screening of Useful Organisms

To find a strain of mold that could produce more penicillin, Florey and Chain depended on the same general approach that breeders have always used, originally choosing from the best wild strains they could find. Researchers then screened for still more robust producers after treatment with mutagens.

Researchers did not know in advance which mutations would increase the yield of penicillin. In other cases, however, geneticists can seek a mutant strain with a well-defined biochemical characteristic. An example of such a directed search is the use of special strains to produce the amino acid lysine.

Lysine is an essential amino acid for animals, meaning that animals (including humans) cannot make it themselves and must obtain it from properly balanced diets. Grains, including wheat and corn, are generally low in lysine, so the feed given to farm animals usually contains a lysine supplement. Feed makers are therefore especially eager to have an inexpensive source of lysine—the cheapest now being from genetically manipulated bacteria.

The bacteria used for lysine production have a specific enzyme deficiency that leads to the accumulation of lysine, as shown in Figure 17-2. The missing enzyme, called homoserine dehydrogenase, converts a precursor of lysine called aspartate semialdehyde (AS) into homoserine (HS), which can then be converted only into threonine, methionine, and isoleucine. Bacteria that cannot make HS therefore accumulate more lysine because AS is not diverted to HS synthesis but contributes to lysine production. Such a mutation frees the bacteria from normal constraints that prevent the accumulation of more than needed amounts of a single amino acid. The task, then, was to find a mutant that lacked the ability to produce HS.

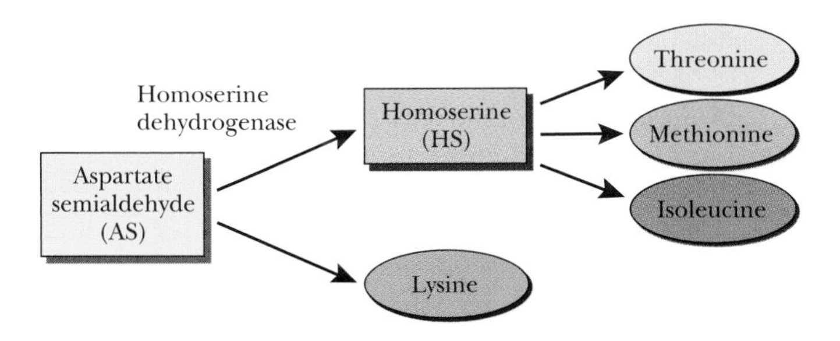

Figure 17-2 Bacteria produce lysine from aspartate semialdehyde (AS). They also produce threonine, methionine, and isoleucine from AS, via an intermediate called homoserine (HS). Homoserine dehydrogenase catalyzes the formation of HS from AS. In mutant bacteria that lack this enzyme, all of the AS can be converted to lysine.

Microbiologists Can Easily Test the Abilities of Individual Mutants to Grow Under Different Culture Conditions

To find desirable mutants, microbiologists often use a chemical mutagen (or x-rays) to induce mutations in a bacterial culture. They then dilute the suspension and spread the bacteria on a glass dish containing a growth medium. If the suspension is dilute enough, the bacteria grow into well-separated colonies, each of which is a clone of a single ancestor, as shown in Figure 17-3a. To keep the descendants of individual bacteria separate from one another, experimenters grow the bacteria on a semisolid surface made of agar (a seaweed extract), usually placed within a covered round glass or plastic dish called a Petri dish or, more commonly, a "plate."

The microbiologist's task is to find the colonies with particular genetic properties; in the case of bacteria to be used for lysine production, for example, the goal was to find bacteria that could not produce HS. In order to find such colonies, researchers use a technique, illustrated in Figure 17-3b, called **replica plating,** in which a part of each colony on a single bacterial plate is transferred, in place, to other plates that contain a different medium (for example, a medium that contains HS). Because the bacterial clones grow in the same positions in the two replicas, microbiologists can easily screen for mutants that grow as desired in the different media. For example, geneticists might screen for mutants that grow in the presence of HS but not in its absence; these mutant bacteria would be found on plates containing HS but not on replicas that do not contain HS (Figure 17-3b). (Ordinarily, bacteria are grown on a "rich" medium that contains plenty of all amino acids; selection for HS dependence uses a minimal medium, which requires the bacteria to make their own amino acids.) Researchers could then test individual clones of these mutants to find the one that produces the most lysine. This mutant could then be grown in larger fermenters in the industrial production of lysine.

Industrial microbiologists have used specific knowledge of biochemical pathways to select hundreds of commercially important strains of bacteria and fungi. Some of the molecules produced by these strains include amino acids such as glutamic acid, whose sodium salt is the widely used flavor enhancer monosodium glutamate (MSG); vitamins such as riboflavin; and even industrial chemicals such as ethanol and acetone. Microorganisms also produce commercially important enzymes, such as those used

(a)

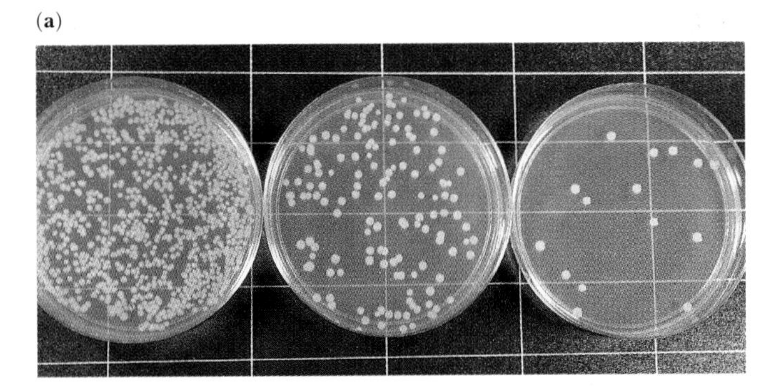

(b)

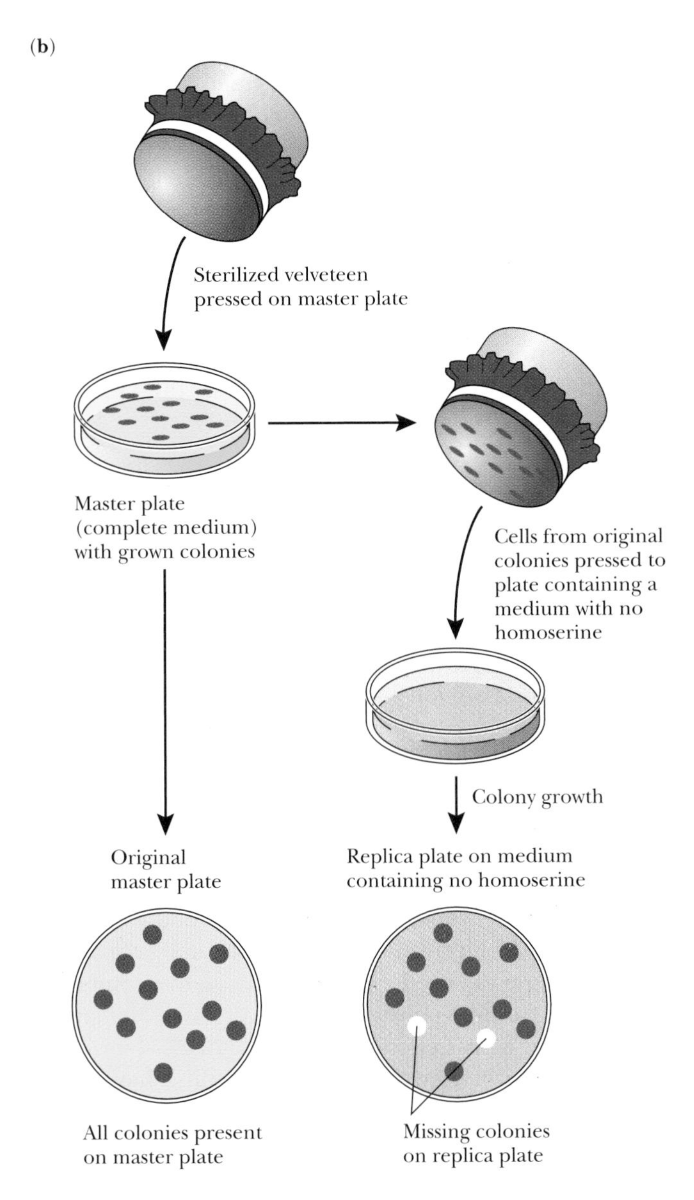

Figure 17-3 Microbiologists usually grow bacteria in glass or plastic plates, called Petri dishes, which contain a jelly-like substance called agar, a product of sea weed. The use of agar came from a suggestion made by the American-born wife of a 19th-century German bacteriologist; as a girl, she had watched her mother use agar to make fruit jellies. **(a)** Serial dilutions of a bacterial stock; each separated colony on a plate represents a clone, the descendants of a single cell; the plate on the left has so many colonies that they are not easily distinguished from one another; the plate in the middle has 1/10 as many colonies as that on the left; and the plate on the right has 1/100 as many colonies. **(b)** Replica plating allows researchers to study the growth of individual colonies (clones) under different conditions, here in the presence and absence of homoserine. *(a, Cytographics, Inc./Visuals Unlimited)*

to convert starch into fructose for use as a sweetener or the protein-digesting enzymes used in laundry detergents.

RECOMBINANT DNA TECHNIQUES ALLOW THE PRODUCTION OF GENETIC VARIANTS THAT DO NOT OCCUR IN NATURE

The realization in the early 1950s that genes are nothing more than linear sequences of nucleotides in DNA led directly to our present ability to manipulate individual genes essentially at will. By the early 1970s, molecular biologists had learned to cut and paste pieces of DNA from different sources, and by the early 1980s they had learned how to change selected individual nucleotides. Studies of viruses and plasmids had already established the means for propagating small pieces of DNA in bacteria and other host cells, and it was not difficult to modify these molecular parasites (that is, the viruses and plasmids) to carry along and reproduce pieces of DNA that were engineered in a test tube.

Recombinant DNAs are DNA molecules consisting of two or more DNA segments that are not found together in nature. Molecular biologists can now produce recombinant DNAs that fulfill particular requirements—scientific, medical, or commercial. Genes manipulated in this way have facetiously been called "designer genes."

Recombinant DNAs provide scientists with large amounts of material with which to study the structure, regulation, and function of individual genes and to unravel the molecular basis of genetic diseases. These genes also provide genetic engineers with a powerful means of reprogramming microorganisms (and even plants and animals) to make them better producers of needed products.

The first commercially important product made by recombinant DNA techniques was **insulin,** a polypeptide hormone that regulates the metabolism of carbohydrates and fats. (A **hormone** is a chemical made by one organ and secreted into the blood, which carries it to another organ, where it influences function; see Chapter 19.)

Insulin deficiency leads to **diabetes mellitus** [Greek, *diabetes,* syphon, running through (referring to the large amount of urine) + *mellitus,* sweet (referring to the presence of glucose in the urine)], a disease in which people cannot properly absorb glucose from the blood. Many people suffering from diabetes mellitus must take injections of insulin every day. Until recently almost all of this insulin came from the pancreases of animals. Pig insulin has a slightly different structure than human insulin, however, and some diabetics cannot tolerate this difference With the isolation of DNA sequences for human insulin, it has become possible to produce human insulin in bacteria. This human insulin is already widely used to treat diabetics.

The development of recombinant DNA technology depended on the coming together of many different lines of research, including work on (1) antiviral defenses of bacteria, (2) DNA replication and repair, (3) replication of viruses and plasmids, and (4) the chemical synthesis of specific nucleotide sequences. We now look at how information and techniques from these disparate fields all contributed to the ability to make recombinant DNAs.

Bacterial Enzymes Important for Antiviral Protection Cut DNA Molecules at Specific Sequences

Recall (from Chapter 3) that the nucleotides of DNA and RNA are linked together by phosphodiester bonds. Each such bond consists of a phosphate group that connects to the 3′-carbon atom of one nucleotide and to the 5′-carbon atom of the next nucleotide in the polynucleotide chain. A **nuclease** is an enzyme that breaks a phosphodiester linkage in DNA or RNA. A nuclease that specifically acts on DNA is called a **DNase,** and one that acts specifically on RNA is called an **RNase.** Nucleases are common enzymes, and most of them break a polynucleotide essentially at random.

A randomly acting DNase (such as that found in the human gut) usually cuts DNA into short fragments, as shown diagrammatically in Figure 17-4. By limiting the time that the nuclease can act, an experimenter may obtain larger fragments of DNA. Even in a collection of originally identical molecules, however, such as the DNA from a single bacterial clone, the fragments of digested DNA are extremely heterogeneous: They are cut in different places, have different sizes, and overlap with one another. There is little hope of finding a specific sequence in such a mess.

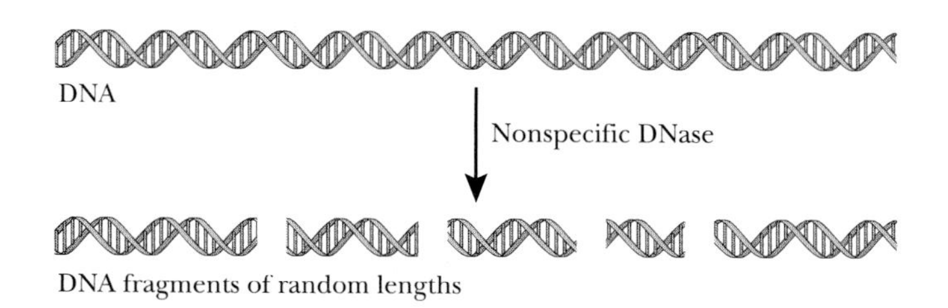

Figure 17-4 Digestion of DNA by DNase produces DNA fragments of random lengths.

The first requirement for the isolation of a specific piece of DNA is the ability to cut DNA at specific sites, using nucleases that act only at certain sequences.

Restriction enzymes are DNases that act as such specialized scissors—cutting DNA only at particular sequences. A **restriction site,** the sequence recognized by a particular restriction enzyme, typically consists of four or six nucleotides, although some restriction sites are much larger. After a restriction enzyme binds to its restriction site in DNA, it hydrolyzes a single phosphodiester bond in each strand of the double helix. In some cases, the cut is in the same place in each strand, so that the resulting fragment of DNA has "blunt" ends, as illustrated in Figure 17-5a. Other restriction enzymes make staggered cuts, which result in "sticky" ends, as shown in Figure 17-5b. That is, the protruding nucleotides resulting from a staggered cut can form base pairs with a complementary sticky end from another fragment cut with the same enzyme. DNA segments with the same type of sticky ends are particularly easy to stick together, making them especially useful in the production of recombinant DNAs.

Many kinds of bacteria produce restriction enzymes, and researchers have already prepared restriction enzymes from over 200 different bacterial strains. Each of these restriction enzymes recognizes a specific nucleotide sequence. More than 90 restriction sites have now been described.

What Good Are Restriction Enzymes to the Bacteria That Produce Them?

Restriction enzymes help protect bacteria from takeover by foreign genes. For example, when most bacteriophages infect *E. coli* strain RY13, a restriction enzyme rapidly degrades the phage's DNA, thereby saving the cell. Researchers have isolated this restriction enzyme, called *EcoRI* [after the name of the bacterial strain], and have studied its action in the test tube. It binds to DNA wherever it finds the sequence of six nucleotides—GAATTC, and it cuts the phosphodiester bond between G and A (Figure 17-5b).

Although a restriction enzyme can protect *E. coli* (or another bacterial strain) from infection by a bacteriophage, such an enzyme would be self-destructive if it also digested *E. coli*'s own DNA. To prevent such molecular suicide, *E. coli* modifies its DNA by adding a methyl group to each target site of the restriction enzyme. An enzyme called *EcoRI* methylase adds a methyl group (CH_3) to an A residue of the GAATTC restriction site whenever this sequence appears in the bacterial DNA. The methyl group protects the bacterial DNA from digestion with *EcoRI,* as illustrated in Figure 17-5c. In contrast, the newly entering viral DNA is not methylated and is cut by *EcoRI* wherever GAATTC appears.

A bacteriophage that can infect such a protected strain must be able to circumvent its defense. Some kinds

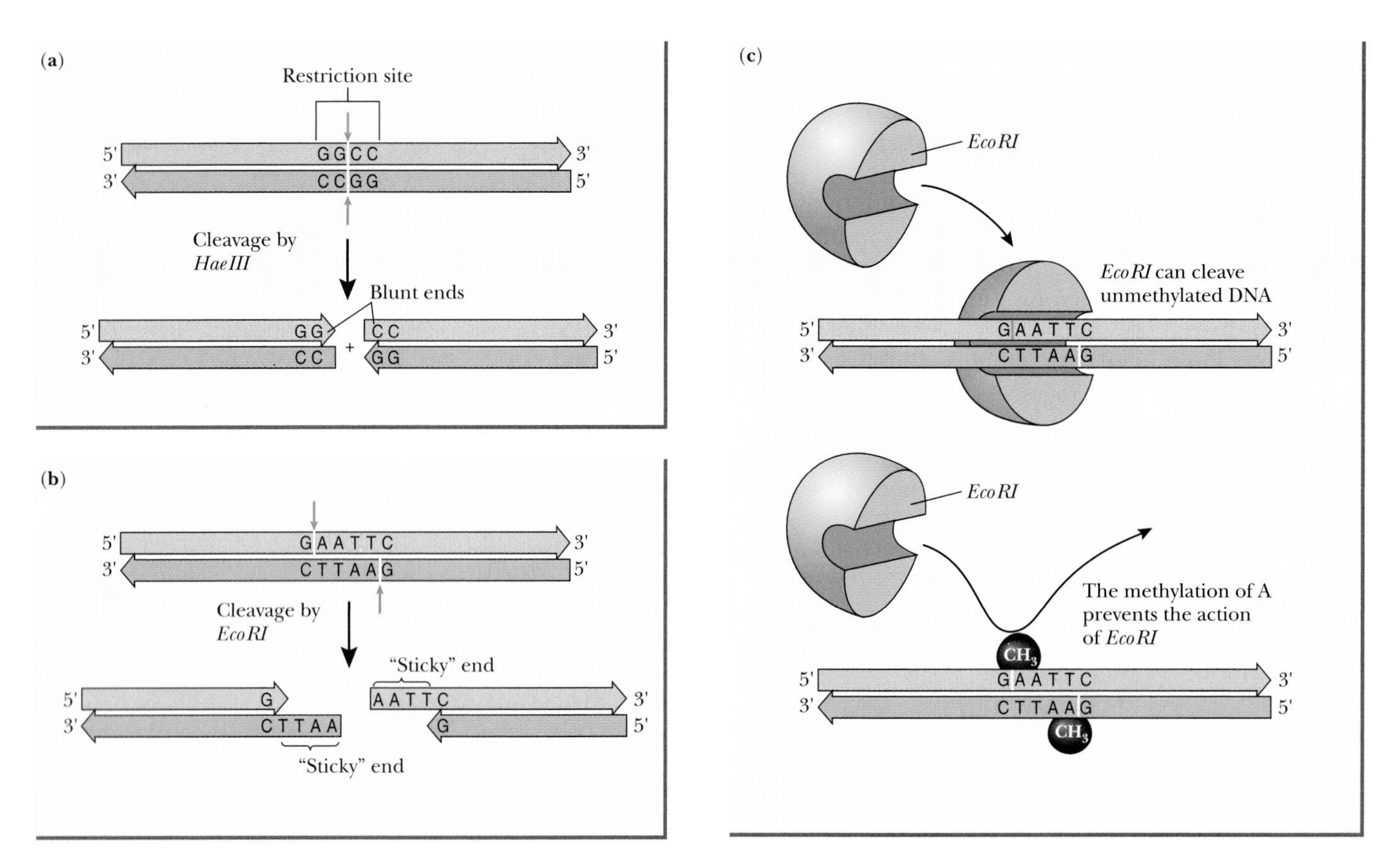

Figure 17-5 Restriction enzymes are specialized molecular scissors that cut DNA at specific sequences. **(a)** Some restriction enzymes, like *HaeIII,* produce blunt ends; **(b)** other restriction enzymes, like *EcoRI,* produce "sticky" ends by cutting at different places in the two strands; **(c)** methylation of the *EcoRI* restriction site prevents the restriction enzyme from cutting the DNA.

of bacteriophages avoid DNA degradation by having DNA that completely lacks restriction sites for the host strain's restriction enzyme. The phage T7, for example, does not contain a single recognition sequence for *EcoRI* among its more than 40,000 nucleotides, and T7 can therefore infect and kill *E. coli* strain RY13. Natural selection has evidently allowed the evolution of a virus that can escape the surveillance system of its host.

Restriction Enzymes Allow the Preparation of Defined DNA Fragments

A restriction enzyme like *EcoRI* can cut DNA every place a particular sequence appears. The result of this digestion is a set of **restriction fragments,** segments of DNA that begin and end with a restriction site. After cutting, each end contains only a partial restriction site, as illustrated in Figure 17-5b.

Electrophoresis in a gel made of agarose or polyacrylamide can separate the resulting restriction fragments according to size, as shown in Figure 17-6a, with larger fragments moving more slowly than smaller fragments. Under each set of conditions, each fragment moves at a rate determined only by its size, as long as the DNA remains as a double helix. By comparing the sizes of restriction fragments from incomplete and complete restriction digestions, one can establish a **restriction map,** a record of the distribution of restriction sites within a piece of DNA. This information allows researchers to cut out and isolate particular DNA fragments. The first restriction map, similar to that shown in Figure 17-6b, was determined in 1971 for the circular tumor virus, SV40. In some cases (when the mixture of restriction fragments is not too complex and the fragments have distinctive sizes), electrophoresis permits the straightforward isolation of individual restriction fragments, which can then be sequenced or combined with pieces of DNA from other sources to make recombinant DNAs.

DNA Ligase Can Link Together Two Pieces of DNA from Different Sources to Produce Recombinant DNA

During DNA replication, DNA ligase links together the discontinuous pieces of the lagging strand within the replication fork. (See Chapter 13, p. 345.) The production of

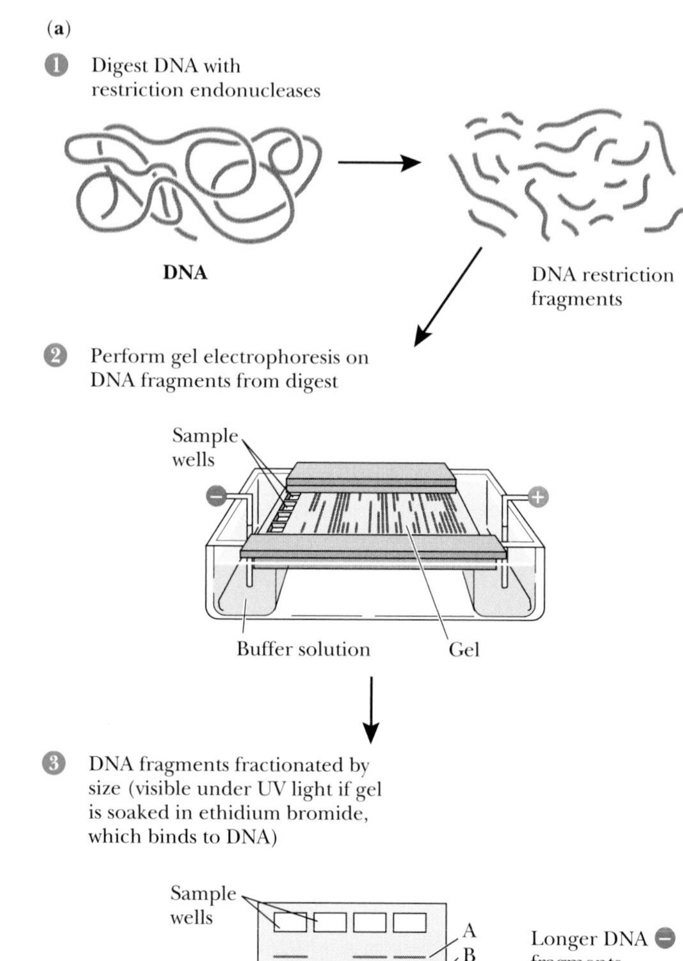

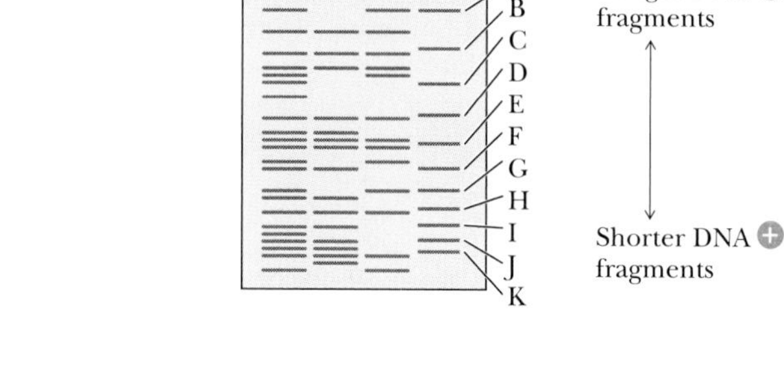

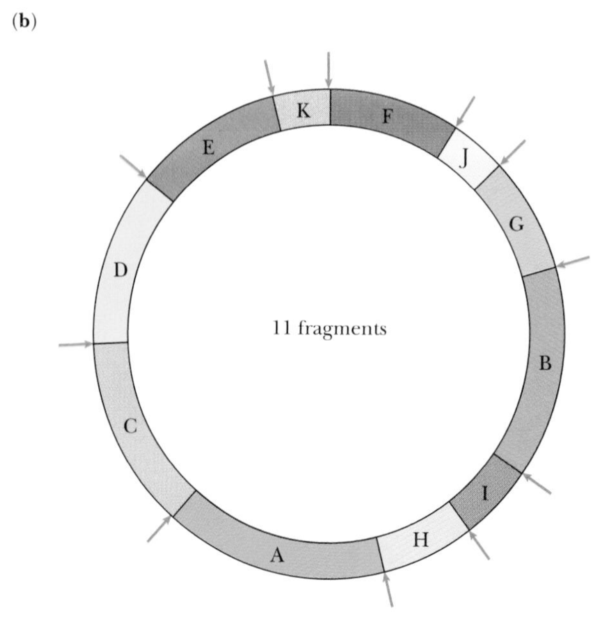

Figure 17-6 Electrophoresis separates DNA restriction fragments according to size. **(a)** Technique for analyzing the sizes of DNA fragments; **(b)** the restriction map of the circular DNA of a DNA tumor virus.

recombinant DNAs has depended largely on the ability to use DNA ligase to link restriction fragments, as shown in Figure 17-7a.

Figure 17-7b shows an example of a recombinant DNA produced by DNA ligase. Researchers put a sequence coding for insulin next to the operator and promoter for β-galactosidase. The recombinant DNA contains a piece of a bacterial regulatory gene (the operator and promoter of the *lac* operon) and of a mammalian structural gene (the coding sequence for insulin). (See Chapter 15.) This recombinant DNA, when put back into *E. coli,* produces insulin, not β-galactosidase, when the bacteria are induced with lactose. It programs the bacteria to produce a mammalian hormone in response to a signal that would ordinarily prepare the bacteria for growth on lactose.

DNA Sequences from Viruses and Plasmids Allow the Propagation of Recombinant DNAs in Bacteria and Other Hosts

Once scientists could produce recombinant DNAs using DNA ligase and isolated restriction fragments, the next goal was to produce large quantities of these recombinant DNAs. In order to achieve this goal, researchers had to find appropriate carriers for propagating foreign genes. Such a carrier, called a **vector,** contains DNA sequences that allow the passenger DNA to replicate within an appropriate host cell.

Recall (from Chapter 16) that R (resistance) factors are naturally occurring plasmids containing genes for antibiotic resistance. By the early 1970s, researchers had isolated and studied many such R-factors. They then used restriction enzymes and DNA ligase to make streamlined plasmid vectors. At a minimum, each plasmid vector contains (1) an origin of replication, the DNA sequence needed to start DNA synthesis, and (2) one or more genes for antibiotic resistance, which allow researchers to distinguish bacteria with the plasmid from those without a plasmid.

With an R-factor as a vector for foreign DNA, bacteria can propagate essentially any DNA sequence. An antibiotic resistance gene, such as amp^R, within the plasmid protects the host bacteria against being killed by antibiotic, for example, ampicillin. Growing bacteria in ampicillin then selects those bacteria that contain a plasmid. The antibiotic kills those bacteria that lack a plasmid.

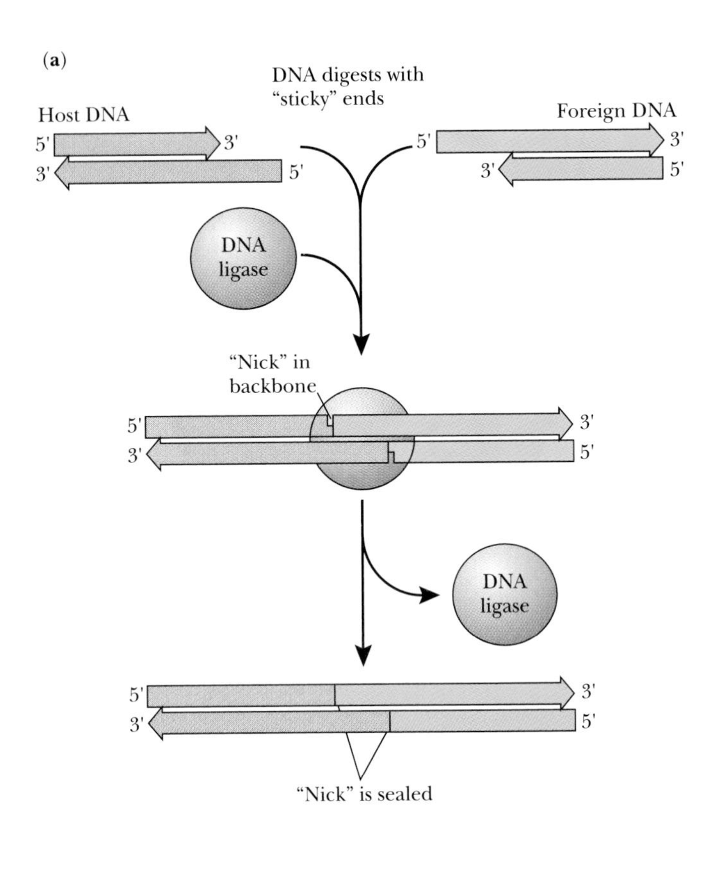

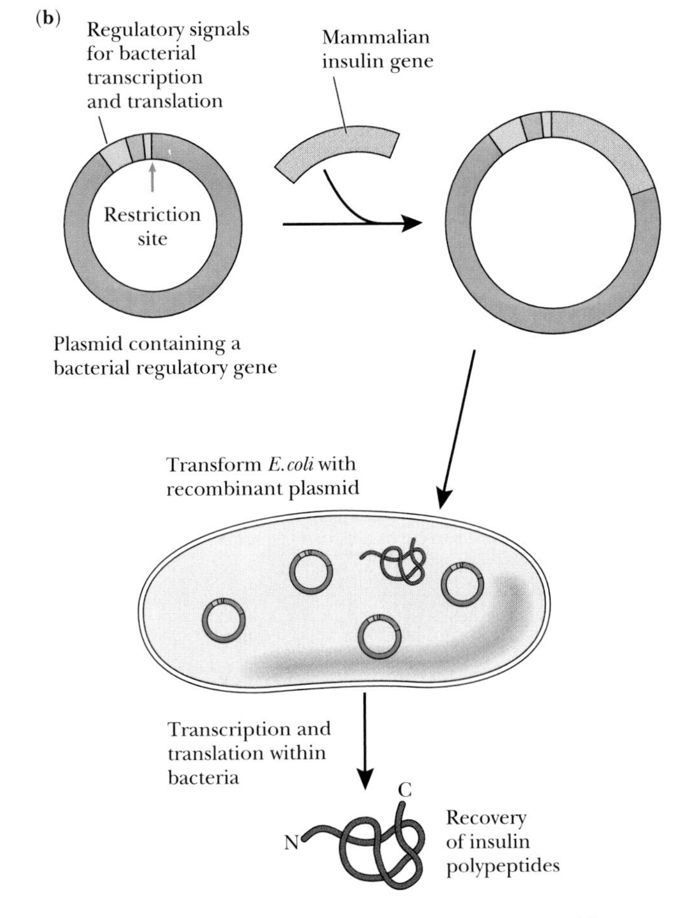

Figure 17-7 Recombinant DNA. **(a)** Use of DNA ligase to join DNAs from different sources, here at sticky ends; **(b)** production of a bacterial plasmid containing a gene encoding mammalian insulin; insulin DNA is placed into a recombinant plasmid containing a bacterial regulatory gene (such as the *lac* promoter and operator); induction with lactose will induce the transcription of insulin DNA into an mRNA that is translated into an insulin polypeptide by bacterial ribosomes.

How Do Experimenters Distinguish Between Recombinant and Nonrecombinant Vectors?

Researchers have engineered many plasmid vectors to allow the rapid distinction between plasmids with passenger DNA and those without such additional sequences. For example, in the widely used plasmid vector diagrammed in Figure 17-8a, each plasmid contains the β-galactosidase gene from *E. coli* (in addition to an origin of replication and a gene for antibiotic resistance). Within the β-galactosidase gene is a restriction site (or a set of restriction sites) into which a DNA segment can be inserted. To use the plasmid as a vector, the researcher inserts the passenger (foreign) DNA into the middle of the β-galactosidase gene, thereby preventing the production of active β-galactosidase. Experimenters can then use a specific assay to distinguish between nonrecombinants (which make β-galactosidase and produce a blue product from a compound called X-gal) and recombinants (which remain white), as illustrated in Figure 17-8b. Even with continual advances in the engineering of recombinant DNAs, some nonrecombinant plasmids manage to persist, so the separation of recombinants from nonrecombinants is sometimes an important step.

Clonal Growth of Host Cells Allows the "Cloning" of Recombinant DNAs

The major problem in characterizing and manipulating genes is that each gene is present as such a small fraction of the total DNA in a cell. The human insulin gene, for example, contains about 0.00005% of total human DNA, and each human cell contains approximately 3×10^{-18} g of insulin DNA. Even if one extracted all the DNA from an entire human (containing 10^{14} cells) and could separate the insulin gene from all the other DNA, one could still obtain only 300 μg of insulin DNA. This amount is less than a researcher can now obtain from a small laboratory culture, for example, in a 100-ml flask. Even more important than the low total yield that would be obtained if one started with nonrecombinant DNA (for example, from an entire human) is the near impossibility of using

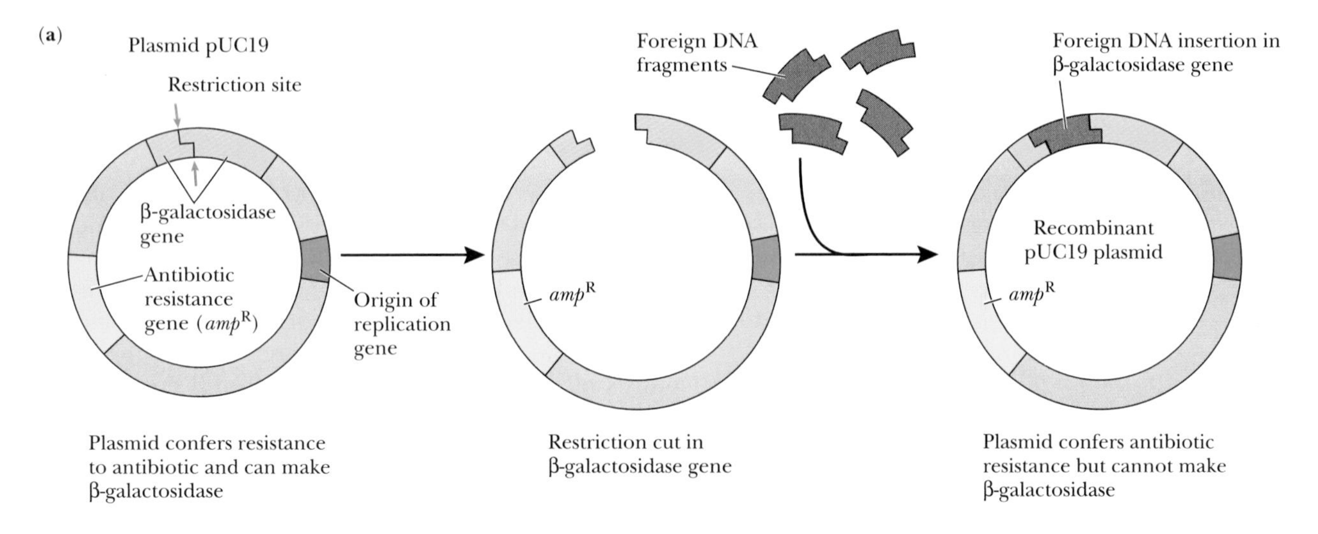

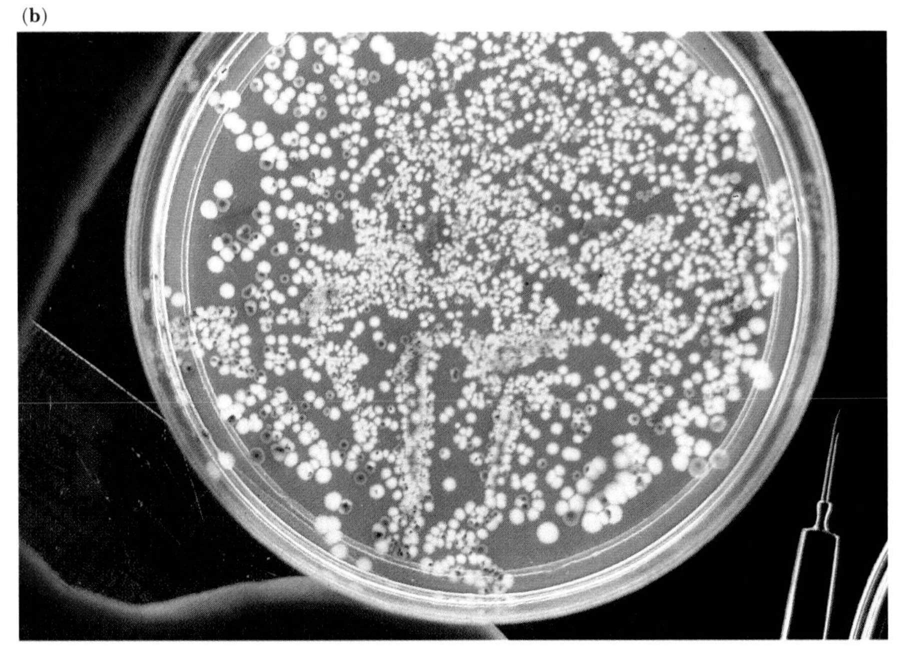

Figure 17-8 Distinguishing recombinant and nonrecombinant DNAs. **(a)** The plasmid pUC19 contains a bacterial origin of replication (which allows the plasmid to replicate within a bacterial host), a gene that conveys ampicillin resistance (amp^R), and a gene for β-galactosidase; if foreign DNA is put into a site within the β-galactosidase gene, the plasmid cannot direct the synthesis of the enzyme. **(b)** In a bacterial strain that lacks its own β-galactosidase, plasmids without any interrupting foreign DNA will make enzyme and produce a blue color from the special substrate, X-gal; plasmids with interrupting foreign DNA will not produce active enzyme, and the colonies will be white rather than blue. *(b, Michael Gabridge/Visuals Unlimited)*

chemical or physical techniques to isolate pure individual genes. Until recently, the only practical way to isolate a particular DNA sequence was to propagate single recombinant DNA molecules in clones of some host cell.

The polymerase chain reaction (PCR) now also allows the preparation of large amounts of individual DNA sequences by using DNA polymerase to amplify DNA spanned by a pair of specific oligonucleotide primers. (See Chapter 13, p. 348.) The lengths of the amplified DNA have generally been limited to 1000 or 2000 nucleotide pairs (1–2 kb), although new methods allow amplification of specific DNA segments up to 25–30 kb. Cloning methods, however, allow the propagation of DNA fragments that are more than 1 million nucleotide pairs long.

To clone a piece of DNA, the experimenter must ensure that each bacterial cell contains no more than one recombinant molecule (passenger plus vector) and that each host cell develops into a physically separated clone. The growth of each bacterial colony then results in the amplification of the single type of recombinant DNA. Because each bacterial colony is a clone and each founder bacterium has only a single plasmid, the plasmids isolated from a whole colony are also a clone.

Each bacterial clone may grow into 10^{12} or more bacteria, each of which carries exactly the same recombinant plasmid. Under some growth conditions, each bacterial cell can contain a thousand or more copies of this plasmid, so such a culture may contain 10^{15} copies, the number that could theoretically be isolated from all the cells of 5–10 humans. To obtain the cloned recombinant DNA, the experimenter's task is far more straightforward: He or she harvests the host bacteria, usually in a centrifuge, and purifies the plasmids.

In addition to plasmid vectors, researchers have used

a variety of other vectors. These include specially modified bacteriophages, animal and plant viruses, and transposable elements. In recent years, researchers have also developed vectors that can propagate huge pieces of DNA (more than 1 million nucleotides long). These vectors are called **yeast artificial chromosomes** (or **YACs**), because they function as tiny chromosomes—replicating and undergoing mitotic movements—in yeast cells, which serve as their hosts. Each YAC has a yeast origin of replication, as well as DNA sequences that serve as a centromere and as the two chromosomal ends (telomeres).

What Is the Best Source of DNA for Making Recombinant DNAs?

The most natural starting material for making recombinant DNA would at first appear to be the genomic DNA of an organism or a virus. For the genes of viruses and microorganisms, genomic DNA is the usual source, but genomic DNAs from plants and animals pose a special problem—the interruption of protein-coding sequences by introns. Because the goal is often to express a eukaryotic gene in bacteria, genomic DNA is often useless because bacteria are unable to remove the introns after transcription and so cannot make a translatable mRNA. In addition, the presence of introns may increase the size of the DNA by a factor of 10 or more, making it even more difficult to propagate the desired recombinant DNA.

A widely used alternative, then, is DNA that has been copied from mRNA (because mRNAs do not contain introns). To prepare such a copy, molecular biologists take advantage of the enzyme **reverse transcriptase** (RNA-dependent DNA polymerase), which copies RNA into DNA and which is present in RNA tumor viruses. (See Chapter 16, p. 433.) Reverse transcriptase copies almost any RNA into DNA as long as an appropriate primer is available. Because almost all eukaryotic mRNAs have a stretch of A residues at their 3′-ends, a chemically synthesized oligonucleotide that consists of a stretch of T residues ("oligo T" or "oligo dT" to show that the oligonucleotide consists of deoxyribonucleotides) can form complementary base pairs and serve as a primer for reverse transcriptase, as shown in Figure 17-9. Alternatively, researchers can use a primer that is specific for a particular RNA sequence or a diverse, random mixture of primers which primes every possible RNA sequence.

The result of the reverse transcriptase reaction is a single-stranded DNA molecule, called **complementary DNA** or **cDNA,** which is complementary to the mRNA from which it was copied. Another enzymatic reaction copies this single-stranded cDNA into double-stranded cDNA, the second strand of which corresponds to the sequence in the original mRNA, as shown in Figure 17-9.

The next step is to link the double-stranded cDNA to vector DNA. Because many vectors (including plasmids and some viruses) are circular DNA molecules, the experimenter must first open the circle by cutting with an appropriate restriction enzyme, as illustrated in Figure 17-10. The most common technique for joining the cDNA

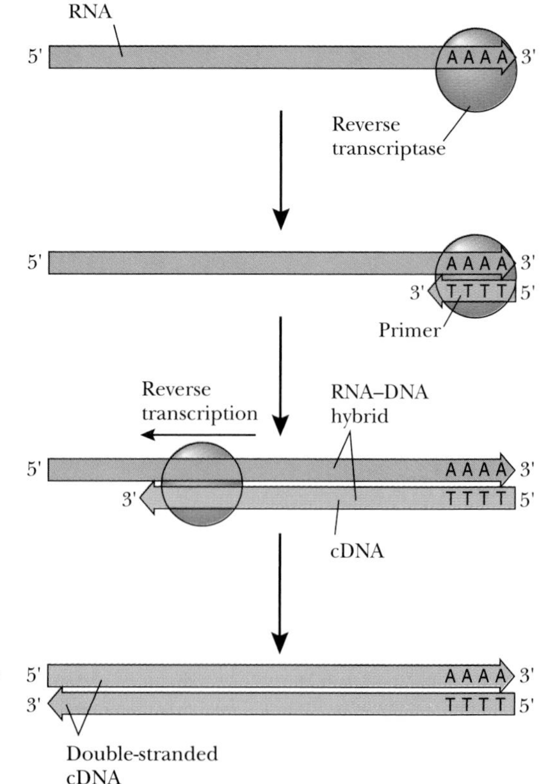

1. Reverse transcriptase copies RNA into DNA. A chemically synthesized oligonucleotide that consists of a stretch of T residues ("oligo T" or "oligo dT") can form complementary base pairs and serve as a primer for reverse transcriptase.

2. The result of reverse transcriptase reaction is a single-stranded DNA molecule, called complementary DNA or cDNA, complementary to the mRNA from which it was copied. The RNA template and the cDNA form a hybrid.

3. Another enzymatic reaction copies this single-stranded cDNA into double-stranded cDNA. The second strand corresponds to that of the original mRNA.

Figure 17-9 Reverse transcriptase (RNA-dependent DNA polymerase) can copy any RNA molecule into DNA. Like other DNA polymerases, reverse transcriptase works by extending a primer at its 3′-end.

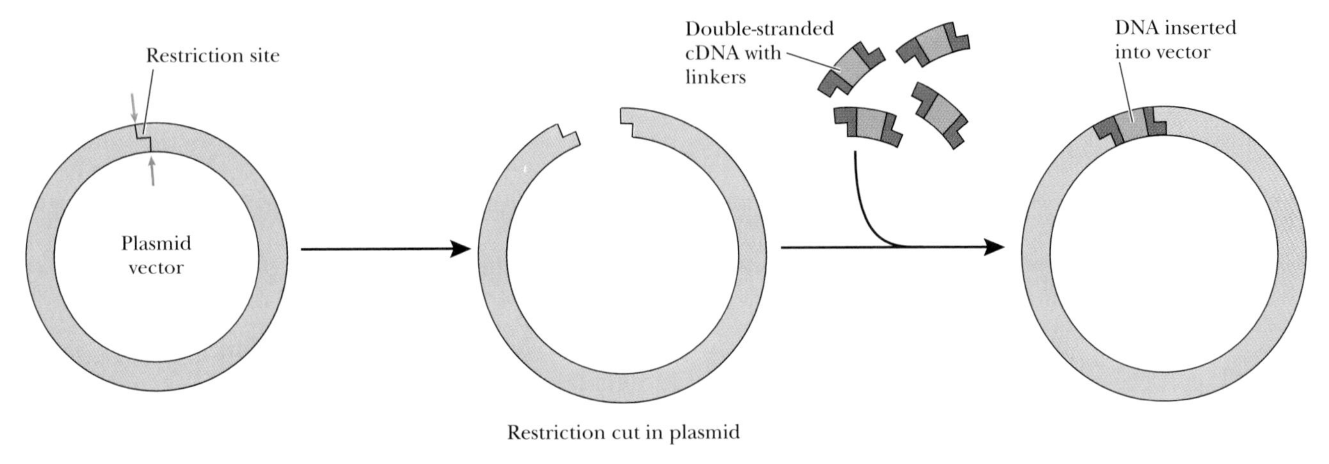

Figure 17-10 Ligation of double-stranded cDNAs, joined to linkers with sticky ends, into the restriction site of a plasmid vector.

ESSAY

RECOMBINANT DNA TECHNOLOGY POSES NEW MEDICAL AND ETHICAL QUESTIONS

Recombinant DNA techniques have already had an important impact on the diagnosis and treatment of human disease. For example, diagnostic laboratories can now detect the presence of disease-causing mutations, such as those responsible for sickle cell disease, muscular dystrophy, Huntington's disease, and cystic fibrosis. Clear diagnoses are important generally because they allow physicians to distinguish similar disorders that require different treatments. And pharmaceutical companies are already using recombinant DNA technology to produce insulin, growth hormone, and other products that are or may be useful therapies for human and animal diseases. Finally, the ability to reprogram cells to produce large amounts of particular proteins is allowing pharmaceutical companies to devise new screening methods. These techniques can allow a more rapid determination of the potential therapeutic effectiveness of newly synthesized chemicals.

In the face of the incredible power of recombinant DNA technology, many people worry that future applications may include the genetic engineering of humans themselves. For example, as we learn more about genetic diseases, some researchers may well try to eliminate diseases by replacing defective genes even before conception. Such a genetic cure has been achieved in mice for a neurological disease characterized by abnormal brain development.

Totalitarian societies have often designated people with certain characteristics as "undesirable" and have often embraced arguments that these characteristics are genetic. If biologists provide the capacity to perform gene replacement in humans, will genetic engineers also try to modify genes that affect characteristics other than those responsible for disease? Will future societies attempt to produce more brown-eyed (or blue-eyed) citizens, or more (or less) intelligent ones, or less (or more) aggressive ones?

While many aspects of physical appearance and behavior are strongly genetic, it seems unlikely in principle that gene replacement could ever produce the kind of uniform society imagined by retrograde totalitarians. Nonetheless, all the people in a society need to discuss the issues raised by the future possibilities of genetic engineering. As in the case of any new advance (such as nuclear weapons and nuclear power), the use of technology is a concern for the society as a whole, not just for scientists.

We do not yet know all the limitations of the new technology. Society must carefully evaluate the risks, benefits, and moral implications that arise from the ability to produce new or previously unavailable products, especially engineered plants and animals.

and the cut vector is to attach the double-stranded cDNA to a chemically synthesized DNA **linker,** a sequence of nucleotides (usually about 15–50) which can then be ligated into a cut restriction site. A widely used method uses a linker with an *EcoRI* restriction site. After treatment with *EcoRI,* the linker converts each end of the cDNA into a "sticky end" that can specifically bind to the complementary sticky ends of the cut vector.

HOW CAN A DESIRED DNA SEQUENCE BE FOUND IN A COLLECTION OF RECOMBINANT DNAs?

The methods we have described can produce thousands, even millions, of clones, each of which contains a different piece of foreign DNA. A collection of clones—called a **gene library**—may contain at least one representative of every sequence in a genome (making it "a complete genomic library") or of every type of mRNA in a particular tissue ("a complete cDNA library"). The problem is to isolate a piece that codes for a particular polypeptide, a task similar to finding a particular book in a university library. But while a university library may occupy a large building, a gene library can fit into a small test tube.

The problem is that there is no card catalog for a gene library: All the recombinant bacteriophages or plasmids are usually jumbled together in no particular order. Finding the desired sequence, then, is really more like finding a needle in a haystack than a book in a library. Recently, however, some molecular geneticists have begun to use robots to make ordered arrays of individual clones, as illustrated in Figure 17-11, keeping "card catalogs" of tens of thousands of clones in computers.

Just as a magnet can help find a needle even in a haystack, molecular biologists often use two kinds of "magnets" to find particular pieces of DNA. The first of these depends on probes for specific DNA sequences and the second on the use of antibodies to detect proteins specified by the recombinant DNAs.

Synthetic Nucleotide Sequences Can Serve as Molecular Probes

The general strategy for finding a specific nucleotide sequence is to use relatively short pieces of single-stranded DNA, called **probes,** usually with a custom-made sequence that forms Watson-Crick duplexes with only one recombinant DNA in a gene library. These duplexes are **hybrid** DNAs; their complementary strands come from different sources. **Molecular hybridization,** the technique of detecting the presence of a nucleotide sequence with a complementary sequence, allows investigators to find (and then to propagate) desired clones. The first problem, however, is to obtain the single-stranded "probe" with the desired sequence.

Automated chemical methods now can routinely produce short segments of nucleic acids (called **oligonucleotides** [Latin, *oligo* = a few]) with any specified sequence up to 60 residues long. (See Chapter 13, p. 348.) But how does one choose the right oligonucleotide?

Suppose we want to find a recombinant DNA that specifies a particular polypeptide, say insulin, whose amino acid sequence we already know. (We actually do not have to know the entire sequence; 5–10 amino acid residues will often do.) The Genetic Code tells us all the possible nucleotide sequences that encode a particular amino acid sequence. Consider, for example, the sequence of the pentapeptide (a peptide five amino acid residues long) shown in Figure 17-12, taken from the sequence of human insulin. Because each amino acid requires a three-

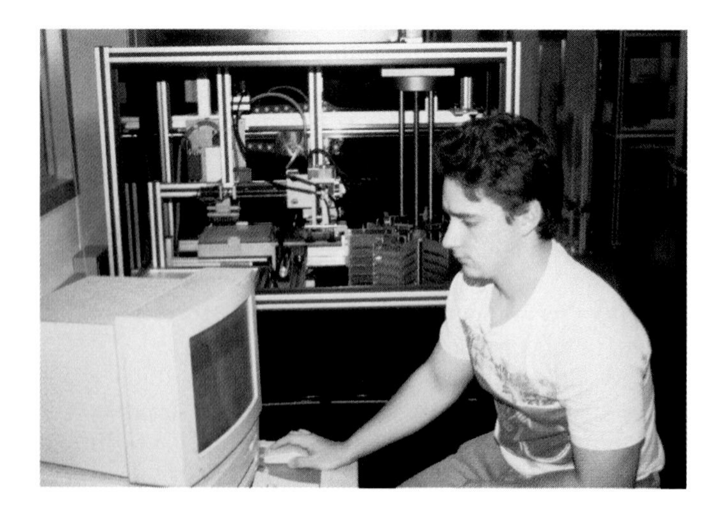

Figure 17-11 Researchers are increasingly relying on computer-controlled robots to produce many replicas of ordered arrays of individual clones. *(Glen Evans, University of Texas, Dallas)*

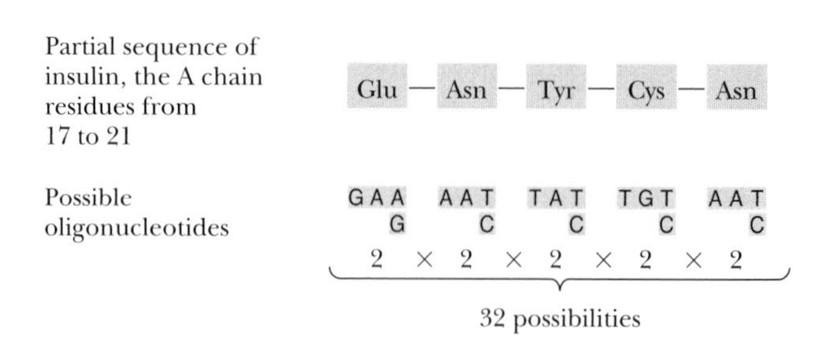

Figure 17-12 Producing an oligonucleotide that will recognize a DNA sequence that encodes a five-residue segment of insulin, Glu-Asn-Tyr-Cys-Asn. Each of these five amino acids can be specified by two different codons, so there are 32 possible coding sequences. Researchers sometimes use a mixture of all 32 possibilities; sometimes they make a best guess at the sequence, realizing that even if they are partly wrong, the oligonucleotide is still likely to hybridize to the actual coding sequence.

nucleotide codon, the DNA segment encoding this pentapeptide must be 15 nucleotides long.

How many possible 15-nucleotide sequences are there? Because each residue has four possibilities, there are $4^{15} = 10^9$ total possibilities. But only a small number of these could actually encode the pentapeptide we want. After looking up these five amino acids in a Genetic Code dictionary (such as that on p. 362), we learn that each of these amino acids could be specified by two different codons. Remember, some amino acids can be specified by as many as six codons, while others have just one codon. So the total number of possible DNA sequences that encode our particular pentapeptide is $2 \times 2 \times 2 \times 2 \times 2 = 32$. We cannot know in advance which one of these 32 nucleotide sequences is actually present in human DNA, so we need to decide which of the 32 to look for.

Molecular biologists have adopted two different strategies for making molecular probes to search libraries for specific coding sequences. One strategy uses a mixture of all possible coding sequences. A mixture of all 32 possible sequences is certain to contain the same sequence as that in one of the clones of the DNA library, although only one of the 32 oligonucleotides will be completely correct.

The second strategy involves making a single oligonucleotide "guess-mer," representing a guess of the most likely nucleotide sequence. The guess-mer strategy depends on two experimental findings: (1) In any species, one codon for a particular amino acid is used more frequently than other equivalent codons for the same amino acid, and (2) oligonucleotides with a few mismatched nucleotides can still form hybrids with the complementary DNAs within the library, allowing the identification of the sought-after recombinant DNA even when the probe and resulting hybrid are not perfect.

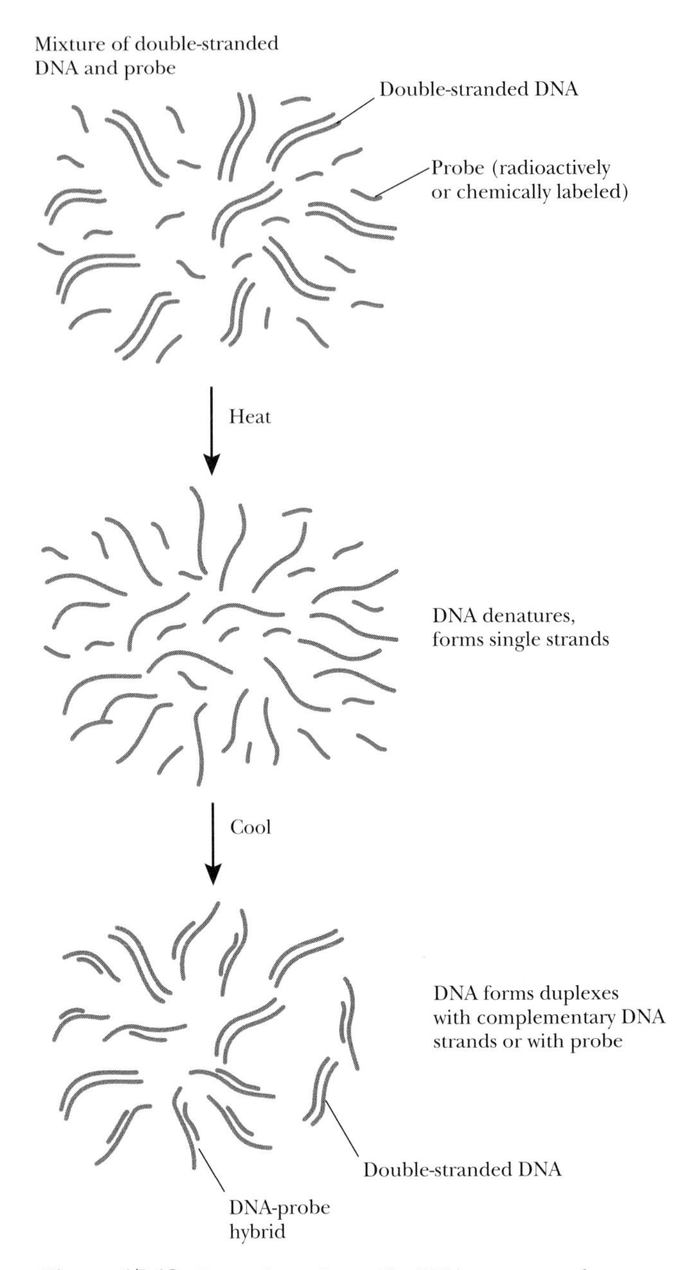

Figure 17-13 Detection of specific DNA sequences by molecular hybridization with a molecular probe. The probe may be a chemically synthesized oligonucleotide or a specific PCR product, labeled with radioactive atoms or with a chemical tag. The probe is allowed to form Watson-Crick duplexes with a mixture of DNAs, and the duplexes are detected by autoradiography or with an antibody.

The Polymerase Chain Reaction Provides Powerful New Approaches to Screening Recombinant DNA Libraries

PCR allows the rapid production of hybridization probes for screening recombinant DNA libraries. These probes may be as long as 1000–2000 nucleotides, far longer (and therefore far more specific) than the oligonucleotide probes (usually 20–50 nucleotides long) described earlier in this chapter. For example, an experimenter may use two oligonucleotides, corresponding to amino acid sequences from different parts of a polypeptide, as primers for a PCR on a cDNA template. This PCR product can then be used as a molecular probe to screen a recombinant DNA library, or it can itself be ligated to a vector and propagated further as a recombinant DNA.

PCR can also be used to verify the presence of a particular known sequence in a recombinant DNA clone. In the case of arrayed libraries, PCR allows the rapid screening of recombinant clones for a specific sequence.

Molecular Hybridization Allows the Detection of Specific Sequences in Recombinant DNA Clones, in Cell Extracts, and in Cells

Figure 17-13 diagrammatically shows how researchers detect hybrids between molecular probes and recombinant DNA clones. A recombinant DNA that encodes the pentapeptide of Figure 17-12 can form a hybrid with the proper oligonucleotide probe. The hybrid can be a perfect hybrid with the exact coding sequence or an imperfect hybrid, with some mismatches that do not form Watson-Crick base pairs, with other closely related probes. To form these hybrids, the experimenter generally mixes the double-stranded DNA with the single-stranded oligonucleotide probe and heats the mixture until the strands of the double-stranded DNA come apart, as in Figure 17-13. Upon cooling, base pairs can again form, both with the original complementary strand and with the added probe (which is generally present at higher concentration than the second strand of the original DNA). The synthetic oligonucleotide or one strand from the denatured PCR product, the original DNA, can then form hydrogen bonds with the complementary strand of insulin DNA.

How Can One Detect Hybrid Molecules in Recombinant DNA Clones?

The molecular probes used to detect hybrids are usually labeled, either with radioactive nucleotides (which can be detected photographically) or with a nucleotide analogue that can be detected immunologically. In either case, the trick is to distinguish hybridized from unhybridized probe. Experimenters commonly use two general approaches: (1) keeping the hybrids attached to some solid support while allowing the unhybridized probe to be rinsed away; and (2) digesting the unhybridized probe with a nuclease that is specific for single-stranded molecules and does not digest hybrids. The immobilized hybrids are then located (by photographic or immunological methods) or, alternatively, are measured in solution.

Researchers use the first of these approaches (immobilization of hybrids on a solid support) to detect specific DNA sequences in individual bacterial clones. Suppose we want to screen a library of recombinant plasmids for one that corresponds to the insulin pentapeptide of Figure 17-12. We can grow bacteria so that each dish has several thousand colonies, with each colony containing a single kind of plasmid. Our task is to find the DNA sequence that encodes the insulin pentapeptide within the thousands of bacterial colonies. We can immobilize a replica of every clone on each plate by a technique almost identical to replica plating. (See Figure 17-3b.) By lightly touching a paper-like sheet of nylon or nitrocellulose onto the bacteria-containing plate, we can pick up a bit of each colony while preserving the arrangement of colonies on the original plate. We can now survey the array of sampled colonies with a labeled hybridization probe to find the ones that correspond to the insulin pentapeptide. Then we can go back to the culture dish to find the original colony.

But how do we release the recombinant DNA from each clone so that it can hybridize to the probe? The technique is surprisingly simple—treating the samples of the arrayed bacterial colonies, blotted onto nitrocellulose or nylon, with high pH. This treatment releases the DNA from the bacteria, separates the DNA strands, and causes the separated strands to stick to the sheet. The sheet is then exposed to the labeled probe, allowed to form hybrids, and washed to remove unhybridized probe. The label is associated with only a few spots, each of which corresponds to a particular clone on the original plate. We can now go back to the original plate, pick the hybridizing clones, and isolate the recombinant DNAs that hybridized.

Molecular Hybridization Can Also Detect Specific Nucleotide Sequences in Cell Extracts and Intact Cells

The general technique for DNA transfer (usually referred to as "blotting") and hybridization was developed by Edward Southern, a molecular biologist working in Scotland. Southern first used this technique to identify specific DNA restriction fragments after electrophoresis, as shown in Figure 17-14a. Molecular biologists now routinely use this technique (called **Southern blotting** or **DNA blotting**).

Other experimenters modified the original technique to detect RNAs with specific nucleotide sequences after electrophoresis. Most molecular biologists call this technique **Northern blotting** (to emphasize its close relationship but distinctness from Southern blotting), although **RNA blotting** is its more serious name.

Both Southern and Northern blotting identify hybridizing nucleic acid molecules in cellular extracts. A further modification of molecular hybridization, called ***in situ* hybridization**, can detect specific nucleotide sequences within intact cells—in chromosomal DNA (as shown in Figure 17-14b) or in cellular mRNA.

Antibodies Can Detect Specific Proteins in Cells, Cell Extracts, and Cells Containing Recombinant DNA Clones

Recall (from Chapters 3 and 5) that antibodies are blood proteins whose surfaces recognize the shapes of foreign molecules, thereby giving animals an important defense against infection. (See Chapter 22 for more about immunological defenses.) When an experimental animal is injected with a foreign protein, it responds by making specific antibodies. Antibodies that specifically recognize molecules present in individual cellular components are valuable tools for cell biologists, allowing them to detect (by immunohistochemistry) the distribution of specific molecules among different types of cells and even within an individual cell. (See Chapter 5.)

Antibodies also provide the means of identifying, within host cells, the products specified by a recombinant DNA molecule. To achieve this detection, the recombinant DNA must contain signals that allow the host cells to read the passenger DNA's genetic information as if it were the

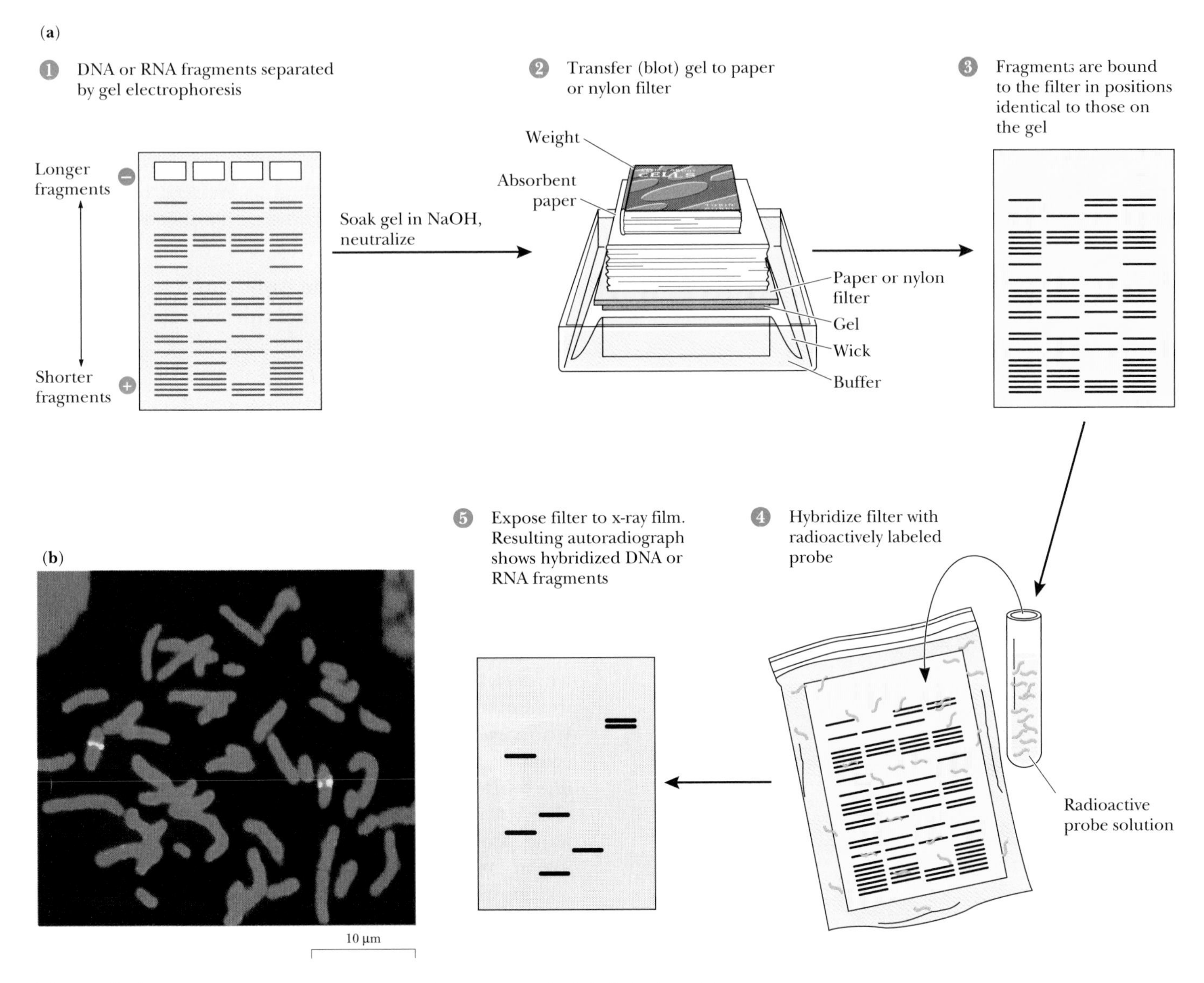

Figure 17-14 Detection of specific nucleotide sequences by hybridization: **(a)** to a blot of DNA fragments (in a Southern blot) or of RNAs (in a Northern blot) separated by electrophoresis; **(b)** to a chromosome spread, detecting the hybridization with a fluorescently labeled antibody to a chemically modified nucleotide. *(b, Dorothy Warburton, Ph.D./Phototake)*

host's own. The production of a polypeptide encoded by passenger DNA is feasible because hosts and passenger DNA use the same genetic code to specify protein sequence.

Some difficulty arises, however, because many genes of interest are from eukaryotes, while the most convenient hosts for cloning are bacteria. A major problem is that a prokaryotic host (unlike a eukaryotic cell) cannot remove introns from an mRNA precursor. To obtain expression of a eukaryotic gene in bacteria, molecular biologists therefore usually start with a coding sequence, free of introns. Usually, as discussed previously, this comes from cDNA rather than genomic DNA.

An **expression vector** is a vector that allows the expression, within the host cell, of the polypeptide encoded in the passenger DNA. Bacterial expression vectors use prokaryotic regulatory signals, as shown in Figure 17-15, (such as those of the *lac* operon) so that the passenger gene's expression is regulated exactly as if it were, for example, the gene for β-galactosidase.

Researchers can use antibodies to screen an "expression library"—a library of cDNAs in an expression vector. To do such screening, the experimenter exposes replicas of bacterial colonies (on nylon or paper sheets) to antibodies and then detects the bound antibodies with a second antibody (or another indicator protein), as shown in Figure 17-16a. The methods for locating bound antibodies are essentially identical to those used to detect the

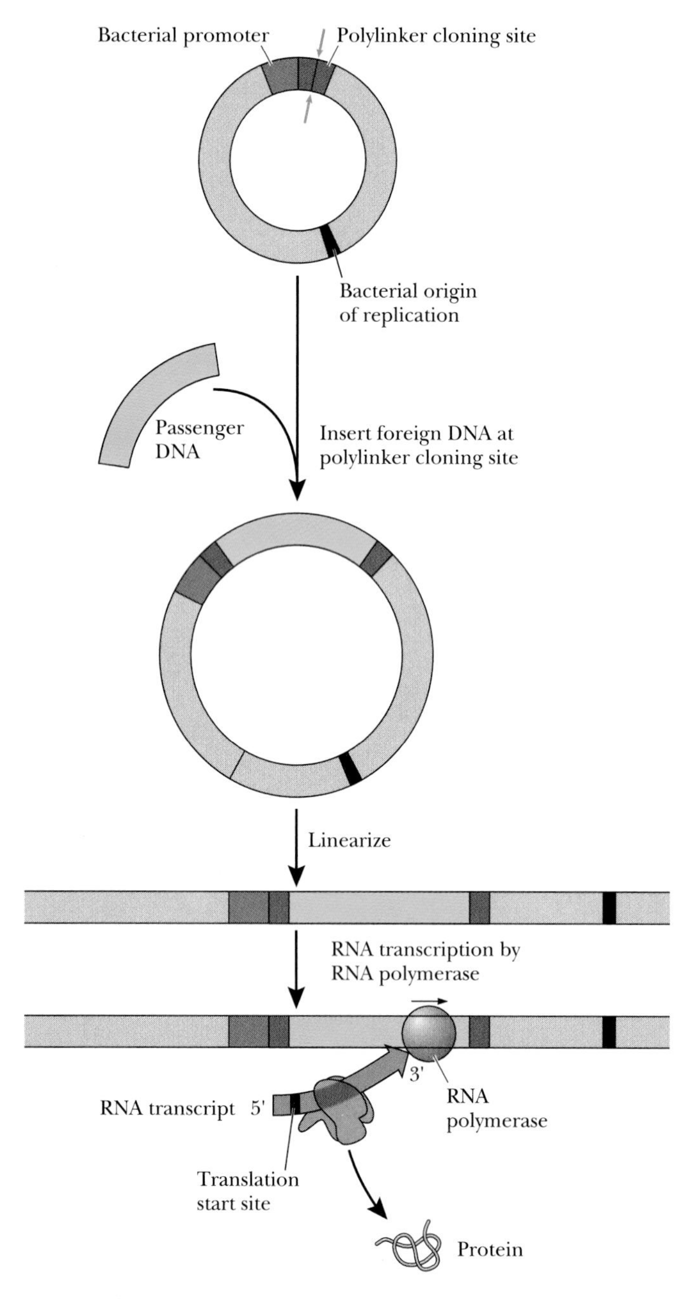

Figure 17-15 Bacterial expression vectors contain signals for bacterial DNA replication, bacterial transcription, and bacterial translation.

cellular sites of antibody binding by immunohistochemistry, as described in Chapter 5.

Eukaryotic expression vectors often use regulatory signals from viruses. One of the most useful eukaryotic expression vectors is derived from Baculovirus, a virus that infects insect cells. Expression vectors are useful not only for finding specific protein-coding sequences in recombinant DNA libraries, but also for programming host cells to make large amounts of proteins that are difficult to obtain in pure form by other means. (More about this later in this chapter.)

Antibodies may also be used to detect specific proteins in cell extracts, as illustrated in Figure 17-16b. The strategy here is the same as in the detection of specific nucleotide sequences in DNA and RNA blots. In fact, almost everyone refers to this method as **Western blotting,** further extending the play on Edward Southern's name.

Functional Assays Test the Ability of a Recombinant DNA to Direct Specific Protein Synthesis

Molecular hybridization can detect the presence of specific DNA sequences in a recombinant DNA library, and antibodies can demonstrate the production of specific proteins. In many cases, however, the ultimate goal of recombinant DNA technology is the expression of a functionally active protein. Usually, a protein made from an expression vector is present at much higher levels than the same protein made *in vivo.* The higher levels of the protein allow the preparation of large amounts of proteins, either for experimental study or for commercial use.

Usually, functional expression requires that the entire protein-coding sequence be present. Functional expression in a bacterial host also demands that the protein of interest not require post-translational modification (for example, cleavage of a signal sequence or glycosylation) in order to function. Many soluble enzymes are functionally active after synthesis in a bacterial host, but most membrane proteins require modifications that can be made only in eukaryotic hosts.

Despite these difficulties, researchers have used recombinant DNAs to express hundreds of functional proteins, either in bacteria or in eukaryotic cells. Because molecular biologists can systematically vary the amino acid sequence of any protein produced from recombinant DNA, researchers are now able to study the relationships between the structural and functional features of proteins in much greater detail than when such studies depended on the ability to isolate pure proteins from extracts of organisms. In addition, recombinant DNA methods have provided much of our current information about gene regulation because researchers can use the functional expression of an easily detectable gene product (such as β-galactosidase) to study the regulatory ability of a cloned DNA sequence from another gene. (See the discussion of "reporters" in Chapter 15, p. 408.)

RECOMBINANT DNAs CAN REPROGRAM CELLS TO MAKE NEW PRODUCTS

The ability to manipulate genes has already had important practical consequences. Hundreds of new and old biotechnology, chemical, and pharmaceutical companies are

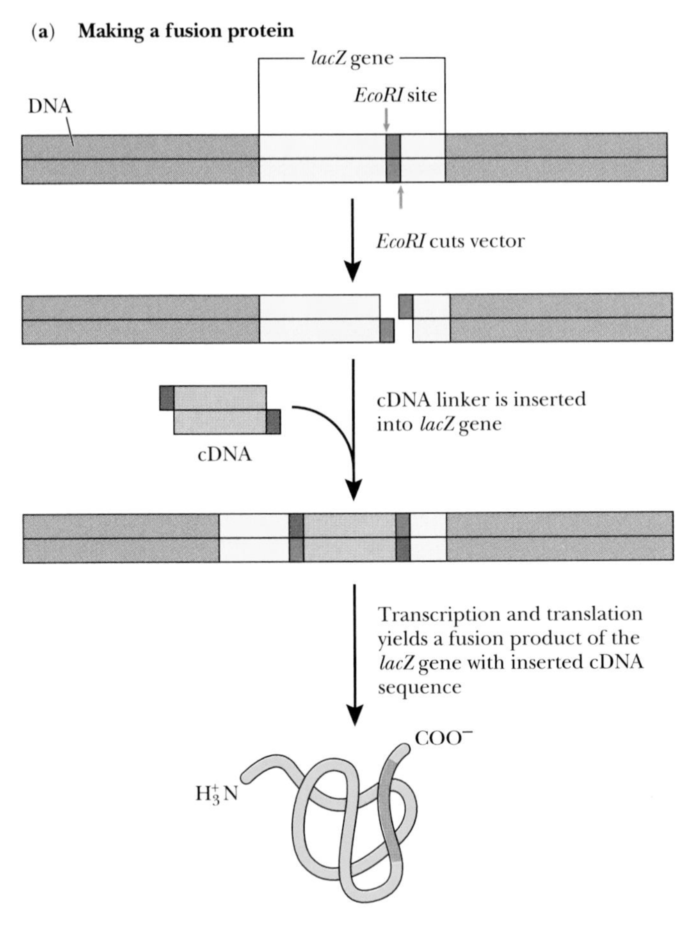

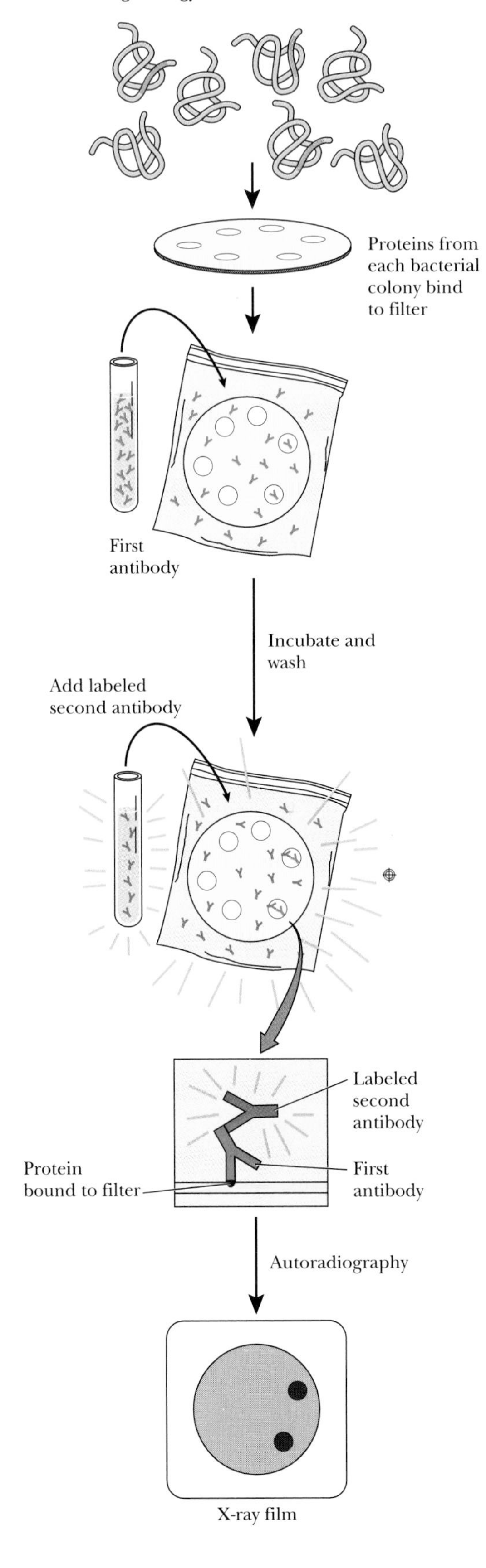

Figure 17-16 Detection of specific proteins with antibodies: **(a)** production of a fusion protein; **(b)** detection of clones making a specific polypeptide by immunoblotting.

applying recombinant DNA methods for commercial purposes. The first products of these companies include proteins already known to be useful to medicine (insulin being the first such product), to agriculture (naturally occurring insecticides), and to industry (enzymes required to synthesize particular chemicals). More recent products include newly engineered forms of old proteins (such as an improved version of an enzyme for laundry detergents), and previously unknown cellular components, such as the protein missing in the lungs of cystic fibrosis patients, or protein factors that may prevent nerve death in several neurological diseases. Finally, recombinant DNA techniques allow the engineering of plants and animals with new characteristics—plants that are resistant to insects or to particular herbicides, sheep that produce a blood-clotting factor in their milk, and pigs that produce human hemoglobin.

Reprogrammed Bacteria and Cultured Eukaryotic Cells Can Make Many Useful Proteins on a Commercial Scale

Insulin and other protein hormones, such as growth hormone, are examples of products whose medical usefulness was clear long before the development of recombinant DNA technology. Another such protein, tissue plasminogen activator, now commercially available from several biotechnology companies, has proved useful in dissolving blood clots in heart attack victims. (See Essay: "Blood Clots and the Therapeutic Uses of Enzymes" in Chapter 4.)

Some proteins, however, are so rare and so difficult to isolate that little was known of them before they could be produced from recombinant DNAs. Among these are a set of proteins—called **cytokines** [Greek, *cyto* = cell + *kinesis* = motion]—responsible for interactions among the cells involved in the immune response. (See Chapter 22.) One of the cytokines, interferon-γ, has long been known to help combat viral infections. But so little pure interferon was available that scientists had only limited opportunity to study its mode of action or its possible medical use.

By 1980, however, the interferon-γ gene had been cloned, and researchers began to study the effectiveness of bacterially produced interferon in preventing and treating viral infections, including those that produce versions of the common cold. Some researchers also initially thought that interferon might also combat certain kinds of cancer. So far, however, the therapeutic applications of interferon-γ have been disappointing. The testing of these substances, however, would have been impossible without the ability to produce large amounts by recombinant DNA techniques. Another cytokine, called IL-2, has been used in the treatment of a type of cancer called malignant melanoma.

Growth factors are another set of rare proteins that keep specific kinds of animal cells alive, stimulate cell division, or trigger particular types of cell specialization. Until recently, however, researchers knew little about the structure of growth factors or how they work. Recombinant DNA techniques have produced relatively large amounts of pure growth factors, and much has already been learned about their structure and mode of action. One of the surprises of this work was the demonstration of several new growth factors, related to but distinct from previously known proteins. Many academic and commercial laboratories are now investigating the possibility that growth factors may be able to prevent nerve degeneration in such devastating diseases as Parkinson's disease, ALS (amyotrophic lateral sclerosis, also called Lou Gehrig's disease), and Huntington's disease.

Recombinant DNA technology allows the reprogramming of cells to make an extraordinary number of products. These include medical treatments, ingredients for processed foods, and enzymes needed to produce valuable small molecules or to destroy pollutants.

Recombinant DNAs can also produce useful vaccines. The production of vaccines against viruses has depended heavily on the ability to propagate viruses in the laboratory. Recall that viruses can grow only inside host cells. In the first half of this century, virologists learned to grow infectious viruses in tissue culture cells or in chick embryos, but to produce a vaccine, they had to inactivate (or "attenuate") the virus so that it would produce an immune response but not disease. Problems remained, however, because even such inactivated vaccines sometimes caused illness. This undesirable response resulted either from the presence of a small amount of virus that remained active or from impurities in the vaccine, such as proteins from the cells in which the virus grew.

Recombinant DNA techniques allow the preparation of uncontaminated vaccines because the immunizing molecules are made from viral DNA rather than from active virus. Instead of growing active virus and then treating to inactivate it, virologists isolate a piece of viral DNA that encodes a single viral protein. A single component of the virus cannot cause disease but can bring about immunity. The application of recombinant DNA technology, then, can produce safer vaccines for many kinds of disease-causing viruses. Such techniques may also produce vaccines for diseases such as herpes and AIDS, for which no vaccines now exist.

Products of Recombinant DNAs Can Be Released Directly from Engineered Somatic Cells

In most cases, the products of recombinant DNAs are made in laboratories and industrial plants. The products are then purified, packaged, and distributed in the same way as other products produced by strictly chemical means. A number of laboratories and companies, however, are now experimenting with a new approach to the delivery of recombinant DNA products—engineering cells within the body to make and deliver the product directly.

One strategy is to take cells out of the body, engineer them to produce the desired product, and then reimplant them in the body. One research group, for example, is engineering the cells that line the blood vessels to produce a clotting factor that is missing in a type of hemophilia.

Another strategy, illustrated in Figure 17-17, is to introduce recombinant genes directly to cells within the body, using a viral vector that allows the expression of a protein that can replace a nonfunctioning protein. This is the approach now being explored for the treatment of cystic fibrosis. A variation of this strategy is to use a viral vector to deliver a recombinant DNA that will make *antisense RNA,* RNA that can form duplexes with a specific mRNA and prevent the expression of a complementary sequence in that mRNA. This approach may be able to prevent the expression of an oncogene responsible for certain cancers.

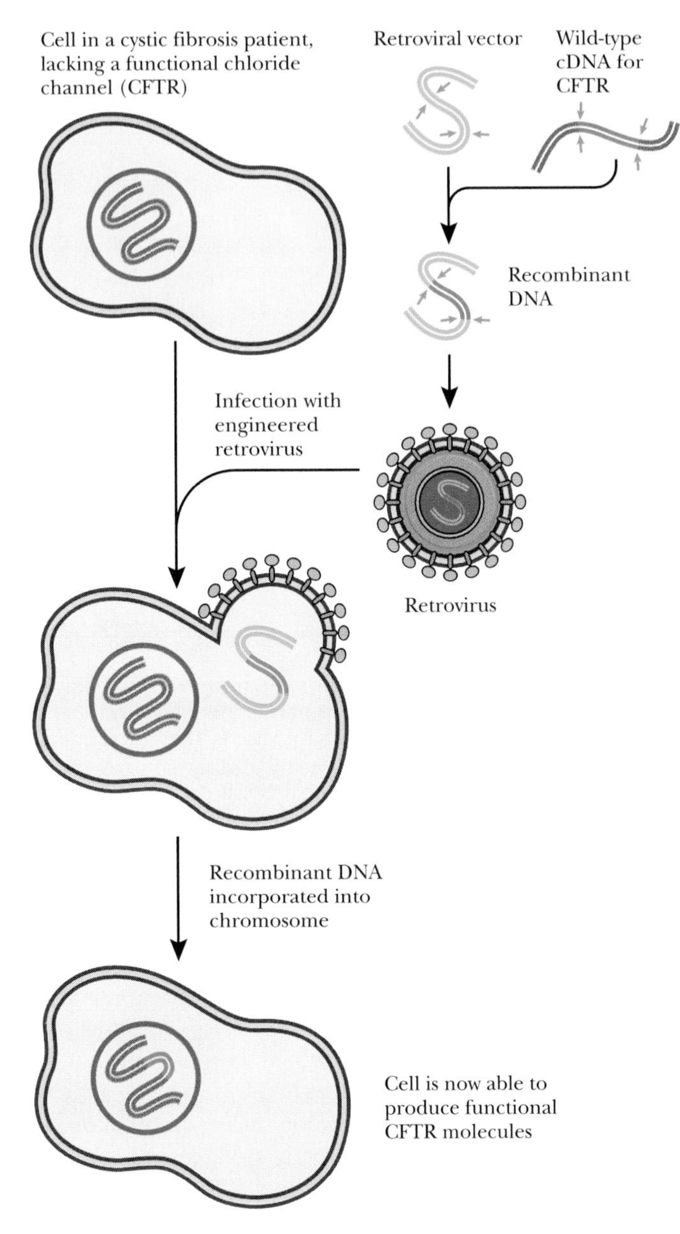

Figure 17-17 Manipulations of recombinant DNA may lead to new therapies for diseases such as hemophilia, cystic fibrosis, and cancer. Cells may be engineered *in vitro,* using a vector derived from a retrovirus, to produce a functional chloride channel (called CFTR), missing in cystic fibrosis patients.

Recombinant DNA Technology Allows the Production of New Proteins Not Present in Nature

New products of recombinant DNA technology range from enzymes deliberately engineered to have altered properties or specific antibodies produced in bacteria to signalling molecules targeted to act only on particular cells. A particularly promising line of research is the attempt to modify antibody structure in such a way that the resulting proteins can serve as enzymes. These engineered enzymes, called "abzymes," can act on synthetic chemicals, including some that now pollute our environment because nothing in nature can break the pollutants down.

RECOMBINANT DNA CAN GENETICALLY ALTER ANIMALS AND PLANTS

Recombinant DNAs can also be used to reprogram whole plants and animals. **Transgenic organisms** carry deliberately engineered DNA in their genomes. The added DNA is called a **transgene.**

To be expressed in a higher organism, a transgene must contain the appropriate control signals. (See Chapter 15.) To identify such regulatory regions, researchers usually couple a suspected segment of DNA to a reporter gene (such as the gene for β-galactosidase) whose cellular distribution can be rapidly determined. Experiments with such recombinant transgenes have already contributed greatly to knowledge about the DNA sequences responsible for specific patterns of expression and developmental regulation of many genes.

How Do Researchers Produce a Transgenic Mammal?

For a gene to be expressed in all the appropriate cells of an animal, researchers must put the transgene into the fertilized egg (the zygote) before the beginning of embryonic development. Then all the cells of the organism contain the engineered DNA. Only some of the cells, however, express that DNA; if the transgene contains *all* the control elements, then the pattern of transgene expression corresponds to that of the endogenous gene.

In the early 1980s, several research groups succeeded in producing transgenic mice. Perhaps the most dramatic of these early experiments produced mice carrying the gene for human growth hormone. At birth, such transgenic mice were already twice the size of their litter mates (see Figure 17-18 photo).

Producing a transgenic mammal with the available techniques requires the surgical removal of eggs from the female, so the number of available eggs is generally small. The experimenter adds sperm to the eggs to accomplish fertilization *in vitro* and then almost immediately injects the engineered gene into the male pronucleus in the resulting zygote. The same technique has produced transgenic pigs, sheep, and frogs, and it seems likely to yield similar results with other species. So far, however, only one transgenic animal is born for every 100 injected zygotes, and (despite tabloid claims to the contrary) no one has made any transgenic human embryos.

It Is Possible to Direct Transgenes to Specific Chromosomal Sites

Most techniques for producing transgenic animals put engineered genes into the host's chromosomes at random,

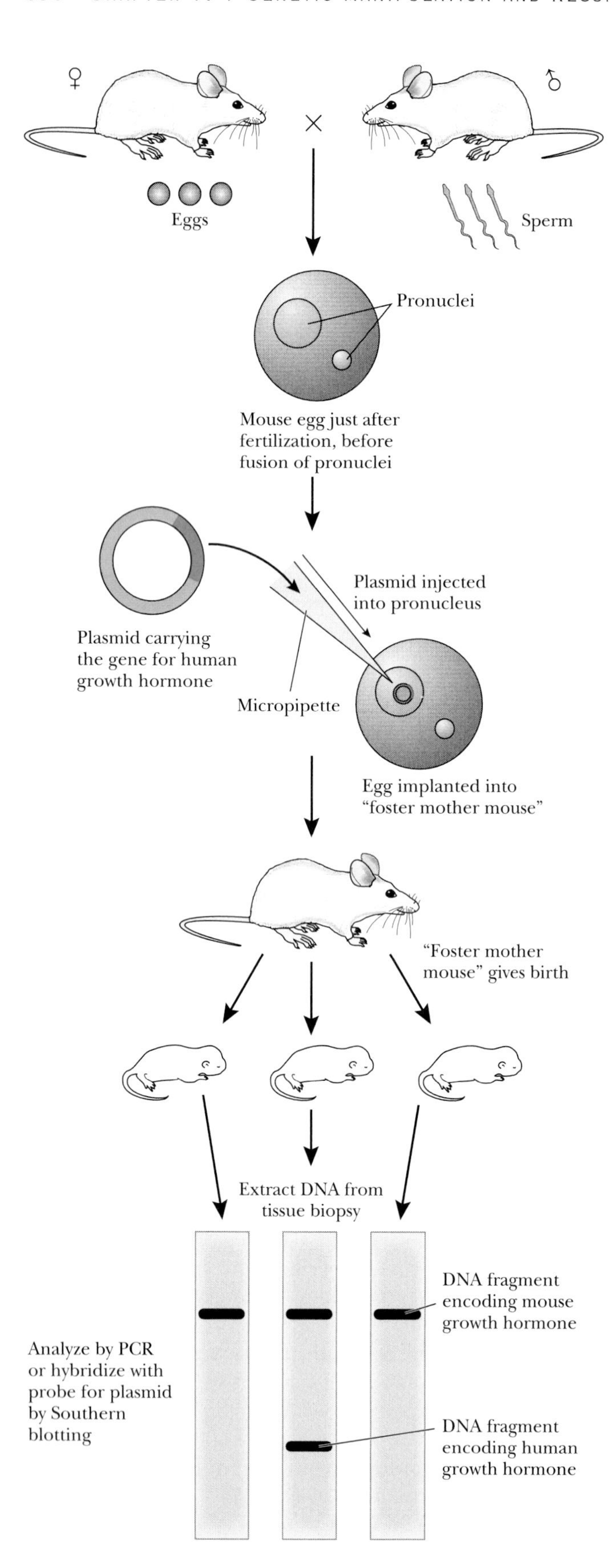

Figure 17-18 Production of transgenic mice. Researchers collect fertilized eggs shortly after mating, after fertilization but before sperm and egg nuclei have fused. They then inject recombinant DNA into the male pronucleus and implant the engineered cells into "foster mothers." Southern blot or PCR analysis of the DNA from the tails of the offspring reveals the presence or absence of the transgenic DNA. Photograph shows two mice from the same litter, one of which contains a transgene encoding human growth hormone. *(Photo courtesy of Ralph L. Brinster, School of Veterinary Medicine, University of Pennsylvania)*

adding to the host's own genes: A transgenic mouse with a gene for human growth hormone also contains two copies of the mouse's own growth hormone genes.

One direction of current research is to try to achieve true gene replacement by **homologous recombination,** in which an engineered gene directly replaces its counterpart, its "homologue," in the animal's chromosomes, as shown in Figure 17-19. Homologous recombination has been possible for some time in yeast and in prokaryotes, but it was not until the 1980s that several laboratories achieved limited success in animal cells. By now, the technique has been applied to dozens of genes, and most mammalian geneticists expect that it will eventually be possible to knock out and replace virtually any gene. The results of one "knockout" experiment are illustrated in Figure 17-19c. (Also see Essay, "Directed Gene Knockouts.")

The Genetic Engineering of Plants Is Easier than That of Animals

In some cases, it has become relatively easy to grow a whole plant from a single cell grown in culture. (See Chapter 21.) Genetic engineers can manipulate cells in culture and then grow them into a new plant, as shown in Figure 17-20 (p. 468). In this way, all the cells of the resulting plant have altered genetic plans.

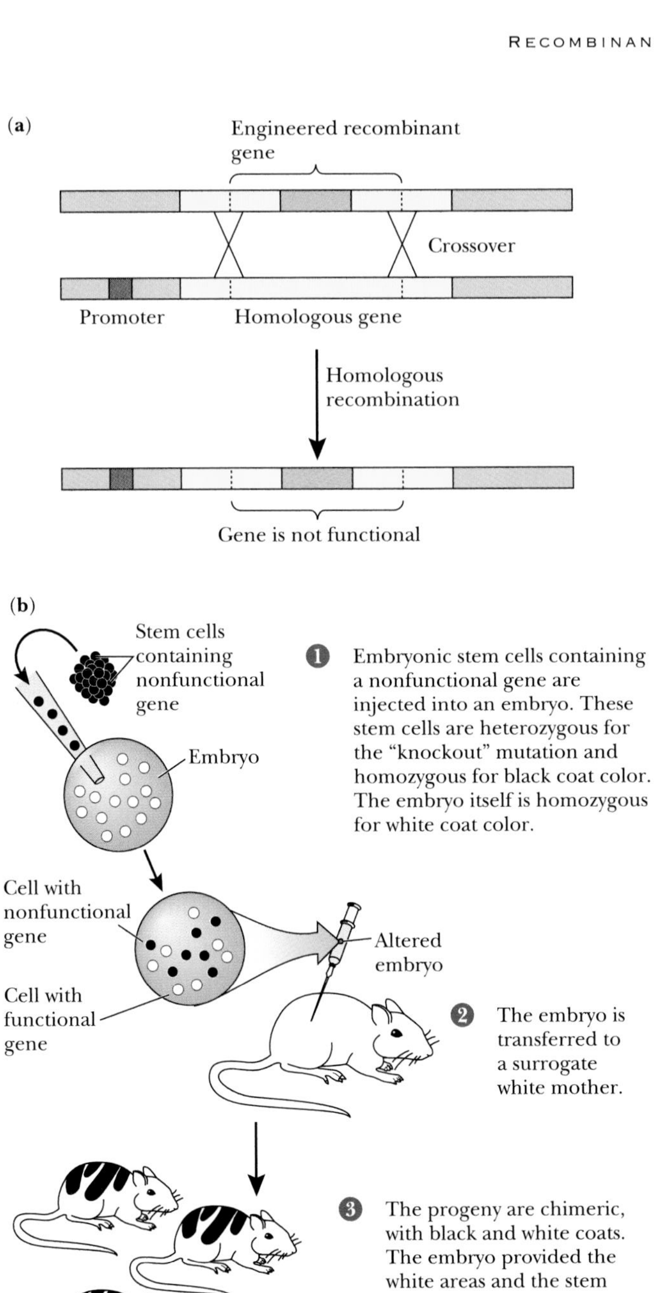

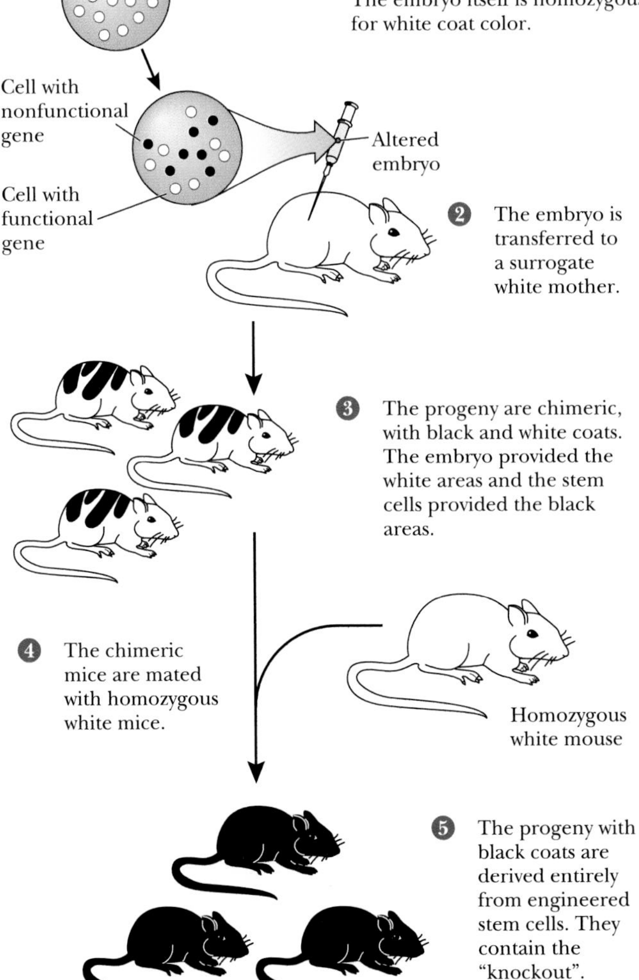

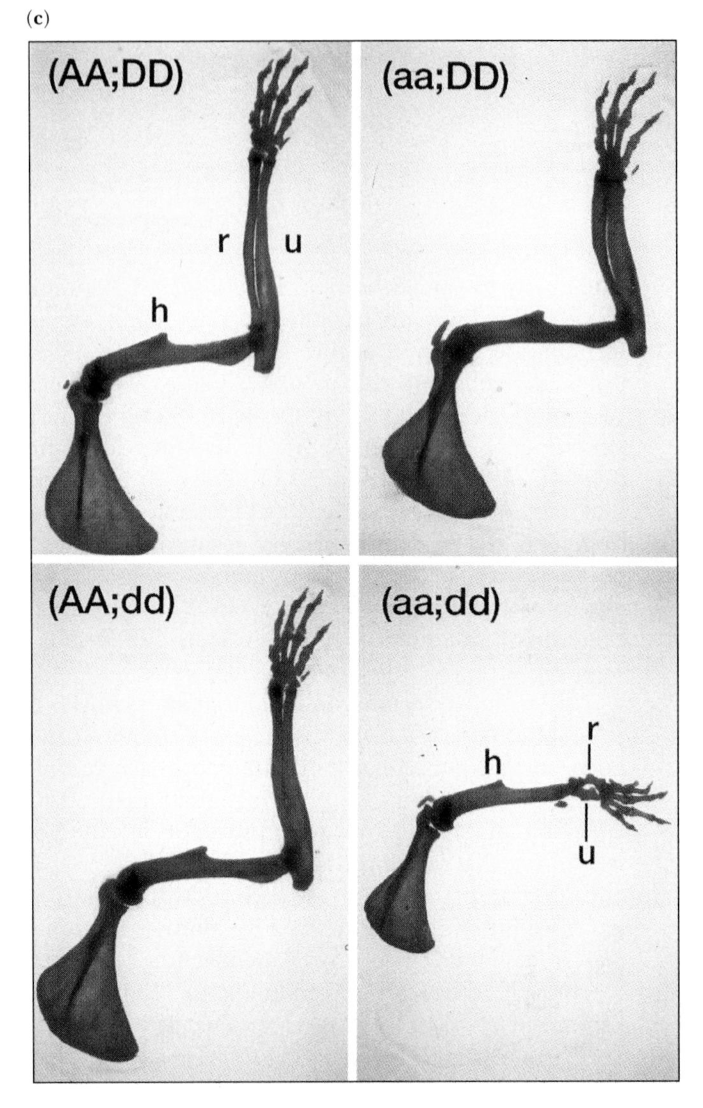

Figure 17-19 Production of "knockout" mice. Researchers manipulate embryonic stem cells to select those in which homologous recombination has occurred, leading to the inactivation of a single disrupted gene. The stem cell is then transplanted into an early mouse embryo, where it contributes to the developing mouse, producing a chimera in which some cells are normal and some are mutant. If cells of the germ line (the precursors of sperm and eggs) derive from mutant cells, the offspring of the chimera will have one disrupted gene in every cell. Breeding of such heterozygotes with each other can produce homozygous knockout mice, which cannot express a particular gene at all. **(a,b)** Strategy for selecting embryonic stem cells in which homologous recombination has occurred with a recombinant DNA; **(c)** forelegs of a normal (AADD) mouse and of three knockout mice, with disruptions in genes that control limb development. *(b, Mario Capecchi, Ph.D., University of Utah)*

To carry genes into plant cells and into plants, many plant molecular biologists use an unconventional genetic system that is involved in causing plant tumors, called crown galls, an example of which is shown in Figure 17-21a. A species of soil bacteria called *Agrobacterium tume-*

(Text continues on page 469.)

ESSAY

DIRECTED GENE KNOCKOUTS

The ability to "knock out" a selected gene by homologous recombination has allowed researchers to determine the function of individual genes not only in individual cells but also in whole plants or animals (meaning mice, the only animal species in which such experiments have been reported). In some cases, the deletion of a particular gene produces a phenotype that immediately reveals the usual function of the missing gene. But in many cases, the result of gene deletion is simply the death of the mutant mouse, often during early embryonic development. Knockouts that cause death in embryonic stages usually give little insight into the function of the deleted gene.

Given this serious limitation of the knockout technique, how can researchers best learn about the function of a particular gene? One approach has involved the development of an ingenious method to knock out genes only in chosen tissues or only at chosen times. This method depends on the ability to induce the cutting and splicing of DNA within specific sets of cells in an adult mouse. Like many of the tools used to engineer DNA *in vitro,* the machinery used to accomplish *in vivo* genetic manipulation comes from a bacterial virus, in this case one called P1.

When bacteriophage P1 infects *E. coli,* its DNA enters the bacterium as a linear molecule. The DNA quickly converts to circular DNA, which persists as a plasmid within the host *E. coli.* The circularization requires just one protein, a "recombinase" enzyme called Cre, which has a molecular size of 38,000. Cre binds to two 34-nucleotide DNA sequences, called *lox* sites, in P1 DNA and catalyzes a recombination event. In the case of P1, the recombination leads to circularization of the linear DNA.

Surprisingly, the bacteriophage-derived Cre can function not only in *E. coli,* but also in eukaryotic cells. When the *cre* gene, which encodes Cre protein, is put into a yeast cell, a plant cell, or an animal cell, it can still catalyze recombination between *lox* sites in DNA. In addition, the source of the DNA between the *lox* sites does not matter; it can equally be from P1, yeast, tobacco, or mice. In all cases, however, recombination does not occur until the cell produces Cre. So, if the *cre* gene is joined to an inducible regulatory element, recombination between *lox* sites occurs only after induction.

With the knowledge that Cre could rearrange eukaryotic DNA that is flanked by *lox* sites, researchers set about designing experiments to rearrange DNA that was already incorporated into a eukaryotic genome. A relatively straightforward application involved the reengineering of transgenic plants. Plants have been the subject of much genetic engineering for agriculturally desirable traits. Scores of laboratories have produced plants with transgenes that provide resistance to insect pests, viruses, or herbicides. But virtually all of these plants also contain genes for antibiotic resistance, which allowed researchers to select for cells with transgenes. The presence of antibiotic-resistance genes in crop plants—particularly those that will end up on the dinner table—is a cause for public concern, because they might contribute to the inactivation of specific antibiotics prescribed by physicians or veterinarians.

With these concerns in mind, researchers have developed a strategy to place *lox* sites on either side of the antibiotic resistance gene. The action of Cre can then remove the selection markers from the plant genome after cells containing the desired transgene have been selected.

More recently, researchers have used the Cre-*lox* system to produce mice with controlled knockouts. Using homologous recombination, they replaced a mouse gene with an engineered version of the same gene, flanked by two *lox* sites, as shown in Figure 1. (Such a gene, flanked by *lox* sites, is said to be a "floxed" gene.) They then bred the engineered *lox*-containing mouse with a transgenic mouse that expresses Cre only in certain tissues or only at certain times. The resulting hybrid mouse contains the floxed gene but does not express Cre in the embryo or in most tissues, and the floxed gene remains as active as it is in the wild-type mouse.

In one case, the floxed gene was DNA polymerase β, which is involved in DNA repair, and the Cre transgene was expressed only in T lymphocytes, cells responsible for cellular immunity. (See Chapter 22.) Unlike more conventional knockout mice with a deleted DNA polymerase β, they readily survived to adulthood. But they did not make DNA polymerase β in their T lymphocytes. Researchers could then begin to evaluate the importance of DNA polymerase β in cellular immunity.

In another case, a Cre transgene was under the control of a promoter for a gene called *Mx1,* whose product contributes to defense against viral infection.

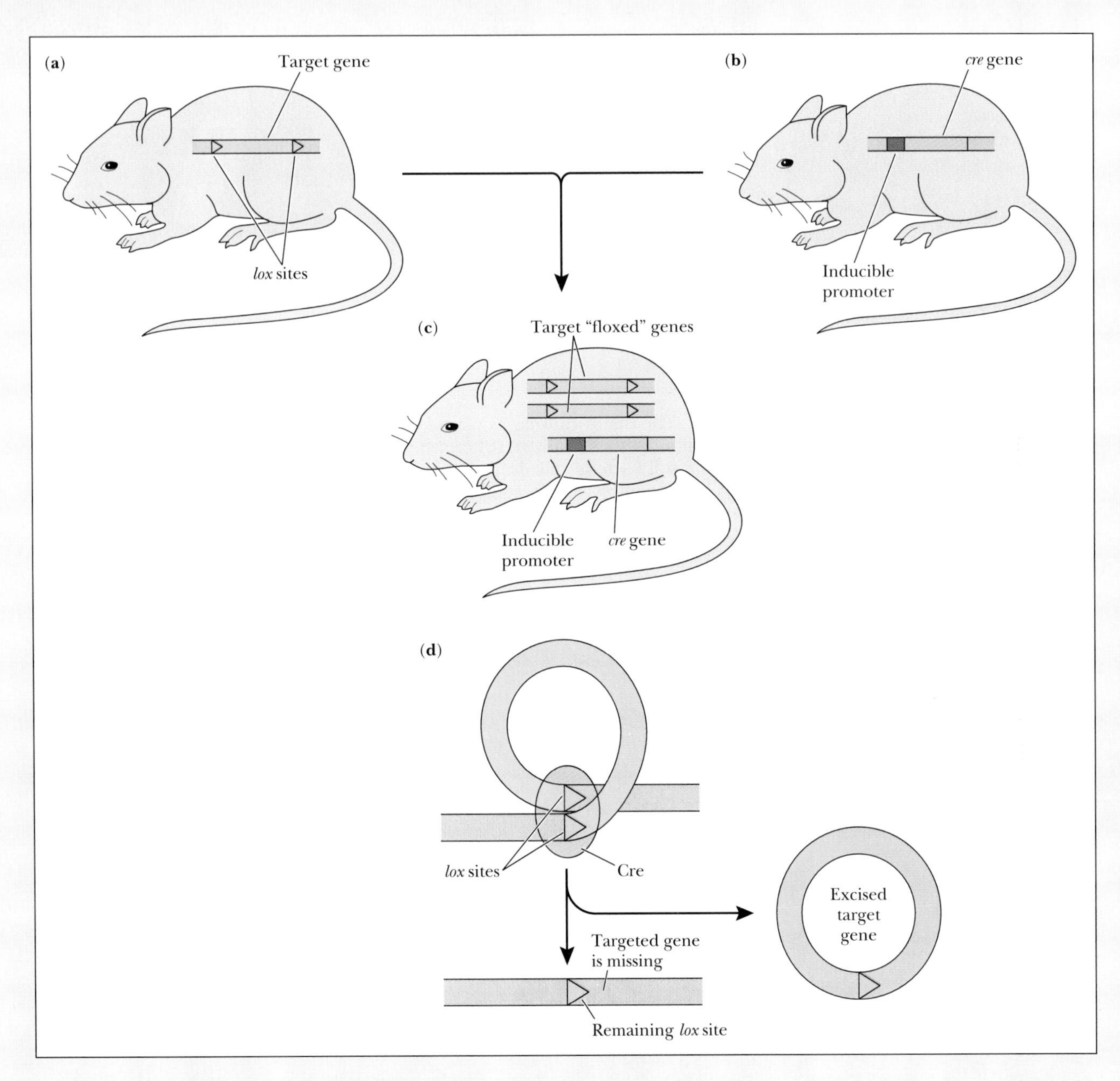

Figure 1 Induced deletion of a "floxed" gene. **(a)** A mouse, genetically engineered by homologous recombination, with a targeted gene flanked by two *lox* sites; the mouse does not contain a *cre* gene; **(b)** a mouse, genetically engineered to contain a Cre transgene, under the control of an inducible promoter; **(c)** a mouse that results from at least two generations of breeding (a) and (b), chosen to be homozygous for the "floxed" gene and to contain an inducible *cre* gene; **(d)** induction of *cre* deletes the targeted gene.

Mx1 is normally silent, but it is activated by *interferon,* whose synthesis is triggered by a viral infection. Injection of interferon into the transgenic *Mx1*-Cre mouse stimulates Cre production, which in turn causes deletion of the floxed gene. So, in these mice, deletion of a floxed gene can be induced at will.

The Cre-*lox* technology should allow the inactivation of any appropriately engineered gene, allowing researchers to learn the role of specific genes in intact animals as well as in individual cells.

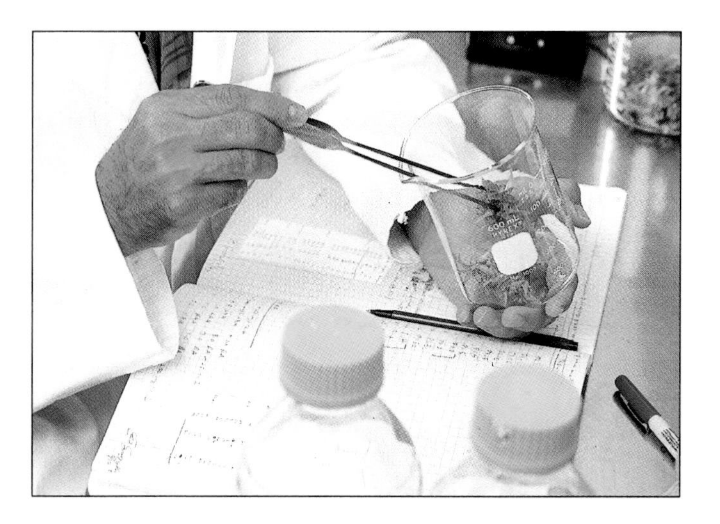

Figure 17-20 Individual plant cells can produce a whole plant much more easily than in animals. Genetic engineering of a single cell can produce a plant with altered genetic properties, as in the case of this genetically engineered tomato plant.
(left, Robert Holmgren/Peter Arnold; right, Richard Nowitz/Phototake NYC)

(a)

Figure 17-21 The tumor-causing Ti plasmid can be used to engineer plant cells. **(a)** A crown gall on a tomato plant; **(b)** crown gall is caused by a plasmid, called the Ti plasmid, within soil bacteria called *Agrobacterium tumefaciens;* **(c)** T-DNA within the Ti plasmid may be modified using recombinant DNA techniques to produce a transgenic plant. *(a, Jack M. Bostrack/Visuals Unlimited)*

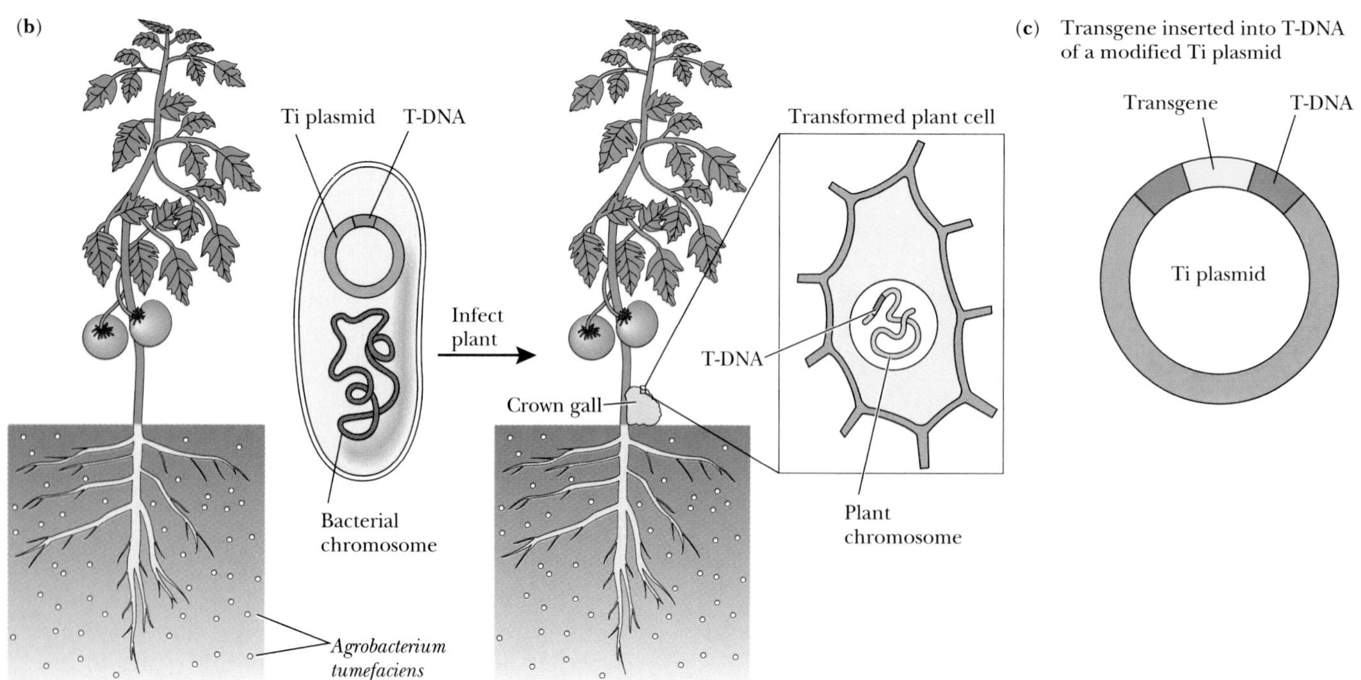

faciens long appeared to be responsible for crown galls, but the actual tumor-causing agent is a plasmid called the **Ti plasmid** [for *t*umor-*i*nducing], which resides within the bacteria (Figure 17-21b). A tumor results from the movement of a piece of plasmid DNA into the genome of the host.

Several plant molecular biologists—including Mary-Del Chilton (then at Washington University in St. Louis) and Jeffrey Schell (then at the University of Ghent in Belgium)—developed recombinant DNA techniques to produce recombinant plasmids containing foreign genes that the plasmid inserts into plant DNA. In addition to inserting genes exactly as they occur in nature, researchers are also able to engineer genes to alter the properties or the distribution of their products.

Many plant scientists have now used the methods shown in Figure 17-21 to produce plants with new genetic properties. Among the products of these experiments are plants that can resist the action of some herbicides. Weeding a crop of such plants could involve simply spraying the herbicide on a field; the herbicide would kill the weeds but not the genetically resistant plants. While some agricultural experts look forward to the use of such methods, many critics have expressed their reservations about the ecological effects of increased herbicide use. Less controversial is the production of new flower varieties.

■ SUMMARY

At least since the beginning of agriculture, humans have practiced the genetic manipulation of organisms. Until the development of recombinant DNA techniques, genetic manipulation consisted mostly of the selection of varieties of organisms with desired properties. Such selection worked more easily for plants and microorganisms because they could be propagated as genetically identical clones, all derived from a single parent.

Standard techniques allow geneticists to grow clones of bacteria separate from one another on agar that contains a growth medium. The technique of replica plating transfers all the colonies on a single plate to other plates, where they may grow in the same relative positions on different media.

Recombinant DNAs consist of two or more DNA segments that are not found together in nature. For example, the manipulation of DNA pieces from humans and *E. coli* allowed researchers to make a recombinant DNA molecule with the regulatory region of β-galactosidase and the structural gene for insulin. Bacteria that contain this DNA make insulin in response to the presence of lactose, the inducer of β-galactosidase production.

Making recombinant DNA depends on the use of enzymes that can cut, link, and modify DNA. Restriction enzymes cut DNA at particular sites, where specific sequences occur, allowing molecular biologists to isolate individual DNA fragments with defined ends. To prepare recombinant DNA, an experimenter can then use DNA ligase, an enzyme used in DNA replication, to link two pieces of DNA from different sources.

DNAs derived from viruses and plasmids can serve as vectors for the propagation of recombinant DNAs within host cells. Some of the most widely used vectors are plasmids that carry antibiotic resistance genes. Other vectors include bacteriophages, plant and animal viruses, and yeast artificial chromosomes.

The starting materials for recombinant DNA may come from the genomes of viruses or organisms or from DNA copies of mRNA. Most plant and animal genes contain introns, which would interfere with their expression in bacteria. So many recombinant DNAs coding for plant and animal proteins are derived from complementary DNAs (cDNAs), copied from mRNA by the reverse transcriptase of a retrovirus.

Collections of recombinant DNAs, called gene libraries, may contain thousands or millions of different DNAs. Finding a desired DNA sequence usually depends on the use of a hybridization probe, often a piece of DNA made in the laboratory, that can form specific hydrogen bonds with one strand of DNA. By labeling the probe with radioactive nucleotides or a chemical marker, the experimenter can identify recombinant DNAs on blots. Antibodies, which have surfaces that recognize the shapes of specific molecules, can permit an experimenter to find bacteria producing a particular protein.

Recombinant DNAs can reprogram organisms to make new products. Some products made by reprogrammed bacteria, such as insulin and tissue plasminogen activator, are already commercially important as pharmaceuticals. In many cases, however, researchers do not know the function of a particular protein until recombinant DNA techniques make enough available to study. Recombinant DNAs are likely to be particularly useful in the preparation of safe vaccines for use in disease prevention.

Recombinant DNA can be used to alter genetically animals and plants. Transgenic organisms carry deliberately added DNA in their genomes. The present method for making transgenic mammals involves the injection of an engineered gene into a zygote produced by *in vitro* fertilization. A new technique, which depends on the ability of cells to perform homologous recombination, may allow the direct replacement of chromosomal genes.

Making transgenic plants is easier than making transgenic animals because experimenters can accomplish gene addition or gene replacement in single cells in culture. The manipulated cells can then be grown in such a way as to produce a whole plant, in which all of the cells derive from the genetically programmed cell. One widely used technique makes use of a vector derived from a tumor-inducing (or Ti) plasmid, which normally resides in bacteria that cause plant tumors called crown galls.

KEY TERMS

biotechnology
complementary DNA (cDNA)
cytokine
diabetes mellitus
DNA blotting
DNase
expression vector
gene library
growth factor
homologous recombination
hormone
hybrid
in situ hybridization
insulin
linker
molecular hybridization
Northern blotting
nuclease
oligonucleotide
recombinant DNA
replica plating
restriction enzyme
restriction fragment
restriction map
restriction site
reverse transcriptase
RNA blotting
RNase
screening
Southern blotting
Ti plasmid
transgene
transgenic organism
vector
Western blotting
yeast artificial chromosome (YAC)

QUESTIONS FOR REVIEW AND UNDERSTANDING

1. What is a clone?
2. How is breeding different from cloning when selecting for a desirable trait?
3. What is recombinant DNA?
4. What is a gene library? What is a cDNA library? What is an expression library?
5. Compare and contrast the following terms:
   (a) DNase vs. RNase
   (b) blunt end vs. sticky end
   (c) Southern blotting vs. Northern blotting
   (d) plasmid vs. yeast artificial chromosome
6. Give examples of some commercially produced recombinant products.
7. What is a transgene?
8. What are the essential genetic elements of a plasmid vector? Explain why each of these elements is important.
9. When selecting for a trait, why is it important to dilute bacteria so that each colony is well separated?
10. Why are restriction enzymes so useful to: (a) bacteria; (b) scientists?
11. Why is the universality of the Genetic Code so essential for the expression of recombinant genes? Give an example to illustrate this concept.
12. Describe three unique methods researchers use to select for bacteria containing a recombinant plasmid.
13. Describe two unique strategies researchers use to obtain and clone a DNA fragment.
14. Propose a model to explain why mutagenesis would cause a strain of mold to produce 100 times more penicillin than the original Peoria strain. (*Hint:* Refer to Chapter 14.)
15. How might a researcher today go about solving the problem of generating large amounts of penicillin?
16. Why is it so difficult to characterize and manipulate a gene in a cell? How have researchers solved this problem?
17. Describe how you would go about generating a cDNA copy of a mammalian gene.
18. Why is it necessary to generate a cDNA copy of a mammalian gene in order for it to be properly expressed in bacteria?
19. Bacteria are not suitable hosts for producing certain recombinant proteins. Explain why.
20. Why is it useful that complete complementarity is not required for molecular hybridization to occur?
21. Give two reasons why recombinant DNAs in general, and expression vectors in particular, are so useful for the study of proteins.
22. Why are recombinant vaccines safer than attenuated virus vaccines?
23. Describe the two strategies being developed that use viral vectors to deliver medical treatment.
24. Describe how a transgenic mammal is produced.
25. Why is it desirable to be able to insert DNA by homologous recombination when generating transgenic animals?

■ SUGGESTED READINGS

Alberts, B., D. Bray, J. Lewis, M. Raff, K. Roberts, and J. D. Watson, *Molecular Biology of the Cell,* 3rd edition, Garland, New York, 1994. Chapter 7 deals with recombinant DNA technology.

Lodish, H., D. Baltimore, A. Berk, S. L. Zipursky, P. Matsudaira, and J. Darnell, *Molecular Cell Biology,* 3rd edition, Scientific American Books, New York, 1995. Chapter 7 deals with recombinant DNA technology.

Watson, J. D., M. Gilman, J. Witkowski, and M. Zoller, *Recombinant DNA,* 2nd edition, Scientific American Books, New York, 1992. This whole book represents an exceptionally clear compendium of most current techniques of molecular biology, as well as of the conclusions drawn from their application.

PART V

Cell Specialization, Integration, and Evolution

SEM of a nerve cell growing on a silicon chip.

LAB TOUR

NAME: JAMES TOWNSEL
TITLE: PROFESSOR OF PHYSIOLOGY
ADDRESS: MEHARRY MEDICAL COLLEGE, NASHVILLE, TN 37208

■ ***What is the major general question that you and your laboratory are asking?***

We are trying to learn how cells regulate the trafficking of proteins that are part of the plasma membrane. The protein we are studying is the choline cotransporter, an integral membrane protein that binds to extracellular choline and moves it across the plasma membrane. Many neurons—especially the motor neurons that drive muscle contraction—use acetylcholine as a neurotransmitter, and the movement of choline across the plasma membrane is the rate-limiting step in acetylcholine synthesis.

■ ***How did you come to choose this question?***

It has been known for some time that acetylcholine synthesis increases after rapid stimulation of neurons. This increase depends on an increase in the number of choline transporters, so I was intrigued by the connection between neural activity and this biochemical event.

I began by asking what were the distinguishing biochemical components necessary to make a cholinergic cell cholinergic; that is, what were the molecules that made a particular neuron produce acetylcholine as a neurotransmitter rather than some other transmitter or no transmitter at all? To study this question, one needs reliable markers of cholinergic neurons. Using the organism that I had long been studying—the horseshoe crab, I surveyed a number of possible markers, including the acetylcholine-synthesizing enzyme, choline acetyltransferase (CAT) and the high affinity choline uptake system.

■ ***Why has the overall question remained important to you?***

An earlier theory had said that the activity of the choline transporter depended upon the levels of choline within the nerve terminal. It was postulated that nerve stimulation caused a "disinhibition" of the choline transporter by a release of acetylcholine (ACh) and the utilization of inter-terminal choline in the subsequent synthesis and replacement of the released ACh. An unexpected finding in our laboratory convinced us that this theory was flawed. We found a high affinity choline transporter system in the *Limulus* cardiac ganglion which was regulated similarly to cholinergic systems.

We wondered if the effects of nerve activity on choline transporters depended on intracellular signals called "second messengers." We started off by asking whether depolarization may have triggered a cascade of second messengers. Studies from a group in Germany started us looking at second messengers that affected protein kinase C (PKC). We found that anything we did to activate PKC led to reduced choline transport, and any inhibition of PKC led to an increase in transport. Our results paralleled what others had found with insulin-dependent glucose transport.

We were able to show that horseshoe crab neurons contained an occluded population of transporter. These occluded transporters were not on the cell surface.

■ ***What other people work with you in your research program?***

During the academic year, my group consists of three graduate students working toward a Ph.D., one student working toward a combined M.D.–Ph.D., a technician, and a postdoctoral fellow. During the summer, we are joined by two undergraduates as well as several visiting faculty from other institutions.

■ ***What organisms do you use in your research? Why those organisms?***

All of our work involves the horseshoe crab, an organism I met as a

Dr. Townsel adjusting the Brandell Cell Harvester for binding studies being carried out on brain hemi-slices.

graduate student at Purdue. In my thesis work I showed that the inhibitory neurotransmitter that regulated the heart beat was serotonin, a widely distributed derivative of the amino acid tryptophan.

The horseshoe crab has a simple nervous system, with hardy cells. The organism and the cells are physiologically durable. Isolated cells continue to work for hours without any apparent defect. In contrast, mammalian cells are much more fragile and begin to show physiological defects after 30 minutes. The horseshoe crab is most famous in neuroscience largely because of Hartline's work on its visual system. I began to worry about other questions, however, particularly about the ways in which particular nerve cells come to use acetylcholine as a neurotransmitter rather than, say, GABA or serotonin.

Every summer I go to the Marine Biological Laboratory at Woods Hole, Massachusetts, where horseshoe crabs are plentiful. The rest of the year, live animals are shipped to me in Nashville.

A favorite starting material for the laboratory is a preparation known as synaptosomes, which is an isolated, resealed, and thus intact, nerve terminal. We were the first to prepare synaptosomes from the horseshoe crab, and we showed that they are almost purely presynaptic.

■ *What instruments do you use in your research, and how do they contribute to the experimental pursuit of your question?*

To prepare synaptosomes, we rely on a special kind of tissue homogenizer, called a Potters homogenizer, which allows us to grind tissue, at a controlled temperature, and with specific tolerances resulting in the production of intact and functioning nerve endings. We also depend on centrifuges, including the ultracentrifuge, to isolate subcellular fractions; scintillation counters; spectrophotometers; chromatography; gel electrophoresis; and a computer-based gel documentation system.

A typical experiment involves the preparation of synaptosomes, after homogenization and centrifugation, and the measurement of choline uptake, using radioactive choline. We use a special apparatus, called a Brandell cell harvester, which allows us to study the binding of particular molecules to these membrane sacs.

■ *Please describe a day in which something happened in your laboratory that was especially exciting.*

The best "eureka" feeling was the day we got our PCMPS (*p*-chloromercuriphenyl sulfonic acid) experiment to work. The result showed—as we had hypothesized—that a substantial fraction of the choline transporters were not on the cell surface.

Other experiments didn't come out exactly as we expected, however. For example, we thought that elevated K^+ triggers increased choline transporter on the cell surface through the inhibition of PKC. However, it appears that the activity of phospholipase A_2 (PLA_2) is far more vital to K^+-stimulated increase in choline transport.

■ *How do new questions arise?*

New questions certainly arise from day to day and from week to week, and each answer leads to more questions.

A lot of questions arise in my laboratory's weekly two-hour meetings. During this weekly "show & tell," we look at the week's activity and plot future studies. Like many other laboratories, we operate with specific aims that may take 2–3 months to complete.

Members of Townsel's laboratory and the object of experimentation—Limulus polyphemus, *the horseshoe crab.*

■ *How does your work fit into the broad context of cell biology?*

The general questions that we're asking have appeal to all kinds of cell biologists. I tell students that while we're working on choline transport, we are asking basic questions about the regulation of integral membrane proteins. Cell biologists do not yet fully understand how cells handle this problem. In the past years, neurobiologists have made great strides in understanding the trafficking of proteins involved in neurotransmissions. It may be that these same processes play a major role in trafficking of integral membrane proteins.

■ *In addition to your scientific work, you are deeply involved in educational projects. What have been your goals in this work?*

My goal is to get young people turned on to the beauty and orderliness of science and the scientific approach. I want them to learn to formulate intellectually interesting questions and to be critical regarding the standards of acceptable evidence. Young people must learn to become critical thinkers.

It is particularly attractive to me to be in an environment with a lot of young people whose backgrounds did not lead them to think they could choose a career in science. It's thrilling to see them become excited about scientific discovery and to pursue new career paths. I find it thrilling to watch their self confidence grow.

CHAPTER 18

Cells and Movement

KEY CONCEPTS

- Muscles are highly organized molecular assemblies that convert the energy of ATP into the relative movement of two types of protein filaments.
- Muscle contraction occurs after a stimulated increase in the availability of calcium ions.
- Proteins like those in muscle filaments are present in many nonmuscular cells and contribute to cellular movement and shape changes.
- Some types of cell movement and shape changes depend on the regulated assembly and movements of microtubules.
- Bacterial movements depend on molecules different from those used by eukaryotes.

OUTLINE

ESSAYS

Movement is so closely associated with being alive that its absence suggests death. Biological movement depends on the first five characteristics of life listed in Chapter 1—organization, energy transformations, chemical reactions, changes with time, and responses to the environment. And movement is required for the sixth characteristic—reproduction. No area of biological inquiry better illustrates the successes and problems of understanding function in terms of cellular and molecular form than does movement.

Biologists distinguish three categories of movement—the familiar movement of whole multicellular organisms, the movement of individual cells, and the movement of subcellular components. In this chapter, we look at how microscopic observations, physiological measurements, and biochemical experiments together led to a detailed molecular model of how organisms, cells, and cell components move. We begin with the most familiar and best understood—our own voluntary movements, accomplished by the muscles attached to our skeletons.

A basketball game, a ballet, or even a walk across campus demonstrates the beauty, the complexity, and the utility of multicellular movements. In humans, movements depend on more than 600 skeletal muscles. Together, the skeletal muscles are the largest tissue in the vertebrate body, making up more than 40% of its weight. Some individual muscles are tiny, like those that move our eyelids, while others are much larger, like those that move our legs. The size, organization, and arrangement of each of these muscles establish which movements we can and cannot perform.

Muscles exert force in only one direction: They can *pull* but not *push*. Most body movement depends on the arrangement of opposing pairs of muscles, which produce movements in opposing directions. For example (as discussed in Chapter 1, p. 18), the contraction of the biceps flexes the forearm, while the opposing action of the triceps extends it.

HOW DID MICROSCOPIC STUDIES LEAD TO A MODEL OF MUSCLE MOVEMENT?

The light microscope and the electron microscope have revealed the organization of contractile proteins in skeletal muscle fibers and have suggested a model for their action. (See Chapter 1.) A skeletal muscle contains many parallel **muscle fibers.** As shown in Figure 18-1a, a muscle fiber is a giant cell that contains the proteins responsible for contraction. Each fiber may have as many as 100 nuclei, all surrounded by the same plasma membrane, the **sarcolemma.** A typical fiber is 10–50 μm in diameter and up to 4 cm long. As can be seen in Figure 18-1b, skeletal muscles and the muscles of the heart show **striations,** cross stripes that lie perpendicular to the muscle's long axis, spaced about 2–3 μm apart. Understanding movement ul-

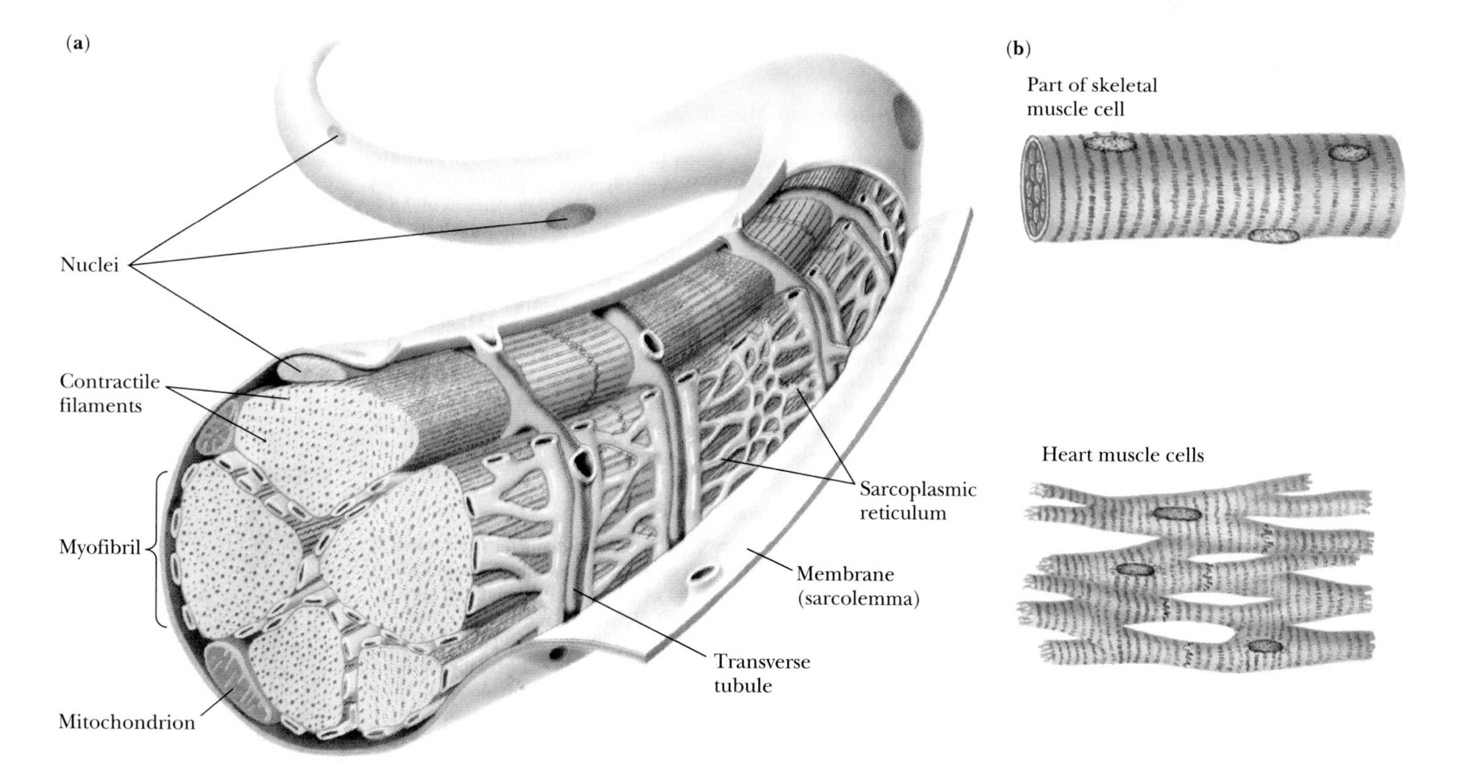

Figure 18-1 A skeletal muscle fiber is a giant cell with as many as 100 nuclei. **(a)** A cut-away of a muscle fiber; **(b)** skeletal muscle and heart muscle are striated, with cross-stripes spaced 2–3 μm apart.

timately depended on knowing the molecular nature of these striations and their changes during muscle contraction.

The Cellular Organization of Muscle Fibers Reflects Their Function

The individual fibers of a skeletal muscle are not all the same. Thicker fibers, called **glycolytic fibers,** derive most of their energy from glycolysis and have few mitochondria. Thinner fibers, called **oxidative fibers,** derive most of their energy from respiration and have many mitochondria.

Glycolytic fibers are also called "fast fibers" because they can deliver lots of power in a short time. They cannot sustain their activity for long, however, because of the excess lactic acid that they accumulate in regenerating NAD^+. (See Chapter 7, p. 210.)

Oxidative fibers, on the other hand, continuously regenerate NAD^+ by oxidative phosphorylation, as described in Chapter 7. These fibers depend on a continuing supply of oxygen and are therefore usually close to blood capillaries. Oxidative fibers contain high levels of myoglobin, an iron-containing protein that helps pull oxygen from the blood. Myoglobin gives these fibers and the muscles that contain them a characteristic red color, while glycolytic muscles have little myoglobin and are white. The "white meat" of a chicken breast consists of glycolytic fast fibers that can be used only for short bursts of activity. In contrast, the breast muscles of birds that actually fly, such as those of a duck or a hummingbird, are red and consist of oxidative fibers. Figure 18-2 compares the structures of oxidative and glycolytic fibers from two different mammalian muscles.

Unused muscles gradually atrophy; they become smaller, losing contractile proteins as well as other cellular components. Anyone who has had a broken arm or leg in a cast knows how weak muscles become from disuse. They also know that exercise can restore the muscles to their previous fullness.

Exercise also influences the state of normal muscle. Brief, high-intensity exercise—like body building—increases the diameter of the glycolytic fibers. The muscles of a trained body builder bulge because the number of myofibrils increases, along with the content of muscle proteins, as shown in Figure 18-3a.

The effect of other forms of exercise—like swimming or long-distance running—is different (see Figure 18-3b). The muscle fibers do not increase in diameter, and the muscles do not bulge. Instead, the oxidative fibers acquire more mitochondria, and the capacity of the blood to carry oxygen to the muscle also increases.

(a) 1 μm

(b) 1 μm

Figure 18-2 (**a**) The leg muscles of a rabbit are composed of glycolytic fast fibers and are fast-fatiguing, while (**b**) the leg muscles of a dog are composed of oxidative fibers and are slow to fatigue. *(Mary Reedy, Duke University Medical Center)*

(a) (b)

Figure 18-3 Different types of training have differing effects on muscle. (**a**) High intensity training increases the number of myofibrils and the diameter of the trained muscles.
(**b**) Sustained aerobic training increases the number of mitochondria and the capacity of the blood to deliver oxygen, but does not increase muscle diameter. *(a, Bruce Curtis/Peter Arnold; b, William Weber/Visuals Unlimited)*

The Visible Striations in Skeletal Muscles Reflect a Regular Molecular Structure

The striated appearance of skeletal muscles, as seen in the light microscope, suggests that they are also highly organized at the molecular level. Skeletal muscle was therefore among the very first tissues examined in the 1950s when biologists began to use the transmission electron microscope to study subcellular structure. Researchers have now come to understand much about the organization of muscle proteins and how they accomplish muscle contraction. Much of this understanding came from the work of Hugh Huxley, a British biologist who, in the mid-1950s, began to use an early electron microscope to study skeletal muscle structure.

A muscle fiber from a skeletal muscle consists of many **myofibrils** [Greek, *myos* = muscle + Latin, *fibrilla* = little fiber], threads each about 1–2 μm in diameter and running the length of a fiber, as shown in Figures 18-1 and 18-4. The myofibrils contain the muscle's contractile proteins. A myofibril consists of a series of smaller units, called **sarcomeres** [Greek, *sarx* = flesh + *meros* = part], small repeating cylinders each about 2.5 μm long. A 2.5 cm-long myofibril, then, consists of about 10,000 end-to-end sar-

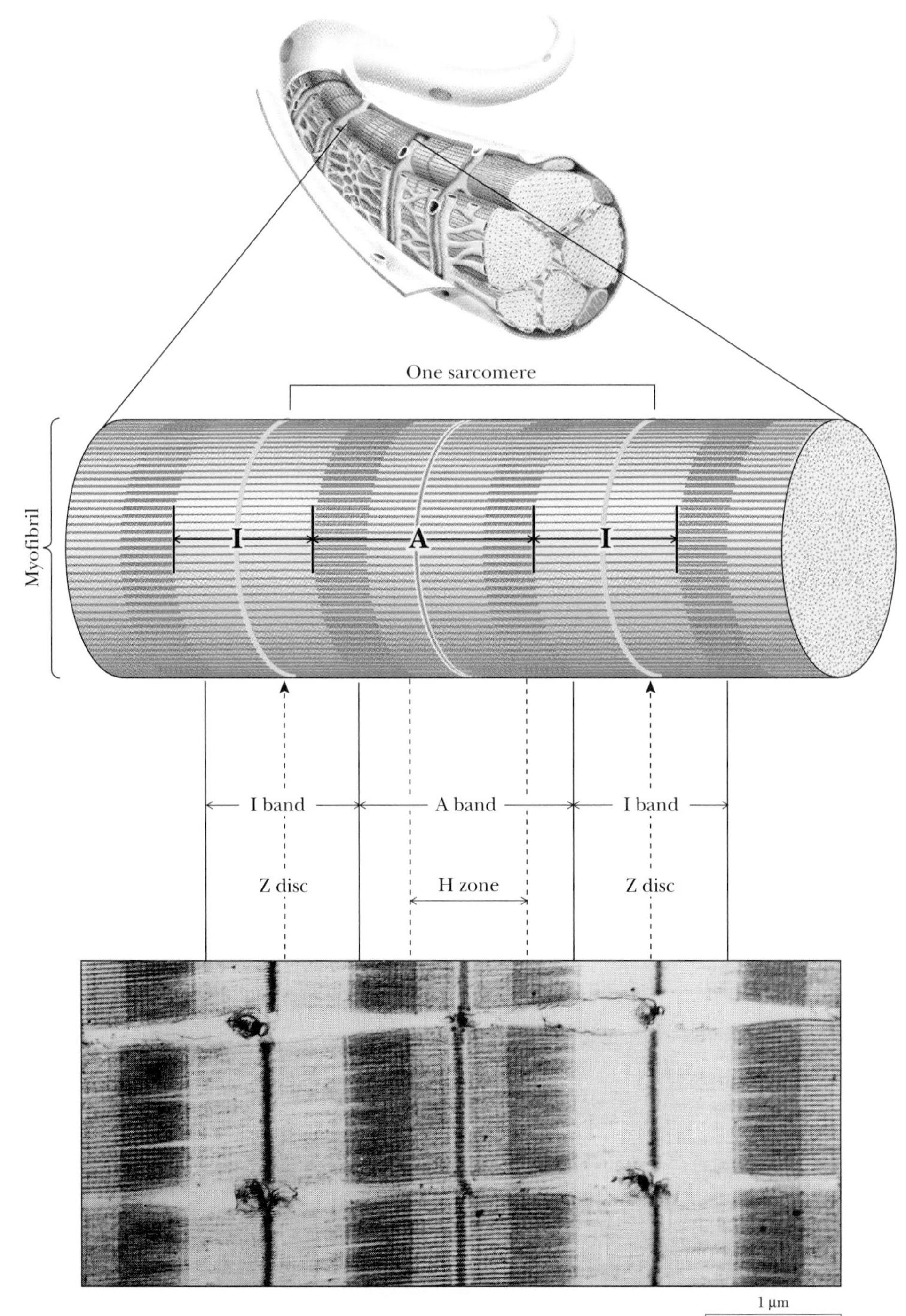

Figure 18-4 A myofibril from a skeletal muscle. *(Hugh Huxley, Brandeis University)*

comeres. Within a fiber, the ends of parallel sarcomeres are aligned, as illustrated in Figure 18-1. This alignment is responsible for the fiber's striations.

Under the light microscope, a stained section of skeletal muscle fiber shows alternating light and dark bands. The darker bands are called **A bands** [for *anisotropic,* meaning that they change the direction of polarized light], and the lighter bands are called **I bands** [for *isotropic*]. A thin dark band, called a **Z disc,** or Z line, interrupts each I band and is the boundary of a sarcomere.

How Does a Muscle Contract?

When a muscle fiber contracts, so do the sarcomeres. The distance between the Z discs changes in the same way as the length of a muscle. As shown in Figure 18-5, the distance between Z discs shortens from about 2.5 μm to about 1.5 μm. The question "how does a muscle contract?" therefore becomes "how does a sarcomere contract?"

We can see that a lighter region (called the H zone) interrupts each A band. After contraction, the A band stays the same size, but the H zone region is smaller. The I band also gets smaller, as shown in Figure 18-5. By analyzing these changes with the electron microscope, researchers were able to figure out what was going on. At even higher magnifications they could see that the changing relationships of the different bands and zones resulted from the sliding of protein filaments.

In ribbon-like sections cut parallel to the axis of the muscle fiber, the electron microscope reveals the basis of the banding pattern. Both the light I bands and dark A bands of each sarcomere contain hundreds of tiny filaments running parallel to the axes of the myofibrils. These filaments come in two sizes—**thin filaments,** 7 nm in diameter, later shown to contain the protein actin, and **thick filaments,** 14 nm in diameter, later shown to contain the protein myosin. (See Chapter 6.) As diagrammed in Figure 18-6, the I band contains only thin filaments, while the A band contains both thick and thin filaments. The light H zone (in the middle of the A band) contains only thick filaments.

One possible explanation for muscle contraction, which we may call the "contracting filament hypothesis" (and which we now know to be incorrect), is that the filaments themselves contract like little coiled springs. Hugh Huxley and his collaborators, however, had another idea. Their electron microscopic observations showed that the filaments themselves did not change length. Although the I zone contracted, the thin filaments had the same length

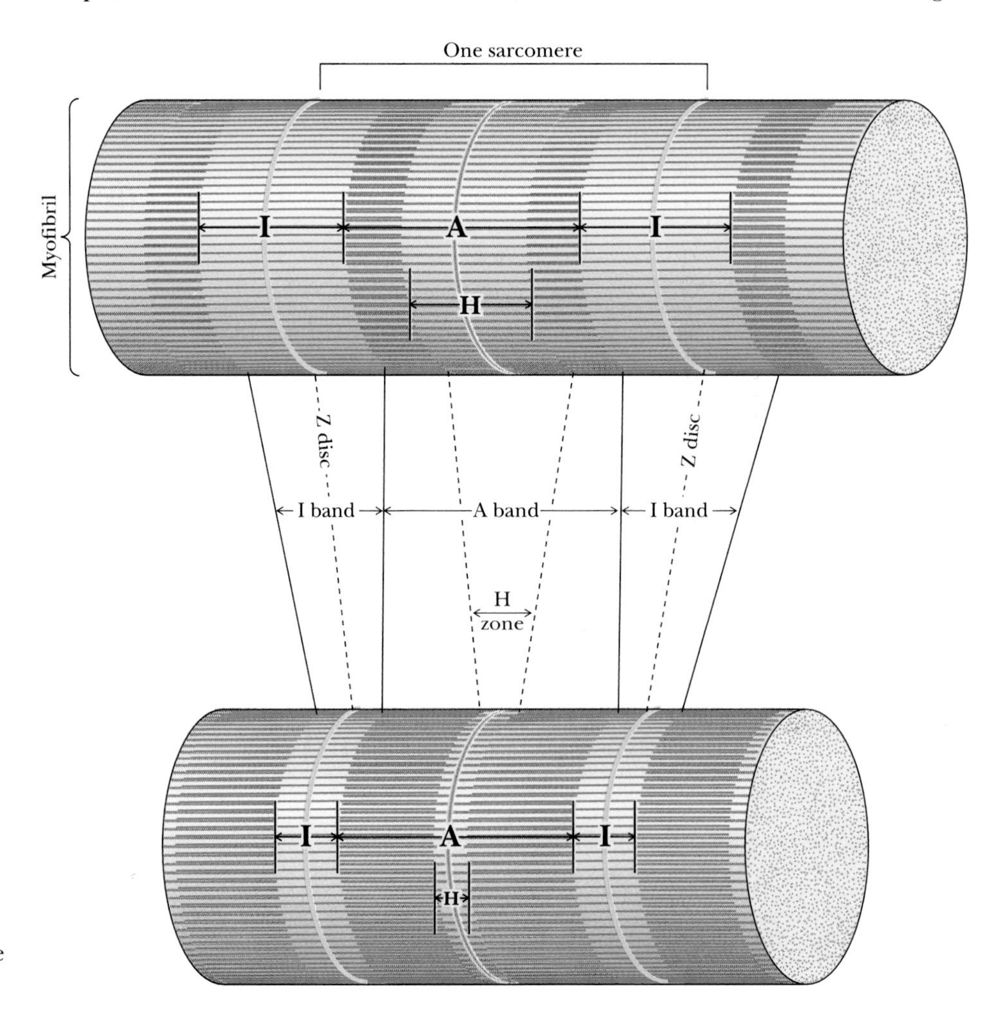

Figure 18-5 The contraction of a myofibril depends on the contraction of the sarcomeres. Each A band stays the same width, while I and H become narrower.

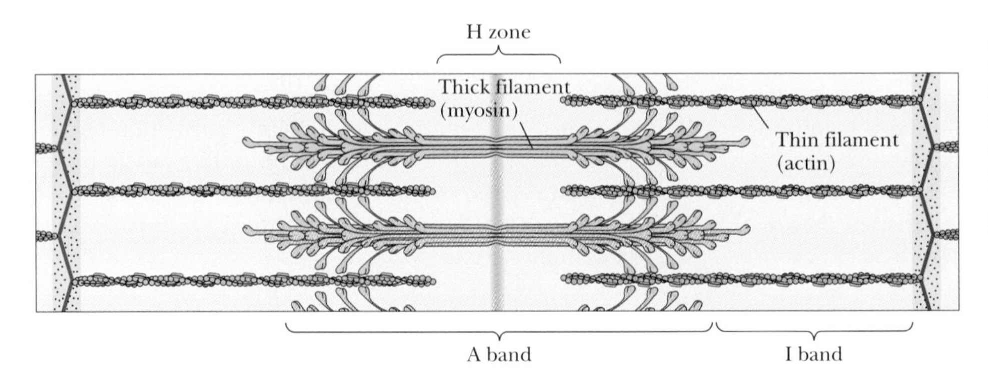

Figure 18-6 The molecular basis of the banding pattern seen in the microscope. The I band consists of thin (actin) filaments, and the A band of thick (myosin) filaments, as well as overlapping thick and thin filaments. The H zone is the region where there is no overlap.

before and after contraction. Furthermore, neither the A band nor the thick filaments changed their length. All that changed was the degree of overlap between the thin and the thick filaments. Huxley therefore proposed the **Sliding Filament Model,** shown in Figure 18-7, in which muscle movement results from changing the relative positions of thick and thin filaments as they slide along one another.

Cross-Bridges Between Thick and Thin Filaments Are Responsible for Muscle Contraction

The electron microscopic studies also suggested a mechanism for moving the filaments. Particularly good photographs showed that the thick and thin filaments were linked to each other at regular intervals by cross-bridges, which, on early electron micrographs, appeared as periodic fuzzy regions and, we now know, are actually extensions of the myosin molecules that make up the thick filaments. With newer techniques, the cross-bridges are far more obvious, as shown in Figure 18-8. Perhaps, Huxley suggested, the cross-bridges pull the thin and thick filaments over one another. Later work showed that this is the case: The cross-bridges actually perform the work of muscle contraction, somehow causing the filaments to move.

Huxley's hypothesis—that the cross-bridges between thick and thin filaments are responsible for generating force and performing work—led to a testable prediction: The force generated by a muscle should depend on the degree of overlap between the thick and thin filaments.

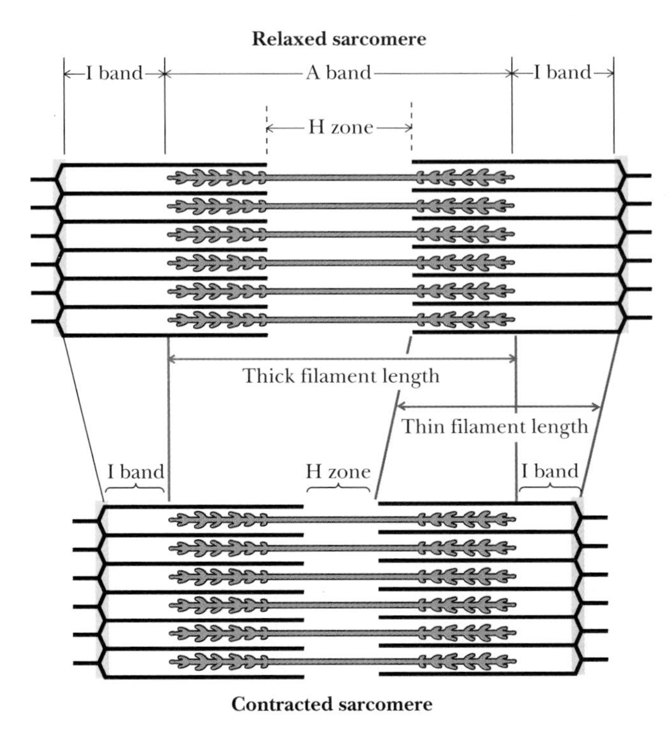

Figure 18-7 The Sliding Filament Model for muscle contraction. The lengths of thick and thin filaments do not change, but their overlap does, reducing the sizes of the I band and the H zone.

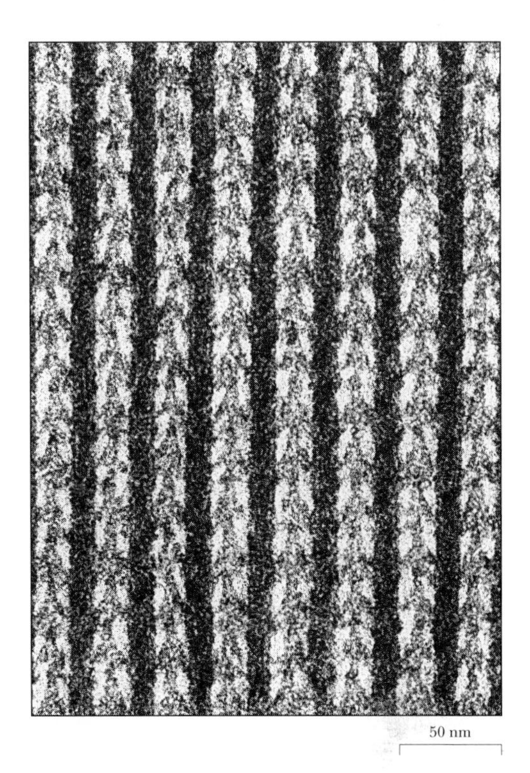

Figure 18-8 Cross-bridges between actin and myosin filaments. *(Mary Reedy, Duke University Medical Center)*

That is, if a muscle is sufficiently stretched, the thick and thin filaments would have little overlap and few cross-bridges would be available to move the filaments. Researchers tested this prediction by measuring the amount of force (by having the muscle pull against a spring) as they stretched a muscle to different lengths. The results were exactly as predicted by the Sliding Filament Model, supporting the importance of the cross-bridges in accomplishing muscle movement.

WHAT ARE THE THICK AND THIN FILAMENTS?

In the late 1940s Albert Szent-Gyorgyi, the Hungarian biochemist who also contributed to the discovery of the citric acid cycle, found a way to treat muscle fibers so that he could remove membranes and soluble materials, leaving behind only the proteins of the thick and thin filaments. These filaments, biochemists later found, consisted largely of two proteins, actin and myosin. Remarkably, the filaments still slid when provided with ATP. (See Chapter 1, p. 19.) So the mystery of muscle contraction must lie in the interactions of specific proteins with ATP.

Thin Filaments Are Composed of Actin

Actin is a relatively small polypeptide, containing about 375 amino acid residues, with a molecular size of about 40,000. (See Chapter 6.) At low salt concentrations actin is roughly spherical with a diameter of about 4 nm; in this state it is called "G-actin" [*g*lobular]. At higher salt concentrations, like those in cells, however, pure actin forms long filaments, called "F-actin" [*f*ilamentous], as shown in Figure 18-9. These actin filaments are 7 nm in diameter, suggesting that they make up the thin filaments of the myofibrils.

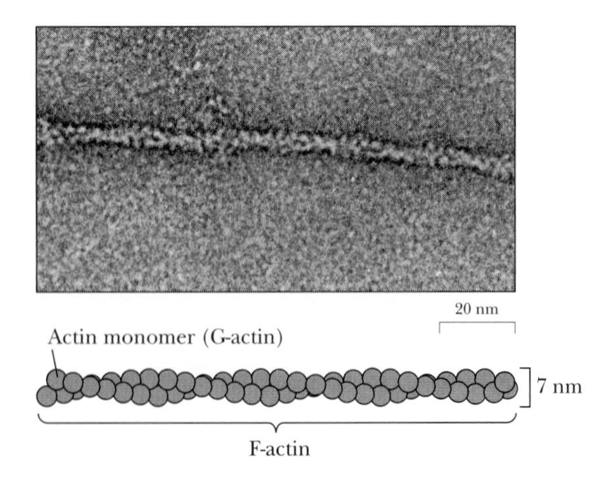

Figure 18-9 An actin filament. *(Hugh Huxley, Brandeis University)*

Immunocytochemistry ultimately confirmed that the thin filaments consist of actin. Researchers injected purified actin into rabbits to evoke the production of antibodies that form complexes with actin. They then tagged these antibodies with a fluorescent dye. The labeled antibodies against actin bind to the I bands and part of the A bands within each sarcomere, showing that the thin filaments contain actin.

Myosin Is the Principal Protein of the Thick Filaments

Myosin, the other major protein of muscle filaments, is much larger than actin: A single myosin molecule consists of six polypeptide chains and has a molecular size of about 500,000—about 4500 amino acid residues, compared with 375 for each actin molecule. Under the electron microscope, each myosin molecule has a long rod-like section, called a tail, and two little heads at one end, as sketched in Figure 18-10a. At higher salt concentrations, hundreds of myosin molecules assemble to form the bipolar structures diagrammed in Figure 18-10b. Each myosin assembly formed *in vitro* resembles the thick filaments of a muscle fiber with projecting cross-bridges. Each assembly contains about 300 myosin molecules and is about 1.5 μm long.

A myosin assembly has a smooth central zone about 0.15 μm long. Beyond this smooth zone are little heads that look like the cross-bridges that attach myosin to actin *in vivo*. By treating purified myosin with a protein-digesting enzyme, researchers can isolate head fragments. These fragments bind to purified actin filaments, just as we would expect for the cross-bridges.

How Does a Myofibril Use the Chemical Energy of ATP?

Muscle contraction accomplishes work and therefore requires energy. In cells this energy comes from ATP, produced mostly in the mitochondria. Szent-Gyorgyi's muscle protein preparation also required ATP for contraction. A good model of muscle contraction must explain how ATP causes the actin and myosin filaments to slide. Again, biochemical experiments with purified actin and myosin suggested an answer.

Myosin is an ATPase; that is, it can split ATP into ADP and phosphate (P_i). This activity is located within the myosin heads (which are the cross-bridges to actin filaments). The ATPase activity of myosin (or of isolated myosin heads) is low, however, unless the myosin first associates with actin. Interaction with actin evidently changes the conformation of the myosin so that the ATPase activity increases dramatically. The ability to split ATP lies entirely in the cross-bridges but depends on the proper interaction of actin and myosin.

(a)

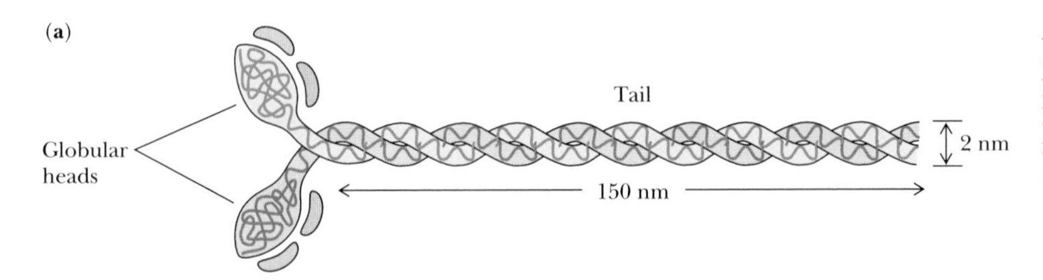

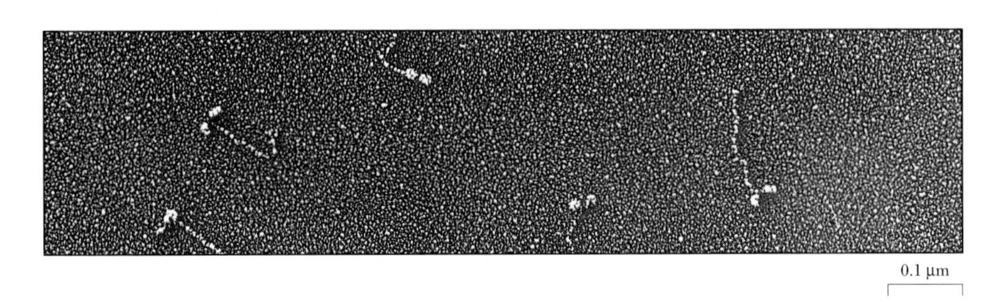

(b)

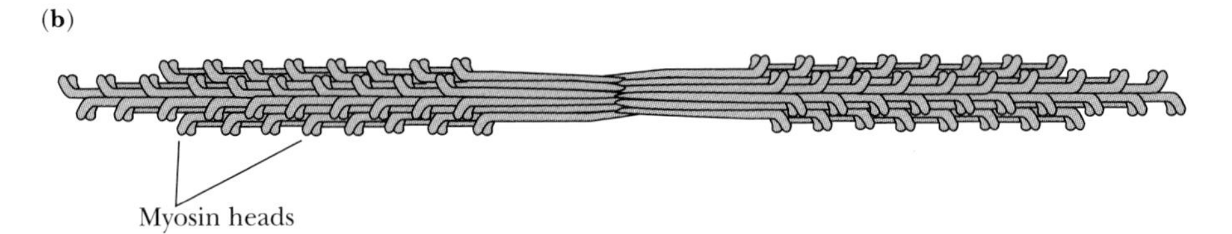

Figure 18-10 A myosin filament. **(a)** A single myosin molecule, showing long rod-like tail attached to two heads; **(b)** the assembly of myosin molecules forms a myosin filament. *(a, Dr. Roger Craig/Peter Arnold)*

The splitting of ATP ultimately provides the energy needed for the cross-bridges to pull the actin and myosin filaments past one another, acting rather like the hands of a rope climber. But the actual mechanism of this energy transduction is still a matter of debate and experiment. As expected, the addition of ATP to a mixture of pure actin and myosin also changes the way they interact, but no one yet knows the precise way in which ATP hydrolysis causes a conformational change that directly drives muscle contraction.

The story, as it is most generally understood, is illustrated in Figure 18-11: When ATP binds to a myosin head, it stimulates the head's detachment from the adjoining actin filament. ATP hydrolysis immediately follows, but the ADP and P_i remain attached to the myosin head, and there is little change in free energy. The "cocked" myosin head (in a high-energy state) now binds to the actin filament, following which the P_i is released and the myosin head changes its conformation, providing a power stroke that pulls the attached actin filament along by about 12 nm, the length of two actin monomers. The cycle begins again when another ATP replaces the attached ADP molecule.

The failure to find biochemical evidence for a conformational change associated with the power stroke (as P_i is released from the actin-myosin complex) has prompted an alternative view of muscle contraction. According to this view, the movement of the myosin head depends on random thermal motion, with the ATP somehow contributing to a ratcheting mechanism that allows thermal motion to drive the filaments in one direction rather than the other.

According to either of these models of filament movement, the depletion of ATP leads to a locking of myosin heads to the actin filaments in a low-energy position (illustrated in Figure 18-11). This arrangement is associated with *rigor mortis,* the rigidity of the body that occurs in the hours following death. Rigor mortis disappears after cross-bridges detach as a result of the fall in calcium ion concentration (as discussed later in this chapter).

Tropomyosin and Troponin Regulate the Interactions of Actin and Myosin Filaments

In intact muscle cells, the trigger for contraction to begin is an increase in the calcium ion concentration. Even isolated muscle proteins (such as those present in Szent-Gyorgyi's preparation) contract only if calcium ions are present. Our next question, then, concerns the role of calcium ions in the sliding of actin and myosin filaments.

Two proteins are attached to the actin filament and regulate its interaction with myosin. These are **tropomyosin,** a rigid, rod-like protein about 40 nm long that stiffens the actin filament, and **troponin,** a large protein that binds both to tropomyosin and to free calcium ions. Experiments with purified tropomyosin, troponin, and actin have revealed the role of these proteins and of calcium ions in regulating muscle movement.

When no calcium ions are present, tropomyosin pre-

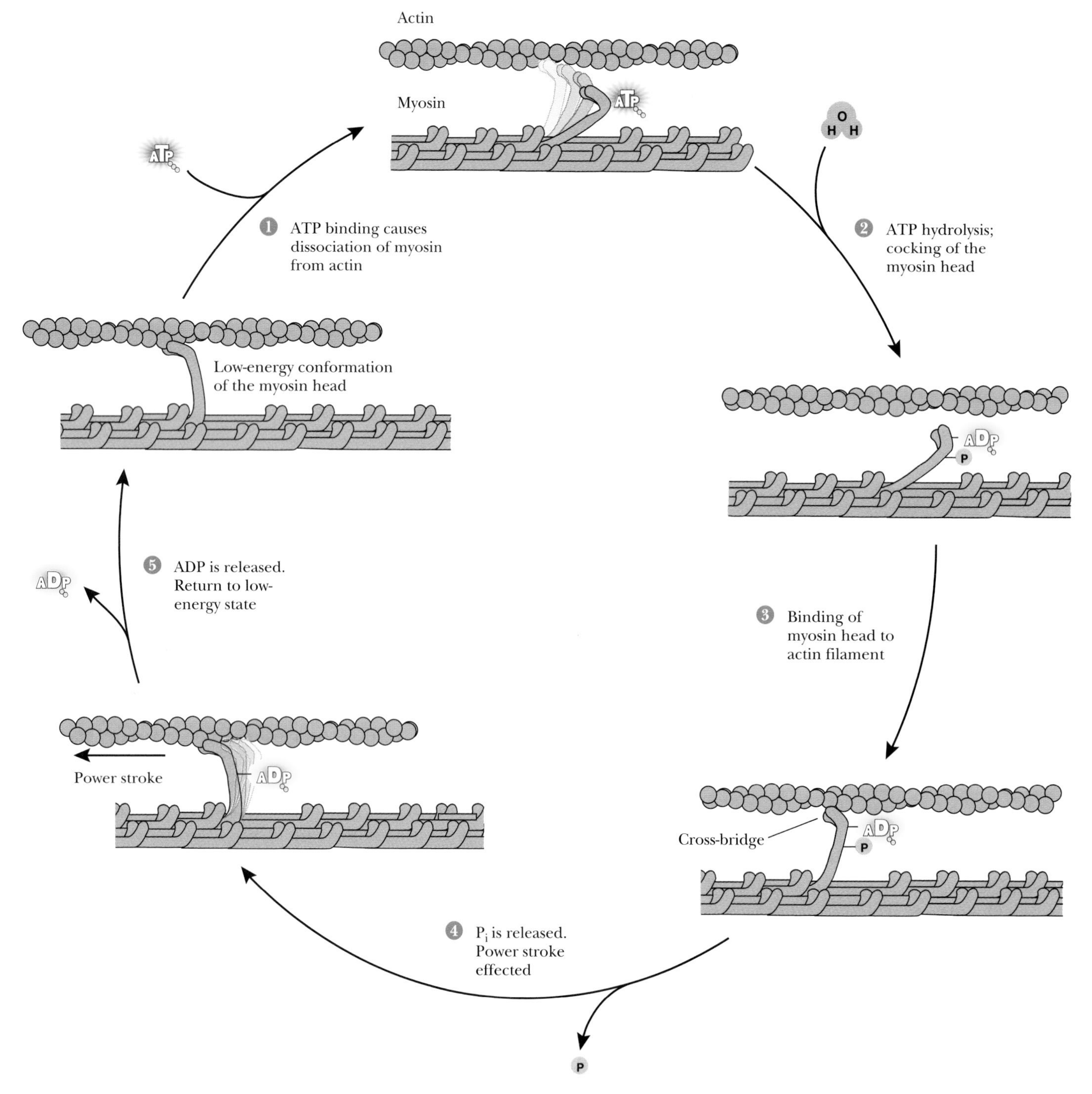

Figure 18-11 A widely accepted model of how ATP causes the sliding of actin and myosin filaments. Each power stroke moves the actin filament by about 12 nm.

vents the myosin heads from binding to actin filaments, as diagrammed in Figure 18-12. Addition of calcium, however, allows actin and myosin to interact.

The action of calcium appears to depend on troponin. Troponin consists of three polypeptides, one of which (TnC) binds to calcium ions. The binding of calcium changes the shape of the troponin molecule. Troponin in turn acts as a wedge to push tropomyosin away from the myosin-binding regions of the actin filaments. In this way, calcium indirectly but nonetheless powerfully controls the sliding of the filaments.

As long as calcium ions are present in high enough concentration, the molecular machinery of the myofibril can use the energy in ATP to drive contractions. When calcium concentrations are too low ($<10^{-7}$ M), the cross-bridges cannot form and no contraction occurs.

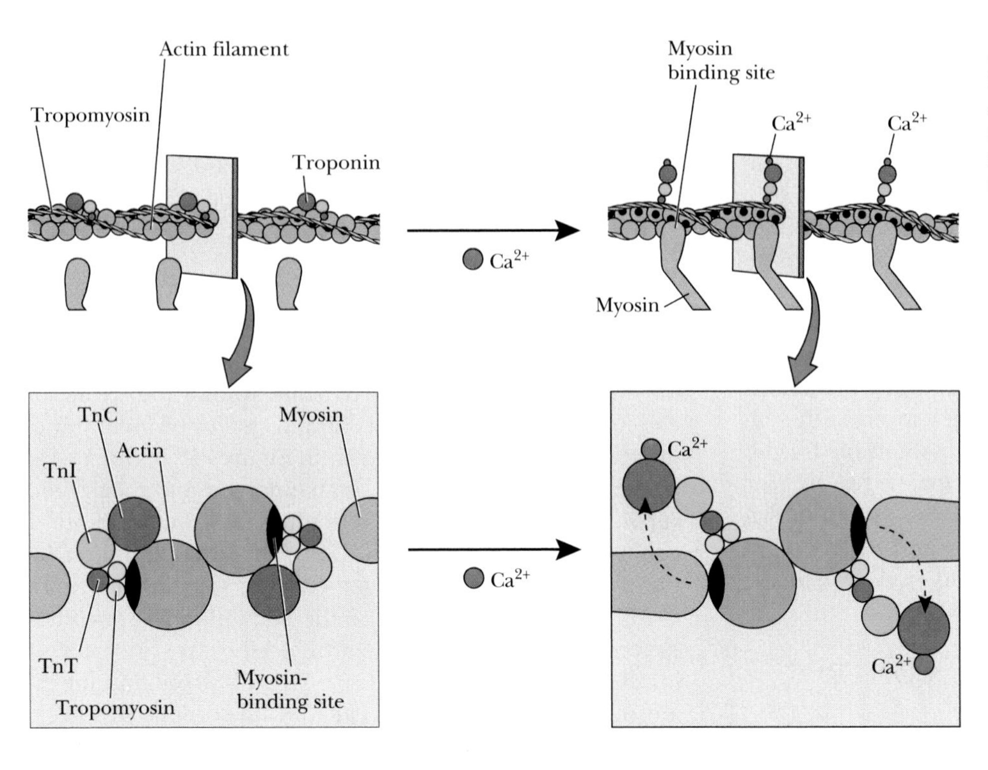

Figure 18-12 Interaction of troponin, tropomyosin, and Ca^{2+} ions. Addition of Ca^{2+} ions to troponin alters the position of tropomyosin along actin, allowing actin to bind to the myosin cross-bridges.

Other Muscle Proteins Anchor Actin and Myosin Filaments to Cellular Structures

The ability of sliding filaments of actin and myosin to power the macroscopic contraction of a skeletal muscle depends on the attachment of the filaments to the structures within each sarcomere and on the linking of parallel myofibrils in register. This linkage depends on a number of proteins whose structures and interactions have been studied *in vitro* and whose intracellular locations have been determined by immunocytochemistry.

One set of proteins anchors the ends of the actin filaments to the Z disc. The anchored ends are the *plus* ends of the actin filaments. (See Chapter 6, p. 180.) As shown in Figure 18-13, a major component of the attachment complex is α-actinin, which is also found in nonmuscle cells, where it acts as a bundling protein, holding together adjacent actin filaments. Another set of proteins includes

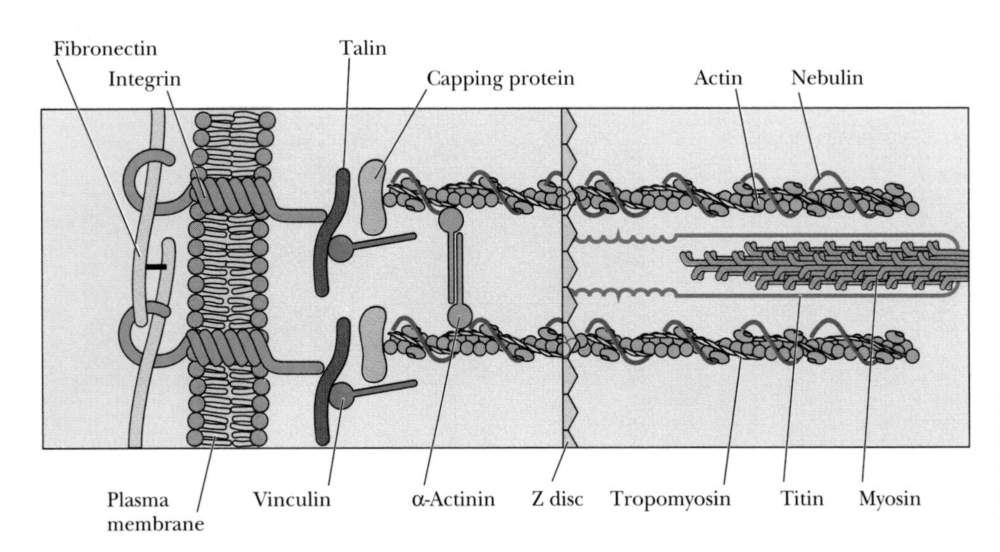

Figure 18-13 Additional muscle proteins, titin and nebulin, stabilize thick and thin filaments and keep them from sliding apart, while α-actinin helps anchor actin filaments to the Z-disc.

spectrin, ankyrin, and vinculin—all of which are also found in nonmuscle cells—and anchors the Z discs to the plasma membrane and to an intermediate filament protein called **desmin.** (See Chapter 6, p. 181.) **Dystrophin,** a protein whose altered structure or absence causes muscular dystrophy, is also involved in this anchoring.

Two distinct sets of proteins stabilize the thick and thin filaments, while another set is found in the central region of the thick filaments. In addition to the anchoring proteins, the Z disc also contains high levels of the enzyme creatine phosphokinase, which is responsible for the production of phosphocreatine, the major reserve for high-energy phosphate bonds in muscle. As shown in Figure 18-13, another set of proteins—*titin* and *nebulin*—appears to keep the filaments from sliding entirely apart.

HOW DO LIVING MUSCLES CONTROL CONTRACTIONS?

Szent-Gyorgyi's preparation of muscle protein (discussed in Chapter 1) shows the dependence of contraction on calcium ions and on ATP. When no plasma membranes are present, an experimenter can easily regulate the concentrations of calcium ions and ATP. But how does an intact cell regulate calcium ion concentration?

An Increase in Calcium Ion Concentration Triggers Muscle Contraction

Researchers have studied rapid changes in calcium ion concentrations by taking advantage of specialized molecules that emit light when they bind calcium. One such molecule is the protein **aequorin,** isolated from a luminescent jellyfish. An experimenter can inject aequorin into a muscle fiber and then use a light detector to measure the amount of emitted light and thereby the concentration of calcium ions, as illustrated in Figure 18-14. Calcium concentrations go up inside the cells (from less than 10^{-7} M to more than 10^{-6} M) just at the beginning of a contraction. This temporary increase stimulates the muscle to contract.

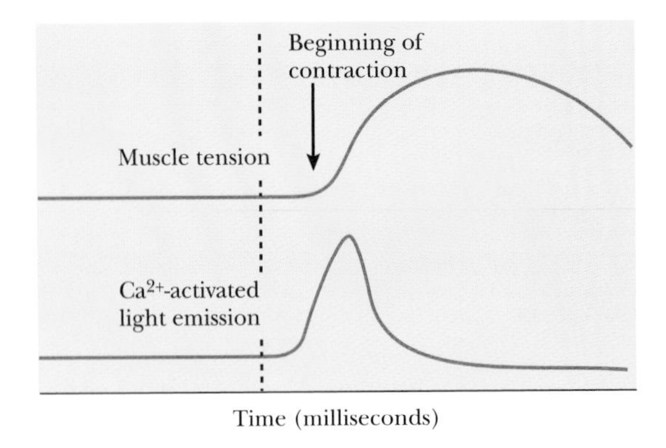

Figure 18-14 An increase in free Ca^{2+} ions, here measured by aequorin fluorescence, triggers muscle contraction.

Where Do Calcium Ions Come From and What Triggers Their Release?

The calcium ions that trigger skeletal muscle contraction do not appear and disappear from thin air, but from an extensive **sarcoplasmic reticulum,** a system of membrane-lined channels and sacs, shown in Figures 18-1 and 18-15. The sarcoplasmic reticulum is an enlarged and specialized version of the endoplasmic reticulum found in most other cell types. (See Chapter 5, p. 135.) The sarcoplasmic reticulum serves as a reservoir for calcium ions in each muscle fiber. The membranes of the sarcoplasmic reticulum have active calcium pumps, which take up free calcium ions from around the myofibrils. They also contain a mechanism for rapidly releasing calcium on command.

The command to release calcium comes from another system of membrane channels, called the transverse tubules, also shown in Figures 18-1 and 18-15. These channels are continuous with the plasma membrane that surrounds the whole fiber. (Remember, the entire fiber is a single cell.) The transverse tubules contact each sarcomere at the Z disc. They signal the sarcoplasmic reticulum to release calcium ions.

Figure 18-15 The sarcoplasmic reticulum, a system of channels that serves as a reservoir for Ca^{2+} ions, and the transverse tubules. *(Don W. Fawcett/Photo Researchers)*

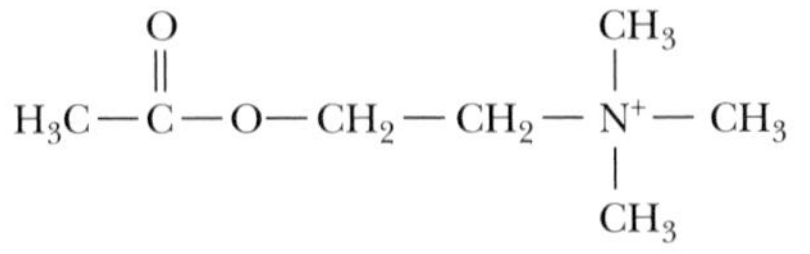

$$H_3C-\overset{\overset{\displaystyle O}{\|}}{C}-O-CH_2-CH_2-\underset{\underset{\displaystyle CH_3}{|}}{\overset{\overset{\displaystyle CH_3}{|}}{N^+}}-CH_3$$

Figure 18-16 The chemical structure of acetylcholine, produced by motor neurons and released into the neuromuscular junction. Acetylcholine stimulates ions to flow through the plasma membrane, ultimately leading to contraction.

Chemical Commands from Nerve Cells Stimulate the Release of Calcium in Muscle Cells

Each muscle fiber receives chemical commands from a motor neuron, a nerve cell that carries information from the nervous system to the muscle. A motor neuron triggers a contraction by releasing **acetylcholine,** whose structure is shown in Figure 18-16. (See Chapters 1 and 23.) The motor neuron does not bathe a muscle cell in acetylcholine, however. Instead, it releases a small puff at the neuromuscular junction, a specialized region of contact between a motor neuron and a muscle fiber.

The acetylcholine diffuses from the motor neuron to the muscle cell's plasma membrane. There it stimulates a change in flow of ions across the membrane. This alteration temporarily changes the voltage that exists across the muscle cell's plasma membrane, as discussed in some detail in Chapter 23. The plasma membrane quickly propagates this voltage change from the neuromuscular junction to the entire plasma membrane within a few thousandths of a second. The voltage change spreads not only over the cell's surface but also through the transverse tubules. These electrical changes bring about an increase in free calcium ions and a consequent contraction of the muscle fiber.

In a condition called **myasthenia gravis,** the patient's immune system inappropriately produces antibodies that recognize and block the receptors for acetylcholine on muscle cells. The result is a weakness that results from the inability to stimulate proper muscle contraction. This tragic illness illustrates the importance of acetylcholine in initiating muscle movement.

Physiologists can mimic the effects of acetylcholine on a muscle cell by electrical stimulation, an experiment first performed by the Italian physiologist Luigi Galvani in 1798. (See Chapter 1, p. 19.) Electrical changes in the transverse tubules directly stimulate the release of calcium from the adjoining sarcoplasmic reticulum. The released calcium in turn triggers the action of actin and myosin.

We may now trace the command to contract from the motor neuron to the myofibril. The motor neuron issues its order in the form of a chemical, acetylcholine. The muscle cell's response is to change the flow of ions across the membrane, creating a change in membrane voltage. This electrical change moves along the cell surface and into the system of transverse tubules and triggers the release of another chemical (called inositol trisphosphate), which quickly diffuses to the sarcoplasmic reticulum. (See Chapter 19.) As a result, the sarcoplasmic reticulum releases some of its accumulated calcium into the myofibrils. The calcium then binds to troponin, causing a shape change that displaces tropomyosin from its position on the actin filaments. The movement of the tropomyosin allows the myosin cross-bridges to bind to actin. Once the cross-bridges form, they use the energy of ATP to pull the myosin filaments along the parallel actin filaments. This process, which occurs in a few milliseconds, is summarized in Figure 18-17.

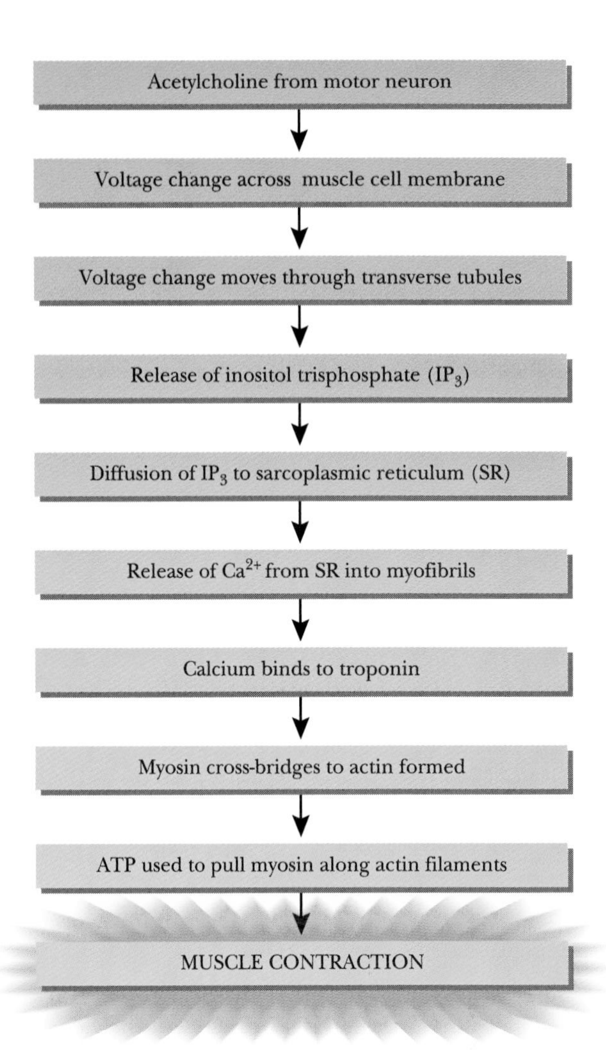

Figure 18-17 The sequence of events initiated by acetylcholine, culminating in muscle contraction.

In vertebrates, motor neurons themselves receive commands from the spinal cord; in most but not all cases, the signal to contract ultimately comes from the brain. In Chapter 23, we examine some of the ways the nervous system coordinates information that affects the activity of motor neurons. However complex this information processing may be, the commands of the motor neuron to the muscle fibers represent the "final common pathway" for the performance of any movement.

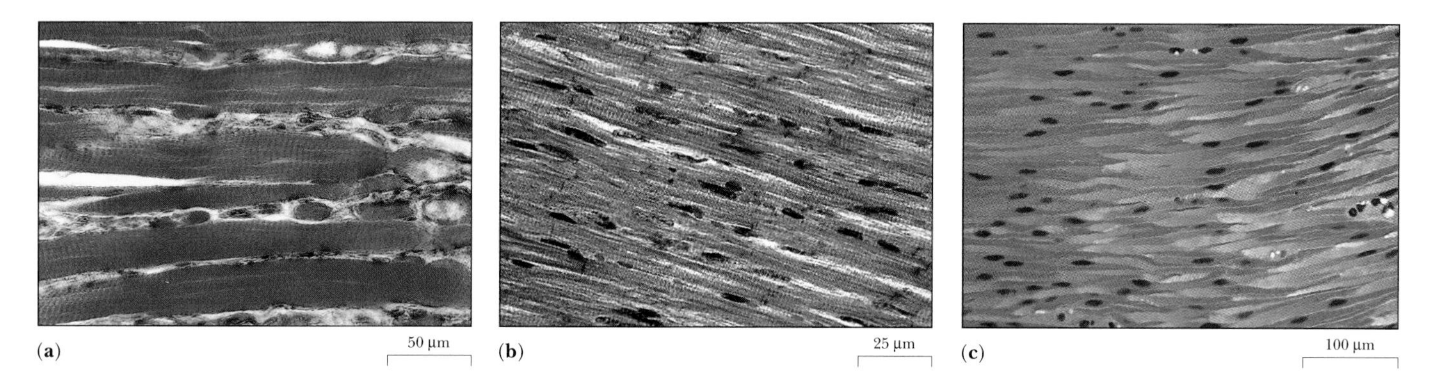

Figure 18-18 Cardiac muscle, like skeletal muscle, is striated, while smooth muscle is not. **(a)** Skeletal muscle; **(b)** cardiac muscle; **(c)** smooth muscle. *(a, David M. Phillips/Visuals Unlimited; b, Biophoto Associates/Science Source/Photo Researchers; c, M. I. Walker/Science Source/Photo Researchers)*

SMOOTH MUSCLES AND CARDIAC MUSCLES ALSO USE ACTIN AND MYOSIN TO ACCOMPLISH MOVEMENT

Vertebrates have other muscles as well as those that move the skeleton. These include **cardiac muscle,** which pumps blood through the heart, and **smooth muscle,** which lines the walls of hollow internal organs such as intestines and blood vessels. Cardiac muscles (like skeletal muscles) are striated, as shown in Figure 18-18b. In contrast, smooth muscles (as their name suggests) do not have striations, as shown in Figure 18-18c.

At least to some extent, humans and other animals control their body movements, so skeletal muscles are often called **voluntary muscles.** In contrast, smooth muscles are **involuntary muscles.** Cardiac muscle, however, is a special case: It is striated but by no means voluntary.

Cardiac and smooth muscle accomplish contraction by using the energy of ATP to drive sliding filaments of actin and myosin, but they differ from skeletal muscle in several ways. For example, each has a distinct type of actin, closely related but not identical to the actin of skeletal muscles. And, as shown in Figure 18-19 for smooth muscle, the arrangements of actin and myosin filaments differ from those of skeletal myofibrils.

Smooth Muscle Consists of Tightly Connected Cells, Each with a Single Nucleus

Smooth muscle consists of individual spindle-shaped cells each of which has a single nucleus. In the human intestine, for example, these cells are about 200 μm long and about 5 μm in diameter, as shown in Figure 18-18c.

Far smaller than the multinucleate fibers of striated muscle, the cells of smooth muscle have no transverse tubules, and their sarcoplasmic reticulum consists mostly of flat vesicles just under the plasma membrane. Smooth muscle cells, however, are closely connected to one another by gap junctions (as shown in Figure 18-20), which allow the passage of materials between the cytoplasm of adjacent cells. (See Chapter 6, p. 174.) Electrical signals pass from cell to cell over distances slightly smaller than those of a skeletal muscle fiber. Because of this electrical coupling, many individual cells can function as a single unit, releasing calcium ions simultaneously in response to stimulation.

Muscle contractions last much longer in smooth muscle than in skeletal muscle, with maximum tension reached in about 5 seconds, in contrast to less than 0.1 second for a skeletal muscle. In addition, smooth muscles have no defined "resting" length, and a smooth muscle can work over a much larger range of lengths than can a skeletal muscle. Stretching a smooth muscle, such as one from the intestinal tract, triggers contraction.

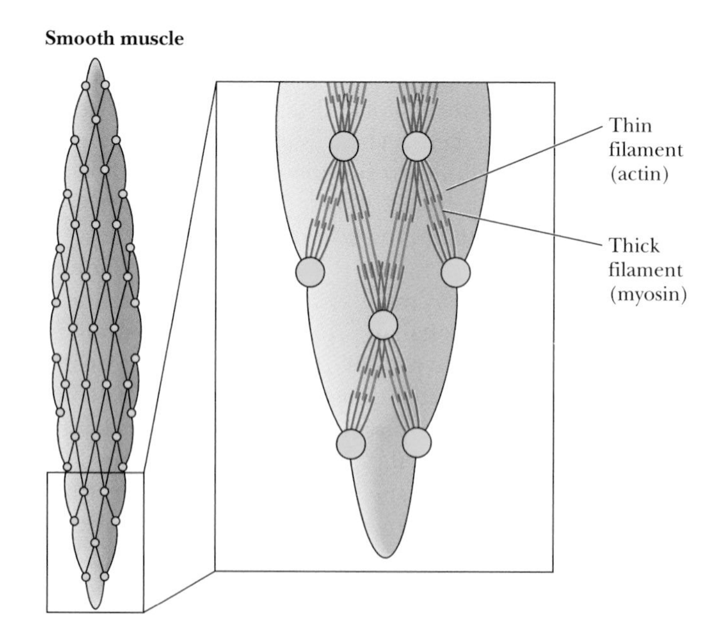

Figure 18-19 The arrangement of actin and myosin filaments in smooth muscle.

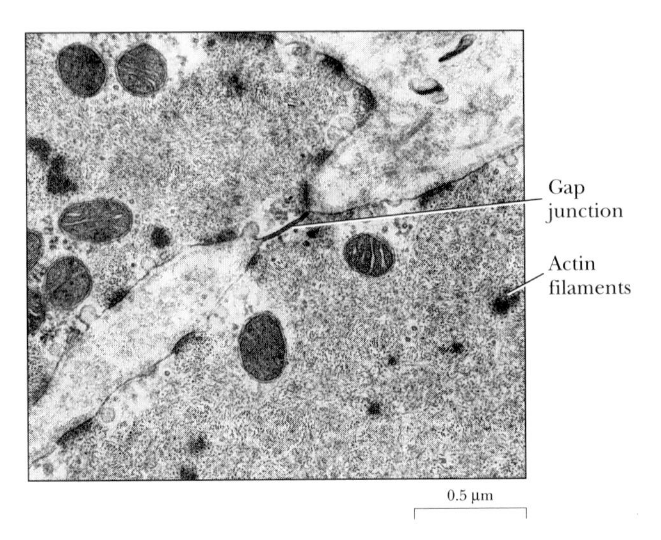

Figure 18-20 Gap junctions between smooth muscle cells allow passage of ions and permit electrical coupling of many cells within a single smooth muscle. *(E. E. Daniel, McMaster University)*

The organization of actin and myosin filaments in smooth muscle is more variable than in skeletal muscle. In smooth muscle, most actin filaments are relatively short and lie in bundles parallel to the long axis of the muscle cell. These bundles attach to the plasma membrane in dense bands that contain vinculin, one of the proteins in the Z-disc complexes of striated muscle. Actin filaments in smooth muscle can also form obliquely arranged complexes with a type of desmin (an intermediate filament). The oblique arrangement of the actin bundles allows smooth muscles to shorten much more than skeletal muscle—to as little as 20% of the starting length, compared with about 60% in skeletal muscle.

The contraction of smooth muscle, like that of skeletal muscle, depends on the concentration of free calcium ions in the cytoplasm. The response to calcium occurs by the mechanism illustrated in Figure 18-21. This mechanism differs from the mechanism in skeletal muscle outlined in Figure 18-17. In smooth muscle, both the binding of myosin to actin filaments and ATPase activity depend on the prior attachment of a phosphate to one of the myosin polypeptides, called the **myosin regulatory light chain.** The enzyme that transfers a phosphate group from ATP to the myosin light chain is called **myosin light-chain kinase.** This enzyme is active only when it complexes, in the presence of calcium ions, with **calmodulin,** a ubiquitous calcium-binding protein. The sequence of regulatory events is slower than that in skeletal muscles: Calcium ions first bind to calmodulin, the complex then binds to myosin light-chain kinase, which then catalyzes the transfer of a phosphate group from ATP to a myosin light chain, which then changes the conformation of myosin so that it can bind to actin and harness energy from ATP hydrolysis. Nonmuscle cells also use myosin light-chain kinase and calmodulin to regulate myosin-dependent cellular movement.

Another calcium-sensitive mechanism (illustrated in Figure 18-21) can also control the contraction of smooth muscles. This regulation involves **caldesmon,** an actin-binding protein that interferes with the binding of actin and myosin. This interference is prevented by the binding of caldesmon to a calcium-calmodulin complex, allowing

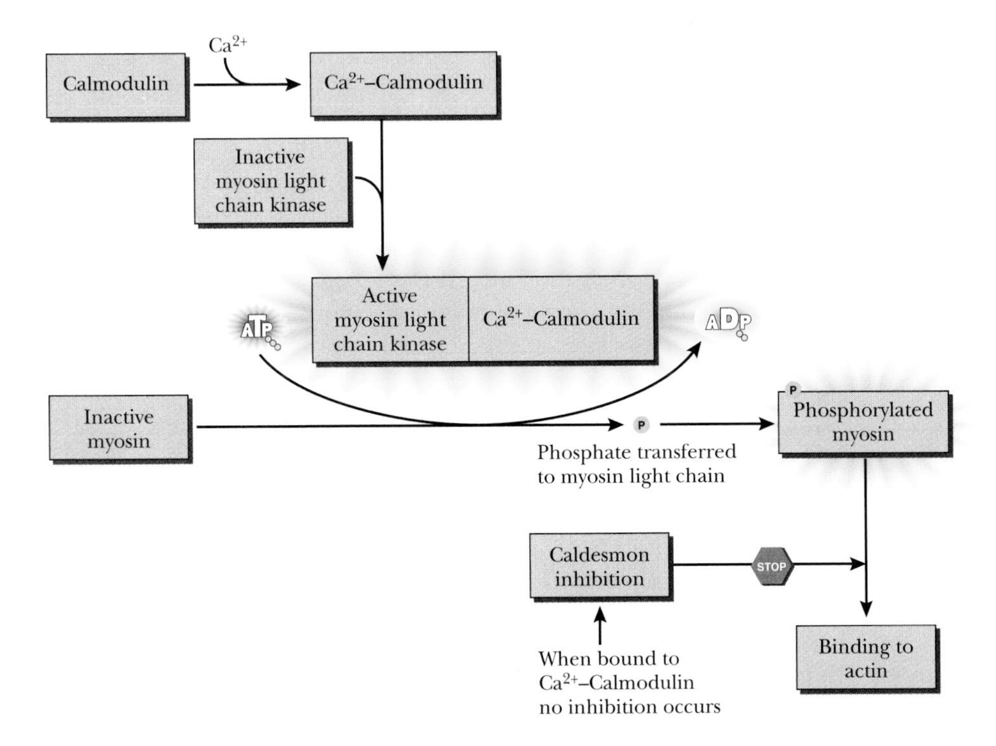

Figure 18-21 As in skeletal muscle, an increase in Ca^{2+} initiates contraction in smooth muscle, but by a different mechanism.

ESSAY

REGULATION OF MYOSIN ACTION BY CALCIUM AND CALMODULIN

Many cellular movements—from muscle filaments to protozoan pseudopods—depend on myosin motors, which propel themselves and their attached passengers along actin tracks. Calcium (Ca^{2+}) ions regulate all these movements. In skeletal muscle, for example, movement requires an increase in free Ca^{2+} ions; skeletal muscle myosin binds actin only after Ca^{2+}-bound troponin displaces interfering tropomyosin from actin filaments.

Cell biologists and biochemists have studied about 20 different myosin motor molecules whose sources range from the epithelia of mammalian intestines to the eyes of *Drosophila* to migrating slime molds. In all cases studied, Ca^{2+} ions regulate myosin's interactions with actin, though by widely differing mechanisms. In contrast to skeletal muscle myosin, for example, the movement of smooth muscle myosin depends on the phosphorylation of a myosin polypeptide. Ca^{2+} ions regulate the activity of the responsible enzyme, myosin light chain kinase. Ca^{2+} ions do not bind directly to myosin, however, but to the ubiquitous regulatory protein, calmodulin. (See Chapter 19.) Ca^{2+} binding changes the three dimensional structure of calmodulin and alters the interaction between calmodulin and the kinase molecule.

Comparing the structures of the many myosins reveals common structural features. These common features suggest some of the ways by which Ca^{2+} ions can regulate myosin movement. All myosins contain at least one huge polypeptide with three distinct domains—head, neck, and tail. As in skeletal muscle myosin (myosin II), whose structure is shown schematically in Figure 18-23, the globular heads contain the catalytic site for ATP hydrolysis, which provides the energy for the molecular motor.

In skeletal muscle myosin, the long tail regions of two molecules interact to form a dimer. In many other myosins, the tails are much shorter and do not form dimers. In some myosins, however, the tails tether myosin to a membrane or, in some cases, regulatory proteins involved in cytoskeletal reorganization.

Each neck region binds two small polypeptides called myosin light chains. In skeletal muscle the light chains do not bind Ca^{2+} but their amino acid sequences have strong resemblance to calmodulin. Calmodulin itself consists of a single polypeptide, about 150 amino acid residues long, roughly the size of skeletal muscle light chains. Calmodulin binds four Ca^{2+} ions, with each Ca^{2+} ion binding to N or O atoms in a loop between two perpendicular α-helices.

The neck region of every known myosin molecule contains a sequence pattern, called an "IQ motif" because each pattern starts with isoleucine (I) and glutamine (Q). Each IQ motif binds calmodulin or another light chain. The brush borders of intestinal epithelia, which contain microvilli rich in actin and myosin, have been a particularly valuable source of calmodulin-binding myosins.

For brush border myosin, interactions between actin and myosin depend on the binding of calmodulin to myosin. At low Ca^{2+} concentrations (below 10^{-7} M) each myosin molecule binds to three or four calmodulin molecules. At high Ca^{2+} concentrations (above 10^{-6} M), calmodulin binds to Ca^{2+} and dissociates from myosin. The calmodulin-free myosin can no longer move along actin filaments.

The number of different myosin molecules is far greater than was expected only a few years ago. Despite their diversity, all the myosins resemble one another in their overall organization and in their ability to associate with Ca^{2+}-binding proteins.

---

the smooth muscle to contract when intracellular calcium increases and reinforcing the action of myosin light-chain kinase.

Nerve connections are also different in smooth muscle. In general, we cannot consciously control the contractions of smooth muscles. Their stimulation comes from nerves of the autonomic nervous system, responsible for the regulation of the internal organs. (One exception to this is the partially conscious control of the smooth muscles of the urinary bladder.) Smooth muscle can generate

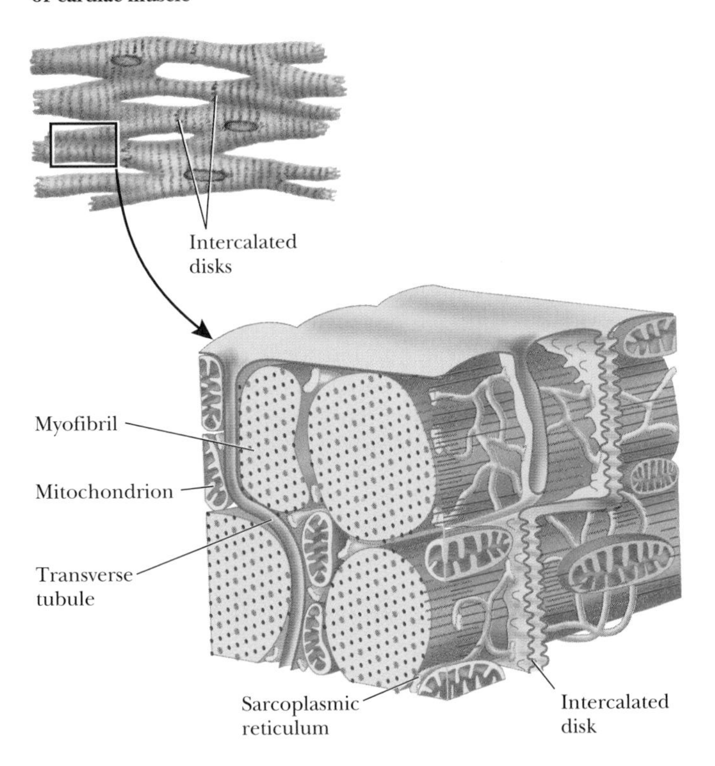

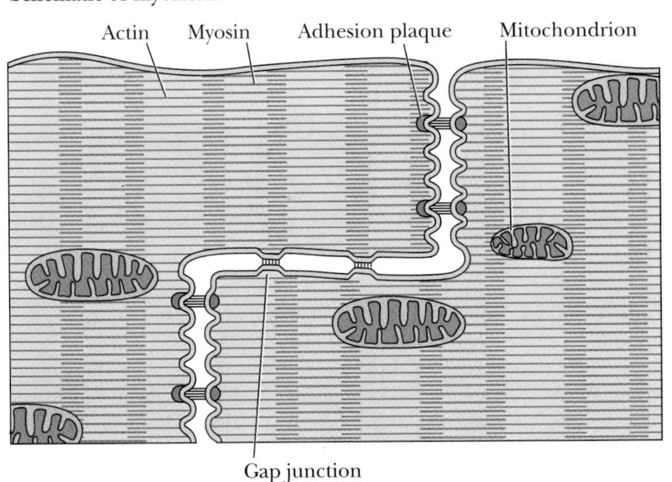

Figure 18-22 Cardiac muscle, showing intercalated discs. *(Photo from Bloom and Fawcett, 11th edition/Photo Researchers)*

its own signals for contraction, even without neural input. This is the case for many of the movements of the intestine.

Cardiac Muscle Consists of Tightly Connected Individual Cells

Cardiac muscle, like smooth muscle, consists of individual cells, each with its own nucleus and cell membranes, as shown in Figure 18-22. The cellular organization of cardiac muscle is unique, with branched fibers interrupted every 100 μm by structures called **intercalated discs,** as illustrated in Figure 18-22.

Like smooth muscle, cardiac muscle cells are electrically coupled through gap junctions. Also like smooth muscle, cardiac muscle can generate its own signals for contraction. Chemical signals from the autonomic nervous system only modulate the intrinsic contractions of the heart.

Some Invertebrate Muscles Show Distinctive Molecular Organizations

Some muscles in invertebrates resemble skeletal muscle, whereas others resemble smooth muscle. A scallop, for example, has two sets of muscles that it uses to open and close its shell—a striated set for the rapid movements of swimming and a smooth set that can hold the shell closed for days or weeks at a time. The smooth muscles, called **catch muscles,** have thick filaments that consist of a protein called **paramyosin,** which is thought to be responsible for maintaining the tension for so long with very little expenditure of energy.

Insect flight muscles are the most rapidly contracting muscles known, with as many as 900 contractions every second. These contractions, however, may represent only 1%–2% of the resting muscle length. The stretching of a flight muscle immediately triggers the contraction of the muscle that pulls the wing in the opposing direction. Studies of the structure of insect flight muscle reveal filaments that resemble titin and nebulin of vertebrate muscles. These filaments may contribute to the flight muscle's ability to contract so rapidly.

ACTIN AND MYOSIN ARE ALSO RESPONSIBLE FOR MANY NONMUSCULAR MOVEMENTS

Actin is a major component of the cytoskeleton of essentially all eukaryotic cells, comprising 5%–10% of a cell's protein. As discussed in Chapter 6, actin filaments contribute to establishing and changing cell shape. For example, during animal development, cells move about and change shape as a single-celled zygote becomes a multicellular embryo and cell specialization begins. These movements depend heavily on the assembly of actin filaments, and they stop when actin assembly is prevented. (See Chapter 20.) Changes in the interactions between actin filaments and actin-binding proteins are also largely responsible for local changes in consistency and viscosity of the cytoplasm, which contribute importantly to cell movement (see later in this chapter).

The actin in nonmuscle cells differs slightly from muscle actin. In mammals, two additional actin genes, distinct from muscle actins, encode nonmuscle actins. The nonmuscle actins are less similar to the muscle actins than the muscle actins are to each other, but no one has yet found the functional significance of structural differences among the six known actins.

The Assembly and Disassembly of Actin Filaments Contribute Both to Cell Movements and to Changes of Cell Shape

The assembly of actin monomers (G-actin) into actin filaments (F-actin) and their disassembly are responsible for most changes in the shapes of cells. Actin monomers can add to either end of a filament, but they are more likely to add to the plus end than to the minus end. (See Figure 6-32, p. 180.) The growth of an actin filament is faster when the monomers are attached to ATP and slower after the actin-ATP is split to actin-ADP. The preferential addition of subunits at the plus end and the preferential subtraction at the minus end bias the movement in a single direction.

Cell biologists often use two compounds produced by fungi to assess the role of actin polymerization and depolymerization in cell movements, such as the movement of amebas, and cell shape changes, such as those that occur during cytokinesis. **Cytochalasin** binds to the plus end of actin filaments and prevents their further growth, while **phalloidin,** another fungal product, binds along the sides of polymerized actin and prevents its dissociation into monomers.

Cells treated with cytochalasin cannot perform cytokinesis, the division of the cytoplasm following mitosis (as described in Chapter 9), although they can accomplish the chromosome movements of mitosis. Cytochalasin treatment also prevents **cell locomotion** (also called **cell crawling** or **ameboid movement),** the movement of cells over a surface, and the changes in cell shape that occur during development. (See Chapter 20.)

Phalloidin cannot pass through the plasma membrane and must be directly injected into the cytoplasm. Such phalloidin-injected cells, like cytochalasin-treated cells, cannot accomplish cytokinesis or cell locomotion. This result supports the view that actin filament depolymerization, as well as polymerization, is also required for cell movements, helping to establish the biological importance of the growth and dissociation of actin filaments. Phalloidin is one of the lethal poisons in the "destroying angel" mushroom, *Amanita phalloides.*

Many Cell Movements Depend on Interactions Between Myosin and Actin

Almost all eukaryotic cells contain myosin as well as actin. Two kinds of myosin contribute to cell movements—myosin II (or *conventional myosin*), which resembles the myosin in muscle filaments, and myosin I, a related but much smaller protein. Like myosin II, myosin I has an actin-binding head and a tail (as shown in Figure 18-23), but myosin I's tail does not promote association with other myosin molecules into thick filaments. The tail of myosin I, however, can bind to membranes and vesicles. Myosin I appears to be responsible for **cytoplasmic streaming,** the movements of membrane-bounded organelles along actin filaments, as shown in Figure 18-24a for *Nitella,* a freshwater alga. Myosin I also mediates the movement of the plasma membrane over the actin-containing cell cortex, as occurs during cell locomotion.

The regular arrangement of actin filaments in *Nitella* has provided a valuable opportunity to study the interactions of myosin and actin. The cells of *Nitella* are enormous—about 5 cm long and 1 mm thick. Just under the plasma membrane are regular arrays of cables, each con-

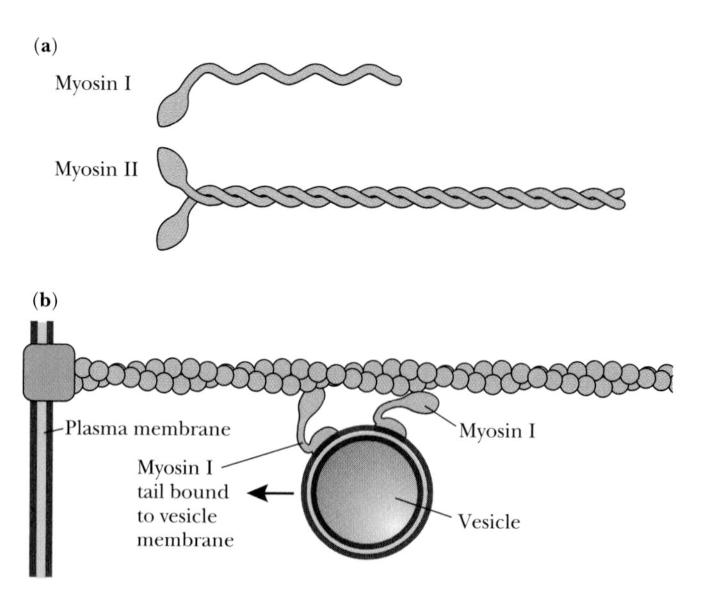

Figure 18-23 Myosin I is a much smaller protein than myosin II, the myosin of skeletal muscle. Myosin I binds to actin and to membranes. **(a)** Comparison of the sizes of myosin I and myosin II; **(b)** myosin I binding to the membrane of a vesicle, allowing the vesicle to move along the attached actin filament.

(a) Cytoplasmic streaming

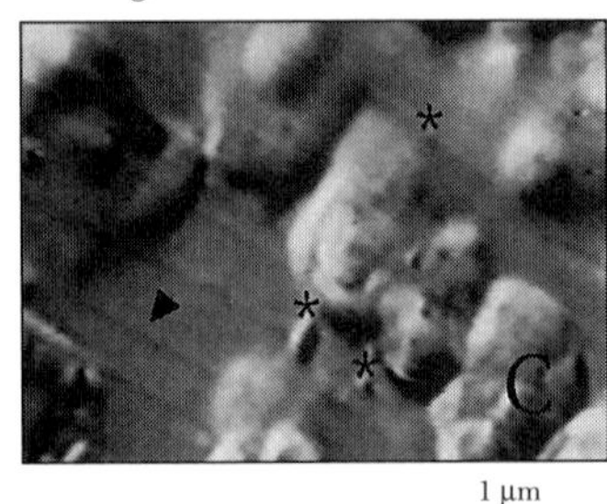

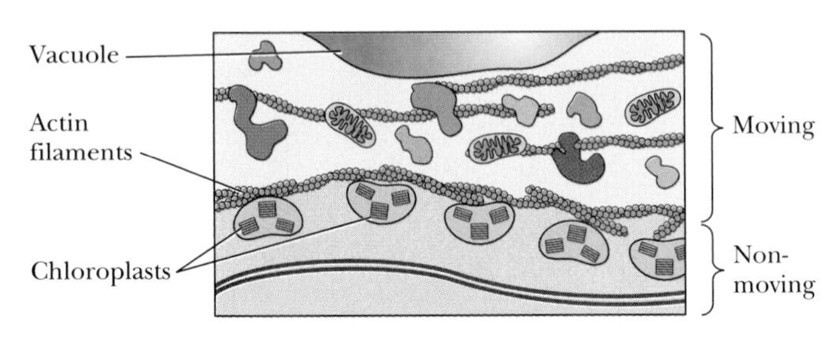

(b)

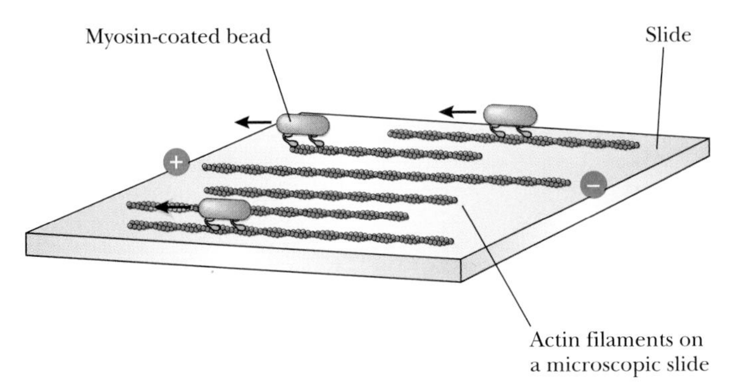

Figure 18-24 Interactions between myosin I, actin, and membrane-bounded vesicles are responsible for cytoplasmic streaming in *Nitella,* a freshwater alga. **(a)** Video micrograph, with schematic drawing, of a small section from an internodal cell of *Nitella,* illustrating the presence of actin filaments (arrow) attached to stationary chloroplasts (C); particles (starred) were moving along these filaments at a rate of 78 μm/sec; **(b)** an experimental system derived from *Nitella,* for studying the interactions of myosin I and actin; in the presence of ATP, myosin-coated beads will walk along the actin cables attached to the slide. *(a, Nina Strömgren Allen, North Carolina State University)*

sisting of hundreds of parallel actin filaments with identical plus-to-minus polarity. In the motility assay illustrated in Figure 18-24b, a *Nitella* cell has been split open and laid out on a microscope slide, exposing the actin cables. The experimenter then adds a suspension of myosin molecules attached to fluorescent beads and measures the movement of the fluorescent beads along the actin carpet. The fastest movement occurs with myosin from skeletal muscle, which requires the splitting of ATP. Beads attached to smooth muscle myosin move more slowly, also in an ATP-dependent manner.

Cell Locomotion Depends on Actin Polymerization, Gel-Sol Transformations, and Myosin-Actin Interactions

As illustrated in Figure 18-25b, cell crawling, or ameboid movement, involves at least three stages: (1) extension of the cell's leading edge, (2) attachment of the leading edge to the surface over which the movement occurs, and (3) contraction of the cell and withdrawal of its trailing edge. Each of these stages involves actin filaments.

As shown in Figure 18-26, extension of the leading edge involves the assembly of actin into **filopodia,** or **microspikes,** hair-like extensions 100–200 nm in diameter and up to 20 μm long, and **lamellipodia,** sheets of cytoplasm about 100–200 nm thick. Microspikes and lamellipodia extend forward, in a process called **ruffling,** during which new actin filaments appear and the existing actin filaments appear to move backward, as diagrammed in Figure 18-25b.

Cytochalasin prevents the formation of new actin filaments and stops the cell's forward movement, showing the importance of actin polymerization. Myosin I probably contributes to ruffling because it is present almost entirely in the tips of the lamellipodia.

Chemotaxis, the movement of a cell toward a higher (or, in some cases, a lower) concentration of a particular chemical, depends (in many eukaryotic cells) on the stimulation of actin filament formation. The migrating amebas of the slime mold, *Dictyostelium,* for example, move toward higher concentrations of cyclic AMP, a widely used intracellular signal (discussed in Chapter 19), used as an extracellular signal by this species.

Cytoplasmic streaming allows the cytoplasm to follow the forward movements of the leading edge. In the giant *Ameba proteus,* for example, researchers can observe streams of fluid cytoplasm ("sol") moving forward, behind the extended membrane, to form a **pseudopod** [Greek, *pseudo* = false + *podos* = little foot], a retractable extension of a moving cell, as shown in Figure 18-27. The cytoplasm then becomes a more viscous "gel" and begins to move backward along the sides of the still-extending pseudopod.

Cytoplasmic streaming occurs even in extracts of *Ameba proteus,* with gel-sol transformation occurring *in vitro* in the presence of calcium ions and ATP. Extracts of motile mammalian cells (such as skin fibroblasts) also show gel-sol transformations *in vitro.* Biochemical studies have established the importance of actin filaments, actin-binding proteins, and myosin in these movements.

Proteins such as α-*actinin,* which cross-links actin filaments (as shown in Figure 18-28a), promote the formation of a three-dimensional gel. On the other hand, proteins such as *gelsolin* promote the dissociation of actin filaments and the formation of a fluid sol, as shown in Figure 18-28b. The transition from cross-linked gel to dissociated sol depends in part on the concentration of calcium

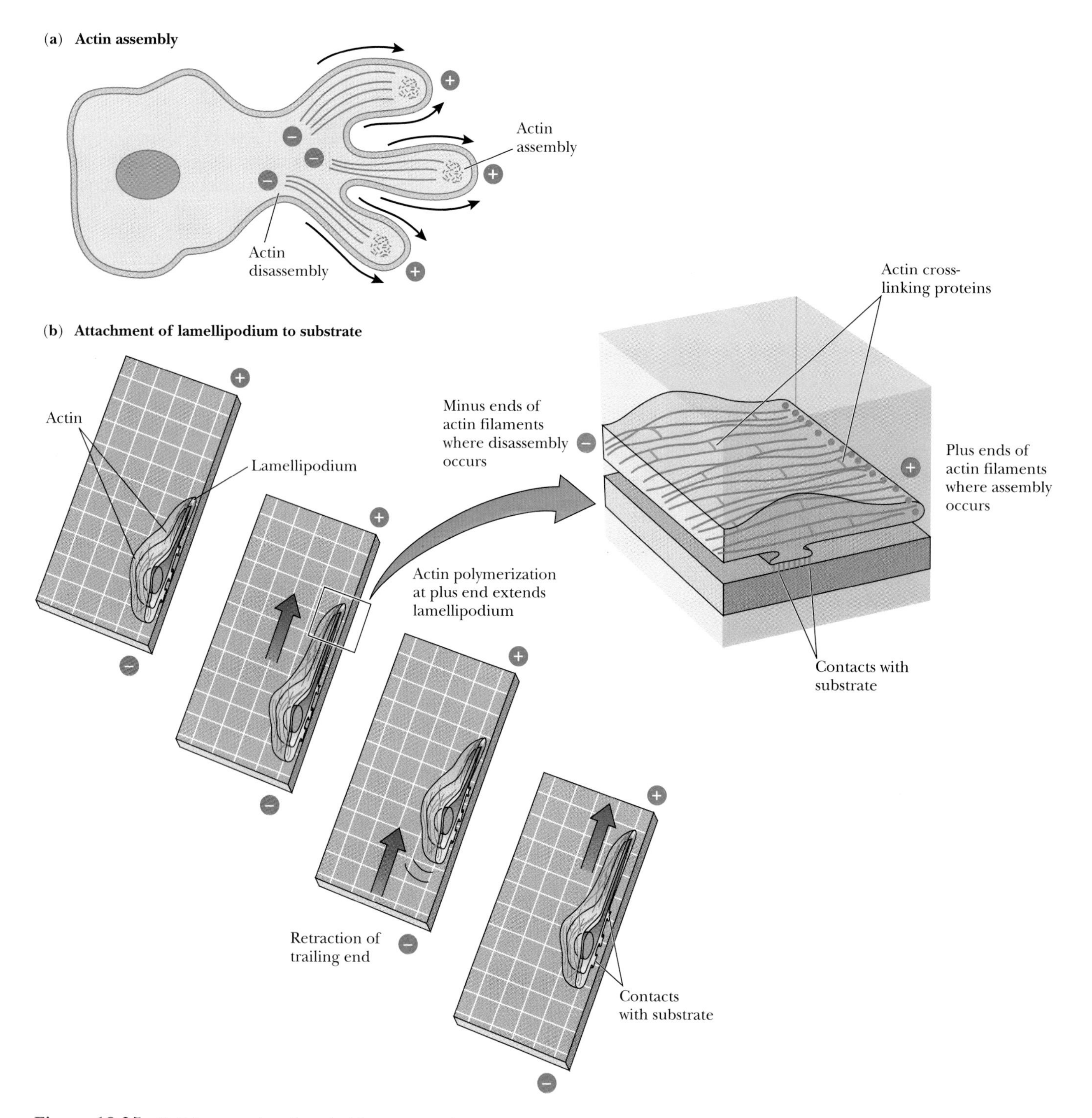

Figure 18-25 Cell locomotion (ameboid movement).

ions. When the concentration is greater than 10^{-7} M, gelsolin binds to actin monomers and to the plus ends of actin filaments. The gelsolin-bound actin monomers no longer add to the growing filaments, whose plus ends are, in any case, also blocked with bound gelsolin. Gelsolin binding stops filament growth, but dissociation continues, allowing transition to the less viscous "sol" state.

The extended edge of a migrating cell attaches to the underlying surface, which cell biologists call the **substratum.** The points, on the exterior of the plasma membrane, at which the cell attaches to the substratum are about 1 μm in diameter and are called **focal contacts,** as shown in Figure 18-29. On the cytoplasmic side of the plasma membrane, focal contacts are the end points of actin bundles.

As shown in Figure 18-29a, the plus ends of these actin bundles associate with a large membrane protein called **integrin** through a protein complex that includes

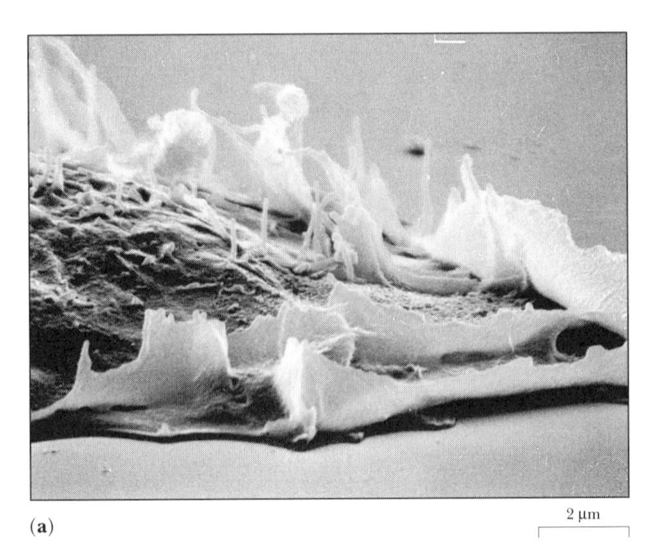

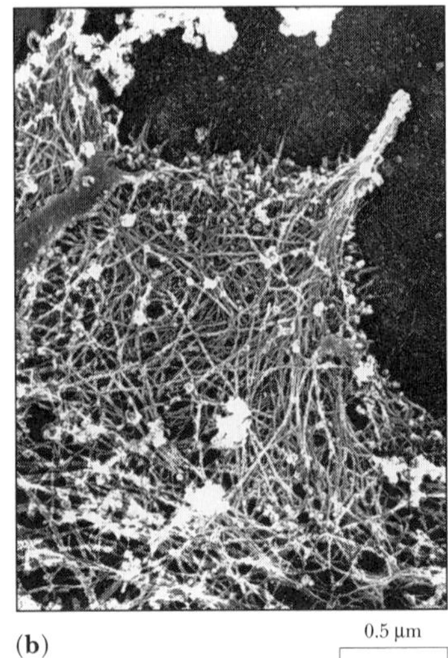

Figure 18-26 Filopodia and lamellipodia extend forward during the movement of a fibroblast, the result of actin filament formation. **(a)** Scanning electron micrograph, showing filopodia, lamellipodia, and ruffles; **(b)** electron micrograph showing actin filaments in a lamellipodium, some of which are bundled into a microspike. *(a, b, Dr. Julian Heath, Baylor College of Medicine; b, reprinted from J. Heath and B. Holifield,* Soc. Exptl. Biol. Symp., *47:35–56, 1993)*

α-actinin and vinculin. Integrin, which consists of two distinctive polypeptides, extends to the outside surface of the membrane, where it binds to **fibronectin,** a component of the extracellular matrix. The specific interactions between integrin and fibronectin are largely responsible for the strong attachment of cells to substratum, although other

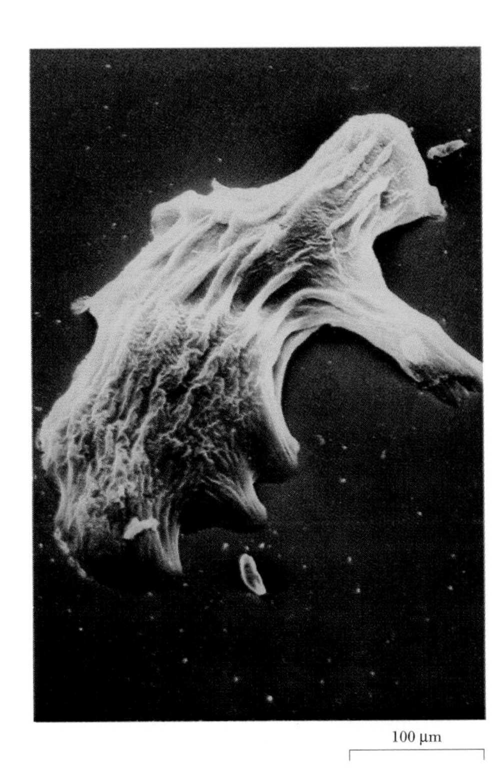

Figure 18-27 Formation of a pseudopod in this *Ameba proteus* depends on transformations between viscous gel ("ectoplasm") and fluid sol ("endoplasm"). *(K. W. Jeon/Visuals Unlimited)*

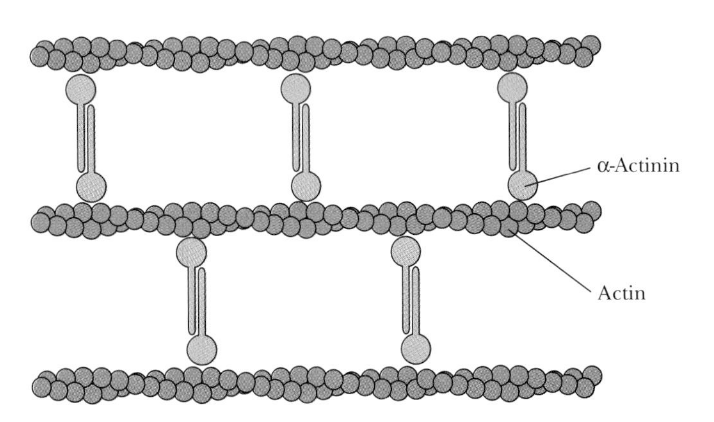

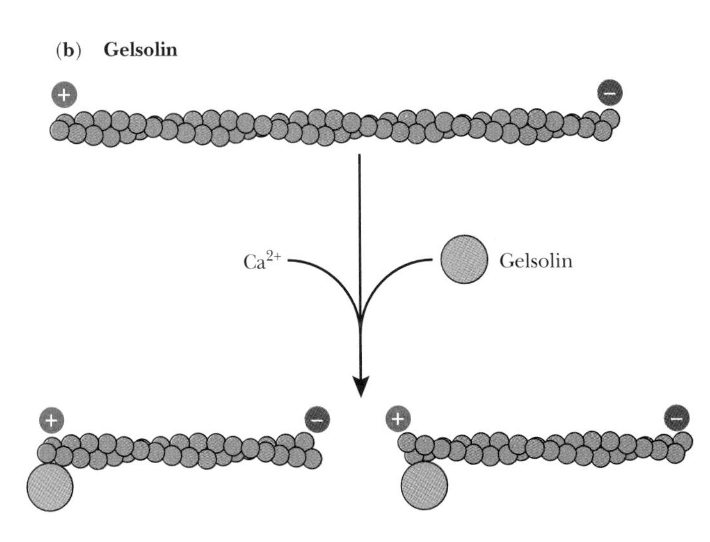

Figure 18-28 Crosslinking of actin filaments by α-actinin promotes gel formation, while interactions with gelsolin promotes sol formation. **(a)** α-Actinin crosslinks actin filaments; **(b)** gelsolin binds to the plus end of actin filaments at Ca^{2+} concentrations above 10^{-7}M.

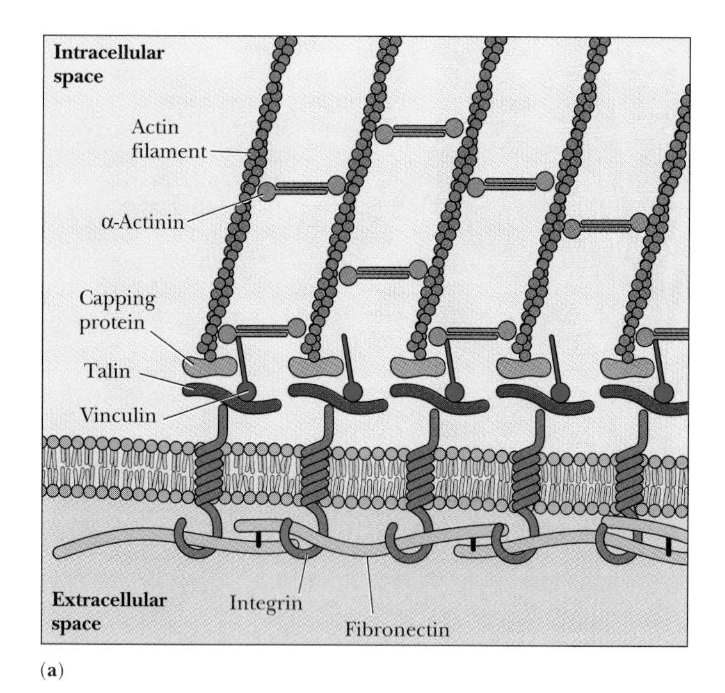

(a)

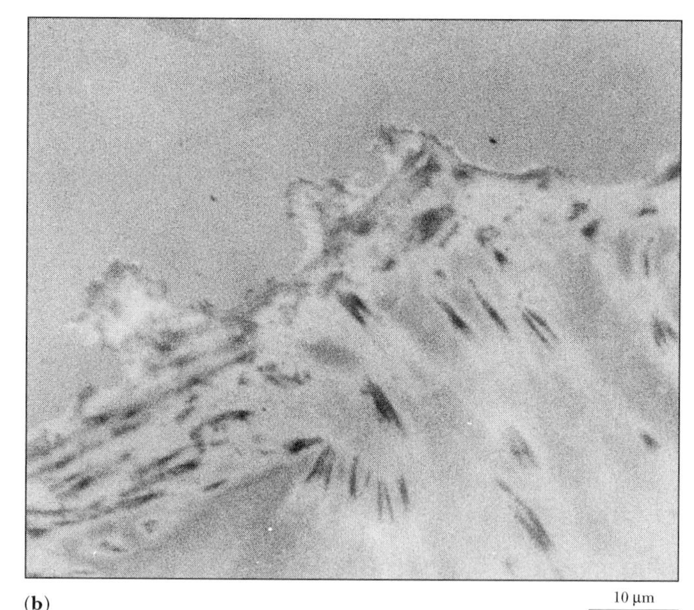

(b)

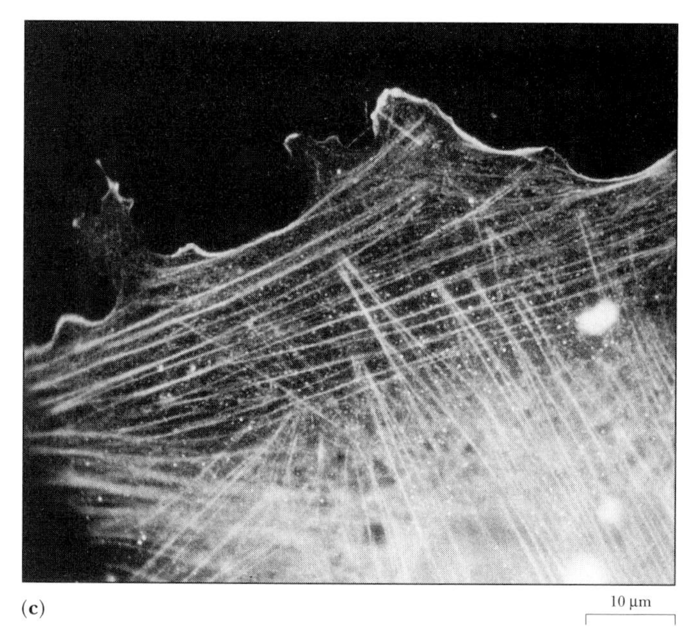

(c)

Figure 18-29 Actin filaments attach to focal contacts, where cells attach to the underlying surface ("substratum"). **(a)** Organization of actin bundles and integrin at a focal contact. **(b)** Region of a live Swiss mouse 3T3 cell visualized using interference reflection optics. The dark streaks represent the focal contacts or points of cell attachment to the cover slip. **(c)** The same region of the 3T3 cell after fixation with an antibody. Many of the actin bundles, or stress fibers, are seen to terminate at the focal contacts. *(b, c, courtesy of Grenham Ireland, University of Manchester)*

interactions between plasma membrane proteins and components of the extracellular matrix also contribute to cell attachment and cell movement.

Cytoplasmic streaming and cell movement also require the participation of myosin and the splitting of ATP. The last phase of the net movement of cytoplasm requires contraction of the cell's contents, particularly of the "tail" region, pulling the whole cell toward the points of attachment. This depends on ATP-driven movements of myosin II and actin filaments.

Increased understanding of the mechanisms of cell movement has depended heavily on information about muscle movement. The similarities and differences illustrate how cells may use the same molecules (or their close relatives) for a wide variety of processes.

■ HOW DO MICROTUBULES CONTRIBUTE TO INTRACELLULAR MOVEMENT AND CELL SWIMMING?

Microtubules, like actin filaments, are part of the cytoskeleton in almost all eukaryotic cells. (See Chapter 6.) Like actin filaments, microtubules are polar helical assemblies of small proteins, but the repeating unit is a dimer (of α and β tubulin) rather than a monomer (of G-actin).

Microtubules are **dynamically unstable,** meaning that their lengths constantly fluctuate, even leading to the complete disappearance of a given microtubule. Like actin, tubulin binds to and hydrolyzes a nucleoside triphosphate, but the nucleotide is GTP, not ATP. (See Chapter 6, p. 176.) Each microtubule in the cytoskeleton has an average lifetime of only about 10 minutes. Microtubules, such as those in Figure 18-30, are constantly changing their form and their component tubulin molecules: They assemble (and disassemble) preferentially at their plus ends, with the minus ends originating in a centrosome (that is, in the cell center). In migrating cells the centrosome lies

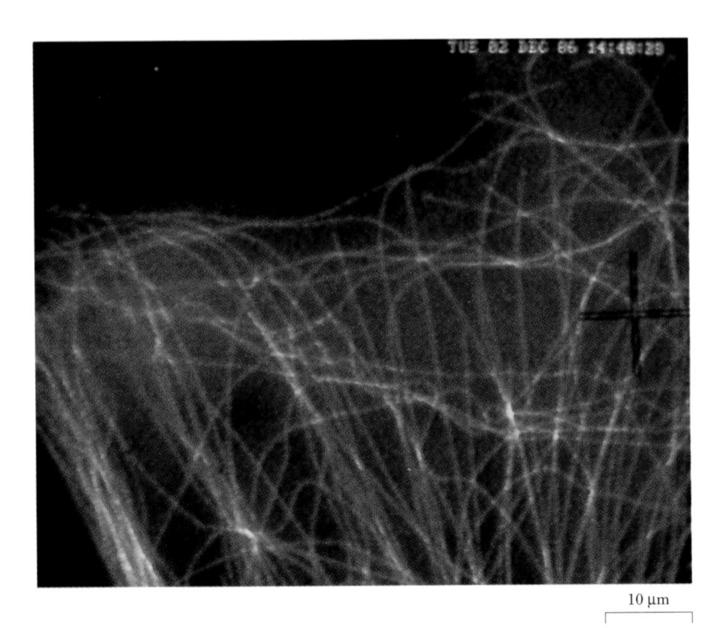

Figure 18-30 Microtubules are constantly changing. *(Paul J. Sammack, University of Minnesota)*

on the side of the nucleus that faces the direction of chemotaxis. Microtubule assembly can therefore reinforce the actin-driven extensions of plasma membrane, which are associated with cell movement.

Cell Survival and Reproduction Depend on the Dynamic Instability of Microtubules

Microtubule assembly is the major driving force in chromosome movement during mitosis. Chemicals like colchicine and colcemid, which preferentially bind to tubulin monomers and promote the disassembly of microtubules, prevent many mitotic movements. (See Chapter 6, p. 176, and Chapter 9, p. 261.) **Taxol,** a compound extracted from the bark of yew trees, increases the stability of microtubules. Treatment with taxol leads to an accumulation of microtubules that blocks mitosis, an effect that has led to the selective destruction of cancer cells in ovarian cancer and breast cancer. The toxic effects of taxol (like those of phalloidin) lend support to the importance of the dynamic instability of cytoskeletal elements: Normal cell life requires depolymerization as well as polymerization.

Kinesin and Dynein Are Microtubular Motor Molecules

As is the case for actin filaments, polymerization and depolymerization are not the whole story of microtubule-dependent movements. Microtubules may also be associated with "motor" proteins that generate movement, using energy obtained by splitting ATP. (See Chapter 6, p. 178.) Two types of motor molecule can bind to microtubules and move along them, using energy provided by the hydrolysis of ATP.

The motor molecule that usually moves toward the plus end of microtubules is called **kinesin.** It consists of four polypeptide chains arranged in a manner that resembles myosin, with two globular heads, a fibrous stalk, and a fan-like tail, as shown in Figure 18-31a. The heads, which have ATPase activity, can bind to microtubules, while the tail may attach to membrane-surrounded vesicles. ATP hydrolysis causes conformational changes that allow the kinesin to walk along the microtubule, toward the plus end, carrying a tail-attached passenger, as illustrated in Figure 18-31b. In this micrograph, kinesin tails have been attached to polystyrene beads. The movement of these beads is away from the centrosome, toward the plus end of the microtubules.

The other motor molecule, **dynein,** is a microtubule-associated motor that moves toward the minus end of microtubules, as also illustrated in Figure 18-31a. Like kinesin, dynein (a much larger molecule with a molecular size of 400,000–500,000) has an ATPase-containing head and a tail that binds to particular membrane-bounded organelles. Kinesin and dynein carry different passengers. Many nerve cells have long extensions called axons that carry signals to other nerve cells or to muscle or hormone-producing cells. Within axons, microtubules extend with their plus ends pointing away from the cell body. Kinesin moves vesicles down the axons, taking newly made protein and membrane away from the Golgi apparatus, as illustrated in Figure 18-31c. In contrast, dynein carries larger vesicles back up the axon to the cell body, where the vesicles' contents undergo degradation and recycling. Despite the usual distinction in the direction of movement by kinesin and dynein motors, however, the directions of preferred movement can be changed by phosphorylation of the proteins.

Cilia and Flagella Have Permanent Microtubular Structures with Dynein Motors

While microtubules within most cells are unstable, many motile cells have permanent microtubular structures—cilia and flagella. Cilia and flagella are each surrounded by extensions of the plasma membrane. They differ in length: Cilia are usually 2–10 μm long, while flagella may be much longer, up to 1 mm or even larger, although they are usually no longer than 200 μm. Cilia usually move like oars, exerting a force parallel to the cell surface and perpendicular to their own axes. (See Chapter 6.) In contrast, a flagellum beats like a whip, generating a force parallel to its own axis and perpendicular to the cell surface.

In many protozoa, rows of cilia move the cell through its aqueous surroundings. Cilia on the surface of animal epithelia, such as those that line the respiratory tract or the oviduct (as shown in Figure 18-32a), push fluids over

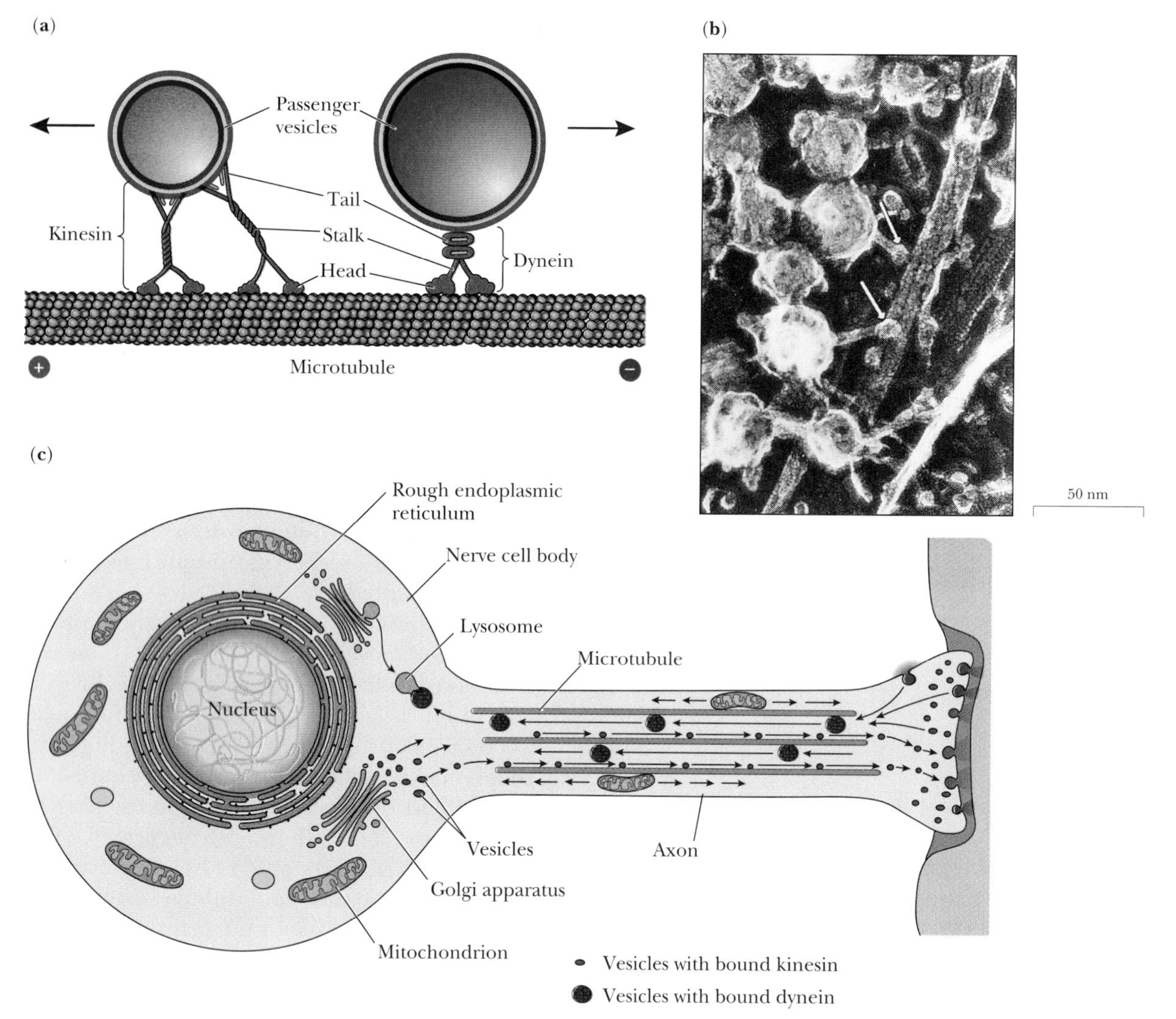

Figure 18-31 Motor-mediated transport uses microtubules as tracks. **(a)** Kinesin motors carry vesicles usually toward the plus end of microtubules, while dynein motors carry vesicles toward the minus end. **(b)** An electron micrograph of a kinesin molecule carrying a polystyrene bead along a microtubule. **(c)** In a nerve axon, kinesin and dynein carry passengers in different directions; the directions can change in response to phosphorylation of the motor molecules. *(b, Hirokawa et al.,* Cell, *56:867–878, Fig. 7)*

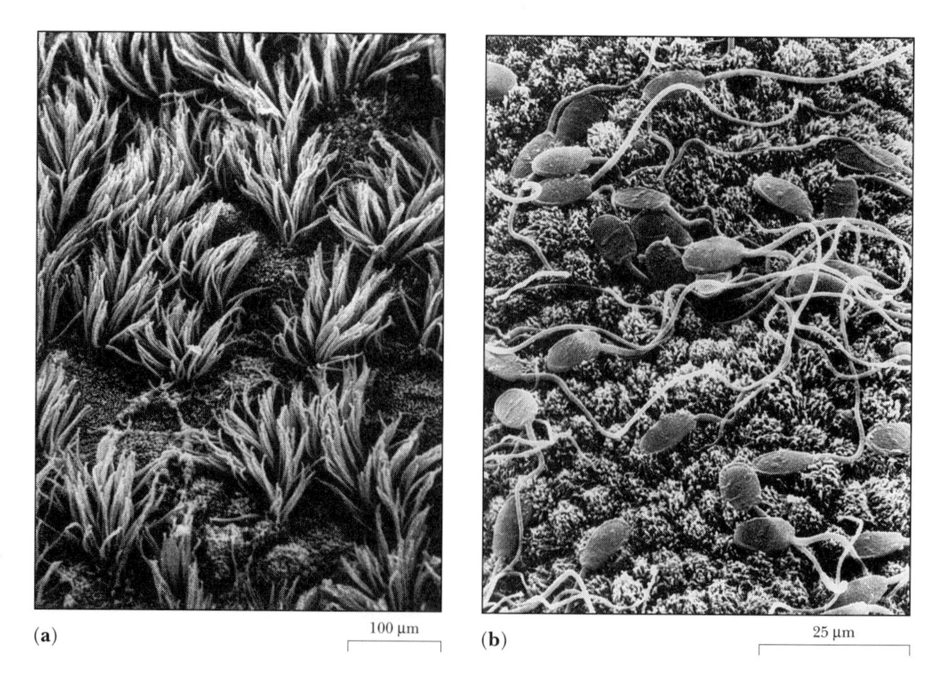

Figure 18-32 Cilia and flagella in reproduction: **(a)** cilia on the epithelium of the human oviduct; **(b)** flagella on rabbit sperm. *(a, Dr. Ellen Dirksen/Photo Researchers; b, David M. Phillips/Photo Researchers)*

ESSAY

HOW DO KINESIN-LIKE PROTEINS CONTRIBUTE TO THE ASSEMBLY OF THE MITOTIC SPINDLE?

The movement of chromosomes at mitosis depends on the spindle apparatus, a highly organized assembly of microtubules. (See Chapter 9.) During prometaphase, each chromosome becomes attached to spindle microtubules; at metaphase, the chromosomes move to the equatorial metaphase plate; and at anaphase, the two chromatids of each chromosome separate and move away from each other along microtubule tracks.

In the 1970s and 1980s, cell biologists realized that microtubules were not unchanging but were "dynamically unstable," alternately growing and shrinking. Studies of dynamic instability of microtubules in general and of spindle microtubules in particular convinced many cell biologists that the chromosomes themselves were only passive participants in the process of mitosis. One description, for example, compared the role of chromosomes during mitosis to the role of the corpse during a funeral; they are the reason for the proceedings, but do not play an active part.

More recent research has shown that it is not correct to view chromosomes as passive objects. Instead, chromosomes play active roles, particularly through motor proteins that assemble at their kinetochores, the structures to which the spindle microtubules bind. As mentioned in Chapter 9, motor molecules within the kinetochores propel chromatids along the spindle fibers at anaphase. Motor molecules increasingly appear to have a major role in the initial assembly of the spindle apparatus.

According to current views, the dynamic instability of spindle microtubules is fundamental to mitosis. Spindle assembly begins with rapidly alternating growth and shrinkage of microtubules from the spindle poles. The growing microtubules serve as searching devices for chromosomes, with a given microtubule becoming stable only after associating with a kinetochore. If no kinetochore attaches, the spindle microtubule is likely to shrink and disappear, with tubulin molecules recycled into other microtubules. Dynamic instability allows microtubules to search efficiently for chromosomes, and the interaction of spindle microtubules with kinetochore proteins is crucial to their stabilization.

The ability to watch mitosis in living cells has also contributed to the realization that kinetochores play an important role. Close analysis of chromosome movements revealed, surprisingly, that chromosomes constantly oscillate on the mitotic spindle. These rapid changes in the direction of chromosome movement suggested that the kinetochore coordinates a switch in the microtubule growth or shrinkage. Analysis of other chromosome movements, for example, at anaphase, suggested that motor proteins may also play a role in moving chromosomes along relatively stable microtubule tracks.

Researchers have identified several motor proteins in the kinetochore, and these proteins are thought to contribute to chromosome translocation along the spindle fibers. But the same motor proteins also contribute to the stability and instability of the fibers themselves. For example, a kinesin-related protein from frog eggs also promotes microtubule depolymerization at the kinetochore. Another kinesin-like protein associated with chromosome arms appears to be involved in assembling spindle microtubules at a common pole, while yet another motor protein is present at the spindle poles and contributes to the moving apart of the spindle poles. While the properties of the microtubules themselves explain many mitotic movements, the contributions of motor proteins are also crucial.

the surface. Flagellated cells (such as the rabbit sperm illustrated in Figure 18-32b) usually have one flagellum whose snake-like movements move the cell through the surrounding medium.

The basic structure in both cilia and eukaryotic flagella, called an **axoneme,** is an array, about 0.25 μm in diameter, of parallel microtubules in a characteristic **"9 + 2 pattern,"** shown in Figure 18-33. Around the outside of the axoneme lie nine *doublet* microtubules, each consisting of two microtubules fused along their length. In the

(a)

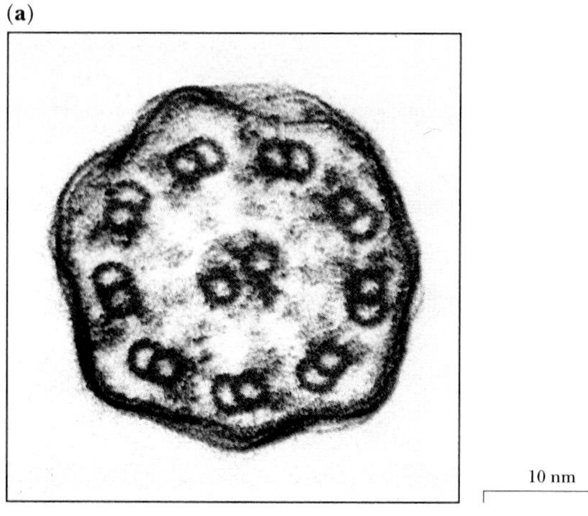

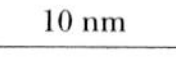

(b)

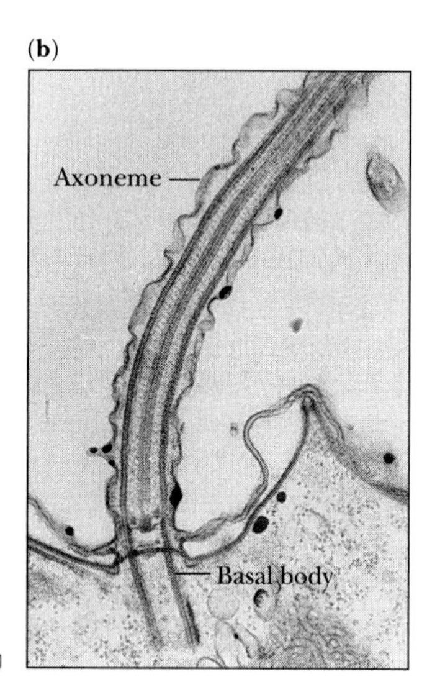

(c)

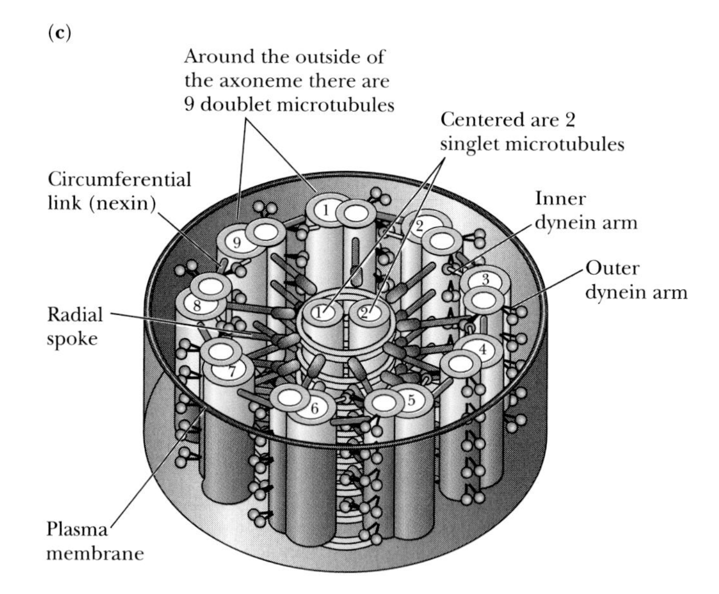

Figure 18-33 The "9 + 2" organization of an axoneme, found in cilia and eukaryotic flagella. **(a)** An electron micrograph of an axoneme, in cross section; **(b)** an electron micrograph of an axoneme, in longitudinal section; **(c)** current view of the structure of an axoneme. *(a, Dr. Gopal Murti/Science Photo Library/Photo Researchers; b, Photo Researchers)*

center of each axoneme are two *singlet* microtubules. Essentially all eukaryotic cilia and all eukaryotic flagella—from *Paramecium* to porcupine—contain the same 9 + 2 pattern, indicating an ancient evolutionary origin of the axoneme design and strong selection pressures for its maintenance. In contrast, the flagella of prokaryotes are entirely different (as discussed later in this chapter).

Axonemes are responsible for the swimming action of cilia and flagella. Movements persist even after the experimental removal of the plasma membranes (with a detergent) and even in the absence of calcium ions. The proteins within the axoneme itself can convert the energy stored in ATP into ciliary or flagellar movement.

Electron micrographs suggest that proteins other than tubulin must contribute to the structure of an axoneme. Biochemical and genetic experiments suggest that each axoneme may contain as many as 200 different types of protein, of which about a dozen have been well characterized. The arms that extend like little ears from each doublet, for example, are dynein molecules and are responsible for the beating of the axonemes.

Other identified axonemal proteins include *nexin,* which links doublets together on the axoneme's circumference, and *tektin,* a protein related to intermediate filament proteins, which forms filaments along the length of the microtubules.

A favorite organism for the analysis of microtubule structure and function is the single-celled green alga *Chlamydomonas*. Each *Chlamydomonas* contains two flagella. Nonmotile mutants of *Chlamydomonas* can be isolated easily. Biochemical and structural analysis of nonmotile mutants has allowed researchers to identify many axonemal proteins and to study their roles in cell movement. One type of nonmotile mutant, for example, lacks dynein. Another class of mutants lacks the radial spokes seen in the wild-type axonemes, while still another class lacks the central sheath and the two singlet microtubules.

The cross-linking proteins of the axoneme—the radial spokes and the circumferential links—make it relatively rigid. Dynein cannot move progressively along the microtubule, as it does along cytoplasmic microtubules. Instead, the ATP-driven changes in dynein structure causes axoneme microtubules to slide against one another, leading to a bending movement, as illustrated in Figure 18-34.

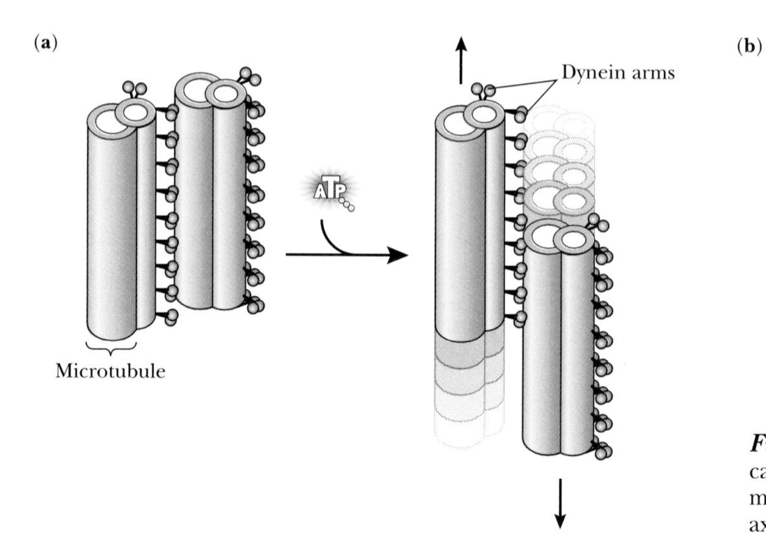

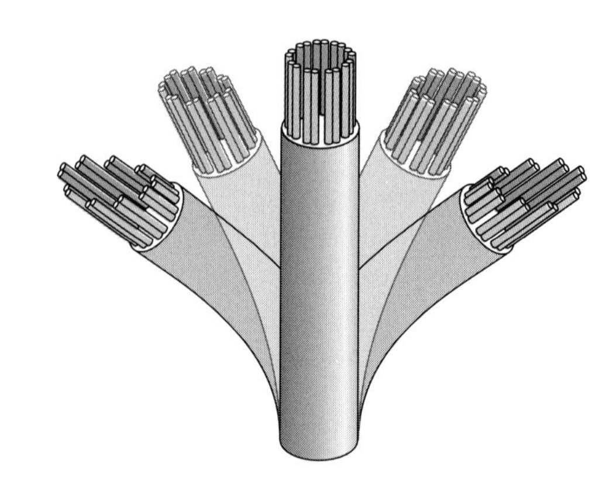

Figure 18-34 ATP-driven sliding of microtubule filaments causes axonemes to bend, causing ciliary and flagellar movement. **(a)** Microtubule movement; **(b)** bending of axoneme.

As in striated muscles, ciliary movement results from the ATP-driven sliding of constrained filaments.

One of the most intriguing problems in contemporary cell biology concerns the assembly and the evolutionary origins of cilia and flagella. An axoneme grows by the addition of tubulin and other proteins to the end farthest from the cell. The base of the axoneme is always anchored in a structure called a **basal body,** a cylinder about 500 nm long that resembles a centriole (Figure 18-35). Compare the electron micrograph of a basal body, Figure

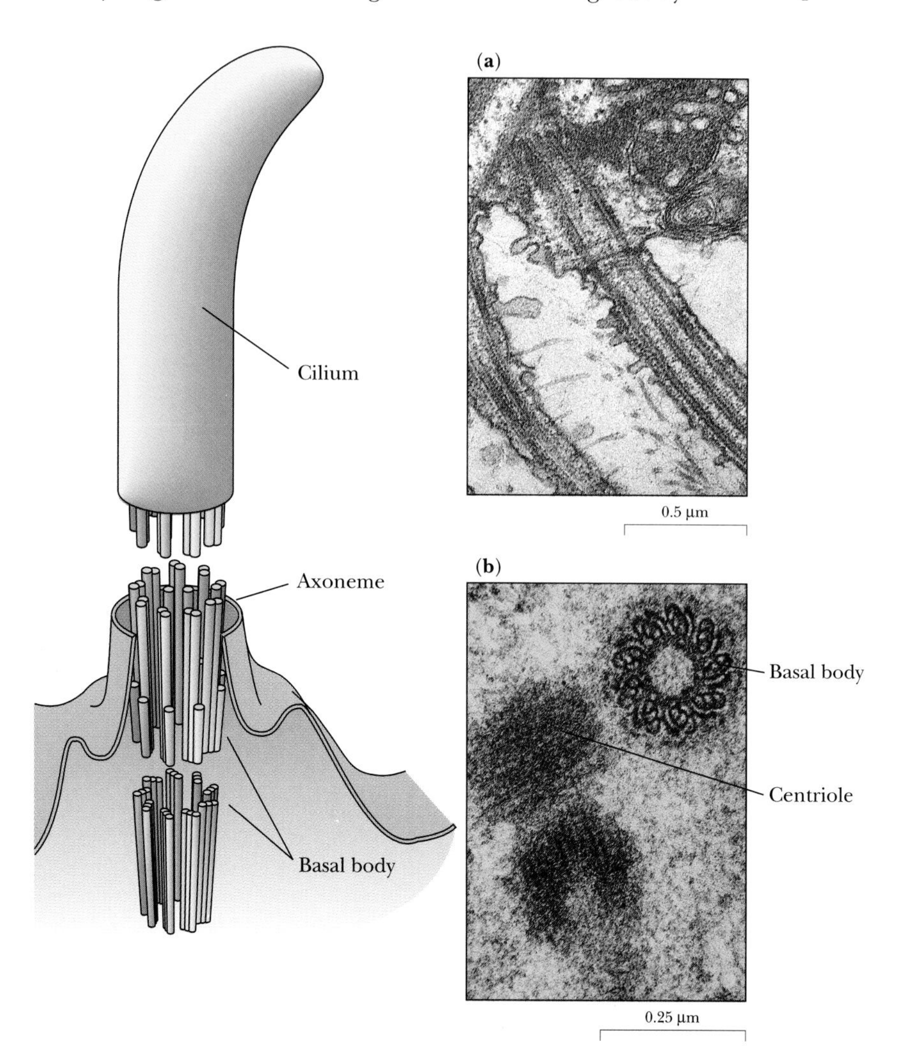

Figure 18-35 A basal body forms the base of a cilium or flagellum. **(a)** A longitudinal section of protozoan flagella; **(b)** resemblance of a centriole to a basal body. *(a, Benjamin Bouck/Photo Researchers; b, David M. Phillips/Visuals Unlimited)*

18-35a, with that of a centriole, Figure 18-35b. In fact, centrioles can convert into basal bodies during cellular differentiation, for example, in the life cycle of the protozoan *Naegleria gruberi.*

Nine groups of microtubule *triplets* lie around the circumference of the basal body. These triplets give rise to the nine doublets in the 9 + 2 pattern of the axoneme, but the basal body has no visible structures that correspond to the two central single microtubules.

Basal bodies and centrioles almost always arise from other basal bodies and centrioles. Their ultimate origin has been the subject of much evolutionary speculation, as discussed in Chapter 24.

■ BACTERIAL FLAGELLA, STRUCTURES UNRELATED TO EUKARYOTIC STRUCTURES, ARE RESPONSIBLE FOR THE MOVEMENTS IN BACTERIAL CHEMOTAXIS

The flagella of a prokaryote, illustrated in Figure 18-36, consist mostly of **flagellin,** a globular protein unrelated to tubulin with a molecular size of about 51,000. The flagellin molecules form a hollow cylinder about 14 nm in diameter and typically 10 μm long—many times the length of the bacterial cell itself. Unlike eukaryotic cilia and flagella, bacterial flagella are not surrounded by a membrane but are in direct contact with the surrounding fluid.

Bacterial flagella provide the only known example of a natural wheel. These flagella do not whip or wave but they rotate, in some cases up to 6000 rpm. As illustrated in Figure 18-36, at the base of the flagellum is the mechanism that drives the rotation: (1) a **flagellar hook,** which connects the flagellum to the rest of the motor; (2) an **S** (or static) **ring** (or *stator*), embedded within the plasma membrane and anchored to the peptidoglycan layer of the bacterial cell wall; and (3) an **M** (or motor) **ring,** a disc of protein that rotates within the S ring. A cylindrical protein rod serves as the shaft of this propeller, attaching the flagellar hook to the motor ring. The rotation of the flagellar motor does not require ATP but is driven by a proton gradient, such as those described in connection with oxidative phosphorylation and photosynthesis in Chapters 7 and 8.

Rotation of bacterial flagella are reversible. They can rotate in either a clockwise or a counterclockwise direction. For *E. coli,* whose flagella are left-handed helices, counterclockwise rotation results in the bundling of the bacteria's flagella, as shown in Figure 18-37a, and the coordinated pushing of the cell forward. Clockwise flagellar rotation pulls rather than pushes. The many flagella pull in different directions, resulting in a random tumbling motion, as illustrated in Figure 18-37b.

The direction of flagellar rotation alternates several times per second. The result is that the bacterium swims smoothly, in a straight line, for a few tenths of a second, a period called a *"run,"* following which the bacterium tumbles for a few tenths of a second. The bacterium's direction at the end of the period of random tumbling determines the next run direction.

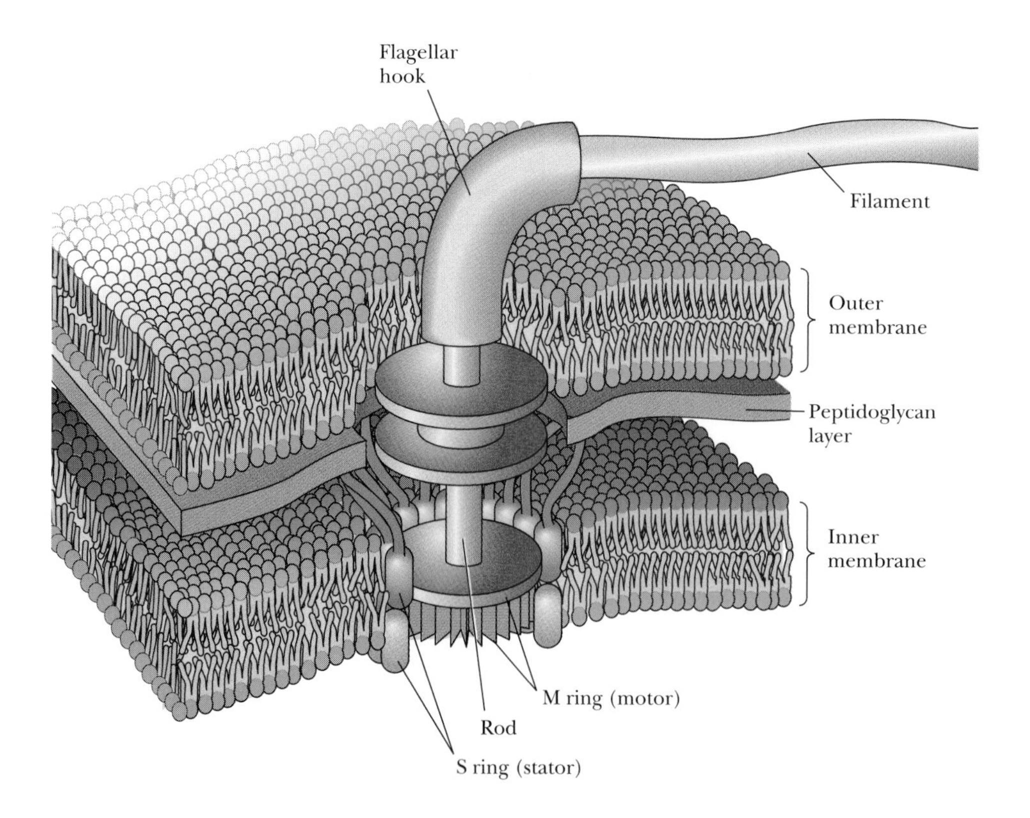

Figure 18-36 A bacterial flagellum, a structure unrelated to the eukaryotic flagellum.

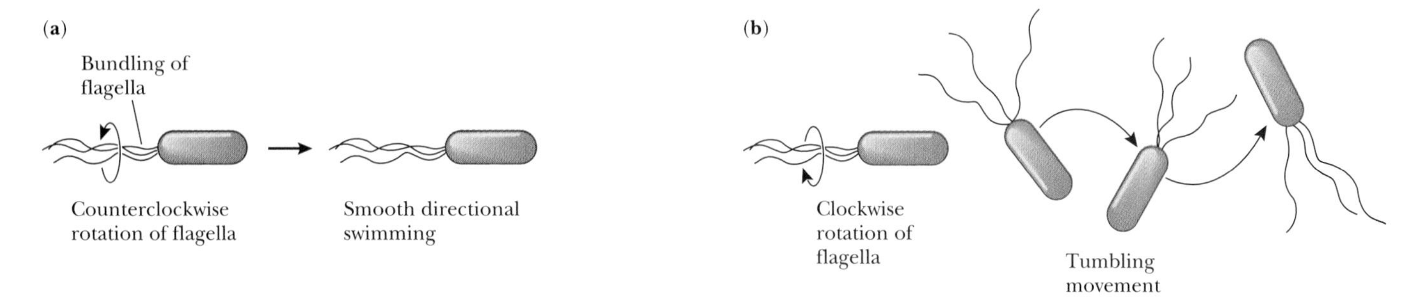

Figure 18-37 The reversible rotation of a bacterial flagellum. **(a)** Counterclockwise rotation pushes the bacterium forward; **(b)** clockwise rotation causes the bacterium to tumble.

A number of chemicals (about 30 in *E. coli*), called **chemattractants** (or repellents), influence the relative lengths of runs and tumbles. A chemattractant suppresses tumbling, while a repellant increases tumbling. As the cell moves in the direction of increasing concentration of a chemattractant, its runs in that direction become longer than the runs in all other directions. The result is net movement in the direction of increasing attractant concentration or away from the direction of increasing repellant concentration.

Genetic analysis of bacterial chemotaxis has allowed the identification of the cellular components responsible for chemotaxis. Attractants and repellents bind to four different sensors, which ultimately act on a cascade of regulatory molecules (called *Che proteins*) to stimulate counterclockwise rotation of the flagellar motor.

SUMMARY

Biological movements depend on filaments of specialized proteins. At a molecular level, the best-understood movements are those of striated muscle, which depends on the sliding of two types of permanent filaments, actin and myosin.

A muscle fiber is a huge multinucleated cell, each of which consists of thousands of repeating units called sarcomeres. Thick filaments, composed largely of myosin, and thin filaments, composed largely of actin, lie parallel to each other within the sarcomeres and are linked together by molecular cross-bridges. ATP causes muscles to contract by acting on the cross-bridges and providing power for filaments to slide.

Filaments slide only when a muscle cell temporarily increases the concentration of free calcium ions. Two proteins, tropomyosin and troponin, are responsible for the calcium dependence of the filament sliding. The calcium ions come from a system of membrane-lined channels and sacs, called the sarcoplasmic reticulum. Electrical signals initiate the release of calcium ions and the contraction of the muscle.

The cardiac muscle of heart and the smooth muscle of internal organs also use actin and myosin to accomplish movement, but the arrangement of the actin and myosin filaments differs from that in striated muscles. Gap junctions allow electrical signals to pass between cells and to trigger the release of calcium ions. The regulation of contraction by calcium ions, however, depends on proteins other than tropomyosin and troponin.

Forms of actin and myosin are also present in nonmuscle cells and are often responsible for cell movement and changes in cell shape. Some cell movements, such as those of cytoplasmic streaming, require the interactions between myosin and actin. Many cell movements, such as the familiar crawling movements of amebas, depend on the association of actin monomers into actin filaments, as well as on the cross-linking of actin filaments to produce a reversible gel-like consistency of the surrounding cytoplasm.

Microtubules, like actin and myosin filaments, are protein filaments formed from the association of smaller proteins. The repeating units of microtubules are tubulin molecules, which continuously add to and remove themselves from microtubules. Microtubules have plus and minus ends, with addition of new tubulin preferentially occurring at the plus end.

Microtubules are associated with motor proteins, kinesin and dynein, which generate movement using energy from the splitting of ATP. Kinesin usually moves along a microtubule carrying attached vesicles toward the plus end, while dynein carries subcellular passengers toward the minus end.

Cilia and eukaryotic flagella are permanent structures composed of microtubules almost always arranged in a characteristic pattern, with nine paired microtubules around the circumference of a cylinder with two single microtubules running down the center. Like muscle fibers, cilia and flagella produce movement by the constrained sliding of molecular filaments. Instead of producing contraction, however, the sliding of microtubules produces bending movements, which cilia use to provide oar-like movements and flagella to produce whip-like movements. The power for bending comes from the splitting of ATP by dynein molecules that are permanently attached to the ciliary or flagellar microtubules.

Bacterial flagella are unrelated to eukaryotic flagella and operate by a totally different mechanism, the only known example of a biological wheel. The rotation of the bacterial flagellum depends on a proton gradient rather than on the splitting of ATP. The direction of flagellar rotation alternates, with counterclockwise rotation pushing the cell forward in a single direction and clockwise rotation pulling the cell into a tumbling mode. Changes in the times of running versus tumbling result in chemotaxis, the directed movement toward specific chemical cues.

■ KEY TERMS

A band
acetylcholine
aequorin
ameboid movement
axoneme
basal body
caldesmon
calmodulin
cardiac muscle
catch muscle
cell crawling
cell locomotion
chemattractant
chemotaxis
cytochalasin
cytoplasmic streaming
desmin
dynamic instability
dynein
dystrophin
fibronectin
filopodia
flagellar hook
flagellin
focal contact
glycolytic fiber
I band
integrin
intercalated disc
involuntary muscle
kinesin
lamellipodia
M ring
microspike
muscle fiber
myasthenia gravis
myofibril
myosin regulatory light chain
myosin light-chain kinase
9 + 2 pattern
oxidative fiber
paramyosin
phalloidin
pseudopod
ruffling
S ring
sarcolemma
sarcomere
sarcoplasmic reticulum
Sliding Filament Model
smooth muscle
striation
substratum
taxol
thick filament
thin filament
tropomyosin
troponin
voluntary muscle
Z disc

■ QUESTIONS FOR REVIEW AND UNDERSTANDING

1. Define and contrast the following terms:
   (a) glycolytic fibers vs. oxidative fibers
   (b) muscle fiber vs. myofibril vs. sarcomere
   (c) actin vs. myosin
   (d) G-actin vs. F-actin
   (e) myosin I vs. myosin II
   (f) troponin vs. calmodulin
   (g) bacterial flagella vs. eukaryotic flagella
   (h) cytochalasin vs. colchicine
2. What is the experimental observation supporting the Sliding Filament Model of muscle contraction and refuting the contracting filament model?
3. What is the experimental evidence suggesting that the small head regions of myosin form cross-bridges with actin?
4. Why are the head regions of myosin thought to be the site of ATP hydrolysis required for muscle contraction?
5. Describe in your own words how tropomyosin and troponin act antagonistically to regulate the association (binding) between actin and myosin.
6. What is the primary role of the sarcoplasmic reticulum?
7. Describe the role of the neuromuscular junction in muscle contraction.
8. In the events leading to muscle contraction, why can inositol triphosphate be thought of as a chemical "second messenger"? (See Chapter 19 for a definition of this term.) Why might calcium be thought of as a "third messenger"?
9. Contrast the properties of skeletal muscle, cardiac muscle, and smooth muscle, taking into consideration structure, control, and function.

10. Muscle cells are highly specialized and employ many "unique" structural components in addition to components commonly found in other types of cells. What components present in skeletal muscle cells and involved in muscle function are also found in other cells? Also, discuss whether there are any truly "unique" components.
11. Describe the role calmodulin plays in smooth muscle contraction.
12. What experimental evidence indicates a role for actin in cell movement and shape changes?
13. How is the tail region of myosin I suited for its function in non-muscle cells? How does it differ from myosin II?
14. Describe, in general terms, why gel-sol transformations are important for cell movement. How is gelsolin involved in this transformation?
15. Why are focal contacts required for cell movement? What is the role of integrin in a focal contact?
16. Why are kinesin and dynein called "motor proteins"? How are these two similar? How are they different?
17. Explain how a chemattractant, by indirectly affecting the frequency of reversing rotation of flagella, allows for bacterial chemotaxis.
18. The protein dystrophin is thought to play a role in anchoring actin microfilaments to the plasma membrane of striated muscle, reinforcing the plasma membrane during contraction. Speculate as to why the absence of this protein, as in individuals afflicted with Duchenne muscular dystrophy, may lead to eventual destruction of the muscle.

SUGGESTED READINGS

ALBERTS, B., D. Bray, J. Lewis, M. Raff, K. Roberts, and J. D. Watson, *Molecular Biology of the Cell,* 3rd edition, Garland, New York, 1994. Chapter 16 deals with the cytoskeleton, including actin filaments, microtubules, motors, and muscles.

BRAY, D., *Cell Movements,* Garland Publishing, New York, 1992. This whole book represents a particularly straightforward treatment of cell movements, including actin-based motility, muscles, microtubules, and motors.

LODISH, H., D. Baltimore, A. Berk, S. L. Zipursky, P. Matsudaira, and J. Darnell, *Molecular Cell Biology,* 3rd edition, Scientific American Books, New York, 1995. Chapter 22 deals with actin filaments, muscle contraction, and cell motility. Chapter 23 deals with microtubules, mitosis, and motors.

CHAPTER 19

Chemical Signaling: Hormones and Receptors

KEY CONCEPTS

- Hormones are chemical signals that move throughout an organism and affect processes in target cells and organs.
- Hormones and other signaling molecules bind to specific receptors in target cells and trigger characteristic responses.
- Most signals that act on the outside of a cell act by stimulating the synthesis of an intracellular signal, called a "second messenger."

OUTLINE

ESSAYS

Both cells and whole organisms use chemicals to communicate. Female silk moths, for example, find mates with an airborne attractor called *bombykol,* while yeast cells use a short, water-soluble polypeptide to trigger a mating response.

A substance secreted by one organism that influences the behavior or physiology of another organism of the same species is called a **pheromone.** Pheromones play a particularly important role in animal behavior; for example, they mark the territories of dogs and wolves, establish the trails of ants, and influence the reproductive cycles of rodents. Extraordinarily low concentrations are effective in inducing behavioral changes: A male moth responds to bombykol when it detects only 40 molecules per second.

In multicellular organisms, chemical signals are responsible for almost all intercellular communication and are especially important in coordinating the responses of different organs to changes in the external or internal environment. Many of these chemical signals are **hormones**—substances that are made and released by cells in a well-defined organ or structure and that move throughout the organism and exert specific effects on specific cells in other organs or structures. Animals and plants use hormones to coordinate the actions of physically distinct tissues and organs. The structures upon which hormones act are called **target organs.** Target organs respond to hormones by changing their metabolism, the activity of their enzymes, or the expression of specific genes.

In animals, hormones are produced by **endocrine** glands, organs that are specialized for secretion into the general circulation (as illustrated in Figure 19-1a), in contrast to **exocrine** organs, whose products are carried to specific targets by ducts, as illustrated in Figure 19-1b. In addition to endocrine hormones, which circulate throughout the organism, animals also produce chemical signals, called **paracrine** hormones, which act only in the immediate region of their production, as illustrated in Figure 19-1c.

Many animals also use the electrical activity of nerve cells (discussed in Chapter 23) to coordinate events in physically separated organs, as illustrated in Figure 19-1d. In most cases, however, the actual signaling between cells depends on a chemical signal called a neurotransmitter, discussed later in this chapter and in Chapter 23.

■ CHEMICAL SIGNALS COORDINATE THE ACTIVITIES OF CELLS AND ORGANISMS

As an example of chemical signals, we discuss the regulation of blood glucose levels in mammals. This regulation depends on separate biochemical events in the liver, the pancreas, and the small intestine.

Shortly after a meal, cells in many organs begin to take up needed fuel and building blocks, and the body is said to be in an **absorptive state,** in which cells take up glucose, make glycogen, and increase the synthesis of fats and proteins. The result of these processes is a drop in the blood concentrations of glucose and other fuel molecules. The body then goes into a **postabsorptive state,** in which cells derive energy and building blocks by breaking down stored glycogen, fats, and proteins.

Several signals contribute to the coordination of absorptive and postabsorptive states. The most important of these is **insulin,** a protein produced by specialized cells of the pancreas. When these cells sense high levels of glucose, characteristic of the absorptive state, they increase the production of insulin, which has been called the "hormone of plenty." Insulin in turn stimulates the uptake of glucose by most cells, decreasing its concentration in the blood. It also stimulates all the biochemical processes characteristic of the absorptive state, in the liver, the muscles, and the adipose tissue. Among the responses in the target organs are the increased synthesis of fats, proteins, and glycogen.

When the intestine runs out of glucose to absorb, blood glucose decreases, and insulin secretion stops. This lowered insulin concentration triggers the postabsorptive state. Glucose uptake slows, and cells begin to break down glycogen, fats, and proteins. **Glucagon,** another protein hormone produced in the pancreas, has the opposite effect on glycogen-glucose interconversions. While insulin promotes the synthesis of glycogen and inhibits its breakdown, glucagon inhibits glycogen synthesis and stimulates its breakdown into glucose.

The failure to produce or to respond to insulin keeps the body in a postabsorptive state. One result is an excessive concentration of glucose in the blood, which leads to its presence in the urine and to excess production of urine. This relatively common condition is called **diabetes mellitus,** meaning the overproduction of sweet (sugar-containing) urine. About 10%–20% of diabetes patients fail to make insulin at all. They depend on daily injections of insulin to lead a relatively normal life.

Hormones Resemble Broadcast Messages, Which Evoke Different Responses in Individual Targets

Hormone and nerve signals have much in common; the major difference between the two types of signals is that hormone signals reach essentially all the tissues of the body, while nerves carry signals only to particular targets. This difference resembles the difference between a message broadcast on the radio and one carried by a cable. Anyone whose radio is tuned to the right station can hear the broadcast message, while the cable message is available only where the wires lead. Another similarity of hormones and broadcast messages is that the same message can evoke different responses in different recipients. Consider, for example, the possible responses of people hearing a weather forecast that predicts a heavy snowfall:

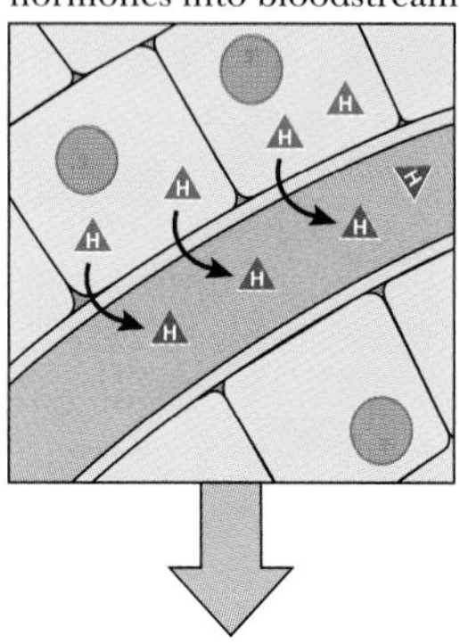

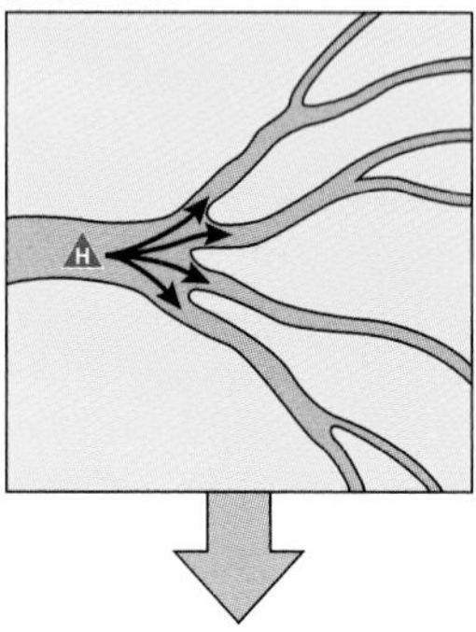

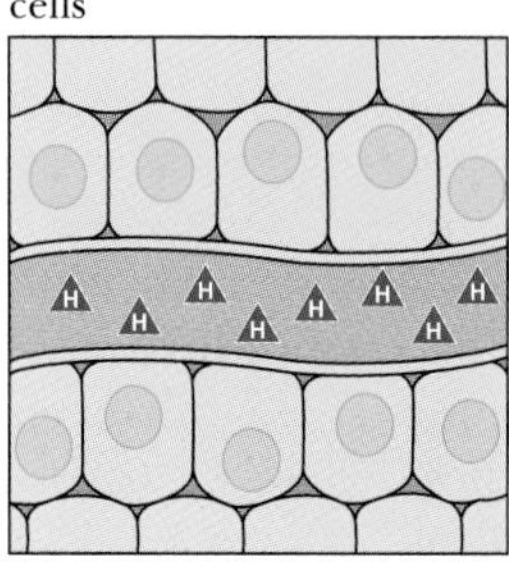

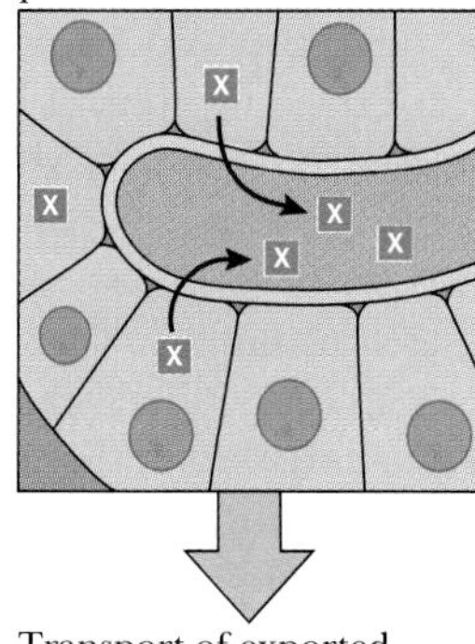

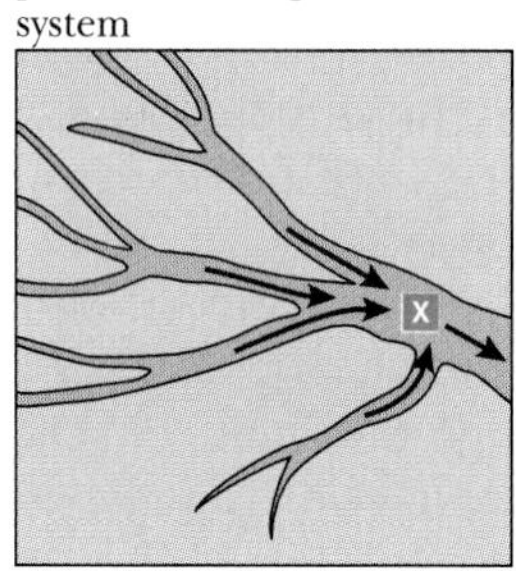

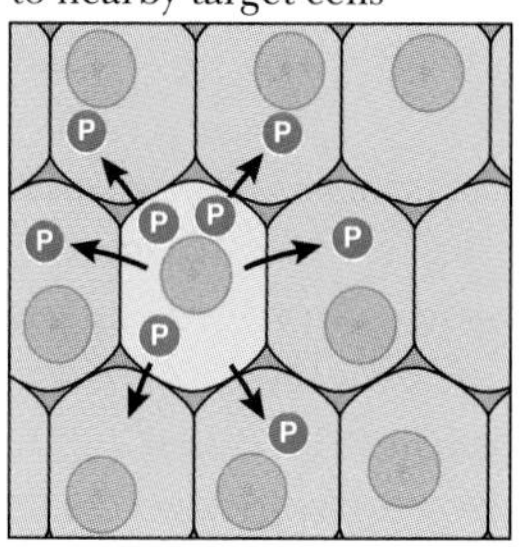

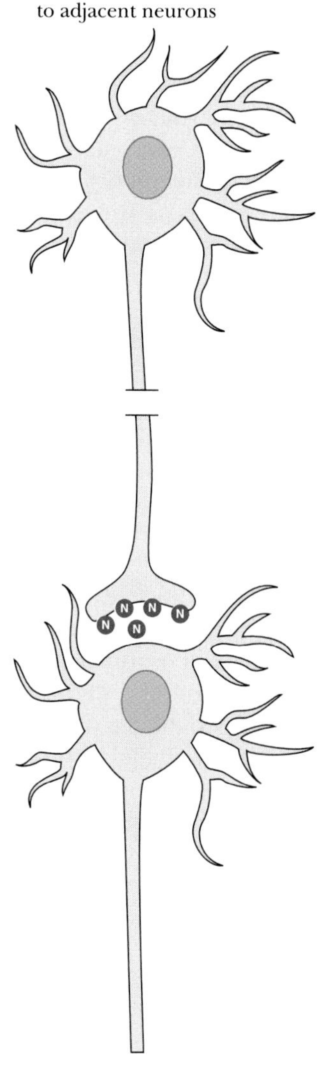

Figure 19-1 Chemical signaling through the circulation depends on hormones made in endocrine glands; chemical signaling over short distances, called paracrine signaling, does not require the circulatory system. **(a)** Endocrine glands secrete hormones into the circulation; hormones then travel to their targets, where they exert their effects. **(b)** Endocrine glands differ from exocrine organs, which export products, such as digestive enzymes, into ducts; the ducts carry the exported products to other locations. **(c)** Paracrine signaling takes place over short distances, without the chemical signals ever entering the circulation. **(d)** Nerve cells carry signals for large distances by taking advantage of the electrical properties of plasma membranes; the signals between nerve cells and their targets, however, are usually specific molecules, called neurotransmitters.

Skiers react differently from merchants or from snowplow operators.

Similarly, cells within individual organs often react differently to the same hormone. For example, the hormone **epinephrine** (or adrenaline), which is made in the adrenal medulla, speeds the heart, dilates the blood vessels, and increases the liver's production of glucose from glycogen, as illustrated in Figure 19-2. All these responses to epinephrine contribute to the "fight or flight" reaction that prepares an animal for immediate action and energy expenditure in the face of stress or danger.

How Do Researchers Identify Hormones and the Glands That Make Them?

Endocrinology, the study of hormones, has traditionally employed a fairly standard approach to the study of each hormone. The goals of this approach are based on the fulfillment of the definition of a hormone. (See p. 507.) The first question is whether a product of one organ affects the state of other organs. To address this question, a researcher destroys or removes a suspected endocrine organ and notes the resulting changes. Removing the testes of

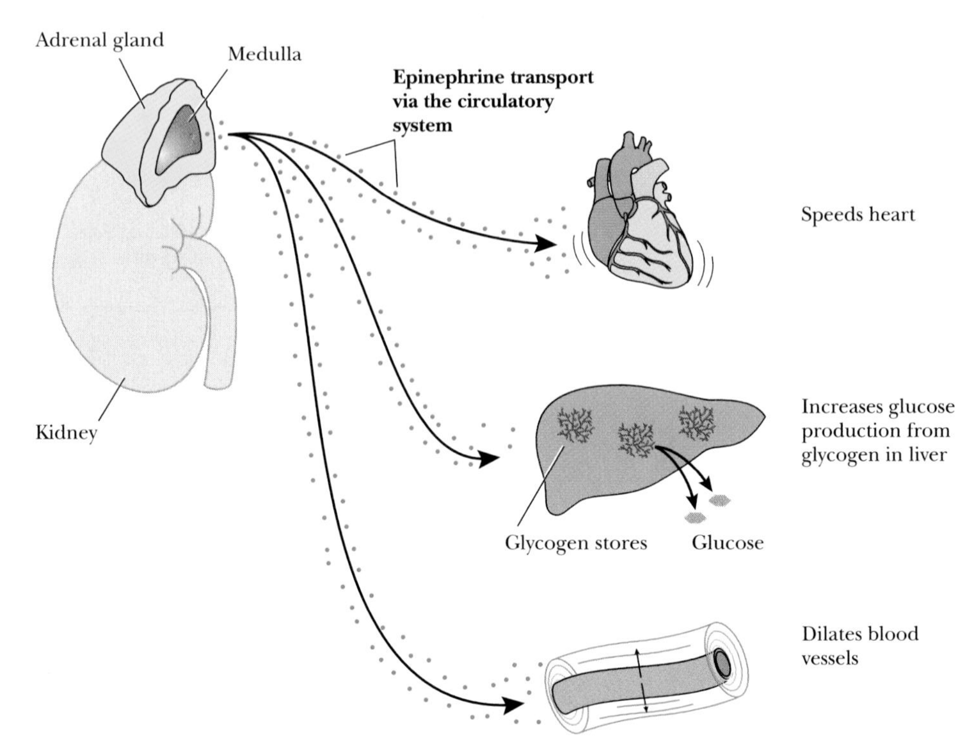

Figure 19-2 The hormone epinephrine, produced by the medulla of the adrenal gland, is responsible for coordinating an animal's "fight or flight reaction" to stress or danger; responses include speeding the heart, dilating the blood vessels, and increasing glucose production by the liver.

an experimental animal, for example, leads to the loss of many male hallmarks—the secondary sexual characteristics that appear at sexual maturity. In roosters, removal of the testes leads to the withering of the comb. A fully formed comb is illustrated in Figure 19-3.

The question addressed in a standard endocrinological study is whether the suspected effects of hormone action depend on a signal that moves through the whole organism, rather than, for example, through a specific duct. To determine this, researchers attempt to restore the normal function by replacing the suspected endocrine gland to another site.

The final question is the chemical identity of the hormone. First, researchers must demonstrate that an extract of the endocrine gland can produce the same effect as the intact gland. Once replacement is accomplished, the task is to isolate and identify the active compound and to determine its chemical structure. Finally, researchers must be able to synthesize the suspected hormone in the laboratory and to show that the chemically produced substance gives the same effect as the suspected hormone. This step rules out the possibility that previous research had isolated the wrong substance from the extracted endocrine organ. For example, the active substance could be a minor contaminant within the extract. Even more importantly, the ability to mimic a hormone's effect with a pure chemical produced in the laboratory shows that the observed action depends on a single hormone, rather than on two or more hormones.

This more or less standard approach has not always been straightforward. Endocrine organs may make more than one hormone, so that a number of substances must be replaced to reverse the effects of organ removal. In vertebrates, for example, the **pituitary gland,** a pea-sized structure at the base of the brain, releases at least nine hormones, as shown in Figure 19-4. The pancreas, as discussed

Figure 19-3 The comb of a rooster is an example of a secondary sexual characteristic. *(Spencer Grant/Photo Researchers)*

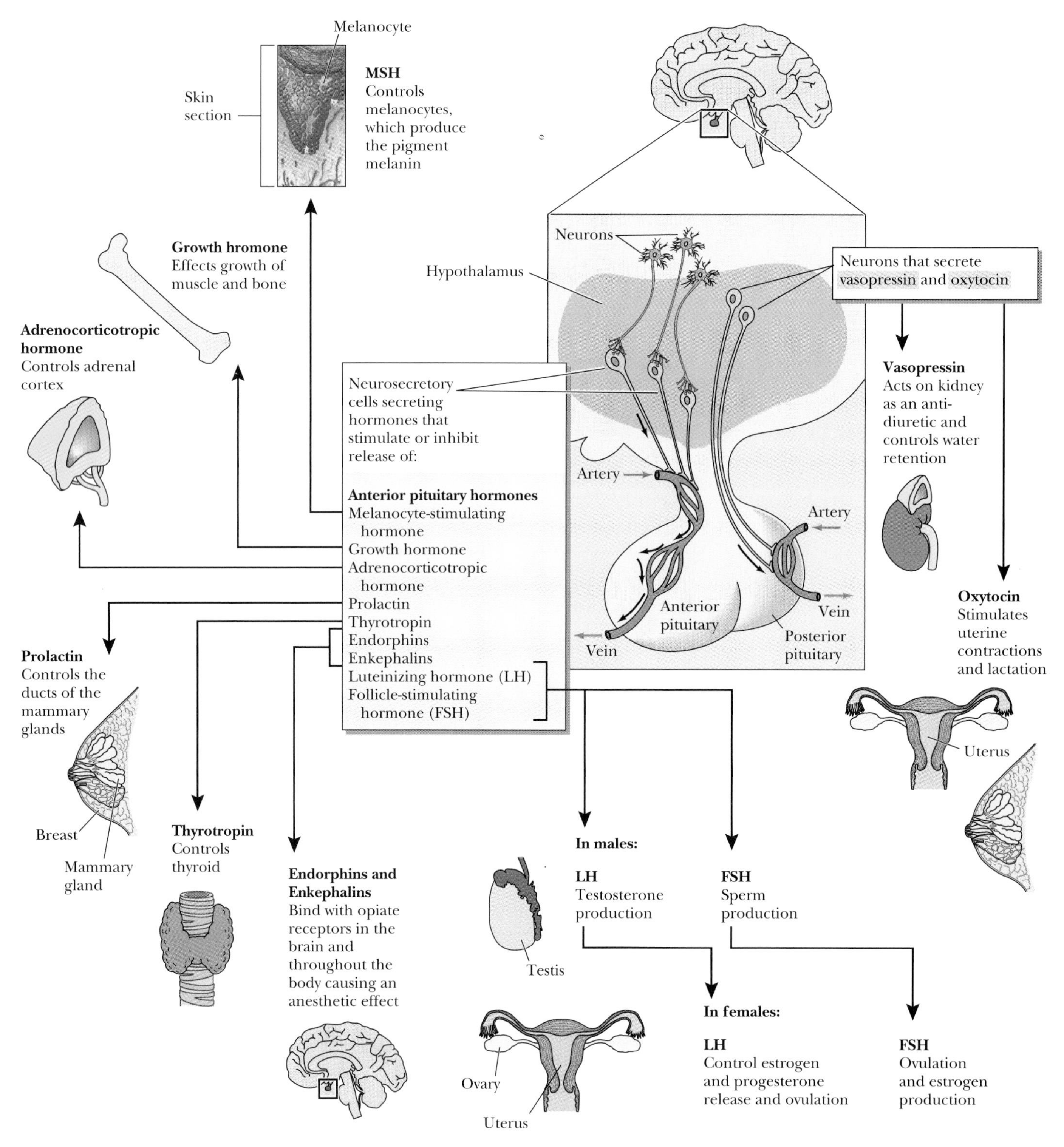

Figure 19-4 The pituitary gland consists of an anterior part, in which cells secrete at least eight different hormones into the circulation, and a posterior part, which secretes two hormones into the circulation. Secretion from the posterior pituitary is actually from the terminals of cells whose cell bodies lie in the hypothalamus, a part of the brain; the secreting cells in the anterior pituitary lie entirely within the pituitary, but they are the targets of hormones released from the hypothalamus.

above, not only produces two hormones with opposing effects but also serves as an exocrine organ, secreting digestive enzymes into the gut. The standard approach outlined above would usually lead to the animal's death and a most confusing picture.

In view of these challenges, researchers have had to develop other ways for identifying and studying hormones. This has meant devising and exploring simpler experimental systems than surgically altered animals. Most contemporary work uses *in vitro* cultures of target organs or target cells to identify signaling molecules and to study their effects. Using both the traditional methods with whole organisms and the more recent approaches with cultured cells and organs, endocrinologists have identified more than 50 hormones and more than 15 hormone-producing structures in vertebrates and similar numbers in a variety of invertebrates.

What Kinds of Molecules Are Hormones?

Studies of the mode of action of hormones show that the major distinction among hormones is between those that are **lipid-soluble signals** (like steroids), which can enter a target cell by passing directly through the plasma membrane, and **water-soluble signals** (like peptides and some amines), which cannot pass through the plasma membrane and must act on its surface. Table 19-1 lists some important vertebrate hormones.

Among the most important lipid-soluble signaling molecules are **steroids,** nonpolar molecules that derive from cholesterol, as illustrated in Figure 19-5. Among the best-studied steroid hormones are the **glucocorticoids** (such as corticosterone), which stimulate the production of carbohydrate from protein and fat; the **mineralocorticoids** (such as aldosterone), which regulate the ion flows in the kidney; and the **sex steroids** (such as testosterone, progesterone, and estrogen), which stimulate, maintain, and regulate reproductive organs and secondary sexual characteristics. The sex steroid testosterone also contributes to the growth of muscles. Chemically synthesized steroids that resemble testosterone have been abused by some athletes to build muscle mass, but these steroids also greatly increase the risks of sterility, cancer, and psychiatric disease.

Water-soluble chemical signals include amines, the simplest chemicals that act as hormones, although some amines (like thyroid hormone) are lipid-soluble hormones. Amines, as illustrated in Figure 19-6, derive from amino acids, usually by the removal of a carboxyl group for the production of epinephrine from tyrosine. Under most biological conditions, the amino group (NH_2) associates with an additional hydrogen ion to form NH_3^+, so

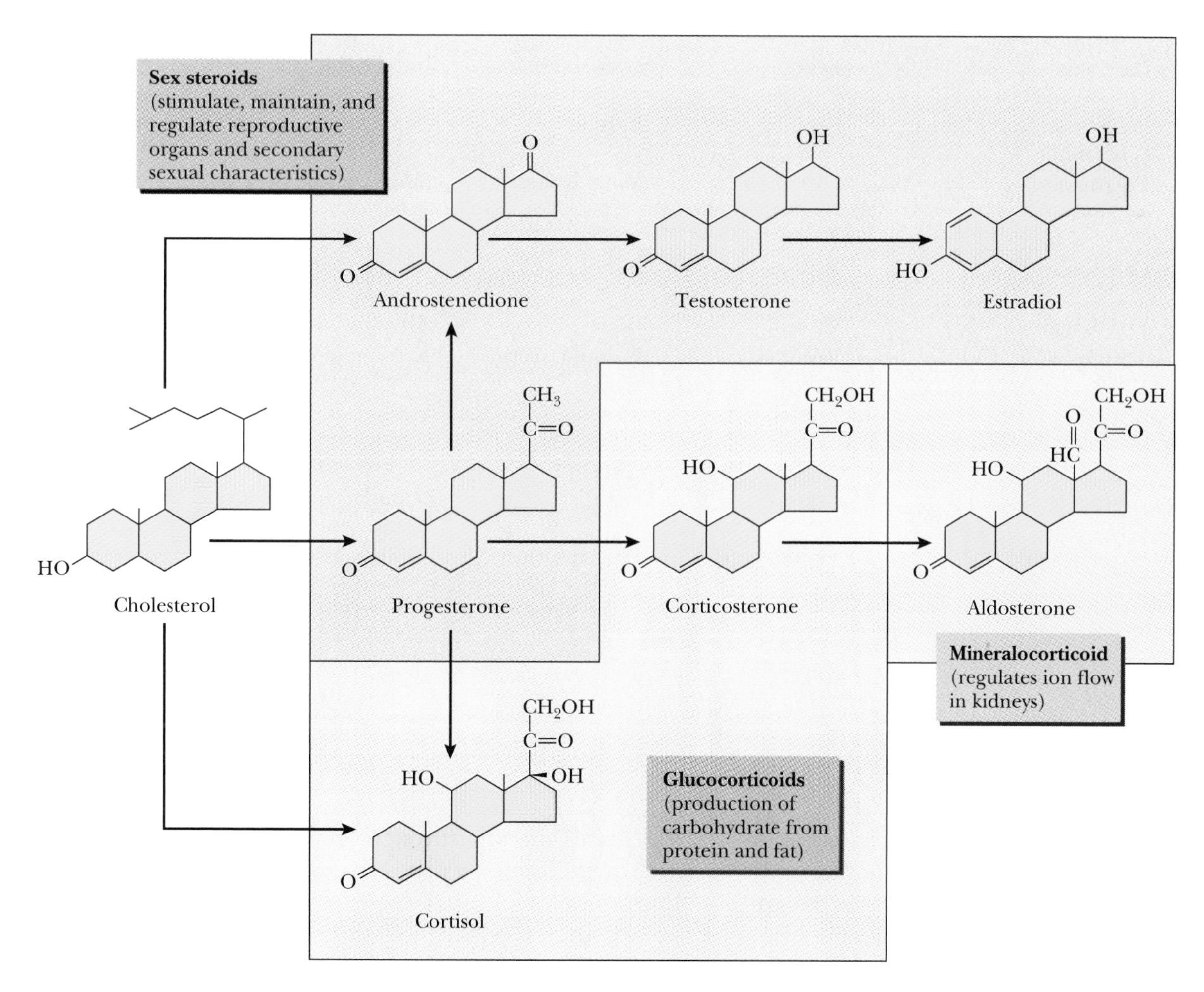

Figure 19-5 Steroid hormones all derive from cholesterol. Although all steroids contain the same characteristic four-ring structure, individual steroids differ in their side groups and have widely differing effects on their targets.

Table 19-1 ***Principal Vertebrate Hormones***

Secreting tissue or gland	Hormone	Chemical nature	Targets	Important properties or actions
Hypothalamus	Releasing and release-inhibiting hormones	Peptides	Anterior pituitary	Control secretion of hormones of anterior pituitary
	Oxytocin, vasopressin	Peptides	(See Posterior pituitary)	Stored and released by posterior pituitary
Anterior pituitary: Tropic hormones	Thyrotropin	Glycoprotein	Thyroid gland	Stimulates synthesis and secretion of thyroxine
	Adrenocorticotropin	Polypeptide	Adrenal cortex	Stimulates release of hormones from adrenal cortex
	Luteinizing hormone	Glycoprotein	Gonads	Stimulates secretion of sex hormones from ovaries and testes
	Follicle-stimulating hormone	Glycoprotein	Gonads	Stimulates growth and maturation of eggs in females; stimulates sperm production in males
Anterior pituitary: Other hormones	Growth hormone	Protein	Bones, liver, muscles	Stimulates protein synthesis and growth
	Prolactin	Protein	Mammary glands	Stimulates milk production
	Melanocyte-stimulating hormone	Peptide	Melanocytes	Controls skin pigmentation
	Endorphins and enkephalins	Peptides	Spinal cord neurons	Decreases painful sensations
Posterior pituitary	Oxytocin	Peptide	Uterus, breasts	Induces birth by stimulating labor contractions; causes milk flow
	Vasopressin (antidiuretic hormone)	Peptide	Kidneys	Stimulates water reabsorption
Thyroid	Thyroxine	Iodinated amino acid derivative	Many tissues	Stimulates and maintains metabolism necessary for normal development and growth
	Calcitonin	Peptide	Bones	Stimulates bone formation; lowers blood calcium
Parathyroids	Parathormone	Protein	Bones	Absorbs bone; raises blood calcium
Thymus	Thymosins	Peptides	Immune system	Activate immune response of T cells in the lymphatic system
Pancreas	Insulin	Protein	Muscles, liver, fat, other tissues	Stimulates uptake and metabolism of glucose; increases conversion of glucose to glycogen and fat
	Glucagon	Protein	Liver	Stimulates breakdown of glycogen and raises blood sugar
	Somatostatin	Peptide	Digestive tract; other cells of the pancreas	Inhibits insulin and glucagon release; decreases secretion, motility, and absorption in the digestive tract

most amine hormones are charged. Other amine hormones—norepinephrine and the thyroid hormones—also derive from tyrosine. While epinephrine and norepinephrine are charged and cannot pass the plasma membrane of their target cells, the thyroid hormone thyroxine keeps its carboxyl group and has no net charge. Thyroxine passes directly through the plasma membrane and acts in the same way as steroids and other lipid-soluble hormones.

Peptides, which are water soluble, are by far the largest class of hormones. A peptide hormone may contain as few as three amino acid residues or well over 100.

Secreting tissue or gland	Hormone	Chemical nature	Targets	Important properties or actions
Adrenal medulla	Adrenaline, noradrenaline	Modified amino acids	Heart, blood vessels, liver, fat cells	Stimulate "fight-or-flight" reactions: increase heart rate, redistribute blood to muscles, raise blood sugar
Adrenal cortex	Glucocorticoids (cortisol)	Steroids	Muscles, immune system, other tissues	Mediate response to stress; reduce metabolism of glucose, increase metabolism of proteins and fats; reduce inflammation and immune responses
	Mineralocorticoids (aldosterone)	Steroids	Kidneys	Stimulates excretion of potassium ions and reabsorption of sodium ions
Stomach lining	Gastrin	Peptide	Stomach	Promotes digestion of food by stimulating release of digestive juices; stimulates stomach movements that mix food and digestive juices
Lining of small intestine	Secretin	Peptide	Pancreas	Stimulates secretion of bicarbonate solution by ducts of pancreas
	Cholecystokinin	Peptide	Pancreas, liver, gall bladder	Stimulates secretion of digestive enzymes by pancreas and other digestive juices from liver; stimulates contractions of gallbladder and ducts
	Enterogastrone	Polypeptide	Stomach	Inhibits digestive activities in the stomach
Pineal	Melatonin	Modified amino acid	Hypothalamus	Involved in biological rhythms
Ovaries	Estrogens	Steroids	Breasts, uterus, other tissues	Stimulate development and maintenance of female characteristics and sexual behavior
	Progesterone	Steroid	Uterus	Sustains pregnancy; helps to maintain secondary female sexual characteristics
Testes	Androgens Testosterone	Steroids	Various tissues	Stimulate development and maintenance of male sexual behavior and secondary male sexual characteristics; stimulate sperm production
Heart	Atrial natriuretic hormone	Peptide	Kidneys	Increases sodium ion excretion

The longer peptide hormones, which are classified as proteins, include insulin and the hormones of the anterior pituitary, which regulate growth, lactation, and the production of hormones by other endocrine organs. Among the shorter peptide hormones are those made by the hypothalamus (a part of the brain) as **releasing factors,** which regulate the release of specific hormones by the anterior pituitary; vasopressin, which stimulates water reabsorption by the kidneys; and oxytocin, which stimulates contractions of the uterus during childbirth. The amino acid sequences of the peptide hormones insulin, vasopressin, and oxytocin are illustrated in Figure 19-7. Nei-

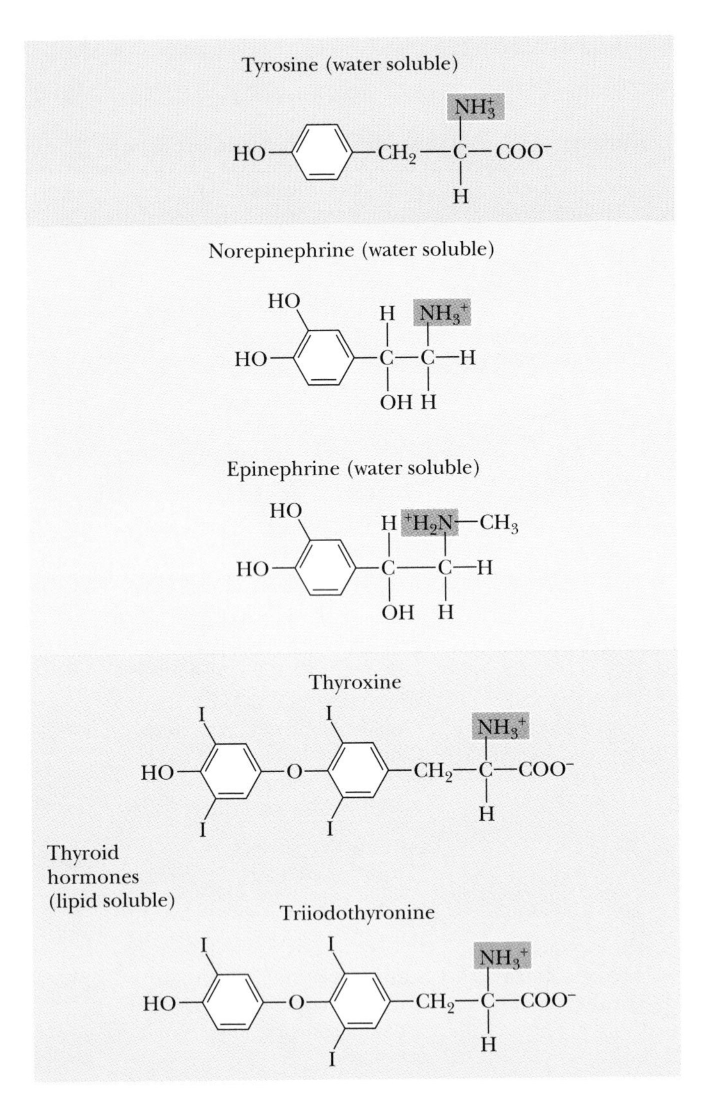

Figure 19-6 The simplest hormones are amines derived from the amino acid tyrosine. Epinephrine and norepinephrine act at cell surfaces, as do other water-soluble hormones. In contrast, the thyroid hormones are lipid soluble, and their actions resemble those of the steroids.

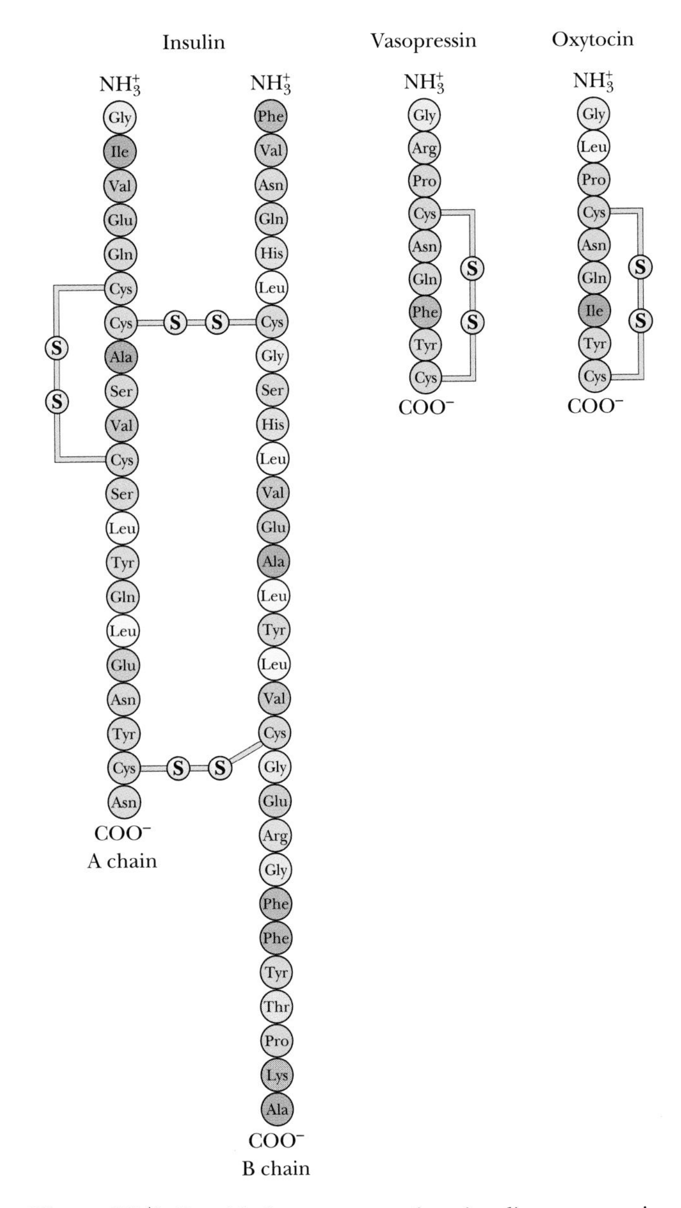

Figure 19-7 Peptide hormones, such as insulin, vasopressin and oxytocin, consist of polypeptide chains that are produced on ribosomes like other polypeptides.

ther the small peptides nor the much larger proteins are soluble in plasma membranes, and none crosses the plasma membrane and enters their target cells. The action of these hormones depends on their acting on specific receptors on the plasma membrane; these receptors in turn stimulate the synthesis or release of intracellular "second messengers" that influence the inner workings of their target cells. (See later in this chapter.)

HOW ARE HORMONES MADE AND RELEASED?

Intercellular signaling in general and hormone action in particular involve six steps: (1) synthesis of the signaling molecule, (2) release from the producing cells, (3) transport to the target cells, (4) detection of signaling molecules by target cells, (5) responses within the target cells, and (6) termination of signaling by the destruction of the signaling molecule.

The synthesis of steroids and amines requires specific series of enzymatic reactions. Making testosterone from cholesterol, for example, requires eight enzymes, while the conversion of tyrosine to epinephrine (illustrated in Figure 19-6) requires five. The pathway of epinephrine synthesis always produces two other important signaling molecules along the way—dopamine, used in a number of important neural circuits in the brain, and

ESSAY

NITRIC OXIDE: AN UNUSUAL SIGNALING MOLECULE

All of the signaling molecules discussed in this chapter have molecular weights ranging from about 100 (for the amines) to 10,000 or more (for the protein hormones). In the last five years, however, cell biologists have come to recognize the tiny nitric oxide molecule (NO) as a potent natural signal. NO consists of just two atoms, nitrogen and oxygen; it has a molecular weight of 30 and is a gas, with the chemical properties of a free radical.

Cells produce NO from the amino acid arginine, using an enzyme called NO synthase (NOS). The chemically reactive NO contributes to the death of tumors and bacteria. In the case of a bacterial infection, a component of the bacterial cell wall activates NOS in macrophages, the scavenger cells that destroy foreign organisms. Increased NO production is necessary for macrophage-mediated killing. Blocking NOS with *N*-methyl-arginine, a specific NOS inhibitor, prevents bacteriocide.

NO also has powerful effects on blood vessels. Signals known to produce vasodilation (relaxation of the blood vessels) do so by triggering the release of NO by the endothelium, the cells that immediately line the blood vessels. The released NO diffuses to the surrounding smooth muscle, where it elicits relaxation. Too much NO release, however, may be involved in an often fatal condition called *septic shock,* and NOS inhibitors are effective treatments for the symptoms of septic shock.

Neurons also produce NO, both in the brain and outside the brain. In the gastrointestinal tract, NO mediates relaxation that is crucial to the normal peristaltic movements of digestion. In the penis, NO produced by pelvic neurons is the major, if not the only, signal for erection.

In the brain, researchers have demonstrated that NO serves as a neurotransmitter, sometimes carrying information in the opposite direction from that of a more conventional neurotransmitter. NO may also serve as the mediator of abnormal cell death in the brain in the course of neurodegenerative diseases such as Huntington's disease. Because the lifetime of any NO molecule is only about 5 seconds, it has been difficult to document all its effects, but the molecule has become a subject of intense research by cell biologists, neurobiologists, physiologists, and biochemists.

norepinephrine, used both in the nervous system and as a circulating hormone.

Cells make peptide hormones on ribosomes associated with the rough endoplasmic reticulum, just as they make exported proteins, using information contained in messenger RNA. (See Chapter 14.) As in the case of exported proteins, peptide hormones are subject to extensive post-translational modifications within the Golgi complex and the lumen of the endoplasmic reticulum. For example, insulin, as illustrated in Figure 19-8, consists of two polypeptides, held together by disulfide bonds. The two chains derive from a single longer polypeptide, called **preproinsulin.** The insulin precursor is cut twice by a proteolytic enzyme, and the intermediate peptide, the C peptide, is removed. Still smaller signaling peptides, like thyroid-hormone releasing hormone (TRH), arise similarly by the action of proteases on longer precursors.

Molecular Biological Studies Have Revealed That a Single Gene Can Encode Several Different Signaling Peptides

The application of molecular biological techniques—in particular, cDNA cloning and sequencing—has allowed comparisons of mRNA structure with the sequences of the resulting peptide hormones. This work has led to two major surprises: (1) A single gene can yield mRNAs for different signaling molecules, according to the pattern of RNA splicing; and (2) a single polypeptide precursor can yield several different peptide hormones, according to the pattern of proteolytic action.

The first surprise was alternative RNA splicing. A single gene encodes both **calcitonin,** which regulates calcium retention by the kidney, and **calcitonin-gene–related peptide** (CGRP), which plays a role in sensory nerve pathways.

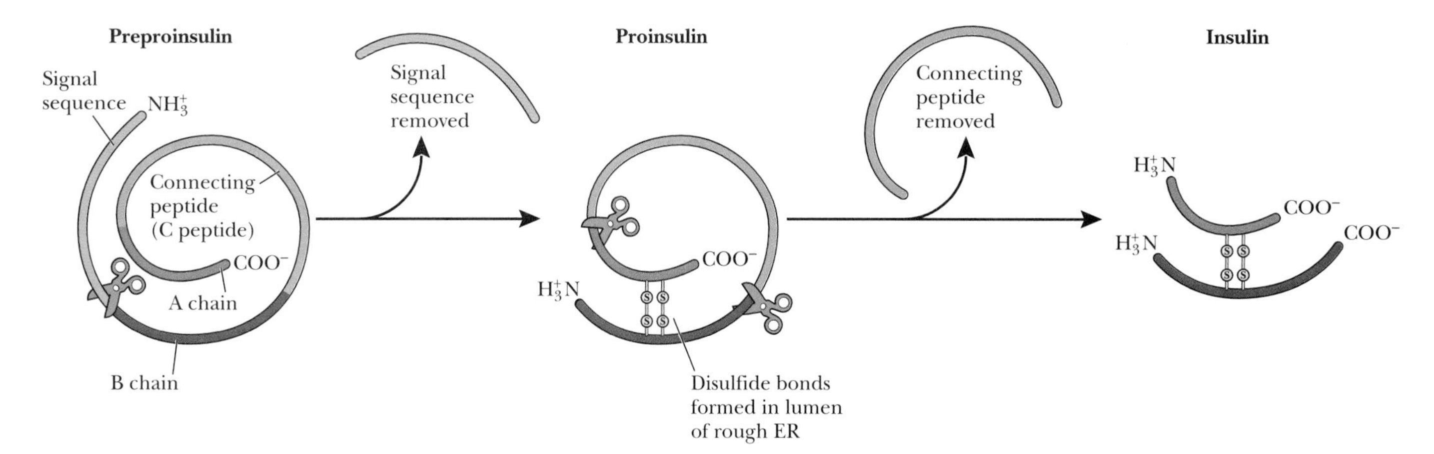

Figure 19-8 Many peptide hormones arise, as does insulin, from a larger precursor. Preproinsulin contains a signal sequence and a peptide segment (the C peptide) that is not part of the mature insulin molecule; as translation occurs on the rough ER, the signal sequence directs the newly made polypeptide into the ER lumen, where disulfide bond formation occurs; peptide cleavage then occurs, releasing the C peptide and resulting in a mature insulin molecule of two polypeptide chains held together with disulfide bonds.

In the thyroid gland, enzymes convert the precursor RNA into an mRNA encoding calcitonin. But in the nervous system, nuclear enzymes convert the same precursor into a different mRNA, which encodes CGRP. (See Chapter 15, p. 418.) In addition to their importance in understanding the synthesis of specific signaling molecules, these findings were of great general interest because they required modification of a previous definition of a gene, "one-gene, one polypeptide chain." (See Chapter 12.) In this case, as well as in many others, a single gene can produce more than one kind of polypeptide chain.

The second surprise—alternative post-translational processing—is well illustrated in the anterior pituitary. There, specific proteases can convert the same polypeptide precursor (called **pro-opiomelanocortin,** or POMC) into hormones with very different actions. As illustrated in Figure 19-9, the products of POMC range from **β-endorphin** [from *endo*genous m*orphin*e], which suppresses pain, to **corticotropin** (also called adrenocorticotrophic hormone or ACTH), which stimulates the production of corticosteroids in the adrenal cortex.

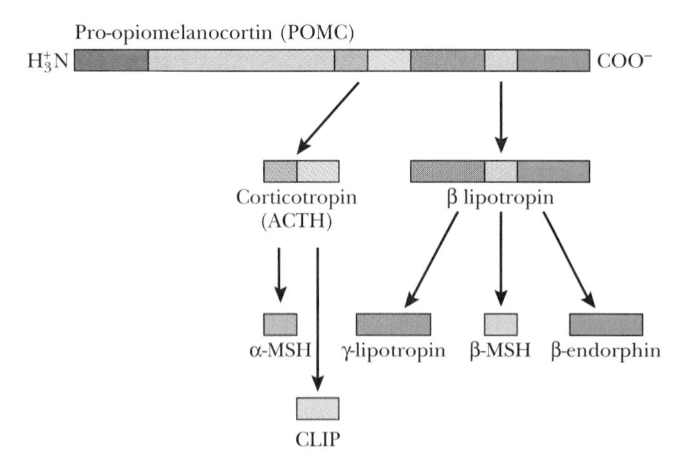

Figure 19-9 A single polypeptide precursor can produce more than one signaling molecule. Pro-opiomelanocortin (POMC) produces at least seven different signaling molecules by different patterns of proteolytic cleavage.

Cells Release Water-Soluble Signals by Regulated Exocytosis

Endocrine cells release lipid-soluble hormones such as steroids directly across their plasma membranes. The amount of hormone released depends on the rate of enzymatic synthesis. This rate, in turn, depends on the amount and activity of the enzymes that convert the hormone precursors to the final product.

In contrast, water-soluble hormones, including peptides and most amines, are packaged into membrane-bounded vesicles. Release of these signals depends on exocytosis, as described in Chapter 6. The rate of release of water-soluble hormones depends on the regulation of exocytosis, which usually depends—among other things—on the concentration of free calcium ions within the hormone-secreting cell. Cells can regulate both the production and the release of hormones. The anterior pituitary and the hypothalamus, illustrated in Figure 19-4, demonstrate the highly regulated nature of hormone release.

The Hypothalamus Regulates Hormone Production and Release by the Cells of the Anterior Pituitary

No set of cells better illustrates the regulation of hormone synthesis than the cells of the anterior pituitary and the

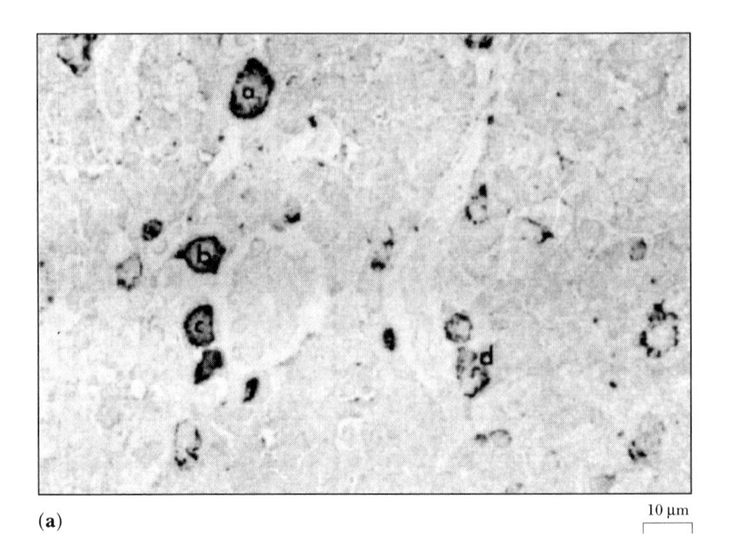

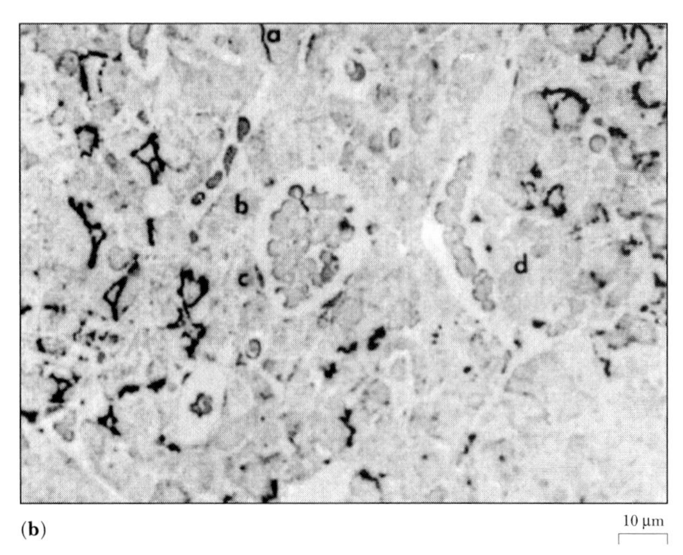

Figure 19-10 Individual cells within the anterior pituitary are specialized for the synthesis of specific hormones. **(a)** Immunohistochemical localization of LH in four cells of the anterior pituitary (designated a, b, c, and d); **(b)** immunohistochemical localization of ACTH in the same section of the anterior pituitary; cells a, b, c, d do not contain ACTH, but other cells do. *(G.V. Childs, University of Texas Medical Branch at Galveston)*

hypothalamus. The anterior pituitary produces and releases seven peptide hormones: corticotrophin (ACTH), which (as mentioned above) stimulates the production of steroids in the adrenal cortex; endorphin, also mentioned above, a natural pain suppressor; thyroid-stimulating hormone (TSH), which stimulates the production of thyroxin in the thyroid gland; growth hormone (GH), which stimulates tissue and skeletal growth; prolactin, which stimulates milk production; melanocyte-stimulating hormone (MSH), which stimulates pigment production in specialized skin cells; follicle-stimulating hormone (FSH), which in females promotes the maturation of the follicle during the menstrual cycle and in males stimulates testosterone production; and luteinizing hormone (LH), which in females induces ovulation and stimulates estrogen production and in males increases testosterone production. The hypothalamus produces at least seven hormones that control the release of hormones by the anterior pituitary.

Distinct sets of cells within the anterior pituitary specialize in the production of a single hormone, as illustrated in Figure 19-10. Some cells, for example, are specialized to produce either growth hormone, ACTH, or TSH. Other cells, however, can produce two hormones—FSH and LH, both of which act on gonadal tissue (and are therefore collectively called gonadotropins).

Because its hormones stimulate so many target organs to produce other hormones, many endocrinologists have referred to the anterior pituitary as the "master gland." In fact, however, the release of hormones by the anterior pituitary depends on other hormones, produced by the hypothalamus. The hypothalamus, not the anterior pituitary, is the true master gland.

The hypothalamus exerts its control of the anterior pituitary by at least five **releasing hormones,** each of which stimulates the release of specific hormones, and two **inhibiting hormones,** which inhibit hormone release. For example, gonadotropin-releasing hormone (GnRH) stimulates the secretion of FSH and LH by the anterior pituitary. Hypothalamic hormones like GnRH enter the general circulation through capillaries in a special structure called the **median eminence.** The median eminence also contains specialized veins that do not flow directly into the rest of the venous circulation but flow into the anterior pituitary, where they divide and form a second capillary bed. This diversion gives the hypothalamic hormones more direct access to their targets in the anterior pituitary.

The regulation of hormone release can involve both negative and positive feedback. In males, for example, FSH and LH stimulate testosterone production in the testes. Testosterone enters the blood from the testes. Circulating testosterone acts on the hypothalamus to inhibit production of GnRH. The lowered GnRH in turn reduces the release of FSH and LH by the anterior pituitary, a state that reduces further testosterone production by the testes. The result is a negative feedback loop: Increased FSH and LH ultimately lead to a reduction in FSH and LH, while increased testosterone leads to a reduction in testosterone synthesis. The result is a relatively stable concentration of testosterone.

In mammalian females, the regulation of GnRH release and gonadotropin release is more complicated: High concentrations of estrogen inhibit GnRH release, ultimately leading to a lowered production of estrogen. But lower estrogen concentrations actually stimulate GnRH release, leading to a surge of LH. During the menstrual cycle, this surge of LH triggers ovulation.

The hypothalamus is a part of the brain, and it also responds to neural information concerning physical and emotional states of the body. Stress—produced, for example, by blood loss, fright, or even a cell biology final—increases the production of corticotropin-releasing hormone (CRH). Increased CRH leads to increases in ACTH and corticosteroids. High corticosteroid levels in turn heighten the body's reaction to stress: Corticosteroids increase heart output, lung ventilation, blood flow to the muscles, and glycogen production in the liver. A small increase in CRH production (0.1 μg) leads to a larger increase in ACTH (1 μg), a still larger increase in corticosteroid (40 μg), and a still larger effect on glycogen synthesis (5600 μg). Unfortunately, in people with stressful lives, the maintenance of high corticosteroid levels may lead to a higher incidence of cardiovascular disease.

HOW DO TARGET CELLS DETECT CHEMICAL SIGNALS?

A chemical signal acts on its target tissues by first binding to a specialized protein, called a **receptor.** Most water-soluble signals (amines and peptides) do not cross the plasma membrane, and their receptors lie on the membranes of target cells. In contrast, lipid-soluble signals, such as steroid hormones, easily cross the plasma membrane, and their receptors are in the cytoplasm. In this section we discuss receptors for water-soluble signals, using receptors for epinephrine as our primary examples.

Most of the same concepts and experimental approaches are useful for the receptors of lipid-soluble signals (such as the steroids). The major differences between the two classes of receptors are that (1) receptors for lipid-soluble signals are in the cytoplasm rather than on the plasma membrane; (2) receptors for lipid-soluble signals act directly to affect transcription, as discussed in Chapter 15, while receptors for water-soluble signals act through intermediates, called **second messengers,** discussed later in this chapter.

Signaling Molecules Bind to Specific Receptors by Noncovalent Bonds

Signaling molecules, such as hormones, form specific complexes with their receptors in much the same way that a substrate associates with an enzyme. (See Chapter 4.) The signal molecule forms noncovalent bonds with the peptide backbone or with amino acid side chains in the region of the receptor called the binding site, as illustrated in Figure 19-11a. Other molecules can mimic the ability of a natural signal to bind to a receptor, forming noncovalent interactions with the same binding site. Such mimics may be natural products (often toxins) or compounds synthesized in the laboratory (and not found in nature).

A molecule that binds to a specific binding site in a protein is called a **ligand** [Latin, *ligare* = to bind]. Some ligands, called **agonists,** affect the target cells in the same way as the natural signal molecule itself, while other ligands, called **antagonists,** prevent the natural signal molecule from binding and triggering its characteristic responses in the target cells. As illustrated in Figure 19-11b and c, the synthetic compounds isoproteronol, propranolol, and alprenolol all share structural features with epinephrine and bind to the same binding site. Isoproteronol is an agonist that mimics the effects of epinephrine, while propranolol and alprenolol are antagonists that bind to epinephrine receptors but do not activate target cell responses.

Ligands Differ in Their Affinity for Receptors

Ligands labeled with radioactive isotopes allow researchers to study the binding of ligands to receptor molecules and to study receptor characteristics. We may describe the binding of a ligand to its receptor as an equilibrium:

$$R + L \rightleftharpoons RL$$

where R is the receptor, L the ligand, and RL the ligand-receptor complex.

The formation of the ligand-receptor complex depends on ligand concentration in the same way as the formation of the substrate-enzyme complex in the Michaelis-Menten model described in Chapter 4. K_D, the dissociation constant of the ligand-receptor complex, is a measure of the affinity of the ligand for the receptor:

$$K_D = \frac{[R][L]}{[RL]}$$

where [R] is the concentration of receptor, [L] the concentration of ligand, and [RL] the concentration of receptor-ligand complex. As in the case of the Michaelis-Menten model, the relationship between the ligand concentration and the formation of ligand-receptor complex is given by a hyperbolic curve. To show this relationship, we use the same algebra as in Chapter 4.

$$[R_T] = [R] + [RL]$$

where $[R_T]$ is the total concentration of receptor. So

$$[R] = [R_T] - [RL]$$

So we may now write,

$$[RL] = \frac{[R][L]}{K_D} = \frac{([R_T] - [RL])[L]}{K_D}$$

Structure	Compound
(3,4-dihydroxyphenyl)–CH(OH)–CH_2–NH–CH_3	Epinephrine
Agonist (3,4-dihydroxyphenyl)–CH(OH)–CH_2–NH–CH(CH_3)–CH_3	Isoproterenol

Structure	Compound
Antagonist CH_2=CH–CH_2–(phenyl)–O–CH_2–CH(OH)–CH_2–NH–CH(CH_3)–CH_3	Alprenolol
Antagonist (naphthyl)–O–CH_2–CH(OH)–CH_2–NH–CH(CH_3)–CH_3	Propranolol

(a) **Epinephrine**

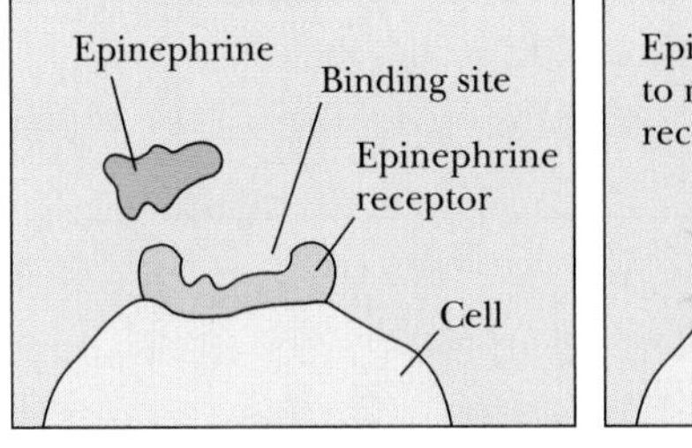

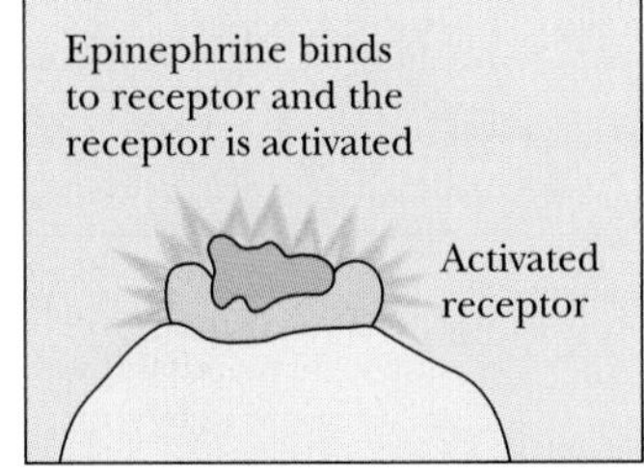

(b) **Epinephrine agonist (isoproterenol)**

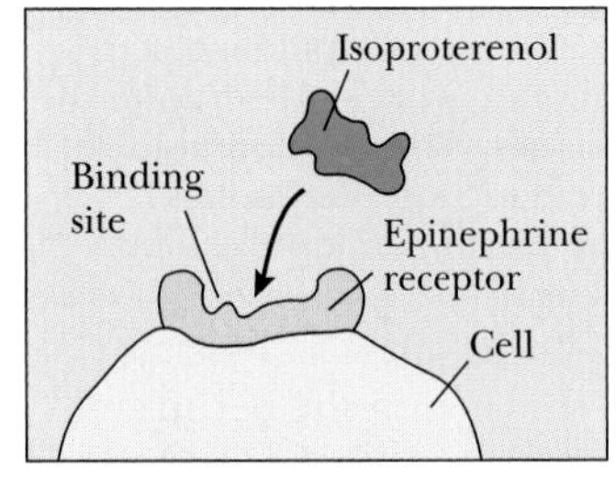

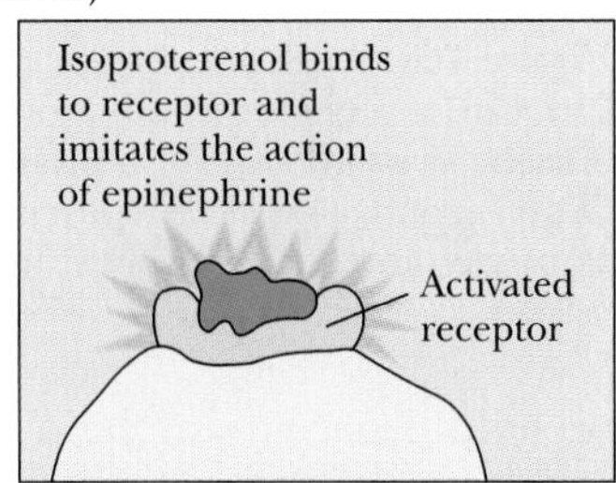

(c) **Epinephrine antagonists (alprenolol and propanolol)**

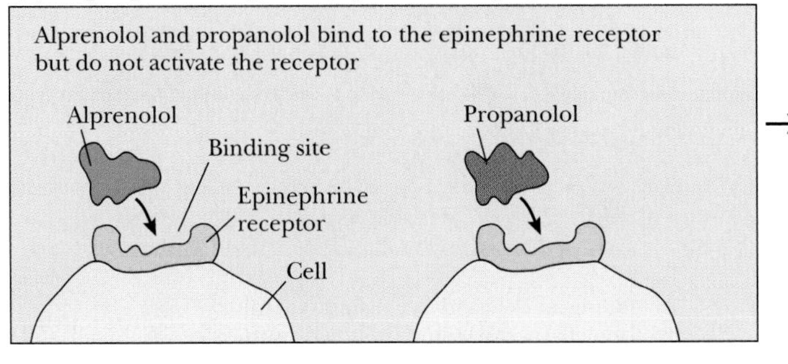

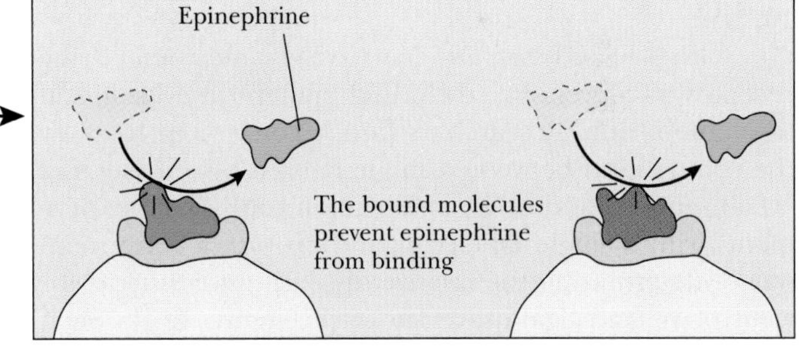

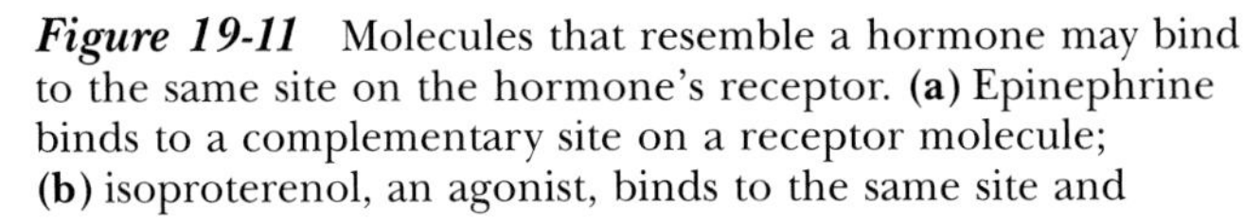

Figure 19-11 Molecules that resemble a hormone may bind to the same site on the hormone's receptor. **(a)** Epinephrine binds to a complementary site on a receptor molecule; **(b)** isoproterenol, an agonist, binds to the same site and imitates the hormone's action; **(c)** alprenolol and propanolol are antagonists; they bind to the same site but do not imitate epinephrine; instead they prevent the hormone itself from binding.

and

$$[RL] = \frac{[R_T][L]}{K_D \dfrac{1}{(1 + [L]/K_D)}} = \frac{[R_T][L]}{[L] + K_D}$$

or

$$\frac{[RL]}{R_T} = \frac{1}{1 + K_D/[L]} = \frac{[L]}{[L] + K_D}$$

When a ligand is present at a concentration that equals its K_D, $[L] = K_D$, $[RL]/R_T = 0.5$; that is, the receptor is half-saturated with ligand. In contrast, when a ligand is present at a concentration that is $1/10$ its K_D, then $[L] = 0.1\ K_D$, $[RL]/R_T = 1/11$, and the receptor is 9% saturated. Figure 19-12 shows the binding of ^3H-labeled alprenolol to the membranes of frog red blood cells, which have specific receptors for epinephrine. Alprenolol is an epinephrine antagonist, as mentioned previously. As an experimenter increases the concentration of alprenolol, the epinephrine receptor binds more and more until the receptor becomes saturated. No matter how much the experimenter then increases the concentration of al-

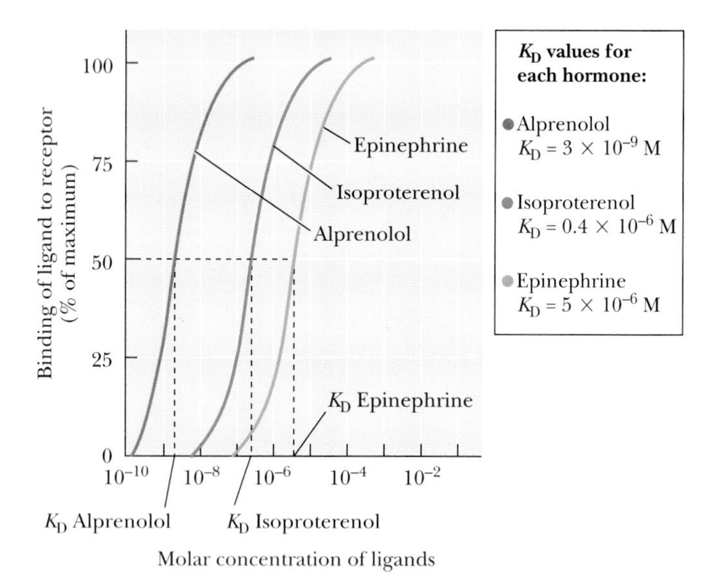

Figure 19-12 By measuring the binding of a radiolabeled ligand to a receptor population, researchers can determine the binding constant K_D for a given ligand, here ^{3}H-alprenolol. Other ligands—such as epinephrine or isoproterenol—compete more or less well with the binding of alprenolol.

prenolol, no more can bind to the membrane because the receptor molecules are already saturated. K_D is the concentration at which the receptors are 50% saturated. For alprenolol in this experiment, the K_D is about 3.4×10^{-9} M.

Other ligands can also bind to the same epinephrine receptor as alprenolol, including epinephrine itself and norepinephrine. Researchers find it convenient to study the competition between a single labeled ligand (such as ^{3}H-alprenolol) and other unlabeled ligands. Such experiments with a single labeled ligand provide a convenient way of determining the relative dissociation constants for a set of related ligands, as shown in Figure 19-12. Such measurements reveal that the receptor's natural ligand, epinephrine, binds to receptors in frog red cells with K_D of 5×10^{-6} M, that is, much less strongly than alprenolol. The synthetic agonist isoproteronol, however, has a K_D of 0.4×10^{-6} M, meaning that it binds more tightly than epinephrine. In general, naturally occurring hormones stimulate responses at extremely low concentrations, typically 10^{-6} to 10^{-9} M, while artificial agonists and antagonists may act at even lower concentrations.

How Do Cell Biologists Study Receptor Molecules?

Because the receptors for water-soluble hormones like epinephrine are associated with membranes, the first step in receptor isolation is usually the removal of the receptor from the membranes. Researchers usually do this by first isolating membranes from the receptor-containing target cells and then treating them with a detergent to make the receptor soluble in water.

A receptor interacts with a ligand.

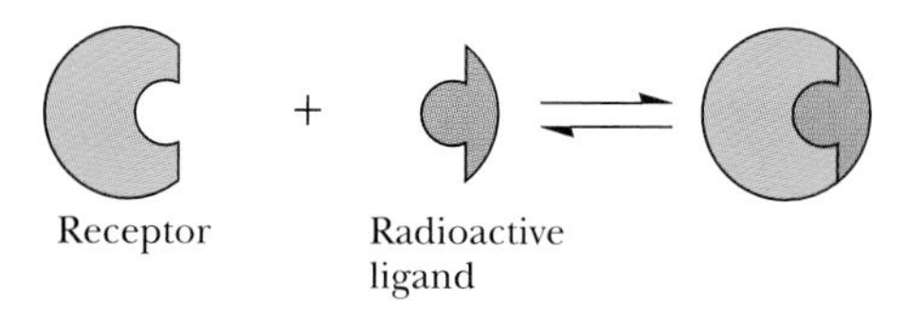

The ligand can be immobilized by coupling it to an insoluble matrix. Cell extracts containing many individual proteins may be passed through the matrix.

Specific receptor protein binds to ligand. All other unbound material is washed out of the matrix.

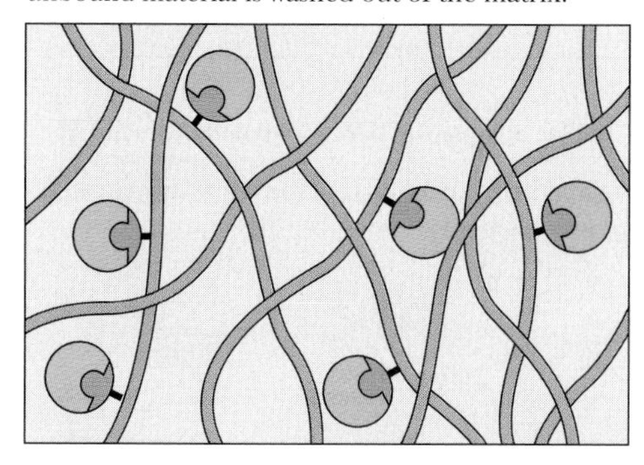

Adding an excess of free ligand that will compete for the bound protein dissociates the protein from the chromatographic matrix. The receptor protein passes out of the column with bound ligand.

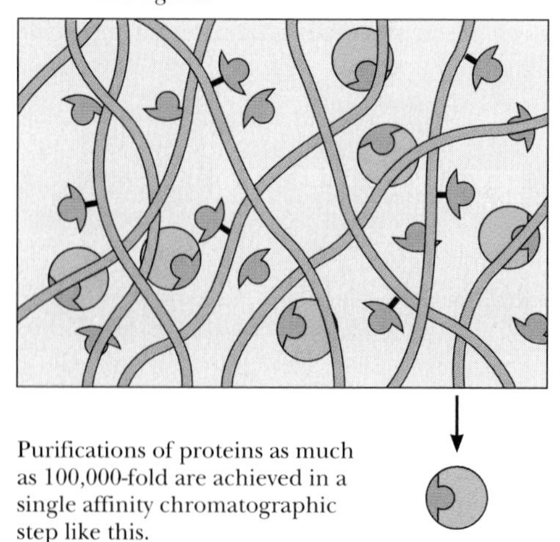

Purifications of proteins as much as 100,000-fold are achieved in a single affinity chromatographic step like this.

Figure 19-13 Affinity chromatography provides a powerful tool for isolating receptor molecules.

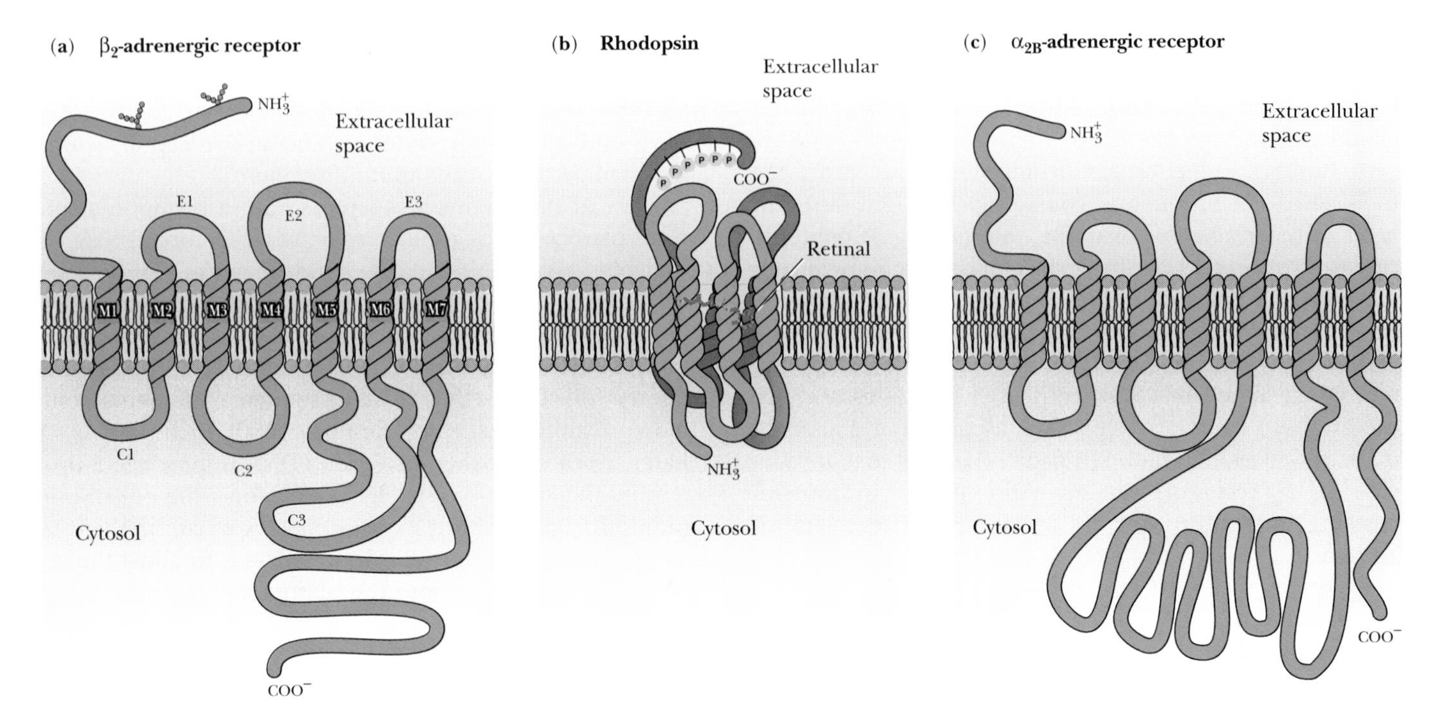

Figure 19-14 Epinephrine binds to different receptors in different tissues, but all epinephrine receptors are membrane proteins with seven transmembrane segments. **(a)** The organization of one epinpehrine receptor, the β_2-adrenergic receptor, showing seven transmembrane α-helices, three intracellular loops, three extracellular loops, an amino terminal extracellular segment, and a carboxy terminal intracellular segment; **(b)** rhodopsin, the light-sensing protein of the retina, also contains seven transmembrane segments but with the amino-terminal end of the polypeptide on the cytosolic side of the plasma membrane; **(c)** another epinephrine receptor, the α_{2B}-adrenergic receptor, also contains seven transmembrane segments with the same orientation as the β_2 receptor.

The avid binding of synthetic ligands, like alprenolol, has allowed the identification of the receptor molecules by a technique called **affinity labeling,** in which a radioactive ligand is chemically attached to the receptor's binding site, as in Figure 19-13. Extending this idea, biochemists have purified many types of receptors by **affinity chromatography,** in which a tightly binding ligand (again, like alprenolol) is attached to an insoluble matrix, as illustrated in Figure 19-13. (See Chapter 3, p. 70.) Solubilized receptor molecules then bind to the matrix, while other membrane proteins do not. This differential binding allows a specific type of receptor to be separated, in a single step, from even a 100,000-fold excess of other proteins.

The isolation of specific receptor molecules allowed researchers to determine their molecular composition—the number and types of polypeptides—and their amino acid sequences. Knowledge of the amino acid sequences in turn allowed the preparation of recombinant DNAs that contained the DNA sequences encoding individual types of receptors. In the case of the epinephrine receptors, studies with recombinant DNAs yielded two surprises: (1) The amino acid sequence of epinephrine-binding receptors is different in different target tissues; and (2) all epinephrine receptors—in fact, all receptors for amine and peptide hormones—share the same structural organization, illustrated in Figure 19-14, as that previously determined for rhodopsin, the protein that detects light in the retina.

The Family of Receptors That Binds Epinephrine Is Part of a Still Larger Family of Membrane Receptors

All receptors that respond to epinephrine are called **adrenergic receptors** [from Adrenaline (as in *adrenal* gland), the patented British trade name for epinephrine]. Tissue-specific differences in adrenergic receptors were already known before the receptor proteins were isolated because tissues differed in their relative responses to epinephrine and norepinephrine. The differing responses led to a distinction between "α-adrenergic receptors" and "β-adrenergic receptors": In general, α-adrenergic receptors respond more effectively to norepinephrine, while β-adrenergic receptors respond more effectively to epinephrine. Smooth muscle cells that line the blood vessels in the intestinal tract, for example, have α-adrenergic receptors and respond more strongly to norepinephrine than to epinephrine, while liver cells, in which receptor action triggers glycogen breakdown and glucose release, have β-adrenergic receptors and respond more strongly to epinephrine. Hormone binding to α-adrenergic receptors triggers a cellular response distinct from that of hormone binding to β-adrenergic receptors, as we see later in this chapter.

Both α and β classes of adrenergic receptors are themselves heterogeneous, consisting of several different types of receptors. For example, the binding of epinephrine to β_1 receptors on the surface of cardiac muscle cells increases the heart rate, while binding to β_2 receptors on smooth muscle cells in the lungs dilates the small airways (bronchioles). Antagonists that are specific for β_1 receptors ("beta blockers") slow cardiac contractions and are used to treat heart problems such as angina and arrhythmias, as well as high blood pressure. Agonists that are specific for β_2 receptors can open the airways without increasing heart rate and are used in the treatment of asthma.

Much of our current understanding of adrenergic receptors comes from studies using recombinant DNA. The isolation and characterization of the first pure adrenergic receptor allowed the determination of its amino acid sequence. This information, in turn, allowed researchers to isolate cDNAs that contained the mRNA sequences for adrenergic receptors for a variety of tissues, using the methods described in Chapter 17.

Comparison of the nucleotide sequences of adrenergic receptor mRNAs from a variety of tissues showed that these receptors are not identical in all tissues but that all adrenergic receptors are part of the same family. Each receptor consists of single polypeptide chains about 450 amino acid residues long. Within each polypeptide are seven segments of 22–24 consecutive hydrophobic amino acid residues. These sequences are thought to be **transmembrane segments** because they are just long enough to span the plasma membrane as an α helix. Figure 19-14 shows the presumed organization of the seven transmembrane helices, designated M1–M7 in Figure 19-14a, and the loops that connect them. Three of these loops (designated C1–C3) lie within the cytoplasm, while three (designated E1–E3) lie in the extracellular space. The largest loop is intracellular loop C3, between M5 and M6.

While the amino acid sequences of adrenergic receptors vary greatly, all share this common plan with seven transmembrane segments. This same plan is also present in the visual pigment rhodopsin, in the receptors for a wide variety of peptide and protein hormones, as well as in a large class of **olfactory receptors,** which bind to odorant molecules and begin to inform the brain about smell.

The Common Properties of Adrenergic Receptors Led to the Identification of Other Receptors

As we've said, despite wide differences in amino acid sequence, adrenergic receptors, rhodopsin, olfactory receptors, and receptors for a variety of peptide hormones, all share a common plan, with seven transmembrane segments. The simplest explanation for the similarities among all these receptors is that all are *homologous;* that is, they have evolved from a common ancestral receptor, which also had seven transmembrane segments.

If all these known receptors have a common structural plan, perhaps other, unknown receptors also share that plan. With this idea in mind, many researchers set about finding other related receptors by a strategy called **homology cloning,** which identifies cDNAs that share common sequence elements. The first step is to screen a cDNA library for cDNAs that contain sequence elements common to known receptor families. With such cDNAs in hand, researchers then use each cDNA to produce a protein in a eukaryotic cell. The reason for using eukaryotic rather than prokaryotic expression—especially in the case of receptors for water-soluble signals—is to allow the receptor to be inserted into the plasma membrane.

Investigators then examine the ability of the encoded protein to act as a hormone receptor, for example by binding to specific ligands. This strategy has been remarkably fruitful, not only for the identification of previously unknown receptors but also for the identification of new members of gene families. Among the fruits of homology cloning are a family of dopamine receptors, each of which has seven transmembrane segments and whose action in the brain is particularly important in pathways affected by Parkinson's disease and schizophrenia.

In another strategy for finding receptor cDNAs, **expression cloning,** researchers identify a cDNA by recognizing the protein that it encodes. In this approach, the experimenter places each cDNA into a vector that contains signals necessary for transcription and translation. (See also Chapter 17, p. 460.)

In a "bacterial expression library," these signals allow each bacterium to express the cDNA it contains, and the researcher may identify the protein product with an antibody. For expression cloning of receptors, however, the cDNAs are inserted into vectors that contain eukaryotic signals, allowing their expression in eukaryotic cells, which can properly process the polypeptide products—targeting them to a membrane address through the Golgi complex and the rough endoplasmic reticulum, as described in Chapter 14. The cells that express the receptor of interest may then be identified by their ability to bind a ligand attached to the surface of the culture dish or to a fluorescent molecule. The attached or fluorescent cells can then be selectively propagated and their cDNAs isolated, sequenced, and further manipulated.

The application of recombinant DNA techniques has allowed the molecular characterization of receptors for a large number of signaling molecules. These studies have shown that many receptors fall into a small number of categories: (1) Receptors for amines and many other water-soluble peptides are cell surface proteins that consist of a single polypeptide with seven transmembrane domains;

(2) receptors for steroids and many lipophilic signaling molecules are soluble proteins that can bind both to signaling molecules and to DNA. (See Chapter 15 for a discussion of steroid receptors and transcription.)

MOST SIGNALS THAT ACT ON THE OUTSIDE OF A CELL REQUIRE A SECOND MESSENGER TO STIMULATE INTRACELLULAR CHANGES

Water-soluble signaling molecules—including amine and peptide hormones—do not enter their target cells but bind to receptors on the outer surface. The resulting receptor-ligand complex does not itself change the internal characteristics of the cells. Instead the receptor acts as part of a relay team in which the signaling molecule is just the first runner, or "first messenger." The action on the receptor of the extracellular signal stimulates the production, within the target cell, of a **second messenger,** whose intracellular action stimulates changes inside the target cell. The second messenger relays—to the cell's interior—the information that a signal has arrived at the cell's surface. In this section, we discuss how signaling molecules can trigger the production of four second messengers—cyclic AMP, calcium ions, diacyl glycerol, and inositol trisphosphate.

Many Cells Use Cyclic AMP as a Second Messenger

In animals, the most widely used second messenger is **cyclic AMP (cAMP),** whose structure is shown in Figure 19-15. Bacteria and some protists, such as the slime mold *Dictyostelium discoideum,* use cAMP as a "hunger" signal. As long as sufficient nutrients are present, *Dictyostelium* lives as single-celled amebas. But when food is scarce the amebas aggregate; the signal to aggregate is cAMP, which the amebas begin to produce as food runs out. *Dictyostelium,* unlike animal cells or *E. coli,* uses cAMP as an extracellular rather than an intracellular signal. Remember, cAMP also signals "hunger" in *E. coli,* stimulating transcription of the *lac* operon and other genes whose products contribute to bacterial catabolism. (See Chapter 15, p. 399.)

In some cases, addition of cAMP to target cells triggers the same cellular responses as epinephrine itself. In such experiments, researchers usually use a modified cAMP, dibutyryl cAMP, because cAMP itself does not penetrate the plasma membrane.

Cyclic AMP synthesis depends on **adenylate cyclase,** the enzyme that catalyzes the conversion of ATP into cAMP, as shown in Figure 19-16a. Adenylate cyclase lies on the inner surface of the plasma membrane. When ATP from the cytosol reaches the enzyme's active site, it is converted to cAMP. The newly formed cAMP then moves from the enzyme back into the cytosol. Another enzyme, **cAMP phosphodiesterase,** is present in the cytosol and hydrolyzes cAMP to AMP, which does not act as a second messenger. The concentration of cAMP within the cytoplasm depends on the activity both of adenylate cyclase, which makes cAMP, and phosphodiesterase, which destroys it, as shown in Figure 19-16b.

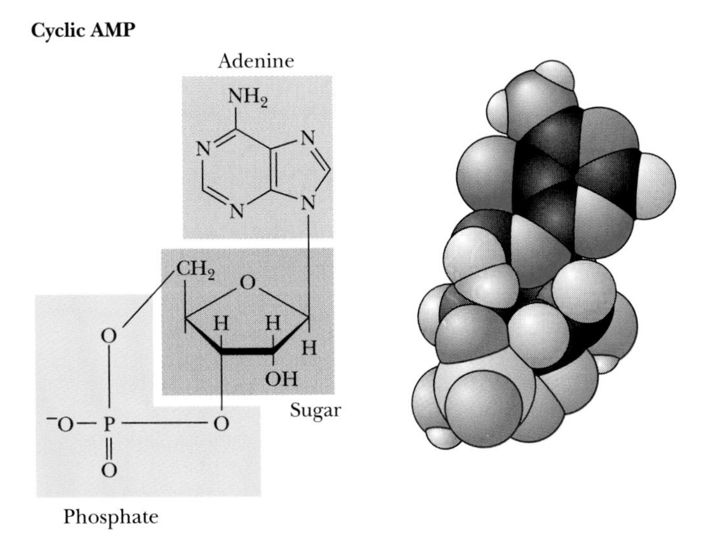

Figure 19-15 The structure of cyclic AMP (cAMP), a widely used intracellular "second messenger" in the action of hormones.

Cyclic AMP-Dependent Phosphorylation Activates Some Enzymes and Inhibits Others

Probably the most important of the enzymes activated by cAMP **is cAMP-dependent protein kinase** (also called protein kinase A), which transfers phosphate groups from ATP to sites in particular proteins. For example, as part of the "flight or fight" response, epinephrine increases the release of glucose by liver cells both by stimulating the breakdown of glycogen and inhibiting the synthesis of glycogen. In both cases, the effects depend on the epinephrine's stimulation of cAMP formation by adenylate cyclase. The higher levels of cAMP after epinephrine action lead to increased activity of protein kinase. Cyclic AMP–dependent protein kinase acts to phosphorylate glycogen synthetase, as shown in Figure 19-17, causing it to lose its enzymatic activity. The net result is that less glucose is sequestered in glycogen and more is available to sustain "fight or flight" activities. Notice that, as in bacteria and slime molds, cAMP effectively acts as a "hunger signal" and stimulates the increased availability of glucose.

In addition to decreasing the incorporation of free glucose into glycogen, cAMP also increases glucose availability by stimulating the activity of **glycogen phosphorylase,** the enzyme that breaks down glycogen into glucose. But cAMP-dependent protein kinase does not act directly on glycogen phosphorylase. Instead it phosphorylates another enzyme **(phosphorylase kinase),** which in turn phos-

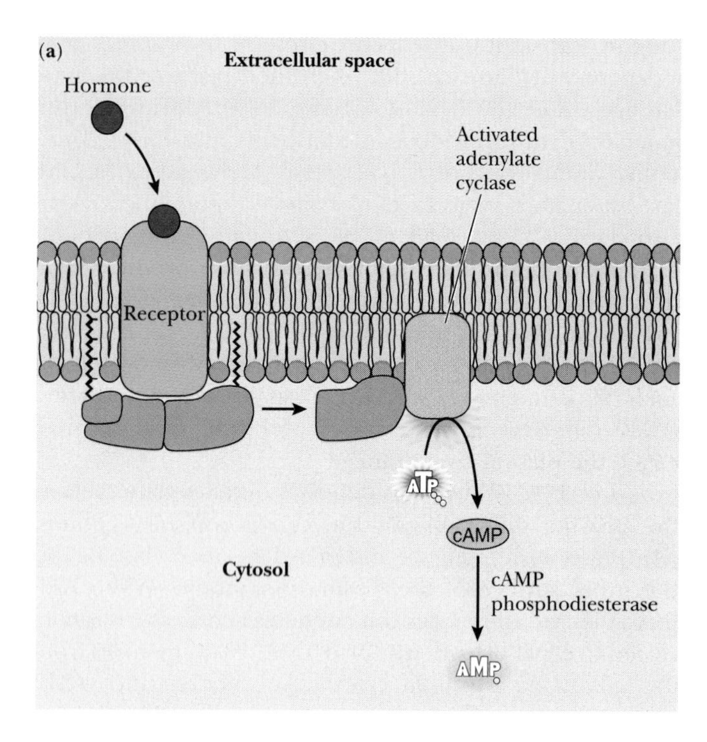

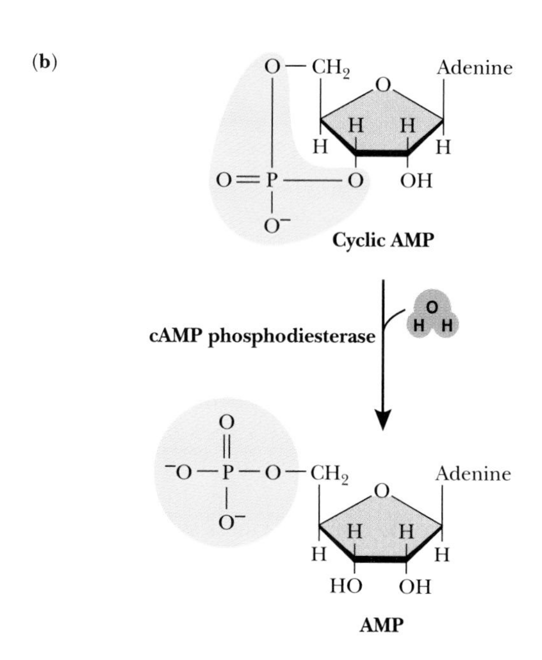

Figure 19-16 The synthesis and degradation of cAMP. **(a)** Binding of a hormone, such as epinephrine, to certain membrane receptors, such as the β-adrenergic receptor, activates adenylate cyclase; adenylate cyclase, another membrane protein, catalyzes the conversion of ATP to cAMP; **(b)** cAMP phosphodiesterase catalyzes the hydrolysis of cAMP to AMP.

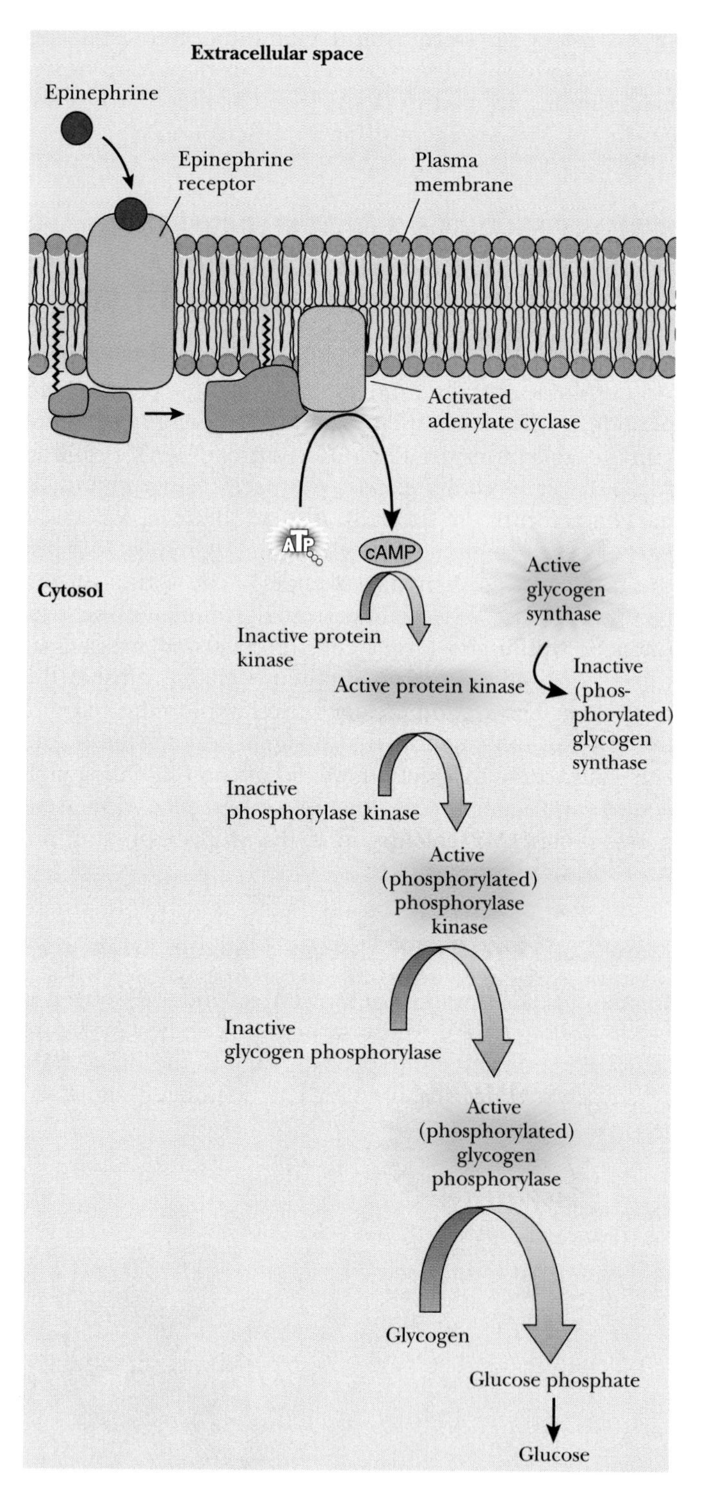

Figure 19-17 A regulatory cascade amplifies the effects of a hormone. There epinephrine stimulates cAMP production, as in Figure 19-16. cAMP activates a protein kinase, which phosphorylates another kinase, which phosphorylates glycogen phosphorylase, the enzyme that breaks down glycogen to glucose phosphate; intracellular glucose phosphate is converted to glucose within cells and then can move into the blood.

phorylates glygogen phosphorylase, as illustrated in Figure 19-17. Epinephrine, adenylate cyclase, cAMP, protein kinase, phosphorylase kinase, and phosphorylase all contribute to a **regulatory cascade,** a set of reactions that *amplify* a signal from a relatively small number of molecules into a response that affects many more molecules. A single epinephrine molecule temporarily activates perhaps 10–100 adenylate cyclase molecules, each of which may produce 10–100 cAMP molecules during the activated pe-

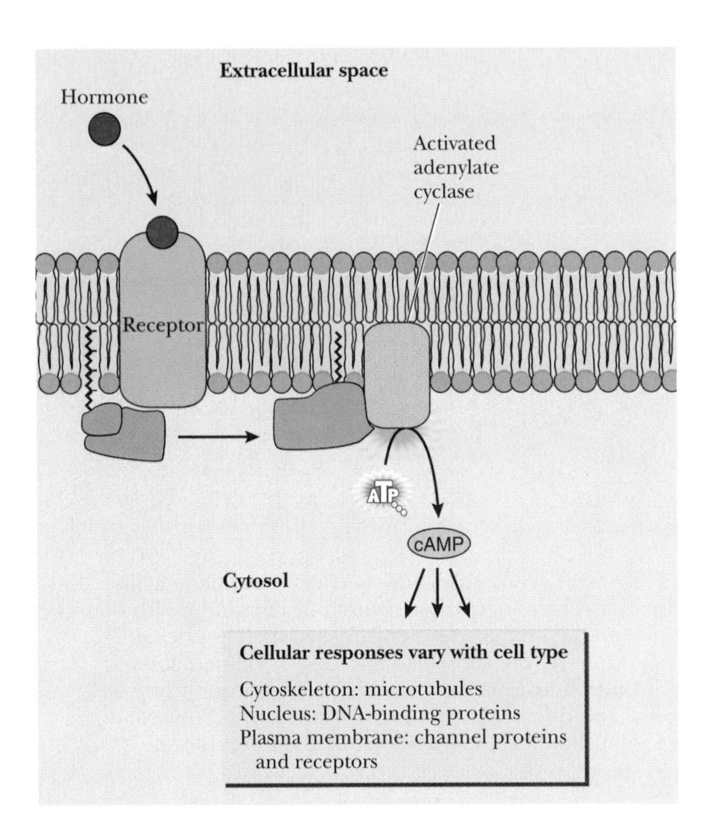

Figure 19-18 cAMP can act at many intracellular targets, which may vary with cell type.

riod. Each cAMP molecule can then stimulate a protein kinase molecule that may affect hundreds of molecules of phosphorylase and stimulate the formation of thousands of glucose molecules.

Cyclic AMP-Dependent Protein Kinases Have Many Protein Targets Within Cells

The protein targets of cAMP-dependent protein kinases vary according to cell type, as shown in Figure 19-18. These target proteins—whose cellular roles can dramatically change after phosphorylation—include microtubule proteins that can change the shape of the target cells, DNA-binding proteins that affect gene expression, and channels and pumps that regulate the passage of ions and molecules through the target cell's plasma membrane.

Many hormones use cAMP as a second messenger, including glucagon, epinephrine, vasopressin, and thyroid stimulating hormone. These hormones all employ the same second messenger in different types of target cells, but the effects of the second messenger differ from cell to cell. A target cell's response to a hormone is an important aspect of its specialized state.

The stimulated increase in cAMP does not last long—usually seconds or minutes. Cyclic AMP phosphodiesterase quickly breaks down cAMP. But the effects of cAMP may still persist in the modified activities of phosphorylated proteins, which may be stable for hours or days. Even these effects ultimately disappear, however, as other enzymes—called **phosphatases**—remove phosphate groups and still other enzymes recycle the building blocks of all the cell's proteins.

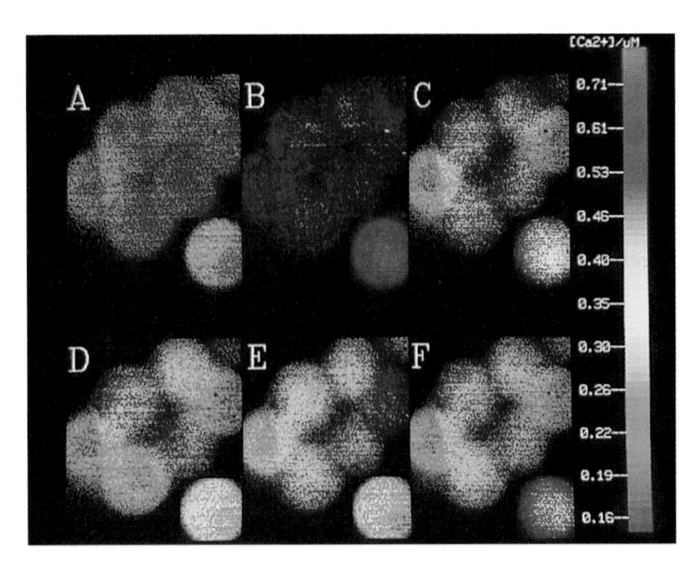

Figure 19-19 Intracellular concentrations of free Ca^{2+} ions can be monitored with the dye *fura*-2, which fluoresces when it binds Ca^{2+} ions. Such experiments demonstrate that hormones often use Ca^{2+} as a second messenger. **(a)** Lack of fluorescence before the stimulation of α_1-adrenergic receptors; **(b–f)** stimulation of α_1-adrenergic receptors increases intracellular Ca^{2+}, as revealed by increased *fura*-2 fluorescence. Redder colors indicate more fluorescence. *(Roger Y. Tsien, Howard Hughes Medical Institute Research Laboratories, University of California, San Diego)*

Calcium Ions Serve as Second Messengers in Many Cells

The binding of many water-soluble signals—including some neurotransmitters and some amines—to their membrane receptors stimulates an increase of free calcium ions (Ca^{2+}) within the cytoplasm. Researchers can observe this increase using the fluorescence of a Ca^{2+}-binding compound called *fura-2,* as illustrated in Figure 19-19. In some cases the increased calcium ion concentration results from a flow through channels in the plasma membrane; in most cases, however, increased cytoplasmic Ca^{2+} comes from cellular stores within the endoplasmic reticulum.

Free Ca^{2+} ions then serve as a second messenger whose effects are mimicked by experimental treatments that increase intracellular free Ca^{2+}. A common way of elevating cytosolic Ca^{2+} experimentally is to use an **ionophore,** a compound that spans the plasma membrane and allows specific ions to pass. A23187 is a commonly used Ca^{2+} ionophore, which fools cells into responding as if they had received a specific extracellular stimulus.

Other compounds (such as EGTA), lower the concentration of Ca^{2+} ions by acting as a **chelator** [Greek, *chela* = claw], a molecule in which several covalently connected atoms bind to the same ion. When Ca^{2+} serves as

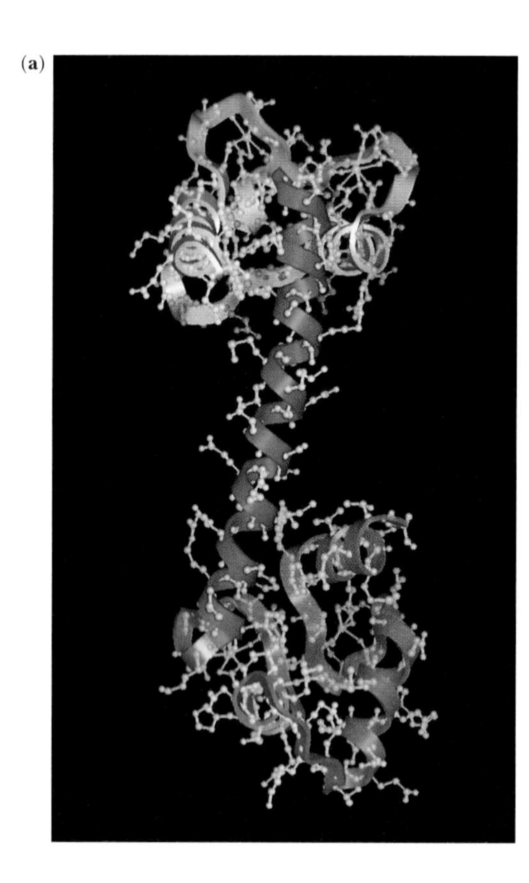

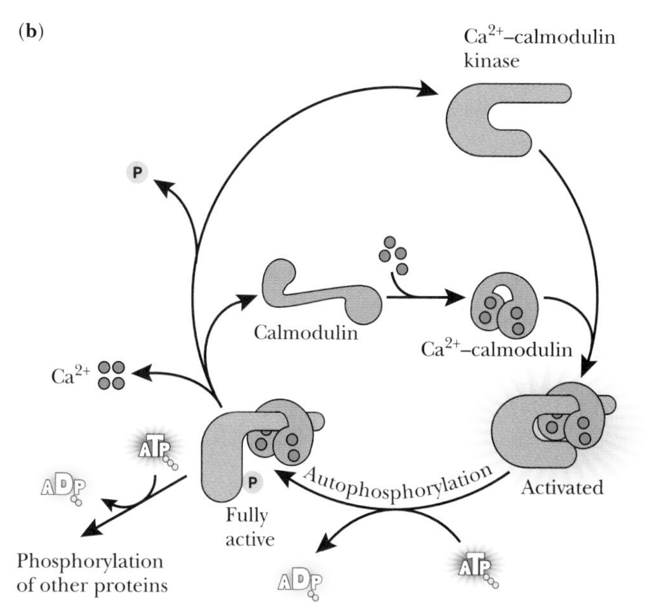

Figure 19-20 Ca^{2+} ions can act as second messenger in hormone action after binding to calmodulin. **(a)** The molecular structure of calmodulin, with bound Ca^{2+} ions. **(b)** Ca^{2+}-calmodulin binds to other proteins, such as calcium-calmodulin kinase, shown here; the complex consisting of calmodulin and calmodulin-kinase becomes fully active after gaining a phosphate group by "autophosphorylation," and the fully active kinase can then phosphorylate other protein targets, altering their functional activities within the cell. *(a, Melinda M. Whaley)*

a second messenger, EGTA added to the outside of cells prevents the effects of the signaling molecule by preventing Ca^{2+} entry into the cell.

Calcium ions bind to specific effector proteins, just as cAMP binds to a specific effector kinase. Like cAMP, Ca^{2+} stimulates a kinase, but only after it binds to **calmodulin,** the cytoplasmic protein that is the most widely distributed Ca^{2+} effector. Calmodulin changes its shape when it binds Ca^{2+}. The activated calmodulin then interacts with other effectors, including a kinase called **calcium/calmodulin kinase.** Figure 19-20 illustrates the interaction of Ca^{2+}, calmodulin, and calcium/calmodulin kinase.

Calmodulin is related to a Ca^{2+}-binding protein in muscle—troponin, discussed in Chapter 18. Recall that the binding of Ca^{2+} to troponin is the immediate stimulus for muscle contraction. Only after troponin changes its shape can the filaments slide. Similarly, only after calmodulin changes it shape (after the binding of Ca^{2+}) can it activate calcium/calmodulin kinase.

While troponin is present only in muscle cells, however, calmodulin is present in cells throughout the animal and plant kingdoms. A single animal cell may contain as many as 10 million molecules of calmodulin, and calmodulin may interact with a large number of cellular components. Among these are the contractile proteins of smooth muscle, the proteins of the cytoskeleton, a variety of protein kinases, and the enzymes that make and degrade cAMP. So one second messenger (Ca^{2+}) can directly affect the operation of another (cAMP).

As was the case for cAMP, the increase in Ca^{2+} ion concentration is only transient. After Ca^{2+} ions flow into the cell, they quickly disappear, entering intracellular stores such as the endoplasmic reticulum. As in the case of cAMP, the effects of the transient concentration increase outlast the return to normal. The effects of increased Ca^{2+} can be particularly prolonged because one of the targets of Ca^{2+}-induced phosphorylation by Ca^{2+}/calmodulin kinase is Ca^{2+}/calmodulin kinase itself. Biochemists call this process **autophosphorylation,** even though each Ca^{2+}/calmodulin kinase molecule acts not on itself but on another calmodulin kinase molecule. After phosphorylation, Ca^{2+}/calmodulin kinase is even more active. The autophosphorylated kinase then continues to phosphorylate other cellular proteins, providing a "molecular memory" of the earlier signal, even after cytosolic Ca^{2+} has returned to a low level.

Phosphatidylinositol Is the Precursor for Two Second Messengers That Affect Both Membranes and Cytosol

In the last decade, researchers have discovered yet another second messenger system that transmits information from the cell surface to the cell's interior. This system uses not one but two second messengers, both of which are derived by breaking down a lipid component of the membrane. Among the hormones that use this membrane-derived lipid messenger system are vasopressin and thyrotropin-releasing hormone.

The source of these second messengers is a phospholipid called **phosphatidylinositol (PI),** illustrated in Figure 19-21a. Two successive phosphorylations convert a

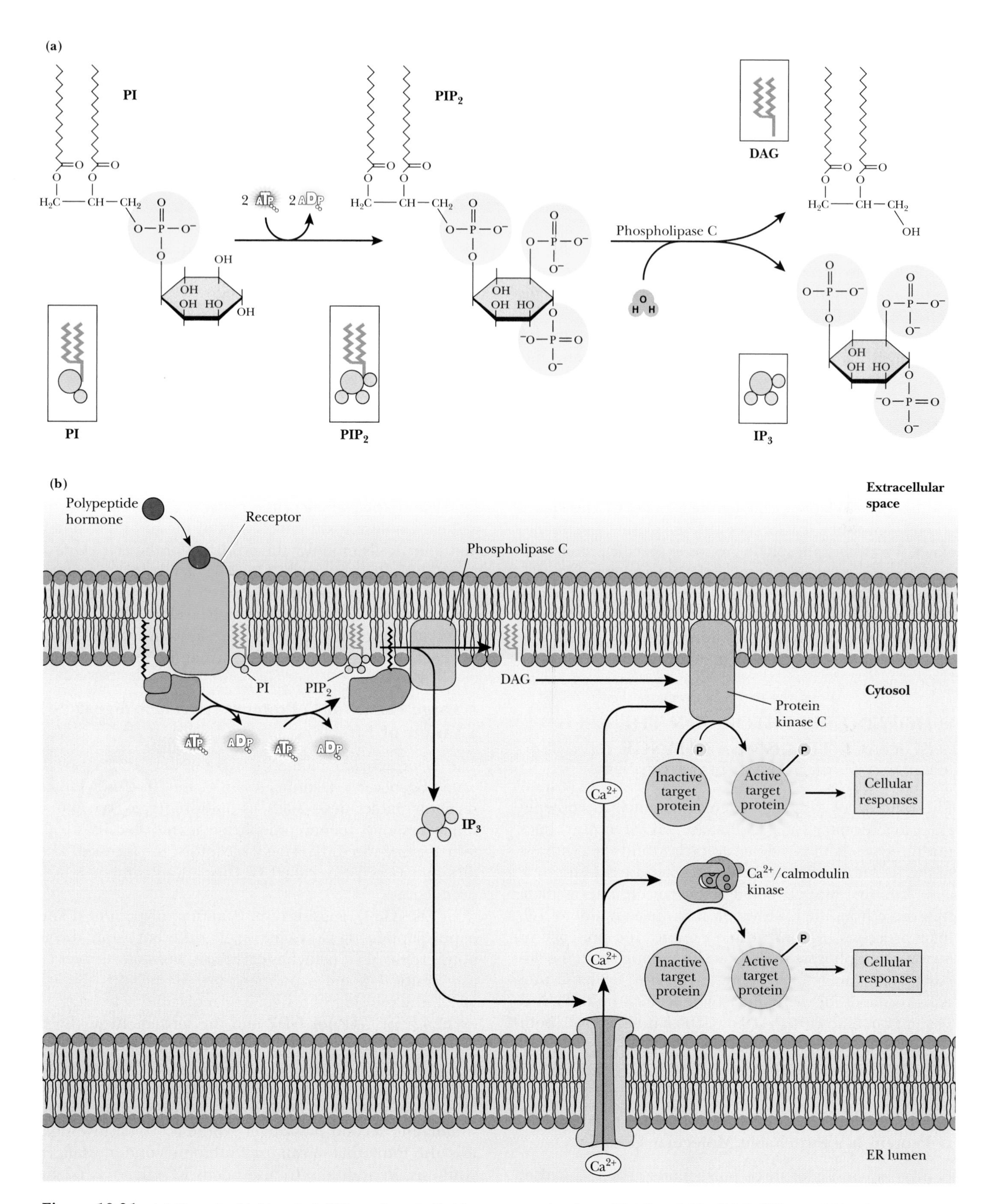

Figure 19-21 **(a)** Phosphatidyl inositol (PI), a phospholipid, can be phosphorylated to produce PIP_2, the precursor of two important second messengers. Phosphoplipase C converts PIP_2 to DAG and IP_3. **(b)** The activity of phospholipase C is controlled by a hormone receptor; binding of hormone to the receptor activates phospholipase C and stimulates the production of DAG and IP_3; DAG diffuses within the membrane to activate protein kinase C, which can act on targets both within the membrane and in the cytosol; IP_3 stimulates the flow of Ca^{2+} from the ER lumen into the cytosol; the increased Ca^{2+} can help activate both calcium-calmodulin kinase and protein kinase C.

small fraction of the PI in the membrane into **phosphatidylinositol 4,5-bisphosphate (PIP_2).** PIP_2 represents less than 1% of all the phospholipid in the membrane, but it plays a crucial role in chemical signaling.

The binding of a chemical signal to one of the receptors that act through the membrane lipid messenger system activates a membrane-bound enzyme, **phospholipase C,** which cleaves PIP_2 into two smaller molecules: **inositol trisphosphate (IP_3),** the phosphate and inositol groups, and **diacylglycerol (DAG),** the glycerol backbone plus the two linked fatty acid chains.

IP_3 is highly charged and immediately enters the cytoplasm. DAG, on the other hand, is nonpolar and diffuses within the plasma membrane. Each of these second messengers, IP_3 and DAG, has a distinctive action.

IP_3 stimulates the endoplasmic reticulum to release Ca^{2+} ions. Calcium ions in turn act as a third messenger, binding to calmodulin and extending the messenger relay.

DAG together with the increased Ca^{2+} ions, stimulates another protein kinase, called **protein kinase C.** This kinase is active only when it associates with the plasma membrane, where it transfers phosphate groups to specific proteins. In the brain, the proteins phosphorylated by protein kinase C include membrane proteins that serve as channels for specific ions, so DAG changes the electrical properties of cells. In other cells, the phosphorylation of specific proteins by protein kinase C leads to the regulation of gene expression.

HOW DO RECEPTORS ON THE CELL SURFACE TRIGGER A CHANGE IN SPECIFIC SECOND MESSENGERS?

The binding of a signaling molecule (such as epinephrine) to a receptor (such as a β-adrenergic receptor) does not directly stimulate adenylate cyclase and the synthesis of cAMP. An intermediate coupling step is required.

The first hint about the nature of the intermediate step came from the observation that the activation of cellular processes by epinephrine requires the presence of guanosine triphosphate (GTP). This finding led to the discovery of a family of membrane proteins, called **G proteins** (G stands for guanyl-nucleotide binding protein). A G protein can bind either GTP or GDP; it interacts with both receptor molecules and adenylate cyclase. A G protein also has an enzymatic function: It catalyzes the hydrolysis of bound GTP to GDP, as illustrated in Figure 19-22.

G Protein Is a Switchable Molecular Coupler

G protein consists of three polypeptide chains, called α, β, and γ. The α polypeptide contains the binding site for GTP or GDP. When no signaling molecule is present, most of the G protein in the membrane of a target cell is bound to GDP and is not associated with a receptor molecule. When the receptor molecule is bound to its ligand (such as epinephrine), however, the receptor-ligand complex binds to G protein, as diagrammed in Figure 19-22. This binding triggers the exchange of GTP for GDP, leading to a multimolecular complex that contains hormone, activated receptor, Gα-GTP, Gβ, and Gγ. This complex immediately begins to dissociate, however, with Gα-GTP diffusing away and binding to adenylate cyclase. The interaction between the Gα-GTP and adenylate cyclase stimulates the synthesis of cAMP. This interaction is ended, however, by the hydrolysis of GTP to GDP, catalyzed by the Gα polypeptide.

Because G protein acts to terminate its own action, it not only relays the signal from the receptor to adenylate cyclase but also serves as a clock that limits the duration of that signal. Derivatives of GTP that cannot be hydrolyzed lead to the permanent activation of adenylate cyclase.

The dangers of such permanent activation are evident in the effects of **cholera toxin,** a protein produced by *Vibrio cholerae,* which causes massive, life-threatening diarrhea. Cholera toxin enters cells of the intestinal lining and specifically modifies the α subunit of the cell's G protein. This alteration locks Gα in its active form, even in the absence of hormone. The result is an excessive stimulation of cAMP synthesis. The high levels of cAMP in turn stimulate the pumping of chloride ions and water out of the cells, leading to severe diarrhea and dehydration.

A Large Family of G Proteins Transduce Signals to a Variety of Effectors

The G protein coupled to the β-adrenergic receptor is called $\mathbf{G_s}$, because it *s*timulates an *increase* in cAMP. Other receptor molecules—such as α_2-adrenergic receptors, which respond to norepinephrine in nerve cells—lead to a *decrease* in cAMP. These receptors are coupled to a different G protein, called $\mathbf{G_i}$ (because it *i*nhibits cAMP production).

Like G_s, G_i consists of α, β, and γ polypeptides. The α polypeptide of G_i, called α_i, is different from the α polypeptide of G_s (which is called α_s), but both G_s and G_i use identical β and γ polypeptides. As shown in Figure 19-23, the binding of a signaling molecule to G_i triggers exchange of GTP for GDP and the movement of the α subunit away from the G_i protein–receptor complex, leaving behind $G_{\beta\gamma}$. The $G_{\beta\gamma}$ complex (which can interact with either α_s or α_i) is present at higher levels than α_i. The $\beta\gamma$ complex then binds to free α_s, preventing it from stimulating second messenger synthesis. Some investigators also think that α_i can itself inhibit second messenger synthesis. In addition, G_i also opens K^+ channels in the plasma membrane.

The understanding of G_i came originally from the analysis of a bacterial toxin made by *Bordetella pertussis,* which causes whooping cough. **Pertussis toxin** causes its damage by covalently modifying the α subunit of G_i, lock-

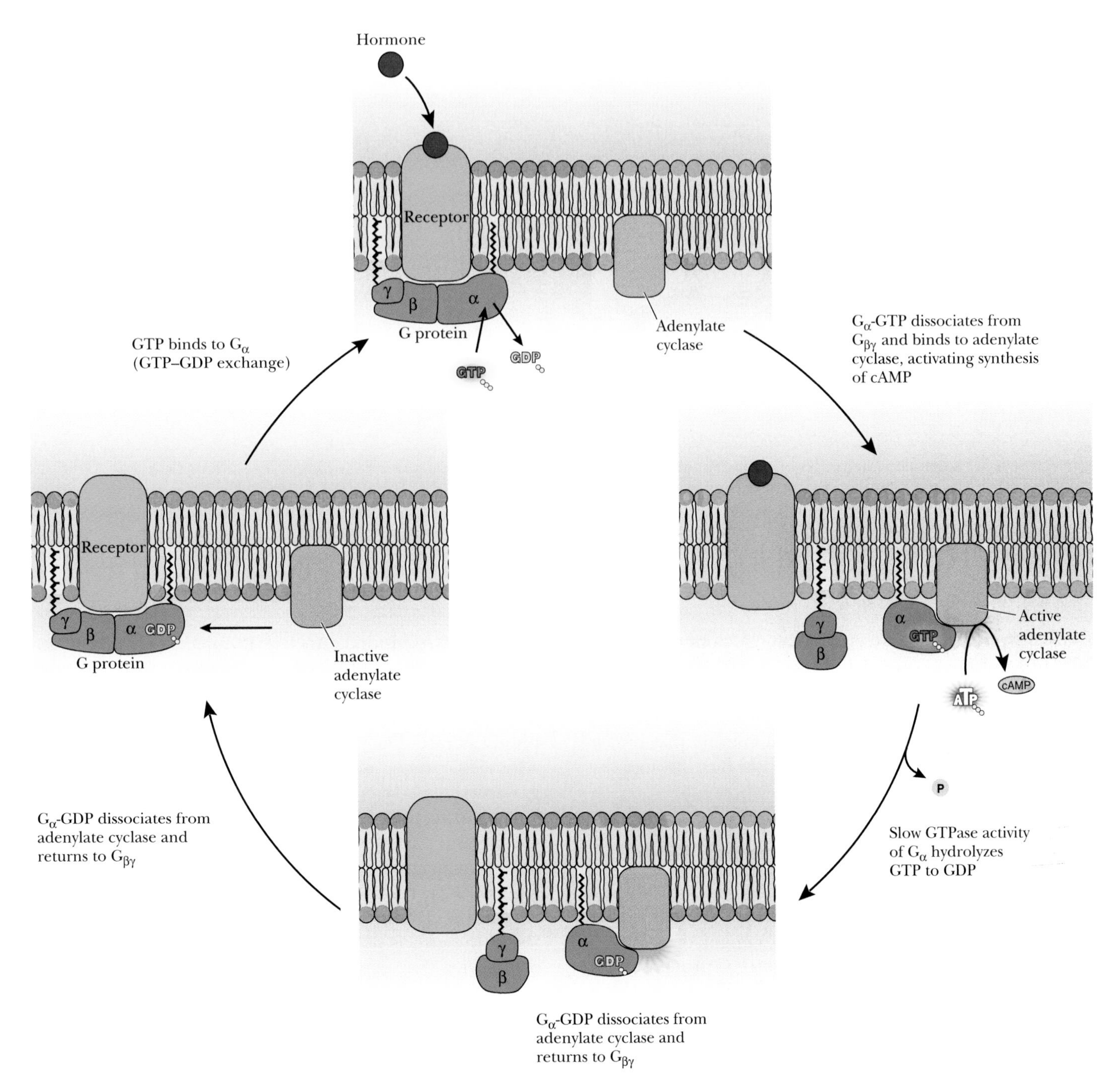

Figure 19-22 A G protein can act as a switch and a clock. The ability to stimulate the production of a second messenger (here cAMP) depends on the initial binding of hormone to its G-protein-coupled receptor. The dissociation of the G protein's three polypeptides depends on the binding of GTP, which displaces a bound GDP on the G protein's α subunit. Even after hormone is no longer present, the stimulated production of second messenger continues until the G protein has hydrolyzed the bound GTP to GDP. Then the GDP-bound α subunit dissociates from adenylate cyclase and reassociates with the β and γ subunits. The duration of cAMP synthesis depends on the GTPase activity of the G protein's α subunit.

ing it in the GDP-binding form. The locked G_i cannot interact with receptors and does not inhibit adenylate cyclase or open K^+ channels in the plasma membrane.

Other G proteins interact with still other members of the family of cell surface receptors with seven transmembrane domains. These receptors include not only those that bind to hormones, but also olfactory receptors, which bind to odorant molecules, and even rhodopsin, the light-sensing molecule of the retina, which changes its shape when it captures a photon.

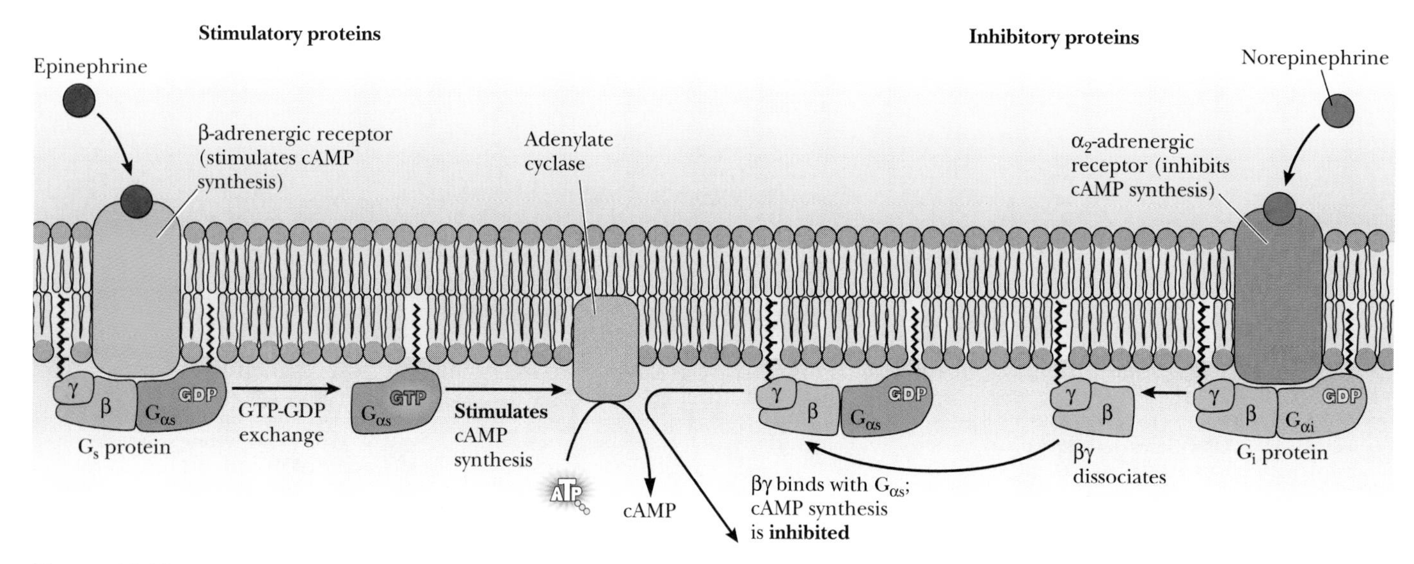

Figure 19-23 Some G proteins, like the one coupled to the β-adrenergic receptor, stimulate the production of cAMP. Other G proteins, like the one coupled to the α_2-adrenergic receptor, inhibit the production of cAMP. The two types of G proteins have identical β and γ subunits but differ in their α subunits. G_i is present at higher levels than G_s, so the norepinephrine-stimulated $\beta\gamma$ dissociation leads to the complexing of α_s with $\beta\gamma$, thereby preventing α_s from stimulating adenylate cyclase.

G proteins may affect cellular effectors other than adenylate cyclase. Some activate phospholipase C, which catalyzes the production of IP_3 and DAG from PIP_2. Transducin, the G protein in the retina that interacts with rhodopsin, changes the activity of an enzyme that breaks down cGMP, which in turn regulates ion flux through the photoreceptor cell membrane.

Which G Protein Does a Given Receptor Bind?

Both α and β adrenergic receptors act via second messengers whose synthesis depends upon interactions with G proteins. The effect on a second messenger depends on which receptors are present and on which G proteins those receptors bind. Depending on the cell type, epinephrine can trigger an increase in cAMP, a decrease in cAMP, or the production of IP_3 and DAG.

Recombinant DNA techniques have allowed cell biologists to determine the structural basis for the different effects of the same signaling molecule. Using a strategy called **domain swapping,** peptide segments ("domains") of two receptors are exchanged and the properties of the resulting mosaic (or "chimeric") receptors are studied. In this experiment, researchers first prepared cDNAs for α_2 and β_2 adrenergic receptors in a vector containing signals that allowed efficient transcription *in vitro* to produce α_2 and β_2 mRNAs. These mRNAs were then injected into large oocytes, the immediate precursors of egg cells, from the frog *Xenopus.* The protein-synthesizing machinery of the oocyte translates the mRNAs into receptors and inserts the receptors into the oocyte membrane.

The *Xenopus* oocyte has its own G proteins (but no adrenergic receptors). An oocyte injected with receptor mRNA can respond to epinephrine (or similar compounds) by increasing or decreasing cAMP, using already present G proteins, adenylate cyclase, and cAMP phosphodiesterase. Oocytes that are programmed to make β_2 adrenergic receptors increase cAMP, while those programmed to make α_2 adrenergic receptors decrease cAMP.

The techniques described in Chapter 17 allow the cutting and pasting of two DNAs as if they were two pieces of film or magnetic tape. In the experiment illustrated in Figure 19-24, researchers exchanged individual segments of α and β cDNA to produce mosaic receptors. Then they tested the effects of epinephrine and related compounds on oocytes that expressed the chimeric receptors. They found that cAMP increased or decreased depending on the source of a segment that contained transmembrane segments 5 and the long cytoplasmic loop between transmembrane segments 5 and 6. The specificity for different signaling molecules, on the other hand, depended on the source of the extracellular loop preceding transmembrane segment 7. These experiments showed that the part of the receptor molecule responsible for binding the chemical signal is distinct from the part of the receptor responsible for the response—an increase or decrease in cAMP.

Some Water-Soluble Signals Do Not Work Through Second Messengers but Directly Stimulate Membrane-Bound Kinases

Not all water-soluble signaling molecules act through second messengers. For example, insulin, like other peptide and protein signals, binds to a receptor on the outside sur-

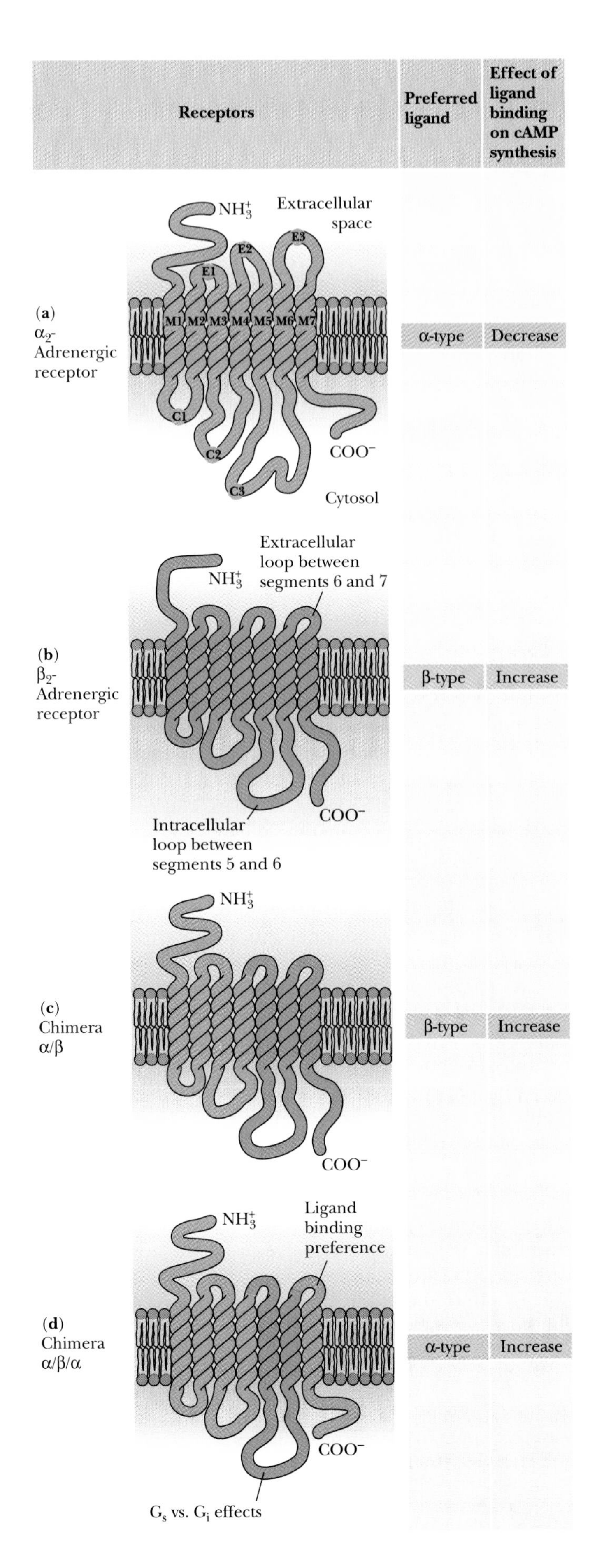

Figure 19-24 Domain swapping experiments reveal the polypeptide segments responsible for the specificity of ligand binding and for the specificity of G protein binding (G_s vs. G_i). Using recombinant DNA techniques, researchers prepared **(a)** a cDNA for the α_2-adrenergic receptor, **(b)** a cDNA for the β_2-adrenergic receptor, **(c)** a chimeric cDNA that encodes the first four transmembrane segments of the α_2-adrenergic receptor and the carboxy-terminal three transmembrane segments of the β_2-adrenergic receptor, and **(d)** a chimeric cDNA that is all α_2-adrenergic receptor except for a segment that contains the extracellular loop between transmembrane segments 4 and 5, transmembrane segments 5 and 6, and the intracellular loop between segments 5 and 6. The researchers then produced mRNAs from each of these cDNAs, injected the mRNAs into frog oocytes, and determined the ligand specificity of each receptor and its effect on cAMP synthesis. The conclusions from this and similar experiments were (1) that the type of ligand binding—preference for α ligands like norepinephrine vs. preference for β ligands like epinephrine—depends on the extracellular loop between segments 6 and 7; and (2) that preference for G_s vs. G_i binding depends on the intracellular loop between segments 5 and 6.

face of its target cells. Instead of changing the concentration of a soluble second messenger, insulin stimulates enzymatic activity of the insulin receptor itself.

The insulin receptor consists of two α and two β chains, all linked together by disulfide bonds, as illustrated in Figure 19-25. The insulin receptor itself functions as a

(Text continues on page 534.)

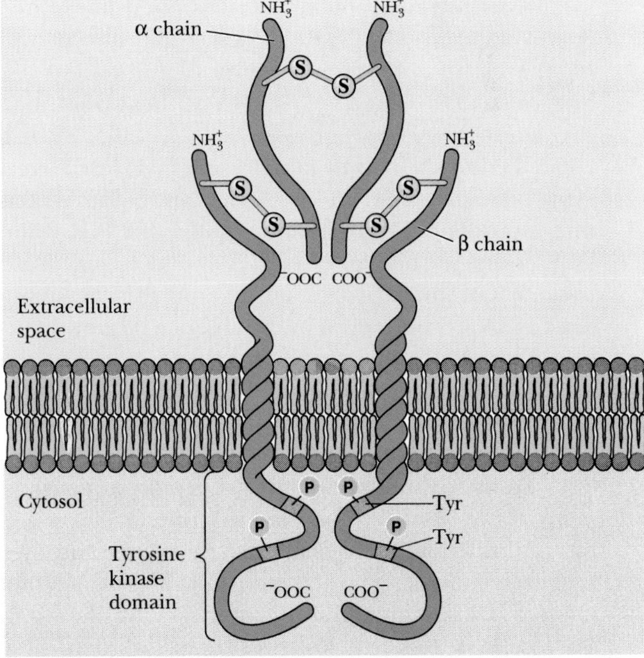

Figure 19-25 The insulin receptor is a tyrosine kinase. It catalyzes the transfer of phosphate groups from ATP to other proteins, including other insulin receptors. By using site-directed mutagenesis, researchers could replace the highlighted tyrosine residues, which are ordinarily autophosphorylated by the insulin receptor. Such mutated insulin receptors can bind insulin but did not produce any biological effects.

ESSAY

THE RAS CASCADE

Some of the most powerful insights into the mechanisms of cell signaling came from a combination of studies of a gene implicated in human bladder cancer and of several genes involved in the development of the *Drosophila* eye. These studies tie together the regulation of cell division and the response of developing cells to extracellular signaling. Both processes, we now realize, involve a cascade of kinases, controlled by a GTP-binding protein that resembles the G proteins discussed in this chapter.

About 20% of human bladder cancers contain an oncogene, called *ras,* that differs from its normal counterpart by a single mutation. The normal counterpart of *ras* (called *c-ras* or the *ras* proto-oncogene) encodes a protein, called Ras, with 170 amino acid residues. The sequence and three dimensional structure of Ras resemble those of a part of the sequence of the larger α_s polypeptide of G_s. Like α_s, Ras is a GTP-binding protein that serves as a switchable molecular coupler that initiates an enzymatic cascade. Also like α_s, Ras hydrolyzes GTP to GDP and is active only when GTP, rather than GDP, is bound.

A simple experiment first illustrated the role of Ras in noncancer cells. Researchers had shown that, under certain conditions of culture, a fibroblast cell line, called 3T3, would begin DNA synthesis only upon the addition of two growth factors, called PDGF (platelet-derived growth factor) and EGF (epidermal growth factor). They then injected antibodies to Ras into individual fibroblasts and showed that PDGF and EGF no longer had an effect. They also found that a mutant Ras, which hydrolyzes GTP only slowly, stimulated DNA synthesis even in the absence of growth factors. They concluded that Ras played a necessary intracellular role in transducing the effects of the growth factors, which act extracellularly. That role depends on Ras's association with GTP.

Studies of mutations that affected eye development in *Drosophila* also pointed to a role for Ras in signal transduction. *Drosophila* eyes are compound eyes that each contain some 800 individual eyes, called ommatidia. Each ommatidium contains eight photosensitive cells, one of which is missing in flies that carry a mutation called *sevenless.* Genetic studies of *sevenless* mutants has revealed several other genes whose products collaborate with the Sevenless protein to mediate cell signaling during the development of the *Drosophila* eye. One of these genes encodes *Drosophila* Ras, which is 80% identical to human Ras. As in the case of fibroblast growth, a mutant Ras with reduced capacity to hydrolyze GTP can overcome a deficiency of extracellular signaling, for example in a *sevenless* mutant. Other members of the *sevenless* signaling pathway are similarly involved in development and cell-to-cell signaling not only in *Drosophila* eyes and rat fibroblasts, but in many other cells and species as well.

The starting point for the pathway is a growth factor receptor (such as the EGF receptor or the insulin receptor) that binds to an extracellular signal. These receptors, called *receptor tyrosine kinases,* are plasma membrane proteins with a single transmembrane domain and a kinase catalytic site on the cytosolic site of the membrane. Ligand binding triggers a dimerization of receptors and the transfer of a phosphate group from ATP to a tyrosine residue of the second chain. The result is receptor *autophosphorylation.*

As shown in Figure 1, the resulting phosphotyrosines then bind to adaptor proteins that have a special binding site (called SH2) for certain phosphotyrosine-containing sequences in the receptor. The adapter then binds to another protein, called Sos (encoded by a gene called *son of sevenless*), which in turn binds to Ras. Sos stimulates the exchange of Ras-bound GDP for GTP, thereby activating Ras and concentrating it on the cytosolic surface of the plasma membrane.

Active Ras then binds to another protein, called Raf, that is a serine/threonine kinase. Raf then binds and phosphorylates another kinase, called MEK, which can phosphorylate at either serine or tyrosine. MEK phosphorylates and activates yet another kinase, called MAP kinase (for microtubule-associate protein or mitogen-activated protein). And, finally, MAP kinase phosphorylates a number of cellular proteins, including transcription factors that regulate gene expression during the cell cycle and cell differentiation.

As in the case of glycogen breakdown in response to epinephrine, the action of a growth factor stimulates a cascade of kinases that vastly amplifies an extracellular signal. Much remains to be learned, both with regard to alternative pathways and alternative targets of phosphorylation. But as complicated as it is, the Ras cascade seems to be involved in widely differing examples of cell signaling.

Figure 1 Binding of a growth factor, such as EGF, to a receptor tyrosine kinase triggers dimerization and autophosphorylation. The resulting phosphotyrosines bind to an adapter molecule, called GRB2, which in turn binds to Sos. Sos binds to Ras and catalyzes the replacement of GDP by GTP. The activated Ras then triggers a cascade of additional kinases, some of which phosphorylate serine and threonine, and at least one of which phosphorylates tyrosine.

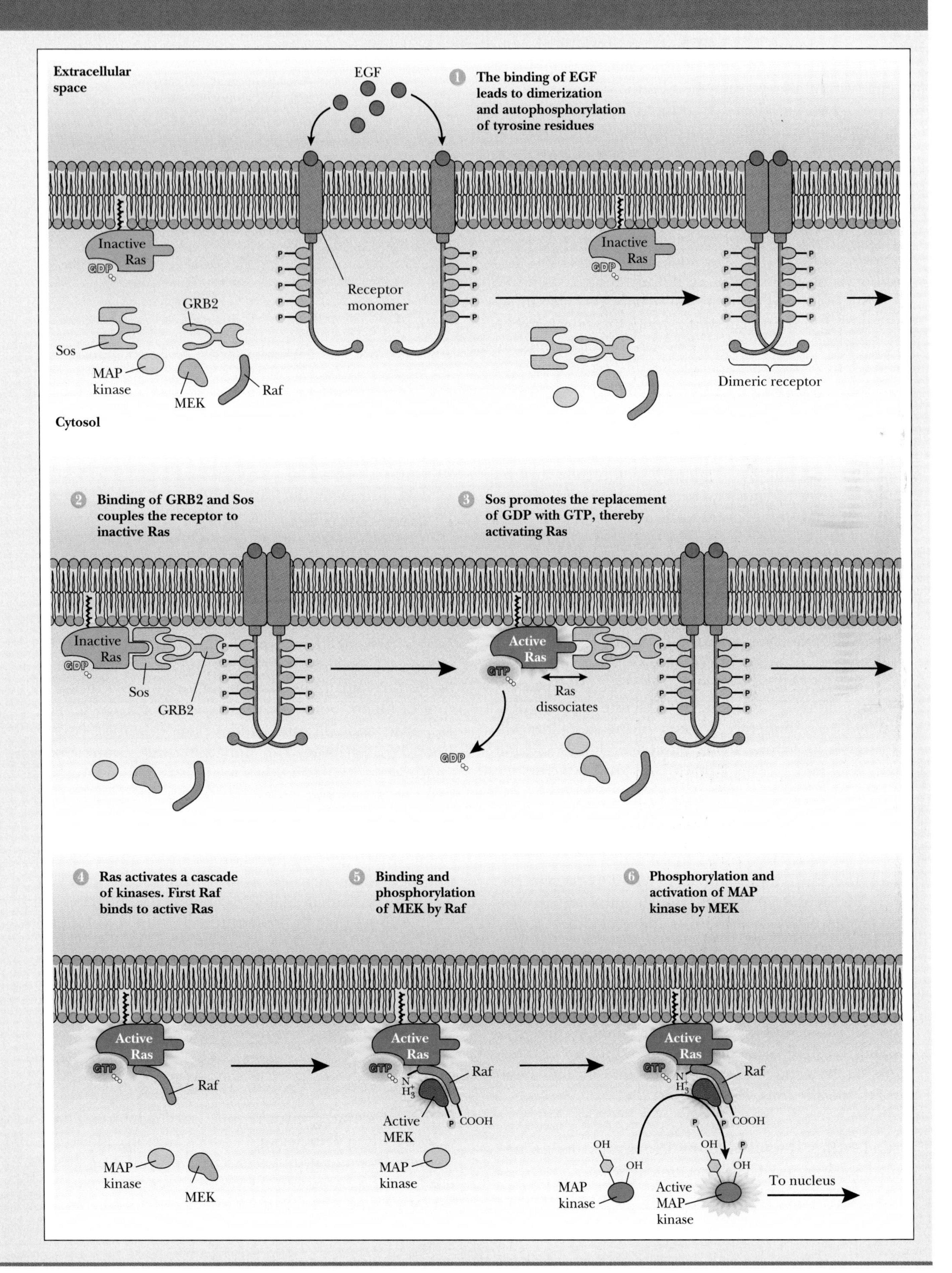
Extracellular space
EGF
1 The binding of EGF leads to dimerization and autophosphorylation of tyrosine residues
Inactive Ras
GDP
Receptor monomer
GRB2
Sos
MAP kinase
MEK
Raf
Cytosol
Dimeric receptor
2 Binding of GRB2 and Sos couples the receptor to inactive Ras
3 Sos promotes the replacement of GDP with GTP, thereby activating Ras
Active Ras
GTP
Ras dissociates
4 Ras activates a cascade of kinases. First Raf binds to active Ras
5 Binding and phosphorylation of MEK by Raf
6 Phosphorylation and activation of MAP kinase by MEK
Active MEK
COOH
OH
Active MAP kinase
To nucleus

kinase, transferring a phosphate group from ATP specifically to particular amino acid residues in a small group of proteins within the plasma membrane. The insulin receptor is a **tyrosine kinase,** transferring phosphate to tyrosine side chains, in contrast to cAMP-dependent protein kinase, which transfers phosphate to the side chains of serine or threonine.

The binding of insulin stimulates the insulin receptor to form a dimer and also activates the kinase activity.

The proteins phosphorylated by insulin receptors are mostly unknown, except that insulin receptors can phosphorylate other insulin receptors, another example of autophosphorylation. Once an insulin receptor molecule is itself phosphorylated, its ability to phosphorylate other proteins is enhanced. In one set of experiments, insulin receptor cDNA was specifically altered to eliminate the phosphorylation targets within the insulin receptor—two tyrosine residues. This cDNA was then used to direct the synthesis of altered insulin receptors in cells that ordinarily lack them. The altered receptors could not autophosphorylate, but they could still bind insulin. Insulin binding, however, did not lead to further biological effects, and the cells did not respond to insulin.

PARACRINE SIGNALS ACT OVER SHORT DISTANCES

Some chemical signals between cells do not travel through the blood but act at much shorter distances. Such interactions are called *paracrine* signaling (in contrast to *endocrine* signaling by hormones). Paracrine signals affect only cells in the immediate vicinity of the signaling cell, as illustrated in Figure 19-1c. Paracrine signals include neurotransmitters, prostaglandins, and growth factors.

Neurotransmitters Are Chemical Signals by Which Nerve Cells Communicate with One Another

Perhaps the most important and varied paracrine signals are the **neurotransmitters,** the chemical signals produced by nerve cells. Neurotransmitters include a variety of amines and peptides—some of which (like norepinephrine and thyrotropin-releasing hormone) are also hormones. Like hormones, neurotransmitters bind to specialized receptors in the target cells. Some neurotransmitters act via second messengers, including cAMP, Ca^{2+} ions, and the lipid messenger system. Other neurotransmitters act by changing the permeability of the membranes to specific ions. (See Chapter 23.)

Prostaglandins Regulate Many Processes, Including Blood Clotting and Uterine Contractions

Another class of paracrine signals includes at least 16 compounds called **prostaglandins,** which all derive from a 20-carbon fatty acid called **arachidonic acid,** as illustrated in Figure 19-26. The enzymes that convert arachidonic acid to prostaglandins are collectively called the **cyclooxygenase pathway.**

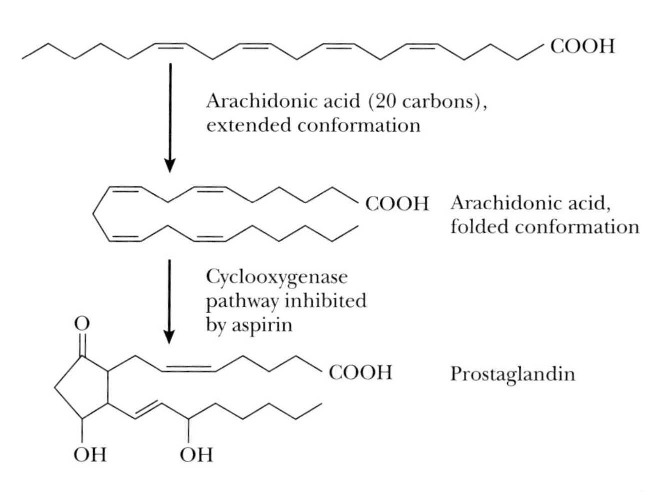

Figure 19-26 Prostaglandins derive from arachidonic acid via the cyclooxygenase pathway, which is inhibited by aspirin.

Many cell types in both vertebrates and invertebrates make prostaglandins. Enzymes that degrade prostaglandins are also widely distributed, so prostaglandins never get far from their sources before they are destroyed.

The effects of prostaglandins depend on the exact variety present. They apparently exert their effects via second messengers, including cAMP and Ca^{2+} ions. Some prostaglandins play an important role in the clotting of blood, others help bring about the inflammatory response, and still others induce smooth muscles to contract. Prostaglandins can initiate uterine contractions in childbirth—a fact that led to their initial discovery and naming as a component of semen, produced in the prostate gland.

Arachidonic acid is a minor component of membrane lipids, and the cyclo-oxygenase pathway produces prostaglandins. Because aspirin inhibits this pathway, it acts as an anti-inflammatory agent and also interferes with the clotting of blood. Recent research has suggested that men over 40 have a reduced risk of heart attack if they take one aspirin every 2 days. Apparently the aspirin reduces the spontaneous incidence of blood clots that would otherwise interfere with the blood supply to the heart.

Growth Factors Act on Membrane-Bound Receptors That Function as Protein Kinases

Cell survival and growth require particular combinations of ions and small molecules—including salts, glucose, and vitamins. Even after many exhaustive studies of the composition of such growth media, however, researchers consistently found that cells grow much better in the presence of blood serum, the liquid part of the blood that remains

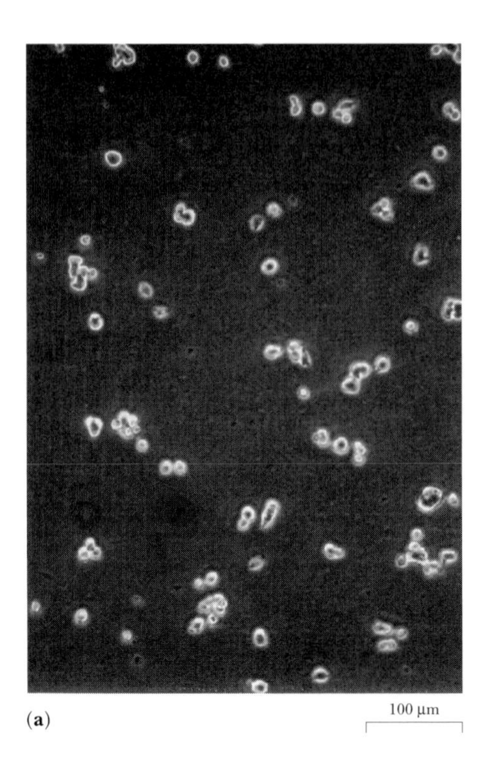

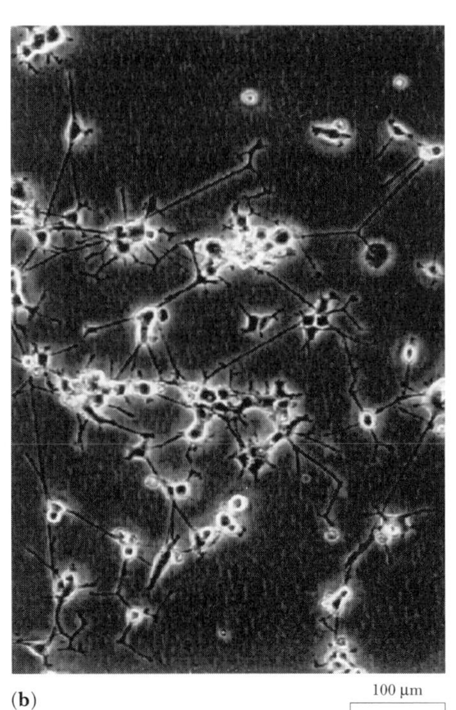

Figure 19-27 Growth factors influence the growth and survival of cells, both in artificial culture and in animals. **(a)** PC12 cells, rat cells derived from a tumor of the adrenal medulla, grown in a standard medium, without added growth factors; **(b)** the same cells grown in the presence of nerve growth factor (NGF) differentiate into cells that have the properties of neurons. *(Eric Shooter, Stanford University Medical School)*

after clotting. Serum is a complex mixture of proteins and small molecules. Which of these is responsible for the success of a cell culture?

Research revealed that serum and tissue extracts contain specific protein molecules, called **growth factors,** which stimulate cell division and cell survival. Every type of cell requires a particular combination of factors in order to grow in the laboratory. For some cells, a mixture of already identified factors is enough. For example, a continuing culture of mouse skin cells (fibroblasts), called 3T3, can grow in a completely artificial medium as long as three growth factors are present. One of the required factors comes from platelets and is called platelet-derived growth factor (PDGF). The others are called epidermal growth factor (EGF) and insulin-like growth factor (IGF). More about these later. Another factor that slightly resembles insulin, called nerve growth factor (NGF), is necessary for the growth and differentiation of particular nerve cells in culture (those of the sympathetic nervous system, which is responsible for part of the "fight or flight" reaction). Figure 19-27a, for example, shows the appearance of a tumor cell line called PC12, derived from the adrenal medulla. In the presence of NGF, PC12 cells take on the appearance of nerve cells, with long processes, as shown in Figure 19-27b.

For some cell types, concoctions of known growth factors are still not enough to allow growth in a defined artificial medium. In these cases researchers must still add some tissue extract of unknown chemical composition, implying that other growth factors remain to be discovered.

Each growth factor is only a small fraction of the total protein in an extract of serum or tissue. The characterization of the factors and their modes of action has required enormous energy and ingenuity. In recent years, biochemical and recombinant DNA techniques have revealed the structures of many growth factors.

The definition of each growth factor initially depended on its effects in an artificial culture system. In many cases, however, researchers have also shown that the same factors are also important *in vivo.* For example, as shown in Figure 19-28, transgenic mice that produce large amounts of NGF have increased numbers of nerve cells in the sympathetic ganglia, which lie next to the spinal cord.

Once growth factor researchers had determined the sequences of a number of growth factors, they began to notice similarities in structure to other factors involved in gene regulation. We have already mentioned the relationship of insulin, IGF, and NGF. EGF also has interesting relatives, including two genes that control the development of invertebrates, one in *Drosophila* and one in the nematode *C. elegans.* Still more amazingly, PDGF is closely related to an oncogene called *sis,* which was originally identified in viruses that cause cancers in cats and monkeys. Evidently, these cancer-causing viruses propagate themselves by causing the overproduction of a growth factor by infected cells.

The mechanism of growth factor action resembles that of peptide hormones. Some appear to act on membrane receptors that involve the lipid second messenger system, leading to protein phosphorylation by protein kinase C. Other growth factor receptors, however, appear to act as kinases themselves (like the insulin receptor). The

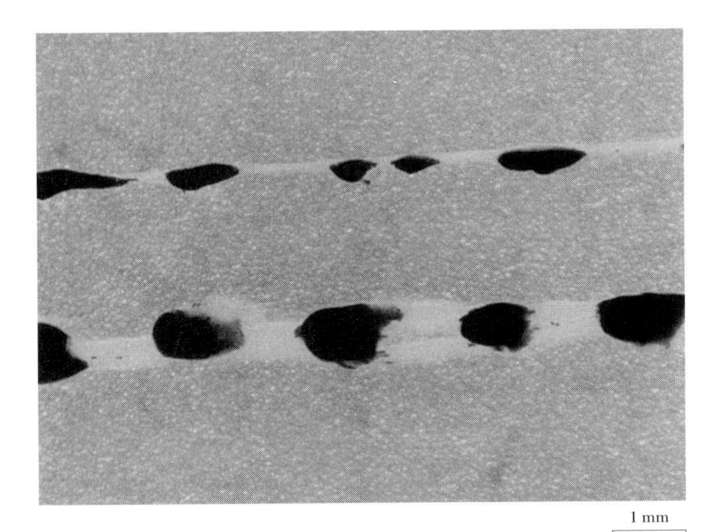

Figure 19-28 A transgenic mouse that overexpresses NGF has increased numbers of neurons in the sympathetic ganglia. The spinal cord on the top shows the sympathetic ganglia (stained blue) from a normal mouse; the spinal cord on the bottom shows the enlarged sympathetic ganglia from the transgenic mouse. *(Gary Hoyle, Tulane University Medical Center)*

EGF receptor, for example, consists of three distinct domains—an extracellular domain, which binds EGF; a membrane domain, which extends across the plasma membrane; and an intracellular domain, which functions as a protein kinase. Like the insulin receptor, the EGF receptor transfers phosphate groups to tyrosine residues instead of serine or threonine residues.

Now another surprise: At least eight different oncogenes, originally identified in tumor viruses, encode protein kinases that transfer phosphates to tyrosine side chains. In fact, one of these oncogenes—originally isolated from a virus that causes the overproduction of red cell precursors in chickens—encodes a protein that is closely related in sequence to the EGF receptor. Another viral oncogene encodes a protein similar to the receptor for a growth factor that stimulates the growth of white blood cells in culture.

The story of oncogenes, then, is closely related to that of growth factors and of hormones. Of the 30 oncogenes that have been studied in some detail, almost half encode protein kinases, one encodes a protein whose sequence resembles a growth factor, two encode proteins whose sequences resemble growth factor receptors, two produce proteins that resemble proteins involved in the stimulation of cAMP production, and one produces a protein that is virtually identical to the receptor for thyroid hormone. Information about chemical signaling among cells also allows us new insights into what may be going wrong in cancer cells.

■ SUMMARY

Multicellular organisms use chemical signals both to coordinate the action of individual organs over long distances and for intracellular communication over short distances. A hormone is a chemical signal, made and released by a specific organ, that travels through the circulation to its target. The hormone insulin, for example, coordinates the uptake of glucose throughout the body, and the hormone epinephrine coordinates the "fight or flight" reaction.

Researchers have identified specific molecules that are responsible for the biological action of many hormones and other chemical signals. These signals fall into two broad categories, lipid-soluble signals and water-soluble signals. Lipid-soluble signals, such as the steroids, enter target cells and affect their patterns of gene expression. Water-soluble signals, such as epinephrine and peptide hormones, act on the surfaces of target cells, usually by stimulating the production of intracellular second messengers.

Peptide hormones form the largest class of water-soluble signals. These peptides often derive from larger polypeptide precursors that are made by ribosomes on the rough endoplasmic reticulum. In some cases, a single gene may encode several different signaling peptides, which arise by alternative pathways of RNA splicing or proteolytic cleavage. Peptide hormones and other water-soluble signals are released by regulated exocytosis. The anterior pituitary, under the hormonal regulation of the hypothalamus, produces and releases seven peptide hormones.

Signaling molecules bind to specific protein receptors either in the plasma membrane, in the case of water-soluble signals, or in the cytosol, in the case of lipid-soluble signals. In many instances, researchers have isolated the proteins that serve as receptors and have determined their primary structures. The proteins that bind epinephrine and many other water-soluble signals are part of a large family of membrane receptors, which contain seven transmembrane α-helices.

Among the intracellular second messengers produced after stimulation by water-soluble signals are cyclic AMP and Ca^{2+} ions. Two other second messengers, diacylglycerol and inositol trisphosphate, derive from a membrane lipid, phosphatidyl inositol. Cyclic AMP, Ca^{2+} ions, and diacylglycerol all exert part of their effects by stimulating the activity of protein kinases.

Receptors on the cell surface trigger a change in specific second messengers through coupling molecules called G proteins, each of which binds to an intracellular domain of a membrane receptor. After the receptor binds to a signal, such as epinephrine, the G protein dissociates and activates an enzyme that stimulates the synthesis of cyclic AMP or of diacylglycerol. G protein action may either increase or decrease the level of a second messenger, depending on the particular G protein coupled to a particular receptor in a particular cell. Experiments with recombinant DNAs have allowed researchers to distinguish the part of a receptor that binds to ligand and the part that binds to a particular G protein.

Paracrine signals do not travel through the blood but diffuse short distances between cells. Neurotransmitters are paracrine signals that act across small distances on specialized regions of nerve and muscle cells. Prostaglandins are a set of paracrine signals that derive from arachidonic acid, a 20-carbon fatty acid. Growth factors are proteins that stimulate cell division and enhance cell survival. The action of many growth factors resembles that of peptide hormones.

KEY TERMS

absorptive state
adenylate cyclase
adrenergic receptor
affinity chromatography
affinity labeling
agonist
antagonist
arachidonic acid
autophosphorylation
β-endorphin
calcitonin
calcitonin-gene-related peptide
calcium/calmodulin kinase
calmodulin
cAMP-dependent protein kinase
cAMP-phosphodiesterase
chelator
cholera toxin
corticotropin
cyclic AMP (cAMP)
cyclo-oxygenase pathway
diabetes mellitus
diacylglycerol (DAG)
domain swapping
endocrine
epinephrine
exocrine
expression cloning
G protein
G_i
glucagon
glucocorticoid
glycogen phosphorylase
growth factor
G_s
homology cloning
hormone
inhibiting hormone
inositol trisphosphate (IP_3)
insulin
ionophore
ligand
lipid-soluble signal
median eminence
mineralocorticoid
neurotransmitter
olfactory receptor
paracrine
pertussis toxin
pheromone
phosphatase
phosphatidylinositol (PI)
phosphatidylinositol 4,5-bisphosphate (PIP_2)
phospholipase C
phosphorylase kinase
pituitary gland
postabsorptive state
preproinsulin
proopiomelanocortin
prostaglandin
protein kinase C
receptor
regulatory cascade
releasing factor
releasing hormone
second messenger
sex steroid
steroid
target organ
transmembrane segment
tyrosine kinase
water-soluble signal

QUESTIONS FOR REVIEW AND UNDERSTANDING

1. Define and contrast the following terms:
   (a) intercellular vs. intracellular
   (b) pheromone vs. hormone
   (c) endocrine hormone vs. paracrine hormone
   (d) insulin vs. glucagon
   (e) agonist vs. antagonist
   (f) binding site vs. ligand
   (g) α-adrenergic receptor vs. β-adrenergic receptor
   (h) homology cloning vs. expression cloning
   (i) adenylate cyclase vs. phosphodiesterase
   (j) kinase vs. phosphatase
   (k) ionophore vs. chelator
2. Where must the receptors for water-soluble signals be located? Are receptors for lipid-soluble signals limited in the same manner? How does this allow the function of the two types of receptors to differ? (See Chapter 15 for a description of steroid receptors.)
3. What are the six processes involved in intercellular signaling?
4. Describe in your own words the two mechanisms by

which a single gene can give rise to more than one peptide hormone.

5. Why is the hypothalamus now considered the "master gland"? What signals influence the activity of the hypothalamus?
6. What is meant by K_D? List the order of affinity (highest to lowest) of the following ligands for their receptor given the following values for K_D:

   L_1: 0.8 μM L_2: 0.008 μM L_3: 1.0 μM

7. Considering the technique of affinity chromatography used to purify receptors, why might this technique be more effective when isolating receptors of large peptide ligands than small (simple) ligands?
8. Explain why the presence of seven transmembrane segments in a diverse group of receptors leads us to suggest that these receptors are homologous.
9. List the four "second messengers." What is the common process promoted by these molecules?
10. Why is it significant that cAMP-dependent protein kinase, calcium/calmodulin-dependent kinase, and protein kinase C do not phosphorylate the same set of proteins?
11. Explain how the same second messenger can elicit different responses in different types of target cells.
12. How does autophosphorylation of calcium/calmodulin-dependent kinase act to perpetuate intracellular signaling in cells?
13. What is a regulatory cascade? Why is it advantageous for reactions or processes that must be changed quickly and in a pronounced way?
14. Why are G proteins considered "molecular couplers"? How are they switched between active and inactive states?
15. How can domain swapping be used to determine the function of a specific region (domain) of a receptor or other protein?
16. How do receptors known as tyrosine kinases differ from receptors such as the adrenergic receptor?
17. How is the epidermal growth factor receptor like the insulin receptor?
18. Given that cancer is the abnormal, uncontrolled growth of cells, why is it logical that so many oncogenes (cancer-causing genes) encode proteins involved in chemical signaling, particularly growth factors, their receptors, and proteins comprising the associated intracellular signal transduction pathways?
19. Recent investigations of signal transduction pathways in cells suggest that there is "crosstalk" between different pathways. That is, when one pathway is activated by hormone binding, parts of other signal transduction pathways will also be activated (or inhibited). From your reading of the chapter, can you think of some examples where crosstalk might occur between the signal transduction pathways discussed?

SUGGESTED READINGS

ALBERTS, B., D. Bray, J. Lewis, M. Raff, K. Roberts, and J. D. Watson, *Molecular Biology of the Cell,* 3rd edition, Garland, New York, 1994. Chapter 15 deals with cell signaling.

LODISH, H., D. Baltimore, A. Berk, S. L. Zipursky, P. Matsudaira, and J. Darnell, *Molecular Cell Biology,* 3rd edition, Scientific American Books, New York, 1995. Chapter 20 deals with cell-to-cell signaling.

WATSON, J. D., M. Gilman, J. Witkowski, and M. Zoller, *Recombinant DNA,* 2nd edition, Scientific American Books, New York, 1992. Chapter 17 deals with cell signaling. Chapter 18 deals with oncogenes, including *ras.*

CHAPTER 20

Cellular Contributions to Animal Development

KEY CONCEPTS

- During development, an animal changes from a single cell to a complex organized structure through processes that involve cell division, cell movement, cell specialization, and pattern formation.
- Sperm and egg cells are highly specialized to accomplish fertilization and to initiate development.
- Each cell contains the genetic information to produce an entire animal, but, as development proceeds, cells become increasingly limited to a particular developmental pathway.
- The pattern of gene expression of individual cells and their participation in the formation of organs and body parts depend on diverse chemical signals, many of which act by regulating transcription of specific genes.
- Virtually all animals appear to use the same mechanisms to achieve pattern formation and programmed cell death.

OUTLINE

ESSAYS

Of all of the wonders of life, none is so amazing as development. Each of us, after all, has developed from an apparently homogeneous single cell less than 100 μm in diameter.

The central question of this chapter is how this development can occur. We begin with the development of vertebrates because they are familiar and they illustrate the general problems of the development of multicellular organisms. We then look at the powerful insights that have come from the study of invertebrates, particularly *Drosophila*.

HOW DO COMPLEX ORGANISMS DEVELOP FROM A FERTILIZED EGG?

For almost all multicellular organisms, biologists have long known that the production of offspring usually requires two parents. But until the late 19th century, they were unable to deduce what each parent actually contributed. For example, Aristotle, for almost 1500 years the most influential thinker in European science, believed that the female merely provides the substance of a new offspring. The male, he argued, somehow molds its form, just as a potter shapes clay into a bowl or a mason uses clay bricks to build a house. Having said this, Aristotle and his followers had little further explanation of how the male accomplished the needed crafting.

Is the Adult Preformed in the Sperm or the Unfertilized Egg?

The question was—and remains—how does *complexity,* for example, a whole human being, arise from *simplicity,* a tiny, apparently homogeneous egg? The easiest way to answer this question has been to deny that simplicity ever existed at all. According to this view, called **preformation,** all the parts of the adult organism already preexist at the earliest stages of life. That is, the adult must already be *preformed* in the sperm or the egg. This view received some support from Leeuwenhoek's discovery of sperm in the microscope. Fanciful drawings of sperm, such as that shown in Figure 20-1, showed a little adult curled up in the sperm head. Other biologists argued that preformed adults existed not in sperm, but in eggs.

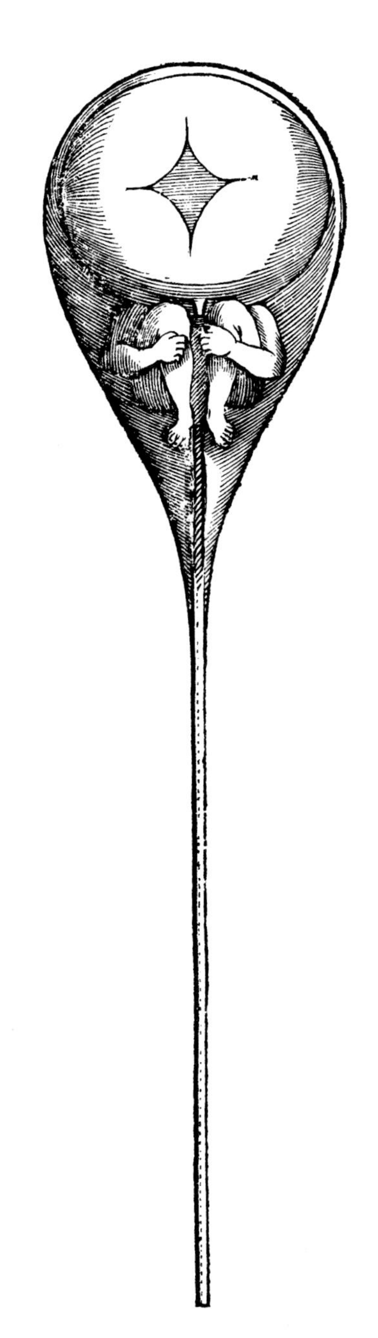

Figure 20-1 A fanciful drawing of human sperm, based on microscopic observations, suggesting that a miniature adult (a "homunculus") is already present, ready to be unfolded. *(Wellcome Institute Library, London)*

Not only did the doctrine of preformation avoid the difficult problem of explaining development, it also provided a simple view of genetics: Children resemble their parents because they were already preformed in the sperm of their father or in the ova of their mother. A problem immediately arises: If either the "ovist" or the "spermist" view is correct, how does the other parent contribute to the genetic makeup of the next generation? The theory of preformation necessarily precludes the participation of one parent. In addition, preformation is uncomfortably deterministic, leaving little room for evolution or even for the appearance of abnormal forms.

According to preformation, future generations would be contained within past generations, like a set of Russian nesting dolls. Only a limited number of generations could be contained, however. But, as geologists discovered that life was far more ancient than suspected and chemists discovered that matter was not infinitely divisible, the theory of preformation began to crumble.

How Did Experiments with Embryos Disprove Preformation?

Notwithstanding the philosophical objections to the theory of preformation, its appeal continued to the end of the 19th century. One of its most appealing aspects was that—unlike the totally abstract theory of Aristotle—scientists could actually test the theory of preformation. Near the end of the 19th century, two German biologists—Wilhelm Roux and Hans Driesch—each performed an important test of the theory.

The experiments were based on the same kind of analysis. Preformation said that the whole organism resulted from the growth of a predetermined miniature, much as a bud unfolds into a leaf. Both Roux and Driesch reasoned that if only half of the embryo were allowed to develop, the result would be an adult (or later embryo) with half its structures missing. Both Roux and Driesch intervened during early development, immediately after the first cleavage division, when the embryo consists of only two cells. If the theory of preformation were correct, then one cell of the two-cell stage should give rise to only half the normal structure.

Working with a frog embryo, Roux pierced one of the two cells with a hot needle. He then watched the other cell continue to develop. As predicted by the theory of preformation, he observed the development of only half an embryo.

But Driesch, working with sea urchin embryos, obtained a different result. Driesch used a different method from Roux: Instead of killing one cell and leaving it in place, Driesch shook the two-celled embryos until the cells came apart. He then watched as each of the two cells grew into a complete sea urchin. This unexpected result disproved the idea, at least in sea urchins, that all the structures of the adult were already preformed in the original fertilized egg.

Driesch's work also suggested that the development of a cell depends on the environment provided by other cells. If Driesch had not disturbed the embryo, the two cells of a two-cell embryo would have produced a single adult; that is, each cell would have given rise to only half an adult. But after he separated the cells, each had the capacity to produce a whole organism!

The contrasting results of Roux and Driesch could have reflected a difference between frogs and sea urchins or a difference in the two techniques; for example, Driesch separated two live cells, but Roux did not remove the killed cell. Some years later, Hans Spemann, another German embryologist, repeated Driesch's sea urchin experiment with the embryos of an amphibian. Spemann used a fine baby hair to separate the two cells of an early salamander embryo. His result confirmed Driesch's view that each cell of the two-cell embryo could give rise to a whole adult. This conclusion is generally true of vertebrates, but it does not apply to all species.

The experiments of Driesch and of Spemann laid to rest the theory of preformation. But the problem remained: How does the complexity of multicellular organisms arise? Driesch was so bothered by this question that he resorted to philosophical speculation. He fell into scientific disrepute: At one point Haeckel, Driesch's former professor and one of the most influential scientists in 19th century Germany, suggested that he should perhaps take some time off in a mental hospital.

We will return to the problem of the origin of complexity later in this chapter. First, however, we outline the main events of animal development.

■ HOW DO CELLS OF THE EMBRYO GIVE RISE TO CELLS OF THE ADULT?

Developmental biologists have a much simplified way of viewing the body plan of a vertebrate. They describe it as "a tube within a tube," actually three concentric tubes, as illustrated in Figure 20-2. Each of these tubes derives from one of three **germ layers,** the embryonic structures that first show a three-layered organization. The outermost tube, which derives from the embryonic **ectoderm** [Greek,

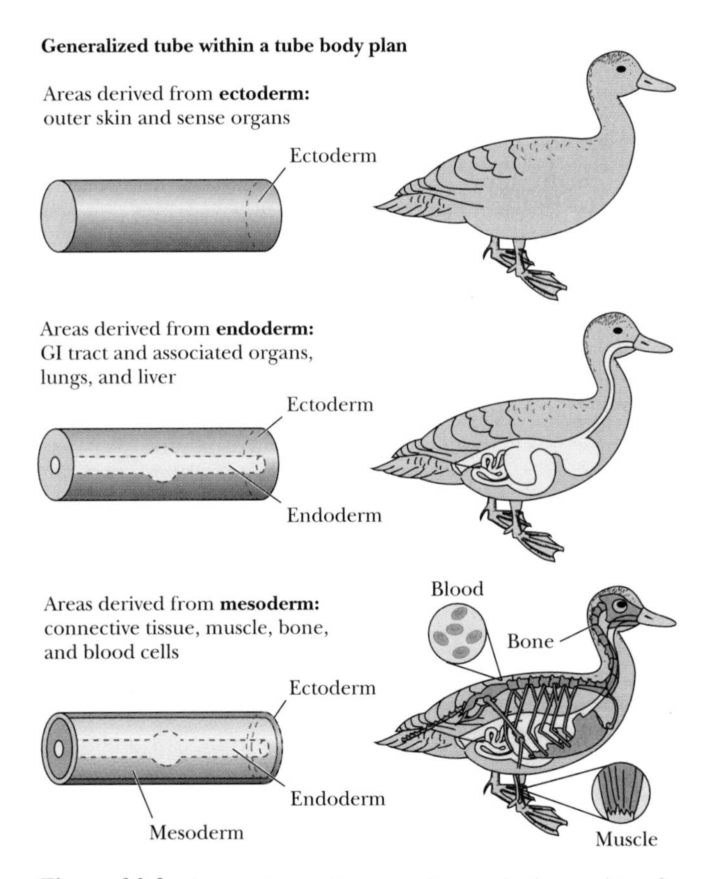

Figure 20-2 An adult vertebrate, such as a duck, consists of a "tube within a tube." The body plan, which includes endoderm, mesoderm, and ectoderm, first arises early in development.

ecto = outside + *derma* = skin], consists of the part of the animal that is in contact with the outside world—the **epidermis,** or outer skin layer, as well as the sense organs and the central nervous system. The innermost tube, which derives from the embryonic **endoderm** [Greek, *endo* = inside + *derma* = skin], is the gastrointestinal tract, together with associated organs like the lungs, pancreas, and liver.

The tissues and organs between the outer and inner tubes, which derive from the embryonic **mesoderm** [Greek, *mesos* = middle + *derma* = skin], consist of muscles, blood, and connective tissues such as bones and tendons. A number of organs including the heart and kidneys also derive from the mesoderm.

The three layers of the adult vertebrate ultimately arise from a single cell—the fertilized egg. A major goal of embryologists was to see how the three-layered organization is accomplished. They studied the early events of development and asked when the germ layers first appear.

The three-layered body plan of vertebrates is also present in the sea urchin, an invertebrate, as are many of the same developmental processes. Because of their relative simplicity and the ability to produce thousands of embryos in the laboratory at the same time by mixing collected sperm and eggs, sea urchins have been much studied by developmental biologists, and we use them in this chapter to illustrate many points about vertebrate development.

Biologists Describe Vertebrate Development in Eight Stages

Although the process of development is continuous, we can conveniently divide it into eight more or less separate stages, as diagrammed in Figure 20-3:

Gamete formation—the production of sperm and eggs; this process involves both meiosis, which produces haploid germ cells, and cell specialization.

Fertilization—the coming together of haploid sperm and haploid egg to form a diploid **zygote;** fertilization also starts the program of gene expression that eventually leads to the formation of many different kinds of specialized cells, all organized in a genetically programmed way.

Cleavage—the division of the zygote into many smaller cells accomplished by a series of often rapid mitotic cell divisions; in many vertebrates the resulting cells form a **blastula** [Greek, little sprout], a sphere with cells on the surface and fluid inside. In mammals, the endpoint of cleavage is a **blastocyst,** a modified blastula, in which cells do not lie within a single layer.

Germ layer formation—the movement of embryonic cells to give the three germ layers described above (ectoderm, mesoderm, and endoderm); the resulting structure is called a **gastrula** [Greek, little stomach] and the process is called **gastrulation.**

Organ formation—the movement and specialization of cells to produce functioning organs such as heart, kidneys, and the nervous system; in mammals, the end point of organ formation is a **fetus.**

Growth and maturation—the increase in size and the change in form of an organism after the organs and basic body plan are established. Growth involves an increase in cell number through cell division. It also involves the production of extracellular materials such as bone, cartilage, and hair. The net result is to bring the organism to a self-reliant form, either a larva or an adult.

Metamorphosis—in many species, a series of dramatic changes in form; in these species, organ formation and growth give rise to a **larva,** a feeding form distinct from the adult. The metamorphosis of a tadpole into a frog is a familiar example, as is that of an initially green caterpillar into a black and orange Monarch butterfly. The tadpole-frog metamorphosis involves a change not only in form but also in the sources of oxygen and food. Not all organisms undergo metamorphosis.

Aging—further development, eventually leading to death. Aging involves the death of cells, changes in the extracellular environment, and decreased function of many organs. Aging occurs at different rates in different species and among individuals within a single species.

We begin our discussion of development with the cleavage stage because the events after fertilization are more or less similar in most animals. In contrast, gamete formation and fertilization, which we discuss later in the chapter, are fundamentally more varied.

Cleavage Converts a Single Cell (the Zygote) into Many Cells

The first process that occurs in a newly fertilized egg is cleavage, a series of rapid mitotic cell divisions, illustrated for the sea urchin in Figure 20-4. These divisions are not accompanied by growth, and the embryo is about the same size after cleavage as it was immediately after fertilization. The results of cleavage are the subdivision of the zygote's cytoplasm and a large increase in the number of nuclei per embryo.

The zygote is invariably much larger than the average somatic cell of an adult and often contains large stores of *yolk,* a mixture of proteins, lipids, and carbohydrates that nourishes the embryo until it can feed itself. As discussed in Chapter 1, large cells—with a lower ratio of surface to volume than small cells—have relatively more difficulty taking in nutrients and excreting wastes. At the end of the cleavage stage, the cells of the embryo are similar in size, and in surface-to-volume ratio, to those of an adult.

The nucleus of a large cell has a much larger "administrative domain" than that of a small cell—challenging its capacity to provide mRNAs (because there are only

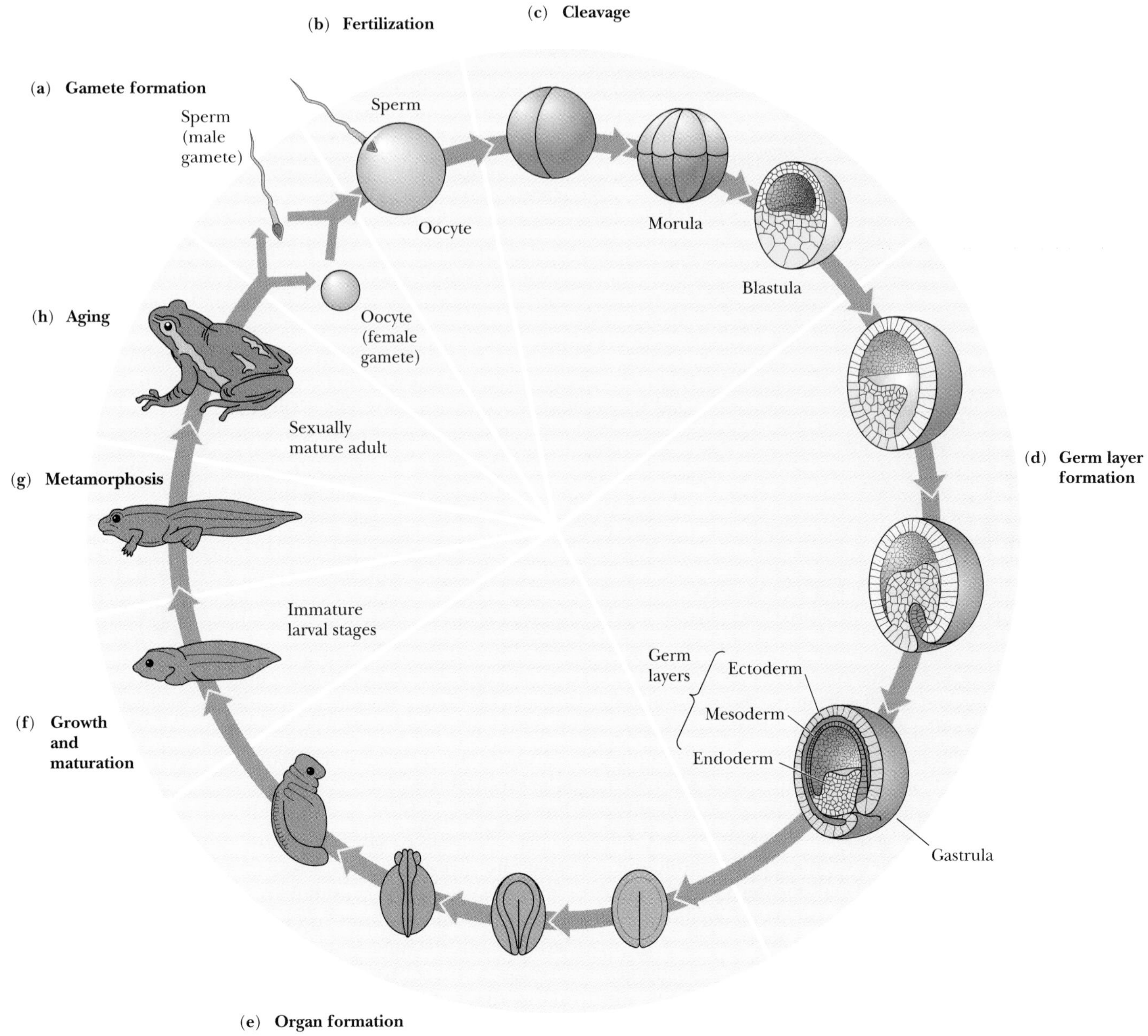

(c) **Cleavage**

(b) **Fertilization**

(a) **Gamete formation**

Sperm (male gamete)

Sperm

Oocyte

Morula

Blastula

Oocyte (female gamete)

(h) **Aging**

Sexually mature adult

(d) **Germ layer formation**

(g) **Metamorphosis**

Immature larval stages

Germ layers

Ectoderm

Mesoderm

Endoderm

(f) **Growth and maturation**

Gastrula

(e) **Organ formation**

Figure 20-3 Stages of vertebrate development: **(a)** gamete formation, leading to sperm and eggs; **(b)** fertilization, leading to a zygote; **(c)** cleavage, leading to a blastula; **(d)** germ layer formation, leading to a gastrula; **(e)** organ formation, leading to a fetus; **(f)** growth and maturation, leading to a larva or an adult; **(g)** metamorphosis, transforming a larva into an adult; **(h)** aging, leading ultimately to death.

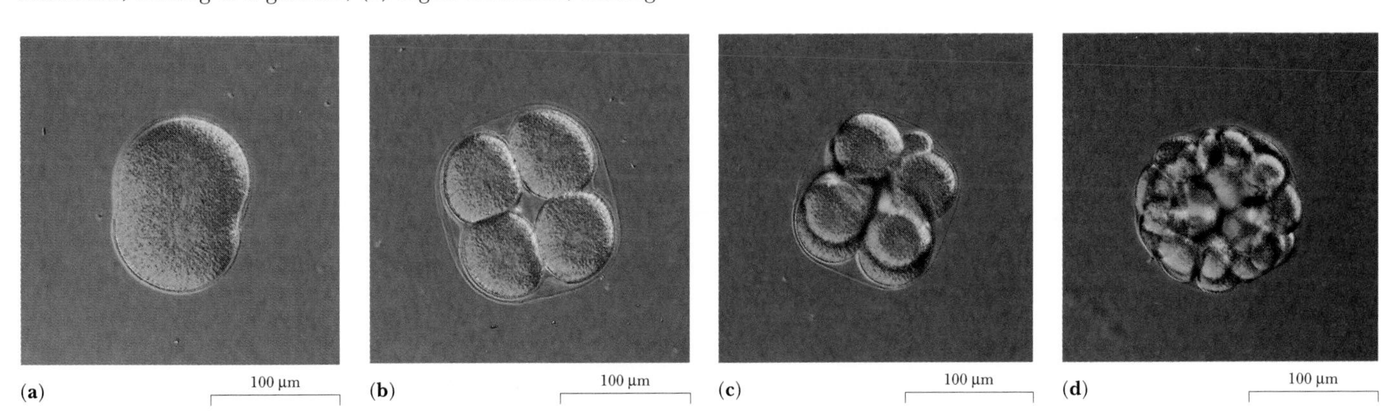

Figure 20-4 Cleavage in the sea urchin embryo, leading to the formation of the blastula, a hollow ball of cells. **(a)** First cell division; **(b)** 4-cell stage; **(c)** 8–12 cell stage; **(d)** 32-cell stage. *(David Fromson, California State University, Fullerton)*

(a) **Bird**

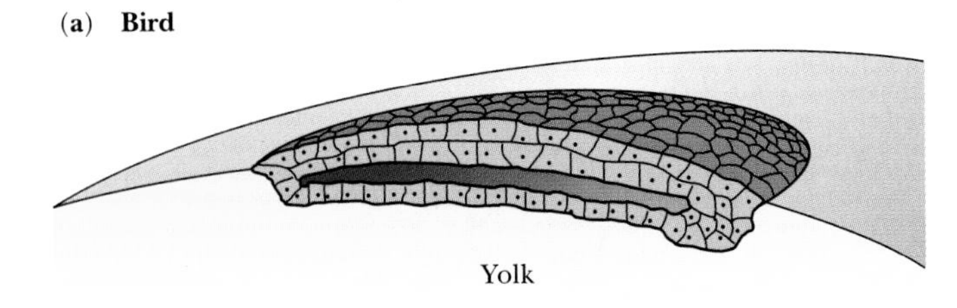

(b) **Sea urchin**

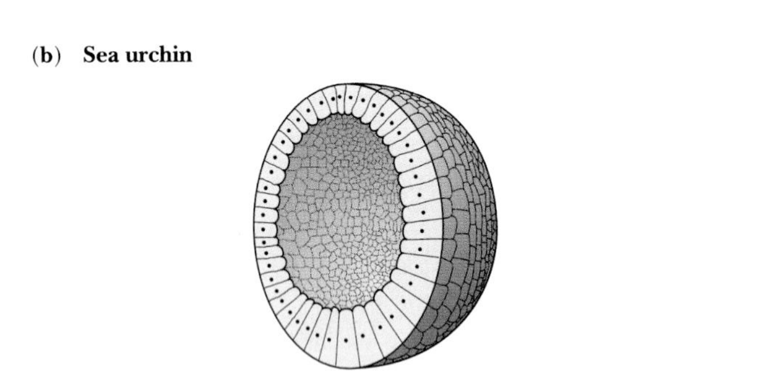

Figure 20-5 Cleavage in **(a)** a yolky bird embryo has a different pattern from that in **(b)** the less yolky sea urchin embryo.

two copies of each gene) and limiting its ability to set up differences in the spatial distribution of those mRNAs. Such spatial differences in mRNA distribution exist in the zygote, but they result from processes that took place before fertilization. The multiplication of nuclei during cleavage not only provides a more abundant source of mRNA but also allows each nucleus to work independently. The independence of individual cells means that different cells can produce different sets of mRNAs, make different proteins, and specialize for different roles.

Despite the common occurrence of rapid cleavage divisions after fertilization, species differ dramatically in the pattern and pace of these divisions. The different cleavage patterns depend both on the orientation of the mitotic spindles in each division and on the distribution of yolk within the egg. Eggs that contain lots of yolk either have relatively slow cleavage divisions or do not completely divide the yolk. Bird eggs, loaded with yolk, confine their cleavage divisions to a disc of incompletely separated cells floating on the surface of the yolk, as shown in Figure 20-5a.

Eggs with little yolk, such as those of the sea urchin, divide more or less equally and symmetrically. In the sea urchin, each division takes less than 1 hour. After seven such divisions the embryo becomes a **blastula,** a hollow, fluid-filled ball of cells, as illustrated in Figure 20-5b. The sea urchin blastula remains a one-layered structure, eventually consisting of 1000–2000 cells.

Cleavage in mammals differs from that in sea urchins. The pace is much slower, each division taking 12–24 hours. In contrast to the sea urchin embryo, whose cleavage divisions occur synchronously, the individual cells of the mammalian embryo divide at different times. After a mammalian embryo has divided to form about eight cells, the cells suddenly huddle together to form a compact structure that develops into the **morula** [Latin = mulberry], a solid ball of cells, as shown in Figure 20-6.

The morula soon forms a **blastocyst** [Greek, *blastos* = sprout + *cystos* = cavity], a modified blastula in which the cells do not lie within a single layer but do enclose an internal cavity. The blastocyst contains two types of cells—the **trophoblast** [Greek, *trephein* = to nourish + *blastos* = germ], the prominent outer cell layer, and the *inner cell mass.* The inner cell mass of the mammalian blastocyst, like the sea urchin blastula, reorganizes into a three-layered structure. This small group of cells will eventually grow into the embryo itself and subsequently into the adult. Cells derived from the trophoblast are responsible for the connections between the embryo and the mother: They attach the embryo to the wall of the uterus and later become part of the placenta, through which mother and fetus exchange nutrients and wastes.

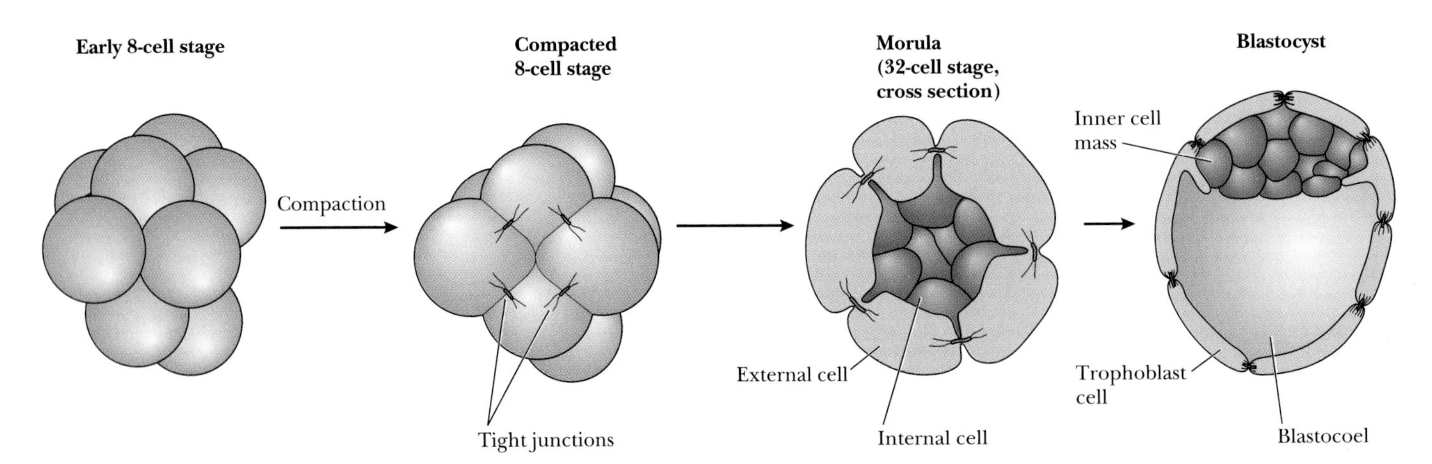

Figure 20-6 Cleavage in a mammalian embryo leads to the formation of a solid morula and then a blastocyst.

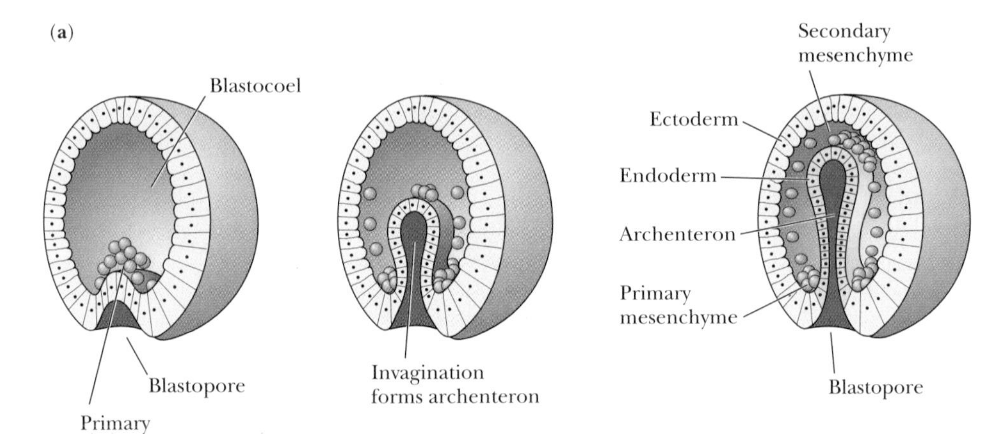

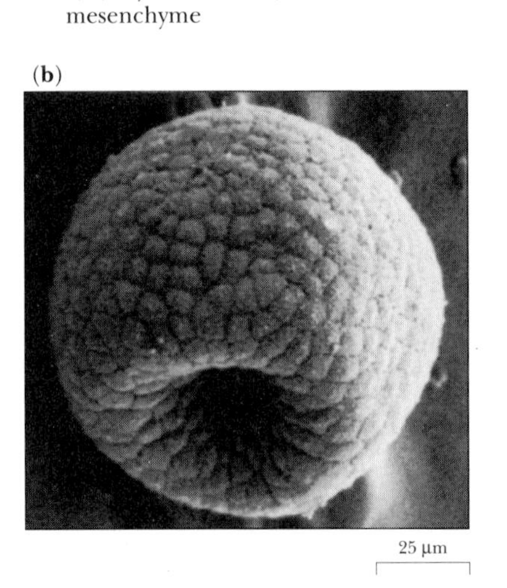

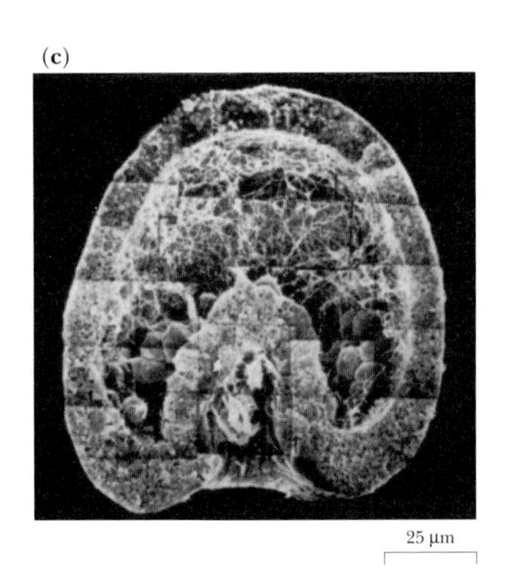

Figure 20-7 Gastrulation in sea urchins. Cells from the bottom of the blastula move into the blastocoel, some individually, others pulling along tightly attached cells to form the endoderm. **(a)** Diagram of gastrulation; **(b)** scanning electron micrograph of early sea urchin gastrula; **(c)** cross section of sea urchin gastrula. *(b, c, from Morrill and Santos, 1985)*

Gastrulation Sets Up the Three-Layered Structure of the Adult Vertebrate

During gastrulation, a vertebrate embryo first sets up the germ layers and becomes "a tube within a tube." The most visible aspect of this process is the formation of the **archenteron** [Greek, *arche* = beginning + *enteron* = gut] or "primitive gut," as shown in Figure 20-7a, the innermost tube, which is lined with endoderm and becomes the digestive tract.

In sea urchins and in all vertebrates, the digestive tract has two openings—the mouth and the anus. The anus is formed from the original site of invagination during germ layer formation. The mouth develops later, as an opening at the forward end of the primitive gut.

Gastrulation in Sea Urchin Embryos Depends on the Coordinated Behavior of Cellular Sheets and Individual Cells

The sea urchin blastula starts out as an almost symmetrical sphere. Soon, however, it changes shape. One side flattens, and a few small cells move—as single cells—into the **blastocoel,** the interior of the blastula, as illustrated in Figure 20-7b and c. These independently moving cells are called the **primary mesenchyme** [Greek, *mesos* = middle + *enchyma* = infusion]. They attach to the interior wall of the blastocoel and eventually produce a calcium-containing "skeleton."

The main business of gastrulation now gets under way—**invagination,** local folding of a cell layer to form an enclosed space with an opening to the outside. In sea urchin gastrulation, the cells on the bottom of the embryo invaginate to form the beginnings of the archenteron. The **blastopore,** the opening of the archenteron, eventually develops into the anus of the mature gut. As invagination proceeds, the cells at the tip of the archenteron detach to form the **secondary mesenchyme.** These cells form contacts with the wall of the blastocoel and actually pull the archenteron the rest of the way to the wall. After the archenteron has attached to the wall, the outer layer forms an opening, which eventually becomes the mouth. The cells of the secondary mesenchyme disperse into the blastocoel and proliferate to form the mesoderm.

Underlying the process of gastrulation are changes in the properties of individual cells. These changes include alterations in the attachments of cells to one another within the single-layered cell sheet and transformations in shape, brought about by the cytoskeleton. The cells on the bottom of the embryo, for example, constrict at the base,

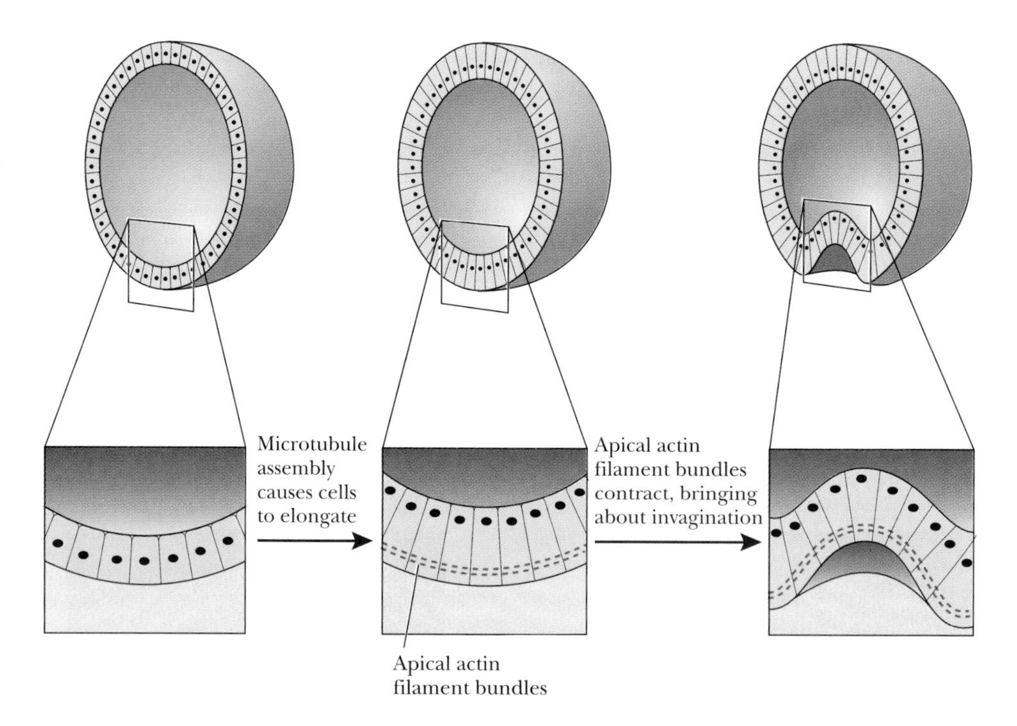

Figure 20-8 Changes in cell shape underlie cell movement during gastrulation: diagram of the relationship between the shape of a cell layer and the shapes of individual cells.

as shown in Figure 20-8, and let go of each other at the inside edge. While the complicated movements of sea urchin gastrulation depend most heavily on the invagination and repacking of cells that are held together in sheets, the precursors of the mesoderm move as individual cells rather than as parts of a continuous sheet.

What Determines Whether a Cell Stays Part of a Sheet or Moves Independently?

The answer to this question lies in the properties of individual cells. Cells that stick together in epithelial sheets do so because of specialized junctions. (See Chapter 6.) Similarly, the folding of cell sheets depends on shape changes in individual cells These changes in turn depend on changes in the cytoskeleton. The characteristics of individual cells account for the complex movements of gastrulation (and for other cell movements during development). We still do not fully know, however, how the embryo coordinates changes in the characteristics of individual cells.

Gastrulation in Amphibian Embryos Depends on the Coordinated Behavior of Cellular Sheets and on the Interactions of Individual Cells with Extracellular Matrix

The embryos of amphibians, especially frogs and salamanders, continue to be a favorite experimental subject for developmental biologists, partly because of their relatively large sizes (with eggs up to 1–2 mm in diameter) and their ability to develop in a simple salt solution (rather than within an egg shell or a uterus). Amphibian gastrulation, however, is more complicated than that of sea urchins because of the presence of considerable yolk.

During the cleavage stage of the amphibian embryo, cells near the top of the embryo (called the "animal pole") divide more rapidly than the large, yolk-laden cells near the bottom (the "vegetal pole"), as illustrated in Figure 20-9. In the cells near the vegetal pole, the cleavage furrow takes longer to get through the yolky cytoplasm, slowing cell division. The resulting blastula, illustrated in Figure 20-9a, is several cell layers thick and is much more asymmetrical than the sea urchin blastula.

Because the yolk-filled cells near the vegetal pole can move only sluggishly, an amphibian embryo accomplishes gastrulation with mechanisms that differ from those in sea urchin embryos. As illustrated in Figure 20-9b and c, germ layers form after cells move through a blastopore that lies near the embryo's equator, rather than at its vegetal pole. Gastrulation begins with an invagination, accompanied by shape changes in *bottle cells* at the dorsal blastopore lip, illustrated in Figure 20-9d. Shortly after the initial invagination, adjacent cells follow the bottle cells inward, in a process called *involution.* The result is the formation of an archenteron, lined by endodermal cells that formerly were located on the surface of the blastula. Some cells detach from the invaginating layers to form the mesoderm.

The future mesodermal cells also play a role in powering gastrulation. These cells, which are within the earliest vanguard of invaginating cells, crawl over the extracellular matrix on the roof of the blastocoel. This crawling depends on the interaction of fibronectin in the extracellular matrix with integrin molecules on the cell surface, as illustrated in Figure 20-10. (See Chapter 18, p. 496.) The binding of fibronectin to integrin depends on

(a) **Blastula**

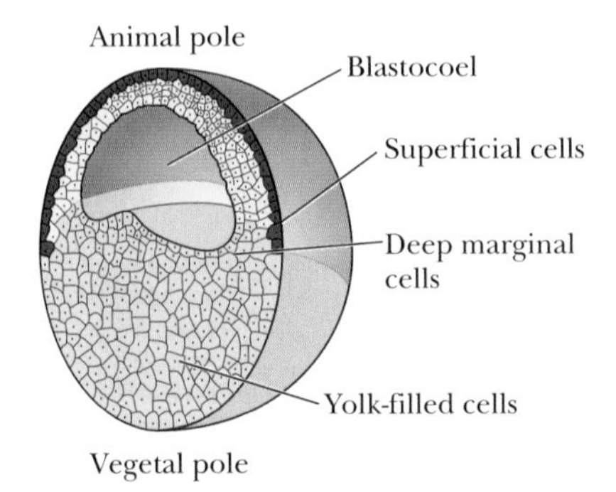

(b) **Germ layer formation**

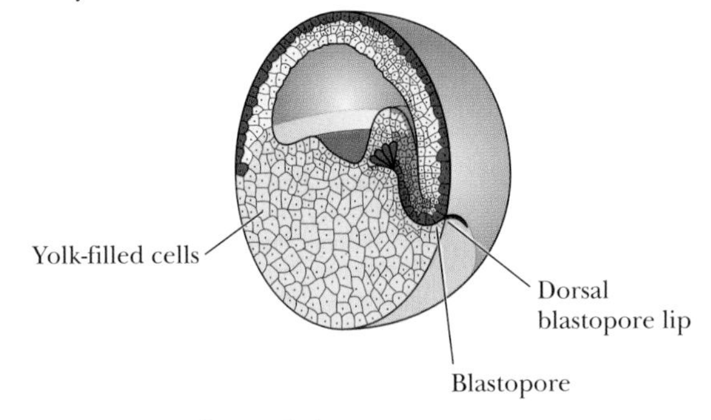

(c) **Successive stages of gastrulation**

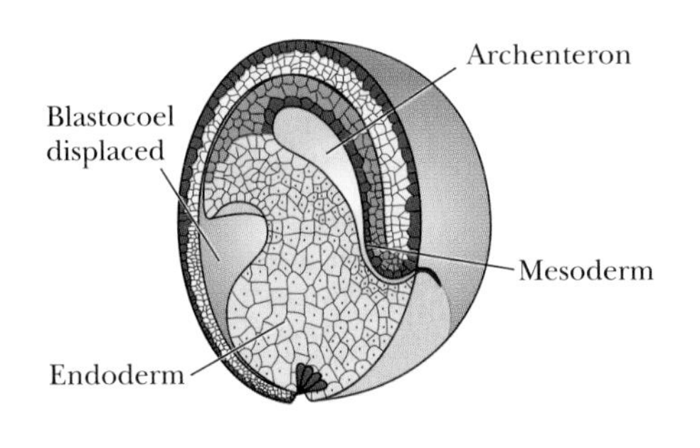

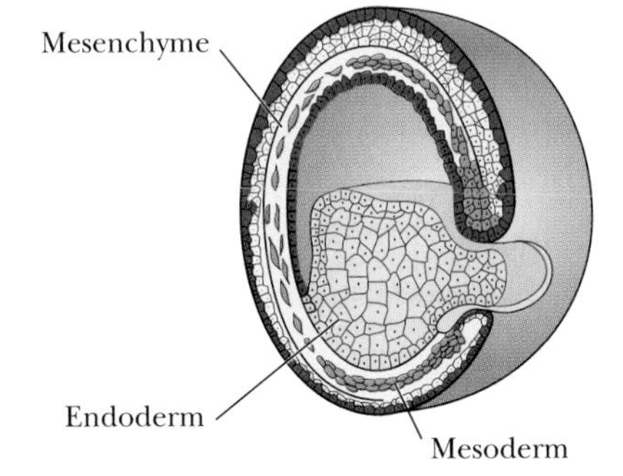

(d) **Shape changes in the cells of the blastopore lip**

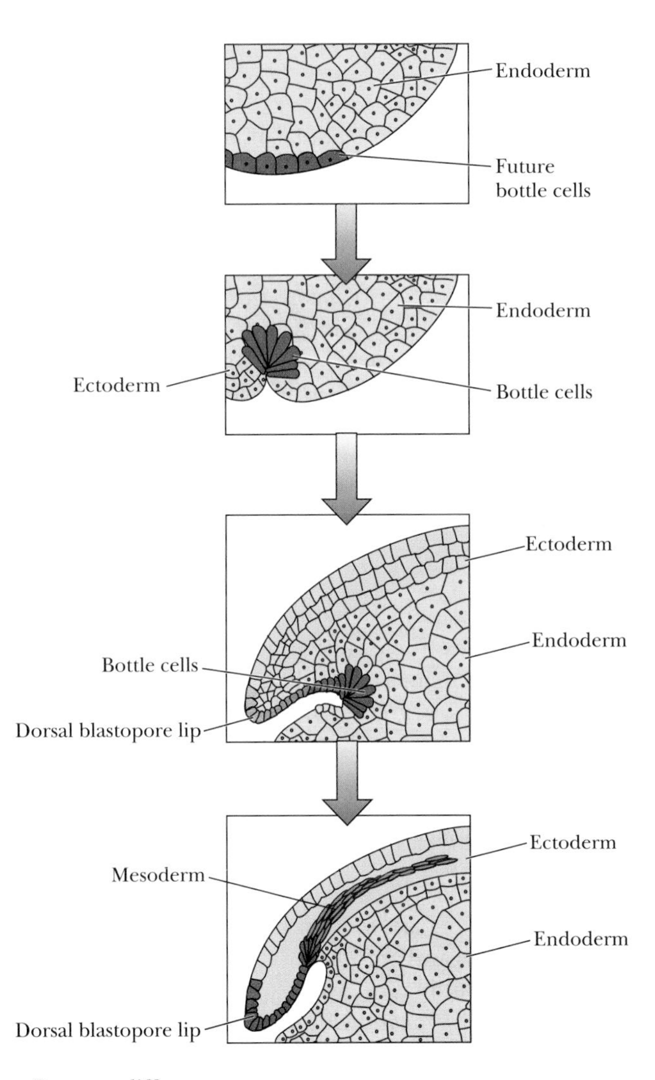

Figure 20-9 Cleavage and gastrulation in an amphibian embryo. Patterns differ from those of sea urchin embryos because of the large amount of yolk. **(a)** Blastula; **(b)** initial cell movements that begin germ layer formation; **(c)** successive stages of gastrulation; **(d)** shape changes in the cells of the blastopore lip.

a stretch of three amino acid residues in the fibronectin molecule—arginine-glycine-aspartic acid (called RGD, using the one-letter abbreviations for the amino acids). Injection of free RGD peptide into a salamander embryo prevents gastrulation from occurring, presumably by preventing the interaction between the fibronectin on the in-

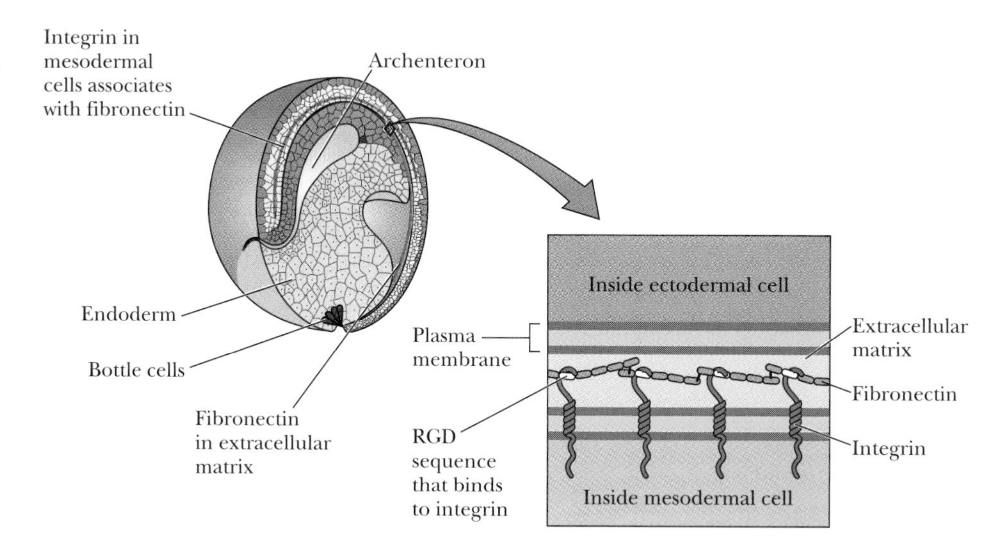

Figure 20-10 The movement of mesodermal cells over the roof of the blastocoel depends on interactions between fibronectin in the extracellular matrix and integrin on the cell surface. The peptide RGD (arginine-glycine-aspartic acid) binds to the fibronectin-binding site in integrin where the same sequence in fibronectin would have bound. Injection of RGD into a salamander embryo prevents germ layer formation.

vaginating cells and the extracellular matrix on the roof of the blastocoel.

The properties of individual cells also stabilize the final arrangement of cells in the gastrula. In one experiment, for example, researchers dissociated pieces of ectoderm and endoderm into suspensions of single cells. They then mixed the two cell suspensions and observed that ectodermal cells aggregated specifically with other ectodermal cells and endodermal cells with other endodermal cells. Even more strikingly, mixing suspensions of ectoderm, mesoderm, and endoderm led to the spontaneous aggregation of the dissociated cells into germ layers, with endodermal cells on the inside, ectodermal cells on the outside, and mesodermal cells in between. These experiments show that some components of the cell surface establish specific interactions between cells. Some of these cell surface molecules, for example, the *cadherins,* which are responsible for calcium-dependent cell-cell adhesion, have been isolated. Researchers continue to study their role both in cell migration and in the stabilization of the transient arrangements of cells in the developing embryo.

The Early Development of a Mammalian Embryo Follows a Pattern Similar to That in Birds

Germ layer formation in mammals, unlike the same process in sea urchins and amphibians, depends mostly on the separation ("delamination") of parallel sheets of cells followed by movements of individual cells. Recall that the mammalian blastocyst consists of a trophoblast and an inner cell mass. The first separation of cells within the inner cell mass produces two layers, called the *hypoblast* and the *epiblast,* as illustrated in Figure 20-11. The cells of the hypoblast enclose what will become the primitive gut, but these cells are later replaced by endoderm cells that are derived from the epiblast. Part of the hypoblast stays just below the epiblast (separated from it by the blastocoel) to form the *blastodisc,* the equivalent of a blastula. The structures of the adult ultimately derive only from cells that originate in the epiblast.

The formation of the germ layers now takes place. Cells of the blastodisc move toward a central line, leading to the formation of the *primitive streak,* the functional equivalent of the blastopore. Individual cells then move through a groove in the streak into the blastocoel. These cells advance independently into the blastocoel, where they later separate into endoderm and mesoderm.

The mammalian pattern of gastrulation is also found in birds and reptiles. This pattern is well suited to the yolky eggs of birds and reptiles, because only the relatively yolk-free cells of the epiblast move. This pattern appears to have evolved early in the history of the land-dwelling vertebrates, when encased, yolky eggs allowed higher vertebrates to become independent of water. The common pattern of embryonic development in reptiles, birds, and mammals supports the view that mammals evolved from an ancestor with an enclosed yolky egg.

Organ Formation Requires Both the Creation of Form and Biochemical Specialization of Individual Cells

Once the embryo has established the three germ layers, it next begins to form organs. An organ may derive from a single germ layer or from two different germ layers. In each case organ formation consists of two major processes—**morphogenesis**, the creation of the organ's form, and **cytodifferentiation,** the production of specialized cells. The formation of the brain in higher vertebrates illustrates many of the main features of organ formation.

At the end of gastrulation, mesoderm has separated from ectoderm and endoderm. In vertebrates, part of the mesoderm forms the **notochord,** an epithelial rod that runs from the front to the rear of the embryo, beneath its back (dorsal) surface. The notochord participates in the

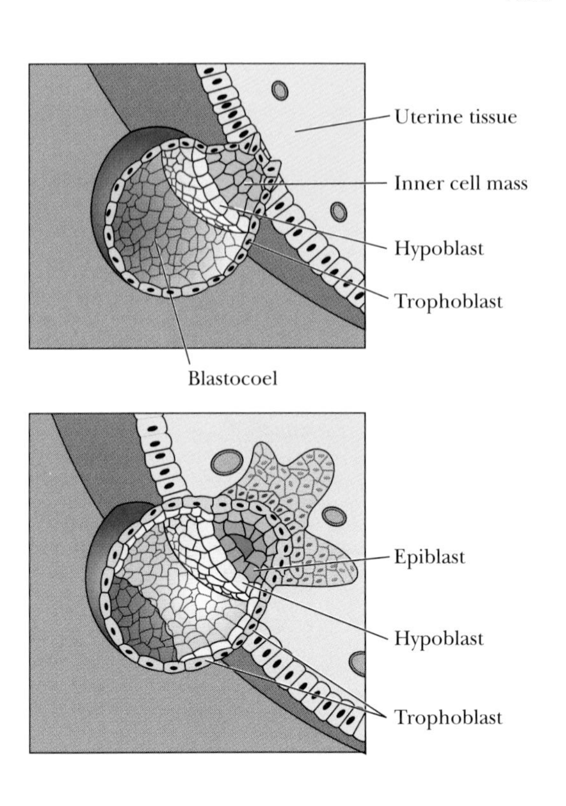

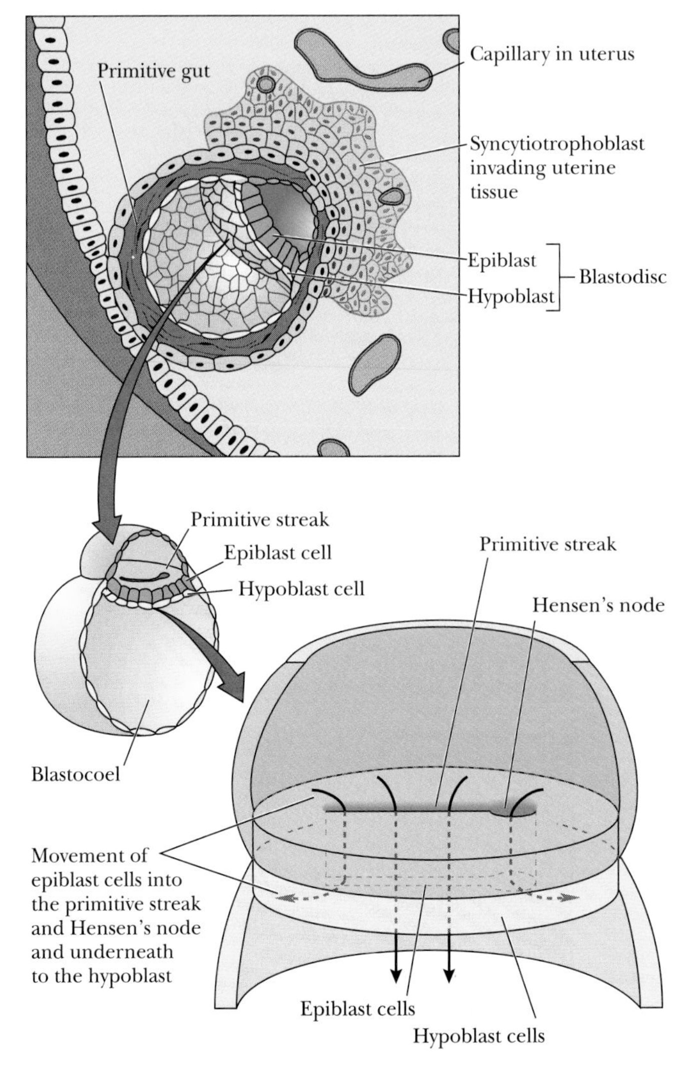

Figure 20-11 Germ layer formation in a mammalian blastocyst. Despite the absence of yolk, a mammalian embryo shows the same discoidal pattern as in a bird embryo, supporting the view that mammals evolved from progenitors with yolky embryos.

elongation of the embryonic axis and in the organization of further embryonic development.

Just above (dorsal to) the notochord, the ectoderm rearranges to form the nervous system, as illustrated in Figure 20-12a. The cells along the central line of the embryo thicken to form the **neural plate**, a flat plate above the notochord, as seen from above in Figure 20-12b. As in the case of the movements of gastrulation, the formation of the neural plate depends on changes in cell shape brought about by the action of the cytoskeleton—here elongation of the cells depends on assembly of microtubules.

The neural plate then curls, again along the embryo's axis, as individual cells in the neural plate constrict at their apical ends, with pinching contractions that depend on actin filaments. The elongated invaginated structure then pinches away from the surface (as seen in Figure 20-12c and d) to form a hollow tube, called the **neural tube,** the precursor structure for the entire central nervous system.

The cells left on the surface, above the neural tube, become skin ectoderm (epidermis), while the cells that had connected the tube to the surface—the **neural crest** cells—migrate away. Descendants of the neural crest develop into different types of cells, depending on where they go after they leave the neural tube behind. Some give rise to the pigmented cells of the skin, others to cells of the peripheral nervous system, and still others to the adrenal medulla. The entire process of forming the neural tube (and neural crest) is called **neurulation.**

The neural tube then undergoes crimping, folding, and cell division to form the beginnings of the different regions of the brain and spinal cord. The front of the tube enlarges and then bulges on each side to form the beginnings of the cerebral hemispheres. Just behind these bulges are the *optic vesicles,* which become the eyes.

The morphogenesis of the eye, illustrated in Figure 20-13, provides a beautiful example of the formation of an incredibly complicated and important organ from a relatively simple set of cell movements. After the optic vesicle comes into contact with the outside surface of the head, the overlying skin ectoderm thickens to form a plate that develops into the lens. Following the appearance of the lens thickening, the optic vesicle folds back on itself, to form a double-walled *optic cup.* At about the same time the lens plate curls and pinches away from the skin ectoderm to form the *lens vesicle.* The stalk that attaches the cup to the rest of the brain becomes the *optic nerve* and eventually carries visual information from the retina to the brain.

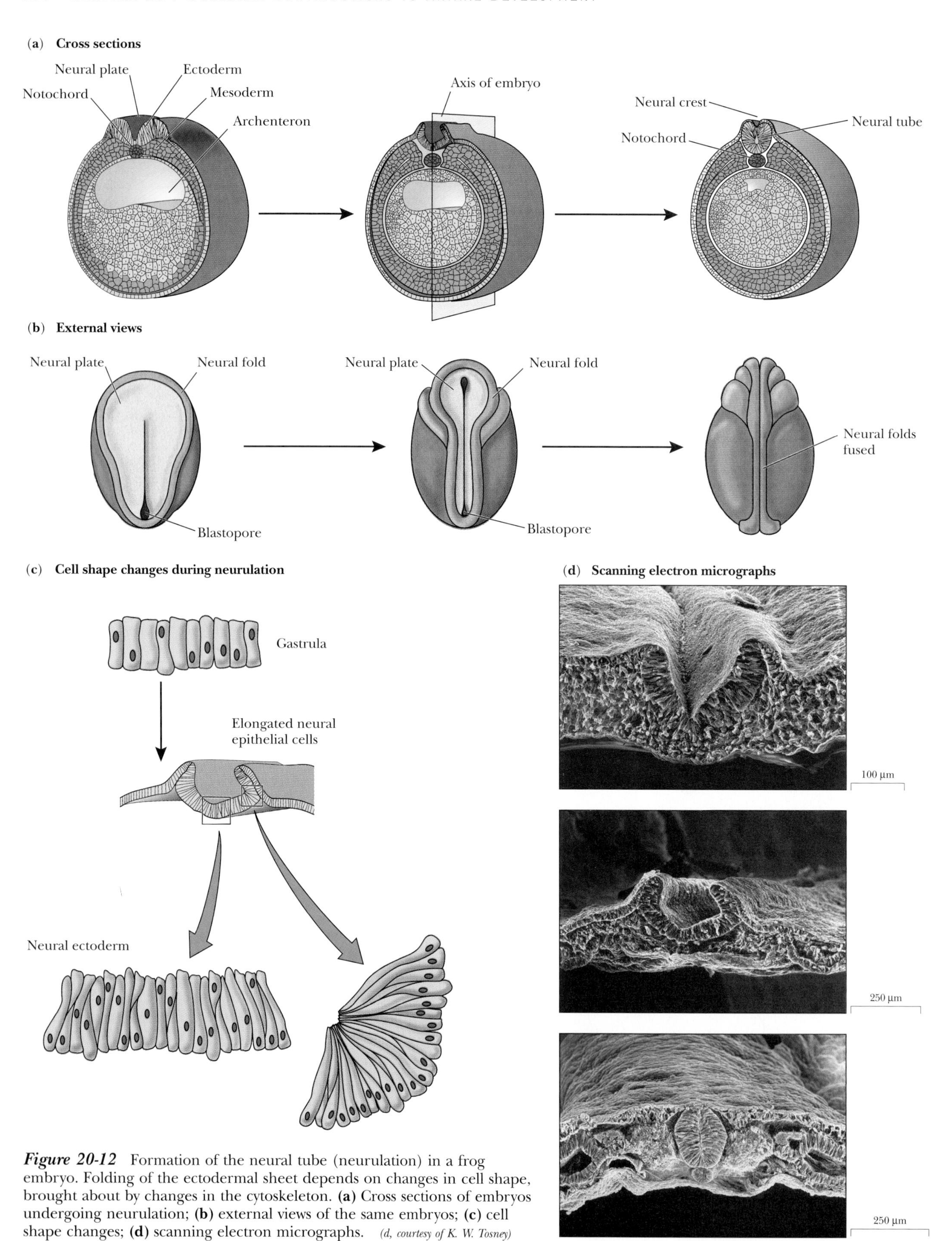

Figure 20-12 Formation of the neural tube (neurulation) in a frog embryo. Folding of the ectodermal sheet depends on changes in cell shape, brought about by changes in the cytoskeleton. **(a)** Cross sections of embryos undergoing neurulation; **(b)** external views of the same embryos; **(c)** cell shape changes; **(d)** scanning electron micrographs. *(d, courtesy of K. W. Tosney)*

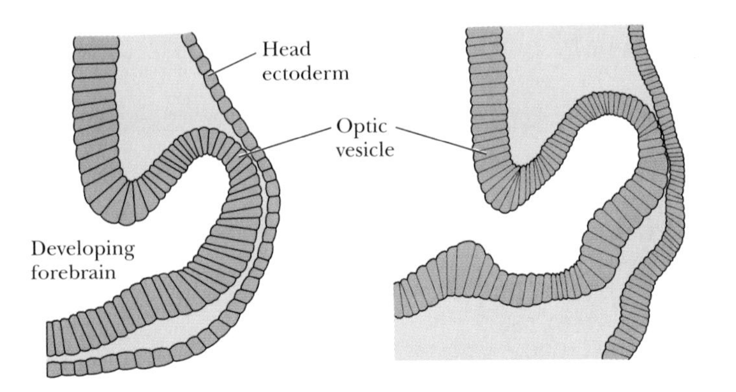

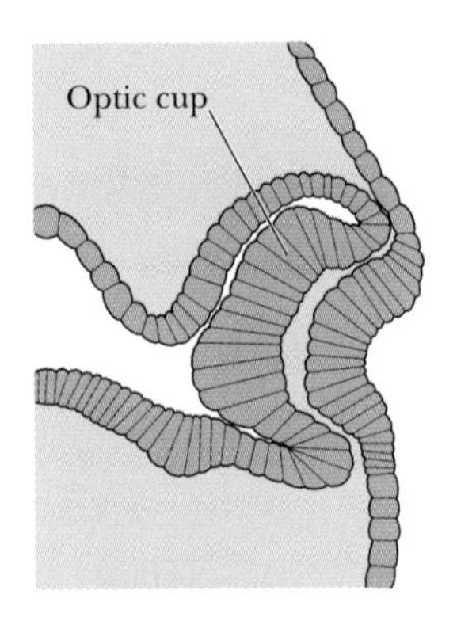

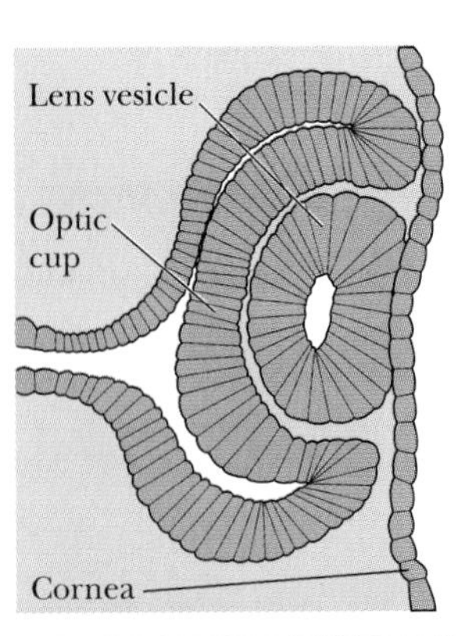

Figure 20-13 Formation of the eye. The folding of ectodermal sheets forms the first optic vesicle, then the optic cup, then the lens and the cornea. The retina later develops from the optic cup. *(Photo courtesy of K. W. Tosney)*

250 μm

After these thickenings, foldings, pouchings, and pinchings, the cells of the lens, the retina, and the overlying skin ectoderm begin to acquire differentiated properties. The outside layer of the cup (the one closest to the lens) begins to develop into several different kinds of cells that detect light (rods and cones). The lens cells start making specialized proteins that give the lens its optical properties. The cells overlying the lens change character to become the *cornea,* the eye's transparent covering. Through these events, an apparently homogeneous layer of cells in the neural tube develops into one of our most prized contacts with our surroundings.

The development of the eye illustrates several generalizations about organ formation: (1) Morphogenesis precedes cytodifferentiation; only after cells have distinctive positions do they start acquiring their specialized properties. (2) Morphogenesis often involves complex folding of cell sheets. Like the foldings of gastrulation, organ formation depends on changes in the cytoskeleton and in the junctions between cells. (3) Organ formation depends on new interactions among cells, brought about by morphogenetic movements.

Growth Requires Cell Division and Even Cell Death

Neurulation is one of many sets of morphogenetic movements. Relatively early in development, each adult (or larval) organ is present as a recognizable **rudiment** [Latin, *rudimentum* = beginning], or initial stage, from which the final form develops. Some embryologists refer to a rudiment as an *anlage* [German = beginning]. The final form of each organ requires cell specialization and growth.

Also early in development, the embryo establishes the proper connections among nerves, muscles, bones, and connective tissue. Each human hand and arm, for example, consists of 43 muscles, 29 bones, and hundreds of nerve pathways. In humans, if all goes well, all these components and others as well are in their proper places by the end of the second month of embryonic life. The embryo can then be called a fetus. The fetus is so tiny that it can live only in the protective environment of the uterus. An extended period of growth, both inside and outside of the uterus, is required for the already formed embryo to become an adult.

"Growth" usually means both the increase in size of individual cells and the increase in cell numbers due to cell division. Paradoxically, however, we can even consider the death of cells as part of this phase of development. Just as a sculptor chips away marble to produce a finished form, **programmed cell death**—also called **apoptosis** [Greek, *apo* = away from + *ptosis* = fall]—contributes to the molding of the body plan, as illustrated in Figure 20-14. Early in development, for example, our fingers and toes are connected like those of ducks, but the cells of our

(Text continues on page 554.)

(a) Duck leg (little cell death)

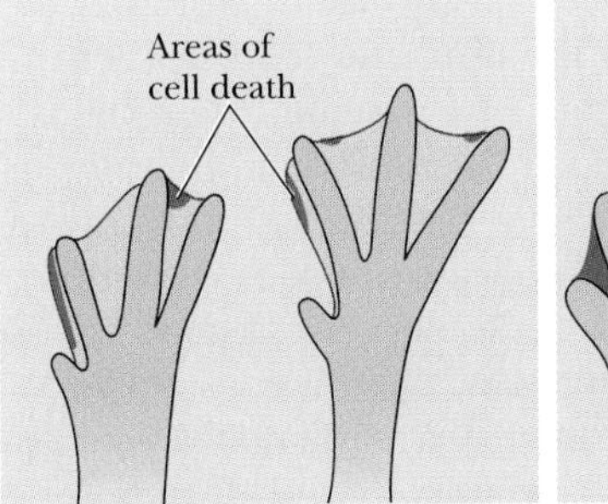

(b) Human (much cell death)

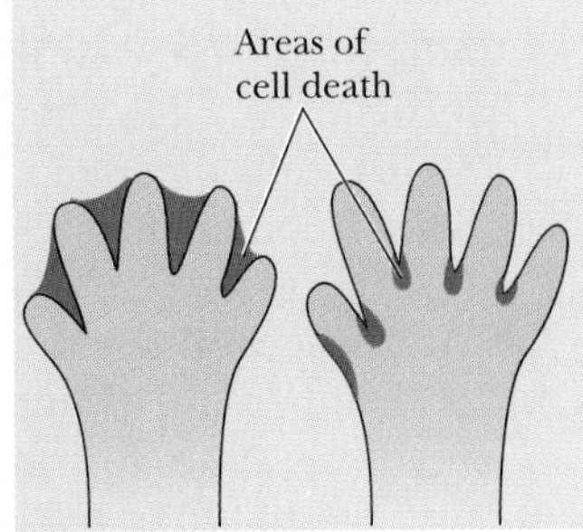

Figure 20-14 Cell death contributes to the generation of form during the growth and maturation of a developing animal. The digits of the human hand are initially connected by a web, as in a duck's foot. The death of the cells in the web occurs during fetal life.

ESSAY

PROGRAMMED CELL DEATH CONTRIBUTES BOTH TO NORMAL DEVELOPMENT AND TO CANCER PROTECTION

Until recently, most developmental biologists asked questions about how cells divide, migrate, and become different—how they organize harmoniously into individual organs and body parts, how they accumulate specific structures and proteins, and how they respond to extracellular signals that stimulate gene expression or promote cell survival. In the last decade, however, researchers have increasingly realized that development also requires *programmed cell death* or *apoptosis,* the death and removal of cells as components in normal developmental pathways, in contrast to *necrosis,* or "accidental" cell death, which occurs as a result of mutation or chemical damage and leads to cell lysis. Unlike necrosis, apoptosis is a tidy process that occurs quickly and does not leave cellular remnants.

One of the best-studied examples of apoptosis in vertebrates is the death of cells between the digits of the embryonic limbs. In the early embryo, a thin tissue lies between the digits. This tissue persists as the familiar webbing of a duck's hind limbs but is destroyed during development in chickens and in other nonaquatic vertebrates, including humans.

Apoptosis occurs in many developmental pathways in both vertebrates and invertebrates—among cells of the immune system that would otherwise attack the body's own cells, among cells of vertebrate nervous systems that arise in development but do not connect to a target, and among the cells of the nematode worm, *Caenorhabditis elegans (C. elegans).* Cells undergoing apoptosis are distinctive both visibly and biochemically: Their nuclei condense, their DNA is systematically digested, and they are methodically engulfed by other cells. The process of apoptosis probably serves to prevent the release of toxic cell products that might perturb other developmental processes in the vicinity of the dying cells.

C. elegans has provided a particularly nice opportunity to study the genetic control of apoptosis. The adult *C. elegans,* shown in Figure 1, is a transparent tube, about 1 mm long, with 959 somatic cells. Because of the simplicity and transparency of *C. elegans* and its embryo, researchers have been able to track the precise pathway of each cell that arises during development. As is the case for many (but not all) invertebrates, the normal developmental pathways of *C. elegans* are invariant: Almost every cell in the embryo has a precisely determined fate. With a few exceptions, destruction of a particular cell of the embryo with a fine laser beam results in an adult that lacks the cells and structures that would have derived from that cell.

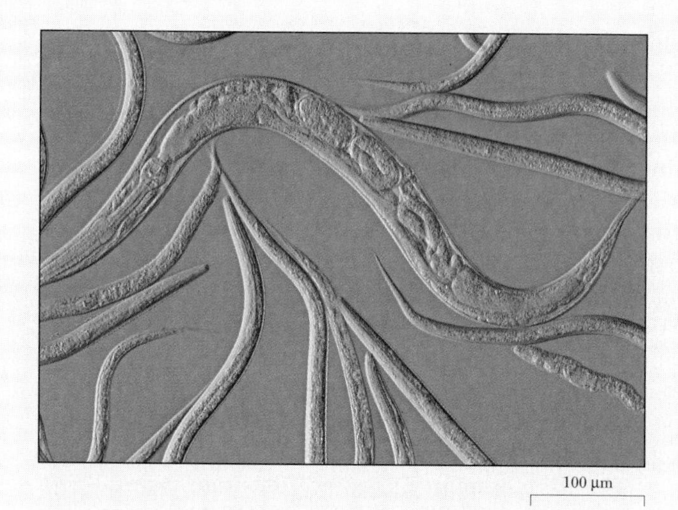

Figure 1 *Caenorhabditis elegans,* a nematode worm, has become a favorite experimental organism for developmental biologists, especially those studying cell death. *(Sinclair Stammers/Science Photo Library/Photo Researchers)*

A number of cells that arise in the embryo, however, have an unusual but invariant fate—programmed cell death. These cells are the sisters of cells that develop into nerve cells of the adult worm. A number of mutations prevent programmed cell death, and these mutations lead to abnormal nervous systems, with too many neurons, as illustrated in Figure 2. Other mutations allow programmed cell death to occur but prevent the engulfing of the cellular corpses. The existence of these mutations demonstrates that the normal functions of the mutated genes contribute to the normally occurring process of apoptosis and engulfment.

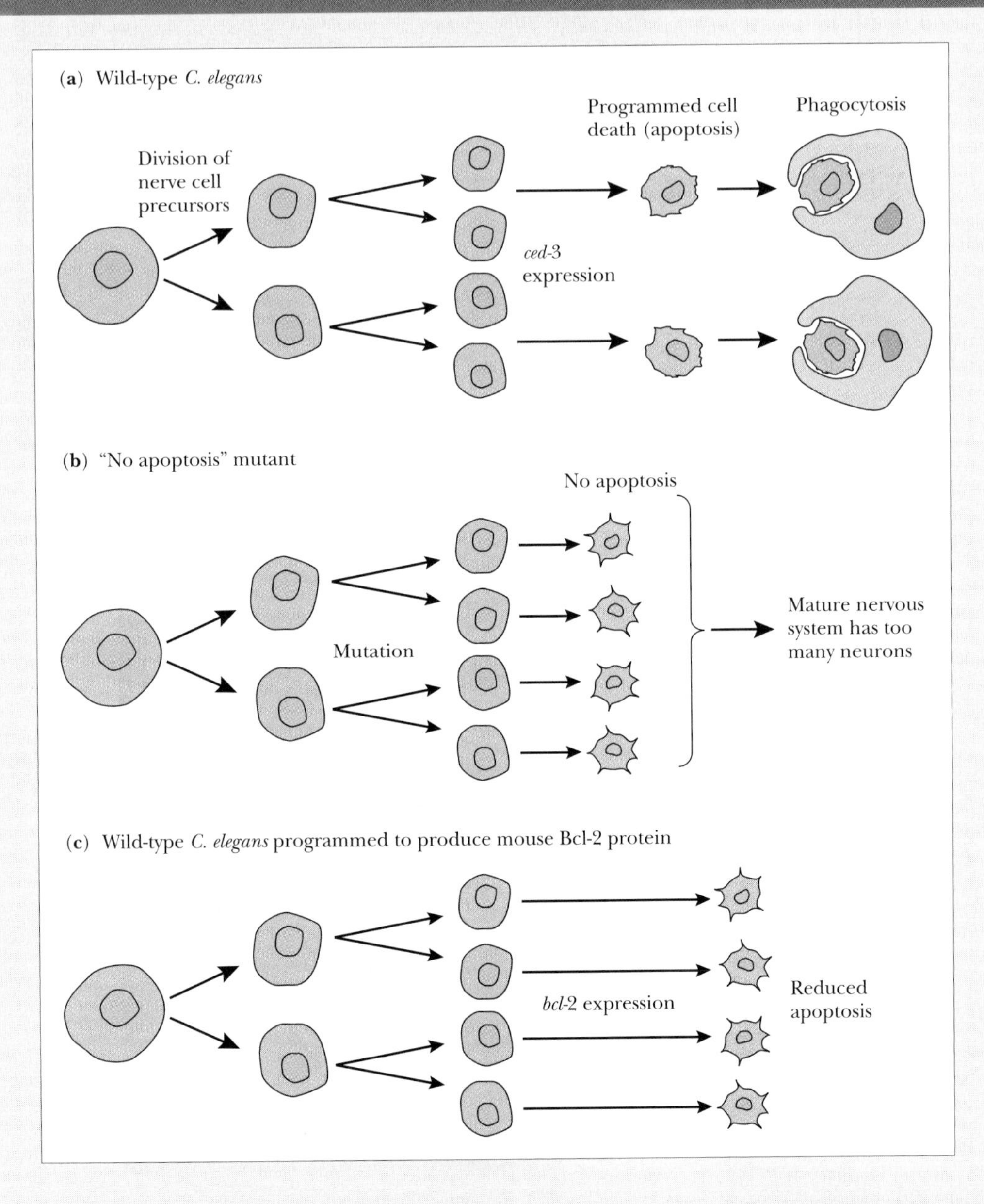

Figure 2 The death of nerve cells during the development of *C. elegans*: **(a)** in wild-type; **(b)** in the *ced*-3 ("cell death abnormal") mutant; **(c)** restoration of normal cell death after injection of *bcl*-2 mRNA; the *C. elegans ced*-3 gene is a distant relative of the *bcl*-2 gene in mammals.

Apoptosis in vertebrates also depends on gene action, at least one of which, called *bcl*-2, corresponds to one of the cell death genes in *C. elegans*. The protein encoded by *bcl*-2 prevents apoptosis of mammalian cells in culture. Surprisingly, however, *C. elegans* that has been genetically engineered to produce mammalian Bcl-2 protein also shows a reduction in programmed cell death. As in the case of the homeo domain genes, the genes that regulate programmed cell death appear to have evolved more than 600 million years ago and serve similar functions in mammals and worms.

Many of the genes that activate programmed cell death are also involved in regulating the cell cycle. One protein, p53, for example, accumulates in cells whose DNA has been damaged. The accumulated p53 stops the cell cycle in G_1, before it enters S, the DNA-synthesizing phase. The p53 also stimulates the cell to undergo apoptosis, preventing the further division of a cell with damaged DNA. Because changes in DNA underlie most cancers, the action of p53 serves as an emergency brake on potential cancer cells, directing them to the apoptosis pathway instead of allowing them to proliferate. Mutations in the *p53* gene make this braking system fail, often leading to unregulated cell growth. Such mutations are present in more than 50% of all human cancers.

webs die, according to a species-specific genetic program, so human digits are separated from one another. In the nervous system, too, programmed cell death helps establish the final "wiring."

After birth, development seldom involves the kind of cell movements that occur during organ formation. Cells do, however, continue to grow, multiply, and specialize. Most postnatal development depends on the increase in size in already formed structures. Typically, an organism's growth begins slowly, accelerates, then slows and stops. Different species have different growth characteristics.

In addition to the differences in growth among species, parts of the body differ in rates of growth. A baby has a proportionally larger head than an adult. Genes somehow control the growth rates of individual body parts.

The patterns of organ formation and embryonic organization are similar in all vertebrates. Early embryos of different species are difficult to tell apart. To a large extent, subsequent differential growth and cell death distinguish vertebrates from one another, as in the case of the loss of webbing between the digits. These immensely important differences, however, happen only in the later stages of development.

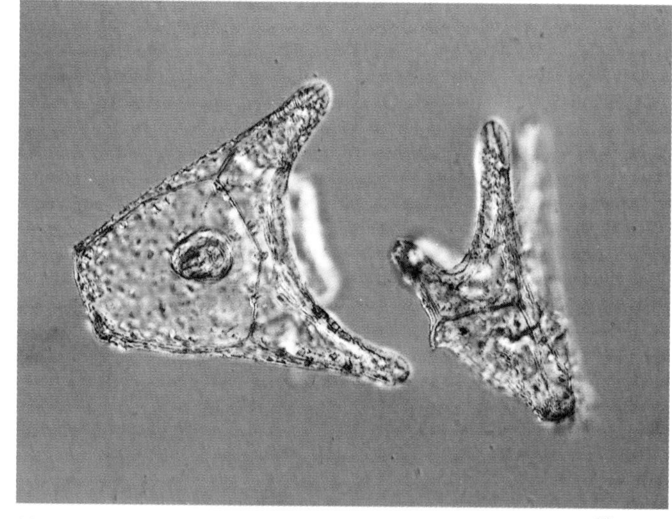
(a)

(b)

Figure 20-15 The pluteus is the larval stage of a sea urchin. **(a)** Pluteus and **(b)** adult of the common sea urchin, *Strongylocentrotus purpuratus.* *(a, David Fromson, California State University, Fullerton; b, Gregory Ochocki/Photo Researchers)*

Metamorphosis Converts a Larva into an Adult

A larva is an immature form that does not resemble the sexually mature adult. The larva is usually the first stage at which an animal has an independent life, able to feed on its own. A frog embryo, for example, develops into a larva called a tadpole, which is aquatic and has a fish-like appearance. It propels itself with a large tail that acts as a fin, it obtains dissolved oxygen from water through its gills, and it subsists on a vegetarian diet. The adult, in contrast, is distinctly frog-like. It has two large hind legs, suitable for jumping; it has no gills and breathes air with its lungs. It is not a vegetarian but consumes flies and other attractive bits of flesh, using its long tongue and big mouth.

Larvae are so different from adults that biologists have sometimes thought that they are different species. The fish-like axolotl, for example, inhabits lakes near Mexico City and has been considered a table delicacy since Mayan times. Only in 1920, when biologists induced metamorphosis in the laboratory, did they realize that the axolotl was actually the larval form of a salamander.

Metamorphosis, the change of form from larva to adult, is more common among invertebrates than vertebrates, especially among the insects. In the sea urchin, the first independent form is a larva, with an appearance and life style totally different from those of the adult. The sea urchin larva, illustrated in Figure 20-15a, is called a *pluteus* [Latin = easel, which it slightly resembles]. The pluteus is mobile and elongated, moving easily on ocean currents. The adult sea urchin (Figure 20-15b) is spiny and more symmetrical, slowly grazing on the bottom of the sea.

The changing of a caterpillar into a butterfly is one of the most spectacular examples of metamorphosis. Many insects, including the beautiful Monarch butterfly, the more prosaic *Drosophila,* as well as fleas, flies, beetles, wasps, moths, and other butterflies, are said to undergo **complete metamorphosis,** meaning that there are distinct larval, pupal, and adult forms. Few adult tissues arise from the differentiated cells of larval tissue. Instead, most of the body parts of the adult (or **imago**) derive from 19 initially unspecialized **imaginal discs,** illustrated in Figure 20-16.

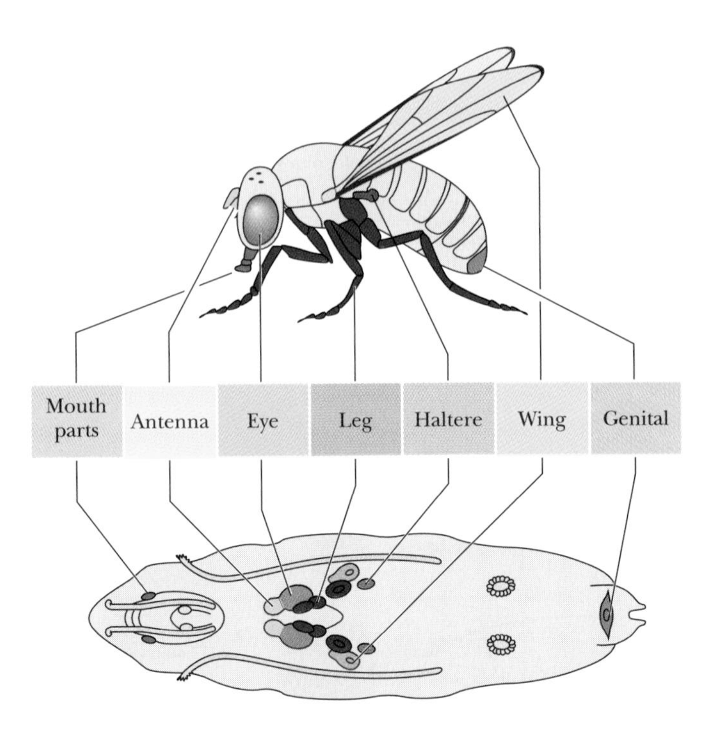

Figure 20-16 The imaginal discs of *Drosophila*. Each of the discs give rise to specific body parts, as indicated.

These groups of larval cells are set aside for later development. Other precursor cells in the larva give rise to muscles, nerves, and other adult structures.

During the development of insects that have complete metamorphosis, the embryo hatches into a wormlike larva (commonly called a grub, a maggot, or a caterpillar). The larva may go through several stages of increasing size before entering an inactive phase, called a **pupa,** which may be enclosed in a *cocoon*. During the pupal stage, almost all the tissues of the larva are digested. The cells of the imaginal discs multiply, move about, and differentiate to form the tissues and organs of the adult. Finally, the pupa emerges from its cocoon as a sexually mature adult.

Aging Depends on the Action of Both Genes and Environment

Human aging is a major focus of medical concern. Many measures of vitality decline with age—fertility, heart output, brain weight, lung capacity. In addition, extracellular proteins of connective tissues lose their earlier resilience. Arteries harden, skin gets less elastic, and joints stiffen.

Aging also occurs at the level of cells. Cells taken from a young organism and grown in tissue culture can divide more times than can cells taken from an older organism. As they get older, cells appear to lose their capacity for proliferation.

Part of the aging process must depend on genes. For example, genetically different strains of laboratory mice may differ radically in their lifespans. Environmental factors, such as nutrition, may also contribute to aging. Genetically determined differences in aging may involve differences in responses to environmental factors, such as diet. Like other aspects of development, aging depends on the interaction of genes and environment.

SPERM AND EGGS ARE HIGHLY SPECIALIZED TO ACCOMPLISH FERTILIZATION AND THEN TO BEGIN EMBRYONIC DEVELOPMENT

We began our discussion of development with the cleavage of the zygote. Now we go back and look at the events that led to the formation of the zygote—the formation of gametes (sperm and eggs) and fertilization. Our discussion illustrates the importance of cell specialization in setting up a program for development.

Gametes are formed in specialized organs called **gonads** [Latin, *gonas* = a primary sex organ], the *testes* in males, the *ovaries* in females. **Primordial germ cells,** the precursors of both sperm and eggs, usually arise early in development. They and their descendants remain separate from the complex changes that produce all of the cells, the somatic cells, in the rest of the body, the **soma.** (See Chapter 10, p. 279.) The primordial germ cells migrate to the developing gonads, where they undergo meiosis and cell specialization. Sperm and eggs both promote the eventual fusion of two haploid genomes, but they do so in strikingly different ways. Except for meiosis, the development of germ cells into sperm cells is radically different from their development into eggs.

Sperm Cells Are Specialized for Rapid Movement

Primordial germ cells in males undergo several mitotic divisions in the testes before beginning meiosis. As illustrated in Figure 20-17, **spermatogonia** are cells, descended from the primordial germ cells, that have become committed to forming sperm cells but can still undergo mitosis. Cells that can no longer undergo mitosis and that enter meiosis are called *primary spermatocytes*. After the first division of meiosis, primary spermatocytes become *secondary spermatocytes*, which have the haploid number of chromosomes but with two chromatids per chromosome. After the second meiotic division, each chromosome consists of a single chromatid, and the cells are called *spermatids*. The development of spermatogonia into mature sperm is called **spermatogenesis.**

Spermatids then become further specialized (differentiated) to become sperm (or **spermatozoa**). They become highly elongated, with widths of a few micrometers and lengths up to 100 μm (depending on species). The nucleus becomes highly condensed, and it occupies only a small fraction of the length of the mature sperm. Most

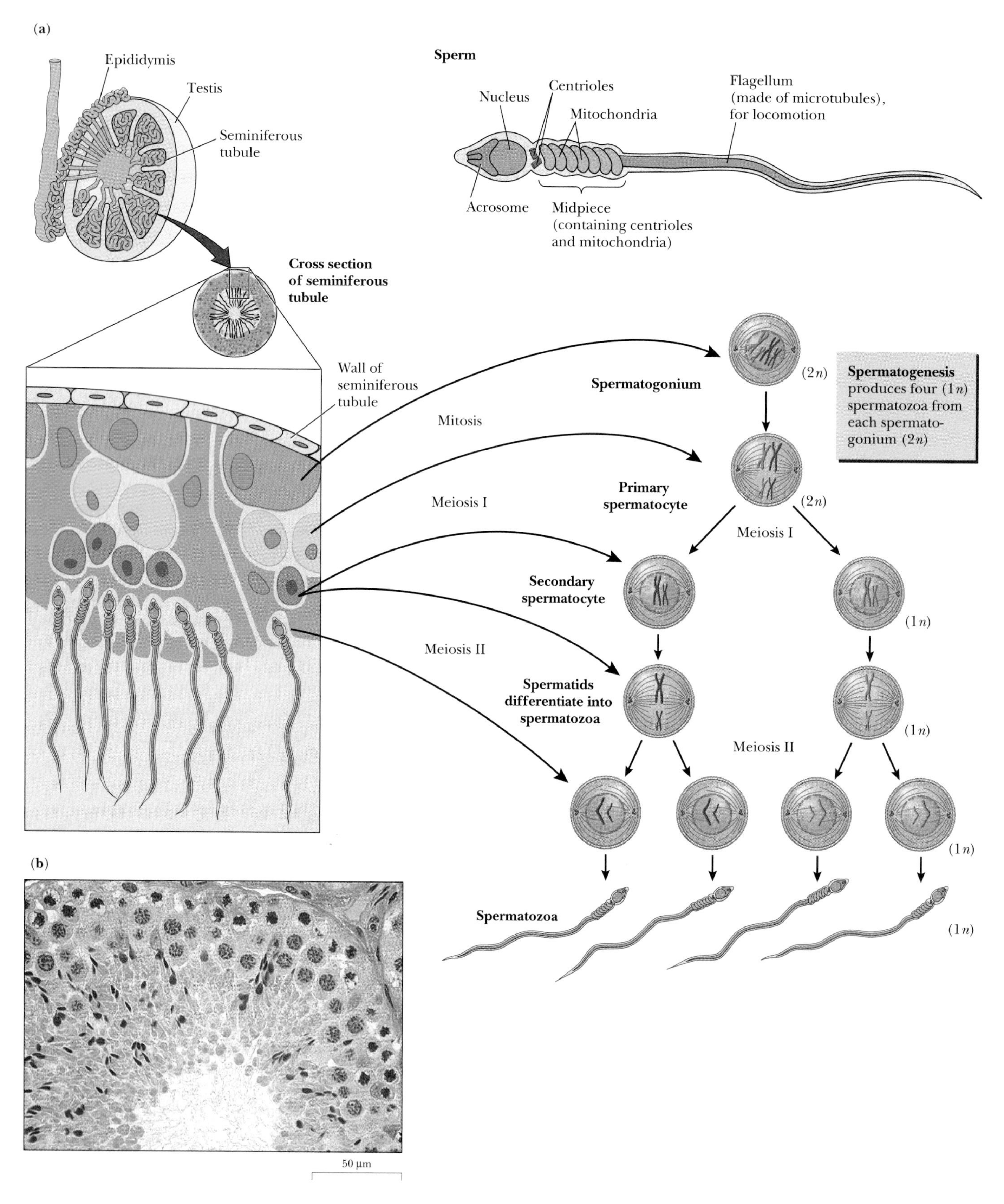

Figure 20-17 **(a)** Spermatogenesis, the development of sperm in the seminiferous tubule of human testis, showing mitosis of spermatogonia, meiosis, spermatocytes, spermatids, and spermatozoa. **(b)** Cross section of testis, showing the different stages of spermatogenesis. *(b, Fred Hossler/Visuals Unlimited)*

of the length is taken up by a powerful flagellum, which propels the sperm in its journey to the egg. Between the nucleus and the flagellum is the *midpiece*, a region that contains densely packed mitochondria, the sources of ATP for the active swimming toward the egg. The sperm contains a special structure at its front end—the **acrosome,** a specialized lysosome whose enzymes eventually allow the sperm to enter the egg. (See later in this chapter.)

Spermatogonia continue to undergo mitosis, forming more spermatogonia, so they provide a continuing supply of sperm from puberty on. In humans, the process of development from spermatogonia to fully differentiated sperm takes about 2.5 months. Each hour about 200 million sperm mature. Each ejaculation releases about 200 million sperm. Unejaculated sperm are either absorbed by other cells in the testes or passed into the urine.

The Specializations of Eggs Allow Them to Provide Energy, Building Materials, and Machinery for the Developing Embryo

At a molecular level, eggs are even more complex than sperm, and in some ways more important in establishing the new embryo. Eggs not only carry the genetic contributions of mothers but also provide new embryos with building materials, energy sources, cell machinery, messenger RNAs, and organizational guides. **Oogenesis,** the production of mature eggs, like spermatogenesis, involves mitosis, meiosis, and cell specialization. In contrast to spermatogenesis, however, cell specialization begins during meiosis, rather than after its completion, as illustrated in Figure 20-18.

Like sperm, eggs derive from primordial germ cells that have made their way to the gonads, in this case the future ovaries, which will provide the environment necessary for oogenesis. After the germ cells arrive in the ovary and become committed to the production of eggs, they are called **oogonia.** Oogonia, like spermatogonia, can undergo mitosis to make more oogonia, or they can differentiate into a *primary oocyte,* which can undergo meiosis, but not mitosis. After the first division of meiosis, primary oocytes become *secondary oocytes.* After the second meiotic division, the egg precursor is called an *ootid.* The ootid then develops into a mature egg, or **ovum** (pl., **ova**).

In sea urchins and many other species, a female produces ova from oogonia throughout her reproductive life. On the other hand, in a human female, oogonia stop dividing during the eighth month of gestation. Most of the 7 million oogonia then present die, and all the rest become primary oocytes and begin meiosis. They stop, however, in the middle of prophase I, where they stay until the onset of puberty, some 12 years or so later. Several million primary oocytes are still present at birth, but only about 500,000 survive to adolescence. Most of these do not complete meiosis. Some, however, complete meiosis and cell specialization, at an average rate of one per 28 days, beginning at puberty. The maturation of ova ceases at *menopause,* usually at about 50 years of age, so that a woman typically produces fewer than 400 mature ova in her entire lifetime.

In mammals, the maturation of an ovum is closely tied to events in the uterus. Hormones secreted by the ovaries act on the lining of the uterus to prepare it to receive and nourish the embryo that would result from fertilization. Hormones made in the ovaries and in the pituitary gland coordinate the maturation of the ovum and the specialization of the uterus during the menstrual cycle. When fertilization does not occur, the uterus sheds, in the menstrual flow, the "nest" built in its lining.

Ova are always relatively large cells, but they vary more widely in size than any other cell type. The ova of sea urchins and humans are each about 100 μm, in diameter. The ova of frogs are about 1 mm in diameter—meaning that their volume is 1000 times that of sea urchins or humans. The largest egg known is that of the ostrich, some 10 cm in diameter. Most of the volume of these large eggs is occupied by yolk. Because mammals provide food for the developing embryo through the placenta, the eggs of mammals have little yolk.

Oogenesis is a particularly important example of cell specialization and the continuation of the species depends on its success. In *Drosophila,* for example, about 10% of the genes are concerned directly with oogenesis. Among the processes that must occur are the accumulation of yolk, the production of ribosomal proteins and RNAs to be used after fertilization, the synthesis of mRNAs to be translated after fertilization, and the distribution of regulatory factors that cause different patterns of gene expression in different regions of the embryo.

Fertilization, Both a Genetic and a Developmental Event, Depends on the Specialized Properties of Both Sperm and Egg

The genetic significance of fertilization is that haploid ovum and haploid sperm come together to form a diploid zygote, bringing together genetic information from the maternal and paternal genomes. (See Chapter 10.) Fertilization is also the prelude to the life of the new organism. After fertilization, the previously quiescent egg begins intense metabolic activity, using energy at a much higher rate than the egg before it is fertilized. The embryo soon begins to make new proteins and to enter the cleavage stage, as we have discussed.

Sperm and Eggs Have Special Adaptations That Contribute to Successful Fertilization

Sea urchins, like many other marine invertebrates, shed their sperm and eggs into tidepools. Successful fertilization requires that the sperm find the eggs of the right species, penetrate the eggs' gelatinous coating, and inject

Figure 20-18 Oogenesis, the development of eggs, showing oogonium, mitosis, primary oocyte, meiosis, secondary oocyte, polar bodies, and ovum.

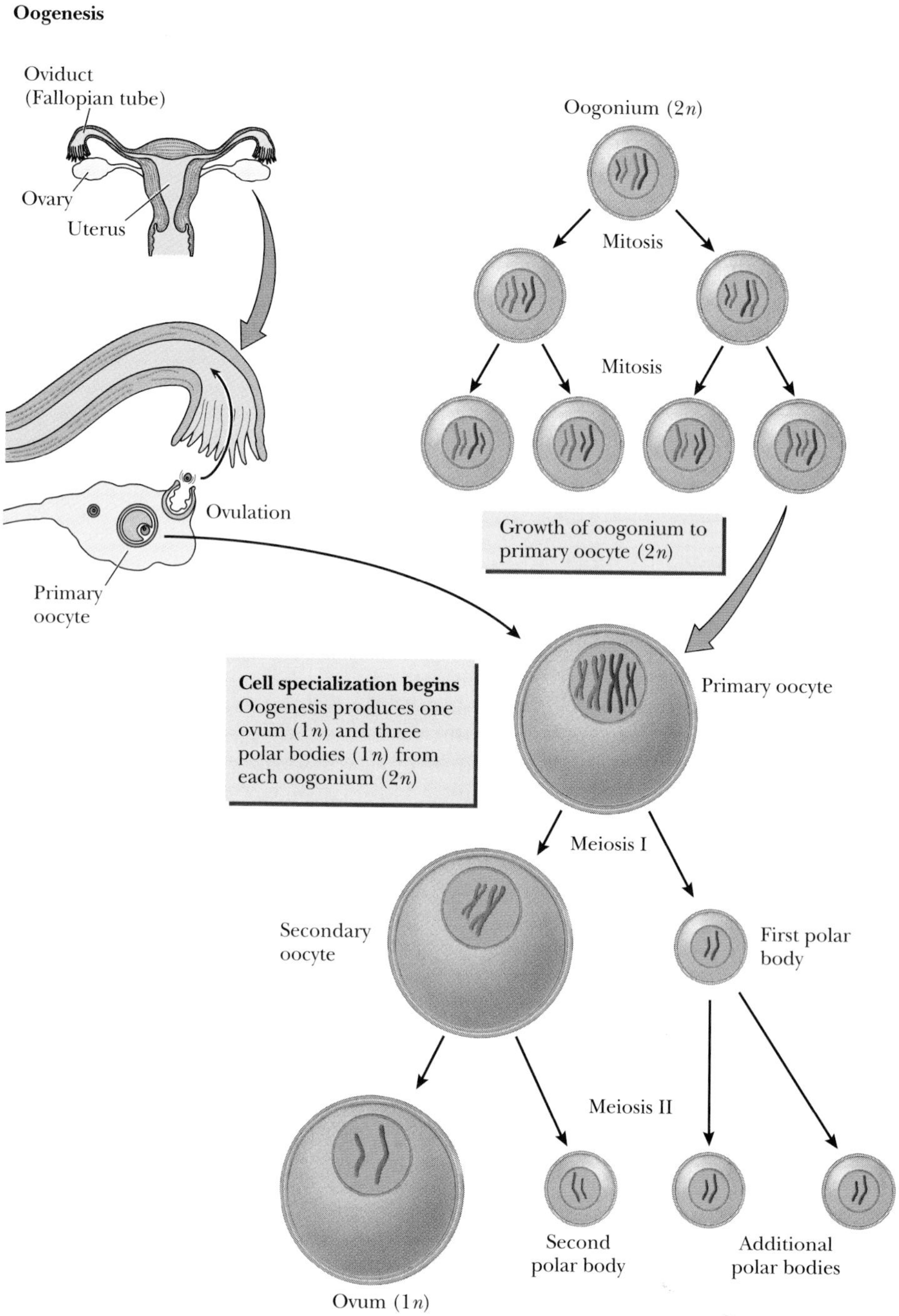

Oogenesis

Oviduct (Fallopian tube)

Ovary

Uterus

Ovulation

Primary oocyte

Oogonium ($2n$)

Mitosis

Mitosis

Growth of oogonium to primary oocyte ($2n$)

Cell specialization begins
Oogenesis produces one ovum ($1n$) and three polar bodies ($1n$) from each oogonium ($2n$)

Primary oocyte

Meiosis I

Secondary oocyte

First polar body

Meiosis II

Second polar body

Additional polar bodies

Ovum ($1n$)

their genetic material. In sea urchins, eggs actually produce a substance that attracts sperm of the same species. Also in sea urchins (and other species as well), contact with the jelly surrounding the egg activates the acrosome of the sperm. Two processes then follow rapidly, as illustrated in Figure 20-19a: (1) The acrosome breaks down, releasing enzymes that digest a path for the sperm through the jelly to the egg itself; and (2) the sperm produces a long extension, composed of actin filaments, which makes initial contact with the surface of the ovum. On the surface of this extension are protein molecules that bind specifically to the ova of the same species.

After the sperm binds, the surface of the egg forms a *fertilization cone,* an extension that surrounds the sperm head, as illustrated in Figure 20-20b. The fertilization cone pulls the sperm into the egg, after which the sperm membrane fuses with that of the egg. The mitochondria and the flagellum of the sperm stay at the egg's surface, where they eventually disintegrate. Meanwhile, the sperm's nucleus and centriole move toward the egg's own nucleus. Finally the two nuclei fuse to form the diploid nucleus of the zygote.

Each egg must have mechanisms to prevent **polyspermy,** fertilization by more than one sperm. Because

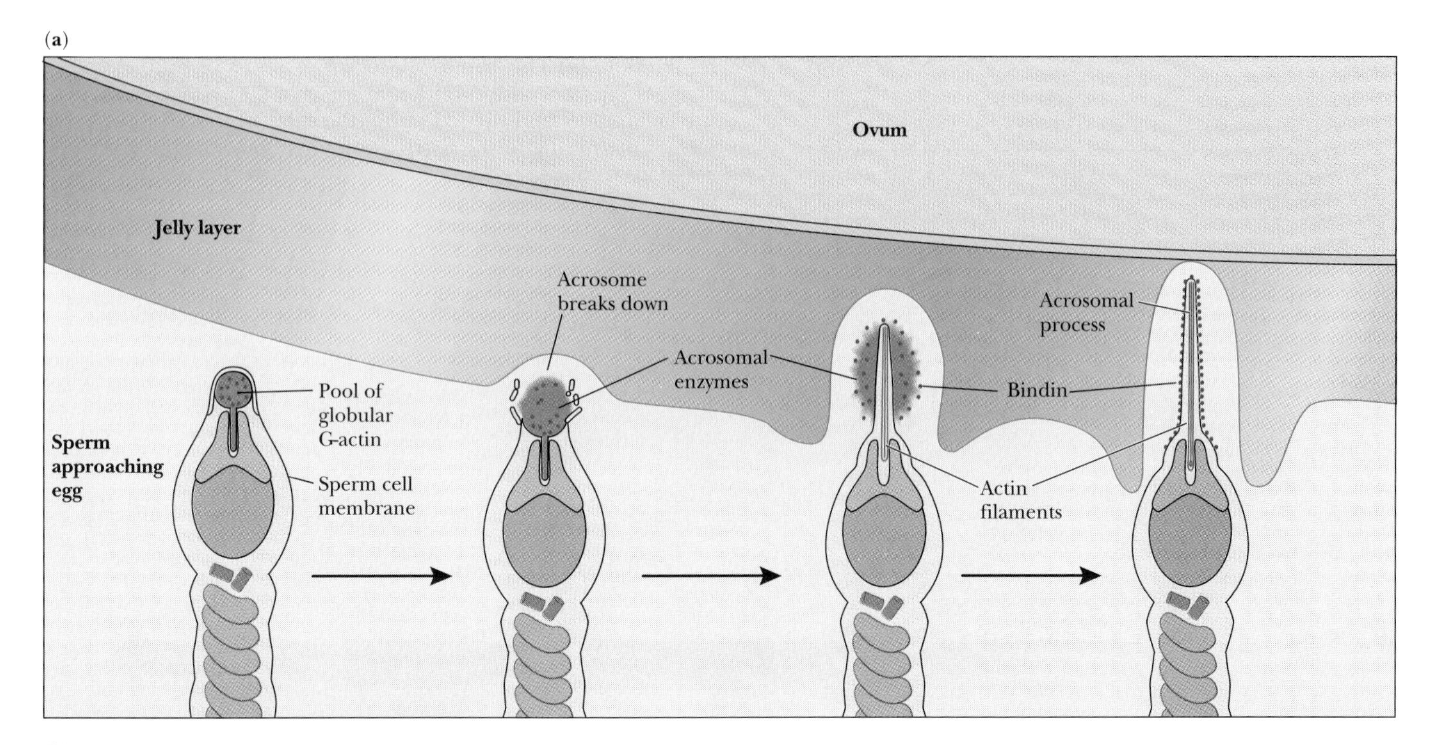

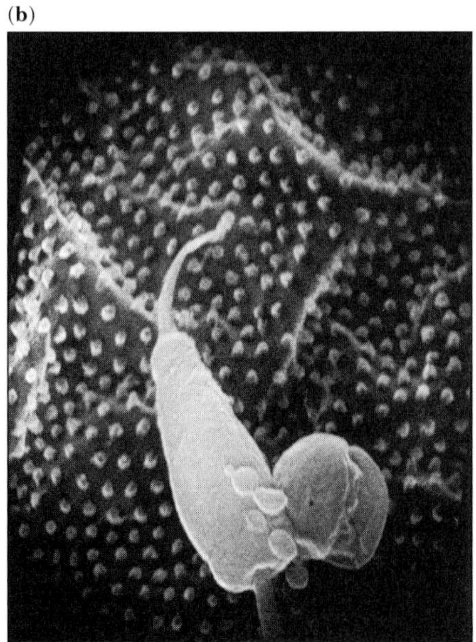

Figure 20-19 Fertilization in sea urchins: **(a)** contact initiates the sperm's acrosome reaction and the extension of actin filaments; **(b)** the beginning of sperm entry. *(b, Gerald Schatten)*

polyspermy leads to an abnormal number of chromosomes, it results in embryonic death or abnormal development. Sea urchin eggs have two ways of preventing polyspermy, illustrated in Figure 20-20. The first of these, the *fast block* to polyspermy, begins within a second of fertilization, as shown in Figure 20-20a. The fusion of the sperm and egg membranes triggers a change in the voltage across the membrane. This change prevents a second sperm from fusing and injecting a nucleus.

The membrane changes associated with the fast block also initiate a change in pH and a wave of increased free calcium ions within the egg cytosol, which propagates across the egg after fertilization. The initial increase in calcium ions is localized to the point of penetration and depends on the influx of calcium. The wave of increased calcium ion concentration, however, depends on calcium ions stored within the egg's endoplasmic reticulum. The release of these calcium stores depends on the coordinated stimulus of calcium ions and inositol trisphosphate. (See Chapter 19, p. 528.)

The wave of calcium ions initiates the slow block to polyspermy, which begins about 20 seconds after fertilization. This block, illustrated in Figure 20-20b consists of raising a **fertilization membrane**, which prevents other sperm from reaching the surface of the egg.

Maternal mRNAs Direct Protein Synthesis in the Newly Fertilized Sea Urchin Egg

Even before the sperm and egg nuclei fuse, profound changes begin within the egg. Within a minute after the binding of a sperm, the egg increases its rate of oxygen consumption, and within ten minutes protein synthesis increases dramatically, as illustrated in Figure 20-21.

One interpretation of the increased protein synthesis is that the newly formed zygote nucleus has begun to transcribe new mRNAs. But the same burst of protein synthesis occurs even in the presence of a drug that prevents

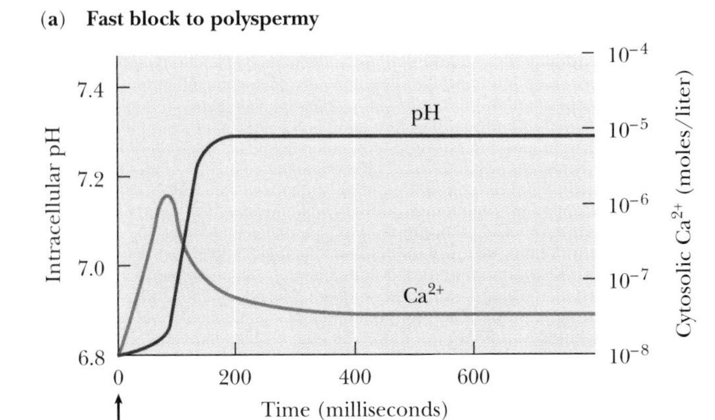

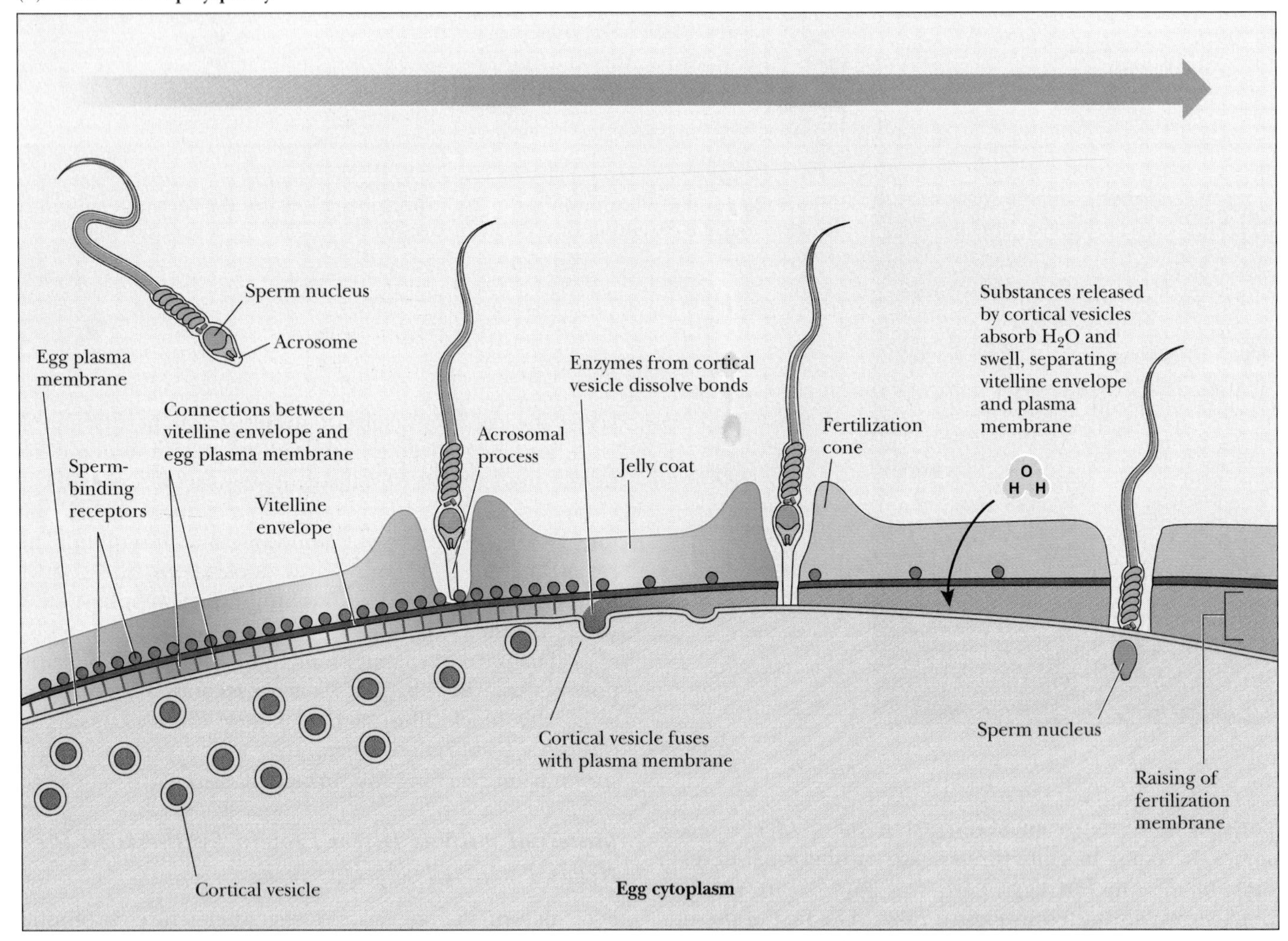

Figure 20-20 Blocks to polyspermy in sea urchin fertilization. **(a)** The "fast block" occurs within a second and changes the Ca^{2+} concentration in the cell, the voltage across the membrane, and the pH within the cell; **(b)** the "slow block," the raising of the fertilization membrane, begins after about 20 seconds.

transcription. Because sperm do not carry mRNAs, the mRNAs that direct this early burst of synthesis must be **maternal mRNAs** that were already present in the egg before fertilization but "masked" by proteins so that they are not translated. These mRNAs were transcribed during oogenesis from genes in the maternal genome. Fertilization not only activates the egg's metabolism but also "unmasks" previously stored messenger RNAs.

Embryos in which transcription has been blocked develop remarkably well through cleavage but stop before germ layer formation. Gastrulation requires the synthesis of new mRNAs, transcribed from the embryo's own DNA.

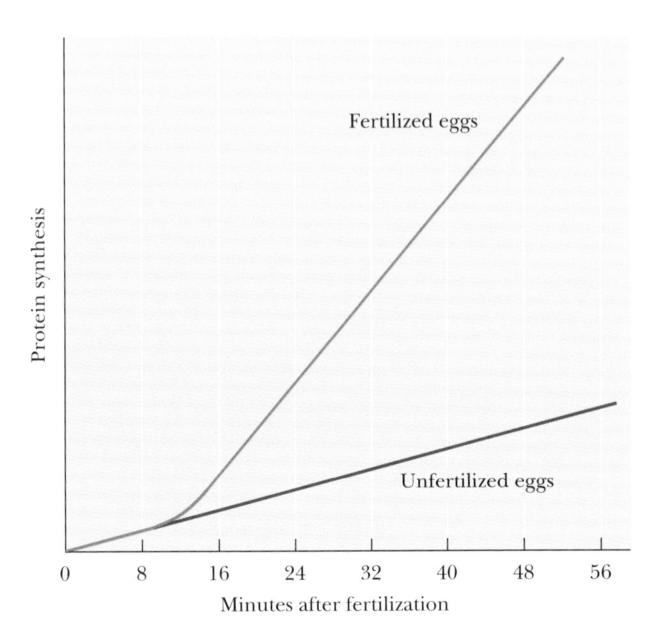

Figure 20-21 Protein synthesis in sea urchin embryos increases immediately after fertilization.

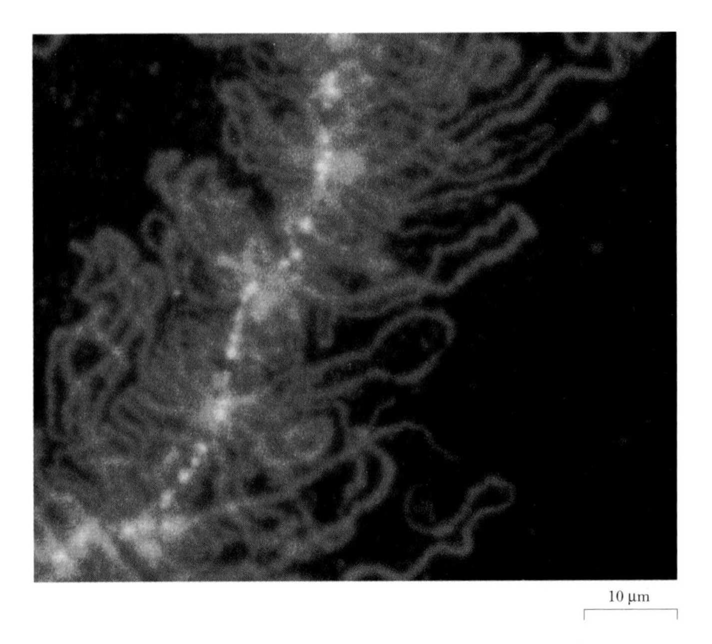

Figure 20-22 Oocytes accumulate mRNAs during oogenesis. This accumulation is particularly dramatic in amphibian oocytes, in which the actively transcribed chromosomes resemble the "lampbrushes" once used to clean gas lamps. *(M. B. Roth and J. G. Gall)*

Germ layer formation requires the onset of distinctive gene activity.

The eggs of other species also contain maternal mRNAs. Amphibians, with their relatively huge eggs, have a particularly large store of mRNAs, which accumulate during the extended period of prophase I, as illustrated in Figure 20-22. Mammalian eggs, too, contain maternal messenger RNAs, but embryos begin to depend on new mRNAs, produced by the embryo itself, sooner than in amphibians and sea urchins.

Mammalian Fertilization Shares Many Characteristics with Fertilization in Sea Urchins

Both sea urchins and mammals share the need for preventing polyspermy, although the context of fertilization differs greatly. The internal fertilization of mammals requires fewer gametes, with a smaller but still significant risk of polyspermy. Mammals appear to rely primarily on the slow rather than the fast block to polyspermy, with the egg also raising a membrane after fertilization. Similarly, mammalian sperm undergo an acrosome reaction roughly similar to that of sea urchins. As in sea urchins, a component of the egg's protective layer activates this reaction. The enzymes released help digest a route for entry to the ovum.

The actual event of fertilization is orchestrated differently in mammals than in invertebrates. In humans only one egg usually matures per month. The precursor of the mature egg is stopped in the middle of prophase of meiosis I, as a primary oocyte. Just before it is to be released, the primary oocyte completes the meiosis. As in many other species, this division is extremely unequal in humans, with most of the materials of the oocyte distributed to a single secondary oocyte, the cell that will become the ovum. The smaller daughter cell becomes a **polar body**, which cannot be fertilized and which soon disintegrates. The second division of meiosis does not occur until after contact of sperm within the reproductive tract. The second division, also asymmetric, produces an ootid and a second polar body, like the first polar body, a cell with no genetic future.

In mammals, the female reproductive tract plays at least two important roles in fertilization: (1) Cilia on its inner surface move the oocyte down toward the approaching sperm; and (2) something within the tract causes the sperm to mature so that they can undergo the acrosome reaction. Sperm not exposed to fluids from the female reproductive tract cannot fertilize an egg.

Fertilization is the beginning of development. Successful fertilization depends on many adaptations of sperm and eggs, as well as of the reproductive tract and of the behavior of the animals.

CELLS BECOME INCREASINGLY RESTRICTED IN POTENCY AS DEVELOPMENT PROCEEDS

Driesch's experiment (described on p. 541) showed that each cell of a two-cell embryo has the ability to develop into a whole (rather than a half) sea urchin. With this and similar results in mind, embryologists have come to distinguish the **fate** of a cell—what it becomes during development—from its **potency**—what it could become if allowed to develop in another environment. In the case of the two cells of Driesch's sea urchin embryos, the fate of each of the cells was to become the right or left half of the sea urchin, but each had the potency to become a whole organism. We may also say that each cell of the two-cell embryo was **totipotent,** meaning that it could—under the right conditions—develop into a whole organism.

What Kinds of Experiments Distinguish Potency and Fate?

At early stages of development and for relatively simple organisms, we may follow the fates of individual cells just by watching and photographing. For example, by observing the movements of cells during sea urchin development, developmental biologists have determined which cells of the blastula give rise to endoderm, which to mesoderm, and which to ectoderm.

For organisms with more complicated cellular movements, however, looking alone is not enough. Gastrulation in a frog, a bird, or a mammal is too complicated to follow by visual observation alone. Embryologists therefore devised ways of marking cells at an early stage so that they could be identified later. These identification tags include particles of carbon, dyes, and radioactively labeled molecules. By following the movements of cells after such marking, embryologists have been able to construct *fate maps* of the early embryo. These maps show what happens to each of the cells, say of the blastula, during normal development.

Defining the potency of a cell is much more complicated. It requires that a cell with a known fate be put into a new environment. Growing such identified cells in different tissue culture conditions provides one way of testing potency because the new, artificial environment may allow cells to develop differently from their normal fate in the embryo.

Another way of examining the potency of embryonic cells is to transplant groups of cells from one embryonic environment to another. This is a direct extension of the strategy, used by both Roux and Driesch, of manipulating the environment of the embryo itself. Because amphibian embryos are relatively large and accessible, frogs and newts have been the subject of many studies of developmental fate and developmental potency. In 1918 Hans Spemann at the University of Freiburg in Germany performed a most influential analysis of developmental potency.

Spemann was studying the development of the newt, using two closely related species with different pigmentation. By transplanting pieces of the pigmented embryo into an unpigmented embryo, he could distinguish easily between structures formed from the host cells and those derived from the transplanted cells. Spemann knew which cells of the blastula and gastrula would eventually give rise to neural plate and which to skin; naming these cells according to their future developmental fate, he called them **presumptive** neural plate and presumptive skin.

Spemann transplanted presumptive neural plate into the area of presumptive skin, as sketched in Figure 20-23. If he performed the transplantation at the early gastrula stage, he found that presumptive neural tissue instead formed skin. That is, transplanted cells from the early gastrula developed according to their new environment, meaning that their developmental potency was greater than their developmental fate. At the early gastrula stage, the potency of the transplanted cells was not yet restricted.

If, however, Spemann waited until the late gastrula stage to perform the transplantation, he obtained a different result. Then the presumptive neural plate still developed a neural plate, even in the new environment. At this later stage of development, the potency of the cells had become committed to their normal developmental fate. Although the cells do not at this point appear any different from presumptive ectoderm, they are already **determined;** that is, they can no longer develop in accordance with new environmental signals.

Nuclear Transplantation Experiments Have Supported the Hypothesis That Nuclei Are Totipotent, Even When Cells Are Not

Spemann's experiment raises two important questions: (1) How does a cell's environment influence its development? (2) How does a cell become determined, so that its developmental fate becomes independent of its environment?

Modern biologists think of determination as the process of establishing which genes will be expressed and which will not. Similarly, biologists think of morphogenesis as achieving the appropriate arrangement of the molecules of the cytoskeleton and differentiation as the process of accumulating a particular set of gene products.

Both the process of determination and that of differentiation are influenced by environmental cues—signaling molecules that influence cell specialization and gene expression. Spemann's transplantation experiment showed that the cues necessary for the determination of the neural plate depend on cell rearrangements during gastrulation. Subsequent changes in cellular environment lead to further changes in cell shape and gene expression during neural development.

An easy explanation for determination would be that, during development, cells progressively lose genes. If this hypothesis were true, then the inability of presumptive neural cells to form skin would be due to their loss of genes needed to make skin. In fact, some organisms do lose chromosomes during development. The question is whether the progressive loss of chromosomes, or of indi-

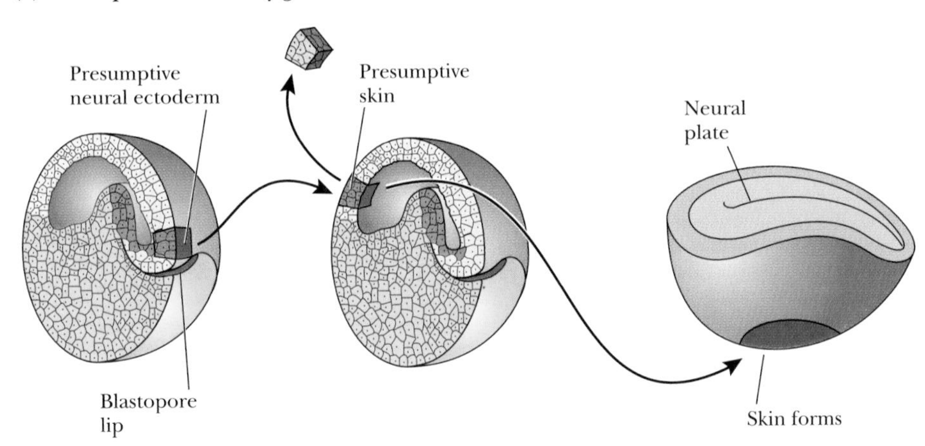

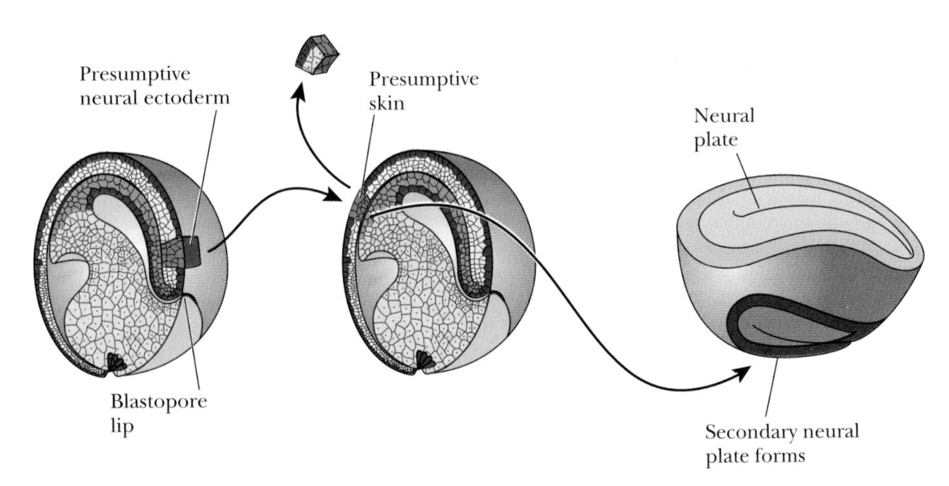

Figure 20-23 Developmental fate and developmental potency, established by marking experiments. Hans Spemann established a fate map for the newt embryo by transplanting tissue from a heavily pigmented species into the same place in a nonpigmented embryo. **(a)** Spemann transplanted presumptive neural tissue from an early gastrula into the region of presumptive skin and found that the transplanted tissue developed, according to its new environment, into skin. **(b)** He transplanted presumptive neural tissue from a late gastrula into the region of presumptive skin and found that the transplanted tissue developed, according to its old environment, into neural tissue.

vidual genes, is the mechanism for determination and cell specialization.

To approach this question experimentally, developmental biologists have asked the general question, "Do cells lose some of their genetic capacity as their developmental potency decreases, that is, when they become determined?" If they do, we would have a ready explanation for determination and differentiation. If they do not, but maintain their genetic capacity, then we would have to think about the problems of determination and overall development in terms of the regulation of gene activity. For the most part, differentiated cells of an organism have the same genes. Ultimately, then, we have to understand development in terms of the activation and repression of genes.

In the 1950s, developmental biologists began to approach the question of gene loss experimentally. Robert Briggs and Thomas King devised a method, illustrated in Figure 20-24a, for **nuclear transplantation,** in which they removed the nucleus of a frog egg and replaced it with the nucleus of another frog cell from a later developmental stage. The egg then developed under the control of the transplanted nucleus. This experiment, similar to the transplantation experiments performed by Spemann, addressed the issue of developmental potency—but now of the nucleus rather than of a group of embryonic cells.

Briggs and King soon discovered that nuclei from a blastula were totipotent; that is, each could direct the development of a whole frog. This finding meant that, by using many nuclei from the same blastula, Briggs and King could produce a group of genetically identical frogs. Because all the cells of the donor blastula were derived from a single cell (the zygote), all the frogs derived from a single blastula are a clone! This experiment has already stimulated a rash of science fiction based on its possible (but unlikely) application to humans.

At least to the blastula stage, frog nuclei are totipotent. But Spemann's experiment suggests that determination may begin only in late gastrulation, so Briggs and King next examined the ability of nuclei at later stages to direct

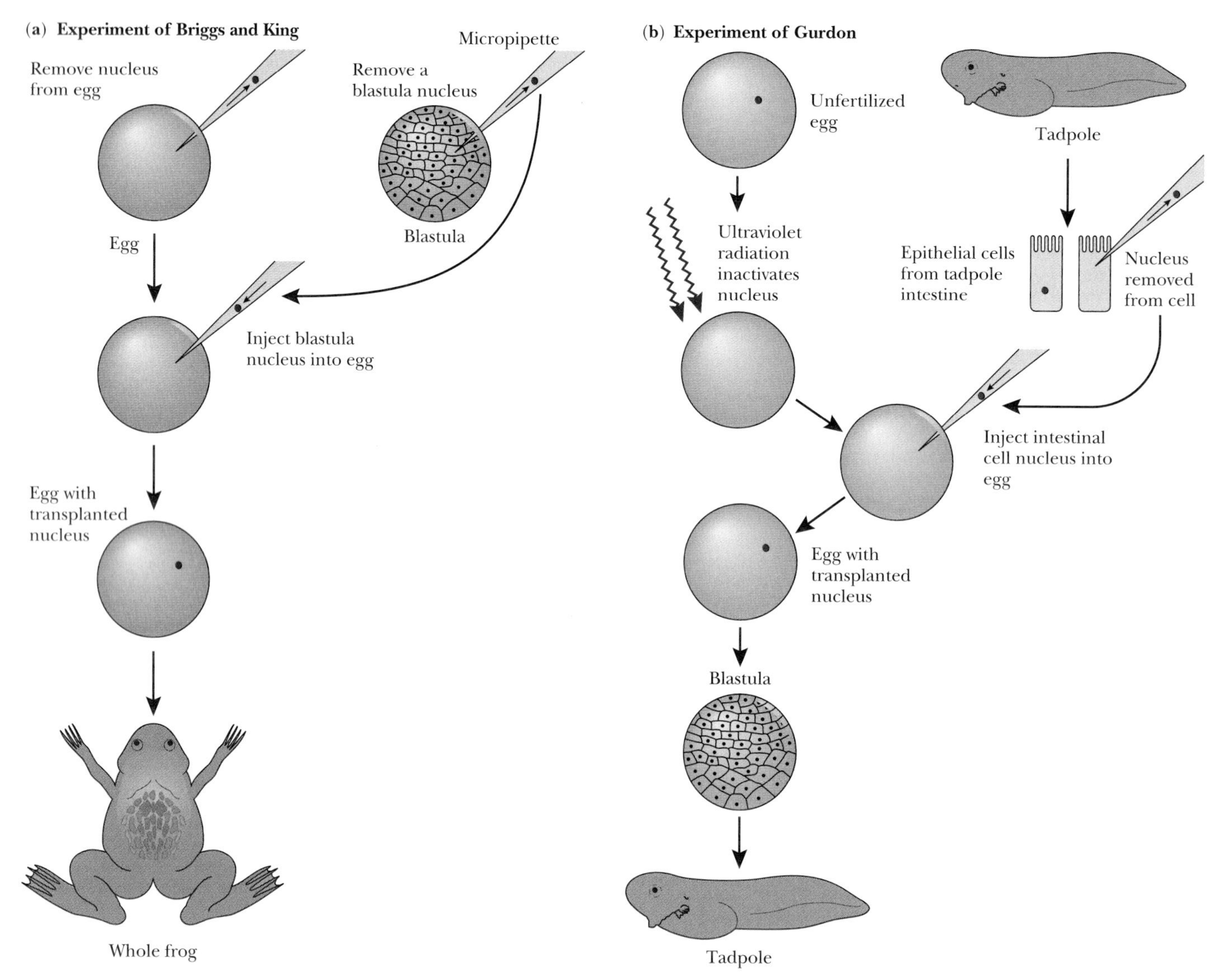

Figure 20-24 Nuclear transplantation experiments have shown that selective gene loss does not underlie determination and development. **(a)** Removal of the egg nucleus and replacement by the nucleus from a blastula cell; the transplanted nucleus can direct the development of an entire frog. **(b)** An alternative method of nuclear transplantation inactivates the egg's nucleus with ultraviolet light; transplantation of an epithelial nucleus can direct the development of a swimming tadpole.

frog development. They found that the potency of the nuclei declined with development; that is, the nuclei became increasingly restricted in what they could do. On the other hand, the developmental potency of a nucleus was always greater than its normal developmental fate.

Briggs and King, working with the American leopard frog, *Rana pipiens,* never produced a swimming tadpole from a nucleus beyond the neurula stage, with the exception of nuclei from the primordial germ cells. They concluded that somatic nuclei gradually lose their capacity to direct complete development.

On the other hand, the English biologist John Gurdon, working with *Xenopus laevis,* the South African clawed frog, and using a slightly different method of nuclear transplantation, found that nuclei from the intestines of tadpoles could generate all types of cells, as illustrated in Figure 20-24b. In some cases, the transplanted intestinal nuclei gave rise to a fertile adult. Gurdon therefore contended that the nuclei of differentiated cells continued to be totipotent. The apparent decrease in potency in the experiments of Briggs and King, Gurdon argued, resulted from the experimental procedure rather than from a molecular change.

The argument is at present unresolved: Biologists do not know whether nuclei remain truly totipotent. We do know, however, that even the nuclei of differentiated cells retain the capacity to express more genes than they normally do. We must therefore consider development to be at least partly a question of gene regulation.

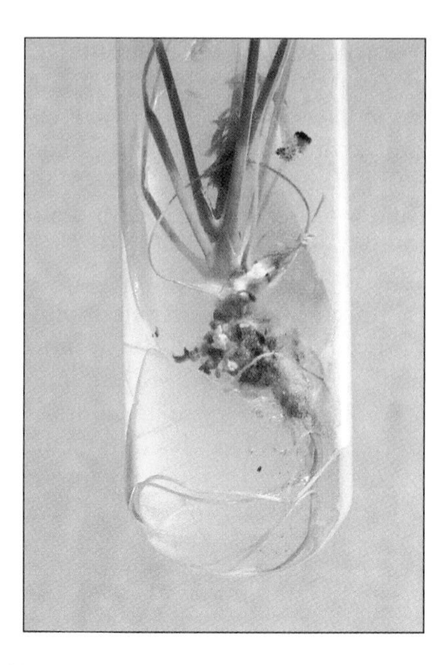

Figure 20-25 A single plant cell, here taken from a carrot root, has given rise to a whole plant, with all types of specialized cells. *(Bruce Iverson)*

A Single Carrot Cell Can Generate a Whole Plant

While this chapter is about animal development, plant cells are wonders of developmental potency, and a particularly dramatic plant experiment sheds light on the subject. Almost anyone who has a houseplant knows how to produce whole new plants from cuttings, and many plants of commercial importance are propagated entirely from cuttings and graftings. (See Chapter 21.) Even a single cell from a fully developed plant can give rise to a whole new plant.

The cloning of a carrot was achieved by F. C. Steward and his colleagues at Cornell University in 1958. From a slice of carrot root, Steward grew colonies of disorganized cells in cultures nourished by coconut milk. Single cells derived from these colonies gave rise to groups of cells that resembled little roots. Steward transferred some of these to an agar plate and found that they could actually produce young plantlets, as illustrated in Figure 20-25. These, in turn, could grow into whole new carrots. The developmental fate of the progenitor cell of these new plants was to be part of the original plant's root transport system. By putting the cell into successively new environments, however, Steward succeeded in demonstrating that it was totipotent. The totipotency of a single cell again points to the dependence of development upon cues from the cellular environment.

Many Cells Maintain Their Determined State Even After Propagation in Artificial Culture

In contrast to the totipotent cells of Steward's carrots, most animal cells have only a limited capacity for differentiation after they are put into culture. When an experimenter puts a piece of tissue into a culture medium, the cells most likely to proliferate are those that have not yet begun their terminal differentiation. Often, the specialization of cells is accompanied by a withdrawal from the cell cycle. Nonetheless, cells from virtually any stage of development generally have only limited developmental potency in culture.

The development of mammalian muscle cells in a culture medium is striking and unusual. During muscle formation, **myoblasts,** the precursors of muscle fibers, fuse to make large multinucleated cells. These fused cells make muscle proteins and soon even begin to contract in the culture dish. For example, myoblasts from the leg of a rat embryo, such as those shown in Figure 20-26, grow for a while in culture as dividing single cells; then they special-

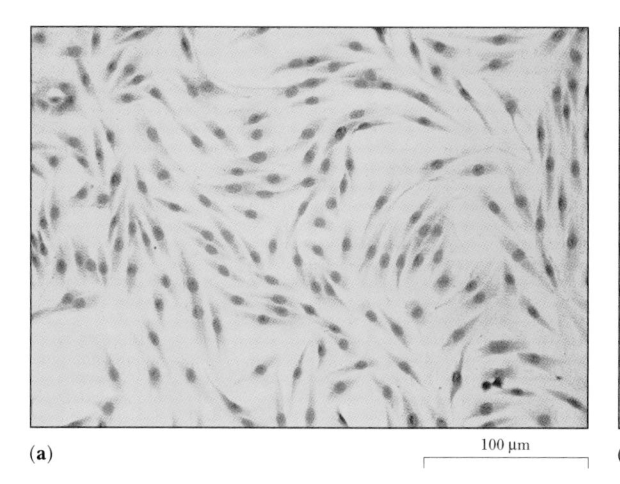

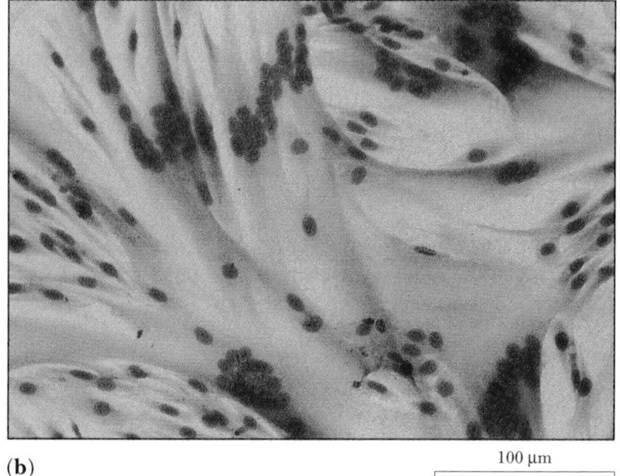

Figure 20-26 Animal cells often maintain their determined state in culture. **(a)** Myoblasts (muscle precursors) from a rat embryo, growing in culture as single, dividing cells; **(b)** fusion of cultured myoblasts to form multinucleate muscle fibers. *(Dr. Daina Z. Ewton)*

ize and fuse into multinucleate muscle cells.

At the time they are first taken from the embryo, the myoblasts are not visibly muscle-like, but they are already determined to be muscle cells. By manipulating the conditions of cell culture, researchers can change the time at which differentiation begins, but they cannot change the conditions so that the cells develop into nerve or skin cells. The cells, then, are determined, and their determined state is stable; that is, they always develop into muscle cells, even after several cycles of division as apparently undifferentiated cells.

Stem Cells Maintain a Determined State While Producing More Stem Cells

Stem cells are cells that can either produce more of themselves by cell division or undergo differentiation to one or more specialized cell types. Spermatogonia are examples of determined stem cells: They are able to divide to produce more spermatogonia, but when they stop dividing they can develop along only a single pathway. Adult vertebrates contain precursor stem cells for cells that must regenerate during adult life—such as the cells of the skin, the linings of intestines, and the blood. Similarly, most plants contain stem cells that can generate even whole new plants. Usually, these are part of the plant's *meristems,* the undifferentiated tissues in which active cell division occurs. (See Chapter 21.) The developmental fate of stem cells—whether they proliferate to make more stem cells or differentiate to produce specialized cells—depends on cues from the environment.

■ DEVELOPMENTAL FATE OFTEN DEPENDS ON CHEMICAL SIGNALING

Because (with a few exceptions) every cell in an animal has the same genes, differences in cell fates must result from the selective expression of genes. Differences in the patterns of gene expression must be influenced by cellular environment. The question then is, how environmental cues influence the developmental fates of cells.

Can the Extraordinary Actions of the Primary Organizer in Amphibian Development Be Explained in Terms of Cells and Molecules?

Arguably the most spectacular experiment in the history of developmental biology was one performed by Spemann's student, Hilde Mangold. Recall (from p. 562) that Spemann established that cells become determined during gastrulation. Mangold and Spemann then wanted to find out what caused this determination.

They correctly guessed that determination of the ectoderm into neural tissue or skin depends on the contact between the ectoderm and the underlying mesoderm. The mesoderm derives from cells that invaginate during gastrulation (as illustrated in Figure 20-9). Mangold wanted to find out what would happen if the invaginating zone—the dorsal lip of the blastopore—were transplanted to another region of the embryo. To her surprise, the transplanted dorsal lip not only invaginated, but, as shown in Figure 20-27, also induced the tissues around it to form a nearly complete second embryo. Spemann was so astonished by this result that he termed the dorsal lip the **primary organizer,** meaning that its action established the organization of the entire early embryo, and he compared the action of the primary organizer to the workings of the mind—suggesting that it could not really be understood in chemical and physical terms. Subsequent work, however, demonstrated that chemical extracts of the dorsal lip can also induce the differentiation of ectoderm, so researchers are certain that they will ultimately understand the chemical basis of the primary organizer action.

Developmental biologists are actively working to identify the molecules responsible for changes in cell fate that occur during early development. The most pressing questions concern the mechanisms of **induction,** the process by which one cell population influences the development of neighboring cells. Much of this work uses *Xenopus laevis,* the same species that Gurdon used in his nuclear transplantation experiments, and employs an *in vitro* assay for induction, the *animal-cap assay.* The *animal cap* refers to the region, near the animal pole of the *Xenopus* blastula, that ordinarily develops into epidermis, both in the intact embryo and in an artificial culture. When an animal cap is cultured in the presence of tissue from the vegetal part of the embryo, however, mesodermal cells (like muscle) develop, along with mesoderm-specific mRNAs.

Studies with animal cap assays have identified four types of secreted proteins that induce mesodermal differentiation, at least partly mimicking the action of the primary organizer. The identified inducing factors include both proteins that have been previously characterized and novel proteins. Among the previously known proteins are (1) transforming growth factor-β (TGF-β), originally identified as the factor responsible for transforming (inducing a cancer-like appearance) of cells in culture; (2) activin, a member of the TGF-β gene family; (3) fibroblast growth factor, originally identified as a factor responsible for stimulating cell division of fibroblasts in culture; and (4) Wnt ("wingless") proteins, transcription factors that represent the vertebrate versions of a *Drosophila* transcription factor involved in wing development. Another, previously unknown secreted protein, called Noggin, consists of about 240 amino acids and appears to be a particularly powerful inducer.

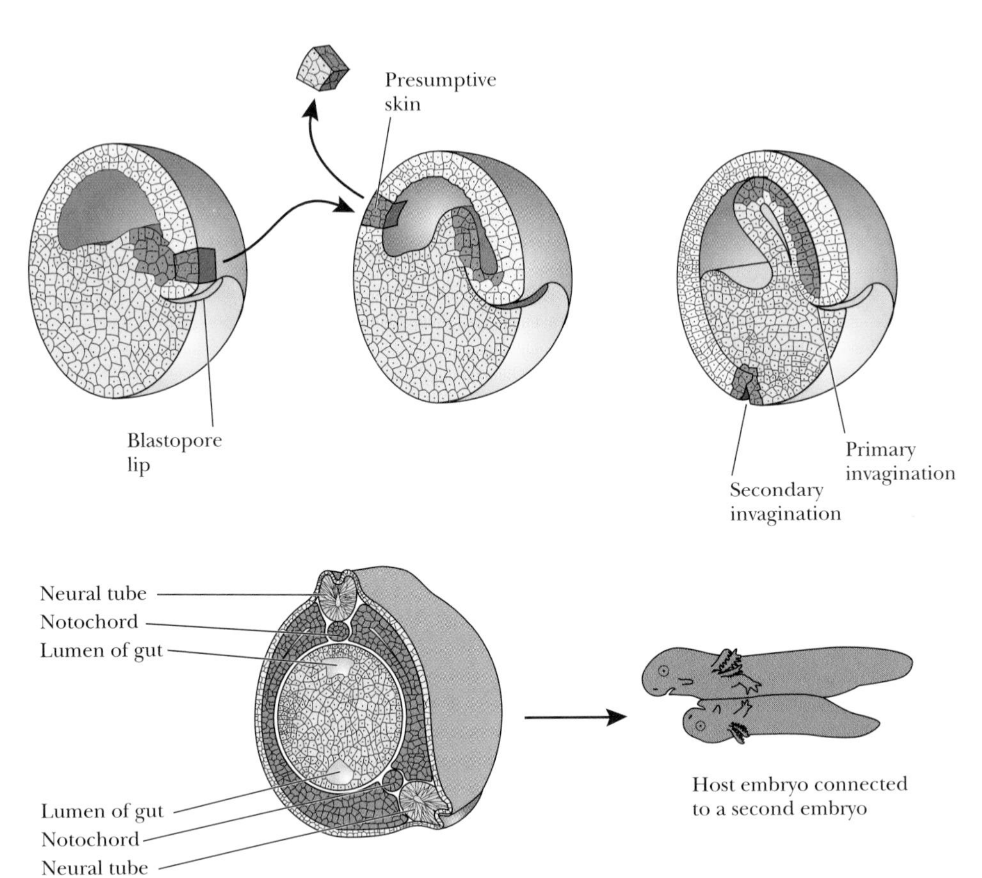

Figure 20-27 Transplantation of the dorsal lip of the blastopore from one newt embryo to the blastocoel of another embryo leads to the organization of an entire second embryo. This astonishing result led Mangold and Spemann to call the dorsal lip "the primary organizer."

Pattern Formation Requires That Cells Know Both Where They Are and What to Do

Development requires both that individual cells express appropriate genes and that cells be organized into functioning tissues and organs. The development of an arm or a wing, for example, requires the cytodifferentiation of muscles, skin, and bone into a characteristic pattern, distinct from the organization of the same cell types in a foot. How does the developing organism achieve this pattern?

The formation of patterns during development depends on each cell's "knowing" *where* it is. It also depends on the interpretation of this information: Each cell must also "know" *what* it is. At the tip of a developing leg bone cells must develop into digits, while cells nearer the body must develop into the other bones (such as the femur or tibia).

Part of pattern formation depends on the interactions between layers of cells. For example, the skin ectoderm of a chicken can give rise to several different kinds of structures—fully tufted feathers, partly tufted feathers, scales, and claws. The developmental fate of the ectoderm depends upon the source of the underlying mesoderm. If wing ectoderm is placed over foot mesoderm, claws develop instead of feathers.

The overall pattern of a limb, however, does not depend on short-range interactions between adjacent cell layers. Instead special mechanisms specify **positional information,** chemical cues that establish the position of a cell in a pattern. The concentration of particular substances, called **morphogens,** specifies the contribution of each cell to a pattern, in the same way that a seat number can specify the contribution of each member of a cheering section to creating the image of a team's mascot.

What molecules serve as morphogens? After decades of unsuccessful attempts to identify morphogens, developmental biologists have identified a number of proteins that serve as inducers during limb development. Among the most interesting of these are fibroblast growth factor, also important in the action of the primary organizer, and a protein called Sonic Hedgehog, because its sequence is specified by a vertebrate gene that is closely related to the *hedgehog* gene of *Drosophila.* In *Drosophila* the Hedgehog protein acts as an extracellular signal that helps organize the early embryo (see later in this chapter). In chick limb development, the engineered production of Sonic Hedgehog in a limb bud leads to a mirror-image duplication of the developing limb. These studies have provided new insights into **pattern formation,** the creation of a spatially organized structure during development.

(a) **Wild-type** (b) ***Antennapedia***

Eye

Antenna

Mouth parts

Figure 20-28 The *Antennapedia* mutant of *Drosophila.* Legs appear were antennae should be. **(a)** Wild-type head; **(b)** *Antennapedia* head. *(Photos courtesy Thomas C. Kaufman)*

Studies of *Drosophila* Development Have Identified Many Genes That Contribute to Pattern Formation

Over the years *Drosophila* geneticists have identified many mutants with weird patterns of development, such as *Antennapedia* (illustrated in Figure 20-28), in which antennae develop into legs, and *bithorax,* in which the fly develops two sets of wings rather than one. *Antennapedia* and *bithorax,* are examples of **homeotic mutations,** which cause the cells that give rise to adult structures to develop into harmonious structures, but with the wrong program (for example, legs instead of antennae). The nature of these mistakes suggests that many structures use the same chemicals to designate position. If the same morphogens are used in each structure, the developing structures must be able to give differing interpretations to the same positional information. In the last decade, *Drosophila* geneticists have identified a number of genes responsible for these specific interpretations. Most of these genes encode proteins that are transcription factors. The Antennapedia protein, for example, is a helix-turn-helix type of transcription factor, distantly related to regulators of bacterial transcription.

A systematic search for mutants that affect early development in *Drosophila* revealed several distinct classes of genes that affect pattern formation in the early embryo. To understand how these genes act, one first needs to understand the normal pattern of early development in

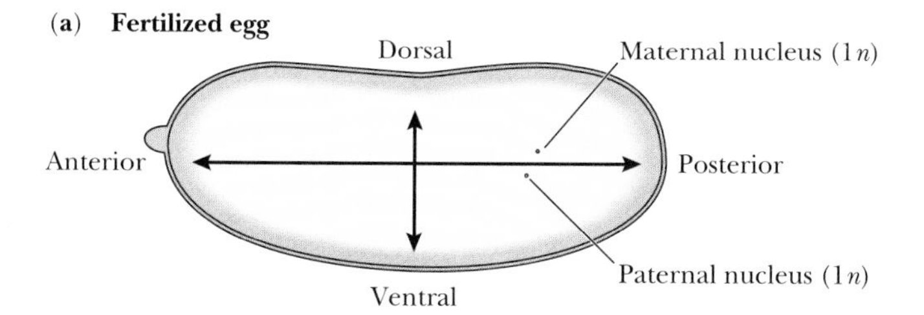

(b) The two 1n parental nuclei fuse to form a 2n zygote

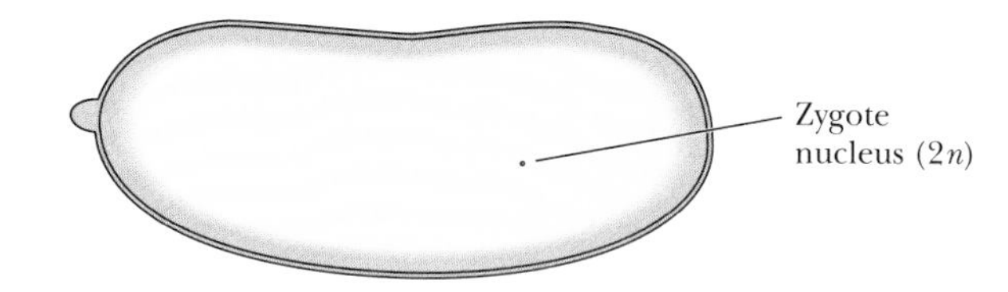

(c) Egg undergoes a series of mitoses without cytokinesis. The result is a syncytium, a structure that contains many nuclei within the same cytoplasm.

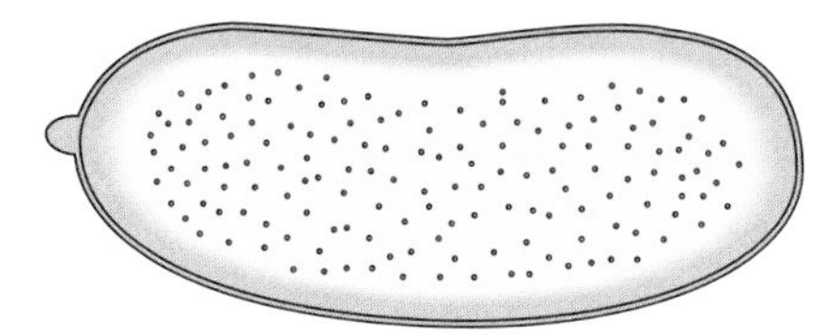

(d) After nine divisions, most of the nuclei migrate toward the embryo's surface, where they form a syncytial blastoderm.

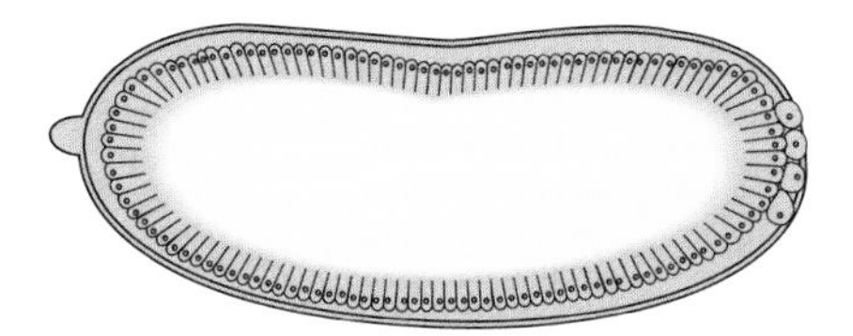

(e) The nuclei then undergo four more divisions following which plasma membranes form to create a cellular blastoderm, consisting of about 5000 cells and corresponding to the blastula of a sea urchin or frog embryo.

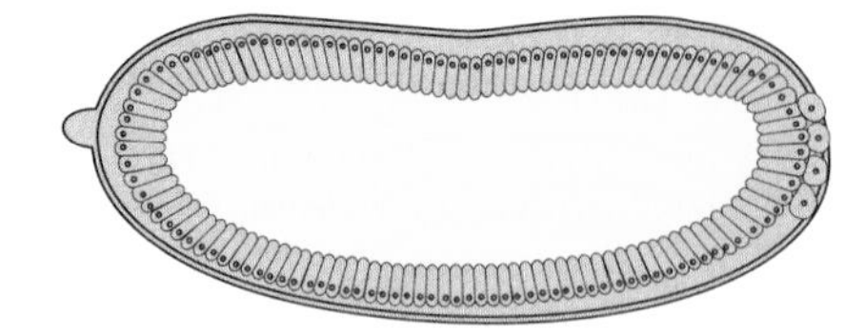

Figure 20-29 Early development of *Drosophila.* **(a)** Fertilized egg; **(b)** zygote with a single nucleus; **(c)** syncytium; **(d)** syncytial blastoderm; **(e)** cellular blastoderm.

Drosophila, which is illustrated in Figure 20-29. After fertilization (shown in Figure 20-29a and b), the egg (which is about 400 μm long and 160 μm in diameter) undergoes a series of mitoses without cytokinesis. The result (shown

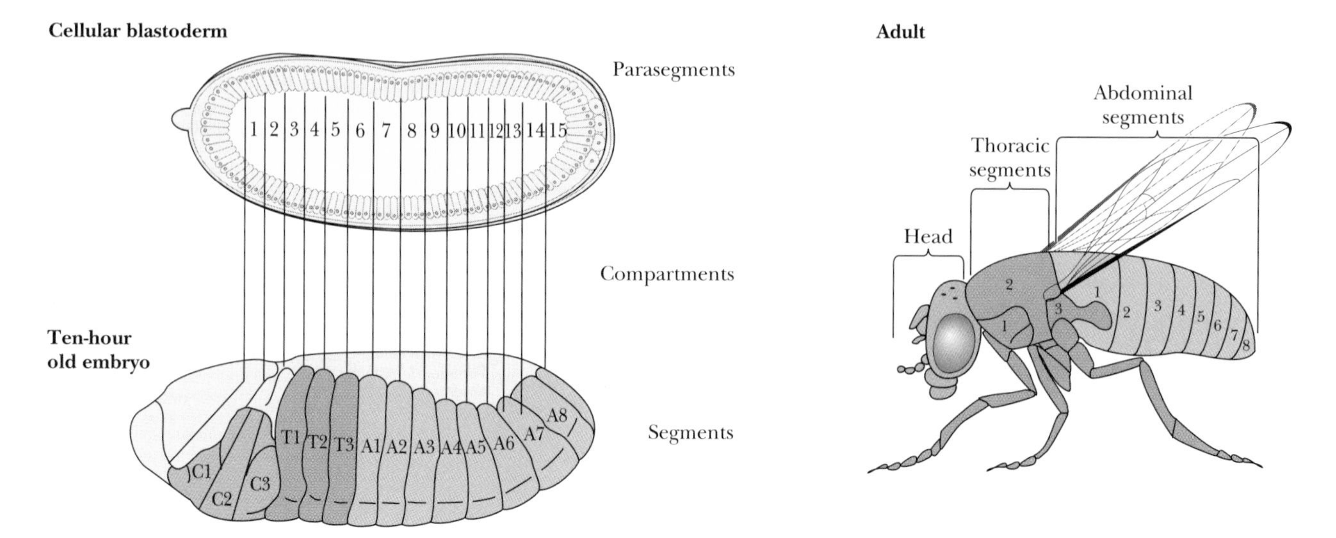

Figure 20-30 The developmental fates of cells in the *Drosophila* embryo are established by the cellular blastoderm stage.

in Figure 20-29c) is a *syncytium,* a structure that contains many nuclei within the same cytoplasm. After nine divisions (which produce 512 nuclei), most of the nuclei migrate toward the embryo's surface, where they form a *syncytial blastoderm* one layer thick, as shown in Figure 20-29d. The nuclei then undergo four more divisions, following which plasma membranes form, creating the *cellular blastoderm,* consisting of about 5000 cells equivalent to the blastula stage of a sea urchin or frog embryo (as pictured in Figure 20-29e).

Germ layer formation soon follows, giving rise to future endoderm, mesoderm, and ectoderm. Even before germ layer formation, however, the developmental fate of each blastoderm cell is well determined with the ultimate fates illustrated in Figure 20-30. While the cells of the cellular blastoderm look identical by conventional microscopy, examination of the pattern of gene expression reveals the existence of highly ordered patterns, consisting of 15 discrete bands, called *parasegments.* In fact, differences in the pattern of gene expression are present even before the embryo constructs its cell boundaries. The ultimate developmental fate of each cell depends on the pattern of gene expression both during oogenesis (as the egg itself is forming) and during the development of the zygote (after fertilization).

Mutations that affect early development may be in either **maternal effect genes,** expressed in the mother during oogenesis, or **zygotic effect genes,** which are expressed only after fertilization. Maternal effect genes are important in establishing the *polarity* of the embryos—which end is front (anterior) and which back (posterior), which side is up (dorsal) and which down (ventral). Mutations in the maternal effect gene *bicoid,* for example, lead to an embryo that lacks anterior structures. Molecular analysis of the *bicoid* gene showed that the *bicoid* mRNA is present in high concentrations in the anterior tip of the unfertilized egg and persists there during embryonic development; this mRNA was produced during oogenesis and laid down in the egg in the anterior tip of the embryo. The *bicoid* mRNA is translated into Bicoid protein after fertilization. The Bicoid protein serves as a morphogen that allows the cells of the cellular blastoderm to know their position.

A set of about 20 known zygotic effect genes, called **segmentation genes,** regulates the development of segments within the early embryo. The segmentation genes fall into the three classes illustrated in Figure 20-31: gap genes, pair rule genes, and segment polarity genes. Mutations in *gap genes* cause the loss of several segments, leading to a gap in the embryo; mutations in *pair rule genes* cause the loss of part of each segment; and mutations in *segment polarity genes* lead to the loss of half of each segment and the duplication of the other half.

Many of the Genes Responsible for Pattern Formation in *Drosophila* Encode Transcription Factors

Recombinant DNA techniques have allowed researchers to identify and study a number of genes responsible for pattern formation in *Drosophila.* The first of those identified were those mutated in *bithorax* and *Antennapedia.* Much to everyone's surprise, these two genes contained sequences that closely resembled one another, a 180-nucleotide pair region now called the **homeo box** [because it appears in a number of genes responsible for homeotic mutants].The homeo box specifies a polypeptide fragment called the *homeo domain,* whose sequence (as mentioned above) resembles the helix-turn-helix type of transcription factor, described in Chapter 15 (p. 411). Later studies re-

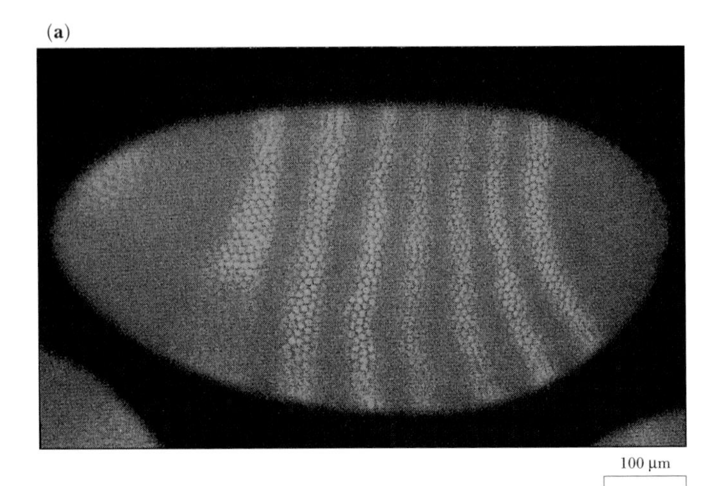

Figure 20-31 Molecular probes reveal segmentation stripes in the early *Drosophila* embryo, before the formation of cellular boundaries. A number of mutations alter this segmentation pattern. **(a)** A normal embryo, showing bands revealed by a probe for the mRNA for *fushi tarazu;* **(b)** altered banding resulting from a gap gene defect, *Krüppel;* **(c)** altered banding resulting from a pair rule gene defect, *fushi tarazu;* **(d)** altered banding resulting from a segment polarity defect, *engrailed.*
(a, Steve Paddock, Jim Langeland, Sean Carroll, Howard Hughes Medical Institute, University of Wisconsin)

Altered banding patterns in segmentation mutants

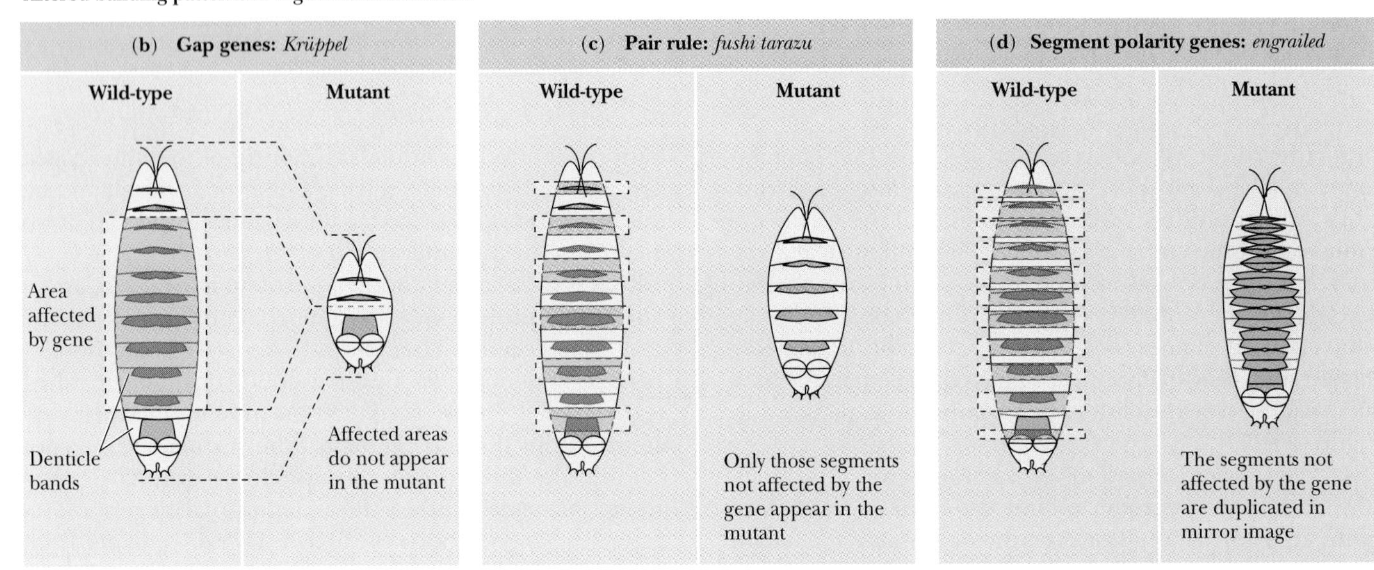

vealed that the proteins encoded by *bithorax* and *Antennapedia* do act as transcription factors, as their sequences suggested.

Other genes responsible for early *Drosophila* development also encode transcription factors. The Bicoid protein, required for the normal development of the early embryo, for example, also contains a homeo domain.

Removing cytoplasm from the anterior pole region of a wild-type *Drosophila* embryo leads to the production of an embryo that lacks anterior structures—exactly the same phenotype as a *bicoid* mutant. Once the *bicoid* gene was cloned, experimenters were able to produce the mRNA encoding the wild-type product *in vitro* and to inject it into a mutant *bicoid* embryo. The result was stunning: The wild-type *bicoid* mRNA rescued the mutant embryos. The injected embryos developed normal anterior structures while the uninjected *bicoid* embryos did not make any anterior structures. These experiments suggested that the wild-type Bicoid protein, normally present in the cytoplasm near the anterior pole, regulates the expression of other genes as embryonic development proceeds.

Among the genes whose expression is regulated by the *bicoid* gene is the gap gene *hunchback.* As shown in Figure 20-32, the mRNA encoded by *hunchback* is normally present at high levels near the embryo's anterior pole, in the same region where the Bicoid protein is normally concentrated. In a *bicoid* embryo, however, *hunchback* mRNA is absent from the anterior (although *hunchback* mRNA is present near the posterior end, having accumulated there during oogenesis).

Researchers have used a variety of recombinant DNA techniques to show that Bicoid protein interacts with the promoter region of the *hunchback* gene. They showed, for example, that different DNA segments within the promoter region can bind the Bicoid protein and activate transcription of the Hunchback protein. The greater the concentration of the Bicoid protein, the greater the production of *hunchback* mRNA and, ultimately, of Hunchback protein.

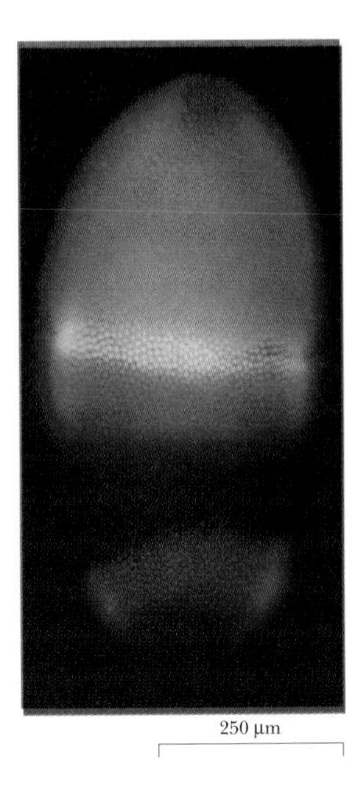

Figure 20-32 *Hunchback* mRNA accumulates near the anterior pole of the early *Drosophila* embryo. This accumulation requires the normal expression of the *bicoid* gene. *(C. Rushlow and M. Levine)*

Gap genes, including *hunchback*, also encode transcription factors of the zinc finger family. (See Chapter 15, p. 411.) Among the regulated targets of these factors are other gap genes. For example, Hunchback protein represses the expression of another gap gene, *Krüppel*. When *hunchback* is underexpressed (for example, in *bicoid* flies) *Krüppel* mRNA is present in more anterior regions than usual. The interpretation of this finding is that Hunchback protein ordinarily prevents *Krüppel* expression in the anterior region, helping to establish the normal domain of the *Krüppel* gene. Gap gene proteins like Hunchback and Krüppel also regulate the expression of pair rule genes, such as *hairy*.

The pair rule genes, such as *hairy* and *fushi tarazu*, are the next genes active in the determination of embryonic patterns. They encode transcription factors that in turn, regulate the transcription both of other pair rule genes and of homeotic genes, such as *bithorax* and *Antennapedia*.

Developmental biologists now view pattern formation in early *Drosophila* development as the result of a cascade of transcriptional regulation, as diagrammed in Figure 20-33. The mRNAs of maternal effect genes lie in gradients in the unfertilized egg. After fertilization the mRNAs are translated into transcription factors that stimulate (or repress) the activity of gap genes. Gap gene products then regulate the expression of pair rule genes and of other gap genes to produce the striped patterns in the syncytial blastoderm, as revealed by *in situ* hybridization or immunohistochemistry. By the time the embryo reaches the cellular blastoderm stage, each nucleus has a unique set of transcription factors that stimulate the expression of appropriate homeotic genes, ultimately leading to the development of an adult with all the proper body parts.

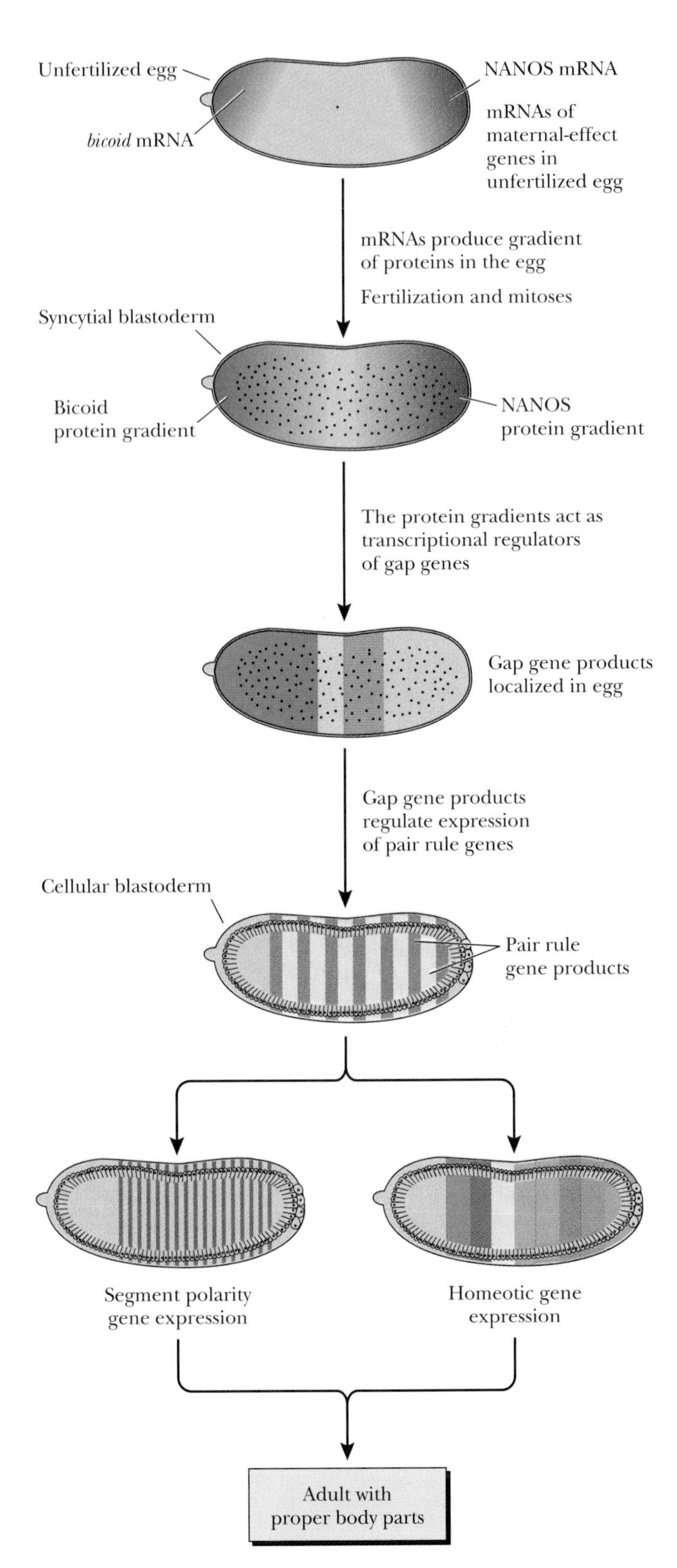

Figure 20-33 Positional information in the early *Drosophila* embryo depends on the sequential expression of genes. The characteristic spatial restrictions of each gene's mRNA ultimately depend on the action of the maternal gene, *bicoid*.

ESSAY

EXTRA BODY PARTS

Since the beginning of this century, experimental embryologists have known that cells of the developing embryo have a greater potency than they normally express. In 1918, for example, Ross Harrison showed that the mesoderm of a limb bud could induce overlying ectoderm to develop into a leg rather than just into the skin of the trunk. Until the last few years, however, no one knew what molecules were responsible for the induction of specific body parts. The clues have come from a variety of sources, with the most striking discoveries often coming from the study of *Drosophila* development.

The photograph that accompanies this essay attests to the level of present understanding of the development of body parts. Figure 1 shows a leg from a *Drosophila* that was genetically manipulated to produce an extra eye. The extra eye was induced by the action of a gene called *eyeless.* In related experiments in chick embryos, a protein, fibroblast growth factor, has been shown to induce the development of an extra leg.

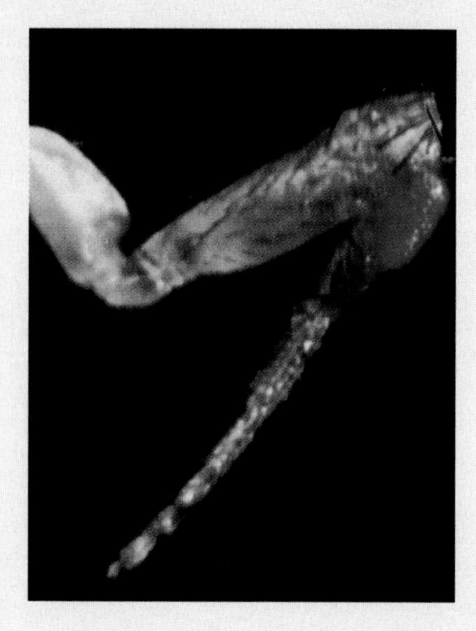

Figure 1 The activation of a normal allele of the *eyeless* locus resulted in the formation of a compound eye on this *Drosophila* leg. *(BIOZENTRUM/University of Basel/Switzerland/Courtesy of Professor W. H. Gehring)*

As the name implies, geneticists first identified the *eyeless* gene as the site of mutations that caused eyeless flies. Painstaking genetics and molecular biology cloning led to the identification and sequencing of the wild-type *eyeless* gene. Not surprisingly by now, the *eyeless* gene has a homeo domain and appears to act as a transcription factor. Moreover, the sequence of the *eyeless* gene of *Drosophila* is related to the *small eye* gene in mice and the *aniridia* ("no iris") gene of humans, again showing the conservation of developmental signals through evolution. The poly-eyed fly in the photograph resulted from the genetically engineered expression of the wild-type *eyeless* gene in developing wings, legs, or antennae.

The chick embryo's extra leg is more complicated and involves not only fibroblast growth factor but also two other diffusible signals, the lipid-soluble retinoic acid and the protein Sonic hedgehog, as well as two transcription factors—the retinoic acid receptor and a Hox gene, Hoxb-8. Researchers now think limb bud induction begins with the action of retinoic acid, produced in the mesoderm, inducing the expression of Hoxb-8 in the overlying ectoderm.

Retinoic acid has long been known to regulate the pattern of digit formation in the developing limb bud of the chick embryo and in regenerating limbs of newts. In the chick limb bud, added retinoic acid can induce the formation of extra digits, with the number of extra digits depending on the concentration of retinoic acid. In the regenerating newt limb, researchers have demonstrated that retinoic acid is present in a concentration gradient, meaning a continuous ("graded") variation in concentration, with the highest concentration in the *proximal* limb (the part that is nearest to the body) and the lowest concentration in the *distal* limb (the part farthest from the body). The retinoic acid receptor is a transcription factor, a close relative of the steroid receptors discussed in Chapter 15.

According to the current model, the underlying mesoderm then produces fibroblast growth factor, which induces proliferation of the limb bud ectoderm and the expression of the Sonic hedgehog protein, which establishes the polarity of the limb. Hoxb-8 allows the induction of Sonic hedgehog and may also contribute to establishing the identity of the developing limb—whether it is to be a leg or a wing.

This model of limb development is likely to become much more refined as new experiments are done. But the ability of specific molecules to induce extra limbs, like the ability of other molecules to induce extra eyes, suggests that researchers are well on their way to understanding the molecular and cellular bases of development.

How the products of the homeotic genes stimulate the production of individual body parts is not yet understood. An example of the subtlety and power of the homeotic genes comes from an experiment in which the DNA-binding homeo domain of the *Deformed* gene (a member of the *Antennapedia* complex) was switched with the corresponding region of a *bithorax* gene. The result was the formation of a thorax where the head should have been. This dramatic change resulted from a difference of only 22 amino acid residues in a single transcription factor.

Mammals and Flies Use Many of the Same Transcription Factors to Establish Patterns During Development

Information about the structure and action of genes that regulate *Drosophila* development led immediately to questions about the role of similar genes during mammalian development. Researchers almost immediately asked whether mammalian genomes had counterparts of the amazing *Drosophila* genes. The needed experiments were straightforward—to use recombinant *Drosophila* DNAs as probes to find corresponding sequences in mammalian DNA. The mammalian counterparts were easy to find. *Drosophila* homeo domain genes, for example, have mammalian homologs, called **Hox genes,** whose sequences are remarkably similar to the *Drosophila* counterparts. A particularly striking example of this similarity is reflected in a comparison of the mouse gene Hox 1.1 and the *Drosophila* gene *Antennapedia:* The homeo domain of mouse Hox 1.1 shares 59 of 60 amino acid residues with the homeo domain of *Antennapedia.* Even more amazingly, the genetically engineered expression of two vertebrate Hox genes (Hox 2.2 and Hox 4.2) in flies led to the same defects in development as mutations in the corresponding *Drosophila* genes (*Antennapedia* and *Deformed*).

The biggest surprise in the study of the Hox genes was that the corresponding genes in flies and mice share the same organizational principle, with the order of genes on the chromosome matching the pattern of action within the embryo. The arrangement of homeotic genes on *Drosophila* chromosome 3 parallels the pattern of expression of those genes on the embryo's axis: Genes that affect head development lie at one end of the cluster, while those that affect tail lie at the other. The same pattern holds for the arrangement of the Hox genes on the mouse chromosome and their pattern of expression in the primitive nervous system of the mouse embryo.

The most recent common ancestor of *Drosophila* and mice lived about 600 million years ago. Flies and vertebrates have evidently managed to use—with very few modifications—a system of specifying positional information that must have been already well established 600 million years ago.

SUMMARY

Animal development begins with a zygote that appears simple and homogeneous. While early embryologists imagined that sperm or egg contained preformed adults, experiments in the late 19th and early 20th centuries showed that most complexity resulted from interactions among cells during development.

Each species undergoes a characteristic sequence of developmental events, but all vertebrates pass through the same stages: gamete formation, fertilization, cleavage, germ layer formation, organ formation, growth, metamorphosis (in many but not all species), and aging. Development starts with fertilization, following which cleavage divisions lead to the formation of a single layer of cells. Cell movements then lead to the formation of a gastrula, whose three germ layers are precursors of distinctive adult structures. Organ formation depends both on cytodifferentiation, the acquisition of specific proteins and subcellular structures, and on morphogenesis, the creation of form.

Sperm and eggs are highly specialized cells that accomplish fertilization and initiate embryonic development. Sperm are specialized for rapid movement, while eggs are specialized to allow fertilization by a single, but not multiple, sperm. The egg also contains materials that provide energy (as yolk granules) and information (as mRNAs) to the developing organism.

The zygote and some cells of the early cleavage embryo are able, under the proper conditions, to develop into a whole organism. As development proceeds, however, cells become increasingly determined, committed to particular developmental fates. Even after a cell becomes determined, its DNA still contains all the information necessary to produce a whole organism, as demonstrated by nuclear transplantation experiments.

Chemical signaling between different parts of the embryo is responsible for cells' sensing where they are (positional information), leading to the formation of harmonious patterns of organs and body parts. Studies of mutants that affect the early development of *Drosophila* have revealed that the establishment of pattern depends on the production and distinctive distribution of a cascade of transcription factors. Many of the proteins responsible for pattern formation in *Drosophila* are virtually identical to mammalian proteins.

KEY TERMS

acrosome
apoptosis
archenteron
blastocoel
blastocyst
blastopore
blastula
cleavage
complete metamorphosis
cytodifferentiation
ectoderm
endoderm
epidermis
fertilization
fertilization membrane
fetus
gamete formation
gastrula
gastrulation
germ layer
germ layer formation
gonad
homeo box
homeotic mutation
Hox gene
imaginal disc
induction
invagination
larva
maternal effect gene
maternal mRNA
mesoderm
metamorphosis
morphogen
morphogenesis
morula
neural crest
neural plate
neural tube
neurulation
notochord
nuclear transplantation
oogenesis
oogonia
ovum
pattern formation
polar body
polyspermy
positional information
potency
preformation
primary mesenchyme
primary organizer
primordial germ cell
programmed cell death
pupa
rudiment
secondary mesenchyme
soma
spermatogenesis
spermatogonia
spermatozoa
stem cell
totipotent
trophoblast
zygote
zygotic effect gene

QUESTIONS FOR REVIEW AND UNDERSTANDING

1. What does yolk consist of? What is its function?
2. What are some of the unique properties of sperm? Of eggs?
3. How are polar bodies generated? What is their function?
4. List and briefly describe the eight stages of vertebrate development.
5. Describe the organization and function of cells making up the blastocyst.
6. Describe the organization and function of cells making up the gastrula.
7. What is a stem cell?
8. Compare necrosis and apoptosis.
9. Describe the hallmarks of a cell undergoing apoptosis.
10. How does programmed cell death play a role in development?
11. What is metamorphosis?
12. Describe the role morphogens play in development.
13. What is a totipotent cell? Can a differentiated cell be totipotent?
14. Describe the experimental evidence that showed that specialized cells of the gastrula can interact and respond to their relative positions.
15. What are the two major processes involved in organ formation? Describe both. Make an argument as to why one process must precede the other.
16. Describe the process of fertilization of the sea urchin.
17. Why is polyspermy not desirable biologically?
18. Describe two mechanisms which prevent polyspermy in sea urchins.
19. Explain how a fate map is constructed.
20. Why is *C. elegans* such a useful organism for studying cell fate?
21. What is the difference between a cell's fate and its potency? Give an example to illustrate this difference.
22. What is pattern formation? What information do cells undergoing this process require?
23. Explain what is meant by the term "induction"? Give three examples of induction.
24. Is cell determination driven by the progressive loss of genetic material? Briefly describe the types of experiments that support your answer.

25. Describe the theory of preformation. How and when was this theory disproved?
26. What is the function of maternal effect genes?
27. What is the function of zygotic effect genes?
28. How would you identify a DNA sequence that encoded a homeo box? How are genes that contain homeo boxes evolutionarily significant?
29. Describe the experiment that demonstrated the function of *bicoid*.
30. Discuss why the study of apoptosis may be important to the study of cancer.

SUGGESTED READINGS

ALBERTS, B., D. Bray, J. Lewis, M. Raff, K. Roberts, and J. D. Watson, *Molecular Biology of the Cell*, 3rd edition, Garland, New York, 1994. Chapter 21 deals with cellular mechanisms of plant and animal development.

BROWDER, L. W., C. A. Erickson, and W. R. Jeffery, *Developmental Biology*, 3rd edition, Saunders College Publishing, Philadelphia, 1991. An excellent text on developmental biology.

GILBERT, S. F., *Developmental Biology*, 3rd edition, Sinauer, Sunderland, MA, 1991. Another excellent text on developmental biology.

LODISH, H., D. Baltimore, A. Berk, S. L. Zipursky, P. Matsudaira, and J. Darnell, *Molecular Cell Biology*, 3rd edition, Scientific American Books, New York, 1995. Chapter 24 deals with multicellularity, focusing on cell-cell interactions and the extracellular matrix.

WATSON, J. D., M. Gilman, J. Witkowski, and M. Zoller, *Recombinant DNA*, 2nd edition, Scientific American Books, New York, 1992. Chapter 18 deals with genes that control the development of *Drosophila*.

CHAPTER 21

Growth and Development of Flowering Plants

KEY CONCEPTS

- The growth and development of plants depend on the repeated generation of standard modules at the ends of the shoot and the root.
- Specific genes control the development of flowering parts.
- Flowering plants undergo alternation of generations, with diploid sporophytes producing haploid gametophytes, which then produce haploid gametes.
- Plant embryos establish a body plan soon after fertilization and pause after seed production.
- Hormones direct development and coordinate cellular and biochemical activities in many cells throughout the plant.
- Many responses to light and to seasonal changes depend on a pigment called phytochrome.

OUTLINE

ESSAYS

The German poet Goethe was said to have soothed himself to sleep by thinking about a bud unfolding into a leaf. Perhaps this process calmed Goethe because he considered it so predictable. Each bud develops into a leaf or flower characteristic of the species.

Little else in the development of a plant is so predictable, however. Plants change their patterns of growth throughout their life cycles. We may compare the life cycle of an animal to a play: It has a script with a beginning, a middle, and an end. In contrast, the life cycle of a plant resembles an improvisation, whose form, content, and length depend on cues from the audience.

Unlike animals, plants cannot move to seek desirable or avoid undesirable environments. Instead they must make the most of the conditions where they are. Environmental cues dictate the pattern of growth, so that the same cells can give rise to leaves, stem, or flowers. The direction of growth responds to light and gravity, and in many flowering plants, the time of flowering depends on the relative lengths of day and night. The responses to environment are even more impressive than those of animals because plant cells, enclosed in a cellulose wall, are more rigid and more tightly glued to one another than those of animals. Movement toward light, for example, depends on the elongation of cells farther from the light compared with those closer to the light.

In this chapter we discuss the development of flowering plants, starting with a description of their general anatomy. We use as an example *Arabidopsis thaliana,* a small weed and a member of the mustard family, which is widely used for genetic studies of plant development. Later in this chapter we examine the molecular and cellular bases of the responses of plants to environmental cues.

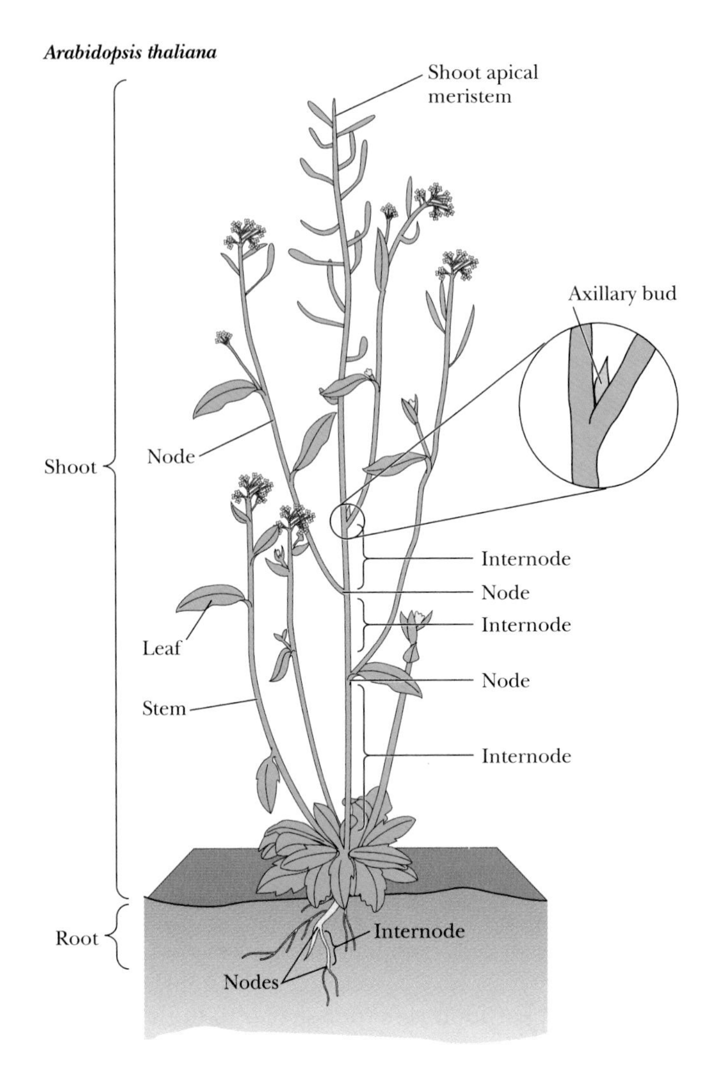

Figure 21-1 The general anatomy and modular organization of a flowering plant as exhibited by *Arabidopsis thaliana.* Note the regular repetition of meristem-derived modules.

HOW DO PLANTS REPEATEDLY GENERATE STRUCTURES AT THE ENDS OF SHOOT AND ROOT?

The most important organizing principle of a plant is the polarity between the **shoot,** the part of the plant above the ground, and the **root,** the part below the ground, as illustrated in Figure 21-1. The shoot-root polarity, roughly parallel to the head-tail polarity of an animal, is established early in the plant embryo.

Just behind the tip of both shoot and root are **apical meristems** [Latin, *apex* = top + Greek, *meristos* = divided], self-renewing groups of undifferentiated cells that generate the differentiated structures of shoots and roots. Like animal stem cells, discussed in Chapter 20 (p. 566), meristem cells can either divide to give rise to more meristem cells or can ultimately undergo cytodifferentiation.

An apical meristem can repeatedly generate more or less the same set of plant parts, which serves as a kind of construction module, as illustrated in Figure 21-1. This module may be repeated an undetermined number of times, depending on the conditions of plant growth. In contrast, animal development usually provides a fixed number of modules (segments) in the embryo, as discussed in Chapter 20 for both *Drosophila* and mammals.

A shoot meristem can generate a stem and leaf. The region of the stem to which the leaf attaches is called a **node,** and the region between nodes is an **internode.** As the meristem produces new cells, the new cells form files that push the meristem forward, so the newly made module also has an apical meristem just behind the tip. The shoot apical meristem resembles a brick factory that is constantly being raised higher on the bricks it has deposited below.

Looking from node to node in a plant, then, we see repeating modules, each consisting of a segment of stem (the internode), a leaf, and a meristem. In both shoot and root, the action of the apical meristems is responsible for the plant's **primary growth,** the extension of its length.

Each species of plant has a characteristic pattern of modular repetition. Modular organization and fixed repetition rules underlie the branching patterns of stems and leaves, as well as the pleasing symmetry of many flowers and fruits.

Morphogenesis Depends on Oriented Cell Division and Cell Expansion

In both shoot and root modules, meristem cells themselves are relatively small, with thin walls, prominent nuclei, and small vacuoles. Away from the meristem, cells no longer divide, but they continue to expand, mostly by taking water into a large central vacuole. The enlarged cells, already arranged in columns, are up to 50 times larger than those of the meristem itself. These cells ultimately differentiate into a variety of cell types.

Unlike animal cells, plant cells cannot move around and form new associations during development. After division ceases, they can only expand and sometimes bend. Plant development therefore depends heavily on orderly cell division, followed by expansion and differentiation. As cells divide within the meristem, the planes of cell division establish their future spatial relationships. Plant biologists distinguish three directions of division, as shown in Figure 21-2. Length increases depend on *transverse* divisions; increases in girth and surface area depend on *periclinal* and *anticlinal* divisions.

The files of cells visible in the tips of shoot and root arise from a series of transverse cell divisions, in which each new cell partition is perpendicular to the axis of the column. The placement of the new cell wall in turn depends on the orientation of microtubules, called **cortical microtubules,** that encircle the dividing cell just inside the plasma membrane, as shown in Figure 21-3. The cortical microtubules somehow determine the orientation of the spindle-derived microtubules that contribute to forming the phragmoplast, as described in Chapter 9 (p. 266).

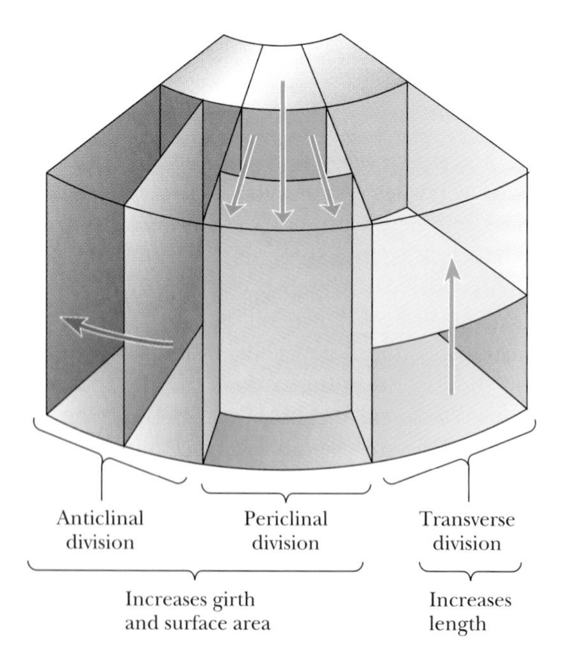

Figure 21-2 The planes of cell division.

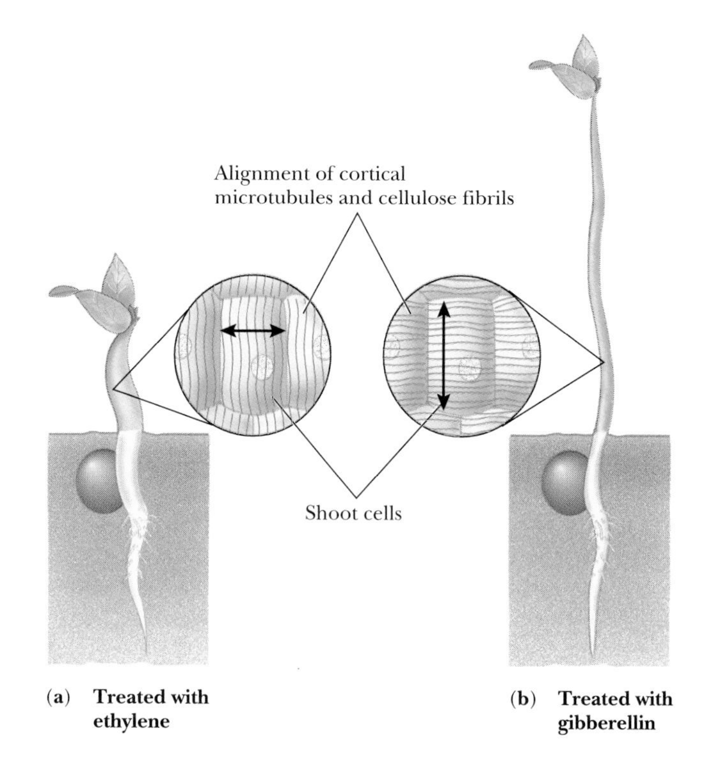

Figure 21-3 The effects of plant hormones on the pattern of microtubule alignment and cellulose deposition. **(a)** Treatment with ethylene produces shortened cells; **(b)** treatment with gibberellin produces elongated cells.

Cell expansion depends mainly on the osmotic influx of water into the central vacuole. While the expansion of the vacuole has no intrinsic direction, the cellulose cell walls constrain the direction of expansion, so the net effect is to elongate the cell in a particular direction.

Understanding the basis of oriented expansion requires understanding how cells establish the pattern of cellulose deposition. Studies of cellulose deposition show that cellulose microfibrils follow a pattern parallel to that established by the cortical microtubules, as illustrated in Figure 21-3. The placement of the cortical microtubules is critical to establishing both the plane of cell division and the direction of cell expansion, but no one yet knows exactly how this is accomplished.

The placement of the cortical microtubules responds to a number of plant hormones, which we discuss later in this chapter. Because of their effects on the patterns of cell division and expansion, plant hormones can dramatically influence growth of an entire plant. The hormone **ethylene,** for example, causes a longitudinal orientation of the cortical microtubules. The parallel cellulose fibrils constrain cell expansion to the horizontal direction, as shown in Figure 21-3a, leading to the production of short, fat seedlings. In contrast, **gibberellin,** another plant hormone, stimulates the transverse orientation of cortical microtubules and cellulose fibrils, leading to tall, thin shoots, as shown in Figure 21-3b.

Although most of this discussion has centered on the growth and differentiation of cells in the shoot, the same general considerations apply to the growth of the root. Nonetheless, the growth of roots and the growth of shoots have several distinctive characteristics.

Growing Root Tips Have Three Distinct Zones

Primary growth of roots occurs almost exclusively near their tips. As shown in Figure 21-4, the growth zone has three separate regions—the root cap, the apical meristem, and an elongation zone. At the very tip is the **root cap,** whose cells are full of Golgi complexes and vesicles. These organelles contribute to the synthesis of mucigel, a polysaccharide slime that lubricates the path of the growing root through the soil. The root cap also serves as a mechanical shield for the apical meristem, which lies just behind.

The apical meristem contains most of the root's dividing cells. If we allow growing roots to take up a radioactive precursor of DNA (^{3}H-thymidine), almost all of the labeled cells lie in the meristem. The cells within the meristem have a small uniform size, but the pattern of cell division produces files of cells whose organization already resembles that of the mature root.

Farther back along the root's axis is the **elongation zone.** The cells in each file are increasingly large. Cells farther from the tip are longer than those nearest the meristem, which have arisen more recently. The farther cells in the elongation zone are from the meristem, the more differentiated they are. Many botanists distinguish between the "zone of elongation," in which cells lengthen, and the "zone of differentiation," in which most of the cells of the primary tissues differentiate into mature cells. The appearance of abundant root hairs marks the end of the differentiation zone.

Primary Growth of Shoots Is More Complicated than That of Roots

In the shoot, the apical meristem must produce stem, leaves, branches, and flowers, so primary growth is more complicated than in the root. The region of primary growth may, in some plants, extend for as much as 10–15 cm behind the shoot's tip and may include several nodes. As in the root, the stem apical meristem contains the dividing cells. Cells produced in the meristem form files of cells that become longer, pushing the meristem upward. The farther cells are from the meristem, the more they have differentiated. As in the root, these columns of cells form concentric cylinders.

Each leaf originates as a **leaf primordium,** a tiny extension of the apical meristem, shown in Figure 21-5a. Each primordium contains a meristem that divides to produce the leaf's differentiated cells. A **bud** consists of sev-

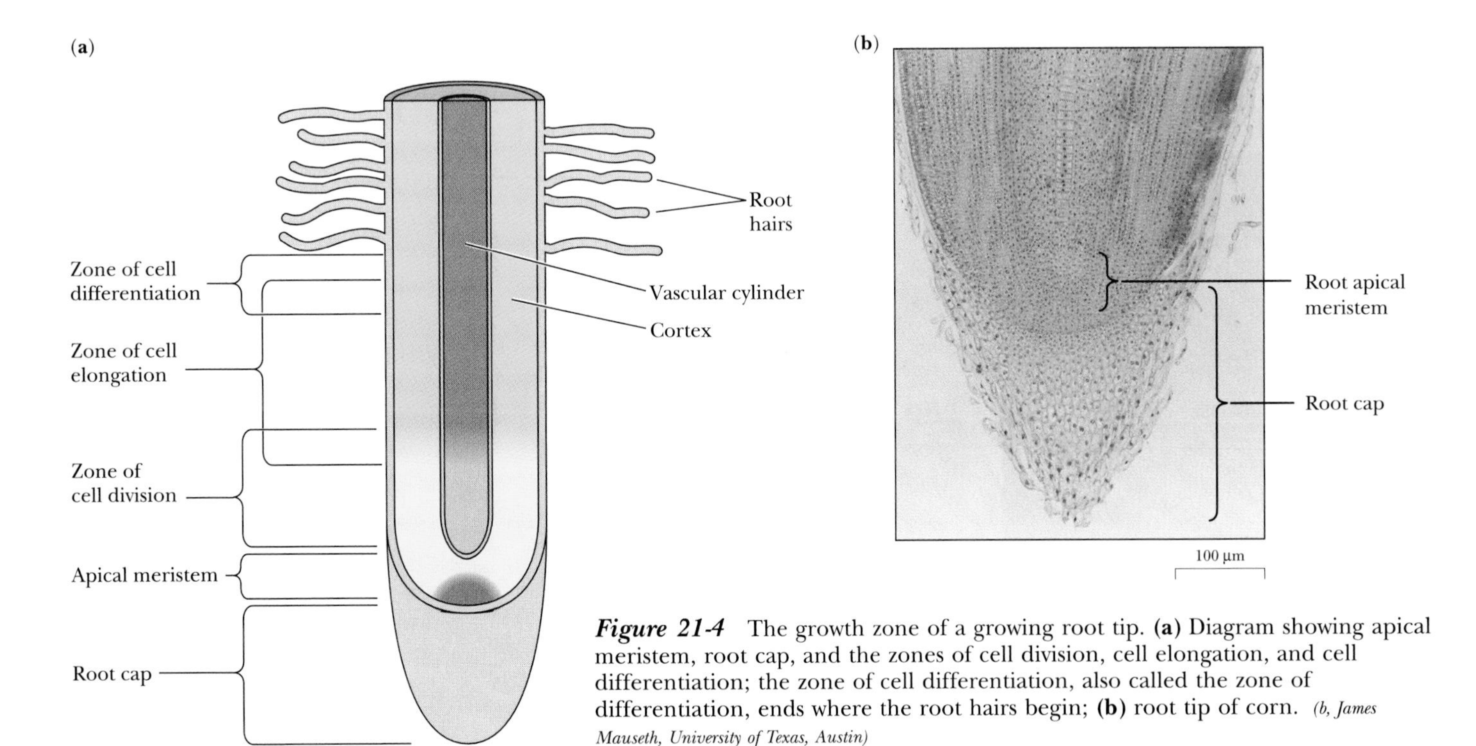

Figure 21-4 The growth zone of a growing root tip. **(a)** Diagram showing apical meristem, root cap, and the zones of cell division, cell elongation, and cell differentiation; the zone of cell differentiation, also called the zone of differentiation, ends where the root hairs begin; **(b)** root tip of corn. *(b, James Mauseth, University of Texas, Austin)*

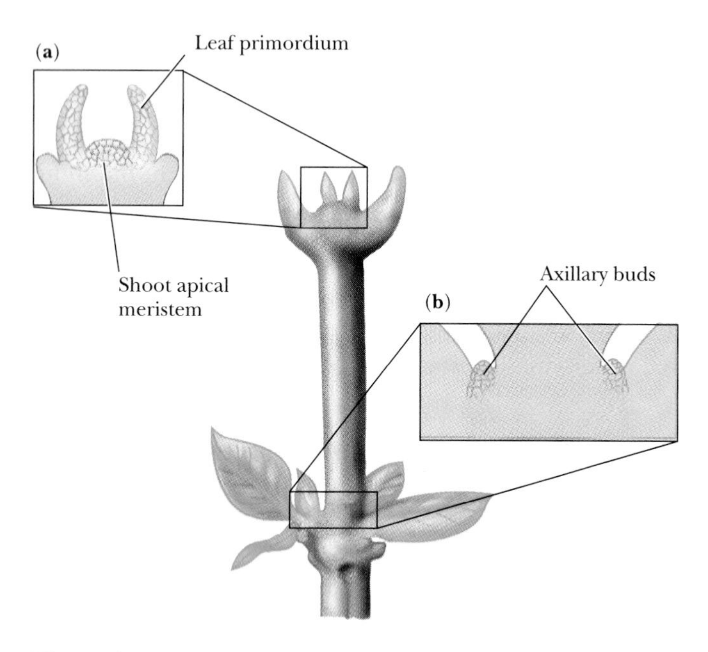

Figure 21-5 A bud contains several leaf primordia and several internodes. **(a)** A leaf primordium, an extension of the shoot apical meristem; **(b)** axillary buds.

eral leaf primordia, separated by internodes whose cells have not elongated.

In addition to the **terminal buds,** which lie at the tip of a shoot, most plants have **axillary** [Greek, *axilla* = armpit] or **lateral buds,** as illustrated in Figure 21-5b, which form just above the points where leaves join the stem and which can develop into branches. Like the terminal buds, the axillary buds contain apical meristems. Many species show **apical dominance,** meaning that the terminal buds suppress the activity of lateral buds. Lateral buds close to the tip of a plant are more inhibited than those farther away. Removal of a terminal bud allows lateral buds to grow, and many gardeners routinely produce bushier plants by pinching off tops, as shown in Figure 21-6. Apical dominance depends on the action of the plant hormone, auxin, which we discuss later in this chapter.

How Does a Flower Derive from a Specialized Bud?

Some buds are specialized to form flowers or a flowering shoot. In most species, flowers contain both **stamens,** male reproductive organs, and **carpels** (or pistils), female reproductive structures. Each flower consists of up to four sets of parts arranged in **whorls,** concentric circles of

Figure 21-6 Apical dominance. **(a)** A tomato plant; **(b)** plant whose terminal bud has been pinched off; the removal of the terminal bud allows the growth of lateral buds.
(a, © Patricia Agre 1984/Photo Researchers)

(a) *Antirrhinum majus*

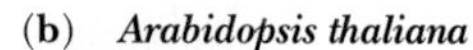

(b) *Arabidopsis thaliana*

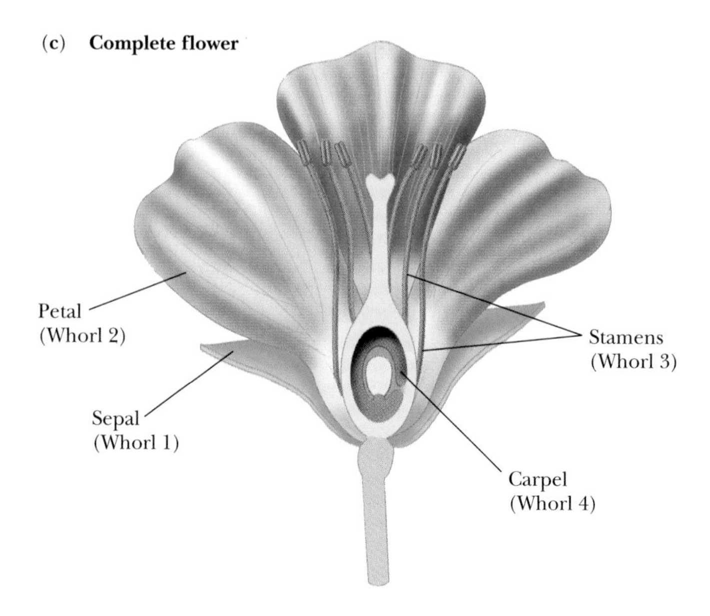

Figure 21-7 Flowers of **(a)** *Antirrhinum majus* (snapdragon) and **(b)** *Arabidopsis thaliana.* The *Arabidopsis* flower is only 1/10 the size (and 1/1000 the volume) of the *Antirrhinum* flower. **(c)** Both have the same plan, with four whorls. *(a, Lefer/Grushow from Grant Heilman; b, Dr. Jeremy Burgess/Science Photo Library/Photo Researchers)*

flower parts at the end of a specialized stem, as shown in Figure 21-7. The two inner whorls are said to be fertile, because they produce gametes. The outer two whorls, 1 and 2 in Figure 21-7c, are the **sepals,** the individual green parts at the base of a flower, and the **petals,** the showy, usually colored, parts of the flower. These whorls do not produce gametes and are said to be sterile. Whorl 3 are the stamens, and whorl 4, the innermost whorl, are the carpels.

Botanists have long noted that not all flowers have all four whorls. A flower that contains all four whorls is said to be a *complete* flower, while an *incomplete* flower is missing one or more whorls. A flower that contains both carpels and stamens (and produces both gametes of both sexes) is said to be *perfect,* while if either is missing it is said to be *imperfect.* The snapdragon and *Arabidopsis* flowers in Figure 21-7 are complete (and therefore perfect) flowers.

In contrast, the corn flowers shown in Figure 21-8 are incomplete and imperfect. The tassel-like corn flowers, at the ends of stems, lack carpels but have stamens and produce male gametes. The lateral flowers, within the "ears," lack stamens but have carpels and produce female gametes. Corn is an example of a *monoecious* [Greek, *monos* = single + *oikos* = house] species, in which both male and female flowers appear on the same plant. Willow trees and date palms are *dioecious* [Greek, *di* = two], with male and female flowers on different plants.

Among the most interesting questions being addressed by contemporary plant biologists are those dealing with the control of flower development: What molecules and cellular processes govern the development of complete versus incomplete flowers, or of male versus female flowers? In the last few years, major insights have come from the genetic analysis of plant mutants, particularly of *Arabidopsis thaliana.*

MOLECULAR ANALYSIS OF MUTATIONS PROVIDES INSIGHTS INTO PLANT DEVELOPMENT

Genetic studies of several plant species, especially of corn and *Arabidopsis,* have identified a number of mutations that dramatically affect the development of plants. By combining genetic analysis with molecular biology, plant biologists have begun to unravel many of the cellular and molecular mechanisms responsible for establishing the identity of meristems and their derivatives. Just as the genetic analysis of *Drosophila* and of *Caenorhabditis elegans* has revolutionized the understanding of animal development,

(a) (b)

Figure 21-8 **(a)** Ears of corn are really large groups (inflorescences) of carpellate flowers surrounded by protective leaflike bracts (the husks). The corn "silks" are long stigmas. **(b)** Corn tassels are inflorescences of staminate flowers. *(James Mauseth, University of Texas, Austin)*

so, it seems, will similar studies of *Arabidopsis* lead to a new view of plant development. Many of the cellular decisions in plant development surprisingly resemble corresponding decisions in animal development, depending, for example, on regulation of gene expression by specific transcription factors.

Arabidopsis thaliana is Particularly Well Suited for Genetic and Molecular Studies

Arabidopsis, the small common weed pictured in Figure 21-1, also called "wall cress," is so well suited for genetic studies that even in 1946 it was called a "botanical *Drosophila.*" Only in the 1980s, however, did it become a major experimental organism for plant biologists. Figure 21-7 shows a view of the *Arabidopsis* flower.

The major advantage of *Arabidopsis* is its rapid generation time, 5–6 weeks (compared with less than 2 weeks for *Drosophila*). Unlike *Drosophila,* however, *Arabidopsis* can self-fertilize, making it much easier to isolate homozygous strains. *Arabidopsis* is small, usually less than 30 cm tall and thin enough that several plants can be grown per square centimeter. Plants can be grown easily and inexpensively in simple soils or in sterile media. Its seeds are tiny; 35,000 seeds occupy only about 1 ml (1 cm^3).

The genome of *Arabidopsis* contains about 70 million nucleotide pairs, only about half as much as the *Drosophila* or *C. elegans* genome and about one-fortieth as much as a mammalian genome. Molecular geneticists have produced an ordered collection of all of the *Arabidopsis* genome, greatly simplifying the task of finding new genes that have been mapped genetically or interrupted with a transposon. Like Steward's carrot cells discussed in Chapter 20, individual *Arabidopsis* cells can regenerate a whole plant. Finally, molecular biologists can readily engineer *Arabidopsis* with recombinant DNAs in Ti-derived vectors, as discussed in Chapter 17 (p. 464).

Geneticists have identified well over 100 mutations in *Arabidopsis.* Some of these mutations lead to specific enzyme deficiencies, others to alterations in response to plant hormones, and still others to alterations in pattern. For example, one mutation leads to a missing shoot, another to a missing stem, another to a missing root. Mutations affect mature plants as well as seedlings. One mutant completely lacks a shoot meristem, both in the intact plant and in tissue culture. Other *Arabidopsis* mutations affect the formation of specific types of tissue, while still others prevent the proper establishment of the embryo shoot-root axis.

Once a mutation has been found, it is now a relatively straightforward—if time-consuming—task to identify and sequence the gene in which the mutation lies. The function of the gene can then be studied by putting it back into a wild-type or mutant *Arabidopsis* using Ti-based transformation, as described in Chapter 17. Researchers have performed such studies for a relatively small number of genes but have already learned much about *Arabidopsis* development.

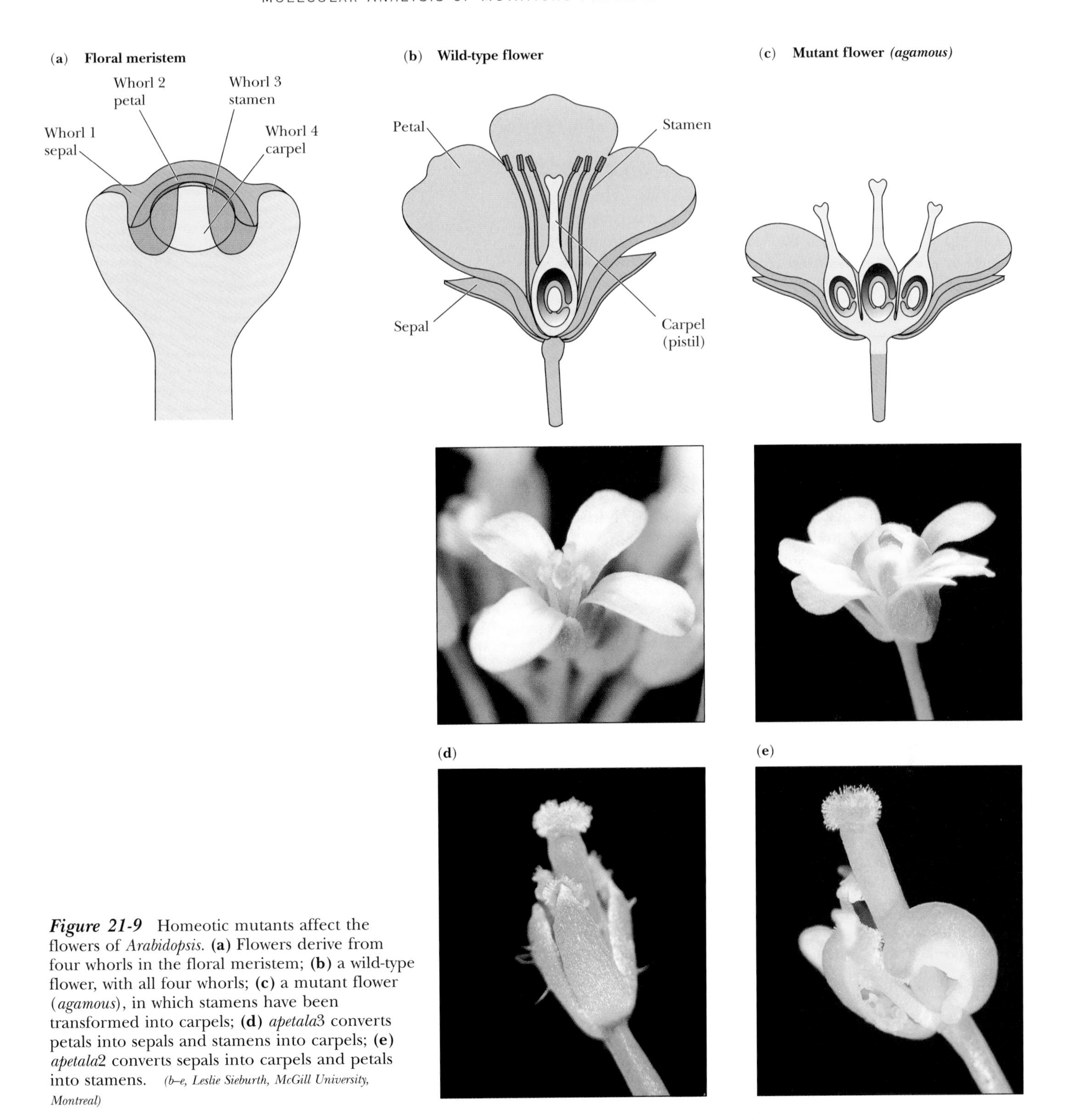

Figure 21-9 Homeotic mutants affect the flowers of *Arabidopsis*. **(a)** Flowers derive from four whorls in the floral meristem; **(b)** a wild-type flower, with all four whorls; **(c)** a mutant flower (*agamous*), in which stamens have been transformed into carpels; **(d)** *apetala3* converts petals into sepals and stamens into carpels; **(e)** *apetala2* converts sepals into carpels and petals into stamens. *(b–e, Leslie Sieburth, McGill University, Montreal)*

Homeotic Selector Genes Establish the Fates of Flower Parts

Figure 21-9 shows abnormal *Arabidopsis* flowers resulting from mutations that resemble homeotic mutations in *Drosophila*. Just as *Antennapedia* transforms *Drosophila* antennae into legs (see Figure 20-28), *agamous* transforms *Arabidopsis* stamens into carpels (Figure 21-9c). Similarly, the *apetala3* mutation converts petals into sepals and stamens into carpels (Figure 21-9d), and *apetala2* converts sepals into carpels and petals into stamens (Figure 21-9e). Each of these **homeotic selector genes** changes the fate of one or more whorls, much as a homeotic gene in *Drosophila* changes the fate of an imaginal disc.

Researchers have identified a number of the homeotic selector genes that participate in the "war of the whorls," that is, the specification of the fate of the four whorls in the floral meristem. Similar genes serve the same roles in flower development in snapdragons. The sequences of the homeotic selector genes reveal a close relationship to one another, just as the homeotic gene in *Drosophila* are closely related. And, just as the homeotic genes in *Drosophila* are clearly related to the Hox genes of mammals, the homeotic selector genes of snapdragon and *Arabidopsis* genes are strikingly similar to one another. The sequence match between these flowering plants, which diverged within the last 125 million years, is less astonishing than that between *Drosophila* and mammals, which diverged at least 600 million years ago. But it is remarkable nonetheless.

The homeotic selector genes are also related to genes previously identified as transcription factors in yeast and animals. Despite the great differences in the mechanism of pattern formation, the development of both plants and animals shares common regulatory strategies and even common molecular designs.

All of the homeotic selector genes discovered so far fall into one of three classes. In order to understand the action of these genes, researchers created a triple mutant that does not express any of the wild-type proteins known to participate in the designation of flower parts. The result is a flower in which all the parts have become leaves. Researchers interpret this result to mean that leaf development is the default pathway (or "ground state") of meristem development. If one or two or three of the homeotic selector genes are expressed, then the whorls instead develop into flower parts.

The homeotic selector genes responsible for flower development provide striking examples of the influence of genes on plant development. The next questions concern the targets of the proteins specified by homeotic selector genes, with a view to understanding the molecular mechanisms of flower development.

HOW DO PLANT EMBRYOS ESTABLISH A BODY PLAN?

The life cycle of a flowering plant consists of two phases, summarized in Figure 21-10: the diploid sporophyte, which produces haploid spores by meiosis; and the haploid gametophyte, which produces haploid gametes by mitosis. The cycling of sporophyte and gametophyte phases is called **alternation of generations.** Most of our discussion here concerns the diploid, sporophyte generation, the dominant phase of the life of a flowering plant. (In nonflowering plants, the gametophyte is often the dominant phase in the life cycle, but we do not discuss these plants

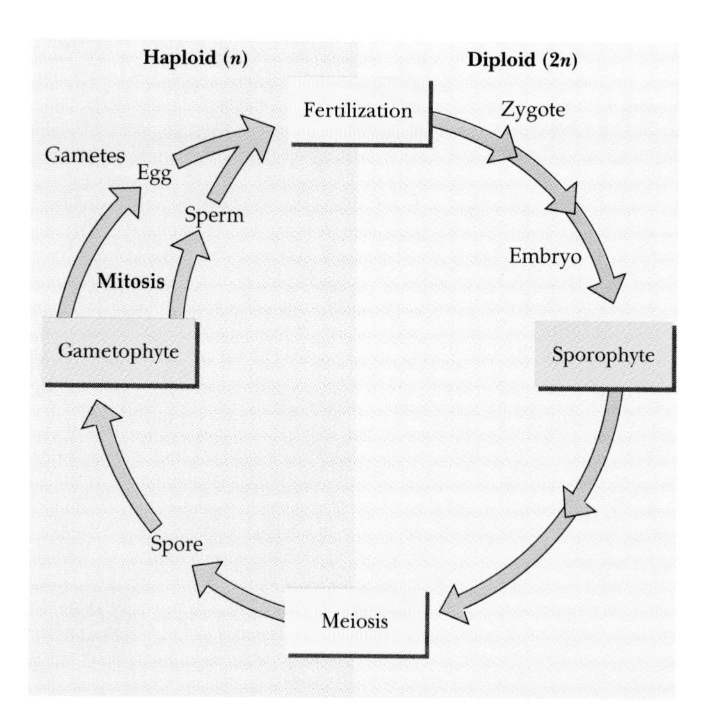

Figure 21-10 Alternation of generations in a flowering plant.

at all.) We begin by describing the development of the gametophytes, the production of gametes, and their union to produce a diploid zygote. Our current understanding of these processes began with careful studies by microscopists in the late 19th century.

Stamens Produce Haploid Microspores, Which Develop into Male Gametophytes

The anther is the pollen-producing part of the stamen, attached to the flower's base with a filament. Within the anther, as illustrated in Figure 21-11a, are four pollen sacs, in which diploid cells undergo meiosis to form haploid microspores [Greek, *mikros* = small], each of which can develop into a male gametophyte. The name "microspore" distinguishes them from the usually larger spores that develop into the female gametophytes.

Each microspore develops into a male gametophyte, which is actually a small, two-celled haploid plant. First the microspore undergoes mitosis to produce two haploid cells, called the generative cell and the vegetative cell, enclosed within a common and often elaborate wall to make up a pollen grain. The nucleus of the generative cell di-

Figure 21-11 Production of male and female gametophytes and gametes. **(a)** Anthers produce male gametophytes within pollen sacs; **(b)** carpels produce female gametophytes within the ovary. *(anther, Astrid & Hanns-Frieder Wichler/Science Photo Library/Photo Researchers; ovule, Ray Evert; embryo sac, Susan Eichhorn)*

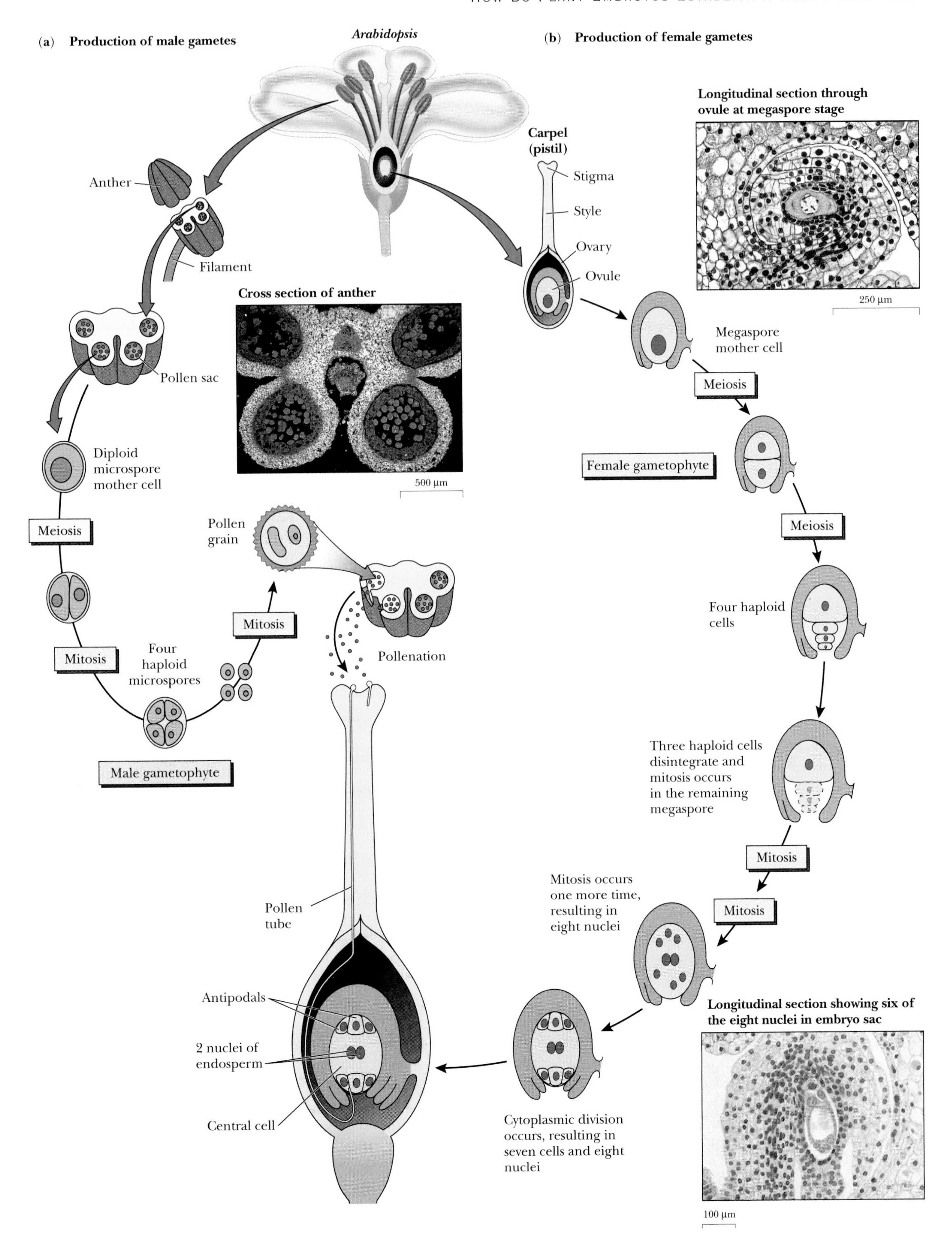
(a) Production of male gametes
Arabidopsis
(b) Production of female gametes
Anther
Filament
Carpel (pistil)
Stigma
Style
Ovary
Ovule
Longitudinal section through ovule at megaspore stage
250 μm
Cross section of anther
500 μm
Pollen sac
Megaspore mother cell
Meiosis
Diploid microspore mother cell
Female gametophyte
Meiosis
Pollen grain
Meiosis
Four haploid cells
Mitosis
Mitosis
Four haploid microspores
Pollenation
Male gametophyte
Three haploid cells disintegrate and mitosis occurs in the remaining megaspore
Mitosis
Mitosis occurs one more time, resulting in eight nuclei
Mitosis
Pollen tube
Antipodals
Longitudinal section showing six of the eight nuclei in embryo sac
2 nuclei of endosperm
Central cell
Cytoplasmic division occurs, resulting in seven cells and eight nuclei
100 μm

vides once again, giving two sperm nuclei surrounded by a little cytoplasm, as shown in Figure 21-11a. The anthers release pollen grains, which may contain either two or three haploid nuclei at the time of their release. Insects, wind, or other agents may then carry a pollen grain to a receptive carpel.

Carpels Produce Haploid Megaspores, Which Develop into Female Gametophytes

In the broad base of the carpel, illustrated in Figure 21-11b, is the **ovary** [Latin, *ovum* = egg], which produces the female gametophytes. A projection, called the *style,* connects the ovary to the *stigma,* the structure that receives the pollen grains.

The wall of the ovary produces one or more **ovules** [Latin, *ovulum* = little egg], structures (illustrated in Figure 21-11b) which, after fertilization, form a seed. One diploid cell in each ovule undergoes meiosis to form four haploid cells. One of these, called the **megaspore** [Greek, *megas* = large] develops into the embryo sac, the mature female gametophyte, while the other three haploid cells degenerate. The diploid tissue from which the megaspore develops and which continues to surround the megaspore and the mature gametophyte is called the *nucellus.* The megaspore nucleus undergoes three rounds of mitosis to form eight haploid nuclei. These eight nuclei, however, form only seven cells. One cell, called the **central cell,** contains two nuclei. One of the other six cells of the embryo sac is the egg. The seven cells together make up a haploid plant, the female gametophyte.

While the haploid gametophyte is developing, the surrounding diploid cells of the ovule are also dividing, as also shown in Figure 21-11b. The expanded ovule encloses the gametophyte in the *integument,* a fold of tissue that eventually becomes the seed coat.

Flowering Plants Undergo Double Fertilization

The pollen grain develops further after it lands on the stigma, producing an extension called a **pollen tube,** derived from the vegetative nucleus, which carries the sperm nuclei into the ovule. As shown in Figure 21-12, the pollen tube grows through a pore within the integument to reach the embryo sac.

Flowering plants differ from all other organisms in undergoing **double fertilization,** the simultaneous fusion of two sperm nuclei from the pollen tube with two cells of the embryo sac, as shown in Figure 21-12. One sperm nucleus fuses with the egg nucleus to form the diploid zygote, which develops into an embryo and eventually into a new sporophyte plant. Later in this chapter we discuss how the embryo develops its shoot-root polarity and its apical meristems.

The other sperm nucleus fuses with the two nuclei of the central cell to form a triploid nucleus. This *endosperm nucleus* gives rise to the seed's *endosperm,* a triploid tissue that provides nourishment and hormones for the growing embryo and for germination (as discussed later in this chapter). Both the zygote and the endosperm nucleus undergo mitosis, starting shortly after fertilization. The seed consists of the embryo, the endosperm, the nucellus, and the integument.

Seeds Contain Embryos Whose Development Has Paused

The pattern of cell divisions in the early embryo permanently fixes the relative positions of cells and the shape of the embryo. During the formation of a mature seed a typical dicot embryo, for example, changes its shape from a sphere to a heart to a torpedo, as shown in Figure 21-12.

The first cell division in a dicot embryo divides the zygote into two unequal cells, called the basal cell and the terminal cell. The larger basal cell divides several times to form a narrow column of cells called a *suspensor* (shown in Figure 21-12), which attaches the embryo to surrounding tissue and nourishes the embryo. The suspensor does not develop further but functions only within the seed.

The smaller terminal cell divides to form a spherical embryo (shown in Figure 21-12). The part of the embryo attached to the suspensor develops into the root apical meristem, while the part farthest from the suspensor develops into a shoot apical meristem. The fundamental polarity of the embryo is established early in plant development, as it is in animal development.

Continuing division of the spherical embryo leads to the formation of three cell layers, which roughly correspond to the three germ layers of vertebrate embryos. These layers produce the three layers of the mature plant —the **epidermis,** the outermost layer of the embryo and the plant; the **ground tissue,** which occupies most of the interior of the embryo and the plant; and the **vascular system,** which serves as a transport system of the plant.

Cells soon begin to divide more rapidly at the end of the embryo away from the suspensor, producing an embryo with a heart-like shape, as shown in Figure 21-12. By this time the embryo has developed apical meristems for both root and shoot, as shown in Figure 21-12.

The continued division of cells in the shoot end produces "seed leaves," or **cotyledons** [Greek, *kotyle* = a deep cup]. The number of cotyledons divides flowering plants into two huge categories, dicots and monocots. **Dicots**—like *Arabidopsis* or peas—have two cotyledons, while **monocots**—like corn or grass—have one. The cotyledons can absorb nutrients from the endosperm and later provide them to the developing plant. As the cotyledons grow longer, they form a stage that in many species resembles a torpedo, as shown in Figure 21-12.

Most of the embryo's cells come to resemble the large, differentiated cells of a mature plant, except that their plastids do not yet contain chlorophyll. The meri-

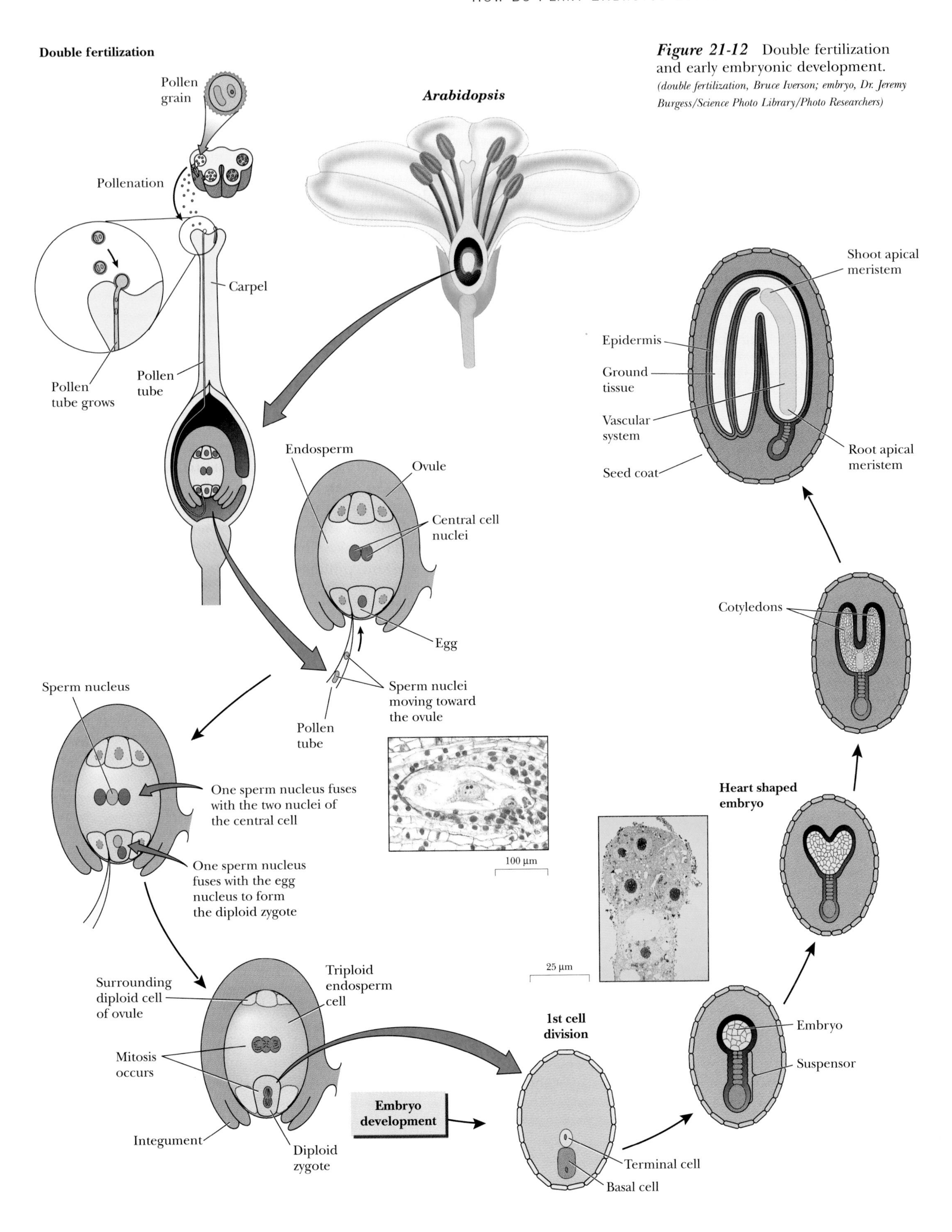

Figure 21-12 Double fertilization and early embryonic development. *(double fertilization, Bruce Iverson; embryo, Dr. Jeremy Burgess/Science Photo Library/Photo Researchers)*

stem cells remain small and undifferentiated. The shoot apical meristem is nestled between the cotyledons, just above their points of attachment, while the root apical meristem lies near the point of attachment of the suspensor. Many plants, especially woody plants, later develop secondary meristems, which are responsible for secondary growth, the increase in the thickness of roots and shoots.

After the torpedo stage, the seed loses most of its water. The seed coat, derived from the integument, surrounds the embryo and the endosperm. Development all but stops, and the embryo enters a **quiescent** stage, in which growth is suspended until restarted by environmental cues. The resumption of growth is called **germination** [Latin, *germinare* = to sprout].

Before describing the process of germination, we must first define the terms used to describe the parts of the seed and the embryo. We do so for a bean, a dicot seed.

Figure 21-13 shows a bean germinating to form a seedling. The two cotyledons, which occupy most of the space inside the seed coat, are attached to a stem-like axis. Above the attachment point of the cotyledons, the axis is called the **epicotyl** [Greek, *epi* = above], and below it is called the **hypocotyl** [Greek, *hypo* = under]. In the soybean and in many (but not all) other plants, the epicotyl also contains the *plumule* [Latin, *plumula* = a small feather], which develops into the first foliage leaves. At the end of the hypocotyl is the *radicle,* or primordial root, which contains the root's apical meristem.

Species vary widely in the total amounts of food reserves and the final sizes of their seeds. Lettuce and tomato seeds are about 1–2 mm in diameter, while peas, beans, and corn are about 10 times larger. The smallest known seeds are the microscopic seeds of orchids, and the largest are the "double coconuts" of the Seychelles nut palm, about half a meter across.

How Does the Plant Resume Development?

Germination begins with **imbibition,** the taking in of water, during which the seed may double or triple in size. Imbibition restarts development: The quiescent embryo awakes, with a burst of metabolic activity, cell division, and growth.

The first part of the embryo to emerge from the seed coat is the radicle and the attached hypocotyl, as shown in Figure 21-13. The radicle immediately turns downward and quickly forms a functioning root system that provides the growing plant with water and minerals.

In the soybean and many other dicots, the top of the hypocotyl forms a hook that pushes through the soil toward the sun, pulling the cotyledons and epicotyl behind,

Seed germination and growth of a bean plant, a dicot

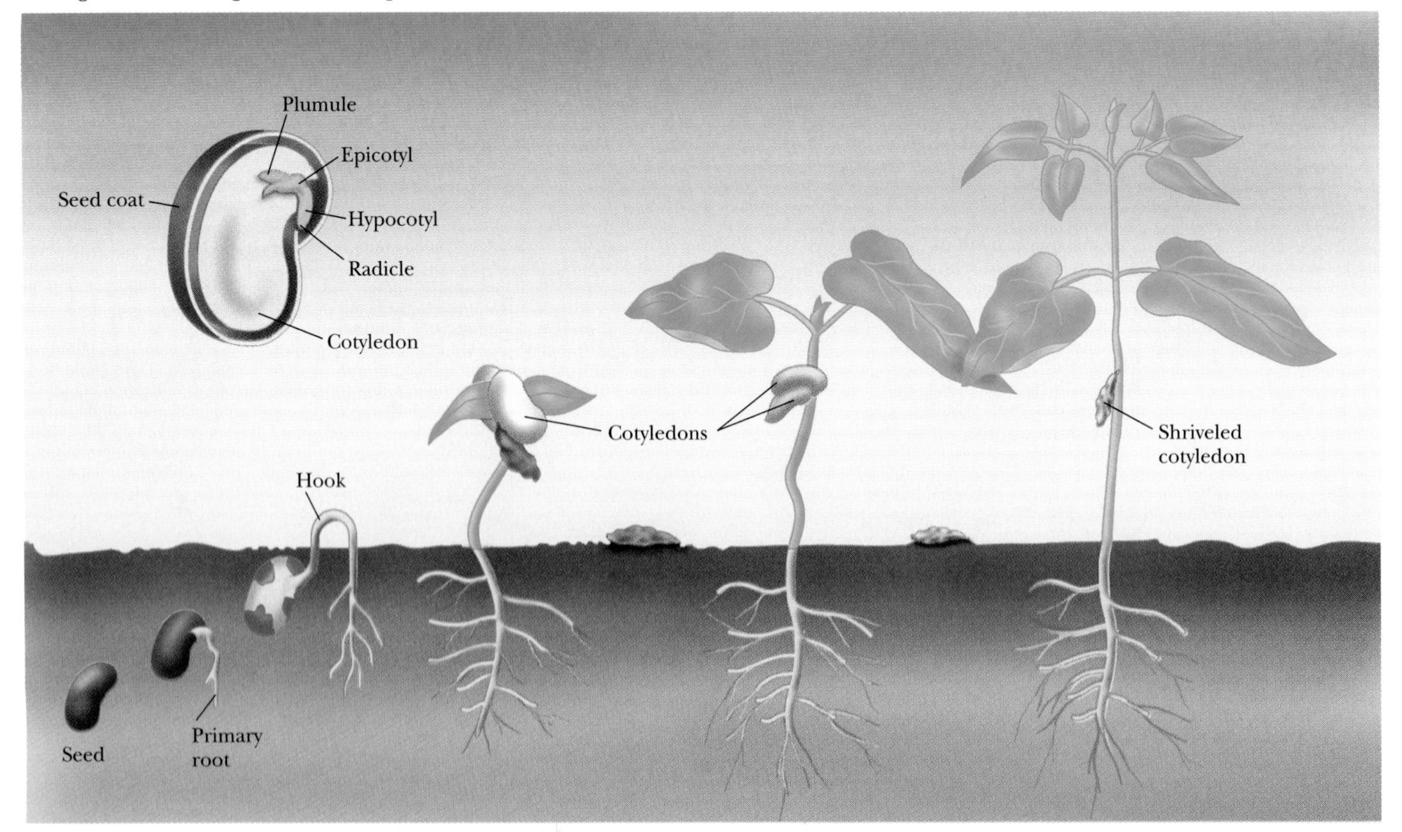

Figure 21-13 Germination in a bean, a dicot.

as shown in Figure 21-13. The arching hypocotyl hook then straightens, the cotyledons spread, and the epicotyl points upward and resumes its growth. In these plants, most of the stem comes from the epicotyl, with only the first centimeter or so (below the cotyledons) deriving from the hypocotyl.

We can see the effect of environment on plant growth by keeping a germinating seedling in the dark. Instead of straightening, the epicotyl arch keeps growing upward. The plant displays a condition called **etiolation** [French, *etioler* = to blanch or whiten], with a thin, spindly appearance, poor leaf development, and no chlorophyll production—familiar in bean sprouts. The abnormal appearance of such sprouts shows that light is important to the normal development of a plant.

In a seed, a plant embryo may suspend its growth for days, months, years, decades, even centuries, depending on the species. The record for proven seed longevity comes from lotus in a former swamp bed in Manchuria. Radioactive dating of these seeds showed that they were about 1000 years old, yet they still germinated. Still more amazingly, researchers have succeeded in growing plants from lupine seeds that were found in 10,000-year-old frozen sediments in Alaska, but the ages of the seeds themselves are unknown. Botanists have concluded that seeds do not have a built-in clock that determines the time of germination but instead depend on cues from the environment.

Plants use a wide variety of mechanisms to prevent premature germination. In many legumes (although not in domesticated and edible ones such as beans and peas), for example, a hard seed coat prevents the entrance of water or oxygen needed for germination. In these and other species, germination often requires *scarification,* harsh treatment to break the seed coat barrier. In nature, scarification conditions include fire, the acid and enzymes of an animal's digestive system, scraping against sand or rocks, successive freezing and thawing, and even the action of a fungus. In the laboratory, researchers may accomplish the scarification of such seeds with sandpaper, alcohol, or even sulfuric acid. Farmers also use scarification to promote the germination of crop seeds.

Other species prevent premature germination with chemical inhibitors. Rainfall then stimulates germination by washing away the inhibitors. Different species use widely varying inhibitory chemicals, including sodium chloride, cyanide, mustard oils, and other organic compounds.

The most widespread compound known to inhibit germination is the plant hormone **abscisic acid,** whose structure is shown in Figure 21-14a. Researchers have shown that abscisic acid inhibits the translation of the messenger RNAs in seeds, preventing the synthesis of enzymes necessary for germination. Abscisic acid is important in preventing precocious germination during seed development. Some mutant corn plants, for example, fail to produce abscisic acid, and their seeds germinate while the ear

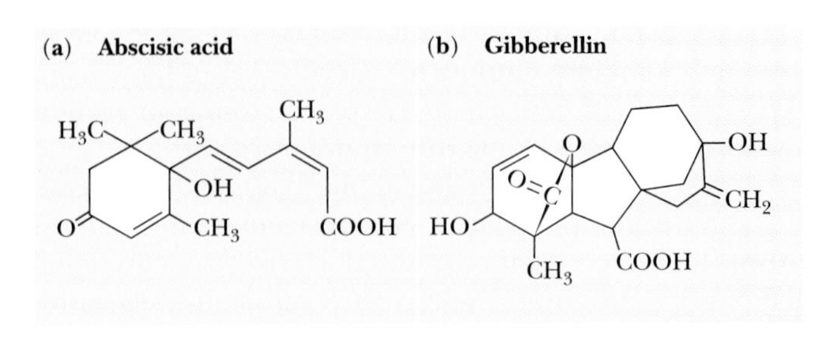

Figure 21-14 Plant hormones help regulate germination. **(a)** Abscisic acid inhibits germination; **(b)** gibberellin induces the expression of enzymes that digest foodstores within the endosperm.

is still attached to the plant. We discuss other actions of abscisic acid later in this chapter.

Cells in the quiescent embryo are not dead, and they always maintain a low level of metabolism. Germination, however, triggers a huge increase in biochemical activity, with energy and building blocks derived from the breakdown of stored starches, oils, and proteins. The presence of these stores also means that seeds are rich foods for a wide variety of animals, including ourselves.

The breakdown of these stores requires the action of enzymes that are not active, not present, or present in much lower amounts in the ungerminated seed. In barley, wheat, and other cereal grains, these enzymes are secreted by the **aleurone layer,** a border of cells that surrounds the endosperm. After germination begins, the aleurone layer secretes enzymes that digest the starches and other storage molecules of the endosperm. One of these enzymes, for example, is α-amylase, which helps to break down starches.

When researchers remove a barley embryo from endosperm and aleurone layer, the endosperm stays undigested. This experiment shows that the embryo itself must stimulate the aleurone to make digestive enzymes. Plant physiologists used isolated aleurone to identify the exact chemical signal that acts on the aleurone layer: It is gibberellin, the same hormone that stimulates the elongation of seedlings, as mentioned earlier in this chapter. Figure 21-14b shows the chemical structure of one form of gibberellin. As little as 10^{-11} gram (one 100 billionth of a gram) of pure gibberellin induces the expression of the gene encoding α-amylase. We discuss other actions of gibberellin later in this chapter.

HORMONES DIRECT DEVELOPMENT AND COORDINATE ACTIVITIES IN CELLS THROUGHOUT THE PLANT

Meristems not only can produce repeating modules of shoot, flower, and root but can even produce whole new plants. **Vegetative reproduction** is the process of producing a new plant without fertilization or seed production. Two fairly familiar examples of vegetative reproduction are the formation of new strawberry plants from runners and the growth of new plants from houseplant cuttings.

Vegetative reproduction allows plants with a successful combination of genes to produce genetically identical offspring, while sexual reproduction leads to new combinations of genes from the two parental plants. Vegetative reproduction generally avoids the most fragile stage in the sexual cycle—the newly germinated seedling. In addition to its widespread occurrence in nature, vegetative reproduction has allowed farmers, gardeners, horticulturists, and researchers to propagate plants with desirable properties.

Vegetative reproduction depends on the ability of meristems to produce the whole range of cell types of a mature plant. That is, shoot meristems can form roots, and vice versa. A rootless cutting of a plant put into a glass of water can generate new roots. A root meristem forms from a group of thin-walled (parenchyma) cells, the **callus,** undifferentiated tissue that forms at the cut surface. Such new meristems are important for vegetative reproduction, both in nature and in the laboratory or garden. The differentiation and growth of a callus depend strongly on the action of plant hormones.

As discussed in Chapter 19, a **hormone** is an organic compound produced in a tissue or organ and transported to another tissue or organ, called the **target,** where it produces one or more specific effects. Plant hormones move from source to target both by diffusion from cell to cell and by transport in the vascular system. Unlike most animal hormones, however, plant hormones not only affect distant targets but also influence the very cells that make them.

Plant physiologists have identified five major plant hormones: auxin, gibberellin, cytokinin, ethylene, and abscisic acid. In fact, auxin, gibberellin, and cytokinin are not single compounds, but each is a small family of related compounds. There are three naturally occurring auxins, at least 60 gibberellins, and several cytokinins. Other related compounds can trigger the same biological responses but are not made by plants. These include compounds isolated from animals, fungi, and bacteria, as well as the products of laboratory synthesis.

In the laboratory, plant tissue placed on a nutrient medium can develop into a callus. The development of the callus into a plantlet depends on its chemical environment. Because a callus usually does not perform photosynthesis, an artificial culture medium must provide the growing cells with an energy source, usually sucrose, as well as with needed minerals.

As shown in Figure 21-15, two plant hormones—auxin, already mentioned, whose structure is shown in Figure 21-16a, and **cytokinin**, whose structure is shown in Figure 21-16b—also influence the growth and differentiation of a callus culture. For example, when the ratio of auxin to cytokinin is low, the callus develops shoot parts—buds, stems, and leaves—but no roots. At higher ratios of auxin to cytokinin, however, roots develop as well, regenerating an entire plant. Auxin and other plant hormones influence meristem development in culture, as they do in intact plants. Later in this chapter we discuss the ways in which these hormones coordinate other responses of plants to their environments.

Auxin Coordinates Many Aspects of Plant Growth and Development

The first naturally occurring plant hormone to be recognized was the most common natural auxin, indoleacetic acid (IAA), a derivative of the amnio acid tryptophan. The discoverer of auxin was a young Dutch researcher named Frits Went. In 1926, when he discovered IAA, Went was serving his compulsory military service during the day and spending his evenings and nights working as a graduate student in his father's plant physiology laboratory at the University of Utrecht.

The question that occupied Went and his father was that of **phototropism,** the bending of plants toward light. Some 45 years earlier, another father and son, Charles and Francis Darwin, had shown that something made in the tip of oat seedlings somehow caused the seedlings to bend toward the light, as illustrated in Figure 21-17a. When the Darwins masked the tips of the seedlings with a light-proof foil, the seedlings no longer bent toward the light. The same mask applied below the tip had no effect on bending.

Went set out to isolate the growth-promoting substance from the tip, using the strategy illustrated in Figure 21-17b. On a tiny cube of gelatin, he placed as many tips as he could fit. After an hour, he removed the tips, reasoning that the hypothetical substance would have entered the block. He then applied the block to one side of a seedling stump that was growing in the dark and waited to see what would happen. At first nothing happened. Hours later, at 3 AM, however, the seedling curved away from the gelatin cube. Excitedly, Went ran home, burst into his parents' bedroom, and declared, "Father, come and see, I've got the growth substance!" His father sleepily told him to repeat the experiment during the day. "If it is any good," the elder Went said, "it will work again, and then I can see it." The younger Went named the growth substance auxin. The isolated substance turned out to be IAA.

Many other researchers have repeated Went's experiment with oat seedlings, using a variety of natural and synthetic auxins. A fixed amount of auxin gives a reproducible amount of curvature. Went's experiment established a **bioassay,** a method that estimates the concentration of a substance by measuring its biological activity. This bioassay allowed plant physiologists to estimate auxin concentrations by measuring seedling curvature with a pro-

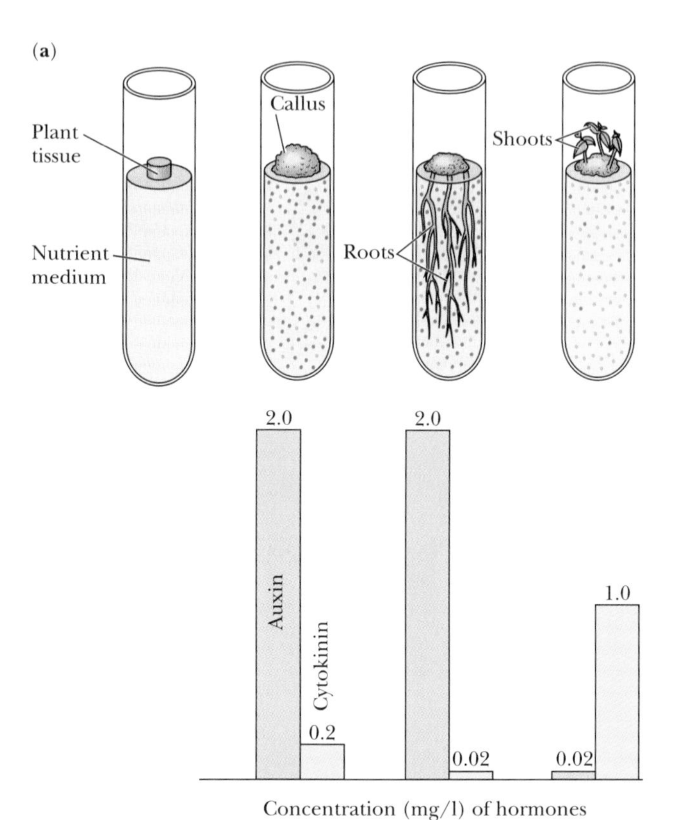

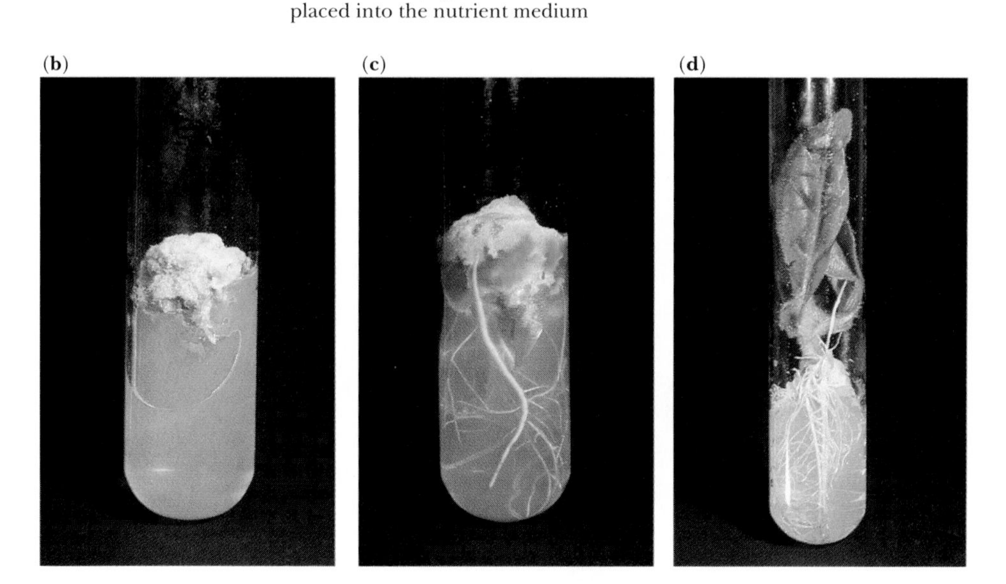

Figure 21-15 The cut surface of a plant can generate a callus, from which a new plant can grow. **(a)** in tissue culture, the growth and differentiation of a callus respond to differing concentrations of the plant hormones auxin and cytokinin; **(b)** an undifferentiated callus; **(c)** a callus differentiating to produce roots; **(d)** a callus differentiating to produce a shoot. *(b–d, Carolina Biological Supply Company)*

(a) Indoleacetic acid (IAA, an auxin)

$H_2C-COOH$

(b) Zeatin (a cytokinin)

$HN-CH_2-CH=C(CH_3)(CH_2OH)$

(c) Ethylene

$H_2C=CH_2$

Figure 21-16 Other plant hormones: **(a)** an auxin, indoleacetic acid (IAA); **(b)** zeatin (a cytokinin); **(c)** ethylene.

tractor. More recently, however, they have measured the concentration of auxin and other plant hormones by chromatography, which depends on chemical rather than biological properties. Other bioassays have been useful for the identification of hormones and other functionally important molecules not only in plants, but in other types of organisms.

Went's experiments suggested that auxin could mediate phototropism if light somehow caused more auxin to be concentrated on the side of the tip away from the light. In the early 1960s, Winslow Briggs, a plant physiologist at Stanford University, tested this prediction in the experiment shown in Figure 21-17c. First, Briggs asked whether light affected the total amount of auxin in the growing tip

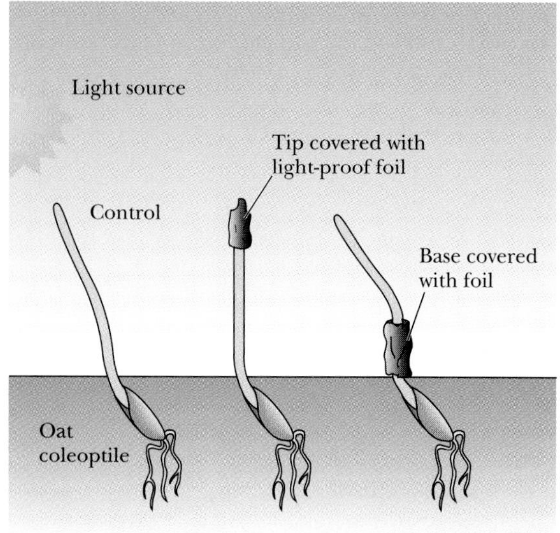

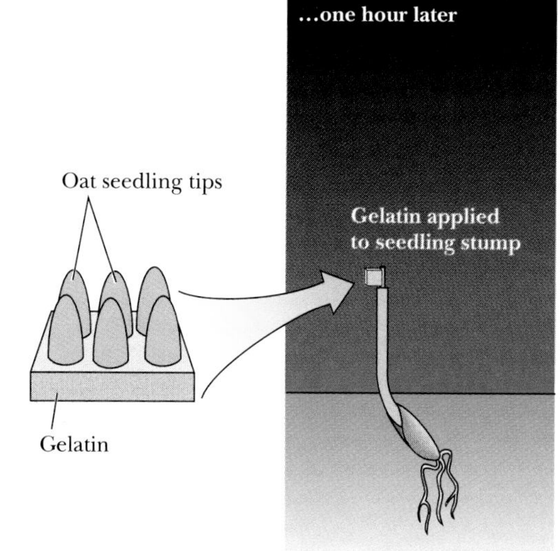

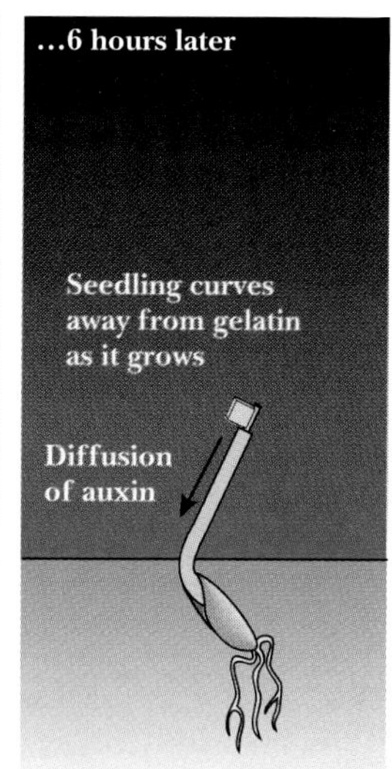

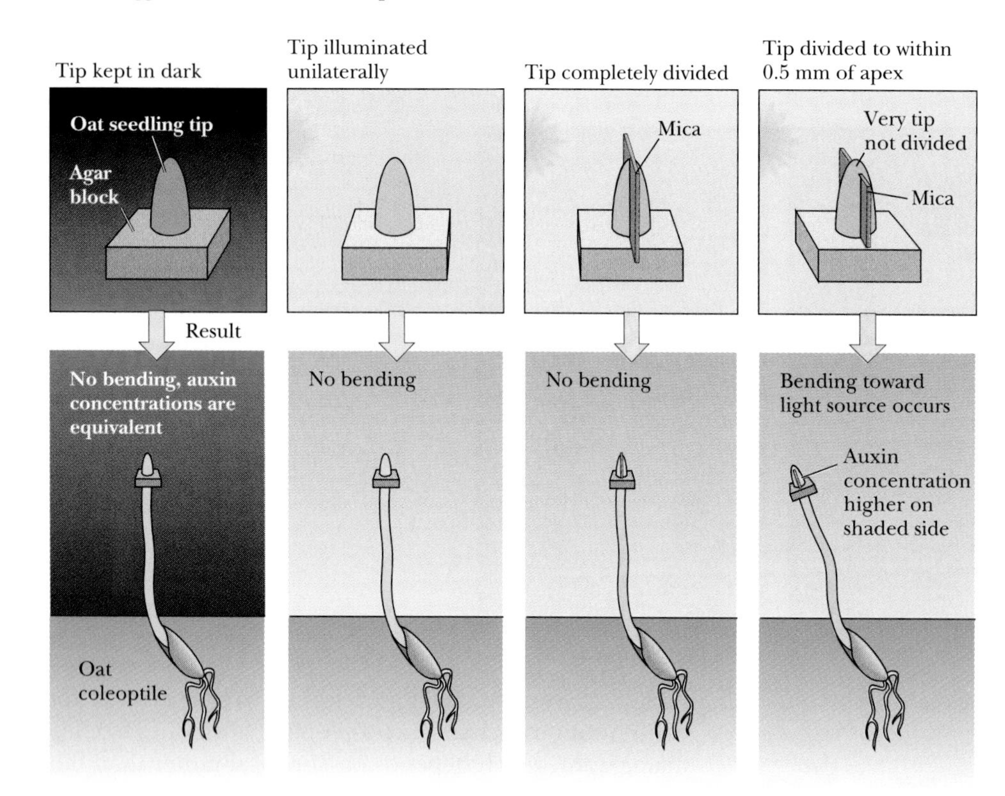

Figure 21-17 Successive attempts to understand how plants accomplish phototropism. **(a)** The experiment of Charles and Francis Darwin showed that phototropism in oat seedlings depends on the action of cells in the tip; **(b)** the experiment of Frits Went showed that a diffusible substance (auxin) is responsible for differential growth; **(c)** the experiment of Winslow Briggs showed that differential growth depends on the cellular distribution of auxin within the tip.

of an oat seedling. After placing tips on blocks of agar, he illuminated some tips and kept others in the dark. He then used the curvature bioassay to estimate the amount of auxin in each tip and found no difference in the amounts of bending, therefore the amount of auxin in the two sets of tips was the same. Phototropism, then, must depend not on the total amount of auxin in an illuminated tip but on the distribution within the tip.

Briggs therefore looked for differences in auxin concentrations between the side of the tip nearer the light and the side away from the light. He inserted a thin piece of mica between the two sides. The mica completely separated the lighted side from the shaded side, and Briggs found no difference in the auxin levels of the two sides. When Briggs left a gap of 0.5 mm near the very end of the tip, however, the auxin concentration on the shaded side was nearly twice that on the illuminated side. The light somehow caused the transport of auxin from the illuminated to the shaded side of the tip. The higher concentration of auxin on the shaded side then caused that side to elongate more rapidly, leading to a curve in the direction of the light.

Briggs' experiments, like those of Went, suggested that auxin acts at the level of individual cells, stimulating cell elongation in both shoots and roots. Elongation requires that a cell's rigid wall must loosen, its contents expand, and the cell wall reform around the expanded cell. Most plant cells exert **turgor pressure,** osmotic pressure against their cell walls resulting from the higher concentration of dissolved molecules and ions in cytoplasm than in the surrounding fluid. (See Chapter 6, p. 161.) The primary event in cell elongation is the selective breaking of the bonds that make the wall rigid.

In addition to its effects on cell elongation, auxin also affects the gene expression in target cells. It stimulates the synthesis of several messenger RNAs and proteins. While researchers do not yet know the identities or functions of the affected proteins, some of these changes occur rapidly, within 10–20 minutes after exposure to auxin. It seems likely that such changes underlie the action of auxin in stimulating new patterns of plant growth, such as the stimulation of adventitious roots.

Besides controlling cell elongation and stimulating the formation of adventitious roots, auxin affects the growth of many other plant tissues. For example, the relative concentrations of auxin and gibberellin in vascular tissue help establish the developmental fate of individual cells—whether the cells develop into xylem, which transports water and minerals from the roots, or phloem, which transports sucrose and other materials from the leaves. Auxin also affects the course of development of plant cells in tissue culture. The mechanism of these hormonal influences is a subject of current investigation, particularly at the level of transcriptional regulation.

In addition to its effects on meristems, auxin stimulates the development of fruit, sometimes even in the absence of fertilization. Unfertilized tomato flowers treated with auxin, for example, develop into seedless tomatoes.

Besides its growth-promoting activity, auxin also inhibits the further development of lateral buds. In plants that show apical dominance, lateral buds do not develop into branches unless the terminal bud is removed. Even after removing the terminal bud, however, lateral buds do not grow if one places auxin on the cut surface. (See later, Figure 21-19.)

Finally, high concentrations of auxin are effective as herbicides; that is, they kill plants, although no one knows exactly why. Synthetic auxins are especially effective because plants cannot break down these substances, and so they persist for a long time. Three synthetic auxins—2,4-D [2,4-dichlorophenoxyacetic acid], 2,4,5-T [2,4,5-trichlorophenoxyacetic acid], and MCPA [methylchlorophenoxyacetic acid]—have been particularly popular because they are especially toxic to dicots and not to monocots. Grasses, corn, and wheat are monocots, so farmers and lawn keepers can selectively kill dicot weeds in fields of corn or wheat, and ranchers can kill sagebrush and mesquite, both dicots, in grassy pastures or rangelands.

How Do Gibberellins Act on Target Cells?

Plants have more different kinds of gibberellins than of any other plant hormone. All the gibberellins are complicated organic molecules with 19 or 20 carbon atoms grouped into four or five ring structures. Most plant species have several different kinds of gibberellin, and some contain as many as 15. Figure 21-14b shows the chemical structure of just one gibberellin, called GA_3. Some researchers think that most of the different forms are precursors or degradation products of a few active forms. Others think that different forms may have different biological effects.

Gibberellins generally affect the overall growth of intact plants far more than the other plant hormones. In fact, their discovery arose from studies of a disease of rice plants caused by a fungus called *Gibberella fujikuroi.* The disease, first described in Japan in the 1890s, causes rice plants to grow so tall that they cannot support their own weight. The Japanese researchers who studied this condition, shown in Figure 21-18, called it "foolish seedling" disease. In the 1930s two Japanese scientists identified the chemical compound, made by the fungus, that is responsible for the extravagant growth of the seedlings. The active compound is the gibberellin GA_3. In the 1950s, plant physiologists showed that gibberellins are also present in uninfected plants and are likely to play a role in the normal regulation of plant growth.

In one experiment, the effect of an added gibberellin on the elongation of segments of oat stem was studied. Gibberellin and sucrose (to provide energy to the nonphotosynthesizing tissue) stimulated a 15-fold increase in length without any increase in cell division. This experiment suggests that gibberellins, like auxin, may promote cell elongation by loosening the attachment of cell walls and protoplasts.

Gibberellins also stimulate the breakdown of starches and sucrose to monosaccharides. The increased glucose and fructose concentrations not only increase the availability of energy but also lower the water potential of the

Figure 21-18 Foolish seedling disease in rice plants, caused by a fungus, *Gibberella fujikuroi.* In an infected plant (left), the parasite releases gibberellin, which causes the stems to elongate rapidly and grow weak and spindly. A healthy, uninfected plant is shown on the right. *(International Rice Research Institute, Manila, Philippines)*

Figure 21-19 Witches' broom disease, caused by cytokinin produced by bacteria. *(Bruce Iverson)*

protoplasts. More water then flows into cells and contributes to their elongation.

In addition to loosening cell walls and increasing hydrolysis of starches, gibberellins can, in some cases, stimulate cell division—for example, in the apical meristem of the shoot. Finally, as already mentioned, gibberellins can act to change the pattern of expression of specific genes, as in the case of α-amylase induction in the germinating barley seed.

Some plant mutants—of peas, corn, beans, rice, and other crops— lack gibberellins altogether and are much shorter than their normal counterparts. When these dwarf mutants are treated with gibberellins, in some cases with as little as 10^{-9} gram per plant, they grow to normal size. The increased growth of dwarf peas has provided a valuable bioassay for the gibberellins.

Gibberellins appear to be important in the normal growth of shoots, the germination of seeds, the flowering of plants, and the mobilization of food reserves from the endosperm and cotyledon of monocots. Their precise role, however, has been difficult to understand because there are many of them, they are present in many tissues, and they are active at extremely low concentrations.

Cytokinins Stimulate Cell Division and Growth

Since the early years of this century, researchers have known that some substances, now called cytokinins, stimulate cytokinesis (cell division) in plant tissue culture. Pieces of vascular tissue and extracts of coconut milk can produce this stimulation. In all cases, however, these materials stimulated cell division only in the presence of appropriate levels of auxin.

In addition to stimulating cell division, cytokinin and auxin together establish the developmental fate of plant tissue, both in artificial culture and in intact plants. For example, the plant tissue culture experiment shown in Figure 21-15 shows that the relative levels of cytokinin and auxin establish whether a callus culture will continue to divide without differentiation or will differentiate into shoots or roots.

Cytokinin and auxin also have contrasting effects on the growth of lateral buds. Auxin, applied to a cut stem, suppresses the growth of lateral buds. On the other hand, cytokinin, applied directly to a lateral bud, stimulates cell division and growth. A plant infection called "witches' broom disease," illustrated in Figure 21-19, provides an extreme example of the effects of cytokinin. The disease-causing bacteria secrete cytokinin, which stimulates the increased lateral branching. Although experimentally applied cytokinin and auxin affect the pattern of terminal and lateral branching, researchers are still not certain of the precise roles of these hormones in the normal regulation of lateral branching.

When a leaf is removed from a plant, it usually begins to lose chlorophyll and to turn yellow. This yellowing is part of the process of **senescence,** the breakdown of cellular components leading to cell death. Application of cytokinin to a cut leaf, however, delays senescence.

During the normal life cycle of a plant, cytokinin transported to leaves by the xylem appears to prevent senescence from occurring prematurely. Several fungi, and even two species of caterpillars, keep up their food supply in leaves by secreting cytokinins to create "green islands" on otherwise yellowing leaves.

Researchers usually determine the chemical structure of a hormone by using a bioassay to detect the hormone as they purify it by chromatography or other methods of chemical separation. In the case of cytokinin, however, this approach was extremely difficult. First, plant tissues contain only tiny amounts of cytokinins, limiting the number of different purification procedures that could be tried. And, second, the bioassay for cytokinins was the stimulation of cell division, which often took as long as 2 weeks to observe. So trying each new purification scheme required large amounts of starting material and a long time. Nonetheless, Folke Skoog, a plant physiologist at the University of Wisconsin, achieved a 1000-fold enrichment of cytokinin activity from an extract of coconut milk. In spite of this accomplishment, however, the cytokinin was still not pure enough to determine its chemical structure.

Skoog's laboratory eventually succeeded in determining the structure of a cytokinin, but only after using a different approach. Carlos Miller, then a postdoctoral fellow, began to test known chemicals (from the laboratory's shelves) for cytokinin activity, hoping to find support for a proposal that cytokinins were somehow involved in RNA or DNA metabolism. The material that was most effective in supplying cytokinin activity came from an old bottle labeled "herring sperm DNA." Miller and Skoog, together with others from Skoog's laboratory, finally succeeded in identifying the chemical structure of the active substance. They called this compound "kinetin" and demonstrated that it was a derivative of adenine (a nitrogenous base in DNA and RNA). Kinetin was present in the bottle because the original DNA had degraded with time to produce a variety of nucleotide derivatives.

Kinetin itself does not occur in plants. Other cytokinins, however, do, and all of these are also derivatives of adenine, with side chains rich in carbon and hydrogen. As expected from the early experiments, the naturally occurring cytokinins are present at exceedingly low levels. The first cytokinin isolated from plant tissue, for example, required the extraction of 60 kg of corn kernels to obtain 1 mg of pure compound. Researchers have found only a few cytokinins in plant tissues, but a number of other compounds, including natural products of several fungi and bacteria, have cytokinin activity.

Modern techniques (such as high-pressure liquid chromatography) have made it possible to determine the amounts of specific cytokinins at levels as low as 1 ng (10^{-9} g). This sensitivity has allowed the measurement of cytokinin concentrations in many plant tissues. At this time, however, researchers have not been able to demonstrate changes in cytokinin concentrations that may account for the response of plants to changed environmental cues. Some plant scientists therefore think that it is incorrect to call cytokinins hormones. Nor does anyone yet know exactly how cytokinins act on plant cells, although it seems likely that they affect the pattern of gene expression.

How Did Researchers Identify Ethylene as a Plant Hormone?

Ethylene, whose structure is shown in Figure 21-16c, is a much simpler molecule than the other plant hormones, consisting of just two carbon atoms and four hydrogen atoms. Unlike the other plant hormones, ethylene is a gas.

The most dramatic, delicious, and economically important effect of ethylene is on fruit ripening. The ripening of a fleshy fruit (like an apple) is a complex process. Among the chemical changes that occur during ripening are the breaking down of chlorophyll, changing the fruit's color. Ripening also involves the breakdown of starches, increasing the fruit's sweetness, and the breakdown of cellulose, making it softer. Ethylene starts or speeds up these changes.

Ethylene also triggers the increased production of more ethylene, so that once ripening starts it spreads rapidly, both within a single fruit and from fruit to fruit. This acceleration of ripening has practical consequences for fruit lovers. One can hasten the ripening of green fruit by keeping them in an enclosed bag to trap the ethylene. It also explains why "one rotten apple spoils the barrel."

Fruit growers and shippers knew how to hasten ripening long before they knew the chemical cause. The ancient Chinese ripened fruit in a room with burning incense, and Puerto Rican growers built bonfires near their crops to ripen their pineapple crops. As we now know, both these procedures worked because the incense and wood fires produced ethylene.

The identification of ethylene as a plant hormone came from a practical problem first noticed in Germany in 1864. Before electrical lights lined city streets, gas lines carrying "illuminating gas" supplied fuel to street lights. Leaks developed in several such lines, causing shade trees to lose their leaves. The damage was due to the ethylene in the leaking gas because another effect of ethylene is to promote the dropping off of leaves as well as fruit. The demonstration that the active component of illuminating gas was ethylene came, however, only in 1901. A Russian graduate student named Dimitry Neljubov showed that illuminating gas stimulated the horizontal growth of pea plants and that the active component of the gas was ethylene.

Ethylene is present almost everywhere in a plant. In addition to its effect on fruit, it participates in many other responses to environmental changes. For example, as shown in Figure 21-20, ethylene causes a *triple response* in the stems of tomato seedlings—stems thicken, stop elongating, and begin to grow horizontally. Tomato seedlings grow in a similar way if they encounter a barrier as they emerge through the soil. The production of ethylene, primarily in the hook of the seedling's epicotyl, appears to coordinate the ability to grow around an interfering object. When the seedlings find their way to light, the amount of ethylene decreases, and the seedlings resume their up-

(a)

(b)

Figure 21-20 The triple response to ethylene in seedlings: the stem thickens, stops elongating, and begins to grow horizontally. **(a)** Response of wild-type tomato seedlings; **(b)** response of tomato mutants that are insensitive to ethylene. *(a,b, Yen, Lee, Tanksley, Klee, and Giovannoni,* Plant Physiology *(1995) 107: 1343–1353)*

ward growth. Tomato mutants that are unresponsive to ethylene do not show any aspect of the triple response, as illustrated in Figure 21-20b.

Recent studies of mutant plants that do not respond normally to ethylene have begun to reveal the molecular mechanisms that underlie ethylene action. One such mutant, the ethylene-insensitive tomato mutant *Never-ripe,* ripens more slowly than usual—an advantage in transporting tomatoes to market. Seedlings of *Never-ripe* do not respond normally to ethylene, meaning that they do not produce the classic triple response.

Arabidopsis researchers have also isolated many ethylene-insensitive mutants and have identified a number of genes responsible for detecting ethylene and bringing about its effects on cell elongation, fruit ripening, and the establishment of cell fate. Two of the genes involved in ethylene's signal transduction pathway resemble genes involved in transcriptional regulation in both prokaryotes and eukaryotes. One of the genes is strikingly similar to the floral homeotic gene *apetala2*. Current research involves further analysis of genes involved in the response to ethylene.

Abscisic Acid Promotes Dormancy in Buds and Seeds

Abscisic acid [Latin, *abscissus* = to cut off], whose structure is shown in Figure 21-14a, takes its name from its ability to stimulate the abscission (dropping) of leaves and fruit. Despite its name, abscisic acid plays only a minor role in abscission, but it does play a major role in the suspension of development in buds and seeds.

Unlike the other identified plant hormones, abscisic acid is usually an inhibitor, not a stimulator. It antagonizes the effect of gibberellins and auxins. Abscisic acid slows the growth of oat seedlings, inhibits the germination of wheat embryos, and prevents the synthesis of α-amylase by the aleurone layer of barley seeds. At a cellular level, it inhibits the synthesis of both proteins and RNA. Abscisic acid also coordinates the responses to a variety of environmental stresses, including drought, excess salt, waterlogging, cold, and mineral deficiency.

Are There Other Plant Hormones?

Research in the last decade has shown that a number of other chemicals can coordinate cellular activities in different parts of a plant. For example, wounding a plant leaf can evoke both local and long-distance effects, suggesting the involvement of one or more plant hormones. (See Essay: "Can Research on the Response to Plant Pathogens Lead to Improved Disease Resistance in Crops?") Plants can also develop resistance to specific viruses, bacteria, or fungi, not only at the site of infection but also far away. In both cases, plant biologists have identified signaling molecules that may mediate both local and long-range effects.

The two molecules shown in Figure 21-21—salicylic acid and jasmonic acid—are often proposed to be new

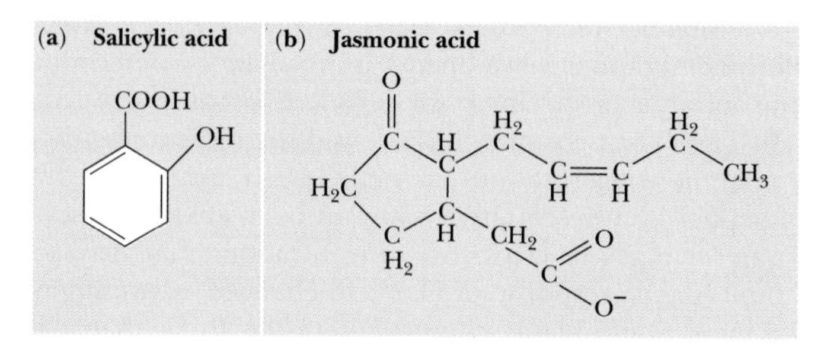

Figure 21-21 Resistance to pathogens in plants depends on salicylic acid, jasmonic acid, and a polypeptide hormone called systemin (not shown). **(a)** Structure of salicylic acid; **(b)** structure of jasmonic acid.

ESSAY

CAN RESEARCH ON THE RESPONSE TO PLANT PATHOGENS LEAD TO IMPROVED DISEASE RESISTANCE IN CROPS?

Plant pathogens—bacteria, fungi, and viruses—destroy about 12% of all crops worldwide. Improving the resistance of crop plants to disease could make a significant contribution to feeding the world's population. Modern molecular and genetic approaches have given plant biologists new information about the ways that plants respond to pathogens. This research suggests new strategies for engineering plants to resist disease.

Plants use three types of mechanism to resist pathogens. First, many plants continuously make molecules that resist pests and pathogens. The waxy cuticle that covers plant parts is the first defense. In addition, plants make toxic substances that help them resist infection. In some plants, up to 10% of the dry weight consists of such chemicals, and some chemists have estimated that plants may produce as many as 10,000 different molecules that participate in the resistance to pests and pathogens. Prominent among these are the *phenolics,* aromatic compounds related to tyrosine, which not only are toxic themselves but also polymerize to make *lignin,* which coats the conductive tissues of plants and reduces the ability of bacteria and fungi to infect the plant. Other classes of molecules important in plant defenses are nitrogen-containing compounds such as the *alkaloids* and *terpenes.* Chemists have already determined the structures of about 5000 different types of alkaloids, including such molecules as caffeine, cocaine, and nicotine, which can have devastating effects on animals (including humans) that eat plants that contain them.

Terpenes are lipid molecules which, like steroids, are synthesized from a 5-carbon building block called *isoprene.* Terpenes are often aromatic, as in the case of the oils of peppermint, basil, sage, and pine trees. In addition to small molecules like the phenolics, alkaloids, and terpenes, many plants constitutively make proteins that are toxic to invading microorganisms or to animal pests.

The second line of plant defense involves the local reactions at the site of infection or wounding. The local response may include the strengthening of cell walls and cuticle, the increased synthesis of specific small molecule toxins, and the induction of a class of proteins, called *pathogen-related* (PR) proteins. Among the PR proteins are protease inhibitors, which disrupt the digestive system of insect pests. The functions of other PR proteins are generally unknown but are subjects of active investigation. Even without knowing the function of all the PR proteins, however, researchers can observe a plant's response to a pathogen by the appearance of identifiable PR mRNAs.

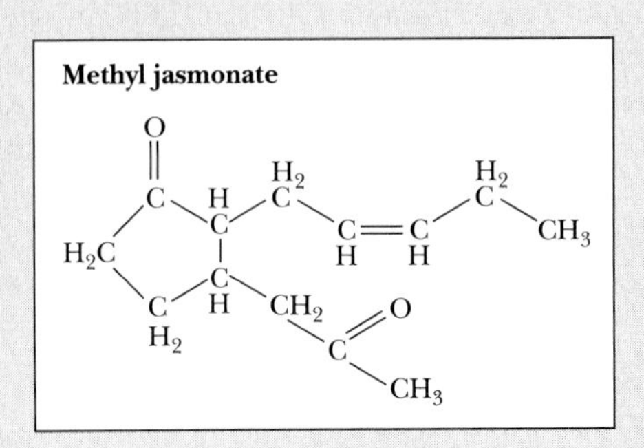

Figure 1 The structure of methyl jasmonate, an additional signaling molecule implicated in the systemic response to infection and injury.

A plant's third line of defense to pathogens and pest is called *systemic acquired resistance,* or SAR, which occurs not only locally but throughout the entire plant. SAR involves the production of proteins and simpler molecules at long distances from the site of infection or wounding. Some of the proteins produced during SAR are the same as the PR proteins produced locally.

How does a plant transmit the information that coordinates systemic acquired resistance? Researchers have identified five signaling molecules, three of whose structures are shown in Figure 21-21 and Figure 1—salicylic acid, systemin, jasmonic acid and its sister molecule methyl jasmonate, and ethylene. In each case, the concentration of the signaling molecule increases after infection or wounding, prior to the development of SAR. And application of low concentrations of each signaling molecule to plants can induce protein synthesis characteristic of SAR. Researchers are still unsure, however, which of these molecules serve as the primary physiological signal in promoting SAR and which are merely consequences of SAR. Signaling molecules may serve different roles after different types of stimuli; for example, salicylic acid appears to activate SAR after pathogen infection but not after tissue damage by feeding insects. Unraveling the pathways of signal transduction certainly suggests new ways for humans to encourage pathogen resistance in important crop plants.

plant hormones, as is a polypeptide called systemin, currently the only known peptide signaling molecule in plants. **Salicylic acid,** closely related to aspirin (acetylsalicylic acid), derives its name from the willow tree [Latin = *salix*], whose bark was used for centuries by American Indians and ancient Greeks to treat aches and fevers. Remarkably, salicylic acid not only helps cure human ills but also helps prevent plant infections. In animals, aspirin and salicylic acid act on prostaglandin synthesis, as discussed in Chapter 19. In plants, salicylic acid promotes flowering and inhibits the production of ethylene.

One of salicylic acid's most striking actions is on reproductive structures of a number of heat-generating (thermogenic) flowers, such as those of skunk cabbage and the voodoo lily. In these flowers, salicylic acid stimulates the production of a mitochondrial electron carrier whose action—unlike those of the other mitochondrial electron carriers described in Chapter 8—does *not* produce a proton gradient. The result is a mitochondrial "short circuit": In this condition, the net result of glycolysis and respiration is the production of heat, not the generation of ATP. The heat production is so dramatic that, on the day of flowering, the flower's temperature increases by 14°C (25°F). The high temperature stimulates the vaporization of foul-smelling organic chemicals that attract pollinating insects. Salicylic acid appears to be the most important signal in stimulating the production of heat and scent in these unusual plants.

In view of the discovery of these other signaling molecules, many plant scientists argue that plant hormones should be defined differently from animal hormones. Plant hormones can act both at long and short distances, like both endocrine and paracrine signals in animals. They can also act on the same cells that produce them, like growth regulators or "autocrine" signals in animals. While researchers have no clear consensus on the definition, all agree about the importance of identifying endogenous substances that control plant development, gene expression, and the responses of a plant to its environment. One researcher has proposed that a plant hormone should be defined simply as "an endogenous plant substance that acts at low concentration to affect physiological processes."

HOW DO PLANTS DETECT ENVIRONMENTAL CHANGES AND HOW DO THEY COORDINATE THEIR RESPONSES TO THOSE CHANGES?

Environmental cues that govern the lives of plants include gravity, light, and season. In the rest of this chapter we examine the mechanisms by which plants detect and respond to these factors.

For example, plants respond to gravity by changing their pattern of growth. The bending or curving toward or away from a stimulus is called a **tropism** [Greek, *trope* = turning]. The response of a plant to gravity, for example, is called **gravitropism.** Roots, especially primary roots, grow down; shoots, especially primary shoots, grow up; branches are usually horizontal. Gravitropism is an active process, as can be seen by turning a potted plant on its side. Soon both root and shoot turn and grow, respectively, downward and upward. Our first question, then, is exactly how does a plant "know" which way is up?

Gravitropsim in Roots Involves Detection by Cells of the Root Cap and Growth Coordination with Several Plant Hormones

A simple experiment demonstrated the location of a root's gravity detector: Removal of the root cap abolishes the root's gravitropism. The response of the root must depend on the cells of the root cap. But which component of the root cap cells detects gravity?

As shown in Figure 21-22, large cells in the root cap contain organelles called **amyloplasts,** each of which contains several starch grains. (See Chapter 5, p. 148.) The amyloplasts are denser than other organelles, and they settle on the lower surface of the cell. Plant physiologists have long suspected that the amyloplasts are the root's **statoliths** [Greek, *statos* = standing + *lithos* = stone], or gravity detectors.

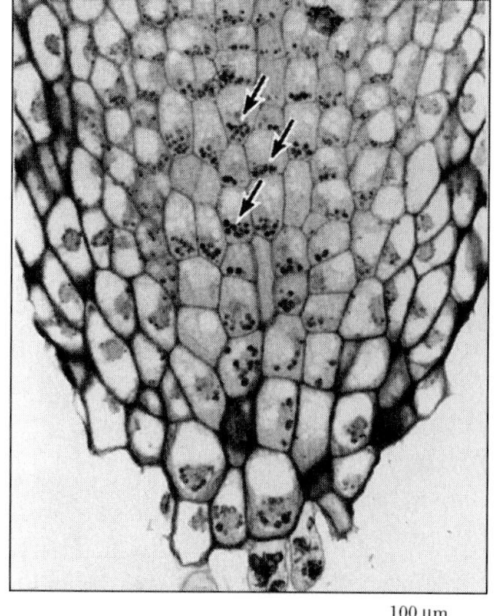

Figure 21-22 Gravitropism depends on amyloplasts in the root cap. *(Randy Moore, University of Akron, Ohio)*

ESSAY

HOW DO RESEARCHERS LOOK FOR A NEW PLANT HORMONE?

A plant wounded by an insect or even by a mechanical stimulus responds by making protease inhibitors, first locally at the site of the wound and then systemically, as discussed in the accompanying Essay, "Can Research on the Response to Plant Pathogens Lead to Improved Disease Resistance in Crops?" The newly produced proteinase inhibitors are adaptive for the plant because they disrupt the digestive systems of feeding insects.

The systemic response depends on a signal that moves from the wound site to the rest of the plant. In a tomato plant, for example, inhibitor synthesis begins first in the wounded leaf and then in more distant leaves. Removing the wounded leaf soon after injury prevents inhibitor synthesis by other leaves. In order to ask the question, "what is the signal for the induction of protease inhibitors in the distant leaves?" researchers have developed methods for the rapid assay of induction. In some cases, they have measured inhibitor activity; in other cases, they have determined the levels of inhibitor mRNAs; and in still other cases, they have used genetically engineered plants in which the promoter of an inhibitor gene is linked to a reporter molecule whose level can be more easily measured than the inhibitor itself. (See Chapter 15.)

Two experimental setups have been useful. In one type of experiment, the researcher sprays suspected signaling molecules onto the leaves of an intact unwounded plant and then measures the induction of inhibitor. In the other type of experiment, the researcher uses a cut rather than an intact plant, applying the suspected signal through the cut stem in an aqueous solution and then examines the plant's responses. The suspected molecules are salicylic acid, systemin, jasmonic acid, and methyl jasmonate.

Methyl jasmonate, the most volatile of the suspected signaling molecules, has the most robust effect in the intact plant assay. When sprayed onto a plant, methyl jasmonate induces protease inhibitors. Even a 30-minute exposure to low concentration (about 10 parts per billion) of methyl jasmonate in a sealed chamber induced inhibitor formation. Methyl jasmonate also induces inhibitor production when applied through a cut stem. The question still remains, however, whether methyl jasmonate is the natural signal for the long-distance response: Does a wounded plant produce it in sufficient amounts to account for the distant induction of inhibitor synthesis?

To test this hypothesis, researchers found conditions that altered the amyloplasts and then examined the effect on the gravitropic response. For example, treatment of roots with gibberellin and cytokinin induces the digestion of the starch grains. With no starch grains, the amyloplasts no longer settled when the root was turned, and the root itself no longer exhibited gravitropism. Later, the root made new starch grains and recovered its gravitropism. These experiments support the hypothesis that the amyloplasts are the gravity sensors in the root cap. Some cells in the stem also contain amyloplasts. These cells, just outside the vascular bundles, are the best candidates for the stem's gravity sensors.

Changes in the orientation of shoot or root cause large differences in the rate of elongation of cells on top and bottom. For example, the cells on the bottom of a horizontal stem may grow 10 times faster than those on the top. This difference is at least partly responsible for the upward bending of the stem. The bending response is similar to that of phototropism. It is not surprising, then, that auxin appears to be necessary. But differences in auxin concentrations between top and bottom are not high enough to account for the difference in the growth response. Other substances must also participate. Among the suspected signals are ethylene and abscisic acid. Calcium ions may also contribute—they inhibit cell elongation and are present in higher concentration at the top than at the bottom of a bending stem.

The mechanism of gravitropism is an active research area. Not only is it a long-standing issue in plant physiology, but interplanetary travel may ultimately depend on the ability to raise crops in the weightless environment of space.

Many Light-Dependent Processes Involve a Pigment Other than Chlorophyll

A plant's survival depends on its ability to get enough light to support photosynthesis, and plants respond to light in many ways. Among these are phototropism, solar tracking (the turning of leaves to follow the sun), and germination.

As in the case of gravitropism, we would first like to know the nature of the light detector responsible for these processes. One way of approaching this question is to study the effect of light of different wavelengths (which we perceive as different colors).

Every light-dependent process has a characteristic *action spectrum,* the effectiveness of different wavelengths in triggering a specific response, shown for the germination of lettuce seeds in Figure 21-23a. (See Chapter 8, p. 230.) Because the effect of light on a process depends on its absorption by some molecule, the action spectrum should correspond to that molecule's *absorption spectrum,* the relative absorption of different wavelengths of light, shown for chlorophyll in Figure 8-8. If chlorophyll were the detector for germination, for example, the action spectrum should correspond to the absorption spectrum of chlorophyll. But this is not the case. The light-absorbing molecule responsible for phototropism and solar movements is therefore not chlorophyll, but some compound that absorbs red but not blue or yellow light.

Light absorption somehow stimulates the transport of auxin from the illuminated to the shaded side of a growing shoot tip, as discussed on page 591. The absorbing pigment, however, is not necessarily in the tip. Analyses of action spectra for many developmental processes point to the importance of pigments other than chlorophyll. Studies of photoperiodism, described in the next section, have identified the most prominent of these pigments.

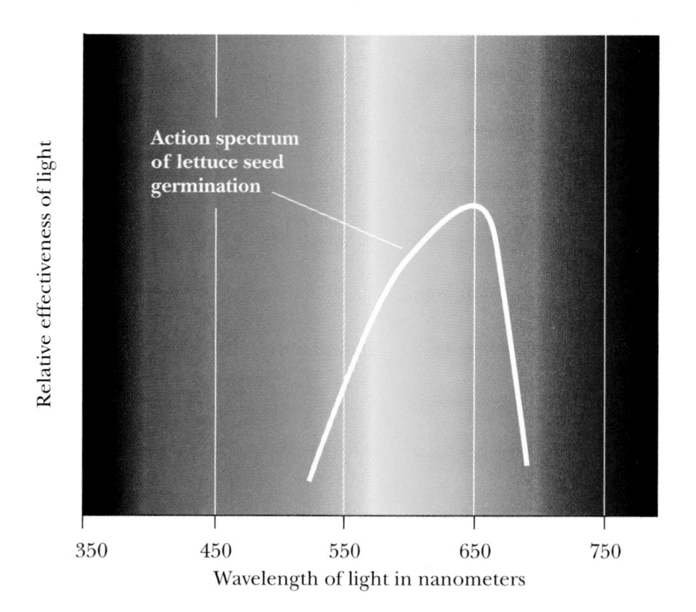

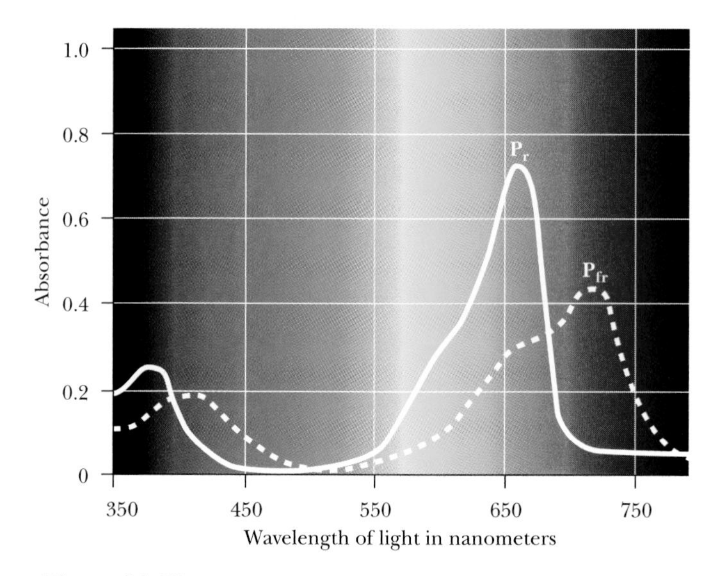

Figure 21-23 The identification of phytochrome, the pigment responsible for many adaptive responses of plants to light. **(a)** A comparison of the action spectrum of lettuce seed germination with the absorption spectrum of chlorophyll suggests that chlorophyll is not the responsible transducer; **(b)** the absorption spectrum of the two forms of phytochrome, P_r and P_{fr}, suggests that light absorption by P_r may regulate germination.

Many Responses to Seasonal Changes Depend on a Pigment Called Phytochrome

Every gardener, farmer, and hiker knows that different plants flower in different seasons. The flowers of crocuses and daffodils appear in the spring, and those of carnations and black-eyed susans in the summer. In many cases, appropriate timing of flower and seed production determines whether new plants have enough warmth and water and what other organisms are its competitors, consumers, and pollinators. The question addressed in this section is how plants detect and respond to seasonal changes.

In some species (like cucumbers, peas, and tomatoes), the time of flowering depends only on maturity or size. Flowering may occur after a plant has grown for weeks, months, or even years. In other species, flowering always happens around a certain date, even if the plant reaches maturity much earlier.

Understanding the seasonal control of flowering is a practical as well as a scientific issue. Soybean farmers, for example, once tried to stagger their harvests by planting fields at 2-week intervals. But they found that (for the variety they were using) all the flowers appeared at once, late in the summer, and the whole harvest was ready at the same time regardless of the planting time.

For any species, the date of flowering varies with latitude, so it cannot result from the operation of an internal annual clock, in the way that sleep movements depend on an internal daily clock. The flowering response turns out to be an example of **photoperiodism,** the response to the relative lengths of day and night in their daily cycle. *Short-day plants* flower when days are short and nights are long, in the late summer, fall, or even winter. Examples include chrysanthemum, cocklebur, and corn. In contrast, *long-day plants* require long days and short nights. These generally flower in the late spring or early summer and include spinach, sugar beets, and black-eyed susans. Finally, the flowering of *day-neutral plants,* such as tomatoes and dandelions, does not depend on day length. Plants from tropical areas are often day-neutral, while many from regions closer to the poles are long-day plants.

One of the first questions that researchers asked was whether flowering depended on the length of the day or the length of the night. Their approach was to grow short-day plants—cocklebur in one laboratory, a soybean variety in another—in artificial cycles of light and dark. The plants flowered, as they would in the field, whenever "days" were shorter than the "nights," regardless of the total duration of the cycle.

In 1938, Karl Hamner and James Bonner, plant physiologists at the University of Chicago, introduced a new kind of experiment, illustrated in Figure 21-24. They interrupted the light period with a period of darkness or the dark period with a period of light. For both short-day and long-day plants, interrupting the artificial day made no difference in flowering. For short-day plants (Figure 21-24a) breaking up the night prevented flowering, and for long-day plants (Figure 21-24b) breaking up the night promoted flowering. The interruption did not need to be large—an ordinary light bulb turned on for half a minute was enough to stop the flowering of a short-day plant. The conclusion of these experiments was that both short-day and long-day plants measure the length of the night, not the day. Short-day plants should really be called "long-night" plants, and long-day plants should be called "short-night" plants.

Other work on the basis of photoperiodism was going on at the US Department of Agriculture Laboratory in Beltsville, Maryland, where photoperiodism had first been discovered in the 1920s. The work, indeed, immediately helped agriculture: Farmers were better able to control when their crops flowered and set seed, breeders could test new varieties under controlled light-dark conditions, and florists began to provide year-round supplies of popular flowers that had once been seasonal. But still no one understood the chemical basis of photoperiodism. In 1940, however, researchers at Beltsville realized that they could learn about how plants detect darkness by determining the action spectrum of the light that most effectively interrupted the long nights required for soybeans to flower.

Determining the action spectrum for a complex process in a whole organism was no easy matter. The researchers first demonstrated that the light effect was in the leaves and that illuminating a single leaf was enough to suppress flowering of a short-day plant. They then had to build a large instrument that would light whole leaves in separate groups of plants with light of different colors. This experiment showed that red light and violet light were most effective in promoting flowering. The researchers concluded that the light detector must be a green or blue pigment, perhaps even chlorophyll. After further measurements, however, they realized that it was another pigment, which they named **phytochrome** [Greek, *phyton* = plant]. Later work showed that phytochrome is the pigment responsible for many processes that depend on the timing of dark and light, including flowering, germination, and leaf formation.

The Beltsville group discovered that they could reverse the effect of red light with a light flash of longer wavelength. The most effective wavelength for the red light effect was about 660 nm, and the most effective wavelength for the reversing flash was about 730 nm, in the "far-red" region of the spectrum. Sterling Hendricks, the plant physiologist who had built the instrument to determine the action spectrum, made a radical hypothesis to explain the far-red reversal: Phytochrome must exist in two interconvertible forms, one absorbing red light and one absorbing far-red light. He called the red-absorbing form P_r and the far-red absorbing form P_{fr}. When P_r absorbs red light it converts to P_{fr}, and far-red light converts P_{fr} back to P_r.

Hendricks and his colleagues proposed the following explanation (illustrated in Figure 21-24c) for the effect of light and dark on flowering: (1) Sunlight contains more energy in the red than in the far-red part of the spectrum; (2) during the day, the red component in sunlight converts P_r to P_{fr}; and (3) in the dark, P_{fr} slowly reverts back to P_r. In short-day plants, flowering occurs only when P_{fr} levels are sufficiently low. Red light supplied at night decreases P_r and increases P_{fr}, thus suppressing flowering. High levels of P_{fr} inhibit flowering in short-day plants and promote flowering in long-day plants. Other plant physi-

Figure 21-24 The flowering of long-day and short-day plants depends on the conversion of phytochrome from P_r to P_{fr}. **(a)** The interruption experiments of Hamner and Bonner with a short-day (long-night) plant; **(b)** interruption experiments with a long-day (short-night) plant; **(c)** a model for the flowering of long-day and short-day plants.

ologists were initially skeptical about the existence of phytochrome. One doubter waggishly termed it "a pigment of the imagination."

Hendricks' hypothesis, however, made a strong prediction: Purified phytochrome should be a single compound whose absorption spectrum can switch from one form to another. This prediction stimulated a biochemical search that was long and difficult for two reasons: (1) Phytochrome acts catalytically, so plants do not contain great amounts; and (2) P_{fr} is unstable, so researchers had to work in the dark to try to keep phytochrome in the P_r form.

After many years of work, several research groups isolated pure phytochrome. It is a protein with a molecular size of about 120,000 atomic mass units, linked to a much smaller organic compound that absorbs light. Purified phytochrome, however, behaved like the pigment imagined by Hendricks and his colleagues almost 40 years earlier. (Under natural sunlight, which contains both red and far-red light, both forms of phytochrome exist, with about 40% P_r and about 60% P_{fr}. In darkness, the P_{fr} level declines, either by reversion to P_r or by enzymatic destruction.

The action spectrum for other light-regulated changes suggested that phytochrome is important in many developmental processes. These processes vary from species to species but include seed germination, the elongation of new seedlings, the beginning of chlorophyll synthesis, and the production of enzymes needed for photosynthesis. For example, in mustard seedlings that have been grown in the dark and transferred to the light, phytochrome controls the following changes: opening of the hypocotyl hook, development of primary leaves, differentiation of xylem, degradation of storage protein, chlorophyll synthesis, protein synthesis, and RNA synthesis.

In almost every case, the developmental change controlled by phytochrome depends on the accumulation of P_{fr}. Red light (or daylight) promotes these processes and far-red light inhibits them. Phytochrome permits the sun, or the plant physiologist, to turn a molecular switch on or off. Some of the effects of phytochrome may depend on induced changes in the synthesis, degradation, or transport of plant hormones. Some effects, however, depend on a more direct influence on the production of particular mRNAs.

Genetic approaches, especially using *Arabidopsis,* have considerably revised previous ideas about phytochrome action. The response of etiolated *Arabidopsis* seedlings to light, for example, involves the straightening of the hypocotyl hook, the separation and growth of cotyledons, and the development of chloroplasts. Studies of the response of mutant seedlings to red and far-red light have revealed the involvement of at least two forms of phytochrome, designated phyA and phyB, which have opposing responses to red and far-red light. Together, the two phytochromes modulate the response to light in the process of seedling development, a process that involves the transcriptional regulation of a number of genes, including that for ribulose bisphosphate carboxylase (Rubisco), the enzyme, discussed in Chapter 8, that catalyzes the first reaction in the fixation of carbon dioxide.

Analysis of phytochrome sequences has not revealed any similarity to molecules involved in other signal transduction pathways, and researchers still do not know exactly how phytochrome affects cellular processes. Studies of mutant *Arabidopsis* seedlings, however, have revealed that phytochrome action depends on a pathway that involves the protein products of a number of genes.

Arabidopsis seedlings that have been grown in the dark respond to light in characteristic ways: They develop chloroplasts, their cotyledons separate and expand, and their hypocotyl hook straightens. Some mutants, however, develop in the dark as if they were grown in the light. Seedlings that contain a mutation in both copies of the gene called *cop,* for example, develop identically in both light and dark, suggesting that the *cop* gene ordinarily suppresses development in the dark. In the light, the *cop* gene product—somehow altered by phytochrome action—no longer suppresses development. Using molecular techniques to localize the *cop* protein, researchers have shown that, in the dark, the *cop* protein is in the nucleus, but in the light it moves to the cytoplasm, consistent with the protein's serving as a regulator of transcription in the dark but not in the light. The sequence of the *cop* protein suggests that it acts as a zinc-finger type of DNA-binding transcriptional regulator, similar to other transcription factors discussed in Chapter 14.

The *cop* protein, and other genetically identified proteins as well, serves as a molecular switch whose activity is somehow controlled by phytochrome. The link between phytochrome action and these switches now appears to involve second messengers that have been previously identified in animal cells—particularly Ca^{2+} and cyclic GMP. (See Chapter 19.) The questions now being addressed by phytochrome researchers concern the molecules that are directly and indirectly affected by phytochrome: How do they talk to each other? And how do they distinguish among the different versions of P_r and P_{fr}?

SUMMARY

Meristems are self-renewing groups of undifferentiated cells that generate the differentiated structures of a plant. The apical meristems at the end of shoots and roots repeatedly generate standard modules that are characteristic of each species of plant. Differentiation involves oriented cell division and the growth of nondividing cells away from the meristems.

A flower develops from a specialized bud and consists of four sets of parts, arranged in concentric circles called whorls. The outer two whorls are sepals and petals, the inner two are the male and female reproductive structures, stamens and carpels.

A small common weed, called *Arabidopsis,* has become a favorite organism for plant biologists, providing many opportunities for genetic and molecular studies. Mutations in flower development, for example, have revealed genes that establish the fates of flower parts, similar in action to the homeotic genes of *Drosophila.*

Stamens produce haploid microspores, which develop into male gametophytes, pollen grains. Carpels produce haploid megaspores, which develop into female gametophytes, the embryo sacs. The male and female gametophytes in turn produce male and female gametes.

Upon pollination, a pollen grain develops into a pollen tube with two sperm nuclei. Flowering plants undergo double fertilization, in which the two sperm nuclei fuse with two cells of the embryo sac within the carpel. One fertilization event produces the zygote, which grows into an embryo. The other gives rise to the endosperm, a triploid tissue that nourishes the growing embryo.

The zygote divides to form an embryo that first resembles a sphere, then a heart, then a torpedo. Significant cell differentiation has occurred by the torpedo stage, but meristem cells remain small and undifferentiated. Development stops soon thereafter, as the embryo loses most of its water and becomes quiescent. A seed consists of a quiescent embryo and its adjacent endosperm, surrounded by a seed coat.

The resumption of growth is called germination, a process that begins with imbibition, the taking in of water. The first part of the embryo to emerge from the seed is the radicle, which contains the root's apical meristem and quickly forms a functioning root system. The shoot apical meristem lies above the point of attachment of the cotyledons, or seed leaves.

A seed may maintain its dormant embryo for many years, until the proper environmental changes bring about germination. Germination triggers a huge increase in biochemical activity, for which the growing plants derive energy from stored starches, oils, and proteins.

The primary growth of both roots and shoots depends on apical meristems. The growing root tip has three distinct zones—the root cap, the apical meristem, and an elongation zone. The growing tip of a stem, however, must be able to produce stem, leaves, branches, and flowers.

Plant physiologists have identified five plant hormones that influence development—auxin, gibberellin, cytokinin, ethylene, and abscisic acid. The first plant hormone to be recognized was auxin, which (among other effects) coordinates phototropism, the growth of a plant toward light. Illumination of one side of the growing tip of an oat seedling causes the transport of auxin away from the lighted side. The higher concentration in the shaded side causes it to grow more rapidly, causing the plant to curve toward the light.

Other plant hormones have effects distinct from those of auxin. Gibberellins promote germination, trigger the mobilization of food reserves in germinating seeds, and stimulate growth in mature plants. Cytokinins stimulate cell division and growth and prevent leaf senescence. Ethylene promotes leaf senescence and fruit ripening and inhibits elongation of stems and roots. And abscisic acid promotes dormancy in buds and seeds. Recent research on the responses of plants to wounds and infection has suggested the existence of other plant hormones—salicylic acid, jasmonic acid, and systemin.

A plant's response to environmental cues depends on its ability to detect such cues. Plants use different detection mechanisms for different kinds of stimuli. Specialized cells in the root cap detect gravity. One kind of unidentified pigment detects light and brings about phototropism. Another kind of pigment, called phytochrome, serves as a molecular switch for many seasonal processes, including flowering and seed production.

Absorption of red light converts phytochrome into a form that no longer absorbs red light, but only far-red light. At night, this form, called P_{fr}, spontaneously switches back to a form that absorbs red light. Most of the processes controlled by phytochrome depend on the accumulation of enough far-red absorbing form. Studies of *Arabidopsis* mutants are leading to a new picture of how phytochrome transmits information to change cellular processes.

KEY TERMS

abscisic acid
aleurone layer
alternation of generations
amyloplast
apical meristem
apical dominance
auxin
axillary bud
bioassay
bud
callus
carpel
central cell
cortical microtubule
cotyledon
cytokinin
dicot
double fertilization
elongation zone
epicotyl
epidermis
ethylene
etiolation
germination
gibberellin
gravitropism
ground tissue
homeotic selector gene
hormone
hypocotyl
imbibition
internode
lateral bud
leaf primordium
megaspore
meristem
monocot
node
ovary
ovules
petal
photoperiodism
phototropism
phytochrome
pollen tube
primary growth
quiescent
root
root cap
salicylic acid
sepal
senescence
shoot
stamen
statolith
target
terminal bud
tropism
turgor pressure
vascular system
vegetative reproduction
whorl

QUESTIONS FOR REVIEW AND UNDERSTANDING

1. Compare and contrast the following terms:
   (a) meristem vs. node vs. internode
   (b) terminal bud vs. axillary bud
   (c) monoecious vs. dioecious
   (d) epicotyl vs. hypocotyl vs. radicle
2. Discuss why *Arabidopsis* is well-suited for experimental studies.
3. Explain how a single gene mutation may result in dramatically different flower morphology.
4. Why is leaf development considered the default pathway of meristem development? How was this demonstrated?
5. What are the two products of double fertilization?
6. Describe the relationship between the embryo, aleurone layer, and endosperm in promoting germination and growth.
7. Describe how the hormones auxin and cytokinin influence differentiation of callus tissue.
8. How does a bioassay differ from chemical analysis (such as chromatography) for determining the concentration of a substance?
9. What are the two primary effects of auxin?
10. How are mutant plants used in a bioassay for gibberellins?
11. How is the process of senescence affected by cytokinins?
12. What is the triple response induced by ethylene? How does ethylene's effect contrast with that of cytokinin? What is the experimental evidence suggesting that ethylene alters gene expression?
13. Why might you expect the level of abscisic acid to increase in a plant subjected to environmental stress?
14. How can the comparison of an action spectrum and an absorbtion spectrum assist in identifying the light "receptor" for a given light-regulated process?
15. Why can phytochrome be thought of as a molecular switch?
16. What is the evidence suggesting that novel hormones are involved in the development of resistance to infectious agents?
17. Speculate as to why some bacteria and fungi produce plant hormones.

■ SUGGESTED READINGS

ALBERTS, B., D. Bray, J. Lewis, M. Raff, K. Roberts, and J. D. Watson, *Molecular Biology of the Cell,* 3rd edition, Garland, New York, 1994. Chapter 21 deals with cellular mechanisms of plant and animal development.

COEN, E. S., and E. M. Meyerowitz, "The war of the whorls: Genetic interactions controlling flower development," *Nature* 353, 31–37, 1991. A review of work that defined the genetics of flower development.

MAUSETH, J. D., *Botany: An Introduction to Plant Biology,* 2nd edition, Saunders College Publishing, 1991. Chapter 14 deals with development and morphogenesis.

MEYEROWITZ, E. M., "Arabidopsis: A Useful Weed," *Cell* 56, 263–269, 1989. A somewhat specialized introduction to the molecular genetics of plant development.

RAVEN, P. H., R. F. Evert, and S. E. Eichhorn, *Biology of Plants,* 5th edition, Worth, New York, 1992. Chapters 20–25 deal with plant development and plant hormones.

CHAPTER 22

Defense Mechanisms: Inflammation and Immunity

KEY CONCEPTS

- Inflammation, complement, interleukins, and interferon provide nonspecific internal defenses against infection.
- The immune system consists of two components—circulating antibodies and cytotoxic cells—which recognize foreign molecular shapes and participate in their elimination.
- The immune response demonstrates specificity, memory, diversity, and the ability to distinguish self from nonself.
- The immune system's ability to recognize so many different kinds of molecules depends on the extraordinary ability of specific gene segments to rearrange during the development of individual cells.
- The descendants of each immune cell express the same rearranged genes as the progenitor. These cells are subject to severe selection during development and to large expansion after exposure to specific infections.
- Cells interact with one another in producing an immune response, relying on cell surface molecules and secreted signaling molecules called cytokines.

OUTLINE

ESSAYS

Many of the viruses and microorganisms that teem throughout the biosphere represent great potential harm. The evolutionary success of animals has depended heavily on the ability to prevent and combat invasions by potentially harmful microorganisms. In this chapter we discuss cellular defense systems, focusing on the mammalian immune system, which arguably provides the most flexible and most responsive defenses in the animal kingdom.

HOW DOES INFLAMMATION HELP THE BODY RESIST INFECTION?

The skin provides the first line of defense against infection or invasion, and relatively few microorganisms breach its barrier. Those that do must confront a second line of more active defenses.

The first internal defense against infection is **inflammation,** a set of responses to local injury. The name comes from the familiar characteristics of injured tissue—redness, heat, swelling, and pain—as if there really were a flame within. Inflammation, unlike the immune response, does not depend on the specific properties of the invading organisms.

Inflammation Depends on a Series of Cellular and Biochemical Events

Microscopic observations have revealed a sequence of cellular events that contribute to the inflammatory response: (a) dilation of the capillaries in the injured region, (b) the attachment of white blood cells, called *neutrophils,* to the capillary walls, (c) movement of neutrophils through the capillary wall into the damaged tissue, and (d) phagocytosis of microorganisms within the tissue. (See Figure 22-1.)

The dilated blood vessels accelerate the delivery of cells and plasma proteins that contribute to both the inflammatory and the immune response. Among the specific cells that move into the inflamed region are cells that counterattack invading organisms.

In addition to these cellular responses, plasma proteins also participate in the body's general defenses. A biochemical cascade, similar to the cascades of the clotting and anticlotting systems described in Chapter 4, leads to the production—by the liver and by widely distributed cells called macrophages—of **complement,** a set of pro-

(a)

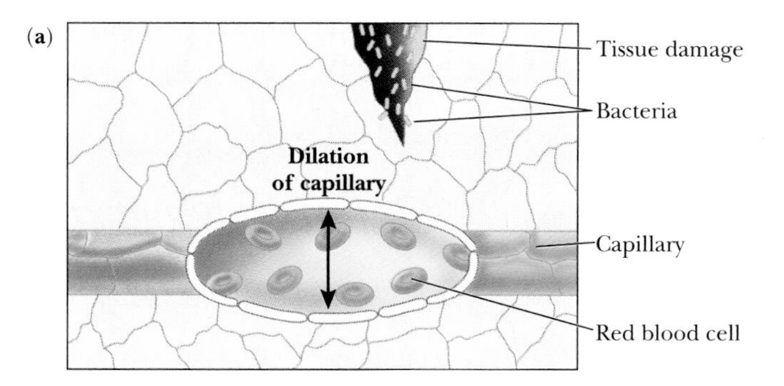

(b)

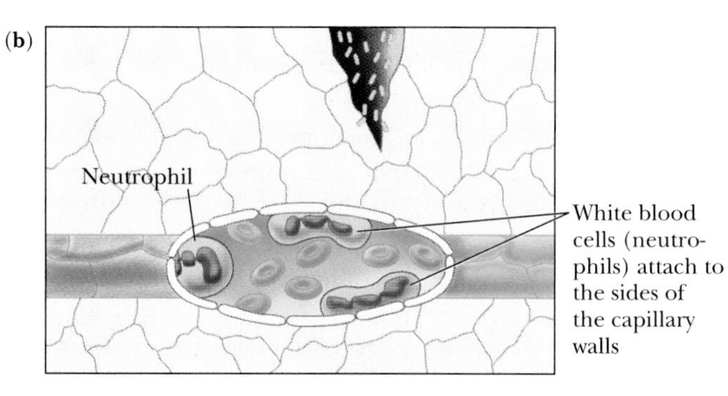

(c)

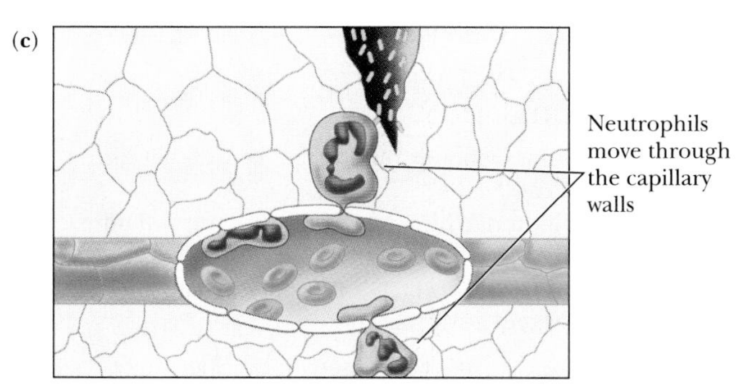

(d)

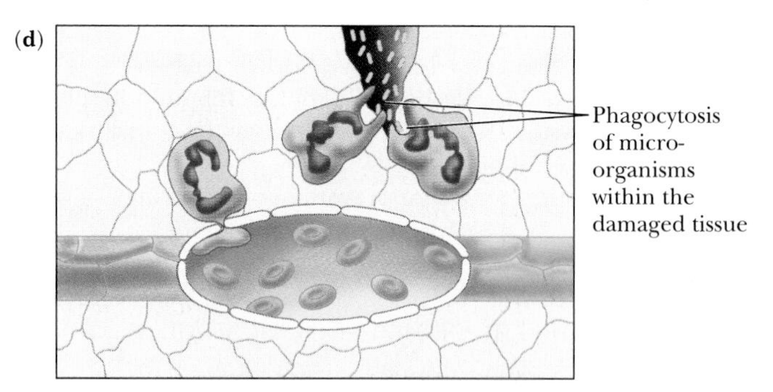

(e)

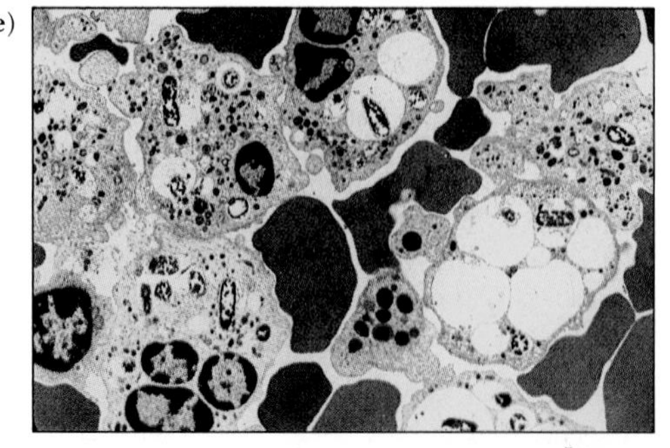

Figure 22-1 Microscopic studies of inflammation have revealed the processes illustrated in these drawings. **(a)** The dilation of blood capillaries; **(b)** the attachment of white blood cells (neutrophils) to capillary walls; **(c)** the movement of neutrophils through the capillary walls; **(d)** phagocytosis of microorganisms within the damaged tissue; **(e)** white blood cells with ingested bacteria. *(e, David M. Phillips/Visuals Unlimited)*

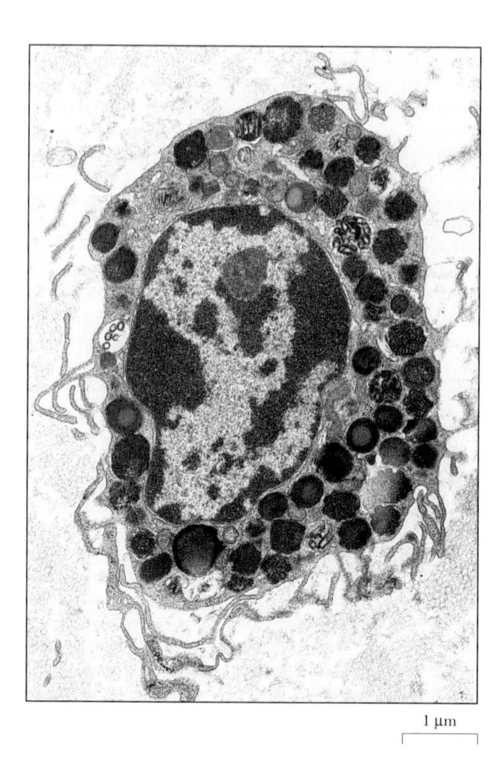

Figure 22-2 A mast cell, packed with histamine-containing granules. Tissue damage stimulates exocytosis of the granules and the release of histamine. Mast cells also release cytokines, which attract migrating neutrophils and help initiate the immune response. *(Barry King/Biological Photo Services)*

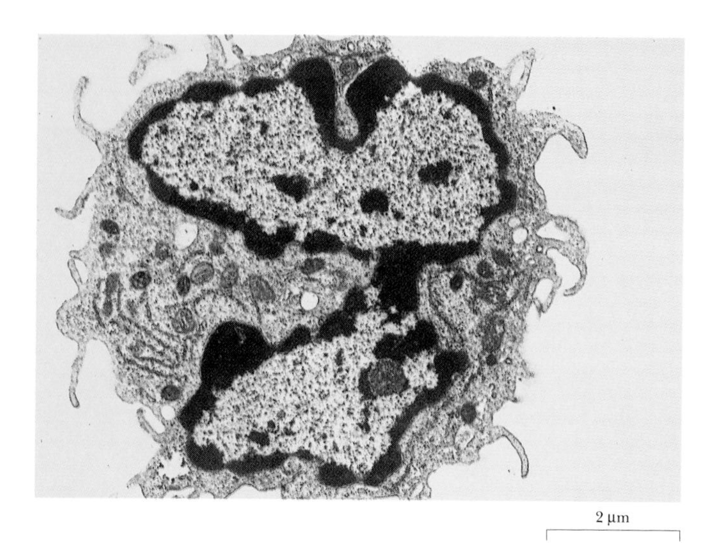

Figure 22-3 A macrophage has an extensive Golgi apparatus and rough endoplasmic reticulum. Macrophages produce and secrete a wide variety of proteins, including complement and interleukins. *(Don Fawcett/Visuals Unlimited)*

teins that attack microbiological invaders. Some of these proteins form a **membrane attack complex,** which kills foreign organisms by boring holes through their membranes. Other components of the complement system coat the microbes' surfaces and identify them as targets for phagocytosis.

Which Cells and Which Molecules Initiate Inflammation?

Mast cells, such as the one illustrated in Figure 22-2, are distributed throughout the body's connective tissues. These cells are large and round, 15–20 μm in diameter, and they are filled with small vesicles, also called "granules." Tissue damage stimulates exocytosis of these vesicles and the release of **histamine** (the amino acid histidine minus the carboxyl group), a major stimulus for the inflammatory response. Mast cells also release other chemicals that attract migrating neutrophils to the site of injury.

Macrophages [Greek, *makros* = large + *phagein* = to eat], illustrated in Figure 22-3, are also large phagocytic cells, widely distributed throughout the body. Immature macrophages (*monocytes*) make up about 5% of the cells in the blood, and mature macrophages are present in connective tissue, as well as in the liver, brain, spleen, lungs, and lymph nodes. Macrophages play an active role in consuming microorganisms and cellular debris, and they also have a major role in initiating the inflammatory response and the immune response.

Macrophages in the area of inflammation produce at least four **cytokines,** small proteins that regulate proliferation and protein synthesis in cells that participate in inflammation and in the immune response, as shown in Figure 22-4. Four cytokines produced early in the inflammatory response are called tumor necrosis factor-α (TNF-α), interleukin-1 (IL-1), interleukin 6 (IL-6), and interferon (IFN). The interleukins derive their name from the fact that they are produced by **leukocytes** (the generic term for all white blood cells; Greek, *leukos* = clear, white) and act on other leukocytes, binding to specific receptors and initiating specific biochemical responses.

Interleukin-1, or IL-1, produced early during inflammation, stimulates responses both in leukocytes and in organs distant from the site of infection. Among its targets, diagrammed in Figure 22-4, are (1) cells in the hypothalamus, a region of the brain that, among other things, regulates body temperature; the effect of IL-1 is to produce a fever, which interferes with microbial growth and promotes the body's immune defenses; (2) macrophages (mostly neutrophils) in the damaged tissue; the effect of IL-1 is to attract more phagocytic cells to the site of injury; (3) endothelial cells in the capillary walls; the effect of IL-1 action is to stimulate the expression of certain cell surface proteins; passing neutrophils in the blood bind to these surface proteins, and the neutrophils soon move through the capillary wall and accumulate near the site of injury; and (4) lymphocytes, discussed later in this chapter, which initiate the immune response to specific molecules of the invading microorganisms.

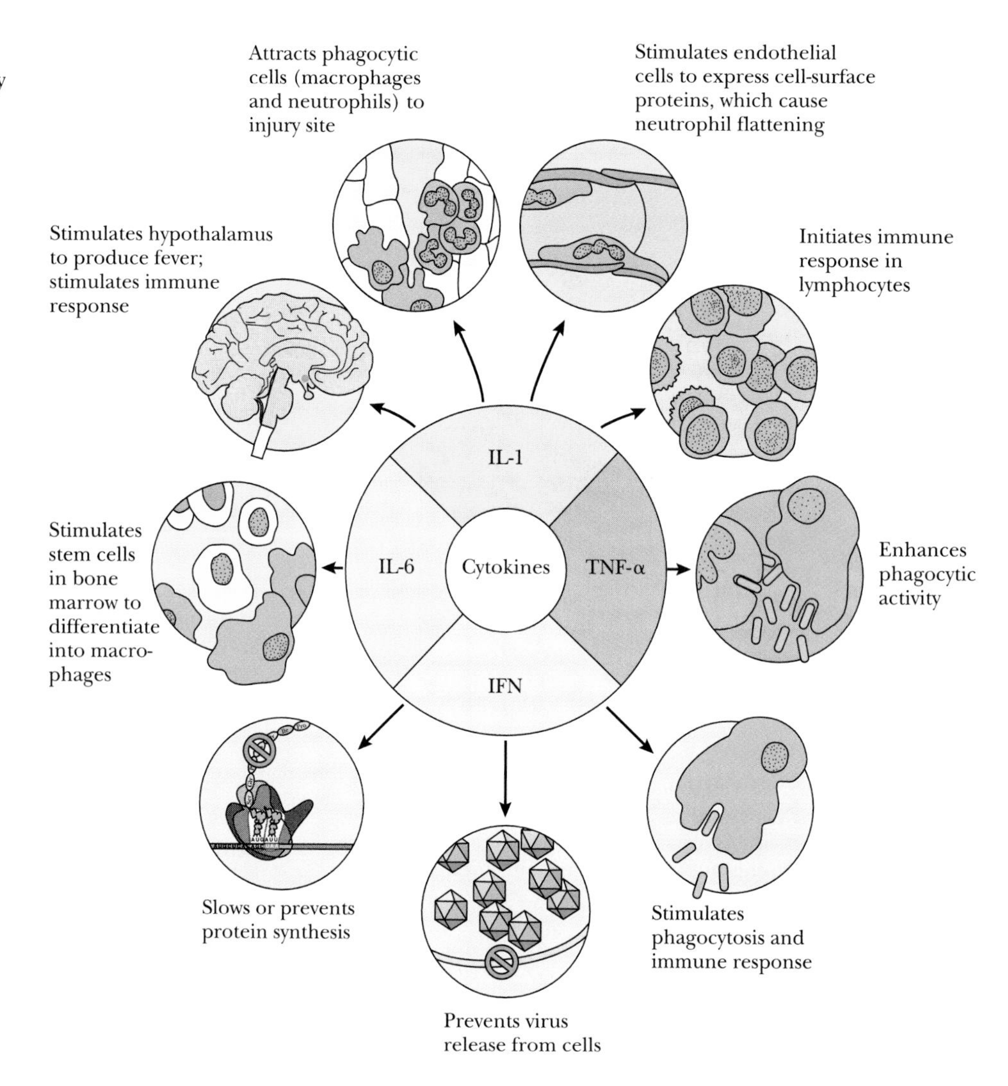

Figure 22-4 During inflammation, cytokines trigger responses in a variety of cells.

As also shown in Figure 22-4, IL-6 acts on stem cells in the bone marrow, stimulating them to produce more macrophages, while TNF-α acts locally on macrophages and neutrophils to enhance their phagocytic activity.

Viral infection stimulates the infected cells to produce another set of cytokines, called **interferons** because they interfere with further spread of the virus. The induction of interferon synthesis depends on double-stranded RNAs, which are common during the life cycle of many viruses, but which are not present in normal cells. Within the infected cells, interferons induce cells to slow or prevent protein synthesis, as diagrammed in Figure 22-4. The shut-down of protein synthesis may be fatal to the cell itself, but it prevents the release of more virus, decreasing the viral threat to the rest of the organism. Interferons also stimulate phagocytosis and the immune response.

HOW DOES THE IMMUNE SYSTEM RECOGNIZE MOLECULES?

The most powerful and the most specific defenses against infection are those of the immune system. The immune system produces two distinct responses to invading organisms: (1) **humoral immunity** [Latin, *umor* = fluid], the production of **antibodies,** blood proteins that form complexes with foreign molecules such as those on the surfaces of microorganisms; and (2) **cellular immunity,** the proliferation of specific cells that attack foreign cells, including the body's own cells that have become cancerous or have been infected with a virus. The cells responsible for the immune response are the **lymphocytes,** a class of white blood cells that develop within the lymphoid tissues (including lymph nodes, spleen, thymus, and tonsils). As discussed later in this chapter, the two types of immunity depend on two distinct sets of cells, as diagrammed in Figure 22-5: B lymphocytes mediate humoral immunity, while T lymphocytes mediate cellular immunity.

The Immune Response Demonstrates Specificity, Memory, Diversity, and the Ability to Distinguish Self from Nonself

Most of us have had contact with someone infected with chickenpox, a common childhood disease. If we have not had chickenpox, we are susceptible to infection. Once we

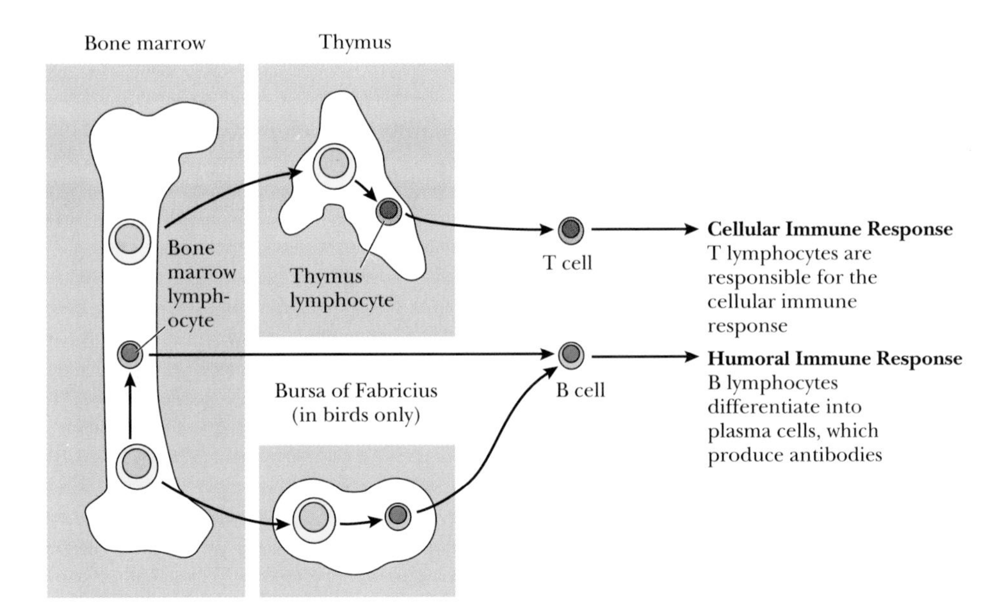

Figure 22-5 Two types of lymphocytes are responsible for two distinct defensive responses, humoral and cellular immunity. B lymphocytes differentiate into plasma cells, which produce antibodies, the agents of humoral immunity. T lymphocytes are responsible for the cellular immune response. In mammals, B lymphocytes mature in the bone marrow; in birds, they mature in an organ called the bursa of Fabricius. T lymphocytes mature in the thymus.

have suffered its itchy ravages, however, we are no longer susceptible; we then say that we are *immune* to chickenpox. Immunologists speak about four characteristics of immunity: specificity, memory, diversity, and self-nonself recognition.

Immunity is **specific:** Cells of the immune system recognize a single type of invader at a time. If we are immune to chickenpox, we are still just as likely to get other diseases to which we are not immune. The specificity of the immune response distinguishes it from the nonspecific responses of inflammation and interferon production.

The immune system has **memory:** Once the immune system has developed defenses against a particular invader, it can attack that molecule, organism, or virus whenever it appears, even decades later. The reappearance of the invader stimulates a **secondary immune response,** which is usually faster and more robust than the primary response, when the invader was first encountered.

The immune system is capable of incredible **diversity:** It can respond to virtually any challenge, including bacteria, viruses, purified proteins or carbohydrates, and even to molecules that are not natural products but have been synthesized by organic chemists. This diversity serves the body well. For example, some viruses such as influenza rapidly evolve and escape from a host population's growing immunity. Each new strain of influenza, for example, has a different protein on the outer coat, allowing influenza epidemics to occur again and again in the same human population. After infection with a new virus strain, however, a person can develop new immunity against that strain, an illustration of the diversity of the immune response.

The diversity of the immune response poses one of the immune system's greatest challenges—**self-nonself recognition,** the ability to distinguish the body's own components from others. In almost all cases the immune system demonstrates **tolerance,** a state of induced unresponsiveness to self components. Tolerance prevents immune attacks on the body's own components.

Until the 1950s, explaining these four properties of the immune response—specificity, memory, diversity, and self-nonself recognition—seemed almost an impossible challenge to biologists. Perhaps the most demanding aspect of the problem was the tremendous diversity of the immune response: How can an animal produce proteins and cells that specifically recognize substances that neither the animal itself nor any of its ancestors could ever have encountered in all of evolutionary history? And how can an animal have specific responses to so many different challenges?

Researchers first asked these questions about humoral immunity, but the same issues arose for cellular immunity. The answer to these questions depends on the realization that, while the number of possible atomic arrangements is impossibly large, there are many fewer shapes. **The immune system produces a huge but not limitless array of proteins that can bind to nonself molecules; each such molecule recognizes a characteristic shape and charge distribution rather than a specific arrangement of particular atoms.** Different arrangements of atoms "look" the same to the protein molecules responsible for immunity.

Antibody Molecules Recognize Specific Shapes

Just as an enzyme can bind to molecules such as competitive inhibitors, whose shapes resemble that of the substrate, an antibody molecule (the agent of humoral immunity) can recognize a limited set of similar molecules, which form equivalent sets of noncovalent bonds. Similarly, the cellular immune response depends on the ability of protein receptors on the surface of T lymphocytes to recognize specific shapes rather than arrangements of particular atoms.

An **antigen** [an *anti*body *gen*erator] is a molecule that stimulates the production of an antibody or of another immune response. Among the most biologically important antigens are those that signal an infection—the coat proteins of viruses and the surface proteins, carbohydrates, and glycoproteins of bacteria.

Each antigen contains a number of **antigenic determinants,** or **epitopes** [Greek, *epi* = upon + *topos* = place], the specific shapes and charge distributions (composed of clusters of atoms) to which antibodies or other recognition proteins can bind. For a protein antigen reacting with an antibody, a single epitope typically consists of five to eight amino acid residues. It is impossible to count the number of actual epitopes recognized by the immune system, but immunologists estimate that the humoral immune system has the capability of making antibodies that recognize more than 10^8 different epitopes.

Antibodies are most likely to recognize epitopes on the accessible outside surface of their antigens. When the antigen is a protein, the epitopes are most likely to consist of hydrophilic or charged amino acid residues and most likely to come from segments that bend away from regular structures, such as the α-helices in the myoglobin molecule illustrated in Figure 22-6.

Lymphocytes Are Responsible for the Immune Response

With a standard microscope, lymphocytes are almost indistinguishable from one another, but they vary greatly at a biochemical level. Altogether the human immune system consists of about 10^{12} lymphocytes, about the same number of cells as are in your brain.

The lymphocytes responsible for the humoral response are called **B lymphocytes** (shown in Figure 22-7a).

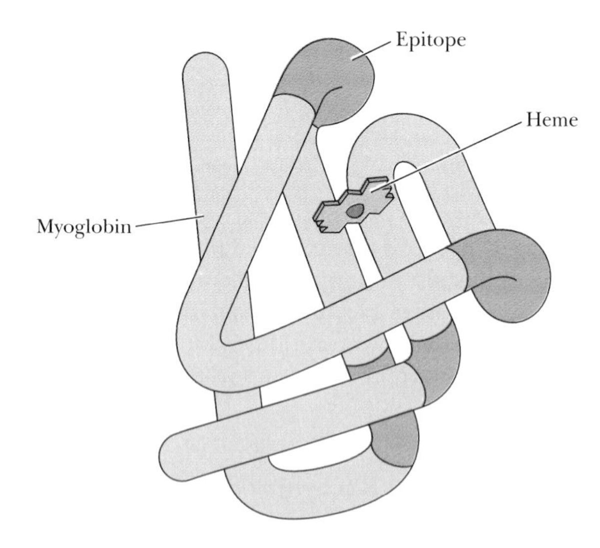

Figure 22-6 The four sequential epitopes shown here lie at the bends between α-helices in the oxygen-binding protein myoglobin. Epitopes recognized by antibodies may be sequential, consisting of adjacent amino acid residues, or nonsequential, consisting of residues brought together only when the protein adopts its three-dimensional conformation.

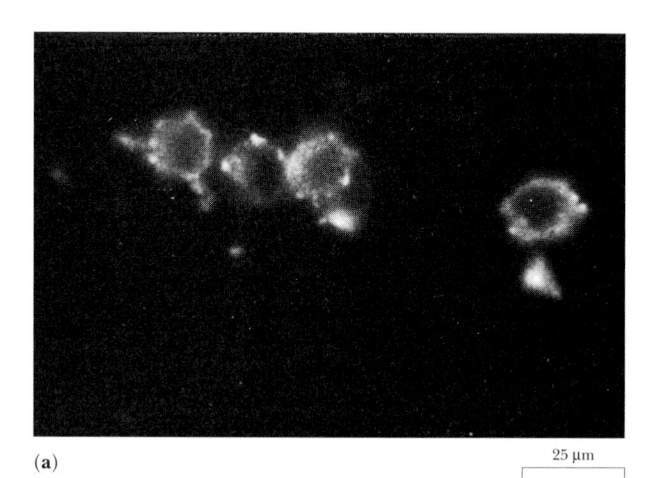

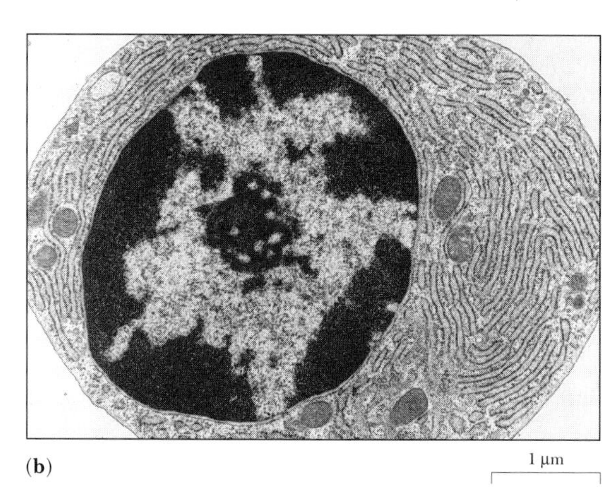

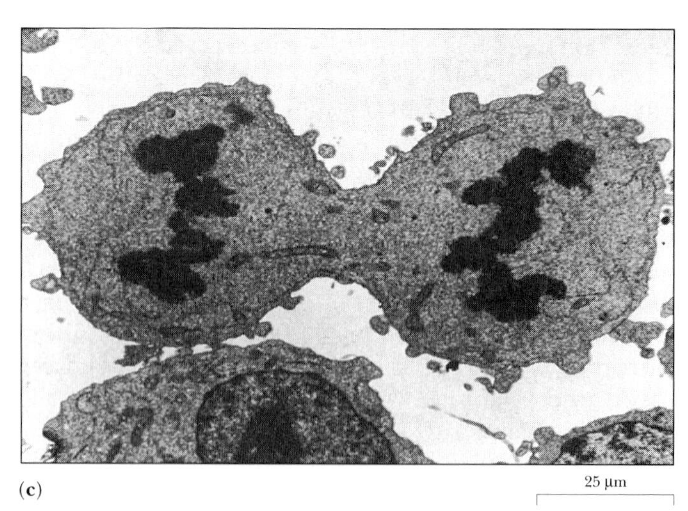

Figure 22-7 Electron micrographs of cells involved in the immune response: **(a)** B lymphocytes in the bone marrow; **(b)** plasma cells, specialized for the production of antibodies; **(c)** T lymphocytes in the thymus. *(a, courtesy of Dr. D. H. Lewis; b, D. W. Fawcett/Photo Researchers; c, David M. Phillips/Visuals Unlimited)*

The letter "B" refers to the fact that, in birds, these lymphocytes mature in an organ called the bursa of Fabricius, an outgrowth of the cloaca. Mammals do not have a bursa, and the B lymphocytes mature in the bone marrow, as indicated in Figure 22-5. Upon stimulation with an appropriate antigen, certain B lymphocytes can develop into **plasma cells** (illustrated in Figure 22-7b), specialized factories for the production and secretion of antibodies.

Other lymphocytes—**T lymphocytes**—mature in the thymus gland, located in the upper chest cavity above the heart, as illustrated in Figure 22-5. The T lymphocytes include cells responsible solely for cellular immunity, as well as other cells that participate in both humoral and cellular immunity.

Clones Derived from Individual B Lymphocytes Produce a Single Kind of Antibody

Lymphocytes ordinarily divide in culture for only a few divisions. In 1975, however, Cesar Milstein and Georges Kohler succeeded in "immortalizing" lymphocytes so that they would grow in culture indefinitely. Such immortal lymphocytes forever maintain the ability to produce specific antibodies. They did this by fusing B lymphocytes from a mouse to a cell line derived from a lymphoma (a cancer of lymphocytes) to obtain a **hybridoma** (a hybrid cell line derived from a lymphoma). They could then isolate and grow individual hybridomas so that the descendants of each original lymphocyte stayed separate from the descendants of other lymphocytes.

The set of descendants from a single cell is a clone. Milstein and Kohler showed that each such clone made only one kind of antibody, which they therefore called a **monoclonal antibody.** Each monoclonal antibody has a defined chemical structure and the ability to recognize a single epitope, as illustrated in Figure 22-8. In contrast, the set of antibodies ordinarily produced against a particular antigen in an intact responding animal is said to be **polyclonal,** meaning that the antibodies are made in the descendants of *many* B cells. Each of the many B-cell clones produces an antibody that recognizes a different epitope on the antigen.

Researchers have been able to study both the chemical and the three-dimensional structures of monoclonal antibodies far more easily than they could ever study the mix of structures in the huge array of antibodies in blood plasma. Studies of monoclonal antibodies confirmed and extended previous knowledge, most of which had come from studies of antibodies produced in cancers of the immune system.

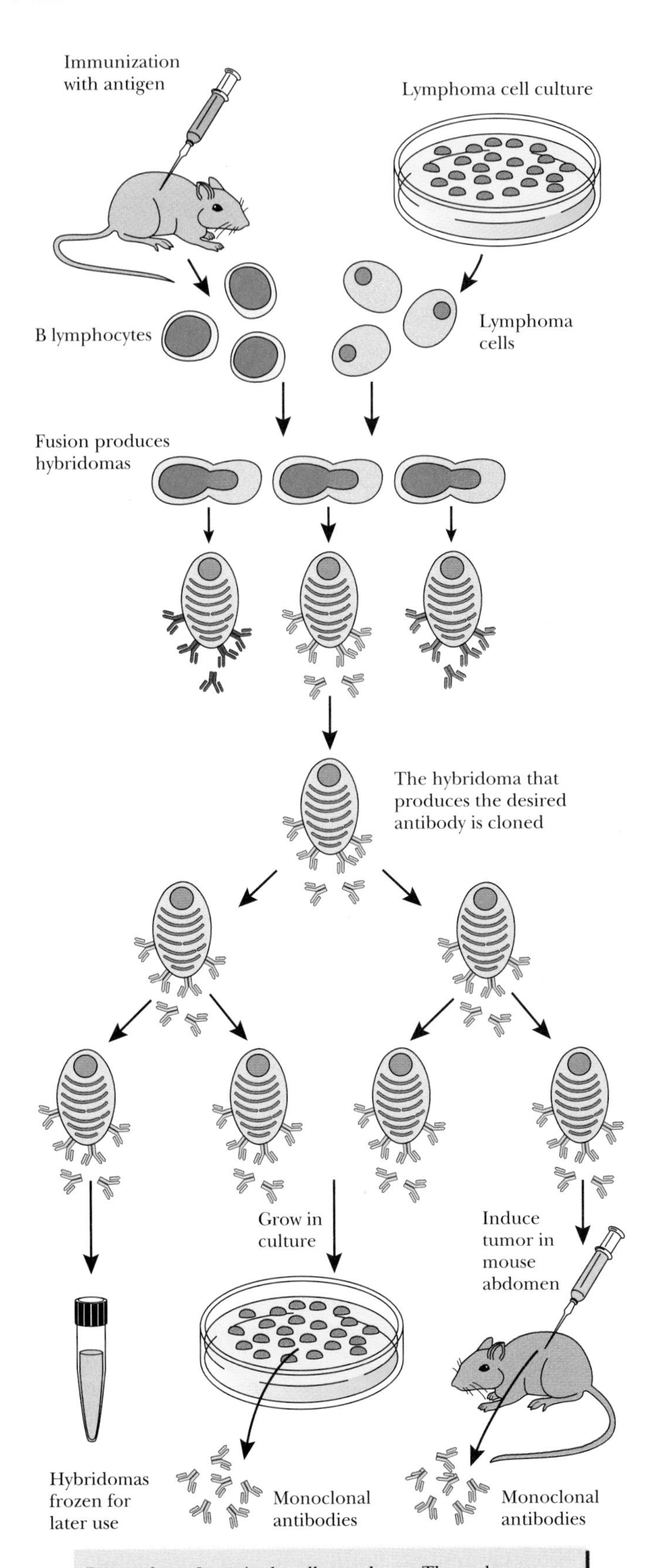

Figure 22-8 The production of monoclonal antibodies from hybridoma cells. Each hybridoma clone makes only one type of antibody, which recognizes a single epitope.

The Chemical Structure of Antibodies Explains Their Ability to Bind Molecules of Specific Shapes

All antibodies are proteins called **immunoglobulins** because they are members of a group of globular plasma proteins called globulins. There are five classes of immunoglobulins—called IgG, IgM, IgA, IgD, and IgE—which differ in size, amino acid sequence, site of synthesis, and biological function. All the immunoglobulins, however, contain antigen-recognition sites made from two distinct polypeptide chains, called **light chain** and **heavy chain.** The five classes of immunoglobulins differ in their heavy chains. All classes can use the same kind of light chain, which come from one of two classes, called κ (kappa) and λ (lambda).

The most abundant class of immunoglobulin, IgG, has a molecular size of about 150,000 and consists of four polypeptide chains—two identical light chains, each containing about 220 amino acid residues, and two identical heavy chains, each containing about 440 amino acid residues. Disulfide bonds between cysteine side chains connect the heavy chains to the light chains and the two heavy chains to each other. The resulting structure, shown in Figure 22-9, is shaped like a Y and contains two binding sites for antigens.

Figure 22-10 compares the structures of the five major types of immunoglobulins. IgG, IgE, and IgD all have two binding sites, while IgA has four and IgM has 10.

The presence of two or more binding sites on each antibody molecule helps explain the effectiveness of antibodies in combating infection. The surface of a microorganism may contain thousands of identical protein or carbohydrate molecules. An individual antibody molecule can bind to a site on one bacterium and to another site on another bacterium. The result of many antibodies binding to many bacteria, then, is an aggregate that is an easy target for scavenger cells, such as macrophages and neutrophils.

Each of the Five Classes of Antibody Makes Distinctive Contributions to the Immune Response

Both IgG and IgM activate complement, a group of serum proteins that form an enzyme cascade that leads to the lysis of targeted cells. (See Essay, "Blood Clots and the Therapeutic Uses of Enzymes," p. 108.) The other classes of antibodies do not activate complement. IgM molecules are already present on the surface of B cells before stimulation by antigen, and IgM molecules are the first to be secreted into the blood. They are the first to appear after exposure to antigen, while IgG molecules are made in much greater amounts during a secondary response.

IgA molecules are especially abundant in tears, saliva, and milk, as well as in the gastrointestinal and respiratory tracts. Secreted IgA plays an important role on mucous membranes, binding to the surfaces of bacteria and viruses. More IgA is produced every day than any other class of antibody.

IgE molecules are worth a special comment. These antibodies are able to mobilize a distinct set of cellular and chemical responses. Some antigens (such as ragweed pollen) appear to be especially effective in evoking the production of IgE. The IgE molecules circulate through the blood and attach themselves to mast cells. When antigen binds to the IgE on the surface of a mast cell, it stimulates the cell to release chemicals that attract a relatively rare type of white blood cell, called *eosinophils.* IgE binding also stimulates the mast cell to produce histamine, which triggers the inflammation reaction. This set of defenses appears to provide defenses particularly important against parasitic worms.

Unfortunately, however, some people mount this complex response inappropriately—to antigens that are not dangerous at all. The IgE–mast cell–eosinophil–inflammation response to a harmless antigen is called an **allergy.** Such antigens include dusts, pollen, and components of foods and medicines. When one of these antigens enters the respiratory passages of a sensitized person, histamine release stimulates increased mucus secretion and contraction of the airways as well as dilation of the blood vessels and an increase in their permeability. Typical allergy symptoms include congestion, sneezing, a runny nose, and difficulty in breathing. In some cases, the response is limited to humoral immunity and happens rapidly. In other cases—commonly with antigens in the leaves of poison ivy or poison oak—the reaction involves T cells and does not appear for several days after exposure.

X-Ray Diffraction Studies Have Revealed the Three-Dimensional Structure of Several Antibody Molecules and Their Interactions with Antigen

X-ray diffraction studies of antigen-antibody complexes have shown that, as illustrated in Figure 22-11a (and as expected from previous biochemical studies), antigen binds at the top of each of the two arms of the antibody, forming specific noncovalent interactions.

The x-ray diffraction studies show that the light and heavy chains of an immunoglobulin molecule consists of two to five **domains,** polypeptide segments that fold as independent units. The light chain consists of two domains, each about 110 amino acid residues long, while the heavy chain consists of four or five domains, each also about 110 amino acid residues long, with the number of domains depending on the class of immunoglobulin, as illustrated in Figure 22-10.

Each domain contains a characteristic compact structure, called the **immunoglobulin fold,** which contains two β structures, each with three or four antiparallel strands. (See Figure 22-11b.) The two β sheets are held together both by hydrophobic amino acid side chains and by a disul-

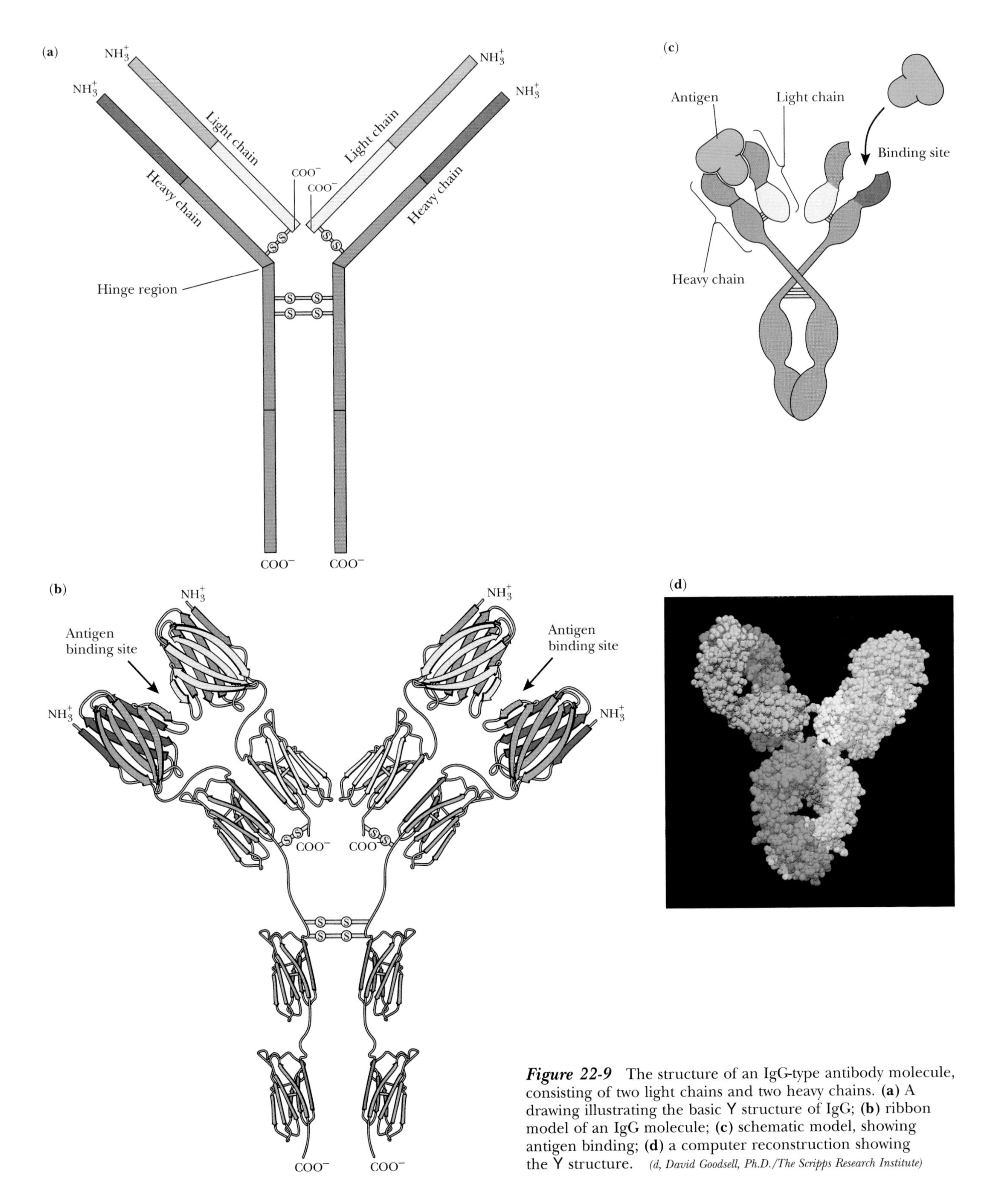

Figure 22-9 The structure of an IgG-type antibody molecule, consisting of two light chains and two heavy chains. **(a)** A drawing illustrating the basic Y structure of IgG; **(b)** ribbon model of an IgG molecule; **(c)** schematic model, showing antigen binding; **(d)** a computer reconstruction showing the Y structure. *(d, David Goodsell, Ph.D./The Scripps Research Institute)*

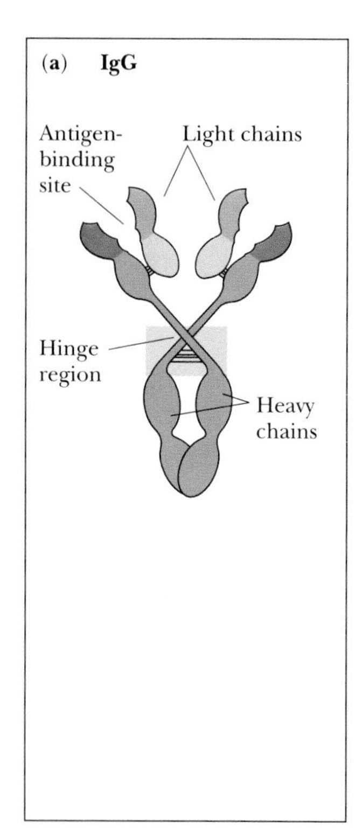

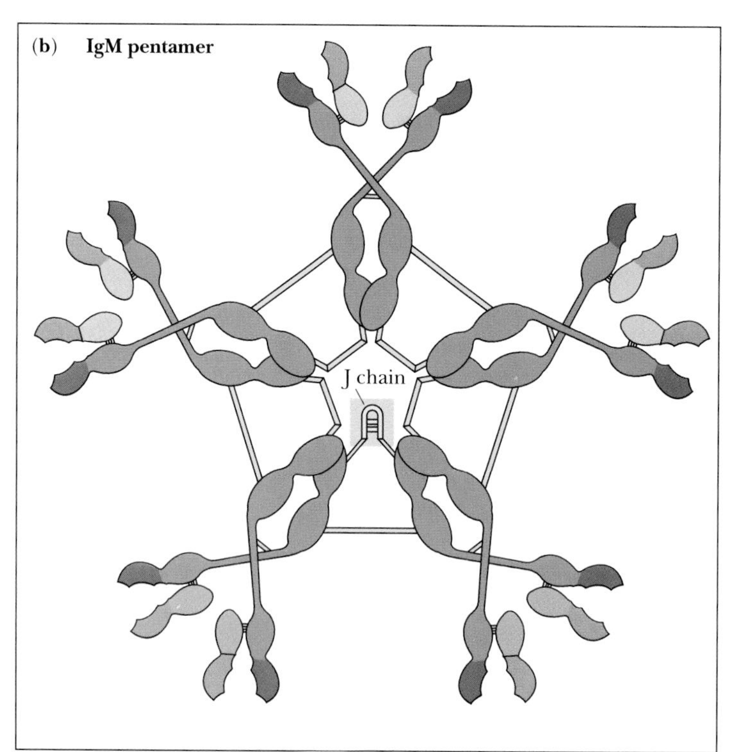

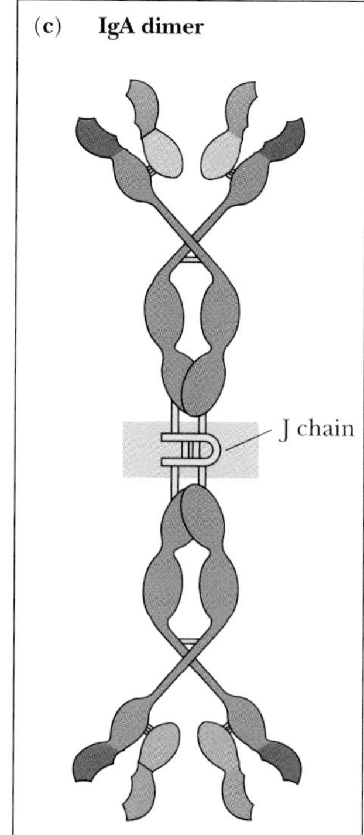

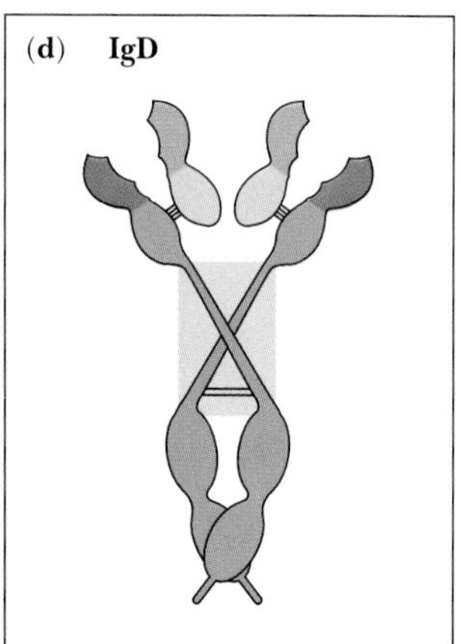

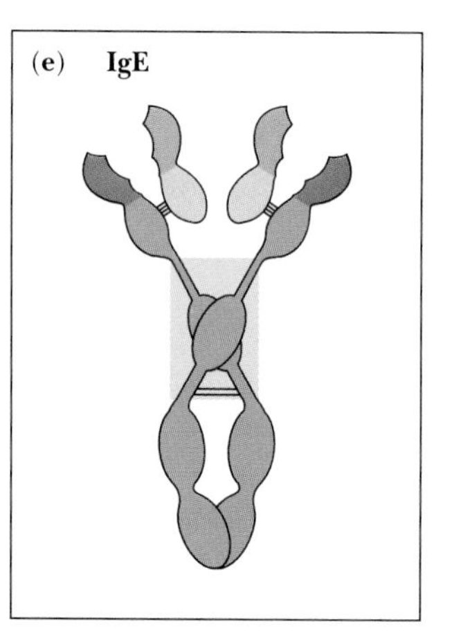

Figure 22-10 The structures of the five major classes of immunoglobulins, which differ in their heavy chains. **(a)** IgG, showing the regions important for complement activation and cell binding; **(b)** IgM, showing its pentameric structure and an additional polypeptide, called the J chain, which participates in the association of heavy chains; **(c)** IgA, showing its dimeric structure and the additional J chain; **(d)** IgD, showing the long hinge region; and **(e)** IgE, with no hinge region and an extra domain.

fide bond between two cysteine side chains, so that each domain resembles a bread-and-butter sandwich with a toothpick through the middle. In this popular description, the bread represents the β sheets, the butter the oily side chains, and the toothpick the disulfide bond.

Both the heavy and the light chains participate in forming noncovalent bonds with the antigen. The amino acid residues that participate in antigen binding all lie in the amino-terminal domains of both heavy and light chains. The segments most involved in binding antigen, shown in Figure 22-11a and b, are called the **complementarity determining regions,** or **CDRs,** because their shape is complementary to that of the antigen. The ability to recognize an antigen depends on the detailed three-dimensional structure of the immunoglobulin molecule.

HOW ARE ANTIBODIES ABLE TO RECOGNIZE SO MANY DIFFERENT MOLECULAR SHAPES?

The diversity of the humoral immune response depends on an animal's ability to produce many different antibodies with a wide variety of three-dimensional structures. Recall (from Chapter 3) that the three-dimensional structure of a protein depends upon the sequence of amino acid residues, which in turn is specified by a gene. But the three-dimensional structure also depends on the conditions in which the protein folds, including pH and the presence of an allosteric effector. (See Chapter 4, p. 112.)

Most vertebrates have only enough DNA to make about 1 million polypeptides, and most geneticists now

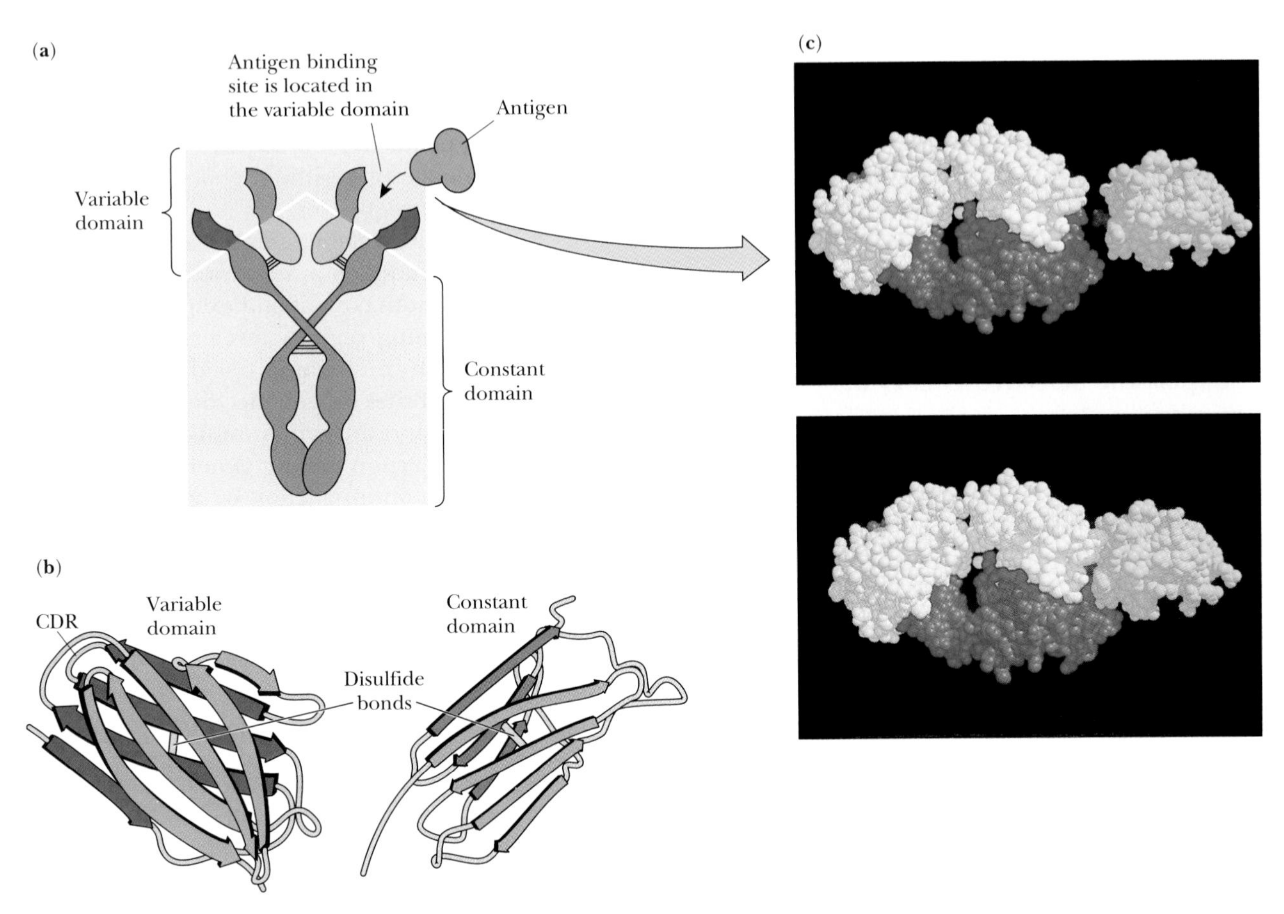

Figure 22-11 The three-dimensional structure of an antibody molecule, determined by x-ray diffraction, showing its interaction with an antigen. **(a)** Model of IgG molecule, showing the antigen binding site; **(b)** ribbon models of the immunoglobulin folds of the variable and constant regions of IgG; each fold contains two β sheets with antiparallel strands; the two β sheets are held together by a disulfide bond and by hydrophobic interactions; **(c)** computer representations of antigen-antibody interaction. *(c, Prof. Simon E. V. Phillips)*

think that the human genome contains only about 100,000 genes. How, then, can we account for the diversity of the immune response, which recognizes at least 1 million different epitopes (and possibly as many as 10^8)?

Immunologists once visualized two ways by which organisms could generate antibody diversity: (1) from antibodies with different amino acid sequences, with each sequence producing a distinct three-dimensional structure able to bind a distinct epitope; or (2) from different conformations of relatively few antibody molecules, which fold under different environmental conditions. In particular, many immunologists thought that the folding of an antibody might depend on the presence of the antigen.

Does Antibody Diversity Result from Selection or Instruction?

The first hypothesis requires that antibodies have an enormous number of different amino acid sequences—at least a million, but perhaps as many as 10^8. According to this **selective theory,** the immune response depends on the animal's ability to select the set of antibodies to make in response to the presence of a particular infection. If every different antibody sequence requires a separate gene, however, the selective theory would appear to require more genes than most vertebrates have in all their DNA.

According to the second possibility, called the **instructive theory,** an antigen would, by its presence, determine how each antibody molecule folds. Because the folding would differ according to which antigens are present, the instructive theory would not require a large number of genes. The dependence of folding on the presence of antigen could also provide an explanation for immunological memory: Perhaps enough antigen persists in the body to accelerate the future production of specific antibodies.

Despite the attractions of the instructive theory, it turned out to be wrong. The ability of an antibody to recognize an antigen depends primarily on the amino acid sequences of the antibody's polypeptides. The evidence is that, even in the absence of antigen, immunoglobulin polypeptides fold to form three-dimensional structures that bind particular antigens. The diversity of antigen recognition therefore depends on the diversity of polypeptides, not on variations in the folding of a single polypeptide.

Differences in the Amino Acid Sequences Among Antibodies Support the Selective Theory of Antibody Diversity

The development of automated methods for determining amino acid sequences has made it possible to compare the sequences of many immunoglobulin polypeptides. Antibody-producing hybridomas provide a source of individual species of immunoglobulin polypeptides. Each hybridoma makes a single kind of antibody molecule, consisting of a single kind of heavy chain and a single kind of light chain. As predicted by the selective theory, each of the immunoglobulin polypeptides from different antibodies, which recognize different epitopes, have different amino acid sequences.

The first big surprise that came from comparing amino acid sequences of immunoglobulin polypeptides was that differences in amino acid sequences among antibodies are not spread throughout the polypeptide chains; instead, they are concentrated in a few regions, as illustrated in Figure 22-12. All of the sequence variation in both heavy and light chains is confined to the domains closest to the amino terminal of each polypeptide, which is called the **variable region.** The remaining domains—one in each light chain and three in the IgG heavy chain—are called **constant regions,** because they are identical in all antibodies of a particular class. The existence of variable and constant regions within a single polypeptide poses a perplexing paradox: How did many different variable region sequences evolve without generating diversity in the constant region sequences?

Within the variable regions some segments, the **hypervariable regions,** are much more variable than others. The hypervariable regions correspond exactly to the complementarity determining regions revealed by x-ray diffraction studies. The diversity of amino acid sequences and the correspondence of sites of sequence diversity with epitope binding sites support the hypothesis that antibody diversity depends on selection among genetically specified antibodies rather than on instruction by antigens.

How Can a Limited Number of Immunoglobulin Genes Encode 10^6 to 10^8 Different Antibodies?

The realization that an antibody molecule consists of two different polypeptides (heavy and light chains) makes the question of antibody diversity slightly less troublesome. If every possible light chain can combine with every possible heavy chain, then 1000 different light chains and 1000 different heavy chains could make 10^6 different kinds of antibody. This would require only 2000 genes, still a substantial number, but a lot less than 1 million. If the number of different antibodies is actually 10^8, then 20,000 genes would be needed. Recent work has shown that still fewer genes are needed to make still more types of antibodies, as discussed later in this chapter.

The question remained whether these genes are part of the germ line, the genes passed from parent to child. Immunologists debated two theories, the **germ line the-**

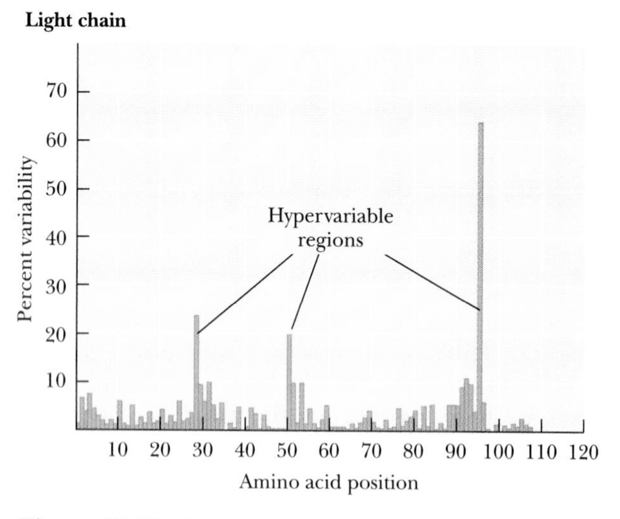

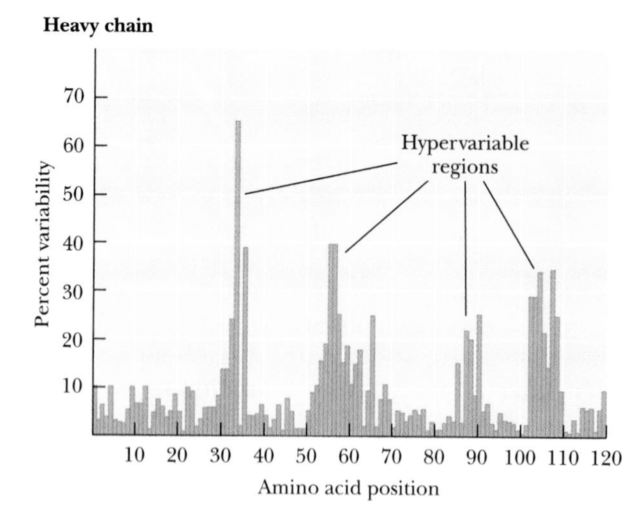

Figure 22-12 Sequence variation among immunoglobulin molecules, determined from monoclonal antibodies. Only the amino-terminal domain (the V region) is shown. Most of the sequence variation is confined to a few regions. The hypervariable regions correspond to the complementarity determining region (CDR) shown in Figure 22-11.

ory, which held that all the needed genes were inherited, and the **somatic theory,** which held that antibody diversity arose sometime during development.

Many geneticists argued that the required number of genes was just too great to be contained in a limited genome, and they supported the somatic theory. A smaller number of genes, these theorists argued, could produce more genes, perhaps by some mutational process or perhaps by gene rearrangements. In a prophetic 1965 paper, for example, Dreyer and Bennett argued that separate genes, encoding variable and constant regions of immunoglobulin polypeptides, might somehow recombine during somatic development to produce the huge antibody repertoire.

The techniques of molecular biology permitted the direct study of immunoglobulin genes themselves, allowing immunologists to distinguish between the germ line and somatic theories of antibody diversity. When molecular biologists began to study the organization and expression of the genes for immunoglobulins, they found that these genes were different from any others that they knew about. Studies of the genes responsible for the immune response have produced surprise after surprise.

The biggest surprise came in 1976, when Susumu Tonegawa and his colleagues discovered that the arrangement of immunoglobulin genes in the germ line was different from that in lymphocytes. Tonegawa's experiment showed that immunoglobulin genes actually rearrange during the differentiation of the immune system. This rearrangement shows the importance of a somatic process in the generation of diversity. But at least part of the diversity of antibody structure is already present in the germ line.

The gene rearrangement consists of the joining of DNA fragments that are separated in germ line DNA. In Tonegawa's first experiment, one of the movable fragments contained information for the variable region of a light chain and the other for the constant region. The joining of these two pieces of DNA produces a gene that codes for a single light chain polypeptide. The discovery was revolutionary: Instead of "one gene—one polypeptide," Tonegawa demonstrated "two genes—one polypeptide." We now know that the situation is even more complicated.

How Many Immunoglobulins from How Many Genes?

In the germ line, the genes for almost all the immunoglobulin light chains (the κ light chains) are present in three groups of DNA sequences on the same chromosome, as illustrated in Figure 22-13a. Another, much smaller group of λ light chain genes lie on another chromosome. The genes for heavy chains are on yet another chromosome (see Figure 22-13b). One group of adjacent sequences has the information for about 300 different variable regions. Some distance away on the same chromosome—at least 10^5 nucleotides from the variable region genes—is a single DNA sequence that encodes the constant region of the light chain. The production of a single light chain involves the splicing of a DNA segment containing a particular variable region sequence to a DNA segment specifying a constant region sequence.

Light Chains Derive from Genes That Have Undergone Two Rearrangements

Another small group of DNA sequences lies between the V and C regions of the κ light chain genes. These are the **J segments** (or **joining region segments**), which code for the part of the variable region closest to the constant region. During the DNA splicing of V and C regions, one of these J segments is moved next to a particular V region segment. The joining process deletes all the DNA between the selected V and J segments. The rearranged gene may still contain extra J segments, but each light chain contains only one J region. Removal of this extra information occurs during RNA splicing, which treats the extra J regions as a disposable intron. (See Chapter 15.)

The production of light chains is a "mix and match" process, involving three sets of sequences—the mouse κ light chain gene contains about 300 V segments, 4 J segments, and a single C segment. Altogether, there are about 1200 possible combinations.

The DNA sequences adjacent to those encoding the V, J, and C segments provide a clue to the mechanism of gene rearrangement. In every case, characteristic sequences lie next to segments that can be joined. Each such sequence, called a **recombination signal sequence,** consists of either 28 or 39 nucleotides, organized (as shown in Figure 22-14a) in three segments that consist of a fixed seven-base-pair sequence and a fixed nine-base-pair sequence, separated by a variant spacer sequence of either 12 or 23 base pairs. Rearrangement always involves the joining of a sequence containing a 12-base-pair spacer with a sequence containing a 23-base-pair spacer, as shown in Figure 22-14b. The machinery for recognizing and joining immunoglobulin DNA segments involves at least two identified proteins, the products of adjacent genes called RAG-1 and RAG-2 [*r*ecombination *a*ctivating *g*enes].

Additional diversity arises because of flexibility in the joining of V and J segments. Because the joining is not exact, two thirds of the time the rearranged DNA contains a frame shift and does not code for a functional immunoglobulin. But this flexibility allows an additional diversity at the VJ junction and increases the diversity of light chains by a factor of 3. As indicated in Table 22-1, the number of possible κ light chains in mice is estimated to be about 3600.

Heavy Chains Derive from Genes That Have Undergone Three Rearrangements

Heavy chain genes also consist of separated clusters of V

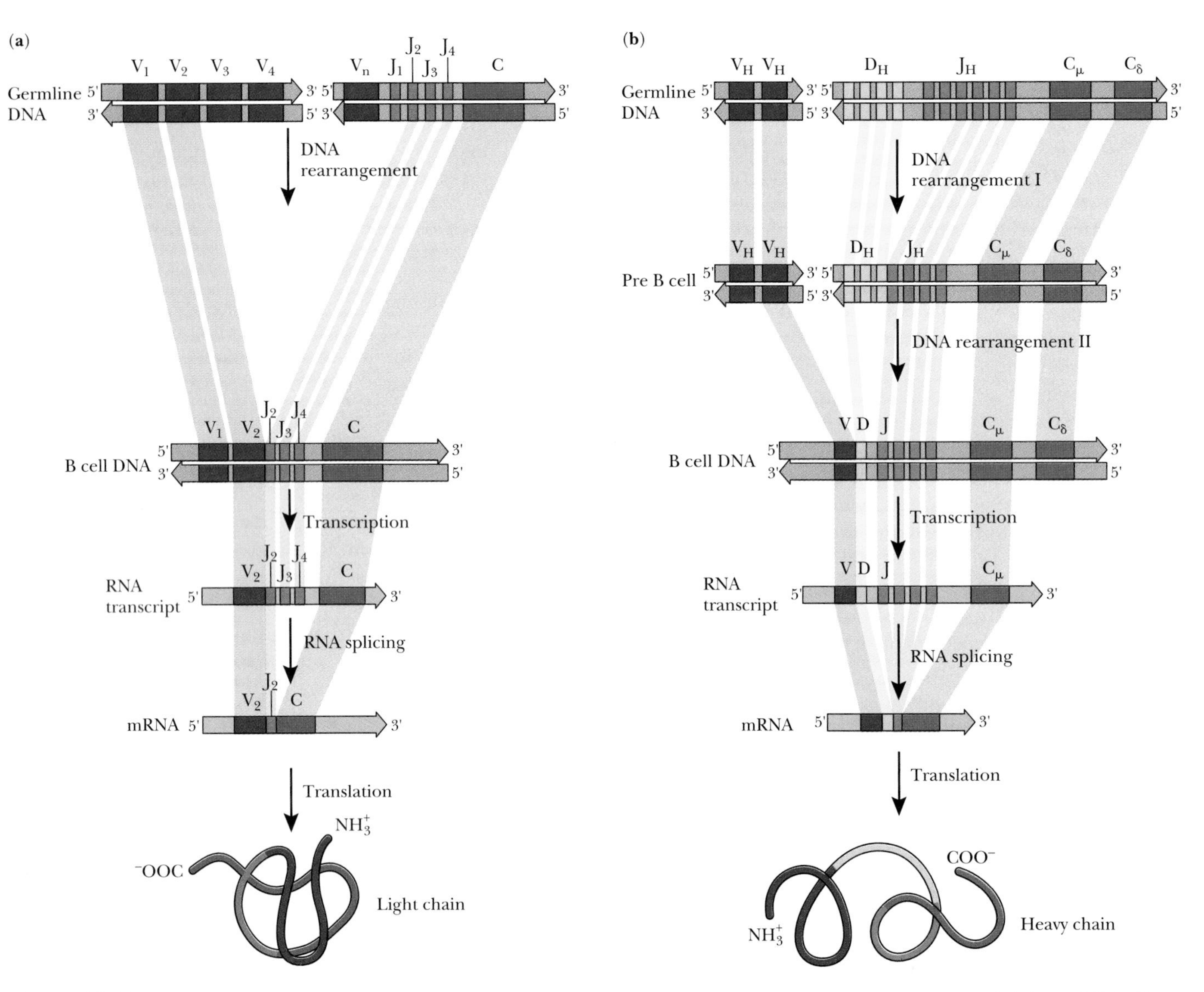

Figure 22-13 The organization of immunoglobulin. **(a)** The organization of κ light chain genes on mouse chromosome 6 and their rearrangement to produce an active gene; note the joining of V, J, and C regions; **(b)** the organization of heavy chain genes on mouse chromosome 12 and their rearrangement to produce an active gene; note the joining of V, D, J, and C regions; also note the adjacent constant region segments corresponding to the different classes of antibodies. Only the first two (for IgM and IgD) are shown.

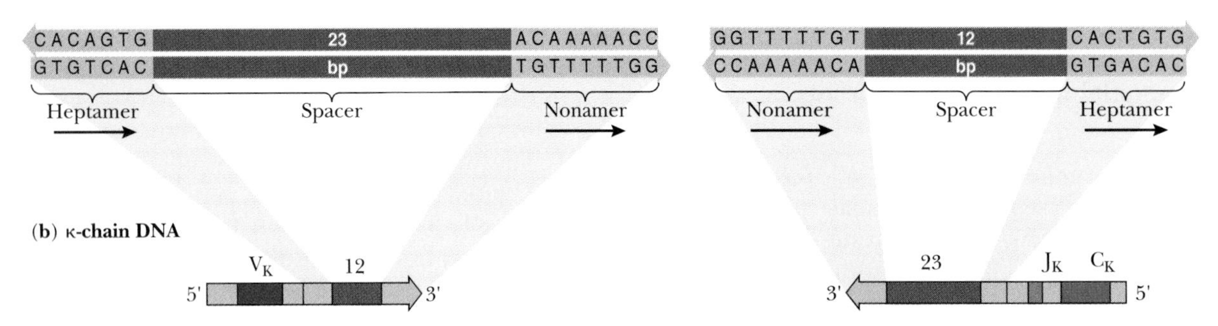

Figure 22-14 Immunoglobulin gene rearrangements depend on the ability of the DNA splicing machinery to recognize recombination signal sequences. **(a)** The nucleotide sequences of recombination signal sequences; **(b)** the placement of signals adjacent to V and J segments in κ chain genes.

Table 22-1 ***Potential Diversity of Immunoglobulins in Mouse***

Mechanism of Diversity	κ Light Chain	λ Light Chain	Heavy Chain
Germ line gene segments			
V	300	2	300
D	0	0	12
J	4	3	4
Combinations of VJ and VDJ joining	$300 \times 4 =$ 1200	$2 \times 3 =$ 6	$300 \times 12 \times 4 =$ ~ 14,000
Junctional flexibility (three different rearrangements possible at each junction)	3	3	9
N-region nucleotide addition	–	–	+
Somatic mutation	+	+	+
Possible diversity	$> 3{,}600 =$ $> 3.6 \times 10^3$	> 18	$> 9 \times 14{,}000$ $= > 1.3 \times 10^5$
Associations of heavy and light chains	$>> 4.7 \times 10^8$		

From J. Kuby, *Immunology*, W. H. Freeman and Company, New York, 1992, Table 8-3, p. 181.

and J segments. In the mouse, the number of heavy chain V segments is similar to that of κ light chain V segments, about 300, and the number of heavy chain J segments is also four. So, including junctional flexibility, we would expect about 3600 different heavy chains. But heavy chain genes contain yet another set of DNA segments—the D [for *d*iversity] segments, which multiplies the number of possible heavy chains by another factor of 36 = 12 (the number of D segments) × 3 (from junctional flexibility at a second junction). So with 3600 different possible light chains and about 100,000 different heavy chains, we can expect more than 10^8 different antibodies.

Still more heavy chain diversity derives from the addition of nucleotides during the joining of V and D segments and of D and J segments. The added nucleotides, which are not present in the genome but are added at random during the joining process, are termed the N region, and the addition process is called **N-region nucleotide addition.**

A single B cell and its descendants contain a single rearranged gene for the light chain and a single rearranged gene for the heavy chain. Every other B cell contains another unique combination of rearranged V and J light chain segments and of V, J, and D heavy chain segments.

Somatic Mutations Increase Antibody Diversity

As if this were not diversity enough, mutations in immunoglobulin genes occur at high rates during the production of lymphocytes. No one yet knows what causes these additional changes to occur specifically in variable region segments, but immunologists estimate that mutations can add another factor of 10 to antibody diversity. Mammals appear to be able to make on the order of 10^8 to 10^9 different kinds of antibody molecules.

Heavy Chain Gene Rearrangements Can Couple the Same Variable Region to Different Constant Regions to Produce Antibodies of Different Classes

Different heavy chains account for the individual properties of five different classes of immunoglobulin molecules discussed above. Studies of the organization and expression of the heavy chain immunoglobulin gene have revealed how these different classes arise. The information for the five types of heavy chain is contained in five adjacent gene segments, as diagrammed in Figure 22-13b. The choice of antibody class depends on the selective expression of one of these segments. In most cases, the choice involves a splicing of DNA similar to that responsible for the joining of V, D, and J segments. The existence of different heavy chains, however, does not add to antibody diversity but only allows antibodies to perform different functions.

THE CLONAL SELECTION THEORY EXPLAINS IMMUNOLOGICAL MEMORY AND SELF-NONSELF RECOGNITION

Studies of the organization of the immunoglobulin genes help explain the specificity and diversity of the immune response. To understand memory and self-nonself recognition, however, we need to understand the cellular basis of the immune response. In this section, we deal with the cellular basis of humoral immunity, focusing on the development of B lymphocytes and their responses to antigens. Below, we discuss the roles of T lymphocytes in both humoral and cellular immunity.

Every B lymphocyte contains just two rearranged genes—one that constitutes an active heavy chain gene and one that constitutes an active light chain gene. Just as every hybridoma clone makes only one kind of antibody, so every B cell and its descendants can make only a single kind of antibody. The particular gene arrangements within a B cell determine the structure of that antibody.

Gene Rearrangements Activate the Transcription of Rearranged Genes by Bringing Together Transcriptional Regulatory Elements

The same gene rearrangements that determine the structure of the expressed immunoglobulin peptides also activate transcription. This transcriptional activation occurs because gene rearrangement moves transcriptional regulatory elements into close proximity with one another. The rearrangement of a single heavy chain gene, randomly chosen, is an early event in B-lymphocyte development. A heavy chain gene rearrangement leads to the subsequent rearrangement of a light chain gene and the inhibition of further heavy chain gene rearrangements. The result is that each mature B lymphocyte expresses only one kind of immunoglobulin.

During embryonic life precursors of B lymphocytes —in the absence of any antigen—form a **lymphocyte library,** with each lymphocyte in the library containing a unique random rearrangement of immunoglobulin genes. Each member of the B-lymphocyte library—and all its descendants—then can make only a single type of antigen recognition site. As we discuss later in this chapter, similar rearrangements of T-cell receptor genes produce a corresponding T-lymphocyte library, each member of which also contains a single type of antigen recognition site. The next question is how to call on the proper cells to produce antibody in the face of an immunological challenge, such as a bacterial infection.

How Do B Lymphocytes Respond to Antigen?

Immunological defenses depend on the ability of lymphocytes, already programmed, to respond only when their particular abilities are needed. The stimulus to which lymphocytes respond is a particular antigen. B lymphocytes respond to particular antigens because they have **B-cell receptors,** cell surface proteins with the same antigen recognition sites as the antibody that they can produce. In fact, B-cell receptors are just membrane-bound forms of IgM or IgD.

Binding of antigen to the B-cell receptors activates two pathways of B-lymphocyte cytodifferentiation, as shown in Figure 22-15. One path leads to the production of **effector cells,** which carry out the appropriate immune function, and the other to the production of **memory cells,** which can later be stimulated to produce more effector cells. The effector cells for humoral immunity are the plasma cells, packed with rough endoplasmic reticulum, which make and secrete antibodies.

Memory cells are indistinguishable in appearance from unstimulated B cells, except that there are many more of them after exposure to antigen. These cells have the same genetic rearrangements as the parent B cell, but they do not immediately participate in the immune response. They become effectors when the animal next encounters the same antigen. The second exposure to an antigen therefore evokes a much stronger and faster response than the first.

The proliferation of memory cells underlies the immunity to childhood diseases after a first exposure. Because lymphocytes respond to small antigenic sites (epitopes), rather than to intact molecules, the initial proliferation does not depend on exposure to exactly the same antigen. Antigens that mimic disease-causing bacteria or viruses are often as effective in producing immunity as the pathogens themselves. A heat-killed polio virus can induce immunity to polio, or a mild infection of cowpox can produce immunity against the devastating smallpox virus.

An Experimental Test of the Clonal Selection Theory Also Suggested a Mechanism for Self-Nonself Recognition

The **Clonal Selection Theory** summarizes the role of selective proliferation of responding lymphocytes, already programmed, into clones of cells that contain specific gene rearrangements. This selective expansion of particular cells was recognized even before the molecular basis of antibody diversity was understood. This understanding resulted mostly from the insights of F. MacFarlane Burnett and Niels Jerne (working independently in Australia and Switzerland). The clonal selection theory makes a strong testable prediction: Removal of all the cells able to respond to a given antigen prevents a further response to that antigen.

In one test of this prediction, diagrammed in Figure 22-16a, researchers removed lymphocytes from mice and mixed them with a specific antigen that was highly radioactive. When lymphocytes bound this antigen, the radiation killed them and the mice were then no longer able to respond to the antigen. The researchers then gave lymphocytes from these treated mice to other mice whose own immune response they had previously destroyed by treatment with x-rays. The mice that received the lymphocytes could then respond to many antigens, but not to the antigen that had been used in radioactive form.

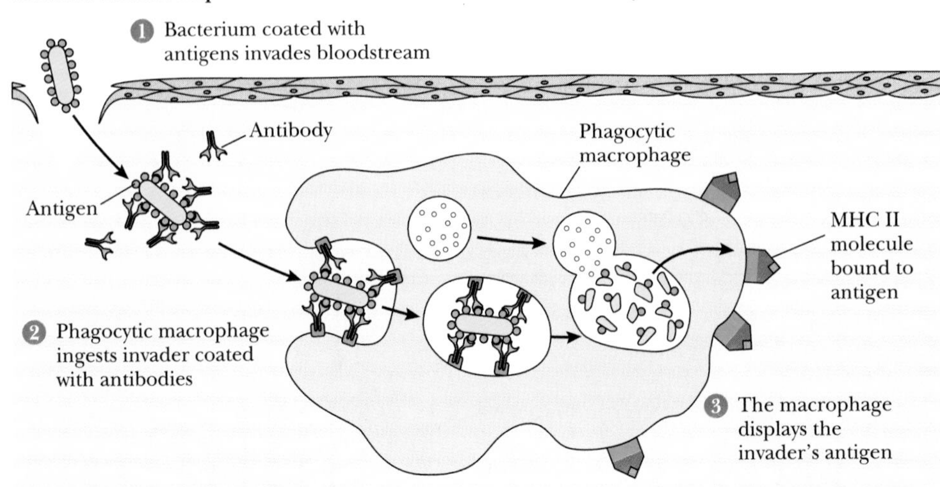

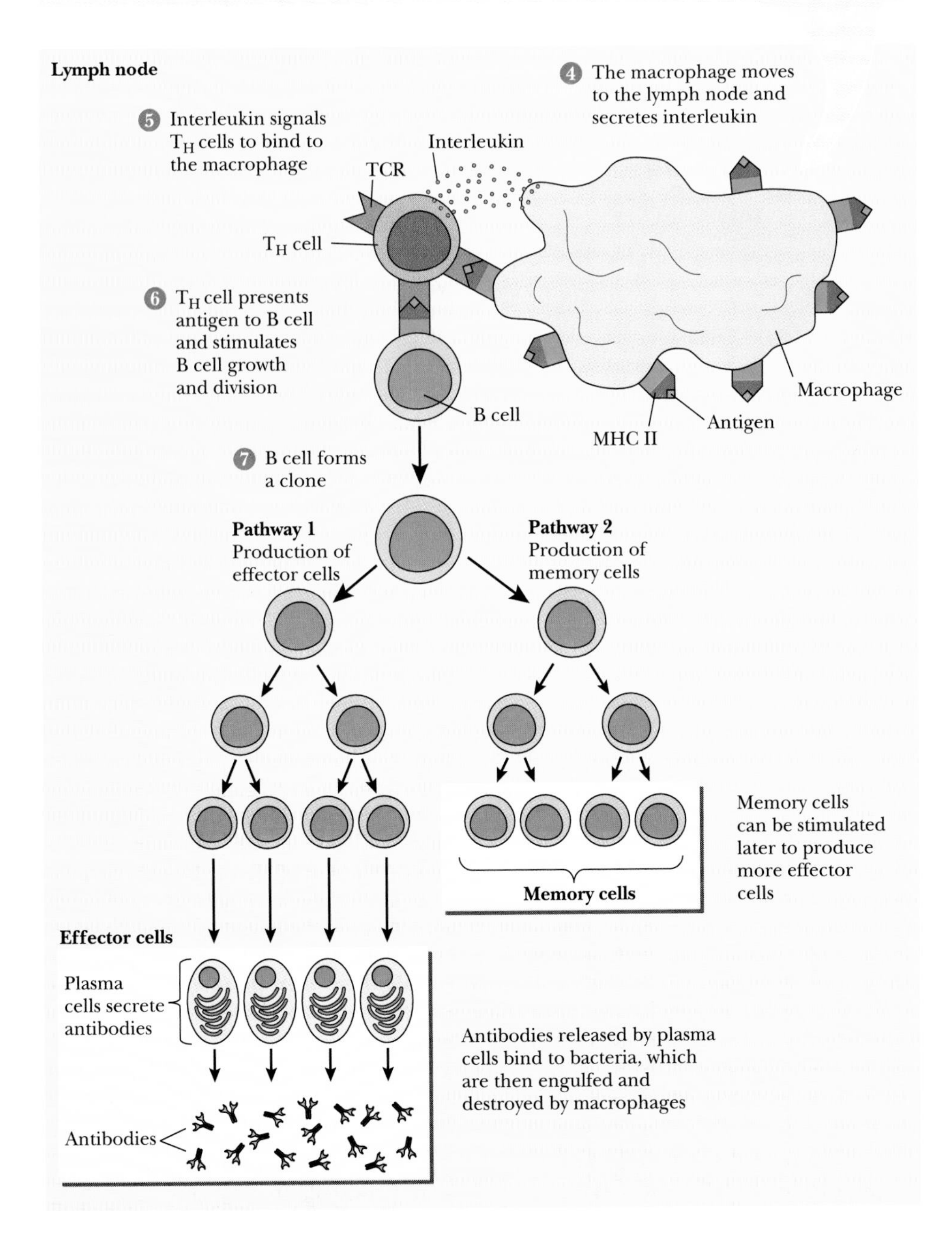

Figure 22-15 Exposure to antigen stimulates a B lymphocyte to produce both effector cells and memory cells. The persistence of the memory cells means that a second exposure to the same antigen evokes a much greater and faster response.

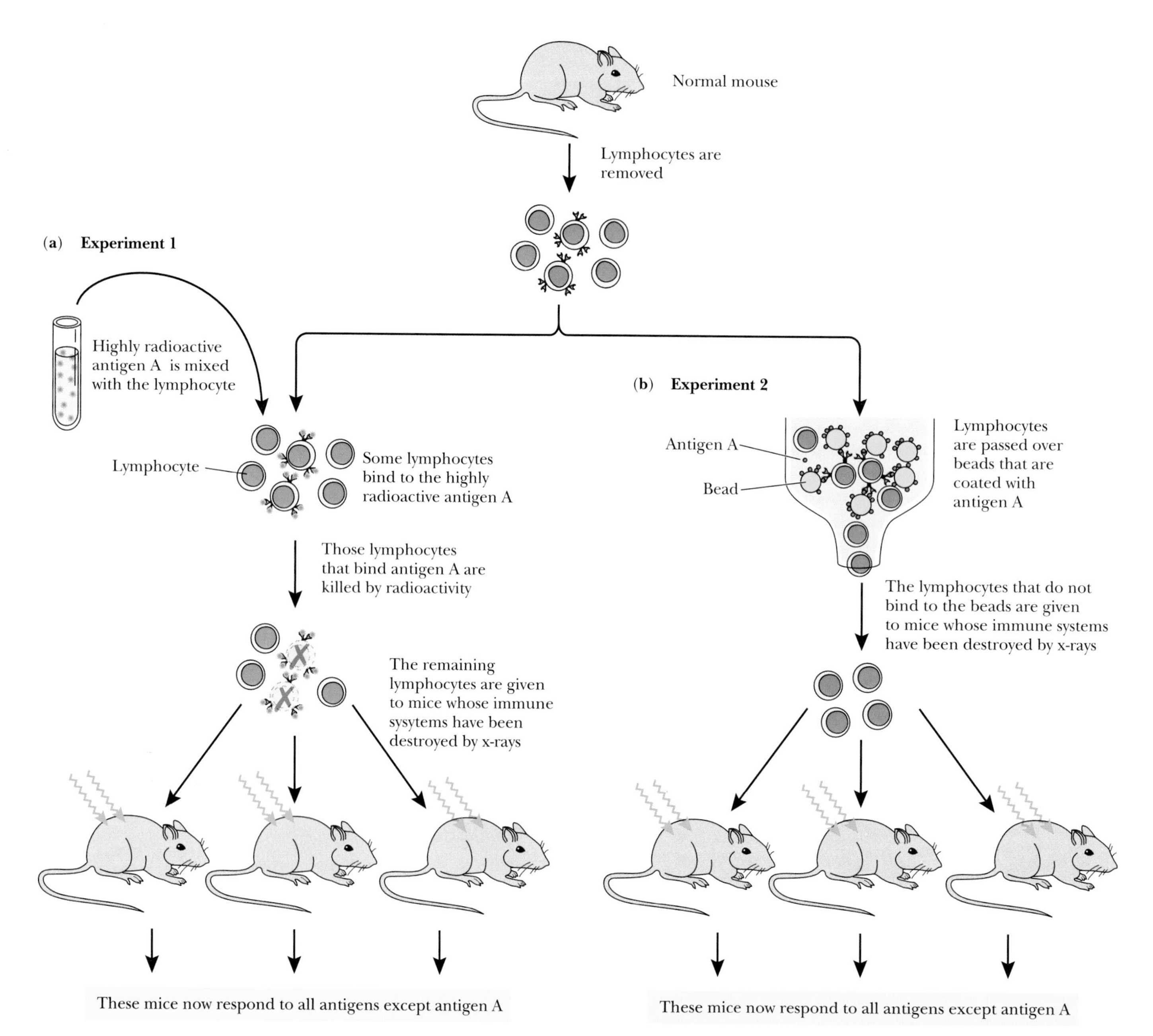

Figure 22-16 Two experiments demonstrating clonal selection. **(a)** B lymphocytes that bind the specific, radioactively labeled antigen are destroyed by radioactivity; **(b)** B lymphocytes that bind to the specific antigen are removed from the blood by exposing them to beads coated with the antigen. Once the specific B lymphocytes are removed, the mice do not respond to that antigen but can respond to other antigens.

In another version of the experiment, illustrated in Figure 22-16b, B lymphocytes that react with a specific antigen were removed by passing lymphocytes from one mouse over beads that had been coated with the antigen. The lymphocytes were then injected into mice whose own immune response had been destroyed. Again, the transfused mice could not respond to the particular antigen but could respond to other antigens.

The clonal selection theory suggests a mechanism for self-nonself recognition and immunological tolerance. After the establishment of the lymphocyte library, an animal could somehow eliminate or suppress the action of all the lymphocytes programmed to react against the animal's own tissues. The mechanism of this recognition depends on T lymphocytes. (See later in this chapter.)

T LYMPHOCYTES ARE RESPONSIBLE FOR CELLULAR IMMUNITY AND ALSO CONTRIBUTE TO HUMORAL IMMUNITY

The humoral immune system produces antibodies that can recognize bacteria and bacterial products and target them for destruction by complement or by phagocytosis. But the humoral immune system cannot provide defenses against intracellular pathogens, such as viruses, nor can it effectively distinguish the body's own tissues from transplants or tumors. The cellular immune system provides another set of specific defenses against agents that lead to altered cells. These same defenses allow the immune system to reject skin grafts or organ transplants and often to recognize and eliminate cancer cells.

Figure 22-17 A *nude* mouse, which lacks a thymus and T lymphocytes. (*Custom Medical Stock Photo*)

The main mediators of cellular immunity are T lymphocytes. Because T cells mature in the thymus, a child born without a thymus (a condition called DiGeorge syndrome) lacks T cells and has no cellular immunity. A DiGeorge syndrome child can withstand most bacterial infections but is continually beset by infections of viruses, intracellular bacteria, and fungi. The *nude* mouse, like a DiGeorge human, lacks a thymus and T cells. (The name of this mutation comes from the fact that the mice are also hairless, as shown in Figure 22-17.) Unlike wild-type mice, which reject skin grafts from genetically different mouse strains, a *nude* mouse can readily accept grafts from other mouse strains or even other species.

Our current understanding of cellular immunity has derived much from studies of skin grafts among different strains of mice. The strains employed are highly inbred, meaning that all the mice of a particular strain are genetically identical; that is, they are homozygous for the same alleles for essentially every gene. Mouse geneticists have also bred *congenic* mice, which are genetically identical except for a single gene or closely linked set of genes. When experimenters study the ability of congenic mice to accept or reject skin grafts, they found that graft rejection (mediated by cellular immunity) depends on genetic differences in the **major histocompatibility complex (MHC)** a single set of 40–50 closely linked genes, some of which encode proteins on the surface of every somatic cell. In mice this complex lies on chromosome 17; in humans the corresponding genes similarly lie in a single complex on chromosome 6.

T Lymphocytes Do Not Recognize Antigens by Themselves but Only as Altered Self

The experiment shown in Figure 22-18 illustrates the role of MHC in cellular autoimmunity. Researchers induced immunity to a virus (called LCM virus) in a strain of mouse that expressed a type of MHC called H-2^k. T cells were then taken from the spleen of the immunized mouse and allowed to attack cells, from several different mouse strains, infected with the same virus. Before exposure to the T cells, the target cells were incubated with radioactive chromium ions (^{51}Cr), which they concentrate intracellularly and release only upon lysis. Release of radioac-

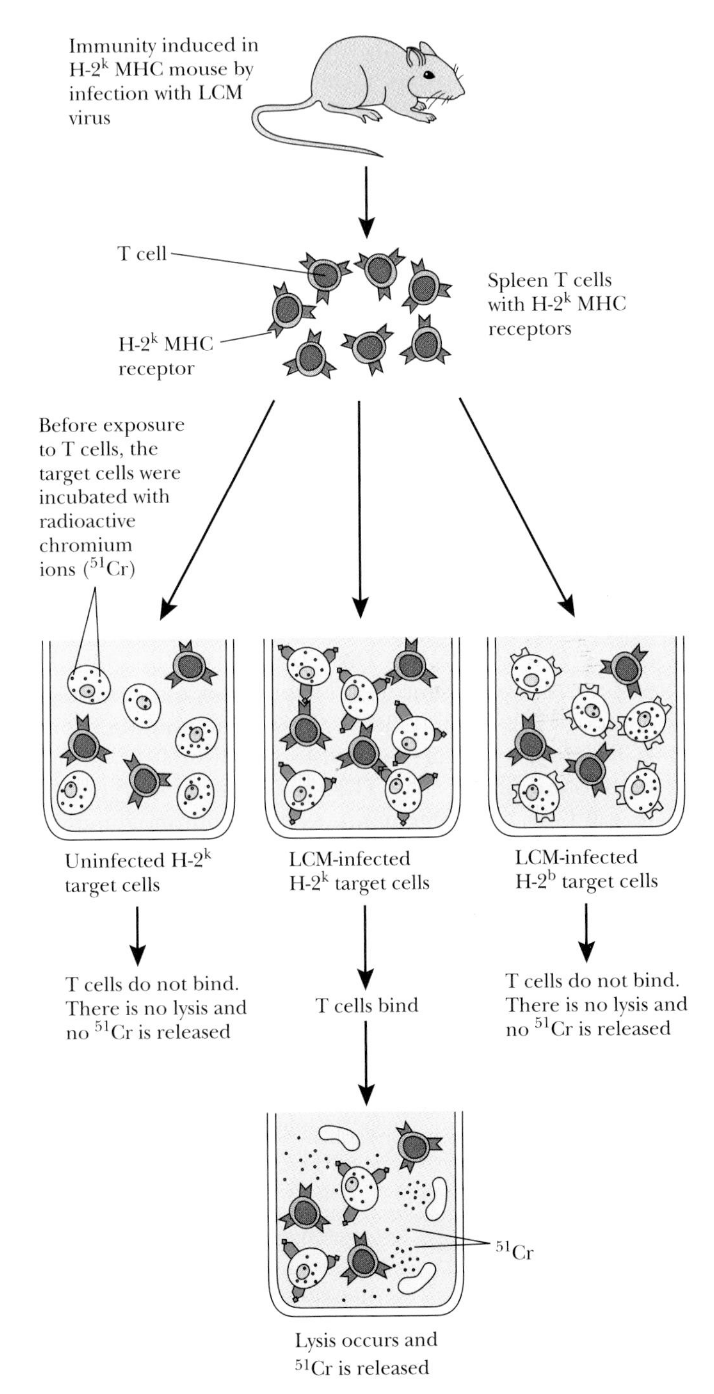

Figure 22-18 Certain T lymphocytes kill cells infected with LCM virus, but only if the viral antigens are presented in the context of the proper MHC molecule. T cells taken from LCM-infected mice with type H-2^k MHC can attack other LCM-infected cells, but only if they come from an H-2^k mouse. LCM-infected cells from an H-2^b mouse are not attacked.

tivity could then indicate cell lysis. The T cells lysed virally infected ^{51}Cr-loaded cells from the same strain, but not uninfected cells or infected cells from another strain, with a different MHC type, such as H-2^b.

The experiment with LCM virus illustrates two important properties of cellular immunity: **cytotoxicity,** the ability of certain T lymphocytes to kill cells with altered surfaces, and **MHC restriction,** the inability of T cells to recognize foreign antigens except in cells with the same MHC type as the responding animal. Another way of stating this conclusion is that T cells recognize only **altered self,** nonself antigens in a self context, somehow specified by MHC genes.

(a) **TCRαδ**

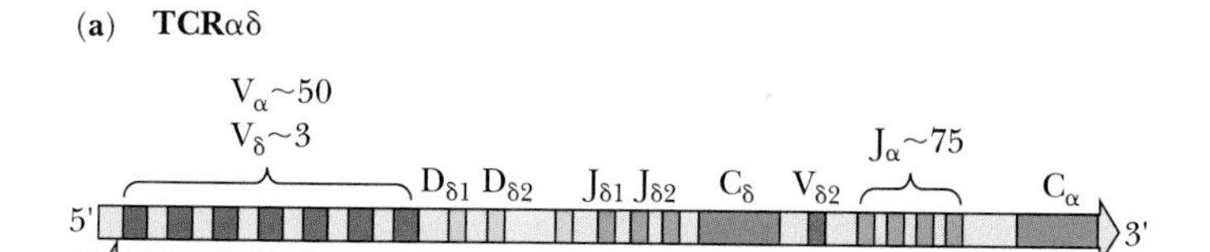

(b) **TCRβ**

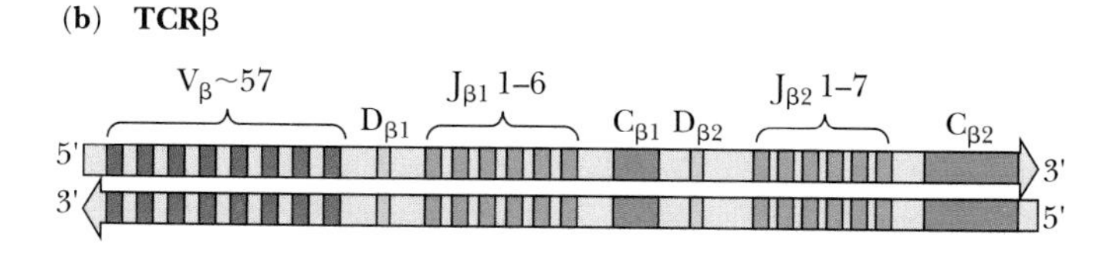

Figure 22-19 The organization of TCR genes. **(a)** Clustering of V and J regions with the C region of the intermingled α and δ TCR gene on mouse chromosome 14; **(b)** clustering of V, D, and J regions with the C region of the β TCR gene on mouse chromosome 6.

Antigen Recognition by T Lymphocytes Depends on T-Cell Receptors, Which Are Even More Diverse than Antibodies

The surfaces of T lymphocytes contain **T-cell receptors (TCRs),** recognition molecules that can recognize foreign antigens as altered self. While immunologists thought for many years that T-cell receptors might be members of a special class of immunoglobulins, this hypothesis turned out to be incorrect.

An important step in establishing the molecular identity of the T-cell receptors was the isolation of a recombinant DNA encoding a TCR peptide. TCR DNA was first isolated as a cDNA, a DNA copy of the mRNA encoding a protein present in T lymphocytes but not in B lymphocytes. Having this cDNA allowed the identification of TCR genes, which are related to but distinct from immunoglobulin genes. The TCR genes can undergo the same kind of rearrangements as the immunoglobulin genes and can generate an even more diverse set of recognition proteins.

Each TCR contains two polypeptide chains, either α and β, or (more rarely) γ and δ. As shown in Figure 22-19a, the α genes, like that of the κ light chain immunoglobulin, consists of clusters of V and J segments with a single constant region segment. The gene for the TCR β chain, shown in Figure 22-19b, like that of the immunoglobulin heavy chain, also contains D-region segments.

As in the case of the immunoglobulins, tremendous diversity arises from gene rearrangements, junctional flexibility, N-region nucleotide addition, and the random association of α and β (or γ and δ) polypeptides. There is no evidence, however, for somatic mutation in the TCR genes as there is in immunoglobulin genes. Nonetheless, the calculations summarized in Table 22-2 suggest that T-cell receptor diversity is even greater than immunoglobulin diversity.

Table 22-2 ***Potential Diversity of αβ T-Cell Receptors in Mouse***

Mechanism of Diversity	α	β
Germ line gene segments		
V	75	25
D	0	2
J	50	12
Combinations of VJ and VDJ joining	$75 \times 50 =$ 3750	$25 \times 2 \times 12 =$ 600
Alternative joining of D gene segments	–	+ (some)
Junctional flexibility	+	+
N-region nucleotide addition	+	+
Somatic mutation	–	–
Possible diversity	$\sim 10^{15}$	

From J. Kuby, *Immunology,* W. H. Freeman and Company, New York, 1992, Table 10-1, p. 222.

T-Cell Receptors Recognize Only Antigens That Are Bound to MHC Proteins

The antigens recognized by T cells differ from those recognized by antibodies in two respects: (1) While antibodies recognize foreign antigens in solution as well as on cell surfaces, TCRs bind foreign antigens only on the surfaces of cells that are recognizably self; and (2) while antibodies recognize the three-dimensional shapes on the outside of proteins and carbohydrates, TCRs recognize only linear segments of polypeptide chains. The antigenic determinants (epitopes) recognized by TCRs are typically peptides about 20 amino acid residues long.

T lymphocytes fall into two general classes: **T-cytotoxic (T_C)** cells, which are responsible for the killing of cells recognized as nonself, and **T-helper (T_H)** cells, which activate both the humoral and cellular immune responses. Upon stimulation with a foreign antigen, some T_C cells differentiate into effector cells, called **cytotoxic T lymphocytes (CTL),** which actually kill the target cells, as illustrated in Figure 22-20. (See Essay, "How Do Cytotoxic T Lymphocytes Kill Target Cells?")

In contrast, antigen recognition by a T_H cell leads not to cell killing but to the production of chemical signals (cytokines) that can activate both T and B lymphocytes. While some cytokines (such as IL-1, mentioned earlier in this chapter) can act systemically, most cytokines are paracrine signals, which activate only cells in the immediate vicinity. For example, IL-2, a cytokine made by T_H cells, stimulates the differentiation of T_C cells into CTLs.

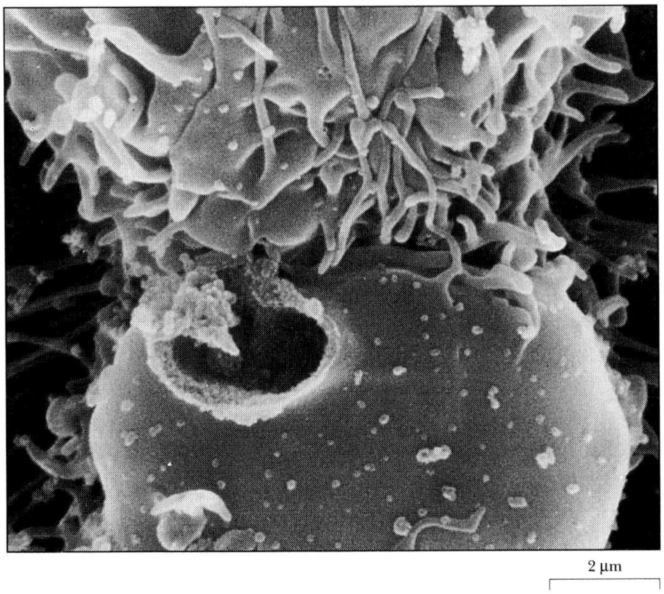

Figure 22-20 After stimulation with a foreign antigen, some T cytotoxic (T_C) cells differentiate into effector cells, called cytotoxic T lymphocytes (CTL). Here a CTL has attacked a tumor cell and produced a hole in the target cell membrane. *(John D. E. Young, Rockefeller University)*

While both T_C and T_H cells use T-cell receptors to recognize nonself antigens, the two cell types differ biochemically and serve different defensive needs. Among the biochemical differences are the presence of a protein called CD8 on T_C cells and of a protein called CD4 on T_H cells. Immunologists use antibodies to CD4 and to CD8 to distinguish the two types of cells (although the correlation of these proteins with T cell type is not absolute). T_C cells and their effectors (CTLs) recognize **endogenous** proteins, made by the body's own cells. CTLs destroy cells that produce nonself endogenous proteins, such as those of a virus or the mutated versions of cell proteins present in a cancer cell. In contrast, T_H cells recognize **exogenous antigens,** such as those of a microorganism, which have been taken up by phagocytosis. T_H cells then mobilize either a humoral or a cellular response, as discussed later in this chapter. Before addressing the functional differences of T_C and T_H cells, however, we must first examine the nature of the proteins specified by the MHC complex.

MHC Class I and Class II Genes Specify Membrane Proteins That Bind Antigens and Present Them to T Cells

In the last decade, molecular biologists and immunologists have learned much about the nature of MHC genes and the proteins they encode. These genes fall into three classes, two of which—Class I and Class II genes—are important for our discussion here.

MHC Class I Molecules Are Present on the Surface of Essentially Every Nucleated Cell (Except Sperm)

MHC Class I molecules present endogenous antigens, binding to polypeptide segments of proteins made within the cells. **MHC Class I** molecules are single polypeptide chains about 340 amino acid residues long. As illustrated in Figure 22-21a, each MHC Class I molecule contains three domains (called α_1, α_2, and α_3), each of about 90 residues, which lie outside of the cell, a transmembrane domain of about 40 residues, and a cytoplasmic domain of about 30 residues. The α_3 domain contains a sequence similar to that of an immunoglobulin fold, suggesting that MHC Class I is a member of the immunoglobulin superfamily. As shown in Figure 22-21a, each Class I molecule associates tightly with **β_2-microglobulin**, a small protein about the size of a single α domain, also a member of the immunoglobulin superfamily.

X-ray diffraction studies of the external portion of one MHC Class I molecule have yielded a detailed model of its three-dimensional structure and, unexpectedly, of its interactions with an antigenic peptide to which the crystallized MHC protein was surprisingly bound. The antigen-

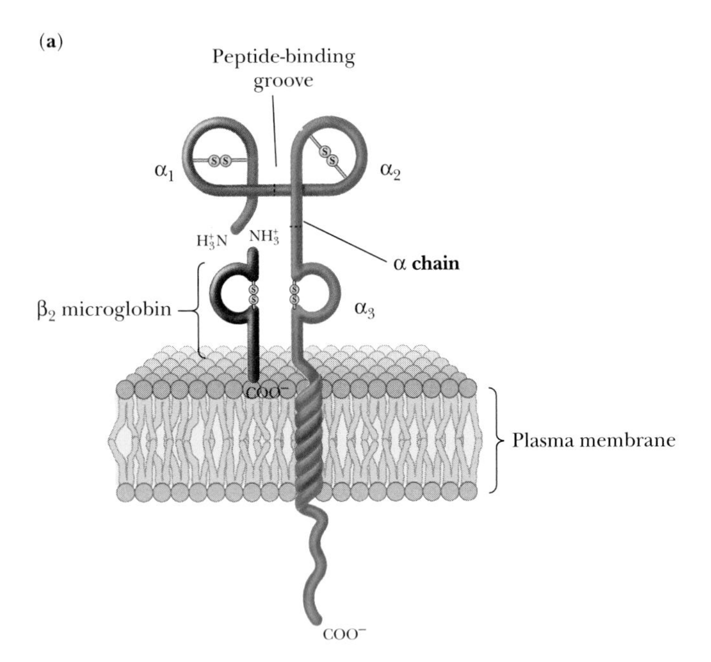

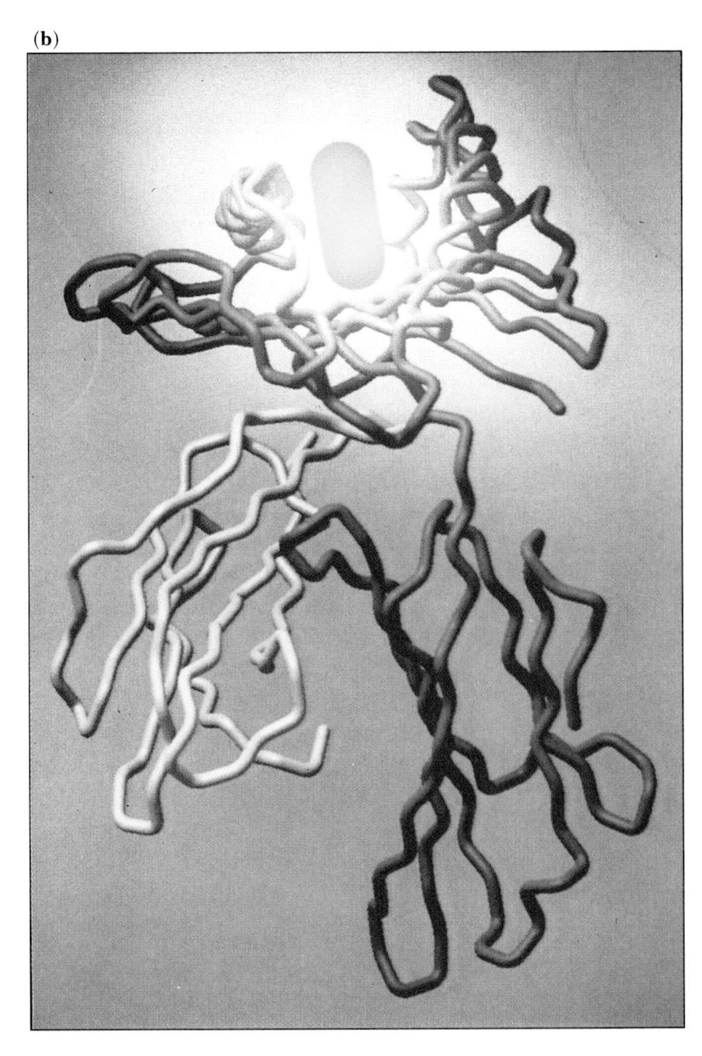

Figure 22-21 The structure of an MHC Class I protein: **(a)** each MHC Class I molecule consists of a single polypeptide, with three extracellular domains (labeled α_1, α_2, and α_3) each about 100 amino acid residues long; each MHC Class I molecule associates with a molecule of β_2-microglobulin; **(b)** model of MHC Class I molecule based on x-ray diffraction studies, showing the cleft where an antigen peptide binds. *(b, Arthur J. Olson, Ph. D., Research Institute of Scripps Clinic)*

binding site consists of a deep groove or cleft, as shown in Figure 22-21b, large enough to accommodate a peptide of 10–20 amino acid residues. The floor of the cleft consists of a β structure with eight antiparallel peptide segments, while the sides of the cleft are two α helices.

MHC Class II Molecules Are Present on the Surfaces of Specialized Antigen-Presenting Cells

They present exogenous antigens, polypeptide segments almost always derived from viruses or microorganisms taken up by phagocytosis. **MHC Class II** molecules consist of two polypeptides, called α and β, each of which can derive from one of several genes. Like MHC Class I molecules, Class II molecules are members of the immunoglobulin superfamily and are membrane proteins that bind antigen peptides, as illustrated in Figure 22-22a. Each peptide has two extracellular domains (α_1 and α_2, or β_1 and β_2), a transmembrane domain, and a cytoplasmic anchor. Antigen binding involves the α_1 and β_1 domains, as illustrated in the three-dimensional structure shown in Figure 22-22b.

Every Human Differs from Every Other (Except for an Identical Twin) in MHC Type

Recombinant DNA studies have established the organization of the MHC genes in mice and humans; their organization on human chromosome 6 is illustrated in Figure 22-23. Every human and every mouse expresses at least six MHC Class I genes, a paternally derived and a maternally derived copy of three different Class I genes. Studies of gene structure have revealed even more potential Class I genes in the MHC region, but their expression pattern is still uncertain. Similarly, every human and every mouse expresses at least six MHC Class II genes. The MHC genes are extremely polymorphic, with each gene existing in as many as 100 different alleles, so (excepting identical twins) every human is likely to differ from every other in MHC type and, therefore, in the manner of presentation of antigens.

Studies of inbred mouse strains, however, have revealed much about the manner in which antigen presentation depends on interactions between antigen peptides and MHC structure. Peptides differ in their ability to serve as antigens, in large part because of the differing affinities of individual peptides for MHC binding.

(a)

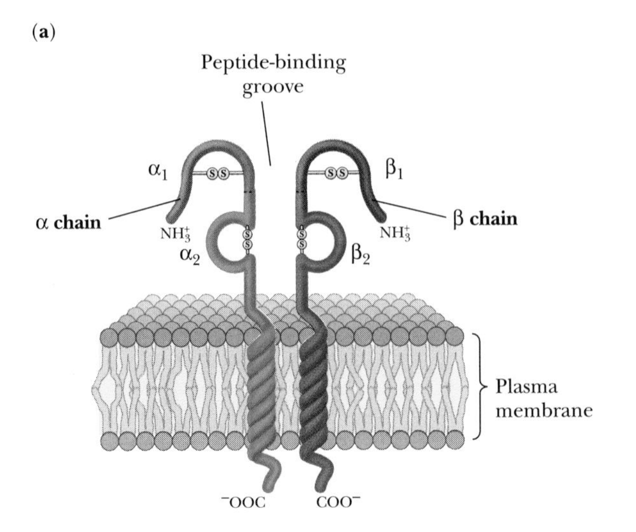

(b)

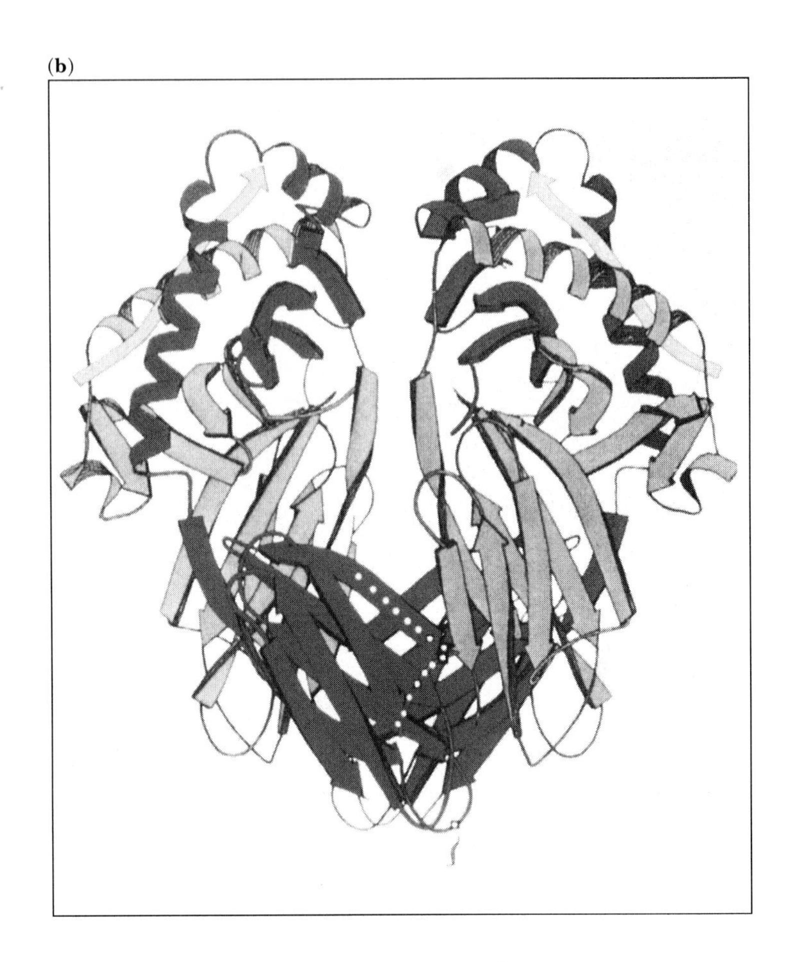

Figure 22-22 The structure of an MHC Class II protein: **(a)** association of Class II MHC polypeptides; **(b)** model of MHC Class II molecule based on x-ray diffraction studies, showing the cleft where an antigen peptide binds. *(b, from Brown, J. H., et al.,* Nature, *364: 38, 1993. With permission.)*

During Maturation in the Thymus, T Lymphocytes Undergo Both Positive and Negative Selection

Cells of the immune system undergo stringent selection as they develop. For T cells, the selected cells can recognize antigens bound to the individual's own MHC molecules. Otherwise, neither T_C nor T_H cells could recognize antigens at all, because antigens must be bound to MHC molecules.

During the maturation of T_C cells in the thymus, surviving precursor cells are those that can recognize antigens within the individual's own MHC Class I molecules. Similarly, the surviving T_H cell precursors are those that can recognize antigens bound to the individual's own MHC Class II molecules. The vast majority of T-cell precursors do not fulfill these criteria. They undergo apoptosis and disappear.

Slightly later during T-cell differentiation, most of the T-cell precursors that recognize self antigens bound to self MHC receptors also undergo apoptosis, leaving only T cells that can recognize nonself antigens bound to self MHC molecules. Altogether, 90–99% of T-cell precursors die within the thymus. The T cells that leave the thymus are highly selected to recognize altered self antigens—nonself antigen in association with self MHC. In principle, all surviving cells recognize nonself antigens in the context of self MHC molecules.

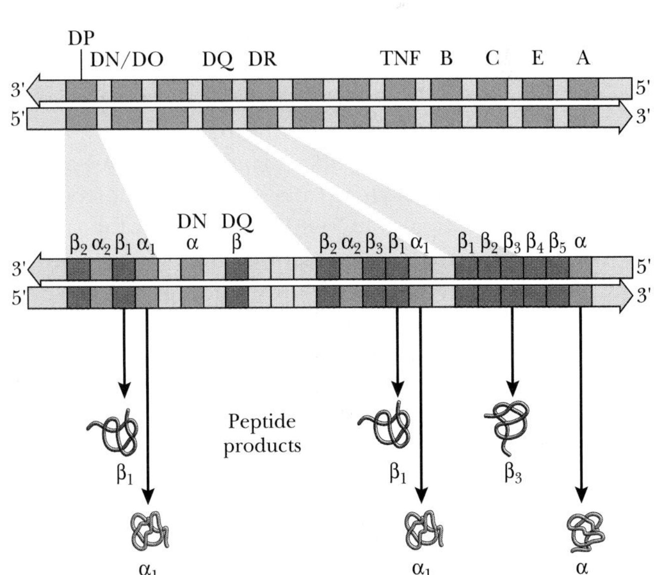

Figure 22-23 The genes of the human major histocompatibility complex (MHC), showing the three major classes of genes. DP, DQ, and DR all encode class II MHC polypeptides; the DP region contains two α and two β genes, of which one α and one β are expressed from each chromosome; similarly, one α and one β are expressed from the DQ and the DR regions. The genes are all highly polymorphic, meaning that individuals in a population can have different versions (alleles) of each gene.

(a)

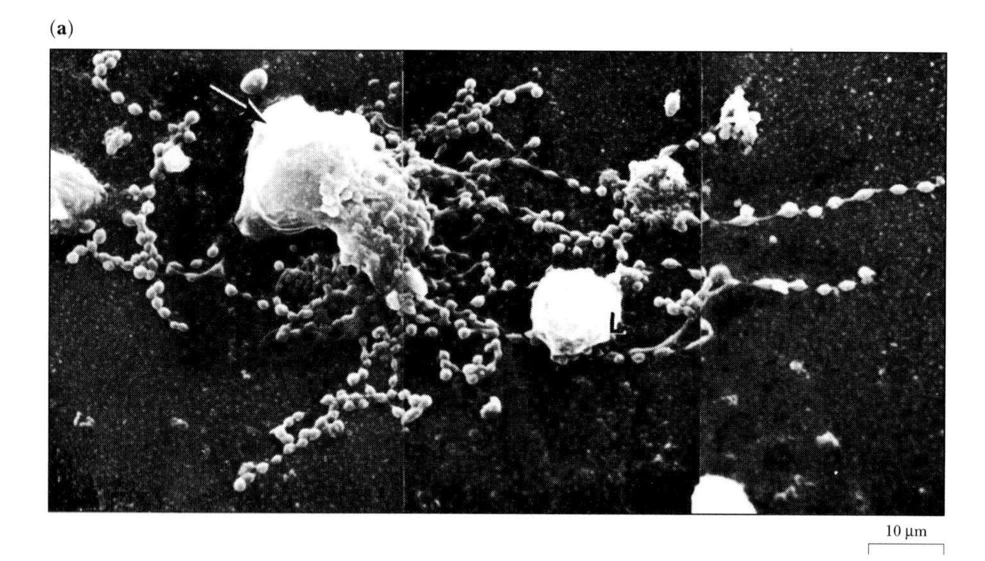

(b)

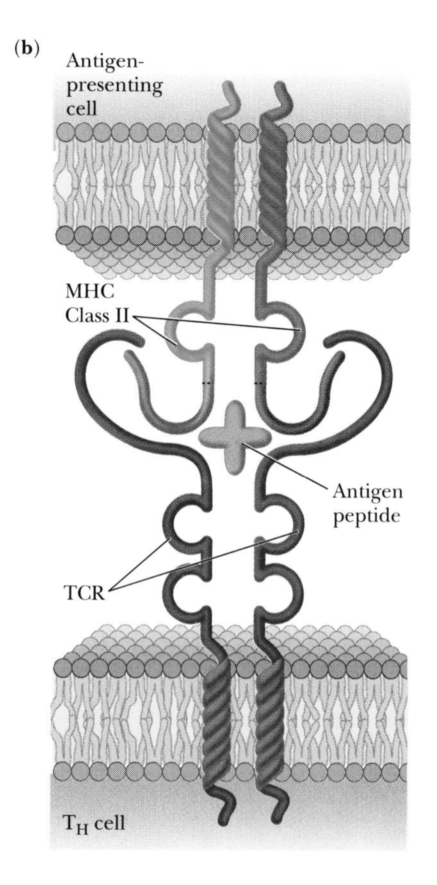

Figure 22-24 Antigen-presenting cells, such as dendritic cells, present exogenous antigens to T_H cells. **(a)** SEM of dendritic cells; **(b)** a diagram of a dendritic cell presenting antigen to a T_H cell. *(a, from Szakal, A. K., et al.,* J. Immunol., *134: 1353–1354, 1985, © 1985 American Association of Immunologists. Reprinted with permission.)*

T_C and T_H Cells Recognize Peptide Determinants in Different MHC Contexts Presented by Distinct Cell Types

T_C cells recognize endogenous antigens, molecules that are synthesized within the cells targeted for killing. T_C cells are especially effective at eliminating cells infected with virus, because animal viruses almost always modify the surfaces of their hosts. In contrast, T_H cells recognize exogenous antigens, such as those of viral polypeptides.

In both cases, the T-cell receptor recognizes a peptide bound to an MHC molecule—a Class I molecule in the case of T_C cells and a Class II molecule in the case of T_H cells. Before an antigenic protein can be recognized by a T lymphocyte, however, it must undergo **antigen processing,** degradation into small peptides and association with an MHC molecule.

Most cells in the body can process antigens and present them to T_C cells in the context of an MHC Class I molecule. Most cells are therefore capable of presenting nonself antigens that they themselves synthesize, such as viral proteins or proteins associated with a cancerous state.

Only certain cells, called **antigen-presenting cells (APCs),** produce MHC Class II molecules and are capable of presenting exogenous antigens, such as those present in grafts, transplants, or extracellular parasites. APCs include macrophages, B lymphocytes, and **dendritic cells** (shown in Figure 22-24a), highly extended cells present in many organs and tissues throughout the body. APCs engulf foreign antigens by endocytosis or phagocytosis (as described in Chapter 6, p. 165), degrade them into peptides, and then present them to T_H cells bound to MHC Class II molecules. The peptide antigen is bound to an MHC Class II molecule on the surface of the APC and to a TCR molecule on the surface of the T_H cell, as shown in Figure 22-24b.

Binding of Processed Antigen to Class II MHC Molecules on the Surface of a T_H Cell Initiates Cytokine Production in the Antigen-Presenting Cell and in the T_H Cell

As summarized in Figure 22-25, the formation of the MHC–antigen–T-cell receptor complex stimulates cytokine synthesis. The APC in the complex produces IL-1, which stimulates the T_H cell. The T_H cell produces IL-2 and IL-2 receptors, leading to the proliferation of the T_H cell. In the cellular immune response, illustrated in Figure 22-25a, IL-2 produced by the expanded clone of T_H cells stimulates T_C cells to differentiate into effector cells (cytotoxic T lymphocytes) and to destroy the target cells.

In the humoral immune response, illustrated in Figure 22-25b, the APC is a B cell, and the T_H cell responds by producing not only IL-2 but also IL-4, IL-5, IL-6, and interferon-γ. These cytokines stimulate the B cell to pro-

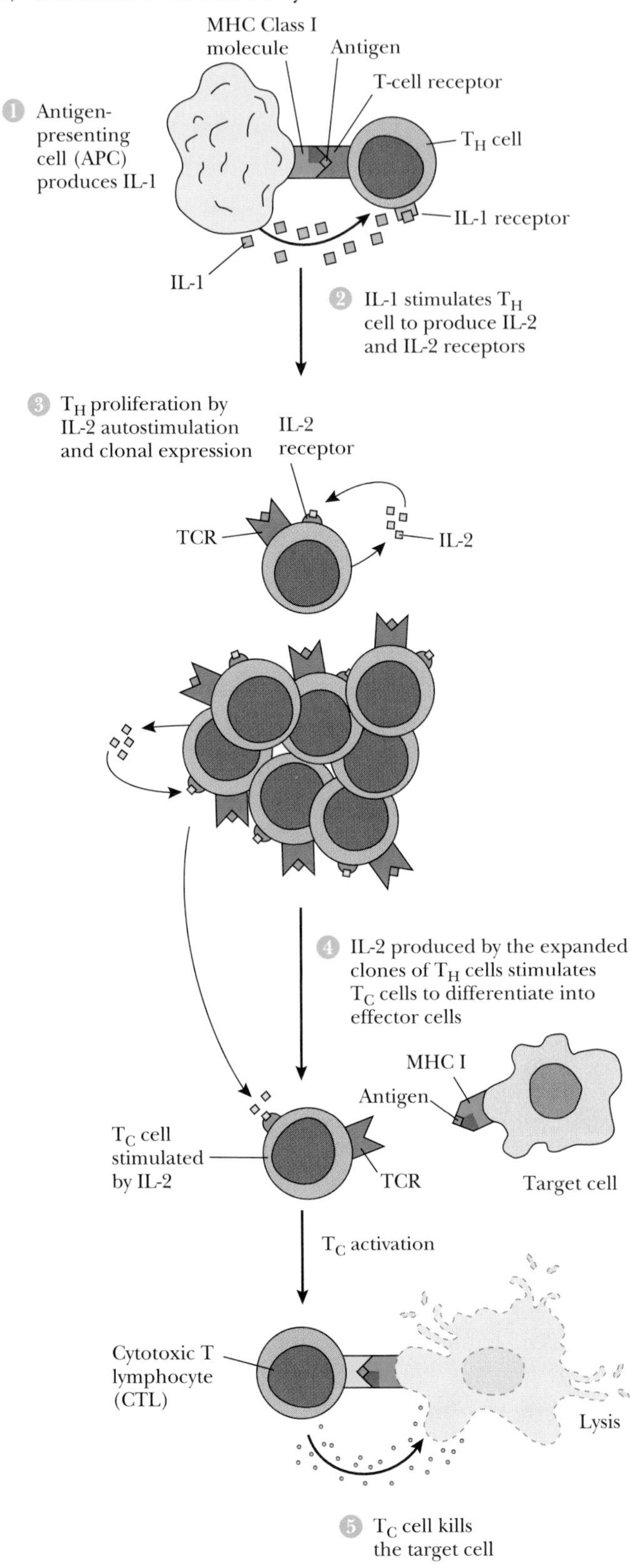

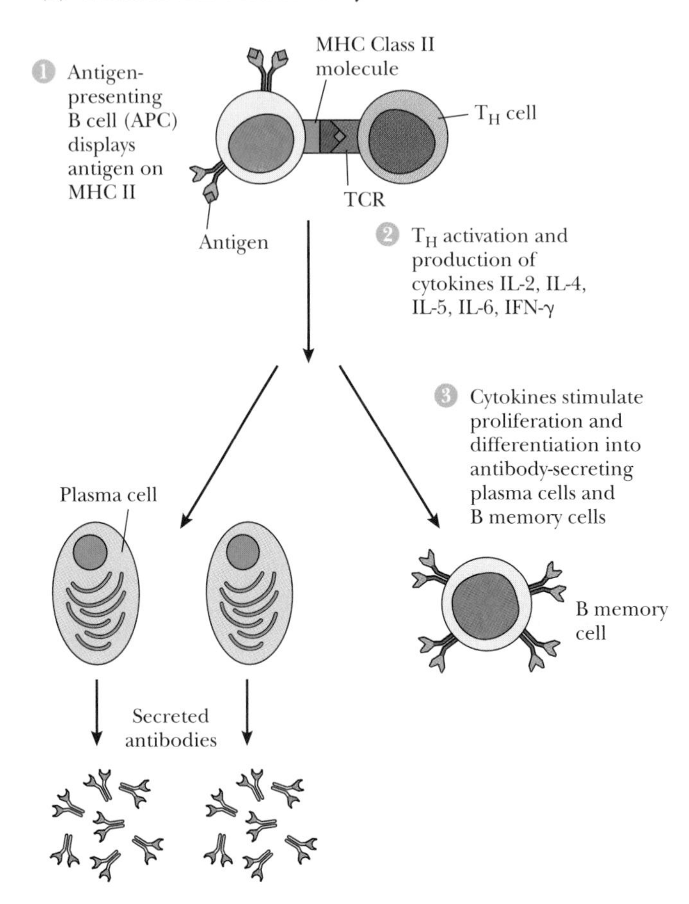

Figure 22-25 The binding of an antigen-presenting cell to a T_H cell stimulates cytokine production, initiating cellular or humoral immunity. **(a)** Initiation of cellular immunity: IL-1 produced by the APC stimulates IL-2 production by T_H, leading to the proliferation of T_H cells and the differentiation of T_C cells into CTLs; **(b)** initiation of humoral immunity: T_H cells produce several cytokines, including IL-2, IL-4, IL-5, and interferon-γ; these signals stimulate the proliferation and differentiation of B cells into plasma cells and memory cells.

liferate and to differentiate into effector cells (plasma cells) that produce antibodies specific for the original antigen.

Immunological researchers are now able to determine the production of cytokines from individual T cells after challenge by different antigens. Studies of cytokine production have revealed distinctive subpopulations of T_H cells, which respond to different types of antigens by producing different sets of cytokines. The different sets of cytokines stimulate distinctive immune responses. T_H1 cells characteristically produce IL-2 and interferon-γ, which stimulate cell-mediated immunity and do not stimulate a specific humoral response. In contrast, T_H2 cells secrete IL-4 and IL-5, which stimulate B-cell proliferation and antibody synthesis. Different types of immunological challenges stimulate different subsets of T_H cells; leprosy-causing bacteria, for example, stimulate a T_H1-mediated response, whereas certain intestinal parasites (such as tapeworms) evoke a T_H2 response.

ESSAY

HOW DO CYTOTOXIC LYMPHOCYTES KILL TARGET CELLS?

Antibodies and complement can attack and destroy many free-living organisms, but what about microorganisms and viruses that grow inside of cells? Defending against such hidden invaders is the task of the cellular immune system, which recognizes and destroys cells whose surfaces hold foreign antigens bound to MHC proteins. The cellular immune system also helps eliminate tumor cells that express viral antigens (such as oncogene peptides) or altered self antigens. In all these cases, the effector cells are cytotoxic T lymphocytes (CTLs). As demonstrated in the experiment shown in Figure 22-18, CTLs attack only cells that have self MHC molecules bound to nonself antigens. Because CTLs are so important in the immune response to viruses, tumors, and transplants, researchers have been particularly concerned about the mechanism by which they ultimately kill their target cells.

Knowledge of the cell biology and biochemistry of CTLs has depended on the relatively recent ability to grow individual CTL clones in culture. The trick was to grow these CTLs along with their target cells in tiny cultures with high concentrations of IL-2. Recall from Figure 22-25 that IL-2 produced by T_H cells stimulates the differentiation of T_C cells in CTLs. In culture, IL-2 promotes the proliferation of the CTLs and has allowed researchers to study the events following their attachment to their target cells.

These studies showed that CTL precursors attach to their target cells after recognizing a specific antigen bound to an MHC Class I molecule. Following attachment, the precursors mature into CTLs, acquiring characteristic cytoplasmic granules, as shown in Figure 1. These granules are secretory granules that contain a protein called *perforin,* which is related to a cell-lysing component of the complement system, as well as a number of proteases.

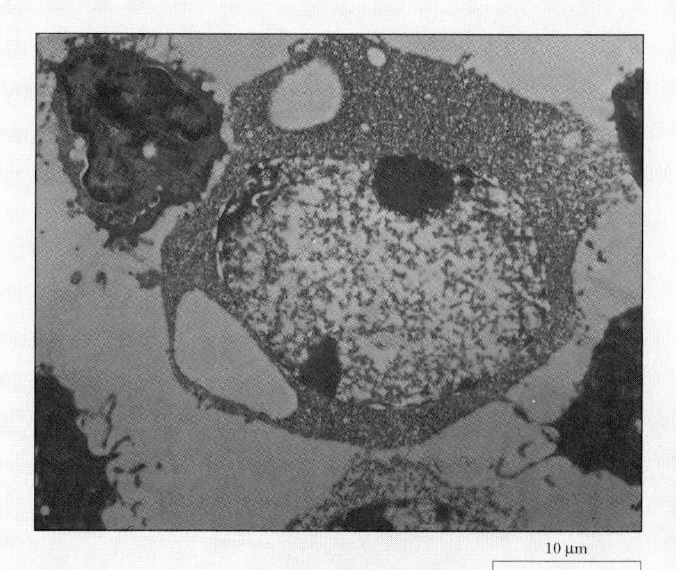

Figure 1 *A cytotoxic T lymphocyte (CTL) attacking a larger target cell. This transmission electron micrograph shows the exocytotic vesicles ("granules") in the mature CTL; these granules contain perforin and proteases.* (Dr. J. Ortaldo/Peter Arnold)

At some point in the killing process, the CTL's cytoskeleton reorients; microtubules emanate from the point of contact. After this reorientation, the secretory granules undergo exocytosis. Perforin moves onto the target cell. Perforin molecules associate within the target cell's plasma membrane to form transmembrane channels, as illustrated in Figure 2. Other proteins from the secretory granules also move into the target cell and contribute to the death of the target cell. Until the early 1990s, most researchers thought that the collaboration of perforin and the proteolytic enzymes explained the killing ability of CTLs.

NEW IMMUNOLOGICAL KNOWLEDGE HAS IMPORTANT PRACTICAL CONSEQUENCES

The ability to manipulate the immune response with modern cellular and molecular techniques has had a major impact on medical technology and has resulted in the formation of hundreds of new biotechnology companies. Monoclonal antibodies are now widely used in diagnostic tests—to detect drugs, hormones, specific pathogens, and antibody responses to pathogens, and cell surface markers. Recombinant DNAs can now be used to produce vac-

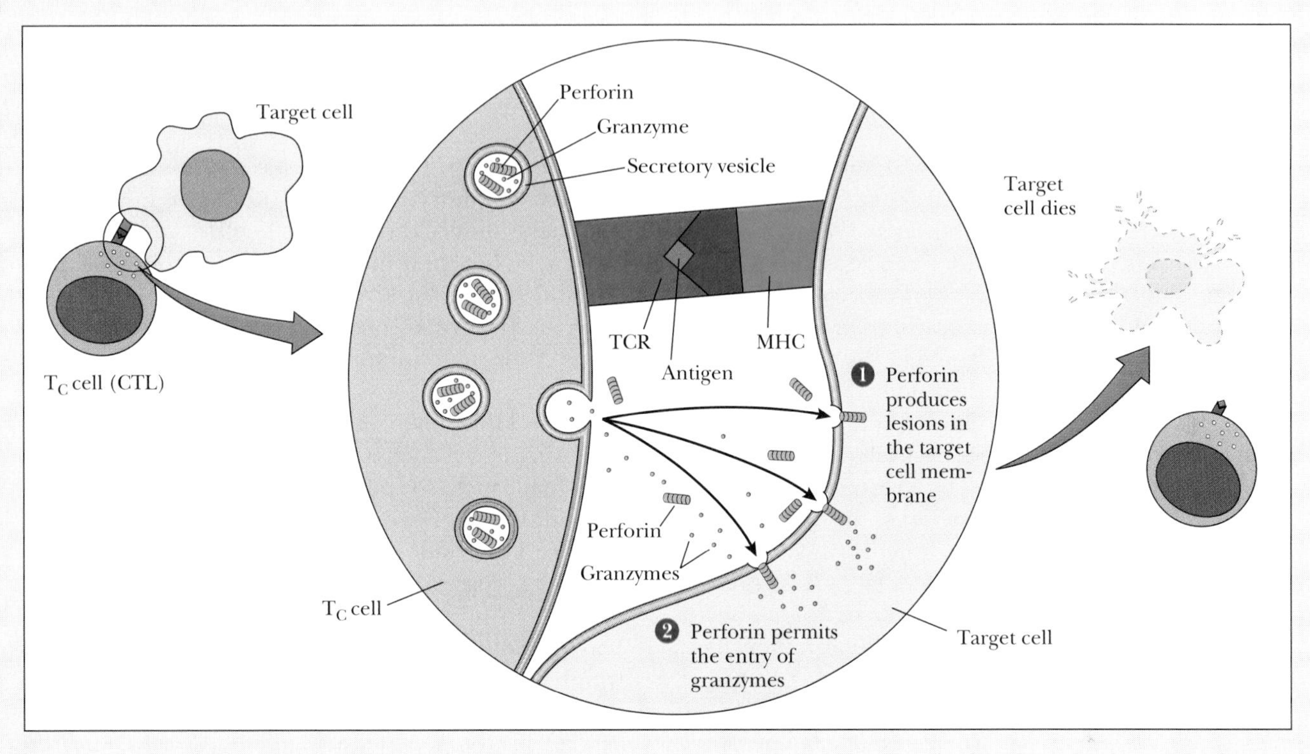

Figure 2 *The killing action of perforin and proteases ("granzymes") released by exocytosis of the secretory vesicles (granules) of CTLs after they attach to their target cells.*

Two experiments, however, showed that CTLs kill their target cells by another mechanism as well. The first was the demonstration that, in some cases, target cell killing was independent of exocytosis and occurred in the absence of Ca^{2+} (which is required for exocytosis). The second type of confounding experiment involved "knock-out" mice that had been genetically engineered so that they did not contain a functional perforin gene. (See Chapter 17 for a discussion of knock-out mice made by homologous recombination.) Surprisingly, CTLs that could not make perforin nonetheless killed their targets.

Immunologists have had to re-examine their ideas about CTL-mediated cell death. Current research indicates that CTL binding to target cells initiates a cell-death program, or apoptosis. (See Essay, "Programmed Cell Death Contributes Both to Normal Development and to Cancer Protection," in Chapter 20.) The trigger for apoptosis appears to be a protein, called Fas, on the surface of the target cell. Fas stimulates apoptosis when it binds a protein, called FasL, on the surface of the CTL. Destruction of a target cell depends not only on the CTL's bags of destruction but also on the preprogrammed ability of the target cell to self destruct. The CTL's "kiss of death" requires the participation of both parties.

cines to particular antigens, rather than using entire pathogens.

Ideas and tests for new therapeutic uses of immunological knowledge abound. Among these are the use of monoclonal antibodies to target cancer cells for toxic compounds, the enhancement of the immune response to cancer cells by adding cells engineered to produce particular cytokines, and the suppression of autoimmunity by inducing tolerance to the antigenic determinant involved in molecular mimicry. (See Essay, "Autoimmunity: Why Does the Immune System Sometimes Attack Self?")

ESSAY

AUTOIMMUNITY: WHY DOES THE IMMUNE SYSTEM SOMETIMES ATTACK SELF?

Although the immune system is largely successful in distinguishing self from nonself, about 5% of the population suffers from some *autoimmune disease,* in which an individual's response to a self antigen causes tissue damage or malfunction. In *myasthenia gravis,* for example, antibodies block the muscle cell receptors that normally respond to acetylcholine, the chemical signal that triggers muscle contraction. (See Chapter 1, p. 22, and Chapter 23.) When autoantibody levels are high, the muscles cannot be activated, resulting in weakness or even paralysis. One treatment for myasthenia gravis involves *plasmapheresis,* the recycling of blood cells after removing the proteins (including antibodies) from the blood plasma.

In *insulin-dependent diabetes mellitus,* which affects about 0.5% of the US population, cellular rather than humoral autoimmunity causes the disease. T lymphocytes attack and destroy the insulin-producing cells of the pancreas, as illustrated in Figure 1. Patients survive by injecting themselves with insulin purified from pig pancreas or made in bacteria from recombinant DNA.

Until the early 1970s, immunologists believed that such autoimmune diseases resulted from a failure of a mechanism that should have eliminated all self-reactive lymphocytes early in life. Now, however, researchers understand that all healthy adults contain some self-reactive lymphocytes, but that, in most people and for almost all antigens, the immune system prevents autoreactive cells from attacking self. In patients with autoimmune disease, something goes wrong.

We know only some of the circumstances that can bring about such a misdirected defense. The bacteria that cause "strep throat," for example, have antigens on their surface that resemble antigens in heart muscle. The result of this similarity is *molecular mimicry,* the recognition by antibodies and T cells of the physically similar antigens in a pathogen and in the body's own tissues. In the case of rheumatic fever, the lymphocytes and antibodies that recognize streptococcal antigens attack and damage the heart. In the case of insulin-dependent diabetes, T cells that recognize an antigen of a common virus (Coxsackie virus) also recognize a peptide segment of an enzyme called GAD (glutamic acid decarboxylase) found in the insulin-producing cells, as illustrated in Figure 2. The working hypothesis is that specific T-cell clones,

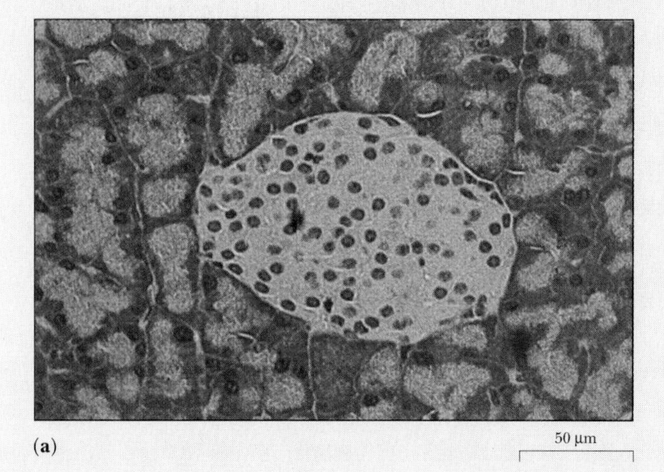

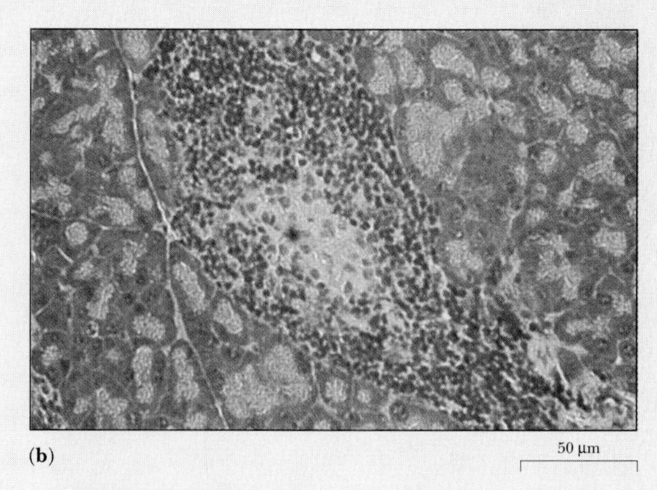

Figure 1 *In insulin-dependent diabetes mellitus, lymphocytes infiltrate into the pancreas and destroy the insulin-producing cells in the islets of Langerhans.* ***(a)*** *A pancreatic islet in a normal mouse;* ***(b)*** *a pancreatic islet in a mouse with insulin-dependent diabetes; notice the large number of white blood cells.* (From M. A. Atkinson and N. K. MacLaren, 1990, Sci. Am. 263(1), 62)

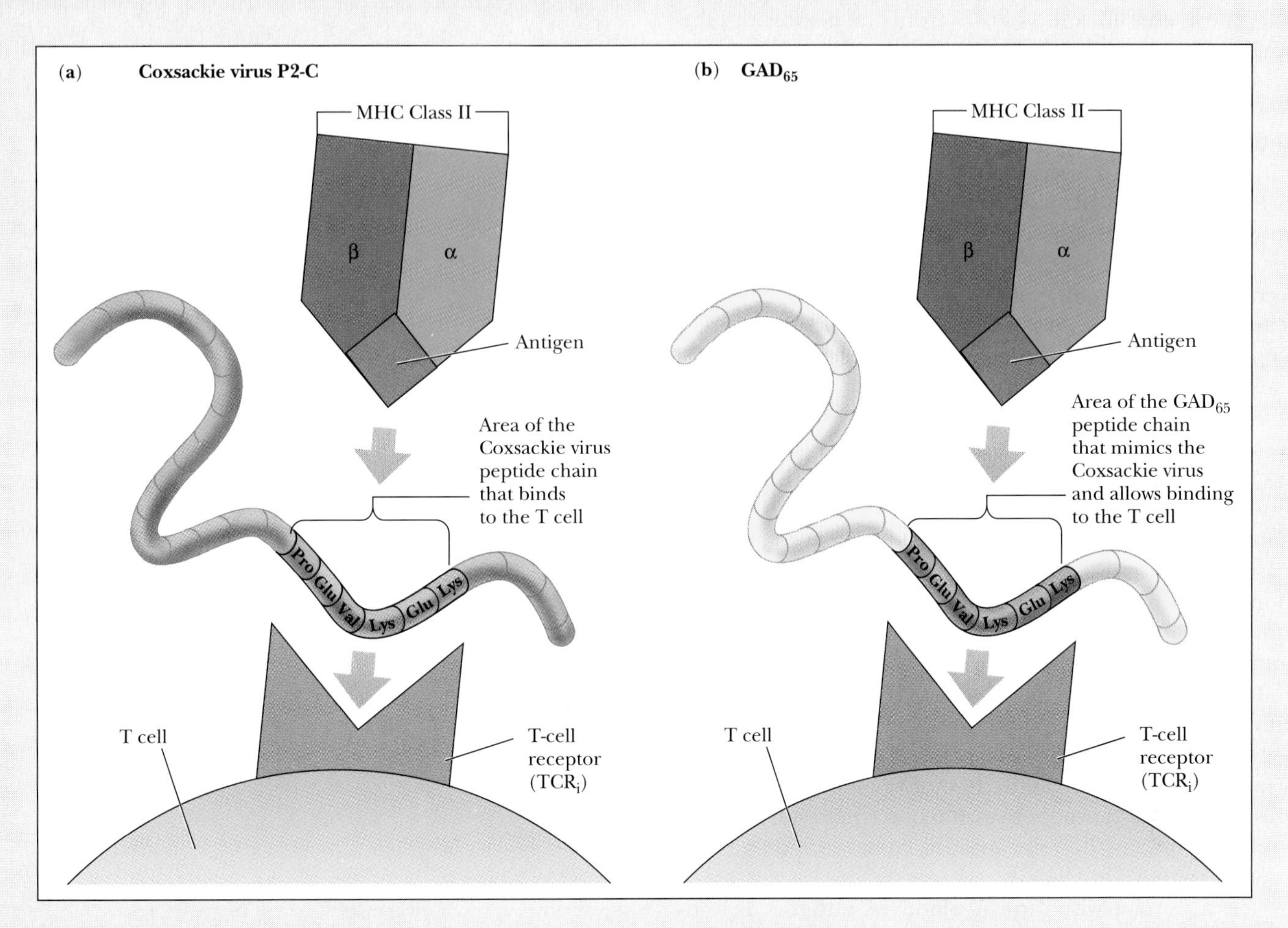

Figure 2 *Presumed molecular mimicry between glutamic acid decarboxylase and a protein from Coxsackie virus. Of the 24 amino acid residues in the mimicking region, 19 are either identical or similar.*

which have proliferated in response to a Coxsackie virus infection, later go on to attack pancreatic cells and to cause diabetes. Supporting this view, many individuals who are just beginning to show signs of diabetes have T_H cells that react with the peptide sequence shown in Figure 2.

Because mimicking antigens behave like other exogenous antigens, they are presented to the immune system as part of a complex with MHC molecules. It is not surprising, then, that individuals with different MHC alleles differ in their susceptibility to particular autoimmune diseases. Individuals that express both the DR3 and DQw8 MHC Class II genes are 100 times more likely than the population as a whole to develop insulin-dependent diabetes.

If autoimmunity to a mimicking antigen is indeed responsible for insulin-dependent diabetes or for any autoimmune disease, then the disease might be prevented by preventing the immune response, for example, with immunosuppressants used in organ transplantation. Another approach is to induce tolerance to the offending antigen—somehow reintroducing the antigen as self. This approach has worked for animal models of several autoimmune diseases, including insulin-dependent diabetes. A novel approach to inducing tolerance comes from the observation that most people are immunologically tolerant to the foods that they eat. By feeding the appropriate mimicking protein to people at risk for insulin-dependent diabetes and other autoimmune diseases, researchers hope to prevent a later autoimmune attack.

Individuals Have Distinctive Antigens on Blood Cell Surfaces

The diversity and individuality of MHC proteins also allow a definitive identification of cells from one person and the tracing of paternity in genetic studies and legal disputes. Other surface antigens besides the MHC proteins, however, can direct an immune response to transplanted cells. Antigens on the surfaces of blood cells are especially important because of the common use of blood transfusions to treat patients who have lost blood in an injury or in surgery. Individuals differ genetically in the kinds of antigens they produce, and each individual may be classified according to a number of "blood groups." Before physicians understood the nature of the antigenic differences, blood transfusions were dangerous procedures.

The most important classification is that of the ABO blood groups, illustrated in Figure 22-26. Individuals of blood group A have an antigen called A on the surfaces of their red blood cells; those of blood group B have another antigen, called B; those of group AB have both A and B; those of blood group O have neither A nor B. The reason that the ABO grouping is so important is the common presence of antibodies to blood ABO antigens in the human population. An individual who does not make A antigen (that is, someone with type O or B blood) has antibodies to A antigen. They have these antibodies even if they have never been exposed to type A blood, probably because similar epitopes are present on the surfaces of common gut bacteria. Similarly, individuals who do not express the B antigen (people with blood types O or A) have antibodies against B. A person with type AB blood has antibodies to neither A nor B.

If a person with type A blood (A antigen, B antibodies) receives a transfusion of type B blood (B antigen, A antibodies), his or her antibodies attack the red cells, link them together in big aggregates, and clog the blood vessels. Such an occurrence could easily be fatal. (The transfused blood itself contains antibodies against the A antigens in the recipient's blood, but these are not so dangerous, because the transfusion dilutes them with the recipient's blood.)

Another antigen on the surface of red cells of some individuals is called the Rh factor—after the rhesus monkeys, in which the factor was first identified. Unlike the situation with the A and B antigens, individuals who lack the Rh antigen do not generally make antibodies against it unless they are first exposed to Rh-positive blood. This would happen, for example, after transfusions of Rh-positive blood into an Rh-negative recipient—a situation that modern medicine seeks to avoid. Another situation, far more common, exposes Rh-negative women to Rh-positive blood—pregnancy. If a fetus receives the Rh factor gene from the father it becomes Rh positive. The mother may then develop antibodies against Rh factor, if she becomes exposed to fetal blood during delivery. Such antibodies are not usually present at high concentrations after one pregnancy with an Rh-positive fetus. Subsequent pregnancies, however, can provoke a strong secondary response in the mother. In some cases, the mother's antibodies can cross the placenta near the end of pregnancy and attack the fetus's red cells. Recent immunological knowledge has allowed physicians to prevent the immune response to Rh factors by treating the mothers with anti-Rh antibodies to remove memory cells for the Rh factor.

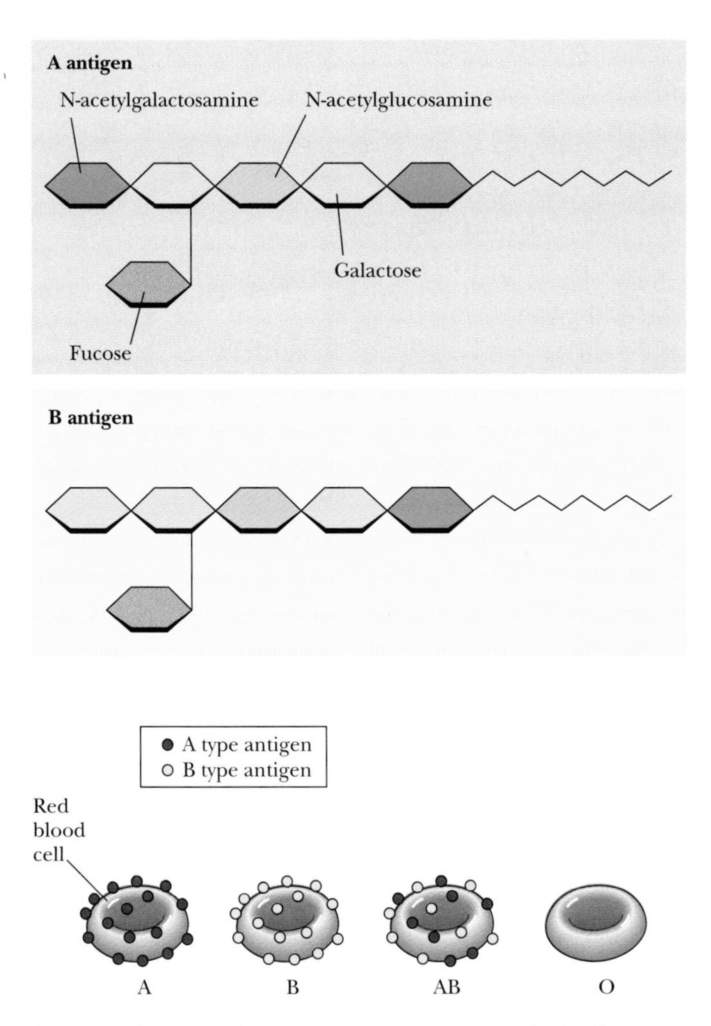

Figure 22-26 ABO blood groups result from distinctive sugars attached to the surface proteins and lipids of red blood cells.

Failures of the Immune System Can Lead to Infection or Cancer

The immune response depends on the ability of cells to proliferate when stimulated by specific antigens. Anything that interferes with cell division in one or more kinds of lymphocytes—for example, high levels of x-rays or other radiation can compromise the immune response. Residents of Hiroshima and Nagasaki who survived the atomic bomb explosions during the Second World War suffered from such radiation-induced damage. They were especially susceptible to infection and were especially likely to develop cancers. The increase in cancers may reflect an increase in the number of failures in the regulation of cell growth or a decreased effectiveness of T cells in eliminating cells that represent altered self, as occurs in cancer.

Mice or other experimental animals treated with high levels of radiation fail to develop normal immune responses. Immunologists can restore the immune response of these irradiated animals by injecting them with lymphocytes from healthy mice. Lymphocyte precursors can also restore the immune responses of irradiated mice. The appropriate precursors, however, lie in two separate locations—in the bone marrow, where the B-cell precursors develop, and in the thymus gland, where the T-cell precursors reside.

The ability to mount an effective immune response may also depend on an individual's general state of health. Every infectious disease has its own pathogen, but generally healthy people are far more successful in defending themselves than generally unhealthy people. Considering the human population as a whole, we may say that malnutrition and poverty are the most prevalent "causes" of infectious disease. Researchers are still unsure whether psychological state—happiness, stress, or depression—also affects the immune response and resistance to disease.

The immunological deficiency that has become the focus of the greatest public attention in the last decade is **AIDS,** or **acquired immunodeficiency syndrome.** AIDS is caused by a retrovirus, called **human immunodeficiency virus (HIV).** HIV infects and kills T_H lymphocytes by first binding to the CD4 protein on the cell surface. The lack of T_H cells means that AIDS patients are unable to mount effective immune responses to infections or to certain kinds of cancers. AIDS is a major public health problem, with an estimated 10 million people (as of 1995) worldwide carrying HIV.

■ SUMMARY

Inflammation, a set of responses to local injury, helps an animal to resist infection by increasing circulation, attracting and mobilizing phagocytic cells, and inducing signaling molecules. One signaling molecule, interleukin-1, stimulates responses in phagocytic cells, in the endothelial cells of the capillaries, and in the brain cells that control body temperature.

The immune system recognizes specific foreign molecular shapes as foreign and participates in their elimination. The humoral immune system depends on the binding of circulating proteins, called antibodies, to specific challenging molecules, called antigens. Humoral immunity depends on white blood cells called B lymphocytes, which develop into antibody-producing plasma cells. Cellular immunity depends on the ability of other white blood cells, called T lymphocytes, to recognize antigens on the surfaces of the body's own cells.

Unlike inflammation, the immune response is specific and relies on the binding of foreign shapes to specific immune molecules. The immune system also shows memory, diversity, and the ability to distinguish between self and nonself.

The simplest antibody molecules, called immunoglobulin G or IgG, consist of four polypeptide chains, two heavy chains and two light chains. Both heavy chains and light chains contain a variable region, extending approximately 110 amino acid residues from their amino-terminal ends, as well as a constant region, extending to the carboxyl-terminal end. Antibody diversity depends on the ability to combine many different light chains with many different heavy chains.

While the immune system can produce as many as 10^8 antibodies, a single antibody-producing cell can produce only one type. At one time, researchers thought that antibody diversity depended on the ability of the same polypeptide chains to fold into many different shapes, with the particular shape determined by the pattern of folding in the presence of antigen. Evidence from immunoglobulin sequencing, however, shows that individual antibodies have different amino acid sequences. The immune response depends on an animal's ability to select among an existing repertoire of cells, each able to make a single type of antibody.

Both light and heavy chains of immunoglobulin are encoded by genes that undergo rearrangement. To form a functional gene, light chain gene regions undergo two rearrangements and heavy chain gene regions undergo three rearrangements. Somatic mutations can further increase the diversity of immunoglobulin sequences.

During embryonic life, precursors of B lymphocytes form a lymphocyte library, with each lymphocyte containing a unique random rearrangement of immunoglobulin genes. According to the clonal selection theory, an individual B lymphocyte proliferates when exposed to an antigen that binds to its particular antibody. Its descendants include effector cells, called plasma cells, which produce and export antibody, and memory cells, which can develop into effector cells after a subsequent challenge to produce an even more robust response to antigen. The clonal selection theory also suggests a mechanism for self-nonself recognition and immunological tolerance—cells in the lymphocyte library that recognize self antigens are eliminated or suppressed.

Cellular immunity depends on antigen recognition

by molecules, called T-cell receptors, on the surface of T lymphocytes. T-cell receptors are even more diverse than antibodies. Like antibodies, they are encoded by the rearrangement of gene segments to form unique combinations within individual cells of a lymphocyte library.

T-cell receptors do not recognize free antigens, but only short polypeptides bound by the proteins of the major histocompatibility complex (MHC). Two classes of MHC proteins can present antigens to T cells. MHC Class I molecules, which are on the surface of all cells in the body, present endogenous antigens to cytotoxic T cells, which are responsible for killing cells recognized as nonself, including virus-infected cells and tumor cells. The antigens presented by MHC Class I are produced within the cells themselves.

MHC Class II molecules present exogenous antigens on the surfaces of specialized antigen-presenting cells, which have taken up foreign organisms by phagocytosis. These cells interact with another class of T lymphocytes called T-helper cells. These latter cells produce cytokines that can activate either the cellular or the humoral response. The antigens presented by MHC Class II cells are those of foreign proteins, taken up by phagocytosis by the antigen-presenting cells and processed for presentation.

KEY TERMS

acquired immunodeficiency syndrome (AIDS)
allergy
altered self
antibody
antigen
antigen-presenting cell (APC)
antigen processing
antigenic determinant
β_2-microglobulin
B-cell receptor
B lymphocyte
cellular immunity
clonal selection theory
complement
complementarity determining region (CDR)
constant region
cytokine
cytotoxic T lymphocyte (CTL)
cytotoxicity
dendritic cell
domain
effector cell
endogenous antigen
epitope
exogenous antigen
germ line theory
histamine
human immunodeficiency virus (HIV)
humoral immunity
hybridoma
hypervariable region
immunoglobulin
immunoglobulin fold
immunoglobulin heavy chain
immunoglobulin light chain
immunological diversity
immunological memory
immunological specificity
inflammation
instructive theory
interferon
interleukin-1
J segment (joining region segment)
leukocyte
lymphocyte
lymphocyte library
macrophage
major histocompatibility complex (MHC)
mast cell
membrane attack complex
memory cell
MHC Class I
MHC Class II
MHC restriction
monoclonal antibody
N-region nucleotide addition
plasma cell
polyclonal
recombination signal sequence
secondary immune response
selective theory
self-nonself recognition
somatic theory
T-cell receptor (TCR)
T-cytotoxic (T_C) cell
T-helper (T_H) cell
T lymphocyte
tolerance
variable region

QUESTIONS FOR REVIEW AND UNDERSTANDING

1. Define and contrast the following pairs of terms:
   (a) monoclonal antibody vs. polyclonal antibody
   (b) humoral immunity vs. cellular immunity
   (c) primary immune response vs. secondary immune response
   (d) antigen vs. epitope
   (e) IgG vs. IgM
   (f) variable region vs. constant region
   (g) selective theory vs. instructive theory
2. What are the four characteristics of inflamed tissue?
3. Describe the four cellular and biochemical events that occur during the inflammatory response.
4. Give some examples of lymphoid tissue.
5. What are cytokines?
6. List and describe the four major characteristics of the immune system.

7. What are the two mechanisms used by complement to attack microbial invaders?
8. What are the two mechanisms that are mediated by antibodies that cause the destruction of bacteria?
9. Why is the secondary immune response more robust than the primary immune response?
10. What is antigen processing?
11. How do the molecular interactions between proteins of the immune response and antigens allow the immune system to recognize so many different antigens?
12. Do all classes of immunoglobulins perform the same functions? Give an example to support your answer.
13. What are B cell receptors?
14. How are the discontinuous genetic elements of the immunoglobulin molecule brought together to encode a functional polypeptide?
15. What are the four mechanisms of diversity shared by immunoglobulin and the T cell receptor? What is the one mechanism that they do not share?
16. How does the genetic organization of immunoglobulin genes promote diversity?
17. What was the testable prediction suggested by the clonal selection theory? Describe the two experiments, and their results, that were done to test that prediction.
18. How does the recognition of antigen by the receptors of T cells and B cells differ?
19. What are congenic mice?
20. What is the relationship between the major histocompatibility complex and graft rejection?
21. What is "altered" self?
22. What are some of the functional differences between MHC I and MHC II?
23. What are some of the functional differences between T helper and T cytotoxic cells?
24. What are the ways that apoptosis figures prominently in the development and action of the immune response?

SUGGESTED READINGS

ALBERTS, B., D. Bray, J. Lewis, M. Raff, K. Roberts, and J. D. Watson, *Molecular Biology of the Cell*, 3rd edition, Garland, New York, 1994. Chapter 23 deals with the immune system.

KUBY, J., *Immunology*, W. H. Freeman and Company, New York, 1992. An excellent immunology textbook.

LODISH, H., D. Baltimore, A. Berks, S. L. Zipursky, P. Matsudaira, and J. Darnell, *Molecular Cell Biology*, 3rd edition, Scientific American Books, New York, 1995. Chapter 27 deals with immunity.

TIZARD, I. R., *Immunology: An Introduction*, 4th edition, Saunders College Publishing, Philadelphia, 1995. An excellent and up-to-date immunology textbook.

WATSON, J. D., M. Gilman, J. Witkowski, and M. Zoller, *Recombinant DNA*, 2nd edition, Scientific American Books, New York, 1992. Chapter 16 deals with the molecules of immune recognition.

CHAPTER 23

The Cells of the Nervous System

KEY CONCEPTS

- The nervous system consists of neurons, cells that carry electrical signals, and glial cells, which provide metabolic and structural support.
- Neurons contain specialized proteins that allow them to produce and carry electrical signals, called action potentials, over long distances.
- Connections between neurons may be either electrical, mediated by gap junctions, or chemical, mediated by the release of signaling molecules called neurotransmitters.
- Neurotransmitters act on target neurons either by directly changing the properties of ion channels or by altering the levels of second messengers like cyclic AMP.
- Learning in simple systems can be understood in terms of the properties of a small number of neurons.

OUTLINE

ESSAYS

No question has more puzzled thinking people than the nature of thinking itself. We now take for granted that the brain is the organ responsible for thought and emotion, but this was not always the case. Even now, we ascribe strong emotions to the heart—as evidenced by popular music, Valentine's Day cards, and bumper stickers.

The earliest recorded evidence for the brain's importance is a treatise of an ancient Egyptian physician, who noted—with surprise—that a wound in the head could interfere with walking or other movements by the legs. More recent physicians and researchers have also learned much about the functions of the brain by studying the effects of specific injuries. Related information comes from observations during brain surgery. Mechanical or electrical stimulation of a particular part of the brain may lead to movements, perceptions, memories, and even emotions.

Even apparently simple sensory stimuli can evoke complex responses. Among the best documented are those written almost 100 years ago by the French novelist Marcel Proust, who provided an extraordinary account of his responses—thoughts, memories, and fantasies—to the tastes of tea and cake. All the functions of the nervous system ultimately depend on its ability to collect and process information about the outside world.

THE NERVOUS SYSTEM IS A CELLULAR NETWORK THAT RESPONDS TO CHANGES IN AN ANIMAL'S EXTERNAL AND INTERNAL ENVIRONMENT

Neurologists—physicians who specialize in disorders of the nervous system—have charted the relationships between the sites of brain injuries and the symptoms reported by their patients. Patients with damage to the rear part of the brain often report strange visual phenomena, such as flashes of light. And patients with damage to the side of the brain complain of buzzes or whines. The correlation of specific symptoms with damage to particular regions has allowed researchers to make a functional "map" of the brain. For example, the *motor cortex,* a region of the brain that lies between two folds on the surface near the front of the brain, is responsible for movements, with each area specifying a particular part of the body. Stimulating one area of the motor cortex, for example, causes a leg to move, while another area commands the tongue. The control of a given body part may require more or less of the motor cortex: For example, much more of the motor cortex is devoted to commanding the hand than the rest of the arm.

One of the main functions of the nervous system is to serve as an input-output computer in the service of homeostasis. The nervous system as a whole receives reports on the state both of the exterior world—determined by vision, touch, hearing, smell, and taste—and of the interior world—determined mostly by chemical receptors that monitor the internal environment.

Figure 23-1 provides an overview of the human nervous system and the locations of structures that perform specific functions. External and internal sensors report to the central nervous system in messages encoded in ion currents. During evolution, natural selection has favored adaptations for the transmission and interpretation of external and internal information important for survival and reproduction.

Modern Neuroscience Has Fine Tools for Probing the Function of Particular Components of the Nervous System

The electrical properties of nerve cells allow neuroscientists to probe individual cells and the interconnections of cells within the nervous system. Researchers can record currents from the surface of the brain, within individual cells, and even in minuscule patches of nerve cell membranes. Such observations can show directly how the brain responds to external stimuli.

Researchers can electrically stimulate the brain of a human patient or of an experimental animal and then record what happens. Such tests on humans are common during neurosurgery, when the patient is usually awake and aware. Researchers can study both the final response—such as the movement of a limb—and the intermediate steps along a neural pathway, as the currents in one region trigger electrical activity elsewhere. In this way, researchers can begin to define the pathways of information transfer within the nervous system. These pathways involve communication between **neurons,** nerve cells that are specialized for the conduction of electrical and chemical signals.

Other methods allow neuroscientists to visualize neural pathways directly in experimental animals. Nerve cells can transport some dyes in the same direction in which they transmit information. When such a dye is injected into the motor cortex, it moves through a lower brain region called the *medulla,* and from there into the spinal cord. Nerve cells in the spinal cord—called **motor neurons** (or **motoneurons**)—directly stimulate muscles to contract. The nerve fibers from the motor cortex cross from left to right (or from right to left) as they pass through the medulla, so that the left side of the brain controls muscles on the right side of the body and vice versa.

Other dyes move in the direction opposite to that in which neural commands flow. These dyes allow neuroscientists to determine the sources of information processed by a particular set of nerve cells. Such studies show that cells of the motor cortex receive information from the spinal cord, from other regions of the cortex, and from brain structures that lie below the cortex. Each part of the nervous system—and almost every nerve cell—receives information from many sources.

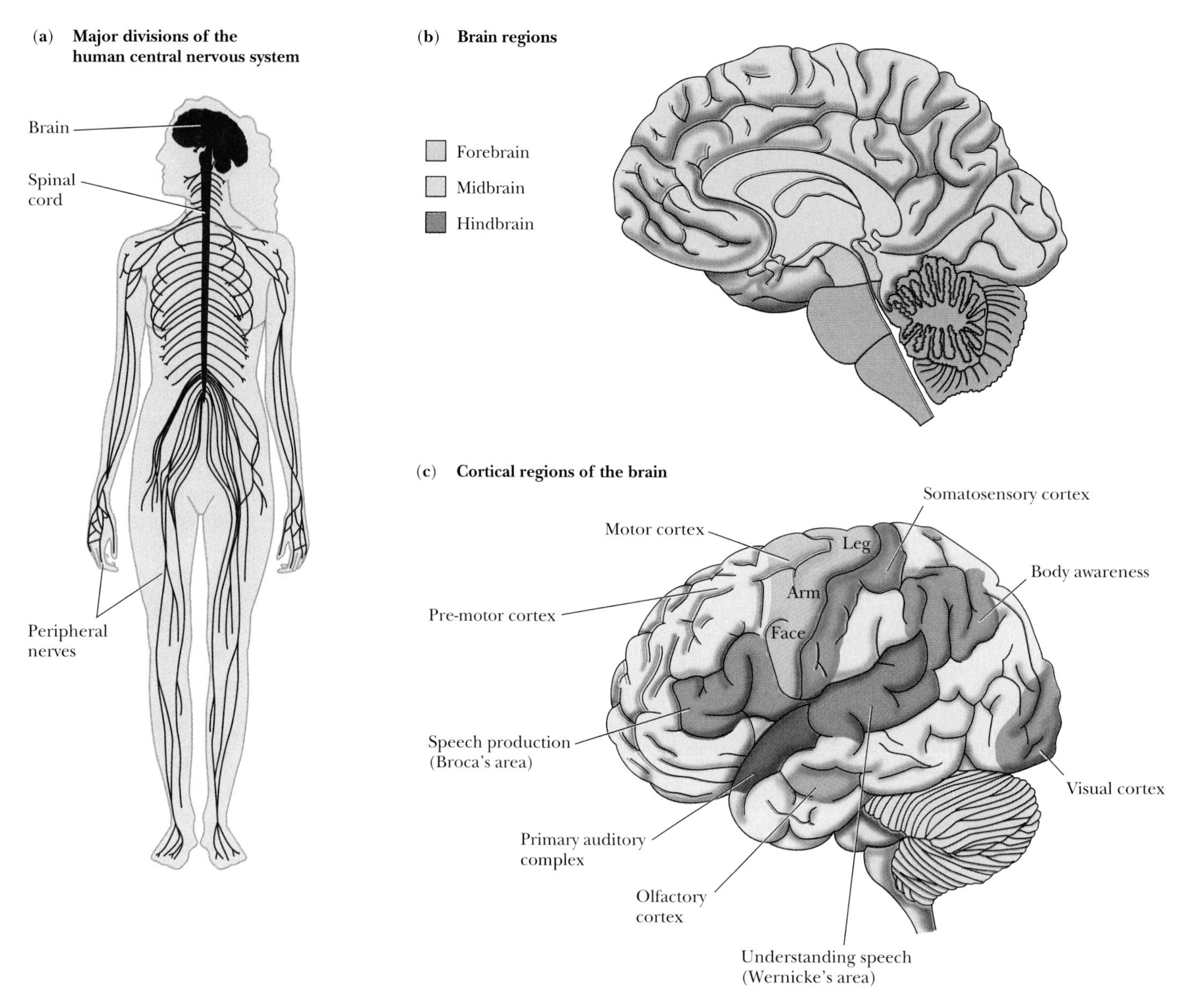

Figure 23-1 **(a)** The human nervous system consists of the brain and spinal cord, which together make up the central nervous system, and peripheral nerves, which carry information to and from the central nervous system. **(b)** The brain contains a forebrain, midbrain, and hindbrain. The outer layer of the forebrain, the cerebral cortex, is particularly convoluted in humans, greatly increasing its surface area compared even with other primates. **(c)** The motor cortex initiates voluntary movements, with each region of the motor cortex commanding a particular body part. Similarly, the somatosensory cortex receives touch information from individual body parts. Other regions of the cerebral cortex are highly involved in language and other higher functions.

The nervous system is the ultimate homeostatic organ, coordinating not only internal regulatory mechanisms but also simple and complex behaviors. A pain or a smell or a sound or a visual image can send a signal that directs an appropriate homeostatic response—withdrawing a hand from a hot stove or running away from an enemy. The connections of nerve cells determine how animals respond to stimuli.

The Vertebrate Nervous System Includes Brain, Spinal Cord, and Peripheral Nerves

Nerve networks exist in all animal phyla except the parazoa (sponges). The simplest organisms to have a nerve network are cnidarians, which include the hydra and jellyfish illustrated in Figure 23-2. Hydras have a web of interconnecting neurons that carry information to all parts of the

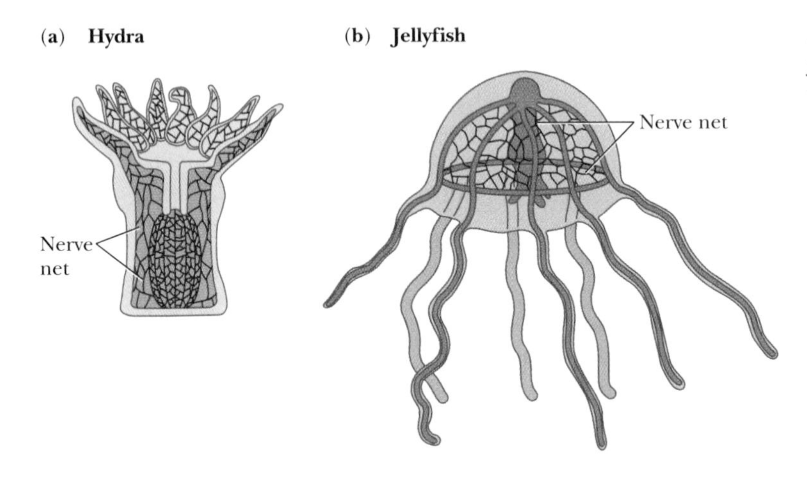

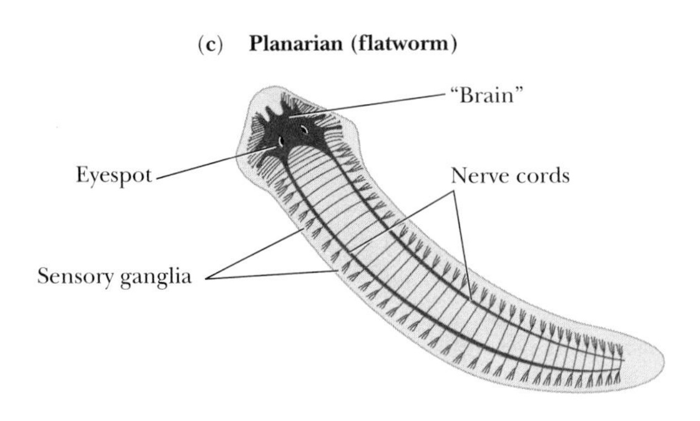

Figure 23-2 Simple nervous systems. **(a)** Hydra; **(b)** jellyfish; **(c)** flatworm, showing two parallel nerve cords and sensory ganglia; peripheral nerves (not shown) extend outward to the muscles.

body. A stimulus anywhere on a hydra's body can trigger reflex muscle contractions, much as a physician's hammer stimulates the knee-jerk reflex. The nerve nets of other cnidarians, such as the jellyfish shown in Figure 23-2b, coordinate complex swimming movements. But, as far as anyone can tell, neither a hydra nor a jellyfish can learn associatively, and their behaviors do not appear to be influenced by experience.

Flatworms (Figure 23-2c) are the simplest animals capable of learning. The nervous system of a flatworm resembles those of more complicated species. The flatworm's nervous system consists of two parallel nerve cords, with peripheral nerves extending outward to the muscles. Like other bilaterally symmetrical animals, flatworms have a well-defined direction of movement. They have eyespots at the front ends, and they process sensory information in *ganglia* (singular, *ganglion*), clusters of nerve cells, also shown in Figure 23-2c. The two nerve cords also converge near the front of the body. Flatworms provide the simplest example of the expansion of nervous systems at an animal's anterior end into something that resembles a "brain." This convergence of nerves even in a simple brain provides many opportunities to generate complex patterns of neural activity and resulting behavior. More complex animals show greater degrees of **cephalization**—the concentration of sense organs and ganglia in the anterior end.

The nervous systems of vertebrates show the highest degree of cephalization, the most sophisticated sense organs, and the greatest capacity for learning. As in other animals with complicated behaviors, most neurons in vertebrates are concentrated in the **central nervous system (CNS),** which consists of brain and spinal cord, as illustrated in Figure 23-1. In vertebrates, the CNS lies above (dorsal to) the central body axis, while the CNS of invertebrates is ventral.

In vertebrates, the **peripheral nervous system (PNS),** which consists of nerve cells outside the CNS, carries information from the CNS to muscles and organs and to the CNS from sense organs. **Afferent** nerves carry *sensory* information *to* the CNS, while **efferent** nerves carry commands concerning *motor* activity *from* the CNS to the muscles and other effector organs, including the endocrine glands.

The efferent nerves that carry motor commands to the striated muscles make up the *voluntary* or *somatic nervous system.* Other efferent nerves are part of the *involuntary* or **autonomic nervous system,** which coordinates the responses of smooth muscles, cardiac muscles, and other effector organs—including those of the endocrine, digestive, excretory, respiratory, and cardiovascular systems.

The autonomic nervous system consists of two distinct sets of nerves—called the **sympathetic** and the **parasympathetic** nerves. Sympathetic and parasympathetic nerves often, but not always, have opposite effects. The sympathetic nervous system brings about the responses of the "fight or flight" reaction. These responses involve the mobilization of energy stores and the preparation for vigorous action often critical to the survival of an individual in a crisis. The sympathetic nervous system collaborates with the adrenal medulla, which stimulates "fight or flight" responses with the hormone epinephrine. (See Chapter 19.) In contrast, the parasympathetic nervous system generally acts to husband resources, for example by slowing the heart and increasing intestinal absorption.

How Are Nerve Cells Special?

Even a generation after the general acceptance of the cell theory, biologists were still not at all certain whether the nervous system really contained conventional cells. Until the late 19th century, anatomists studied the brain in more or less the same ways that they studied other tissues. They fixed and cut and stained, just as we described in Chapter 5. But when researchers examined the brain with a microscope, they did not see distinct cells. Instead, they saw dots, blurs, and dark spots that resemble nuclei, as seen in Figure 23-3a. The problem, as we now know, was that there are so many cells and so many intertwined processes. The trick to seeing individual cells was to devise means of staining one cell while not staining its neighbors. One such procedure, called the Golgi method, developed in the 1880s by Camillo Golgi, stains only a small number of cells in each sample, as seen in Figure 23-3b.

About 100 years ago, the great Spanish neuroanatomist, Santiago Ramon y Cajal, used the Golgi method to study nerve cells in the nervous systems of many species. By his careful and insightful work, Ramon y Cajal provided neuroscience with many insights about the organization and the development of the nervous system. The most basic of his contributions was convincing evidence that the nervous system—like other functional systems in the body—consists of cells. These cells come in a variety of sizes and shapes, as illustrated in Figure 23-4. In general, nerve cells are distinctive in appearance—most are far longer than other types of cells. The cells that command the legs of a giraffe, for example, may be several meters long.

Neurons, examples of which are illustrated in Figure 23-4a, are the nerve cells that carry information in the form of electrical and chemical signals. The nervous system also contains **glial cells** (illustrated in Figure 23-4b), cells that do not themselves transmit electrical and chemical signals but provide metabolic and structural support for neurons. Some glial cells serve as guides and scaffolding during neural development. Other glial cells, called **oligodendrocytes** (or, in the peripheral nervous system, **Schwann cells**) wrap around nerve cell extensions in a **myelin sheath,** as illustrated in Figure 23-4b. (Also see Figure 6-3, p. 154.)

Looking at the cells of Figure 23-4a, we can see that neurons exhibit many variations of one basic plan. Most neurons have a distinct center, called the cell body, which is about the same size as most other cells. The most distinctive features of neurons are the processes that extend away from the cell body. Some of these processes, called **dendrites,** carry signals to the cell body. Others, called **ax-**

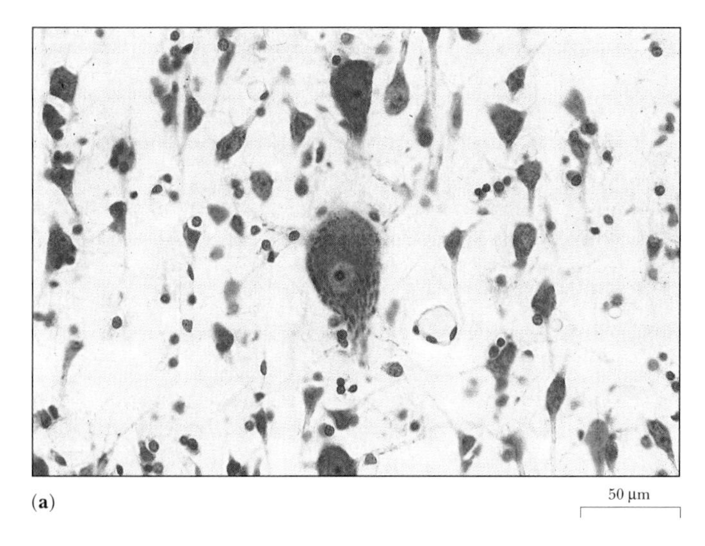

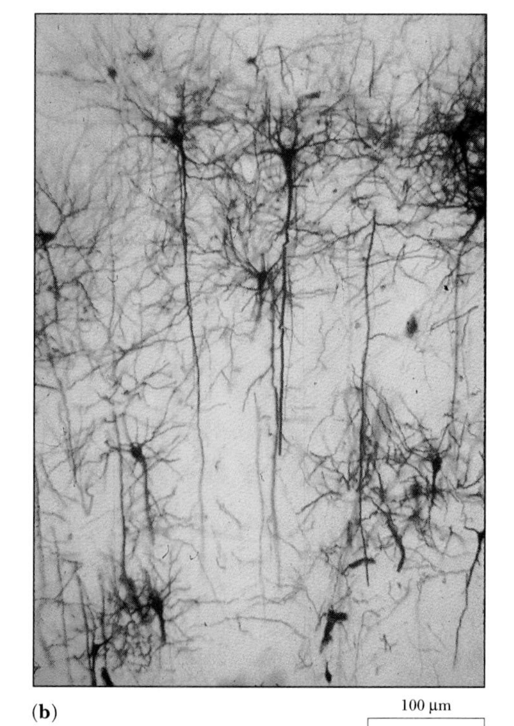

Figure 23-3 There are so many cells in the brain that it is hard to discern where one begins and another ends. Only special staining techniques allow one to see discrete cells. **(a)** A section of human cerebral cortex, treated with a standard histological stain (hematoxylin and eosin); **(b)** human cerebral cortex treated with the Golgi method, which stains only a small fraction of cells. *(John D. Cunningham/Visuals Unlimited)*

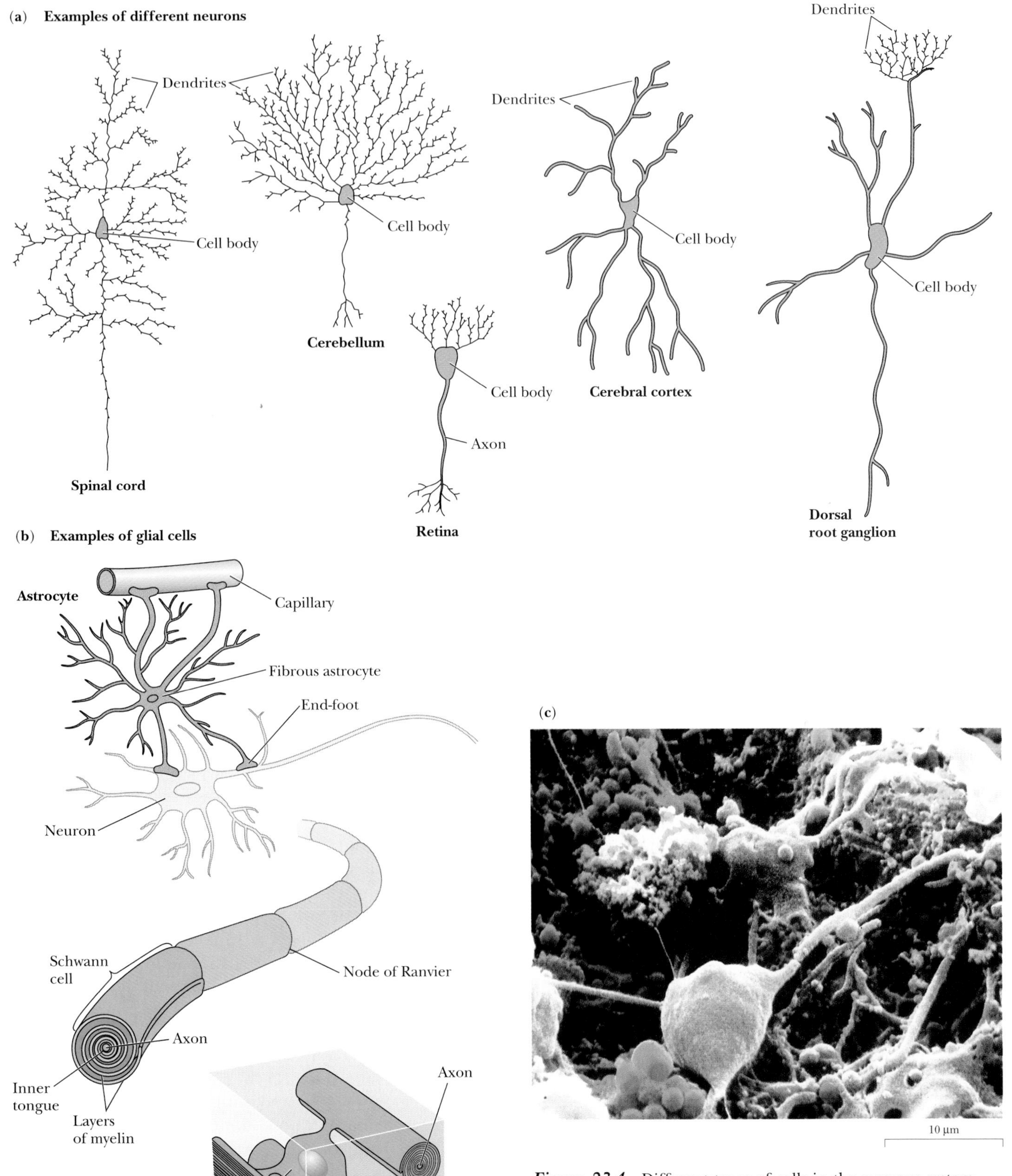

Figure 23-4 Different types of cells in the nervous system. **(a)** Neurons may have few or many processes; **(b)** glial cells include astrocytes, the most numerous cells in the central nervous system, and oligodendrocytes and Schwann cells, which participate in wrapping the long processes of neurons with a sheath of myelin; **(c)** scanning electron micrograph of cerebral cortex using false colors to distinguish neurons (in gray) and glial cells (in orange). *(c, Prof. P. Motta, Dept. of Anatomy, University "La Sapienza," Rome/Science Photo Library/Custom Medical Stock Photo)*

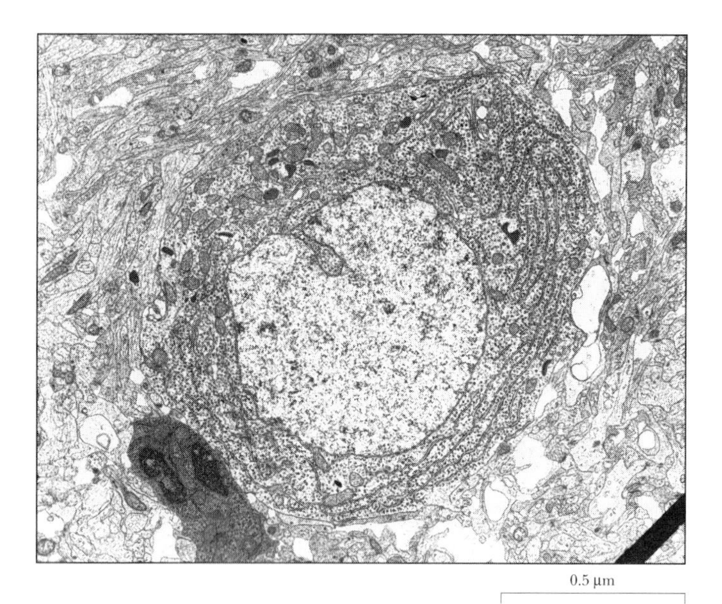

Figure 23-5 An electron micrograph of the cell body of a neuron in a rabbit brain. Note the abundance of mitochondria, ribosomes, and rough endoplasmic reticulum. *(Dr. Dennis Kunkel/Phototake)*

ons, usually carry signals away from the cell body and connect to other cells. Neurons usually have more dendrites than axons; one type of cell in the cerebellum appears to have dendrites that carry inputs from over 50,000 other cells. While dendrites may be more numerous, axons usually extend over a longer distance. Axons may branch at their ends to make contact with many target cells.

Despite their unusual characteristics, neurons have exactly the same kinds of organelles as other cells, as can be seen in Figure 23-5. They have particularly high levels of mitochondria and ribosomes, suggesting that they are active users of energy and synthesizers of proteins. Studies of brain metabolism support this conclusion: The human brain, for example, makes up only about 2% of the body's weight but uses about 20% of all the energy.

The specialized structures in neurons are involved in carrying information over long distances and in making specific contacts with other neurons. Axons and dendrites contain high levels of cytoskeletal proteins, with dendrites rich in microtubules and axons full of microtubules and neurofilaments, neuron-specific intermediate filaments. The contacts between nerve cells—called synapses—are highly specialized, as discussed in detail later in this chapter.

How Do the Neurons Responsible for the Knee-Jerk Reflex Illustrate the Principles of Neural Connections?

Movements and behavior ultimately depend on coordinated signals from motoneurons. These motoneurons represent the final common pathway of the instructions that direct behavior.

Complex behaviors—such as threading a needle, performing a laboratory experiment, or taking a midterm examination—originate in commands from the brain, particularly in the cerebral cortex. Less complex behaviors—such as withdrawing a hand from a flame—often do not depend on the brain at all, but on signal processing in the spinal cord. Behavioral activity mediated by the spinal cord tends to be the most automatic, that is, the least influenced by thought or previous experience.

A **reflex** is an automatic behavior pattern, with motor activity arising directly in response to a sensory stimulus. A reflex familiar to anyone who has had a medical examination is the jerking of the leg after a physician taps your knee with a little hammer. Let us see exactly how this happens.

As illustrated in Figure 23-6, the hammer hits the patellar tendon, which stretches over the knee. This tendon connects a muscle in the thigh, the quadriceps, to a bone in the lower leg. The physician's hammer pulls the tendon and stretches the attached muscle. The muscle contains *stretch receptors,* modified muscle cells whose electrical output depends on the degree to which they are stretched. Information from these receptors flows to **sensory neurons** in the dorsal part of the spinal cord. These sensory neurons make direct contact with motoneurons in the ventral part of the spinal cord, which in turn command the muscles to contract. The knee jerks. The information passes through just one synapse, so the stretch reflex is said to be *monosynaptic.* The normal function of this reflex is to restore the position of the lower leg after a rapid dislocation.

Information from the stretch receptors also travels to a second group of spinal cord neurons, as illustrated in Figure 23-6. These neurons are inhibitory and act on the antagonistic muscles, the hamstrings, which act against the knee jerk. So not only does the signal from the stretch receptor trigger one set of motoneurons to fire, but it also inhibits another set. Yet another set of neurons in the spinal cord carries information about the status of the leg muscle to brain areas that coordinate movements and posture.

The connections involved in the stretch reflex form the simplest known neural circuit in vertebrates. Other circuits are more complicated, with at least one **inhibitory interneuron,** a neuron whose stimulation by one neuron prevents the firing of another neuron in the same area. Interneurons can integrate information from many sources—from stretch receptors, from other somatic receptors, and from elsewhere in the nervous system, including the brain. Circuits in which such integration occurs are called *convergent.*

An interneuron can also participate in a *divergent* circuit, in which a single interneuron may coordinate the action of many motoneurons. Walking, for example, requires

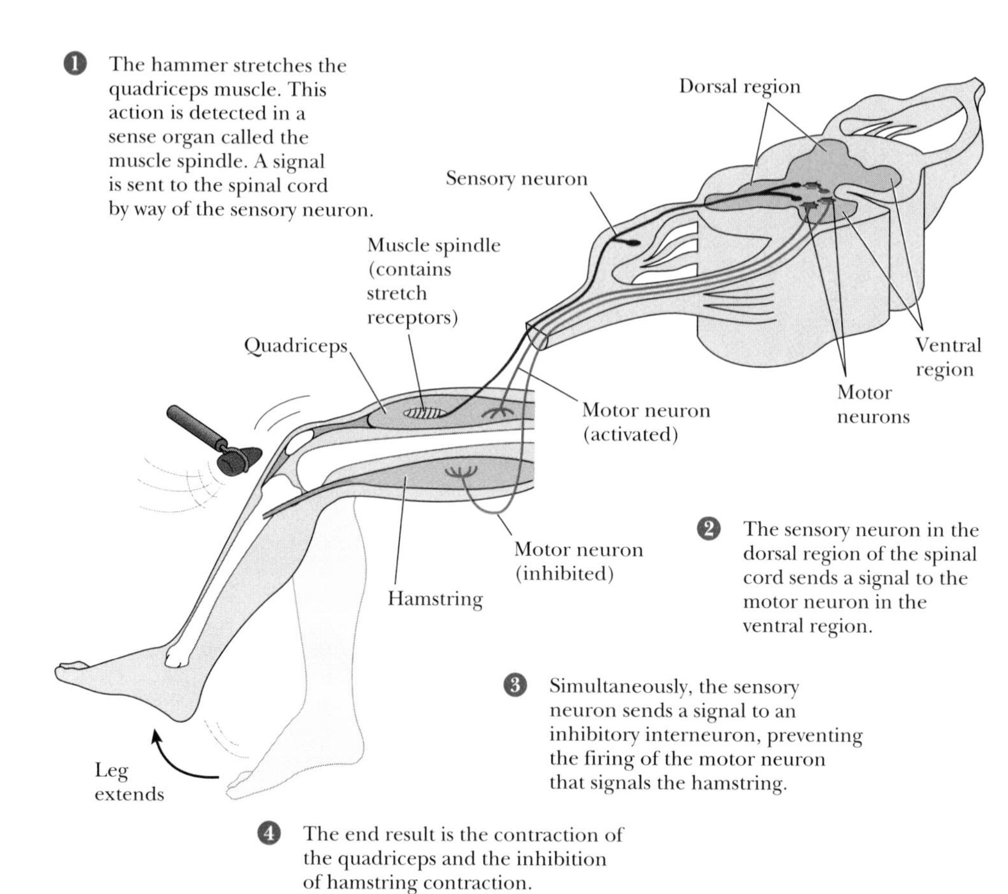

Figure 23-6 The knee jerk is an example of a monosynaptic reflex. The physician's hammer stretches the quadriceps muscle, which is detected in sense organs called muscle spindles, which send signals to the spinal cord. Sensory neurons in the dorsal region of the spinal cord send signals to motor neurons in the ventral region. The result is the contraction of the quadriceps muscle, jerking the lower leg upward. At the same time, sensory neurons send signals to inhibitory interneurons, which in turn inhibit the firing of the motor neurons that signal the hamstring. The result is the contraction of the quadriceps and the inhibition of the antagonistic motoneurons.

the cyclical activity of different groups of motoneurons. The needed coordination does not require the involvement of the brain, however; the interneurons of the spinal cord are enough. Input from the brain nonetheless modifies the activity of spinal cord motoneurons.

Neurons communicate with each other by chemical and electrical signaling. Before we can understand the mechanisms responsible for signaling, however, we must first discuss the electrical properties of cells in general.

■ MOST CELLS HAVE ELECTRICAL POTENTIALS ACROSS THEIR MEMBRANES

Every cell has a chemical composition that is different from its surroundings. These differences result mostly from the properties and activities of the plasma membrane. As discussed in Chapter 6, plasma membranes allow some molecules and ions to pass freely while serving as impermeable boundaries to others. Particular membrane proteins serve as **channels,** which allow the passage of specific molecules and ions. Some of these channel proteins have **gates,** which open and close the channels. The gates respond to environmental signals and regulate the passage of ions or molecules through the membrane. Other membrane proteins serve as *pumps,* using energy from ATP (or from other sources) to bring about specific concentration differences between the internal and external solutions.

The operation of channels, gates, and pumps results in a cell's interior having different concentrations of many molecules and ions from its exterior. As illustrated in Figure 23-7, differences in the concentrations of molecules and ions across the plasma membrane represent a potential source of energy, whose amount we can calculate. The free energy resulting from a concentration difference of an uncharged molecule across a membrane is

$$\Delta G = RT \ln \frac{[C_{out}]}{[C_{in}]}$$

where C_{out} and C_{in} are the concentrations of a particular molecule outside and inside the cell. (See Chapter 4, p. 94.)

When an ion, a charged particle, is present in different concentrations outside and inside a cell, as in Figure 23-7b, a concentration gradient produces an **electrical potential (voltage),** a measure of the energy change associated with a **current** (the movement of charges). The greater the voltage, the greater the work that can be performed and the greater the tendency for a reaction to happen spontaneously. The magnitude of the electrical potential *(E)* is defined as the free energy divided by the charge. For a mole of ions—the charge is $Z\mathcal{F}$, where Z is

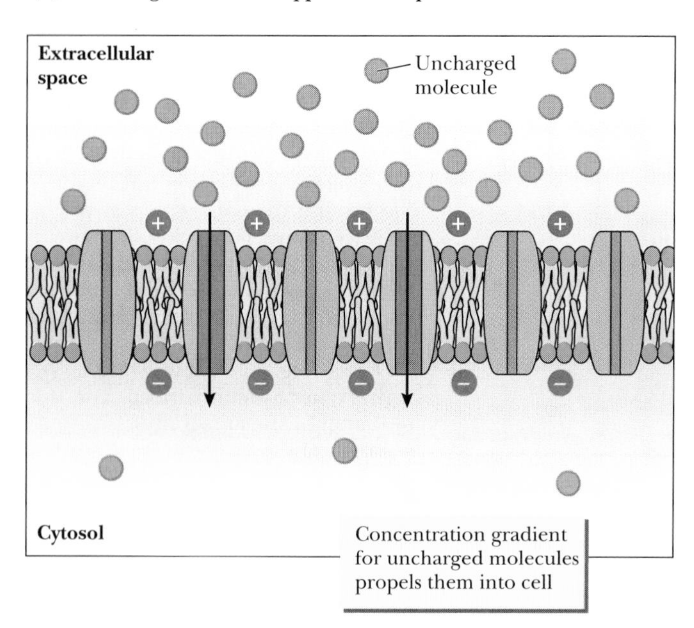

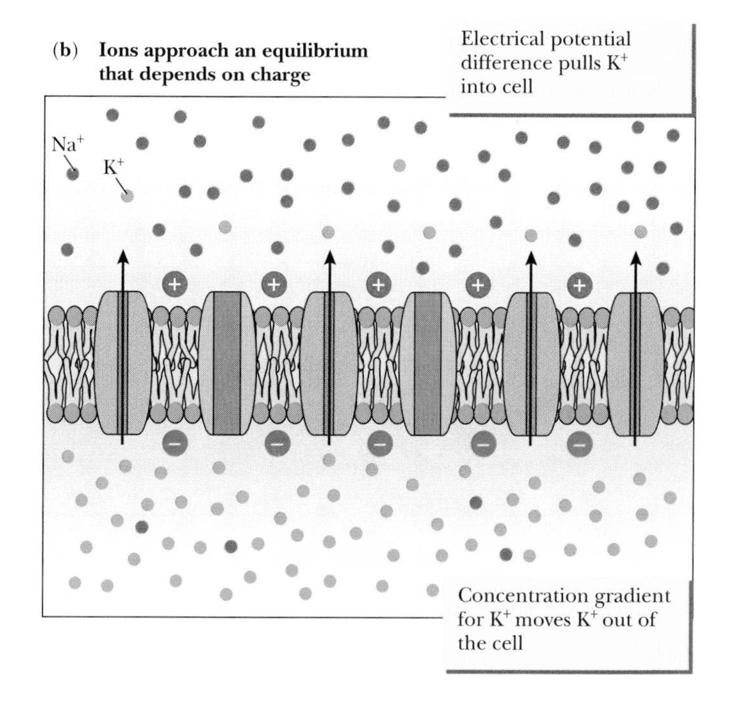

Figure 23-7 Differences in the concentration of a molecule or an ion across a membrane represent potential energy because useful work could be performed as molecules or ions spontaneously move toward equilibrium. **(a)** Uncharged molecules approach an equilibrium with equal concentrations on the two sides of the membrane; **(b)** charged ions approach an equilibrium with a higher concentration of anions (negatively charged ions) on the negative side of the membrane and a higher concentration of cations (positively charged ions) on the positive side of the membrane.

the charge on each ion and $\mathscr{F}$ (Faraday's constant) is the charge on a mole of electrons. The electrical potential is given by an equation called the **Nernst equation:**

$$E = \frac{RT}{Z\mathscr{F}} \ln \frac{[C_{out}]}{[C_{in}]}$$

or

$$E \text{ (in millivolts)} = 58 \log_{10} \frac{[C_{out}]}{[C_{in}]}$$

where C_{out} and C_{in} are the concentrations of a particular ion outside and inside the cell, ln is the natural logarithm, and $\log_{10}$ is the logarithm to the base 10. Moving a positively charged ion to a region with an excess of positive charges requires energy, while moving a negatively charged ion in the same direction releases energy. Let us now see how the Nernst equation applies to a resting nerve cell.

The Resting Potential of a Neuron Depends on the Selectivity of Ion Channels and Pumps

The outside of a neuron (or of any cell) generally has a greater concentration of positive ions than its inside. The main reasons for this excess of positive charge outside the cell are (1) the character of the membrane's sodium-potassium pump; and (2) the limited and selective permeability of the membrane to sodium (Na^+) and potassium (K^+) ions. Recall (from Chapter 6) that the sodium-potassium pump uses the energy of ATP to push Na^+ out and pull K^+ in. For each ATP spent, 3 Na^+ ions leave and 2 K^+ ions enter. The results are (1) Na^+ ion concentration is higher out than in; (2) K^+ ion concentration is higher in than out; and (3) more positive charges move out than in. By restricting the passage of Na^+ and K^+ ions, the membrane maintains the uneven concentrations of ions generated by the sodium-potassium pump. The result is an electrical potential, a voltage, across the membrane.

Electricians measure voltage with a voltmeter, a device that determines the tendency of electrons to flow between two contacts, called *electrodes*. Neurophysiologists can measure voltage in much the same way, but they usually couple their measuring device to cells with **microelectrodes,** electrodes that contain salt solutions and are small enough to insert into or to maintain contact with a single cell, as illustrated in Figure 23-8.

The actual voltages across plasma membranes are much smaller than those in your wall sockets or even in your transistor radio. A standard radio battery, for example, generates a voltage of about 1.5 volts, while most cells have voltages measured in millivolts. One millivolt is 1/1000 of a volt, so a radio battery generates 1500 millivolts. A resting nerve cell has a voltage across its membrane of about 70 millivolts. A neurobiologist's measuring

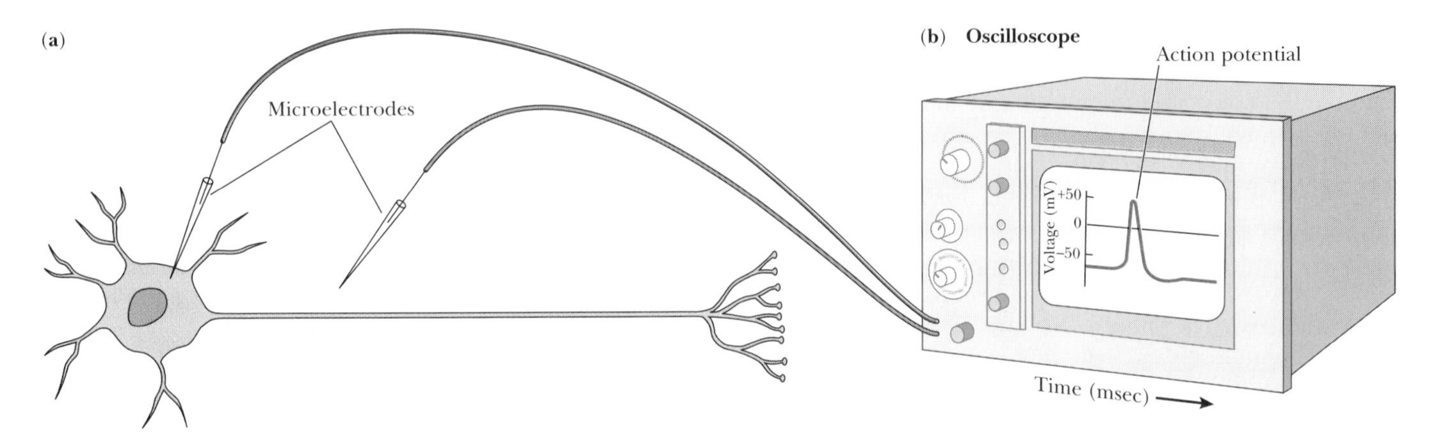

Figure 23-8 Neurobiologists often measure voltages across the plasma membrane using a glass microelectrode inserted into a cell. The difference in electrical potential across the membrane is recorded on an oscilloscope, which displays the voltage as a function of time. **(a)** A pair of microelectrodes, one on the inside and one on the outside of a cell; **(b)** an oscilloscope showing the changes in voltage as a nerve cell produces an action potential.

devices must be much more sensitive than the electrician's. The measuring instrument usually used to measure cellular voltages is an **oscilloscope,** illustrated in Figure 23-8b, which displays voltages on a television-like screen, showing variations with time.

When a physiologist pokes a microelectrode into a nerve cell, the oscilloscope registers a negative voltage of about 70 millivolts. Other cells give similar readings. The big difference is that the nerve cell can suddenly and repeatedly change its electrical potential. Such a sudden change is called an **action potential.** Before we try to understand an action potential, we must first explain the **resting potential,** the voltage across the plasma membrane when the cell is not "firing" (producing action potentials).

To understand the origin of the resting potential, we can first consider three simplified, fictional membranes, as diagrammed in Figure 23-9. Suppose we put a solution of potassium chloride on each side, with a more concentrated solution, say 100 mM, on the left and a less concentrated solution, say 10 mM, on the right. Given enough time, both K^+ ions and Cl^- ions should come to equal concentrations in the two compartments. But—in the fictional case shown in Figure 23-9a — neither ion can cross the membrane, and the voltage across the membrane is

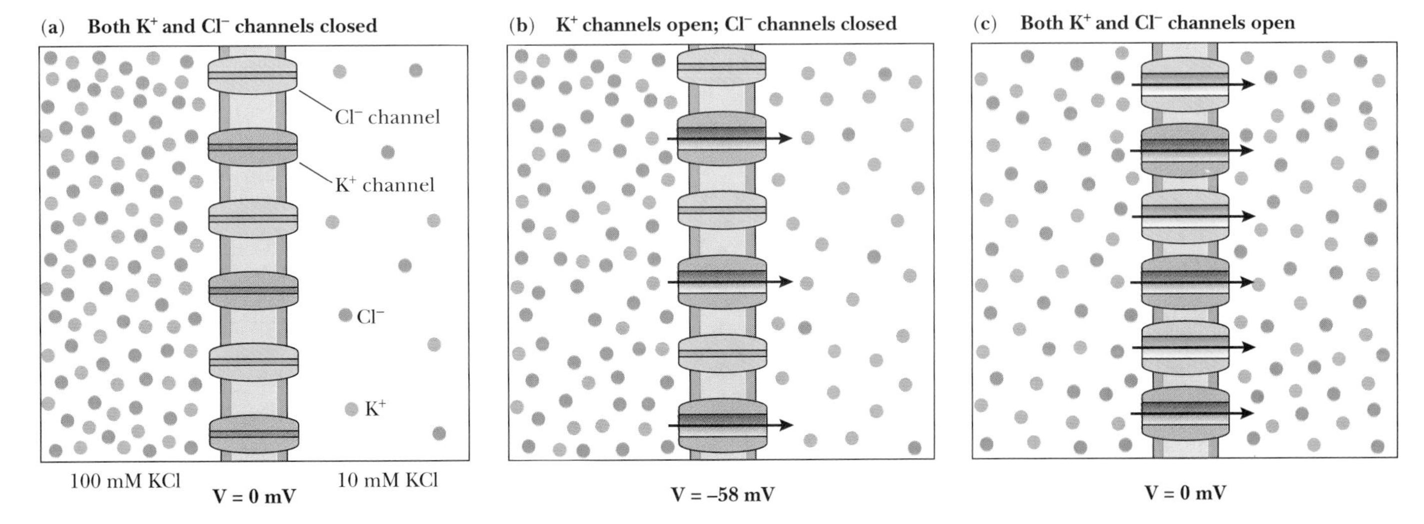

Figure 23-9 An idealized membrane, separating a solution of 100 mM KCl, on the left, from 10 mM KCl, on the right. The membrane contains two types of channels, one that allows the passage of K^+ ions and one that allows the passage of Cl^- ions. **(a)** Both types of channel are closed, and no ions can move. The voltage across the membrane is zero. **(b)** The K^+ channels are open, and the Cl^- channels closed. K^+ ions move to the right, and the voltage across the membrane, given by the Nernst equation, is −58 millivolts. **(c)** Both types of channel are open; Cl^- ions move to the right, and the voltage across the membrane returns to zero.

zero. In the fictional membrane of Figure 23-9b, the membrane contains channels that allow K^+ ions to cross but not Cl^- ions. K^+ ions cross the membrane in both directions, but the net movement is toward the right, down their concentration gradient.

As K^+ ions move to the right, they leave behind a net negative charge on the left side. As this charge accumulates, it slows the further flow of K^+ ions to the right. Finally, the system comes to equilibrium, at which point there is no more net flow of K^+ ions. At equilibrium, the chemical potential due to the difference in concentration exactly balances the electrical potential due to the difference in charge. We can now use the Nernst equation to calculate the voltage across the membrane when the concentration of K^+ ions is 10 times greater on the left than on the right:

$$E \text{ (in millivolts)} = 58 \log_{10} [C_{\text{right}}]/[C_{\text{left}}] = 58 \log_{10} 10/100 = -58 \text{ millivolts}$$

This voltage, −58 millivolts, is the **equilibrium potential,** the voltage at which the electrical force driving the ion in one direction exactly balances the tendency of the ion to move down a concentration gradient.

In this fictional example, K^+ ions lie at their equilibrium potential, but Cl^- ions do not. This means that, given the opportunity, Cl^- ions move from right to left. In doing so, they would move down a concentration gradient, from more to less concentrated. And, because they have a negative charge, the Cl^- ions are also moving down an electrical gradient (from negative to positive).

Imagine now that we could suddenly open a gate that would allow Cl^- ions to pass, as illustrated in Figure 23-9c. If we kept the gate open long enough, Cl^- ions would flow until they came to equal concentrations on both sides. But if we kept the gate open for just a bit, only a small fraction of the Cl^- ions would move from left to right, creating a transient electrical current.

In principle, we could harness this electrical current to perform some kind of work. Consideration of this imaginary, idealized membrane shows, then, that the electrical potential across a membrane—formed by the uneven distributions of ions—has the capacity to do work and that temporary changes in a membrane's permeability to specific ions can lead to changes in voltage and the generation of electrical currents.

Sodium Ions Are Far from Equilibrium but Do Not Contribute Much to a Neuron's Resting Potential

Differences in the charge distribution across nerve plasma membranes are only slightly more complicated than our fictional example. To explain the resting potential in a real neuron, we need to consider three ions instead of two—Cl^-, K^+, and Na^+. Figure 23-10 shows the concentrations of these three ions in an actual neuron, the giant neuron of a squid. Other cellular ions differ in concentration across the membrane, such as the negatively charged organic ions and proteins that cannot cross the membrane. But these ions do not contribute to the resting potential because they can never cross the membrane. (Recall that the Cl^- ions do not contribute to the potential across the fictional membrane of Figure 23-9b, which has closed Cl^- channels, but do contribute to the potential across the membrane of Figure 23-9c, which has open Cl^- channels.)

The resting potential of a neuron is about −70 millivolts, with the inside of the cell more negative than the outside. The potential difference comes mostly from the difference in the concentration of K^+ ions across the membrane. The difference in the sodium ion concentrations across the membrane contributes little to the resting potential because the membrane allows hardly any Na^+

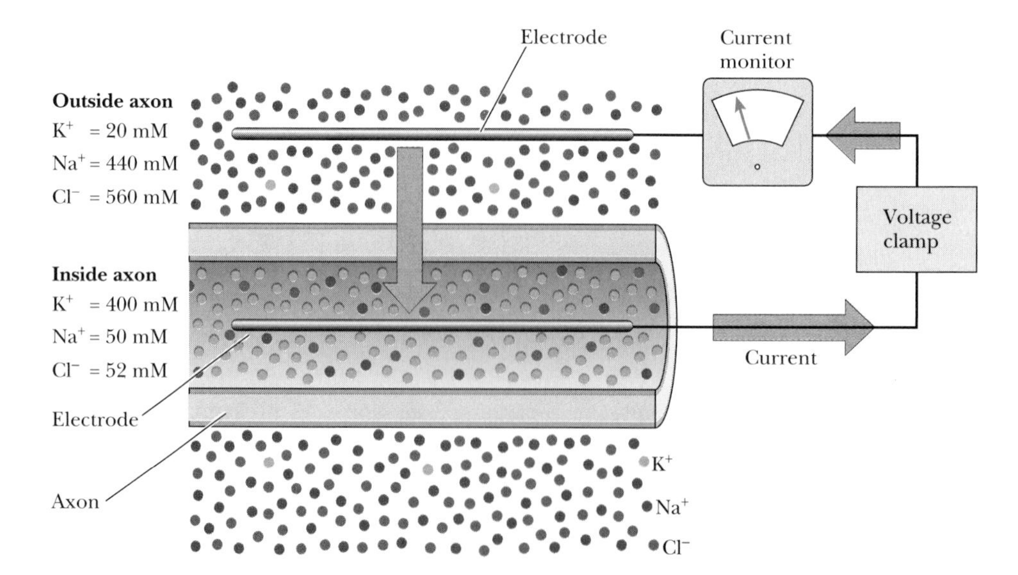

Figure 23-10 Alan Hodgkin, Andrew Huxley, and Bernard Katz first developed an understanding of changing ionic conductances in neurons by studying the giant axon of a squid. A squid giant axon, showing the concentrations of K^+, Na^+, and Cl^- in the cytoplasm and extracellular fluid; the resting potential depends almost entirely upon the K^+ ion gradient. The inserted electrodes allowed Hodgkin and Huxley to perform voltage clamp experiments, altering the voltage across the membrane and measuring the resulting currents.

ions to cross. Actually, some Na^+ ions do pass, and the K^+ ions don't pass completely freely, so the measured potential is a little less than the equilibrium potential calculated from the Nernest equation. Cl^- is also near its equilibrium potential, with more outside than inside, because the interior of the cell is negative.

The main difference between the idealized membrane of Figure 23-9 and the actual membrane of Figure 23-10 is that Na^+ ions are far from equilibrium across nerve membranes. A temporary opening of a sodium channel in the membrane should therefore allow Na^+ ions to flow down both a concentration gradient (higher concentration outside than inside) and an electrical gradient (more positive charge outside than inside), in the same way that Cl^- ions would have flowed in our hypothetical example. Such a temporary opening is exactly what happens during an action potential.

HOW DO CHANGES IN ION PERMEABILITY BRING ABOUT ACTION POTENTIALS?

Understanding how nerve cells generate resting potentials and action potentials must come from experiments, not from idealized models and equations. Such experiments were impossible for a long time, however, because nerve cells are so small. Researchers were unable to put electrodes into cells, to measure internal ion concentrations, or to vary the internal ion concentrations. Doing the proper experiments would require either a giant nerve cell or techniques for dealing with very small cells.

Eventually, neuroscientists developed ways of studying both large and small nerve cells. But most of our current appreciation of how a nerve cell fires—how it creates an action potential—came from experiments done on the giant nerve cell from the squid, illustrated in Figure 23-10. These neurons are responsible for the signals that cause the squid to produce a jet of water, allowing the animal to escape from predators. The axons of these cells are extraordinarily large, up to 1 mm in diameter, and until 1936, biologists thought that they were blood vessels. With the discovery of their real nature, physiologists soon began to use them to understand how neurons generate action potentials.

The most penetrating and influential work on action potentials was that of Alan Hodgkin and Andrew Huxley, which began in the late 1930s in Plymouth, England, and continued after World War II at Cambridge University. Hodgkin and Huxley and their collaborators took advantage of the size of the squid axon to insert electrodes into the interior and to measure the voltage changes as they varied the ion concentrations both inside and outside the cell. By varying the concentration of K^+, they showed its importance in determining the level of the resting potential, just as predicted by our discussion above. They could also measure the ion concentrations within the axon, after pressing out its contents with a rubber roller.

Hodgkin and Huxley then showed that an action potential, unlike the resting potential, depends on the flow of Na^+ into the cell. In one experiment, for example, they removed all Na^+ ions from the extracellular fluid, replacing them with a larger ion, choline, which could not pass through a Na^+ channel in the membrane even when the channel was open. The result was that the axon could no longer produce an action potential at all. This experiment underscored the conclusion that understanding the action potential requires an understanding of how the membrane temporarily changes its permeability to Na^+ ions.

How Does a Membrane Change Its Permeability to Sodium?

Hodgkin and Huxley developed a method, called **voltage clamping,** which allowed control of the voltage across a membrane. Changes that make the inside of the cell *less* negative are said to be **depolarizing,** while changes that make the inside of the cell *more* negative are **hyperpolarizing.** Using a voltage clamp, Hodgkin and Huxley could study currents (that is, the flow of ions) as they varied the voltage. They could tell which ions were contributing to the measured currents by varying the salt composition outside the cell—for example, by removing Na^+ ions or by replacing them with choline ions.

These experiments confirmed the importance of Na^+ ions in the action potential and revealed much about the behavior of the sodium channels. The most important characteristic of the sodium channels is that their opening depends on the voltage across the membrane. At the normal resting potential, sodium channels are effectively closed. Depolarization, which makes the inside of the membrane more positive (less negative), causes the channels to open briefly and to admit a small horde of Na^+ ions. Because their ability to admit Na^+ depends on the membrane potential, sodium channels are said to be **voltage-gated,** meaning that they are open or closed according to the voltage across the membrane.

Hodgkin and Huxley were able to explain the generation of action potentials in terms of the voltage dependence of sodium channels. When a nerve cell is sufficiently depolarized, the membrane reaches a *threshold* voltage, at which point sodium channels open and Na^+ ions flow inward. The flow soon stops, however, within about 2 msec, and the sodium channels become less likely to open again. During this time, called the **refractory period,** the K^+ ions flow out of the cell and restore the original resting potential. During each action potential, then, some Na^+ ions move into the cell and some K^+ ions move out of the cell. The flow of K^+ ions outward increases during the action potential, because the membrane also con-

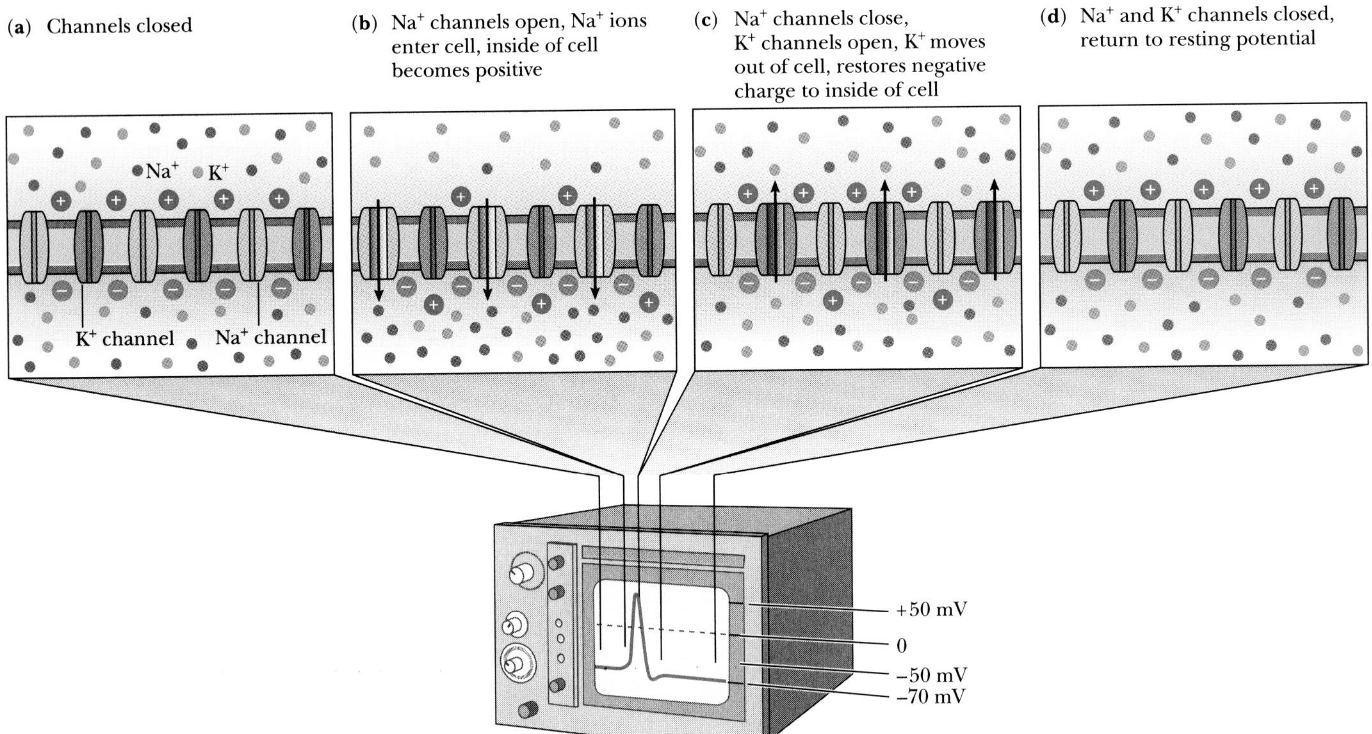

Figure 23-11 A brief current into the cell depolarizes the cell, making the inside of the cell less negative than the resting potential. **(a)** Before depolarization, both Na^+ and K^+ channels are closed, and the inside of the cell is negative with respect to the outside. **(b)** Voltage-gated Na^+ channels briefly open, admitting Na^+ ions into the cell, causing still further depolarization. **(c)** The Na^+ channels then close again, despite the continuing depolarization; voltage-gated K^+ channels open, allowing K^+ to flow outward and restoring the transmembrane voltage to the resting potential. **(d)** Both Na^+ and K^+ channels are closed.

tains voltage-gated potassium channels. Figure 23-11 shows the sequence of channel openings in response to a change in voltage across the membrane. The number of ions that exchange is small compared with the total inside the cell, and the sodium-potassium pump soon restores the original distribution.

In the late 1970s, researchers developed a method, illustrated in Figure 23-12a, called **patch clamping,** which allows the measurement of currents across a tiny piece of membrane (rather than across a whole cell membrane). The patch is so small that it may contain only a single channel (so the method is also called **single-channel recording**). Patch clamping allows researchers to measure the ability of individual channels to admit ions as a function of the voltage imposed across the membrane. Using this technique, researchers demonstrated that individual sodium channels switch back and forth between open and closed states, as illustrated in Figure 23-12b. Each sodium channel stays open for about 1 msec at a time, allowing about 10,000 Na^+ ions to pass through the membrane; the channel then spontaneously closes.

The total amount of open and closed time for each channel depends on the voltage across the membrane. At the resting potential of the cell, the channels are mostly closed. The number of times they open each second increases as the voltage inside the cell becomes more positive. An actual nerve cell membrane contains thousands of such channels, and the current of the action potential results from summing the activity of these channels.

Biochemical and Molecular Biological Studies Have Identified the Voltage-Dependent Sodium Channel Protein

Many poisons and stimulants affect the nervous system profoundly by binding to specific proteins important to neural function. One such poison, **tetrodotoxin,** was isolated from the Japanese pufferfish *("fugu")*, illustrated in Figure 23-13. Tetrodotoxin blocks the passage of Na^+ ions through the voltage-gated channels. Tetrodotoxin and similar toxins allowed biochemists to isolate the protein that serves as the voltage-gated sodium channel, using affinity chromatography.

The electric organ of the electric eel served as a particularly rich source of the sodium channel protein. The isolated protein consists of a single polypeptide with a mol-

(a) **Technique for making patches**

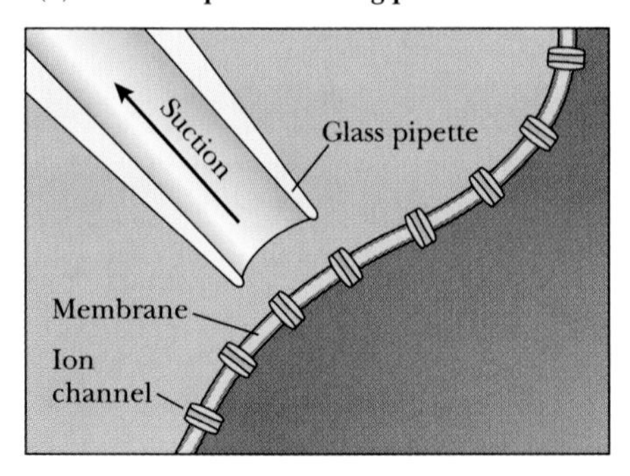

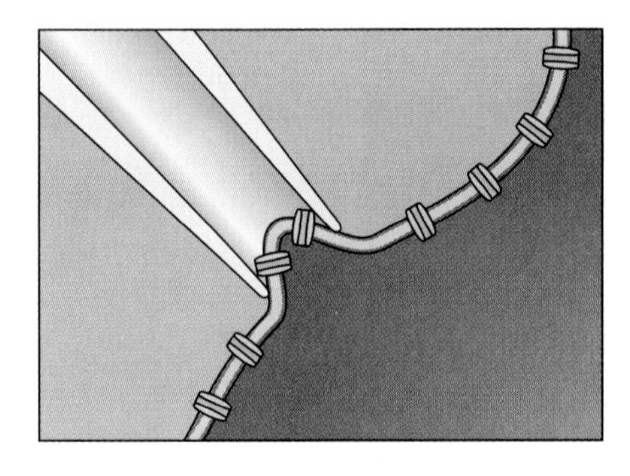

(b) **Depolarization stimulates Na^+ flow through ion channel**

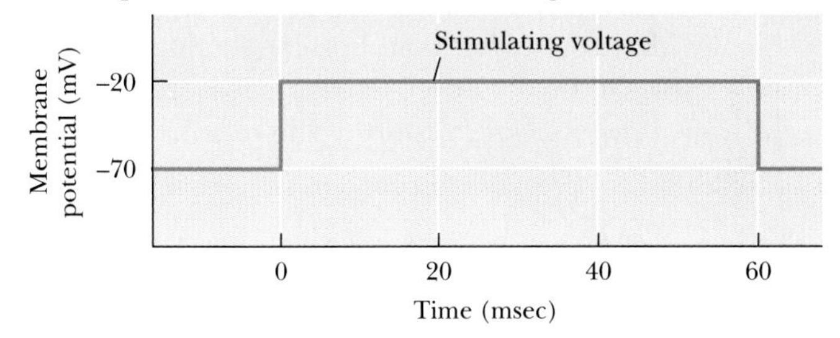

Current through a single Na^+ channel

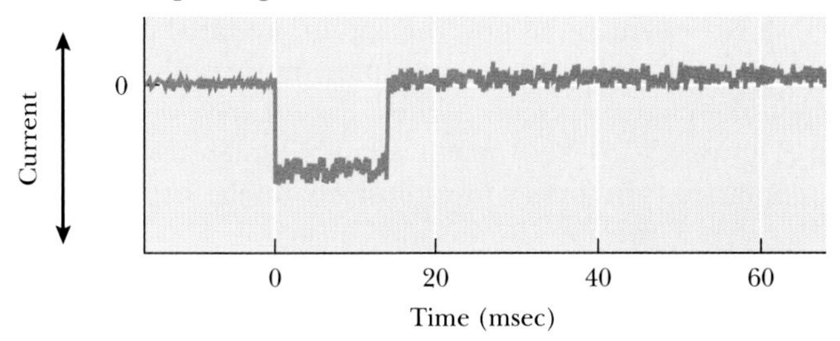

Aggregate (summed) current through many Na^+ channels

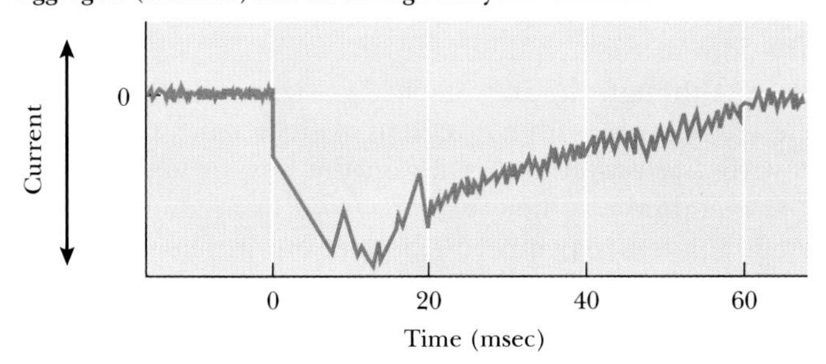

Figure 23-12 To make patch clamp recordings, a researcher uses a micropipette that forms a tight seal with a neuronal membrane and pulls the membrane away from the rest of the cell. The detached patch contains only a few channels, sometimes only a single channel, which can then be studied electrically. **(a)** Technique for making patches; **(b)** measurements of Na^+ flow through a single voltage-gated sodium channel.

ecular size of about 250,000. When this protein was combined with artificial membrane vesicles, researchers could demonstrate a voltage-dependent sodium permeability, showing that they had isolated the right protein.

The ability to isolate the channel protein allowed a group of Japanese scientists led by S. Numa to isolate a recombinant DNA encoding the channel protein. They determined the amino acid sequence of the sodium channel polypeptide. They also produced sodium channel mRNA from the recombinant DNA and injected the mRNA into a frog *(Xenopus)* oocyte. Patch clump experiments showed that voltage-gated sodium channels were present in injected but not uninjected oocytes.

Knowledge of the amino acid sequence of the sodium channel allowed researchers to make models for how it might work. The polypeptide consists of about 2000 amino acid residues, which consists of four similar domains, as illustrated in Figure 23-14. Each of these domains in turn appears to contain six transmembrane segments, each of which consists of mostly nonpolar amino acid residues. Ongoing research on the sodium channel is seeking to understand the relationship between the protein's

Figure 23-13 Tetrodotoxin, a poison isolated from the Japanese puffer fish *(fugu)*, specifically binds to voltage-gated sodium channels. Affinity chromatography with tetrodotoxin and similar toxins allowed the biochemical isolation of the sodium channel. Served in small amounts that tingle the taste buds, *fugu* is considered a delicacy in Japan. *(Jeffrey L. Rotman/Peter Arnold)*

structure and its function. Some researchers have suggested, for example, that four transmembrane segments, one from each domain, assemble to form a pore through which Na^+ ions can flow, as shown in Figure 23-14b.

Other research concerns the nature of the gate. The sequence provided an important hint: In the fourth transmembrane segment (called S4) of each domain, every third residue is positively charged. Researchers have suggested that this positively charged membrane segment is a **gating helix,** responsible for the voltage gate. According to the model shown in Figure 23-14c, the accumulation of negative charges on the exterior surface of the membrane, following depolarization, pulls on the helix and deforms it. The result is that the whole protein moves a bit, opening the pore and allowing Na^+ ions to pass.

To test this hypothesis, researchers used the techniques of site-directed mutagenesis, as described in Chapter 17. They substituted neutral amino acids for the positively charged amino acids in the fourth transmembrane segment. The resulting protein, expressed in injected oocytes, required much more depolarization before Na^+ ions could pass, supporting the idea that the suspected segment contributes to the voltage-sensitive gate.

How Does an Action Potential Move Down an Axon?

The properties of the voltage-gated sodium channel explain not only how action potentials occur but also how they can move for long distances along nerve cell membranes. First, we can list six consecutive events in the production of an action potential in a single spot in the membrane:

1. Something causes a local depolarization of the membrane, large enough to exceed the threshold.
2. Sodium channels open, and Na^+ ions flow into the cell.
3. As Na^+ ions flow inward, the inside of the membrane becomes more positive at that point of the cell's surface.
4. The decreased polarization of the membrane causes more channels to open, increasing the positive charge of the membrane still further.
5. Finally (that is, after less than 1 msec), the sodium channels close (spontaneously).
6. K^+ ions flow outward, restoring the membrane to the resting potential.

Notice that the pumping of Na^+ and K^+ ions does not directly enter into the events of the action potential. This pumping serves only to reestablish and maintain the distribution of ions responsible for the resting potential.

Now we can see how the action potential propagates itself. As the membrane becomes more positive at one spot, the charge spreads to the adjacent spot, as illustrated in Figure 23-15. That patch of membrane also becomes depolarized and opens its sodium channels, leading to further depolarization. In this way the action potential propagates itself along the axon, regenerating itself as it goes. The action potential moves in just one direction, however, because once sodium channels have opened, they cannot immediately open again. So only the sodium channels on the "downstream" side can open.

Although the movement of action potentials superficially resembles the movement of electrical current down a wire, the movement relies on the flow of ions through the membrane rather than electrons from atom to atom. Action potentials move much more slowly. The thicker the axon, the more rapid the ion movement, so larger axons conduct more rapidly than small. Axons can conduct impulses in this way at speeds of 1–10 meters per second.

Nerves Wrapped in Myelin Conduct Action Potentials More Rapidly

Most nerve cells have axons much thinner than the squid giant axon. Given the expected rate of conduction in the nerves that carry information from and to our feet, for example, it would take a painfully long time for us to realize that we had stepped on a nail or come too close to a campfire. Most vertebrate nerves have a special arrangement that allows a much more rapid conduction of nerve impulses. The rapid rate of conduction depends on myelin, a specialized, glistening sheath that surrounds nerve axons.

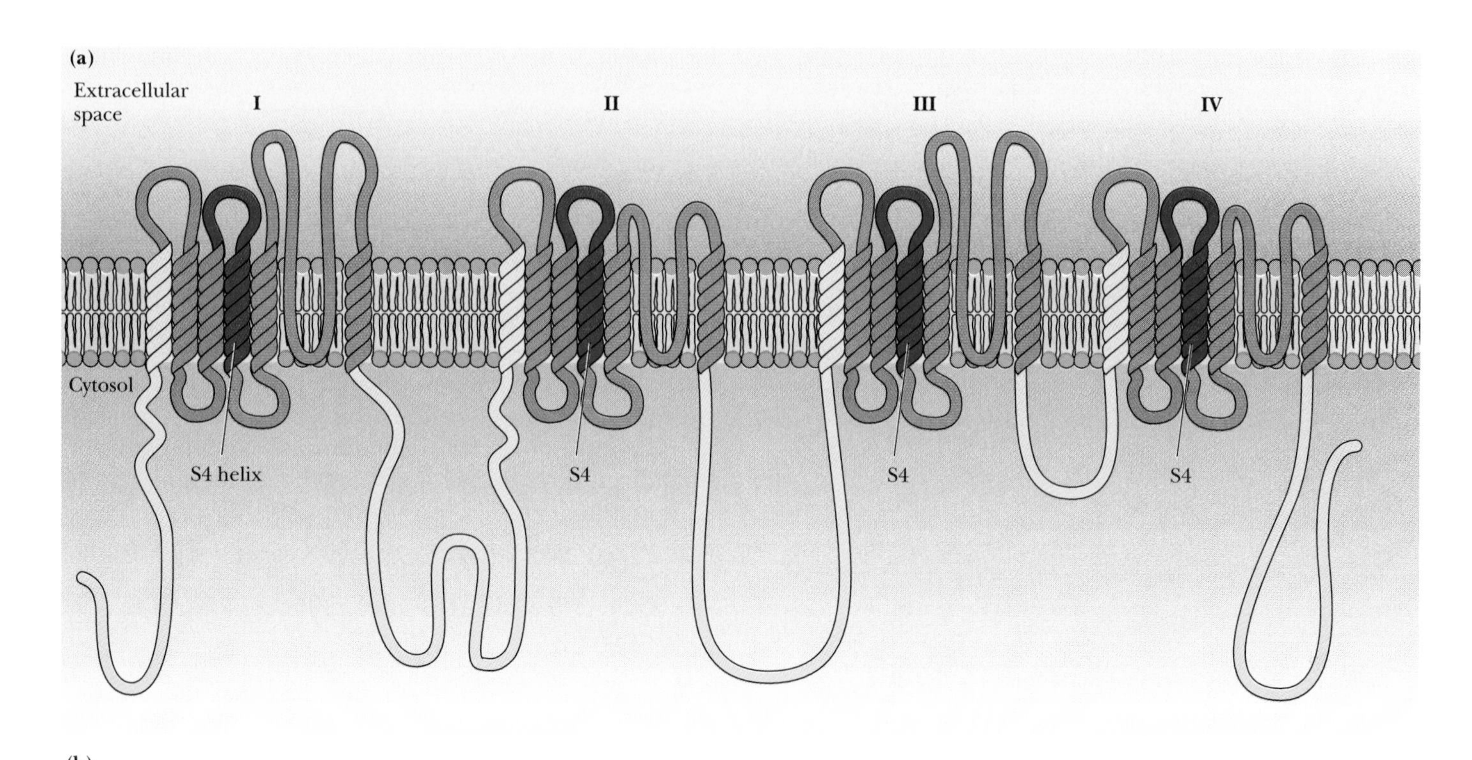

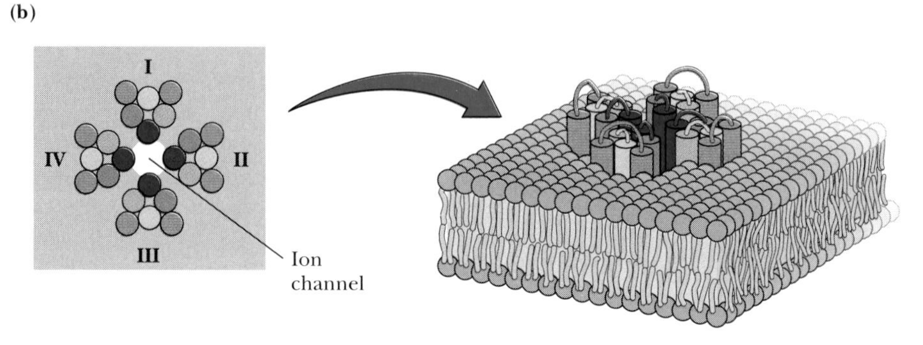

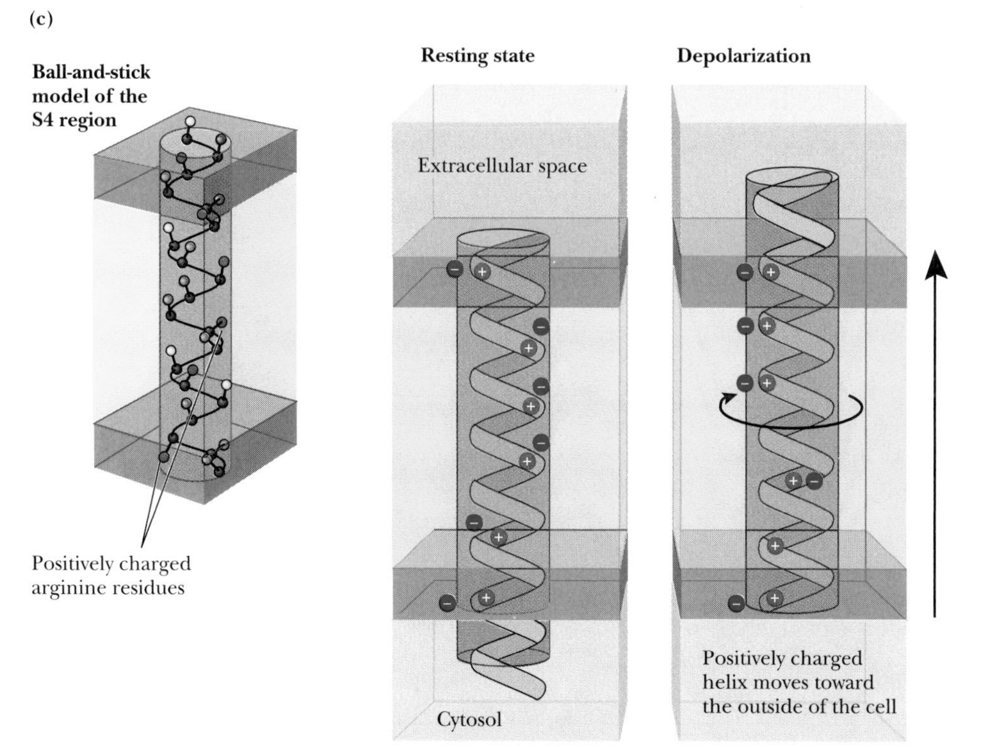

Figure 23-14 The amino acid sequence of the voltage-gated sodium channel suggests how it lies in the neuronal membrane, how it forms an ion channel, and how it changes in response to depolarization. **(a)** The sequence of the receptor suggests that it consists of four domains, designated I, II, III, and IV, each with six transmembrane segments; **(b)** one helix from each of the four domains is thought to form an ion channel; **(c)** the fourth transmembrane segment of each domain contains six positively charged arginine residues; when the membrane is depolarized, the positively charged helix is thought to move toward the outside; this movement may be the trigger for the opening of the ion channel.

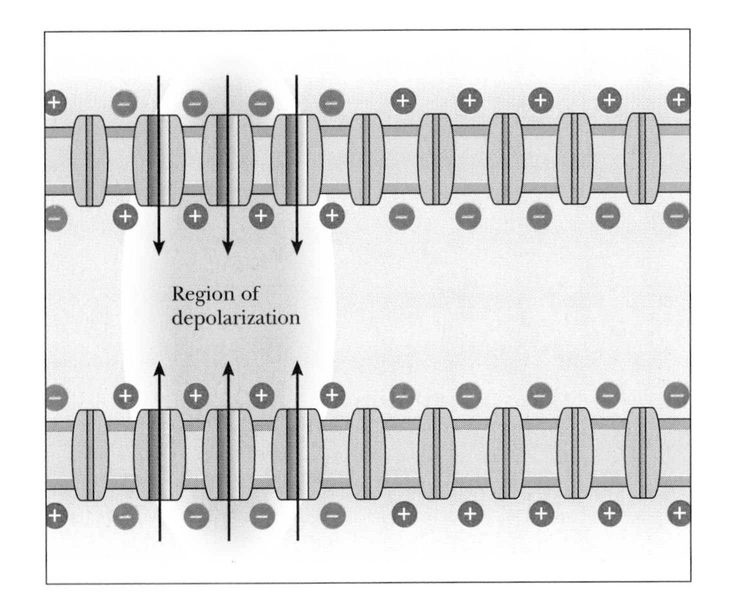

Figure 23-15 The propagation of an action potential along an axon. Na^+ channels open in the region of depolarization. The flow of Na^+ ions causes the depolarization of the adjacent membrane; the action potential moves in only one direction because the sodium channels on the trailing side are temporarily inactivated after they reclose.

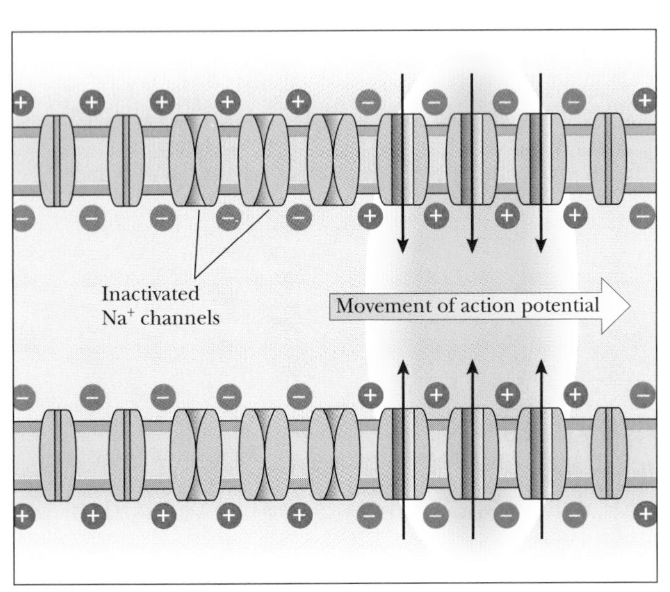

The myelin sheath insulates most of the axon's membrane, so that it cannot allow the passage of ions, as shown in Figure 23-16. Depolarization and ion flow are possible only in the **nodes of Ranvier,** gaps between the myelin wrappings, spaced about every millimeter along the axon. Instead of running down the axon continuously, the nerve impulse moves by **saltatory conduction,** jumping down a myelinated axon at rates up to 100 times faster than down an unmyelinated axon, up to about 120 m/sec.

HOW DO NEURONS COMMUNICATE WITH EACH OTHER?

Ramon y Cajal's studies demonstrated that the nervous system consists of discrete cells. More recently, studies with the electron microscope revealed that there is a distinct boundary—the **synapse**—between most communicating neurons. Figure 23-17 shows a particularly well studied example of a synapse, in this case between a motoneuron and a muscle cell. We have already seen how a neuron in the brain can send information over long distances down an axon. But we still need to understand how a **presynaptic neuron,** lying "upstream" from the synapse, influences the electrical activity of a **postsynaptic neuron,** to which it connects through a synapse.

The easiest type of synapse to understand is the **electrical synapse,** which joins presynaptic and postsynaptic

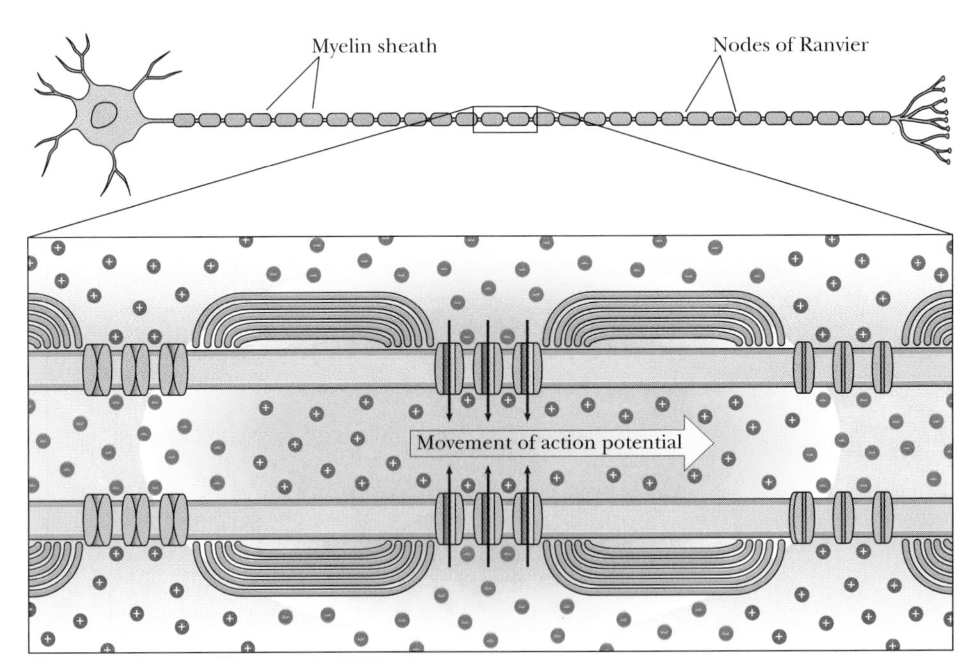

Figure 23-16 The propagation of an action potential along a myelinated axon. The myelin insulates the axon, so current flow occurs only at the places where the sheath is interrupted, the nodes of Ranvier.

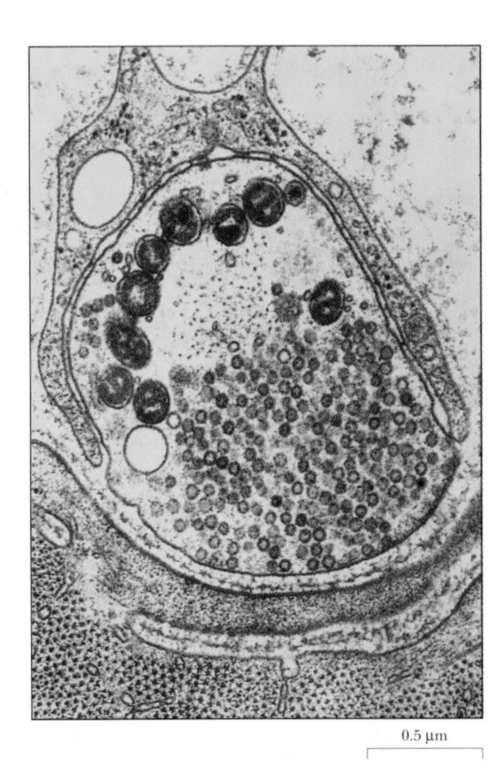

Figure 23-17 Synapses are specialized regions that allow signaling between neurons. The neuromuscular junction, a chemical synapse between a motoneuron and a muscle cell; the fuzzy dark zones are thought to be the docking sites for acetylcholine-containing vesicles. *(T. Reese & D. W. Fawcett/Visuals Unlimited)*

neurons through gap junctions. (Recall from Chapter 6, p. 174, that gap junctions are channels through the membranes of adjacent cells; they allow ions and small molecules to pass freely from one cell to the next.) An electrical synapse allows an action potential to continue traveling to the postsynaptic cell in the same manner and at about the same rate as it traveled down the presynaptic axon. All electrical synapses are **excitatory,** meaning that action potentials in the presynaptic cell stimulate action potentials in the postsynaptic cell.

The more common type of synapse is the **chemical synapse,** of which the neuromuscular junction of Figure 23-17 is a prime example. In a chemical synapse, presynaptic and postsynaptic membranes do not join closely. Instead, the presynaptic and postsynaptic cells are separated by the **synaptic cleft,** a space of about 20 nm. Communication across this gap requires the diffusion of one or more **neurotransmitters,** signaling molecules made in the presynaptic cell that affect the electrical activity of the postsynaptic cell. Chemical synapses may be either excitatory or **inhibitory.**

The electrical activity of the presynaptic neuron triggers the release of a neurotransmitter, which then diffuses across the synaptic cleft. A neurotransmitter can either excite or inhibit the activity of the postsynaptic neuron, but the effect is always delayed by the time needed to diffuse across the gap, generally about 0.5 msec. No such delay occurs in an electrical synapse. On the other hand, chemical synapses have two important advantages over electrical synapses: They may be either excitatory or inhibitory, and they can greatly amplify the signal of a small presynaptic neuron, which may represent only a small ion current.

Most synapses in vertebrate central nervous systems are chemical rather than electrical. Electrical synapses, however, occur in the vertebrate heart, where they coordinate the synchronous contraction of heart muscle cells, and in a variety of invertebrate and vertebrate nerve circuits, where fast conduction between cells is advantageous (for example in an escape reflex).

How Does a Chemical Synapse Work?

Understanding how chemical synapses work is important not only to our understanding of the brain, but also to significant medical and social problems. Essentially all the medications that affect the functioning of the brain act on synapses. Among these substances are sleeping pills, tranquilizers, antipsychotics, pain killers, and illegal drugs like heroin and cocaine. Each of these chemicals acts by mimicking or interfering with the production, release, or action of some neurotransmitter.

Let us look more closely at the structure of a chemical synapse. The best studied chemical synapse is the vertebrate **neuromuscular junction,** the synapse between a motor neuron and a voluntary muscle cell, shown in Figure 23-17. These synapses are similar to those between neurons, but they have been much easier to study.

On the presynaptic side, the electron microscope reveals tens of thousands of tiny vesicles, each about 50 nm in diameter and bounded by a membrane. Cell biologists have isolated these vesicles and shown that they are full of the neurotransmitter acetylcholine. These vesicles resemble the secretory vesicles that we discussed in Chapters 6, and they release their contents by exocytosis, as illustrated in Figure 23-18.

The next questions are (1) what triggers the release of transmitter and (2) how does the released transmitter cause electrochemical changes in the postsynaptic cell?

As in other examples of exocytosis, the trigger for vesicular release appears to be an increase in intracellular Ca^{2+} ions. Recall that the concentration of Ca^{2+} ions in the cytoplasm (generally less than 10^{-7} M) is much less than in the extracellular fluid (about 10^{-3} M). Ca^{2+} ions flow into the cell only if there is a route.

During an action potential, Ca^{2+} ions flow into a presynaptic neuron, but, as illustrated in Figure 23-18, they do so in a localized manner—through the membrane next to the synapse. The Ca^{2+} ions move through distinct channels that open in response to the changed voltage during

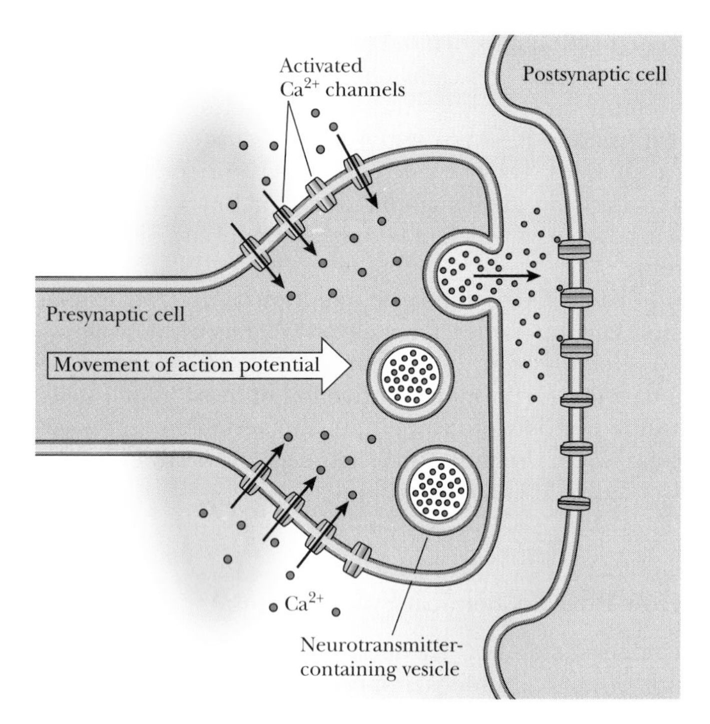

Figure 23-18 Neurotransmitter-containing vesicles release their contents into the synaptic cleft. The arrival of an action potential triggers the voltage-dependent influx of Ca^{2+}, which stimulates exocytosis.

an action potential. Like the sodium channels, these calcium channels are voltage-gated.

Biochemists and molecular biologists have been able to identify the protein that serves as the voltage-gated calcium channel. The voltage-gated calcium channel is remarkably similar to the voltage-gated sodium channel, with about 2000 amino acid residues organized into four domains, each of which contains six transmembrane segments. As in the case of the sodium channel, the fourth transmembrane segment of each domain appears to serve as the voltage-sensitive gate.

A neuron has many fewer calcium channels than sodium channels. There are so few Ca^{2+} ions in the cytoplasm, however, that even a small flow is enough to cause a big change in concentration and to trigger exocytosis and transmitter release.

Many Neurotransmitter Receptors Are Ligand-Gated Ion Channels

In the case of acetylcholine, we know a lot about the action on the postsynaptic side of the synaptic cleft, especially in the neuromuscular junction. When acetylcholine arrives at the postsynaptic membrane, it binds to a receptor molecule. This acetylcholine receptor is itself an ion channel that allows the passage of Na^+ and K^+ ions. It is a gated channel, like the voltage-gated sodium and calcium channels, but acetylcholine rather than voltage unlocks the gate. A molecule that specifically binds to another molecule is called a ligand, so the acetylcholine receptor is an example of a **ligand-gated channel.** (See Chapter 19.)

The acetylcholine receptor was the first ligand-gated channel to be studied in detail, because it was available in large amounts in the electric organs of electric eels and of sting rays. The electric organs of these fish are modified muscles, and they contain many of the small molecules present in neuromuscular junctions. The density of acetylcholine receptors in the electric organs is so great that they can be readily visualized by electron microscopy. Figure 23-19 shows computer reconstructions of the acetylcholine receptor, showing that the pore changes upon binding acetylcholine.

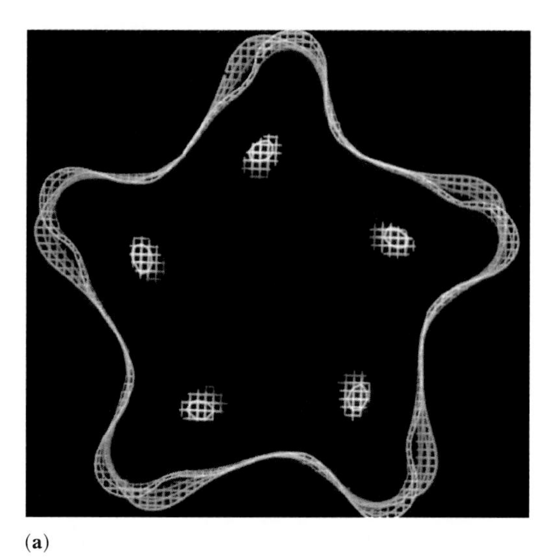

(a)

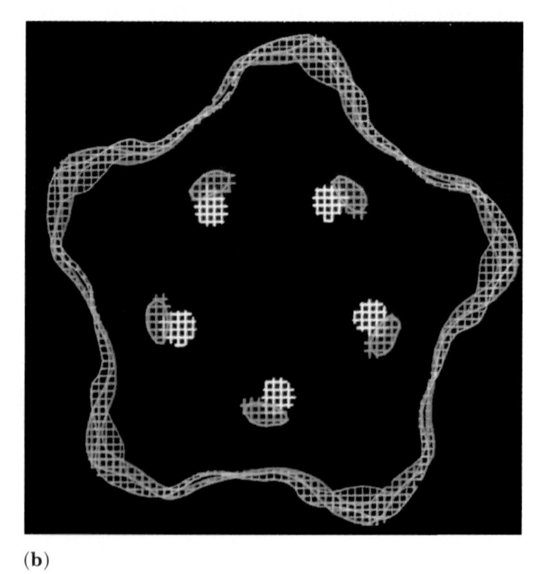

(b)

Figure 23-19 Computer reconstructions of the structure of the acetylcholine receptor from the electric fish, *Torpedo californica.* **(a)** In the "closed" state, **(b)** in the "open" state, in the presence of acetylcholine. *(Nigel Unwin, MRC Laboratory of Molecular Biology, Cambridge, UK)*

The acetylcholine receptor was also the first neurotransmitter receptor for which recombinant DNAs became available, allowing researchers to study the function not only of the naturally occurring receptor but also of receptors whose structures have been altered by site-directed mutagenesis. Even so, many questions remain. We still do not know, for example, the full three-dimensional structure of the receptor, although researchers are actively testing models predicted from the amino acid sequence.

The current model of the ligand-gated acetylcholine receptor is shown in Figure 23-20. The receptor has a molecular size of about 250,000 and consists of five polypep-

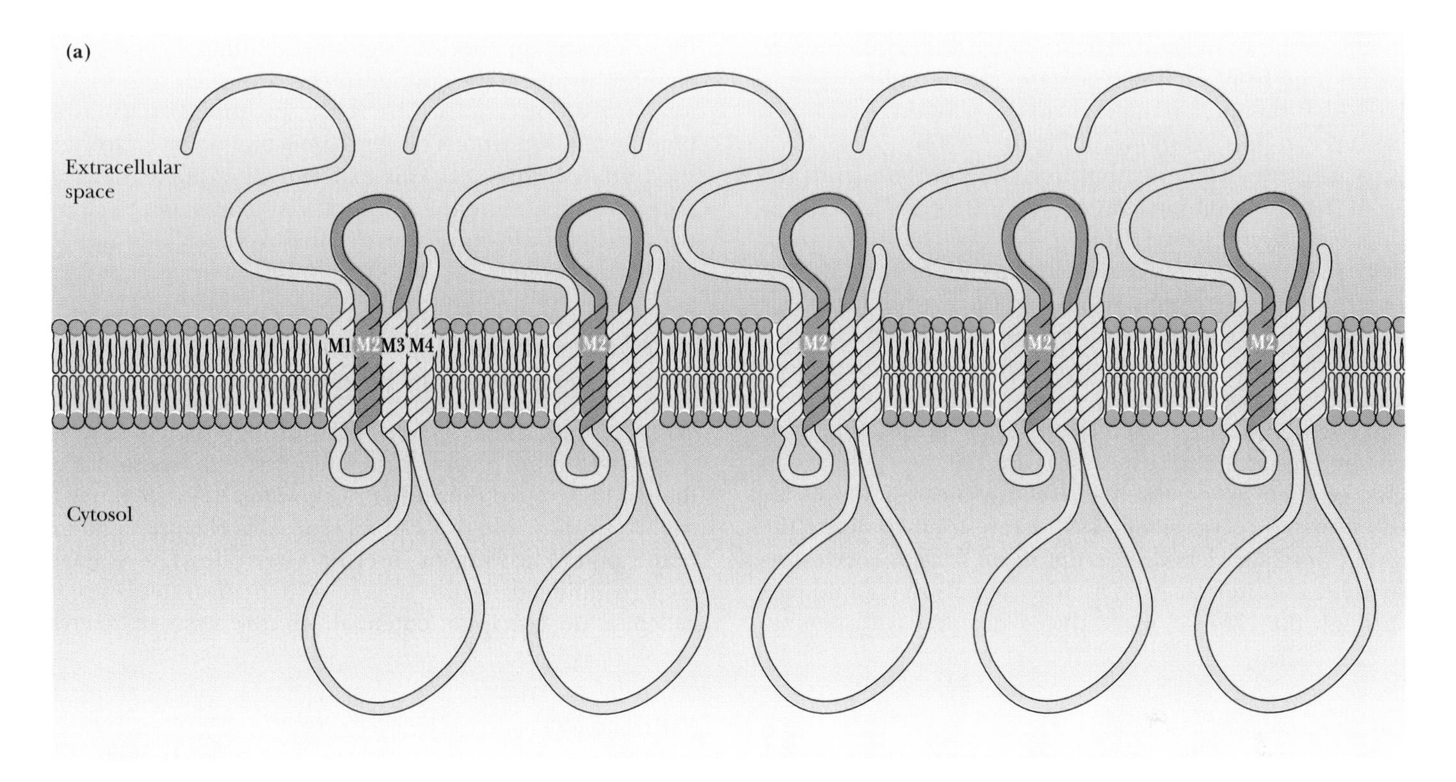

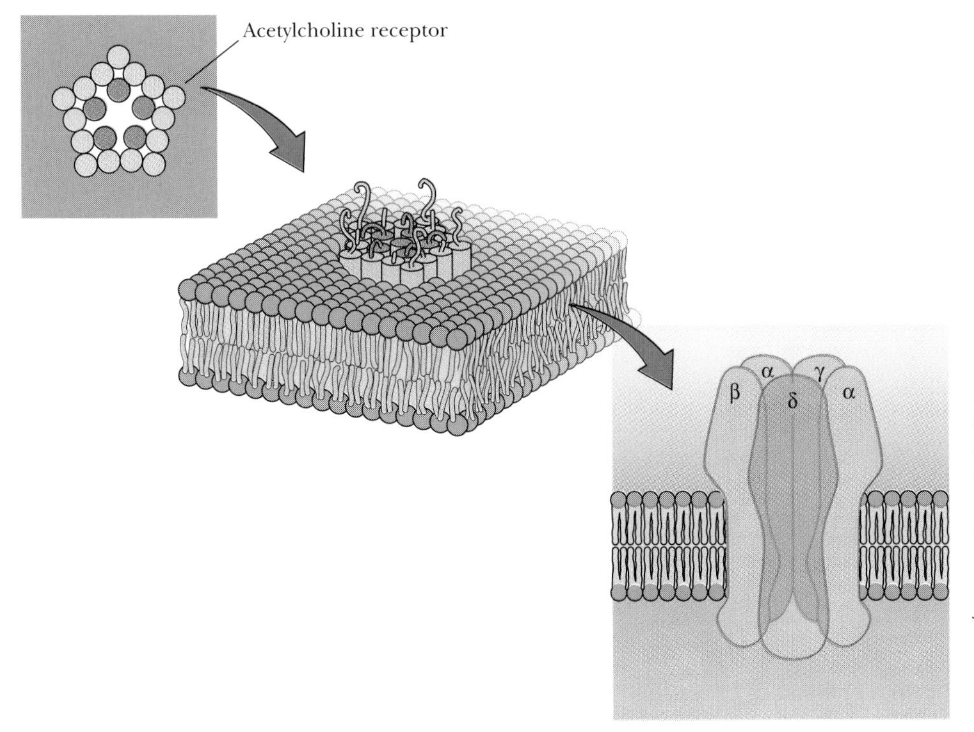

Figure 23-20 Model of the acetylcholine receptor. **(a)** Each subunit contains four transmembrane segments; **(b)** the second transmembrane segment (M2) of each subunit lines the wall of an ion channel that, when open, allows the passage of Na^+ and K^+ ions; changing just three amino acid residues in M2 changes the nature of the channel, so that it now allows the passage of anions like Cl^-, rather than cations like Na^+.

tides: two α chains (each with a molecular size of about 40,000), and one each of three other chains, called β, γ, and δ (with molecular sizes of 49,000, 57,000, and 65,000). The four polypeptides are recognizably similar to one another and almost certainly evolved from a common ancestry.

Each polypeptide of the neuromuscular junction's acetylcholine receptor contains four transmembrane segments, as shown in Figure 23-20a. Five transmembrane segments, one from each polypeptide, are thought to assemble within the membrane to form the pore through which cations can pass, as shown in Figure 23-20b.

The five segments that line the pore are from the second transmembrane segment (M2) of each polypeptide. Negatively charged amino acid side chains (of glutamate or aspartate residues) lie at the ends of each M2 segment. These side chains are thought to form a negatively charged ring at the top and bottom of the ion channel, easing the entry of the positively charged ions that move through the pore.

Researchers tested the role of these negatively charged residues by using site-directed mutagenesis to change them to positively charged lysine residues. After the mutated recombinant DNAs were used to direct the synthesis of acetylcholine receptors in *Xenopus* oocytes, researchers showed that the number of ions that could pass through the channel was reduced during each opening. This experiment supported the hypothesis that the negatively charged side chains are important in determining the properties of the channel.

An even more dramatic demonstration of the importance of the M2 transmembrane segments came from an experiment in which researchers used site-directed mutagenesis to change three residues of the M2 segment of the acetycholine receptor. They changed these residues to those present in the M2 region of another ligand-gated channel protein, the glycine receptor, which admits anions instead of cations. The result was that acetylcholine binding allowed anions rather than cations to pass through the mutated channel. This experiment showed that the part of the acetylcholine receptor that makes up the channel is different from the part of the protein that binds to neurotransmitter.

Neurotransmitters May Produce Either Excitatory or Inhibitory Effects in Postsynaptic Neurons

When cations, like Na^+ ions, pass through the channel of the acetylcholine receptor, they depolarize the membrane; that is, they make the inside of the membrane less negative. Neurophysiologists can measure this change in membrane potential with an inserted electrode. The voltage change induced by the action of a neurotransmitter is called a **postsynaptic potential.** In the case of acetyl-

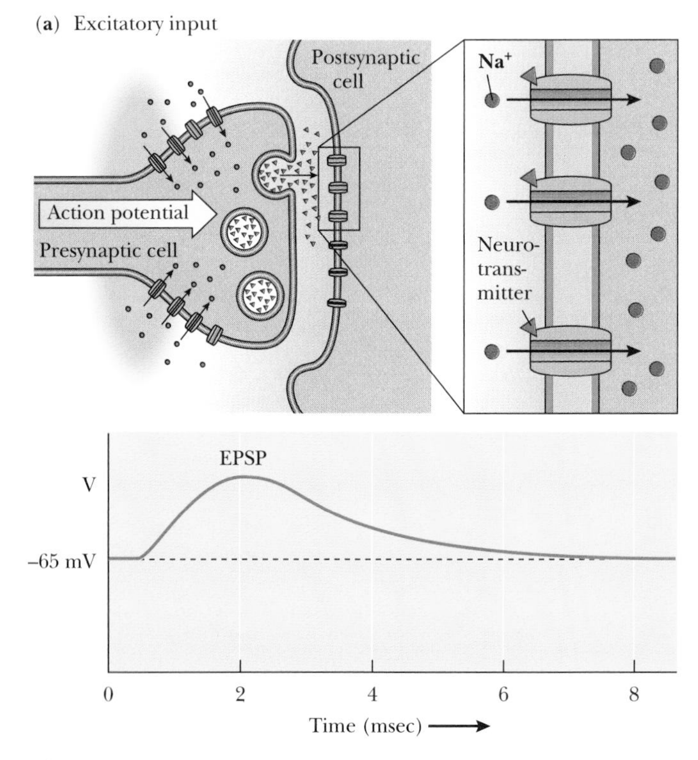

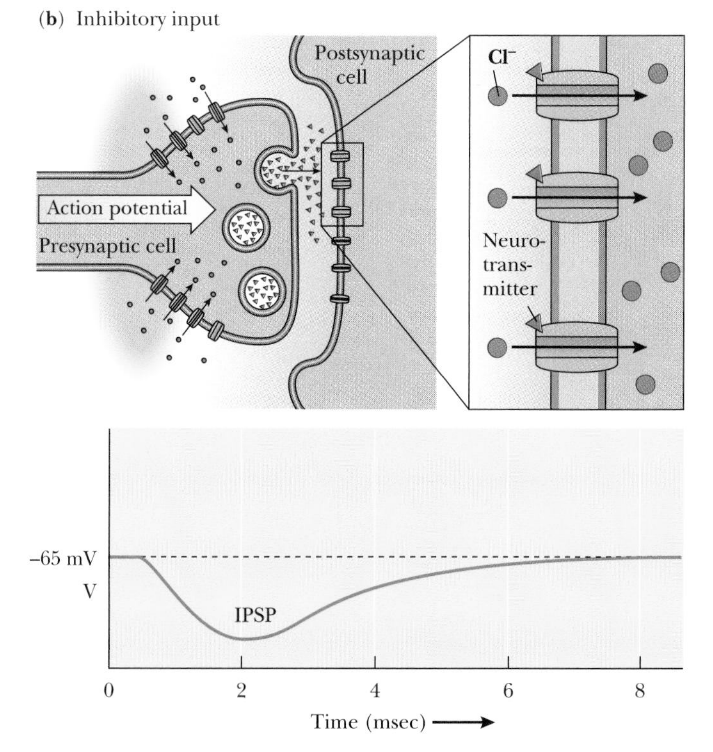

Figure 23-21 A postsynaptic neuron can integrate excitatory and inhibitory inputs. **(a)** An excitatory presynaptic cell evokes an excitatory postsynaptic potential (EPSP); **(b)** an inhibitory presynaptic cell evokes an inhibitory postsynaptic potential (IPSP); because the anion channels opened by the inhibitory signal short circuit the cation channels opened by the excitatory signal, the postsynaptic neuron's response depends on the temporal and spatial relationship between all its presynaptic inputs.

choline, the postsynaptic potential excites the cell and is therefore called an **excitatory postsynaptic potential** (or **EPSP**). (See Figure 23-21a.)

If the EPSP is sufficiently depolarizing, the postsynaptic cell fires an action potential. When the postsynaptic cell is a muscle cell, the action potential triggers contraction, as discussed in Chapter 18. When the postsynaptic cell is a neuron, an induced action potential can then travel through the cell body and axons to reach the next synapse in a neural pathway.

Whether the postsynaptic neuron reaches the threshold for firing often depends not just on one synapse, however. Many presynaptic neurons can converge on a single postsynaptic neuron. Some may be excitatory, while others may be inhibitory. Inhibitory neurotransmitters often increase the negative charge on the inside of a membrane. These neurotransmitters trigger an **inhibitory postsynaptic potential** (or **IPSP**), illustrated in Figure 23-21b.

The activity of a postsynaptic cell depends on the summing of the effects of EPSPs and IPSPs. A single postsynaptic neuron may intergrate inputs from as many as 100,000 different presynaptic cells. Each neuron effectively serves as a microcomputer that determines whether to fire or not to fire according to its inputs, its history, and its computational program. (See Essay, "How Does a Doubly Gated Channel Contribute to Memory?")

The inhibitory neurotransmitter most widely used in the brain is **γ-aminobutyric acid (GABA),** whose structure is shown in Figure 23-22b. About 20% of all neurons in the brain make GABA, and virtually every neuron can respond to GABA. Most of these responses depend on the binding of GABA to the **$GABA_A$ receptor,** a ligand-gated chloride channel that, like the acetylcholine receptor of the neuromuscular junction, consists of five related polypeptide chains.

Binding of GABA to a $GABA_A$ receptor opens a channel that admits Cl^- ions through a channel that is probably surrounded by M2 transmembrane segments from the receptor's five polypeptides. The opening of the chloride channel usually results in the inward movement of Cl^- ions, hyperpolarizing the cell (that is, making the inside of the cell more negative). Because opening a chloride channel also allows negatively charged Cl^- ions to flow inward at the same time as positively charged Na^+ ions may flow through voltage-gated channels, the $GABA_A$ receptor also "short-circuits" the inward current carried by Na^+ ions and decreases the chance that the postsynaptic cell will fire an action potential.

Neurons May Respond to the Same Transmitters in Different Ways

By the mid-1970s, researchers had discovered and studied a handful of neurotransmitters. Some of these, like acetylcholine, were excitatory, while others were inhibitory. By studying the action of these neurotransmitters, neurobiologists came to realize that the same transmitter could have different effects in different postsynaptic neurons.

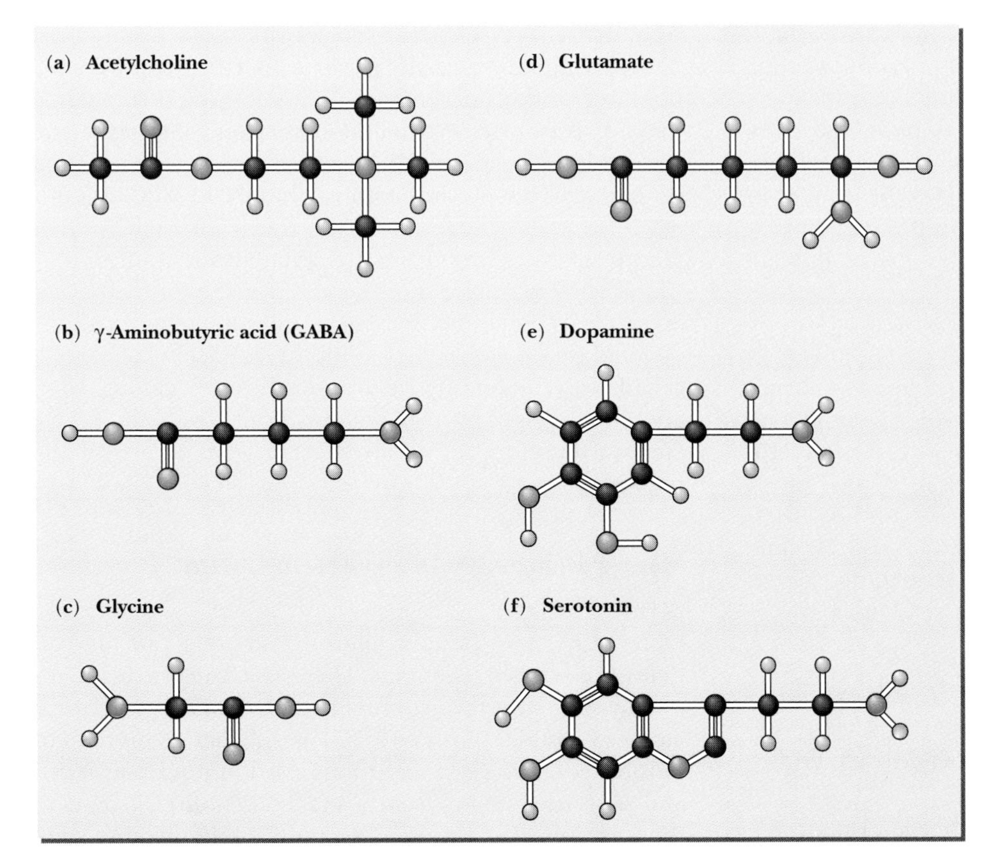

Figure 23-22 Some small molecules that serve as neurotransmitters. **(a)** Acetylcholine; **(b)** GABA (γ-aminobutyric acid); **(c)** glycine; **(d)** glutamate; **(e)** dopamine; **(f)** serotonin (5-hydroxytryptamine).

These differences result from the existence of more than one type of receptor, each with distinct responses to neurotransmitters and other compounds. For example, the compound nicotine, found in tobacco, mimics the effects of acetylcholine in skeletal muscle and in some neurons. On the other hand, muscarine, a poison found in certain mushrooms, stimulates heart muscle and other neurons that respond to acetylcholine but not to nicotine.

So acetylcholine receptors come in two varieties, called nicotinic and muscarinic receptors. The **nicotinic acetylcholine receptor,** found in the neuromuscular junction and elsewhere as well, is a ligand-gated channel. In contrast, the **muscarinic acetylcholine receptor** responds to acetylcholine by inhibiting the synthesis of cyclic AMP (cAMP); its action resembles that of membrane-bound receptors for epinephrine, particularly that of the α_2-adrenergic receptor discussed in Chapter 19 (p. 528).

The decrease in cAMP, triggered by the muscarinic acetylcholine receptor, leads to a decrease in the activity of cAMP-dependent protein kinase, as also discussed in Chapter 19. The result is that cells with muscarinic receptors respond to acetylcholine differently from those with nicotinic receptors. When smooth muscle and cardiac muscle, for example, receive acetylcholine from a neuron, they are less likely to produce an action potential and to contract. Acetylcholine reduces the heart rate and slows the peristaltic contractions of the intestines. These changes occur much more slowly than the altered ion flows produced by ligand-gated channels. The response to acetylcholine via the muscarinic acetylcholine receptor may take seconds instead of milliseconds.

In the last 15 years, researchers have discovered a large number of compounds that can trigger postsynaptic changes. These include other amino acids and their derivatives, as well as many polypeptides. Some of these are certainly neurotransmitters, while others are only suspected. For example, a compound may be found in synapses, and it may trigger responses in the postsynaptic cell, but the appearance of the compound may not reflect the activity of the presynaptic cell.

The receptors for all the identified neurotransmitters act in one of two ways—they are either ligand-gated ion channels or they affect the production of a second messenger. The second messenger may be cAMP or one of the membrane-lipid second messengers (diacylglycerol or phosphotidyl inositol). Neurotransmitters that affect cAMP levels may act either to increase or to inhibit the synthesis of cAMP. In sum, neurotransmitters use the same kind of mechanisms to exert their effects as do the hormones and other signaling molecules discussed in Chapter 19.

The Action of Neurotransmitters Depends on Synthetic Enzymes, Releasing Mechanisms, Postsynaptic Receptors, Uptake Mechanisms, and Degradative Enzymes

We can now see what is needed to understand the influence of a neuron on postsynaptic neurons. We want to know both what controls the synthesis and secretion of neurotransmitters in the presynaptic cell, how the binding of the neurotransmitter to its receptor triggers changes in the postsynaptic membrane, and what terminates the action of the neurotransmitter. For example, two related enzymes, both called **glutamic acid decarboxylase (GAD),** catalyze the formation of GABA from glutamic acid, as illustrated in Figure 23-23a. The two GADs differ slightly in size and amino acid sequence and are independently regulated and separately targeted to axon terminals. Most of the GABA produced in a neuron moves into synaptic vesicles, as illustrated in Figure 23-23b. The movement into vesicles depends on a **vesicular transporter,** a protein within the vesicular membrane that couples neurotransmitter accumulation to the energy stored in the vesicle's pH gradient. The accumulated vesicular neurotransmitter is released by exocytosis.

Recent biochemical and molecular biological experiments have begun to identify the molecules and the mechanisms of Ca^{2+}-dependent vesicular release. The biggest surprise in this work has been the discovery that the proteins involved in transmitter release from mammalian neurons are virtually identical to proteins involved in the trafficking of vesicles in other types of cells, for example in the transport of vesicles containing secreted proteins from the rough endoplasmic reticulum complex to the Golgi complex. Figure 23-24 shows a current model for neurotransmitter release.

Finally, as in the case of the signaling molecules discussed in Chapter 19, neurotransmitters must be cleared from the synaptic cleft to make way for the next set of signals. Termination usually involves specific **reuptake,** the pumping of transmitter from the synaptic cleft either into the presynaptic neuron or into surrounding glial cells. The recovered transmitter is then either recycled into vesicles or degraded by specific enzymes. In the case of GABA, for example, an enzyme called GABA transaminase (shown in Figure 23-23), present in GABA neurons and in surrounding glial cells, catalyzes the conversion of GABA into an inactive compound that eventually enters the citric acid cycle. In the case of acetylcholine, an extracellular enzyme called **acetylcholinesterase** destroys the acetylcholine soon after it is released into the synapse. Both GABA and acetylcholine therefore have only short periods to act on the postsynaptic cell.

Many Psychoactive Drugs Act on Chemical Synapses

Other molecules besides neurotransmitters can bind to neurotransmitter receptors. These substances can either mimic or disrupt the action of a neurotransmitter. For example, nicotine (the neurobiologically active ingredient in tobacco), curare (a toxin long used in poison arrows), and α-bungarotoxin (from a snake's venom) all bind to the nicotinic acetylcholine receptor. While nicotine stim-

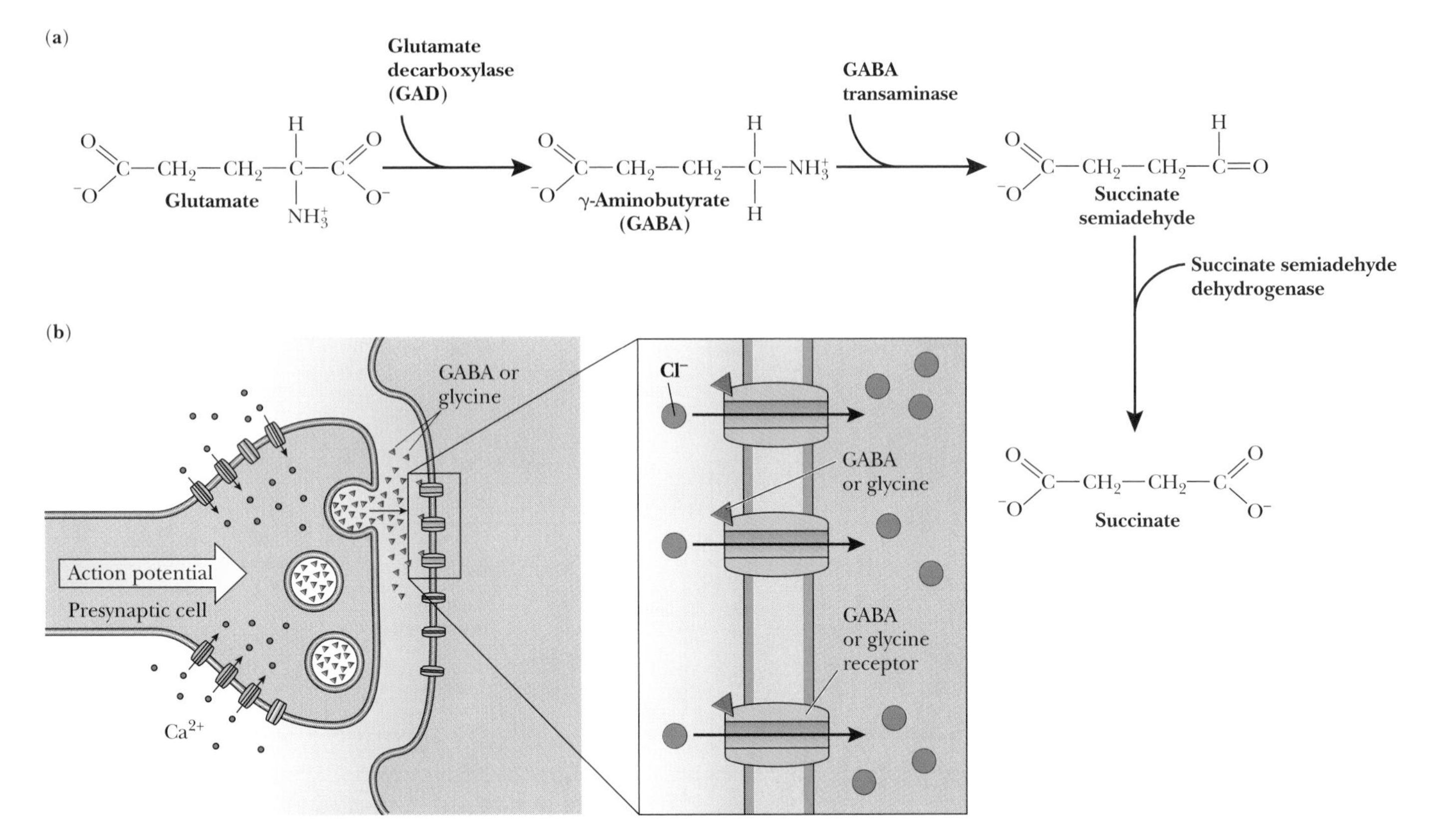

Figure 23-23 GABA is the most prominent inhibitory neurotransmitter in the brain. **(a)** Production of GABA from glutamate, catalyzed by glutamic acid decarboxylase (GAD); **(b)** like other neurotransmitters, GABA is concentrated into synaptic vesicles and released by exocytosis following the entry of Ca^{2+} into the axon terminal.

ulates action potentials, however, the others block the receptor. Similarly, atropine, from the belladonna plant (whose Latin name is *Atropa*), inhibits the muscarinic acetylcholine receptor.

Many compounds that bind to neurotransmitter receptors have proved to be useful medications. For 2000 years, for example, physicians prescribed belladonna extracts to combat diarrhea. We now know that the extracts worked because the atropine blocks the effects of acetylcholine in the smooth muscles of the intestines.

Studying the noxious effects of drugs can provide insights into the brain and its diseases. Too much belladonna, for example, is deadly, giving the belladonna genus its name *(Atropa),* after the Greek goddess Atropus, who cut the thread of life after it had been spun by the Fates. Before a victim of atropine poisoning succumbs, physicians note a loss of memory and general disorientation—symptoms similar to those of **Alzheimer's disease,** a disease usually associated with aging, in which patients suffer massive memory loss. This resemblance stimulated modern researchers to examine the state of acetylcholine neurons in the brains of people who have died with Alzheimer's disease. Remarkably, Alzheimer's patients appear to have a massive degeneration of a part of the brain that is rich in acetylcholine and in muscarinic acetylcholine receptors (See Essay, "Gene-Based Therapies for Neurodegenerative Diseases.")

Other naturally occurring or chemically synthesized compounds affect other neurotransmitter receptors. For example, **benzodiazepines,** commonly prescribed antianxiety drugs, bind to $GABA_A$ receptors and increase the effectiveness of GABA in opening the anion channel.

Still other substances influence the release, reuptake, or inactivation of specific neurotransmitters. Some of these are effective as pain killers, others are tranquilizers, and still others are stimulants. Some drugs bring about mood changes or visions similar to those of psychosis; other drugs provide a means of regulating depression and schizophrenia. The widely used antidepressant Prozac, for example, inhibits the reuptake of serotonin.

Drugs that act on the brain have had enormous social consequences as well. Morphine and heroin, long used as pain killers, have become major people killers, destroying life both directly (by acting on the brain and body) and indirectly (by stimulating crime). Similarly, abuse of cocaine and its derivative "crack" has become widespread. Many people feel that the abuse of these and other neuroactive drugs represents the major social problem of the 1990s.

In spite of the social seriousness of drug abuse, re-

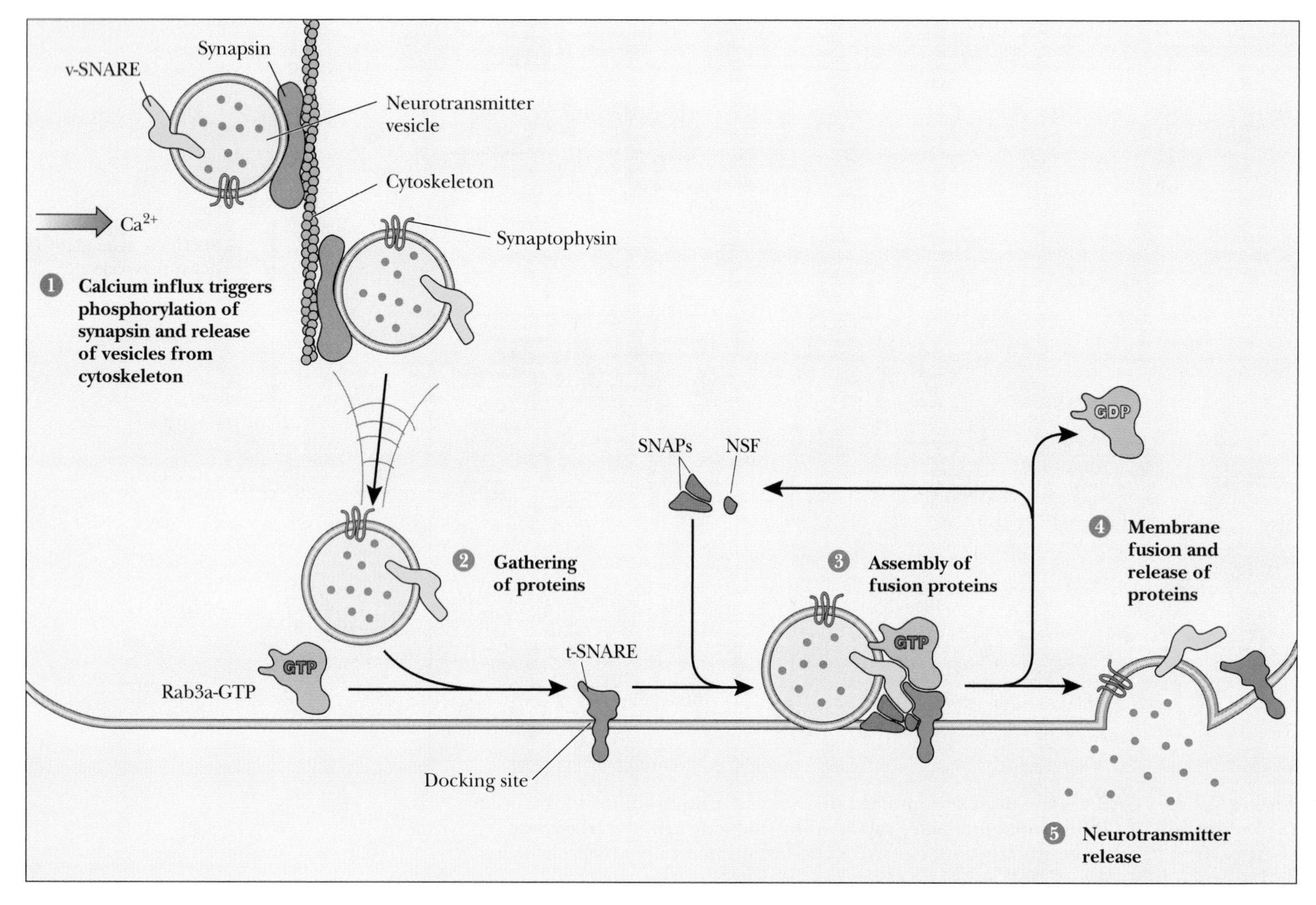

Figure 23-24 Neurotransmitter release depends on at least four sets of proteins, many of which also participate in vesicle fusion in other types of cells. Synapsins link the vesicle membrane to the cytoskeleton, stabilizing the vesicles in the absence of a signal; an increase in Ca^{2+}, for example during an action potential, triggers phosphorylation of synapsins and the release of vesicles from the cytoskeleton. Soluble proteins, called NSF and SNAP, link proteins on the surface of the vesicles with proteins at the docking site on the surface of the plasma membrane. Proteins on the surface of the vesicles include v-SNARE, the vesicular receptor for SNAP (consisting of synaptobrevin and synaptotagmin) and another transmembrane protein called synaptophysin; proteins at the docking site on the plasma membrane include t-SNARE, the target receptor for SNAP (consisting of SNAP25, syntaxin, and neurexin); Rab3a is also involved in the docking of the vesicles and in the trafficking of vesicles within the cell.

search into the way that neuroactive drugs work has already led to important basic discoveries about the brain. One of the most startling of these discoveries occurred in the late 1970s, as researchers tried to understand why **opiates**—morphine, heroin, and related compounds—are such powerful drugs. The result of this research was the realization that the nervous system communicates information about pain using neurotransmitters that were previously unknown.

The route to this insight was as follows: (1) If opiates act powerfully on the brain, they must bind to specific receptor molecules; (2) the binding of radioactively labeled opiates to the brain should reveal the distribution of opiate receptors; (3) the specific distribution and properties of these receptors suggest that they must ordinarily bind to some endogenous brain compound, as well as to the extracts of poppies and the products of chemists. In the 1970s, several research groups identified three classes of such natural opiates, called the **enkephalins,** the **endorphins,** and the **dynorphins**—polypeptides that act as natural pain suppressors. (See Figure 23-25). The natural opiates also appear to be involved in the regulation of mood.

In the early 1990s, molecular biologists, after more than a decade of trying, were able to identify recombinant DNAs that specified the structures of opiate receptors, as illustrated in Figure 23-25b. Understanding how these receptors work may allow us to understand the neural basis of pain and to develop ways of lessening it.

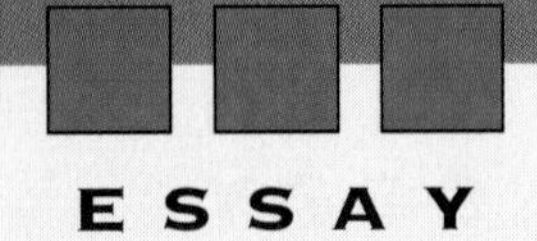

ESSAY

GENE-BASED THERAPIES FOR NEURODEGENERATIVE DISEASES

Alzheimer's disease involves a tragic loss of memory and cognitive ability. Because of the medical and social importance of this disease, which affects millions of people in the United States alone, neuroscientists have expended much effort to define what goes wrong in the brain. The first approach has been to examine the brains of patients who have died of Alzheimer's disease, to see if their brains show any characteristic pathology. Among the most prominent changes in the brain are the presence of extracellular depositions, called *amyloid plaques,* seen in Figure 1a, and of intracellular depositions, called *neurofibrillary tangles,* which consist of certain cytoskeletal proteins (Figure 1b).

Besides the plaques and tangles, neuropathologists have also shown that Alzheimer's disease is a *neurodegenerative disease,* a disorder in which specific nerve cells die prematurely. In Alzheimer's disease the neurons that die most prominently lie in a region called the basal forebrain, as shown in Figure 1c. Among the most characteristic properties of the dying neurons are their use of acetylcholine as a neurotransmitter. Somewhat surprisingly, these cells also possess receptors for *nerve growth factor (NGF),* a protein long implicated in the survival of sensory neurons and sympathetic neurons in the peripheral nervous system.

NGF is one of a family of proteins, called the *neurotrophins,* which sustain the survival and development of specific types of neurons in tissue culture and in intact animals. Neurotrophins act by binding to receptors that lie on the surfaces of particular sets of neurons. The neurotrophin receptors are tyrosine kinases, and many are related to an oncogene called *trk,* to whose protein NGF binds. The immediate effect of neurotrophin binding is the phosphorylation of target proteins within the plasma membrane; the long-term effect is to promote cell survival. In cultures of neurons, withdrawal of specific neurotrophins often leads to cell death by apoptosis.

The role of neurotrophins in keeping neurons alive suggests a way of retarding cell death in neu-

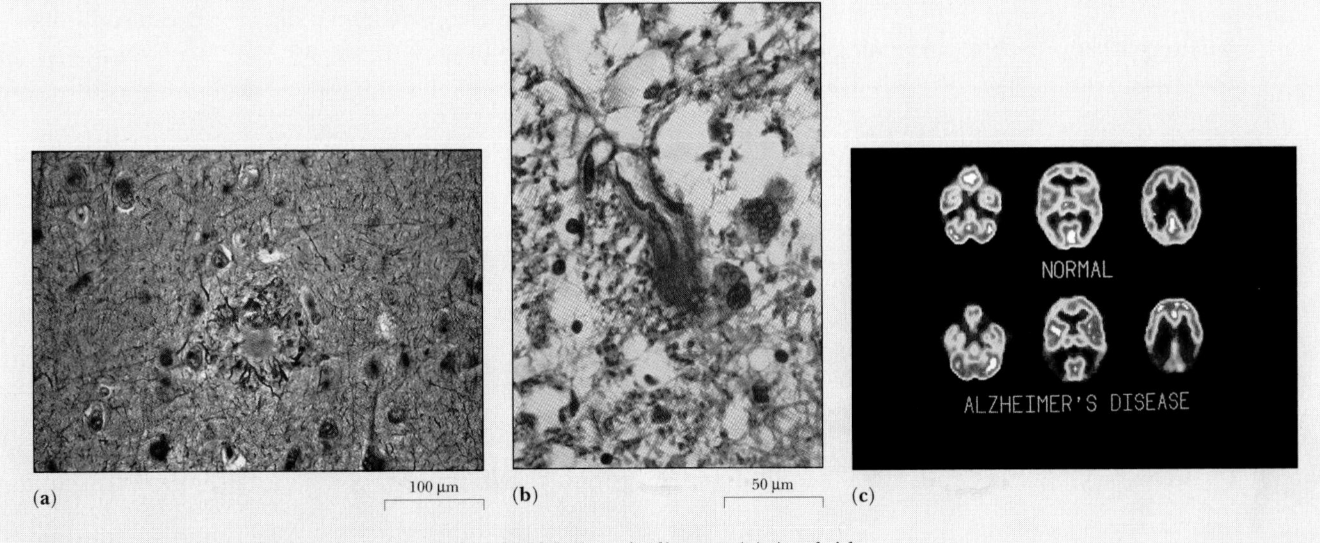

Figure 1 Pathological changes associated with Alzheimer's disease. **(a)** Amyloid plaques, consisting of extracellular depositions of polypeptides derived from a protein called amyloid precursor protein; **(b)** neurofibrillary tangles, intracellular depositions of cytoskeletal proteins; **(c)** cell loss in the basal forebrain revealed by positron emission tomography (PET). *(a, Martin M. Rotker/Science Source/Photo Researchers; b, G. Musil/Visuals Unlimited; c, Science VU/Visuals Unlimited)*

(continued)

ESSAY *continued*

rodegenerative diseases—by supplying extra neurotrophin in the area of cell loss. To evaluate this therapeutic approach to Alzheimer's disease, however, researchers have first had to develop a model of the disease in an experimental animal. Several groups have produced transgenic mice that overexpress the proteins present in plaques (amyloid peptides) or in tangles (the cytoskeletal protein, tau), but researchers do not yet know how well these genetic models mimic the pattern of neuron loss in Alzheimer's disease.

Another widely used model of Alzheimer's disease produces Alzheimer-like cell loss by making a surgical cut of a structure in the rat brain called the fimbria fornix. After the surgery, acetylcholine-producing neurons begin to degenerate in just the region that corresponds to the basal forebrain in humans. With this model in hand, researchers can evaluate possible therapeutic interventions. Does anything prevent or slow neuronal destruction?

Two therapies seem promising—at least for rats. The first is the direct application of NGF, produced from recombinant DNA using methods such as those described in Chapter 17. In one experiment, researchers injected NGF into the fluid-filled ventricles of the rat brain. Enough NGF diffuses from the ventricles into the brain itself that some of the cell loss is prevented.

The second approach to preventing cell loss has been to produce NGF locally. Researchers have taken cultured rat fibroblasts (skin cells) and programmed them with appropriate recombinant DNAs to make and secrete NGF. They then transplant the engineered fibroblasts into the region of the damage. The transplanted cells produce NGF and slow the degeneration that would otherwise occur after the surgical cut.

In the last few years, neuroscientists and virologists have begun to use another approach to neural cell engineering—the use of modified animal viruses to carry genes into cells that are already part of a functioning brain. Neural engineering—either of cultured cells for transplantation or of cells already in the brain—now represents a major research effort.

In addition to producing cells that make trophic factors, researchers have made cells that produce acetylcholine, GABA, and dopamine. These cells, transplanted into the proper place, could eventually help replace lost neuronal function. The ultimate—and still far-off—goal will be to engineer cells also to make the proper synaptic connections. In the meantime, neuroscientists are using engineered cells to investigate the roles of neurotrophins and neurotransmitters in the function and maintenance of neural circuits. The hope is that such engineering experiments will lead to treatments for neurodegenerative diseases such as Alzheimer's disease, Parkinson's disease, and Huntington's disease.

(a)

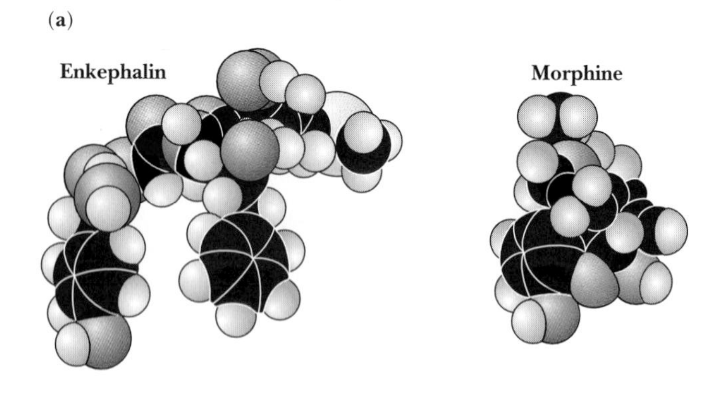

(b) **Structure of enkephalin receptor**

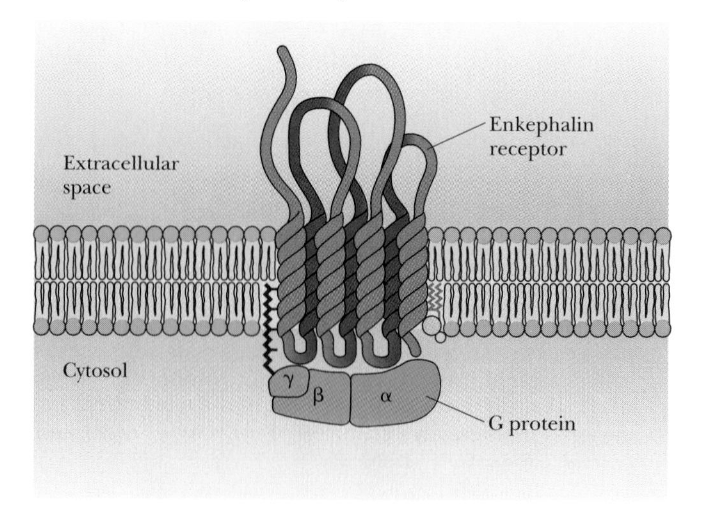

Figure 23-25 The opiates within the brain are polypeptides, while morphine is a small organic compound. Yet morphine mimics the peptides, relieving pain and inducing sleep. **(a)** Similarities in the three-dimensional structures of morphine and enkephalin; **(b)** the structure of an enkephalin receptor.

HOW DOES EXPERIENCE MODIFY NEURONAL CIRCUITS?

Many neural networks are capable of **learning,** modification of neural activity and behavior as the result of experience. We humans are exquisitely aware of our learning, for example as we see a child acquire ever more sophisticated language skills.

Learning researchers have distinguished and studied two types of learning—nonassociative learning and associative learning. In **nonassociative learning,** the subject changes sensitivity to a stimulus after repeated exposure. **Habituation** refers to the decrease in a behavioral response following repeated exposure to a harmless stimulus: Responses to a loud but not deafening noise, for example, decrease as the noise is repeated. **Sensitization** refers to the increased behavioral response to a noxious stimulus: If a painfully loud sound is repeated, the subject is more likely to move away after repetition.

Associative learning refers to the pairing of stimuli, so that the subject learns to respond to a stimulus not obviously related to the primary stimulus, called the **conditioned stimulus,** because it is associated with the **unconditioned stimulus,** which leads to an appropriate physiological response. The classic example of associative learning comes from the work of the Russian physiologist Pavlov, who trained dogs to salivate at the sound of a bell. Pavlov's trick was first to ring the bell (the conditioned stimulus) upon presenting the dog with dinner (the unconditioned stimulus). The dog would salivate at the sight and smell of the meat, which was accompanied by the ringing of the bell. Gradually the dog salivated at the sound of the bell alone.

Experience Can Modify the Gill Withdrawal Reflex of the Sea Hare *Aplysia*

Experimenters have studied both nonassociative and associative learning in the "sea hare" *Aplysia,* a large, shell-less aquatic snail, shown in Figure 23-26a. The advantage of studying *Aplysia* is the simplicity of the nervous system, which consists of eight paired and one unpaired ganglia. Each ganglion consists of fewer than 2000 neurons whose large sizes (up to 1 mm in diameter) and invariant positions allow their ready identification for electrophysiological recording.

The most studied behavior in *Aplysia* is the withdrawal of the gill in response to stimuli that range from a

(Text continues on page 670.)

(a)

(b) **Gill withdrawal reflex**

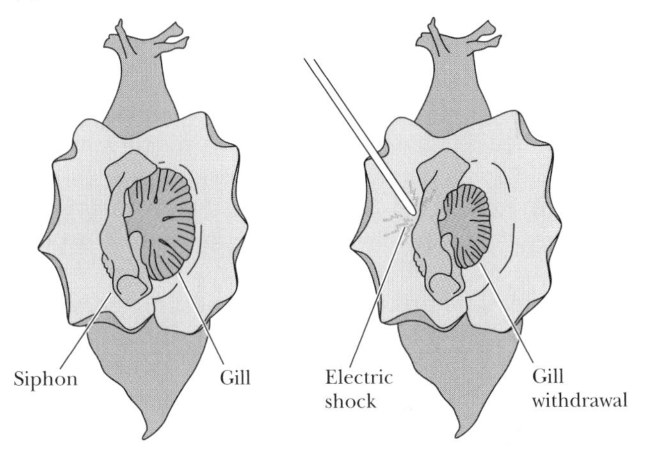

(c) **L7 neuron in abdominal ganglion**

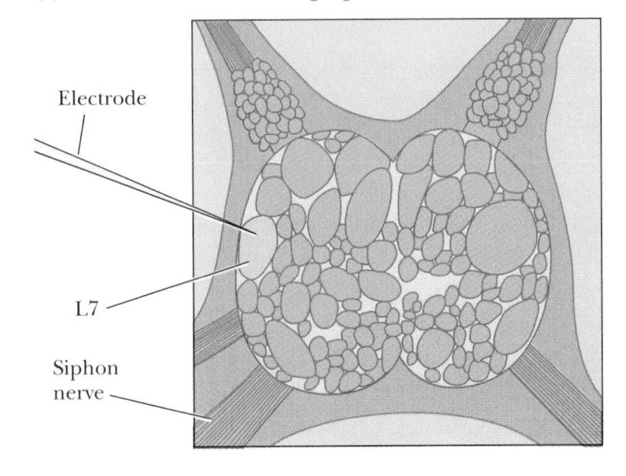

(d) **Circuit diagram**

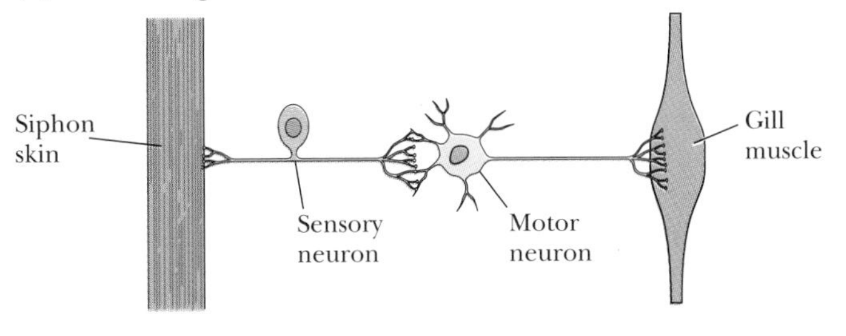

Figure 23-26 The marine snail (or "sea hare") *Aplysia* has a simple nervous system that is nonetheless able to learn. **(a)** *Aplysia;* **(b)** gill withdrawal, a reflex response to an electric shock; **(c)** an L7 neuron in the abdominal ganglion; **(d)** a circuit diagram of the gill withdrawal reflex. *(a, Dan Gotshall/Visuals Unlimited)*

ESSAY

HOW DOES A DOUBLY GATED CHANNEL CONTRIBUTE TO MEMORY?

One of the most studied structures in the mammalian brain is the hippocampus, whose placement and organization in the rat brain are shown in Figure 1. The hippocampus has an important role in the formation of long-term memory and is also an important site in the generation and regulation of epileptic seizures. One of the most forceful illustrations of the role of the hippocampus came from a patient, called H. M., whose hippocampus was removed to stop persistent and otherwise untreatable seizures. After his operation, H. M. could still speak and understand language; he maintained memories from before his surgery, and he could remember for seconds or minutes. But he lost the capacity to form new long-term memories. Studies of H. M., of other patients, and of experimental animals all point to the hippocampus as a crossroads of short-term and long-term memory.

The cells of the hippocampus form a relatively simple neural circuit, as illustrated in Figure 1b. Input comes through the *perforant fiber pathway* to the *dentate region,* where the fibers form synapses on the dendrites of large cells called *granule cells.* The granule cells send axons, called *mossy fibers,* to a set of large cells, called *pyramidal neurons,* in a region called CA3. CA3 pyramidal neurons send information along axons called *Schaffer collaterals* to another set of pyramidal neurons in the CA1 region. Neuroscientists have performed many studies of the relationship between the input and output of each set of neurons in this pathway. These studies have been particularly convenient because the three sets of neurons and their axons all lie roughly within the same plane, so researchers can study input-output relationships not only in intact animals but also in isolated slices of hippocampus in a culture dish.

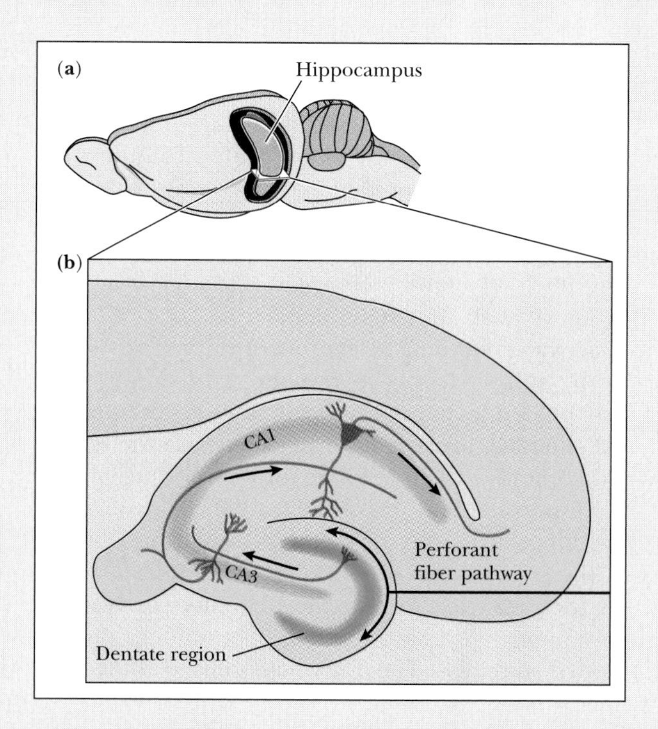

Figure 1 The hippocampus, a C-shaped structure on each side of the forebrain, plays a crucial role in converting short-term memory into long-term memory. **(a)** The position of the hippocampus in a rat brain; the front end of the brain is on the left, and the posterior end and spinal cord are on the right; **(b)** a cross section of the hippocampus, showing the pathway by which neural input moves into the dentate region, from the dentate region to CA3 and from CA3 to CA1.

Figure 2a shows a typical EPSP in a CA1 pyramidal cell, after a single stimulus of an axon within the Schaffer collaterals. Remarkably, however, the size of the EPSP in CA1 neurons increases dramatically if the Schaffer collateral is first subjected to a brief, high-frequency train of impulses, say 100 stimuli each 10 msec long. After just 1 or 2 seconds of such treatment, the increased response of the CA1 cell is changed for days or even weeks. Because of the persistence of this response, neurobiologists refer to this effect as *long-term potentiation (LTP).* Understanding the cellular and molecular basis of LTP has been a major challenge for the last decade.

The starting point for any understanding of this circuit is the synaptic signals and their receptors. The CA3 cells whose axons are stimulated use glutamate as a neurotransmitter. Glutamate in turn can act on at least three different types of receptors—a *metabotropic receptor,* which activates second messengers, and at least two types of *ionotropic receptors,* which are ligand-gated ion channels.

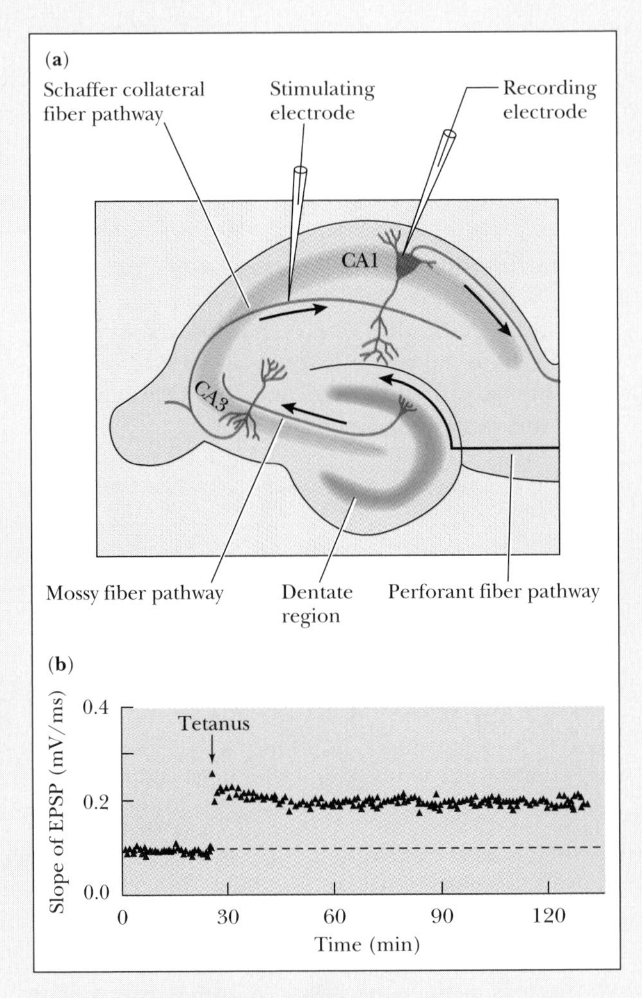

Figure 2 Long-term potentiation (LTP): A train of pulses, just 1 second long, changes the response of postsynaptic neurons. The rate of rise of an excitatory postsynaptic potential (EPSP), recorded from a CA1 neuron after a single stimulation of a CA1 axon, is about 0.1 mV per msec. Immediately after a train of pulses, the EPSP rises about twice as rapidly. **(a)** Cross section of a rat hippocampus, showing the placement of recording and stimulating electrodes; **(b)** the rate of rise of the EPSP before and after a train of pulses (tetanus).

The metabotropic glutamate receptor responds to glutamate by stimulating the activity of phospholipase C, leading to the formation of inositol trisphosphate (IP3) and diacylglycerol (DAG). (See Chapter 19) These second messengers then contribute to the regulation of ion channels in the membrane. The period over which metabotropic receptors regulate ion flow is generally longer than the period over which ligand-gated receptors are regulated.

The ionotropic glutamate receptors respond to glutamate by opening ion channels. Ionotropic glutamate receptors are particularly diverse. Like the ionotropic receptors for GABA and acetylcholine, each receptor consists of multiple polypeptides, probably five per receptor, and each polypeptide may derive from one of many members of a gene family. Different combinations of polypeptides produce channels with different properties.

While there are many types of ionotropic glutamate receptors, the major distinction among them is between receptors that most actively bind an artificial ligand called N-methyl-D-aspartate (NMDA) and those that do not bind NMDA. NMDA-type glutamate receptors have properties (and structures) that distinguish them from other types of ionotropic receptors, while the non-NMDA glutamate receptors are more or less conventional ligand-gated channels. Both types of receptor are gated channels that allow Na^+ and K^+ to move through the membrane. The NMDA receptor (and some non-NMDA receptors) also admit Ca^{2+} ions under some conditions.

The NMDA-type glutamate receptor has several unusual features, which allow it to play a unique role in LTP and other processes: (1) It admits Ca^{2+} as well as Na^+ and K^+; (2) the channel opens only in the presence of both glutamate and glycine; and (3) at the cell's resting potential, the ion channel is blocked by an Mg^{2+} ion; the channel opens only when the neuron is depolarized and when glutamate and glycine are bound. Because of this last property, the NMDA-type glutamate receptor is called a *doubly gated receptor,* meaning that channel opening depends on both voltage and ligand binding.

Most neurobiologists believe that the key to understanding LTP is the ability of NMDA-type glutamate receptors to admit, into already depolarized hippocampal neurons, Ca^{2+} ions, which can then serve as a second messenger. Even before all the properties of NMDA receptors were known, researchers knew that

(continued)

ESSAY *continued*

Ca^{2+} ions were important in LTP because they could prevent LTP by preventing Ca^{2+} influx; they could also initiate LTP by injecting Ca^{2+} ions into hippocampal neurons. Proof that NMDA receptors are important came from an experiment in which researchers specifically blocked NMDA receptors: The result was the prevention of LTP.

Figure 3 shows a recent formulation of LTP induction: (1) Glutamate released from CA3 neurons activates non-NMDA glutamate receptors, depolarizing the CA1 neurons; (2) depolarization of the CA3 neurons removes the Mg^{2+} block from the NMDA-type glutamate receptors, allowing Ca^{2+} to enter the cell; and (3) Ca^{2+} acts as a second messenger, stimulating long-term changes in the postsynaptic cell.

But even more is required to understand LTP because it involves changes in the presynaptic as well as the postsynaptic neuron. Analysis of the pattern of transmitter release from the CA3 neurons has suggested that LTP increases the amount of glutamate released in response to stimulation. Researchers now believe that this presynaptic effect depends on the action of a *retrograde messenger,* a signal from the postsynaptic cell that acts on the presynaptic cell, whose production is triggered by the Ca^{2+} influx. The identity and mode of action of the retrograde messengers are still matters of debate, but popular candidates include two gases, nitric oxide and carbon monoxide.

LTP has been a provocative model of how neurons can learn. Studies of LTP have raised questions not only about the role of the doubly gated NMDA receptor but also about the conversations between presynaptic and postsynaptic cells.

Figure 3 A model of LTP. **(a)** Glutamate released by a single stimulus acts mostly on metabotropic and non-NMDA type glutamate receptors; NMDA type glutamate receptors are blocked by Mg^{2+} ions; **(b)** depolarization caused by a train of impulses activates the NMDA-type glutamate receptor so that Ca^{2+} ions flow through the channel, activating a second messenger system and ultimately leading to the production of a retrograde messenger to the presynaptic neuron.

simple touch on the siphon to a sharp blow or electrical shock, as shown in Figure 23-26b. This behavior occurs as a reflex that can be modified by experience.

The neurons responsible for the gill withdrawal reflex lie in the abdominal ganglion, shown in Figure 23-26c. A sensory neuron from the siphon makes a monosynaptic connection to the motor neuron (called L7) that drives gill withdrawal, as shown schematically in Figure 23-26d. Another neuron, called a facilitating interneuron, synapses with the axon terminal of the sensory neuron, also illustrated in Figure 23-26d.

A single stimulation of the sensory neuron leads to the production of an action potential in L7. Repeated stimulation of the sensory neuron leads to depression, a decrease in the size of the evoked action potential in the postsynaptic neuron. The synaptic depression, which underlies the behavioral habituation, results from a reduction in **synaptic strength,** the response of the postsynaptic neuron to a single stimulation of the presynaptic neuron. The observed reduction in synaptic strength results from a decrease in the amount of released neurotransmitter, which in turn depends on a decrease in the number of Ca^{2+} ions that enter the axon terminal.

How Does an *Aplysia* Learn?

Stimulation of the facilitating interneuron reverses the synaptic depression. This synaptic facilitation underlies the

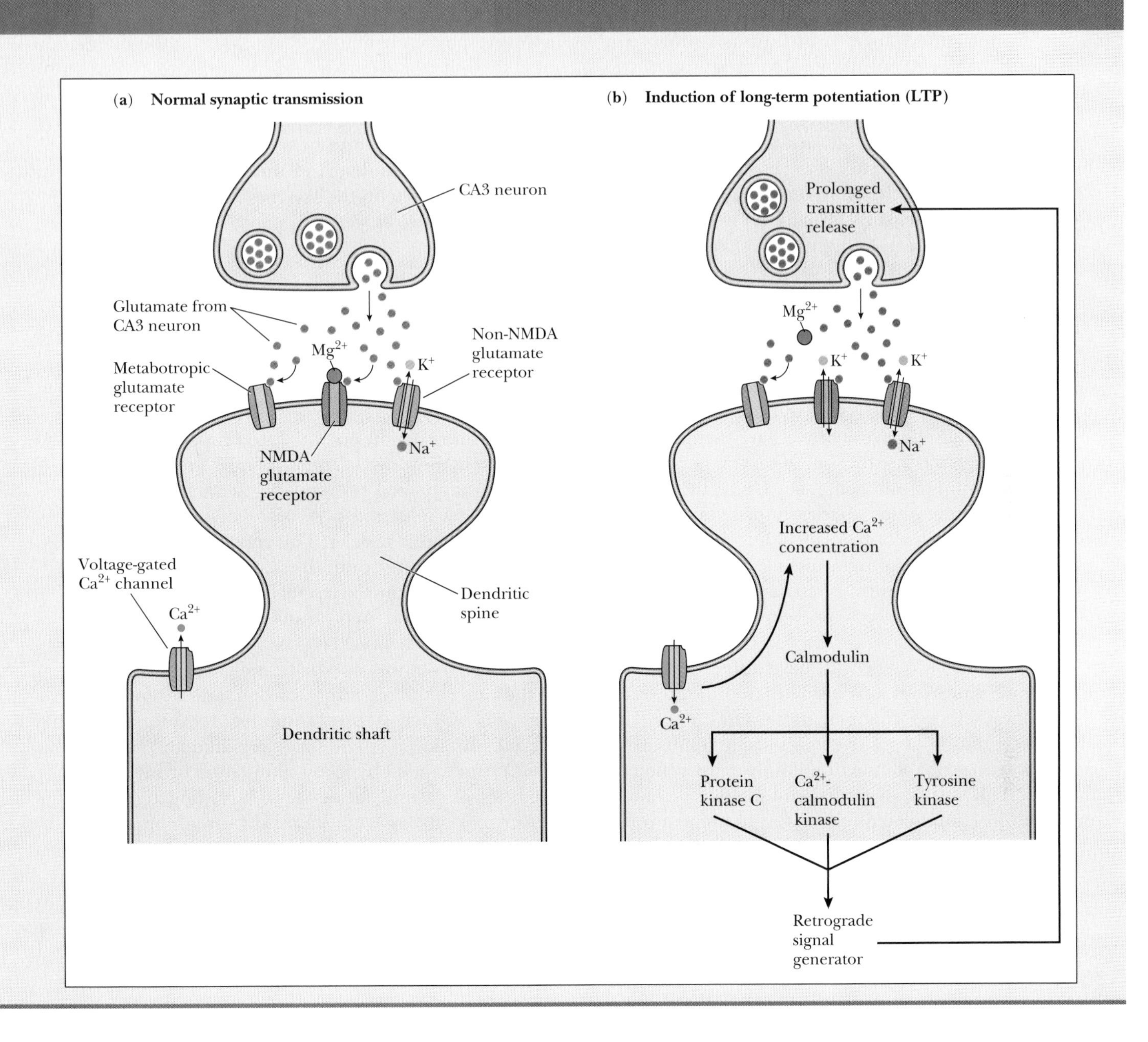

behavioral sensitization. Researchers at Columbia University, led by Eric Kandel, have determined the mechanism of synaptic facilitation by a combination of electrophysiological and biochemical studies. Stimulation of the facilitating interneuron leads to a release of a neurotransmitter called serotonin (whose structure is shown in Figure 23-22f). Serotonin released from the facilitating neuron acts on G-protein–coupled receptors on the axon terminals of the sensory neuron, stimulating the synthesis of cAMP (as discussed for the β-adrenergic receptor in Chapter 19). The increased cAMP activates a cAMP-dependent protein kinase, which in turn inactivates a potassium channel that participates in the repolarization of the sensory neuron after firing. The result of serotonin action is therefore a lengthening of the period during which the axon terminal is depolarized. The extended depolarization allows more Ca^{2+} ions to enter the axon terminal through the voltage-dependent calcium channels. The greater Ca^{2+} concentration leads to a larger release of neurotransmitter and a stronger response by the postsynaptic motor neuron.

A single stimulus to the tail (or to the facilitating neuron) sensitizes the gill withdrawal reflex for several minutes. Repeated stimulation of the tail, however, brings about a much longer period of sensitization, lasting for days or even weeks. This long-term sensitization is a form of associative learning, with the tail shocks serving as conditioned stimulus and the stimulation of the siphon the unconditioned stimulus. Like short-term sensitization, long-term sensitization depends on the inactivation of a repolarizing potassium current. But, unlike short-term sensitization, long-term sensitization requires protein synthesis: Inhibition of protein synthesis during the training period prevents learning.

Both long-term and short-term sensitization depends on increased cAMP that occurs after serotonin release by the facilitating neuron. In short-term sensitization, increased cAMP acts principally through its effect on cAMP-dependent protein kinase. As discussed in Chapter 15 (p. 411), however, cAMP also stimulates the transcription of certain genes—those that contain the eight-nucleotide cAMP response element (CRE). Blocking the interaction of CRE with its specific transcription factor (CREB) prevents the learning, implicating transcriptional regulation as a necessary step in learning. By combining molecular, electrophysiological, and behavioral techniques, neurobiologists hope to understand the mechanism of learning not only in invertebrates like *Aplysia* but in mammals (including humans) as well.

SUMMARY

The operation of the nervous system depends on the properties of individual nerve cells and the way in which they connect with one another. Vertebrates have the most complex nervous systems, consisting of the central nervous system, made up of brain and spinal cord, and the peripheral nervous system, which carries information to and from muscles and other organs, including sense organs. The cells of the nervous system include neurons, which carry information in the form of chemical and electrical signals, and glial cells, which provide mechanical and metabolic support.

Neurons, like all other cells, have different concentrations of molecules and ions inside than out, the result of the action of selective channels, gates, and pumps. The greater concentration of positive ions outside results in an electrical potential of about -70 millivolts across the cell membrane. This resting potential can suddenly change when a neuron is stimulated, giving rise to an action potential, which can travel over large distances. Action potentials move more rapidly through axons that are surrounded by myelin.

The generation and propagation of action potential depend largely on the voltage dependence of sodium channels in the neuronal membrane. The sodium channel is a large protein that consists of four domains, each of which spans the membrane many times. Recombinant DNA techniques are permitting researchers to study the relationship between the structure and function of the sodium channel, as well as of other proteins important in nerve cells.

Neurons communicate with one another at synapses, which may be either electrical or chemical. Electrical synapses allow action potentials to propagate in much the same way as they travel down axons. At a chemical synapse, the presynaptic cell responds to the arrival of an action potential by releasing a store of neurotransmitter contained in synaptic vesicles. This release depends on the influx of Ca^{2+} ions into the axon terminal through a voltage-gated calcium channel.

The released neurotransmitter diffuses across the synaptic cleft to act on receptors on the postsynaptic neuron. These receptors may be ligand-gated ion channels, as in the case of the nicotinic acetylcholine receptor in muscle cells. Other neurotransmitter receptors resemble the membrane receptors for adrenaline and bring about their postsynaptic effects by stimulating or inhibiting the synthesis of second messengers, including cAMP. Many psychoactive drugs act by mimicking or inhibiting the production, release, degradation, or action of neurotransmitters.

Even simple nerve circuits can mediate learning, as in the case of the sea hare, *Aplysia.* The characteristics of three neurons—a sensory neuron, a motor neuron, and a facilitating interneuron—are responsible for the modification of the gill withdrawal reflex. Short-term learning in this system depends on cAMP-dependent protein phosphorylation, while long-term learning depends on cAMP-dependent changes in protein synthesis.

KEY TERMS

acetylcholinesterase
action potential
afferent
Alzheimer's disease
associative learning
autonomic nervous system
axon
benzodiazepine
central nervous system (CNS)
cephalization
channel
chemical synapse
conditioned stimulus
current
dendrite

depolarizing
dynorphin
efferent
electrical potential (voltage)
electrical synapse
endorphin
enkephalin
equilibrium potential
excitatory postsynaptic potential (EPSP)
excitatory synapse
γ-aminobutyric acid (GABA)
$GABA_A$ receptor
gate
gating helix
glial cell
glutamic acid decarboxylase (GAD)
habituation
hyperpolarizing
inhibitory interneuron
inhibitory postsynaptic potential (IPSP)
inhibitory synapse
learning
ligand-gated channel
microelectrode
motor neuron (motoneuron)
muscarinic acetylcholine receptor
myelin sheath
Nernst equation
neuromuscular junction
neuron
neurotransmitter
nicotinic acetylcholine receptor
node of Ranvier
nonassociative learning
oligodendrocyte
opiate
oscilloscope
parasympathetic
patch clamping
peripheral nervous system (PNS)
postsynaptic neuron
postsynaptic potential
presynaptic neuron
reflex
refractory period
resting potential
reuptake
saltatory conduction
Schwann cell
sensitization
sensory neuron
single-channel recording
sympathetic
synapse
synaptic cleft
synaptic strength
tetrodotoxin
unconditioned stimulus
vesicular transporter
voltage clamping
voltage-gated channel

QUESTIONS FOR REVIEW AND UNDERSTANDING

1. Define and contrast the following terms:
   (a) central nervous system vs. peripheral nervous system
   (b) afferent nerves vs. efferent nerves
   (c) axons vs. dendrites
   (d) action potential vs. resting potential
   (e) electrical synapse vs. chemical synapse
   (f) nonassociative learning vs. associative learning
2. What is cephalization? Why do you think evolution would support a trend toward cephalization?
3. Describe how a functional map of the human brain has been constructed.
4. What is an inhibitory interneuron?
5. How do sensors transmit information to the central nervous system?
6. Describe a reflex response at the level of cellular communication?
7. Describe two methods researchers use to monitor the direction of information flow in neural pathways?
8. Contrast a voltage-gated channel with a ligand-gated channel and give examples of each.
9. Describe how the structure of a gating helix allows it to perform its function.
10. The vertebrate brain is highly active metabolically. Why do you think the human brain consumes so much energy? (Consider active transport.)
11. Describe the functions of the autonomic nervous system.
12. Describe three mechanisms which allow molecules and ions to move through a plasma membrane.
13. Explain how the ionic disequilibrium that exists across the membrane of neurons is generated. What occurs as a result of this ionic disequilibrium?
14. Why is a neuron, or any cell, at equilibrium unable to do work? Explain.
15. An action potential is generated by the flow of what ion species? How was this shown?
16. Draw an action potential as you would see it on an oscilloscope. Describe the ionic events at each stage of the action potential.
17. How does an action potential propagate itself? What causes the action potential to travel in only one direction?
18. Describe the two experiments that demonstrated that the M2 segments of the acetylcholine receptor are critical for ion movement.
19. How can acetylcholine induce different kinds of responses?
20. Discuss why it is adaptive that neurotransmitters be short-lived in the synaptic cleft.
21. How does the study of receptors and transmitters assist researchers in designing drugs such as pain killers and mood enhancers?

SUGGESTED READINGS

ALBERTS, B., D. Bray, J. Lewis, M. Raff, K. Roberts, and J. D. Watson, *Molecular Biology of the Cell*, 3rd edition, Garland, New York, 1994. Chapter 11 deals with membrane excitability, including action potentials and synapic transmission; Chapter 15 deals with cell signaling.

KANDEL, E. R., J. H. Schwartz, and T. M. Jessell, *Essentials of Neural Science and Behavior*, Appleton & Lange, Norwalk, Connecticut, 1995. An outstanding introduction to neuroscience, including all the topics covered in this chapter.

LODISH, H., D. Baltimore, A. Berk, S. L. Zipursky, P. Matsudaira, and J. Darnell, *Molecular Cell Biology*, 3rd edition, Scientific American Books, New York, 1995. Chapter 21 deals with nerve cells.

WATSON, J. D., M. Gilman, J. Witkowski, and M. Zoller, *Recombinant DNA*, 2nd edition, Scientific American Books, New York, 1992. Chapter 21 deals with neural development.

CHAPTER 24

How Did the First Cells Evolve?

KEY CONCEPTS

- Cells appeared soon after the earth's origin.
- Scientists can construct a reasonable scenario for the evolution of the first biological molecules.
- The first cells had to develop a means of coupling protein synthesis to genetic instructions.
- Eukaryotic cells probably arose by the association of prokaryotic cells.

OUTLINE

ESSAYS

Understanding life requires an appreciation of its history as well as of life's ongoing mechanisms. But explanations of historical events—even of such recent events as the Vietnam War or the last presidential election—are always incomplete. Even when we have amassed more facts than we can possibly absorb, people differ on which facts are the most important. Someone always argues that the most important facts are not yet known. If we cannot satisfy ourselves that we understand historical events of the last few decades, then think how much more difficult it is to understand those of the last 3.5 billion years!

If frustration is inevitable, why bother at all? We can never know the identity of the first cell, much less give a recipe for making another just like it. Should we therefore abandon evolutionary inquiry?

The answer is no, for two reasons: (1) We still want a plausible explanation of how present cells came to be the way they are, and this means knowing biological history, even if it is incomplete; and (2) we want to see whether currently understood biological and chemical processes could conceivably explain the evolutionary processes that we think must have occurred. In this chapter we examine the way in which biologists have asked questions about the origin of life. Some of our current understanding depends on the collection of rare early fossils. Much of our current view, however, derives from a kind of informed speculation. Evolutionary biologists have inferred part of the story from the shared properties of current organisms—the common building blocks of all macromolecules, the common organelles, the universal use of ATP as energy currency.

But much of the story of early evolution must be based on guesswork. The point of the exercise is to show that life could have arisen by some series of steps, each of which is reasonable in terms of physics and chemistry as we can now study them. By showing that they can tell such a reasonable story, biologists have refuted the idea of an uncrossable gulf between living and nonliving matter.

FOSSILS CAN PROVIDE DIRECT EVIDENCE FOR TESTING EVOLUTIONARY HYPOTHESES

Fossils [Latin, *fossilis* = dug up] are objects, usually found in the ground, that are the remains or imprints of past life. Fossils allow evolutionary biologists to test hypotheses about the evolution of species and of cells. **Paleontology** [Greek, *palaios* = ancient + *on* = being + *logos* = discourse], the study of ancient life, relies heavily on fossils as evidence for the structures and environments of past organisms.

The most direct evidence for evolution comes from the fossil record. The most spectacular fossils are the intact skeletons of extinct animals, such as dinosaurs or mammoths. Complete skeletons, however, are relatively rare compared with the hundreds of thousands of other kinds of fossils, collected by both professionals and amateurs. These fossils include not only preserved bones and teeth, but other traces of life as well, such as those shown in Figure 24-1—including whole organisms frozen in Arctic glaciers or preserved in hardened sap, a surface impression of a body, footprints, or even a bit of dung.

An organism becomes a fossil only if it dies at a place and time that preserves its integrity. By far the majority of dead plants and animals either decay or are eaten. Fossils persist only if they are quickly buried, for example, in a bog or in the bottom of a sea or lake. Then, protected from decay and scavengers, fossils become more deeply buried with time as successive layers of mud and sand settle upon them. Rocks that form from the compression of such settling debris are called sedimentary rocks.

The key to reading the fossil record was the realization that fossils are not all the same age. In general, the older the fossils, the more deeply they are buried. For a fossil to be discovered, the older and deeper layers must be laid bare by the pushing, pulling, and erosion of the earth's surface.

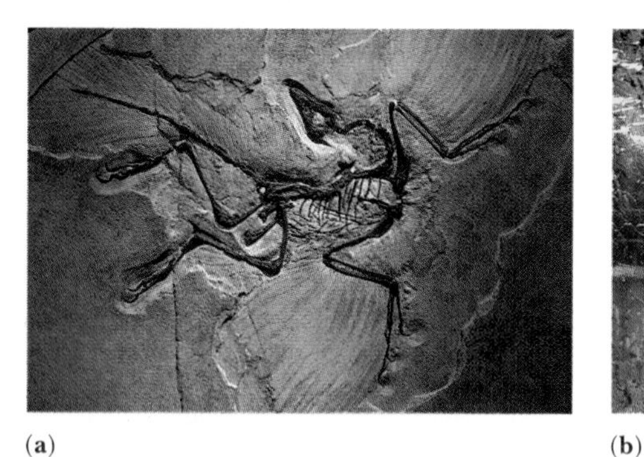

(a)

(b)

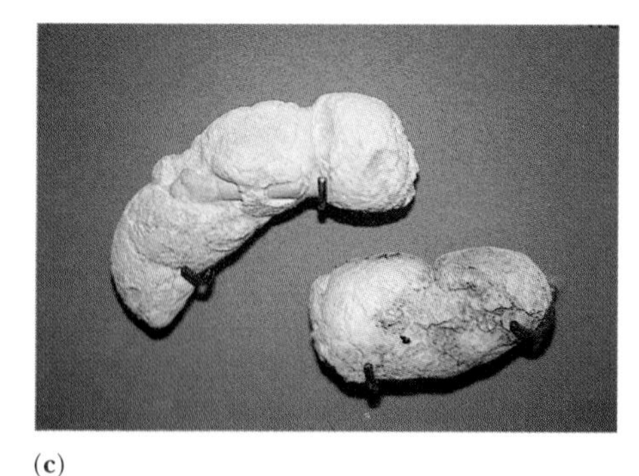

(c)

Figure 24-1 Different kinds of fossils have contributed to our current understanding of the history of life. **(a)** A fossilized *Archaeopteryx,* a 150 million-year-old flying feathered reptile that was a precursor of birds; **(b)** an insect trapped in amber about 35 million years ago; **(c)** coprolites, bits of preserved dung about 20 million years old. *(a, Oxford Scientific Films/Photo Researchers; b, Vaughan Fleming/Science Photo Library/Photo Researchers; c, A. J. Copley/Visuals Unlimited)*

Figure 24-2 The Grand Canyon. The deeper layers represent more ancient times, ranging from 200 million years ago, near the top of the canyon, to more than 2 billion years ago, near the bottom. *(Paraskevas Photography)*

An extraordinary illustration of the historical information in the fossil record lies in the Grand Canyon, a mile-deep slice through sandstone and limestone, shown in Figure 24-2. The rocks in the Grand Canyon, like sedimentary rocks elsewhere, lie in strata, layers formed under ancient oceans.

Paleontologists have determined the ages of fossils in two distinct ways. First, the layering of the rocks gives their relative ages: Within any set of undisturbed strata, the deeper rocks are older. Second, radioactive dating can give the actual ages of many rocks. By observing the layers in a formation like the Grand Canyon as well as the ages of rocks in those layers from radioactive dating, geologists are able to read the layers of rock as if they were pages in a history book. By the mid-19th century, paleontologists established names for the chapters in our planet's history and showed that the same chapters, often with the same stories, lie in the rocks and fossils from many places of the earth.

The rocks at the top of the Grand Canyon are about 200 million years old, while those at the bottom are 2 billion years old. Geologists and paleontologists have helped each other establish the history of the earth's lands, seas, and organisms. For example, although the Grand Canyon has been a rich source of fossils (Figure 24-3), even the youngest rocks near the top contain no fossils of mammals. Because these rocks are more than 200 million years old, the absence of mammalian fossils indicates that mammals must be relatively recent newcomers on our planet. Other rocks of the same age, in other parts of our planet, also lack mammalian fossils, reinforcing the conclusion that mammals evolved relatively recently.

The fossils in the rocks near the top of the canyon have traces of reptiles. Deeper, in rocks some 400 million years old, there are no more reptiles but fossils of extinct fish are found. Still deeper, there are no traces of fish or any other vertebrate. In the rocks older than about 570 million years, we cannot find any fossils that are visible to the naked eye. However, the microscope reveals fossils of still more ancient life, single-celled eukaryotes.

LIFE APPEARED SOON AFTER THE EARTH'S ORIGIN

Since the 19th century, scientists have divided the earth's history into sections, chapters, and subchapters. Paleontologists usually divide the history of life into four **eras,** periods of time that range in length from 65 million to several billion years. As shown in Figure 24-4, each era contains distinctive forms of life in its fossil record. More re-

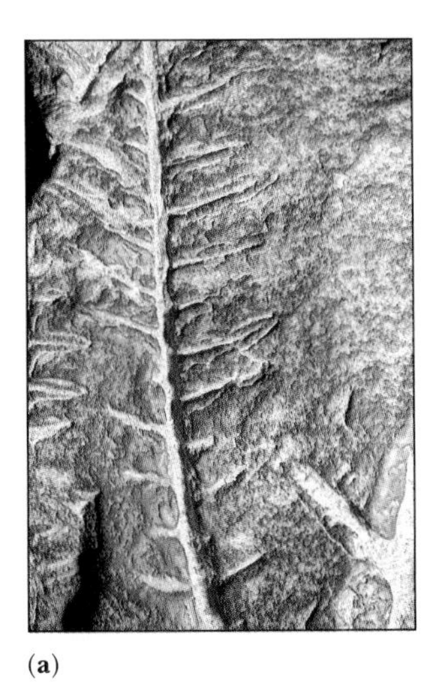

(a)

(b)

(c)

Figure 24-3 Different kinds of fossils have been found in the Grand Canyon. **(a)** Fossil fern; **(b)** sandstone showing the fossil track of a small dinosaur's hind leg; **(c)** rows of fossil sponges and other marine animals. *(a, b, William Ferguson; c, Frank T. Aubrey/Visuals Unlimited)*

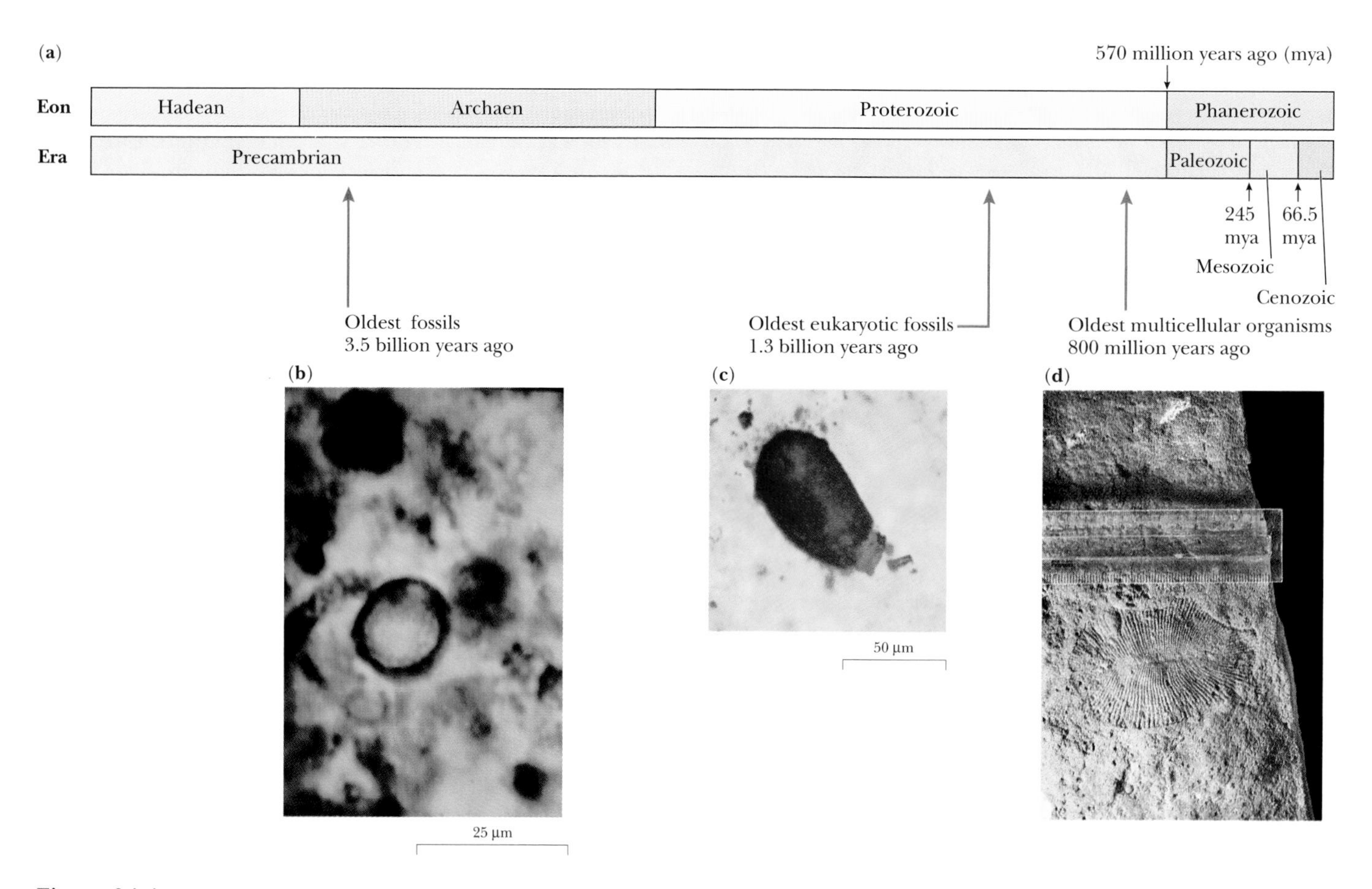

Figure 24-4 The eras of the earth's history, with distinctive fossils from each. **(a)** The division of historical time into eons and eras; **(b)** prokaryotic fossil cells, 3.5 billion years old; **(c)** an early eukaryotic fossil, 700 million years old; **(d)** a fossil specimen of *Dickinsonia,* which has been interpreted as a segmented worm. *(b, Science VU-USM/Visuals Unlimited; c, Andrew H. Knoll, Knoll and Calder,* Palaeontology *26: 467, 1983; d, Biological Photo Service)*

cently, however, geologists and paleontologists have also spoken of four **eons,** spans of time still longer than eras, as also shown in Figure 24-4.

Before 1950, no one had seen a fossil older than about 590 million years (the beginning of the *Cambrian era*). If we were to rely only on visible fossils, then we would have had to say that many forms of life burst forth almost at once, in geological terms. But microscopic fossils tell quite a different story.

Life First Appeared in the Precambrian Era

In the 1950s, paleontologists began to find **Precambrian** fossils—meaning fossils that date from more than 590 million years ago, that is, before the beginning of the Cambrian era. The reason that no one had previously found Precambrian fossils was that they had not looked in the right places. Professional and amateur paleontologists found later fossils in sedimentary rocks—shales, sandstones, and limestones—in which the hard parts of organisms became trapped. But, as we now realize, Precambrian organisms had no hard parts. As the remains of these early forms settled, their substance dissolved or decayed. The remains of soft-bodied organisms persisted only when caught in the beginnings of a relatively rare hard rock called *chert.* Only by examining thin slices of cherts in the microscope did paleontologists find Precambrian fossils, some of which, from the 2.1 billion-year-old Gunflint Chert, are shown in Figure 24-5.

Precambrian rocks are mostly far below the earth's surface, covered by later deposits. They are accessible at only a few spots. Fossils from these few rocks, however, provide the best record of early evolution.

The oldest fossils known are about 3.5 billion years old. These fossils, shown in Figure 24-6, were embedded in two cherts—the Warrawoona rocks of Western Australia and the Fig Tree chert, in southern Africa. These earliest forms appear to be prokaryotic.

Some of the earliest fossils, such as those shown in Figure 24-4b, are small spheres; others, as shown in Figure 24-6, are filamentous structures that resemble modern filamentous cyanobacteria. Still other ancient fossils appear as banded structures within the cherts, which appear to be stromatolites, deposits of limestone formed around

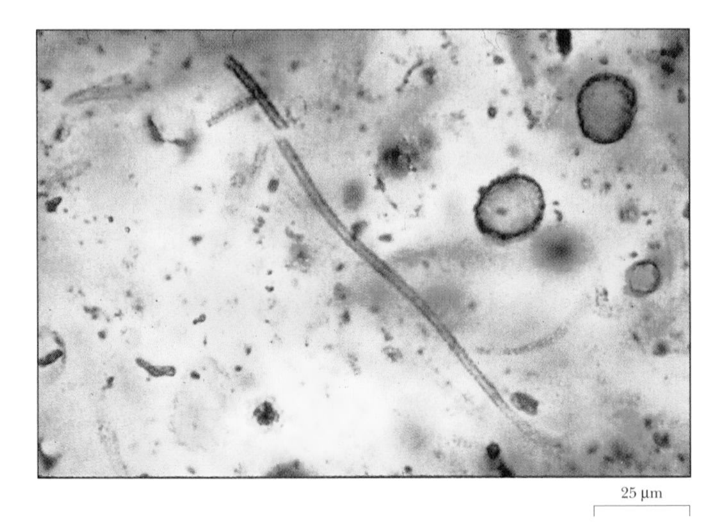

Figure 24-5 Precambrian microfossils from a formation called the Gunflint chert. Chert is silicon dioxide (SiO_2). It does not form by orderly crystalline growth, as does quartz (which has the same chemical composition). Instead, many cherts seem to have arisen on the bottom of shallow seas from jelly-like precipitates. The Precambrian fossils are the remains of organisms trapped in these precipitates. *(J. Robert Waaland/Biological Photo Service)*

10 μm

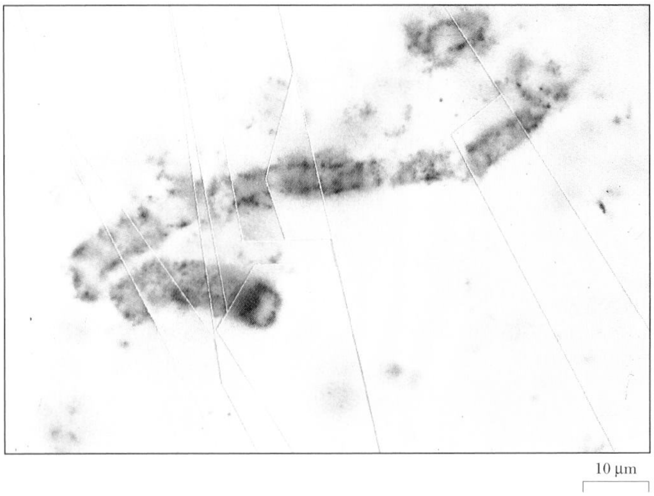

Figure 24-6 A filamentous alga from the Warrawoona rocks of Western Australia, almost 3.5 billion years old. *(Stanley Awramik/Biological Photo Service)*

huge colonies of cyanobacteria. Stromatolites still exist today, as illustrated in Chapter 1, p. 8.

The dates of the Fig Tree and Warrawoona cherts are important. They are the oldest rocks that could possibly have preserved evidence of life. The earth formed about 4.6 billion years ago, and its crust solidified only about 4 billion years ago. The oldest known rocks are about 3.8 billion years old. The oldest fossils are about 3.5 billion years old. Life on earth, then, probably originated rather quickly—within a few hundred million years—after the earth had cooled enough to sustain it.

Fossils of eukaryotes first appear in rocks 1.3 billion years old, as seen in Figure 24-4c. About 630 million years ago, at the beginning of the Phanerozoic eon, and near the end of the traditional Precambrian era, an amazing variety of multicellular organisms appeared, including animals that resemble modern jellyfish, corals, and segmented worms, as well as animals unlike any organisms living today. Fossils of some of these animals, shown in Figure 24-4d, are relatively rare, probably because they were soft-bodied and were preserved only under exceptional circumstances.

Where Did the First Cells Come From?

There are two general explanations for the "sudden" appearance of prokaryotic cells soon after the earth had cooled: (1) They arose spontaneously from nonliving organic molecules in the primitive environment; or (2) they arrived from elsewhere in the universe. It is almost impossible to imagine how we could experimentally address either of these explanations. We can, however, examine the reasons for thinking either of them reasonable.

The second hypothesis appears to be the simpler explanation for the appearance of life. According to this view, the present diversity of life derives from a huge **adaptive radiation,** the derivation of many species from a single ancestor, with each species specialized to a distinct way of life.

An appealing feature of this hypothesis is that the universe is much older than the earth. The universe is thought to have originated 12–16 billion years ago, 8–12 billion years before the cooling of the earth. This longer history would loosen the difficult requirement that life arose spontaneously within a few hundred million years. Furthermore, it could explain some otherwise unexpected chemical properties of organisms, for example, the requirement for molybdenum—an extremely rare element on this planet—in key reactions of photosynthesis and nitrogen fixation.

Still, even if all life on earth derived from an ancient colonization from outer space, we would still want to understand how those organisms first arose. If this explanation is true, the problem of the origin of life would then be still harder to approach, because we would not even know what planet to contemplate. Biologists have therefore accepted the challenge of showing that life *could* have arisen from nonliving molecules here on our own planet. Most biologists believe that life must at some point have come from nonliving molecules on the early earth's surface. The development of organisms from nonliving matter is called **prebiotic evolution.**

CAN SCIENTISTS CONSTRUCT A REASONABLE SCENARIO FOR PREBIOTIC EVOLUTION?

By the 1930s biochemists realized that all organisms are made from the same building blocks—the sugars, lipids, amino acids, and nucleotides that make up the "biochemical alphabets" that we described in Chapter 2. Two influential scientists—Alexander Ivanovich Oparin, a Russian biochemist, and J. B. S. Haldane, an English evolutionary biologist—proposed that these building blocks had been formed in the primitive environment. The original synthesis of these materials, they argued, was prebiotic—before life.

This seems a reasonable starting point for a model of the long-term evolution of organisms. The next question to address is whether we can imagine a reasonable scenario for the "spontaneous" production of the building blocks and whether laboratory experiments support the steps in the presumed process.

What Was the Earth Like When It Was Young?

The conditions during which life must have first appeared were dramatically different from those that exist today. In particular, the earth's atmosphere certainly contained no free oxygen, which would quickly have oxidized the starting materials. Nor was there the plethora of modern organisms that find and consume every source of energy. The origins of life probably extended over at least 100 million years. How this happened is the subject of experiments and debate.

Scientists calculate that the earth formed about 4.6 billion years ago from dust and gas attracted by the sun's gravity. At first, continual collisions with debris in its orbit and the decay of radioactive elements kept earth a molten mass. After 500–600 million years, earth's surface cooled enough to form a thin crust. Gases formed in the interior and escaped through cracks in this crust to form the primitive atmosphere. Among these gases was water vapor, which then condensed, in millions of years of torrential rains, to form the oceans.

The composition of the primitive atmosphere has been a matter of much debate. Scientists agree that the early atmosphere did not contain free oxygen; this came later, almost entirely as a by-product of photosynthesis. (See Chapter 8.) They agree that the primordial atmosphere must have contained hydrogen (H_2), ammonia (NH_3), and methane (CH_4), but they disagree on the relative proportions of these gases and how these proportions changed during the earth's history. Other gases thought to be present in the early atmosphere include carbon dioxide (CO_2) and hydrogen sulfide (H_2S).

Besides these raw materials in the atmosphere, prebiotic synthesis of building blocks required energy because the synthetic reactions are almost all thermodynamically uphill. (See Chapter 4.) The most dramatic energy source in these times was the lightning storms that probably accompanied the filling of the oceans. Still more energy was available from the direct radiation of the sun. Ultraviolet light carries energy in the right amounts to make and break many chemical bonds, and it would have been particularly effective in promoting chemical reactions. Other energy sources depended on the continuing generation of heat in the earth's interior.

Is Prebiotic Synthesis Plausible?

By the early 1950s, scientists had some idea of what the earth was like when the first cells appeared: It had both the raw materials for organic synthesis and sources of energy. But no one knew whether the conditions of the earth's early history would allow synthesis of biologically important molecules.

The researcher who first demonstrated the plausibility of prebiotic synthesis was Stanley Miller, a graduate student at the University of Chicago in the laboratory of Harold Urey, a chemist interested in the early atmosphere. Miller undertook to simulate the primitive conditions of the earth inside a glass apparatus built mostly from the standard equipment of a sophomore organic chemistry laboratory. Miller's simulation of the early earth, pictured in Figure 24-7, consisted of water (the ocean) in a boiler and a chamber fitted with electrodes (the atmosphere, with lighting). In one such experiment, Miller admitted to the atmosphere a mixture of hydrogen (H_2), methane (CH_4), and ammonia (NH_3). As the electric discharges passed through this atmosphere, these simple molecules reacted with one another and with water.

After a week, Miller analyzed the contents of his ocean. He found that more than 10% of the carbon (from the methane) was now in organic molecules. These included at least five amino acids, as well as compounds that are precursors of other amino acids and of the bases of nucleic acids. Among these precursors is hydrogen cyanide (HCN), five molecules of which can directly form adenine, as shown in Figure 24-7b.

Miller and others have repeated this basic experiment, with different atmospheres and different sources of energy. As long as free oxygen (O_2) is absent, the same building blocks appear. And other sources of energy also work—most notably ultraviolet light, the principal energy source available on the earth's surface before the earth developed its protective ozone (O_3) layer. It seems likely, then, that —within a few million years—the primitive oceans became loaded with the organic molecules that were to become the building blocks of life. Scientists think that in many places this "primordial soup" contained as much as 1% organic molecules, about the same as a weak bouillon.

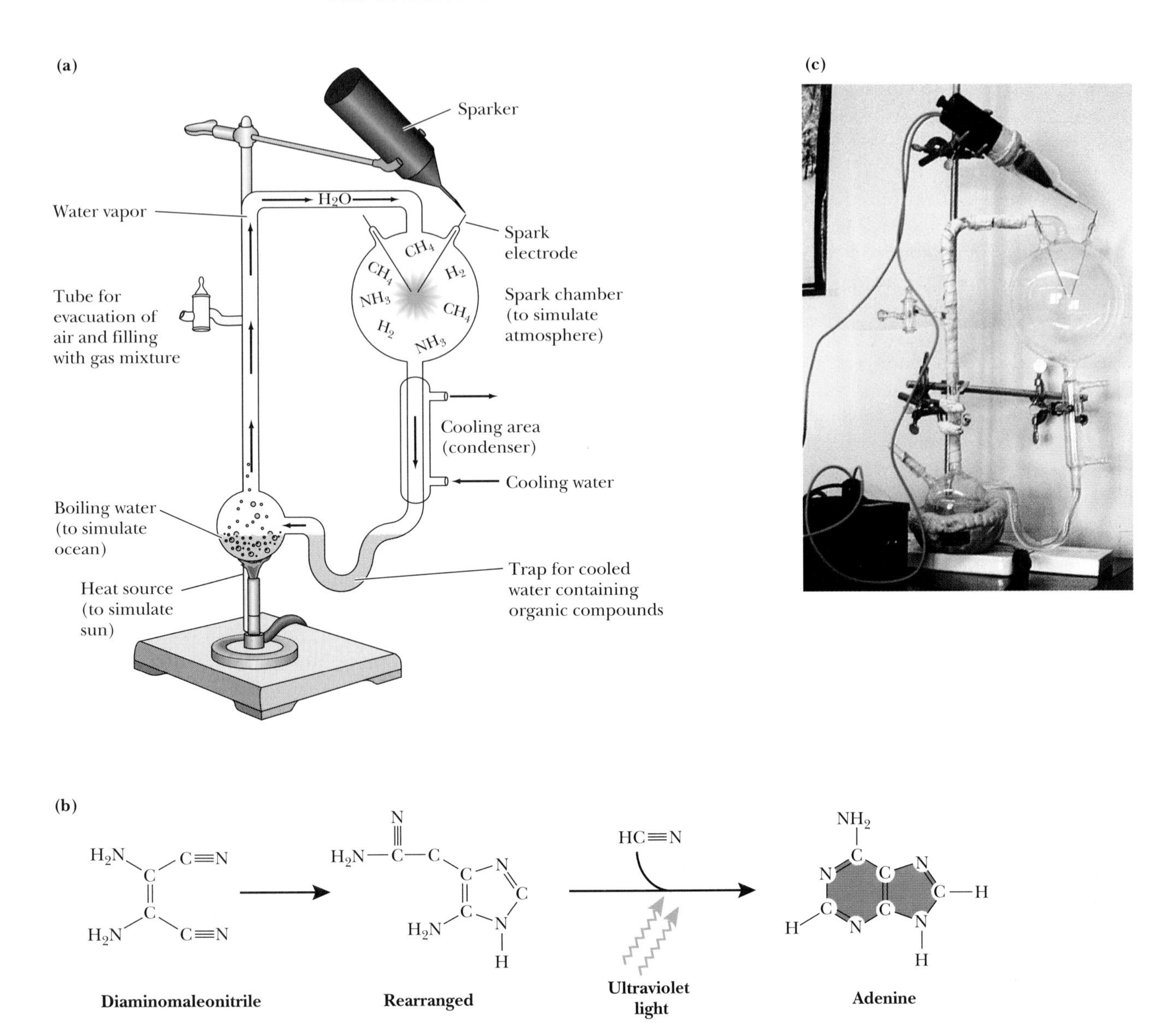

Figure 24-7 In the 1950s, scientists began to simulate the prebiotic environment in the laboratory. **(a)** Stanley Miller's apparatus contained water, a mixture of gases, and artificial lightning; **(b)** among the complex products of Miller's experiment was adenine, the nitrogenous base in ATP. *(c, courtesy of Dr. Stanley Miller)*

How Can Biological Building Blocks Assemble Outside of Cells?

The ability to produce building blocks in laboratory simulations removes the mystery from the first steps of prebiotic evolution. The next question concerns the assembly of building blocks into macromolecules. Can this process also be simulated in the laboratory?

As mentioned above and shown in Figure 24-7, adenine can form from five molecules of hydrogen cyanide (HCN), which was almost certainly present in the primitive atmosphere and ocean. The other nucleic acid bases require more complex but nonetheless straightforward reactions. Adenine, however, was probably the first nitrogenous base to form. It is not surprising that its activated product, ATP, became the universal energy currency of life.

How did ATP form? In laboratory experiments, mixtures of adenine, ribose, and phosphate can actually form ATP when exposed to ultraviolet light. This reaction is faster in the presence of apatite, a common form of calcium phosphate likely to have been present on the primitive earth. Other triphosphates may have arisen in the same way.

The First Biological Polymers Were Probably RNA

As biologists began to think about the ways in which the first biological polymers were formed, they attempted to find laboratory conditions that would stimulate the formation of polypeptides and polynucleotides. They achieved only limited success with polypeptides, using activated amino acids, but the order of the amino acid residues was always random.

Many researchers think that ATP in the primordial broth may have somehow contributed to the synthesis of activated amino acids. (See Chapter 14.) Solutions of such activated amino acids can spontaneously give polypeptide chains up to 50 residues long. These polymerizations occur only slowly, but certain clays speed them up considerably.

Polypeptides also form when researchers heat dry mixtures of unactivated amino acids. Sidney Fox, a biochemist at the University of Miami, has suggested that such "proteinoids" could have formed in volcanic cinder cones and washed into the sea. Other researchers argue that polymerization first took place in the sea or in pond water. In view of the changing environment of early earth, an attractive suggestion is that dissolved amino acids dried onto the surface of clay particles, polymerized, and then redissolved. A series of such cycles could lead to the production of long polypeptides. Whether they formed in the seas or on clays, however, the earliest polypeptides almost certainly did not have consistent sequences of amino acid residues, as do cellular proteins.

In the 1970s, biochemists discovered that nucleotides can combine in the laboratory to form short RNAs. These RNAs, moreover, form double-stranded molecules that follow the Watson-Crick rules for base pairing. So the prebiotic replication of nucleic acids is not at all far fetched. The suggestion that RNA formation was an early event in prebiotic evolution is also consistent with the diverse roles (discussed in Chapter 14) that RNA plays in protein synthesis—as template (messenger RNA), as adaptor (transfer RNA), as workbench (ribosomal RNA), and as a facilitator of splicing (small nuclear RNA).

Even more compelling were the demonstrations that RNA can act as an enzyme. The first two examples of RNA enzymes were the cleavage of a transfer RNA from *Escherichia coli* and the self-splicing of a ribosomal RNA precursor from the protozoan *Tetrahymena thermophila*. (See Essay in Chapter 4, "Almost All Enzymes Are Proteins, but Some Are RNA.")

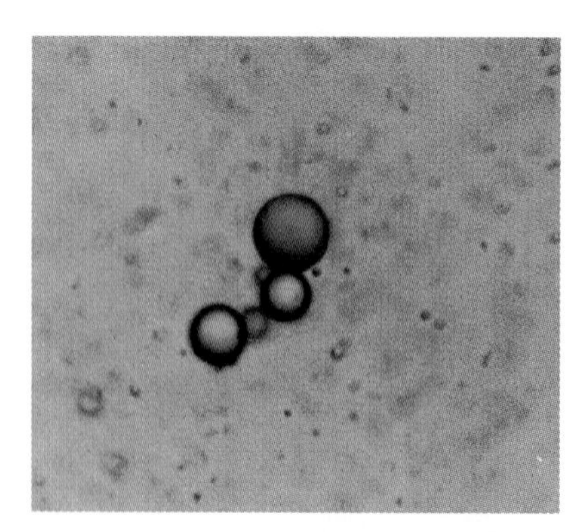

Figure 24-8 Solutions of proteins and other macromolecules can form cell-like associations: Fox's microspheres. Each protenoid microsphere is 1 to 2 micrometers in diameter. *(Steven Brooke and Richard LeDuc)*

The evolution of life depended on establishing an ordered relationship between the self-replicating nucleic acids and reproducible sequences of amino acid residues. This meant the development both of the genetic code and of a mechanism of translation. Before this could happen, however, polypeptides and other molecules probably began to come together into organized associations. **Protobionts,** primitive cells, could concentrate organic molecules and begin to evolve the first metabolic pathways.

HOW MIGHT THE FIRST CELLS HAVE FORMED?

Some variant of the chemical reactions described above may well have filled the open seas, or even small pools and puddles, with the building blocks of life, perhaps as long ago as 4 billion years. But the materials of life are not the same as life. The question is: How did the materials of life organize into structures that accomplish the ordered transformations of matter and energy, responsiveness to environment, and reproduction?

Proteins, Polysaccharides, and Lipids Can Form Cell-Like Structures

Proteins and polysaccharides can spontaneously concentrate themselves into discrete tiny droplets, which Oparin called **coacervates** [Latin, *coacervatus* = heaped up]. Other mixtures can also give rise to isolated, cell-like structures. Fox, for example, found that the protein-like polymers formed from dry amino acids could form tiny spheres about 2 μm in diameter. These structures could even absorb more polymers from solution, forming buds or even a new generation of spheres, as shown in Figure 24-8. Finally, a mixture of phospholipids and proteins can form yet another kind of enclosed structure. These self-enclosed, self-organizing structures are not cells. Their existence, however, shows that cell-like structures could have formed spontaneously in the primordial soup.

These structures also allowed rudimentary metabolism. Oparin showed, for example, that coacervates containing the enzyme phosphorylase would take up glucose-1-phosphate from the surrounding medium, convert it to starch, and concentrate it inside the droplet, as shown in Figure 24-9a. Oparin then made coacervates containing both phosphorylase and amylase, an enzyme that breaks down starch into the disaccharide maltose. These coacervates would convert one component of their environment, glucose-1-phosphate, into another, maltose.

Self-contained, nonliving assemblies show the beginnings of reproduction, responsiveness to their envi-

(a) "Metabolism" within a coacervate drop

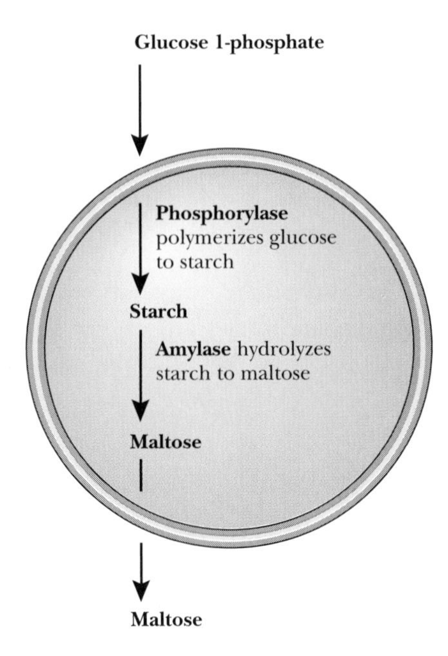

(b) Evolution of a metabolic pathway

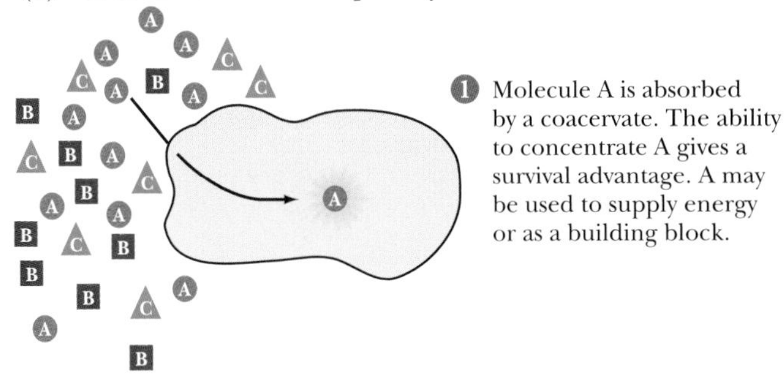

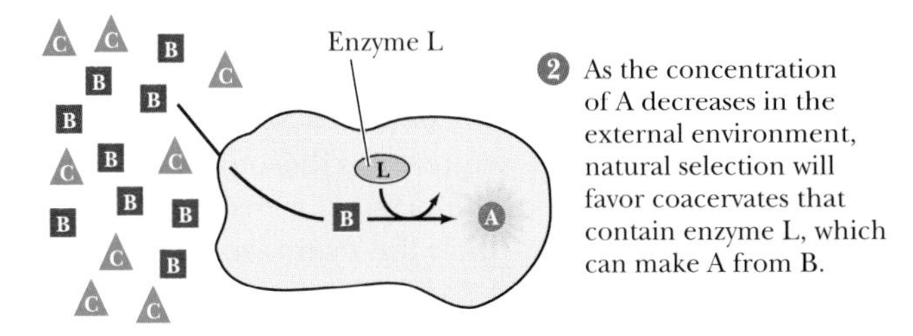

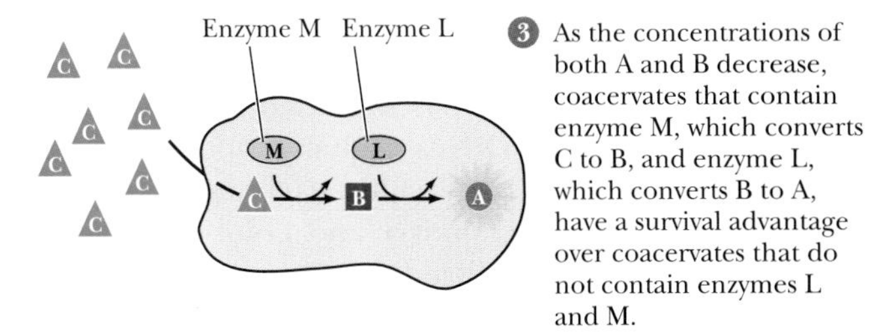

Figure 24-9 How did metabolic pathways evolve? **(a)** Oparin's coacervates were able to convert glucose-phosphate into starch, which they could concentrate inside the droplet; **(b)** possible schemes for the evolution of a metabolic pathway.

ronment, and metabolism. We can even see how such protobionts could evolve. More stable coacervates would persist, as less stable associations disappeared. This means that the composition and organization of the coacervates must have developed relatively quickly.

Similarly, individual coacervates would compete for the limited amount of organic material in the primordial soup. Selection would favor those coacervates with some catalytic ability that would speed the accumulation of those materials. One can easily imagine the "discovery" of catalysts that speed up specific needed chemical reactions.

Finally, we can envision the evolution of primitive metabolic pathways. As coacervates used up a given compound in the environment, those coacervates able to make that compound from another component of the soup would be more successful. Once that precursor was exhausted, further growth of protobionts would depend on the ability to produce the needed product from still another precursor. According to this scheme, a number of pathways would have evolved—one backward step at a time, as diagramed in Figure 24-9b.

Living Cells Need to Couple Protein Synthesis to Genetic Instructions

Even with the evolution of primitive metabolic pathways, protobiont "life" was severely limited. Despite their ability to grow and to bud, protobionts really could not reproduce. Even if they had accumulated catalysts for specific chemical reactions, protobionts could not specifically produce more of those catalysts. To do so required the invention of a genetic system—a way of storing, replicating, and executing the information for the building of catalysts. Because scientists can demonstrate the nonbiological replication of RNA in a test tube, we now think that the original genetic material was RNA. The question is how the sequence of nucleotides in RNA becomes coupled to the sequence of amino acid residues in a polypeptide. Such a coupling would have given its possessor an enormous selective advantage as it led to the capacity to inherit increased abilities for growth and reproduction.

How can we even guess how the first translation system arose? We must rely on comparative studies of the translation machinery in present-day organisms. The first thing to notice is the central role of RNA in translation, again supporting the idea that RNA was the primitive genetic material. Indeed, all organisms (and all viruses) use messenger RNA to direct the ordered synthesis of proteins. Second, all organisms share three other features of protein synthesis:

1. All organisms use essentially the same triplet genetic code.
2. Amino acids are assembled after linkage to transfer RNA adaptors.

3. All organisms use ribosomes to carry out translation. All ribosomes contain RNAs and proteins.

What Does the Character of the Genetic Code Suggest About its Origins

The first thing we can say about the genetic code is that most of the information for each codon is in the first two nucleotides: For 7 of the 20 amino acids, the third "letter" is irrelevant. (See Figure 14-3, p. 362.) Does this mean that the code was originally a doublet?

Almost certainly not. If the code had started out as a doublet, then all previously evolved genes would have become useless when the code became a triplet. Instead, the present characteristics of the code were almost certainly fixed after its initial success. As Francis Crick put it, the code is a "frozen-accident." Information in mRNA was therefore probably read three bases at a time from the beginning. But, at first, two letters of each codon contained all the meaning.

The code contains other interesting patterns. For example, whenever the second position is occupied by U, the corresponding amino acid is almost always hydrophobic. Charged amino acids use codons rich in A and G. These patterns suggest some kind of chemical affinity between codons and amino acids. But chemists have not demonstrated any such associations. It seems more likely that the genetic code evolved slowly, gradually refining its specification. At first, perhaps, the trick was to specify an ordering of nonpolar and polar amino acids. Only later did organisms discover how to distinguish between two similar amino acids, say, between leucine and valine.

The characteristics of the modern code, then, suggests a gradual evolution and refinement of the code. Because the code is universal among both prokaryotes and eukaryotes, it must have completed its refinement near the time of the common origin of all organisms.

How Did tRNAs Evolve?

As discussed in Chapter 14, all cells attach amino acids to transfer RNAs (tRNAs) before assembling a polypeptide. All present-day tRNAs have similar "stem and loop" structures, as shown in Figure 24-10, and all attach to an amino acid at their 3′ ends. In addition, all tRNAs contain an anticodon with common surrounding elements.

All tRNAs are likely to have diverged from a single tRNA present early in the history of life. The first adaptor used in translation, however, must have been much simpler than current tRNAs. The primitive adaptor could have contained the triplet anticodon and a few additional bases to stabilize its interaction with mRNA. We can easily imagine that such a simple adaptor, say seven bases long, may have arisen at random. The ability to couple RNA replication to the production of a functional protein would have given a strong selective advantage to a protobiont that could use an RNA sequence in this way.

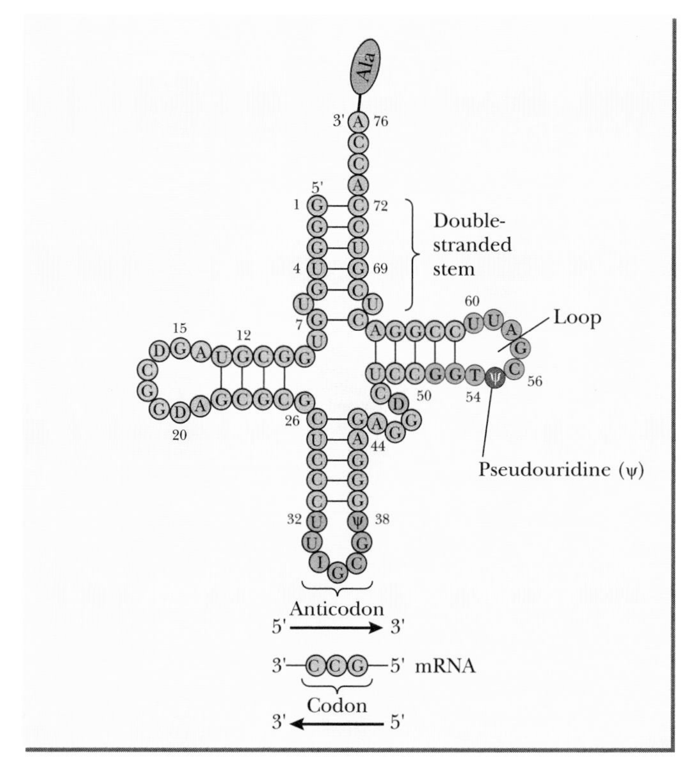

Figure 24-10 All contemporary tRNAs have a similar "stem and loop" structure. All have a similar loop containing a modified base called pseudouridine (ψ), and all contain their anticodon in a similar context.

How Did Ribosomes Evolve?

Ribosomes today contain more than 50 different proteins as well as ribosomal RNAs (rRNAs). Because each protein has a defined amino acid sequence, ribosomal proteins could not have been part of the original translational machinery. It seems more likely that the primitive machinery depended primarily on RNA to facilitate the interaction of amino acids, adaptors, and mRNAs. In prokaryotes, mRNAs can form specific base pairs with part of the smaller rRNA.

The comparative anatomy of rRNAs also supports the idea that rRNAs directly participate in protein synthesis. For example, researchers have determined the nucleotide sequences of the smaller rRNAs of many widely diverged organisms, ranging from bacteria to vertebrates, as well as of ribosomes from mitochondria and chloroplasts. For each rRNA researchers predicted the pattern of stems and loops that could be formed by the formation of duplexes of complementary sequences, as shown in Figure 24-11. While the sequences of the rRNAs are widely diverged, the patterns of stems and loops are highly conserved. Such

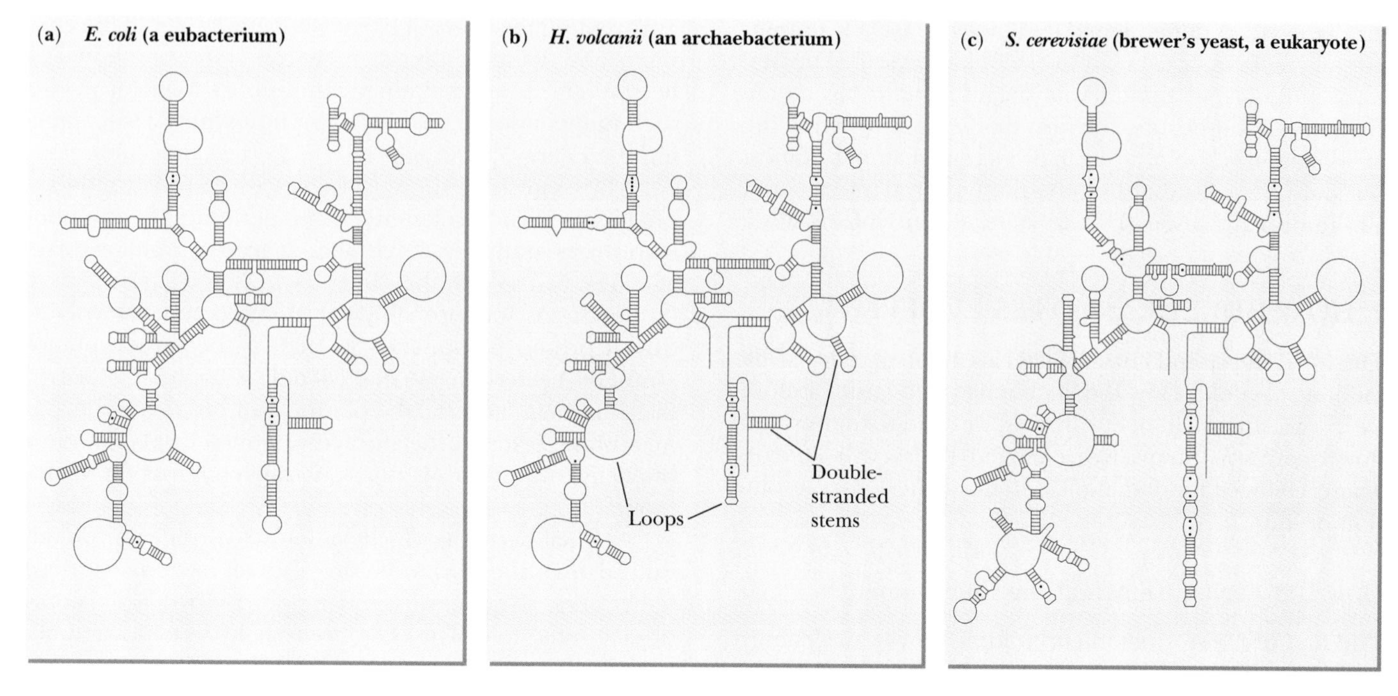

Figure 24-11 All contemporary ribosomal RNAs have a similar "stem and loop" structure.

conservation suggests a function for the pattern of stems and loops, probably as a catalyst for directed protein synthesis.

How Did the First Cells Use Genetic Information?

From laboratory simulations and from comparisons of protein synthesis in modern organisms, biologists have put together a picture of the first true cells. These cells probably had the following properties:

1. Some kind of boundary, such as that in artificially produced coacervates
2. Primitive enzymes that allowed the processing of chemicals in their environment to derive energy
3. Energy storage in ATP
4. RNA as the genetic material that specified the structure of specific enzymes
5. RNA replication, using the Watson-Crick rules of base pairing
6. Specification of each amino acid by a triplet codon, with most of the information conveyed by one or two bases
7. Small adaptor RNAs for protein synthesis, possibly as short as seven bases
8. Translational machinery that consisted of RNAs with specific stem and loop structures

The first enzymes probably arose by the random assembly of amino acids into short polypeptide segments. Such primitive enzymes were probably neither as active nor as stable as modern enzymes. But if a polypeptide segment could form the active site of a catalyst, then its possessor would have enormous advantages over its competitors.

Similarly, the first genetically useful RNAs probably had only short stretches of information. Once the first genetic system had evolved, however, natural selection (as opposed to mere "chemical selection") would begin to work. Selection would favor cells with increased capacity to reproduce, as well as cells better able to extract energy from chemicals in the environment. Reproduction, protein synthesis, and metabolism must therefore have evolved together. The most successful organisms would be those that best coupled their genetic and metabolic systems—those best able to pass on useful catalysts to their descendants.

Among the early improvements in the capacity to reproduce was introduction of DNA as the genetic material. DNA is much more stable than RNA, but its use requires a new set of enzymes for replication and transcription into RNA. Separating replication and translation was perhaps the first step in the division of labor—compartmentalization—that is characteristic of modern cells, especially of eukaryotes.

Scientists think that DNA was originally copied from RNA by the enzyme reverse transcriptase. Reverse transcriptase (RNA-dependent DNA polymerase) is present not only in RNA viruses but in essentially all cells. The universal presence of the enzyme suggests that it has an ancient origin, supporting the idea that it was involved in the introduction of DNA as the new genetic material. Almost certainly, the cells that became the universal ancestors of all life on earth used DNA to store genetic information.

HOW DID EUKARYOTES EVOLVE?

The fossil shown in Figure 24-4c, clearly eukaryotic, is 700 million years old. How did such eukaryotes arise? Before we discuss the origin of eukaryotes, we need to understand how organisms themselves changed their environment. These changes fundamentally affected the future evolution of life.

How Did Early Life Affect the Environment?

The first organisms obtained building blocks and energy from free molecules in their environment. The first nutrients were the components of the primordial soup. Later, however, organisms themselves must have contributed their waste products. Gradually early predator organisms must have arisen to feed on other organisms. As each kind of nutrient was exhausted, evolutionary success depended on the invention of new metabolic pathways—to make new molecules from precursors or to break down the more complex molecules of other organisms.

The earliest cells were **heterotrophs** [Greek, *heteros* = other + *trophe* = nourishment], organisms that cannot derive energy from sunlight or from inorganic chemicals but obtain energy by degrading organic molecules. Modern heterotrophs obtain the needed organic molecules by eating other organisms (or their products or remains). The earliest cells, however, depended on the original supply of organic molecules present in the primordial soup. Because glycolysis is common to all present organisms, it seems likely that this pathway had already evolved in early heterotrophs—at least those whose descendants survived.

But what happened when this food supply was exhausted? The only way that life could have continued was for organisms to begin producing nutrients from inorganic molecules in their environment. One way organisms achieved such independence was to oxidize inorganic chemicals, such as sulfur, in their environment and use the derived energy to build organic molecules from CO_2. Such cells were probably the first **autotrophs,** [Greek, *autos* = self], organisms able to derive energy from inorganic chemicals or from sunlight. The present-day bacteria, *Sulfolobus,* for example, gets its energy by converting sulfur to sulfur dioxide.

Most of the organic molecules in the primordial soup stored solar energy, carried to earth by ultraviolet light. But making the soup may have taken millions of years, and organisms could quickly deplete it of nutritive value. Only by inventing new ways of harnessing energy could life continue, and the most abundant energy source continued to be sunlight. Overwhelmingly, the most successful way to such independence was through photosynthesis—capturing the energy of sunlight in energy-rich molecules.

By examining the variety of ways in which modern organisms harness solar energy, we may guess at how photosynthesis gradually developed, at least 2.2 billion years ago. The first step in the direct harnessing of solar energy by organisms was probably the development of a light-driven proton pump, such as that in the purple photosynthetic bacterium we discussed in Chapter 7. (See p. 204.) In these bacteria, the pumping of protons across the membrane leads to the storage of energy in ATP and to the active transport of other substances across the membrane.

Other bacteria, which biologists consider more advanced than the purple photosynthetic bacteria, can use light energy to generate reducing power (hydrogen atoms or electrons) to convert CO_2 to carbohydrate. Some photosynthetic bacteria obtain the needed electrons from such donors as hydrogen gas or H_2S. (See Chapter 8, p. 228.) Water is much harder to oxidize, and it seems likely that it was not the first used. But water is much more abundant than anything else, and some organisms eventually developed the capacity to use water as a source of reducing power. The ability to produce organic molecules from water and CO_2 represented an enormous advantage for photosynthetic organisms. But it produced a global toxic waste problem. Besides making carbohydrates, photosynthesis also produced molecular oxygen (O_2). Sensitive molecules, previously stable in the anaerobic environment, could not survive.

On the other hand, the availability of O_2 allowed more energy to be derived by complete oxidation of organic compounds to H_2O and CO_2. The evolution of the pathway of respiration, then, gave a tremendous selective advantage to a new class of heterotrophic organisms. The advantages of life in oxygen were so great that today anaerobic organisms persist in only a few environments.

The First Eukaryotes Probably Possessed Mitochondria

Almost all eukaryotes—protists, fungi, plants, and animals—have mitochondira and can use respiration to obtain energy. Only plants and some protists, however, have chloroplasts. To many biologists, it seems likely that the first eukaryotes had mitochondria but no chloroplasts.

Mitochondria, which produce most of a cell's energy, depend on oxygen to perform. They therefore must have appeared only after photosynthetic prokaryotes had produced lots of oxygen. The ancient origin of mitochondria may explain the amazing fact that the genetic code in mitochondria is different from the "universal" genetic code, as shown in Figure 24-12.

	Codon	UGA	AUA	CUA	AGA/AGG
	Universal code	STOP	Ile	Leu	Arg
Mitochondrial codes	Human mitochondria	Trp	Met	Leu	STOP
	Drosophila mitochondria	Trp	Met	Leu	Ser
	Yeast mitochondria	Trp	Met	Thr	Arg
	Plant mitochondria	STOP	Ile	Leu	Arg

Figure 24-12 The genetic code in mitochondria differs from the "universal" code. There are three known variants, which have been determined in the mitochondria of humans, yeast, and *Drosophila.*

How Did Mitochondria and Chloroplasts Arise?

The sizes and internal organization of chloroplasts are similar to those of cyanobacteria. Such similarities have long suggested that chloroplasts derived from previously free-living organisms. Indeed, there are many examples today of photosynthetic organisms living within animals or protists, one example of which is shown in Figure 24-13. This arrangement is called **endosymbiosis** [Greek, *endon* = within + *syn* = together + *bios* = life] because it is a close association of two organisms, one of which lives inside the other. The endosymbiotic association of a cyanobacterium and a protist, for example, is mutually beneficial because it provides an internal source of food for the host and mobility and protection for the cyanobacteria.

Most biologists now favor the view that both chloroplasts and mitochondria are the endosymbiotic descendants of free-living organisms. One of the most forceful proponents of this view has been Lynn Margulis, an evolutionary biologist now at the University of Massachusetts. In her 1970 book, *The Origin of Eukaryotic Cells,* Margulis went one step further than many other biologists have been willing to go. She speculated that the microtubules of the eukaryotic flagellum arose from a symbiosis with a spirochaete (a kind of bacteria). It would be but a short step, she argued, to construct the mitotic spindle apparatus, once cells acquired the ability to produce a flagellum. The actual origin of eukaryotic cells can never be proven, but Margulis's arguments—some of which remain controversial—have galvanized thinking on this subject. Figure 24-14 shows one model for how eukaryotic cells could have evolved.

Mitochondria and Chloroplasts Process Genetic Information Similarly to Prokaryotes

Both mitochondria and chloroplasts have independent genetic systems. (See Chapter 16.) In each case the system includes components similar to those of present-day free-living prokaryotes:

1. A circular DNA molecule not complexed with histones
2. Ribosomes smaller than those in the cytoplasm of eukaryotes
3. Protein synthesis that is inhibited by antibiotics that are specific for prokaryotes
4. tRNA structures similar to those of prokaryotes
5. Small rRNAs with stem and loop structures similar to those of prokaryotes

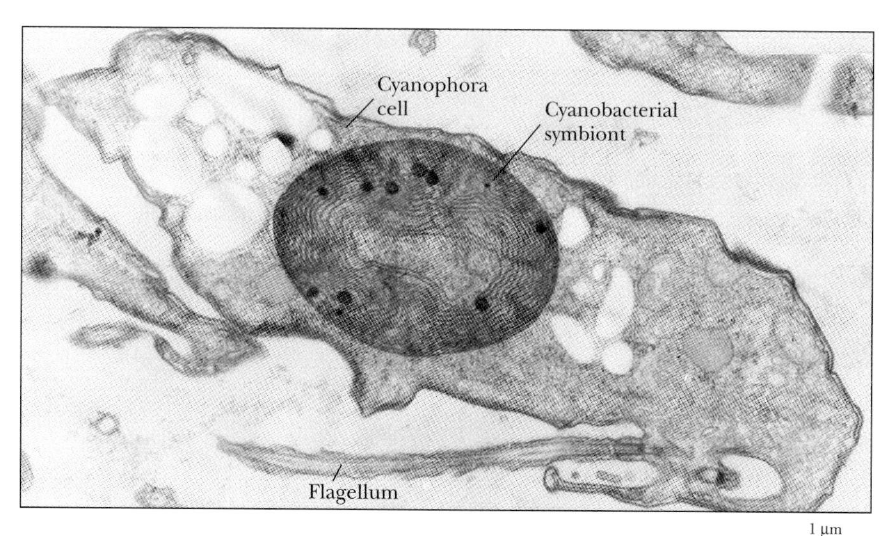

Figure 24-13 A contemporary example of endosymbiosis. The protist *Cyanophora* with an endosymbiotic cyanobacterium. *(Biophoto Associates)*

The small rRNAs of chloroplasts have stem and loop structures similar to those of cyanobacteria, while those of plant mitochondria resemble those of purple sulfur bacteria. The mitochondria of plants, yeasts, and humans, however, seem in many ways distinct from one another, and it seems likely that mitochondria arose more than once.

The analysis of the stem and loop structures of the small rRNAs supports the endosymbiotic origin of chloroplasts and mitochondria. Comparisons of rRNAs from many species have suggested a new evolutionary tree to describe the radiations of the descendants of the "universal" ancestor, as diagrammed in Figure 24-15. For example, bacteria with unusual metabolism and habitat, such as *Sulfolobus,* have rRNA structures distinct from either eukaryotes or prokaryotes that have more conventional metabolism. These bacteria seem to represent an ancient and diverse group, and many biologists think that they even comprise a separate kingdom, called the **Archaebacteria.**

Some evolutionary researchers, notably Carl Woese at the University of Illinois, have convincingly argued that—on a molecular level—the Archaebacteria are as different from conventional bacteria (eubacteria) as they are from eukaryotes. Woese has proposed to classify all organisms into three huge "domains," the *Bacteria* (eubacteria), the *Archaea* (archaebacteria), and the *Eucarya* (eukaryotes). Each domain consists of two or more kingdoms, with the Eucarya including at least four kingdoms—

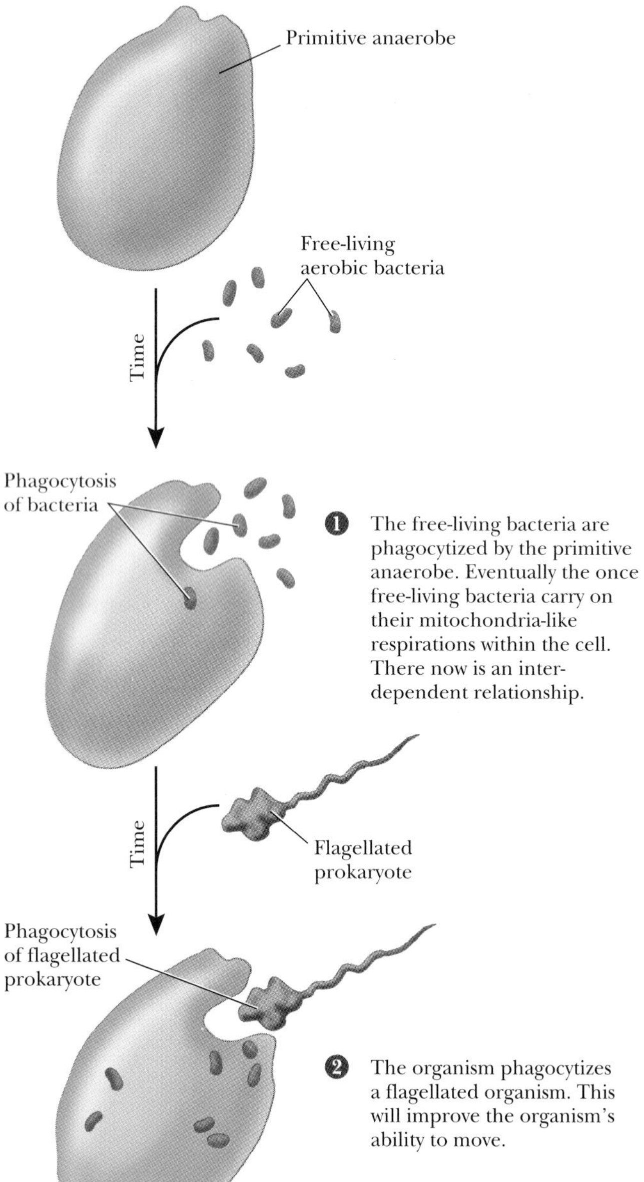

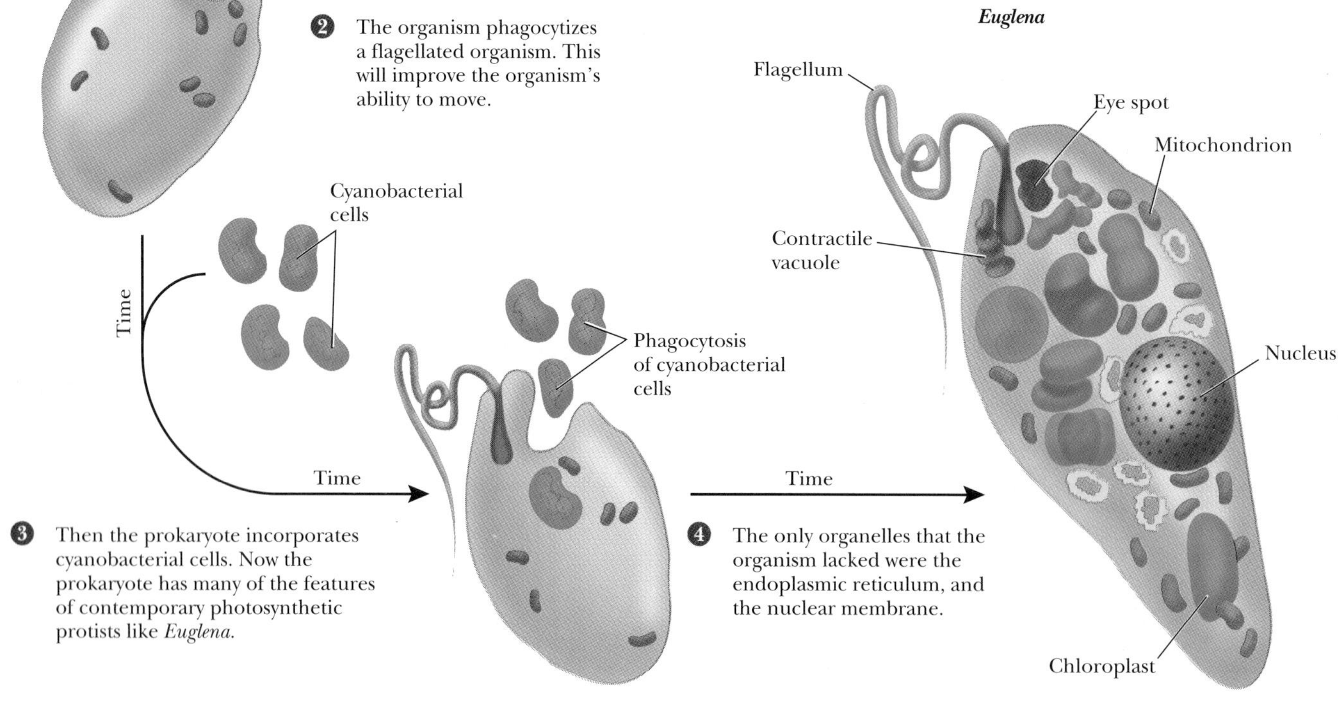

Figure 24-14 Did flagella also arise by endosymbiosis? Here is a possible scheme for the evolution of a photosynthetic protist.

ESSAY

ARCHAEBACTERIA AND BIOTECHNOLOGY

The Archaebacteria (now usually called the Archaea) include three broad groups of single-celled organisms that live under extreme environmental conditions. Biologists have found and studied three broad groups of archaea: the *hyperthermophiles* (which live at temperatures above 70°C), the *halophiles* (which live at extremely high salt concentrations), and the *methanogens* (which live only in the absence of oxygen and produce methane). As microbiologists have learned to grow these unusual organisms in the laboratory, they have learned more and more about their unusual biochemical properties. Some of the questions originally asked for purely scientific reasons have led to practical and profitable applications.

Most enzymes from conventional organisms denature above 60°C, and biochemists have long been interested in the chemical adaptations that allow the enzymes of hyperthermophilic archaea to resist denaturation and chemical destruction. In pursuit of answers to this set of questions, microbiologists collected and cultured archaea from environments that include industrial wastes, ore tailings, hot springs, and deep-sea hydrothermal vents. These studies have led to some new insights about protein stability: that enzymes of hyperthermophilic archaea have much lower levels of certain amino acids (cysteine, asparagine, and glutamine), which are chemically unstable at high temperatures, and that these enzymes have much more compact structures than their conventional counterparts.

The hyperthermophiles have provided more than insights about protein structure and activity at high temperature, however. One enzyme, the DNA polymerase from *Thermus aquaticus* (a eubacterium)—called *Taq* polymerase—is widely used in the commercial application of PCR technology. But while *Taq* polymerase is much more stable than *E. coli* DNA polymerase, it is barely stable enough for PCR, which requires high temperature to separate the strands of the DNA duplex between each round of amplification. Moreover, *Taq* polymerase cannot proofread, and it makes more mistakes than users would like. Many researchers are therefore turning to a hyperthermophilic archaeon (archaebacteria), called *Pyrococcus furiosus,* whose DNA polymerase, called *Pfu* polymerase, has an error frequency 14-fold less than *Taq* polymerase.

Other enzymes from hyperthermophilic archaea are beginning to provide catalytic power for high-temperature commercial operations. Such applications include the production of glucose from starch and the synthesis of specialized chemicals in organic solvents and even in gaseous environments. In one pilot operation, miners are using iron-oxidizing and sulfur-oxidizing thermophiles to extract uranium ore.

The second group of archaea—the halophiles—contain enzymes that can work at high concentrations of salt. In one traditional application, protein hydrolysis by naturally occurring halophilic enzymes contributes to the characteristic flavor of fish paste; on the other hand, some naturally occurring halophiles contribute to putrefaction of salted meat, hides, and dried fish. A more recent, and more deliberate, application of halophile biotechnology has been to the production of biodegradable plastics, such as polyhydroxybutyrate, already mentioned in Chapter 8, p. 243. One plastic-producing halophilic archaeon, *Haloferax mediterranei,* may provide economic opportunities for countries with hot, saline deserts, where brackish ponds either occur naturally or could be created.

Researchers have also produced electricity from artificial membranes that contain bacteriorhodopsin from the halophilic archaeon *Halobacterium halobium,* already mentioned in Chapter 7, p. 204. Other researchers have been trying to develop bacteriorhodopsin-containing membranes in "biochips" that could store and process information.

Finally, the most widely distributed archaea, the methanogens, have highly complex and diverse pathways for generating methane from simple organic compounds, including carbon dioxide, methanol, acetic acid, and carbon monoxide. Of all the archaea, the methanogens are the most widely used at a practical level, with many countries (especially China and India) using methanogens to generate energy-rich "biogas" from human and animal wastes.

The biotechnological applications of archaea appear to be increasing rapidly. The next decade will almost certainly witness applications that we have not yet even considered.

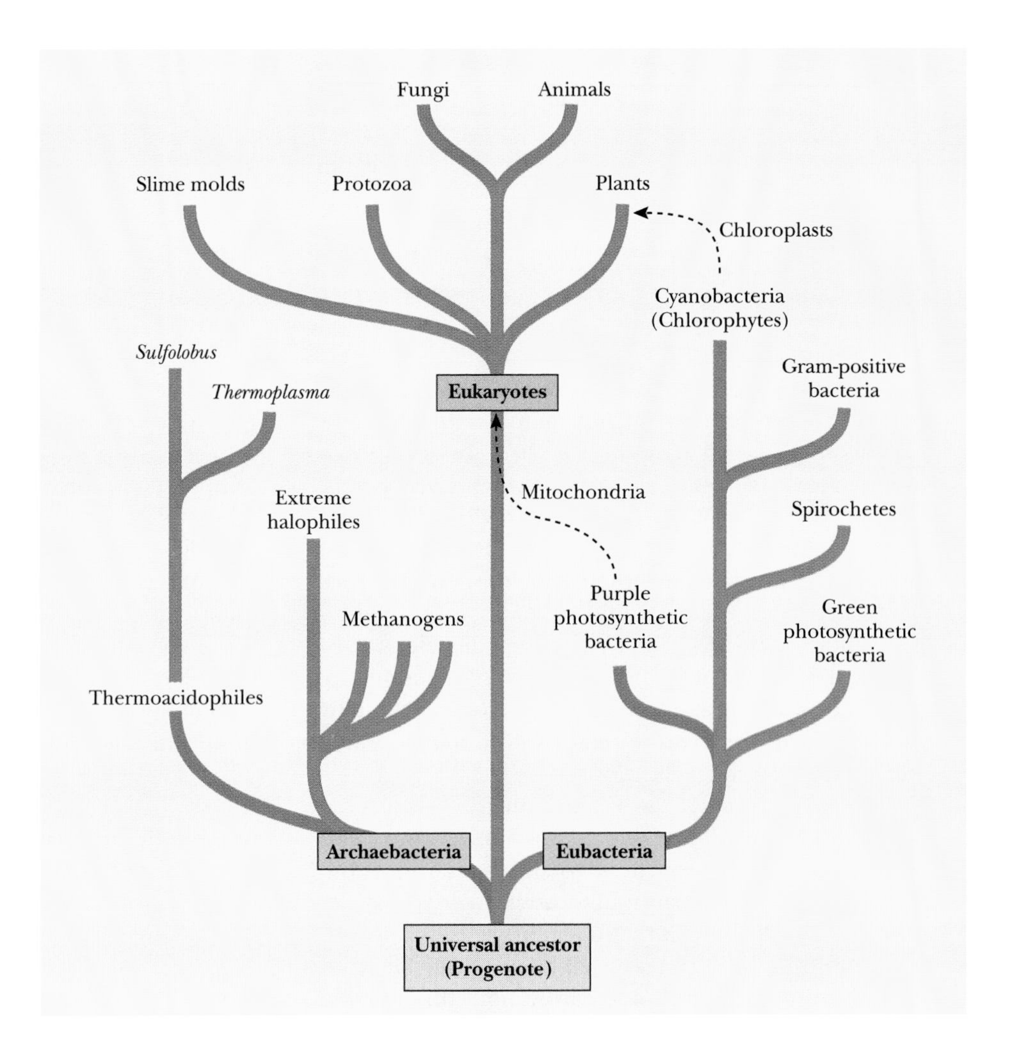

Figure 24-15 A contemporary evolutionary tree based on comparisons of rRNAs. This tree emphasizes the three "domains" of organisms. Archaebacteria, also called Archaea, are as distinct from Eubacteria as they are from Eukaryotes.

protists, fungi, plants, and animals. In the new classification, the definition of eukaryotes depends on molecular structures (such as those of rRNAs) rather than on cellular structures (such as subcellular organelles). Some biologists argue that the earliest eukaryotes existed before the rise of atmospheric O_2 and did not have mitochondria.

Some Mitochondrial and Chloroplast Genes Contain Introns, Like Current Eukaryotes

Most eukaryotic genes are split. Exons, DNA sequences that specify the amino acid sequences of segments of a polypeptide, alternate with introns, or intervening sequences, which do not specify sequences in that polypeptide. In contrast, most genes in prokaryotes are continuous.

The apparent absence of split genes in mitochondria and chloroplasts was once taken as further support of the endosymbiotic theory. But, as it turned out, nature was not so simple. Some genes in mitochondria and in chloroplasts are split. To make things still more confusing, the same gene may be continuous in the mitochondria of one species and split in another.

The strange distribution of split genes not only confuses the question of the origin of eukaryotic cells. It also makes us wonder more urgently about whether the first genes were split or continuous. Originally, many people thought that because split genes were present only in eukaryotes, they must have arisen only after the origin of eukaryotes. But many biologists now think that the first genes were split and that the continuous genes of prokaryotes were the evolutionary novelty. Here is their argument:

1. The first functional coding sequences in RNA must have specified only short segments of polypeptide.
2. Some RNAs spontaneously cut and splice themselves, as in *Tetrahymena.* Ancient split genes might similarly have brought together separately evolved domains.
3. The physical separation of smaller domains would have allowed the mixing and matching of different polypeptide segments. Parts of the sequence of two distinct enzymes of glycolysis, for example, appear to derive from the same ancestral sequence, as shown in Figure 24-16. Walter Gilbert, a molecular biologist at Harvard University, has called this process **exon shuffling.**

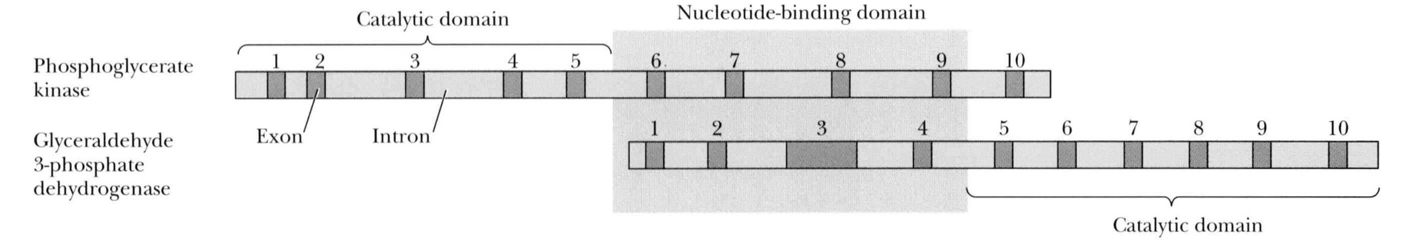

Figure 24-16 Parts of two genes—those encoding the glycolytic enzymes phosphoglycerate kinase and glyceraldehyde 3-phosphate dehydrogenase—appear to have a common origin. The evolutionary mixing of gene segments may have resulted from exon shuffling. Exons are represented in green, introns in yellow.

4. RNA splicing appears to be an ancient process. It occurs in the synthesis of tRNAs in eukaryotes and in prokaryotes. Even the archaebacterium *Sulfolobus* removes introns present in tRNA precursors.
5. Similar sequences and secondary structures appear in introns of genes encoded by nuclei and organelles.
6. Intron loss can occur. Rats, for example, have two genes for the protein hormone insulin. One of these genes contains two introns, like the insulin genes of other mammals. The second gene contains no introns. Because the introns occur in the same place in all the split insulin genes, the intron-less insulin gene seems to have arisen by intron loss. A number of similar examples have convinced most researchers that introns can be lost.
7. The continuous genes of prokaryotes could have arisen by similar intron loss. Prokaryotes divide much more rapidly than eukaryotes and have undergone far more cell divisions than eukaryotes, especially eukaryotes that reproduce sexually. The prokaryotes thus have had a far greater opportunity—and far greater selection pressure—to streamline their genes by losing introns and other excess DNA.
8. The genes of mitochondria and chloroplasts have also evolved. Some, for example, have lost introns; other genes have been transferred from these organelles to the nucleus. But mitochondrial DNAs, in particular, retain many ancestral characteristics, including long sequences of apparent nonsense.

Most biologists now agree that mitochondria and chloroplasts derive from endosymbiotic prokaryotes. Some mitochondrial and chloroplast genes have split genes because mitochondria and chloroplasts have different evolutionary histories from the free-living prokaryotes.

Some Biologists Argue That Eukaryotes Arose in a Different Manner, by the Proliferation of Internal Membranes

Some biologists still consider the endosymbiotic theory too speculative. Some of these critics reject Margulis's argument for the endosymbiotic origin of flagella and the mitotic apparatus. Others believe that chloroplasts may well have originated as free-living cyanobacteria but that the case for mitochondria is much weaker. And most biologists think that there is as yet no convincing argument for the origin of eukaryotic nuclei, although some evolutionary biologists have argued that the thermoacidophile, *Thermoplasma acidophilum,* which contains histones and actin and lacks a cell wall, may resemble the organism that provided the first nucleus. *Cryptomonas,* a photosynthetic organism, may contain an organelle similar to the primordial nucleus.

Opponents of the endosymbiotic theory argue that eukaryotes arose by the proliferation of membranes to form new cellular compartments. Lysosomes and the endoplasmic reticulum, for example, form from cellular membranes—not from free living forms. Perhaps mitochondria and chloroplasts formed when membranes surrounded pieces of mobile DNA (plasmids or transposons) that detached from nuclear genes.

The endosymbiotic theory produces a more complicated evolutionary tree than the traditional view. With this complexity, however, comes a consistency with current views of prebiotic evolution. For the moment, at least, most biologists seem to find endosymbiosis the most convincing general explanation for the origin of at least some subcellular organelles—especially mitochondria and chloroplasts—within eukaryotic cells.

PROTEINS AND NUCLEIC ACIDS PROVIDE A MOLECULAR CLOCK FOR EVOLUTIONARY BIOLOGISTS

The ability to determine the amino acid sequences of proteins and the nucleotide sequences of DNA allows biologists to compare the molecular properties of living species (and, in some cases, even of fossils in which some DNA has been preserved). Such studies, like studies of visible structures, have led to hypotheses concerning the evolution of species in general, and particularly of the **primates,** the group (or "order") that includes humans, monkeys,

(a) Cytochrome c sequences

Species	Sequence (positions −9 … −1, 1 … 40+)
Human/chimpanzee	a G D V E K G K K I F I M K C S Q C H T V E K G G K H K T G P N L H G L F G R K T G Q A
Rhesus monkey	a G D V E K G K K I F I M K C S Q C H T V E K G G K H K T G P N L H G L F G R K T G Q A
Cow, pig, sheep	a G D V E K G K K I F V Q K C A Q C H T V E K G G K H K T G P N L H G L F G R K T G Q A
California gray whale	a G D V E K G K K I F V M K C A Q C H T V E K G G K H K T G P N L H G L F G R K T G Q A
Chicken, turkey	a G D I E K G K K I F V Q K C S Q C H T V E K G G K H K T G P N L H G L F G R K T G Q A
Rattlesnake	a G D V E K G K K I F T M K C S Q C H T V E K G G K H K T G P N L H G L F G R K T G Q A
Tuna	a G D V A K G K K T F V Q K C A Q C H T V E N G G K H K V G P N L W G L F G R K T G Q A
Tobacco hornworm moth	h G V P A G N A D N G K K I F V Q R C A Q C H T V E A G G K H K V G P N L H G F F G R K T G Q A
Drosophila (fruit fly)	h G V P A G D V E K G K K L F V Q R C A Q C H T V E A G G K H K V G P N L H G L I G R K T G Q A
Baker's yeast	h T E F K A G S A K K G A T L F K T R C L Q C H T V E K G G P H K V G P N L H G I F G R H S G Q A
Neurospora crassa (a mold)	h G F S A G D S K K G A N L F K T R C A Q C H T L E E G G G N K I G P A L H G L F G R K T G S V
Wheat germ	a A S F S E A P P G N P D A G A K T F K T K C A Q C H T V D A G A G H K Q G P N L H G L F G R Q S G T T

(b) Family tree based on cytochrome c

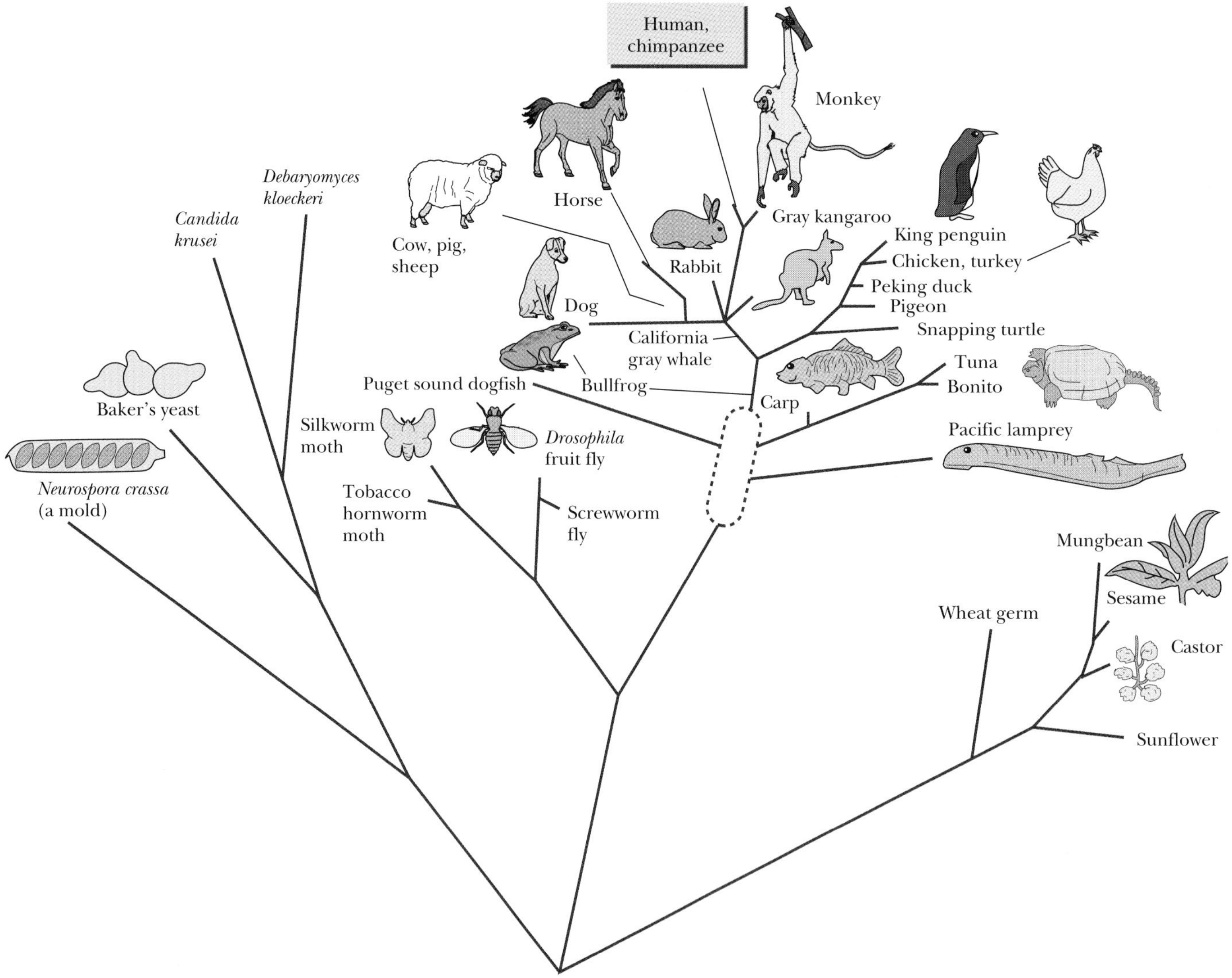

Human, chimpanzee
Monkey
Horse
Cow, pig, sheep
Rabbit
Gray kangaroo
King penguin
Chicken, turkey
Peking duck
Pigeon
Dog
Snapping turtle
California gray whale
Bullfrog
Tuna
Bonito
Puget sound dogfish
Carp
Pacific lamprey
Debaryomyces kloeckeri
Candida krusei
Baker's yeast
Neurospora crassa (a mold)
Silkworm moth
Drosophila fruit fly
Tobacco hornworm moth
Screwworm fly
Mungbean
Sesame
Wheat germ
Castor
Sunflower

Figure 24-17 The use of sequence information, here from cytochrome c, as a molecular clock; cytochrome c of humans is identical to that of chimpanzees, but has 12 differences from the cytochrome c of horses and 44 from the bread mold *Neurospora*. Averaging the differences of many species, fish differ from vertebrates at 19 positions and insects at 27. **(a)** Sequence alignments of cytochrome c from several species; **(b)** an evolutionary tree of all kingdoms, based on cytochrome c sequences; **(c)** a cytochrome c molecule. *(c, Irving Geis)*

45	50	55	60	65	70	75	80	85	90	95	100 (Difference) – 104
P G Y S Y	T A A N K	N K G I I	W G E D T	L M E Y L	N P K K Y	I P G T K	M I F V G	I K K K E	E R A D L	I A Y L K	K A **T** N E
P G Y S Y	T A A N K	N K G I I	W G E D T	L M E Y L	N P K K Y	I P G T K	M I F V G	I K K K E	E R A D L	I A Y L K	K A **A** N E
P G F S Y	T D A N K	N K G I T	W G E E T	L M E Y L	N P K K Y	I P G T K	M I F A G	I K K K G	E R E D L	I A Y L K	K A A N E
V G F S Y	T D A N K	N K G I T	W G E E T	L M E Y L	N P K K Y	I P G T K	M I F A G	I K K K G	E R A D L	I A Y L K	K A A N E
E G F S Y	T D A N K	N K G I T	W G E D T	L M E Y L	N P K K Y	I P G T K	M I F V G	I K K K S	E R V D L	I A Y L K	D A T S K
V G Y S Y	T A A N K	N K G I T	W G D D T	L M E Y L	N P K K Y	I P G T K	M V F V G	L S K K K	E R T N L	I A Y L K	E K T A A
E G Y S Y	T D A N K	S K G I T	W N N D T	L M E Y L	N P K K Y	I P G T K	M I F A G	I K K K G	E R Q D L	V A Y L K	S A T S
P G F S Y	S N A N K	A K G I T	W F E D T	L M E Y L	N P K K Y	I P G T K	M V F A G	L K K A N	E R A D L	I A Y L K	Q A T K
A G F A Y	T N A N K	A K G I T	W Q D D T	L F E Y L	N P K K Y	I P G T K	M I F A G	L K K P N	E R G D L	I A Y L L	K S A T K
Q G Y S Y	T D A N I	K K N V N	W D E N N	M S E Y L	N P X K Y	I P G T K	M A F G G	L K K E K	D R N D L	I T Y L K	K A A C E
D G Y A Y	T D A N K	Q K G I T	W D E N T	L F E Y L	N P X K Y	I P G T K	M A F V G	L K K D K	D R N D I	I T F M K	E A T A
A G Y S Y	S A A N K	N K A V T	W E E N T	L Y D Y L	N P X K Y	I P G T K	M V F P G	L X K P Q	D R A D L	I A Y L K	K A T S S

(c) **Cytochrome c**

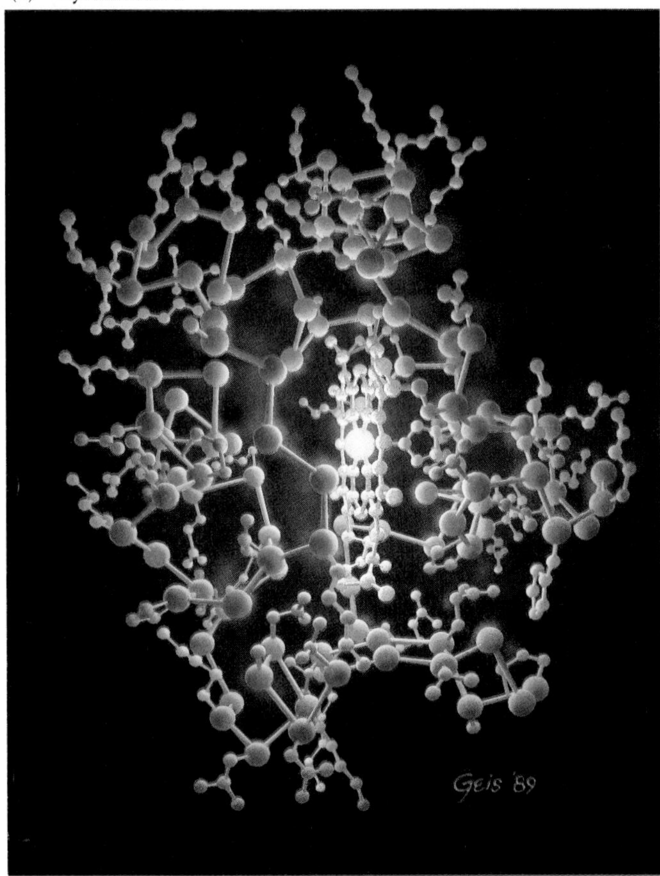

apes, and the lemurs of Madagascar. The distinguishing characteristics of the primates are feet (or hands) with five digits and a grasping thumb, relatively large brains, flexible shoulder joints, and eyes adapted for binocular vision.

The Comparative Anatomy of the Primates Suggests Their Family Tree

Monkeys, apes, and humans—the **anthropoids**—are more like each other than we are like lemurs; we are generally bigger and have shorter snouts. New World monkeys, which inhabit South and Central America, are distinct in two ways from humans, apes, and the Old World monkeys, which inhabit Asia and Africa: The New World monkeys have prehensile (grasping) tails, and their nostrils are more separated and point sideways rather than down. Our group of anthropoids is therefore called the *Catarrhines* [Greek, *kata* = down + *rina* = nose].

Finally, humans and apes—together called the *hominoids* [Latin, *homo* = man] have characteristics not shared by other catarrhines: We have large skulls and long arms and tend to walk at least partially erect. The living hominoids include the gibbons, orangutans, gorillas, chimpanzees, and humans. The similarities and differences, together with fossil data, provide the basis for much of our current understanding of early human evolution.

Family Trees Derived from Molecular Sequence Comparisons Are Consistent with Comparative Anatomy and the Fossil Record

Just as the comparison of molecular structures has led to a new view of early evolution, similar (and simpler) comparisons of amino acid and nucleotide sequences have led to an understanding of later evolutionary relationships. One of the first macromolecules whose structure biochemists compared in many organisms was cytochrome c, a participant in the electron transport chain. (See Chapter 7, p. 199). Cytochrome c is a single polypeptide chain with 104 amino acid residues. Its exact amino acid sequence has been determined in dozens of species.

In all of the species examined, 35 of the 104 residues are always identical. The remaining 69 residues vary from species to species, with fewer differences between species that visibly resemble each other and more differences between species that biologists had already thought to be more distantly related. The similarities and differences among the amino acid sequences appear to reflect evolutionary history in the same way as more visible comparisons. Researchers have used these comparisons to draw a family tree, shown in Figure 24-17, that includes animals,

(Text continues on page 697.)

ESSAY

MITOCHONDRIAL MOLECULAR CLOCKS AND HUMAN EVOLUTION

Molecular analysis predicts that the common ancestors of gorillas, chimpanzees, and humans diverged from the orangutan's ancestors about 8 million years ago, while the common ancestors of humans and chimpanzees may have lived within the last 3–4 million years. The fossil evidence generally supports this view, although the study of human evolution suffers from having so few fossil specimens.

Distinctively human species, called *hominids,* have two defining characteristics: (1) They were bipedal; that is, they walked on two rather than four feet; and (2) they had rounded, rather than rectangular jaws. More recent hominids also had brain cases much larger than those of other anthropoids, including the older hominids.

The earliest fossils that are clearly hominids are the *australopithecines* [Latin, *australis* = south + Greek, *pithekos* = ape], which date from 3.6 million to 1.4 million years ago. These fossils, like all other hominid fossils older than about 1.5 million years, come from Africa, south of the Sahara desert. The australopithecines appeared to fall into four distinct species, which lived at slightly different but overlapping times. During a 2-million-year period australopithecine species became taller and acquired bigger brains.

The first fossils recognized as members of our genus, *Homo,* date from about 2 million years ago. These hominids are called *Homo habilis* [Latin, *habilis* = able] because their fossils lie next to stones that had been shaped into crude tools. The fossils were also adjacent to the bones of large animals, which *Homo habilis* had apparently butchered. *Homo habilis* stood about 1.7 meters (5 feet) tall and had an average brain size of about 750 cubic centimeters. Perhaps the larger brains allowed the greater level of planning needed to make and use tools. Fossils of this species extend over a period of 500,000 years.

Another distinctive hominid species, *Homo erectus,* first appeared about 1.6 million years ago. *Homo erectus* appears to have been the first hominid to leave Africa, with 1–1.2 million-year-old fossils found in China and Indonesia. *Homo erectus* skeletons are larger than those of *Homo habilis;* the brain case is larger, ranging to more than 1000 cubic centimeters; and *Homo habilis's* tools are far more sophisticated. Hand axes of the same design occur at many sites, suggesting that *Homo habilis* could both teach and learn. *Homo habilis* seems also to have learned, about 500,000 years ago, to control fire. This ability would not only have helped warm *Homo habilis* in colder, non-African climates, but would also have increased the digestibility and energy yield from food.

Our own species, *Homo sapiens,* with an average brain size of 1500 cubic centimeters, first appeared about 300,000 years ago. Paleontologists can distinguish at least two groups of *Homo sapiens,* an early group called archaic *Homo sapiens* and a later group (including ourselves) called *Homo sapiens sapiens,* which first appeared about 90,000 years ago. The two groups differ in some physical features as well as in the types of tools that they used.

Anthropologists now agree that during evolution hominids achieved upright posture before increased brain size. Various anthropologists and paleontologists, however, have espoused two distinct views (summarized in Figure 1) concerning recent human evolution: (1) that *Homo erectus* migrated from Africa, starting about 1.5 million years ago, with subsequent waves of migration evolving separately into varieties, or races, that differ from one another; or (2) that *Homo erectus* in its various sites was replaced first by archaic *Homo sapiens* and finally by *Homo sapiens sapiens,* each of which migrated separately from Africa.

Current evidence supports the latter view because all hominids of the last 30,000 years are *Homo sapiens sapiens.* No fossils have characteristics that are intermediate between the two varieties of *Homo sapiens,* nor do any known fossils have properties intermediate between *Homo sapiens* and *Homo erectus.* In addition to the fossil evidence, however, recent molecular analyses suggest that all humans derive from a relatively recent common ancestor.

Rebecca Cann and Allan Wilson, molecular biologists then at the University of California, Berkeley, compared sequences in the mitochondrial DNA from the cells of people from around the world. Sequences in mitochondrial DNA, like sequences in chromosomal DNA, accumulate mutations at a relatively constant rate, so they can serve as a molecular clock. The mitochondrial clock ticks faster (accumulates muta-

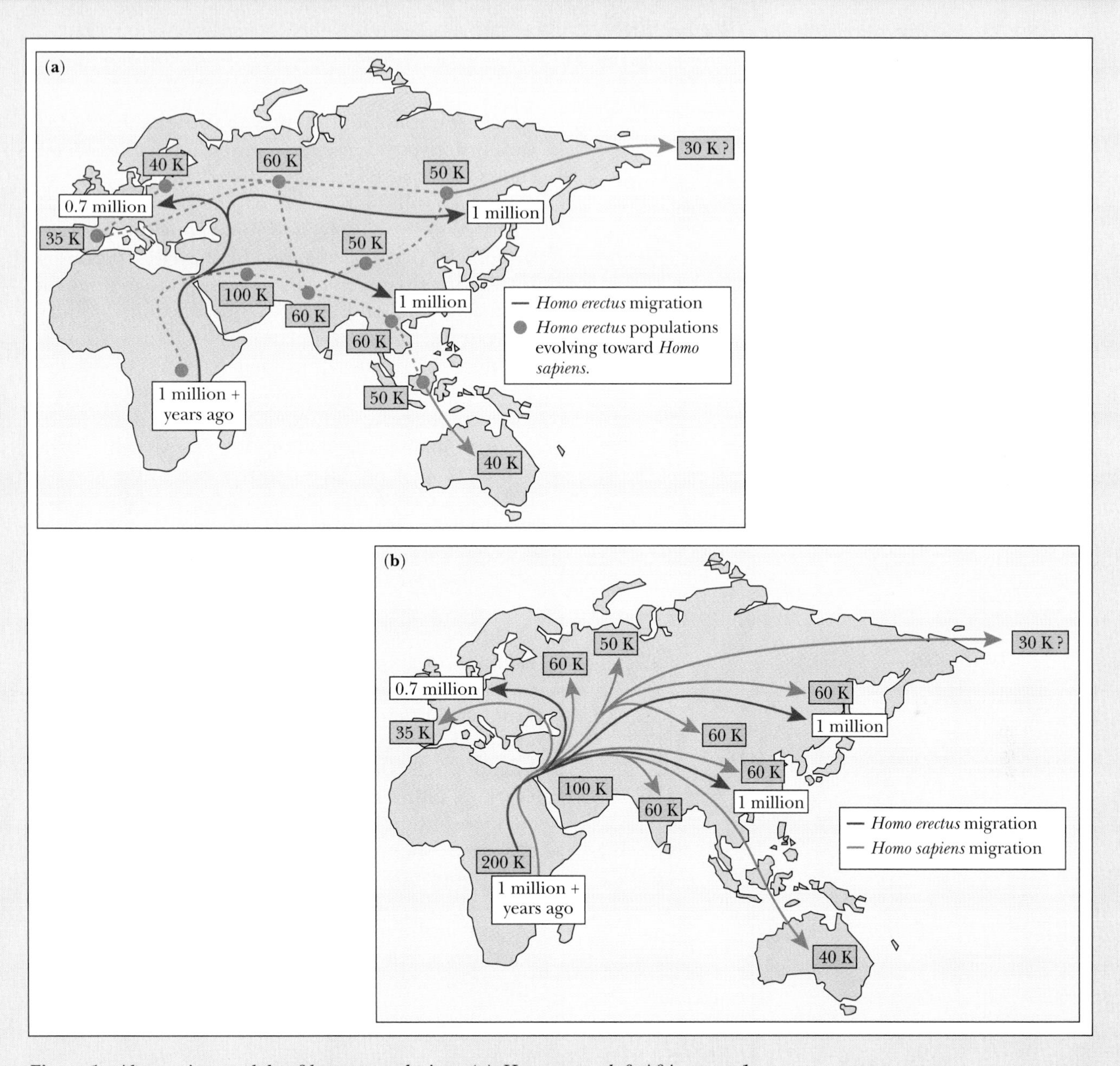

Figure 1 Alternative models of human evolution: **(a)** *Homo erectus* left Africa over 1 million years ago and evolved into *Homo sapiens* in many different places; **(b)** different hominid species replaced one another; *Homo erectus* migrated from Africa and was later replaced by *Homo sapiens,* who migrated from Africa only 200,000 years ago.

tions more rapidly) than most other DNA sequences, so it was especially useful for the study of recent human evolution.

Cann and Wilson found many fewer sequence differences among people than anyone had expected. This suggests that only a short time—about 200,000 years—has elapsed since the common ancestor of all humans was alive. As illustrated in Figure 2, non-Africans have fewer sequence differences than Africans, suggesting a still more recent origin. Cann and Wilson suggested that descendants of the original ancestor left Africa about 100,000 years ago, beginning the colonization of the rest of the planet. This

(continued)

ESSAY *(continued)*

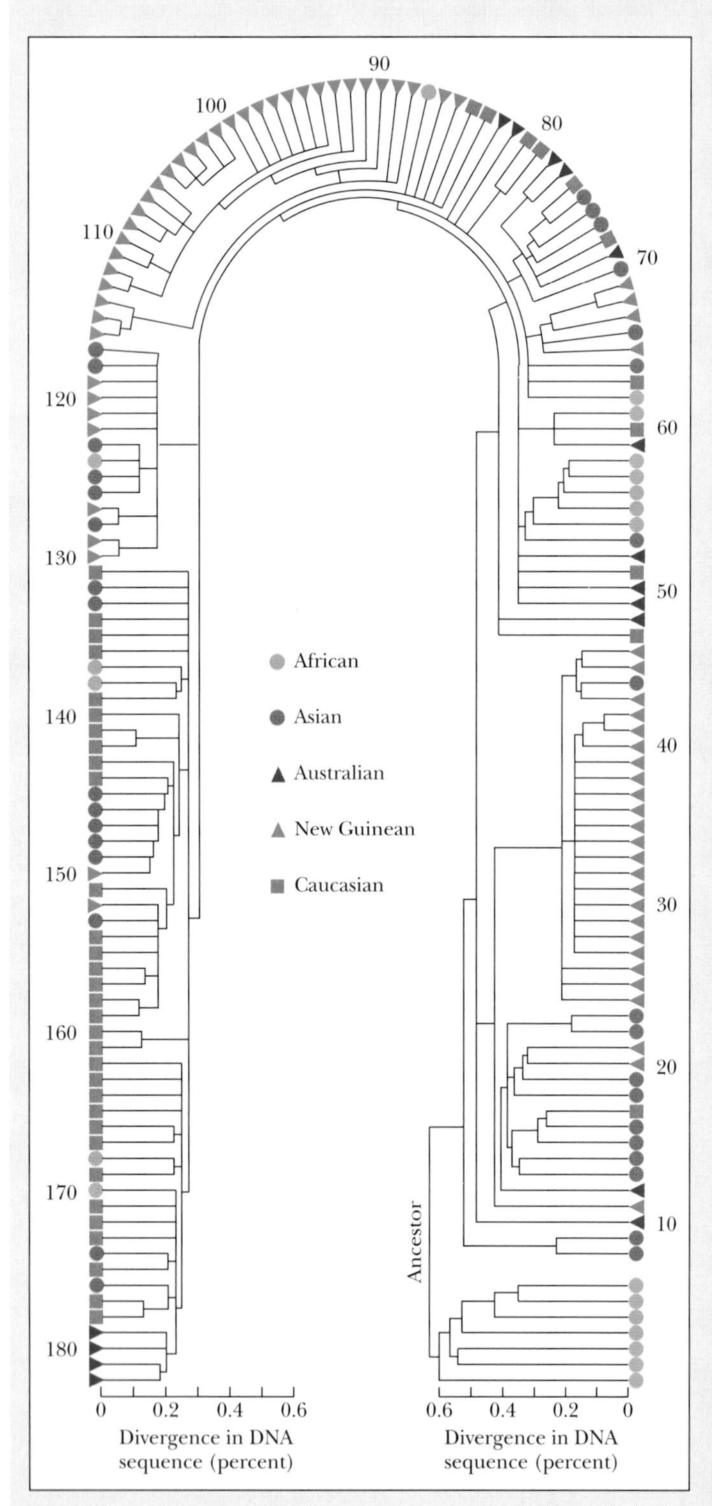

conclusion is not consistent with the idea that human variety derives from separate evolution from different stocks of *Homo erectus*.

At fertilization, each parent provides a set of chromosomes, so each contributes equally to the chromosome of the offspring. But, because the sperm's mitochondria do not enter the egg, only the mother contributes mitochondrial DNA. All the mitochondrial DNA in the human population thus derives from that of one woman, now affectionately known as the *mitochondrial Eve*. This conclusion does not mean that at some recent point in human history, there was only a single woman engaged in reproductive activity. Rather it means, that, by chance, only one mitochondrial DNA was passed on, perhaps from a population of hundreds or thousands. Still, the story is a surprising one: All humans appear to be more closely related than most researchers had previously suspected.

Like other arguments about human evolution, the use of mitochondrial DNA data to support a recent common origin for all humans has evoked much discussion and debate. In particular, a number of researchers criticized the conclusions of Cann and Wilson on statistical grounds. Their original analysis, for example, relied on the analysis of sites for restriction enzymes, rather than on DNA sequences. More recent comparisons, now based on DNA sequences for a specific gene, support the conclusion that the entire human population had a common origin sometime between 220,000 and 370,000 years ago. Some researchers argue that the mitochondrial Eve lived in Asia rather than Africa, but most of the available evidence is still consistent with the "Out-of-Africa" hypothesis.

Figure 2 Analysis of mitochondrial DNA suggests that all humans originated within the last 200,000–400,000 years, probably in Africa. The greater sequence variation of mitochondrial DNA among contemporary Africans and the ability to construct an evolutionary tree for other mitochondrial DNAs are both consistent with a single African origin. *(Adapted from "The Recent African Genesis of Humans," by Allan C. Wilson and Rebecca L. Cann, © April, 1992 by* Scientific American, Inc. *All rights reserved.)*

plants, and fungi. This tree is essentially identical to a tree made on the basis of other comparative studies. This correspondence helped convince more traditional evolutionary biologists of the power of such molecular comparisons.

The cytochrome c sequences of humans and chimpanzees are identical, and there is only one difference between human cytochrome c and that of the rhesus monkey (a macaque), an Old World monkey. Researchers have determined the amino acid sequences of many other proteins, as well as the nucleotide sequences of many genes. Such molecular comparisons led to the same suggestions concerning anthropoid evolution as the comparisons of visible structures. In addition, however, the molecular data suggested—more clearly than other kinds of comparisons—that humans and chimpanzees are more closely related to each other than either species is to gorillas or orangutans.

By comparing the sequences of proteins and DNA, molecular biologists have proposed many hypotheses about the order in which different groups of organisms diverged from each other during evolution. By and large, these hypotheses are consistent with evidence from the fossil record. Consistent with what we can infer from comparative anatomy, for example, molecular analysis predicts that the common ancestor of humans and monkeys lived after the common ancestor of mammals and reptiles. The fossil record is totally consistent with this: The first anthropoid fossils date from about 30 million years ago, while fossils of the first mammal-like reptiles date from at least 180 million years ago.

By using dates from the fossil record, evolutionary biologists have discovered a somewhat unexpected rule: The number of differences in a given molecule in two species is proportional to the time since their last common ancestor. The changes appear to accumulate in each type of protein or in each corresponding piece of DNA like ticks of a "molecular clock." Each gene or protein, however, changes at a different rate, so each clock needs a separate calibration. Counting such differences allows researchers to predict when two species diverged, even if they have no fossil record of a common ancestor.

■ SUMMARY

Some of the current understanding of the origin of life depends on fossils from the Precambrian era and some from comparisons of the properties of current organisms. Much of the discussion of early life, however, depends on the demonstration that plausible chemical reactions could have produced molecules and organized structures similar to those now seen in organisms.

Precambrian rocks contain fossils as old as 3.5 billion years, only 300 million years after the formation of the oldest known rocks. While some evolutionary biologists have argued that life may have arrived from elsewhere in the universe, most researchers think that life evolved from nonliving organic molecules formed on earth.

During the earth's early history, the atmosphere contained no free oxygen but probably contained hydrogen, ammonia, and methane, as well as other gases. In the laboratory, a mixture of these molecules, exposed to electrical discharges, produces many organic molecules, including amino acids and precursors of nucleotides. Scientists think that the primitive oceans became loaded with such organic molecules. These organic molecules, either within the oceans themselves or after drying onto solid surfaces, can form polymers that resemble proteins and nucleic acids.

Mixtures of different kinds of molecules can assemble into organized structures. Some simple structures can take up molecules from their environments and convert them into other compounds. Living cells, however, must be able to couple protein synthesis to genetic instructions.

By comparing the characteristics of protein synthesis in contemporary organisms, scientists have inferred that the first organisms used RNA as the genetic material, with an early version of the triplet genetic code. Protein synthesis probably used a primitive type of tRNA, perhaps with as few as seven nucleotides. RNAs, rather than ribosomes, probably catalyzed the formation of peptide bonds.

Eukaryotes evolved long after prokaryotes, probably after the appearance of oxygen in the atmosphere. Most atmospheric oxygen derived, then as now, from organisms that perform photosynthesis. Photosynthetic organisms probably evolved after early life had depleted the primitive oceans of energy-rich organic molecules.

Most biologists believe that mitochondria and chloroplasts derived from previously free-living organisms. For the most part, mitochondria and chloroplasts carry and process genetic information in ways that resemble those of present-day prokaryotes. In one respect, however, some mitochondrial and chloroplast genes resemble those of eukaryotic nuclei: Their protein coding sequences are interrupted by introns. Some researchers interpret the presence of such split genes to show that the first genes were themselves split and that the unsplit genes of present prokaryotes evolved later. Others suggest that mitochondria and chloroplasts may not have derived from prokaryotes, but from membrane-enclosed pieces of mobile DNA from eukaryotic nuclei.

The fossil record of evolution is incomplete because the chances are small that any given organism will become a fossil. In spite of its limitations, however, the fossil record gives a clear outline of the history of life. Fossils also provide evidence for the hypotheses suggested by anatomical and molecular comparisons of living species. By using dates from the fossil record, evolutionary biologists have discovered that the number of differences between the sequences of amino acids in a particular protein (or of nucleotides in a particular gene) are proportional to the time since the last common ancestor of the species being compared. Counting such molecular differences allows researchers to predict when two species diverged.

KEY TERMS

adaptive radiation
anthropoid
Archaebacteria
autotroph
coacervate
endosymbiosis
eon
era
exon shuffling
fossil
heterotroph
paleontology
prebiotic evolution
Precambrian era
primate
protobiont

QUESTIONS FOR REVIEW AND UNDERSTANDING

1. How did the discovery of microscopic Precambrian fossils contribute to our understanding of early evolution?
2. What are two explanations to account for the appearance of cells on early earth?
3. What conditions on early earth would have facilitated prebiotic evolution?
4. What were the results of Miller's experiments? What conclusions could be drawn from them?
5. How could polypeptides have been formed prebiotically? How would they differ from polypeptides found in cells?
6. What characteristics of RNA molecules led us to suggest a primary role for these molecules in early evolution?
7. Why are compartmented structures such as coacervates thought to be important in allowing life to evolve?
8. Why is it thought that metabolic pathways "evolved backwards"?
9. Why was the development of a genetic system critical to evolution of cells?
10. All present-day organisms use the same genetic code. What does this tell us about the evolution of cells?
11. Why is the evolution of tRNA molecules to carry amino acids particularly difficult to conceptualize?
12. Why is it logical that ribosomal RNA plays such a critical role in translation?
13. Why is it logical to suggest that heterotrophs preceded autotrophs?
14. What were the contributions of early photosynthetic organisms to the evolutionary opportunities for heterotrophic species?
15. Outline the evidence supporting the endosymbiotic origin of chloroplasts and mitochondria.
16. Summarize the arguments favoring the view that the first genes contained introns.
17. How does the ability to compare sequences of proteins and nucleic acids allow for a better understanding of evolutionary relationships among organisms?

SUGGESTED READINGS

ALBERTS, B., D. Bray, J. Lewis, M. Raff, K. Roberts, and J. D. Watson, *Molecular Biology of the Cell,* 3rd edition, Garland, New York, 1994. Chapter 1 deals with the evolution of the cell.

LEWIN, R., *The Origin of Modern Humans,* Scientific American Library, 1993. A nice introduction to current issues concerning human evolution.

SIMPSON, G. G., *Fossils and the History of Life,* Scientific American Books, New York, 1983. A wonderful introduction to paleontology by a master.

WOESE, C. R., O. Kandler, and M. L. Wheelis, "Towards a natural system of organisms: Proposal for the domains Archaea, Bacteria, and Eucarya," *Proceedings of the National Academy of Sciences, U.S.A.* 87, 4576–4579, 1990. A powerful argument for revising the five-kingdom and six-kingdom classification, highly relevant to understanding the nature of the first cells.

GLOSSARY

A band [for *anisotropic,* meaning that it changes the direction of polarized light] In muscle, the region of a sarcomere that contains myosin filaments.

abscisic acid A plant hormone.

absorption spectrum The relative amounts of light of different wavelengths that a substance or solution absorbs.

absorptive state The state of an animal shortly after a meal, during which cells in many organs begin to take up needed fuel and building blocks; cells take up glucose, make glycogen, and increase the synthesis of fats and proteins.

acetyl CoA A compound that consists of the two-carbon acetic acid (acetate) linked to a larger molecule called coenzyme A; acetyl CoA is the end point of glycolysis and enters the citric acid cycle.

acetylcholine A neurotransmitter used in the neuromuscular junction to stimulate muscle contraction.

acetylcholinesterase An extracellular enzyme that destroys acetylcholine soon after it is released into a synapse.

acid A molecule (or part of a molecule) that can give up a hydrogen ion.

acquired immunodeficiency syndrome (AIDS) A disease characterized by a lack of T_H cells; AIDS patients cannot mount effective immune responses to infections or to certain kinds of cancers.

acridine dye A chemical mutagen that causes mutations by inserting itself into the backbone of the DNA double helix.

acrocentric Refers to a chromosome in which the centromere is near one end; such a chromosome looks like a bent capital L.

acrosome A specialized lysosome in a spermatozoan; contains enzymes that allow the sperm to enter the egg.

actin The protein that makes up actin filaments; closely related to the protein first identified as making up the thin filaments of muscle.

actin filaments, also called **microfilaments** The most flexible elements of the cytoskeleton; formed by the association of actin molecules into long filaments about 7 nm in diameter.

action potential, also called a **nerve impulse** A rapid, transient, and self-propagating change of voltage across the membrane of a neuron or a muscle cell; allows long-distance signaling in the nervous system.

action spectrum The relative effectiveness of different wavelengths in promoting a specific light-dependent process.

activation energy The minimum energy needed for a process to occur.

active site A groove or cleft on an enzyme's surface to which a substrate binds.

active transport Movement of a substance across a membrane in a manner that does not occur spontaneously but requires an expenditure of energy.

adaptive radiation The derivation of many species from a single ancestor, with each species specialized to a distinct way of life.

adenosine diphosphate (ADP) [Greek, *di* = two] A nucleotide that consists of adenosine, ribose, and two phosphate groups; the product of the hydrolysis of a single phosphate from ATP.

adenosine triphosphate (ATP) A molecule that provides energy for many biochemical processes in all organisms.

adenylate cyclase The enzyme that catalyzes the conversion of ATP into cyclic AMP.

adhering junction A molecular assembly on the surface of an animal cell that connects the cell surface to actin filaments on the inner surface of the plasma membrane.

adrenal virilism The development of male secondary sexual characteristics in women with excessive adrenal gland activity.

adrenergic receptor [from Adrenaline (as in *adrenal* gland), the patented British trade name for epinephrine] One of a set of membrane proteins that binds to epinephrine and initiates a biological response.

aequorin A protein, isolated from a luminescent jellyfish, that emits light when it binds Ca^{2+} ions.

aerobic In oxygen.

afferent Carrying sensory information from peripheral nerves or sense organs to the central nervous system.

affinity chromatography A type of chromatography in which the stationary phase is a protein or small molecule bound to an insoluble material and packed into a column.

affinity labeling The chemical attachment of a radioactive (or fluorescent) ligand to a binding site within a protein.

aging Progressive developmental changes in an adult organism; includes cell death, changes in the extracellular environment, and the decreased function of many organs.

agonist A ligand that affects a target cell in the same way as the natural signal itself.

aleurone layer In a seed, a border of cells that surrounds the endosperm.

allele One of several variant versions of the same gene.

allergy An inflammatory response to a harmless antigen; involves the production of IgE antibodies and the release of histamine from mast cells.

allosteric effector [because it acts other than sterically] A molecule or ion that changes the activity of an enzyme by binding to a site different from the active site.

α helix A common secondary structure in proteins in which every carbonyl group of the polypeptide backbone is hydrogen bonded to the amino group four amino acids farther down the polypeptide chain.

altered self In the immune response, nonself antigens presented in the context of an MHC molecule.

alternation of generations A sexual life cycle in which haploid and diploid phases alternate.

Alzheimer's disease A disease usually associated with aging, in which patients suffer massive memory loss.

ameboid movement, also called **cell locomotion** or **cell crawling** The movement of cells over a surface.

amine A chemical signal derived from an amino acid, usually by the removal of a carboxyl group.

amino A functional group that consists of a nitrogen and two hydrogen atoms (NH_2).

amino acid A small molecule that contains both amino and carboxyl groups; the building block of polypeptides and proteins.

aminoacyl tRNA synthetase A "decoding enzyme" that links a transfer RNA to its corresponding amino acid to form a "charged" (aminoacyl) tRNA.

aminoacyl-tRNA A transfer RNA molecule linked at its 3′ end to the appropriate amino acid.

amniocentesis [Greek, *centes* = puncture] Sampling of cells in the amniotic fluid, which is withdrawn with a syringe.

amniotic fluid [Greek, *amnion* = membrane around a fetus] The fluid that surrounds the fetus.

amphipathic [Greek, *amphi* = both + *pathos* = feeling] Having both a hydrophilic and a hydrophobic region.

amphipathic helix A class of transcription factors that contain α-helices with nonpolar side chains extending from one side, allowing the formation of dimers.

amyloplast In a plant, a plastid that contains large granules of starches.

anabolism [Greek, *ana* = up + *ballein* = to throw] The synthesis of complex molecules from smaller ones.

anaerobic Without oxygen.

anaphase [Greek, *ana* = up, again] The stage of mitosis during which sister chromatids separate.

anaphase A The part of anaphase during which the chromosomes move.

anaphase B The part of anaphase during which the poles of the spindle move apart.

anastral Having neither centrioles nor asters; a characteristic of mitosis in vascular plants.

anchoring junctions In animal cells, molecular assemblies that attach cells to each other or to extracellular matrix; anchoring junctions help provide strength to resist mechanical stress while not affecting the passage of molecules between them.

aneuploid Having an abnormal number of chromosomes.

Animalia The kingdom that includes animals, heterotrophic multicellular organisms that undergo embryonic development.

anion [so called because it will move toward a positively charged electrode, or anode] A negatively charged ion.

antagonist A ligand that prevents a chemical signal from binding to its receptor and triggering its characteristic response in target cells.

antagonistic pair Two muscles that accomplish opposing movements.

antenna complex In a chloroplast, an association of chlorophyll and carotenoids that traps light and transfers the energy to the chlorophyll molecules that actually participate in photosynthesis.

anthropoid A monkey, ape, or human.

antibiotic A substance produced by a microorganism that kills (or interferes with) other microorganisms.

antibody A blood protein that forms complexes with molecules (antigens) such as those on the surfaces of microorganisms.

anticodon A sequence of three nucleotides in transfer RNA that forms specific base pairs with the corresponding codon sequence in mRNA.

antigen [an *antibody* generator] A molecule that stimulates the production of an antibody, or of another immune response; a molecule that binds to an antibody or other recognition molecule within an animal's immune system.

antigen presenting cell (APC) In the immune system, a cell that produces MHC Class II molecules and can present exogenous antigens, such as those present in grafts, transplants, or extracellular parasites.

antigen processing In the cellular immune response, degradation of an antigen into small peptides that associate with an MHC molecule.

antigenic determinant, also called an **epitope** A specific shape and charge distribution to which antibodies or other recognition proteins can bind.

antiterminator A protein factor that can allow elongation of a transcribed RNA molecule to continue through DNA sequences that would otherwise serve as termination sites.

apical dominance In plant development, the suppression of lateral bud development by terminal buds.

apical meristem [Latin, *apex* = top + Greek, *meristos* = divided] In a plant, a self-renewing group of undifferentiated cells, just behind the tip of the shoot or root, that generates differentiated structures.

apoptosis [Greek, *apo* = away from + *ptosis* = fall] Programmed cell death.

aqueous Watery; dissolved in water.

arachidonic acid A 20-carbon fatty acid that is the precursor of prostaglandins.

Archaea [Greek, *archein* = to begin], also called the Archaebacteria The kingdom (or "domain") of single-celled organisms that live under extreme environmental conditions and have distinctive biochemical features.

Archaebacteria [Greek, *archein* = to begin], also called the Archaea The kingdom (or "domain") of single-celled organisms that live under extreme environmental conditions and have distinctive biochemical features.

archenteron [Greek, *arche* = beginning + *enteron* = gut] The "primitive gut," the innermost tube of an animal embryo; lined with endoderm, it will become the digestive tract.

aromatic Having a benzene-like planar ring of atoms.

artifact A remnant of what the experimenter does, rather than of what exists in nature.

asexual reproduction A process that produces offspring with genes from a single parent.

associative learning A type of learning in which the subject learns to respond to a stimulus (called the conditioned stimulus) that is not obviously related to the primary stimulus; the response to the conditioned stimulus arises because of its association with a stimulus that leads directly to a physiological response.

aster [Latin, = star] A star-like object visible in most dividing eukaryotic cells (other than those of vascular plants); contains the microtubule organizing center.

astral Having an aster and centrioles that participate in mitosis; characteristic of mitosis in animals and in nonvascular plants.

astral microtubules Microtubules that extend from each pole of the mitotic spindle without attaching to any other visible structure.

asymmetric Refers to a carbon atom surrounded by four different groupings of atoms.

ATP synthase An enzyme responsible for making ATP.

ATPase An enzyme that hydrolyzes ATP into ADP and phosphate.

attenuation The regulated premature termination of transcription, best studied in the *trp* operon of *E. coli.*

attenuator site A sequence in the *trp* operon of *E. coli,* where transcription terminates even before the gene is fully transcribed into RNA.

autonomic nervous system Involuntary nervous system; coordinates the responses of smooth muscles, cardiac muscles, and other effector organs—including those of the endocrine, digestive, excretory, respiratory, and cardiovascular systems.

autonomous replication Propagation of a virus or plasmid in which the replicating DNA remains separate from the DNA of the host cell.

autophosphorylation The transfer of a phosphate group by a kinase to an identical kinase molecule.

autoradiography A method that detects radioactive compounds by their ability to expose a photographic film or a photographic emulsion.

autosomes The chromosomes that do not differ between males and females.

autotroph [Greek, *autos* = self + *trophos* = feeder] An organism that can make its own organic molecules from simple inorganic compounds (like carbon dioxide water and ammonia).

auxotroph An individual that cannot synthesize all the compounds needed to grow on a minimal medium; or an individual that requires a specific compound to grow.

axillary bud [Greek, *axilla* = armpit], also called a **lateral bud** A bud that forms just above the point where a leaf joins the stem and which can develop into a branch.

axon A type of neuronal process that usually carries signals away from the cell body and connects to other cells.

axoneme The basic structure in cilia and eukaryotic flagella; consists of parallel microtubules in a characteristic "9 + 2 pattern".

B lymphocyte ["B" refers to the fact that, in birds, these lymphocytes mature in an organ called the *b*ursa of Fabricius, an outgrowth of the cloaca] One of the lymphocytes responsible for the humoral response; when stimulated, can develop into an antibody-producing plasma cell.

B-cell receptor A protein, on the surface of a B lymphocyte, with the same antigen recognition sites as the antibody that that cell can produce.

bacterial flagellum [Latin, = whip; plural **flagella**] A long structure, about 10–20 nm thick and up to 10 μm long, whose rotation propels a bacterium through a liquid.

bacteriophage [Greek, *bakterion* = little rod + *phagein* = to eat], or **phage** A virus that infects bacteria.

basal body The base of an axoneme; a cylinder about 500 nm long that resembles a centriole; the microtubule organizing center of a cilium or a eukaryotic flagellum.

basal lamina In epithelial tissue, extracellular matrix that forms a relatively thin molecular carpet to which the epithelial sheet attaches.

base A molecule (or part of a molecule) that can accept a hydrogen ion.

base substitution A type of point mutation in which one base (nucleotide) is replaced by another.

benzodiazepine One of a number of commonly prescribed anti-anxiety drugs that bind to $GABA_A$ receptors and increase the effectiveness of GABA in opening the chloride channel.

β sheet A β structure consisting of two to five parallel or antiparallel sections of a single polypeptide chain.

β structure A common secondary structure in proteins in which the amino and carbonyl groups of the polypeptide chain are hydrogen bonded to other polypeptide chains or to distant regions of the same chain folded back on itself.

β-endorphin [from *endo*genous m*orphin*e] A peptide signaling molecule; suppresses pain.

β_2-microglobulin A small protein that associates with a Class I molecule specified by the major histocompatibility complex (MHC); a member of the immunoglobulin superfamily.

biceps The major muscle on the front side of the upper arm.

binary fission The process of cell division (in prokaryotes) in which a cell pinches in two, distributing its materials and molecular machinery more or less evenly to the two daughter cells.

binding assay A quantitative measurement of a molecule's ability to bind to another molecule.

binding site The region of a receptor molecule to which a chemical signal binds.

bioassay A method that estimates the concentration of a substance by measuring its biological activity.

biochemistry, also called **biological chemistry** The study of the structures and reactions that actually occur in living organisms.

bivalent, also called a **tetrad** A chromosome pair visible during meiosis I.

blastocoel In an animal embryo, the interior of the blastula.

blastocyst [Greek, *blastos* = sprout + *cystos* = cavity] A modified blastula in which the cells do not lie within a single layer, but do enclose an internal cavity.

blastopore The opening of the archenteron; eventually develops into the anus of the mature gut.

blastula [Greek, = little sprout] A stage of an animal embryo that consists of a sphere with cells on the surface and fluid inside.

branch point The temporary point of attachment of the 5′ end of an intron during the process of its removal; attachment at the branch point forms a lariat structure.

Brownian motion [after Robert Brown, a 19th-century British surgeon and botanist] Jerky movements of small particles, reflecting the random thermal motion of molecules.

bud Several leaf primordia, separated by internodes whose cells have not elongated.

budding yeast *Saccharomyces cerevisiae* (also called baker's yeast and brewer's yeast), a widely used experimental organism that divides by budding rather than by binary fission.

buffer A molecule that easily converts between acidic and basic forms by donating or accepting one or more hydrogen ions.

calcitonin A peptide hormone, made by the thyroid, that regulates calcium retention by the kidney.

calcitonin-gene-related peptide (CGRP) A peptide signaling molecule, made in the nervous system and derived from the same gene as calcitonin; CGRP plays a role in sensory nerve pathways.

calcium/calmodulin kinase An enzyme that, when bound to Ca^{2+}-calmodulin, transfers phosphate groups from ATP to sites in particular proteins.

caldesmon An actin-binding protein that interferes with the binding of actin and myosin in smooth muscle.

callus Undifferentiated tissue that forms at the cut surface of a plant.

calmodulin In eukaryotes, a ubiquitous calcium-binding protein which regulates many cellular processes that depend on changes in Ca^{2+} concentration.

calorie Measure of energy; the amount of energy needed to raise the temperature of one gram of water 1°C.

cAMP-dependent protein kinase, also called **A-kinase** An enzyme that, when bound to cyclic AMP, transfers phosphate groups from ATP to sites in particular proteins.

cAMP phosphodiesterase The enzyme that hydrolyzes cAMP to AMP.

capsid A coat of protein on the outside surface of a virus.

catabolite activator protein (CAP) A positive regulator of transcription in *E. coli;* CAP binds to the promoter region of the *lac* operon and increases transcription whenever cyclic AMP is present at sufficiently high levels.

carbohydrate A compound that contains the equivalent of one water molecule (one oxygen atom and two hydrogen atoms) for every carbon atom; carbohydrates include sugars and polysaccharides.

carbonyl A functional group that consists of a carbon atom attached to an oxygen atom by a double bond (C=O).

carboxyl A functional group that contains one carbon atom and two oxygen atoms; the carbon atom forms a double bond with one oxygen atom and a single bond with the other.

cardiac muscle The muscles that pump blood through the heart.

carotenoids Yellow or orange plant pigments.

carpel, also called a **pistil** In a flower, a female reproductive structure.

catabolism [Greek, *cata* = down + *ballein* = to throw] The reactions of metabolism that break down complex molecules, such as those in food.

catalyst A substance that lowers the activation energy of a reaction and is not itself consumed in that reaction.

catch muscle In a mollusc, a smooth muscle that holds the shell closed.

cation [so called because it will move toward a negatively charged electrode or *cat*hode] A positively charged ion.

cell body The center of a neuron, as distinct from its axon or dendrites.

cell cortex A meshwork of actin filaments just below the surface of a eukaryotic cell.

cell center, also called the **centrosome** A microtubule organizing center; a complex of proteins that in animal cells includes a pair of centrioles.

cell crawling, also called **cell locomotion** or **ameboid movement** The movement of cells over a surface.

cell cycle The orderly sequence of events that accomplish cell reproduction.

cell division The process by which a parent cell gives rise to two daughter cells that carry the same genetic information as the parent cell.

cell division cycle (cdc) mutants Mutant organisms (generally yeasts) that cannot complete the cell division cycle because of alterations (mutations) in certain genes.

cell locomotion, also called **cell crawling** or **ameboid movement** The movement of cells over a surface.

cell plate The precursor of a new cell wall between two daughter plant cells.

Cell Theory (1) All organisms are composed of one or more cells; (2) cells themselves are alive and are the basic living unit of function and organization of all organisms; (3) all cells come from other cells.

cell wall An external rigid structure surrounding all plant cells and most prokaryotes.

cellular immunity The proliferation of specific cells that attack foreign cells, including the body's own cells that have become cancerous or have been infected with a virus.

cellular oncogene, also called a **proto-oncogene** The cellular counterpart of a viral oncogene; participates in the normal control of growth and differentiation.

central cell A cell within the ovule of a flower; contains two haploid nuclei, derived from the megaspore; develops into endosperm after fertilization.

Central Dogma "DNA specifies RNA, which specifies proteins."

central nervous system (CNS) Brain and spinal cord.

centrifugation A method for separating macromolecules and subcellular structures according to size (and shape) by subjecting them to high gravitational fields in a spinning tube.

centriole A pair of small cylindrical structures each about 0.2 μm in diameter and 0.4 μm long, that lie at right angles to one another; present at each pole of the mitotic spindle in animal cells and in some other eukaryotes.

centromere The point at which the two chromatids of a single chromosome are joined.

centrosome A microtubule organizing center that (in many cells) also contains a distinctive organelle called the centriole; the centrosome plays an important role in cell division.

cephalization The tendency to concentrate sense organs and ganglia in the anterior end of an animal.

channel A membrane protein that allows the passage of specific molecules or ions.

chaperonin A protein in mitochondria that helps stabilize the unfolded state of mitochondrial proteins, allowing proper folding once a protein arrives at its target.

chelator [Greek, *chela* = claw] A molecule in which several covalently connected atoms bind to the same ion.

chemattractant A chemical to which cells are attracted in chemotaxis; in bacterial chemotaxis, a chemical that influences the relative lengths of runs and tumbles.

chemical synapse Distinct boundary between two neurons through which neurotransmitters diffuse.

chemical work The formation of chemical bonds.

chemiosmosis [Greek, *osmos,* to push] The linking of chemical and transport processes.

chemistry The study of the properties and the transformations of matter.

chemotaxis The movement of a cell toward a higher (or, in some cases, a lower) concentration of a particular chemical.

chiasma [Greek, = cross; plural, **chiasmata**] The sites of exchange of DNA between homologous chromosomes during meiosis; visible during prophase of meiosis I.

chlorophyll The green pigment of plant leaves; responsible for absorbing light for photosynthesis.

chloroplast A large, green organelle that performs photosynthesis.

cholera toxin A protein produced by the bacterium *Vibrio cholerae;* causes massive, life-threatening diarrhea; specifically modifies the α subunit of G_s protein.

cholesterol A small molecule that consists of a pattern of four interconnected rings of carbon atoms; the starting point in the synthesis of steroids.

chorionic villus sampling A procedure for sampling cells of the early embryo; uses fetal cells present in the placenta.

chromatic aberration The failure of light of different colors (wavelengths) to focus at the same point.

chromatid One of the two separate but connected bodies that make up a chromosome at the beginning of mitosis, when the chromosomes first become visible.

chromatin [Greek, *chroma* = color] A diffuse material within the nucleus of a nondividing eukaryotic cell; consists of DNA and proteins.

chromatography A method for separating molecules according to their relative affinities for a stationary support, called the stationary phase, and a moving solution, called the mobile phase.

chromoplast A plastid that contains the pigments of fruits and flowers.

chromosomal mutation A change in a relatively large region of a chromosome.

chromosome [Greek, *chroma* = color + *soma* = body, because it is stained by certain dyes] A discrete complex of DNA and proteins, visible with a light microscope within a dividing cell; originally used only for eukaryotic cells, but now also used to mean a single large molecule of DNA that contains the genes of a bacterium or a virus.

chromosome banding The distinctive pattern, visible in a microscope, that results from the selective binding of certain dyes to individual chromosomes.

cilium [Latin, = eyelid, from the hairlike appearance of a cilium; plural, **cilia**] A protein assembly, consisting of microtubules, that can move a cell through a liquid medium (or a liquid medium over a cellular surface); a single cell usually contains many cilia, often arranged in rows; cilia have the same organizational plan as eukaryotic flagella but cilia are much shorter.

circular For DNA, forming a continuous loop; examples of circular DNAs include those of bacteria, energy-producing organelles, and many viruses.

***cis, medial,* and *trans* Golgi** Regions of the Golgi apparatus that make distinctive contributions to protein processing.

***cis*-regulatory element** [Latin, *cis* = on the same side, because the regulatory element is on the same piece of DNA that is being regulated], also called a **response element** or **enhancer** A DNA sequence that regulates transcription by binding to a specific protein called a transcription factor.

citric acid cycle, also called the **Krebs cycle** or the **tricarboxlyic acid cycle** A set of reactions that converts the carbon atoms of acetyl CoA into carbon dioxide.

clathrin The protein that lines a coated pit, an intermediate structure in receptor-mediated endocytosis.

cleavage A series of rapid cell divisions following fertilization in many early embryos; cleavage divides the embryo without increasing its mass.

cleavage furrow A groove formed from the cell membrane in a dividing cell as the contractile ring tightens.

Clonal Selection Theory Summarizes the role of selective proliferation of responding lymphocytes, already programmed, into clones of cells that contain specific gene rearrangements.

clone [Greek, *klon* = twig] A population of genetically identical individuals or cells descended from a single ancestor.

coacervate [Latin, *coacervatus* = heaped up] Discrete tiny droplet into which proteins and polysaccharides can spontaneously concentrate.

coated pit A depression in the plasma membrane, lined with clathrin molecules; an intermediate structure in receptor-mediated endocytosis.

codominant Refers to two alleles that each contribute to the phenotype of a heterozygote.

codon A group of three nucleotides that specifies a single amino acid residue.

coenzyme An organic molecule (but not a protein) that is a necessary participant in an enzyme reaction.

collagen helix A regular structure found principally in the structural protein collagen; the collagen helix consists of three polypeptide chains.

column chromatography A type of chromatography in which the stationary phase is an insoluble material packed into a glass or metal column.

commensalism An association between individuals of two species in which one organism benefits without harming the other one.

communicating junctions Membrane-associated structures that allow small molecules to pass freely between two adjacent cells.

competitive inhibitor A molecule whose inhibitory effects on an enzyme can be overcome by increased substrate concentration.

complement A set of blood proteins that attack microbial invaders.

complementarity determining region (CDR) The segment of an immunoglobulin molecule most involved in binding antigen; so called because its shape is complementary to that of the antigen.

complementary Fitting together to make a unified whole; may refer to the two strands of DNA or to the surface of a protein and a substrate, ligand, or antigen.

complete metamorphosis Pattern of development in which there are distinct larval, pupal, and adult forms; characteristic of many insects, including *Drosophila,* fleas, flies, beetles, wasps, moths, and butterflies.

complex transposon A transposable element that contains other genes besides transposase, for example, a gene for antibiotic resistance.

compound microscope A light microscope that contains several lenses.

concentration gradient [because the concentration changes in a *graded* manner] A graded difference in the concentration of a substance.

concentration work Bringing a substance to a different concentration in one region from that in the surroundings.

condensation (or **dehydration condensation**) **reaction** The linking of two building blocks, accompanied by the removal of a water molecule.

condenser lens A lens in a microscope that allows the even illumination of the specimen.

conditional mutation A change in DNA that alters a protein so that it can function under some conditions but not under others.

conditioned stimulus In associative learning, the arbitrary stimulus that is paired with a stimulus (the "unconditioned stimulus") that directly evokes a physiological response.

confocal scanning microscope A microscope that uses a computer controlled laser beam to selectively stimulate fluorescence in a single plane of a section, providing thin "optical sections" of a biological specimen.

conformation The three-dimensional arrangement of a polypeptide chain, the equivalent of its tertiary structure.

congenic Animals that are genetically identical except for a single gene or closely linked set of genes.

conjugation A process in which two temporarily attached bacteria exchange genetic material.

connective tissue A relatively sparse population of cells within a bed of extracellular matrix; includes bone and cartilage.

consensus sequences also called "sequence motifs" or "boxes" [because of the way they are marked in diagrams] Characteristic nucleotide arrangements that are common to many genes.

conservative A possible pattern of DNA replication in which each parent molecule would remain intact and both strands of the descendant molecules would be newly assembled; actual DNA replication is not conservative, but semiconservative.

constant region The part of an immunoglobulin light chain or heavy chain that is identical in all antibodies of a particular class.

constitutive Made at the same level, independent of the presence of an inducer.

contact inhibition of cell division The ability of cells to stop dividing when neighboring cells all touch each other.

contractile ring A bundle of actin filaments that surrounds a dividing cell.

contrast An object's ability to absorb more or less light than its surroundings.

control A situation in which there has been no experimental manipulation; comparing measurements made in a control versus an experimental situation allows the evaluation of an hypothesis.

coordinate regulation The simultaneous expression of several genes.

cortical microtubules Microtubules that encircle a dividing plant cell; determine the placement of a new cell wall.

corticotropin, also called **adrenocorticotrophic hormone** or **ACTH** A peptide hormone, made in the anterior pituitary, that stimulates the production of corticosteroids in the adrenal cortex.

cotranslational Occurring while translation is still in progress.

cotransport The coupled transport of two substances across a membrane; depends on specific transmembrane protein molecules.

cotransporter A transmembrane protein that binds to two or more molecules or ions and transports them across a membrane.

cotyledon [Greek, *kotyle* = a deep cup] In the embryos of seed plants, the "seed leaves," in which nutrients are stored for use after germination.

coupling factor, also called the **F_0-F_1 complex** A protein complex that uses the flow of protons to drive the synthesis of ATP.

covalent bonds Shared arrangements of electrons that hold atoms together in molecules.

CREB [*c*yclic-AMP *r*esponse *e*lement *b*inding protein] The transcription factor that binds to the cyclic AMP response element; a member of the amphipathic helix family.

cristae Elaborate folds of the inner membrane of a mitochondrion.

cro protein A protein that represses the expression of the λ repressor thereby allowing the expression of the genes within λ DNA.

cross-breeding (or **crossing**) The breeding of two genetically distinct organisms.

crossing over One type of genetic recombination, involves the breakage and rejoining of single chromatids of homologous chromosomes.

crystal A solid that is enclosed by geometrically regular faces; the starting material point for x-ray diffraction analysis.

current The movement of charges.

cyclic AMP (cAMP) Adenosine monophosphate in which the same phosphate group is linked to carbon atoms 3 and 5; used as a "hunger" signal in bacteria and protists and as a second messenger in mammalian cells.

cyclic AMP response element (CRE) A sequence in DNA recognized by CREB, a eukaryotic transcription factor that is responsive to cyclic AMP.

cyclic photophosphorylation The production of ATP from light energy by a series of electron transfers that regenerate the absorbing chlorophyll; in plants, cyclic photophosphorylation depends on the flow of electrons from excited P_{700} in a cycle that regenerates P_{700}, which is then able to absorb another photon of light.

cyclin A protein that regularly increases and decreases in concentration during the cell cycle; a component of MPF.

cyclooxygenase pathway The enzymes that convert arachidonic acid to prostaglandins.

cytochalasin A poison that interferes with the growth of an actin filament.

cytochrome [Greek, *kytos* = hollow vessel + *chroma* = color] One of a set of heme-containing electron carrier proteins that change color as they accept or donate electrons.

cytodifferentiation The development of a specialized cell.

cytokine One of a number of small proteins that regulate proliferation and protein synthesis in cells that participate in inflammation and in the immune response.

cytokinesis [Greek, *kytos* = hollow vessel + *kinesis* = movement] The division of the cytoplasm and formation of two separate plasma membranes.

cytokinin A plant hormone that influences the rate of division and differentiation.

cytoplasm In eukaryotes, the part of the cell outside the nucleus but inside the plasma membrane.

cytoplasmic streaming The movements of membrane-bounded organelles along actin filaments.

cytoskeleton A network of protein fibers that runs through the cytosol of eukaryotic cells; it consists of microtubules, actin filaments, and intermediate filaments, along with other associated proteins.

cytosol In eukaryotic cells, the part of the cytoplasm not contained in membrane-bounded organelles.

cytosolic face The half of a membrane's lipid bilayer that touches the cytosol.

cytotoxic T lymphocyte (CTL) An effector cell, derived from a T_C cell, that actually kills a target cell.

cytotoxicity In the immune response, the ability of certain T lymphocytes to kill cells with altered surfaces.

daughter cells The cells produced from a single parent cell after cell division.

deamination The removal of an amino group from adenine, guanine, or cytosine.

degenerate [a term borrowed from quantum physics] Meaning that several codons are equivalent; that is, that they specify the same amino acid.

deletion A type of mutation; the removal of one or more nucleotides.

denatured A form of an enzyme or other protein that lacks functional activity.

dendrite A type of neuronal process that usually carries signals to the cell body.

dendritic cell In the immune response, a type of antigen-presenting cell; a highly extended cell present in many organs and tissues throughout the body.

density gradient A changing concentration of a dissolved substance—usually sucrose, glycerol, or cesium chloride—with the highest concentration at the bottom and the lowest at the top.

depolarizing A voltage change across an excitable membrane that makes the inside of the cell less negative and more apt to produce an action potential.

depression A decrease in the size of the evoked action potential in a postsynaptic neuron that results from the repeated stimulation of a sensory neuron.

depurination The loss of a purine base from deoxyribose in DNA.

desmin An intermediate filament protein in muscle.

desmosome An anchoring junction that consists of transmembrane proteins that attach to a cell's intermediate filaments; serves as rivets between cells.

detergent An amphipathic molecule that interacts both with water and with hydrophobic molecules.

determined In development, having a limited potency, so that it can no longer develop in accordance with new environmental signals.

diabetes mellitus A disease characterized by the overproduction of sweet (sugar-containing) urine.

diacylglycerol (DAG) Glycerol linked to two fatty acid chains; used in animal cells as a nonpolar second messenger that stimulates protein kinase C in the plasma membrane.

dicot One of a large group of plants—including *Arabidopsis* and peas—with two cotyledons.

dideoxynucleotide triphosphate A deoxynucleotide triphosphate that lacks both a 2′ and a 3′-hydroxyl group.

differential centrifugation Successive centrifugations at increasing speeds; allows an experimenter to obtain pellets with distinct subcellular components.

differential gene expression The production of differing amounts of individual proteins (and RNAs) in different types of cells and at different times of development.

differential interference contrast microscopy, also called **Nomarski optics** A special optical arrangement (distinct from phase contrast microscopy) that reveals phase differences among different cell components without the use of stains.

differentiation The process by which tissues and cells become specialized and different from one another.

diffraction The scattering of electromagnetic waves by regular structures so that their interference produces a pattern of lines or spots; the diffraction of x-rays by a crystal produces a pattern of spots that can be analyzed to reveal the structure of the molecules within the crystal.

diffusion [Latin, *diffundere* = to pour out] Random movements that lead to a uniform distribution of molecules both within a solution and on the two sides of a membrane.

digestion The process of hydrolyzing large molecules into smaller units.

diploid [Greek, *di* = double + *ploion* = vessel] Having two sets of chromosomes.

diplontic Having the diploid phase dominate the life cycle.

disjunction In meiosis, the moving apart of two homologous chromosomes during anaphase I and of two sister chromatids in anaphase II.

dispersive A possible pattern of DNA replication in which each daughter molecule contains interspersed pieces of old (parental) DNA and newly polymerized DNA; actual DNA replication is not dispersive but semiconservative.

disulfide A covalent bond between two sulfhydryl groups; common cross link between two cysteine side groups within a single polypeptide or between two polypeptides.

DNA affinity chromatography A method for purifying a DNA-binding protein by virtue of its ability to bind to a specific DNA sequence.

DNA gyrase An enzyme that contributes to DNA replication by preventing the accumulation of kinks at the stems of the replication forks.

DNA ligase An enzyme that links fragments of DNA together by catalyzing the formation of a phosphodiester bond.

DNA polymerase The enzyme that strings together nucleotides into DNA.

DNA-binding domain A structural domain within a transcription factor that binds to a specific DNA regulatory sequence.

docking protein, also called the **SRP receptor** A membrane protein that pulls a growing polypeptide into the lumen of the endoplasmic reticulum.

dolichol A long-chain lipid to which an oligosaccharide attaches before it is transferred to a newly made polypeptide.

domain A polypeptide segment that folds as an independent unit; usually consists of 50–350 amino acid residues.

domain swapping A strategy for making recombinant DNAs in which peptide segments ("domains") of two molecules are exchanged and the properties of the resulting mosaic (or "chimeric") receptors are studied.

dominant Refers to an allele that alone determines the phenotype of a heterozygote.

dormant [French, *dormire* = to sleep] Referring to a stage in the development of a seed, in which growth is suspended until restarted by environmental cues.

dosage compensation The adjustment of gene expression as a function of the number of copies of a specific gene; in *Drosophila,* for example, both X chromosomes are active in females, but each gene is only half as active as it would be in a male.

double fertilization In flowering plants, the simultaneous fusion of two sperm nuclei from the pollen tube with two cells of the embryo sac.

double covalent bond A covalent bond in which two atoms share two pairs of electrons.

Down syndrome, also called **Down's syndrome** A disorder, resulting from a particular chromosomal abnormality, that leads to mental retardation and the abnormal development of the face, heart, and other parts of the body; almost always caused by trisomy of chromosome 21.

downstream In DNA or RNA, toward the 3′ end.

duplication A mutation in which part of a chromosome appears twice.

dynamically unstable Meaning that a microtubule or an actin filament loses tubulin or actin monomers and shrinks if it is not actively growing.

dynein A microtubule-associated motor molecule that moves toward the minus end of microtubules.

dynorphin One of three classes of polypeptides that act as natural pain suppressors.

dystrophin In muscle, the protein whose altered structure or absence causes muscular dystrophy; involved in anchoring the Z disc to the plasma membrane.

early cell plate A small, flattened disc, formed in mitotic telophase in a plant cell, that is the beginning of a new cell wall.

ectoderm [Greek, *ecto* = outside + *derma* = skin] The outermost tube of a vertebrate embryo, including the future skin, sense organs, and central nervous system.

effector Any molecule or ion that changes the activity of an enzyme.

effector cell In the immune system, a cell that carries out a particular immune function.

effector domain, also called a **transactivating domain** A structural domain within a transcription factor that can interact with the transcription machinery or with another transcription factor.

efferent Carrying commands concerning motor activity from the central nervous system to the muscles and other effector organs, including the endocrine glands.

egg, also called an **ovum** A female gamete.

electrical potential, also called **voltage** A measure of the potential energy that results from a charge separation.

electrical synapse A connection between two neurons through gap junctions.

electrical work Energy change caused by changing the separation of charges.

electrochemical gradient A double gradient composed of a *chemical* gradient (the difference in hydrogen ion concentration, or pH) and an *electrical* gradient (the difference in charge).

electrogenic Leading to the accumulation of charge on one side of a membrane.

electromagnetic radiation A form of energy—including light—transmitted through space as periodically changing electrical and magnetic forces.

electron carriers A series of other molecules that carry electrons from high-energy, reduced compounds (NADH, NADPH, and $FADH_2$).

electron microscope Powerful microscope that uses electrons instead of light waves to reveal structures; has much higher resolution than a light microscope.

electron transport chain The pathway of electrons in oxidative phosphorylation or photophosphorylation.

electronegativity A measure of the tendency of an atom to gain electrons.

electrophoresis A method for separating charged molecules (such as proteins) according to their ability to move in an electric field.

electrostatic interaction The attraction or repulsion of charges.

elongation The adding of additional nucleotide or amino acid residues to a growing polynucleotide or polypeptide.

elongation zone A region of the growth zone of a plant's root.

embryo [Greek, *en* = in + *bryein* = to be full of] A set of early developmental stages in which a plant or animal differs from its mature form.

emergent properties Characteristics that arise only at more complex levels of organization.

endergonic Referring to a process in which free energy increases.

endocrine gland An organ that is specialized for secretion of a hormone into the general circulation.

endocrinology The study of hormones.

endocytosis [Greek, *endon* = within] The process of taking in materials from outside a cell in vesicles that arise by the inward folding ("invagination") of the plasma membrane.

endoderm [Greek, *endo* = inside + *derma* = skin] The innermost tube of a vertebrate embryo, including the future gastrointestinal tract and associated organs such as pancreas and liver.

endoplasmic reticulum (ER) [Greek, *endon* = within + *plasmein* = to mold + Latin, *reticulum* = network] An extensive and convoluted network of membranes within a eukaryotic cell.

endorphin One of three classes of polypeptides that act as natural pain suppressors.

endosymbiosis [Greek, *endon* = within + *syn* = together + *bios* = life] The close association of two organisms, one of which lives inside the other.

endosymbiotic theory The generally accepted view that present day energy organelles are descended from prokaryotes that once lived within the early eukaryotic cells.

energy The capacity to do work, to move an object against an opposing force.

enhancer [because it stimulates the transcription of neighboring DNA, without itself serving as a promoter], also called a **response element** or ***cis*-regulatory element** A DNA sequence that regulates transcription by binding to a specific regulatory protein called a transcription factor.

enkephalin One of three classes of polypeptides that act as natural pain suppressors.

enthalpy (H) The heat content of a system, roughly a measure of total energy stored in chemical bonds.

entropy (S) A formal measure of disorder; entropy has a high value when objects are disordered or distributed at random, and a low value when they are ordered.

enzyme A large molecule, almost always a protein, that accelerates the rate of a specific chemical reaction.

enzyme assay A quantitative measurement of an enzyme's ability to catalyze a particular chemical reaction.

enzyme-substrate complex The association of enzyme and substrate that forms in the course of catalysis.

eon One of four periods of time into which geologists and paleontologists have divided the history of life; eons are much longer than eras.

epicotyl [Greek, *epi* = above] The axis of a developing plant above the attachment point of the cotyledons.

epidermal growth factor (EGF) A protein paracrine signal that was first isolated from epidermal cells.

epidermis In animals, the outer layer of the skin; in plants, the outermost layer of the embryo and the plant.

epinephrine, also called **Adrenaline** An amine hormone, made in the adrenal medulla, that speeds the heart, dilates the blood vessels, and increases the liver's production of glucose from glycogen.

episome A virus or other genetic system that can propagate either autonomously or as an integrated part of the host's chromosome.

epistatic gene A type of modifier gene that limits the expression of another gene.

epithelial tissue [Greek, *epi* = upon + *thele* = nipple] Cells tightly linked together to form a sheet with little extracellular matrix.

equilibrium The point at which no further net conversion of reactants and products takes place.

equilibrium constant (K_{eq}) The concentration ratio (the concentration of product divided by the concentration of reactant) at equilibrium.

equilibrium potential The voltage at which the electrical force driving an ion in one direction exactly balances the tendency of that ion to move down a concentration gradient.

era One of four periods of time, ranging in length from 65 million to several billion years, into which paleontologists usually divide the history of life.

Escherichia coli (E. coli) A common eubacterial resident of the human gut; a favorite experimental organism for thousands of research and industrial biologists.

ethylene A plant hormone, a two-carbon molecule containing a double bond.

etiolated [French, *etioler* = to blanch or whiten] Having a thin, spindly appearance, poor leaf development, and no chlorophyll production.

Eubacteria [Greek, *eu* = true] The commonly occurring prokaryotes that live in water or soil, or within larger organisms; Archaebacteria and Eubacteria differ from each other in their metabolic abilities, the composition of their membranes, and the structure of their ribosomes.

eukaryotic [Greek, *eu* = true + *karyon* = nucleus] Referring to cells that contain a nucleus and other membrane-bounded organelles.

eukaryotic flagellum [Latin, = whip; plural **flagella**] A protein assembly, consisting of microtubules, that can move a cell through a liquid medium (or a liquid medium over a cellular surface); a single cell usually contains only one or two flagella; eukaryotic flagella have the same organizational plan as cilia but flagella are much longer.

euploid Having the correct number of chromosomes.

excitatory In the nervous system, leading to the production of action potentials.

excitatory postsynaptic potential (EPSP) A depolarizing response in a postsynaptic cell resulting from the action of an excitatory neurotransmitter.

exergonic Referring to a process in which free energy decreases.

exocrine organ An organ whose products are carried to specific targets by ducts.

exocytosis The transport of materials contained within a vesicle to the outside of a cell.

exon A segment of a gene (or of a pre-mRNA) that is also present in mature mRNA; most exon sequences encode polypeptide segments.

exon shuffling During evolution, the mixing and matching of exons from different ancestral genes.

exoplasmic face The half of a membrane's lipid bilayer that contacts the outside of the cell.

expression cloning A strategy for isolating recombinant DNAs by recognizing the protein that each encodes.

expressivity The range of phenotypes associated with a given genotype.

extension The forceful unbending of a limb.

extracellular matrix A complex network of extracellular molecules, both carbohydrates and proteins, that surrounds many animal cells.

F-factor [for *f*ertility factor] A kind of episome in bacteria that can replicate either autonomously or in integrated form; can move from one bacterium to another during conjugation.

F1, or **first filial generation** [Latin, = son] The initial progeny of a cross.

F2, or **second filial generation** The progeny of the F1 generation.

facilitated diffusion An increased rate of passive transport; depends on the action of specific transporter molecules within the membrane.

facultative anaerobe A microorganism that can live either anaerobically (by fermentation) or aerobically (using oxidative phosphorylation).

FAD [*f*lavin *a*denine *d*inucleotide] An electron acceptor in oxidative phosphorylation.

$FADH_2$ The reduced form of FAD.

familial hypercholesterolemia A disease characterized by high blood cholesterol and increased susceptibility to heart attacks; caused by alterations in the structure of the membrane receptor for LDL.

fat A triacylglycerol that is solid at room temperature; the component fatty acids are usually saturated.

fate What a cell or a tissue becomes during development.

fatty acid A small molecule consisting of a hydrocarbon chain ending in a carboxyl group; a component of phospholipids and triacylglycerides.

feedback inhibition The inhibition of an enzyme reaction by the product of that reaction or of the end-product of an entire pathway; a homeostatic mechanism.

fermentation [Latin, *fervere* = to boil] The anaerobic extraction of energy from organic compounds.

fertilization The union of two haploid gametes to form a diploid cell or zygote.

fetal masculinization The development of male characteristics in a fetus carried by a woman treated with male hormones during pregnancy.

fetus In mammals, a stage of development in which all organs have formed.

fibronectin A protein of the extracellular matrix.

fibrous protein A protein with an elongated shape; includes most structural proteins.

filament A small protein fiber.

First Law of Thermodynamics The statement that the total amount of energy stays constant in any process; that is, energy is neither lost nor gained—it only changes form.

fission yeast *Schizosaccharomyces pombe,* a widely used experimental organism that divides by binary fission.

5′-end The end of a polynucleotide at which the 5′ carbon of the nucleotide residue is not attached to another residue, but to a phosphate or a hydroxyl group.

fixation A process that preserves cells for microscopic examination; fixation keeps the sections from falling apart by binding cellular components together; often involves treatments that make covalent bonds between cellular molecules.

flagellar hook In a prokaryotic flagellum, the structure that connects the flagellum to the rest of the flagellar motor.

flagellin A globular protein that is the main component of prokaryotic flagella; unrelated to tubulin, the major protein in eukaryotic flagella.

flexion The forceful bending of a limb.

Fluid Mosaic Model The accepted model of membrane structure; stresses that proteins and phospholipid molecules can move within each leaf of the lipid bilayer unless they are restricted by special interactions.

fluidity A measure of the ability of substances to move within a membrane.

focal contact In an animal cell, a specialized membrane region at which an adhering junction anchors actin filaments within a cell to extracellular matrix or to a substratum.

follicle-stimulating hormone (FSH) A peptide hormone, made in the anterior pituitary, that, in females, promotes the maturation of the follicle during the menstrual cycle and, in males, stimulates testosterone production.

footprinting A method for detecting proteins bound to DNA by showing that a bound protein prevents the chemical or enzymatic cleavage of a specific DNA sequence.

fossil [Latin, *fossilis* = dug up] An object, usually found in the ground, that represents the remains or imprint of past life.

fraction collector A device that collects sequential samples from a chromatographic column or other source.

frame-shift mutation An insertion or deletion mutation that alters the groupings of nucleotides into codons.

free energy A measure of available energy under the conditions of a biochemical reaction; abbreviated ***G***, after the American thermodynamicist J. Willard *G*ibbs.

free ribosomes Ribosomes present in the cytosol, responsible for the synthesis of soluble proteins.

functional group A standard small grouping of atoms that contributes to the characteristics of an organic molecule.

Fungi The kingdom that includes heterotrophic organisms, both multicellular and single-celled organisms.

G protein [G stands for *g*uanyl-nucleotide binding protein] One of a number of proteins that mediate the action of water-soluble chemical signals; a G protein can bind either GTP or GDP, and it interacts both with receptor molecules and with an enzyme that produces (or degrades) a second messenger.

G_0 The state of a cell that has withdrawn from the cell cycle.

G_1 The period of the cell cycle that represents the *g*ap between the completion of mitosis and the beginning of DNA replication; also called the first growth phase.

G_2 The period of the cell cycle that represents the *g*ap between the completion of DNA synthesis and the beginning of mitosis (of the next cell cycle).

$GABA_A$ receptor A ligand-gated channel that responds to GABA by opening a chloride channel.

gamete [Greek *gamos* = marriage] A specialized reproductive cell through which sexually reproducing parents pass chromosomes to their offspring; a sperm or an egg.

gamete formation The production of sperm and eggs.

gametophyte The haploid form of a life cycle characterized by alternation of generations.

γ-aminobutyric acid (GABA) The inhibitory neurotransmitter most widely used in the brain.

gap junctions Protein assemblies that form channels between adjacent animal cells.

gastrula [Greek, = little stomach] A stage of an animal embryo in which the three germ layers have just formed.

gastrulation The process of forming a gastrula.

gate The part of a channel protein that opens and closes the channel in response to environmental signals.

gating helix The polypeptide segment of a voltage-gated channel responsible for the voltage gate.

gel retardation assay A method that detects protein binding to DNA by comparing the electrophoretic mobility of a specific DNA fragment in the presence and absence of protein.

gene A DNA sequence that is transcribed as a single unit and encodes a single polypeptide or a set of closely related polypeptides.

gene amplification The specific multiplication of genes; known to occur in the genes encoding ribosomal RNA during amphibian development.

gene rearrangement The cutting and splicing of segments of specific genes; known to occur within the immune system.

Genetic Code The relationship between nucleotide sequence in mRNA and amino acid sequence in polypeptides.

genetic linkage The tendency of two or more genes to segregate together.

genetic map A summary of the genetic distances between genes; a linkage map.

genetic recombination Associations of genes that did not exist in the parents.

genetics The study of inheritance.

genome The collection of all the DNA in an organism.

genotype The genes present in a particular organism or cell.

geometrical isomers Two molecules that differ only in the arrangement of groups around a double bond.

germ cells or **germ line** Gametes and the cells from which they arise.

germ layer One of the three tubes of the vertebrate embryo—ectoderm, mesoderm, and endoderm.

germ layer formation The movement of embryonic cells to give the three germ layers (ectoderm, mesoderm, and endoderm); gastrulation.

germ line theory The view that antibody diversity results from genetic information that is already present in the zygote.

germination [Latin, *germinare* = to sprout] The resumption of growth by a seed.

G_i [because it *i*nhibits cAMP production] A G protein whose binding to a receptor-ligand complex leads to an inhibition of adenyl cyclase.

gibberellin A plant hormone.

glial cell A cell within the nervous system that does not itself transmit electrical and chemical signals, but which provides metabolic and structural support for neurons.

globular proteins Relatively compact proteins that are roughly spherical in shape.

glucagon A protein hormone produced in the pancreas; a signal for the postabsorptive state; glucagon inhibits glycogen synthesis and stimulates its breakdown into glucose.

glucocorticoid A steroid hormone that stimulates the production of carbohydrate from protein.

glutamic acid decarboxylase (GAD) One of two related enzymes that catalyze the formation of GABA from glutamic acid.

glycerol A polar three-carbon molecule with three hydroxyl groups; a starting compound for triacylglycerides and phospholipids.

glycocalyx [Greek, *glykos* = sweet + Latin, *calix* = cup], also called the **cell coat** A densely staining zone just outside most eukaryotic cells.

glycogen phosphorylase The enzyme that breaks down glycogen into glucose.

glycolipid A molecule that consists of a lipid attached to carbohydrates.

glycolysis [Greek, *glykys* = sweet (referring to sugar) + *lyein* = to loosen] A set of ten chemical reactions that is the first stage in the metabolism of glucose.

glycolytic fiber A large muscle fiber that derives most of its energy from glycolysis.

glycoprotein [Greek, *glykis* = sweet] A protein that contains covalently attached carbohydrates.

glycosidase An enzyme that catalyzes the digestion of glycosidic bonds; used to digest starch or glycogen.

glycosidic bond The covalent bond between two sugar molecules in a polysaccharide or oligosaccharide.

glycosylation The process of adding sugars to a newly made protein; it takes place in the ER lumen and in the Golgi apparatus.

glyoxisome A specialized peroxisome present in the seeds of some plants; glyoxisomes provide energy for the growing plant embryo by breaking down stored fats.

Golgi apparatus, also called the **Golgi complex** In eukaryotic cells, a set of flattened discs, usually near the nucleus, involved in the processing and export of proteins.

gonad [Greek, *gonos* = seed] A gamete-producing organ; an ovary or testis.

gonadotrophin One of two hormones (FSH and LH), made in the anterior pituitary, that act on gonadal tissue.

gonadotropin releasing hormone (GnRH) A hormone, made by the hypothalamus, that stimulates the secretion of FSH and LH by the anterior pituitary.

granum [Latin, = grain; plural, **grana**] A stack of thylakoids within a chloroplast, bounded by the thylakoid membrane.

gravitropism The response of a plant to gravity.

ground tissue In plants, the cells that occupy most of the interior of the embryo and the plant.

growth control Regulatory mechanism that prevents delinquent division by allowing the cell cycle to proceed under some conditions and to stop under others.

growth and maturation Increase in size and the change in form of an organism after the organs and basic body plan are established.

growth factor One of a number of protein paracrine signals that stimulate cell division and cell survival.

growth hormone (GH) A peptide hormone, made in the anterior pituitary, that stimulates tissue and skeletal growth.

G_s [because it *s*timulates an increase in cAMP] A G protein that stimulates adenyl cyclase after binding to a receptor-ligand complex.

habituation The decrease in a behavioral response following repeated exposure to a harmless stimulus.

haploid [Greek, *haploos* = single] Having a single set of chromosomes.

haplontic Having the haploid phase dominate the life cycle; in a haplontic life cycle, the only diploid cell is the zygote, which itself undergoes meiosis to produce haploid spores.

heat The form of kinetic energy contained in moving molecules.

heat shock A sudden increase in temperature that leads to the production of specific proteins both in bacteria and in eukaryotes.

heavy chain One of the polypeptides present in an immunoglobulin molecule.

helicase An enzyme that uses energy from ATP to untwist the DNA helix.

helix-turn-helix A class of transcription factors that contain three α-helices separated by short turns.

heme An iron-containing organic molecule that gives hemoglobin and the cytochromes their red color; heme may donate or accept electrons, as its iron atom changes its charge between Fe^{2+} and Fe^{3+}.

hemidesmosome An anchoring junction that attaches the intermediate filaments of a cell to extracellular matrix.

hemizygous In male mammals, having one (instead of two) copies of a gene because it lies on the X chromosome.

hemoglobin The red oxygen-binding protein of the blood.

heterogeneous nuclear RNA (hnRNA) The collection of pre-mRNAs in the nucleus of a eukaryotic cell.

heterokaryon [Greek, *heteros* = different + *karyon* = kernel, nucleus] A cell that has two nuclei from different sources.

heterotroph [Greek, *heteros* = other + *trophe* = nourishment] An organism that cannot derive energy from sunlight or from inorganic chemicals but must obtain energy by degrading organic molecules.

heterozygous Having two different alleles for a single gene (in a diploid organism).

hexose [Greek, *hex* = six] A sugar that contains six carbon atoms; glucose is a hexose.

Hfr bacterium ["*h*igh *f*requency of *r*ecombination," so called because an integrated F-factor can stimulate the transfer of genetic information and produce new combinations of genes in the receiving cell] An *E. coli* cell that contains an integrated F-factor.

high performance liquid chromatography (HPLC) A chromatography technique in which high pressure drives the mobile phase over a finely divided stationary phase in a metal column.

high-energy bonds Relatively unstable chemical bonds that give up energy as new, more stable, bonds form.

histamine The amino acid histidine minus the carboxyl group, a major stimulus for the inflammatory response.

histone [Greek, *histos* = web] One of a set of small, positively charged proteins that bind to DNA in eukaryotic cells.

homeo box [since it appears in a number of genes responsible for homeotic mutants] A DNA sequence that specifies a polypeptide fragment called the homeo domain, which functions as a helix-turn-helix type of transcription factor.

homeostasis [Greek, *homeo* = like, similar + stasis = standing] The process of achieving a relatively stable internal environment.

homeotic mutation A mutation that causes the cells of an embryo to give rise to an inappropriate structure in the adult, for example, to legs instead of antennae.

homeotic selector gene In a plant, a gene that establishes the fate of one or more whorls of a developing flower.

homologous chromosomes The two matching chromosomes that align during meiosis I.

homology cloning A strategy for isolating recombinant DNAs that share common sequence elements.

homozygous Having two copies of the same allele (in a diploid organism).

hormone A substance, made and released by cells in a specific organ or structure, that moves throughout the organism and exerts specific effects on specific cells in other organs or structures.

host range For a virus, the set of hosts that a particular virus can infect.

***Hox* gene** Mammalian counterpart of a *Drosophila* homeo domain gene.

human immunodeficiency virus (HIV) The retrovirus that causes AIDS; HIV infects and kills T_H lymphocytes by first binding to the CD4 protein on the cell surface.

humoral immunity [Latin, *umor* = fluid] The production of antibodies.

hybrid The progeny of a cross of two genetically distinct organisms.

hybridoma A hybrid cell line derived from a lymphoma.

hydrocarbon chain A chain of connected carbon atoms, with hydrogen atoms sharing other available outer shell electrons.

hydrogen bond A weak attraction between a hydrogen atom in one molecule that has a slight positive charge and a negatively charged atom in another molecule.

hydrogen ion (H^+) The result of a dissociation of a water molecule; also called a **proton,** though it is actually a **hydronium ion.**

hydrolysis [Greek, *hydro* = water + *lysis* = breaking] Breaking the bond between two building blocks by adding a water molecule, reversing the dehydration-condensation reaction.

hydronium ion (H_3O^+) A water molecule that has acquired an extra proton and a charge of +1; the proper name for a **hydrogen ion**.

hydrophobic [Greek, *hydro* = water + phobos = fear] Avoiding associations with water; nonpolar.

hydrophobic interaction The association of nonpolar molecules.

hydroxyl The —OH functional group, which allows molecules that contain it to form hydrogen bonds.

hyperpolarizing A voltage change across a membrane that makes the inside of the cell more negative and less apt to produce an action potential.

hypertonic [Greek, *hyper* = above] Having a total concentration of solutes higher than that within a cell.

hypervariable region The polypeptide segments, within the variable region of an immunoglobulin light chain or heavy chain, in which the most sequence variation occurs among antibodies; the hypervariable regions of an immunoglobulin molecule correspond exactly to the complementarity determining regions, where antigens bind.

hypocotyl [Greek, *hypo* = under] The axis of a developing plant below the attachment point of the cotyledons.

hypothesis An informed guess about the way a process works or a structure is organized.

hypotonic [Greek, *hypo* = under] Having a total concentration of solutes lower than that within a cell.

I band [for *isotropic*, meaning that it does not change the direction of polarized light] In muscle, the region of a sarcomere that contains only actin filaments and no myosin filaments.

imaginal disc In insect development, a group of larval cells from which adult structures later develop.

imago Adult stage of an insect.

imbibition The taking in of water at the time of seed germination.

immune response The defense system by which animals resist foreign invaders.

immunoaffinity chromatography A type of chromatography in which an antibody is attached to the stationary phase, and the specifically bound protein is eluted by changing the solution and weakening the attachment to the antibody.

immunocytochemistry [Greek, *immunis* = free (referring to the immune response) + *cyto* = cell], also called **immunohistochemistry** A method that uses specific antibodies to reveal the cellular locations of particular molecules of known structure or function.

immunoglobulin An antibody; one of the members of a group of globular blood proteins called globulins.

immunoglobulin fold A characteristic compact structure, within an immunoglobulin molecule; contains two β sheets held together both by hydrophobic amino acid side chains and by a disulfide bond between two cysteine side chains.

immunohistochemistry [Greek, *immunis* = free (referring to the immune response) + *histos* = tissue], also called **immunocytochemistry** A method that uses specific antibodies to reveal the cellular locations of particular molecules of known structure or function.

impermeable junctions, also called **occluding junctions** Molecular assemblies that connect epithelial cells tightly together; they also prevent molecules from leaking between them.

in vivo [Latin, = in life] Present in organisms.

in vitro [Latin, = in glass] In a test tube.

induced fit A change in the conformation of an enzyme brought about by the binding of the substrate.

inducible Made at higher levels in the presence of a specific inducer.

induction In development, the process by which one cell population influences the development of neighboring cells.

inductive reasoning Establishing a generalization from a pattern of specific examples.

inflammation A set of responses to local injury; characteristics of injured tissue include redness, heat, swelling, and pain.

inhibiting hormone A peptide hormone, made by the hypothalamus, that inhibits the release of a specific hormone.

inhibitory In the nervous system, tending to prevent the production of action potentials.

inhibitory interneuron A neuron whose stimulation by one neuron prevents the firing of another neuron in the same area.

inhibitory postsynaptic potential (IPSP) A hyperpolarizing response in a postsynaptic cell resulting from the action of an inhibitory neurotransmitter.

initiation The start of synthesis.

initiation site The first nucleotide actually transcribed from DNA into RNA.

inner cell mass In a mammalian embryo, a small group of cells within a blastocyst that will eventually grow into the embryo itself and subsequently into the adult.

inositol trisphosphate (IP_3) Inositol plus the three phosphate groups derived from phosphatidyl inositol 4,5-bisphosphate (PIP_2); a second messenger.

insertion A type of mutation; the addition of one or more nucleotides.

insertion sequence, also called a **simple transposon** The simplest transposable element, a short length of DNA (up to a few thousand nucleotide pairs long) that can move from place to place in the genome.

instructive theory The now abandoned view that an antigen would, by its presence, determine how each antibody molecule folds.

insulin A protein hormone, produced by specialized cells of the pancreas, that regulates glucose uptake; a signal for the absorptive state; promotes the synthesis of glycogen and inhibits its breakdown.

insulin-like growth factor (IGF) A protein paracrine signal that resembles insulin.

integral protein also called a transmembrane protein A membrane protein that spans the lipid bilayer.

integrated replication A method for viral propagation in which the virus's DNA becomes integrated into the host cell's DNA and is replicated along with that of the host.

integrin A transmembrane protein that connects a cell to the extracellular matrix.

intercalated disc Structures that join heart muscle cells end-to-end.

interferon A cytokine that interferes nonspecifically with the reproduction of viruses.

interleukin-1 (IL-1) A cytokine produced early during inflammation; stimulates responses both in leukocytes and in organs distant from the site of infection.

intermediate filament In eukaryotic cells, a component of the cytoskeleton that consists of filaments 8–10 nm in diameter, thinner than microtubules but thicker than actin filaments.

internode In a plant, the region of a stem between nodes.

interphase The part of the cell cycle in which the chromosomes are not condensed and the cytoplasm is not dividing.

intron (or **intervening sequence**) A segment of a gene (or of a pre-mRNA) that is transcribed into RNA but excised before the primary transcript matures into functional mRNA.

invagination Local folding of a cell layer to form an enclosed space with an opening to the outside.

inversion A mutation in which a segment of a chromosome is turned 180° from its normal orientation.

involuntary muscle A muscle that cannot be consciously controlled; smooth muscle.

ion An atom (or a molecule) with a net electrical charge, the result of a different number of electrons and protons.

ion exchange chromatography A frequently used type of chromatography in which the stationary phase has a net charge.

ionophore A compound that spans the plasma membrane and allows specific ions to pass.

iron-sulfur center The electron-accepting and electron-donating part of an iron-sulfur protein; it contains 2–4 iron atoms and an equal number of sulfur atoms.

iron-sulfur protein One of at least six proteins that participate in the electron transport chain; each contains an iron-sulfur center with 2–4 iron atoms and an equal number of sulfur atoms.

irreversible inhibitor A molecule that forms stable covalent bonds with an enzyme and reduces its catalytic ability.

isoelectric focusing A variation of electrophoresis in which molecules migrate in an electric field in such a way that they concentrate ("focus") at a point where they cease to have a net electrical charge.

isolated system A region that does not exchange matter or energy with its surroundings.

isomers Molecules that contain the same atoms arranged differently.

isotonic [Greek, *isos* = equal + *tonos* = tension] Having a total concentration of solutes that is the same as a cell's interior.

joining region segment, also called a **J segment** The part of the variable region of an immunoglobulin light or heavy chain that is closest to the constant region.

karyotype [Greek, *karyon* = kernel or nucleus + *typos* = stamp] The chromosomal makeup of a cell.

kinesin A microtubule-associated motor molecule that usually moves toward the plus end of a microtubule.

kinetic energy The energy of moving objects.

kinetics [Greek, *kinetikos* = moving] The study of the rates of reactions.

kinetochore A specialized disc-shaped structure that attaches the mitotic spindle to the centromere.

kinetochore microtubules A subset of polar microtubules that run from a pole of the mitotic spindle to a kinetochore.

***lac* repressor** A protein that binds to the *lac* operator and prevents the expression of the *lac* operon.

lactose operon (or ***lac* operon**) A region of *E. coli* DNA that encodes three proteins used to derive energy from lactose: the enzyme β-galactosidase (which splits lactose into galactose and glucose), and two other proteins, called permease and acetylase.

lagging strand In replicating DNA, the newly made strand that is extended discontinuously.

lambda repressor, also called **cI protein** A protein that prevents the transcription of λ genes.

lamellipodia Sheets of cytoplasm that extend forward from a cell engaged in ameboid movement.

lariat structure [because it resembles a rope-twirler's lariat] A transient structure formed during RNA splicing.

larva A feeding form of an animal distinct from the later adult.

lateral bud, also called an **axillary bud** A bud that forms just above the point where a leaf joins the stem and which can develop into a branch.

leader RNA A sequence transcribed from the region between the promoter and the *trp E* structural gene, which is responsible for attenuation.

leading strand In replicating DNA, the newly made strand that extends continuously, with DNA polymerase adding nucleotides to its 3′ end.

leaf primordium A tiny extension of the apical meristem; grows into a leaf.

learning Modification of neural activity and behavior as the result of experience.

leucine zipper A series of leucine side chains that extend in a single direction from two α helices; mediates the dimerization of several transcription factor polypeptides.

leukocyte [Greek, *leukos* = clear, white] White blood cell.

ligand [Latin, *ligare* = to bind] A molecule that binds to a specific binding site in a protein.

ligand-binding domain A structural domain within a transcription factor that can bind to a small molecule such as a steroid hormone; the binding of the small hormone changes the conformation and the activity of the factor.

light chain One of the polypeptides present in an immunoglobulin molecule.

light source The part of a microscope that illuminates a specimen.

linkage group A set of genes that do not assort independently because they are physically close to one another on the same chromosome.

linkage map A summary of the genetic distances between genes; a genetic map.

lipid A compound that is less soluble in water than in nonpolar solvents like olive oil.

lipid-soluble signal A chemical signal that can enter a target cell by passing directly through the plasma membrane.

liposomes Artificially produced vesicles that are surrounded by phospholipid bilayers.

locus [plural, **loci;** Latin, = place] The position of a gene on a chromosome.

low density lipoprotein (LDL) A carrier protein in the blood that binds to cholesterol.

low energy bonds Relatively stable chemical bonds.

lumen [Latin, = light, an opening] An enclosed space, bounded either by membranes (as in the ER lumen) or by an epithelium (as in the lumen of the gut).

luteinizing hormone (LH) A peptide hormone, made in the anterior pituitary, that, in females, induces ovulation and stimulates estrogen production and, in males, increases testosterone production.

lymphocyte One of a class of white blood cells that develop within the lymphoid tissues (including lymph nodes, spleen, thymus, and tonsils); the cells responsible for the immune response.

lymphocyte library The collection of B lymphocytes, each containing a unique random rearrangement of immunoglobulin genes.

lysis [Greek, *lysis* = loosening] The breaking down of a plasma membrane.

lysogenic cycle [Greek, *lysis* = loosen + *genos* = offspring, because it can *gen*erate the destructive *lyt*ic cycle] The reproduction of a virus along with its host without causing the host cell to lyse.

lysogenic virus A virus that reproduces either in a lytic cycle, in which it destroys its host, or along with the host; the virus can lie in a dormant (lysogenic) state and can be activated to enter a lytic cycle.

lysosome A small membrane-bounded organelle that contains hydrolytic enzymes which break down proteins, nucleic acids, sugars, lipids, and other complex molecules.

lytic cycle Viral reproduction that destroys the host.

lytic virus A virus that destroys its host cell by lysing it.

macromolecules [Greek, *macro* = large] Large molecules formed by the polymerization of smaller building blocks.

macrophage [Greek, *makros* = large + *phagein* = to eat] A large phagocytic cell, widely distributed throughout the body.

magnification The ratio of the size of an image to the size of the object itself.

major histocompatibility complex (MHC) A set of 40–50 closely linked genes, some of which encode proteins on the surface of every somatic cell; MHC proteins present antigens to the immune system.

map unit, also called a **centiMorgan (cM)** The genetic distance between two genes that produces 1% recombinant gametes and 99% parental gametes.

mast cell A large round cell, distributed throughout the connective tissues; filled with small histamine-containing vesicles, which are released at the sites of tissue damage.

maternal mRNA An mRNA that is already present in the egg before fertilization.

maternal effect genes Genes that are expressed in the mother during oogenesis.

matrix (of a mitochondrion) The interior compartment within the inner mitochondrial membrane.

mechanical work Energy change resulting from the movement of an object against a force.

median eminence A structure that carries hypothalamic hormones to the anterior pituitary.

megaspore [Greek, *megas* = large] In a flower, the haploid cell within an ovule that develops into the embryo sac, the mature female gametophyte.

meiosis [Greek, *meioun* = to make smaller] The process by which haploid gametes arise from diploid cells; meiosis distributes chromosomes so that each of four daughter cells receives one chromosome from each homologous pair.

meiosis I The first of the two divisions of meiosis, during which homologous chromosomes pair and are distributed into two daughter cells.

meiosis II The second of the two divisions of meiosis, during which sister chromatids are distributed to daughter cells.

melanocyte stimulating hormone (MSH) A peptide hormone, made in the anterior pituitary, which stimulates pigment production in specialized skin cells.

membrane attack complex A protein complex that kills foreign organisms by boring holes through their membranes.

membrane envelope A membrane that surrounds a virus particle.

membrane-bound ribosomes Ribosomes associated with a cell's internal membranes, responsible for the synthesis of secreted proteins as well as many membrane-associated proteins.

membrane-spanning segment A sequence of 20–25 hydrophobic amino acid residues that can form an α helix extending through the membrane in an integral membrane protein.

memory cell In the immune system, a cell that can later be stimulated to produce effector cells with particular antigen specificity.

mesoderm [Greek, *mesos* = middle + *derma* = skin] The middle tube of a vertebrate embryo, including the future connective tissues, bones, muscles, and tendons, and the cells of the blood.

messenger RNA (mRNA) The RNA molecules that carry information from DNA to the ribosomes, where the mRNA is translated into a polypeptide.

metabolism The complex network of biochemical conversions within a cell.

metacentric Refers to a chromosome in which the centromere is at or near the middle; such a chromosome looks like a V.

metallothionein A small protein that tightly binds several metals ions, including zinc, copper, and mercury.

metamorphosis In many species, a series of dramatic changes in form leading from a larva to an adult.

metaphase The stage of mitosis or meiosis during which chromosomes move halfway between the two poles of the spindle, where they accumulate in the metaphase plate.

metaphase plate A disc formed during metaphase in which all of a cell's chromosomes lie in a single plane at right angles to the spindle fibers.

MHC restriction In the immune response, the inability of T cells to recognize foreign antigens except in cells with the same MHC type as the responding animal.

micelle A cluster of amphipathic molecules.

Michaelis constant (K_M) In an enzyme reaction that obeys Michaelis-Menten kinetics, the concentration of substrate at which the reaction velocity is half of its maximum.

Michaelis-Menten kinetics An analysis of the velocity of an enzymatic reaction based on the necessity for the formation of an enzyme-substrate complex.

microelectrode An electrode that is small enough to maintain contact with a single cell.

microsomes Vesicles derived from fragments of the endoplasmic reticulum after experimental disruption of eukaryotic cells.

microspike Hair-like extension at the leading edge of a cell engaged in ameboid movement.

microsurgery The physical manipulation of subcellular structures.

microtubule organizing center (MTOC) The region of a eukaryotic cell from which microtubules emanate.
microtubule-associated protein (MAP) A protein that binds to microtubules and influences their organization.
microtubules In eukaryotic cells, the largest elements of the cytoskeleton; consist of tubulin molecules assembled into hollow rods, about 25 nm in diameter and of variable lengths, up to several μm.
midbody The thin connection between daughter cells that persists until the end of cytokinesis; the midbody is packed with microtubules from the spindle apparatus.
mineralocorticoid A steroid hormone that regulates ion flow in the kidney.
missense mutation A change in DNA that alters the codon for one amino acid residue into a codon for another amino acid residue.
mitochondrial matrix The compartment surrounded by the inner mitochondrial membrane.
mitochondrion [Greek, *mitos* = thread + *chondrion* = a grain; plural, **mitochondria**] One of the most prominent organelles in most eukaryotic cells; the site of oxidative phosphorylation; responsible for the production of most of a eukaryotic cell's ATP.
mitosis [Greek, *mitos* = thread] The process of the equal distribution of chromosomes during cell division.
mitotic spindle An elongated structure that develops outside the nucleus during early mitosis; contains the microtubular machinery that moves the chromatids apart.
mobile gene A gene whose chromosomal address changes.
model A simplified view of how the components of a structure operate.
modifier gene A gene that regulates the expression of another, separate gene.
molarity The concentration of a dissolved substance in moles per liter of solution.
mole The amount of a substance in grams equal to its molecular mass.
molecular genetics The branch of genetics that studies how DNA carries genetic instructions and how cells carry out these instructions.
molecule A specific combination of individual atoms held together by covalent bonds.
Monera The kingdom that, until recently, was considered to include all prokaryotes.
monoclonal antibody An antibody produced from a single hybridoma clone; each monoclonal antibody has a defined chemical structure and has the ability to recognize a single epitope.
monocot One of large group of plants—including corn and grass—whose embryos have one cotyledon.
monomer [Greek, *mono* = single] Each of the component parts of a polymer.
monosaccharide [Greek, *mono* = one + *saccharine* = sugar] A simple sugar with three to nine carbon atoms; the building block for polysaccharides.
monosomy The presence of only a single copy of a chromosome in a diploid cell.
morphogen A substance that specifies the position of a cell within a pattern.
morphogenesis The development of form.
morula [Latin, = mulberry] A solid ball of cells in an early mammalian embryo.
motor cortex A region of the brain responsible for movements, with each area specifying a particular part of the body.
motor neuron, also called a **motoneuron** A neuron that directly connects with a muscle and commands muscle contraction; the final common pathway of the instructions that direct movement.
motor ring, also called an **M ring** In a prokaryotic flagellum, a disc of protein that rotates within the stator.
motor A microtubule-associated protein that generates movement, using energy obtained by splitting ATP.
MPF, for **mitosis promoting factor**[originally called "*m*aturation *p*romoting *f*actor," because of its effect on the maturation of frog eggs] A protein factor that triggers the events of mitosis.
muscarinic acetylcholine receptor A G-protein coupled receptor that responds to acetylcholine by inhibiting the synthesis of cyclic AMP.
muscle fiber A giant muscle cell that contains the proteins responsible for contraction.
mutagen An agent that increases the rate of mutation; a mutagen can be a chemical or a form of radiation.
mutation A change in the nucleotide sequences of DNA.
mutation hot spot A DNA sequence where a mutation is much more likely to occur than in other DNA sequences.
mutualism An association between organisms of two species from which both organisms benefit.
myasthenia gravis A muscle disease in which the patient's immune system inappropriately produces antibodies that recognize and block the receptors for acetylcholine on muscle cells.
myelin An insulating structure around a nerve fiber; consists of extensions of the plasma membrane of a glial cell.
myelin sheath Extensions of the plasma membranes of a Schwann cell or an oligodendrocyte, wrapped around a nerve fiber (axon).
myoblast A cell that is a precursor of a muscle fiber.
myofibril [Greek, *myos* = muscle + Latin, *fibrilla* = little fiber] A thread, each about 1–2 μm in diameter, that runs the length of a muscle fiber.
myosin A long, two-headed protein that interacts with actin and generates movement in an ATP-dependent manner.
myosin light-chain kinase The enzyme that transfers a phosphate group from ATP to the myosin light chain, thereby triggering smooth muscle contraction.
myosin regulatory light chain A polypeptide whose phosphorylation triggers the contraction of smooth muscle.

N-region nucleotide addition Process of nucleotide addition that contributes to the diversity of immunoglobulin heavy chains.
Na^+-K^+ ATPase, also called the **sodium-potassium pump** An important active transporter that simultaneously transports sodium ions (Na^+) out of cells and potassium ions (K^+) into cells, using the energy of ATP.
NAD^+ [*n*icotinamide *a*denine *d*inucleotide] The major electron acceptor in oxidative phosphorylation.
NADH The reduced form of NAD^+
native The form of a protein that has biological activity.

negative feedback The process of neutralizing external changes.

negative regulator A protein that reduces transcription of a particular gene or operon.

Nernst equation An equation that predicts the relationship between the ion concentration and electric potential.

nerve growth factor (NGF) A protein paracrine signal, first isolated from salivary glands; necessary for the growth and differentiation of specific nerve cells in culture.

neural crest A set of embryonic cells derived from the roof of the neural tube; neural crest cells migrate to different locations in the embryo and develop into a variety of different cell types in the adult.

neural plate A flat plate above the notochord in a vertebrate embryo; the future nervous system.

neural tube A hollow tube that forms from the neural plate, above the notochord in a vertebrate embryo; the precursor structure for the central nervous system.

neuromuscular junction The chemical synapse between a motor neuron and a voluntary muscle cell.

neuron A nerve cell specialized for the conduction of electrical and chemical signals.

neurotransmitter A signaling molecule, made in a presynaptic cell, that affects the electrical activity of the postsynaptic cell.

neurulation The process of forming the neural tube and neural crest.

nicotinic acetylcholine receptor A ligand gated cation channel that responds to acetylcholine; found in the neuromuscular junction and elsewhere.

9 + 2 pattern The pattern of microtubules in an axoneme; consists of nine doublet microtubules, each consisting of two microtubules fused along their length, plus two singlet microtubules in the center.

nitrogenous base [so called because it contains nitrogen and can accept hydrogen ions] A small molecule that contains one or two aromatic rings of carbon and nitrogen atoms, with attached hydrogen atoms; attached to these rings are polar functional groups that can form hydrogen bonds with water or with other nitrogenous bases.

node The region of the stem to which the leaf attaches.

node of Ranvier A gap between the myelin wrappings, spaced about every millimeter along the axon.

nonassociative learning Learning in which the subject changes sensitivity to a stimulus after repeated exposure.

noncompetitive inhibitor A molecule whose inhibitory effects on an enzyme cannot be overcome by increased substrate concentration; a noncompetitive inhibitor usually binds to an enzyme at a location other than the active site.

noncovalent interaction Attraction (or repulsion) between separate molecules or ions; noncovalent interactions include electrostatic interactions, hydrogen bonds, van der Waals interactions, and hydrophobic interactions.

noncyclic photophosphorylation The production of ATP from light energy by a series of electron transfers that do not directly regenerate the absorbing chlorophyll; in plants noncyclic photophosphorylation occurs within photosystem II, with electron flow ultimately depending on both photosystems.

nondisjunction The failure of homologous chromosomes or sister chromatids to move apart during meiosis; nondisjunction results in a gamete having too many or too few chromosomes.

noninducible Not made at higher levels in the presence of an inducer that would normally increase production.

nonpolar Having an approximately uniform charge distribution.

nonsense Not specifying any amino acid; a nonsense codon is also called a "stop" codon.

nonsense mutation A change in DNA that changes the codon for an amino acid residue into a nonsense or termination codon and thereby leads to a truncated polypeptide.

notochord A rod that runs from the front to the rear of a vertebrate embryo, beneath its back (dorsal) surface.

nuclear envelope The boundary of a nucleus; it consists of a double membrane separated by about 20–40 nm.

nuclear lamina A fibrous sheath of intermediate filament proteins, just below the nuclear envelope.

nuclear pores Interruptions of the nuclear envelope that form channels between the contents of the nucleus and the cytosol.

nuclear transplantation A technique for moving a nucleus from one cell to another.

nuclease An enzyme that catalyzes the hydrolysis of a nuclei acid.

nucleic acid A macromolecule formed by the polymerization of nucleotides.

nucleoid In prokaryotic cells, the restricted part of a cell that contains the cell's DNA; not surrounded by a membrane.

nucleolar organizer A region of one or more chromosomes that contains the genes for ribosomal RNAs.

nucleolus [Latin, = a small nucleus; plural, **nucleoli**] A conspicuous structure within the nucleus, in which ribosomal RNAs are made.

nucleoplasm The contents of the nucleus.

nucleoside A nitrogenous base attached to a sugar by a covalent bond between a carbon atom of the sugar and a nitrogen atom of the base.

nucleosome A DNA-histone complex, about 11 nm in diameter; each nucleosome contains a 146-nucleotide long stretch of DNA and eight histone molecules (two each of H2A, H2B, H3, and H4).

nucleotide A small molecule that consists of a nitrogen-containing aromatic ring compound, a sugar, and one or more phosphate groups.

nucleus In eukaryotic cells, the membrane-enclosed structure that contains most of a cell's genetic instructions, in the form of DNA.

null mutation A change in DNA that leads to the failure to produce a polypeptide or the rapid degradation of a polypeptide.

objective In a microscope, the lens nearest the sample.

obligate anaerobe A microorganism that can grow only in the absence of oxygen.

Octet Rule The generalization that an atom is particularly stable and chemically unreactive when its outermost shell is full, meaning (usually) that it contains eight electrons.

ocular The lens of a microscope nearest the eye.

oil A triacylglycerol that is liquid at room temperature; the component fatty acids are usually unsaturated.

Okazaki fragment [after its discoverer Reijii Okazaki] In DNA synthesis, a stretch of DNA that is to be added to the lagging strand.

olfactory receptor A protein, in the membrane of the olfactory epithelium, that binds to odorant molecules; binding is the first step in informing the brain about smell.

oligodendrocyte A glial cell in the central nervous system; the oligodendrocyte plasma membrane wraps around nerve cell extensions in a myelin sheet.

oncogene [Greek, *onkos* = bulk, tumor] A gene that can change normal cells into cancer-like cells.

oocyte A precursor of an egg cell.

oogenesis The production of mature eggs.

oogonium A cell that is committed to forming eggs, but which can still undergo mitosis.

open system A region that exchanges both materials and heat with its surroundings.

operator The DNA sequence in an operon to which a repressor protein binds.

operon In prokaryotes, a set of genes transcribed into a single mRNA.

opiates Morphine, heroin, and related compounds.

optical isomers Two molecules that differ only in the arrangement of four different groupings of atoms attached to a single carbon atom.

orbital [so called to distinguish from a planet's restricted "orbit"] A limited portion of an atom's space through which an electron moves.

organelle A small structure within a cell that performs a specialized task; in eukaryotic cells, many organelles are enclosed by membranes, which isolate the contents of the organelle from the rest of the cytoplasm.

organ formation The movement and specialization of tissues and cells to produce functioning organs such as heart, kidneys, and the nervous system.

organic Carbon-containing; refers to all carbon-containing compounds, even when they have nothing to do with organisms.

organic chemistry The study of the structures and reactions of carbon compounds.

origin of replication A DNA sequence at which replication begins.

oscilloscope A measuring instrument that displays voltages on a television-like screen, showing variations with time.

osmosis [Greek, *osmos* = push, thrust] The flow of water across a selectively permeable membrane as a result of concentration differences.

ovary [Latin, *ovum* = egg] The organ (in an animal or a flower) that produces female germ cells (or gametophytes); a female gonad.

overlapping genes A single stretch of DNA that contains the information for distinct polypeptides in different reading frames.

ovule [Latin, *ovulum* = little egg] Structure in a carpel that, after fertilization, forms a seed.

ovum [Latin, = egg; plural, **ova**] A female gamete; an egg.

oxaloacetate A four-carbon compound that can combine with the acetyl group from acetyl CoA to produce citric acid, beginning the citric acid cycle.

oxidative fiber A thin muscle fiber that derives most of its energy from respiration.

oxidative phosphorylation The process that couples the oxidation of NADH and $FADH_2$ to the production of high-energy phosphate bonds in ATP.

oxidizing agent The electron acceptor in a redox reaction.

oxygen evolving complex A manganese-containing protein that directly participates in the electron transfers that convert coordinates of water to oxygen in photosynthesis.

oxytocin A peptide hormone, released by the posterior pituitary, that stimulates contractions of the uterus during childbirth.

paleontology [Greek, *palaios* = ancient + *on* = being + *logos* = discourse] The study of ancient life.

palindrome A sequence of letters or nucleotides that reads the same in both directions; a famous verbal palindrome is "Madam, I'm Adam"; in DNA, a symmetrical sequence to which a symmetrical protein can bind.

paracrine signal A chemical signal that acts only in the immediate region of its production.

paramyosin The specialized myosin of a mollusc's catch muscle; helps to maintain tension for a long time with little energy expenditure.

parasitism An association between organisms of two species that benefits only one participant.

parasympathetic nervous system A division of of the autonomic nervous system; generally acts to husband resources, for example by slowing the heart and increasing intestinal absorption.

parenchymal cells Plant cells with thin cell walls, chloroplasts, large vacuoles, and the machinery to perform photosynthesis.

parental, or **P**, generation The original parents in a cross.

passive transport Movement of a substance across a membrane that occurs spontaneously, without the expenditure of energy.

patch clamping A method that allows the measurement of currents across a tiny piece of membrane rather than across a whole cell membrane.

pattern formation The creation of a spatially organized structure during development.

pedigree A family tree; often used to show the inheritance of a disease within a family.

pellet The part of a sample that, during centrifugation, moves to the bottom of the centrifuge tube.

penetrance The fraction of individuals with a particular genotype that show a corresponding phenotype.

pentose [Greek, *pente* = five] A sugar that contains five carbon atoms; ribose and deoxyribose, which are components of nucleotides, are pentoses.

peptide A water-soluble chemical signal composed of amino acid residues linked by peptide bonds.

peptide bond The covalent bond between two amino acid molecules in a polypeptide; formed by the carboxy group of one amino acid attaching to the amino group of another amino acid.

peptide bond formation During protein synthesis, the transfer of an amino acid from aminoacyl-tRNA to the growing polypeptide chain.

peripheral nervous system (PNS) Nerve cells outside the central nervous system (CNS); carries information from the CNS to muscles and organs and to the CNS from sense organs.

peripheral proteins Membrane proteins located on the outer and inner surfaces of the plasma membrane.

permissive temperature Usually relatively low, a temperature at which a temperature-sensitive mutation has no phenotypic effect.

peroxisome In eukaryotic cells, a kind of membrane-bounded vesicle; peroxisomes contain enzymes that use oxygen to break down molecules by pathways that produce hydrogen peroxide (H_2O_2).

pertussis toxin A protein produced by the bacterium *Bordetella pertussis;* causes whooping cough; specifically modifies the α subunit of G_i, locking it in the GDP-binding form.

petal One of the showy, usually colored, parts of a flower.

pH The logarithm (to the base 10) of the molar hydrogen ion concentration; $pH = \log_{10} 1/[H^+] = -\log [H^+]$.

phagocytosis [Greek, *phagein* = to eat + *kytos* = hollow vessel] A type of cellular ingestion in which the cell's membrane surrounds a relatively large solid particle, such as a microorganism or cell debris.

phalloidin A poison that prevents the dissociation of an actin filament into monomers.

phase contrast microscopy A special optical arrangement (distinct from differential interference contrast microscopy) that reveals phase differences among different cell components without the use of stains.

phenotype The set of an organism's observable properties.

phenotypic trait A single aspect of phenotype in which individuals may vary.

phenylketonuria (PKU) A human disease that results from the absence of the enzyme phenylalanine hydroxylase, which converts phenylalanine to tyrosine.

pheromone A substance secreted by one organism that influences the behavior or physiology of another organism of the same species.

phosphatase An enzyme that hydrolyzes phosphate esters to produce free phosphate ions.

phosphate The ion (PO_4^{3-}) formed by the dissociation of hydrogen ions from phosphoric acid (H_3PO_4).

phosphatidyl inositol (PI) A phospholipid that is phosphorylated to form phosphatidyl inositol 4,5-bisphosphate (PIP_2), which is the precursor of two second messenger molecules, diacylglycerol (DAG) and inositol trisphosphate (IP_3).

phosphatidyl inositol 4,5-bisphosphate (PIP_2) The precursor of two second messenger molecules, diacylglycerol (DAG) and inositol trisphosphate (IP_3).

phosphodiester bond The links between nucleotides formed by the phosphate group of one nucleotide attaching to a carbon atom in the sugar component of another nucleotide.

phospholipase C The enzyme that cleaves PIP_2 into two smaller molecules that serve as second messengers, diacylglycerol (DAG) and inositol trisphosphate (IP_3).

phospholipid An amphipathic derivative of glycerol in which two hydroxyl groups attach to fatty acids and the third to a phosphate ester; a principal component of biological membranes.

phosphoric acid ester A molecule in which a hydroxyl group attached to a carbon atom replaces an oxygen atom in phosphoric acid.

phosphorylase kinase An enzyme that transfers phosphate groups from ATP to glycogen phosphorylase.

photochemical reaction center A complex of the chlorophyll molecules and proteins that convert captured light energy to chemical energy.

photon A package of energy; a light particle.

photoperiodism The response of a plant to the relative lengths of day and night.

photorespiration In plants, an oxygen-dependent process that does not produce ATP or NADPH; photorespiration converts ribulose bisphosphate into CO_2 and serine.

photosynthesis [Greek, *photos* = light + *syntithenai* = to put together] The process that converts light energy into the energy of chemical bonds.

photosystem I One of two distinct but interacting sets of electron transfer reactions responsible for storing light energy in high-energy chemical bonds; photosystem I best absorbs and uses light with wavelengths of about 700 nm.

photosystem II One of two distinct but interacting sets of electron transfer reactions responsible for storing light energy in high-energy chemical bonds; photosystem II best absorbs and uses light with wavelengths of about 680 nm.

phototropism The bending of plants toward light.

phragmoplast [Greek, *phragmos* = fence + *plasma* = mold, form] A set of microtubules that extends between two dividing plant cells at right angles to the cell plate.

phytochrome [Greek, *phyton* = plant] A plant pigment involved in many processes that depend on the timing of dark and light, including flowering, germination, and leaf formation.

pigment A molecule that absorbs visible light and has color to human eyes.

pilus In bacteria, a long appendage that serves as the means of attachment of conjugating bacteria and a conduit for the transfer of DNA.

pinocytosis [Greek, *pinein* = to drink] A type of endocytosis; the nonspecific uptake of bits of liquid and dissolved molecules.

pistil, also called a **carpel** In a flower, a female reproductive structure.

pituitary gland A pea-sized structure at the base of the brain that releases at least nine hormones.

Plantae The kingdom that includes plants, autotrophic multicellular organisms that undergo embryonic development.

plasma cell A cell specialized for the production of antibodies.

plasma membrane A cell's boundary.

plasmid A circular DNA that can replicate autonomously.

plasmodesmata (singular, **plasmodesma**) [Greek, *plassein* = to mold + *desmos* = to bond] Fine intercellular channels between plant cells, derived from vesicles trapped in the growing cell plate.

plasmolysis [Greek, *plasma* = form + *lysis* = loosening] In a plant cell, the flow of water out of the vacuole in a hypertonic solution and the resulting separation of the plasma membrane from the cell wall.

plastid A plant organelle surrounded by a double membrane; plastids include chloroplasts, chromoplasts, and amyloplasts.

platelet derived growth factor (PDGF) A protein paracrine signal that was first isolated from platelets.

pleiotropy [Greek, *pleios* = more + *trope* = turning] The capacity of a single gene to affect many aspects of phenotype.

point mutation, also called a **single-base mutation** A change in a single nucleotide pair of DNA; in some cases, a change in a few nucleotide pairs still qualifies as a point mutation.

polar Having uneven distributions of electrical charge; having positive and negative ends (or poles).

polar body The smaller daughter cell that results from the unequal division of an oocyte.

polar microtubules Microtubules that run from each pole of the mitotic spindle toward the equator.

pollen tube An extension of a pollen grain, which carries the sperm nuclei into the ovule.

polycistronic [Greek, *poly* = many + *cistron,* an invented synonym for "gene" based on a technique called the "*cis-trans* test," not discussed in this book] An mRNA that encodes more than one polypeptide.

polyclonal Referring to antibodies made in the descendants of many B cells, with each B-cell clone producing an antibody that recognizes a different epitope.

polygenic Dependent on more than one gene.

polymer [Greek, *poly* = many + *meros* = part] A molecule that consists of smaller molecules linked together.

polymerase chain reaction (PCR) A method that specifically and repetitively copies a segment of DNA between two defined nucleotide sequences.

polymerization [Greek, *polys* = many + *meros* = part] The assembly of many small molecules into a larger one.

polynucleotide A chain of nucleotide residues held together by phosphodiester bonds.

polypeptide A chain of amino acid residues held together by peptide bonds.

polyribosome, also called a **polysome** Several ribosomes attached to a single mRNA.

polysaccharide A carbohydrate macromolecule formed by the polymerization of simple sugars (monosaccharides).

polyspermy Fertilization by more than one sperm.

polytene chromosomes [Greek, *polys* = many + *tainia* = ribbon] Giant chromosomes, commonly studied in the larvae of *Drosophila* and other flies, which consist of about 1000 chromatids aligned in parallel

polyunsaturated Referring to a hydrocarbon chain (or fatty acid) with many double bonds.

positional information Chemical cues that establish the position of a cell in a pattern.

positioning During protein synthesis, the binding of the next aminoacyl-tRNA to the A site in the ribosome-mRNA complex; the first step of elongation.

positive regulator A protein that increases the rate of transcription.

post-transcriptional processing A series of chemical modifications that convert the primary transcript of a gene to a mature RNA.

postabsorptive state The state of an animal in which cells derive energy and building blocks by breaking down stored glycogen, fats, and proteins.

postsynaptic Refers to the neuron that is "downstream" from a synapse; a postsynaptic neuron responds to the presynaptic neuron.

postsynaptic potential A voltage change induced by the action of a neurotransmitter.

posttranslational Occurring after translation is completed.

potency What a cell or tissue could become during development if it were allowed to develop in another environment.

potential energy A general term for energy that can ultimately be converted into kinetic energy.

pre-mRNA The primary transcript of a protein-coding gene; also used to refer to the partly modified, not yet mature mRNA precursor.

prebiotic evolution The evolution of organisms from nonliving matter.

Precambrian Referring to fossils that date from more than 590 million years ago, that is, before the beginning of the Cambrian Era.

preformation The theory, now discredited, that all the parts of the adult organism already preexist at the earliest stages of life.

preproinsulin A single polypeptide that is the precursor of the two chains of insulin.

presumptive, having a specified developmental fate.

presynaptic Refers to the neuron that is "upstream" from a synapse; an action potenital in the presynaptic neuron can trigger the release of neurotransmitter into the synapse.

Pribnow box [after its discoverer] A consensus sequence in bacterial promoters.

primary growth In a plant, the extension of length.

primary lysosomes Nearly spherical lysosomes that have not yet fused with endosomes.

primary mesenchyme [Greek, *mesos* = middle + *enchyma* = infusion] The first cells to move into the interior of a blastula.

primary organizer The term used by Spemann to describe the dorsal lip of the blastopore, meaning that its action established the organization of the entire early embryo.

primary structure The linear sequence of amino acid residues in each polypeptide chain of a protein.

primary transcript The RNA first transcribed from a particular gene; the precursor of a mature RNA.

primase An RNA polymerase that copies short stretches (fewer than 10 nucleotides) of DNA into RNA; this RNA serves as a primer for DNA polymerase.

primate One of a group (or "order") that includes humans, monkeys, apes, and the lemurs of Madagascar.

primer An already existing polynucleotide to which additional nucleotides are added; all known DNA polymerases add nucleotides to a primer.

primordial germ cells The precursors of both sperm and eggs, which usually arise early in development.

Principle of Independent Assortment, also called **Mendel's Second Law** The generalization that the alleles for one gene segregate independently of the alleles of another gene.

Principle of Segregation, also called **Mendel's First Law** The generalization that a sexually reproducing organism has two "determinants" (or genes, in modern terms) for each characteristic, and these two copies segregate (or separate) during the production of gametes.

pro-opiomelanocortin (POMC) A polypeptide, made in the anterior pituitary, that is cut by specific proteases to produce at least seven biologically active peptides, including adrenocorticotropic hormone; β-endorphin; β and γ-lipotropic hormones; and α-, β-, and γ-melanocyte-stimulating hormones.

probability The chance that a given event will happen; calculated as the number of times an event has actually occurred divided by the number of opportunities it could have occurred; in the most familiar example, the probability of a tossed coin turning up "heads" is 1 (the number of sides that are heads) divided by 2 (the total number of sides).

product rule The statement that the probability of two independent events taking place together is the product of their probabilities.

programmed cell death The death of specific cells as a normal part of development.

prokaryotic [Greek, *pro* = before] Cells that contain neither a nucleus nor other membrane-bounded organelles.

prolactin A peptide hormone, made in the anterior pituitary, that stimulates milk production.

prometaphase [Greek, *meta* = middle], previously called early metaphase, the stage of mitosis during which the nuclear membrane disappears and the chromosomes attach to the spindle fibers.

promoter A DNA sequence that specifies the starting point and direction of transcription; where RNA polymerase first binds as it begins transcription.

prophage [Greek, *pro* = before] The dormant form of a bacteriophage in which the phage DNA has integrated into that of the host cell.

prophase [Greek, *pro* = before] The first phase of mitosis, when the diffusely stained chromatin resolves into discrete chromosomes, each consisting of two chromatids joined together at the centromere.

proplastid Common precursor of all plastids.

prostaglandin One of 16 paracrine signals derived from a 20-carbon fatty acid called arachidonic acid.

protease, also called a **peptidase** A digestive enzyme that catalyzes the hydrolysis of peptide bonds; used to digest proteins.

protein A macromolecule consisting of one or more polypeptides.

protein kinase An enzyme that transfers the phosphate group from an ATP to a protein.

protein kinase A An enzyme that, when stimulated by cyclic AMP, transfers a phosphate group from ATP to specific proteins.

protein kinase C An enzyme that, when bound to diacylglycerol, transfers phosphate groups from ATP to sites in particular proteins.

Protista The kingdom that consists mostly of single-celled organisms, but that also contains some related multicellular species; protists include algae, water molds, slime molds, and protozoa.

proto-oncogene, also called a **cellular oncogene** The cellular counterpart of a viral oncogene; participates in the normal control of growth and differentiation.

protobiont A primitive cell that could concentrate organic molecules and begin to evolve the first metabolic pathway.

proton channel The route that hydrogen ions follow as they flow down an electrochemical gradient.

proton gradient The difference in H^+ concentration across a membrane.

provirus [Greek, *pro* = before] The dormant form of a virus, in which the viral DNA has integrated into that of the host cell.

pseudopod [Greek, *pseudo* = false + *pous* = foot] A retractable extension of a cell engaged in ameboid movement.

pupa An inactive phase, following a larval stage, of insect development; may be enclosed in a cocoon.

purine A nitrogenous base that contains a particular nine-membered double ring, with five carbon and four nitrogen atoms; adenine and guanine are purines.

pyrimidine A nitrogenous base that contains a particular six-membered ring, with four carbon and two nitrogen atoms; thymine, uracil, and cytosine are pyrimidines.

pyrophosphate A molecule that consists of two phosphate groups linked together in a high energy bond.

pyruvate A three-carbon compound that is the end point of the first stage of glycolysis.

quantitative character A phenotypic trait that may be described numerically within a range of values rather than as clear cut alternatives.

quaternary structure The relationship among separate polypeptide chains in a protein.

reading frame The grouping of nucleotides into codons that specify an amino acid sequence.

receptor A protein to which a chemical signal first binds.

receptor-mediated endocytosis The uptake of specific substances, which are recognized by receptor proteins on the cell membrane.

recessive Refers to an allele that does not contribute to the phenotype of a heterozygote.

recombinant Containing a combination of genes not found in nature.

recombination nodules Large assemblies of proteins within the synaptonemal complex that assist in the formation of chiasmata between corresponding segments of the two homologs.

recombination signal sequence A characteristic DNA sequence in an immunoglobulin gene that lies next to segments that can be joined.

redox potential The differing tendencies of two redox reactions to move electrons from electron donor (reducing agent) to electron acceptor (oxidizing agent); comparison of the redox potentials of two reactions predict the direction of electron flow.

reducing agent The electron donor in a redox reaction.

reduction-oxidation (redox) reaction A reaction that transfers electrons from one molecule (the reducing agent, or electron donor) to another (the oxidizing agent, or electron acceptor).

reductionism The effort to understand the whole by understanding the parts.

reflex The most automatic behavior pattern, with motor activity directly responding to a sensory stimulus.

refractory period A period during which a particular channel remains closed.

regulatory cascade A set of reactions that amplify a signal from a relatively small number of molecules into a response that affects many more molecules.

regulatory mutation A change in DNA that affects the amount (rather than the structure) of a polypeptide.

release factor A cytoplasmic protein that a binds an mRNA-ribosome-polypeptide complex and releases the newly made polypeptide.

releasing factor, also called **releasing hormone** A peptide hormone, made by the hypothalamus, that regulates the release of specific hormones by the anterior pituitary.

replication unit In eukaryotic DNA replication cells, a group of 20–50 origins of replication that form replication forks at the same time.

replication [Latin, *replicare* = to fold back] Copying of a single DNA molecule (or a single set of DNA molecules) into two copies.

replication fork A Y-shaped region of DNA where the two strands of the helix have come apart during DNA replication.

reporter A protein (or gene) whose activity serves as a measure of the ability of a test sequence to regulate expression in a particular type of cell.

residue A building block of a macromolecule from which a water molecule has been removed in a dehydration-condensation reaction.

resistance factor, also called an **R factor** A plasmid that contains a gene for an enzyme that inactivates an antibiotic.

resolution The minimum distance between two objects that allows them to form distinct images.

respiration The oxygen-dependent extraction of energy from food molecules.

respiratory chain The pathway of electrons in oxidative phosphorylation.

response element, also called a ***cis*-regulatory element** or **enhancer** A DNA sequence that regulates transcription by binding to a specific protein, a transcription factor.

resting potential The voltage across the cell membrane when a cell is not producing action potentials.

restriction enzyme An enzyme that cuts DNA at a particular sequence.

restrictive temperature Generally higher than the permissive temperature, a temperature at which a temperature-sensitive mutation has no phenotypic effect.

retrovirus A virus that uses reverse transcriptase for viral replication.

reuptake The pumping of transmitter from the synaptic cleft either into the presynaptic neuron or into surrounding glial cells.

reverse transcriptase An enzyme that copies RNA into DNA; RNA-dependent DNA polymerase; present in RNA tumor viruses.

reversion A mutation that restores the previous version of the sequence.

rho (ρ) A protein factor that participates in the termination of transcription.

rhodopsin The protein that detects light in the retina.

ribbon model A way of depicting the tertiary structure of a protein in way that stresses the secondary structures of a globular protein; in the ribbon model an α helix is depicted as a coil, while β sheets are sets of arrows.

ribosomes [Latin, *soma* = body + "ribo" because they contain *ribo*nucleic acid (RNA)] Complex assemblies of RNAs and proteins, about 15–30 nm in diameter, that are responsible for carrying out protein synthesis.

RNA polymerase The enzyme responsible for transcribing DNA into RNA.

root The part of a plant below the ground.

root cap A region of the growth zone of a plant's root; produces a polysaccharide slime that lubricates the path of the growing root through the soil and serves as a mechanical shield for the apical meristem, which lies just behind.

rough ER Endoplasmic reticulum that is associated with ribosomes.

rudiment [Latin, *rudimentum* = beginning] Initial embryonic stage of an organ, from which the final form will develop.

ruffling In a cell engaged in ameboid movement, the extension of processes at the leading edge of the cell and their subsequent movement away from the leading edge.

S The period of the cell cycle during which DNA replicates; the period of DNA synthesis, when a cell doubles its DNA content.

saltatory conduction The jumping movement of an action potential down a myelinated axon, from one node of Ranvier to the next; occurs at rates up to 100 times faster than down an unmyelinated axon.

sarcomere [Greek, *sarx* = flesh + *meros* = part] In muscle, a small repeating cylinder within a myofibril; each sarcomere is about 2.5 μm long.

sarcoplasmic reticulum In a muscle fiber, a system of membrane-lined channels and sacs; a specialization of the endoplasmic reticulum where Ca^{2+} ions are stored.

saturated A hydrocarbon chain (or fatty acid) in which all the bonds between carbon atoms are single bonds; so called because it contains the maximum number of hydrogen atoms.

scanning electron microscope (SEM) A type of electron microscope in which electrons are reflected from the surface of the observed object.

Schwann cell A glial cell in the peripheral nervous system; its plasma membrane wraps around nerve cell extensions in amyelin sheet.

scientific method A formal manner of formulating, testing, and eliminating hypotheses.

Second Law of Thermodynamics The statement that, while the total energy in the universe does not change, less and less energy remains available to do work.

second messenger An intracellular molecule whose synthesis and degradation depends on the action of an extracellular signal and which stimulates changes inside a target cell.

secondary immune response A robust immune response triggered by the reappearance of a previously encountered antigen or microbe.

secondary lysosomes Irregular lysosomes that result from the fusion of a primary lysosome and an endosome; they contain material that is in the process of being digested.

secondary mesenchyme Cells that become mesoderm as gastrulation proceeds.

secondary sexual characteristics Hallmarks of sexual differentiation that appear at sexual maturity; in humans these include an increase in body size, a deepening of the voice, hair growth, and the development of sexual drive.

secondary structure Regular local structures resulting from regular hydrogen bonding within adjacent stretches of a polypeptide backbone; includes α helices and β structures.

secretory vesicles In eukaryotic cells, small vesicles, formed on *trans* side of the Golgi apparatus; they contain glycoproteins that are to be secreted.

section A thin slice of tissue prepared for microscopy.

segmentation gene A gene that regulates the development of segments within the early embryo.

segregation The separation of two homologous chromosomes during meiosis.

selective theory The view that the immune response depends on an animal's ability to select the set of antibodies to make in response to the presence of a particular infection.

selectively permeable Allowing the passage of some ions and molecules (especially of water) much more rapidly than others.

self-splicing Removal of intervening sequences from an RNA without the participation of other molecules.

selfish gene A sequence of DNA whose only role is to reproduce itself.

semiconservative The pattern of DNA replication in which half of each parent molecule (one strand) is present in each daughter molecule.

senescence In plant cells, the breakdown of cellular components leading to cell death.

sensitization The increased behavioral response to a noxious stimulus.

sepal One of the green parts at the base of a flower.

sex chromosomes The chromosomes that differ between males and females; in humans, the 23rd pair of chromosomes.

sex steroid A steroid hormone that stimulates, maintains, and regulates reproductive organs and secondary sexual characteristics.

sex-determining region In male mammals, a gene on the Y chromosome that encodes a regulatory protein that promotes the sexual differentiation of the embryonic gonads into testes.

sex-linked gene Having a different pattern of inheritance in males and females because the gene is contained on a chromosome (the X chromosome in mammals) that also determines the gender of the offspring.

sexual reproduction A process that produces offspring that have inherited genetic information from two parents rather than one; because the genes from each parent are likely to differ, sexual reproduction provides new combinations of genes.

shell A group of orbitals whose electrons have nearly equal energy.

shoot The part of a plant above the ground.

sickle cell disease, also called **sickle cell anemia** A blood disorder that gets its name from the curled appearance of red blood cells in sickle cell patients; a genetic disease caused by a mutation in the gene encoding the β-polypeptide of hemoglobin.

sigma (σ) An accessory protein that allows bacterial RNA polymerase to bind strongly to promoter regions in DNA.

signal hypothesis The statement that a hydrophobic signal peptide directs a growing polypeptide chain through the membrane of the endoplasmic reticulum, into the ER lumen.

signal peptidase An enzyme, confined to the ER lumen, that removes the signal peptide from a polypeptide.

signal peptide A polypeptide segment of 16–30 amino acid residues at the amino terminal end of the newly made protein; a signal peptide directs the newly made protein into the lumen of the endoplasmic reticulum.

signal recognition particle (SRP) A complex of proteins and RNA that recognizes the signal peptide and shuttles the marked polypeptide to the ER membrane.

signal sequence receptor A protein on the ER membrane that binds the signal sequence after its release by SRP and the SRP receptor.

silent mutation A change in DNA that has no effect on phenotype.

simple transposon, also called an **insertion sequence** The simplest transposable element, a short length of DNA (up to a few thousand nucleotide pairs long) that can move from place to place in the genome.

single-channel recording An application of the patch clamp method that measures currents that pass through a single channel.

single-strand binding protein In DNA replication, a protein that binds to the unwound DNA and keeps open the jaws of the replication fork.

sister chromatids The two chromatids that make up a single chromosome; the sister chromatids are duplicate copies of the same genetic information.

site-directed mutagenesis A technique that specifically alters a chosen nucleotide sequence.

Sliding Filament Model The generally accepted view of muscle movement, which holds that movement results from changing the relative positions of thick and thin filaments rather than from filament contraction.

smooth ER Endoplasmic reticulum that is not associated with ribosomes.

smooth muscle The involuntary muscles that line the walls of hollow internal organs such as intestines and blood vessels.

snRNP [small nuclear ribonucleoprotein particle; pronounced "snirp"] Within a spliceosome, one of several complexes of proteins and small nuclear RNAs that catalyze the removal of introns and the splicing of exons.

solute A substance that dissolves within a solvent.

solvent A substance in which other substances dissolve.

soma The somatic cells of an organism; in contrast to germ cells, or germ line.

somatic cells [Greek, *soma* = body], also called the **soma** In a multicelled organism, the cells that do not give rise to germ cells or gametes.

somatic theory The view that antibody diversity arises during development, rather than being inherited.
specific activity The amount of enzymatic or binding activity per gram of total protein.
sperm, also called a **spermatozoan** A male gamete.
spermatogenesis The development of spermatogonia into mature sperm.
spermatogonium A cell that is committed to forming sperm cells, but can still undergo mitosis.
spermatozoan [Greek, *sperma* = seed + *zoos* = living], also called a **sperm** A fully differentiated male germ cell.
spliceosome A complex of pre-mRNA, proteins and small nuclear RNAs that catalyzes the removal of introns and the splicing of exons.
spontaneous A reaction that occurs without any external input of energy.
spontaneous abortion, also called **miscarriage** In mammalian development, embryonic death.
spore A cell that divides by mitosis to produce new individuals.
sporophyte The diploid form of a life cycle characterized by alternation of generations; produces haploid spores that give rise to haploid gametophytes by meiosis.
SRP receptor, or **docking protein** A membrane protein that pulls a growing polypeptide into the lumen of the endoplasmic reticulum.
stage On a microscope, a moveable platform that holds the observed specimen.
stain A dye that binds differently to cell components and thereby increases contrast for microscopy.
stamen In a flower, a male reproductive organ.
Start, also called the **restriction (R) point** The "point of no return" in the G_1 phase of the cell cycle; once a cell proceeds beyond Start, it proceeds through the rest of the cycle, including mitosis and cytokinesis.
statistics The science of collecting and analyzing numerical data.
statolith [Greek, *statos* = standing + *lithos* = stone] Gravity detector.
stator, also called an **S (or static) ring** In a prokaryotic flagellum, the structure that holds the motor ring; embedded within the plasma membrane and anchored to the peptidoglycan layer of the cell wall.
stem cell A cell that can either produce more of itself by cell division or undergo differentiation to one or more specialized cell types.
stereoisomers Molecules with the same atoms and functional groups; two stereoisomers differ only in the spatial arrangements of their atoms.
steric inhibition Inhibition of an enzyme by a molecule whose shape resembles that of the substrate.
steroid A lipid-soluble molecule derived from cholesterol.
stoma [Greek, *stoma* = mouth; plural, **stomata**], also called a **stomate** A minute pore in a plant leaf that allows air to pass to the interior of the leaf.
stop codon A codon in mRNA that signals the end of a polypeptide chain.
striations Cross stripes, 2–3 μm apart, that lie perpendicular to the long axis of a skeletal or cardiac muscle.
stroma [Latin, = mattress] The region within a chloroplast bounded by the inner chloroplast membrane.
structural isomers Molecules that contain the same atoms, grouped in different ways to produce different functional groups.
structural mutation A change in DNA that leads to an alteration in the amino acid sequence of a polypeptide.
subcellular fractionation The bulk isolation of specific subcellular structures, usually by differential centrifugation.
substrate A reacting molecule in an enzyme reaction; usually (but not always) much smaller than an enzyme.
substrate level phosphorylation A reaction in which a redox reaction is coupled to the production of a high energy phosphate.
sugar A simple carbohydrate; a molecule that has the equivalent of one molecule of water (that is two hydrogen and one oxygen atom) for every atom of carbon.
sulfhydryl A functional group that contains a sulfur atom bonded to a hydrogen atom.
sum rule Statement that the probability that either of two separate events will occur is the sum of their individual probabilities.
supernatant The part of a sample that, during centrifugation, remains in suspension.
suppressor mutation A mutation in one gene that prevents the phenotypic changes caused by a mutation in another gene.
surface-to-volume ratio The amount of surface area for each bit of volume.
sympathetic nervous system A division of of the autonomic nervous system; initiates the "fight or flight" reaction.
symport [Greek, *sym* = together with] The coupled transport of two substances across a membrane, in the same direction.
synapse The distinct boundary between two communicating neurons.
synapsis [Greek, = union] The pairing of homologous chromosomes in prophase I.
synaptic cleft A space of about 20 nm that separates the presynaptic and postsynaptic cells in a chemical synapse.
synaptic strength The response of the postsynaptic neuron to a single stimulation of the presynaptic neurons.
synaptonemal complex A structure, visibile during prophase of meiosis I, in which proteins hold homologous chromosomes together.
synchronous cell populations Cells that are all at the same stage of the cell cycle.

T lymphocyte [because it matures in the *t*hymus gland] One of the cells responsible solely for cellular immunity or of the cells that participate in both humoral and cellular immunity.
T cytotoxic cell (T_C cell) A T lymphocyte that is responsible for the killing of cells recognized as nonself.
T-cell receptor A protein, on the surface of a T lymphocyte, that can bind to a complex of an antigen and MHC molecule.
T helper (T_H) cell A T lymphocyte that activates both the humoral and cellular immune responses.
target organ A structure upon which a hormone acts.
TATA box [after the first 4 nucleotides in the sequence] A consensus sequence in eukaryotic promoters.
taxol A compound extracted from the bark of yew trees; increases the stability of microtubules.

telocentric Refers to a chromosome in which the centromere is at one end.

telophase [Greek, *telos* = end] The last phase of mitosis, during which the mitotic apparatus (including kinetochore, polar, and astral microtubules) disperses and the chromosomes lose their distinct identities.

temperature-sensitive mutation A change in DNA that alters a protein so that it functions normally at one temperature (called the permissive temperature, usually relatively low), but not at a second temperature (the restrictive temperature, generally higher than the permissive temperature).

template A guide for the assembly of a complementary shape; in the context of DNA or RNA synthesis, one strand acts as a template for the assembly of a complementary sequence.

terminal bud A bud that lies at the tip of a shoot.

termination The ending of chain growth.

termination signal A DNA sequence that determines the end of transcription.

tertiary structure The complete arrangement of all the atoms of a polypeptide.

testcross A cross between an individual that is homozygous for a recessive allele of a particular gene and an individual whose genotype (homozygote or heterozygote) for that gene is unknown; the phenotype of the progeny reveals the presence of the recessive allele in the parent.

testicular feminization A condition in which a genetically male mammal develops a female phenotype because of the absence of a functional receptor for testosterone.

testis [plural, **testes**] A male gonad, which produces sperm.

tetrad, also called a **bivalent** A united chromosome pair visible during meiosis I; a tetrad consists of four chromatids.

tetrodotoxin A poison, isolated from the Japanese pufferfish ("*fugu*"), that blocks the passage of Na^+ ions through voltage-gated channels.

theory A system of statements and ideas that explain a group of facts or phenomena.

thermodynamics [Greek, *thermo* = heat + *dynamis* = power, force] The study of the relationships among different forms of energy.

thick filament In muscle, a myosin filament, 14 nm in diameter.

thin filament In muscle, an actin filament, 7 nm in diameter.

3′-end The end of a polynucleotide at which the 3′ carbon of the nucleotide residue is not attached to another residue but to a phosphate or a hydroxyl group.

thylakoid [Greek, *thylakos* = sac + *oides* = like] A flattened disk surrounded by the innermost membranes of a chloroplast.

thylakoid membrane A membrane, within the stroma of a chloroplast, that delineates stacked vesicles, called grana.

thymine dimers Two adjacent T residues linked together in the same DNA strand; formed as a result of exposure to ultraviolet light.

thyroid stimulating hormone (TSH) A peptide hormone, made in the anterior pituitary, that stimulates the production of thyroxin in the thyroid gland.

tight junction The main kind of impermeable junction in vertebrates; formed by the fusion of the membranes of adjacent cells.

tolerance In the immune system, a state of induced unresponsiveness to molecules that represent self; tolerance prevents immune attacks on the body's own components.

topogenic sequence A hydrophobic polypeptide segment that determines the orientation of a newly made polypeptide within a membrane.

totipotent Able to develop into a whole organism.

***trans* RNA splicing** The linking together of two separately made RNAs to produce a single mRNA.

***trans*-acting factor** [Latin, *trans* = across, because they regulate the expression of other molecules], also called a **transcription factor** A protein that binds to DNA and regulates transcription.

transactivating domain, also called an **effector domain** A structural domain within a transcription factor that can interact with the transcription machinery or with another transcription factor.

transcribed spacer RNA Within the large common precursor of the ribosomal RNAs, the RNA segments that lie between the mature rRNA sequences; the spacers are rapidly destroyed within the nucleus.

transcription The production of RNA from DNA.

transcription factor, also called a ***trans*-acting factor** A protein that binds to DNA and regulates transcription.

transducin The G-protein in the retina that interacts with rhodopsin; changes the activity of an enzyme that breaks down cyclic GMP, which in turn regulates ion flux through the photoreceptor cell membrane.

transfer RNA (tRNA) A small RNA molecule that serves as an adaptor in protein synthesis; each tRNA contains an anticodon that allows it to bind to a codon in mRNA and each becomes linked to a specific amino acid.

transformation The transfer of one or more genes from one organism to another.

transgenic Containing a particular recombinant DNA.

transit peptide A polypeptide sequence that targets a newly completed polypeptide to a specific organelle.

transition state A distorted form of a substrate that is intermediate between the starting reactant and the final product.

translation Protein synthesis; the conversion of information from an mRNA molecule into a polypeptide.

translation repressor protein A protein that can bind to sequences in the 5′ untranslated region of an mRNA and prevent its translation; used to regulate the translation of the iron-binding protein ferritin.

translational control Alterations in the rate at which specific mRNAs are translated.

translocation In chromosomes, a mutation in which part of one chromosome is moved to another chromosome; in protein synthesis, the movement of the growing polypeptide chain with its attached tRNA from the A site to the P site; the third step of the elongation cycle.

transmembrane segment A segment of consecutive hydrophobic amino acid residues that span the plasma membrane.

transmission electron microscope (TEM) A type of electron microscope in which the beam passes through a specimen.

transmission genetics The branch of genetics that deals with patterns of inheritance.

transposable element A mobile gene that can replicate only in integrated form, as part of a host's chromosome.

transposase An enzyme that catalyzes insertion of a transposable element into new sites.

triacylglycerol, also called a **triglyceride** A nonpolar derivative of glycerol in which all three hydroxyl groups are attached to fatty acids.

triceps The major muscle on the back side of the upper arm.

triple covalent bond A covalent bond in which two atoms share three pairs of electrons.

trisomy [Greek, *tri* = three + *soma* = body] The presence of three, rather than two, copies of a chromosome.

trophoblast [Greek, *trephein* = to nourish + *blastos* = germ] In a mammalian embryo, a prominent outer cell layer in a blastocyst.

tropism [Greek, *trope* = turning] The bending or curving toward or away from a stimulus.

tropomyosin In muscle, a rigid, rod-like protein, about 40 nm long, that stiffens the actin filament.

troponin In muscle, a large protein that binds both to tropomyosin and to free calcium ions.

true-breeding Individuals whose offspring have the same phenotype, generation after generation.

tube In a microscope, the part that has lenses at each end.

tubulin One of two globular proteins, called α and β tubulin, from which microtubules are assembled.

tumor suppressor gene, also called an **anti-oncogene** A gene that antagonizes the action of an oncogene; intimately involved in the normal regulation of cell growth.

tumor virus A virus that causes its host cell to lose its normal ability to regulate cell division.

turgor pressure [Latin, *turgor*, swelling] In plant cells, osmotic pressure against their cell walls resulting from the higher concentration of dissolved molecules and ions in cytoplasm than in the surrounding fluid.

two-dimensional electrophoresis A method for separating proteins in a complex mixture that involves the sequential application of isoelectric focusing and SDS electrophoresis.

tyrosine kinase A receptor protein that transfers phosphate groups from ATP to tyrosine side chains within a target protein.

ultracentrifuge A high-speed centrifuge that can spin tubes at speeds up to 80,000 rpm, subjecting their contents to forces up to 500,000 × gravity.

ultrastructure Subcellular structures visible only with an electron microscope.

unconditioned stimulus In associative learning, the stimulus that directly leads to a physiological response.

unconventional gene A gene without a stable cellular address.

universal Used by all species.

unsaturated A hydrocarbon chain (or fatty acid) that contains at least one double bond between two carbon atoms; so called because it could accept two more hydrogen atoms per double bond.

untranscribed spacer DNA Within the cluster of genes encoding ribosomal RNAs, the DNA sequences lying between each of the repeated copies.

upstream In a 5′ direction.

vacuole A membrane-bounded sac, without any obvious internal structure; larger than a vesicle.

variable region The amino-terminal polypeptide segment, of about 110 amino acid residues, an immunoglobulin light chain or heavy chain; contains the sequence variation responsible for antibody diversity.

vascular system In plants, the structures and cells that serve as a transport system; forms the central core of the plant.

vasopressin A peptide hormone, released by the posterior pituitary, that stimulates water reabsorption by the kidneys.

vegetative reproduction The process of producing a new organism without fertilization or seed production.

velocity The rate of appearance of product in an enzymatic reaction.

vesicle A membrane-bounded sac; some vesicles are present inside living eukaryotic cells; others form after a cell is broken open.

vesicular transporter A protein, within the membrane of a vesicle, that couples neurotransmitter accumulation to the energy stored in a vesicle's pH gradient.

viral oncogene A viral gene that affects growth control in a host cell.

virus An assembly of nucleic acid (DNA or RNA) and proteins (and occasionally other components, such as lipids or carbohydrates) that can reproduce only within a living cell; viruses depend on cells to obtain energy and perform chemical reactions.

visible light Electromagnetic radiation, with wavelengths from about 400 to about 750 nm, that can be perceived by human eyes and brains.

voltage, also called **electrical potential** A measure of the potential energy that results from a charge separation.

voltage clamping A method that allows the control of the voltage across a cell membrane.

voltage-gated Open or closed according to the voltage across the membrane.

voluntary muscle The muscles that can be consciously controlled; skeletal muscle.

water potential The potential energy of water per gram (or kilogram).

water-soluble signal A chemical signal that cannot pass through the plasma membrane and must act on its surface.

wavelength The distance between the crests of two successive waves.

Western blotting [after a similar blotting method, called "Southern blotting" because it was invented by Edward Southern] A method for detecting a particular protein after electrophoresis by blotting the contents of the electrophoresis gel onto a paper-like support and incubating this "blot" with a specific antibody.

whorl One of four concentric circles of flower parts at the end of a specialized stem.

wild type Refers to an allele that is most highly represented in wild populations.

wobble Nonstandard base pairing between the first base in the anticodon of transfer RNA and the third base of the codon in mRNA; wobble allows some tRNAs to recognize more than one codon for the same amino acid.

work The movement of an object against a force, or the conversion of energy into a electrical energy, chemical energy, or concentration energy.

yolk A mixture of proteins, lipids and carbohydrates that nourishes an embryo until it can feed itself.

Z band, also called the **Z disc** or **Z line** In muscle, the boundary of a sarcomere, interrupts the I band.

zinc finger A class of transcription factors that contain zinc atoms bound either to the side chains of four cysteine residues or of two cysteine and two histidine residues.

zygote [Greek, = yoke] The first cell of an embryo, formed by fertilization.

zygotic effect gene A gene that is expressed only after fertilization.

INDEX

Note: Boldfaced page number = discussion of a key term; italicized page number = figure or illustration; "t" = table; and "f" = featured in a chapter Essay.